Modern Electrodynamics

An engaging writing style and a strong focus on the physics make this comprehensive, graduate-level textbook unique among existing classical electromagnetism textbooks.

Charged particles in vacuum and the electrodynamics of continuous media are given equal attention in discussions of electrostatics, magnetostatics, quasistatics, conservation laws, wave propagation, radiation, scattering, special relativity, and field theory. Extensive use of qualitative arguments similar to those used by working physicists makes *Modern Electrodynamics* a must-have for every student of this subject.

In 24 chapters, the textbook covers many more topics than can be presented in a typical two-semester course, making it easy for instructors to tailor courses to their specific needs. Close to 120 worked examples and 80 applications boxes help the reader build physical intuition and develop technical skill. Nearly 600 end-of-chapter homework problems encourage students to engage actively with the material. A solutions manual is available for instructors at www.cambridge.org/Zangwill.

Andrew Zangwill is a Professor of Physics at the Georgia Institute of Technology and a Fellow of the American Physical Society. He is the author of the popular monograph *Physics at Surfaces* (Cambridge University Press, 1988).

Modern Electrodynamics

ANDREW ZANGWILL
Georgia Institute of Technology

CAMBRIDGE
UNIVERSITY PRESS

University Printing House, Cambridge CB2 8BS, United Kingdom

Cambridge University Press is part of the University of Cambridge.

It furthers the University's mission by disseminating knowledge in the pursuit of education, learning and research at the highest international levels of excellence.

www.cambridge.org
Information on this title: www.cambridge.org/Zangwill

First published 2013
Reprinted 2015

A catalogue record for this publication is available from the British Library

Library of Congress Cataloguing in Publication data
Zangwill, Andrew.
Modern electrodynamics / Andrew Zangwill.
pages cm
Includes bibliographical references and index.
ISBN 978-0-521-89697-9
1. Electrodynamics – Textbooks. I. Title.
QC631.Z36 2012
537.6 – dc23 2012035054

ISBN 978-0-521-89697-9 Hardback

Additional resources for this publication at www.cambridge.org/Zangwill

There are more things in heaven & earth connected with electromagnetism than are yet dream't of in philosophy.

Joseph Henry, letter to Lewis C. Beck (1827)

The search for reason ends at the shore of the known; on the immense expanse beyond it only the ineffable can glide.

Abraham Joshua Heschel, *Man is Not Alone* (1951)

Why repeat all this? Because there are new generations born every day. Because there are great ideas developed in the history of man, and these ideas do not last unless they are passed purposely and clearly from generation to generation.

Richard Feynman, *The Meaning of It All* (1963)

Contents

Table of Applications — *page* xv
Preface — xix

1 Mathematical Preliminaries — 1
1.1 Introduction — 1
1.2 Vectors — 1
1.3 Derivatives — 7
1.4 Integrals — 9
1.5 Generalized Functions — 11
1.6 Fourier Analysis — 15
1.7 Orthogonal Transformations — 18
1.8 Cartesian Tensors — 20
1.9 The Helmholtz Theorem — 22
1.10 Lagrange Multipliers — 24
Sources, References, and Additional Reading — 24
Problems — 25

2 The Maxwell Equations — 29
2.1 Introduction — 29
2.2 The Maxwell Equations in Vacuum — 33
2.3 Microscopic vs. Macroscopic — 38
2.4 The Maxwell Equations in Matter — 43
2.5 Quantum Limits and New Physics — 46
2.6 SI Units — 50
2.7 A Heuristic Derivation — 51
Sources, References, and Additional Reading — 53
Problems — 55

3 Electrostatics — 58
3.1 Introduction — 58
3.2 Coulomb's Law — 59
3.3 The Scalar Potential — 60
3.4 Gauss' Law and Solid Angle — 68
3.5 Electrostatic Potential Energy — 74
3.6 Electrostatic Total Energy — 76
3.7 The Electric Stress Tensor — 81
Sources, References, and Additional Reading — 84
Problems — 85

4 Electric Multipoles 90

4.1 Introduction 90
4.2 The Electric Dipole 92
4.3 Electric Dipole Layers 98
4.4 The Electric Quadrupole 102
4.5 Spherical Mathematics 106
4.6 Spherical and Azimuthal Multipoles 109
4.7 Primitive and Traceless Multipole Moments 116
Sources, References, and Additional Reading 119
Problems 121

5 Conducting Matter 126

5.1 Introduction 126
5.2 Electrostatic Induction 126
5.3 Screening and Shielding 133
5.4 Capacitance 134
5.5 The Energy of a System of Conductors 142
5.6 Forces on Conductors 143
5.7 Real Conductors 149
Sources, References, and Additional Reading 151
Problems 152

6 Dielectric Matter 158

6.1 Introduction 158
6.2 Polarization 158
6.3 The Field Produced by Polarized Matter 162
6.4 The Total Electric Field 165
6.5 Simple Dielectric Matter 167
6.6 The Physics of the Dielectric Constant 175
6.7 The Energy of Dielectric Matter 178
6.8 Forces on Dielectric Matter 184
Sources, References, and Additional Reading 191
Problems 193

7 Laplace's Equation 197

7.1 Introduction 197
7.2 Potential Theory 198
7.3 Uniqueness 199
7.4 Separation of Variables 201
7.5 Cartesian Symmetry 203
7.6 Azimuthal Symmetry 209
7.7 Spherical Symmetry 212
7.8 Cylindrical Symmetry 215
7.9 Polar Coordinates 218
7.10 The Complex Potential 221
7.11 A Variational Principle 226
Sources, References, and Additional Reading 228
Problems 229

8 Poisson's Equation 236

8.1 Introduction 236
8.2 The Key Idea: Superposition 236
8.3 The Method of Images 237
8.4 The Green Function Method 250
8.5 The Dirichlet Green Function 252
8.6 The Complex Logarithm Potential 260
8.7 The Poisson-Boltzmann Equation 262
Sources, References, and Additional Reading 264
Problems 265

9 Steady Current 272

9.1 Introduction 272
9.2 Current in Vacuum 273
9.3 Current in Matter 275
9.4 Potential Theory for Ohmic Matter 276
9.5 Electrical Resistance 277
9.6 Joule Heating 280
9.7 Electromotive Force 282
9.8 Current Sources 287
9.9 Diffusion Current: Fick's Law 291
Sources, References, and Additional Reading 293
Problems 294

10 Magnetostatics 301

10.1 Introduction 301
10.2 The Law of Biot and Savart 304
10.3 Ampère's Law 307
10.4 The Magnetic Scalar Potential 312
10.5 The Vector Potential 320
10.6 The Topology of Magnetic Field Lines 325
Sources, References, and Additional Reading 328
Problems 329

11 Magnetic Multipoles 336

11.1 Introduction 336
11.2 The Magnetic Dipole 337
11.3 Magnetic Dipole Layers 345
11.4 Exterior Multipoles 346
11.5 Interior Multipoles 353
11.6 Axially Symmetric Magnetic Fields 357
Sources, References, and Additional Reading 359
Problems 361

12 Magnetic Force and Energy 365

12.1 Introduction 365
12.2 Charged Particle Motion 366

12.3 The Force between Steady Currents 368
12.4 The Magnetic Dipole 372
12.5 The Magnetic Stress Tensor 381
12.6 Magnetostatic Total Energy 384
12.7 Magnetostatic Potential Energy 389
12.8 Inductance 394
Sources, References, and Additional Reading 399
Problems 401

13 Magnetic Matter 407

13.1 Introduction 407
13.2 Magnetization 407
13.3 The Field Produced by Magnetized Matter 412
13.4 Fictitious Magnetic Charge 415
13.5 The Total Magnetic Field 419
13.6 Simple Magnetic Matter 421
13.7 The Energy of Magnetic Matter 433
13.8 Forces on Magnetic Matter 435
13.9 Permanent Magnetic Matter 443
Sources, References, and Additional Reading 447
Problems 448

14 Dynamic and Quasistatic Fields 455

14.1 Introduction 455
14.2 The Ampère-Maxwell Law 456
14.3 Faraday's Law 460
14.4 Electromagnetic Induction 462
14.5 Slowly Time-Varying Charge in Vacuum 467
14.6 Slowly Time-Varying Current in Vacuum 470
14.7 Quasistatic Fields in Matter 472
14.8 Poor Conductors: Quasi-Electrostatics 473
14.9 Good Conductors: Quasi-Magnetostatics 475
14.10 The Skin Effect 477
14.11 Magnetic Diffusion 481
14.12 Eddy-Current Phenomena 483
14.13 AC Circuit Theory 486
Sources, References, and Additional Reading 493
Problems 494

15 General Electromagnetic Fields 501

15.1 Introduction 501
15.2 Symmetry 501
15.3 Electromagnetic Potentials 503
15.4 Conservation of Energy 507
15.5 Conservation of Linear Momentum 511
15.6 Conservation of Angular Momentum 516
15.7 The Center of Energy 519
15.8 Conservation Laws in Matter 522

15.9 The Force on Isolated Matter 526
Sources, References, and Additional Reading 529
Problems 531

16 Waves in Vacuum 536
16.1 Introduction 536
16.2 The Wave Equation 537
16.3 Plane Waves 539
16.4 Polarization 545
16.5 Wave Packets 552
16.6 The Helmholtz Equation 557
16.7 Beam-Like Waves 558
16.8 Spherical Waves 565
16.9 Hertz Vectors 569
16.10 Forces on Particles in Free Fields 571
Sources, References, and Additional Reading 575
Problems 577

17 Waves in Simple Matter 584
17.1 Introduction 584
17.2 Plane Waves 584
17.3 Reflection and Refraction 588
17.4 Radiation Pressure 599
17.5 Layered Matter 602
17.6 Simple Conducting Matter 607
17.7 Anisotropic Matter 613
Sources, References, and Additional Reading 616
Problems 617

18 Waves in Dispersive Matter 624
18.1 Introduction 624
18.2 Frequency Dispersion 624
18.3 Energy in Dispersive Matter 627
18.4 Transverse and Longitudinal Waves 629
18.5 Classical Models for Frequency Dispersion 630
18.6 Wave Packets in Dispersive Matter 641
18.7 The Consequences of Causality 649
18.8 Spatial Dispersion 656
Sources, References, and Additional Reading 657
Problems 659

19 Guided and Confined Waves 666
19.1 Introduction 666
19.2 Transmission Lines 667
19.3 Planar Conductors 672
19.4 Conducting Tubes 675
19.5 Dielectric Waveguides 687
19.6 Conducting Cavities 693

19.7 Dielectric Resonators 704
Sources, References, and Additional Reading 706
Problems 707

20 Retardation and Radiation 714

20.1 Introduction 714
20.2 Inhomogeneous Wave Equations 715
20.3 Retardation 719
20.4 The Time-Dependent Electric Dipole 727
20.5 Radiation 730
20.6 Thin-Wire Antennas 737
20.7 Cartesian Multipole Radiation 743
20.8 Spherical Multipole Radiation 755
20.9 Radiation in Matter 762
Sources, References, and Additional Reading 765
Problems 767

21 Scattering and Diffraction 775

21.1 Introduction 775
21.2 The Scattering Cross Section 776
21.3 Thomson Scattering 777
21.4 Rayleigh Scattering 782
21.5 Two Exactly Solvable Problems 783
21.6 Two Approximation Schemes 790
21.7 The Total Cross Section 793
21.8 Diffraction by a Planar Aperture 797
21.9 Generalized Optical Principles 807
Sources, References, and Additional Reading 812
Problems 814

22 Special Relativity 822

22.1 Introduction 822
22.2 Galileo's Relativity 823
22.3 Einstein's Relativity 825
22.4 The Lorentz Transformation 826
22.5 Four-Vectors 834
22.6 Electromagnetic Quantities 839
22.7 Covariant Electrodynamics 848
22.8 Matter in Uniform Motion 858
Sources, References, and Additional Reading 863
Problems 865

23 Fields from Moving Charges 870

23.1 Introduction 870
23.2 The Liénard-Wiechert Problem 870
23.3 Radiation in the Time Domain 880
23.4 Radiation in the Frequency Domain 886
23.5 Synchrotron Radiation 891

23.6 Radiation Reaction 899
23.7 Cherenkov Radiation 906
Sources, References, and Additional Reading 910
Problems 912

24 Lagrangian and Hamiltonian Methods 916

24.1 Introduction 916
24.2 Hamilton's Principle 916
24.3 Lagrangian Description 918
24.4 Invariance and Conservation Laws 927
24.5 Hamiltonian Description 931
Sources, References, and Additional Reading 940
Problems 941

Appendix A List of Important Symbols 945
Appendix B Gaussian Units 949
Appendix C Special Functions 953
Appendix D Managing Minus Signs in Special Relativity 959

Index 964

Table of Applications

Number	Name	Section
1.1	Two Identities for $\nabla \times \mathbf{L}$	1.2.7
1.2	Inversion and Reflection	1.8.1
2.1	Moving Point Charges	2.1.3
3.1	Field Lines for a Point Charge in a Uniform Field	3.4.4
3.2	The Ionization Potential of a Metal Cluster	3.6.2
4.1	Monolayer Electric Dipole Drops	4.3.2
4.2	Nuclear Quadrupole Moments	4.4.2
4.3	The Potential Produced by $\sigma(\theta) = \sigma_0 \cos\theta$	4.6.3
4.4	The Liquid Drop Model of Nuclear Fission	4.6.4
4.5	The Dielectric Polarization $\mathbf{P}(\mathbf{r})$	4.7.1
5.1	$\sigma(\mathbf{r}_S)$ for a Conducting Disk	5.2.2
5.2	Coulomb Blockade	5.5
6.1	A Uniformly Polarized Sphere	6.3
6.2	Refraction of Field Lines at a Dielectric Interface	6.5.4
6.3	A Classical Model for Quark Confinement	6.7.2
6.4	The Electric Force on an Embedded Volume	6.8.5
7.1	A Conducting Duct	7.5
7.2	Going Off the Axis	7.6.1
7.3	The Unisphere	7.7
7.4	An Electrostatic Lens	7.8.2
8.1	The Electrostatics of an Ion Channel	8.3.4
8.2	A Point Charge Outside a Grounded Tube	8.5.5
8.3	A Wire Array Above a Grounded Plane	8.6
9.1	Contact Resistance	9.5
9.2	The Electric Field Outside a Current-Carrying Wire	9.7.4
9.3	The Four-Point Resistance Probe	9.8
9.4	The Resting Potential of a Cell	9.9

Number	Name	Section
10.1	Irrotational Current Sources	10.2
10.2	Magnetic Resonance Imaging	10.4.2
10.3	Chaotic Lines of **B**	10.6.1
11.1	The Point Magnetic Monopole	11.2.3
11.2	Parity Violation and the Anapole Moment	11.4.1
11.3	The Helmholtz Anti-Coil	11.4.6
11.4	Strong Focusing by Quadrupole Magnetic Fields	11.5.2
11.5	The Principle of the Electron Microscope	11.6
12.1	The Magnetic Mirror	12.4.1
12.2	Magnetic Trapping	12.4.2
12.3	Magnetic Bacteria in a Rotating Field	12.4.4
12.4	The Magnetic Virial Theorem	12.5.1
12.5	The Solar Corona	12.6.1
13.1	A Uniformly Magnetized Sphere	13.4.1
13.2	Magnetic Shielding	13.6.4
13.3	The Magnetic Force on an Embedded Volume	13.8.5
13.4	Refrigerator Magnets	13.9.1
14.1	The Dielectric Constant of a Plasma	14.2.2
14.2	Faraday's Disk Generator	14.4.1
14.3	The Quadrupole Mass Spectrometer	14.5.1
14.4	A Solenoid with a Time-Harmonic Current	14.6.1
14.5	The Power Dissipated in a Metal by a Passing Charge	14.8
14.6	Shielding of an AC Magnetic Field by a Cylindrical Shell	14.10.1
14.7	Thomson's Jumping Ring	14.13.3
15.1	Uniqueness of Solutions and Boundary Conditions	15.4.1
15.2	$\mathbf{P}_{\rm EM}$ for a Capacitor in a Uniform Magnetic Field	15.5.3
15.3	The Hidden Momentum of a Static System	15.7
16.1	The Importance of Being Angular	16.8.2
16.2	Whittaker's Theorem Revisited	16.9.2
17.1	Alfvèn Waves	17.2.1
17.2	Bragg Mirrors	17.5.2
17.3	Plane Wave Propagation in a Uniaxial Crystal	17.7.1
18.1	Charge Relaxation in an Ohmic Medium	18.5.3
18.2	Reflection of Radio Waves by the Ionosphere	18.5.5
18.3	Lorentz Model Velocities	18.6.1
18.4	Whistlers	18.6.2
18.5	Sum Rules	18.7.4
19.1	Slow Waves for Charged Particle Acceleration	19.4.8
19.2	Quantum Billiards in a Resonant Cavity	19.6.4
19.3	Whispering Gallery Modes	19.7
20.1	The Fields of a Point Charge in Uniform Motion I	20.2
20.2	Schott's Formulae	20.3.4
20.3	The Classical Zeeman Effect	20.7.1
21.1	The Polarization of Cosmic Microwave Radiation	21.3
21.2	Approximate Mie Scattering	21.7.1
21.3	Sub-wavelength Apertures and Near-Field Optics	21.9.1

Number	Name	Section
22.1	The Global Positioning System	22.3.1
22.2	Heavy Ion Collisions and the Quark Gluon Plasma	22.4.2
22.3	Charged Particle Motion in a Plane Wave	22.5.2
22.4	Reflection from a Moving Mirror	22.6.5
22.5	The Electric Field Outside a Pulsar	22.8.2
23.1	The Fields of a Point Charge in Uniform Motion II	23.2.4
23.2	A Collinear Acceleration Burst	23.4.4

Preface

A textbook, as opposed to a treatise, should include everything a student must know, not everything the author does know.
Kenneth Johnson, quoted by Francis Low (1997)

In his *Lectures on Physics*, Richard Feynman asserts that "ten thousand years from now, there can be little doubt that the most significant event of the 19th century will be judged as Maxwell's discovery of the laws of electrodynamics". Whether this prediction is borne out or not, it is impossible to deny the significance of Maxwell's achievement to the history, practice, and future of physics. That is why electrodynamics has a permanent place in the physics curriculum, along with classical mechanics, quantum mechanics, and statistical mechanics. Of these four, students often find electrodynamics the most challenging. One reason is surely the mathematical demands of vector calculus and partial differential equations. Another stumbling block is the non-algorithmic nature of electromagnetic problem-solving. There are many entry points to a typical electromagnetism problem, but it is rarely obvious which lead to a quick solution and which lead to frustrating complications. Finally, Freeman Dyson points to the "two-level" structure of the theory.[1] A first layer of linear equations relates the electric and magnetic fields to their sources and to each other. A second layer of equations for force, energy, and stress are quadratic in the fields. Our senses and measurements probe the second-layer quantities, which are determined only indirectly by the fundamental first-layer quantities.

Modern Electrodynamics is a resource for graduate-level readers interested to deepen their understanding of electromagnetism without minimizing the role of the mathematics. The book's size was dictated by two considerations: first, my aim that it serve both as a classroom text and as a reference volume; second, my struggle to apply the epigraph at the top of the page.[2] Physicists are a prickly and opinionated bunch, so it is not surprising that there is very little agreement about "everything a student must know" about electromagnetism at the graduate level. Beyond a very basic core (the main content of undergraduate texts), the topics which appear in graduate-level textbooks and (electronically) published lecture notes depend strongly on the research background of the writer and whether he or she is a theorist or an experimenter. Some instructors view the subject as a convenient setting to illustrate the methods of mathematical physics and/or computational physics. Others see it as an opportunity to introduce topics (optics, plasma physics, astrophysics, biophysics, etc.) into a curriculum which might otherwise not include them. Still others teach electromagnetism for the main purpose of introducing the methods of relativistic field theory to their students.

Given the many uses of this foundational course, *Modern Electrodynamics* purposely contains much more material than can be comfortably covered in a two-semester course. Presentations with quite

[1] F.J. Dyson, "Why is Maxwell's theory so hard to understand?", in *James Clerk Maxwell Commemorative Booklet* (The James Clerk Maxwell Foundation, Edinburgh, 1999). Available at www.clerkmaxwellfoundation.org/DysonFreemanArticle.pdf.
[2] From the preface to F.E. Low, *Classical Field Theory* (Wiley, New York, 1997), p. xi.

different emphases can be constructed by making different choices from among the many topics offered for discussion. All instructors will omit various sections and probably entire chapters. Consistent with this point of view, I do not offer a single, idiosyncratic "vision" of electromagnetism. Rather, I have aimed to present what seemed (to me) to be the pedagogically soundest approach for students coming to this material after a first serious exposure at the junior/senior undergraduate level. In many cases, the same issue is examined from more than one point of view. The mathematics of the subject is given its proper due, but the qualitative and physical arguments I provide may ultimately remain with the reader longer.

The organization of this book reflects my personal experience as an instructor. After experimenting with relativity-first, Lagrangian-first, and radiation-first approaches, I concluded that the majority of students grasped the subject matter best when I used a traditional arrangement of topics. The text is purposely repetitive. This is done both to reinforce key ideas and to help readers who do not read the text in chapter order. My background as a condensed matter physicist appears in various places, including an emphasis on the practical (rather than the formal) aspects of microscopic averaging, a discussion of the limitations of the Lorentz model of dielectric and magnetic matter, and the presence of an entire chapter devoted to the experimentally important subject of quasistatics.

Every chapter of *Modern Electrodynamics* contains worked examples chosen either to develop problem-solving skills or to reveal subtleties of the subject which do not appear when one's exposure is limited to a few standard examples. Every chapter also contains several "applications" drawn from all the major subfields of physics. By and large, these are topics I was unwilling to relegate to the end-of-chapter homework for fear many readers would never see them. About half the chapters include a boxed excursion into a issue (often historical) where words serve better than equations, and every chapter ends with an annotated list of *Sources, References, & Additional Reading* to acknowledge my debt to others and to stimulate inquisitive readers. Finally, every chapter contains a large number of homework problems. These range from undergraduate-type drill problems to more challenging problems drawn directly from the research literature. Like most textbook authors, I emphasize that active engagement with the homework problems is an important part of the learning process. This is particularly important for electromagnetism where the struggle with difficult problems has somehow (wrongly) been elevated to a rite of passage. My desired outcome is a reader who, after completion of a course based on this book, can comfortably read and understand (if not necessarily reproduce in detail) a non-trivial electromagnetic argument or calculation which appears in the course of his or her research or reading.

The modernity of the text indicated by its title is *not* associated with the use of particularly "modern" mathematical methods. Rather, it derives from the inclusion of topics which have attracted new or renewed attention in recent decades. Examples include the electrostatics of ion channels, the modern theory of electric polarization, magnetic resonance imaging, the quantum Hall effect, optical tweezers, negative refraction, the time-domain approach to radiation, the polarization anisotropy of the cosmic microwave background, near-field optics, and relativistic heavy ion collisions. To keep the text finite, some familiar special topics from other texts have been omitted or barely touched upon. Examples include collisions and energy exchange between charged particles, the method of virtual quanta, transition radiation, energy loss in matter, and classical models of the electron. On the other hand, *Modern Electrodynamics* includes an overview of Dirac's Hamiltonian approach to electrodynamics and an update on certain "perpetual" problems like the correctness of the Lorentz-Dirac equation of motion for classical point particles and the Abraham-Minkowski controversy over the electromagnetic energy-momentum tensor in matter. All of these are illustrative of the self-refreshing nature of a subject which is re-invented by every new generation to meet its needs.

Finally, two choices I have made may give pause to some readers. One is my use of SI units throughout. The other is my use of the imaginary number i to impose the metric in special relativity. The technical rationale for using SI units is given in Section 2.6. An equally good reason is simply that this system has become the worldwide standard and nearly all undergraduate textbooks use it without apology. However, because the physics literature is replete with books and research papers which use Gaussian units, Appendix B discusses this system and provides an algorithm to painlessly convert from SI to Gaussian and vice versa.

My reason for using the "old-fashioned" Minkowski metric is purely pedagogical and cannot be stated more clearly than Nobel prize winner Gerard 't Hooft did in the preface to his *Introduction to General Relativity* (2001), namely: "In special relativity, the i has considerable practical advantage: Lorentz transformations are orthogonal, and all inner products only come with + signs. No confusion over signs remains". Although he switches to a metric tensor to discuss general relativity (as he must), 't Hooft further champions his use of i in special relativity with the remark, "I see no reason to shield students against the phenomenon of changes in convention and notation. Such transitions are necessary whenever one switches from one research field to another. They better get used to it". That being said, Appendix D outlines the use of the metric tensor $g_{\mu\nu}$ in special relativity.

It would have been impossible for me to write this book without help. At the top of the list, I gratefully acknowledge the contributions of Wayne Saslow (Texas A&M University) and Glenn Smith (Georgia Tech). These colleagues read and commented on many chapters and were always willing to talk electromagnetism with me. My colleague Brian Kennedy (Georgia Tech) repeatedly put aside his research work and helped on the many occasions when I managed to confuse myself. Special thanks go to Michael Cohen (University of Pennsylvania), who tried to teach me this subject when I was a graduate student and who generously shared many of his insights and thoughtful homework problems. I regret that Mike never found the time to write his own book on this subject.

I am happy to acknowledge David Vanderbilt (Rutgers University) and Stephen Barnett (University of Strathclyde) who provided essential help for my discussions of dielectric polarization and radiation pressure, respectively. Olivier Darrigol (CNRS, Laboratoire SPHERE) answered my questions about the history of Lagrangians in electrodynamics and Andrew Scherbakov (Georgia Tech) gave assistance with the homework problems. I also thank the brave souls who provided feedback after testing some of this material in their classrooms: Michael Pustilnik, Roman Grigoriev, and Pablo Laguna (Georgia Tech), Peter McIntyre (Texas A&M University), Brian Tonner (University of Central Florida), Jiang Xiao (Fudan University), and Kapil Krishan (Jawaharlal Nehru University). Finally, I am delighted to thank my editor Simon Capelin and his first-rate staff at Cambridge University Press. Simon never failed to show enthusiasm for this project, now over more years than either of us would care to remember.

I dedicate this work to my wife Sonia and daughter Hannah. They suffered, but not in silence.

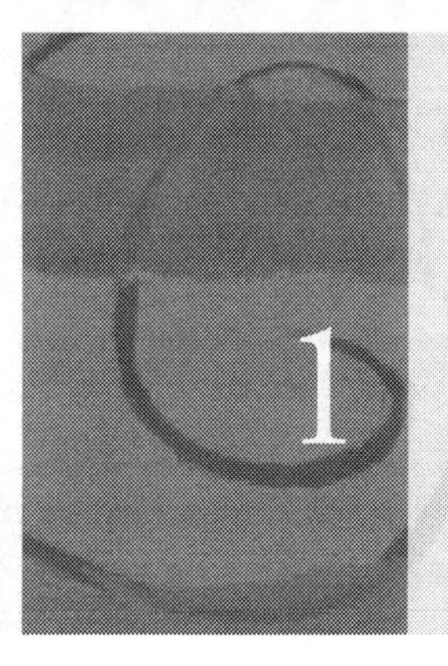

1 Mathematical Preliminaries

The enormous usefulness of mathematics in the natural sciences is something bordering on the mysterious.
Eugene Wigner (1960)

1.1 Introduction

This chapter presents a collection of mathematical notation, definitions, identities, theorems, and transformations that play an important role in the study of electromagnetism. A brief discussion accompanies some of the less familiar topics and only a few proofs are given in detail. For more details and complete proofs, the reader should consult the books and papers listed in Sources, References, and Additional Reading at the end of the chapter. Appendix C at the end of the book summarizes the properties of Legendre polynomials, spherical harmonics, and Bessel functions.

1.2 Vectors

A vector is a geometrical object characterized by a magnitude and direction.[1] Although not necessary, it is convenient to discuss an arbitrary vector using its components defined with respect to a given coordinate system. An example is the right-handed coordinate system with orthogonal unit basis vectors $(\hat{\mathbf{e}}_1, \hat{\mathbf{e}}_2, \hat{\mathbf{e}}_3)$ shown in Figure 1.1, where

$$\hat{\mathbf{e}}_1 \cdot \hat{\mathbf{e}}_1 = 1 \qquad \hat{\mathbf{e}}_2 \cdot \hat{\mathbf{e}}_2 = 1 \qquad \hat{\mathbf{e}}_3 \cdot \hat{\mathbf{e}}_3 = 1 \tag{1.1}$$

$$\hat{\mathbf{e}}_1 \cdot \hat{\mathbf{e}}_2 = 0 \qquad \hat{\mathbf{e}}_2 \cdot \hat{\mathbf{e}}_3 = 0 \qquad \hat{\mathbf{e}}_3 \cdot \hat{\mathbf{e}}_1 = 0 \tag{1.2}$$

$$\hat{\mathbf{e}}_1 \times \hat{\mathbf{e}}_2 = \hat{\mathbf{e}}_3 \qquad \hat{\mathbf{e}}_2 \times \hat{\mathbf{e}}_3 = \hat{\mathbf{e}}_1 \qquad \hat{\mathbf{e}}_3 \times \hat{\mathbf{e}}_1 = \hat{\mathbf{e}}_2. \tag{1.3}$$

We express an arbitrary vector $\mathbf{V}$ in this basis using components $V_k = \hat{\mathbf{e}}_k \cdot \mathbf{V}$,

$$\mathbf{V} = V_1\hat{\mathbf{e}}_1 + V_2\hat{\mathbf{e}}_2 + V_3\hat{\mathbf{e}}_3. \tag{1.4}$$

A vector can be decomposed in any coordinate system we please, so

$$\sum_{k=1}^{3} V_k\hat{\mathbf{e}}_k = \sum_{k=1}^{3} V'_k\hat{\mathbf{e}}'_k. \tag{1.5}$$

[1] A more precise definition of a vector is given in Section 1.8.

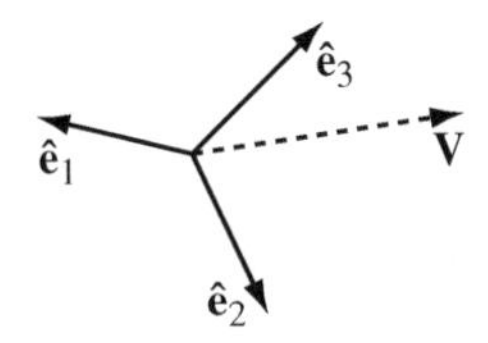

Figure 1.1: An orthonormal set of unit vectors $\hat{\mathbf{e}}_1, \hat{\mathbf{e}}_2, \hat{\mathbf{e}}_3$. $\mathbf{V}$ is an arbitrary vector.

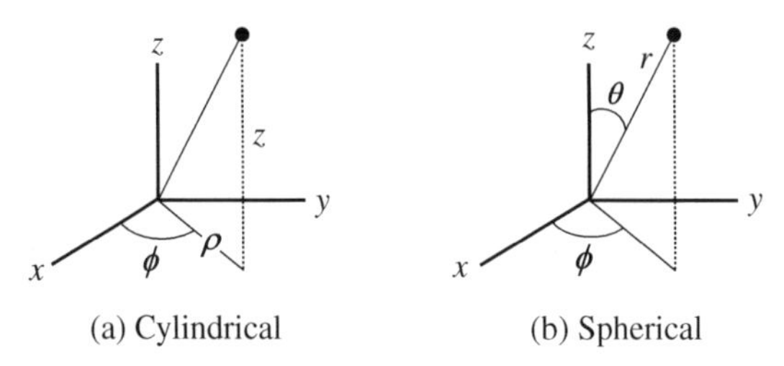

Figure 1.2: Two curvilinear coordinate systems.

1.2.1 Cartesian Coordinates

Our notation for Cartesian components and unit vectors is

$$\mathbf{V} = V_x\hat{\mathbf{x}} + V_y\hat{\mathbf{y}} + V_z\hat{\mathbf{z}}. \tag{1.6}$$

In particular, r_k always denotes the Cartesian components of the position vector,

$$\mathbf{r} = x\hat{\mathbf{x}} + y\hat{\mathbf{y}} + z\hat{\mathbf{z}}. \tag{1.7}$$

It is not obvious geometrically (see Example 1.7 in Section 1.8), but the gradient operator is a vector with the Cartesian representation

$$\nabla = \hat{\mathbf{x}}\frac{\partial}{\partial x} + \hat{\mathbf{y}}\frac{\partial}{\partial y} + \hat{\mathbf{z}}\frac{\partial}{\partial z}. \tag{1.8}$$

The divergence, curl, and Laplacian operations are, respectively,

$$\nabla \cdot \mathbf{V} = \frac{\partial V_x}{\partial x} + \frac{\partial V_y}{\partial y} + \frac{\partial V_z}{\partial z} \tag{1.9}$$

$$\nabla \times \mathbf{V} = \left(\frac{\partial V_z}{\partial y} - \frac{\partial V_y}{\partial z}\right)\hat{\mathbf{x}} + \left(\frac{\partial V_x}{\partial z} - \frac{\partial V_z}{\partial x}\right)\hat{\mathbf{y}} + \left(\frac{\partial V_y}{\partial x} - \frac{\partial V_x}{\partial y}\right)\hat{\mathbf{z}} \tag{1.10}$$

$$\nabla^2 A = \frac{\partial^2 A}{\partial x^2} + \frac{\partial^2 A}{\partial y^2} + \frac{\partial^2 A}{\partial z^2}. \tag{1.11}$$

1.2.2 Cylindrical Coordinates

Figure 1.2(a) defines cylindrical coordinates (ρ, ϕ, z). Our notation for the components and unit vectors in this system is

$$\mathbf{V} = V_\rho\hat{\boldsymbol{\rho}} + V_\phi\hat{\boldsymbol{\phi}} + V_z\hat{\mathbf{z}}. \tag{1.12}$$

The transformation to Cartesian coordinates is

$$x = \rho\cos\phi \qquad y = \rho\sin\phi \qquad z = z. \tag{1.13}$$

The volume element in cylindrical coordinates is $d^3r = \rho d\rho d\phi dz$. The unit vectors $(\hat{\boldsymbol{\rho}}, \hat{\boldsymbol{\phi}}, \hat{\mathbf{z}})$ form a right-handed orthogonal triad. $\hat{\mathbf{z}}$ is the same as in Cartesian coordinates. Otherwise,

$$\hat{\boldsymbol{\rho}} = \hat{\mathbf{x}}\cos\phi + \hat{\mathbf{y}}\sin\phi \qquad \hat{\mathbf{x}} = \hat{\boldsymbol{\rho}}\cos\phi - \hat{\boldsymbol{\phi}}\sin\phi \tag{1.14}$$

$$\hat{\boldsymbol{\phi}} = -\hat{\mathbf{x}}\sin\phi + \hat{\mathbf{y}}\cos\phi \qquad \hat{\mathbf{y}} = \hat{\boldsymbol{\rho}}\sin\phi + \hat{\boldsymbol{\phi}}\cos\phi. \tag{1.15}$$

The gradient operator in cylindrical coordinates is

$$\nabla = \hat{\boldsymbol{\rho}}\frac{\partial}{\partial\rho} + \frac{\hat{\boldsymbol{\phi}}}{\rho}\frac{\partial}{\partial\phi} + \hat{\mathbf{z}}\frac{\partial}{\partial z}. \tag{1.16}$$

The divergence, curl, and Laplacian operations are, respectively,

$$\nabla\cdot\mathbf{V} = \frac{1}{\rho}\frac{\partial(\rho V_\rho)}{\partial\rho} + \frac{1}{\rho}\frac{\partial V_\phi}{\partial\phi} + \frac{\partial V_z}{\partial z} \tag{1.17}$$

$$\nabla\times\mathbf{V} = \left[\frac{1}{\rho}\frac{\partial V_z}{\partial\phi} - \frac{\partial V_\phi}{\partial z}\right]\hat{\boldsymbol{\rho}} + \left[\frac{\partial V_\rho}{\partial z} - \frac{\partial V_z}{\partial\rho}\right]\hat{\boldsymbol{\phi}} + \frac{1}{\rho}\left[\frac{\partial(\rho V_\phi)}{\partial\rho} - \frac{\partial V_\rho}{\partial\phi}\right]\hat{\mathbf{z}} \tag{1.18}$$

$$\nabla^2 A = \frac{1}{\rho}\frac{\partial}{\partial\rho}\left(\rho\frac{\partial A}{\partial\rho}\right) + \frac{1}{\rho^2}\frac{\partial^2 A}{\partial\phi^2} + \frac{\partial^2 A}{\partial z^2}. \tag{1.19}$$

1.2.3 Spherical Coordinates

Figure 1.2(b) defines spherical coordinates (r, θ, ϕ). Our notation for the components and unit vectors in this system is

$$\mathbf{V} = V_r\hat{\mathbf{r}} + V_\theta\hat{\boldsymbol{\theta}} + V_\phi\hat{\boldsymbol{\phi}}. \tag{1.20}$$

The transformation to Cartesian coordinates is

$$x = r\sin\theta\cos\phi \qquad y = r\sin\theta\sin\phi \qquad z = r\cos\theta. \tag{1.21}$$

The volume element in spherical coordinates is $d^3r = r^2\sin\theta dr d\theta d\phi$. The unit vectors are related by

$$\hat{\mathbf{r}} = \hat{\mathbf{x}}\sin\theta\cos\phi + \hat{\mathbf{y}}\sin\theta\sin\phi + \hat{\mathbf{z}}\cos\theta \qquad \hat{\mathbf{x}} = \hat{\mathbf{r}}\sin\theta\cos\phi + \hat{\boldsymbol{\theta}}\cos\theta\cos\phi - \hat{\boldsymbol{\phi}}\sin\phi \tag{1.22}$$

$$\hat{\boldsymbol{\theta}} = \hat{\mathbf{x}}\cos\theta\cos\phi + \hat{\mathbf{y}}\cos\theta\sin\phi - \hat{\mathbf{z}}\sin\theta \qquad \hat{\mathbf{y}} = \hat{\mathbf{r}}\sin\theta\sin\phi + \hat{\boldsymbol{\theta}}\cos\theta\sin\phi + \hat{\boldsymbol{\phi}}\cos\phi \tag{1.23}$$

$$\hat{\boldsymbol{\phi}} = -\hat{\mathbf{x}}\sin\phi + \hat{\mathbf{y}}\cos\phi \qquad \hat{\mathbf{z}} = \hat{\mathbf{r}}\cos\theta - \hat{\boldsymbol{\theta}}\sin\theta. \tag{1.24}$$

The gradient operator in spherical coordinates is

$$\nabla = \hat{\mathbf{r}}\frac{\partial}{\partial r} + \frac{\hat{\boldsymbol{\theta}}}{r}\frac{\partial}{\partial\theta} + \frac{\hat{\boldsymbol{\phi}}}{r\sin\theta}\frac{\partial}{\partial\phi}. \tag{1.25}$$

The divergence, curl, and Laplacian operations are, respectively,

$$\nabla\cdot\mathbf{V} = \frac{1}{r^2}\frac{\partial(r^2 V_r)}{\partial r} + \frac{1}{r\sin\theta}\frac{\partial(\sin\theta\, V_\theta)}{\partial\theta} + \frac{1}{r\sin\theta}\frac{\partial V_\phi}{\partial\phi} \tag{1.26}$$

$$\begin{aligned}\nabla\times\mathbf{V} &= \frac{1}{r\sin\theta}\left[\frac{\partial(\sin\theta\, V_\phi)}{\partial\theta} - \frac{\partial V_\theta}{\partial\phi}\right]\hat{\mathbf{r}} \\ &\quad + \frac{1}{r}\left[\frac{1}{\sin\theta}\frac{\partial V_r}{\partial\phi} - \frac{\partial(rV_\phi)}{\partial r}\right]\hat{\boldsymbol{\theta}} + \frac{1}{r}\left[\frac{\partial(rV_\theta)}{\partial r} - \frac{\partial V_r}{\partial\theta}\right]\hat{\boldsymbol{\phi}}\end{aligned} \tag{1.27}$$

$$\nabla^2 A = \frac{1}{r^2}\frac{\partial}{\partial r}\left(r^2\frac{\partial A}{\partial r}\right) + \frac{1}{r^2\sin\theta}\frac{\partial}{\partial\theta}\left(\sin\theta\frac{\partial A}{\partial\theta}\right) + \frac{1}{r^2\sin^2\theta}\frac{\partial^2 A}{\partial\phi^2}. \tag{1.28}$$

1.2.4 The Einstein Summation Convention

Einstein (1916) introduced the following convention. An index which appears exactly twice in a single term of a mathematical expression is implicitly summed over all possible values for that index. The range of this *dummy index* must be clear from context and the index cannot be used elsewhere in the same expression for another purpose. In this book, the range for a roman index like i is from 1 to 3, indicating a sum over the Cartesian indices x, y, and z. Thus, $\mathbf{V}$ in (1.6) and its dot product with another vector $\mathbf{F}$ are written

$$\mathbf{V} = \sum_{k=1}^{3} V_k \hat{\mathbf{e}}_k \equiv V_k \hat{\mathbf{e}}_k \qquad \mathbf{V} \cdot \mathbf{F} = \sum_{k=1}^{3} V_k F_k \equiv V_k F_k. \tag{1.29}$$

In a Cartesian basis, the gradient of a scalar φ and the divergence of a vector $\mathbf{D}$ can be variously written

$$\nabla\varphi = \hat{\mathbf{e}}_k \nabla_k \varphi = \hat{\mathbf{e}}_k \partial_k \varphi = \hat{\mathbf{e}}_k \frac{\partial \varphi}{\partial r_k} \tag{1.30}$$

$$\nabla \cdot \mathbf{D} = \nabla_k D_k = \partial_k D_k = \frac{\partial D_k}{\partial r_k}. \tag{1.31}$$

If an $N \times N$ matrix $\mathbf{C}$ is the product of an $N \times M$ matrix $\mathbf{A}$ and an $M \times N$ matrix $\mathbf{B}$,

$$C_{ik} = \sum_{j=1}^{M} A_{ij} B_{jk} = A_{ij} B_{jk}. \tag{1.32}$$

1.2.5 The Kronecker and Levi-Cività Symbols

The Kronecker delta symbol δ_{ij} and Levi-Cività permutation symbol ϵ_{ijk} have roman indices i, j, and k which take on the Cartesian coordinate values x, y, and z. They are defined by

$$\delta_{ij} = \begin{cases} 1 & i = j, \\ 0 & i \neq j, \end{cases} \tag{1.33}$$

and

$$\epsilon_{ijk} = \begin{cases} 1 & ijk = xyz \;\; yzx \;\; zxy, \\ -1 & ijk = xzy \;\; yxz \;\; zyx, \\ 0 & \text{otherwise.} \end{cases} \tag{1.34}$$

Some useful Kronecker delta and Levi-Cività symbol identities are

$$\hat{\mathbf{e}}_i \cdot \hat{\mathbf{e}}_j = \delta_{ij} \qquad \delta_{kk} = 3 \tag{1.35}$$

$$\partial_k r_j = \delta_{jk} \qquad V_k \delta_{kj} = V_j \tag{1.36}$$

$$[\mathbf{V} \times \mathbf{F}]_i = \epsilon_{ijk} V_j F_k \qquad [\nabla \times \mathbf{A}]_i = \epsilon_{ijk} \partial_j A_k \tag{1.37}$$

$$\delta_{ij} \epsilon_{ijk} = 0 \qquad \epsilon_{ijk} \epsilon_{ijk} = 6. \tag{1.38}$$

A particularly useful identity involves a single sum over the repeated index i:

$$\epsilon_{ijk} \epsilon_{ist} = \delta_{js} \delta_{kt} - \delta_{jt} \delta_{ks}. \tag{1.39}$$

A generalization of (1.39) when there are no repeated indices to sum over is the determinant

$$\epsilon_{ki\ell} \epsilon_{mpq} = \begin{vmatrix} \delta_{km} & \delta_{im} & \delta_{\ell m} \\ \delta_{kp} & \delta_{ip} & \delta_{\ell p} \\ \delta_{kq} & \delta_{iq} & \delta_{\ell q} \end{vmatrix}. \tag{1.40}$$

Finally, let $\mathbf{C}$ be a 3×3 matrix with matrix elements C_{11}, C_{12}, etc. The determinant of $\mathbf{C}$ can be written using either an expansion by columns,

$$\det \mathbf{C} = \epsilon_{ijk} C_{i1} C_{j2} C_{k3}, \tag{1.41}$$

or an expansion by rows,

$$\det \mathbf{C} = \epsilon_{ijk} C_{1i} C_{2j} C_{3k}. \tag{1.42}$$

A closely related identity we will use in Section 1.8.1 is

$$\epsilon_{\ell mn} \det \mathbf{C} = \epsilon_{ijk} C_{\ell i} C_{mj} C_{nk}. \tag{1.43}$$

1.2.6 Vector Identities in Cartesian Components

The Kronecker and Levi-Cività symbols simplify the proof of vector identities. An example is

$$\mathbf{a} \times (\mathbf{b} \times \mathbf{c}) = \mathbf{b}(\mathbf{a} \cdot \mathbf{c}) - \mathbf{c}(\mathbf{a} \cdot \mathbf{b}). \tag{1.44}$$

Using the left side of (1.37), the ith component of $\mathbf{a} \times (\mathbf{b} \times \mathbf{c})$ is

$$[\mathbf{a} \times (\mathbf{b} \times \mathbf{c})]_i = \epsilon_{ijk} a_j (\mathbf{b} \times \mathbf{c})_k = \epsilon_{ijk} a_j \epsilon_{k\ell m} b_l c_m. \tag{1.45}$$

The definition (1.34) tells us that $\epsilon_{ijk} = \epsilon_{kij}$. Therefore, the identity (1.39) gives

$$[\mathbf{a} \times (\mathbf{b} \times \mathbf{c})]_i = \epsilon_{kij} \epsilon_{k\ell m} a_j b_\ell c_m = (\delta_{i\ell}\delta_{jm} - \delta_{im}\delta_{j\ell}) a_j b_\ell c_m = a_j b_i c_j - a_j b_j c_i. \tag{1.46}$$

The final result, $b_i(\mathbf{a} \cdot \mathbf{c}) - c_i(\mathbf{a} \cdot \mathbf{b})$, is indeed the ith component of the right side of (1.44). The same method of proof applies to gradient-, divergence-, and curl-type vector identities because the components of the ∇ operator transform like the components of a vector [see above (1.8)]. The next three examples illustrate this point.

Example 1.1 Prove that $\nabla \cdot (\nabla \times \mathbf{g}) = 0$.

Solution: Begin with $\nabla \cdot (\nabla \times \mathbf{g}) = \partial_i \epsilon_{ijk} \partial_j g_k = \frac{1}{2} \partial_i \partial_j g_k \epsilon_{ijk} + \frac{1}{2} \partial_i \partial_j g_k \epsilon_{ijk}$. Exchanging the dummy indices i and j in the last term gives

$$\nabla \cdot \nabla \times \mathbf{g} = \tfrac{1}{2} \partial_i \partial_j g_k \epsilon_{ijk} + \tfrac{1}{2} \partial_j \partial_i g_k \epsilon_{jik} = \tfrac{1}{2} \{\epsilon_{ijk} + \epsilon_{jik}\} \partial_i \partial_j g_k = 0.$$

The final zero comes from $\epsilon_{ijk} = -\epsilon_{jik}$, which is a consequence of (1.34).

Example 1.2 Prove that $\nabla \times (\mathbf{A} \times \mathbf{B}) = \mathbf{A} \nabla \cdot \mathbf{B} - (\mathbf{A} \cdot \nabla)\mathbf{B} + (\mathbf{B} \cdot \nabla)\mathbf{A} - \mathbf{B} \nabla \cdot \mathbf{A}$.

Solution: Focus on the ith Cartesian component and use the left side of (1.37) to write

$$[\nabla \times (\mathbf{A} \times \mathbf{B})]_i = \epsilon_{ijk} \partial_j (\mathbf{A} \times \mathbf{B})_k = \epsilon_{ijk} \epsilon_{kst} \partial_j (A_s B_t).$$

The cyclic properties of the Levi-Cività symbol and the identity (1.39) give

$$[\nabla \times (\mathbf{A} \times \mathbf{B})]_i = \epsilon_{kij} \epsilon_{kst} \partial_j (A_s B_t) = (\delta_{is}\delta_{jt} - \delta_{it}\delta_{js})(A_s \partial_j B_t + B_t \partial_j A_s).$$

Therefore,

$$[\nabla \times (\mathbf{A} \times \mathbf{B})]_i = A_i \partial_j B_j - A_j \partial_j B_i + B_j \partial_j A_i - B_i \partial_j A_j.$$

This proves the identity because the choice of i is arbitrary.

Example 1.3 Prove the "double-curl identity" $\nabla\times(\nabla\times\mathbf{A}) = \nabla(\nabla\cdot\mathbf{A}) - \nabla^2\mathbf{A}$.

Solution: Consider the ith Cartesian component. The identity on the left side of (1.37) and the invariance of the Levi-Cività symbol with respect to cyclic permutations of its indices give

$$[\nabla\times(\nabla\times\mathbf{A})]_i = \epsilon_{ijk}\partial_j(\nabla\times\mathbf{A})_k = \epsilon_{ijk}\partial_j\epsilon_{kpq}\partial_p A_q = \epsilon_{kij}\epsilon_{kpq}\partial_j\partial_p A_q.$$

Now apply the identity (1.39) to get

$$[\nabla\times(\nabla\times\mathbf{A})]_i = (\delta_{ip}\delta_{jq} - \delta_{iq}\delta_{jp})\partial_j\partial_p A_q = \partial_i\partial_j A_j - \partial_j\partial_j A_i = \nabla_i(\nabla\cdot\mathbf{A}) - \nabla^2 A_i.$$

The double-curl identity follows because

$$\nabla^2\mathbf{A} = \nabla^2(A_x\hat{\mathbf{x}} + A_y\hat{\mathbf{y}} + A_z\hat{\mathbf{z}}) = \hat{\mathbf{x}}\nabla^2 A_x + \hat{\mathbf{y}}\nabla^2 A_y + \hat{\mathbf{z}}\nabla^2 A_z.$$

1.2.7 Vector Identities in Curvilinear Components

Care is needed to interpret the vector identities in Examples 1.2 and 1.3 when the vectors in question are decomposed into spherical or cylindrical components such as $\mathbf{A} = A_r\hat{\mathbf{r}} + A_\theta\hat{\boldsymbol{\theta}} + A_\phi\hat{\boldsymbol{\phi}}$. This can be seen from Example 1.3 where the final step is no longer valid because $\hat{\mathbf{r}}$, $\hat{\boldsymbol{\theta}}$, and $\hat{\boldsymbol{\phi}}$ are not constant vectors. In other words,

$$\nabla^2\mathbf{A} = \nabla\cdot\nabla(A_r\hat{\mathbf{r}} + A_\theta\hat{\boldsymbol{\theta}} + A_\phi\hat{\boldsymbol{\phi}}) \neq \hat{\mathbf{r}}\nabla^2 A_r + \hat{\boldsymbol{\theta}}\nabla^2 A_\theta + \hat{\boldsymbol{\phi}}\nabla^2 A_\phi. \tag{1.47}$$

One way to proceed is to work out the components of $\nabla(A_r\hat{\mathbf{r}})$, $\nabla(A_\theta\hat{\boldsymbol{\theta}})$, and $\nabla(A_\phi\hat{\boldsymbol{\phi}})$. Alternatively, we may simply *define* the meaning of the operation $\nabla^2\mathbf{A}$ when $\mathbf{A}$ is expressed using curvilinear components. For example,

$$[\nabla^2\mathbf{A}]_\phi \equiv \partial_\phi(\nabla\cdot\mathbf{A}) - [\nabla\times(\nabla\times\mathbf{A})]_\phi, \tag{1.48}$$

and similarly for $(\nabla^2\mathbf{A})_r$ and $(\nabla^2\mathbf{A})_\theta$.

Exactly the same issue arises when we examine the last step in Example 1.2, namely

$$[\nabla\times(\mathbf{A}\times\mathbf{B})]_i = A_i\nabla\cdot\mathbf{B} - (\mathbf{A}\cdot\nabla)B_i + (\mathbf{B}\cdot\nabla)A_i - B_i\nabla\cdot\mathbf{A}. \tag{1.49}$$

By construction, this equation makes sense when i stands for x, y, or z. It does *not* make sense if i stands for, say, r, θ, or ϕ. On the other hand, the full vector version of the identity is correct as long as we retain the r, θ, and ϕ variations of $\hat{\mathbf{r}}$, $\hat{\boldsymbol{\theta}}$, and $\hat{\boldsymbol{\phi}}$. For example,

$$(\mathbf{A}\cdot\nabla)\mathbf{B} = \left[A_r\frac{\partial}{\partial r} + \frac{A_\theta}{r}\frac{\partial}{\partial\theta} + \frac{A_\phi}{r\sin\theta}\frac{\partial}{\partial\phi}\right](B_r\hat{\mathbf{r}} + B_\theta\hat{\boldsymbol{\theta}} + B_\phi\hat{\boldsymbol{\phi}}). \tag{1.50}$$

Application 1.1 Two Identities for $\nabla\times\mathbf{L}$

The $\hbar = 1$ version of the quantum mechanical angular momentum operator, $\mathbf{L} = -i\mathbf{r}\times\nabla$, plays a useful role in the analysis of classical spherical systems. In this Application, we prove two operator identities which will appear later in the text:

(A) $\nabla\times\mathbf{L} = -i\mathbf{r}\nabla^2 + i\nabla(1+\mathbf{r}\cdot\nabla)$

(B) $\nabla\times\mathbf{L} = (\hat{\mathbf{r}}\times\mathbf{L})\left(\frac{1}{r}\frac{\partial}{\partial r}r\right) + \hat{\mathbf{r}}\frac{i}{r}L^2.$

Proof of Identity (A):

We use (1.37), (1.39), and the cyclic property of the Levi-Cività symbol to evaluate the kth component of $\nabla \times \mathbf{L}$ acting on a scalar function ϕ:

$$[\nabla \times \mathbf{L}]_k \phi = -i[\nabla \times (\mathbf{r} \times \nabla)]_k \phi = -i\epsilon_{mk\ell}\epsilon_{mst}\partial_\ell r_s \partial_t \phi = -i[\partial_\ell r_k \partial_\ell \phi - \partial_\ell r_\ell \partial_k \phi]. \quad (1.51)$$

Because $\partial_\ell r_\ell = 3$ and $\partial_\ell r_k = \delta_{\ell k}$,

$$\nabla \times \mathbf{L}\phi = [-i\mathbf{r}\nabla^2 + i2\nabla + i(\mathbf{r} \cdot \nabla)\nabla]\phi. \quad (1.52)$$

However,

$$\partial_k [r_\ell \partial_\ell \phi] = \partial_k \phi + r_\ell \partial_\ell \partial_k \phi, \quad (1.53)$$

which is the kth component of $\nabla(\mathbf{r} \cdot \nabla)\phi = \nabla\phi + (\mathbf{r} \cdot \nabla)\nabla\phi$. Substituting the latter into (1.53) gives Identity (A).

Proof of Identity (B):

We decompose the gradient operator into its radial and angular pieces:

$$\nabla = \hat{\mathbf{r}}(\hat{\mathbf{r}} \cdot \nabla) - \hat{\mathbf{r}} \times (\hat{\mathbf{r}} \times \nabla) = \hat{\mathbf{r}}\frac{\partial}{\partial r} - \frac{i}{r}\hat{\mathbf{r}} \times \mathbf{L}. \quad (1.54)$$

Equation (1.54) and the Levi-Cività formalism produce the intermediate result

$$\nabla \times \mathbf{L} = (\hat{\mathbf{r}} \times \mathbf{L})\frac{\partial}{\partial r} - \frac{i}{r}(\hat{\mathbf{r}} \times \mathbf{L}) \times \mathbf{L} = (\hat{\mathbf{r}} \times \mathbf{L})\frac{\partial}{\partial r} - \frac{i}{r}\left[\hat{r}_k \mathbf{L} L_k - \hat{\mathbf{r}} L^2\right]. \quad (1.55)$$

However, the angular momentum operator obeys commutation relations which can be summarized as $\mathbf{L} \times \mathbf{L} = i\mathbf{L}$. Therefore,

$$\hat{\mathbf{r}} \times (\mathbf{L} \times \mathbf{L}) = i\hat{\mathbf{r}} \times \mathbf{L} \quad \Rightarrow \quad \hat{r}_k \mathbf{L} L_k - \hat{r}_k L_k \mathbf{L} = i\hat{\mathbf{r}} \times \mathbf{L}. \quad (1.56)$$

On the other hand, $r_k L_k = 0$ because $\mathbf{L}$ is perpendicular to both $\mathbf{r}$ and ∇. Therefore, $\hat{r}_k \mathbf{L} L_k = i\hat{\mathbf{r}} \times \mathbf{L}$, which we can substitute into (1.55). The result is identity (B) because, for any scalar function ϕ,

$$\frac{1}{r}\left[\frac{\partial}{\partial r}(r\phi)\right] = \frac{\partial \phi}{\partial r} + \frac{\phi}{r}. \quad (1.57)$$

■

1.3 Derivatives

1.3.1 Functions of $\mathbf{r}$ and $|\mathbf{r}|$

The position vector is $\mathbf{r} = r\hat{\mathbf{r}}$ with $r = \sqrt{x^2 + y^2 + z^2}$. If $f(r)$ is a scalar function and $f'(r) = df/dr$,

$$\nabla r = \hat{\mathbf{r}} \qquad \nabla \times \mathbf{r} = 0 \quad (1.58)$$

$$\nabla f = f'\hat{\mathbf{r}} \qquad \nabla^2 f = \frac{(r^2 f')'}{r^2} \quad (1.59)$$

$$\nabla \cdot (f\mathbf{r}) = \frac{(r^3 f)'}{r^2} \qquad \nabla \times (f\mathbf{r}) = 0. \quad (1.60)$$

Similarly, if $\mathbf{g}(r)$ is a vector function and $\mathbf{c}$ is a constant vector,

$$\nabla \cdot \mathbf{g} = \mathbf{g}' \cdot \hat{\mathbf{r}} \qquad \nabla \times \mathbf{g} = \hat{\mathbf{r}} \times \mathbf{g}' \quad (1.61)$$

$$(\mathbf{g} \cdot \nabla)\mathbf{r} = \mathbf{g} \qquad (\mathbf{r} \cdot \nabla)\mathbf{g} = r\mathbf{g}' \quad (1.62)$$

$$\nabla(\mathbf{r}\cdot\mathbf{g}) = \mathbf{g} + \frac{(\mathbf{r}\cdot\mathbf{g}')\mathbf{r}}{r} \qquad \nabla\cdot(\mathbf{g}\times\mathbf{r}) = 0 \tag{1.63}$$

$$\nabla\times(\mathbf{g}\times\mathbf{r}) = 2\mathbf{g} + r\mathbf{g}' - \frac{(\mathbf{r}\cdot\mathbf{g}')\mathbf{r}}{r} \qquad \nabla(\mathbf{c}\cdot\mathbf{r}) = \mathbf{c}. \tag{1.64}$$

1.3.2 Functions of $\mathbf{r} - \mathbf{r}'$

Let $\mathbf{R} = \mathbf{r} - \mathbf{r}' = (x - x')\hat{\mathbf{x}} + (y - y')\hat{\mathbf{y}} + (z - z')\hat{\mathbf{z}}$. Then,

$$\nabla f(R) = f'(R)\hat{\mathbf{R}} \qquad \nabla\cdot\mathbf{g}(R) = \mathbf{g}'(R)\cdot\hat{\mathbf{R}} \qquad \nabla\times\mathbf{g}(R) = \hat{\mathbf{R}}\times\mathbf{g}'(R). \tag{1.65}$$

Moreover, because

$$\nabla = \hat{\mathbf{x}}\frac{\partial}{\partial x} + \hat{\mathbf{y}}\frac{\partial}{\partial y} + \hat{\mathbf{z}}\frac{\partial}{\partial z} \qquad \text{and} \qquad \nabla' = \hat{\mathbf{x}}\frac{\partial}{\partial x'} + \hat{\mathbf{y}}\frac{\partial}{\partial y'} + \hat{\mathbf{z}}\frac{\partial}{\partial z'}, \tag{1.66}$$

it it straightforward to confirm that

$$\nabla' f(R) = -\nabla f(R). \tag{1.67}$$

1.3.3 The Convective Derivative

Let $\phi(\mathbf{r}, t)$ be a scalar function of space and time. An observer who repeatedly samples the value of ϕ at a fixed point in space, $\mathbf{r}$, records the time rate of change of ϕ as the partial derivative $\partial\phi/\partial t$. However, the same observer who repeatedly samples ϕ along a trajectory in space $\mathbf{r}(t)$ that moves with velocity $\boldsymbol{\upsilon}(t) = \dot{\mathbf{r}}(t)$ records the time rate of change of ϕ as the *convective derivative*,

$$\frac{d\phi}{dt} = \frac{\partial\phi}{\partial t} + \frac{dx}{dt}\frac{\partial\phi}{\partial x} + \frac{dy}{dt}\frac{\partial\phi}{\partial y} + \frac{dz}{dt}\frac{\partial\phi}{\partial z} = \frac{\partial\phi}{\partial t} + (\boldsymbol{\upsilon}\cdot\nabla)\phi. \tag{1.68}$$

For a vector function $\mathbf{g}(\mathbf{r}, t)$, the corresponding convective derivative is

$$\frac{d\mathbf{g}}{dt} = \frac{\partial\mathbf{g}}{\partial t} + (\boldsymbol{\upsilon}\cdot\nabla)\mathbf{g}. \tag{1.69}$$

1.3.4 Taylor's Theorem

Taylor's theorem in one dimension is

$$f(x) = f(a) + (x - a)\left.\frac{df}{dx}\right|_{x=a} + \frac{1}{2!}(x - a)^2\left.\frac{d^2 f}{dx^2}\right|_{x=a} + \cdots. \tag{1.70}$$

An alternative form follows from (1.70) if $x \to x + \epsilon$ and $a \to x$:

$$f(x + \epsilon) = f(x) + \epsilon\frac{df}{dx} + \frac{1}{2!}\epsilon^2\frac{d^2 f}{dx^2} + \cdots. \tag{1.71}$$

Equivalently,

$$f(x + \epsilon) = \left[1 + \epsilon\frac{d}{dx} + \frac{1}{2!}\left(\epsilon\frac{d}{dx}\right)^2 + \cdots\right] f(x) = \exp\left(\epsilon\frac{d}{dx}\right) f(x). \tag{1.72}$$

This generalizes for a function of three variables to

$$f(x + \epsilon_x, y + \epsilon_y, z + \epsilon_z) = \exp\left(\epsilon_x\frac{\partial}{\partial x}\right)\exp\left(\epsilon_y\frac{\partial}{\partial y}\right)\exp\left(\epsilon_z\frac{\partial}{\partial z}\right) f(x, y, z), \tag{1.73}$$

or

$$f(\mathbf{r} + \boldsymbol{\epsilon}) = \exp(\boldsymbol{\epsilon}\cdot\nabla)\, f(\mathbf{r}) = \left[1 + \boldsymbol{\epsilon}\cdot\nabla + \frac{1}{2!}(\boldsymbol{\epsilon}\cdot\nabla)^2 + \cdots\right] f(\mathbf{r}). \tag{1.74}$$

1.4 Integrals

1.4.1 Jacobian Determinant

The determinant of the Jacobian matrix $\mathbf{J}$ relates volume elements when changing variables in an integral. For example, suppose $\mathbf{x}$ and $\mathbf{y}$ are N-dimensional space vectors in two different coordinate systems, e.g., Cartesian and spherical. The volume elements $d^N x$ and $d^N y$ are related by

$$d^N x = |\mathbf{J}(\mathbf{x}, \mathbf{y})|\, d^N y = \begin{vmatrix} \dfrac{\partial x_1}{\partial y_1} & \dfrac{\partial x_1}{\partial y_2} & \cdots & \dfrac{\partial x_1}{\partial y_N} \\ \vdots & \vdots & \vdots & \vdots \\ \dfrac{\partial x_N}{\partial y_1} & \dfrac{\partial x_N}{\partial y_2} & \cdots & \dfrac{\partial x_N}{\partial y_N} \end{vmatrix} d^N y. \tag{1.75}$$

1.4.2 The Divergence Theorem

Let $\mathbf{F}(\mathbf{r})$ be a vector function defined in a volume V enclosed by a surface S with an outward normal $\hat{\mathbf{n}}$. If $d\mathbf{S} = dS\hat{\mathbf{n}}$, the divergence theorem is

$$\int_V d^3r\, \nabla \cdot \mathbf{F} = \int_S d\mathbf{S} \cdot \mathbf{F}. \tag{1.76}$$

Special choices for the vector function $\mathbf{F}(\mathbf{r})$ produce various integral identities based on (1.76). For example, if $\mathbf{c}$ is an arbitrary constant vector, the reader can confirm that the choices $\mathbf{F}(\mathbf{r}) = \mathbf{c}\psi(\mathbf{r})$ and $\mathbf{F}(\mathbf{r}) = \mathbf{A}(\mathbf{r}) \times \mathbf{c}$ substituted into (1.76) respectively yield

$$\int_V d^3r\, \nabla\psi = \int_S d\mathbf{S}\psi \tag{1.77}$$

$$\int_V d^3r\, \nabla \times \mathbf{A} = \int_S d\mathbf{S} \times \mathbf{A}. \tag{1.78}$$

1.4.3 Green's Identities

The choice $\mathbf{F}(\mathbf{r}) = \phi(\mathbf{r})\nabla\psi(\mathbf{r})$ in (1.76) leads to *Green's first identity*,

$$\int_V d^3r\, [\phi\nabla^2\psi + \nabla\phi \cdot \nabla\psi] = \int_S d\mathbf{S} \cdot \phi\nabla\psi. \tag{1.79}$$

Writing (1.79) with the roles of ϕ and ψ exchanged and subtracting that equation from (1.79) itself gives *Green's second identity*,

$$\int_V d^3r\, [\phi\nabla^2\psi - \psi\nabla^2\phi] = \int_S d\mathbf{S} \cdot [\phi\nabla\psi - \psi\nabla\phi]. \tag{1.80}$$

The choice $\mathbf{F} = \mathbf{P} \times \nabla \times \mathbf{Q}$ in (1.76) and the identity $\nabla \cdot (\mathbf{A} \times \mathbf{B}) = \mathbf{B} \cdot \nabla \times \mathbf{A} - \mathbf{A} \cdot \nabla \times \mathbf{B}$ produces a vector analog of Green's first identity:

$$\int_V d^3r\, [\nabla \times \mathbf{P} \cdot \nabla \times \mathbf{Q} - \mathbf{P} \cdot \nabla \times \nabla \times \mathbf{Q}] = \int_S d\mathbf{S} \cdot (\mathbf{P} \times \nabla \times \mathbf{Q}). \tag{1.81}$$

Writing (1.81) with **P** and **Q** interchanged and subtracting that equation from (1.81) gives a vector analog of Green's second identity:

$$\int_V d^3r\,[\mathbf{Q}\cdot\nabla\times\nabla\times\mathbf{P}-\mathbf{P}\cdot\nabla\times\nabla\times\mathbf{Q}]=\int_S d\mathbf{S}\cdot[\mathbf{P}\times\nabla\times\mathbf{Q}-\mathbf{Q}\times\nabla\times\mathbf{P}]. \quad (1.82)$$

1.4.4 Stokes' Theorem

Stokes' theorem applies to a vector function $\mathbf{F}(\mathbf{r})$ defined on an open surface S bounded by a closed curve C. If $d\boldsymbol{\ell}$ is a line element of C,

$$\int_S d\mathbf{S}\cdot\nabla\times\mathbf{F}=\oint_C d\boldsymbol{\ell}\cdot\mathbf{F}. \quad (1.83)$$

The curve C in (1.83) is traversed in the direction given by the right-hand rule when the thumb points in the direction of $d\mathbf{S}$. As with the divergence theorem, variations of (1.83) follow from the choices $\mathbf{F}=\mathbf{c}\psi$ and $\mathbf{F}=\mathbf{A}\times\mathbf{c}$:

$$\int_S d\mathbf{S}\times\nabla\psi=\oint_C d\boldsymbol{\ell}\,\psi \quad (1.84)$$

$$\oint_C d\boldsymbol{\ell}\times\mathbf{A}=\int_S dS_k\nabla A_k-\int_S d\mathbf{S}(\nabla\cdot\mathbf{A}). \quad (1.85)$$

1.4.5 The Time Derivative of a Flux Integral

Leibniz' Rule for the time derivative of a one-dimensional integral is

$$\frac{d}{dt}\int_{x_1(t)}^{x_2(t)} dx\,b(x,t)=b(x_2,t)\frac{dx_2}{dt}-b(x_1,t)\frac{dx_1}{dt}+\int_{x_1(t)}^{x_2(t)} dx\,\frac{\partial b}{\partial t}. \quad (1.86)$$

This formula generalizes to integrals over circuits, surfaces, and volumes which move through space. Our treatment of Faraday's law makes use of the time derivative of a surface integral where the surface $S(t)$ moves because its individual area elements move with velocity $\boldsymbol{\upsilon}(\mathbf{r},t)$. In that case,

$$\frac{d}{dt}\int_{S(t)} d\mathbf{S}\cdot\mathbf{B}=\int_{S(t)} d\mathbf{S}\cdot\left[\boldsymbol{\upsilon}(\nabla\cdot\mathbf{B})-\nabla\times(\boldsymbol{\upsilon}\times\mathbf{B})+\frac{\partial\mathbf{B}}{\partial t}\right]. \quad (1.87)$$

Proof: We calculate the change in flux from

$$\delta\left[\int\mathbf{B}\cdot d\mathbf{S}\right]=\int\delta\mathbf{B}\cdot d\mathbf{S}+\int\mathbf{B}\cdot\delta(\hat{\mathbf{n}}dS). \quad (1.88)$$

The first term on the right comes from time variations of **B**. The second term comes from time variations of the surface. Multiplication of every term in (1.88) by $1/\delta t$ gives

$$\frac{d}{dt}\int\mathbf{B}\cdot d\mathbf{S}=\int\frac{\partial\mathbf{B}}{\partial t}\cdot d\mathbf{S}+\frac{1}{\delta t}\int\mathbf{B}\cdot\delta(\hat{\mathbf{n}}dS). \quad (1.89)$$

We can focus on the second term on the right-hand side of (1.89) because the first term appears already as the last term in (1.87). Figure 1.3 shows an open surface $S(t)$ with local normal $\hat{\mathbf{n}}(t)$ which moves and/or distorts to the surface $S(t+\delta t)$ with local normal $\hat{\mathbf{n}}(t+\delta t)$ in time δt.

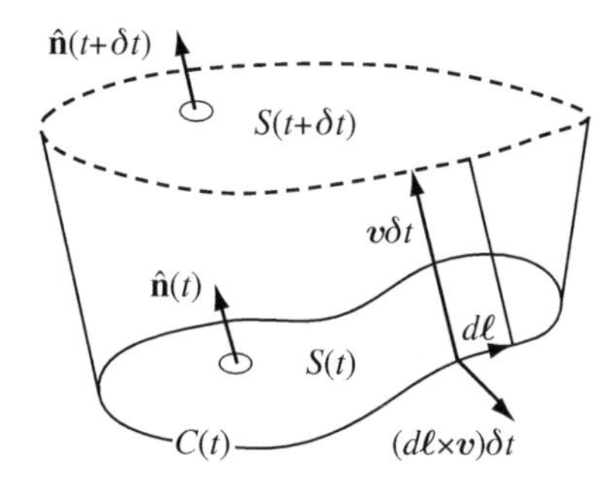

Figure 1.3: A surface $S(t)$ [bounded by the solid curve labeled $C(t)$] changes to the surface $S(t+\delta t)$ (dashed curve) because each element of surface moves by an amount $\boldsymbol{v}\delta t$.

Our strategy is to integrate $\nabla \cdot \mathbf{B}$ over the volume V bounded by $S(t)$, $S(t+\delta t)$, and the ribbon-like surface of infinitesimal width that connects the two. Figure 1.3 shows that an area element of the latter is $d\boldsymbol{\ell} \times \boldsymbol{v}\delta t$. Therefore, using the divergence theorem,

$$\int_V d^3r\, \nabla \cdot \mathbf{B} = \int_{S(t+\delta t)} \mathbf{B} \cdot \hat{\mathbf{n}} dS - \int_{S(t)} \mathbf{B} \cdot \hat{\mathbf{n}} dS + \oint_{C(t)} \mathbf{B} \cdot (d\boldsymbol{\ell} \times \boldsymbol{v}\delta t). \tag{1.90}$$

The minus sign appears in (1.90) because the divergence theorem involves the *outward* normal to the surface bounded by V.

The volume integral on the left side of (1.90) can be rewritten as a surface integral over $S(t)$ because $d^3r = \hat{\mathbf{n}}\, dS \cdot \boldsymbol{v}\delta t$. In the circuit integral over $C(t)$, we use the fact that $\mathbf{B} \cdot (d\boldsymbol{\ell} \times \boldsymbol{v}) = d\boldsymbol{\ell} \cdot (\boldsymbol{v} \times \mathbf{B})$. Finally, the two surface integrals on the right side of (1.90) can be combined into one. Putting all this together transforms (1.90) to

$$\delta t \int_{S(t)} d\mathbf{S} \cdot \boldsymbol{v}(\nabla \cdot \mathbf{B}) = \int_{S(t)} \mathbf{B} \cdot \delta(\hat{\mathbf{n}} dS) + \delta t \oint_{C(t)} d\boldsymbol{\ell} \cdot (\boldsymbol{v} \times \mathbf{B}). \tag{1.91}$$

Finally, we use Stokes' theorem to write the circuit integral in (1.91) as a surface integral. This gives

$$\int_{S(t)} \mathbf{B} \cdot \delta(\hat{\mathbf{n}} dS) = \delta t \int_{S(t)} d\mathbf{S} \cdot \boldsymbol{v}(\nabla \cdot \mathbf{B}) - \delta t \int_{S(t)} d\mathbf{S} \cdot \nabla \times (\boldsymbol{v} \times \mathbf{B}). \tag{1.92}$$

Substitution of (1.92) into (1.89) produces (1.87).

1.5 Generalized Functions

1.5.1 The Delta Function in One Dimension

The one-dimensional generalized function $\delta(x)$ is defined by its "filtering" action on a smooth but otherwise arbitrary test function $f(x)$:

$$\int_{-\infty}^{\infty} dx f(x)\delta(x - x') = f(x'). \tag{1.93}$$

An informal definition consistent with (1.93) is

$$\delta(x) = 0 \quad \text{for} \quad x \neq 0 \qquad \text{but} \qquad \int_{-\infty}^{\infty} dx\, \delta(x) = 1. \tag{1.94}$$

If the variable x has dimensions of length, the integrals in these equations make sense only if $\delta(x)$ has dimensions of inverse length. Note also that the integration ranges in (1.93) and (1.94) need only be large enough to include the point where the argument of the delta function vanishes.

The delta function can be understood as the limit of a sequence of functions which become more and more highly peaked at the point where its argument vanishes. Some examples are

$$\delta(x) = \lim_{m\to\infty} \frac{\sin mx}{\pi x} \tag{1.95}$$

$$\delta(x) = \lim_{m\to\infty} \frac{m}{\sqrt{\pi}} \exp(-m^2x^2) \tag{1.96}$$

$$\delta(x) = \lim_{\epsilon\to 0} \frac{\epsilon/\pi}{x^2+\epsilon^2}. \tag{1.97}$$

We prove the correctness of any of these proposed representations by showing that it possesses the filtering property (1.93). The same method is used to prove delta function identities like

$$\delta(ax) = \frac{1}{|a|}\delta(x), \qquad a \neq 0 \tag{1.98}$$

$$\int_{-\infty}^{\infty} dx f(x) \frac{d}{dx}\delta(x-x') = -\left.\frac{df}{dx}\right|_{x=x'} \tag{1.99}$$

$$\delta[g(x)] = \sum_n \frac{1}{|g'(x_n)|}\delta(x-x_n) \qquad \text{where} \qquad g(x_n)=0, \;\; g'(x_n)\neq 0 \tag{1.100}$$

$$\delta(x-x') = \frac{1}{2\pi}\int_{-\infty}^{\infty} dk\, e^{ik(x-x')}. \tag{1.101}$$

Formula (1.101) may be read as a statement of the completeness of plane waves labeled with the continuous index k:

$$\psi_k(x) = \frac{1}{\sqrt{2\pi}} e^{-ikx}. \tag{1.102}$$

The general result for a complete set of normalized basis functions $\psi_n(x)$ labeled with the discrete index n is[2]

$$\delta(x-x') = \sum_{n=1}^{\infty} \psi_n^*(x)\psi_n(x'). \tag{1.103}$$

Example 1.4 Prove the identity (1.101) in the form

$$\delta(x) = \frac{1}{2\pi}\int_{-\infty}^{\infty} dk\, e^{ikx}.$$

[2] $\psi_n^*(x)$ is the complex conjugate of $\psi_n(x)$.

Solution: By direct calculation,

$$\int_{-\infty}^{\infty} dk\, e^{ikx} = \int_0^{\infty} dk\, e^{ikx} + \int_0^{\infty} dk\, e^{-ikx} = \lim_{\epsilon\to 0}\left[\int_0^{\infty} dk\, e^{ik(x+i\epsilon)} + \int_0^{\infty} dk\, e^{-ik(x-i\epsilon)}\right].$$

The convergence factors make the integrands zero at the upper limit, so

$$\int_{-\infty}^{\infty} dk\, e^{ikx} = \lim_{\epsilon\to 0}\left[\frac{i}{x+i\epsilon} - \frac{i}{x-i\epsilon}\right] = \lim_{\epsilon\to 0}\frac{2\epsilon}{x^2+\epsilon^2} = 2\pi\delta(x).$$

The final equality on the far right follows from (1.97).

1.5.2 The Principal Value Integral and the Plemelj Formula

The Cauchy principal value is a generalized function defined by its action under an integral with an arbitrary function $f(x)$, namely,

$$\mathcal{P}\int_{-\infty}^{\infty} dx\frac{f(x)}{x-x_0} = \lim_{\epsilon\to 0}\left[\int_{-\infty}^{x_0-\epsilon} dx\frac{f(x)}{x-x_0} + \int_{x_0+\epsilon}^{\infty} dx\frac{f(x)}{x-x_0}\right]. \tag{1.104}$$

An important application where the principal value plays a role is the *Plemelj formula*:

$$\lim_{\epsilon\to 0}\frac{1}{x-x_0\pm i\epsilon} = \mathcal{P}\frac{1}{x-x_0} \mp i\pi\delta(x-x_0). \tag{1.105}$$

This expression is symbolic in the sense that it gains meaning when we multiply every term by an arbitrary function $f(x)$ and integrate over x from $-\infty$ to ∞.

The correctness of (1.105) can be appreciated from Figure 1.4 and the identity

$$\frac{1}{x-x_0\pm i\epsilon} = \frac{x-x_0}{(x-x_0)^2+\epsilon^2} \mp i\frac{\epsilon}{(x-x_0)^2+\epsilon^2}. \tag{1.106}$$

The real part of (1.106) generates the principal value in (1.105) because it is a symmetrically cut-off version of $1/(x-x_0)$. The imaginary part of (1.106) generates the delta function in (1.105) by virtue of (1.97).

1.5.3 The Step Function and Sign Function

The Heaviside step function $\Theta(x)$ is defined by

$$\Theta(x) = \begin{cases} 0 & x < 0, \\ 1 & x > 0. \end{cases} \tag{1.107}$$

The delta function is the derivative of the theta function,

$$\frac{d\Theta(x)}{dx} = \delta(x). \tag{1.108}$$

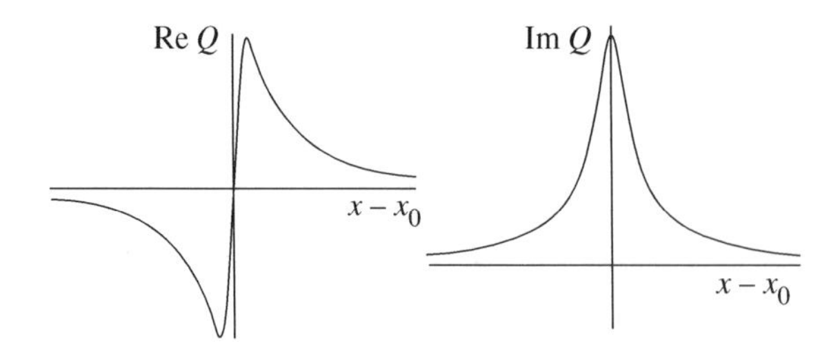

Figure 1.4: The real and imaginary parts of $Q(x) = 1/(x - x_0 - i\epsilon)$.

A useful representation is

$$\Theta(x) = \lim_{\epsilon \to 0} \frac{i}{2\pi} \int_{-\infty}^{\infty} dk \frac{1}{k + i\epsilon} e^{-ikx}. \tag{1.109}$$

The sign function sgn(x) is defined by

$$\text{sgn}(x) = \frac{d}{dx}|x| = \begin{cases} -1 & x < 0, \\ 1 & x > 0. \end{cases} \tag{1.110}$$

A convenient representation is

$$\text{sgn}(x) = -1 + 2 \int_{-\infty}^{x} dy\, \delta(y). \tag{1.111}$$

1.5.4 The Delta Function in Three Dimensions

The definition (1.93) leads us to define a three-dimensional delta function using an integral over a volume V and a smooth but otherwise arbitrary "test" function $f(\mathbf{r})$:

$$\int_V d^3r\, f(\mathbf{r})\delta(\mathbf{r} - \mathbf{r}') = \begin{cases} f(\mathbf{r}') & \mathbf{r}' \in V, \\ 0 & \mathbf{r}' \notin V. \end{cases} \tag{1.112}$$

A less formal definition consistent with (1.112) is

$$\delta(\mathbf{r}) = 0 \quad \text{for} \quad \mathbf{r} \neq 0 \qquad \text{but} \qquad \int_V d^3r\, \delta(\mathbf{r}) = \begin{cases} 1 & \mathbf{r} = 0 \in V, \\ 0 & \mathbf{r} = 0 \notin V. \end{cases} \tag{1.113}$$

These definitions tell us that $\delta(\mathbf{r})$ has dimensions of inverse volume. In Cartesian coordinates,

$$\delta(\mathbf{r}) = \delta(x)\delta(y)\delta(z). \tag{1.114}$$

In curvilinear coordinates, the constraint on the right side of (1.113) and the form of the volume elements for cylindrical and spherical coordinates imply that

$$\delta(\mathbf{r} - \mathbf{r}') = \frac{\delta(\rho - \rho')\delta(\phi - \phi')\delta(z - z')}{\rho} = \frac{\delta(r - r')\delta(\theta - \theta')\delta(\phi - \phi')}{r^2 \sin\theta}. \tag{1.115}$$

The special case $\mathbf{r}' = 0$ requires that we *define* the one-dimensional radial delta function so

$$\int_0^{\infty} dr\, \delta(r) = 1. \tag{1.116}$$

More generally, if $\mathbf{x}_0$ and $\mathbf{y}_0$ represent the same point in two different N-dimensional coordinate systems, we can use

$$\int d^N x\, \delta(\mathbf{x}-\mathbf{x}_0) = \int d^N y\, \delta(\mathbf{y}-\mathbf{y}_0) \tag{1.117}$$

and the Jacobian determinant result (1.75) to deduce that

$$\delta(\mathbf{x}-\mathbf{x}_0) = \frac{1}{|\mathbf{J}(\mathbf{x},\mathbf{y})|}\delta(\mathbf{y}-\mathbf{y}_0). \tag{1.118}$$

1.5.5 Some Useful Delta Function Identities

$$\delta(\mathbf{r}-\mathbf{r}') = \frac{1}{(2\pi)^3}\int d^3k\, e^{i\mathbf{k}\cdot(\mathbf{r}-\mathbf{r}')} \tag{1.119}$$

$$\int d^3r\, f(\mathbf{r})\delta[g(\mathbf{r})] = \int_S dS \frac{f(\mathbf{r}_S)}{|\nabla g(\mathbf{r}_S)|} \qquad \text{where } g(\mathbf{r}_S)=0 \text{ defines } S \tag{1.120}$$

$$\nabla^2 \frac{1}{|\mathbf{r}-\mathbf{r}'|} = -4\pi\delta(\mathbf{r}-\mathbf{r}') \tag{1.121}$$

$$\frac{\partial}{\partial r_k}\frac{\partial}{\partial r_m}\frac{1}{r} = \frac{3r_k r_m - r^2\delta_{km}}{r^5} - \frac{4\pi}{3}\delta_{km}\delta(\mathbf{r}). \tag{1.122}$$

Example 1.5 Use the divergence theorem to prove (1.121) in the form

$$\nabla^2\frac{1}{r} = -4\pi\delta(\mathbf{r}).$$

Solution: In spherical coordinates,

$$\nabla^2\frac{1}{r} = \frac{1}{r^2}\frac{\partial}{\partial r}r^2\frac{\partial}{\partial r}\frac{1}{r} = 0 \quad \text{when} \quad r\neq 0.$$

To learn the behavior at $r=0$, we integrate $\nabla^2(1/r)$ over a tiny spherical volume V centered at the origin. Since $d\mathbf{S} = r^2\sin\theta d\theta d\phi\hat{\mathbf{r}}$ and $\nabla(1/r) = -\hat{\mathbf{r}}/r^2$, the divergence theorem gives

$$\int_V d^3r\,\nabla^2\frac{1}{r} = \int_V d^3r\,\nabla\cdot\nabla\frac{1}{r} = \int_S d\mathbf{S}\cdot\left(-\frac{\hat{\mathbf{r}}}{r^2}\right) = -\int_0^{2\pi} d\phi\int_0^{\pi} d\theta\,\sin\theta = -4\pi.$$

In light of (1.113), these two facts taken together establish the identity.

1.6 Fourier Analysis

Every periodic function $f(x+L) = f(x)$ has a Fourier series representation

$$f(x) = \sum_{m=-\infty}^{\infty} \hat{f}_m e^{i2\pi mx/L}. \tag{1.123}$$

The Fourier expansion coefficients in (1.123) are given by

$$\hat{f}_m = \frac{1}{L}\int_0^L dx f(x) e^{-i2\pi mx/L}. \tag{1.124}$$

For non-periodic functions, the sum over integers in (1.123) becomes an integral over the real line. When the integral converges, we find the *Fourier transform* pair:

$$f(x) = \frac{1}{2\pi}\int_{-\infty}^{\infty} dk\, \hat{f}(k)\, e^{ikx} \tag{1.125}$$

$$\hat{f}(k) = \int_{-\infty}^{\infty} dx\, f(x)\, e^{-ikx}. \tag{1.126}$$

If $f(x)$ happens to be a real function, it follows from these definitions that

$$f(x) = f^*(x) \Rightarrow \hat{f}(k) = \hat{f}(-k) \tag{1.127}$$

In the time domain, it is conventional to write

$$g(t) = \frac{1}{2\pi}\int_{-\infty}^{\infty} d\omega\, \hat{g}(\omega)\, e^{-i\omega t} \qquad \hat{g}(\omega) = \int_{-\infty}^{\infty} dt\, g(t)\, e^{i\omega t}. \tag{1.128}$$

Thus, our convention for the Fourier transform and inverse Fourier transform of a function $f(\mathbf{r}, t)$ of time and all three spatial variables is

$$f(\mathbf{r}, t) = \frac{1}{(2\pi)^4}\int d^3k \int_{-\infty}^{\infty} d\omega\, \hat{f}(\mathbf{k}|\omega)\, e^{i(\mathbf{k}\cdot\mathbf{r}-\omega t)} \tag{1.129}$$

$$\hat{f}(\mathbf{k}|\omega) = \int d^3r \int_{-\infty}^{\infty} dt\, f(\mathbf{r}, t)\, e^{-i(\mathbf{k}\cdot\mathbf{r}-\omega t)}. \tag{1.130}$$

1.6.1 Parseval's Theorem

$$\int_{-\infty}^{\infty} dt g_1^*(t)\, g_2(t) = \frac{1}{2\pi}\int_{-\infty}^{\infty} d\omega\, \hat{g}_1^*(\omega)\, \hat{g}_2(\omega). \tag{1.131}$$

The proof follows by substituting the left member of (1.128) into (1.131) for $g_1^*(t)$ and $g_1(t)$ and using the representation (1.101) of the delta function.

1.6.2 The Convolution Theorem

A function $h(t)$ is called the *convolution* of $f(t)$ and $g(t)$ if

$$h(t) = \int_{-\infty}^{\infty} dt'\, f(t-t')\, g(t'). \tag{1.132}$$

The convolution theorem states that the Fourier transforms $\hat{h}(\omega)$, $\hat{f}(\omega)$, and $\hat{g}(\omega)$ are related by

$$\hat{h}(\omega) = \hat{f}(\omega)\hat{g}(\omega). \tag{1.133}$$

We prove this assertion by using the left side of (1.128) to rewrite (1.132) as

$$h(t) = \int_{-\infty}^{\infty} dt' \left[\frac{1}{2\pi} \int_{-\infty}^{\infty} d\omega \hat{f}(\omega) e^{-i\omega(t-t')} \right] \left[\frac{1}{2\pi} \int_{-\infty}^{\infty} d\omega' \hat{g}(\omega') e^{-i\omega' t'} \right]. \tag{1.134}$$

Rearranging terms gives

$$h(t) = \frac{1}{2\pi} \int_{-\infty}^{\infty} d\omega\, e^{-i\omega t} \hat{f}(\omega) \int_{-\infty}^{\infty} d\omega'\, \hat{g}(\omega') \left[\frac{1}{2\pi} \int_{-\infty}^{\infty} dt'\, e^{-i(\omega'-\omega)t'} \right]. \tag{1.135}$$

The identity (1.101) identifies the quantity in square brackets as the delta function $\delta(\omega - \omega')$. Therefore,

$$h(t) = \frac{1}{2\pi} \int_{-\infty}^{\infty} d\omega\, e^{-i\omega t} \hat{f}(\omega)\, \hat{g}(\omega). \tag{1.136}$$

Comparing (1.136) to the definition on the left side of (1.128) completes the proof. ■

1.6.3 A Time-Averaging Theorem

Let $A(\mathbf{r}, t) = a(\mathbf{r}) \exp(-i\omega t)$ and $B(\mathbf{r}, t) = b(\mathbf{r}) \exp(-i\omega t)$, where $a(\mathbf{r})$ and $b(\mathbf{r})$ are complex-valued functions. If $T = 2\pi/\omega$, it is useful to know that

$$\langle \mathrm{Re}\,[A(\mathbf{r}, t)]\, \mathrm{Re}\,[B(\mathbf{r}, t)] \rangle = \frac{1}{T} \int_0^T dt\, \mathrm{Re}\,[A(\mathbf{r}, t)]\, \mathrm{Re}\,[B(\mathbf{r}, t)] = \tfrac{1}{2} \mathrm{Re}\,[a(\mathbf{r}) b^*(\mathbf{r})]. \tag{1.137}$$

We prove (1.137) by writing $\mathrm{Re}\,[A] = \frac{1}{2}(A + A^*)$ and $\mathrm{Re}\,[B] = \frac{1}{2}(B + B^*)$ so

$$\langle \mathrm{Re}\,[A]\, \mathrm{Re}\,[B] \rangle = \frac{1}{4T} \int_0^T dt \left\{ ab e^{-2i\omega t} + a^* b^* e^{2i\omega t} + ab^* + ba^* \right\}. \tag{1.138}$$

The time-dependent terms in the integrand of (1.138) integrate to zero over one full period. Therefore,

$$\langle \mathrm{Re}\,[A]\, \mathrm{Re}\,[B] \rangle = \frac{1}{4} \left[ab^* + a^* b \right] = \frac{1}{2} \mathrm{Re} \left[ab^* \right] = \frac{1}{2} \mathrm{Re} \left[a^* b \right]. \tag{1.139}$$

Example 1.6 Find the Fourier series which represents the periodic function

$$f(x) = \sum_{p=-\infty}^{\infty} \delta(x - 2\pi p).$$

Solution: This function is periodic in the interval $-\pi \le x < \pi$. Therefore, using (1.124),

$$\hat{f}_m = \frac{1}{2\pi} \sum_{p=-\infty}^{\infty} \int_{-\pi}^{\pi} dx\, \delta(x - 2\pi p) e^{-imx} = \frac{1}{2\pi}.$$

Substituting $\hat{f}_m$ into (1.123) gives

$$f(x) = \frac{1}{2\pi} \sum_{m=-\infty}^{\infty} e^{imx} = \frac{1}{2\pi} + \frac{1}{\pi} \sum_{m=1}^{\infty} \cos mx.$$

1.7 Orthogonal Transformations

Let $(\hat{\mathbf{e}}_1, \hat{\mathbf{e}}_2, \hat{\mathbf{e}}_3)$ and $(\hat{\mathbf{e}}'_1, \hat{\mathbf{e}}'_2, \hat{\mathbf{e}}'_3)$ be two sets of orthogonal Cartesian unit vectors. Each is a complete basis for vectors in three dimensions, so

$$\hat{\mathbf{e}}'_i = A_{ij} \hat{\mathbf{e}}_j. \tag{1.140}$$

The set of scalars A_{ij} are called *direction cosines*. Using the unit vector properties from Section 1.2,

$$\delta_{ij} = \hat{\mathbf{e}}'_i \cdot \hat{\mathbf{e}}'_j = A_{ik} A_{jk}. \tag{1.141}$$

Equation (1.141) says that the transpose of the matrix $\mathbf{A}$, called $\mathbf{A}^{\mathrm{T}}$, is identical to the inverse of the matrix $\mathbf{A}$, called $\mathbf{A}^{-1}$. This is the definition of a matrix that describes an *orthogonal* transformation,

$$\mathbf{A}\mathbf{A}^{\mathrm{T}} = \mathbf{A}\mathbf{A}^{-1} = 1. \tag{1.142}$$

There are two classes of orthogonal coordinate transformations. These follow from the determinant of (1.142):

$$\det\,[\mathbf{A}\mathbf{A}^{\mathrm{T}}] = \det\mathbf{A}\,\det\mathbf{A}^{\mathrm{T}} = (\det\mathbf{A})^2 = 1. \tag{1.143}$$

A **rotation** has $\det \mathbf{A} = 1$. Figure 1.5(a) shows an example where

$$\mathbf{A} = \begin{bmatrix} \cos\theta & \sin\theta & 0 \\ -\sin\theta & \cos\theta & 0 \\ 0 & 0 & 1 \end{bmatrix}. \tag{1.144}$$

A **reflection** has $\det \mathbf{A} = -1$. Figure 1.5(b) shows an example where

$$\mathbf{A} = \begin{bmatrix} -1 & 0 & 0 \\ 0 & 1 & 0 \\ 0 & 0 & 1 \end{bmatrix}. \tag{1.145}$$

The **inversion** transformation is represented by $A_{ij} = -\delta_{ij}$ so $\det \mathbf{A} = -1$ like a reflection. However, a sequence of reflections can have $\det \mathbf{A} = 1$ like a rotation.

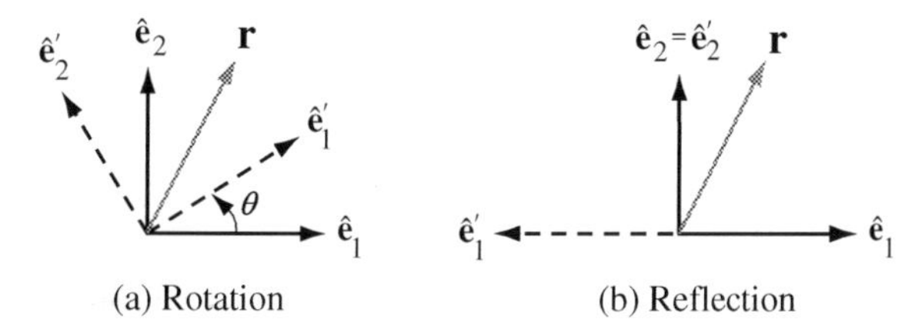

Figure 1.5: Orthogonal transformations from the *passive* point of view. The Cartesian coordinate system transforms. The position vector $\mathbf{r}$ is fixed. (a) a rotation where $\det \mathbf{A} = 1$; (b) a reflection where $\det \mathbf{A} = -1$.

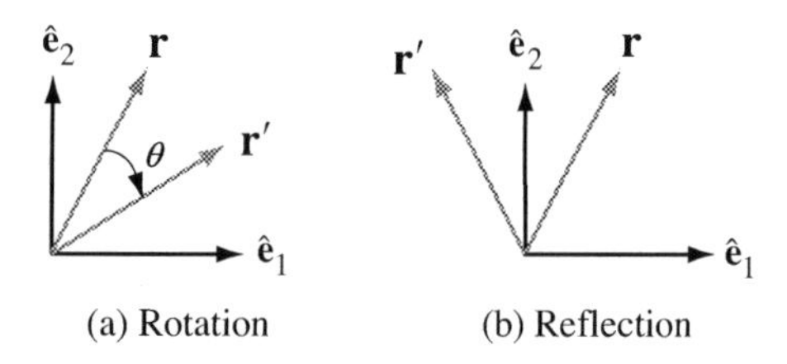

Figure 1.6: Orthogonal transformations from the *active* point of view. The vector $\mathbf{r}$ transforms to the vector $\mathbf{r}'$. The Cartesian coordinate system is fixed. (a) a rotation where $\det \mathbf{A} = 1$; (b) a reflection where $\det \mathbf{A} = -1$.

1.7.1 Passive Point of View

Consider the position vector $\mathbf{r}$ drawn in Figure 1.5. This object can be decomposed using either the $\{\hat{\mathbf{e}}\}$ basis or the $\{\hat{\mathbf{e}}'\}$ basis:

$$\mathbf{r} = r_i \hat{\mathbf{e}}_i = r_j' \hat{\mathbf{e}}_j' = (\mathbf{r})'. \tag{1.146}$$

The notation on the far right side of (1.146) indicates that the vector $\mathbf{r}$ is represented in the primed coordinate system. Substitution of (1.140) in (1.146) shows that the components of $\mathbf{r}$ transform like the unit vectors in (1.140):

$$r_i' = A_{ij} r_j \qquad\qquad r_j = A_{kj} r_k'. \tag{1.147}$$

This description is called the *passive* point of view. The position vector $\mathbf{r}$ is a spectator fixed in space while the coordinate system transforms. The matrix $\mathbf{A}$ is regarded as an operator that transforms the components of $\mathbf{r}$ in the $\{\hat{\mathbf{e}}\}$ basis to the components of $\mathbf{r}$ in the $\{\hat{\mathbf{e}}'\}$ basis. The matrix form of the transformation connects components of the *same* vector in *different* coordinate systems:

$$(\mathbf{r})' = \mathbf{A}\mathbf{r}. \tag{1.148}$$

1.7.2 Active Point of View

The *active* point of view is an alternative (and equivalent) way to think about an orthogonal transformation. Here, the matrix $\mathbf{A}$ is regarded as an operator which transforms $\mathbf{r}$ to a new vector $\mathbf{r}'$ with no change in the underlying coordinate system. The matrix form of the transformation connects the components of *different* vectors in the *same* coordinate system:

$$\mathbf{r}' = \mathbf{A}\mathbf{r}. \tag{1.149}$$

Direct calculation using (1.149) confirms the active point of view illustrated in Figure 1.6 for transformations represented by the rotation matrix (1.144) and the reflection matrix (1.145). Figure 1.6(a) shows that $\mathbf{r}'$ is rotated in the *clockwise* direction with respect to $\mathbf{r}$ as compared to the *counterclockwise* rotation of the set $\{\hat{\mathbf{e}}'\}$ with respect to the set $\{\hat{\mathbf{e}}\}$ in Figure 1.5. Figure 1.6(b) shows that $\mathbf{r}'$ is the image of $\mathbf{r}$ reflected in a mirror at $x = 0$.

1.8 Cartesian Tensors

Tensors are mathematical objects defined by their behavior under orthogonal coordinate transformations. Physical quantities are classified as *rotational tensors* of various ranks depending on how they transform under rotations. In this section, we adopt the passive point of view (Section 1.7.1) where the Cartesian coordinate system alone transforms.

A *tensor of rank 0* is a one-component quantity where the result of a rotational transformation from the passive point of view is

$$f'(\mathbf{r}') = f(\mathbf{r}). \tag{1.150}$$

An ordinary **scalar** is a tensor of rank 0. A *tensor of rank 1* is an object whose three components transform under rotation like the three components of $\mathbf{r}$ in (1.147):

$$V_i'(\mathbf{r}') = A_{ij} V_j(\mathbf{r}). \tag{1.151}$$

A geometrical **vector** is a tensor of rank 1 because (1.151) together with (1.140) and (1.141) guarantees that [*cf.* Equation (1.5)]

$$V_i \hat{\mathbf{e}}_i = V_j' \hat{\mathbf{e}}_j'. \tag{1.152}$$

Many authors use the transformation rule (1.151) to *define* the components of a vector. In light of (1.141), a vector is characterized by the preservation of its length under a change of coordinates:

$$V_i' V_i' = A_{ij} V_j A_{ik} V_k = V_k V_k. \tag{1.153}$$

A *tensor of rank 2* is a nine-component quantity whose components transform under rotation by the rule

$$T_{ij}'(\mathbf{r}') = A_{ik} A_{jm} T_{km}(\mathbf{r}). \tag{1.154}$$

A **dyadic** is a tensor of rank 2 composed of a linear combination of pairs of juxtaposed (not multiplied) vectors. The examples we will encounter in this book all have the form[3]

$$\mathbf{T} \equiv \hat{\mathbf{e}}_i T_{ij} \hat{\mathbf{e}}_j. \tag{1.155}$$

The structure of (1.155) implies that the scalar product (or vector product) of a dyadic with a vector that stands to its left (or to its right) is also a vector. For example, the *unit dyadic* $\mathbf{I}$ has $I_{ij} = \delta_{ij}$ so $\mathbf{I} = \hat{\mathbf{e}}_i \hat{\mathbf{e}}_i$. A brief calculation confirms that the scalar product of a test vector with $\mathbf{I}$ returns the test vector:

$$\mathbf{v} \cdot \mathbf{I} = \mathbf{I} \cdot \mathbf{v} = \mathbf{v}. \tag{1.156}$$

[3] We use boldface type to denote both vectors and dyadics. Context should be sufficient to avoid confusion.

Example 1.7 Show that the gradient operator (1.8) transforms like a vector under an orthogonal coordinate transformation.

Solution: The proof follows from the chain rule and the equation on the right side of (1.147). The latter shows that $A_{kj} = \partial r_j / \partial r'_k$. Therefore, if $\varphi(\mathbf{r})$ is a scalar function,

$$\left(\frac{\partial \varphi}{\partial r_k}\right)' = \frac{\partial \varphi}{\partial r'_k} = \frac{\partial \varphi}{\partial r_j}\frac{\partial r_j}{\partial r'_k} = A_{kj}\frac{\partial \varphi}{\partial r_j}.$$

This is the transformation rule (1.151) for a vector.

1.8.1 Reflection, Inversion and Pseudotensors

We study here the transformation properties of rotational tensors (Section 1.8) under a general orthogonal transformation $\mathbf{A}$. An instructive example is the cross product $\mathbf{m} = \mathbf{p} \times \mathbf{w}$. In component form,

$$m_i = \epsilon_{ijk} p_j w_k. \tag{1.157}$$

By direct calculation using (1.151) and the definition (1.33) of the Kronecker delta,

$$m'_i = \epsilon_{ijk} p'_j w'_k = \epsilon_{ijk} A_{js} p_s A_{k\ell} w_\ell = \epsilon_{pjk} \delta_{ip} A_{js} A_{k\ell} p_s w_\ell. \tag{1.158}$$

Using (1.141) to eliminate δ_{ip} in (1.158) gives

$$m'_i = \epsilon_{pjk} A_{iq} A_{pq} A_{js} A_{k\ell} p_s w_\ell = (\epsilon_{pjk} A_{pq} A_{js} A_{k\ell}) A_{iq} p_s w_\ell. \tag{1.159}$$

In light of (1.43), (1.159) is exactly

$$m'_i = (\det \mathbf{A})\, A_{iq} \epsilon_{qs\ell} p_s w_\ell. \tag{1.160}$$

Therefore, using (1.157),

$$m'_i = (\det \mathbf{A}) A_{iq} m_q. \tag{1.161}$$

Equation (1.161) shows that $\mathbf{m}$ transforms like an ordinary vector under rotations when $\det \mathbf{A} = 1$. An extra minus sign occurs in (1.161) when $\mathbf{A}$ corresponds to a reflection or an inversion where $\det \mathbf{A} = -1$.

More generally, any rotational vector where an explicit determinant factor appears in its transformation rule is called a "pseudovector" or an *axial vector* to distinguish it from an "ordinary" or *polar vector* where no such determinant factor appears in the transformation rule. The nature of the vector produced by the cross product of two other vectors is summarized by

$$\text{axial vector} \times \text{polar vector} = \text{polar vector} \tag{1.162}$$

$$\text{polar vector} \times \text{polar vector} = \text{axial vector} \tag{1.163}$$

$$\text{axial vector} \times \text{axial vector} = \text{axial vector}. \tag{1.164}$$

Application 1.2 Inversion and Reflection

The position vector $\mathbf{r}$ is an ordinary polar vector because the transformation law (1.147) does not include the determinant factor in (1.161). Therefore, if the orthogonal transformation $\mathbf{A}$ corresponds to inversion, the active point of view (Section 1.7.2) tells us that the effect of inversion is

$$\mathbf{r} \to \mathbf{r}' = -\mathbf{r}. \tag{1.165}$$

By definition, any other polar vector $\mathbf{P}$ behaves the same way under inversion:

$$\mathbf{P} \to \mathbf{P}' = -\mathbf{P} \qquad \text{(inversion of a polar vector).} \tag{1.166}$$

Also by definition, an axial vector $\mathbf{Q}$ feels the effect of the determinant in (1.161). This introduces a minus sign for the case of inversion (see Section 1.7), so

$$\mathbf{Q} \to \mathbf{Q}' = \mathbf{Q} \qquad \text{(inversion of an axial vector).} \tag{1.167}$$

This idea generalizes immediately to the case of a *pseudoscalar* or a *pseudotensor* of any rank. For example, if $\mathbf{a}$, $\mathbf{b}$, and $\mathbf{c}$ are polar vectors, the triple product $\mathbf{a} \cdot (\mathbf{b} \times \mathbf{c})$ is a pseudoscalar. The Levi-Cività symbol (1.34) is a third-rank pseudotensor.

The operation of mirror reflection through the x-y plane inverts the z-component of the position vector so

$$(x, y, z) \to (x', y', z') = (x, y, -z). \tag{1.168}$$

A polar vector $\mathbf{P}$ behaves the same way under inversion:

$$(P_x, P_y, P_z) \to (P'_x, P'_y, P'_z) = (P_x, P_y, -P_z) \qquad \text{(reflection of a polar vector).} \tag{1.169}$$

The transformation matrix for reflection satisfies $\det \mathbf{A} = -1$ (see Section 1.7). Therefore, (1.161) dictates that the transformation law for an axial vector $\mathbf{Q}$ includes an overall factor of -1 (compared to a polar vector). In other words,

$$(Q_x, Q_y, Q_z) \to (Q'_x, Q'_y, Q'_z) = (-Q_x, -Q_y, Q_z) \qquad \text{(reflection of an axial vector).} \tag{1.170}$$

■

Example 1.8 Show that the magnetic field $\mathbf{B}$ is a pseudovector.

Solution: By definition, the position vector $\mathbf{r}$ is a polar vector. Therefore, so are the velocity $\mathbf{v} = d\mathbf{r}/dt$ and the current density $\mathbf{j} = \rho\mathbf{v}$. The gradient operator $\nabla = \partial/\partial\mathbf{r}$ transforms like $\mathbf{r}$ under inversion or reflection, so it is also a polar vector. Finally, Ampère's law tells us that $\nabla \times \mathbf{B} = \mu_0 \mathbf{j}$. Since ∇ and $\mathbf{j}$ are polar vectors, we conclude from (1.162) that $\mathbf{B}$ is an axial vector.

1.9 The Helmholtz Theorem

Statement:

An arbitrary vector field $\mathbf{C}(\mathbf{r})$ can always be decomposed into the sum of two vector fields: one with zero divergence and one with zero curl. Specifically,

$$\mathbf{C} = \mathbf{C}_\perp + \mathbf{C}_\parallel, \tag{1.171}$$

where

$$\nabla \cdot \mathbf{C}_\perp = 0 \qquad \text{and} \qquad \nabla \times \mathbf{C}_\parallel = 0. \tag{1.172}$$

An explicit representation of special interest is

$$\mathbf{C}(\mathbf{r}) = \nabla \times \mathbf{F}(\mathbf{r}) - \nabla\Omega(\mathbf{r}). \tag{1.173}$$

When the following integrals over all space converge, $\Omega(\mathbf{r})$ and $\mathbf{F}(\mathbf{r})$ are uniquely given by

$$\Omega(\mathbf{r}) = \frac{1}{4\pi} \int d^3r' \, \frac{\nabla' \cdot \mathbf{C}(\mathbf{r}')}{|\mathbf{r} - \mathbf{r}'|} \tag{1.174}$$

and

$$\mathbf{F}(\mathbf{r}) = \frac{1}{4\pi}\int d^3r' \frac{\nabla' \times \mathbf{C}(\mathbf{r}')}{|\mathbf{r}-\mathbf{r}'|}. \tag{1.175}$$

This result is valid for both static and time-dependent vector fields.

Existence:
The delta function properties (1.112) and (1.121) imply that

$$\mathbf{C}(\mathbf{r}) = \int d^3r' \mathbf{C}(\mathbf{r}')\,\delta(\mathbf{r}-\mathbf{r}') = -\frac{1}{4\pi}\int d^3r'\,\mathbf{C}(\mathbf{r}')\,\nabla^2 \frac{1}{|\mathbf{r}-\mathbf{r}'|}. \tag{1.176}$$

Exchanging $\mathbf{C}(\mathbf{r}')$ and ∇^2 in the last term and using the double-curl identity in Example 1.3 gives

$$\mathbf{C}(\mathbf{r}) = -\frac{1}{4\pi}\nabla\int d^3r'\,\nabla\cdot\left[\frac{\mathbf{C}(\mathbf{r}')}{|\mathbf{r}-\mathbf{r}'|}\right] + \frac{1}{4\pi}\nabla\times\int d^3r'\,\nabla\times\left[\frac{\mathbf{C}(\mathbf{r}')}{|\mathbf{r}-\mathbf{r}'|}\right]. \tag{1.177}$$

Now, since $\nabla f(|\mathbf{r}-\mathbf{r}'|) = -\nabla' f(|\mathbf{r}-\mathbf{r}'|)$, we deduce that

$$\nabla'\cdot\left[\frac{\mathbf{C}(\mathbf{r}')}{|\mathbf{r}-\mathbf{r}'|}\right] = \frac{\nabla'\cdot\mathbf{C}(\mathbf{r}')}{|\mathbf{r}-\mathbf{r}'|} - \mathbf{C}(\mathbf{r}')\cdot\nabla\frac{1}{|\mathbf{r}-\mathbf{r}'|}. \tag{1.178}$$

Moving $\mathbf{C}(\mathbf{r}')$ to the right of ∇ in the last term and rearranging gives

$$\nabla\cdot\left[\frac{\mathbf{C}(\mathbf{r}')}{|\mathbf{r}-\mathbf{r}'|}\right] = \frac{\nabla'\cdot\mathbf{C}(\mathbf{r}')}{|\mathbf{r}-\mathbf{r}'|} - \nabla'\cdot\frac{\mathbf{C}(\mathbf{r}')}{|\mathbf{r}-\mathbf{r}'|}. \tag{1.179}$$

In exactly the same way,

$$\nabla\times\left[\frac{\mathbf{C}(\mathbf{r}')}{|\mathbf{r}-\mathbf{r}'|}\right] = \frac{\nabla'\times\mathbf{C}(\mathbf{r}')}{|\mathbf{r}-\mathbf{r}'|} - \nabla'\times\frac{\mathbf{C}(\mathbf{r}')}{|\mathbf{r}-\mathbf{r}'|}. \tag{1.180}$$

Inserting (1.179) and (1.180) into (1.177) generates four terms:

$$\begin{aligned}\mathbf{C}(\mathbf{r}) = &-\frac{1}{4\pi}\nabla\int d^3r'\,\frac{\nabla'\cdot\mathbf{C}(\mathbf{r}')}{|\mathbf{r}-\mathbf{r}'|} + \frac{1}{4\pi}\nabla\times\int d^3r'\,\frac{\nabla'\times\mathbf{C}(\mathbf{r}')}{|\mathbf{r}-\mathbf{r}'|}\\ &+\frac{1}{4\pi}\nabla\int d^3r'\,\nabla'\cdot\left[\frac{\mathbf{C}(\mathbf{r}')}{|\mathbf{r}-\mathbf{r}'|}\right] - \frac{1}{4\pi}\nabla\times\int d^3r'\,\nabla'\times\left[\frac{\mathbf{C}(\mathbf{r}')}{|\mathbf{r}-\mathbf{r}'|}\right].\end{aligned} \tag{1.181}$$

The last two integrals in (1.181) are *zero* when the first two integrals converge. This follows from (1.76) and (1.78), which show that the divergence theorem transforms the volume integrals in the last two terms in (1.181) into the surface integrals

$$\int d\mathbf{S}'\cdot\frac{\mathbf{C}(\mathbf{r}')}{|\mathbf{r}-\mathbf{r}'|} \qquad \text{and} \qquad \int d\mathbf{S}'\times\frac{\mathbf{C}(\mathbf{r}')}{|\mathbf{r}-\mathbf{r}'|}. \tag{1.182}$$

The surface of integration for both integrals in (1.182) lies at infinity. Therefore, both integrals vanish if $\mathbf{C}(\mathbf{r})$ goes to zero faster than $1/r$ as $r\to\infty$. The same condition guarantees that the integrals in the first two terms in (1.181) converge. The final result is exactly the representation of $\mathbf{C}(\mathbf{r})$ given in the statement of the theorem.

Uniqueness:
Suppose that $\nabla\cdot\mathbf{C}_1 = \nabla\cdot\mathbf{C}_2$ and $\nabla\times\mathbf{C}_1 = \nabla\times\mathbf{C}_2$. Then, if $\mathbf{W} = \mathbf{C}_1 - \mathbf{C}_2$, we have $\nabla\cdot\mathbf{W} = 0$, $\nabla\times\mathbf{W} = 0$. The double-curl identity in Example 1.3 tells us that $\nabla^2\mathbf{W} = 0$ also. With this information, Green's first identity (1.79) with $\phi = \psi = W$ (W is any Cartesian component of $\mathbf{W}$) takes the form

$$\int_V d^3r\,|\nabla W|^2 = \int_S d\mathbf{S}\cdot W\nabla W. \tag{1.183}$$

The surface integral on the right side of (1.183) goes to zero when $\mathbf{C}(\mathbf{r})$ behaves at infinity as indicated above. Therefore, $\nabla W = 0$ or $W = const$. But $W \to 0$ at infinity so $W = 0$ or $\mathbf{C}_1 = \mathbf{C}_2$ as required.

1.10 Lagrange Multipliers

Suppose we wish to minimize (or maximize) a function of two variables $f(x, y)$. The rules of calculus tell us to set the total differential equal to zero:

$$df = \frac{\partial f}{\partial x}dx + \frac{\partial f}{\partial y}dy = 0. \tag{1.184}$$

But dx and dy are arbitrary, so (1.184) implies that

$$\frac{\partial f}{\partial x} = 0 \qquad \text{and} \qquad \frac{\partial f}{\partial y} = 0. \tag{1.185}$$

Equation (1.185) is the correct requirement for an extremum if x and y are independent variables. However, suppose the two variables are constrained by the equation

$$g(x, y) = const. \tag{1.186}$$

Equation (1.186) implies that

$$dg = \frac{\partial g}{\partial x}dx + \frac{\partial g}{\partial y}dy = 0. \tag{1.187}$$

Therefore, (1.184) and (1.187) together tell us that

$$\frac{\partial f/\partial x}{\partial g/\partial x} = \frac{\partial f/\partial y}{\partial g/\partial y} = \lambda. \tag{1.188}$$

The constant λ appears because the two ratios in (1.188) cannot otherwise be equal for all values of x and y. In other words,

$$\frac{\partial f}{\partial x} - \lambda\frac{\partial g}{\partial x} = 0 \qquad \text{and} \qquad \frac{\partial f}{\partial y} - \lambda\frac{\partial g}{\partial y} = 0. \tag{1.189}$$

These are the equations we would have gotten in the first place by trying to extremize, *without constraint*, the function

$$F(x, y) = f(x, y) - \lambda g(x, y). \tag{1.190}$$

The *Lagrange constant* λ is not determined (nor does it need to be) by this procedure. Its value can be adjusted to fix the constant in (1.186) if desired.

Sources, References, and Additional Reading

The quotation at the beginning of the chapter is from

E.P. Wigner, "The unreasonable effectiveness of mathematics in the natural sciences", *Communications on Pure and Applied Mathematics* **13**, 1 (1960).

Section 1.1 Most of the material in this chapter is discussed in general treatments of mathematical physics. Three textbooks with rather different styles are

J. Mathews and R.L. Walker, *Mathematical Methods of Physics* (Benjamin/Cummings, Menlo Park, CA, 1970).

G. Arfken, *Mathematical Methods for Physicists*, 3rd edition (Academic, San Diego, 1985).

M. Stone and P. Goldbart, *Mathematics for Physics* (Cambridge University Press, Cambridge, 2009).

Four textbooks of electromagnetism with particularly complete treatments of the mathematical preliminaries are

B. Podolsky and K.S. Kunz, *Fundamentals of Electrodynamics* (Marcel Dekker, New York, 1969).

W. Hauser, *Introduction to the Principles of Electromagnetism* (Addison-Wesley, Reading, MA, 1971).

R.H. Good, Jr. and T.J. Nelson, *Classical Theory of Electric and Magnetic Fields* (Academic, New York, 1971).

A.M. Portis, *Electromagnetic Fields: Sources and Media* (Wiley, New York, 1978).

Section 1.2 Two readable textbooks of vector analysis and vector calculus are

D.E. Bourne and P.C. Kendall, *Vector Analysis* (Oldbourne, London, 1967).

H.F. Davis and A. D. Snider, *Introduction to Vector Analysis*, 7th edition (William C. Brown, Dubuque, IA, 1995).

Section 1.4 Our proof of the flux theorem in Section 1.4.5 is adapted from

M. Abraham and R. Becker, *The Classical Theory of Electricity and Magnetism* (Blackie, London, 1932), pp. 39-40.

Section 1.5 *Lighthill* and *Barton* discuss the delta function and other generalized functions with clarity and precision.

M.J. Lighthill, *Fourier Analysis and Generalized Functions* (Cambridge University Press, Cambridge, 1964).

G. Barton, *Elements of Green's Functions and Propagation* (Clarendon Press, Oxford, 1989).

Frahm and *Hnizdo* discuss delta function identities with less and more rigor, respectively, in

C.P. Frahm, "Some novel delta function identities", *American Journal of Physics* **51**, 826 (1983).

V. Hnizdo, "Generalized second-order partial derivatives of $1/r$", *European Journal of Physics* **32**, 287 (2011).

Section 1.6 A good, short introduction to Fourier transforms and their applications to physics is

T. Schücker, *Distributions, Fourier Transforms and Some of Their Applications to Physics* (World Scientific, Singapore, 1991).

Section 1.7 The distinction between the passive and active points of view for an orthogonal transformation is made in all editions of

H. Goldstein, *Classical Mechanics* (Addison-Wesley, Cambridge, MA, 1950).

Section 1.8 *Hauser* (see Section 1.1 above) is particularly good on Cartesian tensors. See also

J. Rosen, "Transformation properties of electromagnetic quantities under space inversion, time reversal, and charge conjugation", *American Journal of Physics* **41**, 586 (1973).

Section 1.9 For more on the Helmholtz theorem, see

R.B. McQuistan, *Scalar and Vector Fields: A Physical Interpretation* (Wiley, New York, 1965)

J. Van Bladel, "A discussion of Helmholtz' theorem", *Electromagnetics* **13**, 95 (1993).

Van Bladel discusses a generalization of the Helmholtz decomposition theorem to the case when $\mathbf{C}(\mathbf{r}) \to 0$ does not go to zero more rapidly than $1/r$ as $r \to \infty$.

Problems

1.1 Levi-Cività Practice I

(a) Let $(\hat{\mathbf{e}}_1, \hat{\mathbf{e}}_2, \hat{\mathbf{e}}_2)$ be unit vectors of a right-handed, orthogonal coordinate system. Show that the Levi-Cività symbol satisfies

$$\epsilon_{ijk} = \hat{\mathbf{e}}_i \cdot (\hat{\mathbf{e}}_j \times \hat{\mathbf{e}}_k).$$

(b) Prove that

$$\mathbf{a} \times \mathbf{b} = \begin{vmatrix} \hat{\mathbf{e}}_1 & \hat{\mathbf{e}}_2 & \hat{\mathbf{e}}_3 \\ a_1 & a_2 & a_3 \\ b_1 & b_2 & b_2 \end{vmatrix} = \epsilon_{ijk}\hat{\mathbf{e}}_i a_j b_k.$$

(c) Prove that $\epsilon_{ijk}\epsilon_{ist} = \delta_{js}\delta_{kt} - \delta_{jt}\delta_{ks}$.

(d) In quantum mechanics, the Cartesian components of the angular momentum operator $\hat{\mathbf{L}}$ obey the commutation relation $[\hat{L}_i, \hat{L}_j] = i\hbar\epsilon_{ijk}\hat{L}_k$. Let **a** and **b** be constant vectors and prove the commutator identity

$$[\hat{\mathbf{L}} \cdot \mathbf{a}, \hat{\mathbf{L}} \cdot \mathbf{b}] = i\hbar\hat{\mathbf{L}} \cdot (\mathbf{a} \times \mathbf{b}).$$

1.2 Levi-Cività Practice II Evaluate the following expressions which exploit the Einstein summation convention.

(a) δ_{ii}.
(b) $\delta_{ij}\epsilon_{ijk}$.
(c) $\epsilon_{ijk}\epsilon_{\ell jk}$.
(d) $\epsilon_{ijk}\epsilon_{ijk}$.

1.3 Vector Identities Use the Levi-Cività symbol to prove that

(a) $(\mathbf{A} \times \mathbf{B}) \cdot (\mathbf{C} \times \mathbf{D}) = (\mathbf{A} \cdot \mathbf{C})(\mathbf{B} \cdot \mathbf{D}) - (\mathbf{A} \cdot \mathbf{D})(\mathbf{B} \cdot \mathbf{C})$.
(b) $\nabla \cdot (\mathbf{f} \times \mathbf{g}) = \mathbf{g} \cdot (\nabla \times \mathbf{f}) - \mathbf{f} \cdot (\nabla \times \mathbf{g})$.
(c) $(\mathbf{A} \times \mathbf{B}) \times (\mathbf{C} \times \mathbf{D}) = (\mathbf{A} \cdot \mathbf{C} \times \mathbf{D})\mathbf{B} - (\mathbf{B} \cdot \mathbf{C} \times \mathbf{D})\mathbf{A}$.
(d) The 2×2 Pauli matrices σ_x, σ_y, and σ_z used in quantum mechanics satisfy $\sigma_i\sigma_j = \delta_{ij} + i\epsilon_{ijk}\sigma_k$. If **a** and **b** are ordinary vectors, prove that $(\boldsymbol{\sigma} \cdot \mathbf{a})(\boldsymbol{\sigma} \cdot \mathbf{b}) = \mathbf{a} \cdot \mathbf{b} + i\boldsymbol{\sigma} \cdot (\mathbf{a} \times \mathbf{b})$.

1.4 Vector Derivative Identities Use the Levi-Cività symbol to prove that

(a) $\nabla \cdot (f\mathbf{g}) = f\nabla \cdot \mathbf{g} + \mathbf{g} \cdot \nabla f$.
(b) $\nabla \times (f\mathbf{g}) = f\nabla \times \mathbf{g} - \mathbf{g} \times \nabla f$.
(c) $\nabla \times (\mathbf{g} \times \mathbf{r}) = 2\mathbf{g} + r\dfrac{\partial \mathbf{g}}{\partial r} - \mathbf{r}(\nabla \cdot \mathbf{g})$.

1.5 Delta Function Identities A test function as part of the integrand is required to prove any delta function identity. With this in mind:

(a) Prove that $\delta(ax) = \dfrac{1}{|a|}\delta(x), \quad a \neq 0$.
(b) Use the identity in part (a) to prove that

$$\delta[g(x)] = \sum_m \frac{1}{|g'(x_m)|}\delta(x - x_m), \qquad \text{where} \quad g(x_m) = 0 \text{ and } g'(x_m) \neq 0.$$

(c) Confirm that

$$I = \int_0^\infty dx\, \delta(\cos x)\exp(-x) = \frac{1}{2\sinh(\pi/2)}.$$

1.6 Radial Delta Functions

(a) Show that $\delta(r)/r = -\delta'(r)$ when it appears as part of the integrand of a three-dimensional integral in spherical coordinates. Convince yourself that the test function $f(r)$ does not provide any information. Then try $f(r)/r$.
(b) Show that $\nabla \cdot [\delta(r - a)\hat{\mathbf{r}}] = (a^2/r^2)\delta'(r - a)$ when it appears as part of the integrand of a three-dimensional integral in spherical coordinates.

1.7 A Representation of the Delta Function Show that $D(x) = \lim\limits_{m\to\infty} \dfrac{\sin mx}{\pi x}$ is a representation of $\delta(x)$ by showing that $\int_{-\infty}^{\infty} dx f(x)D(x) = f(0)$.

1.8 An Application of Stokes' Theorem Without using vector identities:

(a) Use Stokes' Theorem $\int d\mathbf{S}\cdot(\nabla\times\mathbf{A}) = \oint d\mathbf{s}\cdot\mathbf{A}$ with $\mathbf{A}=\mathbf{c}\times\mathbf{F}$ where $\mathbf{c}$ is an arbitrary constant vector to establish the equality on the left side of

$$\oint_C d\mathbf{s}\times\mathbf{F} = \int_S dS\,\{\hat{n}_i\nabla F_i - \hat{\mathbf{n}}(\nabla\cdot\mathbf{F})\} = \int_S dS\,(\hat{\mathbf{n}}\times\nabla)\times\mathbf{F}.$$

(b) Confirm the equality on the right side of this expression.

(c) Show that $\oint_C \mathbf{r}\times d\mathbf{s} = 2\int_S d\mathbf{S}$.

1.9 Three Derivative Identities Without using vector identities, prove that

(a) $\nabla f(\mathbf{r}-\mathbf{r}') = -\nabla' f(\mathbf{r}-\mathbf{r}')$.

(b) $\nabla\cdot[\mathbf{A}(r)\times\mathbf{r}] = 0$.

(c) $d\mathbf{Q} = (d\mathbf{s}\cdot\nabla)\mathbf{Q}$ where $d\mathbf{Q}$ is a differential change in $\mathbf{Q}$ and $d\mathbf{s}$ is an element of arc length.

1.10 Derivatives of $\exp(i\mathbf{k}\cdot\mathbf{r})$ Let $\mathbf{A}(\mathbf{r}) = \mathbf{c}\exp(i\mathbf{k}\cdot\mathbf{r})$ where $\mathbf{c}$ is constant. Show that, in every case, the replacement $\nabla\to i\mathbf{k}$ produces the correct answer for $\nabla\cdot\mathbf{A}$, $\nabla\times\mathbf{A}$, $\nabla\times(\nabla\times\mathbf{A})$, $\nabla(\nabla\cdot\mathbf{A})$, and $\nabla^2\mathbf{A}$.

1.11 Some Integral Identities Assume that $\varphi(\mathbf{r})$ and $|\mathbf{G}(\mathbf{r})|$ both go to zero faster than $1/r$ as $r\to\infty$.

(a) Let $\mathbf{F}=\nabla\varphi$ and $\nabla\cdot\mathbf{G}=0$. Show that $\int d^3r\,\mathbf{F}\cdot\mathbf{G}=0$.

(b) Let $\mathbf{F}=\nabla\varphi$ and $\nabla\times\mathbf{G}=0$. Show that $\int d^3r\,\mathbf{F}\times\mathbf{G}=0$.

(c) Begin with the vector with components $\partial_j(P_j\mathbf{G})$ and prove that

$$\int_V d^3r\,\mathbf{P} = -\int_V d^3r\,\mathbf{r}(\nabla\cdot\mathbf{P}) + \int_S dS(\hat{\mathbf{n}}\cdot\mathbf{P})\mathbf{r}.$$

1.12 Unit Vector Practice Express the following in terms of $\hat{\mathbf{r}}$, $\hat{\boldsymbol{\theta}}$, and $\hat{\boldsymbol{\phi}}$:

$$\frac{\partial\hat{\mathbf{r}}}{\partial\theta}\qquad\frac{\partial\hat{\mathbf{r}}}{\partial\phi}\qquad\frac{\partial\hat{\boldsymbol{\theta}}}{\partial\theta}\qquad\frac{\partial\hat{\boldsymbol{\theta}}}{\partial\phi}\qquad\frac{\partial\hat{\boldsymbol{\phi}}}{\partial\theta}\qquad\frac{\partial\hat{\boldsymbol{\phi}}}{\partial\phi}.$$

1.13 Compute the Normal Vector Compute the unit normal vector $\hat{\mathbf{n}}$ to the ellipsoidal surfaces defined by constant values of

$$\Phi(x,y,z) = V\left[\frac{x^2}{a^2}+\frac{y^2}{b^2}+\frac{z^2}{c^2}\right].$$

Check that you get the expected answer when $a=b=c$.

1.14 A Variant of the Helmholtz Theorem I Mimic the proof of Helmholtz' theorem in the text and prove that

$$\varphi(\mathbf{r}) = -\nabla\cdot\frac{1}{4\pi}\int_V d^3r'\,\frac{\nabla'\varphi(\mathbf{r}')}{|\mathbf{r}-\mathbf{r}'|} + \nabla\cdot\frac{1}{4\pi}\int_S d\mathbf{S}'\,\frac{\varphi(\mathbf{r}')}{|\mathbf{r}-\mathbf{r}'|}.$$

1.15 A Variant of the Helmholtz Theorem II A vector function $\mathbf{Z}(\mathbf{r})$ satisfies $\nabla\cdot\mathbf{Z}=0$ and $\nabla\times\mathbf{Z}=0$ everywhere in a simply-connected volume V bounded by a surface S. Modify the proof of the Helmholtz theorem in the text and show that $\mathbf{Z}(\mathbf{r})$ can be found everywhere in V if its value is specified at every point on S.

1.16 Densities of States Let E be a positive real number. Evaluate

(a) $g_1(E) = \int_{-\infty}^{\infty} dk_x\, \delta(E - k_x^2)$.

(b) $g_2(E) = \int_{-\infty}^{\infty} dk_x \int_{-\infty}^{\infty} dk_y\, \delta(E - k_x^2 - k_y^2)$.

(c) $g_3(E) = \int_{-\infty}^{\infty} dk_x \int_{-\infty}^{\infty} dk_y \int_{-\infty}^{\infty} dk_z\, \delta(E - k_x^2 - k_y^2 - k_z^2)$.

1.17 Dot and Cross Products Let $\mathbf{b}$ be a vector and $\hat{\mathbf{n}}$ a unit vector.

(a) Use the Levi-Cività symbol to prove that $\mathbf{b} = (\mathbf{b} \cdot \hat{\mathbf{n}})\hat{\mathbf{n}} + \hat{\mathbf{n}} \times (\mathbf{b} \times \hat{\mathbf{n}})$.
(b) Interpret the decomposition in part (a) geometrically.
(c) Let $\omega = \mathbf{a} \cdot (\mathbf{b} \times \mathbf{c})$ where $\mathbf{a}$, $\mathbf{b}$, and $\mathbf{c}$ are any three non-coplanar vectors. Now let

$$\mathbf{A} = \frac{\mathbf{b} \times \mathbf{c}}{\omega} \qquad \mathbf{B} = \frac{\mathbf{c} \times \mathbf{a}}{\omega} \qquad \mathbf{C} = \frac{\mathbf{a} \times \mathbf{b}}{\omega}.$$

Express $\Omega = \mathbf{A} \cdot (\mathbf{B} \times \mathbf{C})$ entirely in terms of ω.

1.18 S_{ij} and T_{ij}

(a) Show that $\epsilon_{ijk} S_{ij} = 0$ if $\mathbf{S}$ is a symmetric matrix.
(b) Let $\mathbf{b}$ and $\mathbf{y}$ be vectors. The components of the latter are defined by $y_i = b_k T_{ki}$ where $T_{ij} = -T_{ji}$ is an anti-symmetric object. Find a vector $\boldsymbol{\omega}$ such that $\mathbf{y} = \mathbf{b} \times \boldsymbol{\omega}$. Why does it makes sense that $\mathbf{T}$ and $\boldsymbol{\omega}$ could have the same information content?

1.19 Two Surface Integrals Let S be the surface that bounds a volume V. Show that (a) $\int_S d\mathbf{S} = 0$; (b) $\frac{1}{3} \int_S d\mathbf{S} \cdot \mathbf{r} = V$.

1.20 Electrostatic Dot and Cross Products If $\mathbf{a}$ and $\mathbf{b}$ are constant vectors, $\varphi(\mathbf{r}) = (\mathbf{a} \times \mathbf{r}) \cdot (\mathbf{b} \times \mathbf{r})$ is the electrostatic potential in some region of space. Find the electric field $\mathbf{E} = -\nabla\varphi$ and then the charge density $\rho = \epsilon_0 \nabla \cdot \mathbf{E}$ associated with this potential.

1.21 A Decomposition Identity Let $\mathbf{A}$ and $\mathbf{B}$ be vectors. Show that

$$A_i B_j = \frac{1}{2}\epsilon_{ijk}(\mathbf{A} \times \mathbf{B})_k + \frac{1}{2}(A_i B_j + A_j B_i).$$

2 The Maxwell Equations

If you wake up a physicist in the middle of the night and say "Maxwell", he is sure to say "electromagnetic field".
Rudolph Peierls (1962)

2.1 Introduction

All physical phenomena in our Universe derive from four fundamental forces. Gravity binds stars and creates the tides. The strong force binds baryons and mesons and controls nuclear reactions. The weak force mediates neutrino interactions and changes the flavor of quarks. The fourth force, the Coulomb-Lorentz force, animates a particle with charge q and velocity $\boldsymbol{v}$ in the presence of an electric field $\mathbf{E}$ and a magnetic field $\mathbf{B}$:

$$\mathbf{F} = q(\mathbf{E} + \boldsymbol{v} \times \mathbf{B}). \tag{2.1}$$

The subject we call *electromagnetism* concerns the origin and behavior of the fields $\mathbf{E}(\mathbf{r}, t)$ and $\mathbf{B}(\mathbf{r}, t)$ responsible for the force (2.1).

We will come to learn that the electric and magnetic fields are very closely related. However, like many siblings, the *time-independent* quantities $\mathbf{E}(\mathbf{r})$ and $\mathbf{B}(\mathbf{r})$ do not look alike and do not interact. Static electric fields require charge separation. The largest such separations ($\sim 10^5$ m) are associated with electrostatic discharges in the upper atmosphere. Static magnetic fields (apart from magnetic matter) require only charge in steady motion. As far as is known, the maximum size of magnetic field patterns may approach cosmic dimensions ($\sim 10^{20}$ m).

Time-dependent $\mathbf{E}$ and $\mathbf{B}$ fields are more like a newly married couple. Initially, they remain close to their sources. Then, in a moment of subtle reorganization, they break free and—in the form of electromagnetic radiation—race away to an independent, intertwined existence. The two fields are inextricably bound together in the X-rays which reveal the atomic-scale structure of DNA, in the microwaves which facilitate contemporary telecommunications, and in the radio waves which reveal the large-scale structure of the Universe.

The full story of these matters is neither short nor simple. In this chapter, we begin with the primitive concepts of charge and current. A brief review of the history of electromagnetism leads to definitions for $\mathbf{E}(\mathbf{r}, t)$ and $\mathbf{B}(\mathbf{r}, t)$ and to the Maxwell equations which relate the fields to sources of charge and current. We then turn to the relationship between microscopic electromagnetism and macroscopic electromagnetism. This includes a discussion of spatial averaging and a derivation of the matching conditions required by the macroscopic theory. Two short sections discuss the limits of validity of the classical theory and the SI system of units used in this book. The chapter concludes with a heuristic "derivation" of the Maxwell equations.

2.1.1 Electric Charge

The word *electric* derives from the Greek word for amber ($\eta\lambda\varepsilon\kappa\tau\rho o\nu$), a substance which attracts bits of chaff when rubbed. Sporadic and often contradictory reports of this peculiar phenomenon appeared for centuries. Then, in 1600, William Gilbert dismissed all of them as "esoteric, miracle-mongering, abstruse, recondite, and mystical". His own careful experiments showed that many materials, when suitably prepared, produced an "electric force" like amber.

Electrical research was revolutionized in 1751 when Benjamin Franklin postulated that rubbing transfers a tangible electric "fluid" from one body to another, leaving one with a surplus and the other with a deficit. When word of the American polymath's proposal reached Europe, Franz Aepinus realized that the *electric charge* Q was a variable that could be assigned to an electric body. He used it to express Franklin's law of conservation of charge in algebraic form. He also pointed out that the electric force was proportional to Q. Today, we understand charge to be an intrinsic property of matter, like mass. Moreover, no known particle possesses a charge which is not an integer multiple of the minimum value of the electron charge[1]

$$e = 1.602\,177\,33(49) \times 10^{-19}\, C. \tag{2.2}$$

Typically, a neutral atom, molecule, or macroscopic body acquires a net charge only through the gain or loss of electrons, each of which possesses a charge $-e$.

Despite the fundamental discreteness of charge implied by its quantization, electromagnetic theory develops most naturally if we define a continuous charge per unit volume or volume *charge density*, $\rho(\mathbf{r})$. By construction, $dQ = \rho(\mathbf{r})d^3r$ is the amount of charge contained in an infinitesimal volume d^3r. The total charge Q in a finite volume V is

$$Q = \int_V d^3r\, \rho(\mathbf{r}). \tag{2.3}$$

Classically, this is a straightforward definition when $\rho(\mathbf{r})$ is a continuous function of the usual sort. It is equally straightforward in quantum mechanics because the charge density is defined in terms of continuous wave functions. For example, the charge density for a system of N indistinguishable particles (each with charge q) described by the many-particle wave function $\Psi(\mathbf{r}_1, \mathbf{r}_2, \ldots, \mathbf{r}_N)$ is

$$\rho(\mathbf{r}) = Nq \int d^3r_2 \int d^3r_3 \cdots d^3r_N\, |\Psi(\mathbf{r}, \mathbf{r}_2, \ldots, \mathbf{r}_N)|^2. \tag{2.4}$$

It is often useful to imagine continuous distributions of charge which are confined to infinitesimally thin surface layers. This suggests we identify $\mathbf{r}_S$ as a point centered on an infinitesimal element of surface dS and define a charge per unit area, or surface charge density, $\sigma(\mathbf{r}_S)$ so $dQ = \sigma(\mathbf{r}_S)dS$. The total charge associated with a finite surface S is

$$Q = \int_S dS\, \sigma(\mathbf{r}_S). \tag{2.5}$$

A charge per unit length, or linear charge density $\lambda(\boldsymbol{\ell})$, plays a similar role for continuous distributions confined to a one-dimensional filament. In that case, $dQ = \lambda(\boldsymbol{\ell})d\ell$, where $\boldsymbol{\ell}$ points from the origin to to the line element $d\boldsymbol{\ell}$.

Finally, the experimental fact that a free electron has zero size motivates us to define a classical *point charge* as a vanishingly small object which carries a finite amount of charge.[2] The mathematical

[1] We omit quarks, particles with fractional charge, because they cannot be separated from the hadrons they constitute. Equation (2.2) defines the Coulomb (C) as the charge carried by 6.2415096×10^{18} electrons.

[2] Electron-positron scattering experiments judge the electron to be a structureless object with a charge radius less than 1.2×10^{-19} m. See A. Bajo, I. Dymnikova, A. Sakharov, *et al.*, in *Quantum Electrodynamics and Physics of the Vacuum*, edited by G. Cantatore, AIP Conference Proceeding, volume 564 (AIP, Woodbury, NY, 2001), pp. 255-262.

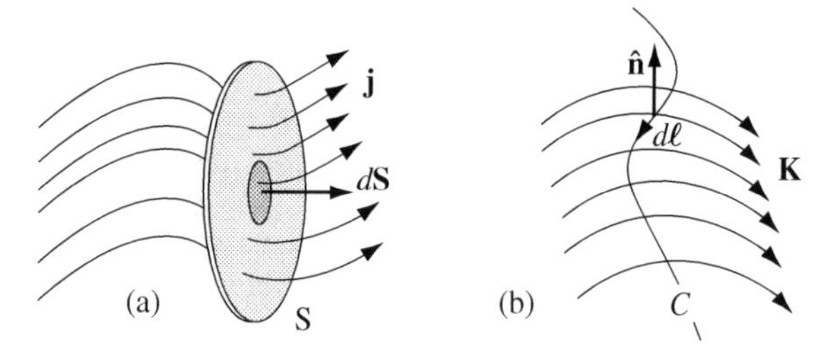

Figure 2.1: (a) The current I is the integral over S of the projection of the volume current density $\mathbf{j}$ onto the area element $d\mathbf{S}$; (b) the current I is the integral over C of the projection of $\mathbf{K} \times \hat{\mathbf{n}}$ (the cross product of the surface current density $\mathbf{K}$ and the surface normal $\hat{\mathbf{n}}$) onto the line element $d\boldsymbol{\ell}$.

properties of the delta function are well suited to this task and we represent the charge density of N point charges q_k located at positions $\mathbf{r}_k$ as

$$\rho(\mathbf{r}) = \sum_{k=1}^{N} q_k \delta(\mathbf{r} - \mathbf{r}_k). \tag{2.6}$$

Substituting (2.6) into (2.3) yields the correct total charge. However, one should be alert for possible (unphysical) divergences in other quantities associated with the singular nature of (2.6).

2.1.2 Electric Current

Electric charge in organized motion is called *electric current*. The identification of electric current as electric charge in continuous motion around a closed path is due to Alessandro Volta. Volta was stimulated by Luigi Galvani and his observation that the touch of a metal electrode induces the dramatic contraction of a frog's leg. A skillful experimenter himself, Volta discovered that the contraction was associated exclusively with the passage of electric charge through the frog's leg. Subsequent experiments designed to generate large electric currents led him to invent the battery in 1800.

We make the concept of electric current quantitative using Figure 2.1(a). By analogy with fluid flow, let $\hat{\mathbf{n}}$ be the local unit normal to an element of surface dS and let $d\mathbf{S} = dS\hat{\mathbf{n}}$. We define a *current density* $\mathbf{j}(\mathbf{r}, t)$ so $dI = \mathbf{j} \cdot d\mathbf{S} = \mathbf{j} \cdot \hat{\mathbf{n}} dS$ is the rate at which charge passes through dS. The total current that passes through a finite surface S is

$$I = \frac{dQ}{dt} = \int_S d\mathbf{S} \cdot \mathbf{j}. \tag{2.7}$$

We can write an explicit formula for $\mathbf{j}(\mathbf{r}, t)$ when a velocity field $\boldsymbol{v}(\mathbf{r}, t)$ characterizes the motion of a charge density $\rho(\mathbf{r}, t)$. In that case, the fluid analogy suggests that the current density is

$$\mathbf{j} = \rho \boldsymbol{v}. \tag{2.8}$$

If the charge is entirely confined to a two-dimensional surface, it is appropriate to replace (2.8) by a surface current density

$$\mathbf{K} = \sigma \boldsymbol{v}. \tag{2.9}$$

Figure 2.1(b) shows that the current that flows past a curve C on a surface can be expressed in terms of the local surface normal $\hat{\mathbf{n}}$ using

$$I = \int_C d\boldsymbol{\ell} \cdot \mathbf{K} \times \hat{\mathbf{n}} = \int_C \mathbf{K} \cdot (\hat{\mathbf{n}} \times d\boldsymbol{\ell}). \tag{2.10}$$

The second integral in (2.10) makes it clear that only the projection of $\mathbf{K}$ onto the normal to the line element $d\boldsymbol{\ell}$ (in the plane of the surface) contributes to I [cf. (2.7)].

2.1.3 Conservation of Charge

As far as we know, electric charge is absolutely conserved by all known physical processes. The only way to change the net charge in a finite volume is to move charged particles into or out of that volume. Chemical reactions create and destroy chemical species, and quantum processes create and destroy elementary particles, but the total charge before and after any of these events is always the same. Indeed, the most stringent tests of this *principle of charge conservation* search for the spontaneous decay of the electron into neutral particles like photons and neutrinos. If this occurs at all, the mean time for decay exceeds 10^{24} years (Belli *et al.* 1999).

For our purposes, the most useful statement of charge conservation begins with the surface integral representation of the current I in (2.7). If we choose the surface S to be closed, the divergence theorem (Section 1.4.2) permits us to express I as an integral over the enclosed volume V:

$$I = \int_V d^3r\, \nabla \cdot \mathbf{j}. \tag{2.11}$$

Because the vector $d\mathbf{S}$ in (2.7) points outward from V, (2.11) is the rate at which the total charge Q *decreases* in the volume V. An explicit expression for the latter is

$$-\frac{dQ}{dt} = -\frac{d}{dt}\int_V d^3r\, \rho = -\int_V d^3r\, \frac{\partial \rho}{\partial t}. \tag{2.12}$$

Equating (2.11) and (2.12) for an arbitrary volume yields a local statement of charge conservation called the *continuity equation*,

$$\frac{\partial \rho}{\partial t} + \nabla \cdot \mathbf{j} = 0. \tag{2.13}$$

The continuity equation says that the total charge in any infinitesimal volume is constant unless there is a net flow of pre-existing charge into or out of the volume through its surface.

Application 2.1 Moving Point Charges

Let N point charges q_k follow trajectories $\mathbf{r}_k(t)$. The charge density of this system of moving point charges is a time-dependent generalization of (2.6):

$$\rho(\mathbf{r}, t) = \sum_{k=1}^{N} q_k \delta(\mathbf{r} - \mathbf{r}_k(t)). \tag{2.14}$$

We can use the continuity equation to derive the corresponding current density. The particle velocities are $\boldsymbol{v}_k(t) = \dot{\mathbf{r}}_k(t)$, so the chain rule gives

$$\frac{\partial \rho}{\partial t} = \sum_k q_k \frac{\partial}{\partial t}\delta(\mathbf{r} - \mathbf{r}_k) = -\sum_k q_k \boldsymbol{v}_k \cdot \nabla \delta(\mathbf{r} - \mathbf{r}_k). \tag{2.15}$$

Since $\nabla \cdot \boldsymbol{v}_k = 0$, this can be written in the form

$$\partial\rho/\partial t = -\nabla \cdot \sum_k q_k \boldsymbol{v}_k \delta(\mathbf{r} - \mathbf{r}_k). \tag{2.16}$$

This formula is consistent with the continuity equation (2.13) if

$$\mathbf{j}(\mathbf{r}, t) = \sum_{k=1}^{N} q_k \boldsymbol{v}_k \delta(\mathbf{r} - \mathbf{r}_k). \tag{2.17}$$

■

2.2 The Maxwell Equations in Vacuum

The unification of electricity, magnetism, and optics was achieved by James Clerk Maxwell with the publication of his monumental *Treatise on Electricity and Magnetism* (1873). Maxwell characterized his theory as an attempt to "mathematize" the results of many different experimental investigations of electric and magnetic phenomena. A generation later, Heinrich Hertz famously remarked that "Maxwell's theory is Maxwell's system of equations." He made this statement because he "not always felt quite certain of having grasped the physical significance" of the arguments given by Maxwell in the *Treatise*. The four "Maxwell equations" Hertz had in mind (independently proposed by Heaviside and Hertz) are actually a concise version of twelve equations offered by Maxwell.

It is traditional to develop the Maxwell equations through a review of the experiments that motivated their construction. We do this here (briefly) for the simple reason that every physicist should know some of the history of this subject. As an alternative, Section 2.7 offers a heuristic "derivation" of the Maxwell equations based on symmetry arguments and minimal experimental input.

2.2.1 Electrostatics

Experimental work by Priestley, Cavendish, and Coulomb at the end of the 18th century established the nature of the force between stationary charged objects. Extrapolated to the case of point charges, the force on a charge q at the point $\mathbf{r}$ due to N point charges q_k at points $\mathbf{r}_k$ is given by the inverse-square *Coulomb's law*:

$$\mathbf{F} = \frac{1}{4\pi\epsilon_0} \sum_{k=1}^{N} q q_k \frac{\mathbf{r} - \mathbf{r}_k}{|\mathbf{r} - \mathbf{r}_k|^3}. \tag{2.18}$$

The pre-factor $1/4\pi\epsilon_0$ reflects our choice of SI units (see Section 2.6). Using the point charge density (2.6), we can restate Coulomb's law in the form

$$\mathbf{F} = q\mathbf{E}(\mathbf{r}), \tag{2.19}$$

where the vector field $\mathbf{E}(\mathbf{r})$ is called the *electric field*:

$$\mathbf{E}(\mathbf{r}) = \frac{1}{4\pi\epsilon_0} \int d^3r'\, \rho(\mathbf{r}') \frac{\mathbf{r} - \mathbf{r}'}{|\mathbf{r} - \mathbf{r}'|^3}. \tag{2.20}$$

Generalizing, we define (2.20) to be the electric field for any choice of $\rho(\mathbf{r})$. This definition makes the *principle of superposition* explicit: the electric field produced by an arbitrary charge distribution is the vector sum of the electric fields produced by each of its constituent pieces. Given (2.20), the mathematical identities

$$\nabla \frac{1}{|\mathbf{r} - \mathbf{r}'|} = -\frac{\mathbf{r} - \mathbf{r}'}{|\mathbf{r} - \mathbf{r}'|^3} \qquad \text{and} \qquad \nabla^2 \frac{1}{|\mathbf{r} - \mathbf{r}'|} = -4\pi\delta(\mathbf{r} - \mathbf{r}') \tag{2.21}$$

are sufficient to show that

$$\nabla \cdot \mathbf{E} = \rho/\epsilon_0 \qquad \text{and} \qquad \nabla \times \mathbf{E} = 0. \tag{2.22}$$

The first equation of (2.22) is *Gauss' law*. It is the first of the four Maxwell equations. The second equation of (2.22) is valid for electrostatics only.

2.2.2 The Field Concept

An important conceptual shift occurs when we pass from (2.18) to (2.19). The first of these conjures up the picture of a non-local force which acts between pairs of charges over arbitrarily large distances. By contrast, Coulomb's law in the form (2.19) suggests the rather different picture of an electric *field* which pervades all of space.[3] A particle experiences a force determined by the local value of the field at the position of the particle. For static problems, the "action-at-a-distance" and field points of view are completely equivalent.

The true superiority of the field approach becomes evident only when we turn to time-dependent problems. In that context, we will see that fields can exist quite independently of the presence or absence of charged particles. We will endow them with properties like energy, linear momentum, and angular momentum and treat them as dynamical objects with the same mechanical status as the particles.

2.2.3 Magnetostatics

William Gilbert was the first person to conduct systematic experiments on the nature of magnetism.[4] In *de Magnete* (1600), Gilbert correctly concluded that the Earth behaves like a giant permanent magnet. But it was not until 1750 that John Michell showed that the force between the ends ("poles") of two rod-like permanent magnets follows an inverse-square law with attraction (repulsion) between unlike (like) poles. Somewhat later, Coulomb and Gauss confirmed these results and Poisson eventually developed a theory of magnetic "charge" and force in complete analogy with electrostatics.

In 1820, Oersted made the dramatic discovery that a current-carrying wire produces effects qualitatively similar to those of a permanent magnet. Biot, Savart, and Ampère followed up quickly with quantitative experiments. A watershed moment occurred when Ampère published his calculation of the force on a closed loop carrying a current I due to the presence of N other loops carrying currents I_k (see Figure 2.2). If $\mathbf{r}$ points to the line element $d\boldsymbol{\ell}$ of loop I and $\mathbf{r}_k$ points to the element $d\boldsymbol{\ell}_k$ of the kth loop, Ampère's formula for the force on I is[5]

$$\mathbf{F} = -\frac{\mu_0}{4\pi} \oint I d\boldsymbol{\ell} \cdot \sum_{k=1}^{N} \oint I_k d\boldsymbol{\ell}_k \frac{\mathbf{r} - \mathbf{r}_k}{|\mathbf{r} - \mathbf{r}_k|^3}. \tag{2.23}$$

The pre-factor $\mu_0/4\pi$ reflects our choice of SI units (see Section 2.6).

With a bit of manipulation, (2.23) can be recast in the form

$$\mathbf{F} = \oint I d\boldsymbol{\ell} \times \mathbf{B}(\mathbf{r}), \tag{2.24}$$

[3] The concept of the field, if not its mathematical expression, is generally credited to Michael Faraday. See McMullin (2002) in Sources, References, and Additional Reading.

[4] See the first paragraph of Section 2.1.1. The word *magnetic* derives from the proper name Magnesia. This is a district of central Greece rich in the naturally magnetic mineral lodestone.

[5] The scalar product in (2.23) is $d\boldsymbol{\ell} \cdot d\boldsymbol{\ell}_k$.

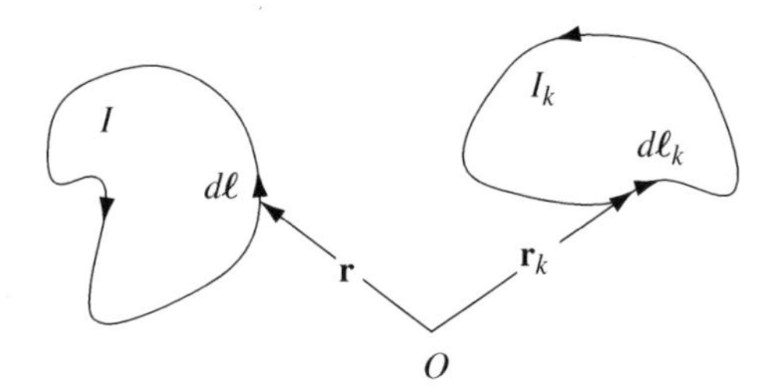

Figure 2.2: Two filamentary loops carry current I and I_k. The vectors $\mathbf{r}$ and $\mathbf{r}_k$ point to the line elements $d\boldsymbol{\ell}$ and $d\boldsymbol{\ell}_k$, respectively.

where the *magnetic field* $\mathbf{B}(\mathbf{r})$ has a form first determined by Biot and Savart in 1820:

$$\mathbf{B}(\mathbf{r}) = \frac{\mu_0}{4\pi}\sum_{k=1}^{N}\oint I_k d\boldsymbol{\ell}_k \times \frac{\mathbf{r}-\mathbf{r}_k}{|\mathbf{r}-\mathbf{r}_k|^3}. \tag{2.25}$$

Later (Section 9.3.1), we will learn that the substitution $I\oint d\boldsymbol{\ell} \to \int d^3r\,\mathbf{j}$ transforms formulae valid for linear circuits into formulae valid for volume distributions of current. Accordingly, we generalize (2.25) and *define* the magnetic field produced by any time-independent current density as

$$\mathbf{B}(\mathbf{r}) = \frac{\mu_0}{4\pi}\int d^3r'\,\frac{\mathbf{j}(\mathbf{r}')\times(\mathbf{r}-\mathbf{r}')}{|\mathbf{r}-\mathbf{r}'|^3}. \tag{2.26}$$

The principle of superposition is again paramount: the magnetic field produced by a steady current distribution is the vector sum of the magnetic fields produced by each of its constituent pieces.

In 1851, William Thomson (later Lord Kelvin) used an equivalence between current loops and permanent magnets due to Ampère to show that the magnetic field produced by both types of sources satisfies

$$\nabla\cdot\mathbf{B} = 0 \qquad \text{and} \qquad \nabla\times\mathbf{B} = \mu_0\mathbf{j}. \tag{2.27}$$

It is a worthwhile exercise to confirm that (2.26) is consistent with both equations in (2.27) as long as the current density satisfies the *steady-current* condition [cf. (2.13)],

$$\nabla\cdot\mathbf{j} = 0. \tag{2.28}$$

The first equation of (2.27) is the second of the Maxwell equations. It has no commonly agreed-upon name. The equation on the right side of (2.27), which is valid for magnetostatics only, is often called *Ampère's law*.

2.2.4 Faraday's Law

Michael Faraday was the greatest experimental scientist of the 19th century.[6] Of particular importance was Faraday's discovery that a transient electric current flows through a circuit whenever the magnetic flux through that circuit *changes* (Figure 2.3). Starting from surprisingly different points of view, the mathematical expression of this fact was achieved by Neumann, Helmholtz, Thomson, Weber, and Maxwell. In modern notation, Faraday's observation applied to a circuit with resistance R means that

$$-\frac{d}{dt}\int_S d\mathbf{S}\cdot\mathbf{B} = IR. \tag{2.29}$$

[6] Every physicist should at least glance through Faraday's *Diary* or his *Experimental Researches in Electricity*. Both are wonderfully readable chronicles of over 40 years (1820-1862) of experimental work.

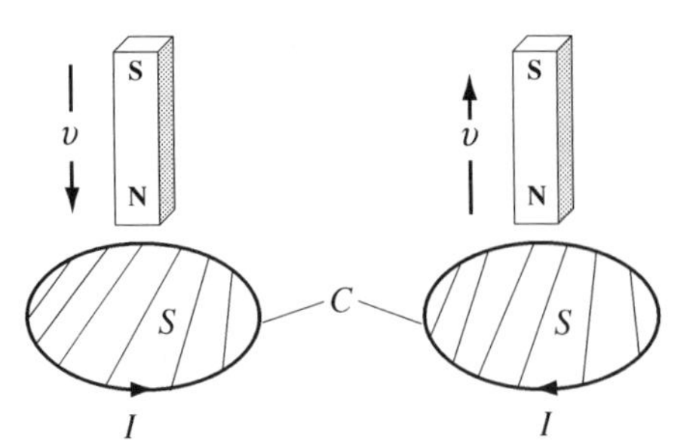

Figure 2.3: A typical experiment which reveals Faraday's law. Current flows in opposite directions in the filamentary wire C when the permanent magnet moves upward or downward. The area S is bounded by the wire C.

The domain of integration S is any surface whose boundary curve coincides with the circuit. Our convention is that the right-hand rule relates the direction of current flow to the direction of $d\mathbf{S}$. In that case, the minus sign in (2.29) reflects *Lenz' law*: the current creates a magnetic field which opposes the original change in magnetic flux.

To complete the story, we need only recognize that scientists of the 19th century understood Ohm's law for current flow in a closed circuit C to mean that

$$IR = \oint_C d\boldsymbol{\ell} \cdot \mathbf{E}. \tag{2.30}$$

For our application, the path C in (2.30) bounds the surface S in (2.29). Therefore, after setting (2.30) equal to (2.29), Stokes' theorem (Section 1.4.4) yields the differential form of *Faraday's law*, the third Maxwell equation:

$$\nabla \times \mathbf{E} = -\frac{\partial \mathbf{B}}{\partial t}. \tag{2.31}$$

2.2.5 The Displacement Current

The *displacement current* is Maxwell's transcendent contribution to the theory of electromagnetism. Writing in 1862, Maxwell used an elaborate mechanical model of rotating "magnetic" vortices with interposed "electric" ball bearings (see Figure 2.4) to argue that the current density $\mathbf{j}$ in Ampère's law must be supplemented by another term when the electric field varies in time. This is the displacement current, $\mathbf{j}_D = \epsilon_0 \partial \mathbf{E}/\partial t$. Inserting this into (2.27), we get the fourth and final Maxwell equation, usually called the *Ampère-Maxwell law*,

$$\nabla \times \mathbf{B} = \mu_0 \mathbf{j} + \frac{1}{c^2}\frac{\partial \mathbf{E}}{\partial t}. \tag{2.32}$$

Today, it is usual to say that the displacement current is absolutely *required* if Ampère's law and Gauss' law are to be consistent with the continuity equation (2.13). This argument was unavailable to Maxwell because he did not associate electric current with electric charge in motion.[7]

Maxwell dispensed entirely with mechanical models when he wrote his *Treatise*. Instead, he introduced $\mathbf{j}_D$ without motivation, remarking only that it is "one of the chief peculiarities" of the theory.

[7] Maxwell (who worked long before the discovery of the electron) did not regard charge as an intrinsic property of matter subject to a law of conservation. To him, Gauss' law did not mean that charge was the source of an electric field. It meant that spatial variations of an electric field were a source of charge. Maxwell's conception of current is not easily summarized. See Buchwald (1985) in Sources, References, and Additional Reading.

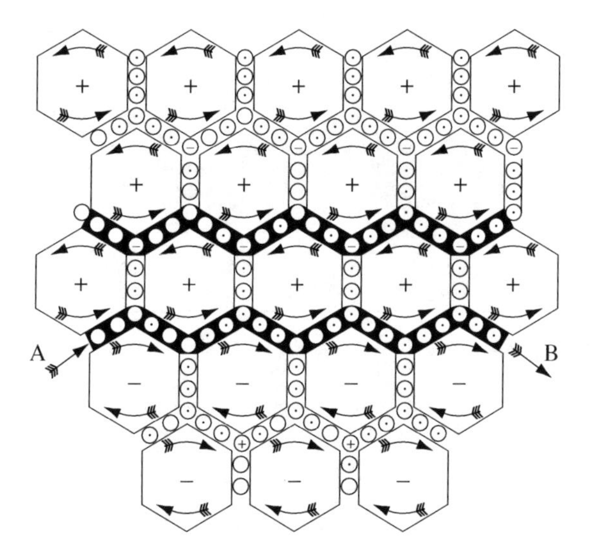

Figure 2.4: Maxwell's sketch of a mechanical model of rotating hexagonal vortices with interposed ball bearings. He inferred the existence of the displacement current from a study of the kinematics of this device. Figure from Maxwell (1861).

Presumably, the fact that the displacement current was essential for wave solutions which propagate at the speed of light had convinced him of its basic correctness. The rest of the world became convinced in 1888 when Hertz discovered electromagnetic waves (in the microwave range) with all the properties predicted by Maxwell's theory.

2.2.6 Putting It All Together

Classical electromagnetism summarizes a vast body of experimental information using the concepts of charge density $\rho(\mathbf{r}, t)$, current density $\mathbf{j}(\mathbf{r}, t)$, electric field $\mathbf{E}(\mathbf{r}, t)$, and magnetic field $\mathbf{B}(\mathbf{r}, t)$. We have seen in this chapter that partial differential equations connect the fields to the sources and vice versa. Two of these are the explicitly time-dependent curl equations (2.31) and (2.32). Two others are the divergence equations in (2.22) and (2.27), which survive the transition from static fields to time-dependent fields without change. Taken together, we get the foundational equations of our subject and the main result of this chapter, the *Maxwell equations*:

$$\nabla \cdot \mathbf{E} = \frac{\rho}{\epsilon_0} \qquad\qquad \nabla \cdot \mathbf{B} = 0 \tag{2.33}$$

$$\nabla \times \mathbf{E} = -\frac{\partial \mathbf{B}}{\partial t} \qquad\qquad \nabla \times \mathbf{B} = \mu_0 \mathbf{j} + \frac{1}{c^2}\frac{\partial \mathbf{E}}{\partial t}. \tag{2.34}$$

The direct connection to experience comes when we specify the force which $\rho(\mathbf{r}, t)$ and $\mathbf{j}(\mathbf{r}, t)$ exert on a charge density $\rho^*(\mathbf{r}, t)$ and a current density $\mathbf{j}^*(\mathbf{r}, t)$:

$$\mathbf{F}(t) = \int d^3r\, [\rho^*(\mathbf{r}, t)\mathbf{E}(\mathbf{r}, t) + \mathbf{j}^*(\mathbf{r}, t) \times \mathbf{B}(\mathbf{r}, t)]. \tag{2.35}$$

The electric part of (2.35) generalizes Coulomb's law of electrostatics to time-dependent situations. The magnetic part of (2.35) was derived by Oliver Heaviside in 1889 by supplementing the Maxwell equations with a postulated expression for the energy of interaction between a current loop and an external magnetic field. An essentially similar derivation was presented somewhat later by Lorentz. Following tradition, we will call (2.35) the "Coulomb-Lorentz" force law.

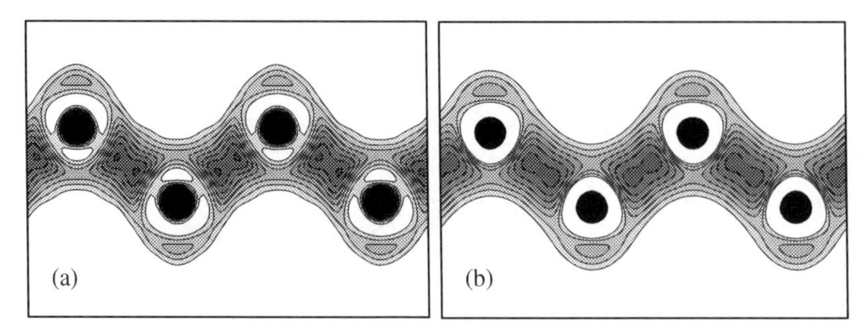

Figure 2.5: Contour plot of the valence charge density of crystalline silicon: (a) $\rho(\mathbf{r})$ extracted from x-ray diffraction data; (b) $\rho(\mathbf{r})$ computed from quantum mechanical calculations and the Maxwell equations. Periodic repetition of either plot in the vertical and horizontal directions generates one plane of the crystal. The rectangular contours reflect "bond" charge between the atoms (black circles). The white regions of very low valence charge density reflect nodes in the sp^3 wave functions. Figure from Tanaka, Takata, and Sakata (2002).

2.3 Microscopic vs. Macroscopic

Hendrik Lorentz had the brilliant insight that the Maxwell equations developed to analyze macroscopic phenomena might apply to microscopic situations as well if the source densities $\rho(\mathbf{r}, t)$ and $\mathbf{j}(\mathbf{r}, t)$ were themselves microscopic. This motivated him to propose a classical model for neutral matter based on point charge densities like (2.6) and their associated current densities (2.17). Lorentz understood that his model would produce electric and magnetic fields which varied rapidly on a microscopic scale. He also understood that only a spatial average over such variations could be relevant to macroscopic measurements. Lorentz therefore proposed to *derive* Maxwell's macroscopic theory by suitably averaging over his microscopic theory.

Lorentz' idea makes sense today if quantum mechanical expressions are used for the charge and current densities. However, before any consideration is given to averaging, it is necessary to prove what Lorentz supposed: namely, that Maxwell's electromagnetism is valid when the sources of charge and current are truly microscopic. Particularly compelling evidence for the microscopic validity of Maxwell's theory comes from the remarkable agreement between experiment and theory for the spectrum of atomic hydrogen.[8] At the level of the single-particle Dirac equation, the bound-state energy levels for an electron moving in the classical Coulomb field of a proton are

$$E(n, j) = mc^2\left[1 + \frac{\alpha^2}{(n-\delta)^2}\right]^{-1/2} \qquad \delta = j + \tfrac{1}{2} - \sqrt{(j+\tfrac{1}{2})^2 - \alpha^2}, \tag{2.36}$$

where m is the reduced mass, n and j are the principal and total angular momentum quantum numbers, and $\alpha = e^2/4\pi\epsilon_0\hbar c \approx 1/137$ is the fine structure constant. Historically, the discrepancies between the measured energy levels and (2.36) stimulated the rapid development of quantum electrodynamics (QED) in the late 1940s. QED effects contribute to the energy levels at order α^5 and thus are very small indeed.

Today, it is commonplace to calculate the spectrum and other properties of atoms, molecules, and solids by combining microscopic Maxwell theory with non-relativistic or relativistic quantum mechanics. The calculations are necessarily approximate because the quantum part of the problem cannot be solved exactly for multi-electron systems. Nevertheless, quantitative agrement with experiment is the norm as long as the spatial variations of the charge density and electrostatic potential are retained in full. This is illustrated by Figure 2.5, which compares the charge density in crystalline silicon

[8] See, e.g., G.W. Erickson, "Energy levels of one-electron atoms", *Journal of Physical and Chemical Reference Data* **6**, 831 (1977).

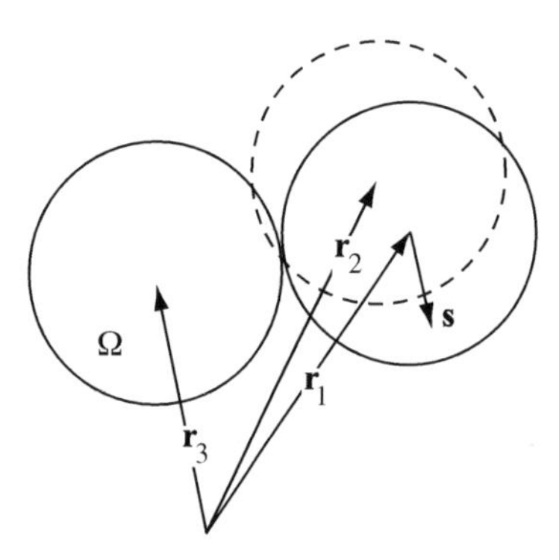

Figure 2.6: Spherical volumes centered at arbitrary points in space $\mathbf{r}_1$, $\mathbf{r}_2$, and $\mathbf{r}_3$.

as measured by x-ray diffraction with quantum mechanical calculations of the same quantity. First-principles calculations for diamagnetic molecules and ferromagnetic solids show similar agreement with experiment when the relevant magnetic fields are computed using microscopic magnetostatics. Indeed, all such evidence suggests that the vacuum Maxwell equations (2.33) and (2.34) are valid for all length scales down to the Compton wavelength of the electron $\lambda_c = h/mc \simeq 2.43 \times 10^{-12}$ m (see Section 2.5). We take this to be the spatial resolution scale for the theory.

In contrast to the spatial degrees of freedom, there is no compelling reason to average over time in the microscopic Maxwell equations. The period of typical electron motion in atoms is 10^{-15}–10^{-16} s. If this motion were averaged away, no consistent Maxwell theory of ultraviolet or x-ray radiation could be contemplated. This is clearly undesirable.

2.3.1 Lorentz Averaging

Lorentz spatial averaging is a mathematical procedure which produces slowly varying *macroscopic* sources and fields from rapidly varying *microscopic* sources and fields. No single averaging scheme applies to every physical situation and the desired resolution scale typically differs from problem to problem. Moreover, the average is almost never carried out explicitly. Nevertheless, it is important to grasp the basic ideas because Lorentz averaging produces certain characteristic features of the macroscopic theory which are absent from the microscopic theory (see Section 2.3.2).

Quantum mechanical calculations show that microscopic quantities like the electric field $\mathbf{E}_{\text{micro}}(\mathbf{r}, t)$ can vary appreciably on the scale of the Bohr radius a_B. These spatial variations average out when viewed at the much larger resolution scale of a macroscopic observer. This suggests the following three-step procedure: (1) center a sphere with volume $\Omega \gg a_B^3$ at every microscopic point $\mathbf{r}$ (see Figure 2.6); (2) carry out a spatial average of $\mathbf{E}_{\text{micro}}(\mathbf{r}, t)$ over the volume of the sphere,

$$\mathbf{E}(\mathbf{r}, t) = \frac{1}{\Omega} \int_{\Omega} d^3s \, \mathbf{E}_{\text{micro}}(\mathbf{s} + \mathbf{r}, t); \tag{2.37}$$

and (3) exploit the fact that $\mathbf{E}(\mathbf{r}, t)$ varies slowly on the resolution scale of the continuous variable $\mathbf{r}$ and replace $\mathbf{E}(\mathbf{r}, t)$ by $\mathbf{E}(\mathbf{R}, t)$ where $\mathbf{R}$ is a low-resolution spatial variable. By "low resolution" we mean that the distance between two "adjacent" points in the continuous $\mathbf{R}$-space is approximately the diameter of the averaging sphere (as measured in $\mathbf{r}$-space). Thus, the "distant" points $\mathbf{r}_1$ and $\mathbf{r}_3$ in Figure 2.6 could serve as adjacent points in $\mathbf{R}$-space. $\mathbf{E}(\mathbf{R}, t)$ is the macroscopic electric field we seek.

For a gas, a good rule of thumb for the averaging sphere is to choose Ω in Figure 2.6 and in (2.37) equal to the inverse density of atoms. For a crystal, Ω is better chosen as the volume of a unit cell. If necessary or desirable, one can perform an additional spatial average of the macroscopic variable

$\mathbf{E}(\mathbf{R}, t)$ over a larger length scale determined by the spatial resolution of an experimental probe or by the spatial extent of density or compositional inhomogeneities.

Some authors offer much more elaborate and mathematically formal discussions of Lorentz averaging than we have done here. However, experience shows that it matters little whether one uses (2.37) or some alternative scheme (e.g., one which employs a smoothly varying weight function rather than a sharp cutoff to the averaging volume) to eliminate the spatial variations that occur within the averaging volume. What matters most is that the spatial averaging algorithm be a *linear* operation. This implies that the space and time derivatives which appear in the Maxwell equations commute with spatial averaging. For example, if we use angle brackets to denote the complete Lorentz averaging procedure, it is not difficult to confirm using (2.37) that

$$\nabla_{\mathbf{R}} \times \mathbf{E}(\mathbf{R}) = \langle \nabla_{\mathbf{r}} \times \mathbf{E}_{\text{micro}}(\mathbf{r}) \rangle \tag{2.38}$$

and

$$\frac{\partial \mathbf{E}}{\partial t} = \left\langle \frac{\partial \mathbf{E}_{\text{micro}}}{\partial t} \right\rangle . \tag{2.39}$$

Similar results apply to $\nabla_{\mathbf{R}} \cdot \mathbf{E}$ and to the space and time derivatives of $\mathbf{B}$. The important conclusion to be drawn from identities like (2.38) and (2.39) is that the Maxwell equations have exactly the same *form* whether they are written in microscopic variables or macroscopic variables. Therefore, to simplify notation, we will generally write simply $\mathbf{E}(\mathbf{r}, t)$ and $\mathbf{B}(\mathbf{r}, t)$ and rely on context to inform the reader whether these symbols refer to microscopic or macroscopic quantities.

A small cloud appears on the Lorentz averaging horizon when we turn to quantities which are bilinear in the fields and sources. An example is the force on the charge density $\rho(\mathbf{r}, t)$ and the current density $\mathbf{j}(\mathbf{r}, t)$ in a volume V due to electromagnetic fields $\mathbf{E}(\mathbf{r}, t)$ and $\mathbf{B}(\mathbf{r}, t)$:

$$\mathbf{F} = \int_V d^3r \, \{\rho \mathbf{E} + \mathbf{j} \times \mathbf{B}\} . \tag{2.40}$$

Interpreted as a microscopic formula, direct substitution of ρ and $\mathbf{j}$ from (2.6) and (2.17) confirms that (2.40) reproduces the Lorentz force law (2.1) for each microscopic particle. However, it is generally the case that $\langle \rho \, \mathbf{E} \rangle \neq \langle \rho \rangle \langle \mathbf{E} \rangle$ and $\langle \mathbf{j} \times \mathbf{B} \rangle \neq \langle \mathbf{j} \rangle \times \langle \mathbf{B} \rangle$. How, then, shall we compute the force on a macroscopic body? The answer, not often stated explicitly, is that we simply *assume* that (2.40) remains valid when all the variables are interpreted macroscopically. No ambiguities arise as long as $\mathbf{F}$ is the total force on an isolated sample of macroscopic matter in vacuum.

2.3.2 The Macroscopic Surface

Lorentz averaging unavoidably produces singularities and discontinuities in macroscopic quantities when the averaging is performed in the immediate vicinity of a surface or interface. This is so because the input microscopic quantities (which are perfectly smooth and continuous at every point in space) exhibit rapid spatial variations near surfaces which are uncharacteristic of the variations which occur elsewhere. To illustrate this, the top panel of Figure 2.7 shows a contour map of the microscopic, ground state, valence electron charge density $\rho_0(\mathbf{r})$ near a flat, crystalline surface of metallic Ag in vacuum. The density was calculated by quantum mechanical methods similar to those used to obtain Figure 2.5(b).

The corrugation of the contour lines of $\rho_0(\mathbf{r})$ adjacent to the vacuum region (to the right) is characteristic of vacuum interfaces. Another characteristic feature—the "spilling-out" of the electron distribution into the vacuum—becomes most apparent when we average $\rho_0(\mathbf{r})$ over planes parallel to

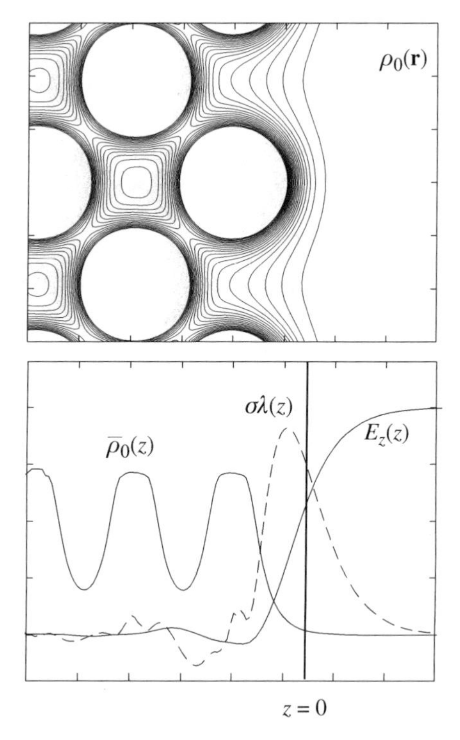

Figure 2.7: Side view of the free surface of crystalline Ag metal. The vacuum lies to the right. Top panel: contour plot of the valence electron charge density $\rho_0(\mathbf{r})$ in one crystalline plane perpendicular to the surface. The white circles are regions of very high electron density centered on the atomic nuclei. Bottom panel: the solid curve labeled $\bar{\rho}_0(z)$ is $\rho_0(\mathbf{r})$ averaged over planes parallel to the surface. The local maxima of this curve are rounded off and do not accurately reflect the electron density near the nuclei; the dashed curve labeled $\sigma\lambda(z)$ is the planar average of the *change* in electron density induced by an electric field directed to the left; the solid curve labeled $E_z(z)$ is the planar average of the electric field perpendicular to the macroscopic surface. The vertical scale is different for each curve. The solid vertical line is the location $z = 0$ of the macroscopic surface. Tick marks on the horizontal scale are separated by $2a_{\rm Bohr} \simeq 1$ Å. Figure adapted from Ishida and Liebsch (2002).

the surface with area A. The function that results,

$$\bar{\rho}_0(z) = \frac{1}{A}\int_A dx\,dy\,\rho_0(\mathbf{r}), \tag{2.41}$$

is plotted in the bottom panel of Figure 2.7. The essential point is that $\bar{\rho}_0(z)$ falls to zero in the vacuum over a distance comparable to the interatomic spacing. The change of scale which accompanies Lorentz averaging in the z-direction destroys the fine resolution needed to see this behavior. This obliges us to define the "edge" of the macroscopic surface (solid vertical line) as a plane where the macroscopic charge distribution falls discontinuously to zero.

Now apply a horizontal electric field perpendicular to the surface and directed to the left. This perturbation induces a distortion of the electron wave functions in the immediate vicinity of the Ag surface. If $\bar{\rho}(z)$ is the planar average of the electronic charge density *in the presence of the field* and σ is the induced charge per unit area of surface, a function $\lambda(z)$ is defined by

$$\bar{\rho}(z) = \bar{\rho}_0(z) + \sigma\lambda(z). \tag{2.42}$$

The dashed curve of $\sigma\lambda(z)$ in the lower panel of Figure 2.7 shows that the main effect of the electric field is to pull some electronic charge away from the last plane of nuclei and farther out into the vacuum. The horizontal scale (about 1 Å between tick marks) shows that the width of the dashed curve cannot be resolved macroscopically. Therefore, after Lorentz averaging in the z-direction, all the induced charge resides in the $z = 0$ plane. Quantitatively, Lorentz averaging replaces the smooth microscopic function $\sigma\lambda(z)$ by a singular macroscopic delta function:[9]

$$\sigma\lambda(z) \to \sigma\delta(z) \qquad \text{(Lorentz averaging).} \tag{2.43}$$

The curve labeled $E_z(z)$ in the lower panel of Figure 2.7 is the planar average of the microscopic electric field (see Section 2.4.2) in the direction normal to the surface. $E_z(z)$ is non-zero in the vacuum, but falls very rapidly to zero in the near-surface region. Therefore, a Lorentz average in the z-direction recovers the familiar result that the macroscopic electric field just *outside* a perfect conductor drops discontinuously to zero just *inside* a perfect conductor (see Section 5.2.2).

2.3.3 Matching Conditions

There is an analytic connection between the discontinuous electric field and the singular distribution of surface charge discussed in the previous section. To discover it, let the plane $z = 0$ separate two regions labeled L and R. $\mathbf{E}_L(\mathbf{r})$ and $\rho_L(\mathbf{r})$ are the macroscopic electric field and charge density in region L. $\mathbf{E}_R(\mathbf{r})$ and $\rho_R(\mathbf{r})$ are the macroscopic electric field and charge density in region R. Now, the step function $\Theta(z)$ is defined (Section 1.5.3) by

$$\Theta(z) = \begin{cases} 0 & z < 0, \\ 1 & z > 0. \end{cases} \tag{2.44}$$

Using this function, we can write the electric field at every point in space as

$$\mathbf{E}(\mathbf{r}) = \mathbf{E}_R(\mathbf{r})\Theta(z) + \mathbf{E}_L(\mathbf{r})\Theta(-z). \tag{2.45}$$

The charge density can be written similarly except that we must allow for the possibility of a surface charge density $\sigma(x, y)$ localized exactly at $z = 0$. In light of (2.43), we write

$$\rho(\mathbf{r}) = \rho_R(\mathbf{r})\Theta(z) + \rho_L(\mathbf{r})\Theta(-z) + \sigma\delta(z). \tag{2.46}$$

Motivated by Gauss' law, $\nabla \cdot \mathbf{E} = \rho/\epsilon_0$, the divergence of (2.45) is[10]

$$\nabla \cdot \mathbf{E} = [\nabla \cdot \mathbf{E}_R]\Theta(z) + [\nabla \cdot \mathbf{E}_L]\Theta(-z) + \hat{\mathbf{z}} \cdot (\mathbf{E}_R - \mathbf{E}_L)\delta(z). \tag{2.47}$$

On the other hand, $\nabla \cdot \mathbf{E}_L = \rho_L/\epsilon_0$ and $\nabla \cdot \mathbf{E}_R = \rho_R/\epsilon_0$. Therefore, if we set (2.46) equal to (2.47) times ϵ_0 and use square brackets to denote the evaluation of $\mathbf{E}_L$ and $\mathbf{E}_R$ at points which are infinitesimally close to one another on opposite sides of $z = 0$, the final result is

$$\hat{\mathbf{z}} \cdot [\mathbf{E}_R - \mathbf{E}_L] = \sigma/\epsilon_0. \tag{2.48}$$

This is a *matching condition* which relates the discontinuity in the normal component of the macroscopic electric field to the magnitude of the singular charge density at the interface. Our derivation based on the differential form of Gauss' law draws explicit attention to the non-analytic nature of the

[9] The centroid of $\lambda(z)$ does not exactly coincide with the macroscopic termination of the zero-field charge distribution. We may locate the macroscopic induced charge density at $z = 0$ nevertheless because the error we make is undetectable at the macroscopic scale.

[10] $\nabla \cdot [\mathbf{f}\Theta(\pm z)] = \Theta(\pm z)\nabla \cdot \mathbf{f} + \mathbf{f} \cdot \nabla\Theta(\pm z) = \Theta(\pm z)\nabla \cdot \mathbf{f} \pm \delta(z)\mathbf{f} \cdot \hat{\mathbf{z}}$.

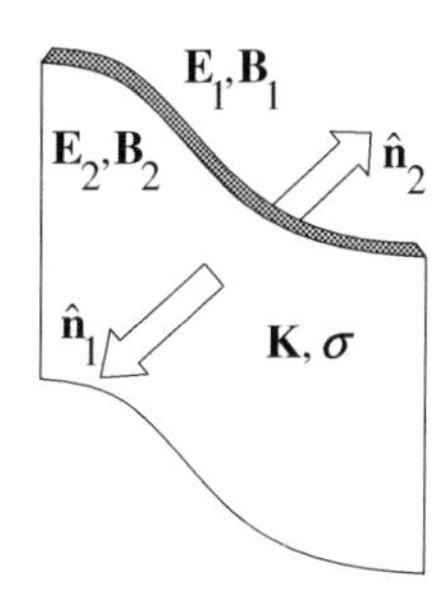

Figure 2.8: An infinitesimally thin interface between two macroscopic regions (labeled 1 and 2) carries a surface charge density σ and a surface current density $\mathbf{K}$. The unit normal $\hat{\mathbf{n}}_1$ points outward from region 1. The unit normal $\hat{\mathbf{n}}_2$ points outward from region 2.

macroscopic fields and sources [exemplified by (2.45) and (2.46)] when surfaces and interfaces are present.

Three additional matching conditions follow similarly from the three remaining Maxwell equations. The general case is an interface with a surface charge density $\sigma(\mathbf{r}_S)$ and a surface current density $\mathbf{K}(\mathbf{r}_S)$. If the unit normal $\hat{\mathbf{n}}_1$ points outward from region 1 and the unit normal $\hat{\mathbf{n}}_2$ points outward from region 2 (Figure 2.8), the full set of matching conditions is

$$\begin{aligned} \hat{\mathbf{n}}_2 \cdot [\mathbf{E}_1 - \mathbf{E}_2] &= \sigma/\epsilon_0 \\ \hat{\mathbf{n}}_2 \cdot [\mathbf{B}_1 - \mathbf{B}_2] &= 0 \\ \hat{\mathbf{n}}_2 \times [\mathbf{E}_1 - \mathbf{E}_2] &= 0 \\ \hat{\mathbf{n}}_2 \times [\mathbf{B}_1 - \mathbf{B}_2] &= \mu_0 \mathbf{K}. \end{aligned} \tag{2.49}$$

The two middle equations in (2.49) express the fact that the normal component of $\mathbf{B}(\mathbf{r}, t)$ and the tangential component of $\mathbf{E}(\mathbf{r}, t)$ are continuous across the interface. The last equation in (2.49) indicates that the tangential component of $\mathbf{B}(\mathbf{r}, t)$ suffers a discontinuity proportional to the local surface current density. Later in the text, we will derive the relations (2.49) using more "physical" arguments.

We close with the remark that the stated matching conditions apply only to an interface which is at rest in the frame of reference where the fields are measured. The last two conditions in (2.49) each acquire an extra term when the interface moves with velocity $\mathbf{v}$ namely,[11]

$$\begin{aligned} \hat{\mathbf{n}}_2 \times [\mathbf{E}_1 - \mathbf{E}_2] - (\mathbf{v} \cdot \hat{\mathbf{n}}_2)[\mathbf{B}_1 - \mathbf{B}_2] &= 0 \\ \hat{\mathbf{n}}_2 \times [\mathbf{B}_1 - \mathbf{B}_2] + \frac{(\hat{\mathbf{v}} \cdot \hat{\mathbf{n}}_2)}{c^2}[\mathbf{E}_1 - \mathbf{E}_2] &= \mu_0 \mathbf{K}. \end{aligned} \tag{2.50}$$

2.4 The Maxwell Equations in Matter

The 19th-century founders of electromagnetism understood their subject very differently than we do today. The physical nature of charge and current inside matter was unknown, so attention focused on the response of an experimental sample to external sources of charge or current. As a result, the theory developed somewhat differently from the "history" recounted in Section 2.2. William Thomson (Lord Kelvin) appreciated that the magnetic field $\mathbf{B}$ produced by Faraday induction differed in some essential way from the field (called $\mathbf{H}$ by him) induced in matter by a steady external current. Similarly,

[11] See Namias (1988) in Sources, References, and Additional Reading.

Maxwell appreciated that the electric field **E** responsible for the force on a charged object differed in some essential way from the field (called **D** by him) induced in matter by an external charge.

Today, we understand that the distinction between Thomson's **B** and **H** and Maxwell's **E** and **D** arises when we distinguish the fields produced by charge and current densities *intrinsic* to the matter from the fields produced by sources *extrinsic* to the matter. It nevertheless proves useful to retain the 19th-century language of the founders. First, it is deeply embedded in the literature and developed intuition of the subject. Second, there are many situations where the four-field formalism simplifies calculations. Finally, the constitutive relationships between **D** and **E** (on the one hand) and between **H** and **B** (on the other hand) force us to confront the quantum effects that distinguish real matter from mere distributions of charge and current.[12]

2.4.1 Macroscopic Sources and Fields

The macroscopic charge density is *zero* at every point inside an isolated sample of neutral matter.[13] The same is true of the macroscopic current density if we except ferromagnetic matter. There being no sources, it follows that an isolated sample of matter produces no electromagnetic field. However, suppose we introduce "free" densities of charge and current, $\rho_f(\mathbf{r}, t)$ and $\mathbf{j}_f(\mathbf{r}, t)$, which are entirely extrinsic to the matter. The fields produced by the extrinsic sources induce charge reorganization and current flow in dielectric matter and current flow in magnetic matter. The fields produced by these induced sources contribute to the total field both inside and outside the matter.

It is traditional to use a vector field $\mathbf{P}(\mathbf{r}, t)$ to characterize the polarization of a dielectric. A vector field $\mathbf{M}(\mathbf{r}, t)$ is used similarly to characterize the magnetization of a magnet. These quantities enter Maxwell's theory by writing

$$\rho(\mathbf{r}, t) = \rho_f(\mathbf{r}, t) - \nabla \cdot \mathbf{P}(\mathbf{r}, t) \tag{2.51}$$

and

$$\mathbf{j}(\mathbf{r}, t) = \mathbf{j}_f(\mathbf{r}, t) + \nabla \times \mathbf{M}(\mathbf{r}, t) + \frac{\partial \mathbf{P}(\mathbf{r}, t)}{\partial t}. \tag{2.52}$$

Later chapters will explain why (2.51) and (2.52) have the forms they do.

We now define the auxiliary macroscopic fields introduced by Maxwell and Thomson as

$$\mathbf{D}(\mathbf{r}, t) = \epsilon_0 \mathbf{E}(\mathbf{r}, t) + \mathbf{P}(\mathbf{r}, t) \tag{2.53}$$

and

$$\mathbf{H}(\mathbf{r}, t) = \mu_0^{-1} \mathbf{B}(\mathbf{r}, t) - \mathbf{M}(\mathbf{r}, t). \tag{2.54}$$

These definitions are natural because, when (2.51) and (2.52) are substituted into (2.33) and (2.34), the resulting equations have a very simple form in terms of the auxiliary fields namely,

$$\nabla \cdot \mathbf{D} = \rho_f \qquad \nabla \cdot \mathbf{B} = 0 \tag{2.55}$$

and

$$\nabla \times \mathbf{E} = -\frac{\partial \mathbf{B}}{\partial t} \qquad \nabla \times \mathbf{H} = \mathbf{j}_f + \frac{\partial \mathbf{D}}{\partial t}. \tag{2.56}$$

[12] Some textbooks refer to **E** as the "electric field", **H** as the "magnetic field", **D** as the "electric displacement", and **B** as the "magnetic induction". In this book, only the *electric field* **E** and the *magnetic field* **B** are fundamental. We give no special names to the auxiliary fields **D** and **H**.

[13] This macroscopic charge density is the Lorentz average (Section 2.3.1) of the microscopic charge density associated with the distribution of protons and electrons in the matter. The microscopic charge density is non-zero and varies rapidly in space.

Comparing (2.55) and (2.56) to the vacuum Maxwell equations (2.33) and (2.34) shows that the first and the last of the matching conditions in (2.49) change to

$$\begin{aligned}\hat{\mathbf{n}}_2 \cdot [\mathbf{D}_1 - \mathbf{D}_2] &= \sigma_f \\ \hat{\mathbf{n}}_2 \times [\mathbf{H}_1 - \mathbf{H}_2] &= \mathbf{K}_f .\end{aligned} \tag{2.57}$$

The two scalar equations in (2.55) and the six components of the two vector equations of (2.56) are not sufficient to determine the twelve components of **E**, **B**, **D**, and **H**. So-called "constitutive relations" of the form $\mathbf{D} = \mathbf{D}\{\mathbf{E}, \mathbf{B}\}$ and $\mathbf{H} = \mathbf{H}\{\mathbf{E}, \mathbf{B}\}$ are needed to close the equation set. When the field strengths are low, a linear approximation is often valid. In the static limit, this defines the dielectric permittivity ϵ and the magnetic permeability μ of macroscopic electrodynamics:

$$\mathbf{D}(\mathbf{r}) = \epsilon \mathbf{E}(\mathbf{r}) \qquad \mathbf{B}(\mathbf{r}) = \mu \mathbf{H}(\mathbf{r}). \tag{2.58}$$

Equivalently, we can work with the electric susceptibility χ and the magnetic susceptibility χ_m:

$$\mathbf{P}(\mathbf{r}) = \epsilon_0 \chi \mathbf{E}(\mathbf{r}) \qquad \mathbf{M}(\mathbf{r}) = \chi_m \mathbf{H}(\mathbf{r}). \tag{2.59}$$

In this book, we will use the term "simple" media to refer to matter where (2.58) and (2.59) are good approximations. We will also discuss "less-simple" media where these formulae break down and require generalization.

2.4.2 Microscopic Fields

Figure 2.5 and Figure 2.7 testify to our ability to calculate charge densities and electric fields in matter at the microscopic scale. On the other hand, not every macroscopic equation in the preceding section remains valid when we pass to the micro-scale. An example is the constitutive relation (2.58), which is known to be incorrect for microscopic applications. Indeed, rather than using $D_z(z) = \epsilon E_z(z)$, or even $D_z(z) = \epsilon(z) E_z(z)$, the field $E_z(z)$ plotted in the bottom panel of Figure 2.7 was calculated using a spatially *non-local* dielectric function $\epsilon(z, z')$ which permits the field at one point in space to influence the field at nearby points in space:[14]

$$D_z(z) = \int dz' \epsilon(z, z') E_z(z'). \tag{2.60}$$

Substituting (2.60) into $\nabla \cdot \mathbf{D} = 0$ gives

$$\frac{d}{dz} \int dz' \epsilon(z, z') E_z(z') = 0. \tag{2.61}$$

This integrates to

$$\int dz' \epsilon(z, z') E_z(z') = \epsilon_0 E_z(\infty), \tag{2.62}$$

where the integration constant $E_z(\infty)$ is the (specified) field far from the surface (in the vacuum) where $\epsilon(z \to \infty, z') = \delta(z - z')$. The hard part of this problem is the quantum mechanical calculation of $\epsilon(z, z')$. Once that is known, (2.62) is an integral equation for $E_z(z)$ which yields the smooth and continuous function plotted in Figure 2.7.

[14] Usually, $\epsilon(z, z') \to 0$ if $|z - z'|$ is larger than one or two atomic spacings.

2.5 Quantum Limits and New Physics

The domain of classical electromagnetism is very large, but not infinite. The known limits of the theory are set by quantum mechanics. The unknown limits of the theory are set by the possible discovery of "new physics" which would render the theory incomplete. This section provides a brief introduction to both these topics.

2.5.1 Quantized Matter

Classical electromagnetism is a theory of the electromagnetic field and its interaction with matter. The entire enterprise works wonderfully well for macroscopic distributions of matter like capacitors, electromagnets, circuit elements, and antennas. Difficulties arise at the micro-scale when quantum effects begin to assert themselves. The photoelectric effect and the absorption and emission of radiation by atoms are historically significant examples where this occurs. Happily, the solution in many cases is a semi-classical description where the matter is treated quantum mechanically and the electromagnetic field is treated classically. Calculations of this kind produce quantitatively accurate results for the rates at which matter absorbs/emits electromagnetic waves and for the rate at which electrons are liberated from matter by photoemission. We will see an example in Chapter 20, where the replacement of a classical particle current by a first-quantized particle current transforms a completely classical expression into a proper quantum mechanical expression.

2.5.2 Vacuum Polarization

Maxwell's theory is the classical limit of quantum electrodynamics (QED), a theory where charged particles and electromagnetic fields are treated on an equal footing as quantum objects. It is the most accurate theory of Nature we possess. Second quantization produces electrodynamic effects which cannot be described by any classical theory. An example is *vacuum polarization*, which is the virtual excitation of electron-positron pairs in the presence of an external electromagnetic field. If the external field is produced by a static point charge q, this relativistic quantum effect modifies Coulomb's law at distances less than the Compton wavelength of the electron, $\lambda_c = h/mc$. To lowest order in the fine structure constant, $\alpha = e^2/2\epsilon_0 hc$, the QED result for the electrostatic potential of a point charge is[15]

$$\varphi(r) = \frac{q(r)}{4\pi\epsilon_0 r} = \frac{q}{4\pi\epsilon_0 r} \times \begin{cases} 1 - \dfrac{2\alpha}{3\pi}\ln(r/\lambda_c) & r/\lambda_c \ll 1, \\ 1 + \dfrac{\alpha}{4\sqrt{\pi}}\left(\dfrac{r}{\lambda_c}\right)^{-3/2} e^{-2r/\lambda_c} & r/\lambda_c \gg 1. \end{cases} \tag{2.63}$$

We can interpret the behavior of $q(r)$ as the quantum analog of the screening of a point charge q by a medium with dielectric permittivity $\epsilon = \epsilon_0\kappa$ (Chapter 6). In a dielectric medium, polarization charge in the immediate vicinity of the source reduces the effective value of the charge from q to q/κ. In the present case, the *divergent* "bare" charge $q(0)$ polarizes the vacuum in its immediate vicinity by attracting virtual electrons and repelling virtual positrons. When observed from distances $r \gg \lambda_c$, the effective magnitude of the source charge is "renormalized" to the finite value q.

[15] See Section 114 of V.B. Berestetskii, E.M. Lifshitz, and L.P. Pitaevskii, *Quantum Electrodynamics*, 2nd edition (Pergamon, Oxford, 1980).

The effects of vacuum polarization become significant when the external field strengths approach

$$
\begin{aligned}
E_c &\sim \frac{mc^2}{e\lambda_c} = \frac{m^2c^3}{eh} \sim 10^{18}\ \mathrm{V/m} \\
B_c &\sim \frac{m^2c^2}{eh} \sim 10^9\ \mathrm{T}.
\end{aligned} \tag{2.64}
$$

Detailed calculations predict a dramatic effect: the breakdown of the linearity of the vacuum Maxwell equations (and thus the principle of superposition) when the field strengths approach (2.64). This QED result (valid for slowly varying fields) can be cast in the form of non-linear constitutive relations for the vacuum. To lowest order in α, virtual pair production generates a vacuum polarization $\mathbf{P}$ and a vacuum magnetization $\mathbf{M}$ given by[16]

$$
\begin{aligned}
\mathbf{P} &= \frac{2\epsilon_0\alpha}{E_c^2}\left\{2(E^2 - c^2B^2)\mathbf{E} + 7c^2(\mathbf{E}\cdot\mathbf{B})\mathbf{B}\right\} \\
\mathbf{M} &= -\frac{2\alpha}{\mu_0 E_c^2}\left\{2(E^2 - c^2B^2)\mathbf{B} + 7(\mathbf{E}\cdot\mathbf{B})\mathbf{E}\right\}.
\end{aligned} \tag{2.65}
$$

At the time of this writing, these effects have not yet been detected.

2.5.3 Quantum Fluctuations

An entirely different restriction on the validity of classical electrodynamics arises because $\mathbf{E}(\mathbf{r}, t)$ and $\mathbf{B}(\mathbf{r}, t)$ are non-commuting vector operators in QED rather than c-number vector fields. We infer from the uncertainty principle that the electric field and the magnetic field cannot take on sharp values simultaneously. Quantum fluctuations of the field amplitudes and phases are always present, even in the vacuum state. On the other hand, the fields produced by macroscopic sources like a light bulb, a laser, a microwave generator, or a blackbody radiator invariably exhibit (much larger) non-quantum fluctuations also. The non-classical regime of *quantum optics* emerges when the classical fluctuations are suppressed to reveal the quantum fluctuations.

Glauber (1963) pointed out that it is possible to distinguish a classical electromagnetic field from a quantum electromagnetic field by focusing on the (time-averaged) field intensity operator $\hat{I}$ and expectation values like

$$
g^{(2)} = \frac{\langle \hat{I}^2 \rangle}{\langle \hat{I} \rangle^2}. \tag{2.66}
$$

Field intensity is a positive quantity, so $g^{(2)} \geq 0$ whether the fluctuations are classical or quantum. However, for fields described by classical electrodynamics, the sharper inequality $g^{(2)} \geq 1$ holds.

Quantum effects generally reveal themselves when we pass from the macroscopic limit of many atoms to the microscopic limit of one or a few atoms. This suggests that a single atom should be a good source of non-classical radiation. The data for $g^{(2)}$ shown in Figure 2.9 illustrate this for low-intensity laser light directed through a cavity containing about a dozen Rb atoms. The parameter $\Delta = 2(\omega_L - \omega_A)\tau$ is the difference between the laser frequency ω_L and the frequency ω_A of a Rb atomic transition, normalized by the radiative lifetime τ of the transition. When $\Delta \simeq 0$, the atoms resonantly scatter the laser light and non-classical values $0 \leq g^{(2)} < 1$ are seen. This means that the operator character of the field variables plays an essential role in the description of the transmitted

[16] See, for example, M. Soljačić and M. Segev, "Self-trapping of electromagnetic beams in vacuum supported by QED nonlinear effects", *Physical Review A* **62**, 043817 (2000) and references therein.

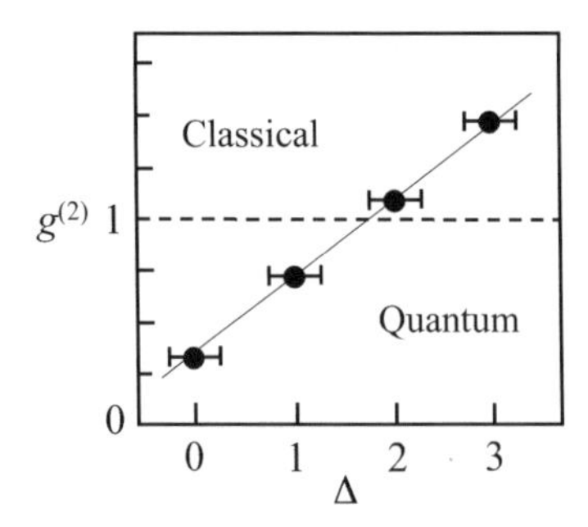

Figure 2.9: Cavity QED data for $g^{(2)}$ as defined in (2.66). The horizontal axis is the laser detuning Δ. The classical theory is valid only when $g^{(2)} \geq 1$. The solid line is a guide to the eye. Figure adapted from Foster, Mielke, and Orozco (2000).

light. As the detuning Δ increases, the atoms scatter less and less. Eventually, $g^{(2)}$ exceeds unity and classical theory suffices to describe the transmitted light as well as the incident light.[17]

2.5.4 New Physics

The Maxwell equations are the mathematical expression of the known facts of experimental electromagnetism. They are subject to modification if any future experiment gives convincing evidence for deviations from any of these "facts". Conversely, one can speculate that the laws of physics are not precisely as we usually imagine them and posit that the Maxwell equations differ (albeit minutely) from their usual form. If so, the modified theory will predict heretofore unobserved phenomena which can be sought in the laboratory. If not seen in experiment, such measurements set limits on the size of the presumed deviations from the orthodox theory.

Two examples of very long standing are the possibilities that (i) Coulomb's law is not precisely inverse-square and (ii) magnetic charge exists and is a source for magnetic fields. The Coulomb question is often addressed by supposing that the force between two point charges varies as $1/r^{2+\epsilon}$ rather than as $1/r^2$. This proposal—which we interpret today as a signature of a non-zero mass for the photon—has a variety of experimental implications. An electrostatic test was performed by Maxwell himself, who concluded that $|\epsilon| \leq 5 \times 10^{-5}$. Contemporary experiments of the same basic design give $|\epsilon| \leq 6 \times 10^{-17}$.[18] Magnetic charge is the subject of Section 2.5.5 below.

Apart from the massive photon and magnetic charge, a variety of other "new physics" scenarios have been suggested which alter the Maxwell equations in various ways. These speculations include (a) charge is not exactly conserved; (b) electromagnetism is not exactly the same in every inertial frame; and (c) electromagnetism violates rotational and/or inversion symmetry. There is no experimental evidence for any of these at the present time, but it is necessary to keep an open mind.

2.5.5 Magnetic Charge

As first-rate physical theories go, the Maxwell equations are embarrassingly asymmetrical. $\nabla \cdot \mathbf{E}$ is proportional to an electric charge density ρ, but $\nabla \cdot \mathbf{B}$ is *not* proportional to a magnetic charge

[17] In this experiment, the average number of photons in the cavity is always much less than one. That is, $\epsilon_0 E^2 V/\hbar\omega \ll 1$, where V is the cavity volume. The data for $\Delta > 2$ in Figure 2.9 show that very small mean photon number alone is not sufficient to guarantee that an electromagnetic field is non-classical.

[18] See Sources, References, and Additional Reading for references to the experimental and theoretical literature on this subject.

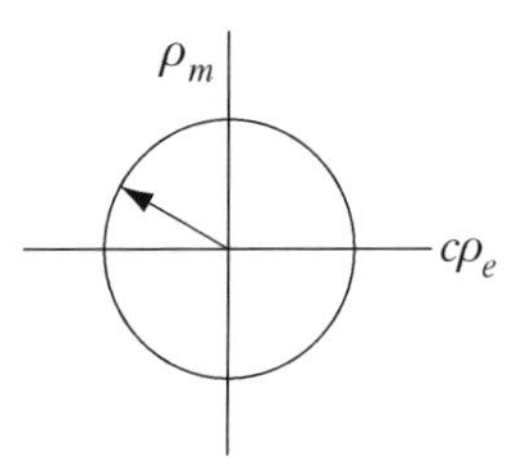

Figure 2.10: Allowed values of ρ_m and $c\rho_e$ lie on the circle. The radius vector indicates the ratio of magnetic charge to electric charge for a hypothetical elementary particle.

density ρ_m. Similarly, an electric current density $\mathbf{j}$ appears in the Ampère-Maxwell law, but no magnetic current density $\mathbf{j}_m$ appears in Faraday's law. To the extent that we associate symmetry with mathematical beauty, the Maxwell equations violate Dirac's dictum that "physical laws should have mathematical beauty".[19]

This state of affairs has led many physicists to *symmetrize* the Maxwell equations by supposing that (i) magnetic charge exists and (ii) the motion of particles with magnetic charge produces a magnetic current density $\mathbf{j}_m$ which satisfies $\nabla \cdot \mathbf{j}_m + \partial\rho_m/\partial t = 0$. If we temporarily let ρ_e and $\mathbf{j}_e$ stand for the usual electric charge density and current density, these assumptions generalize (2.33) and (2.34) to[20]

$$\nabla \cdot \mathbf{E} = \frac{\rho_e}{\epsilon_0} \qquad \nabla \cdot \mathbf{B} = \mu_0 \rho_m \tag{2.67}$$

$$\nabla \times \mathbf{E} = -\mu_0 \mathbf{j}_m - \frac{\partial \mathbf{B}}{\partial t} \qquad \nabla \times \mathbf{B} = \mu_0 \mathbf{j}_e + \frac{1}{c^2}\frac{\partial \mathbf{E}}{\partial t}. \tag{2.68}$$

The new terms acquire meaning from a similarly generalized Coulomb-Lorentz force density,[21]

$$\mathbf{f} = (\rho_e \mathbf{E} + \mathbf{j}_e \times \mathbf{B}) + (\rho_m \mathbf{B} - \mathbf{j}_m \times \mathbf{E}/c^2). \tag{2.69}$$

For present purposes, the most interesting property of (2.67), (2.68), and (2.69) is that they are invariant to a *duality* transformation of the fields and sources parameterized by an angle θ:

$$\mathbf{E}' = \mathbf{E}\cos\theta + c\mathbf{B}\sin\theta \qquad c\mathbf{B}' = -\mathbf{E}\sin\theta + c\mathbf{B}\cos\theta \tag{2.70}$$

$$c\rho'_e = c\rho_e \cos\theta + \rho_m \sin\theta \qquad \rho'_m = -c\rho_e \sin\theta + \rho_m \cos\theta \tag{2.71}$$

$$c\mathbf{j}'_e = c\mathbf{j}_e \cos\theta + \mathbf{j}_m \sin\theta \qquad \mathbf{j}'_m = -c\mathbf{j}_e \sin\theta + \mathbf{j}_m \cos\theta. \tag{2.72}$$

This means that $\mathbf{E}'$, $\mathbf{B}'$, ρ'_e, $\mathbf{j}'_e$, ρ'_m, and $\mathbf{j}'_m$ satisfy exactly the same equations as their unprimed counterparts. The only constraints are those imposed by the transformation itself:

$$c^2\rho_e^2 + \rho_m^2 = c^2{\rho'_e}^2 + {\rho'_m}^2. \tag{2.73}$$

Duality implies that it is strictly a matter of convention whether we say that a particle has electric charge only, magnetic charge only, or some mixture of the two. To see this, let the circle in Figure 2.10 be the locus of values of $c\rho_e$ and ρ_m permitted by (2.73). The radius vector specifies the ratio $\rho_m/c\rho_e$ for a hypothetical elementary particle with, say, electric charge $e < 0$ and magnetic charge $g > 0$. However, if the same ratio applies to every other particle in the Universe, no electromagnetic prediction changes if we exploit dual symmetry and rotate the radius vector (choose θ) to make $\rho_m = 0$ for every particle.

[19] Famously written on a blackboard at Moscow State University in 1955.

[20] A particle with no electric charge and one unit of magnetic charge is called a *magnetic monopole*.

[21] We have chosen the SI unit of magnetic charge as the A·m. Some authors choose this unit as the weber, in which case $\mu_0\rho_m \to \rho_m$ and $\mu_0\mathbf{j}_m \to \mathbf{j}_m$ in our extended Maxwell equations and the magnetic charge and current contributions to (2.69) should be divided by μ_0.

This brings us back to the original Maxwell-Lorentz equations, which are consistent with all known experiments. On the other hand, if an elementary particle is ever discovered where the intrinsic ratio g/ce differs from the value shown in Figure 2.10, the option to simultaneously "rotate away" magnetic charge for all particles disappears. In that case, (2.67), (2.68), and (2.69) become the fundamental electromagnetic laws of Nature. This exciting possibility keeps searches for magnetic monopoles an active part of experimental physics.

2.6 SI Units

Maxwell's theory of electromagnetism forced a merger of two mature disciplines (electricity and magnetism) which had developed their own terminologies and practical systems of units. To design a common system of units, practitioners adhered to two principles: self-consistency and convenience. The first was easy to achieve. But, because convenience is in the eye of the beholder, it was inevitable that many different systems of electromagnetic units would be developed and put into use. Most prominent among these today are SI (Système International) units and Gaussian (G) units. This section gives a brief introduction to the SI units used exclusively in this book. Appendix B describes the Gaussian system and explains how to convert from SI to Gaussian and vice-versa.

The SI system is designed to ensure that mechanical energy and electrical energy are measured in exactly the same units. To that end, the meter (m), kilogram (kg), and second (s) are *defined* as independent *base units* for length, mass, and time. The derived unit of force is the newton ($\mathrm{N} = \mathrm{kg{\cdot}m/s^2}$). We begin by applying these choices to the force on a point charge q_1 due to the presence of a point charge q_2 at a distance r. If k_e is a constant to be determined below, this force is

$$F_e = k_e \frac{q_1 q_2}{r^2} = q_1 E. \tag{2.74}$$

The right side of (2.74) *defines* $E = k_e q_2/r^2$ as the magnitude of the radial electric field at a distance r from a point charge q_2. Another application is the force on a length L_2 of straight wire carrying a current I_2 due to the presence of a parallel wire with current I_1 at a distance d. If k_m is a second constant to be determined below, this force is

$$F_m = k_m \frac{2 I_1 I_2 L_2}{d} = I_2 L_2 B. \tag{2.75}$$

The right side of (2.75) *defines* $B = 2k_m I_1/d$ as the magnitude of the circumferential magnetic field at a perpendicular distance d from an infinite wire carrying a current I_1.

The SI system defines the ampere (A) as a fourth independent base unit of current. The derived unit of charge is the Coulomb ($\mathrm{C} = \mathrm{A{\cdot}s}$) because $I = dq/dt$ relates charge and current. With this information, (2.74) and (2.75) imply that E/B has dimensions of velocity and that k_e/k_m is a constant with dimensions of (velocity)^2. These facts are all we need to write dimensionally correct forms for the Lorentz force law,

$$\mathbf{F} = q(\mathbf{E} + \boldsymbol{v} \times \mathbf{B}), \tag{2.76}$$

and the Maxwell equations,

$$\nabla \cdot \mathbf{E} = 4\pi k_e \rho \qquad\qquad \nabla \cdot \mathbf{B} = 0 \tag{2.77}$$

$$\nabla \times \mathbf{E} = -\frac{\partial \mathbf{B}}{\partial t} \qquad\qquad \nabla \times \mathbf{B} = 4\pi k_m \mathbf{j} + \frac{k_m}{k_e}\frac{\partial \mathbf{E}}{\partial t}. \tag{2.78}$$

The factors of 4π are needed to guarantee that these equations reproduce the electric field for a point charge and the magnetic field of an infinite straight wire as defined in (2.74) and (2.75).

To make further progress, we put $\rho = \mathbf{j} = 0$ and take the curl of one of the two equations in (2.78). We then use the vector identity $\nabla \times (\nabla \times \mathbf{F}) = -\nabla^2 \mathbf{F} + \nabla(\nabla \cdot \mathbf{F})$ and substitute in from the other curl equation in (2.78). When these steps are carried out, separately, for each equation in (2.78), the result is

$$\nabla^2 \mathbf{E} - \frac{k_m}{k_e} \frac{\partial^2 \mathbf{E}}{\partial t^2} = 0 \quad \text{and} \quad \nabla^2 \mathbf{B} - \frac{k_m}{k_e} \frac{\partial^2 \mathbf{B}}{\partial t^2} = 0. \tag{2.79}$$

These equations have wave-like solutions which propagate at the speed of light c (as seen in experiment) if

$$\frac{k_e}{k_m} = c^2. \tag{2.80}$$

Until 1983, the value of c was determined from experiment. Since that time, the speed of light has been defined (by the General Conference on Weights and Measures) as exactly

$$c = 299\,792\,458 \text{ m/s}. \tag{2.81}$$

This legislation demotes the meter to a derived unit. It is the distance traveled by light in 1/299792458 seconds.

We have said that the *raison d'être* of the SI system is to make the joule (J = N·m) the natural unit of electrical energy, just as it is for mechanical energy. In 1901, Giorgi pointed out that this will be the case if the constant in (2.75) is chosen to be

$$k_m = 10^{-7} \frac{\text{N}}{\text{A}^2}. \tag{2.82}$$

This choice fixes the definition of the ampere. It is the constant current which, if maintained in two straight parallel conductors of infinite length and negligible circular cross section, produces a force of 2×10^{-7} N per meter of length when the two wires are separated by a distance of 1 m.

Finally, SI eliminates ("rationalizes") the factors of 4π in the Maxwell equations by introducing a magnetic constant μ_0 where

$$\mu_0 = 4\pi k_m = 4\pi \times 10^{-7} \frac{\text{N}}{\text{A}^2}, \tag{2.83}$$

and, using (2.80), an electric constant ϵ_0 where

$$\epsilon_0 = 1/\mu_0 c^2 = 1/4\pi k_e. \tag{2.84}$$

With these definitions, (2.77) and (2.78) take the forms quoted in (2.33) and (2.34). No particular physical meaning attaches to either μ_0 or ϵ_0.

The polarization **P** has dimensions of a volume density of electric dipole moment. The magnetization **M** has dimensions of a volume density of magnetic dipole moment. This is enough information to see that (2.51) and (2.52) are dimensionally correct. Dimensional consistency similarly dictates the appearance of the factors ϵ_0 and μ_0 in (2.53) and (2.54). This leads to the form of the in-matter Maxwell equations written in (2.55) and (2.56). On the other hand, these definitions imply that in vacuum there is a field **D** and a field **E** which describe exactly the same physical state, but which are measured in different units. The same is true for **B** and **H** in vacuum.

2.7 A Heuristic Derivation

We have emphasized that the principles of electromagnetism can only be revealed by experiment. Nevertheless, it is an interesting intellectual exercise to try to deduce the Maxwell equations using only symmetry principles, minimal theoretical assumptions, and (relatively) minimal input from experiment.

The most profound discussions of this type exploit the symmetries of special relativity.[22] Here, we use a heuristic argument based on inversion and rotational symmetry, translational invariance, and four pieces of experimental information: the existence of the Lorentz force, charge conservation, superposition of fields, and the existence of electromagnetic waves.

We begin with experiment and infer the existence of the electric field $\mathbf{E}(\mathbf{r}, t)$ and the magnetic field $\mathbf{B}(\mathbf{r}, t)$ from trajectory measurements on charged particles. These reveal the Lorentz force,

$$\mathbf{F} = q(\mathbf{E} + \boldsymbol{v} \times \mathbf{B}). \tag{2.85}$$

We will assume that electromagnetism respects the symmetry operations of rotation and inversion. This means that (2.85) must be unchanged when these orthogonal transformations (Section 1.7) are performed.

Rotational invariance is guaranteed by the fact that $\mathbf{F}$, $\mathbf{E}$, and $\boldsymbol{v} \times \mathbf{B}$ are all first-rank tensors (vectors). As for inversion, we deduce from the discussion in Section 1.8.1 that $\mathbf{F}$, $\mathbf{E}$, and $\boldsymbol{v} \times \mathbf{B}$ must all be polar vectors or they must all be axial vectors.[23] The position vector $\mathbf{r}$ is a polar vector. Therefore, since $\boldsymbol{v} = d\mathbf{r}/dt$ and $\mathbf{F} = d(m\boldsymbol{v})/dt$, the force $\mathbf{F}$ is also a polar vector. This means that $\mathbf{E}$ and $\boldsymbol{v} \times \mathbf{B}$ must be polar vectors as well. We have just seen that $\boldsymbol{v}$ is a polar vector. Hence, (1.162) shows that $\mathbf{B}$ must be an axial vector (see also Example 1.8).

Our task now is to guess an equation of motion for each field that respects the same symmetries. For the moment, we neglect sources of charge and current. The experimental fact of superposition of fields restricts us to linear equations and, for simplicity, we consider only first derivatives of space and time. The most straightforward guesses, that $\partial\mathbf{E}/\partial t \propto \mathbf{E}$ and $\partial\mathbf{B}/\partial t \propto \mathbf{B}$, cannot be correct because they lead to (unphysical) exponential growth or decay of the fields as a function of time.[24] The more interesting guesses, that $\partial\mathbf{E}/\partial t \propto \mathbf{B}$ and $\partial\mathbf{B}/\partial t \propto \mathbf{E}$, cannot be accepted either because they mix polar vectors and axial vectors on different sides of the same equation. The guesses $\partial\mathbf{E}/\partial t \propto \mathbf{r} \times \mathbf{B}$ and $\partial\mathbf{B}/\partial t \propto \mathbf{r} \times \mathbf{E}$ repair this problem, but violate the reasonable requirement that the dynamical equations should not change if $\mathbf{r} \to \mathbf{r} + \mathbf{c}$ where $\mathbf{c}$ is a constant vector. That is, the field equations should be invariant to uniform translations.

The gradient operator $\nabla \equiv \partial/\partial\mathbf{r}$ changes sign under inversion, but is unchanged when $\mathbf{r} \to \mathbf{r} + \mathbf{c}$. This suggests the use of $\nabla \cdot \mathbf{B}$ and $\nabla \cdot \mathbf{E}$. Unfortunately, these are scalars and thus unacceptable for the right side of a vector equation of motion. On the other hand, if k_1 and k_2 are constants, viable candidate equations of motion which respect the symmetry operations of rotation, inversion, and translation are

$$\frac{\partial\mathbf{E}}{\partial t} = k_1 \nabla \times \mathbf{B} \qquad \text{and} \qquad \frac{\partial\mathbf{B}}{\partial t} = k_2 \nabla \times \mathbf{E}. \tag{2.86}$$

We now conduct an experiment which monitors charge and field in two adjacent, infinitesimal cubical volumes of space (Figure 2.11). Our first observation, at $t = 0$, finds both boxes empty. The next observation, at $t = dt$, reveals equal and opposite static charges in the boxes and a negatively directed electric field everywhere on their common wall. No magnetic field is detected. By charge conservation, an electric current must have flowed through the wall at $t = \frac{1}{2}dt$ and separated charges which were spatially coincident at $t = 0$. This current is the only possible source of the electric field because the charges and the field appeared simultaneously at $t = dt$. A logical inference is[25]

$$\frac{\partial\mathbf{E}}{\partial t} = k_3 \mathbf{j} \qquad \text{at} \quad t = \tfrac{1}{2}dt. \tag{2.87}$$

[22] See, for example, L.D. Landau and E.M. Lifshitz, *The Classical Theory of Fields*, 2nd edition (Addison-Wesley, Reading, MA, 1962), Chapter 4.

[23] We assume that the electric charge q is unchanged by spatial inversion.

[24] Alternatively, we may demand that both sides of the equation transform identically under time-reversal. See Table 15.1 of Section 15.1.

[25] The alternative $\mathbf{E} \propto -\mathbf{j}$ contradicts the static experimental results at $t = dt$.

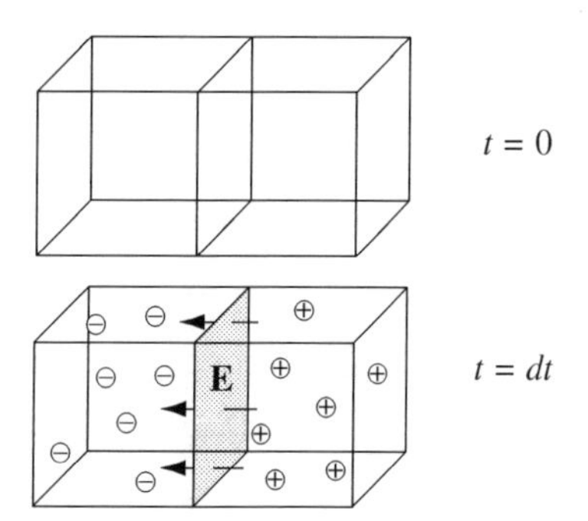

Figure 2.11: A charge-separation experiment. At $t = 0$, no charge or fields are detected in two adjacent volume elements. At $t = dt$, positive charge appears in the right box, negative charge appears in the left box, and an electric field is detected on their common face.

Combining (2.87) with (2.86) yields

$$\frac{\partial \mathbf{E}}{\partial t} = k_1 \nabla \times \mathbf{B} + k_3 \mathbf{j} \qquad \text{and} \qquad \frac{\partial \mathbf{B}}{\partial t} = k_2 \nabla \times \mathbf{E}. \tag{2.88}$$

These are the desired equations of motion.

As a final step, take the divergence of both members of (2.88) and use the continuity equation (2.13). This gives

$$\frac{\partial}{\partial t}(\nabla \cdot \mathbf{B}) = 0 \qquad \text{and} \qquad \frac{\partial}{\partial t}(\nabla \cdot \mathbf{E} + k_3 \rho) = 0. \tag{2.89}$$

If we take as "initial conditions" the vanishing of $\nabla \cdot \mathbf{B}$ and $\nabla \cdot \mathbf{E} + k_2 \rho$, the two equations in (2.89) guarantee that these conditions will remain in place for all time:

$$\nabla \cdot \mathbf{E} = -k_3 \rho \qquad \text{and} \qquad \nabla \cdot \mathbf{B} = 0. \tag{2.90}$$

The structures of (2.88) and (2.90) are now completely determined. It remains only to determine the constants. This depends on the choice of units, as discussed in Section 2.6.

Sources, References, and Additional Reading

The quotation at the beginning of the chapter is from Chapter 2 of

C. Domb, *Clerk Maxwell and Modern Science* (Athlone Press, Bristol, 1963). Printed with permission from the Continuum International Publishing Group © C. Domb, 1963.

Section 2.1 For more on nearly static atmospheric electric fields and nearly static cosmic magnetic fields, see

E.R. Williams, "Sprites, elves, and glow discharge tubes", *Physics Today*, November 2001.

R.M. Kulsrud and E.G. Zweibel, "On the origin of cosmic magnetic fields", *Reports on Progress in Physics* **71**, 046901 (2008).

The experimental test of conservation of charge mentioned in the text is

P. Belli, R. Bernabei, C.J. Dai, *et al.*, "New experimental limit on electron stability", *Physics Letters B* **460**, 236 (1999).

Section 2.2 Two well-regarded histories of electromagnetism are

E.T. Whittaker, *A History of the Theories of Aether and Electricity* (Philosophical Library, New York, 1951).

O. Darrigol, *Electrodynamics from Ampère to Einstein* (University Press, Oxford, 2000).

A thought-provoking essay by a philosopher and historian of science is

E. McMullin, "The origins of the field concept in physics", *Physics in Perspective* **4**, 13 (2002).

To learn how Maxwell and his contemporaries thought about electromagnetism in the years before the discovery of the electron, see

M.S. Longair, *Theoretical Concepts in Physics* (Cambridge University Press, Cambridge, 1984).

J.Z. Buchwald, *From Maxwell to Microphysics* (Cambridge University Press, Chicago, 1985).

B.J. Hunt, *The Maxwellians* (University Press, Cornell, 1991).

Figure 2.4 was taken from the first reference below. The second reference is Maxwell's *Treatise*:

J.C. Maxwell, "On physical lines of force", *Philosophical Magazine* **21**, 281 (1861) by permission of Taylor and Francis Ltd.

J.C. Maxwell, *A Treatise on Electricity and Magnetism* (Clarendon Press, Oxford, 1873).

Our discussion of duality and magnetic charge was inspired by Section 6.11 of

J.D. Jackson, *Classical Electrodynamics*, 3rd edition (Wiley, New York, 1999).

Section 2.3 Figure 2.5 was taken from

H. Tanaka, M. Takata, and M. Sakata, "Experimental observation of valence electron density by maximum entropy method", *Journal of the Physical Society of Japan* **71**, 2595 (2002).

An excellent discussion of the passage from microscopic electromagnetism to macroscopic electromagnetism (and much else) is contained in

F.N.H. Robinson, *Macroscopic Electromagnetism* (Pergamon, Oxford, 1973).

A formal treatment of Lorentz averaging which uses the cell-averaging method of the text is

C. Brouder and S. Rossano, "Microscopic calculation of the constitutive relations", *European Physical Journal B* **45**, 19 (2005).

Microscopic electromagnetic fields near surfaces are discussed in

A. Zangwill, *Physics at Surfaces* (Cambridge University Press, Cambridge, 1988).

A. Liebsch, *Electronic Excitations at Metal Surfaces* (Plenum, New York, 1997).

Figure 2.7 was taken from

H. Ishida and A. Liebsch, "Static and quasistatic response of Ag surfaces to a uniform electric field", *Physical Review B* **66**, 155413 (2002) by permission of the American Physical Society.

The delta function approach to deriving the field matching relations has been re-invented many times over the years. A particularly complete treatment is

V. Namias, "Discontinuity of the electromagnetic fields, potentials, and currents at fixed and moving boundaries", *American Journal of Physics* **56**, 898 (1988).

Section 2.4 An interesting discussion of the origin and use of non-local dielectric functions like the one which appears in Section 2.4.2 is

U. Ritschel, L. Wilets, J.J. Rehr, and M. Grabiak, "Non-local dielectric functions in classical electrostatics and QCD models", *Journal of Physics G: Nuclear and Particle Physics* **18**, 1889 (1992).

Section 2.5 An excellent introduction to non-classical light, quantum optics, and semi-classical radiation theory is

R. Loudon, *The Quantum Theory of Light*, 3rd edition (University Press, Oxford, 2000).

The original proposal to use correlation measurements to distinguish classical from non-classical light was

R. Glauber, "The quantum theory of optical coherence", *Physical Review* **130**, 2529 (1963).

Figure 2.9 was taken from

G.T. Foster, S.L. Mielke, and L.A. Orozco, "Intensity correlations in cavity QED", *Physical Review A* **61**, 53821 (2000) by permission of the American Physical Society.

Electromagnetic tests for "new physics" are the subject of

L.-C. Tu, J. Luo, and G.T. Gilles, "The mass of the photon", *Reports on Progress in Physics* **68**, 77 (2005).

Q. Bailey and A. Kostelecky, "Lorentz-violating electrostatics and magnetostatics", *Physical Review D* **70**, 76006 (2004).

A.S. Goldhaber and W.P. Trower, "Magnetic monopoles", *American Journal of Physics* **58**, 429 (1990).

Section 2.6 The superiority of either the Gaussian or the SI unit system is self-evident to the passionate advocates of each. We draw the reader's attention to two articles which inject some levity into this dreary debate:

W.F. Brown, Jr., "Tutorial paper on dimensions and units", *IEEE Transactions on Magnetics* **20**, 112 (1984).

H.B.G. Casimir, "Electromagnetic units", *Helvetica Physica Acta* **41**, 741 (1968).

Section 2.7 This section was constructed from arguments given in

A.B. Midgal, *Qualitative Methods in Quantum Theory* (W.A. Benjamin, Reading, MA, 1977).

P.B. Visscher, *Fields and Electrodynamics* (Wiley, New York, 1988).

Problems

2.1 Measuring B Let $\mathbf{F}_1$ and $\mathbf{F}_2$ be the instantaneous forces that act on a particle with charge q when it moves through a magnetic field $\mathbf{B}(\mathbf{r})$ with velocities $\boldsymbol{v}_1$ and $\boldsymbol{v}_2$, respectively. Without choosing a coordinate system, show that $\mathbf{B}(\mathbf{r})$ can be determined from the observables $\boldsymbol{v}_1 \times \mathbf{F}_1$ and $\boldsymbol{v}_2 \times \mathbf{F}_2$ if $\boldsymbol{v}_1$ and $\boldsymbol{v}_2$ are appropriately oriented.

2.2 The Coulomb and Biot-Savart Laws The electric and magnetic fields for time-independent distributions of charge and current which go to zero at infinity are

$$\mathbf{E}(\mathbf{r}) = \frac{1}{4\pi\epsilon_0}\int d^3r'\,\rho(\mathbf{r}')\frac{\mathbf{r}-\mathbf{r}'}{|\mathbf{r}-\mathbf{r}'|^3} \qquad \mathbf{B}(\mathbf{r}) = \frac{\mu_0}{4\pi}\int d^3r'\mathbf{j}(\mathbf{r}')\times\frac{\mathbf{r}-\mathbf{r}'}{|\mathbf{r}-\mathbf{r}'|^3}.$$

(a) Calculate $\nabla\cdot\mathbf{E}$ and $\nabla\times\mathbf{E}$.

(b) Calculate $\nabla\cdot\mathbf{B}$ and $\nabla\times\mathbf{B}$. The curl calculation exploits the continuity equation for this situation.

2.3 The Force between Current Loops Let $\mathbf{r}_1$ ($\mathbf{r}_2$) point to a line element $d\mathbf{s}_1$ ($d\mathbf{s}_2$) of a closed loop C_1 (C_2) which carries a current I_1 (I_2). Experiment shows that the force exerted on I_1 by I_2 is

$$\mathbf{F}_1 = -\frac{\mu_0}{4\pi}\oint_{C_1} I_1 d\mathbf{s}_1\cdot\oint_{C_2} I_2 d\mathbf{s}_2\frac{\mathbf{r}_1-\mathbf{r}_2}{|\mathbf{r}_1-\mathbf{r}_2|^3}.$$

(a) Show that $\displaystyle\oint_{C_1} d\mathbf{s}_1\cdot\frac{\mathbf{r}_1-\mathbf{r}_2}{|\mathbf{r}_1-\mathbf{r}_2|^3}=0.$

(b) Use (a) to show that $\mathbf{F}_1 = \displaystyle\oint_{C_1} I_1 d\mathbf{s}_1\times\mathbf{B}_2(\mathbf{r}_1)$ where $\mathbf{B}_2(\mathbf{r})$ is the magnetic field produced by loop C_2.

2.4 Necessity of Displacement Current The magnetostatic equation $\nabla\times\mathbf{B}=\mu_0\mathbf{j}$ is not consistent with conservation of charge for a general time-dependent charge density. Show that consistency can be achieved using $\nabla\times\mathbf{B}=\mu_0\mathbf{j}+\mathbf{j}_\mathrm{D}$ and a suitable choice for $\mathbf{j}_\mathrm{D}$.

2.5 Prelude to Electromagnetic Angular Momentum A particle with charge q is confined to the x-y plane and sits at rest somewhere away from the origin until $t=0$. At that moment, a magnetic field $\mathbf{B}(x,y)=\Phi\delta(x)\delta(y)\hat{\mathbf{z}}$ turns on with a value of Φ which increases at a constant rate from zero. During the subsequent motion of the particle, show that the quantity $\mathbf{L}+q\boldsymbol{\Phi}/2\pi$ is a constant of the motion where $\mathbf{L}$ is the mechanical angular momentum of the particle with respect to the origin and $\boldsymbol{\Phi}=\Phi\hat{\mathbf{z}}$.

2.6 Time-Dependent Charges at Rest Consider a collection of point particles fixed in space with charge density $\rho(\mathbf{r},t)=\sum_k q_k(t)\delta(\mathbf{r}-\mathbf{r}_k)$. Suppose that $\mathbf{E}(\mathbf{r},t=0)=\mathbf{B}(\mathbf{r},t=0)=0$ and

$$\mathbf{E}(\mathbf{r},t)=\frac{1}{4\pi\epsilon_0}\sum_k q_k(t)\frac{\mathbf{r}-\mathbf{r}_k}{|\mathbf{r}-\mathbf{r}_k|^3}.$$

(a) Construct a simple current density which satisfies the continuity equation.

(b) Find $\mathbf{B}(\mathbf{r},t)$ and show that this field and $\mathbf{E}(\mathbf{r},t)$ satisfy all four Maxwell equations.

(c) Describe the flow of charge predicted by the current density computed in part (a).

2.7 Rotation of Free Fields in Vacuum Let θ be a parameter and define "new" electric and magnetic field vectors as linear combinations of the usual electric and magnetic field vectors:

$$\mathbf{E}' = \mathbf{E}\cos\theta + c\mathbf{B}\sin\theta$$
$$c\mathbf{B}' = -\mathbf{E}\sin\theta + c\mathbf{B}\cos\theta.$$

(a) Show that $\mathbf{E}'$ and $\mathbf{B}'$ satisfy the Maxwell equations without sources ($\rho = \mathbf{j} = 0$) if $\mathbf{E}$ and $\mathbf{B}$ satisfy these equations.

(b) Discuss the implications of this result for source-free solutions of the Maxwell equations where $\mathbf{E} \perp \mathbf{B}$ everywhere in space.

2.8 A Current Density Which Varies Linearly in Time An infinitely long cylindrical solenoid carries a spatially uniform but time-dependent surface current density $\mathbf{K}(t) = K_0(t/\tau)\hat{\mathbf{z}}$. K_0 and τ are constants. Find the electric and magnetic fields everywhere in space. Hint: Begin with a simple guess for the magnetic field outside the solenoid.

2.9 A Charge Density Which Varies Linearly in Time If α is a real constant, the continuity equation is satisfied by the charge and current distributions

$$\rho(\mathbf{r}, t) = \alpha t \qquad\qquad \mathbf{j}(\mathbf{r}, t) = -\frac{\alpha}{3}\mathbf{r}.$$

Show by explicit calculation that the net flux of charge into a volume element $dV = r^2 dr d\Omega$ through its bounding surfaces takes the expected value.

2.10 Coulomb Repulsion in One Dimension A point particle with charge q and mass m is fixed at the origin. An identical particle is released from rest at $x = d$. Find the asymptotic ($x \to \infty$) speed of the released particle.

2.11 Ampère-Maxwell Matching Conditions A surface current density $\mathbf{K}(\mathbf{r}_S, t)$ flows in the $z = 0$ plane which separates region 1 ($z > 0$) from region 2 ($z < 0$). Each region contains arbitrary, time-dependent distributions of charge and current.

(a) For fields $\mathbf{B}(\mathbf{r}, t)$ and $\mathbf{E}(\mathbf{r}, t)$, use the theta function method of the text to derive a matching condition based on the Ampère-Maxwell law,

$$\nabla \times \mathbf{B} = \mu_0\mathbf{j} + \frac{1}{c^2}\frac{\partial \mathbf{E}}{\partial t}.$$

(b) Explain how to adapt this result to an interface which is not flat.

2.12 A Variation of Gauss' Law In 1942, Boris Podolsky proposed a generalization of electrostatics that eliminates the divergence of the Coulomb field for a point charge. His theory retains $\nabla \times \mathbf{E} = 0$ but replaces Gauss' law by

$$(1 - a^2\nabla^2)\nabla \cdot \mathbf{E} = \rho/\epsilon_0.$$

(a) Find the electric field predicted by this equation for a point charge at the origin by writing $\mathbf{E} = -\nabla\varphi$ and integrating the equation over an infinitesimal spherical volume. It will be convenient at some point to write

$$\varphi(r) = \frac{q}{4\pi\epsilon_0 r}u(r).$$

(b) Suggest a physical meaning for the parameter a.

2.13 If the Photon Had Mass ... If the photon had a mass m, Gauss' law with $\mathbf{E} = -\nabla\varphi$ changes from $\nabla^2\varphi = -\rho/\epsilon_0$ to an equation which includes a length $L = \hbar/mc$:

$$\nabla^2\varphi = -\frac{\rho}{\epsilon_0} + \frac{\varphi}{L^2}.$$

Experimental searches for m use a geometry first employed by Cavendish where a solid conducting sphere and a concentric, conducting, spherical shell (radii $r_1 < r_2$) are maintained at a common potential Φ by an infinitesimally thin connecting wire. When $m = 0$, all excess charge resides on the outside of the outer shell; no charge accumulates elsewhere.

(a) Use the substitution $\varphi(r) = u(r)/r$ to solve the equation above in the space between the conductors. Find also the electric field in this region.

(b) Use the generalization of Gauss' law for $\mathbf{E}$ implied by the foregoing equation for φ to find the charge Q of the conducting ball.

(c) Show that, to leading order when $L \to \infty$,

$$Q \approx \frac{2\pi\epsilon_0}{3} r_1 \Phi \left(\frac{r_2}{L}\right)^2 \left(1 + \frac{r_1}{r_2}\right).$$

2.14 A Variation of Coulomb's Law Suppose that the electric potential of a point charge at the origin were

$$\varphi(\mathbf{r}) = \frac{q}{4\pi\epsilon_0} \frac{1}{r^{1+\eta}}, \qquad 0 < \eta < 1.$$

Evaluate a superposition integral to find φ inside and outside a spherical shell of radius R with uniform surface charge density $\sigma = Q/4\pi R^2$. Check the $\eta \to 0$ limit.

3 Electrostatics

The subject I am going to recommend to your attention almost terrifies me. The variety it presents is immense, and the enumeration of facts serves to confound rather than to inform. The subject I mean is electricity.
Leonhard Euler (1761)

3.1 Introduction

Maxwell's field theory tells us that every time-independent charge distribution $\rho(\mathbf{r})$ is the source of a vector field $\mathbf{E}(\mathbf{r})$ which satisfies the differential equations

$$\nabla \times \mathbf{E}(\mathbf{r}) = 0 \tag{3.1}$$

and

$$\epsilon_0 \nabla \cdot \mathbf{E}(\mathbf{r}) = \rho(\mathbf{r}). \tag{3.2}$$

The *electric field* $\mathbf{E}(\mathbf{r})$ demands our attention because the force $\mathbf{F}$ and torque $\mathbf{N}$ exerted by $\rho(\mathbf{r})$ on a second charge distribution $\tilde{\rho}(\mathbf{r})$ are

$$\mathbf{F} = \int d^3r\, \tilde{\rho}(\mathbf{r})\mathbf{E}(\mathbf{r}) \tag{3.3}$$

and

$$\mathbf{N} = \int d^3r\, \mathbf{r} \times \tilde{\rho}(\mathbf{r})\mathbf{E}(\mathbf{r}). \tag{3.4}$$

If it happens that neither $\mathbf{F}$ nor $\mathbf{N}$ is of direct interest, the *energy* associated with $\mathbf{E}(\mathbf{r})$, $\rho(\mathbf{r})$, and $\tilde{\rho}(\mathbf{r})$ usually is. We will derive several equivalent expressions for electrostatic total energy and potential energy later in the chapter.

3.1.1 The Scope of Electrostatics

Sommerfeld (1952) divides electrostatics into *summation problems* and *boundary value problems*. In a summation problem, $\rho(\mathbf{r})$ is specified once and for all at every point in space and the problem to find $\mathbf{E}(\mathbf{r})$ reduces to performing an integral. Boundary value problems arise when $\rho(\mathbf{r})$ *cannot* be specified once and for all at every point in space. This occurs when matter of any kind is present because the Coulomb force (3.3) induces the charge density inside the matter to redistribute itself until mechanical equilibrium is established. For historical reasons, this charge rearrangement is called *electrostatic induction* when conductors are involved and *electric polarization* when non-conductors are involved. Remarkably, $\mathbf{E}(\mathbf{r})$ can still be uniquely determined everywhere, provided we specify a model for the polarizable matter. A complete model fixes both the behavior of $\mathbf{E}(\mathbf{r})$ inside the matter and boundary or matching conditions for the Maxwell differential equations. This chapter and the next deal primarily with summation problems. Chapter 7 and Chapter 8 focus on electrostatic boundary value problems.

3.2 Coulomb's Law

It is no accident that the electrostatic Maxwell equations specify exactly the curl of $\mathbf{E}(\mathbf{r})$ and the divergence of $\mathbf{E}(\mathbf{r})$. The Helmholtz theorem (Section 1.9) guarantees that these two quantities provide just enough information to determine $\mathbf{E}(\mathbf{r})$ uniquely. When the integrals converge, the theorem provides the explicit formula

$$\mathbf{E}(\mathbf{r}) = -\nabla \int d^3r' \frac{\nabla' \cdot \mathbf{E}(\mathbf{r}')}{4\pi |\mathbf{r} - \mathbf{r}'|} + \nabla \times \int d^3r' \frac{\nabla' \times \mathbf{E}(\mathbf{r}')}{4\pi |\mathbf{r} - \mathbf{r}'|}. \tag{3.5}$$

Substituting (3.1) and (3.2) into (3.5) gives

$$\mathbf{E}(\mathbf{r}) = -\nabla \frac{1}{4\pi\epsilon_0} \int d^3r' \frac{\rho(\mathbf{r}')}{|\mathbf{r} - \mathbf{r}'|}. \tag{3.6}$$

The gradient operator in (3.6) can be moved inside the integral because it does not act on the variable $\mathbf{r}'$. Then, because

$$\nabla \frac{1}{|\mathbf{r} - \mathbf{r}'|} = -\frac{\mathbf{r} - \mathbf{r}'}{|\mathbf{r} - \mathbf{r}'|^3}, \tag{3.7}$$

we arrive at a superposition integral for the electric field produced by $\rho(\mathbf{r}')$:

$$\mathbf{E}(\mathbf{r}) = \frac{1}{4\pi\epsilon_0} \int d^3r' \, \rho(\mathbf{r}') \frac{\mathbf{r} - \mathbf{r}'}{|\mathbf{r} - \mathbf{r}'|^3}. \tag{3.8}$$

Coulomb's law is the name given to the force that $\rho(\mathbf{r}')$ exerts on $\tilde{\rho}(\mathbf{r})$ when we substitute (3.8) into (3.3):

$$\mathbf{F} = \frac{1}{4\pi\epsilon_0} \int d^3r \int d^3r' \tilde{\rho}(\mathbf{r})\rho(\mathbf{r}') \frac{\mathbf{r} - \mathbf{r}'}{|\mathbf{r} - \mathbf{r}'|^3}. \tag{3.9}$$

It follows immediately from (3.9) that the force that $\tilde{\rho}$ exerts on ρ is $\tilde{\mathbf{F}} = -\mathbf{F}$. This means that $\mathbf{F} = 0$ if $\rho = \tilde{\rho}$. No distribution of charge can exert a net force on itself.[1]

Example 3.1 (a) Find $\mathbf{E}$ on the symmetry axis of a ring with radius R and uniform charge per unit length λ. (b) Use the results of part (a) to find $\mathbf{E}$ on the symmetry axis of a disk with radius R and uniform charge per unit area σ. (c) Use the results of part (b) to find $\mathbf{E}$ for an infinite sheet with uniform charge density σ. Discuss the matching condition at $z = 0$.

Solution: (a) By symmetry, points on opposite sides of the ring in Figure 3.1 contribute equally to $|\mathbf{E}|$ at points $(0, 0, z)$ on the symmetry axis. Moreover, the components of $\mathbf{E}$ transverse to $\hat{\mathbf{z}}$ cancel. The total charge of the ring is $Q = 2\pi R\lambda$. Therefore, since R, z, and α are constant for all points on the ring,

$$\mathbf{E}_{\text{ring}}(z \geq 0) = \frac{\lambda}{4\pi\epsilon_0} \oint d\ell \frac{\cos\alpha}{z^2 + R^2} \hat{\mathbf{z}} = \frac{\lambda}{4\pi\epsilon_0} \frac{2\pi Rz}{(z^2 + R^2)^{3/2}} \hat{\mathbf{z}} = \frac{Q}{4\pi\epsilon_0} \frac{z}{(z^2 + R^2)^{3/2}} \hat{\mathbf{z}}.$$

[1] In practice, $\tilde{\rho}(\mathbf{r})$ is often a *portion* of $\rho(\mathbf{r})$. The force (3.9) is generally non-zero in that case.

We note in passing that $\mathbf{E}_{\text{ring}}(z=0)=0$ because contributions to the field from opposite sides of the ring cancel. $\mathbf{E}_{\text{ring}}(z\to\infty)=0$ is consistent with the Helmholtz theorem and is generally true for fields produced by finite-sized sources.

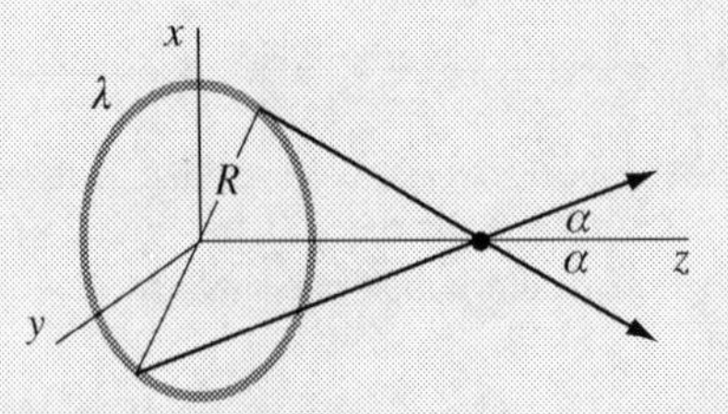

Figure 3.1: A ring with uniform charge per unit length λ.

(b) We superpose the fields produced by rings of radius r with $Q(r)=2\pi r dr\sigma$:

$$\mathbf{E}_{\text{disk}}(z>0)=\frac{z}{4\pi\epsilon_0}\int_0^R dr\frac{2\pi r\sigma}{(z^2+r^2)^{3/2}}\hat{\mathbf{z}}=\frac{\sigma}{2\epsilon_0}\left[1-\frac{z}{\sqrt{z^2+R^2}}\right]\hat{\mathbf{z}}.$$

(c) The limit $R\to\infty$ of $\mathbf{E}_{\text{disk}}$ gives the field of an infinite sheet:

$$\mathbf{E}_{\text{sheet}}(z>0)=\frac{\sigma}{2\epsilon_0}\hat{\mathbf{z}}.$$

By symmetry, $\mathbf{E}(z<0)=-\mathbf{E}(z>0)$ in all three cases. For the disk and the sheet at $z=0$,

$$\mathbf{E}(0^+)-\mathbf{E}(0^-)=\frac{\sigma}{\epsilon_0}\hat{\mathbf{z}}.$$

This agrees with the general matching condition in (2.49).

3.3 The Scalar Potential

The electric field integral (3.8) is difficult to evaluate for all but the simplest choices of $\rho(\mathbf{r})$. However, a glance back at (3.6) shows that we can define a function called the *electrostatic scalar potential*,

$$\varphi(\mathbf{r})=\frac{1}{4\pi\epsilon_0}\int d^3r'\frac{\rho(\mathbf{r}')}{|\mathbf{r}-\mathbf{r}'|},\tag{3.10}$$

and write the electric field in the form

$$\mathbf{E}(\mathbf{r})=-\nabla\varphi(\mathbf{r}).\tag{3.11}$$

This is a great simplification because the scalar integral (3.10) is almost always easier to evaluate than the vector integral (3.8) for $\mathbf{E}(\mathbf{r})$. Two special cases are worth noting. First, if all the charge is confined to a two-dimensional surface and $\sigma(\mathbf{r}_S)$ is the charge density per unit area at a surface point $\mathbf{r}_S$, (3.10) reduces to

$$\varphi(\mathbf{r})=\frac{1}{4\pi\epsilon_0}\int dS\frac{\sigma(\mathbf{r}_S)}{|\mathbf{r}-\mathbf{r}_S|}.\tag{3.12}$$

Similarly, if all the charge is confined to a one-dimensional filament and $\boldsymbol{\ell}$ points to a filamentary element where the charge per unit length is $\lambda(\boldsymbol{\ell})$, (3.10) takes the form[2]

$$\varphi(\mathbf{r}) = \frac{1}{4\pi\epsilon_0}\int d\ell \frac{\lambda(\boldsymbol{\ell})}{|\mathbf{r}-\boldsymbol{\ell}|}. \tag{3.13}$$

Finally, we note for future reference that Gauss' law (3.2) combined with (3.11) gives

$$\nabla^2\varphi(\mathbf{r}) = -\rho(\mathbf{r})/\epsilon_0. \tag{3.14}$$

This is *Poisson's equation*. It can be used as a matter of choice to find $\varphi(\mathbf{r})$ when $\rho(\mathbf{r})$ is specified once and for all. It is used as a matter of necessity to find $\varphi(\mathbf{r})$ when polarizable matter is present (see Chapters 7 and 8).

Example 3.2 The distributions of charge in different atomic nuclei look very similar apart from a change of scale.[3] In figure 3.2, $\rho_A(\mathbf{r})$ is the charge density of nucleus A and $\rho_B(\mathbf{r})$ is the charge density of nucleus B. The two are related by $\rho_B(\lambda\mathbf{r}) = \rho_A(\mathbf{r})$, where λ is a constant. Find the relation between the potentials $\varphi_A(\mathbf{r})$ and $\varphi_B(\mathbf{r})$ and between the electric fields $\mathbf{E}_A(\mathbf{r})$ and $\mathbf{E}_B(\mathbf{r})$.

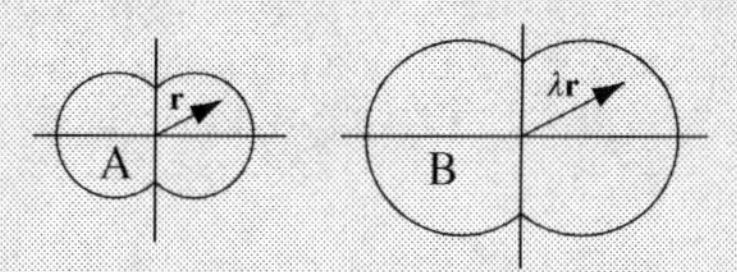

Figure 3.2: Two nuclei related to one another by a change of scale.

Solution: Using the stated information,

$$\varphi_A(\mathbf{r}) = \frac{1}{4\pi\epsilon_0}\int d^3r' \frac{\rho_A(\mathbf{r}')}{|\mathbf{r}-\mathbf{r}'|} = \frac{1}{4\pi\epsilon_0}\int d^3r' \frac{\rho_B(\lambda\mathbf{r}')}{|\mathbf{r}-\mathbf{r}'|}.$$

Changing integration variables to $\lambda\mathbf{r}'$ gives

$$\varphi_A(\mathbf{r}) = \frac{1}{4\pi\epsilon_0}\frac{1}{\lambda^2}\int d^3(\lambda r')\frac{\rho_B(\lambda\mathbf{r}')}{|\lambda\mathbf{r}-\lambda\mathbf{r}'|} = \frac{1}{\lambda^2}\varphi_B(\lambda\mathbf{r}).$$

Then, because $\mathbf{E} = -\nabla\varphi$,

$$\mathbf{E}_B(\lambda\mathbf{r}) = -\frac{\partial}{\partial(\lambda\mathbf{r})}\varphi_B(\lambda\mathbf{r}) = -\frac{\partial}{\partial(\lambda\mathbf{r})}\lambda^2\varphi_A(\mathbf{r}) = \lambda\mathbf{E}_A(\mathbf{r}).$$

3.3.1 The Coulomb Force is Conservative

The gradient formula (3.11) implies that the mechanical work done by $\mathbf{F} = q\mathbf{E}$ when a point charge q moves from $\mathbf{r}_1$ to $\mathbf{r}_2$ depends only on the numerical values of $\varphi(\mathbf{r}_1)$ and $\varphi(\mathbf{r}_2)$:

$$W_E\{\mathbf{r}_1 \to \mathbf{r}_2\} = q\int_{\mathbf{r}_1}^{\mathbf{r}_2} d\boldsymbol{\ell}\cdot\mathbf{E}(\mathbf{r}) = -q\int_{\mathbf{r}_1}^{\mathbf{r}_2} d\boldsymbol{\ell}\cdot\nabla\varphi = -q\left[\varphi(\mathbf{r}_2) - \varphi(\mathbf{r}_1)\right]. \tag{3.15}$$

[2] A one-dimensional charged filament is a good first approximation to a DNA molecule *in vivo* because the phosphate groups of its backbone become de-protonated at pH values typical of physiological environments. The negative linear charge density acquired is $\lambda \approx -|e|/1.7$ Å.

[3] See, for example, B. Frois and C.N. Papanicolas, "Electron scattering and nuclear structure", *Annual Reviews of Nuclear and Particle Science* **37**, 133 (1987).

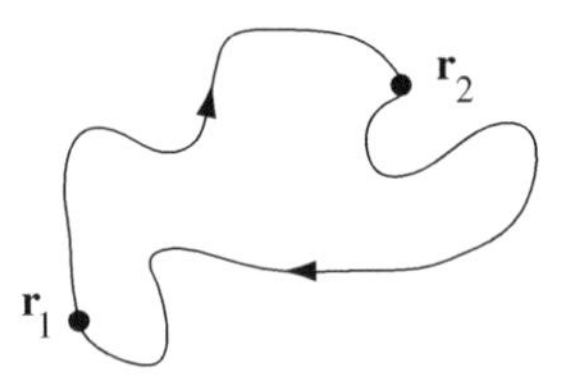

Figure 3.3: A closed path C that can be used in (3.16) to demonstrate that the work done by the Coulomb force is path-independent.

Put another way, the work done by the electric force does not depend on the path taken between the end points. This is the defining characteristic of a *conservative* force. Alternatively, we can compute the work done when q traverses a closed path C through any simply connected portion of space (see Figure 3.3). This work is *zero* because Stokes' theorem (Section 1.4.4) and (3.1) give

$$W_{\text{cycle}} = q \oint_C d\boldsymbol{\ell} \cdot \mathbf{E} = q \int_S d\mathbf{S} \cdot \nabla \times \mathbf{E} = 0. \tag{3.16}$$

This shows that the work done along the path from $\mathbf{r}_1$ to $\mathbf{r}_2$ in Figure 3.3 is the negative of the work done along the remainder of the closed circuit. Reversing the direction of the latter shows that the work done is path-independent. This amounts to a demonstration that (3.11) is correct without the use of the Helmholtz theorem.[4]

The work (3.15) is unchanged if $\varphi(\mathbf{r}) \to \varphi(\mathbf{r}) + \bar{\varphi}$ where $\bar{\varphi}$ is an arbitrary constant. This freedom may be used to fix the zero of potential at any point we choose. Therefore, if we choose $\varphi(\mathbf{r}_0) = 0$, the potential difference

$$\varphi(\mathbf{r}) - \varphi(\mathbf{r}_0) = -\int_{\mathbf{r}_0}^{\mathbf{r}} d\boldsymbol{\ell} \cdot \mathbf{E} \tag{3.17}$$

simplifies to

$$\varphi(\mathbf{r}) = -\int_{\mathbf{r}_0}^{\mathbf{r}} d\boldsymbol{\ell} \cdot \mathbf{E}. \tag{3.18}$$

We note in passing that (3.10) implicitly assumes that the zero of potential is at infinity when $\rho(\mathbf{r}')$ is confined to a finite volume of space.

3.3.2 Matching Conditions for $\varphi(\mathbf{r})$

The macroscopic matching conditions for the electrostatic potential follow from the behavior of the electric field in the vicinity of an interface with surface charge density $\sigma(\mathbf{r}_S)$ (Section 2.3.3). Specifically, if $\hat{\mathbf{n}}_2$ is the outward normal to region 2 and we define the directional derivative as $\partial/\partial n_2 \equiv \hat{\mathbf{n}}_2 \cdot \nabla$, the condition $\hat{\mathbf{n}}_2 \cdot (\mathbf{E}_1 - \mathbf{E}_2) = \sigma/\epsilon_0$ in (2.49) and $\mathbf{E} = -\nabla\varphi$ give

$$\left[\frac{\partial \varphi_2}{\partial n_2} - \frac{\partial \varphi_1}{\partial n_2}\right] = \frac{\sigma}{\epsilon_0}. \tag{3.19}$$

Otherwise, consider the integral on the right side of (3.17) when $\mathbf{r}_0$ and $\mathbf{r}$ are infinitesimally close together but on opposite sides of a surface point $\mathbf{r}_S$. Since $|d\boldsymbol{\ell}| \to 0$, the integral can be non-zero only

[4] Equation (3.16) shows that the Coulomb force is conservative because it is a *central* force [see the rightmost identity in (1.65)], not because it is an inverse-square force.

if the normal component of $\mathbf{E}$ diverges. But (3.19) shows that $\mathbf{E}$ is at most discontinuous. Therefore, the integral is zero and $\varphi(\mathbf{r})$ is *continuous* when we pass from one side of a charged interface to the other:[5]

$$\varphi_1(\mathbf{r}_S) = \varphi_2(\mathbf{r}_S). \tag{3.20}$$

Since the mere existence of the potential depends on the fact that $\nabla \times \mathbf{E} = 0$, it is implicit in the discussion just above that (3.20) carries the same information as the matching condition $\hat{\mathbf{n}}_2 \times [\mathbf{E}_1 - \mathbf{E}_2] = 0$ in (2.49).

3.3.3 Earnshaw's Theorem

Statement: The scalar potential $\varphi(\mathbf{r})$ in a finite, charge-free region of space R takes its maximum or minimum values on the boundary of R.

Proof: Suppose, to the contrary, that $\varphi(\mathbf{r})$ has a local minimum at a point P in R. This means that $\hat{\mathbf{n}} \cdot \nabla\varphi > 0$ at every point on a tiny closed surface S that encloses P. It follows that

$$\int_S dS\, \hat{\mathbf{n}} \cdot \nabla\varphi > 0. \tag{3.21}$$

Using $\mathbf{E} = -\nabla\varphi$ and the divergence theorem, we get an equivalent statement in terms of an integral over the volume V that bounds S:

$$\int_V d^3r\, \nabla \cdot \mathbf{E} < 0. \tag{3.22}$$

But $\nabla \cdot \mathbf{E} = \rho/\epsilon_0$ and $\rho = 0$ everywhere in V. This contradicts (3.22). Therefore, the original assumption that P is a local minimum cannot be true. A similar argument shows that $\varphi(\mathbf{r})$ has no local maxima inside R.

3.3.4 Equipotential Surfaces and Electric Field Lines

The equation

$$\varphi(\mathbf{r}) = \varphi_0 \tag{3.23}$$

defines a family of two-dimensional surfaces in three-dimensional space. According to the left-hand equality in (3.15), no work is required to move a charge along such an *equipotential surface*. Now, consider any path between $\mathbf{r}_1$ and $\mathbf{r}_2$ which lies entirely on an equipotential surface. From the right-hand equality in (3.15),

$$\int_{\mathbf{r}_1}^{\mathbf{r}_2} d\boldsymbol{\ell} \cdot \mathbf{E} = 0. \tag{3.24}$$

This shows that either $\mathbf{E} \equiv 0$ along the path or the direction of $\mathbf{E}$ is perpendicular to the equipotential surface at every point.

As a tool for visualization, we define an *electric field line* as a continuous curve drawn so that the differential element of arc length $d\boldsymbol{\ell}$ points in the direction of $\mathbf{E}(\mathbf{r})$ at every point. In other words, if λ is a constant,

$$d\boldsymbol{\ell} = \lambda\mathbf{E}. \tag{3.25}$$

[5] We will see in Section 4.3.2 that $\varphi(\mathbf{r})$ suffers a jump discontinuity at an interface endowed with a distribution of point electric dipoles oriented perpendicular to the interface.

From the argument just above, we conclude that every electric field line is locally normal to some equipotential surface (except at "null" points where $\mathbf{E} = 0$). This is consistent with the usual geometrical interpretation of the gradient definition $\mathbf{E} = -\nabla\varphi$. Two electric field lines cannot cross (except at a null point) because $\mathbf{E}(\mathbf{r})$ has a unique direction at every point in space.

The electric field line construction is straightforward for a point charge at the origin where (3.10) with $\rho(\mathbf{r}) = q\delta(\mathbf{r})$ gives

$$\varphi(\mathbf{r}) = \frac{q}{4\pi\epsilon_0}\frac{1}{r}. \tag{3.26}$$

The equipotential surfaces are concentric spherical shells and the electric field lines are radial outward for $q > 0$ and radial inward for $q < 0$. More generally, we get a set of differential equations for the field lines by writing out the components of (3.25). In Cartesian coordinates, say, where $\mathbf{E}(\mathbf{r}) = E_x\hat{\mathbf{x}} + E_y\hat{\mathbf{y}} + E_z\hat{\mathbf{z}}$, the equation set is

$$\frac{dx}{E_x} = \frac{dy}{E_y} = \frac{dz}{E_z} = \lambda. \tag{3.27}$$

Analytic solutions of (3.27) are rare, so, with Maxwell (1891), "we cannot afford to despise the humbler method of actually drawing tentative figures on paper and selecting that which appears least unlike the figure we require". Figure 3.4 shows some electric field lines and equipotential curves for a uniform, horizontal field (directed to the right) superposed with the field of a point charge.

Maxwell drew the number density of field lines that pass through any differential element of equipotential surface proportional to $|\mathbf{E}(\mathbf{r}_S)|$ at that element. This implies that the net number of lines that pass through an arbitrary surface S is proportional to the *electric flux*

$$\Phi_E = \int_S dS\,\hat{\mathbf{n}}\cdot\mathbf{E}. \tag{3.28}$$

For the special case of a *closed* surface S, the divergence theorem and $\nabla\cdot\mathbf{E} = \rho/\epsilon_0$ show that the electric flux (3.28) is proportional to the total charge Q_V enclosed by V:

$$\Phi_E = \int_V d^3r\,\nabla\cdot\mathbf{E} = \frac{1}{\epsilon_0}\int_V d^3r\,\rho(\mathbf{r}) = \frac{Q_V}{\epsilon_0}. \tag{3.29}$$

In words, the net number of electric field lines that leave (enter) S is proportional to the net positive (negative) charge enclosed by S. If $Q_V = 0$, every electric field line that enters V must also leave V.

Field line drawings can be a great aid to building physical intuition. Unfortunately, topological distortions are inevitable when a two-dimensional diagram is used to represent a three-dimensional vector field. For example, the total field plotted in Figure 3.4 is dominated by the field produced by the point charge itself at points sufficiently close to the charge. This means that the density of lines adjacent to q should be uniformly distributed around the charge. However, if we were to "fill in" extra field lines to the left of the point charge to do so, the continuation of these lines as they bend around to the right would incorrectly give the impression of a non-uniform field very far to right of the point charge.[6]

3.3.5 A Charged Line Segment

Figure 3.5 shows a one-dimensional line segment of length $2L$ which carries a uniform charge per unit length λ. This is a situation where the electrostatic potential, equipotential surfaces, and electric field

[6] See Wolf *et al.* (1996) in Sources, References, and Additional Reading for further discussion of this point.

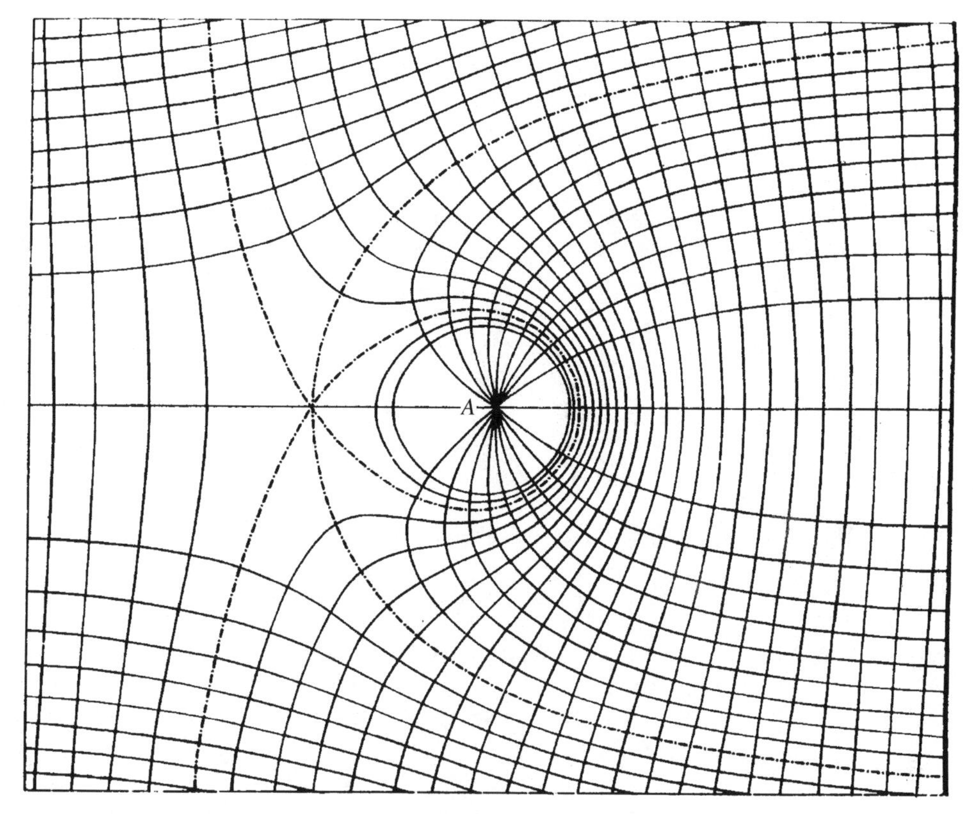

Figure 3.4: Field lines and equipotential curves for a uniform, horizontal electric field (directed to the right) superposed with the electric field of a point charge at A. The plane of the figure passes through the position of the point charge. Two of the depicted equipotentials encircle the point charge and close on themselves. All the other equipotential curves intersect both the top and bottom boundaries of the diagram. The two classes of equipotentials are separated by the single equipotential which intersects itself (dashed-dot). A second dashed-dot curve separates the electric field lines into two classes: those that begin on the point charge and those that do not. Figure reproduced from Maxwell (1891).

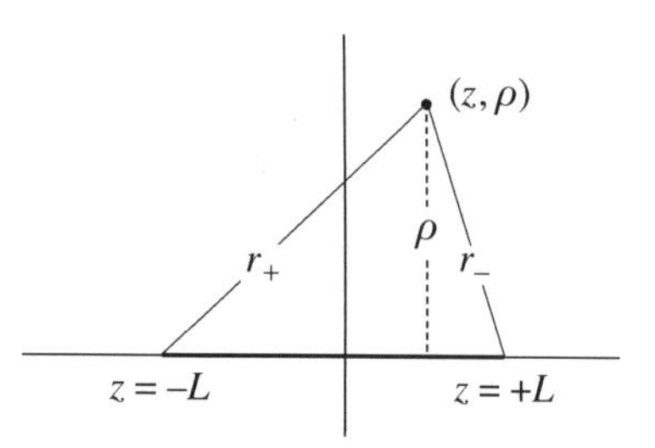

Figure 3.5: A line segment with uniform linear charge density λ.

lines can be calculated exactly. We present it here because it illustrates several of the ideas presented earlier and because the results will be used in Chapter 5 to study the electrostatics of a conducting disk. Unfortunately, the geometrical method of solution does not generalize to other distributions of charge.

The line segment is cylindrically symmetric. Therefore, the integral (3.13) for an observation point (z, ρ) is

$$\varphi(z,\rho) = \frac{\lambda}{4\pi\epsilon_0}\int_{-L}^{+L}\frac{dz'}{\sqrt{(z'-z)^2+\rho^2}} = \frac{\lambda}{4\pi\epsilon_0}\ln\left[\frac{\sqrt{(L-z)^2+\rho^2}+L-z}{\sqrt{(L+z)^2+\rho^2}-L-z}\right]. \tag{3.30}$$

Equation (3.30) simplifies considerably because $r_\pm = \sqrt{\rho^2+(L\pm z)^2}$ are the distances from the line segment end points to the observation point (see Figure 3.5). The key is to define new coordinates (u, t) using

$$\begin{aligned} u &= \tfrac{1}{2}(r_- + r_+) \\ t &= \tfrac{1}{2}(r_- - r_+). \end{aligned} \tag{3.31}$$

In these variables, (3.30) reads

$$\varphi = \frac{\lambda}{4\pi\epsilon_0}\ln\left(\frac{u+t+L-z}{u-t-L-z}\right). \tag{3.32}$$

The identity $ut = -zL$ permits us to eliminate both z and t from (3.32). Specifically,

$$\varphi = \frac{\lambda}{4\pi\epsilon_0}\ln\left[\frac{(u+L)(1+t/L)}{(u-L)(1+t/L)}\right] = \frac{\lambda}{4\pi\epsilon_0}\ln\left(\frac{u+L}{u-L}\right). \tag{3.33}$$

The function on the far right side of (3.33) does not depend on t. Therefore, the equipotentials are simply surfaces of constant u.

Using (3.31), the equipotential condition reads

$$r_+ + r_- = const. \tag{3.34}$$

For a given value of the constant, the set of points in the plane that satisfies (3.34) is the geometrical definition of an ellipse.[7] The foci of the ellipse are the end points of the charged line segment. The curves of constant t are hyperbolae with the same foci (Figure 3.6). The latter may be identified with electric field lines because they are everywhere orthogonal to the equipotentials. More generally, the equipotentials for this problem are prolate ellipsoidal surfaces of revolution. This is the result we will exploit in Application 5.1 to calculate the charge density on the surface of a perfectly conducting disk.

We now return to (3.30) and study the potential at points that are either very far from, or very near to, the charged line segment. For example, if $z \gg L$, the potential at an observation point that lies at a distance $R = \sqrt{\rho^2+z^2}$ from the origin of coordinates is approximately

$$\varphi \approx \frac{\lambda}{4\pi\epsilon_0}\ln\left(\frac{1-z/R+L/R}{1-z/R-L/R}\right). \tag{3.35}$$

If, in addition, $R \gg z$ and $R \gg L$, we can use the fact that $\ln(1+x)\approx x$ when $x \ll 1$ and the fact that $Q = 2L\lambda$ is the total charge of the rod to deduce that (3.35) simplifies to

$$\varphi \approx \frac{1}{4\pi\epsilon_0}\frac{Q}{R}. \tag{3.36}$$

This illustrates a general phenomenon: any finite, charged object "looks" like a point charge when viewed from a sufficiently large distance. The precise meaning of "sufficiently large distance" will be made clear in the next chapter.

[7] See, for example, Section 2.66 of I.N. Bronshtein and K.A. Semendyayev, *Handbook of Mathematics* (Van Nostrand Reinhold, New York, 1985).

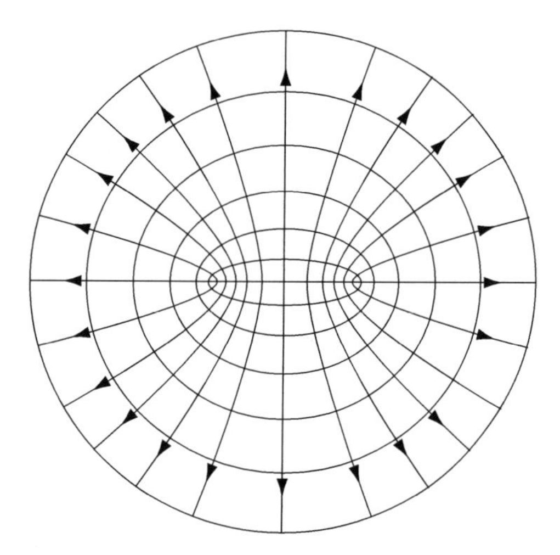

Figure 3.6: The equipotential surfaces of a charged line segment are ellipsoids of revolution. The corresponding electric fields lines are hyperboloids of revolution.

Consider next an observation point that lies very near the origin of coordinates. In that limit, $z \ll L$ and $\rho \ll L$ so

$$\varphi(\rho) \approx \frac{\lambda}{4\pi\epsilon_0} \ln\left(\frac{\sqrt{L^2+\rho^2}+L}{\sqrt{L^2+\rho^2}-L}\right) \approx -\frac{\lambda}{2\pi\epsilon_0}\ln\rho + \frac{\lambda}{2\pi\epsilon_0}\ln(2L). \tag{3.37}$$

The first term on the far right side of (3.37) is the scalar potential for an infinitely long line with uniform charge density λ. This must be the case because, to an observer very close to the segment but far away from its end points, the finite line segment "looks" as if it were infinitely long. The second term on the far right side of (3.37) is a constant that plays no role when we compute the electric field:

$$\mathbf{E} = -\nabla\varphi = \frac{\lambda}{2\pi\epsilon_0\rho}\hat{\boldsymbol{\rho}}. \tag{3.38}$$

The fact that the constant diverges when $L \to \infty$ has no observable consequences. It is an artifact of the unphysical extension of the charge density to infinity.

Example 3.3 An origin-centered spherical shell of infinitesimal thickness has uniform surface charge density $\sigma = Q/4\pi a^2$. A small hole of radius $b \ll a$ is drilled in the shell at the point $\mathbf{R}$ (see Figure 3.7). Find the electric field at all observation points $\mathbf{r}$ where $|\mathbf{r}-\mathbf{R}| \gg b$.

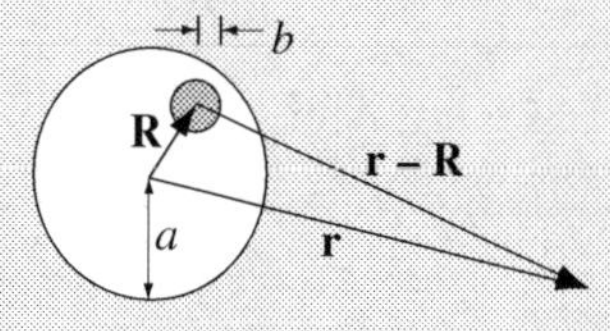

Figure 3.7: A charged spherical shell of radius a with a hole of radius b.

Solution: We exploit the fact that any distribution of charge can be represented as the sum or *superposition* of two or more other charge distributions. Here, we regard the spherical shell with a hole drilled out as the sum of a complete spherical shell and a spherical cap (which just covers the hole) with a surface charge density equal to the negative of the surface charge density of the shell. Because $b \ll a$, the spherical cap is essentially a disk of radius b. Because $|\mathbf{r} - \mathbf{R}| \gg b$, the electric field of the disk is essentially the same as the electric field of a point charge with the same total charge. From what we have said, the latter is $Q' = -\pi b^2 \sigma = -Qb^2/4a^2$. Therefore, using a step function $\Theta(r - a)$, which is equal to one if $r > a$ and is zero otherwise, the electric field observed at $\mathbf{r}$ is

$$\mathbf{E}(\mathbf{r}) = \frac{Q}{4\pi\epsilon_0} \frac{\hat{\mathbf{r}}}{r^2} \Theta(r - a) + \frac{Q'}{4\pi\epsilon_0} \frac{\mathbf{r} - \mathbf{R}}{|\mathbf{r} - \mathbf{R}|^3}.$$

3.4 Gauss' Law and Solid Angle

Equations (3.28) and (3.29) combine to give the integral form of Gauss' law,

$$\varepsilon_0 \int_S d\mathbf{S} \cdot \mathbf{E}(\mathbf{r}) = Q_V. \tag{3.39}$$

This section explores two aspects of this fundamental law. We begin with the role of symmetry to make (3.39) a useful tool to find $\mathbf{E}(\mathbf{r})$. This leads to a derivation of the matching conditions for $\mathbf{E}(\mathbf{r})$ at a charged surface and to an expression for the electrostatic force exerted on such a surface. We then introduce the geometrical concept of *solid angle* and combine this with Coulomb's law for a point charge to provide a derivation of (3.39) that does not make use of $\nabla \cdot \mathbf{E} = \rho/\epsilon_0$.

3.4.1 The Importance of Symmetry

The invariance of a charge distribution to spatial symmetries like rotation, translation, and reflection in a plane is essential to direct calculations of the electric field using (3.39). The invariance is used to determine the direction of the field everywhere and to identify the components of $\mathbf{r}$ upon which $\mathbf{E}(\mathbf{r})$ truly depends. Both of these are needed to identify a surface S where $d\mathbf{S} \cdot \mathbf{E}$ reduces to a scalar constant. For example, the charge distribution $\rho_0(\mathbf{r}) = \rho_0(r)$ in Figure 3.8(a) is invariant to rotations around its origin. This implies that $\mathbf{E}(r, \theta, \phi) = E(r)\hat{\mathbf{r}}$. Therefore, if $Q(r)$ is the net charge enclosed by a Gaussian sphere of radius r, (3.39) shows that

$$\mathbf{E}(\mathbf{r}) = E(r)\hat{\mathbf{r}} \;\Rightarrow\; E(r) = \frac{Q(r)}{4\pi\varepsilon_0 r^2}. \tag{3.40}$$

When $Q(r) = Q$, the integral (3.18) yields the point charge potential

$$\varphi(r) = \frac{Q}{4\pi\epsilon_0 r}. \tag{3.41}$$

An infinitely long cylinder of charge [Figure 3.8(b)] is invariant to translations along the cylinder axis and rotations around the cylinder axis. Therefore, if $\lambda(\rho)$ is the charge per unit length enclosed by a Gaussian cylinder of radius ρ,

$$\mathbf{E}(\mathbf{r}) = E(\rho)\hat{\boldsymbol{\rho}} \;\Rightarrow\; E(\rho) = \frac{\lambda(\rho)}{2\pi\epsilon_0 \rho}. \tag{3.42}$$

For a uniformly charged line, (3.18) yields

$$\varphi(\rho) = -\frac{\lambda}{2\pi\varepsilon_0} \ln \rho, \tag{3.43}$$

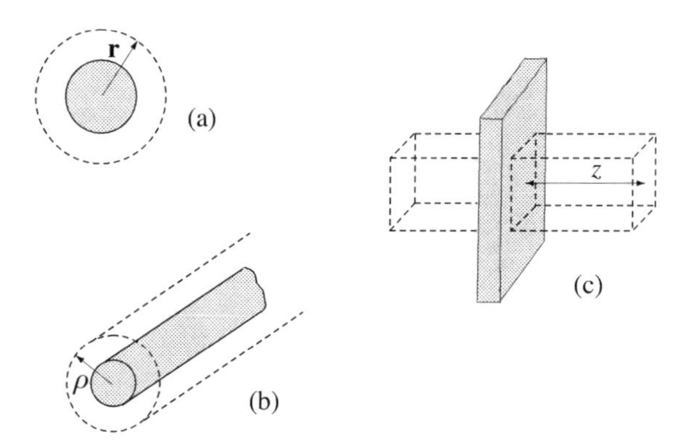

Figure 3.8: Shaded areas are highly symmetrical distributions of charge: (a) $\rho_0(r, \theta, \phi) = \rho_0(r)$; (b) $\rho_0(\rho, \phi, z) = \rho_0(\rho)$; (c) $\rho_0(x, y, z) = \rho_0(z)$. Dashed lines outline possible choices for Gaussian surfaces.

plus a divergent constant of integration as discussed for (3.37). Finally, an infinite slab of charge [Figure 3.8(c)] is invariant to translations along the x- and y-axes. If $\sigma(z)$ is the charge per unit area enclosed by a parallelepiped of width $2z$, Gauss' law tells us that

$$\mathbf{E}(\mathbf{r}) = E(z)\hat{\mathbf{z}} \;\Rightarrow\; E(z) = \frac{\sigma(z)}{2\varepsilon_0}\frac{z}{|z|}. \tag{3.44}$$

Up to another (infinite) constant, the potential for $\sigma(z) = \sigma$ is

$$\varphi(z) = -\frac{\sigma}{2\varepsilon_0}|z|. \tag{3.45}$$

It is important to remark that the symmetry of a charge distribution, i.e., its invariance with respect to spatial operations like rotation, translation, and reflection, tells us only that these symmetry operations transform one solution for $\mathbf{E}(\mathbf{r})$ into another. The Gauss' law solution is *the* solution because it is unique by Helmholtz' theorem.[8]

The potentials (3.41), (3.43), and (3.45) are also worthy of comment. The first of these—the point charge potential—is the building block we used to derive the superposition integral appropriate to situations where the source charge density possesses no particular symmetry:

$$\varphi(\mathbf{r}) = \frac{1}{4\pi\epsilon_0}\int d^3r' \frac{\rho(\mathbf{r}')}{|\mathbf{r} - \mathbf{r}'|}. \tag{3.46}$$

Two-dimensional charge densities are invariant to translations along a fixed direction, say, $\hat{\mathbf{z}}$. To find the corresponding potential, we can superpose line charge potentials like (3.43), each weighted by its own charge per unit length $\Lambda(x, y)dxdy$. This produces the potential

$$\varphi(x, y) = -\frac{1}{2\pi\epsilon_0}\int dx' \int dy' \ln\sqrt{(x - x')^2 + (y - y')^2}\,\Lambda(x', y'). \tag{3.47}$$

The charge density shown in Figure 3.8(b) is a special case of this situation where $\Lambda(x, y) = \Lambda(\sqrt{x^2 + y^2})$. A similar argument applies to one-dimensional charge distributions that are invariant to translations along x and y as in Figure 3.8(c). Superposing sheet potentials like (3.45) with charge per unit area $\gamma(z)dz$ gives the potential

$$\varphi(z) = -\frac{1}{2\epsilon_0}\int dz' |z - z'|\,\gamma(z'). \tag{3.48}$$

[8] The Helmholtz theorem (Section 1.9) can fail if the charge density does not go to zero fast enough at infinity. We can establish uniqueness for infinite lines and planes of charge as the limit of a sequence of charged lines and planes with finite extent.

Example 3.4 Begin with the electric field and derive an expression for the potential $\varphi(r)$ produced by a spherically symmetric charge distribution $\rho(r)$.

Solution: Let

$$Q(r) = 4\pi \int_0^r ds\, s^2 \rho(s)$$

be the charge contained within a ball of radius r centered at the origin. Equation (3.40) gives the electric field due to $Q(r)$ as

$$\mathbf{E}(r) = \frac{Q(r)}{4\pi\epsilon_0 r^2}\hat{\mathbf{r}}.$$

Using (3.18) with $d\boldsymbol{\ell} = dr'\hat{\mathbf{r}}$ and $\mathbf{r}_0$ at infinity, the potential at r is

$$\varphi(r) = -\int_\infty^r d\boldsymbol{\ell}\cdot\mathbf{E}(\mathbf{r}') = -\frac{1}{\epsilon_0}\int_\infty^r \frac{dr'}{r'^2}\int_0^{r'} ds\, s^2\rho(s) = \int_\infty^r d\left(\frac{1}{r'}\right)u(r')$$

where

$$u(r') = \frac{1}{\epsilon_0}\int_0^{r'} ds\, s^2 \rho(s).$$

Integrating by parts gives

$$\varphi(r) = \left.\frac{u(r')}{r'}\right|_\infty^r - \int_\infty^r \frac{dr'}{r'}\frac{du}{dr'}$$

or

$$\varphi(r) = \frac{1}{\epsilon_0 r}\int_0^r dr' r'^2 \rho(r') + \frac{1}{\epsilon_0}\int_r^\infty dr' r' \rho(r').$$

3.4.2 Matching Conditions for $\mathbf{E}(\mathbf{r})$

We can use the field (3.44) to re-derive the matching conditions (2.49) for the electric field near an arbitrarily shaped surface S that carries a surface charge density $\sigma(\mathbf{r}_S)$ (Figure 3.9). The key is to let $\mathbf{E}_1$ and $\mathbf{E}_2$ be the electric fields at points that are infinitesimally close to each other, but on opposite sides of S.

$\mathbf{E}_1$ and $\mathbf{E}_2$ may each be decomposed into two contributions: (1) the field $\mathbf{E}_{\rm disk}$ produced by the tiny, flat disk of charge (centered on $\mathbf{r}_S$) shown shaded in Figure 3.9; and (2) the field $\mathbf{E}_S$ produced by the surface S with the tiny disk at $\mathbf{r}_S$ removed. When viewed at very close range, $\mathbf{E}_{\rm disk}$ is indistinguishable from the field (3.44) produced by an infinite, flat sheet of charge with uniform charge density $\sigma = \sigma(\mathbf{r}_S)$. This field changes sign when the observation point passes through the disk. Conversely, $\mathbf{E}_S$ is smooth and continuous when the observation point passes through the hole created by the removed disk. Hence, by superposition,

$$\mathbf{E}_1 = \mathbf{E}_S - \frac{\sigma}{2\epsilon_0}\hat{\mathbf{n}}_1 \qquad \text{and} \qquad \mathbf{E}_2 = \mathbf{E}_S - \frac{\sigma}{2\epsilon_0}\hat{\mathbf{n}}_2. \tag{3.49}$$

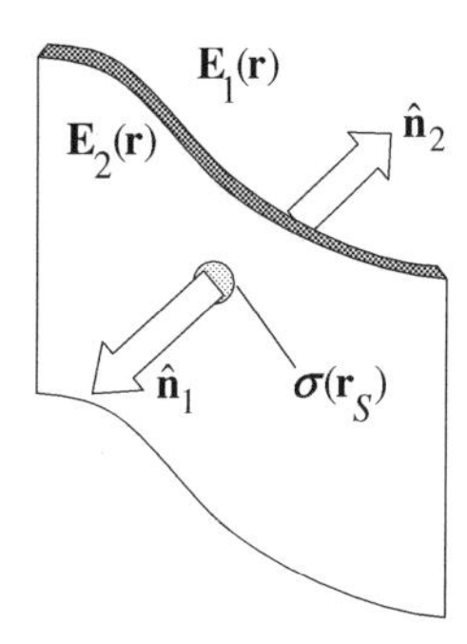

Figure 3.9: A surface S endowed with charge density $\sigma(\mathbf{r}_S)$. The unit normal vector $\hat{\mathbf{n}}_1$ points outward from region 1. The unit normal vector $\hat{\mathbf{n}}_2$ points outward from region 2.

Subtracting the two equations in (3.49) gives $\mathbf{E}_1 - \mathbf{E}_2 = \hat{\mathbf{n}}_2\sigma/\epsilon_0$. Taking the dot product and cross product of this expression with $\hat{\mathbf{n}}_2$ produces the expected matching conditions,

$$\hat{\mathbf{n}}_2 \cdot [\mathbf{E}_1 - \mathbf{E}_2] = \sigma(\mathbf{r}_S)/\epsilon_0 \tag{3.50}$$

and

$$\hat{\mathbf{n}}_2 \times [\mathbf{E}_1 - \mathbf{E}_2] = 0. \tag{3.51}$$

3.4.3 The Force on a Charged Surface

It is not immediately obvious how to calculate the Coulomb force on a charged surface S because (3.50) says that $\hat{\mathbf{n}} \cdot \mathbf{E}$ is discontinuous there. On the other hand, no surface element can exert a force on itself. Therefore, the quantity $\mathbf{E}_S$ in (3.49) must be solely responsible for the force per unit area felt at $\mathbf{r}_S$. Since $\hat{\mathbf{n}}_1 = -\hat{\mathbf{n}}_2$, the sum of the two equations in (3.49) gives $\mathbf{E}_S$ as the simple average of $\mathbf{E}_1$ and $\mathbf{E}_2$. Therefore, the desired force density is

$$\mathbf{f} = \sigma \mathbf{E}_S = \tfrac{1}{2}\sigma(\mathbf{E}_1 + \mathbf{E}_2). \tag{3.52}$$

3.4.4 Solid Angle

This section introduces the geometrical concept of *solid angle* and uses it (together with Coulomb's law for a point charge) to derive the integral form of Gauss' law (3.39). The solid angle will return later when we discuss magnetostatics.

Figure 3.10 shows an origin of coordinates O' and vectors $\mathbf{r}$ and $\mathbf{r}_S$ that label an observation point O and a point on a surface S, respectively. A set of rays fan out radially from O to every point of S. By definition, the *solid angle* Ω_S subtended by S at O is that part of the area of a sphere of unit radius centered at O that is cut off by the rays. In other words, we project every surface element $d\mathbf{S} = dS\hat{\mathbf{n}}$ onto the direction of $\mathbf{r}_S - \mathbf{r}$, divide by the square of this distance (because we are interested in projection onto a unit sphere), and sum over all such elements. This gives

$$\Omega_S(\mathbf{r}) = \int_S d\Omega_S(\mathbf{r}) = \int_S d\mathbf{S} \cdot \frac{\mathbf{r}_S - \mathbf{r}}{|\mathbf{r}_S - \mathbf{r}|^3}. \tag{3.53}$$

It is worth noting that $\Omega_S(\mathbf{r})$ *changes sign* when the observation point $\mathbf{r}$ passes through the surface S. This happens because $\mathbf{r}_S - \mathbf{r}$ changes direction (relative to S) while $d\mathbf{S}$ does not. The direction of the latter is fixed (by convention) so that (3.53) is positive when the curvature of S tends to enclose

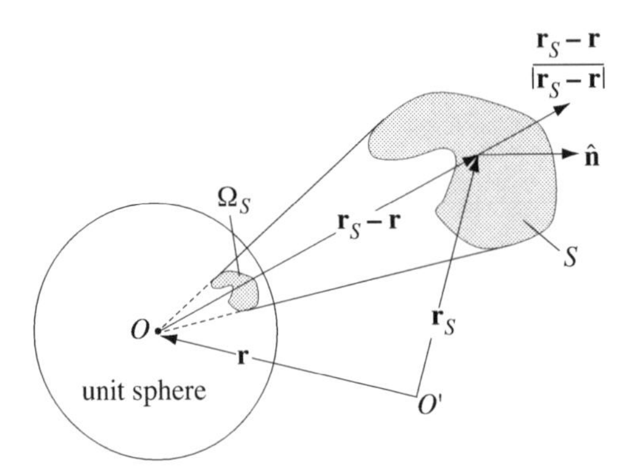

Figure 3.10: Ω_S is the solid angle subtended at the observation point O by the surface S. See text for discussion.

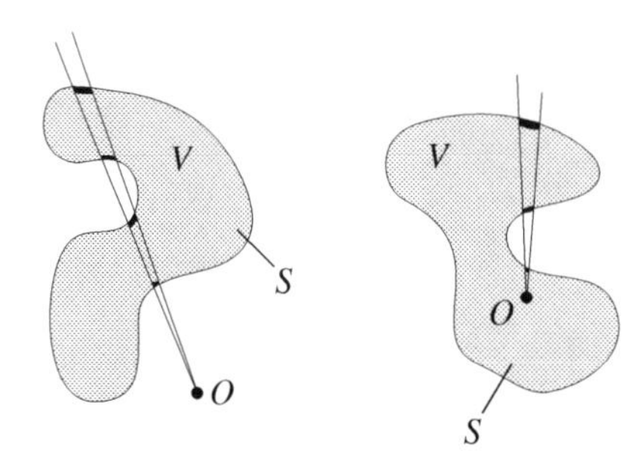

Figure 3.11: A closed surface S that encloses a volume V. The solid angle $\Omega_S = 0$ if the observation point O is outside V. $\Omega_S = 4\pi$ if O is inside V.

the observation point. We note in passing that if O' coincides with O (so $\mathbf{r} = 0$) and S is a spherical surface centered at O (so $\hat{\mathbf{n}} = \hat{\mathbf{r}}_S$) we get the familiar result

$$d\Omega_S = dS\,\frac{\hat{\mathbf{n}}\cdot\hat{\mathbf{r}}_S}{r_S^2} = \frac{dS}{r_S^2} = \sin\theta d\theta d\phi. \tag{3.54}$$

We now put $O' = O$ and specialize to a *closed* surface S that encloses a volume V. The left side of Figure 3.11 shows that every ray from O intersects S an even number of times if O is outside of V. Adjacent intersections where the ray enters and exits V make the same projection onto a unit sphere centered at O but cancel one another in the sum (3.53) because $\hat{\mathbf{n}}\cdot\hat{\mathbf{r}}_S$ changes sign. Conversely, if O is inside of V (right side of Figure 3.11), every ray makes an odd number of intersections and the set of intersections closest to O project onto the complete unit sphere with surface area 4π. With a slight change in notation, this determines our final geometrical result:

$$\Omega_S = \int_S d\mathbf{S}\cdot\frac{\hat{\mathbf{r}}}{r^2} = \begin{cases} 4\pi & O \in V \\ 0 & O \notin V. \end{cases} \tag{3.55}$$

In light of Coulomb's formula for the electric field of a point charge at O, (3.55) is exactly Gauss' law for a point charge:

$$\Phi_E(S) = \int_S d\mathbf{S}\cdot\mathbf{E} = \frac{q}{4\pi\varepsilon_0}\Omega_S. \tag{3.56}$$

The general result (3.39) follows by superposition. The reader should note the crucial role played in this demonstration by the inverse-square nature of the electric force. Gauss' law is not valid for other laws of force.

Application 3.1 Field Lines for a Point Charge in a Uniform Field

A non-trivial application of the solid angle concept permits us to find an equation for the field lines plotted in Figure 3.4 for a point charge $q > 0$ in a uniform electric field $\mathbf{E} = E\hat{\mathbf{z}}$. Figure 3.12 illustrates a few representative field lines redrawn from that figure.

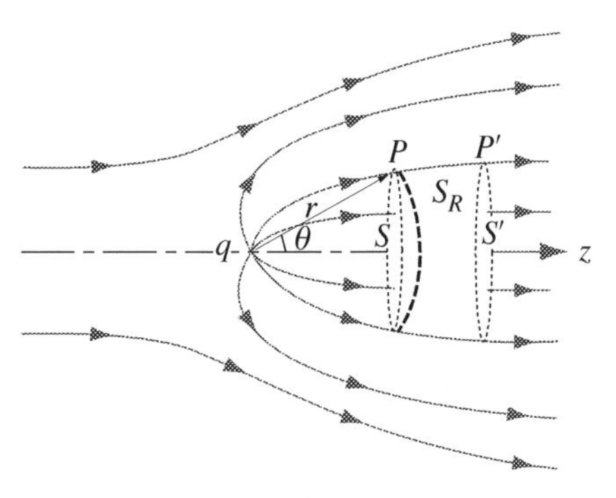

Figure 3.12: Representative electric field lines redrawn from Figure 3.4. The polar coordinates (r, θ) label a point P. One electric field line passes through both P and P'. The complete field line pattern is rotationally symmetric around the z-axis. The same is true of the Gaussian surface composed of the disks S and S' (dotted) and the surface of revolution S_R. The heavy dashed arc indicates a cap-shaped portion of a spherical surface centered at q that passes through P.

Choose one field line and one point P on that line. By the rotational symmetry of the diagram with respect to the z-axis, P defines the perimeter of a circular disk S which lies perpendicular to the plane of the diagram. Let us compute the electric flux Φ through S. We can use (3.56) to compute the contribution from the point charge. Adding the contribution from the uniform field gives

$$\Phi_E(S) = q\Omega_S/4\pi\epsilon_0 + E\pi r^2 \sin^2\theta. \tag{3.57}$$

By construction, the solid angle Ω_S subtended by the disk S is the same as the solid angle subtended by the dashed spherical cap of radius r shown in Figure 3.12. For the cap, $\hat{\mathbf{n}} = \hat{\mathbf{r}}_S$ and $dS = r_S^2 \sin\theta d\theta d\phi$. Therefore,

$$\Omega_S = \int_S dS \frac{\hat{\mathbf{n}} \cdot \hat{\mathbf{r}}_S}{r_S^2} = \int_0^{2\pi} d\phi \int_0^{\theta} \sin\theta d\theta = 2\pi(1 - \cos\theta). \tag{3.58}$$

Now, choose any other point P' which lies on the same electric field line as P and let S' be the axis-centered disk (parallel to S) which passes through P'. Our interest is the closed Gaussian surface G formed by S, S', and the surface of revolution S_R generated by the electric field line segment $\overline{PP'}$. By construction, the electric flux that passes through S_R is zero. Moreover, there is no charge inside G. Therefore, if $\Phi_E(S')$ is the flux through S', the total flux condition (3.29) tells us that

$$\Phi_E(S) + \Phi_E(S') = 0. \tag{3.59}$$

The point P' is arbitrary except that it shares a field line with P. Therefore, for each choice of r and θ, $\Phi_E(S')$ is a fixed real constant (independent of P') which we may combine with the constant $q/2\epsilon_0$

that appears when (3.58) is substituted into (3.57). If we let $a = 2\pi\epsilon_0 E/q$ and $b = 1 + 2\epsilon_0 \Phi_E(S')/q$, the final result is an equation for one field line in Figure 3.12:

$$ar^2 \sin^2\theta - \cos\theta = -b. \tag{3.60}$$

To create the field line pattern, we make a choice for (r, θ) and calculate the corresponding value of $\Phi_E(S')$ [as the negative of $\Phi_E(S)$ from (3.59)]. This produces a value for b. The latter is common to the entire field line, which we plot by varying θ and using (3.60) to find r. Thereafter, every choice for (r, θ) that produces a distinct value for b produces a distinct field line. ∎

3.5 Electrostatic Potential Energy

In classical mechanics, we define the potential energy V of a system acted upon by a conservative force $\mathbf{F}$ so that a small displacement $\delta\mathbf{s}$ of the system induces the change

$$\delta V = -\mathbf{F} \cdot \delta\mathbf{s}. \tag{3.61}$$

Equivalently,

$$\mathbf{F} = -\frac{\delta V}{\delta\mathbf{s}} = -\nabla V. \tag{3.62}$$

Since the Coulomb force $\mathbf{F} = q\mathbf{E}$ is conservative (Section 3.3.1) and $\mathbf{E} = -\nabla\varphi$, we deduce from (3.62) that the *electrostatic potential energy* of a point charge in an electric field is

$$V_E = q\varphi(\mathbf{r}). \tag{3.63}$$

This identifies the electrostatic potential $\varphi(\mathbf{r})$ as a potential energy per unit charge, or specific potential energy. By superposition, the generalization of (3.63) for a charge distribution $\rho_2(\mathbf{r})$ in the field of an electrostatic potential $\varphi_1(\mathbf{r})$ is

$$V_E = \int d^3r\, \rho_2(\mathbf{r})\varphi_1(\mathbf{r}). \tag{3.64}$$

A quick application of (3.63) uses Earnshaw's theorem (Section 3.3.3) to conclude the following: no set of isolated charges can be held in stable equilibrium by electrostatic forces alone. This is so because mechanical equilibrium of, say, q_1 is possible only if the potential energy $q_1\varphi(\mathbf{r})$ (produced by the other charges) has a local minimum at $\mathbf{r}_1$. But Earnshaw's theorem shows that such a local minimum cannot exist anywhere in the free space not occupied by charges. Hence, any point charge inserted into a pre-existing electrostatic field cannot be in stable equilibrium. This argument shows why no classical model of matter composed of positively charged nuclei and negatively charged point electrons can be correct. Quantum mechanics is essential for the stability of matter.

3.5.1 Coulomb Force from Variation of Potential Energy

To gain confidence in (3.64), let us confirm that the expected Coulomb force results from the change in V_E induced by an infinitesimal displacement $\delta\mathbf{s}$ of $\rho_2(\mathbf{r})$. We assume that the source of $\varphi_1(\mathbf{r})$ is fixed in space. The key observation is that the displacement $\delta\mathbf{s}$ induces a charge density change[9]

$$\delta\rho_2(\mathbf{r}) = \rho_2(\mathbf{r} - \delta\mathbf{s}) - \rho_2(\mathbf{r}). \tag{3.65}$$

[9] Figure 3.13 shows a rigid displacement $\delta\mathbf{s}$ of a charge distribution from an initial location with center of mass $\mathbf{R}$ (solid outline) to a final location with center of mass $\mathbf{R} + \delta\mathbf{s}$ (dotted outline). We are interested in the change in charge at the point $\mathbf{r}$ in space. This is $\delta\rho(\mathbf{r}) = \rho_{\text{final}}(\mathbf{r}) - \rho_{\text{initial}}(\mathbf{r})$. Or, using the position of the center of mass as a parameter, $\delta\rho(\mathbf{r}) = \rho(\mathbf{r}, \mathbf{R} + \delta\mathbf{s}) - \rho(\mathbf{r}, \mathbf{R})$. On the other hand, $\rho(\mathbf{r} + \delta\mathbf{s}, \mathbf{R} + \delta\mathbf{s}) = \rho(\mathbf{r}, \mathbf{R})$. Therefore, $\rho(\mathbf{r}, \mathbf{R} + \delta\mathbf{s}) = \rho(\mathbf{r} - \delta\mathbf{s}, \mathbf{R})$. Using this in the last formula for $\delta\rho$ establishes (3.65).

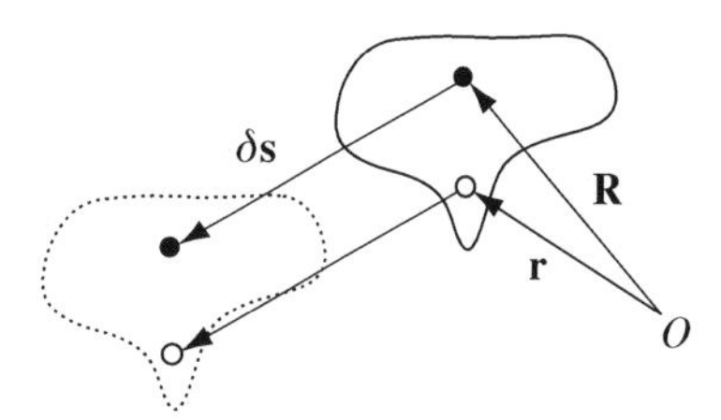

Figure 3.13: Demonstration that $\delta\rho(\mathbf{r}) = \rho(\mathbf{r} - \delta\mathbf{s}) - \rho(\mathbf{r})$.

The corresponding change in the presumptive potential energy (3.64) is

$$\delta V_E = \int d^3r\, \varphi_1(\mathbf{r})\delta\rho_2(\mathbf{r}). \tag{3.66}$$

Therefore, to first order in the small quantity $\delta\mathbf{s}$,

$$\delta V_E = \int d^3r\, \varphi_1(\mathbf{r})\left[\rho_2(\mathbf{r} - \delta\mathbf{s}) - \rho_2(\mathbf{r})\right] = -\int d^3r\, \varphi_1(\mathbf{r})\nabla\rho_2(\mathbf{r})\cdot\delta\mathbf{s}. \tag{3.67}$$

Integrating the last term in (3.67) by parts gives

$$\delta V_E = \int d^3r\, \rho_2(\mathbf{r})\nabla\varphi_1(\mathbf{r})\cdot\delta\mathbf{s} = -\int d^3r\, \rho_2(\mathbf{r})\mathbf{E}_1(\mathbf{r})\cdot\delta\mathbf{s}. \tag{3.68}$$

Comparing (3.68) to (3.61) shows that the force exerted on $\rho_2(\mathbf{r})$ by the field $\mathbf{E}_1 = -\nabla\varphi_1$ is exactly the Coulomb force,

$$\mathbf{F} = -\frac{\partial V_E}{\partial\mathbf{s}} = \int d^3r\, \rho_2(\mathbf{r})\,\mathbf{E}_1(\mathbf{r}). \tag{3.69}$$

3.5.2 Green's Reciprocity Relation

The definition (3.46) of the electrostatic potential permits us to rewrite (3.64) in an important way:

$$\int d^3r\rho_2(\mathbf{r})\varphi_1(\mathbf{r}) = \frac{1}{4\pi\epsilon_0}\int d^3r \int d^3r' \frac{\rho_2(\mathbf{r})\rho_1(\mathbf{r}')}{|\mathbf{r}-\mathbf{r}'|} = \int d^3r'\varphi_2(\mathbf{r}')\rho_1(\mathbf{r}'). \tag{3.70}$$

The equality of the first and last terms in (3.70) is not obvious. It says that the potential energy of $\rho_2(\mathbf{r})$ in the field produced by $\rho_1(\mathbf{r})$ is equal to the potential energy of $\rho_1(\mathbf{r})$ in the field produced by $\rho_2(\mathbf{r})$. This is called *Green's reciprocity relation.*[10]

Reciprocity is often used to solve electrostatic problems that are difficult to analyze by other means. An example is the force between two non-overlapping, spherically symmetric charge distributions with total charges Q_1 and Q_2 (Figure 3.14). Our strategy is to use (3.62) and the potential energy (3.70). We begin with the fact that the Gauss' law electric field produced by $\rho_2(r')$ outside of itself is exactly the same as the electric field produced by a point charge Q_2 located at the center of $\rho_2(r')$ (black dot in Figure 3.14). The electrostatic potential $\varphi_2(r')$ is similarly identical to the point charge potential $\varphi_P(r') = Q_2/(4\pi\epsilon_0 r')$ everywhere outside $\rho_2(r')$.

Using first this point charge observation and then the reciprocity relation (3.70), we can write the potential energy of $\rho_1(r)$ in the field produced by $\rho_2(r')$ in terms of the charge density $\rho_P(\mathbf{r})$ of the Q_2 point charge:

$$V_E = \int d^3r\, \rho_1(\mathbf{r})\varphi_2(\mathbf{r}) = \int d^3r\, \rho_1(\mathbf{r})\varphi_P(\mathbf{r}) = \int d^3r\, \rho_P(\mathbf{r})\varphi_1(\mathbf{r}). \tag{3.71}$$

[10] Some authors refer to (3.70) as Green's reciprocity theorem or Green's reciprocal relation.

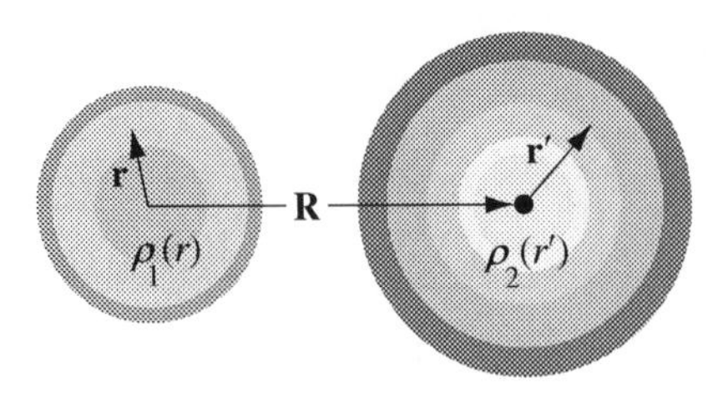

Figure 3.14: Two spherical charge distributions with total charges Q_1 and Q_2. The two distributions do not overlap and the vector **R** points from the center of $\rho_1(r)$ to the center of $\rho_2(r')$.

Since $\rho_P(\mathbf{r}) = Q_2\delta(\mathbf{r} - \mathbf{R})$, the potential energy we seek is

$$V_E = Q_2 \int d^3r\, \delta(\mathbf{r} - \mathbf{R})\varphi_1(\mathbf{r}) = Q_2\varphi_1(\mathbf{R}) = \frac{Q_1 Q_2}{4\pi\epsilon_0 R}. \tag{3.72}$$

The final equality in (3.72) follows from the fact that $\varphi_1(r)$, like $\varphi(r')$, has the form of a point charge potential outside of itself. We conclude that the potential energy (and hence the force) between $\rho_1(r)$ and $\rho_2(r')$ is the same as if all the charge of each was concentrated at a point at its center.

Example 3.5 Use Green's reciprocity relation and an empty sphere with charge spread uniformly over its surface to prove this "mean value theorem": the average of $\varphi(\mathbf{r})$ over a spherical surface S that encloses a charge-free volume is equal to the potential at the center of the sphere:

$$\frac{1}{4\pi R^2}\int_S dS\, \varphi(\mathbf{r}) = \varphi(0).$$

Solution: We use (3.70) and let $\varphi(\mathbf{r})$ and its source $\rho(\mathbf{r})$ play the roles of $\varphi_1(\mathbf{r})$ and $\rho_1(\mathbf{r})$. We choose the "reciprocal system" as suggested in the statement of the problem, i.e.,

$$\rho_2(\mathbf{r}) = \frac{q}{4\pi R^2}\delta(r - R) \qquad \varphi_2(\mathbf{r}) = \frac{q}{4\pi\epsilon_0}\begin{cases} \dfrac{1}{R} & r \le R, \\ \dfrac{1}{r} & r \ge R. \end{cases}$$

Now,

$$\int d^3r\, \rho_2(\mathbf{r})\varphi_1(\mathbf{r}) = \frac{q}{4\pi R^2}\int_S dS\varphi(\mathbf{r}) = q\langle\varphi\rangle_S.$$

On the other hand, since $\rho(\mathbf{r})$ is non-zero only outside the sphere of radius R,

$$\int d^3r'\, \rho_1(\mathbf{r}')\varphi_2(\mathbf{r}') = \frac{q}{4\pi\epsilon_0}\int_{r'>R} d^3r'\, \frac{\rho(\mathbf{r}')}{r'} = q\varphi(0).$$

This proves the theorem because q is arbitrary.

3.6 Electrostatic Total Energy

The *total electrostatic energy* U_E is defined as the total work required to assemble a charge distribution from an initial state where all the charge is dispersed at spatial infinity. We imagine that the work is performed quasistatically by an external agent in such a way that no dissipative effects occur. This ensures that the process is reversible in the thermodynamic sense. U_E is important for many reasons, not least because it achieves its minimum value for the ground state of an electrostatic system. An example is a model charge density $\rho(\mathbf{r})$ which depends on one or more variational parameters. Minimizing U_E with respect to these parameters produces an approximation of the ground state charge density.

It is not difficult to compute $U_{\rm E}$ for a collection of N stationary point charges q_k located at positions $\mathbf{r}_k$. No work is required to bring the first charge into position from infinity because $\mathbf{E} = 0$ everywhere. The work W_{12} required to bring the second charge into position is the *negative* of the work calculated in (3.15) because the external agent does work *against* the Coulomb force exerted by the first charge. Therefore, if $\varphi_1(\mathbf{r})$ is the electrostatic potential (3.26) produced by q_1,

$$W_{12} = q_2[\varphi_1(\mathbf{r}_2) - \varphi_1(\infty)] = q_2\varphi_1(\mathbf{r}_2) = \frac{1}{4\pi\epsilon_0}\frac{q_1 q_2}{|\mathbf{r}_1 - \mathbf{r}_2|}. \tag{3.73}$$

The third charge interacts with both q_1 and q_2 so additional work $W_{13} + W_{23}$ must be done. This generalizes to

$$U_E = W = \frac{1}{4\pi\epsilon_0}\sum_{j=1}^{N}\sum_{i>j}^{N}\frac{q_i q_j}{|\mathbf{r}_i - \mathbf{r}_j|}, \tag{3.74}$$

or

$$U_E = \frac{1}{4\pi\epsilon_0}\frac{1}{2}\sum_{i=1}^{N}\sum_{j\neq i}^{N}\frac{q_i q_j}{|\mathbf{r}_i - \mathbf{r}_j|} = \frac{1}{2}\sum_{i=1}^{N} q_i\varphi(\mathbf{r}_i). \tag{3.75}$$

The factor of $\frac{1}{2}$ in (3.75) corrects for overcounting; the double sum counts each distinct pair of charges twice rather than once as in (3.74). Recalling (3.10), the analog of (3.75) for a continuous distribution of charge is

$$U_E = \frac{1}{8\pi\epsilon_0}\int d^3r \int d^3r' \frac{\rho(\mathbf{r})\rho(\mathbf{r}')}{|\mathbf{r} - \mathbf{r}'|} = \frac{1}{2}\int d^3r\, \rho(\mathbf{r})\varphi(\mathbf{r}). \tag{3.76}$$

The total energy cannot depend on exactly how a charge distribution is assembled. In practice, the symmetry of a problem often suggests a convenient method of assembly to find U_E. An example is a ball of charge with uniform charge density $\rho = Q/\frac{4}{3}\pi R^3$. Rather than evaluate (3.76), we can compute the work required to "build" the ball by successively adding spherical layers of uniform charge and thickness dr. At an intermediate stage of assembly, the ball has radius r and (3.73) identifies the work done against the Coulomb force to add an increment of charge $dq = 4\pi r^2 dr\rho$ as $dW = \varphi_S dq$, where φ_S is the electrostatic potential at the surface of the ball. The method of Example 3.4 gives this potential as $\varphi_S = \rho r^2/3\epsilon_0$. Therefore, the energy we seek is

$$U_E = \frac{4\pi}{3\epsilon_0}\rho^2\int_0^R dr r^4 = \frac{3}{5}\frac{Q^2}{4\pi\epsilon_0 R}. \tag{3.77}$$

A variation of the method used to obtain (3.77) can be exploited to derive (3.76) without passing through the point charge result (3.75). If $\rho(\mathbf{r})$ and $\varphi(\mathbf{r})$ are the final charge density and electrostatic potential, the idea is to let the charge distribution at any intermediate stage of the assembly process be $\lambda\rho(\mathbf{r})$, where λ is a real number which increases continuously from 0 to 1. By linearity, the scalar potential at the same intermediate stage is $\lambda\varphi(\mathbf{r})$. Therefore, as in the preceding charged-ball problem, the work done against the Coulomb force when we add an infinitesimal bit of charge (whereupon λ changes to $\lambda + \delta\lambda$) to any volume element is $[(\delta\lambda)\rho(\mathbf{r})][\lambda\varphi(\mathbf{r})] = [(\delta\lambda)\lambda][\rho(\mathbf{r})\varphi(\mathbf{r})]$. In agreement with (3.76), the total work done is

$$U_E = \int_0^1 \delta\lambda\lambda \int d^3r\, \rho(\mathbf{r})\varphi(\mathbf{r}) = \frac{1}{2}\int d^3r\, \rho(\mathbf{r})\varphi(\mathbf{r}). \tag{3.78}$$

Finally, it is worth remarking that the electrostatic total energy U_E is a quantity which clearly illustrates the fundamental limitations of the point charge concept in classical electrodynamics. $U_{\rm E}$ diverges for a single point charge, which is apparent from the $R \to 0$ limit of (3.77). This unphysical behavior is the price we pay for the computational advantages of writing charge densities as a sum of delta functions as in (2.6). There are circumstances where this divergence can be hidden (see Section 23.6.3), but its presence manifests itself in other ways.

3.6.1 U_E is Positive-Definite

The electrostatic total energy is a non-negative quantity. To see this, it is simplest to eliminate $\rho(\mathbf{r})$ from the far right side of (3.76) using Gauss' law and then integrate by parts. The result is

$$U_E = -\frac{1}{2}\epsilon_0 \int d^3r\, \mathbf{E}\cdot\nabla\varphi + \frac{1}{2}\epsilon_0 \int d^3r\, \nabla\cdot(\varphi\mathbf{E}). \tag{3.79}$$

The divergence theorem applied to the last term in (3.79) gives zero because $\mathbf{E}(\mathbf{r}) \to 0$ at infinity for any realistic source. Then, because $\mathbf{E} = -\nabla\varphi$ in electrostatics,

$$U_E = \frac{1}{2}\epsilon_0 \int d^3r\, |\mathbf{E}|^2 \geq 0. \tag{3.80}$$

$U_E \geq 0$ appears to contradict (3.75) for, say, two equal and opposite point charges. There is no contradiction because the latter explicitly excludes the positive but unphysically infinite self-energy of each point charge [computed by inserting $\mathbf{E} = q\hat{\mathbf{r}}/4\pi\epsilon_0 r^2$ into (3.80)]. This characteristic pathology (remedied only by quantum electrodynamics) shows that the point charge concept must be used with care in classical calculations. By contrast, the self-energy part of (3.80) is integrable and makes a positive (but finite) contribution to the total energy for smooth and non-singular distributions $\rho(\mathbf{r})$ of either sign.

Equation (3.80) invites us to ascribe a positive-definite electrostatic energy density to every point in space,

$$u_E(\mathbf{r}) = \tfrac{1}{2}\epsilon_0|\mathbf{E}(\mathbf{r})|^2. \tag{3.81}$$

This turns out to be correct (see Chapter 15), but a deduction based on (3.80) is unwarranted because (3.76) would lead us to make the same claim for $\frac{1}{2}\rho(\mathbf{r})\varphi(\mathbf{r})$, which is numerically a very different quantity.

Thomson's Problem

The classical "plum pudding" model of the atom was devised by J.J. Thomson in 1904. A contemporary variation of this model asks: what mechanically stable arrangement of N negative point charges q_k has the lowest energy when all the charges are constrained to lie on the surface of a spherical shell of uniform compensating positive charge? The presence of the positive charge vitiates Earnshaw's theorem but, by symmetry, has no other effect on the spatial arrangement of the point charges. The problem thus reduces to minimizing the total energy (3.74) subject to the constraint that each vector $\mathbf{r}_k$ has fixed length.

Surprisingly, the solutions do *not* always correspond to configurations of maximal symmetry. Thus, for $N = 8$, the minimum energy does not occur when the charges are placed at the corners of a cube. Instead, U_E is minimized when the charges are placed at the corners of a twisted, rectangular parallelepiped. The search for the ground state is non-trivial because the number of configurations that correspond to *local* energy minima increases exponentially with N. For large N, the charges arrange themselves locally onto a triangular lattice with six nearest neighbors per charge. Many, but not all, of the ground states are found to possess exact or distorted icosahedral symmetry. If $q_k = |\mathbf{r}_k| = 1$, numerical studies show that the energy of the minimum energy configuration is very well approximated by

$$U_E(N) \simeq \frac{1}{4\pi\epsilon_0}\frac{1}{2}(N^2 - N^{3/2}).$$

Roughly speaking, the first term is the average energy of N charges distributed randomly over the surface of the sphere. The second term corrects for the fact that the charges are not truly randomly distributed but arrange themselves to maximize their distances from each other.

Thomson's problem has a direct bearing on a remarkable phenomenon called *charge inversion* that occurs when large spherical ions with charge Zq are placed in solution with small, mobile particles with charge $-q$. It is reasonable to suppose that Z mobile particles would attach to the surface of each macro-ion to neutralize its charge. However, the form of $U_E(N)$ above actually produces a lower energy when $N > Z$ particles attach!

3.6.2 Interaction Total Energy is Potential Energy

Finally, it is instructive to compute U_E for a charge distribution that is the sum of two parts so $\rho(\mathbf{r}) = \rho_1(\mathbf{r}) + \rho_2(\mathbf{r})$. Using (3.76), the total energy is

$$U_E[\rho_1 + \rho_2] = U_E[\rho_1] + U_E[\rho_2] + \frac{1}{4\pi\epsilon_0}\int d^3r \int d^3r' \frac{\rho_1(\mathbf{r})\rho_2(\mathbf{r}')}{|\mathbf{r}-\mathbf{r}'|}. \tag{3.82}$$

The first two terms on the right side of (3.82) are the total energies of ρ_1 and ρ_2 in isolation. We identify the third term as the *interaction energy* between $\rho_1(\mathbf{r})$ and $\rho_2(\mathbf{r})$:

$$V_E = \frac{1}{4\pi\epsilon_0}\int d^3r \int d^3r' \frac{\rho_1(\mathbf{r})\rho_2(\mathbf{r}')}{|\mathbf{r}-\mathbf{r}'|}. \tag{3.83}$$

We use the symbol V_E in (3.83) because the interaction total energy is exactly the potential energy (3.70). We make this nearly self-evident point here because the same statement will *not* be true when we turn to magnetostatic energy in Chapter 12.

We note also a special case of (3.82) that arises when $\rho_2(\mathbf{r})$ is unspecified but serves to create an external potential $\varphi_{\rm ext}(\mathbf{r})$. In that case, the term $U_E[\rho_2]$ is absent from (3.82) and we deduce that the total electrostatic energy of a charge distribution $\rho_1(\mathbf{r})$ in an external potential is

$$U_E = \frac{1}{8\pi\epsilon_0}\int d^3r \int d^3r' \frac{\rho_1(\mathbf{r})\rho_1(\mathbf{r}')}{|\mathbf{r}-\mathbf{r}'|} + \int d^3r \rho_1(\mathbf{r})\varphi_{\rm ext}(\mathbf{r}). \tag{3.84}$$

Application 3.2 The Ionization Potential of a Metal Cluster

Experiments show that I_N, the ionization potential of a molecular cluster composed of N atoms of a monovalent metallic material like potassium is well represented by the formula [Bréchignac *et al.* (1989)]

$$I_N = U_E^{\rm final} - U_E^{\rm initial} = W + aN^{-1/3}. \tag{3.85}$$

In this formula, a is a constant and W is the work function of a macroscopic ($N \to \infty$) sample of the metal. We will estimate W and a using a model which replaces the cluster by a sphere of radius

R and each atom in the cluster by a sphere of radius $R_S < R$. Conservation of volume implies that $R = R_S N^{1/3}$. The volume of each atomic sphere is chosen as $\frac{4}{3}\pi R_S^3 = n^{-1}$ where n is the valence electron density of the metal.

Figure 3.15 shows models for the cluster before and after the ionization event. The initial state diagram shows the atomic sphere of the one atom that will lose its valence electron by ionization. The electron is modeled as a negative point charge at the center of the sphere (black dot). The neutralizing positive charge of the ionic core is shown smeared out over the sphere volume into a uniform density $\rho_+ = |e|n$ (gray sphere). In the final state diagram, the point electron is gone (ionized) and the total ionic positive charge $+|e|$ left behind is shown distributed over the cluster surface. This simulates the attraction of the ionic charge to the departed electron. Finally, in both the initial and final state, we smear out the total charge of the $N - 1$ neutral atoms that do not participate in the ionization event uniformly over the volume of the entire cluster. This produces the white regions in Figure 3.15 where $\rho(\mathbf{r}) = 0$.

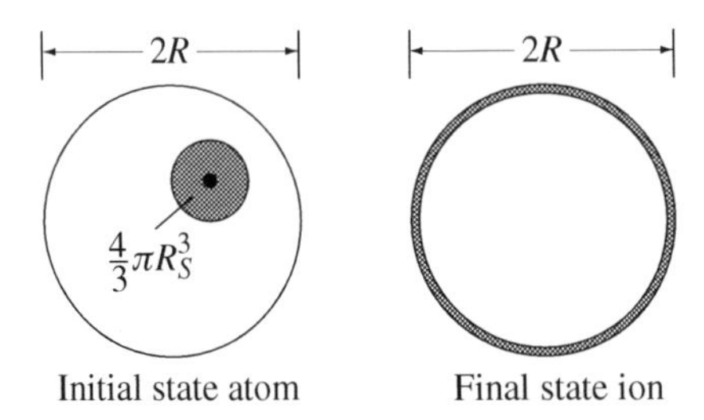

Figure 3.15: Classical models of the charge distribution of a molecular cluster before (left) and after (right) an ionization event. Regions of uniform positive charge density are shaded. Regions of zero charge density are unshaded. The positive charge of the shaded sphere on the left is compensated by the negative charge of the electron to be ionized (black dot) at its center.

U_E^{initial} is the sum of the self-energy of the shaded positive sphere on the left side of Figure 3.15 and the potential energy of interaction between that sphere and the electron at its center. There is no contribution from the non-shaded volume because $\rho(\mathbf{r}) = 0$ there. With respect to the position of the electron, Gauss' law gives the electric field of the sphere as

$$\mathbf{E}_{\text{i}} = \begin{cases} \dfrac{e\mathbf{r}}{4\pi\epsilon_0 R_S^3} & r \le R_S, \\ \dfrac{e\mathbf{r}}{4\pi\epsilon_0 r^3} & r \ge R_S. \end{cases} \tag{3.86}$$

The scalar potential of the point electron is $\varphi_{-e}(r) = -e/4\pi\epsilon_0 r$. Therefore,[11]

$$U_E^{\text{initial}} = \frac{1}{2}\epsilon_0 \int d^3r|\mathbf{E}_{\text{i}}|^2 + \int d^3r \rho_+ \varphi_{-e} = \frac{1}{4\pi\epsilon_0}\frac{e^2}{R_S}\left[\frac{3}{5} - \frac{3}{2}\right]. \tag{3.87}$$

By assumption, U_E^{final} is the electrostatic energy of a sphere with total charge $+e$ uniformly distributed over its surface. From Gauss' law, the electric field is zero inside the sphere and $\mathbf{E}_{\text{f}} = \hat{\mathbf{r}}e/4\pi\epsilon_0 r^2$ outside the sphere. Therefore,

$$U_E^{\text{final}} = \frac{1}{2}\epsilon_0 \int\limits_{r>R} d^3r|\mathbf{E}_{\text{f}}|^2 = \frac{e^2}{8\pi\epsilon_0 R}. \tag{3.88}$$

[11] The first term in (3.87) is an alternative calculation of (3.77).

Using $R = R_S N^{1/3}$, the predicted ionization potential has the form anticipated in (3.85):

$$I_N = U_E^{\text{final}} - U_E^{\text{initial}} \simeq \frac{e^2}{4\pi\epsilon_0 R_S}\left[\frac{9}{10} + \frac{1}{2}N^{-1/3}\right]. \tag{3.89}$$

3.7 The Electric Stress Tensor

Let $\mathbf{E}(\mathbf{r})$ be the electric field produced by a charge density $\rho(\mathbf{r})$. The force exerted on the charge in a volume V (due to the charge that is not in V) is

$$\mathbf{F} = \int_V d^3r\, \rho(\mathbf{r})\mathbf{E}(\mathbf{r}). \tag{3.90}$$

For many applications, it proves useful to eliminate ρ from (3.90) using $\epsilon_0 \nabla \cdot \mathbf{E} = \rho$. To do this easily, we adopt the Einstein convention and sum over repeated Cartesian indices (see Section 1.2.4). In that case, $\nabla \cdot \mathbf{E} = \partial_i E_i$, and

$$F_j = \epsilon_0 \int_V d^3r\, E_j\, \partial_i E_i = \epsilon_0 \int_V d^3r\, \left[\partial_i(E_j E_i) - E_i \partial_i E_j\right]. \tag{3.91}$$

Now, $\nabla \times \mathbf{E} = 0$ implies that $\partial_i E_j = \partial_j E_i$. Therefore,

$$F_j = \epsilon_0 \int_V d^3r\, \left[\partial_i(E_i E_j) - E_i \partial_j E_i\right] = \epsilon_0 \int_V d^3r\, \partial_i(E_i E_j - \tfrac{1}{2}\delta_{ij}\mathbf{E}\cdot\mathbf{E}). \tag{3.92}$$

The final member of (3.92) motivates us to define the Cartesian components of the Maxwell *electric stress tensor* as

$$T_{ij}(\mathbf{E}) = \epsilon_0(E_i E_j - \tfrac{1}{2}\delta_{ij} E^2). \tag{3.93}$$

With this definition and the divergence theorem, (3.92) takes the form

$$F_j = \int_V d^3r\, \partial_i T_{ij}(\mathbf{E}) = \int_S dS\, \hat{n}_i T_{ij}(\mathbf{E}). \tag{3.94}$$

Often, the stress tensor components in (3.93) are used to define the dyadic (Section 1.8)

$$\mathbf{T}(\mathbf{E}) = \hat{\mathbf{e}}_i T_{ij}(\mathbf{E})\hat{\mathbf{e}}_j = \hat{\mathbf{x}}T_{xx}(\mathbf{E})\hat{\mathbf{x}} + \hat{\mathbf{x}}T_{xy}(\mathbf{E})\hat{\mathbf{y}} + \hat{\mathbf{x}}T_{xz}(\mathbf{E})\hat{\mathbf{z}} + \hat{\mathbf{y}}T_{yx}(\mathbf{E})\hat{\mathbf{x}} + \cdots. \tag{3.95}$$

In this language, the net force on the charge inside the closed surface S is

$$\mathbf{F} = \int_S dS\, \hat{\mathbf{n}} \cdot \mathbf{T}(\mathbf{E}). \tag{3.96}$$

Direct substitution from (3.93) confirms that an explicit vector form of (3.96) is

$$\mathbf{F} = \epsilon_0 \int_S dS\, \left[(\hat{\mathbf{n}}\cdot\mathbf{E})\mathbf{E} - \tfrac{1}{2}(\mathbf{E}\cdot\mathbf{E})\hat{\mathbf{n}}\right]. \tag{3.97}$$

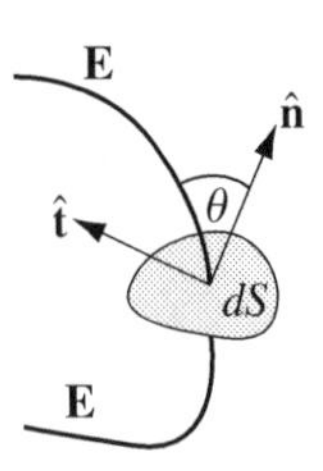

Figure 3.16: The electric field line is coplanar with the surface unit normal $\hat{\mathbf{n}}$ and the unit tangent vector $\hat{\mathbf{t}}$. The angle θ lies between $\hat{\mathbf{n}}$ and the electric field line at the surface element dS.

3.7.1 Applications of the Electric Stress Tensor

Let us apply (3.97) to a spherical balloon which carries a uniform charge per unit area σ. The Gauss' law electric field is zero everywhere inside the balloon and takes the value $\mathbf{E} = \hat{\mathbf{r}}\sigma/\epsilon_0$ just outside the balloon's surface. If we choose the balloon surface as S, the two terms in the integrand of (3.97) combine to give a force per unit area $\mathbf{f} = \hat{\mathbf{r}}\sigma^2/2\epsilon_0$. This agrees with (3.52) where this situation yields $\mathbf{f} = \frac{1}{2}\sigma[0 + \hat{\mathbf{r}}\sigma/\epsilon_0]$. This force density tends to expand the size of the balloon (so each individual bit of charge gets farther apart from every other bit) while keeping the position of the center of mass fixed ($\mathbf{F} = 0$). More generally, Figure 3.16 shows that the electric field $\mathbf{E}$ at any surface element dS can be decomposed into components along the surface normal $\hat{\mathbf{n}}$ and a unit vector $\hat{\mathbf{t}}$ that is tangent to the surface and coplanar with $\mathbf{E}$ and $\hat{\mathbf{n}}$:

$$\mathbf{E} = E(\hat{\mathbf{n}}\cos\theta + \hat{\mathbf{t}}\sin\theta). \tag{3.98}$$

Figure 3.16 does not indicate the direction of $\mathbf{E}$ because (3.98) transforms (3.97) into

$$\mathbf{F} = \tfrac{1}{2}\epsilon_0 \int_S dS\, E^2(\hat{\mathbf{n}}\cos 2\theta + \hat{\mathbf{t}}\sin 2\theta). \tag{3.99}$$

The great virtue of (3.97) and (3.99) is that they replace the volume integral (3.91) by a surface integral, which is often simpler to evaluate. We also gain the view that the net force on the charge inside V is transmitted through each surface element $\hat{\mathbf{n}}\,dS$ by a vector force density $\mathbf{f}$ where $f_j = \hat{n}_i T_{ij}(\mathbf{E})$. It is important to appreciate that the surface in question need not be coincident with the boundary of the charge distribution. Indeed, the fact that $T_{ij}(\mathbf{E})$ can be evaluated at any point in space leads very naturally to the idea that the vacuum acts as a sort of elastic medium capable of supporting stresses (measured by $\mathbf{E}$) which communicate the Coulomb force. Faraday, Maxwell, and their contemporaries believed strongly in the existence of this medium, which they called the "luminiferous aether".

We can use (3.99) to extract mechanical information from the field line patterns shown in Figure 3.17 for two equal-magnitude charges. The key is to choose the perpendicular bisector plane to be part of a surface S that completely encloses one charge when viewed from the *other* charge (see Figure 3.18). The charges are both positive in the left panel and the field lines bend toward tangency as they approach the bisector plane. This corresponds to $\theta = \pi/2$ in Figure 3.16 and a force density $-\frac{1}{2}\epsilon_0 E^2\hat{\mathbf{n}}$ in (3.99). Since $\hat{\mathbf{n}}$ is the *outward* normal, this tells us that the net force on the bisector plane tends to push the bisector plane away from the viewing charge. This is consistent with Coulomb repulsion between the charges. By contrast, the field lines are normal to the bisector plane when the charges have opposite sign (right panel of Figure 3.17). This corresponds to $\theta = 0$ in Figure 3.16 and a force per unit area $+\frac{1}{2}\epsilon_0 E^2\hat{\mathbf{n}}$. For this case, the net force tends to pull the bisector toward the viewing charge. This is consistent with Coulomb attraction between the charges.

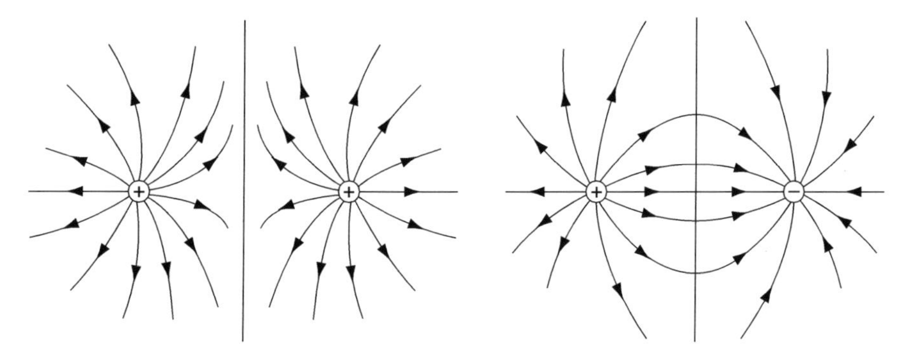

Figure 3.17: The electric field lines associated with two identical, positive point charges (left panel) and two equal but opposite point charges (right panel).

Example 3.6 Confirm that the stress tensor formalism reproduces the familiar Coulomb force law between two identical point charges separated by a distance $2d$. Hint: Integrate (3.97) over a surface S that includes the perpendicular bisector plane between the charges.

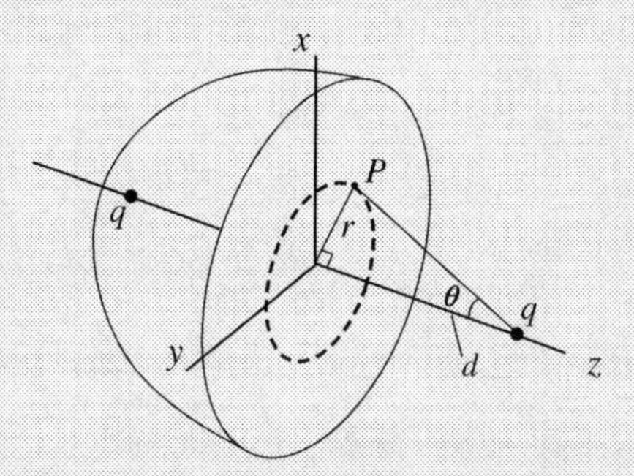

Figure 3.18: The force between two identical charges may be calculated by integrating the electric stress tensor over the bowl-shaped surface (shaded) in the limit when the bowl radius goes to infinity.

Solution: Locate the charges at $\pm d$ on the z-axis. To find the force on the charge at $z = -d$ we choose S as the surface of the solid bowl-shaped object shown in Figure 3.18. This surface encloses the half-space $z < 0$ in the limit when the bowl radius goes to infinity. In the same limit, the hemispherical portion of the bowl makes no contribution to the stress tensor surface integral (3.97) because $dSE^2 \to 0$ as $r \to \infty$. Therefore, it is sufficient to integrate over the bisector plane $z = 0$ where the outward normal $\hat{\mathbf{n}} = \hat{\mathbf{z}}$.

The charge distribution has mirror symmetry with respect to $z = 0$. Therefore, the field $\mathbf{E}_P$ at any point P on the bisector must lie entirely in that plane. That is, $\hat{\mathbf{n}} \cdot \mathbf{E}_P = 0$. The distance between P and either charge is $d/\cos\theta$ where θ is the angle between the z-axis and the line which connects either charge to P. Counting both charges, the component of the electric field in the plane of the bisector at the point P is

$$E_{\|} = 2 \times \frac{q}{4\pi\epsilon_0 d^2} \cos^2\theta \sin\theta.$$

Since $r = d\tan\theta$, the differential element of area on the bisector is

$$dS = r dr d\phi = d^2 \frac{\sin\theta}{\cos^3\theta} d\theta\, d\phi.$$

Only the second term in (3.97) contributes, so the force on the charge enclosed by S is

$$\mathbf{F} = -\frac{\epsilon_0}{2}\hat{\mathbf{z}} \int_{z=0} dS\, E_{\parallel}^2 = -\frac{\epsilon_0}{2}\hat{\mathbf{z}} \left[\frac{q}{2\pi\epsilon_0 d}\right]^2 \int_0^{2\pi} d\phi \int_0^{\pi/2} d\theta\, \cos\theta \sin^3\theta = -\hat{\mathbf{z}}\frac{1}{4\pi\epsilon_0}\frac{q^2}{(2d)^2}.$$

This is indeed Coulomb's law.

Sources, References, and Additional Reading

The quotation at the beginning of the chapter is taken from Letter XXIII of Volume II of

L. Euler, *Letters on Different Subjects in Natural Philosophy Addressed to a German Princess* (Arno Press, New York, 1975).

Section 3.1 The most complete treatment of electrostatics is the three-volume treatise

E. Durand, *Électrostatique* (Masson, Paris, 1964).

Volume I is devoted to basic theory and the potential and field produced by specified distributions of charge. Readers unfamiliar with the French language will still benefit from the beautiful drawings of electric field line patterns and equipotential surfaces for many electrostatic situations. Four textbooks which distinguish summation problems from boundary value potential theory are

A. Sommerfeld, *Electrodynamics* (Academic, New York, 1952).

M.H. Nayfeh and M.K. Brussel, *Electricity and Magnetism* (Wiley, New York, 1985).

L. Eyges, *The Classical Electromagnetic Field* (Dover, New York, 1972).

W. Hauser, *Introduction to the Principles of Electromagnetism* (Addison-Wesley, Reading, MA, 1971).

Section 3.3 See Section 15.3.1 for a classical argument due to Eugene Wigner which relates the freedom to choose the zero of the electrostatic potential to the conservation of electric charge. Example 3.2 is an electrostatic version of a scaling law derived by Newton for the gravitational force. See Section 86 of the thought-provoking monograph

S. Chandrasekhar, *Newton's Principia for the Common Reader* (Clarendon Press, Oxford, 1995).

Figure 3.4 is taken from this reprint of the 3rd edition (1891) of Maxwell's *Treatise*:

J.C. Maxwell, *A Treatise on Electricity and Magnetism* (Clarendon Press, Oxford, 1998). By permission of Oxford University Press.

The problems associated with projecting three-dimensional field line patterns onto two-dimensional diagrams are discussed in

A. Wolf, S.J. Van Hook, and E.R. Weeks, "Electric field lines don't work", *American Journal of Physics* **64**, 714 (1996).

Section 3.4 Application 3.1 is taken from

T.M. Kalotas, A.R. Lee, and J. Liesegang, "Analytical construction of electrostatic field lines with the aid of Gauss' law", *American Journal of Physics* **64**, 373 (1996).

Our discussion of symmetry and Gauss' law is adapted from

R. Shaw, "Symmetry, uniqueness, and the Coulomb law of force", *American Journal of Physics* **33**, 300 (1965).

Section 3.6 Entry points into the literature of Thomson's problem and charge inversion are, respectively,

E.L. Altschuler and A. Pérez-Garrido, "Defect-free global minima in Thomson's problem of charges on a sphere", *Physical Review E* **73**, 036108 (2006).

A.Yu. Grosberg, T.T. Nguyen, and B.I. Shklovskii, "The physics of charge inversion in chemical and biological systems", *Reviews of Modern Physics* **74**, 329 (2002).

The empirical formula (3.85) in Application 3.2 comes from the paper

C. Bréchignac, Ph. Cahuzak, F. Carlier, and J. Leygnier, "Photoionization of mass-selected ions: A test for the ionization scaling law", *Physical Review Letters* **63**, 1368 (1989).

Figure 3.17 and the discussion of the electric stress tensor in Section 3.7 are based on the treatment in

E.J. Konopinski, *Electromagnetic Fields and Relativistic Particles* (McGraw-Hill, New York, 1981).

Problems

3.1 Charged Particle Refraction

(a) A point charge $q > 0$ with total energy E travels through a region of constant potential V_1 and enters a region of potential $V_2 < V_1$. Show that the trajectory bends so that the angles θ_1 and θ_2 in the diagram below obey a type of "Snell's law" with a characteristic "index of refraction" for each medium.

(b) Describe the charge distribution which must exist at the V_1-V_2 interface.

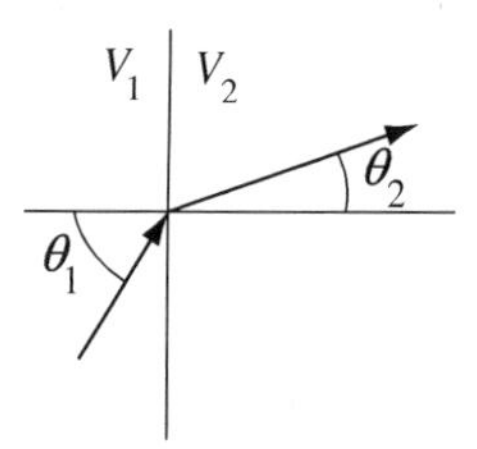

3.2 Symmetric and Traceless The Cartesian components of the electric field in a charge-free region of space are $E_k = C_k + D_{jk}r_j$, where C_k and D_{jk} are constants.

(a) Prove that D_{jk} is symmetric ($D_{jk} = D_{kj}$) and traceless ($\sum_k D_{kk} = 0$).

(b) Find the most general electrostatic potential that generates the electric field in part (a).

3.3 Practice Superposing Fields This problem exploits the ring and disk electric fields calculated in Example 3.1.

(a) Find $\mathbf{E}(\mathbf{r})$ inside and outside a uniformly charged spherical shell by superposing the electric fields produced by a collection of charged rings.

(b) Find $\mathbf{E}(\mathbf{r})$ inside and outside a uniformly charged spherical volume by superposing the electric fields produced by a collection of uniformly charged disks.

3.4 Five Charges in a Line Draw the electric field line pattern for a line of five equally spaced charges with equal magnitude but alternating algebraic signs, as sketched below.

Be sure to choose the scale of your drawing and the number of lines drawn so *all* salient features of the pattern are obvious.

3.5 Gauss' Law Practice Use Gauss' law to find the electric field when the charge density is:

(a) $\rho(x) = \rho_0 \exp\left\{-\kappa\sqrt{x^2}\right\}$. Expresses the answer in Cartesian coordinates.

(b) $\rho(x, y) = \rho_0 \exp\left\{-\kappa\sqrt{x^2 + y^2}\right\}$. Express the answer in cylindrical coordinates.

(c) $\rho(x, y, z) = \rho_0 \exp\left\{-\kappa\sqrt{x^2 + y^2 + z^2}\right\}$. Express the answer in spherical coordinates.

3.6 General Electrostatic Torque Show that the torque exerted on a charge distribution $\rho(\mathbf{r})$ by a distinct charge distribution $\rho'(\mathbf{r}')$ is

$$\mathbf{N} = -\frac{1}{4\pi\epsilon_0}\int d^3r \int d^3r' \frac{\mathbf{r}\times\mathbf{r}'}{|\mathbf{r}-\mathbf{r}'|^3}\rho(\mathbf{r})\rho'(\mathbf{r}').$$

3.7 Field Lines for a Non-Uniformly Charged Disk The z-axis coincides with the symmetry axis of a flat disk of radius a in the x-y plane. The disk carries a uniform charge per unit area $\sigma < 0$. The rim of the disk carries an additional uniform charge per unit length $\lambda > 0$. Use a side (edge) view and sketch the electric field lines everywhere assuming that the total charge of the disk is positive. Your sketch must have enough detail to reveal any interesting topological features of the field line pattern.

3.8 The Electric Field of a Charged Slab and a Charged Sheet

(a) Find $\mathbf{E}(\mathbf{r})$ if $\rho(x, y, z) = \sigma_0\,\delta(x) + \rho_0\Theta(x) - \rho_0\Theta(x-b)$.
(b) Show by explicit calculation that $\rho(x, y, z)$ does not exert a net force on itself.

3.9 The Electric Flux Through a Plane A charge distribution $\rho(\mathbf{r})$ with total charge Q occupies a finite volume V somewhere in the half-space $z < 0$. If the integration surface is $z = 0$, prove that

$$\int_{z=0} dS\,\hat{\mathbf{z}}\cdot\mathbf{E} = \frac{Q}{2\epsilon_0}.$$

3.10 Two Electrostatic Theorems

(a) Use Green's second identity, $\int_V d^3r\,(f\nabla^2 g - g\nabla^2 f) = \int_S d\mathbf{S}\cdot(f\nabla g - g\nabla f)$, to prove that the potential $\varphi(0)$ at the center of a charge-free spherical volume V is equal to the average of $\varphi(\mathbf{r})$ over the surface S of the sphere. We proved this theorem in the text using Green's reciprocity relation.
(b) Use the result of part (a) to provide an alternative to the derivation of Earnshaw's theorem given in the text.

3.11 Potential, Field, and Energy of a Charged Disk A two-dimensional disk of radius R carries a uniform charge per unit area $\sigma > 0$.

(a) Calculate the potential at any point on the symmetry axis of the disk.
(b) Calculate the potential at any point on the rim of the disk. Hint: Use a point on the rim as the origin.
(c) Sketch the electric field pattern everywhere in the plane of the disk.
(d) Calculate the electrostatic total energy U_E of the disk.

3.12 A Charged Spherical Shell with a Hole The figure below shows a circular hole of radius b (white) bored through a spherical shell (gray) with radius R and uniform charge per unit area σ.

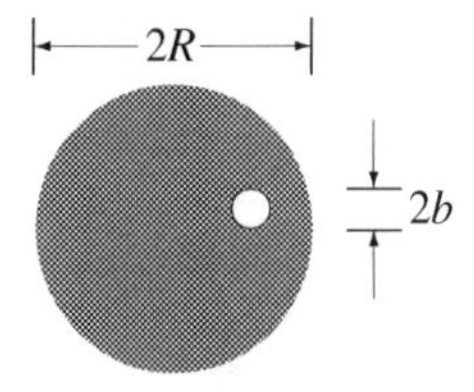

(a) Show that $\mathbf{E}(P) = (\sigma/2\epsilon_0)[1 - \sin(\theta_0/2)]\hat{\mathbf{r}}$, where P is the point at the center of the hole and θ_0 is the opening angle of a cone whose apex is at the center of the sphere and whose open end coincides with the edge of the hole. Perform the calculation by summing the vector electric fields produced at P by all the other points of the shell.
(b) Use an entirely different argument to explain why $\mathbf{E}(P) \approx (\sigma/2\epsilon_0)\hat{\mathbf{r}}$ when $\theta_0 \ll 1$.

3.13 A Uniformly Charged Cube The figure below shows a cube filled uniformly with charge. Determine the ratio φ_0/φ_1 of the potential at the center of the cube to the potential at the corner of the cube. Hint: Think of the cube as formed from the superposition of eight smaller cubes.

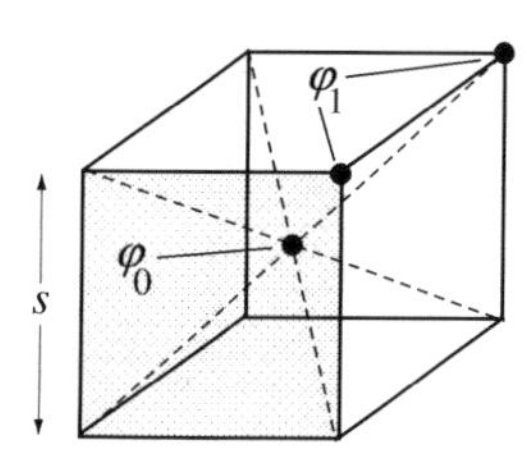

3.14 A Variation on Coulomb's Law Suppose the electrostatic potential of a point charge were $\varphi(r) = (1/4\pi\epsilon_0)r^{-(1+\epsilon)}$ rather than the usual Coulomb formula.

(a) Find the potential $\varphi(r)$ at a point at a distance r from the center of a spherical shell of radius $R > r$ with uniform surface charge per unit area σ. Check the Coulomb limit $\epsilon = 0$.

(b) To first order in ϵ, show that

$$\frac{\varphi(R) - \varphi(r)}{\varphi(R)} = \frac{\epsilon}{2}\left[\frac{R}{r}\ln\frac{R+r}{R-r} - \ln\frac{4R^2}{R^2 - r^2}\right].$$

Since the time of Cavendish, formulae like this one have been used in experimental tests of the correctness of Coulomb's law.

3.15 Practice with Electrostatic Energy Let the space between two concentric spheres with radii a and $R \geq a$ be filled uniformly with charge.

(a) Calculate the total energy U_E in terms of the total charge Q and the variable $x = a/R$. Check the $a = 0$ and $a = R$ limits.

(b) Minimize U_E with respect to x (keeping the total charge Q constant). Identify the physical system which achieves the minimum you find.

3.16 Interaction Energy of Spheres

(a) Evaluate the relevant part of the integral $U_E = \frac{1}{2}\int d^3r\, \rho(\mathbf{r})\varphi(\mathbf{r})$ to find the interaction energy V_E between two identical insulating spheres, each with radius R and charge Q distributed uniformly over their surfaces. The center-to-center separation between the spheres is $d > 2R$. Do *not* assume that $d \gg R$.

(b) Produce a physical argument to explain the dependence of V_E on R.

3.17 Ionization Energy of a Model Hydrogen Atom A model hydrogen atom is composed of a point nucleus with charge $+|e|$ and an electron charge distribution

$$\rho_-(\mathbf{r}) = -\frac{|e|}{\pi a^2 r}\exp(-2r/a).$$

Show that the ionization energy (the energy to remove the electronic charge and disperse it to infinity) of this atom is

$$I = \frac{3}{8}\frac{e^2}{\pi\epsilon_0 a}.$$

Hint: Ignore the (divergent) self-energy of the point-like nucleus.

3.18 Two Spherical Charge Distributions A spherical charge distribution $\rho_1(r)$ has total charge Q_1 and a second, non-overlapping spherical charge distribution $\rho_2(r)$ has total charge Q_2. The distance between the centers of the two distributions is R. Use the stress tensor formalism to prove that the interaction energy between the two is

$$V_E = \frac{1}{4\pi\epsilon_0}\frac{Q_1 Q_2}{R}.$$

Hint: It is not necessary to evaluate any integrals explicitly.

3.19 Two Electric Field Formulae

(a) Show that the electric field produced by a uniform charge density ρ confined to the volume V enclosed by a surface S can be written

$$\mathbf{E}(\mathbf{r}) = \frac{\rho}{4\pi\epsilon_0}\int_S \frac{d\mathbf{S}'}{|\mathbf{r}-\mathbf{r}'|}.$$

(b) Show that the electric field due to an arbitrary but localized charge distribution can always be written in the form

$$\mathbf{E}(\mathbf{r}) = -\frac{1}{4\pi\epsilon_0}\int d^3r' \frac{\nabla'\rho(\mathbf{r}')}{|\mathbf{r}-\mathbf{r}'|}.$$

3.20 The Potential of a Charged Line Segment The line segment from P to P' in the diagram below carries a uniform charge per unit length λ. The vector $\mathbf{a}$ is coincident with the segment. The vectors $\mathbf{c}$ and $\mathbf{b}$ point from the observation point $\mathbf{r}$ to the beginning and end of $\mathbf{a}$, respectively.

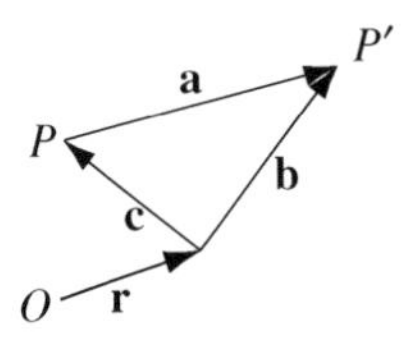

Evaluate the integral for the potential in a coordinate-free manner by parameterizing the line source using a variable s' which is zero at the point on the charged segment that is closest to $\mathbf{r}$. Show thereby that

$$\varphi(\mathbf{r}) = \frac{\lambda}{4\pi\epsilon_0}\ln\frac{\left|\dfrac{\mathbf{b}\cdot\mathbf{a}}{a} + \sqrt{\left(\dfrac{\mathbf{b}\cdot\mathbf{a}}{a}\right)^2 + \dfrac{|\mathbf{b}\times\mathbf{a}|^2}{a^2}}\right|}{\left|\dfrac{\mathbf{c}\cdot\mathbf{a}}{a} + \sqrt{\left(\dfrac{\mathbf{c}\cdot\mathbf{a}}{a}\right)^2 + \dfrac{|\mathbf{b}\times\mathbf{a}|^2}{a^2}}\right|}.$$

3.21 A Variation on Coulomb's Law Suppose that the electrostatic potential produced by a point charge was not Coulombic, but instead varied with distance in a manner determined by a specified scalar function $f(r)$:

$$\varphi(\mathbf{r}) = \frac{q}{4\pi\epsilon_0} f(|\mathbf{r}|).$$

(a) Calculate the potential produced by an infinite flat sheet at $z = 0$ with uniform charge per unit area σ.
(b) Use the result of part (a) to show that the associated electric field is

$$\mathbf{E}(z) = \hat{\mathbf{z}}\frac{\sigma}{2\epsilon_0} z f(z).$$

3.22 A Non-Uniform Charge Distribution on a Surface Let d and s be two unequal lengths. Assume that charge is distributed on the $z = 0$ plane with a surface density

$$\sigma(\rho) = \frac{-qd}{2\pi(\rho^2 + s^2)^{3/2}}.$$

(a) Integrate σ to find the total charge Q on the plane.

(b) Show that the potential $\varphi(z)$ produced by $\sigma(\rho)$ on the z-axis is identical to the potential produced by a point with charge Q on the axis at $z = -s$.

3.23 The Energy outside a Charged Volume The potential takes the constant value φ_0 on the closed surface S which bounds the volume V. The total charge inside V is Q. There is no charge anywhere else. Show that the electrostatic energy contained in the space outside of S is

$$U_E(\text{out}) = \frac{1}{2} Q \varphi_0.$$

3.24 Overcharging A common biological environment consists of large macro-ions with charge $Q < 0$ floating in a solution of point-like micro-ions with charge $q > 0$. Experiments show that N micro-ions adsorb onto the surface of each macro-ion. Model one macro-ion as a sphere with its charge uniformly distributed over its surface. As explained in the boxed discussion of "Thomson's Problem" in the text, the minimum energy configuration of $N \gg 1$ point charges on the surface of a sphere of radius R has total energy

$$E(N) = \frac{q^2}{4\pi\epsilon_0 R} \frac{1}{2} \left[N^2 - N^{3/2} \right].$$

(a) Derive the N^2 term in $E(N)$ by smearing out the micro-ion charge over the surface of the macro-ion.
(b) Give a qualitative argument for the $N^{3/2}$ dependence of the second term in $E(N)$ using the fact that the first term does not account for the fact that the charges try to avoid one another on the sphere's surface.
(c) Find N by minimizing the sum of $E(N)$ and the interaction energy between the micro-ions and the macro-ion. Show thereby that the micro-ions do not simply neutralize the charge of the macro-ion.

4 Electric Multipoles

The electrical phenomena outside the sphere are identical with those arising from an imaginary series of singular points all at the center of the sphere.

James Clerk Maxwell (1891)

By permission of Oxford University Press

4.1 Introduction

Human beings are very large compared to the size of many of the charge distributions that interest them. Sub-atomic particles, atoms, molecules, and biological cells all produce electrostatic potentials which humans sample only at points that lie very far from the source charge density. An atomic electron (with non-zero angular momentum) similarly samples the electrostatic potential of its parent nucleus only at points very far removed from the nucleus itself. Examples like these motivate us to seek an approximation to the Coulomb integral,

$$\varphi(\mathbf{r}) = \frac{1}{4\pi\epsilon_0}\int d^3r' \frac{\rho(\mathbf{r}')}{|\mathbf{r}-\mathbf{r}'|}, \tag{4.1}$$

which isolates the physics most important to a distant observer. Such a scheme adds value even when (4.1) can be evaluated exactly.

This chapter develops the *electrostatic multipole expansion* as a systematic approximation to $\varphi(\mathbf{r})$ when $\mathbf{r}$ lies far from any point within a localized charge distribution. By "localized" we mean that the distribution is confined (or nearly so) to a finite volume of space, e.g., a sphere of radius R like the one shown in Figure 4.1. By "far", we mean that $r \gg R$, where $\mathbf{r}$ is measured from the same point as R. For distant observation points, the ratio R/r is a small parameter which justifies truncation of the multipole expansion after a small number of terms. The expansion coefficients, called *multipole moments*, carry information about the source. In what follows, we present the expansion in both Cartesian and spherical coordinates because both appear widely in the research literature. We also distinguish an "exterior" multipole expansion (applicable to Figure 4.1, where the observation point lies entirely outside the charge distribution) from an "interior" multipole expansion (where the observation point lies inside an empty volume and all the charge lies outside that volume).

4.1.1 The Electric Multipole Expansion

We begin with an exterior multipole expansion that is valid for observation points which lie far outside the sphere drawn in Figure 4.1. This implies that $r' \ll r$ in (4.1) and it is reasonable to replace the factor $|\mathbf{r}-\mathbf{r}'|^{-1}$ by the first few terms of the Taylor series (see Section 1.3.4)

$$\frac{1}{|\mathbf{r}-\mathbf{r}'|} = \frac{1}{r} - \mathbf{r}'\cdot\nabla\frac{1}{r} + \frac{1}{2}(\mathbf{r}'\cdot\nabla)^2\frac{1}{r} - \cdots. \tag{4.2}$$

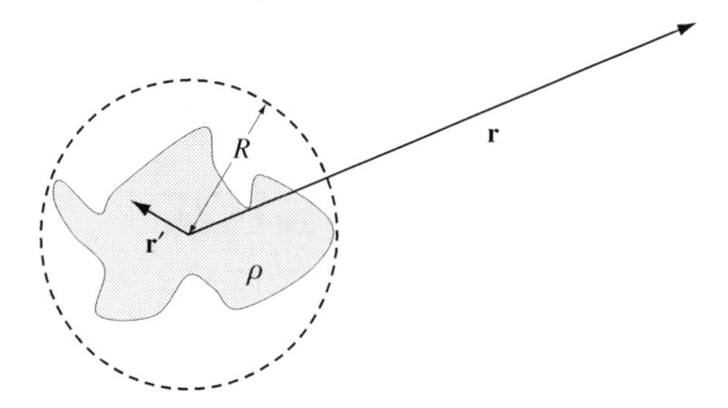

Figure 4.1: A localized charge distribution with total charge Q is confined entirely within a sphere of radius R. The observation point is $\mathbf{r}$ where $r \gg R$.

Substitution of (4.2) into (4.1) produces a series (convergent for $r > R$) known as the primitive Cartesian electric multipole expansion. Using the Einstein convention where repeated indices are summed over x, y, and z (Section 1.2.4),

$$\varphi(\mathbf{r}) = \frac{1}{4\pi\epsilon_0}\left\{\left[\int d^3r' \rho(\mathbf{r}')\right]\frac{1}{r} - \left[\int d^3r' \rho(\mathbf{r}')r_i'\right]\nabla_i\frac{1}{r}\right. \tag{4.3}$$

$$\left. + \left[\frac{1}{2}\int d^3r' \rho(\mathbf{r}')r_i'r_j'\right]\nabla_i\nabla_j\frac{1}{r} - \cdots\right\}.$$

The charge density $\rho(\mathbf{r})$ appears in the integrals enclosed by square brackets in (4.3). For later convenience, we give these integrals special names and evaluate them using the charge density $\rho(\mathbf{r}) = \sum_\alpha q_\alpha \delta(\mathbf{r} - \mathbf{r}_\alpha)$ for a collection of point objects. The *electric monopole moment* is the total charge

$$Q = \int d^3r\, \rho(\mathbf{r}) = \sum_\alpha q_\alpha. \tag{4.4}$$

The *electric dipole moment* is the vector

$$\mathbf{p} = \int d^3r\, \rho(\mathbf{r})\mathbf{r} = \sum_\alpha q_\alpha \mathbf{r}_\alpha. \tag{4.5}$$

The *electric quadrupole moment* is a tensor defined by the nine scalars

$$Q_{ij} = \tfrac{1}{2}\textstyle\int d^3r\, \rho(\mathbf{r})r_i r_j = \tfrac{1}{2}\sum_\alpha q_\alpha r_{\alpha i} r_{\alpha j}. \tag{4.6}$$

Using these definitions and working out the derivatives simplifies (4.3) to

$$\varphi(\mathbf{r}) = \frac{1}{4\pi\epsilon_0}\left\{\frac{Q}{r} + \frac{\mathbf{p}\cdot\mathbf{r}}{r^3} + Q_{ij}\frac{3r_ir_j - r^2\delta_{ij}}{r^5} + \cdots\right\}. \tag{4.7}$$

The charge density in Figure 4.1 is confined to a sphere of radius R. Order-of-magnitude estimates for the dipole moment (4.5) and the quadrupole moment (4.6) based on this observation are $p_i \sim QR$ and $Q_{ij} \sim QR^2$, respectively. Applying this information to the series (4.7) shows that each successive term is smaller than the one that precedes it by a factor of $R/r \ll 1$. This implies that the asymptotic (long-distance) behavior of the electrostatic potential from a localized charge distribution is determined by the first non-zero term of the multipole expansion. Accordingly, the first term in (4.7) dominates the long-distance behavior of the electrostatic potential for objects with a net charge $Q \neq 0$. This confirms

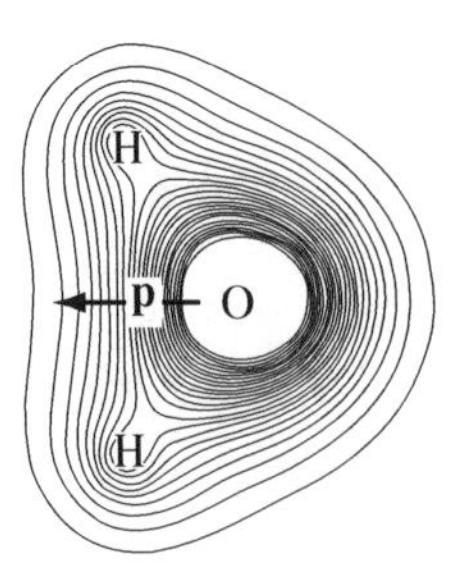

Figure 4.2: Contours of constant charge density for an H_2O molecule. This object has a non-zero electric dipole moment $\mathbf{p}$ because chemical bonding draws negative (electron) charge toward the oxygen atom from the hydrogen atoms. Figure adapted from Bader (1970).

the observation made in Section 3.3.5 that all such objects behave like point charges when viewed from a sufficiently great distance.

4.2 The Electric Dipole

The second term in (4.7) describes the distant electrostatic potential produced by most electrically neutral ($Q = 0$) objects like neutrons, atoms, molecules, plasmas, and ordinary matter. This is the *electric dipole* potential,

$$\varphi(\mathbf{r}) = \frac{1}{4\pi\epsilon_0}\frac{\mathbf{p}\cdot\mathbf{r}}{r^3} \qquad r \gg R. \tag{4.8}$$

The integral that defines the electric dipole moment $\mathbf{p}$ in (4.5) is non-zero for any spatially extended object where the "centers" of positive charge and negative charge do not coincide. We say that a system has a *permanent* electric dipole moment when $\mathbf{p} \neq 0$ is a property of its ground state charge density. A polar molecule like water is a good example (Figure 4.2). We speak of an *induced* dipole moment when $\mathbf{p} \neq 0$ is produced by an external electric field. This is the situation with essentially all conductors and dielectrics.

Whatever the physical origin of the dipole moment, (4.8) *approximates* the true potential far from the distribution. This statement is well defined because the numerical value of $\mathbf{p}$ is independent of the choice of origin when $Q = 0$. To see this, let $\mathbf{p}'$ denote the new dipole moment when we shift the origin by a vector $\mathbf{d}$ so $\mathbf{r} = \mathbf{r}' + \mathbf{d}$. Then, because $d^3r = d^3r'$ and $\rho(\mathbf{r}') = \rho(\mathbf{r})$,

$$\mathbf{p}' = \int d^3r'\,\rho(\mathbf{r}')\mathbf{r}' = \int d^3r\,\rho(\mathbf{r})(\mathbf{r} - \mathbf{d}) = \mathbf{p} - Q\mathbf{d}. \tag{4.9}$$

Hence, $\mathbf{p}' = \mathbf{p}$, if $Q = 0$. Conversely, the electric dipole moment is not uniquely defined for any system with a net charge.

Since $\mathbf{E} = -\nabla\varphi$, the moment $\mathbf{p}$ also completely characterizes the asymptotic *dipole electric field*,

$$\mathbf{E}(\mathbf{r}) = \frac{1}{4\pi\epsilon_0}\frac{3\hat{\mathbf{r}}(\hat{\mathbf{r}}\cdot\mathbf{p}) - \mathbf{p}}{r^3} \qquad r \gg R. \tag{4.10}$$

This formula takes a simple form in polar coordinates because $\mathbf{E}(\mathbf{r})$ must be unchanged when the vector $\mathbf{p}$ rotates about its own axis. Thus, the choice $\mathbf{p} = p\hat{\mathbf{z}}$ and the identity $\hat{\mathbf{z}} = \hat{\mathbf{r}}\cos\theta - \hat{\boldsymbol{\theta}}\sin\theta$ reduce (4.10) to the azimuthally symmetric form

$$\mathbf{E}(r,\theta) = \frac{p}{4\pi\epsilon_0 r^3}\left[2\cos\theta\,\hat{\mathbf{r}} + \sin\theta\,\hat{\boldsymbol{\theta}}\right]. \tag{4.11}$$

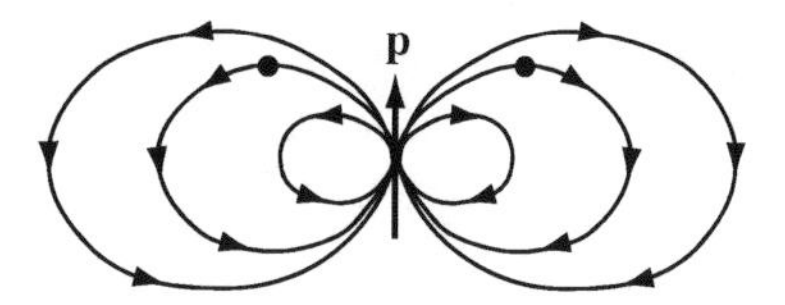

Figure 4.3: Lines of $\mathbf{E}(\mathbf{r})$ for an electric dipole $\mathbf{p} = p\hat{\mathbf{z}}$ located at the center of the diagram. The localized charge distribution responsible for the field is too small to be seen on the scale used in the diagram. The black dots are two points on a circular orbit around the z-axis. See text for details.

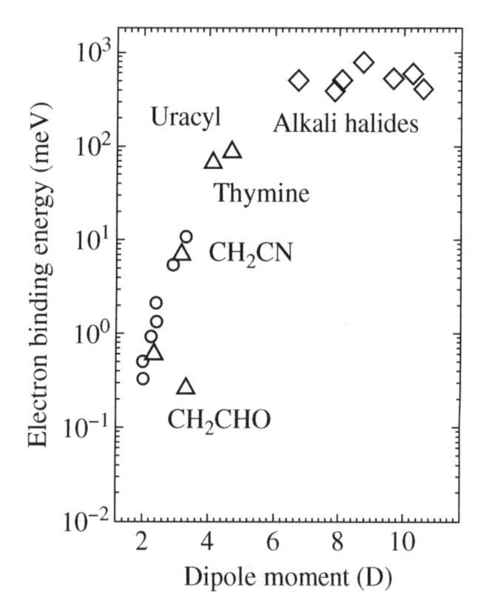

Figure 4.4: The binding energy of an electron captured by a polar molecule. One debye (D) is = 3.36×10^{-30} C·m. Figure adapted from Desfrançois *et al.* (1994).

An equation for the electric field lines follows from the polar analog of the Cartesian formula (3.27):

$$\frac{dr}{E_r} = \frac{rd\theta}{E_\theta} \;\Rightarrow\; \frac{1}{r}\frac{dr}{d\theta} = 2\cot\theta \;\Rightarrow\; r = k\sin^2\theta. \tag{4.12}$$

The integration constant k parameterizes the family of electric field lines (solid curves) shown in Figure 4.3.

The characteristic $1/r^3$ dependence of the dipole field (4.11) differs from the $1/r^2$ behavior of the point charge electric field. A more important difference is that (4.11) is angle-dependent. This has many consequences. Not least of these is that the force $\mathbf{F} = q\mathbf{E}$ that an electric dipole exerts on a point charge q is non-central and leads to non-trivial orbital dynamics. For example, the black dots in Figure 4.3 are two points on a circular orbit around the z-axis in a plane perpendicular to that axis. This can be an orbit for an electron because the electric force is always centripetal (the vertical component of $\mathbf{E}$ is zero). Figure 4.3 implies that an entire family of such circular orbits exists, each one requiring a particular orbital speed to satisfy Newton's second law.

Formal classical mechanics confirms this qualitative prediction. Moreover, the main result of such an analysis—bound states exist if the electric dipole moment is large enough—survives the transition to quantum mechanics. Figure 4.3 thus provides a qualitative rationalization for the experimental fact that a neutral polar molecule can capture an incoming electron to form a stable negative ion. The data plotted in Figure 4.4 indicate that the ionic bound state does indeed disappear when the dipole moment of the host molecule is too small.

Example 4.1 Let $\mathbf{E}(\mathbf{r})$ be the electric field produced by a charge density $\rho(\mathbf{r})$ which lies entirely inside a spherical volume V of radius R. Show that the electric dipole moment of the distribution is given by

$$\mathbf{p} = -3\epsilon_0 \int_V d^3r\, \mathbf{E}(\mathbf{r}).$$

Solution: Assume first that $\rho(\mathbf{r})$ does *not* lie entirely inside V. If we place the origin of coordinates at the center of V, Coulomb's law gives

$$\frac{1}{V}\int_V d^3r\, \mathbf{E}(\mathbf{r}) = \frac{1}{V}\int_V d^3r \frac{1}{4\pi\epsilon_0}\int d^3s\, \rho(\mathbf{s})\frac{\mathbf{r}-\mathbf{s}}{|\mathbf{r}-\mathbf{s}|^3}.$$

Reversing the order of integration and extracting a minus sign produces

$$\frac{1}{V}\int_V d^3r\, \mathbf{E}(\mathbf{r}) = -\frac{1}{V}\int d^3s \left[\frac{1}{4\pi\epsilon_0}\int_V d^3r\, \rho(\mathbf{s})\frac{\mathbf{s}-\mathbf{r}}{|\mathbf{s}-\mathbf{r}|^3}\right].$$

The factor $\rho(\mathbf{s})$ is a constant as far as the integration over r is concerned. Therefore, the quantity in square brackets is the electric field $\mathbf{E}(\mathbf{s})$ produced by a sphere of volume $V = 4\pi R^3/3$ with constant and uniform charge density $\rho(\mathbf{s})$. From Gauss' law, the latter is

$$\mathbf{E}(\mathbf{s}) = \begin{cases} \dfrac{\rho(\mathbf{s})}{3\epsilon_0}\mathbf{s} & s < R, \\ \dfrac{V}{4\pi\epsilon_0}\dfrac{\rho(\mathbf{s})}{s^3}\mathbf{s} & s > R. \end{cases}$$

Therefore, if $\mathbf{p}_{\text{in}}$ is the dipole moment due to the part of ρ which lies inside V and $\mathbf{E}_{\text{out}}(0)$ is the electric field at the origin due to the part of ρ which lies outside V,

$$\frac{1}{V}\int_V d^3r\, \mathbf{E}(\mathbf{r}) = -\frac{1}{3\epsilon_0 V}\int_{s<R} d^3s\, \rho(\mathbf{s})\mathbf{s} - \frac{1}{4\pi\epsilon_0}\int_{s>R} d^3s\, \rho(\mathbf{s})\frac{\mathbf{s}}{s^3} = -\frac{1}{3\epsilon_0}\frac{\mathbf{p}_{\text{in}}}{V} + \mathbf{E}_{\text{out}}(0).$$

We get the stated result if all of $\rho(\mathbf{r})$ is contained in V so $\mathbf{E}_{\text{out}}(0) = 0$ and $\mathbf{p}_{\text{in}} = \mathbf{p}$. On the other hand, if *none* of the charge is contained in V,

$$\mathbf{E}(0) = \frac{1}{V}\int d^3r\, \mathbf{E}(\mathbf{r}).$$

Parity and The Dipole Moment

A well-known theorem of quantum mechanics states that the electric dipole moment is zero for any microscopic system described by a wave function with definite parity:

$$\mathbf{p} = \int d^3r\, \rho(\mathbf{r})\mathbf{r} \;\; = <\psi|\,\mathbf{r}\,|\psi>= 0.$$

The theorem applies to isolated electrons, atoms, and molecules because the Hamiltonian for each of these is invariant under the parity operation. How, then, do we understand the dipole moment of the water molecule indicated in Figure 4.2 and tables of permanent electric dipole moments found in every handbook of molecular properties?

Handbooks write $\mathbf{p} = \mathbf{p}_n + \mathbf{p}_e$, where $\mathbf{p}_n$ is a contribution from the nuclei treated as point charges and $\mathbf{p}_e$ is computed from the integral above using electron wave functions computed in a body-fixed frame of reference. The latter are *not* eigenstates of the parity operator applied to the electron coordinates alone. The quantum mechanical theorem applies to the total molecular wave function in the laboratory frame, including the nuclear coordinates. We get $\mathbf{p} = 0$ in that case because all orientations of the body frame dipole moment are equally probable.

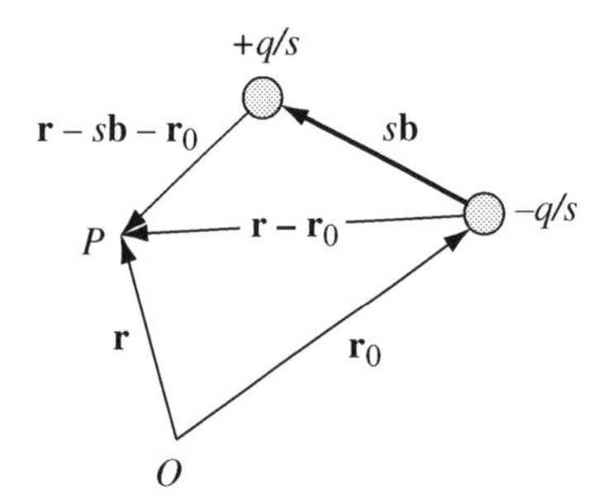

Figure 4.5: A dipole with total charge $Q = 0$ and dipole moment $\mathbf{p} = q\mathbf{b}$. The potential is measured at the observation point P. O is the origin of coordinates.

4.2.1 The Point Electric Dipole

The word "dipole" comes from a geometrical construction due to Maxwell. It begins with the placement of two ("di") equal and opposite charges ("poles") $\pm q/s$ at opposite ends of a vector $s\mathbf{b}$ (Figure 4.5). Now let $s \to 0$. If $\mathbf{p} = q\mathbf{b}$, the electrostatic potential of the resulting "point dipole" located at $\mathbf{r}_0$ is

$$\varphi(\mathbf{r}) = \lim_{s\to 0} \frac{1}{4\pi\epsilon_0}\left[\frac{q/s}{|\mathbf{r}-\mathbf{r}_0 - s\mathbf{b}|} - \frac{q/s}{|\mathbf{r}-\mathbf{r}_0|}\right] = -\frac{1}{4\pi\epsilon_0}\mathbf{p}\cdot\nabla\frac{1}{|\mathbf{r}-\mathbf{r}_0|}. \tag{4.13}$$

This formula is exact because the higher-order terms in the Taylor expansion of $|\mathbf{r}-\mathbf{r}_0 - s\mathbf{b}|^{-1}$ all vanish in the $s \to 0$ limit. The potential (4.13) is identical to (4.8) except that (4.13) is valid at every point in space (except $\mathbf{r}_0$). This is reasonable because $r \gg R$ is always valid when the source size $R \to 0$.

We can derive an analytic formula for the source charge density that produces the point dipole potential (4.13) by applying Maxwell's limiting process directly to the charge density of two point charges. An alternative method exploits (4.13), Poisson's equation (3.14), and the delta function identity (1.121) to get

$$\rho_D(\mathbf{r}) = -\epsilon_0\nabla^2\varphi(\mathbf{r}) = \frac{\mathbf{p}}{4\pi}\cdot\nabla\nabla^2\frac{1}{|\mathbf{r}-\mathbf{r}_0|} = -\mathbf{p}\cdot\nabla\delta(\mathbf{r}-\mathbf{r}_0). \tag{4.14}$$

As far as we know, point electric dipoles with the singular charge density (4.14) do not exist in Nature. Nevertheless, we will find $\rho_D(\mathbf{r})$ a useful tool for computation when the size of a charge distribution is very small compared to all other characteristic lengths in an electrostatic problem.

4.2.2 The Singularity at the Origin

The electric field of a point electric dipole must be given by (4.10) when $\mathbf{r} \neq \mathbf{r}_0$. But what happens precisely at $\mathbf{r} = \mathbf{r}_0$? Given that the point dipole was constructed from two delta function point charges, it is reasonable to guess that a delta function lies at the heart of the electric field of the dipole as well. To check this, it is simplest to use Example 4.1 and a spherical integration volume V centered at $\mathbf{r}_0$. In that case,

$$\int_V d^3r\,\mathbf{E}(\mathbf{r}) = -\frac{\mathbf{p}}{3\epsilon_0}. \tag{4.15}$$

On the other hand, it is clear from the symmetry of Figure 4.3 that (4.11) integrates to *zero* over the volume of any origin-centered sphere. Therefore, if this behavior persists as $V \to 0$, (4.15) will be true if the total electric field of a point dipole is[1]

$$\mathbf{E}(\mathbf{r}) = \frac{1}{4\pi\epsilon_0}\left[\frac{3\hat{\mathbf{n}}(\hat{\mathbf{n}}\cdot\mathbf{p}) - \mathbf{p}}{|\mathbf{r}-\mathbf{r}_0|^3} - \frac{4\pi}{3}\mathbf{p}\,\delta(\mathbf{r}-\mathbf{r}_0)\right]. \tag{4.16}$$

The unit vector $\hat{\mathbf{n}} = (\mathbf{r}-\mathbf{r}_0)/|\mathbf{r}-\mathbf{r}_0|$. A useful way to think about the delta function in (4.16) will emerge in Application 6.1 of Chapter 6, devoted to dielectric matter.

4.2.3 The Dipole Force

The electric dipole moment emerges in a natural way when we calculate the force exerted on a neutral charge distribution $\rho(\mathbf{r}')$ by an external electric field $\mathbf{E}(\mathbf{r}')$ which changes slowly in space. By "slowly" we mean that, over the entire spatial extent of the distribution, $\mathbf{E}(\mathbf{r}')$ is well approximated by a two-term Taylor series expansion around a reference point $\mathbf{r}$ located somewhere inside $\rho(\mathbf{r}')$. Since $\nabla' f(\mathbf{r}')|_{\mathbf{r}'=\mathbf{r}} \equiv \nabla f(\mathbf{r})$, the expansion in question is

$$\mathbf{E}(\mathbf{r}') = \mathbf{E}(\mathbf{r}) + \left[(\mathbf{r}'-\mathbf{r})\cdot\nabla\right]\mathbf{E}(\mathbf{r}) + \cdots. \tag{4.17}$$

The total charge (4.4) is zero by assumption. Therefore, (4.17) gives the first non-zero contribution to the force on $\rho(\mathbf{r})$ as

$$\mathbf{F} = \int d^3r'\,\rho(\mathbf{r}')\mathbf{E}(\mathbf{r}') = \int d^3r'\rho(\mathbf{r}')(\mathbf{r}'\cdot\nabla)\mathbf{E}(\mathbf{r}). \tag{4.18}$$

Using the definition of $\mathbf{p}$ in (4.5), we learn that the force on $\rho(\mathbf{r})$ is entirely characterized by its dipole moment:

$$\mathbf{F} = (\mathbf{p}\cdot\nabla)\mathbf{E}(\mathbf{r}). \tag{4.19}$$

Equation (4.19) is completely general. When $\mathbf{p}$ is a constant vector, the fact that $\nabla\times\mathbf{E} = 0$ and the identity $\nabla(\mathbf{p}\cdot\mathbf{E}) = \mathbf{p}\times(\nabla\times\mathbf{E}) + \mathbf{E}\times(\nabla\times\mathbf{p}) + (\mathbf{p}\cdot\nabla)\mathbf{E} + (\mathbf{E}\cdot\nabla)\mathbf{p}$ transform (4.19) to[2]

$$\mathbf{F} = \nabla(\mathbf{p}\cdot\mathbf{E}). \tag{4.20}$$

We emphasize that an electric field *gradient* is needed to generate a force.

An alternative derivation of (4.19) exploits the fact that the force depends only on $\mathbf{p}$. The idea is to calculate the force that $\mathbf{E}$ exerts on a *point* dipole density (4.14) with the same dipole moment $\mathbf{p}$ as $\rho(\mathbf{r})$. After an integration by parts, the result of this second calculation for a constant $\mathbf{p}$ is

$$\mathbf{F} = \int d^3r'\,\mathbf{E}(\mathbf{r}')\rho_D(\mathbf{r}') = \mathbf{p}\cdot\int d^3r'\,\delta(\mathbf{r}'-\mathbf{r})\nabla'\mathbf{E}(\mathbf{r}'). \tag{4.21}$$

Equation (4.21) reproduces (4.19) because the delta function changes ∇' to ∇ as well as $\mathbf{E}(\mathbf{r}')$ to $\mathbf{E}(\mathbf{r})$.

[1] Equation (4.16) also follows directly from (1.122).

[2] Equation (4.20) is *not* valid for induced electric moments where $\mathbf{p} \propto \mathbf{E}$ and thus generally varies in space.

4.2.4 The Dipole Torque

We can mimic the force calculation in (4.21) to find the Coulomb torque (3.4) exerted on $\rho(\mathbf{r})$ by a field $\mathbf{E}$ that varies slowly in space:

$$\mathbf{N} = \int d^3r' \, \mathbf{r}' \times \mathbf{E}(\mathbf{r}')\rho_D(\mathbf{r}') = \int d^3r' \, \delta(\mathbf{r}' - \mathbf{r})(\mathbf{p} \cdot \nabla')(\mathbf{r}' \times \mathbf{E}). \tag{4.22}$$

The delta function integration gives $(\mathbf{p} \cdot \nabla)(\mathbf{r} \times \mathbf{E})$. Therefore, because $\nabla_i r_\ell = \delta_{i\ell}$, the kth component of the torque is[3]

$$N_k = p_i \nabla_i \epsilon_{k\ell m} r_\ell E_m = \epsilon_{k\ell m} p_i \{\delta_{i\ell} E_m + r_\ell \nabla_i E_m\}. \tag{4.23}$$

Using the dipole force $\mathbf{F}$ in (4.19), the vector form of (4.23) is

$$\mathbf{N} = \mathbf{p} \times \mathbf{E} + \mathbf{r} \times \mathbf{F}. \tag{4.24}$$

The first term in (4.24) is a torque which tends to rotate $\mathbf{p}$ around its center of mass into the direction of $\mathbf{E}$. The second term in (4.24) is a torque which tends to rotate the center of mass of $\mathbf{p}$ around an origin defined by the tail of the position vector $\mathbf{r}$.

4.2.5 The Dipole Potential Energy

The potential energy of interaction between $\rho(\mathbf{r})$ and a slowly varying field $\mathbf{E}(\mathbf{r})$ can be calculated using V_E from (3.64) and the singular charge density (4.14). In detail,

$$V_E(\mathbf{r}) = \int d^3r' \, \varphi(\mathbf{r}')\rho_D(\mathbf{r}') = -\int d^3r' \, \varphi(\mathbf{r}')\mathbf{p} \cdot \nabla'\delta(\mathbf{r}' - \mathbf{r}). \tag{4.25}$$

When $\mathbf{p}$ is constant, the (by now) familiar integration by parts yields

$$V_E(\mathbf{r}) = -\mathbf{p} \cdot \mathbf{E}(\mathbf{r}). \tag{4.26}$$

A quick check uses (4.26) to compute the force and torque on $\mathbf{p}$. We reproduce (4.20) immediately because

$$\mathbf{F} = -\nabla V_E = \nabla(\mathbf{p} \cdot \mathbf{E}). \tag{4.27}$$

To find the torque, we ask how (4.26) changes when the dipole moment $\mathbf{p}$ rotates rigidly by an infinitesimal amount $\delta\boldsymbol{\alpha}$ about its center.[4] The usual rules of mechanics give the change in the dipole moment as

$$\delta\mathbf{p} = \delta\boldsymbol{\alpha} \times \mathbf{p}. \tag{4.28}$$

Therefore, the change in V_E is

$$\delta V_E = -\delta\mathbf{p} \cdot \mathbf{E} = -(\delta\boldsymbol{\alpha} \times \mathbf{p}) \cdot \mathbf{E} = -(\mathbf{p} \times \mathbf{E}) \cdot \delta\boldsymbol{\alpha}. \tag{4.29}$$

This confirms (4.24) because $\delta V_E = -\mathbf{N} \cdot \delta\boldsymbol{\alpha}$ defines the torque.

[3] Recall from (1.37) that $a_k = \epsilon_{k\ell m} b_\ell c_m$ is the kth Cartesian component of $\mathbf{a} = \mathbf{b} \times \mathbf{c}$.

[4] The right-hand rule determines the rotation sense by an angle $|\delta\boldsymbol{\alpha}|$ with respect to an axis pointed in the direction of $\delta\boldsymbol{\alpha}$.

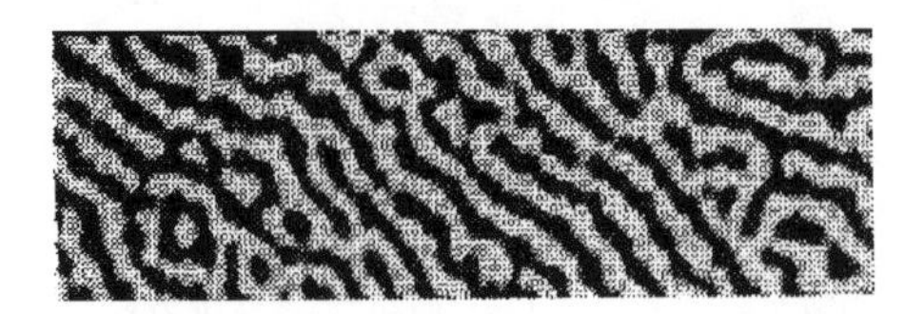

Figure 4.6: Fluorescence microscope image (40 μm × 160 μm) looking down onto a liquid mixture of cholesterol and a phospholipid trapped at an air-water interface parallel to the image. The black regions are long, thin "drops" of the liquid, with a thickness of exactly one molecule. Figure from Seul and Chen (1993).

4.2.6 The Dipole-Dipole Interaction

We can apply (4.26) to compute the total electrostatic energy U_E of a collection of N pre-existing point dipoles located at positions $\mathbf{r}_1, \mathbf{r}_2, \ldots, \mathbf{r}_N$. As in the corresponding point charge problem (Section 3.6), it costs no energy to bring the first dipole $\mathbf{p}_1$ into position. The work done by *us* to bring $\mathbf{p}_2$ into position is exactly the interaction energy (4.25), with the electric field (4.10) of $\mathbf{p}_1$ playing the role of $\mathbf{E}_{\text{ext}}$:

$$W_{12} = -\mathbf{p}_2 \cdot \mathbf{E}_1(\mathbf{r}_2) = \frac{1}{4\pi\epsilon_0}\left\{\frac{\mathbf{p}_1 \cdot \mathbf{p}_2}{|\mathbf{r}_2 - \mathbf{r}_1|^3} - \frac{3\mathbf{p}_1 \cdot (\mathbf{r}_2 - \mathbf{r}_1)\mathbf{p}_2 \cdot (\mathbf{r}_2 - \mathbf{r}_1)}{|\mathbf{r}_2 - \mathbf{r}_1|^5}\right\}. \tag{4.30}$$

Repeating the logic of the point charge example leads to the total energy:

$$U_E = W = \frac{1}{4\pi\epsilon_0}\frac{1}{2}\sum_{i=1}^{N}\sum_{j\neq i}^{N}\left\{\frac{\mathbf{p}_i \cdot \mathbf{p}_j}{|\mathbf{r}_i - \mathbf{r}_j|^3} - \frac{3\mathbf{p}_i \cdot (\mathbf{r}_i - \mathbf{r}_j)\mathbf{p}_j \cdot (\mathbf{r}_i - \mathbf{r}_j)}{|\mathbf{r}_i - \mathbf{r}_j|^5}\right\}. \tag{4.31}$$

A more compact form of (4.31) makes use of the electric field $\mathbf{E}(\mathbf{r}_i)$ at the position of the ith dipole produced by all the *other* dipoles:

$$U_E = -\frac{1}{2}\sum_{i=1}^{N}\mathbf{p}_i \cdot \mathbf{E}(\mathbf{r}_i). \tag{4.32}$$

4.3 Electric Dipole Layers

A distribution of point electric dipoles confined to a surface may be characterized by a dipole moment per unit area. If $\mathbf{r}_S$ is a point on the surface, we use the symbol $\boldsymbol{\tau}(\mathbf{r}_S)$ for this quantity. Consider, for example the collection of polar molecules trapped at an air-water interface shown in Figure 4.6. The microscope image does not resolve individual molecules, but it is known that the permanent electric dipole moment $\mathbf{p}$ of each molecule points directly at the reader (and thus perpendicular to the trapping interface). The molecules in the image have condensed into liquid "drops" (the black regions) which are one molecule thick and shaped like long thin rectangles. Thermal disorder is responsible for the observed meandering and deviations from rectangularity. We will explain at the end of the section (see Application 4.1) why each drop has essentially the same width. Here, we focus on the potential produced by such a system.

Let the dipole moment of each molecule in Figure 4.6 be $\mathbf{p} = q\mathbf{b}$, where b is the molecule's length and q is an effective charge. If each molecule occupies an area S of the interface, it is natural to characterize the closely spaced molecules in each drop by a uniform electric dipole moment per unit area $\boldsymbol{\tau}$, where

$$\boldsymbol{\tau} = \frac{\mathbf{p}}{S} = \frac{q\mathbf{b}}{S} = \sigma\mathbf{b}. \tag{4.33}$$

The uniform surface charge density $\sigma = q/S$ defined by (4.33) suggests a crude representation where we model each drop in Figure 4.6 by two oppositely charged rectangular surfaces separated by the

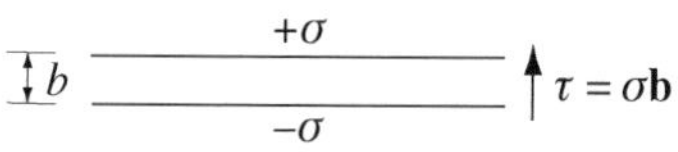

$$\updownarrow b \quad \frac{+\sigma}{-\sigma} \quad \uparrow \tau = \sigma \mathbf{b}$$

Figure 4.7: Side view of a dipole layer: two oppositely charged surfaces with areal charge densities $\pm\sigma$ separated by a distance b. In the macroscopic limit, $\mathbf{b} \to 0$ and $\sigma \to \infty$ with the dipole moment density $\boldsymbol{\tau} = \sigma \mathbf{b}$ held constant.

microscopic distance b (see Figure 4.7). This abstraction makes no attempt to model the actual charge density of each molecule. Rather, we pass to the macroscopic limit ($\mathbf{b} \to 0$ and $\sigma \to \infty$ with $\boldsymbol{\tau} = \sigma \mathbf{b}$ held constant) where the size of the molecules (and the separation between them) is unresolved as in Figure 4.6. This makes the surface electric dipole density $\boldsymbol{\tau}$ a macroscopic property exactly like the surface charge density σ.

The construction illustrated in Figure 4.7 is called a *dipole layer*.[5] In our example, the macroscopic surface dipole density of the dipole layer derives from the intrinsic dipole moment of the constituent molecules. More commonly, dipole layers arise (in the absence of intrinsic dipoles) when mobile charges exist near surfaces or boundary layers. This occurs frequently in plasma physics, condensed matter physics, biophysics, and chemical physics. An example is the surface of a metal (see the graph of $\rho_0(\mathbf{r})$ in Figure 2.7) where the electrons "spill out" beyond the last layer of positive ions. After Lorentz averaging (Section 2.3.1), the electrostatics of this situation is well described by a surface distribution of electric dipole moments.

4.3.1 The Potential of a Dipole Layer

The most general dipole layer is a charge-neutral macroscopic surface S (not necessarily flat) endowed with a dipole moment per unit area that is not necessarily uniform or oriented perpendicular to the surface. This leads us to generalize (4.33) to

$$\boldsymbol{\tau} = \frac{d\mathbf{p}}{dS}. \tag{4.34}$$

The electrostatic potential produced by such an object is computed by superposing point dipole potentials as given by (4.13). Therefore,

$$\varphi(\mathbf{r}) = -\frac{1}{4\pi\epsilon_0}\int_S d\mathbf{p}(\mathbf{r}_S)\cdot\nabla\frac{1}{|\mathbf{r}-\mathbf{r}_S|} = -\frac{1}{4\pi\epsilon_0}\int_S dS\,\boldsymbol{\tau}(\mathbf{r}_S)\cdot\nabla\frac{1}{|\mathbf{r}-\mathbf{r}_S|}. \tag{4.35}$$

A judicious rewriting of (4.35) reveals the singular charge density $\rho_L(\mathbf{r})$ associated with a dipole layer. The key is to convert the surface integral (4.35) to a volume integral using a delta function. For a dipole layer which coincides with $z = 0$, we write $\mathbf{r}'_S = x'\hat{\mathbf{x}} + y'\hat{\mathbf{y}}$ and $\mathbf{r}' = \mathbf{r}'_S + z'\hat{\mathbf{z}}$ to get

$$\varphi(\mathbf{r}) = -\frac{1}{4\pi\epsilon_0}\int d^3r'\delta(z')\boldsymbol{\tau}(\mathbf{r}'_S)\cdot\nabla\frac{1}{|\mathbf{r}-\mathbf{r}'|}. \tag{4.36}$$

Now change the argument of the gradient operator from $\mathbf{r}$ to $\mathbf{r}'$ and integrate (4.36) by parts. Two minus signs later, we get the desired result,

$$\varphi(\mathbf{r}) = -\frac{1}{4\pi\epsilon_0}\int d^3r'\frac{\nabla'\cdot\{\boldsymbol{\tau}(\mathbf{r}'_S)\delta(z')\}}{|\mathbf{r}-\mathbf{r}'|}. \tag{4.37}$$

Comparing (4.37) with (4.1) shows that a surface dipole layer at $z = 0$ is equivalent to the volume charge density[6]

$$\rho_L(\mathbf{r}) = -\nabla\cdot\{\boldsymbol{\tau}(\mathbf{r}_S)\delta(z)\}. \tag{4.38}$$

[5] Some authors reserve the term "double layer" for a dipole layer where every dipole points normal to the surface.

[6] Comparing (4.38) to (2.51) shows that the polarization of this dipole layer is $\mathbf{P}(\mathbf{r}) = \boldsymbol{\tau}(x, y)\delta(z)$.

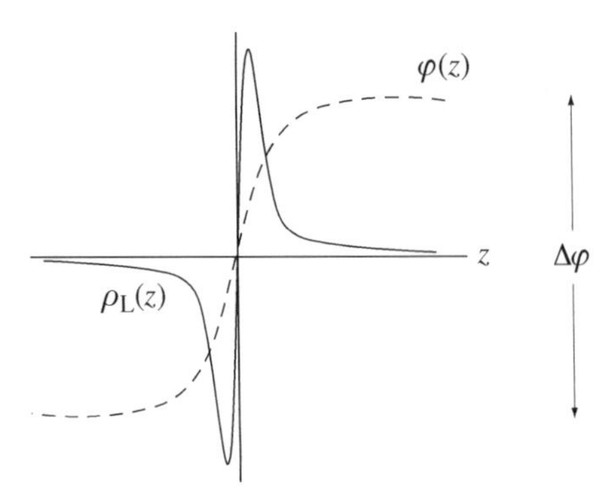

Figure 4.8: The charge density (solid curve) and electrostatic potential (dashed curve) associated with a microscopic dipole layer of finite thickness. In the macroscopic limit, φ is discontinuous by an amount $\Delta\varphi$ and $\rho_{\rm L}$ is singular at $z = 0$.

To interpret (4.38), we define a "surface gradient",

$$\nabla_S \equiv \hat{\mathbf{x}}\,\frac{\partial}{\partial x} + \hat{\mathbf{y}}\,\frac{\partial}{\partial y}, \tag{4.39}$$

and an effective surface charge density,

$$\sigma(\mathbf{r}_S) = -\nabla_S \cdot \boldsymbol{\tau}(\mathbf{r}_S). \tag{4.40}$$

In this language, the singular charge density (4.38) associated with an arbitrary dipole layer is

$$\rho_{\rm L}(\mathbf{r}) = -\tau_z(\mathbf{r}_S)\delta'(z) + \sigma(\mathbf{r}_S)\delta(z). \tag{4.41}$$

The first term in (4.41) is the charge density associated with the component of the layer dipole moment perpendicular to the surface. To see this, compare the capacitor-like charge density in Figure 4.7 with the solid curve in Figure 4.8 labeled $\rho_{\rm L}(z)$.

The latter is a smooth microscopic charge density which Lorentz averages to the macroscopic delta function derivative in (4.41). The associated electrostatic potential (dashed curve in Figure 4.8) is similarly reminiscent of the potential of a parallel-plate capacitor. In detail, the Poisson equation relates the curvature of the potential directly to the charge density:

$$\epsilon_0 \frac{d^2\varphi(z)}{dz^2} = -\rho_{\rm L}(z). \tag{4.42}$$

The surface charge density $\sigma(\mathbf{r}_S)$ in (4.41) arises exclusively from the components of $\boldsymbol{\tau}(\mathbf{r}_S)$ parallel to the surface plane. Qualitatively, the negative end of each in-plane point dipole exactly cancels the positive end of the immediately adjacent in-plane dipole *unless* there is a variation in the magnitude or direction of the dipoles along the surface. The resulting incomplete cancellation is the origin of (4.40).

4.3.2 Matching Conditions at a Dipole Layer

The matching condition, $\epsilon_0\hat{\mathbf{n}}_2 \cdot [\mathbf{E}_1 - \mathbf{E}_2] = \sigma(\mathbf{r}_S)$, tells us that the second term in (4.41) generates a jump in the normal component of the electric field at $z = 0$. To discover the effect of the first term, we write out the Poisson equation (4.42) using only the first term on the right-hand side of (4.41) as the source charge:

$$\epsilon_0 \frac{d^2}{dz^2}\varphi(\mathbf{r}_S, z) = \tau_z(\mathbf{r}_S)\delta'(z). \tag{4.43}$$

Up to an additive constant, integration of (4.43) over z implies that

$$\epsilon_0 \frac{d}{dz}\varphi(\mathbf{r}_S, z) = \tau_z(\mathbf{r}_S)\delta(z). \tag{4.44}$$

The constant disappears when we integrate (4.44) from $z = 0^-$ to $z = 0^+$. What remains is a matching condition for the potential:

$$\varphi(\mathbf{r}_S, z = 0^+) - \varphi(\mathbf{r}_S, z = 0^-) = \tau_z(\mathbf{r}_S)/\epsilon_0. \tag{4.45}$$

The condition (4.45) shows that the electrostatic potential suffers a jump discontinuity at a dipole-layer surface. The variation in $\varphi(z)$ sketched as the dashed curve in Figure 4.8 makes this entirely plausible. The potential rises from left to right across this microscopic dipole layer because the electric field inside the layer points from the positive charge on the right to the negative charge on the left.

The discontinuity (4.45) implies that the tangential component of the electric field is not continuous if the dipole-layer density varies along the surface. In the language of Section 2.3.3, the reader can show that

$$\hat{\mathbf{n}}_2 \times [\mathbf{E}_1 - \mathbf{E}_2] = \nabla \times \{\hat{\mathbf{n}}_2(\hat{\mathbf{n}}_2 \cdot \boldsymbol{\tau})\}/\epsilon_0. \tag{4.46}$$

Under certain conditions, (4.46) generates corrections to the Fresnel formulae (Section 17.3.2) which describe the reflection and transmission of electromagnetic waves from interfaces.[7]

Application 4.1 Monolayer Electric Dipole Drops

Apart from thermal fluctuations, the width w of every rectangular drop in Figure 4.6 is nearly the same, independent of the drop length L. To see why, we ignore interactions between different drops and assume $L \gg w \gg a$, where a is the lateral distance between nearest-neighbor molecules. The energy of each drop is the sum of an electrostatic energy U_E and a "perimeter" energy U_C of chemical origin. The equilibrium drop width we seek is the value of w that minimizes the total energy $U_E + U_C$ keeping the drop area $A = Lw$ fixed. The chemical energy is straightforward. If λ is the chemical energy per unit length of drop perimeter, we use $L \gg w$ to write

$$U_C \approx 2L\lambda = \frac{2\lambda A}{w}. \tag{4.47}$$

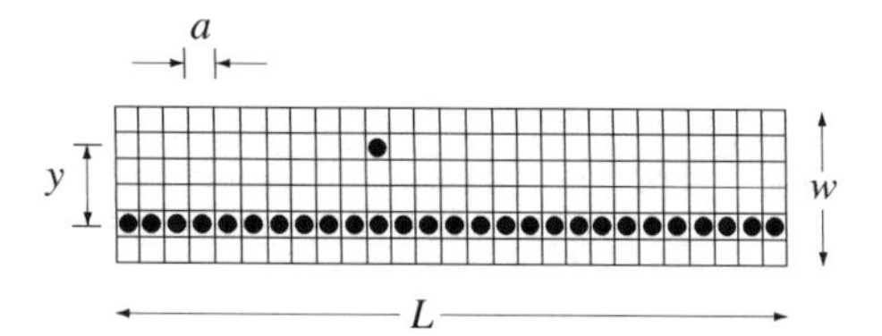

Figure 4.9: Top view looking down onto one monolayer-height rectangular "drop" of area $A = Lw$. A polar molecule pointing out of the page occupies every $a \times a$ square of the indicated checkerboard. Black dots highlight one complete row of molecules and one molecule at a distance y from the row.

To compute U_E, focus on the top view of one rectangular drop shown in Figure 4.9 and imagine every square of the inscribed checkerboard occupied by one polar molecule with its dipole moment pointed out of the page. The black dots draw attention to one row of molecules and one molecule sitting at a distance y from that row. For electrostatic purposes, the rectangle consists of w/a parallel rows (length L) of parallel dipoles. The lateral separation between molecules is a, so (4.33) implies that each row carries a dipole moment per unit length τa. The assumed orientation of the dipoles

[7] See D.C. Langreth, "Macroscopic approach to the theory of reflectivity", *Physical Review B* **39**, 10020 (1989).

implies that the electrostatic interaction energy of two such rows separated by a distance y comes entirely from the first term in the curly brackets in (4.31):

$$U(y) = \frac{1}{4\pi\epsilon_0}\int_0^L dx_1 \int_0^L dx_2 \frac{\tau^2 a^2}{[(x_1 - x_2)^2 + y^2]^{3/2}} = \frac{\tau^2 a^2}{4\pi\epsilon_0}\frac{2}{y^2}\left\{\sqrt{L^2+y^2} - y\right\}. \tag{4.48}$$

Since $L \gg y$, a valid approximation to (4.48) is $U(y) \simeq \tau^2 a^2 L/2\pi\epsilon_0 y^2$. The total electrostatic energy follows by summing over every pair of lines. There are $w/a - 1$ lines separated by a distance a, $w/a - 2$ lines separated by a distance $2a$, etc. Therefore, when $w \gg a$,

$$U_E = \frac{\tau^2 L}{2\pi\epsilon_0}\sum_{n=1}^{w/a}\frac{(w/a - n)}{n^2} \approx \frac{\tau^2 L}{2\pi\epsilon_0}\int_1^{w/a} dn \frac{(w/a - n)}{n^2} \approx \frac{\tau^2 L}{2\pi\epsilon_0}\left\{\frac{w}{a} - \ln\frac{w}{a}\right\}. \tag{4.49}$$

The total energy is the sum of (4.49) and the chemical perimeter energy (4.47):

$$U_T = U_C + U_E = \frac{2A}{w}\left[\lambda + \frac{\tau^2}{4\pi\epsilon_0}\left(\frac{w}{a} - \ln\frac{w}{a}\right)\right]. \tag{4.50}$$

Keeping A fixed and minimizing U_T with respect to w gives the equilibrium width we seek,

$$w_0 \approx a\exp(4\pi\epsilon_0\lambda/\tau^2). \tag{4.51}$$

The width w_0 is independent of the drop length L, as claimed in the statement of the problem. The exponential dependence of w_0 on material parameters is characteristic of two-dimensional dipole systems. Experiments show that the addition of more molecules does not increase the width of any drop beyond w_0. Instead, new drops form with a width w_0. ■

4.4 The Electric Quadrupole

The third (electric quadrupole) term dominates the multipole expansion (4.7) when the charge distribution of interest has zero net charge and zero dipole moment ($Q = \mathbf{p} = 0$). All atomic nuclei produce a quadrupole potential[8], as do all homopolar diatomic molecules. The asymptotic ($r \to \infty$) potential is

$$\varphi(\mathbf{r}) = \frac{1}{4\pi\epsilon_0}Q_{ij}\frac{3r_i r_j - \delta_{ij}r^2}{r^5}. \tag{4.52}$$

Repeating (4.6), the components of the electric quadrupole tensor are

$$Q_{ij} = \tfrac{1}{2}\int d^3r\, \rho(\mathbf{r})r_i r_j. \tag{4.53}$$

Mimicking (4.9), it is not difficult to show that these scalars are uniquely defined and do not depend on the choice of the origin of coordinates if $Q = \mathbf{p} = 0$. Note also that the explicit appearance of Cartesian factors like xy or z^2 in the integrand does not compel us to *evaluate* the integral (4.53) using Cartesian coordinates.

The name "quadrupole" comes from a geometrical construction similar to the dipole case of Figure 4.5. This time, we place two oppositely oriented point dipoles $\pm\mathbf{p}/s$ (thus *four* charges in total) at opposite ends of a vector $s\mathbf{c}$ (Figure 4.10).[9] In the limit $s \to 0$, a calculation similar to (4.13) gives

[8] Of course, the charge Q of a nucleus produces a potential $Q/4\pi\epsilon_0 r$.

[9] The multiple derivative structure of (4.3) implies that each term in the multipole expansion can be represented by point charge configurations constructed from suitable pairings in space of the configurations which describe the previous term. This guarantees that each configuration is electrically neutral and that only 2^n-poles appear.

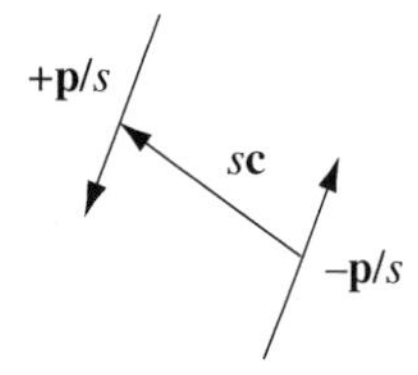

Figure 4.10: A quadrupole composed of two oppositely directed dipoles.

the electrostatic potential of a "point quadrupole" located at the origin:

$$\varphi(\mathbf{r}) = \frac{1}{4\pi\epsilon_0}(\mathbf{c}\cdot\nabla)(\mathbf{p}\cdot\nabla)\frac{1}{r}. \tag{4.54}$$

This is identical to (4.52) with

$$Q_{ij} = \tfrac{1}{2}(c_i p_j + c_j p_i). \tag{4.55}$$

The manifestly symmetric definition (4.55) is needed because $Q_{ij} = Q_{ji}$ is explicit in (4.53). By direct construction, or by mimicking (4.14), the reader can confirm that the corresponding singular charge density for a point quadrupole at the origin is

$$\rho_Q(\mathbf{r}) = Q_{ij}\nabla_i\nabla_j\delta(\mathbf{r}). \tag{4.56}$$

The symmetry $Q_{ij} = Q_{ji}$ implies that only six (rather than nine) independent numbers are needed to characterize a general quadrupole tensor:[10]

$$\mathbf{Q} = \begin{bmatrix} Q_{xx} & Q_{xy} & Q_{xz} \\ Q_{xy} & Q_{yy} & Q_{yz} \\ Q_{xz} & Q_{yz} & Q_{zz} \end{bmatrix}. \tag{4.57}$$

This information can be packaged more efficiently if we adopt the *principal axis* coordinate system (x', y', z') to evaluate the integrals in (4.53). By definition, the quadrupole tensor $\mathbf{Q}'$ is *diagonal* in this system, with components

$$Q'_{ii} = \tfrac{1}{2}\int d^3r'\,\rho(\mathbf{r}')r'_i r'_i. \tag{4.58}$$

In the general case, familiar theorems from linear algebra tell us that

$$\mathbf{Q}' = \mathbf{U}^{-1}\mathbf{Q}\mathbf{U}, \tag{4.59}$$

where $\mathbf{U}$ is the unitary matrix formed from the eigenvectors of $\mathbf{Q}$. These eigenvectors can always be found because $\mathbf{Q}$ is a symmetric matrix. We conclude that the three numbers Q'_{xx}, Q'_{yy}, and Q'_{zz} are sufficient to completely characterize the quadrupole potential (4.52) in the principal axis system. The reader will recognize the direct analogy between the electric quadrupole moment tensor and the moment of inertia tensor used in classical mechanics.

4.4.1 The Traceless Quadrupole Tensor

It is usually not convenient (or even possible) to measure the components of $\mathbf{Q}$ directly in the principal axis system defined by (4.59). Luckily, a strategy exists that economizes the description

[10] Recall from Section 1.8 that we use boldface symbols for tensor quantities like $\mathbf{Q}$. Context should prevent confusion with the total charge Q.

of the quadrupole potential even if we are forced to work in an arbitrary coordinate system. The calculation begins by using (4.53) to write out (4.52):

$$\varphi(\mathbf{r}) = \frac{1}{4\pi\epsilon_0}\left\{\frac{1}{2}\int d^3r'\,\rho(\mathbf{r}')r'_i r'_j\right\}\frac{3r_i r_j - r^2\delta_{ij}}{r^5}. \tag{4.60}$$

The key step uses the identity

$$r'_i r'_j r^2 \delta_{ij} = r^2 r'^2 = r_i r_j r'^2 \delta_{ij} \tag{4.61}$$

to write (4.60) in the form

$$\varphi(\mathbf{r}) = \frac{1}{4\pi\epsilon_0}\left\{\frac{1}{2}\int d^3r'\,\rho(\mathbf{r}')\left(3r'_i r'_j - \delta_{ij}r'^2\right)\right\}\frac{r_i r_j}{r^5}. \tag{4.62}$$

Therefore, if we define a symmetric, *traceless quadrupole tensor*,

$$\Theta_{ij} = \frac{1}{2}\int d^3r'\,\rho(\mathbf{r}')(3r'_i r'_j - r'^2\delta_{ij}) = 3Q_{ij} - Q_{kk}\delta_{ij}, \tag{4.63}$$

the quadrupole potential (4.60) simplifies to

$$\varphi(\mathbf{r}) = \frac{1}{4\pi\epsilon_0}\Theta_{ij}\frac{r_i r_j}{r^5}. \tag{4.64}$$

As its name implies, the point of this algebraic manipulation is that the tensor $\boldsymbol{\Theta}$ has zero trace:

$$\mathrm{Tr}[\boldsymbol{\Theta}] = \delta_{ij}\Theta_{ij} = \frac{1}{2}\int d^3r\,\rho(\mathbf{r})(3r_i r_i - r^2\delta_{ii}) = 0. \tag{4.65}$$

This is important because the constraint

$$\mathrm{Tr}[\boldsymbol{\Theta}] = \Theta_{xx} + \Theta_{yy} + \Theta_{zz} = 0 \tag{4.66}$$

reduces the number of independent components of $\boldsymbol{\Theta}$ by one. Hence, a complete characterization of the quadrupole potential in an arbitrary coordinate system requires specification of only *five* numbers (the independent components of the *traceless* Cartesian quadrupole tensor $\boldsymbol{\Theta}$) rather than six numbers (the components of the *primitive* Cartesian quadrupole tensor $\mathbf{Q}$). Equivalently, it always possible to choose the vectors $\mathbf{c}$ and $\mathbf{p}$ in (4.54) so $|\mathbf{c}| = |\mathbf{p}|$. This similarly reduces the number of independent Cartesian components from six to five.

4.4.2 Force and Torque on a Quadrupole

The force and torque exerted on a point quadrupole in an external electric field can be calculated using the method of Section 4.2.3. The results are

$$\mathbf{F} = Q_{ij}\nabla_i\nabla_j\mathbf{E} \tag{4.67}$$

and

$$\mathbf{N} = 2(\mathbf{Q}\cdot\nabla)\times\mathbf{E} + \mathbf{r}\times\mathbf{F}. \tag{4.68}$$

The corresponding electrostatic interaction energy is

$$V_E(\mathbf{r}) = -Q_{ij}\nabla_i E_j(\mathbf{r}) = -\tfrac{1}{3}\Theta_{ij}\nabla_i E_j(\mathbf{r}). \tag{4.69}$$

A glance back at (4.63) shows that the two expressions in (4.69) differ by a term proportional to $\delta_{ij}\partial_i E_j = \nabla\cdot\mathbf{E}$. This is *zero* because the source charge for the external field $\mathbf{E}$ is assumed to be far from the quadrupole. Following the logic of Section 4.2.3, it not difficult to show that (4.67), (4.68), and (4.69) are valid for any neutral charge distribution with $\mathbf{p} = 0$ if the electric field does not vary too rapidly over the volume of the distribution.

Application 4.2 Nuclear Quadrupole Moments

Most atomic nuclei are rotationally symmetric with respect to one principal axis. If we call this the z-axis, the symmetry and the trace condition (4.66) imply that only *one* independent component of the quadrupole tensor is needed to characterize these nuclei:

$$\Theta_{zz} = -2\Theta_{xx} = -2\Theta_{yy}. \tag{4.70}$$

The algebraic sign of the single parameter $Q \equiv \Theta_{zz}$ characterizes the shape of the proton charge distribution.[11] A cigar-shaped nucleus ($Q > 0$) is called *prolate*; a pumpkin-shaped nucleus ($Q < 0$) is called *oblate* (Figure 4.11). These deviations from nuclear sphericity were first inferred by Schüler and Schmidt (1935) from systematic energy shifts in experimental hyperfine spectra which were interpretable only if the total atomic energy included an electrostatic term like (4.69).

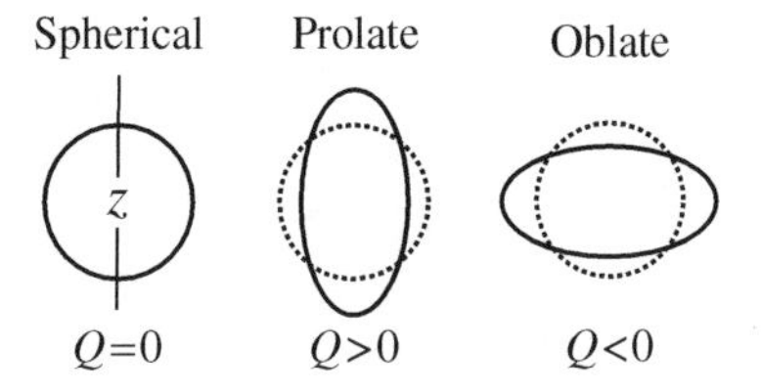

Figure 4.11: Quadrupole distortions of a sphere characterized by $Q = \boldsymbol{\Theta}_{zz}$.

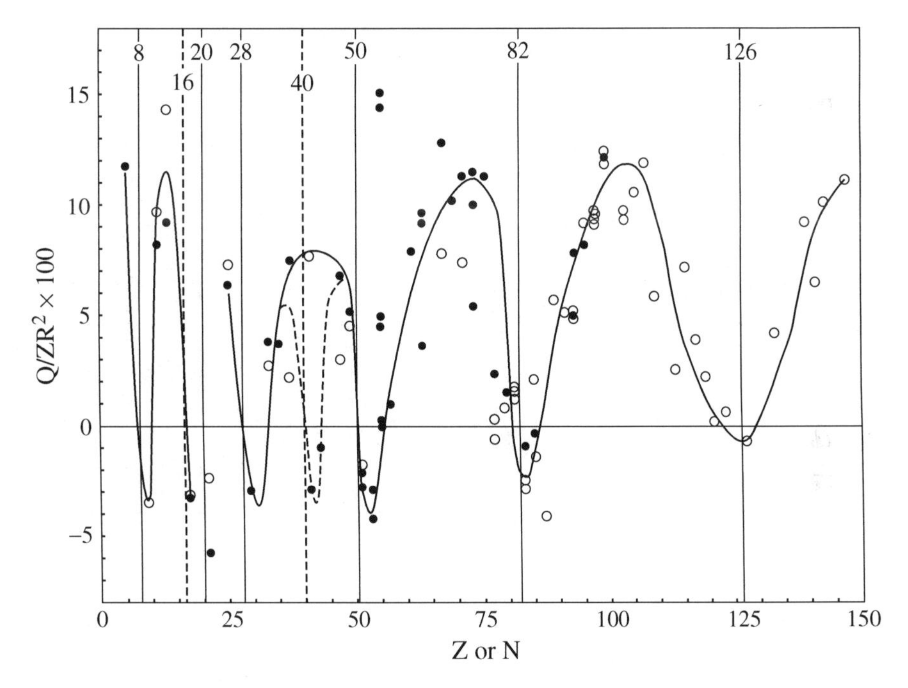

Figure 4.12: Ground state electric quadrupole parameter Q for odd-mass nuclei. Solid (open) circles are nuclei with odd proton (neutron) number $Z(N)$. The data are scaled by ZR^2 where R is the nuclear radius. The vertical lines label "magic" numbers where nuclear shells or subshells close. The curves are guides to the eye. Figure from Rowe and Wood (2010).

Figure 4.12 summarizes quadrupole moment data for odd-mass nuclei. Closed-shell nuclei are spherical, so $Q \approx 0$ when the nuclear-shell model predicts major shell or subshell closings at certain

[11] There should be no confusion with the total charge Q defined in (4.4).

"magic" numbers (vertical lines in Figure 4.12). Oblate shapes are observed for nuclei with very few nucleons in excess of a shell closing. This can be understood qualitatively from Θ_{zz} in (4.63) if we imagine nucleons orbiting the closed shell in the $z = 0$ plane. It is an open question of nuclear physics to explain (in simple terms) the predominance of prolate shapes in Figure 4.12. ■

Example 4.2 Evaluate the traceless quadrupole tensor $\mathbf{\Theta}$ in the principal axis system for an ellipsoid with volume V, uniform charge density ρ, and semi-axes a_x, a_y, and a_z. Choose the center of the ellipsoid as the origin of coordinates.

Solution: Only the diagonal elements of $\mathbf{\Theta}$ are non-zero in the principal axis system. We compute, say,

$$\Theta_{xx} = \tfrac{1}{2}\rho \int_V d^3r' \left\{3x'^2 - r'^2\right\}.$$

A change of variables to $r_i = r'_i/a_i$ transforms the $\mathbf{r}'$ integration over the ellipsoid volume to an $\mathbf{r}$ integration over the volume of a unit sphere. Θ_{xx} depends on the choice of origin because the total charge

$$\rho V = (4\pi/3)\rho\, a_x a_y a_z \neq 0.$$

The suggested choice of the sphere center leads us to compute

$$\Theta_{xx} = \tfrac{1}{2}\rho a_x a_y a_z \int d^3r (2a_x^2 x^2 - a_y^2 y^2 - a_z^2 z^2).$$

However, because the integration volume is a sphere of radius one,

$$\int d^3r\, x^2 = \int d^3r\, y^2 = \int d^3r\, z^2 = \tfrac{1}{3}\int d^3r\, r^2 = (4\pi/3)\int_0^1 dr r^4 = 4\pi/15.$$

Using this, we conclude that

$$\Theta_{xx} = \tfrac{1}{10}\rho V(2a_x^2 - a_y^2 - a_z^2).$$

Then, by symmetry,

$$\Theta_{yy} = \tfrac{1}{10}\rho V(2a_y^2 - a_z^2 - a_x^2)$$

and

$$\Theta_{zz} = \tfrac{1}{10}\rho V(2a_z^2 - a_x^2 - a_y^2).$$

Note that $\Theta_{xx} + \Theta_{yy} + \Theta_{zz} = 0$.

4.5 Spherical Mathematics

The Cartesian multipole expansion (4.3) is not particularly well suited to systems with natural spherical symmetry. Therefore, despite the discussion of nuclear quadrupole moments in the previous section, most treatments of multipole electrostatics in atomic physics and nuclear physics use an alternative, "spherical" expansion. The mathematics needed to derive this expansion begins by replacing the Taylor series used in Section 4.1.1,

$$\frac{1}{|\mathbf{r} - \mathbf{r}'|} = \frac{1}{r} - \mathbf{r}' \cdot \nabla \frac{1}{r} + \frac{1}{2}(\mathbf{r}' \cdot \nabla)^2 \frac{1}{r} - \cdots \qquad r' < r, \tag{4.71}$$

with the binomial expansion

$$\frac{1}{|\mathbf{r}-\mathbf{r}'|}=\frac{1}{\sqrt{r^2-2\mathbf{r}\cdot\mathbf{r}'+r'^2}}=\frac{1}{r}\left[1-2(\hat{\mathbf{r}}\cdot\hat{\mathbf{r}}')\frac{r'}{r}+\frac{r'^2}{r^2}\right]^{-1/2}\qquad r'<r. \tag{4.72}$$

4.5.1 Legendre Polynomials

It is tedious to evaluate the right side of (4.72) using

$$\frac{1}{\sqrt{1+z}}=1-\frac{1}{2}z+\frac{1\cdot 3}{2\cdot 4}z^2-\cdots\qquad -1<z\le 1. \tag{4.73}$$

On the other hand, the expansion we get by doing so is precisely the one used to define the real-valued Legendre polynomials, $P_\ell(x)$, which are well known and tabulated. Specifically,

$$\frac{1}{\sqrt{1-2xt+t^2}}\equiv\sum_{\ell=0}^{\infty}t^\ell P_\ell(x)\qquad |x|\le 1,\ 0<t<1. \tag{4.74}$$

Appendix C summarizes the most important properties of these well-studied functions of mathematical physics. We note here only the orthogonality relation,

$$\int_{-1}^{1}dx\,P_\ell(x)P_{\ell'}(x)=\frac{2}{2\ell+1}\delta_{\ell\ell'}, \tag{4.75}$$

the completeness relation,

$$\sum_{\ell=0}^{\infty}\frac{2\ell+1}{2}P_\ell(x)P_\ell(x')=\delta(x-x'), \tag{4.76}$$

and the first few Legendre polynomials,

$$\begin{aligned}P_0(x)&=1\\ P_1(x)&=x\\ P_2(x)&=\tfrac{1}{2}(3x^2-1).\end{aligned} \tag{4.77}$$

The main result we need here comes from applying (4.74) to (4.72) with $t=r'/r$ and $x=\hat{\mathbf{r}}\cdot\hat{\mathbf{r}}'$. This gives the very useful identity

$$\frac{1}{|\mathbf{r}-\mathbf{r}'|}=\frac{1}{r}\sum_{\ell=0}^{\infty}\left(\frac{r'}{r}\right)^\ell P_\ell(\hat{\mathbf{r}}\cdot\hat{\mathbf{r}}')\qquad r'<r. \tag{4.78}$$

Example 4.3 Use the expansion (4.78) to find the force between two non-overlapping, spherically symmetric charge distributions $\rho_1(r)$ and $\rho_2(r')$ whose centers are separated by a distance R (Figure 4.13).

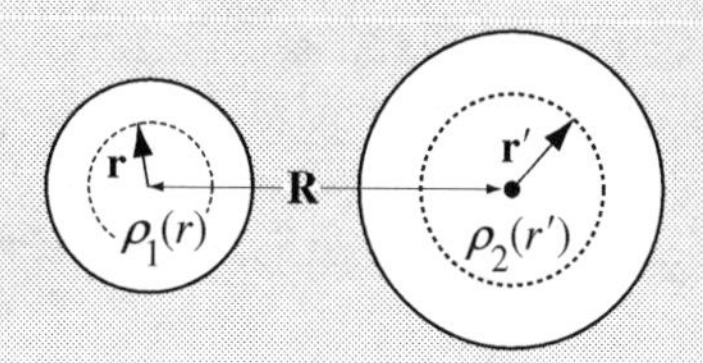

Figure 4.13: Two spherical charge distributions that do not overlap in space.

Solution: We will compute $\mathbf{F} = -\nabla V_E$ where V_E is the interaction energy. By Gauss' law, the electric field produced by $\rho_2(r')$ outside of itself is just the field produced by a point charge $Q_2 = \int d^3r' \rho_2(r')$ placed at $\mathbf{r}' = 0$. The potential produced by $\rho_2(r')$ outside itself is similarly a point charge potential. Hence, the potential energy of interaction is

$$V_E = \int d^3r \, \rho_1(\mathbf{r})\varphi_2(\mathbf{r}) = \frac{Q_2}{4\pi\epsilon_0}\int d^3r \frac{\rho_1(r)}{|\mathbf{r}-\mathbf{R}|}.$$

$\mathbf{R}$ is a fixed vector which we align along the z-axis. In that case, $\hat{\mathbf{r}} \cdot \hat{\mathbf{R}} = \cos\theta$ where θ is the usual polar angle. Therefore, because $P_0(\cos\theta) = 1$ and $r < R$, (4.78) gives

$$V_E = \frac{Q_2}{4\pi\epsilon_0}\int d^3r \, \rho_1(r)\frac{1}{R}\sum_{\ell=0}^{\infty}\left(\frac{r}{R}\right)^{\ell} P_\ell(\cos\theta)P_0(\cos\theta).$$

Only the $\ell = 0$ term survives the sum because the orthogonality relation (4.75) shows that the integral over θ gives $2\delta_{\ell,0}$. Using this information,

$$V_E = \frac{Q_2}{4\pi\epsilon_0 R}\int d^3r \, \rho_1(r) = \frac{Q_1 Q_2}{4\pi\epsilon_0 R}.$$

We conclude that the potential energy between $\rho_1(r)$ and $\rho_2(r')$ is the same as if the total charge of each distribution was concentrated at its center. Hence, the force which $\rho_1(r)$ exerts on $\rho_2(r')$ is

$$\mathbf{F} = -\nabla_{\mathbf{R}} V_E = \frac{Q_1 Q_2}{4\pi\epsilon_0}\frac{\hat{\mathbf{R}}}{R^2}.$$

4.5.2 Spherical Harmonics

The argument of the Legendre polynomials in (4.78) is $\hat{\mathbf{r}} \cdot \hat{\mathbf{r}}' = \cos\gamma$. Figure 4.14 shows that $\cos\gamma$ is an awkward variable because neither $\mathbf{r}$ nor $\mathbf{r}'$ is a constant vector. It would be much more convenient if the right side of (4.78) were expressed explicitly in terms of the spherical coordinates $\mathbf{r} = (r, \theta, \phi)$ and $\mathbf{r}' = (r', \theta', \phi')$ defined with respect to a fixed set of coordinate axes in space. This is the role of the *spherical harmonic addition theorem*,

$$P_\ell(\hat{\mathbf{r}} \cdot \hat{\mathbf{r}}') = \frac{4\pi}{2\ell+1}\sum_{m=-\ell}^{+\ell} Y^*_{\ell m}(\theta', \phi')Y_{\ell m}(\theta, \phi). \tag{4.79}$$

The complex-valued *spherical harmonics* $Y_{\ell m}(\theta, \phi)$ are also well-studied functions of mathematical physics. Appendix C summarizes their most important properties. As implied by (4.79), the index m takes on the $2\ell + 1$ integer values, $-\ell, \ldots, 0, \ldots +\ell$. With the notation $\Omega \equiv (\theta, \phi)$, we mention here only that the spherical harmonics are orthonormal,

$$\int d\Omega \, Y^*_{\ell m}(\Omega)Y_{\ell' m'}(\Omega) = \delta_{\ell\ell'}\delta_{mm'}, \tag{4.80}$$

and complete,[12]

$$\sum_{\ell=0}^{\infty}\sum_{m=-\ell}^{\ell} Y_{\ell m}(\Omega)Y^*_{\ell m}(\Omega') = \frac{1}{\sin\theta}\delta(\theta-\theta')\delta(\phi-\phi'). \tag{4.81}$$

[12] See Section 7.4.1 for the definition of *completeness*.

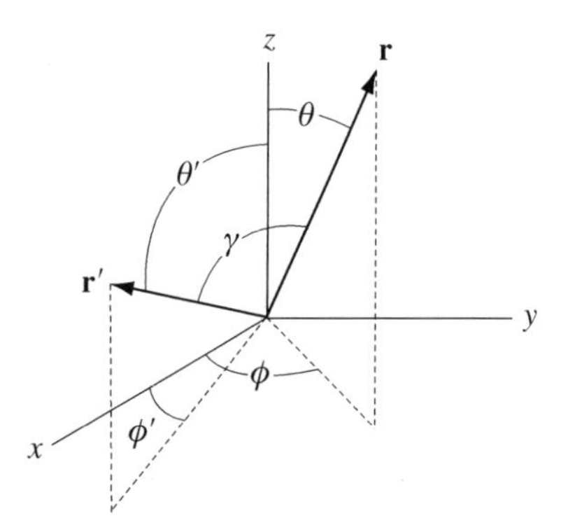

Figure 4.14: Relation between $\mathbf{r} = (r, \theta, \phi)$ and $\mathbf{r}' = (r', \theta', \phi')$ where $\hat{\mathbf{r}} \cdot \hat{\mathbf{r}}' = \cos\gamma$.

The first few spherical harmonics are

$$\begin{aligned}
Y_{00}(\Omega) &= \frac{1}{\sqrt{4\pi}} \\
Y_{10}(\Omega) &= \sqrt{\frac{3}{4\pi}}\cos\theta = \sqrt{\frac{3}{4\pi}}\frac{z}{r} \\
Y_{1\pm1}(\Omega) &= \mp\sqrt{\frac{3}{8\pi}}\sin\theta\exp(\pm i\phi) = \mp\sqrt{\frac{3}{8\pi}}\frac{x \pm iy}{r}.
\end{aligned} \tag{4.82}$$

4.5.3 The Inverse Distance

Substitution of (4.79) into (4.78) produces a spherical expansion for the inverse distance referred to a single origin of coordinates:

$$\frac{1}{|\mathbf{r} - \mathbf{r}'|} = \frac{1}{r}\sum_{\ell=0}^{\infty}\frac{4\pi}{2\ell+1}\left(\frac{r'}{r}\right)^{\ell}\sum_{m=-\ell}^{\ell} Y^*_{\ell m}(\Omega')Y_{\ell m}(\Omega) \qquad r' < r. \tag{4.83}$$

Even better, the alert reader will notice that an exchange of the primed and unprimed variables in (4.83) produces a formula for $1/|\mathbf{r} - \mathbf{r}'|$ that is valid when $r' > r$. This leads us to define $r_<$ as the *smaller* of r and r' and $r_>$ as the *larger* of r and r' in order to write a formula which covers both cases:

$$\frac{1}{|\mathbf{r} - \mathbf{r}'|} = \frac{1}{r_>}\sum_{\ell=0}^{\infty}\frac{4\pi}{2\ell+1}\left(\frac{r_<}{r_>}\right)^{\ell}\sum_{m=-\ell}^{\ell} Y^*_{\ell m}(\Omega_<)Y_{\ell m}(\Omega_>). \tag{4.84}$$

A similar generalization applies to the Legendre identity (4.78).

4.6 Spherical and Azimuthal Multipoles

We derive multipole expansions of the electrostatic potential $\varphi(\mathbf{r})$ appropriate to systems with natural spherical symmetry by substituting (4.84) into

$$\varphi(\mathbf{r}) = \frac{1}{4\pi\epsilon_0}\int d^3r' \frac{\rho(\mathbf{r}')}{|\mathbf{r} - \mathbf{r}'|}. \tag{4.85}$$

The version of (4.84) with $r > r'$ produces an *exterior* multiple expansion analogous to (4.3) that is valid for observation points that lie outside the charge distribution. The version of (4.84) with $r < r'$ gives an *interior* multipole expansion whose character and usefulness we will explore below.

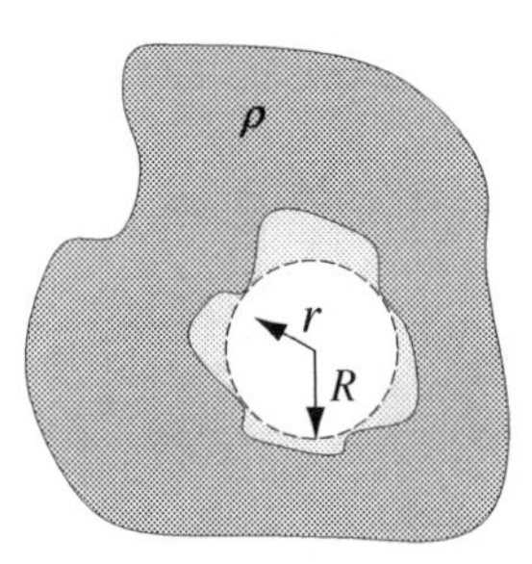

Figure 4.15: A charge distribution $\rho(\mathbf{r})$ (shaded) lies completely outside an origin-centered sphere of radius R. An interior multipole expansion is valid at observation points $\mathbf{r}$ inside the sphere.

4.6.1 The Exterior Spherical Expansion

When the charge density is localized and confined to the sphere of radius R in Figure 4.1, the spherical analog of the Cartesian expansion (4.3) which converges for observation points *exterior* to the sphere is

$$\varphi(\mathbf{r}) = \frac{1}{4\pi\epsilon_0}\sum_{\ell=0}^{\infty}\sum_{m=-\ell}^{\ell} A_{\ell m}\frac{Y_{\ell m}(\Omega)}{r^{\ell+1}} \qquad r > R. \tag{4.86}$$

The complex numbers $A_{\ell m}$ are the *exterior spherical multipole moments* of the charge distribution:

$$A_{\ell m} = \frac{4\pi}{2\ell+1}\int d^3r'\,\rho(\mathbf{r}')r'^{\ell}\,Y^*_{\ell m}(\Omega'). \tag{4.87}$$

The order-of-magnitude estimate $A_{\ell m} \sim QR^{\ell}$ shows that, like the Cartesian multipole expansion, (4.86) is an expansion in the small parameter R/r. Moreover, there is a simple relation between $A_{\ell m}$ and $A_{\ell -m}$ because the spherical harmonics obey

$$Y_{\ell -m}(\theta,\phi) = (-1)^m Y^*_{\ell m}(\theta,\phi). \tag{4.88}$$

The coefficient of $1/r^{\ell+1}$ in (4.86) must be identical to the corresponding coefficient in the Cartesian expansion (4.7). This means that the information encoded by the three spherical moments A_{10}, A_{11}, and $A_{1\,-1}$ is exactly the same as the information encoded by the three components of the electric dipole moment vector $\mathbf{p}$. Put another way, each A_{1m} can be written as a linear combination of the p_k and vice versa. Similarly, the five spherical moments A_{2m} carry the same information as the five components of the traceless quadrupole tensor $\boldsymbol{\Theta}$.[13] When higher moments are needed (as in some nuclear, atomic, and molecular physics problems), the relatively simple analytic properties of the spherical harmonics usually make the spherical multipole expansion the method of choice for approximate electrostatic potential calculations.

4.6.2 The Interior Spherical Expansion

The expansion (4.86) is only valid for observation points that lie *outside* a bounded distribution $\rho(\mathbf{r})$. Another common situation is depicted in Figure 4.15. This shows a region of distributed charge $\rho(\mathbf{r})$ which completely surrounds a charge-free, origin-centered sphere of radius R. To study $\varphi(\mathbf{r})$ inside the sphere, we choose the version of (4.84) that applies when $r < r'$ and substitute this into (4.85).

[13] It is not obvious that the traceless tensor $\boldsymbol{\Theta}$ is relevant here rather than the primitive tensor $\mathbf{Q}$. This point is explored in Section 4.7.

The result is a formula for the potential that is valid at observation points in the *interior* of the sphere in Figure 4.15. The potential is real, so complex conjugation gives

$$\varphi(\mathbf{r}) = \frac{1}{4\pi\epsilon_0}\sum_{\ell=0}^{\infty}\sum_{m=-\ell}^{\ell} B_{\ell m}\, r^{\ell}\, Y_{\ell m}(\Omega) \qquad r < R. \tag{4.89}$$

The complex numbers $B_{\ell m}$ in (4.89) are the *interior spherical multipole moments* of the charge distribution:

$$B_{\ell m} = \frac{4\pi}{2\ell+1}\int d^3r'\, \frac{\rho(\mathbf{r}')}{r'^{\ell+1}}\, Y^*_{\ell m}(\Omega'). \tag{4.90}$$

The Cartesian analog of the interior expansion (4.89) is a power expansion of $\varphi(x, y, z)$ around the origin of coordinates.

Example 4.4 Use the multipole expansion method to find $\varphi(\mathbf{r})$ for a spherically symmetric but otherwise arbitrary charge distribution.

Solution: Figure 4.16 shows an observation point $\mathbf{r}$ at a distance r from the center of the distribution. We use an exterior expansion to find the contribution to $\varphi(r)$ from the charge inside the volume V and an interior expansion to find the contribution to $\varphi(r)$ from the charge outside the volume V.

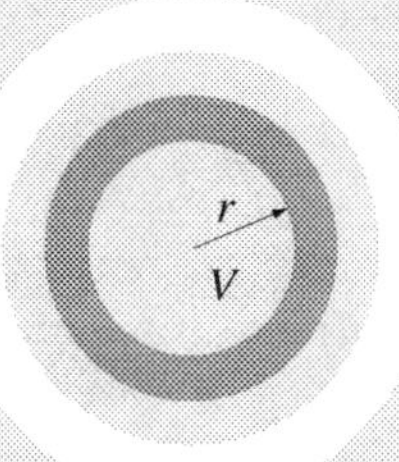

Figure 4.16: A spherically symmetric charge distribution of infinite extent.

Since $Y_{00} = 1/\sqrt{4\pi}$ is a constant, there is no loss of generality if we write $\rho(\mathbf{r}) = \rho(r)Y_{00} = \rho(r)Y^*_{00}$. Then, the exterior and interior multipole moments defined in (4.87) and (4.90) are

$$A_{\ell m}(r) = \frac{4\pi}{2\ell+1}\int d\Omega' Y_{00}(\Omega')Y^*_{\ell m}(\Omega')\int_0^r dr' r'^2\rho(r') = 4\pi\,\delta_{\ell,0}\delta_{m,0}\int_0^r dr' r'^2\rho(r')$$

$$B_{\ell m}(r) = \frac{4\pi}{2\ell+1}\int d\Omega' Y^*_{00}(\Omega')Y_{\ell m}(\Omega')\int_r^{\infty} dr' \frac{\rho(r')}{r'^{\ell-1}} = 4\pi\,\delta_{\ell,0}\delta_{m,0}\int_r^{\infty} dr' r'\rho(r').$$

Since only the A_{00} and B_{00} terms survive we drop the constant factor Y_{00} from the definition of $\rho(\mathbf{r})$ and add the potentials (4.86) and (4.89) to get the desired final result,

$$\varphi(r) = \frac{1}{4\pi\epsilon_0}\left\{\frac{A_{00}(r)}{r} + B_{00}(r)\right\} = \frac{1}{\epsilon_0}\left\{\frac{1}{r}\int_0^r ds\, s^2\rho(s) + \int_r^{\infty} ds\, s\rho(s)\right\}.$$

This agrees with the formula derived for this quantity in Example 3.4.

4.6.3 Azimuthal Multipoles

Specified charge densities that are *azimuthally symmetric* with respect to a fixed axis in space occur quite frequently. For that reason, it is worth learning how (4.86) and (4.89) simplify when $\rho(r,\theta,\phi)=\rho(r,\theta)$. We will need two properties of the spherical harmonics:

$$Y_{\ell 0}(\theta,\phi)=\sqrt{\frac{2\ell+1}{4\pi}}\,P_\ell(\cos\theta) \tag{4.91}$$

and

$$\frac{1}{2\pi}\int_0^{2\pi} d\phi\, Y_{\ell m}(\theta,\phi)=\sqrt{\frac{2\ell+1}{4\pi}}\,P_\ell(\cos\theta)\delta_{m,0}. \tag{4.92}$$

Using (4.92), the expansion coefficients (4.87) and (4.90) become $A_{\ell m}=A_\ell\,\delta_{m0}\sqrt{4\pi/(2\ell+1)}$ and $B_{\ell m}=B_\ell\,\delta_{m0}\sqrt{4\pi/(2\ell+1)}$, where[14]

$$A_\ell=\int d^3r'\, r'^\ell\, \rho(r',\theta')P_\ell(\cos\theta') \tag{4.93}$$

and

$$B_\ell=\int d^3r'\,\frac{1}{r'^{\ell+1}}\rho(r',\theta')P_\ell(\cos\theta'). \tag{4.94}$$

Substituting $A_{\ell m}$ and $B_{\ell m}$ into (4.86) and (4.89), respectively, and using (4.91) yields exterior and interior *azimuthal multipole expansions* in the form

$$\varphi(r,\theta)=\frac{1}{4\pi\epsilon_0}\sum_{\ell=0}^{\infty}\frac{A_\ell}{r^{\ell+1}}P_\ell(\cos\theta) \qquad r>R, \tag{4.95}$$

$$\varphi(r,\theta)=\frac{1}{4\pi\epsilon_0}\sum_{\ell=0}^{\infty}B_\ell\, r^\ell\, P_\ell(\cos\theta) \qquad r<R. \tag{4.96}$$

Application 4.3 The Potential Produced by $\sigma(\theta)=\sigma_0\cos\theta$

A situation which occurs often in electrostatics is an origin-centered spherical shell with radius R and a specified surface charge density

$$\sigma(\theta)=\sigma_0\cos\theta. \tag{4.97}$$

We can represent the potential produced by (4.97) using the exterior azimuthal multipole expansion (4.95) for the space outside the shell and the interior azimuthal multipole expansion (4.96) for the space inside the shell. The fact that $P_1(x)=x$ and the orthogonality relation (4.75) for the Legendre polynomials show that the expansion coefficients (4.93) and (4.94) are

$$\begin{aligned}A_\ell&=\int dS\,R^\ell P_\ell(\cos\theta)\sigma(\theta)=\sigma_0R^{\ell+2}\int_0^{2\pi}d\phi\int_0^{\pi}d\theta\,\sin\theta\,P_\ell(\cos\theta)\cos\theta=\frac{4\pi}{3}R^3\sigma_0\delta_{\ell,1}\\ B_\ell&=\int dS\,\frac{1}{R^{\ell+1}}P_\ell(\cos\theta)\sigma(\theta)=\sigma_0R^{1-\ell}\int_0^{2\pi}d\phi\int_0^{\pi}d\theta\,\sin\theta\,P_\ell(\cos\theta)\cos\theta=\frac{4\pi}{3}\sigma_0\delta_{\ell,1}.\end{aligned} \tag{4.98}$$

[14] Notice that (4.93) and (4.94) include an integration over the azimuthal angle ϕ.

The corresponding potential is

$$\varphi(r,\theta) = \frac{\sigma_0}{3\epsilon_0} \begin{cases} r\cos\theta & r < R, \\ \dfrac{R^3}{r^2}\cos\theta & r > R. \end{cases} \tag{4.99}$$

Note that the two expansions agree at $r = R$. This reflects the continuity of the electrostatic potential as the observation point passes through a layer of charge. The fact that only the $\ell = 1$ term survives tells us that the electric field outside the shell is exactly dipolar. The interior "dipole" potential is proportional to $r\cos\theta = z$. This gives a constant electric field $\mathbf{E} \parallel \hat{\mathbf{z}}$ inside the shell because $\mathbf{E} = -\nabla\varphi$. ■

4.6.4 Preview: The Connection to Potential Theory

The multipole calculation outlined in Application 4.3 exploited the fact that all the source charge lay on the surface of a spherical shell. Another approach to this problem, called *potential theory*, uses the same information to infer that $\varphi(\mathbf{r})$ satisfies Laplace's equation both inside the shell and outside the shell:

$$\nabla^2\varphi(\mathbf{r}) = 0 \qquad r \neq R. \tag{4.100}$$

From this point of view, (4.95) and (4.96) are linear combinations of elementary solutions of Laplace's equation with azimuthal symmetry. The coefficients A_ℓ and B_ℓ are chosen so the potentials inside and outside the sphere satisfy the required matching conditions (Section 3.3.2) at $r = R$. If the source charge on the shell is not azimuthally symmetric, similar remarks apply to a potential given by the simultaneous use of (4.86) and (4.89). We will explore potential theory in detail in Chapters 7 and 8.

Example 4.5 Find the electrostatic potential produced by a uniformly charged line bent into an origin-centered circular ring in the x-y plane. The charge per unit length of the line is $\lambda = Q/2\pi R$.

Solution: The potential $\varphi(r,\theta)$ is determined completely by interior and exterior azimuthal multipole expansions with respect to an origin-centered sphere of radius R. In spherical coordinates, the volume charge density of the ring is

$$\rho(\mathbf{r}) = (\lambda/r)\delta(r-R)\delta(\cos\theta).$$

Note that the right side of this equation is dimensionally correct and satisfies the necessary condition $Q = \int d^3r\rho(\mathbf{r}) = 2\pi R\lambda$. Using (4.93) and (4.94), the expansion coefficients are $A_\ell = QR^\ell P_\ell(0)$ and $B_\ell = QR^{-(\ell+1)}P_\ell(0)$. Therefore,

$$\varphi(r,\theta) = \frac{Q}{4\pi\epsilon_0 r}\sum_{\ell=0}^{\infty}\left(\frac{R}{r}\right)^\ell P_\ell(0)P_\ell(\cos\theta) \qquad r \geq R,$$

$$\varphi(r,\theta) = \frac{Q}{4\pi\epsilon_0 R}\sum_{\ell=0}^{\infty}\left(\frac{r}{R}\right)^\ell P_\ell(0)P_\ell(\cos\theta) \qquad r \leq R.$$

As it must be, this potential is continuous as the observation point passes through the ring.

Application 4.4 The Liquid Drop Model of Nuclear Fission

Bohr and Wheeler (1939) famously discussed a "liquid drop" model of the nucleus where a sphere of uniform charge density $\varrho = Q/(4\pi R_0^3/3)$ undergoes a small amplitude shape distortion as the first step toward fission. Surface energy opposes fission because the distortion increases the nuclear surface area. Electrostatics favors fission because the distortion lowers the Coulomb self-energy of the system. The latter dominates when the nuclear charge is large enough. To see this, we will show the following.

Claim A: Up to quadrupole order, the most general azimuthally symmetric distortion of a uniform sphere with charge q and radius R_0 which preserves both the volume and the position of the center of mass has a surface described by

$$R(\theta) = R_0 \left\{ 1 - \frac{\alpha^2}{5} + \alpha P_2(\cos\theta) \right\} \qquad \alpha \ll 1. \tag{4.101}$$

Claim B: If U_0 is the self-energy of the undistorted nucleus, the change in electrostatic energy induced by the shape distortion (4.101) is

$$\Delta U_E = -\tfrac{1}{5} U_0 \alpha^2. \tag{4.102}$$

Proof of Claim A: The most general azimuthally symmetric form for the nuclear shape (up to quadrupole order) is

$$R(\theta) = R_0 \{1 + \alpha_0 P_0(\cos\theta) + \alpha_1 P_1(\cos\theta) + \alpha_2 P_2(\cos\theta)\}. \tag{4.103}$$

The α_1 term can be discarded because it corresponds to a rigid displacement of the center of mass of the undistorted sphere along the z-axis. To see this, note first that $P_1(\cos\theta) = \cos\theta = z$ on the unit sphere. Then, to first order in $\alpha_1 \ll 1$, the presumptive shape equation $r = 1 + \alpha_1 z$ implies that $r^2 = 1 + 2\alpha_1 z$. On the other hand $r^2 = x^2 + y^2 + z^2$. Equating these two expressions for r^2 implies that $x^2 + y^2 + (z - \alpha_1)^2 = 1$ (to first order in α_1). This is the equation of the displaced unit sphere shown in Figure 4.17(a).

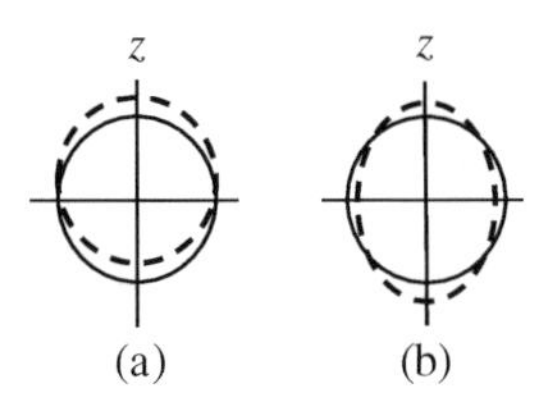

Figure 4.17: Small-amplitude distortions of a unit sphere: (a) $\Delta R(\theta) = \alpha_1 P_1(\cos\theta)$; (b) $\Delta R(\theta) = \alpha_2 P_2(\cos\theta)$.

To preserve the sphere volume, we use $P_0(\cos\theta) = 1$ and demand that

$$\frac{4\pi}{3} R_0^3 = 2\pi \int_0^{\pi} d\theta \sin\theta \int_0^{R(\theta)} dr r^2 = \frac{2\pi}{3} R_0^3 \int_{-1}^{1} dx \, [1 + \alpha_0 + \alpha_2 P_2(x)]^3 . \tag{4.104}$$

The integral on the far right is performed using the orthogonality relation (4.75) for the Legendre polynomials. By omitting terms that are higher order than quadratic in the alpha parameters, we find

$$\frac{4\pi}{3} R_0^3 = \frac{4\pi}{3} R_0^3 + 4\pi R_0^3 \left(\alpha_0 + \alpha_0^2 + \tfrac{1}{5}\alpha_2^2 \right). \tag{4.105}$$

To second order in the alpha parameters, we must put $\alpha_0 = -\alpha_2^2/5$ to make (4.105) an equality. This proves Claim A if we set $\alpha = \alpha_2$ in (4.101).

Proof of Claim B: Let $\varphi_0(\mathbf{r})$ be the electrostatic potential of the undistorted nucleus and use a step function to write the undistorted nuclear charge density as $\rho_0(r) = \varrho\Theta(R_0 - r)$. The same quantities for the shape-distorted nucleus are $\varphi(\mathbf{r})$ and $\rho(r,\theta) = \varrho\Theta(R(\theta) - r)$. With this notation, the change in total Coulomb energy (3.78) upon distortion is

$$\Delta U_E = \tfrac{1}{2}\int d^3r\,[\rho(\mathbf{r})\varphi(\mathbf{r}) - \rho_0(\mathbf{r})\varphi_0(\mathbf{r})]\,. \tag{4.106}$$

To evaluate the first integral in (4.106), we focus on $\rho(\mathbf{r})$ and expand $\Theta(R(\theta) - r)$ to second order in α and write $\delta'(R_0 - r)$ for the radial derivative of $\delta(R_0 - r)$. This gives

$$\begin{aligned}\rho(r,\theta) &\simeq \rho_0(r) + \alpha\varrho P_2(\cos\theta)R_0\delta(R_0 - r)\\ &\quad -\alpha^2\left[\tfrac{1}{5}\varrho R_0\delta(R_0 - r) + \tfrac{1}{2}\varrho R_0^2 P_2^2(\cos\theta)\delta'(R_0 - r)\right] + \cdots\end{aligned} \tag{4.107}$$

It is convenient to write (4.107) as $\rho \simeq \rho_0 + \rho_1 + \rho_2$ where the subscript indicates the power of the small parameter α in each term. The corresponding potential is $\varphi \simeq \varphi_0 + \varphi_1 + \varphi_2$. This means that the product $\rho(\mathbf{r})\varphi(\mathbf{r})$ contains nine terms. One of these terms is $\rho_0\varphi_0$. Three other terms are of order α^3 and can be neglected. The remaining five terms can be reduced to three because Green's reciprocity relation (Section 3.5.2) implies that

$$\int d^3r\rho_0(\mathbf{r})\varphi_1(\mathbf{r}) = \int d^3r\rho_1(\mathbf{r})\varphi_0(\mathbf{r}) \tag{4.108}$$

and

$$\int d^3r\rho_0(\mathbf{r})\varphi_2(\mathbf{r}) = \int d^3r\rho_2(\mathbf{r})\varphi_0(\mathbf{r}). \tag{4.109}$$

We conclude that (to second order in α) the distortion-induced change in Coulomb energy (4.106) is

$$\Delta U_E \simeq \int d^3r\{\rho_0\varphi_1 + \tfrac{1}{2}\rho_1\varphi_1 + \rho_2\varphi_0\} = \Delta U_0 + \Delta U_1 + \Delta U_2. \tag{4.110}$$

To find φ_0, it is simplest to evaluate the formula derived in Example 3.4 at the end of of Section 3.4.1:

$$\varphi_0(r) = \frac{1}{\epsilon_0 r}\int_0^r ds s^2\rho_0(s) + \frac{1}{\epsilon_0}\int_r^\infty ds s\rho_0(s). \tag{4.111}$$

The charge density $\rho_0(r)$ equals ϱ when $r < R_0$ and zero otherwise. Therefore,

$$\varphi_0(r) = \begin{cases}\dfrac{(3R_0^2 - r^2)\varrho}{6\epsilon_0} & r < R,\\[2ex] \dfrac{\varrho R_0^3}{3\epsilon_0 r} & r > R_0.\end{cases} \tag{4.112}$$

To find φ_1, we note from (4.107) that $\rho_1(\mathbf{r})$ has the form $\rho_1 = \sigma\delta(R_0 - r)$ with effective surface charge density $\sigma = \alpha\varphi P_2(\cos\theta)R_0$. This means that the matching condition (3.19) must be imposed at $r = R_0$:

$$\left.\frac{\partial\varphi_1}{\partial r}\right|_{r=R_0^-} - \left.\frac{\partial\varphi_1}{\partial r}\right|_{r=R_0^+} = \frac{\sigma}{\epsilon_0} = \frac{\alpha\varrho R_0 P_2(\cos\theta)}{\epsilon_0}. \tag{4.113}$$

The potential itself has an exterior multipole expansion (4.95) for $r > R_0$ and an interior multipole expansion (4.96) for $r < R_0$. However in light of (4.113), it should be sufficient to keep only the $\ell = 2$ term in each. Therefore, imposing the continuity of φ_1 at $r = R_0$, we write

$$\varphi_1(r,\theta) = \begin{cases}\dfrac{1}{4\pi\epsilon_0}C_2\left(\dfrac{r}{R_0}\right)^2 P_2\cos(\theta) & r < R_0,\\[2ex] \dfrac{1}{4\pi\epsilon_0}C_2\left(\dfrac{R_0}{r}\right)^3 P_2\cos(\theta) & r > R_0.\end{cases} \tag{4.114}$$

Substituting (4.114) into (4.113) determines the coefficient C_2. The final result is

$$\varphi_1(r,\theta) = \begin{cases} \dfrac{\alpha \varrho r^2 P_2(\cos\theta)}{5\epsilon_0} & r < R_0, \\ \dfrac{\alpha \varrho R_0^5 P_2(\cos\theta)}{5\epsilon_0 r^3} & r > R_0. \end{cases} \tag{4.115}$$

With ρ_0, ρ_1, ρ_2, φ_0, and φ_1 in hand, it is straightforward to evaluate the three energy integrals in (4.110) which determine ΔU_E using the Legendre orthogonality relation (4.75) and (for the $\rho_2\varphi_0$ piece) the fact that $\varphi_0'(r)$ is continuous at $r = R_0$. The results are

$$\Delta U_0 = 0, \qquad \Delta U_1 = \frac{2\pi}{25\epsilon_0}\varrho^2 R_0^5 \alpha^2, \qquad \Delta U_2 = -\frac{2\pi}{15\epsilon_0}\varrho^2 R_0^5 \alpha^2. \tag{4.116}$$

Since $\varrho = Q/(4\pi R_0^3/3)$, the sum of the three energies in (4.116) gives the total change in electrostatic energy produced by a quadrupole shape distortion of a spherical nucleus with constant charge density as

$$\Delta U_E = -\left\{\frac{3}{5}\frac{Q^2}{4\pi\epsilon_0 R_0}\right\}\frac{\alpha^2}{5}. \tag{4.117}$$

This proves Claim B because, using (3.77), the total energy of the undistorted nucleus is

$$U_0 = \frac{1}{2}\int d^3r \rho_0(\mathbf{r})\varphi_0(\mathbf{r}) = \frac{3}{5}\frac{Q^2}{4\pi\epsilon_0 R_0}. \tag{4.118}$$

■

4.7 Primitive and Traceless Multipole Moments

There is a deep connection between the spherical multipole expansion derived in Section 4.6 and a traceless version of the Cartesian multiple expansion derived in Section 4.1.1. Namely that the number of independent traceless Cartesian multipole moments $T^{(\ell)}_{ij\ldots m}$ is equal to the number of spherical multipole moments $A_{\ell m}$. Beginning at $\ell = 2$, the number in question is smaller (often much smaller) than the number of independent primitive Cartesian multipole moments, $C^{(\ell)}_{ij\ldots m}$. Our proof of these statements begins with an explicit formula for the entire primitive Cartesian multiple expansion. A convenient form is

$$\varphi(\mathbf{r}) = \frac{1}{4\pi\epsilon_0}\sum_{\ell=0}^{\infty} C^{(\ell)}_{ij\ldots m} N^{(\ell)}_{ij\ldots m}(\mathbf{r}), \tag{4.119}$$

where the primitive electric moment of order ℓ is

$$C^{(\ell)}_{ij\ldots m} = \frac{1}{\ell!}\int d^3r\, \rho(\mathbf{r})\, \underbrace{r_i r_j \cdots r_m}_{\ell \text{ terms}} \tag{4.120}$$

and

$$N^{(\ell)}_{ij\ldots m}(\mathbf{r}) = (-1)^\ell \underbrace{\nabla_i \nabla_j \cdots \nabla_m}_{\ell \text{ terms}} \frac{1}{r}. \tag{4.121}$$

A complete traceless Cartesian multipole expansion is obtained by (i) substituting the expansion (4.78) of $|\mathbf{r} - \mathbf{r}'|^{-1}$ into Coulomb's formula (4.85) for the electrostatic potential; and (ii) writing out the Legendre polynomials (4.77) explicitly using $\hat{\mathbf{r}} \cdot \hat{\mathbf{r}}' = r_i r_i'/rr'$ and $r^2 = r_i r_j \delta_{ij}$. The first few terms

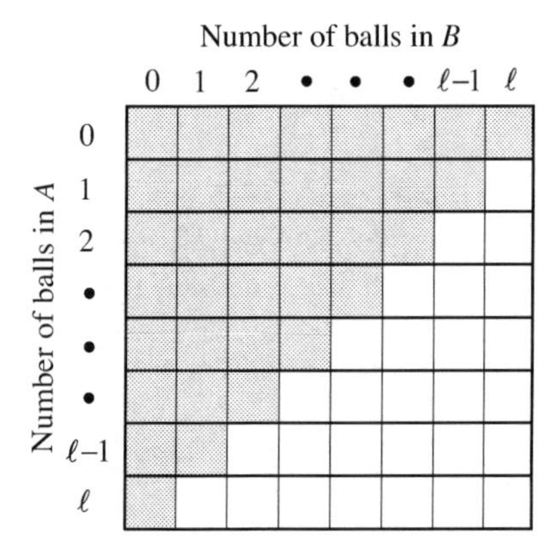

Figure 4.18: $(\ell+1)\times(\ell+1)$ matrix for calculation of the number of independent components of a completely symmetric ℓ-rank tensor.

generated by this procedure are

$$\varphi(\mathbf{r}) = \frac{1}{4\pi\epsilon_0}\left\{\frac{Q}{r} + \frac{\mathbf{p}\cdot\mathbf{r}}{r^3} + \Theta_{ij}\frac{r_i r_j}{r^5} + \cdots\right\}. \tag{4.122}$$

The monopole ($\ell = 0$) and dipole ($\ell = 1$) terms in (4.122) are the same as those in the primitive multiple expansion (4.7). However, the $\ell = 2$ term is not the quadrupole potential (4.52) expressed in terms of the primitive moments Q_{ij}. Instead, it is the quadrupole potential (4.64) expressed in terms of the traceless moments Θ_{ij}. The complete series (4.122) is

$$\varphi(\mathbf{r}) = \frac{1}{4\pi\epsilon_0}\sum_{\ell=0}^{\infty} T^{(\ell)}_{ij\ldots m}\frac{\overbrace{r_i r_j\cdots r_m}^{\ell \text{ terms}}}{r^{2\ell+1}}, \tag{4.123}$$

where the scalars $T^{(\ell)}_{ij\ldots m}$ are the components of an ℓth-order *traceless* Cartesian multipole moment:

$$T^{(\ell)}_{ij\cdots m} = \frac{(-1)^\ell}{\ell!}\int d^3y\rho(\mathbf{y})y^{2\ell+1}\frac{\partial}{\partial y_i}\frac{\partial}{\partial y_j}\cdots\frac{\partial}{\partial y_m}\frac{1}{|\mathbf{y}|}. \tag{4.124}$$

The meaning of "traceless" is that we get *zero* when the trace operation is performed over any pair of indices:

$$\delta_{ij}T^{(\ell)}_{ij\ldots m} = \delta_{im}T^{(\ell)}_{ij\ldots m} = \cdots\cdots = \delta_{mj}T^{(\ell)}_{ij\ldots m} = 0. \tag{4.125}$$

The representation of the electrostatic potential by this traceless Cartesian expansion (4.123) is completely equivalent to the primitive Cartesian expansion (4.119). However, as we will now show, the traceless representation is much more efficient.

4.7.1 Counting Multipole Moments

The primitive Cartesian moments $C^{(\ell)}_{ij\cdots m}$ defined in (4.120) are symmetric with respect to the interchange of any two indices. This means that the number of independent components of the tensor $\mathbf{C}^{(\ell)}$—call this number M—is equal to the number of ways that each of ℓ indices takes one of the three values (x, y, z) regardless of their order. This is equivalent to counting the number of ways to place ℓ indistinguishable balls in three distinguishable urns, A, B, and C. Figure 4.18 represents this problem in matrix form. Each row indicates the number of balls put into urn A. Within any row, each column indicates the number of balls put into urn B. The remaining balls are put into urn C. Each shaded box of the matrix thus represents one possible placement of the balls in the three urns. This means that the number of shaded boxes in Figure 4.18 is the number M we seek. Hence,

$$M = 1 + 2 + \cdots + (\ell + 1) = \tfrac{1}{2}(\ell + 2)(\ell + 1). \tag{4.126}$$

The traceless tensor $T^{(\ell)}_{ij\ldots m}$ defined by (4.123) is also symmetric with respect to index interchange. This tensor would have M independent components also, except that each of the trace conditions (4.125) reduces this number by one. Moreover, each trace represents a choice of two indices to be made equal out of a total of ℓ indices. This means that the number of traces is the binomial coefficient

$$\binom{\ell}{2} = \frac{\ell!}{2!(\ell - 2)!} = \tfrac{1}{2}\ell(\ell - 1). \tag{4.127}$$

Therefore, the number of independent components of $T^{(\ell)}_{ij\ldots m}$ is

$$M' = \tfrac{1}{2}(\ell + 2)(\ell + 1) - \tfrac{1}{2}\ell(\ell - 1) = 2\ell + 1. \tag{4.128}$$

Comparing (4.126) to (4.128) shows that $M' \ll M$ when $\ell \gg 1$. Beginning with the quadrupole term, the traceless moments represent the potential with increasing efficiency.

The number $2\ell + 1$ can been seen in another way using Maxwell's construction of multipole moments. The geometrical constructions shown in Figures 4.5 and 4.10 do not immediately generalize to higher *primitive* moments because there is no unique way to choose and orient two quadrupoles, two octupoles, etc. so they are "equal and opposite" as we did with two charges and two dipole vectors. However, for *traceless* moments, the dipole and quadrupole potential formulae (4.13) and (4.54) *can* be generalized. It turns out[15] that the potential of a "point 2^ℓ-pole" located at the origin can be written in the form

$$\varphi(\mathbf{r}) = \frac{(-)^\ell}{4\pi\epsilon_0}\overbrace{(\mathbf{a}\cdot\nabla)(\mathbf{b}\cdot\nabla)\cdots(\mathbf{p}\cdot\nabla)}^{\ell\text{ terms}}\frac{1}{r}. \tag{4.129}$$

This representation is correct only if $|\mathbf{a}| = |\mathbf{b}| = \cdots = |\mathbf{p}|$. Thus, the potential (4.129) is characterized by two Cartesian components for each vector and one (common) magnitude, i.e., $2\ell + 1$ independent parameters.

It is very satisfying to find that the spherical expansion (4.86) and the traceless Cartesian expansions (4.123) and (4.129) all involve a multiple moment with $2\ell + 1$ components for each spatial factor $r^{-(\ell+1)}$. This means we can generalize the remarks made at the end of Section 4.6.1 for the dipole and quadrupole and assert that each $A_{\ell m}$ can be written as a linear combination of the $2\ell + 1$ components of $\mathbf{T}^{(\ell)}$ and vice versa. In the language of group theory, both representations of the potential are *irreducible*, i.e., maximally efficient in our language.

Application 4.5 The Dielectric Polarization $\mathbf{P}(\mathbf{r})$

In Section 2.4.1, we asserted without proof that a general charge density in the presence of matter can always be written in the form

$$\rho(\mathbf{r}, t) = \rho_c(\mathbf{r}, t) - \nabla\cdot\mathbf{P}(\mathbf{r}, t). \tag{4.130}$$

[15] See G.F. Torres del Castillo and A. Méndez-Garrido, "Differential representation of multipole fields", *Journal of Physics A* **37**, 1437 (2004).

We can support this claim (in the static limit) and generalize (4.14) and (4.56) also by producing a multipole expansion for the charge density of a collection of N point charges q_α located at positions $\mathbf{r}_\alpha$:

$$\rho(\mathbf{r}) = \sum_{\alpha=1}^{N} q_\alpha \delta(\mathbf{r} - \mathbf{r}_\alpha). \tag{4.131}$$

Our strategy is to write a Taylor series (Section 1.3.4) for each delta function in (4.131) with respect to a common origin where $\mathbf{r}_\alpha = 0$. This gives

$$\rho(\mathbf{r}) = \sum_{\alpha} q_\alpha \left\{ 1 - \mathbf{r}_\alpha \cdot \nabla + \frac{1}{2}(\mathbf{r}_\alpha \cdot \nabla)^2 + \cdots \right\} \delta(\mathbf{r}). \tag{4.132}$$

The total charge, dipole moment, and quadrupole moment components of (4.131) are

$$Q = \sum_{\alpha} q_\alpha, \qquad \mathbf{p} = \sum_{\alpha} q_\alpha \mathbf{r}_\alpha, \qquad Q_{ij} = \tfrac{1}{2}\textstyle\sum_\alpha q_\alpha r_{\alpha\, i} r_{\alpha\, j}. \tag{4.133}$$

Therefore, the expansion (4.132) reads

$$\rho(\mathbf{r}) = \left\{ Q - \mathbf{p} \cdot \nabla + Q_{ij} \nabla_i \nabla_j - \cdots \right\} \delta(\mathbf{r}). \tag{4.134}$$

An exercise in integration by parts confirms that substituting (4.134) into (4.1) reproduces the entire multipole expansion (4.7) or (4.119). More to the point, (4.134) has the advertised form

$$\rho(\mathbf{r}) = Q\delta(\mathbf{r}) - \nabla \cdot \mathbf{P}(\mathbf{r}), \tag{4.135}$$

with

$$P_i(\mathbf{r}) = \sum_{\ell=1}^{\infty} (-1)^{\ell-1} C^{(\ell)}_{ij\ldots m} \underbrace{\nabla_j \cdots \nabla_m}_{\ell-1 \text{ terms}} \delta(\mathbf{r}). \tag{4.136}$$

The highly singular nature of the polarization function (4.136) is a direct consequence of the singular nature of the charge density (4.131). We will learn to construct well-behaved versions of $\mathbf{P}(\mathbf{r})$ in Chapter 6. ■

Sources, References, and Additional Reading

The quotation at the beginning of the chapter is from Section 144(b) of

J.C. Maxwell, *A Treatise on Electricity and Magnetism*, 3rd edition (Clarendon Press, Oxford, 1891). By permission of Oxford University Press.

Section 4.1 Textbooks of electromagnetism usually do not treat the different electric multipole expansions which appear in the research literature. Our discussion is modeled on the excellent treatment in

L. Eyges, *The Classical Electromagnetic Field* (Dover, New York, 1972).

Other textbooks with better-than-average discussions of electric multipoles are

W.K.H. Panofsky and M. Phillips, *Classical Electricity and Magnetism*, 2nd edition (Addison-Wesley, Reading, MA, 1962).

R.K. Wangsness, *Electromagnetic Fields*, 2nd edition (Wiley, New York, 1986).

M.A. Heald and J.B. Marion, *Classical Electromagnetic Radiation*, 3rd edition (Saunders, Fort Worth, TX, 1995).

Gray and Gubbins discuss static electric multipoles of all types (primitive Cartesian, traceless Cartesian, azimuthal, and spherical). *Wikswo and Swinney* use pictorial representations of electric multipoles to good effect:

C.G. Gray and K.E Gubbins, *Theory of Molecular Fluids* (Clarendon Press, Oxford, 1984), Chapter 2.

J.P. Wikswo, Jr. and K.R. Swinney, "A comparison of scalar multipole expansions", *Journal of Applied Physics* **56**, 3039 (1984).

Section 4.2 Figure 4.2, Figure 4.4, and Example 4.1 were taken, respectively, from

R.F.W. Bader, *An Introduction to the Electronic Structure of Atoms and Molecules* (Clarke, Irwin & Co., Toronto, 1970).

C. Desfrançois, H. Abdoul-Carime, N. Khelifa, and J.P. Schermann, "From $1/r$ to $1/r^2$ potentials: Electron exchange between Rydberg atoms and polar molecules", *Physical Review Letters* **73**, 2436 (1994).

D.R. Frankl, "Proof of a theorem in electrostatics", *American Journal of Physics* **41**, 428 (1973).

Searches for an intrinsic electric dipole moment for the electron and neutron are discussed in

B.C. Regan, E.D. Commins, C.J. Schmidt, and D. DeMille, "New limit on the electron electric dipole moment", *Physical Review Letters* **88**, 071805 (2002).

S.K. Lamoreaux and R. Golub, "Experimental searches for the neutron electric dipole moment", *Journal of Physics G* **36**, 1004002 (2009).

Section 4.3 Older discussions of electric dipole layers frame the issue in the context of the *contact potential* difference that develops between two metals with different work functions. See, e.g., Chapter II of Part 1 of

M. Planck, *Theory of Electricity and Magnetism*, 2nd edition (Macmillan, London, 1932).

Figure 4.6 and Example 4.1 come, respectively, from

M. Seul and V.S. Chen, "Isotropic and aligned stripe phases in a monomolecular organic film", *Physical Review Letters* **70**, 1658 (1993).

D.J. Keller, H.M. McConnell, and V.T. Moy, "Theory of superstructures in lipid monolayer phase transitions", *Journal of Physical Chemistry* **90**, 2311 (1986).

Section 4.4 The torque exerted on an electric quadrupole by an electric field is the basis for one method to detect buried land mines and other hidden explosives. A good review is

A.N. Garroway, M.L. Buess, J.B. Miller, *el al.*, "Remote sensing by nuclear quadrupole resonance", *IEEE Transactions on Geoscience* **39**, 1108 (2001).

The paper mentioned in Application 4.2 where nuclear quadrupole moments were first inferred is

H. Schüler and Th. Schmidt, "On the deviation of nuclei from spherical symmetry" (in German), *Zeitschrift für Physik* **94**, 457 (1935).

Figure 4.12 was taken from

D.J. Rowe and J.L. Wood, *Fundamentals of Nuclear Models* (World Scientific, Hackensack, NJ, 2010).

Section 4.6 The electrostatic calculation outlined in Application 4.4 was presented first in

N. Bohr and J.A. Wheeler, "The mechanism of nuclear fission", *Physical Review* **56**, 426 (1939).

Our discussion of the electrostatics of nuclear fission is taken from a paper dedicated to the teaching of Julian Schwinger:

J. Bernstein and F. Pollock, "The calculation of the electrostatic energy in the liquid drop model of nuclear fission: A pedagogical note", *Physica A* **96**, 136 (1979).

Section 4.7 The connection between dielectric polarization $\mathbf{P}(\mathbf{r})$ and a multipole expansion of the singular charge density of a collection of classical point charges is part of a much larger discussion of multipole electrodynamics in

D.P. Craig and T. Thirunamachandran, *Molecular Quantum Electrodynamics* (Academic, London, 1984).

Problems

4.1 Dipole Moment Practice Find the electric dipole moment of:

(a) a ring with charge per unit length $\lambda = \lambda_0 \cos\phi$ where ϕ is the angular variable in cylindrical coordinates.
(b) a sphere with charge per unit areas $\sigma = \sigma_0 \cos\theta$ where θ is the polar angle measured from the positive z-axis.

4.2 Smolochowski's Model of a Metal Surface A surprisingly realistic microscopic model for the charge density of a semi-infinite metal (with $z = 0$ as its macroscopic surface) consists of a positive charge distribution

$$n_+(z) = \begin{cases} \bar{n} & z < 0, \\ 0 & z > 0, \end{cases}$$

and a negative charge distribution

$$n_-(z) = \begin{cases} \bar{n}\left\{1 - \frac{1}{2}\exp(\kappa z)\right\} & z < 0, \\ \frac{1}{2}\bar{n}\exp(-\kappa z) & z > 0. \end{cases}$$

(a) Sketch $n_+(z)$ and $n_-(z)$ on the same graph and give a physical explanation of why they might be reasonable.
(b) Calculate the dipole moment $\mathbf{p}$ per unit area of surface for this system.
(c) Calculate the electrostatic potential $\varphi(z)$ and sketch it in the interval $-\infty < z < \infty$.
(d) Calculate and explain the value of $\varphi(\infty) - \varphi(-\infty)$.
(e) Calculate the total electrostatic energy per unit area of this system.

4.3 The Charge Density of a Point Electric Dipole The text used Poisson's equation to show that the charge density of a point electric dipole with moment $\mathbf{p}$ located at the point $\mathbf{r}_0$ is $\rho_D(\mathbf{r}) = -\mathbf{p}\cdot\nabla\delta(\mathbf{r} - \mathbf{r}_0)$.

(a) Derive the given formula for $\rho_D(\mathbf{r})$ using a limiting process analogous to the one used in the text to find its electrostatic potential.
(b) Show that the formula given above for $\rho_D(\mathbf{r})$ is correct by demonstrating that it produces the expected dipole potential when inserted into the Coulomb integral for $\varphi(\mathbf{r})$.

4.4 Stress Tensor Proof of No Self-Force Use the electric stress tensor formalism to prove that no isolated charge distribution $\rho(\mathbf{r})$ can exert a net force on itself. Distinguish the cases when $\rho(\mathbf{r})$ has a net charge and when it does not.

4.5 Point Charge Motion in an Electric Dipole Field Place a point electric dipole $\mathbf{p} = p\hat{\mathbf{z}}$ at the origin and release a point charge q (initially at rest) from the point $(x_0, y_0, 0)$ in the x-y plane away from the origin. Show that the particle moves periodically in a semi-circular arc.

4.6 The Energy to Assemble a Point Dipole Show that $dW = -\mathbf{E}(\mathbf{r})\cdot d\mathbf{p}$ is the work needed to assemble charges $\pm dq$ into a dipole $d\mathbf{p}$ in an external field $\mathbf{E}(\mathbf{r})$. Ignore the dipole self-energy.

4.7 Dipoles at the Vertices of Platonic Solids Identical point electric dipoles are placed at the vertices of the regular polyhedra shown below. All the dipoles are parallel but the direction they point is arbitrary. Show that the electric field at the center of each polyhedron is zero.

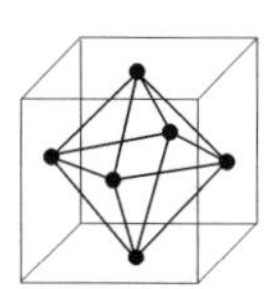
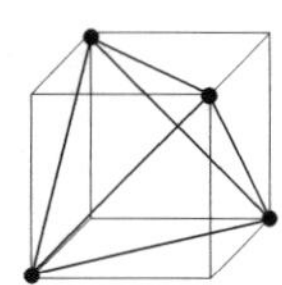
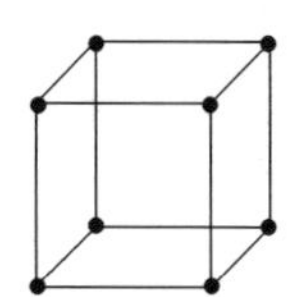

4.8 Two Coplanar Dipoles Two coplanar dipoles are oriented as shown in the figure below.

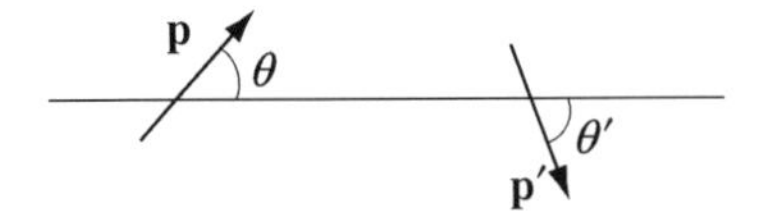

Find the equilibrium value of the angle θ' if the angle θ is fixed.

4.9 Potential of a Double Layer

(a) Show that the potential due to a double-layer surface S with a dipole density $\tau(\mathbf{r}_S)\hat{\mathbf{n}}$ is

$$\varphi(\mathbf{r}) = -\frac{1}{4\tau\epsilon_0}\int_S d\Omega\,\tau(\mathbf{r}_S),$$

where $d\Omega$ is the differential element of solid angle as viewed from $\mathbf{r}$.

(b) Use this result to derive the matching condition at a double-layer surface.

4.10 A Spherical Double Layer A soap bubble (an insulating, spherical shell of radius R) is uniformly coated with polar molecules so that a dipole double layer with $\boldsymbol{\tau} = \tau\hat{\mathbf{r}}$ forms on its surface. Find the potential at every point in space. Check that the matching condition is satisfied at $r = R$.

4.11 The Distant Potential of Two Charged Rings The z-axis is the symmetry axis for an origin-centered ring with charge Q and radius a which lies in the x-y plane. A coplanar and concentric ring with radius $b > a$ has charge $-Q$. Calculate the lowest non-vanishing Cartesian multipole moment to find the asymptotic ($r \to \infty$) electrostatic potential if both charge distributions are uniform. Hint: Use cylindrical coordinates (s, ϕ, z) to perform the integrations.

4.12 The Potential Far from Two Neutral Disks The diagram shows two identical, charge-neutral, origin-centered disks. One disk lies in the x-z plane. The other is tipped away from the first by an angle α around the z-axis. The charge density of each disk depends only on the radial distance from its center. Find the

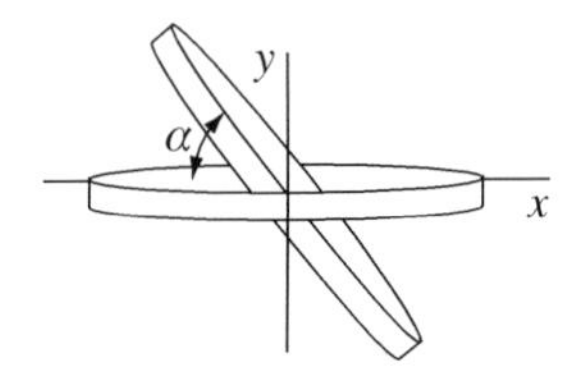

angle α at which the asymptotic electrostatic potential in the x-y plane has the form $\varphi(x, y) = A/s^3$, where A is a constant and $s = \sqrt{x^2 + y^2}$.

4.13 Interaction Energy of Adsorbed Molecules Molecules adsorbed on the surface of a solid crystal surface at low temperature typically arrange themselves into a periodic arrangement, *e.g.*, one molecule lies at the center of each $a \times a$ square of a two-dimensional checkerboard formed by the surface atoms of the crystal. For diatomic molecules which adsorb with their long axis parallel to the surface, the *orientation* of each molecule is determined by the lowest-order electrostatic interaction between nearby molecules.

(a) The CO molecule has a small electric dipole moment $\mathbf{p}$. The sketch below shows a portion of the complete checkerboard where the arrangement of dipole moments is parameterized by an angle α. Treat these as point dipoles and consider the interaction of each dipole with its eight nearest neighbors *only*. Find the angle α that minimizes the total energy and show that the energy/dipole is

$$U = \frac{1}{8\pi\epsilon_0}\frac{p^2}{a^3}\left\{\frac{1}{\sqrt{2}} - 6\right\}.$$

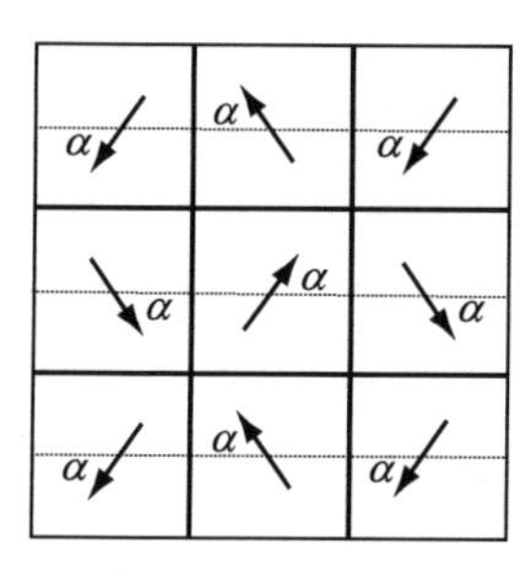

(b) The N_2 molecule has a small quadrupole moment because covalent bonding builds up some negative (electron) charge in the bond region between the atomic nuclei. Use this fact and a *qualitative* argument to help you make a sketch which illustrates the orientational order for a checkerboard of N_2 molecules.

4.14 Practice with Cartesian Multipole Moments Find the primitive, Cartesian monopole, dipole, and quadrupole moments for each of the following charge distributions. Use the geometrical center of each as the origin.

(a) Two charges $+q$ at two diagonal corners of a square $(\pm a, \pm a, 0)$ and two minus charges $-q$ at the two other diagonals of the square $(\pm a, \mp a, 0)$.
(b) A line segment with uniform charge per unit length λ which occupies the interval $-\ell \leq z \leq +\ell$.
(c) An origin-centered ring in the x-y plane with uniform charge per unit length λ and radius R.

4.15 The Many Faces of a Quadrupole

(a) Place two charges $+q$ at two diagonal corners of a square $(\pm a, \pm a, 0)$ and two minus charges $-q$ at the two other diagonals of the square $(\pm a, \mp a, 0)$. Evaluate the primitive quadrupole moment components $Q_{ij} = \frac{1}{2}\int r_i r_j \rho(\mathbf{r}) dV$ and use the result to write down the asymptotic electrostatic potential in Cartesian coordinates.
(b) Write the primitive electric quadrupole tensor explicitly in the form $\mathbf{Q} = \hat{\mathbf{e}}_i Q_{ij} \hat{\mathbf{e}}_j$, where $\hat{\mathbf{e}}_1 = \hat{\mathbf{x}}$, $\hat{\mathbf{e}}_2 = \hat{\mathbf{y}}$, $\hat{\mathbf{e}}_3 = \hat{\mathbf{z}}$, and Q_{ij} are the corresponding matrix elements. Do not write down the vanishing components for the situation of part (a). The remaining parts of this problem exploit the fact that your formula is written in Cartesian coordinates, but gives a valid representation of the quadrupole tensor in any coordinate system.
(c) Express the Cartesian unit vectors $\hat{\mathbf{x}}, \hat{\mathbf{y}}, \hat{\mathbf{z}}$ in terms of the spherical polar unit vectors $\hat{\mathbf{r}}, \hat{\boldsymbol{\theta}}, \hat{\boldsymbol{\phi}}$. Substituting these expressions into the expression for the quadrupole tensor you wrote down in part (b), determine all nine matrix elements of Q in spherical polar coordinates, i.e., Q_{rr}, $Q_{r\theta}$, etc.
(d) Write down the expression for the corresponding electric potential in spherical polar coordinates (where now $\hat{\mathbf{e}}_1 = \hat{\mathbf{r}}$, $\hat{\mathbf{e}}_2 = \hat{\boldsymbol{\theta}}$, $\hat{\mathbf{e}}_3 = \hat{\boldsymbol{\phi}}$) without substituting the matrix elements of Q. Write down only the non-vanishing terms.
(e) Combining the results of parts (c) and (d), compute the electric potential and compare with the answer from part (a).
(f) Explain why you cannot begin in spherical coordinates and use the formula for Q_{ij} quoted in part (a) to reproduce the results of part (d).

4.16 Properties of a Point Electric Quadrupole

(a) Show that the charge density of a point quadrupole is $\rho(\mathbf{r}) = Q_{ij} \nabla_i \nabla_j \delta(\mathbf{r} - \mathbf{r}_0)$.
(b) Show that the force on a point quadrupole in a field $\mathbf{E}(\mathbf{r})$ is $Q_{ij} \partial_i \partial_j \mathbf{E}(\mathbf{r}_0)$.
(c) Show that the torque on a point quadrupole in a field $\mathbf{E}(\mathbf{r})$ is $\mathbf{N} = 2(\mathbf{Q} \cdot \nabla) \times \mathbf{E} + \mathbf{r} \times \mathbf{F}$ where $(\mathbf{Q} \cdot \nabla)_i = Q_{ij} \nabla_j$.
(d) Show that the potential energy of a point quadrupole in a field $\mathbf{E}(\mathbf{r})$ is $V_E = -Q_{ij} \partial_i E_j(\mathbf{r}_0)$.

4.17 Interaction Energy of Nitrogen Molecules How does the leading contribution to the electrostatic interaction energy between two nitrogen molecules depend on the distance R between them?

4.18 A Black Box of Charge A charge distribution is contained entirely inside a black box. Measurements of the electrostatic potential outside the box reveal that all of the exterior multipole moments for $\ell = 1, 2, \ldots$ are zero in a coordinate system with its origin at the center of the box. This does *not* imply that the charge distribution is spherically symmetric. Prove this by constructing a counter-example.

4.19 Foldy's Formula The low-energy Born approximation to the amplitude for electron scattering from a neutron is proportional to the volume integral of the potential energy of interaction between the electron and the neutron,

$$f(0) = -\frac{m}{\pi \hbar^2} \int d^3 r \, V_E(\mathbf{r}).$$

(a) Write a formula for $V_E(\mathbf{r})$ if $\rho_N(\mathbf{s})$ is the charge density of the neutron and $\varphi(\mathbf{s})$ is the electrostatic potential of the electron.

(b) Suppose that $\rho_N(\mathbf{s}) = \rho_N(|\mathbf{s}|)$ and that $\varphi(\mathbf{s})$ varies slowly over the region of space occupied by the neutron's charge distribution. Show that the scattering amplitude gives information about the second radial moment of the neutron charge density:

$$\int d^3 r \, V_E(\mathbf{r}) = -\frac{1}{6} \frac{e}{\epsilon_0} \int d^3 s \, s^2 \rho_N(s).$$

4.20 Practice with Spherical Multipoles

(a) Evaluate the exterior spherical multipole moments for a shell of radius R which carries a surface charge density $\sigma(\theta, \phi) = \sigma_0 \sin\theta \cos\phi$.

(b) Write $\varphi(r > R, \theta, \phi)$ in the form $\varphi(x, y, z, r)$.

(c) Evaluate the interior spherical multipole moments for the shell of part (a).

(d) Write $\varphi(r < R, \theta, \phi)$ in the form $\varphi(x, y, z, r)$.

(e) Check the matching conditions for φ and $\mathbf{E}$ at $r = R$.

(f) Extract the dipole moment $\mathbf{p}$ of the shell from your answer to part (b).

4.21 Proof by Interior Multipole Expansion Let V be a charge-free volume of space. Use an interior spherical multipole expansion to show that the average value of the electrostatic potential $\varphi(r)$ over the surface of any spherical sub-volume inside V is equal to the potential at the center of the sub-volume.

4.22 The Potential outside a Charged Disk The z-axis is the symmetry axis of a disk of radius R which lies in the x-y plane and carries a uniform charge per unit area σ. Let Q be the total charge on the disk.

(a) Evaluate the exterior multipole moments and show that

$$\varphi(r, \theta) = \frac{Q}{4\pi \epsilon_0 r} \sum_{\ell=0}^{\infty} \left(\frac{R}{r}\right)^{\ell} \frac{2}{\ell + 2} P_\ell(0) P_\ell(\cos\theta) \qquad r > R.$$

(b) Compute the potential at any point on the z-axis by elementary means and confirm that your answer agrees with part (a) when $z > R$. Note: $P_\ell(1) = 1$.

4.23 Exterior Multipoles for a Specified Potential on a Sphere

(a) Let $\varphi(R, \theta, \phi)$ be specified values of the electrostatic potential on the surface of a sphere. Show that the general form of an exterior, spherical multipole expansion implies that

$$\varphi(\mathbf{r}) = \sum_{\ell=0}^{\infty} \sum_{m=-\ell}^{\ell} \left(\frac{R}{r}\right)^{\ell+1} Y_{\ell m}(\Omega) \int d\Omega' \varphi(R, \Omega') Y^*_{\ell m}(\Omega') \qquad r > R.$$

(b) The eight octants of a spherical shell are maintained at alternating electrostatic potentials $\pm V$ as shown below in perspective view (a) and looking down the z-axis from above in (b).

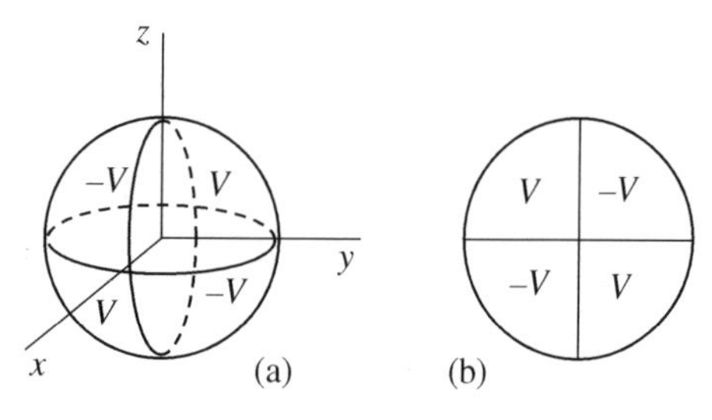

Use the results of part (a) to find the asymptotic ($r \to \infty$) form of the potential produced by the shell.

4.24 A Hexagon of Point Charges Six point charges form an ideal hexagon in the $z = 0$ plane as shown below. The absolute values of the charges are the same, but the signs of any two adjacent charges are opposite.

(a) What is the first non-zero electric multipole moment of this charge distribution? You need not compute its value.

(b) The electrostatic potential far from this distribution varies as $\varphi(\mathbf{r}) \propto r^{-N}$. What is N?

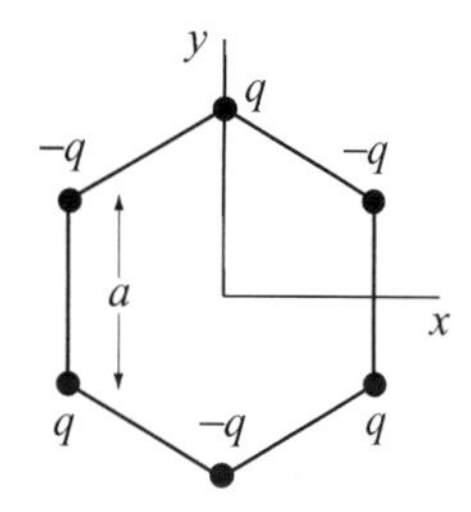

4.25 Analyze This Potential An asymptotic (long-distance) electrostatic potential has the form

$$\varphi(r, \theta, \phi) = \frac{A}{r} + \frac{Bx^2}{r^5} + \text{higher-order terms.}$$

(a) Use a *traceless* Cartesian multipole expansion to show that no *localized* charge distribution exists which can produce an asymptotic potential of this form.

(b) Repeat part (a) using a *primitive* Cartesian multipole expansion.

(c) Show that a suitable $\rho(\mathbf{r})$ can be found which produces the given potential if we remove the restriction that the charge distribution is localized.

5 Conducting Matter

It is evident that the electric fluid in conductors may be considered as moveable.
Henry Cavendish (1771)

5.1 Introduction

A *perfect conductor* is a macroscopic model for real conducting matter with the property that static electric fields are completely excluded from its interior. Since $\epsilon_0 \nabla \cdot \mathbf{E}(\mathbf{r}) = \rho(\mathbf{r})$, a precise definition of a perfect conductor[1] is that both the macroscopic electric field and the macroscopic charge density vanish everywhere inside its volume V:

$$\mathbf{E}(\mathbf{r}) = \rho(\mathbf{r}) = 0 \qquad \mathbf{r} \in V. \tag{5.1}$$

This definition implies that all excess charge accumulates on the surface of a conductor in the form of a surface charge density $\sigma(\mathbf{r}_S)$. The simplest case is an isolated conducting sphere with net charge Q and radius R. By symmetry, the surface charge density is uniform so $\sigma = Q/4\pi R^2$. The Gauss' law electric field outside such a sphere is identical to the electric field of a point charge. In light of (5.1), the continuous electrostatic potential of the sphere is

$$\varphi(r) = \begin{cases} \dfrac{Q}{4\pi\epsilon_0 R} & r \leq R, \\[2ex] \dfrac{Q}{4\pi\epsilon_0 r} & r \geq R. \end{cases} \tag{5.2}$$

5.2 Electrostatic Induction

Cavendish (1771) conceived the idea of a perfect conductor to rationalize the electrostatic properties of metals. In the presence of an external field $\mathbf{E}_{\text{ext}}$, his idea was that the Coulomb force rearranges charge inside the metal until the condition (5.1) is satisfied. This rearrangement process is called *polarization* in the general case and *electrostatic induction* when applied to conductors. The equilibrium state corresponds to zero net force on every particle. In a metal, the charge density $\rho(\mathbf{r})$ of "mobile" conduction electrons is associated with quantum mechanical wave functions which spread out over every atom in the sample. The force density $\rho(\mathbf{r})\mathbf{E}_{\text{ext}}(\mathbf{r})$ distorts the wave functions in such a way that charges with opposite signs are displaced in opposite directions. This force competes with potential energy

[1] Henceforth, the word "conductor" will always mean "perfect conductor".

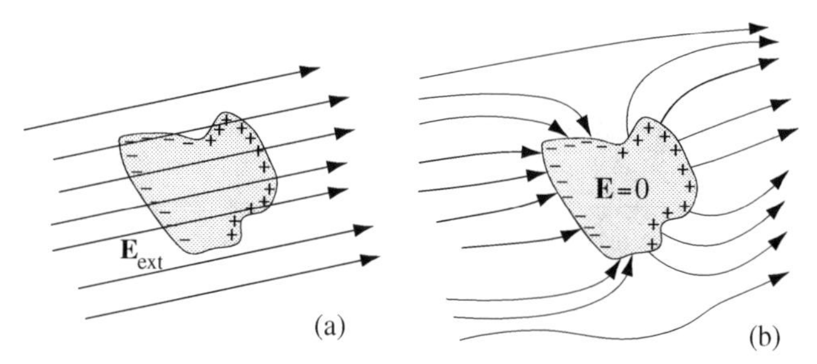

Figure 5.1: Electrostatic induction in a neutral conductor. (a) External field $\mathbf{E}_{\text{ext}}(\mathbf{r})$ induces charge rearrangement; (b) The charge distribution that produces mechanical equilibrium creates a field $\mathbf{E}_{\text{self}}(\mathbf{r})$ (not shown) with the property that $\mathbf{E}(\mathbf{r}) = \mathbf{E}_{\text{ext}}(\mathbf{r}) + \mathbf{E}_{\text{self}}(\mathbf{r}) = 0$ at every point inside the conductor. $\mathbf{E}_{\text{self}}(\mathbf{r})$ is generally dipolar far outside the conductor. Minus (plus) signs indicate a local surfeit (deficit) of electrons.

effects that favor chemical bond formation. Metals are easier to polarize than dielectrics (Chapter 6) because their electronic wave functions feel a relatively smoother potential energy landscape. A perfect conductor is maximally polarizable because, by definition, it possesses a perfectly flat potential energy landscape which offers no barriers to electrostatic induction.

Electrostatic induction in a real metal does *not* involve any long-range displacement of charge. Equilibrium is re-established by tiny perturbations of the conduction electron wave functions at every point in the conductor. When summed over all occupied states, the corresponding perturbed charge density Lorentz averages to zero at all interior points. This is the meaning of (5.1). At all surface points, the perturbed charge density amounts to tiny excesses or deficits of electronic charge which Lorentz average to the macroscopic surface charge density $\sigma(\mathbf{r}_S)$ (cf. Section 2.3.2).

The electric field *outside* a conductor can be quite complicated because it depends on the shape of the conductor, its net charge, and whether or not an external field $\mathbf{E}_{\text{ext}}(\mathbf{r})$ is present. Conversely, the field *inside* a conductor always satisfies (5.1). Therefore, if $\mathbf{E}_{\text{self}}(\mathbf{r})$ labels the electric field produced by $\sigma(\mathbf{r}_S)$, a more explicit version of (5.1) is

$$\mathbf{E}(\mathbf{r}) = \mathbf{E}_{\text{ext}}(\mathbf{r}) + \mathbf{E}_{\text{self}}(\mathbf{r}) = \mathbf{E}_{\text{ext}}(\mathbf{r}) + \frac{1}{4\pi\epsilon_0}\int_S dS'\,\sigma(\mathbf{r}')\frac{\mathbf{r}-\mathbf{r}'}{|\mathbf{r}-\mathbf{r}'|^3} = 0 \qquad \mathbf{r} \in V. \tag{5.3}$$

Figure 5.1 illustrates the phenomenon of electrostatic induction for a neutral conductor immersed in a uniform external field. Panel (a) shows the polarization of charge induced by $\mathbf{E}_{\text{ext}}$. This charge creates a field which induces additional charge rearrangement. The latter alters the induced field, which alters the induced charge distribution, and so on until mechanical equilibrium is achieved. The final surface charge distribution $\sigma(\mathbf{r}_S)$ creates the field $\mathbf{E}_{\text{self}}(\mathbf{r})$ in (5.3). Inside the conductor, $\mathbf{E}_{\text{self}}(\mathbf{r}) = -\mathbf{E}_{\text{ext}}(\mathbf{r})$. Far outside the conductor, $\mathbf{E}_{\text{self}}(\mathbf{r})$ is a dipole electric field parameterized by the macroscopic dipole moment

$$\mathbf{p} = \int_S dS\,\mathbf{r}\,\sigma(\mathbf{r}). \tag{5.4}$$

Figure 5.1(b) shows the total field (5.3) after mechanical equilibrium has been re-established. The field outside the conductor is recognizably distorted from a uniform field by $\mathbf{E}_{\text{self}}(\mathbf{r})$. We will see in Example 5.1 that the correction to the exterior field is exactly dipolar for a spherical conductor. Quadrupole and higher multipole contributions are present for non-spherical conductors.

5.2.1 Thomson's Theorem of Electrostatics

All physical systems in mechanical equilibrium arrange their charge density to minimize their total energy. Physical systems differ only in the degree to which quantum mechanical contributions to the energy compete with electrostatic contributions to the energy to dictate the arrangement of charge.[2] By definition, a perfect conductor is a system where quantum mechanics contributes negligibly to the total energy (except to fix the size and shape of the body). This implies that the fundamental origin of electrostatic induction and the screening and shielding properties of conductors must come from minimizing the classical electrostatic energy alone. This is the content of Thomson's theorem of electrostatics, which minimizes U_E with respect to $\rho(\mathbf{r})$ subject only to holding the total charge, shape, and volume of the sample fixed.

Statement: The electrostatic energy of a body of fixed shape and size is minimized when its charge Q distributes itself to make the electrostatic potential *constant* throughout the body. Since $\mathbf{E} = -\nabla\varphi$, this is the $\mathbf{E} = 0$ condition for the interior of a perfect conductor.

Proof: Let $\rho(\mathbf{r})$ be the charge density of a body of volume V. If $\varphi_{\text{ext}}(\mathbf{r})$ is the potential due to any fixed external charges, the total electrostatic energy is given by (3.84):

$$U_E[\rho] = \frac{1}{8\pi\epsilon_0}\int_V d^3r \int_V d^3r' \frac{\rho(\mathbf{r})\rho(\mathbf{r}')}{|\mathbf{r}-\mathbf{r}'|} + \int_V d^3r\, \rho(\mathbf{r})\varphi_{\text{ext}}(\mathbf{r}). \tag{5.5}$$

Our aim is to minimize U_E subject to the constraint that $Q = \int_V d^3r\, \rho(\mathbf{r})$. The method of Lagrange multipliers (Section 1.10) achieves this constrained minimization by introducing a constant λ and performing an *unconstrained* minimization of

$$I[\rho] = U_E[\rho] - \lambda \int_V d^3r\, \rho(\mathbf{r}). \tag{5.6}$$

This means that, to first order in $\delta\rho$, we insist that

$$\delta I = I[\rho + \delta\rho] - I[\rho] = 0. \tag{5.7}$$

Both $\rho(\mathbf{r})$ and $\rho(\mathbf{r}')$ appear in $I[\rho]$. Therefore, to calculate δI, we let $\rho(\mathbf{r}) \to \rho(\mathbf{r}) + \delta\rho(\mathbf{r})$ and $\rho(\mathbf{r}') \to \rho(\mathbf{r}') + \delta\rho(\mathbf{r}')$ in (5.6). Neglecting terms of order $(\delta\rho)^2$, the result is

$$\begin{aligned}\delta I &= \frac{1}{8\pi\epsilon_0}\int_V d^3r \int_V d^3r' \left[\frac{\rho(\mathbf{r})\delta\rho(\mathbf{r}')}{|\mathbf{r}-\mathbf{r}'|} + \frac{\delta\rho(\mathbf{r})\rho(\mathbf{r}')}{|\mathbf{r}-\mathbf{r}'|}\right] \\ &\quad + \int_V d^3r\, \delta\rho(\mathbf{r})\, \varphi_{\text{ext}}(\mathbf{r}) - \lambda \int_V d^3r\, \delta\rho(\mathbf{r}).\end{aligned} \tag{5.8}$$

Exchanging the variables $\mathbf{r}$ and $\mathbf{r}'$ in the first term in square brackets in (5.8) gives

$$\delta I = \int_V d^3r\, \delta\rho(\mathbf{r}) \left\{ \frac{1}{4\pi\epsilon_0}\int_V d^3r' \frac{\rho(\mathbf{r}')}{|\mathbf{r}-\mathbf{r}'|} + \varphi_{\text{ext}}(\mathbf{r}) - \lambda \right\} = 0. \tag{5.9}$$

Finally, the variation $\delta\rho(\mathbf{r})$ is arbitrary, so the quantity in curly brackets must vanish:

$$\frac{1}{4\pi\epsilon_0}\int_V d^3r' \frac{\rho(\mathbf{r}')}{|\mathbf{r}-\mathbf{r}'|} + \varphi_{\text{ext}}(\mathbf{r}) = \lambda \qquad \mathbf{r} \in V. \tag{5.10}$$

[2] The energy gained by the formation of chemical bonds is an example of a quantum mechanical contribution to the total energy.

The left side of (5.10) is $\varphi_{\rm tot}(\mathbf{r})$, the total electrostatic potential at a point $\mathbf{r} \in V$. The right side of (5.10) is the constant Lagrange parameter λ. Therefore, the energy extremum occurs when $\varphi_{\rm tot}(\mathbf{r})$ assumes the constant value λ throughout V. The fact that the neglected terms of order $(\delta\rho)^2$ are positive definite tells us that (5.10) corresponds to an energy minimum rather than an energy maximum.

Example 5.1 Find the potential produced by a conducting sphere of radius R in a uniform electric field $\mathbf{E}_0$. Show also that the sphere acquires a dipole moment $\mathbf{p} = \alpha\epsilon_0\mathbf{E}_0$ where $\alpha = 4\pi R^3$. The constant α is called the *polarizability*.

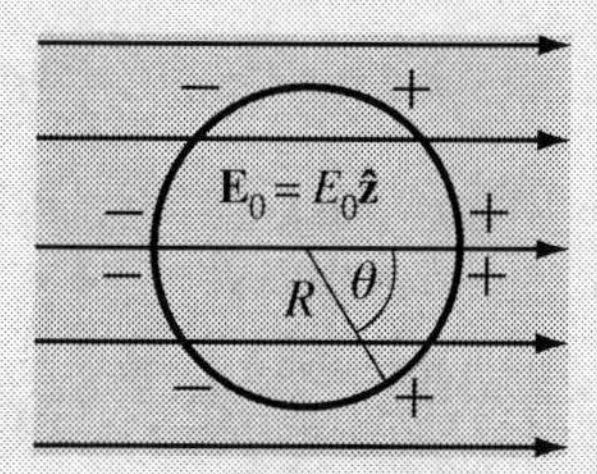

Figure 5.2: A conducting sphere in a uniform external field $E_0\hat{\mathbf{z}}$. Plus and minus signs indicate the induced surface charge density $\sigma(\theta)$.

Solution: By symmetry, a uniform external field induces an azimuthally symmetric charge density $\sigma(\theta)$ on the surface of the sphere. This is shown in Figure 5.2. Inside the conducting sphere, $\sigma(\theta)$ must produce an electric field $\mathbf{E}_{\rm self} = -\mathbf{E}_0$. Since $\mathbf{E}_0 = E_0\hat{\mathbf{z}}$, the corresponding potential is

$$\varphi_{\rm self}(r < R, \theta) = E_0 z = E_0 r \cos\theta.$$

Outside the sphere, $\sigma(\theta)$ produces a potential that can be represented using an exterior azimuthal multipole expansion (Section 4.6.3). The first few terms of such an expansion are

$$\varphi_{\rm self}(r > R, \theta) = \frac{A}{r} + \frac{B}{r^2}\cos\theta + \frac{C}{r^3}P_2(\cos\theta) + \cdots.$$

The potential must be continuous at $r = R$. Therefore, given $\varphi_{\rm self}(r < R, \theta)$ above, the only possibility for the exterior potential is

$$\varphi_{\rm self}(r > R, \theta) = \frac{E_0 R^3}{r^2}\cos\theta.$$

On the other hand, the potential of a point electric dipole with dipole moment $\mathbf{p} = p\,\hat{\mathbf{z}}$ is

$$\varphi(r, \theta) = \frac{1}{4\pi\epsilon_0}\frac{p}{r^2}\cos\theta.$$

Comparing the two preceding potentials shows that $\varphi_{\rm self}(r > R, \theta)$ is *exactly* dipolar (true only for a sphere) and that the induced electric dipole moment is indeed $\mathbf{p} = \alpha\epsilon_0\mathbf{E}_0$ with polarizability $\alpha = 4\pi R^3$. This conforms to a general rule that the polarizability of a compact polarizable object scales with its *volume*.

Example 5.2 A single conductor is formed by two uncharged metal spheres connected by an infinitesimally thin wire; see Figure 5.3. The sphere with radius R_1 is centered at $\mathbf{r}_1$ and the sphere with radius R_2 is centered at $\mathbf{r}_2$. Find an approximate expression for the dipole moment induced in the conductor by a uniform external field $\mathbf{E}_0$. Assume that the wire cannot support a net charge and that the spheres are separated from one another by a distance that is large compared to both R_1 and R_2.

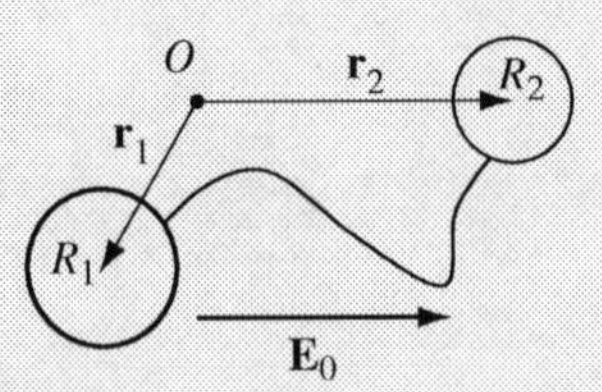

Figure 5.3: An external field $\mathbf{E}_0$ polarizes two distant metal spheres connected by a thin conducting wire.

Solution: The electric field polarizes the system but leaves no charge on the connecting wire. This means that sphere 1 acquires a net charge $-Q$ if sphere 2 acquires a net charge Q. If the separation between the spheres is large enough, the charge density on each may be taken as approximately uniform. This is the crucial approximation needed to write the potential at the center of each sphere as the sum of the external potential and its own monopole potential as given by (5.2):

$$\varphi_1 = -\mathbf{E}_0 \cdot \mathbf{r}_1 - \frac{Q}{4\pi\epsilon_0 R_1},$$

$$\varphi_2 = -\mathbf{E}_0 \cdot \mathbf{r}_2 + \frac{Q}{4\pi\epsilon_0 R_2}.$$

But, $\varphi_1 = \varphi_2$ because the two spheres are connected by a conducting wire. This fixes the value of Q at

$$Q = 4\pi\epsilon_0 \frac{R_1 R_2}{R_1 + R_2} \mathbf{E}_0 \cdot (\mathbf{r}_2 - \mathbf{r}_1).$$

The induced electric dipole moment is

$$\mathbf{p} = Q(\mathbf{r}_2 - \mathbf{r}_1) = Q\mathbf{d} = 4\pi\epsilon_0 \frac{R_1 R_2}{R_1 + R_2} (\mathbf{E}_0 \cdot \mathbf{d})\mathbf{d}.$$

The reader can check that the condition $\mathbf{E}_0 \cdot (\mathbf{r}_2 - \mathbf{r}_1) \gg E_0 R_1, E_0 R_2$, implies that the non-uniformity of the charge density on each sphere produces a negligible correction to the self-monopole potentials used in our solution.

5.2.2 The Surface Charge Density

The charge density $\sigma(\mathbf{r}_S)$ at the surface of a conductor enters the electric field matching conditions (written in terms of the outward normal $\hat{\mathbf{n}}$ to the conductor surface),

$$\hat{\mathbf{n}} \times [\mathbf{E}_{\text{out}} - \mathbf{E}_{\text{in}}] = 0 \tag{5.11}$$

$$\hat{\mathbf{n}} \cdot [\mathbf{E}_{\text{out}} - \mathbf{E}_{\text{in}}] = \sigma(\mathbf{r}_S)/\epsilon_0. \tag{5.12}$$

Because $\mathbf{E}_{\text{in}}(\mathbf{r}) = 0$, (5.11) tells us that the tangential component of $\mathbf{E}_{\text{out}}(\mathbf{r})$ is zero just outside a conducting surface:

$$\mathbf{n} \times \mathbf{E}_{\text{out}}|_S = 0. \tag{5.13}$$

With this information, (5.12) implies that the electric field just outside a conductor is strictly normal to the conductor surface with a magnitude proportional to the local surface charge density:

$$\mathbf{E}_{\text{out}}(\mathbf{r}_S) = \hat{\mathbf{n}}\,\sigma(\mathbf{r}_S)/\epsilon_0. \tag{5.14}$$

Equation (5.14) tells us that the charge density at the surface of a conductor is determined by the electric field immediately outside itself:

$$\sigma(\mathbf{r}_S) = \epsilon_0 \hat{\mathbf{n}} \cdot \mathbf{E}_{\text{out}}(\mathbf{r}_S). \tag{5.15}$$

A quick application of (5.15) is to compute the dipole moment of the polarized sphere in Example 5.1 directly from the integral (5.4). To find the surface charge density, we use the total potential (the sum of the external potential and the induced self-potential) outside the sphere found in Example 5.1:

$$\sigma(\theta) = -\epsilon_0 \frac{\partial}{\partial r}\left[-E_0 r \cos\theta + \frac{E_0 R^3}{r^2}\cos\theta\right]_{r=R} = 3\epsilon_0 E_0 \cos\theta. \tag{5.16}$$

Since $\mathbf{r} = r\sin\theta\cos\phi\hat{\mathbf{x}} + r\sin\theta\sin\phi\hat{\mathbf{y}} + r\cos\theta\hat{\mathbf{z}}$, only the z-component survives the angular integration and we confirm the result found in Example 5.1 that

$$\mathbf{p} = \int dS\,\sigma(\theta)\mathbf{r} = 2\pi R^3 \int_0^{\pi} d\theta\,\sin\theta\cos\theta\,\sigma(\theta)\hat{\mathbf{z}} = 4\pi\epsilon_0 R^3 E_0 \hat{\mathbf{z}}. \tag{5.17}$$

The surface charge density (5.16) is non-uniform because the external field breaks the symmetry of the sphere. Non-uniform surface charge density is similarly the rule for an isolated conductor with a non-spherical shape and a non-zero net charge. An unexpected result is that $\sigma(\mathbf{r}_S)$ *diverges* at the edge of conductors with sharp, knife-like edges. An example is a conducting disk with radius R, net charge Q, and a vanishingly small thickness, where[3]

$$\sigma(\rho) = \frac{Q}{4\pi R\sqrt{R^2-\rho^2}}. \tag{5.18}$$

We derive (5.18) in the next section for interested readers. The generality of the divergence phenomenon for conductors with sharp edges will emerge in Chapter 7.

Application 5.1 $\sigma(\mathbf{r}_S)$ for a Conducting Disk

There is no truly simple way to calculate the surface charge density (5.18) for a charged, conducting disk. In this Application, we use a method which regards the disk as the limiting case of a squashed ellipsoid. The role of the latter becomes clear when we recall the example studied in Section 3.3.5 of a uniformly charged line segment ($-L < z < L$) with total charge Q. In terms of the variables $r_\pm = \sqrt{\rho^2 + (z \pm L)^2}$ (see Figure 5.4) and

$$u = \tfrac{1}{2}(r_+ + r_-), \tag{5.19}$$

our main result from (3.33) was that the segment produced an electrostatic potential

$$\varphi(u) = \frac{Q/2L}{4\pi\epsilon_0}\ln\frac{u+L}{u-L}. \tag{5.20}$$

Equation (5.20) tells us that $u = b$ defines a family of equipotential surfaces parameterized by the constant b. On the other hand, $u = b$ written using (5.19) is the geometric definition of a prolate

[3] The thickness cannot be literally zero because an "interior" must be definable where (5.1) is valid.

ellipsoid of revolution:

$$\tfrac{1}{2}(r_+ + r_-) = b. \tag{5.21}$$

Therefore, by superimposing a conducting ellipsoid on any of the equipotential surfaces of the charged line segment, we can find the surface charge density of the ellipsoid using (5.20) and $\sigma = -\epsilon_0 \nabla\varphi \cdot \hat{\mathbf{n}}|_S$.

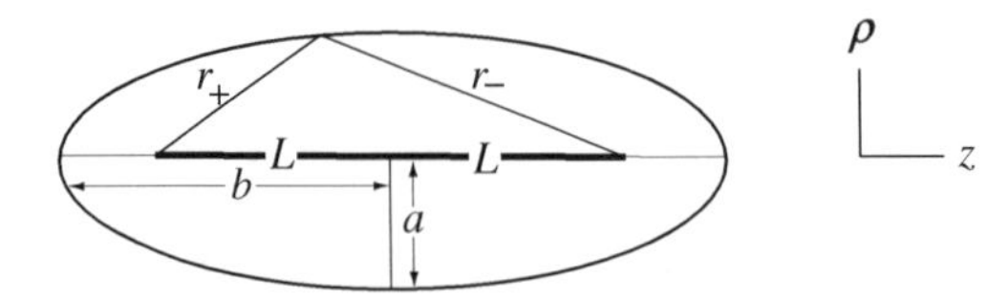

Figure 5.4: A prolate ellipsoid of revolution with semi-minor axis a and semi-major axis b. The focal length $L = \sqrt{b^2 - a^2}$. The line segments $r_\pm$ begin at opposite foci and end at the same point on the ellipsoid surface. The surface is the locus of points where $r_+ + r_- = 2b$.

The electric field associated with (5.20) is guaranteed to be normal to the surface of the ellipsoid. Therefore, since $\varphi = \varphi(\rho, z)$ and $\mathbf{E} = -\nabla\varphi$, (5.14) gives the charge density we seek as ($Q > 0$)

$$\sigma(\rho, z) = \epsilon_0 |\nabla\varphi|_S = \epsilon_0 \left.\sqrt{\left(\frac{\partial\varphi}{\partial z}\right)^2 + \left(\frac{\partial\varphi}{\partial\rho}\right)^2}\right|_S = \epsilon_0 \left|\frac{\partial\varphi}{\partial u}\right|_{u=b} \left.\sqrt{\left(\frac{\partial u}{\partial z}\right)^2 + \left(\frac{\partial u}{\partial\rho}\right)^2}\right|_S . \tag{5.22}$$

Now, $du = (\partial u/\partial\rho)d\rho + (\partial u/\partial z)dz = 0$ on S because $u = b$ on S. This transforms (5.22) to

$$\sigma(\rho, z) = \epsilon_0 \left|\frac{\partial\varphi}{\partial u}\right|_{u=b} \left|\frac{\partial u}{\partial\rho}\right|_S \left.\sqrt{1 + \left(\frac{d\rho}{dz}\right)^2}\right|_S . \tag{5.23}$$

The identity $L^2 = b^2 - a^2$ and straightforward differentiation give the first quantity we need to evaluate (5.23):

$$\left|\frac{\partial\varphi}{\partial u}\right|_{u=b} = \frac{Q}{4\pi\epsilon_0 a^2}. \tag{5.24}$$

Similarly,

$$\left|\frac{\partial u}{\partial\rho}\right|_S = \frac{1}{2}\left[\frac{\partial r_+}{\partial\rho} + \frac{\partial r_-}{\partial\rho}\right]_S = \left.\frac{b\rho}{r_+ r_-}\right|_S . \tag{5.25}$$

We evaluate (5.25) using two more relations from Section 3.3.5:

$$t = \frac{1}{2}(r_- - r_+) \qquad \text{and} \qquad ut = -zL. \tag{5.26}$$

Then, in light of (5.19) and the fact that $u = b$ defines S,

$$r_+ r_-|_S = u^2 - t^2\big|_S = \frac{1}{b^2}\left[b^4 - z^2(b^2 - a^2)\right]. \tag{5.27}$$

Multiplying the factor of b^4 in (5.27) by the left side of the equation for the surface of the ellipsoid,

$$\frac{\rho^2}{a^2} + \frac{z^2}{b^2} = 1, \tag{5.28}$$

and substituting the result into (5.25) gives

$$\left|\frac{\partial u}{\partial\rho}\right|_S = \frac{\rho}{a^2 b}\left[\frac{\rho^2}{a^4} + \frac{z^2}{b^4}\right]^{-1}. \tag{5.29}$$

Finally, because (5.28) guarantees that $\rho d\rho/a^2 + zdz/b^2 = 0$ on the surface of the ellipsoid,

$$\left.\frac{d\rho}{dz}\right|_S = -\frac{a^2}{b^2}\frac{z}{\rho}. \tag{5.30}$$

Substituting (5.24), (5.29), and (5.30) into (5.23) gives the formula we seek for the density of charge on the surface of a conducting prolate ellipsoid:

$$\sigma(\rho, z) = \frac{Q}{4\pi a^2 b}\frac{1}{\sqrt{\rho^2/a^4 + z^2/b^4}}. \tag{5.31}$$

This formula turns out to be valid for an oblate ($b < a$) ellipsoid as well [Stratton (1941)]. Therefore, since the limit $b \to 0$ flattens the ellipsoid into a flat circular disk, we use (5.28) to rewrite (5.31) as

$$\sigma(\rho, z) = \frac{Q}{4\pi a^2}\frac{1}{\sqrt{\rho^2 b^2/a^4 + (1 - \rho^2/a^2)}}. \tag{5.32}$$

The $b \to 0$ limit yields the surface charge density of a flat conducting disk of radius a:

$$\sigma(\rho) = \frac{Q}{4\pi a\sqrt{a^2 - \rho^2}}. \tag{5.33}$$

Geometrically, $\sigma(\rho)$ is the charge per unit area that results when a hemisphere with uniform charge density $Q/4\pi a^2$ is projected onto its equatorial plane. As advertised, $\sigma(\rho)$ diverges at the perimeter of the disk. We will see in Chapter 7 that this behavior is neither exceptional nor particularly unexpected for conductors with macroscopic knife-edges. The divergence reduces to a simple maximum for a real metal disk with finite rounding. ■

5.3 Screening and Shielding

Conductors have the unique ability to *screen* or *shield* a suitably placed sample from the effects of an electric field. By this we mean that a conductor interposed between a sample and a source of electric field generally reduces (and ideally eliminates) the field at the position of the sample. The canonical example of this phenomenon is a neutral conductor with a vacuum cavity "scooped" out of its interior. Figure 5.5(a) shows such a conductor together with a point charge external to its volume. By the definition of a conductor, the electric field is zero at every point in or on the conductor body. The key to shielding is that $\mathbf{E}(\mathbf{r}) = 0$ inside the cavity also.

We prove this assertion by supposing that the electric field inside the cavity is not zero. Every line of $\mathbf{E}_{\text{cavity}}(\mathbf{r})$ must begin on positive charge at some particular point A on the cavity boundary and end on negative charge at a distinct point B on the cavity boundary. We now perform a line integral around the dashed closed path in Figure 5.5(a) which begins at A, follows the path of the presumptive field line inside the cavity to B, and then returns to A by a path that lies entirely inside the body of the conductor. This line integral is zero for any electrostatic field (Section 3.3.1). Therefore, since the field is zero in the body of the conductor,

$$0 = \oint d\boldsymbol{\ell} \cdot \mathbf{E} = \int_A^B d\ell\, |\mathbf{E}_{\text{cavity}}| + \int_B^A d\boldsymbol{\ell} \cdot \mathbf{E}_{\text{conductor}} = \int_A^B d\ell\, |\mathbf{E}_{\text{cavity}}|. \tag{5.34}$$

We conclude from this that $\mathbf{E}_{\text{cavity}} = 0$ because the integrand on the far right side of (5.34) is positive definite.

The fact that $\mathbf{E}_{\text{cavity}} = 0$ implies that any object placed inside the cavity is completely shielded from the electrostatic effects of the exterior point charge q. The field responsible for the shielding

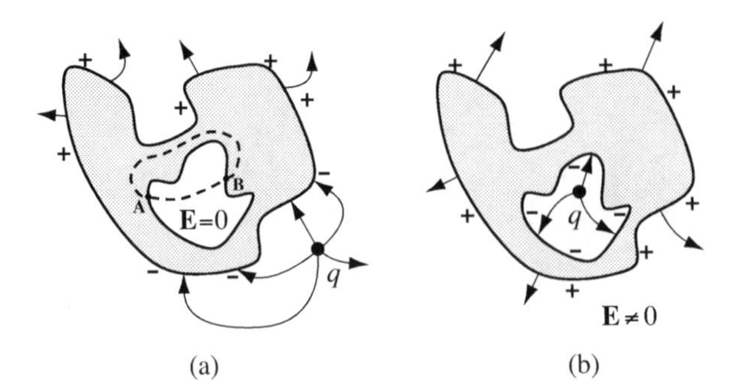

Figure 5.5: A perfect conductor with a cavity. (a) The electric field inside an empty cavity is always zero. If there is source charge outside the conductor, the outer surface of the conductor develops a surface charge density. See text for discussion of the dashed closed loop. (b) The electric field outside the conductor is not zero when there is source charge inside the cavity. The conductor develops a surface charge density on the cavity surface and on the outer surface of the conductor.

is produced by a surface charge density which develops on the outer surface of the conductor by electrostatic induction. Every electric field line that leaves q terminates on the outer surface of the conductor or goes off to infinity. No charge develops on the boundary of the cavity because $\mathbf{E}_{\text{cavity}}$ plays the role of $\mathbf{E}_{\text{out}}$ in (5.15). This type of screening has tremendous practical importance, particularly so because walls made from a metallic mesh shield electrostatic fields almost as effectively as solid metal walls (see Section 7.5.1).

Conversely, a conductor does *not* screen the space outside of itself from charge placed *inside* a scooped-out cavity. This is illustrated by Figure 5.5(b) where the point charge in the cavity induces a surface charge density on both the walls of the cavity and on the exterior surface of the conductor. The latter is necessary because Gauss' law requires non-zero electric flux to pass through any Gaussian surface that encloses the conductor. It is not obvious, but the field outside the conductor does not depend on the exact position of the point charge inside the cavity.[4]

5.4 Capacitance

Conductors are used to store electric charge because their surfaces (where the charge resides) are easily accessible. The concept of *capacitance* measures the quantitative capacity of any particular conductor to store charge, whether in isolation or in the presence of other conductors.

5.4.1 Self-Capacitance

Let V be the potential of an isolated conductor with volume Ω and surface S. Thomson's theorem (Section 5.2.1) and Coulomb's law ensure that the electrostatic potential produced by the conductor is $\varphi(\mathbf{r}) = V\tilde{\varphi}(\mathbf{r})$ where $\tilde{\varphi}(\mathbf{r} \in \Omega) = 1$ and $\tilde{\varphi}(|\mathbf{r}| \to \infty) \to 0$. Gauss' law and $\mathbf{E} = -\nabla\varphi$ give the total charge on the conductor as

$$Q = -\epsilon_0 V \int_S d\mathbf{S} \cdot \nabla\tilde{\varphi}. \tag{5.35}$$

[4] It may amuse the reader to try to prove this statement using only the information provided so far. We give a proof in Section 7.3.1.

The proportionality of Q to V in (5.35) leads us to define a purely geometrical quantity called the *self-capacitance*:

$$C = \frac{Q}{V} = -\epsilon_0 \int_S d\mathbf{S} \cdot \nabla \tilde{\varphi}. \tag{5.36}$$

The self-capacitance of a sphere follows immediately from (5.36) because (5.2) gives the potential at its surface as $Q/4\pi\epsilon_0 R$. Hence,

$$C = 4\pi\epsilon_0 R \qquad \text{(sphere of radius } R\text{)}. \tag{5.37}$$

For an origin-centered conducting disk, we can use the surface charge density (5.33) to compute the potential at the center of the disk from

$$\varphi(0) = \frac{1}{4\pi\epsilon_0} \int_S dS' \frac{\sigma(\mathbf{r}'_S)}{|\mathbf{r}'_S|} = V. \tag{5.38}$$

Bearing in mind that both sides of the disk carry a surface charge $\sigma(\rho)$, (5.38) reduces to

$$V = \frac{2}{4\pi\epsilon_0} \frac{Q}{4\pi R} \int_0^{2\pi} d\phi \int_0^R \frac{d\rho}{\sqrt{R^2 - \rho^2}} = \frac{Q}{8\epsilon_0 R}. \tag{5.39}$$

Using (5.36), the self-capacitance of the disk is

$$C = 8\epsilon_0 R \qquad \text{(disk of radius } R\text{)}. \tag{5.40}$$

On dimensional grounds alone, the self-capacitance must be proportional to $\epsilon_0 \times$ length. The charge on a conductor is confined to its surface, so we might guess that $C \propto \epsilon_0 \sqrt{A}$ where A is the surface area. Indeed, the simple formula

$$C \simeq \epsilon_0 \sqrt{4\pi A} \tag{5.41}$$

turns out to be a remarkably good approximation, even for open structures.[5] Figure 5.6 illustrates the quality of (5.41) for the case of an ellipsoid of revolution.

5.4.2 What Does it Mean to "Ground" a Conductor?

The Earth is a very large and spherical conductor whose self-capacitance (5.37) is extraordinarily large compared to any laboratory conductor. Therefore, because the net charge of the Earth (due to lightning activity) is finite, we can use (5.36) to put $C_{\text{Earth}} \to \infty$ and $\varphi_{\text{Earth}} \to 0$ as a first approximation. This amounts to treating the Earth as a charge reservoir because finite amounts of charge may be added to or taken from its surface without appreciably changing its potential from zero.

We "ground" a conductor (fix its potential at zero) by connecting it to Earth using a fine conducting wire. Figure 5.7 illustrates two common situations. If the conductor initially has a net charge Q, the connection to ground causes this charge to spontaneously flow to the Earth. This lowers the potential energy of every charge on the conductor and leads to a final equilibrium state of $Q = \varphi = 0$ for the conductor. Conversely, an initially uncharged conductor will draw charge "up from ground" due to Coulomb attraction if a charge q is placed near the conductor. This example of electrostatic induction similarly lowers the energy of the connected system of conductor + ground.

[5] See Chow and Yovanovich (1982) in Sources, References, and Additional Reading.

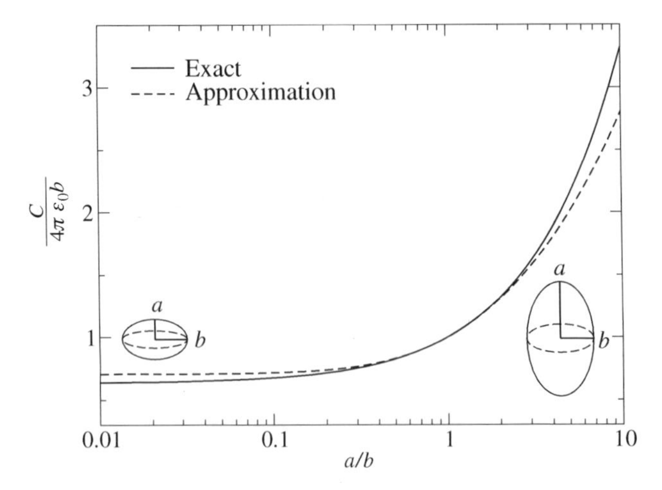

Figure 5.6: Normalized self-capacitance of an ellipsoid of revolution as a function of the aspect ratio a/b (note logarithmic scale). The dimension a is varied while b is held constant. Dashed curve is (5.41). Solid curve is the exact result from Landau and Lifshitz (1960).

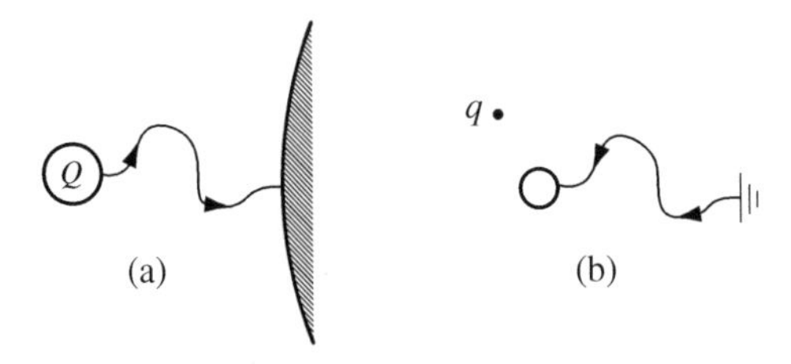

Figure 5.7: A spherical conductor connected to the Earth by a fine conducting wire. The arrows on the wire show that (a) a conductor with initial charge Q spontaneously loses its charge to Earth when grounded; (b) an initially uncharged conductor draws charge up from ground in the presence of a nearby charge q. The symbol for ground is used in panel (b).

5.4.3 The Capacitance Matrix

A modern integrated circuit contains thousands of metallic contacts. A *capacitance matrix* **C** describes how the contacts influence one another electrostatically. The matrix elements C_{ij} (which must be measured or calculated) generalize (5.36) by relating the conductor charges Q_i to the conductor potentials φ_j for a collection of N conductors:

$$Q_i = \sum_{j=1}^{N} C_{ij}\varphi_j. \tag{5.42}$$

To see that (5.42) is correct, Figure 5.8 shows a conductor with potential φ_1 in the presence of $N-1$ other conductors held at zero potential by connection to ground. The charge on conductor 1 associated with φ_1 induces each zero-potential conductor to acquire a net charge of its own, drawn up from ground.[6] Specifically,

$$Q_1^{(1)} = C_{11}\varphi_1 \qquad Q_2^{(1)} = C_{21}\varphi_1 \qquad \cdots \qquad Q_N^{(1)} = C_{N1}\varphi_1, \tag{5.43}$$

[6] Technically, we need a theorem from Section 7.3 to guarantee that the charge drawn up to each grounded conductor is uniquely determined by the potential φ_1.

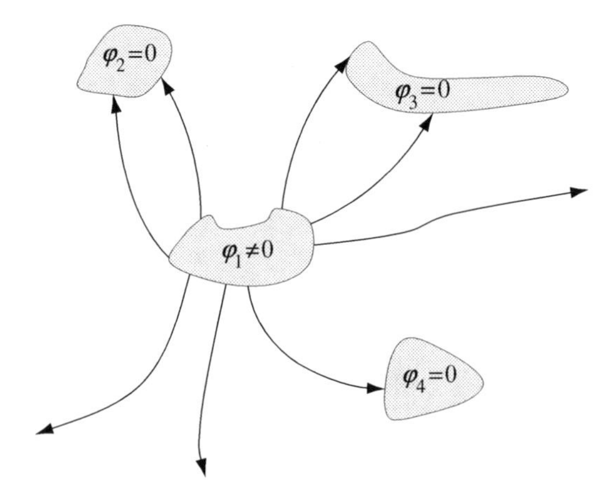

Figure 5.8: Sketch of four conductors. Conductor 1 is held at a non-zero potential φ_1. The other three conductors are held at zero potential. A few representative electric field lines are indicated for the case when $\varphi_1 > 0$.

where the *coefficients of capacitance* $C_{11}, C_{21} \ldots, C_{N1}$ depend only on the shapes and geometrical arrangement of the conductors. If the wires used to ground the conductors are sufficiently thin, the charges and potentials in (5.43) do not change if we remove these wires after static equilibrium has been reached.

Now consider a second situation where conductor 2 is held at potential $\varphi_2 \neq 0$ while all the other conductors are held at zero potential by connection to ground. The same considerations apply as above (including the removal of the physical connections to ground at the end) and we infer that the charges on each conductor are

$$Q_1^{(2)} = C_{12}\varphi_2 \qquad Q_2^{(2)} = C_{22}\varphi_2 \qquad \ldots. \qquad Q_N^{(2)} = C_{N2}\varphi_2. \tag{5.44}$$

Repeating this scenario $N-2$ times, each time grounding only one conductor, produces $N-2$ more expressions like (5.43) and (5.44) for the charge on each conductor. The final step is to superpose all of the foregoing solutions to produce a situation where conductor 1 has potential φ_1, conductor 2 has potential φ_2, etc. We get the desired formula (5.42) by straightforward addition because the charge on the kth conductor is

$$Q_k = Q_k^{(1)} + Q_k^{(2)} + \cdots + Q_k^{(N)}. \tag{5.45}$$

The diagonal elements C_{kk} of the capacitance matrix are *not* simply the self-capacitances discussed in Section 5.4. This can be seen from Figure 5.8 where the charge induced on the grounded conductors influences the potential of the non-grounded conductor. The off-diagonal elements C_{kj} are called *mutual* or *cross capacitances*. The most important property of these quantities is that they are symmetric in their indices:

$$C_{ij} = C_{ji}. \tag{5.46}$$

We prove (5.46) using Green's reciprocity relation (Section 3.5.2), which relates a charge distribution $\rho(\mathbf{r})$ and its potential $\varphi(\mathbf{r})$ to an entirely different charge distribution $\rho'(\mathbf{r})$ and its potential $\varphi'(\mathbf{r})$:

$$\int d^3r\, \varphi(\mathbf{r})\rho'(\mathbf{r}) = \int d^3r\, \varphi'(\mathbf{r})\rho(\mathbf{r}). \tag{5.47}$$

For a collection of N conductors, reciprocity relates a situation where the conductors possess charges Q_i and potentials φ_i to an entirely different situation where the *same* set of conductors possess charges

Q_i' and potentials φ_i'. This reduces (5.47) to

$$\sum_{i=1}^{N} \varphi_i Q_i' = \sum_{i=1}^{N} \varphi_i' Q_i. \tag{5.48}$$

On the other hand, (5.42) applies to both Q_i and Q_i' in (5.48). Therefore,

$$\sum_{i=1}^{N}\sum_{j=1}^{N} \varphi_i C_{ij} \varphi_j' = \sum_{i=1}^{N}\sum_{j=1}^{N} \varphi_i' C_{ij} \varphi_j. \tag{5.49}$$

Exchanging the dummy indices on one side of (5.49) proves that (5.46) is correct.

The elements of the capacitance matrix satisfy the *Maxwell inequalities*:

$$\begin{aligned} C_{kk} &> 0 \\ C_{kj} &< 0 \\ \sum_j C_{kj} &\geq 0. \end{aligned} \tag{5.50}$$

These inequalities can be understood from a version of Figure 5.8 where the kth conductor is held at unit positive potential and all the other conductors are held at zero potential. For this situation, all the electric field lines must begin on conductor k and end on one of the other conductors (or at infinity). In particular, no field lines directly connect any other pair of conductors. The inequalities (5.50) follow from the facts that $Q_m = C_{mk}$ and that the number of field lines that leave each conductor is proportional to the total charge on that conductor (Gauss' law). We get $\sum_j C_{kj} = 0$ only when no field lines escape to infinity.

Example 5.3 Figure 5.9(a) shows a point charge q a perpendicular distance z_0 from one of two parallel, infinite, and grounded conducting plates separated by a distance d. Use Green's reciprocity to find the amount of charge drawn up from ground by each plate.

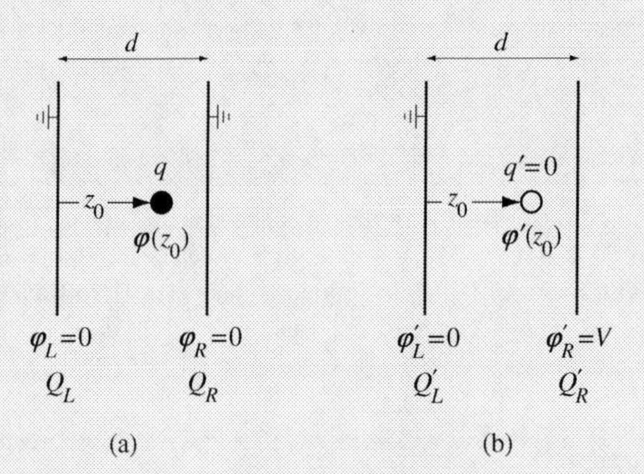

Figure 5.9: Application of Green's reciprocity relation to find the charges Q_L and Q_R induced on two infinite, grounded plates by an interposed point charge q.

Solution: Treat the point charge in Figure 5.9(a) as an infinitesimal spherical conductor placed at $z = z_0$ with potential $\varphi(z_0)$ and charge q. The planes at $z = 0$ and $z = d$ have charges Q_L and Q_R and potentials φ_L and φ_R, respectively. If we use the same notation for the "primed" charges and potentials, the reciprocity relation (5.48) with $N = 3$ reads

$$\varphi_L Q'_L + \varphi_R Q'_R + \varphi(z_0)q' = \varphi'_L Q_L + \varphi'_R Q_R + \varphi'(z_0)q.$$

The trick to using reciprocity is to choose physically sensible values for the primed variables in order to isolate the desired quantities and eliminate the unwanted quantities from the preceding equation. Since $\varphi(z_0)$ is unknown, we choose $q' = 0$ and fix the potentials on the plates at $\varphi'_L = 0$ and $\varphi'_R = V$ [see Figure 5.9(b)]. Since $\varphi_L = \varphi_R = 0$, these choices reduce the preceding equation to

$$0 = VQ_R + \varphi'(z_0)q.$$

By translational invariance, the charge density is uniform on the primed plates. This means that $\mathbf{E}'$ is constant and $\varphi'(z) = Az + B$ between the plates. The boundary values $\varphi'_L(0) = 0$ and $\varphi'_R(d) = V$ fix the constants so $\varphi'(z_0) = z_0 V/d$. This determines Q_R. A second comparison system with $\varphi''_L = V$ and $\varphi''_R = 0$ determines Q_L. Altogether, we find

$$Q_L = -\left(1 - \frac{z_0}{d}\right)q \qquad \text{and} \qquad Q_R = -\frac{z_0}{d}q.$$

The charge induced on each plate is directly proportional to the distance between q and the other plate. The total charge drawn up by both plates is $-q$. This implies that every electric field line that begins on the point charge ends on one of the plates. We will explore the generality of this result in Chapter 7.

Example 5.4 $N + 1$ identical conductors labeled $m = 0, 1, \ldots, N$ are arranged in a straight line. The capacitance matrix is

$$C_{ij} = \begin{cases} C_0 & i = j, \\ -C & |i - j| = 1, \\ 0 & |i - j| > 1. \end{cases}$$

Suppose conductor $m = 0$ at the end of the line has charge Q and all the other conductors are uncharged. Find an approximate expression for the potential φ_m on every conductor in the limit $C \ll C_0$.

Solution: The charge-potential relations (5.42) are

$$\begin{aligned} Q &= C_0\varphi_0 - C\varphi_1 \\ 0 &= C_0\varphi_m - C(\varphi_{m+1} + \varphi_{m-1}) \qquad m \neq 0, N \\ 0 &= C_0\varphi_N - C\varphi_{N-1}. \end{aligned}$$

The last equation is $\varphi_N = (C/C_0)\varphi_{N-1}$. Substituting this into the middle equation with $m = N - 1$ and using $C \ll C_0$ produces

$$\left(\frac{C_0}{C} - \frac{C}{C_0}\right)\varphi_{N-1} = \varphi_{N-2} \qquad \text{or} \qquad \varphi_{N-1} \approx (C/C_0)\varphi_{N-2}.$$

Repeating this argument gives $\varphi_m = (C/C_0)\varphi_{m-1}$. Therefore, $\varphi_m = (C/C_0)^m\varphi_0$. But $Q = C_0\varphi_0$ in the same limit, so

$$\varphi_m \approx \frac{Q}{C_0}\left(\frac{C}{C_0}\right)^m = \frac{Q}{C_0}\exp\{-m\ln(C_0/C)\}.$$

The Triode

Before semiconductor technology made low-power vacuum tubes obsolete, the *triode* performed many of the same functions as the transistor. In this device, a current in the form of a stream of electrons flows through an enclosed vacuum between a heated *cathode* and an *anode*. The figure below is the circuit diagram symbol for a triode. The essential feature is that a voltage applied to a metal *grid* controls the magnitude of the current.

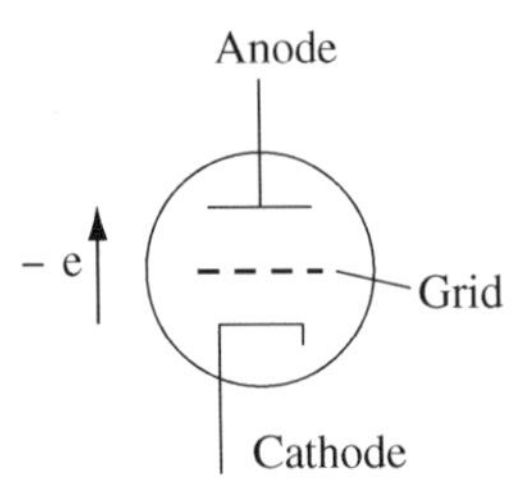

The operation of the triode can be understood qualitatively as follows. Let φ_C, φ_G, and φ_A be the potential (voltage) of the cathode, grid, and anode, respectively. According to (5.42), the charge Q_C on the cathode is

$$Q_C = C_{CC}\varphi_C + C_{CG}\varphi_G + C_{CA}\varphi_A.$$

This equation shows that the grid voltage modulates the charge on the cathode Q_C regardless of the potential difference between the cathode and the anode. From (5.14), Q_C is proportional to the field strength $\mathbf{E}_C$ just outside the cathode. This field, in turn, determines the acceleration (and hence the current) of electrons toward the anode. Note that the physical placement of the grid *between* the cathode and the anode in the figure above is irrelevant to this argument.

5.4.4 The Two-Conductor Capacitor

Figure 5.10 shows a *capacitor* composed of two conductors with equal and opposite charge. To analyze this arrangement—so familiar from elementary circuit theory—we use the inverse $\mathbf{P} = \mathbf{C}^{-1}$ of the capacitance matrix[7] to write

$$\varphi_i = \sum_{j=1}^{N} P_{ij}\, Q_j. \tag{5.51}$$

The P_{ij} are called *coefficients of potential*. Since $Q_1 = -Q_2$, (5.51) for a capacitor simplifies to

$$\begin{aligned} \varphi_1 &= P_{11}Q_1 + P_{12}Q_2 = (P_{11} - P_{12})Q \\ \varphi_2 &= P_{21}Q_1 + P_{22}Q_2 = (P_{12} - P_{22})Q. \end{aligned} \tag{5.52}$$

Subtracting the second equation in (5.52) from the first gives the potential difference $\varphi_1 - \varphi_2 = (P_{11} + P_{22} - 2P_{12})Q$. This leads us to define the *capacitance* C of the capacitor[8] as

$$C = \frac{Q}{\varphi_1 - \varphi_2} = \frac{1}{P_{11} + P_{22} - 2P_{12}} = \frac{C_{11}C_{22} - C_{12}^2}{C_{11} + C_{22} + 2C_{12}}. \tag{5.53}$$

[7] The Maxwell inequalities (5.50) guarantee that this inverse exists.

[8] Older literature speaks more descriptively of the *capacity* C of a *condenser*.

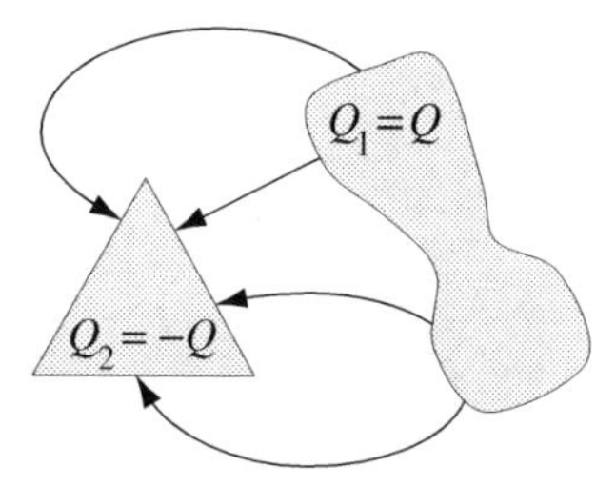

Figure 5.10: The electric field near a two-conductor capacitor.

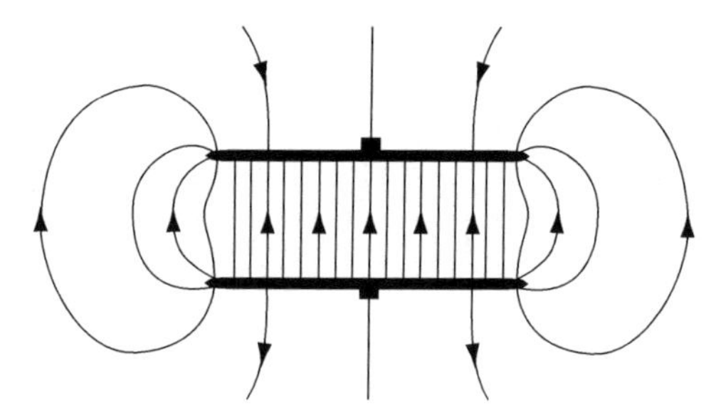

Figure 5.11: The fringe electric field near a parallel-plate capacitor.

The last equality in (5.53) follows from the matrix inverse,

$$\mathbf{P} = \mathbf{C}^{-1} = \begin{bmatrix} C_{11} & C_{12} \\ C_{12} & C_{22} \end{bmatrix}^{-1} = \frac{1}{C_{11}C_{22} - C_{12}^2} \begin{bmatrix} C_{22} & -C_{12} \\ -C_{12} & C_{11} \end{bmatrix}. \tag{5.54}$$

The limit $C_{12} \ll C_{11}, C_{22}$ is noteworthy because it simplifies (5.53) to

$$\frac{1}{C} = \frac{1}{C_{11}} + \frac{1}{C_{22}}. \tag{5.55}$$

If the same limit reduces C_{kk} to the self-capacitance of conductor k, (5.55) identifies the two-conductor capacitance C as the two constituent self-capacitances added in parallel.

More generally, if S is any surface that encloses the positive conductor,

$$C = \frac{\epsilon_0 \int_S d\mathbf{S} \cdot \mathbf{E}}{\int_1^2 d\boldsymbol{\ell} \cdot \mathbf{E}}. \tag{5.56}$$

The line integral in the denominator of (5.56) follows any path from the positive conductor to the negative conductor in Figure 5.10. The most familiar example of a two-conductor capacitor is the "parallel-plate" geometry where a distance d separates two flat conducting plates with uniform surface charge densities $\pm Q/A$. If $d \ll \sqrt{A}$, we use (5.14) and a uniform field approximation to write $|\mathbf{E}| = Q/A\epsilon_0 = (\varphi_1 - \varphi_2)/d$. The capacitance follows from the leftmost equality in (5.53),

$$C_0 = \frac{\epsilon_0 A}{d}. \tag{5.57}$$

Figure 5.11 shows $\mathbf{E}(\mathbf{r})$ for a parallel-plate capacitor where d is *not* very small compared to $\sqrt{A}$. The presence of "fringe" electric field lines which begin and end on the outer surfaces of the plates shows that some of the charge of each plate resides on those outer surfaces. This implies that $|\mathbf{E}|$

between the plates is slightly less than the infinite-plate value $Q/A\epsilon_0$. The correspondingly smaller value of the denominator in (5.56) tells us that the capacitance $C_{\rm exact}$ of a real parallel-plate capacitor *exceeds* (5.57).

5.5 The Energy of a System of Conductors

The total electrostatic energy of a collection of N conductors is straightforward to calculate because the charge of the kth conductor Q_k is distributed with density $\sigma_k(\mathbf{r}_S)$ over its surface S_k where the potential φ_k is constant. This gives

$$U_E = \tfrac{1}{2}\int d^3r\,\rho(\mathbf{r})\varphi(\mathbf{r}) = \tfrac{1}{2}\sum_{k=1}^{N}\varphi_k\int_{S_k} dS\,\sigma_k(\mathbf{r}_S) = \tfrac{1}{2}\sum_{k=1}^{N} Q_k\,\varphi_k. \tag{5.58}$$

Substituting (5.42) and (5.51) into (5.58) produces the alternative expressions

$$U_E = \tfrac{1}{2}\sum_{i=1}^{N}\sum_{j=1}^{N}\varphi_i\,C_{ij}\,\varphi_j \tag{5.59}$$

and

$$U_E = \tfrac{1}{2}\sum_{i=1}^{N}\sum_{j=1}^{N} Q_i\,P_{ij}\,Q_j. \tag{5.60}$$

For a two-conductor capacitor, we use the results of Section 5.4.4 to evaluate the sum in (5.60). The result is

$$U_E = \tfrac{1}{2}Q^2(P_{11}+P_{22}-2P_{12}) = \frac{Q^2}{2C} = \frac{1}{2}C(\varphi_1-\varphi_2)^2. \tag{5.61}$$

This proves that $C > 0$ in (5.53) because we have proved previously (Section 3.6.1) that $U_E \geq 0$. The same formula gives the charging energy of a single isolated conductor in terms of its self-capacitance C:

$$U_E = \frac{Q^2}{2C}. \tag{5.62}$$

Application 5.2 Coulomb Blockade

It is not difficult to transfer a macroscopic number of electrons to and from a macroscopic conductor because the charging energy (5.62) gained or lost is extremely small. This is no longer true for a "quantum dot" (a conducting object whose size scale R is microscopic rather than macroscopic) because (5.41) and (5.62) show that the charging energy $U_E \propto R^{-1}$ is large. In practice, it is possible to interpose a microscopically thin layer of insulating material between a quantum dot ($R \sim 10$ nm) and a much larger conductor which functions as an electron reservoir (Figure 5.12). The insulator creates a potential energy barrier between the dot and the reservoir. At very low temperature, electrons tunnel through the barrier to and from the dot, one by one, if it is energetically favorable to do so. Our task is to find the charge Q on the dot as a function of an external potential φ applied to the dot.

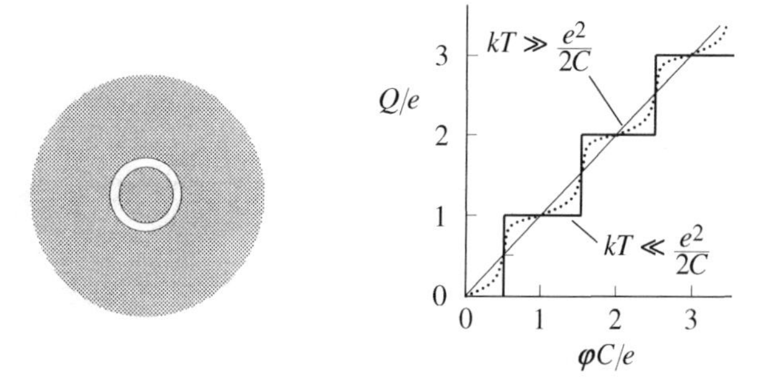

Figure 5.12: Left: two-dimensional quantum dot (shaded circle) separated from an electron reservoir (shaded annulus) by an insulating ring (white). Right: charge on the quantum dot as a function of an external potential φ applied to the dot. See text for discussion.

If N is the number of electrons on the dot and C is the capacitance of the dot-reservoir capacitor, the total electrostatic energy of the system is

$$U_E(N) = (Ne)^2/2C - Ne\varphi. \tag{5.63}$$

In fact, $C \approx C_{11}$, the self-capacitance of the dot, because $C_{22} \gg C_{11}, C_{12}$ in (5.53). When $\varphi = 0$, U_E is smallest when $N = 0$. Otherwise, the energy difference

$$U_E(N + M) - U_E(N) = \{(N + M/2)e/C - \varphi\}\, Me \tag{5.64}$$

first becomes negative when $\varphi = e(N + M/2)/C$. This quantity is smallest for $M = 1$ so, as φ increases, the charge on the dot jumps from N to $N + 1$ when $\varphi C/e = N + 1/2$. This implies that the dot charge $Q(\varphi)$ is the staircase function plotted in Figure 5.12. The horizontal segments in this figure are regions of "Coulomb blockade" where the charge on the dot does not increase as the applied potential increases due to the large cost in electrostatic energy. This phenomenon can be used to establish a capacitance standard simply by counting electrons.

When kT is much larger than the charging energy, electrons are thermally excited over the insulating barrier in copious numbers. This phenomenon washes out the quantum staircase and we obtain the macroscopic result $Q = C\varphi$. This is the straight line plotted in Figure 5.12. The dotted curve is for an intermediate temperature. ■

5.6 Forces on Conductors

In Section 3.4.3, we derived $\mathbf{f} = \frac{1}{2}\sigma(\mathbf{E}_{\text{in}} + \mathbf{E}_{\text{out}})$ as the electrostatic force per unit area on an element of surface. Since $\mathbf{E} = 0$ inside a conductor, the net force on the center of mass of a conductor involves only the field just outside the conductor surface. Using (5.14), we find

$$\mathbf{F} = \frac{1}{2}\int dS\, \sigma \mathbf{E}_{\text{out}} = \frac{1}{2\epsilon_0}\int dS\, \sigma^2 \hat{\mathbf{n}}. \tag{5.65}$$

The same force expression follows from the Maxwell stress tensor formula (3.97) because the electric field just outside a conductor is $\mathbf{E} = \hat{\mathbf{n}}\sigma/\epsilon_0$. The outward normal $\hat{\mathbf{n}}$ in the integrand of (5.65) shows that an outward *pressure* acts on every element of surface. This pressure (which is opposed by quantum mechanical forces of cohesion) reflects the tendency of infinitesimal bits of charge of the same sign to repel each other on each element of surface. Of course, a conductor cannot exert a net force on itself, so the integral (5.65) must be zero for a conductor in isolation.

In Chapter 8, we calculate the force (5.65) when a charge distribution $\rho(\mathbf{r})$ is responsible for $\mathbf{E}_{\text{out}}$. Here, we exploit the total energy of a collection of conductors to compute the force exerted on a given conductor by other conductors. Our strategy is to mimic Section 3.5.1 where we calculated the Coulomb force $\mathbf{F} = -\delta V_E/\delta\mathbf{s}$ from the change in electrostatic potential energy δV_E that occurs when a charge distribution is displaced by an infinitesimal amount $\delta\mathbf{s}$. A subtlety is that we can perform the displacement in one of two ways: holding the charge of the other conductors constant or holding the potential of the other conductors constant. Equations (5.42) and (5.51) remind us that these quantities are not independent. Therefore, since partial derivatives will be involved, it is important to work with an energy function whose natural independent variables are the same as the variables we propose to hold constant. The Legendre transformation mathematics needed to do this is familiar to students of classical mechanics and thermodynamics.

5.6.1 Charges Held Constant

Consider a collection of N conductors with center-of-mass positions $\mathbf{R}_k$, charges Q_k, and potentials φ_k. If the conductors are *isolated*, conservation of charge guarantees that the charge of any one of them cannot change when its center of mass suffers an infinitesimal displacement. To find the force exerted on any one of them, the proper quantity to vary must be a natural function of the conductor charges. We assert this is the total electrostatic energy,

$$U_E = U_E(Q_1, \ldots, Q_N, \mathbf{R}_1, \ldots, \mathbf{R}_N). \tag{5.66}$$

Equation (5.66) follows from the definition of U_E as the total work done to assemble a charge distribution. The work-energy theorem identifies $dU_E = \varphi_k\, dQ_k$ as the change in energy when a bit of charge dQ_k is brought from infinity (zero potential) to the surface of a conductor with potential φ_k. For our problem, this generalizes to

$$dU_E = \sum_{k=1}^{N} \varphi_k\, dQ_k - \sum_{k=1}^{N} \mathbf{F}_k \cdot d\mathbf{R}_k. \tag{5.67}$$

The negative sign in the second term ensures that the force $\mathbf{F}_k$ on the kth conductor moves the system toward lower energy.

On the other hand, it is a matter of calculus that the total differential of (5.66) is

$$dU_E = \sum_{k=1}^{N} \left(\frac{\partial U_E}{\partial Q_k}\right)_{Q',\mathbf{R}} dQ_k + \sum_{k=1}^{N} \left(\frac{\partial U_E}{\partial \mathbf{R}_k}\right)_{Q,\mathbf{R}'} \cdot d\mathbf{R}_k. \tag{5.68}$$

The subscript Q in (5.68) stands for the entire set of conductor charges. The subscript Q' stands for the entire set of conductor charges except for the one involved in the derivative. A similar notation is used for other variables. Comparing (5.68) to (5.67), we conclude that

$$\varphi_k = \left(\frac{\partial U_E}{\partial Q_k}\right)_{Q',\mathbf{R}} \qquad \text{and} \qquad \mathbf{F}_k = -\left(\frac{\partial U_E}{\partial \mathbf{R}_k}\right)_{Q,\mathbf{R}'}. \tag{5.69}$$

All of this is precisely analogous to the situation in thermodynamics where the relation $dU = TdS - pdV$ confirms that the internal energy $U = U(S, V)$ is a natural function of entropy and volume with $T = (\partial U/\partial S)_V$ and $p = -(\partial U/\partial V)_S$. In this sense, the partial derivative on the left side of (5.69) shows that φ_k and Q_k are *conjugate variables* like pressure and volume.

Because the Q_k are held constant, the partial derivative on the right side of (5.69) is simplest to evaluate using (5.60) as our expression for U_E. This expresses the force exerted on the kth conductor in terms of the coefficients of potential P_{ij},[9]

$$\mathbf{F}_k = -\frac{1}{2}\sum_{i,j} Q_i Q_j \left(\frac{\partial P_{ij}}{\partial \mathbf{R}_k}\right)_{Q,\mathbf{R}'}. \tag{5.70}$$

5.6.2 Potentials Held Constant

Electrostatic potential is not a conserved quantity like charge. Therefore, physical connection to a battery (or to ground) is required to maintain the potential of a conductor at a fixed value. Moreover, to calculate forces, we are obliged to replace the total energy U_E in (5.66) with an energy-like quantity that is a natural function of the potentials rather than of the charges.

We showed in the previous section that φ_k and Q_k are conjugate variables. Therefore, following a procedure familiar from thermodynamics and classical mechanics[10] we perform a *Legendre transformation* on U_E and define

$$\hat{U}_E = U_E - \sum_{i=1}^{N} Q_i\varphi_i. \tag{5.71}$$

To check that this is correct, we compute the differential $d\hat{U}_E$ and use (5.67). The result is

$$d\hat{U}_E = dU_E - \sum_{i=1}^{N}\varphi_i dQ_i - \sum_{i=1}^{N} Q_i d\varphi_i = -\sum_{i=1}^{N} Q_i d\varphi_i - \sum_{i=1}^{N}\mathbf{F}_i\cdot d\mathbf{R}_i. \tag{5.72}$$

The usual rules of calculus then confirm that

$$\hat{U}_E = \hat{U}_E(\varphi_1,\ldots,\varphi_N,\mathbf{R}_1,\ldots,\mathbf{R}_N), \tag{5.73}$$

with

$$Q_i = -\left(\frac{\partial \hat{U}_E}{\partial \varphi_i}\right)_{\varphi',\mathbf{R}} \qquad \text{and} \qquad \mathbf{F}_i = -\left(\frac{\partial \hat{U}_E}{\partial \mathbf{R}_i}\right)_{\varphi,\mathbf{R}'}. \tag{5.74}$$

To make further progress, we combine (5.71) with (5.58) to get the explicit formula

$$\hat{U}_E = -\tfrac{1}{2}\sum_i \varphi_i Q_i. \tag{5.75}$$

Comparing (5.75) with (5.58) shows that $\hat{U}_E$ happens to equal $-U_E$. Therefore, since the potentials φ_k are now held constant, it is simplest to evaluate the right side of (5.74) using the negative of (5.59). This expresses the force exerted on the kth conductor in terms of the coefficients of capacitance:

$$\mathbf{F}_k = \frac{1}{2}\sum_{i=1}^{N}\sum_{j=1}^{N}\varphi_i\varphi_j\left(\frac{\partial C_{ij}}{\partial \mathbf{R}_k}\right)_{\varphi,\mathbf{R}'}. \tag{5.76}$$

[9] See, however, Section 5.6.3.

[10] We have noted that the thermodynamic internal energy $U(S,V)$ is a natural function of entropy S and volume V. To get a natural function of temperature and volume, we use $F = U - TS$ to define the Helmholtz free energy $F(T,V)$. In classical mechanics, the Lagrangian $L(q,\dot{q})$ is a natural function of position q and velocity $\dot{q}$. To get a natural function of position and momentum p, we use $H = p\dot{q} - L$ to define the Hamiltonian $H(p,q)$. The overall minus sign in the latter example compared to (5.71) is conventional.

Example 5.5 Calculate the force acting on a metal sheet of thickness t inserted a distance x into a capacitor with parallel-plate separation $d > t$ and voltage difference V; see Figure 5.13. The fringe fields can be neglected if an energy method is used to find $\mathbf{F}$.

Solution: By symmetry, the force is in the x-direction only. Since the potential is held constant, we calculate the force from (5.74) in the form

$$F_x = -\frac{\partial \hat{U}_E}{\partial x} = \frac{\partial U_E}{\partial x} = \lim_{\Delta \to 0} \frac{U_E(x+\Delta) - U_E(x)}{\Delta}.$$

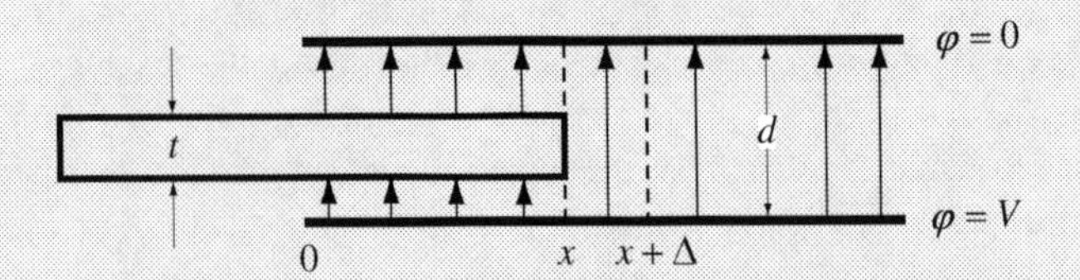

Figure 5.13: A conducting sheet of thickness t partially inserted into a capacitor with plate separation $d > t$.

The energy in this formula is

$$U_E = \frac{1}{2}\epsilon_0 \int d^3r\, |\mathbf{E}|^2.$$

Moreover, there is no contribution to the force formula from the energy stored in the capacitor to the right of the point $x + \Delta$. Since $\mathbf{E} = 0$ inside the slab and

$$V = \int_0^d d\boldsymbol{\ell} \cdot \mathbf{E},$$

the magnitude of the electric field is $V/(d-t)$ above and below the slab. The electric field magnitude is V/d between x and $x + \Delta$. Therefore, if w is width of the capacitor in the direction perpendicular to the paper,

$$U_E(x+\Delta) - U_E(x) = \frac{1}{2}\epsilon_0 \left(\frac{V}{d-t}\right)^2 w(x+\Delta)(d-t)$$
$$- \left[\frac{1}{2}\epsilon_0 \left(\frac{V}{d-t}\right)^2 wx(d-t) + \frac{1}{2}\epsilon_0 \left(\frac{V}{d}\right)^2 w\Delta d\right].$$

We conclude that

$$\mathbf{F} = \frac{1}{2}\epsilon_0 V^2 w \left[\frac{1}{d-t} - \frac{1}{d}\right]\hat{\mathbf{x}}.$$

This force tends to draw the conductor into the capacitor. The same conclusion follows from (5.65) *only* if we take account of the non-uniformity of the electric field near the end of the slab. The contributions to this force integral are equal, opposite, and vertical on the top and bottom faces of the slab. The contributions to (5.65) in the $+x$-direction come from charge induced by electric field lines which bend around and terminate on the vertical edge of the slab that lies inside the capacitor.

5.6.3 Why U_E Differs From $\hat{U}_E$

A naive force calculation using $\mathbf{F}_k = -\partial U_E/\partial \mathbf{R}_k$ and (5.59) gives the *negative* of (5.76). This wrong answer reflects the fact that U_E is not a natural function of the φ_k. Physically, the sign change

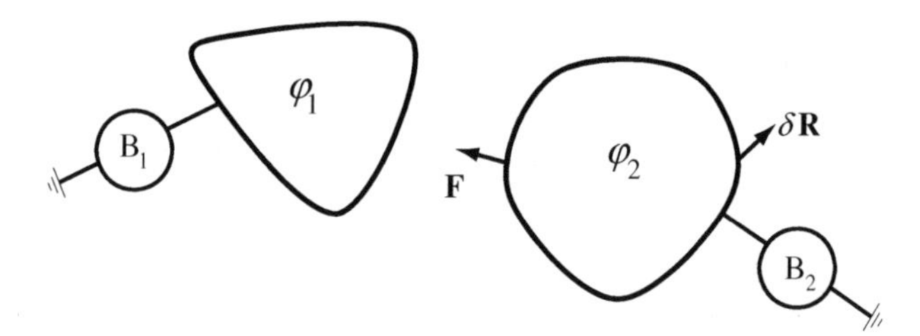

Figure 5.14: Two conductors maintained at fixed potentials by batteries.

occurs because the fixed-potential conductors do not form a closed system. Instead, each conductor is connected to a charge reservoir (a battery) which supplies or extracts charge from the conductor to maintain its potential during an infinitesimal displacement.

Figure 5.14 shows two conductors held at fixed potentials φ_1 and φ_2 by batteries B_1 and B_2. The vector **F** indicates the mechanical force exerted on conductor φ_2 by conductor φ_1. When the conductor φ_2 suffers a rigid displacement $\delta\mathbf{R}$, the charge on φ_1 changes from $Q_1^{(0)}$ to Q_1 and the charge on φ_2 changes from $Q_2^{(0)}$ to Q_2. Our strategy is to compute the energy change $\Delta\hat{U}_E$ which occurs during this process from the change ΔE_{sys} that occurs in the total energy:

$$\Delta E_{\text{sys}} = \mathbf{F}\cdot\delta\mathbf{R} + \Delta U_E + \Delta E_{\text{batt}} = 0. \tag{5.77}$$

In (5.77), $\mathbf{F}\cdot\delta\mathbf{R}$ is the work done by **F**, ΔU_E is the change in the total electrostatic energy associated with the conductors, and ΔE_{batt} is the change in the total energy of the batteries. The zero on the far right side of (5.77) is a statement of the conservation of energy for the isolated system shown in Figure 5.14.

The fixed potential relation on the far right side of (5.74) permits us to eliminate $\mathbf{F}\cdot\delta\mathbf{R}$ in (5.77) in favor of $\Delta\hat{U}_E$. Consequently,

$$\Delta\hat{U}_E = \Delta U_E + \Delta E_{\text{batt}}. \tag{5.78}$$

The change in electrostatic energy follows from (5.58):

$$\Delta U_E = U_E(Q_1, Q_2) - U_B(Q_1^{(0)}, Q_2^{(0)}) = \frac{1}{2}\sum_{k=1}^{N}\varphi_k(Q_k - Q_k^{(0)}). \tag{5.79}$$

As for ΔE_{batt}, the batteries are external to the conductors, so the work done by them is calculable using (5.67). Using $N = 2$ and the fact that positive work done to maintain the potentials on the conductors reduces the total energy of the batteries,

$$\Delta E_{\text{batt}} = -W_{\text{batt}} = -\sum_{k=1}^{N}\int_{Q_k^{(0)}}^{Q_k}\varphi_k dQ_k = -\sum_{k=1}^{N}\varphi_k(Q_k - Q_k^{(0)}). \tag{5.80}$$

Substituting (5.79) and (5.80) into (5.78) gives

$$\Delta\hat{U}_E = -\frac{1}{2}\sum_{k=1}^{N}\varphi_k(Q_k - Q_k^{(0)}). \tag{5.81}$$

Hence, in agreement with (5.75),

$$\hat{U}_E = -\frac{1}{2}\sum_{k=1}^{N}\varphi_k Q_k. \tag{5.82}$$

Given all these remarks, the reader may be surprised to learn that the charge-fixed formula (5.70) and potential-fixed formula (5.76) give exactly the same force for a static collection of conductors. This must be the case because the displacements imagined in Sections 5.6.1 and 5.6.2 are not real, but merely virtual. Consequently, the same force must result whether (in our minds) we hold the charges fixed or hold the potentials fixed. We can illustrate this using the attractive force which acts between the two conductors in Figure 5.10. From the fixed-charge point of view, we use (5.61) and (5.69) to get the force on conductor 1 as

$$\mathbf{F}_1 = -\frac{\partial U_E}{\partial \mathbf{R}_1} = -\frac{Q^2}{2}\frac{\partial}{\partial \mathbf{R}_1}\frac{1}{C} = \frac{U_E}{C}\frac{\partial C}{\partial \mathbf{R}_1}. \tag{5.83}$$

From the fixed-potential point of view, (5.61), (5.74), and the fact that $\hat{U}_E = -U_E$ give the same result:

$$\mathbf{F}_1 = -\frac{\partial \hat{U}_E}{\partial \mathbf{R}_1} = \frac{1}{2}(\varphi_1 - \varphi_2)^2\frac{\partial C}{\partial \mathbf{R}_1} = \frac{U_E}{C}\frac{\partial C}{\partial \mathbf{R}_1}. \tag{5.84}$$

Example 5.6 (a) Calculate the total energy change when the parallel-plate separation increases from s to $s + \delta s$ for the capacitor C in Figure 5.15. Use this to find the force between the plates of capacitor C. (b) Repeat the calculation when the switches are closed. Ignore the effect of the connecting wires and assume that s^2 is very small compared to the capacitor plate area A.

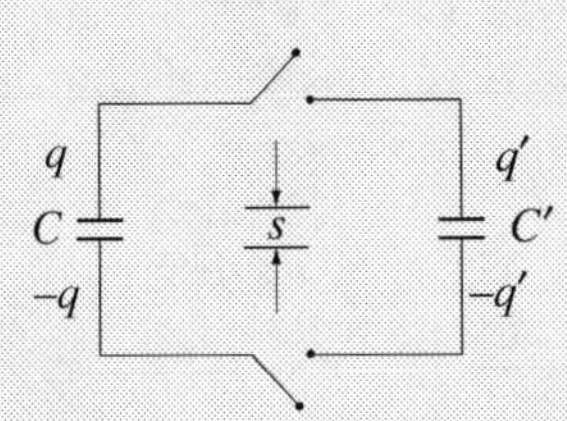

Figure 5.15: Two parallel-plate capacitors.

Solution:

(a) With the switches open, C is an isolated system with total energy $U_E = q^2/2C$. From (5.57), the capacitance $C = \epsilon_0 A/s$ and $\sigma = q/A$ so

$$\delta U_E = -\frac{1}{2}\left(\frac{q}{C}\right)^2 \delta C = \frac{1}{2}\left(\frac{q}{C}\right)^2 \frac{\epsilon_0 A\delta s}{s^2} = \frac{\sigma^2}{2\epsilon_0}A\delta s.$$

The energy increases as δs increases so the force per unit area is attractive. This is consistent with the negative sign in $F = -\delta U_E/\delta s = -\sigma^2 A/2\epsilon_0$ because the direction of the force is opposite to the direction of δs.

(b) With the switches closed, the total energy of the system is $U_E = q^2/2C + q'^2/2C'$. When we change s, there is a change δC and a flow of charge between the capacitors. Therefore, because $\delta q = -\delta q'$,

$$\delta U_E = -\frac{1}{2}\left(\frac{q}{C}\right)^2 \delta C + \frac{q}{C}\delta q + \frac{q'}{C'}\delta q' = -\frac{1}{2}\left(\frac{q}{C}\right)^2 \delta C + \left(\frac{q}{C} - \frac{q'}{C'}\right)\delta q.$$

The last term in parenthesis is *zero* because $q = VC$ and $q' = VC'$ where V is the (common) potential difference between the plates. The term that remains is the energy change computed earlier, so the force is the same. In the reservoir limit $C' \to \infty$, it is not difficult to check that the energy change above is $-\delta\hat{U}_E$ as it should be.

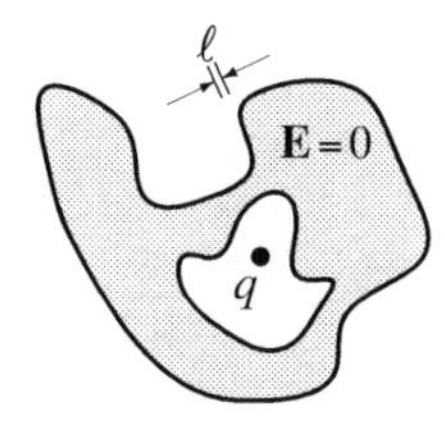

Figure 5.16: A real conductor with a point charge q inside a scooped-out cavity. The electric field is zero in the bulk of the conductor (grey) except within a screening layer (black) of thickness ℓ at the surfaces.

5.7 Real Conductors

Section 5.3 discussed the screening properties of a perfect conductor with a cavity scooped out of its interior. Among other things, we learned that a point charge q inside the cavity induces charge on the surfaces of both the cavity and the conductor. Figure 5.16 shows that real conducting matter behaves similarly except that the infinitesimally thin layers of induced surface charge spread out into diffuse layers of thickness ℓ. This *screening length* ℓ varies from $\sim 10^{-10}$ m for a good metal to $\sim 10^{-8}$ m for a biological plasma, $\sim 10^{-5}$ m for a laboratory plasma, and ~ 10 m for an astrophysical plasma.

To understand the origin of the screening layer, consider an infinite system with three sources of charge: (1) a positive point charge at the origin; (2) a uniform distribution of positive charge with density en_0; and (3) a distribution of mobile negative charge with density $-en(\mathbf{r})$. This model approximates a point impurity in a metal where the positive ions are immobile. It also approximates a point impurity in an astrophysical, biological, or laboratory plasma where the positive ions move very slowly compared to the electrons.

The Poisson equation (2.14) for this model is

$$\epsilon_0 \nabla^2 \varphi(\mathbf{r}) = -e[n_0 - n(\mathbf{r})] - q\delta(\mathbf{r}). \tag{5.85}$$

When $q = 0$, translational invariance implies that $n(\mathbf{r}) = n_0$ and $\varphi(\mathbf{r}) = 0$ if the system is overall charge-neutral. When $q \neq 0$, a spatially varying electrostatic potential arises and a spatially varying charge density develops in response. Intuitively, we expect negative charge to move from points of low potential to points of high potential. To describe this behavior, recall first that the *chemical potential* $\mu(\mathbf{r})$ is the control parameter for particle number density in near-equilibrium situations.[11] This means that $n_0 = n_0(\mu)$. Moreover, $n_0(\mu)$ is a non-decreasing function of its argument.[12] These facts suggest a simple way to control the migration of mobile negative charge: replace $n = n_0(\mu)$ by

$$n(\mathbf{r}) = n_0(\mu + e\varphi(\mathbf{r})). \tag{5.86}$$

This local approximation captures the dominant response of mobile charge at the point $\mathbf{r}$ to the electrostatic potential at the point $\mathbf{r}$.

To make progress, expand (5.86) to first order in $\varphi(\mathbf{r})$ and use the result to approximate the first term on the right-hand side of (5.85)) as

$$-e[n_0(\mu) - n_0(\mu + e\varphi)] \approx e^2 \frac{\partial n_0}{\partial \mu} \varphi. \tag{5.87}$$

[11] Position-dependent chemical potentials are discussed in, e.g., P.M. Morse, *Thermal Physics*, 2nd edition (Benajmin/Cummings, Reading, MA, 1969).

[12] This follows from the thermodynamic identity $(\partial n/\partial \mu)_{T,V} = -n^2 V^{-1}(\partial V/\partial P)_T = n^2\kappa \geq 0$, where κ is the isothermal compressibility.

Next, define

$$\frac{1}{\ell^2} = \frac{e^2}{\epsilon_0}\frac{\partial n_0}{\partial \mu} \tag{5.88}$$

and substitute (5.87) into (5.85) to get

$$\nabla^2\varphi(\mathbf{r}) - \frac{\varphi(\mathbf{r})}{\ell^2} = -q\delta(\mathbf{r}). \tag{5.89}$$

It is not difficult to solve (5.89) if we focus on points away from the origin, where (5.89) simplifies to

$$\nabla^2\varphi(\mathbf{r}) - \frac{\varphi(\mathbf{r})}{\ell^2} = 0. \tag{5.90}$$

A point charge is spherically symmetric. Therefore, $\varphi(\mathbf{r}) = \varphi(r)$ and a brief calculation confirms that (5.90) simplifies to

$$\frac{d^2(r\varphi)}{dr^2} - \frac{r\varphi}{\ell^2} = 0. \tag{5.91}$$

The solution to (5.91) which satisfies the physical boundary conditions $\varphi(r \to \infty) = 0$ and $\varphi(r \to 0) = q/4\pi\epsilon_0 r$ is

$$\varphi(r) = \frac{q}{4\pi\epsilon_0 r}\exp(-r/\ell). \tag{5.92}$$

Using (5.85), the electron density implied by (5.92) satisfies

$$\int d^3r\, n(\mathbf{r}) = -q. \tag{5.93}$$

We conclude from (5.93) that the mobile negative charge attracted toward the origin exactly compensates the positive charge of the point impurity. The decaying exponential in (5.92) makes it clear that the electric field is essentially zero at any distance greater than about ℓ from the point charge.[13] For this reason, ℓ is called the *screening length*.

For a sample of finite size, Gauss' law provides a boundary condition at the conductor surface that can be used with (5.90) to confirm our assertion that the positively charged layer in Figure 5.16 extends a distance ℓ into the conductor. Similarly, if we replace the point charge by a uniform, static, external electric field, one finds from (5.90) that the electric field penetrates a distance ℓ into the bulk of the conductor. All of this shows that, for a sample with a characteristic size D, the ratio ℓ/D measures how nearly "perfectly" the conductor responds to electrostatic perturbations.

5.7.1 Debye-Hückel and Thomas-Fermi Screening

The qualitative physics of the screening layer lies in the proportionality between $\partial n_0/\partial\mu$ and the compressibility quoted in footnote 12. By definition, a perfect conductor has zero screening length. Therefore, according to (5.88), this system is infinitely compressible in the sense that it costs no energy to squeeze all the screening charge into an infinitesimally thin surface layer. By contrast, the screening charge in a classical thermal plasma (like the cytoplasm of a cell or a doped semiconductor) has finite compressibility because the charges gain configurational entropy by spreading out in space. The screening electrons in a metal similarly resist compression because, as the quantum mechanical particle-in-a-box problem teaches us, the kinetic energy of the electrons increases as the volume of the box decreases. For these reasons, the infinitesimally thin surface distributions of a perfect conductor broaden into the screening layers illustrated in Figure 5.16.

[13] Compare the exponential in (5.92) with the exponential in (2.63).

Quantitatively, we need the function $n_0(\mu)$ to compute ℓ from (5.88). For our "classical" examples, it is a standard result of Boltzmann statistics that the particle density depends on chemical potential as

$$n_0(\mu) = \frac{(2\pi mkT)^{3/2}}{h^3} \exp\{\mu/kT\}. \tag{5.94}$$

The associated screening length calculated from (5.88) is

$$\ell_{\rm DH} = \sqrt{\frac{\epsilon_0 kT}{e^2 n}}. \tag{5.95}$$

The subscript "DH" in (5.95) honors P. Debye and E. Hückel, who identified this quantity in their analysis of screening in electrolytes. Typically, $\ell_{\rm DH} \sim 10 - 100$ Å in such solutions and also in doped semiconductors. For a metal, Fermi statistics at $T = 0$ gives the dependence of the particle density on chemical potential as

$$n_0(\mu) = \frac{8\pi}{3}\left(\frac{2m\mu}{h^2}\right)^{3/2}. \tag{5.96}$$

Using (5.96) in (5.88) gives the screening length,

$$\ell_{\rm TF} = \sqrt{\frac{\pi a_B}{4k_F}}, \tag{5.97}$$

where $\mu = \hbar^2 k_F^2/2m$ and $a_B = 4\pi\epsilon_0\hbar^2/me^2$. Here, the subscript "TF" remembers L.H. Thomas and E. Fermi because the approximations of this section were used by them in a statistical theory of atomic structure. In a good metal, $\ell_{\rm TF} \sim 1$ Å.

Sources, References, and Additional Reading

The Cavendish quotation at the beginning of the chapter is taken from Section 98 of "An attempt to explain some of the principal phenomena of electricity by means of an elastic fluid", *Philosophical Transactions* **61**, 584 (1771). The article is reprinted (with a penetrating commentary) in

J.C. Maxwell, *The Electrical Researches of the Honourable Henry Cavendish* (Frank Cass, London, 1967).

Section 5.1 Textbooks that discuss the physics of conductors without an immediate plunge into boundary value potential theory include

V.C.A. Ferraro, *Electromagnetic Theory* (Athlone Press, London, 1954).

W.T. Scott, *The Physics of Electricity and Magnetism*, 2nd edition (Wiley, New York, 1966).

R.K. Wangsness, *Electromagnetic Fields*, 2nd edition (Wiley, New York, 1986).

In the course of his study of thermal fluctuations, Einstein considered the effect of a heat bath on the charge carried by the surface of a conductor. He proposed to use amplification by successive steps of electrostatic induction to observe the tiny changes in charge predicted by his theory. The story is told in detail in

D. Segers and J. Uyttenhove, "Einstein's 'little machine' as an example of charging by induction", *American Journal of Physics* **74**, 670 (2006).

Section 5.2 Application 5.1 is taken from the surprisingly readable

G. Green, *An Essay on the Application of Mathematical Analysis to the Theories of Electricity and Magnetism* (1828). Facsimile edition by Wezäta-Melins Aktieborg, Göteborg, Sweden (1958), Section 12.

Techniques used to find $\sigma(\mathbf{r}_S)$ for a conducting disk different from the one used in Application 5.1 include (i) solving Laplace's equation in ellipsoidal coordinates; (ii) solving an integral equation; and (iii) a purely geometric approach. These methods are discussed, respectively, in

J.A. Stratton, *Electromagnetic Theory* (McGraw-Hill, New York, 1941).

L. Egyes, *The Classical Electromagnetic Field* (Dover, New York, 1972).

A.M. Portis, *Electromagnetic Fields: Sources and Media* (Wiley, New York, 1978).

Section 5.3 A pedagogical discussion of electrostatic shielding which complements the one given in the text is

S.H. Yoon, "Shielding by perfect conductors: An alternative approach", *American Journal of Physics* **71**, 930 (2003).

Section 5.4 The approximate self-capacitance formula (5.41) and the exact formulae graphed in Figure 5.6 are discussed, respectively, in

Y.L. Chow and M.M. Yovanovich, "The shape factor for the capacitance of a conductor", *Journal of Applied Physics* **53**, 8470 (1982).

L.D. Landau and E.M. Lifshitz, *The Electrodynamics of Continuous Media* (Pergamon, Oxford, 1960).

Example 5.4 is a simplified version of a problem analyzed in

K.K. Likharev, N.S. Bakhvalov, G.S. Kazacha, and S.I. Serdyukova, "Single-electron tunnel junction array", *IEEE Transactions on Magnetics* **25**, 1436 (1989).

The boxed material after Example 5.4 comes from

A. Marcus, "The theory of the triode as a three-body problem in electrostatics", *The American Physics Teacher* **7**, 196 (1939).

The capacitance matrix is important to the design of very large scale integrated (VLSI) circuits. A typical example is

Z.-Q. Ning, P.M. Dewilde, and F.L. Neerhoff, "Capacitance coefficients for VLSI multilevel metallization lines", *IEEE Transactions on Electron Devices* **34**, 644 (1987).

Section 5.5 Entry points to learn about Coulomb blockade and quantum dots are

H. Grabert and M.H. Devoret, *Coulomb Blockade Phenomena in Nanostructures*, NATO ASI Series B, volume 294 (Plenum, New York, 1992).

M.W. Keller, A.L. Eichenberger, J.M. Martinis, and N.M. Zimmerman, "A capacitance standard based on counting electrons", *Science* **285**, 1707 (1999).

Section 5.6 Our "thermodynamic" approach to the forces exerted on conductors was inspired by the masterful but characteristically terse discussion in Section 5 of *Landau and Lifshitz* (see Section 5.4 above).

Examples 5.5 and 5.6 were adapted, respectively, from

O.D. Jefimenko, *Electricity and Magnetism* (Appleton-Century-Crofts, New York, 1966).

W.M. Saslow, *Electricity, Magnetism, and Light* (Academic, Amsterdam, 2002).

Section 5.7 An excellent introduction to the screening response of "real" conductors (both classical and quantum) to static and non-static perturbations is

J.-N. Chazalviel, *Coulomb Screening by Mobile Charges* (Birkhäuser, Boston, 1999).

Problems

5.1 A Conductor with a Cavity A solid conductor has a vacuum cavity of arbitrary shape scooped out of its interior. Use Earnshaw's theorem to prove that $\mathbf{E} = 0$ inside the cavity.

5.2 Two Spherical Capacitors A spherical conducting shell with radius b is concentric with and encloses a conducting ball with radius a. Calculate the 2×2 capacitance matrix for this geometry and show that C_{22} can be interpreted as the ordinary capacitance of two capacitors in parallel.

5.3 Concentric Cylindrical Shells A capacitor is formed from three very long, concentric, conducting, cylindrical shells with radii $a < b < c$. Find the capacitance per unit length of this structure if a fine wire connects the inner and outer shells and λ_b is the uniform charge per unit length on the middle cylinder.

5.4 A Charged Sheet between Grounded Planes Two infinite conducting planes are held at zero potential at $z = -d$ and $z = d$. An infinite sheet with uniform charge per unit area σ is interposed between them at an arbitrary point.

(a) Find the charge density induced on each grounded plane and the potential at the position of the sheet of charge.
(b) Find the force per unit area which acts on the sheet of charge.

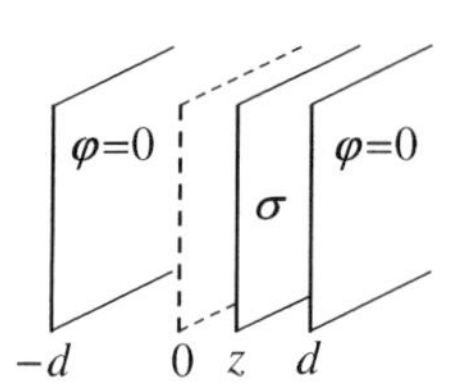

5.5 The Charge Distribution Induced on a Neutral Sphere A point charge q lies a distance $r > R$ from the center of an uncharged, conducting sphere of radius R. Express the induced surface charge density in the form

$$\sigma(\theta) = \sum_{\ell=1}^{\infty} \sigma_\ell P_\ell(\cos\theta)$$

where θ is the polar angle measured from a positive z-axis which points from the sphere center to the point charge.

(a) Show that the total electrostatic energy is

$$U_E = \frac{1}{\epsilon_0} \sum_{\ell=1}^{\infty} \frac{\sigma_\ell}{2\ell+1} \left[\frac{R^3 \sigma_\ell}{2} \frac{4\pi}{2\ell+1} + \frac{q\, R^{\ell+2}}{r^{\ell+1}} \right].$$

(b) Use Thomson's theorem to find $\sigma(\theta)$.

5.6 Charge Transfer between Conducting Spheres A metal ball with radius R_1 has charge Q. A second metal ball with radius R_2 has zero charge. Now connect the balls together using a fine conducting wire. Assume that the balls are separated by a distance R which is large enough that the charge distribution on each ball remains uniform. Show that the ball with radius R_1 possesses a final charge

$$Q_1 = \frac{QR_1}{R_1 + R_2} \left[1 + \frac{(R_1 - R_2)R_2}{(R_1 + R_2)R} \right].$$

5.7 Concentric Spherical Shells Three concentric spherical metallic shells with radii $c > b > a$ have charges e_c, e_b, and e_a, respectively. Find the change in potential of the outermost shell when the innermost shell is grounded.

5.8 Don't Believe Everything You Read in Journals A research paper published in the journal *Applied Physics Letters* describes experiments performed with three identical spherical conductors suspended from above by insulating wires so a (fictitious) horizontal plane passes through the center of all three spheres. It was reported that a large voltage V applied to one sphere induced equal and opposite rotation in the two isolated spheres (see top view below). The authors suggested that the isolated spheres were set into motion by electrostatic torque. Show, to the contrary, that this torque is zero.

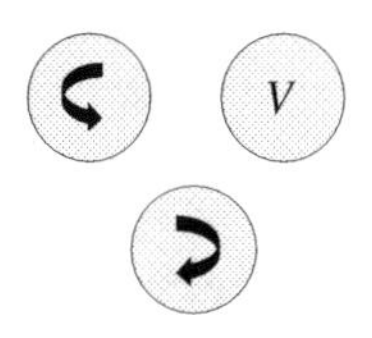

5.9 A Dipole in a Cavity A point electric dipole with moment $\mathbf{p}$ is placed at the center of a hollow spherical cavity scooped out of an infinite conducting medium.

(a) Find the surface charge density induced on the surface of the cavity.
(b) Show that the force on the dipole is zero.

5.10 Charge Induction by a Dipole A point dipole $\mathbf{p}$ is placed at $\mathbf{r} = \mathbf{r}_0$ outside a grounded conducting sphere of radius R. Use Green's reciprocity (and a comparison system with zero volume charge density) to find the charge drawn up from ground onto the sphere.

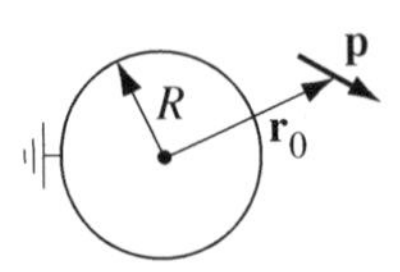

5.11 Charge Induction by a Potential Patch The square region defined by $-a \le x \le a$ and $-a \le y \le a$ in the plane $z = 0$ is a conductor held at potential $\varphi = V$. The rest of the plane $z = 0$ is a conductor held at potential $\varphi = 0$. The plane $z = d$ is also a conductor held at zero potential. Use Green's reciprocity relation to find the total charge induced on the entire $z = 0$ plane.

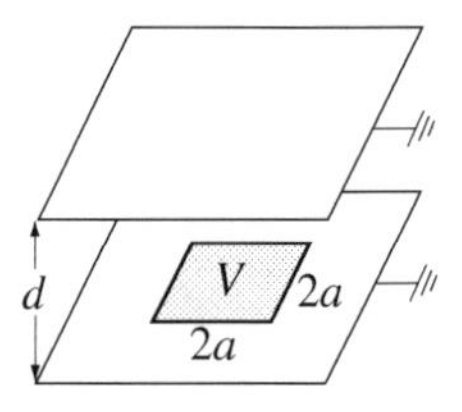

5.12 Charge Sharing among Three Metal Balls Four identical conducting balls are attached to insulating supports that sit on the floor as shown below. One ball has charge Q; its support is fixed in space. The other three balls are uncharged but their supports can be moved around. Describe a procedure (that involves only moving and/or bringing balls into contact) that will leave the $+Q$ ball with its full charge and give the three originally uncharged balls charges q, $-q/2$, and $-q/2$. You may assume that $q < Q$.

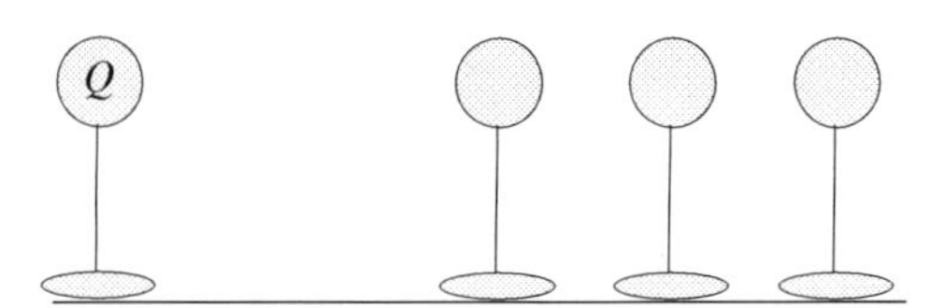

5.13 A Conducting Disk
A conducting disk of radius R held at potential V sits in the x-y plane centered on the z-axis.

(a) Use the charge density for this system calculated in the text to find the potential everywhere on the z-axis.
(b) Ground the disk and place a unit point charge q_0 on the axis at $z = d$. Use the results of part (a) and Green's reciprocity relation to find the amount of charge drawn up from ground to the disk.

5.14 The Capacitance of Spheres

(a) What is the self-capacitance (in farads) of the Earth? How much energy is required to add one electron to the (neutral) Earth?
(b) What is the self-capacitance (in farads) of a conducting nanosphere of radius 10 nm? How much energy (in electron volts) is required to add one electron to the (neutral) sphere?

(c) Two conducting spheres with radii R_A and R_B are separated by a distance R and carry net charges Q_A and Q_B. Find the potential matrix **P** and the capacitance matrix **C** assuming that the two spheres influence one another but that $R \gg R_A, R_B$ so that the charge density on each remains spherical.

(d) Compare the diagonal elements of **C** computed at the end of part (c) with the self-capacitances of the spheres.

5.15 Practice with Green's Reciprocity The text derived Green's reciprocity theorem for a set of conductors as a special case of a more general result. For conductors with charges and potentials (q_k, φ_K) and the same set of conductors with charges and potentials $(\tilde{q}_k, \tilde{\varphi}_k)$, the theorem reads

$$\sum_{i=1}^{N} q_i \tilde{\varphi}_i = \sum_{i=1}^{N} \tilde{q}_i \varphi_i .$$

(a) Use the symmetry of the capacitance matrix to prove the theorem directly.

(b) Three identical conducting spheres are placed at the corners of an equilateral triangle. When the sphere potentials are $(\phi, 0, 0)$, their charges are (q, q_0, q_0). What is the charge q' on each sphere when the potential of each sphere is ϕ'?

(c) What is the potential on each sphere when their charges are $(q'', 0, 0)$?

5.16 Maxwell Was Not Always Right A non-conducting square has a fixed surface charge distribution. Make a rectangle with the same area and total charge by cutting off a slice from one side of the square and gluing it onto an adjacent side. The energy of the rectangle is lower than the energy of the square because we have moved charge from points of high potential to points of low potential. An even lower energy results if we let the charge of the rectangle rearrange itself in any manner that keeps the total charge fixed. By definition, the rectangle is now a conductor. The electrostatic energy $U_E = Q^2/2C$ of the rectangle is lower, so the capacitance of the rectangle is larger than the capacitance of the square.

Maxwell made this argument in 1879 in the course of editing the papers of Henry Cavendish. Much later, the eminent mathematician György Polya observed that the conclusion is correct but that "Maxwell's proof is amazingly fallacious."

(a) Find the logical error in Maxwell's argument.

(b) Make a physical argument which shows that $C_{\text{rect}} > C_{\text{sq}}$. Hint: Think about the electrostatic energy cost to add a bit of charge δQ to either the square or the rectangle.

5.17 Two-Dimensional Electron Gas Capacitor Let d be the separation between two infinite, parallel, perfectly conducting plates. The lower surface of the upper plate has charge per unit area $\sigma_1 > 0$. The upper surface of the lower plate has charge per unit area $\sigma_2 > 0$. At a distance L above the lower plate, there is an infinite sheet of charge with charge per unit area $\sigma_0 = -(\sigma_1 + \sigma_2)$. If we approximate the latter by a two-dimensional electron gas, elementary statistical mechanics tells us that the energy per unit area of the sheet is $u_0 = \pi \hbar^2 \sigma_0^2 / 2me^2$.

Fix σ_1 and show that the total energy per unit area of the system is minimized when

$$\sigma_2 = -\sigma_1 \frac{C_2}{C_2 + C_0},$$

where C_2 is a characteristic geometric capacitance and C_0 is the "quantum capacitance" of the two-dimensional electron gas. Discuss the classical limit of C_0 and confirm that the corresponding value for σ_2 makes sense.

5.18 Two Pyramidal Conductors Two pyramid-shaped conductors each carry a net charge Q.

(a) Transfer charge δQ from pyramid 2 to pyramid 1. Derive a condition on the coefficients of potential P_{ij} which guarantees that this charge transfer *lowers* the total energy of the system.

(b) Show that the condition in part (a) implies a condition on the coefficients of capacitance.

(c) Under what circumstances can the condition in part (b) be used to determine that pyramid 1 is larger than pyramid 2?

5.19 Capacitance Matrix Practice A grounded metal plate is partially inserted into a parallel-plate capacitor with potential difference $\varphi_2 - \varphi_1 > 0$ as shown in the diagram below. Find the elements of the capacitance matrix. Assume that all plates extend a distance d in the direction perpendicular to the paper. Ignore fringing fields.

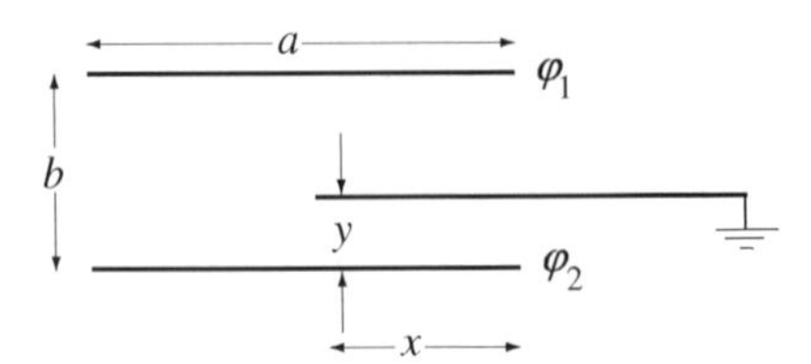

5.20 Bounds on Parallel-Plate Capacitance Let C be the capacitance of capacitor formed from two identical, flat conductor plates separated by a distance d. The plates have area A and arbitrary shape. When $d \ll \sqrt{A}$, we know that the capacitance approaches the value $C_0 = A\epsilon_0/d$.

(a) If $\delta\mathbf{E} = \mathbf{E} - \mathbf{E}_0$, prove the identity

$$\int_V d^3r\, |\mathbf{E}|^2 - \int_V d^3r\, |\mathbf{E}_0|^2 = \int_V d^3r\, |\delta\mathbf{E}|^2 + 2\int_V d^3r\, \mathbf{E}_0 \cdot \delta\mathbf{E}.$$

(b) Let $\mathbf{E} = -\nabla\varphi$ be the actual field between the finite-area plates and let $\mathbf{E}_0 = -\nabla\varphi_0$ be the uniform field that would be present if A were infinite. Use the identity in part (a) to prove that $C > C_0$, using V as the volume between the finite-area plates. Assume that the potentials φ and φ_0 take the same (constant) values on the plates.

5.21 A Two-Wire Capacitor A long, straight wire has length L and a circular cross section with area πa^2. Arrange two such wires so they are parallel and separated by a distance d. You may assume that $L \gg d \gg a$ and ignore end effects.

(a) Graph the electrostatic potential along the straight line which connects the center of one wire to the center of the other wire for the case when the two wires have equal and opposite charge per unit length.
(b) Derive an approximate expression for the capacitance of this two-conductor system.

5.22 An Off-Center Spherical Capacitor A battery maintains the potential difference V between the spheres of a spherical capacitor with capacitance C. Move the center of the inner sphere away from the center of the outer sphere by an amount Δ and call the new capacitance C'.

(a) Use a symmetry argument to show that $C' = C$ to first order in Δ.
(a) Use a force/energy argument to show that $C' = C$ to first order in Δ.

5.23 The Force between Conducting Hemispheres

(a) A spherical metal shell is charged to an electrostatic potential V. Cut this shell in half and pull the halves infinitesimally apart. Find the force with which one hemisphere of the shell repels the other hemisphere.
(b) A spherical capacitor is formed from concentric metal shells with charges $\pm Q$ and radii $b > a$. Cut this capacitor in half and pull the halves infinitesimally apart. Find the force with which the two halves repel one another.

5.24 Holding a Sphere Together A conducting shell of radius R has total charge Q. If sawed in half, the two halves of the shell will fly apart. This can be prevented by placing a point charge Q' at the center of the shell.

(a) What value of Q' just barely holds the shell together?

(b) How does the answer to part (a) change for the case of an insulating sphere with uniform charge density $\sigma = Q/4\pi R^2$?

5.25 Force Equivalence Confirm the assertion made in the text that the inverse relation between the matrix of capacitive coefficients and the matrix of potential coefficients implies the equivalence of these two expressions for the force on the kth conductor of a collection of N conductors:

$$\mathbf{F}_k = \tfrac{1}{2}\sum_{i=1}^{N}\sum_{j=1}^{N}\varphi_i\varphi_j\left(\frac{\partial C_{ij}}{\partial \mathbf{R}_k}\right)_{\mathbf{R}',\varphi} \qquad \text{and} \qquad \mathbf{F}_k = -\tfrac{1}{2}\sum_{i=1}^{N}\sum_{j=1}^{N}Q_iQ_j\left(\frac{\partial P_{ij}}{\partial \mathbf{R}_k}\right)_{\mathbf{R}',Q}.$$

6 Dielectric Matter

If we develop only a macroscopic description of matter in an electric field, we shall find it hard to answer some rather obvious sounding questions.
Edward M. Purcell (1965)

6.1 Introduction

A *dielectric* is a medium that cannot completely screen a static, external, macroscopic electric field from its interior. This property of incomplete screening is a consequence of chemical bonding and other quantum mechanical effects which constrain the rearrangement of its internal charge density when an external field is applied. The same constraints are responsible for the fact that dielectrics do not conduct (or poorly conduct) electric current.

Like a conductor, a dielectric responds to an external field $\mathbf{E}_{\text{ext}}(\mathbf{r})$ by distorting its ground state charge density to produce a field $\mathbf{E}_{\text{self}}(\mathbf{r})$. The total electric field is the sum of these two fields. Unlike a conductor, the total macroscopic field is *non-zero* both inside and outside the volume of the dielectric:

$$\mathbf{E}_{\text{tot}}(\mathbf{r}) = \mathbf{E}_{\text{self}}(\mathbf{r}) + \mathbf{E}_{\text{ext}}(\mathbf{r}) \neq 0. \tag{6.1}$$

The charge distortions that produce $\mathbf{E}_{\text{self}}(\mathbf{r})$ are small for gases and liquids because individual electrons are strongly bound to individual nuclei. Charge rearrangement is greater in solid dielectrics where electrons are *not* necessarily bound to specific nuclei. Nevertheless, chemical bonding effects produce potential energy maxima and minima which inhibit the electron wave functions from exploring some parts of configuration space. We make contact with Chapter 5 with the observation that a perfect conductor is a special case of a dielectric where the non-electrostatic potential energy landscape for charge rearrangement is perfectly flat.

6.2 Polarization

The word *polarization* is used in two ways in the theory of dielectric matter. First, polarization refers to the rearrangement of internal charge that occurs when matter is exposed to an external field. Second, polarization is the name given to a function $\mathbf{P}(\mathbf{r})$ used to characterize the details of the rearrangement. We begin with the intuitive idea of polarization and specify that the source of $\mathbf{E}_{\text{ext}}(\mathbf{r})$ in (6.1) is a charge density $\rho_f(\mathbf{r})$ which is wholly extraneous to the dielectric. A long tradition refers to this as *free charge*. Examples of free charge include the charge on the surface of capacitor plates and point charges we might place inside or outside the body of a dielectric.

The source of $\mathbf{E}_{\text{self}}(\mathbf{r})$ in (6.1) is often called *bound charge*. A more descriptive term is *polarization charge* and we will use the symbol $\rho_{\text{P}}(\mathbf{r})$ for its density. To understand the origin of $\rho_{\text{P}}(\mathbf{r})$, we remind the reader that the macroscopic charge density $\rho(\mathbf{r})$ is *zero* at every point inside a neutral dielectric when $\mathbf{E}_{\rm ext} = 0$ (see Section 2.4.1). When $\mathbf{E}_{\rm ext}$ first appears, positive charge is pushed in one direction and negative charge is pushed in the opposite direction. Charge rearrangement continues until mechanical equilibrium is re-established, and we identify $\rho_P(\mathbf{r})$ as the macroscopic charge density that makes the Coulomb force density $\rho_P(\mathbf{r})\mathbf{E}_{\rm tot}(\mathbf{r})$ equal and opposite to the force density produced by chemical bonding and other non-electrostatic effects. The total charge density that enters Maxwell's theory is the sum of the "free" and "bound" charge densities:

$$\rho(\mathbf{r}) = \rho_f(\mathbf{r}) + \rho_P(\mathbf{r}). \tag{6.2}$$

We use a model-independent approach to introduce the polarization $\mathbf{P}(\mathbf{r})$. The first step is to separate the macroscopic polarization charge density into a surface part, $\sigma_P(\mathbf{r}_S)$, and a volume part, $\rho_P(\mathbf{r})$. Next, we recognize that a neutral dielectric with volume V and surface S remains a neutral dielectric in the presence of free charge of any kind. In that case, the polarization charge densities satisfy the constraint

$$\int_V d^3r\, \rho_P(\mathbf{r}) + \int_S dS\, \sigma_P(\mathbf{r}_S) = 0. \tag{6.3}$$

A neutral conductor satisfies (6.3) with $\rho_P(\mathbf{r}) = 0$ and $\sigma_P(\mathbf{r}_S) \neq 0$. A dielectric uses the polarization $\mathbf{P}(\mathbf{r})$ to satisfy (6.3) with $\rho_P(\mathbf{r}) \neq 0$ and $\sigma_P(\mathbf{r}_S) \neq 0$. The key observation is that the left side of (6.3) is identically *zero* if the divergence theorem is used after substituting

$$\rho_P(\mathbf{r}) = -\nabla \cdot \mathbf{P}(\mathbf{r}) \qquad \mathbf{r} \in V, \tag{6.4}$$

$$\sigma_P(\mathbf{r}_S) = \mathbf{P}(\mathbf{r}_S) \cdot \hat{\mathbf{n}}(\mathbf{r}_S) \qquad \mathbf{r}_S \in S, \tag{6.5}$$

$$\mathbf{P}(\mathbf{r}) = 0 \qquad \mathbf{r} \notin V. \tag{6.6}$$

Three remarks are germane. First, the vector $\hat{\mathbf{n}}(\mathbf{r}_S)$ in (6.5) is the outward unit normal to the surface S at the point $\mathbf{r}_S$. Second, $\mathbf{P}(\mathbf{r}) = 0$ outside the sample in (6.6) because polarization is associated with matter and there is no matter outside V.[1] Third, the equations (6.4), (6.5), and (6.6) do not determine $\mathbf{P}(\mathbf{r})$ uniquely. This follows from Helmholtz' theorem (Section 1.9) and the fact that we did not specify $\nabla \times \mathbf{P}(\mathbf{r})$ *inside* the sample volume V.

To summarize, the macroscopic electrostatic field of a dielectric sample is produced by macroscopic polarization charge densities $\rho_P(\mathbf{r})$ and $\sigma_P(\mathbf{r})$. These, in turn, are determined by the polarization $\mathbf{P}(\mathbf{r})$ of the sample. The latter is fundamental to the theory and our attention must now turn to its physical meaning and methods that can be used to calculate it.

6.2.1 The Volume Integral of P(r)

An important clue to the physical meaning of the polarization $\mathbf{P}(\mathbf{r})$ is that its volume integral is the total electric dipole moment $\mathbf{p}$ of a dielectric sample. To see this, we integrate the $k^{\rm th}$ Cartesian component of $\mathbf{P}(\mathbf{r})$ over the sample volume V. Because $\mathbf{P} \cdot \nabla r_k = P_k$,

$$\int_V d^3r\, P_k = \int_V d^3r\, \nabla \cdot (r_k \mathbf{P}) - \int_V d^3r\, r_k (\nabla \cdot \mathbf{P}). \tag{6.7}$$

[1] Some authors build (6.6) into their definition of $\mathbf{P}(\mathbf{r})$. This generates the surface density (6.5) as a singular piece of the volume density (6.4). An example is a dielectric which occupies the half-space $z \leq 0$ with polarization $\mathbf{P}(\mathbf{r}) = \mathbf{P}_0(\mathbf{r})\Theta(-z)$. In that case, $\rho_P = -\nabla \cdot \mathbf{P}(\mathbf{r}) = -\Theta(-z)\nabla \cdot \mathbf{P}_0 + \delta(z)\mathbf{P}_0 \cdot \hat{\mathbf{z}}$. The last term is σ_P. Our preference is to display the volume and surface parts of the polarization separately.

Using the divergence theorem and the definitions in (6.4) and (6.5),

$$\int_V d^3r\, P_k = \int_S dS\, r_k \sigma_P(\mathbf{r}_S) + \int_V d^3r\, r_k\, \rho_P(\mathbf{r}). \tag{6.8}$$

The right-hand side of (6.8) defines the kth Cartesian component of the electric dipole moment $\mathbf{p}$ of the sample. We conclude that the integral of the polarization $\mathbf{P}(\mathbf{r})$ over the volume of a dielectric is equal to the total dipole moment of the dielectric:

$$\int_V d^3r\, \mathbf{P}(\mathbf{r}) = \mathbf{p}. \tag{6.9}$$

6.2.2 The Lorentz Model

Following Lorentz (1902), many authors use a microscopic version of (6.9) to identify $\mathbf{P}(\mathbf{r})$ as an "electric dipole moment per unit volume". The physical idea is to regard a polarized dielectric as a collection of atomic or molecular electric dipoles. The mathematical prescription defines the polarization at a macroscopic point $\mathbf{r}$ as the electric dipole moment of a microscopic cell with volume Ω labeled by $\mathbf{r}$:[2]

$$\mathbf{P}(\mathbf{r}) = \frac{1}{\Omega}\int_\Omega d^3s\, \mathbf{s}\, \rho_{\text{micro}}(\mathbf{s}) = \frac{\mathbf{p}(\mathbf{r})}{\Omega}. \tag{6.10}$$

For finite Ω, (6.10) replaces the true charge distribution in each cell by a point electric dipole. In the limit $\Omega \to 0$, the Lorentz approximation replaces the entire dielectric by a continuous distribution of point electric dipoles with a density $\mathbf{P}(\mathbf{r})$ computed from (6.10).

Despite its widespread use, the Lorentz formula (6.10) is usually a poor approximation to the true polarization (see Section 6.2.3). The exceptions which prove the rule are dielectric gases, non-polar liquids, and molecular solids where the constituent atoms and molecules interact very weakly. The inaccuracies of the Lorentz model appear consistently for all dielectrics where chemical bonding is important and "bond charge" is present on the boundaries of the averaging cell Ω. Worse, (6.10) generally gives different values for $\mathbf{p}(\mathbf{r})/\Omega$ when one uses different (but equally sensible) choices for Ω. The simple model of an ionic crystal shown in Figure 6.1 shows that this can be true even when no bond charge is present. The two panels show two choices for the Lorentz cell Ω. Using (6.10), they lead to oppositely directed $\mathbf{P}(\mathbf{r})$ vectors. Other cell choices give other values, including $\mathbf{P}(\mathbf{r}) = 0$.

6.2.3 The Modern Theory of Polarization

In the 1990s, an unambiguous theory of the polarization $\mathbf{P}(\mathbf{r})$ was developed which abandons the Lorentz point of view. To get a flavor for this approach, we recall from (6.4) that the dielectric charge density $\rho(\mathbf{r})$ is the (negative) divergence of the polarization. This implies that $\mathbf{P}(\mathbf{r})$ contains *more* information than $\rho(\mathbf{r})$. However, if the polarization were simply the dipole moment per unit volume of the charge density, $\mathbf{P}(\mathbf{r})$ would contain *less* information than $\rho(\mathbf{r})$. We resolve this paradox by recalling from (2.4) that quantum mechanics defines $\rho(\mathbf{r})$ as the absolute square of a system's wave function. $\mathbf{P}(\mathbf{r})$ contains more information than $\rho(\mathbf{r})$ because the polarization subtly encodes information about the *phase* of the system wave function. This is the essential insight provided by the modern theory of polarization.[3]

[2] In this chapter, $\mathbf{r}$ is a macroscopic variable and $\mathbf{s}$ is a microscopic variable. In Section 2.3.1 on Lorentz averaging, the macroscopic variable was called $\mathbf{R}$ and the microscopic variable was called $\mathbf{r}$.

[3] See Resta and Vanderbilt (2007) in Sources, References, and Additional Reading.

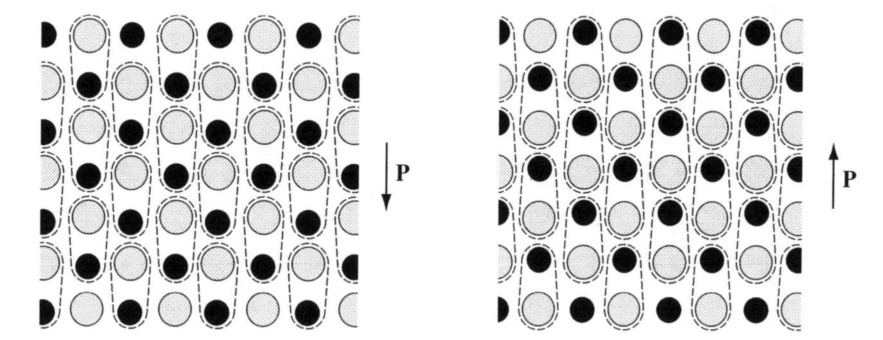

Figure 6.1: Cartoon view of an ionic crystal. White spheres are negative ions. Black spheres are positive ions. The arrows indicate the local polarization $\mathbf{P}(\mathbf{r})$ computed from Lorentz' formula (6.10). The figure, reproduced from Purcell (1965), repeats periodically like a checkerboard.

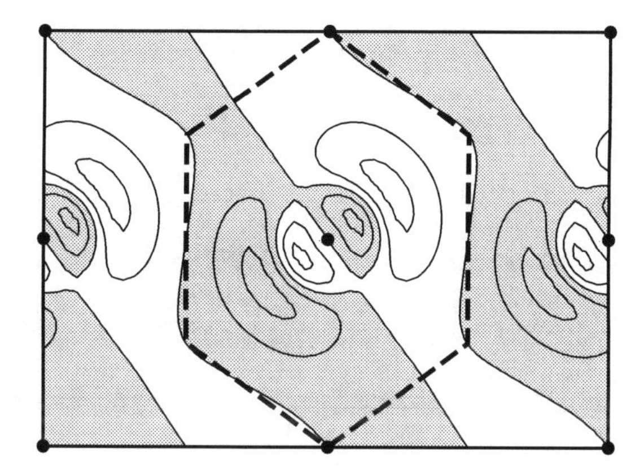

Figure 6.2: Contour plot of the charge density induced in a NaBr crystal by an electric field that points from the lower left to the upper right. Shaded and unshaded regions correspond, respectively, to an excess or deficit of negative (electron) charge compared to the unpolarized state. The dots on the upper and lower borders are Na nuclei. The three dots across the middle are Br nuclei. The figure repeats periodically like a checkerboard. Reproduced from Umari *et al.* (2001).

We indicated in Section 2.4.1 that a current density $\mathbf{j}_P = \partial\mathbf{P}/\partial t$ is associated with time variations of the polarization. Without going into details, we use this formula to state the entirely plausible answer: replace (6.10) by the Lorentz cell average of the local, time-integrated current density of polarization charge which flows when an external electric field is switched on to polarize a dielectric:[4]

$$\mathbf{P}(\mathbf{r}) = \frac{1}{\Omega}\int_{\Omega} d^3s \int_{-\infty}^{\infty} dt\, \mathbf{j}_{P,\mathrm{micro}}(\mathbf{s}, t). \tag{6.11}$$

A quantum mechanical calculation is generally required to find the microscopic current density $\mathbf{j}_{P,\mathrm{micro}}(\mathbf{s}, t)$.

An important application of (6.11) is to test the quality of the Lorentz approximation. Figure 6.2 shows a contour plot of the (quantum mechanically) calculated microscopic charge density $\rho_{\mathrm{micro}}(\mathbf{s})$ induced in a NaBr crystal by an external electric field $\mathbf{E}$. This is a favorable case where the Lorentz

[4] This definition assumes that $\mathbf{P} = 0$ at $t = -\infty$.

picture that motivates (6.10) might be expected to be correct. Therefore, it is comforting to see that the field does indeed displace charge in a dipole-like fashion inside the cell defined by the dashed lines. A smaller, oppositely directed dipole displacement occurs very close to each nucleus. However, the macroscopic $\mathbf{P}(\mathbf{r})$ calculated from the Lorentz formula (6.10) (using the dashed cell as Ω) turns out to account for less than half of the true polarization calculated from (6.11). The majority of $\mathbf{P}(\mathbf{r})$ comes from polarization currents which flow *between* averaging cells. These currents play no role in the Lorentz approximation. Similar calculations performed for covalent dielectrics like silicon produce highly complex plots for $\rho(\mathbf{r})$ and $\mathbf{P}(\mathbf{r})$ which completely rule out an interpretation of the polarization in terms of cell dipole moments. The Lorentz picture is qualitatively incorrect for these dielectrics.

To summarize, Lorentz' physical model of a dielectric composed of polarized atoms or molecules is realistic only when those entities retain their individual integrity as quantum mechanical objects. Gases, simple liquids, and van der Waals-bonded molecular solids satisfy this criterion, but the vast majority of dielectrics do not. For these latter cases, the Lorentz formula (6.10) is not reliable and more sophisticated methods are needed to calculate $\mathbf{P}(\mathbf{r})$. However, once $\mathbf{P}(\mathbf{r})$ is known—by whatever means—it turns out that Lorentz' notion that a dielectric behaves like a continuous distribution of point electric dipoles with density $\mathbf{P}(\mathbf{r})$ is rigorously correct. We delay the proof of this assertion until Section 6.3.1 in order to gain some introductory appreciation of the potential and field produced by polarized matter.

6.3 The Field Produced by Polarized Matter

This section focuses on the potential φ_P and field $\mathbf{E}_P$ produced by polarized matter, regardless of how the polarization $\mathbf{P}(\mathbf{r})$ is produced. Section 6.4 combines $\mathbf{E}_P$ with the field produced by other sources to get the electric field $\mathbf{E}$ of the Maxwell equations.

We assume that $\mathbf{P}(\mathbf{r})$ is specified once and for all. In that case, the macroscopic charge densities (6.4) and (6.5) inserted into the general expressions (3.10) and (3.12) produce the scalar potential

$$\varphi_P(\mathbf{r}) = \frac{1}{4\pi\epsilon_0}\int_V d^3r' \, \frac{-\nabla'\cdot\mathbf{P}(\mathbf{r}')}{|\mathbf{r}-\mathbf{r}'|} + \frac{1}{4\pi\epsilon_0}\int_S d\mathbf{S}'\cdot\frac{\mathbf{P}(\mathbf{r}')}{|\mathbf{r}-\mathbf{r}'|}. \tag{6.12}$$

The corresponding Coulomb's law electric field is $\mathbf{E}_P = -\nabla\varphi_P$, or

$$\mathbf{E}_P(\mathbf{r}) = \frac{1}{4\pi\epsilon_0}\int_V d^3r' \, \frac{-\nabla'\cdot\mathbf{P}(\mathbf{r}')}{|\mathbf{r}-\mathbf{r}'|^3}(\mathbf{r}-\mathbf{r}') + \frac{1}{4\pi\epsilon_0}\int_S d\mathbf{S}'\cdot\mathbf{P}(\mathbf{r}')\frac{\mathbf{r}-\mathbf{r}'}{|\mathbf{r}-\mathbf{r}'|^3}. \tag{6.13}$$

The integrals in (6.12) and (6.13) converge and give $\varphi_P(\mathbf{r})$ and $\mathbf{E}_P(\mathbf{r})$ correctly everywhere, including points that lie inside V.

The special case of a *uniformly* polarized sample simplifies (6.13) considerably because $\nabla\cdot\mathbf{P}=0$ leaves only the surface integral to evaluate. The divergence theorem transforms the latter to

$$\mathbf{E}_P(\mathbf{r}) = \frac{1}{4\pi\epsilon_0}\int_V d^3r' \, \nabla'_k\left[P_k\frac{\mathbf{r}-\mathbf{r}'}{|\mathbf{r}-\mathbf{r}'|^3}\right]. \tag{6.14}$$

Now, move the constant P_k outside the integral and replace ∇'_k by $-\nabla_k$ using the fact that the gradient acts only on a function of $\mathbf{r}-\mathbf{r}'$. The final result, known as *Poisson's formula*, expresses $\mathbf{E}_P(\mathbf{r})$ in terms of the electric field $\mathcal{E}(\mathbf{r})$ produced by the same object with the uniform polarization replaced by a uniform charge density with unit magnitude:

$$\mathbf{E}_P(\mathbf{r}) = -\mathbf{P}\cdot\nabla\frac{1}{4\pi\epsilon_0}\int_V d^3r' \, \frac{\mathbf{r}-\mathbf{r}'}{|\mathbf{r}-\mathbf{r}'|^3} = -(\mathbf{P}\cdot\nabla)\mathcal{E}(\mathbf{r}). \tag{6.15}$$

Application 6.1 A Uniformly Polarized Sphere

The electric field produced by a sphere of radius R with uniform polarization $\mathbf{P}$ plays a role in many applications. We derive it here by two different methods.

Method I: A direct application of Poisson's formula (6.15) requires the electric field of a sphere with uniform charge density $\rho = 1$. This is a standard problem solved using Gauss' law in integral form. The field is

$$\mathcal{E}(\mathbf{r}) = \begin{cases} \dfrac{\mathbf{r}}{3\epsilon_0} & r < R, \\ \dfrac{\mathbf{r}V}{4\pi\epsilon_0 r^3} & r > R. \end{cases} \tag{6.16}$$

Then, because $\nabla_i r_j = \delta_{ij}$ and $\nabla_i(r_j/r^3) = \nabla_j(r_i/r^3)$, (6.15) gives

$$\mathbf{E}_P(\mathbf{r}) = \begin{cases} -\dfrac{\mathbf{P}}{3\epsilon_0} & r < R, \\ \dfrac{V}{4\pi\epsilon_0}\left\{\dfrac{3(\hat{\mathbf{r}}\cdot\mathbf{P})\hat{\mathbf{r}} - \mathbf{P}}{r^3}\right\} & r > R. \end{cases} \tag{6.17}$$

Figure 6.3 is a plot of $\mathbf{E}_P(\mathbf{r})$ as given by (6.17). Every electric field line begins at a point where the surface polarization charge density $\mathbf{P}\cdot\hat{\mathbf{n}}$ is positive and ends at a point where σ_P is negative. Outside the sphere, $\mathbf{E}_P(\mathbf{r})$ is identical to the electric field of a point dipole at the origin with moment $\mathbf{p} = V\mathbf{P}$. This is the dipole moment of the entire sphere [see (6.9)]. Inside the sphere, the electric field is constant[5] and anti-parallel to $\mathbf{P}$. Moreover, a glance back at (4.16) shows that this constant electric field is identical to the singular (delta function) part of the electric field of a point electric dipole. We conclude that a point electric dipole may sensibly be regarded as the $R \to 0$ limit of a uniformly polarized sphere.

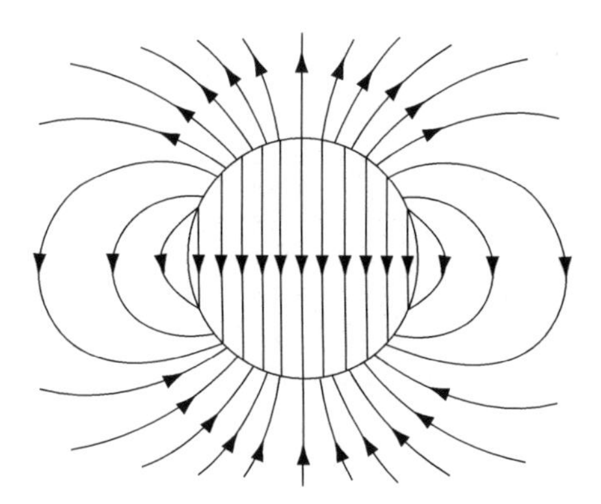

Figure 6.3: The electric field of a uniformly polarized sphere. Inside the sphere, $\mathbf{E}_P$ is anti-parallel to the polarization density vector $\mathbf{P}$. Outside the sphere, $\mathbf{E}_P(\mathbf{r})$ is exactly dipolar.

Method II: If $\mathbf{P} = P\hat{\mathbf{z}}$, there is no volume polarization charge density and the surface polarization charge density is

$$\sigma_P = \mathbf{P}\cdot\hat{\mathbf{r}} = P\cos\theta. \tag{6.18}$$

[5] The electric field inside a polarized volume is constant if and only if the volume shape is ellipsoidal. $\mathbf{E}_P$ generally points in a different direction than $\mathbf{P}$. See Brownstein (1987) in Sources, References, and Additional Reading.

In Application 4.3, we used a multipole method to find the electrostatic potential produced by a charge density of the form (6.18) confined to the surface of a sphere. Writing that result in the language of the current problem gives

$$\varphi(r,\theta) = \frac{P}{3\epsilon_0}\begin{cases} z & r < R, \\ \dfrac{R^3}{r^2}\cos\theta & r > R. \end{cases} \tag{6.19}$$

The electric field $\mathbf{E} = -\nabla\varphi$ computing using (6.19) reproduces (6.17).

6.3.1 Polarized Matter as a Superposition of Point Dipoles

In this section, we identify the fundamental physics behind $\varphi_P(\mathbf{r})$ in (6.12) and explain why the Lorentz model has enduring value, despite its deficiencies as a physical model for dielectric matter (see Section 6.2.3). The essential step is to rewrite the surface integral in (6.12) using the divergence theorem. This gives

$$\varphi_P(\mathbf{r}) = \frac{1}{4\pi\epsilon_0}\int_V d^3r' \,\frac{-\nabla'\cdot\mathbf{P}(\mathbf{r}')}{|\mathbf{r}-\mathbf{r}'|} + \frac{1}{4\pi\epsilon_0}\int_V d^3r' \,\nabla'\cdot\left[\frac{\mathbf{P}(\mathbf{r}')}{|\mathbf{r}-\mathbf{r}'|}\right]. \tag{6.20}$$

Writing out the divergence explicitly,

$$\nabla'\cdot\left[\frac{\mathbf{P}(\mathbf{r}')}{|\mathbf{r}-\mathbf{r}'|}\right] = \frac{\nabla'\cdot\mathbf{P}(\mathbf{r}')}{|\mathbf{r}-\mathbf{r}'|} + \mathbf{P}(\mathbf{r}')\cdot\nabla'\frac{1}{|\mathbf{r}-\mathbf{r}'|}, \tag{6.21}$$

and substituting back into (6.20) produces a cancellation of two terms. The final result is the convergent integral

$$\varphi_P(\mathbf{r}) = \frac{1}{4\pi\epsilon_0}\int_V d^3r' \,\mathbf{P}(\mathbf{r}')\cdot\nabla'\frac{1}{|\mathbf{r}-\mathbf{r}'|}. \tag{6.22}$$

Equation (6.22) is a key result. Because $\nabla'|\mathbf{r}-\mathbf{r}'|^{-1} = -\nabla|\mathbf{r}-\mathbf{r}'|^{-1}$, comparing (6.22) with (4.13) or (4.35) shows that $\varphi_P(\mathbf{r})$ is the electrostatic potential produced by collection of *point* electric dipoles with moments $d\mathbf{p}(\mathbf{r}) = \mathbf{P}(\mathbf{r})d^3r$. In other words, (6.22) is the electrostatic potential produced by a volume distribution of point electric dipole moments with density $\mathbf{P}(\mathbf{r})$. What does this mean?

We know from Section 4.2.1 that point electric dipoles do not exist in Nature. We also know (from Section 6.2.2) that the Lorentz model of molecular dipoles is only occasionally applicable to real dielectric matter. Nevertheless, (6.22) shows that the electrostatic potential of a dielectric is indistinguishable from the potential produced by a continuous distribution of *fictitious point electric dipoles* with a dipole moment per unit volume equal to $\mathbf{P}(\mathbf{r})$. This shows that Lorentz' idea to represent a polarized solid by a volume distribution of point dipoles is perfectly valid. We simply must provide a better approximation for $\mathbf{P}(\mathbf{r})$ than Lorentz could offer.

The point dipole representation of the polarization qualitatively rationalizes the dependence of the polarization charge densities $\rho_P(\mathbf{r})$ and $\sigma_P(\mathbf{r})$ on $\mathbf{P}(\mathbf{r})$. Consider the one-dimensional representation of $\mathbf{P}(\mathbf{r})$ shown in Figure 6.4. In the interior, the positively charged end of one dipole compensates the negatively charged end of an adjacent dipole unless there is some variation in either the magnitude or the direction of the dipole density. This is the origin of the volume polarization charge density $\rho_P = -\nabla\cdot\mathbf{P}$. Similarly, the projection onto the surface normal of those point dipoles which reside just at the sample surface leaves some charge uncompensated in the sense just described. This is the origin of the surface charge density $\sigma = \mathbf{P}\cdot\hat{\mathbf{n}}$.

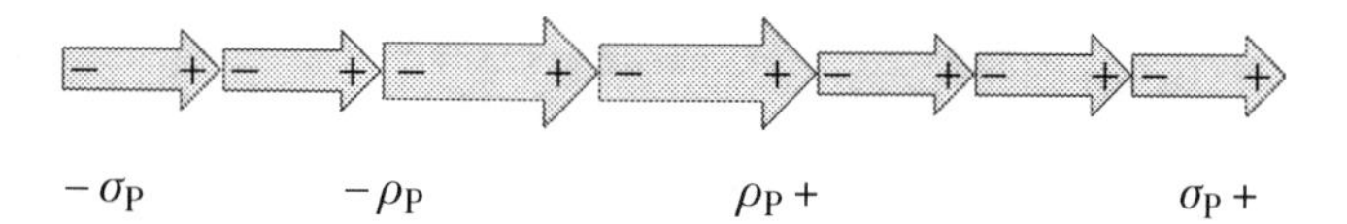

Figure 6.4: One-dimensional point dipole representation of the polarization $\mathbf{P}(\mathbf{r})$ with $\rho_P = -\nabla \cdot \mathbf{P}$ and $\sigma_P = \mathbf{P} \cdot \hat{\mathbf{n}}$. The size of each arrow is proportional to the magnitude of $\mathbf{P}(\mathbf{r})$ at that point.

The point of view offered in this section leads some authors to construct an alternative to the electric field (6.13) by superposing point dipole electric fields (Section 4.2.1) suitably weighted by $\mathbf{P}(\mathbf{r})$. This gives

$$\mathbf{E}_P^\dagger(\mathbf{r}) = \frac{1}{4\pi\epsilon_0} \int_V d^3r' \left\{ \frac{3(\mathbf{r}-\mathbf{r}')\mathbf{P}(\mathbf{r}')\cdot(\mathbf{r}-\mathbf{r}')}{|\mathbf{r}-\mathbf{r}'|^5} - \frac{\mathbf{P}(\mathbf{r}')}{|\mathbf{r}-\mathbf{r}'|^3} \right\}. \tag{6.23}$$

We have been careful to label this field $\mathbf{E}_P^\dagger(\mathbf{r})$ to distinguish it from the field $\mathbf{E}_P(\mathbf{r})$ in (6.13). The two are identical for observation points $\mathbf{r}$ that lie *outside* the dielectric volume V. Awkwardly, (6.23) *diverges* for observation points that lie *inside* the polarized sample. The divergence is handled by scooping out an infinitesimal vacuum cavity around $\mathbf{r}$ and studying the limit as the volume of the cavity goes to zero. Unfortunately, the value for the integral obtained by doing this is not unique – it depends on the *shape* of the cavity.[6] This leads us to eschew (6.23) in favor of (6.13) when $\mathbf{E}_P(\mathbf{r})$ is needed at interior points.

6.4 The Total Electric Field

The total electric field $\mathbf{E}$ in the Maxwell equations is the sum of the electric field $\mathbf{E}_P$ produced by the polarization charge $\rho_P(\mathbf{r}) = -\nabla \cdot \mathbf{P}$ and the electric field produced by all charges not associated with dielectric bodies. We designated the latter "free charge" with density $\rho_f(\mathbf{r})$ at the beginning of Section 6.2. Gauss' law involves the total charge, which is the sum of the polarization charge and the free charge. Therefore,

$$\epsilon_0 \nabla \cdot \mathbf{E}(\mathbf{r}) = \rho(\mathbf{r}) = \rho_P(\mathbf{r}) + \rho_f(\mathbf{r}) = -\nabla \cdot \mathbf{P}(\mathbf{r}) + \rho_f(\mathbf{r}). \tag{6.24}$$

6.4.1 The Auxiliary Field $\mathbf{D}(\mathbf{r})$

Motivated by (6.24), it is traditional to define an auxiliary vector field $\mathbf{D}(\mathbf{r})$ which combines the electric field with the polarization,[7]

$$\mathbf{D}(\mathbf{r}) = \epsilon_0 \mathbf{E}(\mathbf{r}) + \mathbf{P}(\mathbf{r}). \tag{6.25}$$

The fundamental electrostatic condition remains valid in matter:

$$\nabla \times \mathbf{E} = 0. \tag{6.26}$$

Therefore, the preceding three equations imply that

$$\nabla \cdot \mathbf{D} = \rho_f \tag{6.27}$$

and

$$\nabla \times \mathbf{D} = \nabla \times \mathbf{P}. \tag{6.28}$$

[6] Older textbooks speak of "cavity definitions" for the field inside matter in this context.

[7] $\mathbf{D}(\mathbf{r})$ is often called the "electric displacement" in the older literature of electromagnetism.

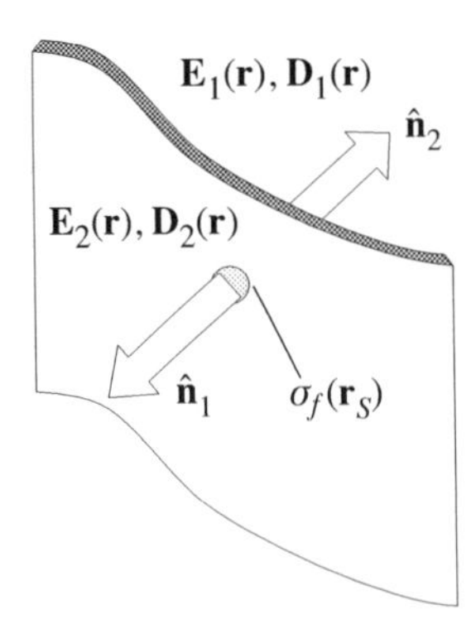

Figure 6.5: A boundary layer with free charge density σ_f separates two dissimilar dielectrics. The unit normal $\hat{\mathbf{n}}_1$ points outward from region 1. The unit normal $\hat{\mathbf{n}}_2$ points outward from region 2.

If Q_f is the free charge enclosed by a surface S, it is often useful to exploit the integral form of (6.27),

$$\int_S d\mathbf{S} \cdot \mathbf{D} = Q_f. \tag{6.29}$$

Elementary applications of (6.29) sometimes create the false impression that free charge is the sole source of the auxiliary field $\mathbf{D}$. The truth emerges when we use the Helmholtz theorem (Section 1.9) to express $\mathbf{D}(\mathbf{r})$ explicitly in terms of its divergence (6.27) and curl (6.28). This gives

$$\mathbf{D}(\mathbf{r}) = -\nabla \int d^3r' \, \frac{\rho_f(\mathbf{r}')}{4\pi|\mathbf{r}-\mathbf{r}'|} + \nabla \times \int d^3r' \, \frac{\nabla' \times \mathbf{P}(\mathbf{r}')}{4\pi|\mathbf{r}-\mathbf{r}'|}. \tag{6.30}$$

The second term in (6.30) tells us that most spatial variations in $\mathbf{P}(\mathbf{r})$ (including abrupt changes at macroscopic surfaces and interfaces) contribute to $\mathbf{D}(\mathbf{r})$. It is interesting that this term has the same structure as a magnetic field $\mathbf{B} = \nabla \times \mathbf{A}$ in the Coulomb gauge.

6.4.2 Matching Conditions

Following the method of Section 2.3.3 (or otherwise), the matching conditions implied by (6.27) and (6.28) at an interface between two regions of space endowed with a surface density of free charge σ_f are (Figure 6.5)

$$\hat{\mathbf{n}}_2 \cdot (\mathbf{D}_1 - \mathbf{D}_2) = \sigma_f \tag{6.31}$$

and

$$\hat{\mathbf{n}}_2 \times (\mathbf{D}_1 - \mathbf{D}_2) = \hat{\mathbf{n}}_2 \times (\mathbf{P}_1 - \mathbf{P}_2). \tag{6.32}$$

In light of (6.25), the latter is simply the familiar

$$\hat{\mathbf{n}}_2 \times (\mathbf{E}_1 - \mathbf{E}_2) = 0. \tag{6.33}$$

6.4.3 Constitutive Relations

The equations (6.25), (6.26), and (6.27) cannot be solved simultaneously unless one (i) specifies $\mathbf{P}(\mathbf{r})$ once and for all, or (ii) invokes a *constitutive relation* that relates $\mathbf{P}(\mathbf{r})$ to $\mathbf{E}(\mathbf{r})$. It is here where we distinguish real dielectric matter from mere charge by using experiment, theory, or phenomenology to inject quantum mechanical and/or statistical mechanical information into classical electrodynamics.

For example, a *ferroelectric* is a form of matter where **P** can exist in the absence of an external field. We will not pause to treat this case because it is rather uncommon. Instead, we focus on the vast majority of systems which are unpolarized in their ground state but which acquire a macroscopic polarization in the presence of a uniform external electric field. The general rule revealed by experiment is

$$P_i = \epsilon_0 \chi_{ij} E_j + \epsilon_0 \chi^{(2)}_{ijk} E_j E_k + \cdots . \tag{6.34}$$

The tensor character of the constants χ_{ij} and $\chi^{(2)}_{ijk}$ allows for the possibility that **P** is not parallel to **E**. This is realized in spatially anisotropic matter. The second-order and higher-order terms in (6.34) allow for the possibility that the polarization depends non-linearly on the field. This is realized in all matter when the electric field strength is large enough.

6.5 Simple Dielectric Matter

The first term on the right-hand side of (6.34) is sufficient to describe the polarization of a *linear* dielectric. In this book, a dielectric that is both linear and spatially isotropic will be called *simple*. A simple dielectric obeys the constitutive relation

$$\mathbf{P} = \epsilon_0 \chi \mathbf{E}. \tag{6.35}$$

The constant χ is called the *electric susceptibility*. We also introduce the *permittivity* ϵ and the dimensionless *dielectric constant* κ through

$$\mathbf{P} = (\epsilon - \epsilon_0)\mathbf{E} = \epsilon_0(\kappa - 1)\mathbf{E}. \tag{6.36}$$

Statistical mechanical arguments show that $\chi \geq 0$.[8] Therefore, $\epsilon \geq \epsilon_0$ and $\kappa \geq 1$ as well. This is plausible because a dielectric incompletely screens an external field. In a simple medium, the auxiliary field defined in (6.25) becomes

$$\mathbf{D} = \epsilon \mathbf{E} = \kappa \epsilon_0 \mathbf{E} = \epsilon_0(1 + \chi)\mathbf{E}. \tag{6.37}$$

Experiments show that (6.35), (6.36), and (6.37) apply equally well to macroscopic fields that vary with position in a simple medium.

Example 6.1 A point dipole $\mathbf{p}_0$ is embedded at the center of a dielectric sphere with volume V and dielectric constant κ. Find the total dipole moment $\mathbf{p}_{\rm tot}$ of the entire system.

Solution: From (6.9) and (6.36), the dipole moment of the polarizable dielectric is

$$\mathbf{p} = \int_V d^3r\, \mathbf{P} = \epsilon_0(\kappa - 1)\int_V d^3r\, \mathbf{E}.$$

On the other hand, we proved in Example 4.1 that

$$\int_V d^3r\, \mathbf{E} = -\frac{\mathbf{p}_{\rm tot}}{3\epsilon_0}.$$

Therefore,

$$\mathbf{p}_{\rm tot} = \mathbf{p}_0 + \mathbf{p} = \mathbf{p}_0 - \tfrac{1}{3}(\kappa - 1)\mathbf{p}_{\rm tot},$$

[8] See, for example, T.M. Sanders, Jr., "On the sign of the static susceptibility", *American Journal of Physics* **56**, 448 (1988).

or

$$\mathbf{p}_{\text{tot}} = \frac{3}{2+\kappa}\mathbf{p}_0.$$

Since $\kappa \geq 1$, a dielectric medium generally screens (reduces the magnitude of) the embedded dipole.

6.5.1 Fields and Sources in Simple Dielectric Matter

We begin our study of simple dielectrics by inserting (6.37) into the Maxwell equation (6.27) to get

$$\epsilon \nabla \cdot \mathbf{E} + \mathbf{E} \cdot \nabla \epsilon = \rho_f. \tag{6.38}$$

The general problem posed by (6.38) is difficult, particularly when the dielectric "constant" varies smoothly with position. In this book, we restrict ourselves to situations no more complicated than Figure 6.7 in Section 6.5.3 below, where the dielectric constant takes (different) constant values in distinct regions of space separated by sharp boundaries. In each region, (6.38) simplifies to

$$\epsilon \nabla \cdot \mathbf{E} = \rho_f. \tag{6.39}$$

The global electric field is constructed from the fields calculated in each region by enforcing the matching conditions (6.31) and (6.33) at each sharp boundary. Alternately, because $\nabla \times \mathbf{E} = 0$ is always true in electrostatics, we can insert $\mathbf{E} = -\nabla\varphi$ into (6.39) and solve a Poisson equation,

$$\epsilon \nabla^2 \varphi = -\rho_f, \tag{6.40}$$

in each region where the permittivity is constant. Section 6.5.5 explores this potential-theory approach to simple dielectric matter.

Inserting (6.35) and (6.37) into the Maxwell equation (6.28) gives zero on both sides because $\nabla \times \mathbf{E} = 0$. On the other hand, writing (6.28) in the form

$$\nabla \times \mathbf{P} = 0 \tag{6.41}$$

seems to imply that the second integral in (6.30) vanishes, from which we might conclude that free charge is the only source of the field $\mathbf{D}(\mathbf{r})$. This is not true (in general) because a contribution to the polarization integral survives from every surface S_k where the dielectric constant changes discontinuously. We leave it as an exercise for the reader to show that (6.30) reduces in this case to

$$\mathbf{D}(\mathbf{r}) = -\nabla \int d^3r' \frac{\rho_f(\mathbf{r}')}{4\pi|\mathbf{r}-\mathbf{r}'|} + \sum_k \nabla \times \int dS'_k \frac{\mathbf{P}(\mathbf{r}') \times \hat{\mathbf{n}}(\mathbf{r}')}{4\pi|\mathbf{r}-\mathbf{r}'|}. \tag{6.42}$$

Finally, we have learned that polarization charge is the fundamental source of the field produced by a polarization $\mathbf{P}(\mathbf{r})$. Using (6.36) and (6.39), the volume and surface polarization charge densities produced by a simple dielectric with permittivity $\epsilon_0\kappa$ are

$$\rho_P = -\nabla \cdot \mathbf{P} = \left(\frac{1}{\kappa} - 1\right)\rho_f \tag{6.43}$$

and

$$\sigma_P = \mathbf{P} \cdot \hat{\mathbf{n}}|_S = \epsilon_0(\kappa - 1)\mathbf{E} \cdot \hat{\mathbf{n}}|_S. \tag{6.44}$$

Typically, there is a contribution to (6.44) from both sides of an interface between two simple dielectrics.

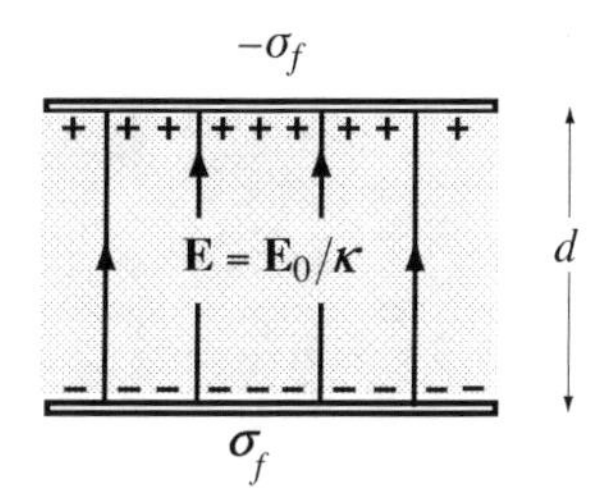

Figure 6.6: A capacitor with fixed plate charge density σ_f. Plus and minus signs denote polarization charge at the surface of the dielectric. A few lines of the electric field $\mathbf{E} = E\hat{\mathbf{z}}$ are indicated.

6.5.2 Simple Dielectric Response to Free Charge

A point charge embedded in a medium with dielectric constant κ is an example of a volume distribution of free charge $\rho_f(\mathbf{r})$. According to (6.43), the point charge induces a polarization charge which occupies the same point in space as the point charge itself. This macroscopic statement is a consequence of Lorentz averaging (Section 2.3). In microscopic reality, the point charge polarizes nearby matter and partially neutralizes itself by attracting nearby charges in the dielectric (of opposite sign) to itself. Because $\kappa \geq 1$, this partial neutralization appears on the macroscopic level when we use (6.43) to calculate the total volume charge density,

$$\rho(\mathbf{r}) = \rho_P(\mathbf{r}) + \rho_f(\mathbf{r}) = \frac{\rho_f(\mathbf{r})}{\kappa}. \tag{6.45}$$

The screening physics of (6.45) is built in when we solve Gauss' law (6.39) for a point charge q_0 embedded at $\mathbf{r}_0$ in a simple dielectric medium:

$$\mathbf{E}(\mathbf{r}) = \frac{q_0}{4\pi\epsilon}\frac{\mathbf{r}-\mathbf{r}_0}{|\mathbf{r}-\mathbf{r}_0|^3} = \frac{\mathbf{E}_{\text{vac}}(\mathbf{r})}{\kappa}. \tag{6.46}$$

The corresponding solution of the Poisson equation (6.40) for the electrostatic potential is

$$\varphi(\mathbf{r}) = \frac{q_0}{4\pi\,\epsilon}\frac{1}{|\mathbf{r}-\mathbf{r}_0|} = \frac{\varphi_{\text{vac}}(\mathbf{r})}{\kappa}. \tag{6.47}$$

It is not always appreciated that the field and potential produced by the embedded point charge are screened (reduced) by a factor of κ compared to the case of a point charge in vacuum, whether the observation point lies within the dielectric medium or not.[9] This is a consequence of the spatially *local* character of (6.45).

A Parallel-Plate Capacitor with Fixed Charge

Figure 6.6 shows a parallel-plate capacitor with vacuum capacitance $C_0 = \epsilon_0 A/d$. What happens when a dielectric is inserted to fill the space between the plates and the system is isolated so the plate charge Q remains fixed? The charge density $\sigma_f = Q_f/A$ is fixed also, so Gauss' law (6.29) and a pillbox-shaped Gaussian surface with one face inside the lower conducting plate and one face inside the dielectric tell us that $\mathbf{D} = \sigma_f\hat{\mathbf{z}}$.

Therefore,

$$\mathbf{E} = \frac{\mathbf{D}}{\epsilon} = \frac{\mathbf{E}_0}{\kappa}. \tag{6.48}$$

The electric field is reduced (screened) compared to its value $\mathbf{E}_0 = \hat{\mathbf{z}}\sigma/\epsilon_0$ in the absence of the dielectric. This is consistent with (6.46) because the free charge at the surface of each metal plate

[9] If the observation point lies outside the medium, there are generally other contributions to the total field besides the field from the screened point charge itself. See Section 6.5.3.

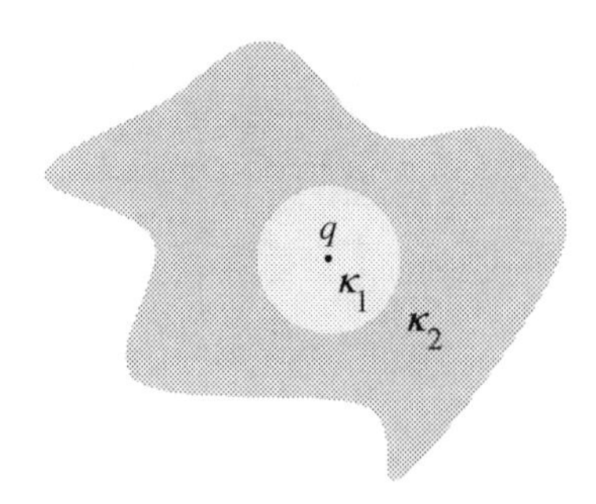

Figure 6.7: A point charge q embedded at the center of a sphere with dielectric constant κ_1. The sphere is itself embedded in an infinite volume with dielectric constant κ_2.

is partially neutralized by polarization charge which appears at the dielectric surfaces. At the lower surface, say, the definition (6.5) of σ_P gives

$$\sigma_P = -\mathbf{P} \cdot \hat{\mathbf{z}} = -\epsilon_0(\kappa - 1)\mathbf{E} \cdot \hat{\mathbf{z}} = \frac{1-\kappa}{\kappa}\sigma_f. \tag{6.49}$$

This gives the total charge density at the metal/dielectric interface (which is the source of $\mathbf{E}$) as a special case of (6.45):

$$\sigma = \sigma_f + \sigma_P = \sigma_f/\kappa. \tag{6.50}$$

The potential difference between the plates is

$$\Delta\varphi = \int_0^d d\boldsymbol{\ell} \cdot \mathbf{E} = \frac{1}{\kappa}\int_0^d d\boldsymbol{\ell} \cdot \mathbf{E}_0. \tag{6.51}$$

We have assumed that the charge Q of the plate does not change. Therefore, the capacitance of the dielectric-filled capacitor is

$$C = \frac{Q}{\Delta\varphi} = \kappa C_0. \tag{6.52}$$

6.5.3 Polarization Charge at a Simple Interface

The polarization charge induced at the common interface between two polarized dielectrics is the source of an electric field at every point in space. This observation resolves the following "paradox". Figure 6.7 shows a point charge q embedded in a sphere of radius R and dielectric constant κ_1. The sphere is itself embedded in an infinite medium with dielectric constant κ_2. We begin with the spherical symmetry of the problem and use Gauss' law in the form (6.29) to get

$$\mathbf{D}(r) = \frac{q}{4\pi}\frac{\hat{\mathbf{r}}}{r^2}. \tag{6.53}$$

Since $\mathbf{D} = \epsilon\mathbf{E}$, we conclude that $\mathbf{E}_1(r) = q\hat{\mathbf{r}}/4\pi\epsilon_1 r^2$ in medium 1 and $\mathbf{E}_2(r) = q\hat{\mathbf{r}}/4\pi\epsilon_2 r^2$ in medium 2. On the other hand, (6.46) tells us that the electric field produced by the embedded point charge *at every point in space* is

$$\mathbf{E}_q(r) = \frac{q}{4\pi\epsilon_1}\frac{\hat{\mathbf{r}}}{r^2}. \tag{6.54}$$

If (6.54) holds true in medium 2, how does the Gauss' law field quoted just above for $\mathbf{E}_2(r)$ arise?

There is no true paradox here because $\mathbf{E}_2(r)$ is the sum of two terms: the field (6.54) produced by q and a field $\mathbf{E}_\sigma(r)$ produced by a uniform density of surface polarization charge σ_P at the $r = R$

dielectric interface. The spherical symmetry of the boundary guarantees that there is no comparable contribution to $\mathbf{E}_1(r)$. By Gauss' law,

$$\mathbf{E}_\sigma(r) = \frac{\sigma_P R^2}{\epsilon_0}\frac{\hat{\mathbf{r}}}{r^2} \qquad r > R. \tag{6.55}$$

We get σ_P itself from (6.5), noting that both dielectrics make a contribution:

$$\sigma_P(r) = \hat{\mathbf{r}}\cdot(\mathbf{P}_1 - \mathbf{P}_2)|_{r=R} = \hat{\mathbf{r}}\cdot[\epsilon_0(\kappa_1 - 1)\mathbf{E}_1 - \epsilon_0(\kappa_2 - 1)\mathbf{E}_2]_{r=R}\,. \tag{6.56}$$

However, $\mathbf{D}_1(R) = \mathbf{D}_2(R)$ because there is no free charge at the interface. Therefore

$$\sigma_P = \hat{\mathbf{r}}\cdot(\mathbf{D}_1 - \mathbf{D}_2)_{r=R} - \epsilon_0\hat{\mathbf{r}}\cdot(\mathbf{E}_1 - \mathbf{E}_2)_{r=R} = -\epsilon_0\hat{\mathbf{r}}\cdot\mathbf{D}(R)\left[\frac{1}{\epsilon_1} - \frac{1}{\epsilon_2}\right]. \tag{6.57}$$

Inserting $\mathbf{D}(R)$ from (6.53) into (6.57) gives

$$\mathbf{E}_\sigma(r) = \frac{q}{4\pi\epsilon_0}\frac{\hat{\mathbf{r}}}{r^2}\left[\frac{1}{\kappa_2} - \frac{1}{\kappa_1}\right] \qquad r > R. \tag{6.58}$$

Combining (6.54) with (6.55) produces the elementary Gauss' law result,

$$\mathbf{E}_2(r) = \mathbf{E}_q(r) + \mathbf{E}_\sigma(r) = \frac{q}{4\pi\epsilon_2}\frac{\hat{\mathbf{r}}}{r^2} \qquad r > R. \tag{6.59}$$

Example 6.2 A point charge q sits at $z = -d$ on the z-axis (Figure 6.8). Find the polarization charge density $\sigma_P(x, y)$ induced on the plane $z = 0$ when the half-space $z > 0$ is filled with a medium with dielectric constant κ_R and the half-space $z < 0$ is filled with a medium with dielectric constant κ_L.

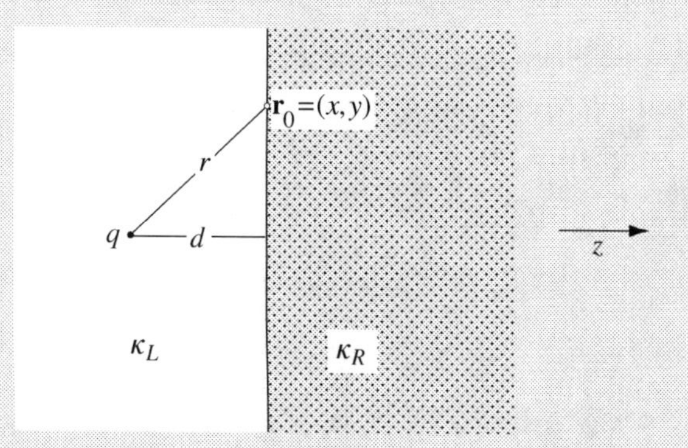

Figure 6.8: Two semi-infinite dielectrics meet at a planar interface. A point charge q sits a distance d to the left of the interface. The point $\mathbf{r}_0 = (x, y)$ lies in the $z = 0$ plane.

Solution: This problem is typically solved using the method of images (Section 8.3.3). The alternative method offered here is less elegant, but will deepen the reader's understanding of the electrostatics of dielectrics.

Let $\mathbf{E}_L$ and $\mathbf{E}_R$ be the electric fields at points infinitesimally close to, but on opposite sides of, an interface point $\mathbf{r}_0 = (x, y)$. The surface polarization charge density at $\mathbf{r}_0$ is

$$\sigma_P(\mathbf{r}_0) = \hat{\mathbf{z}}\cdot(\mathbf{P}_L - \mathbf{P}_R) = \epsilon_0(\chi_L\mathbf{E}_L - \chi_R\mathbf{E}_R)\cdot\hat{\mathbf{z}}.$$

The sources of $\mathbf{E}_L$ and $\mathbf{E}_R$ are the screened point charge q/κ_L and the surface polarization charge $\sigma_P(\mathbf{r}_0)$ we are trying to find. Following Section 3.4.2, the field produced by the surface charge on either side of (but very near) the surface itself is the sum of two terms: a contribution $\pm\hat{\mathbf{z}}\sigma_P(\mathbf{r}_0)/2\epsilon_0$ from the charge infinitesimally near to $\mathbf{r}_0$ and a contribution $\mathbf{E}_S$ from the surface charge away from $\mathbf{r}_0$. The latter has no component along $\hat{\mathbf{z}}$. Therefore, using the trigonometric factor d/r to project out the z-component of the point charge field,

$$\epsilon_0 \mathbf{E}_L \cdot \hat{\mathbf{z}} = \frac{1}{4\pi\kappa_L}\frac{q}{r^2}\frac{d}{r} - \frac{\sigma_P(\mathbf{r}_0)}{2} \qquad \epsilon_0 \mathbf{E}_R \cdot \hat{\mathbf{z}} = \frac{1}{4\pi\kappa_L}\frac{q}{r^2}\frac{d}{r} + \frac{\sigma_P(\mathbf{r}_0)}{2}.$$

There is no free charge at the interface, so the matching condition (6.31) is

$$\kappa_L \mathbf{E}_L \cdot \hat{\mathbf{z}} = \kappa_R \mathbf{E}_R \cdot \hat{\mathbf{z}}.$$

Combining all these equations gives the polarization surface charge density as

$$\sigma_P(x, y) = \frac{1}{2\pi\kappa_L}\frac{\kappa_L - \kappa_R}{\kappa_L + \kappa_R}\frac{qd}{(x^2 + y^2 + d^2)^{3/2}}.$$

We will use this result in Example 6.7 to find the force exerted on q by the dielectric.

6.5.4 Simple Dielectric Response to Fixed Fields

In the absence of free charge, the Maxwell equations (6.26) and (6.27) simplify to

$$\nabla \times \mathbf{E} = 0 \qquad \text{and} \qquad \nabla \cdot \mathbf{D} = 0. \tag{6.60}$$

The most interesting aspect of these equations is the matching conditions they imply, namely,

$$\hat{\mathbf{n}} \times [\mathbf{E}_1 - \mathbf{E}_2]_S = 0 \qquad \text{and} \qquad \hat{\mathbf{n}} \cdot [\mathbf{D}_1 - \mathbf{D}_2]_S = 0. \tag{6.61}$$

The tangential component of **E** and the normal component of **D** are continuous at a dielectric interface with no free charge.

A Parallel-Plate Capacitor with Fixed Potential

Figure 6.9 shows a parallel-plate capacitor with vacuum capacitance $C_0 = \epsilon_0 A/d$. What happens when a dielectric is inserted to fill the space between the plates and the potential difference V between the plates is held constant? Because the left side of

$$V = \int_0^d d\boldsymbol{\ell} \cdot \mathbf{E} \tag{6.62}$$

is fixed, the field between the plates is

$$\mathbf{E} = \mathbf{E}_0 = \frac{V}{d}\hat{\mathbf{z}}. \tag{6.63}$$

In contrast to (6.48), the electric field is *not* screened by the dielectric.

The polarization charge density induced at the lower surface of the dielectric is

$$\sigma_P = -\mathbf{P} \cdot \hat{\mathbf{z}} = -\epsilon_0(\kappa - 1)V/d. \tag{6.64}$$

On the other hand, $\mathbf{D} = \epsilon_0 \kappa \mathbf{E}$, so the charge density induced on the lower metal plate is

$$\sigma_f = \mathbf{D} \cdot \hat{\mathbf{z}} = \epsilon_0 \kappa \frac{V}{d}. \tag{6.65}$$

This shows that the capacitance C with fixed potential is the same as the capacitance (6.52) we found earlier for fixed charge:

$$C = \frac{\sigma_f A}{V} = \kappa C_0. \tag{6.66}$$

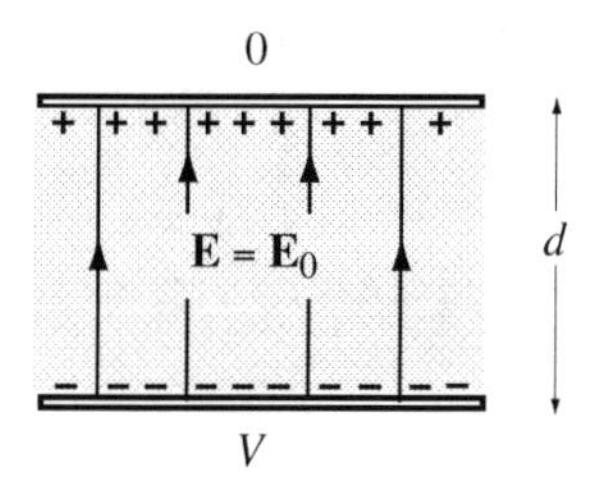

Figure 6.9: A capacitor with fixed plate potentials. Plus and minus signs denote polarization charge at the surface of the dielectric. A few lines of the electric field $\mathbf{E} = E\hat{\mathbf{z}}$ are indicated.

The sum of (6.65) and (6.64) is exactly the free charge density $\sigma_0 = \epsilon_0 V/d$ that produces the field $\mathbf{E}_0$ when the dielectric is absent. In other words, $\mathbf{E} = \mathbf{E}_0$ when the dielectric is present because charge flows from the battery (or whatever maintains the plates at fixed potential) to the surface of the plates to exactly cancel the polarization charge on the adjacent dielectric surfaces. Faraday exploited (6.65) to determine κ for many materials. His method was to measure the change in the amount of charge drawn onto the metal plates of a (spherical) capacitor when a dielectric was interposed between them. The SI unit of capacitance is called the *farad* to honor this aspect of Faraday's work.

Macroscopically, the charge densities (6.65) and (6.64) are spatially coincident. This is not so microscopically, and (6.65) implies that the local electric field immediately adjacent to both metal plates is *larger* in magnitude than the electric field produced by the vacuum capacitor. If so, there must also be regions inside the dielectric where the electric field is smaller than (or oppositely directed from) the field of the vacuum capacitor. We will return to this point at the end of Section 6.6.2.

Application 6.2 Refraction of Field Lines at a Dielectric Interface

The matching conditions (6.61) imply that the lines of $\mathbf{E}(\mathbf{r})$ *refract* when they pass from a medium with permittivity ϵ_1 to a medium with permittivity ϵ_2. Figure 6.10 shows a flat interface between these media and uses the angles α_1 and α_2 to label the angle between the interface normal and the electric field vector in each medium. The continuity of the normal component of $\mathbf{D} = \epsilon\mathbf{E}$ and the continuity of the tangential component of $\mathbf{E}$ read

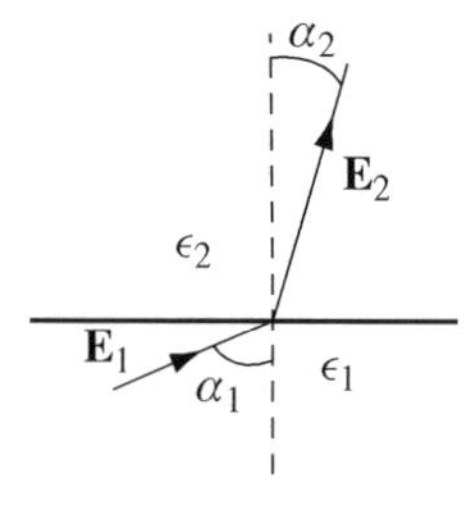

Figure 6.10: The refraction of $\mathbf{E}(\mathbf{r})$ at a flat interface between two simple dielectrics.

$$\begin{aligned} \epsilon_1 E_1 \cos\alpha_1 &= \epsilon_2 E_2 \cos\alpha_2 \\ E_1 \sin\alpha_1 &= E_2 \sin\alpha_2. \end{aligned} \tag{6.67}$$

Dividing one equation in (6.67) by the other gives the desired law of refraction:

$$\epsilon_1 \cot\alpha_1 = \epsilon_2 \cot\alpha_2. \tag{6.68}$$

The case $\epsilon_2 \gg \epsilon_1$ is interesting because both $\alpha_1 \approx 0$ and $\alpha_2 \approx \pi/2$ satisfy the law of refraction. The first solution is consistent with $\mathbf{E} = (\sigma/\epsilon_0)\hat{\mathbf{n}}$ for the electric field just outside the surface of a perfect conductor. The second solution says that the lines of $\mathbf{D}$ and $\mathbf{E}$ in the high-permittivity material are nearly parallel to the interface with the low-permittivity material. The actual solution adopted by Nature depends on the detailed geometry of the problem, including the positions of free charges away from the interface. ■

6.5.5 Potential Theory for a Simple Dielectric

Gauss' law for simple dielectric matter (6.39) combined with $\mathbf{E} = -\nabla\varphi$ tells us that the electrostatic potential inside a simple dielectric with permittivity $\epsilon = \epsilon_0\kappa$ satisfies Poisson's equation with $\rho(\mathbf{r})$ replaced by $\rho_f(\mathbf{r})/\kappa$:

$$\epsilon_0\nabla^2\varphi = -\rho_f/\kappa. \tag{6.69}$$

As discussed in Section 3.3.2, the matching conditions (6.33) and (6.31) can be expressed entirely in potential language, namely,

$$\varphi_1(\mathbf{r}_S) = \varphi_2(\mathbf{r}_S) \tag{6.70}$$

and

$$\left[\kappa_2\frac{\partial\varphi_2}{\partial n_2} - \kappa_1\frac{\partial\varphi_1}{\partial n_2}\right]_S = \frac{\sigma_f}{\epsilon_0}. \tag{6.71}$$

When there is no free charge anywhere, the Poisson-like equation (6.69) reduces everywhere to *Laplace's equation*,

$$\nabla^2\varphi = 0. \tag{6.72}$$

The matching condition (6.71) simplifies similarly to

$$\kappa_1\left.\frac{\partial\varphi_1}{\partial n}\right|_S = \kappa_2\left.\frac{\partial\varphi_2}{\partial n}\right|_S. \tag{6.73}$$

Chapters 7 and 8 discuss methods to solve the Laplace and Poisson equations, respectively, when simple dielectric matter is present.

Example 6.3 Figure 6.11 shows a spherical cavity scooped out of an infinite medium with dielectric constant κ. Capacitor plates at infinity produce a fixed and uniform external field $\mathbf{E}_0 = E_0\hat{\mathbf{z}}$. Find the potential and field everywhere.

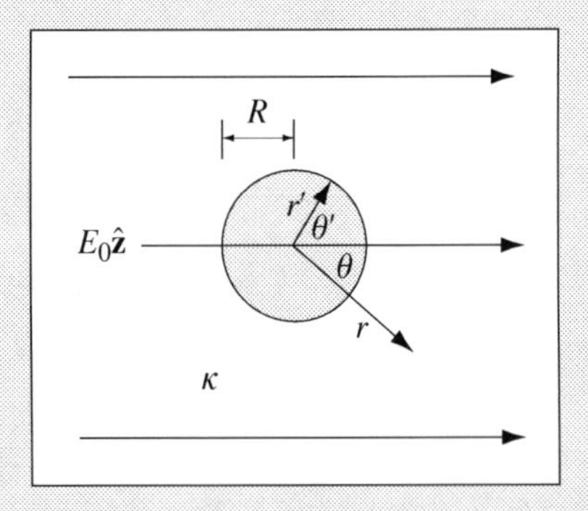

Figure 6.11: A sphere of radius R scooped out of a medium with dielectric constant κ. The field lines shown are for the external field $\mathbf{E}_0 = E_0\hat{\mathbf{z}}$ only.

Solution: The field $\mathbf{E}_0$ induces a uniform polarization $\mathbf{P}_0 = P_0\hat{\mathbf{z}}$ in the original infinite uniform medium. After the cavity is created, a polarization charge density $\sigma(\theta) = \mathbf{P}_0 \cdot \hat{\mathbf{n}} = -P_0 \cos\theta$ develops on its surface.[10] This charge distribution produces an electric field which supplements $\mathbf{E}_0$. This is the situation analyzed in Application 4.3, where we showed that a charge density $\sigma(\theta) \propto \cos\theta$ confined to a sphere produces a potential which varies as $r\cos\theta$ inside the sphere and $\cos\theta/r^2$ outside the sphere. Consequently, the total potential has the form

$$\varphi(r,\theta) = \begin{cases} Ar\cos\theta & r \le R, \\ \left(-E_0 r + \dfrac{B}{r^2}\right)\cos\theta & r \ge R. \end{cases}$$

The contribution $-E_0 r\cos\theta$ to $\varphi(r,\theta)$ is associated with the *unscreened* $\mathbf{E}_0$. This is consistent with the discussion immediately above of the parallel-plate capacitor with fixed plate potential.

The matching conditions (6.70) and (6.73) applied at $r = R$ determine the coefficients A and B:

$$A = -\frac{3\kappa}{2\kappa+1}E_0 \qquad \text{and} \qquad B = -\frac{\kappa-1}{2\kappa+1}E_0 R^3.$$

The total field $\mathbf{E} = -\nabla\varphi$ outside the cavity is the sum of $\mathbf{E}_0$ and a pure dipole field. The total field inside the cavity is uniform, points in the same direction as $\mathbf{E}_0$, and is larger in magnitude than $\mathbf{E}_0$:

$$E_{\text{in}} = -A = \frac{3\kappa}{2\kappa+1}E_0 > E_0.$$

Physically, this occurs because the polarization charge density is positive (negative) on the left (right) hemisphere of the cavity in Figure 6.11:

$$\sigma_P(\theta) = \mathbf{P}\cdot\hat{\mathbf{n}} = -\epsilon_0(\kappa-1)\mathbf{E}_{\text{out}}\cdot\hat{\mathbf{r}}.$$

It is worth noting that $\sigma_P(\theta) = \mathbf{P}\cdot\hat{\mathbf{n}}$ is *not* the same as $\sigma(\theta) = \mathbf{P}_0\cdot\hat{\mathbf{n}}$ introduced at the beginning of the problem because the final electric field (and hence the final polarization) in the medium is not uniform. This fact does not invalidate our solution because $\sigma_P(\theta)$ and $\sigma(\theta)$ are both proportional to $\cos\theta$, which is the only information we used to help determine the form of $\varphi(r,\theta)$ above.

6.6 The Physics of the Dielectric Constant

Table 6.1 lists the experimental dielectric constants for two gases (helium and nitrogen), two liquids (methane and water), and two solids (silicon dioxide and silicon). The electrostatics of the physical models which predict these numbers is interesting. All make essential use of the *local electric field* which exists inside a polarized dielectric.

6.6.1 The Electric Polarizability

Example 5.1 of Section 5.2 demonstrated that a uniform electric field $\mathbf{E}_0$ induces an electric dipole moment $\mathbf{p} = 4\pi\epsilon_0 R^3\mathbf{E}_0$ in a conducting sphere of radius R. More generally, we define the *polarizability* α of any small dielectric body as the proportionality constant between a uniform field $\mathbf{E}_0$ and the dipole moment $\mathbf{p}$ induced in the body by that field:[11]

$$\mathbf{p} = \alpha\epsilon_0\mathbf{E}_0. \tag{6.74}$$

[10] There is a negative sign because $\hat{\mathbf{n}}$ is the *outward* unit normal to the dielectric surface.

[11] Compare (6.74) with (6.35).

Table 6.1. Dielectric constants at 10^5 Pa (1 atm) and 20° C.

Substance	κ
He	1.000065
N_2	1.00055
CH_4	1.7
SiO_2	4.5
Si	11.8
H_2O	80

Equation (6.74) generalizes to microscopic bodies like atoms and molecules where the value of the induced dipole moment depends on the value of a microscopic electric field at the position of the atom or molecule:

$$\mathbf{p}(\mathbf{r}) = \alpha\epsilon_0\mathbf{E}_{\text{micro}}(\mathbf{r}). \tag{6.75}$$

Hundreds of research papers have been devoted to calculations and measurements of electric polarizabilities.

6.6.2 The Clausius-Mossotti Formula

The Lorentz model for polarizable matter (Section 6.2.2) relates the macroscopic dielectric constant κ [see (6.36)] of a dielectric gas to the microscopic polarizability α [see (6.75)] of the constituent atoms or molecules. The main ingredient is the spatial integral over a suitable averaging volume Ω of the microscopic electric field on the right side of (6.75):[12]

$$\mathbf{E}(\mathbf{R}) = \frac{1}{\Omega}\int_{\Omega} d^3r\, \mathbf{E}_{\text{micro}}(\mathbf{r}). \tag{6.76}$$

We also need an alternative definition of the cell-averaged dipole moment $\mathbf{p}_\mathbf{R}$ first defined in (6.10), namely,

$$\mathbf{p}_\mathbf{R} = \frac{1}{\Omega}\int_{\Omega} d^3r\, \mathbf{p}(\mathbf{r}). \tag{6.77}$$

Finally, because $n = 1/\Omega$ is the density of molecules in the gas, (6.10), (6.76), and (6.77) combine to generate an expression for the polarization $\mathbf{P}$ which appears in the macroscopic equation (6.25):

$$\mathbf{D}(\mathbf{R}) = \epsilon_0\mathbf{E}(\mathbf{R}) + \mathbf{P}(\mathbf{R}) = \epsilon_0\mathbf{E}(\mathbf{R}) + n\alpha\epsilon_0\mathbf{E}(\mathbf{R}). \tag{6.78}$$

Comparing (6.78) with (6.37) shows that the static dielectric constant is

$$\kappa = 1 + n\alpha. \tag{6.79}$$

Our derivation of (6.79) implicitly assumes that the dimensionless parameter $n\alpha$ is much smaller than one. To see this, note first that the field in (6.75) which polarizes the molecule in a given cell cannot include the microscopic field $\mathbf{E}_{\text{self}}(\mathbf{r})$ produced by the molecule itself. On the other hand, the macroscopic field $\mathbf{E}(\mathbf{R})$ in (6.78) is the average of the field in a given cell from *all* sources, including

[12] This is the Lorentz averaging procedure discussed in Section 2.3.1. We follow the notation of that section where $\mathbf{r}$ denotes a microscopic spatial variable and $\mathbf{R}$ denotes a macroscopic spatial variable.

$\mathbf{E}_{\rm self}(\mathbf{r})$. Therefore, it is necessary to subtract the Lorentz average of the self-field when we calculate the polarization:

$$\mathbf{P}(\mathbf{R}) = n\alpha\epsilon_0 \left\{ \mathbf{E}(\mathbf{R}) - \frac{1}{\Omega}\int_\Omega d^3s\, \mathbf{E}_{\rm self}(\mathbf{s}) \right\} = n\alpha\epsilon_0 \mathbf{E}_{\rm local}(\mathbf{R}). \tag{6.80}$$

The *local field* $\mathbf{E}_{\rm local}(\mathbf{R})$ defined by (6.80) is a better approximation to the macroscopic field that polarizes each molecule. We have performed the integral in (6.80) already for the case of a spherical cell (Example 4.1). The result given there is a good approximation for non-spherical cells also, particularly if the cell is large compared to the size of the molecular charge distribution:

$$\frac{1}{\Omega}\int_\Omega d^3s\, \mathbf{E}_{\rm self}(\mathbf{s}) = -\frac{\mathbf{p}}{3\epsilon_0\Omega} = -\frac{\mathbf{P}(\mathbf{R})}{3\epsilon_0}. \tag{6.81}$$

Substituting (6.81) into (6.80) gives an algebraic equation for $\mathbf{P}(\mathbf{R})$. The solution,

$$\mathbf{P}(\mathbf{R}) = \frac{n\alpha\epsilon_0}{1 - n\alpha/3}\mathbf{E}(\mathbf{R}), \tag{6.82}$$

together with the leftmost equation in (6.78) gives the desired result for the dielectric constant:

$$\kappa = 1 + \frac{n\alpha}{1 - n\alpha/3}. \tag{6.83}$$

Only when $n\alpha \ll 1$ does (6.83) reproduce (6.79). In that limit, the averaging volume Ω is large and the self-field makes a negligible contribution to the Lorentz average. Experimental tests of (6.83) usually employ an equivalent form known as the *Clausius-Mossotti formula*:

$$\frac{\kappa - 1}{\kappa + 2} = \frac{n\alpha}{3}. \tag{6.84}$$

Non-polar gases and a few simple liquids like He, N_2, and C_6H_6 obey formula (6.84) rather well. The polarizability is either measured independently or estimated from $\alpha = 3V$.[13]

We can also combine (6.80) with (6.82) to discover that (cf. Example 6.1)

$$\mathbf{E}_{\rm local}(\mathbf{R}) = \frac{\mathbf{E}(\mathbf{R})}{1 - n\alpha/3} = \frac{\kappa + 2}{3}\mathbf{E}(\mathbf{R}). \tag{6.85}$$

This formula confirms the remark made at the end of Section 6.5.4 that the local electric field typically exceeds the average macroscopic field inside a dielectric. $\mathbf{E}_{\rm local}(\mathbf{R})$ is due to all the *other* molecules, so it is characteristic of the volume *between* gas molecules. This may be contrasted with the volume *inside* each molecule where (the reader may wish to confirm) the dipole self-field is very large and directed oppositely to $\mathbf{E}_{\rm local}(\mathbf{R})$.

6.6.3 Polar Liquids and Solids

Table 6.1 reports the dielectric constants for liquid methane and water. Experiments show that CH_4 obeys the Clausius-Mossotti relation (6.84) while H_2O does not. The physical reason for this is that methane is a non-polar molecule while water is a polar molecule with a permanent electric dipole moment $\mathbf{p}_0$ (see Figure 4.2). Onsager (1936) realized that the presence of $\mathbf{p}_0$ dramatically affects the local field needed to calculate κ. He derived a generalization of the Clausius-Mossotti formula using, in part, the cavity calculation of Example 6.3 generalized to include a dipole $\mathbf{p}_0$ at the center of the cavity.

[13] See the remark at the end of Example 5.1 of Section 5.2.1.

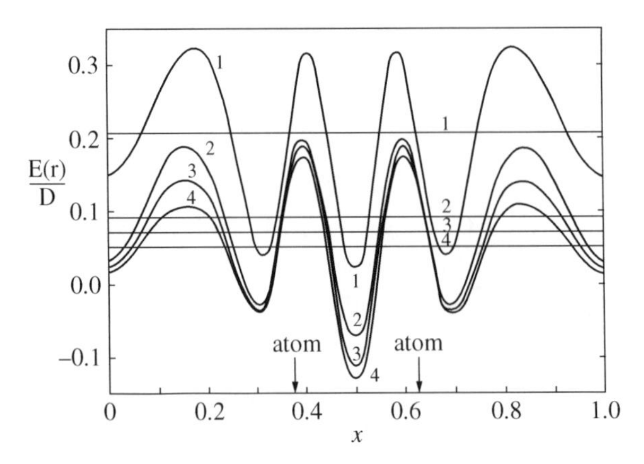

Figure 6.12: The local field $\mathbf{E}_{\text{micro}}(\mathbf{r})$ (solid curves) and the macroscopic field $\mathbf{E}$ (horizontal lines) for four Group IV crystals: 1 = C, 2 = Si, 3 = Ge, 4 = Sn. The horizontal axis is a line (parallel to $\mathbf{D}$) that passes through two nuclei in the averaging cell. The latter repeats periodically in the horizontal direction. Figure adapted from Baldereschi, Car, and Tosatti (1979).

No simple formula predicts the dielectric constants of the solids SiO_2 (quartz) and Si in Table 6.1. Reliable estimates for such systems require quantum mechanical perturbation theory. The relevant perturbation is a uniform field, which it will be convenient to call $\mathbf{D}$. The response is the total electric field $\mathbf{E}_{\text{micro}}(\mathbf{r})$. Given the latter, we compute the macroscopic electric field from (6.76). Every averaging cell is the same in a perfect crystal so $\mathbf{E}(\mathbf{R}) = \mathbf{E}$. The static dielectric constant follows immediately from $\epsilon_0\kappa = \mathbf{D}/\mathbf{E}$.

Figure 6.12 shows $\mathbf{E}_{\text{micro}}(\mathbf{r})$ for a sequence of Group IV dielectric solids. For each material, the horizontal line is the macroscopic field (6.76). The calculation uses $|\mathbf{D}| = 1$, so the intercept of each line with the vertical axis is $1/\kappa$. Notice that κ increases and the intracell variations in $\mathbf{E}_{\text{micro}}(\mathbf{r})$ decrease as one proceeds down the Group IV column of the periodic table. This corresponds to increasing spreading out (delocalization) of the electronic wave functions.

6.6.4 The Limit $\kappa \to \infty$

The discussion just above implies that $\kappa \to \infty$ produces $\mathbf{E}(\mathbf{R}) = 0$ inside a dielectric exposed to a static external field. The latter is the definition of a perfect conductor. For that reason, the $\kappa \to \infty$ limit of a formula derived using dielectric theory often reproduces the results of a perfect conductor calculation. This can be a useful strategy to check calculations and to build intuition. Some caution is required because there is no reason to expect perfect-conductor behavior in the $\kappa \to \infty$ limit when there is free charge in intimate contact with the dielectric. This occurs, *e.g.*, when a point charge is embedded in the bulk of a dielectric or when free charge accumulates on the surface of a dielectric (see Section 6.5.4).

6.7 The Energy of Dielectric Matter

The total energy is a fundamental property of dielectric matter. It contributes essentially to the thermodynamics of a dielectric and greatly facilitates the calculation of forces that act on and in dielectric matter. In this section, we compute (i) the total energy of a polarized dielectric, and (ii) the change in energy which occurs when a dielectric becomes polarized. Both calculations are done in two ways.

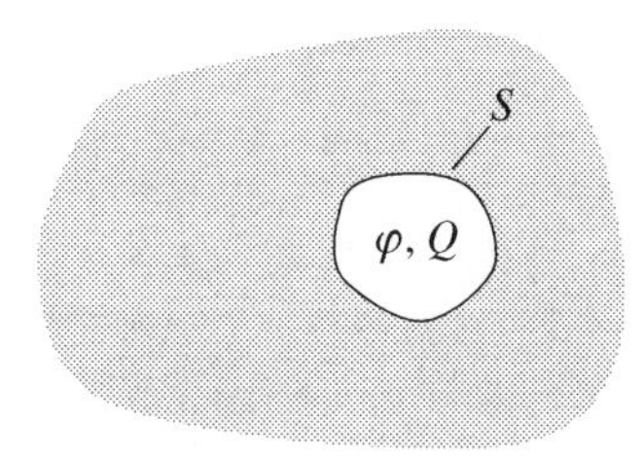

Figure 6.13: A conductor with charge Q and potential φ embedded in a dielectric medium.

6.7.1 The Total Energy U_E

Figure 6.13 shows a charged conductor embedded in an infinite dielectric medium. The charge on the conductor is the source of the field in the dielectric. Therefore, the mechanical work $\varphi\delta Q$ required to transport a charge δQ from infinity to the conductor surface (see Section 5.6.1) is identical to the energy δU_E required to change the field in the dielectric. Our goal is to write δU_E entirely in terms of the field.[14] Accordingly, if S is the conductor surface and V is the infinite volume *not* occupied by the conductor, Gauss' law and the divergence theorem give

$$\delta U_E = \varphi\delta Q = \int_S d\mathbf{S}\cdot\delta\mathbf{D}\varphi = -\int_V d^3r\,\nabla\cdot(\delta\mathbf{D}\varphi). \tag{6.86}$$

We assume there is no free charge in V. This makes $\nabla\cdot(\delta\mathbf{D}) = 0$ and simplifies (6.86) to the desired final expression,

$$\delta U_E = -\int_V d^3r\,\nabla\varphi\cdot\delta\mathbf{D} = \int d^3r\,\mathbf{E}\cdot\delta\mathbf{D}. \tag{6.87}$$

The final integral in (6.87) is over all of space because $\mathbf{E} = 0$ in the volume occupied by the conductor.

An alternative derivation of (6.87) identifies the different contributions to the total energy. We learned in Section 6.2.3 that polarization is created in a dielectric when an electric field $\mathbf{E}(\mathbf{r})$ induces a flow of polarization charge $\rho_P(\mathbf{r})$ with a current density $\mathbf{j}_P = \partial\mathbf{P}/\partial t$. Put another way, the Coulomb force density $\rho_P(\mathbf{r})\mathbf{E}(\mathbf{r})$ does quasistatic work on an element of matter with velocity $\boldsymbol{v}(\mathbf{r})$ at a rate

$$\rho_P(\mathbf{r})\mathbf{E}(\mathbf{r})\cdot\boldsymbol{v}(\mathbf{r}) = \mathbf{j}_P(\mathbf{r})\cdot\mathbf{E}(\mathbf{r}). \tag{6.88}$$

This work is done against the cohesive forces of matter which hold the dielectric together and oppose the creation of the polarization charge. Therefore, for an isolated dielectric, the work done by the electric field increases the internal energy of the matter at a rate per unit volume

$$\frac{\partial u_{\rm mat}}{\partial t} = \frac{\partial\mathbf{P}}{\partial t}\cdot\mathbf{E}. \tag{6.89}$$

To get the change in total energy density, we add to (6.89) the rate of change of the electric field energy density $u_{\rm field} = \frac{1}{2}\epsilon_0 E^2$.[15] Using $\mathbf{D} = \epsilon_0\mathbf{E} + \mathbf{P}$, we get a change in total internal energy density which agrees with (6.87):

$$du_E = du_{\rm field} + du_{\rm mat} = d\left\{\tfrac{1}{2}\epsilon_0 E^2\right\} + \mathbf{E}\cdot d\mathbf{P} = \mathbf{E}\cdot d\mathbf{D}. \tag{6.90}$$

[14] This will make the result applicable to cases where the source of the field inside a dielectric is not a charged conductor.

[15] It is an *assumption* that the field energy density in the dielectric is the same as the field energy density in the vacuum. See Section 9.14 of Purcell (1965) and Section 4.2 of Bobbio (2000) in Sources, References, and Additional Reading.

The total energy is the net work required to establish the field **D**. Our final result is the desired expression for the total energy of a macroscopic polarized object:

$$U_E = \int d^3r \int_0^{\mathbf{D}} \mathbf{E} \cdot \delta\mathbf{D}. \tag{6.91}$$

It is important to appreciate that the integration from zero to **D** in (6.91) cannot be performed until we have specified a constitutive relation which relates **D**(**r**) to **E**(**r**) in every volume element of a sample with arbitrary shape.

Our use of the symbol U_E indicates that (6.91) is the dielectric analog of the function used in Section 5.6.1 to compute the force on isolated conductors. That earlier quantity was a natural function of the free charges Q_k carried by conductors labeled $k = 1, 2, \ldots, N$. The same is true here because (6.90) shows that U_E is a natural function of **D** and free charge determines the latter from $\nabla \cdot \mathbf{D} = \rho_f$. Hence, if **R** labels the center of mass of the dielectric,

$$U_E = U_E(\mathbf{D}, \mathbf{R}) \qquad \text{or} \qquad U_E = U_E(Q_k, \mathbf{R}). \tag{6.92}$$

The analog of (5.69) is

$$\mathbf{E} = \frac{1}{V}\left(\frac{\partial U_E}{\partial \mathbf{D}}\right)_{\mathbf{R}} \qquad \text{and} \qquad \mathbf{F} = -\left(\frac{\partial U_E}{\partial \mathbf{R}}\right)_{\mathbf{D}} \qquad \text{or} \qquad \mathbf{F} = -\left(\frac{\partial U_E}{\partial \mathbf{R}}\right)_Q. \tag{6.93}$$

6.7.2 U_E for a Simple Dielectric

The integral in (6.91) is tractable only for simple choices of the dielectric constitutive relation. By far the most common situation is a simple dielectric where $\mathbf{D} = \epsilon \mathbf{E}$ and

$$U_E[\mathbf{D}] = \int d^3r \int_0^{\mathbf{D}} \frac{\mathbf{D} \cdot \delta\mathbf{D}}{\epsilon} = \frac{1}{2}\int d^3r \, \frac{|\mathbf{D}|^2}{\epsilon} = \frac{1}{2}\int d^3r \, \mathbf{E} \cdot \mathbf{D}. \tag{6.94}$$

Equation (6.94) predicts that free charge is attracted to regions where ϵ is large. This is so because the **D**-field produced by free charge is largest near to itself and the volume integral of $|\mathbf{D}|^2/\epsilon$ in (6.94) is minimized when ϵ is large in the same regions of space where $|\mathbf{D}|$ is large. This is sensible because the Coulomb energy is lowered when q polarizes the medium to draw charge of the opposite sign toward itself. If we except situations constrained by symmetry, a more general statement is that $|\mathbf{E}|$ tends to be large where ϵ is large and small where ϵ is small. This is so because the polarization created by **E** in a simple medium amounts to a collection of point electric dipoles aligned with **E** with magnitudes proportional to ϵ. Therefore, the potential energy $-\mathbf{p} \cdot \mathbf{E}$ [see (4.26)] gained by each dipole is maximized when **E** is large in regions where ϵ is large.

Application 6.3 A Classical Model for Quark Confinement

Quantum chromodynamics (QCD) assigns a "color" degree of freedom to quarks. For some purposes, the QCD vacuum may be regarded as a dielectric medium for color charge with a vanishingly small dielectric constant κ. Because $\kappa < 1$, this fictitious dielectric medium "anti-screens" free charge in the sense that (6.43) tells us that the polarization charge has the same sign as the free charge, thereby enhancing the effect of the latter. On the other hand, Coulomb repulsion prevents the free charge and the polarization charge from getting too close together. This suggests that a tiny sphere with color charge q and radius a digs itself a vacuum hole of radius R in the dielectric. This is indicated in both panels of Figure 6.14.

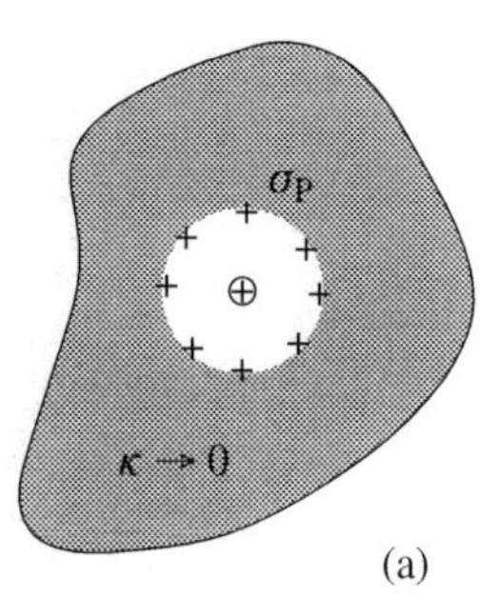

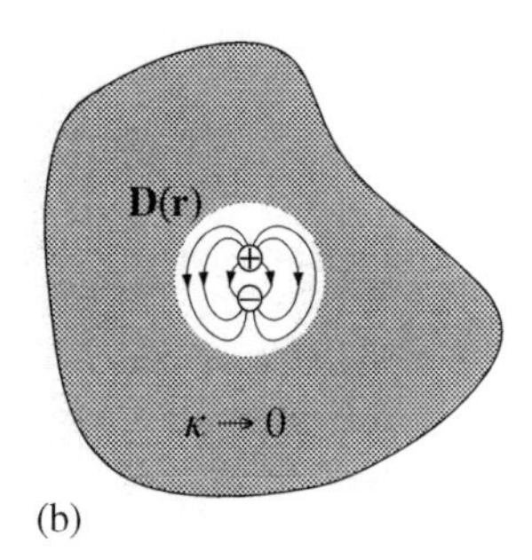

Figure 6.14: Classical models for (a) a free quark and (b) a quark-antiquark pair (meson). The QCD vacuum is modeled as an infinite dielectric medium with a dielectric constant $\kappa \to 0$: (a) a tiny sphere (radius a) with "color" charge q at the center of a spherical cavity (radius R) scooped out of the dielectric induces surface polarization charge of the same sign; (b) a tiny dipole at the center of a spherical cavity produces lines of $\mathbf{D}$ which do not leave the cavity when $\kappa \to 0$. Figure from Lee (1980).

We focus first on the "free quark" situation depicted in Figure 6.14(a) and minimize the sum of the electrostatic energy (6.94) and a surface energy U_S given by the product of the cavity area and an energy per unit area γ. The fields inside and outside the cavity satisfy

$$\epsilon_0 \mathbf{E}_{\text{in}} = \mathbf{D}_{\text{in}} = \frac{q}{4\pi r^2}\hat{\mathbf{r}} = \mathbf{D}_{\text{out}} = \kappa\epsilon_0 \mathbf{E}_{\text{out}}. \tag{6.95}$$

The large value of $\mathbf{E}_{\text{out}}$ implied by (6.95) when $\kappa \to 0$ arises from the large density of surface polarization charge σ_P indicated in Figure 6.14(a). We split the integral (6.94) into $U_E = U_{\text{in}} + U_{\text{out}}$ and focus on the limit $a \ll R$. A straightforward calculation gives

$$U_{\text{tot}} = U_{\text{in}} + U_{\text{out}} + U_S = \frac{q^2}{8\pi\epsilon_0}\left\{\frac{1}{a} + \frac{1}{\kappa R}\right\} + \gamma 4\pi R^2. \tag{6.96}$$

Minimizing U_{tot} with respect to R shows that (6.96) has a stable minimum at a cavity radius $R^* \propto \kappa^{-1/3}$. On the other hand, the energy $U_{\text{tot}}(R^*) \propto \kappa^{-2/3}$. The divergence of this energy when $\kappa \to 0$ "explains" why free quarks are never seen in isolation.

Figure 6.14(b) illustrates a classical model for a meson (a quark-antiquark pair) composed of two tiny spheres with equal and opposite charge. The energetic preference to eliminate $\mathbf{D}$ from regions of space with a vanishingly small dielectric constant [see the commentary following (6.94)] is easily accommodated by the creation of a cavity and a slight distortion of the natural field line pattern of a dipole. This geometry has $U_{\text{out}} = 0$ because $\mathbf{D}_{\text{out}} = 0$ when $\kappa = 0$. In the same limit, dimensional analysis tells us that U_{in} must contain a term proportional to $p^2/\epsilon_0 R^3$, where p is the dipole moment of the two charges. This implies that a cavity does indeed form around the charges because $U_{\text{tot}} = U_{\text{in}} + U_S$ has an absolute minimum at a κ-independent value of R. The total energy of this classical meson is finite in the $\kappa \to 0$ limit appropriate to QCD. ■

6.7.3 The Energy to Polarize Simple Matter

Let $\mathbf{E}_0$ be an electric field in vacuum. It is an interesting task to calculate the change in energy that occurs when a simple dielectric is inserted into the field, holding the charges which created $\mathbf{E}_0$ fixed. Because the dielectric polarizes, the electric field $\mathbf{E}_0(\mathbf{r})$ changes to $\mathbf{E}(\mathbf{r})$ and the auxiliary field $\mathbf{D}_0(\mathbf{r}) = \epsilon_0 \mathbf{E}_0(\mathbf{r})$ changes to $\mathbf{D}(\mathbf{r})$. Using (6.94), the change in total energy is

$$\Delta U_E = \frac{1}{2}\int d^3r \, [\, \mathbf{E}\cdot\mathbf{D} - \mathbf{E}_0\cdot\mathbf{D}_0 \,]. \tag{6.97}$$

By adding and subtracting $\frac{1}{2}(\mathbf{E}\cdot\mathbf{D}_0 - \mathbf{E}_0\cdot\mathbf{D})$, we can write (6.97) in the form

$$\Delta U_E = \frac{1}{2}\int d^3r\ [\mathbf{E}\cdot(\mathbf{D}-\mathbf{D}_0) + (\mathbf{D}-\mathbf{D}_0)\cdot\mathbf{E}_0 + \mathbf{E}\cdot\mathbf{D}_0 - \mathbf{D}\cdot\mathbf{E}_0\,]. \tag{6.98}$$

This complication is temporary because, as we shall now prove, only the last two integrals in (6.98) are non-zero.

The trick is to adopt a geometry like Figure 6.13 and follow the steps from (6.86) to (6.87) in reverse order. This shows that

$$\int d^3r\,\mathbf{E}\cdot(\mathbf{D}-\mathbf{D}_0) = \varphi(Q-Q_0). \tag{6.99}$$

The right-hand side of (6.99) is zero because we have held the charge fixed. By the same logic,

$$\int d^3r\,(\mathbf{D}-\mathbf{D}_0)\cdot\mathbf{E}_0 = (Q-Q_0)\varphi_0 = 0. \tag{6.100}$$

The identity $\mathbf{D} = \epsilon_0(\mathbf{E}+\mathbf{P})$ transforms the two remaining terms in (6.98) into the desired final expression,

$$\Delta U_E = -\frac{1}{2}\int d^3r\,\mathbf{P}\cdot\mathbf{E}_0. \tag{6.101}$$

6.7.4 The Physical Origin of ΔU_E

The derivation of (6.101) in the preceding section does not address the physical origin of ΔU_E. To correct this, we begin with the special case of a linear dielectric with a very small size and use a different method to calculate ΔU_E. For such an object, we know from Section 6.6.1 that an external electric field $\mathbf{E}_0$ induces an electric dipole moment $\mathbf{p}$. One contribution to ΔU_E is the interaction energy (4.26) between $\mathbf{E}_0$ and the moment $\mathbf{p}$ in the final state:

$$U_{\rm ext} = -\mathbf{p}\cdot\mathbf{E}_0 = -pE_0. \tag{6.102}$$

The energy gain (6.102) is opposed by the energy $U_{\rm int}$ required to create the dipole in the first place. This is the work done by the field against internal cohesive forces which oppose the creation of the dipole moment. $U_{\rm int}$ must be (at least) quadratic in $\mathbf{p}$ to produce a total energy with an absolute minimum. Hence, we write

$$\Delta U_E = U_{\rm ext} + U_{\rm int} = -pE_0 + \zeta p^2, \tag{6.103}$$

and determine the positive constant ζ by minimizing ΔU_E with respect to p. Using (6.102), the final result can be written

$$\Delta U_E = -pE_0 + \frac{1}{2}pE_0 = -\frac{1}{2}\mathbf{p}\cdot\mathbf{E}_0 = -\frac{1}{2}\epsilon_0\alpha|\mathbf{E}_0|^2. \tag{6.104}$$

The similarity of (6.104) to (6.101) is apparent.

A finite-sized sample differs from the "point" object to which (6.104) applies because the field at any point in a finite dielectric is the sum of the external field $\mathbf{E}_0$ and the field $\mathbf{E}_{\rm self}$ created by the dielectric itself. Taking this into account, our strategy is to derive (6.101) by following the derivation of (6.104) as closely as possible. Thus, we write

$$\Delta U_E = U_{\rm ext} + U_{\rm int} + U_{\rm self}, \tag{6.105}$$

where $U_{\rm ext}$ is the interaction energy between $\mathbf{P}(\mathbf{r})$ and the external field $\mathbf{E}_0(\mathbf{r})$ and $U_{\rm int}$ is the energy required to overcome the internal cohesive forces and polarize the dielectric. The new term $U_{\rm self}$ is the interaction energy between $\mathbf{P}(\mathbf{r})$ and the field $\mathbf{E}_{\rm ind}(\mathbf{r})$ produced by the dielectric itself. We calculate each of these in turn.

The first term on the right side of (6.105) is a generalization of the interaction energy $-\mathbf{p}\cdot\mathbf{E}_0$ in (6.102) to the case of a continuous distribution of point dipoles with density $\mathbf{P}(\mathbf{r})$; namely,

$$U_{\rm ext} = -\int d^3r\, \mathbf{P}(\mathbf{r})\cdot\mathbf{E}_0(\mathbf{r}). \tag{6.106}$$

$U_{\rm self}$ follows similarly from the self-energy (4.32) of a collection of point dipoles:

$$U_{\rm self} = -\frac{1}{2}\int d^3r\, \mathbf{P}(\mathbf{r})\cdot\mathbf{E}_{\rm ind}(\mathbf{r}). \tag{6.107}$$

We compute $U_{\rm int}$ using the fact that (6.89) is the rate of change of energy required to *create* a bit of polarization $\delta\mathbf{P}$. Since $\mathbf{E} = \mathbf{E}_0 + \mathbf{E}_{\rm ind}$, the linearity between $\mathbf{P}$ and $\mathbf{E}$ allows us to perform the integral exactly as we did in (6.94):

$$U_{\rm int} = \int d^3r \int_0^{\mathbf{P}} \mathbf{E}\cdot\delta\mathbf{P} = \frac{1}{2}\int d^3r\, \mathbf{P}(\mathbf{r})\cdot[\mathbf{E}_0(\mathbf{r}) + \mathbf{E}_{\rm ind}(\mathbf{r})]\,. \tag{6.108}$$

The terms which involve $\mathbf{E}_{\rm ind}$ cancel when we sum (6.106), (6.107), and (6.108) to get ΔU_E in (6.105). The result is exactly the same as ΔU_E in (6.101).

Example 6.4 A parallel-plate capacitor has a fixed surface charge density Q_0/A. The electric field is $\mathbf{E}_0$ when the potential difference between the plates is φ_0. The potential difference changes to φ when a linear, isotropic dielectric sample with volume V is placed between the plates. Show that the magnitude of the dipole moment induced in the dielectric is $p = \epsilon_0 A\Delta\varphi$, where $\Delta\varphi = \varphi - \varphi_0$.

Solution: Since $|\mathbf{E}_0| = Q_0/A\epsilon_0$ is uniform, and the dipole moment of the sample is $\mathbf{p} = \int_V d^3r\,\mathbf{P}$,

$$\int_V d^3r\,\mathbf{P}\cdot\mathbf{E}_0 = \frac{p\,Q_0}{A\epsilon_0}.$$

$\mathbf{P} = 0$ outside the dielectric. Therefore, nothing changes if we expand V to a volume V' which includes the positively charged capacitor plate. Then, if $\mathbf{E}$ is the electric field between the plates after the insertion of the dielectric,

$$\int_V d^3r\,\mathbf{P}\cdot\mathbf{E}_0 = \int\int_{V'} d^3r\ [(\mathbf{D} - \epsilon_0\mathbf{E}_0)\cdot\mathbf{E}_0 + \epsilon_0(\mathbf{E}_0 - \mathbf{E})\cdot\mathbf{E}_0]\,.$$

If Q is the charge on the positive plate after the insertion of the sample, the method of integration used to derive (6.87) and (6.101) gives, in the present case,

$$\frac{p\,Q_0}{A\epsilon_0} = (Q - Q_0)\varphi_0 + (\varphi_0 - \varphi)Q_0.$$

This gives the desired result because $Q = Q_0$, by assumption.

6.7.5 The Energy Functional $\hat{U}_E$

We conclude our discussion of energy and dielectric matter by mimicking the logic used for conductors in Section 5.6.2 to derive an energy function for a dielectric in the presence of N conductors held at fixed potentials φ_k. Accordingly, we use a Legendre transformation to define an energy function $\hat{U}_E$

from

$$\hat{U}_E = U_E - \int d^3r\, \mathbf{D}\cdot\mathbf{E} = -\frac{1}{2}\int d^3r\, \mathbf{D}\cdot\mathbf{E}. \tag{6.109}$$

Since $U_E = U_E[\mathbf{D}]$, variation of (6.109) gives

$$\delta\hat{U}_E = \delta\left[U_E - \int d^3r\, \mathbf{D}\cdot\mathbf{E}\right] = -\int d^3r\, \mathbf{D}\cdot\delta\mathbf{E}. \tag{6.110}$$

This shows that $\hat{U}_E$ is a natural function of **E** rather than **D**. The notation is again the same as we used for conductors because [cf. (6.86) and (6.87)],

$$\int d^3r\, \mathbf{D}\cdot\delta\mathbf{E} = \sum_k Q_k\, \delta\varphi_k. \tag{6.111}$$

As anticipated, (6.111) shows that $\hat{U}_E$ may be regarded as a natural function of the fixed potentials applied to any conductors in the problem. Therefore, when we carry over from U_E the implicit dependence of the energy on the center of mass coordinate **R** of the dielectric,

$$\hat{U}_E = \hat{U}_E(\mathbf{E},\mathbf{R}) \qquad \text{or} \qquad \hat{U}_E = \hat{U}_E(\varphi_k,\mathbf{R}). \tag{6.112}$$

Our aim is to derive the analog of the partial derivative relations in (6.93). The first step uses the rules of calculus and (6.112) to conclude that

$$\delta\hat{U}_E = \left.\frac{\partial\hat{U}_E}{\partial\mathbf{E}}\right|_{\mathbf{R}}\cdot\delta\mathbf{E} + \left.\frac{\partial\hat{U}_E}{\partial\mathbf{R}}\right|_{\mathbf{E}}\cdot\delta\mathbf{R}. \tag{6.113}$$

Next, compare (6.113) to (6.110) and insist that the mechanical force on a dielectric always acts to reduce its energy. The formulae that result are the relations we seek:

$$\mathbf{D} = -\frac{1}{V}\left(\frac{\partial\hat{U}_E}{\partial\mathbf{E}}\right)_{\mathbf{R}} \qquad \text{and} \qquad \mathbf{F} = -\left(\frac{\partial\hat{U}_E}{\partial\mathbf{R}}\right)_{\mathbf{E}} \qquad \text{or} \qquad \mathbf{F} = -\left(\frac{\partial\hat{U}_E}{\partial\mathbf{R}}\right)_{\varphi}. \tag{6.114}$$

We will use the force formula on the far right side of (6.114) in Example 6.8 at the end of the chapter.

6.8 Forces on Dielectric Matter

The electric force that acts on dielectric matter is straightforward to calculate when we ask for the net force on an entire isolated object in vacuum. Subtleties arise when we ask for the force density needed to calculate the force on a sub-volume of a dielectric sample. For that reason, we treat the two cases separately.

6.8.1 An Isolated Dielectric Body

Figure 6.15 shows an isolated sample of dielectric matter with polarization $\mathbf{P}(\mathbf{r})$ in an external electric field $\mathbf{E}_0(\mathbf{r})$. It is immaterial whether **P** is due to $\mathbf{E}_0$ or not. The electromagnetic force on this object is given by Coulomb's law applied to (i) the volume and surface polarization charge densities, $\rho_P = -\nabla\cdot\mathbf{P}$ and $\sigma_P = \mathbf{P}\cdot\hat{\mathbf{n}}$; and (ii) whatever free charge density ρ_f happens to be present inside the dielectric. Thus,

$$\mathbf{F} = \int d^3r\, \left[\rho_f(\mathbf{r}) - \nabla\cdot\mathbf{P}(\mathbf{r})\right]\mathbf{E}_0(\mathbf{r}) + \int d\mathbf{S}\cdot\mathbf{P}(\mathbf{r}_S)\mathbf{E}_0(\mathbf{r}_S). \tag{6.115}$$

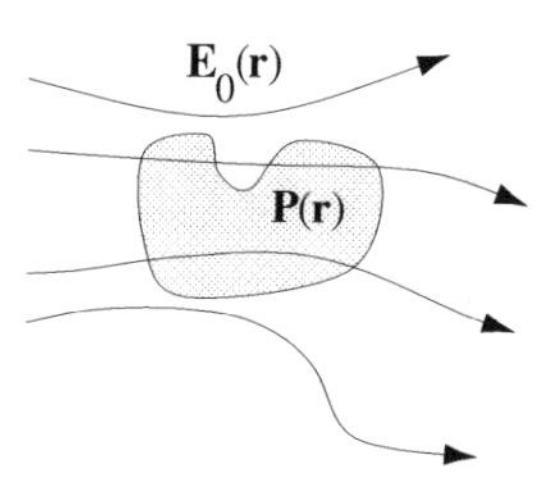

Figure 6.15: Polarized matter in vacuum exposed to an external electric field $\mathbf{E}_0(\mathbf{r})$. The displayed field lines represent the external field only.

An alternative point of view recalls from Section 6.3.1 that $\mathbf{P}(\mathbf{r})$ behaves exactly like a volume distribution of point electric dipoles. Therefore, a generalization of the point dipole results of Section 4.2.1 suggests that the force on the sample volume is

$$\mathbf{F} = \int d^3r\, [\rho_f(\mathbf{r}) + \mathbf{P}(\mathbf{r}) \cdot \nabla\,]\mathbf{E}_0(\mathbf{r}). \tag{6.116}$$

The force formulae (6.115) and (6.116) are equal by virtue of the vector identity

$$\int dS(\hat{\mathbf{n}} \cdot \mathbf{a})\mathbf{b} = \int d^3r\, (\nabla \cdot \mathbf{a})\, \mathbf{b} + \int d^3r\, (\mathbf{a} \cdot \nabla)\, \mathbf{b}. \tag{6.117}$$

The total electric field, $\mathbf{E}(\mathbf{r}) = \mathbf{E}_0(\mathbf{r}) + \mathbf{E}_{\text{self}}(\mathbf{r})$, is the sum of the external field and the field produced by the dielectric. However, because a dielectric cannot exert a net force on itself, (6.115) does not change if we replace $\mathbf{E}_0(\mathbf{r})$ by $\mathbf{E}(\mathbf{r})$ in the volume integral and $\mathbf{E}_0(\mathbf{r}_S)$ by $\mathbf{E}_{\text{avg}}(\mathbf{r}_S) = \frac{1}{2}[\mathbf{E}_{\text{in}}(\mathbf{r}_S) + \mathbf{E}_{\text{out}}(\mathbf{r}_S)]$ in the surface integral. The average is required because a surface charge distribution generally creates a discontinuity in the field right at the surface (see Section 3.4.3). In other words, a valid expression for the force on an isolated dielectric is

$$\mathbf{F} = \int d^3r\, \left[\rho_f(\mathbf{r}) - \nabla \cdot \mathbf{P}(\mathbf{r})\right] \mathbf{E}(\mathbf{r}) + \int d\mathbf{S} \cdot \mathbf{P}(\mathbf{r}_S)\mathbf{E}_{\text{avg}}(\mathbf{r}_S). \tag{6.118}$$

We leave it as an exercise for the reader to transform (6.118) into the form of (6.116) that can be used in conjunction with the total field:

$$\mathbf{F} = \int d^3r\, [\rho_f(\mathbf{r}) + \mathbf{P}(\mathbf{r}) \cdot \nabla\,]\mathbf{E}(\mathbf{r}) + \frac{1}{2\epsilon_0} \int d\mathbf{S}\, [\hat{\mathbf{n}}(\mathbf{r}_S) \cdot \mathbf{P}(\mathbf{r}_S)]^2. \tag{6.119}$$

We note in closing that the bilinear character of the integrands in this section raises the same questions about the macroscopic validity of our force formulae as arose in the Lorentz averaging discussion of Section 2.3.1. For that reason, it is necessary to treat the validity of (6.118) and (6.119) as logically independent assumptions of the macroscopic theory subject to verification by experiment.

Example 6.5 How does the force between a charged object and an uncharged simple dielectric depend on the distance r between them? Assume that r is large compared to the size of either object.

Solution: Because the object sizes are small compared to r, the variation of the electric field at the position of the dielectric can be taken as $\mathbf{E}_0 = (Q/4\pi\epsilon_0)\hat{\mathbf{r}}/r^2$. For the same reason, the force (6.116) on the dielectric can be written in terms of the dipole moment $\mathbf{p} = \int d^3r\, \mathbf{P}(\mathbf{r})$ of the dielectric body:

$$\mathbf{F} = \int d^3r\, (\mathbf{P} \cdot \nabla)\mathbf{E}_0 \simeq (\mathbf{p} \cdot \nabla)\mathbf{E}_0.$$

The dielectric is linear, so $\mathbf{p} = \gamma \, \mathbf{E}_0$ where γ is a constant. Therefore, because $\nabla \times \mathbf{E}_0 = 0$,

$$\mathbf{F} = \gamma(\mathbf{E}_0 \cdot \nabla)\mathbf{E}_0 = \tfrac{1}{2}\gamma \, \nabla|\mathbf{E}_0|^2 \propto -Q^2 \frac{\hat{\mathbf{r}}}{r^5}.$$

This force is always attractive, independent of the sign of Q. The charged object attracts polarization charge of unlike sign and repels polarization charge of like sign. The former is nearer to Q than the latter and thus dominates the net Coulomb interaction.

With other choices for the non-uniform external field $\mathbf{E}_0(\mathbf{r})$, the force $\mathbf{F} = \tfrac{1}{2}\gamma \, \nabla|\mathbf{E}_0|^2$ has been successfully used to sort, position, and transport many different types of minute dielectric bodies, including living cells. This phenomenon is known as *dielectrophoresis*, from the Greek verb meaning "to bear" or "to carry".[16]

6.8.2 A Simple Dielectric Sub-Volume

The formulae derived in the previous section from Coulomb's law give the net force on an isolated dielectric body in vacuum. They *cannot* be used to find the force on a dielectric *sub-volume*. To see this, consider the force per unit area $\mathbf{f} = f\hat{\mathbf{z}}$ which acts on the infinitesimally thin sub-volume enclosed by the dashed lines in Figure 6.16. This is the force per unit area which acts on the abrupt interface where the dielectric constant changes from κ_1 to κ_2. More precisely, it is the force per unit area that acts on the polarization charge density σ_P at the $z = 0$ interface.

Let us calculate $\mathbf{f}$ using Coulomb's law (6.115) applied to the sub-volume bounded by the surfaces $z = \epsilon$ and $z = -\epsilon$ in the limit when $\epsilon \to 0$. There is no contribution from the volume integrals. Therefore,

$$\mathbf{f} = \sigma_P \mathbf{E}_0 = [\mathbf{P}_1 \cdot \hat{\mathbf{z}} + \mathbf{P}_2 \cdot (-\hat{\mathbf{z}})]\mathbf{E}_0 = \left[\epsilon_0 \frac{\kappa_1 - 1}{\kappa_1} E_0 - \epsilon_0 \frac{\kappa_2 - 1}{\kappa_2} E_0\right] E_0 \hat{\mathbf{z}}. \tag{6.120}$$

Gauss' law gives the external field applied to the dielectric as $E_0 = \sigma_f/\epsilon_0$. The final result is the following *incorrect* formula for the force per unit area on the dielectric interface:

$$\mathbf{f} = \frac{\sigma_f^2}{\epsilon_0}\left(\frac{1}{\kappa_2} - \frac{1}{\kappa_1}\right)\hat{\mathbf{z}} \qquad \text{(incorrect)}. \tag{6.121}$$

The problem with (6.121) becomes clear when we take the limit $\kappa_1 \to \infty$ to get

$$\mathbf{f} \to \frac{\sigma_f^2}{\kappa_2 \epsilon_0}\hat{\mathbf{z}}. \tag{6.122}$$

According to Section 6.6.4, $\kappa_1 \to \infty$ amounts to replacing the $z < 0$ matter in Figure 6.16 by a perfect conductor. Therefore, (6.122) should be the force per unit area which the upper plate exerts on the lower plate of Figure 6.16 in that limit. If the plate area is A and the plate separation is d, (6.52) tells us that the capacitance is $C = \kappa_2 \epsilon_0 A/d$. Then, using (5.83) with $\mathbf{R} = -\hat{\mathbf{z}}d$ to evaluate the force on the lower plate, the force density in question is

$$\mathbf{f} = \frac{Q^2}{2C}\frac{\partial C}{\partial \mathbf{R}} = -\frac{1}{A}\frac{Q^2}{2C^2}\frac{\partial C}{\partial d}\hat{\mathbf{z}} = \frac{\sigma_f^2}{2\kappa_2\epsilon_0}\hat{\mathbf{z}}. \tag{6.123}$$

This differs from (6.122) by a factor of 2. We will derive a correct version of (6.121) in Example 6.6.

[16] See Voldman (2006) in Sources, References, and Additional Reading for more information about dielectrophoresis.

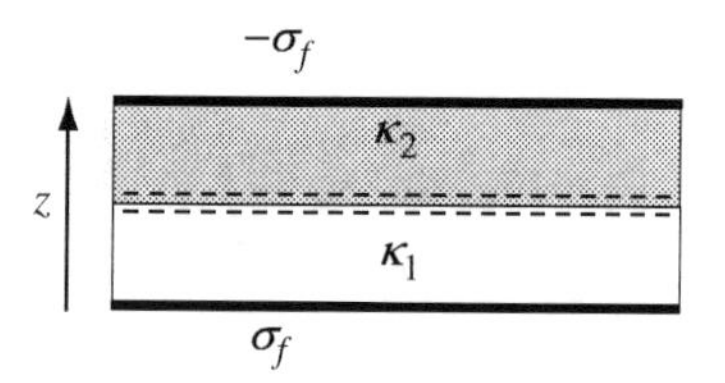

Figure 6.16: A fixed-charge capacitor where $\epsilon(z)$ changes abruptly from ϵ_1 to ϵ_2 at $z = 0$. The force on the interface polarization charge is the force exerted on the dielectric sub-volume enclosed by the dotted lines.

6.8.3 The Role of Short-Range Forces

The force formulae derived in Section 6.8 fail when applied to a dielectric *sub-volume* because they omit the effect of short-range, non-electrostatic forces which act on the sub-volume boundary. These quantum mechanical forces (which are responsible for elasticity) contribute to the field-induced force of interest because their magnitude and direction are polarization-dependent. On the other hand, internal forces of this kind cancel out when we sum over all sub-volumes to find the net force on an entire body.

Brown (1951) calculated forces in dielectric matter using the theory of elastic stress generalized to take account of dielectric polarization. This direct and explicit method highlights three important points. First, the surface terms needed to correct (6.115) and (6.116) for a sub-volume are not the same. Second, the requirements of local mechanical equilibrium make it possible to express the final force entirely in terms of macroscopic quantities. Third, a force density expression exists (due to Helmholtz) for which the associated surface term may be ignored because it integrates to zero (at least for a dielectric fluid). The goal of the next section is to derive Helmholtz' force density formula and apply it to the interface problem of Figure 6.16.

6.8.4 Force from Variation of Energy for a Simple Dielectric

It is instructive to calculate the electrostatic force on a simple dielectric medium from the change in energy U_E which accompanies a virtual displacement $\delta\mathbf{r}$ of the dielectric:

$$\delta U_E = -\mathbf{F} \cdot \delta\mathbf{r}. \tag{6.124}$$

This generalizes the derivation in Section 3.5.1 of the Lorentz force and produces $\mathbf{F}$ in the form of an integral over all of space. Moreover, the integrand of this integral agrees with the force density obtained by other methods where one can check that the effects of short-range forces cancel out when integrated over the surface of any sub-volume (see Section 6.8.3).

Beginning with $U_E = \frac{1}{2}\int d^3r\, |\mathbf{D}|^2/\epsilon$ from (6.94), the variation of U_E has two terms:

$$\delta U_E = \int d^3r\, \frac{\mathbf{D} \cdot \delta\mathbf{D}}{\epsilon} - \frac{1}{2}\int d^3r\, \frac{\mathbf{D} \cdot \mathbf{D}}{\epsilon^2}\delta\epsilon = \int d^3r\, \mathbf{E} \cdot \delta\mathbf{D} - \frac{1}{2}\int d^3r\, E^2 \delta\epsilon. \tag{6.125}$$

The integrations in (6.125) are taken over all of space. Moreover, if the variations $\delta\mathbf{D}$ and $\delta\epsilon$ are induced by a displacement $\delta\mathbf{r}$ of the matter, the analog of Figure 3.13 gives [see (3.65)]

$$\delta\mathbf{D} = \mathbf{D}(\mathbf{r} - \delta\mathbf{r}) - \mathbf{D}(\mathbf{r}) = -(\delta\mathbf{r} \cdot \nabla)\mathbf{D} \tag{6.126}$$

and

$$\delta\epsilon = \epsilon(\mathbf{r} - \delta\mathbf{r}) - \epsilon(\mathbf{r}) = -(\delta\mathbf{r} \cdot \nabla)\epsilon. \tag{6.127}$$

Substituting these into (6.125) gives

$$\delta U_E = -\int d^3r\, \mathbf{E}\cdot(\delta\mathbf{r}\cdot\nabla)\mathbf{D} + \frac{1}{2}\int d^3r\, E^2\nabla\epsilon\cdot\delta\mathbf{r}. \tag{6.128}$$

The second term in (6.128) already has the desired form (6.124). To transform the first term, we use the vector identity

$$\nabla\times(\delta\mathbf{r}\times\mathbf{D}) = \delta\mathbf{r}(\nabla\cdot\mathbf{D}) - (\delta\mathbf{r}\cdot\nabla)\mathbf{D}, \tag{6.129}$$

integration by parts, and the Maxwell equations $\nabla\cdot\mathbf{D} = \rho_f$ and $\nabla\times\mathbf{E} = 0$. The final result is called the *Helmholtz formula*:

$$\mathbf{F} = -\frac{\delta U_E}{\delta\mathbf{r}} = \int d^3r\, \rho_f(\mathbf{r})\mathbf{E}(\mathbf{r}) - \frac{1}{2}\int d^3r\, E^2(\mathbf{r})\nabla\epsilon(\mathbf{r}). \tag{6.130}$$

When the domain of integration is restricted to a sub-volume, (6.130) gives the mechanical force exerted on that sub-volume.

It is worth noting that (6.130) does not contain the polarization $\mathbf{P}$ or the polarization charge density ρ_P explicitly. From this point of view, the electric field exerts a force only on free charge and at points in space where the dielectric "constant" varies in space. The physically correct force at the abrupt dielectric interface in Figure 6.16 comes precisely from this latter contribution (see Example 6.6).

Example 6.6 Use (6.130) to calculate the force per unit area on the ϵ_1/ϵ_2 interface ($z=0$) of the parallel-plate capacitor shown in Figure 6.16. Check the limit $\kappa_1 \to \infty$.

Solution: The force (6.130) is non-zero only in the immediate vicinity of the $z=0$ interface where $\epsilon(z)$ changes abruptly. Therefore, with $\mathbf{E} = E_z\hat{\mathbf{z}}$, the force per unit area is $\mathbf{f} = f\hat{\mathbf{z}}$ where

$$f = -\frac{1}{2}\int_{0^-}^{0^+} E_z^2 \frac{d\epsilon}{dz}dz.$$

Formally, $\epsilon(z) = \epsilon_1\Theta(-z) + \epsilon_2\Theta(z)$ so $d\epsilon/dz = (\epsilon_2 - \epsilon_1)\delta(z)$. This makes the integral ill-defined because E_z is not continuous at $z=0$. However, D_z *is* continuous [see (6.31)] and Gauss' law gives $D_z = \sigma_f$ everywhere between the capacitor plates. Therefore,

$$f = -\frac{1}{2}\int_{0^-}^{0^+} \frac{D_z^2}{\epsilon^2}\frac{d\epsilon}{dz}dz = \frac{1}{2}\int_{0^-}^{0^+} D_z^2 \frac{d}{dz}\left(\frac{1}{\epsilon}\right)dz = \frac{\sigma_f^2}{2\epsilon_0}\left(\frac{1}{\kappa_2} - \frac{1}{\kappa_1}\right).$$

Independent of the direction of the electric field, this force tends to increase the volume of the medium with large ϵ and decrease the volume of the medium with small dielectric constant.

Since $E_0 = \sigma_f/\epsilon_0$ is the electric field between the plates of a vacuum capacitor, the $\kappa_1 \to \infty$ limit of the force density can be written in the form

$$f \to \frac{1}{2}\frac{E_0}{\kappa_2}\sigma_f.$$

As discussed in the paragraph following (6.122), this is the expected force of attraction per unit area between the plates of a capacitor filled with matter with dielectric constant κ_2.

6.8.5 The Electric Stress Tensor for a Simple Dielectric

The subtleties of the Helmholtz formula (6.130) revealed by Example 6.6 motivate us to rewrite the force in a form that avoids $\nabla\epsilon$. To do this, we use $\frac{1}{2}E^2\partial_j\epsilon = \frac{1}{2}\partial_j(D_kE_k) - D_k\partial_jE_k$ and replace the last term in this identity by $D_k\partial_k E_j$ because $\nabla \times \mathbf{E} = 0$. Then, because $\nabla \cdot \mathbf{D} = \rho_f$, (6.130) takes the form

$$\mathbf{F} = \int d^3r\,(\nabla \cdot \mathbf{D})\mathbf{E} - \int d^3r\,(\mathbf{D} \cdot \nabla)\mathbf{E} - \frac{1}{2}\int d^3r\,\nabla(\mathbf{D} \cdot \mathbf{E}). \tag{6.131}$$

Now convert the last term in (6.131) to a surface integral and use the identity (6.117) with $\mathbf{a} = \mathbf{D}$ and $\mathbf{b} = \mathbf{E}$ to do the same with the first two terms. The result is a very convenient formula for calculations that should remind the reader of a similar formula derived in Section 3.7 for the force on a distribution of pure charge in vacuum:

$$\mathbf{F} = \int_S dS\left\{(\hat{\mathbf{n}} \cdot \mathbf{D})\mathbf{E} - \tfrac{1}{2}\hat{\mathbf{n}}(\mathbf{D} \cdot \mathbf{E})\right\}. \tag{6.132}$$

This expression has two important virtues compared to the Helmholtz formula (6.130): (i) it contains no space derivatives; and (ii) it requires only the evaluation of a surface integral rather than a volume integral. A non-trivial use of (6.132) is the subject of Application 6.4 below.

Finally, the structure of the integrand in (6.132) motivates us to define an *electric stress tensor* for simple dielectric matter as

$$\mathcal{T}_{ij}(\mathbf{D}) \equiv D_i E_j - \tfrac{1}{2}\delta_{ij}\mathbf{D} \cdot \mathbf{E}. \tag{6.133}$$

This object is symmetric because $D_i E_j = \epsilon E_i E_j$. Moreover, in vacuum where $\mathbf{D} = \epsilon_0\mathbf{E}$, (6.133) reduces to the Maxwell electric stress tensor defined in Section 3.7. Using (6.133), (6.132) takes the form

$$\mathbf{F} = \int_S dS\,\hat{\mathbf{n}} \cdot \mathcal{T}(\mathbf{D}). \tag{6.134}$$

A volume integral version of (6.134) follows from the divergence theorem; namely, the jth component of the force (6.132) on a dielectric sub-volume Ω is

$$F_j = \int_\Omega d^3r\,\partial_i \mathcal{T}_{ij}(\mathbf{D}). \tag{6.135}$$

Application 6.4 The Electric Force on an Embedded Volume

Figure 6.17 shows a simple dielectric with permittivity $\epsilon = \kappa\epsilon_0$. Embedded in this dielectric is a point charge q and a second dielectric with permittivity $\epsilon' = \kappa'\epsilon_0$. A point charge q' is embedded in the second dielectric. Our goal is to find the force exerted on the embedded dielectric and q'. In principle, this force can be computed using the Helmholtz formula (6.130). We proceed differently and use the stress tensor method to emphasize the failure of a formula like (6.115) when applied to a dielectric sub-volume.

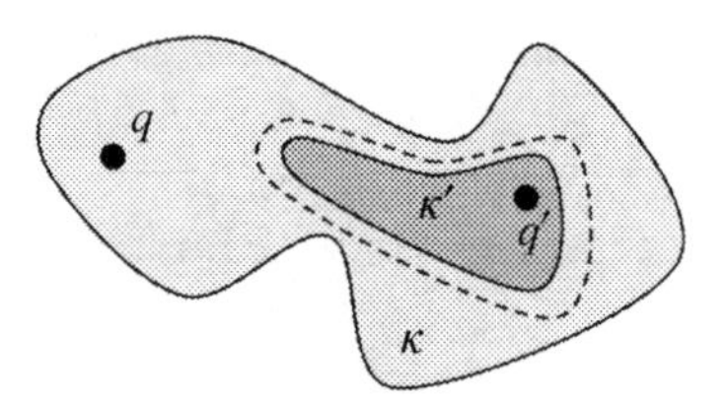

Figure 6.17: The dashed line is the boundary S of a volume Ω which is infinitesimally larger than the volume of a dielectric with permittivity κ'. The latter is embedded in a larger dielectric with permittivity κ.

Our strategy is to use (6.132) and choose S as a surface (dashed in Figure 6.17) which bounds a volume Ω which is infinitesimally larger than the physical volume of the embedded dielectric. Since $\mathbf{D} = \epsilon\mathbf{E}$ everywhere on S, we can pull ϵ outside the integral and then transform the surface integral into a volume integral as we did to get (6.135). The result is

$$F_j = \epsilon \int_\Omega d^3r\, \partial_i (E_i E_j - \tfrac{1}{2}\delta_{ij}\mathbf{E}\cdot\mathbf{E}). \tag{6.136}$$

Repeating the algebraic steps used in Section 3.7 (in reverse order) shows that (6.136) reduces to

$$\mathbf{F} = \kappa \int_\Omega d^3r\, \rho(\mathbf{r})\mathbf{E}(\mathbf{r}). \tag{6.137}$$

The charge $\rho(\mathbf{r})$ in (6.137) is the sum of the free and polarization charge densities in Ω. This makes (6.137) superficially similar to the volume integral in (6.118). The difference is the multiplicative factor κ which (not obviously) comes from the field-dependent short-range force (see Section 6.8.3) exerted on the κ' dielectric by the κ dielectric. ■

Example 6.7 Find the force exerted on the point charge q in Example 6.2 of Section 6.5.3.

Solution: The force on q is given by (6.137) or (equivalently) the first term in (6.130). $\mathbf{E}(\mathbf{r})$ is the field produced by the surface charge distribution σ_P computed in Example 6.2. By symmetry, we need only the z-component of this field evaluated at the position of the point charge. Therefore, since $\mathbf{s} = x\hat{\mathbf{x}} + y\hat{\mathbf{y}}$ labels points on the interface and $r = \sqrt{s^2 + d^2}$ is the distance to q (see Figure 6.8),

$$\mathbf{F} = -\hat{\mathbf{z}} \int d^2s \frac{q}{4\pi\epsilon_0} \frac{\sigma_P(\mathbf{s})(d\hat{\mathbf{z}} + \mathbf{s})}{|d\hat{\mathbf{z}} + \mathbf{s}|^3} = -\hat{\mathbf{z}}\, 2\pi \int_0^\infty ds\, s \frac{q}{4\pi\epsilon_0} \left\{ \frac{1}{2\pi\kappa_L} \frac{\kappa_L - \kappa_R}{\kappa_L + \kappa_R} \frac{qd}{r^3} \right\} \frac{d}{r^3}.$$

The integral is straightforward so

$$\mathbf{F}_{\text{on } q} = -\hat{\mathbf{z}} \frac{1}{\kappa_L} \frac{\kappa_L - \kappa_R}{\kappa_L + \kappa_R} \frac{1}{4\pi\epsilon_0} \left(\frac{q}{2d}\right)^2.$$

The force between q and the interface is attractive when $\kappa_R > \kappa_L$ and repulsive when $\kappa_R < \kappa_L$. This agrees with Section 6.7.2 where we concluded that free charge is attracted to regions where the dielectric constant is large. We will return to this example in Section 8.3.3.

Example 6.8 Figure 6.18 shows a slab of simple dielectric matter partially inserted into the gap d between the square ($L \times L$) plates of a capacitor. If the fixed potential difference between the plates is V, show that the slab is drawn into the capacitor by a force $\mathbf{F} = \hat{\mathbf{x}}\epsilon_0(\kappa - 1)LV^2/2d$. Ignore fringe field effects.

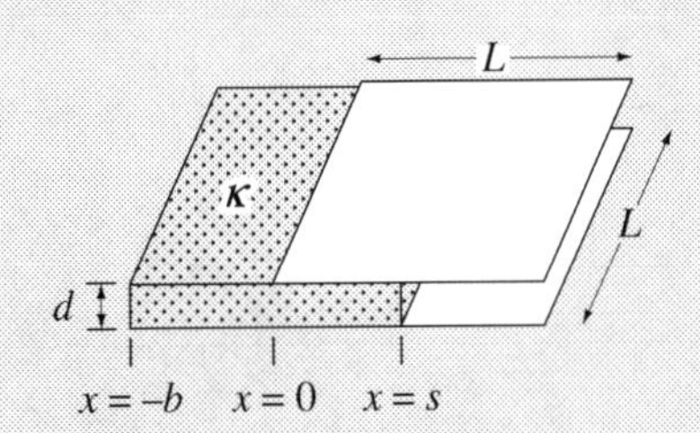

Figure 6.18: A dielectric slab partially inserted into a capacitor.

Method I: If $\mathbf{E}$ is the actual electric field inside the polarized volume V of the dielectric, (6.101) gives the total energy gain of the system as

$$\Delta\hat{U}_E = -\frac{1}{2}\int_V d^3r\,\mathbf{P}\cdot\mathbf{E}_0 = -\frac{1}{2}\epsilon_0(\kappa-1)\int_V d^3r\,\mathbf{E}\cdot\mathbf{E}_0.$$

For fixed potential difference, $|\mathbf{E}_0| = |\mathbf{E}| = V/d$, so

$$\Delta\hat{U}_E = -\frac{1}{2}\epsilon_0(\kappa-1)\frac{V^2}{d^2}\times(Lds).$$

This gives the force $\mathbf{F} = -\partial\hat{U}_E/\partial\mathbf{s} = \hat{\mathbf{x}}\epsilon_0(\kappa-1)LV^2/2d$ as required.

Method II: There is no free charge in the dielectric so (6.130) gives the force on the dielectric as

$$\mathbf{F} = -\frac{1}{2}\int d^3r E^2\nabla\epsilon.$$

If we ignore fringing effects, $E = V/d$ inside the capacitor and $E = 0$ outside the capacitor. Therefore, we can ignore the $x = -b$ end of the dielectric and write $\epsilon(x) = \epsilon_0\Theta(x-s) + \epsilon\Theta(s-x)$. This gives $\nabla\epsilon = \epsilon_0(1-\kappa)\delta(x-s)\hat{\mathbf{x}}$. The electric field is continuous at $x = s$ and the cross sectional area of the dielectric is Ld. Therefore,

$$\mathbf{F} = -\frac{1}{2}Ld\left(\frac{V}{d}\right)^2\epsilon_0(1-\kappa)\hat{\mathbf{x}} = \epsilon_0(\kappa-1)L\frac{V^2}{2d}\hat{\mathbf{x}}.$$

Remark: Unfortunately, neither Method I nor Method II clearly identifies the true physical origin of the force which attracts the dielectric into the capacitor. We are interested in the net force on the entire volume V of the dielectric, so it is permissible to use the fundamental law (6.116)

$$\mathbf{F} = \int_V d^3r\,[\mathbf{P}(\mathbf{r})\cdot\nabla]\mathbf{E}_0(\mathbf{r}).$$

This formula shows that the force can only come from those portions of the dielectric where the capacitor electric field $\mathbf{E}_0(\mathbf{r})$ deviates from uniformity. In other words, the force arises entirely from the fringing field near the edges of the capacitor plates.

Sources, References, and Additional Reading

The quotation at the beginning of the chapter is taken from Chapter 9 of

E.M. Purcell, *Electricity and Magnetism* (McGraw-Hill, New York, 1965).

Section 6.1 The book cited above by *Purcell* contains the best elementary discussion of the physics of dielectrics. Two intermediate-level textbooks distinguished by the instructive examples they present are

O.D. Jefimenko, *Electricity and Magnetism* (Appleton-Century-Crofts, New York, 1966).

D.J. Griffiths, *Introduction to Electrodynamics*, 3rd edition (Prentice-Hall, Upper Saddle River, NJ, 1999).

Section 6.2 Our approach to volume and surface polarization charge follows

L.D. Landau and E.M. Lifshitz, *Electrodynamics of Continuous Media* (Pergamon, Oxford, 1960).

The original discussion of Lorentz and a well-presented version of this traditional theory of polarized matter are

H.A. Lorentz, "The fundamental equations for electromagnetic phenomena in ponderable bodies deduced from the theory of electrons", *Proceedings of the Royal Academy of Amsterdam* **5**, 254 (1902).

C.J.F. Böttcher, *The Theory of Electric Polarization* (Elsevier, Amsterdam, 1952).

A good description of the modern theory of polarized matter by its originators is

R. Resta and D. Vanderbilt, "Theory of polarization: A modern approach", in *Physics of Ferroelectrics: A Modern Perspective*, edited by K.M. Rabe, C.H. Ahn, and J.-M. Triscone (Springer, New York, 2007), pp. 31-68.

Figure 6.1 comes from *Purcell* (see Section 6.1 above). Figure 6.2 in Section 6.2.3 is taken from

P. Umari, A. Dal Corso, and R. Resta, "Inside dielectrics: Microscopic and macroscopic polarization", in *Fundamental Physics of Ferroelectrics*, edited by H. Krakauer (AIP, Melville, NY, 2001), pp. 107-117. Copyright 2001. American Institute of Physics.

Section 6.3 The theorem quoted in Application 6.1 is the subject of

K.R. Brownstein, "Unique shape of uniformly polarizable dielectrics", *Journal of Mathematical Physics* **28**, 978 (1987).

Section 6.5 The examples in this section were drawn from

W.T. Scott, *The Physics of Electricity and Magnetism*, 2nd edition (Wiley, New York, 1966).

L. Eyges, *The Classical Electromagnetic Field* (Dover, New York, 1972).

A.M. Portis, *Electromagnetic Fields: Sources and Media* (Wiley, New York, 1978).

Section 6.6 There are many ways to derive the Clausius-Mossotti formula (6.84), not all of them completely legitimate. Three useful discussions of this subject are

B.R.A. Nijboer and F.W. De Wette, "The internal field in dipole lattices", *Physica* **25**, 422 (1958).

M. Bucher, "Re-evaluation of the local field", *Journal of the Physics and Chemistry of Solids* **51**, 1241 (1990).

D.E. Aspnes, "Local field effects and effective medium theory: A microscopic perspective", *American Journal of Physics* **50**, 704 (1982).

Onsager's calculation of the dielectric constant for water appears in

L. Onsager, "Electric moments of molecules in liquids", *Journal of the American Chemical Society* **58**, 1486 (1936).

The source for Figure 6.12 is

A. Baldereschi, R. Car, and E. Tosatti, "Microscopic local fields in dielectrics", *Solid State Communications* **32**, 757 (1979), with permission from Elsevier.

Section 6.7 Our treatment of energy in dielectric matter benefitted from *Landau and Lifshitz* (see Section 6.2 above) and *Egyes* (see Section 6.5 above). Two discussions which make contact with the statistical mechanical (Hamiltonian) point of view are

R. Balian, *From Microphysics to Macrophysics* (Springer, Berlin, 2007), Volume I, Section 6.6.5.

O. Narayan and A.P. Young, "Free energies in the presence of electric and magnetic fields", *American Journal of Physics* **73**, 293 (2005).

Authors who recognize that it is an assumption to set the field energy density in a dielectric equal to the field energy in vacuum include *Purcell* (see Section 6.1 above) and

S. Bobbio, *Electrodynamics of Materials* (Academic, San Diego, 2000).

Figure 6.14 and the classical model of quark confinement in Application 6.3 come from the article by *Lee*. Example 6.4 is taken from the book by *Robinson*.

T.D. Lee, "Is the physical vacuum a medium?", *Transactions of the New York Academy of Sciences*, Series II, **40**, 111 (1980).

F.N.H. Robinson, *Macroscopic Electromagnetism* (Pergamon, Oxford, 1973).

Section 6.8 The subject of forces in and on dielectric matter is subtle and complex. Our discussion is consistent with, but less complete than, those in *Landau and Lifshitz* (see Section 6.2 above), *Robinson* and *Bobbio* (see Section 6.7 above), and

J.A. Stratton, *Electromagnetic Theory* (McGraw-Hill, New York, 1941).

Example 6.5 explains the origin of dielectrophoresis. A review of this subject in a biophysical context is

J. Voldman, "Electrical forces for micro-scale cell manipulation", *Annual Reviews of Biomedical Engineering* **8**, 425 (2006).

An instructive discussion of the role of short-range forces in dielectric force calculations is

D.L. Webster, "Unscrambling the dielectric constant", *The American Physics Teacher* **2**, 149 (1934).

Example 6.6 was familiar to Maxwell. Our method of solution follows

J. Schwinger, L.L. DeRaad, Jr., K.A. Milton, and W-Y. Tsai, *Classical Electrodynamics* (Perseus, Reading, MA, 1998).

Application 6.4 is taken from

W.F. Brown, Jr., "Electric and magnetic forces: A direct calculation", *American Journal of Physics* **19**, 290 (1951).

Example 6.8 has been discussed by generations of physicists. Our treatment is based on

S. Margulies, "Force on a dielectric slab inserted into a parallel-plate capacitor", *American Journal of Physics* **52**, 515 (1984).

Problems

6.1 Polarization by Superposition Two spheres with radius R have uniform but equal and opposite charge densities $\pm\rho$. The centers of the two spheres fail to coincide by an infinitesimal displacement vector $\boldsymbol{\delta}$. Show by direct superposition that the electric field produced by the spheres is identical to the electric field produced by a sphere with a suitably chosen uniform polarization $\mathbf{P}$.

6.2 How to Make a Uniformly Charged Sphere Find a polarization $\mathbf{P}(r)$ which produces a polarization charge density in the form of an origin-centered sphere with radius R and uniform volume charge density ρ_P.

6.3 The Energy of a Polarized Ball Find the total electrostatic energy of a ball with radius R and uniform polarization $\mathbf{P}$.

6.4 A Hole in Radially Polarized Matter The polarization in all of space has the form $\mathbf{P} = P\Theta(r - R)\hat{\mathbf{r}}$, where P and R are constants. Find the polarization charge density and the electric field everywhere.

6.5 The Field at the Center of a Polarized Cube A cube is polarized uniformly parallel to one of its edges. Show that the electric field at the center of the cube is $\mathbf{E}(0) = -\mathbf{P}/3\epsilon_0$. Compare with $\mathbf{E}(0)$ for a uniformly polarized sphere. Hint: Recall the definition of solid angle.

6.6 Practice with Poisson's Formula Confirm Poisson formula (derived in Section 6.3) for the case when the volume V is a rectangular slab which is infinite in the x and y directions and occupies the interval $-t \le z \le t$ otherwise. Keep the direction of $\mathbf{P}_0$ arbitrary.

6.7 Isotropic Polarization The electrostatic polarization inside an origin-centered sphere is $\mathbf{P}(\mathbf{r}) = \mathbf{P}(r)$.

(a) Show that $\varphi(\mathbf{r})$ outside the sphere is equal to the potential of a point electric dipole at the origin with a moment equal to the electric dipole moment of the entire sphere.

(b) What is $\varphi(\mathbf{r})$ inside the sphere?

6.8 E and D for an Annular Dielectric

(a) The entire volume between two concentric spherical shells is filled with a material with uniform polarization $\mathbf{P}$. Find $\mathbf{E}(\mathbf{r})$ everywhere.

(b) The entire volume inside a sphere of radius R is filled with polarized matter. Find $\mathbf{D}(\mathbf{r})$ everywhere if $\mathbf{P} = P\hat{\mathbf{r}}/r^2$.

6.9 The Correct Way to Define E Students are often told that $\mathbf{E} = \mathbf{F}_q/q$ defines the electric field at a point if $\mathbf{F}_q$ is the measured force on a tiny charge q placed at that point. More careful instructors let $q \to 0$ to avoid the polarization of nearby matter due to the presence of q. Unfortunately, this experiment is impossible to perform. A better definition uses $\mathbf{F}_q$ and the force $\mathbf{F}_{-q}$ measured when $-q$ sits at the same point. There is no need to let $q \to 0$, even if conductors or (linear) dielectric matter is present. Derive an expression that relates $\mathbf{E}$ to $\mathbf{F}_q$ and $\mathbf{F}_{-q}$.

6.10 Charge and Polarizable Matter Coincident A polarizable sphere of radius R is filled with free charge with uniform density ρ_c. The dielectric constant of the sphere is κ.

(a) Find the polarization $\mathbf{P}(\mathbf{r})$.
(b) Confirm explicitly that the total volume polarization charge and the total surface polarization charge sum to the expected value.

6.11 Cavity Field A uniform electric field $\mathbf{E}_0$ exists throughout a homogeneous dielectric with permittivity ϵ. What is the electric field inside a vacuum cavity cut out of the interior of the dielectric in the shape of a rectangular pancake with dimensions $L \times L \times h$? Assume that $h \ll L$ and express $\mathbf{E}_{\text{cav}}$ entirely in terms of $\mathbf{E}_0$, $\hat{\mathbf{n}}$, ϵ, and ϵ_0.

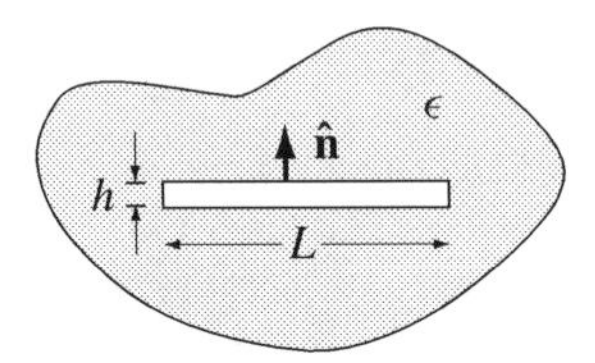

6.12 Making External Fields Identical An origin-centered sphere with permittivity ϵ and radius a is placed in a uniform external electric field $\mathbf{E}_0$. What radius $b < a$ should an origin-centered, perfectly conducting sphere similarly placed in a similar field $\mathbf{E}_0$ have so that the two situations produce identical electric fields for $r > a$?

6.13 The Capacitance Matrix for a Spherical Sandwich Two concentric, spherical, conducting shells have radii $R_2 > R_1$ and charges q_2 and q_1. The volume between the shells is filled with a linear dielectric with permittivity $\epsilon = \kappa\epsilon_0$. Determine the elements of the capacitance matrix for this system.

6.14 A Spherical Conductor Embedded in a Dielectric A spherical conductor of radius R_1 is surrounded by a polarizable medium which extends from R_1 to R_2 with dielectric constant κ.

(a) The conductor has charge Q. Find $\mathbf{E}$ everywhere and confirm that the total polarization charge is zero.
(b) The conductor is grounded and the entire system is placed in a uniform electric field $\mathbf{E}_0$. Find the electrostatic potential everywhere and determine how much charge is drawn up from ground to the conductor.

6.15 A Parallel-Plate Capacitor with an Air Gap An air-gap capacitor with parallel-plate area A discharges by the electrical breakdown of the air between its parallel plates (separation d) when the voltage between its plates exceeds V_0. Lay a slab with dielectric constant κ and thickness $t < d$ on the surface of the lower plate and maintain a potential difference V between the plates.

(a) Find the capacitance of this structure with the dielectric slab present.
(b) Show that the value of V where electric breakdown occurs in the air portion of the capacitor gap is

$$V' = V_0 \left[1 - \frac{t}{d}\left(1 - \frac{1}{\kappa}\right)\right].$$

6.16 Helmholtz Theorem for D(r) Write the Helmholtz theorem expression for $\mathbf{D}(\mathbf{r})$ and eliminate $\mathbf{D}$ itself from the integrals you write down. How does this formula simplify (if at all) for simple dielectric matter?

6.17 Electrostatics of a Doped Semiconductor A semiconductor with permittivity ϵ occupies the space $z \geq 0$. One "dopes" such a semiconductor by implanting neutral, foreign atoms with uniform density N_D in the near-surface region $0 \le z \le d$. Assume that one electron from each dopant atom ionizes and migrates to the free surface of the semiconductor. The final result (illustrated by the diagram) is a region with uniform positive charge density eN_D and a layer of negative charge with density σ localized at $z = 0$.

(a) Find and sketch the electric field $E_+(z)$ at every point in space produced by the volume charge.

(b) Find σ and the electric field $E_-(z)$ produced by σ. Sketch E_- on the same graph used to sketch E_+ in part (a).

(c) Sketch the total electric field and check that your graph is consistent with integrating Gauss' law from $z = -\infty$ to $z = \infty$.

6.18 Surface Polarization Charge Point charges $q_1, q_2, \ldots, q_N$ are embedded in a body with permittivity κ_{in}. The latter is itself embedded in a body with permittivity κ_{out}. Find the total polarization charge Q_{pol} induced on the boundary between the two dielectrics.

6.19 An Elastic Dielectric The parallel-plate capacitor shown below is made of two identical conducting plates of area A carrying charges $\pm q$. The capacitor is filled with a compressible dielectric solid with permittivity ϵ and elastic energy

$$U_e = \frac{1}{2}k(d - d_0)^2.$$

(a) Find the equilibrium separation between the plates $d(q)$.

(b) Sketch the potential difference between the plates $V(q)$. Comment on any unusual behavior of the differential capacitance $C_d(q) = dq/dV$.

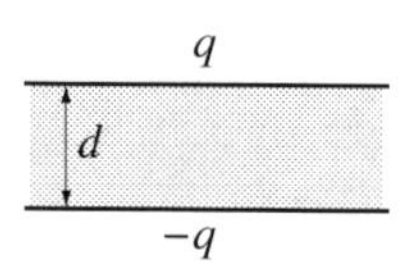

6.20 A Dielectric Inclusion A dielectric body with permittivity ϵ_{in} is embedded in an infinite volume of dielectric matter with permittivity ϵ_{out}. The entire system is polarized by an external electric field $\mathbf{E}_{\text{ext}}$. If φ is the exact electrostatic potential and S is the surface of the embedded body, show that the dipole moment of the system can be written in the form

$$\mathbf{p} = (\epsilon_{\text{out}} - \epsilon_{\text{in}}) \int_S dS\,\hat{\mathbf{n}}\varphi(\mathbf{r}_S).$$

6.21 A Classical Meson Application 6.3 modeled a meson (a quark-antiquark pair) as a finite dipole placed at the center of a spherical cavity with radius R and unit dielectric constant scooped out of an infinite medium with dielectric constant $\kappa \to 0$. For this problem, we replace the finite dipole by a point dipole $\mathbf{p}$.

(a) Find $\mathbf{D}$ and $\mathbf{E}$ everywhere for finite κ.

(b) Confirm the statements made in the Application regarding $\mathbf{D}$ and U_E when $\kappa = 0$. Assume a cutoff distance $a \ll R$ to simulate the size of the original dipole.

6.22 An Application of the Dielectric Stress Tensor A metal ball with charge Q sits at the center of a thin, spherical, conducting shell. The shell has charge Q' and the space between the shell and the ball is filled with matter with dielectric constant κ. Use the stress tensor method to prove that if the shell were split into two hemispheres, the two halves stay together only if Q' has the opposite sign to Q and

$$1-\frac{1}{\sqrt{\kappa}} \leq \left|\frac{Q'}{Q}\right| \leq 1+\frac{1}{\sqrt{\kappa}}.$$

6.23 Two Dielectric Interfaces The figure shows two fixed-potential capacitors filled with equal amounts of two different types of simple dielectric matter. Use the stress tensor method to compare the force per unit area which acts on the two dielectric interfaces. Express your answer in terms of the electric field E_0 which would be present if the dielectric matter were absent.

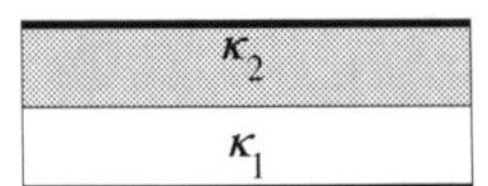

6.24 The Force on an Isolated Dielectric The text proved that the force on an isolated dielectric is

$$\mathbf{F} = \int d^3r \left[\rho_f(\mathbf{r}) - \nabla\cdot\mathbf{P}(\mathbf{r})\right]\mathbf{E}(\mathbf{r}) + \int d\mathbf{S}\cdot\mathbf{P}(\mathbf{r}_S)\mathbf{E}_{\text{avg}}(\mathbf{r}_S),$$

where $\mathbf{E}(\mathbf{r})$ is the total field at an interior point $\mathbf{r}$ and $\mathbf{E}_{\text{avg}}(\mathbf{r}_S)$ is the average of the total field just inside and just outside the dielectric at the surface point $\mathbf{r}_S$. Show that this expression can be rewritten in the form

$$\mathbf{F} = \int d^3r\,[\rho_f(\mathbf{r}) + \mathbf{P}(\mathbf{r})\cdot\nabla\,]\mathbf{E}(\mathbf{r}) + \frac{1}{2\epsilon_0}\int d\mathbf{S}\,[\hat{\mathbf{n}}(\mathbf{r}_S)\cdot\mathbf{P}(\mathbf{r}_S)]^2.$$

6.25 Minimizing the Total Energy Functional Use the method of Lagrange multipliers to show that, among all functions $\mathbf{D}(\mathbf{r})$ which satisfy $\nabla\cdot\mathbf{D} = \rho_f$, the minimum (not merely the extremum) of $U_E = \frac{1}{2}\int d^3r\,|\mathbf{D}|^2/\epsilon$ occurs when $\mathbf{D}(\mathbf{r}) = \epsilon\mathbf{E}(\mathbf{r})$. Assume that $\hat{\mathbf{n}}\cdot\mathbf{D}$ takes specified values on the boundary surface of the volume V of integration. Hint: Because $\nabla\cdot\mathbf{D}(\mathbf{r}) = \rho_f(\mathbf{r})$ is a constraint at every $\mathbf{r}$, it is necessary to use a Lagrange *function* $\varphi(\mathbf{r})$ (rather than a single Lagrange constant) to enforce the constraint throughout V.

7 Laplace's Equation

The problem of finding the solution to any electrostatic problem
is equivalent to finding a solution of Laplace's equation
throughout the space not occupied by conductors.
Sir James Jeans (1925)

7.1 Introduction

A recurring theme in electrostatics is the inevitable rearrangement of charge which occurs inside matter when electric fields are present. The Coulomb integral,

$$\varphi(\mathbf{r}) = \frac{1}{4\pi\epsilon_0}\int d^3r' \frac{\rho(\mathbf{r}')}{|\mathbf{r}-\mathbf{r}'|}, \tag{7.1}$$

loses its usefulness for these situations because we cannot specify $\rho(\mathbf{r})$ once and for all. Luckily, a different approach is available for finding the potential produced by conductors and linear dielectrics. This is possible because we have introduced explicit models for the behavior of these materials in an electrostatic field.

$\mathbf{E} = 0$ inside a perfect conductor (Section 5.1) and the potential inside a material with dielectric constant κ satisfies Poisson's equation (Section 6.5.5):

$$\epsilon_0 \nabla^2 \varphi(\mathbf{r}) = -\rho_f(\mathbf{r})/\kappa. \tag{7.2}$$

Matching conditions describe how the potential behaves when an abrupt interface with free charge density σ_f separates a material with dielectric constant κ_1 from a material with dielectric constant κ_2 (Figure 7.1). If $\varphi_1(\mathbf{r}_S)$ and $\varphi_2(\mathbf{r}_S)$ are the potentials in the two regions infinitesimally close to an interface point $\mathbf{r}_S$, the matching conditions are

$$\varphi_1(\mathbf{r}_S) = \varphi_2(\mathbf{r}_S) \tag{7.3}$$

and

$$\kappa_1 \left.\frac{\partial \varphi_1}{\partial n_1}\right|_S + \kappa_2 \left.\frac{\partial \varphi_2}{\partial n_2}\right|_S = \frac{\sigma_f}{\epsilon_0}. \tag{7.4}$$

When combined with suitable boundary conditions, the three preceding equations lead to a strategy to solve electrostatics problems for conductors and linear dielectrics known as *potential theory*.

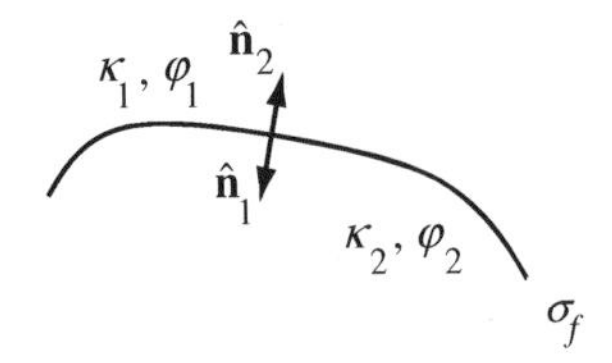

Figure 7.1: An interface where dielectric properties change abruptly: $\hat{\mathbf{n}}_1$ is the outward normal from region 1; $\hat{\mathbf{n}}_2$ is the outward normal from region 2.

7.2 Potential Theory

Potential theory—the branch of mathematics devoted to the study of the Poisson and Laplace equations—is a mature subject which has been in continuous development for over 200 years. For that reason, physics students of the 21st century are entitled to ask whether two chapters devoted to 19th-century mathematics are really necessary when numerical solvers of the Poisson and Laplace equations are available on every desktop computer. Users of these packages know that the ability to rapidly change geometries and boundary conditions is an invaluable aid to building physical insight. Sophisticated graphical interfaces make it possible to study the details with unprecedented ease.

The case for analytic methods is self-evident when a particular Laplace or Poisson problem admits a closed-form solution where both global behavior and limiting cases lay in plain view. The argument is less straightforward when the solution takes the form of sums of products of functions which are typically evaluated numerically anyway. Nevertheless, physicists have exploited this approach for generations and the contemporary student must acquire familiarity (if not absolute fluency) just to make sense of the literature. Moreover, complicated-looking analytic formulae often simplify dramatically in various limits of physical interest. If so, the intuitive virtues of the simple analytic solution return. Finally, despite the assurances of software developers, it is always worthwhile to use an analytic solution (no matter how complicated) as a check on the correctness of a purely numerical solution.

In this chapter, we begin by demonstrating that the solution to Poisson's equation is *unique* in (vacuum or linear dielectric) spaces bounded by conductors. We then limit ourselves to situations where one or more surface charge densities $\sigma(\mathbf{r}_S)$ account for all the charge in the system. This special case simplifies our task to solving Laplace's equation at all points that do not lie on the charged surfaces:

$$\nabla^2\varphi(\mathbf{r}) = 0. \tag{7.5}$$

If $\sigma(\mathbf{r}_S)$ happens to be known, it is a matter of taste and computational convenience whether we evaluate the integral (7.1) or use the matching conditions to knit together the global potential $\varphi(\mathbf{r})$ from the separate solutions of (7.5) that are valid on opposite sides of each interface. The solution for $\varphi(\mathbf{r})$ obtained in this way is identical to (7.1) and thus is unique by Helmholtz' theorem (Section 1.9). The usual case is that $\sigma(\mathbf{r}_S)$ is *not* known and we use the methods of potential theory to find it. More precisely, we use potential theory to find $\varphi(\mathbf{r})$ and then calculate $\sigma(\mathbf{r}_S)$ *post factum* using the matching condition (7.4). By uniqueness, the calculated $\sigma(\mathbf{r}_S)$ and (7.1) reproduce the Laplace equation solution.

The main task of this chapter is to explore the methods that have been developed to solve (7.5) for different geometries. We place special emphasis on *separation of variables* and *analytic function theory*. The next chapter deals with Poisson's equation in full generality.

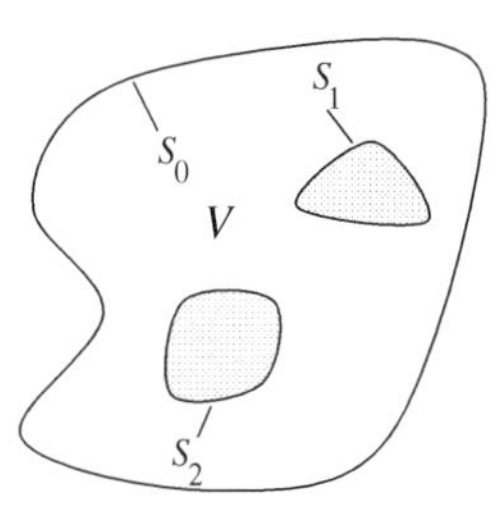

Figure 7.2: A volume V bounded by a closed exterior surface S_0 and two closed interior surfaces S_1 and S_2.

7.3 Uniqueness

The diverse methods of potential theory often lead to representations for $\varphi(\mathbf{r})$ that look remarkably different. Nevertheless, if the physical problem is well posed, all such representations are strictly equivalent at every point in the solution domain. This is guaranteed by the fact that the solutions to the Poisson equation are unique when suitable conditions are satisfied. Therefore, if a valid solution is obtained by any means—including guesswork—it is *the* solution.

We will prove uniqueness for the solution of Poisson's equation in the connected volume V shown in Figure 7.2. The volume is bounded by any number of non-overlapping closed surfaces $S_1, S_2, \ldots$ and an outer enclosing surface S_0. These surfaces need *not* coincide with matter-vacuum interfaces. There may also be a volume density of source charge $\rho(\mathbf{r})$ in V. The proof proceeds by contradiction. Thus, we suppose that V supports *two* distinct solutions of the Poisson equation with the same charge density $\rho(\mathbf{r})$. In other words, $\nabla^2\varphi_1(\mathbf{r}) = -\rho(\mathbf{r})/\epsilon$ and $\nabla^2\varphi_2(r) = -\rho(\mathbf{r})/\epsilon$ for all $\mathbf{r} \in V$.

Our strategy is to construct the difference $\Phi(\mathbf{r}) = \varphi_1(\mathbf{r}) - \varphi_2(\mathbf{r})$ and observe that $\nabla^2\Phi = 0$ in V. Therefore, Green's first identity (Section 1.4.3),

$$\int_V d^3r \left(f\nabla^2 g + \nabla f \cdot \nabla g\right) = \int_S d\mathbf{S} \cdot f\nabla g, \tag{7.6}$$

with $f = g = \Phi$ gives

$$\int_V d^3r|\nabla\Phi|^2 = \sum_i \int_{S_i} d\mathbf{S} \cdot \Phi\,\nabla\Phi. \tag{7.7}$$

The sum is over all the bounding surfaces S_i and each vector $d\mathbf{S}_i$ points *outward* from V. The key observation is that the integrand on the left side of (7.7) is non-negative. This means that $\nabla\Phi(\mathbf{r}) = 0$ if boundary conditions are chosen so that the right side of (7.7) vanishes. In that case, $\varphi_1(\mathbf{r}) = \varphi_2(\mathbf{r})$ (up to an inessential constant) and the potential that solves Poisson's equation is unique.

Each term in the sum on the right side of (7.7) has the same form. Therefore, we focus on making the following integral equal to zero:

$$I = \int_S d\mathbf{S} \cdot \Phi\,\nabla\Phi. \tag{7.8}$$

There are three ways to do this:

1. <u>Dirichlet</u> boundary conditions specify the value of the potential at each surface point. $I = 0$ because the constraint $\varphi_1(\mathbf{r}_S) = \varphi_2(\mathbf{r}_S)$ implies that $\Phi(\mathbf{r}_S) = 0$.
2. <u>Neumann</u> boundary conditions specify the value of the normal component of the gradient of the potential at each surface point. $I = 0$ in this case because the corresponding constraint $\hat{\mathbf{n}} \cdot \nabla\varphi_1(\mathbf{r}_S) = \hat{\mathbf{n}} \cdot \nabla\varphi_2(\mathbf{r}_S)$ implies that $d\mathbf{S} \cdot \nabla\Phi(\mathbf{r}_S) = 0$.

3. Mixed boundary conditions specify $\varphi(\mathbf{r}_S)$ over a portion of S and specify $\hat{\mathbf{n}} \cdot \nabla\varphi(\mathbf{r}_S)$ over the remainder. This gives $I = 0$ as well.

More detailed considerations[1] show that these choices are mutually exclusive. That is, we cannot specify *both* $\varphi(\mathbf{r}_S)$ and $\hat{\mathbf{n}} \cdot \nabla\varphi(\mathbf{r}_S)$ simultaneously at any point on the bounding surface. If we do, the problem becomes over-determined and no solution exists.

7.3.1 Uniqueness When Conductors are Present

The epigraph at the beginning of this chapter considers the bounding surfaces S_k in Figure 7.2 to be perfectly conducting.[2] In that case, it is straightforward to prove that the solution to Poisson's equation in V is unique if, for every conductor, we specify either

$$\begin{aligned} &\text{(A) the conductor potential } \varphi_S, \quad \text{or} \\ &\text{(B) the conductor total charge } Q_S. \end{aligned} \tag{7.9}$$

Choice A is a Dirichlet condition. Choice B is related to a Neumann condition, but provides less information. It is nevertheless sufficient to guarantee uniqueness because, using Gauss' law, (7.8) for a conductor surface is

$$\int_S d\mathbf{S} \cdot \Phi\,\nabla\Phi = -(\varphi_1 - \varphi_2)\int_S d\mathbf{S} \cdot (\mathbf{E}_1 - \mathbf{E}_2) = (\varphi_1 - \varphi_2)(Q_1 - Q_2)/\epsilon_0. \tag{7.10}$$

We note in passing that choice B explains why the electric field outside a perfect conductor is independent of how charge is distributed inside a vacuum cavity scooped out of the body of the conductor (see Section 5.3). Only the total amount of such charge matters to the unique solution.

The alternatives in (7.9) are mutually exclusive. Suppose we compute $\varphi(\mathbf{r})$ everywhere from φ_S data. Since $\mathbf{E} = \hat{\mathbf{n}}\,\sigma/\epsilon_0$ at the surface of a conductor, its total charge is

$$Q_S = \int_S dS\,\sigma(\mathbf{r}_S) = -\epsilon_0 \int_S dS\,\hat{\mathbf{n}} \cdot \nabla\varphi(\mathbf{r}_S). \tag{7.11}$$

Conversely, we can compute $\varphi(\mathbf{r})$ in terms of a set of unspecified constants $\varphi_S = \varphi(\mathbf{r}_S)$. Then, given Q_S data for each conductor, (7.11) written out for every conductor is a set of linear equations that determine the φ_S. Indeed, they are just the capacitance equations

$$Q_i = \sum_j C_{ij}\,\varphi_j. \tag{7.12}$$

As a practical matter, Neumann boundary conditions do not arise when only dielectric matter and/or perfect conductors are present. They occur instead in waveguide theory (Chapter 19) and when a steady current flows through an ohmic medium (Chapter 9). Mixed boundary conditions occur occasionally, but the mathematics needed to apply them is rather exotic.[3] For these reasons, all the potential theory problems worked out in this chapter have Dirichlet boundary conditions.

[1] See, e.g., P.M. Morse and H. Feshbach, *Methods of Theoretical Physics* (McGraw-Hill, New York, 1953), Section 6.1.

[2] The enclosing surface S_0 need not be a conductor if it is placed at infinity. The condition $\varphi(\mathbf{r}) \to 0$ as $r \to \infty$ is a Dirichlet condition.

[3] See Sneddon (1966) and Fabrikant (1991) in Sources, References, and Additional Reading.

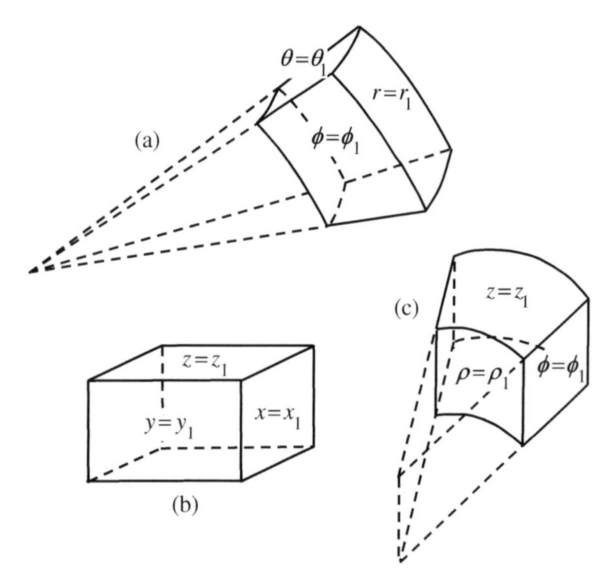

Figure 7.3: Volumes bounded by surfaces where one coordinate variable is constant: (a) spherical coordinates; (b) Cartesian coordinates; (c) cylindrical coordinates.

Example 7.1 An uncharged and isolated conductor of arbitrary shape has a cavity of arbitrary shape scooped out of its interior. A point dipole $\mathbf{p}$ sits inside the cavity. Show that $\mathbf{E}(\mathbf{r}) = 0$ is the unique solution outside the conductor. Contrast with the case of a point charge in the cavity.

Solution: Let the conductor fill all of space except for the cavity. The dipole induces a surface charge density $\sigma^*(\mathbf{r}_S)$ on the cavity wall S. By the definition of a conductor, the field produced by $\sigma^*(\mathbf{r}_S)$ exactly cancels the field due to $\mathbf{p}$ everywhere outside the cavity. In other words, $\mathbf{p}$ and $\sigma^*(\mathbf{r}_S)$ together produce $\mathbf{E} = 0$ at every point outside the cavity. For that reason, the same will be true if we apply $\sigma^*(\mathbf{r}_S)$ to the cavity wall of the original finite-size conductor. Unlike the case of a point charge in a scooped-out cavity, this can be done without inducing charge on the outside surface of the conductor because Gauss' law guarantees that $\sigma^*(\mathbf{r}_S)$ integrates to zero over the cavity wall. Moreover, $\mathbf{E}(\mathbf{r})$ is still zero everywhere outside the cavity, including the space outside the conductor. The total conductor charge is zero, so the exterior solution $\mathbf{E}(\mathbf{r}) = 0$ is unique by choice B of (7.9).

7.4 Separation of Variables

Many potential problems have a symmetry that makes one particular coordinate system most natural for its description. Consider the three empty volumes shown in Figure 7.3. Each has been constructed so every bounding surface is defined by a constant value for one spherical, Cartesian, or cylindrical coordinate. Semi-infinite and infinite volumes are included by expanding the range of the relevant coordinates beyond what is shown in the figure. Our task is to solve $\nabla^2\varphi = 0$ in each volume subject to specified conditions on each boundary surface.

In an orthogonal coordinate system where (u, v, w) labels a point in space, the method of *separation of variables* assumes a product solution of the form

$$\varphi(u, v, w) = A(u)B(v)C(w). \tag{7.13}$$

This guess separates Laplace's partial differential equation into three ordinary, second-order differential equations in 13 different coordinate systems.[4] The separation process generates three undetermined "separation constants" (not all independent) which serve as parameters in the differential equations and labels for their solutions. Thus, if α is the sole separation constant in the differential equation for $A(u)$, the most general solution of the latter is an arbitrary sum of its two linearly independent solutions: $a_\alpha A_\alpha(u) + \tilde{a}_\alpha \tilde{A}_\alpha(u)$. Similarly, if β and γ are the sole separation constants in the differential equations for $B(v)$ and $C(w)$, respectively, the most general solution to Laplace's equation is a sum over all possible values of the separation constants of products of the form (7.13):

$$\varphi(u, v, w) = \sum_{\alpha\beta\gamma} \left[a_\alpha A_\alpha(u) + \tilde{a}_\alpha \tilde{A}_\alpha(u)\right] \left[b_\beta B_\beta(v) + \tilde{b}_\beta \tilde{B}_\beta(v)\right] \left[c_\gamma C_\gamma(w) + \tilde{c}_\gamma \tilde{C}_\gamma(w)\right]. \tag{7.14}$$

The art to solving a potential problem by separation of variables is to determine the expansion coefficients, a_α, $\tilde{a}_\alpha$, b_β, $\tilde{b}_\beta$, c_γ, and $\tilde{c}_\gamma$ and the nature (real, imaginary, integer) and values of the separation constants α, β, and γ that must be retained in the sums. If the retained values of a separation constant turn out to take on continuous (rather than discrete) values, the corresponding sum in (7.14) should be read as an integral. The various constants are chosen so the total potential (i) satisfies the boundary conditions, (ii) remains finite throughout the solution volume, and (iii) respects the symmetries of the physical problem. The method is best understood by the study of examples, so the next several sections provide a variety of these for problems with Cartesian, azimuthal, spherical, cylindrical, and polar symmetry. In all but the simplest cases, it turns out to be essential to arrange matters so that some of the separated differential equations have boundary conditions which produce a Sturm-Liouville eigenvalue problem. This ensures that the associated eigenfunctions are complete and orthonormal in the manner described in the next section. The boundary conditions needed to do this are either homogeneous Dirichlet ($\varphi = 0$), homogeneous Neumann ($\partial\varphi/\partial n = 0$), periodic, or boundedness at infinity.

7.4.1 Complete and Orthonormal Sets of Functions

A set of eigenfunctions, $\psi_k(v)$, defined on the interval $a \le v \le b$ is said to be *complete* if any function $F(v)$ defined on the same interval can be written as a linear combination of the $\psi_k(v)$. In other words, coefficients F_k exist such that

$$F(v) = \sum_k F_k \psi_k(v) \qquad a \le v \le b. \tag{7.15}$$

A convenient statement of completeness is the *closure relation*,

$$\sum_k \psi_k(v)\psi_k^*(v') = \delta(v - v'). \tag{7.16}$$

The index k takes discrete values when the range of v is finite and continuous values when the range of v is semi-infinite or infinite. In the latter case, the sum becomes an integral in (7.16).

To show that closure implies completeness, we write

$$F(v) = \int_a^b dv' \delta(v - v') F(v') \qquad a \le v, v' \le b. \tag{7.17}$$

[4] Besides Cartesian, spherical, and cylindrical coordinates, Laplace's equation is separable in conical, bispherical, ellipsoidal, parabolic cylindrical, elliptic cylindrical, oblate spheroidal, prolate spheroidal, parabolic, paraboloidal, and toroidal coordinates.

Substituting (7.16) into (7.17) gives

$$F(v) = \sum_k \int_a^b dv' \psi_k(v)\psi_k^*(v')F(v'). \tag{7.18}$$

This confirms (7.15) with

$$F_k = \int_a^b dv' \psi_k^*(v')F(v'). \tag{7.19}$$

Finally, consider $F(v) = \psi_j(v)$ in (7.15). This forces $F_k = \delta_{kj}$ and substitution into (7.19) gives the *orthonormality* relation

$$\int_a^b dv\, \psi_k^*(v)\psi_j(v) = \delta_{kj}. \tag{7.20}$$

7.5 Cartesian Symmetry

For potential problems with natural rectangular boundaries, the trial solution $\varphi(x, y, z) = X(x)Y(y)Z(z)$ converts Laplace's equation,

$$\nabla^2\varphi = \frac{\partial^2\varphi}{\partial x^2} + \frac{\partial^2\varphi}{\partial y^2} + \frac{\partial^2\varphi}{\partial z^2} = 0, \tag{7.21}$$

into

$$\frac{X''(x)}{X(x)} + \frac{Y''(y)}{Y(y)} + \frac{Z''(z)}{Z(z)} = 0. \tag{7.22}$$

Each of the three ratios in (7.22) is a function of only one variable. Therefore, their sum can be zero only if each is separately equal to a distinct constant. This gives

$$\frac{d^2X}{dx^2} = \alpha^2 X, \qquad \frac{d^2Y}{dy^2} = \beta^2 Y, \qquad \text{and} \qquad \frac{d^2Z}{dz^2} = \gamma^2 Z, \tag{7.23}$$

where the separation constants α^2, β^2, and γ^2 are real[5] and satisfy

$$\alpha^2 + \beta^2 + \gamma^2 = 0. \tag{7.24}$$

The methodology outlined in Section 7.4 directs us to identify all the elementary solutions of the differential equations in (7.23). The separation constants can be zero or non-zero, so

$$X_\alpha(x) = \begin{cases} A_0 + B_0 x & \alpha = 0, \\ A_\alpha e^{\alpha x} + B_\alpha e^{-\alpha x} & \alpha \neq 0, \end{cases} \tag{7.25}$$

$$Y_\beta(y) = \begin{cases} C_0 + D_0 y & \beta = 0, \\ C_\beta e^{\beta y} + D_\beta e^{-\beta y} & \beta \neq 0, \end{cases} \tag{7.26}$$

$$Z_\gamma(z) = \begin{cases} E_0 + F_0 z & \gamma = 0, \\ E_\gamma e^{\gamma z} + F_\gamma e^{-\gamma z} & \gamma \neq 0. \end{cases} \tag{7.27}$$

[5] The possibility that α, β, and γ are neither purely real nor purely imaginary is precluded by the boundary conditions for all the problems we will encounter in this book.

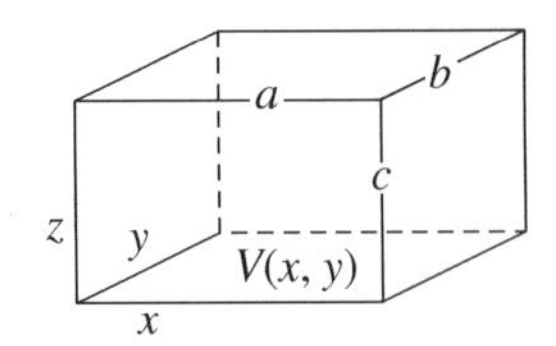

Figure 7.4: An empty box with five walls maintained at zero potential and the $z = 0$ bottom wall maintained at potential $V(x, y)$.

The linearity of Laplace's equation permits us to superpose the products of these elementary solutions. Therefore, using a delta function to enforce (7.24), the general solution reads

$$\varphi(x, y, z) = \sum_{\alpha}\sum_{\beta}\sum_{\gamma} X_\alpha(x)Y_\beta(y)Z_\gamma(z)\delta(\alpha^2 + \beta^2 + \gamma^2). \tag{7.28}$$

The form of (7.25) shows that α can be restricted to positive real values if $\alpha^2 > 0$ and to positive imaginary values if $\alpha^2 < 0$. Similar remarks apply to β and γ.

As an example, let us use (7.28) to find the electrostatic potential inside the rectangular box shown in Figure 7.4. We assume that all the walls are fixed at zero potential except for the $z = 0$ wall, where the potential takes specified values $V(x, y)$.[6] The homogeneous Dirichlet boundary conditions on the vertical side walls are not difficult to satisfy if we write $\alpha = i\alpha'$ and $\beta = i\beta'$ in (7.25) and (7.26). We then choose α', β', and the expansion coefficients to make $X_\alpha(x)$ and $Y_\beta(y)$ sine functions which vanish at $x = a$ and $y = b$, respectively. Bearing in mind the delta function constraint, (7.28) takes the form

$$\varphi(x, y, z) = \sum_{m=1}^{\infty}\sum_{n=1}^{\infty} \sin\left(\frac{m\pi x}{a}\right)\sin\left(\frac{n\pi y}{b}\right)\left[E_{mn}\exp(\gamma_{mn}z) + F_{mn}\exp(-\gamma_{mn}z)\right], \tag{7.29}$$

where

$$\gamma_{mn}^2 = \left(\frac{m\pi}{a}\right)^2 + \left(\frac{n\pi}{b}\right)^2. \tag{7.30}$$

Our next task is to choose E_{mn} and F_{mn} so the potential vanishes at $z = c$. If V_{mn} are coefficients still to be determined, a convenient way to write the result is

$$\varphi(x, y, z) = \sum_{m=1}^{\infty}\sum_{n=1}^{\infty} V_{mn}\sin\left(\frac{m\pi x}{a}\right)\sin\left(\frac{n\pi y}{b}\right)\frac{\sinh[\gamma_{mn}(c - z)]}{\sinh(\gamma_{mn}c)}. \tag{7.31}$$

It remains only to impose the final boundary condition that $\varphi(x, y, 0) = V(x, y)$. This gives

$$V(x, y) = \sum_{m=1}^{\infty}\sum_{n=1}^{\infty} V_{mn}\sin\left(\frac{m\pi x}{a}\right)\sin\left(\frac{n\pi y}{b}\right), \tag{7.32}$$

which is a double Fourier sine series representation of $V(x, y)$. To find the coefficients V_{mn}, multiply both sides of (7.32) by $\sin(m'\pi x/a)\sin(n'\pi y/b)$ and integrate over the intervals $0 \le x \le a$ and $0 \le y \le b$. This completes the problem because the orthogonality integral

$$\int_0^\pi ds\,\sin(ms)\sin(ns) = \frac{\pi}{2}\delta_{mn} \tag{7.33}$$

[6] We assume that a thin strip of insulating material isolates the bottom wall from the others.

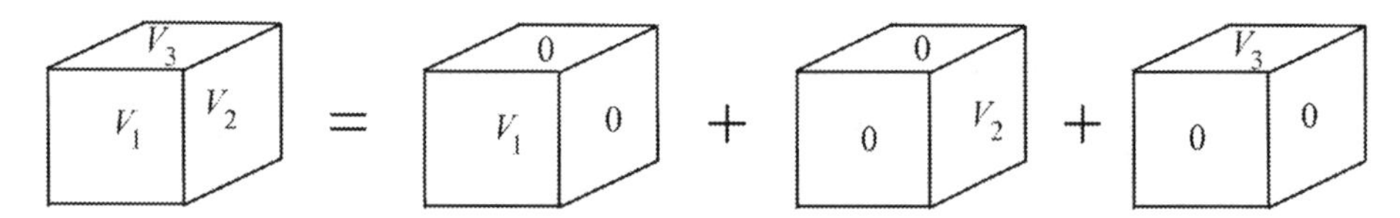

Figure 7.5: The potential in a box with three $\varphi = 0$ walls and three $\varphi \neq 0$ walls represented as the sum of three box potentials, each with five $\varphi = 0$ walls and one $\varphi \neq 0$ wall.

shows that

$$V_{mn} = \frac{4}{ab}\int_0^a\int_0^b dxdy\, V(x, y)\sin\left(\frac{m\pi x}{a}\right)\sin\left(\frac{n\pi y}{b}\right). \tag{7.34}$$

It was not an accident that a Fourier series appeared in (7.32) when the need arose to represent the arbitrary boundary data $V(x, y)$. The key was the homogeneous Dirichlet (zero potential) boundary condition imposed on each of the vertical walls in Figure 7.4. This transformed the differential equations for $X(x)$ and $Y(y)$ into eigenvalue problems with complete sets of orthogonal sine functions as their eigenfunctions.

Our example raises the question of how to "arrange" a complete set of eigenfunctions if we had specified non-zero potentials on any (or all) of the vertical side walls. The solution (indicated schematically in Figure 7.5) is to superpose the separated-variable solutions to several independent potential problems, each like the one we have just solved but with a different wall held at a non-zero potential. This general approach works for other coordinate systems also. Application 7.1 illustrates another method.

Application 7.1 A Conducting Duct

Figure 7.6 is a cross sectional view of an infinitely long, hollow, conducting duct. The walls are maintained at the constant potentials indicated and our task is to find the electrostatic potential everywhere inside the duct. A straightforward approach to this problem mimics Figure 7.5 and superposes the solutions of three different potential problems, each with only one wall held at a non-zero potential. The reader can confirm that only two problems actually need be superposed: one with $V' = 0$ and another with $V = 0$. In this Application, we follow a third path and use the $\beta = 0$ solution in (7.26) to remove the inhomogeneous boundary condition $\varphi(x, L) = V'$. More generally, we use a systematic "inspection" method which retains every elementary solution in (7.25), (7.26), and (7.27) until a boundary condition or other physical consideration forces us to remove it. Uniqueness guarantees that the solutions obtained by these three approaches will agree numerically at every point, despite their different analytic appearances.

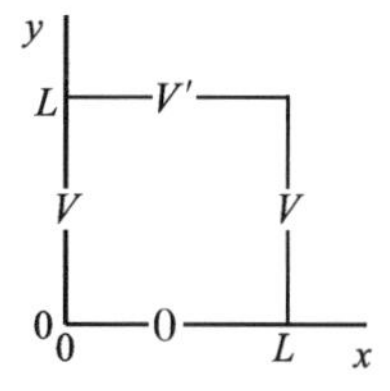

Figure 7.6: Cross sectional view of a conducting duct with Dirichlet boundary conditions.

The translational invariance of the duct implies that $\varphi(x, y, z) = \varphi(x, y)$. This means that $E_\gamma = F_\gamma = 0$ in (7.27) when $\gamma \neq 0$. The $\gamma = 0$ contribution survives with $E_0 = 1$ and $F_0 = 0$. Using (7.24), $\gamma = 0$ implies that $\beta^2 = -\alpha^2$, which in turn implies that the $\alpha = 0$ solution $X_0(x)$ pairs exclusively with the $\beta = 0$ solution $Y_0(y)$ in (7.28). Otherwise, we choose $\alpha^2 > 0$ so the $\beta \neq 0$ basis functions in (7.26) are complex exponentials (or sines and cosines). The general solution for the potential at this juncture is

$$\varphi(x, y) = (a_0 + b_0 x)(c_0 + d_0 y) + \sum_\alpha (a_\alpha e^{\alpha x} + b_\alpha e^{-\alpha x})(c_\alpha \sin \alpha y + d_\alpha \cos \alpha y). \tag{7.35}$$

The boundary condition at $y = 0$ simplifies (7.35) to

$$\varphi(x, 0) = 0 = (a_0 + b_0 x)c_0 + \sum_\alpha (a_\alpha e^{\alpha x} + b_\alpha e^{-\alpha x})d_\alpha. \tag{7.36}$$

This shows that $c_0 = d_\alpha = 0$. We are free to choose $d_0 = c_\alpha = 1$ in the resulting expression for $\varphi(x, y)$, so the boundary condition $\varphi(x, L) = V'$ reads

$$V' = (a_0 + b_0 x)L + \sum_\alpha (a_\alpha e^{\alpha x} + b_\alpha e^{-\alpha x}) \sin \alpha L. \tag{7.37}$$

Equation (7.37) becomes an identity when we choose $b_0 = 0$, $a_0 = V'/L$, and $\alpha = m\pi/L$ where m is a positive integer. In other words, we can use the $\beta = 0$ linear solution in (7.26) to account for the inhomogeneous boundary condition at $y = L$. The potential inside the duct now takes the form

$$\varphi(x, y) = \frac{V'}{L} y + \sum_{m=1}^{\infty} (a_m e^{m\pi x/L} + b_m e^{-m\pi x/L}) \sin \frac{m\pi y}{L}. \tag{7.38}$$

The constants a_m and b_m in (7.38) are determined by the two boundary conditions that remain,

$$\varphi(0, y) = V = \frac{V'}{L} y + \sum_{m=1}^{\infty} (a_m + b_m) \sin \frac{m\pi y}{L} \tag{7.39}$$

and

$$\varphi(L, y) = V = \frac{V'}{L} y + \sum_{m=1}^{\infty} (a_m e^{m\pi} + b_m e^{-m\pi}) \sin \frac{m\pi y}{L}. \tag{7.40}$$

The key step is to multiply both (7.39) and (7.40) by $\sin(n\pi y/L)$ and exploit the orthogonality integral (7.33). For each value of m, this gives two equations for the two unknowns, a_m and b_m, namely,

$$a_m + b_m = \Omega_m = a_m e^{m\pi} + b_m e^{-m\pi}, \tag{7.41}$$

where

$$\Omega_m = \frac{2}{L} \int_0^L dy \sin \frac{m\pi y}{L} \left(V - \frac{V'}{L} y \right) = \frac{2}{m\pi} \left[V + (-1)^m (V' - V) \right]. \tag{7.42}$$

It remains only to solve (7.41) for a_m and b_m. This bit of algebra gives

$$a_m = \frac{\Omega_m (1 - e^{-m\pi})}{2 \sinh(m\pi)} \qquad \text{and} \qquad b_m = \frac{\Omega_m (e^{m\pi} - 1)}{2 \sinh(m\pi)}. \tag{7.43}$$

Substituting these coefficients back into (7.38) completes the problem. ■

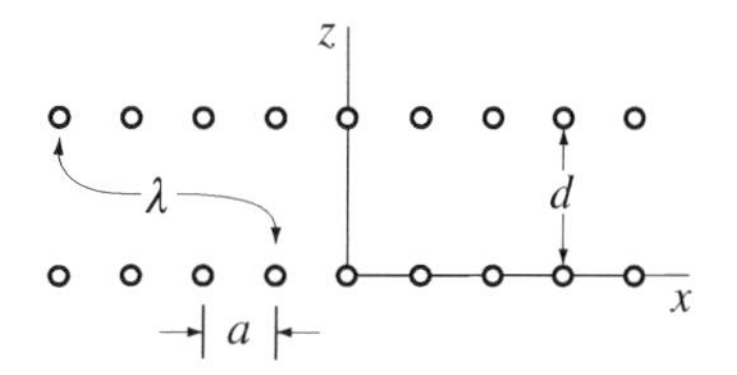

Figure 7.7: Edge view of a "parallel-plate" Faraday cage composed of two parallel arrays of infinitely long charged lines aligned parallel to the y-axis. Each line has a uniform charge per unit length λ. The diagram repeats periodically out to $x = \pm\infty$. The two walls of the cage are separated by a distance d. Within each wall, the charged lines are separated by a distance a.

Faraday's *diary*, January 15 1836

Have been for some days engaged in building up a cube of 12 feet in the side. It consists of a slight wooden frame held steady by diagonal ties of cord; the whole being mounted on four glass feet. The sides, top and bottom are covered in paper. The top and bottom have each a cross framing of copper wire connected by copper wires passing down the four corner uprights. The sheets of paper which constitute the four sides have each two slips of tin foil pasted on their inner surface. These are connected below with copper wire so that all the metallic parts are in communication. Access to the inside was made by cutting a flap in the paper. Connecting this cube by a wire with the Electrical Machine, I can quickly and well electrify the whole. I now went in the cube standing on a stool and [my assistant] Anderson worked the machine until the cube was fully charged. I could by no appearance find any traces of electricity in myself or the surrounding objects. In fact, the electrification without produced no consequent effects within.

7.5.1 Faraday's Cage

The diary entry just above describes a wire mesh "cage" which effectively shields its interior from an electrostatic potential applied to its surface. The physics of the shielding can be understood using the "parallel-plate" cage shown in Figure 7.7, where an infinite number of infinitely long and uniformly charged lines lie in two parallel planes. This is a good model for a grid of conducting wires because, if every line carries a charge per unit length λ, the potential at a distance $s \ll a$ from any particular line (regarded as the origin) is

$$\varphi(s) = -\frac{\lambda}{2\pi\epsilon_0}\ln|s| + const. \tag{7.44}$$

The equipotentials of (7.44) are cylinders so the charged lines are electrostatically equivalent to an array of conducting wires held at a common potential.

To find the potential between the "plates" of the cage, we begin with the potential produced by the lower grid of wires alone. The reflection symmetry of the grid implies that this two-dimensional potential must obey $\varphi(-x, -z) = \varphi(x, z)$. Moreover, $\varphi(x, z) \to 0$ as $|z| \to \infty$ because the source charge is localized in the z-direction. This tells us to choose the sign of the separation constant so $Z(z)$ in (7.27) decays exponentially away from the grid in both directions. Putting all this information together yields a separated-variable solution for the lower grid of wires with the form

$$\varphi(x, z) = A + B|z| + \sum_{\gamma} C_\gamma \cos(\gamma x)\exp(-\gamma|z|). \tag{7.45}$$

It is essential that we retain the term proportional to $|z|$ [the $\gamma = 0$ term in (7.27)] because, when $|z| \gg a$, the grid is indistinguishable from an infinite charged sheet. Indeed, precisely this fact tells us that $A = 0$ and

$$B = -\lambda/2a\epsilon_0. \tag{7.46}$$

The geometry of the wires imposes the *periodicity* condition $\varphi(x + a, z) = \varphi(x, z)$. This means that the separation constant γ takes the discrete values $\gamma = 2\pi m/a$ with $m = 1, 2, \ldots$ Accordingly,

$$\varphi(x, z) = -\frac{\lambda}{2a\epsilon_0}|z| + \sum_{m=1}^{\infty} C_m \cos(2\pi mx/a)\exp(-2\pi m|z|/a). \tag{7.47}$$

We fix the expansion coefficients C_m by imposing the matching condition (7.4) in the form

$$\left.\frac{\partial\varphi}{\partial z}\right|_{z=0^-} - \left.\frac{\partial\varphi}{\partial z}\right|_{z=0^+} = \frac{\sigma(x)}{\epsilon_0}. \tag{7.48}$$

The relevant $\sigma(x)$ is the surface charge density of the array of charged lines,

$$\sigma(x) = \lambda \sum_{p=-\infty}^{\infty} \delta(x - pa). \tag{7.49}$$

A brief calculation using the Fourier expansion[7]

$$\sum_{p=-\infty}^{\infty} \delta(x - 2\pi p) = \frac{1}{2\pi} + \frac{1}{\pi}\sum_{m=1}^{\infty} \cos mx \tag{7.50}$$

yields the potential created by the $z = 0$ wire mesh:

$$\varphi(x, z) = -\frac{\lambda}{2a\epsilon_0}|z| + \frac{\lambda}{2\pi\epsilon_0}\sum_{m=1}^{\infty} \frac{1}{m}\cos(2\pi mx/a)\exp(-2\pi m|z|/a). \tag{7.51}$$

The reader can confirm that (7.51) reproduces (7.44) in the plane of the grid.

The upper grid of wires contributes a potential identical to (7.51) except that $z \to z - d$ where d is the separation between the two grids. When we add the two together, the first term in (7.51) is replaced by a position-independent *constant* when $0 < z < d$. The remaining, mesh-induced, contributions are *exponentially small* if $z \gg a$ and $d - z \gg a$. Therefore, the electric field inside the cage is essentially *zero* at all observation points that lie farther away from the cage walls than the spacing between the wires of the cage. The condition $d \gg a$ is a design prerequisite for all practical Faraday cages.

7.5.2 The Electrostatic Potential Has Zero Curvature

The duct potential (7.38) and the cage potential (7.51) are both composed of products of sinusoidal and exponential functions in orthogonal directions. This is a consequence of the constraint (7.24) imposed on the separation constants. Since Laplace's equation (7.5) is the origin of the constraint, we can use geometrical language and ascribe this behavior to the requirement that the total *curvature* of $\varphi(\mathbf{r})$ vanish at every point where $\nabla^2\varphi = 0$. This insight provides a qualitative understanding of Earnshaw's theorem[8] because it says that the one-dimensional curvatures of the potential along each of the three Cartesian directions cannot all have the same algebraic sign.

[7] See Example 1.6.

[8] Earnshaw's theorem states that $\varphi(\mathbf{r})$ cannot have a local maximum or a local minimum in a charge-free volume of space. See Section 3.3.3.

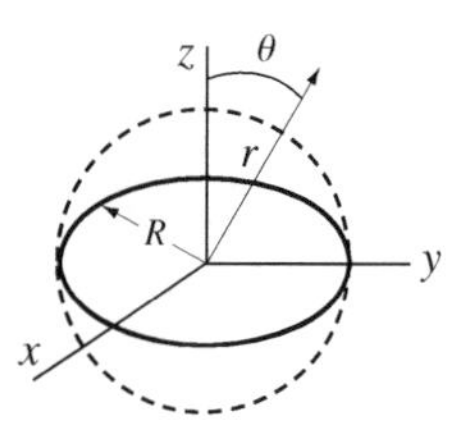

Figure 7.8: An origin-centered ring in the $z = 0$ plane with uniform charge per unit length $\lambda = Q/2\pi R$. The dashed sphere separates space into a region with $r < R$ and a region with $r > R$.

The zero-curvature condition guarantees that the solutions of Laplace's equation are not bounded in at least one Cartesian direction. Nevertheless, as both the duct and cage examples demonstrate, true divergence never occurs because a steadily increasing potential along some direction always signals the eventual appearance of a region of source charge where Laplace's equation is no longer valid. The only possible solution to $\nabla^2\varphi = 0$ in completely empty space is $\varphi = const.$, which corresponds to no electric field at all.

7.6 Azimuthal Symmetry

In Example 4.5, we found the potential of a uniformly charged ring (Figure 7.8) by evaluating all of its electrostatic multipole moments. Another way to solve this problem exploits separation of variables in spherical coordinates. The azimuthal symmetry of the ring implies that $\varphi(r, \theta, \phi) = \varphi(r, \theta)$. Therefore, for this problem (and any problem with azimuthal symmetry), the Laplace equation reduces to

$$\nabla^2\varphi = \frac{1}{r^2}\frac{\partial}{\partial r}\left(r^2\frac{\partial\varphi}{\partial r}\right) + \frac{1}{r^2\sin\theta}\frac{\partial}{\partial\theta}\left(\sin\theta\frac{\partial\varphi}{\partial\theta}\right) = 0. \tag{7.52}$$

With a change of variable to $x = \cos\theta$, the trial solution $\varphi(r, x) = R(r)M(x)$ separates (7.52) into the two ordinary differential equations

$$\frac{d}{dr}\left(r^2\frac{dR}{dr}\right) - \kappa R = 0 \tag{7.53}$$

and

$$\frac{d}{dx}\left[(1-x^2)\frac{dM}{dx}\right] + \kappa M = 0. \tag{7.54}$$

If we write the real separation constant as $\kappa = \nu(\nu+1)$, it is straightforward to verify that (7.53) is solved by

$$R_\nu(r) = A_\nu r^\nu + B_\nu r^{-(\nu+1)}. \tag{7.55}$$

The same substitution in (7.54) produces Legendre's differential equation:

$$(x^2-1)\frac{d^2M}{dx^2} + 2x\frac{dM}{dx} - \nu(\nu+1)M = 0. \tag{7.56}$$

The linearly independent solutions of (7.56) are called *Legendre functions* of the first and second kind. We denote them by $P_\nu(x)$ and $Q_\nu(x)$, respectively. This yields the general solution to Laplace's equation for a problem with azimuthal symmetry as

$$\varphi(r,\theta) = \sum_\nu \left[A_\nu r^\nu + B_\nu r^{-(\nu+1)}\right]\left[C_\nu P_\nu(\cos\theta) + D_\nu Q_\nu(\cos\theta)\right]. \tag{7.57}$$

For arbitrary values of ν, the Legendre functions have the property that $P_\nu(-1) = Q_\nu(\pm 1) = \infty$. These divergences are unphysical, so (7.57) in full generality applies only to problems where the domain of interest does not include the z-axis. An example is the space between two coaxial cones which open upward along the positive z-axis with their common vertex at the origin of coordinates. If we ask for the potential inside *one* such cone, (7.57) can include all the $P_\nu(x)$—but none of the $Q_\nu(x)$.

It remains only to provide a representation of the potential that can be used for problems where the entire z-axis is part of the physical domain. The answer turns out to be (7.57), provided we exclude the $Q_\nu(x)$ functions and restrict the values of ν to the non-negative integers, $\ell = 0, 1, 2, \ldots$ The latter choice reduces the $P_\nu(\cos\theta)$ functions to the Legendre polynomials $P_\ell(\cos\theta)$ (see Section 4.5.1), which are finite and well behaved over the entire angular range $0 \le \theta \le \pi$. The $Q_\ell(\cos\theta)$ functions remain singular at $\theta = \pi$.

7.6.1 A Uniformly Charged Ring

The charged ring in Figure 7.8 is a problem where the potential is required over the full angular range $0 \le \theta \le \pi$. In such a case, the preceding paragraph tells us that only the Legendre polynomials may appear in (7.57). Accordingly,

$$\varphi(r,\theta) = \sum_{\ell=0}^{\infty} \left[A_\ell r^\ell + B_\ell r^{-(\ell+1)}\right] P_\ell(\cos\theta). \tag{7.58}$$

On the other hand, neither r^ℓ nor $r^{-(\ell+1)}$ is finite throughout the entire radial domain $0 \le r < \infty$. This suggests a divide-and-conquer strategy: construct separate, regular solutions to Laplace's equation in the disjoint regions $r < R$ and $r > R$ and match them together at the surface of the ring ($r = R$). We will do this for a general (but specified) charge density $\sigma(\theta)$ applied to a spherical surface at $r = R$ and then specialize to the uniformly charged ring.

A representation which ensures that $\varphi(r,\theta)$ is both regular at the origin and goes to zero as $r \to \infty$ is

$$\varphi(r,\theta) = \begin{cases} \displaystyle\sum_{\ell=0}^{\infty} c_\ell \left(\frac{r}{R}\right)^\ell P_\ell(\cos\theta) & r \le R, \\ \displaystyle\sum_{\ell=0}^{\infty} c_\ell \left(\frac{R}{r}\right)^{\ell+1} P_\ell(\cos\theta) & r \ge R. \end{cases} \tag{7.59}$$

Notice that (7.59) builds in the continuity of the potential at $r = R$ as specified by (7.3). Using an obvious notation, the matching condition (7.4) is[9]

$$\epsilon_0 \left[\frac{\partial\varphi_<}{\partial r} - \frac{\partial\varphi_>}{\partial r}\right]_{r=R} = \sigma(\theta). \tag{7.60}$$

When applied to (7.59), this gives

$$\epsilon_0 \sum_{\ell=0}^{\infty} \frac{c_\ell}{R}(2\ell+1)P_\ell(\cos\theta) = \sigma(\theta). \tag{7.61}$$

[9] The representation (7.59) is valid for any value of R. We choose the radius of the ring in order to exploit the matching condition (7.60).

The final step is to multiply (7.61) by $P_m(\cos\theta)$ and integrate over the interval $0 \le \theta \le \pi$ using the orthogonality relation for the Legendre polynomials,

$$\int_{-1}^{1} dx\, P_\ell(x) P_m(x) = \frac{2}{2\ell+1}\delta_{\ell m}. \tag{7.62}$$

This yields the expansion coefficients in the form

$$c_m = \frac{R}{2\epsilon_0}\int_0^\pi d\theta\, \sin\theta\, \sigma(\theta) P_m(\cos\theta). \tag{7.63}$$

We infer from Example 4.5 that the surface charge density of the uniformly charged ring sketched in Figure 7.8 is $\sigma(\theta) = (\lambda/R)\,\delta(\cos\theta)$. Plugging this into (7.63) gives $c_m = QP_m(0)/4\pi\epsilon_0 R$. Therefore, the electrostatic potential (7.59) of the ring is

$$\varphi(r,\theta) = \begin{cases} \dfrac{Q}{4\pi\epsilon_0}\displaystyle\sum_{\ell=0}^{\infty}\frac{r^\ell}{R^{\ell+1}}P_\ell(0)P_\ell(\cos\theta) & r \le R, \\ \dfrac{Q}{4\pi\epsilon_0}\displaystyle\sum_{\ell=0}^{\infty}\frac{R^\ell}{r^{\ell+1}}P_\ell(0)P_\ell(\cos\theta) & r \ge R. \end{cases} \tag{7.64}$$

This agrees with the multipole solution obtained in Example 4.5.

Application 7.2 Going Off the Axis

It is not absolutely necessary to use the matching condition (7.60) to find the expansion coefficients c_ℓ in (7.59). An alternative method exploits the uniqueness of the solutions to Laplace's equation. The idea is to compare the general formula (7.59) with an easily computable special case. For the latter, we let ϕ be the azimuthal variable, specialize to the charged ring, and use the Coulomb integral (7.1) to find the potential *on the z-axis*:

$$\varphi(z) = \frac{\lambda}{4\pi\epsilon_0}\int_0^{2\pi}\frac{R\,d\phi}{\sqrt{R^2+z^2}} = \frac{Q}{4\pi\epsilon_0}\frac{1}{\sqrt{R^2+z^2}}. \tag{7.65}$$

This expression can be rewritten using the generating function for the Legendre polynomials (see Section 4.5.1):

$$\frac{1}{\sqrt{1-2xt+t^2}} = \sum_{\ell=0}^{\infty} t^\ell P_\ell(x) \qquad |x| \le 1,\ 0 < t < 1. \tag{7.66}$$

When $|z| < R$, use of (7.66) with $x = 0$ in (7.65) gives

$$\varphi(z) = \frac{Q}{4\pi\epsilon_0}\frac{1}{\sqrt{R^2+z^2}} = \frac{Q}{4\pi\epsilon_0 R}\sum_{\ell=0}^{\infty}(z/R)^\ell P_\ell(0). \tag{7.67}$$

Comparing this with (7.59) specialized to the z-axis (where $r = z$ and $\theta = 0$) gives $c_\ell = QP_\ell(0)/4\pi\epsilon_0 R$ as before because $P_\ell(1) = 1$. Given the c_ℓ, we can now use (7.59) to "go off the axis" and find the potential everywhere. This procedure shows that *any* azimuthally symmetric potential is uniquely determined by its values on the symmetry axis. ■

Example 7.2 In 1936, Lars Onsager constructed a theory of the dielectric constant for a polar liquid using a model of a point dipole **p** placed at the center of a spherical cavity of radius a scooped out of an infinite medium with dielectric constant κ. Find the electric field that acts on the dipole if the entire system is exposed to a uniform external electric field $\mathbf{E}_0 \parallel \mathbf{p}$.

Solution: Let the polar axis in spherical coordinates point along $\mathbf{E}_0$ and $\mathbf{p}$. The presence of the external field guarantees that $\varphi(r \to \infty) \to -E_0 r \cos\theta$. The presence of the point dipole at the center of the cavity guarantees that $\varphi(r \to 0) = p\cos\theta/(4\pi\epsilon_0 r^2)$. Everywhere else, the potential satisfies Laplace's equation. Therefore, because both sources behave as $P_1(\cos\theta) = \cos\theta$, only $\cos\theta$ terms can be present in the general solution (7.59). In other words,

$$\varphi(r) = \begin{cases} \left[Ar + \dfrac{p}{4\pi\epsilon_0 r^2} \right] \cos\theta & r \le a, \\ \left[-E_0 r + \dfrac{B}{r^2} \right] \cos\theta & r \ge a. \end{cases}$$

There is no free charge at the cavity boundary, so the matching conditions (7.3) and (7.4) are

$$\varphi_{\rm in}(a,\theta) = \varphi_{\rm out}(a,\theta) \qquad \left.\frac{\partial \varphi_{\rm in}}{\partial r}\right|_{r=a} = \kappa \left.\frac{\partial \varphi_{\rm out}}{\partial r}\right|_{r=a}.$$

Therefore, we find without complication that

$$A = -\frac{3\kappa}{2\kappa+1}E_0 - \frac{2p}{4\pi\epsilon_0 a^3}\frac{\kappa-1}{2\kappa+1} \qquad \text{and} \qquad B = -a^3 E_0 \frac{\kappa-1}{2\kappa+1} + \frac{p}{4\pi\epsilon_0}\frac{3}{2\kappa+1}.$$

The electric field that acts on the dipole is $\mathbf{E} = -\nabla\varphi$ for $r \le a$ minus the contribution from **p** itself:

$$\mathbf{E}(0) = \frac{3\kappa}{2\kappa+1}\mathbf{E}_0 + \frac{2\mathbf{p}}{4\pi\epsilon_0 a^3}\frac{\kappa-1}{2\kappa+1}.$$

7.7 Spherical Symmetry

In this section, we solve Laplace's equation for problems with natural spherical boundaries that lack full azimuthal symmetry. These situations require the complete Laplacian operator in spherical coordinates:

$$\nabla^2\varphi = \frac{1}{r^2}\frac{\partial}{\partial r}\left(r^2 \frac{\partial \varphi}{\partial r}\right) + \frac{1}{r^2 \sin\theta}\frac{\partial}{\partial\theta}\left(\sin\theta \frac{\partial\varphi}{\partial\theta}\right) + \frac{1}{r^2\sin^2\theta}\frac{\partial^2\varphi}{\partial\phi^2} = 0. \tag{7.68}$$

The trial solution $\varphi(r,\theta,\phi) = R(r)Y(\theta,\phi)$ separates (7.68) into the ordinary differential equation

$$\frac{d}{dr}\left(r^2\frac{dR}{dr}\right) = \ell(\ell+1)R \tag{7.69}$$

and the partial differential equation

$$-\frac{1}{\sin\theta}\frac{\partial}{\partial\theta}\left(\sin\theta\frac{\partial Y}{\partial\theta}\right) - \frac{1}{\sin^2\theta}\frac{\partial^2 Y}{\partial\phi^2} = \ell(\ell+1)Y. \tag{7.70}$$

The choice of the separation constant as $\ell(\ell+1)$ allows us to borrow the solution of the radial equation (7.69) from Section 7.6:

$$R_\ell(r) = A_\ell r^\ell + B_\ell r^{-(\ell+1)}. \tag{7.71}$$

The notation indicates that we have already specialized to the case where ℓ is a non-negative integer. In practice, this case applies to the vast majority of non-contrived electrostatics problems with spherical boundaries. Moreover, when $\ell = 0, 1, 2, \ldots$, (7.70) is exactly the eigenvalue equation for the (dimensionless) quantum mechanical orbital angular momentum operator $\hat{L}^2$:

$$\hat{L}^2 Y_{\ell m}(\theta, \phi) = \ell(\ell+1) Y_{\ell m}(\theta, \phi). \tag{7.72}$$

The complex-valued eigenfunctions $Y_{\ell m}(\theta, \phi)$ are the *spherical harmonics* introduced in Section 4.5.2. From the point of view of separation of variables, the constant m^2 separates (7.70) into two ordinary differential equations when we write $Y(\theta, \phi) = B(\theta)G(\phi)$. The choice of m as an integer in the range $-\ell \le m \le \ell$ guarantees that the spherical harmonics are finite and well behaved when $0 \le \theta < \pi$ and $0 \le \phi < 2\pi$. Appendix C gives a table of spherical harmonics and lists a few of their properties. Here, we exhibit only the orthogonality integral,[10]

$$\int d\Omega \; Y^*_{\ell' m'}(\Omega) Y_{\ell m}(\Omega) = \delta_{\ell \ell'} \delta_{mm'}, \tag{7.73}$$

and the phase relation

$$Y_{\ell, -m}(\theta, \phi) = (-)^m Y^*_{\ell m}(\theta, \phi). \tag{7.74}$$

Combining all the above, the general solution to Laplace's equation in spherical coordinates is

$$\varphi(r, \theta, \phi) = \sum_{\ell=0}^{\infty} \sum_{m=-\ell}^{\ell} \left[B_{\ell m} r^{\ell} + A_{\ell m} r^{-(\ell+1)} \right] Y_{\ell m}(\theta, \phi). \tag{7.75}$$

A typical problem requires a partition of the radial space as in (7.59) to ensure that the solution is regular at the origin and at infinity. Thus, (7.75) shows why the exterior multipole expansion (4.86) represents the potential for $r > R$ when charge occurs only *inside* an origin-centered sphere of radius R:

$$\varphi(r, \theta, \phi) = \sum_{\ell=0}^{\infty} \sum_{m=-\ell}^{\ell} A_{\ell m} \frac{Y_{\ell m}(\theta, \phi)}{r^{\ell+1}} \qquad r > R. \tag{7.76}$$

The solution (7.75) also shows why the interior multipole expansion (4.89) represents the potential for $r < R$ when charge occurs only *outside* an origin-centered sphere of radius R:

$$\varphi(r, \theta, \phi) = \sum_{\ell=0}^{\infty} \sum_{m=-\ell}^{\ell} B_{\ell m} r^{\ell} Y^*_{\ell m}(\theta, \phi) \qquad r < R. \tag{7.77}$$

These expansions are valid only in regions of space *free* from the source charge which defines the multipole moments $A_{\ell m}$ and $B_{\ell m}$.

Application 7.3 The Unisphere

The stainless steel "Unisphere" is the largest representation of the Earth ever constructed (Figure 7.9). Let this object be a model for a spherical conducting shell from which finite portions of the surface have been removed. The real Unisphere is grounded for safety. Here, we assume the shell is charged to a potential φ_0 and show that the *difference* in the surface charge density inside and outside the shell is a *constant* over the entire surface.

[10] $d\Omega \equiv \sin\theta d\theta d\phi$.

Figure 7.9: The "Unisphere" was the symbol of the 1964 World's Fair. The sphere radius is $R \approx 18$ m. Photograph from www.fotocommunity.de.

Place the origin of coordinates at the center of the shell. A representation that guarantees that $\varphi(\mathbf{r})$ is regular, continuous, and satisfies Laplace's equation everywhere off the shell is

$$\varphi(r,\theta,\phi) = \begin{cases} \sum_{\ell m} A_{\ell m} \left(\frac{r}{R}\right)^{\ell} Y_{\ell m}(\theta,\phi) & r \le R, \\ \sum_{\ell m} A_{\ell m} \left(\frac{R}{r}\right)^{\ell+1} Y_{\ell m}(\theta,\phi) & r \ge R. \end{cases} \tag{7.78}$$

The Dirichlet boundary condition is

$$\varphi_0 = \sum_{\ell m} A_{\ell m}\, Y_{\ell m}(\theta,\phi)|_{\text{on}}, \tag{7.79}$$

where the subscript "on" indicates that the equality holds only for angles (θ,ϕ) which coincide with the conducting surface of the shell. Therefore, the charge density difference $\Delta\sigma(\theta,\phi) = \sigma_{\text{out}}(\theta,\phi) - \sigma_{\text{in}}(\theta,\phi)$ between the outer and inner surfaces of the shell is

$$\Delta\sigma(\theta,\phi) = \epsilon_0 \left[\frac{\partial\varphi_>}{\partial r} + \frac{\partial\varphi_<}{\partial r}\right]_{\text{on}} = -\frac{\epsilon_0}{R}\sum_{\ell m} A_{\ell m}\, Y_{\ell m}(\theta,\phi)|_{\text{on}} = -\frac{\epsilon_0\,\varphi_0}{R}. \tag{7.80}$$

As advertised, this is indeed a constant, independent of (θ,ϕ). ■

Example 7.3 An origin-centered sphere has radius R. Find the volume charge density $\rho(r,\theta,\phi)$ (confined to $r < R$) and the surface charge density $\sigma(\theta,\phi)$ (confined to $r = R$) which together produce the electric field given below. Express the answer using trigonometric functions.

$$\mathbf{E} = -\frac{2V_0 x}{R^2}\hat{\mathbf{x}} + \frac{2V_0 y}{R^2}\hat{\mathbf{y}} - \frac{V_0}{R}\hat{\mathbf{z}} \qquad x^2+y^2+z^2 \le R^2.$$

Solution: Integrating each component of $\mathbf{E} = -\nabla\varphi$ gives

$$\hat{x}: \qquad \varphi = \frac{V_0}{R^2}x^2 + f(y,z)$$

$$\hat{y}: \qquad \varphi = -\frac{V_0}{R^2}y^2 + g(x,z)$$

$$\hat{z}: \qquad \varphi = \frac{V_0}{R}z + h(x,y).$$

Therefore,

$$\varphi_{\rm in}(x,y,z)=\frac{V_0}{R^2}\left(x^2-y^2\right)+\frac{V_0}{R}z+const. \qquad x^2+y^2+z^2\leq R^2.$$

Direct computation in Cartesian coordinates shows that $\varphi_{\rm in}$ satisfies Laplace's equation. Since $\rho=-\epsilon_0\nabla^2\varphi$, we conclude that there is no volume charge inside the sphere. On the other hand, in spherical coordinates, we know that solutions of Laplace's equation take the form (7.78). This means that the x^2-y^2 term in $\varphi_{\rm in}$ is (at worst) a linear combination of $\ell=2$ terms. The z term in $\varphi_{\rm in}$ is an $\ell=1$ term. Therefore, because $x=r\sin\theta\cos\phi$, $y=r\sin\theta\sin\phi$, and $z=r\cos\theta$,

$$\varphi(r,\theta,\phi)=\begin{cases} V_0\dfrac{r}{R}\cos\theta+V_0\left(\dfrac{r}{R}\right)^2\sin^2\theta\cos 2\phi & r\leq R, \\[2ex] V_0\left(\dfrac{R}{r}\right)^2\cos\theta+V_0\left(\dfrac{R}{r}\right)^3\sin^2\theta\cos 2\phi & r\geq R. \end{cases}$$

The charge density follows from the matching condition

$$\sigma(\theta,\phi)=\epsilon_0\left[\frac{\partial\varphi_{\rm in}}{\partial r}-\frac{\partial\varphi_{\rm out}}{\partial r}\right]_{r=R}=\epsilon_0\frac{V_0}{R}\left(3\cos\theta+5\sin^2\theta\cos 2\phi\right).$$

7.8 Cylindrical Symmetry

Laplace's equation in cylindrical coordinates is

$$\nabla^2\varphi=\frac{1}{\rho}\frac{\partial}{\partial\rho}\left(\rho\frac{\partial\varphi}{\partial\rho}\right)+\frac{1}{\rho^2}\frac{\partial^2\varphi}{\partial\phi^2}+\frac{\partial^2\varphi}{\partial z^2}=0. \tag{7.81}$$

For problems with cylindrical boundaries or, more generally, for problems with a unique preferred axis, the trial solution $\varphi(\rho,\phi,z)=R(\rho)G(\phi)Z(z)$ separates (7.81) into three ordinary differential equations with two real separation constants α^2 and k^2:

$$\rho\frac{d}{d\rho}\left(\rho\frac{dR}{d\rho}\right)+(k^2\rho^2-\alpha^2)R=0 \tag{7.82}$$

$$\frac{d^2G}{d\phi^2}+\alpha^2G=0 \tag{7.83}$$

$$\frac{d^2Z}{dz^2}-k^2Z=0. \tag{7.84}$$

Boundary and regularity (finiteness) conditions may or may not decide for us whether to choose α^2 and k^2 positive or negative. If both are chosen positive, the elementary solutions for $G(\phi)$ and $Z(z)$ are

$$G_\alpha(\phi)=\begin{cases} x_0+y_0\phi & \alpha=0, \\ x_\alpha\exp(i\alpha\phi)+y_\alpha\exp(-i\alpha\phi) & \alpha\neq 0, \end{cases} \tag{7.85}$$

and

$$Z_k(z)=\begin{cases} s_0+t_0z & k=0, \\ s_k\exp(kz)+t_k\exp(-kz) & k\neq 0. \end{cases} \tag{7.86}$$

7.8.1 Bessel Functions

The radial equation (7.82) is Bessel's differential equation. The elementary solutions of this equation are

$$R_\alpha^k(\rho) = \begin{cases} A_0 + B_0 \ln \rho & k = 0, \quad \alpha = 0, \\ A_\alpha \rho^\alpha + B_\alpha \rho^{-\alpha} & k = 0, \quad \alpha \neq 0, \\ A_\alpha^k J_\alpha(k\rho) + B_\alpha^k N_\alpha(k\rho) & k^2 > 0, \\ A_\alpha^k I_\alpha(\kappa\rho) + B_\alpha^k K_\alpha(\kappa\rho) & k^2 < 0. \end{cases} \tag{7.87}$$

$J_\alpha(x)$ and $N_\alpha(x)$ are called *Bessel functions* of the first and second kind, respectively. If we let $k = i\kappa$, the *modified Bessel functions* $I_\alpha(x)$ and $K_\alpha(x)$ are Bessel functions with pure imaginary arguments:

$$I_\alpha(\kappa\rho) = i^{-\alpha} J_\alpha(i\kappa\rho) \qquad K_\alpha(\kappa\rho) = \frac{\pi}{2} i^{\alpha+1}[J_\alpha(i\kappa\rho) + iN_\alpha(i\kappa\rho)]. \tag{7.88}$$

We define $J_\alpha(x)$, $N_\alpha(x)$, $I_\alpha(x)$, and $K_\alpha(x)$ for $x \geq 0$ only and refer the reader to Appendix C for more information. However, when α is real, it is important to know that $J_\alpha(x)$ is regular everywhere, $N_\alpha(x)$ diverges as $x \to 0$, and the asymptotic behavior ($x \gg 1$) of both is damped oscillatory:

$$\begin{aligned} J_\alpha(x) &\to \sqrt{\frac{2}{\pi x}} \cos\left(x - \alpha\frac{\pi}{2} - \frac{\pi}{4}\right) \\ N_\alpha(x) &\to \sqrt{\frac{2}{\pi x}} \sin\left(x - \alpha\frac{\pi}{2} - \frac{\pi}{4}\right). \end{aligned} \tag{7.89}$$

The modified Bessel function $I_\alpha(x)$ is finite at the origin and diverges exponentially as $x \to \infty$. $K_\alpha(x)$ diverges as $x \to 0$ but goes to zero exponentially as $x \to \infty$.

The general solution of Laplace's equation in cylindrical coordinates is a linear combination of the elementary solutions

$$\varphi(\rho, \phi, z) = \sum_\alpha \sum_k R_\alpha^k(\rho) G_\alpha(\phi) Z_k(z). \tag{7.90}$$

The results of the previous paragraph show that the choice $k^2 > 0$ produces a solution (7.90) which pairs oscillatory Bessel function behavior for $R(\rho)$ with real exponential functions for $Z(z)$. Conversely, the choice $k^2 < 0$ pairs a Fourier representation for $Z(z)$ with modified Bessel functions and thus real exponential behavior for $R(\rho)$. The fact that $\varphi(\rho, \phi, z)$ always exhibits simultaneous oscillatory and exponential behavior in its non-angular variables is the cylindrical manifestation of the "zero-curvature" property of the solutions to Laplace's equation discussed in Section 7.5.2. Finally, the matching conditions produce two natural constraints on the angular variation in (7.90). The first, $G(0) = G(2\pi)$, is a consequence of the continuity of the potential (7.3). The second, $G'(0) = G'(2\pi)$, is a consequence of (7.4) if the full angular range $0 \leq \phi \leq 2\pi$ is free of charge. Using (7.85), both conditions together force $\alpha = n$ where $n = 0, 1, 2, \ldots$ is a non-negative integer.

7.8.2 Fourier-Bessel Series

An interesting class of potential problems asks us to solve $\nabla^2\varphi = 0$ inside a cylinder of radius R when the potential is specified on two cross sections, say, $\varphi(\rho \leq R, \phi, z = z_1) = f_1(\rho, \phi)$ and $\varphi(\rho \leq R, \phi, z = z_2) = f_2(\rho, \phi)$. In the spirit of the problem defined by Figure 7.4, this calls for complete sets of eigenfunctions in the variables ρ and ϕ. Periodic boundary conditions are appropriate for the angular variable and a glance at (7.85) shows that the functions $G_m(\phi) = \exp(im\phi)$ suffice if $m = 0, \pm 1, \pm 2, \ldots$ This forces $\alpha = m$ in (7.87) and we are further restricted to functions $R_m^k(\rho)$ which are finite at the origin. Accordingly, the radial eigenfunctions are the set of functions which satisfy the homogeneous boundary condition $J_m(kR) = 0$. This condition fixes the allowed values of the separation constant at $k_{mn} = x_{mn}/R$ where x_{mn} is the nth zero of the Bessel function $J_m(x)$. Invoking completeness, we conclude it must be possible to construct a *Fourier-Bessel series* where, say,

$$f_1(\rho, \phi) = \sum_{m=-\infty}^{\infty} \sum_{n=1}^{\infty} c_{mn} J_m(x_{mn}\rho/R) \exp(im\phi). \tag{7.91}$$

To compute the c_{mn}, multiply both sides of (7.91) by $\rho J_m(k_{mn'}\rho)\exp(-im'\phi)$ and integrate over the intervals $0 \leq \rho \leq R$ and $0 \leq \phi < R$. This is sufficient because the exponential and Bessel functions satisfy the orthogonality relations

$$\frac{1}{2\pi} \int_0^{2\pi} d\phi \, e^{i(m-m')\phi} = \delta_{mm'} \tag{7.92}$$

and

$$\int_0^R d\rho \, \rho J_m(k_{mn}\rho) J_m(k_{mn'}\rho) = \delta_{nn'} \frac{1}{2} R^2 [J_{m+1}(k_{mn}R)]^2. \tag{7.93}$$

Application 7.4 An Electrostatic Lens

In 1931, Davisson and Calbick discovered that a circular hole in a charged metal plate focuses electrons exactly like an optical lens focuses light.[11] In fact, all electrostatic potentials with cylindrical symmetry have this property. Figure 7.10 shows another common electron lens: two adjacent and coaxial metal tubes of radius R separated by a small gap d. A potential difference $V_R - V_L$ is maintained between the tubes. In this Application, we calculate the potential inside the tubes (in the $d \to 0$ limit) and briefly discuss their focusing properties.

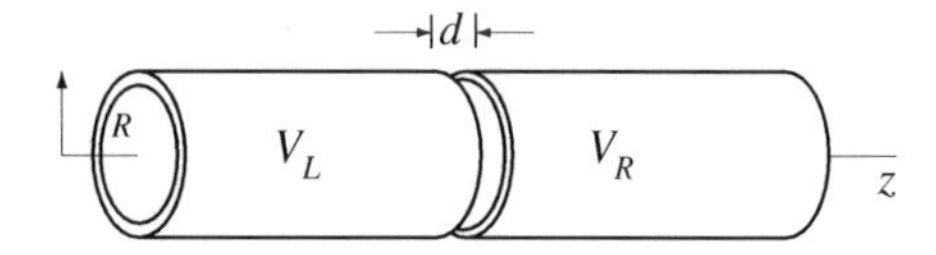

Figure 7.10: A two-tube electron lens.

If we separate variables in Laplace's equation in cylindrical coordinates, the rotational symmetry of the tubes fixes the separation constant in (7.83) at $\alpha = 0$. The choices $x_0 = 1$ and $y_0 = 0$ in (7.85) reduce $\varphi(\rho, \phi, z)$ to $\varphi(\rho, z)$. Here, we write $k = i\kappa$ and show (by construction) that the problem can be solved using radial functions with $k^2 < 0$ only. We invite the reader to show that a solution which looks different (but which is numerically equal by uniqueness) can be constructed using only radial functions with $k^2 > 0$. The potential must be finite at $\rho = 0$. Therefore, the discussion in Section 7.8.1 tells us that the most general form of the potential at this point is

$$\varphi(\rho, z) = \frac{1}{2\pi} \int_{-\infty}^{-\infty} d\kappa \, A(\kappa) I_0(|\kappa|\rho) e^{i\kappa z}. \tag{7.94}$$

The extracted factor of 2π emphasizes that (7.94) is a Fourier integral. Therefore, an application of Fourier's inversion theorem (i.e., the orthogonality of the complex exponential functions) gives

$$A(\kappa) = \frac{1}{I_0(|\kappa|\rho)} \int_{-\infty}^{\infty} dz \, \varphi(\rho, z) e^{-i\kappa z}. \tag{7.95}$$

[11] C.J. Davisson and C.J. Calbick, "Electron lenses", *Physical Review* **38**, 585 (1931).

To evaluate (7.95), we let $d \to 0$ in Figure 7.10 so $\varphi(R, z) = V_L$ when $z < 0$ and $\varphi(R, z) = V_R$ when $z > 0$. This gives

$$A(\kappa) = \frac{1}{I_0(|\kappa|R)} \left[V_L \int_{-\infty}^{0} dz\, e^{-i\kappa z} + V_R \int_{0}^{\infty} dz\, e^{-i\kappa z} \right]. \tag{7.96}$$

Regularizing the integrals and using the Plemelj formula (1.105) gives[12]

$$\varphi(\rho, z) = \frac{V_R + V_L}{2} + \frac{V_R - V_L}{2\pi i} \int_{-\infty}^{\infty} \frac{d\kappa}{\kappa} \frac{I_0(|\kappa|\rho)}{I_0(|\kappa|R)} e^{i\kappa z}. \tag{7.97}$$

The real part of the integrand in (7.97) is an odd function of κ. Therefore, the potential inside the tube is

$$\varphi(\rho, z) = \frac{V_R + V_L}{2} + \frac{V_R - V_L}{\pi} \int_{0}^{\infty} \frac{d\kappa}{\kappa} \frac{I_0(\kappa\rho)}{I_0(\kappa R)} \sin \kappa z. \tag{7.98}$$

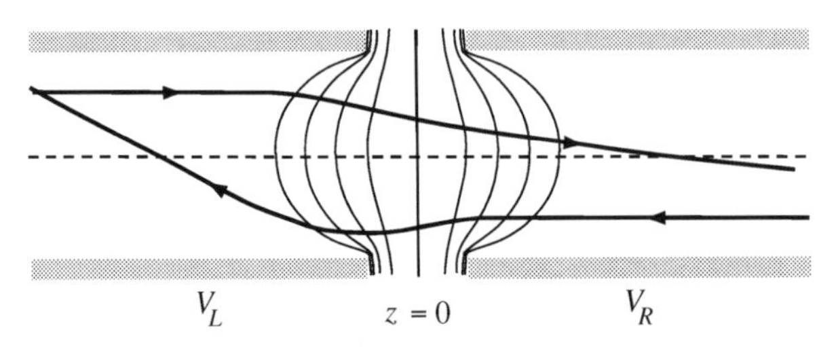

Figure 7.11: A two-tube electron lens with the symmetry axis (dashed) and a few equipotentials (light solid lines) indicated. The heavy solid lines are two particle trajectories drawn for $V_R > V_L$. Figure adapted from Heddle (1991).

Figure 7.11 is a cut-open view of the two tubes with a few equipotentials of (7.98) drawn as light solid lines. The dark solid lines are the trajectories of two charged particles. Although they move in opposite directions, both initially follow straight-line paths parallel to the z-axis in the $\mathbf{E} = 0$ region inside one tube. Both particles bend in the vicinity of the gap and then cross the z-axis during subsequent straight-line motion in the $\mathbf{E} = 0$ region of the other tube. In the language of optics, the $\mathbf{E} \neq 0$ regions of space on opposite sides of $z = 0$ deflect particles moving from left to right in Figure 7.11 first like a converging lens and then like a diverging lens. Particles moving from right to left are deflected first like a diverging lens and then like a converging lens. The diverging effect is weaker than the converging effect.[13] Therefore, particles are always focused toward the symmetry axis on the far side of the gap. ■

7.9 Polar Coordinates

There are many physical situations where the electrostatic potential is effectively a function of two (rather than three) spatial variables. The conducting duct (Application 7.1) and the Faraday cage (Section 7.5.1) are examples we solved in Cartesian coordinates. When the symmetry of the problem

[12] We use the regularization $\int_0^{\pm\infty} dz\, e^{-i\kappa z} = \lim_{\alpha \to 0} \int_0^{\pm\infty} dz\, e^{-i\kappa z} e^{\mp \alpha z}$.

[13] We leave it as an exercise for the reader to show that this is a generic feature of charged particle motion near the symmetry axis of a cylindrical potential.

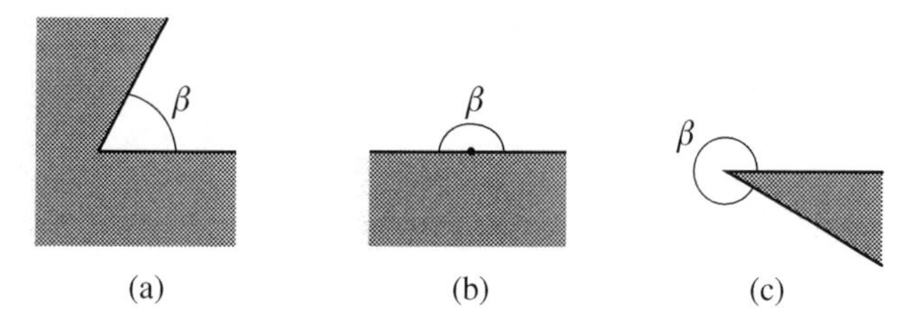

Figure 7.12: Side view of a two-dimensional wedge ($0 \leq \phi \leq \beta$) cut out of perfectly conducting (shaded) matter. The conductor is held at zero potential and otherwise fills all of space.

warrants it, it may be more natural to study separated-variable solutions of the two-dimensional Laplace's equation written in polar coordinates,

$$\nabla^2 \varphi = \frac{1}{\rho}\frac{\partial}{\partial \rho}\left(\rho \frac{\partial \varphi}{\partial \rho}\right) + \frac{1}{\rho^2}\frac{\partial^2 \varphi}{\partial \phi^2} = 0. \tag{7.99}$$

Alternatively, a general, separated-variable solution to (7.99) follows from the cylindrical-coordinates solution (7.90) with $Z(z) = 1$ and $k = 0$. If we switch from exponential functions to sinusoidal functions of the polar angle ϕ, the result can be written

$$\varphi(\rho, \phi) = (A_0 + B_0 \ln \rho)(x_0 + y_0 \phi) + \sum_{\alpha \neq 0} [A_\alpha \rho^\alpha + B_\alpha \rho^{-\alpha}][x_\alpha \sin \alpha\phi + y_\alpha \cos \alpha\phi]. \tag{7.100}$$

If, say, the potential far from the origin were due to a uniform electric field $\mathbf{E} = E_0 \hat{\mathbf{x}}$, it would be necessary to put $B_0 = A_\alpha = 0$ except for the single term $E_0 \rho \cos \phi$.

7.9.1 The Electric Field near a Sharp Corner or Edge

The electric field $\mathbf{E}(\rho, \phi)$ inside a two-dimensional wedge ($0 \leq \phi \leq \beta$) bounded by a grounded perfect conductor (Figure 7.12) provides insight into the nature of electric fields near sharp conducting corners. The potential cannot be singular as $\rho \to 0$ so (7.100) simplifies to

$$\varphi(\rho, \phi) = A + B\phi + \sum_{\alpha > 0} C_\alpha \rho^\alpha \sin(\alpha\phi + \delta_\alpha). \tag{7.101}$$

We get $A = B = 0$ because $\varphi(0, \phi) = 0$. Moreover, $\varphi(\rho, 0) = \varphi(\rho, \beta) = 0$ implies that $\delta_\alpha = 0$ and $\alpha = m\pi/\beta$ where m is a positive integer. The coefficients C_m are determined by boundary or matching conditions far from $\rho = 0$ which we do not specify. On the other hand, the $m = 1$ term dominates the sum as $\rho \to 0$. Therefore, up to a multiplicative constant, the potential very near the apex is

$$\varphi(\rho, \phi) \approx \rho^{\pi/\beta} \sin(\pi\phi/\beta). \tag{7.102}$$

The associated electric field is

$$\mathbf{E} = -\nabla\varphi = -\frac{\pi}{\beta}\rho^{\pi/\beta - 1}\left\{\hat{\boldsymbol{\rho}} \sin(\pi\phi/\beta) + \hat{\boldsymbol{\phi}} \cos(\pi\phi/\beta)\right\}. \tag{7.103}$$

Equation (7.103) correctly gives $\mathbf{E}$ as a vertically directed constant vector when $\beta = \pi$ [Figure 7.12(b)]. Otherwise, $|\mathbf{E}| \to 0$ as $\rho \to 0$ when $\beta < \pi$ [Figure 7.12(a)] but $|\mathbf{E}| \to \infty$ as $\rho \to 0$ when $\beta > \pi$ [Figure 7.12(c)]. This singular behavior is not new: we saw in (5.33) that the surface charge density of a conducting disk has a square-root singularity at its edge. This is consistent with the electric field here in the limit $\beta \to 2\pi$. A related singularity is often invoked to explain the behavior of a lightning rod.[14]

Analogy with Gas Diffusion

Imagine a collection of gas particles that *diffuse* from infinity toward the wedge in Figure 7.12. Each particle sticks to the wedge at the point of impact. The space and time dependence of the gas density $n(\mathbf{r}, t)$ is governed by the diffusion constant D and the diffusion equation, $D\nabla^2 n = \partial n/\partial t$. If we continuously supply gas from infinity, the density quickly reaches a time-independent steady state governed by $\nabla^2 n(\mathbf{r}) = 0$. The boundary condition for this equation is $n(\mathbf{r}_S) = 0$ at the wedge surface because "sticking" irreversibly removes particles from the gas. This establishes a one-to-one correspondence between the steady-state gas diffusion problem and the electrostatic problem solved just above. The analog of the electric field for the gas problem is the particle number current density $\mathbf{j}(\mathbf{r}) = -D\nabla n$. Now recall that diffusing particles execute a random walk in space. When $\beta > \pi$, such particles are very likely to encounter the convex apex before any other portion of the wedge, i.e., there is a large flux of particles to the apex. Conversely, when $\beta < \pi$, diffusing particles are very *unlikely* to reach the concave corner before striking another portion of the wedge. The net particle flux to the corner is very small. These are precisely the behaviors exhibited by $\mathbf{E}(\mathbf{r})$ near the apex.

Example 7.4 Find $\varphi(\rho, \phi)$ in the region bounded by the two arcs $\rho = \rho_1$ and $\rho = \rho_2$ and the two rays $\phi = \phi_1$ and $\phi = \phi_2$. All the boundaries are grounded except that $\varphi(\rho, \phi_2) = f(\rho)$. How does the nature of the separation constant change in the limit $\rho_1 \to 0$?

Solution: The homogeneous boundary conditions $\varphi(\rho_1, \phi) = \varphi(\rho_2, \phi) = 0$ eliminate the $\alpha = 0$ terms in the general solution (7.100). On the other hand, it seems impossible to satisfy these boundary conditions using the functions ρ^α and $\rho^{-\alpha}$ until we realize that $\rho^{\pm\alpha} = \exp(\pm\alpha \ln \rho)$. The choice $\alpha = i\gamma$ then turns the ρ dependence of (7.100) into a Fourier series in the variable $\ln \rho$. The general solution at this point is

$$\varphi(\rho, \phi) = \sum_{\gamma \neq 0} (A_\gamma e^{i\gamma \ln \rho} + B_\gamma e^{-i\gamma \ln \rho})(x_\gamma \sinh \gamma\phi + y_\gamma \cosh \gamma\phi).$$

We satisfy $\varphi(\rho_2, \phi) = 0$ by choosing the ratio A_γ / B_γ so the radial functions are $\sin\{\gamma \ln(\rho/\rho_2)\}$. We satisfy $\varphi(\rho_1, \phi) = 0$ by choosing $\gamma = m\pi / \ln(\rho_1/\rho_2)$ where m is an integer. The condition $\varphi(r, \phi_1) = 0$ leads to

$$\varphi(\rho, \phi) = \sum_{m=-\infty}^{\infty} A_m \sin\left\{\frac{m\pi \ln(\rho/\rho_2)}{\ln(\rho_1/\rho_2)}\right\} \sinh\left\{\frac{m\pi(\phi - \phi_1)}{\ln(\rho_1/\rho_2)}\right\}.$$

Finally, let $\theta = \pi \ln(\rho/\rho_2)/\ln(\rho_1/\rho_2)$, multiply both sides of the foregoing by $\sin(n\theta)$, and integrate over the interval $0 \le \theta \le \pi$. This completes the solution because the orthogonality of the sine functions uniquely determines the A_m.

[14] The field is very large but not truly singular near a rounded edge or near the tip of a lighting rod. Our analysis is approximately valid as long as the edge or tip is spatially isolated from other parts of the conductor and the radius of curvature of the edge or tip is small compared to any other length scale in the problem.

As $\rho_1 \to 0$, the interval between successive values of the separation constant γ goes to zero in the sum above because $\Delta\gamma = \Delta m\pi/\ln(\rho_1/\rho_2) = \pi/\ln(\rho_1/\rho_2)$. This means that the sum on α should be replaced by an integral over α. If, in addition, we absorb some constants into α,

$$\varphi(\rho, \phi) = \int_{-\infty}^{\infty} d\gamma\, A(\gamma) \sin\{\gamma \ln(\rho/\rho_2)\} \sinh\{\gamma(\phi - \phi_1)\}.$$

The Fourier inversion theorem expresses the function $A(\alpha)$ as an integral over the boundary function $f(\rho)$.

7.10 The Complex Potential

The theory of analytic functions is uniquely suited to finding solutions to the Laplace equation in two dimensions:

$$\frac{\partial^2\varphi}{\partial x^2} + \frac{\partial^2\varphi}{\partial y^2} = 0. \tag{7.104}$$

To see why, recall first that all electrostatics problems satisfy

$$\nabla \times \mathbf{E} = 0 \quad \Longrightarrow \quad \mathbf{E} = -\nabla\varphi. \tag{7.105}$$

In charge-free regions of space, it is also true that

$$\nabla \cdot \mathbf{E} = 0 \quad \Longrightarrow \quad \mathbf{E} = \nabla \times \mathbf{C}. \tag{7.106}$$

Therefore, because $\mathbf{E}(x, y) = -\nabla\varphi(x, y)$ is not a function of z, there is no loss of generality if we choose $\mathbf{C}(x, y) = -\psi(x, y)\hat{\mathbf{z}}$. In that case, (7.105) and (7.106) give

$$E_x = -\frac{\partial\varphi}{\partial x} = -\frac{\partial\psi}{\partial y} \tag{7.107}$$

$$E_y = -\frac{\partial\varphi}{\partial y} = \frac{\partial\psi}{\partial x}. \tag{7.108}$$

The equations on the far right sides of (7.107) and (7.108) are called the *Cauchy-Riemann* equations. Using them, it is straightforward to confirm that both $\varphi(x, y)$ and $\psi(x, y)$ satisfy Laplace's equation:

$$\nabla^2\varphi = 0 = \nabla^2\psi. \tag{7.109}$$

More importantly, if $w = x + iy$ is a complex variable, the Cauchy-Riemann equations also imply that

$$f(w) = \varphi(x, y) + i\,\psi(x, y) \tag{7.110}$$

is an *analytic* function. This means that the derivative

$$\frac{df}{dw} = \lim_{\Delta w \to 0} \frac{f(w + \Delta w) - f(w)}{\Delta w} \tag{7.111}$$

is independent of the direction of Δw in the complex plane. For example, $\Delta w = \Delta x$ when we compute the limit along a path parallel to the real axis. This gives

$$\frac{df}{dw} \to \frac{\partial f}{\partial x} = \frac{\partial\varphi}{\partial x} + i\frac{\partial\psi}{\partial x}. \tag{7.112}$$

The choice $\Delta w = i\Delta y$ means that we compute (7.111) along a path parallel to the imaginary axis. The result in this case is

$$\frac{df}{dw} \to \frac{\partial f}{\partial (iy)} = \frac{\partial \psi}{\partial y} - i\frac{\partial \varphi}{\partial y}. \tag{7.113}$$

Equating the real and imaginary parts of (7.112) and (7.113) reproduces the Cauchy-Riemann equations (7.107) and (7.108).

The Cauchy-Riemann equations also guarantee that

$$\nabla\varphi \cdot \nabla\psi = \frac{\partial \varphi}{\partial x}\frac{\partial \psi}{\partial x} + \frac{\partial \varphi}{\partial y}\frac{\partial \psi}{\partial y} = -\frac{\partial \psi}{\partial y}\frac{\partial \psi}{\partial x} + \frac{\partial \psi}{\partial x}\frac{\partial \psi}{\partial y} = 0. \tag{7.114}$$

This shows that the curves $\varphi(x, y) = const.$ are everywhere perpendicular to the curves $\psi(x, y) = const.$ Since (7.105) identifies $\varphi(x, y)$ as the electrostatic potential, we infer that the curves $\psi(x, y) = const.$ correspond to electric field lines. Finally, it is convenient for calculations to combine the electric field components in (7.107) and (7.108) into a single complex number E and use (7.112) to write

$$E = E_x - iE_y = -\frac{df}{dw}. \tag{7.115}$$

The foregoing sketch of the theory of analytic functions applies to two-dimensional electrostatic boundary value problems if we can find an analytic function $f(w)$—known as the *complex potential*—where either its real part or its imaginary part satisfies the boundary conditions. Conversely, we generate a storehouse of useful information by systematically surveying analytic functions and identifying the two-dimensional potential problems they describe.

7.10.1 A Uniform Electric Field

The simplest complex potential is

$$f(w) = -E_0 w = -E_0(x + iy) = \varphi + i\psi. \tag{7.116}$$

The physical potential is $\mathrm{Re}\, f = \varphi(x) = -E_0 x$, which we identify as the potential of a uniform electric field $\mathbf{E} = E_0\hat{\mathbf{x}}$. The straight electric field lines correspond to constant values of $\mathrm{Im}\, f = -E_0 y$.

7.10.2 A Quadrupole Potential

A quadrupole potential can be represented by the complex potential

$$f(w) = -\tfrac{1}{2}w^2 = -\tfrac{1}{2}(x + iy)^2 = -\tfrac{1}{2}(x^2 - y^2) - ixy = \varphi + i\psi. \tag{7.117}$$

The solid lines in Figure 7.13 show equipotentials where $\varphi(x, y) = const.$ The dashed lines are electric field lines where $\psi(x, y) = const.$ From (7.115), the field itself is

$$\mathbf{E} = x\hat{\mathbf{x}} - y\hat{\mathbf{y}}. \tag{7.118}$$

The figure also shows portions of four conducting rods (pointed out of the plane of the paper) that can produce this field because their surfaces are coincident with equipotentials.

7.10.3 A Conducting Strip in an External Field

An interesting example is

$$f(w) = \sqrt{w^2 - a^2} = \varphi + i\psi. \tag{7.119}$$

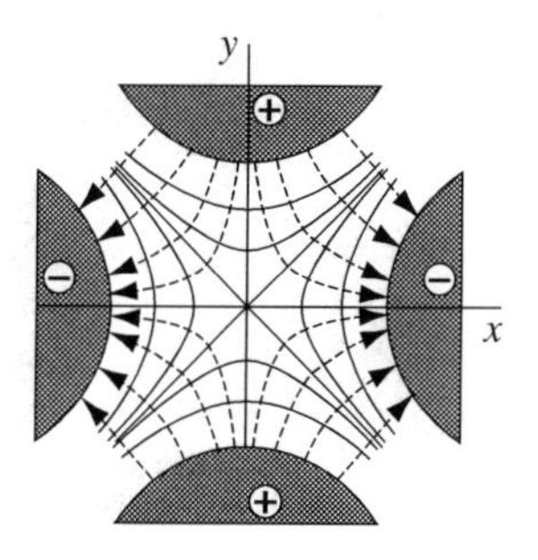

Figure 7.13: Equipotentials (solid curves) and electric field lines (dashed curves) for a pure quadrupole potential. Electrodes shaped like equipotentials are indicated. Figure from Klemperer (1972).

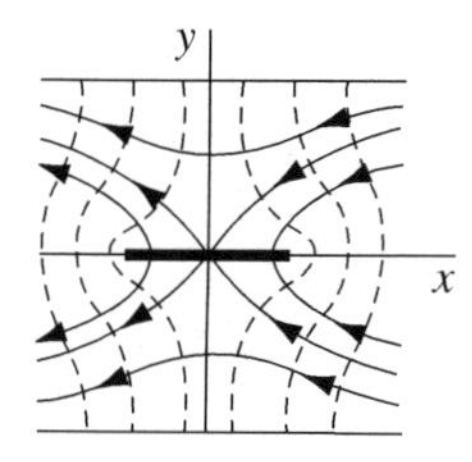

Figure 7.14: Electric field lines (solid curves) and equipotentials (dashed curves) near a conducting plate in a uniform electric field parallel to the x-axis. Figure from Durand (1966).

In (7.119), $\varphi = \mathrm{Re} f$ is a constant (equal to zero) when $w = x$ and $|x| < a$. Moreover, $f(w) \to w$ when $|w| \gg a$. This suggests that $\varphi(x, y)$ is the electrostatic potential of a unit strength electric field $\mathbf{E}_0 = -\hat{\mathbf{x}}$ perturbed by a two-dimensional conducting plate $(-a < x < a)$ oriented parallel to the field in the $y = 0$ plane.

We confirm this assignment by squaring the equation on the far right side of (7.119), setting $w = x + iy$, and equating real and imaginary parts. The two equations that result are

$$x^2 - y^2 - a^2 = \varphi^2 - \psi^2 \qquad \text{and} \qquad xy = \psi\varphi. \tag{7.120}$$

Eliminating ψ from (7.120) gives an equation (parameterized by different constant values of φ) for the equipotentials shown as dashed lines in Figure 7.14. Eliminating φ from (7.120) similarly gives an equation for the (solid) electric field lines in Figure 7.14. These are parameterized by different constant values of ψ. The electric field lines are not normal to the conductor at $x = y = 0$ because $x = 0$ and $y = 0$ are $\varphi = 0$ equipotentials which intersect at that point.

It is instructive to calculate the charge density induced on the plate's surface. By symmetry, the charge distribution is the same on the top and bottom, but it is worth seeing how this comes out of the mathematics. Specifically,

$$\sigma = \epsilon_0 \mathbf{E} \cdot \hat{\mathbf{n}}|_S = \pm\epsilon_0 E_y|_{y=0}, \tag{7.121}$$

where the upper (lower) sign refers to the upper (lower) surface of the plate. In light of (7.115), the surface charge density is

$$\sigma(x) = \pm\epsilon_0 \left[\frac{-ix}{\sqrt{w^2 - a^2}}\right]_{y \to 0}. \tag{7.122}$$

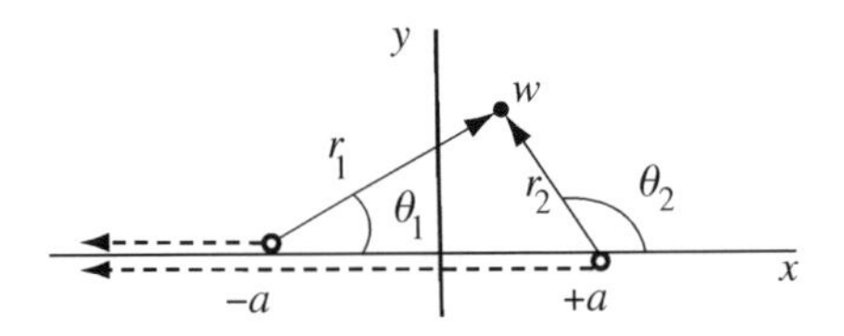

Figure 7.15: The function $f(w) = \sqrt{w^2 - a^2}$ is single-valued in the $w = x + iy$ plane if $w + a = r_1 \exp(i\theta_1)$ and $w - a = r_2 \exp(i\theta_2)$ with $-\pi \le \theta_1, \theta_2 < \pi$.

The $y \to 0$ step must be taken carefully because the factor $\sqrt{w^2 - a^2}$ in (7.122) is not single-valued.

We proceed by introducing the branch cuts shown as dashed lines in Figure 7.15. One line runs from $x = +a$ to $x = -\infty$, the other from $x = -a$ to $x = -\infty$. Branch cuts cannot be crossed, so we use angles θ_1 and θ_2 restricted to $-\pi \le \theta_1, \theta_2 \le \pi$ to write two polar forms for an arbitrary point in the $w = x + iy$ plane:

$$w = -a + r_1 \exp(i\theta_1) \qquad \text{and} \qquad w = a + r_2 \exp(i\theta_2). \tag{7.123}$$

The parameterizations in (7.123) have the effect of making the square root in (7.122) a single-valued function

$$\frac{1}{\sqrt{w^2 - a^2}} = \frac{1}{\sqrt{r_1 r_2}} \exp\{-i\tfrac{1}{2}(\theta_1 + \theta_2)\}. \tag{7.124}$$

The angle θ_2 is π on the top of the plate and $-\pi$ on the bottom of the plate. In both cases, $\theta_1 = 0$, $r_1 = a + x$, and $r_2 = a - x$. Using this information in (7.124) to evaluate (7.122) shows that

$$\sigma(x) = \pm\epsilon_0 \left[\frac{-ix}{\sqrt{a^2 - x^2}}\right]\left[-i \sin\left(\pm\frac{\pi}{2}\right)\right] = -\epsilon_0 \frac{x}{\sqrt{a^2 - x^2}}. \tag{7.125}$$

The negative sign is consistent with Figure 7.14 because the unperturbed electric field points in the $-x$-direction. The square-root singularity in (7.125) is the same one we found in (5.33) for the conducting disk and in (7.103) with $\beta = 2\pi$.

7.10.4 Conformal Mapping

The search for analytic functions which satisfy prescribed boundary conditions for any particular two-dimensional potential problem would be hopeless if not for the fact that analytic functions generate *conformal mappings* of the complex plane into itself. For example, $g(w) = u(x, y) + iv(x, y)$ describes a mapping from the complex $w = x + iy$ plane to the complex $g = u + iv$ plane. Conformal mappings transform curves into curves with the property that the angle of intersection between two curves is not changed by the transformation. In other words, a conformal map preserves the electrostatic property (7.114) that equipotentials and electric field cross at right angles.

We are now in a position to suggest a solution method for two-dimensional Laplace problems with awkward-shaped boundaries: find a conformal map which deforms the boundaries of the posed problem into the boundaries of a problem where Laplace's equation is easy to solve. The inverse mapping then deforms the equipotentials and field lines of the easy problem into the equipotentials and field lines of the posed problem. There are many tools available to aid in the search for suitable conformal transformations. For example, the Riemann mapping theorem states that *any* simply-connected domain in the w-plane can be mapped into a circular disk in the g-plane. Riemann's theorem is not constructive, but a great many explicit mapping functions for other situations have been analyzed and catalogued for ready application.

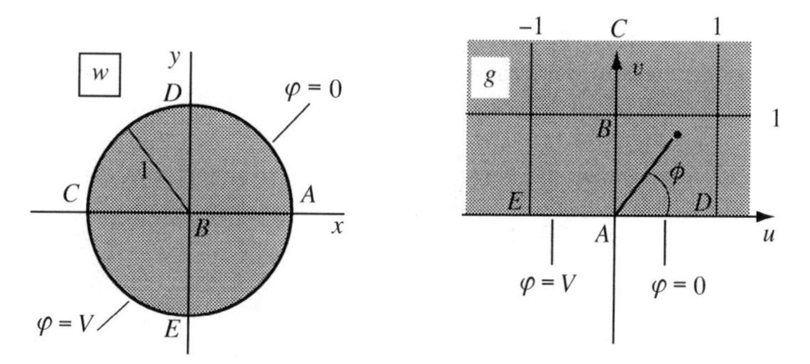

Figure 7.16: Left side: a cylinder with unit radius in the $w = x + iy$ plane. Right side: Equation (7.126) maps the interior of the cylinder to the upper half $g = u + iv$ plane. The point C is at ∞ on the u-axis.

7.10.5 The Potential inside a Split Cylinder

The left side of Figure 7.16 is a cross sectional view of a hollow conducting cylinder with unit radius. The upper half of the cylinder is grounded. The (infinitesimally displaced) lower half of the cylinder is held at potential V. To find the potential everywhere inside the cylinder, we consider the mapping

$$g(w) = -i\frac{w-1}{w+1}. \tag{7.126}$$

The points labeled A-E in and on the cylinder in the w-plane map to the corresponding points labeled A-E in the g plane. The upper semi-circle in the w-plane maps to the positive u-axis and the lower semi-circle in the w-plane maps to the negative u-axis. The point C maps to $g = i\infty$.

It is not hard to find the potential everywhere in the upper half g-plane when the negative real axis is held at potential V and the positive real axis is held at zero potential. If $\arg[g]$ is the argument of the complex number g, the separated-variable solution of Laplace's equation (7.100) that satisfies the boundary conditions is

$$\varphi(g) = \frac{V}{\pi}\phi = \frac{V}{\pi}\arg[g]. \tag{7.127}$$

We now use (7.126) to express (7.127) in the original variables

$$\varphi(x, y) = \frac{V}{\pi}\arg\left[-i\frac{w-1}{w+1}\right] = \frac{V}{\pi}\tan^{-1}\left[\frac{1-x^2-y^2}{2y}\right]. \tag{7.128}$$

The argument of the inverse tangent is a fixed constant on each equipotential. This defines a family of equipotential curves parameterized by a constant c:

$$x^2 + (y+c)^2 = 1 + c^2. \tag{7.129}$$

These are portions of circles centered on $x = 0$ that pass through both $(1, 0)$ and $(-1, 0)$ in the x-y plane. The solid lines in Figure 7.17 are a few of these equipotentials. The dashed lines represent electric field lines.

7.10.6 The Fringing Field of a Capacitor

In his *Treatise on Electricity and Magnetism*, Maxwell famously used conformal mapping to study the fringing field at the edges of a finite-area parallel-plate capacitor. If $g = u + iv$ is a typical point

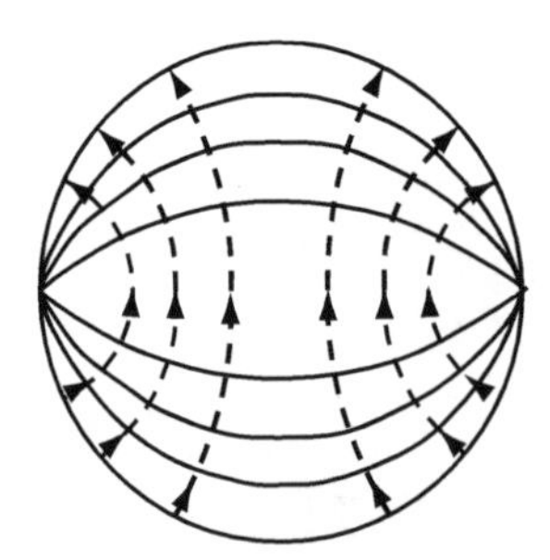

Figure 7.17: Equipotentials (solid curves) and electric field lines (dashed curves) for the two-dimensional potential problem defined by the left side of Figure 7.16.

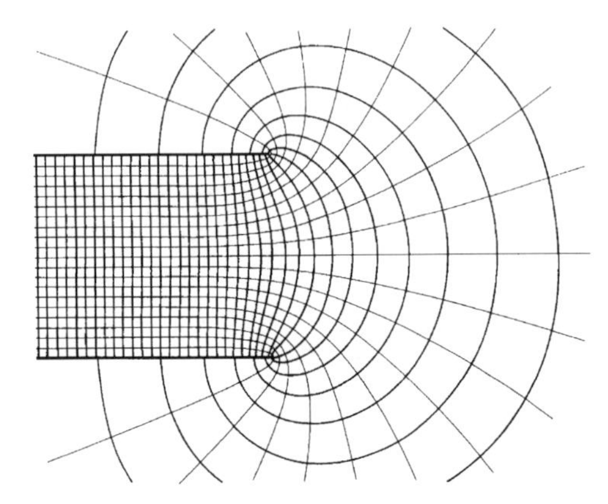

Figure 7.18: The equipotentials and electric field lines near the edge of a semi-infinite parallel-plate capacitor obtained by conformal mapping. Figure from Maxwell (1873).

in the complex g-plane, the mapping function he used for this purpose is

$$w = \frac{d}{2\pi}\left\{1 + g + \exp(g)\right\}. \tag{7.130}$$

Separating (7.130) into real and imaginary parts gives

$$x = \frac{d}{2\pi}\{1 + u + e^u \cos v\} \qquad \text{and} \qquad y = \frac{d}{2\pi}\{v + e^u \sin v\}. \tag{7.131}$$

The choice $v = \pm\pi$ gives $y = \pm d/2$. However, as u varies linearly from $-\infty$ to $+\infty$, x varies non-linearly from $-\infty$ to zero (when $u = 0$) and then back to $-\infty$. In other words, the two plates of an infinite parallel-plate capacitor in the g-plane ($v = \pm i\pi$ for all u) map to the two plates of a semi-infinite parallel-plate capacitor in the w-plane ($y = \pm d/2$ with $x < 0$).

In light of (7.116), we conclude that constant values of the real and imaginary parts of the complex potential $f(u, v) = u + iv$ in the g-plane map to the electric field lines and equipotentials of a semi-infinite capacitor in the w-plane. Figure 7.18 shows the result of doing this with a close-up view in the vicinity of the capacitor edge.

7.11 A Variational Principle

Thomson's theorem for electrostatics (Section 5.2.1) established that, among all charge densities $\rho(\mathbf{r})$ with total charge Q in a volume V, the electrostatic energy is lowest for the particular charge density that produces a constant value for the electrostatic potential throughout V. We prove a related theorem in this section: among all potential functions $\varphi(\mathbf{r})$ which take constant values on a set of (conductor) surfaces, the electrostatic energy is lowest for the particular potential function which satisfies Laplace's equation in the volume bounded by the surfaces. When this theorem is applied to candidate potential functions with adjustable parameters, we have a *variational principle* that can be used to find approximate solutions to Laplace's equation.

To begin, let $\varphi(\mathbf{r})$ be the exact solution to an electrostatic Dirichlet problem with specified constant values for the potential on a fixed set of (conductor) surfaces. Now let $\psi(\mathbf{r})$ be an arbitrary function which takes the *same* specified values on the surfaces. Physical intuition suggests that the electrostatic energy $U_E[\psi]$ associated with ψ is greater than the electrostatic energy $U_E[\varphi]$ associated with φ. To prove this, we must show that $\delta U_E = U_E[\psi] - U_E[\varphi] > 0$, where

$$U_E[\varphi] = \frac{\epsilon_0}{2}\int d^3r\, \nabla\varphi \cdot \nabla\varphi. \tag{7.132}$$

There is no loss of generality if we write

$$\psi(\mathbf{r}) = \varphi(\mathbf{r}) + \delta\varphi(\mathbf{r}), \tag{7.133}$$

as long as the variation $\delta\varphi(\mathbf{r})$ satisfies $\delta\varphi(\mathbf{r}_S) = 0$. This ensures that $\varphi(\mathbf{r})$ and $\psi(\mathbf{r})$ satisfy the same boundary conditions.

The change in energy associated with the variation (7.133) is

$$\delta U_E = U_E[\varphi + \delta\varphi] - U_E[\varphi] = \epsilon_0 \int d^3r\, \nabla\varphi \cdot \nabla\delta\varphi + \frac{\epsilon_0}{2}\int d^3r\, |\nabla\delta\varphi|^2. \tag{7.134}$$

Now apply $\nabla \cdot (\delta\varphi\, \mathbf{E}) = \mathbf{E} \cdot \nabla\delta\varphi + \delta\varphi\, \nabla \cdot \mathbf{E}$ to the first integral and use the divergence theorem. This gives

$$\delta U_E = -\epsilon_0 \int dS\, \delta\varphi\, \hat{\mathbf{n}} \cdot \nabla\varphi - \epsilon_0 \int d^3r\, \delta\varphi\, \nabla^2\varphi + \frac{\epsilon_0}{2}\int d^3r\, |\nabla\delta\varphi|^2. \tag{7.135}$$

The first term on the right side of (7.135) vanishes because $\delta\varphi$ vanishes at all boundary points. The second term is zero because $\varphi(\mathbf{r})$ satisfies Laplace's equation. The expression that remains establishes the result we seek,

$$\delta U_E = \frac{\epsilon_0}{2}\int d^3r\, |\nabla\delta\varphi|^2 > 0. \tag{7.136}$$

The lowest electrostatic energy is achieved by the unique potential function that satisfies the boundary conditions *and* solves Laplace's equation.

The result (7.136) is actually a bit more general than we have indicated. First, it is not necessary that $\varphi(\mathbf{r}_S)$ assume only constant values. If the bounding surfaces are not conductors, we still have $\delta\varphi(\mathbf{r}_S) = 0$ if more general Dirichlet conditions are specified. The surface integral in (7.135) also vanishes if the special Neumann condition $\mathbf{n} \cdot \nabla\varphi(\mathbf{r}_S) = 0$ is imposed on some or all of the bounding surfaces.

With the information just gained, the following procedure immediately suggests itself to find approximate solutions to boundary value potential theory problems. Construct a trial solution to Laplace's equation $\psi(\mathbf{r})$ that satisfies the boundary conditions exactly but otherwise depends on some number of adjustable parameters. Minimize $U_E[\psi]$ with respect to those parameters. The resulting function does not satisfy Laplace's equation exactly but the total energy differs from the exact answer $U_E[\varphi]$ by an amount that is *quadratic* in the difference between $\varphi(\mathbf{r})$ and $\psi(\mathbf{r})$. This is significant because most applications demand accuracy in the energy (or force) rather than accuracy in the potentials themselves. Of course, δU_E can be made as small as we like simply by adding more and more variational parameters to the candidate function $\psi(\mathbf{r})$.

Example 7.5 Use $\psi(x,y) = \varphi_0 x(L-x)\exp(-py)$ as a trial solution with a variational parameter p to find the electrostatic potential inside a rectangular slot with boundary conditions $\psi(x,0) = \varphi_0 x(L-x)$, $\psi(0,y) = \psi(L,y) = 0$, and $\psi(x, y\to\infty) = 0$.

Solution: The trial potential satisfies the boundary conditions but not Laplace's equation. We find p by minimizing $U_E(p)$. The trial electric field is $\mathbf{E}(x,y) = \varphi_0\left[(2x-L)\hat{\mathbf{x}} + px(L-x)\hat{\mathbf{y}}\right]\exp(-py)$. Therefore,

$$U_E(p) = \tfrac{1}{2}\epsilon_0 \int_0^L dx \int_0^\infty dy\, \mathbf{E}\cdot\mathbf{E} = \tfrac{1}{12}\epsilon_0\varphi_0^2 L^3[p^{-1} + pL^2/10].$$

The minimum of $U_E(p)$ occurs when $p = \sqrt{10}/L$. This gives the optimal solution as

$$\psi(x,y) = \varphi_0 x(L-x)\exp\left[-\sqrt{10}y/L\right].$$

For comparison, the exact separated-variable solution is

$$\varphi(x,y) = \sum_{n=1}^{\infty} A_n \sin\frac{n\pi x}{L} e^{-n\pi y/L} \qquad \text{with} \qquad A_n = \frac{2\varphi_0}{L}\int_0^L dx\, x(L-x)\sin\frac{n\pi x}{L}.$$

The largest contribution to the sum comes from the $n=1$ term:

$$\varphi_1(x,y) = (8\varphi_0 L^2/\pi^3)\sin(\pi x/L)e^{-\pi y/L}.$$

This compares very well with the variational solution in both magnitude and shape.

Sources, References, and Additional Reading

The quotation at the beginning of the chapter is taken from Section 232 of

J.H. Jeans, *The Mathematical Theory of Electricity and Magnetism*, 5th edition (Cambridge University Press, Cambridge, 1925).

Section 7.1 Textbooks of electromagnetism where solution methods for Laplace's equation are discussed in detail include

W.R. Smythe, *Static and Dynamic Electricity* (McGraw-Hill, New York, 1939).

E. Durand, *Électrostatique* (Masson, Paris, 1966), Volume II.

Section 7.2 An informative essay on the history of potential theory is

L. Gårding, "The Dirichlet problem", *The Mathematical Intelligencer* **2**, 43 (1979).

Potential problems with mixed boundary conditions are discussed at length in

I.N. Sneddon, *Mixed Boundary Value Problems in Potential Theory* (North-Holland, Amsterdam, 1966).

V.I. Fabrikant, *Mixed Boundary Value Problems of Potential Theory and Their Applications in Engineering* (Kluwer, Dordrecht, 1991).

Section 7.4 The introductory text by *Asmar* and the more sophisticated treatment by *Zauderer* are typical of discussions which place Laplace's equation in the more general setting of partial differential equations:

N.H. Asmar, *Partial Differential Equations with Fourier Series and Boundary Value Problems*, 2nd edition (Pearson, Upper Saddle River, NJ, 2005).

E. Zauderer, *Partial Differential Equations of Applied Mathematics*, 2nd edition (Wiley, New York, 1989).

Section 7.5 Application 7.1 was taken from

S. Hassani, "Laplace's equation with physically uncommon boundary conditions", *American Journal of Physics* **59**, 470 (1991).

The boxed material just above Section 7.5.1 comes from

M. Faraday, *Faraday's Diary of Experimental Investigation 1820-1862* (G. Bell and Sons, London, 1936). Courtesy and copyright by The Royal Institution of Great Britain & HR Direct.

Section 7.6 Onsager's discussion of the dielectric constant for polar liquids is

L. Onsager, "Electric moments of molecules in liquids", *Journal of the American Chemical Society* **43**, 189 (1936).

Section 7.7 Application 7.3 is adapted from an exercise in *Jeans*, cited above as the source for the quotation at the beginning of this chapter.

Section 7.8 Figure 7.11 and the electrostatic lens discussion in Section 7.4 come, respectively, from

D.W.O. Heddle, *Electrostatic Lens Systems* (Adam Hilger, Bristol, 1991).

M. Szilagyi, *Electron and Ion Optics* (Plenum, New York, 1988).

Section 7.10 Two textbooks of electromagnetism with good discussions of complex variable methods for two-dimensional electrostatics are

J. Vanderlinde, *Classical Electromagnetic Theory* (Wiley, New York, 1993).

R.H. Good, Jr. and T.J. Nelson, *Classical Theory of Electric and Magnetic Fields* (Academic, New York, 1971).

Figure 7.13, Figure 7.14, and Figure 7.18 come, respectively, from

O. Klemperer, *Electron Physics*, 2nd edition (Butterworths, London, 1972).

E. Durand, *Électrostatique* (Masson, Paris, 1966), Volume II.

J.C. Maxwell, *A Treatise on Electricity and Magnetism* (Clarendon Press, Oxford, 1873). By permission of Oxford University Press.

Problems

7.1 Two Electrostatic Theorems Use the orthogonality properties of the spherical harmonics to prove the following identities for a function $\varphi(\mathbf{r})$ which satisfies Laplace's equation in and on an origin-centered spherical surface S of radius R:

(a) $\displaystyle\int_S dS\,\varphi(\mathbf{r}) = 4\pi R^2\varphi(0).$

(b) $\displaystyle\int_S dS z\varphi(\mathbf{r}) = \frac{4\pi}{3}R^4\left.\frac{\partial\varphi}{\partial z}\right|_{\mathbf{r}=0}.$

7.2 Green's Formula Let $\hat{\mathbf{n}}$ be the normal to an equipotential surface at a point P. The principal radii of curvature of the surface at P are R_1 and R_2. A formula due to George Green relates normal derivatives ($\partial/\partial n \equiv \hat{\mathbf{n}}\cdot\nabla$) of the potential $\varphi(\mathbf{r})$ (which satisfies Laplace's equation) at the equipotential surface to the mean curvature of that equipotential surface $\kappa = \frac{1}{2}(R_1^{-1} + R_2^{-1})$:

$$\frac{\partial^2\varphi}{\partial n^2} + 2\kappa\frac{\partial\varphi}{\partial n} = 0.$$

Derive Green's equation by direct manipulation of Laplace's equation.

7.3 Poisson's Formula for a Sphere The Poisson integral formula

$$\varphi(\mathbf{r}) = \frac{(R^2 - r^2)}{4\pi R}\int_{|\mathbf{r}'_S|=R} dS'\frac{\bar{\varphi}(\mathbf{r}'_S)}{|\mathbf{r}-\mathbf{r}'_S|^3} \qquad |\mathbf{r}| < R$$

gives the potential at any point $\mathbf{r}$ inside a sphere if we specify the potential $\bar{\varphi}(\mathbf{r}_S)$ at every point on the surface of the sphere. Derive this formula by summing the general solution of Laplace's equation inside the sphere using the derivatives (with respect to r and R) of the identity

$$\frac{1}{|\mathbf{r}-\mathbf{r}'_S|} = \sum_{\ell=0}^{\infty} \frac{r^\ell}{R^{\ell+1}} P_\ell(\hat{\mathbf{r}} \cdot \hat{\mathbf{r}}'_S).$$

7.4 The Potential inside an Ohmic Duct The z-axis runs down the center of an infinitely long heating duct with a square cross section. For a real metal duct (not a perfect conductor), the electrostatic potential $\varphi(x, y)$ varies *linearly* along the side walls of the duct. Suppose that the duct corners at $(\pm a, 0)$ are held at potential $+V$ and the duct corners at $(0, \pm a)$ are held at potential $-V$. Find the potential inside the duct beginning with the trial solution

$$\varphi(x, y) = A + Bx + Cy + Dx^2 + Ey^2 + Fxy.$$

7.5 The Near-Origin Potential of Four Point Charges Four identical positive point charges sit at (a, a), $(-a, a)$, $(-a, -a)$, and $(a, -a)$ in the $z = 0$ plane. Very near the origin, the electrostatic potential can be written in the form

$$\varphi(x, y, z) = A + Bx + Cy + Dz + Exy + Fxz + Gyz + Hx^2 + Iy^2 + Jz^2.$$

(a) Deduce the non-zero terms in this expansion and the algebraic signs of their coefficients. Do not calculate the exact value of the non-zero coefficients.

(b) Sketch electric field lines and equipotentials in the $z = 0$ plane everywhere inside the square and a little bit outside the square.

7.6 The Microchannel Plate The parallel plates of a *microchannel plate* electron multiplier are segmented into conducting strips of width b so the potential can be fixed on the strips at staggered values. We model this using infinite-area plates, a finite portion of which is shown below. Find the potential $\varphi(x, y)$ between the plates and sketch representative field lines and equipotentials. Note the orientation of the x- and y-axes.

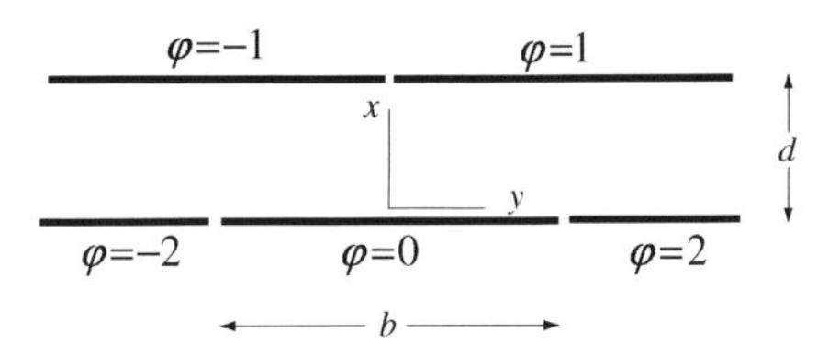

7.7 A Potential Patch by Separation of Variables The square region defined by $-a \le x \le a$ and $-a \le y \le a$ in the $z = 0$ plane is a conductor held at potential $\varphi = V$. The rest of the $z = 0$ plane is a conductor held at potential $\varphi = 0$. The plane $z = d$ is also a conductor held at zero potential.

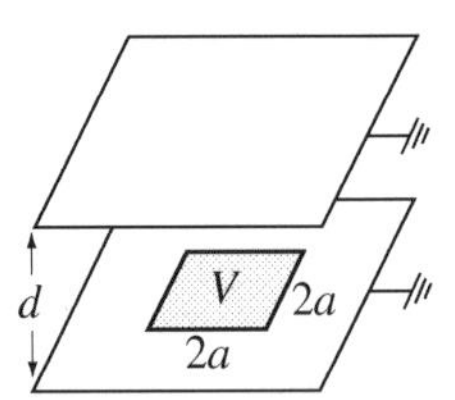

(a) Find the potential for $0 \le z \le d$ in the form of a Fourier integral.

(b) Find the total charge induced on the upper surface of the lower ($z = 0$) plate. The answer is very simple. Do not leave it in the form of an unevaluated integral or infinite series.

(c) Sketch field lines of $\mathbf{E}(\mathbf{r})$ between the plates.

7.8 A Conducting Slot The figure shows an infinitely long and deep slot formed by two grounded conductor plates at $x = 0$ and $x = a$ and a conductor plate at $z = 0$ held at a potential φ_0. Find the potential inside the slot and determine its asymptotic behavior when $z \gg a$.

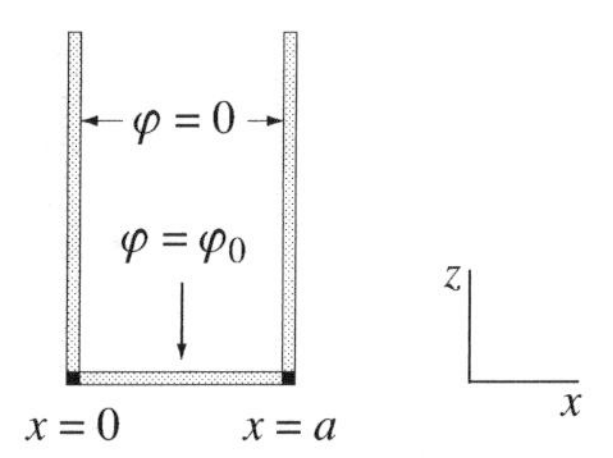

7.9 A Two-Dimensional Potential Problem in Cartesian Coordinates Two flat conductor plates (infinite in the x- and y-directions) occupy the planes $z = \pm d$. The $x > 0$ portion of both plates is held at $\varphi = +\varphi_0$. The $x < 0$ portion of both plates is held at $\varphi = -\varphi_0$. Derive an expression for the potential between the plates using a Fourier integral to represent the x variation of $\varphi(x, z)$.

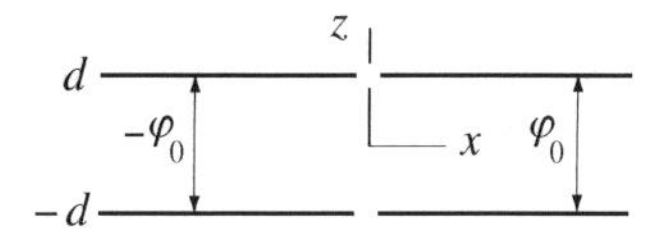

7.10 An Electrostatic Analog of the Helmholtz Coil A spherical shell of radius R is divided into three conducting segments by two very thin air gaps located at latitudes θ_0 and $\pi - \theta_0$. The center segment is grounded. The upper and lower segments are maintained at potentials V and $-V$, respectively. Find the angle θ_0 such that the electric field inside the shell is as nearly constant as possible near the center of the sphere.

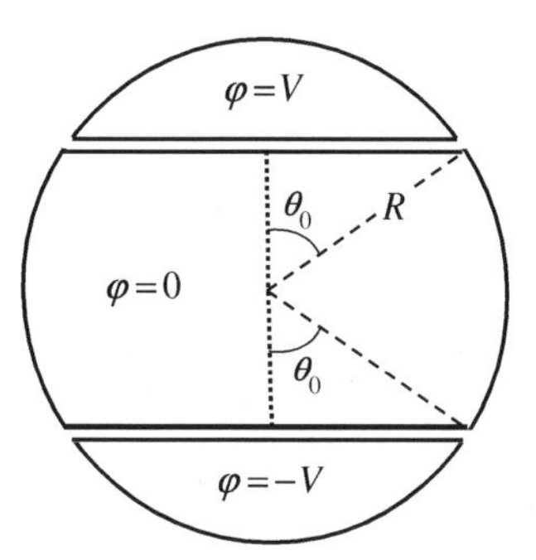

7.11 Make a Field inside a Sphere Find the volume charge density ρ and surface charge density σ which must be placed in and on a sphere of radius R to produce a field inside the sphere of

$$\mathbf{E} = -2V_0 \frac{xy}{R^3}\hat{\mathbf{x}} + \frac{V_0}{R^3}(y^2 - x^2)\hat{\mathbf{y}} - \frac{V_0}{R}\hat{\mathbf{z}}.$$

There is no other charge anywhere. Express your answer in terms of trigonometric functions of θ and ϕ.

7.12 The Capacitance of an Off-Center Capacitor A spherical conducting shell centered at the origin has radius R_1 and is maintained at potential V_1. A second spherical conducting shell maintained at potential V_2 has radius $R_2 > R_1$ but is centered at the point $s\hat{\mathbf{z}}$ where $s \ll R_1$.

(a) To lowest order in s, show that the charge density induced on the surface of the inner shell is

$$\sigma(\theta) = \epsilon_0 \frac{R_1 R_2 (V_2 - V_1)}{R_2 - R_1} \left[\frac{1}{R_1^2} - \frac{3s}{R_2^3 - R_1^3} \cos\theta \right].$$

Hint: Show first that the boundary of the outer shell is $r_2 \approx R_2 + s\cos\theta$.

(b) To lowest order in s, show that the force exerted on the inner shell is

$$\mathbf{F} = \int dS \frac{\sigma^2}{2\epsilon_0} \hat{\mathbf{n}} = \hat{\mathbf{z}} 2\pi R_1^2 \int_0^\pi d\theta \sin\theta \frac{\sigma^2(\theta)}{2\epsilon_0} \cos\theta = -\frac{Q^2}{4\pi\epsilon_0} \frac{s\hat{\mathbf{z}}}{R_2^3 - R_1^3}.$$

(c) Integrate the force in (b) to find the capacitance of this structure to second order in s.

7.13 The Plane-Cone Capacitor A capacitor is formed by the infinite grounded plane $z = 0$ and an infinite, solid, conducting cone with interior angle $\pi/4$ held at potential V. A tiny insulating spot at the cone vertex (the origin of coordinates) isolates the two conductors.

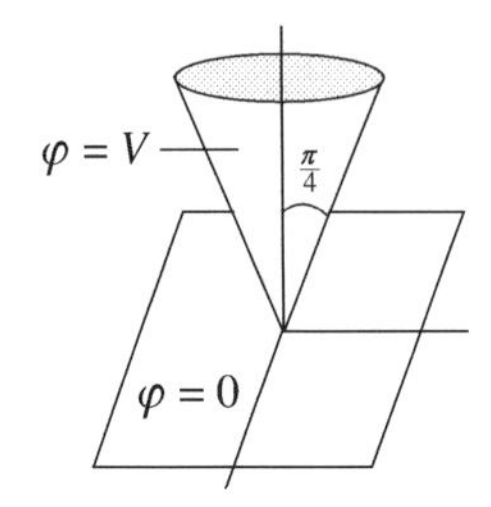

(a) Explain why $\varphi(r, \theta, \phi) = \varphi(\theta)$ in the space between the capacitor "plates".

(b) Integrate Laplace's equation explicitly to find the potential between the plates.

7.14 A Conducting Sphere at a Dielectric Boundary A conducting sphere with radius R and charge Q sits at the origin of coordinates. The space outside the sphere above the $z = 0$ plane has dielectric constant κ_1. The space outside the sphere below the $z = 0$ plane has dielectric constant κ_2.

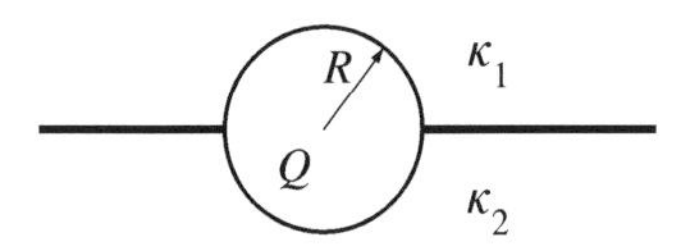

(a) Find the potential everywhere outside the conductor.

(b) Find the distributions of free charge and polarization charge wherever they may be.

7.15 The Force on an Inserted Conductor A set of known constants α_n parameterizes the potential in a volume $r < a$ as

$$\varphi_{\text{ext}}(r, \theta) = \sum_{n=1}^{\infty} \alpha_n \left(\frac{r}{R}\right)^n P_n(\cos\theta).$$

Let $\hat{\mathbf{z}}$ point along $\theta = 0$ and insert a solid conducting sphere of radius $R < a$ at the origin. Show that the force exerted on the sphere when it is connected to ground is in the z-direction and

$$F_z = 4\pi\epsilon_0 \sum_{n=1}^{\infty} (n+1)\alpha_n \alpha_{n+1}.$$

Hint: The Legendre polynomials satisfy $(n+1)P_{n+1}(x) + nP_{n-1}(x) = (2n+1)xP_n(x)$.

7.16 A Segmented Cylinder The figure below is a cross section of an infinite, conducting cylindrical shell. Two infinitesimally thin strips of insulating material divide the cylinder into two segments. One segment is held at unit potential. The other segment is held at zero potential. Find the electrostatic potential inside the cylinder.

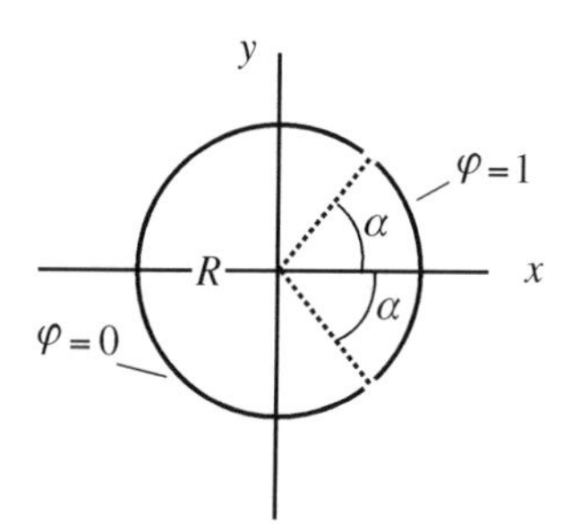

7.17 An Incomplete Cylinder The figure below shows an infinitely long cylindrical shell from which a finite angular range has been removed. Let the shell be a conductor raised to a potential corresponding to a charge per unit length λ. Find the fraction of charge which resides on the inner surface of the shell in terms of λ and the angular parameter p. Hint: Calculate $Q_{\text{in}} - Q_{\text{out}}$.

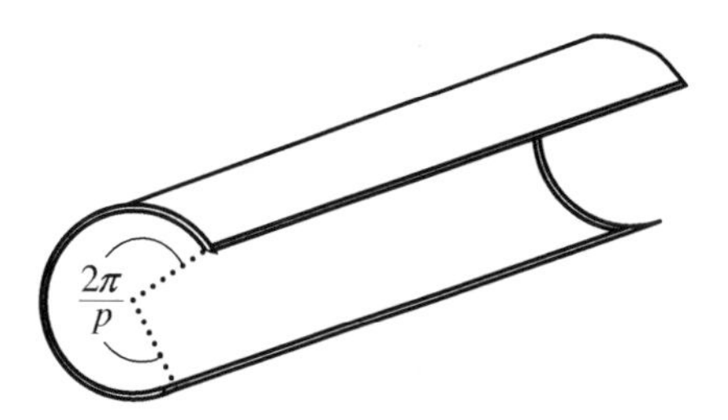

7.18 The Two-Cylinder Electron Lens Two semi-infinite, hollow cylinders of radius R are coaxial with the z-axis. Apart from an insulating ring of thickness $d \to 0$, the two cylinders abut one another at $z = 0$ and are held at potentials V_L and V_R. Find the potential everywhere inside both cylinders. You will need the integrals

$$\lambda \int_0^1 ds\, s\, J_0(\lambda s) = J_1(\lambda) \qquad \text{and} \qquad 2\int_0^1 ds\, s\, J_0(x_n s) J_0(x_m s) = J_1^2(x_n)\delta_{nm}.$$

The real numbers x_m satisfy $J_0(x_m) = 0$.

7.19 A Periodic Array of Charged Rings Let the z-axis be the symmetry axis for an infinite number of identical rings, each with charge Q and radius R. There is one ring in each of the planes $z = 0$, $z = \pm b$, $z = \pm 2b$, etc. Exploit the Fourier expansion in Example 1.6 to find the potential everywhere in space. Check that your solution makes sense in the limit that the cylindrical variable $\rho \gg R, b$. Hint: If $I_\alpha(y)$ and K_α are modified Bessel functions,

$$I'_\alpha(y)K_\alpha(y) - I_\alpha(y)K'_\alpha(y) = 1/y.$$

7.20 Axially Symmetric Potentials Let $V(z)$ be the potential on the axis of an axially symmetric electrostatic potential in vacuum. Show that the potential at any point in space is

$$V(\rho, z) = \frac{1}{\pi}\int_0^\pi d\zeta\; V(z + i\rho\cos\zeta).$$

Hint: Show that the proposed solution satisfies Laplace's equation and exploit uniqueness.

7.21 Circular-Plate Capacitor Consider a parallel-plate capacitor with circular plates of radius a separated by a distance $2L$.

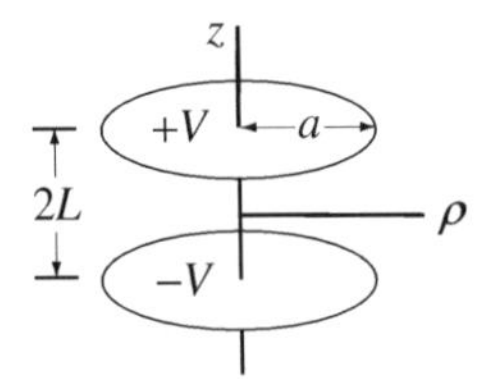

A paper published in 1983 proposed a solution for the potential for this situation of the form

$$\varphi(\rho, z) = \int_0^\infty dk\, A(k) f(k, z) J_0(k\rho),$$

where J_0 is the zero-order Bessel function and

$$A(k) = \frac{2V}{1 - e^{-2kL}} \frac{\sin(ka)}{\pi k}.$$

(a) Find the function $f(k, z)$ so the proposed solution satisfies the boundary conditions on the surfaces of the plates. You may make use of the integral

$$\int_0^\infty dk \frac{\sin(ka)}{k} J_0(k\rho) = \begin{cases} \pi/2 & 0 \le \rho \le a \\ \sin^{-1}(a/\rho) & \rho \ge a. \end{cases}$$

(b) Show that the proposed solution nevertheless fails to solve the problem because the electric field it predicts is not a continuous function of z when $\rho > a$.

7.22 A Dielectric Wedge in Polar Coordinates Two wedge-shaped dielectrics meet along the ray $\phi = 0$. The opposite edge of each wedge is held at a fixed potential by a metal plate. The system is invariant to translations perpendicular to the diagram.

(a) Explain why the potential $\varphi(\rho, \phi)$ between the plates does not depend on the polar coordinate ρ.
(b) Find the potential everywhere between the plates.

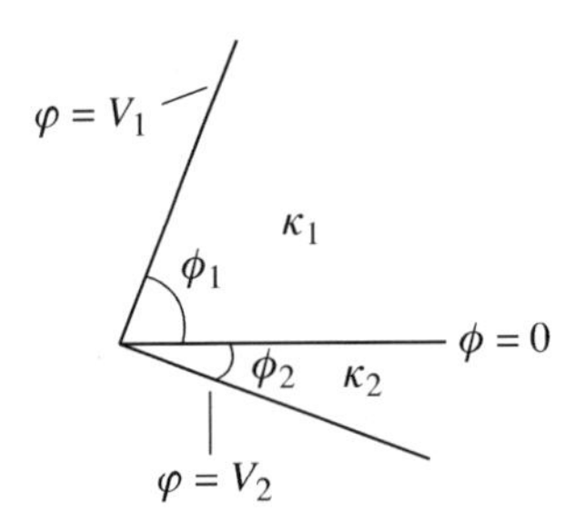

7.23 Contact Potential The $x > 0$ half of a conducting plane at $z = 0$ is held at zero potential. The $x < 0$ half of the plane is held at potential V. A tiny gap at $x = 0$ prevents electrical contact between the two halves.

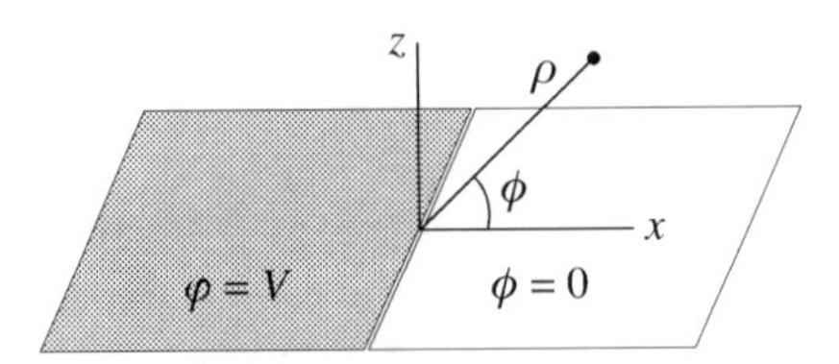

(a) Use a change-of-scale argument to conclude that the $z > 0$ potential $\varphi(\rho, \phi)$ in plane polar coordinates cannot depend on the radial variable ρ.
(b) Find the electrostatic potential in the $z > 0$ half-space.
(c) Make a semi-quantitative sketch of the electric field lines and use words to describe the most important features.

7.24 A Complex Potential Give a physical realization of the electrostatic boundary value problem whose solution is provided by the complex potential

$$f(w) = i\frac{V_1 + V_2}{2} + \frac{V_1 - V_2}{2} \ln\left[\frac{R + iw}{R - iw}\right].$$

7.25 A Cylinder in a Uniform Field by Conformal Mapping An infinitely long conducting cylinder (radius a) oriented along the z-axis is exposed to a uniform electric field $E_0\hat{\mathbf{y}}$.

(a) Consider the conformal map $g(w) = w + a^2/w$, where $g = u + iv$ and $w = x + iy$. Show that the circle $|w| = a$ and the parts of the x-axis that lie outside the circle map onto the entire u-axis.
(b) Let the potential on the cylinder be zero. What is the potential on the x-axis? Use this potential and the mapping in part (a) to solve the corresponding electrostatic problem in the g-plane. Find a complex potential $f(u, v)$ which satisfies the boundary conditions.
(c) Map the complex potential from part (b) back into the w-plane. Find the physical electrostatic potential $\varphi(x, y)$ and the electric field $\mathbf{E}(x, y)$. Sketch the electric field and the equipotentials.

8 Poisson's Equation

It may not be amiss to give a general idea of the method that has enabled us to arrive at results, remarkable in their simplicity and generality, which it would be very hard if not impossible to demonstrate in the ordinary way.

George Green (1828)

8.1 Introduction

Poisson's equation for the electrostatic potential in vacuum $\varphi(\mathbf{r})$ is

$$\nabla^2\varphi(\mathbf{r}) = -\rho(\mathbf{r})/\epsilon_0. \tag{8.1}$$

When $\rho(\mathbf{r})$ is completely specified, $\varphi(\mathbf{r})$ can be calculated by direct integration or by evaluating Coulomb's integral (7.1). These options are not available when (i) polarizable matter is present and (ii) at least some portion of $\rho(\mathbf{r})$ consists of true volume charge. This state of affairs obliges us to solve (8.1) as a *boundary value problem* in potential theory. Most often, the volume part of $\rho(\mathbf{r})$ is specified and the task at hand is to find the field produced by the charge induced on the surface of nearby conductors and/or isotropic linear dielectrics.

The most effective solution strategies for (8.1) exploit the principle of superposition. We construct $\varphi(\mathbf{r})$ as the sum of a particular solution of Poisson's equation and a general solution of Laplace's equation. The latter is chosen so the total potential satisfies the specified boundary conditions. The method of images exploits exactly the same idea for the case of point and line charges in the presence of conductors and/or dielectrics. These special solutions are valuable because the solutions to (8.1) for more complex volume distributions are constructed by superposing the potentials from point or line sources with suitable weights. The Green function method brings this point of view to its greatest level of generality and sophistication.

8.2 The Key Idea: Superposition

Superposition is the key idea used to solve Poisson's equation in electrostatics. Consider the prototype problem of a point charge q fixed at $\mathbf{r}_0$ in a volume V with specified values for the potential $\varphi(\mathbf{r})$ on the boundary surface of V. The unique solution to this problem can always be written in the form

$$\varphi(\mathbf{r}) = \frac{q}{4\pi\epsilon_0|\mathbf{r}-\mathbf{r}_0|} + \text{a solution to Laplace's equation in } V. \tag{8.2}$$

The first term in (8.2) is the potential of a point charge in infinite space. It is a *particular* solution of the Poisson equation (8.1) with $\rho(\mathbf{r}) = q\delta(\mathbf{r}-\mathbf{r}_0)$. By superposition, the potential (8.2) is a solution of the same Poisson equation because any solution of Laplace's equation contributes zero to the right

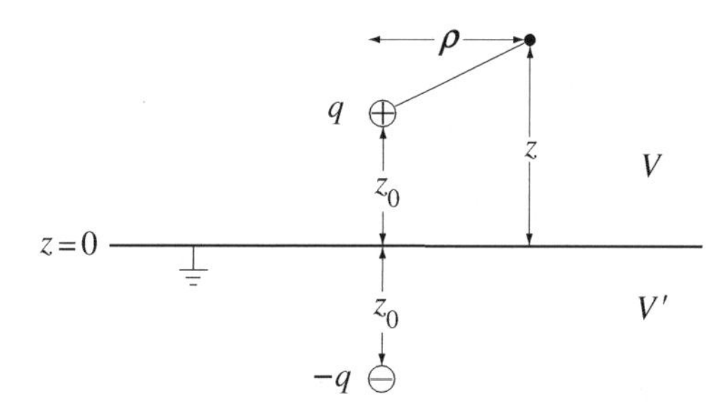

Figure 8.1: A point charge q at a distance z_0 above an infinite, grounded, conducting plane at $z = 0$. The potential in V ($z \geq 0$) is unchanged if the conductor is replaced by an "image" charge $-q$ at a distance z_0 below the $z = 0$ plane.

side of Poisson's equation. The role of the second term in (8.2) is to guarantee that the total potential satisfies the Dirichlet conditions imposed on the boundary of the volume V. It is always possible to do this because, by completeness (Section 7.4), the necessary function can be represented by the most general separated-variable solution of Laplace's equation. Put another way, the point charge q induces charge on the boundary of V and the field produced by this charge satisfies Laplace's equation in V. We will see in Section 8.4 how this scenario generalizes for the case of an arbitrary but specified $\rho(\mathbf{r})$.

The science of solving Poisson-type boundary value problems in electrostatics is to find at least one representation where (8.2) satisfies the boundary conditions. By uniqueness (Section 7.3), all other representations are (numerically) equivalent. The art to solving these problems is to find a form for (8.2) that is as simple as possible. The method described in the next section is unsurpassed in this regard.

8.3 The Method of Images

The *method of images* solves a small but important class of Poisson equation problems where a planar, spherical, or cylindrical boundary separates space into a volume V and a complementary volume V'. If a point (or line) charge is present in V, the method delivers the electrostatic potential in V by introducing fictitious "image" charges into V'. The boundary is usually assumed to be grounded and the volumes may be vacuum or filled with a simple dielectric.

8.3.1 A Point Charge and a Conducting Plane

Figure 8.1 shows a point charge q which lies a distance z_0 above an infinite, grounded, conducting plane at $z = 0$. We choose V as the half-space $z \geq 0$ and V' as the half-space $z < 0$. The method exploits the fact that the plane $z = 0$ remains an equipotential at $\varphi = 0$ if we replace the conductor by a fictitious "image" point charge $-q$ at a distance $2z_0$ below q.

To be consistent with (8.2), it is crucial that the potential due to the image charge in V' satisfies Laplace's equation in V. This guarantees that the potential produced by q and its image satisfies the same Poisson equation in V. Therefore, by uniqueness, the potential $\varphi(\mathbf{r} \in V)$ is exactly the same for the two problems. The method tells us nothing about the potential in V'.

Directly from Figure 8.1, the potential for $z \geq 0$ in cylindrical coordinates is

$$\varphi(\rho, z \geq 0) = \frac{q}{4\pi\epsilon_0}\left[\frac{1}{\sqrt{\rho^2 + (z - z_0)^2}} - \frac{1}{\sqrt{\rho^2 + (z + z_0)^2}}\right]. \tag{8.3}$$

The first term on the right side of (8.3) is a particular solution of Poisson's equation for $z \geq 0$. The second (image) term is a solution of Laplace's equation for $z \geq 0$ which ensures that the total potential satisfies the boundary condition $\varphi(z=0)=0$. This shows that (8.3) has exactly the structure of the solution envisioned in (8.2). The associated electric field in the solution volume V is the superposition of two inverse-square Coulomb fields, one from the real charge and one from the image charge:

$$\mathbf{E}(\rho, z \geq 0) = \frac{q}{4\pi\epsilon_0}\left[\frac{\rho\hat{\boldsymbol{\rho}} + (z-z_0)\hat{\mathbf{z}}}{[\rho^2+(z-z_0)^2]^{3/2}} - \frac{\rho\hat{\boldsymbol{\rho}} + (z+z_0)\hat{\mathbf{z}}}{[\rho^2+(z+z_0)^2]^{3/2}}\right]. \tag{8.4}$$

By construction, the field line pattern for (8.4) is identical to the pattern on the positive charge side of the planar boundary on the right side of Figure 3.17.

The true source of the second term in (8.3)—the "image potential"—is the charge induced on the conductor surface by the point charge q. Using (8.4), that charge is distributed according to

$$\sigma(\rho) = \epsilon_0\hat{\mathbf{z}}\cdot\mathbf{E}(\rho, z=0) = -\frac{qz_0}{2\pi(\rho^2+z_0^2)^{3/2}}. \tag{8.5}$$

The total amount of charge drawn up from ground is

$$q_{\text{ind}} = \int\limits_{z=0} dS\,\sigma = -q\int_0^{2\pi}\frac{d\phi}{2\pi}\int_0^{\infty}\frac{s\,ds}{(s^2+1)^{3/2}} = -q. \tag{8.6}$$

The fact that $q_{\text{ind}} = -q$ means that every electric field line that leaves q terminates somewhere on the infinite conducting plane. No electric flux escapes to infinity.

The force exerted on the plane by the point charge is

$$\mathbf{F}(z_0) = \frac{\hat{\mathbf{z}}}{2\epsilon_0}\int\limits_S dS\,\sigma^2 = \hat{\mathbf{z}}\frac{q^2}{4\pi\epsilon_0 z_0^2}\int_0^{\infty}\frac{s\,ds}{(s^2+1)^3} = \hat{\mathbf{z}}\frac{q^2}{4\pi\epsilon_0}\frac{1}{(2z_0)^2}. \tag{8.7}$$

The equal and opposite force $\mathbf{F}_q$ exerted on the point charge by the plane is precisely the Coulomb force that the fictitious image charge exerts on the real charge. To check this, we subtract the field $\mathbf{E}_{\text{self}}$ created by q itself from (8.4) and compute

$$\mathbf{F}_q(z_0) = q[\mathbf{E}-\mathbf{E}_{\text{self}}](\rho, z_0) = -\hat{\mathbf{z}}\frac{q^2}{4\pi\epsilon_0}\frac{1}{(2z_0)^2}. \tag{8.8}$$

This wonderful and quite general result is true because the image charge produces the physically correct electric field at the position of q.

A more subtle question is the interaction potential energy $V_E(z)$ between the point charge q and the grounded plane. By definition, $\mathbf{F}_q = -\nabla V_E$ where the gradient acts on the coordinates of q. Therefore,

$$V_E(z_0) = -\int_{\infty}^{z_0} d\hat{\mathbf{z}}\cdot\mathbf{F}_q(z) = -\frac{q^2}{16\pi\epsilon_0}\frac{1}{z_0}. \tag{8.9}$$

$V_E(z_0)$ may be compared with the interaction potential energy $\tilde{V}_E(z_0)$ between q and the image point charge $-q$. The electrostatic potential produced by the latter is the second term in (8.3), so

$$\tilde{V}_E(z_0) = q\times\left[-\frac{q}{4\pi\epsilon_0}\frac{1}{\sqrt{\rho^2+(z+z_0)^2}}\right]_{(0,z_0)} = -\frac{q^2}{8\pi\epsilon_0}\frac{1}{z_0}. \tag{8.10}$$

The energies (8.9) and (8.10) differ because the interaction potential energy is part of the total energy of assembly (Section 3.6.2). In the presence of q, no work is required to transport charge from infinity to the grounded plane (because the charges move on an equipotential) whereas work must be done to bring the image charge from infinity to its final position.

Image-Potential States

Direct evidence for the existence of the image force comes from measured binding energies of electrons trapped just outside flat metal surfaces. If (8.8) is correct, the binding energies should be the eigenvalues of a one-dimensional Schrödinger equation with the potential energy function (8.9):

$$V(z) = -\frac{1}{4\pi\epsilon_0}\frac{e^2}{4z}.$$

If we interpret $V(z)$ as the Coulomb energy between an electron with charge $e/2$ and a "proton" with charge $e/2$, the predicted eigenvalues are $E(n) = \epsilon(n)/16$, where $\epsilon(n)$ is Bohr's formula for the energy levels of a hydrogen atom:

$$\epsilon(n) = -(me^4/2\hbar^2)/n^2 = -(13.6\,\text{eV})/n^2.$$

This prediction is borne out quantitatively by photoelectric effect measurements of electrons bound to noble metal surfaces.[a]

8.3.2 Image Theory from Potential Theory

Not everyone would think to solve the boundary value problem in Figure 8.1 by superposing the potential of a true charge with the potential of a fictitious charge. Therefore, it is reassuring to learn that the image solution (8.3) can be derived more or less directly using (8.2) and separation of variables. We are looking to supplement the first term in (8.2) with a solution of Laplace's equation that makes the total potential for $z \geq 0$ vanish at $z = 0$. A key observation is that the charge induced on the flat surface (which produces the Laplace potential) has cylindrical symmetry around the z-axis. Looking back at this class of solutions of Laplace's equation (Section 7.8) and choosing the one that goes to zero at $z \to \infty$, we conclude that

$$\varphi(\rho, z \geq 0) = \frac{q}{4\pi\epsilon_0}\frac{1}{\sqrt{\rho^2 + (z - z_0)^2}} + \int_0^\infty dk\, A(k) J_0(k\rho) e^{-kz}. \tag{8.11}$$

The expansion coefficients $A(k)$ in (8.11) are chosen to make $\varphi(\rho, z = 0) = 0$. This would be difficult to do if not for the Bessel function identity[1]

$$\frac{1}{\sqrt{\rho^2 + z^2}} = \int_0^\infty dk\, J_0(k\rho) e^{-k|z|}. \tag{8.12}$$

Using (8.12) to represent the first term in (8.11), we see by inspection that $\varphi(\rho, z = 0) = 0$ if we choose $A(k) = -q\exp(-kz_0)/4\pi\epsilon_0$. This means that the potential we seek is

$$\varphi(\rho, z \geq 0) = \frac{q}{4\pi\epsilon_0}\frac{1}{\sqrt{\rho^2 + (z - z_0)^2}} + \frac{-q}{4\pi\epsilon_0}\int_0^\infty dk\, J_0(k\rho) e^{-k(z+z_0)}. \tag{8.13}$$

[a] See Fauster (1994) in Sources, References, and Additional Reading.

[1] We know that $(\rho^2 + z^2)^{-1/2} = \int_0^\infty dk B(k) J_0(k\rho) e^{-k|z|}$ because both sides are bounded and cylindrically symmetric solutions of Laplace's equation away from the origin. Now let $\rho = 0$. This gives $B(k) = 1$ because $J_0(0) = 1$ and $\int_0^\infty dk e^{-k|z|} = 1/|z|$.

Using (8.12) again to rewrite the second term in (8.13) shows that the latter is exactly the image solution (8.3).

We leave it as an exercise for the reader to confirm that the method of images solves the two-dimensional Poisson problem that results when the point charge in Figure 8.1 is replaced by an infinite line of uniform charge density oriented perpendicular to the plane of the figure.

Example 8.1 A point dipole $\mathbf{p}$ sits a distance d from the flat surface of a grounded metal sample. Calculate the work required to rotate the dipole from a perpendicular orientation (pointed directly at the metal) to a parallel orientation.

Solution: Figure 8.2 shows the dipole $\mathbf{p}$ (not to scale) misaligned by an angle α from the perpendicular orientation. The plus and minus charges which constitute the dipole form images of themselves of opposite sign a distance d behind the surface of the conductor. Therefore, if the real dipole is $\mathbf{p} = -p\cos\alpha\hat{\mathbf{x}} + p\sin\alpha\hat{\mathbf{y}}$ in the coordinate system shown, the image dipole is

$$\mathbf{p}' = -p\cos\alpha\hat{\mathbf{x}} - p\sin\alpha\hat{\mathbf{y}}.$$

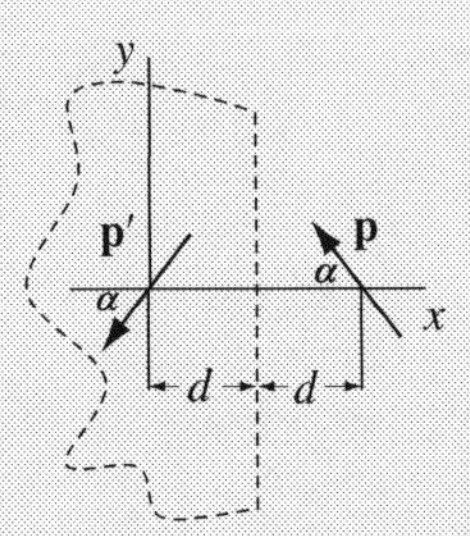

Figure 8.2: A point dipole in vacuum and its image inside a grounded metal sample.

The electric field produced by the image at the position $\mathbf{r} = 2d\hat{\mathbf{x}}$ of the real dipole is

$$\mathbf{E}' = \frac{3\mathbf{r}(\mathbf{p}' \cdot \mathbf{r}) - \mathbf{p}'r^2}{4\pi\epsilon_0 r^5} = \frac{-2p\cos\alpha\hat{\mathbf{x}} + p\sin\alpha\hat{\mathbf{y}}}{32\pi\epsilon_0 d^3}.$$

The torque exerted by the image dipole on the real dipole is

$$\mathbf{N} = \mathbf{p} \times \mathbf{E}' = \frac{p^2}{32\pi\epsilon_0 d^3}\sin\alpha\cos\alpha\hat{\mathbf{z}}.$$

Therefore, the work required to rotate the dipole from perpendicular to parallel orientation is

$$W = \int_0^{\pi/2} d\alpha \cdot \mathbf{N} = \frac{p^2}{32\pi\epsilon_0 d^3}\int_0^{\pi/2} d\alpha \sin\alpha\cos\alpha = \frac{p^2}{64\pi\epsilon_0 d^3}.$$

8.3.3 Dielectric Boundaries

The planar interface in Figure 8.3 separates space into two semi-infinite simple dielectrics. The method of images can be used to find the total electric field $\mathbf{E}^*$ at the position of the point charge q embedded in the medium with dielectric constant κ_L. We may then apply (6.130) to find the force $\mathbf{F} = q\mathbf{E}^*$ which the dielectric exerts on q.

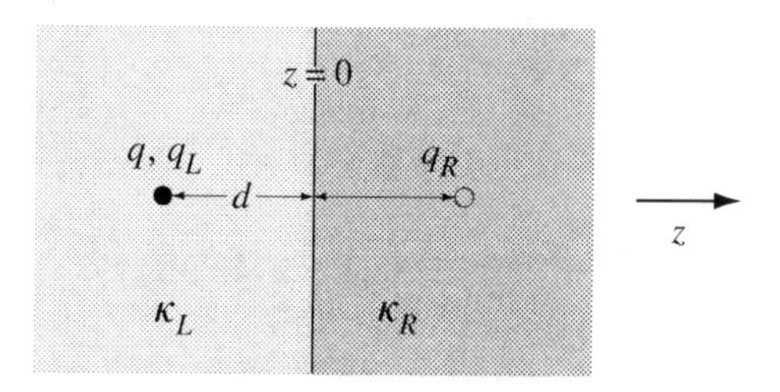

Figure 8.3: A point charge q lies a distance d from the $z = 0$ interface between two semi-infinite linear dielectrics.

We learned in Section 6.5.5 that the electrostatic potential for this type of problem must satisfy two matching conditions at the dielectric interface:

$$\varphi_L(\mathbf{r}_S) = \varphi_R(\mathbf{r}_S) \tag{8.14}$$

and

$$\kappa_L \frac{\partial \varphi_L}{\partial n}\bigg|_S = \kappa_R \frac{\partial \varphi_R}{\partial n}\bigg|_S . \tag{8.15}$$

To find the potential $\varphi_L(\mathbf{r})$ in the $z < 0$ half-space, we fill *all* space with dielectric constant κ_L and supplement the potential of the charge q at $z = -d$ with the potential of an image charge q_R at $z = d$. To find the potential $\varphi_R(\mathbf{r})$ in the $z > 0$ half-space, we fill all space with dielectric constant κ_R and place an image charge q_L at $z = -d$.[2] Section 6.5.2 taught us that the potential of an embedded point charge is screened by the dielectric constant of its host medium. Therefore, the proposed image system generates the potentials

$$\varphi_L(\rho, z) = \frac{1}{4\pi\epsilon_L}\left[\frac{q}{\sqrt{\rho^2 + (z+d)^2}} + \frac{q_R}{\sqrt{\rho^2 + (z-d)^2}}\right] \tag{8.16}$$

and

$$\varphi_R(\rho, z) = \frac{1}{4\pi\epsilon_R}\frac{q_L}{\sqrt{\rho^2 + (z+d)^2}}. \tag{8.17}$$

A few lines of algebra shows that (8.16) and (8.17) do indeed satisfy (8.14) and (8.15) if the image charges take the values

$$q_R = \frac{\kappa_L - \kappa_R}{\kappa_L + \kappa_R} q \qquad \text{and} \qquad q_L = \frac{2\kappa_R}{\kappa_L + \kappa_R} q. \tag{8.18}$$

The force on q comes entirely from the field produced by q_R evaluated at the position of q. This is the electric field associated with the second term in (8.16) evaluated at $(0, -d)$. Therefore,

$$\mathbf{F} = -\frac{1}{4\pi\epsilon_L}\frac{q q_R}{4d^2}\hat{\mathbf{z}}. \tag{8.19}$$

This agrees with the result obtained in Example 6.7 by another method. It is an instructive exercise to use the stress tensor formalism (Section 6.8.5) to check that the force exerted on the dielectric interface by the point charge is exactly the negative of (8.19).

Figure 8.4 shows the lines of $\mathbf{E}(\mathbf{r})$ for this problem. In the κ_R matter, the straight lines reflect the point charge electric field from q_L alone. In the κ_L material, the curved lines reflect the superposed Coulomb fields from q and q_R. Apart from the "refraction" of the field lines at the dielectric boundary

[2] The placement of q_R and q_L at $z = \pm d$ is *not* obvious for this problem.

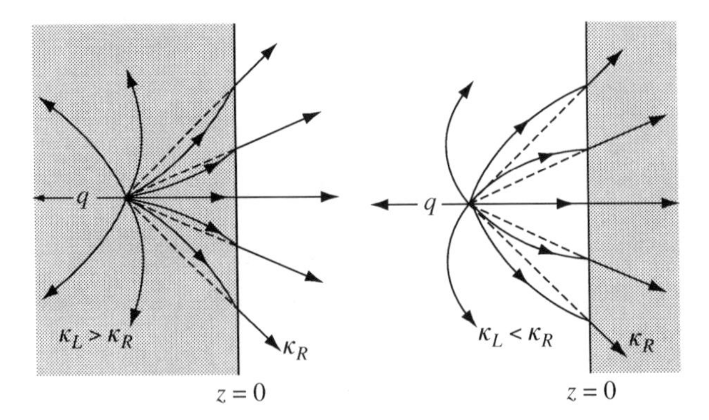

Figure 8.4: Lines of $\mathbf{E}(\mathbf{r})$ for a point charge near a flat dielectric interface for $\kappa_L > \kappa_R$ (left panel) and $\kappa_L < \kappa_R$ (right panel). Figure from Durand (1966).

(see Application 6.2), the most interesting feature is the change of sign of the curvature of the field lines near the point charge for $\kappa_L > \kappa_R$ compared to $\kappa_L < \kappa_R$. This can be understood because (8.18) shows that q_R is positive in the first case and negative in the second case. Therefore, a glance back at Figure 3.17 shows that the field in medium κ_L resembles the $z < 0$ part of the field between two point charges of opposite sign when $\kappa_L < \kappa_R$ and the $z < 0$ part of the field between two point charges of like sign when $\kappa_L > \kappa_R$.

When $\kappa_R \to \infty$, the image charge $q_R = -q$ and the field in medium κ_L reduces to the $z < 0$ part of the field of an electric dipole. The field lines in this case are strictly perpendicular to the dielectric interface at $z = 0$. Since this is the behavior of field lines at the surface of a perfect conductor, we confirm our earlier deduction (Section 6.6.4) that a linear dielectric without embedded free charge behaves like a perfect conductor when its permittivity goes to infinity.

8.3.4 Multiple Images

Poisson problems with multiple planar boundaries sometimes can be solved using multiple images. The only restriction is that the solution volume remain image-free. Consider the potential produced by a point charge q which lies on the z-axis midway between two infinite, parallel, grounded, conducting plates (Figure 8.5). Our previous experience tells us that an image charge $-q$ placed on the z-axis at $z = 3d/2$ makes the plane $z = d$ a zero equipotential. The plane $z = 0$ will be a zero equipotential also if we add an image $-q$ at $z = -d/2$ and another charge q at $z = -3d/2$. Unfortunately, these two images destroy the zero equipotential at $z = d$. Additional images at $z = 5d/2$ and $z = 7d/2$ repair the damage at $z = d$, but at the cost of spoiling the equipotential at $z = 0$. One quickly realizes that an infinite sequence of images is needed: positive charges at $z = 2dm + d/2$ and negative charges at $z = 2dm - d/2$ where $m = 0, \pm 1, \pm 2, \ldots$

As Figure 8.5 shows, the final image system is an infinite set of collinear and equally spaced point charges with equal magnitude and alternating sign. Any intuition we may have about the potential produced by this set of charges may be applied immediately to the potential in the volume $0 \leq z \leq d$. For example, the multiple point charge problem has cylindrical symmetry around the z-axis where the charges lay. This means that the potential satisfies Laplace's equation everywhere away from the z-axis:

$$\nabla^2 \varphi = \frac{\partial^2 \varphi}{\partial \rho^2} + \frac{1}{\rho}\frac{\partial \varphi}{\partial \rho} + \frac{\partial^2 \varphi}{\partial z^2} \approx \frac{\partial^2 \varphi}{\partial \rho^2} + \frac{\partial^2 \varphi}{\partial z^2} = 0. \tag{8.20}$$

The approximation indicated just above applies when $\rho \gg d$, the latter being the only length scale in the problem. The final equation on the far right side of (8.20) is a two-dimensional, Cartesian-type

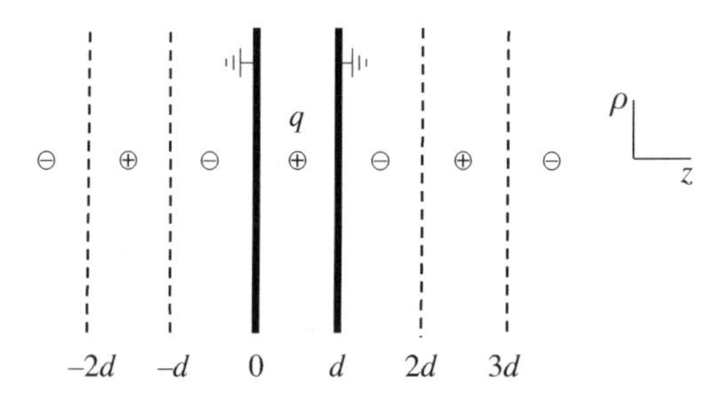

Figure 8.5: A positive point charge q on the z-axis lies midway between two grounded conducting plates at $z = 0$ and $z = d$. The other charges shown are images.

Laplace equation to which the method of separation of variables applies. The general solution is (7.35) with x replaced by ρ and y replaced by z. But the line of point charges has periodicity $L = 2d$ in the z-direction. Therefore, when $\rho \gg d$, the general solution will be dominated by the $\alpha = 2\pi/L$ term in (7.35). Dimensional analysis determines the pre-factor (up to a factor of order one), so the expected asymptotic behavior of the potential is[3]

$$\varphi(\rho, z) \sim \frac{q}{\epsilon_0 d} \exp(-\pi\rho/d) \sin(\pi z/d) \qquad \rho \gg d, \quad 0 \le z \le d. \tag{8.21}$$

Applying this result to the volume between the grounded plates in Figure 8.5 shows that the charge induced on the plates screens the $1/r$ potential of the point charge very effectively.

Application 8.1 The Electrostatics of an Ion Channel

Figure 8.6 shows protein tubules embedded in the wall of a cell. The hollow channel of each tubule is a pathway for ions to travel back and forth between the cell interior and the cell exterior. The cell exterior, the channel, and the cell interior are mostly water with a dielectric constant $\kappa_W \approx 80$. The tubule itself has a dielectric constant $\kappa_L \approx 2$. In this Application, we estimate the energy of a positive ion in such a channel and show that an energy barrier must be overcome for an ion to pass entirely through the channel.

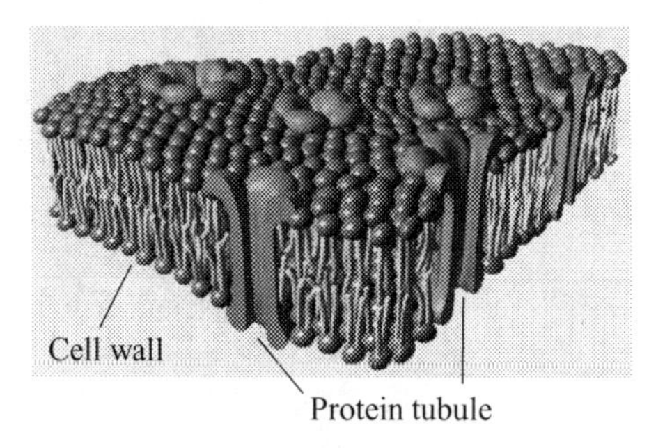

Figure 8.6: Protein tubules embedded in wall of a cell. Figure from Gonzales and Carrasco (2003).

Figure 8.7 shows the electric field line pattern when the ion sits at the midpoint of a model ion channel with length L and radius $a \ll L$. The field lines bend parallel to the channel walls to avoid

[3] Example 8.4 confirms this expectation.

the low-permittivity tubule. This is consistent with the discussion in Application 6.2 and with the left panel of Figure 8.4 in the limit $\kappa_L \gg \kappa_R$ when $q_R \approx q$ in (8.18). It also means that, except very

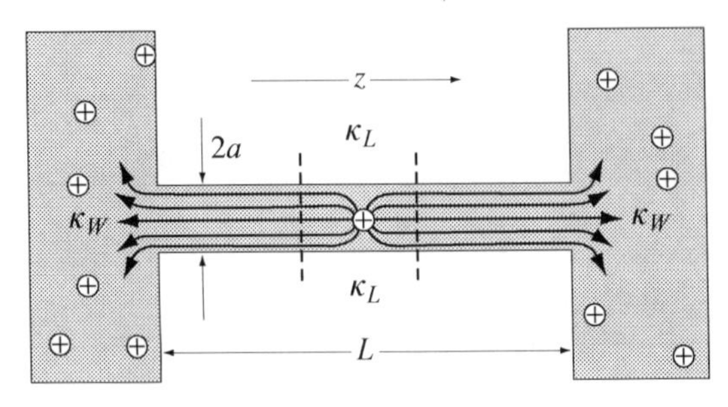

Figure 8.7: A positive ion q sits at the midpoint of a channel which connects two reservoirs of ion-filled water. The protein walls of the channel have a dielectric constant κ_L that is much less than the dielectric constant κ_W of water.

near the point charge, the field strength in Figure 8.7 may be calculated using $\epsilon_0 \nabla \cdot \mathbf{E} = q/\kappa_W$ and a cylindrical Gaussian surface formed from the channel walls and the two vertical caps indicated by the vertical dashed lines in Figure 8.7. This gives the constant field

$$E_0 = \frac{q}{2\pi \epsilon_0 a^2 \kappa_W}. \tag{8.22}$$

E_0 is exactly the field produced by an infinite plane at the center of the channel with uniform surface charge density $\sigma = q/\pi a^2$.

The electrostatic energy (6.94) is a function of the position z of the ion, where $z = -L/2$ and $z = L/2$ label the left and right ends of the channel, respectively:

$$U_E(z) = \frac{1}{2} \int d^3r\, \mathbf{E} \cdot \mathbf{D}. \tag{8.23}$$

To estimate (8.23) for our channel of volume $V = \pi a^2 L$, we replace the highly conducting reservoirs by grounded conducting planes at the entrance and exit of the channel. Based on (8.22), we similarly replace q by a plane with charge density σ. Figure 8.8 shows this equivalent electrostatic problem when q is off-center. The presence of the grounded planes implies that $E_L \neq E_R$, where E_L is the constant field to the left of σ and E_R is the constant field to the right of σ. The voltage drop $\int d\boldsymbol{\ell} \cdot \mathbf{E}$ across the channel is zero and the total charge drawn up from ground onto the two grounded planes is $-\sigma$. These two conditions can be written in the form

$$\begin{aligned} E_L(L/2 + z) &= E_R(L/2 - z) \\ E_L + E_R &= \frac{q}{\epsilon_0 \pi a^2 \kappa_W}. \end{aligned} \tag{8.24}$$

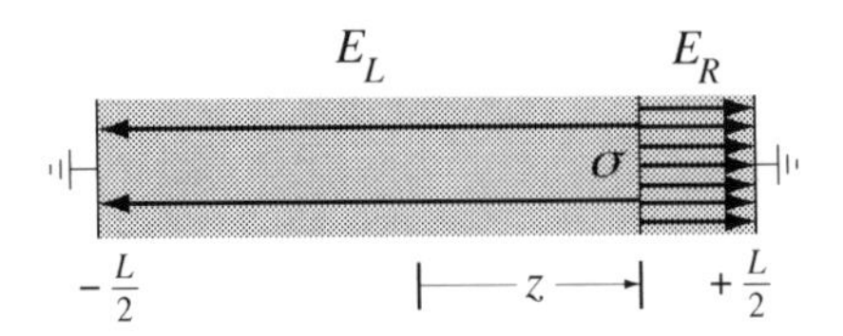

Figure 8.8: A model for the electrostatics of Figure 8.7 replaces the point charge at z by a plane with charge density $\sigma = q/\pi a^2$ and replaces the reservoirs by grounded conducting planes at the ends of the channel. The midpoint of the channel is $z = 0$.

It is straightforward to solve (8.24) for E_L and E_R and evaluate (8.23) using the fact that E_L occupies a volume $V(\frac{1}{2}+z/L)$ and E_R occupies a volume $V(\frac{1}{2}-z/L)$. The final result shows that the ion must surmount a quadratic potential energy barrier to pass through the channel:

$$U_E(z) = 2\pi\epsilon_0 E_0^2 V\left(\frac{1}{4}-\frac{z^2}{L^2}\right). \tag{8.25}$$

The energy barrier at $z=0$ predicted by (8.25) is much larger than kT at ambient temperature. Entropy effects and a better treatment of the interaction of the ion with the molecules of the tubule must be taken into account to lower the barrier sufficiently far to make ion passage through a cell wall plausible *in vivo*.[4]

8.3.5 A Point Charge outside a Conducting Sphere

The method of images can be used to find the electrostatic potential outside (inside) a grounded, spherical, conducting shell when a point charge is placed outside (inside) the shell. Figure 8.9 shows a point charge q outside such a shell and, by symmetry, the presumptive image charge q' must lie somewhere on the z-axis defined by q and the center of the sphere. Therefore, if θ is the angle defined in Figure 8.9, the equipotential requirement on the surface of the sphere is

$$\varphi(R,\theta) = \frac{1}{4\pi\epsilon_0}\left\{\frac{q}{\sqrt{s^2+R^2-2sR\cos\theta}}+\frac{q'}{\sqrt{b^2+R^2-2bR\cos\theta}}\right\} = 0. \tag{8.26}$$

Two special cases of (8.26) are

$$q^2(b^2+R^2) = q'^2(s^2+R^2) \qquad \theta = \pi/2, \tag{8.27}$$

and

$$q^2(b-R)^2 = q'^2(s-R)^2 \qquad \theta = 0. \tag{8.28}$$

Subtracting (8.28) from (8.27) gives $q^2 b = q'^2 s$. With this information, both (8.27) and (8.28) become

$$sb^2-(s^2+R^2)b+sR^2 = 0. \tag{8.29}$$

The physically sensible root of this quadratic equation gives the position and magnitude of the image charge as

$$b = \frac{R^2}{s} \qquad \text{and} \qquad q' = -\frac{R}{s}q = -\frac{b}{R}q. \tag{8.30}$$

The image charge lies inside the sphere because the left equation in (8.30) defines R as the geometric mean of s and b.

Substituting (8.30) into (8.26) correctly produces $\varphi(R,\theta)=0$ for *all* values of θ. By uniqueness, we conclude that a point charge q located at any point $\mathbf{s}$ outside a grounded, conducting sphere of radius R produces the following electrostatic potential at any other point $\mathbf{r}$ outside the sphere:

$$\varphi(\mathbf{r}) = \frac{q}{4\pi\epsilon_0}\left\{\frac{1}{|\mathbf{r}-\mathbf{s}|}-\frac{R/s}{|\mathbf{r}-\mathbf{s}R^2/s^2|}\right\} \qquad s,r>R. \tag{8.31}$$

Gauss' law applied to the image system tells us that q' is the integrated charge drawn up to the conducting sphere from ground. Because $|q'|<q$, not all the electric field lines which begin at q terminate on the conducting sphere. The remainder go off to infinity. This "loss" of flux to infinity differs from the case of the conducting plane in Figure 8.1 because the surface area of a sphere is finite.

[4] See Kamenev *et al.* (2006) in Sources, References, and Additional Reading.

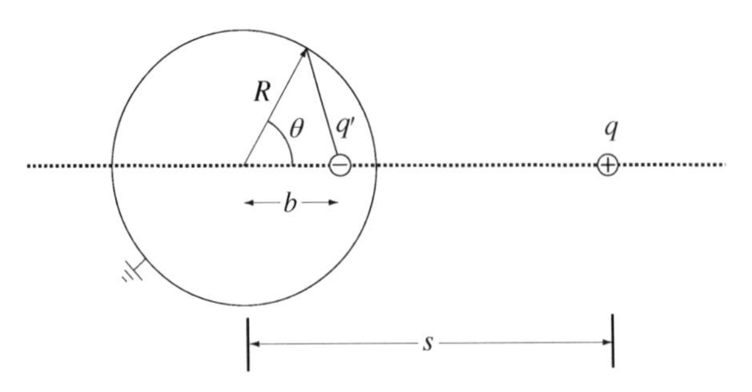

Figure 8.9: A point charge q outside a grounded, conducting sphere. An image system with $q' \neq -q$ makes $r = R$ an equipotential surface.

The force exerted by q on the grounded conducting sphere is $-q\mathbf{E}_S$, where $\mathbf{E}_S$ is the electric field produced by the sphere at the position of q. This can be computed from $\mathbf{F} = q\nabla\varphi(\mathbf{r})|_{\mathbf{r}=\mathbf{s}}$ or directly from Coulomb's inverse-square force law using the position and magnitude of the image charge determined above. In other words, the force exerted on the sphere is

$$\mathbf{F} = -\frac{qq'}{4\pi\epsilon_0}\frac{\hat{\mathbf{s}}}{(s-b)^2} = \frac{q^2}{4\pi\epsilon_0}\frac{R\mathbf{s}}{(s^2-R^2)^2}. \tag{8.32}$$

We note that this force is always attractive and that its asymptotic ($s \gg R$) behavior with distance is inverse-*cube* rather than the inverse-square behavior seen in (8.7).

Example 8.2 Let $\Phi(r, \theta, \phi)$ be any function that satisfies Laplace's equation *inside* a sphere of radius $r = R$. Show that $\Psi(r, \theta, \phi) = (R/r)\Phi(R^2/r, \theta, \phi)$ satisfies Laplace's equation *outside* the sphere. Choose Φ as the potential of a point charge q at $\mathbf{s}$ with $s > R$ and show that (8.31) is identical to

$$\varphi(r \geq R, \theta, \phi) = \Phi(r, \theta, \phi) - \Psi(r, \theta, \phi).$$

This example exploits the *method of inversion* to derive a solution of Laplace's equation in one domain from a known solution of Laplace's equation in a complementary domain.

Solution: Let the operator L^2 stand for all the angle-dependent pieces of the Laplacian operator in spherical coordinates. By the assumption of the problem,

$$\nabla^2\Phi = \frac{1}{r}\frac{\partial^2}{\partial r^2}(r\Phi) + \frac{L^2}{r^2}\Phi = 0 \qquad r < R.$$

Now, substitute the proposed solution Ψ into this equation, let $u = R^2/r$, and use the chain rule. The result is

$$\nabla^2\Psi(r, \theta, \phi) = \frac{R^5}{r^5}\left[\frac{1}{u}\frac{\partial^2}{\partial u^2}(u\Phi) - \frac{L^2}{u^2}\Phi\right] = \frac{R^5}{r^5}\nabla^2\Phi(u, \theta, \phi).$$

The rightmost term of this expression is zero when $u = R^2/r < R$. Following the chain of equalities to the left tells us that $\Psi(r, \theta, \phi)$ satisfies Laplace's equation when $r > R$. Now choose

$$\Phi(\mathbf{r}) = \frac{q}{4\pi\epsilon_0|\mathbf{r}-\mathbf{s}|}.$$

This function satisfies Poisson's equation *outside* the sphere and Laplace's equation *inside* the sphere. We have just shown that the associated inverse function $\Psi(\mathbf{r})$ satisfies Laplace's equation outside the sphere. Therefore, since $\Psi(R,\theta,\phi)=\Phi(R,\theta,\phi)$, the unique potential in $r\geq R$ that vanishes at $r=R$ is

$$\varphi(r\geq R,\theta,\phi)=\Phi(r,\theta,\phi)-\Psi(r,\theta,\phi).$$

It remains only to show that $-\Psi(\mathbf{r})$ is the image term in (8.31). To that end, let $s\hat{\mathbf{z}}$ be the position of q in Figure 8.9 so $\mathbf{r}=(r,\theta,\phi)$. The Legendre polynomial expansion (4.78) of the true point charge potential when $r>s$ is

$$\Phi(r,\theta,\phi)=\frac{1}{4\pi\epsilon_0}\frac{q}{r}\sum_{\ell=0}^{\infty}\left(\frac{s}{r}\right)^{\ell}P_\ell(\cos\theta).$$

Therefore, since $R^2/r<s$ when $r>s>R$,

$$-\frac{R}{r}\Phi(R^2/r,\theta,\phi)=\frac{1}{4\pi\epsilon_0}\frac{-qR/s}{r}\sum_{\ell=0}^{\infty}\left(\frac{R^2/s}{r}\right)^{\ell}P_\ell(\cos\theta).$$

This is the potential of a point charge $q'=-qR/s$ on the positive z-axis at a distance $R^2/s<R$ from the origin—in other words, the potential of the image charge in Figure 8.9.

8.3.6 Other Sphere Problems

It is straightforward to generalize the grounded-sphere problem to the case of a conducting sphere held at a fixed potential φ_0. Just add to (8.31) the potential from a second image with charge $q_0=4\pi\epsilon_0 R\varphi_0$ placed at the center of the sphere. This raises the potential of the sphere from zero to φ_0. The unique solution is

$$\varphi(\mathbf{r})=\frac{q}{4\pi\epsilon_0}\left\{\frac{1}{|\mathbf{r}-\mathbf{s}|}-\frac{R/s}{|\mathbf{r}-\mathbf{s}R^2/s^2|}\right\}+\frac{R\varphi_0}{r}\qquad s,r>R. \tag{8.33}$$

Nearly identical reasoning produces the potential of a point charge q in the presence of an isolated conducting sphere which carries a net charge Q. We use image charges at the center to replace the net charge of the grounded sphere by Q. This gives a potential outside the sphere of

$$\varphi(\mathbf{r})=\frac{q}{4\pi\epsilon_0}\left\{\frac{1}{|\mathbf{r}-\mathbf{s}|}-\frac{R/s}{|\mathbf{r}-\mathbf{s}R^2/s^2|}\right\}+\frac{1}{4\pi\epsilon_0}\frac{Q+qR/s}{r}\qquad s,r>R. \tag{8.34}$$

When $Q=0$, (8.34) is the potential of a point charge outside an isolated, neutral, spherical conductor. The force between these two objects is (8.32) plus the repulsive force between q and the image charge qR/s at the center of the sphere. A moment's reflection shows that the net force is always attractive. This raises the question: is the same conclusion true for a conductor with an arbitrary shape? When the distance s is very large, the force is surely attractive and shape-independent because the conductor may be replaced by a dipole that points away from q. Otherwise, we can identify at least one class of shapes where the force is repulsive for some distances. We prove this in a more general setting as follows.

Let $\rho(\mathbf{r})$ be a finite, but otherwise arbitrary, charge distribution. This distribution possesses a collection of equipotential surfaces, including a zero equipotential at infinity. Now consider an uncharged conductor whose shape, in part or in full, matches any one of the equipotential surfaces of $\rho(\mathbf{r})$. The electrostatic interaction energy V_E between the conductor and $\rho(\mathbf{r})$ is zero at infinity and (because $Q=0$) zero again when the conductor overlays the shape-matching equipotential surface. Moreover, V_E decreases from zero as the conductor moves in from infinity because (see above) $\mathbf{F}=-\nabla V_E$ is attractive when the conductor is far away from $\rho(\mathbf{r})$. Therefore, the force must become repulsive as the conductor approaches the special equipotential and V_E returns to zero. Figure 8.10 shows a point

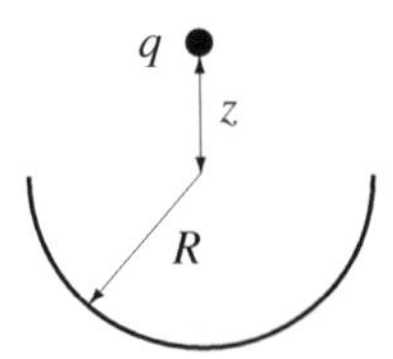

Figure 8.10: At $z = 0$, the isolated, conducting, hemispherical shell is coincident with one of the equipotentials of the point charge q. Figure from Levin and Johnson (2011).

charge on the symmetry axis of a neutral, conducting, hemispherical, shell. The interaction energy $V_E = 0$ when $z = 0$, and our argument shows that the force between the two objects is initially repulsive as z increases and then turns attractive. The reader is encouraged to interpret this result in terms of the angular distribution of charge induced on the surface of the shell.

Finally, the experience of Section 8.3 suggests that the method of images could be applied to find the potential produced by a point charge near a *dielectric* sphere. This turns out to be true, but the image is (surprisingly) not a simple point charge. It is a finite line segment with a continuous but non-uniform linear charge density.[5]

8.3.7 A Line Charge outside a Conducting Cylinder

Let the circle in Figure 8.9 be the cross section of an infinitely long conducting cylinder. We define a two-dimensional Poisson problem by re-interpreting the point q in Figure 8.9 as the cross section of an infinitely long line charge at $\mathbf{s} = s\hat{\mathbf{x}}$ with uniform charge per unit length λ. Up to a constant, the potential of this line charge is

$$\varphi_0(\mathbf{r}) = -\frac{\lambda}{2\pi\epsilon_0}\ln|\mathbf{r}-\mathbf{s}|. \tag{8.35}$$

We seek a solution for $\varphi(x, y)$ outside the cylinder as the sum of the potentials from the true line charge and from a parallel image line charge with uniform density λ' placed at the position of q'. Using (8.35), our ansatz for the potential is

$$\varphi(x, y) = -\frac{\lambda}{2\pi\epsilon_0}\ln\sqrt{(x-s)^2+y^2} - \frac{\lambda'}{2\pi\epsilon_0}\ln\sqrt{(x-b)^2+y^2} \tag{8.36}$$

or

$$\varphi(x, y) = -\frac{1}{4\pi\epsilon_0}\ln\frac{[(x-s)^2+y^2]^{+\lambda}}{[(x-b)^2+y^2]^{-\lambda'}}, \qquad x^2+y^2 > R^2. \tag{8.37}$$

Our challenge is to find the equipotential surfaces of (8.37) by finding conditions where the numerator and denominator of the logarithm have a constant ratio C. A moment's reflection shows that this is only possible if $\lambda' = -\lambda$. Writing out this condition and collecting terms gives

$$x^2 + 2\left[\frac{bC-s}{1-C}\right]x + y^2 = \frac{Cb^2-s^2}{1-C}. \tag{8.38}$$

The choice $C = s/b$ produces our solution because it reduces (8.38) to the equation of an origin-centered cylinder with radius $R = \sqrt{sb}$. In other words, the Poisson problem outside the cylinder is solved by an image line charge inside the cylinder with position and charge density

$$b = \frac{R^2}{s} \qquad \text{and} \qquad \lambda' = -\lambda. \tag{8.39}$$

[5] See P. Bussemer, "Comment on image theory for electrostatic and magnetostatic problems involving a material sphere", *American Journal of Physics* **62**, 657 (1994).

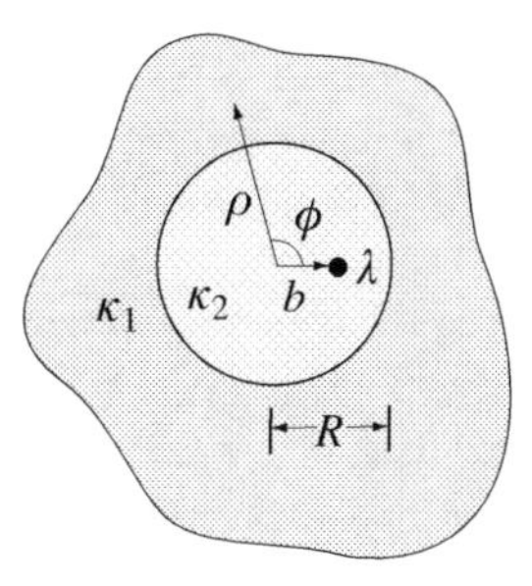

Figure 8.11: A line charge λ embedded in a dielectric cylinder κ_2. The cylinder is embedded in an infinite dielectric space κ_1.

Comparing these values to (8.30) shows that the point and line images sit at the same distance from the origin but that the source and image lines have the same strength. Unlike the point charge and the sphere, every electric field line that leaves the line charge terminates on the infinitely long cylinder.

It is worth noticing that the choices in (8.39) do *not* give zero for the constant value of the potential on the cylinder surface. To make it zero, we must add a constant to the right side of (8.36). This simple expedient is not possible for the sphere problem because the charge and potential of a conducting sphere cannot be specified independently. The difference arises because the point charge potential (8.2) goes to zero at infinity while the line charge potential (8.35) diverges at infinity.

8.3.8 A Dielectric Cylinder

Figure 8.11 shows an infinite line with charge density λ embedded at a point $\mathbf{b}$ inside a cylinder with radius R and dielectric constant κ_2. The cylinder is itself embedded in an infinite space with dielectric constant κ_1. Figure 8.12 shows image systems that determine the potential both inside and outside the cylinder. To find $\varphi_{\rm in}(\rho,\phi)$, we fill all space with dielectric constant $\kappa_2 = \epsilon_2/\epsilon_0$ and supplement the source line charge with a line with charge density λ' at the "inverse point" that lies a distance $s = R^2/b$ from the origin of the cylinder [cf. (8.39)].

To find $\varphi_{\rm out}(\rho,\phi)$, we fill all space with dielectric constant $\kappa_1 = \epsilon_1/\epsilon_0$ and place an image line with charge density λ'' at the position of the original line. The cylinder is net neutral, so it is necessary to add an image line somewhere inside the cylinder with compensating charge density $\lambda - \lambda''$. The only point that suggests itself is the origin. Accordingly, if $C_{\rm in}$ and $C_{\rm out}$ are constants, the potential created by these image systems is

$$\varphi_{\rm in} = -\frac{\lambda}{2\pi\epsilon_2}\ln\rho_2 - \frac{\lambda'}{2\pi\epsilon_2}\ln\rho_1 + C_{\rm in} \tag{8.40}$$

$$\varphi_{\rm out} = -\frac{\lambda''}{2\pi\epsilon_1}\ln\rho_2 - \frac{\lambda-\lambda''}{2\pi\epsilon_1}\ln\rho + C_{\rm out}. \tag{8.41}$$

The potentials (8.40) and (8.41) must be equal when ρ, ρ_1, and ρ_2 meet at any point on the cylinder boundary as shown in Figure 8.12. Our choice of the inverse point for λ' has the effect of making the smallest and largest of the three triangles in Figure 8.12(a) geometrically *similar*. This implies that $\rho_1/R = \rho_2/b$. Using this fact to eliminate ρ_1 from (8.40) we demand that the coefficients of $\ln\rho_2$ agree when we set $\varphi_{\rm in} = \varphi_{\rm out}$ on the boundary where $\rho = R$. This implies that

$$\kappa_1(\lambda + \lambda') = \kappa_2\lambda''. \tag{8.42}$$

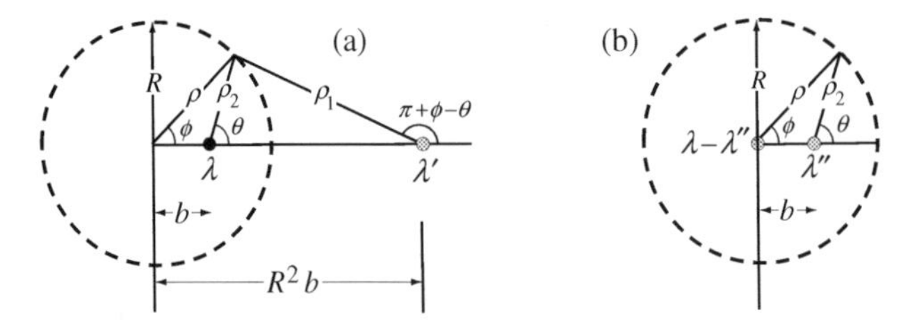

Figure 8.12: Line charges used to solve the Poisson problem defined by Figure 8.11 for (a) $\varphi_{\rm in}$: the source line λ inside the cylinder plus an image line λ' outside the cylinder; (b) $\varphi_{\rm out}$: an image line λ'' at the position of the source line and an image line $\lambda - \lambda''$ at the center of the cylinder.

It is algebraically awkward to enforce the scalar potential matching condition (8.15) derived from the continuity of the normal component of $\mathbf{D}(\mathbf{r})$. The calculation simplifies considerably if we use a trick and write the $\mathbf{D}$-field produced by a line charge using a "vector potential" rather than the scalar potential (8.35). Specifically, with K a constant, let

$$\psi(\phi) = \frac{\lambda}{2\pi}\phi + K, \tag{8.43}$$

so

$$\mathbf{D}(\rho) = \nabla \times (\psi \hat{\mathbf{z}}) = \frac{1}{\rho}\frac{\partial \psi}{\partial \phi}\hat{\boldsymbol{\rho}} = \frac{\lambda}{2\pi\rho}\boldsymbol{\rho}. \tag{8.44}$$

This representation is useful because the continuity of (8.44) implies that ψ is continuous at any boundary.

Using the angles defined in Figure 8.12, we superpose the ψ-potentials from the true and image line charges to write the total potentials inside and outside the cylinder boundary:

$$\psi_{\rm in} = \frac{\lambda}{2\pi}\theta + \frac{\lambda'}{2\pi}(\pi - \theta + \phi) + K_{\rm in} \tag{8.45}$$

$$\psi_{\rm out} = \frac{\lambda''}{2\pi}\theta + \frac{\lambda - \lambda''}{2\pi}\phi + K_{\rm out}. \tag{8.46}$$

Setting the coefficients of θ and ϕ equal in (8.45) and (8.46) gives

$$\lambda - \lambda'' = \lambda'. \tag{8.47}$$

Combining (8.47) with (8.42) fixes the image charge densities at

$$\lambda' = \frac{\kappa_2 - \kappa_1}{\kappa_2 + \kappa_1}\lambda \qquad \text{and} \qquad \lambda'' = \frac{2\kappa_1}{\kappa_2 + \kappa_1}\lambda. \tag{8.48}$$

These values are exactly the same as those used to solve the corresponding problem with a planar dielectric interface in Section 8.3.3.

8.4 The Green Function Method

In this section, we derive a general solution to the Poisson equation in a volume V,

$$\nabla^2\varphi(\mathbf{r}) = -\rho(\mathbf{r})/\epsilon_0 \qquad \mathbf{r} \in V, \tag{8.49}$$

with either Dirichlet or Neumann conditions specified on the volume's boundary surface S. The solution exploits a function of two variables called the *Green function*, $G(\mathbf{r}, \mathbf{r}')$, which satisfies (8.49) when $\mathbf{r} \in V$ and $\rho(\mathbf{r})$ is a point charge density located at $\mathbf{r} = \mathbf{r}' \in V$:

$$\nabla^2 G(\mathbf{r}, \mathbf{r}') = -\delta(\mathbf{r} - \mathbf{r}')/\epsilon_0 \qquad \mathbf{r}, \mathbf{r}' \in V. \tag{8.50}$$

The Green function is unique if the boundary conditions are satisfied when $\mathbf{r}$ lies on S.

The connection between the potential and the Green function becomes clear when we substitute $f(\mathbf{r}') = \varphi(\mathbf{r}')$ and $g(\mathbf{r}') = G(\mathbf{r}', \mathbf{r})$ into Green's second identity (1.80),

$$\int_V d^3r' (f\nabla'^2 g - g\nabla'^2 f) = \int_S dS'\, \hat{\mathbf{n}}' \cdot (f\nabla' g - g\nabla' f). \tag{8.51}$$

The unit normal $\hat{\mathbf{n}}'$ points outward from V. Using (8.49) and (8.50) explicitly, the notation $\partial\varphi/\partial n \equiv \hat{\mathbf{n}} \cdot \nabla\varphi$ permits us to write (8.51) in the form

$$\varphi(\mathbf{r} \in V) = \int_V d^3r'\, \rho(\mathbf{r}')G(\mathbf{r}', \mathbf{r}) - \epsilon_0 \int_S dS'\, \varphi(\mathbf{r}') \frac{\partial G(\mathbf{r}', \mathbf{r})}{\partial n'} + \epsilon_0 \int_S dS'\, G(\mathbf{r}', \mathbf{r}) \frac{\partial \varphi(\mathbf{r}')}{\partial n'}. \tag{8.52}$$

Equation (8.52) is an integral equation for $\varphi(\mathbf{r})$ which involves the boundary values of both φ and $\partial\varphi/\partial n$. The key to the Green-function method is to choose boundary conditions for $G(\mathbf{r}, \mathbf{r}')$ which transform (8.52) into an explicit formula for $\varphi(\mathbf{r})$. There are two natural choices.

The Miller of Nottingham

The theory developed in this section first appeared in an 1828 memoir by George Green (1793-1841) entitled *An Essay on the Application of Mathematical Analysis to the Theories of Electricity and Magnetism*. The *Essay* is remarkable, not least because the young Green received only one year of schooling before he was apprenticed to his father's flour mill. Only at the age of 30 did Green join the Nottingham Subscription Library and discover higher mathematics through the works of Laplace, Lagrange, Legendre, and Poisson. Greatly stimulated, Green worked out his ideas and published his 72-page *Essay* (privately) for a subscription list of local notables. It received little attention and the discouraged Green returned to milling. A few years later, he followed the advice of friends and entered Cambridge as a 40-year-old undergraduate. Green graduated in 1837 and quickly published six papers on hydrodynamics, acoustics, and optics before returning abruptly to Nottingham in 1841. He died there at the age of 47, almost entirely unknown. William Thomson (later Lord Kelvin) found a copy of the *Essay* just after his own graduation from Cambridge in 1845. He arranged for its publication (in Germany) in 1850. Riemann coined the term "Green function" soon thereafter. The first widely seen edition in English was published in 1871.

8.4.1 Dirichlet Boundary Conditions

The *Dirichlet* Green function satisfies (8.50) and the Dirichlet boundary condition

$$G_D(\mathbf{r}_S, \mathbf{r}') = 0 \qquad \mathbf{r}_S \in S, \quad \mathbf{r}' \in V. \tag{8.53}$$

This choice makes the last integral in (8.52) zero. Therefore, once we have solved for $G_D(\mathbf{r}, \mathbf{r}')$ and supplied the boundary data $\varphi_S(\mathbf{r}_S)$, the unique solution for the potential is

$$\varphi(\mathbf{r} \in V) = \int_V d^3r'\, \rho(\mathbf{r}')G_D(\mathbf{r}', \mathbf{r}) - \epsilon_0 \int_S dS'\, \varphi_S(\mathbf{r}') \frac{\partial G_D(\mathbf{r}', \mathbf{r})}{\partial n'}. \tag{8.54}$$

8.4.2 Neumann Boundary Conditions

The *Neumann* Green function satisfies (8.50) and a Neumann boundary condition which involves the area A of the surface S which bounds V:

$$\frac{\partial G_N(\mathbf{r}_S, \mathbf{r}')}{\partial n} = -\frac{1}{\epsilon_0 A} \qquad \mathbf{r}_S \in S, \quad \mathbf{r}' \in V. \tag{8.55}$$

This choice[6] reduces the first surface integral in (8.52) to the average value of the potential over S,

$$\langle \varphi \rangle_S = \frac{1}{A} \int_S dS\, \varphi(\mathbf{r}). \tag{8.56}$$

Therefore, once we have solved for $G_N(\mathbf{r}, \mathbf{r}')$ and supplied the boundary data $\partial\varphi_S(\mathbf{r}_S)/\partial n$, (8.52) simplifies to a formula for the unique potential in V:

$$\varphi(\mathbf{r} \in V) = \langle \varphi \rangle_S + \int_V d^3r'\, \rho(\mathbf{r}') G_N(\mathbf{r}', \mathbf{r}) + \epsilon_0 \int_S dS'\, \frac{\partial \varphi_S(\mathbf{r}')}{\partial n'} G_N(\mathbf{r}', \mathbf{r}). \tag{8.57}$$

Neumann boundary conditions do not occur naturally in electrostatics problems with stationary charge.[7] It is true that $\sigma(\mathbf{r}_S) = -\epsilon_0 \partial\varphi(\mathbf{r}_S)/\partial n$ relates the normal derivative of the potential to the surface charge density on a perfect conductor. Nevertheless, $\partial\varphi/\partial n$ can never be prescribed in advance; it is *determined* by the solution of the boundary value problem. We can imagine specifying the surface charge density on the surface of a non-conductor, but, in that case, $\sigma(\mathbf{r}_S)$ is almost always associated with a matching condition, not a boundary condition.

Mixed boundary conditions (Neumann and Dirichlet conditions on different parts of S) arise when, say, a conducting shell occupies a finite portion of a boundary surface. Finite-sized capacitor plates and apertures in conducting walls fall into this category also. Unfortunately, the mathematics needed to treat this class of problems is fairly exotic.[8]

8.5 The Dirichlet Green Function

The Dirichlet Green function has a simple physical interpretation. According to (8.50) and (8.53), $G_D(\mathbf{r}, \mathbf{r}')$ is the electrostatic potential at any point $\mathbf{r}$ in a volume V bounded by grounded conducting walls (or infinity) due to a unit-strength point charge placed at any point $\mathbf{r}'$ in the same volume. Remarkably, $G_D(\mathbf{r}, \mathbf{r}')$ is *also* the electrostatic potential at any point $\mathbf{r}' \in V$ due to a unit-strength point charge placed at any point $\mathbf{r} \in V$. The latter statement is not obvious, but is a consequence of the *reciprocal property* of the Dirichlet Green function,

$$G_D(\mathbf{r}, \mathbf{r}') = G_D(\mathbf{r}', \mathbf{r}). \tag{8.58}$$

We leave the proof of (8.58) as an exercise for the reader.

Two further points are worth noting. First, the usual Coulomb potential satisfies (8.50) and (8.53) when the bounding surface S recedes to infinity. In this context, the Coulomb potential is called the *free-space* Green function,

$$G_0(\mathbf{r}, \mathbf{r}') = \frac{1}{4\pi\epsilon_0} \frac{1}{|\mathbf{r} - \mathbf{r}'|}. \tag{8.59}$$

[6] $\partial G_N(\mathbf{r}_S, \mathbf{r}')/\partial n = 0$ cannot be used as a boundary condition for (8.50) because integrating the latter over V and using the divergence theorem shows that $\int d\mathbf{S}\cdot\nabla G_N(\mathbf{r}, \mathbf{r}') = -1/\epsilon_0$.

[7] Neumann boundary conditions occur in the theory of steady current flow (Chapter 9) and in the theory of conducting waveguides (Chapter 19).

[8] See Sneddon (1966) and Fabrikant (1991) in Sources, References, and Additional Reading.

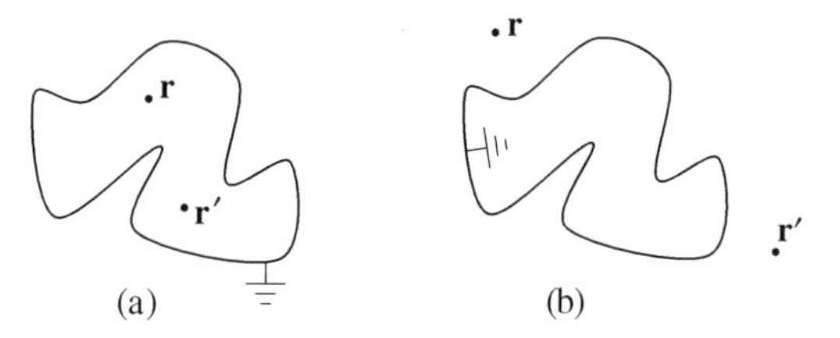

Figure 8.13: (a) $\mathbf{r}$ and $\mathbf{r}'$ lie inside a grounded shell of finite volume; (b) $\mathbf{r}$ and $\mathbf{r}'$ lie outside a grounded shell of finite volume.

Second, two distinct Green functions can be defined for a zero-potential surface which encloses a finite volume (Figure 8.13). $G_D(\mathbf{r}, \mathbf{r}')$ is an *interior* Green function when $\mathbf{r}$ and $\mathbf{r}'$ both lie inside the finite volume. $G_D(\mathbf{r}, \mathbf{r}')$ is an *exterior* Green function when $\mathbf{r}$ and $\mathbf{r}'$ both lie outside the finite volume. The reciprocity relation (8.58) applies to both the interior and exterior Green functions.

8.5.1 The Magic Rule

Using (8.58), it is convenient to write (8.54) in the form

$$\varphi(\mathbf{r} \in V) = \int_V d^3r'\, G_D(\mathbf{r}, \mathbf{r}')\rho(\mathbf{r}') - \epsilon_0 \int_S dS'\, \varphi_S(\mathbf{r}') \frac{\partial G_D(\mathbf{r}, \mathbf{r}')}{\partial n'}. \tag{8.60}$$

Barton (1989) refers to (8.60) as the "magic rule" for Poisson's equation. The name is apt because (8.60) uses superposition and a Green function calculated once (for a given V) to deliver $\varphi(\mathbf{r})$ for arbitrary choices of $\varphi_S(\mathbf{r}_S)$ and $\rho(\mathbf{r})$. On the other hand, it may be difficult to actually compute $G_D(\mathbf{r}, \mathbf{r}')$ for all values of $\mathbf{r} \in V$ and $\mathbf{r}' \in V$. For that reason, (8.60) is particularly well suited to "production-line" calculations where the potential is desired for many different choices of $\varphi_S(\mathbf{r}_S)$ and $\rho(\mathbf{r})$.

The general features of the magic rule can be appreciated without any detailed knowledge of the Green function. First, we should check that $\varphi(\mathbf{r}_S)$ calculated from (8.60) actually produces the specified boundary value $\varphi_S(\mathbf{r}_S)$. In light of (8.53), only the second term in (8.60) contributes. Moreover, the curved integration surface S appears flat when the observation point approaches $\mathbf{r}_S$. Therefore, we are justified in replacing S by the x-y plane with $\mathbf{r}_S$ at the origin. If we orient the positive z-axis to point toward V,

$$\varphi(\mathbf{r}_S) = \epsilon_0 \lim_{z \to 0^+} \int_{z'=0} dS'\, \varphi_S(\mathbf{r}') \left. \frac{\partial G_D(0, 0, z|\mathbf{r}')}{\partial z'} \right|_{z'=0}. \tag{8.61}$$

A key insight is that the Dirichlet Green function in (8.61) is exactly the potential in the $z' > 0$ half-space above a flat, grounded conductor due to a unit point charge on the positive z-axis. But this is exactly the problem we solved in Section 8.3 using the method of images. Therefore, making use of (8.3), we find that

$$\lim_{z \to 0^+} \epsilon_0 \left. \frac{\partial G_D}{\partial z'} \right|_{z'=0} = \lim_{z \to 0^+} \frac{1}{2\pi} \frac{z}{[x'^2 + y'^2 + z^2]^{3/2}}. \tag{8.62}$$

Finally, the presumed correctness of (8.61) implies that the right side of (8.62) is a representation of the two-dimensional delta function $\delta(x')\delta(y')$. This is so because (i) it vanishes when $z \to 0$ unless $x' = 0$ and $y' = 0$ and (ii) its integral over a disk of radius R centered at the origin is unity. The latter is best carried out in polar coordinates with $\rho' = \sqrt{x'^2 + y'^2} = z\sigma$ because

$$\lim_{z\to 0} \int_0^R d\rho'\, \rho' \frac{z}{(\rho'^2 + z^2)^{3/2}} = \lim_{z\to 0} \int_0^{R/z} d\sigma\, \frac{\sigma}{(\sigma^2+1)^{3/2}} = 1. \tag{8.63}$$

We now turn to the first term—the volume integral—in the magic rule (8.60). The meaning of this term is best appreciated if we choose the boundary surface S to be a perfect conductor held at potential φ_0. In that case, the divergence theorem and (8.50) reduce the surface integral in (8.60) to the constant value φ_0. We then mimic (8.2) and decompose the Green function according to

$$G_D(\mathbf{r}, \mathbf{r}') = \frac{1}{4\pi\epsilon_0} \frac{1}{|\mathbf{r} - \mathbf{r}'|} + \Lambda(\mathbf{r}, \mathbf{r}') \qquad \mathbf{r}, \mathbf{r}' \in V. \tag{8.64}$$

The point charge potential on the right side of (8.64) is a particular solution of (8.50). Therefore, the function $\Lambda(\mathbf{r}, \mathbf{r}')$ must (i) solve Laplace's equation in V,

$$\nabla^2 \Lambda(\mathbf{r}, \mathbf{r}') = 0 \qquad \mathbf{r}, \mathbf{r}' \in V, \tag{8.65}$$

and (ii) ensure that the sum (8.64) satisfies the boundary condition (8.53). Substituting (8.64) into the (remaining) volume term in (8.60) gives

$$\varphi(\mathbf{r} \in V) = \frac{1}{4\pi\epsilon_0} \int_V d^3r' \frac{\rho(\mathbf{r}')}{|\mathbf{r} - \mathbf{r}'|} + \int_V d^3r'\, \Lambda(\mathbf{r}, \mathbf{r}')\rho(\mathbf{r}') + \varphi_0. \tag{8.66}$$

The first term is the *direct* effect of $\rho(\mathbf{r})$. The second term is the *indirect* effect of $\rho(\mathbf{r})$. That is, $\Lambda(\mathbf{r}, \mathbf{r}')$ is the electrostatic potential at $\mathbf{r}$ due to the charge *induced* on the conductor surface by a unit strength point charge at $\mathbf{r}'$.

A quick application of (8.66) is to find the force on a point charge q located at a point $\mathbf{s}$ inside (or outside) an oddly shaped, perfectly conducting shell. Because $\rho(\mathbf{r}') = q\delta(\mathbf{r}' - \mathbf{s})$, and the charge cannot exert a force on itself,

$$\mathbf{F}(\mathbf{s}) = q\mathbf{E}_{\text{ind}}(\mathbf{s}) = -q^2 \nabla \Lambda(\mathbf{r}, \mathbf{s})\big|_{\mathbf{r}=\mathbf{s}}. \tag{8.67}$$

8.5.2 Calculation of Dirichlet Green Functions

We have seen that the Dirichlet Green function is the electrostatic potential of a unit-strength point charge in the presence of a grounded boundary. For planar and spherical boundaries, the method of images is sufficient to find $G_D(\mathbf{r}, \mathbf{r}')$. In the sections to follow, we outline three other methods used to calculate Dirichlet Green functions. They are discussed in order of increasing explicitness of their treatment of the delta function singularity in (8.50).

8.5.3 The Method of Eigenfunction Expansion

This approach to $G_D(\mathbf{r}, \mathbf{r}')$ is best understood by analogy with the quantum mechanics of a free particle trapped in a volume V by impenetrable walls. The eigenfunctions of such a particle satisfy the Schrödinger equation (in appropriate units),

$$-\nabla^2 \psi_n(\mathbf{r}) = \lambda_n \psi_n(\mathbf{r}) \qquad \mathbf{r} \in V, \tag{8.68}$$

and the boundary condition,

$$\psi_n(\mathbf{r}_S) = 0. \tag{8.69}$$

The eigenvalues λ_n are real and positive. The normalized eigenfunctions are complete (Section 7.4) so

$$\sum_n \psi_n(\mathbf{r})\psi_n^*(\mathbf{r}') = \delta(\mathbf{r} - \mathbf{r}'). \tag{8.70}$$

Using this information, only two steps are needed to prove that the Dirichlet Green function for the volume V is

$$G_D(\mathbf{r}, \mathbf{r}') = \frac{1}{\epsilon_0} \sum_n \frac{\psi_n(\mathbf{r})\psi_n^*(\mathbf{r}')}{\lambda_n}. \tag{8.71}$$

First, $G_D(\mathbf{r}, \mathbf{r}')$ satisfies the differential equation (8.50). This follows from (8.70) and, using (8.68), the fact that

$$\nabla^2 G_D(\mathbf{r}, \mathbf{r}') = \frac{1}{\epsilon_0} \sum_n \frac{\nabla^2 \psi_n(\mathbf{r})\psi_n^*(\mathbf{r}')}{\lambda_n} = -\frac{1}{\epsilon_0} \sum_n \psi_n(\mathbf{r})\psi_n^*(\mathbf{r}'). \tag{8.72}$$

Second, $G_D(\mathbf{r}, \mathbf{r}')$ satisfies the boundary condition (8.53). This is a consequence of (8.69):

$$G_D(\mathbf{r}_S, \mathbf{r}') = \frac{1}{\epsilon_0} \sum_n \frac{\psi_n(\mathbf{r}_S)\psi_n^*(\mathbf{r}')}{\lambda_n} = 0. \tag{8.73}$$

Example 8.3 Find the charge density that must be glued to the surface of an insulating, cubical box ($0 \le x, y, z \le a$) so the electric field everywhere outside the box is identical to the field produced by a (fictitious) point charge Q located at the center of the box.

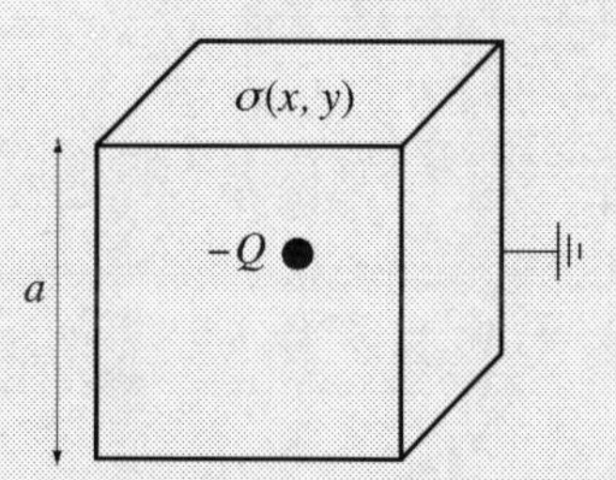

Figure 8.14: A grounded box with a point charge $-Q$ at its center.

Solution: Replace the box in question by an identical box with perfectly conducting walls. Ground the box and put a point charge $-Q$ at its center (Figure 8.14). The charge drawn up from ground must exactly annul the field of the point charge at every point outside the box. In other words, at every exterior point, the field produced by the charge induced on the inside surface of the box is identical to the field of a point charge $+Q$ at the center of the box. This is the charge density we should glue to the surface of the insulating box.

By symmetry, it is sufficient to consider only the $z = a$ surface of the box. Since the interior Dirichlet Green function $G_D(x, y, z; x', y', z')$ is exactly the electrostatic potential at (x, y, z) inside the box due to a unit point charge at (x', y', z') inside the box, the charge density of interest is

$$\sigma(x, y) = -\epsilon_0 Q \frac{\partial}{\partial z} G_D(x, y, z; \tfrac{1}{2}a, \tfrac{1}{2}a, \tfrac{1}{2}a)\Big|_{z=a}.$$

A straightforward application of separation of variables to (8.68) yields the normalized eigenfunctions indexed by the positive integers n, ℓ, and m:

$$\psi_{n\ell m}(x, y, z) = \sqrt{\frac{8}{a^3}} \sin\left[\frac{n\pi x}{a}\right] \sin\left[\frac{\ell\pi y}{a}\right] \sin\left[\frac{m\pi z}{a}\right].$$

The eigenvalues are $\lambda_{n\ell m} = (\pi^2/a^2)(n^2+\ell^2+m^2)$ so, by (8.71), the Green function is

$$G_D(\mathbf{r},\mathbf{r}') = \frac{8}{\epsilon_0\pi^2 a}\sum_{n,\ell,m=1}^{\infty}\frac{\sin\left[\frac{n\pi x}{a}\right]\sin\left[\frac{\ell\pi y}{a}\right]\sin\left[\frac{m\pi z}{a}\right]\sin\left[\frac{n\pi x'}{a}\right]\sin\left[\frac{\ell\pi y'}{a}\right]\sin\left[\frac{m\pi z'}{a}\right]}{n^2+\ell^2+m^2}.$$

Using the first equation above, this gives the surface charge density in the form of the infinite series

$$\sigma(x,y) = \frac{8Q}{\pi a^2}\sum_{n,\ell,m\ \mathrm{odd}}(-1)^{\frac{1}{2}(n+\ell+m-3)}\frac{m}{n^2+\ell^2+m^2}\sin\left[\frac{n\pi x}{a}\right]\sin\left[\frac{\ell\pi y}{a}\right].$$

An end-of-chapter exercise will invite the reader to derive an entirely different representation for the Green function calculated here and compare the numerical convergence of the corresponding formula for $\sigma(x,y)$ with the one obtained here.

8.5.4 The Method of Direct Integration

The direct-integration method proceeds in four steps. First, write $\delta(\mathbf{r}-\mathbf{r}')$ in (8.50) as the product of three one-dimensional delta functions. Second, use completeness relations to represent *two* of the three delta functions. Third, use the same relations to motivate an ansatz for a Green function which transforms (8.50) into an inhomogeneous, ordinary differential equation. Finally, integrate this differential equation taking proper account of the point charge singularity and the boundary conditions.

As an example, let us find a cylindrical representation of the free-space Green function,

$$G_0(\mathbf{r},\mathbf{r}') = \frac{1}{4\pi\epsilon_0}\frac{1}{|\mathbf{r}-\mathbf{r}'|}. \tag{8.74}$$

This problem amounts to solving (8.50) in the form

$$\nabla^2 G_0(\rho,\phi,z;\rho',\phi',z') = -\frac{1}{\epsilon_0}\frac{\delta(\rho-\rho')\delta(\phi-\phi')\delta(z-z')}{\rho}, \tag{8.75}$$

with the boundary condition that $G_0(\mathbf{r},\mathbf{r}')\to 0$ when $|\mathbf{r}|\to\infty$.

Our strategy calls for replacing two of the delta functions in (8.75) with completeness relations constructed from orthonormal solutions to Laplace's equation in cylindrical coordinates (Section 7.8). There are no boundaries in the z- or ϕ-directions, so it is appropriate to use completeness relations constructed from "plane wave" exponential functions like

$$\delta(z-z') = \frac{1}{2\pi}\int_{-\infty}^{\infty} dk\, e^{ik(z-z')} = \frac{1}{\pi}\int_0^{\infty} dk\, \cos k(z-z') \tag{8.76}$$

and

$$\delta(\phi-\phi') = \frac{1}{2\pi}\sum_{m=-\infty}^{\infty} e^{im(\phi-\phi')}. \tag{8.77}$$

Guided by (8.76) and (8.77), we look for a Green function of the form

$$G_0(\mathbf{r},\mathbf{r}') = \frac{1}{2\pi^2}\sum_{m=-\infty}^{\infty}\int_0^{\infty} dk\, e^{im(\phi-\phi')}\cos k(z-z')G_m(\rho,\rho'|k). \tag{8.78}$$

Substituting (8.78) into (8.75) eliminates the delta functions (8.76) and (8.77) from the right side of (8.75) and leaves an ordinary differential equation for $G_m(\rho, \rho'|k)$:

$$\frac{1}{\rho}\frac{d}{d\rho}\left(\rho\frac{dG_m}{d\rho}\right) - \left(k^2 + \frac{m^2}{\rho^2}\right)G_m = -\frac{1}{\epsilon_0}\frac{\delta(\rho-\rho')}{\rho}. \tag{8.79}$$

When $\rho \neq \rho'$, (8.79) is Bessel's equation (7.82) with $\alpha = m$ and $k^2 < 0$. Looking back at Section 7.8.1, we see that the solution of (8.79) must be a linear combination of the modified Bessel functions $I_m(k\rho)$ and $K_m(k\rho)$. $I_m(x)$ is regular at the origin and diverges as $x \to \infty$, while $K_m(x)$ diverges as $x \to 0$ and goes to zero as $x \to \infty$. Our solution must also behave properly at $\rho = 0$ and as $\rho \to \infty$. Therefore, if A is a constant, $\rho_> = \max[\rho, \rho']$, and $\rho_< = \min[\rho, \rho']$, the Green function we need is

$$G_m(\rho, \rho'|k) = AI_m(k\rho_<)K_m(k\rho_>). \tag{8.80}$$

This formula builds in the fact that $G_m(\rho, \rho'|k)$ is continuous at $\rho = \rho'$. If it were not, we would have $dG_m/d\rho \propto \delta(\rho - \rho')$ and $d^2G_m/d\rho^2 \propto \delta'(\rho - \rho')$, which contradicts (8.79).

The constant A is determined by a "jump condition" obtained by integrating (8.79) over an infinitesimal interval from $\rho = \rho' - \delta$ to $\rho = \rho' + \delta$:

$$\left.\frac{dG_m}{d\rho}\right|_{\rho=\rho'+\delta} - \left.\frac{dG_m}{d\rho}\right|_{\rho=\rho'-\delta} = -\frac{1}{\epsilon_0\rho'}. \tag{8.81}$$

Physically, (8.81) is nothing more or less than the matching condition $\hat{\mathbf{n}}_2 \cdot [\mathbf{E}_1 - \mathbf{E}_2] = \sigma/\epsilon_0$ for the case of a point charge on the cylindrical surface $\rho = \rho'$. Finally, we use (8.80) to evaluate (8.81). The result is

$$A\{K_m(k\rho')I'_m(k\rho') - K'_m(k\rho')I_m(k\rho')\} = \frac{1}{\epsilon_0 k\rho'}. \tag{8.82}$$

The quantity in curly brackets is the *Wronskian* of the modified Bessel functions $K_m(x)$ and $I_m(x)$ (see Appendix C.3.1):

$$W[K_m(x), I_m(x)] = K_m(x)I'_m(x) - K'_m(x)I_m(x) = \frac{1}{x}. \tag{8.83}$$

This fixes $A = 1/\epsilon_0$ in (8.82).

We conclude that the desired cylindrical representation of the free-space Green function is

$$G_0(\mathbf{r}, \mathbf{r}') = \frac{1}{2\pi^2\epsilon_0}\sum_{m=-\infty}^{\infty}\int_0^{\infty} dk\, e^{im(\phi-\phi')}\cos k(z-z')I_m(k\rho_<)K_m(k\rho_>). \tag{8.84}$$

This formula is the cylindrical analog of the spherical representation (4.84) derived in connection with the multipole expansion:

$$\frac{1}{4\pi\epsilon_0}\frac{1}{|\mathbf{r}-\mathbf{r}'|} = \frac{1}{4\pi\epsilon_0 r_>}\sum_{\ell=0}^{\infty}\frac{4\pi}{2\ell+1}\left(\frac{r_<}{r_>}\right)^{\ell}\sum_{m=-\ell}^{\ell} Y^*_{\ell m}(\Omega_<)Y_{\ell m}(\Omega_>). \tag{8.85}$$

Indeed, the reader may wish to derive (8.85) using the method of direct integration and the completeness relation for the spherical harmonics,

$$\sum_{\ell=0}^{\infty}\sum_{m=-\ell}^{\ell} Y^*_{\ell m}(\theta', \phi')Y_{\ell m}(\theta, \phi) = \delta(\cos\theta - \cos\theta')\delta(\phi - \phi'). \tag{8.86}$$

Example 8.4 Find the potential between two infinite, grounded plates coincident with $z = 0$ and $z = d$ when a point charge q is interposed between them at $z = z_0$. Find the charge induced on the $z = d$ plate.

Solution: The Dirichlet Green function for this problem (the potential between the plates) satisfies (8.75) with the boundary condition that $G_D(\mathbf{r}, \mathbf{r}')$ vanishes at $z = 0$ and $z = d$. The construction algorithm is same as the one that led to (8.84) except that we replace the delta function representation (8.76) by a completeness relation constructed from solutions of Laplace's equation that satisfy the zero-potential boundary conditions at $z = 0$ and $z = d$:

$$\frac{2}{d}\sum_{n=1}^{\infty} \sin\left[\frac{n\pi z}{d}\right] \sin\left[\frac{n\pi z'}{d}\right] = \delta(z - z').$$

Proceeding exactly as before, the Green function for the volume between the plates is

$$G_D(\mathbf{r}, \mathbf{r}') = \frac{1}{\epsilon_0 \pi d} \sum_{n=1}^{\infty} \sum_{m=-\infty}^{\infty} e^{im(\phi-\phi')} \sin\left[\frac{n\pi z}{d}\right] \sin\left[\frac{n\pi z'}{d}\right] I_m\left(\frac{n\pi\rho_<}{d}\right) K_m\left(\frac{n\pi\rho_>}{d}\right).$$

The potential is rotationally symmetric around an axis perpendicular to the plates which passes through the point charge position (x_0, y_0, z_0). Therefore, if we choose the origin of coordinates so $x_0 = y_0 = 0$, we can put $m = 0$ and $\rho' = 0$ in the Green function. Because $I_0(0) = 1$, we conclude that the potential between the plates is

$$\varphi(\rho, z) = \frac{q}{\epsilon_0 \pi d} \sum_{n=1}^{\infty} \sin\left[\frac{n\pi z}{d}\right] \sin\left[\frac{n\pi z_0}{d}\right] K_0\left(\frac{n\pi\rho}{d}\right).$$

We note in passing that the asymptotic behavior of the modified Bessel function is

$$K_m(x \gg 1) \simeq \sqrt{\frac{\pi}{2x}}\, e^{-x}.$$

This implies that the $n = 1$ term dominates the sum when $\rho \gg d$. Up to a slowly varying pre-factor, it also confirms the correctness of the long-distance exponential decay we predicted in (8.21) for this problem. The charge induced on the $z = d$ plate is[9]

$$Q(d) = \epsilon_0 \int_0^{2\pi} d\phi \int_0^{\infty} d\rho\, \rho \left.\frac{\partial\varphi}{\partial z}\right|_{z=d} = \frac{2q}{\pi} \sum_{n=1}^{\infty} (-1)^n \frac{1}{n} \sin\left[\frac{n\pi z_0}{d}\right] \int_0^{\infty} dy\, y\, K_0(y) = -\frac{z_0}{d} q.$$

This confirms the result found in Example 5.3 (Section 5.4.3) using Green's reciprocity theorem.

8.5.5 The Method of Splitting

We conclude with a calculational method which splits the Green function into two terms. One term is the Coulomb potential of a point charge at $\mathbf{r} = \mathbf{r}'$. The other term is a linear combination of solutions to Laplace's equation chosen to guarantee that the boundary conditions are satisfied. This was the strategy used in (8.64), which we repeat here for convenience:

$$G_D(\mathbf{r}, \mathbf{r}') = G_0(\mathbf{r}, \mathbf{r}') + \Lambda(\mathbf{r}, \mathbf{r}') = \frac{1}{4\pi\epsilon_0} \frac{1}{|\mathbf{r} - \mathbf{r}'|} + \Lambda(\mathbf{r}, \mathbf{r}'). \tag{8.87}$$

[9] The integral over y is one and the sum over n is $-\pi z_0/2d$. See, e.g., I.S. Gradshteyn and I.M. Ryzhik, *Table of Integrals, Series, and Products* (Academic, New York, 1980).

The symmetry of the problem dictates the forms of both $G_0(\mathbf{r}, \mathbf{r}')$ and $\Lambda(\mathbf{r}, \mathbf{r}')$. For example, to find the exterior Green function of an infinitely long tube centered on the z-axis, we use the cylindrical representation (8.84) for the point charge potential in (8.87) and a sum of solutions of Laplace's equations in cylindrical coordinates for $\Lambda(\mathbf{r}, \mathbf{r}')$. A choice that vanishes at infinity, respects the reciprocal symmetry (8.58), and resembles the point charge potential as much as possible is

$$\Lambda(\mathbf{r}, \mathbf{r}') = \frac{1}{2\pi^2\epsilon_0} \sum_{m=-\infty}^{\infty} \int_0^{\infty} dk\, e^{im(\phi-\phi')} \cos k(z-z') B_m K_m(k\rho) K_m(k\rho'). \tag{8.88}$$

The boundary condition $G_D(\rho = R, \mathbf{r}') = 0$ fixes the values of the expansion coefficients as $B_m = -I_m(kR)/K_m(kR)$. Therefore, the exterior Dirichlet Green function for a cylinder with radius R is

$$G_D(\mathbf{r}, \mathbf{r}') = \frac{1}{2\pi^2\epsilon_0} \sum_{m=-\infty}^{\infty} \int_0^{\infty} dk\, e^{im(\phi-\phi')} \cos k(z-z') \left\{ I_m(k\rho_<) - \frac{I_m(kR)}{K_m(kR)} K_m(k\rho_<) \right\} K_m(k\rho_>). \tag{8.89}$$

The first term in the curly brackets in (8.89) is $G_0(\mathbf{r}, \mathbf{r}')$ in (8.87). The second term in the curly brackets is $\Lambda(\mathbf{r}, \mathbf{r}')$ in (8.87).

Application 8.2 A Point Charge outside a Grounded Tube

We can use the Green function (8.89) to find the attractive force between a grounded cylindrical tube of radius R and a point charge q which lies at a point $\mathbf{r}' = (\varrho, 0, 0)$ far outside the tube. This is the analog of the image force (8.7) exerted on a charge by a grounded plane. We choose this example because it is difficult to obtain the answer *without* appeal to a Green function. Symmetry demands that $\mathbf{F} = F\hat{\boldsymbol{\rho}}$ and a direct application of (8.67) using (8.89) gives

$$F(\varrho) = \frac{q^2}{2\pi^2\epsilon_0 R^2} \sum_{m=-\infty}^{\infty} \int_0^{\infty} dx\, x \frac{I_m(x)}{K_m(x)} K_m'(x\varrho/R) K_m(x\varrho/R). \tag{8.90}$$

We are interested in the asymptotic limit of this expression when $\varrho \gg R$. The guess that $F \propto 1/\varrho$ (true for a point charge far from a uniform line of charge) is not correct because the charge density induced on the surface of the cylinder is not uniform along its length.

To make progress, we use the asymptotic form of the modified Bessel function $K_m(x)$ quoted in Example 8.4 to conclude that

$$K_m'(x\varrho/R) K_m(x\varrho/R) \sim -\frac{\pi}{2} \frac{R}{x\varrho} e^{-2x\varrho/R} \qquad \varrho \gg R. \tag{8.91}$$

The corresponding asymptotic formula $I_m(x \gg 1) \sim e^x/\sqrt{2\pi x}$ implies that the integrand of (8.90) goes exponentially to zero at the upper limit of the integral. On the other hand, $K_m(x) \propto 1/x^m$ and $I_m(x) \propto x^m$ as $x \to 0$ while $K_0(x) \sim \ln(2/x)$ and $I_0(x) \sim 1$ as $x \to 0$. With this information, it is not difficult to convince oneself that (8.90) is dominated by the $m = 0$ term in the sum near the lower limit of the integral. Therefore, when $\varrho \gg R$,

$$F \sim \frac{q^2}{2\pi^2\epsilon_0 R^2} \frac{\pi R}{2\varrho} \int_0^{\infty} \frac{dx}{\ln x} e^{-2\varrho x/R}. \tag{8.92}$$

The integral in (8.92) can be evaluated asymptotically[10] with the result that

$$F(\varrho) \sim -\frac{q^2}{8\pi\epsilon_0}\frac{1}{\varrho^2 \ln(2\varrho/R)} \qquad \varrho \gg R. \tag{8.93}$$

As might be expected on geometric grounds, the magnitude of the asymptotic force between a point charge and an infinite grounded cylinder is weaker than the force between a point charge and an infinite grounded plane ($F \sim \varrho^{-2}$) but stronger than the force between a point charge and a finite grounded sphere ($F \sim \varrho^{-3}$). ■

8.6 The Complex Logarithm Potential

Section 7.10 outlined an approach to two-dimensional electrostatics based on the fact that the real and imaginary parts of an analytic function both satisfy Laplace's equation. This method generalizes to two-dimensional Poisson problems because the potential of a line charge at the origin with uniform charge density λ is the real part[11] of the complex potential

$$f(w) = -\frac{\lambda}{2\pi\epsilon_0}\ln w = -\frac{\lambda}{2\pi\epsilon_0}\{\ln r + i\theta\}. \tag{8.94}$$

The function $\ln w$ is analytic everywhere except at $w = 0$. Therefore, we can superpose complex potentials like (8.94) and synthesize virtually any two-dimensional charge distribution of interest. For example, the potential produced by a positive line charge at $w = -a$ and a negative line charge at $w = a$ is the real part of the complex function

$$f_1(w) = -\frac{\lambda}{2\pi\epsilon_0}\ln\frac{w+a}{w-a}. \tag{8.95}$$

We can use (8.95) to derive the complex potential of a *line dipole*. This is the limit of $f_1(w)$ when $\lambda \to \infty$ and $a \to 0$ while holding their product finite. Using $\ln(1+\delta) \simeq \delta$, the result is

$$f_2(w) = -\frac{\lambda a}{\pi\epsilon_0 w}. \tag{8.96}$$

Therefore, since $w = x + iy$ and $r^2 = x^2 + y^2$, the physical potential of the line dipole can be written in terms of the dipole moment per unit length $\mathbf{p} = -2\lambda a\hat{\mathbf{x}}$ as

$$\varphi(x, y) = \operatorname{Re} f_2(w) = \frac{1}{2\pi\epsilon_0}\frac{\mathbf{p}\cdot\mathbf{r}}{r^2}. \tag{8.97}$$

Application 8.3 A Wire Array above a Grounded Plane

In Section 7.5.1, we studied an array of parallel charged lines as a model for a Faraday cage. This situation mimics an array of conducting wires because the equipotential surfaces are the same for observation points close to each wire. A closely related problem with many practical applications[12] is

[10] See Equation (2.20) of B. Wong, *Asymptotic Approximations of Integrals* (Academic, Boston, 1989).

[11] In the notation of Section 7.10, we choose $f = \varphi + i\psi$.

[12] An example is the *wire chamber*, a device used in particle physics to infer the trajectory of rapidly moving particles of ionizing radiation.

an infinite planar array of parallel lines with charge density λ placed at a distance d above a flat, grounded conductor of infinite extent (Figure 8.15). The method of images tells us that the ground plane can be replaced by an array of lines with charge density $-\lambda$ located at a distance d below the ground plane. Therefore, the potential for $y > 0$ can be generated by an array of pairs of oppositely charged lines of the sort described by (8.95).

If a is the distance between adjacent lines, a fruitful line of analysis focuses on the complex potential

$$f_3(w) = -\frac{\lambda}{2\pi\epsilon_0}\ln\left\{\frac{\sin\left[(\pi/a)(w - id)\right]}{\sin\left[(\pi/a)(w + id)\right]}\right\}. \tag{8.98}$$

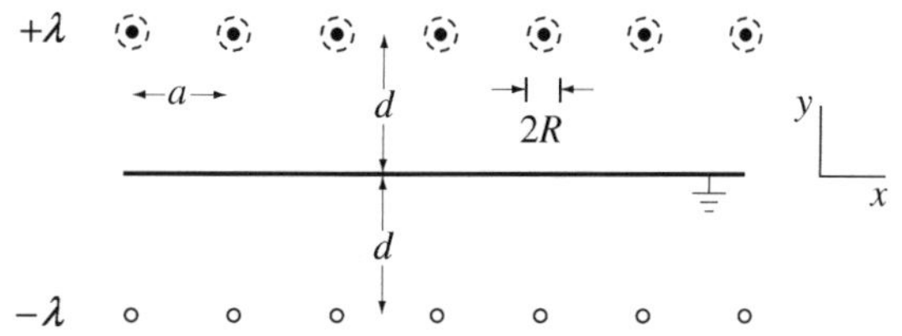

Figure 8.15: An array of identical line charges (solid dots) above a ground plane at $y = 0$. The open dots are image line charges that can replace the ground plane to find the potential for $y > 0$. The dashed circles are equipotentials very close to the line charges that can be regarded as the surfaces of an array of conducting wires.

The complex number $w = id + na + w_0$ is especially interesting if n is an integer and $|w_0| \ll 1$. This identifies w as a point in the immediate vicinity of $w_n = (na, d)$. Moreover,

$$\ln\sin[(\pi/a)(w - id)] = \ln[(-1)^n \sin(\pi w_0/a)] \approx \ln[(-1)^n(\pi w_0/a)]. \tag{8.99}$$

Therefore, up to an inessential constant, the real part of (8.99) is $\ln|w_0|$. Comparison with (8.94) shows that the numerator of (8.98) contributes to singularities in the potential like a collection of positive line charges located at the points (na, d). A similar argument shows that the denominator of (8.98) contributes to singularities in the potential like a collection of negative line charges located at the points $(na, -d)$. That being said, $f_3(w)$ in (8.98) is itself an analytic function of w. This leads us to conclude that the physical potential above the ground plane is

$$\varphi(x, y) = \operatorname{Re} f_3(z) = \frac{\lambda}{4\pi\epsilon_0}\ln\left\{\frac{\sin^2(\pi x/a) + \sinh^2[(\pi/a)(y + d)]}{\sin^2(\pi x/a) + \sinh^2[(\pi/a)(y - d)]}\right\}. \tag{8.100}$$

It is not difficult to calculate the capacitance of the structure in Figure 8.15 if we interpret the dashed equipotentials as the surfaces of conducting wires. Let R be the wire radius and assume that $R \ll a \ll d$. We can compute the potential V at a typical point on the wire array using (8.100) with, say, $x = 0$ and $y = d + R$. The result is

$$V \approx \frac{\lambda}{2\pi\epsilon_0}\ln\frac{a}{2\pi R} + \frac{\lambda d}{\epsilon_0 a}. \tag{8.101}$$

Therefore, the capacitance *per unit area* between the wire array and the ground plane is

$$\frac{C}{A} = \frac{\sigma}{V} = \frac{\lambda/a}{V} \simeq \frac{2\pi\epsilon_0}{a\ln(a/2\pi R) + 2\pi d}. \tag{8.102}$$

This formula correctly recovers the parallel-plate result, $C/A = \epsilon_0/d$, when the spacing a between the wires is small compared to their separation $2d$. ▪

Example 8.5 Find the potential between two grounded plates at $y = 0$ and $y = d$ with an interposed line charge λ at $(0, d/2)$.

Solution: A straightforward application of the method of images produces a solution to this problem in the form of an infinite sequence of images (cf. Section 8.3.4). However, we can reduce our labor to a single image and get a closed-form solution in the bargain if we use the method of conformal mapping (see Section 7.10.4) to map the strip $0 \le y \le d$ into the upper half complex plane. Figure 8.16 illustrates the mapping from $w = x + iy$ to

$$g(w) = u(w) + i\,v(w) = \exp(\pi w/d).$$

As indicated by the capital letters in the figure, the line $y = 0$ from $(-\infty, 0)$ to $(\infty, 0)$ maps to the positive u-axis from $(0, 0)$ to $(0, \infty)$. Similarly, the line $y = d$ from $(-\infty, d)$ to (∞, d) maps to the negative u-axis from $(0, 0)$ to $(0, -\infty)$. The $x = 0$ line segment from $(0, 0)$ through $(0, d/2)$ to $(0, d)$ maps to the upper half of the unit circle $|g| = 1$ from $(1, 0)$ through $(0, 1)$ to $(-1, 0)$. Therefore, a line charge at $w = id/2$ maps to a line charge at $g_0 = i$.

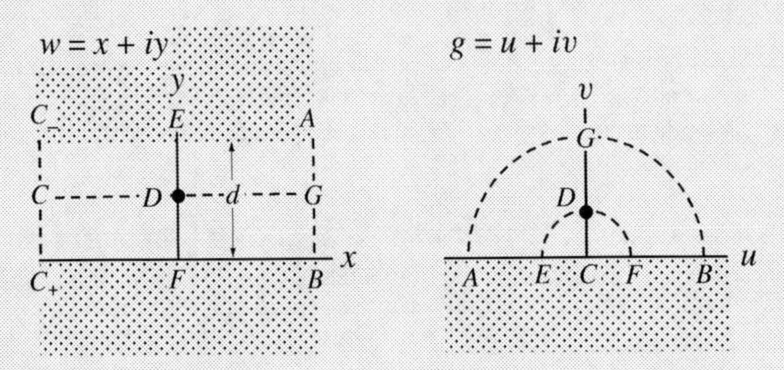

Figure 8.16: The conformal transformation $g = \exp(\pi w/d)$ maps the strip $0 \le y \le d$ of the $w = x + iy$ plane to the upper half $g = u + iv$ plane. The capital letters are points before and after mapping. The dark dot labeled D is the position of the line charge.

The plane $v = 0$ will be at zero potential if an image line charge is placed at the point $g_0^* = -i$. Using (8.94), the complex potential of a positive line charge λ at g_0 and a negative line charge $-\lambda$ at g_0^* is

$$f(g(w)) = -\frac{\lambda}{2\pi\epsilon_0}\ln\left[\frac{g(w) - g_0}{g(w) - g_0^*}\right] = -\frac{\lambda}{2\pi\epsilon_0}\ln\left[\frac{\exp(\pi w/d) - i}{\exp(\pi w/d) + i}\right].$$

The physical electrostatic potential is the real part of $f(w)$:

$$\varphi(x, y) = -\frac{\lambda}{2\pi\epsilon_0}\ln\left\{\frac{e^{2\pi x/d}\cos^2(\pi x/d) + [e^{\pi x/d}\sin(\pi y/d) - 1]^2}{e^{2\pi x/d}\cos^2(\pi x/d) + [e^{\pi x/d}\sin(\pi y/d) + 1]^2}\right\}.$$

This function vanishes at $y = 0$ and $y = d$ as it must.

8.7 The Poisson-Boltzmann Equation

We have focused so far on situations where the volume part of the Poisson charge density is specified and immobile. The resulting polarization of nearby conductors and simple dielectrics leads to induced distributions of charge on the surface of these bodies. However, it often occurs in biophysics, plasma

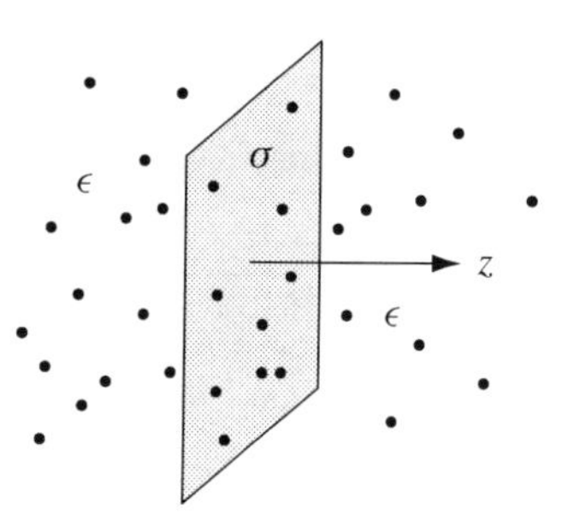

Figure 8.17: A charged sheet at $z = 0$ immersed in a neutral dielectric. Mobile charged particles (black dots) dispersed in the dielectric make the system overall charge-neutral.

physics, and condensed matter physics that immobile *surface* charges redistribute mobile *volume* charge in such a way that $\rho(\mathbf{r}) \neq 0$ in equilibrium.

Consider the situation depicted in Figure 8.17, where a plane ($z = 0$) with uniform charge density σ is embedded in a medium with dielectric permittivity ϵ. This creates an electric field $\mathbf{E}_0 = \hat{\mathbf{z}}\,\mathrm{sgn}(z)\sigma/2\epsilon$. Now neutralize the system by adding to the dielectric medium a number density n_0 of mobile particles, each with charge q. Our interest is to find $\mathbf{E}(z)$ and the equilibrium density of mobile particles $n(z)$. We assume that the latter obey Boltzmann statistics.

If $\varphi(z)$ is the electrostatic potential, the charge density of the system is

$$\rho(z) = \sigma\delta(z) + qn(z) = \sigma\delta(z) + qn_0 \exp\{-q\varphi(z)/kT\}. \tag{8.103}$$

Substituting this into (8.1) (with ϵ_0 replaced by ϵ) produces a special case of the *Poisson-Boltzmann equation*,

$$\frac{d^2\varphi}{dz^2} + \frac{qn_0}{\epsilon}e^{-q\varphi/kT} = -\frac{\sigma}{\epsilon}\delta(z). \tag{8.104}$$

We get one boundary condition by integrating (8.104) over an infinitesimal interval around $z = 0$. Since $\varphi(z) = \varphi(-z)$, this gives

$$\left.\frac{d\varphi}{dz}\right|_{z=0^+} = -\frac{\sigma}{2\epsilon}. \tag{8.105}$$

We will also impose the boundary condition that $\mathbf{E}(z) \to 0$ as $z \to \infty$.

It is convenient to define a dimensionless potential $\psi(z) = q\varphi/kT$ and a length $\ell = q^2/\epsilon kT$ so (8.104) becomes

$$\frac{d^2\psi}{dz^2} + \ell\, n_0\, e^{-\psi} = 0 \qquad (z \neq 0). \tag{8.106}$$

To solve (8.106), multiply the entire equation by $d\psi/dz$, integrate over z between z_1 and z_2, and use the Boltzmann expression for $n(z)$. The result is

$$\left(\frac{d\psi}{dz}\right)^2_{z=z_1} - \left(\frac{d\psi}{dz}\right)^2_{z=z_2} = 2\ell\,[n(z_1) - n(z_2)]. \tag{8.107}$$

We can integrate (8.107) again if $n(z_2) = 0$, $\partial\psi/\partial z|_{z_2} = 0$, and $z_2 = \infty$. This produces the reduced potential

$$\psi(z) = 2\ln\left\{\sqrt{\frac{\ell n_0}{2}}(z+b)\right\}. \tag{8.108}$$

Bearing in mind that q and σ have opposite signs, the choice $b = 4|q|/\ell\sigma$ ensures that $\psi(z)$ satisfies the $z = 0$ boundary condition. This solves the problem.

The charge density of the mobile particles is

$$qn(z) = qn_0 \exp(-\psi) = \frac{2q}{\ell}\frac{1}{(z+b)^2}. \tag{8.109}$$

As expected, (8.109) integrates to $-\sigma/2$ on both sides of the interface. We also find that the electric field $\mathbf{E} = -(kT/q)\nabla\psi$ in the medium is

$$\mathbf{E}(z) = \frac{b}{|z|+b}\mathbf{E}_0. \tag{8.110}$$

This formula shows that mobile charges screen the uniform electric field of an oppositely charged interface by a factor that varies *algebraically* with z. This may be compared with the situation studied in Section 5.7 where a neutral medium composed of mobile charges of one sign and immobile charges of opposite sign screens the Coulomb field of a point charge by a factor that varies *exponentially* with distance from the point charge.

Sources, References, and Additional Reading

The quotation at the beginning of the chapter is taken from the introduction to

G. Green, *An Essay on the Application of Mathematical Analysis to the Theories of Electricity and Magnetism* (1828). Facsimile edition by Wezäta-Melins Aktieborg, Göteborg, Sweden (1958).

Section 8.1 The mathematical physics books listed here discuss Poisson's equation with particular clarity and precision. The last includes a large number of worked examples.

G. Barton, *Elements of Green's Functions and Propagation* (Clarendon Press, Oxford, 1989).

P. Morse and H. Feshbach, *Methods of Theoretical Physics* (Feshbach Publishing, Minneapolis, MN, 2005).

N.N. Lebedev, I.P. Skal'skaya, and Ya.S. Uflyand, *Problems in Mathematical Physics* (Pergamon, Oxford, 1966).

Textbooks of electromagnetism where the subject matter of this chapter is treated in detail include

W.R. Smythe, *Static and Dynamic Electricity* (McGraw-Hill, New York, 1939).

E. Weber, *Electromagnetic Fields* (Wiley, New York, 1950).

W. Hauser, *Introduction to the Principles of Electromagnetism* (Addison-Wesley, Reading, MA, 1971).

J.D. Jackson, *Classical Electrodynamics*, 3rd edition (Wiley, New York, 1999).

Section 8.3 The method of images was first described by William Thomson in an 1845 letter to Joseph Liouville. Thomson was unaware that Newton had invented the method over 150 years earlier to solve gravitational problems. See Section 81 of

S. Chandrasekhar, *Newton's Principia for the Common Reader* (Clarendon Press, Oxford, 1995).

Experiments showing that image forces trap electrons near metal surfaces are reviewed in

Th. Fauster, "Quantization of electronic states on metal surfaces", *Applied Physics A* **59**, 479 (1994).

Example 8.1 and Figure 8.4 come, respectively, from

R.H. Good, Jr. and T.J. Nelson, *Classical Theory of Electric and Magnetic Fields* (Academic, New York, 1971).

E. Durand, *Électrostatique* (Masson, Paris, 1966), Volume III.

Figure 8.6 in Application 8.1 comes from *Gonzales and Carrasco*. Our discussion of the electrostatics of an ion channel was adopted from *Kamenev et al.*:

M.E. Gonzalez and L. Carrasco, "Viroporins", *FEBS Letters* **552**, 28 (2003). With permission from Elsevier.

A. Kamenev, J. Zhang, A.I. Larkin, and B.I. Shklovskii, "Transport in one-dimensional Coulomb gases", *Physica A* **359**, 129 (2006).

Example 8.2 comes from

L.G. Chambers, *An Introduction to the Mathematics of Electricity and Magnetism* (Chapman and Hall, London, 1973).

Group theoretical arguments are used to deduce the set of all closed volumes bounded by planar surfaces that can be analyzed using image theory in

R. Terras and R.A. Swanson, "Electrostatic image problems with plane boundaries", *American Journal of Physics* **48**, 531 (1980).

Figure 8.10 and its discussion were adapted from

M. Levin and G. Johnson, "Is the electric force between a point charge and a neutral metallic object always attractive?", *American Journal of Physics* **79**, 843 (2011).

The textbook by *Smythe* (see Section 8.1 above) and the three textbooks below each present a different way to deduce the image solution for the dielectric cylinder problem (Section 8.3.8 above). Our discussion follows the characteristically simple and elegant approach of *Landau and Lifshitz*.

L.D. Landau and E.M. Lifshitz, *Electrodynamics of Continuous Media* (Pergamon, Oxford, 1960).

P.C. Clemmow, *An Introduction to Electromagnetic Theory* (Cambridge University Press, Cambridge, 1973).

J. Vanderlinde, *Classical Electromagnetic Theory* (Wiley, New York, 1993).

Section 8.4 Two excellent books which discuss the use of Green functions to solve the Poisson equation are *Barton* (see Section 8.1 above) and

M.D. Greenberg, *Application of Green Functions in Science and Engineering* (Prentice-Hall, Englewood Cliffs, NJ, 1971).

Julian Schwinger was the master of using Green functions to solve a wide variety of inhomogeneous problems in field theory. His virtuosity in the context of Poisson's equation is evident in

J. Schwinger, L.L. DeRaad, Jr., K.A. Milton, and W.-Y. Tsai, *Classical Electrodynamics* (Perseus, Reading, MA, 1998).

A biography of the enigmatic George Green is

D.M. Cannell, *George Green* (Athlone Press, London, 1993).

Poisson problems with apertures lead to mixed boundary condition problems of the sort discussed in

I.N. Sneddon, *Mixed Boundary Value Problems in Potential Theory* (North-Holland, Amsterdam, 1966).

V.I. Fabrikant, *Mixed Boundary Value Problems of Potential Theory and Their Applications in Engineering* (Kluwer, Dordrecht, 1991).

Section 8.5 Our approach to Dirichlet Green functions draws heavily on *Barton* (see Section 8.1 above).
Example 8.3 was devised by Prof. Michael Cohen of the University of Pennsylvania. Application 8.2 was drawn from

B.E. Granger, P. Král, H.R. Sadeghpour, and M. Shapiro, "Highly extended image states around nanotubes", *Physical Review Letters* **89**, 135506 (2002).

Section 8.6 Application 8.3 comes from *Morse and Feshbach* (see Section 8.1 above).

Section 8.7 For more information about the Poisson-Boltzmann equation, see

J.-N. Chazalviel, *Coulomb Screening by Mobile Charges* (Birkhäuser, Boston, 1999).

D. Andelman, "Electrostatic properties of membranes", in *Handbook of Biological Physics*, edited by R. Lipowsky and E. Sackmann (Elsevier, Amsterdam, 1995), Volume 1.

Problems

8.1 The Image Force and Its Limits The text showed that the attractive force **F** between an origin-centered, grounded, conducting sphere of radius R and a point charge located at a point $s > R$ on the positive z-axis varies as $1/s^3$ when $s \gg R$. Replace the sphere by a grounded conductor of any shape. Use Green's reciprocity principle to show that the force between the conductor and the charge still varies as $1/s^3$ when s is large compared to the characteristic size of the conductor.

8.2 Point Charge near a Corner Two semi-infinite and grounded conducting planes meet at a right angle as seen edge-on in the diagram. Find the charge induced on each plane when a point charge Q is introduced as shown.

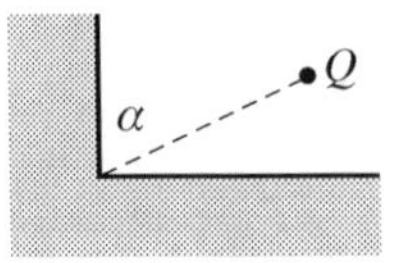

8.3 Rod and Plane The diagram below shows a rod of length L and net charge Q (distributed uniformly over its length) oriented parallel to a grounded infinite conducting plane at the distance d from the plane.

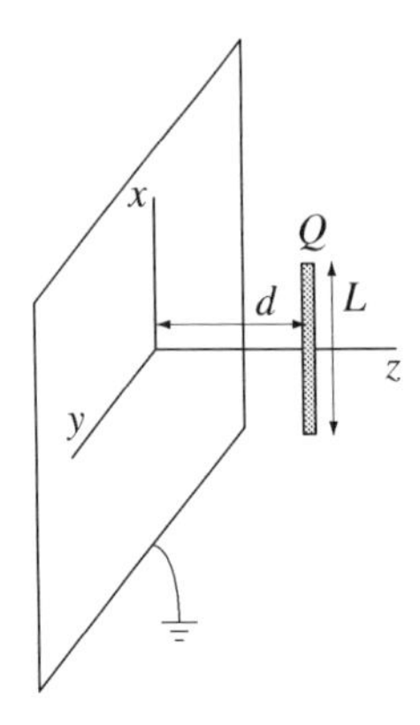

(a) Evaluate a double integral to find the exact force exerted on the rod by the plane.
(b) Simplify your answer in part (a) in the limit $d \gg L$. Give a physical argument for your result.
(c) Find the charge density $\sigma(x, y)$ induced on the conducting plane.
(d) Find the total charge induced on the plane *without* integrating $\sigma(x, y)$.

8.4 A Dielectric Slab Intervenes An infinite slab with dielectric constant $\kappa = \epsilon/\epsilon_0$ lies between $z = a$ and $z = b = a + c$. A point charge q sits at the origin of coordinates. Let $\beta = (\kappa - 1)/(\kappa + 1)$ and use solutions of Laplace's equation in cylindrical coordinates to show that

$$\varphi(z > b) = \frac{q(1-\beta^2)}{4\pi\epsilon_0}\int_0^\infty dk \frac{J_0(k\rho)\exp(-kz)}{1-\beta^2\exp(-2kc)} = \frac{q(1-\beta^2)}{4\pi\epsilon_0}\sum_{n=0}^{\infty}\frac{\beta^{2n}}{\sqrt{(z+2nc)^2+\rho^2}}.$$

Note: The rightmost formula is a sum over image potentials, but it is much more tedious to use images from the start.

8.5 The Force Exerted by a Charge on a Dielectric Interface The half-plane $z < 0$ has dielectric constant κ_L and the half-plane $z > 0$ has dielectric constant κ_R. Embed a point charge q on the z-axis at $z = -d$. The text computed the force on q to be

$$\mathbf{F}_q = -\frac{1}{4\pi\epsilon_L}\frac{q^2}{4d^2}\frac{\kappa_L - \kappa_R}{\kappa_L + \kappa_R}\hat{\mathbf{z}}.$$

(a) Use the stress tensor formalism to show that the force on the $z = 0$ interface is equal in magnitude but opposite in direction to $\mathbf{F}_q$.
(b) Show that the Coulomb force on the interfacial surface polarization charge density,

$$\mathbf{F} = \int dS\sigma_P \frac{\mathbf{E}_L + \mathbf{E}_R}{2},$$

does *not* give the correct force. Hint: See Application 6.2.

8.6 Image Energy and Real Energy Suppose that a collection of image point charges $q_1, q_2, \ldots, q_N$ is used to find the force on a point charge q at position $\mathbf{r}_q$ due to the presence of a conductor held at potential φ_C. Let U_A be the electrostatic potential energy between q and the conductor. Let U_B be the electrostatic energy of q in the presence of the image charges. Find the general relation between U_A and U_B and confirm that $U_A = \frac{1}{2}U_B$ when $\varphi_C = 0$.

8.7 Images in Spheres I A point charge q is placed at a distance $2R$ from the center of an isolated, conducting sphere of radius R. The force on q is observed to be zero at this position. Now move the charge to a distance $3R$ from the center of the sphere. Show that the force on q at its new position is repulsive with magnitude

$$F = \frac{1}{4\pi\epsilon_0}\frac{173}{5184}\frac{q^2}{R^2}.$$

Hint: A spherical equipotential surface remains an equipotential surface if an image point charge is placed at its center.

8.8 Images in Spheres II Positive charges Q and Q' are placed on opposite sides of a grounded sphere of radius R at distances of $2R$ and $4R$, respectively, from the sphere center. Show that Q' is repelled from the sphere if $Q' < (25/144)Q$.

8.9 Debye's Model for the Work Function In 1910, Debye suggested that the work function W of a metal could be computed as the work performed against the electrostatic image force when an electron is removed from the interior of a finite piece of metal to a point infinitely far outside the metal. Model the metal as a perfectly conducting sphere with a macroscopic radius R and suppose that the image force only becomes operative at a microscopic distance d outside the surface of the metal.

(a) Show that

$$W = \frac{e^2}{8\pi\epsilon_0}\left[\frac{2}{R+d} - \frac{R}{(R+d)^2} + \frac{R}{(R+d)^2 - R^2}\right].$$

Hint: Removing an electron from the metal leaves the metal with a net charge.

(b) Let $x = d/R$ and take the limit $R \to \infty$ to find the Debye model prediction for the work function of a semi-infinite sample. It is understood today that the image force plays an insignificant role in the work function.

8.10 Force between a Line Charge and a Conducting Cylinder Let b be the perpendicular distance between an infinite line with uniform charge per unit length λ and the center of an infinite conducting cylinder with radius $R = b/2$.

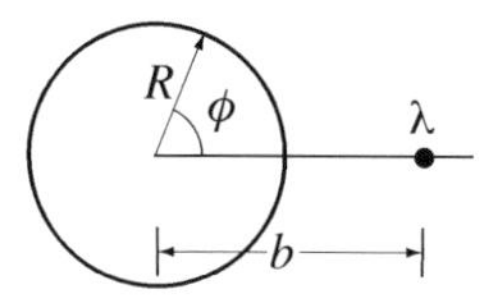

(a) Show that the charge density induced on the surface of the cylinder is

$$\sigma(\phi) = -\frac{\lambda}{2\pi R}\left(\frac{3}{5 - 4\cos\phi}\right).$$

(b) Find the force per unit length on the cylinder by an appropriate integration over $\sigma(\phi)$.

(c) Confirm your answer to (b) by computing the force per unit length on the cylinder by another method.

Hint: Let a single image line inside the sphere fix the potential of the cylinder.

8.11 Point Dipole in a Grounded Shell A point electric dipole with moment $\mathbf{p}$ sits at the center of a grounded, conducting, spherical shell of radius R. Use the method of images to show that the electric field inside the shell is the sum of the electric field produced by $\mathbf{p}$ and a *constant* electric field $\mathbf{E} = \mathbf{p}/4\pi\epsilon_0 R^3$.

Hint: Use the formula for the charge density induced on a grounded plane by a point charge q located a distance z_0 above the plane: $\sigma(\rho) = -qz_0/[2\pi(\rho^2 + z_0^2)^{3/2}]$.

8.12 Inversion in a Cylinder

(a) Let $\Phi(\rho, \phi)$ be a solution of Laplace's equation in a cylindrical region $\rho < R$. Show that the function $\Psi(\rho, \phi) = \Phi(R^2/\rho, \phi)$ is a solution of Laplace's equation in the region $\rho > R$.
(b) Show that a suitable linear combination of the functions Φ and Ψ in part (a) can be used to solve Poisson's equation for a line charge located anywhere inside or outside a grounded conducting cylindrical shell.
(c) Show that a linear combination of the functions Φ and Ψ can be used to solve Poisson's equation for a line charge located inside or outside a solid cylinder of radius R and dielectric constant κ_1 when the cylinder is embedded in a space with dielectric constant κ_2.
(d) Comment on the implications of this problem for the method of images.

8.13 Symmetry of the Dirichlet Green Function Use Green's second identity to prove that $G_D(\mathbf{r}, \mathbf{r}') = G_D(\mathbf{r}', \mathbf{r})$.

8.14 Green Function Inequalities The Dirichlet Green function for any finite volume V can always be written in the form

$$G_D(\mathbf{r}, \mathbf{r}') = \frac{1}{4\pi\epsilon_0}\frac{1}{|\mathbf{r} - \mathbf{r}'|} + \Lambda(\mathbf{r}, \mathbf{r}') \qquad \mathbf{r}, \mathbf{r}' \in V.$$

The function $\Lambda(\mathbf{r}, \mathbf{r}')$ satisfies $\nabla^2\Lambda(\mathbf{r}, \mathbf{r}') = 0$.

(a) Use the physical meaning of the Dirichlet Green function to prove that

$$G_D(\mathbf{r}, \mathbf{r}') < \frac{1}{4\pi\epsilon_0}\frac{1}{|\mathbf{r} - \mathbf{r}'|}.$$

(b) Use Earnshaw's theorem to prove that

$$G_D(\mathbf{r}, \mathbf{r}') > 0.$$

8.15 The Potential of a Voltage Patch The plane $z = 0$ is grounded except for an finite area S_0 which is held at potential φ_0. Show that the electrostatic potential away from the plane is

$$\varphi(x, y, z) = \frac{\varphi_0|z|}{2\pi}\int_{S_0}\frac{d^2r'}{|\mathbf{r} - \mathbf{r}'|^3}.$$

8.16 The Charge Induced by Induced Charge Maintain the plane $z = 0$ at potential V and introduce a grounded conductor somewhere into the space $z > 0$. Use the "magic rule" for the Dirichlet Green function to find the charge density $\sigma(x, y)$ induced on the $z = 0$ plane by the charge $\sigma_0(\mathbf{r})$ induced on the surface S_0 of the grounded conductor.

8.17 Free-Space Green Functions by Eigenfunction Expansion Find the free-space Green function $G_0^{(d)}(\mathbf{r}, \mathbf{r}')$ in $d = 1, 2, 3$ space dimensions by the method of eigenfunction expansion. For $d = 2$, you will need (i) an integral representation of $J_0(x)$; (ii) the regularization $k^{-1} = \lim_{\eta\to 0} k/(k^2 + \eta^2)$; and (iii) an integral representation of $K_0(x)$. For $d = 1$, you will need (1.111) and the fundamental definition of the Green function.

8.18 Free-Space Green Function in Polar Coordinates The free-space Green function in two dimensions (potential of a line charge) is $G_0^{(2)}(\mathbf{r}, \mathbf{r}') = -\ln|\mathbf{r} - \mathbf{r}'|/2\pi\epsilon_0$. Use the method of direct integration to reduce the two-dimensional equation $\epsilon_0 \nabla^2 G(\mathbf{r}, \mathbf{r}') = -\delta(\mathbf{r} - \mathbf{r}')$ to a one-dimensional equation and establish the alternative representation

$$G_0^{(2)}(\mathbf{r}, \mathbf{r}') = -\frac{1}{2\pi\epsilon_0} \ln \rho_> + \frac{1}{2\pi\epsilon_0} \sum_{m=1}^{\infty} \frac{1}{m} \frac{\rho_<^m}{\rho_>^m} \cos m(\phi - \phi').$$

8.19 Using a Cube to Simulate a Point Charge

(a) Use completeness relations to represent $\delta(x - x')\delta(y - y')$ and then the method of direct integration for the inhomogeneous differential equation which remains to find the interior Dirichlet Green function for a cubical box with side walls at $x = \pm a$, $y = \pm a$, and $z = \pm a$.

(b) Use the result of part (a) to find the charge density that must be glued onto the surfaces of an insulating box with sides walls at $x = \pm a$, $y = \pm a$, and $z = \pm a$ so that the electric field everywhere outside the box is identical to the field of a (fictitious) point charge Q located at the center of the box. It is sufficient to calculate $\sigma(x, y)$ for the $z = a$ face. Give a numerical value (accurate to 0.1%) for $\sigma(0, 0)$.

(c) This problem was solved in the text by a different method. Check that both methods give the same (numerical) answer for $\sigma(0, 0)$.

8.20 Green Function for a Sphere by Direct Integration

(a) Use the completeness relation,

$$\sum_{\ell m} Y_{\ell m}(\hat{\mathbf{r}}) Y_{\ell m}^*(\hat{\mathbf{r}}') = \frac{1}{\sin\theta} \delta(\theta - \theta')\delta(\phi - \phi'),$$

and the method of direct integration to show that

$$G(\mathbf{r}, \mathbf{r}') = \frac{1}{4\pi\epsilon_0} \sum_{\ell=0}^{\infty} \left\{ \frac{r_<^\ell}{r_>^{\ell+1}} - \frac{r_<^\ell r_>^\ell}{R^{2\ell+1}} \right\} P_\ell(\hat{\mathbf{r}} \cdot \hat{\mathbf{r}}')$$

is the interior Dirichlet Green function for a sphere of radius R.

(b) Show that $G(\mathbf{r}, \mathbf{r}')$ above is identical to the image solution for this problem.

8.21 The Charge Induced on a Conducting Tube

(a) Derive an integral expression for the charge density $\sigma(\phi, z)$ induced on the outer surface of a conducting tube of radius R when a point charge q is placed at a perpendicular distance $s > R$ from the symmetry axis of the tube.

(b) Confirm that the point charge induces a total charge $-q$ on the tube surface.

(c) Show that the angle-averaged linear charge density falls off extremely slowly with distance along the length of the tube. Specifically, show that, as $z \to \infty$,

$$\lambda(z) = R \int_0^{2\pi} d\phi\, \sigma(\phi, z) \sim -\frac{q \ln(s/R)}{z \ln^2(z/R)}.$$

8.22 Green Function for a Dented Beer Can An empty beer can is bounded by the surfaces $z = 0$, $z = h$, and $\rho = R$. By slamming it against his forehead, a frustrated football fan dents the can into the shape shown below. Our interest is the interior Dirichlet Green function of the dented can.

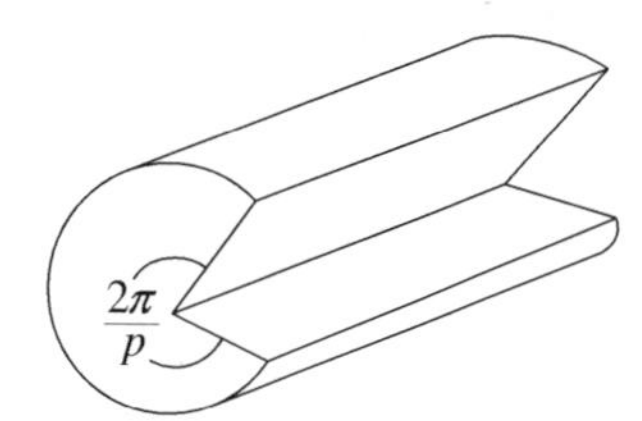

(a) Show that suitable choices for the allowed values of γ and β in the sums makes the ansatz

$$G_D(\mathbf{r},\mathbf{r}') = \sum_\gamma \sum_\beta \sin(\gamma z)\sin(\gamma z')\sin(\beta\phi)\sin(\beta\phi')g_{\gamma\beta}(\rho,\rho')$$

satisfy the boundary conditions at $z=0$, $z=h$, $\phi=0$, and $\phi=2\pi/p$, and thus reduces the defining equation for the Green function to a one-dimensional differential equation for $g_{\alpha\beta}(\rho,\rho')$.

(b) Complete the solution for $G_D(\mathbf{r},\mathbf{r}')$.

8.23 Weyl's Formula Write $\delta(\mathbf{r}-\mathbf{r}') = \delta(\mathbf{r}_\perp - \mathbf{r}'_\perp)\delta(z-z')$ and use direct integration to derive Weyl's formula for the free-space Green function in three dimensions,

$$G_0(\mathbf{r},\mathbf{r}') = \frac{1}{2\epsilon_0}\int \frac{d^2k_\perp}{(2\pi)^2} e^{i\mathbf{k}_\perp\cdot(\mathbf{r}_\perp - \mathbf{r}'_\perp)}\frac{1}{k_\perp}e^{-k_\perp|z-z'|}.$$

8.24 Electrostatics of a Cosmic String A *cosmic string* is a one-dimensional object with an extraordinarily large linear mass density ($\mu \sim 10^{22}$ kg/m) which (in some theories) formed during the initial cool-down of the Universe after the Big Bang. In two-dimensional (2D) general relativity, such an object distorts flat spacetime into an extremely shallow cone with the cosmic string at its apex. Alternatively, one can regard flat 2D space as shown below: undistorted but with a tiny wedge-shaped region removed from the physical domain. The usual angular range $0 \le \phi < 2\pi$ is thus reduced to $0 \le \phi < 2\pi/p$ where $p^{-1} = 1 - 4G\mu/c^2$, G is Newton's gravitational constant, and c is the speed of light. The two edges of the wedge are indistinguishable so any physical quantity $f(\phi)$ satisfies $f(0) = f(2\pi/p)$.

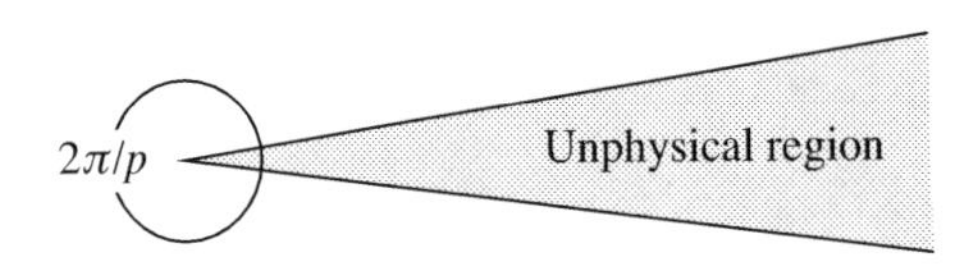

(a) Begin with no string. Show that the free-space Green function in 2D is

$$G_0(\boldsymbol{\rho},\boldsymbol{\rho}') = -\frac{1}{2\pi\epsilon_0}\ln|\boldsymbol{\rho}-\boldsymbol{\rho}'|.$$

(b) Now add the string so $p \neq 1$. To find the modified free-space Green function $G_0^p(\boldsymbol{\rho},\boldsymbol{\rho}')$, a representation of the delta function is required which exhibits the proper angular behavior. Show that a suitable form is

$$\delta(\phi-\phi') = \frac{p}{2\pi}\sum_{m=-\infty}^{\infty} e^{imp(\phi-\phi')}.$$

(c) Exploit the ansatz

$$G_0^p(\rho,\phi,\rho',\phi') = \frac{p}{2\pi}\sum_{m=-\infty}^{\infty} e^{imp(\phi-\phi')}G_m(\rho,\rho')$$

to show that

$$G_0^p(\rho,\phi,\rho',\phi') = \frac{1}{2\pi}\sum_{m=1}^{\infty}\cos[mp(\phi-\phi')]\frac{1}{m}\left(\frac{\rho_<}{\rho_>}\right)^{mp} - \frac{p}{2\pi}\ln\rho_>.$$

(d) Perform the indicated sum and find a closed-form expression for G_0^p. Check that $G_0^1(\boldsymbol{\rho},\boldsymbol{\rho}')$ correctly reproduces your answer in part (a).

(e) Show that a cosmic string at the origin and a line charge q at $\boldsymbol{\rho}$ are attracted with a force

$$\mathbf{F} = (p-1)\frac{q^2\hat{\boldsymbol{\rho}}}{4\pi\epsilon_0\rho}.$$

8.25 Practice with Complex Potentials Show that

$$f(z) = -\frac{\lambda}{2\pi\epsilon_0} \ln \tan \frac{\pi z}{a}$$

can be used as the complex potential for an array of equally spaced, parallel, charged lines in the $y = 0$ plane. Let n be an integer and let $x = na$ and $x = (n + \frac{1}{2})a$ be the positions of the positive and negatively charged lines, respectively. Find the asymptotic behavior ($|y| \to \infty$) of the physical potential.

9 Steady Current

The endless circulation of the electric fluid may appear paradoxical and even inexplicable, but it is no less true.
Alessandro Volta (1800)

9.1 Introduction

We conclude our study of electrostatics with a class of problems where the ultimate source of $\mathbf{E}(\mathbf{r})$ is electric charge in steady motion. In Section 2.1.2, we used a current density $\mathbf{j}(\mathbf{r}, t)$ to characterize the current I through an arbitrary surface S by

$$I = \int_S d\mathbf{S} \cdot \mathbf{j}. \tag{9.1}$$

We applied (9.1) to a closed surface and derived a continuity equation which relates $\mathbf{j}(\mathbf{r}, t)$ to variations in the charge density $\rho(\mathbf{r}, t)$:

$$\frac{\partial \rho}{\partial t} + \nabla \cdot \mathbf{j} = 0. \tag{9.2}$$

In this chapter, we focus on charge densities which do not depend explicitly on time so (9.2) reduces to the *steady-current* condition,

$$\nabla \cdot \mathbf{j}(\mathbf{r}) = 0. \tag{9.3}$$

Steady currents produce time-independent magnetic fields (see Chapter 10). We will see in this chapter that they can also produce time-independent *electric* fields. However, because the Maxwell equations do not mix static electric fields with static magnetic fields, the fields of interest to us here satisfy the conventional equations of electrostatics,

$$\nabla \cdot \mathbf{E} = \frac{\rho}{\epsilon_0} \qquad \text{and} \qquad \nabla \times \mathbf{E} = 0. \tag{9.4}$$

9.1.1 The Steady-Current Condition

The meaning of the steady-current condition (9.3) becomes clear if we adopt a field line representation for $\mathbf{j}(\mathbf{r})$ patterned after the field line representation for $\mathbf{E}(\mathbf{r})$ (Section 3.3.4). Our experience with Gauss' law [left equation in (9.4)] tells us that lines of steady current density are like electric field lines when no charge is present. This means that every line which enters an infinitesimal volume must also exit that volume. In other words, every line of $\mathbf{j}(\mathbf{r})$ either closes on itself or begins and ends at infinity.

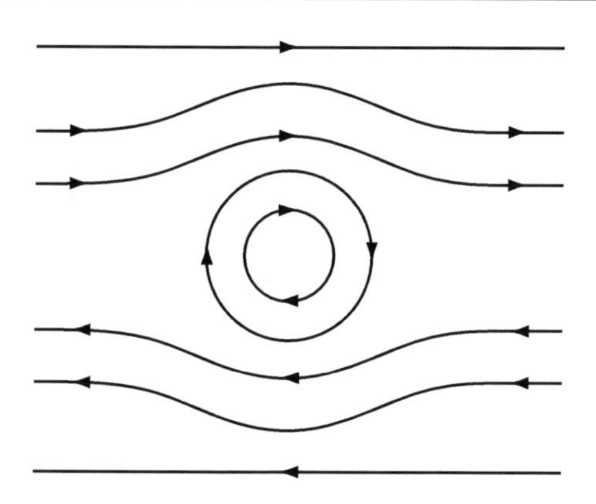

Figure 9.1: Field lines of a current density $\mathbf{j}(\mathbf{r})$ which satisfies the steady-current condition (9.3).

Figure 9.1 shows the field line pattern for a model current density which satisfies the steady-current condition (9.3). To the extent that this sketch reminds the reader of an eddy pool formed by two currents of water flowing in opposite directions, Figure 9.1 illustrates the analogy between electric current flow and ordinary fluid flow.[1] Indeed, we exploited this analogy in Section 2.1.2 to write down a formula for $\mathbf{j}(\mathbf{r})$ when a velocity field $\boldsymbol{v}(\mathbf{r})$ characterizes the motion of a charge density $\rho(\mathbf{r})$:

$$\mathbf{j}(\mathbf{r}) = \rho(\mathbf{r})\boldsymbol{v}(\mathbf{r}). \tag{9.5}$$

The term *convection current density* is used for (9.5) when $\rho(\mathbf{r})$ moves rigidly with uniform velocity $\boldsymbol{v}(\mathbf{r}) = \boldsymbol{v}_0$. A microscopic convection current density often applies to neutral systems like semiconductors and astrophysical and biological plasmas where several types of particles (not all with the same algebraic sign of charge) contribute to the current density. When M species of particles with charges q_k and spatially uniform number densities n_k move with uniform velocities $\boldsymbol{v}_k$, the current density is

$$\mathbf{j} = \sum_{k=1}^{M} q_k n_k \boldsymbol{v}_k. \tag{9.6}$$

9.2 Current in Vacuum

A stream of identical charged particles moving uniformly in vacuum is the simplest example of an electric current. It is an interesting example because the magnitude of the current is limited by the electric field generated by the particles themselves. To see this, we analyze a vacuum diode[2]—a parallel-plate capacitor with one plate heated so it ejects electrons by thermionic emission (Figure 9.2). For simplicity, we assume that every electron is ejected with zero speed.

When the plate temperature is low, the number of emitted electrons is very small and each particle simply accelerates across the gap under the influence of the potential $\varphi(x) = xV/L$. This is the usual solution of Laplace's equation between the plates of a capacitor (curve 1 of Figure 9.3). From (9.1) and (9.5), electrons in the volume AL between the plates with an average speed υ produce an average current per electron

$$I = A \times \frac{1}{AL} \times e\upsilon = \frac{e}{L}\upsilon. \tag{9.7}$$

[1] No less a notable than J.J. Thomson—the discoverer of the electron—was explicit in 1937: "The service of the electric fluid concept to the science of electricity, by suggesting and coordinating research, can hardly be overestimated." The analogy breaks down for phenomenon that depend on the details of the inter-particle forces in the fluid.

[2] This device has disappeared from low-power consumer electronics—a victim of semiconductor technology.

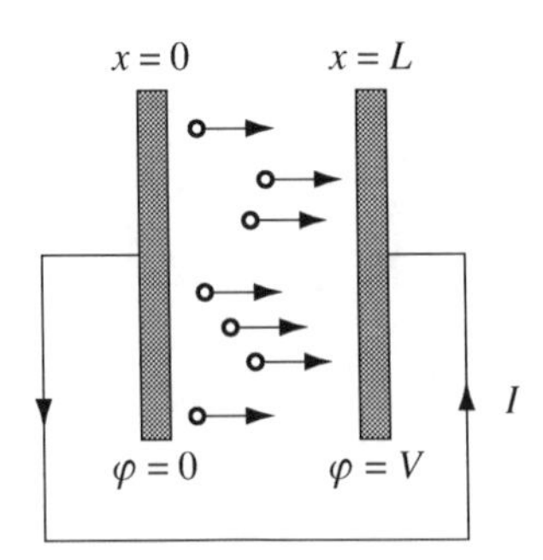

Figure 9.2: Schematic of a vacuum diode. Electrons are accelerated from a hot cathode ($x = 0$) to a cold metal anode ($x = L$).

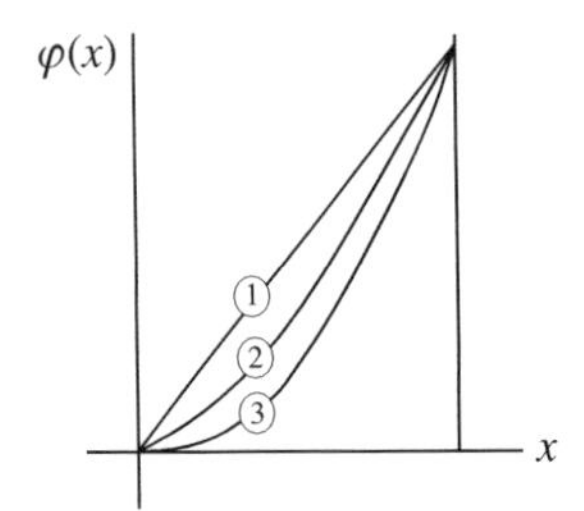

Figure 9.3: Electrostatic potential between the plates of a vacuum diode for (1) zero current; (2) an intermediate value of current; and (3) saturated current.

The same average current also flows through the rest of the circuit. This follows from the fact (see Example 5.3 in Section 5.4.3) that a charge e at a point x between parallel plates induces charges $Q_L = -e(1 - x/L)$ and $Q_R = -ex/L$ on the left and right plates. When e moves, the plate charges vary in time as $\dot{Q}_L = e\dot{x}/L$ and $\dot{Q}_R = -e\dot{x}/L$. Therefore, a current per electron $I = ev/L$ flows through the circuit.[3]

The rate of electron emission into the vacuum gap increases as the temperature of the cathode increases. The current increases proportionally because each electron makes a contribution like (9.7). Eventually, a non-negligible *space charge*[4] density $\rho(\mathbf{r}) < 0$ builds up between the plates. The potential is no longer determined by Laplace's equation but by Poisson's equation,

$$\nabla^2\varphi = -\rho/\epsilon_0. \tag{9.8}$$

This is a signal that the Coulomb repulsion between electrons cannot be ignored. Since $\nabla^2 \equiv d^2/dx^2$ for this problem, (9.8) says that the potential $\varphi(x)$ develops a positive curvature everywhere between the plates as the electric self-field of the electrons begins to retard their acceleration (curve 2 of Figure 9.3). The charge density and current increase with temperature nevertheless (albeit more slowly) until, finally, $d\varphi/dx \to 0$ at $x = 0$. The current saturates at this point because there is no electric field at the cathode to accelerate additional electrons across the gap (curve 3 of Figure 9.3).

The steady, saturated current satisfies (9.3) so the product $j = \rho(x)\upsilon(x)$ is a constant everywhere. Also, because we assume $\upsilon(0) = 0$, the kinetic energy of any electron is

$$\tfrac{1}{2}m\upsilon^2(x) = e\varphi(x). \tag{9.9}$$

[3] This argument implicitly uses a quasistatic approximation. See Chapter 14 for more details.

[4] The term "space charge" generically refers to a volume of uncompensated charge in an otherwise neutral volume.

Combining these facts with (9.8) gives

$$\frac{d^2\varphi}{dx^2} = \frac{|j|}{\epsilon_0}\sqrt{\frac{m}{2e}}\frac{1}{\sqrt{\varphi}}. \tag{9.10}$$

Now, multiply both sides of (9.10) by $2d\varphi/dx$, integrate, and demand that $\varphi(0) = 0$ and $d\varphi/dx|_{x=0} = 0$. The latter is the condition for saturation. The result is

$$\left(\frac{d\varphi}{dx}\right)^2 = \frac{|j|}{\epsilon_0}\sqrt{\frac{8m}{e}}\sqrt{\varphi}. \tag{9.11}$$

Finally, take the square root of (9.11) and integrate again. This gives the space-charge-limited current density in a form known as the *Child-Langmuir law*:

$$|j| = \frac{4}{9}\sqrt{\frac{2e}{m}}\frac{\epsilon_0 V^{3/2}}{L^2}. \tag{9.12}$$

Since $C = \epsilon_0 A/L$ is the capacitance of the diode, we can rationalize the dimensional structure of (9.12) using the time Δt it takes for the maximum capacitor charge ΔQ to traverse the vacuum gap:

$$I \sim \frac{\Delta Q}{\Delta t} \sim \frac{C\Delta\varphi}{\Delta t} \sim \frac{(A\epsilon_0/L)V}{L/\sqrt{2eV/m}}. \tag{9.13}$$

9.3 Current in Matter

Current flow in neutral matter is a complex phenomenon whose proper microscopic description requires arguments from statistical physics and quantum mechanics. Nevertheless, it is well established phenomenologically that the current density in many systems obeys *Ohm's law*,

$$\mathbf{j} = \sigma\mathbf{E}. \tag{9.14}$$

Like $\mathbf{P} = \epsilon_0\chi\mathbf{E}$, Ohm's law is a constitutive relation which describes the material-dependent response of a many-particle system to a specified stimulus. Here, it is the *conductivity* σ which carries the material-dependent information.

Ohm's law describes the motion of charged particles that are accelerated by an electric field but suffer energy and momentum degrading collisions (scattering events) with other particles in the system.[5] The linear dependence of $\mathbf{j}$ on $\mathbf{E}$ in (9.14)—as opposed to the non-linear dependence exhibited by the Child-Langmuir law (9.12)—is one consequence of these collisions. Qualitatively, this can be seen from a classical derivation (due to Drude) which balances the electric Lorentz force on a typical conduction electron in a metal with an effective drag force due to collisions. If τ is the average time between collisions, this balance makes it possible for the particles to achieve a terminal velocity $\boldsymbol{v}$ determined by

$$m\frac{d\boldsymbol{v}}{dt} = -e\mathbf{E} - \frac{m\boldsymbol{v}}{\tau} = 0. \tag{9.15}$$

The steady solution to (9.15) is called the *drift velocity*,[6]

$$\boldsymbol{v}_{\mathrm{d}} = -\frac{e\tau}{m}\mathbf{E}. \tag{9.16}$$

[5] In neutral systems, these collisions typically involve oppositely charged particles of a different species. In a metal, electrons suffer collisions with nearly immobile ions. In a plasma, electrons suffer collisions with much heavier, and thus much slower-moving, ions.

[6] The drift velocity for electrons in the copper wires which carry current to small household appliances is very small, about 10^{-3} m/s.

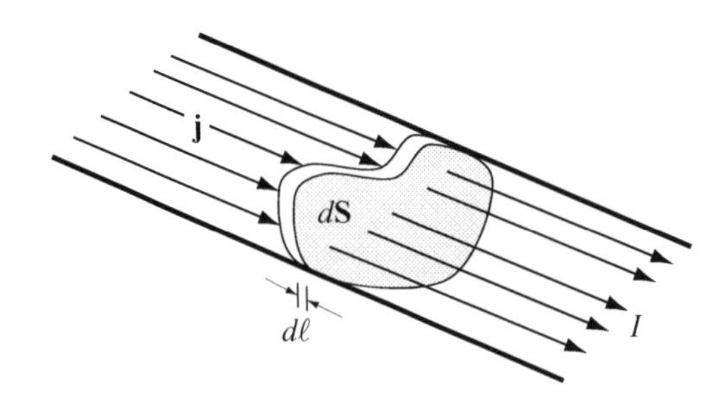

Figure 9.4: Portion of a filamentary ohmic wire where the streamlines of $\mathbf{j}$ and the vector $d\mathbf{S}$ are everywhere parallel to the walls of the wire.

Substituting (9.16) into (9.6) with $M = 1$ gives Ohm's law (9.14) with

$$\sigma = \frac{ne^2\tau}{m}. \tag{9.17}$$

Proper quantum mechanical calculations for a metal give the same result, albeit with a somewhat different meaning for some of the symbols.[7]

9.3.1 The Filamentary-Wire Limit

Very often, we will be concerned with ohmic current flow in a thin filamentary "wire" where the streamlines $d\boldsymbol{\ell}$ of $\mathbf{j}$ are everywhere parallel to the walls of the wire (Figure 9.4). In that case, if $d\mathbf{S} \parallel \mathbf{j}$ is a vector whose magnitude is a differential element of cross sectional area,

$$\int \mathbf{j}\, d^3r = \int\int \mathbf{j}\,(d\mathbf{S}\cdot d\boldsymbol{\ell}) = \int (\mathbf{j}\cdot d\mathbf{S}) \int d\boldsymbol{\ell} = I\int d\boldsymbol{\ell}. \tag{9.18}$$

In all that follows, we will freely use (9.18) to take the filamentary limit of any volume integral which contains $\mathbf{j}(\mathbf{r})$ in the integrand. For example, if $\mathbf{C}(\mathbf{r})$ is an arbitrary vector field,

$$\int d^3r\, \mathbf{j}\times\mathbf{C} \to I\int d\boldsymbol{\ell}\times\mathbf{C}. \tag{9.19}$$

9.4 Potential Theory for Ohmic Matter

Like other constitutive equations, Ohm's law closes the Maxwell equations because it expresses the charges or the currents—the sources of the fields—in terms of the fields themselves. If we generalize (9.14) to account for macroscopic spatial inhomogeneities, the defining equations for current flow in an ohmic medium are

$$\nabla\cdot\mathbf{j}(\mathbf{r}) = 0 \tag{9.20}$$

$$\mathbf{j}(\mathbf{r}) = \sigma(\mathbf{r})\mathbf{E}(\mathbf{r}) \tag{9.21}$$

$$\nabla\times\mathbf{E}(\mathbf{r}) = 0. \tag{9.22}$$

The first two of these combine to give

$$\sigma(\mathbf{r})\,\nabla\cdot\mathbf{E}(\mathbf{r}) + \mathbf{E}(\mathbf{r})\cdot\nabla\sigma(\mathbf{r}) = 0. \tag{9.23}$$

The case when the conductivity is a constant independent of position is very important because the second term on the left side of (9.23) vanishes. Since $\rho(\mathbf{r}) = \epsilon_0\nabla\cdot\mathbf{E}(\mathbf{r})$ and (9.22) implies that

[7] See, for example, J.M. Ziman, *Principles of the Theory of Solids* (Cambridge University Press, Cambridge, 1964).

$\mathbf{E} = -\nabla\varphi$, we conclude that the interior of an ohmic conductor with uniform conductivity is characterized by

$$\sigma(\mathbf{r}) = \sigma \qquad \rho(\mathbf{r}) = 0 \qquad \nabla^2\varphi(\mathbf{r}) = 0. \tag{9.24}$$

This is a significant result. It shows that the powerful machinery of potential theory can be applied to find $\mathbf{E}(\mathbf{r})$ and $\mathbf{j}(\mathbf{r})$ for steady current flow in homogeneous ohmic matter.

9.4.1 Matching Conditions for Ohmic Matter

The matching conditions at an interface between two materials with conductivities σ_1 and σ_2 are derived from (9.20) and (9.22) in the usual way (see Section 2.3.3). They are

$$\hat{\mathbf{n}}_2 \cdot [\mathbf{j}_1 - \mathbf{j}_2] = 0 \qquad \text{or} \qquad \sigma_1 \left.\frac{\partial\varphi_1}{\partial n}\right|_S = \sigma_2 \left.\frac{\partial\varphi_2}{\partial n}\right|_S, \tag{9.25}$$

and

$$\hat{\mathbf{n}}_2 \times [\mathbf{E}_1 - \mathbf{E}_2] = 0 \qquad \text{or} \qquad \varphi_1|_S = \varphi_2|_S. \tag{9.26}$$

These equations are very similar to the matching conditions (6.70) and (6.71) at the interface between two dielectrics. A significant difference is that if no current flows through a boundary surface S which separates an ohmic medium from a non-conducting medium, (9.25) provides the natural Neumann boundary condition,

$$\left.\frac{\partial\varphi}{\partial n}\right|_S = 0. \tag{9.27}$$

Example 9.1 A steady current with uniform density $\mathbf{j}$ flows through a flat interface between a medium with conductivity σ_1 and dielectric permittivity ϵ_1 and a medium with conductivity σ_2 and dielectric permittivity ϵ_2. Find the free charge that accumulates on the interface.

Solution: Let $\hat{\mathbf{n}}_2$ be the outward unit normal from medium 2. The free surface charge density σ_f appears in the matching condition (6.31):

$$\sigma_f = \hat{\mathbf{n}}_2 \cdot (\mathbf{D}_1 - \mathbf{D}_2).$$

On the other hand, (9.25) says that

$$\mathbf{j} = \sigma_1\mathbf{E}_1 = \sigma_2\mathbf{E}_2.$$

Therefore, since $\mathbf{D}_1 = \epsilon_1\mathbf{E}_1$ and $\mathbf{D}_2 = \epsilon_2\mathbf{E}_2$,

$$\sigma_f = \left(\frac{\epsilon_1}{\sigma_1} - \frac{\epsilon_2}{\sigma_2}\right)\hat{\mathbf{n}}_2 \cdot \mathbf{j}.$$

9.5 Electrical Resistance

Figure 9.5 illustrates a current flow boundary value problem where a potential difference $V = \varphi_A - \varphi_B$ between two perfectly conducting electrodes drives a steady current I through an ohmic sample embedded in an insulting material (which could be vacuum). A typical task is to compute the *resistance* R of the sample, defined by

$$R = \frac{\varphi_A - \varphi_B}{I}. \tag{9.28}$$

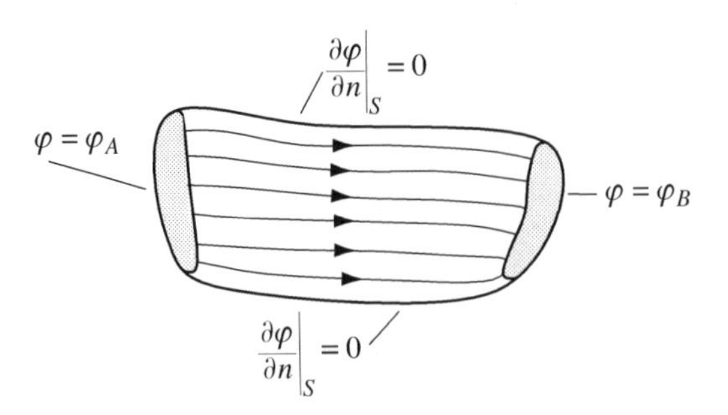

Figure 9.5: Two conducting electrodes attached to the ends of an ohmic conductor embedded in an insulating medium. A potential difference $\varphi_A - \varphi_B$ between the electrodes drives a current as indicated by the lines of $\mathbf{j}(\mathbf{r})$.

The unknown quantity in (9.28) is the current I, which we compute from the flux integral (9.1) of the current density over any cross section of the conductor. To find $\mathbf{j}(\mathbf{r})$, we solve (9.24) using the boundary conditions indicated in Figure 9.5: φ is specified on each electrode and $\partial\varphi/\partial n$ is specified on the boundary with the insulating medium. These mixed Dirichlet/Neumann boundary conditions are sufficient to guarantee a unique solution to the potential problem defined by (9.24) (see Section 7.3).

Since $\mathbf{j} = -\sigma\,\nabla\varphi$, an alternative to fixing the voltage difference is to fix the current I through each electrode surface S using

$$\sigma \int_S dS\, \frac{\partial \varphi}{\partial n} = \pm I. \tag{9.29}$$

The choice of the plus/minus sign depends on whether the current enters/exits the sample through the electrode in question. In this scenario, we calculate the potential difference from the line integral of $\mathbf{E}$ from one electrode to the other. Therefore, if S is any cross section of the conductor, the resistance is[8]

$$R = \frac{\int_A^B d\boldsymbol{\ell}\cdot\mathbf{E}}{\int_S d\mathbf{S}\cdot\mathbf{j}} = \frac{1}{\sigma} \times \frac{\int_A^B d\boldsymbol{\ell}\cdot\mathbf{E}}{\int_S d\mathbf{S}\cdot\mathbf{E}}. \tag{9.30}$$

In practice, one uses (9.30) and the measured resistance to extract the geometry-independent conductivity (or its reciprocal, the *resistivity* $\rho = 1/\sigma$).

It is interesting to apply (9.30) to the case where two conducting electrodes are embedded in an infinite medium with conductivity σ and S is any surface which completely encloses one of the electrodes. The key is to compare (9.30) with the definition of the capacitance of the two-conductor capacitor offered in (5.56). This comparison produces a remarkable formula which relates the capacitance of two conductors in vacuum to the resistance between the same two conductors embedded in ohmic matter:

$$RC = \frac{\epsilon_0}{\sigma}. \tag{9.31}$$

Equation (9.31) has obvious practical value when one is asked to find R for a given geometry when a simple method or clear analogy exists to find C for the same geometry. Conversely, a simple method may exist to compute R, when what is really required is the corresponding value for C.

[8] The boundary condition (9.29) produces a unique solution because it is analogous to specifying the total charge on a perfect conductor.

Example 9.2 (a) A straight wire has length L, resistivity ρ, and cross sectional area A. The shape of the cross section is arbitrary but is uniform over the length of the wire. Prove that the current density in the wire is uniform and find the wire's resistance. (b) Solve Laplace's equation to find the resistance of an ohmic "washer" with inner radius a, outer radius b, and thickness h. Assume that a steady current flows radially from a to b. (c) Check this computation by a judicious application of the answer to part (a).

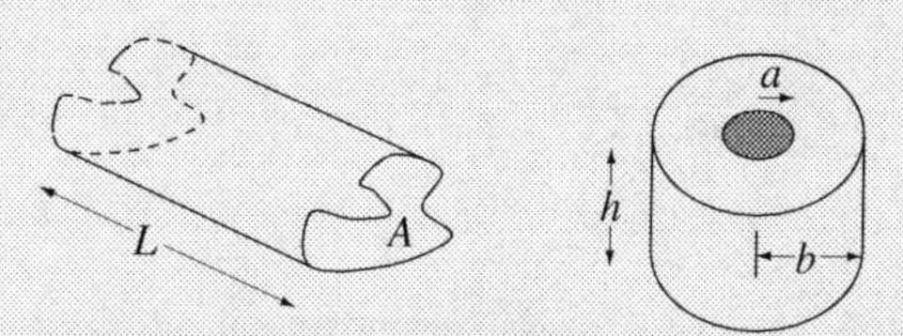

Figure 9.6: A straight wire and a thick washer, both with uniform resistivity $\rho = 1/\sigma$. Current flows longitudinally through the wire and radially through the washer.

Solution:

(a) Begin with a parallel-plate capacitor whose plates are separated by a distance L. The electric field between the infinite-area plates is $\mathbf{E} = \hat{\mathbf{z}}\, V/L$. Now insert the resistive wire between the capacitor plates so that the wire ends are flush with the plates. The electric field everywhere (including inside the wire) remains $\mathbf{E} = \hat{\mathbf{z}}\, V/L$ because the corresponding potential $\varphi = -Vz/L$ satisfies Laplace's equation and the boundary condition (9.27) on the side walls of the wire. $\mathbf{E}$ is uniform in space, so the current density $\mathbf{j} = \sigma\mathbf{E}$ is uniform as well. Finally, inserting $\mathbf{E}$ into (9.30) gives $R = \rho L/A$.

(b) We learned in Section 7.8 that $\varphi(r) = A + B\ln r$ is the only solution of Laplace's equation in cylindrical coordinates which does not depend on the Cartesian coordinate $0 \le z \le h$ or the angular coordinate $0 \le \phi < 2\pi$. The boundary condition (9.29) applied at either $r = a$ or $r = b$ fixes B and the constant A is immaterial. The associated electric field is $\mathbf{E} = \hat{\mathbf{r}} I/2\pi\sigma hr$. Inserting this into (9.30) using any coaxial cylinder with radius $a \le r \le b$ for S gives $R = (\rho/2\pi h)\ln(b/a)$.

(c) We can check this answer using the results of part (a) because radial current flow through the washer is one-dimensional. $dR = \rho dr/2\pi rh$ is the contribution to the total resistance from a cylindrical slice of thickness dr. Each slice is in series with the others, so a sum over all slice resistances gives the desired result,

$$R = \frac{\rho}{2\pi h}\int_a^b \frac{dr}{r} = \frac{\rho}{2\pi h}\ln\frac{b}{a}.$$

Application 9.1 Contact Resistance

The intrinsic roughness of material surfaces ensures that current flow between two macroscopic conductors occurs only at the tiny asperities where the conductors truly touch. A model for this situation is the flow of fluid through an impermeable membrane into which many very small holes have been drilled. To calculate the *contact resistance* between two such conductors, it is important to know the resistance R produced by a single hole of radius a cut into an infinitely large planar baffle of insulating material. The baffle divides an infinite ohmic medium into two parts. Figure 9.7 is a side view which shows the expected lines of current density $\mathbf{j} = \sigma\mathbf{E}$ (dashed) and the corresponding equipotentials (solid) for this situation.

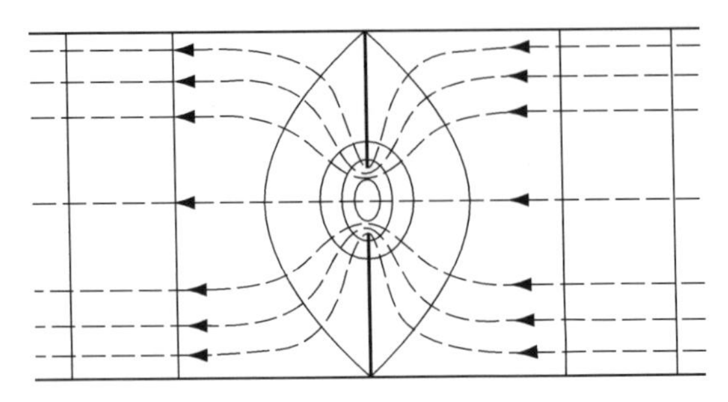

Figure 9.7: Electric field lines (dashed) and equipotentials (solid) for ohmic current flow through a circular orifice viewed edge-on. Figure adapted from Jansen, van Gelder, and Wyder (1980).

In this Application, we find R by exploiting the relation (9.31) between resistance and capacitance. The key step is to replace the hole with a negatively charged conducting disk, eliminate the insulating baffle, and recognize that the pattern of $\mathbf{E}(\mathbf{r})$ produced by the disk is exactly the same as the pattern of $\mathbf{E}(\mathbf{r})$ in Figure 9.7 with the direction of the arrows reversed on the left side of the baffle. Therefore, the quantity of interest is the two-conductor capacitance C between the disk and a spherical conducting shell of infinite radius which serves as the source of the electric field lines which terminate on the disk. In light of (5.55), $C = C_{\text{self}} = 8\epsilon_0 a$ where C_{self} is the self-capacitance of a conducting disk of radius a (see Section 5.4.1).

Actually, only *half* of C is relevant because negative charge accumulates on both sides of the conducting disk while our problem has field lines both entering and leaving the hole. In other words, (9.31) with $C \to C/2$ gives only the contribution to the contact resistance produced by the lines of $\mathbf{j}$ which enter the hole in Figure 9.7:

$$R_{\text{in}} = \frac{\epsilon_0}{\sigma(C/2)} = \frac{1}{4\sigma a}. \tag{9.32}$$

The total contact resistance is R_{in} in series with an identical resistance R_{out} produced by the lines of $\mathbf{j}$ which exit the hole. Hence,

$$R = R_{\text{in}} + R_{\text{out}} = \frac{1}{2\sigma a}. \tag{9.33}$$

The concept of contact resistance becomes very important when the size of electrical contacts shrinks to the nanometer scale. At the same time, the momentum-relaxation approximation behind Ohm's law (Section 9.3) begins to break down and the transport of electrons becomes more nearly ballistic. Theory and experiment reveal a crossover from ohmic to non-ohmic behavior as the size of the contact decreases.[9] ■

9.6 Joule Heating

The inelastic collisions responsible for Ohm's law rob the current-carrying charged particles of their kinetic energy. That energy is dissipated in the rest of the sample in the form of Joule heat. To compute the rate of heat production, we recall first that the mobile charged particles in an ohmic medium move

[9] See Sharvin (1965) and Erts *et al.* (2000) in Sources, References, and Additional Reading.

at the constant drift velocity (9.16) with no change in potential energy. Therefore, from the first law of thermodynamics, the rate of Joule heating is equal to the rate at which the electric field does work on the current-carrying charged particles:

$$\frac{dW}{dt} = \frac{d}{dt}\sum_{k=1}^{N} q_k\, \mathbf{E}(\mathbf{r}_k)\cdot \mathbf{r}_k(t) = \int_V d^3r\, \mathbf{j}\cdot\mathbf{E}. \tag{9.34}$$

Using (9.1) and (9.28), it is straightforward to express the heating rate (9.34) in terms of the total current I which flows through a medium of resistance R between perfectly conducting electrodes A and B (see Figure 9.5):

$$\int_V d^3r\, \mathbf{j}\cdot\mathbf{E} = -\int_V d^3r\, \nabla\cdot(\mathbf{j}\varphi) = -\int_S d\mathbf{S}\cdot\mathbf{j}\,\varphi = (\varphi_A - \varphi_B)I = I^2R. \tag{9.35}$$

9.6.1 The Current Density That Minimizes Joule Heating

The analogy between current flow and fluid flow provides some insight into the behavior of the current density in an ohmic medium. We will exploit the fact that some classes of steady fluid flows correspond to situations of minimum energy dissipation.[10] Since Joule heating is itself a form of energy dissipation, it is reasonable to suppose that the distribution of current in ohmic matter minimizes the rate of Joule heating. To test this idea, we assume (as an empirical matter) that $|\mathbf{j}|^2/\sigma$ is the local heating rate and also that a specified, steady current I flows through the sample between two electrodes. We will *not* assume the validity of Ohm's law (9.21) or the zero-curl condition (9.22).

The physical requirements that $\mathbf{j}(\mathbf{r})$ be steady and correspond to a specified total current direct us perform a constrained minimization using the method of Lagrange multipliers.[11] To ensure that only steady flows are considered, we introduce a Lagrange function $\psi(\mathbf{r})$ to fix $\nabla\cdot\mathbf{j} = 0$. To ensure that a specified amount of current flows out of one electrode and into the other, we introduce two Lagrange constants λ_α $(\alpha = 1, 2)$ to fix $I = \pm\int d\mathbf{S}_\alpha\cdot\mathbf{j}$. Putting all this together, our task is to minimize the functional

$$F[\mathbf{j}] = \frac{1}{2}\int_V d^3r\, \frac{|\mathbf{j}|^2}{\sigma} - \int_V d^3r\, \psi(\mathbf{r})\nabla\cdot\mathbf{j} + \sum_\alpha \lambda_\alpha \int_{S_\alpha} d\mathbf{S}_\alpha\cdot\mathbf{j}. \tag{9.36}$$

The factor of $\frac{1}{2}$ and the minus sign are inserted for convenience. Operationally, we compute $\delta F = F[\mathbf{j}+\delta\mathbf{j}] - F[\mathbf{j}]$ and look for the conditions which make $\delta F = 0$ to first order in $\delta\mathbf{j}$. This extremum is a minimum if $\delta F > 0$ to second order in $\delta\mathbf{j}$.

It is simplest to integrate (9.36) by parts first to get

$$F[\mathbf{j}] = \frac{1}{2}\int_V d^3r\, \frac{|\mathbf{j}|^2}{\sigma} + \int_V d^3r\, \mathbf{j}\cdot\nabla\psi(\mathbf{r}) - \int_S d\mathbf{S}\cdot\mathbf{j}\,\psi + \sum_\alpha \lambda_\alpha \int_{S_\alpha} d\mathbf{S}_\alpha\cdot\mathbf{j}. \tag{9.37}$$

If we assume that no current flows through the non-electrode parts of the surface S (which might be at infinity), a calculation of δF to first order in $\delta\mathbf{j}$ gives

$$\delta F = \int_V d^3r\, \delta\mathbf{j}\cdot\left[\frac{\mathbf{j}}{\sigma} + \nabla\psi\right] - \int_S d\mathbf{S}_\alpha\cdot\delta\mathbf{j}\,(\psi - \lambda_\alpha). \tag{9.38}$$

[10] See, for example, G.K. Batchelor, *Introduction to Fluid Dynamics* (Cambridge University Press, Cambridge, 1967), Section 4.8.

[11] The steps here closely follow the constrained minimization of $U_E[\mathbf{D}]$ performed in Section 6.7.2.

Since the variation $\delta\mathbf{j}$ is arbitrary, we get $\delta F = 0$ if $\mathbf{j}(\mathbf{r}) = -\sigma\,\nabla\psi(\mathbf{r})$ and $\psi(\mathbf{r})$ takes the constant values λ_α on the electrodes. The second-order term is $\frac{1}{2}\int d^3r\,|\delta\mathbf{j}|^2/\sigma$ so these conditions indeed correspond to minimum energy dissipation.

We conclude that $\psi(\mathbf{r})$ is the electrostatic potential $\varphi(\mathbf{r})$ and that Ohm's law is indeed satisfied. Moreover, since $\nabla\cdot\mathbf{j} = 0$ was a constraint, we recover the potential problem (9.24) with mixed Dirichlet and Neumann boundary conditions. Therefore, among all possible $\mathbf{j}(\mathbf{r})$, the distribution of current which flows in an ohmic medium is precisely the one which generates the least Joule heating. Since we have minimized $|\mathbf{j}|^2/\sigma$, the lines of $\mathbf{j}$ will be most concentrated in those parts of V with the highest conductivity.

Example 9.3 A spherical sample of radius R and conductivity σ is embedded in an infinite medium with conductivity σ_0. Far away from the sphere, the current density j_0 is uniform. Find the value of σ which maximizes the rate at which energy is dissipated inside the sphere.

Solution: In light of (9.35), our goal is to maximize the integral $\int d^3r\,\mathbf{j}\cdot\mathbf{E}$ over the volume of the sphere. To find $\mathbf{E}$ inside the sphere, we exploit the matching-condition isomorphism mentioned just after (9.25) and (9.26). In Example 6.3 of Section 6.5.5, we calculated the electric field inside a spherical vacuum cavity scooped out of an infinite medium with dielectric permittivity ϵ in a uniform external field $\mathbf{E}_0$. The answer we found was

$$\mathbf{E}_{\text{in}} = \frac{3\epsilon}{2\epsilon + \epsilon_0}\mathbf{E}_0.$$

This becomes the solution of the present problem if we put $\epsilon \to \sigma_0$, $\epsilon_0 \to \sigma$, and $\mathbf{E}_0 \to \mathbf{j}_0/\sigma_0$:

$$\mathbf{E}_{\text{in}} = \frac{3}{2\sigma_0 + \sigma}\mathbf{j}_0.$$

The rate of Joule heating inside the sphere is

$$\int d^3r\,\mathbf{j}\cdot\mathbf{E} = \tfrac{4}{3}\pi R^3\sigma E_{\text{in}}^2 = 12\pi R^3 j_0^2 \frac{\sigma}{(2\sigma_0 + \sigma)^2}.$$

This rate is largest when $\sigma = 2\sigma_0$.

9.7 Electromotive Force

Direct (steady) current flow occurs in an ohmic medium if a source of energy is present to replenish the energy lost to Joule heating. The electrostatic field $\mathbf{E} = \mathbf{j}/\sigma$ inside the medium is not a candidate because, from (9.22) and Stokes theorem, this field does zero net work when the drifting particles of the medium execute closed circuits in order to satisfy (9.20). Therefore, somewhere in the circuit, there must be one or more non-electrostatic sources of energy (chemical, thermal, gravitational, nuclear, or even electromagnetic) to maintain the current. For historical reasons, we say that these sources impress an *electromotive force* (EMF) on the system. Microscopically, the detailed action of any particular source of EMF may be quite complicated to describe. Macroscopically, we can represent the effect of any source of EMF by a fictitious electric field $\mathbf{E}'$ which modifies Ohm's law to

$$\mathbf{j}(\mathbf{r}) = \sigma[\mathbf{E}(\mathbf{r}) + \mathbf{E}'(\mathbf{r})]. \tag{9.39}$$

Typically, $\mathbf{E}'(\mathbf{r})$ is non-zero only in a very localized region of space (such as between the terminals of a battery) and is specified once and for all. An exception is the true electric field associated with electromagnetic induction (Faraday's law). We will study this important special case later.

9.7.1 Voltage Difference

As a matter of convention, we define the *voltage difference* between two points 1 and 2 along a circuit as

$$V_1 - V_2 = \int_1^2 d\boldsymbol{\ell} \cdot \mathbf{E}. \tag{9.40}$$

Similarly, we define the EMF between the points 1 and 2 as

$$\mathcal{E}_{12} = \int_1^2 d\boldsymbol{\ell} \cdot \mathbf{E}'. \tag{9.41}$$

Finally, because $I = jA$, the resistance R_{12} along the path length L_{12} is

$$R_{12} = \frac{1}{I\sigma} \int_1^2 d\boldsymbol{\ell} \cdot \mathbf{j} = \frac{L_{12}}{A\sigma}. \tag{9.42}$$

Using the three previous equations, (9.39) implies that

$$IR_{12} = V_1 - V_2 + \mathcal{E}_{12} = \varphi_1 - \varphi_2 + \mathcal{E}_{12}. \tag{9.43}$$

The last equality in (9.43) reminds us that when $\nabla \times \mathbf{E} = 0$ (as is the case here), the integral in (9.40) is also the electrostatic potential difference $\varphi_1 - \varphi_2$.[12]

For a closed circuit, $\oint d\boldsymbol{\ell} \cdot \mathbf{E} = 0$, and (9.43) simplifies to

$$\mathcal{E} = IR, \tag{9.44}$$

where R is the resistance of the entire circuit and the EMF of the entire circuit is

$$\mathcal{E} = \oint d\boldsymbol{\ell} \cdot \mathbf{E}' = \frac{1}{q} \oint d\boldsymbol{\ell} \cdot \mathbf{F}. \tag{9.45}$$

The last equality in (9.45) expresses the effective electric field as a force per unit charge. This is a reasonable way to think about electromotive "force" because, by definition, every source of EMF drives a current. We will return to this point of view in Section 14.4.1.

9.7.2 The Physical Meaning of EMF

The physical content of (9.43) is not difficult to appreciate if we locate the points 1 and 2 infinitesimally close to the "terminals" of a source of EMF like a battery.[13] In that case, R_{12} plays the role of an

[12] The definition of voltage (9.40) and the left equality in (9.43) remain valid when the electric field varies slowly in time (AC circuit theory). The right equality in (9.43) is *not* valid in that case because $\mathbf{E} \neq -\nabla\varphi$.

[13] A battery is a combination of fundamental units called *voltaic cells*. The EMF of a voltaic cell is generated by chemical reactions at the terminals of the device. The energy content is fixed by the amount of chemical reactant in the cell. See Saslow (1999) in Sources, References, and Additional Reading.

effective "internal resistance" which accounts for loss mechanisms within the source itself. Multiplying through by the current I gives

$$P = I\mathcal{E}_{12} = I(\varphi_2 - \varphi_1) + I^2 R_{12} \tag{9.46}$$

or

$$dU = dq\,\mathcal{E}_{12} = dq(\varphi_2 - \varphi_1) + I^2 R_{12} dt. \tag{9.47}$$

Equation (9.46) identifies the EMF as the non-electrostatic power P delivered to the circuit per unit current I. Equation (9.47) identifies the EMF as the non-electrostatic energy per unit charge delivered to the circuit. Apart from losses associated with the internal resistance, that energy goes to establish a potential difference between the terminals. In other words, the EMF maintains a potential energy difference $dq(\varphi_1 - \varphi_2)$ between a bit of charge dq at one terminal and a bit of charge dq at the other terminal.

Example 9.4 Find the potential difference $\varphi_2 - \varphi_1$ between the two indicated points in the circuit sketched in Figure 9.8. Confirm that its value for an "open circuit" ($R \to \infty$) is identical to the EMF the battery supplies to a closed circuit. R_{12} is the internal resistance of the battery.

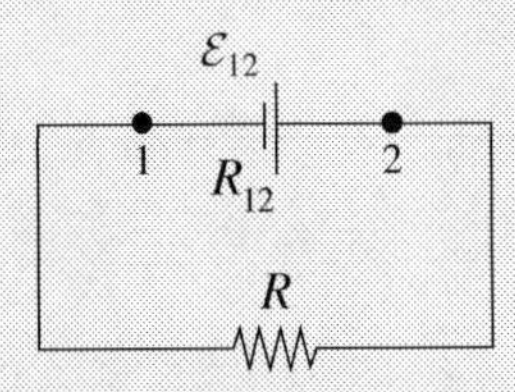

Figure 9.8: A simple circuit to illustrate the relationship between voltage and EMF.

Solution: The expression (9.43) is valid for the path from point 1 to point 2 through the battery. The expression $I(R + R_{12}) = \mathcal{E}$ is valid for the entire circuit. However, $\mathcal{E} = \mathcal{E}_{12}$, because the ohmic wire supplies no EMF. From these facts, a bit of algebra gives

$$\varphi_2 - \varphi_1 = \mathcal{E}\frac{R}{R + R_{12}}.$$

We get the desired result, $V_T = \varphi_2 - \varphi_1 = \mathcal{E}$, when we cut the wire ($R \to \infty$) to create an open circuit. V_T is often called the "terminal voltage" of a battery.

9.7.3 Kirchhoff's Laws

Kirchhoff's laws are a restatement of the differential laws $\nabla \cdot \mathbf{j} = 0$ and $\nabla \times \mathbf{E} = 0$ for filamentary-wire circuits. Figure 9.9 shows a typical direct-current (DC) multiloop circuit with "nodes" (dark circles) at points where the current can split into parts which follow different paths. Typically, one assigns a current I_k to every segment of wire between two nodes and defines several closed loops as shown in the figure. In this language, the steady-current condition $\nabla \cdot \mathbf{j} = 0$ amounts to the statement that currents must flow into and out of each node so no accumulation of charge occurs:

$$\sum_k I_k = 0 \qquad \text{(Kirchhoff's current law).} \tag{9.48}$$

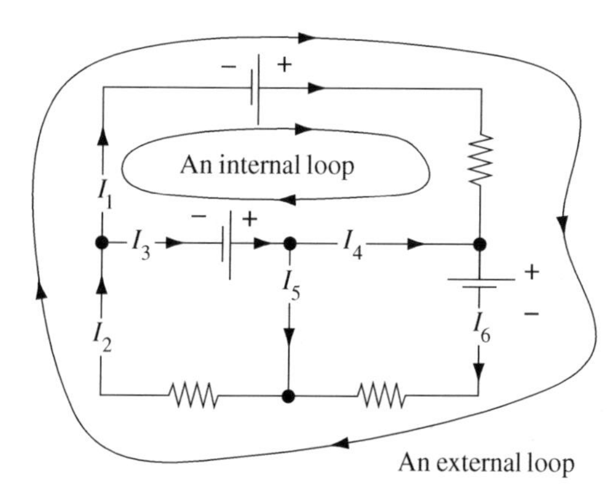

Figure 9.9: A DC circuit. The +/− signs fix the algebraic sign of the EMF for each battery. Black dots indicate nodes. The direction of the current flow arrow between each pair of nodes is arbitrary. Two loops which may be used to evaluate (9.49) are sketched. Other internal loops are not shown.

For each closed loop in Figure 9.9, $\nabla \times \mathbf{E} = 0$ implies that $\oint d\boldsymbol{\ell} \cdot \mathbf{E} = 0$. Therefore, if I_n is the current which flows through resistor R_n, the generalization of (9.44) is

$$\sum_k \mathcal{E}_k = \sum_n I_n R_n \qquad \text{(Kirchhoff's voltage law).} \tag{9.49}$$

The left side of (9.49) is the algebraic sum of the EMFs for one closed loop ($\mathcal{E}_k$ is reckoned positive if the loop passes through the seat of EMF from − to +). The right side of (9.49) is the sum of the voltage drops across each resistive element (reckoned positive if the loop is traversed in the same direction as the direction of I_n). If there are N unknown currents, we apply (9.48) and (9.49) repeatedly until N linearly independent equations are obtained. If the solution produces $I_k < 0$, the actual direction of current flow in the kth wire segment is opposite to the direction of the arrow (assigned arbitrarily) in Figure 9.9.

9.7.4 The Charges That Produce $\mathbf{E} = \mathbf{j}/\sigma$

We have identified the EMF as the agent which maintains a steady current in a circuit. But we have not identified the charge which maintains the static electric field $\mathbf{E} = \mathbf{j}/\sigma$. The relevant charge cannot lie inside the wire because $\rho = \epsilon_0 \nabla \cdot \mathbf{E} \propto \nabla \cdot \mathbf{j} = 0$. Nor can it be associated primarily with the source of EMF. This is so because we can bend the wire at a point far from the EMF and the electric field must rearrange to maintain Ohm's law with the new geometry. This suggests that the charge which is most important is quite nearby to the bend. In fact, it resides on the *surface* of the wire. This means that every circuit (indeed every conductor) which carries a steady current produces a static electric field outside of itself.

Figure 9.10 illustrates the electric field outside a DC circuit where a battery drives a current through a two-dimensional wedge-shaped conductor. By Ohm's law, the electric field inside the conductor (not shown), is everywhere parallel to the conductor walls. The tangential component of the electric field is continuous,[14] so this component must be present just *outside* the conductor as well. There is also a normal component to $\mathbf{E}_{\text{out}}(\mathbf{r})$. Otherwise, there would be no surface charge density on the conductor surface to guide the flow of current. Figure 9.10 shows that the surface charge density changes algebraic sign in the immediate vicinity of the sharp bend. We invite the reader to make a

[14] See the left side of (9.26).

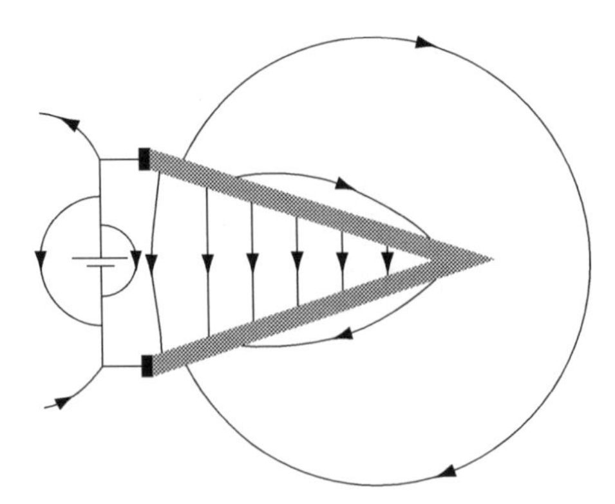

Figure 9.10: Electric field near a circuit where a battery supplies current to a wedge-shaped ohmic conductor. The conductor extends infinitely in the direction perpendicular to the page. The field lines begin and end on charges which reside on the surface of the conductor. Figure from Jefimenko (1966).

simple estimate of the total guiding charge Q at a 90° bend in a wire with conductivity σ which carries a current I. The answer is

$$Q \sim \epsilon_0 I/\sigma. \tag{9.50}$$

Numerically, this is about one electron's worth of charge when 1 A of current flows through a typical copper ($\sigma = 6 \times 10^7$ Ω·m) wire.

It is worth emphasizing that the time-independent surface charge density $\sigma(\mathbf{r}_S) = \epsilon_0\, \hat{\mathbf{n}} \cdot \mathbf{E}_{\text{out}}(\mathbf{r}_S)$ is produced by a collection of immobile charge carriers which are fixed in space. These coexist with the charge carriers that constitute the steady current in the bulk of the wire. The latter move at the average drift velocity (9.16).

Application 9.2 The Electric Field outside a Current-Carrying Wire

It is instructive to find the external electric field $\mathbf{E}_{\text{out}}(\rho, z)$ and the surface charge density $\sigma(z)$ associated with a long straight ohmic wire where a potential difference V drives a steady current from $-\ell/2$ to $\ell/2$ on the z-axis. If the circular cross section of the wire has radius R, we can calculate these quantities approximately in the limit when $R \ll \ell$ and $\rho, z \ll \ell$.

The current is steady and $\mathbf{j} = \sigma\mathbf{E}$, so the potential satisfies $\nabla^2\varphi = 0$ both inside and outside the wire. The finite length of the wire complicates the boundary value problem considerably. However, the restriction $\rho, z \ll \ell$ allows us to treat the wire as infinite as a first approximation. Ohm's law guarantees that $\varphi_{\text{in}} = -E_0 z$ inside the wire where $E_0 = V/\ell$. Outside the wire, the appropriate azimuthally symmetric solution to this Laplace's equation problem is $\varphi_{\text{out}} = Az\ln(\rho/L)$ where A and L are constants (see Section 7.8). There is no term in φ_{out} proportional to $\ln\rho$ itself because we assume the wire has zero net charge.

The matching condition $\varphi_{\text{in}} = \varphi_{\text{out}}$ at $\rho = R$ fixes A so

$$\varphi_{\text{out}}(\rho, z) = -V\frac{z}{\ell}\frac{\ln(\rho/L)}{\ln(R/L)}. \tag{9.51}$$

The constant L is determined by the behavior of φ_{out} when ρ or z is large—a domain where our solution is no longer valid. On the other hand, the physical solution should depend only on dimensionless ratios of the various lengths in the problem. This suggests that $L = \ell$, a guess that can be confirmed explicitly.[15] Therefore,

$$\varphi_{\rm out}(\rho, z) = -V\frac{z}{\ell}\frac{\ln(\rho/\ell)}{\ln(R/\ell)} \qquad \rho, z \ll \ell, \tag{9.52}$$

$$\mathbf{E}_{\rm out}(\rho, z) = -\frac{V}{\ell}\frac{\hat{\mathbf{z}}\ln(\rho/\ell) + \hat{\boldsymbol{\rho}}(z/\rho)}{\ln(R/\ell)} \qquad \rho, z \ll \ell. \tag{9.53}$$

$\mathbf{E}_{\rm in}$ points along $\hat{\mathbf{z}}$ so the charge density on the surface of the wire is

$$\sigma(z) = \epsilon_0[\mathbf{E}_{\rm out}(R, z) - \mathbf{E}_{\rm in}(R, z)] \cdot \hat{\boldsymbol{\rho}} = \epsilon_0\frac{V}{\ell}\frac{z/R}{\ln(R/\ell)} \qquad z \ll \ell. \tag{9.54}$$

$\sigma(z)$ is an odd function of z, so the total charge on the wire surface is zero, as expected. The formula above for $\mathbf{E}_{\rm out}$ approximately describes the field lines which begin and end on the same side of the wedge conductor in Figure 9.10. ■

9.8 Current Sources

The lines of $\mathbf{j}(\mathbf{r})$ shown in Figure 9.5 begin and end at the conducting electrodes attached to the surface of the ohmic sample. A more general situation is shown in Figure 9.11. The left panel is a sketch by Michael Faraday of the pattern of current flow around an electric fish submerged in a tank of water. The right panel shows modern data of the current flow equipotentials produced by a related electric fish. The two patterns are consistent and clearly indicate that the fish acts as a three-dimensional source of biochemical electromotive force.

With this insight, we can combine (9.39) with $\nabla \cdot \mathbf{j} = 0$ and $\mathbf{E} = -\nabla\varphi$ to derive an equation for the potential:

$$\sigma\, \nabla^2\varphi(\mathbf{r}) = -f(\mathbf{r}). \tag{9.55}$$

The function $f(\mathbf{r}) = -\sigma\nabla \cdot \mathbf{E}'(\mathbf{r})$ represents the spatial distribution of current sources and sinks which produce $\mathbf{j}(\mathbf{r})$. When the sources/sinks are point-like, $f(\mathbf{r})$ reduces to one or more delta functions. An example is the current dipole sketched in Figure 9.12(a). This model is often used to discuss the electrical activity of organs like the brain and the heart. Point sources can also model small spherical electrodes embedded in a conventional ohmic medium if the insulated wires which bring the current to and from the electrodes are thin enough that they do not disturb the current distribution in the medium. A tiny hemisphere is used to model an electrode which is very small compared to the size of the conductor to which it is attached [Figure 9.12(b)].

The family resemblance to Poisson's equation points the way to solution methods for (9.55). Thus, the potential of a single point current source $f(\mathbf{r}) = I\delta(\mathbf{r})$ at the origin of an infinite conducting medium is

$$\varphi(r) = \frac{I}{4\pi\sigma r}. \tag{9.56}$$

The hemispherical "point" source in Figure 9.12 expels its current exclusively into the semi-infinite conductor below, so

$$\varphi(r) = \frac{I}{2\pi\sigma r}. \tag{9.57}$$

Note that (9.57) satisfies the boundary condition $\partial\varphi/\partial n = 0$ at the surface.

[15] See Marcus (1941) and Coombes and Laue (1981) in Sources, References, and Additional Reading.

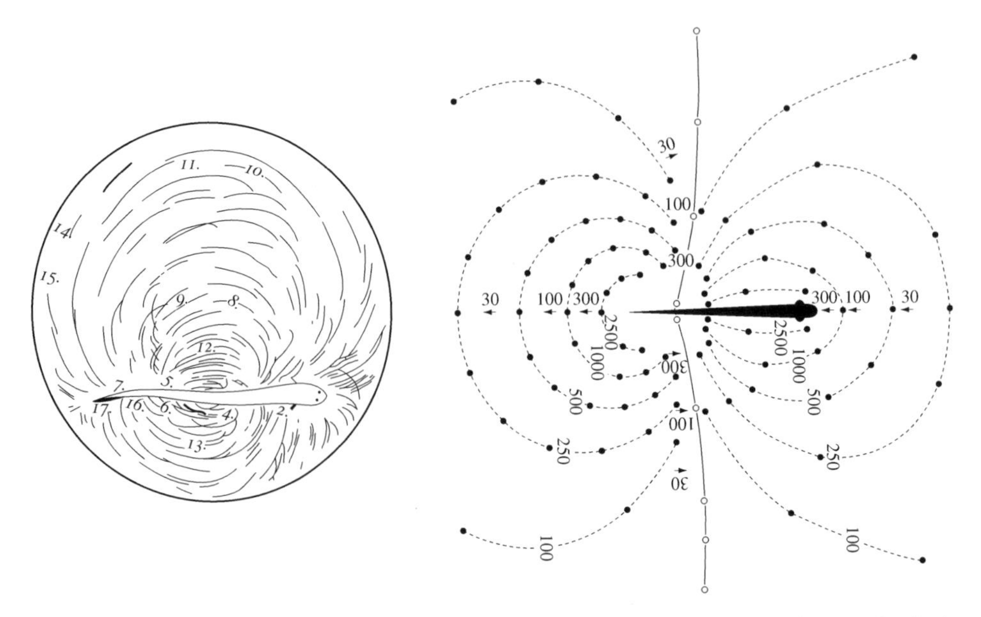

Figure 9.11: Left: Sketch from Faraday's *Diary* entry of 22 October 1838 showing lines of current density in water, produced by the electric fish *Gymnotus*. Right: Measured equipotentials in water produced by the electric fish *Apteronotus*. Data from Knudsen (1975).

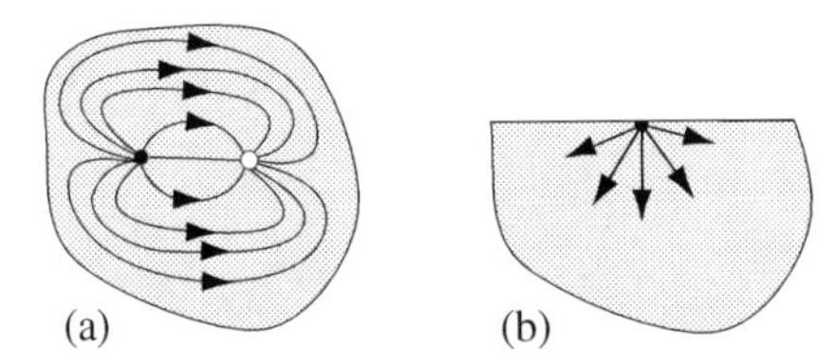

Figure 9.12: Lines of current density produced by (a) an embedded current dipole; and (b) a hemispherical point electrode attached to the surface of a semi-infinite ohmic medium.

When the volume of the conducting medium is finite but has high symmetry, it is sometimes possible to guess or compute a particular solution. To do this, we add a suitable linear combination of solutions of the homogeneous (Laplace) equation to satisfy the boundary conditions. This is one approach to the general Green function method discussed in Section 8.4. Indeed, using results derived there, it is straightforward to confirm that, up to a constant, the potential for any situation (like the current dipole in Figure 9.12) where the Neumann boundary condition (9.27) is imposed everywhere on the surface S of the conducting volume V is

$$\varphi(\mathbf{r} \in V) = \frac{1}{\sigma} \int_V d^3r' \, G_N(\mathbf{r}, \mathbf{r}') f(\mathbf{r}'). \tag{9.58}$$

The boundary condition (8.55) for the Neumann Green function $G_N(\mathbf{r}, \mathbf{r}')$ sometimes complicates matters analytically, but it poses no particular problems for numerical work.

Application 9.3 The Four-Point Resistance Probe

The resistivity $\rho = 1/\sigma$ of geological, biological, and solid-state samples is often measured by contacting four uniformly spaced leads to the sample surface in a straight line (Figure 9.13). Current flows through the two outer leads and the voltage is measured between the two inner leads. As an idealization, this Application establishes the relationship between the measured resistance and the resistivity for a sample which has thickness h but is infinite in the two other directions. We model the contact between each lead and the sample as an infinitesimal hemisphere, as in Figure 9.12(b).

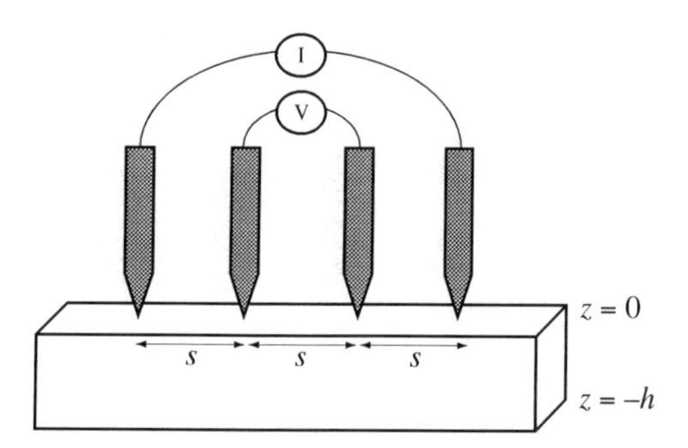

Figure 9.13: Schematic diagram of a four-point probe resistivity measurement.

For a semi-infinite medium, (9.57) tells us that the potential produced by the two current leads would be $\varphi_+(r_+) + \varphi_-(r_-)$ where $\varphi_\pm(r) = \pm\rho I/2\pi r$, r_+ is the distance to the observation point from the "positive" electrode (where the current enters the sample), and r_- is the distance from the observation point to the "negative" electrode (where the current exits the sample). This potential satisfies the Neumann boundary condition $\partial\varphi/\partial z = 0$ at $z = 0$ but not at $z = -h$.

Drawing on the analogy to electrostatics, we appeal to the method of images. Focus first on the positive electrode in Figure 9.14. The potential $\varphi_+(r_+)$ added to the potential from a positive image point current located directly below the positive electrode at $z = -2h$ produces a total potential which satisfies the boundary condition at $z = -h$ but not at $z = 0$. The latter is corrected by adding a second positive image current directly above the original electrode at $z = 2h$. As in the corresponding Dirichlet electrostatics problem (Figure 7.5), this second image has the effect of making the boundary condition no longer valid at $z = -h$. An infinite number of images is needed to satisfy the boundary condition at both surfaces.

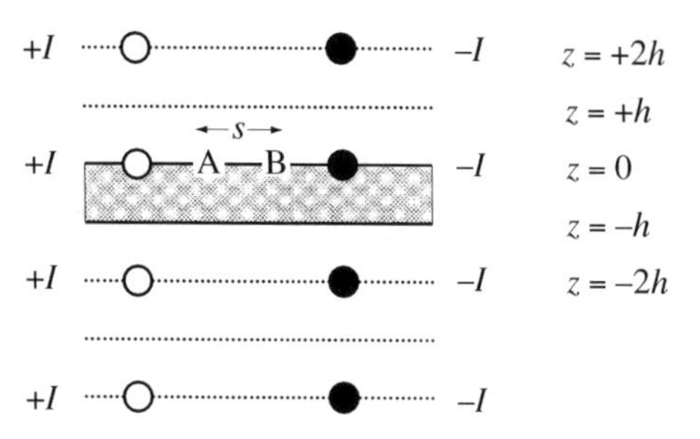

Figure 9.14: Image current system for the four-point probe.

The potential from an infinite array of negative image point currents, added to $\varphi_-(r_-)$, similarly satisfies $\partial\varphi/\partial z = 0$ at both $z = 0$ and $z = -h$. Summing the contribution from all these sources gives the potential at one voltage probe as

$$\varphi_A = \frac{\rho I}{2\pi}\left[\frac{1}{2s} + 2\sum_{n=1}^{\infty}\left\{\frac{1}{\sqrt{s^2 + (2nh)^2}} - \frac{1}{\sqrt{(2s)^2 + (2nh)^2}}\right\}\right]. \tag{9.59}$$

It is not difficult to check that $\varphi_B = -\varphi_A$ so the resistance of the sample is

$$R = \frac{\varphi_A - \varphi_B}{I} = \frac{\rho}{2\pi}\left[\frac{1}{s} + \frac{2}{h}\sum_{n=1}^{\infty}\left\{\frac{1}{\sqrt{(s/2h)^2 + n^2}} - \frac{1}{\sqrt{(s/h)^2 + n^2}}\right\}\right]. \tag{9.60}$$

The limit $h/s \gg 1$ eliminates the effect of all the images so $R = \rho/2\pi s$. When $h/s \ll 1$, the first two terms of the Euler-Maclaurin summation formula[16] give $R = (\rho/\pi h)\ln 2$. As a check on this limit, we can study a truly two-dimensional sample where we replace the point currents used above by a positive and negative *line* current (with current/length Γ) oriented perpendicular to the film. In that case, $\varphi_+(r_+) = (\rho\Gamma/2\pi)\ln r_+$, and similarly for $\varphi_-(r_-)$. This gives

$$\varphi_A = -\varphi_B = (\rho\Gamma/2\pi)[\ln s - \ln(2s)]. \tag{9.61}$$

Therefore, in agreement with the limiting result, the resistance is

$$R = \frac{\varphi_A - \varphi_B}{\Gamma h} = \frac{\rho}{\pi h}\ln 2. \tag{9.62}$$

■

Example 9.5 Electrocardiography gathers information about the electrical activity of the heart from measurements of the electric potential at various points on the surface of the heart. This information can be used to estimate the first moment of the effective current source $f(\mathbf{r})$ in (9.55). Specifically, for a body with uniform conductivity σ whose volume V is bounded by an insulating medium, prove that the first moment $\mathbf{M}$ of the source strength $f(\mathbf{r})$ is

$$\mathbf{M} = \int_V d^3r\, f(\mathbf{r})\mathbf{r} = \sigma \int_S d\mathbf{S}\,\varphi(\mathbf{r}).$$

Thus, measurements of electric potential over the surface of the heart give an indication of the integrated strength of the internal current source.

Solution: Begin with (9.55) and focus on one component of the desired moment. This gives

$$\int_V d^3r\, f\, r_k = -\sigma\int_V d^3r\, r_k\,\nabla^2\varphi = -\sigma\int_V d^3r\,\nabla\cdot(r_k\nabla\varphi) + \sigma\int_V d^3r\,\nabla\varphi\cdot\nabla r_k.$$

Now apply the divergence theorem to the first term on the far right to get

$$M_k = \int_V d^3r\, f\, r_k = -\sigma\int_S d\mathbf{S}\cdot\nabla\varphi\, r_k + \sigma\int_V d^3r\,\nabla\varphi\cdot\nabla r_k.$$

The insulator boundary condition $\partial\varphi/\partial n = 0$ everywhere on S makes the first term on the right vanish. On the other hand, because $\nabla^2 r_k = 0$,

$$\int_V d^3r\,\nabla\varphi\cdot\nabla r_k = \int_V d^3r\,\nabla\cdot(\varphi\nabla r_k) - \int_V d^3r\,\varphi\nabla^2 r_k = \int_S d\mathbf{S}\cdot\varphi\,\nabla r_k.$$

This proves the stated identity because

$$M_k = \sigma\int_S d\mathbf{S}\cdot\varphi\,\nabla r_k = \sigma\int_S dS_k\,\varphi.$$

[16] $\sum_{n=1}^{\infty} f(n) = \int_0^{\infty} dn\, f(n) + \frac{1}{2}[f(0) + f(\infty)] + \cdots$.

9.9 Diffusion Current: Fick's Law

The Drude derivation of Ohm's law sketched in Section 9.3 balanced the Coulomb force $q\mathbf{E}$ against a drag force produced by collisions between unlike particles. However, if the density $n(\mathbf{r})$ of a particle species varies in space, it has been known for over a century that collisions of this sort induce a particle current even when the particles are uncharged. Arguments from kinetic theory[17] show that the associated particle number current density is given by *Fick's law*:

$$\mathbf{j}_n = -D\nabla n. \tag{9.63}$$

The quantity $D > 0$ is called the *diffusion constant*.

For particles which carry a charge q, (9.63) generates a charge current density $\mathbf{j} = q\mathbf{j}_n$ which supplements the Ohm's law current density (9.21). The resulting phenomenological expression for the current density is called the *drift-diffusion equation*:

$$\mathbf{j}(\mathbf{r}) = \sigma\mathbf{E}(\mathbf{r}) - qD\nabla n(\mathbf{r}). \tag{9.64}$$

Some of the physics of the drift-diffusion equation is revealed when we use $\epsilon_0\nabla\cdot\mathbf{E} = \rho$ to eliminate $\rho(\mathbf{r}) = qn(\mathbf{r})$ from (9.64) and use the identity $\nabla\times\nabla\times\mathbf{E} = \nabla(\nabla\cdot\mathbf{E}) - \nabla^2\mathbf{E}$. Since static fields satisfy $\nabla\times\mathbf{E} = 0$, the resulting current density is

$$\mathbf{j}(\mathbf{r}) = \sigma\mathbf{E}(\mathbf{r}) - \epsilon_0 D\nabla^2\mathbf{E}(\mathbf{r}). \tag{9.65}$$

No current flows in static equilibrium. Therefore, putting $\mathbf{j} = 0$ in (9.65) gives an equation for the equilibrium electric field inside the sample:

$$\nabla^2\mathbf{E} - \frac{\sigma}{\epsilon_0 D}\mathbf{E} = 0. \tag{9.66}$$

Dimensional analysis of (9.66) shows that $\sqrt{\epsilon_0 D/\sigma}$ is a characteristic length for this problem. In fact, comparing (9.66) to (5.90) identifies this quantity as the screening length ℓ defined in our discussion (Section 5.7) of the electrostatics of real conductors:

$$\frac{\sigma}{\epsilon_0 D} = \frac{1}{\ell^2}. \tag{9.67}$$

This formula is called the *Einstein relation*. It relates the screening length—an equilibrium quantity which is independent of the details of particle transport—to the ratio of two transport coefficients.

The Einstein relation bears directly on the relative importance of the Ohm's law and Fick's law components of (9.65). Recall from Section 5.7.1 that ℓ is the Thomas-Fermi length $\ell_{\rm TF} = \sqrt{\pi a_{\rm B}/4k_{\rm F}}$ in a good metal and the Debye-Hückel length $\ell_{\rm DH} = \sqrt{\epsilon_0 kT/e^2 n}$ in a thermal plasma. Ohm's law works well for good metals because $l_{\rm TF}$ is microscopically small. According to (9.67), any corrections due to diffusion currents arise only within a microscopic distance of free surfaces or interfaces where the electron density changes abruptly. Conversely, $\ell_{\rm DH}$ can be macroscopically large in a thermal plasma. This means that diffusion currents can be significant over large distances from boundary layers in the ionosphere and from interfaces in doped semiconductors and living cells.

Application 9.4 The Resting Potential of a Cell

A potential difference of about 100 mV exists across the cell wall of nerve and muscle cells. The interior of the cell has a lower potential than the exterior. A simple model for the origin of this *resting*

[17] See, for example, C. Kittel and H. Kroemer, *Thermal Physics*, 2nd edition (W.H. Freeman, New York, 1980), Chapter 14.

potential difference exploits the fact that these cells have evolved over time so their cell walls separate two compositionally different conducting plasmas. The concentration of K^+ ions is relatively high in the cytoplasm of such a cell and relatively low in the plasma which surrounds the cell (top panel of Figure 9.15).

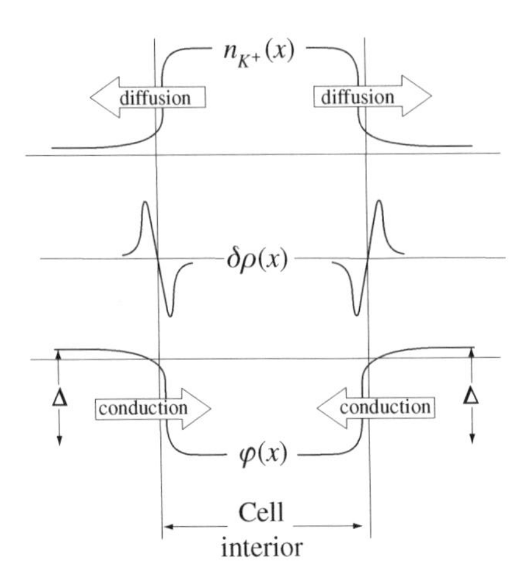

Figure 9.15: Current flow through a cell wall. Top: concentration of K^+ ions. Middle: charge density forms a dipole layer. Bottom: electrostatic potential. Block arrows indicate the direction of diffusion currents and conduction currents.

Fick's law predicts that a diffusion current of K^+ ions flows across the wall from just inside the cell to just outside the cell. This motion of positive ions destroys the local charge neutrality which existed previously and leads to a slight excess of positive (negative) charge just outside (inside) the cell wall. As shown in the middle panel of Figure 9.15, an electric double layer (Section 4.3) forms. Like a parallel-plate capacitor, the electric field of a double layer is only non-zero inside the double layer itself. This field drives a conduction current of K^+ ions back into the cell. In equilibrium, the conduction current and the diffusion current balance one another and the "built-in" electric field which remains is responsible for the change in electrostatic potential across the cell wall (lower panel of Figure 9.15).

For applications like this one, it is usual to define a *mobility* $\hat{\mu} = \sigma/nq$ so the drift velocity (9.16) is $\mathbf{v}_\mathrm{d} = \hat{\mu}\,\mathbf{E}$. In that case, (9.64) is usually called the Nernst-Planck equation:

$$\mathbf{j}(\mathbf{r}) = qn\hat{\mu}\mathbf{E}(\mathbf{r}) - qD\nabla n(\mathbf{r}). \tag{9.68}$$

The total current (9.68) is zero in equilibrium. With $\mathbf{E} = -\nabla\varphi$, this means that $\hat{\mu}d\varphi = -Ddn/n$. Integrating across the membrane gives the resting potential difference:

$$\Delta = \varphi_\mathrm{out} - \varphi_\mathrm{in} = \frac{D}{\hat{\mu}}\ln(n_\mathrm{in}/n_\mathrm{out}) = \frac{kT}{q}\ln(n_\mathrm{in}/n_\mathrm{out}). \tag{9.69}$$

The last equality follows from the Einstein relation (9.67) because the Debye-Hückel screening length quoted in the text just above applies to a biological plasma. The K^+ concentrations for a quiescent neuron membrane are $n_\mathrm{in} \approx 120$ mmol/L and $n_\mathrm{out} \approx 5$ mmol/L. At body temperature ($T = 310$ K), the formula above gives $\Delta \approx 86$ mV, which is not far from experiment. ■

Sources, References, and Additional Reading

The quotation at the beginning of the chapter is taken from

A. Volta, "On the electricity excited by the mere contact of conducting substances of different kinds", *Philosophical Magazine* **7**, 289 (1800). Reproduced by permission of Taylor & Francis Ltd.

An excellent biography of Volta is

G. Pancaldi, *Volta* (University Press, Princeton, 2003).

Section 9.1 Particulary good textbook treatments of the electrostatics of conducting media include

I.E. Tamm, *Fundamentals of Electricity*, 9th edition (Mir, Moscow, 1979).

O.D. Jefimenko, *Electricity and Magnetism* (Appleton Century-Crofts, New York, 1966).

M.H. Nayfeh and M.K. Brussel, *Electricity and Magnetism* (Wiley, New York, 1985).

The quotation in the footnote to Section 9.1.1 comes from

J.J. Thomson, *Recollections and Reflections* (MacMillan, New York, 1937).

Section 9.2 It is well worth reading the paper where the Child-Langmuir law was announced:

I. Langmuir, "The effect of space charge and residual gases on thermionic currents in high vacuum", *Physical Review* **2**, 409 (1913).

Section 9.5 Gustav Kirchhoff was the first to propose the method of Figure 9.5 to extract intrinsic electrical conductivity from measurements of current and potential difference for samples of specified shape and size. Two modern contributions to the Green function method introduced by Kirchhoff are

S. Murashina, "Neumann Green functions for Laplace's equation for a circular cylinder of finite length", *Japanese Journal of Applied Physics* **12**, 1232 (1973).

H. Levine, "On the theory of Kirchhoff's method for the determination of electrical conductivity", *Physica A* **96**, 60 (1979).

The source of Figure 9.7 is *Jansen et al.* This article summarizes the results of Application 9.1 and also those of *Sharvin*. *Erts et al.* report contact measurements of the crossover from ohmic to ballistic transport.

A.G.M. Jansen, A.P. van Gelder, and P. Wyder, "Point-contact spectroscopy in metals", *Journal of Physics C* **13**, 6073 (1980).

Yu.V. Sharvin, "A possible method for studying Fermi surfaces", *Journal of Experimental and Theoretical Physics* **48**, 984, (1965).

D. Erts, H. Olin, L. Ryen, E. Olsson, and A. Thölén, "Maxwell and Sharvin conductance in gold point contacts investigated using TEM-STM", *Physical Review B* **61**, 12725 (2000).

Section 9.6 This section is adapted from

A. Kovetz, *Electromagnetic Theory* (University Press, Oxford, 2000), Section 36.

Section 9.7 The physics of voltaic cells is discussed with care in

W.M. Saslow, "Voltaic cells for physicists", *American Journal of Physics* **67**, 574 (1999).

Figure 9.10 comes from Jefimenko (see Section 9.1 above). Application 9.2 was adapted from

A. Marcus, "The electric field associated with steady current in a long cylindrical conductor", *American Journal of Physics* **9**, 225 (1941).

C.A. Coombes and H. Laue, "Electric fields and charge distributions associated with steady currents", *American Journal of Physics* **49**, 450 (1981).

Section 9.8 The left and right panels in Figure 9.11, respectively, come from

R.T. Cox, "Electric fish", *American Journal of Physics* **11**, 13 (1943).

E.I. Knudsen, "Spatial aspects of the electric fields generated by weakly electric fish", *Journal of Comparative Physiology* **99**, 103 (1975).

Example 9.5 is taken from this paper by Dennis Gabor, who later won a Nobel prize for his invention of holography:

D. Gabor and C.V. Nelson, "Determination of the resultant dipole of the heart from measurements on the body surface", *Journal of Applied Physics* **25**, 413 (1954).

Application 9.3 was adapted from

L.B. Valdes, "Resistivity measurements on germanium for transistors", *Proceedings of the Institute of Radio Engineers* **42**, 420 (1954).

Section 9.9 Figure 9.15 and Application 9.4 come from

M. Uehara, K.K. Sakane, H.S. Maciel, and W.I. Urruchi, "Physics and biology: Bio-plasma physics", *American Journal of Physics* **68**, 450 (2000).

Problems

9.1 A Power Theorem A steady current is produced by a collection of moving charges confined to a volume V. Prove that the rate at which work is done on these moving charges by the electric field produced by a static charge distribution (not necessarily confined to V) is zero.

9.2 A Salt-Water Tank A battery maintains a potential difference V between the two halves of the cover of a tank ($L \times \infty \times h$) filled with salty water. Find the current density $\mathbf{j}(x, y, z)$ induced in the water.

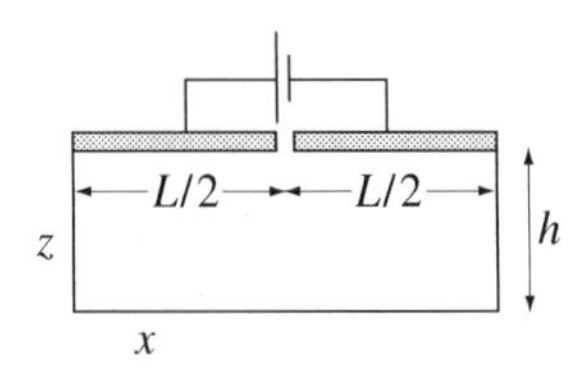

9.3 Radial Hall Effect An infinitely long cylindrical conductor carries a constant current with density $j_z(r)$.

(a) Despite Ohm's law, compute the radial electric field $E_r(r)$ that ensures that the radial component of the Lorentz force is zero for every current-carrying electron.

(b) The source of $E_r(r)$ is $\rho(r) = \rho_+ + \rho_c(r)$ where ρ_+ comes from a uniform distribution of immobile positive ions and $\rho_c(r) = j_z(r)/v$ comes from electrons with velocity v. Show that $\rho_c(r) = \rho_c = -\rho_+/(1 - v^2/c^2)$. Do not use special relativity.

(c) Estimate the potential difference from the center to the surface of a copper wire with circular cross section 1 cm^2 that carries a current of 1 A.

9.4 Acceleration EMF

(a) A long straight rod with cross sectional area A and conductivity σ accelerates parallel to its length with acceleration $\mathbf{a}$. Write down the Drude-like equation of motion for the average velocity $\mathbf{v}$ of an electron of mass m in the rod relative to the motion of the rod itself. Show that, in the absence of an external electric field, there is a steady solution that corresponds to a current $I = A\sigma ma/e$ flowing in the wire. Assume that a flexible, no-loss wire and a perfect ammeter close the circuit.

(b) Argue that this result generalizes to an EMF

$$\mathcal{E} = \oint d\boldsymbol{\ell} \cdot (m\mathbf{a}/e)$$

for the case of electrons in an accelerated closed circuit. Ignore the Drude drag force.

(c) Use the results of part (b) to show that a circular wire ring rotated with angular acceleration $\dot{\Omega}$ carries a current

$$I = \frac{2m\dot{\Omega}S}{eR},$$

where R is the total resistance of the wire and S is the area of the wire circle. Estimate the magnitude of this current for a copper ring of radius 1 cm and cross sectional area 1 mm^2 which oscillates harmonically around the ring axis at 500 Hz, with a maximum angular amplitude of 1° of arc.

9.5 Membrane Boundary Conditions A thin membrane with conductivity σ' and thickness δ separates two regions with conductivity σ.

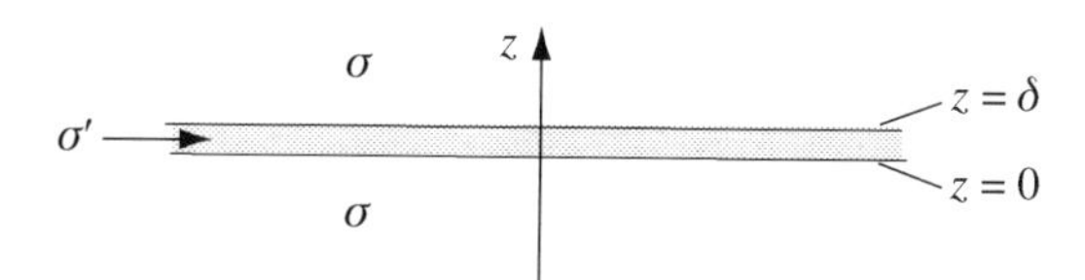

Assume uniform current flow in the z-direction in the figure above. When δ is small, it makes sense to seek "across-the-membrane" matching conditions for the electrostatic potential $\varphi(z)$ defined entirely in terms of quantities defined outside the membrane. Find the potential in all three regions of the figure and prove that suitable matching conditions are

$$\varphi(z=\delta^+) - \varphi(z=0^-) = \delta \frac{\sigma}{\sigma'} \left.\frac{d\varphi}{dz}\right|_{z=0^-}$$

$$\left.\frac{d\varphi}{dz}\right|_{z=\delta^+} - \left.\frac{d\varphi}{dz}\right|_{z=0^-} = 0.$$

9.6 Current Flow to a Bump A voltage difference V_0 causes a steady current to flow from the top conductor to the bottom conductor (in the sketch below) through an ohmic medium with conductivity σ. Find an approximate expression for the current I that flows into the hemispherical bump (radius R) portion of the lower conductor. Assume that $d \gg R$.

Hint: A grounded spherical conductor in an external field $\mathbf{E}_0$ acquires a surface charge density $\Sigma = 3\epsilon_0 E_0 \cos\theta$ where θ is the polar angle measured from $\hat{\mathbf{E}}$.

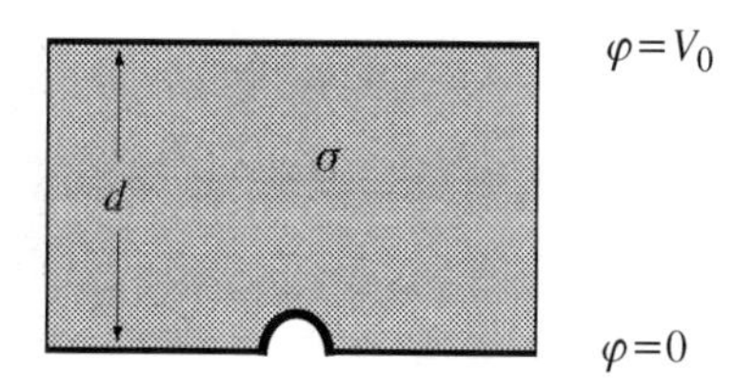

9.7 The Charge at a Bend in a Wire A wire with conductivity σ carries a steady current I. Confirm the statement made in the text that a charge $Q \sim \epsilon_0 I/\sigma$ accumulates on the wire's surface in the immediate neighborhood of a 90° right-angle bend. Make a sketch of the wire indicating the position and sign of the surface charges. Explain the physical origin of the algebraic sign of the charges that you draw.

9.8 Spherical Child-Langmuir Problem The electrodes of a spherical capacitor have radii a and $b > a$. The inner electrode is grounded; the outer electrode is held at potential V. In vacuum diode mode, the thermionic current which flows from the inner cathode to the outer anode increases with temperature until the electric field due to space charge produced between the concentric electrodes compensates the voltage-induced field and $\partial\varphi/\partial r|_{r=a} = 0$ on the surface of the cathode. Assume that the initial velocity of the thermally emitted electrons is zero.

(a) Show that the maximum current I between the electrodes is described by the expression

$$I = \sqrt{\frac{2e}{m}} 4\pi\epsilon_0 \left[\frac{V}{y(x)}\right]^{3/2}, \qquad x = \ln(b/a).$$

The function $y(x)$ is the solution to a differential equation which cannot be integrated in closed form.

(b) Find the maximum current in the limit when $b \gg a$.

9.9 A Honeycomb Resistor Network An infinite, two-dimensional network has a honeycomb structure with one hexagon edge removed. Otherwise, the resistance of every hexagon edge is r. Find the resistance of the network when a current I enters point A and is extracted at the point B. Hint: think about the current flow if the edge between A and B were not missing.

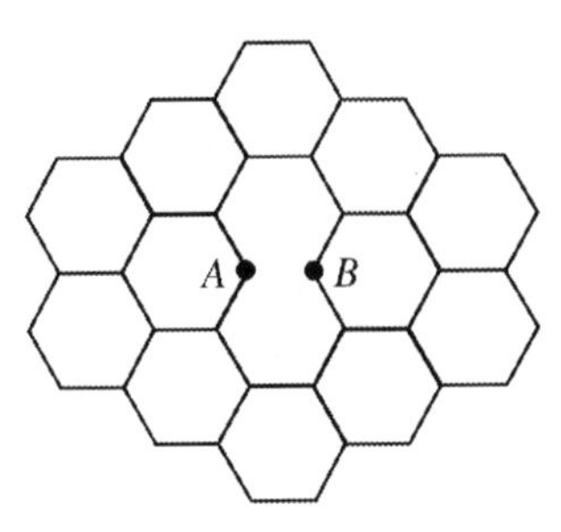

9.10 **Refraction of Current Density** Show that the lines of current density $\mathbf{j}$ obey a "law of refraction" at the flat boundary between two ohmic media with conductivities σ_1 and σ_2. Use the geometry shown below.

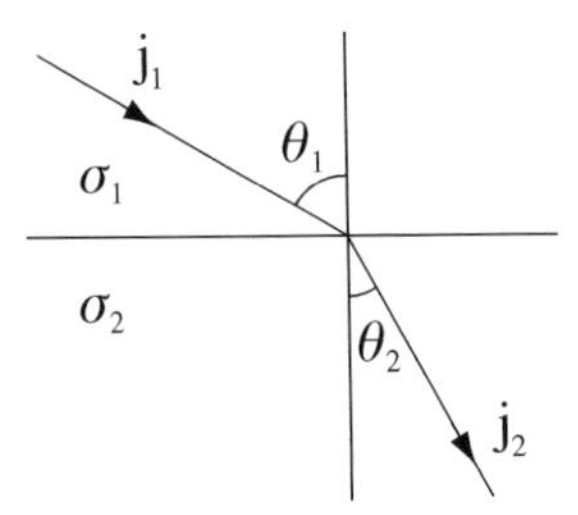

9.11 **Resistance to Ground** The diagram shows a wire connected to the Earth (conductivity σ_E) through a perfectly conducting sphere of radius a which is half-buried in the Earth. The layer of earth immediately adjacent to the sphere with thickness $b - a$ has conductivity σ_2. Find the resistance between the end of the wire and a point deep within the Earth (taken as infinitely large).

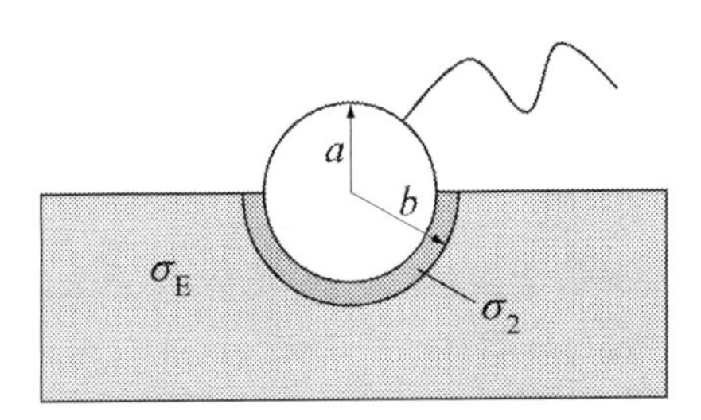

9.12 **A Separation-Independent Resistance** Two highly conducting spheres with radii a_1 and a_2 are used to inject and extract current from points deep inside a tank of weakly conducting fluid. Show that the resistance between the spheres depends very weakly on their separation d when d is large compared to a_1 and a_2. Confirm that the dependence on d disappears entirely if d is large enough. Assume that the electrolyte has permittivity ϵ and conductivity σ.

9.13 **Inhomogeneous Conductivity** Steady current flows in the x-direction in an infinite, two-dimensional strip defined by $|y| < L$. The current density $\mathbf{j}$ is constant everywhere in the strip and the conductivity varies in space as

$$\sigma(x) = \frac{\sigma_0}{1 + a\cos kx} \qquad a < 1.$$

The conductivity is zero outside the strip. Find the electric field $\mathbf{E}$ everywhere.

9.14 A Variable Resistor A square plate of copper metal can be used as a crude variable resistor by making suitable choices of the places to attach leads that carry current to and from the plate.

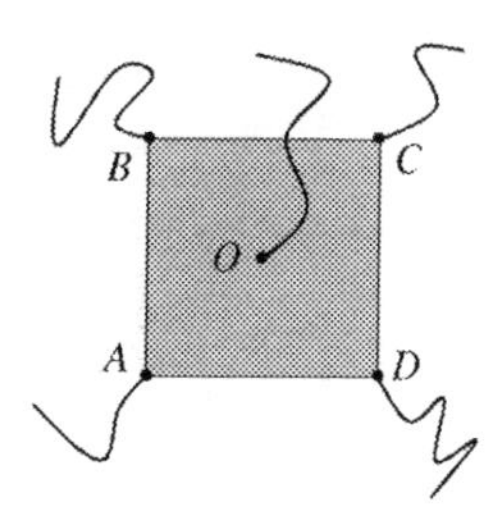

Sketch the lines of current density for the four cases described below. As far as possible, use your sketches to order the effective resistances of the plate for the four cases from smallest to largest. Explain your logic and ignore the contact resistances at the points where the wires enter and exit the plate.

(a) Current enters at A and exits at B.
(b) Current enters at A and exits at C.
(c) Current enters at A and exits at O.
(d) Current enters at A and exits at B and D, with the wires leaving those points joined together.

9.15 The Resistance of an Ohmic Sphere A current I flows up the z-axis and is intercepted by an origin-centered sphere with radius R and conductivity σ. The current enters and exits the sphere through small conducting electrodes which occupy the portion of the sphere's surface defined by $\theta \leq \alpha$ and $\pi - \alpha \leq \theta \leq \pi$. Derive an expression for the resistance of the sphere to the flowing current. Assume that $\alpha \ll 1$ and comment on the limit $\alpha \to 0$. Hint:

$$(2\ell + 1) \int_{x_1}^{x_2} dx \, P_\ell(x) = [P_{\ell+1}(x) - P_{\ell-1}(x)]_{x_1}^{x_2} .$$

9.16 Space-Charge-Limited Current in Matter Consider the vacuum diode problem treated in the text with the space between the plates filled with a poor conductor with dielectric permittivity ϵ. For matter of this kind, $\mathbf{v} = \tilde{\mu}\mathbf{E}$, where the *mobility* $\tilde{\mu}$ is the constant of proportionality between the drift velocity of the electrons and the electric field in the matter. Find the replacement for the Child-Langmuir law for the dependence of the maximum current density on the material constants, the plate separation L, and the plate potential difference V.

9.17 van der Pauw's Formula The diagram below shows an ohmic film with conductivity σ, thickness d, infinite length, and semi-infinite width. A total current I enters the film at the point A through a line contact (modeled as a half-cylinder with negligible radius) and exits the film similarly at the point B. The potential difference $V_D - V_C$ between the contact at C and the contact at D determines the resistance $R_{AB,CD}$. The contact separations are a, b, and c, as indicated.

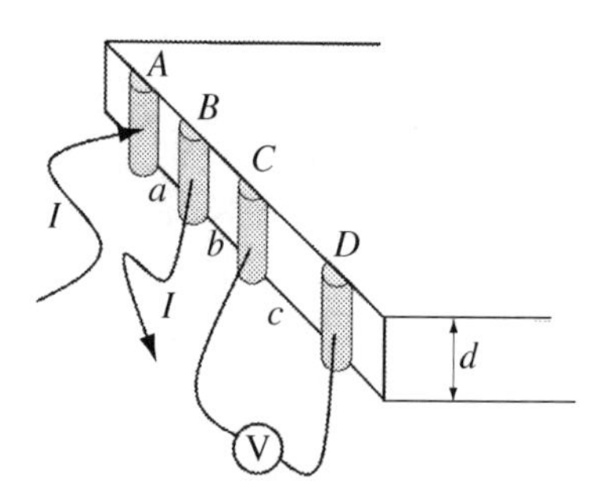

(a) Show that the electrostatic potential produced at point C by the current injected at point A is

$$\varphi_{AC} = -(I/\pi d\sigma)\ln(a+b).$$

(b) Prove that

$$\exp(-\pi d\sigma R_{AB,CD}) + \exp(-\pi d\sigma R_{BC,DA}) = 1.$$

9.18 Rayleigh-Carson reciprocity The diagram below illustrates a reciprocal principle satisfied by an ohmic sample of any shape.

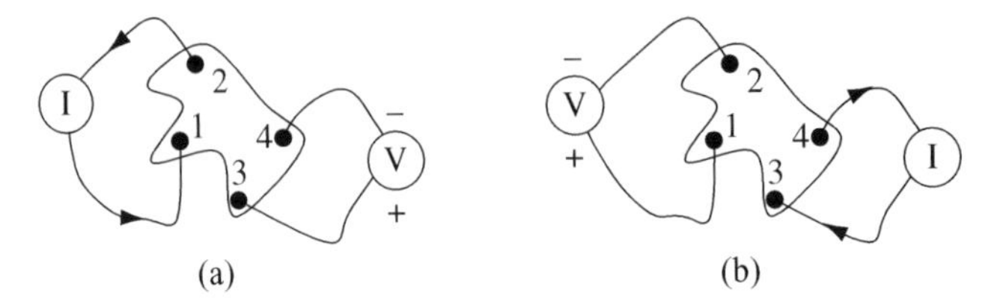

The principle asserts that if the impressed currents satisfy $I_A = I_B$, the measured voltages satisfy $V_A = V_B$. Prove this by evaluating the integral

$$\int d^3r\, [\nabla \cdot (\varphi_A \mathbf{j}_B) - \nabla \cdot (\varphi_B \mathbf{j}_A)]$$

in two different ways. Assume that the current enters and exits the sample at delta function sources and sinks.

9.19 The Electric Field of an Ohmic Tube Roll up an ohmic sheet to form an infinitely long, origin-centered cylinder of radius a. Cut a narrow slot along the length of the cylinder and insert a line source of EMF so the electrostatic potential within the sheet (in polar coordinates) is

$$\varphi(a,\phi) = \frac{V_0}{2\pi}\phi \qquad -\pi < \phi < \pi.$$

(a) Find a separated-variable solution to Laplace's equation in polar coordinates for the electrostatic potential inside and outside the cylinder.

(b) Sum the series in part (a) and show that the equipotentials inside the cylinder are straight lines drawn from the seat of the EMF. Hint:

$$\sum_{k=1}^{\infty} \frac{b^k \sin kx}{k} = \tan^{-1}\left[\frac{b\sin x}{1-b\cos x}\right].$$

(c) Prove that the potential outside the cylinder is dipolar and indicate the direction of the dipole moment. Sketch several equipotentials inside and outside the cylinder.

(d) Show that the surface charge density on both sides of the cylinder wall is $\sigma(\alpha) = (\epsilon_0 V_0/2\pi a)\tan\alpha$, where α is a polar angle defined with respect to the seat of the EMF.

(e) Show that the electric field inside the cylinder is $\mathbf{E} = -(V_0/\pi s)\hat{\boldsymbol{\alpha}}$, where s is the polar distance from the seat of the EMF. Sketch several electric field lines inside and outside the cylinder. Take some care with the angle at which the field lines approach the cylinder.

9.20 Current Density in a Curved Segment of Wire A potential difference V drives a steady current I through an ohmic wire with conductivity σ and a constant circular cross section. One portion of the wire has the shape of a circular arc with inner radius of curvature R_1 and outer radius of curvature R_2. Find the dependence of the current density j on the local radius of curvature r defined in the diagram.

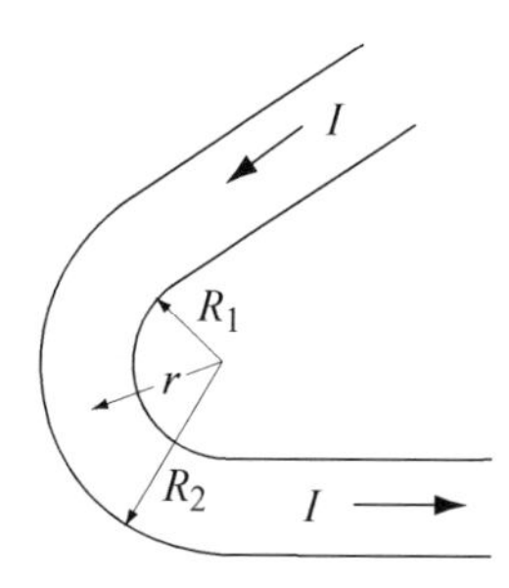

9.21 **The Annulus and the Trapezoid** The annulus shown below is cut from a planar metal sheet with thickness t and conductivity σ.

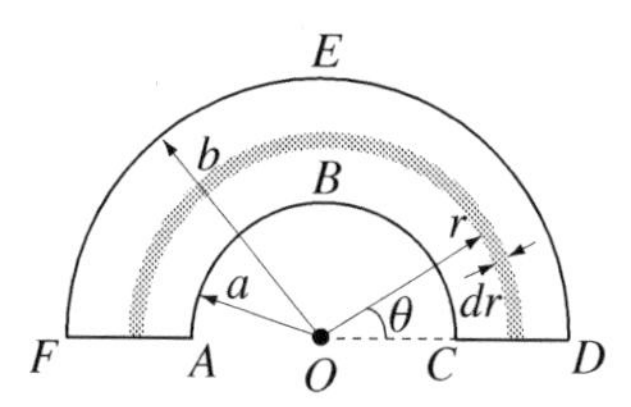

(a) Let V be the voltage between the edge CD and the edge FA. Solve Laplace's equation to find the electrostatic potential, current density, and resistance of the annulus.

(b) Divide the annulus into a sequence of concentric sub-annuli, each with width dr. Show how to combine the resistances of the individual sub-annuli to reproduce the resistance computed in part (a). Use the lines of current density predicted in each case to explain why the two calculations agree.

(c) Let V be the voltage between the edge ABC and the edge DEF of the original annulus. Repeat all the steps of part (a) and part (b).

(d) The trapezoid shown below is cut from a planar metal sheet with thickness t and conductivity σ. Let V be the voltage between the edge AB and the edge CD. Explain why the exact resistance computed by solving Laplace's equation for the entire trapezoid is *not* the same as the resistance computed by summing the resistances for sub-trapezoids like the one indicated by shading in the figure below. Does the summation calculation overestimate or underestimate the exact resistance?

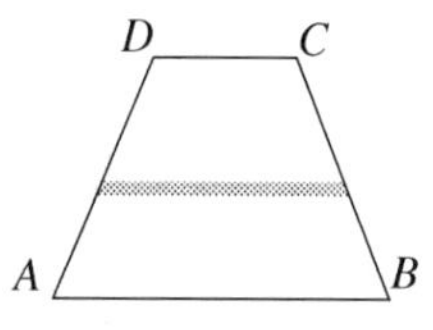

9.22 **Joule Heating of a Shell** Current flows on the surface of a spherical shell with radius R and conductivity σ. The potential is specified on two rings as $\varphi(\theta = \alpha) = V \cos n\phi$ and $\varphi(\theta = \pi - \alpha) = -V \cos n\phi$. Show that the rate at which Joule heat is generated between the two rings is

$$\mathcal{R} = \frac{2\pi n \sigma V^2}{\cos \alpha}.$$

Hint: The substitution $y = \ln[\tan(\theta/2)]$ will be useful.

9.23 **The Resistance of a Shell** A spherical shell with radius a has conductivity σ in the angular range $\alpha_1 < \theta < \pi - \alpha_2$. Otherwise, the shell is perfectly conducting and a potential difference V is maintained between $\theta = 0$ and $\theta = \pi$.

(a) Solve Laplace's equation to find the potential, surface current density, and resistance of the shell between $\theta = 0$ and $\theta = \pi$.

(b) Divide the shell into many thin rings. Find the resistance of each and combine them to find the resistance and confirm the answer derived in part (a).

Hint: The substitution $y = \ln[\tan(\theta/2)]$ will be useful.

9.24 The Resistance of the Atmosphere The conductivity of the Earth's atmosphere increases with height due to ionization by solar radiation. At an altitude of about $H = 50$ km, the atmosphere can be considered practically an ideal conductor. Experiment shows that the height dependence of the conductivity of the atmosphere can be approximated by

$$\sigma(r) = \sigma_0 + A(r - r_0)^2,$$

where $r_0 = 6.4 \times 10^6$ m is the radius of the Earth and r is the distance from the center of the Earth to the observation point. The conductivity at the surface of the Earth is $\sigma_0 = 3 \times 10^{-14}$ S/m and the constant $A = 0.5 \times 10^{-20}$ S/m^3. Experiment also shows that an electric field $E_0 \approx -100$ V/m exists near the Earth's surface and is directed downward. Estimate the resistance of the atmosphere.

9.25 Ohmic Loss in an Infinite Circuit The diagram below shows that the resistance between the terminals A and B is determined by a motif of three resistors $R_2 - R_1 - R_2$ repeated an infinite number of times. Determine the ratio R_1/R_2 such that the rate at which Joule heat is produced by all the R_1 resistors combined is a fraction α of the heat produced by the entire circuit. Find the largest value α can attain.

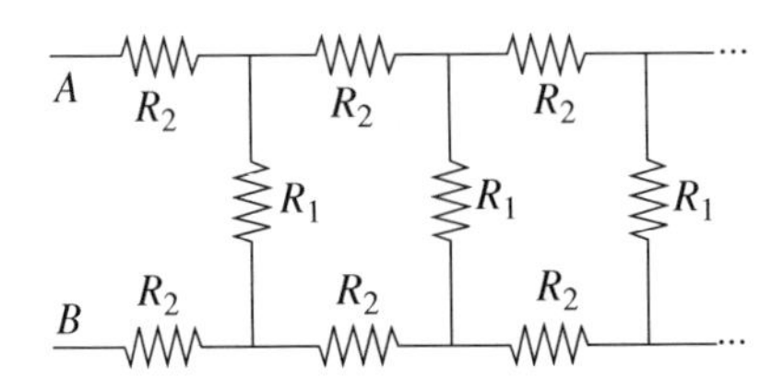

10 Magnetostatics

At every point in space at which there is a finite magnetic force there is . . . a magnetic field.
William Thomson (1851)

10.1 Introduction

Maxwell's field theory tells us that every time-independent current distribution $\mathbf{j}(\mathbf{r})$ is the source of a vector field $\mathbf{B}(\mathbf{r})$ which satisfies the differential equations

$$\nabla \cdot \mathbf{B}(\mathbf{r}) = 0 \tag{10.1}$$

and

$$\nabla \times \mathbf{B}(\mathbf{r}) = \mu_0 \mathbf{j}(\mathbf{r}). \tag{10.2}$$

The *magnetic field* $\mathbf{B}(\mathbf{r})$ demands our attention because the force $\mathbf{F}$ and torque $\mathbf{N}$ exerted by $\mathbf{j}(\mathbf{r})$ on a second current distribution $\tilde{\mathbf{j}}(\mathbf{r})$ are

$$\mathbf{F} = \int d^3r \; \tilde{\mathbf{j}}(\mathbf{r}) \times \mathbf{B}(\mathbf{r}) \tag{10.3}$$

and

$$\mathbf{N} = \int d^3r \; \mathbf{r} \times [\tilde{\mathbf{j}}(\mathbf{r}) \times \mathbf{B}(\mathbf{r})]. \tag{10.4}$$

If it happens that neither $\mathbf{F}$ nor $\mathbf{N}$ is of direct interest, the *energy* associated with $\mathbf{B}(\mathbf{r})$, $\mathbf{j}(\mathbf{r})$, and $\tilde{\mathbf{j}}(\mathbf{r})$ usually is. We will derive several equivalent expressions for magnetostatic total energy and magnetostatic potential energy in Chapter 12.

10.1.1 The Scope of Magnetostatics

Magnetostatics is a more subtle and complex subject than electrostatics. In part, this is so because the current density is a vector and the Lorentz force (10.3) involves a cross product. The most important point of physics is that two quite different types of current produce magnetic fields: *electric* current from moving charge and *magnetization* current from the quantum mechanical spin of point-like particles.[1] This makes magnetizable matter much more varied than polarizable matter in both its fundamental

[1] Magnetization current due to orbiting charge is a special case of electric current. See Chapter 13.

nature and in its response to external fields. Particularly important in this regard is the ubiquity and practical importance of permanent magnetism (ferromagnetism) due to the spontaneous alignment of electron spins. The analogous phenomenon in dielectrics (ferroelectricity) is relatively uncommon.

A quick comparison of the magnetostatic equations in Section 10.1 with the electrostatic equations in Section 3.1 shows that the roles of the divergence and curl operators are reversed in the static Maxwell equations. This leads to methods and results for magnetostatics which are quite different from the methods and results of electrostatics. In practice, it is important to appreciate, ponder, and master these differences to build physical intuition and calculational skill. We will see later that special relativity ascribes the distinction between $\mathbf{E}(\mathbf{r})$ and $\mathbf{B}(\mathbf{r})$ (as a matter of principle) to nothing more than a choice of observer reference frame (Chapter 22).

10.1.2 Magnetostatic Fields Imply Steady Currents

The divergence operation applied to both sides of (10.2) shows that the current densities relevant to magnetostatics satisfy the *steady-current* condition,

$$\nabla \cdot \mathbf{j}(\mathbf{r}) = 0. \tag{10.5}$$

Figure 9.1 of the previous chapter shows lines of $\mathbf{j}(\mathbf{r})$ for a current density of this sort.[2] The same diagram is a possible field line pattern for $\mathbf{B}(\mathbf{r})$. In general, we mimic electrostatics (Section 3.3.4) and use a scalar parameter λ to ensure that every differential element of magnetic field line $d\mathbf{s}$ is parallel to $\mathbf{B}(\mathbf{r})$ itself:

$$d\mathbf{s} = \lambda \mathbf{B}. \tag{10.6}$$

Writing (10.6) out in components produces differential equations for the field lines like[3]

$$\frac{dy}{dx} = \frac{B_y}{B_x} \qquad \text{and} \qquad \frac{dz}{dx} = \frac{B_z}{B_x}. \tag{10.7}$$

Analytic solutions for the magnetic field lines defined by (10.6) are not common. Nevertheless, an essential characteristic shared by all magnetic field patterns reveals itself when we define the *magnetic flux* through a surface S as

$$\Phi_B = \int_S d\mathbf{S} \cdot \mathbf{B}. \tag{10.8}$$

If S encloses a volume V, the divergence theorem (Section 1.4.2) and the Maxwell equation $\nabla \cdot \mathbf{B} = 0$ give

$$\Phi_B = \int_V d^3r \; \nabla \cdot \mathbf{B} = 0. \tag{10.9}$$

The zero flux condition (10.9) is a powerful constraint which cannot be satisfied unless magnetic field lines do not begin or end inside V. Since V can be infinitesimal, we conclude that lines of $\mathbf{B}(\mathbf{r})$ do not begin or end at isolated points in space (Figure 10.1).

10.1.3 Magnetic Monopoles Do Not Exist

The constraint (10.9) implies that magnetic charge does not exist. Magnetic field lines either close on themselves, begin at infinity and end at infinity, or do not close at all (see Section 10.6). Put another

[2] The force (10.3) and torque (10.4) are valid irrespective of the steady-current condition.

[3] See Section 4.2 for an electrostatic example.

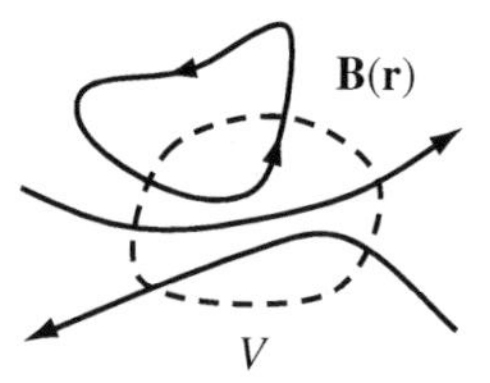

Figure 10.1: Magnetic field lines do not begin or end in any volume V of space.

way, the Maxwell equation $\nabla \cdot \mathbf{B} = 0$ summarizes the fact that no reproducible experiment has ever detected a magnetic monopole—the magnetic analog of the electric point charge. This is significant for many reasons, not least because it robs the Maxwell equations of perfect symmetry between electric and magnetic phenomena.[4]

10.1.4 Thomson's Theorem of Magnetostatics

Earnshaw's theorem of electrostatics (Section 3.3.3) states that the electric scalar potential $\varphi(\mathbf{r})$ has no local maxima or minima in a charge-free volume of space. A related theorem applies to the magnitude of a magnetostatic field.

Theorem: $|\mathbf{B}(\mathbf{r})|$ can have a local minimum, but never a local maximum, in a current-free volume of space.

Proof: Suppose, to the contrary, that $|\mathbf{B}(\mathbf{r})|$ or, equivalently, $\mathbf{B} \cdot \mathbf{B}$ has a local maximum at a point P where $\nabla \cdot \mathbf{B} = 0$ and $\nabla \times \mathbf{B} = 0$. Then, by assumption, if S is the surface of a small spherical volume V centered at P,

$$\int_S d\mathbf{S} \cdot \nabla(\mathbf{B} \cdot \mathbf{B}) < 0. \tag{10.10}$$

On the other hand, using the divergence theorem,

$$\int_S d\mathbf{S} \cdot \nabla(\mathbf{B} \cdot \mathbf{B}) = \int_V d^3r\ \nabla_i \nabla_i (B_j B_j) = 2\int_V d^3r\, (\nabla_i B_j)^2 + 2\int_V d^3r\ B_j \nabla^2 B_j. \tag{10.11}$$

The last term in (10.11) vanishes because $\nabla \times \mathbf{B} = 0$ in V implies that $\nabla_i B_j = \nabla_j B_i$. Specifically,

$$\int_V d^3r\ B_j \nabla^2 B_j = \int_V d^3r\ B_j \nabla_i \nabla_j B_i = \int_V d^3r\ B_j \nabla_j (\nabla \cdot \mathbf{B}) = 0. \tag{10.12}$$

We conclude that

$$\int_S d\mathbf{S} \cdot \nabla(\mathbf{B} \cdot \mathbf{B}) = 2\int_V d^3r\, (\nabla_i B_j)^2 \geq 0. \tag{10.13}$$

This contradicts our original assumption and so proves the assertion. Application 12.2 in Section 12.4.2 explores some practical applications of Thomson's theorem.

[4] See Section 2.5.5 on the "symmetric" Maxwell equations. For the current status of magnetic monopoles in theory and experiment, see Milton (2006) in Sources, References, and Additional Reading.

10.2 The Law of Biot and Savart

Equations (10.1) and (10.2) specify the divergence and curl of $\mathbf{B}(\mathbf{r})$. This is precisely the information required by the Helmholtz theorem (Section 1.9) to give an explicit expression for the unique magnetic field produced by $\mathbf{j}(\mathbf{r})$. Since $\nabla \cdot \mathbf{B} = 0$ and $\nabla \times \mathbf{B} = \mu_0 \mathbf{j}$, the theorem gives

$$\mathbf{B}(\mathbf{r}) = \nabla \times \frac{\mu_0}{4\pi} \int d^3r' \, \frac{\mathbf{j}(\mathbf{r}')}{|\mathbf{r} - \mathbf{r}'|}. \tag{10.14}$$

The integral in (10.14) converges for all well-behaved, spatially localized current distributions. It may not converge for a current source of infinite extent but, even in that case, it is often possible to extract $\mathbf{B}(\mathbf{r})$ as the limit of the field produced by a source of finite size.[5]

A fundamental superposition formula for $\mathbf{B}(\mathbf{r})$ follows from (10.14) when we bring the curl inside the integral and use the fact that $\mathbf{j}(\mathbf{r}')$ is not a function of $\mathbf{r}$. The result is the historically significant *law of Biot and Savart*:

$$\mathbf{B}(\mathbf{r}) = \frac{\mu_0}{4\pi} \int d^3r' \, \frac{\mathbf{j}(\mathbf{r}') \times (\mathbf{r} - \mathbf{r}')}{|\mathbf{r} - \mathbf{r}'|^3}. \tag{10.15}$$

An alternate form which applies when a surface current density $\mathbf{K}(\mathbf{r}_S)$ is specified rather than a volume current density is

$$\mathbf{B}(\mathbf{r}) = \frac{\mu_0}{4\pi} \int dS \frac{\mathbf{K}(\mathbf{r}_S) \times (\mathbf{r} - \mathbf{r}_S)}{|\mathbf{r} - \mathbf{r}_S|^3}. \tag{10.16}$$

When a current I flows in a filamentary (one-dimensional) wire, and $\boldsymbol{\ell}$ is a vector which points to a line element $d\boldsymbol{\ell}$ of the wire, the substitution $\mathbf{j}\, d^3r \to I d\boldsymbol{\ell}$ (see Section 9.3.1) simplifies (10.15) to

$$\mathbf{B}(\mathbf{r}) = \frac{\mu_0 I}{4\pi} \int \frac{d\boldsymbol{\ell} \times (\mathbf{r} - \boldsymbol{\ell})}{|\mathbf{r} - \boldsymbol{\ell}|^3}. \tag{10.17}$$

The next few sections report Biot-Savart results we will exploit in later sections.

Application 10.1 Irrotational Current Sources

Physical current densities where $\nabla \times \mathbf{j} = 0$ produce $\mathbf{B}(\mathbf{r}) = 0$ at every point in space. To prove this, write the Biot-Savart integral (10.15) in the form

$$\mathbf{B}(\mathbf{r}) = \frac{\mu_0}{4\pi} \int d^3r' \, \mathbf{j}(\mathbf{r}') \times \nabla' \frac{1}{|\mathbf{r} - \mathbf{r}'|}. \tag{10.18}$$

Now, integrate by parts and use the integral theorem (1.78). The result is

$$\mathbf{B}(\mathbf{r}) = \frac{\mu_0}{4\pi} \int d^3r' \, \frac{\nabla' \times \mathbf{j}(\mathbf{r}')}{|\mathbf{r} - \mathbf{r}'|} - \frac{\mu_0}{4\pi} \int d\mathbf{S} \times \frac{\mathbf{j}(\mathbf{r}')}{|\mathbf{r} - \mathbf{r}'|}. \tag{10.19}$$

Physically interesting currents are localized in the sense that they vanish at infinity sufficiently fast to ensure that the surface integral in (10.19) is zero. We conclude that localized and curl-free current distributions ($\nabla \times \mathbf{j} = 0$) produce no magnetic field anywhere. ■

10.2.1 A Circular Current Loop

No simple analytic formula gives the magnetic field at every point in space for a circular current loop. However, the Biot-Savart integral is not difficult to evaluate if we restrict ourselves to observation

[5] See Example 10.4.

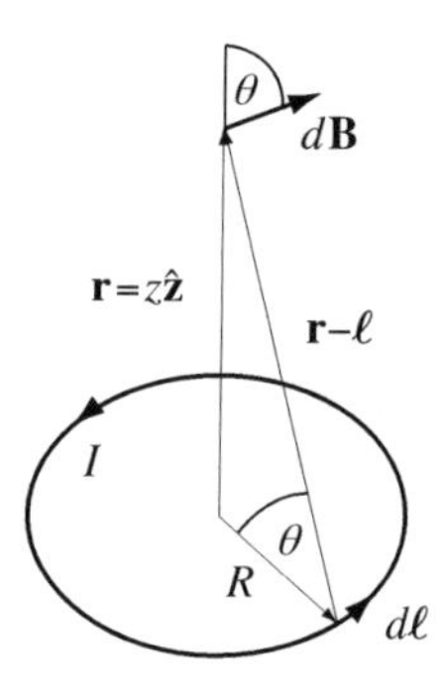

Figure 10.2: Geometry for a Biot-Savart calculation of the magnetic field on the symmetry axis of a circular current loop.

points that lie on the symmetry axis of the loop. The vector $d\mathbf{B}$ in Figure 10.2 is a typical contribution to the integrand of (10.17). The components of $d\mathbf{B}$ perpendicular to the symmetry (z) axis cancel when $d\boldsymbol{\ell}$ traverses the entire loop. The z-components add and have the same magnitude for every $d\boldsymbol{\ell}$. Therefore,

$$\mathbf{B}(z) = \hat{\mathbf{z}}\frac{\mu_0 I}{4\pi}\frac{\cos\theta}{R^2+z^2}\oint d\ell = \hat{\mathbf{z}}\frac{\mu_0 I}{2}\frac{R^2}{(R^2+z^2)^{3/2}}. \tag{10.20}$$

The non-zero value of $\mathbf{B}(z=0)$ predicted by (10.20) differs from the result $\mathbf{E}(z=0)=0$ found in Example 3.1 for a uniformly charged ring. The difference may be traced to the cross product in the Biot-Savart law and its absence in the corresponding electric field formula (3.8). As a result, contributions to $\mathbf{B}(z=0)$ from opposite sides of the ring add rather than subtract as the contributions to $\mathbf{E}(z=0)$ do for the charged ring. On the other hand, $\mathbf{B}(z\to\infty)=0$ is similar to $\mathbf{E}(z\to\infty)=0$ for the uniformly charged ring. This is consistent with the Helmholtz theorem and is a general feature of magnetic fields produced by finite-sized sources.

10.2.2 An Infinitely Long Solenoid

Figure 10.3 shows an azimuthal current $\mathbf{K}$ flowing on the surface of an infinitely long solenoid. The cross sectional shape of the solenoid is arbitrary but uniform along its length. The vector $\mathbf{R}=\mathbf{r}-\mathbf{r}_S$ in the Biot-Savart integral (10.16) is drawn for the case when the observation point P lies outside the body of the solenoid. However, the calculation to be outlined below applies equally well when P lies inside the body of the solenoid.

We exploit the fact that $\mathbf{K}d\ell = K d\boldsymbol{\ell}$ is an azimuthal vector by factoring the surface integral in (10.16) into a z-integral and a line integral around the solenoid's perimeter:

$$\mathbf{B}(P) = \frac{\mu_0}{4\pi}\int dS\,\frac{\mathbf{K}\times\mathbf{R}}{R^3} = \frac{\mu_0 K}{4\pi}\int_{-\infty}^{\infty} dz\oint\frac{d\boldsymbol{\ell}\times\mathbf{R}}{R^3}. \tag{10.21}$$

From the geometry, $\mathbf{R}+\mathbf{R}'=-z\hat{\mathbf{z}}$ and $R^2=R'^2+z^2$. Using this information and $d\boldsymbol{\ell}=d\mathbf{R}'$ in (10.21) gives

$$\mathbf{B}(P) = \frac{\mu_0 K}{4\pi}\oint\left[(\mathbf{R}'\times d\mathbf{R}')\int_{-\infty}^{\infty}\frac{dz}{(R'^2+z^2)^{3/2}} + (\hat{\mathbf{z}}\times d\mathbf{R}')\int_{-\infty}^{\infty}dz\frac{z}{(R'^2+z^2)^{3/2}}\right]. \tag{10.22}$$

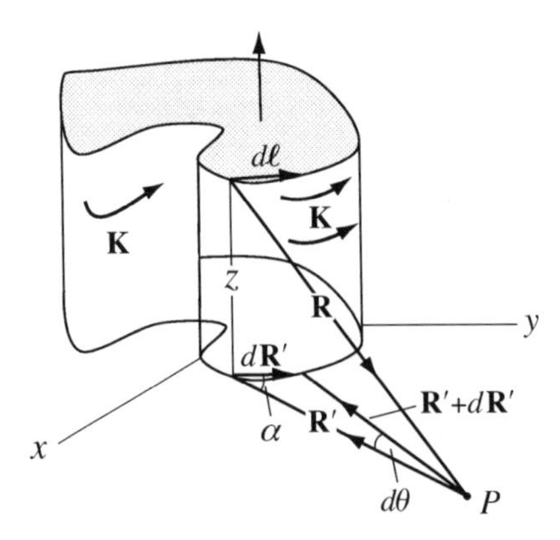

Figure 10.3: An infinitely long solenoid with a uniform cross sectional shape. The surface current density **K** has constant magnitude, but is everywhere parallel to the azimuthal vector $d\boldsymbol{\ell} = d\mathbf{R}'$ tangent to the solenoid surface.

The first integral in square brackets in (10.22) has the value $2/R'^2$. The second integral vanishes because its integrand is an odd function of z. Therefore,

$$\mathbf{B}(P) = \frac{\mu_0 K}{2\pi} \oint \frac{\mathbf{R}' \times d\mathbf{R}'}{R'^2}. \tag{10.23}$$

An important observation is that the vector $\mathbf{R}' \times d\mathbf{R}'$ points in the $-\hat{\mathbf{z}}$-direction when P lies outside the solenoid and points in the $+\hat{\mathbf{z}}$-direction when P lies inside the solenoid. Moreover,

$$|\mathbf{R}' \times d\mathbf{R}'| = R' dR' \sin(\pi - \alpha) = R' dR' \sin\alpha, \tag{10.24}$$

and the law of sines gives

$$dR' \sin\alpha = |\mathbf{R}' + d\mathbf{R}'| \sin(d\theta) \approx R' d\theta. \tag{10.25}$$

Therefore, $|\mathbf{R}' \times d\mathbf{R}'| \approx R'^2 d\theta$, and the magnitude of (10.23) is

$$B(P) = \frac{\mu_0 K}{2\pi} \oint d\theta. \tag{10.26}$$

When P lies outside the solenoid, the vector $\mathbf{R}'$ in Figure 10.3 sweeps out zero net angle θ as its tip traces out the closed circuit of the integral (10.26). When P lies inside the solenoid, $\mathbf{R}'$ sweeps out an angle 2π over the same closed circuit. Hence,

$$\mathbf{B}(P) = \begin{cases} \mu_0 K \hat{\mathbf{z}} & P \text{ inside the solenoid,} \\ 0 & P \text{ outside the solenoid.} \end{cases} \tag{10.27}$$

The magnetic field is uniform and axial everywhere inside the solenoid and vanishes everywhere outside the solenoid.

Example 10.1 It is convenient for some applications to construct a closed path for current flow from a set of straight wire segments connected head-to-tail. (a) Express the Biot-Savart magnetic field produced by the wire segment **a** (shown as a dotted line in Figure 10.4) in terms of its current I and the co-planar vectors **b** and **c** which begin at the observation point P and end at the beginning and end points of the segment, respectively. (b) Use the analysis of part (a) to find the magnetic field produced by an infinitely long filament of current.

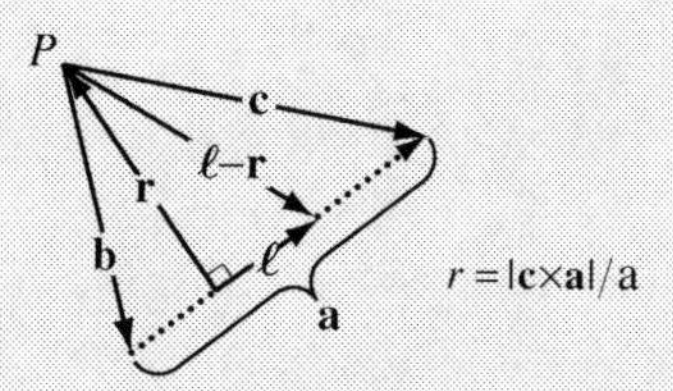

Figure 10.4: The wire segment **a** (dotted) carries a steady current I from the head of the vector **b** to the head of the vector **c**. The observation point is P.

Solution: (a) Choose the origin of the vectors $\mathbf{r}$ and $\boldsymbol{\ell}$ in the Biot-Savart integral (10.17) as the point on the line defined by the wire which lies closest to the observation point P. Since $|\mathbf{r} - \boldsymbol{\ell}| = \sqrt{r^2 + \ell^2}$ and the vectors $d\boldsymbol{\ell} \times \mathbf{r}$ and $\mathbf{c} \times \mathbf{a}$ are parallel,

$$\frac{d\boldsymbol{\ell} \times (\mathbf{r} - \boldsymbol{\ell})}{|\mathbf{r} - \boldsymbol{\ell}|^3} = \frac{d\boldsymbol{\ell} \times \mathbf{r}}{|\mathbf{r} - \boldsymbol{\ell}|^3} = \frac{d\ell\, r}{(r^2 + \ell^2)^{3/2}} \frac{\mathbf{c} \times \mathbf{a}}{|\mathbf{c} \times \mathbf{a}|}.$$

The integral on ℓ runs from $\ell_b = \mathbf{a} \cdot \mathbf{b}/a$ to $\ell_c = \mathbf{a} \cdot \mathbf{c}/a$. Therefore, since $r = |\mathbf{c} \times \mathbf{a}|/a$, the Biot-Savart magnetic field is

$$\mathbf{B} = \frac{\mu_0 I}{4\pi} \frac{\mathbf{c} \times \mathbf{a}}{|\mathbf{c} \times \mathbf{a}|} \frac{1}{r} \frac{\ell}{\sqrt{\ell^2 + r^2}} \Bigg|_{\mathbf{a}\cdot\mathbf{b}/a}^{\mathbf{a}\cdot\mathbf{c}/a} = \frac{\mu_0 I}{4\pi} \frac{\mathbf{c} \times \mathbf{a}}{|\mathbf{c} \times \mathbf{a}|^2} \left\{ \frac{\mathbf{a} \cdot \mathbf{c}}{c} - \frac{\mathbf{a} \cdot \mathbf{b}}{b} \right\}.$$

The direction of $\mathbf{c} \times \mathbf{a}$ is perpendicular to the plane of the diagram so, by rotational symmetry, the lines of **B** form closed circles coaxial with the axis of the current-carrying segment.

(b) For an infinitely long wire, the limits of the Biot-Savart integral are $\ell_b = -\infty$ and $\ell_c = \infty$. Moreover, $\mathbf{c} \times \mathbf{a}$ points along the azimuthal unit vector $\hat{\boldsymbol{\phi}}$ if we adopt the right-hand rule (see next section). Therefore, the magnetic field at a distance r from an infinitely long current-carrying wire is

$$\mathbf{B} = \hat{\boldsymbol{\phi}} \frac{\mu_0 I}{4\pi} \frac{1}{r} \frac{\ell}{\sqrt{\ell^2 + r^2}} \Bigg|_{-\infty}^{\infty} = \frac{\mu_0 I}{2\pi r} \hat{\boldsymbol{\phi}}.$$

10.3 Ampère's Law

The fundamental magnetostatic equation (10.2) is called *Ampère's law*. Elementary discussions focus on its integral form, which we derive by integrating both sides of (10.2) over a surface S and using Stokes' theorem to transform the surface integral of $\nabla \times \mathbf{B}$ into a circuit integral of **B**. The desired result follows because (2.7) defines the current I_C as the integral of $\mathbf{j} \cdot d\mathbf{S}$ over S:

$$\oint_C d\boldsymbol{\ell} \cdot \mathbf{B} = \mu_0 \int_S d\mathbf{S} \cdot \mathbf{j} = \mu_0 I_C. \tag{10.28}$$

The sign of I_C is fixed by the right-hand rule: the thumb points in the direction of positive current when the curl of the fingers follows the direction of integration around C.

Ampère's law in the form (10.28) can sometimes be used to find the magnetic field when $\mathbf{j}(\mathbf{r})$ is highly symmetrical. The key is to use symmetry (or other arguments) to discover (i) which components of the magnetic field vector are non-zero and (ii) how each non-zero component depends on the components of **r**. The calculation of $\mathbf{B}(\mathbf{r})$ reduces to algebra if this information reveals an integration path C where the integrand of (10.28) is a constant which can be taken out of the integral.

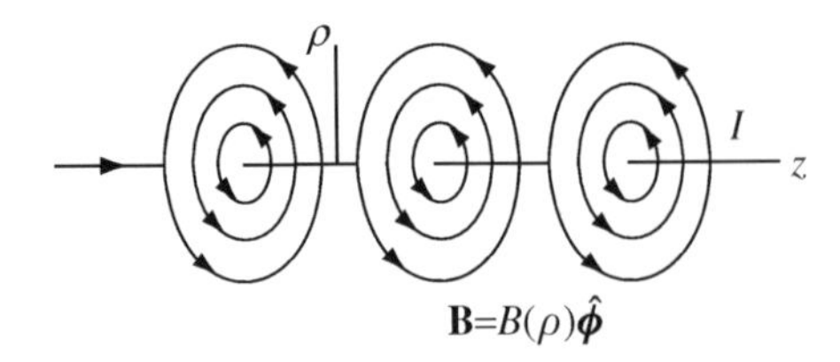

Figure 10.5: The azimuthal magnetic field of a straight and infinite filament of current. The vertical radial vector lies in the $z = 0$ plane.

10.3.1 An Infinite Line of Current

Every physics student learns to use Ampère's law in integral form to find the magnetic field produced by an infinite line of current (see Figure 10.5). One assumes that $\mathbf{B}(\mathbf{r}) = B(\rho)\hat{\boldsymbol{\phi}}$ and uses circles concentric with the line of current as Ampèrian integration paths. Since I in Figure 10.5 points along $\hat{\mathbf{z}}$, the resulting field,

$$\mathbf{B}(\rho) = \frac{\mu_0 I}{2\pi\rho}\hat{\boldsymbol{\phi}}, \tag{10.29}$$

is consistent with the right-hand rule and reproduces the formula derived at the end of Example 10.1. It remains only to provide a convincing argument for the assumed form for the field.

The invariance of the line current to translations along, and rotations around, the z-axis implies that $\mathbf{B}(\rho, \phi, z) = \mathbf{B}(\rho)$. The same symmetries, combined with $\nabla \cdot \mathbf{B} = 0$, tell us that the radial component $B_\rho = 0$. Otherwise, field lines would begin or end at the line of current. Finally, consider reversing the sign of I by reflecting the line of current in the $z = 0$ plane of Figure 10.5. The linearity of the Maxwell equations guarantees that this operation reverses the sign of $\mathbf{B}$ also. However, as we will now show, the z-component of $\mathbf{B}$ does *not* change under mirror reflection through $z = 0$. Therefore, $B_z = 0$ and we confirm the correctness of our Ampère ansatz that $\mathbf{B} = B(\rho)\hat{\boldsymbol{\phi}}$.

By definition, reflection in the $z = 0$ plane transforms the position vector $\mathbf{r}$ to

$$\mathbf{r}' = x'\hat{\mathbf{x}} + y'\hat{\mathbf{y}} + z'\hat{\mathbf{z}} = x\hat{\mathbf{x}} + y\hat{\mathbf{y}} - z\hat{\mathbf{z}}. \tag{10.30}$$

The z-component of $\mathbf{r}'$ is the negative of the z-component of $\mathbf{r}$. The x- and y-components of $\mathbf{r}'$ and $\mathbf{r}$ are the same. The Cartesian components of the gradient $\nabla = \partial/\partial\mathbf{r}$, and the current density $\mathbf{j} = \rho\mathbf{v} = \rho d\mathbf{r}/dt$, behave similarly because they are constructed from $\mathbf{r}$ and scalars which do not change under reflection. To discover how the reflected magnetic field $\mathbf{B}'(x, y, -z)$ is related to the original magnetic field $\mathbf{B}(x, y, z)$, it is sufficient to demand that the transformed equation, $\nabla' \times \mathbf{B}' = \mu_0\mathbf{j}'$, be consistent with the original equation, $\nabla \times \mathbf{B} = \mu_0\mathbf{j}$. Writing out the Cartesian components of each equation and using the transformation properties of ∇ and $\mathbf{j}$ shows immediately that reflection through the $z = 0$ plane leaves the z-component of the magnetic field unaffected and changes the algebraic sign of the x- and y-components:

$$\mathbf{B}' = B'_x\hat{\mathbf{x}} + B'_y\hat{\mathbf{y}} + B'_z\hat{\mathbf{z}} = -B_x\hat{\mathbf{x}} - B_y\hat{\mathbf{y}} + B_z\hat{\mathbf{z}}. \tag{10.31}$$

This discussion is consistent with Application 1.2 at the end of Section 1.8.1 where vectors that transform under reflection like (10.30) were called "ordinary" or "polar" vectors and vectors that transform like (10.31) were called "axial" or "pseudo" vectors. An independent deduction that $\mathbf{B}(\mathbf{r})$ is an axial vector appears at the beginning of Section 2.7.

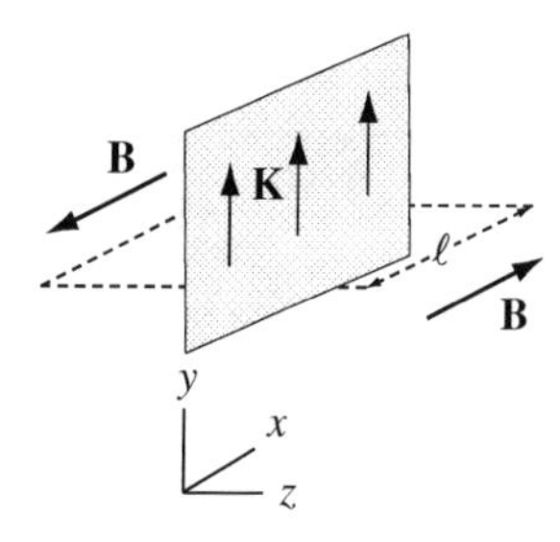

Figure 10.6: An infinite sheet of current with density $\mathbf{K} = K\hat{\mathbf{z}}$ and the Ampèrian loop (dashed) used to find $\mathbf{B}$.

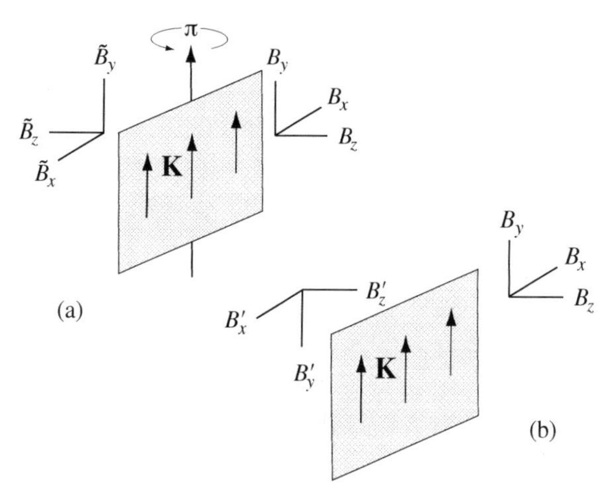

Figure 10.7: Magnetic field of a current sheet at $x = 0$: (a) rotation by π around the z-axis transforms $\mathbf{B}$ to $\tilde{\mathbf{B}}$; (b) reflection through $x = 0$ transforms $\mathbf{B}$ to $\mathbf{B}'$.

10.3.2 A Uniform Sheet of Current

Figure 10.6 shows an infinite sheet with uniform surface current density $\mathrm{K}(\mathbf{r}_S) = K\hat{\mathbf{y}}$ confined to the x-y plane. Two symmetry arguments facilitate a calculation of the sheet's magnetic field using Ampère's law. The first uses the translational invariance of the current sheet along x and y to deduce that $\mathbf{B} = \mathbf{B}(z)$. The second exploits the fact that $\mathbf{B}(z)$ must transform identically under two symmetry operations if both operations leave the current invariant and relocate every observation point identically. The symmetries in question are 180° rotation around the y-axis and reflection in the $z = 0$ plane. Figure 10.7(a) shows how 180° rotation transforms the magnetic field vector $\mathbf{B}(x, y, z)$ to a vector $\tilde{\mathbf{B}}(x, y, -z)$.[6] Analytically,

$$\tilde{\mathbf{B}} = \tilde{B}_x\hat{\mathbf{x}} + \tilde{B}_y\hat{\mathbf{y}} + \tilde{B}_z\hat{\mathbf{z}} = -B_x\hat{\mathbf{x}} + B_y\hat{\mathbf{y}} - B_z\hat{\mathbf{z}}. \tag{10.32}$$

Figure 10.7(b) shows how mirror reflection transforms $\mathbf{B}(x, y, z)$ to a vector $\mathbf{B}'(x, y, -z)$ according to (10.31). The requirement that each Cartesian component of $\tilde{\mathbf{B}}$ in (10.32) equal the corresponding component of $\mathbf{B}'$ in (10.31) shows that $B_y = B_z = 0$ and that

$$\mathbf{B} = B_x(z)\hat{\mathbf{x}} \quad \text{and} \quad B_x(-z) = -B_x(z). \tag{10.33}$$

[6] We use the active interpretation of this transformation. See Section 1.7.2.

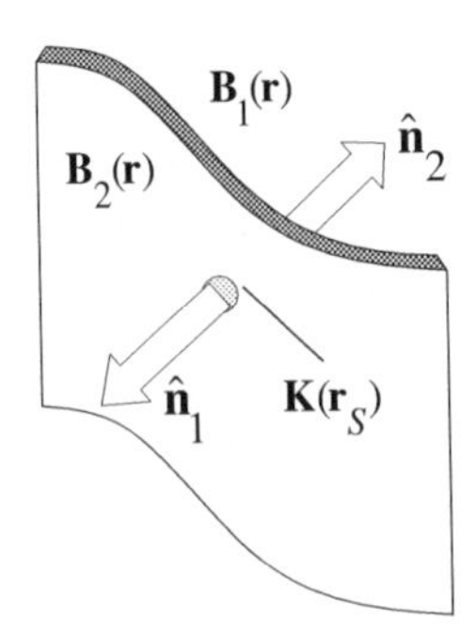

Figure 10.8: A surface that carries an areal current density $\mathbf{K}(\mathbf{r}_S)$. The unit normal vectors $\hat{\mathbf{n}}_k$ points outward from region k.

Using (10.33), the choice of the dashed rectangular loop in Figure 10.6 as the circuit C reduces the integral in (10.28) to the factor $2B_y\ell$. The enclosed current and the surface current density are related by $I_C = K\ell$, so

$$\mathbf{B}(x, y, z) = \begin{cases} \frac{1}{2}\mu_0 K\hat{\mathbf{x}} & z > 0, \\ -\frac{1}{2}\mu_0 K\hat{\mathbf{x}} & z < 0. \end{cases} \tag{10.34}$$

The magnetic field is uniform, but oppositely directed, on opposite sides of an infinite sheet of uniform current density. On each side, the field is parallel to the sheet, but perpendicular to the direction of current flow. If $\hat{\mathbf{s}}$ is a unit normal to the current sheet which points to the observation point, a generally valid formula is

$$\mathbf{B} = \frac{1}{2}\mu_0 \mathbf{K} \times \hat{\mathbf{s}}. \tag{10.35}$$

10.3.3 Matching Conditions for $\mathbf{B}(\mathbf{r})$

We can use (10.35) to re-derive the matching conditions (2.49) for the magnetic field near an arbitrarily shaped surface S which carries a surface current density $\mathbf{K}(\mathbf{r}_S)$ (Figure 10.8). Following Section 3.4.2, let $\mathbf{B}_1(\mathbf{r}_S)$ and $\mathbf{B}_2(\mathbf{r}_S)$ be the magnetic fields at two points which are infinitesimally close to each other, but on opposite sides of S at the point $\mathbf{r}_S$. Each of these fields can be written as a superposition of two fields. One is the field produced by the differential element of surface which contains $\mathbf{r}_S$ itself (shown as a shaded disk in Figure 10.8). This is well approximated by (10.35) at the observation points in question. The second contribution to $\mathbf{B}_1(\mathbf{r}_S)$ and $\mathbf{B}_2(\mathbf{r}_S)$ is the field $\mathbf{B}_S$ produced by all the source current *except* for the current which flows through the disk. The latter is continuous passing through the hole occupied by the disk. Therefore,

$$\mathbf{B}_1 = \mathbf{B}_S + \tfrac{1}{2}\mu_0 \mathbf{K}(\mathbf{r}_S) \times \hat{\mathbf{n}}_2 \qquad \text{and} \qquad \mathbf{B}_2 = \mathbf{B}_S - \tfrac{1}{2}\mu_0 \mathbf{K}(\mathbf{r}_S) \times \hat{\mathbf{n}}_2. \tag{10.36}$$

Subtracting the two equations in (10.36) gives $\mathbf{B}_1 - \mathbf{B}_2 = \mu_0 \mathbf{K}(\mathbf{r}_S) \times \hat{\mathbf{n}}_2$. Taking the dot product and cross product of this equation with $\hat{\mathbf{n}}_2$ produces the matching conditions

$$\hat{\mathbf{n}}_2 \cdot [\mathbf{B}_1 - \mathbf{B}_2] = 0 \tag{10.37}$$

$$\hat{\mathbf{n}}_2 \times [\mathbf{B}_1 - \mathbf{B}_2] = \mu_0 \mathbf{K}(\mathbf{r}_S). \tag{10.38}$$

The normal component of $\mathbf{B}$ is continuous across a current-carrying surface. The tangential component of $\mathbf{B}$ can suffer a discontinuity. We note in passing that (10.38) combined with the fact that $\mathbf{B}_{\text{in}} = 0$ is the static field inside a perfect conductor (see Section 13.6.6) relates the magnetic field just outside a perfect conductor to the current flowing on its surface:

$$\mu_0 \mathbf{K} = \hat{\mathbf{n}} \times \mathbf{B}_{\text{out}}. \tag{10.39}$$

10.3.4 The Force on a Sheet of Current

The tiny disk of current shown shaded in Figure 10.8 cannot exert a force on itself. Therefore, the force per unit area on any particular element of current-carrying surface is due to the net field $\mathbf{B}_S$ produced by all the other surface elements. Adding together the two equations in (10.36) shows that $\mathbf{B}_S$ is the simple average of $\mathbf{B}_1$ and $\mathbf{B}_2$. Hence, the force per unit area $\mathbf{f}(\mathbf{r}_S)$ exerted on a current-carrying surface is $\mathbf{K}(\mathbf{r}_S) \times \mathbf{B}_S$:

$$\mathbf{f}(\mathbf{r}_S) = \tfrac{1}{2}\, \mathbf{K}(\mathbf{r}_S) \times (\mathbf{B}_1 + \mathbf{B}_2). \tag{10.40}$$

We will make use of (10.40) in Chapter 12.

Example 10.2 Figure 10.9 shows a filamentary wire which carries a steady current I up the symmetry axis of a hollow cylindrical can of radius R and height L. The return current flows radially outward along the upper end cap, down the cylindrical wall, and then radially inward along the lower end cap. (a) Argue that the magnetic field everywhere has the form $\mathbf{B}(\mathbf{r}) = B(\rho, z)\hat{\boldsymbol{\phi}}$ in cylindrical coordinates. (b) Find $B(\rho, z)$ at every point in space. (c) Show that the solution satisfies the matching conditions (10.37) and (10.38) on the walls of the can.

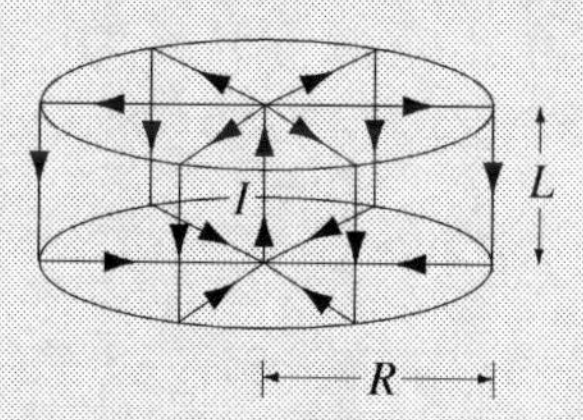

Figure 10.9: A straight filament carries a current I up the center of a hollow cylindrical can. The return current flows radially outward along the top end cap, down the tubular wall, and radially inward along the lower end cap.

Solution: (a) The field $\mathbf{B} = \mathbf{B}(\rho, z)$ because the source current has rotational symmetry around the z-axis. The current density has the form $\mathbf{j}(\mathbf{r}) = j_\rho(\rho, z)\hat{\boldsymbol{\rho}} + j_z(\rho, z)\hat{\mathbf{z}}$, so the three components of Ampère's law in cylindrical coordinates simplify to

$$-\frac{\partial B_\phi}{\partial z} = \mu_0 j_\rho(\rho, z), \qquad \frac{\partial B_\rho}{\partial z} - \frac{\partial B_z}{\partial \rho} = 0, \qquad \frac{1}{\rho}\frac{\partial}{\partial \rho}(\rho B_\phi) = \mu_0 j_z(\rho, z).$$

All three equations are consistent with a magnetic field of the form $\mathbf{B}(\mathbf{r}) = B(\rho, z)\hat{\boldsymbol{\phi}}$. The latter also satisfies $\nabla \cdot \mathbf{B} = 0$.

(b) With the ansatz of part (a), Ampère's law (10.28) with circular Ampèrian integration paths coaxial with the symmetry axis gives

$$\mathbf{B}(\rho, z) = \begin{cases} \dfrac{\mu_0 I}{2\pi\rho}\hat{\boldsymbol{\phi}} & \text{inside the can,} \\ 0 & \text{outside the can.} \end{cases}$$

The fact that the field inside the can is exactly the same as the field (10.29) produced by an infinitely long wire does *not* imply that the field comes entirely from the vertical segment of current enclosed by the Ampèrian loop. All the currents in and on the can contribute to the total calculated field.

(c) If $\hat{\mathbf{n}}$ is the outward normal to any wall of the can, both matching conditions are represented by

$$\mathbf{B}_{\text{out}} - \mathbf{B}_{\text{in}} = \mu_0 \mathbf{K} \times \hat{\mathbf{n}}.$$

The surface current densities on the upper (+) and lower (−) end caps are $\mathbf{K}_{\pm} = \pm\hat{\boldsymbol{\rho}} I/2\pi\rho$. The surface current density on the outside wall is $\mathbf{K}_0 = -\hat{\mathbf{z}} I/2\pi R$. Using these values, we confirm that the matching condition on the side wall is an identity:

$$0 - \frac{\mu_0 I}{2\pi R}\hat{\boldsymbol{\phi}} = -\mu_0 \frac{I}{2\pi R}\hat{\mathbf{z}} \times \hat{\boldsymbol{\rho}}.$$

The same is true for the matching conditions on the end caps:

$$0 - \frac{\mu_0 I}{2\pi\rho}\hat{\boldsymbol{\phi}} = \pm\mu_0 \frac{I}{2\pi\rho}\{\hat{\boldsymbol{\rho}} \times (\pm\hat{\mathbf{z}})\}.$$

Remark: The current density used in the preceding analysis, $\mathbf{j}(\mathbf{r}) = j_\rho(\rho, z)\hat{\boldsymbol{\rho}} + j_z(\rho, z)\hat{\mathbf{z}}$, applies equally well to a *toroidal solenoid* (Figure 10.10) formed by bending a conventional linear solenoid so its two open ends join smoothly together. Therefore, when N turns of a wire with current I are wound to form the toroid shown in Figure 10.10, the Ampère's law analysis just above gives the magnetic field immediately as

$$\mathbf{B}(\rho, z) = \begin{cases} \dfrac{\mu_0 N I}{2\pi\rho}\hat{\boldsymbol{\phi}} & \text{inside the toroidal volume,} \\ 0 & \text{outside the toroidal volume.} \end{cases}$$

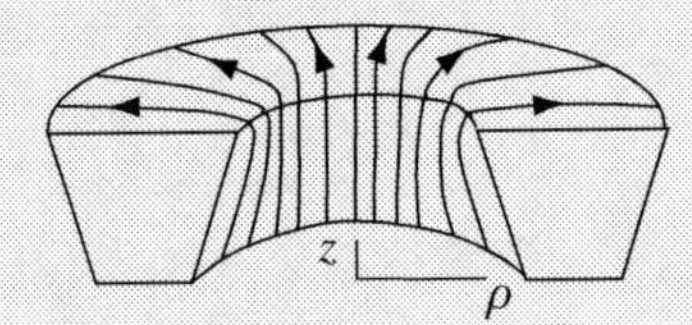

Figure 10.10: Cross section of a toroidal solenoid with rotational symmetry around the z-axis.

10.4 The Magnetic Scalar Potential

Ampère's law (10.2) reads $\nabla \times \mathbf{B}(\mathbf{r}) = 0$ in any volume of space V where the current density $\mathbf{j}(\mathbf{r}) = 0$. Within that volume, the analogy with electrostatics implies that we may define a *magnetic scalar potential* $\psi(\mathbf{r})$ where

$$\mathbf{B}(\mathbf{r}) = -\nabla\psi(\mathbf{r}) \qquad \mathbf{r} \in V. \tag{10.41}$$

Since $\nabla \cdot \mathbf{B}(\mathbf{r}) = 0$ everywhere, $\psi(\mathbf{r})$ satisfies Laplace's equation at the same set of points,

$$\nabla^2\psi(\mathbf{r}) = 0 \qquad \mathbf{r} \in V. \tag{10.42}$$

This is an important result. It shows that the potential-theory methods of Chapter 7 apply to any magnetostatic problem where the current is confined to surfaces which separate space into disjoint, current-free sub-volumes. We need only solve (10.42) in each sub-volume and use the matching conditions (10.37) and (10.38) to stitch the solutions together.

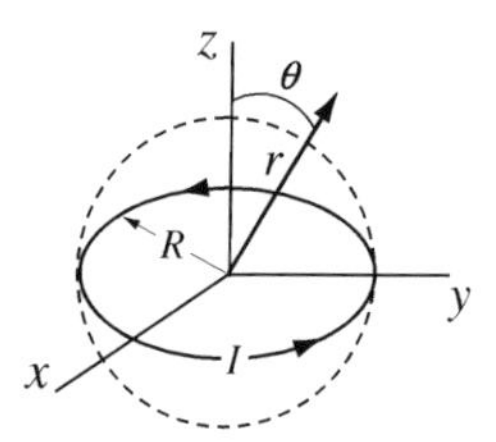

Figure 10.11: A origin-centered ring of radius R carries a steady current I in the $z = 0$ plane.

10.4.1 A Uniform Ring of Current

In Section 7.6.1, we calculated the electric scalar potential of a uniform circular ring with net charge Q. Here, we calculate the magnetic scalar potential of a circular ring which carries a current I. We center the ring on the z-axis in the x-y plane (see Figure 10.11) so $r = R$ and $\theta = \pi/2$ define the position of the ring in polar coordinates. The spherical matching surface $r = R$ divides space into two current-free sub-volumes.

The problem has azimuthal symmetry so the appropriate solutions of Laplace's equation that are regular in each sub-volume are

$$\psi(r, \theta) = \begin{cases} \displaystyle\sum_{\ell=1}^{\infty} A_\ell \left(\frac{r}{R}\right)^\ell P_\ell(\cos\theta) & r < R, \\ \displaystyle\sum_{\ell=1}^{\infty} B_\ell \left(\frac{R}{r}\right)^{\ell+1} P_\ell(\cos\theta) & r > R. \end{cases} \tag{10.43}$$

Notice that the sums in (10.43) begin at $\ell = 1$ rather than at $\ell = 0$ as they did for the charged-ring problem. This follows[7] from the matching condition (10.37) because the continuity of $\hat{\mathbf{r}} \cdot \nabla\psi$ at $r = R$ gives $B_0 = 0$ as a special case of

$$B_\ell = -\frac{\ell}{\ell+1} A_\ell. \tag{10.44}$$

The tangential matching condition (10.38) determines the coefficients A_ℓ in terms of so-called "associated" Legendre functions (Appendix C). To avoid this complication, we "go off the axis" as discussed in Application 7.2 following Section 7.6.1. This requires an independent calculation of the field *on* the axis. We did this in Section 10.2.1 with the result that

$$B_z(z) = \frac{\mu_0 I}{2} \frac{R^2}{(R^2 + z^2)^{3/2}}. \tag{10.45}$$

Using (10.45), we integrate the z-component of (10.41) and set the integration constant equal to zero. This gives

$$\psi(z) = -\frac{\mu_0 I}{2} \frac{z}{\sqrt{R^2 + z^2}}. \tag{10.46}$$

The key step is to rewrite (10.46) using the generating function (4.74) for the Legendre polynomials. When we do this for $z < R$, the result is

$$\psi(z) = -\frac{\mu_0 I}{2} \sum_{\ell=1}^{\infty} \left(\frac{z}{R}\right)^\ell P_{\ell-1}(0). \tag{10.47}$$

[7] A deeper explanation for this fact will be given in the next chapter.

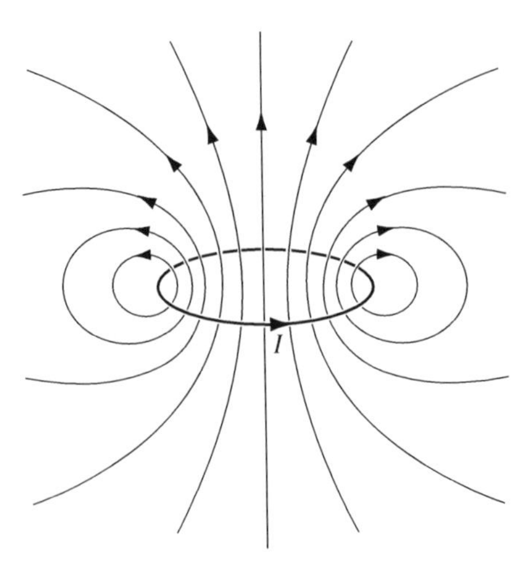

Figure 10.12: Representative magnetic field lines produced by a circular ring of current. The full field line pattern shares the rotational symmetry of the ring.

The expansion (10.47) must be equivalent to the $r < R$ expansion in (10.43) with $\theta = 0$ and $r = z$. Since $P_\ell(1) = 1$, this establishes that $A_\ell = -\frac{1}{2}\mu_0 I P_{\ell-1}(0)$. We then use (10.44) and the facts that $\ell P_{\ell-1}(0) = -(\ell+1)P_{\ell+1}(0)$ and $P_\ell(0) = 0$ when ℓ is odd. The final result is

$$\psi(r,\theta) = \begin{cases} -\dfrac{\mu_0 I}{2} \displaystyle\sum_{\ell=1,3,\ldots}^{\infty} \left(\dfrac{r}{R}\right)^{\ell} P_{\ell-1}(0) P_\ell(\cos\theta) & r < R, \\ -\dfrac{\mu_0 I}{2} \displaystyle\sum_{\ell=1,3,\ldots}^{\infty} \left(\dfrac{R}{r}\right)^{\ell+1} P_{\ell+1}(0) P_\ell(\cos\theta) & r > R. \end{cases} \tag{10.48}$$

It is important to notice that (10.48) is *not* continuous at $r = R$.[8] Unlike the electric scalar potential, $\varphi(\mathbf{r})$, there is no requirement that $\psi(\mathbf{r})$ be continuous at a matching surface.

Figure 10.12 illustrates a few representative field lines of

$$\mathbf{B}(r,\theta) = -\frac{\partial\psi}{\partial r}\hat{\mathbf{r}} - \frac{1}{r}\frac{\partial\psi}{\partial\theta}\hat{\boldsymbol{\theta}}. \tag{10.49}$$

Observers very near the ring see circular field loops which "link" the ring because, at these distances, the ring looks just like a straight segment of current (see Section 10.3.1). Farther away, the circles expand asymmetrically into elliptical loops. Finally, when $r \gg R$, we can approximate $\psi(r,\theta)$ using just the $\ell = 1$ term in (10.48):

$$\psi(r,\theta) \simeq \frac{\mu_0}{4\pi}\frac{\pi R^2 I}{r^2}\cos\theta. \tag{10.50}$$

The $\cos\theta/r^2$ structure of the magnetic scalar potential (10.50) is identical to the electric scalar potential of a point electric dipole.[9] Therefore, since $\mathbf{B} = -\nabla\psi$ and $\mathbf{E} = -\nabla\varphi$, the $\mathbf{B}$ field line structure far from the current ring is identical to the $\mathbf{E}$ field line structure of a point electric dipole (Figure 4.3).

[8] This is so despite the fact that (10.46) *is* continuous at $z = R$. By construction, (10.48) reproduces (10.46) when $z < R$. However, (10.48) goes to zero when $r \to \infty$ while (10.46) approaches equal and opposite constant values when $z \to \pm\infty$. This difference has no effect on the magnetic field because $\mathbf{B} = -\nabla\psi$.

[9] We will show in the next chapter that the product of the ring area πR^2 and the current I in the numerator of (10.50) is the magnitude of the *magnetic dipole moment* of a current ring.

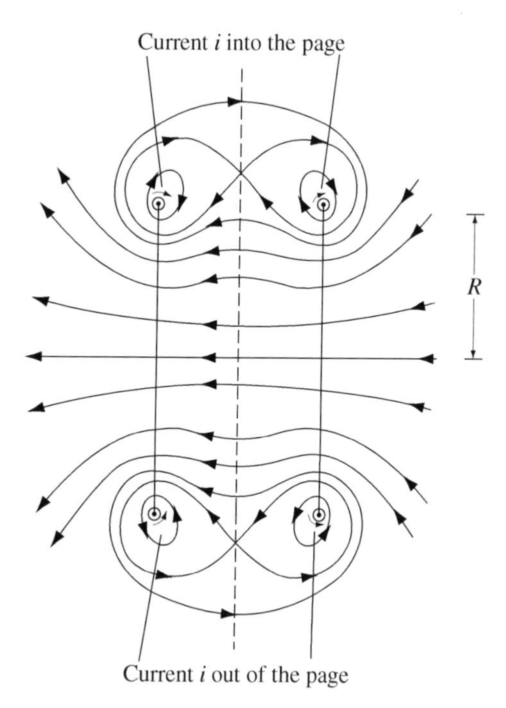

Figure 10.13: Lines of $\mathbf{B}(\mathbf{r})$ near a Helmholtz coil. The parallel rings (radius R) of current are oriented vertically and perpendicular to the plane of the figure. The vertical dashed line is the plane $z = R/2$. Figure from Scott (1966).

10.4.2 The Helmholtz Coil

In 1853, the French scientist Gaugain pointed out that a circular coil of current with radius R in the $z = 0$ plane produces a region of fairly uniform magnetic field near the point $z = R/2$ on the symmetry axis. This follows from (10.45) because $z = R/2$ is an inflection point where the second derivative $B_z''(z)$ is zero. The great German physicist-physiologist, Hermann von Helmholtz, knew that the field is even more uniform in the same region of space if one adds a second, identically oriented current coil in the $z = R$ plane (Figure 10.13).

It is not difficult to appreciate the origin of the field uniformity of the Helmholtz coil. For two coaxial rings of radius R separated by a distance R (one at $z = 0$ and one at $z = R$), the total field on the symmetry axis still satisfies $B_z''(R/2) = 0$ by construction. More importantly, the reflection symmetry of the two-coil configuration with respect to the plane $z = R/2$ implies that every odd derivative of the total field is also zero at $z = R/2$. This means that the first non-zero derivative of $B_z(z)$ at $z = R/2$ is d^4B_z/dz^4. This guarantees that $B_z(z)$ near the midpoint between the two coils will be very uniform indeed.

Helmholtz' Lament

Writing to his friend du Bois-Raymond in May of 1853, Helmholtz sought to establish his priority:

Gaugain has constructed a [coil] according to the same principle as mine, but in any event in a very unadvantageous form. I gave a lecture on the principle of my instrument in 1848 or 1849 at the Berlin

Physical Society and would like to know if enough is noted in the minutes that I could refer back to them.

du Bois-Raymond responded:

> Your priority in the matter of the Gaugain [coil] is irretrievably lost. The minutes of the year 1849, to which your communication belongs, are unfortunately lost . . . This is a shame, but it is a new warning not to hide one's talents. Don't let this hold you back from describing your instrument now, however.

Helmholtz chose not to publish. Nevertheless, the name *Helmholtz coil* became associated with his two-coil magnet, probably because Gustav Wiedemann attributed the design to Helmholtz in the first edition of his encyclopedic *Die Lehre von der Elektricität* (1861).

Application 10.2 Magnetic Resonance Imaging

Magnetic resonance imaging (MRI) is a medical diagnostic technique which exploits nuclear magnetic resonance to excite proton spins in the human body. Image contrast comes from variations in the resonance signal strength due to differences in the density and spin relaxation rate of protons found in different types of tissue (bone, muscle, fat, etc.) In 2003, the physicist Peter Mansfield won a share of the Nobel Prize in Physiology or Medicine for the development of a "snapshot" technique which greatly reduced the time needed for data acquisition. This, in turn, required the invention of "active magnetic shielding" of the fields produced by current-carrying coils (used to define the image plane and its coordinate axes) from the superconducting magnet used to align the spins. We illustrate the idea using the concentric cylinders in Figure 10.14. Given a current density $\mathbf{K}_1(\phi, z)$ on the inner cylinder, the goal is to choose a current density $\mathbf{K}_2(\phi, z)$ on the outer cylinder so the total magnetic field $\mathbf{B} = 0$ when $\rho > R_2$.

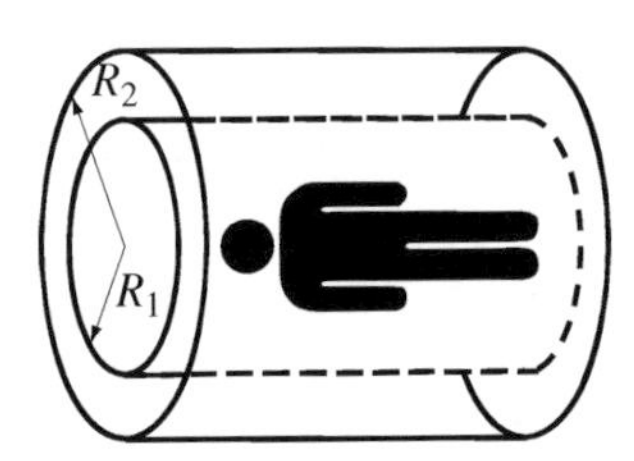

Figure 10.14: Active shielding specifies current on two cylindrical shells to make $\mathbf{B}(\rho, z)$ a desired field inside the inner cylinder and $\mathbf{B} = 0$ outside the outer cylinder.

A preliminary problem considers a magnetostatic scalar potential $\psi(\rho, \phi, z)$ which satisfies Laplace's equation inside and outside a cylindrical surface at $\rho = R$ subject to the matching conditions[10]

$$\hat{\rho} \cdot (\mathbf{B}^{\text{out}} - \mathbf{B}^{\text{in}}) = 0 \qquad \hat{\rho} \times (\mathbf{B}^{\text{out}} - \mathbf{B}^{\text{in}}) = \mu_0 \mathcal{K}. \tag{10.51}$$

Section 7.8 discussed solutions of $\nabla^2 \psi = 0$ in cylindrical coordinates. We choose the separation constant $k^2 = (i\kappa)^2 < 0$ in (7.87) to get a Fourier integral in the z-variable and modified Bessel

[10] For the remainder of this Application, we use the symbol $\mathcal{K}$ for surface current density to distinguish it from the modified Bessel function $K_m(x)$.

functions [see (7.88)] in the ρ-variable. The physical requirements on ψ are that it be bounded everywhere and go to zero as $\rho \to \infty$. It is also convenient to build in the continuity of $B_\rho = -\partial\psi/\partial\rho$ at $\rho = R$ to satisfy the matching condition on the left side of (10.51). The general solution which does all these things is

$$\psi(\rho, \phi, z) = \sum_{m=-\infty}^{\infty} e^{im\phi} \int_{-\infty}^{\infty} d\kappa\, A_\phi(\kappa) e^{i\kappa z} \times \begin{cases} K'_m(|\kappa|R) I_m(|\kappa|\rho) & \rho < R, \\ I'_m(|\kappa|R) K_m(|\kappa|\rho) & \rho > R. \end{cases} \tag{10.52}$$

A strategy to find the expansion functions $A_\phi(\kappa)$ uses the ϕ-component of the matching condition for $\mathcal{K}$, the Bessel function Wronskian, $x[I'(x)K(x) - I(x)K'(x)] = 1$, and the Fourier components $\mathcal{K}^m_\phi(\kappa)$ defined by

$$\hat{\boldsymbol{\phi}} \cdot \mathcal{K}(\phi, z) = \sum_{m-\infty}^{\infty} e^{im\phi} \int_{-\infty}^{\infty} \frac{d\kappa}{2\pi} e^{i\kappa z} \mathcal{K}^m_\phi(\kappa). \tag{10.53}$$

The potential which results is[11]

$$\psi(\rho, \phi, z) = \frac{\mu_0 R}{2\pi} \sum_{m=-\infty}^{\infty} e^{im\phi} \int_{-\infty}^{\infty} d\kappa \frac{|\kappa|}{i\kappa} \mathcal{K}^m_\phi(\kappa) e^{i\kappa z} \times \begin{cases} K'_m(|\kappa R|) I_m(|\kappa|\rho) & \rho < R, \\ I'_m(|\kappa|R) K_m(|\kappa|\rho) & \rho > R. \end{cases} \tag{10.54}$$

To solve the original problem, we put $\rho > R_2 > R_1$ and use (10.54) twice to add the potential produced by $\mathcal{K}_1$ to the potential produced by $\mathcal{K}_2$:

$$\psi(\rho, \phi, z) = \frac{\mu_0}{2\pi} \sum_{m=-\infty}^{\infty} e^{im\phi} \int_{-\infty}^{\infty} d\kappa \frac{|\kappa|}{i\kappa} e^{i\kappa z} K_m(|\kappa|\rho) \left[R_1 \mathcal{K}^m_{1\phi} I'_m(|\kappa|R_1) + R_2 \mathcal{K}^m_{2\phi} I'_m(|\kappa|R_2) \right]. \tag{10.55}$$

The immediate conclusion is that ψ and $\mathbf{B}$ are both zero outside the outer cylinder if the Fourier components of $\mathcal{K}_2$ are related to the Fourier components of $\mathcal{K}_1$ by

$$\mathcal{K}^m_{1\phi} = -\frac{R_1 I'_m(|\kappa|R_1)}{R_2 I'_m(|\kappa|R_2)} \mathcal{K}^m_{2\phi}. \tag{10.56}$$

This approach to magnetic shielding is used in virtually all commercial MRI scanners. ■

10.4.3 Topological Aspects of $\psi(\mathbf{r})$

The magnetic scalar potential differs from its electrostatic counterpart because it is not a single-valued function of its argument. To see this, let V be the vacuum space outside a filamentary current loop (Figure 10.15). The relation $\mathbf{B} = -\nabla\psi$ is valid everywhere in V. Therefore, its integrated form is also correct as long as the path of integration path from A to B lies entirely in V:

$$\psi(A) - \psi(B) = \int_A^B d\boldsymbol{\ell} \cdot \mathbf{B}. \tag{10.57}$$

[11] The reader can check that (10.54) satisfies the z-component of the matching condition on the right side of (10.51). The two components of the matching condition are not independent because $\nabla \cdot \mathcal{K} = 0$.

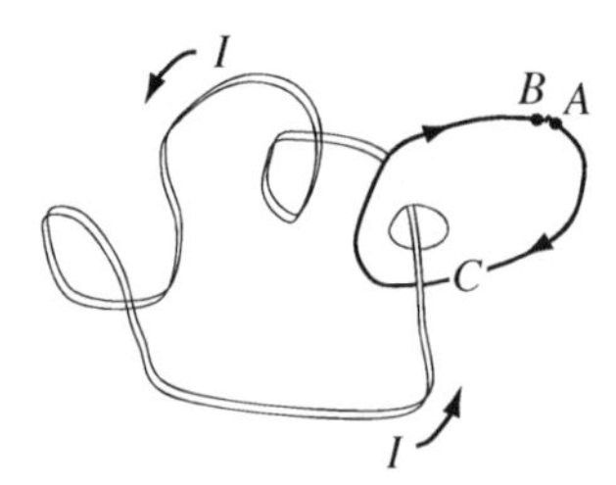

Figure 10.15: The volume V is all of space except for the filamentary current loop I. The curve C "links" the current loop and so cannot be shrunk to a point within V.

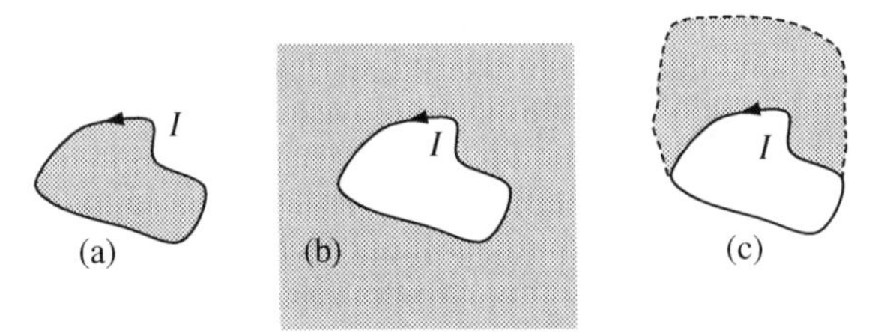

Figure 10.16: Three open "barrier" surfaces (shaded) which bound the current loop I and make V a simply-connected volume: (a) the surface is planar and finite; (b) the surface is planar and extends to infinity; (c) the surface has the shape of a billowing windsock.

However, unlike the line integral of $\mathbf{E}(\mathbf{r})$ over a closed path (which is always zero), the line integral of $\mathbf{B}(\mathbf{r})$ is *not* zero if the closed path encloses a current I. This is the content of Ampère's law:

$$\oint_C d\boldsymbol{\ell}\cdot\mathbf{B} = \mu_0 I. \tag{10.58}$$

To be more precise, let A and B be infinitesimally nearby points which define the beginning and end of a curve C which encircles the current I in Figure 10.15. In the language of (10.57), the Ampère's law constraint (10.58) reads

$$\psi(A) - \psi(B) = \pm\mu_0 I. \tag{10.59}$$

The plus/minus sign applies when the encircled current flows parallel/anti-parallel to the "thumb" when the right-hand rule is applied to C. This shows that $\psi(\mathbf{r})$ is not a single-valued function of position.

The magnetic scalar potential $\psi(\mathbf{r})$ is inevitably multi-valued when V is not *simply connected*, that is, when integration paths exist which encircle (or "link") a closed loop of current as in Figure 10.15. Fortunately, it is not difficult to make $\psi(\mathbf{r})$ single-valued and thereby restore its usefulness. The trick is to insert "barrier" surfaces into V which prevent the occurrence of linking paths. A sufficient condition is that the barrier(s) transform V into a simply-connected domain. Figure 10.16 illustrates several alternatives for the case of a planar current loop I. In each case, we are entitled to solve $\nabla^2\psi = 0$ in the volume which excludes the barrier. From this point of view, (10.59) is a *matching condition* which applies across the barrier.

We recall at this point that the magnetic scalar potential (10.48) for a current ring is not continuous at $r = R$. This discontinuity is *not* a consequence of (10.59). For the ring, $r = R$ is a *closed* matching surface which partitions space into disjoint regions rather than an *open* barrier surface with the current source as its boundary. An alternative method of solution using true barriers must be adopted for the ring problem in order to apply (10.59) to the current-ring problem.

10.4.4 Solid Angle Representation

We conclude this section by deriving a representation of the magnetic scalar potential $\psi(\mathbf{r})$ for a filamentary current loop C using the vector identity (1.85):

$$\oint_C d\boldsymbol{\ell} \times \mathbf{f} = \int_S dS_k \nabla f_k - \int_S d\mathbf{S} \nabla \cdot \mathbf{f}. \tag{10.60}$$

The idea is to identify the left side of (10.60) with the Biot-Savart integral (10.17) for a current circuit C. Then, if S is any surface which bounds C and we use the variable $\mathbf{r}'$ in place of $\boldsymbol{\ell}$, (10.60) reads

$$\mathbf{B}(\mathbf{r}) = \frac{\mu_0 I}{4\pi} \int_S dS_k \nabla' \frac{(\mathbf{r}-\mathbf{r}')_k}{|\mathbf{r}-\mathbf{r}'|^3} - \frac{\mu_0 I}{4\pi} \int_S d\mathbf{S}\, \nabla' \cdot \frac{\mathbf{r}-\mathbf{r}'}{|\mathbf{r}-\mathbf{r}'|^3}. \tag{10.61}$$

The integrand of the second term in (10.61) is proportional to $\delta(\mathbf{r}-\mathbf{r}')$. This term vanishes if we limit ourselves to observation points $\mathbf{r}$ that *do not* lie on the surface S. Moreover, at the cost of a minus sign, we can replace ∇' by ∇ in the integrand of the first term in (10.61). Therefore,

$$\mathbf{B}(\mathbf{r}) = \frac{\mu_0 I}{4\pi} \nabla \int_S d\mathbf{S} \cdot \frac{\mathbf{r}'-\mathbf{r}}{|\mathbf{r}-\mathbf{r}'|^3} = \frac{\mu_0 I}{4\pi} \nabla \Omega_S(\mathbf{r}), \tag{10.62}$$

where $\Omega_S(\mathbf{r})$ is the solid angle (Section 3.4.4) subtended by S at the observation point $\mathbf{r}$. Comparison with (10.41) shows that, up to an irrelevant constant, the magnetic scalar potential for a current loop is

$$\psi(\mathbf{r}) = -\frac{\mu_0 I}{4\pi} \Omega_S(\mathbf{r}). \tag{10.63}$$

It is important to remember that the direction of the vector $d\mathbf{S}$ in (10.62) is fixed by the direction of current flow and the right-hand rule. With that information, the properties of the solid angle when the observation point passes through S provide an alternative proof of the jump condition (10.59). We will return to this point in the next chapter when we discuss dipole layers.

Example 10.3 Use the solid angle representation (10.62) to find $\mathbf{B}(z)$ on the symmetry axis of a current ring.

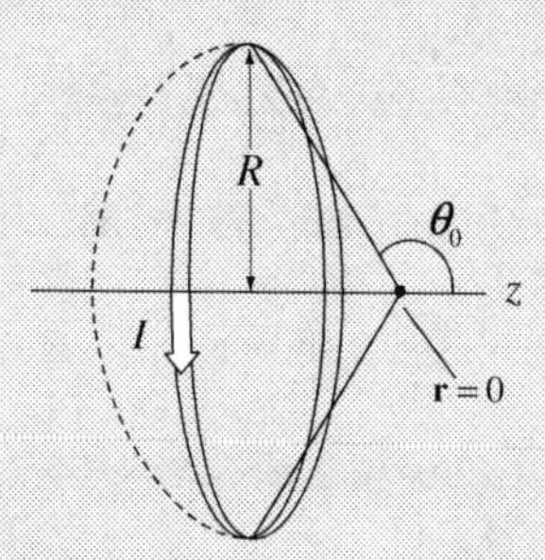

Figure 10.17: A current ring of radius R lies in the plane $z = 0$. The dashed line is part of sphere centered at the observation point $\mathbf{r} = 0$.

Solution: In Figure 10.17, a sphere is centered on the observation point $\mathbf{r} = 0$ such that a portion of that sphere (dashed) forms a spherical cap for the current ring. In that case, the solid angle subtended by the ring is the same as the solid angle subtended by the cap. The direction of current flow fixes $d\mathbf{S} = -a^2 \sin\theta d\theta d\phi\, \hat{\mathbf{s}}$. Therefore, the solid angle is

$$\Omega(z) = -\int_0^{2\pi} d\phi \int_{\theta_0}^{\pi} d\theta \sin\theta = 2\pi(\cos\theta_0 - 1) = 2\pi\left(\frac{z}{\sqrt{R^2+z^2}} - 1\right).$$

In agreement with (10.45), the magnetic field (10.62) is

$$\mathbf{B}(z) = \frac{\mu_0 I}{4\pi}\frac{d}{dz}\Omega(z)\hat{\mathbf{z}} = \frac{\mu_0 I}{2}\frac{R^2}{(z^2+R^2)^{3/2}}\hat{\mathbf{z}}.$$

10.5 The Vector Potential

The magnetic scalar potential formula $\mathbf{B} = -\nabla\psi$ is not valid at points in space where $\mathbf{j}(\mathbf{r}) \neq 0$. A more general approach to $\mathbf{B}(\mathbf{r})$ exploits the zero-divergence condition $\nabla \cdot \mathbf{B} = 0$ to infer that a *vector potential* $\mathbf{A}(\mathbf{r})$ exists such that

$$\mathbf{B}(\mathbf{r}) = \nabla \times \mathbf{A}(\mathbf{r}). \tag{10.64}$$

The alert reader will have appreciated this fact already from our application of the Helmholtz theorem to get (10.14). Here is another proof.

Assume that $\mathbf{B}(x, y, z)$ is given at every point in space. We construct a vector potential which satisfies (10.64) as follows. Let $\mathbf{A} = A_x\hat{\mathbf{x}} + A_y\hat{\mathbf{y}} + A_z\hat{\mathbf{z}}$, where A_z is an arbitrary function of z alone and A_x and A_y are defined by

$$A_x = \int dz\, B_y \qquad \text{and} \qquad A_y = -\int dz\, B_x. \tag{10.65}$$

These conditions imply that

$$\nabla \times \mathbf{A} = \left(B_x, B_y, \frac{\partial A_y}{\partial x} - \frac{\partial A_x}{\partial y}\right). \tag{10.66}$$

On the other hand, $\nabla \cdot \mathbf{B} = 0$, so (10.65) and (10.66) together imply that

$$\hat{\mathbf{z}} \cdot \nabla \times \mathbf{A} = -\int dz \left(\frac{\partial B_x}{\partial x} + \frac{\partial B_y}{\partial y}\right) = \int dz \frac{\partial B_z}{\partial z} = B_z. \tag{10.67}$$

Combining (10.66) with (10.67) confirms (10.64).

An immediate consequence of the existence of the vector potential is a circuit-integral representation of the magnetic flux (10.8). Using Stokes' theorem (Section 1.4.4),

$$\Phi_B = \int_S d\mathbf{S} \cdot \mathbf{B} = \int_S d\mathbf{S} \cdot \nabla \times \mathbf{A} = \int_C d\boldsymbol{\ell} \cdot \mathbf{A}. \tag{10.68}$$

This representation has many uses, not the least being its essential role in the quantum mechanics of particles moving in the presence of a magnetic field.[12]

[12] See, for example, Peshkin and Tonomura (1989) in Sources, References, and Additional Reading.

10.5.1 The Non-Uniqueness of $\mathbf{A}(\mathbf{r})$

Like the electric scalar potential $\varphi(\mathbf{r})$, the magnetic vector potential $\mathbf{A}(\mathbf{r})$ is not uniquely defined. The arbitrariness of $A_z(z)$ in the existence proof just above makes this clear. More generally, if $\chi(\mathbf{r})$ is any scalar function, (10.64) and the identity $\nabla \times \nabla\chi \equiv 0$ show that the magnetic field produced by a vector potential $\mathbf{A}$ is the same as the magnetic field produced by the alternative vector potential

$$\mathbf{A}'(\mathbf{r}) = \mathbf{A}(\mathbf{r}) + \nabla\chi(\mathbf{r}). \tag{10.69}$$

This suggests a powerful idea: choose the *gauge function* $\chi(\mathbf{r})$ to simplify calculations. In practice, it is rare to specify the gauge function itself. Instead, we impose a *constraint* on the vector potential which is equivalent to some choice of $\chi(\mathbf{r})$.

As an example, let us choose the *Coulomb gauge* constraint

$$\nabla \cdot \mathbf{A}(\mathbf{r}) = 0. \tag{10.70}$$

If, to the contrary, an otherwise acceptable vector potential had the property that $\nabla \cdot \mathbf{A} \neq 0$, we could use (10.69) and demand that $\nabla \cdot \mathbf{A}' = 0$. This implies that

$$\nabla^2\chi(\mathbf{r}) = -\nabla \cdot \mathbf{A}(\mathbf{r}). \tag{10.71}$$

Thus, any gauge function $\chi(\mathbf{r})$ which satisfies Poisson's equation with $\nabla \cdot \mathbf{A}(\mathbf{r})$ as its source term leads, via (10.69), to a vector potential $\mathbf{A}'(\mathbf{r})$ which satisfies (10.70).

10.5.2 The Vector Poisson Equation

Substituting $\mathbf{B} = \nabla \times \mathbf{A}$ into $\nabla \times \mathbf{B} = \mu_0\mathbf{j}$ gives

$$\nabla \times (\nabla \times \mathbf{A}) = \mu_0\mathbf{j}. \tag{10.72}$$

This is the magnetostatic equivalent of the Poisson equation, $\nabla \cdot \nabla\varphi = -\rho/\epsilon_0$, which we integrated in Section 8.4 using a Green function method. A related, but somewhat awkward, Green function method can be used to integrate the double-curl equation. Here, we shall proceed differently and note that, if $j_k(\mathbf{r})$ is the kth Cartesian component of $\mathbf{j}(\mathbf{r})$, (10.72) is equivalent to[13]

$$\nabla_k(\nabla \cdot \mathbf{A}) - \nabla^2 A_k = \mu_0 j_k. \tag{10.73}$$

This equation is not easy to analyze because it non-trivially couples together all three Cartesian components of $\mathbf{A}(\mathbf{r})$. However, if we demand that the vector potential satisfy the Coulomb gauge constraint (10.70), (10.73) decouples into three independent Poisson equations, one for each Cartesian component $A_k(\mathbf{r})$. In other words, the vector $\mathbf{A}$ satisfies

$$\nabla^2\mathbf{A}(\mathbf{r}) = -\mu_0\mathbf{j}(\mathbf{r}). \tag{10.74}$$

Our experience with Poisson's equation in electrostatics allows us to immediately write down the solution of (10.74) for each Cartesian component $A_k(\mathbf{r})$:

$$A_k(\mathbf{r}) = \frac{\mu_0}{4\pi}\int d^3r' \, \frac{j_k(\mathbf{r}')}{|\mathbf{r} - \mathbf{r}'|}. \tag{10.75}$$

Re-assembling the components (10.75) into a single vector gives the vector potential in the Coulomb gauge as

$$\mathbf{A}(\mathbf{r}) = \frac{\mu_0}{4\pi}\int d^3r' \, \frac{\mathbf{j}(\mathbf{r}')}{|\mathbf{r} - \mathbf{r}'|}. \tag{10.76}$$

[13] See Section 1.2.7 for a discussion of the equivalence of (10.72) and (10.73).

When a steady current I flows in a filamentary circuit C, the substitution $\mathbf{j}\,d^3r \to I d\boldsymbol{\ell}$ (see Section 9.3.1) shows that (10.76) simplifies to

$$\mathbf{A}(\mathbf{r}) = \frac{\mu_0 I}{4\pi}\int \frac{d\boldsymbol{\ell}}{|\mathbf{r}-\boldsymbol{\ell}|}. \tag{10.77}$$

Similarly, a current distribution confined to a surface with areal density $\mathbf{K}(\mathbf{r}_S)$ produces the vector potential

$$\mathbf{A}(\mathbf{r}) = \frac{\mu_0}{4\pi}\int_S dS\,\frac{\mathbf{K}(\mathbf{r}_S)}{|\mathbf{r}-\mathbf{r}_S|}. \tag{10.78}$$

Two points are worth noting. First, comparing (10.76) with (10.14) shows that the Coulomb gauge choice is implicit in the Helmholtz theorem. Second, (10.78) shows that each Cartesian component of $\mathbf{A}(\mathbf{r})$ is related to each component of $\mathbf{K}(\mathbf{r}_S)$ in exactly the same way that the scalar potential $\varphi(\mathbf{r})$ is related to a surface charge density $\sigma(\mathbf{r}_S)$. Because $\varphi(\mathbf{r})$ is continuous when the observation point $\mathbf{r}$ passes through a layer of charge, we may conclude that $\mathbf{A}(\mathbf{r})$ is continuous when the observation point passes through a layer of current. In other words, the vector potential matching condition is

$$\mathbf{A}_1(\mathbf{r}_S) = \mathbf{A}_2(\mathbf{r}_S). \tag{10.79}$$

Example 10.4 Find $\mathbf{A}(\mathbf{r})$ and then $\mathbf{B}(\mathbf{r})$ for an infinitely long filamentary wire which carries a current I up the z-axis.

Solution: We use (10.77) with $d\boldsymbol{\ell} = dz\hat{\mathbf{z}}$. This tells us that $\mathbf{A}(\mathbf{r}) = A_z(\mathbf{r})\hat{\mathbf{z}}$. The invariance of the source with respect to translations along the z-axis and rotations around the z-axis implies that $A_z = A_z(\rho)$ in cylindrical coordinates. Therefore,

$$A_z(\rho) = \frac{\mu_0 I}{4\pi}\int_{-\infty}^{\infty}\frac{dz}{\sqrt{z^2+\rho^2}}.$$

This integral diverges. To make progress, we let the wire extend from $-L$ to $+L$, compute A_z to lowest order in ρ/L, and then let $L \to \infty$ at the end. The required integral is

$$\int_{-L}^{L}\frac{dz}{\sqrt{z^2+\rho^2}} = \ln\frac{\sqrt{1+(\rho/L)^2}+1}{\sqrt{1+(\rho/L)^2}-1} \simeq \ln 4 + 2\ln(L/\rho).$$

Although the constant $2\ln L$ diverges as $L \to \infty$, this term (and the other constant) drop out of the magnetic field $\mathbf{B} = \nabla\times\mathbf{A}$ for any finite L. Therefore, in agreement with Section 10.3.1, we conclude that

$$A_z(\rho) = -\frac{\mu_0 I}{2\pi}\ln\rho \qquad \text{and} \qquad \mathbf{B}(\rho) = \frac{\mu_0 I}{2\pi\rho}\hat{\boldsymbol{\phi}}.$$

10.5.3 An Instructive Example

Example 10.4 exploited the result of (10.75) that a current flowing in one Cartesian direction produces a vector potential pointed in the same Cartesian direction. By contrast, a current density $\mathbf{j} = j(\mathbf{r})\hat{\boldsymbol{\phi}}$ generally does *not* produce a vector potential $\mathbf{A} = A\hat{\boldsymbol{\phi}}$. This is so because the curvilinear unit vector $\hat{\boldsymbol{\phi}}$ is not a constant vector and we must write

$$\mathbf{j}(\mathbf{r}) = j(\mathbf{r})\hat{\boldsymbol{\phi}} = j(\mathbf{r})\{\cos\phi\hat{\mathbf{y}} - \sin\phi\hat{\mathbf{x}}\} \tag{10.80}$$

before using (10.76) to evaluate the vector potential.

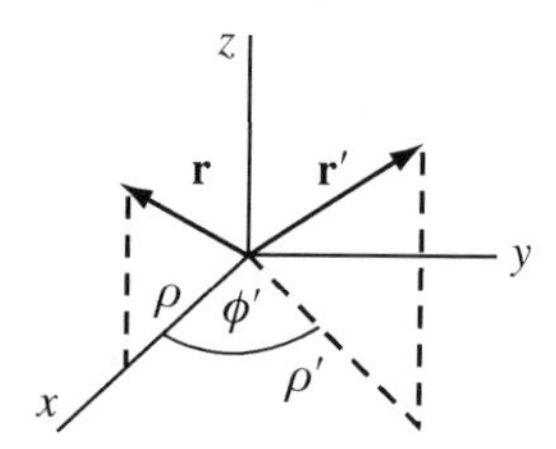

Figure 10.18: Coordinate system used to evaluate (10.76) when $\mathbf{j}(\mathbf{r}) = j(\rho, z)\hat{\boldsymbol{\phi}}$.

An instructive exception is the current density $\mathbf{j}(\mathbf{r}) = j(\rho, z)\hat{\boldsymbol{\phi}}$. If we choose the observation point at $\mathbf{r} = \rho\hat{\mathbf{x}} + z\hat{\mathbf{z}}$, as shown in Figure 10.18, the integral (10.76) takes the form

$$\mathbf{A}(\mathbf{r}) = \frac{\mu_0}{4\pi}\int d^3r' \frac{j(\rho', z')\{\cos\phi'\hat{\mathbf{y}} - \sin\phi'\hat{\mathbf{x}}\}}{\sqrt{(\rho - \rho'\cos\phi')^2 + (\rho'\sin\phi')^2 + (z - z')^2}}. \tag{10.81}$$

The $\hat{\mathbf{x}}$-component of (10.81) is zero because the integrand is an odd function of ϕ'. Therefore, $\mathbf{A}$ points in the $\hat{\mathbf{y}}$-direction, which happens to be the $\hat{\boldsymbol{\phi}}$ direction for an observation point in the x-z plane. On the other hand, the rotational symmetry of the current density tells us that this choice of observation point is not special. Therefore, the vector potential is

$$\mathbf{A}(\rho, z) = \hat{\boldsymbol{\phi}}\,\frac{\mu_0}{4\pi}\int d^3r' \frac{j(\rho, z)\cos\phi'}{|\mathbf{r} - \mathbf{r}'|}, \tag{10.82}$$

and we conclude that

$$\mathbf{j} = j(\rho, z)\hat{\boldsymbol{\phi}} \quad\Longrightarrow\quad \mathbf{A} = A(\rho, z)\hat{\boldsymbol{\phi}}. \tag{10.83}$$

A similar argument in spherical coordinates shows that

$$\mathbf{j} = j(r, \theta)\hat{\boldsymbol{\phi}} \quad\Longrightarrow\quad \mathbf{A} = A(r, \theta)\hat{\boldsymbol{\phi}}. \tag{10.84}$$

The filamentary current ring (Section 10.4.1) is an example where these results apply.

10.5.4 The Double-Curl Equation

Our derivation of the vector Poisson equation (10.74) and the vector potential in the Coulomb gauge (10.76) passed quickly over the fundamental equation (10.72) satisfied by $\mathbf{A}(\mathbf{r})$ before any choice of gauge is made. This *double-curl equation* bears repeating:

$$\nabla \times (\nabla \times \mathbf{A}) = \mu_0 \mathbf{j}. \tag{10.85}$$

When $\mathbf{j}(\mathbf{r})$ has sufficient symmetry that $\mathbf{A}(\mathbf{r})$ has a single vector component, (10.85) simplifies to a single, inhomogeneous, partial differential equation. In favorable cases, this differential equation may be easier to solve than performing the vector potential integral (10.76). Example 10.5 and Example 10.6 illustrate this approach to finding $\mathbf{A}(\mathbf{r})$.

Example 10.5 Integrate $\nabla \times (\nabla \times \mathbf{A}) = \mu_0\mathbf{j}$ to find $\mathbf{A}(\mathbf{r})$ for a long, straight cylindrical wire with radius a which carries a uniform current density j. Find $\mathbf{B}(\mathbf{r})$ from $\mathbf{A}(\mathbf{r})$ for $\rho < a$ and $\rho > a$.

Solution: If $\mathbf{j} = j\hat{\mathbf{z}}$, (10.75) tells us that $\mathbf{A} = (0, 0, A_z)$. By symmetry, $A_z = A_z(\rho)$ in cylindrical coordinates and (10.85) reduces to

$$\frac{1}{\rho}\frac{\partial}{\partial\rho}\left(\rho\frac{\partial A_z}{\partial\rho}\right) = -\mu_0 j.$$

Direct integration produces a particular solution of the inhomogeneous equation and a general solution of the homogeneous equation:

$$A_z(\rho) = -\tfrac{1}{4}\mu_0 j\rho^2 + C\ln\rho + D.$$

The $\ln\rho$ term is absent for $\rho < a$ and the ρ^2 term is absent for $\rho > a$. If D and D' are the constants for $\rho < a$ and $\rho > a$, the matching condition (10.79) applied at $\rho = a$ fixes the value of D'. We find

$$A_z(\rho) = \begin{cases} D_0 - \tfrac{1}{4}\mu_0 j\rho^2 & \rho \le a, \\ D_0 - \tfrac{1}{4}\mu_0 ja^2 + C\ln(\rho/a) & \rho \ge a. \end{cases}$$

Since $\mathbf{B} = -(dA_z/d\rho)\hat{\boldsymbol{\phi}}$, we can put $D_0 = 0$ and the matching condition (10.37) fixes $C = -\tfrac{1}{2}\mu_0 ja^2$. Since $I = j\pi a^2$, the final vector potential and magnetic field are

$$A_z(\rho) = \begin{cases} -\dfrac{\mu_0 I}{4\pi}\dfrac{\rho^2}{a^2} & \rho \le a, \\ -\dfrac{\mu_0 I}{4\pi}\left[1 + 2\ln\left(\dfrac{\rho}{a}\right)\right] & \rho \ge a, \end{cases} \qquad B_\phi(\rho) = \begin{cases} \dfrac{\mu_0 I}{2\pi}\dfrac{\rho}{a^2} & \rho \le a, \\ \dfrac{\mu_0 I}{2\pi\rho}, & \rho \ge a. \end{cases}$$

Example 10.6 Find $\mathbf{A}(\mathbf{r})$ in cylindrical coordinates for a filamentary current ring of radius R. The ring is coaxial with the z-axis and centered at the origin, as in Figure 10.11.

Solution: Our strategy is to solve $\nabla \times \nabla \times \mathbf{A} = 0$ in the space above and below $z = 0$ and match the solutions together using (10.38). The surface current density $\mathbf{K} = I\delta(\rho - R)\hat{\boldsymbol{\phi}}$ at $z = 0$ is an example of (10.83), so the double-curl equation reduces to a partial differential equation for $A_\phi(\rho, z)$ alone. With $\mathbf{A} = A(\rho, z)\hat{\boldsymbol{\phi}}$, we find

$$\frac{\partial^2 A}{\partial z^2} + \frac{\partial^2 A}{\partial \rho^2} + \frac{1}{\rho}\frac{\partial A}{\partial \rho} - \frac{A}{\rho^2} = 0.$$

Separating variables with $A(\rho, z) = R(\rho)Z(z)$ and a separation constant k^2 gives $Z(z) = \exp(\pm kz)$ and the differential equation

$$R'' + \frac{R'}{\rho} - \frac{R}{\rho^2} + k^2 R = 0.$$

This is Bessel's equation (7.82) of order one. Of the two linearly independent solutions (Section 7.8.1), $J_1(k\rho)$ is regular for all ρ, but $N_1(k\rho)$ is not. The vector potential above (+) and below (−) the $z = 0$ plane is a linear combination of elementary regular solutions with all possible values of k:

$$A_\pm(\rho, z) = \int_0^\infty dk\, a(k) J_1(k\rho) e^{\mp kz}.$$

The expansion coefficients $a(k)$ are determined from the jump (10.38) in the tangential component of the magnetic field at the surface of the ring,

$$\left[\frac{\partial A_+}{\partial z} - \frac{\partial A_-}{\partial z}\right]_{z=0} = -\mu_0 I\delta(\rho - R),$$

and the completeness relation for the Bessel functions,

$$\int_0^\infty d\rho\, k\rho J_1(k\rho)J_1(k'\rho) = \delta(k-k').$$

The final result for the vector potential is

$$\mathbf{A}(\rho, z) = \hat{\boldsymbol{\phi}}\frac{1}{2}\mu_0 I R \int_0^\infty dk\, J_1(k\rho)J_1(kR)e^{-k|z|}.$$

10.6 The Topology of Magnetic Field Lines

The magnetic field line patterns produced by currents confined to a straight wire, a circular loop, or a planar sheet are not truly representative of magnetic fields that satisfy $\nabla \cdot \mathbf{B} = 0$. To see this, focus on the magnetic field $\mathbf{B}_1$ produced by a circular ring. As seen in Figure 10.12, the field lines very near the ring form closed circles centered on the current. The set of all such field line circles with a given radius forms a torus. Now add an infinite straight wire which carries a current along the symmetry axis of the current-carrying ring (Figure 10.19). Some of the circular field lines produced by the wire field alone, call it $\mathbf{B}_2$, are tangent to the torus. Therefore, some lines of the field $\mathbf{B}_1 + \mathbf{B}_2$ must be tangent to the torus and spiral around it in a helical manner. Only for certain values of the ring and wire currents does the helix close on itself. A similar argument shows that field lines spiral helically around the straight wire as well.

10.6.1 Magnetic Reconnection

The topology of a static magnetic field pattern is fixed once and for all. However, there are many situations where the sources of the field change so slowly that a "quasistatic" approximation can be used at each position of the source.[14] When that occurs, it is not difficult to imagine source motions which bring magnetic field lines very close together. They can even touch at points where $\mathbf{B} = 0$. These null points are important because they are places where the connectivity of the field lines can change, resulting in a change in the overall topology of the magnetic field. This phenomenon is called *magnetic reconnection.*

There is good experimental evidence that magnetic reconnection occurs inside tokamaks and other fusion research machines. The magnetic fields in question are associated with currents of electrons and ions in the form of a plasma. The hydrodynamics of the plasma complicates the description of reconnection considerably. Regardless, space is the place for plasmas, and there is reason to believe that reconnection is operative in the evolution of essentially all extraterrestrial magnetic fields. The magnetohydrodynamics of the Sun is a good example, particularly in connection with solar flares and solar prominences. A particularly well-studied case is the interaction of the interplanetary magnetic field (IMF) with the dipole magnetic field of the Earth (see Section 11.2.1).

Figure 10.20 is an oversimplified cartoon of the magnetic field line pattern in the vicinity of the Earth. The field lines of the Earth are connected to its poles. The vertical lines on the left are the IMF being "blown" toward the Earth by a plasma called the "solar wind" (white arrows). Magnetic reconnection occurs at a point (indicated by a star) where the IMF is anti-parallel to the field lines connected to the Earth. This transiently connects the IMF to the Earth. The individual identities of the IMF field and

[14] Chapter 14 explores the quasistatic approximation in detail.

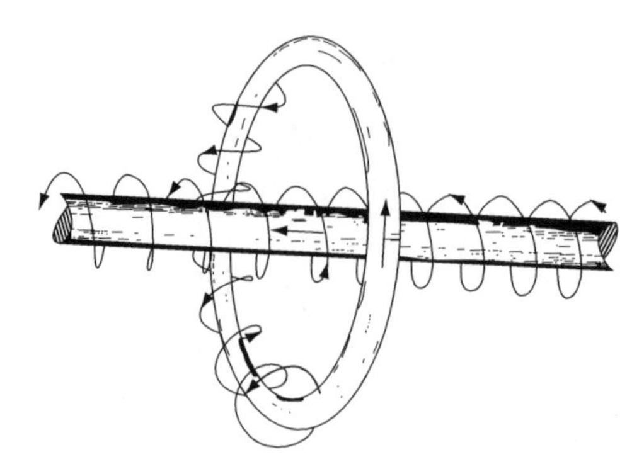

Figure 10.19: A line current which threads a ring current. The magnetic field lines around the ring and around the wire generally do not close on themselves. Figure adapted from McDonald (1954).

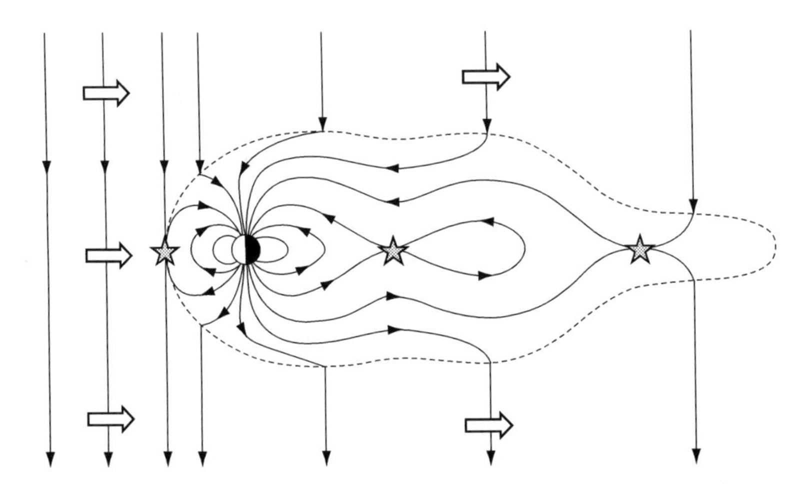

Figure 10.20: Cartoon of magnetic field lines in the vicinity of the Earth. White arrows indicate the solar wind. Dark arrows indicate the direction of magnetic field lines. Stars indicate points of magnetic reconnection. Dashed lines indicate the magnetopause. See text for discussion. Figure courtesy of Jeffrey J. Love.

the Earth's field are re-established downstream beyond additional points of reconnection. The dotted line in the diagram indicates the "magnetopause" boundary between the interplanetary plasma and the plasma of the near-Earth environment.

Application 10.3 Chaotic Lines of B

The complexity of a large class of magnetic field line configurations can be appreciated using a field constructed from a constant B_0 and an arbitrary scalar function $f(\mathbf{r})$:

$$\mathbf{B}(\mathbf{r}) = B_0\hat{\mathbf{z}} + \hat{\mathbf{z}} \times \nabla f(\mathbf{r}). \tag{10.86}$$

This field satisfies $\nabla \cdot \mathbf{B} = 0$ by construction. From (10.6), the equation for the field lines is

$$\frac{B_x}{dx} = \frac{B_y}{dy} = \frac{B_z}{dz} = \lambda, \tag{10.87}$$

or

$$\frac{dx}{dz} = -\frac{1}{B_0}\frac{\partial f}{\partial y} \quad \text{and} \quad \frac{dy}{dz} = +\frac{1}{B_0}\frac{\partial f}{\partial x}. \tag{10.88}$$

Now, change variables in (10.88) so $x = q$, $y = p$, and $z = t$. If, in addition, we let $f = -B_0 H$, the two equations above are exactly Hamilton's equations of classical mechanics,

$$\dot{q} = \frac{\partial H}{\partial p} \qquad \text{and} \qquad \dot{p} = -\frac{\partial H}{\partial x}. \tag{10.89}$$

Therefore, the magnetic field lines are the "time"-dependent trajectories in (p, q) phase space of a "particle" with Hamiltonian $H = -f/B_0$. Since most Hamiltonians are non-integrable and produce chaotic trajectories, the magnetic field line configuration will be very complex indeed. ■

Example 10.7 The *helicity* $h = \int d^3r \mathbf{A} \cdot \mathbf{B}$ is a quantitative measure of the topological complexity of a magnetic field configuration. It is used extensively in solar physics and other situations where dynamo action produces very complicated field line patterns (see Figure 11.4). To illustrate the idea, define a *flux tube* to be a bundle of parallel magnetic field lines where the same magnetic flux $\Phi = \int d\mathbf{S} \cdot \mathbf{B}$ passes through every cross section of the tube. Figure 10.21 shows two flux tubes which close on themselves. The two tubes are unlinked in Figure 10.21(a) and linked in Figure 10.21(b).

(a) Show that $h = 0$ when the tubes are unlinked and $h = 2\Phi_1\Phi_2$ when the tubes are linked.
(b) Find the conditions required to make the definition of h gauge invariant.

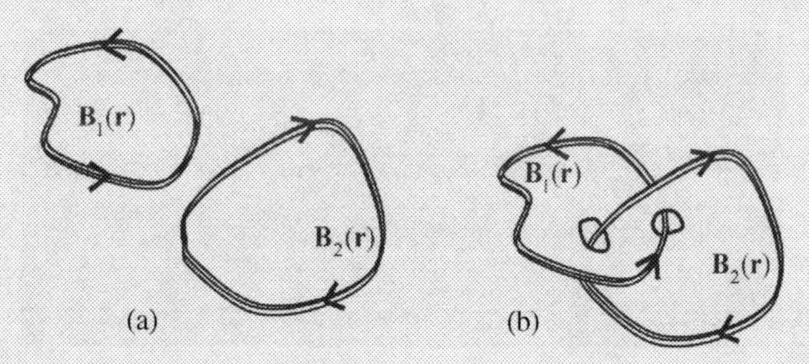

Figure 10.21: Two closed flux tubes made from bundles of magnetic field lines. Panels (a) and (b) show the tubes unlinked and linked, respectively.

Solution: (a) Let $\mathbf{A} = \mathbf{A}_1 + \mathbf{A}_2$. Because $\mathbf{B}_1 = \nabla \times \mathbf{A}_1$ is confined to the volume V_1 of tube 1 and $\mathbf{B}_2 = \nabla \times \mathbf{A}_2$ is confined to the volume of tube 2,

$$h = \int d^3r \, \mathbf{A} \cdot \mathbf{B} = \int_{V_1} d^3r \, \mathbf{A} \cdot \mathbf{B}_1 + \int_{V_2} d^3r \, \mathbf{A} \cdot \mathbf{B}_2.$$

Inside each tube, $d^3r = d\mathbf{S} \cdot d\boldsymbol{\ell}$, where $\mathbf{B} \parallel d\mathbf{S} \parallel d\boldsymbol{\ell}$ and $d\mathbf{S}$ is an element of the tube cross section. Therefore, if C_1 and C_2 are the closed curves which define the tubes longitudinally,

$$h = \int_{C_1} d\boldsymbol{\ell} \cdot \mathbf{A} \int_{S_1} d\mathbf{S} \cdot \mathbf{B}_1 + \int_{C_2} d\boldsymbol{\ell} \cdot \mathbf{A} \int_{S_2} d\mathbf{S} \cdot \mathbf{B}_2 = \Phi_1 \int_{C_1} d\boldsymbol{\ell} \cdot \mathbf{A} + \Phi_2 \int_{C_2} d\boldsymbol{\ell} \cdot \mathbf{A}.$$

Now, from (10.68), the magnetic flux through the open surface S bounded by a curve C is $\Phi = \int_C d\boldsymbol{\ell} \cdot \mathbf{A}$. When the two flux tubes are unlinked as in Figure 10.21(a), no magnetic flux passes through either S_1 or S_2. However, when the two flux tubes are linked as in Figure 10.21(b), the full flux of $\mathbf{B}_1$ passes through $\mathcal{S}_2$ and the full flux of $\mathbf{B}_2$ passes through $\mathcal{S}_1$. Therefore,

$$h = \begin{cases} 0 & \text{tubes are unlinked,} \\ 2\Phi_1\Phi_2 & \text{tubes are linked.} \end{cases}$$

(b) From (10.69), a change of gauge considers $\mathbf{A}' \to \mathbf{A} + \nabla\chi$. We thus compute

$$h' = \int d^3r\, \mathbf{A}' \cdot \mathbf{B} = h + \int d^3r\, \nabla\chi \cdot \mathbf{B}.$$

Since $\nabla \cdot \mathbf{B} = 0$,

$$h' = h + \int d^3r\, \nabla \cdot (\mathbf{B}\chi) = h + \int d\mathbf{S} \cdot \mathbf{B}\chi.$$

Therefore, $h' = h$ if $\mathbf{B}$ vanishes at the boundary of the integration volume or, less restrictively, $\hat{\mathbf{n}} \cdot \mathbf{B} = 0$ everywhere on the boundary.

Sources, References, and Additional Reading

The quotation at the beginning of the chapter is taken from the paper where William Thomson (later Lord Kelvin) first embraces Faraday's concept of "field" to interpret his mathematical results.

W. Thomson, "On the theory of magnetic induction in crystalline and non-crystalline substances", *Philosophical Magazine* **1**, 177 (1851). Reproduced with permission of Taylor & Francis Ltd.

Section 10.1 Two authoritative monographs on the subject matter of this chapter are

E. Durand, *Magnétostatique* (Masson, Paris, 1968).

H.E. Knoepfel, *Magnetic Fields* (Wiley, New York, 2000).

Textbooks of electromagnetism with particulary good treatments of magnetostatics include

W. Hauser, *Introduction to the Principles of Electromagnetism* (Addison-Wesley, Reading, MA, 1971).

M.H. Nayfeh and M.K. Brussel, *Electricity and Magnetism* (Wiley, New York, 1985).

H.A. Haus and J.R. Melcher, *Electromagnetic Fields and Energy* (Prentice Hall, Englewood Cliffs, NJ, 1989).

The current status of magnetic monopoles is reviewed in

K. A. Milton, "Theoretical and experimental status of magnetic monopoles", *Reports on Progress in Physics*, **69**, 1637 (2006).

The Earnshaw-like theorem for $|\mathbf{B}(\mathbf{r})|$ discussed in Section 10.1.4 was proved by William Thomson in 1847 in connection with a study of the magnetic levitation of diamagnetic objects. The theorem has been rediscovered many times over the years. The original discussion and a modern treatment are

William Thomson, *Reprints of Papers on Electrostatics and Magnetism* (Macmillan, London, 1872), Section 665.

M.V. Berry and A.K. Geim, "Of flying frogs and levitrons", *European Journal of Physics* **18**, 307 (1997).

Section 10.2 Example 10.1 is adapted from Section 8.2 of *Haus and Melcher* (see Section 10.1 above). The calculation in Section 10.2.2 is taken from

V. Namias, "On the magnetic field due to a solenoid of arbitrary cross section", *American Journal of Physics* **53**, 588 (1985).

Section 10.3 The symmetry argument used in Section 10.3.2 is adapted from Section 9.8 of *Hauser* (see Section 10.1 above).

Section 10.4 The text and translation into English of the extract from the correspondence between Helmholtz and du Bois-Raymond quoted after Section 10.4.2 was provided by Prof. Kathryn Olesko (Georgetown University). Figure 10.13 comes from

W.T. Scott, *The Physics of Electricity and Magnetism*, 2nd edition (Wiley, New York, 1966).

Generalizations of the Helmholtz coil are discussed in

J.L. Kirschvink, "Uniform magnetic fields and double-wrapped coil systems", *Bioelectromagnetics* **13**, 401 (1992).

A very clear discussion of the use of scalar potential theory to calculate magnetic fields is contained in Part 1, "Generation and Computation of Magnetic Fields", of the monograph

H. Zijlstra, *Experimental Methods in Magnetism* (North-Holland, Amsterdam, 1967).

Application 10.2 on magnetic shielding in MRI scanners was adapted from

P. Mansfield and B. Chapman, "Multishield active magnetic screening of coil structures in NMR", *Journal of Magnetic Resonance* **72**, 211 (1987).

The topological aspects of the magnetic scalar potential are discussed in more detail in

A. Vourdas and K.J. Binns, "Magnetostatics with scalar potentials in multiply connected regions", *IEE Proceedings A* **136**, 49 (1989).

Section 10.5 The role of the vector potential (and its gauge invariance) in quantum mechanics is discussed authoritatively in

M. Peshkin and A. Tonomura, *The Aharonov-Bohm Effect* (Springer, Berlin, 1989).

Our treatment of the vector Poisson equation and the double curl equation draws heavily on

L.D. Landau and E.M. Lifshitz, *Electrodynamics of Continuous Media* (Pergamon, Oxford, 1960), Section 29.

Vector potential topics we do not discuss, like uniqueness and direct integration of the double curl equation, are treated in

J.A. Stratton, *Electromagnetic Theory* (McGraw-Hill, New York, 1941).

Example 10.6 is taken from the Appendix to

W.G. Hurley, "Calculation of self and mutual inductances in planar magnetic structures", *IEEE Transactions on Magnetics* **31**, 2416 (1995).

Section 10.6 The stimulating article from which Figure 10.19 was taken is

K.L. McDonald, "Topology of steady current magnetic fields", *American Journal of Physics* **22**, 586 (1954). Copyright 1954, American Association of Physics Teachers.

An entry point to the literature of Hamiltonian approaches to magnetic field lines is

M. Sita Janaki and G. Ghosh, "Hamiltonian formulation of magnetic field line equations", *Journal of Physics A* **20**, 3679 (1987).

Example 10.7 was taken from

H.K. Moffatt, *Magnetic Field Generation in Electrically Conducting Fluids* (Cambridge University Press, Cambridge, 1978).

Problems

10.1 In-Plane Field of a Current Strip A uniform surface current $\mathbf{K} = K\hat{\mathbf{z}}$ confined to a strip of width b carries a total current I. Find the magnetic field at a point in the plane of the strip that lies a perpendicular distance a from the strip in the $\hat{\mathbf{y}}$-direction.

10.2 Current Flow in a Disk The z-axis coincides with the symmetry axis of a flat disk of radius R in the x-y plane. Sketch and justify in words the pattern of currents that must flow in the disk to produce the magnetic field pattern shown below (as viewed edge-on with the disk). The field pattern has the rotational symmetry of the disk.

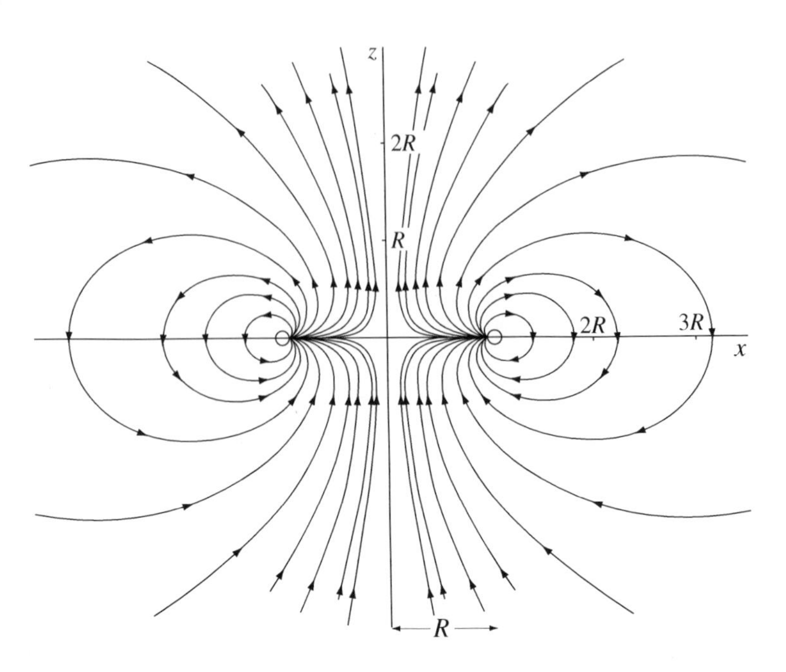

10.3 Finite-length Solenoid I

(a) Consider a semi-infinite and tightly wound solenoid with a circular cross section. Prove that the magnetic flux which passes out through the open end of the solenoid is exactly one-half the flux which passes through a cross section deep inside the solenoid.

(b) A tightly wound solenoid has length L and a circular cross section. Let $L = 5R$, where R is the radius of the cross section. Sketch the magnetic field lines associated with this solenoid. Take special care with the lines near the open ends. Do any field lines penetrate the walls of the solenoid? If not, explain why not. If so, discuss their behavior very near the walls.

10.4 Helmholtz and Gradient Coils

(a) Two rings of radius R, coaxial with the z-axis, are separated by a distance $2b$ and carry a current I in the same direction. Make explicit use of the formula for the magnetic field of a single current ring on its symmetry axis to derive the Helmholtz relation between b and R that makes $B_z(z)$ most uniform in the neighborhood of the midpoint between the rings.

(b) A gradient coil has the same geometry as part (a) except that the rings carry current in opposite directions. Compute $B_z(z)$ in the immediate vicinity of the midpoint between the rings for an arbitrary choice of the ring separation.

10.5 A Step off the Symmetry Axis A circular loop with radius R and current I lies in the x-y plane centered on the z-axis. The magnetic field on the symmetry axis is

$$\mathbf{B}(z) = \frac{1}{2}\mu_0 I \frac{R^2}{(R^2 + z^2)^{3/2}}\hat{\mathbf{z}}.$$

In cylindrical coordinates, $B_\rho(\rho, z) = f(z)\rho$ when $\rho \ll R$. Use only the Maxwell equations to find $f(z)$ and then $B_z(\rho, z)$ when $\rho \ll R$.

10.6 Two Approaches to the Field of a Current Sheet

(a) Use the Biot-Savart law to find $\mathbf{B}(\mathbf{r})$ everywhere for a current sheet at $x = 0$ with $\mathbf{K} = K\,\hat{\mathbf{z}}$.
(b) Check your answer to part (a) by superposing the magnetic field from an infinite number of straight current-carrying wires.

10.7 The Geometry of Biot and Savart Biot and Savart derived their eponymous formula using a current-carrying wire bent as shown below. Find $\mathbf{B}(\mathbf{r})$ in the plane of the wire at a distance d from the bend along the axis of symmetry.

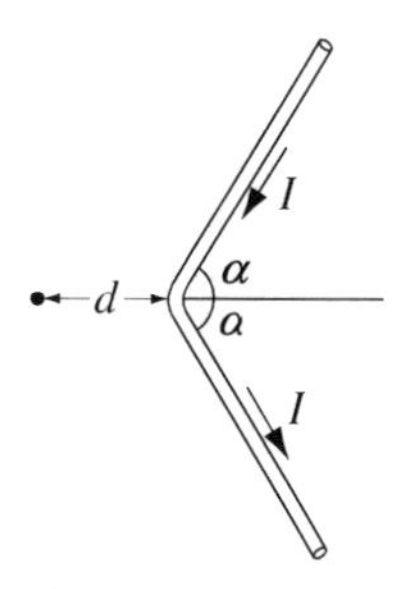

10.8 The Magnetic Field of Planar Circuits

(a) Let I be the current carried by a wire bent into a planar loop. Place the origin of coordinates at an observation point P in the plane of the loop. Show that the magnitude of the magnetic field at the point P is

$$B(P) = \frac{\mu_0 I}{4\pi} \int_0^{2\pi} \frac{d\phi}{r(\phi)},$$

where $r(\phi)$ is the distance from the origin of coordinates at P to the point on the loop located at an angle ϕ from the positive x-axis.
(b) Show that the magnetic field at the center of a current-carrying wire bent into an ellipse with major and minor axes $2a$ and $2b$ is proportional to a complete elliptic integral of the second kind. Show that you get easily understandable answers when $a = b$ and when $a \to \infty$ with b fixed.
(c) An infinitesimally thin wire is wound in the form of a planar coil which can be modeled using an effective surface current density $\mathbf{K} = K\hat{\boldsymbol{\phi}}$. Find the magnetic field at a point P on the symmetry axis of the coil. Express your answer in terms of the angle α subtended by the coil at P.

10.9 Invert the Biot-Savart Law Let $\mathbf{B}(x, z)$ be the magnetic field produced by a surface current density $\mathbf{K}(y, z) = K(z)\hat{\mathbf{y}}$ confined to the $x = x_0$ plane.

(a) Show that the Biot-Savart law for this situation reduces to a one-dimensional convolution integral for each component of $\mathbf{B}$.
(b) Confine your attention to $x < x_0$ and show that

$$\mu_0 K(z) = \frac{1}{\pi} \int_{-\infty}^{\infty} dz' \int_{-\infty}^{\infty} dk \, \exp\{ik(z - z') + |k|(x_0 - x)\} B_z(x, z').$$

(c) Why does the single component $B_z(x, z)$ evaluated at one (arbitrary) value of $x < x_0$ provide enough information to determine $K(z)$?

10.10 Symmetry and Ampère's Law The figure below shows a current I which flows down the z-axis from infinity and then spreads out radially and uniformly to infinity in the $z = 0$ plane.

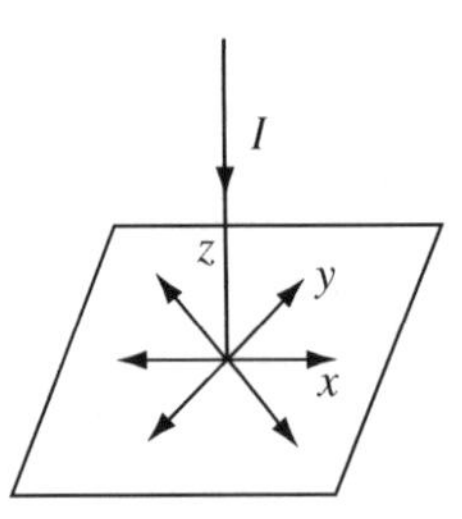

(a) The given current distribution is invariant to reflection through the y-z plane. Prove that, when reflected through this plane, the cylindrical components of the magnetic field transform from $\mathbf{B} = B_\rho\hat{\boldsymbol{\rho}} + B_\phi\hat{\boldsymbol{\phi}} + B_z\hat{\mathbf{z}}$ to

$$\mathbf{B}' = B'_\rho\hat{\boldsymbol{\rho}} + B'_\phi\hat{\boldsymbol{\phi}} + B'_z\hat{\mathbf{z}} = -B_\rho\hat{\boldsymbol{\rho}} + B_\phi\hat{\boldsymbol{\phi}} - B_z\hat{\mathbf{z}}.$$

(b) Compare the results of part (a) to the transformation of $\mathbf{B}$ to $\tilde{\mathbf{B}}$, where the latter is a π rotation around the z-axis that also leaves the current invariant. Use this and any other symmetry argument you need to conclude that $\mathbf{B}(\mathbf{r}) = B_\phi(\rho, z)\hat{\boldsymbol{\phi}}$ everywhere.

(c) Use the results of part (b) and Ampère's law to find the magnetic field everywhere.

(d) Check explicitly that your solution satisfies the magnetic field matching conditions at the $z = 0$ plane.

10.11 Current Flow over a Sphere A current I starts at $z = -\infty$ and flows up the z-axis as a linear filament until its hits an origin-centered sphere of radius R. The current spreads out uniformly over the surface of the sphere and flows up lines of longitude from the south pole to the north pole. The recombined current flows thereafter as a linear filament up the z-axis to $z = +\infty$.

(a) Find the current density on the sphere.

(b) Use explicitly stated symmetry arguments and Ampère's law in integral form to find the magnetic field at every point in space.

(c) Check that your solution satisfies the magnetic field matching conditions at the surface of the sphere.

10.12 Finite-Length Solenoid II

(a) Use superposition and the magnetic field on the symmetry axis of a current ring to find the magnetic field at the midpoint of the symmetry axis of a cylindrical solenoid. The solenoid has radius R, length L, and is wound with n turns per unit length of a wire that carries a current I.

(b) Assume that $L \gg R$ and use the results of part (a) and Ampère's law to estimate the magnetic field $\mathbf{B}_{\rm out}$ just outside the solenoid walls (but far from its edges).

10.13 How the Biot-Savart Law Differs from Ampère's Law A current I_0 flows up the z-axis from $z = z_1$ to $z = z_2$ as shown below.

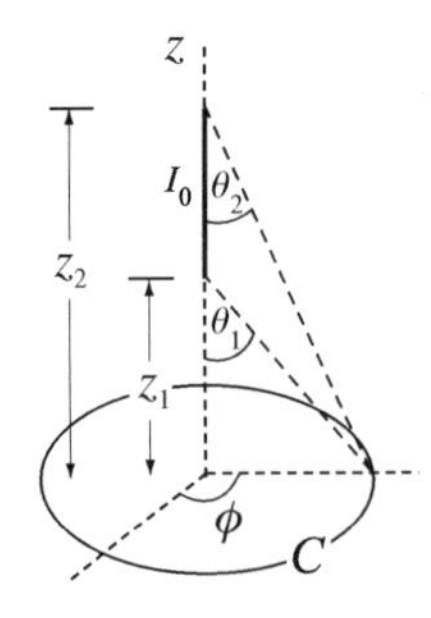

(a) Use the Biot-Savart law to show that the magnetic field in the $z = 0$ plane is

$$\mathbf{B}(\rho, \phi, 0) = \frac{\mu_0}{4\pi}\frac{I_0}{\rho}\{\cos\theta_2 - \cos\theta_1\}\,\hat{\boldsymbol{\phi}}.$$

(b) Symmetry and the Coulomb gauge vector potential show that $\mathbf{A} = A_z(\rho, z)\hat{\mathbf{z}}$ and $\mathbf{B} = \nabla \times \mathbf{A} = B_\phi(\rho, z)\hat{\boldsymbol{\phi}}$. However, an origin-centered, circular Ampèrian loop C in the $z = 0$ plane gives $\mathbf{B} = 0$, rather than the answer obtained in part (a). The reason is that the current segment I_0 does not satisfy $\nabla \cdot \mathbf{j} = 0$. To reconcile Biot-Savart with Ampère, supplement I_0 by the current densities

$$\mathbf{j}_1(\mathbf{r}) = -\frac{I_0}{4\pi} \frac{\mathbf{r} - \mathbf{r}_1}{|\mathbf{r} - \mathbf{r}_1|^3} \qquad \text{and} \qquad \mathbf{j}_2(\mathbf{r}) = \frac{I_0}{4\pi} \frac{\mathbf{r} - \mathbf{r}_2}{|\mathbf{r} - \mathbf{r}_2|^3},$$

where $\mathbf{r}_1 = (0, 0, z_1)$ and $\mathbf{r}_2 = (0, 0, z_2)$. Describe $\mathbf{j}_1(\mathbf{r})$ and $\mathbf{j}_2(\mathbf{r})$ in words and show quantitatively that they, together with I_0, form a closed circuit.

(c) Using I_0, $\mathbf{j}_1$, and $\mathbf{j}_2$ as current sources, apply Ampère's law in integral form to recover the formula derived in part (a).

(d) Show that the addition of $\mathbf{j}_1(\mathbf{r})$ and $\mathbf{j}_2(\mathbf{r})$ does not spoil the Biot-Savart calculation of part (a).

10.14 Find Surface Current from the Field inside a Sphere Find the surface current density $\mathbf{K}(\theta, \phi)$ on the surface of sphere of radius a which will produce a magnetic field inside the sphere of $\mathbf{B}_<(x, y, z) = (B_0/a)(x\,\hat{\mathbf{x}} - y\,\hat{\mathbf{y}})$. Express your answer in terms of elementary trigonometric functions.

10.15 A Spinning Spherical Shell of Charge A charge Q is uniformly distributed over the surface of a sphere of radius R. The sphere spins at a constant angular frequency with $\boldsymbol{\omega} = \omega\hat{\mathbf{z}}$. Use $\mathbf{B} = -\nabla\psi$ to find the magnetic field everywhere. Hint: See Appendix C.1.1.

10.16 The Distant Field of a Helical Coil The figure below shows an infinitely long current filament wound in the form of a circular helix with radius R and pitch ℓ, i.e., ℓ is the distance along the z-axis occupied by one wind of the helix. Find the ρ dependence of the magnetic field $\mathbf{B}(\rho, \phi, z)$ in the limit $\rho \gg \ell$. Does the $\ell \to 0$ limit make sense?

Hint: The magnetic scalar potential obeys $\nabla^2\psi = 0$ for $\rho > R$.

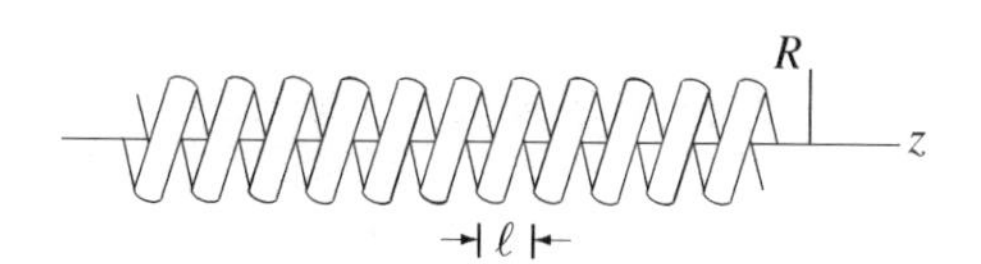

10.17 The Distant Field of Helmholtz Coils The text describes a Helmholtz coil as two parallel, coaxial, and circular current loops of radius R separated by a distance R. Each loop carries a current I in the same direction.

(a) Use the magnetic scalar potential and *both* matching conditions at an appropriate spherical surface to show that each component of the Helmholtz coil magnetic field behaves asymptotically ($r \to \infty$) as

$$\frac{f(\theta)}{r^3} + \frac{g(\theta)}{r^7} + \cdots.$$

(b) Show that a second set of Helmholtz coils arranged coaxially with the first can be used to cancel the dipole field so that only the hexadecapole field remains at long distance.

Hint: Make use of the properties of the associated Legendre functions (Appendix C).

10.18 Solid Angles for Magnetic Fields Use the solid angle representation of the magnetic scalar potential $\psi(\mathbf{r})$ to find $\mathbf{B}(\mathbf{r})$ everywhere for an infinite, straight line of current I. State carefully the surface you have chosen to "cut" the current-free volume to make $\psi(\mathbf{r})$ single-valued.

10.19 A Matching Condition for Vector Potential Show that the normal derivative of the Coulomb gauge vector potential suffers a jump discontinuity at a surface endowed with a current density $\mathbf{K}(\mathbf{r}_S)$.

10.20 Magnetic Potentials Consider a vector potential $\mathbf{A} = \frac{1}{2}C\ln(x^2 + y^2)\hat{\mathbf{z}}$. Find a vector potential $\mathbf{A}' = A'_x\hat{\mathbf{x}} + A'_y\hat{\mathbf{y}}$ which produces the same magnetic field.

10.21 Consequences of Gauge Choices

(a) Show by direct calculation that the Coulomb gauge condition $\nabla \cdot \mathbf{A} = 0$ applies to

$$\mathbf{A}(\mathbf{r}) = \frac{\mu_0}{4\pi} \int d^3r' \frac{\mathbf{j}(\mathbf{r}')}{|\mathbf{r} - \mathbf{r}'|}.$$

(b) Find the choice of gauge where a valid representation of the vector potential is

$$\mathbf{A}(\mathbf{r}) = \frac{1}{4\pi} \int d^3r' \frac{\mathbf{B}(\mathbf{r}') \times (\mathbf{r} - \mathbf{r}')}{|\mathbf{r} - \mathbf{r}'|^3}.$$

10.22 The Magnetic Field of Charge in Uniform Motion Consider a charge distribution $\rho(\mathbf{r})$ in rigid, uniform motion with velocity $\boldsymbol{v}$.

(a) Show that the magnetic field produced by this system is $\mathbf{B}(\mathbf{r}) = (\boldsymbol{v}/c^2) \times \mathbf{E}$, where $\mathbf{E}(\mathbf{r})$ is the electric field produced by $\rho(\mathbf{r})$ at rest.
(b) Use this result to find $\mathbf{B}(\mathbf{r})$ for an infinite line of current and an infinite sheet of current (both uniform) from the corresponding electrostatic problem.

10.23 A Geometry of Aharonov and Bohm

(a) Find the vector potential inside and outside a solenoid that generates a magnetic field $\mathbf{B} = B\hat{\mathbf{z}}$ inside an infinite cylinder of radius R. Work in the Coulomb gauge.
(b) The Aharonov-Bohm effect occurs because the magnetic flux $\Phi_B = \oint d\mathbf{s} \cdot \mathbf{A}$ is non-zero when the integration circuit is, say, the rim of a disk of radius $\rho > R$ which lies perpendicular to the solenoid axis. Show that $\mathbf{A}' = \mathbf{A} + \nabla\chi$ with $\chi = -\Phi\phi/2\pi$ (ϕ is the angle in cylindrical coordinates) leads to identically zero vector potential outside the solenoid.
(c) The result in part (b) implies that we could eliminate the Aharonov-Bohm effect by a gauge transformation. Show, however, that the new magnetic field corresponds to a different physical problem, where

$$\mathbf{B}' = \mathbf{B} - \hat{\mathbf{z}}\frac{\Phi}{2\pi\rho}\delta(\rho).$$

10.24 Lamb's Formula A quantum particle with charge q, mass m, and momentum $\mathbf{p}$ in a magnetic field $\mathbf{B}(\mathbf{r}) = \nabla \times \mathbf{A}(\mathbf{r})$ has velocity $\boldsymbol{v}(\mathbf{r}) = \mathbf{p}/m - (q/m)\mathbf{A}(\mathbf{r})$. This means that a charge distribution $\rho(\mathbf{r})$ generates a "diamagnetic current" $\mathbf{j}(\mathbf{r}) = -(q/m)\rho(\mathbf{r})\mathbf{A}(\mathbf{r})$ when it is placed in a magnetic field.

(a) Show that $\mathbf{A}(\mathbf{r}) = \frac{1}{2}\mathcal{B} \times \mathbf{r}$ is a legitimate vector potential for a uniform magnetic field $\mathcal{B}$.
(b) Let $\rho(\mathbf{r}) = \rho(r)$ be the spherically symmetric charge distribution associated with the electrons of an atom. Expose the atom to a uniform magnetic field $\mathcal{B}$ and show that the ensuing diamagnetic current induces a vector potential

$$\mathbf{A}_{\rm ind}(\mathbf{r}) = \frac{\mu_0}{4\pi}\frac{e\mathcal{B} \times \mathbf{r}}{6m}\left[\frac{1}{r^3}\int_{r'<r} d^3r'\, \rho(r')r'^2 + \int_{r'>r} d^3r' \frac{\rho(r')}{r'}\right].$$

(c) Expand $\mathbf{A}_{\rm ind}(\mathbf{r})$ for small values of r and show that the diamagnetic field at the atomic nucleus can be written in terms of $\varphi(0)$, the electrostatic potential at the nucleus produced by $\rho(r)$:

$$\mathbf{B}_{\rm ind}(0) = \frac{e\varphi(0)}{3mc^2}\mathcal{B}.$$

This formula was obtained by Willis Lamb in 1941. He had been asked by Isidore Rabi to determine whether $\mathbf{B}_{\rm ind}(0)$ could be ignored when nuclear magnetic moments were extracted from molecular beam data.

10.25 Toroidal and Poloidal Magnetic Fields It is true (but not obvious) that any vector field $\mathbf{V}(\mathbf{r})$ which satisfies $\nabla \cdot \mathbf{V}(\mathbf{r}) = 0$ can be written *uniquely* in the form

$$\mathbf{V}(\mathbf{r}) = \mathbf{T}(\mathbf{r}) + \mathbf{P}(\mathbf{r}) = \mathbf{L}\psi(\mathbf{r}) + \nabla \times \mathbf{L}\gamma(\mathbf{r}),$$

where $\mathbf{L} = -i\mathbf{r} \times \nabla$ is the angular momentum operator and $\psi(\mathbf{r})$ and $\gamma(\mathbf{r})$ are scalar fields. $\mathbf{T}(\mathbf{r}) = \mathbf{L}\psi(\mathbf{r})$ is called a *toroidal field* and $\mathbf{P}(\mathbf{r}) = \nabla \times \mathbf{L}\gamma(\mathbf{r})$ is called a *poloidal field*. This decomposition is widely used in laboratory plasma physics.

(a) Confirm that $\nabla \cdot \mathbf{V}(\mathbf{r}) = 0$.
(b) Show that a poloidal current density generates a toroidal magnetic field and vice versa.
(c) Show that $\mathbf{B}(\mathbf{r})$ is toroidal for a toroidal solenoid.
(d) Suppose there is no current in a finite volume V. Show that $\nabla^2\mathbf{B}(\mathbf{r}) = 0$ in V.
(e) Show that $\mathbf{A}(\mathbf{r})$ in the Coulomb gauge is purely toroidal in V when $\psi(\mathbf{r})$ and $\gamma(\mathbf{r})$ are chosen so that $\nabla^2\mathbf{B}(\mathbf{r}) = 0$ in V.

11 Magnetic Multipoles

We therefore have an absolute method for measuring the sign and magnitude of the [magnetic] moment of any system.
Isidore Rabi (1937)

11.1 Introduction

The magnetic field produced by most interesting current distributions cannot be calculated exactly. However, when $\mathbf{j}(\mathbf{r})$ is spatially localized, our experience with electrostatics (Chapter 3) suggests a systematic scheme of approximation based on a *multipole expansion*. In its most familiar form, we calculate $\mathbf{B} = \nabla \times \mathbf{A}$ from an expansion of the factor $|\mathbf{r} - \mathbf{r}'|^{-1}$ in the vector potential

$$\mathbf{A}(\mathbf{r}) = \frac{\mu_0}{4\pi} \int d^3r' \frac{\mathbf{j}(\mathbf{r}')}{|\mathbf{r} - \mathbf{r}'|}. \tag{11.1}$$

More so than in electrostatics, both exterior and interior multipole expansions occur in common magnetic experience. Exterior expansions arise in many atomic and nuclear problems where the observation point lies *outside* a finite volume which contains the source current. Interior expansions arise in many experimental and diagnostic geometries where the observation point lies *inside* a finite volume which excludes the source current. Both expansions apply to regions of space where $\mathbf{j}(\mathbf{r}) = 0$. Analogous expansions exist for the magnetic scalar potential $\psi(\mathbf{r})$ at the same set of points.

11.1.1 The Magnetic Multipole Expansion

We begin with the situation depicted in Figure 11.1 where an observation point $\mathbf{r}$ lies far outside a sphere of radius R which entirely encloses a localized current distribution $\mathbf{j}(\mathbf{r})$.

Since $r \gg R$, the Taylor expansion of $|\mathbf{r} - \mathbf{r}'|^{-1}$ in (11.1) is dominated by the first few terms:

$$\frac{1}{|\mathbf{r} - \mathbf{r}'|} = \frac{1}{r} + \frac{\mathbf{r}' \cdot \mathbf{r}}{r^3} + \cdots. \tag{11.2}$$

This gives the kth Cartesian component of the approximate vector potential as

$$A_k(\mathbf{r}) = \frac{\mu_0}{4\pi} \left[\frac{1}{r} \int d^3r' \, j_k(\mathbf{r}') + \frac{\mathbf{r}}{r^3} \cdot \int d^3r' \, j_k(\mathbf{r}')\mathbf{r}' + \cdots \right]. \tag{11.3}$$

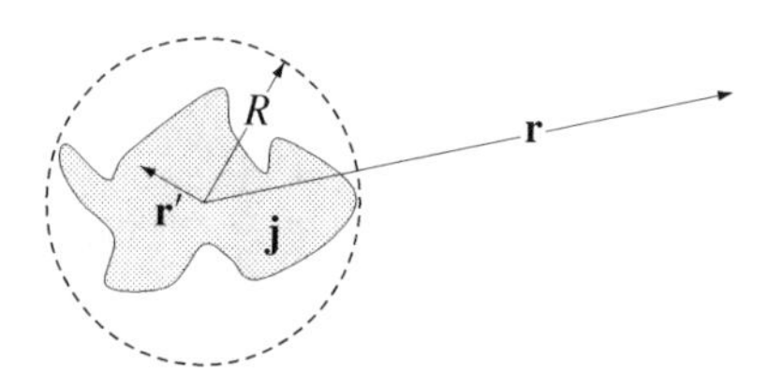

Figure 11.1: A localized current distribution confined to the interior of a sphere of radius R. The observation point $\mathbf{r}$ lies outside the sphere.

11.1.2 The Magnetic Monopole

The first term—the monopole term—in the magnetic multipole expansion (11.3) is always identically zero. The reason for this is *not* the apparent absence of magnetic charge in Nature. It is rather the constraint imposed on the three components of the current density $j_k(\mathbf{r})$ by the steady-current condition,

$$\nabla \cdot \mathbf{j}(\mathbf{r}) = 0. \tag{11.4}$$

To see this, note first that (11.4) implies that

$$\nabla' \cdot (r'_k \mathbf{j}) = r'_k(\nabla' \cdot \mathbf{j}) + \mathbf{j} \cdot \nabla' r'_k = j_k. \tag{11.5}$$

A key point is that any localized current distribution vanishes at infinity. Therefore, integrating the far left and far right sides of (11.5) over all space and using the divergence theorem, we confirm that

$$\int d^3r'\, j_k(\mathbf{r}') = \int d^3r' \nabla' \cdot (r'_k \mathbf{j}) = \int d\mathbf{S}' \cdot (r'_k \mathbf{j}) = 0. \tag{11.6}$$

The equivalent statement for a filamentary current is

$$I \oint d\boldsymbol{\ell} = 0. \tag{11.7}$$

11.2 The Magnetic Dipole

We have just proved that the $1/r$ term vanishes in (11.3) for localized and steady current distributions. Therefore, unless the second term vanishes also, the asymptotic (long-distance) behavior of the vector potential is determined by the nine integrals $T_{k\ell}$ which constitute the *magnetic dipole* term:

$$A_k(\mathbf{r}) = \frac{\mu_0}{4\pi}\left[\int d^3r'\, j_k(\mathbf{r}') r'_\ell\right]\frac{r_\ell}{r^3} = \frac{\mu_0}{4\pi} T_{k\ell}\frac{r_\ell}{r^3}. \tag{11.8}$$

Two identities are needed to simplify (11.8). The first generalizes (11.5) and uses (11.4):

$$\nabla' \cdot (r'_\ell r'_k \mathbf{j}) = r'_\ell r'_k \nabla' \cdot \mathbf{j} + r'_\ell j_k + r'_k j_\ell = r'_\ell j_k + r'_k j_\ell. \tag{11.9}$$

The second,

$$\epsilon_{\ell k i}(\mathbf{r}' \times \mathbf{j})_i = r'_\ell j_k - r'_k j_\ell, \tag{11.10}$$

is seen to be correct by multiplying every term in (11.10) by r_k and summing over the repeated index. The result is the ℓth Cartesian component of the vector identity $\mathbf{r} \times (\mathbf{r}' \times \mathbf{j}) = \mathbf{r}'(\mathbf{r} \cdot \mathbf{j}) - \mathbf{j}(\mathbf{r} \cdot \mathbf{r}')$. Because $\epsilon_{\ell k i} = \epsilon_{k i \ell}$, adding (11.9) to (11.10), integrating over all space, and using the divergence theorem as in (11.6) shows that

$$T_{k\ell} = \int d^3r'\, j_k r'_\ell = \tfrac{1}{2}\epsilon_{k i \ell} \int d^3r'\, (\mathbf{r}' \times \mathbf{j})_i \equiv \epsilon_{k i \ell} m_i. \tag{11.11}$$

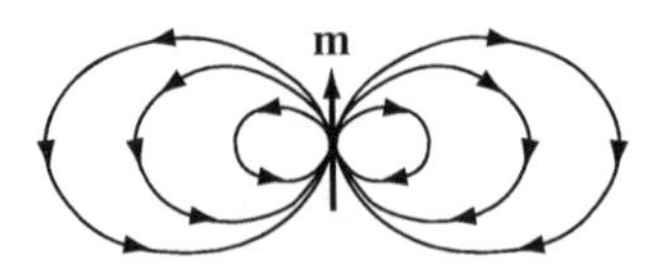

Figure 11.2: Lines of $\mathbf{B}(\mathbf{r})$ produced by the dipole moment $\mathbf{m} = m\hat{\mathbf{z}}$ of a localized current distribution at the center of the diagram. The distribution is too small to be seen on the scale of the diagram.

The last equality in (11.11) defines the *magnetic dipole moment* vector,

$$\mathbf{m} = \frac{1}{2}\int d^3r \; \mathbf{r} \times \mathbf{j}(\mathbf{r}). \tag{11.12}$$

Substituting (11.11) into (11.8) produces the magnetic dipole approximation to the vector potential:

$$\mathbf{A}(\mathbf{r}) = \frac{\mu_0}{4\pi}\frac{\mathbf{m} \times \mathbf{r}}{r^3} \qquad r \gg R. \tag{11.13}$$

The magnetic field associated with (11.13) is

$$\mathbf{B}(\mathbf{r}) = \nabla \times \left\{\frac{\mu_0}{4\pi}\frac{\mathbf{m} \times \mathbf{r}}{r^3}\right\} = \frac{\mu_0}{4\pi}\left[\mathbf{m}\left(\nabla \cdot \frac{\mathbf{r}}{r^3}\right) - (\mathbf{m} \cdot \nabla)\frac{\mathbf{r}}{r^3}\right]. \tag{11.14}$$

The first term in the square brackets in (11.14) produces a delta function at the origin. The second term produces another delta function at the origin and a piece which is non-zero when $\mathbf{r} \neq 0$. Only the latter is relevant here because (11.13) was derived assuming $r \gg R$. Accordingly,

$$\mathbf{B}(\mathbf{r}) = -\frac{\mu_0}{4\pi}(\mathbf{m} \cdot \nabla)\frac{\mathbf{r}}{r^3} = \frac{\mu_0}{4\pi}\frac{3\hat{\mathbf{r}}(\hat{\mathbf{r}} \cdot \mathbf{m}) - \mathbf{m}}{r^3} \qquad r \gg R. \tag{11.15}$$

The dipole field (11.15) dominates the physics when $r \gg R$ because all higher terms in the expansion (11.3) generate contributions to $\mathbf{A}(\mathbf{r})$ which are smaller by additional factors of R/r. The field (11.15) may be compared to the expression (4.10) derived in Section 4.2 for the asymptotic electric field produced by a neutral charge distribution with a non-zero electric dipole moment $\mathbf{p}$:

$$\mathbf{E}(\mathbf{r}) = \frac{1}{4\pi\epsilon_0}\frac{3\hat{\mathbf{r}}(\hat{\mathbf{r}} \cdot \mathbf{p}) - \mathbf{p}}{r^3} \qquad r \gg R. \tag{11.16}$$

The two formulae have exactly the same structure. Hence, the lines of $\mathbf{B}(\mathbf{r})$ plotted in Figure 11.2 for a magnetic dipole are identical to the lines of $\mathbf{E}(\mathbf{r})$ plotted in Figure 4.3 for an electric dipole.

11.2.1 The Magnetic Dipole Moment

The magnetic dipole moment defined by (11.12) characterizes the far field of a steady current very efficiently. We started with nine components of $T_{k\ell}$ in the vector potential (11.8) and ended with three components of $\mathbf{m}$ in the equivalent vector potential (11.13). Moreover, since $T_{k\ell} = -T_{\ell k}$, it is not difficult to confirm that[1]

$$m_i = \tfrac{1}{2}\epsilon_{i\ell k}T_{\ell k}. \tag{11.17}$$

In the language of tensor analysis, (11.17) identifies the magnetic moment as the asymmetric part of the decomposition of an arbitrary second-rank tensor $S_{\ell k}$ into a symmetric traceless part, an asymmetric

[1] The anti-symmetry of $T_{k\ell}$ follows from the equality of the first and last terms in (11.11).

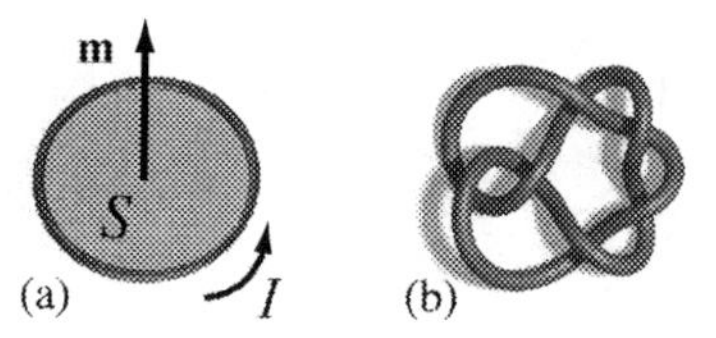

Figure 11.3: (a) A planar loop of current I with its magnetic dipole moment $\mathbf{m}$; (b) a stevedore knot.

part, and a scalar part:

$$S_{\ell k} = \tfrac{1}{2}(S_{\ell k} + S_{k\ell} - \tfrac{2}{3}S_{jj}\delta_{\ell k}) + \tfrac{1}{2}(S_{\ell k} - S_{k\ell}) + \tfrac{1}{3}S_{jj}\delta_{\ell k}. \tag{11.18}$$

As a concrete example, consider a filamentary loop of arbitrary shape which carries a steady current I around its circuit C. Using $\mathbf{j}\,d^3r \to I d\boldsymbol{\ell}$, this idealization simplifies (11.12) to

$$\mathbf{m} = \frac{1}{2} I \oint_C \mathbf{r} \times d\boldsymbol{\ell}. \tag{11.19}$$

A corollary of Stokes' theorem[2] transforms (11.19) to

$$\mathbf{m} = \tfrac{1}{2} I \int_S d\mathbf{S}\, \nabla \cdot \mathbf{r} - \tfrac{1}{2} I \int_S dS_k \nabla r_k = I \int_S d\mathbf{S} \equiv I\mathbf{S}. \tag{11.20}$$

This result for $\mathbf{m}$ is correct for *any* surface S which has C as its boundary.

For the planar loop shown in Figure 11.3(a), (11.20) predicts $\mathbf{m} = IA\hat{\mathbf{n}}$, where A is the area circumscribed by the loop and $\hat{\mathbf{n}}$ is the normal to the plane defined by the direction of the current and the right-hand rule. It is not so obvious how to choose S if the current loop is non-planar like the stevedore's knot shown in Figure 11.3(b). A convenient choice is the *minimal* surface with the smallest geometrical area. This is the surface of the film that clings to the loop when it is dipped into a soap solution. Of course, the vector area element $d\mathbf{S}$ in (11.20) points in different directions as the integration is carried out for any non-planar loop.[3]

Equation (11.20) gives $\mathbf{m} = I\pi R^2\hat{\mathbf{z}}$ for an origin-centered ring of radius R which carries a current I around the z-axis. If our theory is consistent, this magnetic moment should emerge naturally from the $r \to \infty$ limit of the exact magnetic scalar potential calculation for this current source performed in Section 10.4.1. Indeed, a glance back at (10.50) shows that

$$\lim_{r\to\infty} \psi(r,\theta) = \frac{\mu_0}{4\pi}\frac{\pi R^2 I}{r^2}\cos\theta = \frac{\mu_0}{4\pi}\frac{\mathbf{m}\cdot\mathbf{r}}{r^3}. \tag{11.21}$$

Moreover, because $\mathbf{m}$ is a constant vector and $\partial_i(r_k/r^3) = \partial_k(r_i/r^3)$, the magnetic field derived from (11.21) agrees exactly with (11.15):

$$\mathbf{B} = -\nabla\psi = -\frac{\mu_0}{4\pi}\nabla\left[\frac{\mathbf{m}\cdot\mathbf{r}}{r^3}\right] = -\frac{\mu_0}{4\pi}(\mathbf{m}\cdot\nabla)\frac{\mathbf{r}}{r^3}. \tag{11.22}$$

The Magnetic Field of the Earth

The magnetic field outside the surface of the Earth looks very much like Figure 11.2. However, instead of steady currents, the Earth's field is produced by a complex pattern of time-dependent electric currents generated by electromagnetic induction (Faraday's law) and the flow of a conducting liquid iron alloy in the Earth's outer core driven by convection, buoyancy, and the rotation

[2] $\oint_C d\boldsymbol{\ell} \times \mathbf{A} = \int_S dS_k \nabla A_k - \int_S d\mathbf{S}(\nabla \cdot \mathbf{A})$. See Section 1.4.4.

[3] Equation (11.19) may be preferable for calculating $\mathbf{m}$ for some non-planar loops.

of the Earth. A remarkable fact gleaned from volcanic rock studies is that the magnetic dipole moment of the Earth has reversed its direction nearly a hundred times over the past 20 million years. The current moment points 11° away from the South geographic pole and has magnitude $m_E = 8 \times 10^{22}$ A·m^2.

The left panel of Figure 11.4 shows the Earth's magnetic field obtained from a numerical solution of the non-linear equations which result when Ohm's law and a hydrodynamic description of the outer core are combined with Maxwell's equations. The field line pattern is very nearly dipolar at the Earth's surface (large white circle), but is extraordinarily complex in the core region (inside the small white circle). The right panel of Figure 11.4 shows numerical results for the azimuthal component of the electric current density on the spherical surface that separates the Earth's conducting core from its insulating mantle. White (black) identifies regions where $j_\phi(\theta, \phi)$ flows eastward (westward) in a frame of reference which rotates with the Earth. The ellipse in the right panel is a Mollweide projection of the smaller white circle in the left panel.

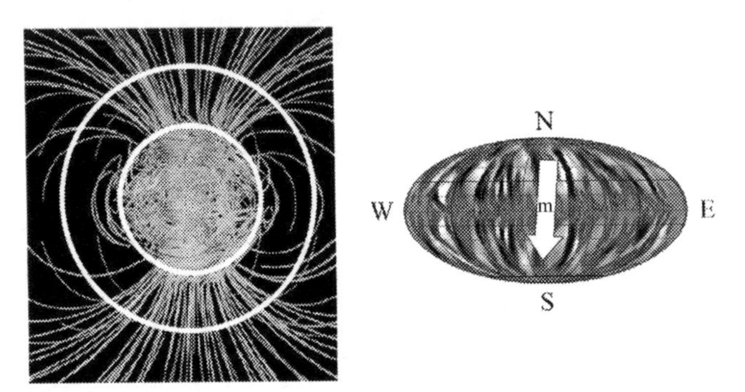

Figure 11.4: Left panel: the magnetic field of the Earth determined from numerical simulations. The large white circle is the surface of the Earth. The small white circle is the core-mantle boundary (CMB). Figure from Glatzmaier and Clune (2000); Copyright 2000, IEEE. Right panel: the magnetic moment of the Earth's dipole field superimposed on the azimuthal component of the electric current density at the CMB. See text for details. Figure adapted from Sakuraba and Hamano (2007).

11.2.2 Orbital and Spin Magnetic Moments

In this section, we discuss the magnetic moments of atoms and molecules. Experiments show that these moments derive from (i) the orbital angular momentum of the electrons and (ii) the spin angular momentum of the electrons, protons, and neutrons. Not obviously, the magnetic moment from the orbital motion can be calculated using a classical model of N charged particles whose motions keep them close to a fixed origin of coordinates. If the particles have charges q_k, velocities $\mathbf{v}_k = d\mathbf{r}_k/dt$, and masses m_k, the associated current density is

$$\mathbf{j}(\mathbf{r}) = \sum_{k=1}^{N} q_k \mathbf{v}_k \delta(\mathbf{r} - \mathbf{r}_k). \tag{11.23}$$

Substituting (11.23) into (11.12) gives the *orbital magnetic moment* $\mathbf{m}_L$ in terms of the orbital angular momentum, $\mathbf{L}_k = m_k \mathbf{r}_k \times \mathbf{v}_k$, of the k th particle with respect to the origin of coordinates:

$$\mathbf{m}_L = \frac{1}{2}\sum_{k=1}^{N} q_k(\mathbf{r}_k \times \mathbf{v}_k) = \sum_{k=1}^{N} \frac{q_k}{2m_k}\mathbf{L}_k. \tag{11.24}$$

If all the particles have the same charge-to-mass ratio, $\mathbf{m}_L$ is proportional to the total angular momentum $\mathbf{L} = \sum_k \mathbf{L}_k$:

$$\mathbf{m}_L = \frac{q}{2m}\mathbf{L}. \tag{11.25}$$

It is remarkable that the classical proportionality between the magnetic moment and the angular momentum survives the transition to quantum mechanics. In other words, (11.25) gives the correct orbital magnetic moment of an atom when $\mathbf{L}$ is the total orbital angular momentum of the atomic electrons.

Even more remarkably, experiments show that quantum particles with spin angular momentum $\mathbf{S}$ also possesses an intrinsic *spin magnetic moment*,

$$\mathbf{m}_S = g\frac{q}{2m}\mathbf{S}. \tag{11.26}$$

The quantum physics of (11.26) is buried in the dimensionless "g-factor". For the electron, g is very slightly greater than 2 and the scale for the moment is set by the *Bohr magneton*,

$$\mu_{\text{B}} = \frac{e\hbar}{2m_e} = 9.3\times 10^{-24}\ \text{A}\cdot\text{m}^2. \tag{11.27}$$

For more complex quantum systems with total angular momentum $\mathbf{J}$, it is common to define a *gyromagnetic ratio* γ so the total magnetic moment is

$$\mathbf{m} = \gamma\mathbf{J}. \tag{11.28}$$

Example 11.1 Let $\mathbf{B}(\mathbf{r})$ be the magnetic field produced by a current density $\mathbf{j}(\mathbf{r})$ that lies entirely inside a spherical volume V of radius R. Show that the magnetic moment of $\mathbf{j}(\mathbf{r})$ is

$$\mathbf{m} = \frac{3}{2\mu_0}\int_V d^3r\,\mathbf{B}(\mathbf{r}).$$

This problem is the magnetic analog of Example 4.1, which expressed the electric dipole moment of a charge distribution using a spherical average of the electric field produced by the charge.

Solution: Assume first that $\mathbf{j}(\mathbf{r})$ does *not* lie entirely inside V. If we place the origin of coordinates at the center of V, the Biot-Savart law (10.15) gives

$$\frac{1}{V}\int_V d^3r\,\mathbf{B}(\mathbf{r}) = -\frac{\mu_0}{4\pi V}\int d^3r'\mathbf{j}(\mathbf{r}')\times\int_V d^3r\frac{(\mathbf{r}'-\mathbf{r})}{|\mathbf{r}'-\mathbf{r}|^3}.$$

The $\mathbf{r}$ integral is exactly the "electric" field $\mathbf{E}(\mathbf{r}')$ due to a uniform charge density $\rho(\mathbf{r}) = 4\pi\epsilon_0$. From Gauss' law or otherwise, this is

$$\mathbf{E}(\mathbf{r}') = \begin{cases} \dfrac{\rho\mathbf{r}'}{3\epsilon_0} = \dfrac{4\pi}{3}\mathbf{r}' & r' < R, \\[2ex] \dfrac{\rho V}{4\pi\epsilon_0}\dfrac{\mathbf{r}'}{r'^3} = V\dfrac{\mathbf{r}'}{r'^3} & r' > R. \end{cases}$$

Therefore,

$$\frac{1}{V}\int_V d^3r\,\mathbf{B}(\mathbf{r}) = \frac{\mu_0}{3V}\int_{r'\le R} d^3r'\,\mathbf{r}'\times\mathbf{j}(\mathbf{r}') - \frac{\mu_0}{4\pi}\int_{r'>R} d^3r'\frac{\mathbf{j}(\mathbf{r}')\times\mathbf{r}'}{r'^3}.$$

From (11.12), the first integral on the right side of this equation is proportional to $\mathbf{m}_{\rm in}$, the magnetic moment due to the part of $\mathbf{j}$ which lies inside V. The second integral is the Biot-Savart magnetic field at the origin, $\mathbf{B}_{\rm out}(0)$, due to that part of $\mathbf{j}$ which lies outside V. Therefore,

$$\frac{1}{V}\int_V d^3r\,\mathbf{B}(\mathbf{r}) = \frac{2}{3}\mu_0\frac{\mathbf{m}_{\rm in}}{V} + \mathbf{B}_{\rm out}(0).$$

This formula reduces to the advertised result if all of $\mathbf{j}(\mathbf{r})$ is contained in V. On the other hand, if *none* of the current is contained in V,

$$\mathbf{B}(0) = \frac{1}{V}\int d^3r\,\mathbf{B}(\mathbf{r}).$$

Example 11.2 Find $\mathbf{j}(\mathbf{r})$ for a hydrogen atom in the eigenstates $\Psi(\mathbf{r}) = \langle \mathbf{r}\,|\,2\,1\,m\rangle$ where $m = 0, \pm 1$. Solve $\nabla\times\nabla\times\mathbf{A} = \mu_0\mathbf{j}$ using an ansatz for $\mathbf{A}(\mathbf{r})$ with the same angular dependence as $\mathbf{j}(\mathbf{r})$. Find the orbital magnetic moment for each state from (11.13) and the asymptotic ($r\to\infty$) form of $\mathbf{A}(\mathbf{r})$.

Solution: The $|2\ 1\ m\rangle$ hydrogenic wave functions are

$$\Psi_{21m}(\mathbf{r}) = R(r)\times\begin{cases}\sqrt{3/4\pi}\cos\theta & m=0,\\ -\sqrt{3/8\pi}\sin\theta\exp(i\phi) & m=1,\\ \sqrt{3/8\pi}\sin\theta\exp(-i\phi) & m=-1,\end{cases}$$

where $R(r) = (6a^3)^{-1/2}(r/2a)\exp(-r/2a)$ and a is the Bohr radius. If $\mu_B = e\hbar/2m$ is the Bohr magneton, the quantum mechanical expression for the current associated with such wave functions is

$$\mathbf{j} = i\mu_B(\Psi^*\nabla\Psi - \Psi\nabla\Psi^*) = -2\mu_B{\rm Im}(\Psi^*\nabla\Psi).$$

Therefore, by direct computation, $\mathbf{j}_{210}(\mathbf{r}) = 0$ and

$$\mathbf{j}_{21\pm1}(\mathbf{r}) = \mp\frac{\mu_B}{32\pi}\frac{r}{a^5}e^{-r/a}\sin\theta\,\hat{\boldsymbol{\phi}} = \mp j_0 r\,e^{-r/a}\sin\theta\,\hat{\boldsymbol{\phi}}.$$

Now, we know from (10.84) of Section 10.5.3 that a current density $\mathbf{j} = j_\phi(r,\theta)\hat{\boldsymbol{\phi}}$ generates a vector potential of the form $\mathbf{A} = A_\phi(r,\theta)\hat{\boldsymbol{\phi}}$. This reduces $\nabla\times(\nabla\times\mathbf{A}) = \mu_0\mathbf{j}$ to

$$\frac{1}{r}\frac{\partial^2}{\partial r^2}(rA_\phi) + \frac{1}{r}\frac{\partial}{\partial\theta}\left\{\frac{1}{r\sin\theta}\frac{\partial}{\partial\theta}(\sin\theta\,A_\phi)\right\} = -\mu_0 j_\phi.$$

The fact that $j_\phi \propto \sin\theta$ motivates us to try $A_\phi(r,\theta) = f(r)\sin\theta/r^2$. This choice simplifies the preceding equation to

$$\frac{d}{dr}\left(\frac{f'}{r^2}\right) = \pm\mu_0 j_0 r\,e^{-r/a}.$$

After two radial integrations, we get $f(r)$ in terms of constants K_1 and K_2:

$$f(r) = \pm\mu_0 a^2 j_0 e^{-r/a}\left\{r^3 + 4ar^2 + 8a^2r + 8a^3\right\} + K_1r^3 + K_2.$$

Therefore,

$$\mathbf{A}(r,\theta) = \pm\frac{\mu_0\mu_B\sin\theta}{32\pi a^2}\times\left\{e^{-r/a}\left(\frac{r}{a} + 4 + \frac{8a}{r} + \frac{8a^2}{r^2}\right) + K_2r + \frac{K_3}{r^2}\right\}\hat{\boldsymbol{\phi}}.$$

A must go to zero at infinity. This fixes $K_2 = 0$. **A** also must not diverge as $r \to 0$. The expansion $\exp(-r/a) = 1 - r/a + \frac{1}{2}r^2/a^2 + \cdots$ shows that the choice $K_3 = -8a^2$ makes this true. We conclude that the dominant term at long distance is

$$\lim_{r\to\infty} \mathbf{A}(r,\theta) = \mp \frac{\mu_0}{4\pi}\frac{\mu_B \sin\theta}{r^2}\hat{\boldsymbol{\phi}}.$$

Comparing this result with (11.13) shows that the $|2\ 1\ m\rangle$ hydrogenic state carries an orbital magnetic dipole moment equal to $-m\mu_B$. This is precisely the standard quantum mechanical result (because the electron charge is negative).

11.2.3 The Point Magnetic Dipole

A point magnetic dipole is a current distribution that produces a pure magnetic dipole field at every point in space. The electron is a near-perfect realization of a point magnetic dipole because it has both zero size and a spin magnetic moment given by (11.26).[4] The significance of this fact cannot be understated, not least because electron spin is responsible for the magnetism of permanent magnets (see Chapter 13). Needless to say, the point magnetic dipole is also an excellent model for the macroscopic magnetic response of microscopic but finite-sized particles like the proton and neutron.

The vector potential of a point magnetic dipole at the origin is given by (11.13) with the restriction to large distances removed. If the dipole sits at $\mathbf{r}_0$,

$$\mathbf{A}(\mathbf{r}) = \frac{\mu_0}{4\pi}\frac{\mathbf{m}\times(\mathbf{r}-\mathbf{r}_0)}{|\mathbf{r}-\mathbf{r}_0|^3}. \tag{11.29}$$

We do not avoid the origin, so (11.14), (11.15), and $\nabla^2|\mathbf{r}-\mathbf{r}_0|^{-1} = -4\pi\delta(\mathbf{r}-\mathbf{r}_0)$ give the magnetic field $\mathbf{B} = \nabla\times\mathbf{A}$ of a point magnetic dipole $\mathbf{m}$ at $\mathbf{r}_0$ as

$$\mathbf{B}(\mathbf{r}) = \mu_0\left[\mathbf{m}\,\delta(\mathbf{r}-\mathbf{r}_0) - \nabla\frac{1}{4\pi}\frac{\mathbf{m}\cdot(\mathbf{r}-\mathbf{r}_0)}{|\mathbf{r}-\mathbf{r}_0|^3}\right]. \tag{11.30}$$

A virtue of (11.30) is that it leads quickly to the current density of a point magnetic dipole. Since $\mathbf{j} = \mu_0^{-1}\nabla\times\mathbf{B}$, we find without trouble that

$$\mathbf{j}_D(\mathbf{r}) = \nabla\times[\mathbf{m}\delta(\mathbf{r}-\mathbf{r}_0)] = -\mathbf{m}\times\nabla\delta(\mathbf{r}-\mathbf{r}_0). \tag{11.31}$$

The gradient term in (11.30) conceals a delta function at $\mathbf{r} = \mathbf{r}_0$. We leave it to the reader to extract the strength of this delta function by direct computation. Here, we find the total strength of the delta function at the heart of a point magnetic dipole by exploiting Example 11.1 and a spherical integration volume V centered at $\mathbf{r}_0$. This procedure tells us that

$$\int_V d^3r\,\mathbf{B}(\mathbf{r}) = \frac{2\mu_0}{3}\mathbf{m}. \tag{11.32}$$

By the symmetry of Figure 11.2, (11.32) vanishes when we use (11.15) for the integrand. Therefore, if this behavior persists as $V \to 0$, the integral will have the correct value only if the total magnetic field for a point magnetic dipole is[5]

$$\mathbf{B}(\mathbf{r}) = \frac{\mu_0}{4\pi}\left[\frac{3\hat{\mathbf{n}}(\hat{\mathbf{n}}\cdot\mathbf{m})-\mathbf{m}}{|\mathbf{r}-\mathbf{r}_0|^3} + \frac{8\pi}{3}\mathbf{m}\delta(\mathbf{r}-\mathbf{r}_0)\right]. \tag{11.33}$$

[4] Unlike point *electric* dipoles, which do not seem to exist in Nature.

[5] Equation (11.33) also follows directly from (11.30) and (1.122).

A useful way to think about the delta function in (11.33) will emerge in Chapter 13. Here, we note only that the coefficient of the delta function in (11.33) differs from the corresponding coefficient in the electric field formula (4.16) for a point electric dipole.

Application 11.1 The Point Magnetic Monopole

There is no experimental evidence for the existence of free magnetic monopoles. Nevertheless, we can synthesize one from a semi-infinite solenoid (N turns/length of wire with current I) in the limit when the solenoid's cross sectional area $S \to 0$ (Figure 11.5).

The construction begins with a planar, circular loop with current I which lies in the x-y plane and is coaxial with the z-axis. The magnetic moment of the loop is $\mathbf{m}_0 = IS\,\hat{\mathbf{z}}$. If $\mathbf{r} = \rho\,\hat{\boldsymbol{\rho}} + z\,\hat{\mathbf{z}}$, the vector potential far from the loop is given by the dipole formula,

$$\mathbf{A} = \frac{\mu_0}{4\pi}\frac{\mathbf{m}_0 \times \mathbf{r}}{r^3} = \frac{\mu_0 m_0}{4\pi}\frac{\rho}{(\rho^2 + z^2)^{3/2}}\hat{\boldsymbol{\phi}}. \tag{11.34}$$

The vector potential of the semi-infinite solenoid follows by superposing contributions of this form from a stack of loops which extends from $z_0 = -\infty$ to $z_0 = 0$ on the negative z-axis. If $g = Nm_0$ is the magnetic dipole moment per unit length, we let $\mathbf{A} \to d\mathbf{A}$ and $m_0 \to Nm_0 dz_0 = g dz_0$, so

$$\mathbf{A} = \int d\mathbf{A} = \frac{\mu_0 g}{4\pi}\int_{-\infty}^{0} dz_0 \frac{\rho\hat{\boldsymbol{\phi}}}{[\rho^2 + (z - z_0)^2]^{3/2}} = \frac{\mu_0 g}{4\pi r}\frac{1 - \cos\theta}{\sin\theta}\hat{\boldsymbol{\phi}}. \tag{11.35}$$

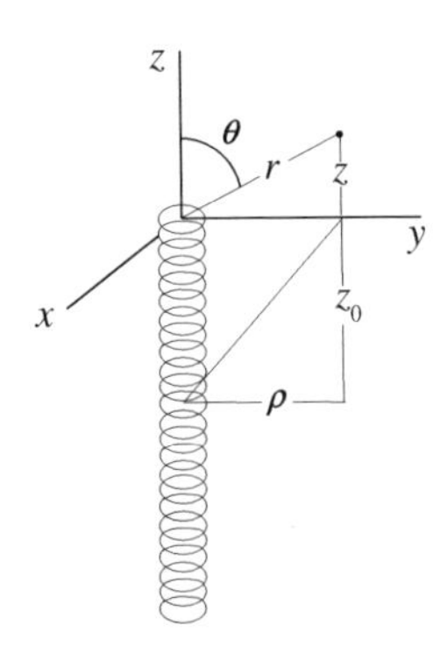

Figure 11.5: A "monopole" at the origin simulated by a semi-infinite solenoid coincident with the negative z-axis.

The associated magnetic field $\mathbf{B} = \nabla \times \mathbf{A}$ is

$$\mathbf{B}(\mathbf{r}) = \frac{1}{r\sin\theta}\frac{\partial}{\partial\theta}\left(\sin\theta A_\phi\right)\hat{\mathbf{r}} - \frac{1}{r}\frac{\partial}{\partial r}\left(rA_\phi\right)\hat{\boldsymbol{\theta}} = \frac{\mu_0 g}{4\pi}\frac{\hat{\mathbf{r}}}{r^2}. \tag{11.36}$$

This Coulomb-type formula is valid at all points that are sufficiently far from the solenoid that the dipole approximation is valid. This domain expands to include all of space (except the negative z-axis) in the limit when $S \to 0$ (so $m_0 \to 0$) and $N \to \infty$ in such a way that g remains constant. The magnetic field above satisfies

$$\nabla \cdot \mathbf{B}(\mathbf{r}) = \mu_0 g\,\delta(\mathbf{r}) \qquad \text{and} \qquad \nabla \times \mathbf{B}(\mathbf{r}) = 0. \tag{11.37}$$

These are the equations we expect for the field of a magnetic monopole at the origin with magnetic charge g. ∎

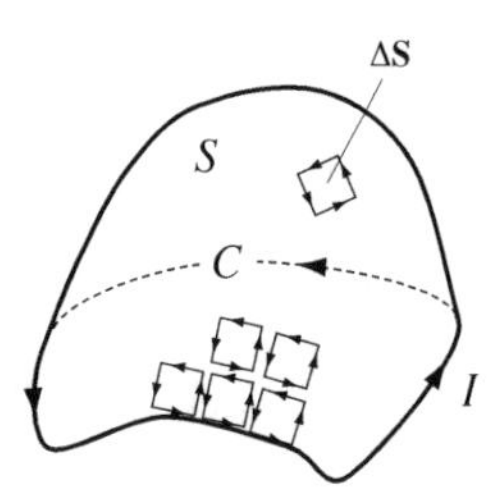

Figure 11.6: A current I flows through a closed circuit C which is the boundary of a surface S. The surface is "tiled" by small current loops with area ΔS. Adjacent small loops carry current in the same direction, so nearest-neighbor parallel legs on adjacent loops carry current in opposite divections.

11.3 Magnetic Dipole Layers

The interesting physics of the electric dipole layer (Section 4.3) motivates us to study the analogous problem of a magnetic dipole layer. This is a surface S which carries a continuous distribution of point magnetic dipoles. The principal result is known as *Ampère's theorem*: if S is open and the dipoles are oriented normal to the surface,[6] the magnetic field produced by S is identical to the magnetic fields produced by a current flowing around the boundary of S.

The qualitative correctness of this result can be appreciated immediately from Figure 11.6. The open surface S (which has C as its boundary) is tiled by small planar current loops, each of which has area ΔS and carries a current I. The tiling is such that two adjacent legs carry current in opposite directions. In the limit $\Delta S \to 0$, the magnetic fields from adjacent internal legs cancel pairwise and only the field from the external legs (which constitute C) contributes to the field. In the same limit, each infinitesimal loop is indistinguishable from a point magnetic dipole with differential dipole moment $d\mathbf{m} = Id\mathbf{S} = IdS\hat{\mathbf{n}}$.

For a quantitative proof of Ampère's theorem, we use the dipole vector potential in (11.29) and sum the contributions to $\mathbf{A}(\mathbf{r})$ from every point on the surface. This gives

$$\mathbf{A}(\mathbf{r}) = \frac{\mu_0}{4\pi}\int_S d\mathbf{m} \times \frac{\mathbf{r}-\mathbf{r}_S}{|\mathbf{r}-\mathbf{r}_S|^3} = \frac{\mu_0 I}{4\pi}\int_S dS\;\hat{\mathbf{n}} \times \nabla_S \frac{1}{|\mathbf{r}-\mathbf{r}_S|}. \tag{11.38}$$

Now, if $\boldsymbol{\xi}$ is a constant vector, the cyclic property of the scalar triple product, and Stokes' theorem, imply that

$$\boldsymbol{\xi}\cdot\mathbf{A}(\mathbf{r}) = \frac{\mu_0 I}{4\pi}\int_S d\mathbf{S}\cdot\nabla_S \times \frac{\boldsymbol{\xi}}{|\mathbf{r}-\mathbf{r}_S|} = \frac{\mu_0 I}{4\pi}\oint_C d\boldsymbol{\ell}\cdot\frac{\boldsymbol{\xi}}{|\mathbf{r}-\mathbf{r}_S|}. \tag{11.39}$$

Since $\boldsymbol{\xi}$ is completely arbitrary, we conclude from (11.39) that

$$\mathbf{A}(\mathbf{r}) = \frac{\mu_0 I}{4\pi}\oint_C \frac{d\boldsymbol{\ell}}{|\mathbf{r}-\mathbf{r}_S|}. \tag{11.40}$$

This is indeed the vector potential of a loop C which carries a current I.

Ampère's equivalence between a current-carrying circuit and any magnetic double layer bounded by that circuit can also be established using the magnetic scalar potential. In that case, we ascribe the jump discontinuity (10.59) in $\psi(\mathbf{r})$ to the passage of the observation point $\mathbf{r}$ through a suitably chosen magnetic double layer. This is analogous to the jump in the electric scalar potential $\varphi(\mathbf{r})$ that

[6] We called this a *double layer* in the electric case.

occurs when $\mathbf{r}$ passes through an electric double layer (Section 4.3.2) and provides an alternative to the interpretation of the jump in terms of the properties of the solid angle (Section 10.4.4).

Example 11.3 A conventional disc drive stores data using tiny magnetic elements with their dipole moments oriented parallel to the surface of the disc. The next generation of magnetic disc drives will use magnetic elements with their dipole moments oriented perpendicular to the disc surface. Thus, a "virgin" sample of magnetic recording tape may be modeled as a thin film uniformly coated with parallel point magnetic dipoles oriented perpendicular to its surface. Find the magnetic field produced by an infinitely long and flat strip of such a tape with width d and negligible thickness (Figure 11.7).

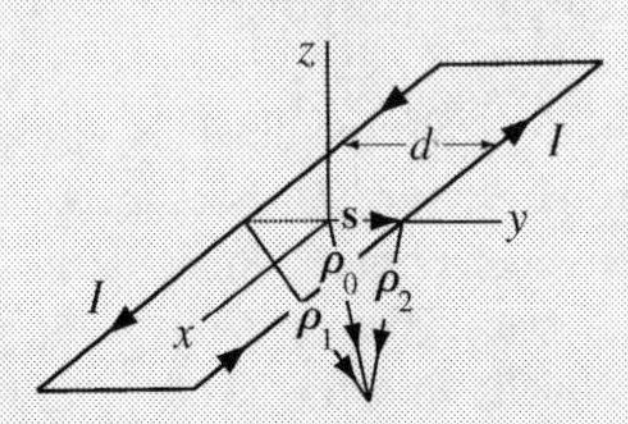

Figure 11.7: An infinitely long strip of magnetic recording tape with width d, negligible thickness, and a uniform distribution of point magnetic dipoles pointed along $+\hat{\mathbf{z}}$. The vector $\mathbf{s} = (d/2)\hat{\mathbf{y}}$. The observation point is the tip of the vector $\boldsymbol{\rho}_0$.

Solution: Let $I\hat{\mathbf{z}}$ be the magnetic moment per unit area of the tape. By Ampère's theorem, the magnetic field of the tape is identical to the field produced by two infinitely long and straight filamentary wires. One wire (coincident with the $y = -d/2$ edge of the tape) carries a current I along $+\hat{\mathbf{x}}$. The other wire (coincident with the $y = d/2$ edge of the tape) carries a current I along $-\hat{\mathbf{x}}$. The magnetic field of a single wire is $\mathbf{B}(\rho) = (\mu_0 I/2\pi\rho)\hat{\boldsymbol{\phi}}$, where ρ is the perpendicular distance form the wire and $\hat{\boldsymbol{\phi}}$ is the azimuthal direction determined by the direction of current flow and the right-hand rule. Therefore, if $\mathbf{s} = (d/2)\hat{\mathbf{y}}$, $\boldsymbol{\rho}_1 = \boldsymbol{\rho}_0 + \mathbf{s}$, and $\boldsymbol{\rho}_2 = \boldsymbol{\rho}_0 - \mathbf{s}$, superposition gives the magnetic field of the tape as

$$\mathbf{B}(\mathbf{r}) = \frac{\mu_0 I}{2\pi}\left\{\frac{\hat{\mathbf{x}} \times \hat{\boldsymbol{\rho}}_1}{\rho_1} - \frac{\hat{\mathbf{x}} \times \hat{\boldsymbol{\rho}}_2}{\rho_2}\right\},$$

or

$$\mathbf{B}(x, y, z) = \frac{\mu_0 I}{2\pi}\left\{\frac{(y - \frac{1}{2}d)\hat{\mathbf{z}} - z\hat{\mathbf{y}}}{(y - \frac{1}{2}d)^2 + z^2} - \frac{(y + \frac{1}{2}d)\hat{\mathbf{z}} - z\hat{\mathbf{y}}}{(y + \frac{1}{2}d)^2 + z^2}\right\}.$$

11.4 Exterior Multipoles

The dipole formula (11.13) is a good approximation to the exact vector potential when the magnetic dipole moment $\mathbf{m} \neq 0$ and the observation point is much farther from a compact source than the spatial extent of the source itself. If either of these conditions is violated, higher-order terms in the multipole expansion are needed. This section treats the general expansion for problems with rectangular, spherical, and azimuthal symmetry.

11.4.1 Cartesian Expansion for A(r)

In Cartesian coordinates, the primitive structure of the complete multipole expansion for $\mathbf{A}(\mathbf{r})$ is

$$A_k(\mathbf{r}) = \frac{\mu_0}{4\pi}\left\{\left[\int d^3s\ j_k(\mathbf{s})\right]\frac{1}{r} - \left[\frac{1}{1!}\int d^3s\ j_k(\mathbf{s})s_\ell\right]\nabla_\ell\frac{1}{r} + \left[\frac{1}{2!}\int d^3s\ j_k(\mathbf{s})s_\ell s_m\right]\nabla_\ell\nabla_m\frac{1}{r} - \cdots\right\}. \tag{11.41}$$

We learned in Section 11.1.2 that the first term in (11.41) is zero. Therefore, if

$$T^{(n)}_{k\ell\ldots m} = \frac{1}{n!}\int d^3s\ j_k(\mathbf{s})\ \underbrace{s_\ell\cdots s_m}_{n\text{ terms}}, \tag{11.42}$$

the vector potential (11.41) takes the compact form

$$A_k(\mathbf{r}) = \frac{\mu_0}{4\pi}\sum_{n=1}^{\infty}(-)^n T^{(n)}_{k\ell\ldots m}\ \underbrace{\nabla_\ell\cdots\nabla_m}_{n\text{ terms}}\frac{1}{r}. \tag{11.43}$$

Our experience with the magnetic dipole term (Section 11.1.1) suggests that a representation of $A_k(\mathbf{r})$ more efficient than (11.43) can be derived if we exploit the steady-current condition. This is indeed the case. Moreover, all the essential steps are already present in the third-order, magnetic quadrupole term. To avoid confusion in what follows, we use different symbols to denote derivatives with respect to the Cartesian components of the variables $\mathbf{r}$ and $\mathbf{s}$, namely,

$$\nabla_k \equiv \frac{\partial}{\partial r_k} \qquad \text{and} \qquad \partial_k \equiv \frac{\partial}{\partial s_k}. \tag{11.44}$$

We begin with a generalization of (11.9):

$$\partial_p(s_k s_\ell s_m j_p) = s_k s_\ell s_m \partial_p j_p + j_\ell s_k s_m + j_m s_k s_\ell + j_k s_\ell s_m. \tag{11.45}$$

The steady-current condition, $\partial_p j_p = 0$, and the identity (11.10) (used twice with different indices) transform (11.45) into

$$\partial_p(s_k s_\ell s_m j_p) = 3 j_k s_\ell s_m + s_\ell \epsilon_{kmi}(\mathbf{s}\times\mathbf{j})_i + s_m \epsilon_{k\ell i}(\mathbf{s}\times\mathbf{j})_i. \tag{11.46}$$

The structure of the third-order term in (11.41) motivates us to insert the operator $\nabla_\ell\nabla_m$ on the right side of every term in (11.46) and sum over repeated indices. This makes the two cross product terms in (11.46) identical. Therefore, integration over all space and the divergence theorem give, for a localized current distribution,

$$\int d^3s\ j_k(\mathbf{s})(\mathbf{s}\cdot\nabla)^2 = \frac{2}{3}\epsilon_{ki\ell}\int d^3s\ (\mathbf{s}\times\mathbf{j})_i(\mathbf{s}\cdot\nabla)\nabla_\ell. \tag{11.47}$$

Exactly similar manipulations show that the general integral in (11.41) is

$$\int d^3s\ j_k(\mathbf{s})(\mathbf{s}\cdot\nabla)^n = \frac{n}{n+1}\epsilon_{ki\ell}\int d^3s\ (\mathbf{s}\times\mathbf{j})_i(\mathbf{s}\cdot\nabla)^{n-1}\nabla_\ell. \tag{11.48}$$

Therefore, taking note of the factorials which appear in the denominators in (11.41), we conclude that the exterior Cartesian multipole expansion for the vector potential can be written in the form

$$A_k(\mathbf{r}) = \frac{\mu_0}{4\pi}\epsilon_{ki\ell}\sum_{n=1}^{\infty}(-1)^n m^{(n)}_{ip\cdots q}\ \underbrace{\nabla_p\cdots\nabla_q}_{n-1\text{ terms}}\nabla_\ell\frac{1}{r}, \tag{11.49}$$

with Cartesian magnetic multipole moments defined by

$$m^{(n)}_{ip\cdots q} = \frac{n}{(n+1)!}\int d^3s\ (\mathbf{s}\times\mathbf{j})_i \underbrace{s_p \cdots s_q}_{n-1 \text{ terms}}. \tag{11.50}$$

The transformation from (11.43) to (11.49) represents a considerable simplification because $T^{(n)}_{k\ell\cdots m}$ is a tensor with $n+1$ indices whereas $m^{(n)}_{ip\cdots q}$ is a tensor with only n indices.

Application 11.2 Parity Violation and the Anapole Moment

We showed in Section 4.2.1 that the electric dipole moment $\mathbf{p} = \int d^3r\ \rho(\mathbf{r})\mathbf{r}$ vanishes for any microscopic system described by a wave function with definite parity because the integrand changes sign when $\mathbf{r} \to -\mathbf{r}$. Because the current density $\mathbf{j} = \rho\boldsymbol{v} = \rho\dot{\mathbf{r}}$ also changes sign when $\mathbf{r} \to -\mathbf{r}$, the same statement is true for the Cartesian components of the magnetic quadrupole moment defined in (11.42),

$$T^{(2)}_{k\ell m} \equiv \frac{1}{2}\int d^3r\ j_k(\mathbf{r}) r_\ell r_m. \tag{11.51}$$

However, soon after the discovery of the parity-violating weak interaction, it was predicted that parity mixing could generate a current in an atomic nucleus with the form of a toroidal solenoid (see Example 10.2) which makes a non-zero contribution to $T_{k\ell m}$ called the *anapole moment* **a**.[7] Forty years passed before high-precision hyperfine spectroscopy provided convincing evidence for **a**.[8]

Our task in this Application is to isolate the anapole piece of the magnetic quadrupole moment. To that end, we use (11.47) and the definition $\mathbf{N} = \mathbf{r}\times\mathbf{j}$ to write

$$T^{(2)}_{k\ell m} = \frac{1}{3}\epsilon_{ki\ell}\int d^3r\ N_i r_m. \tag{11.52}$$

Since $N_i r_m = \frac{1}{2}(N_i r_m + N_m r_i) + \frac{1}{2}(N_i r_m - N_m r_i)$, we define $\Lambda_{im} = \int d^3r\ (N_i r_m + N_m r_i)$ and write (11.52) in the form

$$T^{(2)}_{k\ell m} = \frac{1}{6}\epsilon_{ki\ell}\Lambda_{im} + \frac{1}{6}\epsilon_{ki\ell}\int d^3r\ (r_m\epsilon_{ipq} - r_i\epsilon_{mpq}) r_p j_q. \tag{11.53}$$

Two identities help simplify the integral in (11.53). The first connects the product of two Levi-Cività symbols to a determinant of Kronecker delta symbols (see Section 1.2.5):

$$\epsilon_{ki\ell}\epsilon_{mpq} = \begin{vmatrix} \delta_{km} & \delta_{im} & \delta_{\ell m} \\ \delta_{kp} & \delta_{ip} & \delta_{\ell p} \\ \delta_{kq} & \delta_{iq} & \delta_{\ell q} \end{vmatrix}. \tag{11.54}$$

The second follows from (11.45) with $\partial_p j_p = 0$ by setting $\ell = m$ and integrating over all of space:

$$\int d^3r\ \mathbf{r}(\mathbf{j}\cdot\mathbf{r}) = -\tfrac{1}{2}\int d^3r\ r^2\mathbf{j}. \tag{11.55}$$

After a bit of algebra, we find that

$$T^{(2)}_{k\ell m} = \frac{1}{6}\epsilon_{ki\ell}\Lambda_{im} + \frac{1}{4\pi}(\delta_{km}a_\ell - \delta_{\ell m}a_k), \tag{11.56}$$

[7] Ya.B. Zeldovich, "Electromagnetic interaction with parity violation", *Soviet Physics JETP* **6**, 1184 (1958).

[8] C.S. Wood *et al.*, "Measurement of parity non-conservation and an anapole moment in cesium", *Science* **275**, 1759 (1997).

where the *anapole moment* vector is defined as

$$\mathbf{a} = -\pi \int d^3r\, r^2 \mathbf{j}(\mathbf{r}). \tag{11.57}$$

Now, the quadrupole piece of the vector potential (11.43) is

$$A_k(\mathbf{r}) = \frac{\mu_0}{4\pi} T^{(2)}_{k\ell m} \nabla_\ell \nabla_m \frac{1}{r}. \tag{11.58}$$

But $\nabla^2(1/r) = -4\pi\delta(\mathbf{r})$, so the anapole pieces of (11.56) produce a vector potential

$$\mathbf{A}(\mathbf{r}) = \frac{\mu_0}{4\pi}\left\{\mathbf{a}\,\delta(\mathbf{r}) + \nabla(\mathbf{a}\cdot\nabla)\frac{1}{4\pi r}\right\}. \tag{11.59}$$

The corresponding magnetic field is zero everywhere except at $\mathbf{r} = 0$:

$$\mathbf{B}(\mathbf{r}) = \nabla \times \mathbf{A}(\mathbf{r}) = -\frac{\mu_0}{4\pi}\mathbf{a} \times \nabla\delta(\mathbf{r}). \tag{11.60}$$

Like every multipole moment, the (singular) current density associated with $\mathbf{a}$ lies exactly at the origin. Therefore, (11.60) says that the magnetic field of an anapole moment is zero everywhere outside of its source current density. This property is characteristic of a toroidal solenoid (see Example 10.2). For that reason, the solenoid winding shown in Figure 10.10 is a fair (classical) model for a nuclear current density that produces an anapole moment. ■

11.4.2 Spherical Expansion for $\psi(\mathbf{r})$

In spherical coordinates, it is simplest to generate an exterior spherical multipole expansion for the magnetic scalar potential $\psi(\mathbf{r})$. Our strategy is to use $\mathbf{B} = -\nabla\psi$ and first derive an expansion for

$$\mathbf{r}\cdot\mathbf{B} = -r\frac{\partial\psi}{\partial r}. \tag{11.61}$$

The expansion for $\psi(\mathbf{r})$ itself follows by integrating (11.61). The first step combines $\nabla\cdot\mathbf{B} = 0$ with the curl of $\nabla\times\mathbf{B} = \mu_0\mathbf{j}$ to get (see Example 1.3)

$$\nabla^2\mathbf{B} = -\mu_0\nabla\times\mathbf{j}. \tag{11.62}$$

The next step applies the identity $\mathbf{r}\cdot(\nabla^2\mathbf{B}) = \nabla^2(\mathbf{r}\cdot\mathbf{B}) - 2\nabla\cdot\mathbf{B}$ to (11.62) to get

$$\nabla^2(\mathbf{r}\cdot\mathbf{B}) = -\mu_0\mathbf{r}\cdot\nabla\times\mathbf{j}. \tag{11.63}$$

This is a vector Poisson equation of the sort solved in Section 10.5.2 for the vector potential in the Coulomb gauge. Transcribing the solution derived there to the present problem gives

$$\mathbf{r}\cdot\mathbf{B} = \frac{\mu_0}{4\pi}\int d^3r' \,\frac{\mathbf{r}'\cdot\nabla'\times\mathbf{j}(\mathbf{r}')}{|\mathbf{r}-\mathbf{r}'|}. \tag{11.64}$$

An exterior spherical multipole expansion for $\mathbf{r}\cdot\mathbf{B}$ follows by substituting the inverse-distance formula (4.83) into (11.64):

$$\frac{1}{|\mathbf{r}-\mathbf{r}'|} = \frac{1}{r}\sum_{\ell=0}^{\infty}\sum_{m=-\ell}^{\ell}\frac{4\pi}{2\ell+1}\left(\frac{r'}{r}\right)^{\ell} Y^*_{\ell m}(\Omega')Y_{\ell m}(\Omega) \qquad r' < r. \tag{11.65}$$

The last step is to integrate (11.64) using (11.61) and define *spherical magnetic multipole moments*

$$M_{\ell m} = \frac{1}{\ell+1}\int d^3r\, r^\ell\, Y^*_{\ell m}(\Omega)\,\mathbf{r}\cdot[\nabla\times\mathbf{j}(\mathbf{r})]. \tag{11.66}$$

The expansion of the magnetic scalar potential which results,

$$\psi(\mathbf{r}) = \frac{\mu_0}{4\pi} \sum_{\ell=1}^{\infty} \sum_{m=-\ell}^{\ell} \frac{4\pi}{2\ell+1} M_{\ell m} \frac{Y_{\ell m}(\Omega)}{r^{\ell+1}}, \tag{11.67}$$

is strikingly similar to the spherical exterior multipole expansion (4.86) for the *electric* scalar potential except (as noted earlier) there is no magnetic monopole term.[9]

11.4.3 Alternative Forms for $M_{\ell m}$

There are several equivalent forms for the spherical magnetic multipole moments defined by (11.66). For example, because $\nabla \cdot (\mathbf{r} \times \mathbf{j}) = -\mathbf{r} \cdot (\nabla \times \mathbf{j})$,

$$M_{\ell m} = -\frac{1}{\ell+1} \int d^3r \; r^\ell \, Y^*_{\ell m}(\Omega) \nabla \cdot [\mathbf{r} \times \mathbf{j}(\mathbf{r})]. \tag{11.68}$$

Integrating (11.68) by parts gives another formula,

$$M_{\ell m} = \frac{1}{\ell+1} \int d^3r \; [\mathbf{r} \times \mathbf{j}(\mathbf{r})] \cdot \nabla \, [r^\ell \, Y^*_{\ell m}(\Omega)]. \tag{11.69}$$

An immediate deduction from (11.69) is $M_{00} = 0$. This accounts for the absence of the $\ell = 0$ term in (11.67). We can also substitute the identity

$$\mathbf{r} \cdot (\nabla \times \mathbf{j}) = \epsilon_{ijk} r_i \partial_j j_k = \epsilon_{jki} r_j \partial_k j_i = (\mathbf{r} \times \nabla) \cdot \mathbf{j} \tag{11.70}$$

into (11.66) to get

$$M_{\ell m} = \frac{1}{\ell+1} \int d^3r \; r^\ell \, Y^*_{\ell m}(\Omega) \, (\mathbf{r} \times \nabla) \cdot \mathbf{j}(\mathbf{r}). \tag{11.71}$$

A particularly interesting expression for $M_{\ell m}$ is simplest to derive if we define

$$\mathbf{L} = -i\mathbf{r} \times \nabla, \tag{11.72}$$

and write (11.71) in the form

$$M_{\ell m} = \frac{i}{\ell+1} \int d^3r \; r^\ell \, Y^*_{\ell m}(\Omega) \, \mathbf{L} \cdot \mathbf{j}(\mathbf{r}). \tag{11.73}$$

The reader will recognize that $\hbar\mathbf{L}$ is the quantum mechanical operator for orbital angular momentum. In nuclear physics, particulary, it is common to define a *vector spherical harmonic*,

$$\mathbf{X}_{\ell m} = \frac{1}{\sqrt{\ell(\ell+1)}} \mathbf{L} Y_{\ell m}(\Omega), \tag{11.74}$$

and use the Hermitian property of $\mathbf{L}$ to transform (11.73) to

$$M_{\ell m} = i\sqrt{\frac{\ell}{\ell+1}} \int d^3r \, r^\ell \, \mathbf{X}^*_{\ell m}(\Omega) \cdot \mathbf{j}(\mathbf{r}). \tag{11.75}$$

The applications of this expression (see Section 11.4.6) exploit the fact that[10]

$$\int d\Omega \, \mathbf{X}^*_{\ell m} \cdot \mathbf{X}_{\ell' m'} = \delta_{\ell\ell'} \delta_{mm'}. \tag{11.76}$$

[9] We have put the integration constant equal to zero so $\psi(\infty) = 0$. The integration of (11.61) does not generate an additional arbitrary function of θ and ϕ because $\nabla^2\psi = 0$ (Section 10.4) and (11.67) already contains all the fundamental solutions of Laplace's equation in spherical coordinates that are regular at infinity (Section 7.7).

[10] The orthogonality relation (11.76) is true because $\mathbf{L}$ is Hermitian and the $Y_{\ell m}$ are orthogonal.

11.4.4 Spherical Expansion for $\mathbf{A}(\mathbf{r})$

It is a short step from the spherical multipole expansion (11.67) for the magnetic scalar potential to a spherical multipole expansion for the vector potential $\mathbf{A}(\mathbf{r})$. With the definition of $\mathbf{L}$ in (11.72), the key ingredient is the vector identity[11]

$$\nabla \times \mathbf{L} = -i\mathbf{r}\nabla^2 + i\nabla(1 + \mathbf{r} \cdot \nabla). \tag{11.77}$$

Specifically, because $\nabla^2[Y_{\ell m}(\Omega)/r^{\ell+1}] = 0$, (11.77) tells us that

$$(i\ell\nabla + \nabla \times \mathbf{L})\frac{Y_{\ell m}(\Omega)}{r^{\ell+1}} = 0. \tag{11.78}$$

Therefore, (11.67) and $\mathbf{B} = -\nabla\psi = \nabla \times \mathbf{A}$ imply that

$$\mathbf{A}(\mathbf{r}) = \frac{\mu_0}{4\pi}\sum_{\ell=1}^{\infty}\sum_{m=-\ell}^{\ell}\frac{4\pi}{2\ell+1}\frac{1}{i\ell}M_{\ell m}\mathbf{L}\frac{Y_{\ell m}(\Omega)}{r^{\ell+1}}. \tag{11.79}$$

This is a conventional form of the magnetostatic multipole expansion of the vector potential.

11.4.5 Azimuthal Expansions

In spherical coordinates, the current density of a system with azimuthal symmetry takes the form $\mathbf{j}(\mathbf{r}) = \mathbf{j}(r, \theta)$. For situations like this, the ϕ integration in (11.66) can be done using

$$\frac{1}{2\pi}\int_0^{2\pi} d\phi\, Y_{\ell m}(\theta, \phi) = \sqrt{\frac{2\ell+1}{4\pi}}P_\ell(\cos\theta)\delta_{m0}. \tag{11.80}$$

Then, because $Y_{\ell 0}(\theta, \phi) = \sqrt{(2\ell+1)/4\pi}\, P_\ell(\cos\theta)$, our spherical multipole expansion (11.67) reduces to an *azimuthal multipole expansion*,

$$\psi(r, \theta) = \frac{\mu_0}{4\pi}\sum_{\ell=1}^{\infty} M_\ell \frac{P_\ell(\cos\theta)}{r^{\ell+1}}, \tag{11.81}$$

with azimuthal magnetic multipole moments

$$M_\ell = \frac{1}{\ell+1}\int d^3r\, r^\ell\, P_\ell(\cos\theta)\,\mathbf{r}\cdot(\nabla \times \mathbf{j}). \tag{11.82}$$

11.4.6 Currents That Produce Pure Multipole Fields

It is instructive (and possibly useful in the laboratory) to construct current distributions which produce a multipole field with given values of ℓ and m. A glance at (11.75) and (11.76) shows that a current density proportional to the vector spherical harmonic $\mathbf{X}_{\ell m}$ [defined by (11.74) and (11.72)] has exactly the desired properties. For simplicity, we specialize to axial symmetry and use (11.70) and (11.72) to write the azimuthal moment (11.82) as

$$M_\ell = \frac{i}{\ell+1}\int d^3r\, r^\ell P_\ell(\cos\theta)\mathbf{L}\cdot\mathbf{j}. \tag{11.83}$$

The radial dependence is not important to this argument, so we choose a model current density confined to the surface of a sphere of radius R, with the form

$$\mathbf{j}_L(\theta) = I\frac{\delta(r-R)}{R}(\mathbf{r}\times\nabla)P_L(\cos\theta) = I\frac{\delta(r-R)}{R}i\mathbf{L}P_L(\cos\theta). \tag{11.84}$$

[11] See Application 1.1 at the end of Section 1.2.

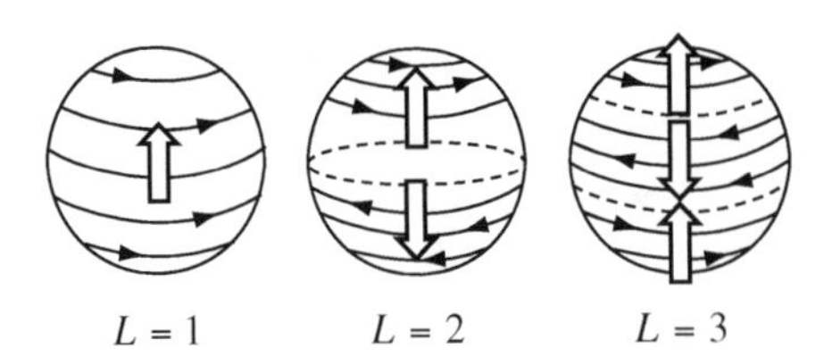

Figure 11.8: Lines of surface current density on a sphere that produce pure dipole ($L = 1$), quadrupole ($L = 2$), and octupole ($L = 3$) magnetic fields outside the sphere. See text for discussion of the superimposed arrows.

If we insert (11.84) into (11.83) and use $\mathbf{L}^2 P_L = L(L+1)P_L$, the orthogonality relation (C.3) of the Legendre polynomials guarantees that the only non-zero magnetic multipole moment generated by $\mathbf{j}_L$ is M_L. In practice, we write out the cross product in (11.84) in spherical coordinates to get

$$\mathbf{j}_L(\theta) = I\frac{\delta(r-R)}{R}\frac{\partial P_L(\cos\theta)}{\partial\theta}\hat{\boldsymbol{\phi}}. \tag{11.85}$$

The cartoons in Figure 11.8 illustrate the surface current density (11.85) for $L = 1$, $L = 2$, and $L = 3$. These currents produce purely dipole, quadrupole, and octupole magnetic fields, respectively, at every point outside the sphere. These assignments can be understood if we use Figure 11.3(a) to assign a magnetic dipole moment to each current loop on each sphere. Summing nearby moments which point in the same direction leads to the partial dipole moments drawn as arrows on the spheres. By analogy with the case of electric multipole moments (Chapter 4), we see that a magnetic quadrupole moment derives from the juxtaposition of two oppositely oriented magnetic dipole moments and that a magnetic octupole moment derives from the juxtaposition of two oppositely oriented magnetic quadrupole moments.

Application 11.3 The Helmholtz Anti-Coil

Figure 11.9 shows two coaxial rings that carry current in opposite directions.[12] To determine the distant magnetic field produced by this object, we focus first on the upper ring at $z = d/2$. The current density of this ring can be written as

$$\mathbf{j}_+(r,\theta) = \hat{\boldsymbol{\phi}}\frac{I}{r}\delta(r - R\csc\alpha)\delta(\theta-\alpha). \tag{11.86}$$

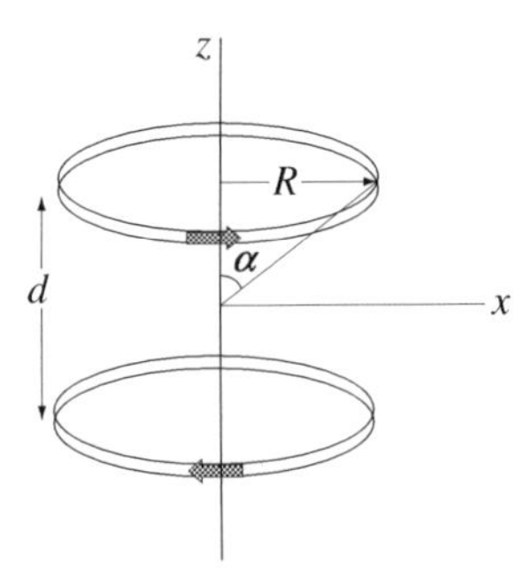

Figure 11.9: A current I flows in opposite directions (arrows) through two coaxial rings (radius R) separated by a distance d.

[12] The coaxial rings of a conventional Helmholtz coil carry current in the same direction. See Section 10.4.2.

Therefore, because $(\nabla \times \mathbf{j})_r = (1/\sin\theta)\partial_\theta(\sin\theta j_\phi)$,

$$\mathbf{r} \cdot (\nabla \times \mathbf{j}_+) = \frac{I}{r\sin\theta}\delta(r - R\csc\alpha)[\cos\theta\delta(\theta-\alpha) + \sin\theta\delta'(\theta-\alpha)]. \tag{11.87}$$

Inserting (11.87) into (11.82) and writing $P'_\ell(x) = (d/dx)P_\ell(x)$ gives the moments (11.82) as

$$M_\ell^+ = \frac{2\pi I R^2}{\ell+1}\left(\frac{R}{\sin\alpha}\right)^{\ell-1} P'_\ell(\cos\alpha). \tag{11.88}$$

The special case $\alpha = \pi/2$ locates the upper ring in the x-y plane. This simplifies (11.88) to

$$M_\ell^+ = \frac{2\pi I R^{\ell+1}}{\ell+1} P'_\ell(0). \tag{11.89}$$

Using the Legendre polynomial identity, $P'_\ell(0) = \ell P_{\ell-1}(0)$, the reader can check that the multipole moments (11.89) inserted into (11.81) exactly reproduce the $r > R$ expansion of the current ring magnetic scalar potential derived in Section 10.4.1.

Return now to a general value for the angle α and focus on the lower ring at $z = -d/2$ in Figure 11.9. The magnetic multipole moments M_ℓ^- for this ring are given by (11.88) with $I \to -I$ and $\alpha \to \pi - \alpha$. Therefore, since $P_\ell(-x) = (-1)^\ell P_\ell(x)$,

$$M_\ell^- = (-1)^\ell M_\ell^+. \tag{11.90}$$

The multipole moments for the two-ring system are $M_\ell = M_\ell^+ + M_\ell^-$ This shows that all the odd-order moments vanish. Moreover, if $m = I\pi R^2$ is the magnitude of the dipole moment of either ring, we find from (11.89) and (11.90) that the magnetic field very far away from this object has the character of a *magnetic quadrupole* with magnetic quadrupole moment

$$M_2 = 4mR\cot\alpha = 2md. \tag{11.91}$$

The appearance of a magnetic quadrupole for this situation is consistent with the analysis of Section 11.3. ■

11.5 Interior Multipoles

Many practical applications demand an approximation to $\mathbf{B}(\mathbf{r})$ when currents *outside* a specified volume create a magnetic field *inside* that volume.[13] The spherical volume shown in Figure 11.10 is an example. One approach to this problem exploits the interior analog of the exterior Cartesian multipole expansion of Section 11.1.1. If the observation point $\mathbf{r}$ lies close to the center of the sphere, we can exchange $\mathbf{r}$ and $\mathbf{r}'$ in the expansion (11.2) and approximate the vector potential by

$$A_k(\mathbf{r}) = \frac{\mu_0}{4\pi}\left[\int d^3r' \frac{j_k(\mathbf{r}')}{r'} + \mathbf{r}\cdot\int d^3r' \frac{j_k(\mathbf{r}')\mathbf{r}'}{r'^3} + \cdots\right]. \tag{11.92}$$

The first term in (11.92) is a constant which does not contribute to the magnetic field $\mathbf{B} = \nabla \times \mathbf{A}$. The second term is linear in $\mathbf{r}$ and we leave it as an exercise for the reader to show that the *constant* magnetic field it produces is

$$\mathbf{B}(0) = \frac{\mu_0}{4\pi}\int d^3r' \frac{\mathbf{r}' \times \mathbf{j}(\mathbf{r}')}{r'^3} \qquad r \ll R. \tag{11.93}$$

[13] The exposure of a small experimental sample to an external magnetic field almost always requires this geometry.

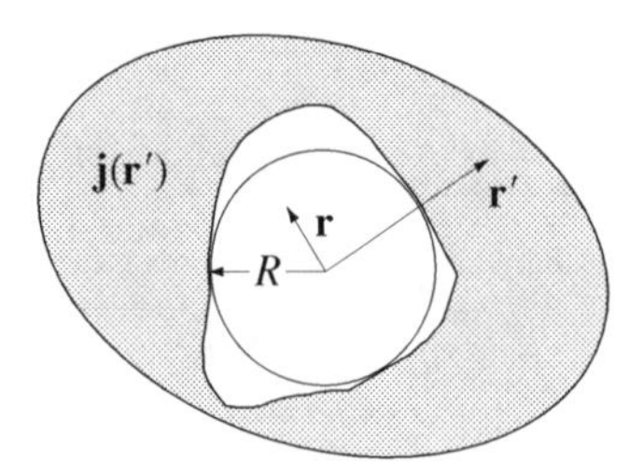

Figure 11.10: The observation point **r** lies inside a spherical volume of radius R which completely excludes a distribution of source current $\mathbf{j}(\mathbf{r})$ (dark shading).

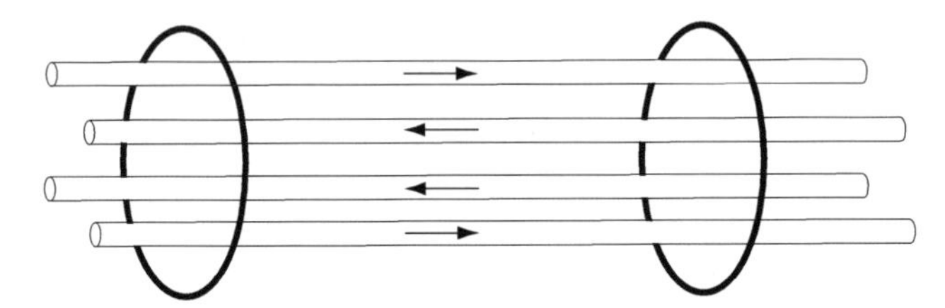

Figure 11.11: Four equally spaced wires arranged on a cylindrical surface. The dark circles are guides to the eye. Adjacent wires carry current in opposite directions.

The notation $\mathbf{B}(0)$ is appropriate because (11.93) is the Biot-Savart magnetic field due to $\mathbf{j}(\mathbf{r}')$ evaluated at the center of the sphere. It is straightforward, but not very common, to continue the Cartesian expansion (11.92) beyond the dipole term studied above. For that reason, we pass on to the more interesting and practically important case of effectively two-dimensional magnetic fields.

11.5.1 Interior Expansion for Two-Dimensional Fields

Two-dimensional magnetic fields of the form $\mathbf{B}(\rho, \phi)$ are ubiquitous in many subfields of physics. A short list of applications includes neutral and charged particle trapping, beam focusing and spectroscopy, plasma confinement, and magnetic resonance imaging. The goal in each case is to expose an experimental sample to a specified magnetic field inside a cylindrical sample volume. Fields of this type are often produced using parallel arrays of straight current-carrying wires. A four-wire example is shown in Figure 11.11. Example 11.3 in Section 11.3 may be regarded as a two-wire example.

To compute $\mathbf{B}(\mathbf{r})$, we take the vector potential $\mathbf{A} = -\hat{\mathbf{z}}(\mu_0 I/2\pi)\ln\rho$ of a single wire,[14] shift the wire position suitably, and superpose the contributions from all the wires. More generally, if $\boldsymbol{\rho} = (\rho, \phi)$ is a two-dimensional vector, the vector potential $\mathbf{A}(\boldsymbol{\rho}) = A_z(\boldsymbol{\rho})\hat{\mathbf{z}}$ associated with the two-dimensional current density $\mathbf{j}(\boldsymbol{\rho}) = j_z(\boldsymbol{\rho})\hat{\mathbf{z}}$ is

$$A_z(\boldsymbol{\rho}) = \frac{\mu_0}{2\pi}\int d^2\rho'\, j_z(\boldsymbol{\rho}')\ln|\boldsymbol{\rho} - \boldsymbol{\rho}'|. \tag{11.94}$$

The corresponding magnetic field lies in the plane perpendicular to the z-axis:

$$\mathbf{B}(\boldsymbol{\rho}) = \nabla \times A_z(\boldsymbol{\rho})\hat{\mathbf{z}} = -\hat{\mathbf{z}} \times \nabla A_z(\boldsymbol{\rho}). \tag{11.95}$$

[14] See Example 10.4.

We note in passing that the magnetic field lines for (11.95) are defined by the condition $A_z(\boldsymbol{\rho}) = const.$ This is so because the equation for the field lines (10.7) is identical to

$$0 = dA_z = \frac{\partial A_z}{\partial x}dx + \frac{\partial A_z}{\partial y}dy = -B_y dx + B_x dy. \tag{11.96}$$

To make progress with (11.94), let $\rho_< = \min(\rho, \rho')$ and $\rho_> = \max(\rho, \rho')$ so

$$|\boldsymbol{\rho} - \boldsymbol{\rho}'|^2 = \rho_>^2 \left[1 - \frac{\rho_<}{\rho_>}e^{i(\phi-\phi')}\right]\left[1 - \frac{\rho_<}{\rho_>}e^{-i(\phi-\phi')}\right]. \tag{11.97}$$

Combining (11.97) with the identity

$$\log(1-z) = -\sum_{n=1}^{\infty}\frac{z^n}{n} \qquad |z| < 1 \tag{11.98}$$

gives

$$\ln|\boldsymbol{\rho} - \boldsymbol{\rho}'| = \ln\rho_> - \sum_{n=1}^{\infty}\left(\frac{\rho_<}{\rho_>}\right)^n \frac{\cos n(\phi-\phi')}{n}. \tag{11.99}$$

If $j_z(\boldsymbol{\rho}')$ is non-zero only when $\rho' \geq R$ and the observation points of interest always satisfy $\rho < R$, substitution of (11.99) into (11.94) leads to the interior multipole expansion

$$A_z(\rho,\phi) = \frac{\mu_0}{2\pi}\sum_{n=1}^{\infty}\rho^n\left\{C_n\cos n\phi + S_n \sin n\phi\right\} \qquad \rho < R, \tag{11.100}$$

with interior multipole moments

$$\begin{Bmatrix} S_n \\ C_n \end{Bmatrix} = \frac{1}{n}\int_0^{2\pi} d\phi \int_0^{\infty} d\rho\, \rho^{1-n} \begin{Bmatrix} \sin n\phi \\ \cos n\phi \end{Bmatrix} j_z(\rho,\phi). \tag{11.101}$$

We have dropped from (11.100) a non-essential constant which derives from the $\ln\rho_>$ term in (11.99).

11.5.2 2N Parallel Current-Carrying Wires

Let us evaluate the multipole moments (11.101) explicitly for the situation depicted in Figure 11.11 generalized to the case of $2N$ equally spaced wires with alternating directions of current flow. If all the wires lie on a cylinder of radius R, and one wire carrying current in the $+z$-direction passes through the point $(R, 0)$, a suitable current density is

$$j_z(\rho,\phi) = \frac{I}{R}\delta(\rho - R)\sum_{p=0}^{2N-1}(-1)^p\delta(\phi - p\pi/N). \tag{11.102}$$

This quantity has a periodicity in ϕ of $2\pi/N$. Therefore, the only non-zero terms in (11.100) have $n = mN$ where m is an integer. This fact, together with the identity

$$\sum_{p=0}^{2N-1}(-1)^p\exp(imp\pi) = \begin{cases} 2N & m \text{ odd}, \\ 0 & m \text{ even}, \end{cases} \tag{11.103}$$

shows that $S_n \equiv 0$ in (11.101). The C_n are not zero and lead to

$$A_z(\rho,\phi) = \frac{\mu_0 I}{\pi}\sum_{m=1,3,5}^{\infty}\frac{1}{m}\left(\frac{\rho}{R}\right)^{mN}\cos mN\phi. \tag{11.104}$$

When $\rho \ll R$, the sum (11.104) is dominated by the $m = 1$ term. This term is called a pure multipole of order $2N$:

$$A_z(\rho, \phi) \propto \rho^N \cos N\phi. \tag{11.105}$$

The fields derived from (11.105) with $N = 1, 2, 3, 4$ are called dipole, quadrupole, sextupole, and octupole fields, which reminds us that $2N$ filamentary currents produce the field in question. This may be contrasted with the three-dimensional electric multipole case where 2^N point charges produce dipole, quadrupole, octupole, and hexadecapole fields for $N = 1, 2, 3, 4$, respectively.

Two parallel wires which carry current in opposite directions corresponds to $N = 1$ in (11.105). We calculated the exact magnetic field for this geometry in Example 11.3. Very near the origin, $A_z \propto \rho \cos\phi = x$, so the associated magnetic field (11.95) is *uniform* and points in the $-\hat{\mathbf{y}}$-direction.[15]

Application 11.4 Strong Focusing by Quadrupole Magnetic Fields

The discovery of *strong focusing* in the early 1950s had a major impact on the evolution of high-energy particle accelerators. To illustrate the principle we note that, very near the origin, the four-wire configuration shown in Figure 11.11 produces a pure quadrupole ($N = 2$) magnetic field. If a is a length, the field has the form

$$\mathbf{B}(x, y) = -(B_0/a)(x\hat{\mathbf{y}} + y\hat{\mathbf{x}}) \qquad x, y \ll R. \tag{11.106}$$

Since $\mathbf{B} = -\nabla\psi$, a magnetic scalar potential $\psi(x, y) = (B_0/a)xy$ produces this field also. Figure 11.12 shows some typical field lines. $\mathbf{B}(x, y)$ deflects a particle with charge q and velocity $\boldsymbol{v} = v\hat{\mathbf{z}}$ with a transverse force

$$\mathbf{F} = qvB_0(x\hat{\mathbf{x}} - y\hat{\mathbf{y}}). \tag{11.107}$$

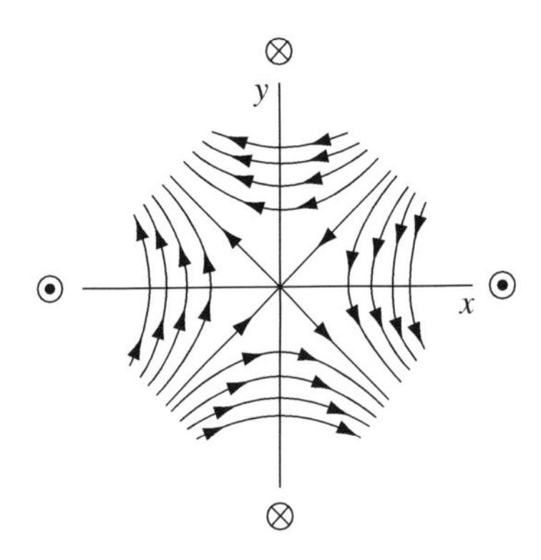

Figure 11.12: Two wires carrying current I toward the reader (solid dots) and two wires carrying current I away from the reader (crosses) as shown produce a pure quadrupole field near the origin of coordinates.

The particle is deflected toward the symmetry axis (focused) in the y-direction and deflected away from the symmetry axis (de-focused) in the x-direction. However, if the particle immediately enters the field of a second quadrupole magnet which is rotated by 90° with respect to the first, the alternating effect of focusing and de-focusing can produce overall focusing in both transverse directions. This phenomenon—known as *strong focusing*—is used in particle accelerators to reduce the dimensions and transverse velocity spread of charged particle beams.

[15] This geometry played an important role in the historical development of molecular beam techniques to measure magnetic moments. See Appendix F of N.F. Ramsey, *Molecular Beams* (Oxford University Press, New York, 1955).

The physics of strong focusing becomes clearer when we recall the classical optics formula for the effective focal length f of two thick lenses separated by a distance d,[16]

$$\frac{1}{f} = \frac{1}{f_1} + \frac{1}{f_2} - \frac{d}{f_1 f_2}. \tag{11.108}$$

A lens focuses when $f > 0$ and de-focuses when $f < 0$. Therefore, overall focusing occurs if we choose $f_2 = -f_1$ as in the lens arrangement in Figure 11.13. It is worth noting that a force like (11.107) occurs when a particle with charge q encounters the electric field $\mathbf{E} \propto x\hat{\mathbf{x}} - y\hat{\mathbf{y}}$. This field, shown in Figure 7.14, derives from the purely quadrupole electrostatic potential $\varphi(x, y) \propto y^2 - x^2$. Compared to the magnetostatic strong focusing used in high-energy electron, proton, and anti-proton accelerators, electrostatic strong focusing is better suited for low-energy atomic and molecular ion accelerators.

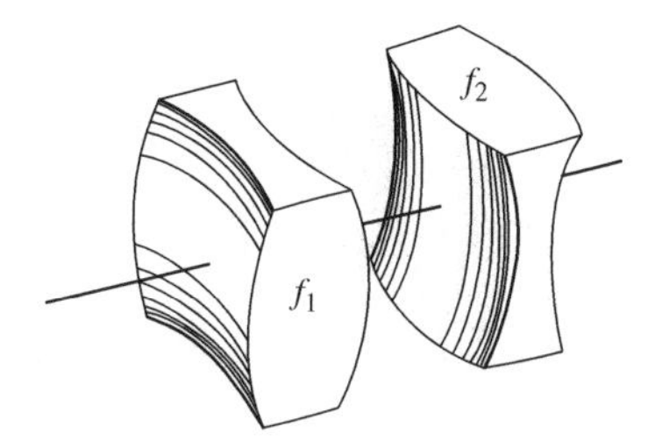

Figure 11.13: Complementary converging and diverging lenses for strong focusing in the two orthogonal directions transverse to the beam direction. Figure from Bullock (1955).

11.6 Axially Symmetric Magnetic Fields

We conclude this chapter with an analysis of magnetic fields with rotational symmetry with respect to a fixed axis. Fields of this kind are important because they function as lenses for charged particles (see Application 11.5 below). We limit our study to the field in the immediate vicinity of the symmetry axis.

In cylindrical coordinates, the magnetic scalar potential of a system with axial symmetry has the property that $\psi = \psi(\rho, z)$. When ρ is small compared to the radius of a cylinder which plays the role of the sphere in Figure 11.10, it makes sense to study an expansion of the form

$$\psi(\rho, z) = \sum_{n=0}^{\infty} a_n(z)\rho^n. \tag{11.109}$$

Inside the cylinder, the scalar potential satisfies Laplace's equation,

$$\nabla^2\psi = \frac{1}{\rho}\frac{\partial}{\partial\rho}\left(\rho\frac{\partial\psi}{\partial\rho}\right) + \frac{\partial^2\psi}{\partial z^2} = 0. \tag{11.110}$$

Substituting (11.109) into (11.110) and rearranging so that terms with common powers of ρ can be compared leads to the condition

$$\sum_{m=-2}^{\infty} a_{m+2}(z)(m+2)^2\rho^m + \sum_{n=0}^{\infty} a_n''(z)\rho^n = 0. \tag{11.111}$$

[16] See, e.g., E. Hecht and H. Zajac, *Optics* (Addison-Wesley, Reading, MA, 1974).

The $m = -2$ term vanishes identically. The $m = -1$ term forces $a_1(z) = 0$ to avoid a divergence at the origin. Otherwise, the coefficient of every power of ρ must vanish:

$$a_{n+2} = -\frac{a_n''(z)}{(n+2)^2} \qquad n \geq 0. \tag{11.112}$$

This shows that only even powers of ρ enter the expansion (11.109).

The magnetic field for this situation is

$$\mathbf{B}(\rho, z) = -\frac{\partial \psi}{\partial \rho}\hat{\boldsymbol{\rho}} - \frac{\partial \psi}{\partial z}\hat{\mathbf{z}}. \tag{11.113}$$

Therefore, the z-component of the magnetic field on the symmetry axis is

$$B_z(z) \equiv B_z(0, z) = -a_0'(z). \tag{11.114}$$

Given (11.112), we conclude that a magnetic field with azimuthal symmetry is determined everywhere by its on-axis piece, $B_z(z)$.[17] In detail,

$$B_z(\rho, z) = B_z(z) - \frac{\rho^2}{4}B_z''(z) + \frac{\rho^4}{64}B_z''''(z) + \cdots \tag{11.115}$$

and

$$B_\rho(\rho, z) = -\frac{\rho}{2}B_z'(z) + \frac{\rho^3}{16}B_z'''(z) + \cdots. \tag{11.116}$$

Application 11.5 The Principle of the Electron Microscope

The electron microscope evolved from the 1931 discovery by Ruska and Knoll that a fast electron beam passing through a solenoid coil focuses toward the symmetry axis of the coil. We will prove here that *every* axially symmetric magnetic field produces focusing as long as the beam stays close to the symmetry axis and the velocity components satisfy

$$v_z \simeq const. \qquad \text{and} \qquad v_z \gg v_\rho, v_\phi. \tag{11.117}$$

Quantitatively, we assume that $\rho \ll |B_z(z)/B_z'(z)|$, so (11.115) and (11.116) imply that

$$B_\rho(\rho, z) \simeq -\frac{\rho}{2}B_z'(\rho, z). \tag{11.118}$$

In cylindrical coordinates, a particle with charge q and mass m has velocity $\boldsymbol{v} = (\dot{\rho}, \rho\dot{\phi}, \dot{z})$ and acceleration $\mathbf{a} = (\ddot{\rho} - \rho\dot{\phi}^2, 2\dot{\rho}\dot{\phi} + \rho\ddot{\rho}, \ddot{z})$. These kinematic formulae[18] and the Lorentz force law $\mathbf{F} = q\boldsymbol{v} \times \mathbf{B}$ give the equations of motion,

$$\ddot{\rho} - \rho\dot{\phi}^2 = \frac{q\rho}{m}B_z\dot{\phi} \tag{11.119}$$

and

$$\ddot{\phi} - 2\dot{\phi}\frac{\rho}{\dot{\rho}} = \frac{q}{m\rho}\left[v_z B_\rho - v_\rho B_z\right]. \tag{11.120}$$

We regard $\rho(z, t)$ as the trajectory of the particle so

$$\dot{\rho} = \frac{d\rho}{dz}\frac{dz}{dt} = \rho'\dot{z}. \tag{11.121}$$

[17] The same conclusion follows from the "going off the axis" calculation for the current ring in Section 10.4.1.

[18] K. Rossberg, *A First Course in Analytic Mechanics* (Wiley, New York, 1983).

Using (11.117),

$$\ddot{\rho} = \ddot{z}\rho' + \dot{z}^2\rho'' \simeq \dot{z}^2\rho''. \tag{11.122}$$

Similarly, $\dot{\phi} = \phi'\dot{z}$ and $\ddot{\phi} \simeq \dot{z}^2\phi''$. These results, together with (11.118) and the definition

$$K(z) = qB_z/mv_z, \tag{11.123}$$

permit us to re-express the equations of motion entirely in terms of spatial derivatives. A few lines of algebra show that the rewritten equations of motion are

$$\rho'' - \rho(\phi')^2 = K\rho\phi' \tag{11.124}$$

and

$$\frac{d}{dz}\left\{\rho^2(\phi' + K/2)\right\} = 0. \tag{11.125}$$

The azimuthal equation (11.125) integrates to

$$\phi'(z) = -\frac{K(z)}{2} + \frac{C}{\rho^2}. \tag{11.126}$$

Our main interest is a particle that enters the magnetic field from a region of zero field (so $K = 0$) along a trajectory which is parallel to the symmetry axis (so $\phi' = 0$). These initial conditions are consistent with (11.126) if we choose the integration constant $C = 0$. Therefore, when it does enter the magnetic field, (11.126) tells us that the particle spirals as it moves along the z-axis. As for the radial motion, substitution of (11.126) into (11.124) gives

$$\frac{d^2\rho}{dz^2} = -\tfrac{1}{4}K^2\rho. \tag{11.127}$$

By assumption, $\rho'' = 0$ before the particle enters the field. Therefore, when the field begins to act, (11.127) says that the particle bends toward the symmetry axis. The trajectory returns to a straight line when the Lorentz force ceases. Therefore, by a judicious choice of the strength and spatial extent of the field, one can control the point where the particle crosses the symmetry axis, i.e., the focal point of this "magnetic lens". Figure 11.14 shows the trajectories of three particles which approach a current ring parallel to the symmetry axis of the ring.

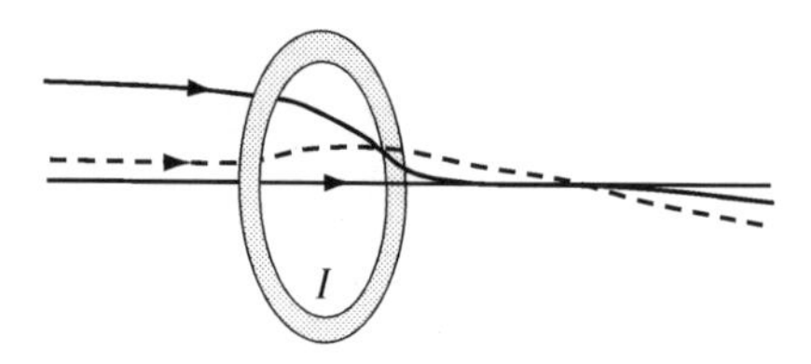

Figure 11.14: Three charged particles with (initially) parallel trajectories are focused by a circular current loop.

Sources, References, and Additional Reading

The quotation at the beginning of the chapter is taken from the abstract to

I.I. Rabi, "Space quantization in a gyrating magnetic field", *Physical Review* **51**, 652 (1937). Copyright 1937 by the American Physical Society.

Section 11.1 Textbooks of electromagnetism with particularly good treatments of magnetic dipole physics and the magnetic multipole expansion include

R.H. Good, Jr. and T.J. Nelson, *Classical Theory of Electric and Magnetic Fields* (Academic, New York, 1971)

J. Konopinski, *Electromagnetic Fields and Relativistic Particles* (McGraw-Hill, New York, 1981).

R.K. Wangsness, *Electromagnetic Fields*, 2nd edition (Wiley, New York, 1986).

V.D. Barger and M.G. Olsson, *Classical Electricity and Magnetism* (Allyn & Bacon, Boston, 1987).

Section 11.2 *Geissman* readably discusses the geomagnetic field and its reversals. The second and third references below are the sources for the left and right panels of Figure 11.4, respectively.

J.W. Geissman, "Geomagnetic flip", *Physics World* **17**(4) 31 (2004).

G.A. Glatzmaier and T. Clune, "Computational aspects of geodynamo simulations", *Computing in Science and Engineering* **2**(3), 61 (2000). © 2000 IEEE.

A. Sakuraba and Y. Hamano, "Turbulent structure in Earth's fluid core inferred from time series of the geomagnetic dipole moment", *Geophysical Research Letters* **34**, L15308 (2007). Reproduced by permission of American Geophysical Union.

Example 11.1 and Example 11.2 are taken from, respectively,

A.M. Portis, *Electromagnetic Fields: Sources and Media* (Wiley, New York, 1978).

W. Gough, "The magnetic field produced by a hydrogen atom", *European Journal of Physics* **17**, 208 (1996).

Section 11.4 The Cartesian magnetic multipole and anapole moment discussions in Section 11.4.1 were drawn, respectively, from

A. Castellanos, M. Panizo, and J. Rivas, "Magnetostatic mulitpoles in Cartesian coordinates", *American Journal of Physics* **46**, 1116 (1978).

I.B. Khriplovich, *Parity Non-Conservation in Atomic Phenomena* (Gordon & Breach, Philadelphia, PA, 1991), Chapter 8.

The spherical magnetic multipole discussion of Section 11.4.2 is adapted from

C.G. Gray and P.J. Stiles, "Spherical tensor approach to multipole expansions: Magnetostatic interactions", *Canadian Journal of Physics* **54**, 513 (1976).

C.G. Gray, "Simplified derivation of the magnetostatic multipole expansion using the scalar potential", *American Journal of Physics* **46**, 582 (1978).

The book by *Rose* is a definitive treatment of multipole fields. Application 11.3 comes from *Smith*, which also contains an elementary discussion of higher-order magnetic multipole moments in nuclei.

M.E. Rose, *Multipole Fields* (Wiley, New York, 1955).

D.G. Smith, "Magnetic multiples in theory and practice", *American Journal of Physics* **48**, 739 (1980).

An interesting treatment of magnetic multipoles different from the one in the text is

J.B. Bronzan, "Magnetic scalar potentials and the multipole expansion in magnetostatics", in *Electromagnetism: Paths to Research*, edited by D. Teplitz (Plenum, New York, 1982).

Figure 11.8 and Section 11.4.6 were inspired by

E. Ley-Koo and M.A. Góngora-Treviño, "Interior and exterior multipole expansions", *Revista Mexicana de Física* **34**, 645 (1988).

Section 11.5 Our discussion of interior multipole expansions for two-dimensional magnetic fields is drawn from

D.E. Lobb, "Properties of some useful two-dimensional magnetic fields", *Nuclear Instruments and Methods* **64**, 251 (1968).

J.P. Boris and A.F. Kuckes, "Closed form expressions for the magnetic field of two-dimensional multipole configurations", *Nuclear Fusion* **8**, 323 (1968).

The first reference below is the source of Figure 11.13. The second reference discusses strong focusing by magnetic quadrupole fields.

M. L. Bullock, "Electrostatic strong-focusing lens", *American Journal of Physics* **23**, 264 (1955). Copyright 1955, American Association of Physics Teachers.

M. S. Livingston and J.P. Blewett, *Particle Accelerators* (McGraw-Hill, New York, 1962).

Section 11.6 The following monographs discuss axial magnetic fields and charged particle motion from several different points of view:

P.S. Farago, *Free-Electron Physics* (Penguin, Harmondsworth, UK, 1970).

O. Klemperer, *Electron Physics*, 2nd edition (Butterworths, London, 1972).

The Nobel Prize winner Ernst Ruska recounts the history of magnetic electron optics in

E. Ruska, "The development of the electron microscope and electron microscopy", *Reviews of Modern Physics* **59**, 627 (1987).

Problems

11.1 Magnetic Dipole Moment Practice A current distribution produces the vector potential

$$\mathbf{A}(r,\theta,\phi) = \hat{\boldsymbol{\phi}}\,\frac{\mu_0}{4\pi}\,\frac{A_0 \sin\theta}{r}\,\exp(-\lambda r).$$

Find the magnetic moment of this current distribution without performing any integrals.

11.2 Origin Independence of Magnetic Multipole Moments Let $\mathbf{j}(\mathbf{r})$ be an arbitrary current distribution.

(a) Show that the components of the magnetic dipole moment $\mathbf{m} = \frac{1}{2}\int d^3r\, \mathbf{r} \times \mathbf{j}$ are invariant to a rigid shift of the origin of coordinates.

(b) Show that the components of the Cartesian magnetic quadrupole moment

$$m^{(2)}_{ij} = \tfrac{1}{3}\int d^3r\,(\mathbf{r} \times \mathbf{j})_i r_j$$

are invariant to a rigid shift of the origin only if the magnetic moment vanishes.

11.3 The Field outside a Finite Solenoid A cylindrical solenoid with length L and cross sectional area $A = \pi R^2$ is formed by wrapping n turns per unit length of a wire that carries a current I. Estimate the magnitude of the magnetic field just outside the solenoid and far away from the ends when $L \gg R$. Do this by integrating $\nabla \cdot \mathbf{B}$ over a hemispherical volume of radius $a \to \infty$ oriented so the disk-like portion of its surface cuts perpendicularly through the solenoid at a point far away from both ends.

11.4 The Magnetic Moment of a Rotating Charged Disk A compact disk with radius R and uniform surface charge density σ rotates with angular speed ω. Find the magnetic dipole moment $\mathbf{m}$ when the axis of rotation is

(a) the symmetry axis of the disk;

(b) any diameter of the disk.

11.5 The Field inside a Semi-Infinite Solenoid A semi-infinite solenoid (concentric with the negative z-axis) has cross sectional area $A = \pi R^2$, n turns per unit length of a wire with current I, and magnetic moment per unit length $\mathbf{m} = nIA\hat{\mathbf{z}}$. When $A \to 0$ and $N \to \infty$ with $\mathbf{m}$ fixed, Application 11.1 showed that the magnetic field outside this solenoid (i.e., excluding the negative z-axis) is

$$\mathbf{B}_{\rm out} = \frac{\mu_0 m}{4\pi}\,\frac{\hat{\mathbf{r}}}{r^2}.$$

Use the field inside an infinite solenoid to approximate the field $\mathbf{B}_{\rm in}$ inside a semi-infinite, finite-area solenoid. Show that when $A \to 0$, $\mathbf{B}_{\rm in}$ is confined to the negative z-axis and $\nabla \cdot \mathbf{B}_{\rm in} = -\mu_0 m \delta(\mathbf{r})$. This guarantees that $\nabla \cdot (\mathbf{B}_{\rm in} + \mathbf{B}_{\rm out}) = 0$ as it must. Hint: Show that $\lim_{A\to 0} \Theta(R-\rho)/A = \delta(x)\delta(y)$.

11.6 A Spinning Spherical Shell of Charge A charge Q is uniformly distributed over the surface of a sphere of radius R. The sphere spins at a constant angular frequency ω.

(a) Show that the current density of this configuration can be written in the form $\mathbf{j} = \nabla \times [\mathbf{M}\Theta(R - r)]$ where $\mathbf{M}$ is a constant vector.
(b) Find the magnetic moment of this current distribution.
(c) Evaluate the Coulomb gauge integral to find $\mathbf{A}(\mathbf{r})$.
(d) Find $\mathbf{B}(\mathbf{r})$ everywhere.

11.7 Magnetic Moment of a Planar Spiral Find the magnetic moment of a planar spiral with inner radius a and outer radius b composed of N turns of a filamentary wire that carries a steady current I.

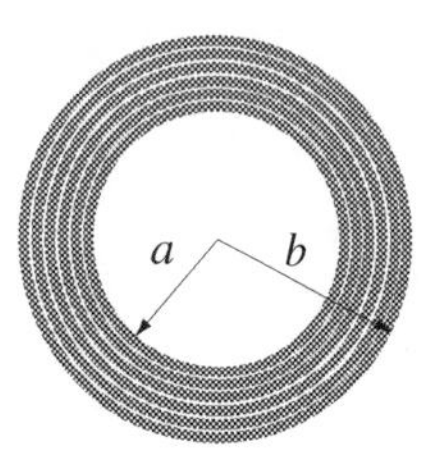

11.8 A Hidden Delta Function The text writes two expressions for the magnetic field of point magnetic dipole at the origin:

$$\mathbf{B}(\mathbf{r}) = \mu_0 \mathbf{m}\delta(\mathbf{r}) - \nabla\psi_0(\mathbf{r})$$

$$\mathbf{B}(\mathbf{r}) = \frac{\mu_0}{4\pi}\frac{3\hat{\mathbf{r}}(\hat{\mathbf{r}} \cdot \mathbf{m}) - \mathbf{m}}{r^3} + \frac{2}{3}\mu_0 \mathbf{m}\delta(\mathbf{r}).$$

$\psi_0(\mathbf{r})$ is the magnetic scalar potential of a point dipole. Prove that the delta function content of these two formulae are the same, at least when used as part of an integration over a volume integral.

11.9 Magnetic Dipole and Quadrupole Moments for $\psi(\mathbf{r})$ The text produced a spherical multipole expansion for the magnetic scalar potential $\psi(\mathbf{r})$ based on the identity

$$-r\frac{\partial \psi}{\partial r} = I = \frac{\mu_0}{4\pi}\int d^3r' \frac{\mathbf{r}' \cdot \nabla' \times \mathbf{j}(\mathbf{r}')}{|\mathbf{r} - \mathbf{r}'|}.$$

A Cartesian expansion for the scalar potential can be developed from the same starting point.

(a) Expand $|\mathbf{r} - \mathbf{r}'|^{-1}$ and confirm that

$$\psi(\mathbf{r}) = \frac{\mu_0}{4\pi}\left[-\mathbf{m} \cdot \nabla\frac{1}{r} + \frac{1}{3}m_{ij}^{(2)}\nabla_i\nabla_j\frac{1}{r} + \cdots\right],$$

where

$$\mathbf{m} = \frac{1}{2}\int d^3r\,(\mathbf{r} \cdot \nabla \times \mathbf{j})\mathbf{r} \qquad m_{ij}^{(2)} = \frac{1}{2}\int d^3r\,(\mathbf{r} \cdot \nabla \times \mathbf{j})r_i r_j.$$

(b) Show that the dipole moment can be written in the familiar form $\mathbf{m} = \frac{1}{2}\int d^3r\,\mathbf{r} \times \mathbf{j}$.
(c) Show that the magnetic quadrupole can be written as

$$m_{ij}^{(2)} = \frac{1}{2}\int d^3r\,(\mathbf{r} \times \mathbf{j}) \cdot \nabla(r_i r_j).$$

(d) Prove that $m_{ij}^{(2)}$ in part (c) is traceless.
(e) Rewrite the formula in part (c) as $\mathbf{m}^{(2)} = \frac{1}{2}\int d^3r\,[(\mathbf{r} \times \mathbf{j})\mathbf{r} + \mathbf{r}(\mathbf{r} \times \mathbf{j})]$.
(f) Prove that the magnetic quadrupole moment $\mathbf{M}^{(2)} = \int d^3r\,(\mathbf{r} \times \mathbf{j})\mathbf{r}$ produces the same quadrupole scalar potential as $\mathbf{m}^{(2)}$ in part (c). $\mathbf{M}^{(2)}$ is the moment which appears in the Cartesian multipole expansion for the *vector* potential derived in the text.

11.10 Biot-Savart at the Origin Show that the first non-zero term in an interior Cartesian multipole expansion of the vector potential can be written in the form $\mathbf{A}(\mathbf{r}) = (\mu_0/4\pi)\mathbf{G} \times \mathbf{r}$ where $\mathbf{G}$ is a constant vector. Show that the associated magnetic field is a Biot-Savart field.

11.11 Purcell's Loop A filamentary current loop traverses eight edges of a cube with side length $2b$ as shown below.

(a) Find the magnetic dipole moment $\mathbf{m}$ of this structure.
(b) Do you expect a negligible or a non-negligible magnetic quadrupole moment? Place the origin of coordinates at the center of the cube as shown.

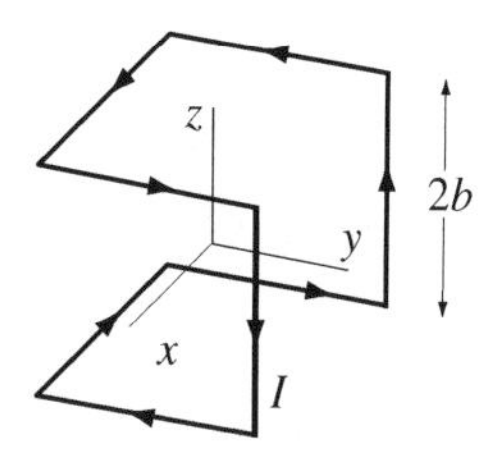

11.12 Dipole field from Monopole Field If such a thing existed, the magnetic field of a point particle with magnetic charge g at rest at the origin would be $\mathbf{B}_{\rm mono}(\mathbf{r}) = (\mu_0 g\mathbf{r}/(4\pi r^3)$. Show that the magnetic field of a point magnetic dipole $\mathbf{m}$ is $\mathbf{B} = -(\mathbf{m}\cdot\nabla)\mathbf{B}_{\rm mono}/g$ at points away from the dipole.

11.13 The Spherical Magnetic Dipole Moment Show that the formula for the magnetic dipole moment derived in Example 11.1,

$$\mathbf{m} = \frac{3}{2\mu_0}\int_{\rm sphere} d^3r\,\mathbf{B}(\mathbf{r}),$$

is consistent with the spherical multipole expansion of the vector potential derived in Section 11.4,

$$\mathbf{A}(\mathbf{r}) = \frac{\mu_0}{4\pi}\sum_{\ell=1}^{\infty}\sum_{m=-\ell}^{\ell}\frac{4\pi}{2\ell+1}\frac{1}{i\ell}M_{\ell m}\mathbf{L}\frac{Y_{\ell m}(\Omega)}{r^{\ell+1}}.$$

Hint: Use (1.78) and $\nabla = \hat{\mathbf{r}}(\partial/\partial r) + \nabla_\perp$ to simplify the required integral to $\int d\Omega\nabla_\perp Y_{\ell m}$. Because $\nabla(r^\ell Y_{\ell m})$ satisfies Laplace's equation, use Example 3.5 to prove that this integral is zero unless $\ell = 1$.

11.14 No Magnetic Dipole Moment Show that a current density with vector potential $\mathbf{A}(\mathbf{r}) = f(\mathbf{r})\mathbf{r}$ has zero magnetic dipole moment.

11.15 A Spherical Superconductor A superconductor has the property that its interior has $\mathbf{B} = 0$ under all conditions. Let a sphere (radius R) of this kind sit in a uniform magnetic field $\mathbf{B}_0$.

(a) Place a fictitious point magnetic dipole $\mathbf{m}$ at the center of the sphere. Find $\mathbf{m}$ from the matching condition on the normal component of $\mathbf{B}$.
(b) In reality, the dipole field in part (a) is created by a current density $\mathbf{K}$ which appears on the surface of the sphere. Find $\mathbf{K}$ from the matching condition on the tangential component of $\mathbf{B}$.
(c) Confirm your answer in part (a) by computing the magnetic dipole moment associated with $\mathbf{K}$ from part (b).

11.16 Azimuthal Moments for Concentric Current Rings Compute the azimuthal multipole moments M_ℓ ($\ell = 1 - 8$) for a Helmholtz coil. Let the rings of the coil have radius R and carry a current I.

11.17 Dipole Field from Biot-Savart Expand the integrand of the Biot-Savart formula for the magnetic field and show that $\mathbf{B}(\mathbf{r})$ very far from a localized source of current is exactly the dipole magnetic field

$$\mathbf{B}(\mathbf{r}) = \frac{\mu_0}{4\pi}\frac{3(\mathbf{m}\cdot\hat{\mathbf{r}})\hat{\mathbf{r}} - \mathbf{m}}{r^3}.$$

11.18 Octupoles from Dipoles Let $\mathbf{m}_1$, $\mathbf{m}_2$, and $\mathbf{m}_3$ be three point dipoles.

(a) Find the constraints that must be imposed on the $\mathbf{m}_\alpha$ and their positions $\mathbf{r}_\alpha$ if the asymptotic magnetic field is octupolar.
(b) Show that one class of solutions places all the dipoles (with suitable magnitudes) on a line (separated by suitable distances) with their moments pointing parallel or anti-parallel to one another. Sketch an example.
(c) Show that another class of solutions has all the dipoles at the corners of a triangle and, if s_α is the length of the side opposite $\mathbf{m}_\alpha$, $m_1 : m_2 : m_2 = s_1 : s_2 : s_3$.

Hint: The quadratic form $x_i A_{ij} x_j = 0$ if $A_{ij} = -A_{ji}$.

11.19 Magnetic Multipoles from Electric Multipoles Consider a static and azimuthally symmetric charge distribution $\rho(\mathbf{r})$ which produces no electric field outside itself except a single, pure, spherical electric multipole field of order ℓ. Show that, when rotated rigidly about its symmetry axis with frequency ω, such a distribution generally produces a magnetic field outside of itself which is a superposition of *two* pure magnetic multipole fields, one of order $l + 1$ and one of order $\ell - 1$.

11.20 A Seven-Wire Circuit A steady current flows through seven wires as indicated in the figure. Find the asymptotic form of the vector potential using Cartesian coordinates and Cartesian unit vectors.

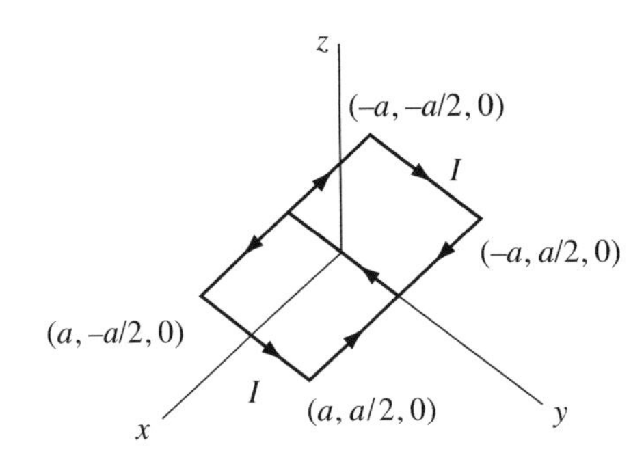

12 Magnetic Force and Energy

I had constructed an expression for the attraction of two infinitely small currents which was, in truth, only a hypothesis, but the simplest one that could be adopted and, consequently, the one that should be tried first.
André-Marie Ampère (1820)

12.1 Introduction

We stated at the beginning of Chapter 10 that a magnetic field $\mathbf{B}(\mathbf{r})$ exerts both a force $\mathbf{F}$ and a torque $\mathbf{N}$ on a current density $\mathbf{j}(\mathbf{r})$:

$$\mathbf{F} = \int d^3r\ \mathbf{j}(\mathbf{r}) \times \mathbf{B}(\mathbf{r}) \tag{12.1}$$

$$\mathbf{N} = \int d^3r\ \mathbf{r} \times [\mathbf{j}(\mathbf{r}) \times \mathbf{B}(\mathbf{r})]. \tag{12.2}$$

In this chapter, we study $\mathbf{F}$ and $\mathbf{N}$ in detail, including the motion of charged particles and the fact that no current density can exert a net force or torque on itelf. We will also study the *energy* associated with current densities and magnetic fields and discover that the total magnetic energy U_B of an isolated magnetic system differs from the potential energy $\hat{U}_B$ of a magnetic system with fixed currents. U_B is most useful for the calculation of equilibrium properties; $\hat{U}_B$ is most useful for the calculation of mechanical forces.[1]

A familiar special case of (12.1) confines a current I to a filamentary wire which traces out a closed curve C. The substitution $d^3r\,\mathbf{j}(\mathbf{r}) \to I d\boldsymbol{\ell}$ (see Section 9.3.1) gives[2]

$$\mathbf{F} = I \int_C d\boldsymbol{\ell} \times \mathbf{B}(\mathbf{r}). \tag{12.3}$$

We showed in Section 10.3.4 that the force density at a point $\mathbf{r}_S$ on a surface S which carries a current density $\mathbf{K}(\mathbf{r}_S)$ is $\mathbf{K}(\mathbf{r}_S) \times \mathbf{B}_{\rm avg}(\mathbf{r}_S)$, where $\mathbf{B}_{\rm avg}(\mathbf{r}_S)$ is the average of the magnetic fields infinitesimally close to $\mathbf{r}_S$ but on opposite sides of S. The net force on the entire surface S is therefore

$$\mathbf{F} = \int_S dS\ \mathbf{K}(\mathbf{r}_S) \times \mathbf{B}_{\rm avg}(\mathbf{r}_S). \tag{12.4}$$

Finally, we can pass to the limit of individual charges q_k which move with velocities $\boldsymbol{v}_k = \dot{\mathbf{r}}_k(t)$. The current density for that situation is

$$\mathbf{j}(\mathbf{r}, t) = \sum_{k=1}^{N} q_k \boldsymbol{v}_k \delta(\mathbf{r} - \mathbf{r}_k). \tag{12.5}$$

[1] The total energy of an isolated system of charged conductors is the same as its potential energy.
[2] It is straightforward to write down the corresponding special case of (12.2) here and below.

Substituting (12.5) into (12.1) recovers the familiar Lorentz force law,

$$\mathbf{F} = \sum_{k=1}^{n} q_k \boldsymbol{v}_k \times \mathbf{B}(\mathbf{r}_k). \tag{12.6}$$

12.2 Charged Particle Motion

Applications 11.4 and 11.5 of the previous chapter demonstrated the ability of quadrupole and axial magnetic fields to focus charged particles. This section examines the effect of magnetic fields on charged particles from a more general point of view. The main conclusions will play an important role in all our subsequent thinking about the subject.

12.2.1 The Lorentz Force Does No Work

The most important fact of life for charged particles in magnetic fields is that the Lorentz force (12.6) *performs zero mechanical work* on such particles. This is so because, when $\boldsymbol{v}$ produces a displacement of the particle $\delta\mathbf{r} = \boldsymbol{v}\delta t$ in time δt, the work done by (12.6) is

$$\delta W = \mathbf{F}\cdot\delta\mathbf{r} = q(\boldsymbol{v}\times\mathbf{B})\cdot\boldsymbol{v}\delta t = 0. \tag{12.7}$$

The implications of this fact for charged particle trajectories will appear throughout the chapter. Section 12.6 will focus on the implications of (12.7) for the calculation of magnetic energy.

The equation of motion for the center of mass of a particle with charge q and mass m in a magnetic field $\mathbf{B}(\mathbf{r})$ is[3]

$$m\frac{d\boldsymbol{v}}{dt} = q\boldsymbol{v}\times\mathbf{B}(\mathbf{r}). \tag{12.8}$$

The cross-product structure of the Lorentz force suggests that we decompose the velocity vector $\boldsymbol{v}$ into a piece parallel to the local magnetic field and a piece perpendicular to the local magnetic field:

$$\boldsymbol{v} = \boldsymbol{v}_{\parallel} + \boldsymbol{v}_{\perp}. \tag{12.9}$$

The zero-work condition (12.7) tells us that the speed υ and kinetic energy T are constants of the motion for a charged particle in a magnetic field. In the language of (12.9),

$$T = \tfrac{1}{2}m\upsilon^2 = \tfrac{1}{2}m\upsilon_{\parallel}^2 + \tfrac{1}{2}m\upsilon_{\perp}^2 = T_{\parallel} + T_{\perp} = const. \tag{12.10}$$

By facilitating the *exchange* of kinetic energy between $T_{\parallel}$ and $T_{\perp}$, a general magnetic field can change the direction of a particle velocity $\boldsymbol{v}$, but it cannot change its magnitude.

12.2.2 Helical Motion in a Uniform Magnetic Field

The equation of motion (12.8) is integrable when $\mathbf{B}(\mathbf{r})$ is a constant vector. If, say, $\mathbf{B} = B\hat{\mathbf{z}}$, the Lorentz force produces no acceleration along the z-direction and

$$\upsilon_z = const. \tag{12.11}$$

Otherwise, the *cyclotron frequency*,

$$\omega_c = \frac{qB}{m}, \tag{12.12}$$

[3] We are concerned here with non-relativistic ($\upsilon \ll c$) motion only. See Chapter 22 for relativistic motion.

appears when we write out the x- and y-components of (12.8):

$$\begin{aligned}\ddot{x} &= \omega_c \dot{y} \\ \ddot{y} &= -\omega_c \dot{x}.\end{aligned} \tag{12.13}$$

Direct substitution into (12.11) and (12.13) confirms that the particle trajectory is parameterized in terms of the constants x_0, y_0, R, and ϕ:

$$\begin{aligned}x(t) &= x_0 + R\cos(\omega_c t - \phi) \\ y(t) &= y_0 - R\sin(\omega_c t - \phi) \\ z(t) &= \upsilon_z t.\end{aligned} \tag{12.14}$$

The first two equations in (12.14) say that the particle executes circular motion in the x-y plane at the cyclotron frequency ω_c. They also fix the speed $\upsilon_\perp$ of this motion because

$$\upsilon_\perp^2 = \dot{x}^2 + \dot{y}^2 = \omega_c^2 R^2. \tag{12.15}$$

Using (12.12) and (12.15), the *cyclotron radius* R is

$$R = \frac{\upsilon_\perp}{|\omega_c|} = \frac{m\upsilon_\perp}{|q|B}. \tag{12.16}$$

Superposing this circular motion with $z(t)$ in (12.14) shows that the general trajectory of a charged particle in a constant magnetic field is a helix. The motion is "guided" by the magnetic field line that coincides with the symmetry axis of the helix.

12.2.3 Larmor's Theorem

In many cases, the effect of a magnetic field is simply to perturb the motion of a charged particle caused by a much larger force **F**. When the effect of **F** is to keep the particle in the vicinity of a fixed origin, we have the setting for a theorem due to Joseph Larmor. The theorem states that the effect of a (weak) magnetic field can be eliminated by transformation to a suitable rotating coordinate system.[4]

Let $(\mathbf{r}_0, \dot{\mathbf{r}}_0, \ddot{\mathbf{r}}_0)$ be the position, velocity, and acceleration of the particle as viewed from a reference frame which rotates with angular velocity $\boldsymbol{\omega}$ around the fixed origin. Textbooks of mechanics[5] show that the laboratory-frame velocity $\boldsymbol{\upsilon} = \dot{\mathbf{r}}$ and acceleration $\dot{\boldsymbol{\upsilon}} = \ddot{\mathbf{r}}$ that enter the equation of motion (12.8) are related to the rotating-frame quantities by

$$\begin{aligned}\dot{\mathbf{r}} &= \dot{\mathbf{r}}_0 + \boldsymbol{\omega} \times \mathbf{r}_0 \\ \ddot{\mathbf{r}} &= \ddot{\mathbf{r}}_0 + 2\boldsymbol{\omega} \times \dot{\mathbf{r}}_0 + \boldsymbol{\omega} \times (\boldsymbol{\omega} \times \mathbf{r}_0).\end{aligned} \tag{12.17}$$

Coriolis and centrifugal effects are responsible for the two extra terms in the acceleration formula. Substituting (12.17) into (12.8) and rearranging gives

$$m\ddot{\mathbf{r}}_0 = \mathbf{F} + \dot{\mathbf{r}}_0 \times (q\mathbf{B} + 2m\boldsymbol{\omega}) + (\boldsymbol{\omega} \times \mathbf{r}_0) \times (q\mathbf{B} + m\boldsymbol{\omega}). \tag{12.18}$$

The second term on the right side of (12.18) disappears if $\boldsymbol{\omega} = \boldsymbol{\omega}_L$, where the *classical Larmor frequency* is defined by

$$\boldsymbol{\omega}_L = -\frac{q}{2m}\mathbf{B}. \tag{12.19}$$

The resulting equation of motion in the rotating frame is

$$m\ddot{\mathbf{r}}_0 = \mathbf{F} + m\boldsymbol{\omega}_L \times (\boldsymbol{\omega}_L \times \mathbf{r}_0). \tag{12.20}$$

[4] Larmor had in mind a classical atom where a negative particle orbits a positive particle because of the Coulomb force **F** between them.

[5] See, for example, Section 4-9 of H. Goldstein, *Classical Mechanics* (Addison-Wesley, Cambridge, MA, 1950).

Therefore, if the magnetic field is weak enough that the final centrifugal term in (12.20) (which is second order in **B**) can be dropped, the particle motion as viewed in the rotating frame is exactly the same as the motion in the original inertial frame without the magnetic field. This is Larmor's theorem.

The centrifugal term in (12.20) cannot be dropped if $\mathbf{F} = 0$. Indeed, if **B** is constant, the action of this term in the Larmor rotating frame simply reproduces the free motion studied in Section 12.2.2 where the particle orbits in a plane perpendicular to **B** at the cyclotron frequency $\omega_c = 2\omega_L$.

Example 12.1 Find the trajectory of a particle with charge q and mass m released at rest from the origin in the crossed fields $\mathbf{B} = B\hat{\mathbf{z}}$ and $\mathbf{E} = E\hat{\mathbf{y}}$. Hint: Let $\boldsymbol{v}(t) = \boldsymbol{v}_0 + \boldsymbol{v}_1(t)$ and make a suitable choice for the constant $\boldsymbol{v}_0$.

Solution: The equation of motion to be analyzed is

$$m\frac{d\boldsymbol{v}}{dt} = q\left(\boldsymbol{v} \times \mathbf{B} + \mathbf{E}\right).$$

Since $\boldsymbol{v}_0$ is a constant, the suggested substitution gives

$$\frac{d\boldsymbol{v}_1}{dt} = \frac{q}{m}\boldsymbol{v}_1 \times \mathbf{B} + \frac{q}{m}\left(\boldsymbol{v}_0 \times \mathbf{B} + \mathbf{E}\right).$$

For the fields given, the last term vanishes if

$$\boldsymbol{v}_0 = \frac{E}{B}\hat{\mathbf{x}} = \frac{\mathbf{E} \times \mathbf{B}}{B^2}.$$

This contribution to the particle velocity is often called "$\mathbf{E} \times \mathbf{B}$ drift". The equation for $\boldsymbol{v}_1$ which remains is exactly (12.8). Therefore, the solution we seek is (12.14) with a term $v_0 t$ added to the expression for $x(t)$. The initial conditions fix all the constants. The trajectory traces out a cycloid in the x-y plane (Figure 12.1):

$$x(t) = \frac{E}{B}t - \frac{E}{B\omega_c}\sin(\omega_c t)$$

$$y(t) = \frac{E}{B\omega_c}[\cos(\omega_c t) - 1].$$

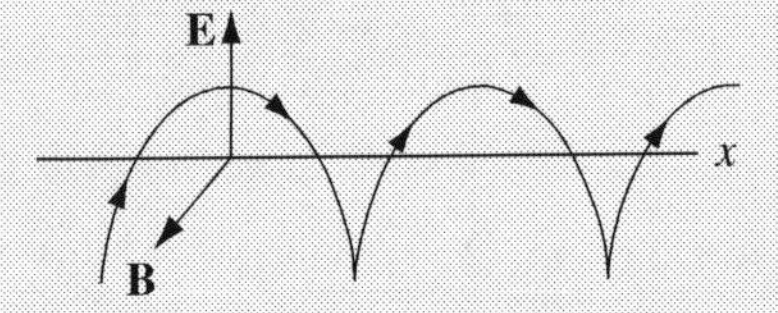

Figure 12.1: The trajectory of a charge $q > 0$ in the crossed fields $\mathbf{B} = B\hat{\mathbf{z}}$ and $\mathbf{E} = E\hat{\mathbf{y}}$. The particle begins at rest from the origin.

12.3 The Force between Steady Currents

The magnetic force between two steady current distributions obeys a vector-variant of Coulomb's law. We derive it by combining the Biot-Savart magnetic field (10.15) produced by a distribution $\mathbf{j}_1(\mathbf{r})$ with the force (12.1) that acts on a distribution $\mathbf{j}_2(\mathbf{r})$. This gives the force of 1 on 2 as

$$\mathbf{F}_2 = \frac{\mu_0}{4\pi}\int d^3r\, \mathbf{j}_2(\mathbf{r}) \times \int d^3r'\, \frac{\mathbf{j}_1(\mathbf{r}') \times (\mathbf{r} - \mathbf{r}')}{|\mathbf{r} - \mathbf{r}'|^3}. \tag{12.21}$$

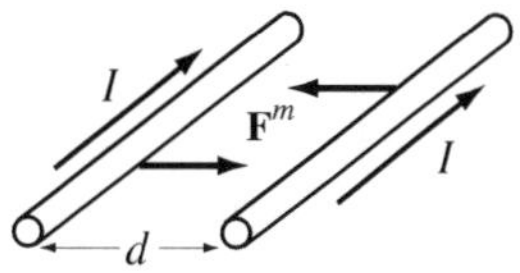

Figure 12.2: Two parallel wires which carry current I in the same direction attract one another.

Expanding the triple cross product temporarily complicates (12.21) to

$$\mathbf{F}_2 = \frac{\mu_0}{4\pi} \int d^3r \int d^3r' \, \frac{\mathbf{j}_1(\mathbf{r}')\,[\mathbf{j}_2(\mathbf{r}) \cdot (\mathbf{r} - \mathbf{r}')] - [\mathbf{j}_1(\mathbf{r}') \cdot \mathbf{j}_2(\mathbf{r})]\,(\mathbf{r} - \mathbf{r}')}{|\mathbf{r} - \mathbf{r}'|^3}. \tag{12.22}$$

We now perform a short side-calculation and note that

$$\int d^3r \, \nabla \cdot \left\{ \frac{\mathbf{j}_2(\mathbf{r})}{|\mathbf{r} - \mathbf{r}'|} \right\} = \int d^3r \frac{\nabla \cdot \mathbf{j}_2(\mathbf{r})}{|\mathbf{r} - \mathbf{r}'|} + \int d^3r \, \mathbf{j}_2(\mathbf{r}) \cdot \nabla \frac{1}{|\mathbf{r} - \mathbf{r}'|}. \tag{12.23}$$

If $\mathbf{j}_2(\mathbf{r}) \to 0$ faster than $1/r$ as $r \to \infty$, the divergence theorem guarantees that the left side of (12.23) is zero. The steady-current condition $\nabla \cdot \mathbf{j}_2(\mathbf{r}) = 0$ makes the first term on the right side of (12.23) zero also. This means that the integral on the far right side of (12.23) is zero as well. The latter is one of the integrals that appears in (12.22). Therefore, the magnetostatic force which $\mathbf{j}_1$ exerts on $\mathbf{j}_2$ is[6]

$$\mathbf{F}_2^m = -\frac{\mu_0}{4\pi} \int d^3r \int d^3r' \, \mathbf{j}_2(\mathbf{r}) \cdot \mathbf{j}_1(\mathbf{r}') \frac{\mathbf{r} - \mathbf{r}'}{|\mathbf{r} - \mathbf{r}'|^3}. \tag{12.24}$$

It is apparent from (12.24) that the magnetic force satisfies Newton's third law,

$$\mathbf{F}_1^m = -\mathbf{F}_2^m. \tag{12.25}$$

The special case of (12.25) when $\mathbf{j}_2(\mathbf{r}) = \mathbf{j}_1(\mathbf{r})$ gives $\mathbf{F}^m = 0$ and thus confirms the assertion made at the beginning of the chapter that no current distribution can exert a force on itself. Finally, as suggested earlier, the magnetic force law (12.24) is indeed similar to the electrostatic force between two charge densities,

$$\mathbf{F}_2^e = \frac{1}{4\pi\varepsilon_0} \int d^3r \int d^3r' \, \rho_2(\mathbf{r})\rho_1(\mathbf{r}') \frac{\mathbf{r} - \mathbf{r}'}{|\mathbf{r} - \mathbf{r}'|^3}. \tag{12.26}$$

For two infinitely long and straight wires which carry current I in the same direction, $\mathbf{j}_2(\mathbf{r}) \cdot \mathbf{j}_1(\mathbf{r}') d^3r d^3r' \to I^2 d\ell d\ell'$ in (12.24). This is the geometry used to define the ampere in the SI system of units (see Section 2.6). The corresponding electrostatic problem puts $\rho_2(\mathbf{r})\rho_1(\mathbf{r}') \to \lambda^2 d\ell d\ell'$ in (12.26) if each wire carries a linear charge density λ. With these substitutions, (12.24) and (12.26) differ only by a numerical constant. Hence, the variation of the force with the distance d between the wires is exactly the same in the two cases. However, as Figure 12.2 indicates, the sign difference between (12.24) and (12.26) tells us that parallel currents attract and anti-parallel currents repel.

More generally, if the moving charges in (12.24) are related to the static charges in (12.26) by $\mathbf{j} = \rho \boldsymbol{v}$, the fact that $\mu_0\epsilon_0 = 1/c^2$ tells us that

$$|\mathbf{F}^m| \sim \frac{v^2}{c^2} |\mathbf{F}^e|. \tag{12.27}$$

For electrons which carry a current of 1 A in a copper wire with a cross sectional area of 1 mm^2, the ratio of the drift velocity (see Section 9.3) to the speed of light is $v/c \sim 10^{-12}$. The magnetic force

[6] Ampère deduced the force (12.24) with the factor $\mathbf{j}_2(\mathbf{r}) \cdot \mathbf{j}_1(\mathbf{r}')$ replaced by $3\mathbf{j}_2(\mathbf{r}) \cdot (\mathbf{r} - \mathbf{r}')\mathbf{j}_1(\mathbf{r}') \cdot (\mathbf{r} - \mathbf{r}')/|\mathbf{r} - \mathbf{r}'|^2 - 2\mathbf{j}_2(\mathbf{r}) \cdot \mathbf{j}_1(\mathbf{r}')$. The two force formulae turn out to be identical. See the quotation at the beginning of the chapter.

would be utterly negligible compared to the electric force except for the immobile nuclei of the copper atoms which constitute the wire. These positive charges make the entire wire charge-neutral. As a result, the only electric force which occurs between current-carrying wires comes from the charge which develops on the surface of such wires (see Section 9.7.4).

The similarity between (12.24) and (12.26) for two parallel wires breaks down when $\mathbf{j}_1$ and $\mathbf{j}_2$ are two localized and widely separated current distributions. The factor $\mathbf{j}_2 \cdot \mathbf{j}_1$ inevitably has both positive and negative contributions which partly cancel. As a result, the asymptotic magnetic force between two current loops (i) depends sensitively on the relative orientation of the two distributions and (ii) falls off more rapidly with the loop separation than the inverse-square force between two distant charge distributions (see Section 12.4 and Example 12.7). A related cancellation produces *zero* net force when a compact current distribution is exposed to a uniform external field $\mathbf{B}_0$. This is most apparent for a closed current loop because $\oint d\boldsymbol{\ell} = 0$ implies that the magnetic force (12.3) vanishes:

$$\mathbf{F} = I\left[\oint d\boldsymbol{\ell}\right] \times \mathbf{B}_0 = 0. \tag{12.28}$$

The corresponding magnetic torque does not vanish (see Section 12.4.4).

Example 12.2 Figure 12.3 illustrates a set of $N \gg 1$ identical wires wound tightly around a cylinder of radius R with a pitch angle θ. Each wire carries a current I and winds around the cylinder once after a distance L. Find the force per unit area $\mathbf{f}$ and the critical pitch angle θ_c at which $\mathbf{f} = 0$. Discuss the cases $\theta < \theta_c$ and $\theta > \theta_c$ physically.

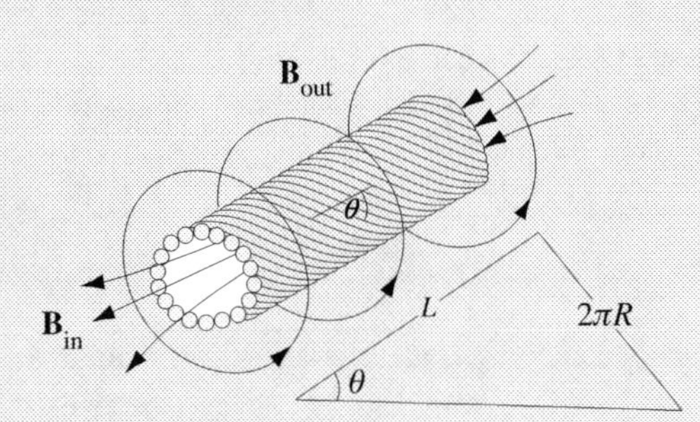

Figure 12.3: N identical, current-carrying wires wound around a cylinder of radius R with a pitch angle θ. The winding pattern repeats after a length L.

Solution: If $\theta = 0$, the surface current density is $\mathbf{K} = \hat{\mathbf{z}}NI/2\pi R$. The integral form of Ampère's law gives zero for the field inside the cylinder and $\mathbf{B}_{\text{wire}} = \hat{\boldsymbol{\phi}}\mu_0 NI/2\pi\rho$ for the field outside the cylinder. If $\theta = \pi/2$, the solenoidal surface current density is $\mathbf{K} = \hat{\boldsymbol{\phi}}NI/L$. According to Section 10.2.2, this source current produces zero field outside the cylinder and $\mathbf{B}_{\text{sol}} = \mu_0 K\hat{\mathbf{z}}$ inside the cylinder.

The current density for a general winding is

$$\mathbf{K}(\theta) = \hat{\mathbf{z}}\frac{NI}{2\pi R}\cos\theta + \hat{\boldsymbol{\phi}}\frac{NI}{L}\sin\theta.$$

Therefore, by superposition, the total magnetic field is

$$\mathbf{B}_{\text{in}} = \hat{\mathbf{z}}\frac{\mu_0 NI}{L}\sin\theta \qquad \mathbf{B}_{\text{out}} = \hat{\boldsymbol{\phi}}\frac{\mu_0 NI}{2\pi\rho}\cos\theta.$$

From (12.4), the force density on the surface is $\frac{1}{2}\mathbf{K} \times (\mathbf{B}_{\rm in} + \mathbf{B}_{\rm out})$:

$$\mathbf{f}(\theta) = \hat{\boldsymbol{\rho}}\frac{\mu_0}{2}\left[\left(\frac{NI\sin\theta}{L}\right)^2 - \left(\frac{NI\cos\theta}{2\pi R}\right)^2\right].$$

The critical pitch angle θ_c is defined by the condition $\mathbf{f}(\theta_c) = 0$. This establishes that $\tan\theta_c = L/2\pi R$. However, Figure (12.3) shows that the geometrical connection between the pitch angle and the winding repeat distance is $\cot\theta = L/2\pi R$. Therefore, $\mathbf{f} = 0$ when $\theta = \theta_c = \pi/4$ and $L = 2\pi R$.

In light of (12.24), the outward radial contribution to $\mathbf{f}$ (the $\sin^2\theta$ term) reflects the repulsion between anti-parallel current elements on opposite sides of a current loop. The inward radial piece of $\mathbf{f}$ (the $\cos^2\theta$ term) reflects the attraction between parallel current elements. The latter (called the *pinch effect* in plasma physics and beam physics) can generate quite large forces. This is illustrated by Figure 12.4, which shows a hollow-tube lightning rod crushed by the pinch effect. For the present problem, the repulsive solenoid-type force dominates when $\theta > \pi/4$. The attractive pinch effect force dominates when $\theta < \pi/4$.

Figure 12.4: A post-strike photograph of a lightning rod fashioned from a hollow copper tube. Figure from Pollock and Barraclough (1905). The authors estimated the current responsible for crushing the tube as $\sim 10^5$ A. This is the same order of magnitude as lightning-strike currents measured by present-day methods.

12.3.1 A Magnetic Work Paradox?

Let a current-carrying wire loop C be exposed to a spatially inhomogeneous magnetic field $\mathbf{B}(\mathbf{r})$ where (12.28) does not apply. When the loop has accelerated to a velocity $\boldsymbol{v}_c$, the work done by the magnetic force (12.3) in a small time interval δt is[7]

$$\delta W = \mathbf{F} \cdot \boldsymbol{v}_c \delta t. \tag{12.29}$$

This raises an apparent paradox: how is the work (12.29) consistent with the fact (see Section 12.2.1) that the Lorentz magnetic force does no work on moving charged particles? The answer lies in Faraday's law, $\nabla \times \mathbf{E} = -\partial \mathbf{B}/\partial t$. The kinetic energy gained by the loop as a whole is compensated by the kinetic energy lost by individual electrons slowed down by an electric field induced inside the moving loop. In other words, the current in the loop decreases.

To be quantitative, we evaluate the total Lorentz force which acts on the wire loop shown in Figure 12.5. The wire is composed of N mobile electrons (each with charge e) and a fixed arrangement of N ions (each with charge $-e$). The ions at instantaneous positions $\mathbf{R}_k$ provide most of the loop's mass and all of its rigidity. Hence, the velocity of every ion in the conductor is $\boldsymbol{v}_c$ during the time interval δt considered just above. The total work done on all the ions by the magnetic field is

$$\delta W_{\rm ion} = -e\sum_{k=1}^{N}[\boldsymbol{v}_c \times \mathbf{B}(\mathbf{R}_k)] \cdot \boldsymbol{v}_c \delta t = 0. \tag{12.30}$$

[7] Equation (12.28) tells us that the field must be non-uniform or the net force on the loop is zero.

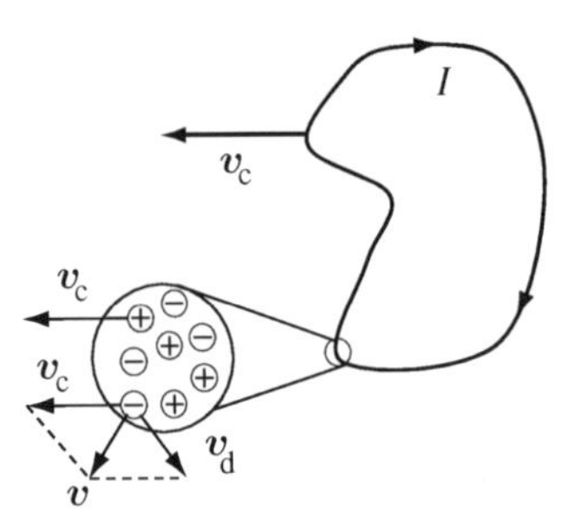

Figure 12.5: A wire loop composed of N positively charged ions and N negatively charged electrons. When the center-of-mass velocity of the loop, $\boldsymbol{v}_c$, is zero, the ions are immobile and the current-carrying electrons move with drift velocity $\boldsymbol{v}_d$.

Each electron in the loop has instantaneous position $\mathbf{r}_k$, contributes to the net current I through its drift velocity [see (9.16)] $\boldsymbol{v}_{d,k}$, and gains an additional velocity $\boldsymbol{v}_c$ due to collisions with the ions. Therefore, the total velocity of the kth electron in the loop is

$$\boldsymbol{v}_k = \boldsymbol{v}_c + \boldsymbol{v}_{d,k}. \tag{12.31}$$

The net work done on all the electrons by the Lorentz force is

$$\delta W_{\text{elec}} = e \sum_{k=1}^{N} [(\boldsymbol{v}_c + \boldsymbol{v}_{d,k}) \times \mathbf{B}(\mathbf{r}_k)] \cdot (\boldsymbol{v}_c + \boldsymbol{v}_{d,k})\delta t. \tag{12.32}$$

Two of the four terms in (12.32) vanish identically. The two terms that remain sum to zero by the cyclic property of the vector triple product:

$$\delta W_{\text{elec}} = e \sum_{k=1}^{N} [\boldsymbol{v}_{d,k} \times \mathbf{B}(\mathbf{r}_k)] \cdot \boldsymbol{v}_c \delta t + e \sum_{k=1}^{N} [\boldsymbol{v}_c \times \mathbf{B}(\mathbf{r}_k)] \cdot \boldsymbol{v}_{d,k} \delta t = 0. \tag{12.33}$$

This calculation confirms our expectation that the Lorentz force does zero work on a current-carrying wire. The resolution of the "paradox" comes when we realize that the first term in (12.33) is exactly the work (12.29) done to move the loop as a rigid whole. This is positive if the loop is accelerating. The second term in (12.33) is the work done on the drifting electrons by the effective electric field $\boldsymbol{v}_c \times \mathbf{B}$.[8] This is *negative* under the present conditions, which means that the drift velocity of each electron decreases. Quite apart from Joule heating (neglected here), the current in the wire decreases to conserve the total kinetic energy of the electrons. A battery (or some other source of energy) is needed to maintain the current when a force like (12.3) acts on a circuit.

12.4 The Magnetic Dipole

This section explores the magnetic dipole approximation to the force, torque, and energy of a current distribution in an external magnetic field $\mathbf{B}_{\text{ext}}(\mathbf{r})$. Our experience with magnetic multipoles (Chapter 11) suggests that this approximation should be very accurate if the field does not change very much over the size of the distribution.

[8] In Section 14.4, we will use the term "motional electromotive force" in connection with this phenomenon.

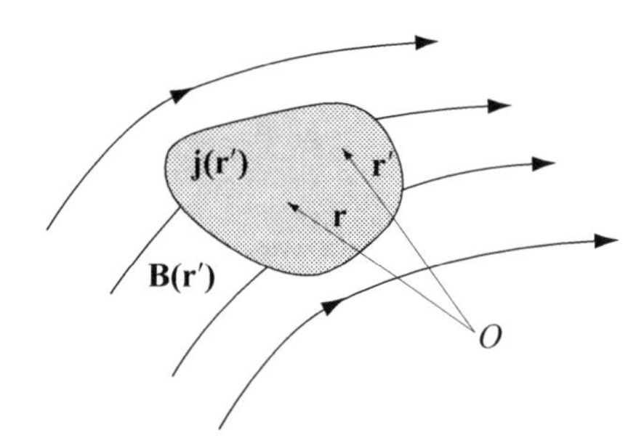

Figure 12.6: A current distribution $\mathbf{j}(\mathbf{r}')$ immersed in a magnetic field $\mathbf{B}(\mathbf{r}')$ which varies slowly in space.

12.4.1 The Dipole Force

Figure 12.6 shows a current distribution $\mathbf{j}(\mathbf{r}')$ in the presence of a magnetic field $\mathbf{B}(\mathbf{r}')$. The Lorentz force that the field exerts on the current is

$$\mathbf{F} = \int d^3r'\, \mathbf{j}(\mathbf{r}') \times \mathbf{B}(\mathbf{r}'). \tag{12.34}$$

If the magnitude and direction of the field do not change very much over the volume of the distribution, we can perform a Taylor expansion[9] of the integration variable $\mathbf{r}'$ around a reference point $\mathbf{r}$ and write

$$\mathbf{B}(\mathbf{r}') = \mathbf{B}(\mathbf{r}) + [(\mathbf{r}' - \mathbf{r}) \cdot \nabla]\mathbf{B}(\mathbf{r}) + \cdots. \tag{12.35}$$

Using (12.35) and the fact that $\int d^3r'\, \mathbf{j}(\mathbf{r}') = 0$ [See (10.5)], the first non-zero contribution to the magnetic force is

$$\mathbf{F} = \int d^3r'\; \mathbf{j}(\mathbf{r}') \times (\mathbf{r}' \cdot \nabla)\mathbf{B}(\mathbf{r}). \tag{12.36}$$

Two ingredients simplify the evaluation of (12.36). The first is the identity (11.11), which we repeat here for convenience:

$$\int d^3r'\; j_k r'_\ell = -\tfrac{1}{2}\epsilon_{k\ell i} \int d^3r'\, (\mathbf{r}' \times \mathbf{j})_i. \tag{12.37}$$

The second is the definition (11.12) of the magnetic dipole moment,

$$\mathbf{m} = \frac{1}{2}\int d^3r\;\; \mathbf{r} \times \mathbf{j}(\mathbf{r}). \tag{12.38}$$

Using (12.38) to rewrite (12.37), and substituting the latter into (12.36), the Levi-Cività identity $\epsilon_{\ell km}\epsilon_{\ell pi} = \delta_{pk}\delta_{im} - \delta_{ik}\delta_{pm}$ gives

$$\mathbf{F} = m_k \nabla B_k - \mathbf{m}\nabla \cdot \mathbf{B}. \tag{12.39}$$

It is always true that $\nabla \cdot \mathbf{B} = 0$. Therefore, because our result depends on the choice of the reference point $\mathbf{r}$, the force exerted on the current density in Figure 12.6 depends only on its magnetic moment:

$$\mathbf{F}(\mathbf{r}) = m_k \nabla B_k(\mathbf{r}). \tag{12.40}$$

Stern and Gerlach famously discovered the quantization of angular momentum in atoms [as reflected in their magnetic moments (see Section 11.2.2)] by exploiting the force (12.40) to deflect atoms in an inhomogeneous magnetic field.

[9] See Section 1.3.4.

When $\mathbf{m}$ is a constant vector, an alternative form of (12.40) is

$$\mathbf{F} = \nabla(\mathbf{m} \cdot \mathbf{B}). \tag{12.41}$$

When the sources of $\mathbf{B}$ are far away, so $\nabla \times \mathbf{B} = 0$, the identity $\nabla(\mathbf{m} \cdot \mathbf{B}) = \mathbf{m} \times (\nabla \times \mathbf{B}) + \mathbf{B} \times (\nabla \times \mathbf{m}) + (\mathbf{m} \cdot \nabla)\mathbf{B} + (\mathbf{B} \cdot \nabla)\mathbf{m}$ transforms (12.41) to[10]

$$\mathbf{F} = (\mathbf{m} \cdot \nabla)\mathbf{B}. \tag{12.42}$$

We emphasize that a magnetic field *gradient* is needed to generate a force. This is consistent with (12.28) for a current loop.

A quick application of (12.41) is to study the dipole approximation to the force on a current distribution $\mathbf{j}_2$ produced by a distant current distribution $\mathbf{j}_1$. If the latter is centered at the origin and has magnetic dipole moment $\mathbf{m}_1$, the distant field it produces is well approximated by the dipole field:

$$\mathbf{B}_1(\mathbf{r}) = \frac{\mu_0}{4\pi} \frac{3\hat{\mathbf{r}}(\hat{\mathbf{r}} \cdot \mathbf{m}_1) - \mathbf{m}_1}{r^3}. \tag{12.43}$$

This field varies slowly over the dimensions of $\mathbf{j}_2$ if the latter lies sufficient far from $\mathbf{j}_1$. In that case, (12.41) shows that the force $\mathbf{F}_2 = (\mathbf{m}_2 \cdot \nabla)\mathbf{B}_1(\mathbf{r})$ exerted by $\mathbf{j}_1$ on $\mathbf{j}_2$ is

$$\mathbf{F}_2 = \frac{3\mu_0}{4\pi r^4} \left[(\mathbf{m}_1 \cdot \mathbf{m}_2)\hat{\mathbf{r}} + (\mathbf{m}_1 \cdot \hat{\mathbf{r}})\mathbf{m}_2 + (\mathbf{m}_2 \cdot \hat{\mathbf{r}})\mathbf{m}_1 - 5(\mathbf{m}_1 \cdot \hat{\mathbf{r}})(\mathbf{m}_2 \cdot \hat{\mathbf{r}})\hat{\mathbf{r}} \right]. \tag{12.44}$$

The dipole-dipole force (12.44) is non-central and falls off as $1/r^4$. This confirms the remark preceding (12.28) that the magnetostatic force (12.24) between two currents falls off more rapidly with separation, and has a more complicated angular dependence, than the corresponding electrostatic Coulomb force between two distant charge distributions.

The four terms which appear in (12.44) are not unexpected: each represents one of the four distinct ways to combine the three vectors $\mathbf{m}_1$, $\mathbf{m}_2$, and $\hat{\mathbf{r}}$ into a single vector. The reader is urged to study the direction of the net force (12.44) for various orientations of these three vectors. These turn out to be consistent with the direction expected from superposition and Coulomb's law if each dipole moment were of electric type rather than magnetic type. The reason for this will become clear when we introduce the concept of "fictitious magnetic charge" in Chapter 13.

Example 12.3 Show that the force exerted on a point magnetic dipole by an arbitrary magnetic field $\mathbf{B}$ is (12.41).

Solution: The current density of a point magnetic dipole at $\mathbf{r}_0$ is $\mathbf{j}_D(\mathbf{r}) = -\mathbf{m} \times \nabla\delta(\mathbf{r} - \mathbf{r}_0)$ (see Section 11.2.3). Using this density, the force (12.1) becomes

$$\mathbf{F} = \int d^3r \; \mathbf{j}_D \times \mathbf{B} = \int d^3r \; \mathbf{B} \times [\mathbf{m} \times \nabla\delta(\mathbf{r} - \mathbf{r}_0)].$$

Expanding the double cross product and integrating by parts gives, first,

$$\mathbf{F} = \int d^3r \; [\mathbf{B} \cdot \nabla\delta(\mathbf{r} - \mathbf{r}_0)\mathbf{m} - (\mathbf{m} \cdot \mathbf{B})\nabla\delta(\mathbf{r} - \mathbf{r}_0)],$$

and then

$$\mathbf{F} = -\int d^3r \; \mathbf{m}\,\delta(\mathbf{r} - \mathbf{r}_0)\nabla \cdot \mathbf{B} + \int d^3r \; \delta(\mathbf{r} - \mathbf{r}_0)\nabla(\mathbf{m} \cdot \mathbf{B}).$$

Since $\nabla \cdot \mathbf{B} = 0$, we get the advertised result,

$$\mathbf{F} = \nabla_0[\mathbf{m} \cdot \mathbf{B}(\mathbf{r}_0)].$$

[10] Notice that (12.41) and (12.42) differ from (12.40) by a factor of 2 for an induced magnetic moment where $\mathbf{m} \propto \mathbf{B}$.

This confirms our expectation (based on the electric dipole results of Section 4.2.1) that the current density of a point magnetic dipole can always be used in place of the true density of a current distribution when $\mathbf{B}(\mathbf{r})$ varies slowly over the spatial extent of the current.

Application 12.1 The Magnetic Mirror

The magnetic dipole force (12.41) plays an important role when we generalize the problem of cyclotron motion (Section 12.2.2) to situations where the magnetic field is not strictly constant. Previously, we found that a charged particle spirals in a circular helix around any straight field line of a constant magnetic field. We show here that helical motion persists when the guiding field line changes direction—provided the change occurs slowly enough. Suppose v is the particle speed and $\omega_c = qB/m$ is the cyclotron frequency (12.12). A sufficient condition for a spiraling particle to follow a field line wherever it leads is that each Cartesian component of the field satisfies a condition of "adiabatic" change:

$$|B_k/\nabla B_k| \gg v/\omega_c. \tag{12.45}$$

The condition (12.45) also produces a characteristic *magnetic mirror effect*, where a particle spirals into a region of increasing field, reverses direction, and then spirals back out the way it came. Figure 12.7 illustrates the first part of this motion for an axially symmetric magnetic field with $B_z > 0$.

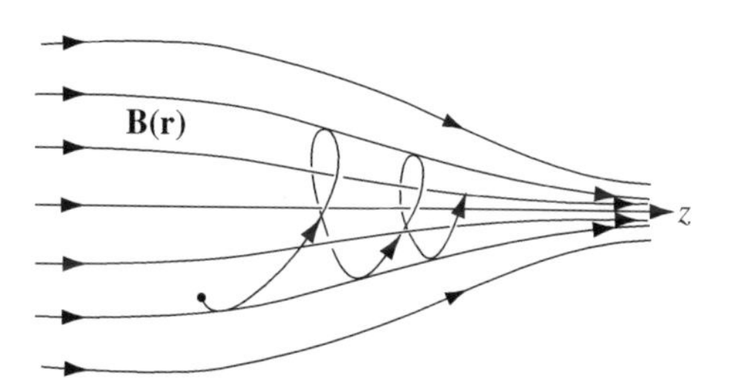

Figure 12.7: The trajectory of a point charge in an axially symmetric magnetic field $\mathbf{B}(\mathbf{r})$ whose magnitude increases as z increases (to the right).

To demonstrate these phenomena, we work in cylindrical coordinates and use (11.25) to deduce that the spiralling charge in Figure 12.7 produces an effective orbital magnetic moment,

$$\mathbf{m}_L = \frac{q}{2m}\mathbf{L} = \frac{q}{2m}(\boldsymbol{\rho} \times m\mathbf{v}_\perp) = -\frac{1}{2}|q|v_\perp \rho \hat{\mathbf{z}} = -\frac{T_\perp}{B_z}\hat{\mathbf{z}}, \tag{12.46}$$

where $T_\perp = \frac{1}{2}mv_\perp^2$ is the "perpendicular" kinetic energy [see (12.10)] and $\rho = mv_\perp/|q|B_z$ [see (12.16)]. Since $T_\perp > 0$, the magnetic moment (12.46) is *anti-parallel* to $\mathbf{B}_\parallel \equiv B_z\hat{\mathbf{z}}$ regardless of the sign of q. Equation (12.46) implies that the magnetic flux through the orbit of the particle is

$$\Phi = \pi R^2 B_z = -\frac{2\pi m}{q^2} m_L. \tag{12.47}$$

We now show that m_L (and therefore Φ) is a constant of the motion when the field changes adiabatically. This will explain why the particle trajectory in Figure 12.7 spirals along the surface of a single flux tube: $\nabla \cdot \mathbf{B} = 0$ guarantees that the same flux passes through any two cross sections of the tube.

The adiabatic condition (12.45) permits us to assume that the derivative $\partial B_z/\partial z$ is approximately constant. In that case, the Maxwell equation,

$$\nabla \cdot \mathbf{B} = 0 = \frac{1}{\rho}\frac{\partial}{\partial \rho}(\rho B_\rho) + \frac{\partial B_z}{\partial z}, \tag{12.48}$$

can be integrated and we find [cf. (11.118)]

$$B_\rho = -\frac{1}{2}\rho\frac{\partial B_z}{\partial z}. \tag{12.49}$$

Using (12.49) and (12.46), Newton's law in the z-direction is

$$m\frac{d\mathbf{v}_\parallel}{dt} = qv_\perp B_\rho\hat{\mathbf{z}} = -\frac{1}{2}q\rho v_\perp\frac{\partial B_z}{\partial z}\hat{\mathbf{z}} = \mathbf{m}_L\frac{\partial B_z}{\partial z}. \tag{12.50}$$

Therefore, the dot product of the first and last members of (12.50) with $\mathbf{v}_\parallel = (dz/dt)\hat{\mathbf{z}}$ implies that

$$\frac{d}{dt}\left(\frac{1}{2}mv_\parallel^2\right) = -m_L\frac{\partial B_z}{\partial t}. \tag{12.51}$$

Finally, because (12.10) ensures that $\dot{T}_\parallel = -\dot{T}_\perp$, (12.51) becomes

$$\frac{dT_\perp}{dt} = m_L\frac{\partial B_z}{\partial t}. \tag{12.52}$$

Equation (12.52) establishes that the orbital magnetic moment is a time-independent constant because, in addition, the equality of the first and last members of (12.46) tells us that

$$\frac{dT_\perp}{dt} = \frac{d}{dt}(m_L B_z). \tag{12.53}$$

More sophisticated treatments refer to our conclusion that $m_L = const.$ as a statement of the *adiabatic invariance* of the magnetic moment.

The magnetic mirror effect emerges from an analysis of the z-component of the force (12.41) exerted on the orbit regarded as a point dipole with moment (12.46):

$$F_z = (\mathbf{m}_L\cdot\nabla)B_z = -m_L\frac{\partial B_z}{\partial z}. \tag{12.54}$$

This force is always negative for the field configuration shown in Figure 12.7. Therefore, a particle that enters from the left with $v_z > 0$ has its longitudinal motion slowed, brought to a stop, and then reversed. The chirality of the spiral motion does not change during the reflection. Clearly, $T_\perp$ increases as T_z decreases (and vice versa) so their sum remains constant. As mentioned following (12.10), this transfer of kinetic energy between longitudinal motion and transverse motion is characteristic of Lorentz force dynamics where no work is done.

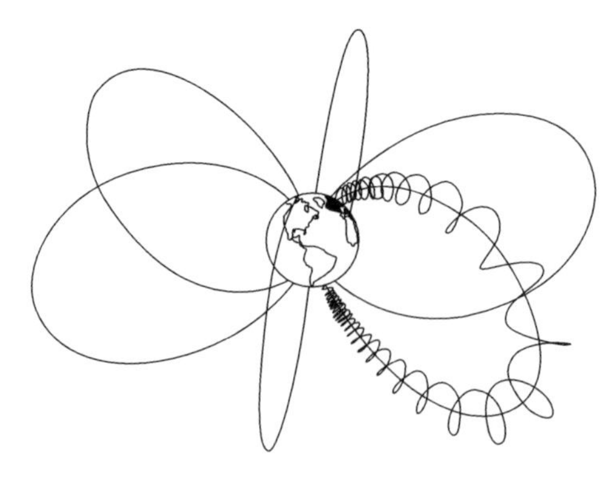

Figure 12.8: Schematic view of the dipolar magnetic field produced by the Earth and the trajectory of a charged particle guided by one field line. The particle spirals back and forth between two magnetic mirrors near the North and South poles.

A dramatic example of the mirror effect occurs when electrons and protons from the Sun are trapped by the dipole magnetic field of the Earth.[11] As the cartoon in Figure 12.8 indicates, the van Allen radiation belts are composed of particles that spiral back and forth between the mirror points of a "magnetic bottle" formed by the field lines of the dipole. A small fraction of these particles collide with and excite atmospheric atoms and molecules. The de-excitation of these species contributes to the dramatic optical phenomenon known as the *aurora*. ■

12.4.2 Dipole Potential Energy

The dipole force $\mathbf{F} = \nabla(\mathbf{m}\cdot\mathbf{B})$ in (12.41) manifestly has zero curl. This means that $\mathbf{F}$ is a conservative force (see Section 3.3.1) that can be derived from a potential energy. Evidently, $\mathbf{F} = -\nabla \hat{V}_B$, where $\hat{V}_B$ is the interaction potential energy function

$$\hat{V}_B(\mathbf{r}) = -\mathbf{m}\cdot\mathbf{B}(\mathbf{r}). \tag{12.55}$$

Superficially, (12.55) is similar to the potential energy of a point electric dipole in an external electric field. From (4.26), the latter is

$$V_E(\mathbf{r}) = -\mathbf{p}\cdot\mathbf{E}(\mathbf{r}). \tag{12.56}$$

The fact that we have written $\hat{V}_B$ in (12.55) rather than V_B is a first indication that magnetostatic potential energy is a more subtle subject than electrostatic potential energy. We will study this issue thoroughly in Section 12.7.

Application 12.2 Magnetic Trapping

Thomson's magnetostatic theorem (Section 10.1.4) states that $|\mathbf{B}(\mathbf{r})|$ has no local maximum in a current-free region of space. An example is

$$\mathbf{B}(\mathbf{r}) = (y\hat{\mathbf{x}} + x\hat{\mathbf{y}})B_1 + \hat{\mathbf{z}}B_0. \tag{12.57}$$

This field is divergence-free, curl-free, and unbounded in magnitude. On the other hand, $|\mathbf{B}(\mathbf{r})|$ takes its minimum value, B_0, at every point on the z-axis:

$$|\mathbf{B}(\mathbf{r})| = \sqrt{B_0^2 + B_1^2(x^2+y^2)}. \tag{12.58}$$

Because (12.58) has no local maximum, the potential energy (12.55) has no local minimum when $\mathbf{m}$ is parallel to the magnetic field (12.57). The force (12.41) merely pushes such a dipole toward regions of space where either $\mathbf{B} = const.$ or $\mathbf{j}(\mathbf{r}) \neq 0$. On the other hand, a dipole with $\mathbf{m}$ anti-parallel to the field (12.57) will be pushed to the z-axis and remain trapped there.

Magnetic traps were studied originally in connection with plasma-confinement schemes for nuclear fusion. More recent experiments exploit Thomson's theorem to study Bose-Einstein condensation and related phenomena in ultra-cold atomic ensembles. For atoms with total angular momentum $\mathbf{J}$ and magnetic moment $\mathbf{m} = \gamma\mathbf{J}$ (Section 11.2.2), static field trapping occurs for atoms prepared and maintained in quantum states where $\gamma\mathbf{J}$ is anti-parallel to the local magnetic field. Typically, trapping is desired at a single point in space, which requires a more complicated field than (12.57). A historically important example (for both plasma physics and cold-atom physics) is the Ioffe-Pritchard geometry shown in Figure 12.9. Application 11.4 showed that the four straight wires in Figure 12.9 produce the B_1 magnetic quadrupole term in (12.57) near the z-axis. Near $z = 0$ on the z-axis, the oppositely

[11] See the box at the end of Section 11.2.1.

wound Helmholtz coils (see Application 11.3) in the figure produce a field where $B_z = B_0 + B_2 z^2$. The sum of these fields leads to an absolute minimum of $|\mathbf{B}(\mathbf{r})|$ at the origin.

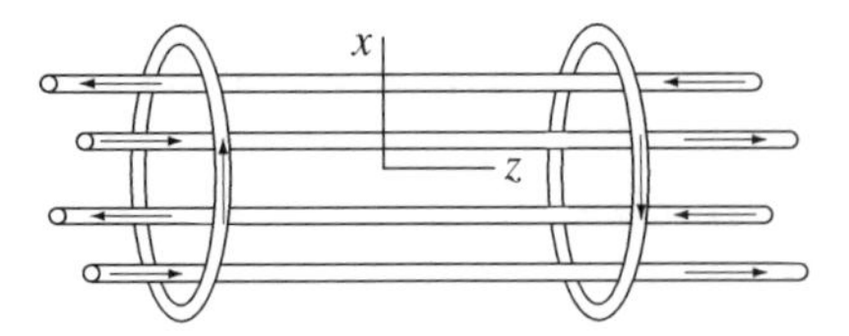

Figure 12.9: The Ioffe-Pritchard configuration produces a minimum of $|\mathbf{B}(\mathbf{r})|$ at its center. Arrows indicate that adjacent wires carry current in opposite directions.

12.4.3 The Dipole–Dipole Interaction Potential Energy

We compute the potential energy of interaction between a point dipole $\mathbf{m}_1$ at $\mathbf{r}_1$ and a point dipole $\mathbf{m}_2$ at $\mathbf{r}_2$ using (12.55) and the full point dipole magnetic field (11.33). If $\hat{\mathbf{n}}_{12} = \mathbf{r}_2 - \mathbf{r}_1/|\mathbf{r}_2 - \mathbf{r}_1|$, this energy is

$$\hat{V}_{12} = \frac{\mu_0}{4\pi}\left[\frac{\mathbf{m}_1 \cdot \mathbf{m}_2 - 3(\hat{\mathbf{n}}_{12} \cdot \mathbf{m}_1)(\hat{\mathbf{n}}_{12} \cdot \mathbf{m}_2)}{|\mathbf{r}_2 - \mathbf{r}_1|^3} - \frac{8\pi}{3}(\mathbf{m}_1 \cdot \mathbf{m}_2)\delta(\mathbf{r}_2 - \mathbf{r}_1)\right]. \tag{12.59}$$

For a collection of N point magnetic dipoles at non-overlapping positions $\mathbf{r}_i$, we sum terms like (12.59) and insert a factor of $\frac{1}{2}$ to avoid double-counting. If $\mathbf{B}(\mathbf{r}_i)$ is the magnetic field at the position of dipole i due to all the other dipoles, the total interaction potential energy stored by the collection is

$$\hat{V}_B = -\frac{1}{2}\sum_{i=1}^{N} \mathbf{m}_i \cdot \mathbf{B}(\mathbf{r}_i). \tag{12.60}$$

The non-singular term in (12.59) is the potential energy of *any* two current distributions separated by a distance $|\mathbf{r}_2 - \mathbf{r}_1|$ which is large compared to the spatial extent of either one. Thus, when $\mathbf{m}_1$, $\mathbf{m}_2$, and $\hat{\mathbf{n}}_{12}$ all point in the same direction, the force $\mathbf{F} = -\nabla \hat{V}_B$ [see (12.44)] is attractive because $\hat{V}_B$ is increasingly negative as the dipoles approach one another. This is the situation for two distant and *coaxial* current rings which carry current in the same direction. Conversely, (12.59) is increasingly positive when two distant and *coplanar* current rings approach one another with currents that flow in the same direction. This corresponds to a repulsive force between the two.

The delta function term in (12.59) cannot play a role in classical physics because two distinct current distributions cannot occupy the same point in space. On the other hand, the delta function applies directly to the theory of the hyperfine interaction in atoms where, say, $\mathbf{m}_1$ is the magnetic moment of the nucleus and $\mathbf{m}_2$ is the magnetic moment of an electron. There is a non-zero probability $|\psi_2(0)|^2$ that the electron and nucleus coincide in space so the delta function, or *Fermi contact term*, contributes to the hyperfine energy of the electron.

12.4.4 The Dipole Torque

Following Example 12.3, we use the point magnetic dipole current density $\mathbf{j}_D(\mathbf{r}) = -\mathbf{m} \times \nabla\delta(\mathbf{r} - \mathbf{r}_0)$ (see Section 11.2.3) to calculate the torque (12.2) exerted by a weakly inhomogeneous magnetic field

$\mathbf{B}(\mathbf{r})$ on a localized current density $\mathbf{j}(\mathbf{r})$. After expanding the double cross product, this is

$$\mathbf{N} = \int d^3r \; \mathbf{r} \times \{\mathbf{B} \cdot \nabla\delta(\mathbf{r} - \mathbf{r}_0)\mathbf{m} - (\mathbf{m} \cdot \mathbf{B})\nabla\delta(\mathbf{r} - \mathbf{r}_0)\} \,. \tag{12.61}$$

Integration by parts yields

$$N_i = \int d^3r \; \delta(\mathbf{r} - \mathbf{r}_0)\epsilon_{ijk} \{m_s \nabla_k - m_k \nabla_s\} r_j B_s. \tag{12.62}$$

The derivatives under the integral give four terms. One term vanishes because $\nabla \cdot \mathbf{B} = 0$. Another term vanishes because $\epsilon_{ijk}\delta_{kj} = 0$. The two non-zero terms give the desired result in terms of the magnetic force $\mathbf{F}(\mathbf{r}_0) = m_k \nabla B_k(\mathbf{r}_0)$ from (12.40):

$$\mathbf{N}(\mathbf{r}_0) = \mathbf{m} \times \mathbf{B}(\mathbf{r}_0) + \mathbf{r}_0 \times \mathbf{F}(\mathbf{r}_0). \tag{12.63}$$

The second term in (12.63) vanishes if we put the origin of coordinates at $\mathbf{r}_0$. The intrinsic magnetic torque that remains is

$$\mathbf{N}(\mathbf{r}) = \mathbf{m} \times \mathbf{B}(\mathbf{r}). \tag{12.64}$$

The torque (12.64) is equal to the time rate of change of the intrinsic angular momentum, $\mathbf{J}$, of the moment-carrying system. For a macroscopic system, $\mathbf{J}$ is dominated by the mechanical angular momentum $\tilde{\mathbf{I}} \cdot \boldsymbol{\omega}$, where $\mathbf{I}$ is the moment of inertia tensor and $\boldsymbol{\omega}$ is the angular velocity vector. In other words,

$$\frac{d\mathbf{J}}{dt} = \tilde{\mathbf{I}} \cdot \frac{d\boldsymbol{\omega}}{dt} = \mathbf{m} \times \mathbf{B}. \tag{12.65}$$

As a matter of principle, (12.65) rotates $\mathbf{m}$ into the direction of $\mathbf{B}$ to minimize the interaction potential energy (12.55). However, to solve (12.65) and learn the details, we must first refer $\mathbf{m}$ to the principal axes of $\tilde{\mathbf{I}}$. This is not difficult for a current-carrying wire ring where the ring geometry fixes the average trajectory of the moving charges. The case of a permanently magnetized compass needle is less straightforward because the *magnetic anisotropy* of the needle plays a key role.[12]

Application 12.3 Magnetic Bacteria in a Rotating Field

Some rod-shaped marine and freshwater bacteria contain magnetic particles which give the bacteria a net magnetic moment parallel to their length. The torque (12.64) causes them to continuously orient themselves parallel to the local direction of the Earth's magnetic field. This evolutionary strategy guides them downward toward the oxygen-poor environment they prefer. Apart from their rotational-orientational motion, the bacteria move at a constant speed v along their length because their self-propulsive force is balanced by the viscous drag of the water. In this Application, we show that trajectories like those shown in Figure 12.10 occur when the bacteria move in the plane of a rotating magnetic field.

[12] Magnetic anisotropy quantifies the shape- and material-dependent energy of an object with respect to the direction of its magnetic moment. See, e.g., K.H.J. Buschow and F.R. De Boer, *Physics of Magnetism and Magnetic Materials* (Kluwer, New York, 2004).

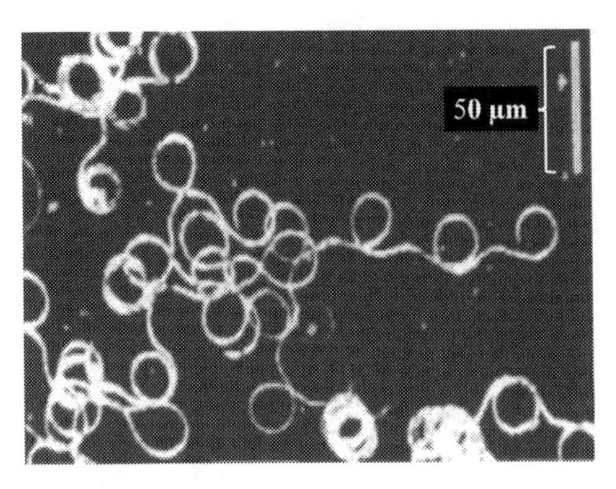

Figure 12.10: Swimming tracks of magnetic bacteria in a magnetic field which rotates in the plane of the diagram. Figure from Steinberger *et al.* (1994).

Let $\mathbf{B}(t)$ be a uniform magnetic field rotating in the x-y plane at frequency ω and let $\theta(t)$ be the angle between a bacterium's magnetic moment $\mathbf{m}$ and the x-axis of that plane. Thus,

$$\mathbf{B}(t) = B(\cos\omega t\hat{\mathbf{x}} + \sin\omega t\hat{\mathbf{y}}) \qquad \text{and} \qquad \mathbf{m} = m(\cos\theta\hat{\mathbf{x}} + \sin\theta\hat{\mathbf{y}}). \tag{12.66}$$

Our statement that the motion of the bacterium in water is non-inertial implies that its rotational equation of motion is $\Gamma\dot{\theta} = N$, where Γ is a drag coefficient and N is the driving torque. For our problem,

$$\mathbf{N} = \mathbf{m} \times \mathbf{B} = mB\sin(\omega t - \theta)\hat{\mathbf{z}}. \tag{12.67}$$

$\Omega = mB/\Gamma$ is a characteristic frequency, so the orientation of the bacterium is determined by the equation

$$\frac{d\theta}{dt} = \Omega\sin(\omega t - \theta). \tag{12.68}$$

The equation of motion (12.68) has both synchronous and non-synchronous solutions. The synchronous solutions have $\theta(t) = \omega t + K$, where K is a constant. Substituting this information back into (12.68) gives

$$\omega = \frac{d\theta}{dt} = \Omega\sin K, \tag{12.69}$$

which shows that these solutions are valid only at low frequency, when $\gamma = \Omega/\omega \geq 1$. When $\gamma < 1$, the bacterium cannot keep up with the rotating field and a direct integration of (12.68) gives

$$\theta(t) = \omega t - 2\tan^{-1}\left[\gamma + \sqrt{1-\gamma^2}\tan\left\{\tfrac{1}{2}\omega t\sqrt{1-\gamma^2}\right\}\right]. \tag{12.70}$$

To discover the bacterium's trajectory, we note that $\mathbf{v} = r\dot{\theta}\hat{\boldsymbol{\theta}}$ and $\mathbf{a} = -r\dot{\theta}^2\hat{\mathbf{r}} = -v\dot{\theta}\hat{\mathbf{r}}$ are the velocity and acceleration in polar coordinates for an object moving at constant speed. On the other hand, the centripetal acceleration is $\mathbf{a} = -(v^2/R)\hat{\mathbf{r}}$, where R is the instantaneous radius of curvature of the trajectory. Equating these two expressions for the acceleration implies that the time-dependent curvature satisfies

$$\frac{1}{R} = \frac{1}{v}\frac{d\theta}{dt} = \frac{\Omega}{v}\sin(\omega t - \theta). \tag{12.71}$$

For the synchronous solutions, (12.71) tells us that the bacterium (and its magnetic moment) remains adiabatically aligned with the magnetic field by moving in a circle of radius $R = v/|\omega|$. Figure 12.10 contains bacterium tracks of this kind.

For the non-synchronous solutions, (12.70) says that $\omega t - \theta$ is a periodic function of time. Using (12.71), we infer that $1/R$ is a periodic function of time also and that this curvature *changes sign* every time $\omega t - \theta$ passes through integer multiples of π. We conclude that, however simple or complex the trajectory may be, the bacterium always alternates between regions of positive and negative curvature as it moves. Figure 12.10 exhibits tracks of this kind also. ■

12.4.5 Larmor Precession for Microscopic Systems

For a microscopic system where $\mathbf{m} = \gamma \mathbf{J}$ is valid (see Section 11.2.2), we can use (12.64) and the classical torque equation $d\mathbf{J}/dt = \mathbf{N}$ to write an equation of motion for the magnetic moment:

$$\frac{d\mathbf{m}}{dt} = \gamma\, \mathbf{m} \times \mathbf{B}. \tag{12.72}$$

The scalar product of (12.72) with $\mathbf{m}$ and the scalar product of (12.72) with $\mathbf{B}$ show that

$$\mathbf{m} \cdot \frac{d\mathbf{m}}{dt} = 0 = \mathbf{B} \cdot \frac{d\mathbf{m}}{dt}. \tag{12.73}$$

The equality on the left side of (12.73) says that $\frac{1}{2}(d/dt)|\mathbf{m}|^2 = 0$ so $|\mathbf{m}|$ is a constant of the motion. The equality on the right side of (12.73) says that the component of $\mathbf{m}$ parallel to $\mathbf{B}$ is also a constant of the motion. This means that the magnetic moment does *not* turn into the direction of $\mathbf{B}$. Instead, it *precesses* around the direction of $\mathbf{B}$ at the *Larmor frequency* $\Omega_L = \gamma B$.[13] If we choose $\mathbf{B} = B\hat{\mathbf{z}}$, this conclusion follows from the fact that (12.72) is solved by

$$m_z = m\cos\theta \qquad m_y = m\sin\theta\cos\Omega_L t \qquad m_x = m\sin\theta\sin\Omega_L t. \tag{12.74}$$

The z-component of the moment is fixed while the vector $\mathbf{m}_\parallel = m_x\hat{\mathbf{x}} + m_y\hat{\mathbf{y}}$ rotates in the x-y plane at the Larmor frequency (Figure 12.11).

Nuclear magnetic resonance (NMR) and other magnetic resonance techniques supplement the static field $\mathbf{B}$ in (12.72) with a smaller, time-dependent magnetic field $\mathbf{B}_1(t) = \mathbf{B}_1\cos\omega t$ which is oriented perpendicular to $\mathbf{B}$. When the tunable frequency ω approaches Ω_L, the precession axis resonantly switches from $\hat{\mathbf{B}}$ to $\hat{\mathbf{B}}_1$ in an easily detectable manner. This phenomenon is widely exploited in many subfields of physics, chemistry, biology, and diagnostic medicine to identify the chemical environment of the precessing moments.

12.5 The Magnetic Stress Tensor

Following the electrostatic example of Section 3.7, the magnetic force density $\mathbf{j} \times \mathbf{B}$ in (12.1) can be expressed entirely in terms of the Cartesian components of the magnetic field. For this purpose, we assume that $\mathbf{j}$ is the sole source of $\mathbf{B}$ and restrict the volume of integration in (12.1) to a subset V of the total volume Ω occupied by the current. Using $\nabla \times \mathbf{B} = \mu_0\mathbf{j}$ and the Levi-Cività identity (1.39), it is not difficult to see that

$$\mathbf{j} \times \mathbf{B} = \mu_0^{-1}\left[(\mathbf{B}\cdot\nabla)\mathbf{B} - B_k\nabla B_k\right] = \mu_0^{-1}\left[(\mathbf{B}\cdot\nabla)\mathbf{B} - \tfrac{1}{2}\nabla B^2\right]. \tag{12.75}$$

[13] This conventional definition of the Larmor frequency Ω_L generalizes the "classical" Larmor frequency ω_L defined in (12.19). The latter applies to a purely orbital magnetic moment when the leftmost equation in (12.46) is relevant.

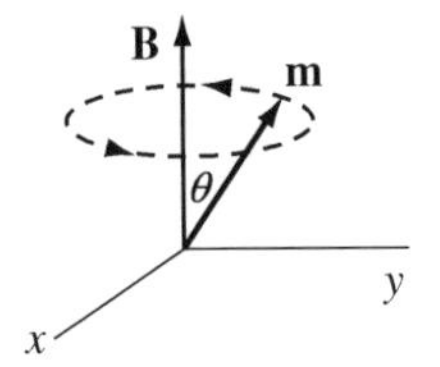

Figure 12.11: The magnetic moment $\mathbf{m} = \gamma \mathbf{J}$ precesses around the magnetic field $\mathbf{B} = B\hat{\mathbf{z}}$ at the Larmor frequency $\Omega_L = \gamma B$.

Moreover, because $\nabla \cdot \mathbf{B} = 0$, (12.75) is identical to

$$(\mathbf{j} \times \mathbf{B})_k = \mu_0^{-1} \partial_i \left\{ B_i B_k - \tfrac{1}{2} \delta_{ik} B^2 \right\} \equiv \partial_i T_{ik}(\mathbf{B}). \tag{12.76}$$

The last equality in (12.76) defines the Cartesian components of the *Maxwell magnetic stress tensor* $T_{ik}(\mathbf{B})$.

One virtue of (12.76) is that it expresses the Lorentz force density as a total divergence. Consequently, the total magnetic force on a volume V can be written as an integral over any surface S which bounds V:

$$F_k = \int_V d^3r \, (\mathbf{j} \times \mathbf{B})_k = \int_V d^3r \, \partial_i T_{ik}(\mathbf{B}) = \int_S dS \, \hat{n}_i \, T_{ik}(\mathbf{B}). \tag{12.77}$$

For computational purposes, it is often preferable to use the surface integral representation of the magnetic force,

$$\mathbf{F} = \frac{1}{\mu_0} \int_S dS \, [(\hat{\mathbf{n}} \cdot \mathbf{B})\mathbf{B} - \tfrac{1}{2} \hat{\mathbf{n}} (\mathbf{B} \cdot \mathbf{B})]. \tag{12.78}$$

It is important to appreciate that the integration surface S in (12.78) need *not* coincide with the physical boundary of $\mathbf{j}(\mathbf{r})$. If this is so, and S lies entirely in the vacuum, one can say that the magnetic force $\mathbf{F}$ is "transmitted" through empty space from the magnetic field at S to the enclosed current density $\mathbf{j}(\mathbf{r})$.

Despite the very different appearances of the electric and magnetic force densities in the Coulomb-Lorentz force law, $\mathbf{F} = \int d\mathbf{r} \, [\rho \mathbf{E} + \mathbf{j} \times \mathbf{B}]$, it is striking that the Cartesian components of the magnetic stress tensor,

$$T_{ij}(\mathbf{B}) = \frac{1}{\mu_0} [B_i B_j - \tfrac{1}{2} \delta_{ij} B^2], \tag{12.79}$$

have exactly the same structure as the components of the electric stress tensor (3.93),

$$T_{ij}(\mathbf{E}) = \varepsilon_0 [E_i E_j - \tfrac{1}{2} \delta_{ij} E^2]. \tag{12.80}$$

This is no accident. The similarity of (12.79) to (12.80) reflects a deep similarity between electric and magnetic fields which becomes most obvious when we analyze electrodynamics from the point of view of special relativity (Chapter 22).

12.5.1 Magnetic Tension and Magnetic Pressure

The two terms on the far right side of (12.75) are often interpreted as forces due to "magnetic tension" and "magnetic pressure", respectively. The pressure interpretation is reasonable because $-\hat{\mathbf{n}} B^2 / 2\mu_0$ is a force per unit area which pushes on every element of the enclosing surface S in (12.78). The use of the word "tension" in this context can be understood using the self-field of a long and straight current-carrying wire. Let the wire have a circular cross section with radius a and a uniform current density $\mathbf{j} = j\hat{\mathbf{z}}$.

The magnetic field produced by this wire has the form $\mathbf{B} = B_\phi(\rho)\hat{\boldsymbol{\phi}}$ where (see Example 10.5)

$$B_\phi(\rho) = \begin{cases} \frac{1}{2}\mu_0 j\rho & \rho < a, \\ \frac{1}{2}\mu_0 j a^2/\rho & \rho > a. \end{cases} \tag{12.81}$$

The first term of the force density (12.75) for the field (12.81) is[14]

$$\mathbf{f}_1 = \frac{1}{\mu_0}(\mathbf{B}\cdot\nabla)\mathbf{B} = \frac{B_\phi^2}{\mu_0\rho}\frac{\partial\hat{\boldsymbol{\phi}}}{\partial\phi} = -\frac{B_\phi^2}{\mu_0\rho}\hat{\boldsymbol{\rho}}. \tag{12.82}$$

This quantity points radially inward at every point in space. Now, if we ascribe to magnetic field lines the ability to "transmit" force to the wire, we can reproduce (12.82) if the radii of the circular lines which describe (12.81) tend to contract. But this is exactly what would happen if the field lines were elastic ropes in a state of tension: each closed circle of rope shrinks to relieve the stress.

Turning now to the magnetic pressure, the second term in (12.75) for this situation is

$$\mathbf{f}_2 = -\frac{1}{2\mu_0}\nabla(B^2) = -\frac{\partial}{\partial\rho}\left(\frac{B_\phi^2}{2\mu_0}\right)\hat{\boldsymbol{\rho}}. \tag{12.83}$$

Using (12.81), we see that $\mathbf{f}_1 + \mathbf{f}_2$ is zero outside the wire (as it must be because $\mathbf{j}\times\mathbf{B} = 0$ there). Inside the wire, the two terms add and the circular field lines exert a radially inward force. This is again the magnetic pinch effect (see Example 12.2).

Application 12.4 The Magnetic Virial Theorem

In the 1950s, physicists interested in generating very intense magnetic fields were mindful of the explosive consequences of passing a very large current through solenoidal coils of wire.[15] For that reason, they sought winding strategies that would produce zero magnetic force density on the surface of the coils. This turns out to be impossible unless the current source extends to infinity, as in Example 12.2. The proof goes as follows.

Let $\hat{\mathbf{f}}(\mathbf{r})$ denote the volume density of external forces of constraint needed to oppose the magnetic forces. Our aim is to show that $\hat{\mathbf{f}} \equiv 0$ is impossible for physical current distributions. We begin with (12.76) and write the condition of local mechanical equilibrium, $\hat{\mathbf{f}} + \mathbf{j}\times\mathbf{B} = 0$, in the form

$$\hat{f}_j + \partial_i T_{ij}(\mathbf{B}) = 0. \tag{12.84}$$

Now, multiply this formula by r_j and sum over the Cartesian index j. Since (12.79) gives the trace of the magnetic stress tensor as $\sum_i T_{ii}(\mathbf{B}) = -B^2/2\mu_0$, we find

$$r_j\hat{f}_j + \partial_i[r_j T_{ij}(\mathbf{B})] + B^2/2\mu_0 = 0. \tag{12.85}$$

Finally, integrate over a volume V and use the divergence theorem. The result is called the *magnetic virial theorem:*

$$\int_V d^3r\,\{r_j\hat{f}_j + B^2/2\mu_0\} = -\int_S dS\,\hat{n}_i r_j T_{ij}(\mathbf{B}). \tag{12.86}$$

[14] See Section 1.2.7 for the meaning of $(\mathbf{B}\cdot\nabla)\mathbf{B}$ in curvilinear components.

[15] See the first term in the final expression for the force density $\mathbf{f}$ in Example 12.2.

For our application, we choose the surface S far outside the finite-sized coils so the asymptotic $1/r^3$ behavior of the dipole magnetic field they produce makes the integral on the right side of (12.86) vanish. This gives

$$\int_V d^3r\, r_j\, \hat{f}_j = -\frac{1}{2\mu_0}\int_V d^3r\, |\mathbf{B}|^2. \tag{12.87}$$

The right side of this equation cannot be zero. Therefore, a non-zero external force density $\hat{\mathbf{f}}(\mathbf{r})$ *must* be present to maintain mechanical equilibrium. The best we can do is spread out the required force over a large region. ■

12.6 Magnetostatic Total Energy

By definition, the total magnetic energy U_B of an isolated current distribution $\mathbf{j}(\mathbf{r})$ is the total reversible work required to create $\mathbf{j}(\mathbf{r})$ and its associated magnetic field $\mathbf{B}(\mathbf{r})$. This quantity is important for many reasons, not least because U_B is a minimum when an isolated magnetostatic system is at equilibrium. If we assume a functional form for $\mathbf{j}(\mathbf{r})$ which depends on one or more variational parameters, minimizing U_B with respect to these parameters produces an approximation to the equilibrium current density.

To derive an expression for U_B, we focus on a closed loop C of filamentary wire and recall that the Lorentz force does no work on the mobile charges in the wire (Section 12.2.1). Therefore, the external work done *against* the Coulomb force on the moving charges is the only plausible candidate for the required work. In a time δt, the displacement of charge q_i with velocity $\boldsymbol{v}_i$ is $\delta\mathbf{r}_i = \boldsymbol{v}_i\delta t$. Therefore, since the current density of the moving charges is $\mathbf{j}(\mathbf{r}) = \sum q_i \boldsymbol{v}_i \delta(\mathbf{r} - \mathbf{r}_i)$,

$$\delta W_{\text{ext}} = -\sum_i q_i \mathbf{E}\cdot\boldsymbol{v}_i\delta t = -\int d^3r\, \mathbf{j}(\mathbf{r})\cdot\mathbf{E}(\mathbf{r})\delta t = -I\oint_C d\boldsymbol{\ell}\cdot\mathbf{E}\delta t. \tag{12.88}$$

An electric field arises in our problem because, no matter how slowly the current is increased, the accompanying change in magnetic field induces an electric field according to Faraday's law,

$$\nabla\times\mathbf{E} = -\frac{\partial\mathbf{B}}{\partial t}. \tag{12.89}$$

Using Stokes' theorem (Section 1.4.4) and (12.89), the work (12.88) can be rewritten as an integral over any surface S with C as its boundary:

$$\delta W_{\text{ext}} = -I\int_S d\mathbf{S}\cdot\nabla\times\mathbf{E}\,\delta t = I\int_S d\mathbf{S}\cdot\frac{\partial\mathbf{B}}{\partial t}\delta t = I\frac{d}{dt}\int_S d\mathbf{S}\cdot\mathbf{B}\,\delta t. \tag{12.90}$$

The transfer of the time derivative from inside to outside the integral in (12.90) is valid as long as the boundary of S is neither moved nor distorted in time δt. The rightmost integral in (12.90) is the magnetic flux Φ through the loop C [see (10.8)]. Therefore, since $(d\Phi/dt)\delta t = \delta\Phi$ is a change in magnetic flux, we conclude that the work increment done in time δt at a moment when the current in the loop is I is[16]

$$\delta W_{\text{ext}} = I\,\delta\Phi. \tag{12.91}$$

[16] This is the analog of the electrostatic work $\varphi\delta Q$ required to add charge δQ to an isolated conductor with potential φ. See Section 5.6.1.

By our definition, the work increment (12.91) is exactly the change in total magnetic energy,

$$\delta U_B = I\delta\Phi. \tag{12.92}$$

An expression for δU_B for a general current density $\mathbf{j}(\mathbf{r})$ follows from the definition of magnetic flux; namely, because

$$I\Phi_B = I\int_S d\mathbf{S}\cdot\mathbf{B} = I\int_S d\mathbf{S}\cdot\nabla\times\mathbf{A} = I\oint_C d\boldsymbol{\ell}\cdot\mathbf{A}, \tag{12.93}$$

the mnemonic device $Id\boldsymbol{\ell} \to \mathbf{j}\,d^3r$ implies that

$$\delta U_B = \int d^3r\;\mathbf{j}\cdot\delta\mathbf{A}. \tag{12.94}$$

12.6.1 Explicit Formulae for U_B

The energy increment (12.92) is all we need to derive an explicit expression for U_B. Our strategy is to quasistatically increase the current in a filamentary loop from zero to a final value of I.[17] At any intermediate stage, the current in the loop is $I(\lambda) = \lambda I$, where $0 \leq \lambda \leq 1$. The linearity of magnetostatics guarantees that the instantaneous flux through the loop is $\lambda\Phi$ and that the change in flux as λ increases infinitesimally is $\delta\Phi(\lambda) = (\delta\lambda)\Phi$. Therefore, the total energy required to raise the current in the loop from zero to I is

$$U_B = \int I(\lambda)\delta\Phi(\lambda) = \int_0^1 [\lambda I][(\delta\lambda)\Phi] = I\Phi\int_0^1 \delta\lambda\lambda = \frac{1}{2}I\Phi. \tag{12.95}$$

Generalizing to multiple loops and mimicking the steps which led from (12.92) to (12.94) shows that

$$U_B = \frac{1}{2}\sum_{k=1}^{N} I_k\Phi_k = \frac{1}{2}\int d^3r\;\mathbf{j}\cdot\mathbf{A}. \tag{12.96}$$

Two additional versions of the last member of (12.96) are worth knowing. The first follows from the Coulomb gauge formula (10.76) for the vector potential:

$$U_B = \frac{\mu_0}{8\pi}\int d^3r\int d^3r'\,\frac{\mathbf{j}(\mathbf{r})\cdot\mathbf{j}(\mathbf{r}')}{|\mathbf{r}-\mathbf{r}'|}. \tag{12.97}$$

The second follows from $\nabla\times\mathbf{B} = \mu_0\mathbf{j}$, the vector identity $\mathbf{A}\cdot(\nabla\times\mathbf{B}) = \mathbf{B}\cdot(\nabla\times\mathbf{A}) - \nabla\cdot(\mathbf{A}\times\mathbf{B})$, and the divergence theorem. The surface integral in the latter vanishes because $\mathbf{A}\times\mathbf{B}$ falls off faster than $1/r^2$ as $r\to\infty$ for a localized current density. Therefore,

$$U_B = \frac{1}{2\mu_0}\int d^3r\,|\mathbf{B}(\mathbf{r})|^2 \geq 0. \tag{12.98}$$

It is not apparent from (12.96) or (12.97), but (12.98) makes it clear that the total magnetic energy U_B is a positive definite quantity. This must be so or circulating currents would appear spontaneously everywhere.

[17] Compare this with the second method used in Section 3.6 to derive the total electrostatic energy U_E.

Application 12.5 The Solar Corona

Figure 12.12 is a not-to-scale cartoon which identifies the two outermost layers of the Sun: the opaque photosphere where sunlight is emitted and the nearly transparent corona. The magnetic field of the photosphere can be measured spectroscopically. The magnetic field of the corona is difficult to measure, but it can be calculated approximately from the known magnetic field of the photosphere. The calculation exploits the fact that the corona is in approximate quasistatic equilibrium. Therefore, its properties can be determined by minimizing the total magnetic energy U_B subject to two boundary conditions: (i) specified values for $\hat{\mathbf{n}} \cdot \mathbf{B}$ at the boundary between the corona and the photosphere, and (ii) $\mathbf{B} = 0$ at a boundary surface very far from the Sun. Under these conditions, we show below that the coronal volume V is characterized by $\mathbf{j}(\mathbf{r}) = 0$ and that magnetostatic potential theory determines the corona magnetic field $\mathbf{B}(\mathbf{r})$.

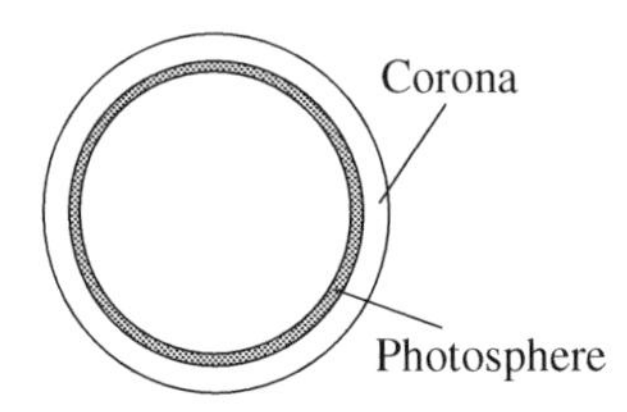

Figure 12.12: The two outermost layers of the Sun are the opaque photosphere (gray) and the nearly transparent corona (white). The layer thicknesses are not to scale.

The equilibrium state of the corona has the property that its energy is stationary (does not change) when the corona magnetic field changes from $\mathbf{B}$ to $\mathbf{B} + \delta\mathbf{B}$ subject to the boundary condition that $\hat{\mathbf{n}} \cdot \delta\mathbf{B} = 0$ on S. In other words, the equilibrium state is characterized by the condition $\delta U_B = 0$. To study the consequences of this constraint, we calculate the change in the total magnetic energy to first order in $\delta\mathbf{B}$ when $\mathbf{B}$ changes as indicated above. Using (12.98),

$$\delta U_B = \frac{1}{\mu_0} \int_V d^3r \; \mathbf{B} \cdot \delta\mathbf{B} = \frac{1}{\mu_0} \int_V d^3r \; \mathbf{B} \cdot (\nabla \times \delta\mathbf{A}). \tag{12.99}$$

The second equality in (12.99) expresses the variation in $\mathbf{B}$ in terms of a variation in the vector potential $\mathbf{A}$. This guarantees that the variation does not spoil the $\nabla \cdot \mathbf{B} = 0$ property of the magnetic field. Now, integrate (12.99) by parts, use the divergence theorem, and apply the identity $\hat{\mathbf{n}} \cdot (\delta\mathbf{A} \times \mathbf{B}) = \mathbf{B} \cdot (\hat{\mathbf{n}} \times \delta\mathbf{A})$ to get

$$\delta U_B = \frac{1}{\mu_0} \int_S dS \, \mathbf{B} \cdot (\hat{\mathbf{n}} \times \delta\mathbf{A}) + \frac{1}{\mu_0} \int_V d^3r \; (\nabla \times \mathbf{B}) \cdot \delta\mathbf{A}. \tag{12.100}$$

The boundary condition $\hat{\mathbf{n}} \cdot \delta\mathbf{B}|_S = 0$ is satisfied automatically if the variations of the vector potential satisfy $\hat{\mathbf{n}} \times \delta\mathbf{A}|_S = 0$.[18] Therefore, $\delta U_B = 0$ implies that $\nabla \times \mathbf{B} = 0$ in V. Ampère's law then gives $\mathbf{j}(\mathbf{r}) = 0$ in V as well. The energy of this state is a minimum because (12.98) is a positive definite functional of $\mathbf{B}(\mathbf{r})$.

We have just performed a constrained minimization of the magnetic total energy to learn that $\nabla \times \mathbf{B} = 0$ in the coronal volume V. Since $\nabla \cdot \mathbf{B} = 0$ everywhere, the results of Section 10.4 tell us that the magnetic field in V is determined by $\mathbf{B} = -\nabla\psi$ with $\nabla^2\psi = 0$. Moreover, $\hat{\mathbf{n}} \cdot \mathbf{B} = -\partial\psi/\partial n$

[18] If $\epsilon_{jik} n_i \delta A_k = 0$ on S, then $\epsilon_{ijk} n_i \delta A_k = 0$ on S as well. This implies, in turn, that $n_i \epsilon_{ijk} \partial_j \delta A_k = \hat{\mathbf{n}} \cdot \nabla \times \delta\mathbf{A} = \hat{\mathbf{n}} \cdot \delta\mathbf{B} = 0$ on S.

is specified on the boundary between the corona and the photosphere and we can let $\psi \to 0$ on the outer boundary of the corona at infinity. Therefore, the uniqueness theorem for Laplace's equation with mixed boundary conditions (Section 7.3) guarantees that $\psi(\mathbf{r})$ is unique and computable in V using the methods of potential theory. Since $\mathbf{B} = -\nabla\psi$, we have achieved our goal of deducing the magnetic field in the corona without direct observations of that quantity. ■

12.6.2 Interaction Total Energy and Reciprocity

We derive the interaction energy between two current densities $\mathbf{j}_1(\mathbf{r})$ and $\mathbf{j}_2(\mathbf{r})$ by inserting $\mathbf{j}(\mathbf{r}) = \mathbf{j}_1(\mathbf{r}) + \mathbf{j}_2(\mathbf{r})$ into (12.97). The result takes the form

$$U_B[\mathbf{j}_1 + \mathbf{j}_2] = U_B[\mathbf{j}_1] + U_B[\mathbf{j}_2] + V_B[\mathbf{j}_1, \mathbf{j}_2], \tag{12.101}$$

where $U_B[\mathbf{j}_1]$ and $U_B[\mathbf{j}_2]$ are the total magnetic energies of $\mathbf{j}_1(\mathbf{r})$ and $\mathbf{j}_2(\mathbf{r})$ in isolation and $V_B[\mathbf{j}_1, \mathbf{j}_2]$ is the energy of interaction between the two distributions:

$$V_B[\mathbf{j}_1, \mathbf{j}_2] = \frac{\mu_0}{4\pi} \int d^3r \int d^3r' \, \frac{\mathbf{j}_1(\mathbf{r}') \cdot \mathbf{j}_2(\mathbf{r})}{|\mathbf{r} - \mathbf{r}'|}. \tag{12.102}$$

Reversing the logic that led from (12.96) to (12.97) permits us to write the interaction energy (12.102) in two equivalent forms and set them equal. This produces a magnetostatic *reciprocity relation* analogous to Green's reciprocity relation (Section 3.5.2):

$$V_B = \int d^3r \, \mathbf{j}_1(\mathbf{r}) \cdot \mathbf{A}_2(\mathbf{r}) = \int d^3r \, \mathbf{j}_2(\mathbf{r}) \cdot \mathbf{A}_1(\mathbf{r}). \tag{12.103}$$

In light of (12.96), we can apply (12.103) twice to a given set of N current loops, once when the currents and fluxes are (I_k, Φ_k) and once again when the currents and fluxes are (I'_k, Φ'_k). The resulting filamentary loop version of the reciprocity relation is

$$\sum_{k=1}^{N} I_k \Phi'_k = \sum_{k=1}^{N} I'_k \Phi_k. \tag{12.104}$$

We leave it as an exercise for the reader to show that, if a field $\mathbf{B}^{\text{ext}}(\mathbf{r})$ (whose source is unspecified) produces a magnetic flux Φ_k^{ext} through the kth filamentary loop, the total energy changes from (12.96) to

$$U_B = \frac{1}{2} \sum_{k=1}^{N} I_k \Phi_k + \sum_{k=1}^{N} I_k \Phi_k^{\text{ext}}. \tag{12.105}$$

12.6.3 V_B is Not a Potential Energy for Constant Currents

It is natural to suppose that the interaction energy V_B defined in (12.102) is related to the mechanical force (12.24) between $\mathbf{j}_1(\mathbf{r})$ and $\mathbf{j}_2(\mathbf{r})$. Unfortunately, V_B is increasingly *positive* when two parallel currents are brought closer together while experiment and (12.24) show that two parallel currents attract one another. This means that $\mathbf{F} = -\nabla V_B$ produces the wrong sign (although the correct magnitude) for the Lorentz force between two current distributions. This is unlike the electrostatic case, where $\mathbf{F} = -\nabla V_E$ produces the correct Coulomb force between two charge distributions.

The origin of this problem is that V_B was defined for an *isolated* system of currents while our force scenario imagines two wires with *specified* currents. These wires cannot be isolated because Section 12.3.1 showed that the current in an isolated loop of wire changes when the force (12.24) acts. To maintain a fixed current, a battery or generator must be connected to the loop. We will see below that taking proper account of the energy required to maintain fixed currents has the effect of fixing the sign problem. More precisely, the interaction *potential* energy $\hat{V}_B$ needed to compute forces for problems with fixed currents turns out to be the *negative* of the interaction *total* energy V_B. The corresponding problem did not arise in electrostatics because the total energy is also the potential energy for problems with fixed charge.

A clear hint that U_B and V_B are not ideally suited to discuss situations of imposed currents is that neither is a natural function of the currents we are interested in fixing. Instead, we show in the next section that U_B and V_B are natural functions of the *magnetic fluxes* which pass through each loop. In Section 12.7, we derive the potential energy function which resolves the issues raised here.

12.6.4 U_B is a Natural Function of Magnetic Flux

In this section, we guess and then prove that the total magnetic energy U_B is a natural function of the vector potential $\mathbf{A}$ or, equivalently, of the magnetic fluxes, Φ_1, Φ_2, ..., Φ_N associated with a collection of N filamentary current loops. We guess, in addition, that U_B is a natural function of the center-of-mass coordinate of each loop, $\mathbf{R}_k$. An explicit statement which identifies the complete set of independent variables is

$$U_B = U_B(\mathbf{A}, \mathbf{R}_1, \ldots, \mathbf{R}_N) \qquad \text{or} \qquad U_B(\Phi_1, \ldots, \Phi_N, \mathbf{R}_1, \ldots, \mathbf{R}_N). \tag{12.106}$$

Now, the expressions for U_B derived earlier in this section depend only on the instantaneous configuration of currents and not on the details of the assembly process. In the language of thermodynamics, this means that U_B is a function of state and dU_B is a perfect differential. Therefore, the ordinary rules of calculus applied to, say, the right member of (12.106) tell us that

$$dU_B = \sum_{k=1}^{N} \left(\frac{\partial U_B}{\partial \Phi_k}\right)_{\Phi',\mathbf{R}} d\Phi_k + \sum_{k=1}^{N} \left(\frac{\partial U_B}{\partial \mathbf{R}_k}\right)_{\Phi,\mathbf{R}'} \cdot d\mathbf{R}_k. \tag{12.107}$$

The primes in (12.107) indicate variables which are held constant except for the one varied to form the derivative.

The essential step which confirms the correctness of (12.106) is that (12.107) agrees exactly with (12.92) and with the usual mechanical connection between force and energy if

$$I_k = \left(\frac{\partial U_B}{\partial \Phi_k}\right)_{\Phi',\mathbf{R}'} \qquad \text{and} \qquad \mathbf{F}_k = -\left(\frac{\partial U_B}{\partial \mathbf{R}_k}\right)_{\Phi,\mathbf{R}'}. \tag{12.108}$$

It is not immediately obvious how to "hold the magnetic flux constant" as the partial derivatives in (12.108) require. As a practical matter, this is possible only for current loops fabricated from superconducting wires with zero electrical resistance. This follows from an application of Faraday's law to such a loop in the form [see (14.42)]

$$IR = -\frac{d}{dt}\int d\mathbf{S} \cdot \mathbf{B} = \frac{d\Phi}{dt}. \tag{12.109}$$

Since $R = 0$ for a superconductor, and the current I in the ring can be non-zero, (12.109) requires that $\Phi = const.$ for the ring.

The Quantum Hall Effect

The left panel below shows a current I flowing through a thin film in the presence of a perpendicular magnetic field $\mathbf{B}$. The Lorentz force, $\mathbf{F} = q\mathbf{v} \times \mathbf{B}$ pushes the current-carrying electrons in the direction transverse to I and $\mathbf{B}$. Eventually, the Lorentz force is balanced by the Coulomb force from the transverse electric field created by the displaced electrons. The Hall voltage V_H is a measure of that electric field.

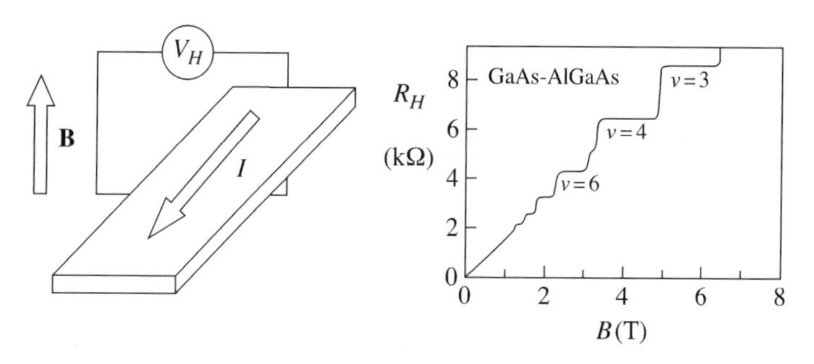

The right panel above shows the Hall resistance, $R_H = V_H/I$, measured by Paalanen, Tsui, and Gossard (1982) for a two-dimensional electron gas trapped at the interface between a GaAs crystal and a GaAlAs crystal. When the magnetic field is large, the data clearly show horizontal steps where $R_H(B)$ takes the quantized values

$$R_H = \frac{1}{\nu}\frac{h}{e^2} = \frac{25.8\ \text{k}\Omega}{\nu}, \qquad \nu = 1, 2, 3, \ldots$$

In 1981, Laughlin presented a "thought experiment" to explain the steps in the Hall resistance using the left member of (12.108), gauge invariance, and the solutions of the Schrödinger equation for a two-dimensional system of free electrons in a perpendicular magnetic field. Extremely briefly, Laughlin showed that the smallest increase in magnetic flux that can be accommodated by the current-carrying two-dimensional electron gas is $\Delta\Phi = h/e$. When this occurs, the system experiences a change in energy $\Delta U_B = \nu e V_H$ due to the effective transfer of ν electrons from one side of the film to the other. The integer ν is the number of highly degenerate quantum energy levels which are fully occupied by electrons at the magnetic field of interest. Therefore, using (12.108),

$$R_H = \frac{V_H}{I} = V_H \frac{\Delta\Phi}{\Delta U_B} = V_H \frac{h/e}{\nu e V_H} = \frac{1}{\nu}\frac{h}{e^2}.$$

It is worth noting that this theoretical analysis holds V_H constant and uses (12.108) to compute I as B changes. In a typical experiment, one holds I constant and measures V_H as B changes.

12.7 Magnetostatic Potential Energy

In this section, we derive an energy function $\hat{U}_B$ whose interaction part $\hat{V}_B$ functions as a potential energy for magnetostatic problems with fixed currents. The ability to fix $\mathbf{j}(\mathbf{r})$ or the set $\{I_k\}$ implies that the latter will be the natural independent variables for $\hat{U}_B$. Having proved in the previous section that $\mathbf{A}(\mathbf{r})$ or the set $\{\Phi_k\}$ are the natural independent variables for the total magnetic energy U_B, our experience with electrostatics (Section 5.6.2) suggests that a Legendre transformation of U_B from flux to current should produce $\hat{U}_B$. The proof will be to confirm that $\mathbf{F} = -\nabla \hat{V}_B$ is the correct magnetic force between fixed currents.

12.7.1 $\hat{U}_B$ is a Natural Function of Current

We define $\hat{U}_B$ using the Legendre transformation,

$$\hat{U}_B = U_B - \sum_{k=1}^{N} I_k \Phi_k. \tag{12.110}$$

By direct calculation,

$$d\hat{U}_B = dU_B - \sum_{k=1}^{N} I_k d\Phi_k - \sum_{k=1}^{N} \Phi_k dI_k. \tag{12.111}$$

Substituting dU_B from (12.92) into (12.111) gives

$$d\hat{U}_B = -\sum_{k=1}^{N} \Phi_k dI_k - \sum_{k=1}^{N} \mathbf{F}_k \cdot d\mathbf{R}_k. \tag{12.112}$$

We have added the second term on the right side of (12.112) to account for the fact that quasistatic equilibrium requires that a work $\delta W_F = -\mathbf{F} \cdot \delta\mathbf{R}$ be performed against the mechanical force $\mathbf{F}$ in (12.3) to prevent acceleration of the center of mass $\mathbf{R}$ of a current loop.

Equation (12.112) tells us that the natural variables of $\hat{U}_B$ are the currents I_k and coordinates $\mathbf{R}_k$:

$$\hat{U}_B = \hat{U}_B(I_1, \ldots, I_N, \mathbf{R}_1, \ldots, \mathbf{R}_N). \tag{12.113}$$

This is so because, if (12.113) is correct, the ordinary rules of calculus tell us that

$$d\hat{U}_B = \sum_{k=1}^{N} \left(\frac{\partial \hat{U}_B}{\partial I_k} \right)_{I',\mathbf{R}} dI_k + \sum_{k=1}^{N} \left(\frac{\partial \hat{U}_B}{\partial \mathbf{R}_k} \right)_{I,\mathbf{R}'} \cdot d\mathbf{R}_k. \tag{12.114}$$

Comparing (12.114) with (12.112) shows that

$$\mathbf{F}_k = -\left(\frac{\partial \hat{U}_B}{\partial \mathbf{R}_k} \right)_{I,\mathbf{R}'} \qquad \text{and} \qquad \Phi_k = -\left(\frac{\partial \hat{U}_B}{\partial I_k} \right)_{I',\mathbf{R}}. \tag{12.115}$$

A similar argument shows that the work done against the magnetic torque (12.2) which tends to rotate a loop with current I_k by an amount $\delta\alpha_k$ around an axis $\hat{\mathbf{n}}_k$ is $-\mathbf{N}_k \cdot \hat{\mathbf{n}}_k \delta\alpha_k$. Hence, the torque can be calculated from

$$\mathbf{N}_k \cdot \hat{\mathbf{n}}_k = -\left(\frac{\partial \hat{U}_B}{\partial \alpha_k} \right)_{I,\mathbf{R}}. \tag{12.116}$$

12.7.2 Explicit Formulae for $\hat{U}_B$

Using (12.96) to evaluate (12.110) shows that

$$\hat{U}_B = -\frac{1}{2} \sum_{k=1}^{N} I_k \Phi_k = -U_B. \tag{12.117}$$

The magnetostatic potential energy is (numerically) the negative of magnetostatic total energy. It is important to be aware that if (12.117) is used for $\hat{U}_B$, (12.113) implies that the fluxes $\{\Phi_k\}$ must be expressed in terms of the currents $\{I_k\}$ before any of the derivatives in (12.115) or (12.116) are evaluated. We will do this explicitly in Section 12.8.2.

Alternative expressions for $\hat{U}_B$ follow straightforwardly by mimicking the manipulations performed in Section 12.6.1 for U_B. We find

$$\hat{U}_B = -\frac{1}{2} \int d^3r\, \mathbf{j} \cdot \mathbf{A} = -\frac{1}{2\mu_0} \int d^3r\, |\mathbf{B}(\mathbf{r})|^2 \tag{12.118}$$

and

$$\hat{U}_B = -\frac{\mu_0}{8\pi} \int d^3r \int d^3r' \frac{\mathbf{j}(\mathbf{r}) \cdot \mathbf{j}(\mathbf{r}')}{|\mathbf{r} - \mathbf{r}'|}. \tag{12.119}$$

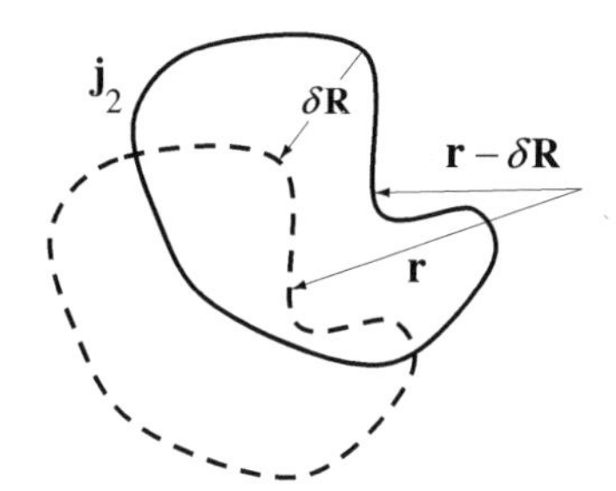

Figure 12.13: Illustration that a rigid shift $\delta\mathbf{R}$ of the current density $\mathbf{j}_2$ produces a change in the current density at a point $\mathbf{r}$ of $\delta\mathbf{j}_2(\mathbf{r}) = \mathbf{j}_2(\mathbf{r}-\delta\mathbf{R}) - \mathbf{j}_2(\mathbf{r})$.

Similarly [cf. (12.102)], the interaction potential energy between two distinct current distributions is

$$\hat{V}_B = -\frac{\mu_0}{4\pi}\int d^3r \int d^3r' \frac{\mathbf{j}_1(\mathbf{r}')\cdot\mathbf{j}_2(\mathbf{r})}{|\mathbf{r}-\mathbf{r}'|}. \tag{12.120}$$

12.7.3 Force from Variation of Potential Energy

We confirm in this section that the force

$$\mathbf{F} = -\left.\frac{\delta\hat{U}_B}{\delta\mathbf{R}}\right|_I \tag{12.121}$$

on the left side of (12.115) correctly reproduces the magnetic force (12.24) which $\mathbf{j}_1(\mathbf{r})$ exerts on $\mathbf{j}_2(\mathbf{r})$ when $\delta\mathbf{R}$ is a virtual displacement of $\mathbf{j}_2(\mathbf{r})$. More precisely, we replace $\hat{U}_B$ by $\hat{V}_B$ in (12.121) to compute the derivative because the displacement $\delta\mathbf{R}$ produces no change in the self-energies $\hat{U}_B[\mathbf{j}_1]$ and $\hat{U}_B[\mathbf{j}_2]$ in the analog of (12.101). Figure 12.13 shows that a rigid shift of $\mathbf{j}_2(\mathbf{r})$ by $\delta\mathbf{R}$ is equivalent to a change[19]

$$\delta\mathbf{j}_2(\mathbf{r}) = \mathbf{j}_2(\mathbf{r}-\delta\mathbf{R}) - \mathbf{j}_2(\mathbf{r}) = -(\delta\mathbf{s}\cdot\nabla)\mathbf{j}_2(\mathbf{r}). \tag{12.122}$$

Inserting (12.122) into (12.120) and integrating by parts gives the change in interaction energy,

$$\delta\hat{V}_B = -\frac{\mu_0}{4\pi}\int d^3r \int d^3r'\, \mathbf{j}_2(\mathbf{r})\cdot\mathbf{j}_1(\mathbf{r}')\,\delta\mathbf{R}\cdot\nabla\frac{1}{|\mathbf{r}-\mathbf{r}'|}. \tag{12.123}$$

Writing (12.123) in the form $\delta\hat{V}_B = -\mathbf{F}_2\cdot\delta\mathbf{R}$ gives the desired result,

$$\mathbf{F}_2 = -\frac{\mu_0}{4\pi}\int d^3r \int d^3r'\, \mathbf{j}_2(\mathbf{r})\cdot\mathbf{j}_1(\mathbf{r}')\frac{\mathbf{r}-\mathbf{r}'}{|\mathbf{r}-\mathbf{r}'|^3}. \tag{12.124}$$

This confirms our expectation that the Lorentz force produces mechanical effects which tend to reduce the magnetic potential energy of a system of fixed currents. On the other hand, (12.117) implies that the very same forces tend to *increase* the magnetic total energy of such systems. This differs from electrostatics, where the Coulomb force reduces the total energy and the potential energy (which are equal) for a system of fixed charges. The physical reason for this difference is the subject of the next section.

[19] See the footnote to (3.65) in Section 3.5.1, where we displaced a charge distribution in a similar manner.

Example 12.4 Use (12.120) to confirm the formula (12.55) for the interaction potential energy between a point magnetic dipole at the origin and a current source which produces a magnetic field $\mathbf{B}(\mathbf{r})$.

Solution: We showed in Section 11.2.3 that the fixed current density of a point magnetic dipole at the origin is $\mathbf{j} = \nabla \times [\mathbf{m}\delta(\mathbf{r})]$. Therefore,

$$\hat{V}_B = -\int d^3r\, \mathbf{j} \cdot \mathbf{A} = -\int d^3r\, \mathbf{A} \cdot \nabla \times [\mathbf{m}\delta(\mathbf{r})] = -\epsilon_{k\ell p} \int d^3r\, A_k \partial_\ell [m_p \delta(\mathbf{r})].$$

The magnetic dipole moment component m_p is a constant. Therefore, integration by parts gives

$$\hat{V}_B = \epsilon_{k\ell p} m_p \int d^3r\, \delta(\mathbf{r}) \partial_\ell A_k = -\int d^3r\, \delta(\mathbf{r}) \mathbf{m} \cdot \nabla \times \mathbf{A}.$$

Therefore, in agreement with (12.55), the potential energy of interaction is

$$\hat{V}_B = -\mathbf{m} \cdot \mathbf{B}(0).$$

Example 12.5 A current-carrying wire is bent into a closed rectangle with dimensions $\ell \times d$. The current I is uniformly distributed across the wire's cross sectional area πa^2. Assume that $a \ll d \ll \ell$. (a) Use an energy method to estimate the force between the two long sides of the rectangle. (b) Do the same to find the force between the two short sides of the rectangle.

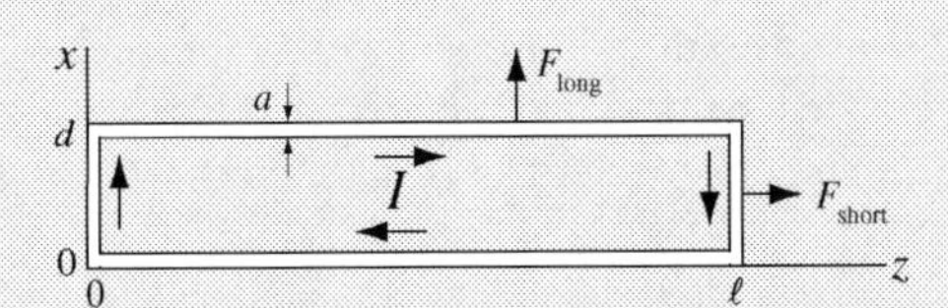

Figure 12.14: A current-carrying wire bent into a rectangle with $a \ll d \ll \ell$.

Solution: (a) We compute the force on the long segment from the derivative of the potential energy $\hat{V}_B$ with respect to the length d. For the stated geometry, we can neglect the energy associated with the two short wire segments and approximate the energy per unit length of the two long segments by the energy per unit length of two straight, infinite wires which carry current in opposite directions.

Let $\mathbf{j}_1 = \hat{\mathbf{z}} I/\pi a^2$ be the current density in the upper wire in Figure 12.14. The interaction potential energy with $\mathbf{j}_1$ involves the vector potential $\mathbf{A}_2$ produced by the lower wire, which carries current in the $-z$-direction. Since $d \gg a$, we can use $\rho \approx d$ in the results of Example 10.5 and approximate $\mathbf{A}_2$ at points on the upper wire as

$$\mathbf{A}_2 \approx \frac{\mu_0 I}{2\pi} \ln(d/a)\hat{\mathbf{z}}.$$

Therefore, if $\Omega = \pi a^2 \ell$ is the volume of the upper wire,

$$\hat{V}_B = -\int_\Omega d^3r\, \mathbf{j}_1 \cdot \mathbf{A}_2 \approx -\frac{\mu_0 I^2}{2\pi^2 a^2} \int_\Omega d^3r\, \ln(d/a) = -\frac{\mu_0 I^2 \ell}{2\pi} \ln(d/a).$$

This potential energy becomes more negative as d increases. Therefore, the upper wire segment is repelled from the lower wire segment as indicated in Figure 12.14:

$$\mathbf{F}_{\text{long}} = -\frac{\partial \hat{V}_B}{\partial d}\hat{\mathbf{x}} = \frac{\mu_0 I^2}{2\pi}\frac{\ell}{d}\hat{\mathbf{x}}.$$

The reader can check that this force agrees with a calculation based on $\mathbf{F} = I\int d\boldsymbol{\ell} \times \mathbf{B}$ and the magnetic field produced by an infinite and straight wire.

(b) The predicted force on the right-hand short segment is more subtle. It does *not* result from a direct interaction with the left-hand short segment. Rather, it emerges as an expansive "tension" in each long segment:

$$\mathbf{F}_{\text{short}} = -\frac{\partial \hat{V}_B}{\partial \ell}\hat{\mathbf{z}} \simeq \frac{\mu_0 I^2}{2\pi}\ln(d/a)\hat{\mathbf{z}}.$$

This example shows how the energy (per unit length) of an infinitely long wire can be used to find the force at the end of a finite wire. Example 6.8 used the same approach with infinitely large plates to find the force needed to insert a dielectric into a parallel-plate capacitor.

12.7.4 The Meaning of the Minus Sign in $\hat{U}_B$

The minus sign in front of $\hat{V}_B$ in (12.120) guarantees that this expression is increasingly *negative* when two parallel filaments are brought closer together. Because two parallel currents attract, this confirms our interpretation of $\hat{V}_B$ as a magnetic potential energy for fixed currents. The force $\mathbf{F} = -\nabla\hat{V}_B$ and potential energy $\hat{V}_B = -\mathbf{m}\cdot\mathbf{B}(0)$ from Example 12.4 similary reproduce (12.41) because a specified current density implies a specified magnetic moment and vice versa.

Our task is to understand why $\hat{U}_B(I) = -U_B(\Phi)$ in (12.117). A similar issue arose in Section 5.6.2. There, we found that $\hat{U}_E(\varphi) = -U_E(Q)$ for a conductor where either the charge Q or the potential φ could be held constant. The energy that accounted for the change of sign there came from the work done by external batteries to hold the potentials fixed. Here, it is similarly the work done by external generators to keep the current fixed which provides the energy needed to change the algebraic sign.

Figure 12.15 shows two loops held at fixed current by generators I_1 and I_2. The vector $\mathbf{F}_2$ indicates the mechanical force exerted on loop I_2 by loop I_1. When loop I_2 suffers a rigid displacement $\delta\mathbf{R}_2$, the flux through loop I_1 changes from $\Phi_1^{(0)}$ to Φ_1 and the flux through loop I_2 changes from $\Phi_2^{(0)}$ to Φ_2. Our strategy is to determine the potential energy change $\Delta\hat{U}_B$ during this process from a computation of the change which occurs in the total system energy, ΔE_{sys}:

$$\Delta E_{\text{sys}} = \mathbf{F}_2\cdot\delta\mathbf{R}_2 + \Delta U_B + \Delta E_{\text{gen}} = 0. \tag{12.125}$$

In (12.125), $\mathbf{F}_2\cdot\delta\mathbf{R}_2$ is the work done by $\mathbf{F}_2$, ΔU_B is the change in the total magnetic energy associated with the loops, and ΔE_{gen} is the change in the total energy of the generators. The zero on the far right side of (12.125) is a statement of the conservation of energy for the isolated system shown in Figure 12.15.

The fixed current relation on the left side of (12.115) permits us to eliminate $\mathbf{F}_2\cdot\delta\mathbf{R}_2$ in (12.125) in favor of $\Delta\hat{U}_B$. Consequently,

$$\Delta\hat{U}_B = \Delta U_B + \Delta E_{\text{gen}}. \tag{12.126}$$

The change in magnetic energy follows directly from (12.96):

$$\Delta U_B = U_B(\Phi_1, \Phi_2) - U_B(\Phi_1^{(0)}, \Phi_2^{(0)}) = \frac{1}{2}\sum_{k=1}^{N} I_k(\Phi_k - \Phi_k^{(0)}). \tag{12.127}$$

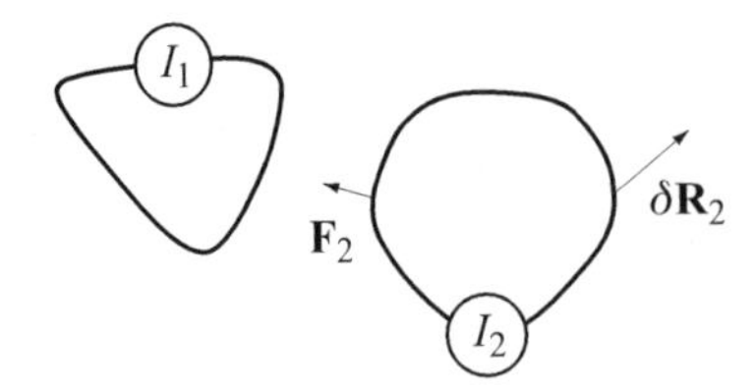

Figure 12.15: Two loops attached to generators which keep the current in each loop fixed.

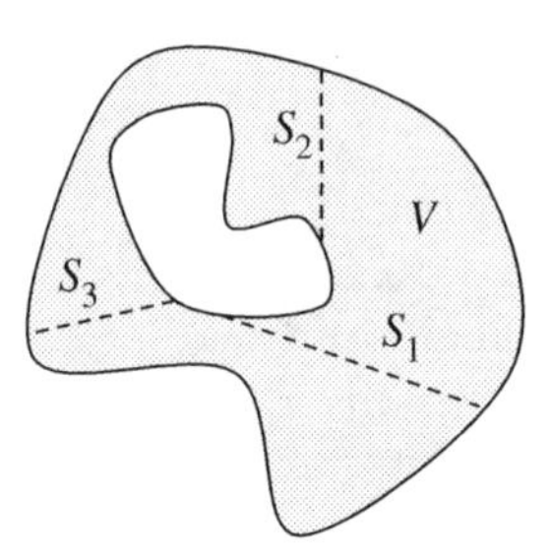

Figure 12.16: If $\nabla \cdot \mathbf{j} = 0$ in the shaded region, a unique current I flows through every dashed cross section.

As for ΔE_{gen}, the generators are external to the loops, so the work done by them is calculable using (12.91). The positive work done to maintain the current in the rings reduces the total energy of the generators, so

$$\Delta E_{\text{gen}} = -W_{\text{gen}} = -\sum_{k=1}^{N} \int_{\Phi_k^{(0)}}^{\Phi_k} I_k d\Phi_k = -\sum_{k=1}^{N} I_k(\Phi_k - \Phi_k^{(0)}). \tag{12.128}$$

Substituting (12.128) and (12.127) into (12.126) gives

$$\Delta \hat{U}_B = -\frac{1}{2}\sum_{k=1}^{N} I_k(\Phi_k - \Phi_k^{(0)}). \tag{12.129}$$

Hence, in agreement with (12.117),

$$\hat{U}_B = -\frac{1}{2}\sum_{k=1}^{N} I_k \Phi_k. \tag{12.130}$$

12.8 Inductance

The magnetostatic constraint $\nabla \cdot \mathbf{j}(\mathbf{r}) = 0$ implies that a unique current I can be defined for any current distribution with the topology of a torus. Figure 12.16 is an example where $\mathbf{j}(\mathbf{r}_S) \cdot \hat{\mathbf{n}} = 0$ everywhere on the closed surface of the shaded region and the planes S_1, S_2, and S_3 are generalized cross sections. Let us check this claim for the volume V bounded by S_1 and S_2 and the walls of the shaded region. Since no current flows through the latter,

$$0 = \int_V d^3r \; \nabla \cdot \mathbf{j} = \int_{S_1} d\mathbf{S} \cdot \mathbf{j} + \int_{S_2} d\mathbf{S} \cdot \mathbf{j} = I_1 + I_2. \tag{12.131}$$

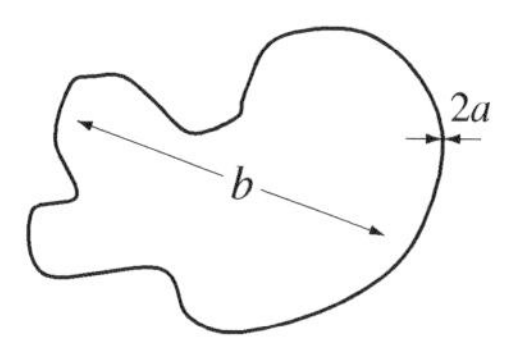

Figure 12.17: A cylindrical current-carrying wire (radius a) bent into a closed loop with characteristic size b. The current density j in the wire is uniform.

We conclude that I_1 and I_2 are equal and opposite currents. This is correct because one enters V and the other exits V. All other choices of cross section lead to exactly the same value of current.

The argument above permits us to uniquely define the *self-inductance* L of a steady current distribution $\mathbf{j}(\mathbf{r})$ from its total magnetic energy as

$$U_B = \tfrac{1}{2} L I^2. \tag{12.132}$$

Comparing (12.132) with (12.95) shows that the self-inductance relates the magnetic flux through a loop to the magnitude of the current in the loop which produces that flux:

$$\Phi = LI. \tag{12.133}$$

Comparing (12.132) with (12.98) shows that

$$L = \frac{1}{\mu_0 I^2} \int d^3r\, |\mathbf{B}(\mathbf{r})|^2 = \frac{1}{I^2} \int d^3r\, \mathbf{j}(\mathbf{r}) \cdot \mathbf{A}(\mathbf{r}). \tag{12.134}$$

Dimensional analysis tells us that inductance has dimensions of $\mu_0 \times$ length. An elementary example is a cylindrical solenoid of radius R and length ℓ tightly wound with N turns of a wire carrying a current. When $\ell \gg R$, the field strength is zero outside the solenoid and is nearly uniform with magnitude $B = \mu_0 IN/\ell$ inside the solenoid. Therefore, the self-inductance defined by the leftmost equality in (12.134) is

$$L = \mu_0 \pi N^2 \frac{R^2}{\ell}. \tag{12.135}$$

This example notwithstanding, it is often quite challenging to calculate self-inductances analytically.

12.8.1 The Self-Inductance of a Wire Loop

Figure 12.17 shows a cylindrical wire with radius a bent into a closed loop with a characteristic size b (Figure 12.17). Our aim is to show that

$$L = \mu_0 b\, [k_1 \ln(b/a) + k_2] \qquad a \ll b, \tag{12.136}$$

where k_1 and k_2 are numerical constants which depend on the details of the geometry. When $a \ll b$, the current density $j = I/\pi a^2$ is uniform and it is reasonable to use the results of Example 10.5 for the magnetic field inside and outside a straight wire. We estimate the contribution to the integral (12.134) from the volume inside the wire as

$$L_{\text{in}} = \frac{1}{\mu_0 I^2} 2\pi b \int_0^a dr\, r\, B_{\text{in}}^2 = \frac{\mu_0}{8\pi} b. \tag{12.137}$$

This gives the second term in the square brackets in (12.136). A similar estimate of the dominant contribution to the field energy outside the wire uses the wire radius a and the size scale b as lower

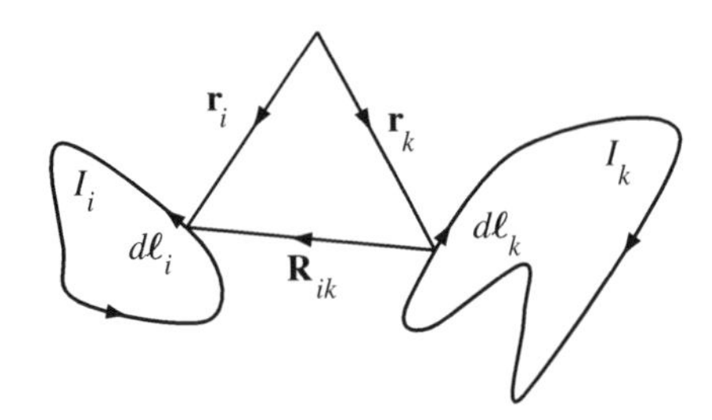

Figure 12.18: Two filamentary loops of current.

and upper limits of the integral, respectively:

$$L_{\text{out}} = \frac{1}{\mu_0 I^2} 2\pi b \int_a^b dr\, r\, B_{\text{out}}^2 = \frac{\mu_0}{2\pi} b \ln(b/a). \tag{12.138}$$

This gives the first term in the square brackets in (12.136).

A more accurate calculation for a circular loop[20] shows that the self-inductance has the form (12.136) with $k_1 = 1$ and $k_2 \approx 1/3$ when we interpret b as the loop radius. For a long, skinny, rectangular loop with side lengths b and $c \gg b$, we should replace the multiplicative pre-factor b in (12.136) by c because this was our estimate of the total wire length. This change makes (12.136) consistent with Example 12.5.

12.8.2 Mutual Inductance

Generalizing (12.132), we can use (12.97) to write the total energy of a system of N distinct current distributions $\mathbf{j}_k(\mathbf{r})$ in the form

$$U_B = \frac{1}{2} \sum_{i=1}^{N} \sum_{k=1}^{N} M_{ik} I_i I_k. \tag{12.139}$$

The real numbers M_{ik} are the elements of an inductance matrix $\mathbf{M}$:

$$M_{ik} = \frac{\mu_0}{4\pi} \frac{1}{I_i I_k} \int d^3r \int d^3r' \frac{\mathbf{j}_i(\mathbf{r}) \cdot \mathbf{j}_k(\mathbf{r}')}{|\mathbf{r} - \mathbf{r}'|}. \tag{12.140}$$

The self-inductances $L_i \equiv M_{ii}$ discussed earlier are joined now by *mutual inductances* M_{ik}. The symmetry

$$M_{ik} = M_{ki} \tag{12.141}$$

is manifest in (12.140) and is a restatement of the reciprocity property (12.103). Moreover, the positivity of (12.139) for non-trivial situations [see (12.98)] tells us that $\mathbf{M}$ is a positive-definite matrix. Among other things, the positive-definiteness of $\mathbf{M}$ ensures that

$$|M_{ik}| \le \sqrt{L_i L_k} \qquad \text{and} \qquad |M_{ik}| \le \frac{L_i + L_k}{2}. \tag{12.142}$$

For a pair of filamentary current loops like those in Figure 12.18, (12.140) simplifies to *Neumann's formula*,

$$M_{ik} = \frac{\mu_0}{4\pi} \oint_{C_i} \oint_{C_k} \frac{d\boldsymbol{\ell}_i \cdot d\boldsymbol{\ell}_k}{|\mathbf{r}_i - \mathbf{r}_k|}. \tag{12.143}$$

[20] See, for example, W.R. Smythe, *Static and Dynamic Electricity* (McGraw-Hill, New York, 1939).

This formula makes it manifest that the inductances depend only on the geometry of the current loops, and not on the values of the currents they carry. Indeed, comparing (12.139) to (12.96) shows that the single-loop self-inductance formula $\Phi = LI$ generalizes to

$$\Phi_i = \sum_{k=1}^{N} M_{ik} I_k. \tag{12.144}$$

According to this formula, we find M_{ik} by calculating the magnetic flux through the ith loop when the kth loop carries unit current and all other loops carry zero current. In this context, it is common to speak of the magnetic flux "linked" to the ith loop.

More generally, the representations of U_B in (12.96), (12.97), and (12.98) and the magnetic reciprocity relation (12.103) lead to various equivalent expressions for the coefficients of mutual inductance. Thus,

$$M_{ik} = \frac{1}{I_i I_k}\int d^3r\, \mathbf{j}_i(\mathbf{r})\cdot\mathbf{A}_k(\mathbf{r}) = \frac{1}{\mu_0 I_i I_k}\int d^3r\, \mathbf{B}_i(\mathbf{r})\cdot\mathbf{B}_k(\mathbf{r}). \tag{12.145}$$

In light of (12.141), these formulae are equally valid when i and k are reversed in the integrands.

Example 12.6 An infinite wire carries a current up the rotational symmetry axis of a toroidal solenoid with N tightly wound turns and a circular cross section. The inner radius of the torioid is a and the outer radus is b. Find the mutual inductance M between the wire and the solenoid

Solution: We will use (12.144) and compute $M = \Phi/I$ where Φ is the flux of the magnetic field of the wire through the solenoid. We center the toroid at the origin of a cylindrical coordinate system and write the magnetic field of the wire as $\mathbf{B} = (\mu_0 I/2\pi\rho)\hat{\boldsymbol{\phi}}$. If A is the cross sectional area of the solenoid, the flux of $\mathbf{B}$ through the entire solenoid is

$$\Phi = N\int_A d\mathbf{S}\cdot\mathbf{B} = \frac{\mu_0 NI}{2\pi}\int_a^b \frac{d\rho}{\rho}\int_{-z_0}^{z_0} dz,$$

where $\pm z_0$ are indicated in Figure 12.19 below.

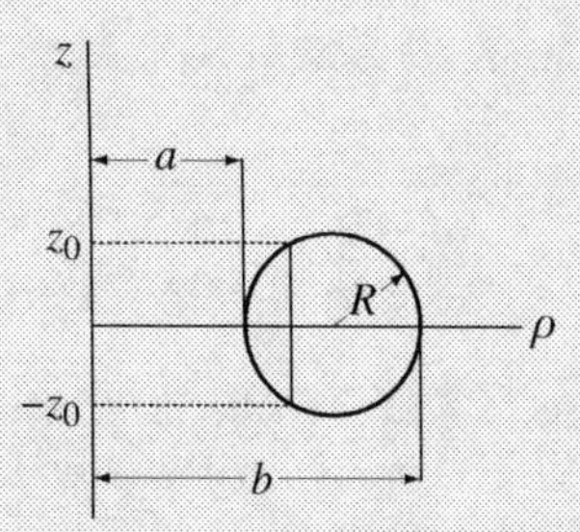

Figure 12.19: A cross sectional view of a toroidal solenoid and a current I which flows up the z-axis.

The limits of integration for the z-integral are determined by the equation for the circle which is the cross section of the solenoid:

$$\left(\rho - \frac{a+b}{2}\right)^2 + z^2 = \left(\frac{b-a}{2}\right)^2.$$

This gives the limits as $\pm z_0$ where $z_0 = \sqrt{(b-\rho)(\rho-a)}$. Therefore,

$$\Phi = \frac{\mu_0 NI}{2\pi}\int_a^b d\rho \frac{\sqrt{(b-\rho)(\rho-a)}}{\rho} = \frac{\mu_0 NI}{2\pi} \times J.$$

A standard reference gives[21]

$$J = \sqrt{(b-\rho)(\rho-a)}\Big|_a^b - \sqrt{ab}\sin^{-1}\left[\frac{(a+b)\rho - 2ab}{\rho(b-a)}\right]_a^b - \left[\frac{a+b}{2}\right]\sin^{-1}\left[\frac{a+b-2\rho}{b-a}\right]_a^b,$$

or $J = \pi\left[(a+b)/2 - \sqrt{ab}\right]$. Therefore, the mutual inductance is

$$M = \frac{1}{2}\mu_0 N\left(\sqrt{b} - \sqrt{a}\right)^2.$$

As a quick check, we eliminate b in this formula in favor of the radius $R = (b-a)/2$ defined by Figure 12.19. Then, when $R \ll a$,

$$M = \frac{1}{2}\mu_0 N\left[\sqrt{2R+a} - \sqrt{a}\right]^2 \approx \frac{1}{2}\mu_0 N \frac{R^2}{a}.$$

This is the expected result because, in this limit,

$$M = \frac{\Phi}{I} = \frac{BNA}{I} \approx \frac{\mu_0 N}{2\pi a}\cdot \pi R^2.$$

12.8.3 Magnetic Force in Terms of Inductance

We are now in a position to calculate the force on any one of a collection of filamentary loops with specified currents. Since $\hat{U}_B = -U_B$ is a natural function of the currents, (12.115) and (12.139) combine to give

$$\mathbf{F}_k = \frac{1}{2}\sum_{i=1}^N\sum_{j=1}^N I_i I_j \left(\frac{\partial M_{ij}}{\partial \mathbf{R}_k}\right)_{I,\mathbf{R}'}. \tag{12.146}$$

Alternatively, we can invert (12.144) to express the currents in terms of the fluxes and compute the force from

$$\mathbf{F}_k = -\left(\frac{\partial U_B}{\partial \mathbf{R}_k}\right)_{\Phi,\mathbf{R}'} = -\frac{1}{2}\sum_{i=1}^N\sum_{j=1}^N \Phi_i\Phi_j\left(\frac{\partial M_{ij}^{-1}}{\partial \mathbf{R}_k}\right)_{\Phi,\mathbf{R}'}. \tag{12.147}$$

The formula (12.147) does not contradict our earlier statement that it is most natural to compute forces using $\hat{U}_B$. The calculation is indeed unnatural (since we are required to hold the flux constant) but not impossible if done properly as we have indicated. This is so because magnetostatic force is an equilibrium property that cannot know whether flux or current is held fixed. Indeed, the leftmost equality in (12.147) follows immediately by identifying the spatial derivatives on the far right side of (12.107) with $-\mathbf{F}_k$ as we have consistently done in this book. The reader should prove explicitly that (12.146) is equal to (12.147).

[21] I.S. Gradshteyn and I.M. Ryzhik, *Table of Integrals, Series, and Products* (Academic, New York, 1980), Section 2.267.

Example 12.7 Use the vector potential from Example 10.6 to find an integral representation of the mutual inductance M and magnetic force between two identical and coaxial current rings (radius R) separated by a distance h. Do the integral in the limit $h \gg R$ and interpret the result.

Solution: We compute M from (12.145) in the form

$$M = \frac{1}{I^2}\int d^3r\, \mathbf{j}_1(\mathbf{r}) \cdot \mathbf{A}_2(\mathbf{r}),$$

where $\mathbf{A}_2(\mathbf{r})$ is the vector potential of a z-axis-centered ring in the x-y plane and $\mathbf{j}_1(\mathbf{r}) = I\delta(z - h)\delta(\rho - R)\hat{\boldsymbol{\phi}}$ is the current density of an identical ring at a height h above $z = 0$. Using $\mathbf{A}_2(\rho, z)$ from Example 10.6, the mutual inductance is

$$M = \mu_0 \pi R^2 \int_0^\infty dk\, J_1^2(kR)\exp(-kh).$$

This integral is known in terms of elliptic integrals, although it is probably simpler to evaluate it numerically. Regardless, since M appears twice in the sum (12.146), the force between the rings is

$$\mathbf{F} = I^2 \frac{dM}{dh}\hat{\mathbf{z}}.$$

When $h \gg R$, the integral is dominated by the $k \to 0$ limit. Since $J_1(x \to 0) = x/2$, it is straightforward to confirm that

$$\mathbf{F} = \frac{\mu_0}{4\pi}(\pi R^2 I)^2 \frac{d^3}{dh^3}\int_0^\infty dk\, \exp(-kh)\hat{\mathbf{z}} = -\frac{3\mu_0}{2\pi}\frac{(\pi R^2 I)^2}{h^4}\hat{\mathbf{z}}.$$

The force is attractive because any current element of one ring is closest to a parallel current element in the other ring. The force would be repulsive if the currents in the two rings circulated in opposite directions. Note also, as indicated in (12.44), that $\mathbf{F}$ varies with the inverse fourth power of the ring separation.

As a check, we can use (12.144) to write $M = \Phi/I$ and estimate the flux as $B_z(h) \times \pi R^2$ where $B_z(h)$ is the $z \gg R$ limit of (10.20). This gives the mutual inductance as

$$M = \frac{\mu_0}{2\pi}\frac{(\pi R^2)^2}{h^3}.$$

The corresponding force is the same as above. Since $m = \pi R^2 I$ is the magnetic moment of each ring,

$$\mathbf{F} = I^2 \frac{dM}{dh}\hat{\mathbf{z}} = -\frac{3\mu_0}{2\pi}\frac{m^2}{h^4}\hat{\mathbf{z}}.$$

This is also the force between two parallel and collinear dipoles calculated from $\mathbf{F} = -\nabla \hat{V}_B$ and (12.59).

Sources, References, and Additional Reading

The Ampère quotation at the beginning of the chapter is taken from Chapter 7 of the biography

J.R. Hoffman, *André-Marie Ampère* (Cambridge University Press, Cambridge, 1995).

Section 12.2 Charged particle motion in spatially varying magnetic fields occurs in many subfields of physics. Readable treatments in the contexts of electron optics, space physics, plasma physics, and accelerator physics are, respectively,

D.A. De Wolf, *Basics of Electron Optics* (Wiley, New York, 1990).

J.W. Chamberlain, *Motion of Charged Particles in the Earth's Magnetic Field* (Gordon & Breach, New York, 1964).

C.L. Longmire, *Elementary Plasma Physics* (Wiley-Interscience, New York, 1963).

S. Humphries, *Principles of Charged Particle Acceleration* (Wiley, New York, 1986).

Section 12.3 Example 12.2 and Section 12.3.1 were inspired, respectively, by

O.D. Jefimenko, *Electricity and Magnetism* (Appleton-Century-Crofts, New York, 1966).

W. Saslow, *Electricity, Magnetism, and Light* (Academic, Amsterdam, 2002), Section 10.8.2.

Figure 12.4 was taken from

J.A. Pollack and S. Barraclough, "Note on a hollow lightning conductor crushed by the discharge", *Proceedings of the Royal Society of New South Wales* **39**, 131 (1905).

Section 12.4 Application 12.1 was adapted from

O. Klemperer, *Electron Physics* 2nd edition (Butterworths, London, 1972).

Figure 12.8 greatly simplifies the trajectories of charged particles trapped by the Earth's magnetic field. For an overview and a detailed analysis, see

M.G. Kivelson and C.T. Russell, *Introduction to Space Physics* (Cambridge University Press, Cambridge, 1995).

S.N. Kuznetsov and B.Yu. Yushkov, "Boundary of adiabatic motion of a charged particle in a dipole magnetic field", *Plasma Physics Reports* **28**, 342 (2002).

Magnetic traps for plasmas and neutral atoms are discussed in detail in

R.F. Post, "The magnetic mirror approach to fusion", *Nuclear Fusion* **27**, 1579 (1987).

T. Bergeman, G. Erez, and H.J. Metcalf, "Magnetostatic trapping fields for neutral atoms", *Physical Review A* **35**, 1535 (1987).

Application 12.3 and Figure 12.10 come, respectively, from

A. Cēbers and M. Ozols, "Dynamics of an active magnetic particle in a rotating magnetic field", *Physical Review E* **73**, 21505 (2006).

B. Steinberger, N. Petersen, H. Petermann, and D.G. Weiss, "Movement of magnetic bacteria in time-varying magnetic fields", *Journal of Fluid Mechanics* **273**, 189 (1994).

Section 12.5 Our discussion of the magnetic stress tensor was adapted from

R.C. Cross, "Magnetic lines of force and rubber bands", *American Journal of Physics* **57**, 722 (1989).

V.O. Jensen, "Magnetic stresses in ideal MHD plasmas", *Physica Scripta* **51**, 490 (1995).

Application 12.4 is based on

E.N. Parker, "Reaction of laboratory magnetic fields against their current coils", *Physical Review* **109**, 1440 (1958).

Section 12.6 The model for our "thermodynamic" approach to magnetic energy is

L.D. Landau and E.M. Lifshitz, *Electrodynamics of Continuous Media* (Pergamon, Oxford, 1960).

Other textbooks that do a good job with magnetic energy are

R.K. Wangsness, *Electromagnetic Fields*, 2nd edition (Wiley, New York, 1986).

A.M. Portis, *Electromagnetic Fields: Sources and Media* (Wiley, New York, 1978).

Application 12.5 appears at the beginning of Section III of

D.W. Longcope, "Separator current sheets: Generic features of minimum-energy magnetic fields subject to flux constraints", *Physics of Plasmas* **8**, 5277 (2001).

Yoshioka gives a clear and well-organized overview of the quantum Hall effect. Our very brief introduction to the integer quantum Hall effect reproduces data from *Paalanen, Tsui, and Gossard* and sketches the original argument by *Laughlin*:

D. Yoshioka, *The Quantum Hall Effect* (Springer, Berlin, 1998).

M.A. Paalanen, D.C. Tsui, and A.C. Gossard, "Quantized Hall effect at low temperatures", *Physical Review B* **25**, 5566 (1982).

R.B. Laughlin, "Quantized Hall conductivity in two dimensions", *Physical Review B* **23**, 5632 (1981).

Section 12.7 This section benefitted from the lucid discussion in

S. Bobbio, *Electrodynamics of Materials* (Academic, San Diego, CA, 2000).

Section 12.8 Analytic calculations of self and mutual inductance can be quite challenging. For that reason, it is common to use approximate formulae which have been developed over the years for various geometries. The classic compendium (reprinted from the original 1946 edition) is

F. Grover, *Inductance Calculations* (Dover, New York, 2004).

Problems

12.1 Bleakney's Theorem A particle with mass m and charge q moves non-relativistically in static fields $\mathbf{E}(\mathbf{r})$ and $\mathbf{B}(\mathbf{r})$. Show that a re-scaling of the magnetic field and the time is sufficient for a particle with mass M and charge q to follow exactly the same trajectory as the original particle. Do the motions of m and M differ at all?

12.2 A Hall Thruster An axial electric field $\mathbf{E} = E\hat{\mathbf{z}}$ and a radial magnetic field $\mathbf{B} = B\hat{\boldsymbol{\rho}}$ coexist in the volume V between two short cylindrical shells concentric with the z-axis. Suppose V is filled with xenon gas and, at a given moment, a discharge ionizes the gas into a plasma composed of n_e electrons per unit volume and $n_i = n_e$ singly charged positive ions per unit volume. This is a model for a propulsion device used on spacecraft.

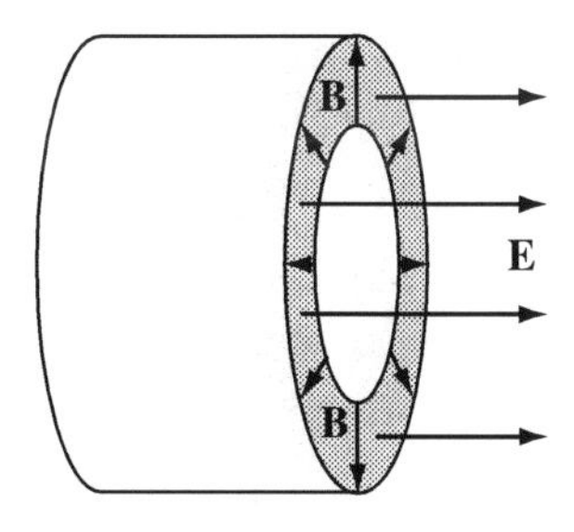

(a) Show that the electrons drift at constant speed in the axial direction because they acquire a velocity $\mathbf{v} = (\mathbf{E} \times \mathbf{B})/B^2$. This defines a *Hall current density* $\mathbf{j}_{\text{Hall}} = -en_e\mathbf{v}$.

(b) Argue that the cylindrical shells experience a net thrust (force) $\mathbf{T} = \mathbf{B} \times \mathbf{j}_{\text{Hall}}$ because the ions are ejected from V. What about the Lorentz force on the ions?

12.3 Charged Particle Motion near a Straight, Current-Carrying Wire Let a steady current I flow up the y-axis and let the initial position and velocity of a particle with mass m and charge q be $\mathbf{r}_0 = (x_0, 0, 0)$ and $\mathbf{v}_0 = (0, v_0, 0)$.

(a) Show that the motion of the particle is confined to the x-y plane.

(b) Prove that $v_y = v_0 + \beta \ln(x/x_0)$ (where β is a constant) and use the fact that the Lorentz force does no work to prove that the particle never leaves the interval $x_0 \le x \le x_0 \exp(-2v_0/\beta)$.

(c) Use (but do not solve) the equation of motion for v_x and the results obtained so far to sketch a typical particle trajectory. Justify your sketch.

(d) Prove that the particle trajectory is given by

$$\frac{dy}{dx} = \frac{v_0 + \beta \ln(x/x_0)}{\sqrt{-\beta \ln(x/x_0)[2v_0 + \beta \ln(x/x_0)]}}.$$

12.4 Anti-Parallel Currents Do Not Always Repel Two long, parallel wires of length L, separation d, and cross sectional radius a are connected by a U-turn at each end to form a closed circuit. Insert a battery with potential difference V at one U-turn and a shunt resistor R at the other U-term. Assume that R is much larger than the ohmic resistance of the wires and the internal resistance of the battery. Show that the wires repel (attract) if R is smaller (greater) than

$$R_0 = \sqrt{\frac{\mu_0}{\epsilon_0}} \frac{1}{\pi} \ln(d/a).$$

12.5 The Mechanical Stability of Concentric Solenoids Two finite-length, concentric, cylindrical solenoids carry current in the same direction. The outer solenoid is very slightly longer and has a very slightly larger radius than the inner solenoid, but their mid-planes are coincident. Determine if the inner solenoid is stable or unstable against a small displacement of its position in the axial direction. Repeat for the case of a small displacement in the transverse direction.

12.6 The Torque between Nested Current Rings Two origin-centered circular rings have radii a and $b \ll a$ and carry currents I_a and I_b. A narrow insulating rod coincident with their common diameter permits the smaller ring to rotate freely inside the larger ring. Show that the torque which must be applied to hold the planes of the rings at a right angle has magnitude

$$N = \frac{\pi}{2} \mu_0 I_a I_b \frac{b^2}{a} \left[1 - \left(\frac{3b}{4a} \right)^2 \right].$$

12.7 Force and Torque Two identical, current-carrying rectangular loops are oriented at right angles, one in the vertical x-y plane, one in the horizontal x-z plane. The horizontal loop moves infinitesimally slowly from $z = -\infty$, through $z = 0$ (where the centers of the two loops coincide), to $z = +\infty$.

(a) Graph (qualitatively) the non-zero component of the force exerted on the vertical loop by the horizontal loop as a function of the position of the latter.
(b) Repeat part (a) for the torque exerted on the vertical loop by the horizontal loop.

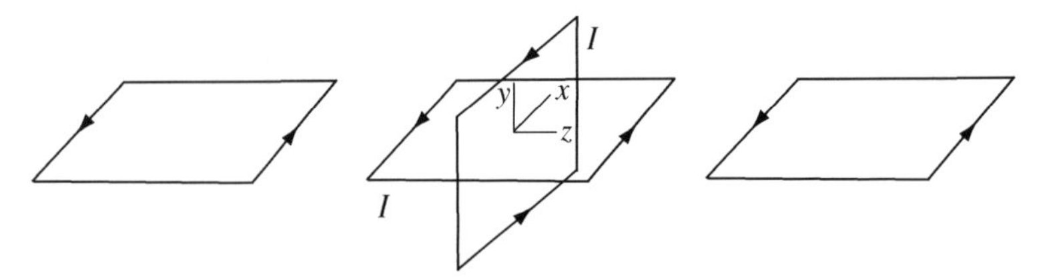

12.8 A General Formula for Magnetostatic Torque Show that the torque exerted on current distribution $\mathbf{j}(\mathbf{r})$ by a distinct current distribution $\mathbf{j}'(\mathbf{r}')$ is

$$\mathbf{N} = \frac{\mu_0}{4\pi} \int d^3r \int d^3r' \left[\frac{\mathbf{j}(\mathbf{r}) \times \mathbf{j}(\mathbf{r}')}{|\mathbf{r} - \mathbf{r}'|} + \frac{\mathbf{r} \times \mathbf{r}'}{|\mathbf{r} - \mathbf{r}'|^3} \mathbf{j}(\mathbf{r}) \cdot \mathbf{j}'(\mathbf{r}') \right].$$

12.9 Force-Free Magnetic Fields Let $\alpha(\mathbf{r})$ be an arbitrary scalar function. A magnetic field which satisfies $\nabla \times \mathbf{B} = \alpha \mathbf{B}$ is called *force-free* because the Lorentz force density $\mathbf{j}(\mathbf{r}) \times \mathbf{B}(\mathbf{r})$ vanishes everywhere. There is some evidence that fields of this sort exist in the Sun's magnetic environment.

(a) Under what conditions is the sum of two force-free fields itself force-free?
(b) Let $\alpha(\mathbf{r}) = \alpha$. Find and sketch the force-free magnetic field where $\mathbf{B}(z) = B_x(z)\hat{\mathbf{x}} + B_y(z)\hat{\mathbf{y}}$ and $\mathbf{B}(0) = B_0\hat{\mathbf{y}}$.
(c) Let $\alpha(\mathbf{r}) = \alpha$. Find and sketch the force-free magnetic field where $\mathbf{B}(\rho) = B_\phi(\rho)\hat{\boldsymbol{\phi}} + B_z(\rho)\hat{\mathbf{z}}$ and $\mathbf{B}(0)$ is finite.
(d) Suppose that $B_z(R) = 0$ in part (c). Find a simple magnetic field $\mathbf{B}^{\text{out}}(\rho)$ in the *current-free* volume $\rho > R$ which matches onto the force-free magnetic field in the $\rho < R$ volume without any current induced on the matching cylinder.

12.10 Nuclear Magnetic Resonance
The chemical diagnostic tool of nuclear magnetic resonance uses a static magnetic field $\mathbf{B}_0 = B_0\hat{\mathbf{z}}$ and a small-amplitude radio-frequency magnetic field $\mathbf{B}_1(t)$ to orient and manipulate nuclear spins in solids and liquids. To get a flavor for the manipulation, let $\mathbf{B}_1(t) = B_1(\hat{\mathbf{x}}\cos\omega t - \hat{\mathbf{y}}\sin\omega t)$ and let $\dot{\mathbf{m}} = \gamma\mathbf{m}\times(\mathbf{B}_0+\mathbf{B}_1)$ be the equation of motion for a nuclear magnetic moment $\mathbf{m}$. Also let $M_x(t)$ and $M_y(t)$ be the transverse components of $\mathbf{m}$ in a coordinate system that rotates with $\mathbf{B}_1$. If $\Omega_L = \gamma B_0$ is the Larmor frequency, show that $\dot{M}_x = (\Omega_L - \omega)M_y$ and $\dot{M}_y = -(\Omega_L-\omega)M_x + \gamma B_1 m_z$ and discuss the trajectory of the moment in the rotating and laboratory frames when $\omega = \Omega_L$.

12.11 Two Dipoles in a Uniform Field Two point dipoles $\mathbf{m}_1$ and $\mathbf{m}_2$ on the x-axis are separated by a distance R and misaligned from the positive x-axis by small angles α and β as shown below. A uniform magnetic field $\mathbf{B}$ points along the negative x-axis. Show that $\alpha=\beta=0$ corresponds to stable equilibrium if

$$B < \frac{\mu_0}{4\pi R^3}\left[m_1+m_2-\sqrt{m_1^2+m_2^2-m_1m_2}\right].$$

12.12 Three Point Dipoles The diagram shows three small magnetic dipoles at the vertices of an equilateral triangle. Moments $\mathbf{m}_\mathrm{B}$ and $\mathbf{m}_\mathrm{C}$ point permanently along the internal angle bisectors. Moment $\mathbf{m}_\mathrm{A}$ is free to rotate in the plane of the triangle. Find the stable equilibrium orientation of the latter and the period of small oscillations around that orientation. Let all the moments have magnitude m and let $\mathbf{m}_\mathrm{A}$ have moment of inertia I about its center of mass.

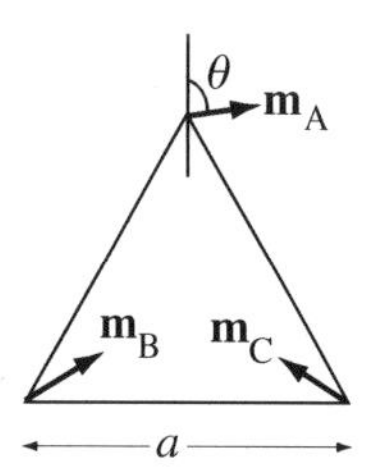

12.13 A Dipole in the Field of Two Dipoles Two identical point dipoles $\mathbf{m} = m\hat{\mathbf{z}}$ sit rigidly at $(\pm a, 0, 0)$. A third point dipole $\mathbf{M}$ is free to rotate at its fixed position $(0, y, z)$. Find the Cartesian components of $\mathbf{M}$ which correspond to stable mechanical equilibrium. Hint: No explicit minimization is required.

12.14 Superconductor Meets Solenoid A superconducting sphere (radius R) placed in a uniform magnetic field $\mathbf{B}$ spontaneously generates currents on its surface which produce a dipole magnetic field. The field is equivalent to that produced by a point dipole at the center of the sphere with magnetic moment $\mathbf{m} = -(2\pi/\mu_0)R^3\mathbf{B}$. Suppose such a sphere (mass M) begins at infinity with speed v_0 and moves toward a solenoid along the symmetry axis of the latter. The field deep inside the solenoid has magnitude B_S. What is the minimum value of v_0 such that the sphere will pass through the solenoid (as opposed to reflecting back)? Assume that $R \ll |\mathbf{B}|/|\nabla\mathbf{B}|$ so the sphere can be treated as a point dipole.

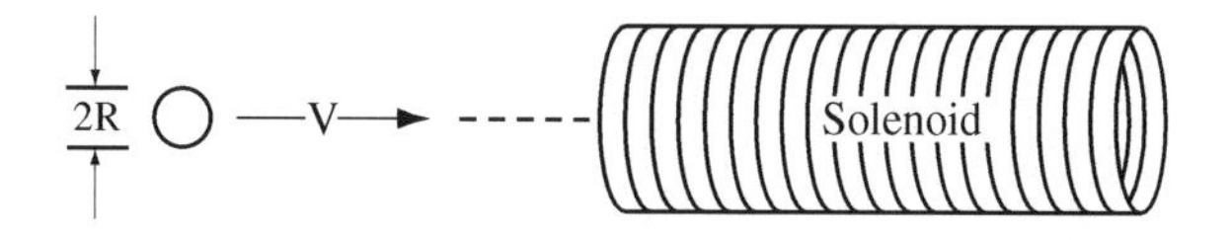

12.15 The Levitron The Levitron™ is a toy in which a spinning magnetic top floats stably in air above a magnetized base. The potential energy of this system is $\hat{V}_B(\mathbf{r}) = Mgz - \mathbf{m}\cdot\mathbf{B}(\mathbf{r})$ if we model the top as a point magnetic dipole with moment $\mathbf{m}$ and mass M moving in the magnetic field of the base. A precession-averaged, adiabatic approximation is valid if the top spins much faster than it precesses and it precesses much faster than the rate of change of the magnetic field experienced by the precessing top. In this approximation, $\hat{V}_B(\mathbf{r}) \approx Mgz - m_B B(\mathbf{r})$, where $B(\mathbf{r}) = |\mathbf{B}(\mathbf{r})|$ and m_B is the (constant) projection of $\mathbf{m}$ on the instantaneous direction of $\mathbf{B}$.

(a) Suppose that $\mathbf{B}(\mathbf{r}) = -\nabla\psi(\mathbf{r})$ is axially symmetric. Show that the magnetic scalar potential above the base and near the z-axis has the form

$$\psi(\rho, z) = \psi_0(z) - \frac{1}{4}\psi_2(z)\rho^2 + \cdots \qquad \text{where} \quad \psi_n(z) = \frac{d^n\psi(0,z)}{dz^n}.$$

(b) Work to second order in ρ for $\psi(\rho, z)$ and show that $m_B < 0$ is a necessary condition for a point of equilibrium to exist on the z-axis that is stable with respect to both vertical and horizontal perturbations. In other words, stable levitation occurs only when the magnetic moment of the top is anti-parallel to the local magnetic field.

(c) If $m_B < 0$, show that the equilibrium and stability conditions require that (i) ψ_1 and ψ_2 have opposite signs; (ii) ψ_1 and ψ_3 have the same signs; and (iii) $\psi_2^2 > 2\psi_3\psi_1$.

12.16 Magnetic Trap I Two infinite, straight, parallel wires, each carrying current I in the same direction, are coincident with the lines $(1, 0, z)$ and $(-1, 0, z)$. In addition, there is a *large* external field $\mathbf{B}_0 = B_0\hat{\mathbf{z}}$. An atom (mass M) whose magnetic dipole moment $\mathbf{m}_0$ is always *anti-parallel* to the local magnetic field direction can be trapped at the origin. What is the approximate frequency of small oscillations in the vicinity of the origin?

12.17 Magnetic Trap II A filamentary wire carries a current along the positive z-axis in the presence of a constant magnetic field $\mathbf{B} = B_0\hat{\mathbf{x}} + B'\hat{\mathbf{z}}$.

(a) Find the straight line in space along which the magnitude of the total magnetic field has an absolute minimum.

(b) Find the frequency of small oscillations in the vicinity of the line found above for an atom (mass M) trapped near the line found in part (a) by the virtue of the fact that its magnetic dipole moment $\mathbf{m}$ remains *anti-parallel* to the local magnetic field direction at all times.

(c) How does the situation change if $\mathbf{B}' = 0$?

12.18 Roget's Spiral A tightly wound, N-turn helical coil (radius R) has unstretched length L.

(a) Use an energy method to determine how much current must be passed through the coil if it is to remain unstretched when a mass m is hung from its bottom end. Neglect stray fields at the ends of the coil and purely mechanical forces which might produce a "spring constant".

(b) Give a simple explanation of the forces at work. Why does this problem not contradict the fact that no system can exert a net magnetic force on itself?

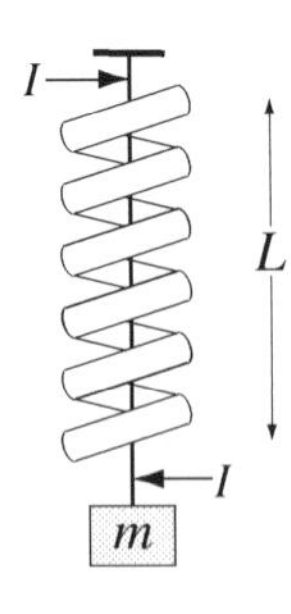

12.19 Equivalence of Force Formulae Consider a collection of current loops where

$$U_B = \frac{1}{2}\sum_{k=1}^{N} I_k \Phi_k.$$

Using the notation of the text for variables that are to be held constant in partial derivatives, prove that the magnetic force exerted on the ith loop satisfies

$$\mathbf{F}_i = -\left.\frac{\partial U_B}{\partial \mathbf{R}_i}\right|_{\Phi'} = -\left.\frac{\partial \hat{U}_B}{\partial \mathbf{R}_i}\right|_{I'}.$$

12.20 The Force between a Current Loop and a Wire A circular loop of radius R carries a current I_1. A straight wire with a current I_2 in the plane of the loop passes a distance d from the loop center.

(a) Use an energy method to show that the force on the loop is

$$\mathbf{F} = \mu_0 I_1 I_2 \left[\frac{d}{\sqrt{d^2 - R^2}} - 1\right]\hat{\mathbf{z}}.$$

(b) Interpret your force formula in the limit $d \gg R$ in the language of magnetic moments.

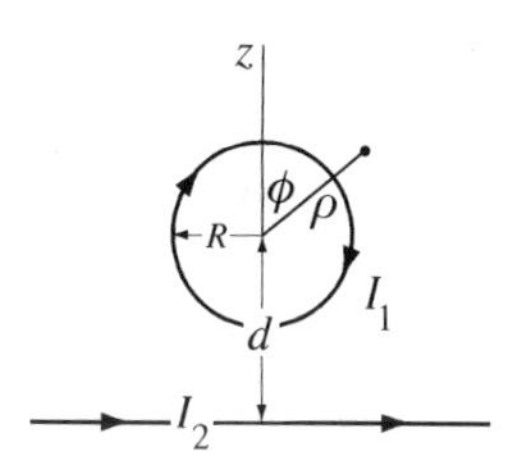

12.21 Toroidal Inductance A straight wire carries a current I_1 down the symmetry axis of a toroidal solenoid with a rectangular cross section of area ab. The solenoid has inner radius R and is composed of N turns of a wire that carries a current I_2.

(a) Find the mutual inductance of the coil with respect to the wire.
(b) Find the mutual inductance of the wire with respect to the coil.

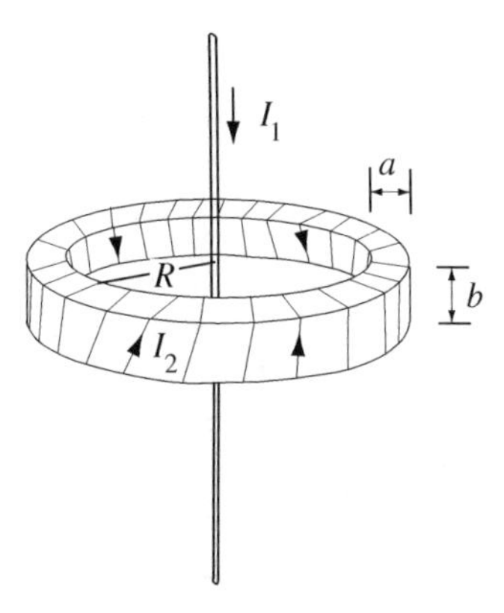

12.22 Force between Square Current Loops The figure shows two square, current-carrying loops with side length a and center-to-center separation c. The currents I_1 and I_2 circulate in the same direction.

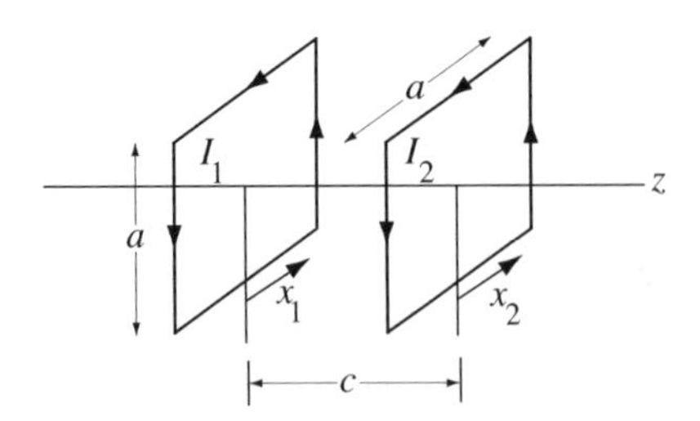

(a) Evaluate Neumann's integral and show that the mutual inductance between the loops is

$$M = \frac{2\mu_0}{\pi}\left[a\sinh^{-1}\left(\frac{a}{c}\right) - a\sinh^{-1}\left(\frac{a}{\sqrt{a^2+c^2}}\right) - 2\sqrt{a^2+c^2} + \sqrt{2a^2+c^2} + c\right].$$

(b) Show that the force of attraction between the loops is

$$F = \frac{2\mu_0 I_1 I_2}{\pi}\left[\frac{c\sqrt{2a^2+c^2}}{a^2+c^2} + 1 - \frac{a^2+2c^2}{c\sqrt{a^2+c^2}}\right].$$

(b) Evaluate the force calculated in part (a) in the limit $c \gg a$ and interpret the formula which results.

12.23 The History of Mutual Inductance The correct form of the interaction energy between two current-carrying circuits was much debated in the 1870s by Maxwell and the German theoretical physicists Carl Neumann, Wilhelm Weber, and Hermann Helmholtz. If $d\mathbf{s}_1$ is a line element of circuit 1 and $d\mathbf{s}_2$ is a line element of circuit 2, all of their proposals can be described as different choices for a constant k in an expression for the mutual inductance of the two line elements:

$$d^2M_{12} = \frac{\mu_0}{4\pi}\left\{\left(\frac{1+k}{2}\right)\frac{d\mathbf{s}_1\cdot d\mathbf{s}_2}{|\mathbf{r}_1-\mathbf{r}_2|} + \left(\frac{1-k}{2}\right)\frac{d\mathbf{s}_1\cdot(\mathbf{r}_1-\mathbf{r}_2)(\mathbf{r}_1-\mathbf{r}_2)\cdot d\mathbf{s}_2}{|\mathbf{r}_1-\mathbf{r}_2|^3}\right\}.$$

Not obviously, the value of k cannot be determined by force measurements on real circuits because the integration of d^2M_{12} over two closed loops produces a formula which does not depend on k. Show this is true by applying Stokes' theorem (twice) to the second term in the curly brackets.

12.24 An Inductance Inequality Consider a two-loop circuit where the total magnetic energy is $U_B = \frac{1}{2}\left(L_1I_1^2 + 2MI_1I_2 + L_2I_2^2\right)$. Prove that $M^2 \le L_1L_2$. Do not simply quote a theorem from the theory of matrices.

12.25 The Self-Inductance of a Spherical Coil
Find the self-inductance of a coil produced by winding a wire $N \gg 1$ times around the surface of sphere of radius R such that the density of turns is uniform along the z-axis perpendicular to the plane of each turn. The wire carries a current I.

13 Magnetic Matter

Plato holds the magnetic virtue to be divine.
William Gilbert (1600)

13.1 Introduction

All matter responds to a static, external magnetic field $\mathbf{B}_{\rm ext}(\mathbf{r})$ by producing a magnetic field of its own, $\mathbf{B}_{\rm self}(\mathbf{r})$. The total magnetic field, both inside and outside the matter, is the sum of the external and induced fields,

$$\mathbf{B}_{\rm tot}(\mathbf{r}) = \mathbf{B}_{\rm ext}(\mathbf{r}) + \mathbf{B}_{\rm self}(\mathbf{r}). \tag{13.1}$$

Far outside the matter, $\mathbf{B}_{\rm self}$ is dipolar and thus may be characterized by a macroscopic magnetic dipole moment $\mathbf{m}$. The origin of this moment is fundamentally quantum mechanical and a long-standing classification scheme reflects characteristic differences in the behavior of $\mathbf{m}$ and $\mathbf{B}_{\rm self}$ for different types of matter. As Figure 13.1 indicates, the magnetic moment of a *paramagnet* points parallel to $\mathbf{B}_{\rm ext}$ while the magnetic moment of a *diamagnet* points anti-parallel to $\mathbf{B}_{\rm ext}$. For most magnets, the induced moment vanishes when the external field is removed. The exception to this rule is a *ferromagnet*, which is a special case of a paramagnet where $\mathbf{m}$ remains non-zero when $\mathbf{B}_{\rm ext} \to 0$. A *superconductor* is a special case of a diamagnet where $\mathbf{B}_{\rm tot} = 0$ inside the volume of the superconductor; it is the magnetic analog of a perfect conductor.

The formal theory of magnetized matter has much in common with the formal theory of polarized matter developed in Chapter 6. However, even as we pursue this commonality, two factors lead to inevitable differences in emphasis and presentation. The first is mathematical and arises from the difference between the curl in $\mathbf{B} = \nabla \times \mathbf{A}$ and the gradient in $\mathbf{E} = -\nabla\varphi$. The second is physical and arises from the technological importance of permanently magnetized matter compared to the relative unimportance of permanently polarized matter.[1] The reader should bear these differences in mind as our story proceeds.

13.2 Magnetization

The word *magnetization* is used in two ways in the theory of magnetic matter. First, magnetization refers to the rearrangement of internal currents that occurs when matter is exposed to an external

[1] Permanent magnets are made from alloys and compounds of the ferromagnetic elements Fe, Co, and Ni. Three common examples are NdFeB, SmCo, and AlNiCo.

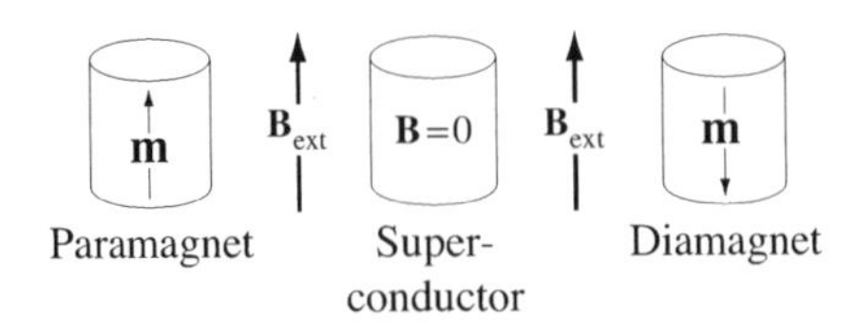

Figure 13.1: The response of a paramagnet, superconductor, and a diamagnet to an external magnetic field. The vector $\mathbf{m}$ is the induced magnetic dipole moment.

magnetic field $\mathbf{B}_{\rm ext}(\mathbf{r})$. Second, magnetization is the word given to a function $\mathbf{M}(\mathbf{r})$ used to characterize the details of the rearrangement. We begin with the external field which initiates the magnetization process and identify its source as a current density $\mathbf{j}_f(\mathbf{r})$ which is wholly extraneous to the magnetizable matter. A long tradition refers to this as *free current*. The most familiar example is a current-carrying wire, perhaps wound into a coil or solenoid.[2]

The source of $\mathbf{B}_{\rm self}(\mathbf{r})$ in (13.1) is often called *bound current*. A more descriptive term is *magnetization current* and we will use the symbol $\mathbf{j}_M(\mathbf{r})$ for its density. The total current density that enters Maxwell's theory is the sum of the "free" and "bound" current densities:

$$\mathbf{j}(\mathbf{r}) = \mathbf{j}_f(\mathbf{r}) + \mathbf{j}_M(\mathbf{r}). \tag{13.2}$$

Experiments show that there are two distinct sources of magnetization current density. At the level of atoms and molecules, we identified these in Section 11.2.2 as (i) the orbital angular momentum of circulating electrons and (ii) the spin angular momentum of electrons, protons, and neutrons. Orbital magnetism is the rule for diamagnets and superconductors. Electron spin is responsible for most of the magnetism in paramagnets and ferromagnets.

Classical electrodynamics treats spin magnetism and orbital magnetism on the same footing by defining a magnetization $\mathbf{M}(\mathbf{r})$ and a magnetization current density $\mathbf{j}_M(\mathbf{r})$ for each. The identification of these functions is the first task of this chapter. The magnetic field $\mathbf{B}_M(\mathbf{r})$ produced by a specified $\mathbf{j}_M(\mathbf{r})$ is our next concern. Thereafter, we add the effect of free current and write the full Maxwell equations for magnetic matter. Subsequent sections treat simple (linear and isotropic) magnets, energy and force in magnetic systems, and, finally, magnetic domains and hysteresis in permanent magnets.

13.2.1 Spin Magnetization

The magnetic fields produced by refrigerator magnets, loudspeaker magnets, and disk drives all arise from the cooperative alignment of electron spins in ferromagnetic matter.[3] This is significant because, without approximation, electrons are known to behave magnetically exactly like point-like magnetic dipoles (Section 11.2.3). Therefore, despite the fact that spin is a non-classical concept, a collection of N electrons with spin magnetic moments $\mathbf{m}_k$ (all with the same magnitude) located at positions $\mathbf{r}_k$ can be characterized by a spin magnetic moment per unit volume, or *spin magnetization*,

$$\mathbf{M}_S(\mathbf{r}) = \sum_{k=1}^{N} \mathbf{m}_k\, \delta(\mathbf{r} - \mathbf{r}_k). \tag{13.3}$$

It is a routine matter to calculate $\mathbf{M}_S(\mathbf{r})$ for ferromagnetic materials using first-principles quantum mechanical methods of the sort used to produce Figure 2.5 and Figure 6.2.

[2] There is no need to specify a source current when a permanent magnet is used to create $\mathbf{B}_{\rm ext}$. See Section 13.6.3.

[3] The contribution to the total magnetic moment from the protons and neutrons in matter can be neglected because their magnetic moments are smaller by a factor of the electron/proton mass ratio.

We know from Section 11.2.3 that the magnetic field produced by a point magnetic moment $\mathbf{m}$ at the origin is identical to the magnetic field produced by the singular current density $\mathbf{j} = \nabla \times [\mathbf{m}\delta(\mathbf{r})]$. Therefore, the magnetic field produced by the spin magnetization (13.3) is exactly the same as the magnetic field produced by the *spin magnetization current density*,

$$\mathbf{j}_S(\mathbf{r}) = \nabla \times \mathbf{M}_S(\mathbf{r}). \tag{13.4}$$

Typically, the spins of a magnetic sample are confined to a closed volume bounded by a sharp surface. The corresponding spin magnetization drops abruptly to zero and one finds that a surface magnetization current density appears. To see this, let a collection of spins occupy the $z \leq 0$ half-space. Using (13.4),

$$\mathbf{j}_S = \nabla \times [\mathbf{M}_S \Theta(-z)] = [\nabla \times \mathbf{M}_S]\Theta(-z) + [\mathbf{M}_S \times \hat{\mathbf{z}}]\delta(z). \tag{13.5}$$

The first term on the far right-hand side of (13.5) is the spin magnetization current density of the bulk. The second term generates a spin magnetization current density at the $z = 0$ surface because

$$\mathbf{K}_S(x, y) = \int dz\, \mathbf{j}_S = \mathbf{M}_S(x, y) \times \hat{\mathbf{z}}. \tag{13.6}$$

The generalization of (13.6) for spins confined to a volume with local outward normal $\hat{\mathbf{n}}(\mathbf{r}_S)$ is

$$\mathbf{K}_S(\mathbf{r}_S) = \mathbf{M}_S(\mathbf{r}_S) \times \hat{\mathbf{n}}(\mathbf{r}_S). \tag{13.7}$$

13.2.2 Orbital Magnetization

Ampère was the first to suggest that microscopic "molecular currents" are responsible for the magnetic fields produced by matter. Today, we understand this to be true for systems where an external field is needed to bias the direction of microscopic or macroscopic circulating currents to produce a non-zero net magnetic moment. In quantum mechanics, every electron in an eigenstate with orbital wave function $\Psi_k(\mathbf{r})$ contributes a steady, dissipationless current density

$$\mathbf{j}_k(\mathbf{r}) = i\mu_B(\Psi_k^* \nabla \Psi_k - \Psi_k \nabla \Psi_k^*). \tag{13.8}$$

Each $\mathbf{j}_k(\mathbf{r})$ describes a closed loop of current and thus a magnetic moment. These moments are randomly oriented in paramagnetic and diamagnetic materials until an external field favors one axis of circulation and thereby induces a non-zero macroscopic magnetic dipole moment.[4]

The foregoing, combined with the general discussion following (13.2), motivates us to avoid classical models of atoms and molecules and use only the existence of closed internal current loops to define an *orbital magnetization current density* $\mathbf{j}_O(\mathbf{r})$ and an *orbital magnetization* $\mathbf{M}_O(\mathbf{r})$. The key observation is that no internal magnetization current loop, or any superposition of such loops, can produce a net current I through any cross sectional surface S' of a finite sample. This is so because a loop that carries magnetization current in one direction through S' inevitably carries the same amount of magnetization current through S' in the opposite direction before it closes on itself.

To make this idea precise, Figure 13.2 shows a bar-shaped magnet with volume V and surface S (dashed lines). The planar cross sectional surface S' is bounded by a perimeter curve C. A prospective volume current density $\mathbf{j}_O(\mathbf{r})$ carries orbital magnetization current through S' and a prospective surface density $\mathbf{K}_O(\mathbf{r})$ carries orbital magnetization current through C. Therefore, our requirement that zero net current flows through any cross section takes the form[5]

$$I = \int_{S'} d\mathbf{S} \cdot \mathbf{j}_O(\mathbf{r}) + \oint_C d\boldsymbol{\ell} \cdot \mathbf{K}_O(\mathbf{r}) \times \hat{\mathbf{n}}(\mathbf{r}_S) = 0. \tag{13.9}$$

[4] The complete story of orbital magnetism lies far outside the scope of this book. A readable introduction is K.H.J. Buschow and F.R. DeBoer, *Physics of Magnetism and Magnetic Materials* (Kluwer, New York, 2004).

[5] Section 2.1.2 discusses the two terms that appear in (13.9).

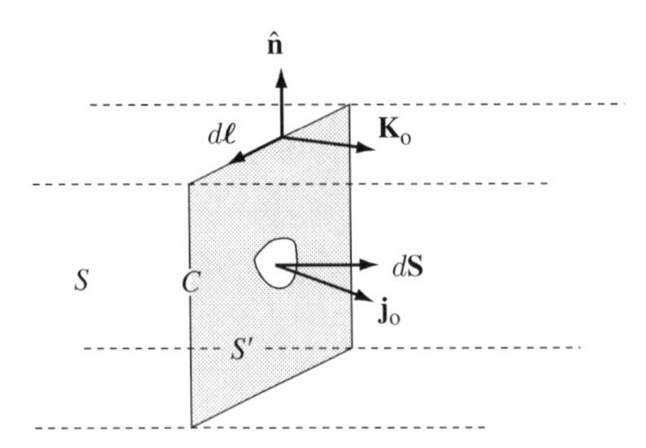

Figure 13.2: The vector $d\mathbf{S}$ is normal to a cross sectional surface S' (shaded) of a bar-shaped magnet that carries an orbital magnetization volume current density $\mathbf{j}_O$ and an orbital magnetization surface current density $\mathbf{K}_O$. The unit vector $\hat{\mathbf{n}}$ is normal to the surface S of the magnet (dashed lines). The line element $d\boldsymbol{\ell}$ is everywhere tangent to the curve C on the sample surface that bounds S'.

Our model-independent strategy is to ask what functions $\mathbf{j}_O(\mathbf{r})$ and $\mathbf{K}_O(\mathbf{r}_S)$ satisfy (13.9) for all choices of sample volume V and all choices of cross section S. Using the hint provided by the spin magnetization current densities (13.4) and (13.7), we write the general answer in terms of an an orbital magnetization function, $\mathbf{M}_O(\mathbf{r})$; namely,

$$\mathbf{j}_O(\mathbf{r}) = \nabla \times \mathbf{M}_O(\mathbf{r}) \qquad \mathbf{r} \in V, \tag{13.10}$$

$$\mathbf{K}_O(\mathbf{r}_S) = \mathbf{M}_O(\mathbf{r}_S) \times \hat{\mathbf{n}}(\mathbf{r}_S) \qquad \mathbf{r}_S \in S, \tag{13.11}$$

$$\mathbf{M}_O(\mathbf{r}) = 0 \qquad \mathbf{r} \notin V. \tag{13.12}$$

It is straightforward to check that (13.10) and (13.11) alone make (13.9) an identity. This is so because

$$\int_{S'} d\mathbf{S} \cdot \mathbf{j}_O = \int_{S'} d\mathbf{S} \cdot \nabla \times \mathbf{M}_O = \oint_C d\boldsymbol{\ell} \cdot \mathbf{M}_O, \tag{13.13}$$

and (because $d\boldsymbol{\ell} \cdot \hat{\mathbf{n}} = 0$)

$$\oint_C d\boldsymbol{\ell} \cdot \mathbf{K}_O(\mathbf{r}_S) \times \hat{\mathbf{n}}(\mathbf{r}_S) = \oint_C d\boldsymbol{\ell} \cdot (\mathbf{M}_O \times \hat{\mathbf{n}}) \times \hat{\mathbf{n}} = -\oint_C d\boldsymbol{\ell} \cdot \mathbf{M}_O. \tag{13.14}$$

Two points are worth noting. First, we set $\mathbf{M}_O(\mathbf{r}) = 0$ outside the sample in (13.12) because magnetization is associated with matter and there is no matter outside V.[6] Second, the equations (13.10), (13.11), and (13.12) do not determine $\mathbf{M}_O(\mathbf{r})$ uniquely. This follows from Helmholtz' theorem (Section 1.9) and the fact that we did not specify $\nabla \cdot \mathbf{M}_O(\mathbf{r})$ *inside* the sample volume V.

13.2.3 Total Magnetization

The *total* magnetization is the sum of the spin and orbital magnetization current densities:

$$\mathbf{M}(\mathbf{r}) = \mathbf{M}_S(\mathbf{r}) + \mathbf{M}_O(\mathbf{r}). \tag{13.15}$$

[6] Some authors build (13.12) into their definition of $\mathbf{M}_O(\mathbf{r})$. This generates the surface density (13.11) as a singular piece of the volume density (13.10). The argument is identical to the one made at the end of Section 13.2.1 for spin magnetization.

With this definition, a summary of our results so far is that the macroscopic magnetic field of a magnetized sample is produced by macroscopic magnetization current densities

$$\mathbf{j}_M(\mathbf{r}) = \nabla \times \mathbf{M}(\mathbf{r}) \qquad \text{and} \qquad \mathbf{K}_M(\mathbf{r}_S) = \mathbf{M}(\mathbf{r}_S) \times \hat{\mathbf{n}}(\mathbf{r}_S). \tag{13.16}$$

The total magnetization $\mathbf{M}(\mathbf{r})$ is fundamental to the theory and our attention must now focus on its physical meaning and methods that can be used to calculate it.

13.2.4 The Volume Integral of M(r)

An immediate consequence of (13.3) is that the volume integral of the spin magnetization $\mathbf{M}_S(\mathbf{r})$ is the spin part of the total magnetic dipole moment of a sample. In this section, we prove an analogous result for the *total* magnetization. Taken together, these two results give an important clue to the physical meaning of the orbital magnetization; namely, the volume integral of $\mathbf{M}_O(\mathbf{r})$ is the orbital part of the total magnetic dipole moment.

The definition of the magnetic moment (11.12) implicitly includes both volume and surface contributions. Therefore, the magnetic moment produced by magnetization current is

$$\mathbf{m} = \frac{1}{2}\int_V d^3r\; [\mathbf{r} \times \mathbf{j}_M] + \frac{1}{2}\int_S dS\; [\mathbf{r} \times \mathbf{K}_M]. \tag{13.17}$$

Substituting the two formulae in (13.16) into (13.17), and using the Levi-Cività representation of the cross product in the integrands to write out the kth Cartesian component of $\mathbf{m}$, gives

$$m_k = \frac{1}{2}\int_V d^3r\; [r_\ell \partial_k M_\ell - r_\ell \partial_\ell M_k] + \frac{1}{2}\int_S dS\; [\hat{n}_\ell M_k r_\ell - \hat{n}_k r_\ell M_\ell]. \tag{13.18}$$

Equation (13.18) can be rewritten in the form

$$m_k = \frac{1}{2}\int_V d^3r\; [\partial_k(r_\ell M_\ell) - M_\ell \partial_k r_\ell] - \frac{1}{2}\int_V d^3r\; [\partial_\ell(r_\ell M_k) - M_k \partial_\ell r_\ell]$$
$$+ \frac{1}{2}\int_S dS\; [\hat{n}_\ell M_k r_\ell - \hat{n}_k r_\ell M_\ell]. \tag{13.19}$$

The integral of the total derivative in the first term of each volume integral in (13.19) produces a surface integral [see (1.77)]. These surface integrals exactly cancel the explicit surface integrals in (13.19). Therefore, using $\partial_k r_\ell = \delta_{k\ell}$ and $\partial_\ell r_\ell = 3$, we conclude that the integral of the total magnetization $\mathbf{M}$ over the volume of a magnetized object is indeed equal to the magnetic dipole moment of the object:[7]

$$\mathbf{m} = \int_V d^3r\, \mathbf{M}. \tag{13.20}$$

13.2.5 The Lorentz Model

Following Lorentz (1902), many authors use a microscopic version of (13.20) to identify $\mathbf{M}(\mathbf{r})$ as a "magnetic dipole moment per unit volume". The physical idea is to regard a magnetized sample as a collection of atomic or molecular magnetic dipoles. The mathematical prescription defines the

[7] In Section 6.2.1, we showed that the integral of the polarization $\mathbf{P}$ over the volume of a polarized object is equal to the electric dipole moment of the object.

magnetization at a macroscopic point $\mathbf{r}$ as the magnetic dipole moment of a microscopic cell with volume Ω labeled by $\mathbf{r}$:[8]

$$\mathbf{M}(\mathbf{r}) = \frac{1}{2\Omega}\int_{\Omega} d^3s\, \mathbf{s} \times \mathbf{j}_{\mathrm{M,micro}}(\mathbf{s}) = \frac{\mathbf{m}(\mathbf{r})}{\Omega}. \tag{13.21}$$

For finite Ω, (13.21) replaces the current density in each averaging cell by a point magnetic dipole $\mathbf{m}$. In the limit $\Omega \to 0$, the Lorentz approximation replaces the entire magnet by a continuous distribution of point magnetic dipoles with a density $\mathbf{M}(\mathbf{r})$ computed from (13.21). This is rigorously true for the spin part of $\mathbf{M}(\mathbf{r})$ and may be approximately true for the orbital part of $\mathbf{M}(\mathbf{r})$. The issue is whether the physical current density in the magnetized sample can be partitioned into a sum of "atomic" current densities which are well localized near the center of each averaging cell. This may be true for a magnet that is an electrical insulator; it may *not* be true for a conducting magnet if there is significant microscopic magnetization current flow *between* averaging cells.

Equation (13.21) and its interpretation are very similar to the Lorentz definition of electric polarization $\mathbf{P}(\mathbf{r})$ and its interpretation presented in Section 6.2.2. For the electric problem, we pointed out the shortcomings of Lorentz' model and described its modern replacement: define $\mathbf{P}(\mathbf{r})$ as the Lorentz average over a cell centered at $\mathbf{r}$ of the time-integral of the microscopically calculable polarization current density $\mathbf{j}_{\mathrm{P,micro}}$. Unfortunately, no similar scheme works in the magnetic case because the orbital part of $\mathbf{M}(\mathbf{r})$ is not uniquely defined. This is the subject of the next section.

13.2.6 The Non-Uniqueness of $\mathbf{M}(\mathbf{r})$

Practitioners who apply first-principles quantum mechanics to magnetic matter use (13.3) to calculate the spin magnetization $\mathbf{M}_S$ and (13.8) to calculate the orbital part of the microscopic magnetization current density $\mathbf{j}_{\mathrm{O,micro}}$. The latter may be Lorentz averaged to get a macroscopic orbital current density, but it is impossible to uniquely invert either the microscopic or the macroscopic versions of $\mathbf{j}_{\mathrm{O}} = \nabla \times \mathbf{M}_{\mathrm{O}}$ to get $\mathbf{M}_{\mathrm{O}}(\mathbf{r})$. Just as there is gauge freedom to choose the vector potential $\mathbf{A}(\mathbf{r})$, the magnetization $\mathbf{M}_{\mathrm{O}}(\mathbf{r})$ and the magnetization $\mathbf{M}_{\mathrm{O}}(\mathbf{r}) + \nabla\Lambda(\mathbf{r})$ correspond to exactly the same orbital magnetization volume current density.[9] For that reason, no observable quantity depends on $\mathbf{M}(\mathbf{r})$ alone. As a prescription to fix $\mathbf{M}(\mathbf{r})$, the physically well-motivated Lorentz approximation (13.21) is one choice among many.[10]

13.3 The Field Produced by Magnetized Matter

This section focuses on the vector potential $\mathbf{A}_{\mathrm{M}}$ and the magnetic field $\mathbf{B}_{\mathrm{M}}$ produced by magnetized matter. We suppose that the magnetization $\mathbf{M}(\mathbf{r})$ is specified in a volume V and do not concern ourselves with how the magnetization was produced. Conventional magnetostatics (Section 10.5.2) tells us that the volume and surface magnetization current densities in (13.16) are the sources of a vector potential

$$\mathbf{A}_{\mathrm{M}}(\mathbf{r}) = \frac{\mu_0}{4\pi}\int_V d^3r' \frac{\nabla' \times \mathbf{M}(\mathbf{r}')}{|\mathbf{r}-\mathbf{r}'|} + \frac{\mu_0}{4\pi}\int_S dS' \frac{\mathbf{M}(\mathbf{r}') \times \hat{\mathbf{n}}'}{|\mathbf{r}-\mathbf{r}'|}. \tag{13.22}$$

[8] In this chapter, $\mathbf{r}$ is a macroscopic variable and $\mathbf{s}$ is a microscopic variable. In Section 2.3.1 on Lorentz averaging, the macroscopic variable was called $\mathbf{R}$ and the microscopic variable was called $\mathbf{r}$.

[9] $\Lambda(\mathbf{r})$ must be constant on S to get the same orbital magnetization surface current density.

[10] See Hirst (1997) in Sources, References, and Additional Reading.

The corresponding Biot-Savart formula with $\mathbf{j}_M(\mathbf{r})$ and $\mathbf{K}_M(\mathbf{r})$ as sources is

$$\mathbf{B}_M(\mathbf{r}) = \frac{\mu_0}{4\pi}\int_V d^3r' \,\frac{[\nabla' \times \mathbf{M}(\mathbf{r}')] \times (\mathbf{r}-\mathbf{r}')}{|\mathbf{r}-\mathbf{r}'|^3} + \frac{\mu_0}{4\pi}\int_S dS' \,\frac{[\mathbf{M}(\mathbf{r}') \times \hat{\mathbf{n}}'] \times (\mathbf{r}-\mathbf{r}')}{|\mathbf{r}-\mathbf{r}'|^3}. \tag{13.23}$$

Equations (13.22) and (13.23) give $\mathbf{A}_M(\mathbf{r})$ and $\mathbf{B}_M(\mathbf{r})$ correctly at every point in space. On the other hand, it may not be necessary to evaluate (13.22) or (13.23) explicitly if we can deduce the field by some other means (say, from an Ampère's law calculation). Example 13.1 is typical of the latter.

Example 13.1 Find the magnetic field $\mathbf{B}_M(\mathbf{r})$ produced by an infinite slab of matter with uniform magnetization as indicated in Figure 13.3.

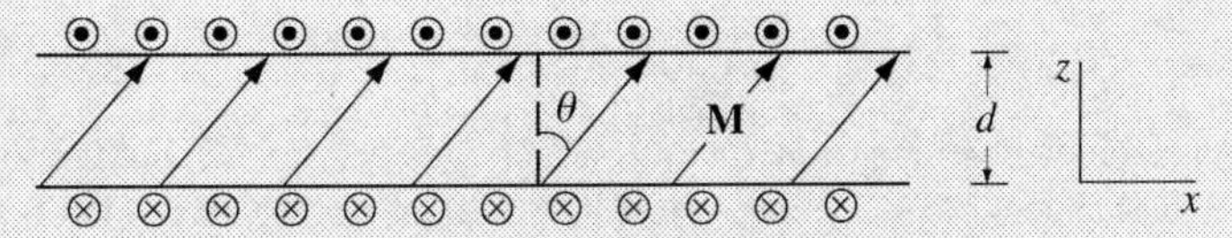

Figure 13.3: Side view of an infinite slab ($0 \le z \le d$) of matter with uniform magnetization **M**. Circles with dots (crosses) indicate magnetization current flowing out of (into) the paper.

Solution: The volume current density (13.10) is zero for uniform magnetization. The surface current density $\mathbf{K} = \mathbf{M} \times \hat{\mathbf{n}}$ flows uniformly out of (into) the page along the upper (lower) surface of the slab with magnitude $K_M = M \sin\theta$. Therefore, the results of Section 10.3.2 tell us that the magnetic field produced by the top sheet of current is $\mathbf{B}_M = \frac{1}{2}\mu_0 K_M \hat{\mathbf{x}}$ below the sheet and the negative of this above the sheet. Superposing the fields from both sheets gives

$$\mathbf{B}_M(\mathbf{r}) = \begin{cases} \mu_0 M \sin\theta\,\hat{\mathbf{x}} & \text{inside the slab,} \\ 0 & \text{outside the slab.} \end{cases} \tag{13.24}$$

13.3.1 Magnetized Matter as a Superposition of Point Dipoles

The physical meanings of $\mathbf{A}_M(\mathbf{r})$ and $\mathbf{B}_M(\mathbf{r})$ reveal themselves when we use Stokes' theorem to write the surface integral in (13.22) as a volume integral:

$$\int_S dS' \,\frac{\mathbf{M}(\mathbf{r}') \times \hat{\mathbf{n}}'}{|\mathbf{r}-\mathbf{r}'|} = -\int_V d^3r' \,\nabla' \times \frac{\mathbf{M}(\mathbf{r}')}{|\mathbf{r}-\mathbf{r}'|}. \tag{13.25}$$

Writing out the curl on the right side of (13.25) and substituting back into (13.22) gives

$$\mathbf{A}_M(\mathbf{r}) = \frac{\mu_0}{4\pi}\int_V d^3r' \,\mathbf{M}(\mathbf{r}') \times \nabla' \frac{1}{|\mathbf{r}-\mathbf{r}'|}. \tag{13.26}$$

Because $\nabla'|\mathbf{r}-\mathbf{r}'|^{-1} = -\nabla|\mathbf{r}-\mathbf{r}'|^{-1}$, comparison of (13.26) with the vector potential (11.13) of a magnetic dipole shows that $\mathbf{A}_M(\mathbf{r})$ is the vector potential of a collection of point magnetic dipoles with moments $d\mathbf{m}(\mathbf{r}) = \mathbf{M}(\mathbf{r})d^3r$. In other words, (13.26) is the vector potential produced by a volume distribution of point magnetic dipoles with density $\mathbf{M}(\mathbf{r})$.

We saw in the previous section that a point dipole description is exact for the spin part of the magnetization but not even uniquely defined for the orbital part of the magnetization. Nevertheless, (13.26) shows that the vector potential of magnetic matter is indistinguishable from the potential

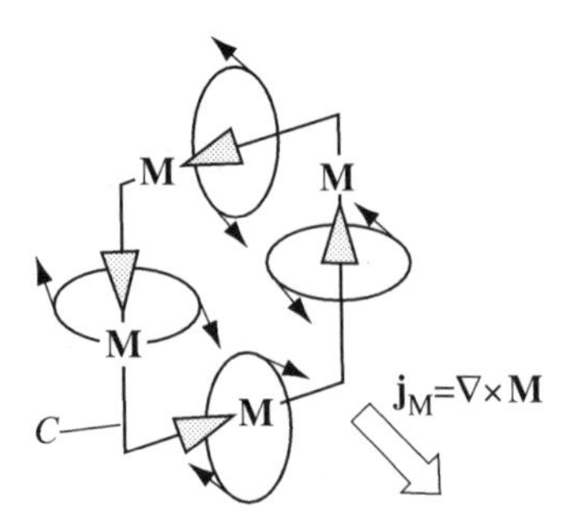

Figure 13.4: Magnetic dipole (big arrows) representation of a point in space (the center of the square loop C) where $\mathbf{j}_M = \nabla \times \mathbf{M} \neq 0$. The equivalent Ampèrian loops shows that a net current (small arrows) flows in the same direction.

produced by a continuous distribution of *fictitious point magnetic dipoles* with a dipole moment per unit volume equal to $\mathbf{M}(\mathbf{r})$. Therefore, to the extent that an expression for $\mathbf{M}(\mathbf{r})$ is available (perhaps only for model purposes), Lorentz' idea to represent a magnetized solid by a volume distribution of point dipoles is perfectly valid.

The point dipole representation of the magnetization qualitatively rationalizes the form of the polarization volume current density $\mathbf{j}_M = \nabla \times \mathbf{M}$. To see this, we recall the geometrical definition of the curl.[11] If C is the boundary of an infinitesimal area element $d\mathbf{S} = dS\hat{\mathbf{n}}$,

$$(\nabla \times \mathbf{M}) \cdot \hat{\mathbf{n}} = \lim_{dS \to 0} \frac{1}{dS} \oint_C d\boldsymbol{\ell} \cdot \mathbf{M}. \tag{13.27}$$

With this definition, the directions of the four point magnetic dipoles drawn on the perimeter C of the square area element in Figure 13.4 guarantee that $\nabla \times \mathbf{M} \neq 0$ at the center of the square. Moreover, we can associate an infinitesimal Ampèrian loop with each of the point dipoles. From the right-hand rule, all four loops contribute to an electric current which flows through the square in the direction of $\nabla \times \mathbf{M}$. This is the volume magnetization current, $\mathbf{j}_M$. We leave it as an exercise for the reader to rationalize the polarization surface current density, $\mathbf{K}_M = \mathbf{M} \times \hat{\mathbf{n}}$, in a similar way.

The superposition point of view offered in this section makes it possible to construct an alternative to the Biot-Savart expression for the magnetic field (13.23) by superposing point dipole magnetic fields suitably weighted by $\mathbf{M}(\mathbf{r})$. Caution is required because, in Section 6.3.1, we superposed dipole electric fields and were led to a field $\mathbf{E}_P^{\dagger}(\mathbf{r})$ that was ill-defined at observation points inside the polarized body. A field $\mathbf{B}_M^{\dagger}(\mathbf{r})$ constructed in the same way is similarly ill-defined at interior points. Happily, the mathematics plays out differently if we use the representation (11.30) derived in Section 11.2.3 for the magnetic field of a point dipole $\mathbf{m}$ located at $\mathbf{r}'$:

$$\mathbf{B}(\mathbf{r}) = \mu_0 \left[\mathbf{m}\,\delta(\mathbf{r} - \mathbf{r}') - \nabla \frac{1}{4\pi} \frac{\mathbf{m} \cdot (\mathbf{r} - \mathbf{r}')}{|\mathbf{r} - \mathbf{r}'|^3} \right]. \tag{13.28}$$

As noted above, we assign a point dipole moment $d\mathbf{m}(\mathbf{r}') = \mathbf{M}(\mathbf{r}')d^3r'$ to every element of the sample volume V. Summing the contribution from every volume element gives

$$\mathbf{B}_M(\mathbf{r}) = \mu_0 \int_V d^3r'\, \mathbf{M}(\mathbf{r}')\delta(\mathbf{r} - \mathbf{r}') - \nabla \frac{\mu_0}{4\pi} \int_V d^3r'\, \mathbf{M}(\mathbf{r}') \cdot \frac{\mathbf{r} - \mathbf{r}'}{|\mathbf{r} - \mathbf{r}'|^3}. \tag{13.29}$$

[11] See, for example, H.M. Schey, *Div, Grad, Curl and All That*, 3rd edition (W.W. Norton, New York, 1997).

The first term in (13.29) contributes only if $\mathbf{r} \in V$. Therefore, the definition

$$\psi_M(\mathbf{r}) = \frac{1}{4\pi} \int_V d^3r' \, \mathbf{M}(\mathbf{r}') \cdot \nabla' \frac{1}{|\mathbf{r} - \mathbf{r}'|} \tag{13.30}$$

permits us to write the field (13.29) in the form

$$\mathbf{B}_M(\mathbf{r}) = \begin{cases} \mu_0 \mathbf{M}(\mathbf{r}) - \mu_0 \nabla \psi_M(\mathbf{r}) & \mathbf{r} \in V, \\ -\mu_0 \nabla \psi_M(\mathbf{r}) & \mathbf{r} \notin V. \end{cases} \tag{13.31}$$

The integral (13.30) converges to a unique value everywhere and thus guarantees that (13.31) reproduces the Biot-Savart field (13.23) both inside and outside V. The reader will appreciate that it is not an accident that a comparison of (13.30) with (11.21) shows that $\psi_M(\mathbf{r})$ is the magnetic scalar potential of a volume distribution of point magnetic dipoles with density $\mathbf{M}(\mathbf{r})$.

Finally, a glance at (13.31) shows that it makes sense to define an auxiliary field $\mathbf{H}_M(\mathbf{r})$ from

$$\mathbf{H}_M(\mathbf{r}) = -\nabla \psi_M(\mathbf{r}). \tag{13.32}$$

Since $\mathbf{M} = 0$ outside the magnetized volume, this definition collapses the pair of equations (13.31) into the fundamental relation of magnetic matter,

$$\mathbf{B}_M(\mathbf{r}) = \mu_0[\mathbf{M}(\mathbf{r}) + \mathbf{H}_M(\mathbf{r})]. \tag{13.33}$$

Using (13.30) and (13.32), it is straightforward to check that the gauge freedom of $\mathbf{M}(\mathbf{r})$ discussed in Section 13.2.6 does not affect $\mathbf{B}_M(\mathbf{r})$ in (13.33). This must be so because $\mathbf{B}_M(\mathbf{r})$ is an observable via the Lorentz force.

13.4 Fictitious Magnetic Charge

A very useful way to think about the functions $\psi_M(\mathbf{r})$ and $\mathbf{H}_M(\mathbf{r})$ defined in the previous section appears when we rewrite (13.30) in the form

$$\psi_M(\mathbf{r}) = \frac{1}{4\pi} \int_V d^3r' \, \nabla' \cdot \left[\frac{\mathbf{M}(\mathbf{r}')}{|\mathbf{r} - \mathbf{r}'|} \right] - \frac{1}{4\pi} \int_V d^3r' \, \frac{\nabla' \cdot \mathbf{M}(\mathbf{r}')}{|\mathbf{r} - \mathbf{r}'|}. \tag{13.34}$$

Using the divergence theorem and exchanging the order of the two terms gives

$$\psi_M(\mathbf{r}) = \frac{1}{4\pi} \int_V d^3r' \, \frac{\rho^*(\mathbf{r}')}{|\mathbf{r} - \mathbf{r}'|} + \frac{1}{4\pi} \int_S dS \, \frac{\sigma^*(\mathbf{r}_S)}{|\mathbf{r} - \mathbf{r}_S|}, \tag{13.35}$$

where

$$\rho^*(\mathbf{r}) = -\nabla \cdot \mathbf{M}(\mathbf{r}) \qquad \text{and} \qquad \sigma^*(\mathbf{r}_S) = \mathbf{M}(\mathbf{r}_S) \cdot \hat{\mathbf{n}}(\mathbf{r}_S). \tag{13.36}$$

The charge density notation used in (13.36) is appropriate because, if $\mathbf{M}(\mathbf{r}) \to \mathbf{P}(\mathbf{r})/\epsilon_0$ in (13.36), the magnetostatic potential (13.35) transforms to the electrostatic potential (6.12) for a sample of polarized dielectric matter. Since $\mathbf{E} = -\nabla\varphi(\mathbf{r})$ is a field produced by volume and surface densities of *true* electric charge, it is natural to interpret $\mathbf{H}_M(\mathbf{r}) = -\nabla\psi_M$ as an "electric-like" field produced by *fictitious magnetic charge* with a volume density $\rho^*(\mathbf{r})$ and a surface density $\sigma^*(\mathbf{r})$.[12] The direct

[12] The magnetic-charge approach to magnetic matter began with Poisson in the early 19th century and dominated textbook treatments of the subject until the middle of the 20th century. The fact that *true* magnetic charge does not exist does not negate the usefulness of *fictitious* magnetic charge as a tool for computation and building intuition.

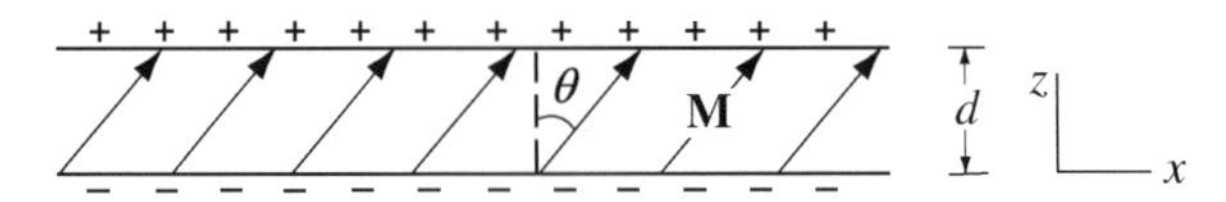

Figure 13.5: Side view of an infinite slab ($0 \le z \le d$) of matter with uniform magnetization $\mathbf{M}$. Plus and minus signs indicate surface distributions of fictitious magnetic charge σ^*.

analogy with Coulomb's law allows us to write down a formula for $\mathbf{H}_{\rm M}(\mathbf{r})$ that is valid at every point in space:

$$\mathbf{H}_{\rm M}(\mathbf{r}) = \frac{1}{4\pi}\int_V d^3r'\,\rho^*(\mathbf{r}')\frac{\mathbf{r}-\mathbf{r}'}{|\mathbf{r}-\mathbf{r}'|^3} + \frac{1}{4\pi}\int_S dS\,\sigma^*(\mathbf{r}_S)\frac{\mathbf{r}-\mathbf{r}_S}{|\mathbf{r}-\mathbf{r}_S|^3}. \tag{13.37}$$

Fictitious magnetic charge (or magnetic "poles") is a powerful idea that makes it possible to exploit electrostatic knowledge to solve problems with magnetized matter. Example 13.1 is a case in point. The magnetization of the slab is uniform, so the volume magnetic charge density $\rho^* = -\nabla\cdot\mathbf{M} = 0$. The top surface of the slab carries a magnetic surface charge density $\sigma^* = \mathbf{M}\cdot\hat{\mathbf{z}} = M\cos\theta$. The bottom surface carries a magnetic charge density $\mathbf{M}\cdot(-\hat{\mathbf{z}}) = -M\cos\theta$. Both distributions are indicated in Figure 13.5, which makes it clear that the slab is isomorphic to a parallel-plate capacitor.

For a capacitor whose plates carry surface charge densities $\pm\sigma$ as shown, we know that $\mathbf{E}_{\rm in} = -(\sigma/\epsilon_0)\hat{\mathbf{z}}$ between the plates and $\mathbf{E}_{\rm out} = 0$ otherwise. Therefore, the analogy between $\mathbf{E}$ and $\mathbf{H}_{\rm M}$ implies that $\mathbf{H}_{\rm M} = -\sigma^*\hat{\mathbf{z}}$ inside the slab and $\mathbf{H}_{\rm M} = 0$ outside the slab. Because $\mathbf{M}_{\rm in} = M\cos\theta\,\hat{\mathbf{z}} + M\sin\theta\,\hat{\mathbf{x}}$ and $\mathbf{M}_{\rm out} = 0$, we exactly reproduce the results of the effective current calculation in Example 13.1:

$$\mathbf{B}_{\rm M}(\mathbf{r}) = \mu_0(\mathbf{H}_{\rm M}+\mathbf{M}) = \begin{cases} \mu_0 M\sin\theta\,\hat{\mathbf{x}} & \text{inside the slab,} \\ 0 & \text{outside the slab.} \end{cases} \tag{13.38}$$

We will see in Section 13.8.1 that the magnetic charge concept can be reliably extended to calculate forces on magnetic bodies.

Finally, the magnetic scalar potential $\psi_{\rm M}(\mathbf{r})$ in (13.35) admits a multipole expansion identical in form to the one developed in Chapter 4 for the electrostatic scalar potential $\varphi(\mathbf{r})$. One simply replaces the true electric charge density in the latter by the fictitious magnetic charge density for the former. Then, because $\mathbf{M} = 0$ outside the matter (where the expansion is valid), the multipole expansion for $\mathbf{B}_{\rm M} = -\mu_0\nabla\psi_{\rm M}$ is identical in form to the multipole expansion for $\mathbf{E} = -\nabla\varphi$.

13.4.1 Potential Theory for $\mathbf{H}_{\rm M}(\mathbf{r})$

The Coulomb-like formula (13.37) suggests that we can construct a potential theory for magnetic matter based on the analogy with electrostatics. Specifically, (13.32), (13.33), and the left side of (13.36) show that

$$\nabla\times\mathbf{H}_{\rm M} = 0 \tag{13.39}$$

and

$$\nabla\cdot\mathbf{H}_{\rm M} = -\nabla\cdot\mathbf{M} = \rho^*. \tag{13.40}$$

The magnetic scalar potential itself satisfies

$$\nabla^2\psi_{\rm M} = \nabla\cdot\mathbf{M} = -\rho^*. \tag{13.41}$$

The Poisson equation (13.41) reduces to Laplace's equation in regions of space where $\mathbf{M}$ is uniform (constant):

$$\nabla^2 \psi_M(\mathbf{r}) = 0. \tag{13.42}$$

We conclude from these equations that all the techniques developed in previous chapters for electrostatic problems become available to solve problems with specified magnetization. The matching conditions for $\psi_M(\mathbf{r})$ are exactly the same as those for $\varphi(\mathbf{r})$ (see Section 3.3.2), namely,

$$\psi_1(\mathbf{r}_S) = \psi_2(\mathbf{r}_S) \tag{13.43}$$

and

$$\left[\frac{\partial \psi_1}{\partial n_1} - \frac{\partial \psi_2}{\partial n_1}\right]_S = [\mathbf{M}_1 - \mathbf{M}_2]_S \cdot \hat{\mathbf{n}}_1. \tag{13.44}$$

The right side of (13.44) is $\sigma_1^* + \sigma_2^*$ because both sides of the interface may contribute to the fictitious surface charge density defined in (13.36).[13] We leave it as an exercise for the reader to prove that any function that satisfies Equations (13.41) through (13.44) uniquely determines $\mathbf{H}_M(\mathbf{r})$ and $\mathbf{B}_M(\mathbf{r})$.

Application 13.1 A Uniformly Magnetized Sphere

Figure 13.6 shows a sphere with uniform magnetization $\mathbf{M} = M\hat{\mathbf{z}}$. The volume densities of both magnetization current $\mathbf{j}_M = \nabla \times \mathbf{M}$ and fictitious magnetic charge $\rho^* = -\nabla \cdot \mathbf{M}$ are zero for this system. Therefore, we may regard the magnetic field inside and outside the sphere as produced *either* by a surface density of fictitious magnetic surface $\sigma^* = \mathbf{M} \cdot \hat{\mathbf{n}}$ (plus and minus signs drawn onto the left-hand sphere) *or* by a surface density of magnetization current $\mathbf{K}_M = \mathbf{M} \times \hat{\mathbf{n}}$ (solid lines drawn onto the right-hand sphere). We will use σ^* to find the magnetostatic potential ψ_M and thus the magnetic field $\mathbf{B}_M$ everywhere.

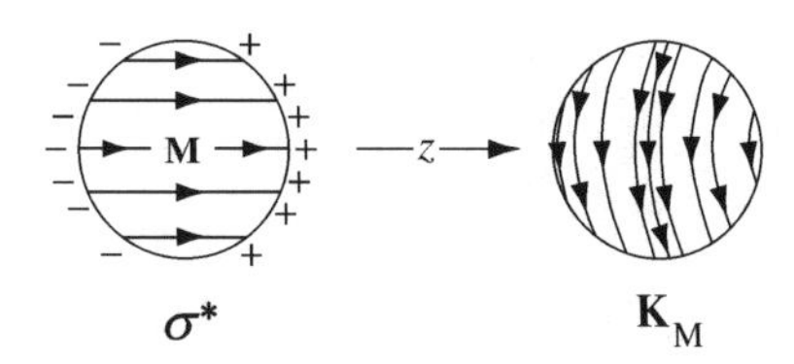

Figure 13.6: A uniformly magnetized sphere. Plus and minus signs on the left indicate the distribution of surface magnetic charge with density $\sigma^* = \mathbf{M} \cdot \hat{\mathbf{n}}$. Solid lines with arrows on the right indicate the distribution of magnetization surface current with density $\mathbf{K} = \mathbf{M} \times \hat{\mathbf{n}}$.

By symmetry, $\psi(\mathbf{r}) = \psi(r, \theta)$. This potential satisfies (13.42) inside and outside the sphere subject to the matching condition (13.44), which here reads

$$\left[\frac{\partial \psi_{\text{in}}}{\partial r} - \frac{\partial \psi_{\text{out}}}{\partial r}\right]_{r=R} = \sigma^* = \mathbf{M} \cdot \hat{\mathbf{r}} = M \cos\theta. \tag{13.45}$$

From the set of all regular, azimuthally symmetric solutions of Laplace's equation (see Section 7.6), the matching condition can be satisfied only if $\psi_{\text{in}} = Ar\cos\theta$ and $\psi_{\text{out}} = B\cos\theta/r^2$. Combining

[13] A direct derivation of (13.44) exploits (13.33) and the continuity of the normal component of $\mathbf{B}_M$ implied by $\nabla \cdot \mathbf{B}_M = 0$

(13.45) with (13.43) gives

$$\psi_{\mathrm{M}}(r,\theta) = \begin{cases} \frac{1}{3}Mz & r < R, \\ \frac{1}{3}MR^3\dfrac{\cos\theta}{r^2} & r > R. \end{cases} \tag{13.46}$$

Using (13.32) and (13.33), we find without difficulty that:

$$\mathbf{H}_{\mathrm{M}}(\mathbf{r}) = \begin{cases} -\dfrac{1}{3}\mathbf{M} & r < R, \\ \dfrac{R^3}{3}\left\{\dfrac{3(\hat{\mathbf{r}}\cdot\mathbf{M})\hat{\mathbf{r}} - \mathbf{M}}{r^3}\right\} & r > R, \end{cases} \tag{13.47}$$

and

$$\mathbf{B}_{\mathrm{M}}(\mathbf{r}) = \begin{cases} \dfrac{2}{3}\mu_0\mathbf{M} & r < R, \\ \mu_0\mathbf{H}_{\mathrm{M}}(\mathbf{r}) & r > R. \end{cases} \tag{13.48}$$

■

Figure 13.7 shows field lines for both $\mathbf{B}_{\mathrm{M}}(\mathbf{r})$ and $\mathbf{H}_{\mathrm{M}}(\mathbf{r})$. Outside the volume V of the sphere, $\mathbf{B}_{\mathrm{M}} = \mu_0\mathbf{H}_{\mathrm{M}}$ is identical to the magnetic field produced by a point magnetic dipole at the center of the sphere with dipole moment $\mathbf{m} = V\mathbf{M}$. From (13.20), this is the dipole moment of the entire sphere. Inside the sphere, $\mathbf{B}_{\mathrm{M}}$ and $\mathbf{H}_{\mathrm{M}}$ are uniform and anti-parallel. This is so because the lines of $\mathbf{B}_{\mathrm{M}}(\mathbf{r})$ must form closed loops while the lines of $\mathbf{H}_{\mathrm{M}}(\mathbf{r})$ must point away (or toward) the surface depending on the sign of the magnetic charge shown in Figure 13.6. A glance back at (11.33) shows that the constant value for $\mathbf{B}_{\mathrm{M}}$ inside the sphere is identical to the singular (delta function) part of the magnetic field of a point magnetic dipole. This shows that a point magnetic dipole may sensibly be regarded as the $R \to 0$ limit of a uniformly magnetized sphere. Otherwise, comparison with Application 6.1 in Section 6.3 shows that the field $\mathbf{H}_{\mathrm{M}}$ for a uniformly magnetized sphere is essentially identical to the field $\mathbf{E}_{\mathrm{P}}$ for a uniformly polarized sphere.

13.4.2 The Demagnetization Field

The field $\mathbf{H}_{\mathrm{M}}$ inside the volume V of a magnetized sample is often called the *demagnetization field* because it tends to point in the opposite direction to the magnetization $\mathbf{M}$. Perusal of Figures 13.5 to 13.7 shows that this happens because the surface charge density $\sigma^* = \mathbf{M}\cdot\hat{\mathbf{n}}$ is the source of the "electric-like" field $\mathbf{H}_{\mathrm{M}}$. In general, $\mathbf{H}_{\mathrm{M}}(\mathbf{r})$ and $\mathbf{B}_{\mathrm{M}}(\mathbf{r})$ vary from point to point inside the volume V of an arbitrary magnetized body. The sole exception is an ellipsoidal-shaped body where $\mathbf{H}_{\mathrm{M}}$ is a constant vector not necessarily parallel to $\mathbf{M}$.[14] This leads us to define a 3×3 *demagnetization tensor* $\mathbf{N}$ and write

$$\mathbf{B}_{\mathrm{M}}(\mathbf{r}\in V) = \mu_0[\mathbf{M} + \mathbf{H}_{\mathrm{M}}] = \mu_0[\mathbf{M} - \mathbf{N}\cdot\mathbf{M}]. \tag{13.49}$$

The obvious virtues of exposing the entire volume of an experimental sample to a uniform field motivates experimenters to exploit (13.49) by fashioning samples in the shape of ellipsoids, particularly limiting cases like thin rods and flat disks. The sphere is a limiting case where we can compute $\mathbf{H}_{\mathrm{M}}$ immediately using the fact that the trace of $\mathbf{N}$ is unity in the diagonal principal axis system, i.e., $N_{xx} + N_{yy} + N_{zz} = 1$.[15] Therefore, the symmetry of a sphere demands that $N_{xx} = N_{yy} = N_{zz} = 1/3$.

[14] See Brownstein (1987) in Sources, References, and Additional Reading.

[15] We leave the proof of this statement as an exercise for the reader.

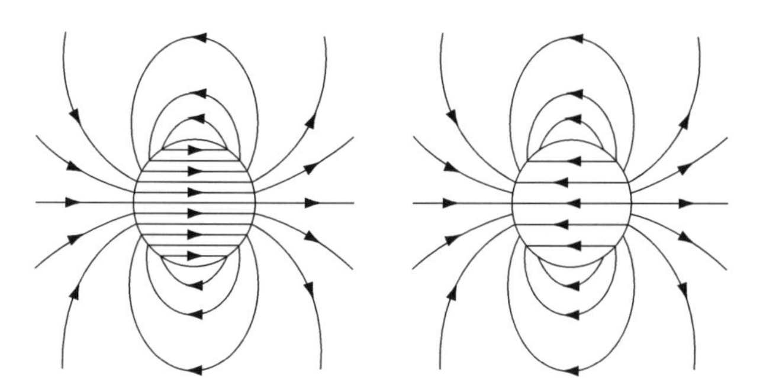

Figure 13.7: $\mathbf{B}_\mathrm{M}(\mathbf{r})$ (left) and $\mathbf{H}_\mathrm{M}(\mathbf{r})$ (right) for a sphere with uniform magnetization $\mathbf{M}$ that points from left to right.

Using this information in (13.49) confirms the result from (13.47) that

$$\mathbf{H}_\mathrm{M} = -\frac{1}{3}\mathbf{M} \qquad \text{(inside a sphere).} \tag{13.50}$$

Magnetic Charge or Circulating Current inside a Proton?

How do we know that the magnetic dipole moment of the proton (and the field it produces) is due to circulating (quark) currents and not due to a suitable distribution of real magnetic charge? The two panels of Figure 13.7 serve as a models for these two situations if we choose the sphere radii equal to the proton radius ($R \sim 10^{-15}$ m). If real magnetic charge existed, we would label the field on the right side of Figure 13.7 $\mathbf{B}_\mathrm{p}^\mathrm{charge}(\mathbf{r})$ to distinguish it from the field $\mathbf{B}_\mathrm{p}^\mathrm{current}(\mathbf{r})$ on left side of the figure. The two are identical outside the proton radius so a probe of the internal magnetic structure of the proton is needed to make a choice between them.

One approach studies the hyperfine splitting of the hydrogen spectrum that occurs when the magnetic moment of its electron $\mathbf{m}_\mathrm{e}$ interacts with the dipole magnetic field of the proton $\mathbf{B}_\mathrm{p}$. The piece of the potential energy $\hat{V}_B = -\mathbf{m}_\mathrm{e}\cdot\mathbf{B}_\mathrm{p}$ (see Section 12.4.2) that comes from the exterior field of the proton averages to zero due to the spherical symmetry of the 1s electron wave function. However, this wave function has non-zero amplitude at the position $\mathbf{r}_0$ of the proton. Therefore, if circulating currents are responsible for the proton's magnetic moment, the relevant magnetic field is the singular part of (11.33):

$$\mathbf{B}_\mathrm{p}^\mathrm{current}(\mathbf{r}) = \frac{2}{3}\mu_0\mathbf{m}_\mathrm{p}\delta(\mathbf{r}-\mathbf{r}_0).$$

If instead the proton is essentially a point magnetic charge dipole, the electrostatic analogy says that the relevant magnetic field is the singular part of the point electric dipole field (4.16) with $\mathbf{p}/\epsilon_0$ replaced by $\mu_0\mathbf{m}$:

$$\mathbf{B}_\mathrm{p}^\mathrm{charge}(\mathbf{r}) = -\frac{1}{3}\mu_0\mathbf{m}_\mathrm{p}\,\delta(\mathbf{r}-\mathbf{r}_0).$$

It is straightforward to compare the hyperfine shift predicted by these two fields with high-precision measurements. This comparison unequivocally favors the circulating current model.

13.5 The Total Magnetic Field

Section 13.2 focused on the magnetization current density $\mathbf{j}_\mathrm{M} = \nabla\times\mathbf{M}$ and the magnetic field $\mathbf{B}_\mathrm{M}$ produced by magnetized matter. No account was taken of any other sources of magnetic field, which were collectively called "free" current with density $\mathbf{j}_f$. In this section, we study the total magnetic field $\mathbf{B}$ produced by the total current density $\mathbf{j} = \mathbf{j}_{\rm M} + \mathbf{j}_f$. These are the quantities which enter Ampère's law, which we now write in the form

$$\nabla \times \mathbf{B} = \mu_0 \mathbf{j} = \mu_0(\mathbf{j}_{\rm M} + \mathbf{j}_f) = \mu_0 \left[\nabla \times \mathbf{M} + \mathbf{j}_f\right]. \tag{13.51}$$

Moving the $\nabla \times \mathbf{M}$ term to the left side of (13.51) motivates us to define the auxiliary vector field $\mathbf{H}(\mathbf{r})$ using

$$\mathbf{B}(\mathbf{r}) = \mu_0[\mathbf{H}(\mathbf{r}) + \mathbf{M}(\mathbf{r})]. \tag{13.52}$$

The field $\mathbf{H}$ in (13.52) is the natural generalization of the field $\mathbf{H}_{\rm M}$ in (13.33) when all sources of magnetic field are taken into account.

It is traditional to use $\mathbf{H}$ to rewrite Ampère's law in (13.51) and the other magnetostatic Maxwell equation,

$$\nabla \cdot \mathbf{B} = 0. \tag{13.53}$$

The two equations which result are the magnetostatic Maxwell equations in matter:

$$\nabla \times \mathbf{H} = \mathbf{j}_f \tag{13.54}$$

and

$$\nabla \cdot \mathbf{H} = -\nabla \cdot \mathbf{M} = \rho^*. \tag{13.55}$$

Equations (13.54) and (13.55) make it clear that both free current and fictitious magnetic charge are sources of $\mathbf{H}(\mathbf{r})$. Indeed, a direct application of the Helmholtz theorem (Section 1.9) generates the magnetic scalar potential as defined by (13.35) in a very natural way:[16]

$$\mathbf{H}(\mathbf{r}) = \frac{1}{4\pi}\int d^3r' \, \frac{\mathbf{j}_f(\mathbf{r}') \times (\mathbf{r} - \mathbf{r}')}{|\mathbf{r} - \mathbf{r}'|^3} - \nabla\psi_{\rm M}(\mathbf{r}). \tag{13.56}$$

13.5.1 Matching Conditions

Referring to Figure 13.8, the matching conditions for $\mathbf{H}(\mathbf{r})$ that derive from (13.54) and (13.55) are

$$\hat{\mathbf{n}}_2 \times [\mathbf{H}_1 - \mathbf{H}_2] = \mathbf{K}_f \tag{13.57}$$

and

$$\hat{\mathbf{n}}_2 \cdot [\mathbf{H}_1 - \mathbf{H}_2] = [\mathbf{M}_2 - \mathbf{M}_1] \cdot \hat{\mathbf{n}}_2. \tag{13.58}$$

Using (13.52), the latter is equivalent to

$$\hat{\mathbf{n}}_2 \cdot [\mathbf{B}_1 - \mathbf{B}_2] = 0. \tag{13.59}$$

These results reduce to those of Section 13.4.1 when there is no free current.

Finally, since it is always possible to write $\mathbf{B} = \nabla \times \mathbf{A}$, we remind the reader that the vector potential is always continuous at interfaces where material properties change discontinuously [see (10.79)]:

$$\mathbf{A}_1(\mathbf{r}_S) = \mathbf{A}_2(\mathbf{r}_S). \tag{13.60}$$

The tangential component of (13.60) is equivalent to (13.59).

[16] The surface integral piece of (13.35) is implicit from this point of view.

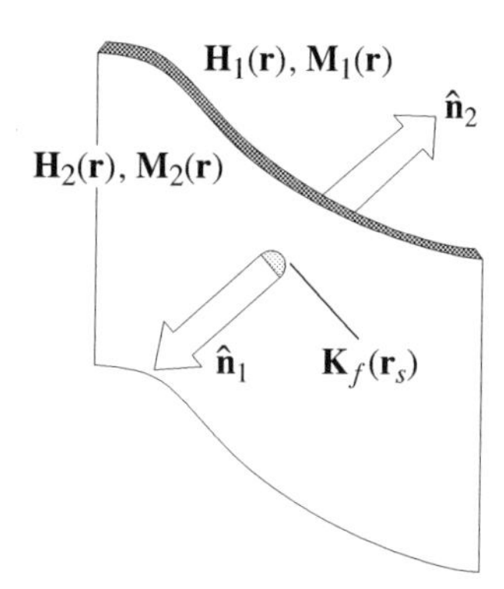

Figure 13.8: A free current density $\mathbf{K}_f$ flows on the boundary surface S between two (possibly magnetized) regions of space. The unit normal $\hat{\mathbf{n}}_1$ points outward from region 1; the unit normal $\hat{\mathbf{n}}_2$ points outward from region 2.

13.5.2 Constitutive Relations

The equations (13.52), (13.53), and (13.54) cannot be solved simultaneously unless (i) we specify $\mathbf{M}(\mathbf{r})$ once and for all or (ii) we invoke a *constitutive relation* which relates $\mathbf{M}$ to $\mathbf{H}$.[17] This is the point where we distinguish real magnetic matter from a collection of magnetic dipole moments by using experiment, theory, or phenomenology to inject quantum mechanical and statistical mechanical information into classical electrodynamics. For example, a large class of systems have no magnetic moment in the absence of a magnetic field, but acquire a macroscopic magnetization in the presence of a uniform external magnetic field. The general rule revealed by experiment for these systems is

$$M_i = \chi_{ij} H_j + \chi^{(2)}_{ijk} H_j H_k \ldots \tag{13.61}$$

The tensor character of the constants χ_{ij} and $\chi^{(2)}_{ijk}$ in (13.61) allows for the possibility that $\mathbf{M}$ is not parallel to $\mathbf{H}$. This is realized in spatially anisotropic matter. The last term in (13.61) allows for the possibility that the magnetization depends non-linearly on the field. This is realized if the magnetic field strength is large enough. For simplicity, we focus most of our attention on a special case of (13.61) which is the magnetic analog of the simple dielectric matter studied in Section 6.5. A brief discussion of permanent magnetism is the subject of Section 13.9 at the end of the chapter.

13.6 Simple Magnetic Matter

The first term on the right-hand side of (13.61) is sufficient to describe the magnetization of a *linear* magnet. In this book, a magnet that is both linear and spatially isotropic will be called *simple*. A simple magnet obeys the constitutive relation

$$\mathbf{M} = \chi_m \mathbf{H}. \tag{13.62}$$

The constant χ_m is called the *magnetic susceptibility*. Using (13.52), it is traditional to define a corresponding *magnetic permeability* μ and relative permeability $\kappa_m = \mu/\mu_0$ from

$$\mathbf{B} = \mu_0(\mathbf{H} + \mathbf{M}) = \mu\mathbf{H} = \kappa_m \mu_0 \mathbf{H} = \mu_0(1 + \chi_m)\mathbf{H}. \tag{13.63}$$

[17] For historical reasons, it is traditional in magnetism to relate $\mathbf{M}$ to the auxiliary field $\mathbf{H}$ rather than to the magnetic field $\mathbf{B}$.

Experiments show that (13.62) and (13.63) apply equally well to macroscopic fields that vary with position in a simple medium.

As stated in the Introduction to this chapter, materials where the field induces a magnetization parallel to **H**, and thus to **B**, are called *paramagnetic* ($\chi_m > 0$). Spin moments make the dominant contribution to **M** in paramagnets. Materials where the field induces a magnetization anti-parallel to **H** are called *diamagnetic* ($\chi_m < 0$). Orbital moments make the dominant contribution to **M** in diamagnets. At room temperature, the elements of the left half of the periodic table are mostly paramagnetic. The elements of the right half of the periodic table are mostly diamagnetic. One can prove[18] that $\mu > 0$ and measured magnetic susceptibilities are typically quite small, $|\chi_m| \sim 10^{-4} - 10^{-6}$. The elements Fe, Ni, and Co in the middle of the periodic table are not simple magnets, but some alloys of these elements (including soft iron and silicon steel) behave effectively as simple magnets with relative permeabilities that are extremely large ($\kappa_m \sim 10^4$).

Example 13.2 A uniform external field $\mathbf{B}_0$ induces a uniform magnetization inside a simple magnetic sphere of radius a and permeability μ.[19] Find the magnetic moment **m** of the sphere.

Solution: The magnetic moment is $\mathbf{m} = \mathbf{M}V$ where $V = (4/3)\pi a^3$ is the sphere volume. Moreover, Application 13.1 showed that a sphere with uniform magnetization **M** produces the fields

$$\mathbf{H}_\mathrm{M}(r < a) = -\frac{1}{3}\mathbf{M} \qquad \text{and} \qquad \mathbf{B}_\mathrm{M}(r < a) = \frac{2}{3}\mu_0\mathbf{M}.$$

Here, **M** is induced by the external field $\mathbf{B}_0$. Therefore, the total fields inside the sphere are

$$\mathbf{B}_\mathrm{in} = \mathbf{B}_0 + \mathbf{B}_\mathrm{M}(r < a) = \mathbf{B}_0 + \frac{2}{3}\mu_0\mathbf{M}$$

and

$$\mathbf{H}_\mathrm{in} = \mathbf{H}_0 + \mathbf{H}_\mathrm{M}(r < a) = \mathbf{H}_0 - \frac{1}{3}\mathbf{M}.$$

Inserting these two equations into $\mathbf{B}_\mathrm{in} = \mu\mathbf{H}_\mathrm{in}$ permits us to solve for **M**. Because $\mathbf{B}_0 = \mu_0\mathbf{H}_0$, the result is

$$\mathbf{M} = \frac{3}{\mu_0}\left(\frac{\mu - \mu_0}{\mu + 2\mu_0}\right)\mathbf{B}_0.$$

The corresponding magnetic moment of the sphere is

$$\mathbf{m} = \mathbf{M}V = 4\pi a^3\left(\frac{\mu - \mu_0}{\mu + 2\mu_0}\right)\mathbf{H}_0.$$

13.6.1 Fields and Sources in Simple Magnetic Matter

We begin our study of simple magnets by inserting (13.63) into the Maxwell equation (13.54) to get

$$\frac{1}{\mu}\nabla \times \mathbf{B} - \mathbf{B} \times \nabla\left(\frac{1}{\mu}\right) = \mathbf{j}_f. \tag{13.64}$$

The general problem posed by (13.64) is difficult, particularly when the permeability varies smoothly with position. In this book, we restrict ourselves to situations no more complicated than Figure 13.10

[18] See, for example, O.V. Dolgov, D.A. Kirzhnitz, and V.V. Losyakov, "On the admissible values of the static magnetic permeability", *Solid State Communications* **46**, 147 (1983).

[19] See Section 13.6.4 for the closely related problem of a magnetizable rod in a transverse external field.

in Section 13.6.2 below, where the permeability takes (different) constant values in distinct regions of space separated by sharp boundaries. Taking account of the second Maxwell equation (13.53), the defining equations in each region,

$$\nabla \times \mathbf{B} = \mu \mathbf{j}_f \qquad \text{and} \qquad \nabla \cdot \mathbf{B} = 0, \tag{13.65}$$

have the same structure as the magnetostatic Maxwell equations in vacuum with $\mu_0 \to \mu$ and $\mathbf{j} \to \mathbf{j}_f$. The global magnetic field is constructed from the fields calculated in each region by enforcing the matching conditions in Section 13.5.1.

We have learned that magnetization current and fictitious magnetic charge are equivalent ways to source the field produced by a magnetization $\mathbf{M} = \chi_m \mathbf{H}$. Using (13.62) and (13.63), the volume and surface densities of magnetization current for a simple magnet are

$$\mathbf{j}_\mathrm{M} = \nabla \times \mathbf{M} = \nabla \times (\chi_m \mathbf{H}) = \chi_m \mathbf{j}_f \tag{13.66}$$

and

$$\mathbf{K}_\mathrm{M} = \mathbf{M} \times \hat{\mathbf{n}} = \chi_m \mathbf{H} \times \mathbf{n}. \tag{13.67}$$

The volume and surface densities of fictitious magnetic charge for a simple magnet are

$$\rho^* = -\nabla \cdot \mathbf{M} = -\chi_m \nabla \cdot \mathbf{H} = 0 \tag{13.68}$$

and

$$\sigma^* = \mathbf{M} \cdot \hat{\mathbf{n}} = \chi_m \mathbf{H} \cdot \hat{\mathbf{n}}. \tag{13.69}$$

Typically, there are contributions to (13.67) and (13.69) from both sides of an interface between two simple magnets.

Substituting (13.68) into (13.55) shows that $\rho^* = \nabla \cdot \mathbf{H} = 0$ in volumes bounded by surfaces or interfaces where σ^* may or may not be zero. If there is no free current in the same volumes, (13.54) shows that $\nabla \times \mathbf{H} = 0$. In these circumstances, $\mathbf{H}$ is derivable from a magnetic scalar potential, $\mathbf{H} = -\nabla \psi$, which satisfies Laplace's equation,

$$\nabla^2 \psi(\mathbf{r}) = 0 \qquad \text{for } \mathbf{r} \text{ in } V \text{ where } \rho^* = 0 \text{ and } \mathbf{j}_f = 0. \tag{13.70}$$

Section 13.6.4 explores the potential theory approach to simple magnetic matter implied by (13.70). If $\rho^* = 0$ and $\mathbf{j}_f \neq 0$, the expression (13.56) is still available for $\mathbf{H}(\mathbf{r})$ except that the $-\nabla \psi_\mathrm{M}$ term now derives entirely from the fictitious surface magnetic charge density σ^*. In light of (13.69), this has the effect of changing an explicit formula for $\mathbf{H}(\mathbf{r})$ into an integral equation for $\mathbf{H}(\mathbf{r})$. For simple geometries, the integral equation becomes an algebraic equation.

13.6.2 Simple Magnetic Response to Free Current

The canonical example of the magnetizing power of a free current is an electromagnet composed of a solenoidal coil wrapped around a cylindrical rod of soft iron (Figure 13.9). We ignore end effects and suppose that n turns per unit length of wire with current I are wrapped around the rod. For a rod coaxial with the z-axis, we can use (13.65) and the infinite-length vacuum ($\mu = \mu_0$) solenoid solution derived in Section 10.2.2 to write down the answer immediately:

$$\mathbf{B} = \mu n I \hat{\mathbf{z}}. \tag{13.71}$$

This is a very large field compared to the vacuum solenoid because soft iron has a very large effective relative permeability $\kappa_\mathrm{m} = \mu/\mu_0$.

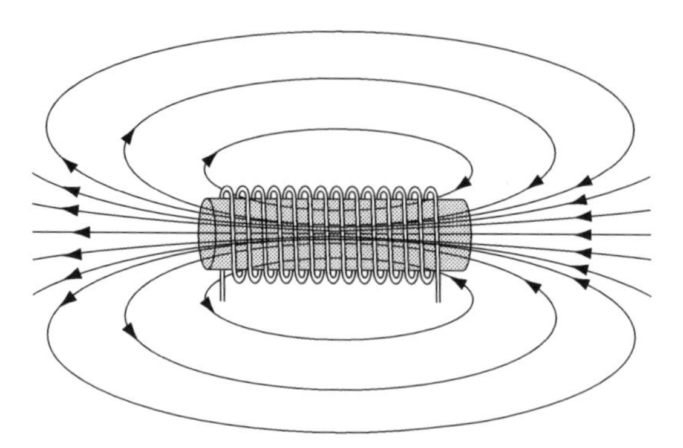

Figure 13.9: Cartoon of an iron-core electromagnet and its magnetic field.

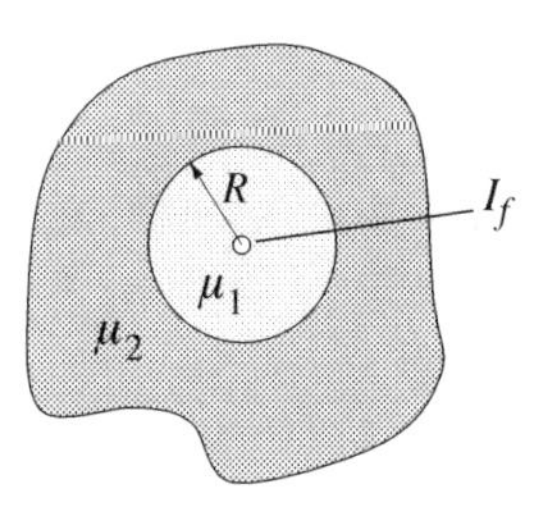

Figure 13.10: A long straight wire embedded in a coaxial cylinder with permeability μ_1 carries a current I_f. The cylinder is itself embedded in a medium with permeability μ_2.

Despite (or perhaps because of) the ease of this solution, it is important to appreciate that it is the contribution of $\mathbf{M} = \chi_m \mathbf{H}$ to $\mathbf{B} = \mu_0(\mathbf{H} + \mathbf{M})$ that one exploits by using an iron core.[20] Here, the only source of $\mathbf{H}$ is the free current supplied by the coil. Complications arise when the rod is finite and a fictitious magnetic charge per unit area $\sigma^* = \mathbf{M} \cdot \hat{\mathbf{n}}$ at each end of the rod introduces a demagnetization field.[21]

The magnetization charge density $\mathbf{j}_{\mathrm{M}}(\mathbf{r})$ in (13.66) is spatially coincident with the free current density $\mathbf{j}_f(\mathbf{r})$ that induces it. This macroscopic statement is a consequence of Lorentz averaging (Section 2.3). In microscopic reality, the current density merely magnetizes matter in its immediate neighborhood. Nevertheless, the strength of the effective current felt far from the source is altered. This underscores the fact that linear magnetic response is intrinsically *local*. The entire scenario is very reminiscent of the analogous situation when free charge polarizes a simple dielectric (see Section 6.5.2), and the example immediately below is correspondingly similar to the dielectric problem solved in Section 6.5.3.

Figure 13.10 shows a cylinder of radius R filled with matter with permeability μ_1 embedded in an infinite medium with permeability μ_2. A filamentary wire carries a free current I_f up the z-axis of the cylinder. In vacuum, the wire produces a magnetic field $\mathbf{B}(\rho) = \hat{\boldsymbol{\phi}}\mu_0 I_f / 2\pi\rho$ at every point in space. However, according to (13.66) and (13.63), the *embedded* wire carries an effective current $I' = I_{\mathrm{M}} + I_f = (\mu_1/\mu_0)I_f$. This source produces a magnetic field at every point in space equal to

$$\mathbf{B}_1(\rho) = \frac{\mu_1 I_f}{2\pi\rho}\hat{\boldsymbol{\phi}}. \tag{13.72}$$

[20] $|\mathbf{M}|$ cannot exceed the "saturation value" achieved when all the spins in the ferromagnet are aligned with $\mathbf{H}$.

[21] Demagnetization effects reduce the field enhancement predicted by (13.71). The latter is valid for a solenoid of length L and radius R only if $L \gg \mu R$.

On the other hand, the integral form of Ampère's law in (13.64) is

$$\oint d\boldsymbol{\ell} \cdot \mathbf{H} = I_f. \tag{13.73}$$

Using $\mathbf{B} = \mu_2 \mathbf{H}$, (13.73) says that the magnetic field in the $r > R$ region is

$$\mathbf{B}(\rho) = \frac{\mu_2 I_f}{2\pi\rho}\hat{\boldsymbol{\phi}}. \tag{13.74}$$

There is no contradiction between (13.72) and (13.74) because $\mathbf{B}_{\rm I}(\rho)$ is not the whole story. To get the Ampère's law result, we must add to (13.72) the magnetic field produced by the induced surface current density $\mathbf{K} = (\mathbf{M}_{\rm I} - \mathbf{M}_2) \times \hat{\mathbf{r}}$ that flows on the cylinder which separates the two regions in Figure 13.10. We leave the details as an exercise for the reader.

13.6.3 Simple Magnetic Response to a Fixed Field

In the absence of free current, external magnetic fields (produced by unspecified sources) must be present to produce magnetization. The field $\mathbf{H}(\mathbf{r})$ is determined by

$$\nabla \times \mathbf{H} = 0 \qquad \text{and} \qquad \nabla \cdot \mathbf{B} = 0. \tag{13.75}$$

In the usual way (see Section 2.3.3), the two equations in (13.75) imply that the tangential component of $\mathbf{H}$ and the normal component of $\mathbf{B}$ are continuous at an interface between media with different magnetic permeability. Ohmic matter is a good example because the (singular) current carried at the surface of a perfect conductor spreads out over the entire sample when the conductivity is finite.

Example 13.3 Show that lines of $\mathbf{B}$ "refract" at the interface as shown in Figure 13.11. Specifically, if α_k is the angle between $\mathbf{B}_k$ and the interface normal in a medium with permeability μ_k, show that

$$\mu_1 \tan\alpha_2 = \mu_2 \tan\alpha_1.$$

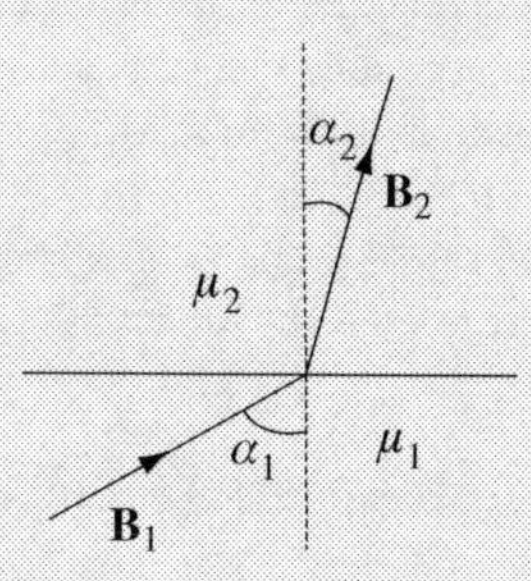

Figure 13.11: Refraction of magnetic field lines near an interface between two magnetizable media.

Solution: The equation on the left side of (13.75) says that the tangential component of $\mathbf{H} = \mathbf{B}/\mu$ is continuous. Therefore,

$$\frac{B_1 \sin\alpha_1}{\mu_1} = \frac{B_2 \sin\alpha_2}{\mu_2}.$$

The equation on the right side of (13.75) says that the normal component of $\mathbf{B}$ is continuous:

$$B_1 \cos\alpha_1 = B_2 \cos\alpha_2.$$

Dividing the first of these equations by the second gives the proposed law of field line refraction. The case $\mu_2 \gg \mu_1$ is interesting because both $\alpha_1 \approx 0$ and $\alpha_2 \approx \pi/2$ satisfy this law. The detailed geometry of the problem (including the locations of free currents away from the interface) determines which solution is adopted by Nature.

13.6.4 Potential Theory for a Simple Magnet

Potential theory applies to simple magnetic matter problems when the free current in (13.64) is confined to two-dimensional surfaces with surface density $\mathbf{K}_f$. Away from those surfaces, $\nabla \times \mathbf{H} = 0$ and the representation $\mathbf{H} = -\nabla\psi$ in terms of a magnetic scalar potential ψ is valid. Moreover, combining (13.68) with (13.55) shows that $\nabla \cdot \mathbf{H} = 0$ also. Therefore,

$$\nabla^2\psi(\mathbf{r}) = 0. \tag{13.76}$$

The matching condition implied by (13.64) comes over unchanged from Section 13.5.1:

$$\hat{\mathbf{n}}_2 \times [\mathbf{H}_1 - \mathbf{H}_2] = \mathbf{K}_f. \tag{13.77}$$

When $\mathbf{K}_f = 0$, the analogy with electrostatics shows that (13.77) simplifies to

$$\psi_1(\mathbf{r}_S) = \psi_2(\mathbf{r}_S). \tag{13.78}$$

Using $\mathbf{B} = \mu\mathbf{H}$, the complementary matching condition that the normal component of $\mathbf{B}$ is continuous is

$$\mu_1 \left.\frac{\partial\psi_1}{\partial n}\right|_S = \mu_2 \left.\frac{\partial\psi_2}{\partial n}\right|_S. \tag{13.79}$$

A Magnetizable Rod in a Transverse External Field

A historically significant application of potential theory to simple magnetic matter asks for the total magnetic field produced when a solid magnetizable rod coaxial with the z-axis is immersed in a uniform transverse external magnetic field $\mathbf{B}_0 = B_0\hat{\mathbf{x}}$ (Figure 13.12). Faraday discovered paramagnetism and diamagnetism using a geometry of this kind. To see how, we begin with the magnetic scalar potential $\psi_0 = -(B_0/\mu_0)x = -H_0\rho\cos\phi$ for the external field. The most general solution of Laplace's equation in polar coordinates (Section 7.9) consistent with both the external field and the matching conditions (13.78) and (13.79) is

$$\psi(r,\theta) = \begin{cases} A\rho\cos\phi & r < R, \\ (C/\rho - H_0\rho)\cos\phi & r > R. \end{cases} \tag{13.80}$$

Applying the matching conditions fixes the values of the coefficients as

$$A = -\frac{2\mu_0}{\mu+\mu_0}H_0 \qquad \text{and} \qquad C = \frac{\mu-\mu_0}{\mu+\mu_0}R^2H_0. \tag{13.81}$$

Outside the rod, $\mathbf{B}(\mathbf{r}) = -\mu_0\nabla\psi(\mathbf{r})$ is the sum of a uniform field and a dipole field. Inside the rod, the field is uniform. In terms of the relative permeability $\kappa_m = \mu/\mu_0$,

$$\mathbf{B}_{\text{in}} = \frac{2\kappa_m}{\kappa_m+1}\mathbf{B}_0. \tag{13.82}$$

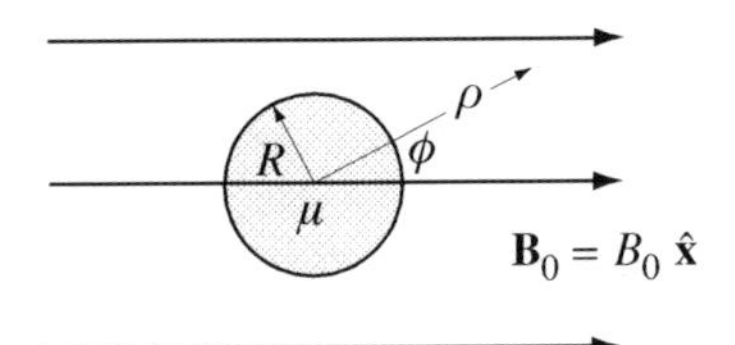

Figure 13.12: View down the z-axis of a rod with magnetic permeability μ immersed in a uniform transverse magnetic field. Only the external field $\mathbf{B}_0$ is shown.

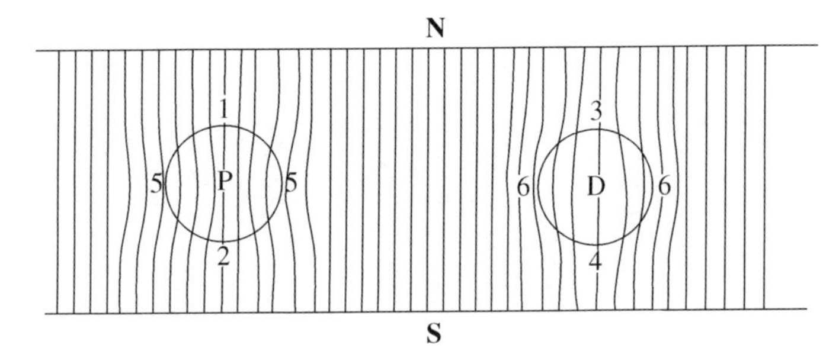

Figure 13.13: A diagram of lines of $\mathbf{B}(\mathbf{r})$ published by Faraday for a uniform field perturbed by magnetizable rods that point out of the plane of the paper. $\mathbf{B}_{\text{in}} > \mathbf{B}_0$ for the paramagnetic (P) rod on the left. $\mathbf{B}_{\text{in}} < \mathbf{B}_0$ for the diamagnetic (D) rod on the right. Faraday did not know that the field inside both rods is uniform [see (13.82)]. The numerical labeling was used by him in his discussion. The diagram appears in an 1851 entry of his *Experimental Researches in Electricity*.

Equation (13.82) shows that $\mathbf{B}_{\text{in}} > \mathbf{B}_0$ when $\kappa_m > 1$ and $\mathbf{B}_{\text{in}} < \mathbf{B}_0$ when $\kappa_m < 1$. In other words, the density of magnetic field lines inside the rod is larger (smaller) than the external field line density for paramagnetic (diamagnetic) matter. This point was appreciated by Faraday when he published Figure 13.13. Without the benefit of mathematics, he correctly understood that paramagnetic matter "attracts" field lines to its interior while diamagnetic matter "expels" magnetic field lines from its interior. The origin of this behavior will become clear in Section 13.7.2.

Faraday created the situation studied above by inserting one end of a magnetizable rod into the inhomogeneous field $\mathbf{B}(z)$ between the poles of a strong magnet. The magnetic dipole moment induced in each volume element dV of the rod is $d\mathbf{m} = dV\mathbf{M} = dV\chi_m\mathbf{H}$. Therefore, if A is the cross sectional area of the rod and B_0 is the field strength at the end of the rod, (12.40) gives the net force on the rod in the z-direction along its length as

$$F_z = \int dm_k \frac{\partial}{\partial z} B_k = \frac{1}{2}\frac{\chi_m}{1+\chi_m}\frac{A}{\mu_0}B_0^2. \tag{13.83}$$

Because χ_m is positive for a paramagnet and negative for a diamagnet, Faraday could distinguish a paramagnetic rod from a diamagnetic rod by the direction of the force (13.83).

An alternative derivation of (13.82) exploits the fact that $\mathbf{H}_0 = H_0\hat{\mathbf{x}}$ induces a magnetization $\mathbf{M} = \chi_m(\mathbf{H}_0 + \mathbf{H}_{\text{ind}})$, and thus a fictitious magnetic surface charge density $\sigma^* = \mathbf{M}\cdot\hat{\boldsymbol{\rho}} = M\cos\phi$ on the cylinder surface. Using the analogy between $\mathbf{H}$ and $\mathbf{E}$, any one of several electrostatic methods shows that a $\cos\phi$ surface charge distribution produces a uniform field $\mathbf{H}_{\text{ind}} = -\frac{1}{2}\mathbf{M}$ inside the cylinder. Therefore,

$$\mathbf{H}_{\text{ind}} = -\frac{1}{2}\chi_m(\mathbf{H}_0 + \mathbf{H}_{\text{ind}}) \qquad \Rightarrow \qquad \mathbf{H}_{\text{ind}} = -\frac{\chi_m}{\chi_m + 2}\mathbf{H}_0. \tag{13.84}$$

Equation (13.82) follows because $\mu = \mu_0(1+\chi_m)$ and $\mathbf{B} = \mu(\mathbf{H}_0 + \mathbf{H}_{\text{ind}})$.

Application 13.2 Magnetic Shielding

The fact that $\mathbf{B}(\mathbf{r})$ tends to concentrate in regions of high permeability is often used to shield isolated regions of space from the unwanted effects of an external field. The potential-theory method of the previous section is well suited to demonstrating this for the case of a cylindrical magnetic shell exposed to a uniform external field $\mathbf{B}_{\text{ext}}$ oriented perpendicular to the cylinder axis. The lines of $\mathbf{B}$ one finds for this problem are sketched in Figure 13.14.

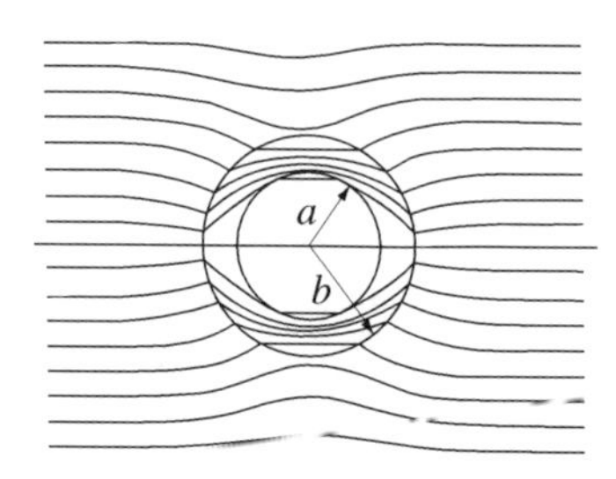

Figure 13.14: Side view of a cylindrical shell of magnetic matter which captures the flux of an external magnetic field. The field inside the shell is so small that only three magnetic field line are visible on the scale used to draw the lines.

If the shell has relative permeability $\kappa_m = \mu/\mu_0$, inner radius a, and outer radius b, the reader can confirm that the magnetic field inside the cylinder is constant:

$$\mathbf{B}_{\text{in}}(\rho < a) = \frac{4\kappa_m b^2}{(\kappa_m + 1)^2 b^2 - (\kappa_m - 1)^2 a^2}\mathbf{B}_{\text{ext}}. \tag{13.85}$$

This formula properly reduces to $\mathbf{B}_{\text{ext}}$ when $\kappa_m = 1$. More interesting is the limit $\kappa_m \gg 1$ where

$$\mathbf{B}_{\text{in}}(\rho < a) \approx \frac{4b^2}{b^2 - a^2}\frac{\mathbf{B}_{\text{ext}}}{\kappa_m} \qquad \kappa_m \gg 1. \tag{13.86}$$

This shows that the interior field goes to zero as $1/\kappa_m$ and that the shielding is most efficient when the thickness of the tube is not very small compared to its radius. ■

13.6.5 The Method of Images for Magnetic Matter

A distribution of magnetization $\mathbf{M}(\mathbf{r})$ or current density $\mathbf{j}_f(\mathbf{r})$ embedded near an interface between two magnetizable media creates fields that can be determined using a magnetic analog of the method of images. Figure 13.15 shows the example of a current loop I embedded in a medium with permeability μ_L. To analyze this situation, we recall from Figure 11.6 that any finite current loop can be decomposed into a sum of infinitesimal current loops. This invites us to treat the loop shown as infinitesimal and thus representable by a point magnetic dipole with moment $\mathbf{m}$ (indicated by the arrow embedded in medium L).

To this information we add the insight gained from the problem of the uniformly magnetized sphere solved in Application 13.1, namely, that the external field of a point magnetic dipole may be regarded as produced by either circulating electric charge or fictitious magnetic charge (indicated by the plus and minus signs embedded in medium L). This suggests the following strategy: calculate the magnetic scalar potential $\psi(\mathbf{r})$ produced by a single magnetic point charge g embedded in medium L and use superposition to treat the cases of embedded $\mathbf{M}(\mathbf{r})$ or embedded $\mathbf{j}_f$.

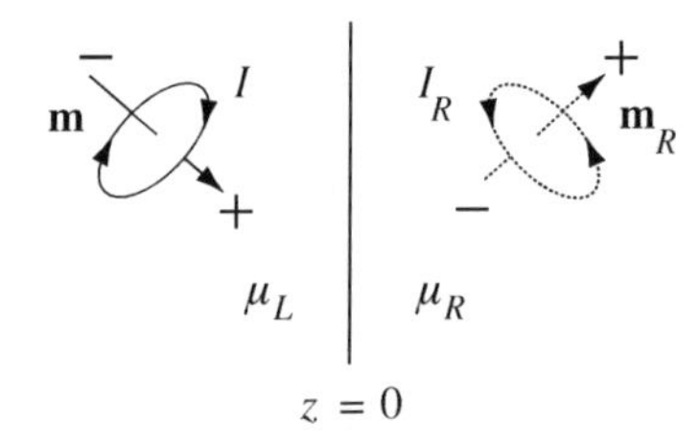

Figure 13.15: A current ring I in magnetic medium L (solid curve) and an image current ring I_R (dotted) in a medium R where $\mu_R > \mu_L$. The associated dipole moment vector in medium L (R) points slightly into (out of) the plane of the paper. Plus and minus signs indicate magnetic charges which produce the same dipole moments. The image moment $\mathbf{m}_R$ reverses direction if $\mu_R < \mu_L$.

From (13.41), the potential of a point magnetic charge with density $\rho^*(\mathbf{r}) = g\delta(\mathbf{r} - \mathbf{r}_0)$ is

$$\psi(\mathbf{r}) = \frac{g}{4\pi} \frac{1}{|\mathbf{r} - \mathbf{r}_0|}. \tag{13.87}$$

Moreover, the magnetic-potential matching conditions (13.78) and (13.79) are exactly the same as the electrostatic-potential matching conditions for the corresponding dielectric problem (Section 8.3.3) with the permittivity ϵ replaced by the permeability μ. Therefore, motivated by the image solution to the dielectric problem, we find that a magnetic charge g embedded at $(\rho, z) = (0, -d)$ in region L generates potentials (in cylindrical coordinates)

$$\psi_L(\rho, z) = \frac{1}{4\pi\mu_L} \left[\frac{g}{\sqrt{\rho^2 + (z+d)^2}} + \frac{g_R}{\sqrt{\rho^2 + (z-d)^2}} \right] \tag{13.88}$$

and

$$\psi_R(\rho, z) = \frac{1}{4\pi\mu_R} \frac{g_L}{\sqrt{\rho^2 + (z+d)^2}}, \tag{13.89}$$

where

$$g_R = \frac{\mu_L - \mu_R}{\mu_L + \mu_R} g \qquad \text{and} \qquad g_L = \frac{2\mu_L}{\mu_L + \mu_R} g. \tag{13.90}$$

The magnetic charge g and its image g_R have opposite signs when $\mu_R > \mu_L$.[22] Focusing on this case, the right side of Figure 13.15 shows the image magnetic charges generated in medium R by the two magnetic charges in medium L. The arrow in medium R shows the orientation of the associated image dipole moment $\mathbf{m}_R$. Notice that the component of $\mathbf{m}_R$ perpendicular (parallel) to interface is parallel (anti-parallel) to the corresponding component of $\mathbf{m}$. The image moment $\mathbf{m}_L$ is strictly parallel to $\mathbf{m}$. The right side of Figure 13.15 also shows the image current loop associated with $\mathbf{m}_R$. By extrapolation, if $\mu_R > \mu_L$, we conclude that a current density $\mathbf{j}(x, y, z)$ in medium L generates a magnetic field in L derivable from $\mathbf{j}$ itself and an image current density in medium R:

$$\mathbf{j}_R(x, y, z) = \frac{\mu_L - \mu_R}{\mu_L + \mu_R} \left[-j_x(x, y, -z), -j_y(x, y, -z), j_z(x, y, -z) \right]. \tag{13.91}$$

The magnetic field in medium R is obtained from the image current density in medium L:

$$\mathbf{j}_L(x, y, z) = \frac{2\mu_L}{\mu_L + \mu_R} \mathbf{j}(x, y, z). \tag{13.92}$$

We will use this result in Application 13.4 to estimate the sticking force of a refrigerator magnet.

[22] This is one way to understand why magnets stick to magnetizable surfaces.

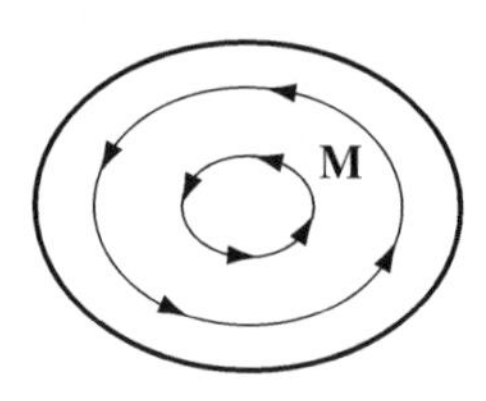

Figure 13.16: A disk-shaped ferromagnet with $\mathbf{M} = M\hat{\boldsymbol{\phi}}$.

13.6.6 Limiting Cases for the Permeability μ

We showed in Section 6.6.4 that a dielectric with permittivity $\epsilon \to \infty$ behaves in many ways like a perfect conductor. It is interesting and useful to study similar limiting cases for the magnetic permeability μ.

(A) The Perfect Ferromagnet ($\mu \to \infty$)

The effective permeability of a ferromagnet is very large (see Section 13.9). This leads us to define a *perfect ferromagnet* as a material described by $\mathbf{B} = \mu\mathbf{H}$ in the limit when $\mu \to \infty$. Since $\mathbf{B} = \mu_0(\mathbf{M} + \mathbf{H})$, the magnetic state of such a system is characterized by

$$\mathbf{H}(\mathbf{r}) = 0 \qquad \text{and} \qquad \mathbf{B}(\mathbf{r}) = \mu_0\mathbf{M}(\mathbf{r}) \qquad \text{(perfect ferromagnet).} \tag{13.93}$$

Figure 13.16 shows an infinitesimally thin disk with magnetization $\mathbf{M} = M\hat{\boldsymbol{\phi}}$. A perfect ferromagnet with this special shape creates no magnetic charge because $\rho^* = -\nabla \cdot \mathbf{M} = 0$ and $\sigma^* = \mathbf{M} \cdot \hat{\mathbf{n}} = 0$. Since the density of free current is also zero, (13.56) gives $\mathbf{H}(\mathbf{r}) = 0$ as advertised.[23]

The ferromagnetic limit can be applied to the refraction of field lines near the interface between two magnetic materials (see Figure 13.11). If medium 2 is a ferromagnet so $\mu_2 \to \infty$, the law of refraction in Example 13.3 is satisfied if the field lines in medium 1 are normal ($\alpha_1 \to 0$) to the surface of the ferromagnet. This is similar to the behavior of electric field lines at the surface of a perfect conductor. Indeed, ferromagnetic pole pieces are machined to match the curvature of desired magnetic equipotential surfaces in the same way that conducting electrodes are machined to match the curvature of desired electric equipotential surfaces. This is illustrated for a magnetic quadrupole field by Figure 13.17, which should be compared with the corresponding electric quadrupole case in Figure 7.13.

The limit $\mu_R \to \infty$ in (13.90) gives an image magnetic charge $g_L = 0$ while $g_R = -g$. This is similar to the behavior of a point charge outside a conducting surface. Combining these two observations leads to the conclusion that a thin sheet of ferromagnetic material screens an external magnetic field in the same way that a thin sheet of conducting material screens an electric field. A caveat is that the refraction law in Example 13.3 is also satisfied if the field lines in the ferromagnet are parallel to the interface ($\alpha_2 \to 90°$). This solution places no particular restriction on α_1. In practice, one finds that the overall geometry of the problem determines whether $\alpha_1 \to 0$ or $\alpha_2 \to 90°$.[24]

(B) The Superconductor ($\mu \to 0$)

A superconductor is a "perfect diamagnet" where $\mathbf{B}(\mathbf{r}) = 0$ at every point inside its volume V. In the presence of an applied magnetic field $\mathbf{B}_{\text{app}}(\mathbf{r})$, a superconductor spontaneously produces dissipation-free, circulating currents on its surface that produce a field $\mathbf{B}_{\text{ind}}(\mathbf{r})$ that cancels the applied field at every interior point:

$$\mathbf{B}(\mathbf{r}) = \mathbf{B}_{\text{app}}(\mathbf{r}) + \mathbf{B}_{\text{ind}}(\mathbf{r}) = 0 \qquad \mathbf{r} \in V. \tag{13.94}$$

[23] In a real ferromagnet, this type of structure rarely occurs because not all directions of $\mathbf{M}$ are energetically degenerate as assumed here. See Section 13.9.1.

[24] See Van Bladel (1961) in Sources, References, and Additional Reading.

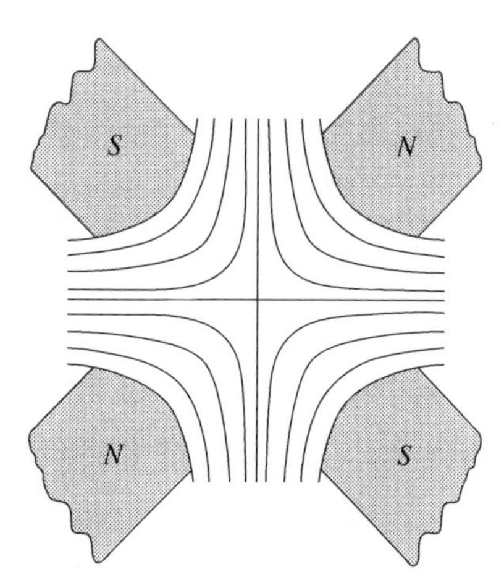

Figure 13.17: Lines of constant magnetic potential ψ for a quadrupole field produced by properly shaped ferromagnetic pole pieces. Figure adapted from Klemperer (1972).

Although perfect diamagnetism is fundamentally a quantum mechanical phenomenon, classical Maxwell theory provides a correct description of the $\mathbf{B}(\mathbf{r}) = 0$ state if a suitable constitutive relation is used for the superconductor.

For limited purposes, a superconductor may be considered a simple magnetic material ($\mathbf{B} = \mu\mathbf{H}$) if $\mu = 0$ is used to enforce (13.94). Combining this with the (always true) continuity of the normal component of the magnetic field produces a boundary condition at the surface of a superconductor for magnetostatics problems. We may write this as a condition on $\mathbf{B}_{\text{out}}$, or as a condition on the magnetic potential ψ_{out} with $\mu_{\text{in}} = 0$ in the matching condition (13.79):

$$\mathbf{B}_{\text{out}} \cdot \hat{\mathbf{n}}|_S = 0 \qquad \text{or} \qquad \left.\frac{\partial \psi_{\text{out}}}{\partial n}\right|_S = 0. \tag{13.95}$$

This boundary condition is consistent with the refraction law in Section 13.3 that tells us that $\mathbf{B}_{\text{out}}$ is always tangential to the surface of a superconductor. The limit $\mu_R \to 0$ applied to the magnetic image discussion of Section 13.6.5 shows that a point magnetic charge g placed outside the flat surface of a superconductor induces an identical image charge g inside the superconductor. It is possible to levitate a magnet (or current loop) above a superconductor because the repulsion of the magnet (or current loop) by its image can be balanced by its weight (see Section 13.8.1). Of course, the true source of the force is the magnetic field produced by currents induced on the surface of the superconductor.

(C) The Perfect Conductor

The limit $\mu \to 0$ can also be used to model the behavior of a perfect conductor. To understand this, we anticipate some dynamical results from Section 14.11, where we show that an external magnetic field penetrates into the bulk of an ohmic conductor at a rate which decreases as the conductivity σ increases. Figure 13.18(a) shows that the penetration is complete in the long-time ($t \to \infty$) limit. Classically, the only exception is the perfect conductor shown in Figure 13.18(b) because $\sigma \to \infty$ and the penetration rate goes to zero. Therefore, for magnetostatic purposes, it is correct to set $\mathbf{B} = 0$ in the interior of a perfect conductor. It is in this sense that the $\mu \to 0$ limit is used and, invoking the continuity of the normal component of $\mathbf{B}$, we get (13.95) as a magnetostatic boundary for a perfect conductor. A classical surface current provides the field $\mathbf{B}_{\text{ind}}$ needed to annul the external field.

These results apply immediately to the image current formula (13.91) derived in connection with Figure 13.15. We put $\mu_R = 0$ to make the right half-space a perfect conductor and conclude that a

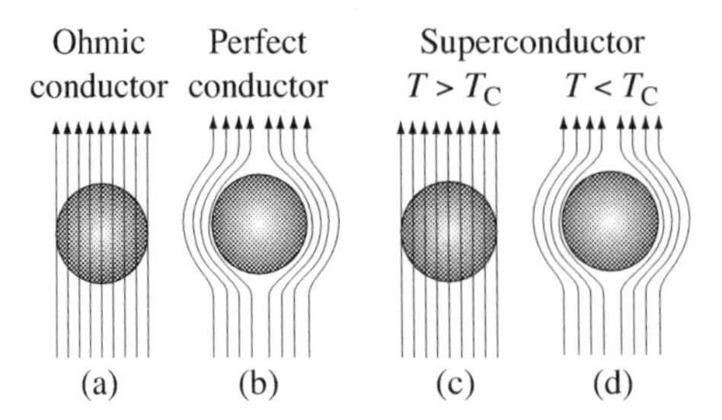

Figure 13.18: Asymptotic ($t \to \infty$) situation after $t = 0$ exposure to a uniform magnetic field: (a) field penetrates an ohmic conductor; (b) field does not penetrate a perfect conductor; (c) field penetrates a superconductor when $T > T_C$; field does not penetrate a superconductor when $T < T_C$.

current flowing in the plane $z = -d$ induces an image current flowing in the opposite direction in the plane $z = d$. Thus, a parallel plane current loop is repelled by a perfect conductor, just as it was by the superconductor in the previous section.

The foregoing suggests that a superconductor and a perfect conductor differ little from the point of view of magnetostatics. To gain a deeper understanding, Figure 13.18(c) shows that field penetration occurs for a superconductor when the temperature T is greater than the superconducting transition temperature T_C. This is so because, when $T > T_C$, a superconductor is just an ordinary ohmic medium. However, when $T < T_C$, quantum mechanical effects induce surface currents which oppose the penetration of an external field into the bulk of a superconductor. The final result, shown in Figure 13.18(d), is $\mathbf{B} = 0$ as discussed above. The difference between a perfect conductor and a superconductor appears when we study Figure 13.18(a) as $\sigma \to \infty$ and Figure 13.18(c) as T drops below T_C. In the first case, nothing happens and the existing magnetic flux remains trapped inside the (now perfectly conducting) sample. In the second case, superconducting surface currents appear and create a field $\mathbf{B}_{\text{ind}}$ which annuls the penetrated flux. In the colorful language of superconductivity, one says that the magnetic flux is "expelled" from the bulk and we return to Figure 13.18(d).

Example 13.4 Find the magnetic dipole moment of a perfectly conducting sphere with radius a.

Solution: We assume a uniform external field $\mathbf{B}_0$. By symmetry, the presumptive magnetic moment $\mathbf{m}$ sits at the center of the sphere and produces a dipole magnetic field $\mathbf{B}_{\text{ind}}$ which, outside the sphere, is identical to the magnetic field produced by the surface current induced on the sphere by $\mathbf{B}_0$. Using the point dipole field (11.33), the boundary condition (13.95) can be written

$$-\hat{\mathbf{r}} \cdot \mathbf{B}_0 = \hat{\mathbf{r}} \cdot \mathbf{B}_{\text{ind}}(r = a) = \frac{\mu_0}{2\pi} \frac{\hat{\mathbf{r}} \cdot \mathbf{m}}{a^3}.$$

Therefore, with $\mathbf{B} = \mu_0 \mathbf{H}_0$, the magnetic moment of the sphere is

$$\mathbf{m} = -2\pi a^3 \mathbf{H}_0.$$

Alternatively, we showed in Example 13.2 that a sphere with permeability μ placed in an external magnetic field $\mathbf{B}_0 = \mu_0 \mathbf{H}_0$ develops a magnetic moment

$$\mathbf{m} = 4\pi a^3 \left(\frac{\mu - \mu_0}{\mu + 2\mu_0} \right) \mathbf{H}_0.$$

The $\mu \to 0$ limit of this expression, $\mathbf{m} = -2\pi a^3 \mathbf{H}_0$, confirms the result above.

13.7 The Energy of Magnetic Matter

The total energy of magnetic matter contributes essentially to its thermodynamic properties. The potential energy of magnetic matter facilitates the calculation of forces that act in and on the matter. We discuss both in this section, beginning with a general magnetized body and then specializing to the special case of a simple magnet.

13.7.1 The Total Energy U_B

We showed in Section 12.2.1 that the Lorentz force does no work on a moving charged particle. Therefore, the internal energy of an element of matter with velocity $\boldsymbol{v}(\mathbf{r})$ changes only because the Coulomb electric force density $\rho(\mathbf{r})\mathbf{E}(\mathbf{r})$ does quasistatic work at a rate

$$\rho(\mathbf{r})\mathbf{E}(\mathbf{r}) \cdot \boldsymbol{v}(\mathbf{r}) = \mathbf{j}(\mathbf{r}) \cdot \mathbf{E}(\mathbf{r}). \tag{13.96}$$

The relevant portion of $\mathbf{j}(\mathbf{r})$ for magnetic matter is the magnetization current density $\mathbf{j}_{\mathrm{M}}(\mathbf{r}) = \nabla \times \mathbf{M}(\mathbf{r})$. This means that the internal energy of magnetic matter increases at a rate per unit volume

$$\frac{\partial u_{\mathrm{int}}}{\partial t} = (\nabla \times \mathbf{M}) \cdot \mathbf{E} = (\nabla \times \mathbf{E}) \cdot \mathbf{M} + \nabla \cdot (\mathbf{M} \times \mathbf{E}). \tag{13.97}$$

The last term vanishes when we integrate over all space. Therefore, using $\nabla \times \mathbf{E} = -\partial \mathbf{B}/\partial t$,

$$\frac{\partial u_{\mathrm{int}}}{\partial t} = -\frac{\partial \mathbf{B}}{\partial t} \cdot \mathbf{M}. \tag{13.98}$$

To this, we add the change in field energy density. Using (12.98) and (13.52), the change in the total internal energy density is then[25]

$$du_B = du_{\mathrm{field}} + du_{\mathrm{int}} = d\left\{B^2/2\mu_0\right\} - \mathbf{M} \cdot d\mathbf{B} = \mathbf{H} \cdot d\mathbf{B}. \tag{13.99}$$

The total energy is the net work required to establish the field $\mathbf{B}$. This leads us to a fundamental expression for the total energy of magnetized matter:

$$U_B = \int d^3r \int_0^{\mathbf{B}} \mathbf{H} \cdot \delta \mathbf{B}. \tag{13.100}$$

We infer from (13.99) that U_B is a natural function of the field $\mathbf{B}$. Therefore, if $\mathbf{R}$ labels the center of mass of the magnet, $U_B = U_B(\mathbf{B}, \mathbf{R})$, and the analog of (12.108) is

$$\mathbf{H} = \frac{1}{V}\left(\frac{\partial U_B}{\partial \mathbf{B}}\right)_{\mathbf{R}} \qquad \text{and} \qquad \mathbf{F} = -\left(\frac{\partial U_B}{\partial \mathbf{R}}\right)_{\mathbf{B}}. \tag{13.101}$$

13.7.2 U_B for a Simple Magnet

The integral (13.100) cannot be done analytically for, say, a ferromagnet where $\mathbf{B}$ depends on $\mathbf{H}$ in a complicated way (see Section 13.9). By contrast, the integration is straightforward for a simple magnet where $\mathbf{B} = \mu \mathbf{H}$. The result is

$$U_B[\mathbf{B}] = \int d^3r \int_0^{\mathbf{B}} \frac{\mathbf{B} \cdot \delta \mathbf{B}}{\mu} = \frac{1}{2}\int d^3r\, \mu |\mathbf{H}|^2 = \frac{1}{2}\int d^3r\, \mathbf{B} \cdot \mathbf{H}. \tag{13.102}$$

[25] It is an *assumption* that the field energy density in the magnet is the same as the field energy density in the vacuum.

An equivalent expression for the magnetic energy follows if we insert $\mathbf{B} = \nabla \times \mathbf{A}$ into the rightmost member of (13.102), integrate by parts, and use $\nabla \times \mathbf{H} = \mathbf{j}_f$:

$$U_B[\mathbf{B}] = \frac{1}{2} \int d^3r \, \mathbf{A} \cdot \mathbf{j}_f. \tag{13.103}$$

Equation (13.102) predicts that free current is attracted to regions where μ is large. This is so because the $\mathbf{H}$ field produced by a free current is largest near itself and the volume integral of $\mu|\mathbf{H}|^2$ in (13.102) is maximized when μ is large where $|\mathbf{H}|$ is large.[26] This explains the attractive force between the current loop and the interface in Figure (13.15) when $\mu_R > \mu_L$.

If we except situations constrained by symmetry, it is generally true that paramagnetic matter "attracts" lines of $\mathbf{B}$ and diamagnetic matter "repels" lines of $\mathbf{B}$. This occurs because the magnetization created by $\mathbf{B}$ in a simple medium amounts to a collection of point magnetic dipoles aligned (anti-aligned) with $\mathbf{B}$ for paramagnetic (diamagnetic) matter. Moreover, the magnitudes of the induced dipoles are proportional to $|\chi_m|$. Therefore, the potential energy $-\mathbf{m} \cdot \mathbf{B}$ [see (12.55)] gained by each dipole in paramagnetic matter is largest when $\mathbf{B}$ is large in regions where $|\chi_m|$ is large. Similarly, the potential energy lost by each dipole in diamagnetic matter is smallest when $\mathbf{B}$ is small in regions where $|\chi_m|$ is large. This provides a natural energetic explanation for the field line patterns in Figure 13.13 and Figure 13.14.

13.7.3 The Energy to Magnetize Simple Matter

Let $\mathbf{B}_0$ be a magnetic field in vacuum. It is instructive to calculate the change in energy that occurs when a sample of simple magnetic matter is inserted into the field, holding the currents which created $\mathbf{B}_0$ fixed. Because the sample magnetizes, the magnetic field $\mathbf{B}_0(\mathbf{r})$ changes to $\mathbf{B}(\mathbf{r})$ and the auxiliary field $\mathbf{H}_0(\mathbf{r}) = \mathbf{B}_0(\mathbf{r})/\mu_0$ changes to $\mathbf{H}(\mathbf{r})$. Using (13.102), the change in energy is

$$\Delta U_B[\mathbf{H}] = \frac{1}{2} \int d^3r \, (\mathbf{H} \cdot \mathbf{B} - \mathbf{H}_0 \cdot \mathbf{B}_0). \tag{13.104}$$

By adding and subtracting $\frac{1}{2}(\mathbf{H} \cdot \mathbf{B}_0 - \mathbf{H}_0 \cdot \mathbf{B})$, we temporarily complicate (13.104) to

$$\Delta U_B = \frac{1}{2} \int d^3r \, [\mathbf{B} \cdot (\mathbf{H} - \mathbf{H}_0) + (\mathbf{H} - \mathbf{H}_0) \cdot \mathbf{B}_0 + \mathbf{B} \cdot \mathbf{H}_0 - \mathbf{H} \cdot \mathbf{B}_0]. \tag{13.105}$$

The key step uses the equality of (13.102) and (13.103) to transform (13.105) to

$$\Delta U_B = \frac{1}{2} \int d^3r \, [\mathbf{A} \cdot (\mathbf{j}_f - \mathbf{j}_{f0}) + (\mathbf{j}_f - \mathbf{j}_{f0}) \cdot \mathbf{A}_0 + (\mathbf{B} - \mu_0 \mathbf{H}) \cdot \mathbf{H}_0]. \tag{13.106}$$

The final result follows from $\mathbf{B} = \mu_0(\mathbf{H} + \mathbf{M})$ and our assumption that $\mathbf{j}_f = \mathbf{j}_{f0}$:

$$\Delta U_B = \frac{1}{2} \int d^3r \, \mathbf{M} \cdot \mathbf{B}_0. \tag{13.107}$$

As discussed in Section 12.7.4 and in the two subsections to follow, ΔU_B is positive because it includes the energy required to maintain the currents. The *negative* of (13.107) is the potential energy gained by the matter during the process of magnetization.

13.7.4 The Potential Energy $\hat{U}_B$

Most magnetic systems operate under conditions of fixed current. In the present context, this makes the free current $\mathbf{j}_f$ the desirable independent variable. Since $\nabla \times \mathbf{H} = \mathbf{j}_f$, the field $\mathbf{H}$ is an equally

[26] Recall from the end of Section 12.7.3 that magnetic forces tend to *maximize* U_B.

good choice for the independent control parameter. Using (13.102) and the method of Legendre transformation (*cf.* Section 6.7.5), it is straightforward to define a potential energy function that is a natural function of **H** rather than **B**. The function of interest is

$$\hat{U}_B = U_B - \int d^3r\, \mathbf{B} \cdot \mathbf{H}. \tag{13.108}$$

To confirm that $\hat{U}_B = \hat{U}_B[\mathbf{H}]$, it is enough to combine (13.100) with the total differential $d\hat{U}_B$ computed from (13.108). The result is

$$d\hat{U}_B = -\int d^3r\, \mathbf{B} \cdot \delta\mathbf{H}. \tag{13.109}$$

This shows that the analog of (13.101) is

$$\mathbf{B} = -\frac{1}{V}\left(\frac{\partial \hat{U}_B}{\partial \mathbf{H}}\right)_{\mathbf{R}} \qquad \text{and} \qquad \mathbf{F} = -\left(\frac{\partial \hat{U}_B}{\partial \mathbf{R}}\right)_{\mathbf{H}}. \tag{13.110}$$

The notation is consistent with that used in Section 12.7.

13.7.5 The Potential Energy of a Simple Magnet

For simple matter, we can use (13.102) for U_B in (13.108). Since $\mathbf{B} = \mu\mathbf{H}$,

$$\hat{U}_B = -\frac{1}{2}\int d^3r\, \mathbf{B} \cdot \mathbf{H} = -\frac{1}{2}\int d^3r\, \mu|\mathbf{H}|^2 = -U_B. \tag{13.111}$$

An equivalent formula follows from (13.103) and the equality of the first and last terms in (13.111):

$$\hat{U}_B = -\frac{1}{2}\int d^3r\, \mathbf{A} \cdot \mathbf{j}_f. \tag{13.112}$$

As noted earlier, (13.111) explains the attraction between the loop of free current and the magnetic interface in Figure 13.15 when $\mu_R > \mu_L$. The potential energy $\hat{U}_B$ is reduced by a force which tends to increase the magnitude of **H** in the medium with the largest permeability. We will make further use of these results in Section 13.8.3.

13.8 Forces on Magnetic Matter

13.8.1 Coulomb's Law for Magnetism

A remarkable and non-obvious feature of the fictitious magnetic charge densities defined in Section 13.4 is that the force between two isolated magnetized objects can be calculated using these densities and a Coulomb-like inverse-square law of force. We demonstrate this here for the case of two very long and very thin magnetized rods of the kind used by Coulomb in his original magnetism experiments. The next section gives a general proof for arbitrary distributions of magnetic matter.

Figure 13.19 shows two magnetic rods in a collinear arrangement separated by a distance d. The force between them is given by Ampère's formula (12.24) evaluated using their respective magnetization currents. The volume current density $\mathbf{j}_{\mathrm{M}} = \nabla \times \mathbf{M}$ is zero because **M** is uniform. The surface current density $\mathbf{K}_{\mathrm{M}} = \mathbf{M} \times \hat{\mathbf{n}}$ is solenoidal and circulates in opposite directions around the circumference of the two rods. There is no surface current density on the end cap surfaces of either rod. Our strategy to calculate the net force between the rods is to add up the forces between a dense set of current rings (representing one rod) and a dense set of coaxial and counter-circulating current rings (representing the other rod).

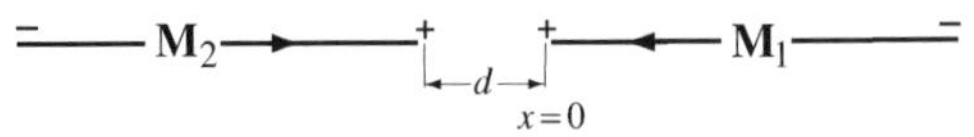

Figure 13.19: Two collinear rods separated by a distance d. The uniform magnetizations of the rods point toward one another as shown. Plus and minus signs indicate fictitious magnetic point charges at the ends of the rods.

In Example 12.7, we calculated the force between two coaxial and counter-circulating current rings with radius R separated by a distance h. The thin rods of interest here correspond to the limit $h \gg R$ where the repulsive force between the rings involves only the magnetic moment of each ring:

$$F = \frac{3\mu_0}{2\pi}\frac{m_1 m_2}{h^4}. \tag{13.113}$$

This shows that our calculation effectively models each rod as a linear array of parallel, point magnetic dipoles. For a rod with magnetization M_k, length L_k, and cross sectional area A_k, the dipole moment per unit length of rod is

$$\frac{m_k}{L_k} = M_k A_k \equiv g_k. \tag{13.114}$$

By superposing forces like (13.113) between every dipole of one rod and every dipole of the other rod, we represent the net repulsive force between the rods in the form

$$F = \frac{3\mu_0}{2\pi}\int\limits_{-L_2-d}^{-d} dx' \int\limits_{0}^{L_1} dx \frac{g_1 g_2}{(x-x')^4}. \tag{13.115}$$

Straightforward integration gives the final result,

$$F = \frac{\mu_0}{4\pi}\frac{g_1 g_2}{d^2}\left[1 - \frac{d^2}{(L_1+d)^2} - \frac{d^2}{(L_2+d)^2} + \frac{d^2}{(L_1+L_2+d)^2}\right]. \tag{13.116}$$

To interpret (13.116), we compute the fictitious magnetic charge density $\sigma^* = \mathbf{M}\cdot\hat{\mathbf{n}}$ associated with each rod. As shown in Figure 13.19, there are two positive point charges at the rod ends nearest to each other and two negative point charges at the rod ends farthest from each other. The magnitudes q_k^* of these charges satisfy

$$\frac{q_k^*}{A_k} = \sigma_k^* = M_k. \tag{13.117}$$

Comparing (13.117) with (13.114) shows that $q_k^* = g_k$. Now, from (10.29) or from (13.37) with $\rho^* = g\delta(\mathbf{r})$, the magnetic field produced by a point magnetic charge g at the origin is

$$\mathbf{B}(\mathbf{r}) = \frac{\mu_0 g}{4\pi}\frac{\hat{\mathbf{r}}}{r^2}. \tag{13.118}$$

This shows that, if $\mathbf{F} = g_0\mathbf{B}(\mathbf{r}_0)$ is the force exerted on g_0 by g, (13.116) is reproduced exactly when we sum this force over four pairs of fictitious magnetic charges at the ends of the rods. We will prove in the next section that this tremendous labor-saving device extends to arbitrary distributions of magnetic charge. The force (13.116) reduces to the net inverse-square law observed by Coulomb in the long-rod limit when $L_1, L_2 \gg d$.

13.8.2 An Isolated Magnetic Body

Figure 13.20 shows an isolated sample of magnetic matter with magnetization $\mathbf{M}(\mathbf{r})$ in an external magnetic field $\mathbf{B}_0(\mathbf{r})$. It is immaterial whether $\mathbf{M}$ is due to $\mathbf{B}_0$ or not. Several formulae are available

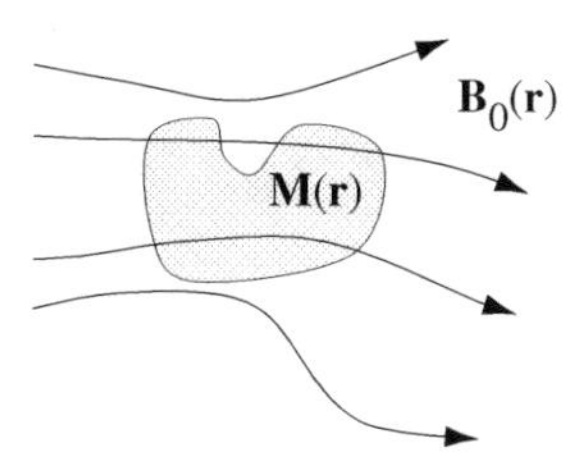

Figure 13.20: An isolated sample of magnetized matter in the presence of an external magnetic field $\mathbf{B}_0(\mathbf{r})$. The field lines represent $\mathbf{B}_0(\mathbf{r})$ only.

to calculate the electromagnetic force exerted by $\mathbf{B}_0(\mathbf{r})$ on this object. The first is the total Lorentz force evaluated using (i) the volume and surface magnetization current densities $\mathbf{j}_M = \nabla \times \mathbf{M}$ and $\mathbf{K}_M = \mathbf{M} \times \hat{\mathbf{n}}$ and (ii) whatever free current density $\mathbf{j}_f$ happens to be present inside the magnet. Thus,

$$\mathbf{F} = \int_V d^3r\,[\mathbf{j}_f(\mathbf{r}) + \nabla \times \mathbf{M}(\mathbf{r})] \times \mathbf{B}_0(\mathbf{r}) + \int_S dS\,[\mathbf{M}(\mathbf{r}_S) \times \hat{\mathbf{n}}(\mathbf{r}_S)] \times \mathbf{B}_0(\mathbf{r}_S). \tag{13.119}$$

A second point of view recalls from Section 13.3.1 that $\mathbf{M}(\mathbf{r})$ behaves exactly like a volume distribution of point magnetic dipoles. Therefore, a generalization of the point dipole result (12.42) suggests that the force on the sample volume is

$$\mathbf{F} = \int_V d^3r\,\mathbf{j}_f(\mathbf{r}) \times \mathbf{B}_0(\mathbf{r}) + \int_V d^3r\,[\mathbf{M}(\mathbf{r}) \cdot \nabla]\mathbf{B}_0(\mathbf{r}). \tag{13.120}$$

The force formulae (13.119) and (13.120) are equal by virtue of the identity

$$\int_V d^3r\,[(\mathbf{M} \cdot \nabla)\mathbf{B}_0 - \mathbf{M}(\nabla \cdot \mathbf{B}_0) + \mathbf{M} \times (\nabla \times \mathbf{B}_0) + \mathbf{B}_0 \times (\nabla \times \mathbf{M})] = \int_S dS\,(\mathbf{M} \times \hat{\mathbf{n}}) \times \mathbf{B}_0, \tag{13.121}$$

and because $\nabla \cdot \mathbf{B}_0 = 0$ (always true) and $\nabla \times \mathbf{B}_0 = 0$ (the sources of $\mathbf{B}_0$ are far away).

Finally, Coulomb's law for magnetism (Section 13.8.1) emerges if we rewrite (13.120) using the identity

$$\int_S dS(\hat{\mathbf{n}} \cdot \mathbf{M})\mathbf{B}_0 = \int_V (\nabla \cdot \mathbf{M})\mathbf{B}_0 + \int_V (\mathbf{M} \cdot \nabla)\mathbf{B}_0. \tag{13.122}$$

Since $\rho^* = \nabla \cdot \mathbf{M}$ and $\sigma^* = \mathbf{M} \cdot \hat{\mathbf{n}}$,

$$\mathbf{F} = \int_V d^3r\,\mathbf{j}_f(\mathbf{r}) \times \mathbf{B}_0(\mathbf{r}) + \int_V d^3r\,\rho^*(\mathbf{r})\mathbf{B}_0(\mathbf{r}) + \int_S dS\,\sigma^*(\mathbf{r}_S)\mathbf{B}_0(\mathbf{r}_S). \tag{13.123}$$

In other words, the magnetic force on an isolated body due to its own magnetization can be computed with no approximation using the idea of fictitious magnetic charge.

The total magnetic field, $\mathbf{B}(\mathbf{r}) = \mathbf{B}_0(\mathbf{r}) + \mathbf{B}_{\text{self}}(\mathbf{r})$, is the sum of the external field and the field produced by the magnet. However, because a magnetic body cannot exert a force on itself, it should be possible to replace the external field $\mathbf{B}_0(\mathbf{r})$ by the total field $\mathbf{B}(\mathbf{r})$ in all the foregoing. This is true for the volume integrals in (13.119) and (13.123). For the corresponding surface integrals, we replace $\mathbf{B}_0(\mathbf{r}_S)$ by $\mathbf{B}_{\text{avg}}(\mathbf{r}_S) = \frac{1}{2}[\mathbf{B}_{\text{in}}(\mathbf{r}_S) + \mathbf{B}_{\text{out}}(\mathbf{r}_S)]$ because the tangential component of $\mathbf{B}$ is generally not continuous at a material interface (see Section 10.3.4). Therefore, if $\mathbf{j}(\mathbf{r})$ and $\mathbf{K}(\mathbf{r}_S)$ are the total volume and surface current densities, valid formulae for the net force are

$$\mathbf{F} = \int_V d^3r\, \mathbf{j}(\mathbf{r}) \times \mathbf{B}(\mathbf{r}) + \int_S dS\, \mathbf{K}(\mathbf{r}_S) \times \mathbf{B}_{\rm avg}(\mathbf{r}_S) \tag{13.124}$$

and

$$\mathbf{F} = \int_V d^3r\, \mathbf{j}_f(\mathbf{r}) \times \mathbf{B}(\mathbf{r}) + \int_V d^3r\, \rho^*(\mathbf{r})\mathbf{B}(\mathbf{r}) + \int_S dS\, \sigma^*(\mathbf{r}_S)\mathbf{B}_{\rm avg}(\mathbf{r}_S). \tag{13.125}$$

The presence of the gradient in (13.120) generates a surface term when we replace $\mathbf{B}_0(\mathbf{r})$ by $\mathbf{B}(\mathbf{r})$. We leave it to the reader to confirm that

$$\mathbf{F} = \int_V d^3r\, \mathbf{j}_f(\mathbf{r}) \times \mathbf{B}(\mathbf{r}) + \int_V d^3r\, [\mathbf{M}(\mathbf{r}) \cdot \nabla]\mathbf{B}(\mathbf{r}) + \frac{\mu_0}{2}\int_S dS\, [\hat{\mathbf{n}}(\mathbf{r}_S) \cdot \mathbf{M}(\mathbf{r}_S)]^2. \tag{13.126}$$

The bilinear character of the integrands in this section raises the same questions about the macroscopic validity of our force formulae as arose in the Lorentz averaging discussion of Section 2.3.1. Therefore, as we did in Section 6.8.1 for dielectric matter, it is necessary to treat the validity of formulae like (13.124), (13.125), and (13.126) as logically independent assumptions of the macroscopic theory, subject to verification by experiment. On the other hand, it is straightforward to check that the gauge freedom of $\mathbf{M}(\mathbf{r})$ (see Section 13.2.5) has no effect on the net force and torque.

13.8.3 A Simple Magnetic Sub-Volume

Section 13.8.2 was devoted to the net force that an external magnetic field exerts on an isolated magnetized body. In this section, we calculate the force on a simple magnetic sub-volume. This class of problems includes the force on a free current distribution embedded in a permeable medium and the force on the interface between two simple magnetic materials. As in the corresponding dielectric problem (Section 6.8.3), this cannot be done without a proper accounting of the short-range, quantum mechanical forces responsible for cohesion and elasticity in the magnet. It is necessary to do this (albeit indirectly through the permeability) because these internal forces depend on the magnetization state of the sample. On the other hand, all internal forces cancel out in pairs when we sum over all sub-volumes to find the net force on an an isolated magnetic body.

13.8.4 Force from Variation of Energy for a Simple Magnet

Following the dielectric example of Section 6.8.4, we calculate force density in simple magnetic matter from the variation

$$\delta \hat{U}_B = -\mathbf{F} \cdot \delta\mathbf{r} \tag{13.127}$$

of the magnetic potential energy induced by a rigid displacement $\delta\mathbf{r}$. Given (13.111), the total variation is

$$\delta \hat{U}_B = -\int d^3r\, \mathbf{B} \cdot \delta\mathbf{H} - \frac{1}{2}\int d^3r\, \delta\mu|\mathbf{H}|^2. \tag{13.128}$$

Without any calculation, the second term on the right side of (13.128) permits us to reiterate a point made slightly differently in Section 13.7.2, namely, that the magnetic potential energy decreases (increases) if a paramagnetic (diamagnetic) material is inserted into a region of empty space where a field is present. This is so because the induced magnetic moment of a paramagnet (diamagnet) is parallel (anti-parallel) to the existing field and the energy of a magnetic dipole moment is lowered (raised) when it is aligned (anti-aligned) with a magnetic field. Hence, a paramagnet is attracted to regions of strong field from regions of weak field. A diamagnet is repelled from regions of strong field toward regions of weak field.

Our strategy is to use (13.103) to rewrite (13.128) in the form

$$\delta \hat{U}_B = -\int d^3r\, \mathbf{A}\cdot\delta\mathbf{j}_f - \frac{1}{2}\int d^3r\, \delta\mu|\mathbf{H}|^2. \tag{13.129}$$

From the analog of Figure 3.13, the variation of free current density is [see (3.65)]

$$\delta\mathbf{j}_f(\mathbf{r}) = \mathbf{j}_f(\mathbf{r}-\delta\mathbf{r}) - \mathbf{j}_f(\mathbf{r}) = -(\delta\mathbf{r}\cdot\nabla)\mathbf{j}_f. \tag{13.130}$$

A similar formula holds for $\delta\mu(\mathbf{r})$. Substituting both into (13.129), we can exploit the steady-current condition $\nabla\cdot\mathbf{j}_f = 0$ and the fact that $\delta\mathbf{r}$ is a constant vector to rewrite the $\mathbf{j}_f$ term using

$$\nabla\times(\delta\mathbf{r}\times\mathbf{j}_f) = [(\nabla\cdot\mathbf{j}_f) + (\mathbf{j}_f\cdot\nabla)]\delta\mathbf{r} - [(\nabla\cdot\delta\mathbf{r}) + (\delta\mathbf{r}\cdot\nabla)]\mathbf{j}_f = -(\delta\mathbf{r}\cdot\nabla)\mathbf{j}_f. \tag{13.131}$$

From here, an integration by parts and (13.127) give *Helmholtz' formula* for the force:

$$\mathbf{F} = -\frac{\delta\hat{U}_B}{\delta\mathbf{r}} = \int d^3r\, \mathbf{j}_f(\mathbf{r})\times\mathbf{B}(\mathbf{r}) - \frac{1}{2}\int d^3r\, H^2(\mathbf{r})\nabla\mu(\mathbf{r}). \tag{13.132}$$

The parallelism between (13.132) and the corresponding formula (6.130) for the force on a dielectric is striking. We draw particular attention to the second term, which shows that there is a force density associated with spatial variations in $\mu(\mathbf{r})$. Example 13.5 below shows that special care is required if $\mu(\mathbf{r})$ changes abruptly from one constant value to another.

13.8.5 The Magnetic Stress Tensor for a Simple Magnet

The analogy with dielectric matter motivates us to manipulate (13.132) into a form that is usually more convenient for calculations. To do this, begin with the identity $\frac{1}{2}H^2\partial_j\mu = \frac{1}{2}\partial_j(B_kH_k) - B_k\partial_jH_k$ and use the Levi-Cività machinery to simplify $\mathbf{j}_f\times\mathbf{B} = (\nabla\times\mathbf{H})\times\mathbf{B}$. Some cancellation occurs and we find

$$\mathbf{F} = \int d^3r\, (\mathbf{B}\cdot\nabla)\mathbf{H} - \frac{1}{2}\int d^3r\, \nabla(\mathbf{B}\cdot\mathbf{H}). \tag{13.133}$$

Now convert the last term in (13.133) to a surface integral and use the identity (13.122) with $\mathbf{a} = \mathbf{B}$ and $\mathbf{b} = \mathbf{H}$ to do the same with the first term. The result is very reminiscent of the formula (12.78) for the force on a distribution of current in vacuum:

$$\mathbf{F} = \int_S dS\, \left\{(\hat{\mathbf{n}}\cdot\mathbf{B})\mathbf{H} - \tfrac{1}{2}\hat{\mathbf{n}}(\mathbf{B}\cdot\mathbf{H})\right\}. \tag{13.134}$$

Indeed, if the analogy with its vacuum counterpart holds up, (13.134) is the expression we seek for the force on a sub-volume Ω of simple magnetic matter enclosed by a surface S.

It is conventional at this point to define the *magnetic stress tensor* for simple magnetic matter as

$$\mathcal{T}_{ij}(\mathbf{H}) = B_iH_j - \tfrac{1}{2}\delta_{ij}\mathbf{B}\cdot\mathbf{H}. \tag{13.135}$$

This object is symmetric because $B_i H_j = \mu H_i H_j$. Combining (13.135) with the divergence theorem, we can write the jth component of the force (13.134) on Ω in the form[27]

$$F_j = \int_\Omega d^3r\, \partial_i \mathcal{T}_{ij}(\mathbf{H}). \tag{13.136}$$

Application 13.3 The Magnetic Force on an Embedded Volume

Figure 13.21 shows a simple magnet with permeability $\mu = \kappa_m \mu_0$. Embedded in this magnet is a free line current I and a second magnet with permeability $\mu' = \kappa'_m \mu_0$. A line current I' is embedded in the second magnet. Our goal is to find the force exerted on the embedded magnet and I'. In principle, this force can be computed using the Helmholtz formula (13.132). The stress tensor method turns out to be simpler.

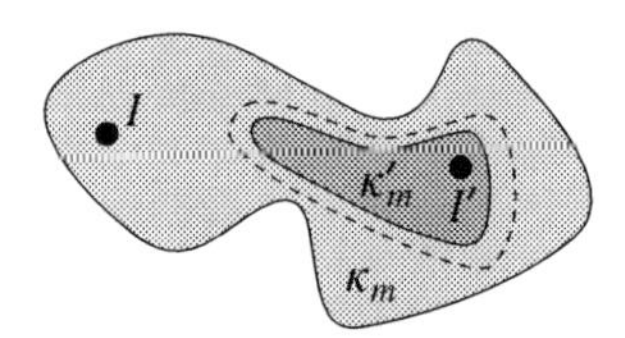

Figure 13.21: The dashed line is the boundary S of a volume Ω which is infinitesimally larger than the volume of a magnet with permeability κ'_m. The latter is embedded in a larger magnet with permeability κ_m.

Our strategy is to evaluate (13.134) using a surface S (dashed in Figure 13.21) that bounds a volume Ω which is infinitesimally larger than the physical volume of the embedded magnet. Because $\mathbf{H} = \mathbf{B}/\mu$ everywhere on S, we can pull $1/\mu$ outside the integral and then transform the surface integral into a volume integral as we did to get (13.136). The result is

$$F_j = \frac{1}{\mu} \int_\Omega d^3r\, \partial_i (B_i B_j - \tfrac{1}{2}\delta_{ij} \mathbf{B}\cdot\mathbf{B}). \tag{13.137}$$

With $\mu = \kappa_m \mu_0$, the algebra in Section 12.5 shows that (13.137) reduces to

$$\mathbf{F} = \frac{1}{\kappa_m} \int_\Omega d^3r\, \mathbf{j}(\mathbf{r}) \times \mathbf{B}(\mathbf{r}). \tag{13.138}$$

Let us apply (13.138) to find the force exerted by a uniform external field $\mathbf{B}_0$ on a filamentary current I that flows up the z-axis of an infinite medium with permeability μ. From (13.66), the only magnetization current in the problem is $\mathbf{j}_M = \chi_m \mathbf{j}_f$, so we can choose Ω as any cylindrical volume coaxial with the z-axis. The total field $\mathbf{B}$ is the sum of $\mathbf{B}_0$ and the field produced by $\mathbf{j} = \mathbf{j}_f + \mathbf{j}_M = \kappa_m \mathbf{j}_f$. But the latter cannot exert a force on Ω itself. Therefore, (13.138) reduces to

$$\mathbf{F} = \int_\Omega d^3r\, \mathbf{j}_f(\mathbf{r}) \times \mathbf{B}_0. \tag{13.139}$$

The force is exactly the same as if the magnetic matter were absent.

[27] See Brown (1951) and Landau and Lifshitz (1960) in Sources, References, and Additional Reading. These authors use entirely different arguments to derive (13.136) for an arbitrary sub-volume. Both consider the possibility of elastic deformation of the sub-volume and the variation of the permeability with density.

A similar argument applies if we replace the wire in the previous example with a current-carrying wire with finite radius and permeability μ'. In this case, $\mathbf{j}(\mathbf{r})$ includes a magnetization current density $\mathbf{K}_M$ induced on the surface of the wire. Nevertheless, the field produced by $\mathbf{j}(\mathbf{r})$ does not exert a force on itself. The field $\mathbf{B}_0$ does not exert any force on the magnetic matter (or the equivalent magnetization currents) because only a non-uniform field produces a net force on one or many point magnetic dipoles. This implies that the force is still given by (13.139). A brute-force calculation using (13.138) confirms this result. ■

Example 13.5 Calculate the force per unit area exerted on the μ_1-μ_2 interface at $z = 0$ in Figure 13.11 in Section 13.6.3.

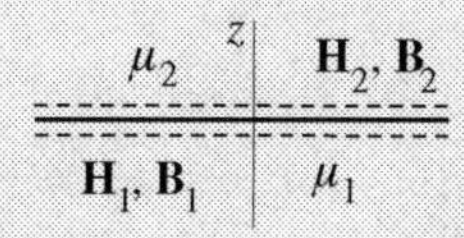

μ_2 | z | $\mathbf{H}_2, \mathbf{B}_2$

$\mathbf{H}_1, \mathbf{B}_1$ | μ_1

Figure 13.22: The interface $z = 0$ separates permeable medium μ_1 from permeable medium μ_2. Dashed lines indicate the integration surface S in (13.134).

Method I: We apply (13.134) to a closed surface S that fits snugly over the $z = 0$ plane (dashed lines in Figure 13.22). $\mathbf{B} = \mu\mathbf{H}$ in each region and it is convenient to decompose all fields into components perpendicular and normal to the interface, e.g., $\mathbf{H} = \mathbf{H}_\perp + H_z\hat{\mathbf{z}}$. Since $\hat{\mathbf{n}} = +\hat{\mathbf{z}}$ and $\hat{\mathbf{n}} = -\hat{\mathbf{z}}$ for the $z > 0$ and $z < 0$ portions of S, respectively, the force per unit area is

$$\begin{aligned} \mathbf{f} &= \mu_2 H_{2z}(\mathbf{H}_{2\perp} + H_{2z}\hat{\mathbf{z}}) - \tfrac{1}{2}\hat{\mathbf{z}}\mu_2(H_{2\perp}^2 + H_{2z}^2) \\ &\quad - \mu_1 H_{1z}(\mathbf{H}_{1\perp} + H_{1z}\hat{\mathbf{z}}) + \tfrac{1}{2}\hat{\mathbf{z}}\mu_1(H_{1\perp}^2 + H_{1z}^2). \end{aligned}$$

From Section 13.5.1, the matching conditions at the interface are

$$\mathbf{H}_{1\perp} = \mathbf{H}_{2\perp}$$

and

$$B_{1z} = \mu_1 H_{1z} = \mu_2 H_{2z} = B_{2z}.$$

This information simplifies the force-density expression to

$$\mathbf{f} = \left[\frac{1}{2}(\mu_1 - \mu_2)H_\perp^2 + \frac{1}{2}B_z^2\left(\frac{1}{\mu_2} - \frac{1}{\mu_1}\right)\right]\hat{\mathbf{z}}.$$

Both terms tend to expand the volume of the medium with large μ and shrink the volume of the medium with small μ.

Method II: We apply (13.132) where the force comes entirely from the second term. This gives the force per unit area in the form $\mathbf{f} = f\hat{\mathbf{z}}$ where

$$f = -\frac{1}{2}\int_{0^-}^{0^+} [H_\perp^2 + H_z^2]\frac{d\mu}{dz}dz.$$

Formally, $\mu(z) = \mu_1\Theta(-z) + \mu_2\Theta(z)$ so $d\mu/dz = (\mu_2 - \mu_1)\delta(z)$. This makes the integral ill-defined because H_z is not continuous at $z = 0$. However, B_z *is* continuous so we rewrite the integral as

$$f = -\frac{1}{2}\int_{0^-}^{0^+} H_\perp^2(\mu_2 - \mu_1)\delta(z)dz + \frac{1}{2}\int_{0^-}^{0^+} B_z^2 \frac{d}{dz}\left(\frac{1}{\mu}\right)dz.$$

Therefore, in agreement with Method I,

$$f = \frac{1}{2}(\mu_1 - \mu_2)H_\perp^2 + \frac{1}{2}B_z^2\left(\frac{1}{\mu_2} - \frac{1}{\mu_1}\right).$$

Example 13.6 Calculate the force that pulls a close-fitting cylindrical rod with permeability μ into a long solenoid. The solenoid in Figure 13.23 has a circular cross sectional area A and is tightly wound with N turns of a wire which carries a current I.

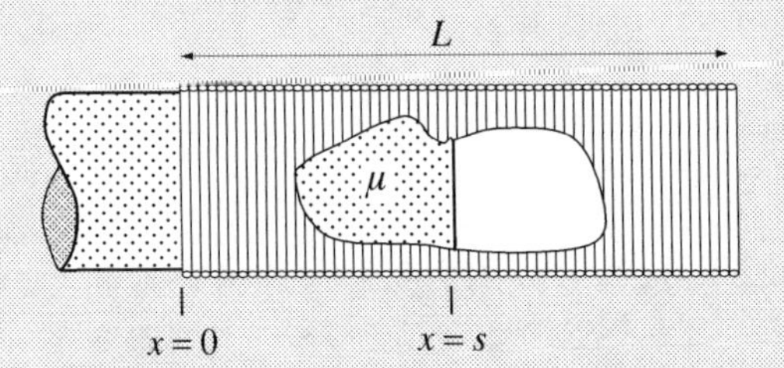

Figure 13.23: Side view of a solenoid winding with a solid, cylindrical rod of magnetizable matter partially inserted. A hole cut into the solenoid reveals how far the rod has been inserted.

Method I: We can compute the force from $\mathbf{F} = -(\partial \hat{U}_B/\partial s)\hat{\mathbf{x}}$ if the rod is inserted by an amount s along the x-axis. Since $\chi_m \ll 1$, the field $\mathbf{H}$ inside the solenoid is very nearly equal to the field $\mathbf{H}_0 = NI/L\hat{\mathbf{x}}$ produced by the solenoid alone. Therefore, ignoring fringing effects, the energy from (13.108) is

$$\hat{U}_B = -\frac{1}{2}\int d^3r\, \mathbf{B}\cdot\mathbf{H}_0 = -\frac{1}{2}A[s\mu + (L-s)\mu_0]H_0^2.$$

This gives the force as

$$\mathbf{F} = -\frac{\partial \hat{U}_B}{\partial s}\hat{\mathbf{x}} = (\mu - \mu_0)\frac{N^2I^2A}{2L^2}\hat{\mathbf{x}}.$$

Method II: We use the stress tensor integral (13.134). The simplest choice for S is the surface of a hollow cylinder (closed at both ends) which barely encloses the entire rod. This is inconvenient because it requires us to estimate $\mathbf{H}$ on those portions of S that lie outside the solenoid. It is better to choose a surface which lies everywhere inside the solenoid, where $\mathbf{B}$ and $\mathbf{H} \approx \mathbf{H}_0$ both point along the x-axis. Only the second term in (13.134) contributes if we do this and choose S as the surface of a cylinder of infinitesimal length (closed at both ends) which barely encloses the end of the rod which lies inside the solenoid. Only the two disk-shaped surfaces which cap the cylinder contribute and these have unit normals which point in opposite directions. Therefore, we find without difficulty that

$$\mathbf{F} = -\frac{1}{2}\int_S dS\,\hat{\mathbf{n}}(\mathbf{B}\cdot\mathbf{H}_0) = \frac{1}{2}A(\mu - \mu_0)H_0^2\hat{\mathbf{x}} = (\mu - \mu_0)\frac{N^2I^2A}{2L^2}\hat{\mathbf{x}}.$$

Remark: Both methods presented here unfortunately obscure the physical origin of the force. Since the rod lies in vacuum, the force on its volume V can be calculated from (13.120):

$$\mathbf{F} = \mu_0 \int_V d^3r \, (\mathbf{M} \cdot \nabla)\mathbf{H}_0.$$

The force must originate from the fringe field at the end of the solenoid because only this part of $\mathbf{H}_0$ is not spatially uniform.

13.9 Permanent Magnetic Matter

The class of materials known as *ferromagnets* have been known since antiquity. A *soft* ferromagnet in a small external field is a simple paramagnet with an effective value of μ that can exceed 10^4. A *hard* ferromagnet—the subject of this section—does not obey the linear constitutive relation (13.61) at all. Instead, a graph of $\mathbf{M}$ versus $\mathbf{H}_{\text{ext}}$ (the latter produced by free currents flowing in nearby coils) typically exhibits the non-reversible and non-single-valued behavior shown in Figure 13.24. This is called *magnetic hysteresis*.

Beginning at $\mathbf{H}_{\text{ext}} = \mathbf{M} = 0$, $\mathbf{M}(\mathbf{H}_{\text{ext}})$ first traces out a highly non-linear "initial magnetization curve".[28] The maximum or *saturation* value of $\mathbf{M}$ occurs when all the magnetic moments in the sample point in the same direction. However, when $\mathbf{H}_{\text{ext}}$ is decreased back to zero, the system does not retrace the initial magnetization curve back to $\mathbf{M} = 0$. A macroscopic *remanent magnetization* $\mathbf{M} \neq 0$ remains when $\mathbf{H}_{\text{ext}} = 0$. This is the property that defines a permanent magnet. The magnetization can be driven to zero and ultimately saturated in the opposite direction by applying $\mathbf{H}_{\text{ext}}$ in the opposite direction. The net result is that $\mathbf{M}(\mathbf{H}_{\text{ext}})$ traces out the closed *hysteresis* curve shown in Figure 13.24.

Why does the curve in Figure 13.24 have the particular shape it does? Why does the magnetization not return to zero when the external field is removed? The details of the answer go far beyond the scope of this book. Nevertheless, for at least some ferromagnets, the most important contributing factor to hysteresis is the creation and motion of *magnetic domains*. This topic merits discussion here because the relevant physics is almost entirely magnetostatics.

13.9.1 Magnetic Domains

For quantum mechanical reasons, every microscopic magnetic moment in a ferromagnet spontaneously aligns itself with the magnetic moments in its neighborhood.[29] This would produce a uniformly magnetized sample except that ferromagnets invariably break up into a collection of macroscopic regions called *magnetic domains*. All the moments are aligned within any given domain, but the direction of $\mathbf{M}$ is not the same in every domain. As we will show, a ferromagnet minimizes its magnetostatic energy by creating magnetic domains and arranging them to produce the smallest (net) amount of fictitious magnetic charge. By this rule, the four-domain arrangement shown in Figure 13.25(a) is energetically preferable to the single-domain state of uniform magnetization shown in Figure 13.25(b).

We consider an isolated ferromagnet in the absence of free currents or other sources of external fields. We also assume that $|\mathbf{M}(\mathbf{r})|$ takes its saturation maximum value M_S at every point in the sample. This means that the internal energy required to create a bit of magnetization is not at issue. The energy

[28] The effective permeability is usually defined as the maximum value of the ratio M/H_{ext} evaluated along the initial magnetization curve.

[29] This cooperative phenomenon disappears above a material-dependent critical temperature.

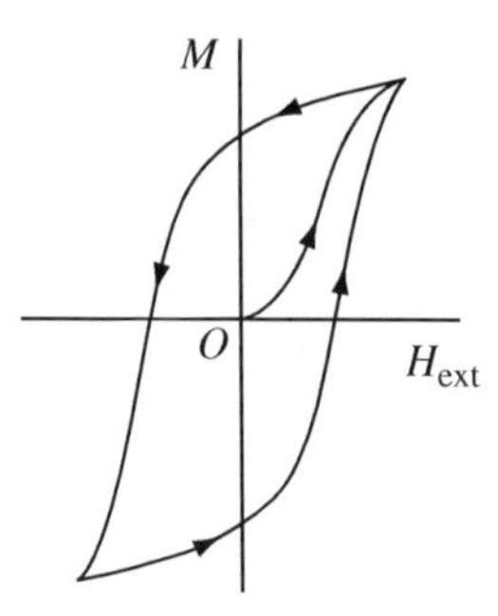

Figure 13.24: A typical hysteresis curve for a ferromagnet like iron. Beginning at O, the arrows indicate the non-reversible variations in M as H_{ext} varies.

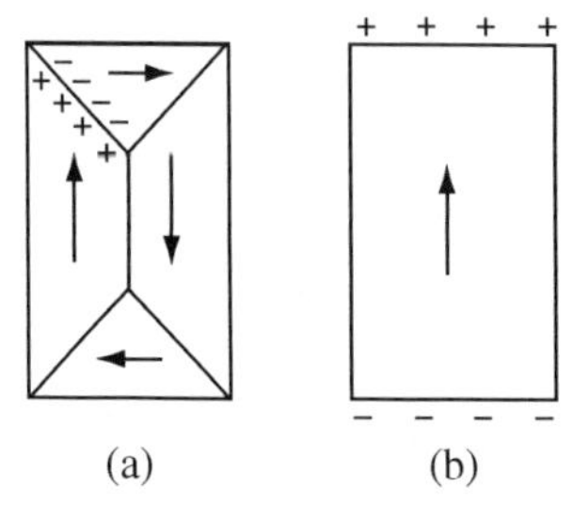

Figure 13.25: Two possible domain patterns for a rectangular sample. Plus and minus signs denote fictitious magnetic charge. The charge distribution is shown for only one of the five internal interfaces of the four-domain configuration.

of interest is the potential energy of each magnetic moment in the field produced by all the other magnetic moments in the sample. This is the magnetostatic analog of the electrostatic self-energy (6.107). Therefore for the present case of zero external field, the energy to be minimized is[30]

$$\hat{U}_{\text{self}} = -\frac{1}{2}\int d^3r\,\mathbf{M}(\mathbf{r})\cdot\mathbf{B}(\mathbf{r}). \tag{13.140}$$

Our first step is to use $\mathbf{B} = \mu_0(\mathbf{H}+\mathbf{M})$ to eliminate $\mathbf{B}$ from (13.140):

$$\hat{U}_{\text{self}} = -\frac{1}{2}\mu_0\int d^3r\,|\mathbf{M}|^2 - \frac{1}{2}\int d^3r\,\mathbf{M}\cdot\mathbf{H}. \tag{13.141}$$

The same equation can be used to eliminate $\mathbf{M}$ from the last term in (13.141). This gives

$$\hat{U}_{\text{self}} = -\frac{1}{2}\mu_0\int d^3r\,|\mathbf{M}|^2 - \frac{1}{2}\int d^3r\,\mathbf{B}\cdot\mathbf{H} + \frac{1}{2}\mu_0\int d^3r\,|\mathbf{H}|^2. \tag{13.142}$$

By assumption, $|\mathbf{M}(\mathbf{r})| = M_S$, so the first term on the right hand-side of (13.142) is a constant that can be dropped. The second term in (13.142) is *zero* because (13.102) and (13.103) are equal and $\mathbf{j}_f = 0$. This leaves the self-energy as

$$\hat{U}_{\text{self}} = \frac{1}{2}\mu_0\int d^3r\,|\mathbf{H}|^2. \tag{13.143}$$

[30] The reader should resist the temptation to minimize (13.111). The latter is valid for simple, linear matter only.

We can now understand why the domain pattern in Figure 13.25(a) is preferred to the single-domain state of Figure 13.25(b). $\hat{U}_{\rm self}$ in (13.143) takes its minimum value when the auxiliary field $\mathbf{H}(\mathbf{r})$ is identically zero. From Section 13.5, we know that the sources of $\mathbf{H}(\mathbf{r})$ are free current and fictitious magnetic charge. Here, $\mathbf{j}_f = 0$ and the volume charge $\rho^* = -\nabla \cdot \mathbf{M}$ is zero inside every magnetic domain. All that remains is the magnetic charge density $\sigma^* = \mathbf{M} \cdot \hat{\mathbf{n}}$ at the surfaces of the domains. In Figure 13.25(a), no magnetic charge is created at any of the exterior sample surfaces and the charge created by adjacent domains at their common interface is equal and opposite because each interface makes an angle of 45° with respect to the sample edges. Therefore, $\mathbf{H}(\mathbf{r}) \equiv 0$ for this arrangement of domains. This is not the case for Figure 13.25(b).

$\hat{U}_{\rm self}$ cannot be reduced to zero for samples of arbitrary shape. First, quantum mechanical effects make some directions of magnetization energetically preferred to others.[31] This prevents the creation of smooth and closed "circles of **M**" like those sketched in Figure 13.16. In addition, there is an energy to be paid at every boundary where dissimilar domains meet. This is because spins on opposite sides of the boundary are not aligned as the ferromagnetic state prefers. Observed domain patterns minimize the sum of (13.143) and these other contributions to the total energy.

Application 13.4 Refrigerator Magnets

A refrigerator magnet is a thin, wafer-like object of ferromagnetic material engineered to have a pattern of magnetic domains like the one shown in Figure 13.26. Such magnets have two important properties: (i) the magnetic field outside the wafer is nearly zero; and (ii) the attraction of the wafer to permeable matter is very strong. The origin of both lies in the behavior of the field $\mathbf{H} = -\nabla\psi$ produced by the wafer.

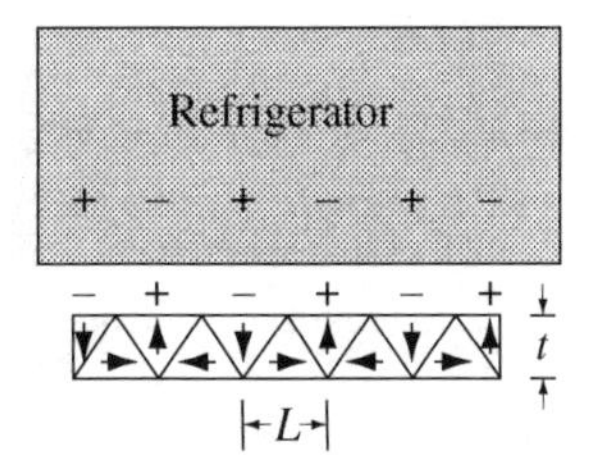

Figure 13.26: Edge view of a thin ($t \ll L$) wafer magnet in vacuum adjacent to a permeable material (refrigerator) with $\mu > \mu_0$. Plus and minus signs are the magnetic charges produced by the domains of the wafer magnet and image charges in the refrigerator.

From Section 13.6.3, we know that the magnetic potential $\psi(\mathbf{r})$ satisfies Laplace's equation outside the wafer. Moreover, the wafer's source magnetic charge alternates in sign with a spatial periodicity L. We may then apply the electrostatic experience gained in Section 7.5.2 and Section 8.3.4 to conclude that the potential produced by such a source falls off as $\exp(-z/L)$ in the direction z perpendicular to the direction of periodicity. The magnetic field has a similarly short range. Therefore, a wafer magnet has essentially no magnetic effect on its environment until it is in intimate contact with another object.

Each domain of the wafer magnet produces a strip of surface magnetic charge with density $\sigma^* = M_S$. The sign of the magnetic charge alternates from one strip to the next. According to Section 13.6.5,

[31] This is the phenomenon of *magnet anisotropy* mentioned in a footnote to Section 12.4.4.

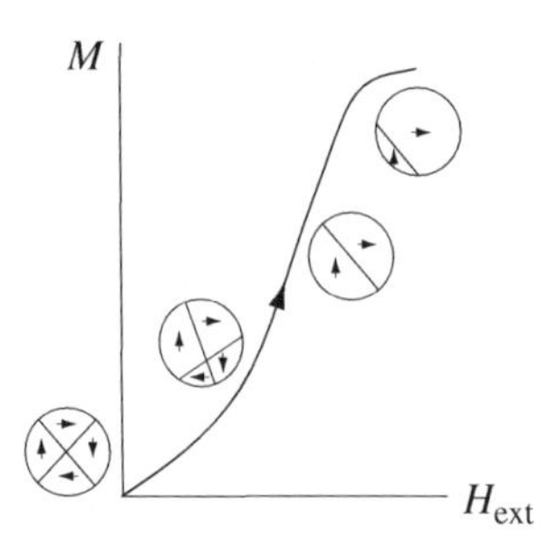

Figure 13.27: Evolution of magnetic domains during the initial magnetization of a ferromagnet. The arrows give the local direction of **M** in each domain.

each strip creates an image strip in the permeable refrigerator with opposite magnetic surface charge density. If $\mu \gg \mu_0$, the surface charge densities are equal and opposite for each strip and its image. Therefore, Coulomb's law of magnetism (Section 13.8.1) tells us that the force between the wafer magnet and the refrigerator is approximately the wafer area A times the attractive force per unit area between the plates of an infinite parallel-plate capacitor with surface charge density σ^*. The field produced by one plate is $B = \frac{1}{2}\mu_0\sigma^*$ so

$$F \approx \frac{1}{2}\mu_0 M_S^2 A. \qquad (13.144)$$

The force per unit area predicted by (13.144) is about six times atmospheric pressure for a typical refrigerator magnet. ■

13.9.2 Magnetic Hysteresis

The shape of the **M** versus $\mathbf{H}_{\rm ext}$ curve in Figure 13.24 is determined largely by the field-induced mobility of the boundaries between adjacent magnetic domains. Consider the initial magnetization curve shown in Figure 13.27. The cartoon on the bottom left side of Figure 13.27 shows the $\mathbf{H}_{\rm ext} = 0$ domain pattern for a disk-shaped sample where only two directions of magnetization are allowed due to quantum mechanical magnetic anisotropy. The *net* magnetization of this state is $\mathbf{M} = 0$.

When a field $H_{\rm ext} > 0$ is turned on, the magnetization curve rises from the origin in Figure 13.24 because the domain boundaries in Figure 13.27 *move*. Consider the boundary between two domains, $\mathbf{M}_1$ and $\mathbf{M}_2$, where $\mathbf{M}_2$ happens to be more nearly parallel to $\mathbf{H}_{\rm ext}$ than $\mathbf{M}_1$. The magnetic potential energy[32] becomes more negative if individual magnetic moments in $\mathbf{M}_1$ right next to the boundary rotate into the direction of $\mathbf{M}_2$, thereby moving the boundary. By this mechanism, the size of $\mathbf{M}_1$ shrinks and the size of $\mathbf{M}_2$ grows. The cartoons in Figure 13.27 show how domain motion driven by $\mathbf{H}_{\rm ext}$ ultimately leads to a state of uniform magnetization. Hysteresis and remanent magnetization occur because, for a variety of material-dependent reasons, the scenario sketched above does not exactly retrace itself when the field is lowered. For a hard ferromagnet, the saturated state remains (nearly) intact all way down to $\mathbf{H} = 0$.

[32] The relevant potential energy, $\hat{U}_B = -\int d^3r\, \mathbf{M}\cdot\mathbf{B}_0$, is the magnetic analog of the electrostatic potential energy (6.101).

Sources, References, and Additional Reading

The quotation at the beginning of the chapter is taken from the Chapter 1 of *de Magnete*, which was published in Latin in 1600. An English-language edition is

W. Gilbert, *On the Magnet* (Dover, Mineola, New York, 1958).

Section 13.1 Three excellent treatments of the classical aspects of the magnetism of matter are

L.D. Landau and E.M. Lifshitz, *Electrodynamics of Continuous Media* (Pergamon, Oxford, 1960).

W.F. Brown Jr., *Magnetostatic Principles in Ferromagnetism* (North-Holland, Amsterdam, 1962).

R.K. Wangsness, *Electromagnetic Fields*, 2nd edition (Wiley, New York, 1986).

Section 13.2 Our approach to volume and surface magnetization current density follows *Landau and Lifshitz* (see Section 13.1 above). The magnetization function $\mathbf{M}_O$ was introduced in *Lorentz. Hirst* discusses the gauge freedom associated with this quantity.

H.A. Lorentz, "The fundamental equations for electromagnetic phenomena in ponderable bodies deduced from the theory of electrons", *Proceedings of the Royal Academy of Amsterdam* **5**, 254 (1902).

L.L. Hirst, "The microscopic magnetization: Concept and application", *Reviews of Modern Physics* **69**, 607 (1997).

Section 13.4 These books illustrate the widespread use of fictitious magnetic charge in applied physics and in older textbooks of electromagnetism.

M.G. Abele, *Structures of Permanent Magnets* (Wiley, New York, 1993).

H.M. Bertram, *Theory of Magnetic Recording* (Cambridge University Press, Cambridge, 1994).

V.C.A. Ferraro, *Electromagnetic Theory* (Athlone Press, London, 1954).

A paper which proves that the demagnetization field is a constant for ellipsoids only is

K.R. Brownstein, "Unique shape of uniformly polarized dielectrics", *Journal of Mathematical Physics* **28**, 978 (1987).

The boxed discussion of circulating currents vs. true magnetic charge as the source of the magnetic dipole moment of the proton was adapted from

W.A. Nierenberg, "The measurement of the nuclear spins and static moments of radioactive isotopes", *Annual Reviews of Nuclear Science* **7**, 349 (1957).

G.I. Opat, "Limits placed on the existence of magnetic charge in the proton by the ground state hyperfine splitting of hydrogen", *Physics Letters B* **60**, 205 (1976).

Section 13.6 Magnetic susceptibility is an intrinsically quantum mechanical phenomenon. Semi-classical models exist, but none has predictive value. See

R.M. White, *Quantum Theory of Magnetism*, 2nd edition (Springer, New York, 1982).

Figure 13.13 comes from this short but excellent biography:

J.M. Thomas, *Michael Faraday and the Royal Institution* (Adam Hilger, Bristol, 1991).

An entry point to the literature of magnetic shielding is

T.J. Sumner, J.M. Pendlebury, and K.F. Smith, "Conventional magnetic shielding", *Journal of Physics D: Applied Physics* **20**, 1095 (1987).

We do not discuss "magnetic circuit theory". The reader should consult *Wangsness* (see Section 13.1 above) for an introduction. This method is used to solve the problem posed by Figure 13.14 in

E. Paperno and I. Sasada, "Magnetic circuit approach to magnetic shielding", *Journal of the Magnetics Society of Japan* **24**, 40 (2000).

Potential theory with ferromagnetic matter is the subject of

S.P. Thompson and M. Walker, "Mirrors of magnetism", *Proceedings of the Physical Society of London* **13**, 310 (1894).

J. Van Bladel, "Magnetostatic fields at an iron-air boundary", *American Journal of Physics* **29**, 732 (1961).

Figure 13.17 was taken from

O. Klemperer, *Electron Physics*, 2nd edition (Butterworths, London, 1972).

The magnetic properties of superconductors are expertly discussed in the book by *Landau and Lifshitz* (see Section 13.1 above). Another good reference is

T.P. Orlando and K.A. Delin, *Foundations of Applied Superconductivity* (Addison-Wesley, Reading, MA, 1991).

Section 13.7 Our treatment of energy in magnetic matter derives from *Brown* and *Landau and Lifshitz* (see Section 13.1 above). Two discussions which make contact with the statistical mechanical (Hamiltonian) point of view are

R. Balian, *From Microphysics to Macrophysics* (Springer, Berlin, 2007), Volume I, Section 6.6.5.

O. Narayan and A.P. Young, "Free energies in the presence of electric and magnetic fields", *American Journal of Physics* **73**, 293 (2005).

Section 13.8 Forces on and in magnetic matter are a subtle subject. See *Landau and Lifshitz* (see Section 13.1 above) for a careful discussion. Application 13.3 comes from a polemical article which eschews the use of energy methods to calculate electromagnetic forces:

W.F. Brown, Jr., "Electric and magnetic forces: a direct calculation", *American Journal of Physics* **19**, 290 (1951).

Example 13.5 was familiar to Maxwell. Our method of solution follows

J. Schwinger, L.L. DeRaad, Jr., K.A. Milton, and W-Y. Tsai, *Classical Electrodynamics* (Perseus, Reading, MA, 1998).

Section 13.9 Besides *Brown* (see Section 13.1 above), two books which provide a modern view of ferromagnetism and magnetic hysteresis are

G. Bertotti, *Hysteresis in Magnetism* (Academic, San Diego, 1998).

É. du Trémolet de Lacheisserie, D. Gignoux, and M. Schlenker, *Magnetism* (Springer, New York, 2005).

Application 13.4 was inspired by an introductory textbook which fully exploits the magnetic pole concept:

W. Saslow, *Electricity, Magnetism, and Light* (Academic, New York, 2002).

Problems

13.1 The Magnetic Field of an Ideal Solenoid Section 10.2.2 of the text evaluated a Biot-Savart integral to show that a surface current $\mathbf{K}$ flowing azimuthally on the surface of an infinitely long solenoid with an arbitrary cross sectional shape produces a magnetic field $\mathbf{B} = \mu_0 K\hat{\mathbf{z}}$ inside the solenoid and $\mathbf{B} = 0$ outside the solenoid. Use an equivalent magnetization argument to deduce the same result.

13.2 Equal and Opposite Magnetization The half-space $z > 0$ has uniform magnetization $\mathbf{M} = -M\hat{\mathbf{z}}$. The half-space $z < 0$ has uniform magnetization $\mathbf{M} = +M\hat{\mathbf{z}}$. Find the magnetic field $\mathbf{B}$ at every point in space using (a) the method of magnetization current and (b) the method of effective magnetization charge.

13.3 Equivalent Currents An insulating sphere with radius R rotates with angular velocity $\boldsymbol{\omega} = \omega\hat{\mathbf{z}}$. The total charge Q of the sphere is uniformly distributed over its surface.

(a) Show that the magnetic field outside the insulating sphere is identical to the magnetic field outside a uniformly magnetized sphere of radius R when the magnetization $\mathbf{M} = M\hat{\mathbf{z}}$ and M are suitably chosen.

(b) Use the result of part (a) to find the magnetic moment of the insulating sphere.

13.4 The Helmholtz Theorem for M

(a) Show that the Helmholtz theorem representation of the magnetization $\mathbf{M}(\mathbf{r})$ is equivalent to the equation $\mathbf{B}_{\text{M}} = \mu_0(\mathbf{H}_{\text{M}} + \mathbf{M})$.

(b) The figure shows a thin film (infinite in the x and y directions) where alternating strips have constant magnetization $\mathbf{M} = \pm M\,\hat{\mathbf{y}}$. Find $\mathbf{B}(\mathbf{r})$ and $\mathbf{H}(\mathbf{r})$ everywhere and discuss the behavior of the effective magnetization current density.

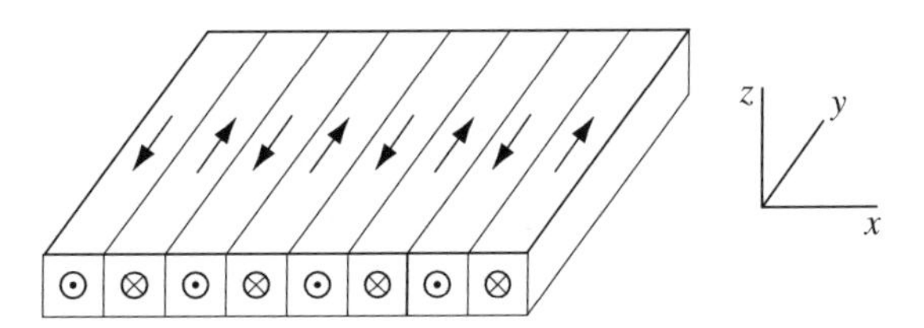

13.5 The Virtues of Magnetic Charge

(a) Show that the magnetic dipole moment of a magnetized body can be written

$$\mathbf{m} = \int d^3r\, \mathbf{r}\rho^*(\mathbf{r}),$$

where $\rho^*(\mathbf{r}) = -\nabla \cdot \mathbf{M}(\mathbf{r})$ is the density of fictitious magnetic charge.

(b) Let $\mathbf{B}_1$ and $\mathbf{B}_2$ be the magnetic fields produced by bodies with magnetizations $\mathbf{M}_1$ and $\mathbf{M}_2$, respectively. Use the fact that magnetization current density is as "real" as electric current density to show that the potential energy of interaction between the two bodies satisfies the reciprocity relation

$$\hat{V}_B = -\int d^3r\, \mathbf{M}_1 \cdot \mathbf{B}_2 = -\int d^3r\, \mathbf{M}_2 \cdot \mathbf{B}_1.$$

(c) If the two bodies in part (b) are solid and cannot overlap in space, show that

$$\hat{V}_B = \frac{\mu_0}{4\pi} \int d^3r \int d^3r' \frac{\rho^*(\mathbf{r})\rho_2^*(\mathbf{r}')}{|\mathbf{r} - \mathbf{r}'|}.$$

13.6 Atom Optics with Magnetic Recording Tape A very long piece of magnetic recording tape has a length L, a width $w \ll L$, and a thickness $t \ll w$. The tape has a magnetization $\mathbf{M}(x) = \hat{\mathbf{x}}\, M \cos kx$.

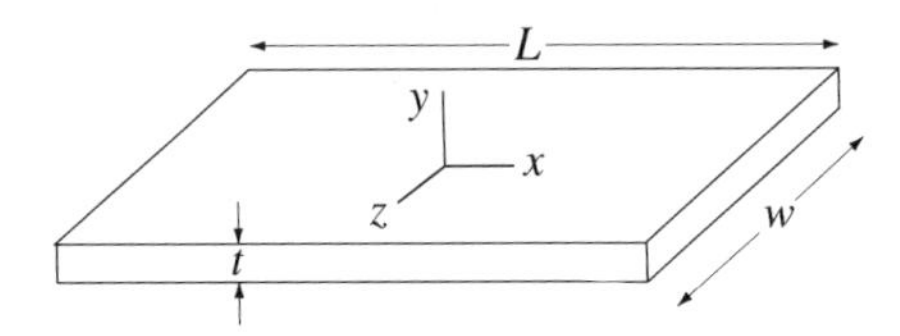

(a) Solve Laplace's equation for the magnetic scalar potential and calculate $\mathbf{B}(x, y, z)$ just above the surface of the tape. For this purpose, it is legitimate to treat the tape as both infinitely wide and infinitely long.

(b) Plot the field lines of $\mathbf{B}(\mathbf{r})$.

(c) Sketch the trajectory of a neutral atom that approaches the tape from above (but not exactly straight down). Assume that the atom's magnetic moment $\mathbf{m}$ remains permanently aligned anti-parallel to $\mathbf{B}$. Hint: Consider the interaction potential energy between the atom and the tape.

13.7 Bitter's Iron Magnet What magnetization $\mathbf{M}(\mathbf{r})$ imposed on an infinite piece of iron produces the largest magnetic field at a given point? Francis Bitter (a pioneer in the design of high-field magnets) posed this problem in 1936. Since the magnetization of a ferromagnet is produced by electron spins, we need first to find the direction of a point magnetic dipole with moment $\mathbf{m}$ located at $\mathbf{r} = (r, \theta, \phi)$ that maximizes the z-component of its magnetic field at the origin.

(a) Show that the maximum of $B_z(0)$ occurs when $2\tan\alpha = \tan\theta$, where $\hat{\mathbf{r}} \cdot \hat{\mathbf{m}} = \cos\alpha$ and $\mathbf{r}$ lies between $\mathbf{m}$ and $\hat{\mathbf{z}}$ in the plane $\phi = 0$.

(b) Show that the maximal condition can be written $\frac{1}{2}\tan\theta = r d\theta/dr$. Use this to establish that the locus of desired spin directions in the ferromagnet is indistinguishable from the field lines of a suitably oriented point magnetic dipole at the origin.

(c) Show that

$$B_z(0) = \frac{\mu_0 m N}{4\pi} \int d^3 r \, \frac{\sqrt{1 + 3\cos^2\theta}}{r^3},$$

where N is the number of spins/volume. Perform the integral for the case of a spherical shell of iron with inner radius r_1 and outer radius r_2.

13.8 Einstein Errs! Einstein published the following argument in 1910. The solid lines with arrows in the figure below show the directions of current flow for a "can-of-current". A constant current I flows up the central z-axis of a hollow cylindrical can of radius R, flows radially outward on the top end-cap with surface density K_+, flows down the can side wall with surface density K_W, and then flows radially inward on the bottom end-cap with surface density K_-. In Example 10.2, we used Ampère's law to show that the magnetic field outside the can is *zero* while the field inside the can is $\mathbf{B}_{\rm in} = \hat{\boldsymbol{\phi}}\mu_0 I/2\pi\rho$.

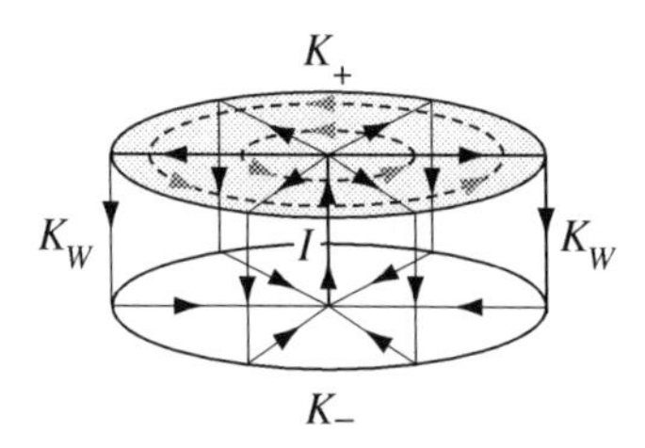

(a) Let the *top* end-cap carry a magnetization $\mathbf{M} = M\hat{\boldsymbol{\phi}}$ as indicated by the gray shading and the dashed lines with arrows. According to the text, this magnetization produces $\mathbf{H} = 0$ everywhere so $\mathbf{B}_{\rm M} = \mu_0\mathbf{M}$ everywhere. Show that $\mathbf{B}_{\rm M}$ exerts a Lorentz force on the end cap which tends to levitate the can. Since the can cannot exert a force on itself, Einstein argued that the Lorentz force density $\mathbf{j} \times \mathbf{B}$ was incorrect and should be replaced by $\mathbf{j} \times \mathbf{H}$.

(b) Let the magnetized end-cap have a tiny thickness. Calculate and sketch the closed lines of magnetization current density.

(c) Use the sketch from part (b) to identify the part of the end-cap magnetization current that feels a force due to $\mathbf{B}_{\rm in}$ in the zero-thickness limit. Show that this force cancels the levitation force calculated in part (a).

13.9 A Hole Drilled through a Permanent Magnet The diagram shows a cylindrical hole of radius R drilled through a permanent magnet which is infinite in the two directions transverse to its thickness t. Let $R \ll t$ and find $\mathbf{B}$ at every point in space that lies very far from both ends of the cylindrical hole.

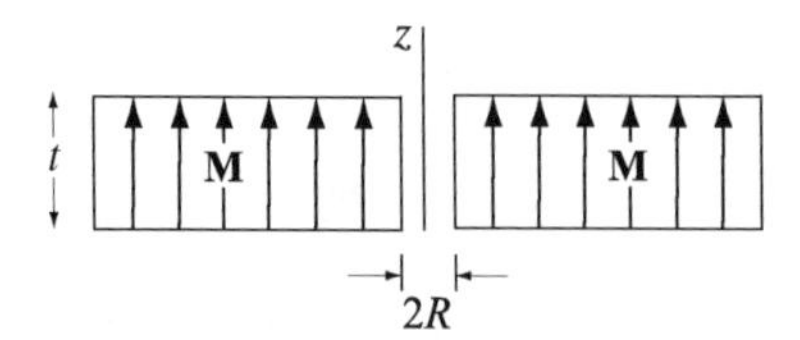

13.10 The Demagnetization Factor for an Ellipsoid Let $\mathbf{N}$ be the demagnetization tensor in the principal axis system of a prolate ($a > b$) ellipsoid of revolution uniformly magnetized along its symmetry axis.

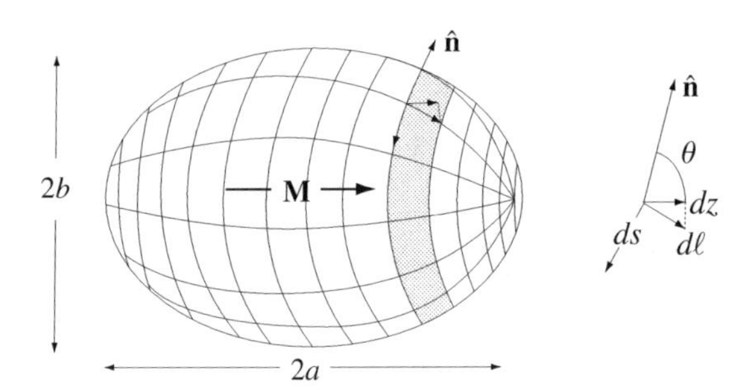

(a) Explain why only N_{zz} need be calculated.

(b) Cut the ellipsoid into circular slices with thickness dz and radius ρ as shown in the figure. The bevelled edge of each slice is perpendicular to the ellipsoid normal $\hat{\mathbf{n}}$ and has edge length $d\ell = dz/\sin\theta$. On the other hand, if $d\boldsymbol{\ell}$ points along the circumference on any slice,

$$\mathbf{M} \times \hat{\mathbf{n}} dS = M \sin\theta \, d\ell \, d\mathbf{s} = M dz d\mathbf{s}.$$

Let $\epsilon = \sqrt{1 - b^2/a^2}$ be the eccentricity of the ellipsoid. Use Ampère's theorem (Section 11.3) and the Biot-Savart result for the on-axis magnetic field of a current ring to show that

$$N_{zz} = \frac{1-\epsilon^2}{\epsilon^2}\left[\frac{1}{2\epsilon}\ln\left(\frac{1+\epsilon}{1-\epsilon}\right) - 1\right] \qquad (\epsilon < 1).$$

(c) Check the spherical limit.

13.11 Lunar Magnetism The Moon has no magnetic field outside of itself, despite the observed permanent magnetization of rocks collected by lunar missions. One explanation supposes that the Moon once had a geodynamo (like the Earth) confined to a core region ($r < b$) that produced a dipole field everywhere outside the core. As a result, the lunar crust ($b \le r \le a$) became permanently magnetized proportional to (and parallel to) the local dipole field. Later, the geodynamo ceased, along with its dipole field. Show that the still-present magnetization of the crust produces zero magnetic field outside the surface of the Moon ($r > a$). Sketch the lines of $\mathbf{B}(\mathbf{r})$ inside the Moon.

13.12 A Dipole in a Magnetizable Sphere A point magnetic dipole is located at the center of a magnetizable sphere with radius R and permeability μ. Find $\mathbf{H}(\mathbf{r})$ everywhere.

13.13 Magnetic Shielding A uniform external field $\mathbf{B}_{\rm ext} = B_{\rm ext}\hat{\mathbf{x}}$ is applied to an infinitely long cylindrical shell with inside radius a, outside radius b, and relative permeability $\kappa_m = \mu/\mu_0$. The rest of space is vacuum. Show that the field inside the shell is screened to the value

$$\mathbf{B}_{\rm in} = \frac{4\kappa_m b^2}{(\kappa_m + 1)^2 b^2 - (\kappa_m - 1)^2 a^2}\mathbf{B}_{\rm ext}.$$

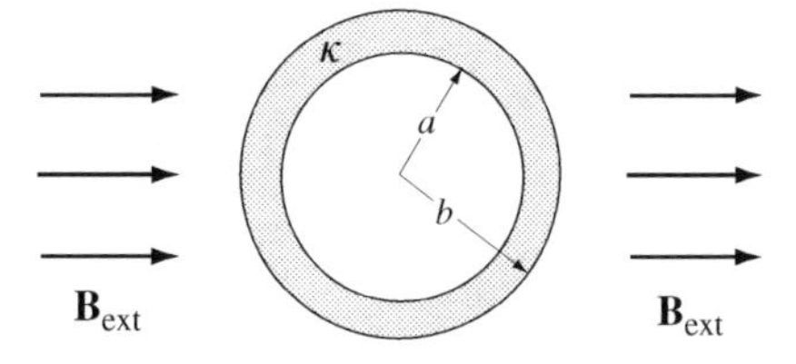

13.14 The Force on a Current-Carrying Magnetizable Wire A straight wire with radius a and magnetic permeability μ carries a conduction current density $\mathbf{j}_0 = j_0\hat{\mathbf{z}} = \hat{\mathbf{z}}I_0/\pi a^2$.

(a) Find the induced magnetization $\mathbf{M}$ and the volume and surface magnetization current densities when the wire is exposed to a uniform external field $\mathbf{B}_0 = B_0\hat{\mathbf{x}}$.

(b) Show that the Lorentz force $\mathbf{F} = \int d^3r\, \mathbf{j} \times \mathbf{B} = I_0 B_0 \hat{\mathbf{y}}$, where $\mathbf{j}$ is the total current density and $\mathbf{B}$ is the total magnetic field. Hint: Do not forget the surface contribution to this integral.

13.15 Active Magnetic Shielding An infinitely long and straight wire carries a steady current I in the $+z$-direction at a distance d from the surface of a perfect conductor that occupies the half-space $x > 0$.

(a) Find the current density induced on the surface of the conductor.

(b) Remove the perfect conductor. Explain how one should locate a collection of N infinitely long and straight wires, and choose the current flowing in each wire, to effectively shield the vacuum space $x > 0$ from the effect of the magnetic field produced by the current-carrying wire at $x = -d$.

13.16 The Role of Interface Magnetization Current An infinite cylinder of radius R filled with matter with permeability μ_1 is embedded in an infinite medium with permeability μ_2. A wire carries a current I_f up the z-axis of the cylinder. Show by explicit calculation that the sum of the fields produced by the free currents and the magnetization currents is equal to the field at every point in space as calculated using Ampère's law directly.

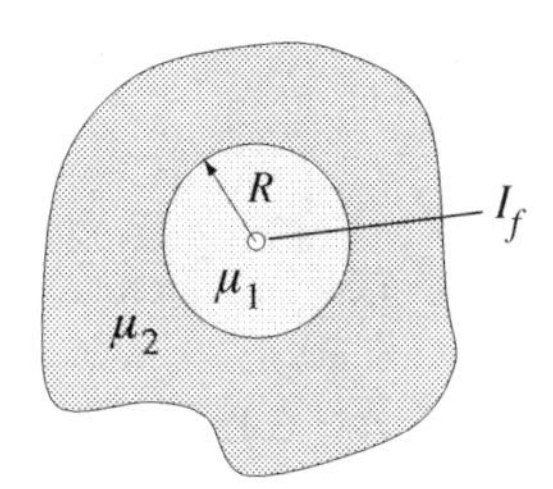

13.17 Magnetic Film and Magnetic Disk

(a) An infinitely large film of insulating magnetic material has permeability μ and thickness h. A uniform external magnetic field $\mathbf{B}_0$ is oriented perpendicular to the plane of the film. Find the magnetic field $\mathbf{B}$ at any point inside the film.

(b) Create a magnetic disk by removing from the film in part (a) all the matter that lies outside a region of radius $R \gg h$ (see diagram below). Find $\mathbf{B}$ at the center of the disk which remains (to first order in h/R) by subtracting, from the answer to part (a), the magnetic field produced at the center of the disk by the matter outside the disk.

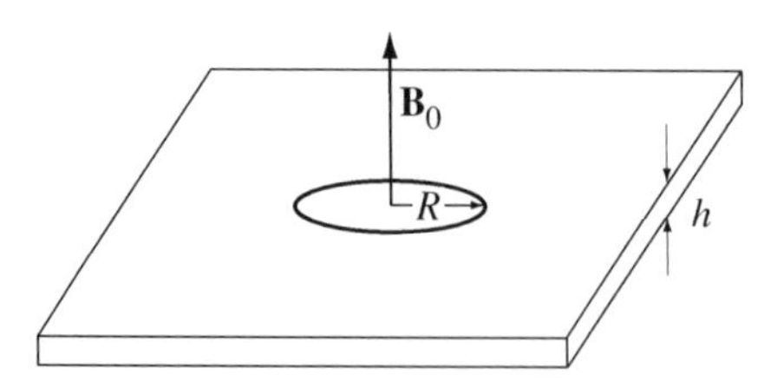

Hint: the field of a point magnetic dipole with moment $\mathbf{m}$ is $\mathbf{B}(\mathbf{r}) = \dfrac{\mu_0}{4\pi}\dfrac{3(\mathbf{m}\cdot\mathbf{r})\mathbf{r} - r^2\mathbf{m}}{r^5}$.

13.18 A Current Loop Levitated by a Bar Magnet The diagram shows a filamentary loop with radius a, steady current I, and mass per unit length ρ. The loop is levitated against its weight at a height h above the north pole of a very long cylindrical bar magnet with radius r. Find the magnetization of the permanent magnet. Assume that $r \ll a$ and $r \ll h$.

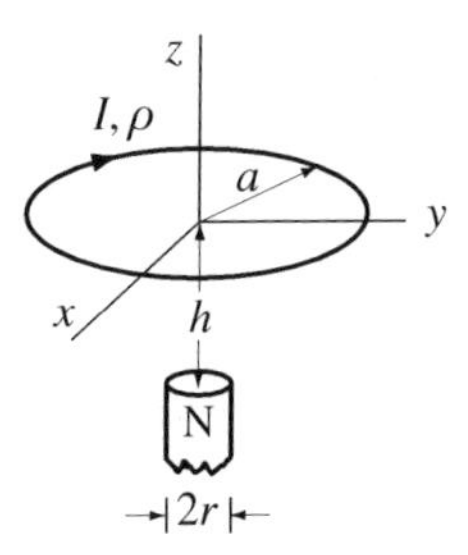

13.19 A Real Electromagnet

(a) A cylindrical solenoid occupies the interval $-L/2 \le z \le L/2$, has radius R, and is wound tightly with N turns of a wire which carries current I. Use superposition and derive an expression for the magnetic field on the symmetry axis in the form $\mathbf{B}_0(z) = (\mu_0 NI/L) f(z)\hat{\mathbf{z}}$. Compute the ratio $B_0(\pm L/2)/B_0(0)$ in the limit when $L \gg R$.

(b) Fill the solenoid with a close-fitting rod of soft iron with effective susceptibility χ_m. Make the approximation that $\mathbf{M}(\mathbf{r}) = M(z)\hat{\mathbf{z}}$, take account of the demagnetization field produced by the (magnetically) charged disks at either end of the rod, but ignore the effect of volume magnetic charge. Show that, inside the solenoid,

$$\mathbf{B}(z) \approx (1 + \chi_m)\{B_0(z) + \mu_0 \sigma^*[f(z) - 1]\}\,\hat{\mathbf{z}},$$

where (when $L \gg R$)

$$\sigma^* \approx \frac{NI}{L}\frac{\chi_m}{\chi_m + 2}.$$

(c) Calculate the amplification of the magnetic field produced by the soft iron at $z = 0$ and at $z = \pm L/2$. Comment on the result when $\chi_m \gg 1$.

13.20 Vector Potential Approach to Image Currents The space $x > 0$ ($x < 0$) is occupied by a medium with magnetic permeability μ_1 (μ_2). A line current I points out of the paper in medium 1 at a distance a from the interface with medium 2.

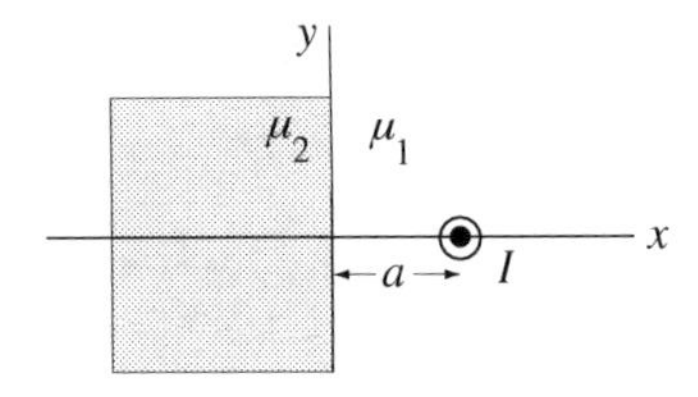

(a) Use the matching conditions for the vector potential $\mathbf{A}$ and the magnetic auxiliary field $\mathbf{H}$ to find the image currents needed to find the magnetic field at every point in space.

(b) Find the force per unit length exerted on I by the $x < 0$ half-space.

13.21 The London Equations for a Superconductor In 1935, the brothers Fritz and Heinz London described superconductivity using a phenomenological constitutive equation where a length $\delta > 0$ relates the current density to the Coulomb vector potential:

$$\mathbf{j} = -\frac{1}{\mu_0 \delta^2}\mathbf{A}.$$

(a) Use the London constitutive equation to derive a differential equation for $\mathbf{B}(\mathbf{r})$.

(b) The London theory predicts that $\mathbf{B}$ is not strictly zero at every point inside a superconductor. To see this, consider a slab of superconductor which is infinite in the x- and y-directions and lies between $z = -d$ and $z = d$. Compute $\mathbf{B}(z)$ inside the superconductor when the slab is placed in a static and uniform magnetic field $\mathbf{B}_0 = B_0\hat{\mathbf{x}}$.

(c) Find the current density inside the superconductor.

13.22 Supercurrent on a Sphere A superconducting sphere of radius R is placed in an external magnetic field $\mathbf{B}_0$. Show that the current which develops on the surface of the superconductor has density $\mu_0\mathbf{K} = (3/2)\hat{\mathbf{r}} \times \mathbf{B}_0$.

13.23 A Cylindrical Refrigerator Magnet A long cylinder has a cross sectional area A of arbitrary shape that is constant over its length $L \gg \sqrt{A}$. The cylinder is longitudinally magnetized with a uniform magnetic moment per unit volume $\mathbf{M}$. Show that the cylinder adheres to a flat, highly permeable surface with a force

$$F = \frac{1}{2}\mu_0 M^2 A - \frac{7}{16\pi}\mu_0 M^2 \frac{A^2}{L^2}.$$

13.24 Magnetic Total Energy The diagram shows two ways to periodically arrange N identical permanent bar magnets. The magnetic moments of the magnets are all parallel in configuration 1. The moments are alternately parallel and anti-parallel in configuration 2.

1 [S N] [S N] [S N] [S N]

2 [S N] [N S] [S N] [N S]

Choose a finite volume V that encloses almost all the magnetic flux produced by either configuration. Which configuration has the larger total magnetic energy in V and why?

13.25 Inductance in a Magnetic Medium Let L_0 be the self-inductance of a free current loop in vacuum. Show that the self-inductance changes to $L = \kappa_m L_0$ when the vacuum is replaced by a magnetic medium with permeability $\mu = \kappa_m \mu_0$.

14 Dynamic and Quasistatic Fields

The fields change but little during the time required for light to travel a distance equal to the maximum dimension of the body.
Wolfgang Pauli (1949)

14.1 Introduction

The remainder of this book is devoted to electric and magnetic fields which vary in time. Unlike their static counterparts, a time-dependent electric field is the source of a magnetic field and a time-dependent magnetic field is the source of an electric field. The coupling between the two in vacuum is dictated by the Maxwell equations,

$$\nabla \cdot \mathbf{E} = \frac{\rho}{\epsilon_0} \qquad \nabla \cdot \mathbf{B} = 0 \tag{14.1}$$

$$\nabla \times \mathbf{E} = -\frac{\partial \mathbf{B}}{\partial t} \qquad \nabla \times \mathbf{B} = \mu_0 \mathbf{j} + \frac{1}{c^2}\frac{\partial \mathbf{E}}{\partial t}. \tag{14.2}$$

In Chapter 6, we wrote $\rho = \rho_f - \nabla \cdot \mathbf{P}$ to distinguish free charge from polarization charge. In Chapter 13, we wrote $\mathbf{j} = \mathbf{j}_f + \nabla \times \mathbf{M}$ to distinguish free current from magnetization current. In this chapter, we introduce the *polarization current density* $\mathbf{j}_{\rm P} = \partial \mathbf{P}/\partial t$ and amend the total current density to $\mathbf{j} = \mathbf{j}_f + \nabla \times \mathbf{M} + \partial \mathbf{P}/\partial t$. Substituting these decompositions of ρ and $\mathbf{j}$ into (14.1) and (14.2) and defining the auxiliary fields $\mathbf{D} = \epsilon_0 \mathbf{E} + \mathbf{P}$ and $\mathbf{H} = \mathbf{B}/\mu_0 - \mathbf{M}$ generates the Maxwell equations in matter:

$$\nabla \cdot \mathbf{D} = \rho_f \qquad \nabla \cdot \mathbf{B} = 0 \tag{14.3}$$

$$\nabla \times \mathbf{E} = -\frac{\partial \mathbf{B}}{\partial t} \qquad \nabla \times \mathbf{H} = \mathbf{j}_f + \frac{\partial \mathbf{D}}{\partial t}. \tag{14.4}$$

Compared to our previous work, the new terms to be reckoned with are $\partial \mathbf{E}/\partial t$ (or $\partial \mathbf{D}/\partial t$) in the Ampère-Maxwell law and $\partial \mathbf{B}/\partial t$ in Faraday's law. This motivates us to begin our discussion with the new physics associated with these time derivatives, most notably the effects of *displacement current* and the phenomenon of *electromagnetic induction*. We then turn to the *quasistatic* limit where the sources change slowly enough in time to justify dropping one or the other of the time derivatives from the Maxwell equations. When the source is a slowly varying charge density $\rho(\mathbf{r}, t)$, we neglect $\partial \mathbf{B}/\partial t$ to get a quasi-electrostatic approximation. When the source is a slowly varying current density $\mathbf{j}(\mathbf{r}, t)$, we neglect $\partial \mathbf{E}/\partial t$ to get a quasi-magnetostatic approximation. Later chapters will focus on the characteristically high-frequency phenomena of radiation and waves where both time derivatives contribute equally to the physics.

A slowly moving point charge in vacuum is a typical quasi-electrostatics problem. The moving charge produces a Coulomb-like electric field and a magnetic field from the time variations of the electric field. A solenoid with a slowly time-varying current is a typical quasi-magnetostatics problem. The solenoid produces a Biot-Savart-like magnetic field and an electric field from the time variations of the magnetic field. AC circuit theory exploits both approximations simultaneously, albeit in different parts of space.

New physics appears when we consider slowly time-varying charge and current in or near polarizable, magnetizable, and conducting matter. Ohmic matter is particularly interesting because the phenomenon of *charge relaxation* eliminates free charge from the interior of a conductor. Quasi-electrostatics applies to poor conductors where charge relaxation is slow. Quasi-magnetostatics applies to good conductors where charge relaxation is fast. Mechanical effects come from the force exerted on distributions of charge and current by fields. It is not obvious, but experiments confirm that the Coulomb-Lorentz force law remains valid when the sources and fields vary in time:

$$\mathbf{F}(t) = \int d^3r \, [\rho(\mathbf{r}, t)\mathbf{E}(\mathbf{r}, t) + \mathbf{j}(\mathbf{r}, t) \times \mathbf{B}(\mathbf{r}, t)] \,. \tag{14.5}$$

14.2 The Ampère-Maxwell Law

The right side of the Ampère-Maxwell law,

$$\nabla \times \mathbf{B} = \mu_0 \mathbf{j} + \frac{1}{c^2}\frac{\partial \mathbf{E}}{\partial t}, \tag{14.6}$$

identifies two sources capable of producing a magnetic field. The electric current density $\mathbf{j}(\mathbf{r}, t)$ is familiar from magnetostatics. The magnetic field produced by a time-varying electric field was completely unknown until Maxwell postulated its existence. In vacuum, the *displacement current* density $\mathbf{j}_D = \epsilon_0 \partial \mathbf{E}/\partial t$ produces magnetic effects indistinguishable from moving charge. The closely related *polarization current* density, $\mathbf{j}_P = \partial \mathbf{P}/\partial t$, accounts for the magnetic field produced by time-dependent charge separation in matter.

14.2.1 Displacement Current

The mechanical model which led Maxwell to introduce the displacement current is no longer a part of physics.[1] From a contemporary perspective, $\mathbf{j}_D$ is best understood as the guarantor that the Maxwell equations respect conservation of charge as dictated by the continuity equation,

$$\nabla \cdot \mathbf{j} + \frac{\partial \rho}{\partial t} = 0. \tag{14.7}$$

Specifically, the time derivative of Gauss' law in (14.1) is

$$\nabla \cdot \epsilon_0 \frac{\partial \mathbf{E}}{\partial t} - \frac{\partial \rho}{\partial t} = 0. \tag{14.8}$$

Adding (14.7) to (14.8) gives

$$\nabla \cdot \left[\mathbf{j} + \epsilon_0 \frac{\partial \mathbf{E}}{\partial t} \right] = 0. \tag{14.9}$$

Comparing (14.9) with the divergence of the Ampère-Maxwell equation shows that $\mathbf{j}_D$ must be present in the latter.

[1] See Chapter 2 for a brief historical sketch.

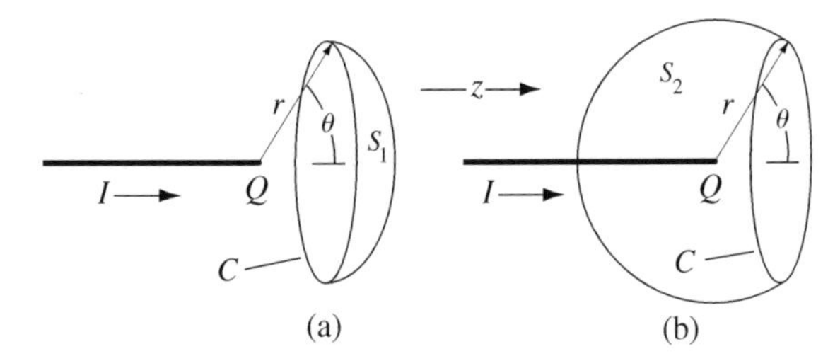

Figure 14.1: A straight-line current I terminates at the origin. Panels (a) and (b) show two choices for an integration surface to evaluate the Ampère-Maxwell equation in integral form.

The displacement current plays a prominent role in situations like the one shown in Figure 14.1, where the steady current I carried by a long and straight wire abruptly terminates at a point of charge accumulation. Subject to later confirmation, we suppose that the sole source of $\mathbf{E}(\mathbf{r}, t)$ is the time-varying charge $Q(t) = It$ at the origin of coordinates. The integral form of Gauss' law then gives

$$\mathbf{E}(r, t) = \frac{Q(t)}{4\pi\epsilon_0 r^2}\hat{\mathbf{r}}. \tag{14.10}$$

Our strategy to find $\mathbf{B}(\mathbf{r}, t)$ exploits the integral form of the Ampère-Maxwell law,

$$\oint_C d\boldsymbol{\ell} \cdot \mathbf{B} = \mu_0 \int_S d\mathbf{S} \cdot \mathbf{j} + \frac{1}{c^2}\int_S d\mathbf{S} \cdot \frac{\partial \mathbf{E}}{\partial t}. \tag{14.11}$$

$\mathbf{B} = \mathbf{B}(r, \theta)$ by symmetry, so we need only the direction of $\mathbf{B}$ to guide our choice for the Ampèrian circuit C. For this purpose, we consider the problem from the Biot-Savart point of view, where the source current is $\mathbf{j}$ from the wire plus $\mathbf{j}_\mathrm{D} = \epsilon_0 \partial \mathbf{E}/\partial t$.[2] The latter has zero curl and thus contributes nothing to the Biot-Savart integral (see Section 10.1). Previous work (see Chapter 10) showed that the magnetic field produced by a straight wire of any length has the form $\mathbf{B} = B_\phi(r, \theta)\hat{\boldsymbol{\phi}}$.

The just-deduced form of $\mathbf{B}$ suggests we choose C as a circle of radius $r \sin\theta$ concentric with the z-axis. Figure 14.1 shows two possible choices for the capping surface S for an observation point with $\theta < \pi/2$. Each is a portion of a sphere of radius r centered on Q. The current in the wire contributes to (14.11) when we use S_2, but not when we use S_1. The displacement current contributes in both cases, but unequally. Carrying out the line integral over C and the two surface integrals over both S_1 and S_2 gives

$$B_\phi 2\pi r \sin\theta = \begin{cases} 0 + \mu_0 \dfrac{I(1-\cos\theta)}{2} & S = S_1, \\ \mu_0 I - \mu_0 \dfrac{I(1+\cos\theta)}{2} & S = S_2. \end{cases} \tag{14.12}$$

The total magnetic field is static (so Faraday's law does not alter our calculation of $\mathbf{E}$) and independent of the choice of S, as it must be:

$$\mathbf{B}(r, \theta) = \frac{\mu_0 I(1-\cos\theta)}{4\pi r \sin\theta}\hat{\boldsymbol{\phi}}. \tag{14.13}$$

A brief calculation shows that (14.13) applies also when $\theta > \pi/2$ and thus correctly reproduces the magnetic field of an infinite wire when $\theta \to \pi$. Figure 14.2 shows representative field lines for both $\mathbf{E}$ and $\mathbf{B}$. The electric field lines emerge radially from the accumulation point. The magnetic field lines

[2] The Biot-Savart integral is a consequence of the Helmholtz theorem, and thus remains valid whether $\nabla \times \mathbf{B} = \mu_0 \mathbf{j}$ or $\nabla \times \mathbf{B} = \mu_0(\mathbf{j} + \mathbf{j}_\mathrm{D})$. However, it is unusual to know $\mathbf{j}_\mathrm{D}$ explicitly, as we do here.

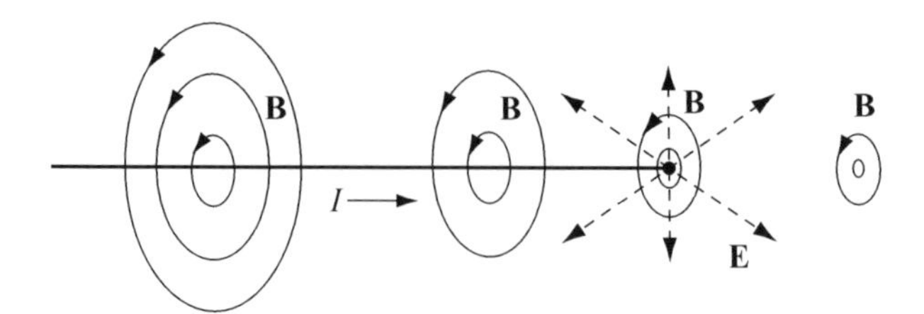

Figure 14.2: A straight-line current I terminates at a point. Closed circles labeled **B** are typical magnetic field lines. Dashed lines labeled **E** are typical electric field lines.

are circles concentric with the symmetry axis of the line current. The latter is true even for points beyond the accumulation point where the circles of **B** do not enclose the conduction current I.

14.2.2 Polarization Current

Polarization current arises whenever bound polarization charge is displaced. We can be quantitative because the polarization current density $\mathbf{j}_P$ is related to the polarization charge density ρ_P by a conservation law no less stringent than (14.7):

$$\frac{\partial \rho_P}{\partial t} + \nabla \cdot \mathbf{j}_P = 0. \tag{14.14}$$

Substituting the explicit form of the polarization charge density, $\rho_P = -\nabla \cdot \mathbf{P}$, into (14.14) gives

$$\nabla \cdot \mathbf{j}_P = \nabla \cdot \frac{\partial \mathbf{P}}{\partial t}. \tag{14.15}$$

A particular solution of (14.15) is

$$\mathbf{j}_P = \frac{\partial \mathbf{P}}{\partial t}. \tag{14.16}$$

This most common form of polarization current density occurs only when $\mathbf{P}(\mathbf{r}, t)$ is an explicit function of time. An example is the charge separation current which accompanies the creation of polarized matter from unpolarized matter (see Section 6.2.3). The boxed material which follows this section and Application 14.1 provide two other examples. In every case, there is an associated magnetic moment,

$$\mathbf{m} = \frac{1}{2}\int d^3r\, \mathbf{r} \times \frac{\partial \mathbf{P}}{\partial t}. \tag{14.17}$$

The general solution of (14.15) supplements (14.16) with a contribution which depends on an arbitrary vector field $\mathbf{M}_P$:

$$\mathbf{j}_P = \frac{\partial \mathbf{P}}{\partial t} + \nabla \times \mathbf{M}_P. \tag{14.18}$$

The notation used for the second term in (14.18) suggests that certain types of polarization produce an effective magnetization. The simplest case is a body with time-independent polarization $\mathbf{P}(\mathbf{r})$ which moves uniformly with velocity $\boldsymbol{v}$. In that case

$$\mathbf{j}_P = \rho_P \boldsymbol{v} = -\boldsymbol{v}\nabla \cdot \mathbf{P}. \tag{14.19}$$

On the other hand, the definition of the convective derivative in Section 1.3.3, and the fact that an observer moving with the body sees no change in polarization at all, means that

$$\frac{d\mathbf{P}}{dt} = \frac{\partial \mathbf{P}}{\partial t} + (\boldsymbol{v} \cdot \nabla)\mathbf{P} = 0. \tag{14.20}$$

Substituting (14.19) and (14.20) into the left and right sides of (14.18), respectively, gives

$$\nabla \times \mathbf{M}_{\mathrm{P}} = (\boldsymbol{\upsilon} \cdot \nabla)\mathbf{P} - \boldsymbol{\upsilon}\nabla \cdot \mathbf{P}. \tag{14.21}$$

The general solution of (14.21) is

$$\mathbf{M}_{\mathrm{P}} = \mathbf{P} \times \boldsymbol{\upsilon}, \tag{14.22}$$

plus the gradient of a scalar function. We choose the latter to vanish[3] and conclude that a polarized body in uniform motion with velocity $\boldsymbol{\upsilon}$ creates the magnetization (14.22). The physical effects of this magnetization are indistinguishable from those produced by orbital or spin magnetic moments.

Polarization Current in Ice

Solid ice conducts electricity by the transport of protons. To understand this, panel (a) below models an ice crystal as a linear chain of water molecules. A proton enters at the left and forms a bond with the O atom of the leftmost water molecule by breaking that oxygen atom's bond with the H atom on the far side of the molecule. The proton released by the broken bond forms a bond with the O atom next in line and this sequence of bond forming/breaking propagates to the end of the chain.

The net result, shown in panel (b), is that the electric dipole moment $\mathbf{p}$ rotates for every water molecule of the chain and the proton exits to the right. Another proton cannot enter from the left until, as panel (c) shows, a random thermal excitation rotates the entire leftmost water molecule. This induces the neighboring molecule to rotate and so on down the chain until the original configuration (a) has been restored. The entire sequence transports one proton from one end of the ice crystal to other with no other change in the system.

In steady state, this process combines a proton conduction current with a proton polarization current $\partial \mathbf{P}/\partial t$. The former comes from protons hopping from molecule to molecule down the chain. The latter comes from the rotating back and forth of the direction of $\mathbf{p}$ for each individual molecule.

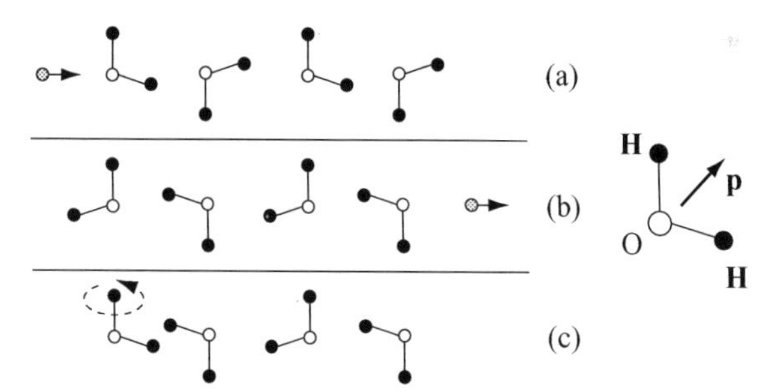

Figure 14.3: Figure adapted from Petrenko and Whitworth (1999).

Application 14.1 The Dielectric Constant of a Plasma

The concept of polarization current leads naturally to a definition for the dielectric constant of a plasma. Example 12.1 analyzed the motion of a particle with mass m and charge q moving in constant fields $\mathbf{E}$ and $\mathbf{B}$ oriented perpendicular to each other. The result was uniform circular motion at the cyclotron

[3] This can be justified using special relativity.

frequency $\omega_c = qB/m$ superimposed on uniform translational motion with velocity $\boldsymbol{v}_0 = \mathbf{E} \times \mathbf{B}/B^2$. This "drift" velocity does not produce a net current in a neutral plasma (with oppositely charged constituents) because $\boldsymbol{v}_0$ does not depend on charge or mass. However, when $\mathbf{E}$ changes with time, we will see below that a second drift velocity $\boldsymbol{v}_P = (m/qB^2)\dot{\mathbf{E}}$ develops because the inertia of the particles prohibits them from adiabatically following the variations of the field. A net current—carried primarily by the most massive particles—arises because oppositely charged particles drift in opposite directions. The resulting separation of charge produces a polarization from which a dielectric constant of the plasma may be defined. In this Application, we offer an energy argument to deduce these facts.

Let a uniform magnetic field $\mathbf{B}$ and a time-varying electric field $\mathbf{E}(t)$ (oriented at right angles to $\mathbf{B}$) act on particle species with mass m, charge q, and number density n. The change in kinetic energy of one particle produced by the changing field is

$$\delta\left(\frac{1}{2}mv_0^2\right) = \delta\left(\frac{1}{2}m\frac{E^2}{B^2}\right) = \frac{m\mathbf{E}\cdot\delta\mathbf{E}}{B^2} = q\mathbf{E}\cdot\delta\mathbf{r}. \tag{14.23}$$

The last equality in (14.23) equates the change in kinetic energy to the work done on the particle by the field during a displacement $\delta\mathbf{r}$. In a time interval δt, the particle acquires a velocity

$$\boldsymbol{v}_P = \frac{\delta\mathbf{r}}{\delta t} = \frac{m}{qB^2}\frac{\partial\mathbf{E}}{\partial t}. \tag{14.24}$$

Therefore, if the plasma contains n particles per unit volume, there is an induced current density

$$\mathbf{j}_P = nq\boldsymbol{v}_P = \frac{nm}{B^2}\frac{\partial\mathbf{E}}{\partial t} = \frac{\partial\mathbf{P}}{\partial t}. \tag{14.25}$$

With $\rho = mn$, the separation of the light particles from the massive particles of opposite charge produces a polarization $\mathbf{P} = (\rho/B^2)\mathbf{E}$ of the plasma.[4] Because

$$\mathbf{D} = \epsilon_0\mathbf{E} + \mathbf{P} = \epsilon\mathbf{E}, \tag{14.26}$$

we conclude that the plasma behaves like a medium with dielectric constant

$$\epsilon = \epsilon_0 + \frac{\rho}{B^2}. \tag{14.27}$$

This result will be relevant later when we study wave propagation in a magnetized plasma (Application 17.1). The propagating wave is the source of the time-varying electric field in that case. ■

14.3 Faraday's Law

The right side of Faraday's law,

$$\nabla \times \mathbf{E} = -\frac{\partial\mathbf{B}}{\partial t}, \tag{14.28}$$

shows that a time-dependent magnetic field is generally the source of an electric field. When the latter varies in time also, the displacement current (see Section 14.2.1) provides a feedback mechanism to create the self-sustaining electric and magnetic fields we call electromagnetic waves. We will have much more to say about these waves in later chapters. Here, we focus narrowly on the effect of Faraday-induced electric fields on charged particles. These effects look rather different if the charged

[4] In a plasma, the kinetic energy of the particles is the beneficiary of the work done to produce charge separation. In conventional dielectric matter, the work to separate charges is done against the potential energy of the cohesive forces of the material.

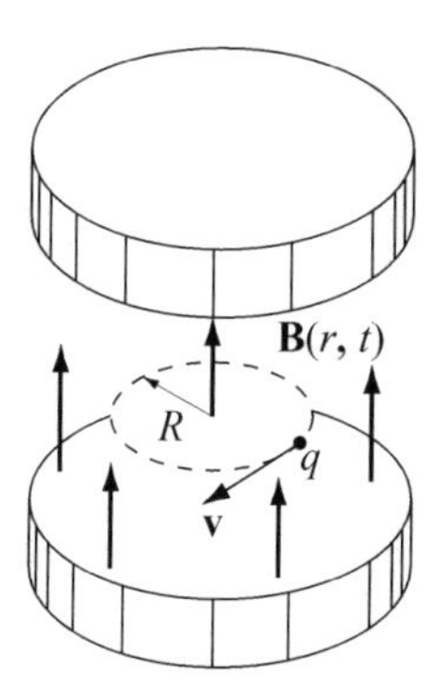

Figure 14.4: The betatron. A particle with charge $q > 0$ moves in a circular orbit of time-independent radius R (dashed line) at velocity $\mathbf{v}$. Heavy solid arrows indicate the magnetic field $\mathbf{B}(r, t)$.

particles move freely in vacuum or if they are confined to wires or extended samples of condensed matter. All of these merit our attention.

14.3.1 The Betatron

A device called the *betatron* exploits a Faraday electric field to accelerate electrons in circular orbits. Betatrons played an important role in the early days of particle physics and radiation oncology.[5] The technical challenge is to choose the space- and time-varying magnetic field so the radius of each particle's orbit remains fixed in time as its speed increases. Figure 14.4 is a cartoon view where the solid arrows denote a magnetic field which points strictly perpendicular to the orbit plane and varies only with the radial distance from the center of the orbit.

The symmetry of the magnetic field guarantees that the tangential electric field is constant at every point on the orbit. Therefore, we evaluate Faraday's law in integral form using the orbit as the circuit C in the line integral in (14.29). Recalling the definition $\Phi_B = \int_S d\mathbf{S} \cdot \mathbf{B}$ of magnetic flux, this gives

$$2\pi R E_\phi(R) = \oint_C d\boldsymbol{\ell} \cdot \mathbf{E} = \int_S d\mathbf{S} \cdot \nabla \times \mathbf{E} = -\int_S d\mathbf{S} \cdot \frac{\partial \mathbf{B}}{\partial t} = -\frac{d\Phi_B}{dt}. \tag{14.29}$$

On the other hand, the radial and tangential components of Newton's second law for circular motion are

$$\frac{mv^2}{R} = qvB(R) \qquad \text{and} \qquad \frac{d}{dt}mv = qE_\phi(R). \tag{14.30}$$

Using (14.29) and (14.30) to eliminate v and $E_\phi(R)$ gives

$$\frac{d\Phi_B}{dt} = -2\pi R^2 \frac{dB(R)}{dt}. \tag{14.31}$$

This equation integrates to the *betatron condition*—a constraint which relates the total magnetic flux through the orbit to the magnitude of the magnetic field on the orbit:

$$|\Phi_B| = 2\pi R^2 B(R). \tag{14.32}$$

The factor of 2 in (14.32) shows that the magnetic field cannot be uniform over the area enclosed by the orbit. The field must be stronger inside the orbit than on the orbit. In detail, $\mathbf{B}(\mathbf{r})$ is chosen to ensure the stability of the orbit against transverse displacements of the particle.

[5] Physicians used the X-rays produced when high-energy electrons from a betatron were directed at a metal target.

14.4 Electromagnetic Induction

In principle, *electromagnetic induction* occurs any time the electric field induced by Faraday's law has a demonstrable effect. In practice, the term is often reserved for situations where the induced electric field drives a current in an ohmic conductor. In this section, we explore this "circuit theory" aspect of Faraday's law and pay special attention to the point in the standard analysis when an explicit low-velocity approximation is made which destroys the exactness (and Lorentz covariance) of the theory.

We begin with the integral form of Faraday's law used already in (14.29). Formally, one integrates (14.28) over an open surface S and uses Stokes' theorem to convert the surface integral of $\nabla \times \mathbf{E}$ into a line integral of $\mathbf{E}$ over the closed curve C which bounds S:

$$\oint_C d\boldsymbol{\ell} \cdot \mathbf{E} = -\int_S d\mathbf{S} \cdot \frac{\partial \mathbf{B}}{\partial t}. \tag{14.33}$$

The key step is to permit the surface S (and hence the boundary C) to move and distort arbitrarily by assigning a velocity $\boldsymbol{\upsilon}_{\rm c}(\mathbf{r}, t)$ to every differential element of S. We can now use the "flux theorem" (See Section 1.4.5),

$$\frac{d}{dt}\int_{S(t)} d\mathbf{S} \cdot \mathbf{B} = \int_{S(t)} d\mathbf{S} \cdot \left\{ \frac{\partial \mathbf{B}}{\partial t} + \boldsymbol{\upsilon}_{\rm c}(\nabla \cdot \mathbf{B}) - \nabla \times (\boldsymbol{\upsilon}_{\rm c} \times \mathbf{B}) \right\}, \tag{14.34}$$

to replace the integral on the right side of (14.33). Using $\nabla \cdot \mathbf{B} = 0$ and Stokes' theorem to rewrite the last term in (14.34) as a line integral shows that (14.33) is equivalent to

$$\oint_C d\boldsymbol{\ell} \cdot (\mathbf{E} + \boldsymbol{\upsilon}_{\rm c} \times \mathbf{B}) = -\frac{d}{dt}\int_S d\mathbf{S} \cdot \mathbf{B}. \tag{14.35}$$

This equation is exact and one could exploit special relativity (Chapter 22) to transform its left side to the rest frame of C for (possibly) easier numerical evaluation.[6]

14.4.1 Faraday Electromotive Force

The choice of C in (14.35) is arbitrary and dictated entirely by convenience. However, for circuit theory and electromechanical applications, C is invariably chosen as a closed path through conducting matter. A loop of wire is the simplest example, but any path through a conducting body which begins and ends at the same point can serve just as well. In both cases, it is important to recognize that $\boldsymbol{\upsilon}_{\rm c}(\mathbf{r}, t)$ in (14.35) is the local velocity of the massive ions which give the matter its mass and rigidity. Current-carrying electrons move with respect to these ions at their *drift velocities* $\boldsymbol{\upsilon}_{\rm d}(\mathbf{r}, t)$.[7] Therefore, in the practically important case when the electrons move much more slowly than the speed of light, the Galilean velocity addition rule tells us that the total velocity of an electron at point $\mathbf{r}$ and time t is

$$\boldsymbol{\upsilon}_{\rm e}(\mathbf{r}, t) = \boldsymbol{\upsilon}_{\rm c}(\mathbf{r}, t) + \boldsymbol{\upsilon}_{\rm d}(\mathbf{r}, t). \tag{14.36}$$

[6] See, for example, H. Gelman, "Faraday's law for relativistic and deformed motion of a circuit", *European Journal of Physics* **12**, 230 (1991).

[7] The drift velocity is the terminal velocity reached by charged carriers in an ohmic medium due to collisions with other particles. See Section 9.3.

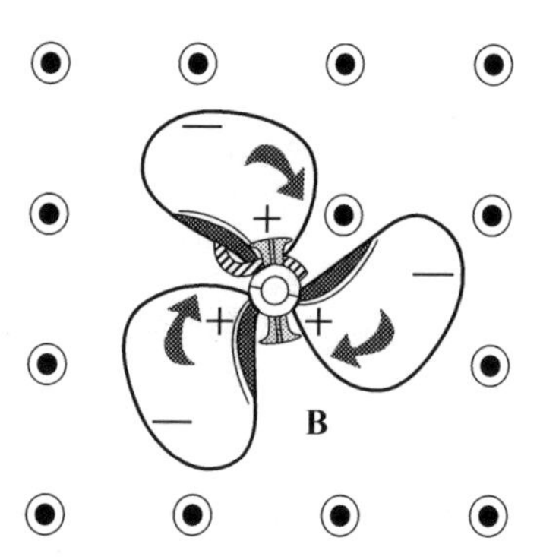

Figure 14.5: The blades of a fan rotate in a uniform magnetic field. Plus and minus signs indicate charge separation in each blade.

Equation (14.36) is not relativistically exact. Therefore, substituting (14.36) into (14.35) gives a low-velocity *approximation* to the latter which is in general use:

$$\oint_C d\boldsymbol{\ell} \cdot (\mathbf{E} + \boldsymbol{v}_\mathrm{e} \times \mathbf{B}) = -\frac{d}{dt}\int_S d\mathbf{S} \cdot \mathbf{B} + \oint_C d\boldsymbol{\ell} \cdot (\boldsymbol{v}_\mathrm{d} \times \mathbf{B}). \tag{14.37}$$

Recall now that $\mathbf{F}_L = q(\mathbf{E} + \boldsymbol{v}_\mathrm{e} \times \mathbf{B})$ is the Coulomb-Lorentz force on an electron with charge q. Therefore, the integral on the left side of (14.37) is a special case of the *electromotive force* (EMF) expression[8]

$$\mathcal{E}_{12} = \frac{1}{q}\int_1^2 d\boldsymbol{\ell} \cdot \mathbf{F}, \tag{14.38}$$

where the points 1 and 2 are coincident and $\mathbf{F} = \mathbf{F}_L$. An example where $\mathbf{F} = \mathbf{F}_L$ and 1 and 2 are not coincident is provided by the fan in Figure 14.5. The metal blades of this fan rotate through a uniform magnetic field. Initially, $\mathbf{E} = 0$ and the Lorentz force on the mobile electrons of the metal drives a separation of charge. The latter causes an electric field to build up until, in steady state, the Coulomb force exactly balances the Lorentz force. The electric part of the line integral (14.38) is the voltage difference between the center of the fan and the tips of the blades.

We have seen that the left side of (14.37) defines a "Faraday" electromotive force,

$$\mathcal{E}_F = \oint_C d\boldsymbol{\ell} \cdot (\mathbf{E} + \boldsymbol{v}_\mathrm{e} \times \mathbf{B}). \tag{14.39}$$

In a circuit context, Ohm's law in the form $\mathcal{E}_F = IR$ relates the Faraday EMF to the current flowing in the circuit. This motivates a search for alternative expressions for $\mathcal{E}_F$ which might be simpler to evaluate for particular situations. One of these follows by substituting (14.39) into (14.37) to get

$$\mathcal{E}_F = -\frac{d}{dt}\int_S d\mathbf{S} \cdot \mathbf{B} + \oint_C d\boldsymbol{\ell} \cdot (\boldsymbol{v}_\mathrm{d} \times \mathbf{B}). \tag{14.40}$$

Substituting (14.36) into (14.40) and using (14.34) produces a third equivalent form,

$$\mathcal{E}_F = -\int_S d\mathbf{S} \cdot \frac{\partial \mathbf{B}}{\partial t} + \oint_C d\boldsymbol{\ell} \cdot (\boldsymbol{v}_\mathrm{e} \times \mathbf{B}). \tag{14.41}$$

[8] In Section 9.7, the symbol **F** in (14.38) stood for the real or effective force associated with any energy source capable of driving a current between points 1 and 2.

Older physics and engineering textbooks use the names "transformer EMF" for the surface integral in (14.41) and "motional EMF" for the circuit integral in (14.41).

Let us return to (14.40). The special case when C is everywhere coincident with a filamentary ohmic wire simplifies this expression because $\boldsymbol{v}_{\rm d}$ is parallel to the integration path element $d\boldsymbol{\ell}$ and $d\boldsymbol{\ell}\cdot(\boldsymbol{v}_{\rm d}\times\mathbf{B})=0$. This reduces (14.40) to a formula which is often called the "flux rule":

$$\mathcal{E}_F = -\frac{d\Phi_B}{dt} = -\frac{d}{dt}\int_S d\mathbf{S}\cdot\mathbf{B}. \tag{14.42}$$

The minus sign in (14.42) reflects *Lenz' law*: the direction of the current driven by $\mathcal{E}_F$ produces a magnetic field which tends to oppose the change in magnetic flux which produced it.[9] Most of the qualitative effects of electromagnetic induction can be understood by systematically applying this law.

We conclude by noting that the drift velocity $\boldsymbol{v}_{\rm d}$ is numerically very small. For that reason, many authors neglect the $\boldsymbol{v}_{\rm d}\times\mathbf{B}$ term in (14.40) altogether. Such a "derivation" of the flux rule must be viewed with caution because situations exist where the two terms in (14.40) have comparable magnitudes. The Hall effect is an example where the drift term alone generates the entire EMF.[10]

Example 14.1 A small circular loop of N turns of wire is oriented so the loop axis is parallel to the local direction of a magnetic field $\mathbf{B}$. A galvanometer (or other current integrating device) measures the total amount of charge Q which flows through the circuit of total resistance R (Figure 14.6) when the coil is flipped so its loop axis reverses direction. Find the relationship between Q and $|\mathbf{B}|$ assuming that $\mathcal{E}_F = IR$ remains valid. This device, known as a *flip coil*, is used for quick measurements of magnetic field strength.

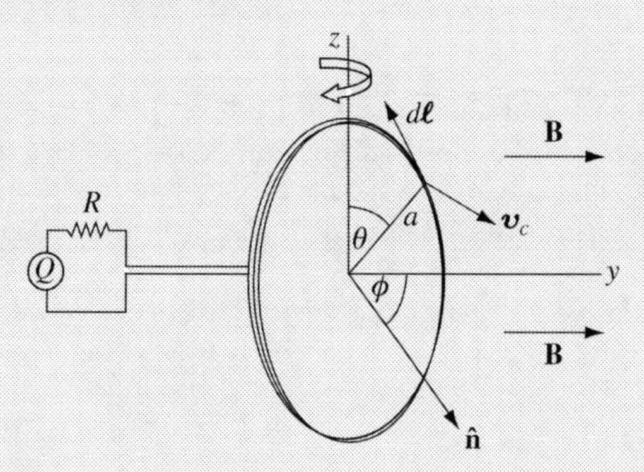

Figure 14.6: A flip coil measurement begins with the normal to the plane of the loop, $\hat{\mathbf{n}}$, pointed along $+y$ and ends with $\hat{\mathbf{n}}$ pointed along $-y$. The induced current flows in the direction of $d\boldsymbol{\ell}$.

Method I: Since $\mathcal{E}_F/R = I = dQ/dt$ relates the induced current to the total charge, we can integrate (14.42) directly to get

$$Q = \int dt\, I = -\frac{1}{R}\int d\Phi_{\rm B} = -\frac{(-N\pi a^2 B)-(N\pi a^2 B)}{R} = \frac{2N\pi a^2 B}{R}.$$

[9] The phrase "back EMF" is used to describe this phenomenon in electric-circuit parlance.

[10] See the first paragraph of the "Quantum Hall Effect" box at the end of Section 12.6.4.

Method II: We compute the EMF using (14.41) and (14.36). The magnetic field does not change in time and there is no contribution from $\boldsymbol{v}_{\rm d}$ for a filamentary wire. Therefore,

$$\mathcal{E}_F = N \oint d\boldsymbol{\ell} \cdot \boldsymbol{v}_{\rm c} \times \mathbf{B}.$$

From Figure 14.6, the velocity of a differential element of the loop (radius a) is $\boldsymbol{v}_{\rm c} = a \sin\theta \dot{\phi}\, \hat{\mathbf{n}}$. Then, because $\mathbf{B} = B\hat{\mathbf{y}}$ and $d\boldsymbol{\ell} = -ad\theta\hat{\boldsymbol{\theta}}$,

$$\mathcal{E}_F = Na^2 B \sin\phi \frac{d\phi}{dt} \int_0^{2\pi} d\theta \sin^2\theta = N\pi a^2 B \sin\phi \frac{d\phi}{dt}.$$

Integrating $I = \mathcal{E}_F / R$ over time (as the coil flips) gives the same result as Method I,

$$Q = \frac{N\pi a^2 B}{R} \int_0^{\pi} d\phi \sin\phi = \frac{2N\pi a^2 B}{R}.$$

Example 14.2 One way to measure a time-varying magnetization $\mathbf{M}(t)$ exploits the EMF $\mathcal{E}_F(t)$ induced by the magnetization in a nearby "pickup" coil of wire (Figure 14.7). Assume that a steady current $I_{\rm coil}$ in the coil produces a magnetic field $\mathbf{B}_{\rm coil}(\mathbf{r})$ and show that

$$\mathcal{E}_F(t) = -\frac{1}{I_{\rm coil}} \frac{d}{dt} \int d^3r\, \mathbf{B}_{\rm coil} \cdot \mathbf{M}(t).$$

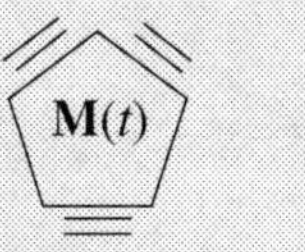

Figure 14.7: A time-dependent magnetization $\mathbf{M}(t)$ and a nearby coil of wire.

Solution: A dynamic magnetization produces an EMF in the pickup coil because the magnetic field $\mathbf{B}_{\rm M}(\mathbf{r}, t)$ produced by $\mathbf{M}(t)$ produces a time-dependent magnetic flux through the coil:

$$\Phi = \int d\mathbf{S} \cdot \mathbf{B}_{\rm M} = \oint d\boldsymbol{\ell} \cdot \mathbf{A}_{\rm M}.$$

The substitution $Id\boldsymbol{\ell} \to \mathbf{j}d^3r$, $\nabla \times \mathbf{B}_{\rm coil} = \mu_0 \mathbf{j}_{\rm coil}$, and an integration by parts permits us to write

$$I_{\rm coil}\Phi = \int d^3r\, \mathbf{j}_{\rm coil} \cdot \mathbf{A}_{\rm M} = \frac{1}{\mu_0} \int d^3r\, \nabla \times \mathbf{B}_{\rm coil} \cdot \mathbf{A}_{\rm M} = \frac{1}{\mu_0} \int d^3r\, \mathbf{B}_{\rm coil} \cdot \mathbf{B}_{\rm M}(t).$$

Now, $\mathbf{B}_{\rm M} = \mu_0(\mathbf{M} + \mathbf{H}_{\rm M})$, so

$$I_{\rm coil}\Phi = \int d^3r\, \mathbf{B}_{\rm coil} \cdot (\mathbf{M} + \mathbf{H}_{\rm M}) = \int d^3r\, \mathbf{B}_{\rm coil} \cdot \mathbf{M}.$$

The volume integral of $\mathbf{B}_{\text{coil}} \cdot \mathbf{H}_{\text{M}}$ vanishes in the preceding equation because $\mathbf{H}_{\text{M}}$ is not produced by free current.[11] Applying the flux rule (14.42), we conclude that

$$\mathcal{E}_F(t) = -\frac{1}{I_{\text{coil}}}\frac{d}{dt}\int d^3r\, \mathbf{B}_{\text{coil}} \cdot \mathbf{M}(t).$$

This formula is used to extract $\mathbf{M}(t)$ from magnetic recording tape (where the coil is scanned over the tape) and from nuclear magnetic resonance data (where the source spins precess in an external field) using a coil with an easily calculable $\mathbf{B}_{\text{coil}}(\mathbf{r})$.

Application 14.2 Faraday's Disk Generator

In 1831, Faraday constructed the device shown in Figure 14.8. A conducting disk of radius a rotates with angular velocity ω in a static and uniform magnetic field $\mathbf{B}$ oriented perpendicular to the plane of the disk. A wire frame GMFA with resistance R makes fixed contact with the center of the disk at point A and sliding contact with the rim of the disk at point G.[12] Our task is to confirm Faraday's experimental result that rotation of the disk generates a current I in the wire frame. For simplicity, we assume that the resistance of the disk is negligible.

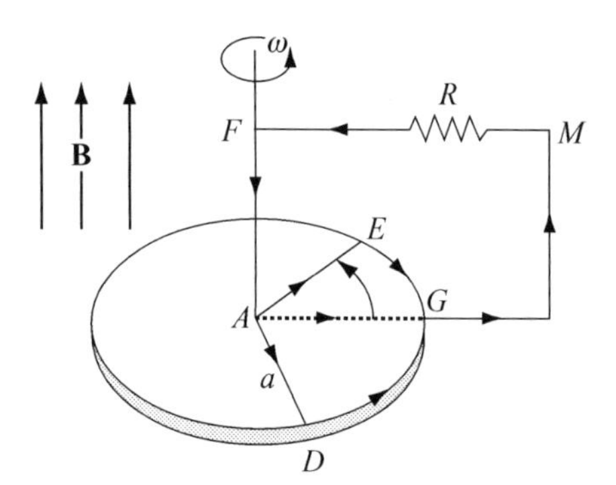

Figure 14.8: The Faraday disk generator. A uniform magnetic field fills the space occupied by the metal disk and the wire frame GMFA. Stick arrowheads indicate various choices for the circuit C discussed in the text.

Method I: We use (14.41) and note that only the motional EMF term contributes because $\mathbf{B}$ does not depend on time. The actual path taken by current-carrying electrons through the disk is not obvious. On the other hand, the velocity component $\boldsymbol{v}_{\parallel}$ which is locally parallel to the integration path is irrelevant because $d\boldsymbol{\ell} \cdot (\boldsymbol{v}_{\parallel} \times \mathbf{B}) = 0$ for any choice of C. Moreover, if C is the fixed rectangle AGMFA in Figure 14.8, the entire electromotive force comes from the perpendicular velocity component $\boldsymbol{v}_{\perp}$ along the path segment AG. We now *assume* that $v_{\perp}(r) = r\omega$, i.e., we assume that the electrons are dragged along rigidly with the ions as the disk rotates. This leads to the steady current $I = \mathcal{E}_F/R$ with

$$\mathcal{E}_F = \int_A^G d\boldsymbol{\ell} \cdot (\boldsymbol{v} \times \mathbf{B}) = \int_0^a dr\, r\omega B = \tfrac{1}{2}\omega a^2 B. \tag{14.43}$$

[11] This follows because (13.102) is equal to (13.103).

[12] The plane of the wire frame is perpendicular to the plane of the disk.

Method II: We use (14.40) and an integration path which, in part, rotates with the disk. This generates a time-dependent area in the flux integral. A suitable choice for C is ADGMFA. This path moves to AGMFA and then to AEGMFA as the disk turns. With the same assumptions as above regarding $\boldsymbol{v}$, this choice yields no contribution from the $\boldsymbol{v}_{\rm d} \times \mathbf{B}$ motional EMF term anywhere along the path. However, since $d\mathbf{S}$ is anti-parallel to $\mathbf{B}$ (for the assumed direction of current flow) during the time t when AG rotates through an angle ωt to AE, we again find $I = \mathcal{E}_F/R$ where

$$\mathcal{E}_F = -\frac{d}{dt}\int\limits_{ADEA} d\mathbf{S}\cdot\mathbf{B} = \frac{d}{dt}\left(\tfrac{1}{2}\omega t a^2 B\right) = \tfrac{1}{2}\omega a^2 B. \tag{14.44}$$

It is interesting to consider how the Faraday disk problem changes if the external magnetic field changes in time. If, say, $B(t) = B\cos\Omega t$, either method used above generates an additional contribution to the EMF, $\mathcal{E}' = \pi a^2 B\Omega \sin\Omega t$. However, the current driven by this EMF is entirely circumferential and confined to the disk itself. Therefore, the current in the external circuit is unchanged from the value computed using just $\mathcal{E}_F$. ■

14.5 Slowly Time-Varying Charge in Vacuum

We turn now to a general discussion of the fields produced by a slowly time-varying charge density $\rho(\mathbf{r}, t)$. Figure 14.9 shows a source of this type in vacuum. We use a length ℓ to characterize its size and a time T to characterize the time needed for $\rho(\mathbf{r}, t)$ to undergo a typical variation of its magnitude. Thus, $T \sim 1/\omega$ for a time-harmonic source which oscillates at frequency ω. These quantities will help us define the meaning of "slowly time-varying" based on the dimensional estimates $\nabla \sim 1/\ell$ and $\partial/\partial t \sim 1/T \sim \omega$.

We apply this scheme first to the continuity equation (14.7). Ignoring the vector character (here and below),

$$\nabla\cdot\mathbf{j} + \frac{\partial\rho}{\partial t} = 0 \quad\Rightarrow\quad \frac{j}{\ell} + \omega\rho = 0 \quad\Rightarrow\quad j \sim \omega\ell\rho. \tag{14.45}$$

To analyze the Maxwell equations similarly, we use Helmholtz' theorem (Section 1.9) and decompose the electric field as $\mathbf{E} = \mathbf{E}_C + \mathbf{E}_F$ where the "Coulomb" and "Faraday" pieces satisfy $\nabla\times\mathbf{E}_C = 0$ and $\nabla\cdot\mathbf{E}_F = 0$. Consequently,

$$\nabla\cdot\mathbf{E}_C = \frac{\rho}{\epsilon_0} \quad\Rightarrow\quad \frac{E_C}{\ell} \sim \frac{\rho}{\epsilon_0} \quad\Rightarrow\quad E_C \sim \frac{\ell}{\epsilon_0}\rho \tag{14.46}$$

and

$$\nabla\times\mathbf{E}_F = \frac{\partial\mathbf{B}}{\partial t} \quad\Rightarrow\quad \frac{E_F}{\ell} \sim \omega B \quad\Rightarrow\quad E_F \sim \omega B\ell. \tag{14.47}$$

We next use (14.46) and (14.47) to form the ratio,

$$\frac{E_F}{E_C} \sim \frac{\varepsilon_0\omega B}{\rho}. \tag{14.48}$$

Now use (14.46) and (14.47) again to estimate the magnetic field in (14.48) using the Ampère-Maxwell law:

$$\nabla\times\mathbf{B} = \mu_0\mathbf{j} + \frac{1}{c^2}\frac{\partial\mathbf{E}}{\partial t} \quad\Rightarrow\quad \frac{B}{\ell} \sim \mu_0 j + \frac{\omega}{c^2}\left(\frac{\ell\rho}{\varepsilon_0} + \omega B\ell\right). \tag{14.49}$$

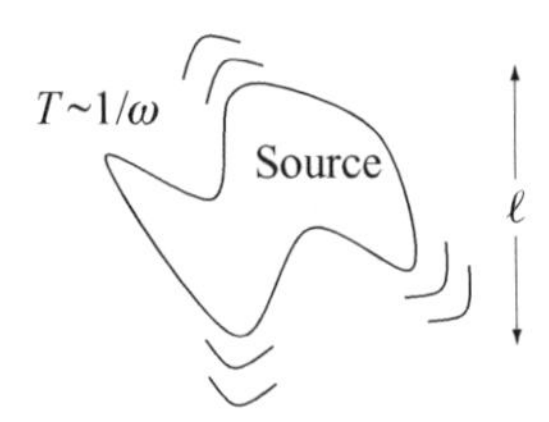

Figure 14.9: A source distribution of size ℓ with maximal time variations that occur over a characteristic time $T \sim 1/\omega$.

A glance back at (14.45) shows that the first and second terms on the far right side of (14.49) are both of order $\mu_0 \omega \ell \rho$. Therefore, solving the rightmost equation in (14.49) for B and substituting into (14.48) gives

$$\frac{E_F}{E_C} \sim \frac{\omega^2 \ell^2 / c^2}{1 - \omega^2 \ell^2 / c^2}. \tag{14.50}$$

Equation (14.50) shows that a sufficiently slowly time-varying charge density produces a negligibly small Faraday electric field compared to its Coulomb electric field. This leads us to characterize a charge distribution as "slowly time-varying" whenever

$$\omega^2 \ll \frac{c^2}{\ell^2}. \tag{14.51}$$

The inequality (14.51) can be read as a condition on ω for fixed system size or as a condition on ℓ for a fixed time variation of the fields. Later, we will see that it is not a coincidence that ℓ/c in (14.51) is the time required for light to travel a distance ℓ.

14.5.1 Quasi-Electrostatic Fields in Vacuum

The estimates of the previous section apply to a source which produces a static electric field in the limit of no time variation. When (14.51) is satisfied, the Faraday electric field is negligible and we may drop the $\partial \mathbf{B}/\partial t$ term in the Maxwell equations to get

$$\nabla \cdot \mathbf{E} = \frac{\rho}{\epsilon_0} \qquad\qquad \nabla \cdot \mathbf{B} = 0 \tag{14.52}$$

$$\nabla \times \mathbf{E} = 0 \qquad\qquad \nabla \times \mathbf{B} = \mu_0 \mathbf{j} + \frac{1}{c^2}\frac{\partial \mathbf{E}}{\partial t}. \tag{14.53}$$

This is a *quasi-electrostatic* limit because $\mathbf{E}(\mathbf{r}, t)$ satisfies the electrostatic Maxwell equations. Helmholtz' theorem (Section 1.9) and the left members of (14.52) and (14.53) provide an explicit formula for the electric field,

$$\mathbf{E}(\mathbf{r}, t) = -\nabla \varphi(\mathbf{r}, t) = -\nabla \frac{1}{4\pi\epsilon_0} \int d^3 r' \, \frac{\rho(\mathbf{r}', t)}{|\mathbf{r} - \mathbf{r}'|}. \tag{14.54}$$

It is *not* obvious, but the magnetic field that satisfies (14.52) and (14.53) is given by the simplest time-dependent generalization of the Biot-Savart law:

$$\mathbf{B}(\mathbf{r}, t) = \nabla \times \mathbf{A}(\mathbf{r}, t) = \nabla \times \frac{\mu_0}{4\pi} \int d^3 r' \, \frac{\mathbf{j}(\mathbf{r}', t)}{|\mathbf{r} - \mathbf{r}'|}. \tag{14.55}$$

The electric field (14.54) is the static Coulomb field produced by the instantaneous value of the charge density. The magnetic field (14.55) is the static Biot-Savart field produced by the instantaneous value of the current density. The two sources are not independent because the continuity condition (14.7) relates the instantaneous value of $\mathbf{j}(\mathbf{r}, t)$ to the instantaneous value of $\rho(\mathbf{r}, t)$.

Application 14.3 The Quadrupole Mass Spectrometer

A quadrupole mass spectrometer is a device which exploits the quasi-electrostatic approximation to separate particles with a specified charge-to-mass ratio from a beam composed of many different mass and charge species. In a typical realization, four metal rods are arranged symmetrically (Figure 14.10) and a time-dependent potential is applied to one pair of non-adjacent rods:

$$\varphi(t) = \varphi_0 + V_0 \cos \omega t. \tag{14.56}$$

A potential $-\varphi(t)$ is applied to the other pair of rods. Our interest is the trajectories of particles with different charges and masses which enter the spectrometer along the z-axis with velocities $\boldsymbol{v} = v_z\hat{\mathbf{z}} + \boldsymbol{v}_\perp$. We assume that $v_\perp \ll v_z$. By symmetry, $E_z = 0$, and the particles do not accelerate along the symmetry axis.

Suppose first that $V_0 = 0$ in (14.56) so the problem is strictly electrostatic. The potential $\varphi(x, y)$ in the volume between the rods satisfies Laplace's equation. The symmetry of the rods ensures that $\varphi(-x, y) = \varphi(x, y)$ and $\varphi(x, -y) = \varphi(x, y)$. Therefore, since $\nabla^2\varphi = 0$, the potential in the immediate vicinity of the z-axis must have the form

$$\varphi(x, y) = \frac{\varphi_0}{R^2}(x^2 - y^2) + \cdots. \tag{14.57}$$

The length R is present for dimensional reasons. The ellipsis indicates that we have written only the lowest-order (quadrupole) term of an interior multipole expansion of the potential. The higher-order terms are needed to satisfy the boundary conditions at the surface of the rods (see Figure 7.13).

In the region of space where (14.57) is valid, Newton's equations of motion for a particle with charge q and mass m are

$$m\ddot{x} + \frac{2q\varphi_0}{R^2}x = 0 \qquad m\ddot{y} - \frac{2q\varphi_0}{R^2}y = 0 \qquad m\ddot{z} = 0. \tag{14.58}$$

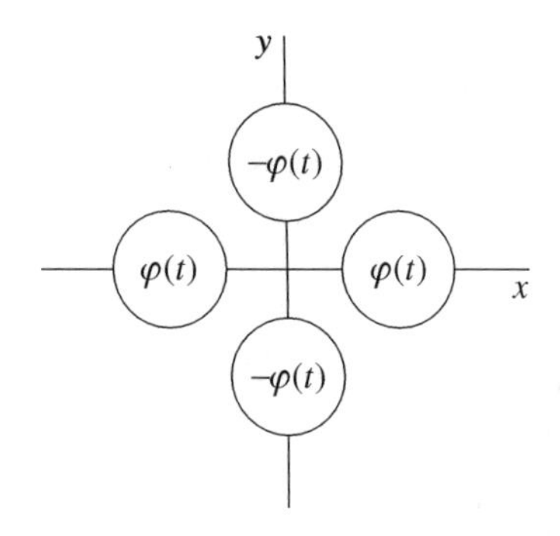

Figure 14.10: Cross section of a quadrupole mass spectrometer. This view looks down the symmetry axis of four perfectly conducting cylindrical rods oriented perpendicular to the page.

There is a restoring force in the x-direction and a repelling force in the y-direction because $x = y = 0$ is a saddle point for the potential energy $q\,\varphi(x, y)$. When $\boldsymbol{v}_\perp \neq 0$, these equations predict uniform motion in the z-direction accompanied by simple harmonic motion in the x-direction and exponential runaway motion in the y-direction. No particle successfully traverses the entire length of the instrument. It either hits one of the rods or escapes sideways.

Now let V_0 in (14.56) be non-zero. If the driving frequency ω satisfies the quasistatic condition (14.51) (with $\ell \sim R$), the equations of motion (14.58) remain valid with φ_0 replaced by $\varphi(t)$.[13] Modulating the potential in this way, we periodically exchange the roles of the x-direction and y-direction as far as restoring and repelling forces are concerned. Thus, with a suitable choice of the parameters ω, φ_0, and V_0, a particle with a given value of q/m that is headed away from the z-axis for one half-cycle of oscillation can be brought back to the axis again in the next half-cycle. This particle enjoys bounded, stable motion in the transverse direction and thus is able to pass completely through the length of the device for collection at the far end.[14] In practice, ω is a radio frequency and one sweeps the values of φ_0 and V_0 (typically in the kV range) to operate the instrument as a mass spectrometer for ions of fixed charge. ■

Example 14.3 Find the electric and magnetic fields produced by a point charge in vacuum which moves with a constant speed $\upsilon \ll c$. Use the latter condition to justify a quasi-electrostatic approximation.

Solution: A point charge has no characteristic size. Therefore, the length ℓ must be the distance from the charge to the observation point. With $\omega = \upsilon/\ell$, the condition $\upsilon \ll c$ is then identical to the quasistatic condition $\omega\ell/c \ll 1$. This tells us we can use (14.54) and (14.55) to compute the fields. The charge and current densities are $\rho(\mathbf{r}, t) = q\delta(\mathbf{r} - \boldsymbol{\upsilon}t)$ and $\mathbf{j}(\mathbf{r}, t) = q\,\boldsymbol{\upsilon}\delta(\mathbf{r} - \boldsymbol{\upsilon}t)$, respectively. Therefore,

$$\mathbf{E}(\mathbf{r}, t) = -\nabla\frac{1}{4\pi\epsilon_0}\int d^3r'\,\frac{q\,\delta(\mathbf{r}' - \boldsymbol{\upsilon}t)}{|\mathbf{r} - \mathbf{r}'|} = \frac{q}{4\pi\epsilon_0}\frac{\mathbf{r} - \boldsymbol{\upsilon}t}{|\mathbf{r} - \boldsymbol{\upsilon}t|^3}$$

$$\mathbf{B}(\mathbf{r}, t) = \nabla\times\frac{\mu_0}{4\pi}\int d^3r'\,\frac{q\,\boldsymbol{\upsilon}\delta(\mathbf{r}' - \boldsymbol{\upsilon}t)}{|\mathbf{r} - \mathbf{r}'|} = \nabla\times\frac{\boldsymbol{\upsilon}}{c^2}\frac{1}{4\pi\epsilon_0}\frac{q}{|\mathbf{r} - \boldsymbol{\upsilon}t|} = \frac{\boldsymbol{\upsilon}}{c^2}\times\mathbf{E}(\mathbf{r}, t).$$

The electric field is the static Coulomb field of the charge rigidly "dragged along" by its motion. The magnetic field reflects the two vectors of the problem ($\mathbf{E}$ and $\boldsymbol{\upsilon}$) combined in the simplest way that yields a vector.

14.6 Slowly Time-Varying Current in Vacuum

The decision as to whether a current density $\mathbf{j}(\mathbf{r}, t)$ in vacuum is "slowly time-varying" or not requires an order-of-magnitude analysis somewhat different from the one performed in Section 14.5 for a time-varying charge density. Using the Ampère-Maxwell law on the right side of (14.53), let $\mathbf{B}_A$ be the magnetic field produced by the Ampère current j and let $\mathbf{B}_d$ be the magnetic field produced by the displacement current

$$\mathbf{j}_d = \epsilon_0\partial\mathbf{E}/\partial t. \tag{14.59}$$

Writing $\nabla \sim 1/\ell$ and $\partial/\partial t \sim \omega$ as described in connection with Figure 14.9 gives the magnitude estimates

$$B_A \sim \mu_0 j\,\ell \qquad \text{and} \qquad B_d \sim \ell\omega E/c^2 \tag{14.60}$$

[13] The estimate $\upsilon B/E \sim (\upsilon/c)(R/cT) \ll 1$ shows that the magnetic Lorentz force is negligible in the quasi-electrostatic limit.

[14] The same physics underlies the principle of strong focusing with quadrupole magnets used in particle accelerators (see Application 11.4 at the end of Section 11.5.2).

However, according to Faraday's law,

$$\nabla \times \mathbf{E} = -\frac{\partial}{\partial t}[\mathbf{B}_A + \mathbf{B}_d] \qquad \Rightarrow \qquad E \sim \mu_0 \omega j \ell^2 + \ell^2 \omega^2 E / c^2. \tag{14.61}$$

Solving (14.61) for E and using this to estimate the displacement current in (14.59) gives the ratio

$$\frac{j_d}{j} \sim \frac{\omega^2 \ell^2 / c^2}{1 - \omega^2 \ell^2 / c^2}. \tag{14.62}$$

Equation (14.62) implies that the Ampère magnetic field produced by a sufficiently slowly varying current density is much greater than the magnetic field produced by the displacement current. Therefore, we characterize a current distribution as "slowly time-varying" whenever

$$\omega^2 \ll \frac{c^2}{\ell^2}. \tag{14.63}$$

Not accidentally, the quasi-magnetostatic condition (14.63) is exactly the same as the quasi-electrostatic condition (14.51). We will see later that this condition ensures that the phenomena of *retardation* and *radiation* do not occur.

14.6.1 Quasi-Magnetostatic Fields in Vacuum

The estimates of the previous section apply to a source which produces a static magnetic field in the limit of no time variation. When (14.63) is satisfied, the displacement current is negligible and we can drop the $\partial \mathbf{E}/\partial t$ term in the Maxwell equations to get

$$\nabla \cdot \mathbf{E} = \rho/\epsilon_0 \qquad \nabla \cdot \mathbf{B} = 0 \tag{14.64}$$

$$\nabla \times \mathbf{E} = -\frac{\partial \mathbf{B}}{\partial t} \qquad \nabla \times \mathbf{B} = \mu_0 \mathbf{j}. \tag{14.65}$$

This is a *quasi-magnetostatic* approximation because $\mathbf{B}(\mathbf{r}, t)$ satisfies the magnetostatic Maxwell equations. Since $\nabla \cdot \nabla \times \mathbf{B} \equiv 0$, the neglect of $\mathbf{j}_D$ also requires that

$$\nabla \cdot \mathbf{j} = 0. \tag{14.66}$$

Comparing (14.66) to the continuity equation (14.7) shows that $\rho(\mathbf{r}, t) = \rho(\mathbf{r})$. The presence of a static charge density—with its associated static electric field—is not precluded by (14.64) and (14.65), but it is not consistent with our premise that the static limit of the source is a steady current. For that reason, there is no loss of generality if we choose

$$\rho(\mathbf{r}, t) = 0. \tag{14.67}$$

The Helmholtz theorem (Section 1.9) and the right members of (14.64) and (14.65) give an explicit formula for $\mathbf{B}(\mathbf{r}, t)$ which is identical to the formula (14.55) for the magnetic field of quasi-electrostatics:

$$\mathbf{B}(\mathbf{r}, t) = \nabla \times \mathbf{A}(\mathbf{r}, t) = \nabla \times \frac{\mu_0}{4\pi} \int d^3 r' \, \frac{\mathbf{j}(\mathbf{r}', t)}{|\mathbf{r} - \mathbf{r}'|}. \tag{14.68}$$

The corresponding quasi-electrostatic electric field is

$$\mathbf{E}(\mathbf{r}, t) = -\frac{\mu_0}{4\pi} \int d^3 r' \, \frac{\partial \mathbf{j}(\mathbf{r}', t)/\partial t}{|\mathbf{r} - \mathbf{r}'|}. \tag{14.69}$$

The reader can check that (14.68) and (14.69) are connected by Faraday's law and that (14.69) has zero divergence when the steady-current condition (14.66) is valid.

Application 14.4 A Solenoid with a Time-Harmonic Current

A familiar magnetostatics problem asks for the magnetic field produced by an infinitely long solenoid formed by tightly wrapping n turns per unit length of wire around a tube of radius a. The wire carries a steady current I. A quasi-magnetostatic generalization seeks the electric and magnetic fields when the current in the wire is $I(t) = I\cos\omega t$. If the tube is coaxial with the z-axis, the solution of the original magnetostatic problem (Section 10.2.2) and (14.68) give the quasistatic magnetic field as

$$\mathbf{B}(t) = \begin{cases} \mu_0 n I(t)\hat{\mathbf{z}} & \rho < a, \\ 0 & \rho > a. \end{cases} \tag{14.70}$$

By symmetry, the electric field everywhere satisfies $\mathbf{E} = E(\rho, t)\hat{\boldsymbol{\phi}}$. This suggests we use a circle of radius ρ coaxial with the z-axis for the curve C in the integral form of Faraday's law,

$$\oint_C d\boldsymbol{\ell}\cdot\mathbf{E} = -\frac{d}{dt}\int_S d\mathbf{S}\cdot\mathbf{B}. \tag{14.71}$$

Using $\mathbf{B}(t)$ from (14.71) gives the electric field everywhere as

$$\mathbf{E}_\phi(\rho, t) = \begin{cases} -\mu_0 n\dfrac{\rho}{2}\dot{I}(t) & \rho < a, \\ -\mu_0 n\dfrac{a^2}{2\rho}\dot{I}(t) & \rho > a. \end{cases} \tag{14.72}$$

The formal solution to the quasi-magnetostatic problem ends here. However, by re-introducing the displacement current density, we can find a first approximation to the (small) magnetic field $\mathbf{B}_{\rm out} = B_{\rm out}(\rho, t)\hat{\mathbf{z}}$ which appears outside the solenoid. Using (14.72) to write out $\nabla\times\mathbf{B}_{\rm out} = c^{-2}\partial\mathbf{E}_{\rm out}/\partial t$ gives

$$\frac{\partial B_{\rm out}}{\partial\rho} = -\frac{\omega^2}{c^2}\frac{\mu_0 n a^2}{2\rho}I(t). \tag{14.73}$$

Therefore, up to a time-dependent "constant" which does not depend on ρ, the magnetic field which appears outside the solenoid due to the time variation of the current is

$$B_{\rm out}(\rho, t) = \int_{c/\omega}^{\rho} d\rho'\frac{\partial B_{\rm out}}{\partial\rho'} = -\frac{1}{2}\mu_0 n I(t)\frac{\omega^2 a^2}{c^2}\ln(\omega\rho/c). \tag{14.74}$$

The lower limit of the integral in (14.74) has been "cut off" at a distance c/ω, which is the natural length scale associated with the time variations of the source and field. The need for such a cutoff is signalled by the fact that the ρ' integral diverges if the usual value of the lower limit (infinity) is retained. The approximate solution (14.74) is not valid at distances greater than c/ω from the solenoid. We will revisit this problem in Chapter 20 when we study propagating fields. ■

14.7 Quasistatic Fields in Matter

When non-conducting dielectric or magnetic matter is present, it makes sense to speak of slowly time-varying distributions of *free* charge and current. When $\mathbf{D} = \epsilon\mathbf{E}$ and $\mathbf{B} = \mu\mathbf{H}$, the order-of-magnitude estimates made in Sections 14.5 and 14.6 can be repeated using the in-matter Maxwell equations (Section 2.4.1). Nothing changes except that $\epsilon_0 \to \epsilon$ and $\mu_0 \to \mu$ everywhere and the condition for slow variation (14.51) is replaced by

$$\omega^2 \ll \frac{1}{\ell^2 \mu \epsilon}. \tag{14.75}$$

14.7.1 Charge Relaxation

Quasistatic behavior in metals and plasmas is dominated by the fact that free electric charge abhors a conducting medium. As a result, the process of *charge relaxation* removes charge from the volume of any system where the current density obeys Ohm's law,

$$\mathbf{j}_f = \sigma \mathbf{E}. \tag{14.76}$$

To see why, substitute (14.76) and then $\epsilon \nabla \cdot \mathbf{E} = \rho_f$ into the continuity equation

$$\nabla \cdot \mathbf{j}_f + \frac{\partial \rho_f}{\partial t} = 0. \tag{14.77}$$

The result is a partial differential equation for the time evolution of the charge density,

$$\frac{\partial \rho_f}{\partial t} + \frac{\sigma}{\epsilon} \rho_f = 0. \tag{14.78}$$

If the conductivity σ is strictly constant, the solution to (14.78) is[15]

$$\rho_f(\mathbf{r}, t) = \rho_f(\mathbf{r}, 0) \exp(-t/\tau_E) \qquad \text{where } \tau_E \equiv \epsilon/\sigma. \tag{14.79}$$

This formula shows that volume charge disappears from the interior of a conductor on a time scale set by an electric time constant τ_E. The greater the conductivity, the faster this process occurs. Since charge is conserved, the disappearance of bulk charge is accompanied by the appearance of charge on the surface of the conductor.

14.8 Poor Conductors: Quasi-Electrostatics

Quasi-electrostatics in conducting matter makes sense when charge relaxation is slow enough to permit a free charge density $\rho_f(\mathbf{r}, t)$ to exert its Coulombic influence before it disappears as dictated by (14.79). For a time-harmonic source that oscillates at frequency ω, this means that the order-of-magnitude estimates made in Section 14.5 remain true when $\omega \tau_E \gg 1$. Switching to dielectric language and making use of (14.75), we conclude that the regime of quasi-electrostatics for poorly conducting simple matter is defined by

$$\frac{\sigma}{\epsilon} \ll \omega \ll \frac{1}{\ell \sqrt{\mu \epsilon}}. \tag{14.80}$$

As a practical matter, this regime exists only for very low-conductivity materials ($\sigma < 10\ \Omega^{-1}\text{m}^{-1}$) which are more naturally classified as lossy dielectrics.

In a poor conductor, the quasi-electrostatic approximation to the Maxwell equations neglects electromagnetic induction and treats the current density as purely ohmic. Therefore, the fields are determined by

$$\nabla \cdot \mathbf{E} = \frac{\rho_f}{\epsilon} \qquad \nabla \cdot \mathbf{B} = 0 \tag{14.81}$$

$$\nabla \times \mathbf{E} = 0 \qquad \nabla \times \mathbf{B} = \mu \sigma \mathbf{E} + \mu \epsilon \frac{\partial \mathbf{E}}{\partial t}. \tag{14.82}$$

[15] See, however, the first paragraph of Section 14.9.

Usually, we solve the electrostatic part of these equations to find $\mathbf{j} = \sigma\mathbf{E}$. If needed, the magnetic field follows as in the vacuum case.

Application 14.5 The Power Dissipated in a Metal by a Passing Point Charge

A point charge q in vacuum travels horizontally with speed υ at a height d over the surface of a flat metal surface with conductivity σ and permittivity ϵ. In the following, we estimate the rate at which power is dissipated in the metal in the limit when $c \gg \upsilon \gg d\sigma/\epsilon$.

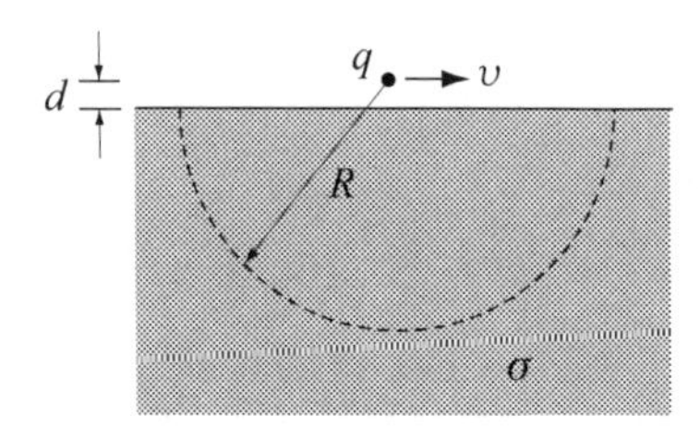

Figure 14.11: A point charge moving parallel to a flat metal surface.

In the $\upsilon \to 0$ limit, the charge polarizes the metal and creates an image electric field which cancels the point charge electric field $\mathbf{E} = q\hat{\mathbf{r}}/4\pi\epsilon r^2$ everywhere inside the metal. When the charge is moving, the metal cannot respond quickly enough to screen out the quasi-electrostatic Coulomb field of q (see Example 14.3) in the immediate vicinity of the charge itself. In other words, $\mathbf{E}$ persists in the metal out to a radius R (see Figure 14.11). Qualitatively, R should increase as the conductivity decreases and go to zero in the static limit. A dimensionally correct guess with this behavior is

$$R \sim \frac{\upsilon\epsilon}{\sigma} \gg d. \tag{14.83}$$

The inequality in (14.83) reiterates the limit assumed in the statement of the problem.

From Section 9.6.1, the rate at which a metal with volume V dissipates energy by Joule heating is $\mathcal{P} = \int_V d^3r\,\sigma E^2$. Since $\mathbf{E} = -\nabla\varphi$ in quasi-electrostatics, a bit of vector calculus shows that

$$\mathcal{P} = \int_V d^3r\,\sigma E^2 = -\sigma\int_V d^3r\mathbf{E}\cdot\nabla\varphi = \sigma\int d^3r\,[\varphi\nabla\cdot\mathbf{E} - \nabla\cdot(\mathbf{E}\varphi)]. \tag{14.84}$$

There is no charge density in the metal, so $\rho = \nabla\cdot\mathbf{E} = 0$ in (14.84). Therefore, using the divergence theorem,

$$\mathcal{P} = -\sigma\int_S d\mathbf{S}\cdot\mathbf{E}\varphi. \tag{14.85}$$

The only portion of the metal surface where the field is not zero is an area of radius πR^2 just below the moving charge. The surface normal and $\mathbf{E}$ are roughly anti-parallel so, using R from (14.83) and a simple estimate of the field and potential at the surface in the integrand, we conclude that

$$\mathcal{P} \sim \sigma(\pi R^2)\left(\frac{q}{4\pi\epsilon d^2}\right)\left(\frac{q}{4\pi\epsilon d}\right) \sim \frac{1}{16\pi}\frac{\upsilon^2 q^2}{\sigma d^3}. \tag{14.86}$$

■

Example 14.4 A battery maintains a voltage V_0 between the parallel plates (separation d) of a capacitor filled with a leaky dielectric with conductivity σ and permittivity ϵ. Find the time evolution of the plate surface charge densities[16] $\Sigma_\pm(t)$ if the battery is abruptly disconnected at $t = 0$.

Solution: Since the usual results of electrostatics apply, $E(t) = V(t)/d$ is the magnitude of the instantaneous electric field between the plates and $\Sigma_\pm(t) = \pm\epsilon E(t)$. Current j flows from the positive plate to the negative plate with no accumulation of charge within the dielectric. Therefore,

$$\frac{d\Sigma_+}{dt} = -j = -\sigma E = -\frac{\sigma}{\epsilon}\Sigma_+ = -\frac{1}{\tau_E}\Sigma_+.$$

The appearance of the electric time constant (14.78) generalized to dielectric matter is expected. The solution of this differential equation that satisfies the initial conditions gives the surface charge density of the positive plate as the capacitor discharges as

$$\Sigma_+(t) = \frac{\epsilon V_0}{d}\exp(-t/\tau_E).$$

14.9 Good Conductors: Quasi-Magnetostatics

The exponential charge-relaxation formula derived in Section 14.7.1 is not strictly valid when the conductivity σ is large. A more careful analysis (Application 18.1) shows that charge relaxation in a good metal ($\sigma \sim 10^8\ \Omega^{-1}\text{m}^{-1}$) occurs less rapidly than (14.79) predicts.[17] Nevertheless, it remains true that charge disappears from the bulk of a good conductor much more rapidly than for a poor conductor—so quickly, in fact, that no long-range Coulomb effects are discernible. This means that, to a very good approximation, we may set

$$\rho_f(\mathbf{r}, t) = 0 \qquad \text{and} \qquad \nabla\cdot\mathbf{j}_f = 0. \tag{14.87}$$

The steady-current condition on the right side of (14.87) follows from $\rho_f = 0$ and the continuity equation (14.77).

To find the quasistatic approximation appropriate to a good conductor, we note first that the current density $\mathbf{j}_f = \sigma\mathbf{E}$ in an ohmic system is always driven by some sort of external source. Sometimes, the latter is an explicit and spatially distinct current density $\mathbf{j}_{\text{ext}}(\mathbf{r}, t)$. At other times, an unspecified current source is presumed to impose a specified electric or magnetic field. This leads us to write the Ampère-Maxwell equation in the form

$$\nabla\times\mathbf{B} = \mu\mathbf{j}_{\text{ext}} + \mu\sigma\mathbf{E} + \mu\epsilon\frac{\partial\mathbf{E}}{\partial t}. \tag{14.88}$$

In (14.88), the ratio of the displacement current density $\mathbf{j}_{\text{D}} = \epsilon\partial\mathbf{E}/\partial t$ to the external current density $\mathbf{j}_{\text{ext}}$ can be estimated exactly as we did in Section 14.6. This gives

$$\frac{j_{\text{D}}}{j_{\text{ext}}} \sim \mu\epsilon\omega^2\ell^2 \ll 1. \tag{14.89}$$

[16] This example retains σ for the conductivity and uses Σ for surface charge density to avoid confusion.

[17] The absurdly small value for τ_E predicted by (14.79) for a good conductor is an indicator that the theory is inadequate for this case.

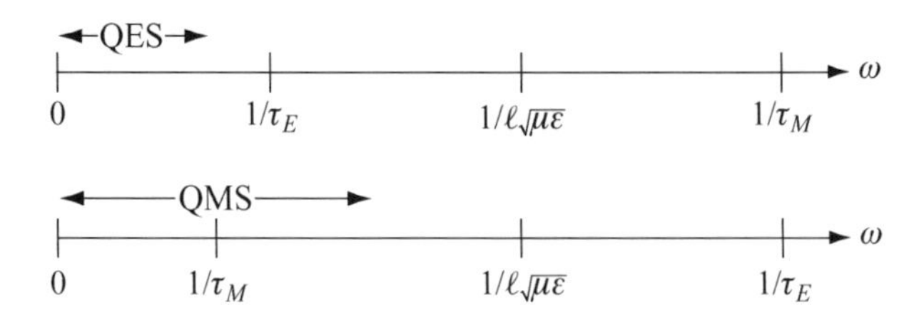

Figure 14.12: Ranges of validity for quasistatic approximations when conducting matter is present. Top: quasi-electrostatics (QES) when $\tau_E \gg \tau_M$. Bottom: quasi-magnetostatics (QMS) when $\tau_M \gg \tau_E$. The horizontal scale is logarithmic in frequency. Figure adapted from Orlando and Devlin (1991).

The imposed inequality in (14.89) is our usual definition of a quasistatic limit. It is valid up to short-wave frequencies (10^7 Hz) for meter-sized objects and up to infrared frequencies (10^{13} Hz) for micron-sized structures.

Since $\tau_E = \epsilon/\sigma$, an order-of-magnitude estimate of the ratio of $\mathbf{j}_\mathrm{D}$ to the conduction current density $\mathbf{j}_f = \sigma\mathbf{E}$ in (14.88) is

$$\frac{j_\mathrm{D}}{j_f} \sim \frac{\epsilon\omega E}{\sigma E} = \omega\tau_E \ll 1. \tag{14.90}$$

The imposed inequality in (14.90) is the opposite of the quasi-electrostatic condition (14.80) and reflects the fact that we are interested in good conductors where (14.87) is true. As a practical matter, (14.90) is satisfied up to ultraviolet frequencies (10^{17} Hz) for high-conductivity materials like metals.

The displacement current may be neglected in (14.88) when both (14.89) and (14.90) are satisfied. This defines the regime of quasi-magnetostatics for good conductors. On the other hand, the physics in this regime depends strongly on the relative importance of "external" versus "induced" fields. For example, following Section 14.6, the field $\mathbf{B}_\mathrm{ext}$ produced by $\mathbf{j}_\mathrm{ext}$ creates a Faraday electric field $E_F \sim \omega\ell B_\mathrm{ext}$. The corresponding conduction current density $\mathbf{j}_F = \sigma\mathbf{E}_F$ produces its own Ampère's law magnetic field $B_F \sim \mu\sigma\ell E_F$. Therefore,

$$\frac{B_F}{B_\mathrm{ext}} \sim \mu\sigma\omega\ell^2 = \omega\tau_M \qquad \text{where } \tau_M \equiv \mu\sigma\ell^2. \tag{14.91}$$

The importance of the magnetic time constant τ_M becomes clear when we realize that the electromagnetic transit time $\sqrt{\mu\epsilon}\ell$ is the geometric mean of τ_E and τ_M. This means that the quasistatic condition (14.89) can be written in the form

$$\mu\epsilon\omega^2\ell^2 = (\omega\tau_E)(\omega\tau_M) \ll 1. \tag{14.92}$$

Given the extraordinary smallness of $\omega\tau_E$ when (14.89) is satisfied, we learn from (14.92) that quasi-magnetostatics makes sense both when $\omega\tau_M \ll 1$ (where electromagnetic induction is negligible) and when $\omega\tau_M \gg 1$ (where electromagnetic induction dominates). This is indicated in the lower scale in Figure 14.12. For comparison, the upper scale of Figure 14.12 shows the more limited range of validity of quasi-electrostatics for poorly conducting matter.

Good conductors exhibit a surprisingly diverse range of quasi-magnetostatic behavior. Section 14.10 and Section 14.11 treat examples where the driving force for current flow is an imposed electric or magnetic field. Section 14.12 treats *eddy-current* phenomena for situations where a specified $\mathbf{j}_\mathrm{ext}(\mathbf{r}, t)$ drives currents in a nearby conductor.

When $\mathbf{j}_{\rm ext}(\mathbf{r}, t) = 0$, the quasi-magnetostatic approximation to the Maxwell equations is

$$\nabla \cdot \mathbf{E} = 0 \qquad\qquad \nabla \cdot \mathbf{B} = 0 \tag{14.93}$$

$$\nabla \times \mathbf{E} = -\frac{\partial \mathbf{B}}{\partial t} \qquad\qquad \nabla \times \mathbf{B} = \mu\sigma\mathbf{E}. \tag{14.94}$$

A straightforward approach to (14.93) and (14.94) substitutes the curl equation on the left side of (14.94) into the curl of the equation on the right side of (14.94). Using $\nabla \cdot \mathbf{B} = 0$ and the vector identity $\nabla \times (\nabla \times \mathbf{B}) = \nabla(\nabla \cdot \mathbf{B}) - \nabla^2\mathbf{B}$, these operations give

$$\mathcal{D}\nabla^2\mathbf{B} = \frac{\partial \mathbf{B}}{\partial t} \qquad \text{with } \mathcal{D} = 1/\mu\sigma. \tag{14.95}$$

Exchanging the role of the two curl equations in (14.94) and repeating the same sequence of steps yields

$$\mathcal{D}\nabla^2\mathbf{E} = \frac{\partial \mathbf{E}}{\partial t} \qquad \text{with } \mathcal{D} = 1/\mu\sigma. \tag{14.96}$$

This analysis shows that $\mathbf{E}(\mathbf{r}, t)$ and $\mathbf{B}(\mathbf{r}, t)$ inside conducting matter satisfy exactly the same equation in the quasi-magnetostatic regime. The two fields differ only in the boundary and/or matching conditions they satisfy. In the frequency domain, we will find the *skin effect*. In the time domain, we will find the phenomenon of *magnetic diffusion*.

14.10 The Skin Effect

The current density is uniform across the cross section of a long straight wire which carries a direct current (see Example 9.2). This is no longer true when the wire carries an alternating current. To see this, let a straight, cylindrical wire with radius R, conductivity σ, and permeability μ carry a time-harmonic current $I(t) = I_0 \exp(-i\omega t)$. Choose the wire parallel to the z-axis so, by symmetry, $\mathbf{E}(\mathbf{r}, t) = \hat{\mathbf{z}}E(\rho)\exp(-i\omega t)$. For this situation, (14.96) in cylindrical coordinates simplifies to

$$\rho\frac{d}{d\rho}\left(\rho\frac{dE}{d\rho}\right) + \kappa^2\rho^2 E = 0, \tag{14.97}$$

where $\kappa = \sqrt{i\mu\sigma\omega} = (1+i)/\delta$ and

$$\delta(\omega) = \sqrt{\frac{2}{\mu\sigma\omega}}. \tag{14.98}$$

The frequency-dependent quantity $\delta(\omega)$ in (14.98) is called the *skin depth*. It is a characteristic length scale for time-harmonic quasi-magnetostatic problems. Indeed, comparing (14.98) to (14.91) shows that[18]

$$\tfrac{1}{2}\omega\tau_M = \frac{R^2}{\delta^2}. \tag{14.99}$$

The linearly independent solutions of (14.97) are the Bessel functions $J_0(\kappa\rho)$ and $N_0(\kappa\rho)$ appropriate to problems with rotational symmetry around the z-axis (Section 7.8.1). The second of these diverges at the center of the wire ($\rho = 0$) so the electric field inside the wire is

$$\mathbf{E}(\rho, t) = \hat{\mathbf{z}}AJ_0(\kappa\rho)\exp(-i\omega t). \tag{14.100}$$

[18] The wire radius is the characteristic length scale for this problem, so $\ell = R$.

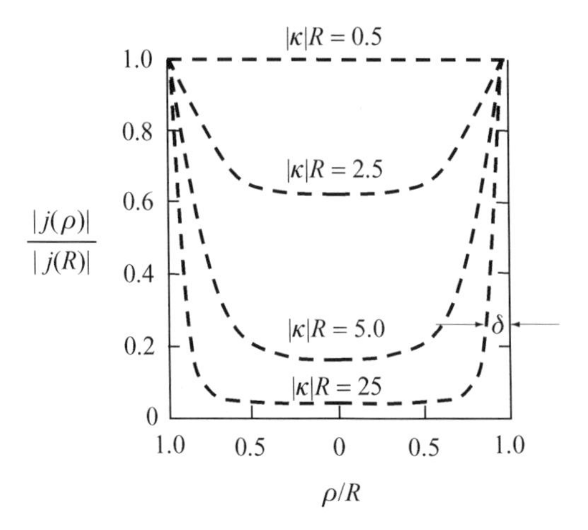

Figure 14.13: The magnitude of the current density $|j(\rho)|$ normalized to its value at the surface of the wire for different values of κR. The thickness of the skin layer for $\kappa R = 25$ is indicated by δ. Figure adapted from King (1945).

The Bessel functions satisfy $J_0'(x) = -J_1(x)$, so Faraday's law in the form $\nabla \times \mathbf{E} = i\omega\mathbf{B}$ gives the accompanying magnetic field,

$$\mathbf{B}(\rho, t) = \hat{\boldsymbol{\phi}}\frac{\kappa A}{i\omega}J_1(\kappa\rho)\exp(-i\omega t). \tag{14.101}$$

The constant A in (14.100) and (14.101) is determined by the continuity of the tangential component of $\mathbf{B}$ at the surface of the wire and the Ampère's law magnetic field $\mathbf{B}(R) = \hat{\boldsymbol{\phi}}\mu I_0/2\pi R$.

The interesting feature of this problem appears when we study the ratio $E(\rho)/E(R)$. Since $\mathbf{j} = \sigma\mathbf{E}$, the graphs of the normalized current density $|j(\rho)|/|j(R)|$ in Figure 14.13 tell the magnitude part of the story. As the magnitude $|\kappa|R = \sqrt{2}R/\delta$ increases, the current becomes increasing confined to a narrow "skin layer" near the wire's surface. This has a dramatic effect on the resistance of the wire because the effective cross sectional area through which the current flows decreases drastically as the driving frequency increases.

From a mathematical point of view, the dashed curves in Figure 14.13 are understandable using (14.100) and the limiting forms of the zero-order Bessel function,

$$J_0(|\kappa|\rho) = 1 - \tfrac{1}{4}|\kappa|^2\rho^2 + \tfrac{1}{64}|\kappa|^4\rho^4 - \cdots \qquad (|\kappa|\rho \ll 1) \tag{14.102}$$

and

$$J_0(|\kappa|\rho) \sim \sqrt{\frac{1}{2\pi|\kappa|\rho}}\cos(|\kappa|\rho - \pi/4) \qquad (|\kappa|\rho \gg 1). \tag{14.103}$$

The keys to the computation are to write the cosine in (14.103) as a sum of complex exponentials, recall that κ is complex, and retain only the dominant terms when computing the electric field ratio of interest. The result is

$$\frac{E(\rho)}{E(R)} \approx \begin{cases} 1 & \delta \gg R, \\ \exp\{(i-1)(R-\rho)/\delta\} & \rho \approx R,\ \delta \ll R. \end{cases} \tag{14.104}$$

The top line of (14.104) confirms that the electric field and current density are uniform throughout the wire's cross section in the low-frequency limit when $\delta \gg R$. In the high-frequency limit, when $\delta \ll R$, the second line in (14.104) shows that the skin depth $\delta(\omega)$ sets the scale for the (initially) exponential

Table 14.1. Skin depth $\delta(\omega)$ for copper metal at several frequencies.

$\delta(\omega)$	ω (Hz)	Band
10 nm	10^{15}	Visible
1 μm	10^{12}	Microwave
0.1 mm	10^{6}	AM radio
1 cm	10	House current

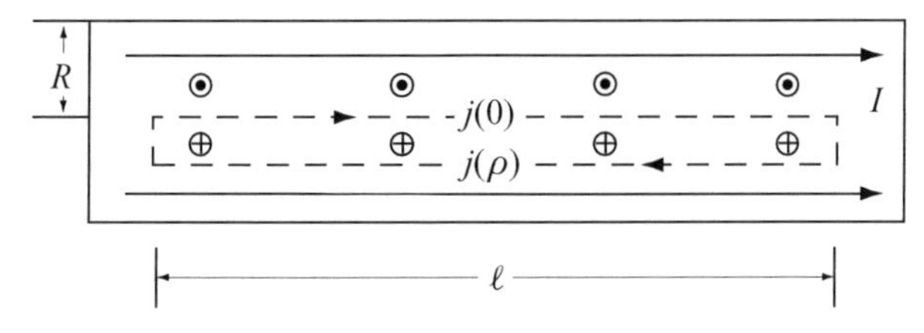

Figure 14.14: A wire carries a uniformly distributed current $I \cos\omega t$. Self-induction induces currents around loops (dashed) which lead to the skin effect. See text for discussion.

decrease of the current density as one moves into the wire from the surface. This exponential behavior is not surprising because, when $\rho \approx R$, Bessel's equation (14.97) simplifies to

$$\frac{d^2E}{d\rho^2} + \kappa^2 E = 0. \tag{14.105}$$

The second line in (14.104) also shows that the current density is an oscillatory function of ρ when $\delta \ll R$. The instantaneous current density in the skin layer thus resembles a set of nested tubes, with adjacent tubes carrying current in opposite directions. Table 14.1 gives a few values for the skin depth of copper.

14.10.1 The Physical Origin of the Skin Effect

The skin effect arises from an interplay between Ohm's law, Ampere's law, and Faraday's law. To gain some insight, it is instructive to derive and analyze an integral equation for the current density $\mathbf{j} = \sigma\mathbf{E}$ using the integral forms,

$$\oint_C d\boldsymbol{\ell}\cdot\mathbf{H} = \int_S d\mathbf{S}\cdot\mathbf{j} \quad \text{and} \quad \oint_{C'} d\boldsymbol{\ell}\cdot\mathbf{E} = -\int_{S'} d\mathbf{S}\cdot\frac{\partial\mathbf{B}}{\partial t}. \tag{14.106}$$

The circular loop C is coincident with, and traversed in the same direction as, a closed magnetic field line of radius ρ defined by the dotted and crossed symbols in Figure 14.14. The rectangular loop C' shown dashed in Figure 14.14 has length ℓ and width ρ. The long leg coincident with the wire's symmetry axis is traversed from left to right, which we define as the direction of positive j in the diagram.

If all quantities vary as $\exp(-i\omega t)$, the integrals in (14.106) simplify to

$$B(\rho) = \frac{\mu}{\rho}\int_0^\rho ds\, s\, j(s) \quad \text{and} \quad j(\rho) = j(0) - i\omega\sigma\int_0^\rho d\rho' B(\rho'). \tag{14.107}$$

These expressions show that the current density in the wire become radially inhomogeneous because Faraday self-induction induces an electric field (and hence a contribution to the current density) which depends on the instantaneous and radially inhomogeneous magnetic field produced by that current density. Inserting the left side of (14.107) into the right side and using (14.98) gives

$$j(\rho) = j(0) - \frac{2i}{\delta}\int_0^{\rho} d\rho' \frac{1}{\rho'}\int_0^{\rho'} ds\, s\, j(s). \tag{14.108}$$

This integral equation can be solved using the iterative method of successive approximations. Only elementary integrals appear and we find

$$\frac{j(\rho)}{j(0)} = 1 - \frac{i}{2}\left(\frac{\rho}{\delta}\right)^2 - \frac{1}{16}\left(\frac{\rho}{\delta}\right)^4 + \cdots \tag{14.109}$$

Because $\kappa = (1+i)/\delta$, the series in (14.109) is identical to the series in (14.102) and the reader can check that the present method reproduces (14.100) exactly because (14.109) is the absolutely convergent series expansion of $J_0(\kappa\rho)$ for all values of ρ/δ.

Young Rutherford

In 1893, the 23-year-old Ernest Rutherford was curious "whether [soft] iron was magnetic in very rapidly oscillating fields or not". To find out, he used a spark gap method to pass alternating currents (up to 10^9 Hz) through iron needles and measured the magnetism of the needles using a torsion balance magnetometer. The experiments were performed in the cellar of Canterbury College, Christchurch, New Zealand. Rutherford was aware of the skin effect (predicted in 1886 by Heaviside and Rayleigh) and confirmed that the magnetization was proportional to the radius, rather than to the cross sectional area, of his needles. To examine their internal state, he exposed similarly magnetized needles to nitric acid for varying amounts of time. The acid dissolved the needles from the outside in, so Rutherford was able to determine the radial variation of the magnetization. Figure 14.15 reproduces the relevant figure from his paper, "Magnetization of iron by high-frequency discharges" [*Transactions of the New Zealand Institute* **27**, 481 (1894)]. The damped oscillatory behavior is consistent with the current density (14.103).

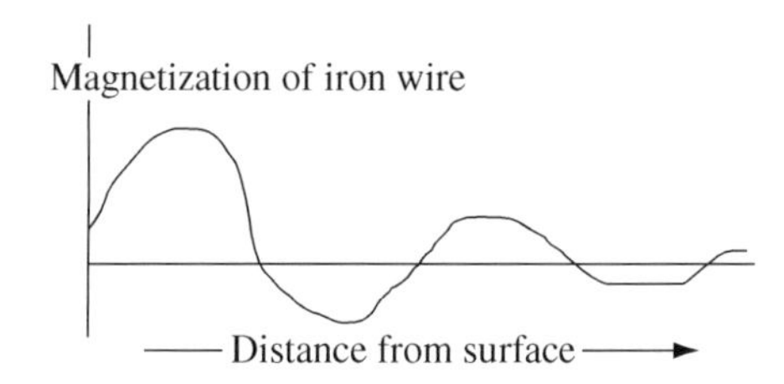

Figure 14.15: Reproduction of a figure from Ernest Rutherford's first research paper.

Application 14.6 Shielding of an AC Magnetic Field by a Cylindrical Shell

A long cylindrical shell coaxial with the z-axis has inner radius R, outer radius $R+d$, and conductivity σ. Unspecified sources generate an external field $\mathbf{B}_0(t) = \hat{\mathbf{z}}B_0\exp(-i\omega t)$ at every point in space (Figure 14.16). Our task is to show that the magnetic field in the cylindrical hollow is negligibly small when $d \ll \delta \ll R$ but $d \gg \delta^2/R$. By symmetry, the magnetic field everywhere has the form

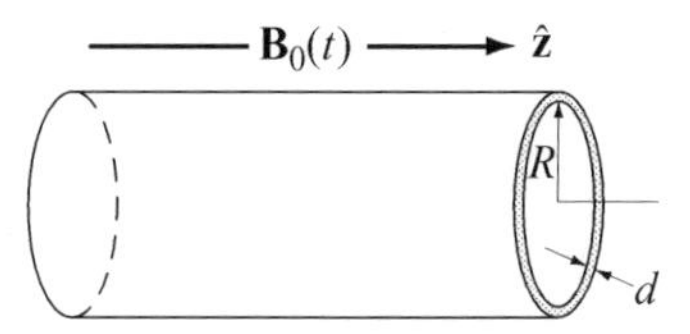

Figure 14.16: Shielding of an AC magnetic field by a cylindrical ohmic shell.

$\mathbf{B}(\mathbf{r}, t) = \hat{\mathbf{z}}B(\rho)\exp(-i\omega t)$. The fields $\mathbf{B}_{\text{in}} = \mathbf{B}(\rho < R)$ and $\mathbf{B}_{\text{out}} = \mathbf{B}_0(\rho > R + d)$ are both uniform in the quasi-magnetostatic limit because both satisfy $\nabla \cdot \mathbf{B} = 0$ and $\nabla \times \mathbf{B} = 0$. The longitudinal and time-varying magnetic field induces an azimuthal electric field by Faraday's law,

$$\oint_C d\boldsymbol{\ell} \cdot \mathbf{E} = -\frac{d}{dt} \int_S d\mathbf{S} \cdot \mathbf{B}. \tag{14.110}$$

Choosing C as a circular circuit with radius R simplifies (14.110) to

$$2\pi R E(R) = i\omega\pi R^2 B_{\text{in}}. \tag{14.111}$$

The limit $d \ll \delta$ implies that the electric field and current density $\mathbf{j} = \sigma\mathbf{E}$ are uniform throughout the ohmic matter of the shell. Indeed, for a sufficiently thin shell, the quasi-magnetostatic matching condition $\hat{\mathbf{n}}_2 \times [\mathbf{B}_1 - \mathbf{B}_2] = \mu\mathbf{K}_f$ reads

$$B_{\text{out}} - B_{\text{in}} = -\mu K_f = \mu j d = \mu\sigma E(R) d. \tag{14.112}$$

Since $\delta^2 = 2/\mu\sigma\omega$, substituting (14.111) into (14.112) yields

$$\frac{B_{\text{in}}}{B_{\text{out}}} = \frac{1}{1 - i\dfrac{dR}{\delta^2}}. \tag{14.113}$$

The inside field is out of phase with the external field but has a very small amplitude when $d \gg \delta^2/R$ because

$$\left|\frac{B_{\text{in}}}{B_{\text{out}}}\right| = \frac{1}{\sqrt{1 + \left(\dfrac{Rd}{\delta^2}\right)^2}}. \tag{14.114}$$

■

14.11 Magnetic Diffusion

We showed in Section 14.9 that the quasistatic magnetic field in a good conductor satisfies

$$\mathcal{D}\nabla^2\mathbf{B} = \frac{\partial \mathbf{B}}{\partial t} \qquad \text{with } \mathcal{D} = 1/\mu\sigma. \tag{14.115}$$

If we replace $\mathbf{B}(\mathbf{r}, t)$ by a density of particles $n(\mathbf{r}, t)$, this equation—called the *diffusion equation*—describes how a small whiff of perfume spreads out to fill an entire room (see boxed material following Section 7.9.1). The diffusion constant $\mathcal{D}$ determines the rate of spreading. To discover the corresponding behavior of $\mathbf{B}(\mathbf{r}, t)$ in a good conductor, imagine a conducting half-space $x > 0$ which is free of all fields for $t < 0$. At $t = 0$, turn on a magnetic field $B_0\hat{\mathbf{z}}$ just at the surface ($x = 0$) and maintain it for

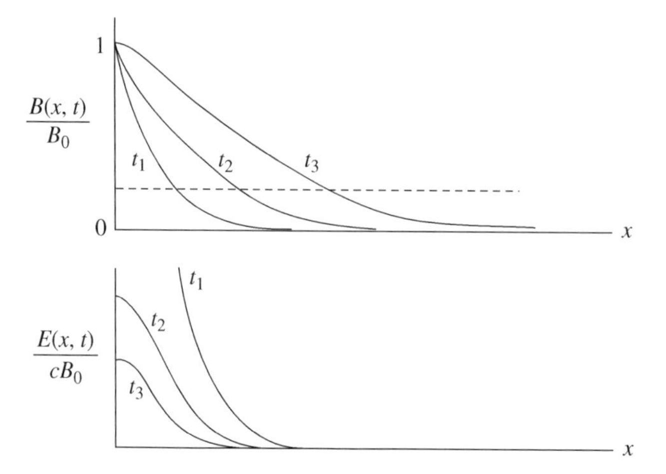

Figure 14.17: Normalized magnetic field $B(x,t)$ (top panel) and electric field $E(x,t)$ (bottom panel) at three different times ($t_1 < t_2 < t_3$) during magnetic diffusion. See text for discussion.

all $t > 0$. We will use the *method of similarity* to solve (14.115) for $\mathbf{B}(x,t) = B(x,t)\hat{\mathbf{z}}$ for all $t > 0$ and then compute $\mathbf{E}(x,t)$ from Ampère's law.

The first step is to check (by direct substitution) that the diffusion equation itself, the boundary condition $B(0,t) = B_0$, and the initial condition $B(x,0) = 0$ are all invariant under the transformation $x \to x/\Delta$ and $t \to t/\Delta^2$ for $\Delta > 0$. This implies that $B(x/\Delta, t/\Delta^2) = B(x,t)$. If we choose the length Δ to have the specific value[19]

$$\Delta = \sqrt{4\mathcal{D}t}, \tag{14.116}$$

this equation reads $B(x,t) = B(x/\Delta,\ 1/4\mathcal{D})$. Notice that the second argument does not depend on x or t. This means that the magnetic field that solves (14.115) for a fixed value of $\mathcal{D}$ is a function of the single variable $s = x/\Delta$:

$$B(x,t) = B_0 f(x/\Delta) = B_0 f(s). \tag{14.117}$$

Changing variables to $g(s) = f'(s)$ simplifies (14.115) to the ordinary differential equation

$$g' + 2sg = 0. \tag{14.118}$$

It is straightforward to integrate (14.118) once to get $g(s)$ and then again to get $f(s)$. Applying the initial and boundary conditions in the form $f(0) = 1$ and $f(\infty) = 0$ produces an integral which defines the complementary error function of mathematical physics, erfc(x). In detail, the magnetic field in the conductor is

$$\mathbf{B}(x,t) = \hat{\mathbf{z}}\frac{2B_0}{\sqrt{\pi}}\int_{x/\Delta}^{\infty} ds\,\exp(-s^2) = \hat{\mathbf{z}}\,B_0\mathrm{erfc}\left(\frac{x}{\sqrt{4\mathcal{D}t}}\right). \tag{14.119}$$

This solution satisfies $\mathbf{B}(0,t) = \hat{\mathbf{z}}B_0$ because erfc(0) = 1. Otherwise, the top panel of Figure 14.17 plots the magnitude of (14.119) as a function of position at three different times. The intersection of each curve with the horizontal dashed line occurs at $x = 2\sqrt{\mathcal{D}t}$. This shows that the time-dependent length (14.116) measures the depth to which the magnetic field penetrates into the ohmic material after a time t.

[19] The factor of 4 in (14.116) simplifies the final multiplicative pre-factor in (14.119).

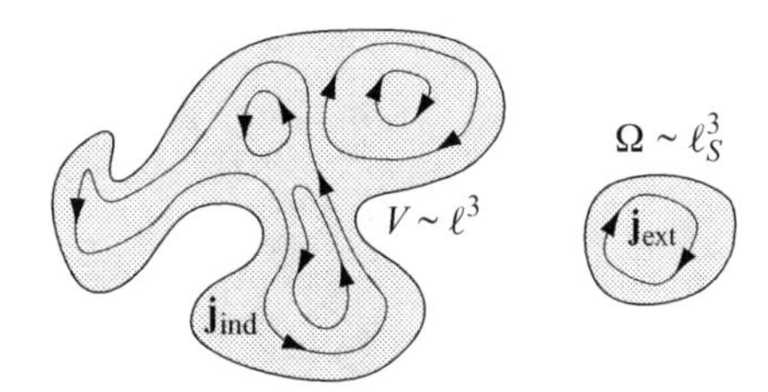

Figure 14.18: By Faraday induction, the closed loops of a time-varying external current density $\mathbf{j}_{\text{ext}}$ induce closed loops of eddy current with density $\mathbf{j}_{\text{ind}}$ in a nearby ohmic medium.

Using the definition of $\mathcal{D}$ in (14.115), we gain the important insight that large conductivity impedes the diffusion of a magnetic field into a conductor. For a perfect conductor, there is no penetration at all. This is the origin of the magnetostatic boundary condition used for perfect conductors (see Section 13.6.6). Note also that (14.116) provides an interpretation of the magnetic time constant τ_M defined on the right side of (14.91). It is the time needed for a magnetic field to diffuse a distance ℓ into a material with conductivity σ and magnetic permeability μ.

The electric field which accompanies (14.119) follows immediately from Ampère's law on the right side of (14.94) and the fact that $(d/dz)\text{erfc}(z) = -(2/\sqrt{\pi})\exp(-z^2)$:

$$\mathbf{E}(x,t) = \hat{\mathbf{y}}\, B_0 \sqrt{\frac{\mathcal{D}}{\pi t}} \exp(-x^2/4\mathcal{D}t). \tag{14.120}$$

This field (bottom panel of Figure 14.17) is zero at $t = 0$ except at $x = 0$ where it is singular. The singularity is a consequence of the abrupt turn-on of the magnetic field at $t = 0$. Thereafter, $\mathbf{E}(x,t)$ first increases exponentially and then decreases as $t^{-1/2}$ as $\mathbf{B}(x,t)$ penetrates into the conductor. The crossover occurs at later times for points in the sample which are further from $x = 0$. In the limit $t \to \infty$, the magnetic field $B_0\hat{\mathbf{z}}$ fills the conductor and the electric field is zero everywhere.

14.12 Eddy-Current Phenomena

Closed loops of current which form in conducting matter due to electromagnetic induction are generically called *eddy currents*. Figure 14.18 shows an external current density $\mathbf{j}_{\text{ext}}(\mathbf{r},t)$ which produces a magnetic field $\mathbf{B}_{\text{ext}}(\mathbf{r},t)$. When this field penetrates a nearby conducting body, Ohm's law guarantees that the electric field induced by Faraday's law, $\mathbf{E}_{\text{ind}}(\mathbf{r},t)$, creates closed loops of eddy current in the body with density $\mathbf{j}_{\text{ind}}(\mathbf{r},t)$.

Eddy currents generate magnetic forces and torques. They also dissipate energy by Joule heating. The forces and torques make possible induction motors, eddy-current braking, and electromagnetic levitation. The energy losses are responsible for the substantial amount of electricity consumed by commercial power transformers and other large electrical machines. Our goal here is to discover the qualitative characteristics of both phenomena.

When all sources and fields oscillate at frequency ω, it is convenient to use the complex exponential $f(\mathbf{r},t) = f(\mathbf{r})\exp(-i\omega t)$ for calculations. Hence, if we choose $\mathbf{j}_{\text{ext}}(\mathbf{r})$ to be a real function, the physical external current density is

$$\text{Re}\left[\mathbf{j}_{\text{ext}}(\mathbf{r},t)\right] = \text{Re}\left[\mathbf{j}_{\text{ext}}(\mathbf{r})\exp(-i\omega t)\right] = \mathbf{j}_{\text{ext}}(\mathbf{r})\cos\omega t. \tag{14.121}$$

A similar formula applies to $\mathbf{B}_{\text{ext}}(\mathbf{r},t)$. The presence of a complex-valued spatial function $f(\mathbf{r})$ is the signal that $f(\mathbf{r},t)$ has a component that oscillates out of phase with the driving source.

The quasi-magnetostatic force $\mathbf{F}$ exerted by $\mathbf{j}_{ext}$ on the resistive medium in Figure 14.18 is equal and opposite to the force exerted on $\mathbf{j}_{ext}$ by the induced magnetic field. Therefore, if $\Omega = \ell_S^3$ is the volume occupied by the source current, the force averaged over one period $T = 2\pi/\omega$ of the current oscillation is

$$\langle \mathbf{F} \rangle = -\int_\Omega d^3r \, \frac{1}{T} \int_0^T dt \, \text{Re}\left[\mathbf{j}_{ext}(\mathbf{r}, t)\right] \times \text{Re}\left[\mathbf{B}_{ind}(\mathbf{r}, t)\right]. \tag{14.122}$$

Similarly, the time-averaged rate of Joule heating in the conducting body with volume $V = \ell^3$ is (Section 9.6)

$$\langle \mathcal{R} \rangle = \int_V d^3r \, \frac{1}{T} \int_0^T dt \, \text{Re}\left[\mathbf{j}_{ind}(\mathbf{r}, t)\right] \cdot \text{Re}\left[\mathbf{E}_{ind}(\mathbf{r}, t)\right]. \tag{14.123}$$

14.12.1 Qualitative Estimates for Forces and Losses

We evaluate (14.122) and (14.123) using $\mathbf{j}_{ind} = \sigma \mathbf{E}_{ind}$, the decomposition $\mathbf{B}_{ind}(\mathbf{r}) = \mathbf{B}'_{ind}(\mathbf{r}) + i\mathbf{B}''_{ind}(\mathbf{r})$, and the time-averaging theorem proved in Section 1.6.3. The results are

$$\langle \mathbf{F} \rangle = -\tfrac{1}{2} \int_\Omega d^3r \, \mathbf{j}_{ext}(\mathbf{r}) \times \mathbf{B}'_{ind}(\mathbf{r}) \tag{14.124}$$

and

$$\langle \mathcal{R} \rangle = \tfrac{1}{2}\sigma \int_V d^3r \, |\mathbf{E}_{ind}(\mathbf{r})|^2. \tag{14.125}$$

To make further progress, we specialize the Maxwell curl equations (14.94) to time-harmonic fields:

$$\nabla \times \mathbf{E} = i\omega \mathbf{B} \qquad \nabla \times \mathbf{B} = \mu\sigma \mathbf{E}. \tag{14.126}$$

A first estimate of $\mathbf{E}_{ind}$ comes from substituting $\mathbf{B}_{ext}$ into the left side of (14.126). Using this field in the right side of (14.126) gives a first estimate of $\mathbf{B}_{ind}$. Following our skin-effect discussion (Section 14.10), we distinguish high-frequency behavior ($\delta \ll \ell$) where the estimate $\nabla \sim 1/\delta$ is appropriate from low-frequency behavior ($\delta \gg \ell$) where our original estimate $\nabla \sim 1/\ell$ is still valid. Since $\delta^2 = 2/\mu\sigma\omega$, this gives

$$E_{ind}^{(1)} \sim \begin{cases} i\omega\delta B_{ext} & \delta \ll \ell, \\ i\omega\ell B_{ext} & \delta \gg \ell, \end{cases} \tag{14.127}$$

and

$$B_{ind}^{(1)} \sim \begin{cases} iB_{ext} & \delta \ll \ell, \\ i\dfrac{\ell^2}{\delta^2} B_{ext} & \delta \gg \ell. \end{cases} \tag{14.128}$$

We can use $\mathbf{E}_{ind}^{(1)}$ directly in (14.125), but $\mathbf{B}_{ind}^{(1)}$ is purely imaginary, while (14.124) requires the real part of $\mathbf{B}_{ind}$. This calls for a second estimate obtained by substituting (14.128) into the left side of (14.126) and substituting the resulting second estimate of $\mathbf{E}_{ind}$ into the right side of (14.126). The result is

$$B_{ind}^{(2)} \sim \begin{cases} -B_{ext} & \delta \ll \ell, \\ -\dfrac{\ell^4}{\delta^4} B_{ext} & \delta \gg \ell. \end{cases} \tag{14.129}$$

Equations (14.128) and (14.129) show that the real and imaginary parts of $\mathbf{B}_{\text{ind}}$ are as large as $\mathbf{B}_{\text{ext}}$ itself when the skin depth is small. This is the fundamental origin of the phenomenon of *magnetic shielding* seen already in Application 14.6 at the end of Section 14.10.

Here, we use (14.127) to estimate the damping rate (14.125) and (14.129) to estimate the force (14.124). When $\delta \ll L$, current only flows through a tiny volume of the conducting body near its surface. This means that V should be replaced by $\ell^2\delta$ when we estimate the integral in (14.125). We will also assume that $\Omega \ll V$ so $\mathbf{B}_{\text{ind}}$ is nearly constant over the volume of the source current. Since $j_{\text{ext}} \sim B_{\text{ext}}/\mu\ell_S$, these approximations give

$$F \sim \begin{cases} \dfrac{\ell^3}{\mu\ell_S} B^2_{\text{ext}} & \delta \ll \ell, \\ \mu\ell_S^2\ell^4\sigma^2\omega^2 B^2_{\text{ext}} & \delta \gg \ell, \end{cases} \tag{14.130}$$

and

$$\mathcal{R} \sim \begin{cases} \sqrt{\dfrac{\omega}{\mu\sigma}} \dfrac{\ell^2}{\mu} B^2_{\text{ext}} & \delta \ll \ell, \\ \sigma\omega^2\ell^5 B^2_{\text{ext}} & \delta \gg \ell. \end{cases} \tag{14.131}$$

Experiments in the $\omega \to 0$ limit confirm that both $\mathbf{F}$ and $\mathcal{R}$ are proportional to ω^2 and that both increase as σ increases. A frequency-independent force and a damping rate $\mathcal{R} \propto \sqrt{\omega}$ are similarly characteristic of the high-frequency, small-skin-depth regime. Application 14.7 at the end of the chapter confirms some of these results in a circuit-theory context.

Example 14.5 A popular lecture demonstration uses a pendulum mechanism to swing a disk of metal between the pole faces of a magnet. Estimate the damping force on the metal at a moment when the pendulum speed is v. What happens when a parallel array of long and narrow slots are cut out of the disk?

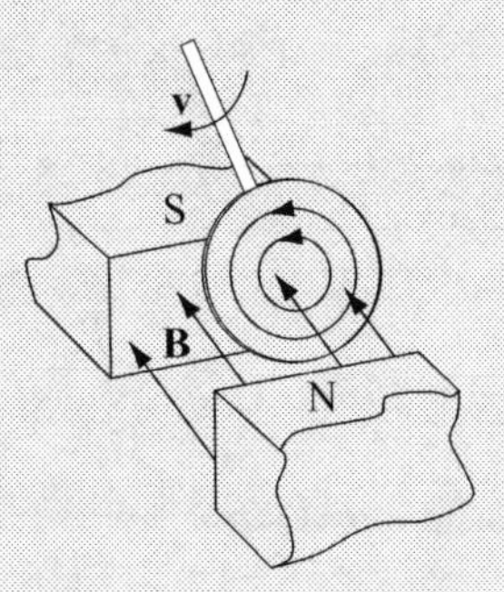

Figure 14.19: An eddy-current pendulum where a conducting disk swings betweens the poles of a magnet.

Solution: Eddy currents appear in the metal as soon as the downward swing brings the leading edge of the metal into the space occupied by the magnetic field $\mathbf{B}$. Let the metal disk in Figure 14.19 have conductivity σ, radius R, and thickness t. At a moment when the disk speed is v, it is simplest to estimate the damping force directly from an integral over the volume $\pi R^2 t$ of the disk:

$$\mathbf{F} = \int d^3r\, \mathbf{j} \times \mathbf{B}.$$

Ohm's law does not seem immediately relevant because there is no source of electric field. However, (14.35) makes clear that a sensible generalization for a conductor in motion with velocity $\mathbf{v}$ is

$$\mathbf{j} = \sigma(\mathbf{E} + \mathbf{v} \times \mathbf{B}).$$

The vectors $\mathbf{v}$ and $\mathbf{B}$ are perpendicular in Figure 14.19. Therefore, when the entire disk is immersed in the field, a good estimate is

$$F \sim \pi R^2 t \sigma v B^2.$$

The direction of this force is such that a sufficiently strong magnet halts the pendulum in mid-swing. This is an effective way to prevent the introduction of magnetic flux into the metal, as Lenz' law dictates. The force disappears if long thin slots are cut out of the metal because closed loops of eddy current cannot form.

14.13 AC Circuit Theory

The quasistatic approximation underlies the most common approach to practical electromagnetic problems—alternating-current (AC) circuit theory. This theory generalizes to time-dependent situations the familiar linear relationship between an electromotive force and a time-independent current,

$$I = \frac{\mathcal{E}}{R}. \tag{14.132}$$

Experiment shows that the relation is still linear, but that $I(t)$ cannot respond instantaneously to $\mathcal{E}(t)$. The inevitability of this time delay leads us to write

$$I(t) = \int_{-\infty}^{t} dt'\, f(t-t')\mathcal{E}(t'). \tag{14.133}$$

The fact that $t' < t$ in (14.133) expresses the principle of *causality*: only past values of the EMF influence the present value of the current.

The practical consequences of (14.133) emerge when both $\mathcal{E}(t)$ and $I(t)$ are expressed in terms of their Fourier frequency components $\hat{\mathcal{E}}(\omega)$ and $\hat{I}(\omega)$:

$$\mathcal{E}(t) = \frac{1}{2\pi}\int_{-\infty}^{\infty} d\omega\, \hat{\mathcal{E}}(\omega)\exp(-i\omega t) \tag{14.134}$$

and

$$I(t) = \frac{1}{2\pi}\int_{-\infty}^{\infty} d\omega\, \hat{I}(\omega)\exp(-i\omega t). \tag{14.135}$$

A similar formula with $f(t)$ and $\hat{f}(\omega)$ always makes sense if we demand that

$$f(t<0) = 0. \tag{14.136}$$

In other words, when (14.136) is valid, we have the Fourier transform partners

$$\hat{f}(\omega) = \int_{-\infty}^{\infty} dt\, f(t)\exp(i\omega t) = \int_{0}^{\infty} dt\, f(t)\exp(i\omega t) \tag{14.137}$$

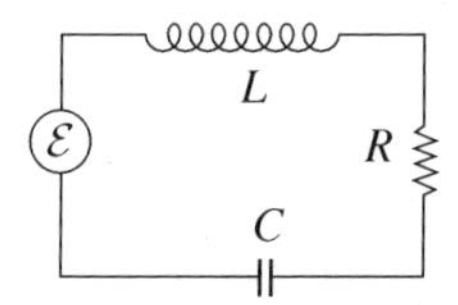

Figure 14.20: A circuit with an ideal capacitor, an ideal inductor, an ideal resistor, and a source of electromotive force.

and

$$f(t)=\frac{1}{2\pi}\int_{-\infty}^{\infty}d\omega\,\hat{f}(\omega)\exp(-i\omega t). \tag{14.138}$$

14.13.1 The Complex Impedance $\hat{Z}(\omega)$

The restriction (14.136) guarantees that the integral (14.133) does not violate causality if we extend its upper limit to $+\infty$:

$$I(t)=\int_{-\infty}^{\infty}dt'\,f(t-t')\mathcal{E}(t'). \tag{14.139}$$

We are now in a position to Fourier transform (14.139) and exploit the convolution theorem (Section 1.6.2). Given (14.134), (14.135), and (14.137), the key result is

$$\hat{I}(\omega)=\hat{f}(\omega)\hat{\mathcal{E}}(\omega). \tag{14.140}$$

The definition $\hat{Z}(\omega)\equiv\hat{f}^{-1}(\omega)$ then produces the natural generalization of (14.132) for time-dependent circuits:

$$\hat{\mathcal{E}}(\omega)=\hat{Z}(\omega)\hat{I}(\omega). \tag{14.141}$$

The complex-valued function $\hat{Z}(\omega)$ is called the *impedance*. The fact that $f(t)$ in (14.133) is real implies that

$$\operatorname{Re}\hat{Z}(\omega)=\operatorname{Re}\hat{Z}(-\omega)\qquad\text{and}\qquad\operatorname{Im}\hat{Z}(\omega)=-\operatorname{Im}\hat{Z}(-\omega). \tag{14.142}$$

Therefore, a low-frequency expansion of the impedance has the form

$$\begin{aligned}\operatorname{Re}\hat{Z}(\omega)&=R+a_1\omega^2+a_2\omega^4+\cdots\\ \operatorname{Im}\hat{Z}(\omega)&=b_1\omega+b_2\omega^3+\cdots.\end{aligned} \tag{14.143}$$

14.13.2 The RLC Circuit: A Damped Oscillator

An important special case of (14.141) emerges from Maxwell theory if we apply quasistatic principles to an RLC circuit. In Figure 14.20, a sinusoidal EMF drives an alternating current through a resistor and an inductor (solenoid) and induces time-varying accumulations of charge on the plates of a capacitor. A quasi-electrostatic approximation applies to the capacitor and a quasi-magnetostatic approximation applies to the inductor. This permits us to exploit conservation of power, *static* expressions for R, L, and C, and static formulae for the electric energy, magnetic energy, and rate of Joule heating.

The rate at which energy is supplied by a source of EMF is equal to the rate of Joule heating plus the rate of change of the stored electric and magnetic field energy. For the circuit shown in Figure 14.20, this power balance reads

$$\mathcal{E}I = RI^2 + \frac{d}{dt}\left\{\frac{1}{2}LI^2 + \frac{Q^2}{2C}\right\} = RI^2 + LI\frac{dI}{dt} + \frac{Q}{C}\frac{dQ}{dt}. \tag{14.144}$$

Dividing (14.144) by $I = dQ/dt$ gives

$$\mathcal{E} = L\frac{d^2Q}{dt^2} + R\frac{dQ}{dt} + \frac{Q}{C}. \tag{14.145}$$

The reader may recognize (14.145) as an application of Kirchhoff's voltage law generalized to time-varying currents when it is written in the form[20]

$$\mathcal{E} - L\frac{dI}{dt} - IR - \frac{Q}{C} = 0. \tag{14.146}$$

The inductive term, in particular, follows from Faraday's flux rule (14.42) and the time derivative of $\Phi = LI$ in (12.133).

We solve the linear equation (14.145) using a Fourier representation of $Q(t)$ and the fact that $\hat{I}(\omega) = -i\omega\hat{Q}(\omega)$. The result is exactly (14.141) with

$$\hat{Z}(\omega) = R - \frac{1}{i\omega C} - i\omega L. \tag{14.147}$$

If $\hat{Z} = |\hat{Z}|\exp(i\phi)$, the physical current driven by a real electromotive force $\hat{\mathcal{E}}(\omega)\cos\omega t$ is

$$I(t) = \mathrm{Re}\left[\hat{I}(\omega)\exp(-i\omega t)\right] = \frac{\hat{\mathcal{E}}(\omega)}{\sqrt{R^2 + (1/\omega C - \omega L)^2}}\cos(\omega t + \phi), \tag{14.148}$$

where

$$\tan\phi = \left(\frac{1}{\omega C} - \omega L\right)\frac{1}{R}. \tag{14.149}$$

The current lags the driving EMF ($\phi < 0$) when the inductance dominates the circuit. This is so because, by Lenz' law, the circuit opposes instantaneous changes in its magnetic state. Conversely, the current leads the driving EMF ($\phi > 0$) when the capacitance dominates. An equivalent (and more physical) statement is that the voltage drop Q/C in (14.145) lags the current because the instantaneous charge $Q(t) = \int^t dt' I(t')$ depends on the current at previous times.

The most striking feature of any RLC circuit is its capacity for unforced self-oscillation. According to (14.141) there is a solution to (14.145) with $\mathcal{E} = 0$ when $Z(\omega) = 0$. Using (14.147), this occurs at the frequency

$$\omega = -i\frac{R}{2L} \pm \sqrt{\frac{1}{LC} - \frac{R^2}{4L^2}}. \tag{14.150}$$

The oscillations are over-damped when $R > 2\sqrt{L/C}$ and under-damped when $R < 2\sqrt{L/C}$. When $R = 0$, undamped oscillations occur at the resonance frequency

$$\omega_T = \frac{1}{\sqrt{LC}}. \tag{14.151}$$

[20] See (14.156) for the complete generalization.

This is called *Thomson's formula*.[21] A glance back at (14.148) shows that ω_T is also the frequency at which a driven RLC circuit responds resonantly. That is, the amplitude of the current is maximal when the EMF is driven at frequency $\omega = \omega_T$.

Since the current and the charge are 90° out of phase, the two terms in the brackets in (14.144) conjure up the image of the total energy of the system (less that lost to ohmic dissipation) sloshing back and forth between the electric energy stored in the capacitor and the magnetic energy stored in the inductor. The under-damped oscillation frequency is the real part of (14.150). The imaginary part of (14.150) gives an estimate of the number of radians of oscillation which occur during the exponential decay time L/R. This is the *quality factor* $\mathcal{Q}$ of the circuit:

$$\mathcal{Q} = \frac{\omega L}{R} \approx \frac{\omega_T L}{R} = \frac{1}{R}\sqrt{\frac{L}{C}}. \tag{14.152}$$

The resistance and inductance which enter the frequency $\omega_M = R/L$ tell us that the oscillator damping is a quasi-magnetostatic effect. To confirm this, it is enough to show that $\omega_M = 2\pi/\tau_M$, where $\tau_M = \mu\sigma\ell^2$ is the magnetic time constant defined on the right side of (14.91). In Figure 14.20, the wire radius a plays the role of ℓ and the inductance of a wire loop of length b is $L \sim \mu b/2\pi$ (Section 12.8). Therefore,

$$\omega_M = \frac{2\pi}{\mu\sigma a^2} = \frac{b/\sigma a^2}{\mu b/2\pi} \sim \frac{R}{L}. \tag{14.153}$$

Example 14.6 Find the complex impedance $\hat{Z}(\omega)$ of the long straight wire studied in Section 14.10 using the surface values of its quasistatic electric and magnetic fields. Use $\hat{Z}(\omega)$ to estimate the low-frequency inductance of the wire.

Solution: If V is the voltage drop along a linear distance ℓ of the wire surface and I is the total current carried by the wire calculated using Ampère's law, the impedance of the wire is sensibly $\hat{Z} = V/I$. Therefore, using $\mathbf{E}(R)$ from (14.100) and $\mathbf{B}(R)$ from (14.101),

$$\hat{Z}(\omega) = \frac{V}{I} = \frac{\mu \int d\boldsymbol{\ell}\cdot\mathbf{E}(R)}{\oint d\boldsymbol{\ell}\cdot\mathbf{B}(R)} = \frac{i\omega\mu\ell J_0(\kappa R)}{2\pi R\kappa J_1(\kappa R)}.$$

Small-argument expansions of the Bessel functions give

$$\frac{J_0(x)}{J_1(x)} = \frac{1 - \dfrac{x^2}{4} + \cdots}{\dfrac{x}{2} - \dfrac{x^3}{16} + \cdots} \approx \frac{2}{x}\left(1 - \frac{x^2}{8}\right).$$

Therefore, since $\kappa^2 = i\mu\sigma\omega = 2i/\delta^2$ and the DC resistance of the wire is $R_{\rm DC} = \ell/\sigma\pi R^2$,

$$\hat{Z}(\omega) \approx \frac{\ell}{\sigma\pi R^2}\left(1 - \frac{i}{8}\mu\sigma\omega R^2\right) = R_{\rm DC}\left(1 - \frac{iR^2}{4\delta^2}\right).$$

We get an estimate of the low-frequency inductance of a length ℓ of wire by comparing this formula for $\hat{Z}(\omega)$ to (14.147). The result is

$$L \approx \frac{\mu\ell}{8\pi}.$$

This agrees with the static result (12.137) calculated in Section 12.8.

[21] Electromagnetism abounds with unrelated scientists named Thomson. The formula (14.151) was derived by William Thomson (Lord Kelvin). The "jumping ring" described at the end of the chapter was invented by Elihu Thomson, co-founder of the General Electric Corporation. Neither should be confused with J.J. Thomson, discoverer of the electron and Nobel laureate.

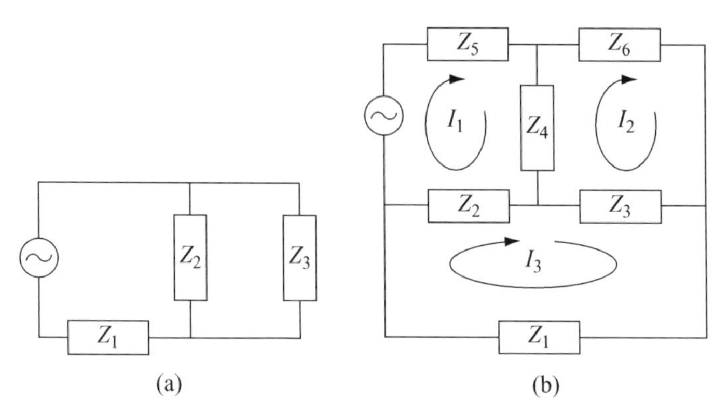

Figure 14.21: Two complex circuits. The small circles with an inscribed sine curve denote a source of time-harmonic EMF.

14.13.3 Network Circuits

The relation $\hat{\mathcal{E}}(\omega) = \hat{Z}(\omega)\hat{I}(\omega)$ also applies to a circuit like the one shown in Figure 14.21(a). This is so because (like resistance in the direct-current case)

$$\hat{Z}(\omega) = \sum_{\alpha} \hat{Z}_{\alpha}(\omega) \tag{14.154}$$

and

$$\hat{Z}^{-1}(\omega) = \sum_{\alpha} \hat{Z}_{\alpha}^{-1}(\omega) \tag{14.155}$$

define $\hat{Z}(\omega)$ for a collection of impedances in series and in parallel, respectively.

To analyze a network circuit like the one shown in Figure 14.21(b), we assign a "loop current" $\hat{I}_k(\omega)$ to every internal closed loop of the circuit. The total current through any point in the circuit is the algebraic sum of the loop currents which pass through that point. Then, if $\hat{\mathcal{E}}_k(\omega)$ is the algebraic sum of the amplitudes of a time-harmonic EMFs associated with the kth loop, the generalization of (14.141) implied by Kirchhoff's voltage law[22] is

$$\sum_{k=1}^{N} \hat{Z}_{jk}(\omega)\hat{I}_k(\omega) = \hat{\mathcal{E}}_j(\omega). \tag{14.156}$$

The elements of the symmetric impedance matrix $\mathbf{Z}$ are

$$\hat{Z}_{jk}(\omega) = R_{jk} - (i\omega C_{jk})^{-1} - i\omega L_{jk}, \tag{14.157}$$

where R_{kk}, C_{kk}, and L_{kk} are the resistance, capacitance, and inductance of the kth loop. An off-diagonal element is generated whenever a resistive, capacitive, or inductive element is shared by two loop currents. For example, when only these effects are included, the circuit of Figure 14.21(b) yields

$$\hat{\mathbf{Z}} = \begin{bmatrix} \hat{Z}_2 + \hat{Z}_4 + \hat{Z}_5 & -\hat{Z}_4 & -\hat{Z}_2 \\ -\hat{Z}_4 & \hat{Z}_3 + \hat{Z}_4 + \hat{Z}_6 & -\hat{Z}_3 \\ -\hat{Z}_2 & -\hat{Z}_3 & \hat{Z}_1 + \hat{Z}_2 + \hat{Z}_3 \end{bmatrix} \tag{14.158}$$

[22] The physical currents and EMFs are the real part of their complex counterparts so Kirchhoff's laws are valid for the complex-valued functions $\hat{I}_k(\omega)$ and $\hat{\mathcal{E}}_j(\omega)$.

with $\hat{\mathcal{E}}_1 = \hat{\mathcal{E}}$ and $\hat{\mathcal{E}}_2 = \hat{\mathcal{E}}_3 = 0$. The off-diagonal elements are negative because two loop currents in a shared branch flow in opposite directions when we choose every loop current to flow clockwise.

Additional contributions to the off-diagonal elements L_{jk} and C_{jk} arise when we generalize (14.144) using expressions for the total electrostatic and magnetostatic energy of a current-carrying network. The magnetic energy yields Neumann's formula (11.101) for the mutual inductance between loop currents j and k:

$$L_{jk} = \frac{\mu_0}{4\pi} \oint_{C_j} \oint_{C_k} \frac{d\boldsymbol{\ell}_j \cdot d\boldsymbol{\ell}'_k}{|\mathbf{x}_j - \mathbf{x}'_k|}. \tag{14.159}$$

This formula applies even when the two loop currents j and k are completely distinct.

There is no volume charge $\rho(\mathbf{r}, t)$ inside a current-carrying wire, so the electrostatic energy generates contributions to C^{-1}_{jk} entirely from the Coulomb interaction between surface charge distributions $\Sigma_j(\mathbf{r}, t)$ and $\Sigma_k(\mathbf{r}, t)$ associated with different current loops. One source of such *stray* or *parasitic* capacitance is the charge on the surface of distinct capacitors. Another is the interaction between current-induced charges on the surfaces of the wires themselves (see Section 9.7.4). These contributions can be important for wires in very close proximity at the highest frequencies where circuit theory is valid.

A familiar example is a tightly wound inductor coil like the one sketched in Figure 14.20. The stray capacitance is not easy to calculate analytically because the induced charge varies over the length of the circuit.[23] In principle, the dominant contribution comes from the Coulomb interaction between charges on adjacent turns of the inductor. In practice, the net effect is often modeled as an effective capacitor in parallel with the coil inductance.

Example 14.7 Find the free-oscillation frequencies of an LC circuit coupled to an $\hat{L}\hat{C}$ circuit by a mutual inductance M. Specialize to the case $LC = \hat{L}\hat{C}$ and show that beating occurs at the frequency

$$\omega = \frac{\omega_T}{\sqrt{1-\kappa}} - \frac{\omega_T}{\sqrt{1+\kappa}}$$

when $\kappa = M/\sqrt{L\hat{L}} \ll 1$. Show that one normal-mode frequency is always very large when $LC \neq \hat{L}\hat{C}$ and $\kappa \approx 1$.

Solution: Unforced oscillation frequencies are determined from (14.156) with $\hat{\mathcal{E}}_j(\omega) = 0$. That is, when $|\hat{\mathbf{Z}}(\omega)| = 0$ or

$$\begin{vmatrix} 1/(\omega^2 C) - L & M \\ M & 1/(\omega^2 C) - \hat{L} \end{vmatrix} = 0.$$

Since $\omega_T = 1/\sqrt{LC}$ and $\hat{\omega}_T = 1/\sqrt{\hat{L}\hat{C}}$, this gives the mode frequencies

$$\omega^2 = \pm \frac{\omega_T^2 + \hat{\omega}_T^2 \pm \sqrt{(\omega_T^2 + \hat{\omega}_T^2)^2 - 4\omega_T^2\hat{\omega}_T^2(1-\kappa^2)}}{2(1-\kappa^2)}.$$

When $\omega_T = \hat{\omega}_T$, $\omega^2 = \omega_T^2(1 \pm \kappa)/(1-\kappa^2)$ or $\omega = \omega_T/\sqrt{1 \pm \kappa}$. If $\kappa \ll 1$, the general solution is a linear combination of sinusoids oscillating at two nearby frequencies. This means that beating

[23] The first reliable calculation of the capacitance of a solenoid was reported in the Ph.D. thesis of the distinguished theoretical physicist Gregory Breit. The published paper, "The distributed capacity of inductance coils", *Physical Review* **17**, 649 (1921), is a tour-de-force of applied mathematics.

occurs at the difference frequency stated above. When $\omega_T \neq \hat{\omega}_T$ and $\kappa \approx 1$, we use $\sqrt{1-\epsilon} \approx 1 - \frac{1}{2}\epsilon$ with $\epsilon = 1 - \kappa^2$ in the general formula for ω^2 above. The two frequencies that result are

$$\omega^2 = \frac{\omega_T^2 + \hat{\omega}_T^2}{1-\kappa} \quad \text{and} \quad \frac{1}{\omega^2} = \frac{1}{\omega_T^2} + \frac{1}{\hat{\omega}_T^2}.$$

The first of these is always much greater than the second when $\kappa \approx 1$.

Application 14.7 Thomson's Jumping Ring

Elihu Thomson's "jumping ring", shown in Figure 14.22, consists of a cylindrical solenoid and a coaxial metal ring with a slightly larger radius a. If the ring rests on a support mounted just above the top end of the solenoid and a current $I_S(t) = I_0 \exp(-i\omega t)$ is applied to the solenoid, the force (14.122) can be sufficient to launch the ring into the air. We will calculate the time-averaged force by treating the ring and solenoid as circuits coupled by a mutual inductance M. An order-of-magnitude estimate should agree with the predictions in (14.130).

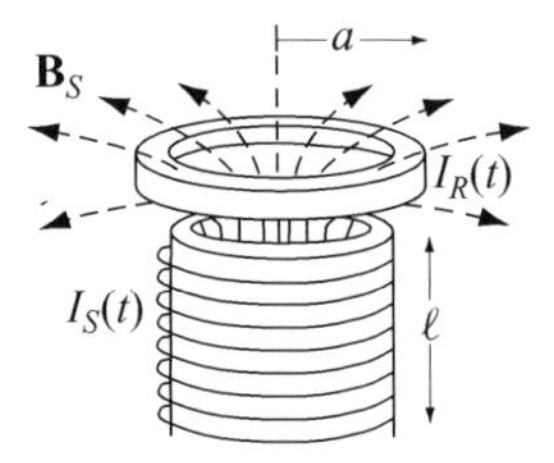

Figure 14.22: Cartoon of Thomson's jumping ring.

The fringing magnetic field $\mathbf{B}_S$ near the top of the solenoid exerts a force on the current $I_R(t)$ induced in the ring by the time variations of $I_S(t)$. If $B_\rho(t)$ is the radial component of $\mathbf{B}_S$, the instantaneous force exerted on the ring in the z-direction is

$$F_z(t) = \hat{\mathbf{z}} \cdot \oint \mathrm{Re}[I_R] d\boldsymbol{\ell} \times \mathrm{Re}[\mathbf{B}_S] = \mathrm{Re}[I_R] 2\pi a \,\mathrm{Re}[B_\rho]. \tag{14.160}$$

If the ring has resistance R and self-inductance L, the linear equation in (14.156) that contains the EMF in the ring (which is zero) is

$$-i\omega M I_S + (R - i\omega L) I_R = \mathcal{E}_R = 0. \tag{14.161}$$

Solving this for the current in the ring gives

$$\mathrm{Re}[I_R(t)] = \frac{\omega M I_0}{R^2 + \omega^2 L^2}(R \sin\omega t - \omega L \cos\omega t). \tag{14.162}$$

Now, $\mathrm{Re}[B_\rho(t)] \approx \mu_0 n \kappa \,\mathrm{Re}[I_S(t)]$, where $n = N/L$ is the number of turns/length of wire wound around the solenoid and κ is a geometrical factor which accounts for the fringing of the field at the position of the ring. Therefore,

$$F_z(t) = \mu_0 n \kappa (2\pi a) \frac{\omega M I_0^2}{R^2 + \omega^2 L^2}(\omega L \cos^2 \omega t - R \sin\omega \cos\omega t). \tag{14.163}$$

Carrying out the time average of this force over one period of the current oscillation explicitly gives

$$\langle F_z \rangle = \frac{1}{2} N \kappa \mu_0 \omega^2 I_0^2 \times \frac{a}{L} \times \frac{ML}{R^2 + \omega^2 L^2}. \tag{14.164}$$

We know from (14.92) and (14.153) that this quasi-magnetostatic analysis makes sense both when $\omega \ll R/L$ and when $\omega \gg R/L$. As in (14.153), we estimate the self-inductance of the ring as $L \sim \mu_0 a/2\pi$. If $\Phi_R = B\pi a^2$ is the magnetic flux through the ring, magnetostatic theory (Section 12.8.2) tells us that the mutual inductance satisfies $\Phi_R = MI_0$. Finally, $R = 2\pi a/\sigma A$ if the ring has cross sectional area A. Substituting this information into the formula just above gives the limiting behaviors

$$\langle F_z \rangle = \begin{cases} \mu_0 \dfrac{I_0^2 M}{L} \dfrac{a}{L} \sim \dfrac{a^2}{\mu_0} B^2 & \omega \gg R/L, \\[2ex] \mu_0 \omega^2 I_0^2 \dfrac{a}{L} \dfrac{ML}{R^2} \sim \mu_0 \omega^2 a^2 A^2 \sigma^2 B^2 & \omega \ll R/L. \end{cases} \tag{14.165}$$

These results agree with (14.130) in detail when we recognize that a measures both the solenoid size ℓ_S and the ring size ℓ in the high-frequency limit. In the low-frequency limit, ℓ is the radius of the wire which constitutes the ring. ■

Sources, References, and Additional Reading

The quotation at the beginning of the chapter is taken from Chapter 3 of

W. Pauli, *Lectures on Physics: Volume 1. Electrodynamics*, edited by C.P. Enz (MIT Press, Cambridge, 1973).

Section 14.1 Textbooks by Russian authors tend to treat quasistatics particularly thoroughly. Some good treatments are

L.D. Landau and E.M. Lifshitz, *Electrodynamics of Continuous Media* (Pergamon, Oxford, 1960).

I.E. Tamm, *Fundamentals of the Theory of Electricity* (Mir, Moscow, 1979).

B.G. Levich, *Theoretical Physics* (North-Holland, Amsterdam, 1970).

The effect of quasistatic fields on the human body is a topic of increasing interest. An entry point to the literature is

R.W.P. King, "Fields and currents in the organs of the human body when exposed to power lines and VLF transmitters", *IEEE Transactions on Biomedical Engineering* **45**, 520 (1998).

Section 14.2 This section benefitted from

E.W. Cowan, *Basic Electromagnetism* (Academic, New York 1968).

A.M. Portis, *Electromagnetic Fields: Sources and Media* (Wiley, New York, 1978).

The sources for Figure 14.3 and Application 14.1 were

V.F. Petrenko and R.W. Whitworth, *The Physics of Ice* (University Press, Oxford, 1999).

C.L. Longmire, *Elementary Plasma Physics* (Wiley-Interscience, New York, 1963).

Section 14.4 This section was adapted from *Tamm* (see Section 14.1 above) and

P.J. Scanlon, R.N. Hendriksen, and J.R. Allen, "Approaches to electromagnetic induction", *American Journal of Physics* **37**, 698 (1969).

G. Giuliani, "A general law for electromagnetic induction", *Europhysics Letters* **81**, 6002 (2008).

Example 14.2 is based on a more general analysis in

N. Smith, "Reciprocity principles for magnetic recording theory", *IEEE Transactions on Magnetics* **23**, 1995 (1987).

Section 14.5 The distinction we make between quasi-electrostatics and quasi-magnetostatics reflects the point of view advocated in

H.A. Haus and J.R. Melcher, *Electromagnetic Fields and Energy* (Prentice-Hall, Englewood Cliffs, NJ, 1989).

T.P. Orlando and K.A. Devin, *Foundations of Applied Superconductivity* (Addison-Wesley, Reading, MA, 1991).

The quadrupole mass spectrometer is discussed in detail in

P.H. Dawson, *Quadrupole Mass Spectrometry and Its Applications* (Elsevier, Amsterdam, 1976).

Section 14.7 For more on charge relaxation, see

H.C. Ohanian, "On the approach to electro- and magneto-static equilibrium", *American Journal of Physics* **51**, 1020 (1983).

Section 14.8 The problem of a point charge moving parallel to an ohmic surface (Application 14.5) has generated a substantial literature since it was posed by Joseph Larmor in 1909. A good entry point is

W.L. Schaich, "Surface response of a conductor: Static and dynamic, electric and magnetic", *American Journal of Physics* **69**, 1267-1276 (2001).

Section 14.9 Complementary discussions of quasi-magnetostatics can be found in

W.R. Smythe, *Static and Dynamic Electricity* (McGraw-Hill, New York, 1939).

H.E. Knoefel, *Magnetic Fields* (Wiley, New York, 2000).

Figure 14.12 was adapted from *Orlando and Devlin* (see Section 14.5 above).

Section 14.10 A master theoretical physicist discusses the skin effect in normal conductors and in superconductors in these three not-quite-consecutive articles:

H.B.G. Casimir and J. Ubbink, *Philips Technical Review* **28**, 271; 300; 366 (1967).

Our treatments of the skin effect, magnetic shielding, magnetic diffusion, and eddy currents were drawn from *Landau and Lifshitz* (see Section 14.1 above) and

G.S. Smith, "A simple derivation for the skin depth in a round wire", *European Journal of Physics* **35**, 025002 (2014).

S. Fahy, C. Kittel, and S.G. Louie, "Electromagnetic screening by metals", *American Journal of Physics* **56**, 989 (1988).

Richard Ghez, *A Primer of Diffusion Problems* (Wiley, New York, 1988).

Figure 14.13 was adapted from

R.W.P. King, *Electromagnetic Engineering* (McGraw-Hill, New York, 1945).

Section 14.12 An engaging discussion of quasistatic forces may be found in

L. Page and N.I. Adams, *Principles of Electricity*, 2nd edition (Van Nostrand, Toronto, 1949).

Section 14.13 Two derivations of AC circuit theory from Maxwell's equations which differ both from each other and from the one given in the text are

R.W.P. King, "Quasistationary and non-stationary currents", in *Encyclopedia of Physics*, edited by S. Flügge, (Springer, Berlin, 1958), vol. XVI, pp. 165-284.

J. Van Bladel, "Circuit parameters from Maxwell's equations", *Applied Scientific Research* **28**, 381 (1973).

Example 14.7 comes from *Landau and Lifshitz* (see Section 14.1 above). For theory and experiment on Thomson's jumping ring (Application 14.7), see

W.M. Saslow, "Electromechanical implications of Faraday's law: A problem collection", *American Journal of Physics* **55**, 986 (1987).

P.J.H. Tjossem and V. Cornejo, "Measurements and mechanisms of Thomson's jumping ring", *American Journal of Physics* **68**, 238 (2000).

Problems

14.1 A Polarized Slab in Motion The text shows that a body with uniform polarization $\mathbf{P}$ and uniform velocity $\boldsymbol{v}$ generates a magnetization $\mathbf{M} = \mathbf{P} \times \boldsymbol{v}$. Confirm this by comparing the convection surface current density to the presumed magnetization surface current density when a slab of matter with polarization $\mathbf{P} = P\hat{\mathbf{z}}$ occupies the volume between the planes $z = 0$ and $z = d$ and the slab moves with velocity $\boldsymbol{v} = v\hat{\mathbf{x}}$.

14.2 Broken Wire? An infinite straight wire with a cross sectional area πa^2 carries a low frequency current $I(t) = I_0 \cos \omega t$. Without close inspection, it is impossible to tell if the wire is broken into two pieces with a tiny gap of width $b \ll a$ separating the two because the displacement current flowing in the gap is exactly equal to $I(t)$. Confirm this by calculating the displacement current $I_d(t)$ from the electric field in the gap.

14.3 Charge Accumulation at a Line A surface current density $\mathbf{K} = -K\hat{\mathbf{x}}$ flows in the half-plane ($x > 0$, $z = 0$). The current accumulates on the line $x = 0$ which bounds the half-plane.

(a) Find $\mathbf{E}(\mathbf{r}, t)$ and $\mathbf{B}(\mathbf{r}, t)$ in the quasi-electrostatic approximation. Hint: Use symmetry and the Ampère-Maxwell law in integral form to find the magnetic field.
(b) Confirm that your solution satisfies the full set of Maxwell equations without approximation.

14.4 Charge Accumulation in a Plane A time-independent surface current with density $\mathbf{K}$ flows in the x-y plane from infinity to the point $\mathbf{r} = 0$ in a radially symmetric manner. As a result, charge accumulates at $\mathbf{r} = 0$ at the rate $dq/dt = I$.

(a) Find the displacement current.
(b) Find the total magnetic field everywhere.

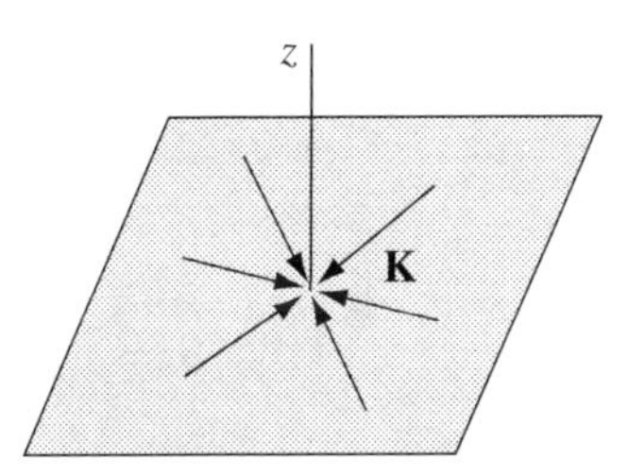

14.5 Rogowski Coil A tightly wound solenoid with n turns per unit length and cross sectional area A is bent into a flexible torus. Two external leads let the current flow in and out. Find the EMF that appears across the leads when a slowly varying, time-dependent current $I(t)$ flows through the hole in the torus as shown below. Conversely, show that no EMF arises from a time-dependent current that does not flow through the hole.

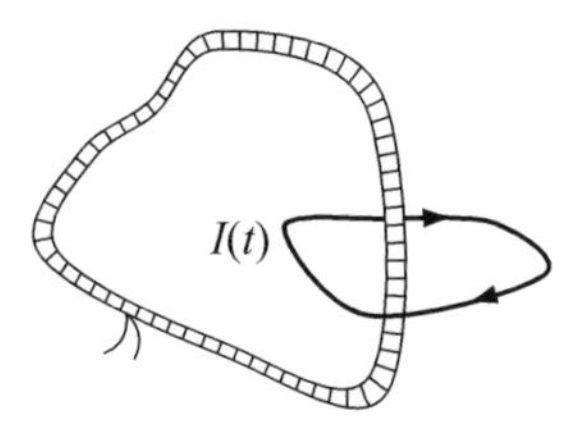

14.6 Magnetic Field of an AC Capacitor A voltage $V(t) = V_0 \sin \omega t$ is applied between the plates of a circular capacitor filled with ohmic matter of conductivity σ. The radius R of the plates is very large compared to the plate separation d. Find the magnetic field between the plates in the quasi-electrostatic approximation.

14.7 A Resistive Ring Comes to Rest An ohmic ring with radius a, mass M, and total resistance R lies in the x-y plane. At $t = 0$, the center of the ring passes by the origin with velocity $\mathbf{v} = v_0\hat{\mathbf{x}}$. How far does the ring travel before stopping if all space is filled with a magnetic field $\mathbf{B} = B_0(x/x_0)\hat{\mathbf{z}}$? Hint: Assume that $a \ll x_0$ and use conservation of energy.

14.8 A Discharging Capacitor A distance d separates the infinitesimally thin and circular plates of a capacitor. The plates have radius $R \gg d$ and instantaneous charges $\pm Q(t)$ as they are slowly discharged by connection to a large resistor (not shown below). Use the non-uniform surface charge distribution of a perfectly conducting circular plate and model the actual current distribution on the right (left) plate using a single surface current density $\mathbf{K}_R$ ($\mathbf{K}_L$).

(a) Integrate the continuity equation to find $K_L(\rho, t)$ and $K_R(\rho, t)$.
(b) Use the Ampère-Maxwell law to find the magnetic field on both sides of both plates when $\rho < R$. Ignore fringe fields.
(c) Show that the fields in (b) satisfy the matching conditions with the current densities in (a).

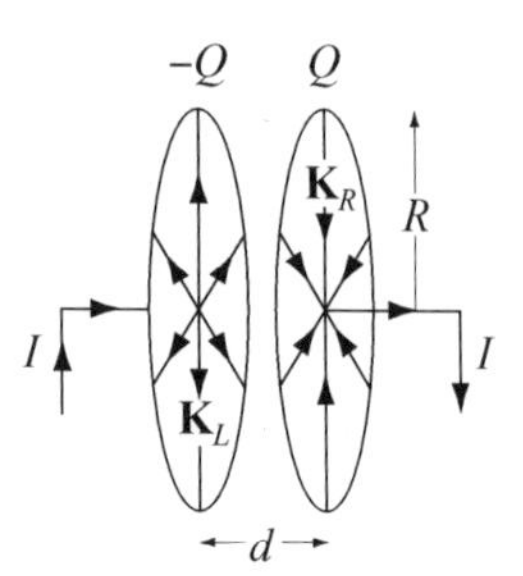

14.9 What Do the Voltmeters Read? The diagram below shows a planar circuit composed of zero-resistance wires, two resistors R_1 and R_2, and two voltmeters V_1 and V_2. A tightly wound solenoid with radius r produces a magnetic field inside itself that points *into* the paper with a time-increasing magnitude $B(t)$. The voltmeters display the value of the line integral of $\mathbf{E}$ along an integration path that passes through the meter from its plus (+) terminal to its minus (−) terminal. What voltage is displayed by voltmeter V_1? What voltage is displayed by voltmeter V_2? Assume that each voltmeter draws negligible current.

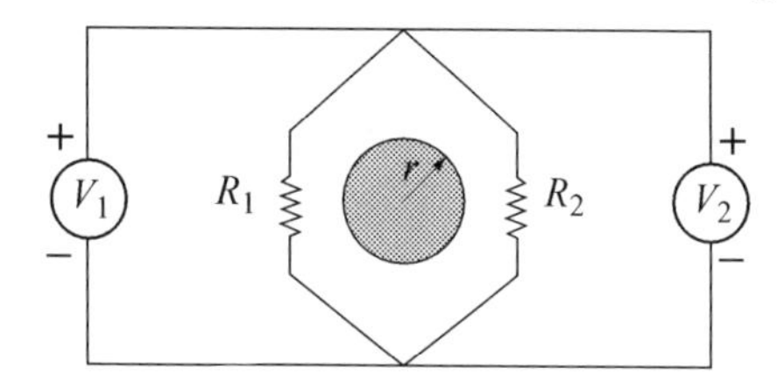

14.10 A Sliding Circuit The side view below shows an iron magnet with two exposed pole faces moving slowly to the right at speed υ. A stationary conducting wire bent into a U-shape (P′Q′QP) is placed at an angle in the gap between the pole faces such that the wire makes electrical contact with the moving magnet at P and P′. Calculate the EMF induced in the wire, both before and after the top edge of the loop Q′Q enters the field $\mathbf{B}$ of the magnet. Make a reasonable assumption about the drift velocity of the electrons in the wire and in the moving magnet.

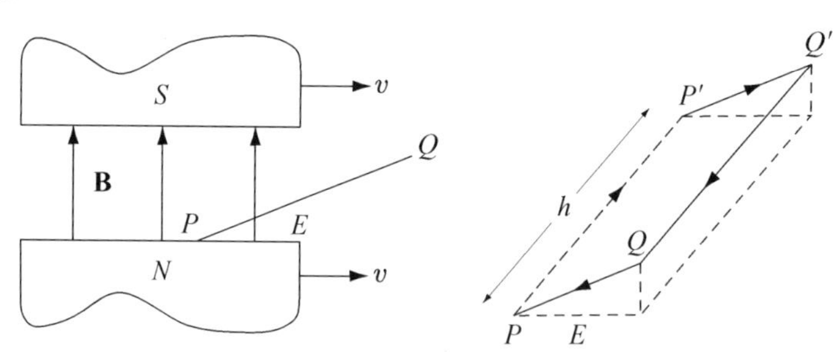

14.11 Townsend-Donaldson Effect A tightly wound solenoid with radius R and length $\ell \gg R$ is composed of N turns of conducting wire. The solenoid carries a slowly varying current $I(t)$ and experiences a voltage drop $V(t)$ over its length. Show that

$$\frac{E_\theta}{E_z} = \frac{\ell/N}{2\pi R},$$

where E_z is the average electric field along the solenoid axis and E_θ is the tangential electric field at the surface of the solenoid.

14.12 A Magnetic Monopole Detector An ohmic wire with resistance R is bent into a closed ring. Suppose a monopole with magnetic charge g approaches Earth from a distant galaxy, passes through the ring, and then continues on its way to another distant galaxy. How much charge flows through a fixed cross section of the wire?

14.13 Corbino Disk An annular disk has thickness t, inner radius R_1, outer radius R_2, and conductivity σ. Let a radial current I_0 flow from the inner periphery to the outer periphery of the disk.

(a) If n is the conduction electron density, use Faraday's law to show that a circular current

$$I = \frac{\sigma B}{2\pi n e} I_0 \ln \frac{R_2}{R_1}$$

flows when a constant magnetic field $\mathbf{B}$ is applied perpendicular to the annular plane.

(b) Show that every origin-centered circle in the disk is an equipotential.

14.14 A Falling Ring and the Lorentz Force

A charge Q is distributed uniformly on a non-conducting ring of radius R and mass M. The ring is dropped from rest from a height h and falls to the ground through a non-uniform magnetic field $\mathbf{B}(\mathbf{r})$. The plane of the ring remains horizontal during its fall.

(a) Explain qualitatively why the ring rotates as it falls.

(b) Use Faraday's flux rule to show that the velocity of the center of mass of the ring when it hits the ground is

$$v_{\text{CM}} = \sqrt{2gh - \frac{Q^2 R^2}{4M^2}[B_z(0) - B_z(h)]^2}.$$

(c) The magnetic Lorentz force in this problem plays the role of static friction when a massive cylinder rolls down a rough incline. Both facilitate the transfer of translational kinetic energy into rotational kinetic energy without performing net work themselves. Thus the rotation in part (a) may be ascribed to the work done by an azimuthal Lorentz force associated with the vertical component of the ring velocity and the horizontal components of the magnetic field. If so, an equal and opposite magnetic work must be done by a vertical Lorentz force which opposes the force of gravity. Confirm that this is the case by an explicit calculation of this force and the net work it performs on the ring. Assume that the magnetic field changes slowly over the area of the ring at any time during its fall.

14.15 Ohmic Dissipation by a Moving Charge A positive point charge q moves with velocity v straight toward an infinitely large, grounded, ohmic plane.

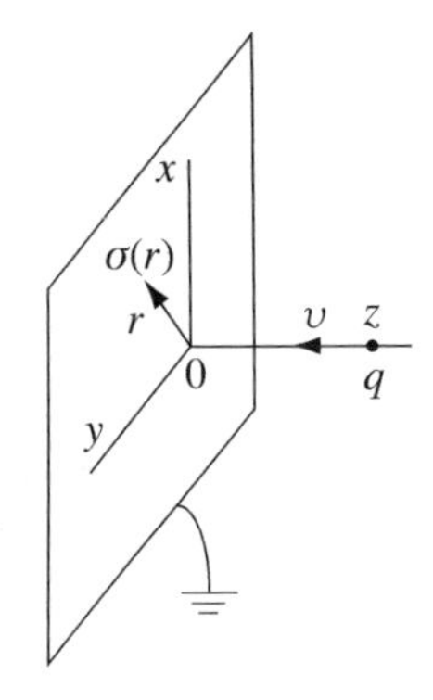

(a) Find the charge density $\sigma(r)$ induced on the plane at the moment when the distance between the point charge and the plane is z.

(b) In the ohmic plane, let the resistance of a ring of radius a and thickness da to a radial current be $dR = R_S(da/2\pi a)$, where R_S is a constant measured in ohms. Show that the rate at which Joule heat is generated in the plane by the moving charge is

$$P = \frac{q^2 v^2 R_S}{8\pi z^2}.$$

14.16 An Unusual Attractive Force The diagram shows a metal sphere (radius a) moving with speed v parallel to a straight wire which carries a current I. The distance between the wire and the center of the sphere is $d \gg a$.

(a) Explain qualitatively why there is an attractive force between the wire and the sphere.

(b) Show that the dependence of the force on dimensional quantities is $F \propto \frac{v^2}{c^2}\frac{a^3}{d^3}\mu_0 I^2$.

Hint: A conducting sphere with radius a develops a dipole moment $\mathbf{p} = 4\pi\epsilon_0 a^3 \mathbf{E}$ in a field $\mathbf{E}$.

14.17 Quasi-Electrostatic Fields Confirm that the formulae below satisfy the four Maxwell equations in the quasi-electrostatic approximation.

$$\mathbf{E}(\mathbf{r}, t) = -\nabla \frac{1}{4\pi\epsilon_0}\int d^3r' \frac{\rho(\mathbf{r}', t)}{|\mathbf{r} - \mathbf{r}'|} \qquad \mathbf{B}(\mathbf{r}, t) = \nabla \times \frac{\mu_0}{4\pi}\int d^3r' \frac{\mathbf{j}(\mathbf{r}', t)}{|\mathbf{r} - \mathbf{r}'|}.$$

14.18 Casimir's Circuit The three parallel, ohmic wires shown below are driven at frequency ω by a common source of EMF. The wires have length L, separation b, and radius a where $a \ll b \ll L$. If δ is the skin depth of the wire material, use Faraday's flux rule to show that

$$\frac{I_2}{I_1} = \frac{I_2}{I_3} = 1 - \frac{\ln 2}{\ln(b/a) + i(\delta^2/a^2)}.$$

Assume that I_1 is real and plot the locus of points traced out by I_2 in the complex I_2 plane as ω varies from 0 to ∞. Use the high-frequency results to rationalize the skin effect for time-harmonic current flow in a thin metal slab with dimensions $L \times 2b$.

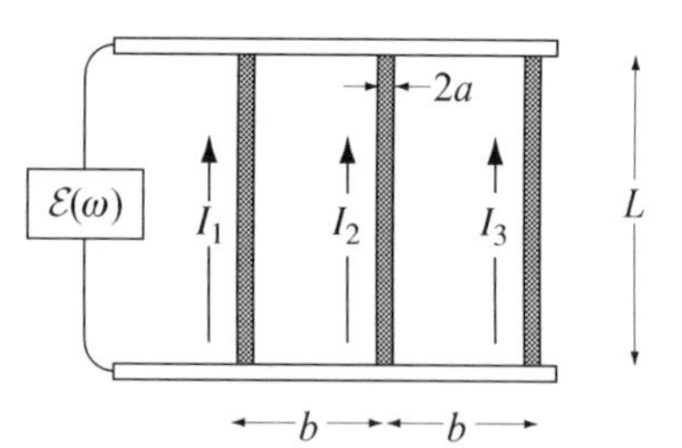

14.19 Inductive Impulse A conducting wire frame with side lengths a and b lies at rest on a frictionless horizontal surface at a distance l from a long straight wire carrying a current I_0 (see figure below). The mass of the frame is m, and its total resistance is R. Use an impulse approximation to find the magnitude and the direction of the velocity of the frame after the current in the long straight wire has been abruptly switched off.

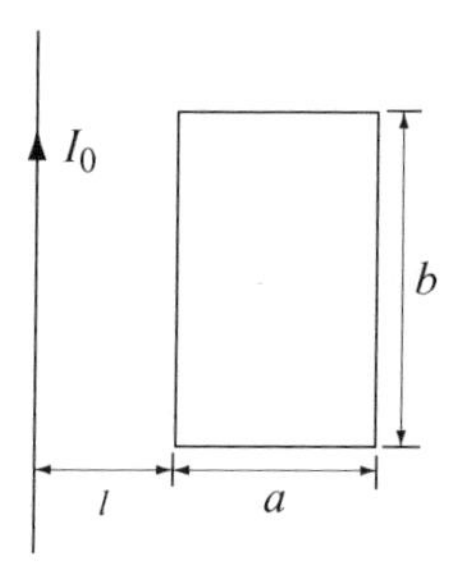

14.20 AC Resistance of an Ohmic Wire Consider an ohmic wire with length L, radius a, and conductivity σ in the *high-frequency* limit of quasi-magnetostatics. Use a simple physical argument to show that (up to a dimensionless constant) the AC resistance $R(\omega)$ of the wire is

$$R(\omega) \propto \frac{L}{a}\sqrt{\frac{\mu_0\omega}{\sigma}}.$$

14.21 A Rotating Magnet The equation of motion for a magnetic dipole moment $\mathbf{m}$ which rotates about its center with an angular velocity $\boldsymbol{\Omega}$ is $d\mathbf{m}/dt = \boldsymbol{\Omega} \times \mathbf{m}$. Find the electric and magnetic fields associated with this object. Neglect the displacement current in the Maxwell equations.

14.22 Magnetic Metal Slab A slab of material with conductivity σ, electric permittivity ϵ, and magnetic permeability μ occupies the infinite volume between the planes $y = \pm d$.

(a) Find the steady-state magnetic field inside the slab if a magnetic field $\mathbf{B}_0 = \hat{\mathbf{z}}B_0 \exp(-i\omega t)$ is present everywhere outside the slab. Assume that the frequency is low enough that the quasi-magnetostatic limit is valid.

(b) Sketch the $t = 0$ field in the limit when the skin depth $\delta \gg d$ and also when $\delta \ll d$.

14.23 Azimuthal Eddy Currents in a Wire A longitudinal AC magnetic field $\mathbf{B}(t) = \hat{\mathbf{z}}B_0 \cos\omega t$ is driven through the interior of an ohmic tube with length L and radius $R \ll L$.

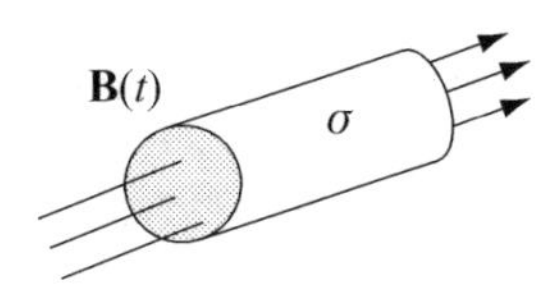

(a) Find the low-frequency eddy-current density inside the tube, neglecting the effects of self-inductance.

(b) Find the correction to the eddy-current density produced by self-inductance (to next order in ω).

(c) Derive the condition where the correction in part (b) can be ignored.

14.24 Eddy-Current Levitation

A wire loop with radius b in the x-y plane carries a time-harmonic current $I_0 \cos\omega t$.

(a) Find the value of I_0 needed to levitate a small sphere with mass m, radius a, and conductivity σ at a height z above the center of the loop. Assume that $a \ll b$ and that $\delta \ll a$, where δ is the skin depth of the sphere.

(b) Compare the time-averaged levitation force you calculate with the qualitative estimate made in the text. Comment if they do not agree.

14.25 Dipole down the Tube A small magnet (weight w) falls under gravity down the center of an infinitely long, vertical, conducting tube of radius a, wall thickness $t \ll a$, and conductivity σ. Let the tube be concentric with the z-axis and model the magnet as a point dipole with moment $\mathbf{m} = m\hat{\mathbf{z}}$. We can find the terminal velocity of the magnet by balancing its weight against the magnetic drag force associated with ohmic loss in the walls of the tube.

(a) At the moment it passes through $z = z_0$, show that the magnetic flux produced by **m** through a ring of radius a at height z' is

$$\Phi_{\rm B} = \frac{\mu_0 m}{2}\frac{a^2}{r_0^3} \qquad \text{where } r_0^2 = a^2 + (z_0 - z')^2.$$

(b) When the speed v of the dipole is small, argue that the Faraday EMF induced in the ring is

$$\mathcal{E} = -\frac{\partial \Phi_{\rm B}}{\partial t} = v\frac{\partial \Phi_{\rm B}}{\partial z'}.$$

(c) Show that the current induced in the thin slice of tube which includes the ring is

$$dI = \frac{3\mu_0 m a v \sigma t}{4\pi}\frac{(z_0 - z')}{r_0^5}dz'.$$

(d) Compute the magnetic drag force **F** on **m** by equating the rate at which the force does work to the power dissipated in the walls of the tube by Joule heating,

$$\mathbf{F}\cdot\mathbf{v} = \int \mathcal{E}\,dI.$$

(e) Find the terminal velocity of the magnet.

15 General Electromagnetic Fields

When we turn our attention to the general case of electrodynamics . . . our first impression is surprise at the enormous complexity of the problems to be solved.
Max Planck (1932)

15.1 Introduction

The most general electromagnetic fields vary in space and time with no restrictions beyond those imposed by the Maxwell equations. The enormity of this subject, and its many different points of entry, motivate us to seek and exploit general principles to help organize and simplify the discussion. Symmetry is one such principle because it connects old solutions of the Maxwell equations to new solutions. A second principle (learned first in electrostatics and magnetostatics) is to exploit suitably defined potential functions to simplify the path to finding these solutions. Finally, using both symmetry and potential functions as tools, we explore the consequences of conservation laws which relate the time variation of a physical quantity at a point to the transport of that quantity from point to point.

The most familiar conservation law in electromagnetism is the continuity equation,

$$\nabla \cdot \mathbf{j} + \frac{\partial \rho}{\partial t} = 0. \tag{15.1}$$

We showed in Section 2.1.3 that (15.1) is an expression of the conservation of electric charge. In this chapter, we derive conservation laws which relate the time variation and transport of energy, linear momentum, and angular momentum. A natural interpretation of these laws assigns each of these mechanical properties to the electromagnetic field itself. This is a profound and liberating idea which has many consequences. Not least, the mechanical effects produced by various configurations of electromagnetic fields become easy to understand when we contemplate the transfer of their energy, linear momentum, and angular momentum to and from nearby distributions of charged particles or matter. The discussion in this chapter is based largely on manipulating the Maxwell equations. Chapter 22 and Chapter 24 revisit the conservation laws from the points of view of special relativity and Hamilton's principle, respectively.

15.2 Symmetry

Symmetry plays many roles in electromagnetism. In Chapters 3 and 10, we used the spatial symmetries of charge and current distributions to help evaluate Coulomb and Biot-Savart integrals. The same symmetries transformed the integral forms of Gauss' law and Ampère's law into useful tools for computation. In Chapters 7 and 8, we used symmetry extensively to choose coordinate systems to

Table 15.1. Behavior of electromagnetic quantities under the discrete operations of space inversion (left column) and time inversion (right column).

$\mathbf{r} \to -\mathbf{r}$	$t \to -t$
$\rho \to \rho$	$\rho \to \rho$
$\mathbf{j} \to -\mathbf{j}$	$\mathbf{j} \to -\mathbf{j}$
$\boldsymbol{v} \to -\boldsymbol{v}$	$\boldsymbol{v} \to -\boldsymbol{v}$
$\mathbf{F} \to -\mathbf{F}$	$\mathbf{F} \to \mathbf{F}$
$\mathbf{E} \to -\mathbf{E}$	$\mathbf{E} \to \mathbf{E}$
$\mathbf{B} \to \mathbf{B}$	$\mathbf{B} \to -\mathbf{B}$
$\varphi \to \varphi$	$\varphi \to \varphi$
$\mathbf{A} \to -\mathbf{A}$	$\mathbf{A} \to -\mathbf{A}$

separate variables for the Laplace and Poisson equations and to fix values for the separation constants. The last section of Chapter 2 exploited symmetry somewhat differently. There, we *postulated* the invariance of the Maxwell equations to the symmetry operations of spatial translation, rotation, and inversion as part of a heuristic "derivation" of these equations.

15.2.1 Discrete Symmetries

A discrete symmetry transformation produces a discontinuous change in a transformed quantity. The discrete symmetries important to electromagnetism are space inversion (or parity), mirror reflection, and time reversal.[1] The operation of space inversion takes the position vector to its opposite: $\mathbf{r} \to -\mathbf{r}$. Other quantities are said to have definite parity under inversion if they either change sign like $\mathbf{r}$ or remain unchanged. Polar vectors and axial vectors are examples of quantities that behave oppositely under inversion (see Application 1.2 at the end of Section 1.8.1).

Following (2.85), we used Newton's second law to establish the polar nature of any force vector and then the Coulomb-Lorentz force $\mathbf{F} = q(\mathbf{E} + \boldsymbol{v} \times \mathbf{B})$ to fix the parity of the particle velocity $\boldsymbol{v} = d\mathbf{r}/dt$, the charge density ρ, the current density $\mathbf{j}$, the electric field $\mathbf{E}$, and the magnetic field $\mathbf{B}$. The left column of Table 15.1 summarizes the results of that discussion, together with results for the potentials derived from $\mathbf{E} = -\nabla\varphi$, $\mathbf{B} = \nabla \times \mathbf{A}$, and the rules summarized in (1.162). Reflection through a mirror is similar to inversion in the sense that the relative minus sign between the transformed $\mathbf{r}$ and $\mathbf{B}$ in Table 15.1 applies on a component-by-component basis [compare (10.30) to (10.31)].

The time-reversal operation $t \to -t$ has no effect on $\mathbf{r}$ or the charge density ρ. On the other hand, $\boldsymbol{v} = d\mathbf{r}/dt$ and hence the linear momentum $\mathbf{p} = m\boldsymbol{v}$ and current density $\mathbf{j} = \rho\boldsymbol{v}$ change sign. From this, we conclude that $\mathbf{F} = d\mathbf{p}/dt$ does not change sign under time reversal. Specializing this result to the Coulomb-Lorentz force shows that $\mathbf{E} \to \mathbf{E}$ under time-reversal while $\mathbf{B} \to -\mathbf{B}$.[2] The magnetic field result may be confirmed from the change in magnetic field that occurs when the direction of current flow reverses in any simple configuration of filamentary wires. The behavior of the potentials follows immediately from the behavior of the fields. The right column of Table 15.1 summarizes these results for time reversal.

[1] The discrete symmetry of charge conjugation (where the electric charge $q \to -q$) is important in quantum mechanics.

[2] This argument assumes that electric charge is unchanged when $t \to -t$.

15.2.2 Dual Symmetry

The name *dual symmetry* is given to the invariance of the source-free ($\rho = \mathbf{j} = 0$) Maxwell equations to the discrete transformation[3]

$$\mathbf{E} \to c\mathbf{B} \qquad \text{and} \qquad \mathbf{B} \to -\mathbf{E}/c. \tag{15.2}$$

In words (and as the reader can check), if $\mathbf{E}$ and $\mathbf{B}$ are solutions to the source-free Maxwell equations, dual symmetry implies that $\mathbf{B}' = \mathbf{E}/c$ and $\mathbf{E}' = -c\mathbf{B}$ are solutions as well. We will use this symmetry in later sections to create new solutions from old solutions.

15.2.3 Continuous Symmetries

A continuous symmetry transformation produces a smooth change in a transformed quantity. Finite changes are regarded as the cumulative effect of a succession of infinitesimal transformations, each of which produces only an infinitesimal change. Familiar continuous symmetries that leave the Maxwell equations invariant include translations in space, rotations in space, and translations in time. Continuous *gauge* transformations of the electromagnetic potentials will occupy our attention in the next section and the continuous *Lorentz transformations* of special relativity are the subject of Chapter 22. For the present, we note only that a homogeneous Lorentz transformation generalizes continuous rotations in three-dimensional space to continuous rotations in four-dimensional space-time.

Continuous symmetries have special interest in theoretical physics because a powerful theorem due to Noether guarantees that each continuous symmetry of a theory generates a conservation law. In the present case, Noether's theorem relates the invariance of the Maxwell equations to translations in space, rotations in space, and translations in time to the conservation laws for linear momentum, angular momentum, and energy, respectively. We delay our discussion of this profound approach to the conservation laws until Chapter 24 when we apply the action principle of Lagrangian mechanics to electrodynamics.

15.3 Electromagnetic Potentials

The scalar potential $\varphi(\mathbf{r})$ and vector potential $\mathbf{A}(\mathbf{r})$ played prominent simplifying roles in electrostatics and magnetostatics. Their time-dependent counterparts do the same for time-varying fields. The starting point, as always, is the Maxwell equations in vacuum,

$$\nabla \cdot \mathbf{B} = 0 \qquad\qquad \nabla \cdot \mathbf{E} = \rho/\epsilon_0 \tag{15.3}$$

$$\nabla \times \mathbf{E} + \frac{\partial \mathbf{B}}{\partial t} = 0 \qquad\qquad \nabla \times \mathbf{B} - \frac{1}{c^2}\frac{\partial \mathbf{E}}{\partial t} = \mu_0 \mathbf{j}. \tag{15.4}$$

The divergence equations carry over without change from statics. In particular, $\nabla \cdot \mathbf{B} = 0$ is a property of *all* magnetic fields. Therefore, the arguments used in Section 10.5 show that

$$\mathbf{B}(\mathbf{r}, t) = \nabla \times \mathbf{A}(\mathbf{r}, t). \tag{15.5}$$

Further progress comes from inserting (15.5) into Faraday's law on the left side of (15.4). This gives

$$0 = \nabla \times \left(\mathbf{E} + \frac{\partial \mathbf{A}}{\partial t}\right). \tag{15.6}$$

Recall now that $\nabla \times \nabla f = 0$ is an identity for any $f(\mathbf{r}, t)$. This means we are justified to set the parenthetical quantity in (15.6) equal to the (negative) gradient of a scalar potential $\varphi(\mathbf{r}, t)$. In other

[3] Equation (15.2) is a physically realizable special case of the transformation (2.70).

words,

$$\mathbf{E}(\mathbf{r},t) = -\nabla\varphi(\mathbf{r},t) - \frac{\partial}{\partial t}\mathbf{A}(\mathbf{r},t). \tag{15.7}$$

The electromagnetic potentials $\mathbf{A}(\mathbf{r},t)$ and $\varphi(\mathbf{r},t)$ serve two immediate purposes. First, the fields (15.5) and (15.7) automatically satisfy the homogeneous (no source) Maxwell equations on the left sides of (15.3) and (15.4). Second, the potentials reduce the number of functions to be determined from six (the scalar components of $\mathbf{E}$ and $\mathbf{B}$) to four (φ and the scalar components of $\mathbf{A}$). The reader may have anticipated this possibility because only four independent functions can result when eight (scalar Maxwell) equations constrain six functions.

Equations of motion for the potentials follow by substituting (15.5) and (15.7) into the inhomogeneous Maxwell equations. These are the equations on the right sides of (15.3) and (15.4) where the charge density and current density appear. Making use of $\nabla\times\nabla\times\mathbf{A} = \nabla(\nabla\cdot\mathbf{A}) - \nabla^2\mathbf{A}$, we find

$$\nabla^2\varphi + \frac{\partial}{\partial t}(\nabla\cdot\mathbf{A}) = -\rho/\epsilon_0 \tag{15.8}$$

$$\nabla^2\mathbf{A} - \frac{1}{c^2}\frac{\partial^2\mathbf{A}}{\partial t^2} - \nabla\left(\nabla\cdot\mathbf{A} + \frac{1}{c^2}\frac{\partial\varphi}{\partial t}\right) = -\mu_0\mathbf{j}. \tag{15.9}$$

The coupled equations (15.8) and (15.9) are not easy to solve. Luckily (as the next section shows), it is never necessary to do so.

15.3.1 Gauge Invariance

Like their static counterparts, $\varphi(\mathbf{r},t)$ and $\mathbf{A}(\mathbf{r},t)$ are not uniquely defined. This has no observable consequences because the non-uniqueness does not affect the electric and magnetic fields that enter the Coulomb-Lorentz law.[4] To confirm this, let $\Lambda(\mathbf{r},t)$ be an arbitrary *gauge function* of space and time and define a new vector potential $\mathbf{A}'$ and a new scalar potential φ' using

$$\mathbf{A}' = \mathbf{A} + \nabla\Lambda \tag{15.10}$$

and

$$\varphi' = \varphi - \frac{\partial\Lambda}{\partial t}. \tag{15.11}$$

The corresponding electric and magnetic fields $\mathbf{E}'$ and $\mathbf{B}'$ follow by substitution of (15.10) and (15.11) into (15.5) and (15.7). Direct calculation shows that all the terms that contain $\Lambda(\mathbf{r},t)$ cancel out. Therefore,

$$\mathbf{E}' = -\nabla\varphi' - \frac{\partial\mathbf{A}'}{\partial t} = -\nabla\varphi - \frac{\partial\mathbf{A}}{\partial t} = \mathbf{E} \tag{15.12}$$

and

$$\mathbf{B}' = \nabla\times\mathbf{A}' = \nabla\times\mathbf{A} = \mathbf{B}. \tag{15.13}$$

By definition, the Maxwell equations are *gauge invariant* because (15.12) and (15.13) show that the physical electric and magnetic fields are independent of the gauge function Λ.

[4] Subtleties arise when quantum mechanics is taken into account. See Sources, References, and Additional Reading.

Wigner's Argument

We will prove in Chapter 24 that the continuity equation (15.1) is a consequence of the gauge invariance of electromagnetism. For the present, the following argument due to Wigner (1949) makes it plausible that the non-uniqueness of $\varphi(\mathbf{r}, t)$ plays an important role in the conservation of electric charge.

Suppose that charge is not conserved and that a charge Q can be created by doing work W. The scale used to measure the electrostatic potential φ is arbitrary, so W cannot depend on the absolute value of φ at the point of creation. Now move Q to a point where the potential on the chosen scale is φ'. Destruction of the charge at this point recovers the work W exactly. The world is now exactly the same as it was before the original creation event except that an energy $Q(\varphi' - \varphi)$ has been gained. This is impossible by conservation of energy. Therefore, as long as no physical process can depend on the scale used to measure electrostatic potential, the assumption that charge is not conserved cannot be true.

From a practical point of view, gauge invariance provides the key to solving (15.8) and (15.9). The idea is to choose a gauge function $\Lambda(\mathbf{r}, t)$ so (15.8) and (15.9) become simple and easy to solve when written in the primed variables. In practice, this "choice of gauge" is rarely made explicitly. An *implicit* choice is made by imposing a constraint on the potential functions. We need only show that the implied $\Lambda(\mathbf{r}, t)$ is calculable in principle (see Example 15.1). With rare exceptions, only two choices of gauge are ever made in classical electromagnetism:

$$\nabla \cdot \mathbf{A} = 0 \qquad \text{(Coulomb gauge)} \tag{15.14}$$

$$\nabla \cdot \mathbf{A} + \frac{1}{c^2}\frac{\partial \varphi}{\partial t} = 0 \qquad \text{(Lorenz gauge).} \tag{15.15}$$

15.3.2 The Coulomb Gauge

The Coulomb gauge choice (15.14) reduces (15.8) and (15.9) to

$$\nabla^2 \varphi_C = -\rho/\epsilon_0 \tag{15.16}$$

and

$$\nabla^2 \mathbf{A}_C - \frac{1}{c^2}\frac{\partial^2 \mathbf{A}_C}{\partial t^2} = -\mu_0 \mathbf{j} + \frac{1}{c^2}\nabla\frac{\partial \varphi_C}{\partial t}. \tag{15.17}$$

The scalar potential obeys Poisson's equation. Therefore, if we specify that $\varphi_C(\mathbf{r}, t) \to 0$ as $|\mathbf{r}| \to \infty$, we know from electrostatics that

$$\varphi_C(\mathbf{r}, t) = \frac{1}{4\pi\epsilon_0}\int d^3r' \,\frac{\rho(\mathbf{r}', t)}{|\mathbf{r} - \mathbf{r}'|}. \tag{15.18}$$

This shows that at least a part of the electric field (15.7) is the familiar, instantaneous Coulomb electric field. The Coulomb gauge is widely used in atomic, molecular, and condensed matter physics precisely because the potential (15.18) binds oppositely charged particles into stable orbits.

Equation (15.18) expresses the scalar potential in the Coulomb gauge as an integral over the charge density alone. It is not obvious from (15.17), but the Coulomb gauge vector potential $\mathbf{A}_C(\mathbf{r}, t)$ can be written similarly as an integral over the current density alone.[5] Unfortunately, the expression for $\mathbf{A}_C$ is awkward, which motivates us to look for the physics in another way. Our strategy is to use (15.18) and

[5] See Jackson (2002) in Sources, References, and Additional Reading.

the continuity equation (15.1) to evaluate the term on the far right side of (15.17). These operations transform (15.17) to

$$\nabla^2 \mathbf{A}_C - \frac{1}{c^2}\frac{\partial^2 \mathbf{A}_C}{\partial t^2} = -\mu_0 \mathbf{j} - \mu_0 \nabla \int \frac{d^3 r'}{4\pi} \frac{\nabla' \cdot \mathbf{j}(\mathbf{r}', t)}{|\mathbf{r} - \mathbf{r}'|}. \tag{15.19}$$

The right-hand side of (15.19) simplifies because the Helmholtz theorem (Section 1.9) decomposes the current density into a longitudinal piece that satisfies $\nabla \times \mathbf{j}_\parallel = 0$ and a transverse piece that satisfies $\nabla \cdot \mathbf{j}_\perp = 0$.[6] Specifically,

$$\mathbf{j}(\mathbf{r}, t) = \mathbf{j}_\parallel(\mathbf{r}, t) + \mathbf{j}_\perp(\mathbf{r}, t) = -\nabla \int \frac{d^3 r'}{4\pi} \frac{\nabla' \cdot \mathbf{j}(\mathbf{r}', t)}{|\mathbf{r} - \mathbf{r}'|} + \nabla \times \int \frac{d^3 r'}{4\pi} \frac{\nabla' \times \mathbf{j}(\mathbf{r}', t)}{|\mathbf{r} - \mathbf{r}'|}. \tag{15.20}$$

Comparing (15.20) to (15.19) shows that the complicated, non-local function of position on the right side of the latter is simply the transverse piece of the current density:

$$\nabla^2 \mathbf{A}_C - \frac{1}{c^2}\frac{\partial^2 \mathbf{A}_C}{\partial t^2} = -\mu_0 \mathbf{j}_\perp. \tag{15.21}$$

One consequence of (15.21) is that a current density with zero curl produces zero magnetic field. This generalizes a previous magnetostatic result [see (10.19)] to arbitrary, time-dependent current densities. The proof proceeds in two steps. First, use $\nabla \times \mathbf{j} = 0$ and $\nabla \times \mathbf{j}_\parallel = 0$ to deduce that $\nabla \times \mathbf{j}_\perp = 0$. Second, use $\nabla \cdot \mathbf{j}_\perp = 0$ and the fact that $\mathbf{j}_\perp \to 0$ at infinity for any localized current distribution to deduce (from the Helmholtz theorem) that $\mathbf{j}_\perp = 0$. Returning to (15.21), we conclude that $\mathbf{A}_C = 0$ and $\mathbf{B} = \nabla \times \mathbf{A}_C = 0$.

Example 15.1 Discuss the gauge functions that transform an arbitrary set of electrodynamic potentials $(\varphi, \mathbf{A})$ to Coulomb gauge potentials $(\varphi', \mathbf{A}')$.

Solution: $\nabla \cdot \mathbf{A} \neq 0$ for an arbitrarily chosen vector potential. We seek $\nabla \cdot \mathbf{A}' = 0$ where (15.10) defines $\mathbf{A}'(\mathbf{r}, t)$. Therefore,

$$\nabla \cdot \mathbf{A}' = \nabla \cdot (\mathbf{A} + \nabla \Lambda) = \nabla \cdot \mathbf{A} + \nabla^2 \Lambda = 0.$$

This shows that the gauge function $\Lambda(\mathbf{r}, t)$ is determined by

$$\nabla^2 \Lambda = -\nabla \cdot \mathbf{A}.$$

We know from Section 7.3 that the solution to this Poisson-like equation is not unique (so some gauge freedom remains) until we choose suitable boundary conditions for $\Lambda(\mathbf{r}, t)$. For example, if we insist that $\Lambda \to 0$ as $r \to \infty$ (and $\nabla \cdot \mathbf{A}$ goes to zero at infinity faster than $1/r$), the unique solution is

$$\Lambda(\mathbf{r}, t) = \frac{1}{4\pi} \int d^3 r' \frac{\nabla \cdot \mathbf{A}(\mathbf{r}', t)}{|\mathbf{r} - \mathbf{r}'|}.$$

[6] The names *longitudinal* and *transverse* refer to the behavior of the spatial Fourier transforms of $\mathbf{j}_\parallel(\mathbf{r}, t)$ and $\mathbf{j}_\perp(\mathbf{r}, t)$, namely, $\mathbf{k} \cdot \hat{\mathbf{j}}_\perp(\mathbf{k}, t) = 0$ and $\mathbf{k} \times \hat{\mathbf{j}}_\parallel(\mathbf{k}, t) = 0$.

15.3.3 The Lorenz Gauge

The Lorenz[7] gauge choice (15.15) uncouples (15.8) and (15.9) to give

$$\nabla^2 \varphi_L - \frac{1}{c^2}\frac{\partial^2 \varphi_L}{\partial t^2} = -\rho/\epsilon_0 \tag{15.22}$$

$$\nabla^2 \mathbf{A}_L - \frac{1}{c^2}\frac{\partial^2 \mathbf{A}_L}{\partial t^2} = -\mu_0 \mathbf{j}. \tag{15.23}$$

These equations have the same structure as (15.17), except that the source terms on the right-hand sides are much simpler. Moreover, the charge density determines φ_L in exactly the same way that the Cartesian components of the current density determine the Cartesian components of $\mathbf{A}_L$. This characteristic makes the Lorenz gauge very popular for problems where the Coulomb potential (15.18) does not simplify the physics. Radiation is a case in point and Chapter 20 is largely devoted to the solution and analysis of (15.22) and (15.23). The Lorenz gauge is also preferred for relativistic calculations because the gauge condition itself is preserved under a Lorentz transformation of the potentials (see Chapter 22).

15.4 Conservation of Energy

Several times in previous chapters we have considered the work done by static electric and magnetic fields on one or more charged particles.[8] The magnetic Lorentz force does no work (Section 12.2.1), so all the work is done by the electric Coulomb force. Specifically, the rate at which $\mathbf{E}(\mathbf{r}, t)$ and $\mathbf{B}(\mathbf{r}, t)$ do mechanical work on a collection of particles with charge density $\rho(\mathbf{r}, t)$ and current density $\mathbf{j}(\mathbf{r}, t) = \rho(\mathbf{r}, t)\boldsymbol{v}(\mathbf{r}, t)$ confined to a volume V is

$$\frac{dW_{\text{mech}}}{dt} = \int_V d^3r\,(\rho\mathbf{E} + \mathbf{j}\times\mathbf{B})\cdot\boldsymbol{v} = \int_V d^3r\,\mathbf{j}\cdot\mathbf{E}. \tag{15.24}$$

Our goal is to rewrite (15.24) in the form of a conservation law. The first step eliminates the current density $\mathbf{j}$ on the far right side of (15.24) using the Ampère-Maxwell equation [right side of (15.4)]. This gives

$$\frac{dW_{\text{mech}}}{dt} = \int_V d^3r \left[\frac{1}{\mu_0}\nabla\times\mathbf{B} - \epsilon_0\frac{\partial \mathbf{E}}{\partial t}\right]\cdot\mathbf{E}. \tag{15.25}$$

Next, use Faraday's law [left side of (15.4)] to eliminate the first term in the integrand of (15.25). The specific identity needed is

$$\nabla\cdot(\mathbf{E}\times\mathbf{B}) = \mathbf{B}\cdot(\nabla\times\mathbf{E}) - \mathbf{E}\cdot(\nabla\times\mathbf{B}) = -\mathbf{B}\cdot\frac{\partial\mathbf{B}}{\partial t} - \mathbf{E}\cdot(\nabla\times\mathbf{B}). \tag{15.26}$$

Combining the result of this substitution with the original equation (15.24) gives the desired law in a form known as *Poynting's theorem*:

$$\int_V d^3r\,\frac{\partial}{\partial t}\frac{1}{2}\epsilon_0\left[\mathbf{E}\cdot\mathbf{E} + c^2\mathbf{B}\cdot\mathbf{B}\right] = -\int_V d^3r\,\mathbf{j}\cdot\mathbf{E} - \int_V d^3r\,\frac{1}{\mu_0}\nabla\cdot(\mathbf{E}\times\mathbf{B}). \tag{15.27}$$

[7] The Danish physicist Ludvig Lorenz introduced the constraint (15.15) in 1867. It is often misattributed to the homonymic Dutch physicist Hendrik Lorentz.

[8] See Section 3.3.1 and Section 12.6.

15.4.1 The Poynting Vector and Field Energy Density

We interpret (15.27) as a power-balance equation based on the physical meaning of (15.24) as the rate at which work is done to change the mechanical energy of the particles, $U_{\rm mech}$. The key step is to identify the left side of (15.27) as the time rate of change of the total energy of the electromagnetic field,

$$U_{\rm EM} = \tfrac{1}{2}\epsilon_0 \int_V d^3r\, [\mathbf{E}\cdot\mathbf{E} + c^2\mathbf{B}\cdot\mathbf{B}]. \tag{15.28}$$

This is very plausible. In the static limit, $U_{\rm EM}$ reduces to the sum of the electrostatic total energy U_E defined in Section 3.6 and the magnetostatic total energy U_B defined in Section 12.6. We note in passing that a (presumptive) term in the integrand of (15.28) proportional to $\mathbf{E}\cdot\mathbf{B}$ vanishes when integrated over all space because it changes sign when $\mathbf{r} \to -\mathbf{r}$ (see Table 15.1).

We now define the *Poynting vector*,

$$\mathbf{S} = \frac{1}{\mu_0}\mathbf{E}\times\mathbf{B}, \tag{15.29}$$

and use the divergence theorem to convert the last term in (15.27) to an integral over the surface S that bounds V. This gives

$$\frac{dU_{\rm tot}}{dt} = \frac{d}{dt}(U_{\rm mech} + U_{\rm EM}) = -\int_S dA\, \mathbf{S}\cdot\hat{\mathbf{n}}. \tag{15.30}$$

The Poynting vector $\mathbf{S}$ has dimensions of (energy/volume) × velocity. This invites us to interpret $\mathbf{S}$ as an *energy current density* by analogy with the usual charge current density $\mathbf{j} = \rho\boldsymbol{\upsilon}$ which has dimensions of (charge/volume) × velocity. In light of our assumption that no charged particles enter or leave the volume V, it is quite reasonable to read (15.30) as a global statement of conservation of energy. The total energy $U_{\rm mech} + U_{\rm EM}$ in a volume changes only if electromagnetic energy flows in or out of V through its surface S. The normal $\hat{\mathbf{n}}$ points outward, so energy flows out of (into) V when $\mathbf{S}$ is parallel (anti-parallel) to $\hat{\mathbf{n}}$.

For unbounded space (modeled as a spherical volume with radius $R \to \infty$), the absolute conservation law

$$U_{\rm mech} + U_{\rm EM} = const. \tag{15.31}$$

is true only if the integral on the right-hand side of (15.30) vanishes as $S \to \infty$. In other words, the magnitude $|\mathbf{S}|$ must decrease more rapidly that $1/r^2$. This is true for all the electric and magnetic fields we have studied so far. It will *not* be true of the fields associated with *radiation* that we will study in Chapter 20.

A spatially local statement of energy conservation follows from (15.27) if we use (15.28) to define an *electromagnetic energy density*,

$$u_{\rm EM} = \tfrac{1}{2}\epsilon_0(\mathbf{E}\cdot\mathbf{E} + c^2\mathbf{B}\cdot\mathbf{B}). \tag{15.32}$$

Then, because the enclosing volume V in (15.27) is arbitrary,

$$\frac{\partial u_{\rm EM}}{\partial t} + \nabla\cdot\mathbf{S} = -\mathbf{j}\cdot\mathbf{E}. \tag{15.33}$$

The analogy between this equation and the continuity equation $\partial\rho/\partial t + \nabla\cdot\mathbf{j} = 0$ reinforces the interpretation of the Poynting vector $\mathbf{S}$ as a current density of electromagnetic energy. From this point of view, $-\mathbf{j}\cdot\mathbf{E}$ is a sink (source) term that transfers energy from (to) the electromagnetic field to (from) the charged particles that interact with the field. The mechanical energy of the particles increases (decreases) accordingly.

Application 15.1 Uniqueness of Solutions and Boundary Conditions

We can use Poynting's theorem to find boundary conditions that guarantee the uniqueness of solutions to Maxwell's equations in a volume V. For this purpose, let $(\mathbf{E}_1, \mathbf{B}_1)$ and $(\mathbf{E}_2, \mathbf{B}_2)$ be two distinct solutions that derive from the same charge and current distributions. By assumption, the two solutions satisfy the same boundary conditions and initial conditions. Because identical sources are involved, the difference fields $\mathbf{E} = \mathbf{E}_1 - \mathbf{E}_2$ and $\mathbf{B} = \mathbf{B}_1 - \mathbf{B}_2$ satisfy the source-free Maxwell equations. Therefore, $\mathbf{E}$ and $\mathbf{B}$ satisfy Poynting's theorem (15.27) without the source term:

$$\frac{d}{dt}\int_V d^3r\ (\mathbf{E}\cdot\mathbf{E} + c^2\mathbf{B}\cdot\mathbf{B}) = -2c^2\int_S dA\,\hat{\mathbf{n}}\cdot\mathbf{E}\times\mathbf{B}. \tag{15.34}$$

Now, $dA\,\hat{\mathbf{n}}\cdot(\mathbf{E}\times\mathbf{B}) = dA\,\mathbf{E}\cdot(\mathbf{B}\times\hat{\mathbf{n}}) = dA\,\mathbf{B}\cdot(\hat{\mathbf{n}}\times\mathbf{E})$. Therefore, if we specify, say, $\hat{\mathbf{n}}\times\mathbf{E}_1$ on the surface S that bounds V, we will have $\hat{\mathbf{n}}\times(\mathbf{E}_1 - \mathbf{E}_2) = \hat{\mathbf{n}}\times\mathbf{E} = 0$ on S. Similarly, if we specify $\hat{\mathbf{n}}\times\mathbf{B}_1$ on S, we will have $\hat{\mathbf{n}}\times(\mathbf{B}_1 - \mathbf{B}_2) = \hat{\mathbf{n}}\times\mathbf{B} = 0$ on S. In either case, the integral on the right-hand side above vanishes. Using (15.28), we see that $\dot{U}_{\text{EM}}(t) = 0$ as well. The total electromagnetic energy associated with $\mathbf{E}$ and $\mathbf{B}$ is a constant.

On the other hand, the initial conditions $\mathbf{E}_1(\mathbf{r}, 0) = \mathbf{E}_2(\mathbf{r}, 0)$ and $\mathbf{B}_1(\mathbf{r}, 0) = \mathbf{B}_2(\mathbf{r}, 0)$ imply that $\mathbf{E}(\mathbf{r}, 0) = \mathbf{B}(\mathbf{r}, 0) = 0$. This fixes the value of the constant at $U_{\text{EM}}(0) = 0$. But $U_{\text{EM}}(t) = 0$ is possible only if $\mathbf{E} \equiv 0$ and $\mathbf{B} \equiv 0$, which means that $\mathbf{E}_1(\mathbf{r}, t) = \mathbf{E}_2(\mathbf{r}, t)$ and $\mathbf{B}_1(\mathbf{r}, t) = \mathbf{B}_2(\mathbf{r}, t)$. We conclude that a solution of the Maxwell equations in V is unique if either $\hat{\mathbf{n}}\times\mathbf{E}$ or $\hat{\mathbf{n}}\times\mathbf{B}$ is specified on the boundary S. When $V \to \infty$, uniqueness is guaranteed if the fields go to zero fast enough as $r \to \infty$ that the integral on the right side of (15.34) vanishes. ■

Example 15.2 Compare the "rest mass" of a uniform spherical shell with radius a and charge q computed from $U_E = m_0c^2$ with the "kinetic mass" computed from $U_B = \frac{1}{2}mv^2$. Assume that the sphere moves with constant speed $v \ll c$. Describe the flow of energy predicted by the associated Poynting vector.

Solution: By Gauss' law, the electric field is zero inside the shell and $\mathbf{E} = \hat{\mathbf{r}}q/4\pi\epsilon_0 r^2$ outside the shell. This gives

$$U_E = \tfrac{1}{2}\epsilon_0\int d^3r|\mathbf{E}|^2 = \frac{q^2}{8\pi\epsilon_0 a} = m_0c^2.$$

When $v \ll c$, the quasi-electrostatic approximation used in Example 14.3 is valid. Therefore, at $t = 0$ when the sphere passes through the origin of coordinates, the electric field is the same as above and the magnetic field is $\mathbf{B} = (\boldsymbol{v}/c^2)\times\mathbf{E}$. If $\boldsymbol{v} = v\hat{\mathbf{z}}$, so $|\boldsymbol{v}\times\mathbf{E}| = vE\sin\theta$, the rest energy U_E is supplemented by

$$U_B = \frac{1}{2\mu_0}\int d^3r|\mathbf{B}|^2 = U_E\frac{v^2}{2c^2}\int_{-1}^{1} d(\cos\theta)\sin^2\theta = \frac{1}{2}\left(\frac{4}{3}m_0\right)v^2 = \tfrac{1}{2}mv^2.$$

The implied mass in this case is a factor of $4/3$ larger than the rest mass.

Since $\hat{\mathbf{r}} = \sin\theta\,\hat{\boldsymbol{\rho}} + \cos\theta\,\hat{\mathbf{z}}$ relates unit vectors in spherical and cylindrical coordinates, the Poynting vector is

$$\mathbf{S} = \frac{1}{\mu_0}\mathbf{E}\times\mathbf{B} = \epsilon_0\mathbf{E}\times(\boldsymbol{v}\times\mathbf{E}) = -\epsilon_0 vE^2\sin\theta\,\hat{\boldsymbol{\theta}}.$$

This formula suggests a unidirectional flow of energy along spherical surfaces from points on the negative z-axis to corresponding points on the positive z-axis. This peculiarity and the factor of

4/3 difference between the rest and kinetic mass found above were of great concern to Abraham, Lorentz, and other physicists who built classical models for the electron immediately after its discovery.

The issue was resolved by Poincaré, who pointed out that a purely electromagnetic electron is not mechanically stable against its own repulsive Coulomb forces unless some internal cohesive forces hold it together. Although mooted by quantum electrodynamics, this aspect of the classical electron problem is solved by adding the contribution to the mass and flow of energy provided by such "Poincaré stresses".

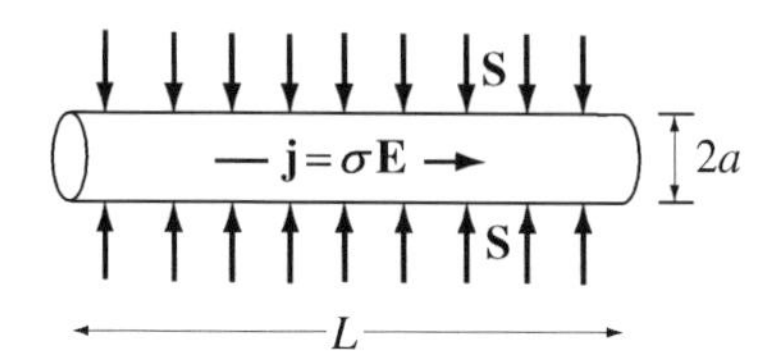

Figure 15.1: The lines of the Poynting vector **S** point radially inward at the surface of a current-carrying ohmic wire.

15.4.2 Energy Flow in Resistive Wires

A long, straight, current-carrying wire with radius a and conductivity σ provides a nice illustration of Poynting's theorem (see Figure 15.1). Let the current I flow in the positive z-direction and let the wire volume be the integration volume V in (15.27). By Ohm's law, $\mathbf{j} = \sigma\mathbf{E}$, and Ampère's law, $\nabla \times \mathbf{B} = \mu_0\mathbf{j}$, the electric and magnetic fields at the wire surface are $\mathbf{E} = \hat{\mathbf{z}}I/\pi a^2\sigma$ and $\mathbf{B} = \hat{\boldsymbol{\phi}}\mu_0 I/2\pi a$. The Poynting vector (15.29) points radially *in* toward the center of the wire. Moreover, because $R = L/\pi a^2\sigma$ is the resistance of the wire, the integral on the right side of (15.30) is

$$-\int_S dA\,\mathbf{S}\cdot\hat{\mathbf{n}} = I^2\frac{L}{\pi a^2\sigma} = I^2 R. \tag{15.35}$$

This confirms the conservation of energy statement (15.30) because $\dot{U}_{\rm EM} = 0$ for steady currents and $\dot{U}_{\rm mech} = I^2 R$ for an ohmic circuit (Section 9.6). It also shows that a constant flow of electromagnetic energy into the wire through its side walls is required to maintain the kinetic energy of the current-carrying particles against the energy they lose to ohmic heating.

The energy delivered to the wire in Figure 15.1 originates from a battery or some other source of electromotive force (see Section 9.7). Figure 15.2 illustrates the global flow of energy for a planar, current-carrying wire loop where the streamlines of the Poynting vector can be calculated exactly. The ohmic loop is drawn as a heavy solid line. Gray arrows indicate the direction of the current. We model the EMF as a vertically oriented point electric dipole located at the position of the upward-pointing gray arrow. The light solid lines of **S** intersect the wire at right angles and deliver energy as indicated by the black solid arrows. An exact calculation of **S** is possible because we endow the ohmic loop with two special characteristics: (i) a shape exactly coincident with one of the electric field lines of the point dipole; and (ii) a conductivity that varies in such a way that $\mathbf{j} = \sigma(\mathbf{r})\mathbf{E}(\mathbf{r})$ is constant everywhere in the wire. With these choices, the presence of the wire does not disturb the electric field pattern of the dipole.

Infinitesimally close to the wire surface, the magnetic field produced by the current forms perfect closed circles concentric with the wire. Therefore, the Poynting vector (15.27) evaluated at the wire surface is everywhere perpendicular to the surface and points toward the wire. A moment's reflection

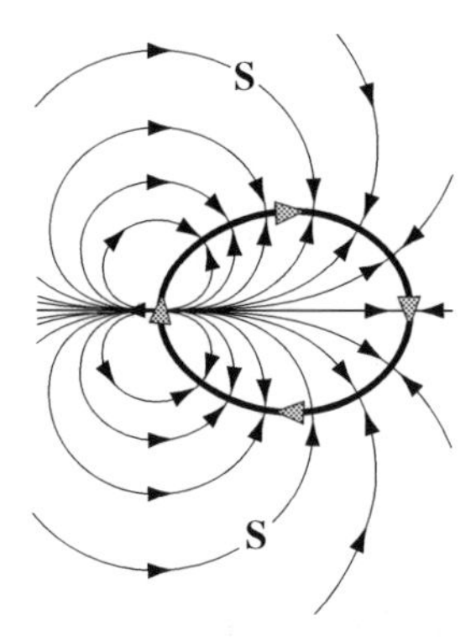

Figure 15.2: Field lines of the Poynting vector $\mathbf{S}$ (light solid lines with black arrows) for an ohmic wire circuit (heavy solid line with gray arrows) with special properties (see text). The magnetic field produced by the loop is not shown.

shows that the lines of $\mathbf{S}(\mathbf{r})$ are exactly coincident with the equipotential curves of the point dipole. It is clear from the diagram that energy flows out of the point dipole source, into the vacuum, and into the wire at every point along its length.

The situation is less simple for an ohmic loop with uniform conductivity and an arbitrary shape. The total electric field is now the sum of the point dipole field and the field produced by charges that appear on the surface of the loop (Section 9.7.4). This implies that some electric field lines intersect the surface of the wire. The normal component of $\mathbf{E}$ at the surface of the wire is proportional to the density of surface electric charge at that point. The tangential component of $\mathbf{E}$ is continuous at the wire surface and contributes to the Poynting vector (which still points locally into the wire).

15.4.3 Non-Uniqueness of the Poynting Vector

It may seem odd that the Poynting vector for a wire circuit does not predict energy flow parallel to the wire itself. This and other unanticipated features of some Poynting flows prompt some authors to define a Poynting vector using

$$\mathbf{S} = \frac{1}{\mu_0}(\mathbf{E} \times \mathbf{B}) + \nabla \times \mathbf{X}. \tag{15.36}$$

The vector field $\mathbf{X}$ is chosen to make (15.36) point in more "natural" directions. The definition (15.36) does not disrupt Poynting's theorem (15.33) because the latter contains only $\nabla \cdot \mathbf{S}$. Relativistic considerations constrain, but do not completely eliminate, this arbitrariness in the definition of $\mathbf{S}$.[9] There is no real problem in any event because the Poynting vector is not an observable.

15.5 Conservation of Linear Momentum

Nothing in our previous work prepares us for the surprising fact that most electromagnetic field configurations carry linear momentum. To make this plausible, we consider two identical particles with charge q released from rest. Figure 15.3(a) shows the motion of the pair (labeled 1 and 2 for clarity) in the center-of-mass frame of reference. By symmetry, the force $\mathbf{F}_1 = q_1[\mathbf{E}_2(\mathbf{r}_1) + \boldsymbol{v}_1 \times \mathbf{B}_2(\mathbf{r}_1)]$ exerted

[9] See U. Backhaus and K. Schäfer, "On the uniqueness of the vector for energy flow density in electromagnetic fields", *American Journal of Physics* **54**, 279 (1986).

Figure 15.3: Two identical charges ($q_1 = q_2 = q$) repel one another. The motion as viewed in (a) the center of mass frame; (b) the frame where q_1 is instantaneously at rest. The distance $R(t) = |\mathbf{r}_1 - \mathbf{r}_2(t)|$.

on q_1 by q_2 must be equal and opposite to the force $\mathbf{F}_2 = q_2[\mathbf{E}_1(\mathbf{r}_2) + \boldsymbol{v}_2 \times \mathbf{B}_1(\mathbf{r}_2)]$ exerted on q_2 by q_1. Newton's third law is satisfied.

We now analyze the same problem in a frame of reference where particle 1 is instantaneously at rest [Figure 15.3(b)]. Neither particle feels a Lorentz magnetic force in this frame. The Coulomb electric force felt by the moving charge due to the field produced by the static charge is

$$\mathbf{F}_2(t) = -q_2 \nabla \varphi_1(\mathbf{r}_2, t). \tag{15.37}$$

The electric force felt by the static charge due to the field produced by the moving charge is

$$\mathbf{F}_1(t) = -q_1 \nabla \varphi_2(\mathbf{r}_1, t) - q_1 \frac{\partial \mathbf{A}_2(\mathbf{r}_1, t)}{\partial t}. \tag{15.38}$$

In the Coulomb gauge, (15.18) shows that the scalar potentials are

$$\varphi_1(\mathbf{r}_2, t) = \frac{1}{4\pi\epsilon_0} \frac{q_1}{|\mathbf{r}_1 - \mathbf{r}_2(t)|} \qquad \text{and} \qquad \varphi_2(\mathbf{r}_1, t) = \frac{1}{4\pi\epsilon_0} \frac{q_2}{|\mathbf{r}_1 - \mathbf{r}_2(t)|}. \tag{15.39}$$

Because $q_1 = q_2$ and $\nabla \varphi_1(\mathbf{r}_2, t) = -\nabla \varphi_2(\mathbf{r}_1, t)$, the electric fields derived from (15.39) that appear in (15.37) and (15.38) have equal magnitude but point in opposite directions.

Newton's second law states that the net force on a system of particles is equal to the time derivative of the system's total mechanical linear momentum, $\mathbf{P}_{\text{mech}}$. Writing this out and using the sum of (15.37) and (15.38) as the net force shows that

$$\frac{d\mathbf{P}_{\text{mech}}}{dt} = \mathbf{F}_1 + \mathbf{F}_2 = -q_1 \frac{\partial \mathbf{A}_2(\mathbf{r}_1, t)}{\partial t}. \tag{15.40}$$

Apparently, the total mechanical linear momentum is *not* a constant of the motion. This is an untenable conclusion for a closed system not acted on by external forces. To save the principle of the conservation of *total* linear momentum, we are forced to hypothesize that the electromagnetic field possesses an intrinsic linear momentum $\mathbf{P}_{\text{EM}}$ and write

$$\frac{d}{dt}(\mathbf{P}_{\text{mech}} + \mathbf{P}_{\text{EM}}) = \mathbf{F}_1 + \mathbf{F}_2 + \frac{d\mathbf{P}_{\text{EM}}}{dt} = 0. \tag{15.41}$$

This equation is consistent with (15.40) if $\mathbf{P}_{\text{EM}} = q_1 \mathbf{A}_2(\mathbf{r}_1, t)$ in the considered frame of reference (see Section 15.5.3).

15.5.1 The Mechanical Force on a Volume

Our strategy to confirm that electromagnetic fields possess an intrinsic linear momentum $\mathbf{P}_{\text{EM}}$ begins with the total Coulomb-Lorentz force that acts on the charge and current densities in a volume V:

$$\mathbf{F}_{\text{mech}} = \int_V d^3r \, (\rho \mathbf{E} + \mathbf{j} \times \mathbf{B}). \tag{15.42}$$

The first step is to eliminate ρ and $\mathbf{j}$ from (15.42) using the Maxwell equations (15.3) and (15.4). Next, add $(\nabla \cdot \mathbf{B})\mathbf{B} = 0$ to the integrand and use

$$\frac{\partial}{\partial t}(\mathbf{E} \times \mathbf{B}) = \mathbf{E} \times \frac{\partial \mathbf{B}}{\partial t} + \frac{\partial \mathbf{E}}{\partial t} \times \mathbf{B}. \tag{15.43}$$

The result is

$$\mathbf{F}_{\text{mech}} = \int_V d^3r \left\{ -\epsilon_0 \frac{\partial}{\partial t}(\mathbf{E} \times \mathbf{B}) + \epsilon_0 \left[(\nabla \cdot \mathbf{E})\mathbf{E} - \mathbf{E} \times (\nabla \times \mathbf{E})\right] + \frac{1}{\mu_0}\left[(\nabla \cdot \mathbf{B})\mathbf{B} - \mathbf{B} \times (\nabla \times \mathbf{B})\right] \right\}. \tag{15.44}$$

The first term on the right side of (15.44) is proportional to the Poynting vector (15.29). Therefore, writing out the curl operations and rearranging gives

$$\mathbf{F}_{\text{mech}} = \int_V d^3r \left\{ -\frac{1}{c^2}\frac{\partial \mathbf{S}}{\partial t} + \epsilon_0 \left[(\nabla \cdot \mathbf{E})\mathbf{E} + (\mathbf{E} \cdot \nabla)\mathbf{E} - \frac{1}{2}\nabla(\mathbf{E} \cdot \mathbf{E})\right] + \frac{1}{\mu_0}\left[(\nabla \cdot \mathbf{B})\mathbf{B} + (\mathbf{B} \cdot \nabla)\mathbf{B} - \frac{1}{2}\nabla(\mathbf{B} \cdot \mathbf{B})\right] \right\}. \tag{15.45}$$

The appearance of (15.45) simplifies considerably if we define the *Maxwell (electromagnetic) stress tensor* as the dyadic[10]

$$\mathbf{T} = \epsilon_0 \left[\mathbf{EE} + c^2\mathbf{BB} - \tfrac{1}{2}\mathbf{I}(E^2 + c^2B^2)\right]. \tag{15.46}$$

The components of this symmetric object are precisely the sum of the corresponding components of the electrostatic stress tensor $T_{ij}(\mathbf{E})$ (Section 3.7) and the magnetostatic stress tensor $T_{ij}(\mathbf{B})$ (Section 12.5)

$$T_{ij} = T_{ij}(\mathbf{E}) + T_{ij}(\mathbf{B}) = \epsilon_0 \left[E_i E_j + c^2 B_i B_j - \tfrac{1}{2}\delta_{ij}(E^2 + c^2B^2)\right]. \tag{15.47}$$

The divergence of the stress tensor, $\nabla \cdot \mathbf{T}$, is a *vector* with components

$$(\nabla \cdot \mathbf{T})_j = \sum_i \frac{\partial}{\partial x_i} T_{ij}. \tag{15.48}$$

Therefore, (15.45) takes the compact form

$$\mathbf{F}_{\text{mech}} = \frac{d\mathbf{P}_{\text{mech}}}{dt} = \int_V d^3r \left\{ -\frac{1}{c^2}\frac{\partial \mathbf{S}}{\partial t} + \nabla \cdot \mathbf{T} \right\}. \tag{15.49}$$

We note in passing that the $\partial \mathbf{S}/\partial t$ term in (15.49) precludes writing the total mechanical force as a surface integral as we did in electrostatics and magnetostatics. Time-harmonic fields are an important exception where the Poynting vector term disappears after averaging over one period of oscillation.

15.5.2 Momentum Density and Momentum Current Density

It is straightforward to interpret (15.49) as a conservation law because a quantity with the dimensions of an *electromagnetic momentum density* is

$$\mathbf{g} = \frac{\mathbf{S}}{c^2} = \epsilon_0(\mathbf{E} \times \mathbf{B}). \tag{15.50}$$

[10] A dyadic is two vectors juxtaposed or a sum of such terms. The unit dyadic is $\mathbf{I} = \hat{\mathbf{x}}\hat{\mathbf{x}} + \hat{\mathbf{y}}\hat{\mathbf{y}} + \hat{\mathbf{z}}\hat{\mathbf{z}}$. See Section 1.8.

The corresponding formula for the total linear momentum of a configuration of electric and magnetic fields is[11]

$$\mathbf{P}_{\text{EM}} = \int d^3r\, \mathbf{g} = \epsilon_0 \int d^3r\, \mathbf{E} \times \mathbf{B}. \tag{15.51}$$

Combining (15.50) and (15.51) with the divergence theorem and $(\hat{\mathbf{n}} \cdot \mathbf{T})_j = \hat{n}_i T_{ij}$ permits us to write (15.49) in the appealing form

$$\frac{d\mathbf{P}_{\text{tot}}}{dt} = \frac{d}{dt}(\mathbf{P}_{\text{mech}} + \mathbf{P}_{\text{EM}}) = \int_S dA\, \hat{\mathbf{n}} \cdot \mathbf{T}. \tag{15.52}$$

Equation (15.52) is the linear momentum analog of the energy conservation law (15.30).

The fact that $\mathbf{T}$ has dimensions of (momentum/volume) × velocity invites us to interpret the stress tensor as a *momentum current density*. One index of $\mathbf{T}$ labels the direction of flow; the other labels the components of the momentum. Thus, T_{ij} is the rate at which the jth component of momentum flows through an area element $dA\,\hat{n}_i$. Not obviously, T_{ij} is also the rate at which the ith component of momentum flows through an area element $dA\,\hat{n}_j$. This is a consequence of the symmetry $T_{ij} = T_{ji}$ of (15.47). For unbounded space, the absolute conservation law

$$\mathbf{P}_{\text{mech}} + \mathbf{P}_{\text{EM}} = const. \tag{15.53}$$

is true only if the integral on the right-hand side of (15.52) vanishes as $S \to \infty$. This is true for all the static and quasistatic fields we have studied so far and makes (15.53) a useful tool for problem-solving provided *all* the contributions to the linear momentum can be identified. This is not usually a great challenge, except when certain "hidden" contributions to $\mathbf{P}_{\text{mech}}$ occur (see Application 15.3 at the end of Section 15.7).

A spatially local statement of linear momentum conservation follows from the foregoing if we glance back at (15.42) and remind ourselves that $\mathbf{F}_{\text{mech}}$ is the volume integral of the Coulomb-Lorentz force density $\mathbf{f}_{\text{mech}} = \rho\mathbf{E} + \mathbf{j} \times \mathbf{B}$. Then, because the integration volume V is arbitrary, substitution of (15.50) into (15.49) produces a local conservation law that is the momentum analog of (15.33):

$$\frac{\partial \mathbf{g}}{\partial t} + \nabla \cdot (-\mathbf{T}) = -\mathbf{f}_{\text{mech}}. \tag{15.54}$$

15.5.3 The Physical Significance of the Vector Potential

We can now calculate $\mathbf{P}_{\text{EM}}$ for the two charged particles shown in Figure 15.3(b). The reference frame there has particle 1 instantaneously at rest and particle 2 in motion. The integral (15.51) involves the total electric field $\mathbf{E}(\mathbf{r}, t)$ and the total magnetic field $\mathbf{B}(\mathbf{r}, t)$. However, if $(\mathbf{E}_1, \mathbf{B}_1)$ and $(\mathbf{E}_2, \mathbf{B}_2)$ are the fields produced by q_1 and q_2, only the "interaction" momentum carried by the cross terms $\mathbf{E}_1 \times \mathbf{B}_2 + \mathbf{E}_2 \times \mathbf{B}_1$ can be relevant to the force between the particles.

$\mathbf{B}_1 = 0$ because particle 1 is at rest. Therefore,

$$\mathbf{P}_{\text{EM}} = \epsilon_0 \int d^3r\, \mathbf{E}_1 \times \mathbf{B}_2. \tag{15.55}$$

Using $(\mathbf{E}_1 \times \mathbf{B}_2)_k = \epsilon_{k\ell m} E_{1\ell} B_{2m}$ and $B_{2m} = \epsilon_{mst}\partial_s A_{2t}$, we find

$$P_{\text{EM},k} = \epsilon_0 \int d^3r\, (E_{1\ell}\,\partial_k A_{2\ell} - E_{1\ell}\,\partial_\ell A_{2k}). \tag{15.56}$$

[11] This formula appeared for the first time in J.J. Thomson, *Notes on Recent Research in Electricity and Magnetism* (Clarendon Press, Oxford, 1893). Not long after, the future Nobel laureate turned his attention to cathode rays and the experiments which led him to discover the electron.

Integrating both terms in (15.56) by parts and assuming that the surface terms vanish gives

$$P_{\mathrm{EM},k} = \epsilon_0 \int d^3r \,(A_{2k} \nabla \cdot \mathbf{E}_1 - A_{2\ell}\, \partial_k E_{1\ell}). \tag{15.57}$$

Now, $\nabla \times \mathbf{E}_1 = 0$, so $\partial_k E_{1\ell} = \partial_\ell E_{1k}$ and another integration by parts yields

$$\mathbf{P}_{\mathrm{EM}} = \epsilon_0 \int d^3r\, \mathbf{A}_2(\nabla \cdot \mathbf{E}_1) + \mathbf{E}_1(\nabla \cdot \mathbf{A}_2). \tag{15.58}$$

Finally, Gauss' law, $\epsilon_0 \nabla \cdot \mathbf{E} = \rho$, and the Coulomb gauge constraint, $\nabla \cdot \mathbf{A}_2 = 0$, reduce (15.58) to

$$\mathbf{P}_{\mathrm{EM}} = \int d^3r\, \rho_1(\mathbf{r})\mathbf{A}_2(\mathbf{r}). \tag{15.59}$$

Equation (15.59) is valid whenever a static or quasistatic charge distribution coexists with a static or quasistatic magnetic field. For the specific problem at hand, $\rho_1(\mathbf{r}) = q_1\delta(\mathbf{r} - \mathbf{r}_1)$ and we conclude that

$$\mathbf{P}_{\mathrm{EM}} = q_1\mathbf{A}_2(\mathbf{r}_1, t). \tag{15.60}$$

This confirms the guess we made at the beginning of this section.[12] It also explains why Maxwell referred to the vector potential $\mathbf{A}(\mathbf{r}, t)$ as "electro-kinetic momentum". However, because $q\varphi(\mathbf{r}, t)$ is the potential energy of a point charge q in the quasistatic limit, it may be more apt to refer to $q\mathbf{A}(\mathbf{r}, t)$ in the Coulomb gauge as the "'potential momentum" in the same limit. In other words, the Coulomb gauge expression $q\mathbf{A}(\mathbf{r}, t)$ may be consistently interpreted as the field momentum available for conversion into particle momentum. The fact that this identification is not gauge invariant does not diminish the usefulness of the Coulomb gauge vector potential to rationalize the non-intuitive ability of quasistatic electric and magnetic fields to store and release linear momentum. This point is not arcane because (15.51) makes clear that $\mathbf{P}_{\mathrm{EM}}$ is non-zero for nearly every situation where electrostatic and magnetostatic fields overlap in space.

Application 15.2 $\mathbf{P}_{\mathrm{EM}}$ for a Capacitor in a Uniform Magnetic Field

Consider a parallel-plate capacitor whose plates have charges $\pm Q$, area A, and separation d. In the presence of a uniform magnetic field $\mathbf{B}_0$, it is common to write $\mathbf{E_0} = (Q/A\epsilon_0)\hat{\mathbf{z}}$ for the magnitude of the electric field between the plates and estimate the linear momentum stored by the fields of this system as

$$\mathbf{P}_{\mathrm{EM}} = \epsilon_0 \int d^3r\, \mathbf{E} \times \mathbf{B} \approx \epsilon_0 A d(\mathbf{E}_0 \times \mathbf{B}_0). \tag{15.61}$$

In this Application, we show that (15.61) is *twice* the correct value due to neglect of the fringing electric field at the edge of the plates. Surprising, this result does not change in the limit $d \ll \sqrt{A}$ when the infinite-area approximation is normally valid. Our strategy is to first compute $\mathbf{P}_{\mathrm{EM}}$ for $\mathbf{B}_0$ in the presence of a finite electric dipole (two equal and opposite point charges separated by d) with moment $\mathbf{p}_0 = -qd\hat{\mathbf{z}}$. We then exploit the fact that $\mathbf{P}_{\mathrm{EM}}$ is linear in $\mathbf{E}$ and use superposition to deduce the field momentum for the capacitor modeled as a dense set of these dipoles distributed uniformly in the x-y plane over an area A.

[12] We will see in Chapter 24 that (15.60) is intimately connected to the fact that $m\boldsymbol{v} + q\mathbf{A}$ is the *canonical momentum* of a particle with mass m and charge q that moves with velocity $\boldsymbol{v}$ in a magnetic field.

Equation (15.59) is available to compute $\mathbf{P}_{\rm EM}$ because the fields are static. Nominally, this requires the charge density of the dipole and a vector potential for $\mathbf{B}_0 = \nabla \times \mathbf{A}$. However, using only

$$\mathbf{A}(\mathbf{r}) = \frac{1}{2}\mathbf{B}_0 \times \mathbf{r} \tag{15.62}$$

and the definition of the electric dipole moment, we conclude that

$$\mathbf{P}_{\rm EM} = \int d^3r\, \rho \mathbf{A} = \frac{1}{2}\mathbf{B}_0 \times \int d^3r\, \rho \mathbf{r} = \frac{1}{2}\mathbf{B}_0 \times \mathbf{p}_0. \tag{15.63}$$

Now, as indicated above, we construct a capacitor by densely arranging parallel dipoles like the fibers of a plush carpet. The dipole moment of this capacitor is $\mathbf{p} = -Qd\hat{\mathbf{z}} = -\epsilon_0 dA\mathbf{E}_0$. Using (15.63), the stored electromagnetic linear momentum is indeed one-half of (15.61):

$$\mathbf{P}_{\rm EM} = -\frac{1}{2}\mathbf{p} \times \mathbf{B}_0 = \frac{1}{2}\epsilon_0 Ad(\mathbf{E}_0 \times \mathbf{B}_0). \tag{15.64}$$

Remark: The failure of (15.61) is another consequence of the linear dependence of $\mathbf{g}_{\rm EM} = \mathbf{E} \times \mathbf{B}$ on $\mathbf{E}$. Small errors in the field produce non-negligible errors in $\mathbf{P}_{\rm EM}$ compared to the force calculation in Example 6.8 where the *quadratic* dependence of the electrostatic field energy on $\mathbf{E}$ made it possible to ignore the effect of fringe fields and use the infinite-area approximation to the electric field of a finite-area capacitor. ■

Example 15.3 A charge q moving slowly through an external magnetic field $\mathbf{B}(\mathbf{r}, t)$ makes a displacement $\delta\mathbf{r}_0$ in a time δt. Derive the force on q from the change in electromagnetic momentum which accompanies the displacement.

Solution: If the particle moves quasi-electrostatically from $\mathbf{r}_0$ to $\mathbf{r}_0 + \delta\mathbf{r}_0$, we can compute the change in the electromagnetic momentum (15.51) using $\mathbf{E} = -\nabla\varphi$:

$$\delta\mathbf{P}_{\rm EM} = -\frac{q}{4\pi}\int d^3r\, \nabla\left[\frac{1}{|\mathbf{r} - \mathbf{r}_0 - \delta\mathbf{r}_0|} - \frac{1}{|\mathbf{r} - \mathbf{r}_0|}\right] \times \mathbf{B}.$$

But $\delta\mathbf{r}_0$ is an infinitesimal constant vector, so

$$\delta\mathbf{P}_{\rm EM} = \frac{q}{4\pi}\int d^3r\, \nabla\left[\delta\mathbf{r}_0 \cdot \nabla\frac{1}{|\mathbf{r} - \mathbf{r}_0|}\right] \times \mathbf{B} = \frac{q}{4\pi}\int d^3r\left[\nabla^2\frac{1}{|\mathbf{r} - \mathbf{r}_0|}\right]\delta\mathbf{r}_0 \times \mathbf{B}.$$

Using $\nabla^2|\mathbf{r} - \mathbf{r}_0|^{-1} = -4\pi\delta(\mathbf{r} - \mathbf{r}_0)$ gives

$$\delta\mathbf{P}_{\rm EM} = -q\delta\mathbf{r}_0 \times \mathbf{B}.$$

Finally, let $\mathbf{P}_{\rm mech}$ be the translational momentum of q and observe that quasistatic fields fall off quickly enough at infinity that (15.52) reads $\mathbf{P}_{\rm mech} + \mathbf{P}_{\rm EM} = const$. In that case, $\delta\mathbf{P}_{\rm mech} = -\delta\mathbf{P}_{\rm EM}$ and the force on the particle is

$$\mathbf{F} = \frac{\delta\mathbf{P}_{\rm mech}}{\delta t} = q\frac{\delta\mathbf{r}_0}{\delta t} \times \mathbf{B} = q\boldsymbol{v} \times \mathbf{B}.$$

This is the familiar Lorentz force.

15.6 Conservation of Angular Momentum

Most configurations of electric and magnetic fields carry angular momentum. Since $\mathbf{L} = \mathbf{r} \times \mathbf{p}$ is the angular momentum of a particle at point $\mathbf{r}$ with linear momentum $\mathbf{p}$, it is natural to suppose that an electromagnetic field with linear momentum density $\mathbf{g} = \epsilon_0 \mathbf{E} \times \mathbf{B}$ also possesses an *electromagnetic angular momentum density* $\mathbf{r} \times \mathbf{g}$. To check this guess, we begin with the mechanical torque and angular momentum exerted on any distribution of charge and current by an electromagnetic field:

$$\mathbf{N}_{\text{mech}} = \frac{d\mathbf{L}_{\text{mech}}}{dt} = \int d^3r\, \mathbf{r} \times \mathbf{f}_{\text{mech}} = \int d^3r\, \mathbf{r} \times (\rho\mathbf{E} + \mathbf{j} \times \mathbf{B}). \tag{15.65}$$

The presence of the force density $\mathbf{f}_{\text{mech}}$ in (15.65) reminds us of the conservation of linear momentum as expressed by (15.54):

$$\frac{\partial \mathbf{g}}{\partial t} - \nabla \cdot \mathbf{T} = -\mathbf{f}_{\text{mech}}. \tag{15.66}$$

The family resemblance becomes stronger when we form the cross product of (15.66) with a time-independent position vector $\mathbf{r}$. This gives

$$\frac{\partial}{\partial t}(\mathbf{r} \times \mathbf{g}) - \mathbf{r} \times \nabla \cdot \mathbf{T} = -\mathbf{r} \times \mathbf{f}_{\text{mech}}, \tag{15.67}$$

where

$$-(\mathbf{r} \times \nabla \cdot \mathbf{T})_i = -\epsilon_{ijk} r_j\, \partial_m T_{mk}. \tag{15.68}$$

However, because

$$\partial_m (T_{mk} r_j) = r_j\, \partial_m T_{mk} + T_{mk}\delta_{mj} = r_j\, \partial_m T_{mk} + T_{jk}, \tag{15.69}$$

we may write

$$-(\mathbf{r} \times \nabla \cdot \mathbf{T})_i = -\varepsilon_{ijk}[\partial_m (T_{mk} r_j) - T_{jk}]. \tag{15.70}$$

On the other hand, the symmetry ($T_{ij} = T_{ji}$) of the electromagnetic stress tensor in (15.47) implies that $\epsilon_{ijk} T_{jk} = 0$. Therefore,

$$-(\mathbf{r} \times \nabla \cdot \mathbf{T})_i = \partial_m \varepsilon_{ikj} T_{mk} r_j = \{\nabla \cdot (\mathbf{T} \times \mathbf{r})\}_i. \tag{15.71}$$

The final step combines (15.67) and (15.71) to get the continuity-like equation

$$\frac{\partial}{\partial t}(\mathbf{r} \times \mathbf{g}) + \nabla \cdot (\mathbf{T} \times \mathbf{r}) = -\mathbf{r} \times \mathbf{f}_{\text{mech}}. \tag{15.72}$$

On dimensional grounds alone, we may conclude that this is a differential expression for conservation of angular momentum analogous to the conservation of linear momentum statement (15.54).

15.6.1 Angular Momentum Current Density

It is conventional at this point to define the dyadic

$$\mathbf{M} = \mathbf{T} \times \mathbf{r} \qquad (M_{ij} = T_{ik} r_\ell \epsilon_{jk\ell}). \tag{15.73}$$

$\mathbf{M}$ has dimensions of (angular momentum/volume) $\times$ velocity and thus may be consistently interpreted as an *angular momentum current density*. Substituting (15.73) into (15.72) and integrating over an arbitrary volume V gives an equation for the mechanical torque that is the analog of (15.49) for the mechanical force:

$$\mathbf{N}_{\text{mech}} = \frac{d\mathbf{L}_{\text{mech}}}{dt} = -\int d^3r \left\{\frac{\partial}{\partial t}(\mathbf{r} \times \mathbf{g}) + \nabla \cdot \mathbf{M}\right\}. \tag{15.74}$$

Equation (15.74) shows that we guessed correctly that $\mathbf{r} \times \mathbf{g}$ is properly interpreted as the angular momentum density of an electromagnetic field. The total field angular momentum in a volume V is

$$\mathbf{L}_{\rm EM} = \epsilon_0 \int d^3r\, \mathbf{r} \times (\mathbf{E} \times \mathbf{B}). \tag{15.75}$$

Combining (15.74) with (15.75) gives the angular momentum analog of the linear momentum conservation law (15.52) and the energy conservation law (15.30):

$$\frac{d\mathbf{L}_{\rm tot}}{dt} = \frac{d}{dt}(\mathbf{L}_{\rm EM} + \mathbf{L}_{\rm mech}) = -\int_V d^3r\, \nabla \cdot \mathbf{M} = -\int_S d\mathbf{S} \cdot \mathbf{M}. \tag{15.76}$$

It is worth repeating and emphasizing the role of the symmetry $T_{ij} = T_{ji}$ of the stress-energy tensor in the derivation of local conservation of angular momentum [see (15.71)]. The same connection occurs in continuum mechanics, general relativity, and other field theories.

Equation (15.76) becomes somewhat less abstract when we construct M_{ij} from the components of $\mathbf{T}$ defined in (15.47) and write out the flux integral. The result is

$$\frac{d\mathbf{L}_{\rm total}}{dt} = \epsilon_0 \int_S d\mathbf{S} \cdot \left\{\mathbf{E}(\mathbf{r} \times \mathbf{E}) + c^2\mathbf{B}(\mathbf{r} \times \mathbf{B})\right\} + \tfrac{1}{2}\epsilon_0 \int_S d\mathbf{S} \times \mathbf{r}(E^2 + c^2B^2). \tag{15.77}$$

The right-hand side of (15.77) goes to zero for static and quasistatic fields when the enclosing surface $S \to \infty$. In that case, we get a strict law of conservation of angular momentum:

$$\frac{d}{dt}(\mathbf{L}_{\rm EM} + \mathbf{L}_{\rm mech}) = 0. \tag{15.78}$$

This is not necessarily the case when radiation fields are present (see Chapter 20).

Generalizing from Section 15.5.3, we may think of $\mathbf{L}_{\rm EM}$ as angular momentum that is potentially available for transfer to matter. Any loss of field angular momentum must be compensated by a gain in mechanical angular momentum with respect to the same origin. For static fields, this idea underlies a famous paradox presented by Feynman in his *Lectures on Physics*.[13] Example 15.4 is a variation of this problem.

Example 15.4 (a) Find $\mathbf{L}_{\rm EM}$ for the ferromagnetic metal sphere shown in Figure 15.4(a) which has radius R, charge Q, and uniform magnetization $\mathbf{M} = M\hat{\mathbf{z}}$. (b) Compare this with the mechanical angular momentum acquired by the sphere when the magnetization is removed [Figure 15.4(b)]. The latter may be accomplished by heating the sphere above the Curie temperature, where ferromagnetism disappears.

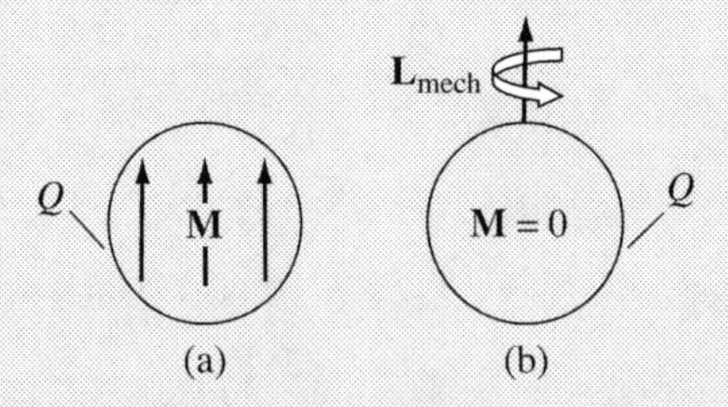

Figure 15.4: A sphere with charge Q begins (a) at rest with uniform magnetization $\mathbf{M}$ and ends (b) demagnetized with a net angular momentum $\mathbf{L}_{\rm mech}$.

[13] See Section 17.4 of R.P. Feynman, R.B. Leighton, and M. Sands, *The Feynman Lectures on Physics*, Volume II (Addison-Wesley, Reading, MA, 1964).

Solution: (a) From Gauss' law, $\mathbf{E}$ is zero inside the sphere and $\mathbf{E} = \hat{\mathbf{r}} Q/4\pi\epsilon_0 r^2$ for $r > R$. We found the magnetic field of a uniformly magnetized sphere in Section 13.4.1. For $r < R$, it is $\mathbf{B}_{\text{in}} = (2/3)\mu_0\mathbf{M}$. For $r > R$ it is the field of a point dipole at the origin with magnetic moment $\mathbf{m} = \hat{\mathbf{z}} 4\pi R^3 M/3$:

$$\mathbf{B}_{\text{out}} = \frac{\mu_0}{4\pi}\frac{m}{r^3}[2\cos\theta\hat{\mathbf{r}} + \sin\theta\hat{\boldsymbol{\theta}}].$$

From (15.75), the electromagnetic angular momentum of the sphere is

$$\mathbf{L}_{\text{EM}} = \epsilon_0 \int d^3r\, \mathbf{r} \times (\mathbf{E} \times \mathbf{B}) = \epsilon_0 \int d^3r\, [(\mathbf{r}\cdot\mathbf{B})\mathbf{E} - (\mathbf{r}\cdot\mathbf{E})\mathbf{B}].$$

By symmetry, only the z-component of the angular momentum is non-zero. Therefore, since $\hat{\mathbf{r}}\cdot\hat{\mathbf{z}} = \cos\theta$ and $\hat{\boldsymbol{\theta}}\cdot\hat{\mathbf{z}} = -\sin\theta$,

$$\mathbf{L}_{\text{EM}} = \hat{\mathbf{z}}\frac{\mu_0}{4\pi}\frac{Qm}{4\pi}\int_0^{2\pi} d\phi \int_{-1}^{1} d(\cos\theta)\sin^2\theta \int_R^{\infty} \frac{dr}{r^2} = \frac{2}{9}\mu_0 Q R^2 \mathbf{M}.$$

(b) A decrease in $\mathbf{M}$ is accompanied by a decrease in $\mathbf{B}$. This change induces an electric field that exerts a torque on the surface charge density $\sigma = Q/4\pi R^2$ of the sphere. By applying Faraday's law to a circular path on the surface of the sphere concentric with the z-axis, we find the electric field:

$$\mathbf{E}(r,\theta) = -\frac{1}{2} r\sin\theta \dot{B}_z \hat{\boldsymbol{\phi}}.$$

B_z is discontinuous at $r = R$, but the magnetic flux integral in Faraday's law (and thus the induced electric field) is not. Therefore, the torque on the sphere is

$$\mathbf{N} = \int_S dS\, \mathbf{r} \times \sigma\mathbf{E} = \hat{\mathbf{z}} \int_S dS\, r\sin\theta\, \sigma E_\phi = -\frac{1}{3} Q R^2 \dot{\mathbf{B}}_{\text{in}}.$$

Since $\mathbf{N} = d\mathbf{L}_{\text{mech}}/dt$ and $\int \dot{\mathbf{B}} dt = -\mathbf{B}_{\text{in}}$, we conclude that

$$\mathbf{L}_{\text{mech}} = \frac{1}{3} Q R^2 \mathbf{B}_{\text{in}} = \frac{2}{9}\mu_0 Q R^2 \mathbf{M}.$$

This is equal to $\mathbf{L}_{\text{EM}}$ calculated in part (a), as expected from conservation of angular momentum.

15.7 The Center of Energy

Newton's laws predict that the center of mass of an isolated system is stationary when its total linear momentum is zero. In this section, we generalize this idea to include electromagnetic fields. By ignoring two surface integrals, we limit the discussion to (i) quasistatic situations and (ii) source-free electromagnetic wave packets confined to a finite volume of space.

The first step is to multiply the differential statement of energy conservation (15.33) by $\mathbf{r}$ to get

$$\left[\frac{\partial u_{\text{EM}}}{\partial t} + \nabla\cdot\mathbf{S}\right]\mathbf{r} = -\mathbf{j}\cdot(\mathbf{E} + \mathbf{v}\times\mathbf{B})\mathbf{r}. \tag{15.79}$$

The second step is to multiply the differential statement of linear momentum conservation (15.54) by $c^2 t$. This gives

$$\left[\frac{\partial \mathbf{g}}{\partial t} - \nabla \cdot \mathbf{T}\right] c^2 t = -(\rho \mathbf{E} + \mathbf{j} \times \mathbf{B}) c^2 t. \tag{15.80}$$

Bearing in mind (15.48), the third step is to subtract (15.80) from (15.79) and exploit the dyadic identity[14]

$$\nabla \cdot (\mathbf{S}\mathbf{r}) = \partial_k (S_k \mathbf{r}) = (\nabla \cdot \mathbf{S})\mathbf{r} + \mathbf{S}. \tag{15.81}$$

Using (15.50), these manipulations produce

$$\frac{\partial}{\partial t}\left(\mathbf{r} u_{\mathrm{EM}} - c^2 t \mathbf{g}\right) + \nabla \cdot \left(\mathbf{S}\mathbf{r} + c^2 t \mathbf{T}\right) = -\mathbf{j} \cdot (\mathbf{E} + \mathbf{v} \times \mathbf{B})\mathbf{r} + (\rho \mathbf{E} + \mathbf{j} \times \mathbf{B}) c^2 t. \tag{15.82}$$

Equation (15.82) simplifies when we specialize to point charges where $\rho = \sum q_i \delta(\mathbf{r} - \mathbf{r}_i)$ and $\mathbf{j} = \sum q_i \mathbf{v}_i \delta(\mathbf{r} - \mathbf{r}_i)$, let $\mathbf{F}_i = q_i(\mathbf{E} + \mathbf{v}_i \times \mathbf{B})$, and integrate over all space. We assume, in addition, that the surface integral produced by the divergence term vanishes. Consequently,

$$\frac{d}{dt}\int d^3 r\, [\mathbf{r} u_{\mathrm{EM}} - c^2 t \mathbf{g}] + \sum_i \left[\mathbf{r}_i \frac{d\mathbf{r}_i}{dt} \cdot \mathbf{F}_i - c^2 t \mathbf{F}_i\right] = 0. \tag{15.83}$$

The usual laws of mechanics tell us that $\mathbf{F}_i = d\mathbf{p}_i/dt$ and $\mathbf{F}_i \cdot d\mathbf{r}_i = d\mathcal{E}_i$ where $\mathbf{p}_i$ and $\mathcal{E}_i$ are the mechanical momentum and total mechanical energy of the ith particle. Therefore,

$$\frac{d}{dt}\int d^3 r\, \mathbf{r} u_{\mathrm{EM}} + \sum_i \mathbf{r}_i \frac{d\mathcal{E}_i}{dt} = c^2 \int d^3 r\, \mathbf{g} + c^2 t \frac{d}{dt}\left[\int d^3 r\, \mathbf{g} + \sum_i \mathbf{p}_i\right]. \tag{15.84}$$

The quantity in square brackets in (15.84) is the total linear momentum $\mathbf{P}_{\mathrm{EM}} + \mathbf{P}_{\mathrm{mech}}$. The time derivative of this quantity vanishes by conservation of linear momentum because we assume the surface integral vanishes in (15.52). Therefore,

$$\frac{d}{dt}\left[\int d^3 r\, \mathbf{r} u_{\mathrm{EM}} + \sum_i \mathbf{r}_i \mathcal{E}_i\right] = c^2 \mathbf{P}_{\mathrm{EM}} + \sum_i \mathbf{v}_i \mathcal{E}_i. \tag{15.85}$$

Now, by exact analogy with the definition of the center of mass, we define the *center of energy* $\mathbf{R}_{\mathrm{E}}$ and the total energy U_{tot} from

$$\left[\int d^3 r\, \mathbf{r} u_{\mathrm{EM}} + \sum_i \mathbf{r}_i \mathcal{E}_i\right] = \mathbf{R}_{\mathrm{E}}\left[\int d^3 r\, u_{\mathrm{EM}} + \sum_i \mathcal{E}_i\right] = \mathbf{R}_{\mathrm{E}} U_{\mathrm{tot}}. \tag{15.86}$$

Conservation of energy tells us that $(d/dt)U_{\mathrm{tot}} = 0$. Therefore, we can combine the time derivative of (15.86) with (15.85) to get

$$U_{\mathrm{tot}} \frac{d\mathbf{R}_{\mathrm{E}}}{dt} = c^2 \left[\mathbf{P}_{\mathrm{EM}} + \sum_i \frac{\mathbf{v}_i}{c^2} \mathcal{E}_i\right]. \tag{15.87}$$

Finally, we anticipate special relativity (Section 22.5.2) and recognize that $\mathbf{p}_i = (\mathbf{v}_i/c^2)\mathcal{E}_i$. This identifies the quantity in square brackets as the total linear momentum of the system and puts (15.87) in a form known as the *center-of-energy theorem*:

$$\mathbf{v}_{\mathrm{E}} = \frac{d\mathbf{R}_{\mathrm{E}}}{dt} = \frac{c^2 \mathbf{P}_{\mathrm{tot}}}{U_{\mathrm{tot}}}. \tag{15.88}$$

[14] See Section 1.8.

Equation (15.88) guarantees that the center of energy of an isolated system remains at rest when its total linear momentum vanishes.

Application 15.3 The Hidden Momentum of a Static System

Figure 15.5 shows a point charge q and a current loop I, both at rest. In this Application, we use the center-of-energy theorem to reveal a "hidden" source of linear momentum in this static system. We begin with the electromagnetic field momentum, and derive a general expression for static fields which exploits $\mathbf{E} = -\nabla\varphi$, $\nabla \times \mathbf{B} = \mu_0 \mathbf{j}$, and an integration by parts where the surface integral vanishes for a localized distribution of charge and current. Specifically,

$$\mathbf{P}_{\text{EM}} = \epsilon_0 \int d^3r\, \mathbf{E} \times \mathbf{B} = -\epsilon_0 \int d^3r\, \nabla\varphi \times \mathbf{B} = \epsilon_0 \int d^3r\, \varphi(\nabla \times \mathbf{B}) = \frac{1}{c^2} \int d^3r\, \varphi \mathbf{j}. \quad (15.89)$$

Figure 15.5: A current loop and a point charge at rest.

If the electric field is nearly constant over the area of the loop, it is a good approximation to insert

$$\varphi(\mathbf{r}) = \varphi(\mathbf{r}_0) - (\mathbf{r} - \mathbf{r}_0) \cdot \mathbf{E}_0 + \cdots \quad (15.90)$$

into the rightmost member of (15.89). We then use (11.6) and (11.11), namely

$$\int d^3r\, j_k = 0 \qquad \text{and} \qquad \int d^3r\, j_k r_\ell = \epsilon_{ki\ell} m_i, \quad (15.91)$$

where m_i is the ith Cartesian component of the magnetic moment of the current distribution. The final result is

$$\mathbf{P}_{\text{EM}} = \frac{\mathbf{E}_0 \times \mathbf{m}}{c^2}. \quad (15.92)$$

The expressions (15.89) and (15.92) are generally non-zero. This is disquieting because our intuition [and the center-of-energy theorem (15.88)] tells us that the total linear momentum (electromagnetic + mechanical) of a static situation like Figure 15.5 should vanish. A sensible place to look for the momentum needed to cancel $\mathbf{P}_{\text{EM}}$ is the moving particles that constitute the current. However, if all the particles have the same charge and mass, the left equation in (15.91) implies that the total particle momentum is also zero:

$$\int d^3r\, \mathbf{j} = \sum_k q\mathbf{v}_k = 0 = \sum_k m\mathbf{v}_k = \mathbf{P}^{\text{NR}}_{\text{mech}}. \quad (15.93)$$

The superscript in $\mathbf{P}^{\text{NR}}_{\text{mech}}$ is a reminder that the momentum calculated in (15.93) is non-relativistic. The source of the "missing" or "hidden momentum" in Figure 15.5 emerges when we recognize that the exact, relativistic momentum expression multiplies $m\mathbf{v}_k$ by the factor $\gamma_k = 1/\sqrt{1 - v_k^2/c^2}$ (see Section 22.5.2). The effect of this change is dramatic, even when all the velocities satisfy $v_k \ll c$. To see this, we focus on *one* particle in the current loop and break its trajectory into infinitesimal segments, each with velocity $\mathbf{v}_\alpha = \delta\boldsymbol{\ell}_\alpha/\delta t$, where $T = N\delta t$ is the total time needed to traverse the loop. The average current contributed by this particle is $I = q/T$, so the time-averaged linear momentum associated with its motion is not zero but

$$\mathbf{P}_{\text{mech}} = \frac{1}{T}\sum_{\alpha=1}^{N} \gamma_\alpha m \mathbf{v}_\alpha \, \delta t = \frac{1}{T}\sum_{\alpha=1}^{N} \gamma_\alpha m \delta \boldsymbol{\ell}_\alpha = \frac{I}{q}\oint \gamma m d\boldsymbol{\ell}. \tag{15.94}$$

Now, at any point on its trajectory, conservation of energy fixes the sum of the kinetic and (electrostatic) potential energies at

$$\gamma mc^2 + q\varphi = \mathcal{E}_{\text{tot}}. \tag{15.95}$$

Hence, using $\oint d\boldsymbol{\ell} = 0$ to add zero to the last term in (15.94) gives

$$\mathbf{P}_{\text{mech}} = -\frac{I}{c^2}\oint d\boldsymbol{\ell}\,\frac{\mathcal{E}_{\text{tot}} - \gamma mc^2}{q} = -\frac{I}{c^2}\oint d\boldsymbol{\ell}\,\varphi. \tag{15.96}$$

Every particle in the current loop contributes similarly and we deduce that (15.96) is the filamentary-current version of a general expression for the (hidden) mechanical momentum, $\mathbf{P}_{\text{hid}}$, of a stationary current distribution:

$$\mathbf{P}_{\text{mech}} = -\frac{1}{c^2}\int d^3r\,\varphi\mathbf{j} = \mathbf{P}_{\text{hid}}. \tag{15.97}$$

Comparing (15.97) with (15.89) proves that the total linear momentum in Figure 15.5 vanishes in accord with the center-of-energy theorem:

$$\mathbf{P}_{\text{EM}} + \mathbf{P}_{\text{mech}} = 0 = \mathbf{P}_{\text{EM}} + \mathbf{P}_{\text{hid}}. \tag{15.98}$$

If the current loop in Figure 15.5 moves with non-relativistic center-of-mass velocity $\mathbf{v}_{\text{CM}}$, $\mathbf{P}_{\text{hid}}$ is still the negative of the electromagnetic momentum computed when $\mathbf{v}_{\text{CM}} = 0$, but (15.98) generalizes to

$$\mathbf{P}_{\text{mech}} = m\mathbf{v}_{\text{CM}} + \mathbf{P}_{\text{hid}}. \tag{15.99}$$

The corresponding force responsible for the motion of the center of mass is

$$\mathbf{F}_{\text{CM}} = m\dot{\mathbf{v}}_{\text{CM}} = \frac{d\mathbf{P}_{\text{mech}}}{dt} - \frac{d\mathbf{P}_{\text{hid}}}{dt} = \mathbf{F}_{\text{mech}} - \frac{d\mathbf{P}_{\text{hid}}}{dt}, \tag{15.100}$$

where $\mathbf{F}_{\text{mech}} = d\mathbf{P}_{\text{mech}}/dt$ is the Coulomb-Lorentz force (15.42). Hence, if the current loop in Figure 15.5 experiences an inhomogeneous magnetic field $\mathbf{B}(\mathbf{r})$, (12.41), (15.92), and the rightmost equation in (15.98) show that the loop accelerates due to the quasistatic force

$$\mathbf{F}_{\text{CM}} = \nabla(\mathbf{m}\cdot\mathbf{B}) - \frac{1}{c^2}\frac{d}{dt}(\mathbf{m}\times\mathbf{E}). \tag{15.101}$$

In this way, the "covert" momentum of the current loop influences its "overt" momentum.[15] ■

15.8 Conservation Laws in Matter

It is a mixed bag to derive conservation laws for arbitrary, time-varying electromagnetic fields in matter. On the one hand, formal manipulations of the in-matter Maxwell equations produce a sensible statement of the conservation of energy in matter and well-accepted expressions for the Poynting

[15] This perspective comes from the very valuable paper by W.H. Furry, "Examples of momentum distributions in the electromagnetic field and in matter", *American Journal of Physics* **37**, 621 (1969).

vector and the electromagnetic energy density in matter. On the other hand, similar manipulations lead to only an *apparent* statement of conservation of linear momentum in matter and there is no single, well-accepted expression for either the momentum of an electromagnetic field in matter or for the time-dependent electromagnetic force density in matter. Part of the problem is familiar from our analysis of static forces in sub-volumes of polarizable and magnetizable matter (Chapter 6 and Chapter 13). The Coulomb-Lorentz force coexists with short-range quantum mechanical forces at the microscopic scale and the latter influences the former at the macroscopic scale in a manner we are obliged to discuss using the dielectric permittivity ϵ and the magnetic permeability μ.

Another problem is the difficulty in distinguishing time-dependent forces in matter from contributions to the transport of electromagnetic field momentum in matter. Decades of theory and experiment addressed to this "Abraham-Minkowksi controversy" have not solved what Nobel laureate Vitaly Ginzburg calls one of the "perpetual problems" of physics.[16] In this chapter, we mimic our discussion of static forces in matter and distinguish carefully between the force exerted by time-dependent fields on an isolated sample of matter and the corresponding force exerted on a sub-volume of matter. The latter lies at the heart of the controversy, which we describe in brief, along with a summary of some recent suggestions for its resolution.

15.8.1 Conservation of Energy in Simple Matter

In Section 15.4, we derived a conservation law for mechanical energy and field energy by straightforward manipulation of the Maxwell equations in vacuum. Here, we apply the same approach to the Maxwell equations in matter:

$$\nabla \cdot \mathbf{D} = \rho_f \qquad\qquad \nabla \cdot \mathbf{B} = 0 \tag{15.102}$$

and

$$\nabla \times \mathbf{E} = -\frac{\partial \mathbf{B}}{\partial t} \qquad\qquad \nabla \times \mathbf{H} = \mathbf{j}_f + \frac{\partial \mathbf{D}}{\partial t}. \tag{15.103}$$

The required algebraic steps are literally the same as for the vacuum case. Therefore, we simply state the analog of Poynting's theorem when matter is present:

$$\int_V d^3r \left[\mathbf{E} \cdot \frac{\partial \mathbf{D}}{\partial t} + \mathbf{H} \cdot \frac{\partial \mathbf{B}}{\partial t} \right] = -\int_V d^3r \, \mathbf{j}_f \cdot \mathbf{E} - \int_V d^3r \, \nabla \cdot (\mathbf{E} \times \mathbf{H}). \tag{15.104}$$

The first term on the right side of (15.104) is unambiguously the mechanical work done by the field on free charges. Therefore, we are entitled to interpret the integral on the left side of (15.104) as the rate of change of electric and magnetic energy. There is similarly no ambiguity if we use the divergence theorem to interpret the last term in (15.104) as the flux of energy through the surface that encloses V. This leads us to generalize the definition of the Poynting vector (15.29) to

$$\mathbf{S} = \mathbf{E} \times \mathbf{H}. \tag{15.105}$$

Simple linear media are defined by $\mathbf{D} = \epsilon \mathbf{E}$ and $\mathbf{B} = \mu \mathbf{H}$. When these formulae are valid, (15.104) implies that the total electromagnetic energy stored in a volume V is

$$U_{\rm EM} = \int_V d^3r \, u_{\rm EM} = \int_V d^3r \, \tfrac{1}{2} \left[\epsilon \, |\mathbf{E}|^2 + \mu \, |\mathbf{H}|^2 \right]. \tag{15.106}$$

[16] See the first paragraph of Chapter 3 of V.L. Ginzburg, *Theoretical Physics and Astrophysics* (Pergamon, Oxford, 1979).

This expression, valid for arbitrary time-dependent fields, is exactly the sum of the electric and magnetic energies we deduced for static fields in linear matter in Section 6.7 and Section 13.7.1, respectively. Then, because the work done on the free charges increases their mechanical energy, we can apply the divergence theorem to the Poynting vector term and combine (15.104), (15.105), and (15.106) to write a global statement for conservation of energy in matter, namely,

$$\frac{d}{dt}(U_{\text{mech}} + U_{\text{EM}}) = -\int_S dA\,\mathbf{S}\cdot\hat{\mathbf{n}}. \tag{15.107}$$

A differential statement of energy conservation in matter follows because (15.104) and (15.107) apply to every possible volume of integration V:

$$\frac{\partial u_{\text{EM}}}{\partial t} + \nabla\cdot\mathbf{S} = -\mathbf{j}_f\cdot\mathbf{E}. \tag{15.108}$$

This is the same as (15.33) except that the definition of each symbol is slightly different when matter is present.

Example 15.5 The AC Stark shift is the change in the energy of an atom produced by an oscillating electric field. Let $\mathbf{E}(t) = \boldsymbol{\mathcal{E}}\cos\omega t$ and compute the Stark shift from the change in the energy of the electric field due to the presence of a gas of atoms.

Solution: Using (15.106), the change in the electric field energy density is

$$\Delta u_E = \frac{1}{2}(\epsilon - \epsilon_0)|\mathbf{E}(t)|^2.$$

The average of $\cos^2(\omega t)$ over one period is $1/2$, so the time-averaged change in the energy density is

$$\langle \Delta u_E \rangle = \frac{1}{4}(\epsilon - \epsilon_0)\mathcal{E}^2.$$

From (6.79) of Section 6.6.2, the dielectric constant of a gas of atoms is

$$\kappa = \frac{\epsilon}{\epsilon_0} = 1 + n_0\alpha,$$

where n_0 is the number density of atoms and α is the polarizability of an atom. Therefore,

$$\langle \Delta u_E \rangle = \frac{1}{4}n_0\alpha\epsilon_0\mathcal{E}^2.$$

By conservation of energy, the change in electric field energy is compensated by a change in the energy of each atom in the gas. Therefore, the expected AC Stark shift per atom is

$$\Delta E_{\text{Stark}} = -\frac{1}{4}\alpha\epsilon_0\mathcal{E}^2.$$

Quantum mechanics predicts the same result.

15.8.2 Conservation of Momentum in Simple Matter

The logic used in Section 15.5.1 to derive a conservation law for linear momentum can be extended to fields in matter by using (15.102) and (15.103) to eliminate ρ_f and $\mathbf{j}_f$ from the Coulomb-Lorentz expression for the force on the free charge and current in any sub-volume V of a sample of matter:

$$\mathbf{F}_{\text{free}} = \int_V d^3r\,[\rho_f\mathbf{E} + (\mathbf{j}_f \times \mathbf{B})] = \int_V d^3r\,(\nabla\cdot\mathbf{D})\mathbf{E} + \left(\nabla\times\mathbf{H} - \frac{\partial\mathbf{D}}{\partial t}\right)\times\mathbf{B}. \tag{15.109}$$

The key trick is to add $(\nabla\cdot\mathbf{B})\mathbf{H} = 0$ and (using Faraday's law) $\mathbf{D}\times(\nabla\times\mathbf{E}) + \mathbf{D}\times\partial\mathbf{B}/\partial t = 0$ to the far right side of (15.109) to get

$$\mathbf{F}_{\text{free}} = \int_V d^3r \left[(\nabla\cdot\mathbf{D})\mathbf{E} + (\nabla\cdot\mathbf{B})\mathbf{H} - \mathbf{B}\times(\nabla\times\mathbf{H}) - \mathbf{D}\times(\nabla\times\mathbf{E}) - \frac{\partial}{\partial t}(\mathbf{D}\times\mathbf{B})\right]. \tag{15.110}$$

The Levi-Cività symbol (1.2.5) facilitates the evaluation of the curl terms and we find that

$$\begin{aligned} F_{\text{free},j} &= -\int_V d^3r\,\frac{\partial}{\partial t}(\mathbf{D}\times\mathbf{B})_j \\ &\quad + \int_V d^3r\,\left[(\nabla\cdot\mathbf{D})E_j + (\mathbf{D}\cdot\nabla)E_j - D_k\partial_j E_k\right] \\ &\quad + \int_V d^3r\,\left[(\nabla\cdot\mathbf{B})H_j + (\mathbf{B}\cdot\nabla)\mathbf{H}_j - B_k\partial_j H_k\right]. \end{aligned} \tag{15.111}$$

We now define the *Minkowski force density*,

$$\mathbf{f}_{\text{M}} = \rho_f\mathbf{E} + \mathbf{j}_f\times\mathbf{B} - \tfrac{1}{2}E^2\nabla\epsilon - \tfrac{1}{2}H^2\nabla\mu, \tag{15.112}$$

and an *electromagnetic stress tensor in matter* $\mathcal{T}$ with components

$$\mathcal{T}_{ij} = D_iE_j + B_iH_j - \frac{1}{2}\delta_{ij}(\mathbf{D}\cdot\mathbf{E} + \mathbf{B}\cdot\mathbf{H}). \tag{15.113}$$

We also restrict ourselves to $\mathbf{D} = \epsilon\mathbf{E}$ and $\mathbf{B} = \mu\mathbf{H}$, so $D_k\partial_jE_k = \frac{1}{2}\left[\partial_j(D_kE_k) - E^2\partial_j\epsilon\right]$ and $B_k\partial_jH_k = \frac{1}{2}\left[\partial_j(B_kH_k) - H^2\partial_j\mu\right]$. This information, together with the identity

$$\int_V d^3r\,\nabla_k(a_k\mathbf{b}) = \int_V d^3r\,(\nabla\cdot\mathbf{a})\,\mathbf{b} + \int_V d^3r\,(\mathbf{a}\cdot\nabla)\,\mathbf{b}, \tag{15.114}$$

permits us to rewrite (15.111) in the form

$$\int_V d^3r\,\mathbf{f}_{\text{M}} = \int_V d^3r\,\nabla\cdot\mathcal{T} - \int_V d^3r\,\frac{\partial}{\partial t}(\mathbf{D}\times\mathbf{B}). \tag{15.115}$$

In 1908, Minkowski interpreted (15.115) as a statement of conservation of linear momentum by asserting that the momentum density associated with an electromagnetic field in matter is

$$\mathbf{g}_{\text{M}} = \mathbf{D}\times\mathbf{B} = \epsilon\mu\mathbf{E}\times\mathbf{H} = \epsilon\mu\mathbf{S}. \tag{15.116}$$

Because V can be chosen arbitrarily small, (15.115) reduces to a continuity equation for the flow of linear momentum:

$$\frac{\partial\mathbf{g}_M}{\partial t} + \nabla\cdot(-\mathcal{T}) = -\mathbf{f}_{\text{M}}. \tag{15.117}$$

The force density $\mathbf{f}_{\text{M}}$ is the sum of the electrostatic and magnetostatic force densities deduced in Section 6.8.4 and Section 13.8.4. Similarly, the stress tensor $\mathcal{T}$ is the sum of the electrostatic and magnetostatic stress tensors defined in Section 6.8.5 and Section 13.8.5. Hence, the static limit of (15.117) reproduces our previous results. Everything seems to be in order, but, as the next section shows, appearances can be deceiving.

15.8.3 The Abraham-Minkowski Controversy

Soon after Minkowski proposed (15.116) as the electromagnetic linear momentum density in matter, Max Abraham suggested that the relationship $\mathbf{g} = \mathbf{S}/c^2$ between the momentum density and the Poynting vector in vacuum [see (15.50)] should remain valid in matter. If this is true, (15.105) implies that the correct expression for the electromagnetic momentum density in matter is

$$\mathbf{g}_{\rm A} = \frac{\mathbf{E} \times \mathbf{H}}{c^2}. \tag{15.118}$$

To incorporate (15.118) into a conservation law, we need only add and subtract the time derivative of $\mathbf{g}_{\rm A}$ on both sides of (15.117) and rearrange the terms slightly. The result is an alternative statement of linear momentum conservation in matter,

$$\frac{\partial \mathbf{g}_A}{\partial t} + \nabla \cdot (-\mathcal{T}) = -\mathbf{f}_{\rm A}, \tag{15.119}$$

where

$$\mathbf{f}_{\rm A} = \mathbf{f}_{\rm M} + \frac{\partial}{\partial t}\left(\mathbf{D} \times \mathbf{B} - \frac{\mathbf{E} \times \mathbf{H}}{c^2}\right) = \mathbf{f}_{\rm M} + \left(\epsilon\mu - \frac{1}{c^2}\right)\frac{\partial \mathbf{S}}{\partial t} \tag{15.120}$$

is an alternative to (15.112) as the electromagnetic force density in matter. The "correction" to the Minkowski force density (15.112) on the far right side of (15.120) vanishes for static fields and also for (time-averaged) harmonic fields because it is a total time derivative.

For more than a century, disputants in the "Abraham-Minkowski controversy" have argued whether (15.116) or (15.118) (or some other expression) is the correct linear momentum density for electromagnetic fields in matter.[17] The controversy lives on because research papers continue to appear (including experiments) which claim to establish the correctness of one particular expression for the momentum density $\mathbf{g}_{\rm EM}$ and the force density $\mathbf{f}$. The truth of the matter is that neither (15.117) nor (15.119) is complete as it stands. To the momentum density, force density, and stress tensor in each, one must add a momentum density, force density, and stress tensor associated solely with the matter. The added terms are not the same in the two cases, but the resulting statement of linear momentum conservation for the total system of field plus matter *is* the same for the two cases.

The controversies in the literature arise mainly from attempts to isolate various pieces of the total conservation law without taking account of the pieces left behind. On the other hand, there is an emerging consensus that both momentum densities are physically meaningful, but that one or the other provides the simplest description of a particular experimental situation. Thus, the transfer of Minkowski momentum (15.116) correctly describes the force exerted by an electromagnetic wave on an object embedded in a dielectric medium while the Abraham momentum (15.118) naturally describes the kinetic momentum of a wave propagating freely through a dielectric medium.[18]

15.9 The Force on Isolated Matter

We derive an expression for the electromagnetic force $\mathbf{F}$ exerted on an isolated sample of matter characterized by a polarization $\mathbf{P}$, a magnetization $\mathbf{M}$, and free charge and current densities ρ_f and $\mathbf{j}_f$ by integrating the Coulomb-Lorentz force density $\rho\mathbf{E}_0 + \mathbf{j} \times \mathbf{B}_0$ over the volume V of the sample

[17] A list of the distinguished contributors to this debate includes Einstein, von Laue, Planck, Pauli, Casimir, Peierls, Shockley, and Ginzburg.

[18] See Sources, References, and Additional Reading at the end of the chapter.

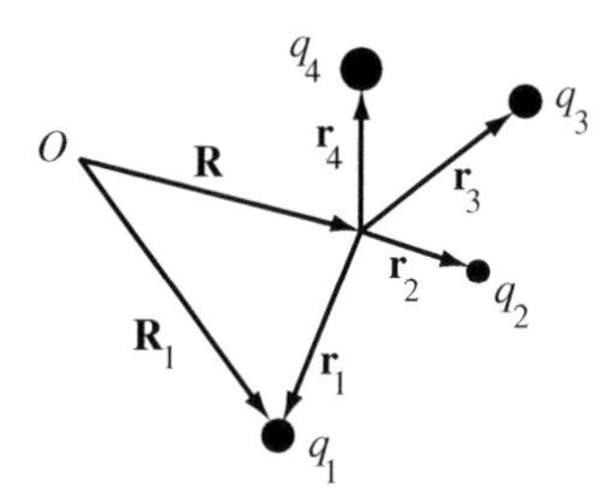

Figure 15.6: A collection of moving point particles with charges q_α which never stray far from their center of mass $\mathbf{R}$. The vector $\mathbf{r}_\alpha$ points from the center of mass to the position of the αth particle.

using $\rho = \rho_f - \nabla \cdot \mathbf{P}$ and $\mathbf{j} = \mathbf{j}_f + \nabla \times \mathbf{M} + \partial \mathbf{P}/\partial t$. In other words,

$$\mathbf{F} = \int_V d^3r \, [\rho_f \mathbf{E}_0 + \mathbf{j}_f \times \mathbf{B}_0] + \int_V d^3r \left\{ -(\nabla \cdot \mathbf{P})\mathbf{E}_0 + \left[\nabla \times \mathbf{M} + \frac{\partial \mathbf{P}}{\partial t} \right] \times \mathbf{B}_0 \right\}. \tag{15.121}$$

The $\nabla \cdot \mathbf{P}$ and $\nabla \times \mathbf{M}$ terms in (15.121) can be transformed using the vector identities

$$(\mathbf{P} \cdot \nabla)\mathbf{E}_0 = \partial_k (P_k \mathbf{E}_0) - (\nabla \cdot \mathbf{P})\mathbf{E}_0 \tag{15.122}$$

$$\left[(\mathbf{M} \times \nabla) \times \mathbf{B}_0 \right]_i = \partial_i (\mathbf{M} \cdot \mathbf{B}_0) - B_{0,k} \partial_i M_k \tag{15.123}$$

$$\left[(\nabla \times \mathbf{M}) \times \mathbf{B}_0 \right]_i = \partial_k (B_k M_i) - (\nabla \cdot \mathbf{B}_0) M_i - B_{0,k} \partial_i M_k. \tag{15.124}$$

The volume integrals of the total derivative terms (the first term on the right-hand side of each of the three preceding equations) vanish because we can expand the volume of integration so its surface lies everywhere in the vacuum (where $\mathbf{P} = \mathbf{M} = 0$) and transform these volume integrals to surface integrals. Therefore, because $\nabla \cdot \mathbf{B}_0 = 0$, the force (15.121) on an isolated sample of matter takes the form

$$\mathbf{F} = \int_V d^3r \, [\rho_f \mathbf{E}_0 + \mathbf{j}_f \times \mathbf{B}_0] + \int_V d^3r \left[(\mathbf{P} \cdot \nabla)\mathbf{E}_0 + \frac{\partial \mathbf{P}}{\partial t} \times \mathbf{B}_0 + (\mathbf{M} \times \nabla) \times \mathbf{B}_0 \right]. \tag{15.125}$$

The $(\mathbf{P} \cdot \nabla)\mathbf{E}_0$ term in (15.125) is familiar from (6.116) as the force on polarized matter regarded as a continuous distribution of point electric dipoles. However, the $(\mathbf{M} \times \nabla) \times \mathbf{B}_0$ term does not resemble the force (13.120) on magnetized matter regarded as a distribution of point magnetic dipoles. The reason for this, and the origin of the force on the polarization current density, can be appreciated from the calculation presented in the next section.

15.9.1 The Force on a Classical Atom

A single atom or molecule is a special case of an isolated sample of matter. In this section, we gain insight into (15.125) by using a low-order multipole expansion to compute the net force on a "classical atom" regarded as an electrically neutral collection of N moving point particles. We assume that none of the particles strays very far from the center of mass $\mathbf{R}$ which moves with velocity $\mathbf{v} = \dot{\mathbf{R}}$ (see Figure 15.6). Each particle has charge q_α, position $\mathbf{R}_\alpha = \mathbf{R} + \mathbf{r}_\alpha$, and velocity $\dot{\mathbf{R}}_\alpha = \mathbf{v} + \dot{\mathbf{r}}_\alpha$.

The Coulomb-Lorentz force on the atom due to external fields $\mathbf{E}_0(\mathbf{r}, t)$ and $\mathbf{B}_0(\mathbf{r}, t)$ is

$$\mathbf{F} = \sum_\alpha q_\alpha \left[\mathbf{E}_0(\mathbf{R}_\alpha) + \dot{\mathbf{R}}_\alpha \times \mathbf{B}_0(\mathbf{R}_\alpha) \right]. \tag{15.126}$$

By assumption, $|\mathbf{r}_\alpha| \ll |\mathbf{R}|$, so it reasonable to perform a Taylor expansion of the external fields $\mathbf{E}_0(\mathbf{R}_\alpha)$ and $\mathbf{B}_0(\mathbf{R}_\alpha)$ around $\mathbf{R}_\alpha = \mathbf{R}$ and keep only the first two terms in each expansion:

$$\mathbf{E}_0(\mathbf{R}_\alpha) = \mathbf{E}_0(\mathbf{R}) + (\mathbf{r}_\alpha \cdot \nabla_\mathbf{R})\mathbf{E}_0(\mathbf{R}) \quad \text{and} \quad \mathbf{B}_0(\mathbf{R}_\alpha) = \mathbf{B}_0(\mathbf{R}) + (\mathbf{r}_\alpha \cdot \nabla_\mathbf{R})\mathbf{B}_0(\mathbf{R}). \tag{15.127}$$

Substituting (15.127) into (15.126) and writing $\mathbf{E}_0$, $\mathbf{B}_0$, and ∇ for $\mathbf{E}_0(\mathbf{R})$, $\mathbf{B}_0(\mathbf{R})$, and $\nabla_\mathbf{R}$ gives

$$\begin{aligned}\mathbf{F} &= \sum_\alpha q_\alpha (\mathbf{E}_0 + \mathbf{v}\times\mathbf{B}_0) + \sum_\alpha q_\alpha(\mathbf{r}_\alpha\cdot\nabla)\mathbf{E}_0 + \sum_\alpha q_\alpha(\mathbf{r}_\alpha\cdot\nabla)(\mathbf{v}\times\mathbf{B}_0)\\ &+ \sum_\alpha q_\alpha \dot{\mathbf{r}}_\alpha \times \mathbf{B}_0 + \sum_\alpha q_\alpha(\mathbf{r}_\alpha\cdot\nabla)(\dot{\mathbf{r}}_\alpha\times\mathbf{B}_0).\end{aligned} \tag{15.128}$$

This expression simplifies if we recall the definitions of total charge q and electric dipole moment vector $\mathbf{p}$ from Section 4.1.1:

$$q = \int d^3r\,\rho = \sum_\alpha q_\alpha \qquad \mathbf{p} = \int d^3r\,\mathbf{r}\rho = \sum_\alpha q_\alpha r_\alpha. \tag{15.129}$$

We have assumed that $q = 0$, so (15.129) reduces (15.128) to

$$\mathbf{F} = (\mathbf{p}\cdot\nabla)\mathbf{E}_0 + \dot{\mathbf{p}}\times\mathbf{B}_0 + (\mathbf{p}\cdot\nabla)(\mathbf{v}\times\mathbf{B}_0) + \sum_\alpha q_\alpha(\mathbf{r}_\alpha\cdot\nabla)(\dot{\mathbf{r}}_\alpha\times\mathbf{B}_0). \tag{15.130}$$

To simplify the last term in (15.130), add

$$(\mathbf{r}_\alpha\times\dot{\mathbf{r}}_\alpha)\times\nabla = \dot{\mathbf{r}}_\alpha(\mathbf{r}_\alpha\cdot\nabla) - \mathbf{r}_\alpha(\dot{\mathbf{r}}_\alpha\cdot\nabla) \tag{15.131}$$

to

$$\frac{d}{dt}(\mathbf{r}_\alpha\mathbf{r}_\alpha)\cdot\nabla = \dot{\mathbf{r}}_\alpha(\mathbf{r}_\alpha\cdot\nabla) + \mathbf{r}_\alpha(\dot{\mathbf{r}}_\alpha\cdot\nabla) \tag{15.132}$$

to get

$$\dot{\mathbf{r}}_\alpha(\mathbf{r}_\alpha\cdot\nabla) = \frac{1}{2}\frac{d}{dt}(\mathbf{r}_\alpha\mathbf{r}_\alpha)\cdot\nabla + \frac{1}{2}(\mathbf{r}_\alpha\times\dot{\mathbf{r}}_\alpha)\times\nabla = (\mathbf{r}_\alpha\cdot\nabla)\dot{\mathbf{r}}_\alpha. \tag{15.133}$$

In addition, recall the definitions of the electric quadrupole moment tensor (Section 4.4) and the magnetic dipole moment vector (Section 11.2),

$$\mathbf{Q} = \frac{1}{2}\int d^3r\,\mathbf{r}\rho\mathbf{r} = \frac{1}{2}\sum_\alpha q_\alpha\mathbf{r}_\alpha\mathbf{r}_\alpha \qquad \mathbf{m} = \frac{1}{2}\int d^3r\,\mathbf{r}\times\mathbf{j} = \frac{1}{2}\sum_\alpha q_\alpha\mathbf{r}_\alpha\times\dot{\mathbf{r}}_\alpha. \tag{15.134}$$

Combining (15.133) and (15.134) with $[\mathbf{Q}\cdot\nabla]_i = Q_{ik}\partial_k$ shows that (15.130) is

$$\mathbf{F} = (\mathbf{p}\cdot\nabla)\mathbf{E}_0 + \dot{\mathbf{p}}\times\mathbf{B}_0 + (\mathbf{p}\cdot\nabla)(\mathbf{v}\times\mathbf{B}_0) + (\mathbf{m}\times\nabla)\times\mathbf{B}_0 + \left[\frac{d\mathbf{Q}}{dt}\cdot\nabla\right]\times\mathbf{B}_0. \tag{15.135}$$

When the center-of-mass velocity $\mathbf{v} = 0$, the electric dipole terms in (15.135) reproduce the electric polarization terms in (15.125) and the magnetic dipole term in (15.135) reproduces the magnetization term in (15.125). We will analyze the electric force in more detail when we discuss the phenomenon of "optical tweezers" in Section 16.10.3. Otherwise, (15.135) is often rewritten using the fact that the dipole moments $\mathbf{p}$ and $\mathbf{m}$ of an atom are functions of time but not space. Moreover, the "atom" in Figure 15.6 moves through the electric and magnetic fields at velocity $\mathbf{v}$. Therefore (see Section 1.3.3),

$$\frac{d\mathbf{B}_0}{dt} = \frac{\partial\mathbf{B}_0}{\partial t} + (\mathbf{v}\cdot\nabla)\mathbf{B}_0. \tag{15.136}$$

With this information, the reader can use a few vector identities and Faraday's law to rewrite the dipole terms in (15.135) in the form

$$\mathbf{F}_{\text{atom}} = \nabla(\mathbf{p}\cdot\mathbf{E}_0) + \nabla(\mathbf{m}\cdot\mathbf{B}_0) + \frac{d}{dt}(\mathbf{p}\times\mathbf{B}_0) + \nabla[\mathbf{p}\cdot(\mathbf{v}\times\mathbf{B}_0)]. \tag{15.137}$$

The first two terms of (15.137) are familiar as the negative gradients of the potential energies of an electric/magnetic dipole in an external electric/magnetic field. The third term is not new either. As a macroscopic density, it appears as the first term in the "Abraham force" in (15.120) because $\mathbf{D} = \epsilon_0\mathbf{E} + \mathbf{P}$ and $\mathbf{B} = \mu_0(\mathbf{H} + \mathbf{M})$ show that

$$\mathbf{D} \times \mathbf{B} - \frac{\mathbf{E} \times \mathbf{H}}{c^2} = \mathbf{P} \times \mathbf{B} + \frac{\mathbf{E} \times \mathbf{M}}{c^2}. \tag{15.138}$$

The rightmost term in (15.138) does not occur in (15.137). However, it would have if we had manipulated the relativistic equation of motion for our classical atom as we did for the current loop to get (15.101).

The final term in (15.137) contributes only when the atom is in motion. As written, it is similar to the first term in (15.137) with a "motional electric field" $\mathbf{v} \times \mathbf{B}_0$. Atomic physicists speak of this as a "Röntgen term" and rewrite it as $\nabla[(\mathbf{p} \times \mathbf{v}) \cdot \mathbf{B}_0]$ to emphasize that a moving electric dipole moment behaves like a magnetic moment.[19] The name honors the 19th-century discoverer of x-rays who (earlier in his career) sought to detect the Maxwell displacement current by detecting the magnetic field produced by a rapidly rotating disk of polarized matter.

Sources, References, and Additional Reading

The quotation at the beginning of the chapter is taken from Section 74 of

M. Planck, *The Theory of Electricity and Magnetism* 2nd edition (Macmillan, London, 1932).

Section 15.1 Three textbooks and a monograph which discuss conservation laws without the use of Lagrangian methods are

R.H. Good, Jr. and T.J. Nelson, *Classical Theory of Electric and Magnetic Fields* (Academic, New York, 1971).

L. Eyges, *The Classical Electromagnetic Field* (Dover, New York, 1972).

D.J. Griffiths, *Introduction to Electrodynamics*, 3rd edition (Prentice-Hall, Upper Saddle River, NJ, 1999).

W.G.V. Rosser, *Interpretation of Classical Electromagnetism* (Kluwer, Dordrecht, 1997).

Section 15.2 The importance of symmetry in physics can hardly be understated. An annotated bibliography which provides an entry point to the vast literature on this subject is

J. Rosen, "Symmetry and group theory in physics", *American Journal of Physics* **49**, 304 (1981).

For more (and then much more) on symmetry in electrodynamics, see

J.D. Jackson, *Classical Electrodynamics*, 3rd edition (Wiley, New York, 1999).

W.I. Fushchich and A.G. Nitikin, *Symmetries of Maxwell's Equations* (Reidel, Dordrecht, 1987).

An exotic symmetry of the Maxwell equations which generalizes the idea of conformal mapping to three space dimensions is discussed in

C. Codirla and H. Osborn, "Conformal invariance and electrodynamics", *Annals of Physics* **260**, 91 (1997).

Section 15.3 These two papers are a review of the concept of gauge invariance and a detailed examination of electromagnetic gauge transformations:

J.D. Jackson and L.B. Okun, "Historical roots of gauge invariance", *Reviews of Modern Physics* **73**, 663 (2001).

J.D. Jackson, "From Lorenz to Coulomb and other explicit gauge transformations", *American Journal of Physics* **70**, 917 (2002).

Wigner's argument in Section 15.3.1 comes from

E. Wigner, "Invariance in physical theory", *Proceedings of the American Philosophical Society* **93**, 521 (1949).

[19] We shall return to this point when we study special relativity. See also Section 14.2.2.

The significance of the electromagnetic potentials in quantum mechanics is reviewed in

S. Olariu and I.I. Popescu, "The quantum effects of electromagnetic fluxes", *Reviews of Modern Physics* **57**, 339 (1985).

Section 15.4 Application 15.1 was adapted from Section 9.2 of

J.A. Stratton, *Electromagnetic Theory* (McGraw-Hill, New York, 1941).

The discussion of Poynting's theorem for resistive wires in Section 15.4.2 was inspired by the original paper on the subject:

J.H. Poynting, "On the transfer of energy in the electromagnetic field", *Philosophical Transactions of the Royal Society of London* **175**, 343 (1884).

The suggestion that the Poynting vector is something other than $\mathbf{S} = \mathbf{E} \times \mathbf{B}/\mu_0$ appears periodically in the literature. See, for example,

U. Backhaus and K. Schäfer, "On the uniqueness of the vector for energy flow density in electromagnetic fields", *American Journal of Physics* **54**, 279 (1986).

Section 15.5 The force and momentum analysis of the situation depicted in Figure 15.3 was adapted from *Labarthe*. Application 15.2 and Example 15.3 come from *McDonald* and *Müller-Kirsten*, respectively:

J.-J. Labarthe, "The vector potential of a moving charge in the Coulomb gauge", *European Journal of Physics* **20**, L31 (1999).

K.T. McDonald, "Electromagnetic momentum of a capacitor in a uniform magnetic field", http://cosmology.princeton.edu/~mcdonald/examples/cap_momentum.pdf

H.J.W. Müller-Kirsten, *Electrodynamics: An Introduction Including Quantum Effects* (World Scientific, Hackensack, NJ, 2004).

Section 15.6 Example 15.4 comes from *Sharma*. Einstein and de Haas famously measured the gyromagnetic ratio of the electron by measuring the angular momentum acquired by an iron cylinder when a current was used to magnetize it. *Galison* engagingly tells the story of how and why they got the wrong answer!

N.L. Sharma, "Field versus action-at-a-distance in a static situation", *American Journal of Physics* **56**, 420 (1988).

P. Galison, *How Experiments End* (University Press, Chicago, 1987).

Section 15.7 This section follows the treatment in

J. Schwinger, L.L. DeRaad, K.A. Milton, and W.-Y. Tsai, *Classical Electrodynamics* (Perseus, Reading, MA, 1998).

Application 15.3 was adapted from *Calkin*. With a clarity characteristic of his many "hidden momentum" papers, *Hnizdo* discusses the recovery of the action-reaction law for quasistatic systems and the general absence of "hidden angular momentum":

M.G. Calkin, "Linear momentum of the source of a static electromagnetic field", *American Journal of Physics* **39**, 513 (1971).

V. Hnizdo, "Conservation of linear momentum and angular momentum and the interaction of a moving charge with a magnetic dipole", *American Journal of Physics* **60**, 242 (1992).

For more about the center-of-energy theorem and its uses see

T.T. Taylor, "Electrodynamics paradox and the center-of-mass principle", *Physical Review* **137**, B467 (1965).

T.H. Boyer, "Illustrations of the relativistic conservation law for the center of energy", *American Journal of Physics* **73**, 953 (2005).

Section 15.8 Example 15.5 comes from

B.J. Sussman, "Five ways to the non-resonant dynamic Stark effect", *American Journal of Physics* **79**, 477 (2011).

Synge is a light-hearted reflection on the Abraham-Minkowski controversy. The other two articles bring the discussion up to date.

J.L. Synge, "On the present status of the electromagnetic energy tensor", *Hermathena* **117**, 80 (1974).

R.N.C. Pfeiffer, T.A. Nieminen, N.R. Heckenberg, and H. Rubinsztein-Dunlop, "Momentum of an electromagnetic wave in dielectric media", *Reviews of Modern Physics* **79**, 1197 (2007).

S.M. Barnett and R. Loudon, "The enigma of optical momentum in a medium", *Philosophical Transactions of the Royal Society of London A* **368**, 927 (2010).

Section 15.9 The multipole treatment of forces on atoms and matter in Section 15.9.1 was inspired by a similar discussion in *Schwinger et al.* (see Section 15.7 above) and in

S.R. DeGroot and L.G. Suttorp, *Foundations of Electrodynamics* (North-Holland, Amsterdam, 1972).

A short and readable account of the role of Wilhelm Röntgen in the history of electromagnetism is

P. Dawson, "Röntgen's other experiment", *The British Journal of Radiology* **70**, 809 (1997).

Problems

15.1 Continuous Creation An early competitor of the Big Bang theory postulates the "continuous creation" of charged matter at a (very small) constant rate R at every point in space. In such a theory, the continuity equation is replaced by

$$\nabla \cdot \mathbf{j} + \frac{\partial \rho}{\partial t} = R.$$

(a) For this to be true, it is necessary to alter the source terms in the Maxwell equations. Show that it is sufficient to modify Gauss' law to

$$\nabla \cdot \mathbf{E} = \rho / \epsilon_0 - \lambda \varphi$$

and the Ampère-Maxwell law to

$$\nabla \times \mathbf{B} = \mu_0 \mathbf{j} + \frac{1}{c^2}\frac{\partial \mathbf{E}}{\partial t} - \lambda \mathbf{A}.$$

Here, λ is a constant and φ and $\mathbf{A}$ are the usual scalar and vector potentials. Is this theory gauge invariant?

(b) Confirm that a spherically symmetric solution of the new equations exists with

$$\mathbf{A}(r,t) = \mathbf{r} f(r,t) \qquad \text{and} \qquad \varphi(\mathbf{r},t) = \varphi_0$$

where $f(r,t)$ is a scalar function and φ_0 is a constant.

(c) Show that the only non-singular solution to the partial differential equation satisfied by $f(r,t)$ is a constant.

(d) Show that the velocity of the charge created by this theory, $\mathbf{v} = \mathbf{j}/\rho$, is a linear function of r. This agrees with Hubble's famous observations.

15.2 Lorenz Gauge Forever

(a) Suppose φ_L and $\mathbf{A}_L$ satisfy the Lorenz gauge constraint. What equation must Λ satisfy to ensure that $\mathbf{A}' = \mathbf{A}_L - \nabla\Lambda$ and $\varphi' = \varphi_L + \dot{\Lambda}$ are Lorenz gauge potentials also?

(b) What equation must Λ satisfy to ensure that φ' satisfies the same inhomogeneous wave equation as φ_L? Show that the same equation for Λ' also ensures that $\mathbf{A}'$ satisfies the same inhomogeneous wave equation as $\mathbf{A}_L$.

15.3 Gauge Invariant Vector Potential The Helmholtz theorem guarantees that the vector potential can be decomposed in the form $\mathbf{A} = \mathbf{A}_\perp + \mathbf{A}_\parallel$ where $\nabla \cdot \mathbf{A}_\perp = 0$ and $\nabla \times \mathbf{A}_\parallel = 0$. Show that a gauge invariant physical observable can be expressed in terms of $\mathbf{A}_\perp$, but not $\mathbf{A}_\parallel$.

15.4 Transverse Current Density in the Coulomb Gauge Show that the Cartesian components of the transverse current density $\mathbf{j}_\perp(\mathbf{r},t)$ used to define the Coulomb gauge vector potential can be written in terms of the total current density $\mathbf{j}(\mathbf{r},t)$ as

$$j_{\perp,k}(\mathbf{r},t) = \int d^3r'\, \delta^\perp_{k\ell}(\mathbf{r}-\mathbf{r}')\, j_\ell(\mathbf{r}',t),$$

where the *transverse delta function* is

$$\delta^\perp_{k\ell}(\mathbf{r}-\mathbf{r}') = -\frac{1}{4\pi}\left(\delta_{k\ell}\nabla^2 - \nabla_k\nabla_\ell\right)\frac{1}{|\mathbf{r}-\mathbf{r}'|}.$$

15.5 Poincaré Gauge

(a) Confirm that $\varphi(\mathbf{r}) = -\mathbf{r}\cdot\mathbf{E}$ and $\mathbf{A} = -\frac{1}{2}\mathbf{r}\times\mathbf{B}$ are acceptable scalar and vector potentials, respectively, for a constant electric field $\mathbf{E}$ and a constant magnetic field $\mathbf{B}$.

(b) By direct computation of $\mathbf{B} = \nabla\times\mathbf{A}$ and $\mathbf{E} = -\nabla\varphi - \partial\mathbf{A}/\partial t$, prove that the generalizations of the formulae in part (a) to arbitrary time-dependent fields are

$$\varphi(\mathbf{r}, t) = -\mathbf{r}\cdot\int_0^1 d\lambda\, \mathbf{E}(\lambda\mathbf{r}, t) \qquad\qquad \mathbf{A}(\mathbf{r}, t) = -\int_0^1 d\lambda\, \lambda\mathbf{r}\times\mathbf{B}(\lambda\mathbf{r}, t).$$

Hint: Prove first that $\dfrac{d}{d\lambda}\mathbf{G}(\lambda\mathbf{r}) = \dfrac{1}{\lambda}\,(\mathbf{r}\cdot\nabla)\,\mathbf{G}(\lambda\mathbf{r})$ for any vector field $\mathbf{G}$.

15.6 First-Order Equations for Numerical Electrodynamics Second derivatives are difficult to calculate numerically with high accuracy. Therefore, if both the fields and the potentials (Lorenz gauge) are of interest, a convenient equation to integrate is

$$\frac{\partial\mathbf{A}}{\partial t} = -\mathbf{E} - \nabla\varphi.$$

(a) Let $\mathcal{C}(\mathbf{r}, t) = \nabla\cdot\mathbf{E} - \rho/\epsilon$ and let the initial conditions satisfy $\mathcal{C}(\mathbf{r}, t = 0) = 0$. If this Gauss' law condition is maintained, show that the equation above combined with the two equations below produces fields which satisfy all four Maxwell equations with properly defined potentials:

$$\frac{1}{c^2}\frac{\partial\mathbf{E}}{\partial t} = \nabla\times(\nabla\times\mathbf{A}) - \mu_0\mathbf{j} \qquad \text{and} \qquad \frac{\partial\varphi}{\partial t} = -c^2\nabla\cdot\mathbf{A}.$$

(b) Show that the three equations above imply that $\partial\mathcal{C}/\partial t = 0$. Hence, any initial differences from zero (due to numerical noise) are frozen onto the computational grid (which is not a good thing).

(c) Show that the two equations in (a) can be replaced by $\dot{\varphi} = -c^2\Gamma$ with

$$\frac{1}{c^2}\frac{\partial\mathbf{E}}{\partial t} = -\nabla^2\mathbf{A} + \nabla\Gamma - \mu_0\mathbf{j} \qquad \text{and} \qquad \frac{\partial\Gamma}{\partial t} = -\rho/\epsilon_0 - \nabla^2\varphi.$$

(d) Show that $\dot{\mathbf{A}} = -\mathbf{E} - \nabla\varphi$ and the three equations in part (c) imply that $\partial^2\mathcal{C}/\partial t^2 = c^2\nabla^2\mathcal{C}$. Hence, any initial differences from zero propagate out of the computational grid at the speed of light. For this reason, set (c) is preferred to set (a) for numerical work.

15.7 Elementary Energy Conservation An ohmic bar with mass m slides without friction on two parallel, perfectly conducting rails. A uniform magnetic field $\mathbf{B}$ points into the page as shown. Let R be the resistance of the bar over the length ℓ. Prove that the initial kinetic energy of the bar, $\frac{1}{2}mv_0^2$, is completely dissipated into Joule heat as $t\to\infty$.

15.8 The Poynting Vector Field A point charge q sits at $(a, 0, 0)$, a point charge $-q$ sits at $(-a, 0, 0)$, and a uniform magnetic field $\mathbf{B} = B\hat{\mathbf{z}}$ fills all of space. (a) Prove that the streamlines of the Poynting vector either close on themselves or begin and end at infinity. (b) Sketch several representative streamlines of the Poynting vector, including the one which passes through the origin of coordinates.

15.9 Poynting Vector Matching Condition How do the normal and tangential components of the Poynting vector in matter behave at an interface between two simple media where no free current flows? Comment on the physical meaning of your result.

15.10 A Poynting Theorem Check A flat and infinitely large sheet with uniform charge density σ moves with constant speed $\boldsymbol{v}$ in a direction parallel to its surface. Confirm the differential form of Poynting's theorem at every point not on the sheet.

15.11 A Charged Particle in Static Fields A particle with charge q and mass m produces fields $\mathbf{E}_q$ and $\mathbf{B}_q$ as it moves in static fields $\mathbf{E}_0(\mathbf{r})$ and $\mathbf{B}_0(\mathbf{r})$. Let $\mathbf{j}_0(\mathbf{r})$ be the current density associated with the static fields and write out Poynting's theorem for the total fields assuming that no radiation is produced (all fields fall off more rapidly that $1/r$).

(a) Argue that the div $\mathbf{S}$ term is absent from the theorem and that only cross terms like $\mathbf{E}_0 \cdot \mathbf{E}_q$, $\mathbf{B}_0 \cdot \mathbf{B}_q$, $\mathbf{j}_0 \cdot \mathbf{E}_q$, and $\mathbf{j}_q \cdot \mathbf{E}_0$ contribute.

(b) Let $\mathbf{E}_0 = -\nabla\varphi_0$, use the work energy theorem, and rewrite the magnetic energy to show that the entire content of the Poynting theorem amounts to the statement, $\frac{1}{2}mv^2 + q\varphi_0(\mathbf{r}_q) = const.$

15.12 Energy Flow in a Coaxial Cable A cable is made from two coaxial cylindrical shells. The outer shell has radius b and charge per unit length λ, and carries a longitudinal current I. The inner cylinder has radius $a < b$ and charge per unit length $-\lambda$, and carries the current I back in the opposite direction.

(a) Integrate the Poynting vector to find the rate at which energy flows through a cross section of the cable.

(b) Show that a resistor R connected between the cylinders dissipates the power calculated in (a).

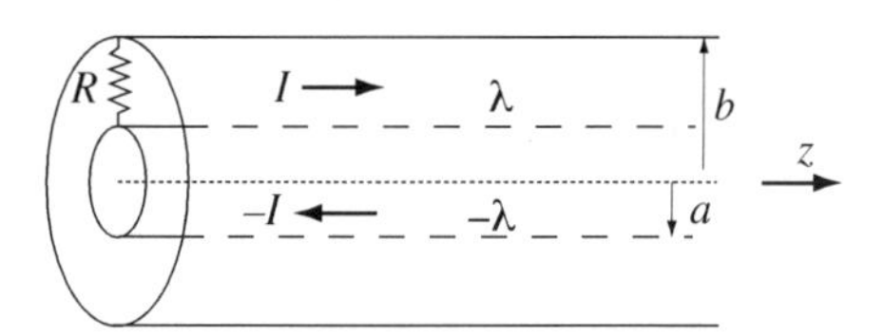

15.13 Energy Conservation for Quasi-Magnetostatic Fields A magnetic field $\mathbf{B}(z,t) = \hat{\mathbf{y}}B_0 \cos\omega t$ is applied just outside the lower ($z = 0$) surface of a semi-infinite slab of ohmic material which extends to $z = \infty$.

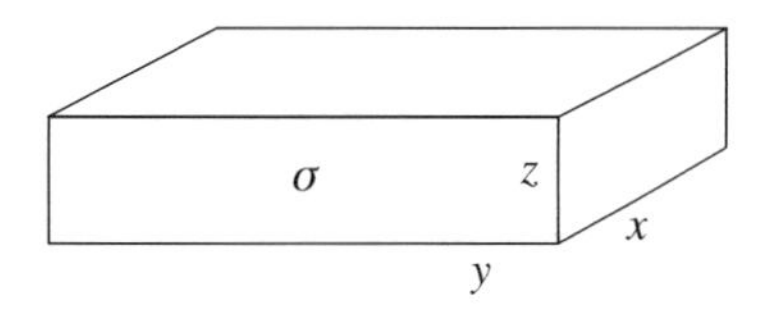

(a) Find the quasi-magnetostatic electric and magnetic fields inside the conductor.

(b) Show by explicit calculation that these fields satisfy the Poynting theorem,

$$\int_V d^3r \left[\frac{\partial u_{\mathrm{EM}}}{\partial t} + \mathbf{j} \cdot \mathbf{E}\right] = -\int_S dA\hat{\mathbf{n}} \cdot \mathbf{S},$$

when averaged over one period of the field oscillation.

15.14 Energy to Spin Up a Charged Cylinder The angular velocity $\omega(t)\hat{\mathbf{z}}$ of the cylindrical shell shown below increases from zero and smoothly approaches the steady value ω_0. The shell has infinitesimal thickness and carries a uniform charge per unit length $\lambda = 2\pi R\sigma$, where σ is a uniform charge per unit area. Assume that the shell radius $R \ll L$ and that the spin-up is very slow so the displacement current may be neglected.

R
λ=2πRσ
ω(t)ẑ
L

(a) Find the static electric field and steady magnetic field when $\omega = \omega_0$.

(b) Find $\mathbf{B}(\mathbf{r},t)$ everywhere during the spin-up and use it to find the time-dependent part of the electric field.

(c) The spin-up is performed by an external agent who supplies power at a rate $-\mathbf{j}\cdot\mathbf{E}$ per unit volume to create the magnetic field. Confirm this by evaluating Poynting's theorem over all of space. Comment on the role of the Poynting vector and the electric field in this calculation.

(d) Evaluate Poynting's theorem using a cylindrical volume with a radius slightly smaller than the shell to study the flow of energy into the interior from the surface of the shell.

15.15 A Momentum Flux Theorem The source of a magnetic field $\mathbf{B}_0$ is a steady current produced by a collection of moving charges confined to a volume V. These particles are exposed to an external electric field $\mathbf{E}_0$ produced by a static charge distribution which is not necessarily confined to V. Prove that the net flux of electromagnetic momentum $\mathbf{g}_{\rm EM} = \epsilon_0(\mathbf{E}_0 \times \mathbf{B}_0)$ through the surface of V is zero.

15.16 An Electromagnetic Inequality Show that the inequality $U_{\rm EM} \geq c\,|\mathbf{P}_{\rm EM}|$ is a general property of electric and magnetic fields. Under what conditions are the two quantities equal?

15.17 Potential Momentum An infinite cylindrical solenoid of radius R is wound tightly with n turns per unit length of a wire that carries a current I. A bead of mass m and charge q slides freely on a non-conducting circular wire of radius $r > R$ that is concentric with the solenoid. Let the bead be at rest (ignore gravity) and then reduce the current in the solenoid winding to zero. The bead begins to slide around the wire.

(a) Use Faraday's law to find the speed v of the bead after the magnetic field in the solenoid has disappeared.

(b) Show that an application of conservation of linear momentum leads to the same answer.

(c) Explain why the electric and magnetic fields produced by the moving charge do not contribute to the momentum balance in part (b).

15.18 $\mathbf{P}_{\rm EM}$ for an Electric Dipole in a Uniform Magnetic Field Compute $\mathbf{P}_{\rm EM}$ for a point electric dipole with moment $\mathbf{p}$ located at the center of a hollow spherical shell (radius R) with uniform surface charge density σ which rotates at frequency ω. Do not assume that the rotation axis is aligned with $\mathbf{p}$. Hint: There are contributions from both inside and outside the sphere.

15.19 $\mathbf{P}_{\rm EM}$ for Electric and Magnetic Dipoles

(a) Show that the electromagnetic linear momentum $\mathbf{P}_{\rm EM}$ for static fields in all of space can be rewritten in terms of the current density $\mathbf{j}(\mathbf{r})$ and the electrostatic potential $\varphi(\mathbf{r})$.

(b) Point charges q and $-q$ sit at points $(0, 0, d)$ and $(0, 0, -d)$, respectively. A point magnetic dipole with moment $\mathbf{m}$ sits at the origin. Show that $\mathbf{P}_{\rm EM} = \mathbf{E}(0) \times \mathbf{m}/c^2$, where $\mathbf{E}(0)$ is the electric field of the charges evaluated at the origin.

(c) A point electric dipole with moment $\mathbf{p}$ sits not far from a steady distribution of current. Express $\mathbf{P}_{\rm EM}$ in terms of $\mathbf{p}$ and the Coulomb gauge vector potential.

15.20 $\mathbf{P}_{\rm EM}$ in the Coulomb Gauge The Helmholtz theorem guarantees that any vector can be decomposed in the form $\mathbf{v} = \mathbf{v}_\perp + \mathbf{v}_\parallel$ where $\nabla\cdot\mathbf{v}_\perp = 0$ and $\nabla\times\mathbf{v}_\parallel = 0$. Use this fact, the identity

$$\int d^3r\,(\mathbf{u}\cdot\nabla)\mathbf{w} = \int (d\mathbf{S}\cdot\mathbf{u})\mathbf{w} - \int d^3r\,(\nabla\cdot\mathbf{u})\mathbf{w},$$

and the Coulomb gauge constraint to prove that the electromagnetic momentum of a charge distribution $\rho(\mathbf{r}, t)$ in the presence of a magnetic field $\mathbf{B}(\mathbf{r}, t)$ can be written in the form

$$\mathbf{P}_{\rm EM} = \int d^3r\,\rho\mathbf{A}_\perp + \epsilon_0 \int d^3r\mathbf{E}_\perp \times \mathbf{B}.$$

15.21 Hidden Momentum in a Bar Magnet? A point charge at rest sits in the magnetostatic field of a stationary permanent magnet with spin magnetization $\mathbf{M}(\mathbf{r})$.

(a) Show that the field momentum is

$$\mathbf{P}_{\rm EM} = \frac{1}{c^2}\int d^3r\,\mathbf{E}\times\mathbf{M}.$$

(b) Is there "hidden momentum" associated with this situation? If so, can you suggest its origin classically or quantum mechanically?

15.22 **$\mathbf{L}_{\rm EM}$ for a Charge in a Two-Dimensional Magnetic Field** A point charge q sits at the origin. A magnetic field $\mathbf{B}(\mathbf{r}) = B(x, y)\hat{\mathbf{z}}$ fills all of space. Show that the field angular momentum is $\mathbf{L}_{\rm EM} = -(q/2\pi)\Phi_B\hat{\mathbf{z}}$, where Φ_B is the flux of $\mathbf{B}$ through the plane $z = 0$.

15.23 **Transformation of Angular Momentum** The figure below shows a cutaway view of an infinite cylindrical solenoid with radius R that creates a magnetic field $\mathbf{B} = B\hat{\mathbf{z}}$ inside itself. An infinitely long cylinder of insulating material with radius $a < R$, permeability μ_0, and permittivity ϵ_0 sits inside (and coaxial with) the solenoid. The cylinder is filled uniformly with charge with density $\bar{\rho} > 0$. A uniform surface charge σ makes the cylinder electrically neutral.

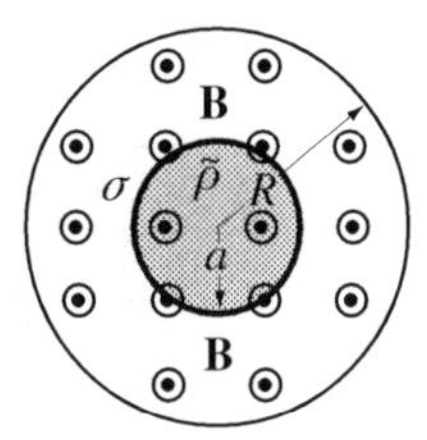

(a) Find the total electromagnetic angular momentum (per unit length) of this system.
(b) Compute the instantaneous torque (per unit length) which acts on the cylinder as the magnetic field is reduced to zero with a time dependence $B(t)$.
(c) Show that the final mechanical angular momentum of the cylinder is equal to the initial angular momentum calculated in part (a).

15.24 **$\mathbf{L}_{\rm EM}$ for Static Fields**

(a) Show that the electromagnetic angular momentum for static fields in the Coulomb gauge is

$$\mathbf{L}_{\rm EM} = \int d^3r\, \rho(\mathbf{r})\mathbf{r} \times \mathbf{A}(\mathbf{r}).$$

(b) Evaluate $\mathbf{L}_{\rm EM}$ when a point electric dipole $\mathbf{p}$ sits at the origin in the presence of a static magnetic field $\mathbf{B}(\mathbf{r})$.
(c) An axially symmetric charge distribution rotates around its symmetry axis with frequency $\boldsymbol{\omega}$. Prove that the magnetic total energy can be expressed as

$$U_B = \frac{1}{2}\boldsymbol{\omega} \cdot \mathbf{L}_{\rm EM}.$$

15.25 **The Dipole Force on Atoms and Molecules** The text showed that the Coulomb-Lorentz force on a classical atom or molecule (a bounded collection of net neutral charges moving with center-of-mass velocity $\mathbf{v}$) due to its electric dipole moment $\mathbf{p}(t)$ and magnetic dipole moment $\mathbf{m}(t)$ is

$$\mathbf{F} = (\mathbf{p} \cdot \nabla)\mathbf{E} + \dot{\mathbf{p}} \times \mathbf{B} + (\mathbf{p} \cdot \nabla)(\mathbf{v} \times \mathbf{B}) + (\mathbf{m} \times \nabla) \times \mathbf{B}.$$

Use the fact that the atom is moving through the fields $\mathbf{E}(\mathbf{r}, t)$ and $\mathbf{B}(\mathbf{r}, t)$ with velocity $\mathbf{v}$ to show that

$$\mathbf{F} = \frac{d}{dt}(\mathbf{p} \times \mathbf{B}) + \nabla\{\mathbf{p} \cdot (\mathbf{E} + \mathbf{v} \times \mathbf{B})\} + \nabla(\mathbf{m} \cdot \mathbf{B}).$$

16 Waves in Vacuum

The theoretical discovery of an electromagnetic wave spreading with the speed of light is one of the greatest achievements in the history of science.
Albert Einstein and Leopold Infeld (1938)

16.1 Introduction

This chapter begins our exploration of the single most important fact of electromagnetic life. The Maxwell equations have wave-like solutions which propagate from point to point through space carrying energy, linear momentum, and angular momentum. Electromagnetic fields of this kind transport life-giving heat from the Sun, reveal the internal structure of the human body, and facilitate communication by radio, television, satellite, and cell phone. The propagating solutions we will study are often called "free fields" because they are not "attached" to distributions of charge or current. Their electric field lines do not terminate on charge and their magnetic field lines do not encircle current. For that reason, electromagnetic waves bear very little relationship (both physically and mathematically) to the electrostatic, magnetostatic, and quasistatic fields we have studied to this point. This chapter focuses on the basic structure and surprising variety of propagating waves in vacuum. Chapter 20 takes up the question of how one produces them.

Electromagnetic waves are solutions of the Maxwell equations in the absence of sources. Such waves are also solutions of a vector wave equation which appears repeatedly through the course of the chapter. We analyze plane wave solutions first because they are simple, important, and provide a convenient setting for discussing polarization. We then superpose plane waves to form wave packets and demonstrate their fundamental properties of complementarity and free-space diffraction. Scalar wave packets occupy most of our attention because (i) it is difficult to superpose vector waves and (ii) scalar waves appear naturally when free fields are derived from electromagnetic potentials. We then turn, successively, to beam-like waves, spherical waves, and the Hertz vector approach to electromagnetic waves. The chapter concludes with a few examples of the behavior of charged and/or polarizable particles in free fields.

The Maxwell equations in vacuum without source terms are

$$\nabla \cdot \mathbf{E} = 0 \qquad\qquad \nabla \cdot \mathbf{B} = 0 \tag{16.1}$$

$$\nabla \times \mathbf{E} = -\frac{\partial \mathbf{B}}{\partial t} \qquad\qquad \nabla \times \mathbf{B} = \frac{1}{c^2}\frac{\partial \mathbf{E}}{\partial t}. \tag{16.2}$$

When read from right to left, the two curl equations are coupled partial differential equations for the time evolution of $\mathbf{E}(\mathbf{r}, t)$ and $\mathbf{B}(\mathbf{r}, t)$. The two divergence equations provide the initial conditions. This follows from the divergence of each curl equation:

$$\frac{\partial}{\partial t}\{\nabla \cdot \mathbf{E}\} = 0 \qquad \text{and} \qquad \frac{\partial}{\partial t}\{\nabla \cdot \mathbf{B}\} = 0. \tag{16.3}$$

Equation (16.3) keeps the fields divergence-free for all time if (16.1) makes them divergence-free at $t = 0$.

16.2 The Wave Equation

The electric and magnetic fields independently satisfy a vector version of the classical wave equation of mathematical physics. To see this, substitute the left equation in (16.2) into the curl of the right equation in (16.2) (and vice versa). The result is

$$\nabla \times (\nabla \times \mathbf{E}) = -\frac{1}{c^2}\frac{\partial^2 \mathbf{E}}{\partial t^2} \qquad \text{and} \qquad \nabla \times (\nabla \times \mathbf{B}) = -\frac{1}{c^2}\frac{\partial^2 \mathbf{B}}{\partial t^2}. \tag{16.4}$$

The vector identity

$$\nabla \times (\nabla \times \mathbf{c}) = \nabla(\nabla \cdot \mathbf{c}) - \nabla^2 \mathbf{c} \tag{16.5}$$

and (16.1) reduce (16.4) to the advertised vector wave equations,

$$\nabla^2 \mathbf{E} - \frac{1}{c^2}\frac{\partial^2 \mathbf{E}}{\partial t^2} = 0 \qquad \text{and} \qquad \nabla^2 \mathbf{B} - \frac{1}{c^2}\frac{\partial^2 \mathbf{B}}{\partial t^2} = 0. \tag{16.6}$$

Equation (16.6) shows that each Cartesian component of $\mathbf{E}(\mathbf{r}, t)$ and $\mathbf{B}(\mathbf{r}, t)$ satisfies the scalar wave equation. The same statement does not apply when these vectors are expressed in spherical or cylindrical components (see Section 1.2.7). It is important to keep in mind that vector functions that happen to satisfy (16.6) do not automatically satisfy the Maxwell equations. Rather, they are candidate solutions which qualify as bona fide electric and magnetic fields only if they can be made divergence-free to satisfy (16.1) and coupled together to satisfy (16.2). These steps are not difficult to carry out for *plane wave* solutions of (16.6). The constraints dictated by the Maxwell equations are less easy to arrange for *beam-like* and *spherical wave* solutions of (16.6). For these (and related) solutions, it is simpler to exploit a set of wave equations based on electromagnetic potentials. The next two subsections show how.

16.2.1 Lorenz Gauge Potentials

In Section 15.3, we wrote the electric and magnetic fields in terms of potential functions:

$$\mathbf{B} = \nabla \times \mathbf{A} \qquad \text{and} \qquad \mathbf{E} = -\nabla\varphi - \frac{\partial \mathbf{A}}{\partial t}. \tag{16.7}$$

The Lorenz gauge condition (Section 15.3.3),

$$\nabla \cdot \mathbf{A}_L + \frac{1}{c^2}\frac{\partial \varphi_L}{\partial t} = 0, \tag{16.8}$$

led us to wave equations for their time evolution:

$$\nabla^2 \varphi_L - \frac{1}{c^2}\frac{\partial^2 \varphi_L}{\partial t^2} = 0 \qquad \text{and} \qquad \nabla^2 \mathbf{A}_L - \frac{1}{c^2}\frac{\partial^2 \mathbf{A}_L}{\partial t^2} = 0. \tag{16.9}$$

Potentials that satisfy (16.8) and (16.9) produce fields (16.7) that satisfy all four Maxwell equations in free space. A key to this approach is that very simple choices for the vector potential generate entire classes of electromagnetic waves. For example, let $\mathbf{s}$ be a *constant* (but otherwise arbitrary) vector and write

$$\mathbf{A}_L(\mathbf{r}, t) = u(\mathbf{r}, t)\mathbf{s}. \tag{16.10}$$

Substituting (16.10) into (16.9) gives

$$\left[\nabla^2 u - \frac{1}{c^2}\frac{\partial^2 u}{\partial t^2}\right]\mathbf{s} = 0. \tag{16.11}$$

This shows that *every* solution of the scalar wave equation

$$\nabla^2 u - \frac{1}{c^2}\frac{\partial^2 u}{\partial t^2} = 0 \tag{16.12}$$

produces a Lorenz gauge vector potential (16.10) that satisfies the right side of (16.9). The linearity of the wave equation ensures that the scalar potential derived from (16.8) and (16.10),

$$\varphi_L(\mathbf{r}, t) = -c^2\mathbf{s}\cdot\int_{-\infty}^{t} dt'\,\nabla u(\mathbf{r}, t'), \tag{16.13}$$

satisfies the left side of (16.9). Thus, every solution of the scalar wave equation generates a set of electromagnetic waves (16.7) parameterized by the vector $\mathbf{s}$ in (16.10).

16.2.2 Coulomb Gauge Potentials

An alternative to (16.8) is the Coulomb gauge constraint (Section 15.3.2),

$$\nabla\cdot\mathbf{A}_C = 0. \tag{16.14}$$

This choice simplifies the equation of motion for the potentials to

$$\begin{aligned}\nabla^2\varphi_C &= 0\\ \nabla^2\mathbf{A}_C - \frac{1}{c^2}\frac{\partial^2\mathbf{A}_C}{\partial t^2} &= \frac{1}{c^2}\nabla\frac{\partial\varphi_C}{\partial t}.\end{aligned} \tag{16.15}$$

In infinite, empty space, we may set $\varphi_C(\mathbf{r}, t) = 0$ without loss of generality because, in that case, the solution of Laplace's equation in (16.15) is at most a constant. Therefore, as long as (16.14) holds, the vector potential satisfies a vector wave equation,

$$\nabla^2\mathbf{A}_C - \frac{1}{c^2}\frac{\partial^2\mathbf{A}_C}{\partial t^2} = 0, \tag{16.16}$$

and all four Maxwell equations are satisfied by

$$\mathbf{B} = \nabla\times\mathbf{A}_C \qquad \text{and} \qquad \mathbf{E} = -\frac{\partial\mathbf{A}_C}{\partial t}. \tag{16.17}$$

The decision to use the Coulomb gauge amounts to weighing the virtues of eliminating the scalar potential against the burden of imposing the Coulomb gauge constraint explicitly on the vector potential.

Whittaker's Theorem

Only two solutions of the scalar wave equation are needed to represent an arbitrary electromagnetic field in empty space.

Proof: The results of this section show that (16.17) describes an arbitrary field if each Cartesian component of $\mathbf{A}_C(\mathbf{r}, t)$ satisfies the scalar wave equation and $\nabla \cdot \mathbf{A}_C = 0$. However, the latter statement implies that only two of the three Cartesian components of $\mathbf{A}_C$ are independent. Therefore, only two solutions of the wave equation are needed. This observation, due to Whittaker (1904), helps rationalize several results we will derive in later sections and chapters.

16.3 Plane Waves

The most important class of solutions to the wave equation are called *plane waves*. To find them, we write a generic vector wave equation,

$$\nabla^2 \mathbf{w} - \frac{1}{c^2}\frac{\partial^2 \mathbf{w}}{\partial t^2} = 0, \tag{16.18}$$

and look for solutions that depend on a single spatial variable. For example, the guess $\mathbf{w} = \mathbf{w}(z, t)$ simplifies (16.18) to

$$\frac{\partial^2 \mathbf{w}}{\partial z^2} - \frac{1}{c^2}\frac{\partial^2 \mathbf{w}}{\partial t^2} = \left(\frac{\partial}{\partial z} + \frac{1}{c}\frac{\partial}{\partial t}\right)\left(\frac{\partial}{\partial z} - \frac{1}{c}\frac{\partial}{\partial t}\right)\mathbf{w} = 0. \tag{16.19}$$

A method of solution due to d'Alembert changes variables to $\xi = z + ct$ and $\eta = z - ct$ and uses the chain rule to compute the derivatives $\partial_z = \partial_\xi + \partial_\eta$ and $\partial_{ct} = \partial_\xi - \partial_\eta$ so that

$$\frac{\partial}{\partial \xi} = \frac{1}{2}\left[\frac{\partial}{\partial z} + \frac{1}{c}\frac{\partial}{\partial t}\right] \quad \text{and} \quad \frac{\partial}{\partial \eta} = \frac{1}{2}\left[\frac{\partial}{\partial z} - \frac{1}{c}\frac{\partial}{\partial t}\right]. \tag{16.20}$$

Comparing (16.19) with (16.20) shows that $\mathbf{w}$ satisfies

$$\frac{\partial^2 \mathbf{w}}{\partial \eta \partial \xi} = 0. \tag{16.21}$$

The variables ξ and η are linearly independent. Therefore, if $\mathbf{f}$ and $\mathbf{g}$ are arbitrary vector functions of one scalar variable, direct integration of (16.21) shows that $\mathbf{w}(\xi) = \mathbf{f}(z + ct)$ and $\mathbf{w}(\eta) = \mathbf{g}(z - ct)$ are linearly independent solutions. We call these *propagating plane waves* because $\mathbf{g}(z - ct)$ takes constant values on planes $z - ct = const.$ which propagate with speed c in the postive z-direction and $\mathbf{f}(z + ct)$ takes constant values on planes $z + ct = const.$ which propagate with speed c in the negative z-direction. The general solution of (16.18) is a linear combination of plane waves propagating in opposite directions:

$$\mathbf{w}(z, t) = \mathbf{g}(z - ct) + \mathbf{f}(z + ct). \tag{16.22}$$

This function typically has *standing-wave* character.

16.3.1 Transverse Electromagnetic Waves

The remarks following (16.6) tell us that plane waves for $\mathbf{E} = \mathbf{E}(z, t)$ and $\mathbf{B} = \mathbf{B}(z, t)$ are candidate constituents of an electromagnetic wave. Our strategy is to construct a valid electric field, and then use

(16.2) to derive the associated magnetic field. The first observation is that $\mathbf{E}(z,t)$ lies entirely in the x-y plane. This follows from (16.1) and (16.2), which give

$$\frac{\partial E_z}{\partial z} = 0 \tag{16.23}$$

and

$$\frac{\partial E_z}{\partial t} = c^2 \hat{\mathbf{z}} \cdot \nabla \times \mathbf{B} = 0. \tag{16.24}$$

Equations (16.23) and (16.24) show that E_z is at most a constant, which we may safely set to zero. Therefore, using the subscript $\perp$ to denote a vector that is perpendicular to $\hat{\mathbf{z}}$, two linearly independent, propagating plane wave solutions for the electric field are

$$\mathbf{E}_+(z,t) = \mathbf{f}_\perp(z+ct) \qquad \text{and} \qquad \mathbf{E}_-(z,t) = \mathbf{g}_\perp(z-ct). \tag{16.25}$$

The magnetic fields $\mathbf{B}_\pm(z,t)$ that accompany the electric fields $\mathbf{E}_\pm(z,t)$ in (16.25) follow from $\nabla \times \mathbf{E} = -\partial \mathbf{B}/\partial t$. Because $E_z = 0$,

$$\frac{\partial}{\partial t}\begin{Bmatrix} \mathbf{B}_+(z,t) \\ \mathbf{B}_-(z,t) \end{Bmatrix} = -\hat{\mathbf{z}} \times \frac{\partial}{\partial z}\begin{Bmatrix} \mathbf{f}_\perp(z+ct) \\ \mathbf{g}_\perp(z-ct) \end{Bmatrix} = -\hat{\mathbf{z}} \times \frac{1}{c}\frac{\partial}{\partial t}\begin{Bmatrix} \mathbf{f}_\perp(z+ct) \\ -\mathbf{g}_\perp(z-ct) \end{Bmatrix}. \tag{16.26}$$

Integration of the first and last terms in (16.26) gives the magnetic field up to a function of z alone. The latter may be dropped because we are only interested in time-varying fields. Therefore,

$$c\mathbf{B}_+(z,t) = -\hat{\mathbf{z}} \times \mathbf{f}_\perp(z+ct) \qquad \text{and} \qquad c\mathbf{B}_-(z,t) = \hat{\mathbf{z}} \times \mathbf{g}_\perp(z-ct). \tag{16.27}$$

The propagating plane waves $(\mathbf{E}_+, \mathbf{B}_+)$ and $(\mathbf{E}_-, \mathbf{B}_-)$ are called *transverse electromagnetic* (TEM) because both $\mathbf{E}$ and $\mathbf{B}$ are perpendicular to $\hat{\mathbf{z}}$. A quick calculation shows that the electric field is orthogonal to the magnetic field for both:

$$\mathbf{E}_+ \cdot \mathbf{B}_+ = 0 \qquad \text{and} \qquad \mathbf{E}_- \cdot \mathbf{B}_- = 0. \tag{16.28}$$

On the other hand, $\mathbf{E} \cdot \mathbf{B} \neq 0$ for a general linear combination of waves propagating in opposite directions like $(\mathbf{E}_+ + \mathbf{E}_-, \mathbf{B}_+ + \mathbf{B}_-)$. Fields of this kind always have at least some standing-wave character (see Example 16.2).

Finally, there is nothing special about the positive or negative z-direction in the foregoing. A plane, transverse, electromagnetic wave that propagates in the $\hat{\mathbf{k}}$ direction can be constructed from an arbitrary vector function of one scalar variable, $\mathbf{E}_\perp(\phi)$, which lies entirely in a plane perpendicular to a constant *wave vector* $\mathbf{k} = k\hat{\mathbf{k}}$:

$$\mathbf{E}(\mathbf{r},t) = \mathbf{E}_\perp(\mathbf{k}\cdot\mathbf{r} - ckt) \qquad c\mathbf{B}(\mathbf{r},t) = \hat{\mathbf{k}} \times \mathbf{E}(\mathbf{r},t). \tag{16.29}$$

The vectors $(\mathbf{E}, \mathbf{B}, \mathbf{k})$ form a right-handed orthogonal triad and, by direct calculation using (16.29),

$$|\mathbf{E}| = c|\mathbf{B}|. \tag{16.30}$$

Figure 16.1 shows two representations of the plane wave (16.29) for a choice of $\mathbf{E}_\perp(\phi)$ that is non-zero over only a limited range of its argument. The upper panel graphs the electric and magnetic field amplitudes (on orthogonal axes) as a function of position at a fixed time. The lower panel shows the corresponding field line pattern, which extends to $\pm\infty$ in the transverse direction. The plane wave shown in Figure 16.1 is unphysical (as is any true plane wave) because no wave with infinite extent in any direction can exist in Nature. This defect does does not deter us from studying plane waves further because (i) many real electromagnetic fields *resemble* a plane wave *locally* over a limited region of space and (ii) we will later superpose plane waves to synthesize real waves with finite extent.

The infinitely long and straight field lines in Figure 16.1 are correct for a plane wave, but obscure the fundamental propagation mechanism for real waves with finite transverse extent. The fields of these waves are not strictly transverse and their field lines form closed loops to satisfy the divergence

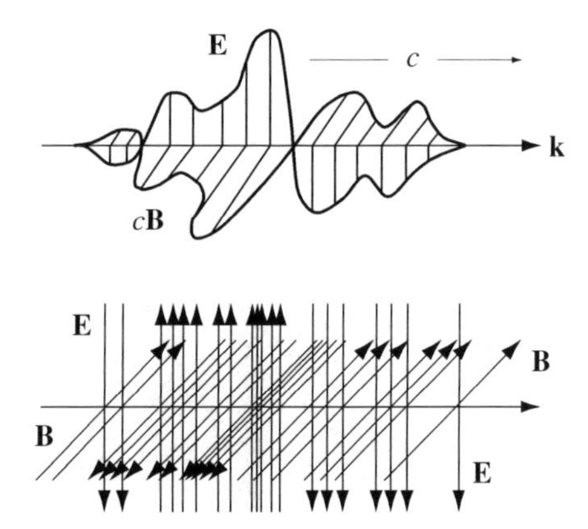
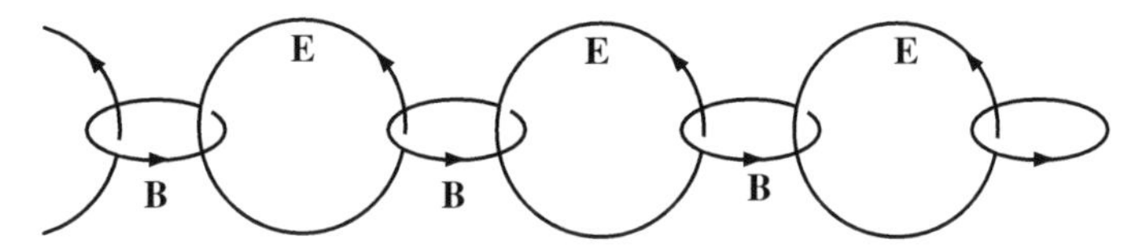

Figure 16.1: Two views of a TEM plane wave in vacuum which propagates at speed c in the $\mathbf{k}$ direction. Upper panel: field amplitudes graphed on orthogonal axes. Lower panel: field lines.

Figure 16.2: Cartoon of some linked electric and magnetic field lines for an electromagnetic wave which propagates to the right. Figure from Born (1924).

equations in (16.2). Moreover, the loops of $\mathbf{E}$ and loops of $\mathbf{B}$ interconnect like the links of a chain (see Figure 16.2). This topology, which Maxwell called the "mutual embrace" of the electric and magnetic fields, ensures that Faraday's law in (16.3) dynamically generates new loops of $\mathbf{E}$ and the Ampère-Maxwell law in (16.3) dynamically generates new loops of $\mathbf{B}$.[1] We will revisit this point when we study beam-like waves in Section 16.7.

16.3.2 Phase Velocity

The *phase* of the wave (16.29) is the scalar function

$$\phi(\mathbf{r}, t) = \mathbf{k} \cdot \mathbf{r} - ckt. \tag{16.31}$$

This definition shows that $\phi(\mathbf{r}, t)$ is constant (at fixed time) everywhere on the plane perpendicular to $\mathbf{k}$ that contains the point $\mathbf{r}$ (Figure 16.3). The amplitudes of $\mathbf{E}$ and $\mathbf{B}$ are correspondingly constant on the same planes. Writing (16.31) as $\phi = \mathbf{k} \cdot [\mathbf{r} - c\hat{\mathbf{k}}t]$ and setting the quantity in brackets equal to $\mathbf{r}_0$ shows that the phase is constant along the trajectory $\mathbf{r}(t) = \mathbf{r}_0 + c\hat{\mathbf{k}}t$. This shows that every plane of constant phase moves perpendicular to itself at the *phase velocity*

$$v_{\rm p} = \frac{d\mathbf{r}}{dt} = c\hat{\mathbf{k}}. \tag{16.32}$$

The fact that the speed of light $c = 1/\sqrt{\epsilon_0\mu_0}$ is the phase velocity for propagating plane waves in vacuum was the key to Maxwell's realization that light is an electromagnetic phenomenon.

We will see later (Application 19.1) that the phase velocity plays an important role in determining the efficiency of energy transfer between an electromagnetic wave and a charged particle. In that

[1] This is best appreciated using the integral form of these laws.

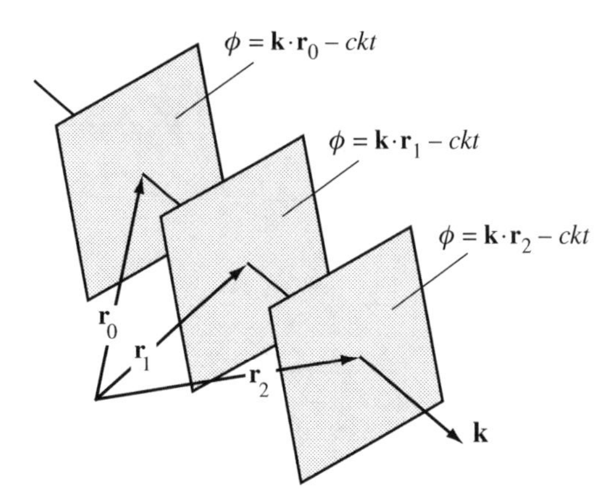

Figure 16.3: The phase (16.31) takes constant values on planes perpendicular to the wave vector $\mathbf{k}$.

context, it is common to write the condition for phase constancy in the form

$$0 = \frac{d\phi(\mathbf{r},t)}{dt} = \frac{\partial\phi}{\partial t} + \frac{\partial\phi}{\partial\mathbf{r}}\cdot\frac{d\mathbf{r}}{dt} = \frac{\partial\phi}{\partial t} + \nabla\phi\cdot\boldsymbol{v}_{\rm p}. \tag{16.33}$$

The phase velocity (16.32) satisfies (16.33) when the phase takes its plane wave value (16.31). Other solutions of the Maxwell equations in free space have more complicated expressions for $\phi(\mathbf{r},t)$. In that case, (16.33) remains true if we use the real part of $\phi(\mathbf{r},t)$ and $\boldsymbol{v}_{\rm p}$ generally varies as a function of position.

Example 16.1 Confirm by direct calculation that the plane wave fields $\mathbf{E}(\mathbf{r},t) = \mathbf{E}_\perp(\mathbf{k}\cdot\mathbf{r} - ckt)$ and $c\mathbf{B}(\mathbf{r},t) = \hat{\mathbf{k}}\times\mathbf{E}(\mathbf{r},t)$ satisfy all four Maxwell equations in vacuum.

Solution: Let $f'(\phi) = df/d\phi$ where the phase ϕ is given by (16.31). By construction, $\mathbf{k}\cdot\mathbf{E}_\perp = 0$. Therefore, the chain rule gives

$$\nabla\cdot\mathbf{E} = \mathbf{E}'_\perp(\phi)\cdot\nabla\phi = \mathbf{k}\cdot\mathbf{E}'_\perp(\phi) = 0.$$

Similarly, $\nabla\times\mathbf{E} = \mathbf{k}\times\mathbf{E}'_\perp$. Therefore, because $\hat{\mathbf{k}}$ is a constant vector,

$$\nabla\cdot c\mathbf{B} = \nabla\cdot(\hat{\mathbf{k}}\times\mathbf{E}) = -\hat{\mathbf{k}}\cdot(\nabla\times\mathbf{E}) = -\hat{\mathbf{k}}\cdot(\mathbf{k}\times\mathbf{E}'_\perp) = 0.$$

We confirm Faraday's law using the expression for $\nabla\times\mathbf{E}$ derived just above,

$$c\dot{\mathbf{B}} = \hat{\mathbf{k}}\times\dot{\mathbf{E}} = \hat{\mathbf{k}}\times\mathbf{E}'_\perp\dot{\phi} = -c\mathbf{k}\times\mathbf{E}'_\perp = -c\nabla\times\mathbf{E}.$$

Finally, $\dot{\mathbf{E}} = -ck\mathbf{E}'_\perp$ and $\hat{\mathbf{k}}$ is a constant vector, so

$$\nabla\times c\mathbf{B} = \nabla\times(\hat{\mathbf{k}}\times\mathbf{E}) = -(\hat{\mathbf{k}}\cdot\nabla)\mathbf{E} + \hat{\mathbf{k}}(\nabla\cdot\mathbf{E}) = -(\hat{\mathbf{k}}\cdot\nabla)\mathbf{E} = -\hat{k}_i\mathbf{E}'_\perp\partial_i\phi = -k\mathbf{E}'_\perp.$$

This shows that $\nabla\times\mathbf{B} = c^{-2}\partial\mathbf{E}/\partial t$, which is the final Maxwell equation in free space.

16.3.3 Mechanical Properties

The energy density (15.32) associated with the plane wave (16.29) is

$$u_{\rm EM} = \frac{1}{2}\epsilon_0\left(\mathbf{E}\cdot\mathbf{E} + c^2\mathbf{B}\cdot\mathbf{B}\right) = \frac{1}{2}\epsilon_0\left[|\mathbf{E}_\perp|^2 + (\hat{\mathbf{k}}\times\mathbf{E}_\perp)\cdot(\hat{\mathbf{k}}\times\mathbf{E}_\perp)\right]. \tag{16.34}$$

Because $\mathbf{E}_\perp$ is transverse to $\mathbf{k}$, (16.34) simplifies to

$$u_{\rm EM}(\mathbf{r}, t) = \epsilon_0 |\mathbf{E}_\perp(\mathbf{k}\cdot\mathbf{r} - ckt)|^2. \tag{16.35}$$

The energy current density $\mathbf{S}$ and linear momentum density $\mathbf{g}$ defined in (15.29) and (15.50), respectively, vary in space and time according to

$$\mathbf{g}(\mathbf{r}, t) = \epsilon_0(\mathbf{E}\times\mathbf{B}) = \frac{\mathbf{S}}{c^2} = \frac{\epsilon_0}{c}|\mathbf{E}_\perp(\mathbf{k}\cdot\mathbf{r} - ckt)|^2\hat{\mathbf{k}}. \tag{16.36}$$

These formulae establish that a plane wave in vacuum carries energy and linear momentum at the speed of light in the direction of propagation of the wave. Specifically,

$$\mathbf{g} = \frac{u_{\rm EM}}{c}\hat{\mathbf{k}} \qquad \text{and} \qquad \mathbf{S} = u_{\rm EM}c\hat{\mathbf{k}}. \tag{16.37}$$

Finally, we showed in Section 15.6.1 that $\mathbf{r}\times\mathbf{g}$ may be interpreted as the angular momentum per unit volume associated with an electromagnetic field. Using (16.37), we see that the electromagnetic plane wave (16.29) carries zero angular momentum along the direction of propagation.[2]

It cannot be stressed too much that the plane electromagnetic waves discussed here possess their full complement of mechanical properties in the complete absence of sources of charge or current. Moreover, the conservation laws discussed in Chapter 15 show that field energy, field linear momentum, and field angular momentum can be exchanged with the energy and momentum of a collection of particles. This leads to the view that $\mathbf{E}(\mathbf{r}, t)$ and $\mathbf{B}(\mathbf{r}, t)$ constitute a mechanical system that is every bit as "real" as any collection of particles.[3]

16.3.4 Monochromatic Plane Waves

A *monochromatic* plane wave with angular frequency ω is a special case of (16.29) where

$$\mathbf{E}_\perp(\phi) = \boldsymbol{\mathcal{E}}_\perp \exp(i\phi) \qquad \text{and} \qquad \omega = c|\mathbf{k}|. \tag{16.38}$$

The vector $\boldsymbol{\mathcal{E}}_\perp$ is generally complex. Therefore, if we define $c\boldsymbol{\mathcal{B}}_\perp = \hat{\mathbf{k}}\times\boldsymbol{\mathcal{E}}_\perp$, the physical fields of a monochromatic plane wave are the *real parts* of the complex vector functions:

$$\mathbf{E}(\mathbf{r}, t) = \boldsymbol{\mathcal{E}}_\perp \exp[i(\mathbf{k}\cdot\mathbf{r} - \omega t)] \qquad \text{and} \qquad \mathbf{B}(\mathbf{r}, t) = \boldsymbol{\mathcal{B}}_\perp \exp[i(\mathbf{k}\cdot\mathbf{r} - \omega t)]. \tag{16.39}$$

The monochromatic plane wave in (16.39) is called *uniform* because the wave amplitude takes constant values on the same planes where the phase is constant. Figure 16.4 shows a typical wave of this kind. It extends infinitely in all three spatial directions with electric and magnetic fields that vary sinusoidally in space and time with wavelength $\lambda = 2\pi/k$ and oscillation period $T = 2\pi/\omega$. The phase difference between two planes in Figure 16.3 separated by a distance λ is exactly 2π for such a wave. These features may be contrasted with the non-monochromatic plane wave shown in the top panel of Fig. 16.1. The latter extends infinitely in any transverse direction, but is spatially localized in the $\pm\mathbf{k}$ directions.

Some care is needed to express the mechanical properties of (16.39) in terms of the complex amplitude $\mathbf{E}_\perp$. For example, the leftmost equation in (16.34) must be written in the form

$$u_{\rm EM} = \tfrac{1}{2}\epsilon_0\left\{|\mathrm{Re}\,\mathbf{E}|^2 + c^2|\mathrm{Re}\,\mathbf{B}|^2\right\}. \tag{16.40}$$

[2] This is related to the infinite extent of a plane wave. Section 16.7.5 discusses $\mathbf{L}_{\rm EM}$ for more realistic electromagnetic waves.

[3] See Mermin (2009) on the "reality" of classical and quantum fields.

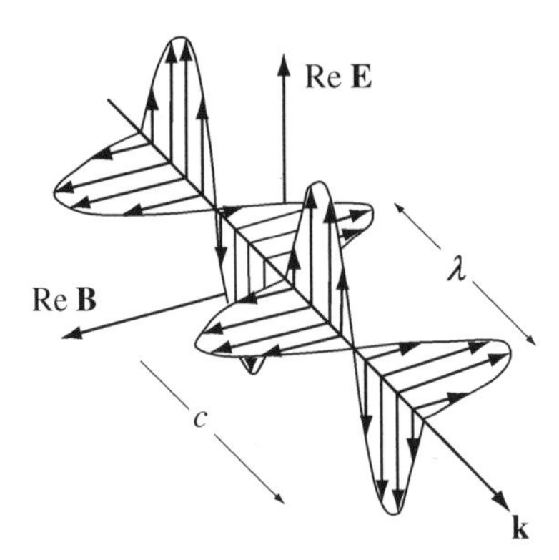

Figure 16.4: A uniform and monochromatic plane wave in vacuum with wavelength λ propagates with phase velocity $\boldsymbol{v} = c\hat{\mathbf{k}}$. This wave has linear polarization.

Writing $\text{Re}\,\mathbf{E} = \frac{1}{2}(\mathbf{E} + \mathbf{E}^*)$ and $\text{Re}\,\mathbf{B} = \frac{1}{2}(\mathbf{B} + \mathbf{B}^*)$ and using (16.39) gives

$$u_{\text{EM}}(\mathbf{r}, t) = \tfrac{1}{4}\epsilon_0 \left[\boldsymbol{\mathcal{E}}_\perp \cdot \boldsymbol{\mathcal{E}}_\perp^* + c^2 \boldsymbol{\mathcal{B}}_\perp \cdot \boldsymbol{\mathcal{B}}_\perp^*\right] + \tfrac{1}{4}\epsilon_0 \text{Re}\left[\left(\boldsymbol{\mathcal{E}}_\perp \cdot \boldsymbol{\mathcal{E}}_\perp + c^2 \boldsymbol{\mathcal{B}}_\perp \cdot \boldsymbol{\mathcal{B}}_\perp\right) \exp(2i\{\mathbf{k} \cdot \mathbf{r} - \omega t\})\right]. \tag{16.41}$$

The two static terms in (16.41) contribute equally [see (16.34)]. The remaining terms vanish when time-averaged over one period of oscillation. Therefore, the time-averaged energy density of the monochromatic wave (16.39) is[4]

$$\langle u_{\text{EM}} \rangle = \tfrac{1}{2}\epsilon_0 |\boldsymbol{\mathcal{E}}_\perp|^2. \tag{16.42}$$

Integration of the constant (16.42) over all space gives a divergent total energy. This is an artifact of the (unphysical) infinite extent of a plane wave.

The time-averaged linear momentum density and energy current density (Poynting vector) of a monochromatic plane wave are similarly

$$\langle \mathbf{g} \rangle = \frac{1}{2}\frac{\epsilon_0}{c} |\boldsymbol{\mathcal{E}}_\perp|^2 \hat{\mathbf{k}} = \frac{\langle u_{\text{EM}} \rangle}{c} \hat{\mathbf{k}} \tag{16.43}$$

and

$$\langle \mathbf{S} \rangle = c^2 \langle \mathbf{g} \rangle = \langle u_{\text{EM}} \rangle c \hat{\mathbf{k}}. \tag{16.44}$$

Rearrangement of (16.44) defines a quantity called the *energy velocity*,

$$\mathbf{v}_{\text{E}} = \frac{\langle \mathbf{S} \rangle}{\langle u_{\text{EM}} \rangle} = c\hat{\mathbf{k}}. \tag{16.45}$$

The energy velocity and the phase velocity (16.32) are identical for a plane wave in vacuum. We will encounter situations in later chapters when this is no longer true.

16.3.5 Wave Intensity

The position-dependent *intensity* of a general electromagnetic wave is defined as the magnitude of the time average of the Poynting vector over a time T that is much larger than any characteristic time scale

[4] The passage from (16.40) to (16.42) recapitulates the theorem proved in Section 1.6.3, namely, that when $A(t) = Ae^{-i\omega t}$ and $B(t) = Be^{-i\omega t}$, the time average $\langle[\text{Re}A(t)][\text{Re}B(t)]\rangle = \frac{1}{2}\text{Re}(A^* B)$.

Figure 16.5: Polarization of the cosmic microwave background (galactic coordinates) as measured by the Wilkinson Microwave Anisotropy Probe (WMAP). The white lines indicate the average direction of the electric field which propagates to the Earth from the indicated part of the sky. Courtesy of NASA and the WMAP Science Team.

associated with the wave:

$$I(\mathbf{r}) = \left| \frac{1}{T} \int_0^T dt\, \mathbf{S}(\mathbf{r}, t) \right|. \tag{16.46}$$

For a monochromatic plane wave, it is sufficient to choose T as the period of the wave and the intensity reduces to the cycle-average. Using (16.43) and (16.44), the resulting position-independent intensity is

$$I = |\langle \mathbf{S} \rangle| = \frac{1}{2}\epsilon_0 c |\boldsymbol{\mathcal{E}}_\perp|^2. \tag{16.47}$$

16.4 Polarization

The *polarization* of an electromagnetic wave characterizes the direction of its field vectors in space and time. This information is important for many reasons, not least because it provides clues to the origin of the wave. A dramatic example is the observed polarization of thermal microwaves which propagate to the Earth from throughout the cosmos (Figure 16.5). Data of this kind provide detailed support for the Big Bang model of the origin of the Universe.[5]

Polarization also provides information about the interactions an electromagnetic wave may have suffered with matter between the time of its creation and the time of its detection. A common example is the polarization of skylight due to scattering from particles in the atmosphere. Practical applications that exploit the polarization of electromagnetic waves are ubiquitous, ranging from terrestrial broadcasting and satellite communications to sunglasses and liquid crystal displays. The human eye turns out to be (weakly) sensitive to polarization and many insects, particularly bees, use the polarization of skylight as a navigational aid.[6]

16.4.1 The Polarization Ellipse

The polarization of a general electromagnetic wave varies as a function of position and may not be the same for the electric and magnetic fields. For simplicity, we restrict ourselves to a monochromatic plane wave propagating in vacuum and set ourselves the task of proving that, as the phase of the wave advances by 2π, the tip of the electric field vector $\mathbf{E}$ traces out an ellipse in the plane perpendicular to

[5] See Application 21.1 for an explanation of the polarization of the cosmic microwave background radiation.

[6] See Sources, References, and Additional Reading.

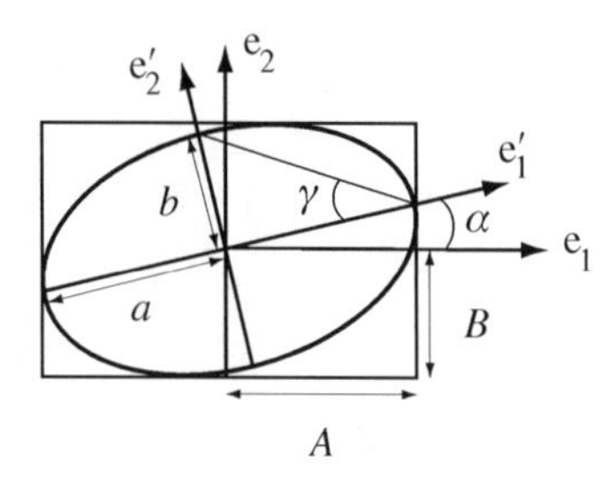

Figure 16.6: The polarization ellipse (heavy solid line) traced out by the tip of the **E** vector for a general monochromatic plane wave.

the propagation direction **k**. For that purpose, we define real unit vectors $\hat{\mathbf{e}}_1$ and $\hat{\mathbf{e}}_2$ so $(\hat{\mathbf{e}}_1, \hat{\mathbf{e}}_2, \hat{\mathbf{k}})$ is a right-handed orthogonal triad of unit vectors:

$$\hat{\mathbf{e}}_1 \cdot \hat{\mathbf{e}}_2 = 0 \qquad \text{and} \qquad \hat{\mathbf{e}}_1 \times \hat{\mathbf{e}}_2 = \hat{\mathbf{k}}. \tag{16.48}$$

The vectors $\hat{\mathbf{e}}_1$ and $\hat{\mathbf{e}}_2$ are a basis for $\boldsymbol{\mathcal{E}}_\perp$ in (16.39), so the electric field may be written

$$\mathbf{E}(\mathbf{r}, t) = [\mathcal{E}_1\hat{\mathbf{e}}_1 + \mathcal{E}_2\hat{\mathbf{e}}_2]\exp\{i(\mathbf{k}\cdot\mathbf{r} - \omega t)\}. \tag{16.49}$$

The complex scalars $\mathcal{E}_1$ and $\mathcal{E}_2$ in (16.49) are conveniently parameterized using the real numbers A, B, δ_1, and δ_2:

$$\mathcal{E}_1 = A\exp(i\delta_1) \qquad \text{and} \qquad \mathcal{E}_2 = B\exp(i\delta_2). \tag{16.50}$$

Using (16.50) and $\phi = \mathbf{k}\cdot\mathbf{r} - \omega t$, the real (physical) part of (16.49) takes the form[7]

$$\mathrm{Re}\,\mathbf{E} = A\cos(\phi + \delta_1)\hat{\mathbf{e}}_1 + B\cos(\phi + \delta_2)\mathbf{e}_2 = E_1\hat{\mathbf{e}}_1 + E_2\hat{\mathbf{e}}_2. \tag{16.51}$$

The key step is to equate the coefficients of $\hat{\mathbf{e}}_1$ and $\mathbf{e}_2$ in the second equality of (16.51) and manipulate the two equations that result to get

$$\begin{aligned}\frac{E_1}{A}\sin\delta_2 - \frac{E_2}{B}\sin\delta_1 &= \sin(\delta_2 - \delta_1)\cos\phi \\ \frac{E_1}{A}\cos\delta_2 - \frac{E_2}{B}\cos\delta_1 &= \sin(\delta_2 - \delta_1)\sin\phi.\end{aligned} \tag{16.52}$$

Squaring and adding the two terms in (16.52) gives

$$\left(\frac{E_1}{A}\right)^2 + \left(\frac{E_2}{B}\right)^2 - 2\left(\frac{E_1}{A}\right)\left(\frac{E_2}{B}\right)\cos\delta = \sin^2\delta, \tag{16.53}$$

where

$$\delta = \delta_2 - \delta_1. \tag{16.54}$$

Equation (16.53) describes an ellipse in the e_1 - e_2 plane (Figure 16.6). Accordingly, one says that the general monochromatic plane wave (16.49) exhibits *elliptical polarization*. The eccentricity and orientation of the ellipse depend on the phase difference δ and the amplitude ratio B/A.

16.4.2 Linear Polarization

The polarization ellipse in Figure 16.6 degenerates to a straight line when the orthogonal electric field components are either in phase or 180° out of phase:

$$\delta = \delta_2 - \delta_1 = m\pi \qquad (m = 0, 1, \ldots). \tag{16.55}$$

[7] The wave vector **k** is real. Later, situations will arise where **k** is complex.

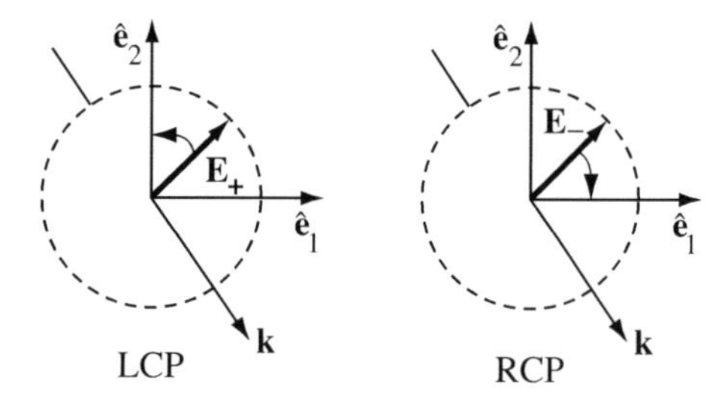

Figure 16.7: When viewed as the wave approaches, the tip of the electric vector of a monochromatic plane wave traces out a clockwise circle for right circular polarization (RCP) and a counterclockwise circle for left circular polarization (LCP).

These situations correspond to *linear polarization* because, up to a sign, $\mathbf{E}(\mathbf{r}, t)$ points in the same fixed direction for all values of $\mathbf{r}$ and t. Specifically, substituting (16.55) into (16.51) gives

$$\mathrm{Re}\,\{\mathbf{E}(\mathbf{r}, t)\} = (A\hat{\mathbf{e}}_1 \pm B\hat{\mathbf{e}}_2)\cos(\mathbf{k}\cdot\mathbf{r} - \omega t + \delta_1), \tag{16.56}$$

where the plus (minus) sign applies when m is even (odd). The monochromatic plane wave sketched in Figure 16.4 is linearly polarized. So are the fields indicated in Figure 16.5.

16.4.3 Circular Polarization

The polarization ellipse in Figure 16.6 simplifies to a circle when the orthogonal electric field components have equal amplitude but are 90° out of phase:

$$A = B = \mathcal{A}/\sqrt{2} \qquad \delta = \delta_2 - \delta_1 = m\pi/2 \qquad (m = \pm 1, \pm 3, \ldots). \tag{16.57}$$

These situations correspond to *circular polarization* because the tip of the electric vector traces out a circle in every fixed plane perpendicular to $\mathbf{k}$. It remains only to determine the direction that the circle is traced out. To that end, specialize to the plane that contains $\mathbf{r} = 0$ and write (16.51) with $\delta_1 = 0$ in the form[8]

$$\mathrm{Re}\,\{\mathbf{E}(0, t)\} = \frac{\mathcal{A}}{\sqrt{2}}\left[\hat{\mathbf{e}}_1 \cos\omega t + \hat{\mathbf{e}}_2 \cos(\omega t - \delta)\right]. \tag{16.58}$$

The condition (16.57) implies that $\cos\delta = 0$ and $\sin\delta = \pm 1$. The choice of sign produces two distinct fields from (16.58):

$$\mathrm{Re}\,\{\mathbf{E}_\pm(0, t)\} = \frac{\mathcal{A}}{\sqrt{2}}\left[\hat{\mathbf{e}}_1 \cos\omega t \pm \hat{\mathbf{e}}_2 \sin\omega t\right]. \tag{16.59}$$

Figure 16.7 illustrates the behavior of (16.59) as viewed from the direction into which the wave is progressing. The tip motion for $\mathbf{E}_+$ is counterclockwise and we speak of a wave with *left circular* polarization (LCP). The tip motion for $\mathbf{E}_-$ is clockwise and we speak of a wave with *right circular* polarization (RCP).[9] Figure 16.8 shows the disposition of $\mathbf{E}(\mathbf{r})$ for a monochromatic RCP plane wave at the fixed time $t = 0$. The tips of the electric field vectors trace a circular spiral. An LCP wave looks similar except that the sense of twist is opposite.

[8] We may set $\delta_1 = 0$ because the polarization depends only on the phase difference (16.54).

[9] There is nothing to be done about the fact that some subfields of physics use our convention for LCP/RCP and other subfields use exactly the reverse. The reader must always check carefully which convention is being used.

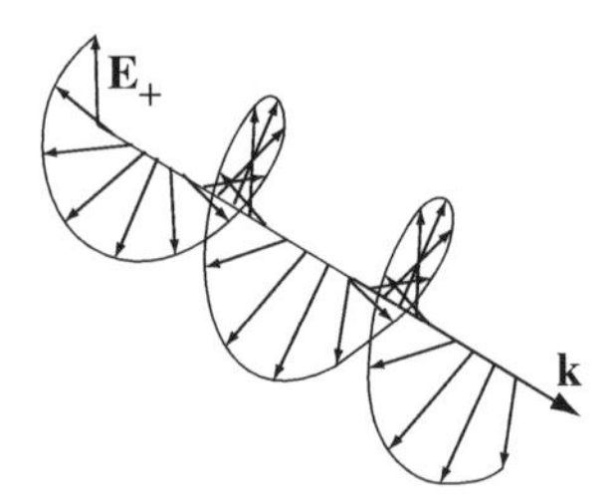

Figure 16.8: A snapshot of $\mathbf{E}(\mathbf{r}, 0)$ for a monochromatic RCP plane wave.

For many applications, it is convenient to replace the Cartesian basis vectors $\hat{\mathbf{e}}_1$ and $\hat{\mathbf{e}}_2$ by two vectors which are complex conjugate pairs:

$$\hat{\mathbf{e}}_+ = \frac{1}{\sqrt{2}}(\hat{\mathbf{e}}_1 + i\,\hat{\mathbf{e}}_2) \qquad \hat{\mathbf{e}}_- = \frac{1}{\sqrt{2}}(\hat{\mathbf{e}}_1 - i\,\hat{\mathbf{e}}_2). \tag{16.60}$$

Direct calculation confirms that these vectors satisfy

$$\hat{\mathbf{e}}_\pm^* \cdot \hat{\mathbf{e}}_\pm = 1 \qquad \text{and} \qquad \hat{\mathbf{e}}_\pm^* \cdot \hat{\mathbf{e}}_\mp = 0. \tag{16.61}$$

The usefulness of (16.60) becomes clear when we recognize that (16.59) is the real part of the following complex electric field (evaluated at $\mathbf{r} = 0$):

$$\mathbf{E}_\pm(\mathbf{r}, t) = \mathcal{A}\hat{\mathbf{e}}_\pm \exp[i(\mathbf{k}\cdot\mathbf{r} - \omega t)] = \mathcal{A}\left[\frac{\hat{\mathbf{e}}_1 \pm i\hat{\mathbf{e}}_2}{\sqrt{2}}\right]\exp[i(\mathbf{k}\cdot\mathbf{r} - \omega t)]. \tag{16.62}$$

This shows that $\hat{\mathbf{e}}_+$ represents a pure LCP wave and $\hat{\mathbf{e}}_-$ represents a pure RCP wave. The vectors (16.60) form a basis in the plane transverse to $\mathbf{k}$. Therefore, a representation fully equivalent to (16.49) is

$$\mathbf{E}(\mathbf{r}, t) = [\mathcal{E}_+\hat{\mathbf{e}}_+ + \mathcal{E}_-\hat{\mathbf{e}}_-]\exp\{i(\mathbf{k}\cdot\mathbf{r} - \omega t)\}. \tag{16.63}$$

Every monochromatic plane wave can be decomposed into components of right circular polarization and left circular polarization as readily as it can be decomposed into components of orthogonal linear polarization.

Complex Vectors

Complex vectors like $\hat{\mathbf{e}}_+$ and $\hat{\mathbf{e}}_-$ in (16.61) do not obey all the usual rules for real vectors. For example, $\mathbf{w}\cdot\mathbf{v} = 0$ and $\mathbf{w}\times\mathbf{v} = 0$ do *not* imply that $\mathbf{w} = 0$ or $\mathbf{v} = 0$ as they do for real vectors. Rather, the magnitude of a complex vector is $|\mathbf{v}|^2 = \mathbf{v}\cdot\mathbf{v}^*$ and two complex vectors $\mathbf{v}$ and $\mathbf{w}$ are *orthogonal* if and only if

$$\mathbf{v}\cdot\mathbf{w}^* = \mathbf{v}^*\cdot\mathbf{w} = 0.$$

The decomposition of a complex vector $\mathbf{v}$ in a complex orthonormal basis $(\mathbf{s}_1, \mathbf{s}_2)$,

$$\mathbf{v} = v_1\mathbf{s}_1 + v_2\mathbf{s}_2 = (\mathbf{v}\cdot\mathbf{s}_1^*)\mathbf{s}_1 + (\mathbf{v}\cdot\mathbf{s}_2^*)\mathbf{s}_2,$$

may look more familiar to students of quantum mechanics if we use Dirac notation:

$$|\mathbf{v}\rangle = |\mathbf{s}_1\rangle\langle\mathbf{s}_1|\mathbf{v}\rangle + |\mathbf{s}_2\rangle\langle\mathbf{s}_2|\mathbf{v}\rangle.$$

The connection between orthogonality and geometric perpendicularity for real vectors is lost for complex vectors. Thus, the vectors $\boldsymbol{\mathcal{E}}$ and $\boldsymbol{\mathcal{B}}$ in (16.38) are not generally orthogonal, despite the fact that the corresponding real electric vector and real magnetic vector are perpendicular.

16.4.4 Elliptical Polarization

The conclusions of Section 16.4.3 for circular polarization generalize to *elliptical polarization* when the phase difference δ and the amplitude ratio A/B in (16.54) take arbitrary values. When $\sin\delta > 0$, the tip of **E** traverses the ellipse in Figure 16.6 in the counterclockwise direction and we speak of left elliptical polarization. When $\sin\delta < 0$, the tip of **E** traverse the ellipse in the clockwise direction and we speak of right elliptical polarization. In both cases, the magnitude of the electric field changes as a function of time as the tip of **E** traces out an elliptical helix in three-dimensional space.

Example 16.2 Find and describe the electromagnetic field produced by a superposition of two equal-amplitude, monochromatic, plane waves that propagate in opposite directions. (a) Let the wave propagating along $+z$ have right circular polarization and the wave propagating along $-z$ have left circular polarization. (b) Repeat when both waves have LCP.

Solution: (a) A possible electric field for an RCP wave propagating along $+z$ and an LCP wave propagating along $-z$ is

$$\mathbf{E}_1 = \tfrac{1}{2}\mathrm{Re}\left\{(\hat{\mathbf{x}} - i\hat{\mathbf{y}})\left[e^{i(kz+\omega t)} + e^{i(kz-\omega t)}\right]\right\} = (\hat{\mathbf{x}}\cos kz + \hat{\mathbf{y}}\sin kz)\cos\omega t.$$

This field is the sum of electric fields like those in (16.25). The associated magnetic fields from (16.27) are

$$c\mathbf{B}_1 = -\hat{\mathbf{z}} \times \tfrac{1}{2}\mathrm{Re}\left\{(\hat{\mathbf{x}} - i\hat{\mathbf{y}})\left[e^{i(kz+\omega t)} - e^{i(kz-\omega t)}\right]\right\} = (\hat{\mathbf{x}}\cos kz + \hat{\mathbf{y}}\sin kz)\sin\omega t.$$

This is a standing wave with $\mathbf{E} \parallel \mathbf{B}$. Since $E_y/E_x = B_y/B_x = \tan kz$, the tips of both field vectors at a fixed time trace out a helix centered on the z-axis.

(b) An electric field where the two counter-propagating waves are both LCP is

$$\mathbf{E}_2 = \mathrm{Re}\left\{(\hat{\mathbf{x}} - i\hat{\mathbf{y}})e^{i(kz+\omega t)} + (\hat{\mathbf{x}} + i\hat{\mathbf{y}})e^{i(kz-\omega t)}\right\} = (\hat{\mathbf{x}}\cos\omega t + \hat{\mathbf{y}}\sin\omega t)\cos kz.$$

The corresponding magnetic field is

$$c\mathbf{B}_2 = \hat{\mathbf{z}} \times \mathrm{Re}\left[(\hat{\mathbf{x}} - i\hat{\mathbf{y}})e^{i(kz+\omega t)} - (\hat{\mathbf{x}} + i\hat{\mathbf{y}})e^{i(kz-\omega t)}\right] = (\hat{\mathbf{x}}\cos\omega t + \hat{\mathbf{y}}\sin\omega t)\sin kz.$$

This is also a standing wave where $\mathbf{E} \parallel \mathbf{B}$. Since $E_y/E_x = B_y/B_x = \tan\omega t$, the entire field pattern lies in a single plane which rotates around the z-axis at frequency ω.

16.4.5 Stokes Parameters and the Poincaré Sphere

The ellipse in Figure 16.6 provides a complete characterization of the polarization of a propagating plane wave. At low frequencies (up to microwave), the electric field vector traces out the polarization ellipse slowly enough to permit the parameters A, B, and δ in (16.53) to be measured more or less directly. At optical and higher frequencies, it is usual to extract A, B, and δ from knowledge of the *Stokes parameters*:

$$\begin{aligned} s_0 &= A^2 + B^2 \\ s_1 &= A^2 - B^2 \\ s_2 &= 2AB\cos\delta \\ s_3 &= 2AB\sin\delta. \end{aligned} \tag{16.64}$$

Only three of the four are needed because

$$s_0^2 = s_1^2 + s_2^2 + s_3^2. \tag{16.65}$$

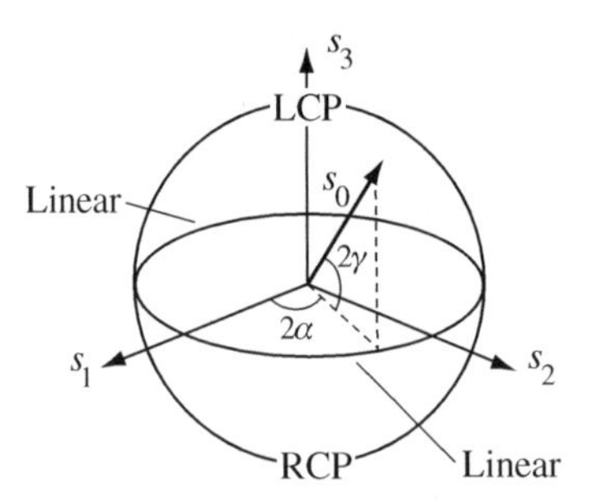

Figure 16.9: A Poincaré sphere with radius s_0. The Cartesian coordinates of the tip of the indicated vector are (s_1, s_2, s_3).

The key to the method is that each Stokes parameter is directly related to wave intensities measured in different orthogonal bases. Specifically, let I_0 and I_{90} be the intensities measured by detectors sensitive to horizontal and vertical linear polarization with respect to the e_1-e_2 coordinate system used in (16.51). Similarly, let I_{+45} and I_{-45} be the intensities measured by detectors sensitive to linear polarizations at angles $+45^\circ$ and -45° with respect to the same axes. Finally, let I_{RCP} and I_{LCP} be the intensities measured by detectors sensitive to right circular and left circular polarization, respectively. With these definitions, we show in Example 16.3 below that

$$
\begin{aligned}
s_0 &= I_0 + I_{90} \\
s_1 &= I_0 - I_{90} \\
s_2 &= I_{+45} - I_{-45} \\
s_3 &= I_{\text{RCP}} - I_{\text{LCP}}.
\end{aligned}
\tag{16.66}
$$

Equation (16.66) establishes that six intensity measurements are sufficient to determine the amplitudes and the relative phase of the two complex electric field components E_1 and E_2 in (16.51).

A notable feature of the Stokes parameters (s_1, s_2, s_3) is that they define a one-to-one correspondence between every state of polarization and a unique point on the surface of sphere of radius s_0 called the *Poincaré sphere* (see Figure 16.9). To see this, note first that the angle α in Figure 16.6 rotates the e_1 - e_2 coordinate system into the principal axis system e_1' - e_2' of the polarization ellipse. The tangent of the angle γ in Figure 16.6 is the ratio of the minor axis to the major axis of the ellipse. By carrying out the rotation explicitly, and writing the field components in the principal axis system as[10]

$$E_1' = a\cos(\mathbf{k}\cdot\mathbf{r} - \omega t + \tau) \quad \text{and} \quad E_2' = \pm b\sin(\mathbf{k}\cdot\mathbf{r} - \omega t + \tau),$$

the reader can use (16.53) and show that

$$\tan 2\alpha = \frac{2AB}{A^2 - B^2}\cos\delta. \tag{16.67}$$

$$\sin 2\gamma = \frac{2AB}{A^2 + B^2}\sin\delta. \tag{16.68}$$

Comparison with (16.64) shows that (16.67) is the ratio s_2/s_1 and that (16.68) is the ratio s_3/s_0. Substituting both of these into (16.65) gives an explicit formula for s_1. Substituting the latter back into

[10] The $\pm$ sign in (16.67) distinguishes right and left elliptical polarization. The common phase angle is τ.

(16.67) gives an explicit formula for s_2. The final results are

$$\begin{aligned} s_1 &= s_0 \cos 2\gamma \cos 2\alpha \\ s_2 &= s_0 \cos 2\gamma \sin 2\alpha \\ s_3 &= s_0 \sin 2\gamma. \end{aligned} \tag{16.69}$$

Equation (16.69) shows that a vector with Cartesian coordinates (s_1, s_2, s_3) is identical to the vector shown in Figure 16.9 with polar coordinates $(s_0, \pi/2 - 2\gamma, 2\alpha)$. This confirms that the Stokes parameters map the original polarization ellipse parameters A, B, and δ onto a unique point on the surface of a sphere. All states of linear polarization (horizontal, vertical, etc.) lie on the sphere's equator, LCP occupies the north pole, and RCP occupies the south pole. All other points on the sphere describe elliptical polarization. For practical work, the Poincaré sphere is a wonderful visualization tool because changes in polarization (produced naturally or by man-made optical elements) are recorded graphically as trajectories on the surface of the sphere.

16.4.6 Partially Polarized and Unpolarized Waves

An electromagnetic field produced by superposing a large number of monochromatic plane waves, all of which have the same propagation vector **k**, is said to be *unpolarized* if the polarization states of the constituent waves are a random mixture of the two orthogonal states of polarization. In such a case, the polarization parameters A, B, and δ in (16.53) are not fixed constants (as they are for a single monochromatic plane wave), but vary randomly as a function of time. The Sun, flames, and man-made incandescent and fluorescent sources all produce electromagnetic waves of this sort. Superpositions of waves where the polarization parameters change in time, but not completely randomly, are said to be *partially polarized.* A virtue of the Stokes parameter formalism is that it applies equally well to these more general situations, provided we perform a suitable time average of the parameters defined in (16.66). An unpolarized wave has average values $\langle s_0 \rangle \neq 0$ and $\langle s_1 \rangle = \langle s_2 \rangle = \langle s_3 \rangle = 0$. The last three parameters take values between zero and s_0 for partially polarized fields.

Example 16.3 Confirm the four formulae quoted in (16.66).

Solution: The exponential factor in (16.49) does not play a role so we write a general state of polarization as

$$\boldsymbol{\mathcal{E}} = \mathcal{E}_1 \hat{\mathbf{e}}_1 + \mathcal{E}_2 \mathbf{e}_2.$$

I_0 and I_{90} are the intensities in the $\hat{\mathbf{e}}_1$ and $\hat{\mathbf{e}}_2$ components, respectively. Therefore, using (16.50), we get immediate agreement with (16.64) that

$$s_0 = I_0 + I_{90} = |\mathcal{E}_1|^2 + |\mathcal{E}_2|^2 = A^2 + B^2$$

and

$$s_1 = I_0 - I_{90} = |\mathcal{E}_1|^2 - |\mathcal{E}_2|^2 = A^2 - B^2.$$

To compute s_2, we rotate the e_1-e_2 coordinate system by $\theta = -45^\circ$ and follow the prescription of Section 1.7. This gives the components in the new basis as

$$\begin{bmatrix} \mathcal{E}_{-45} \\ \mathcal{E}_{+45} \end{bmatrix} = \begin{bmatrix} \cos\theta & \sin\theta \\ -\sin\theta & \cos\theta. \end{bmatrix} \begin{bmatrix} \mathcal{E}_1 \\ \mathcal{E}_2 \end{bmatrix} = \frac{1}{\sqrt{2}} \begin{bmatrix} \mathcal{E}_1 - \mathcal{E}_2 \\ \mathcal{E}_1 + \mathcal{E}_2 \end{bmatrix}.$$

The intensity difference of interest is

$$s_2 = I_{+45} - I_{-45} = |\mathcal{E}_{+45}|^2 - |\mathcal{E}_{-45}|^2 = \frac{1}{2}\left[(\mathcal{E}_1 + \mathcal{E}_2)(\mathcal{E}_1 + \mathcal{E}_2)^* - (\mathcal{E}_1 - \mathcal{E}_2)(\mathcal{E}_1 - \mathcal{E}_2)^*\right].$$

Therefore, in agreement with (16.64),

$$s_2 = \mathcal{E}_1^*\mathcal{E}_2 + \mathcal{E}_1\mathcal{E}_2^* = 2\text{Re}\,[\mathcal{E}_1^*\mathcal{E}_2] = 2AB\cos\delta.$$

Finally, we compute s_3 by rewriting (16.49) in the circular polarization form (16.63). This shows that

$$\mathcal{E}_+ = \frac{\mathcal{E}_1 - i\mathcal{E}_2}{\sqrt{2}} \qquad \text{and} \qquad \mathcal{E}_- = \frac{\mathcal{E}_1 + i\mathcal{E}_2}{\sqrt{2}}.$$

Therefore,

$$s_3 = I_{\text{RCP}} - I_{\text{LCP}} = |\mathcal{E}_+|^2 - |\mathcal{E}_-|^2 = \frac{1}{2}[(\mathcal{E}_1 - i\mathcal{E}_2)(\mathcal{E}_1^* + i\mathcal{E}_2^*) - (\mathcal{E}_1 + i\mathcal{E}_2)(\mathcal{E}_1^* - i\mathcal{E}_2^*)].$$

This also agrees with (16.64) because

$$s_3 = -i[\mathcal{E}_1^*\mathcal{E}_2 - \mathcal{E}_1\mathcal{E}_2^*] = 2\text{Im}\,[\mathcal{E}_1^*\mathcal{E}_2] = 2AB\sin\delta.$$

16.5 Wave Packets

A monochromatic plane wave like (16.49) fills all of space and therefore cannot exist in Nature. However, by superposing plane waves with different wave vectors, one can synthesize solutions of the source-free Maxwell equations that do not extend to infinity in any direction. These physically realizable *wave packets* are, in fact, the only kinds of waves that can be produced by real sources. Using the monochromatic plane wave fields (16.39) as basis functions, the physical fields associated with a general electromagnetic wave packet are[11]

$$\mathbf{E}(\mathbf{r}, t) = \text{Re}\frac{1}{(2\pi)^3}\int d^3k\, \boldsymbol{\mathcal{E}}_\perp(\mathbf{k}) \exp\{i(\mathbf{k}\cdot\mathbf{r} - ckt)\} \tag{16.70}$$

and

$$c\mathbf{B}(\mathbf{r}, t) = \text{Re}\frac{1}{(2\pi)^3}\int d^3k\, \left[\hat{\mathbf{k}} \times \boldsymbol{\mathcal{E}}_\perp(\mathbf{k})\right] \exp\{i(\mathbf{k}\cdot\mathbf{r} - ckt)\}. \tag{16.71}$$

The vector amplitude function $\boldsymbol{\mathcal{E}}_\perp(\mathbf{k})$ in (16.70) satisfies $\mathbf{k}\cdot\boldsymbol{\mathcal{E}}_\perp(\mathbf{k}) = 0$ but is otherwise arbitrary as long as the integrals converge.

16.5.1 The Total Energy

The total energy U_{EM} of an electromagnetic wave packet is finite and time-independent. This differs fundamentally from the infinite and time-varying total energy computed in Section 16.3.4 for any of its plane wave constituents. To show this, we begin by substituting (16.70) and (16.71) into the energy density formula,

$$u_{\text{EM}}(\mathbf{r}, t) = u_E(\mathbf{r}, t) + u_B(\mathbf{r}, t) = \tfrac{1}{2}\epsilon_0(|\text{Re}\mathbf{E}(\mathbf{r}, t)|^2 + c^2|\text{Re}\mathbf{B}(\mathbf{r}, t)|^2). \tag{16.72}$$

[11] The factors of $(2\pi)^{-3}$ are inserted for consistency with the Fourier transform convention in Section 1.6.

The total energy follows by integrating over all space:

$$U_{\rm EM}(t) = U_E(t) + U_B(t) = \int d^3r\, u_{\rm EM}(\mathbf{r}, t). \tag{16.73}$$

Using $\omega = ck$, straightforward algebra shows that the electric energy density $u_E(\mathbf{r}, t)$ contains terms proportional to

$$\boldsymbol{\mathcal{E}}_\perp(\mathbf{k}) \cdot \boldsymbol{\mathcal{E}}^*_\perp(\mathbf{k}')e^{i(\mathbf{k}-\mathbf{k}')\cdot\mathbf{r}}e^{-i(\omega-\omega')t} \quad \text{and} \quad \boldsymbol{\mathcal{E}}_\perp(\mathbf{k}) \cdot \boldsymbol{\mathcal{E}}_\perp(\mathbf{k}')e^{i(\mathbf{k}+\mathbf{k}')\cdot\mathbf{r}}e^{-i(\omega+\omega')t} \tag{16.74}$$

and the complex conjugates of these terms. The reader can confirm that integrating the sum of all these terms over space as indicated in (16.73) and using the delta function identity

$$\int d^3r\, e^{i\mathbf{p}\cdot\mathbf{r}} = (2\pi)^3\delta(\mathbf{p}) \tag{16.75}$$

gives[12]

$$U_E(t) = \frac{\epsilon_0}{8}\frac{1}{(2\pi)^3}\int d^3k\, \left\{|\boldsymbol{\mathcal{E}}_\perp(\mathbf{k})|^2 + \boldsymbol{\mathcal{E}}_\perp(\mathbf{k}) \cdot \boldsymbol{\mathcal{E}}_\perp(-\mathbf{k})e^{-2i\omega t}\right\} + \text{c.c.} \tag{16.76}$$

The corresponding total magnetic energy $U_B(t)$ is identical to (16.76) except that the time-dependent term has the opposite sign. Therefore, the total electromagnetic energy of the wave packet is

$$U_{\rm EM} = \frac{\epsilon_0}{2}\frac{1}{(2\pi)^3}\int d^3k\, |\boldsymbol{\mathcal{E}}_\perp(\mathbf{k})|^2. \tag{16.77}$$

Comparing (16.77) to (16.42) shows that $U_{\rm EM}$ is the sum of the (time-averaged) total energies of the monochromatic plane waves that constitute the packet. The important difference is that $U_{\rm EM}$ for the wave packet is strictly time-independent without any need for time-averaging. This is a consequence of conservation of energy. The total energy of a physically realizable wave packet must be finite and strictly constant in time. If two (or more) wave packets launched from different parts of space propagate and "collide" with one another in free space, the total energy must remain the sum of their individual energies, even as their individual electromagnetic fields superpose during the time they overlap in space. A similar result holds for the total linear momentum of a wave packet, namely,

$$c\mathbf{P}_{\rm EM} = \frac{\epsilon_0}{2}\frac{1}{(2\pi)^3}\int d^3k\, \hat{\mathbf{k}}\, |\boldsymbol{\mathcal{E}}_\perp(\mathbf{k})|^2. \tag{16.78}$$

The foregoing may be contrasted with a superposition of electrostatic fields like $\mathbf{E} = \mathbf{E}_1 + \mathbf{E}_2 + \cdots$ or magnetostatic fields $\mathbf{B} = \mathbf{B}_1 + \mathbf{B}_2 + \cdots$. In these cases, the cross terms in the total energy proportional to the space integrals of $\mathbf{E}_j \cdot \mathbf{E}_k$ and $\mathbf{B}_j \cdot \mathbf{B}_k$ are non-zero and represent the work done to bring the source of field j into the field produced by source k. These terms are zero for free fields because, by definition, they have no sources.

16.5.2 Scalar Wave Packets

Sections 16.2.1 and 16.2.2 taught us to use electromagnetic potentials to generate solutions of the free-space Maxwell equations from solutions of the scalar wave equation. We know also, from experience, that it is easier to add scalars than to add vectors. Therefore, our analysis becomes simpler, with no loss of generality, if we abandon the vector wave packets (16.70) and (16.71) in favor of *scalar wave packets* formed by summing solutions of the scalar wave equation

$$\nabla^2 u - \frac{1}{c^2}\frac{\partial^2 u}{\partial t^2} = 0. \tag{16.79}$$

[12] The abbreviation "c.c." in (16.76) stands for complex conjugate.

A packet composed of monochromatic plane waves with frequency $\omega(\mathbf{k}) = c|\mathbf{k}|$ is

$$u(\mathbf{r}, t) = \frac{1}{(2\pi)^3} \int d^3k \; \hat{u}(\mathbf{k}) \exp\{i[\mathbf{k} \cdot \mathbf{r} - \omega(\mathbf{k})t]\}. \tag{16.80}$$

The amplitude weighting factor $\hat{u}(\mathbf{k})$ in (16.80) is determined by the shape of the wave packet at $t = 0$. Indeed, $\hat{u}(\mathbf{k})$ and $u(\mathbf{r}, 0)$ are precisely Fourier transform partners:[13]

$$u(\mathbf{r}, 0) = \frac{1}{(2\pi)^3} \int d^3k \, \hat{u}(\mathbf{k}) \exp(i\mathbf{k} \cdot \mathbf{r}) \tag{16.81}$$

$$\hat{u}(\mathbf{k}) = \int d^3r \; u(\mathbf{r}, 0) \exp(-i\mathbf{k} \cdot \mathbf{r}). \tag{16.82}$$

The implications of (16.81) and (16.82) are clearest when we choose $\hat{u}(\mathbf{k})$ to be non-zero only for wave vectors that point along the x-axis. Dropping the pre-factor, this reduces (16.81) to the one-dimensional integral

$$u(x, 0) = \int_{\infty}^{\infty} dk_x \; \hat{u}(k_x) \exp(ik_x x). \tag{16.83}$$

To illustrate (16.83), it is convenient to choose $\hat{u}(k_x)$ as a normalized Gaussian with half-width Δk_x centered on k_{0x}:

$$\hat{u}(k_x) = \frac{1}{\sqrt{\pi}\Delta k_x} \exp[-(k_x - k_{0x})^2/\Delta k_x^2]. \tag{16.84}$$

Inserting (16.84) into (16.83) and using

$$\int_{-\infty}^{\infty} ds \exp(as - bs^2) = \sqrt{\frac{\pi}{b}} \exp(a^2/4b) \tag{16.85}$$

produces another Gaussian, this time with half-width $\Delta x = 2/\Delta k_x$:

$$u(x, 0) = \exp(ik_{0x}x) \exp[-x^2/(\Delta x)^2]. \tag{16.86}$$

The real part of (16.86) plotted in Figure 16.10 shows a stationary wave packet composed of a sinusoid with wavelength $2\pi/k_{0x}$ and an amplitude modulated by a Gaussian exponential factor. The Gaussian narrows to a delta function in the limit when $\hat{u}(k_x) = const$. Conversely, the wave packet expands to a single plane wave with infinite width when $\hat{u}(k_x)$ is a delta function.

16.5.3 Complementarity

The inverse relation between Δx and Δk_x quoted following (16.85) is not a special feature of the Gaussian weight function. To see this, consider a packet that extends from x to $x + \Delta x$ composed of plane waves with propagation vectors in the interval $[k, k + \Delta k]$. The wave $\exp(ikx)$ suffers a phase change $k\Delta x$ from one end of the packet to the other. The wave $\exp[i(k + \Delta k)x]$ similarly suffers a phase change $(k + \Delta k)\Delta x$ across the packet. Now, $u(x, 0)$ goes to zero on the left side of the packet in Figure 16.10 because complete destructive interference occurs there. The next point of complete destructive interference is the right side of the packet where $u(x, 0)$ is again zero. Hence, when we add together the two extremal waves considered above,

$$(k + \Delta k)\Delta x - k\Delta x \sim 2\pi \quad \Rightarrow \quad \Delta k \Delta x \sim 2\pi. \tag{16.87}$$

[13] This may be confirmed by substituting one into the other and using (1.119).

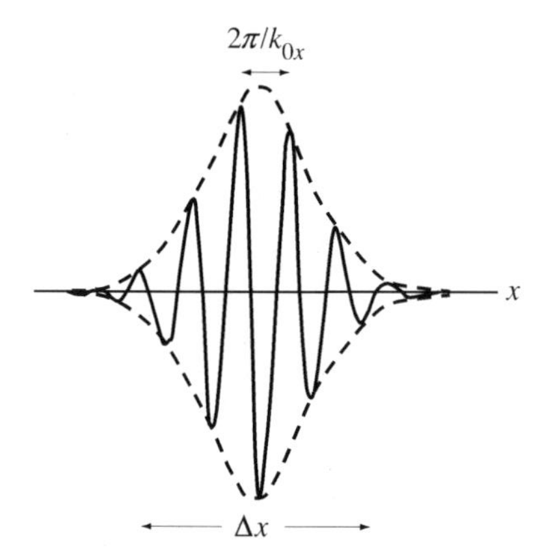

Figure 16.10: A one-dimensional wave packet. An envelope function (dashed lines) of width Δx modulates the amplitude of a carrier wave with wave vector k_{0x} (solid lines).

Equation (16.87) reflects a *complementarity* between Δx and Δk that is fundamental to the superposition of plane waves (and to Fourier analysis in general). A more careful treatment shows that the *variances* of the distributions $\hat{u}(k)$ and $u(x)$ satisfy

$$\Delta x\, \Delta k_x \geq \tfrac{1}{2}. \tag{16.88}$$

For the general scalar wave packet in (16.81), the inequality (16.88) applies together with[14]

$$\Delta y\, \Delta k_y \geq \tfrac{1}{2} \qquad \text{and} \qquad \Delta z\, \Delta k_z \geq \tfrac{1}{2}. \tag{16.89}$$

These results imply that $u(\mathbf{r}, 0)$ can be localized to an arbitrarily small volume of space by including sufficiently many wave vectors $\mathbf{k}$ in the superposition integral (16.81). However, as we shall soon see, the three-dimensional localization of such a wave packet cannot be maintained as the packet propagates.

16.5.4 Group Velocity

The time evolution of (16.80) determines the motion of $u(\mathbf{r}, t)$ through space. The analysis is straightforward when the amplitude function $\hat{u}(\mathbf{k})$ is negligible for all $\mathbf{k}$ except for a small set of wave vectors that do not differ much from a fixed wave vector $\mathbf{k}_0$ in either magnitude or direction. This invites us to keep only the linear term in the Taylor expansion of the *dispersion relation* $\omega(\mathbf{k})$,[15]

$$\omega(\mathbf{k}) = \omega(\mathbf{k}_0) + \left.\frac{\partial \omega}{\partial k_i}\right|_{\mathbf{k}=\mathbf{k}_0} (\mathbf{k} - \mathbf{k}_0)_i + \cdots. \tag{16.90}$$

Now let $\omega_0 = \omega(\mathbf{k}_0)$ and define the *group velocity* as

$$\boldsymbol{v}_g = \nabla_{\mathbf{k}}\omega(\mathbf{k})|_{\mathbf{k}=\mathbf{k}_0} = \left.\frac{\partial \omega}{\partial \mathbf{k}}\right|_{\mathbf{k}=\mathbf{k}_0}. \tag{16.91}$$

[14] These inequalities, combined with the quantum mechanical relation $\mathbf{p} = \hbar\mathbf{k}$ between particle momentum and wave vector, produces the Heisenberg uncertainty relation for momentum and position.

[15] The term "dispersion relation" has its historical origin in optics. A useful mnemonic is that the frequency ω disperses ("changes") as the wave vector $\mathbf{k}$ changes.

In terms of these quantities, substitution of (16.90) into (16.80) gives

$$u(\mathbf{r},t) = e^{i(\mathbf{k}_0\cdot\mathbf{r}-\omega_0 t)}\frac{1}{(2\pi)^3}\int d^3k\ \hat{u}(\mathbf{k})\exp\{i[(\mathbf{k}-\mathbf{k}_0)\cdot(\mathbf{r}-\boldsymbol{v}_{\rm g}t)]\}. \tag{16.92}$$

The meaning of this expression emerges when we define the function

$$\psi(\mathbf{p}) = \frac{1}{(2\pi)^3}\int d^3k\ \hat{u}(\mathbf{k})e^{i(\mathbf{k}-\mathbf{k}_0)\cdot\mathbf{p}} \tag{16.93}$$

and write (16.92) in the form

$$u(\mathbf{r},t) = \exp\{i(\mathbf{k}_0\cdot\mathbf{r}-\omega_0 t)\}\psi(\mathbf{r}-\boldsymbol{v}_{\rm g}t). \tag{16.94}$$

The wave packet (16.94) generalizes (16.86) to a monochromatic *carrier wave* with wave vector $\mathbf{k}_0$ and frequency ω_0 modulated by the *envelope function* $\psi(\mathbf{r}-\boldsymbol{v}_{\rm g}t)$. The argument of ψ shows that the envelope moves without distortion at the group velocity $\boldsymbol{v}_{\rm g}$. For our vacuum problem, $\omega = c|\mathbf{k}|$, and the group velocity (16.91) happens to equal the phase velocity in (16.32):

$$\boldsymbol{v}_{\rm g} = c\frac{\partial}{\partial\mathbf{k}}\sqrt{k_x^2+k_y^2+k_z^2} = c\hat{\mathbf{k}} = \boldsymbol{v}_{\rm p}. \tag{16.95}$$

Later, we will encounter situations (in matter) where $\boldsymbol{v}_{\rm g} \neq \boldsymbol{v}_{\rm p}$. In those cases, the carrier wave train appears to move through the envelope with relative velocity $\boldsymbol{v}_{\rm p} - \boldsymbol{v}_{\rm g}$.

16.5.5 Free-Space Diffraction

Real wave packets in vacuum do not propagate in the undistorted manner predicted by (16.94). Instead, the quadratic and higher-order terms in (16.90) cause such packets to spread out in the directions transverse to the propagation direction.[16] This phenomenon—called *free-space diffraction*—arises from the presence of plane waves in the packet (16.80) with non-zero **k**-vector components transverse to the average propagation direction. The very same plane wave components permit a wave to "bend around" the edge of an opaque obstruction in the context most usually associated with the word "diffraction" (Chapter 21).

The rate at which $u(\mathbf{r},t)$ spreads in the transverse direction is large when the spatial extent of $u(\mathbf{r},0)$ in the transverse direction is small. This is so because (16.88) and (16.89) imply that such a packet can be formed only if (16.80) includes wave vectors whose magnitude and direction differ considerably from a specified vector $\mathbf{k}_0$. This contradicts the assumption made in (16.90) that all terms beyond the first two are negligible. To be more quantitative, we first quote (but do not prove) two general results for scalar wave packet propagation that bear on the question at hand. Then, in Section 16.7, we demonstrate free-space diffraction explicitly for a beam-like wave of the kind produced by a laser or a flashlight in vacuum. The reader may wish to look ahead to Figure 16.11 for an illustration of the spreading phenomenon.

Following Bradford,[17] we consider the time evolution of a scalar wave packet like (16.80) without making any specific choice for either $\hat{u}(\mathbf{k})$ or $\omega(\mathbf{k})$. For any function of position $f(\mathbf{r})$, we define an "expectation value"

$$\langle f\rangle_{\mathbf{r}} = \int d^3r\ u^*(\mathbf{r},t)f(\mathbf{r})u(\mathbf{r},t). \tag{16.96}$$

[16] The exception that proves the rule is a one-dimensional packet composed of vacuum plane waves that all propagate in the same direction. This packet does not distort along the x-direction of propagation because $\omega(\mathbf{k}) = ck_x$ and all the higher-order terms in (16.90) are zero. On the other hand, the packet extends to infinity in every transverse direction.

[17] H.M. Bradford, "Propagation and spreading of a pulse or wave packet", *American Journal of Physics* **44**, 1058 (1976).

For simplicity, we assume that $\hat{u}(\mathbf{k})$ in (16.80) is real so all the plane waves that contribute to the integral are in phase at $\mathbf{r} = 0$ when $t = 0$. In that case, the reader can confirm by direct calculation that the "centroid" of the wave packet satisfies

$$\langle \mathbf{r} \rangle_{\mathbf{r}} = \langle \boldsymbol{v}_{\mathrm{g}} \rangle_{\mathbf{k}} t, \tag{16.97}$$

where

$$\langle \boldsymbol{v}_{\mathrm{g}} \rangle_{\mathbf{k}} = \int d^3k \, |\hat{u}(\mathbf{k})|^2 \boldsymbol{v}_{\mathrm{g}}(\mathbf{k}) = \int d^3k \, |\hat{u}(\mathbf{k})|^2 \nabla_{\mathbf{k}} \omega(\mathbf{k}). \tag{16.98}$$

Equation (16.97) says that the centroid of the wave packet moves at a constant velocity equal to the weighted average of the group velocities that contribute to the packet. For waves in vacuum, this amounts to an average over directions because (16.95) shows that every wave in the packet travels at the speed of light.

More interesting is the second moment of $u(\mathbf{r}, t)$ with respect to the position of the centroid. This gives the time evolution of the wave packet *width* as propagation proceeds. The specific quantity to calculate is the variance,

$$\sigma^2 = \langle (\mathbf{r} - \langle \mathbf{r} \rangle_{\mathbf{r}})^2 \rangle_{\mathbf{r}} = \langle (x - \langle x \rangle_{\mathbf{r}})^2 \rangle_{\mathbf{r}} + \langle (y - \langle y \rangle_{\mathbf{r}})^2 \rangle_{\mathbf{r}} + \langle (z - \langle z \rangle_{\mathbf{r}})^2 \rangle_{\mathbf{r}}. \tag{16.99}$$

Under the same conditions where (16.97) is valid, one finds that

$$\sigma^2(t) = \sigma^2(0) + \langle (\boldsymbol{v}_{\mathrm{g}} - \langle \boldsymbol{v}_{\mathrm{g}} \rangle_{\mathbf{k}})^2 \rangle_{\mathbf{k}} t^2. \tag{16.100}$$

In words, the width of the packet increases as time goes on by an amount that depends on the variance of the group velocities that contribute to the packet. Since all vacuum plane waves travel at the same speed, this means that the rate of spreading is largest for packets composed of plane waves with the widest range of directions.

16.6 The Helmholtz Equation

The harmonic time dependence of the monochromatic plane waves used in the last several sections is the basis for a general approach to time-dependent fields based on Fourier's theorem. Following (1.128), we Fourier analyze an arbitrary function of space and time $u(\mathbf{r}, t)$ and write

$$u(\mathbf{r}, t) = \frac{1}{2\pi} \int_{-\infty}^{\infty} d\omega \, \hat{u}(\mathbf{r}, \omega) e^{-i\omega t}, \qquad \text{where} \qquad \hat{u}(\mathbf{r}, \omega) = \int_{-\infty}^{\infty} dt \, u(\mathbf{r}, t) e^{i\omega t}. \tag{16.101}$$

The Fourier method exploits the linearity of the Maxwell equations to focus on a single frequency component like $\hat{u}(\mathbf{r}, \omega) \exp(-i\omega t)$. Once the behavior of this quantity is known, the integral on the left side of (16.101) determines the behavior of $u(\mathbf{r}, t)$. The method is limited only by our ability to perform the final Fourier synthesis integral.

The Fourier philosophy applied to the scalar wave equation directs us to substitute $\hat{u}(\mathbf{r}, \omega) \exp(-i\omega t)$ into (16.79). The result is a partial differential equation for $\hat{u}(\mathbf{r}, \omega)$ called the *Helmholtz equation*,

$$\left(\nabla^2 + \frac{\omega^2}{c^2} \right) \hat{u}(\mathbf{r}, \omega) = 0. \tag{16.102}$$

A general solution of (16.102) produces a general solution of the wave equation when $\hat{u}(\mathbf{r}, \omega)$ is substituted back into (16.101). In Section 16.8, we used separation of variables (see Chapter 7) to solve the Helmholtz equation. Another method is the subject of the next section.

16.6.1 The Angular Spectrum of Plane Waves

Many scattering, diffraction, and beam problems seek a solution of the Helmholtz equation in the half-space $z > 0$ when information about the solution is supplied on the $z = 0$ plane. The *angular spectrum of plane waves* is very well suited for this task. This superposition method is based on the fact that $\exp(i\mathbf{k}\cdot\mathbf{r})$ is a solution of (16.102) if its wave vector components k_x, k_y, and k_z satisfy

$$k_x^2 + k_y^2 + k_z^2 = \frac{\omega^2}{c^2}. \tag{16.103}$$

Therefore, if $f(\mathbf{k})$ is a weight function, a general solution of the Helmholtz equation is

$$\hat{u}(\mathbf{r},\omega) = \frac{1}{(2\pi)^3}\int d^3k\, f(\mathbf{k})\exp(i\mathbf{k}\cdot\mathbf{r})\delta(\omega - ck). \tag{16.104}$$

Our strategy is to eliminate the delta function by integrating over k_z and write the result in a manner which manifestly converges when $z \geq 0$. If $\omega = ck_0$, and $g(k_x, k_y)$ is another weight function, the integral that remains is

$$\hat{u}(x,y,z,\omega) = \frac{1}{(2\pi)^2}\int\limits_{-\infty}^{\infty} dk_x \int\limits_{-\infty}^{\infty} dk_y g(k_x,k_y) e^{i(k_x x + k_y y)} \times \begin{cases} e^{i\sqrt{k_0^2 - k_x^2 - k_y^2}\,z} & k_x^2 + k_y^2 \leq k_0^2, \\ e^{-\sqrt{k_x^2 + k_y^2 - k_0^2}\,z} & k_x^2 + k_y^2 > k_0^2. \end{cases} \tag{16.105}$$

This weighted sum differs from the plane wave packet (16.80) in Section 16.5.2 because, as the brace in (16.105) indicates, (16.103) forces some values of k_z to be imaginary. Waves of this kind vary exponentially (rather than sinusoidally) as a function of z. Including only the decaying exponentials guarantees that (16.105) converges for $z \geq 0$.

Plane waves with complex wave vectors are called *evanescent*. They occur naturally in the theory of total internal reflection from a dielectric interface.[18] The need to include evanescent waves in (16.105) becomes clear when we suppose that $\hat{u}(\mathbf{r},\omega)$ is known or specified on the plane $z = 0$. In that case, setting $z = 0$ on both sides of (16.105) shows that $g(k_x, k_y)$ is determined from $\hat{u}(x, y, z = 0, \omega)$ by a two-dimensional inverse Fourier transform. In general, all values of k_x and k_y, including those where (16.103) makes k_z imaginary, are needed to represent an arbitrary choice for $\hat{u}(x, y, z = 0, \omega)$. Substituting $g(k_x, k_y)$ back into (16.105) generates a solution to the Helmholtz equation in the space $z \geq 0$ which reduces to the specified boundary values on the $z = 0$ plane. We will exploit the angular spectrum of plane waves when we study the field diffracted by an aperture in a conducting plane (see Chapter 21).

16.7 Beam-Like Waves

Flashlights and lasers produce narrow pencils of illumination called *beams*. All beams have the property that their intensity profile falls off very rapidly in the direction transverse to the direction of propagation. Beams also exhibit free-space diffraction (Section 16.5.5) in the sense that the intensity profile spreads out transversely as propagation proceeds. In this section, we construct and analyze the diffractive properties of *approximate*, beam-like solutions to the scalar wave equation. Using these beam-like scalar waves, it is straightforward to construct approximate, beam-like solutions of the Maxwell equations using the potential functions discussed in Section 16.2.2. An outstanding feature of beam-like Maxwell fields is that they are *not transverse*. In other words, $\mathbf{E}(\mathbf{r}, t)$ and $\mathbf{B}(\mathbf{r}, t)$ generally have non-zero components parallel to the propagation direction.

[18] We will have more to say about evanescent waves in Section 17.3.7.

16.7.1 The Paraxial Approximation

Let the z-axis be the axis of propagation. With $\mathbf{r}_\perp = x\hat{\mathbf{x}} + y\hat{\mathbf{y}}$, we choose a trial solution for the scalar wave equation in the form of a monochromatic plane wave with a spatially modulated amplitude:

$$u(\mathbf{r}, t) = \Psi(\mathbf{r}_\perp, z)e^{i(kz-\omega t)} \qquad \omega = ck. \tag{16.106}$$

This is an ansatz of the form (16.101) with $\hat{u}(\mathbf{r}, \omega) = \Psi(\mathbf{r})\exp(ikz)$. Using this and the definition

$$\nabla^2 = \frac{\partial^2}{\partial x^2} + \frac{\partial^2}{\partial y^2} + \frac{\partial^2}{\partial z^2} \equiv \nabla_\perp^2 + \frac{\partial^2}{\partial z^2} \tag{16.107}$$

simplifies the Helmholtz equation (16.102) to an equation for the modulation function $\Psi(\mathbf{r}_\perp, z)$ alone:

$$\nabla_\perp^2\Psi + 2ik\frac{\partial\Psi}{\partial z} + \frac{\partial^2\Psi}{\partial z^2} = 0. \tag{16.108}$$

We now suppose that $\Psi(\mathbf{r}_\perp, z)$ changes very little in the z-direction over the wavelength $\lambda = 2\pi/k$ of the carrier plane wave. This assumption amounts to what is called the *paraxial approximation*:

$$k\frac{\partial\Psi}{\partial z} \gg \frac{\partial^2\Psi}{\partial z^2}. \tag{16.109}$$

The inequality (16.109) permits us to drop the last term in (16.108) to get the *paraxial wave equation*:

$$-\frac{1}{2k}\nabla_\perp^2\Psi = i\frac{\partial\Psi}{\partial z}. \tag{16.110}$$

It is not difficult to find solutions to (16.110). One quick method notes that, when $z \to t$, (16.110) becomes the time-dependent Schrödinger equation for a free particle in two dimensions with mass $m = \hbar k$. A familiar solution to the latter is a plane wave with two-dimensional wave vector $\mathbf{q}_\perp$ and energy $E = \hbar^2 q_\perp^2/2m$. Hence, an elementary solution of (16.110) is

$$\Psi(\mathbf{r}_\perp, z) = \exp[i(\mathbf{q}_\perp \cdot \mathbf{r}_\perp - q_\perp^2 z/2k)]. \tag{16.111}$$

The general solution is a sum of such waves:

$$\Psi(\mathbf{r}_\perp, z) = \int d^2q_\perp \; \hat{\Psi}(\mathbf{q}_\perp)e^{i(\mathbf{q}_\perp\cdot\mathbf{r}_\perp - q_\perp^2 z/2k)}. \tag{16.112}$$

Therefore, (16.106) takes the form

$$u(\mathbf{r}, t) = \int d^2q_\perp \; \hat{\Psi}(\mathbf{q}_\perp)\exp\left[i\left(\mathbf{q}_\perp \cdot \mathbf{r}_\perp + \left\{k - \frac{q_\perp^2}{2k}\right\}z - \omega t\right)\right]. \tag{16.113}$$

To get beam-like solutions to (16.110), we need only choose $\hat{\Psi}(\mathbf{q}_\perp)$ in (16.113) so $\Psi(\mathbf{r}_\perp, z)$ falls to zero rapidly as $|\mathbf{r}_\perp|$ increases from zero. On the other hand, because $\omega = ck$ was assumed by our derivation, (16.113) is *not* a sum of plane waves that satisfy the scalar wave equation. Section 16.7.3 explores in more detail the physical meaning of the paraxial approximation made here.

16.7.2 The Gaussian Beam

The most important beam-like wave derived from (16.113) exploits the Gaussian weight function

$$\hat{\Psi}(\mathbf{q}_\perp) = w_0^2\exp[-\tfrac{1}{4}w_0^2q_\perp^2]. \tag{16.114}$$

This choice factors (16.112) into two one-dimensional integrals of the form (16.85). The resulting modulation function $\Psi(\mathbf{r}_\perp, z)$, substituted into (16.106), produces an exact, cylindrically symmetric

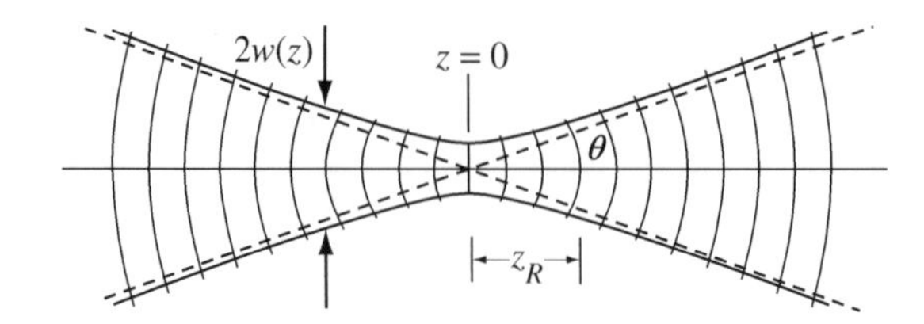

Figure 16.11: A Gaussian beam. The two heavy solid curves indicate the width $2w(z)$. The two dashed lines indicate the asymptotic divergence of the width. The curvature of the surfaces of constant phase (light solid lines) is largest at a distance z_R from the beam waist at $z = 0$ in the center of the drawing.

solution to (16.110) known as the fundamental *Gaussian beam*. In terms of the transverse variable $\rho^2 = x^2 + y^2$, this approximate solution to the original scalar wave equation (16.79) is

$$u(\rho, z, t) = \frac{4\pi w_0^2}{w_0^2 + i2z/k} \exp\left[-\frac{\rho^2}{w_0^2 + i2z/k}\right] \exp[i(kz - \omega t)]. \tag{16.115}$$

The characteristics of the Gaussian beam are best understood if we define

$$z_R = \frac{1}{2} k w_0^2 \tag{16.116}$$

$$R(z) = z + \frac{z_R^2}{z} \tag{16.117}$$

$$w(z) = w_0 \sqrt{1 + (z/z_R)^2} \tag{16.118}$$

$$\alpha(z) = -\tan^{-1}\left[\frac{z}{z_R}\right]. \tag{16.119}$$

Some straightforward (but tedious) algebra shows that (16.115) is identical to

$$u(\rho, z, t) = 4\pi \left[\frac{w_0}{w(z)}\right] \exp\left[-\frac{\rho^2}{w^2(z)}\right] \exp\left\{i\left[\frac{k\rho^2}{2R(z)} + \alpha(z) + kz - \omega t\right]\right\}. \tag{16.120}$$

Equation (16.120) has beam-like properties because the Gaussian factor $\exp[-\rho^2/w^2(z)]$ causes its amplitude to decrease rapidly in the radial direction transverse to the direction of propagation. On the other hand, (16.118) shows that the cross sectional area of the beam, $\pi w^2(z)$, increases from πw_0^2 to $2\pi w_0^2$ when z increases from $z = 0$ (the "beam waist") to $z = z_R$ (the "Rayleigh distance"). This diffractive spreading of the beam comes entirely from the $\mathbf{q}_\perp \neq 0$ terms in (16.113) that propagate in directions other than z.

The heavy solid curves in Figure 16.11 are the locus of points traced out by the width function $w(z)$. These curves define a hyperboloid of revolution around the propagation axis. The asymptotic behavior of $w(z)$ is shown by dashed straight lines inclined at angles θ from the z-axis, where

$$w(z \to \pm\infty) = \frac{w_0}{z_R} z = \theta z. \tag{16.121}$$

The angle $\theta = \lambda/\pi w_0$ is called the *beam divergence*. The inverse relation between θ and w_0 is another example of the complementary nature of free-space diffraction (See Section 16.5.3). The smaller we try to make the beam waist, the more rapidly the beam spreads out as it propagates.

The function $R(z)$ in (16.120) gives the radius of curvature of the surfaces of constant phase at points near the propagation axis (see Example 16.4). The latter are indicated as light solid lines in Fig. 16.11. $R(z)$ is smallest at $z = z_R$ and largest (infinite) at the beam waist and at $z = \pm\infty$. This means that the field of the Gaussian beam approaches a plane wave as $z \to \pm\infty$. A glance at (16.117) shows that $R(z)$ changes sign as z passes through zero. The associated change of sign in (16.120) is responsible for the converging versus diverging effects of diffraction seen in Fig. 16.11 as the beam propagates from left to right through $z = 0$.

The phase $\alpha(z)$ defined in (16.119) is called the *Gouy phase* and arises entirely from the transverse confinement of the beam. The monotonic decrease of this quantity as z goes from $-\infty$ to $+\infty$ can be interpreted simply because the surfaces of constant phase are perpendicular to the curve of $w(z)$. In brief, the beam progressively suffers a phase lag (compared to a plane wave focused to a point at the waist) because the curved "wave optics" path traced by $w(z)$ in Figure 16.11 is shorter than the straight "geometrical optics" path traced by the dashed lines in the figure.

Example 16.4 Produce an argument which supports the interpretation of $R(z)$ in (16.117) as the radius of curvature for a surface of constant phase which passes through the point (ρ, z) near the propagation axis. Neglect the Gouy phase $\alpha(z)$.

Solution: Let the vertical dashed line in Figure 16.12 be a surface of constant phase for the plane wave $\exp[ikz]$. If, instead, the circular arc in the figure were a surface of constant phase, the wave in question would have the form $\exp[ikR]$ with the origin at the point O. Let us try to express this factor, near the z-axis, in cylindrical coordinates (ρ, z).

The Pythagorean theorem applied to either right triangle in Figure 16.12 gives

$$R^2 = (R - \Delta z)^2 + \rho^2 \approx R^2 - 2R\Delta z + (\Delta z)^2 + \rho^2.$$

Near the z-axis, we may neglect the $(\Delta z)^2$ term to get $\Delta z = \rho^2/2R$. Therefore, at points on the arc not far from the axis,

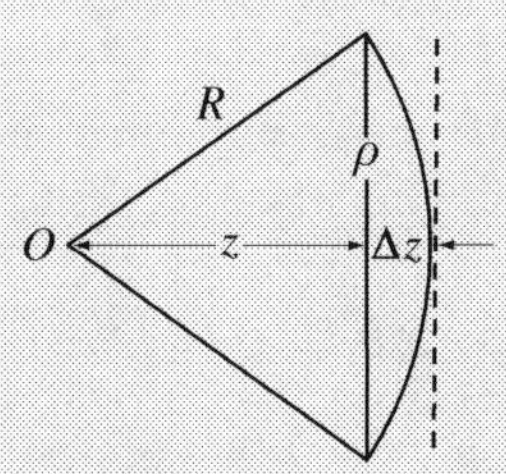

Figure 16.12: A portion of a circle of radius R centered at the point O.

$$\exp\{ikR\} = \exp\{ik(z + \Delta z)\} = \exp\left\{ik\left[z + \frac{\rho^2}{2R}\right]\right\}.$$

The last member of this equation is the complex exponential factor in the Gaussian beam (16.115) when we neglect the Guoy phase $\alpha(z)$. Therefore, at a point (ρ, z) near the propagation axis, the surfaces of constant phase for this beam are portions of spheres of radius $R(z)$.

16.7.3 The Meaning of the Paraxial Approximation

The Rayleigh range z_R in (16.116) is the characteristic longitudinal dimension of the beam. Therefore, we can express the paraxial condition (16.109) used to pass from (16.108) to (16.110) as

$$\frac{k}{z_R} \gg \frac{1}{z_R^2} \qquad \text{or} \qquad k^2 w_0^2 \gg 1. \tag{16.122}$$

The physical meaning of (16.122) becomes clear when we regard our beam-like solution as a wave packet like (16.80). Consider, for example, a packet formed by superposing monochromatic plane

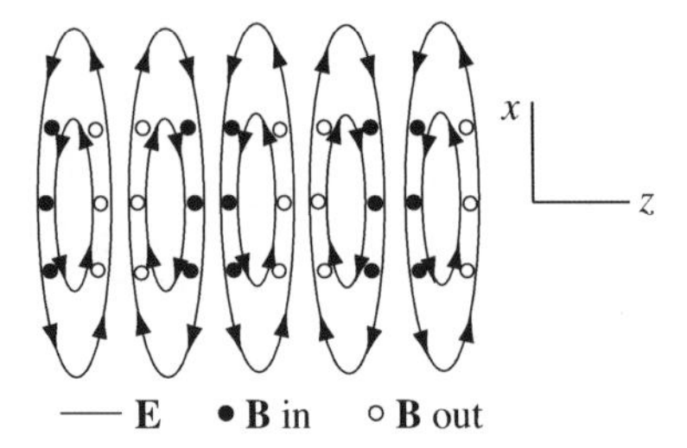

Figure 16.13: Cartoon of the field lines of (16.125) near the waist and in the $y = 0$ plane for a fundamental Gaussian beam propagating in the z-direction. Figure adapted from Davis and Patsakos (1981).

waves with wave vectors $\mathbf{k} = \mathbf{q}_\perp + k_z\hat{\mathbf{z}}$. Since $\omega = c|\mathbf{k}|$, the phase of each wave enters as

$$\exp[i(\mathbf{k}\cdot\mathbf{r} - \omega(\mathbf{k})t)] = \exp(i\mathbf{q}\cdot\mathbf{r}_\perp)\exp(ik_z z)\exp\left(-ict\sqrt{q_\perp^2 + k_z^2}\right). \tag{16.123}$$

Equation (16.113) does *not* contain pieces like (16.123). Rather, it assumes that $q_\perp^2 \ll k^2$ so the second and third exponential factors on the right side of (16.123) can be approximated, respectively, by

$$k_z = \sqrt{k^2 - q_\perp^2} \approx k - \frac{q_\perp^2}{2k} \quad \text{and} \quad \sqrt{q_\perp^2 + k_z^2} \approx k_z \approx k. \tag{16.124}$$

Equation (16.124) shows that only plane waves with wave vectors that deviate very slightly from the propagation axis are included in a beam-like wave packet. Moreover, every wave is taken to propagate with the common frequency $\omega = ck_z$. To connect this physical conclusion with the analytic paraxial approximation made above, it is enough to recall that the waist w_0 is the characteristic transverse dimension of the beam. In light of (16.114), we write $q_\perp \sim 1/w_0$ and thereby confirm that $k^2 \gg q_\perp^2$ reproduces the right side of (16.122). An equivalent statement is that the beam divergence satisfies $\theta^2 \ll 1$.

16.7.4 Paraxial Electromagnetic Waves

Section 16.2.1 described how to use Lorenz gauge potentials and solutions to the scalar wave equation to generate electric and magnetic fields that satisfy the Maxwell equations. The same prescription applied to the scalar Gaussian beam (16.115) generates a beam-like electromagnetic wave which satisfies the Maxwell equations in the paraxial approximation. We leave it as an exercise for the reader to confirm that the fields for a beam propagating in the z-direction can be chosen as

$$\mathbf{E} = u\hat{\mathbf{x}} + \frac{i}{k}\frac{\partial u}{\partial x}\hat{\mathbf{z}} \quad \text{and} \quad \mathbf{B} = -\frac{i}{\omega}\nabla\times\mathbf{E}. \tag{16.125}$$

Figure 16.13 is a sketch of the field lines predicted by (16.125) near the waist of the fundamental Gaussian beam in the $y = 0$ plane. This is a more realistic depiction of the "mutual embrace" of $\mathbf{E}$ and $\mathbf{B}$ than Figure 16.2. The closed loops are consistent with $\nabla\cdot\mathbf{E} = 0$ (up to terms that may be neglected in the paraxial approximation) and make it clear why (16.125) is explicitly not transverse to the z-direction of propagation.

16.7.5 The Angular Momentum of a Paraxial Beam

As early as 1909, Poynting argued that circularly polarized light carries angular momentum, but that linearly polarized light does not.[19] His prediction was confirmed by Beth in 1936, who studied the mechanical torque exerted by polarized light on suspended quartz plates.[20] The source of the torque is the transfer of angular momentum from the field to the matter. This motivates us to analyze the angular momentum stored in a beam-like monochromatic wave that propagates along the z-axis. Thus, with $\mathbf{r}_\perp = (x, y)$, we write

$$\mathbf{E}(x, y, z, t) = \mathbf{E}_0(\mathbf{r}_\perp, z)\exp[i(kz - \omega t)]. \tag{16.126}$$

Using the paraxial approximation (16.109), we will (i) confirm the just-stated experimental phenomenology and (ii) identify the spin and orbital contributions to the total angular momentum per unit area of the beam.

Our strategy is to compute the ratio of L_z (the z-component of the angular momentum per unit length of beam) to U (the total energy per unit length of beam). More precisely, we compute the ratio of the time average[21] of each quantity because the integrals of $\hat{\mathbf{z}} \cdot \mathbf{g}_{\rm EM}$ and $u_{\rm EM}$ over a beam cross section are not time-independent for a monochromatic wave. Therefore, integrating the time averages of the integrands of (15.31) and (15.75) over $d^2r = dxdy$, the ratio in question is

$$\frac{\langle L_z \rangle}{\langle U \rangle} = \frac{{\rm Re}\int d^2r\,[\mathbf{r} \times (\mathbf{E}^* \times \mathbf{B})] \cdot \hat{\mathbf{z}}}{\frac{1}{2}{\rm Re}\int d^2r\,[\mathbf{E}^* \cdot \mathbf{E} + c^2\mathbf{B}^* \cdot \mathbf{B}]}. \tag{16.127}$$

Faraday's law, $\nabla \times \mathbf{E} = i\omega\mathbf{B}$, and the free-field relation $\omega = ck$ transform (16.127) to

$$\frac{\langle L_z \rangle}{\langle U \rangle} = \frac{{\rm Re}\int d^2r\,\{\mathbf{r} \times [\mathbf{E}^* \times (-i\nabla \times \mathbf{E})]\} \cdot \hat{\mathbf{z}}}{\frac{1}{2}\omega{\rm Re}\int d^2r\,[\mathbf{E}^* \cdot \mathbf{E} + k^{-2}(\nabla \times \mathbf{E}^*) \cdot (\nabla \times \mathbf{E})]}. \tag{16.128}$$

In the paraxial approximation (Section 16.7.3), the z-variations of the electric field in (16.126) are carried primarily by the exponential factor rather than by the pre-factor. In other words, $\partial\mathbf{E}_0/\partial z \ll k\mathbf{E}_0$. One consequence of this approximation is that

$$\frac{\partial \mathbf{E}}{\partial z} \approx ik\mathbf{E}. \tag{16.129}$$

Another consequence follows from (16.129) and $\nabla \cdot \mathbf{E} = 0$ if we write $\mathbf{E} = (\mathbf{E}_\perp, E_z)$ and use the paraxial estimate $k^{-1}\nabla_\perp \sim (kw_0)^{-1} \ll 1$ appropriate for the Gaussian beam:

$$E_z \approx \frac{i}{k}\nabla \cdot \mathbf{E}_\perp \approx \frac{i}{kw_0}E_\perp. \tag{16.130}$$

Comparing (16.130) with the right side of (16.122) shows that, in the paraxial limit, quantities that are quadratic in E_z may be neglected compared to quantities that are quadratic in $\mathbf{E}_\perp$.

We are now in a position to express (16.128) entirely in terms of the vector $\mathbf{E}_\perp$. First, the remark made just above tells us that the first term in the denominator of (16.128) satisfies

$$\mathbf{E}^* \cdot \mathbf{E} \approx \mathbf{E}_\perp^* \cdot \mathbf{E}_\perp. \tag{16.131}$$

Similarly, by writing out all the components and using both (16.129) and (16.130), we find the same approximate value for the second term in the denominator of (16.128):

$$k^{-2}(\nabla \times \mathbf{E}^*) \cdot (\nabla \times \mathbf{E}) \approx \mathbf{E}_\perp^* \cdot \mathbf{E}_\perp. \tag{16.132}$$

[19] See Allen, Barnett and Padgett (2003) in Sources, References, and Additional Reading.

[20] R, Beth, "Mechanical detection and measurement of the angular momentum of light", *Physical Review* **50**, 115 (1936).

[21] See the time-averaging theorem in Section 1.6.3.

For the numerator of (16.128), we begin with the fact that

$$\mathbf{r} \times [\mathbf{E}^* \times (\nabla \times \mathbf{E})] = \mathbf{r} \times (E_k^* \nabla E_k) - \mathbf{r} \times (\mathbf{E}^* \cdot \nabla)\mathbf{E}. \tag{16.133}$$

The remark following (16.130) ensures that the substitution $\mathbf{E} \to \mathbf{E}_\perp$ is valid for the $\mathbf{r} \times (E_k^* \nabla E_k)$ term in (16.133). On the other hand, the $\mathbf{r} \times (\mathbf{E}^* \cdot \nabla)\mathbf{E}$ term in (16.133) simplifies in the numerator of (16.128) only after an integration by parts. The reader can confirm that the final result is the gauge invariant expression

$$\frac{\langle L_z \rangle}{\langle U \rangle} = \frac{\text{Re} \int d^2r \, \{\mathbf{E}_\perp^* \cdot [\mathbf{r} \times (-i\nabla)]_z \mathbf{E}_\perp - i[\mathbf{E}_\perp^* \times \mathbf{E}_\perp]_z\}}{\omega \int d^2r \, |\mathbf{E}_\perp|^2}. \tag{16.134}$$

All the integrals in (16.134) converge because the intensity of a beam-like wave goes to zero quickly as $r_\perp$ increases from zero.

16.7.6 Orbital and Spin Angular Momentum

To interpret (16.134), we follow part (c) of Example 16.3 at the end of Section 16.4.5 and use the circular polarization vectors $\hat{\mathbf{e}}_+$ and $\hat{\mathbf{e}}_-$ defined in (16.60) as basis vectors to express the general state of polarization (16.63) as the column vector

$$|\psi\rangle = \begin{pmatrix} \psi_+ \\ \psi_- \end{pmatrix} = \frac{1}{\sqrt{2}} \begin{pmatrix} E_x - iE_y \\ E_x + iE_y \end{pmatrix}. \tag{16.135}$$

In this basis, $\psi_- = 0$ for a left circularly polarized (LCP) wave where $E_y/E_x = +i$. Similarly, $\psi_+ = 0$ for a right circularly polarized (RCP) wave where $E_y/E_x = -i$ (see Section 16.4.3). We also normalize the electric field intensity so

$$\langle \psi | \psi \rangle = \int d^2r \, |\mathbf{E}_\perp|^2 = 1. \tag{16.136}$$

The next step writes the z-components of quantum mechanical orbital and spin-1 angular momentum operators as

$$\ell_z = [\mathbf{r} \times (-i\hbar\nabla)]_z = -i\hbar \frac{\partial}{\partial \phi} \tag{16.137}$$

and

$$s_z = \hbar \begin{pmatrix} 1 & 0 \\ 0 & -1 \end{pmatrix}. \tag{16.138}$$

Using the four preceding equations, (16.134) takes the highly suggestive form[22]

$$\frac{\langle L_z \rangle}{\langle U \rangle} = \frac{\langle \psi | \ell_z | \psi \rangle + \langle \psi | s_z | \psi \rangle}{\hbar\omega}. \tag{16.139}$$

The spin angular momentum in (16.139) is determined entirely by the polarization of the beam. From (16.135) and (16.138),

$$\langle \psi | s_z | \psi \rangle = \hbar \int d^2r \, (|\psi_+|^2 - |\psi_-|^2). \tag{16.140}$$

[22] We insert $\hbar$ into the numerator and denominator of (16.139) to facilitate comparison with familiar quantum mechanical formulae. See Berry (1998) in Sources, References, and Additional Reading for discussion of the spin representation (16.138).

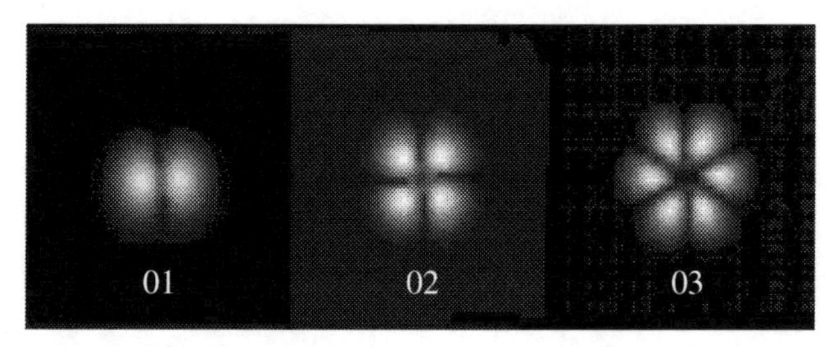

Figure 16.14: Cross sectional intensity profiles of a few (p, ℓ) Laguerre-Gauss beam-like waves defined by (16.141) with non-zero orbital angular momentum per unit length of beam.

The term *helicity* refers to the projection of the spin angular momentum onto the direction of linear momentum (which is $+z$ here). Therefore, an LCP wave has $s_z = +\hbar$ and positive helicity while an RCP wave has $s_z = -\hbar$ and negative helicity. A linearly polarized beam has $\psi_+ = \psi_-$. For that case, $s_z = 0$ and no spin angular momentum is available for transfer to matter.

The derivative on the far right side of (16.137) tells us that the orbital angular momentum in (16.139) is determined by the ϕ dependence of $\mathbf{E}_\perp(\rho, \phi, z, t)$. An immediate consequence is that the fundamental Gaussian beam analyzed in Section 16.7.2 carries zero orbital angular momentum. On the other hand, an important class of solutions of the paraxial wave equation (16.110) *do* possess orbital angular momentum with respect to the beam axis. These solutions—which generalize the fundamental Gaussian solution—are indexed by integers p and ℓ because they involve the associated Laguerre polynomials, $L_p^\ell[x]$.[23] Our notation for these Laguerre-Gauss waves is $u_{p\ell}(\mathbf{r}, t)$, where the fundamental solution (16.115) corresponds to $u_{00}(\mathbf{r}, t)$. The higher-order waves are

$$u_{p\ell}(\rho, \phi, z, t) = \left[\frac{\sqrt{2}\rho}{w(z)}\right]^\ell L_p^\ell\left[\frac{2\rho^2}{w^2(z)}\right] e^{i(2p+\ell)\alpha(z)} e^{i\ell\phi} u_{00}(\rho, z, t). \tag{16.141}$$

Figure 16.14 plots the intensity of the real part of (16.141) for $p = 0$ and $\ell = 1, 2,$ and 3. Such beams have an azimuthal component to their linear momentum density and carry an orbital angular momentum (with respect to the beam axis) proportional to ℓ. They are also "hollow" in the sense that $u_{p\ell}(0, \phi, z, t) = 0$.

We close this section by noting that the elegant separation of the total angular momentum per unit area along the beam direction into orbital and spin components in (16.139) is a special feature of the paraxial approximation. It does not carry over to exact, beam-like solutions of the free-space Maxwell equations. On the other hand, a gauge invariant separation into orbital and spin components *does* occur for true, beam-like electromagnetic waves if one examines the fully longitudinal component, M_{zz}, of the angular momentum current density $\mathbf{M}$ defined in (15.73).[24]

16.8 Spherical Waves

Spherical electromagnetic waves are solutions to the empty-space Maxwell equations where the surfaces of constant phase are spheres rather than planes. Waves of this kind arise naturally in certain guided-wave problems (Chapter 19) and are essential to the theory of radiation (Chapter 20). We introduce these waves with an ansatz for the vector potential:

$$\mathbf{A}_C(\mathbf{r}, t) = \nabla \times [u(\mathbf{r}, t)\mathbf{r}]. \tag{16.142}$$

[23] The associated Laguerre polynomials arise in the Schrödinger problem for the hydrogen atom.

[24] See S.M. Barnett, "'Optical angular momentum flux", *Journal of Optics and Semi-Classical Optics* **4**, S7 (2002).

The subscript indicates that (16.142) satisfies the Coulomb gauge condition, $\nabla \cdot \mathbf{A}_C = 0$. As discussed in Section 16.2.2, we may choose $\varphi_C = 0$ in empty space and $\mathbf{A}_C$ satisfies a vector wave equation. Therefore, our task is to find solutions of

$$\left[\nabla^2 - \frac{1}{c^2}\frac{\partial^2}{\partial t^2}\right][\nabla \times (u\mathbf{r})] = 0. \tag{16.143}$$

Happily, the curl operator commutes with the Laplacian operator,[25] so (16.143) is equivalent to

$$\nabla \times \left[\nabla^2 - \frac{1}{c^2}\frac{\partial^2}{\partial t^2}\right](u\mathbf{r}) = 0. \tag{16.144}$$

On the other hand,

$$\nabla \times [\nabla^2(u\mathbf{r})] = \nabla \times \left\{\mathbf{r}\nabla^2 u + 2\nabla u\right\} = \nabla \times (\mathbf{r}\nabla^2 u). \tag{16.145}$$

Therefore, (16.144) simplifies to

$$\nabla \times \mathbf{r}\left\{\nabla^2 u - \frac{1}{c^2}\frac{\partial^2 u}{\partial t^2}\right\} = 0. \tag{16.146}$$

We conclude from (16.146) that *any* solution $u(\mathbf{r}, t)$ of the scalar wave equation can be used to construct a Coulomb gauge vector potential using (16.142).

16.8.1 TE and TM Vector Waves

The spherical electromagnetic wave fields associated with the vector potential (16.142) are $\mathbf{B} = \nabla \times \mathbf{A}_C$ and $\mathbf{E} = -\partial \mathbf{A}_C/\partial t$. Since $\nabla \times \mathbf{r} = 0$, we find *transverse electric* (TE) waves,

$$\mathbf{E}_{\rm TE} = \mathbf{r} \times \nabla \dot{u} \qquad \text{and} \qquad \mathbf{B}_{\rm TE} = -\nabla \times [\mathbf{r} \times \nabla u]. \tag{16.147}$$

The name is appropriate because the electric field in (16.147) satisfies

$$\mathbf{r} \cdot \mathbf{E}_{\rm TE} = 0. \tag{16.148}$$

It is not obvious, but the fields in (16.147) are perpendicular:

$$\mathbf{E}_{\rm TE} \cdot \mathbf{B}_{\rm TE} = 0. \tag{16.149}$$

To see this, we define the vector operator $\mathbf{L} = -i\mathbf{r} \times \nabla$ and write (16.147) as[26]

$$\mathbf{E}_{\rm TE} = i\mathbf{L}\dot{u} \qquad \text{and} \qquad \mathbf{B}_{\rm TE} = -i\nabla \times \mathbf{L}u. \tag{16.150}$$

The key to the argument is the operator identity[27]

$$\nabla \times \mathbf{L} = (\hat{\mathbf{r}} \times \mathbf{L})\left(\frac{1}{r}\frac{\partial}{\partial r}r\right) + \hat{\mathbf{r}}\frac{i}{r}L^2. \tag{16.151}$$

Equation (16.151) shows that $\mathbf{B}_{\rm TE}$ has a component along $\hat{\mathbf{r}}$ and a component along $\hat{\mathbf{r}} \times \mathbf{L}u$. The latter is perpendicular to $\hat{\mathbf{r}}$, and also to $\mathbf{L}\dot{u}$, because u and $\dot{u}$ have the same angular dependence. This proves (16.149).

To make further progress, we exploit the dual symmetry (Section 15.2.2) of the source-free Maxwell equations (16.1) and (16.2) with respect to the transformation

$$\mathbf{E} \to c\mathbf{B} \qquad \text{and} \qquad \mathbf{B} \to -\mathbf{E}/c. \tag{16.152}$$

[25] A quick proof exploits the Levi-Cività representation of the curl in (1.39).

[26] The operator $\mathbf{L}$ appeared previously in Section 11.4.4.

[27] See Application 1.1 at the end of Section 1.2.

Specifically, an application of (16.152) to the TE solution (16.147) generates a linearly independent, free-field solution of the Maxwell equations. To avoid confusion, we do this using $w(\mathbf{r}, t)$ as the solution to the scalar wave equation in (16.146) and define *transverse magnetic* (TM) waves,

$$c\mathbf{B}_{\text{TM}} = \mathbf{r} \times \nabla \dot{w} \qquad \text{and} \qquad \mathbf{E}_{\text{TM}} = c\nabla \times [\mathbf{r} \times \nabla w]. \tag{16.153}$$

The same arguments used just above for the TE fields show that the fields in (16.153) satisfy

$$\mathbf{r} \cdot \mathbf{B}_{\text{TM}} = 0 \tag{16.154}$$

and

$$\mathbf{E}_{\text{TM}} \cdot \mathbf{B}_{\text{TM}} = 0. \tag{16.155}$$

Finally, we state (but do not prove) an important theorem: every solution of the source-free Maxwell equations can be represented uniquely by a sum of TE and TM fields of the form (16.147) and (16.153).[28] The scalar functions $u(\mathbf{r}, t)$ and $w(\mathbf{r}, t)$ are known as *Debye potentials* in this context and constitute the two functions anticipated by Whittaker's theorem (see Section 16.2.2).

16.8.2 Scalar Waves

With (16.147) and (16.153) in hand, it remains only to find solutions of the scalar wave equation in (16.146) with spherical symmetry. We specialize to time-harmonic waves where $u(\mathbf{r}, t) = \hat{u}(r, \theta, \phi, \omega)\exp(-i\omega t)$ and learn from Section 16.6 that our task is to solve the Helmholtz equation (16.102) in spherical coordinates:

$$\frac{1}{r^2}\frac{\partial}{\partial r}\left(r^2 \frac{\partial \hat{u}}{\partial r}\right) + \frac{1}{r^2 \sin\theta}\frac{\partial}{\partial \theta}\left(\sin\theta \frac{\partial \hat{u}}{\partial \theta}\right) + \frac{1}{r^2 \sin^2\theta}\frac{\partial^2 \hat{u}}{\partial \phi^2} + \frac{\omega^2}{c^2}\hat{u} = 0. \tag{16.156}$$

Separation of variables suggests we try a solution to (16.156) of the form

$$\hat{u}(r, \theta, \phi \,|\, \omega) = R(r)Y(\theta, \phi). \tag{16.157}$$

Mimicking our treatment of Laplace's equation (Section 7.7), we choose $\ell(\ell + 1)$ as a separation constant and set $\omega = ck$. This leads to the ordinary differential equations

$$-\frac{1}{\sin\theta}\frac{\partial}{\partial \theta}\left(\sin\theta \frac{\partial Y}{\partial \theta}\right) - \frac{1}{\sin^2\theta}\frac{\partial^2 Y}{\partial \phi^2} = \ell(\ell + 1)Y \tag{16.158}$$

and

$$\frac{d^2 R}{dr^2} + \frac{2}{r}\frac{dR}{dr} + \left[k^2 - \frac{\ell(\ell + 1)}{r^2}\right] R = 0. \tag{16.159}$$

Equation (16.158) is identical to (7.70), which identifies the angular functions as the spherical harmonics $Y_{\ell m}(\theta, \phi)$ discussed in Section 4.5.2. The radial equation (16.159) is closely related to Bessel's equation of order $\ell + \frac{1}{2}$. We omit the details (see Section C.3.2) and simply state that the solutions to (16.159) of most interest to us here are the *spherical Hankel functions* $h_\ell^{(1)}(kr)$. The first few of these are

$$\begin{aligned}
h_0^{(1)}(kr) &= -i\frac{e^{ikr}}{kr} \\
h_1^{(1)}(kr) &= -\left[1 + \frac{i}{kr}\right]\frac{e^{ikr}}{kr} \\
h_2^{(1)}(kr) &= i\left[1 + \frac{3i}{kr} - \frac{3}{(kr)^2}\right]\frac{e^{ikr}}{kr}.
\end{aligned} \tag{16.160}$$

[28] See Gray and Nickel (1978) in Sources, References, and Additional Reading for a proof and discussion.

When multiplied by $\exp(-i\omega t)$, each $h_\ell^{(1)}(kr)$ represents an *outgoing* spherical wave where the surfaces of constant phase $\phi = kr - \omega t$ expand radially outward from the origin as time goes on. A linearly independent solution is its complex conjugate, $h_\ell^{(2)}(kr)$. When multiplied by $\exp(-i\omega t)$, this represents an *incoming* spherical wave where the surfaces of constant phase $\phi = kr + \omega t$ collapse radially inward toward the origin as time goes on. A general solution of the Helmholtz equation (16.159) is a linear combination of the two:

$$\hat{u}(\mathbf{r}|\omega) = \sum_{\ell=0}^{\infty} \sum_{m=-\ell}^{\ell} [A_\ell(k) h_\ell^{(1)}(kr) + B_\ell(k) h_\ell^{(2)}(kr)] Y_{\ell m}(\theta, \phi). \tag{16.161}$$

Substituting (16.161) into (16.147) generates a complete set of transverse electric (TE) spherical electromagnetic waves. Substituting (16.161) into (16.153) generates a complete set of transverse magnetic (TM) spherical electromagnetic waves. Chapter 20 uses both sets of waves to discuss multipole radiation.

Application 16.1 The Importance of Being Angular

All spherical electromagnetic waves are intrinsically anisotropic, with electric and magnetic fields that depend on at least one angular variable. To prove this, it is sufficient to write out (16.147) and (16.153) using solutions of the scalar wave equation $u(r, t)$ and $w(r, t)$ that depend only on the radial variable r rather than on $\mathbf{r} = (r, \theta, \phi)$. The result is

$$\begin{aligned} \mathbf{E}_{\rm TE}(r, t) &= \mathbf{r} \times \nabla \dot{u}(r, t) = \mathbf{r} \times \left(\frac{\partial^2 u}{\partial r \partial t} \right) \mathbf{r} = 0 \\ \mathbf{B}_{\rm TM}(r, t) &= \frac{1}{c} \mathbf{r} \times \nabla \dot{w}(r, t) = \frac{1}{c} \mathbf{r} \times \left(\frac{\partial^2 w}{\partial r \partial t} \right) \mathbf{r} = 0. \end{aligned} \tag{16.162}$$

The absence of purely radial (and thus isotropic) electromagnetic waves is a consequence of (16.148) and (16.154), namely, $\mathbf{E}_{\rm TE}(\mathbf{r}, t)$ and $\mathbf{B}_{\rm TM}(\mathbf{r}, t)$ are vector waves that must be transverse to $\mathbf{r}$ at every point in space. The formulae (16.147) and (16.153) define continuous vector fields that are everywhere tangent to any origin-centered spherical surface. However, a celebrated theorem of algebraic topology guarantees that this is impossible! More colloquially, this *hairy ball theorem* states that it is impossible to comb a hairy ball so every one of its hairs lays flat.[29] As indicated in Figure 16.15, a cowlick (a point where a hair stands up) or a bald spot must be present somewhere. The transversality conditions (16.148) and (16.154) rule out a cowlick for $\mathbf{E}_{\rm TE}$ and $\mathbf{B}_{\rm TM}$ (which play the role of the hairs). But, neither can $\mathbf{E}_{\rm TE}$ and $\mathbf{B}_{\rm TM}$ have isolated zeroes on any origin-centered sphere (bald spots) when u and w depend only on the radial variable r. Hence, both fields must vanish identically.

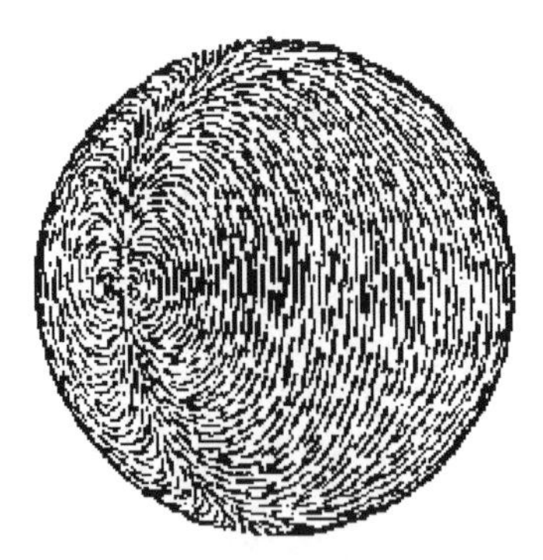

Figure 16.15: A hairy ball must have a bald spot or a cowlick if its hair lays flat at every other point on the surface of the ball.

[29] See Milnor (1978) in Sources, References, and Additional Reading for a proof and discussion.

16.9 Hertz Vectors

A very general approach to electromagnetic fields in vacuum generalizes the "spherical" vector potential (16.142) to define a pair of potential functions known as *Hertz vectors*. In this section, we use Hertz vectors to generalize the idea of transverse electric (TE) and transverse magnetic (TM) waves introduced in Section 16.8.1. Their true value will appear later when we re-introduce the source terms in the Maxwell equations to study radiation and scattering.[30]

16.9.1 The Magnetic Hertz Vector

The *magnetic Hertz vector* $\boldsymbol{\pi}_{\rm m}(\mathbf{r}, t)$ is a potential function defined so the usual vector potential of electromagnetism automatically satisfies the Coulomb gauge constraint $\nabla \cdot \mathbf{A}_C = 0$.[31] This will be the case if

$$\mathbf{A}_C = \nabla \times \boldsymbol{\pi}_{\rm m}. \tag{16.163}$$

Exactly as in the beginning of Section 16.8, we use the facts that (i) $\mathbf{A}_C$ satisfies the vector wave equation (16.16) and (ii) the curl operator commutes with the Laplacian operator to conclude that

$$\nabla \times \left\{ \nabla^2 \boldsymbol{\pi}_{\rm m} - \frac{1}{c^2}\frac{\partial^2 \boldsymbol{\pi}_{\rm m}}{\partial t^2} \right\} = 0. \tag{16.164}$$

In light of our freedom to set the scalar potential $\varphi_C = 0$ when no sources are present (see Section 16.2.2), (16.164) and substitution of (16.163) into (16.17) show that, if

$$\nabla^2 \boldsymbol{\pi}_{\rm m} - \frac{1}{c^2}\frac{\partial^2 \boldsymbol{\pi}_{\rm m}}{\partial t^2} = 0, \tag{16.165}$$

a valid electromagnetic field in empty space is

$$\mathbf{E} = -\nabla \times \frac{\partial \boldsymbol{\pi}_{\rm m}}{\partial t} \qquad \text{with} \qquad \mathbf{B} = \nabla \times \nabla \times \boldsymbol{\pi}_{\rm m}. \tag{16.166}$$

Recycling the trick we have used throughout this chapter, choices for $\boldsymbol{\pi}_{\rm m}$ that reduce the vector wave equation (16.165) to a scalar wave equation turn out to generate practically important electromagnetic fields. Thus, the transverse electric (TE) spherical waves found in Section 16.8 come from the choice $\boldsymbol{\pi}_{\rm m} = u\hat{\mathbf{r}}$. For other purposes, it is sufficient to choose a constant vector $\mathbf{s}$ and form the trial solution

$$\boldsymbol{\pi}_{\rm m}(\mathbf{r}, t) = u(\mathbf{r}, t)\mathbf{s}. \tag{16.167}$$

Substituting (16.167) into (16.165) shows that the latter will be satisfied if $u(\mathbf{r}, t)$ is a solution of the scalar wave equation. Then, making use of

$$\nabla \times \nabla \times \boldsymbol{\pi}_{\rm m} = \nabla(\nabla \cdot \boldsymbol{\pi}_{\rm m}) - \nabla^2 \boldsymbol{\pi}_{\rm m} \tag{16.168}$$

and the three equations that precede it, we find that

$$\mathbf{E}_{\rm TE} = \mathbf{s} \times \nabla \frac{\partial u}{\partial t} \qquad \text{and} \qquad \mathbf{B}_{\rm TE} = (\mathbf{s} \cdot \nabla)\nabla u - \frac{1}{c^2}\frac{\partial^2 u}{\partial t^2}\mathbf{s}. \tag{16.169}$$

The fields (16.169) are transverse electric (TE) because the electric field is always transverse to $\mathbf{s}$.

[30] As the name implies, Hertz vectors have been used with profit in electromagnetic theory for over a century. Some authors do not exploit them because their virtues do not carry over into quantum theory.

[31] The "magnetic" nature of $\boldsymbol{\pi}_{\rm m}$ will emerge when we introduce sources in Chapter 20.

16.9.2 The Electric Hertz Vector

The *electric Hertz vector* $\boldsymbol{\pi}_{\rm e}$ is a potential function defined so the usual scalar and vector potentials of electromagnetism automatically satisfy the Lorenz gauge constraint.[32] The latter is

$$\nabla \cdot \mathbf{A}_L + \frac{1}{c^2}\frac{\partial \varphi_L}{\partial t} = 0. \tag{16.170}$$

Hence, an appropriate defintion of $\boldsymbol{\pi}_{\rm e}$ is

$$\mathbf{A}_L = \frac{1}{c^2}\frac{\partial \boldsymbol{\pi}_{\rm e}}{\partial t} \qquad \text{and} \qquad \varphi_L = -\nabla \cdot \boldsymbol{\pi}_{\rm e}. \tag{16.171}$$

Equation (16.9) reminds us that $\mathbf{A}_L$ and φ_L satisfy a vector and a scalar wave equation, respectively. Substituting the potentials (16.171) into these wave equations, it is straightforward to confirm that both are satisfied if the electric Hertz vector itself satisfies

$$\nabla^2 \boldsymbol{\pi}_{\rm e} - \frac{1}{c^2}\frac{\partial^2 \boldsymbol{\pi}_{\rm e}}{\partial t^2} = 0. \tag{16.172}$$

The associated electromagnetic field follows by inserting (16.171) into (16.7) and making use of (16.168):

$$\mathbf{E} = \nabla \times \nabla \times \boldsymbol{\pi}_{\rm e} \qquad \text{and} \qquad \mathbf{B} = \frac{1}{c^2}\nabla \times \frac{\partial \boldsymbol{\pi}_{\rm e}}{\partial t}. \tag{16.173}$$

Like $\boldsymbol{\pi}_{\rm m}$ in (16.166), there is no restriction or constraint on $\boldsymbol{\pi}_{\rm e}$ in (16.173) except that both must be solutions of the vector wave equation. Indeed, the two sets of fields, (16.166) and (16.173), are dual symmetry partners in the sense that the transformation (16.152) relates one to the other with the choice $\boldsymbol{\pi}_{\rm e} = c\boldsymbol{\pi}_{\rm m}$.

We now mimic (16.167) and use a constant vector $\mathbf{s}$ to write

$$\boldsymbol{\pi}_{\rm e}(\mathbf{r}, t) = w(\mathbf{r}, t)\mathbf{s}. \tag{16.174}$$

In the (by now) familiar way, this ansatz transforms the problem of seeking solutions to the vector wave equation (16.172) into the problem of seeking solutions $w(\mathbf{r}, t)$ to the scalar wave equation. The fields that result when (16.174) is substituted into (16.173) are

$$\mathbf{E}_{\rm TM} = (\mathbf{s}\cdot\nabla)\nabla w - \frac{1}{c^2}\frac{\partial^2 w}{\partial t^2}\mathbf{s} \qquad \text{and} \qquad \mathbf{B}_{\rm TM} = -\mathbf{s} \times \nabla \frac{1}{c^2}\frac{\partial w}{\partial t}. \tag{16.175}$$

The fields (16.175) are transverse magnetic (TM) because the magnetic field is always transverse to $\mathbf{s}$. The paraxial beam-like fields (16.125) are an example which may be regarded as derived from the electric Hertz vector $\boldsymbol{\pi}_{\rm e} = w\hat{\mathbf{x}}$ with $w = u/k^2$.

Application 16.2 Whittaker's Theorem Revisited

Whittaker's theorem (see Section 16.2.2) states that two solutions of the scalar wave equation are sufficient to represent an arbitrary electromagnetic field in vacuum. Since (16.167) defines a one-component $\boldsymbol{\pi}_{\rm m}$ and (16.174) defines a one-component $\boldsymbol{\pi}_{\rm e}$, it is reasonable to suspect that a linear combination of (16.169) and (16.175) will do the job.

To demonstrate this, we write the electric field in the form

$$\mathbf{E}(\mathbf{r}, t) = \frac{1}{2\pi}\int d\omega\, \mathbf{E}(\mathbf{r}|\omega)e^{-i\omega t}, \tag{16.176}$$

[32] The "electric" nature of $\boldsymbol{\pi}_{\rm e}$ will emerge when we introduce sources in Chapter 20.

where

$$\mathbf{E}(\mathbf{r}|\omega) = \frac{1}{(2\pi)^3}\int d^3k\,\mathbf{E}_\perp(\mathbf{k})e^{i\mathbf{k}\cdot\mathbf{r}}\delta(k-\omega/c). \tag{16.177}$$

Comparison with Section 16.6.1 shows that (formally) we are using a plane wave spectrum representation where (see Section 16.3.4) the amplitude function $\mathbf{E}_\perp(\mathbf{k})$ is strictly perpendicular to $\mathbf{k}$. This motivates us to choose a fixed unit vector $\hat{\mathbf{s}}$ and write $\mathbf{E}_\perp(\mathbf{k}) = E_1(\mathbf{k})\hat{\mathbf{e}}_1 + E_2(\mathbf{k})\hat{\mathbf{e}}_2$, where $\hat{\mathbf{e}}_1 = \hat{\mathbf{k}}\times\hat{\mathbf{s}}$ and $\hat{\mathbf{e}}_2 = \hat{\mathbf{k}}\times(\hat{\mathbf{k}}\times\hat{\mathbf{s}})$ are unit vectors that are perpendicular both to $\mathbf{k}$ and to each other. Accordingly,

$$\mathbf{E}(\mathbf{r}|\omega) = \frac{1}{(2\pi)^3}\int d^3k\,\left[(\hat{\mathbf{k}}\times\hat{\mathbf{s}})E_1(\mathbf{k}) + \hat{\mathbf{k}}\times(\hat{\mathbf{k}}\times\hat{\mathbf{s}})E_2(\mathbf{k})\right]e^{i\mathbf{k}\cdot\mathbf{r}}\delta(k-\omega/c). \tag{16.178}$$

When the gradient operator acts on a plane wave, $\nabla \equiv i\mathbf{k}$. Therefore, we can let $\hat{\mathbf{k}} \to -(i/k)\nabla$ in (16.178), pull the gradient operators out of the integral, and use the delta function to write

$$\mathbf{E}(\mathbf{r}|\omega) = i\omega\nabla\times(u\hat{\mathbf{s}}) + \nabla\times\nabla\times(w\hat{\mathbf{s}}), \tag{16.179}$$

where

$$u(\mathbf{r}|\omega) = -\frac{c}{\omega^2}\int\frac{d^3k}{(2\pi)^3}E_1(\mathbf{k})e^{i\mathbf{k}\cdot\mathbf{r}}\delta(k-\omega/c) \tag{16.180}$$

and

$$w(\mathbf{r}|\omega) = -\frac{c^2}{\omega^2}\int\frac{d^3k}{(2\pi)^3}E_2(\mathbf{k})e^{i\mathbf{k}\cdot\mathbf{r}}\delta(k-\omega/c). \tag{16.181}$$

Now, substitute (16.179) into (16.176) and let

$$u(\mathbf{r},t) = u(\mathbf{r}|\omega)e^{-i\omega t} \qquad \text{and} \qquad w(\mathbf{r},t) = w(\mathbf{r}|\omega)e^{-i\omega t}. \tag{16.182}$$

The result is an expression for $\mathbf{E}(\mathbf{r},t)$ that is the sum of the electric field in (16.166) with the magnetic Hertz vector (16.167) and the electric field in (16.173) with the electric Hertz vector (16.174):

$$\mathbf{E}(\mathbf{r},t) = -\frac{\partial}{\partial t}\nabla\times[u(\mathbf{r},t)\hat{\mathbf{s}}] + \nabla\times\nabla\times[w(\mathbf{r},t)\hat{\mathbf{s}}]. \tag{16.183}$$

By construction, the magnetic field associated with (16.183) can only be the sum of the magnetic fields in (16.166) and (16.173):

$$\mathbf{B}(\mathbf{r},t) = \nabla\times\nabla\times[u(\mathbf{r},t)\hat{\mathbf{s}}] + \frac{1}{c^2}\frac{\partial}{\partial t}\nabla\times[w(\mathbf{r},t)\hat{\mathbf{s}}]. \tag{16.184}$$

The fields (16.183) and (16.184) satisfy the conditions of Whittaker's theorem because the two scalar functions defined by (16.182) are plainly solutions of the scalar wave equation. ■

16.10 Forces on Particles in Free Fields

The behavior of particles in time-varying electromagnetic fields is a central issue in accelerator physics, plasma physics, atomic physics, and space physics. In this section, we focus on time-harmonic fields in vacuum and analyze three situations where analytic results are possible: (a) the motion of a charged particle in a monochromatic plane wave; (b) the force that a monochromatic field exerts on a charged particle; and (c) the force that a monochromatic field exerts on an electrically polarizable particle.

16.10.1 Charged Particle Motion in a Plane Wave

Consider a point particle with charge q and mass m in the field of a monochromatic, linearly polarized plane wave with electric field $\mathbf{E} = \hat{\mathbf{x}}E_0\cos(kz - \omega t)$. The electric Coulomb force $q\mathbf{E}$ drives the particle along $\mathbf{E}$. The magnetic Lorentz force drives the particle along $\mathbf{v} \times \mathbf{B}$, where $c\mathbf{B} = \hat{\mathbf{y}}E_0\cos(kz - \omega t)$. Therefore, if $\mathbf{v}$ is in phase with $\mathbf{E}$, this is a scheme for particle acceleration along the z-axis of wave propagation. However, as we will now show, the particle velocity is not in phase with the electric field and there is zero net acceleration in any direction.

Our strategy is to treat the electric force $q\mathbf{E}$ as primary and add the effect of the magnetic force $q\boldsymbol{v} \times \mathbf{B}$ as a perturbation. We choose initial conditions, $x(0) = z(0) = \dot{x}(0) = \dot{z}(0) = 0$. The non-relativistic equation of motion for the Coulomb force alone is[33]

$$m\ddot{x} = qE_0\cos(kz - \omega t) \approx qE_0\cos(\omega t). \tag{16.185}$$

The last term in (16.185) is an approximation which is valid if the particle remains close to the origin. Integrating twice and imposing the initial conditions gives

$$x(t) = \frac{qE_0}{m\omega^2}\left[1 - \cos(\omega t)\right] \equiv d\left[1 - \cos(\omega t)\right]. \tag{16.186}$$

The characteristic length scale $d = qE_0/m\omega^2$ defined in (16.186) measures the field strength of the plane wave.

The Lorentz force may be treated as a perturbation if the particle speed is non-relativistic:

$$\frac{v_x|\mathbf{B}|}{|\mathbf{E}|} = \frac{v_x}{c} \sim \frac{\omega d}{c} = kd \ll 1. \tag{16.187}$$

Accordingly, the equation for the z-motion is

$$\ddot{z} = (q/m)\dot{x}B_y = \tfrac{1}{2}kd^2\omega^2\sin(2\omega t). \tag{16.188}$$

The time averages $\langle\ddot{x}\rangle = \langle\ddot{z}\rangle = 0$ from (16.185) and (16.188) confirm that the particle experiences zero net acceleration. Nevertheless, integrating (16.188) twice and applying the initial conditions shows interesting behavior: oscillatory motion in the z-direction at *twice* the wave frequency superimposed on a steady drift in the same direction,

$$z(t) = \tfrac{1}{8}kd^2\left[2\omega t - \sin(2\omega t)\right]. \tag{16.189}$$

If the drift coordinate is $z_0 = c(kd)^2t/4$, we can eliminate the sinusoidal functions from (16.186) and (16.189) to get

$$16(z - z_0)^2 = k^2x^2(d^2 - x^2). \tag{16.190}$$

Figure 16.16 plots the trajectory (16.190) as a function of d in a (moving) frame of reference centered on z_0. The trajectories are figure-eight Lissajous figures with their lobes on the x-axis. As the field strength parameter $d \to 0$, the motion shrinks to one-dimensional harmonic motion along the x (electric field) axis at frequency ω. The second harmonic 2ω appears in the z-motion because of the non-linear nature of the Lorentz force law: $\boldsymbol{v} \propto \mathbf{E}$ from (16.186) and $c|\mathbf{B}| = |\mathbf{E}|$ for a plane wave so $\mathbf{F} = q\boldsymbol{v} \times \mathbf{B} \propto |\mathbf{E}|^2$. This non-linearity implies that we cannot use the results of this section to deduce the behavior of a particle in a high-intensity field composed of a superposition of plane waves (see the following section).

[33] See Section 22.5.2 for an exact, relativistic solution of this problem.

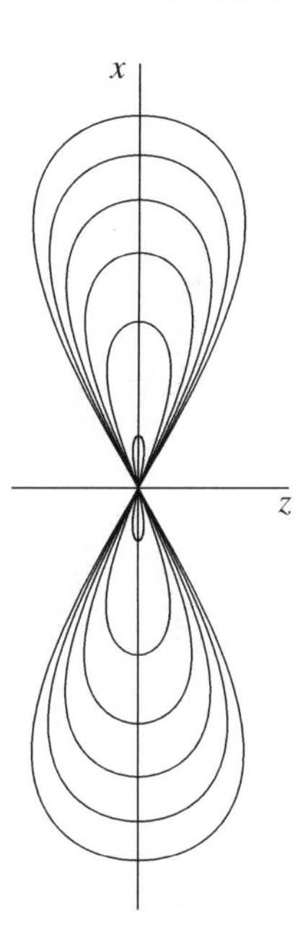

Figure 16.16: Figure-eight trajectories (16.190) (centered on z_0) for a particle with mass m and charge q set into motion by a plane wave propagating in the z-direction with frequency ω and electric field strength E_0. The lobe area increases as $d = qE_0/m\omega^2$ increases.

16.10.2 The Pondermotive Force on a Charged Particle

Unlike the single plane wave studied just above, many time-harmonic electromagnetic waves *are* able to exert a net force on a charged particle. To see this, we again treat the Lorentz magnetic force as a perturbation to the Coulomb electric force. From (16.186), the electric displacement of a particle with charge q away from a suitably chosen origin is

$$\mathbf{r} = -\frac{q}{m\omega^2}\mathbf{E}. \tag{16.191}$$

For small r, we may write $\mathbf{E}(\mathbf{r},t) \approx \mathbf{E}_0 + (\mathbf{r}\cdot\nabla)\mathbf{E}_0$ and $\mathbf{B}(\mathbf{r},t) \approx \mathbf{B}_0$ where $\mathbf{E}_0 = \mathbf{E}(0)e^{-i\omega t}$ and $\mathbf{B}_0 = \mathbf{B}(0)e^{-i\omega t}$. Using these approximations and (16.191) to evaluate the Coulomb-Lorentz force gives

$$\mathbf{F} = q(\mathbf{E} + \dot{\mathbf{r}}\times\mathbf{B}) \approx q\mathbf{E}_0 - \frac{q^2}{m\omega^2}\left[(\mathbf{E}_0\cdot\nabla)\mathbf{E}_0 + \frac{\partial\mathbf{E}_0}{\partial t}\times\mathbf{B}_0\right]. \tag{16.192}$$

The identity $(\mathbf{E}\cdot\nabla)\mathbf{E} = \frac{1}{2}\nabla(\mathbf{E}\cdot\mathbf{E}) - \mathbf{E}\times(\nabla\times\mathbf{E})$ and $\nabla\times\mathbf{E} = -\partial\mathbf{B}/\partial t$ transform (16.192) to

$$\mathbf{F} = q\mathbf{E}_0 - \frac{q^2}{m\omega^2}\left[\tfrac{1}{2}\nabla(\mathbf{E}_0\cdot\mathbf{E}_0) + \frac{\partial}{\partial t}(\mathbf{E}_0\times\mathbf{B}_0)\right]. \tag{16.193}$$

All the fields (and hence the force) in (16.193) are complex, time-harmonic quantities. To find the physical force, we first let $\mathbf{E}_0 \to \text{Re}\,\mathbf{E}_0$ and $\mathbf{B}_0 \to \text{Re}\,\mathbf{B}_0$ and then average over one period of oscillation. When this is done, the $q\mathbf{E}_0$ and total time derivative terms in (16.193) vanish. For the gradient term, we apply the time-averaging theorem for harmonic fields (Section 1.6.3).

The final result applies to any choice of origin, so the net time-averaged force is

$$\langle\mathbf{F}(\mathbf{r})\rangle = -\frac{q^2}{4m\omega^2}\nabla|\mathbf{E}(\mathbf{r})|^2 = -\nabla V_{\text{P}}(\mathbf{r}). \tag{16.194}$$

The name *pondermotive force* is often applied to (16.194) and to any other time-averaged, non-linear force that results from the interaction of matter with an oscillating electromagnetic field.

Equation (16.194) confirms the result of the previous section that a propagating plane wave with $\mathbf{E}(\mathbf{r}) = \exp(i\mathbf{k}\cdot\mathbf{r})$ exerts zero net force on a charged particle. The pondermotive potential $V_P(\mathbf{r})$ defined by (16.194) is manifestly positive, so we expect charged particles (of either sign) to be repelled from regions of space with high field intensity toward regions of space with low field intensity. For example, a low-energy electron incident on the Gaussian beam (16.115) from the side is back-scattered by the beam because the field intensity decreases rapidly in the radial direction from its maximum on the propagation axis. On the other hand, numerical trajectory calculations show that the same electron, given an initial kinetic energy larger than the maximum of $V_P(\mathbf{r})$, can be trapped near the focus of the beam and subsequently accelerated by the beam's longitudinal electric field.[34]

16.10.3 Optical Tweezers: the Force on a Polarizable Particle

Example 6.5 demonstrated that a static electric field exerts an attractive force on a polarizable atom or a molecule. The generalization of this phenomenon to time-dependent fields makes it possible to move and trap tiny polarizable particles with remarkable precision. This phenomenon—called *optical tweezers*—can be analyzed using ray optics when the object size is large compared to the wavelength of the field. In this section, we work in the opposite limit, when the object size is small compared to the wavelength of the field, and replace the object by a point electric dipole $\mathbf{p}(t)$.

The time average of the force (15.135) exerted on a point electric dipole by a field which oscillates at frequency ω is

$$\langle\mathbf{F}\rangle = \tfrac{1}{2}\mathrm{Re}\left\{(\mathbf{p}^*\cdot\nabla)\mathbf{E} + \frac{d\mathbf{p}^*}{dt}\times\mathbf{B}\right\}. \tag{16.195}$$

Using $\nabla\times\mathbf{E} = -\partial\mathbf{B}/\partial t = i\omega\mathbf{B}$ and a complex polarizability $\alpha = \alpha' + i\alpha''$ defined by $\mathbf{p} = \epsilon_0\alpha\mathbf{E}$, (16.195) takes the form

$$\langle\mathbf{F}\rangle = \tfrac{1}{2}\epsilon_0\mathrm{Re}\left\{\alpha^*(\mathbf{E}^*\cdot\nabla)\mathbf{E} + \alpha^*\mathbf{E}^*\times(\nabla\times\mathbf{E})\right\}. \tag{16.196}$$

To simplify (16.196), we assume a real and spatially varying amplitude $\boldsymbol{\mathcal{E}}(\mathbf{r})$ and phase $\phi(\mathbf{r})$ so

$$\mathbf{E}(\mathbf{r}) = \boldsymbol{\mathcal{E}}(\mathbf{r})\exp[i\phi(\mathbf{r})]. \tag{16.197}$$

The Maxwell equation $\nabla\cdot\mathbf{E} = 0$ implies that $\nabla\cdot\boldsymbol{\mathcal{E}} = 0$ and $\boldsymbol{\mathcal{E}}\cdot\nabla\phi = 0$. Substituting (16.197) into (16.196) and using these constraints yields

$$\langle\mathbf{F}\rangle = \tfrac{1}{2}\epsilon_0\mathrm{Re}\left\{\alpha^*(\boldsymbol{\mathcal{E}}\cdot\nabla)\boldsymbol{\mathcal{E}} + \alpha^*[\boldsymbol{\mathcal{E}}\times(\nabla\times\boldsymbol{\mathcal{E}}) + i|\boldsymbol{\mathcal{E}}|^2\nabla\phi]\right\}. \tag{16.198}$$

Therefore, because $\nabla(\boldsymbol{\mathcal{E}}\cdot\boldsymbol{\mathcal{E}}) = 2(\boldsymbol{\mathcal{E}}\cdot\nabla)\boldsymbol{\mathcal{E}} + 2\boldsymbol{\mathcal{E}}\times(\nabla\times\boldsymbol{\mathcal{E}})$,

$$\langle\mathbf{F}\rangle = \mathbf{F}_1 + \mathbf{F}_2 = \tfrac{1}{4}\epsilon_0\alpha'\nabla|\boldsymbol{\mathcal{E}}|^2 + \tfrac{1}{2}\epsilon_0\alpha''|\boldsymbol{\mathcal{E}}|^2\nabla\phi. \tag{16.199}$$

The $\mathbf{F}_1$ term in (16.199) is similar to the pondermotive force (16.194) on a free charged particle except that the polarizability α is generally a frequency-dependent function whose real part can be positive or negative.[35] When $\alpha' > 0$, this "gradient force" pushes a polarizable particle away from field intensity minima and toward field intensity maxima. As a result, $\mathbf{F}_1$ alone is capable of trapping

[34] Y.I. Salamin and C.H. Keitel, "Electron acceleration by a tightly focused laser", *Physical Review Letters* **88**, 095005 (2002).

[35] The frequency response of the "Lorentz atom" studied in Section 18.5.4 implies that $\alpha' > 0$ for frequencies just below the transition frequency of a long-lived atomic resonance and $\alpha' < 0$ for frequencies just above the resonant frequency.

atoms and molecules near the focus of a laser beam. The pondermotive potential associated with $\mathbf{F}_1$ is the time average of the energy $\Delta U_E = -\frac{1}{2}\mathbf{p}\cdot\mathbf{E}$ computed in Section 6.7.4 for a polarizable point object where $\mathbf{p} = \epsilon_0\alpha\mathbf{E}$ and $\alpha'' = 0$ (see also Example 15.5). In detail,

$$\mathbf{F}_1 = -\nabla\langle\Delta U_E\rangle = \frac{1}{2}\epsilon_0\alpha'\nabla\langle|\mathbf{E}|^2\rangle = \frac{1}{4}\epsilon_0\alpha'\nabla|\mathcal{E}|^2. \tag{16.200}$$

The $\mathbf{F}_2$ term in (16.199) is present when the complex polarizability has an imaginary part and the particle can absorb energy from the field (see Example 17.2.2 and Section 18.3). It is often called the "scattering force" because scattering is synonymous with absorption and re-radiation in classical physics. The scattering force points in the direction of the local wave vector $\mathbf{k}(\mathbf{r}) = \nabla\phi(\mathbf{r})$.

Sources, References, and Additional Reading

The quotation at the beginning of the chapter is taken from Chapter 3 of

A. Einstein and L. Infeld, *The Evolution of Physics* (Simon and Schuster, New York, 1938).

Section 16.1 Textbooks with the title "Electromagnetic Waves" are common in the "electromagnetics" sub-discipline of electrical engineering. Three examples of this genre in increasing order of sophistication are

U.S. Inan and A.S. Inan, *Electromagnetic Waves* (Prentice-Hall, Upper Saddle River, NJ, 2000).

D.H. Staelin, A.W. Morgenthaler, and J.A. Kong, *Electromagnetic Waves* (Prentice-Hall, Englewood Cliffs, NJ, 1994).

C.G. Someda, *Electromagnetic Waves* (CRC Press, Boca Raton, FL, 2006).

Section 16.2 *Jones* and *Schelkunoff* prove the theorem of *Whittaker* using arguments different from each other and from the one given in the text.

E.T. Whittaker, "On an expression of the electromagnetic field due to electrons by means of two scalar potential functions", *Proceedings of the London Mathematical Society*, Series 2, **1**, 367 (1904).

D.S. Jones, *The Theory of Electromagnetism* (Macmillan, New York, 1964), Section 1.10.

S.A. Schelkunoff, *Electromagnetic Waves* (Van Nostrand, Princeton, NJ, 1943), Section 10.3.

Section 16.3 Textbooks of electromagnetism where plane waves are treated particularly clearly include

J. Stratton, *Electromagnetic Theory* (McGraw-Hill, New York, 1941).

E.J. Konopinski, *Electromagnetic Fields and Relativistic Particles* (McGraw-Hill, New York, 1981).

Figure 16.2 comes from *Born*. *Wise* provides a historical perspective.

M. Born, *Einstein's Theory of Relativity* (Metheun, London, 1924), Chapter V.

M.N. Wise, "The mutual embrace of electricity and magnetism", *Science* **203**, 1310 (1979).

An entertaining discussion of the "reality" of the classical electromagnetic field is

N.D. Mermin, "What's bad about this habit?", *Physics Today*, May 2009, pp. 8-9.

Section 16.4 Our discussion of wave polarization benefitted from

M.L. Kales, "Elliptically polarized waves and antennas", *Proceedings of the IRE.* **39**, 544 (1951).

D. Goldstein, *Polarized Light*, 2nd edition (Marcel Dekker, New York, 2003).

G.S. Smith, "The polarization of skylight: An example from nature", *American Journal of Physics* **75**, 25 (2007).

Section 16.5 The material of this section was drawn from *Konopinski* (see Section 16.3 above) and

B. Podolsky and K. Kunz, *Fundamentals of Electrodynamics* (Marcel Dekker, New York, 1969).

B.G. Levich, *Theoretical Physics* (North-Holland, Amsterdam, 1970), Volume 1.

Section 16.6 The angular spectrum of plane waves is discussed clearly in

G.S. Smith, *Classical Electromagnetic Radiation* (Cambridge University Press, Cambridge, 1997).

Section 16.7 Beam-like solutions of the Maxwell equations (Section 16.7) are discussed extensively in the literature of lasers. Two points of entry are

A.E. Siegmann, *Lasers* (University Science Books, Sausalito, CA, 1986).

E.J. Galvez, "Gaussian beams in the optics course", *American Journal of Physics* **74**, 355 (2006).

Figure 16.13 was adapted from

L.W. Davis and G. Patsakos, "TE and TM electromagnetic beams in free space", *Optics Letters* **6**, 22 (1981).

The angular momentum calculations in Section 16.7.5 and Section 16.7.6 come from

M. Berry, "Paraxial beams of spinning light", in *Singular Optics*, edited by M.S. Soskin and M.V. Vasnetsov (SPIE, Bellingham, WA, 1998), pp. 6-11.

The preceding paper, as well as the articles by Poynting and Beth cited in the text, are reproduced in the reprint volume,

L. Allen, S.M. Barnett, and M.J. Padgett, *Optical Angular Momentum* (Institute of Physics, Bristol, 2003).

Section 16.8 Two intermediate-level textbooks with brief discussions of spherical electromagnetic waves are

J.R. Reitz and F.J. Milford, *Foundations of Electromagnetic Theory* (Addison-Wesley, Reading, MA, 1960).

M.H. Nayfeh and M.K. Brussel, *Electricity and Magnetism* (Wiley, New York, 1985).

Stratton and *Konopinski* (see Section 16.3 above) discuss spherical electromagnetic waves in detail. The first uses Hertz vectors and also treats cylindrical waves. The use of Debye potentials for free fields is discussed in

C.G. Gray and B.G. Nickel, "Debye potential representation of vector fields", *American Journal of Physics* **46**, 735 (1978).

The "hairy ball theorem" (Application 16.1) is discussed in strictly mathematical terms and then in an electromagnetic context in

J. Milnor, "Analytic proofs of the 'hairy ball theorem' and the Brouwer fixed point theorem", *American Mathematical Monthly* **85**, 521 (1978).

H. Mott, *Polarization in Antennas and Radar* (Wiley, New York, 1986), Appendix B.

Section 16.9 A short and precise description of the use of Hertz vectors in empty space is

F. Kottler, "Diffraction at a black screen, part II: Electromagnetic theory", in *Progress in Optics*, Volume 6, edited by E. Wolf (North-Holland, Amsterdam, 1967), Appendix B.

A Hertz vector study of the fundamental Gaussian beam and its approximants is

P. Varga and P. Torok, "The Gaussian wave solution of Maxwell's equations and the validity of the scalar wave approximation", *Optics Communications* **152**, 108 (1998).

Application 16.2 was adapted from

D.N. Pattanayak and G.P. Agrawal, "Representation of vector electromagnetic beams", *Physical Review A* **22**, 1159 (1980).

Section 16.10 There have been many calculations of the trajectory of a charged point particle in the field of a plane wave. Our treatment simplifies the analysis of

E.S. Sarachik and G.T. Schappert, "Classical theory of the scattering of intense laser radiation by free electrons", *Physical Review D* **1**, 2738 (1970).

An early and very clear treatment of the pondermotive force exerted on a charged particle is

H.A.H. Boot, S.A. Self, and R.B. Robertson-Shersby-Harvie, "Containment of a fully ionized plasma by radio frequency fields", *Journal of Electronics and Control* **4**, 434 (1958).

The paper by *Shimizu and Sasada* is the source of our discussion of forces on polarizable particles. The wave optics approach to optical tweezers (and much else) can be found in the reprint volume of original papers edited by *Padgett et al.*

Y. Shimizu and H. Sasada, "Mechanical force in laser cooling and trapping", *American Journal of Physics* **66**, 960 (1998).

M.J. Padgett, J.E. Molloy, and D. McGloin, *Optical Tweezers: Methods and Applications* (CRC Press, Boca Raton, FL, 2010).

Problems

16.1 Wave Equation vs. Maxwell Equations Let $\mathbf{E}(\mathbf{r}, t)$ be a vector field that satisfies

$$\nabla \cdot \mathbf{E} = 0 \qquad \text{and} \qquad \nabla^2 \mathbf{E} - \frac{1}{c^2}\frac{\partial^2 \mathbf{E}}{\partial t^2} = 0.$$

Use the Helmholtz theorem to write a formula for $\mathbf{B}(\mathbf{r}, t)$ in terms of $\mathbf{E}(\mathbf{r}, t)$. Confirm that $\mathbf{B}(\mathbf{r}, t)$ and $\mathbf{E}(\mathbf{r}, t)$ satisfy all four Maxwell equations in free space.

16.2 No Electromagnetic Bullets Let $f(\xi)$ be an arbitrary scalar function of the scalar variable ξ. We have learned that $f(z - ct)$ is a traveling-wave solution of the *one-dimensional* wave equation. In other words,

$$\left[\frac{\partial^2}{\partial z^2} - \frac{1}{c^2}\frac{\partial^2}{\partial t^2}\right] f(z - ct) = 0.$$

We have also learned that solutions of this equation can be *strictly* localized, i.e., $f(\xi)$ can be exactly zero outside a finite interval of ξ. Now let $\psi(x, y, z - ct)$ be a solution of the *three-dimensional* wave equation.

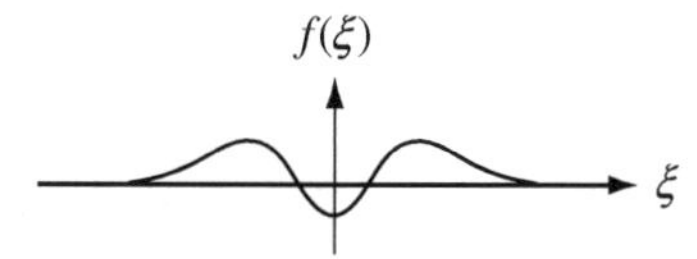

Use the information just given (and your knowledge of electrostatics) to prove that ψ *cannot* be strictly localized in the x, y, and z directions simultaneously.

16.3 An Evanescent Wave in Vacuum Let $\mathbf{E} = \hat{\mathbf{y}} E_0 \exp[i(hz - \omega t) - \kappa x]$ be the electric field of a wave propagating in vacuum.

(a) How are the real parameters h, κ, and ω related to one another?
(b) Find the associated magnetic field $\mathbf{B}$.
(c) Under what conditions is the polarization of the magnetic field close to circular?
(d) Compute the time-averaged Poynting vector.

16.4 Plane Waves from Potentials

(a) Let $\mathbf{A}(x)$ be a vector function of a scalar argument. Find the conditions that make $\mathbf{A}(\mathbf{k} \cdot \mathbf{r} - ckt)$ a legitimate Coulomb gauge vector potential in empty space. Compute $\mathbf{E}(\mathbf{r}, t)$ and $\mathbf{B}(\mathbf{r}, t)$ explicitly and show that both are transverse.
(b) Consider a Coulomb gauge vector potential $\mathbf{A}_C(\mathbf{r}, t) = u(\mathbf{r}, t)\mathbf{a}$ where u is a scalar function and $\mathbf{a}$ is a constant vector. What restrictions must be imposed on $u(\mathbf{r}, t)$ and $\mathbf{a}$, if any? Find the associated electric and magnetic fields.
(c) Specialize (b) to the case when $u(\mathbf{r}, t) = u(\mathbf{k} \cdot \mathbf{r} - ckt)$. Show that $c\mathbf{B} = \hat{\mathbf{k}} \times \mathbf{E}$.
(d) Consider a Lorenz gauge potential $\mathbf{A}_L(\mathbf{r}, t) = u(\mathbf{r}, t)\mathbf{s}$ where u is a scalar function and $\mathbf{s}$ is a constant vector. What restrictions must be imposed on $u(\mathbf{r}, t)$ and $\mathbf{s}$, if any? Find the associated electric and magnetic fields.
(e) Specialize (d) to the case when $u(\mathbf{r}, t)$ is the same plane wave as in part (c). Show that the electric and magnetic fields are exactly the same as those computed in part (c).

16.5 Two Counter-Propagating Plane Waves

(a) Let $\mathbf{E} = E_0 \cos(kz - \omega t)\hat{\mathbf{x}} + E_0 \cos(kz + \omega t)\hat{\mathbf{x}}$. Write $\mathbf{E}(z, t)$ in simpler form and find the associated magnetic field $\mathbf{B}(z, t)$.
(b) For the fields in part (a), find the instantaneous and time-averaged electric and magnetic field energy densities. Find also the instantaneous and time-averaged energy current density.

(c) Let $\mathbf{E} = E_0 \cos(kz - \omega t)\hat{\mathbf{x}} + E_0 \sin(kz + \omega t)\hat{\mathbf{y}}$. Determine the polarization of this field at $z = 0$, $z = \lambda/8$, $z = \lambda/4$, $z = 3\lambda/8$, and $z = \lambda/2$.

(d) Show that the field in part (c) can be written as the sum of a standing wave with polarization $\hat{\mathbf{e}}_+$ and a standing wave with polarization $\hat{\mathbf{e}}_-$.

(e) Find the time-averaged Poynting vector as a function of z for the field in part (c).

16.6 Transverse Plane Waves with E ∥ B Consider transverse plane waves in free space where $\mathbf{E} = \mathbf{E}(z, t)$ and $\mathbf{B} = \mathbf{B}(z, t)$.

(a) Demand that the Poynting vector $\mathbf{S} = 0$ and show that all such solutions must satisfy

$$\frac{\partial u_{\rm EM}}{\partial t} = \frac{\partial u_{\rm EM}}{\partial z} = 0.$$

(b) The constancy of $u_{\rm EM}$ implies that the fields can be parameterized as

$$E_x = \cos\alpha\cos\beta \qquad cB_x = \sin\alpha\cos\gamma$$
$$E_y = \cos\alpha\sin\beta \qquad cB_y = \sin\alpha\sin\gamma,$$

where α, β, and γ are functions of z and t. In particular, show that $\mathbf{S} = 0$ implies that

$$\alpha(z,t) = F(z+ct) + G(z-ct)$$
$$\beta(z,t) = F(z+ct) - G(z-ct)$$
$$\gamma(z,t) = \beta(z,t),$$

where $F(x)$ and $G(x)$ are arbitrary scalar functions.

(c) Let $F = \frac{1}{2}k(z+ct)$ and $G = \pm\frac{1}{2}k(z-ct)$ where $\omega = ck$. For both choices of sign, sketch a snapshot of the fields as a function of z which clearly shows their behavior. Explain why one of these cases corresponds to the superposition of two counter-propagating, circularly polarized waves.

(d) Sketch the fields as above for $F = \frac{1}{2}k(z+ct)$ and $G = 0$.

16.7 Photon Spin for Plane Waves

(a) Show that the angular momentum of an electromagnetic field in empty space (no sources) can be written in the form

$$\mathbf{L}_{\rm EM} = \epsilon_0 \int d^3r\, \mathbf{r}\times(\mathbf{E}\times\mathbf{B}) = \epsilon_0 \int d^3r\, E_k(\mathbf{r}\times\nabla)A_k + \epsilon_0 \int d^3r\, \mathbf{E}\times\mathbf{A} = \mathbf{L}_{\rm orbital} + \mathbf{L}_{\rm spin}.$$

Note any requirements that the fields must satisfy at infinity. The last term is assigned to $\mathbf{L}_{\rm spin}$ because it is a contribution to the angular momentum that does not depend on the "lever arm" $\mathbf{r}$.

(b) Show that the proposed decomposition is not gauge invariant and therefore not physically meaningful.

(c) Despite the foregoing, work in the Coulomb gauge and apply these formulae to a circularly polarized plane wave with electric field

$$\mathbf{E}_\pm = E_0 \frac{\hat{\mathbf{x}} \pm i\hat{\mathbf{y}}}{\sqrt{2}} \exp[i(kz - \omega t)].$$

Show that the time averages obey $\pm\omega\,\hat{\mathbf{z}}\cdot\langle\mathbf{L}_{\rm spin}\rangle = \langle U_{\rm EM}\rangle$. Interpret this formula if $\langle U_{\rm EM}\rangle = \hbar\omega$.

16.8 When Interference Behaves Like Reflection The diagram shows two electromagnetic beams intersecting at right angles. $(\mathbf{E}_{\rm H}, \mathbf{B}_{\rm H})$ propagates in the $+x$-direction. $(\mathbf{E}_{\rm V}, \mathbf{B}_{\rm V})$ propagates in the $+y$-direction. For simplicity, each beam is taken as a pure plane wave (with $\omega = ck = 2\pi c/\lambda$) cut off transversely so its cross section is a perfect square of area λ^2:

$$\mathbf{E}_{\rm H} = -E_0 \exp[i(kx - \omega t)]\hat{\mathbf{z}} \qquad |y| \le \lambda/2, \quad |z| \le \lambda/2,$$
$$c\mathbf{B}_{\rm H} = +E_0 \exp[i(kx - \omega t)]\hat{\mathbf{y}} \qquad |y| \le \lambda/2, \quad |z| \le \lambda/2,$$
$$\mathbf{E}_{\rm V} = E_0 \exp[i(ky - \omega t)]\hat{\mathbf{z}} \qquad |x| \le \lambda/2, \quad |z| \le \lambda/2,$$
$$c\mathbf{B}_{\rm V} = E_0 \exp[i(ky - \omega t)]\hat{\mathbf{x}} \qquad |x| \le \lambda/2, \quad |z| \le \lambda/2.$$

The beams overlap in a cube (volume λ^3) centered at the origin where the total fields are $\mathbf{E} = \mathbf{E}_{\rm H} + \mathbf{E}_{\rm V}$ and $\mathbf{B} = \mathbf{B}_{\rm H} + \mathbf{B}_{\rm V}$.

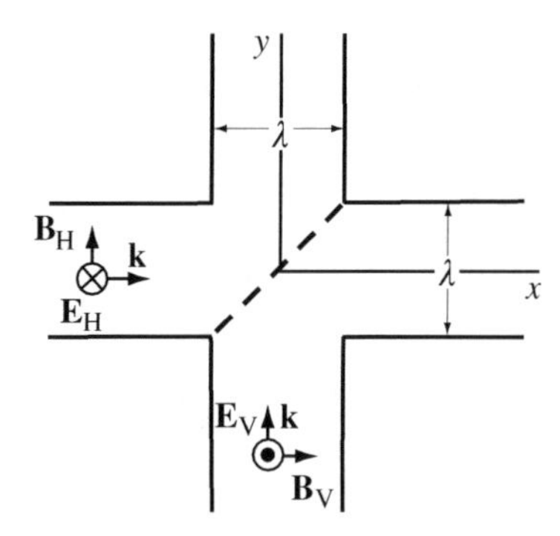

(a) Calculate the time-averaged energy density $\langle u_{\rm EM}(\mathbf{r})\rangle$ for the horizontal (H) beam alone, for the vertical (V) beam alone, and for the total field in the overlap region. Show that the last of these takes its minimum value on the plane $x = y$ shown dashed in the diagram. Compute **E** and **B** on this plane.

(b) Calculate the time-averaged Poynting vector $\langle \mathbf{S}(\mathbf{r})\rangle$ for the H beam, the V beam, and the total field as in part (a). Make a careful sketch of $\langle \mathbf{S}(x, y)\rangle$ everywhere the fields are defined.

(c) Explain why the behavior of both **E** and **B** in the vicinity of the $x = y$ plane is exactly what you would expect if that plane were a perfect conductor. This shows that the interfering beams behave as if they had specularly reflected from each other.

16.9 Zeroes of the Transverse Field It is difficult to visualize time-dependent electromagnetic fields in three-dimensional space. However, for fields that propagate in the z-direction, it is interesting to focus on the locus of points where the transverse (x and y) components of the fields vanish simultaneously. For $\mathbf{E}(\mathbf{r}, t)$, this is the one-dimensional curve where the surface $E_x(x, y, z, t) = 0$ intersects the surface $E_y(x, y, z, t) = 0$. The same condition with $\mathbf{B}(\mathbf{r}, t)$ gives the curve of magnetic zeroes.

(a) Show that the scalar function $\psi(r, t) = k\left\{gx + kx^2 \sin\delta + iy'\right\}\exp i(kz - \omega t)$ solves the wave equation where g and δ are parameters and $y' = y\cos\delta + z\sin\delta$.

(b) Let $\mathbf{E} = \hat{\mathbf{x}}\psi + i\hat{\mathbf{y}}\psi + \hat{\mathbf{z}}E_z$. Use Gauss' law and Faraday's law to deduce $E_z(r, t)$ and $\mathbf{B}(r, t)$ so that **E** and **B** solve the Maxwell equations in free space.

(c) If the parameter $g \gg 1$, there is a large region of space where the x^2 piece of the fields can be dropped with very little loss of accuracy. Do this and show that the curve of electric zeroes is the line $y = -z\tan\delta$ in the $x = 0$ plane.

(d) In the same $g \gg 1$ limit, show that the locus of zeroes for the transverse magnetic field is an elliptic helix that rotates in time around its axis of symmetry. The latter (which we will call the z' axis) is the line of electric zeroes found in part (c). Notice that y' and z' are orthogonal and use polar coordinates (r, θ) defined by $gx = r\cos\theta$ and $y' = r\sin\theta$. Make a sketch that shows where both sets of zeroes occur.

16.10 Superposition and Wave Intensity Let $\mathbf{E} = \mathbf{E}_1 + \mathbf{E}_2$ be the electric field of the sum of two monochromatic plane waves propagating in the z-direction. One wave has frequency ω_1 and is elliptically polarized. The other wave has frequency ω_2 and is elliptically polarized in a different way than the first wave. Derive precise, quantitative conditions that relate the averaging time T to ω_1 and ω_2 so the wave intensities satisfy $I = I_1 + I_2$.

16.11 Antipodes of the Poincaré Sphere Prove that any two antipodes on the Poincaré sphere correspond to orthogonal states of polarization.

16.12 Kepler's Law for Plane Wave Polarization Let $\mathcal{E}(t) = \mathrm{Re}\,\mathbf{E}(0, t) = a_1\cos(\omega t - \delta_1)\hat{\mathbf{e}}_1 + a_2\cos(\omega t - \delta_2)\hat{\mathbf{e}}_2$. Except for the case of linear polarization, this vector sweeps out an ellipse as a function of time. Show that the rate at which $\mathcal{E}$ sweeps out area within the ellipse is

$$\frac{dA}{dt} = \frac{1}{2}\omega a_1 a_2 \sin(\delta_2 - \delta_1).$$

This shows that $\mathcal{E}(t)$ sweeps out equal areas in equal time. This is similar to Kepler's second law except that the origin of $\mathcal{E}(t)$ is the center of the polarization ellipse while the origin of the radius vector in Kepler's law is the focus of an orbital ellipse.

16.13 Elliptical Polarization The electric field of a plane wave is $\mathbf{E}(z,t) = \hat{\mathbf{x}}A\cos(kz - \omega t + \delta_1) + \hat{\mathbf{y}}B\cos(kz - \omega t + \delta_2)$. Show that the principal axes of the polarization ellipse for this field are rotated from the x- and y-axes by an angle α where

$$\tan 2\alpha = \frac{2AB}{A^2 - B^2}\cos(\delta_2 - \delta_1).$$

16.14 A Vector Potential Wave Packet $\mathbf{A}(\mathbf{r},t) = (a\hat{\mathbf{x}} + ib\hat{\mathbf{y}})A_0(\zeta)\exp(i\zeta)$ where $\zeta = k(z + ct)$ is a vector potential wave packet.

(a) Find the (real) electric and magnetic fields in the approximation that the envelope function $A(\zeta)$ varies slowly over one wavelength of the wave.
(b) Find the linear momentum density carried by the wave in part (a). What does the overall algebraic sign of this quantity tell you?

16.15 Fourier Uncertainty Let $f(x)$ and $\hat{f}(k)$ be Fourier transforms pairs so

$$f(x) = \int_{-\infty}^{\infty} \frac{dk}{2\pi}\hat{f}(k)e^{ikx} \qquad \text{and} \qquad \hat{f}(k) = \int_{-\infty}^{\infty} dx f(x)e^{-ikx}.$$

The averages of the operator O with respect to the distributions $f(x)$ and $\hat{f}(k)$ are

$$\langle O\rangle_x = \frac{\int_{-\infty}^{\infty} dx\, f^*(x) O\, f(x)}{\int_{-\infty}^{\infty} dx\, |f(x)|^2} \qquad \text{and} \qquad \langle O\rangle_k = \frac{\int_{-\infty}^{\infty} dk\, \hat{f}^*(k) O\, \hat{f}(k)}{\int_{-\infty}^{\infty} dk\, |\hat{f}(k)|^2}.$$

(a) Show that $\hat{h}(k) = ik\hat{f}(k)$ if $h(x) = df/dx$.
(b) Prove Parseval's theorem,

$$\frac{1}{2\pi}\int_{-\infty}^{\infty} dk\, \hat{f}^*(k)\hat{g}(k) = \int_{-\infty}^{\infty} dx\, f^*(x)g(x).$$

(c) Show that $\langle k\rangle_k = -i\langle d/dx\rangle_x$ and $\langle k^2\rangle_k = -\langle d^2/dx^2\rangle_x$.

(d) Let $\Delta O = \sqrt{\langle O^2\rangle_x - \langle O\rangle_x^2}$ and reproduce a proof of the (Hermitian) operator identity

$$\Delta A\, \Delta B \geq \frac{1}{2}|\langle 2i[A,B]\rangle_x|$$

from any (quantum mechanics) text in enough detail to demonstrate that you understand it.
(e) Let $\Delta k = \sqrt{\langle k^2\rangle_k - \langle k\rangle_k^2}$ and use all of the above to prove that

$$\Delta x\, \Delta k \geq \frac{1}{2}.$$

This shows that the width of a wave packet in real space increases as the wave vector content of the packet decreases.

16.16 Plane Wave Packet from the Helmholtz Equation The scalar wave equation is $c^2\nabla^2 u = \dfrac{\partial^2 u}{\partial t^2}$.

(a) Find a general solution to this equation by separating variables in the form $u(\mathbf{r})T(t)$.
(b) Use Cartesian variables to separate the equation satisfied by $u(\mathbf{r})$. Write the general solution again.
(c) How does your final general solution for $u(\mathbf{r}, t)$ differ from the expression derived in the text for a wave packet composed of monochromatic plane waves? Show that this difference disappears for $\text{Re}\, u(\mathbf{r}, t)$.

16.17 $\mathbf{P}_{\text{EM}}$ **for a Wave Packet** Use a normalization slightly different from the text and write the complex electric field of an electromagnetic wave packet as

$$\mathbf{E}(\mathbf{r}, t) = \frac{1}{(2\pi)^{3/2}} \int d^3k\, \mathbf{E}_\perp(\mathbf{k}) \exp[i(\mathbf{k}\cdot\mathbf{r} - ckt)].$$

(a) Show that the total linear momentum of the wave packet satisfies

$$c\mathbf{P}_{\text{EM}} = \tfrac{1}{2}\epsilon_0 \int d^3k\, \hat{\mathbf{k}}\, |\mathbf{E}_\perp(\mathbf{k})|^2.$$

(b) Produce an argument which shows that

$$U_{\text{EM}} \geq c\,|\mathbf{P}_{\text{EM}}|.$$

(c) When does the equal sign apply in part (b)?

16.18 A Transverse Magnetic Beam Let $\mathbf{A}(\rho, z, t) = U(\rho, z, t)\hat{\mathbf{z}}$ be a Lorenz gauge vector potential where $U(\rho, z, t) = u(\rho, z)\exp(-i\omega t)$ is an azimuthally symmetric, beam-like solution of the scalar wave equation which propagates in the z-direction. Show that the magnetic field $\mathbf{B}(\rho, z, t)$ and electric field $\mathbf{E}(\rho, z, t)$ are both linearly polarized, but that the latter is not transverse.

16.19 Paraxial Fields of the Gaussian Beam Let $u(x, y, z, t)$ be a monochromatic, Gaussian beam solution of the scalar wave equation in the paraxial approximation. The beam propagates in the z-direction. Use a suitably chosen Lorenz gauge vector potential to confirm the statement made in the text that the electric and magnetic fields of this beam in the same approximation can be written

$$\mathbf{E} = u\hat{\mathbf{x}} + \frac{i}{k}\frac{\partial u}{\partial x}\hat{\mathbf{z}} \qquad \text{and} \qquad \mathbf{B} = -\frac{i}{\omega}\nabla\times\mathbf{E}.$$

Confirm also that $\nabla\cdot\mathbf{E} = 0$ and $\nabla\cdot\mathbf{B} = 0$ within the paraxial approximation.

16.20 Physical Origin of the Gouy Phase The function

$$\phi(\rho, z) = f(x, y)\exp\left[i\left(kz - \omega t + k_0\rho^2/2R + \alpha\right)\right]$$

is a beam-like solution to the scalar wave equation (in the paraxial approximation) where

$$\omega = ck, \qquad z_R = \frac{1}{2}k_0 w_0^2, \qquad w(z) = w_0\sqrt{1 + z^2/z_R^2}, \qquad R(z) = z + z_R^2/z, \qquad \alpha(z) = -\tan^{-1}(z/z_R),$$

and

$$f(x, y) = \sqrt{\frac{2}{\pi}}\frac{1}{w}\exp\left[-(x^2 + y^2)/w^2\right].$$

The Gouy phase, $\alpha(z)$, arises from the fact that the beam has a finite size in the transverse direction. To see this, use the fact that $k^2 = k_x^2 + k_y^2 + k_z^2$ to motivate the definition of an effective propagation "constant"

$$\bar{k}_z(z) = \frac{\langle k_z^2\rangle}{k} = k - \frac{\langle k_x^2\rangle}{k} - \frac{\langle k_y^2\rangle}{k}.$$

In this formula, the averages are defined over the distribution of transverse wave vectors that make up the beam. That is,

$$\langle g\rangle \;=\; \int_{-\infty}^{\infty} dk_x \int_{-\infty}^{\infty} dk_y\, g(k_x, k_y)|F(k_x, k_y)|^2$$

where

$$F(k_x, k_y) = \frac{1}{2\pi}\int_{-\infty}^{\infty} dx \int_{-\infty}^{\infty} dy\, f(x, y)\exp[-i(k_x x + k_y y)].$$

Show that $kz + \alpha(z) = \int_0^z dz\, \bar{k}_z$. Confirm that $\alpha(z) = 0$ if there is no localization in the transverse direction.

16.21 D'Alembert Solutions in Two and Three Dimensions If f is a scalar function of one scalar variable, we know that $f(z \pm ct)$ are solutions of the one-dimensional wave equation.

(a) Show that $f(r \pm ct)/r$ are solutions of the three-dimensional wave equation in spherical coordinates.
(b) Show that $f(\rho \pm ct)/\sqrt{\rho}$ are solutions of the two-dimensional wave equation in cylindrical coordinates when ρ is sufficiently large. Make a precise definition of "sufficiently large".

16.22 Wave Interference

(a) Superpose two scalar waves: a plane wave $u_1 \exp(ikx)$ and a spherical wave $(u_2/r)\exp[i(kr + \delta)]$. Show that the locus of points where constructive interference occurs defines a family of parabolas.
(b) Superpose two equal-amplitude, time-harmonic, y-polarized plane waves. Their wave vectors satisfy $|\mathbf{k}_1| = |\mathbf{k}_2|$ and lie in the x-z plane as shown below. Is the sum of these two plane waves itself a plane wave? Is it transverse?
(c) Plot the time-averaged Poynting vector for the wave in part (b) as a function of position for $\theta = 0$, $\theta = \pi/4$, and $\theta = \pi/2$.

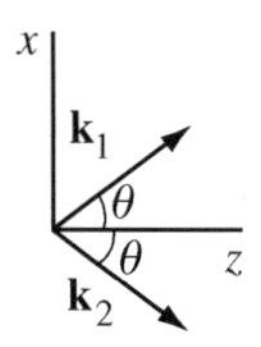

16.23 Phase Velocity of Spherical Waves Consider the outgoing spherical wave solutions of the three-dimensional scalar wave equation in vacuum. Show that the phase velocity of the $\ell = 0$ wave is c but that the phase velocity of the $\ell = 1$ wave is position-dependent and greater than c. What happens as $r \to 0$ and $r \to \infty$? What about the $\ell > 1$ waves?

16.24 Bessel Waves

(a) Separate variables in cylindrical coordinates and find a general time-harmonic solution $\psi(\rho, \phi, z, t)$ of the scalar wave equation which propagates in the z-direction and remains finite on the z-axis.
(b) Find the TE and TM electric and magnetic fields associated with the magnetic Hertz vector $\boldsymbol{\pi}_{\rm m} = \psi_0\hat{\mathbf{z}}$ where ψ_0 is the cylindrically symmetric solution found in part (a).
(c) Interpret $\psi_0(\mathbf{r}, t)$ as a sum of plane waves $\exp[i(\mathbf{q}\cdot\mathbf{r} - \omega t)]$ where the $\mathbf{q}$ vectors have the same magnitude and are distributed uniformly on the surfaced of a cone. Hint: An integral representation of the zero-order Bessel function is

$$J_0(x) = \frac{1}{2\pi}\int_0^{2\pi} d\theta\, \exp(ix\cos\theta).$$

16.25 Charged Particle Motion in a Circularly Polarized Plane Wave A particle with charge q and mass m interacts with a circularly polarized plane wave in vacuum. The electric field of the wave is $\mathbf{E}(z, t) = \mathrm{Re}\,\{(\hat{\mathbf{x}} + i\hat{\mathbf{y}})E_0 \exp[i(kz - \omega t)]\}$.

(a) Let $v_\pm = v_x \pm i v_y$ and $\Omega = 2qE_0/mc$. Show that the equations of motion for the components of the particle's velocity $\boldsymbol{v}$ can be written

$$\frac{dv_z}{dt} = \frac{1}{2}\Omega\left\{v_+ e^{+i(kz-\omega t)} + v_- e^{-i(kz-\omega t)}\right\}$$

$$\frac{dv_\pm}{dt} = \Omega(c - v_z)e^{\mp i(kz-\omega t)}.$$

(b) Let $\ell_\pm = v_\pm e^{\pm i(kz-\omega t)} \pm ic\Omega\omega$ and show that

$$\frac{dv_z}{dt} = \frac{1}{2}\Omega(\ell_+ + \ell_-) = i\frac{\Omega}{2\omega}\frac{d}{dt}(\ell_+ - \ell_-).$$

(c) Let K be the constant of the motion defined by the two $\dot{v}_z$ equations above. Differentiate the equations in part (a) and establish that

$$\frac{d^2 v_z}{dt^2} + \left[\Omega^2 + \omega^2\right] v_z = \omega^2 K.$$

Use the initial conditions $v(0) = 0$ and $v_z'(0) = 0$ to evaluate K and solve for $v_z(t)$. Describe the nature of the particle acceleration in the z-direction.

17 Waves in Simple Matter

It appears that the square of the index of refraction is equal to the product of the specific dielectric capacity and the specific magnetic capacity.
James Clerk Maxwell (1865)

17.1 Introduction

This chapter explores the propagation of monochromatic plane waves in *simple* matter where the electric permittivity ϵ, magnetic permeability μ, and ohmic conductivity σ are all constants. When $\sigma = 0$ this model for matter is *non-dispersive* in the sense that plane waves with different frequencies all have the same phase velocity. This contrasts with real matter, which is *frequency-dispersive* because $\epsilon = \epsilon(\omega)$ is a function of frequency and plane waves with different frequencies propagate with different phase velocities.[1] Nevertheless, by focusing on one frequency at a time—and by not superposing waves with different frequencies—many important effects of wave propagation in real matter can be captured using a non-dispersive model. We will be particularly interested in the reflection, refraction, and interference that occur when waves interact with planar boundaries which separate regions of dissimilar simple matter.

The mathematics of wave propagation in linear and isotropic non-dispersive matter is nearly identical to the mathematics of wave propagation in vacuum. This has the virtue of generating results very quickly (by analogy) and the vice of masking some important physics associated with the matter. In this chapter, we can do little more than name this "hidden" physics; a proper discussion must wait until the reader has acquired an appreciation of retardation and radiation (Chapter 20).

Section 17.6 treats wave propagation in conducting matter where Ohm's law ($\mathbf{j} = \sigma\mathbf{E}$) holds with a constant conductivity σ. Technically, this system is frequency-dispersive and discussion of it could logically be delayed until the next chapter. However, by limiting ourselves to low frequencies, there is pedagogic value in comparing the non-uniform waves in a simple conductor to the non-uniform waves at a dielectric interface. We will also make contact with our previous discussion of quasi-magnetostatics (Section 14.9).

17.2 Plane Waves

The Maxwell equations in matter are

$$\nabla \cdot \mathbf{D} = \rho_f \qquad\qquad \nabla \cdot \mathbf{B} = 0 \tag{17.1}$$

[1] The origins and consequences of frequency dispersion are the subject of Chapter 18.

and

$$\nabla \times \mathbf{E} = -\frac{\partial \mathbf{B}}{\partial t} \qquad \nabla \times \mathbf{H} = \mathbf{j}_f + \frac{\partial \mathbf{D}}{\partial t}. \tag{17.2}$$

In this section, we set $\rho_f = \mathbf{j}_f = 0$ and seek plane wave solutions to (17.1) and (17.2) for non-dispersive, simple matter where the constitutive relations are

$$\mathbf{D} = \epsilon \mathbf{E}, \qquad \text{and} \qquad \mathbf{B} = \mu \mathbf{H}. \tag{17.3}$$

The product and the ratio of the constants ϵ and μ appear so frequently that it is traditional to define the *index of refraction* n and the *intrinsic impedance* Z of a medium as

$$n = c\sqrt{\mu\epsilon} \qquad \text{and} \qquad Z = \sqrt{\mu/\epsilon}. \tag{17.4}$$

The index of refraction is dimensionless and takes the value $n = 1$ in vacuum. The intrinsic impedance has dimensions of resistance and takes the value $Z_0 = \sqrt{\mu_0/\epsilon_0} \approx 377\ \Omega$ in vacuum.

The constitutive relations reduce the number of fields in the Maxwell equations from four to two. We choose $\mathbf{E}$ and $\mathbf{H}$ as the independent fields because the polarization and magnetization of the medium are

$$\mathbf{P} = (\epsilon - \epsilon_0)\mathbf{E} \qquad \text{and} \qquad \mathbf{M} = (\mu/\mu_0 - 1)\mathbf{H}, \tag{17.5}$$

the matching conditions at an interface are the same for $\mathbf{E}$ and $\mathbf{H}$, and the Poynting vector in matter is (see Section 15.8.1)

$$\mathbf{S} = \mathbf{E} \times \mathbf{H}. \tag{17.6}$$

Thus, for an infinite and spatially homogeneous medium defined by (17.3), the source-free Maxwell equations of interest are

$$\nabla \cdot \mathbf{E} = 0 \qquad \nabla \cdot \mathbf{H} = 0 \tag{17.7}$$

and

$$\nabla \times \mathbf{E} = -\mu \frac{\partial \mathbf{H}}{\partial t} \qquad \nabla \times \mathbf{H} = \epsilon \frac{\partial \mathbf{E}}{\partial t}. \tag{17.8}$$

17.2.1 Monochromatic Plane Waves

We are interested in uniform and monochromatic plane waves (Section 16.3.4) where[2]

$$\mathbf{E}(\mathbf{r}, t) = \mathbf{E} \exp[i(\mathbf{k} \cdot \mathbf{r} - \omega t)] \qquad \text{and} \qquad \mathbf{H}(\mathbf{r}, t) = \mathbf{H} \exp[i(\mathbf{k} \cdot \mathbf{r} - \omega t)]. \tag{17.9}$$

Substituting (17.9) into (17.7) and (17.8) generates four constraints:

$$\mathbf{k} \cdot \mathbf{E} = 0 \qquad \mathbf{k} \cdot \mathbf{H} = 0 \tag{17.10}$$

and

$$\mathbf{k} \times \mathbf{E} = \omega\mu\mathbf{H} \qquad \mathbf{k} \times \mathbf{H} = -\omega\epsilon\mathbf{E}. \tag{17.11}$$

An immediate consequence of (17.10) and (17.11) is that the vectors $(\mathbf{E}, \mathbf{H}, \mathbf{k})$ form a right-handed orthogonal triad (Figure 17.1). Therefore, (17.9) is a transverse electromagnetic wave (TEM) to which our discussion of polarization (Section 16.4) applies without change. It also follows from (17.11) that

[2] It should always be clear from context whether the symbols $\mathbf{E}$ and $\mathbf{H}$ refer to the vector functions in (17.9) or to their vector amplitudes.

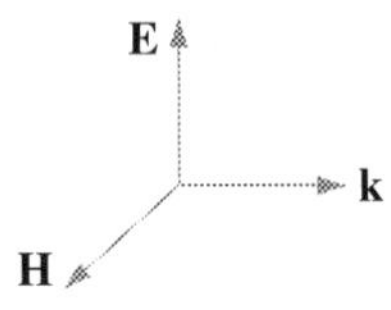

Figure 17.1: E, H, and k for a TEM plane wave in simple matter.

the ratio of the field amplitudes—a scalar quantity called the *wave impedance*—is numerically equal to the intrinsic impedance of the medium defined in (17.4):

$$Z_{\text{wave}} = \frac{E}{H} = \sqrt{\frac{\mu}{\epsilon}} = Z. \tag{17.12}$$

A *dispersion relation* relates the wave frequency ω to the wave vector $\mathbf{k}$. To find it, we form the cross product of $\mathbf{k}$ with the leftmost equation in (17.11) and substitute in from the rightmost equation in (17.11). This gives

$$\mathbf{k} \times (\mathbf{k} \times \mathbf{E}) = -\omega^2 \mu \epsilon \mathbf{E}. \tag{17.13}$$

Expanding the triple cross product and using $\mathbf{k} \cdot \mathbf{E} = 0$ from (17.10) gives

$$\mathbf{k} \cdot \mathbf{k} = \mu \epsilon \omega^2 \qquad \text{or} \qquad \omega(\mathbf{k}) = \frac{c}{n} k. \tag{17.14}$$

Collecting results, an infinite medium defined by (17.3) supports transverse and monochromatic plane waves of the form (17.9) if

$$\mathbf{k} = n\frac{\omega}{c}\hat{\mathbf{k}}, \qquad \mathbf{k} \cdot \mathbf{E} = 0, \qquad \text{and} \qquad Z\mathbf{H} = \hat{\mathbf{k}} \times \mathbf{E}. \tag{17.15}$$

The phase velocity of these waves is

$$\mathbf{v}_{\text{p}} = \frac{\omega}{k}\hat{\mathbf{k}} = \frac{c}{n}\hat{\mathbf{k}}. \tag{17.16}$$

The physical origin of the variation of the phase velocity with the index of refraction is neither obvious nor trivial (see below). On the other hand, using (17.3) to eliminate $\mathbf{D}$ and $\mathbf{H}$ in the original Maxwell equations (17.1) and (17.2), it is not difficult to see that every formula for a vacuum wave derived in Chapter 16—including those for beam-like waves and spherical waves—has a counterpart in matter, obtained by replacing c with c/n everywhere. The phase velocity is just one example. Another example is the energy velocity [see (16.45)],

$$\mathbf{v}_{\text{E}} = \frac{\langle \mathbf{S} \rangle}{\langle u_{\text{EM}} \rangle}, \tag{17.17}$$

which we check using (17.15) to evaluate the ratio of $\mathbf{S} = \mathbf{E} \times \mathbf{H}$ to $u_{\text{EM}} = \frac{1}{2}\epsilon\,|\mathbf{E}|^2 + \frac{1}{2}\mu\,|\mathbf{H}|^2$. In light of the time-averaging theorem,[3]

$$\mathbf{v}_{\text{E}} = \frac{\frac{1}{2}\text{Re}\,\{\mathbf{E} \times \mathbf{H}^*\}}{\frac{1}{2}\text{Re}\left\{\frac{1}{2}\epsilon\mathbf{E} \cdot \mathbf{E}^* + \frac{1}{2}\mu\mathbf{H} \cdot \mathbf{H}^*\right\}} = \frac{2|\mathbf{E}|^2\hat{\mathbf{k}}}{Z(\epsilon\,|\mathbf{E}|^2 + \mu\,|\mathbf{H}|^2)} = \frac{\hat{\mathbf{k}}}{\sqrt{\mu\epsilon}} = \frac{c}{n}\hat{\mathbf{k}}. \tag{17.18}$$

The similarity of an electromagnetic wave inside matter to an electromagnetic wave outside matter was unsurprising to Maxwell and his contemporaries because wave motion in both cases was thought to occur in an all-pervasive medium called the "luminiferous aether". All that happened when passing from vacuum to ordinary matter was a change in constitutive parameters from ϵ_0 and μ_0 to ϵ and μ. This point of view became untenable after Einstein[4] and we have little choice but to conclude that

[3] See Section 1.6.3.

[4] Einstein's special theory of relativity made the aether a superfluous concept. See Chapter 22.

the wave-like variations of the polarization and magnetization implied by (17.5) produce electric and magnetic fields which propagate exactly as they do in vacuum, albeit at a different phase velocity. We prove this explicitly in Chapter 21 in connection with a celebrated result known as the *extinction theorem.*

Application 17.1 Alfvén Waves

A magnetized plasma supports a class of low-frequency electromagnetic waves called *Alfvén waves.* Their behavior is dictated by the dielectric constant $\epsilon = \epsilon_0 + \rho/B_0^2$ computed in Application 14.1 for a plasma with mass density ρ in the presence of a static magnetic field $\mathbf{B}_0$ and a slowly varying electric field $\mathbf{E}_1$ oriented perpendicular to $\mathbf{B}_0$. Here, we associate $\mathbf{E}_1$ with a propagating wave $(E_1\hat{\mathbf{x}}, B_1\hat{\mathbf{y}})$ and choose the magnitude of $\mathbf{B}_0 = B_0\hat{\mathbf{z}}$ large enough that we can neglect the effect of the wave field on the motion of the plasma particles.

Consider the case when $\mathbf{B}_0 \parallel \mathbf{k}$ in Figure 17.1. Typically, $\rho/B_0^2 \gg 1$, so the phase velocity (17.14) is

$$v_A = \frac{1}{\sqrt{\mu_0\epsilon}} \approx \sqrt{\frac{B_0^2}{\mu_0\rho}}. \tag{17.19}$$

It is not an accident that (17.19) is reminiscent of the speed, $v = \sqrt{T/\rho}$, of a transverse wave on a string with tension T and mass density ρ. As shown in Section 12.5.1, B^2/μ_0 is a tension which opposes any attempt to bend a magnetic field line. It is crucial that $\mathbf{k} \parallel \mathbf{B}_0$ so the magnetic field $\mathbf{B}_1$ of the Alfvén wave adds only a sinusoidal ripple to the otherwise straight field lines of $\mathbf{B}_0$ (see Figure 17.2). In other words, the presence of the wave bends the pre-existing field lines of $\mathbf{B}_0$ (against their tension), but does not compress them. These are called *shear type* Alfvén waves.

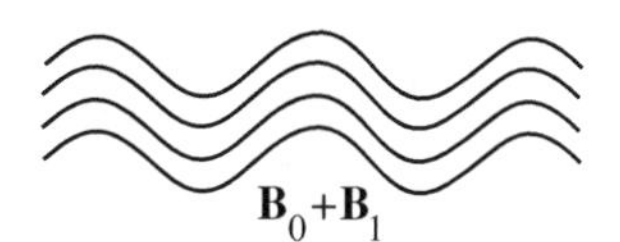

Figure 17.2: Lines of the total magnetic field in a magnetized plasma with a shear Alfvén wave present.

■

Example 17.1 Turpentine is a simple dielectric liquid except that the magnetization is related to the time derivative of the polarization by $\mathbf{M} = s\partial\mathbf{P}/\partial t$, where s is a constant. Prove that $\mathbf{E}\cdot\dot{\mathbf{E}} = 0$ and thereby deduce that only left and right circularly polarized plane waves can propagate in this liquid without the polarization changing.

Solution: In a simple liquid, $\mathbf{D} = \epsilon\mathbf{E}$, $\mathbf{B} = \mu\mathbf{H}$, and the vectors $(\mathbf{E}, \mathbf{B}, \mathbf{k})$ form a right-handed orthogonal triad. This means that $\mathbf{B}\cdot\mathbf{D} = 0$ and $\mathbf{H}\cdot\mathbf{E} = 0$. Combining this with $\mathbf{B} = \mu_0(\mathbf{H}+\mathbf{M})$ and $\mathbf{D} = \epsilon_0\mathbf{E} + \mathbf{P}$ shows that $\mathbf{M}\cdot\mathbf{P} = 0$ also. On the other hand, $\mathbf{M} = s\dot{\mathbf{P}}$ for this specific material. Hence, $\mathbf{P}\cdot\dot{\mathbf{P}} = 0$ and (because the medium is linear) $\mathbf{E}\cdot\dot{\mathbf{E}} = 0$. The latter is a characteristic of circularly polarized waves only (recall that the velocity vector $\mathbf{v} = \dot{\mathbf{r}}$ is always perpendicular to the position vector $\mathbf{r}$ when the tail of the latter is fixed and its head uniformly traces out a circle). We will see in Section 17.7 that turpentine is an example of an *anisotropic* medium where not all plane waves travel with the same phase velocity.

17.2.2 Energy Balance in Simple Matter

Monochromatic fields in matter irreversibly lose energy if the permittivity or permeability of the host matter has an imaginary part. Here, we quantify this statement for the simplest case of a non-magnetic medium with a constant, but complex, permittivity $\epsilon = \epsilon' + i\epsilon''$. There is no free current, so the Poynting theorem (15.104) reads

$$\int_V d^3r \left[\mathbf{E} \cdot \frac{\partial \mathbf{D}}{\partial t} + \mathbf{H} \cdot \frac{\partial \mathbf{B}}{\partial t}\right] = -\int_S d\mathbf{S} \cdot (\mathbf{E} \times \mathbf{H}). \tag{17.20}$$

The fields of interest are

$$\begin{aligned} \mathbf{E} &= \text{Re}[\mathbf{E}_0 \exp(-i\omega t)] = \mathbf{E}_0 \cos(\omega t) \\ \mathbf{D} &= \text{Re}[\epsilon \mathbf{E}_0 \exp(-i\omega t)] = \epsilon' \mathbf{E} + \epsilon'' \mathbf{E}_0 \sin(\omega t) \end{aligned} \tag{17.21}$$

and their magnetic counterparts, with $\mathbf{B} = \mu_0 \mathbf{H}$. Using these, (17.20) becomes

$$\int_S d\mathbf{S} \cdot (\mathbf{E} \times \mathbf{H}) + \frac{d}{dt} \int_V d^3r \left[\frac{\epsilon'}{2}|\mathbf{E}|^2 + \frac{\mu_0}{2}|\mathbf{H}|^2\right] = -\int_V d^3r\, \omega\epsilon''|\mathbf{E}|^2. \tag{17.22}$$

The first term on the left side of (17.22) is the rate at which energy flows out of the volume V of the medium. The second term on the left is the rate of change of the stored energy in the medium. Therefore, by conservation of energy, $\omega\epsilon''|\mathbf{E}|^2$ must be the rate at which the medium absorbs electric field energy per unit volume. Indeed, it can be useful to regard a unit volume of matter with a monochromatic wave present as a damped oscillator with a quality factor

$$Q = \omega \times \frac{\text{maximum energy stored}}{\text{average power loss}}. \tag{17.23}$$

Using (1.137) to compute the time average in the denominator of (17.23) gives

$$Q = \omega \times \frac{\frac{1}{2}\epsilon'|\mathbf{E}_0|^2}{\frac{1}{2}\omega\epsilon''|\mathbf{E}_0|^2} = \frac{\epsilon'}{\epsilon''}. \tag{17.24}$$

We will return to this question in the next chapter when we study waves in frequency-dispersive media.

17.3 Reflection and Refraction

A plane wave incident on a sharp boundary between two dissimilar materials produces a solution to the Maxwell equations in the form of a *reflected* plane wave and a *refracted* plane wave when the wavelength is small compared to the curvature of the boundary. In this section, we give a selective introduction to this subject (familiar from optics) with an emphasis on the use of electromagnetic methods to derive the main results.

In Figure 17.3, $z = 0$ is the boundary between two semi-infinite half-spaces, one characterized by material parameters (ϵ_1, μ_1), the other by material parameters (ϵ_2, μ_2). Everyday experience tells us that an incident plane wave which approaches medium 2 from medium 1 "splits" into a reflected wave confined to medium 1 and a refracted wave confined to medium 2. Thus, the electric field in medium 2 is $\mathbf{E}_2(\mathbf{r}, t) = \mathbf{E}_T \exp[i(\mathbf{k}_T \cdot \mathbf{r} - \omega_T t)]$ and, in medium 1,

$$\mathbf{E}_1(\mathbf{r}, t) = \mathbf{E}_I \exp[i(\mathbf{k}_I \cdot \mathbf{r} - \omega_I t)] + \mathbf{E}_R \exp[i(\mathbf{k}_R \cdot \mathbf{r} - \omega_R t)]. \tag{17.25}$$

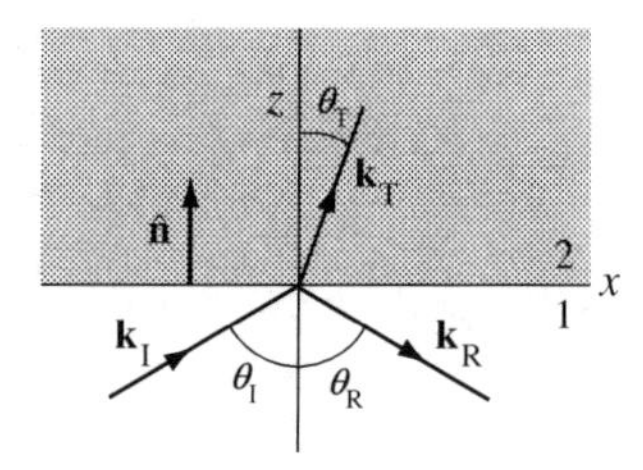

Figure 17.3: Wave vectors for incident (I), reflected (R), and transmitted (T) (or refracted) plane waves at a straight material boundary.

Equation (17.15) completes the specification of these waves, including the magnetic fields with amplitudes $\mathbf{H}_I$, $\mathbf{H}_R$, and $\mathbf{H}_T$. At the interface, the total fields must satisfy the matching conditions which derive from (17.1) and (17.2) when no charge or current is present (see Section 2.3.3):

$$\hat{\mathbf{n}} \cdot [\mathbf{D}_1 - \mathbf{D}_2] = 0 \qquad \hat{\mathbf{n}} \cdot [\mathbf{B}_1 - \mathbf{B}_2] = 0 \tag{17.26}$$

$$\hat{\mathbf{n}} \times [\mathbf{E}_1 - \mathbf{E}_2] = 0 \qquad \hat{\mathbf{n}} \times [\mathbf{H}_1 - \mathbf{H}_2] = 0. \tag{17.27}$$

17.3.1 Specular Reflection and Snell's Law

The matching conditions (17.27) and (17.26) cannot be satisfied for all time and for all points on the $z = 0$ plane unless the phases of the incident, reflected, and refracted waves are all exactly the same. Hence, $\omega_I = \omega_R = \omega_T \equiv \omega$ and $\mathbf{k}_I \cdot \mathbf{r}|_{z=0} = \mathbf{k}_R \cdot \mathbf{r}|_{z=0} = \mathbf{k}_T \cdot \mathbf{r}|_{z=0}$. The last set of equalities implies that

$$k_{Ix} = k_{Rx} = k_{Tx} \qquad \text{and} \qquad k_{Iy} = k_{Ry} = k_{Ty}. \tag{17.28}$$

The three wave vectors $\mathbf{k}_I$, $\mathbf{k}_R$, and $\mathbf{k}_T$ are all coplanar. To see this, orient the Cartesian axes so the *plane of incidence* (the plane that contains both $\mathbf{k}_I$ and the unit normal to the interface $\hat{\mathbf{n}}$) is the x-z plane as shown in Figure 17.3. Then, because $k_{Iy} = 0$, we deduce from the right member of (17.28) that $k_{Ry} = k_{Ty} = 0$ also. The left member of (17.28) generates the laws of reflection and refraction. The incident and reflected waves both lie in medium 1, so the leftmost equation in (17.15) gives

$$k_I = k_R = n_1 \frac{\omega}{c} \equiv k_1. \tag{17.29}$$

Hence, $k_{Ix} = k_{Rx}$ implies that $\sin\theta_I = \sin\theta_R$, which is the *law of reflection*:

$$\theta_I = \theta_R \equiv \theta_1. \tag{17.30}$$

Similarly, in medium 2 we write $k_T = k_2$ and $\theta_T = \theta_2$ so

$$k_{Tx} = k_2 \sin\theta_T \equiv n_2 \frac{\omega}{c} \sin\theta_2. \tag{17.31}$$

Snell's law of refraction comes from $k_{Ix} = k_{Tx}$ in (17.28):

$$n_1 \sin\theta_1 = n_2 \sin\theta_2. \tag{17.32}$$

This relation says that the refracted wave vector $\mathbf{k}_T$ inclines closer to the interface normal than the incident wave vector $\mathbf{k}_I$ when a wave passes from a "fast" medium to a "slow" medium [as reckoned by magnitude of the phase velocity (17.16)], i.e., when $n_1 < n_2$. We note in passing that the laws of reflection and refraction are often said to be "kinematic" because they follow from the wave nature of the fields only. The dynamical laws for the fields (the Maxwell equations) play no role.

Negative-Index Matter

In the late 1990s, it became possible to fabricate artificial materials where the electric permittivity and magnetic permeability are both negative in a limited frequency band (see Section 18.5.6). Although $n = c\sqrt{\epsilon\mu} = c\sqrt{(-\epsilon)(-\mu)}$ suggests that the index of refraction is unaffected, Veselago had pointed out 30 years earlier that the Maxwell equations require that we write $n = -c\sqrt{|\epsilon||\mu|}$. To see this, define a unit vector $\hat{\mathbf{k}}$ so $(\mathbf{E}, \mathbf{H}, \hat{\mathbf{k}})$ is a right-handed orthogonal triad. Substituting the left member of (17.15) into (17.11) gives

$$\frac{n}{c}\hat{\mathbf{k}} \times \mathbf{E} = \mu\mathbf{H} \qquad \text{and} \qquad \frac{n}{c}\hat{\mathbf{k}} \times \mathbf{H} = -\epsilon\mathbf{E}.$$

For both equations to hold simultaneously, we must choose $n < 0$ if $\epsilon < 0$ and $\mu < 0$. On the other hand, $Z\mathbf{H} = \hat{\mathbf{k}} \times \mathbf{E}$ tells us that the wave impedance $Z = \sqrt{\mu/\epsilon} > 0$. The fact that $\mathbf{k} = -|n|(\omega/c)\hat{\mathbf{k}}$ led Veselago to coin the phrase "left-handed materials" because $(\mathbf{E}, \mathbf{H}, \mathbf{k})$ is a left-handed orthogonal triad for wave propagation in these systems.

Negative-index materials have many counter-intuitive properties. One is that the Poynting vector and the phase velocity point in opposite directions. This follows from (17.6) and (17.16) because both n and μ are negative:

$$\mathbf{S} = \frac{n}{c\mu}|\mathbf{E}|^2\hat{\mathbf{k}} \qquad \text{and} \qquad \mathbf{v}_{\rm p} = \frac{c}{n}\hat{\mathbf{k}}.$$

A particularly striking feature is the phenomenon of "negative refraction", which we illustrate by writing Snell's law for wave propagation from vacuum, $n_1 = 1$, into a material with index $n_2 < 0$, namely, $\sin\theta_1 = -|n_2|\sin\theta_2$. This shows that $\theta_2 < 0$, which means that the refracted wave vector $\mathbf{k}_T$ does not point up and to the right in Figure 17.3, but rather up and to the left. This has been confirmed experimentally.

The special case of $n = -1$ has captured special attention because it holds open the possibility of creating a "perfect lens" which is aberration-free and images objects with arbitrarily fine resolution. This may be contrasted with a conventional lens, whose resolution is limited to feature sizes of the order of the wavelength. The main obstacle to realizing such a lens in the laboratory is reducing the dissipative losses which occur when electromagnetic waves pass through matter of any kind, including negative-index material.

17.3.2 The Fresnel Equations

In 1823, Fresnel used a molecular model of the aether and heuristic arguments to predict the intensities of light rays reflected and refracted at a flat interface between two transparent media. Seventy years later, Hertz reproduced Fresnel's results by applying the matching conditions (17.27) and (17.26) to the electric fields $\mathbf{E}_1$ and $\mathbf{E}_2$ (and their magnetic counterparts) defined in (17.25) and the text preceding it. The calculation exploits the fact that an arbitrarily polarized plane wave can be decomposed into the sum of two plane waves with orthogonal polarization.

Figure 17.4 shows the standard choice for these orthogonal waves: a p-polarized wave where $\mathbf{E}$ is strictly parallel to the plane of incidence (which contains $\mathbf{k}_I$ and $\hat{\mathbf{n}}$): and an s-polarized wave where $\mathbf{E}$ is strictly perpendicular to the plane of incidence.[5] The general case follows by superposing the

[5] The label "s" comes from the German word for perpendicular, *senkrecht*. Some authors use $\|$ in place of "p" and $\perp$ in place of "s". We mostly avoid this notation because $\|$ and $\perp$ are used consistently in this book for quantities, e.g., field components, that lie parallel or perpendicular to a surface or interface. Other authors use TM (transverse magnetic) in place of "p" and TE (transverse electric) in place of "s" because $\mathbf{H}$ is strictly perpendicular to the interface normal for a TM wave and $\mathbf{E}$ is strictly perpendicular to the interface normal for a TE wave. We avoid this notation also because,

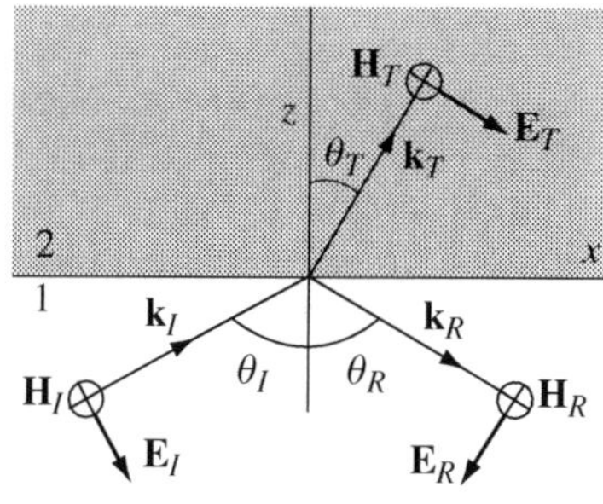

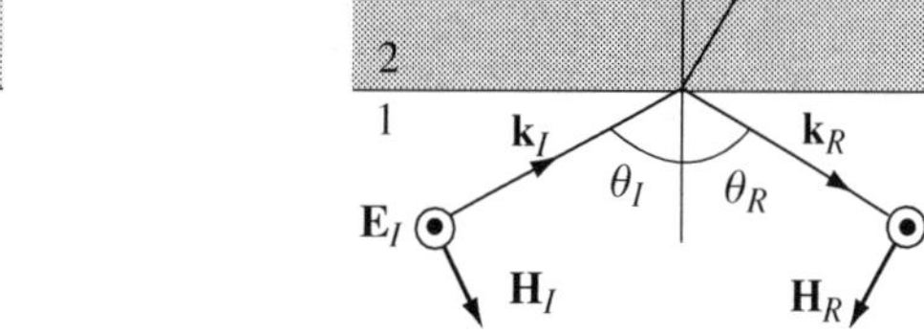

Figure 17.4: Wave vectors and field vectors for incident, reflected, and refracted plane waves at an abrupt interface for two orthogonal polarizations of the incident wave. Circles with crosses (dots) indicate vectors which point into (out of) the plane of the diagram. The plane of the diagrams is the plane of incidence.

field amplitudes obtained by applying the matching conditions to the two component polarizations separately.

The left panel of Figure 17.4, combined with (17.15), shows that the matching conditions (17.27) applied to a p-wave give

$$E_I \cos\theta_1 - E_R \cos\theta_1 = E_T \cos\theta_2 \qquad \text{and} \qquad H_I + H_R = H_T. \tag{17.33}$$

Using (17.15), the second equation in (17.33) becomes $Z_2(E_I + E_R) = Z_1 E_T$. It follows immediately that the reflection and transmission amplitudes are

$$r_\text{p} \equiv \left[\frac{E_R}{E_I}\right]_\text{p} = \frac{Z_1\cos\theta_1 - Z_2\cos\theta_2}{Z_1\cos\theta_1 + Z_2\cos\theta_2} \qquad \text{and} \qquad t_\text{p} \equiv \left[\frac{E_T}{E_I}\right]_\text{p} = \frac{2Z_2\cos\theta_1}{Z_1\cos\theta_1 + Z_2\cos\theta_2}. \tag{17.34}$$

The same matching conditions applied to an s-wave give

$$E_I + E_R = E_T \qquad \text{and} \qquad H_I\cos\theta_1 - H_R\cos\theta_1 = H_T\cos\theta_2. \tag{17.35}$$

Using (17.15), the second equation in (17.35) becomes $Z_2(E_I - E_R)\cos\theta_1 = Z_1 E_T \cos\theta_2$. Therefore,

$$r_\text{s} \equiv \left[\frac{E_R}{E_I}\right]_\text{s} = \frac{Z_2\cos\theta_1 - Z_1\cos\theta_2}{Z_2\cos\theta_1 + Z_1\cos\theta_2} \qquad \text{and} \qquad t_\text{s} \equiv \left[\frac{E_T}{E_I}\right]_\text{s} = \frac{2Z_2\cos\theta_1}{Z_2\cos\theta_1 + Z_1\cos\theta_2}. \tag{17.36}$$

The amplitude formulae in (17.34) and (17.36) are the *Fresnel equations*.

17.3.3 Remarks

1. The matching conditions (17.26) for the normal components of **D** and **B** were not used to derive (17.34) and (17.36) because the information they provide is redundant for this problem.

2. The Fresnel amplitudes (17.34) and (17.36) are not explicit because Snell's law (17.32) is needed to compute the refracted angle θ_2 for a given angle of incidence θ_1.

3. An s-wave cannot be distinguished from a p-wave at normal incidence. Therefore, (17.34) and (17.36) must be identical in that limit. The apparent sign difference between the reflected amplitudes when $\cos\theta_1 = \cos\theta_2 = 1$ occurs because, when $\theta_I = \theta_R = 0$ in Figure 17.4, $\mathbf{E}_I$ and $\mathbf{E}_R$ point in opposite directions for p-polarization but point in the same direction for s-polarization.[6] The latter is

in Section 16.8.1 and in later sections, TE (TM) refers to a wave whose electric (magnetic) field vector, alone, is strictly transverse to the direction of wave propagation.

[6] Some authors avoid this problem by using the ratio H_R/H_I to define r_p. Others choose the directions of the field vectors differently than in Figure 17.4.

most natural, so, at normal incidence,

$$r = \frac{Z_2 - Z_1}{Z_1 + Z_2} \quad \text{and} \quad t = \frac{2Z_2}{Z_1 + Z_2}. \tag{17.37}$$

When both media have the same magnetic properties, $Z = \sqrt{\mu/\epsilon}$ reduces (17.37) to

$$r = \frac{n_1 - n_2}{n_1 + n_2} \quad \text{and} \quad t = \frac{2n_1}{n_1 + n_2}. \tag{17.38}$$

To check (17.38), let $\epsilon_2 \to \infty$ to mimic a perfect conductor (see Section 6.6.4). $Z_2 \to 0$ in this limit, so (17.37) predicts $E_R/E_I = -1$ and $E_T/E_I = 0$. The reflected wave suffers a phase shift of π because the entirely tangential field $\mathbf{E}_1 = \mathbf{E}_I + \mathbf{E}_R$ must vanish at $z = 0$ when medium 2 is a perfect conductor.

4. The incident wave propagates through the interface as if it were absent when the reflection amplitudes in (17.34) and (17.36) vanish. This *impedance matching* phenomenon is fundamental to wave propagation in many different contexts.[7] The name itself is self-explanatory when we specialize to normal incidence so (17.37) is applicable. The no-reflection condition is then precisely a condition that the impedances of the two media be equal:

$$Z_1 = Z_2 \quad \Rightarrow \quad \sqrt{\frac{\mu_1}{\epsilon_1}} = \sqrt{\frac{\mu_2}{\epsilon_2}}. \tag{17.39}$$

5. Z_1 and Z_2 are both real for simple transparent matter because ϵ and μ are both real and positive. This means that reflection and refraction do not alter the linear polarization of a pure s-wave or a pure p-wave. On the other hand, the Fresnel equations imply that a wave with arbitrary linear polarization reflects as a linearly polarized wave with relatively more "s" content than the incident wave and refracts as a linear polarized wave with relatively more "p" content than the incident wave. In other words, the direction of $\mathbf{E}$ turns away from the plane of incidence when reflected and turns toward the plane of incidence when refracted. We discuss the behavior of incident waves with circular and elliptical polarization below.

Example 17.2 (a) Write out the reflection amplitudes r_s and r_p in terms of the permeability μ and the index of refraction n for each half-space. (b) Specialize to non-magnetic matter and derive Fresnel's famous "sine" and "tangent" reflection formulae:

$$r_s = -\frac{\sin(\theta_1 - \theta_2)}{\sin(\theta_1 + \theta_2)} \quad \text{and} \quad r_p = \frac{\tan(\theta_1 - \theta_2)}{\tan(\theta_1 + \theta_2)}.$$

Solution:
(a) We note first that

$$\frac{Z_1}{Z_2} = \frac{\mu_1 n_2}{\mu_2 n_1} = \frac{\epsilon_2 n_1}{\epsilon_1 n_2}.$$

Using the first equality above, the reflection formulae in (17.34) and (17.36) become

$$r_s = \frac{\mu_2 n_1 \cos\theta_1 - \mu_1 n_2 \cos\theta_2}{\mu_2 n_1 \cos\theta_1 + \mu_1 n_2 \cos\theta_2} \quad \text{and} \quad r_p = \frac{\mu_1 n_2 \cos\theta_1 - \mu_2 n_1 \cos\theta_2}{\mu_1 n_2 \cos\theta_1 + \mu_2 n_1 \cos\theta_2}.$$

(b) The permeability drops out of the reflection formulae when we set $\mu_1 = \mu_2 = \mu_0$ for non-magnetic matter. Begin with the "s" expression and multiply the first and second terms in the

[7] See, for example, F.S. Crawford, Jr., *Waves*, (McGraw-Hill, New York, 1968).

numerator and denominator by the left and right sides, respectively, of Snell's law, $n_2 \sin\theta_2 = n_1 \sin\theta_1$. This gives

$$r_s = \frac{n_1\cos\theta_1 - n_2\cos\theta_2}{n_1\cos\theta_1 + n_2\cos\theta_2} = \frac{n_1 n_2}{n_1 n_2} \times \frac{\cos\theta_1 \sin\theta_2 - \cos\theta_2 \sin\theta_1}{\cos\theta_1 \sin\theta_2 + \cos\theta_2 \sin\theta_1} = -\frac{\sin(\theta_1 - \theta_2)}{\sin(\theta_1 + \theta_2)}.$$

Proceeding similarly with the "p" expression, we find

$$r_p = \frac{n_2\cos\theta_1 - n_1\cos\theta_2}{n_2\cos\theta_1 + n_1\cos\theta_2} = \frac{n_1 n_2}{n_1 n_2} \times \frac{\cos\theta_1 \sin\theta_1 - \cos\theta_2 \sin\theta_2}{\cos\theta_1 \sin\theta_1 + \cos\theta_2 \sin\theta_2}.$$

Now multiply the $\cos\theta_1 \sin\theta_1$ terms by $\sin^2\theta_2 + \cos^2\theta_2$ and multiply the $\cos\theta_2 \sin\theta_2$ terms by $\sin^2\theta_1 + \cos^2\theta_1$. This produces four terms in the numerator and four terms in the denominator. Factoring both gives

$$r_p = \frac{\sin\theta_1\cos\theta_2 - \sin\theta_2\cos\theta_1}{\sin\theta_1\cos\theta_2 + \sin\theta_2\cos\theta_1} \times \frac{\cos\theta_1\cos\theta_2 - \sin\theta_1\sin\theta_2}{\cos\theta_1\cos\theta_2 + \sin\theta_1\sin\theta_2}.$$

Hence,

$$r_p = \frac{\sin(\theta_1 - \theta_2)\cos(\theta_1 + \theta_2)}{\sin(\theta_1 + \theta_2)\cos(\theta_1 - \theta_2)} = \frac{\tan(\theta_1 - \theta_2)}{\tan(\theta_1 + \theta_2)}.$$

17.3.4 Energy Transport

Conservation of energy requires that the reflected and refracted waves carry off the energy transported to a flat interface by the incident electromagnetic wave. We confirm this using the Poynting vector (17.6), which measures the energy current density or flow of power in matter. The specific quantity of interest is the energy *reflection coefficient*,

$$R = \left| \frac{\langle \mathbf{S}_R \rangle \cdot \hat{\mathbf{n}}}{\langle \mathbf{S}_I \rangle \cdot \hat{\mathbf{n}}} \right|. \tag{17.40}$$

The time average of the Poynting vector for a wave characterized by (17.9) and (17.15) is

$$\langle \mathbf{S} \rangle = \frac{1}{2}\mathrm{Re}(\mathbf{E} \times \mathbf{H}^*) = \frac{1}{2Z}|\mathbf{E}|^2\hat{\mathbf{k}}. \tag{17.41}$$

The factors $\hat{\mathbf{k}}_I \cdot \hat{\mathbf{n}}$ and $\hat{\mathbf{k}}_R \cdot \hat{\mathbf{n}}$ that appear when we evaluate (17.40) are equal and opposite for the incident and reflected waves, both of which lie entirely in medium 1 of Figure 17.3. Therefore, the reflection coefficient is simply the square of the Fresnel reflection amplitude:

$$R = \left| \frac{E_R}{E_I} \right|^2 = |r|^2. \tag{17.42}$$

Using (17.34) and (17.36), the reflection coefficients for p-waves and s-waves are

$$R_p = \left(\frac{Z_1\cos\theta_1 - Z_2\cos\theta_2}{Z_1\cos\theta_1 + Z_2\cos\theta_2} \right)^2 \quad \text{and} \quad R_s = \left(\frac{Z_2\cos\theta_1 - Z_1\cos\theta_2}{Z_2\cos\theta_1 + Z_1\cos\theta_2} \right)^2. \tag{17.43}$$

The special case of normal incidence reduces both formulae in (17.43) to

$$R = \left(\frac{Z_2 - Z_1}{Z_2 + Z_1} \right)^2 \quad \Longrightarrow \quad \left(\frac{n_2 - n_1}{n_2 + n_1} \right)^2. \tag{17.44}$$

The rightmost expression in (17.44) applies to non-magnetic matter.

The energy *transmission coefficient* is defined similarly except that $\mathbf{k}_T \cdot \hat{\mathbf{n}}$ differs from $\mathbf{k}_I \cdot \hat{\mathbf{n}}$ and Z_1 differs from Z_2. Therefore,

$$T = \left| \frac{\langle \mathbf{S}_T \rangle \cdot \hat{\mathbf{n}}}{\langle \mathbf{S}_I \rangle \cdot \hat{\mathbf{n}}} \right| = \frac{Z_1 \cos\theta_2}{Z_2 \cos\theta_1} \left| \frac{E_T}{E_I} \right|^2 . \tag{17.45}$$

For pure p-waves and pure s-waves, (17.45) reduces to

$$T_{\mathrm{p}} = \frac{4Z_1 Z_2 \cos\theta_1 \cos\theta_2}{(Z_1 \cos\theta_1 + Z_2 \cos\theta_2)^2} \quad \text{and} \quad T_{\mathrm{s}} = \frac{4Z_1 Z_2 \cos\theta_1 \cos\theta_2}{(Z_2 \cos\theta_1 + Z_1 \cos\theta_2)^2}. \tag{17.46}$$

At normal incidence,

$$T = \frac{4Z_1 Z_2}{(Z_1 + Z_2)^2} \quad \Longrightarrow \quad \frac{4n_1 n_2}{(n_1 + n_2)^2}. \tag{17.47}$$

All these reflection and transmission coefficients are consistent with conservation of energy because, as the reader can check,

$$R + T = 1. \tag{17.48}$$

17.3.5 Polarization by Reflection

An unpolarized electromagnetic wave can be polarized by reflection. This occurs because the Fresnel reflection amplitudes for its "s" and "p" components vanish at different angles of incidence. To find the angle where a p-wave does not reflect, we square the p-wave impedance matching condition $Z_1 \cos\theta_1 = Z_2 \cos\theta_2$ derived from (17.34) and add this to the square of Snell's law (17.32). This gives

$$\left(\frac{Z_1}{Z_2} \right)^2 \cos^2\theta_1 + \left(\frac{n_1}{n_2} \right)^2 \sin^2\theta_1 = \cos^2\theta_2 + \sin^2\theta_2 = 1. \tag{17.49}$$

The "s" case produces the same equation—with Z_1 and Z_2 interchanged—when we use the impedance matching condition $Z_1 \cos\theta_2 = Z_2 \cos\theta_1$ derived from (17.36). Therefore, if we replace θ_1 by θ_E for the "s" case and θ_B for the "p" case, these equations imply that

$$\begin{aligned} \mathrm{s}: \qquad & \tan^2\theta_{\mathrm{E}} = \frac{(Z_2/Z_1)^2 - 1}{1 - (n_1/n_2)^2} = \frac{\mu_2}{\mu_1} \cdot \frac{\epsilon_1\mu_2 - \epsilon_2\mu_1}{\epsilon_2\mu_2 - \epsilon_1\mu_1} \\ \mathrm{p}: \qquad & \tan^2\theta_{\mathrm{B}} = \frac{(Z_1/Z_2)^2 - 1}{1 - (n_1/n_2)^2} = \frac{\epsilon_2}{\epsilon_1} \cdot \frac{\epsilon_2\mu_1 - \epsilon_1\mu_2}{\epsilon_2\mu_2 - \epsilon_1\mu_1}. \end{aligned} \tag{17.50}$$

The derivation of (17.50) shows that $\theta_E = \theta_B$ only when $Z_1 = Z_2$, in which case both angles are zero. An arbitrary set of constitutive parameters predicts different values for θ_E and θ_B. Consequently, an unpolarized wave incident at θ_B reflects only its "s"-component and an unpolarized wave incident at θ_E reflects only its "p"-component. This is what is meant by "polarization by reflection". When the two media in question are non-magnetic, $\mu_1 = \mu_2 = \mu_0$ and (17.50) reduces to

$$\tan^2\theta_{\mathrm{E}} = -1 \qquad \text{and} \qquad \tan^2\theta_{\mathrm{B}} = \frac{\epsilon_2}{\epsilon_1}. \tag{17.51}$$

The left equation has no solution. Since $n = c\sqrt{\mu_0\epsilon}$, the right equation shows we can produce a pure s-wave by reflecting an unpolarized wave from the interface at *Brewster's angle* of incidence:

$$\theta_{\mathrm{B}} = \tan^{-1}\left(\frac{n_2}{n_1} \right). \tag{17.52}$$

Combining (17.52) with Snell's law (17.32) shows that

$$\sin\theta_2 = \frac{n_1}{n_2} \sin\theta_B = \cot\theta_B \sin\theta_B = \cos\theta_B \quad \Rightarrow \quad \theta_B + \theta_2 = \frac{\pi}{2}. \tag{17.53}$$

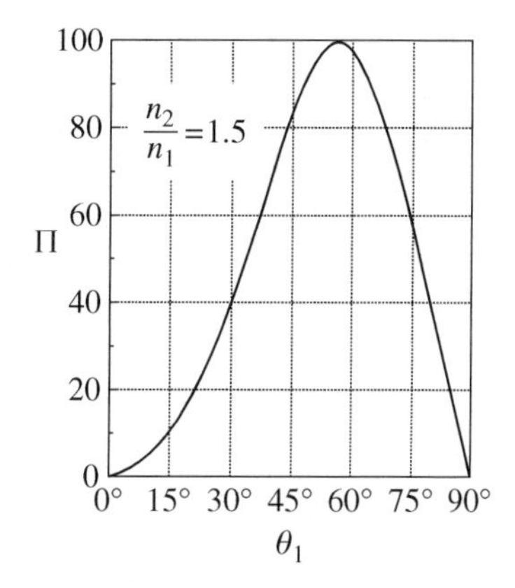

Figure 17.5: Relative polarization by reflection, $\Pi(\theta_1)$.

Brewster's angle occurs when the transmitted wave vector and the putative reflected wave vector are perpendicular.

It is not necessary to operate exactly at the Brewster angle to achieve a significant degree of polarization by reflection. This may be seen from Figure 17.5, which plots the relative polarization,

$$\Pi = \left| \frac{R_s - R_p}{R_s + R_p} \right|, \tag{17.54}$$

as a function of the angle of incidence for visible light in air ($n_1 = 1$) reflected from a glass slide ($n_2 = 1.5$). Complete s-polarization occurs at the Brewster angle, $\theta_B \approx 56°$.

It is worth noting that the "p" reflection coefficient in (17.34) changes sign when the incident angle passes through the Brewster angle. The corresponding "s" reflection coefficient in (17.36) does not.[8] Therefore, because the z-components of $\mathbf{k}_I$ and $\mathbf{k}_R$ have opposite signs, an incident wave with right circular polarization reflects with left elliptical polarization when $\theta_I < \theta_B$ and reflects with right elliptical polarization when $\theta_I > \theta_B$. The refracted wave is also generally elliptical, but retains the rotational sense of the incident wave for all angles of incidence.

Finally, there is the question of the physical origin of the disappearance of the reflected p-wave at the Brewster angle(s). The same question may be asked of the origin of the reflected and refracted waves in the first place. As mentioned following (17.18), the only possibility is that the polarization $\mathbf{P}(\mathbf{r}, t)$ and magnetization $\mathbf{M}(\mathbf{r}, t)$ in both media produce electric fields which (i) interfere destructively with the incident wave in medium 2 and (ii) produce the reflected and refracted waves in medium 1 and medium 2, respectively.[9]

17.3.6 Total Internal Reflection

Snell's law (17.32) predicts that a refracted wave bends away from the interface normal ($\theta_2 > \theta_1$) when a monochromatic plane wave propagates from a high-index material to a low-index material ($n_1 > n_2$). However, when the angle of incidence θ_1 reaches a *critical angle* θ_C defined by

$$\sin\theta_C = n_2/n_1, \tag{17.55}$$

the angle of refraction θ_2 reaches $\pi/2$ and the refracted wave propagates parallel to the interface. The simple geometric meaning of Snell's law breaks down when $\theta_1 > \theta_C$ and a consistent interpretation

[8] See the lower left panel of Figure 17.6.

[9] See Doyle (1985) cited in Sources, References, and Additional Reading.

demands a generalization of our previous analysis. The key is to write out the "p" electric field which refracts into medium 2 in the left panel in Figure 17.4:

$$\mathbf{E}_2(x,z) = \mathbf{E}_T \exp[i(\mathbf{k}_2 \cdot \mathbf{r} - \omega t)] = \mathbf{E}_T \exp[i(k_{2x}x + k_{2z}z - \omega t)]. \tag{17.56}$$

From (17.28) and (17.31), the component of $\mathbf{k}_2$ parallel to the interface is

$$k_{2x} = k_{1x} = k_1 \sin\theta_1 = \frac{\omega}{c} n_1 \sin\theta_1. \tag{17.57}$$

Snell's law gives the component of $\mathbf{k}_2$ normal to the interface as

$$k_{2z}^2 = k_2^2 \cos^2\theta_2 = \left(\frac{\omega}{c}\right)^2 n_2^2 \cos^2\theta_2 = \left(\frac{\omega}{c}\right)^2 n_2^2(1 - \sin^2\theta_2) = \left(\frac{\omega}{c}\right)^2 \left[n_2^2 - n_1^2 \sin^2\theta_1\right]. \tag{17.58}$$

The last term in (17.58) is negative when $n_1 > n_2$ and $\theta_1 > \theta_C$. This motivates us to write $n_2 \cos\theta_2 = i\sqrt{n_1^2 \sin^2\theta_1 - n_2^2}$ and use (17.55) and (17.58) to define a positive real number κ so

$$k_{2z} \equiv i\kappa = i\frac{\omega}{c} n_1 \sqrt{\sin^2\theta_1 - \sin^2\theta_C}. \tag{17.59}$$

The positive square root in (17.59) ensures that the electric field decays (rather than grows) exponentially in medium 2:

$$\mathbf{E}_2(x,z,t) = \mathbf{E}_T e^{-\kappa z} \exp[i(k_1 x \sin\theta_1 - \omega t)]. \tag{17.60}$$

It is appropriate to call (17.60) an *interfacial wave* because it is non-zero only within a mean distance κ^{-1} of the $z=0$ plane in medium 2. The wave propagates in the x-direction (along the interface) with a phase speed

$$v_p = \frac{\omega}{k_{1x}} = \frac{c}{n_1 \sin\theta_1} > \frac{c}{n_1} = v_{p1}. \tag{17.61}$$

However, since $\theta_1 > \theta_C$,

$$v_p = \frac{c}{n_1 \sin\theta_1} = \frac{c/n_2}{(n_1/n_2)\sin\theta_1} = \left(\frac{\sin\theta_C}{\sin\theta_1}\right) v_{p2} < v_{p2}. \tag{17.62}$$

This shows that the refracted wave (17.60) propagates along the interface with a phase speed which lies between the phase speeds characteristic of either of the two bounding media:

$$v_{p1} < v_p < v_{p2}. \tag{17.63}$$

The amplitude $\mathbf{E}_T$ in (17.60) is constrained by the Maxwell law $\nabla \cdot \mathbf{E}_2 = 0$. Using (17.56), (17.59), and the leftmost equation in (17.57), this constraint reads

$$\mathbf{k}_2 \cdot \mathbf{E}_2 = 0 = k_{2x} E_{Tx} + k_{2z} E_{Tz} = k_{1x} E_{Tx} + i\kappa E_{Tz}, \tag{17.64}$$

or

$$\frac{E_{Tx}}{E_{Tz}} = -\frac{i\kappa}{k_1 \sin\theta_1}. \tag{17.65}$$

The important conclusion from (17.60) and (17.65) is that the refracted plane wave field (17.60) is *not uniform* because its amplitude is not constant on planes perpendicular to the x-direction of propagation and it is *not transverse* because it has a non-zero component along the x-direction of propagation. Both features are a consequence of the interface breaking the translational invariance of an infinite medium. The fact that (17.65) is pure imaginary implies that the physical electrical field vector (the real part of $\mathbf{E}_T$) rotates in the x-z plane—a longitudinal version of a conventional transverse plane wave with elliptical polarization.

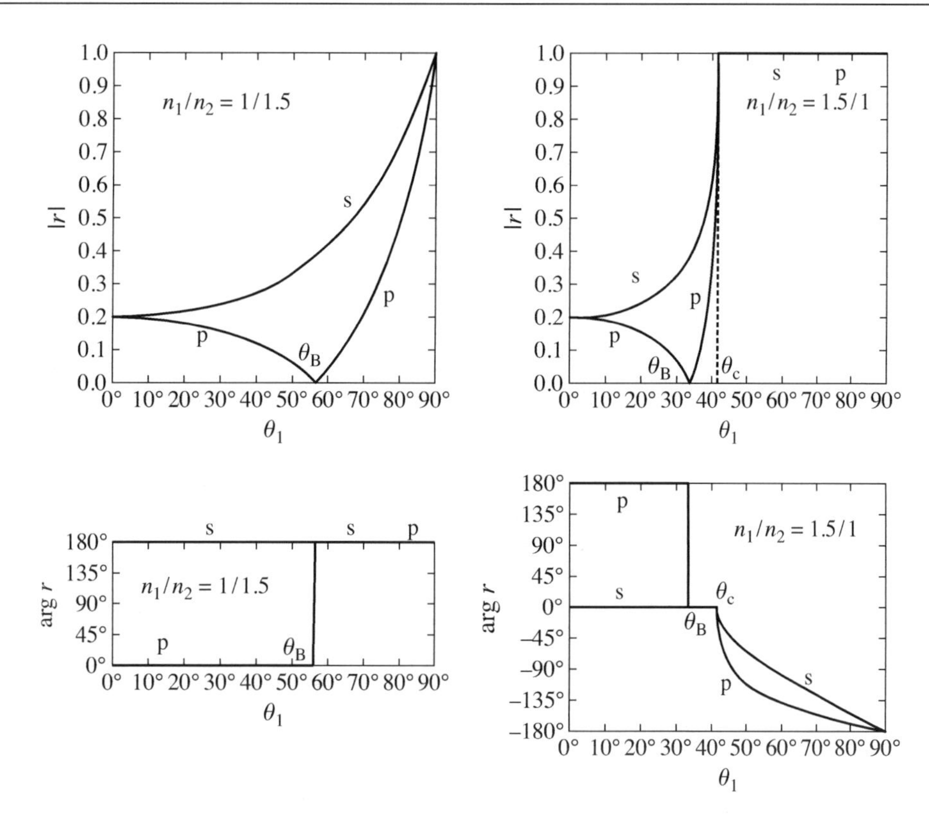

Figure 17.6: Magnitude and phase of the Fresnel reflection amplitude $r = E_R/E_I$ as a function of angle of incidence for $n_1 < n_2$ (left side) and $n_1 > n_2$ (right side). Figure adapted from Zhang and Li (1998).

The phenomenon discussed in this section is called *total internal reflection* because the magnitude of the reflected wave is equal to the magnitude of the incident wave. To see this, set $\mu_1 = \mu_2 = \mu_0$ in the "p" reflection amplitude in part (a) of Example 17.2 and use (17.58) to eliminate $\cos\theta_2$ in favor of k_{2z}. Then use (17.59) to eliminate k_{2z} in favor of κ. Because $k_2 = n_2\omega/c$, the result is

$$r_p = \frac{n_2\cos\theta_1 - n_1\cos\theta_2}{n_2\cos\theta_1 + n_1\cos\theta_2} = \frac{n_2 k_2\cos\theta_1 - i\kappa}{n_2 k_2\cos\theta_1 + i\kappa}. \tag{17.66}$$

The far right side of (17.66) is a complex number of the form $(a - ib)/(a + ib)$, which has magnitude one. In fact,

$$r_p = e^{-i\delta}, \tag{17.67}$$

where $\tan(\delta/2) = b/a$ characterizes the phase of the reflected wave with respect to the incident wave. The latter is shown in the lower right panel of Figure 17.6, which summarizes the behavior of the reflection amplitudes at optical frequencies for air/glass interfaces as a function of angle of incidence. Particularly noteworthy is the non-zero *difference* in the phase acquired by s- and p-waves when each suffers total internal reflection. This means that a totally reflected wave is elliptically polarized. Fresnel exploited this fact himself to convert linearly polarized light into circularly polarized light.

17.3.7 Non-Uniform and Evanescent Plane Waves

The refracted wave (17.60) is an example of an *inhomogeneous* or *non-uniform* plane wave. This is a plane wave-like solution of the source-free Maxwell equations where the amplitude of the wave is *not* constant on planes perpendicular to the direction of propagation. Indeed, an ordinary uniform plane wave propagating along $\mathbf{k}$ may be regarded as a non-uniform plane wave propagating along, say, $\mathbf{p} = k_z\hat{\mathbf{z}}$:

$$\mathbf{E}(\mathbf{r}, t) = \mathbf{E}e^{i(\mathbf{k}\cdot\mathbf{r}-\omega t)} = \mathbf{E}e^{i(k_x x+k_y y)}e^{i(k_z z-\omega t)} = \mathbf{E}'(x, y)e^{i(\mathbf{p}\cdot\mathbf{r}-\omega t)}. \tag{17.68}$$

The refracted wave (17.60) is also an example of an *evanescent* plane wave. The name (from a Latin word meaning "to vanish") is appropriate because the amplitude of an evanescent wave decreases exponentially along some direction in space. A general approach to evanescent waves recognizes that the fields in (17.9) satisfy the vector wave equation even when $\mathbf{k}$ is a complex vector with real and imaginary parts defined by $\mathbf{k} = \mathbf{q} + i\boldsymbol{\kappa}$. The electric field for such a wave is

$$\mathbf{E}(\mathbf{r}, t) = \mathbf{E}\exp[i(\mathbf{q}\cdot\mathbf{r} - \omega t)]\exp(-\boldsymbol{\kappa}\cdot\mathbf{r}). \tag{17.69}$$

We compute the corresponding magnetic field using (17.11) because $Z\mathbf{H} = \hat{\mathbf{k}} \times \mathbf{E}$ in (17.15) is no longer valid when $\mathbf{k}$ is complex and $\mathbf{q}$ and $\boldsymbol{\kappa}$ are not parallel. Accordingly,

$$\mathbf{H}(\mathbf{r}, t) = \frac{1}{\omega\mu}\mathbf{k} \times \mathbf{E}\exp[i(\mathbf{q}\cdot\mathbf{r} - \omega t)]\exp(-\boldsymbol{\kappa}\cdot\mathbf{r}). \tag{17.70}$$

For non-dissipative media, the real and imaginary parts of $\mathbf{k}\cdot\mathbf{k} = \mu\epsilon\omega^2$ from (17.14) give

$$q^2 - \kappa^2 = \mu\epsilon\omega^2 \qquad \text{and} \qquad \mathbf{q}\cdot\boldsymbol{\kappa} = 0. \tag{17.71}$$

The real part of (17.69) diverges as $\boldsymbol{\kappa}\cdot\mathbf{r} \to -\infty$. This unphysical behavior disqualifies it as a possible electric field in an infinite medium. However, as shown by (17.60), (17.69) with $\boldsymbol{\kappa} = \kappa\hat{\mathbf{z}}$ and $\kappa > 0$ is perfectly well behaved in a medium which occupies only the half-space $z > 0$. We leave it as an exercise for the reader to prove that the phase velocity of an evanescent wave is always *less* than the uniform plane wave value c/n.

Several familiar ideas need revision when $\mathbf{k}$ is a complex vector. Not least, the planes of constant phase in (17.69) are perpendicular to the real vector $\mathbf{q}$, while planes of constant amplitude are perpendicular to the real vector $\boldsymbol{\kappa}$. The right member of (17.71) constrains these planes to lie at right angles to each other.[10] The same planes are parallel to one another for an ordinary, uniform plane wave. Equally important, the two Maxwell equations in (17.10) no longer have the meaning of geometrical orthogonality when $\mathbf{E}$, $\mathbf{H}$, and $\mathbf{k}$ are all complex vectors.[11] Similarly, the Maxwell equation $\mathbf{E}\cdot\mathbf{k} = 0$ [see (17.15)] does *not* imply that $\mathbf{E}^*\cdot\mathbf{k} = 0$. Example 17.3 illustrates this explicitly.

Example 17.3 Show that $R_\text{p} = 1$ and $T_\text{p} = 0$ when a p-polarized wave suffers total internal reflection.

Solution: $R_\text{p} = 1$ follows immediately from (17.67) and the formula (17.42) for the reflection coefficient. $T_\text{p} = 0$ is more subtle because the Poynting vector formula (17.41) used to evaluate (17.45) assumes a real wave vector. Instead, (17.57) and (17.59) show that total reflection produces a refracted wave $\mathbf{E}_2$ with a complex wave vector $\mathbf{k}_2 = k_{2x}\hat{\mathbf{x}} + i\kappa\hat{\mathbf{z}}$. Under these conditions, (17.70)

[10] This constraint changes to $\mathbf{q}\cdot\boldsymbol{\kappa} > 0$ in conducting matter. See Section 17.6.

[11] See the box labeled "Complex Vectors" following (16.62).

gives the time averaged Poynting vector as

$$\langle \mathbf{S}_2 \rangle = \frac{1}{2}\mathrm{Re}\left(\mathbf{E}_2^* \times \mathbf{H}_2\right) = \frac{1}{2\omega\mu_2}\mathrm{Re}\left[\mathbf{E}_2^* \times (\mathbf{k}_2 \times \mathbf{E}_2)\right] = \frac{1}{2\omega\mu_2}\mathrm{Re}\left[|\mathbf{E}_2|^2\mathbf{k}_2 - (\mathbf{E}_2^* \cdot \mathbf{k}_2)\mathbf{E}_2\right].$$

The unfamiliar factor here is

$$\mathbf{E}_2^* \cdot \mathbf{k}_2 = E_{2x}^* k_{2x} + i\kappa E_{2z}^*.$$

We eliminate $E_{2x}^* k_{2x}$ from the preceding equation by using $\nabla \cdot \mathbf{E}_2 = 0$ in the form

$$(\mathbf{k}_2 \cdot \mathbf{E}_2)^* = 0 = E_{2x}^* k_{2x} - i\kappa E_{2z}^*.$$

Therefore,

$$\langle \mathbf{S}_2 \rangle \cdot \hat{\mathbf{z}} = \frac{1}{2\omega\mu_2}\mathrm{Re}\left\{i\kappa|\mathbf{E}_2|^2 - 2i\kappa|E_{2z}|^2\right\} = 0.$$

The corresponding transmission coefficient (17.45) vanishes also.

17.4 Radiation Pressure

The force per unit area exerted on a body when it reflects or absorbs an electromagnetic wave is called *radiation pressure*. The ultimate source of this force is a transfer of linear momentum from the wave to the body. Therefore, calculations of radiation pressure exploit the conservation law derived in Section 15.8.2 which relates the mechanical force density $\mathbf{f}$ to the field momentum density, $\mathbf{g}_{\mathrm{EM}}$, and the electromagnetic stress tensor, $\mathcal{T}$:

$$\mathbf{f} = -\frac{\partial \mathbf{g}_{\mathrm{EM}}}{\partial t} + \nabla \cdot \boldsymbol{\mathcal{T}}. \tag{17.72}$$

Integrating (17.72) over a volume V bounded by a surface S and using the divergence theorem, the force that acts on V is

$$\mathbf{F} = -\int_V d^3r \, \frac{\partial \mathbf{g}_{\mathrm{EM}}}{\partial t} + \int_S dS \, \hat{\mathbf{n}} \cdot \boldsymbol{\mathcal{T}}. \tag{17.73}$$

If the incident field is confined to a wave packet, the pressure can be computed entirely from the time rate of change of $\mathbf{g}_{\mathrm{EM}}$ because S can be extended to infinity where the fields of the packet vanish. If the incident field is time-harmonic (and thus extends over all of space), the time-averaged pressure can be computed entirely from $\boldsymbol{\mathcal{T}}$ because the time-average of $\partial \mathbf{g}_{\mathrm{EM}}/\partial t$ is zero at every point in V.

17.4.1 An Incident Time-Harmonic Plane Wave

In this section, we calculate the radiation pressure for the situation shown in Figure 17.7 where a normal-incidence time-harmonic plane wave is partially reflected and partially transmitted by a semi-infinite, non-magnetic dielectric. The "volume" of integration is the plane $z = 0$, so the jth component of the time-averaged force is

$$\langle F_j \rangle = \int_S dA \, n_k \langle \mathcal{T}_{kj} \rangle, \tag{17.74}$$

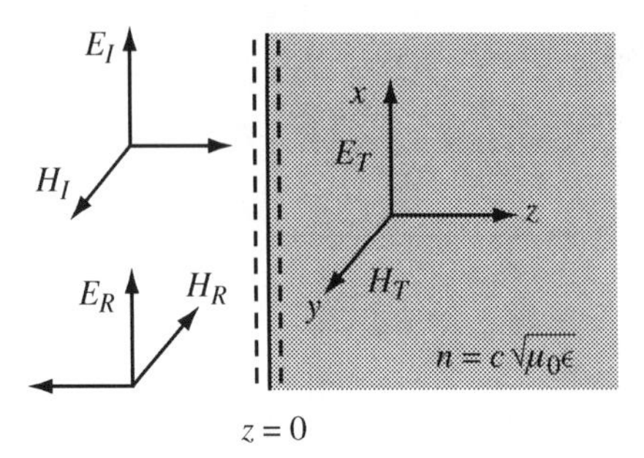

Figure 17.7: The incident, reflected, and transmitted plane waves which produce radiation pressure at the surface of a simple, semi-infinite dielectric.

where the planes $z = 0^-$ and $z = 0^+$ comprise S and the stress tensor is [see (15.113)]

$$\mathcal{T}_{ij} = D_i E_j + B_i H_j - \frac{1}{2}\delta_{ij}(\mathbf{D}\cdot\mathbf{E} + \mathbf{B}\cdot\mathbf{H}). \tag{17.75}$$

For the geometry shown in Figure 17.7, the stress tensor is diagonal and its time average is

$$\langle \mathcal{T} \rangle = \frac{1}{4}\begin{bmatrix} \epsilon|\mathbf{E}|^2 - \mu|\mathbf{H}|^2 & 0 & 0 \\ 0 & \mu|\mathbf{H}|^2 - \epsilon|\mathbf{E}|^2 & 0 \\ 0 & 0 & -\epsilon|\mathbf{E}|^2 - \mu|\mathbf{H}|^2 \end{bmatrix}. \tag{17.76}$$

In light of this diagonal structure, (17.74) gives the time-averaged radiation pressure on an area A of the $z = 0$ plane of Figure 17.7 as[12]

$$\langle \mathcal{P}_{\rm rad} \rangle = \frac{\langle F_z \rangle}{A} = \langle T_{zz}(z = 0^+) - T_{zz}(z = 0^-) \rangle. \tag{17.77}$$

Figure 17.7 and (17.15) give the fields needed to evaluate (17.77) as

$$\mathbf{E} = \begin{cases} (E_I + E_R)\hat{\mathbf{x}} & z = 0^- \\ E_T\hat{\mathbf{x}} & z = 0^+ \end{cases} \qquad \text{and} \qquad \mathbf{H} = \begin{cases} \dfrac{E_I - E_R}{Z_0}\hat{\mathbf{y}} & z = 0^- \\ \dfrac{E_T}{Z}\hat{\mathbf{y}} & z = 0^+. \end{cases} \tag{17.78}$$

Using $Z = \sqrt{\mu/\epsilon}$, we find without difficulty that

$$\langle \mathcal{P}_{\rm rad} \rangle = \frac{1}{2}\epsilon_0 E_I^2 \left[1 + \left(\frac{E_R}{E_I}\right)^2 - \frac{\epsilon}{\epsilon_0}\left(\frac{E_T}{E_I}\right)^2\right]. \tag{17.79}$$

A quick check confirms that $\langle F_x \rangle = \langle F_y \rangle = 0$, as expected for normal incidence.

Equation (17.79) is often written in terms of the incident wave intensity $I_0 = |\langle \mathbf{S}_I \rangle|$, the energy reflection coefficient R, and the energy transmission coefficient T. These quantities are defined by (17.41), (17.42), and (17.45), respectively. Hence, because $\epsilon Z = n\epsilon_0 Z_0$ for our problem, the radiation pressure becomes

$$\langle \mathcal{P}_{\rm rad} \rangle = \frac{I_0}{c}\left[1 + R - nT\right]. \tag{17.80}$$

Inserting the explicit expressions for R and T from (17.44) and (17.47) into (17.80) with $n_1 = 1$ and $n_2 = n$ gives an expression for the radiation pressure on the dielectric surface in Figure 17.7 that was

[12] Notice that $\langle T_{zz} \rangle = -\langle u_{\rm EM} \rangle$ for this problem.

first deduced by Poynting in 1905:

$$\langle \mathcal{P}_{\rm rad} \rangle = -\frac{2I_0}{c}\frac{n-1}{n+1}. \tag{17.81}$$

From the first equality in (17.77), the negative sign in (17.81) means that the radiation force points *away* from a dielectric with $n > 1$. Using (17.79), we conclude that the contribution to the radiation pressure from the transmitted wave exceeds the contributions from the incident and reflected waves.

The analysis of radiation pressure simplifies if the substance that occupies $z > 0$ in Figure 17.7 absorbs all the energy transmitted through its surface. There is no propagating transmitted wave in that case and we can set $T = 0$ in (17.80) to get

$$\langle \mathcal{P}_{\rm rad} \rangle = \frac{I_0}{c}\left[1 + R\right]. \tag{17.82}$$

For a totally absorbing sample, $R = 0$ and $\langle \mathcal{P}_{\rm rad} \rangle = I_0/c$. For a totally reflecting sample, $R = 1$ and $\langle \mathcal{P}_{\rm rad} \rangle = 2I_0/c$. The factor of 2 is explicable qualitatively because twice the momentum must be supplied by the sample to "turn around" the incident beam than is needed to absorb it.

Example 17.4 Use the left side of (17.72) to derive the time-averaged radiation pressure (17.81).

Solution: The Minkowski force density (15.112) is

$$\mathbf{f}_{\rm M} = \rho_f \mathbf{E} + \mathbf{j}_f \times \mathbf{B} - \tfrac{1}{2}E^2\nabla\epsilon - \tfrac{1}{2}H^2\nabla\mu.$$

This quantity produces the same time-averaged force as the Abraham force density (15.120) derived in Section 15.8.3 because the two differ by a total time derivative. It is also possible to use the Lorentz force density recorded in (15.121) of Section 15.9.[13] Using $\mathbf{f}_{\rm M}$, the force on an area A of the free surface at $z = 0$ is

$$\langle F \rangle = \left\langle \int_V d^3r\, \mathbf{f}_{\rm M} \right\rangle = -\hat{\mathbf{z}}\frac{A}{4}\int_{z=0^-}^{z=0^+} E^2 \frac{d\epsilon}{dz} dz.$$

The electric field lies in the x-y plane and thus is continuous passing through $z = 0$. Hence, E^2 comes out of the integral and is equal to the square of the transmitted field [see (17.38)],

$$E_T = \frac{2}{n+1}E_I.$$

Because $n^2 = c^2\epsilon\mu_0$, we find

$$\langle F_z \rangle = -\frac{A}{4}E_T^2(\epsilon - \epsilon_0) = -\frac{A}{4}\epsilon_0\left[\frac{2}{n+1}E_I\right]^2(n^2-1).$$

The incident wave intensity is the magnitude of the incident Poynting vector,

$$I_0 = |\langle \mathbf{S}_I \rangle| = \frac{1}{2}\epsilon_0 c E_I^2.$$

Therefore, we reproduce (17.81),

$$\langle \mathcal{P}_{\rm rad} \rangle = \frac{\langle F_z \rangle}{A} = -\frac{2I_0}{c}\frac{n-1}{n+1}.$$

[13] See, e.g., C.F. Bohren, "Radiation forces and torques without stress tensors", *European Journal of Physics* **32**, 1515 (2011).

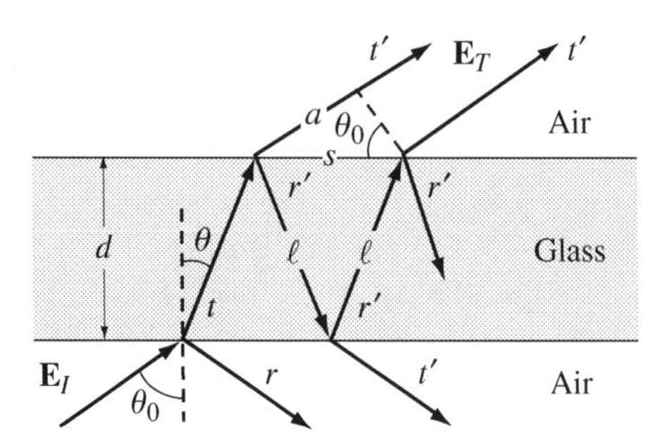

Figure 17.8: A plane wave incident on an embedded layer is imagined to reflect back and forth inside the layer any number of times before it exits.

17.5 Layered Matter

This section generalizes the single-interface problem studied in the previous two sections to the case of layered media characterized by two or more parallel interfaces bounded by regions with uniform dielectric properties. New phenomena occur because reflection and refraction from the interfaces produce wave interference in the matter bounded by the interfaces. This geometry occurs in many natural and man-made settings, and interference phenomena in layered dielectrics have been observed in every spectral range between x-rays and radio waves. Our discussion begins with a two-interface problem, proceeds to a general multilayer structure, and concludes with a practically important structure called the Bragg mirror.

17.5.1 The Fabry-Perot Geometry

A glass microscope slide in air both reflects and transmits light. The net transmission coefficient $\mathcal{T}$ is very sensitive to the interference inside the glass between the forward-going wave transmitted through the front surface of the glass and the backward-going wave reflected from the back surface of the glass. Long before light was identified as an electromagnetic phenomenon, George Airy took account of interference and calculated $\mathcal{T}$ by summing the amplitudes of propagating waves that reflect back and forth any number of times inside the slab before they exit. Matching conditions play no role because each wave (represented by a directed line segment in Figure 17.8) is imagined to propagate interference-free until it suffers reflection or transmission at one of the two interfaces. The calculation requires only the Fresnel amplitudes r and t for a wave incident from the air side of the air/glass interface and the corresponding quantities r' and t' for a wave incident from the glass side of the same interface.

If $\Delta\phi$ is the phase difference between two successive waves that exit the slab at the top of Figure 17.8, the total transmitted electric field takes the form of a geometric series:

$$\mathbf{E}_T = \mathbf{E}_I \left[tt' + tr'r't'e^{i\Delta\phi} + t(r'r')^2 t' e^{i2\Delta\phi} + \cdots \right] = \frac{tt'}{1 - r'r'e^{i\Delta\phi}} \mathbf{E}_I. \tag{17.83}$$

To simplify (17.83), we pause to examine the single-interface refraction event in panel (a) of Figure 17.9 and the time-reversed version of the same event in panel (b). The latter simply reverses the directions of the propagation wave vectors because $(\mathbf{E}, \mathbf{B}) \to (\mathbf{E}, -\mathbf{B})$ under time reversal (see Table 15.1). Both events occur with the same relative amplitudes if there is no absorption of electromagnetic energy. On the other hand, it is easier to understand the event in panel (b) (which produces one outgoing wave from two incoming waves) if we decompose it into the two events shown in panel (c) and panel (d). The requirement that this decomposition preserve the relative amplitudes of the waves produces the

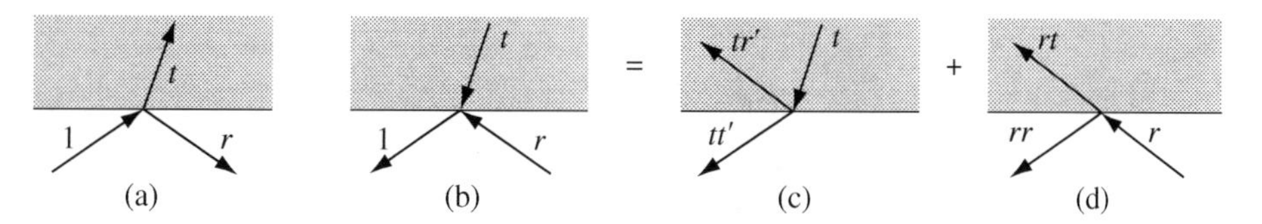

Figure 17.9: Wave-vector diagrams used to prove (17.84). Panel (b) is time-reversed with respect to panel (a). See text for discussion.

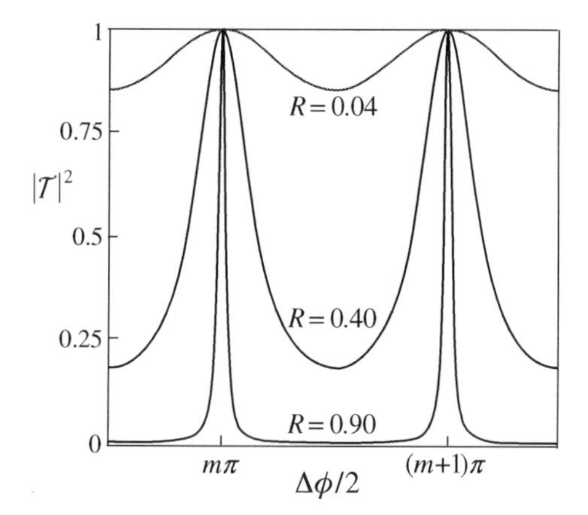

Figure 17.10: The thin-film transmission function (17.85) plotted for several values of the single-interface reflectivity R.

Stokes relations,

$$r^2 + tt' = 1 \qquad \text{and} \qquad tr' + rt = 0. \tag{17.84}$$

Two consequences of (17.84) are $r = -r'$ and $tt' = 1 - r^2 = 1 - R$. Substituting these into (17.83) generates *Airy's formula* for the fraction of energy transmitted through the film:

$$|\mathcal{T}|^2 = \left|\frac{E_T}{E_I}\right|^2 = \left[1 + \frac{4R}{(1-R)^2}\sin^2(\Delta\phi/2)\right]^{-1}. \tag{17.85}$$

Figure 17.8 shows that the phase difference $\Delta\phi = 2k\ell - k_0 a$, where $k = n\omega/c$ and $k_0 = n_0\omega/c$. Also, $\ell = d/\cos\theta$ and $a = s\sin\theta_0$, where $s = 2d\sin\theta/\cos\theta$. Combining these facts with Snell's law, $n_0\sin\theta_0 = n\sin\theta$, gives the argument of the oscillatory factor in (17.85) as

$$\Delta\phi/2 = n\frac{\omega}{c}d\cos\theta. \tag{17.86}$$

Figure 17.10 plots (17.85) as a function of $\Delta\phi/2$ for several values of R. A remarkable feature is that 100% transmission occurs whenever $\Delta\phi/2$ is an integer multiple of π, *independent* of the value of R. At normal incidence, this resonance condition reads $d = m\frac{1}{2}(\lambda/n)$, which shows that complete transmission occurs when the dielectric film supports standing waves with an integer number of half-wavelengths between its surfaces. Even when the transmission amplitude is essentially zero for each individual interface, perfect transmission through the film occurs when all the waves that exit the layer from the top in Figure 17.8 interfere constructively while all the waves that exit from the bottom interfere destructively. Conversely, there is essentially zero transmission at values of ϕ other than $m\pi$ when $R \to 1$. In 1897, Charles Fabry and Alfred Perot realized that this interference-induced

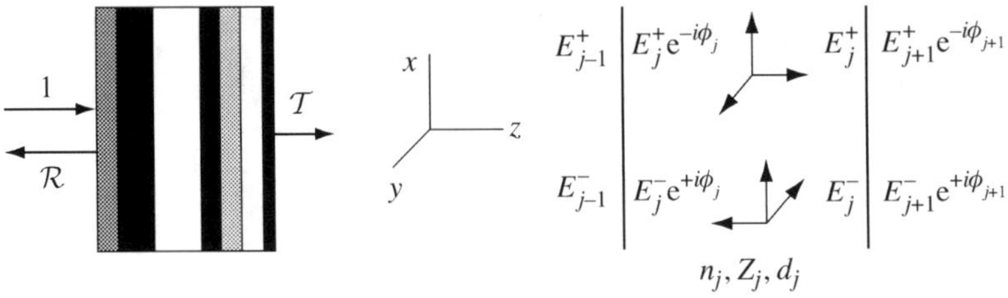

Figure 17.11: Left panel: a cartoon of a multilayer. Right panel: the electric field amplitudes of forward-going and backward-going plane waves in layer j at its boundaries.

contrast could be exploited for interferometry. Today, the Fabry-Perot geometry is widely used in the design of laser cavities and for high-resolution spectroscopy.

17.5.2 Waves in a Multilayer

The repeated-reflection method used in the previous section does not generalize easily to the study of the interplay between reflection, refraction, and interference when a plane wave strikes a multilayered medium like the one shown in the left panel of Figure 17.11. On the other hand, it is straightforward to treat this geometry if we mimic our derivation of the Fresnel equations (Section 17.3.2) and use electromagnetic matching conditions to relate the amplitudes of a minimal set of plane waves. For simplicity, we specialize to normal incidence and calculate the net reflection and transmission amplitudes $\mathcal{R}$ and $\mathcal{T}$ indicated in the diagram.

We treat the multilayer as a collection of $N+1$ layers where layer j has index of refraction n_j, wave impedance Z_j, and thickness d_j. Layer 0 is semi-infinite and extends to $z=-\infty$. Layer N is also semi-infinite and extends to $z=+\infty$. The right panel of Figure 17.11 shows that we decompose the total electric field in layer j, $\mathbf{E}_j(z)=E_j(z)\hat{\mathbf{x}}$, into the sum of a right-going wave $\mathbf{E}_j^+(z)$ and a left-going wave $\mathbf{E}_j^-(z)$. If E_j^+ and E_j^- are the field amplitudes for the right-going and left-going waves at the rightmost boundary of layer j, the right panel of Figure 17.11 also shows that propagation across layer j causes each wave to accumulate a phase

$$\phi_j = n_j d_j \omega / c. \tag{17.87}$$

The fields $\mathbf{E}_j$ and $\mathbf{H}_j$ are tangential and thus continuous everywhere. Imposing this continuity condition at the left boundary of layer j gives

$$\begin{aligned} E_{j-1} &= E_{j-1}^+ + E_{j-1}^- = E_j^+ \exp(-i\phi_j) + E_j^- \exp(i\phi_j) \\ H_{j-1} &= H_{j-1}^+ + H_{j-1}^- = H_j^+ \exp(-i\phi_j) + H_j^- \exp(i\phi_j). \end{aligned} \tag{17.88}$$

Moreover, $H_j = (E_j^+ - E_j^-)/Z_j$ because $Z_j H_j^+ = E_j^+$, $Z_j H_j^- = -E_j^-$, and $E_j = E_j^+ + E_j^-$. Inverting this to get $E_j^+ = \frac{1}{2}(Z_j H_j + E_j)$ and $E_j^- = \frac{1}{2}(E_j - Z_j H_j)$ permits us to write (17.88) in matrix form as

$$\begin{pmatrix} E_{j-1} \\ H_{j-1} \end{pmatrix} = \begin{pmatrix} \cos\phi_j & -iZ_j \sin\phi_j \\ -iZ_j^{-1}\sin\phi_j & \cos\phi_j \end{pmatrix} \begin{pmatrix} E_j \\ H_j \end{pmatrix}. \tag{17.89}$$

This equation relates the wave amplitudes in one layer to those in an adjacent layer. Therefore, repeated application of the *transfer matrix* defined by (17.89) permits us to propagate waves from one side of the multilayer to the other:

$$\begin{pmatrix} E_0 \\ H_0 \end{pmatrix} = \prod_{j=1}^{N-1} \begin{pmatrix} \cos\phi_j & -iZ_j \sin\phi_j \\ -iZ_j^{-1}\sin\phi_j & \cos\phi_j \end{pmatrix} \begin{pmatrix} E_N \\ H_N \end{pmatrix}. \tag{17.90}$$

For our problem, we set $E_0^+ = 1$ and note that there is no left-moving wave in layer N. This information permits us to use $H_j = (E_j^+ - E_j^-)/Z_j$ and

$$E_0^- = \mathcal{R} \qquad Z_0 H_0^+ = 1 \qquad Z_0 H_0^- = -E_0^- \qquad E_N^+ = \mathcal{T} \qquad E_N^- = 0 \tag{17.91}$$

to transform (17.90) to

$$\begin{pmatrix} 1+\mathcal{R} \\ (1-\mathcal{R})Z_0^{-1} \end{pmatrix} = \prod_{j=1}^{N-1} \begin{pmatrix} \cos\phi_j & -iZ_j \sin\phi_j \\ -iZ_j^{-1}\sin\phi_j & \cos\phi_j \end{pmatrix} \begin{pmatrix} \mathcal{T} \\ \mathcal{T}Z_N^{-1} \end{pmatrix}. \tag{17.92}$$

This completes the normal incidence problem because (17.92) generates two equations in the two unknowns $\mathcal{R}$ and $\mathcal{T}$.[14]

Application 17.2 Bragg Mirrors

The multiple interface structures shown in Figure 17.12 are called *Bragg mirrors*. The multilayer in the left panel, composed of Pt metal and amorphous carbon, is used in space telescopes to reflect X-rays. The semiconductor micropillar shown in the middle panel contains two GaAs/AlAs multilayers which reflect near-infrared waves back and forth to form a resonant cavity for a laser. The right panel is a cross section micrograph of alternating layers of high- and low-density chitin in the wing of a Japanese jewel beetle. This reflecting structure, rather than any natural pigmentation, is responsible for the brilliant metallic color of this insect.

Figure 17.12: Left panel: a multilayer which reflects hard X-rays. Transmission electron micrograph from Ohnishi *et al.* (2004). Middle panel: multilayer mirrors define a near-infrared laser cavity. Scanning electron micrograph from Sanvitto *et al.* (2005). Copyright 2005, American Institute of Physics. Right panel: cross section of the wing of a Japanese jewel beetle which reflects very strongly in the visible. Transmission electron micrograph from Kinoshita and Yoshioka (2005).

The physics of a Bragg mirror becomes clear when we glance back at (17.38) and observe that the normal-incidence Fresnel reflection amplitude $r = (n_1 - n_2)/(n_1 + n_2)$ changes sign (but not magnitude) at the successive interfaces of a multilayer composed of alternating layers of two dissimilar

[14] The generalization of (17.92) to an arbitrary angle of incidence requires very little additional work. See, e.g., Lipson, Lipson, and Tannhauser (1995) in Sources, References, and Additional Reading.

materials. Therefore, a net reflective structure will result if we choose the layer thicknesses and indices of refraction so waves advance in phase by the same amount through every layer and all partially reflected waves add constructively. To be more quantitative, we treat the multilayer in Figure 17.11 as a stack of M bilayers where each bilayer consists of a non-magnetic slab with thickness d_A and index of refraction n_A joined to a non-magnetic slab with thickness d_B and index of refraction n_B. Using $2\pi/\lambda = \omega/c$ in (17.87), the physical situation imagined above will be achieved if

$$\phi = 2\pi n_A d_A/\lambda = 2\pi n_B d_B/\lambda = \pi/2, 3\pi/2, \ldots \tag{17.93}$$

For a sufficiently large number M of A/B bilayers, such a multilayer should behave like a highly reflective and wavelength-specific mirror.

The repeated bilayer structure of a Bragg mirror reduces the matrix product in (17.92) to a product of two matrices repeated M times, namely,

$$\left[\begin{pmatrix} \cos\phi & -iZ_A\sin\phi \\ -iZ_A^{-1}\sin\phi & \cos\phi \end{pmatrix}\begin{pmatrix} \cos\phi & -iZ_B\sin\phi \\ -iZ_B^{-1}\sin\phi & \cos\phi \end{pmatrix}\right]^M \tag{17.94}$$

or

$$\begin{pmatrix} \cos^2\phi - \dfrac{Z_A}{Z_B}\sin^2\phi & -i(Z_A+Z_B)\sin\phi\cos\phi \\ -i(Z_A^{-1}+Z_B^{-1})\sin\phi\cos\phi & \cos^2\phi - \dfrac{Z_B}{Z_A}\sin^2\phi \end{pmatrix}^M. \tag{17.95}$$

Our choice of $\phi = \pi/2, 3\pi/2, \ldots$ diagonalizes (17.95), so (17.92) becomes

$$\begin{aligned} 1+\mathcal{R} &= \left(-\frac{Z_A}{Z_B}\right)^M \mathcal{T} \\ 1-\mathcal{R} &= \left(-\frac{Z_B}{Z_A}\right)^M \frac{Z_0}{Z_N}\mathcal{T}. \end{aligned} \tag{17.96}$$

For simplicity, set $Z_N/Z_0 = 1$ and recall that $Z_A/Z_B = n_B/n_A$ for non-magnetic materials. In that case, solving (17.96) for $\mathcal{R}$ gives

$$\mathcal{R} = \frac{\left(-\dfrac{n_B}{n_A}\right)^M - \left(-\dfrac{n_B}{n_A}\right)^{-M}}{\left(-\dfrac{n_B}{n_A}\right)^M + \left(-\dfrac{n_B}{n_A}\right)^{-M}}. \tag{17.97}$$

Whether $n_A/n_B > 1$ or $n_A/n_B < 1$, one term dominates the other in both the numerator and denominator of (17.97) when $M \gg 1$. The result is $\mathcal{R} \to \pm 1$, as anticipated. A numerical example is a GaAs/AlAs multilayer mirror with $M = 25$, $n_{\text{GaAs}} = 3.5$, and $n_{\text{AlAs}} = 3.0$. For these choices, (17.97) gives $|\mathcal{R}| = 0.999$. The name "Bragg mirror" is used for these structures because the condition (17.93) written as $n_A d_A + n_B d_B = \lambda/2$ is reminiscent of the "Bragg condition" for constructive interference when X-rays scatter from parallel planes of atoms in a crystal. ■

Example 17.5 Let a thin layer of a non-magnetic dielectric with index n and thickness d completely cover a substrate material with index of refraction n_s. Find n and d such that a normal incidence electromagnetic plane wave *does not* reflect from the layer. This kind of *anti-reflective coating* is the most common use of a dielectric multilayer.

Solution: We use (17.92) for a single layer and set $\mathcal{R} = 0$. Using $Z_i/Z_j = n_j/n_i$ and $n_0 = 1$ gives

$$1 = (\cos\phi - i\frac{n_s}{n}\sin\phi)\mathcal{T}$$

$$1 = (-in\sin\phi + n_s\cos\phi)\mathcal{T}.$$

Eliminating $\mathcal{T}$ gives $\cos\phi - i(n_s/n)\sin\phi = n_s\cos\phi - in\sin\phi$. The index $n_s \neq 1$, so we must have $\cos\phi = 0$. Therefore, using (17.87) and $\omega/c = 2\pi/\lambda$, we conclude that

$$n = \sqrt{n_s} \qquad \text{and} \qquad nd = \text{odd integer} \times \frac{\pi}{4}\lambda.$$

With this choice for n, the air, coating, and substrate have indices of refraction 1, $\sqrt{n_s}$, and n_s. This implies that the normal-incidence Fresnel reflection amplitude in (17.38) is the *same* for the air/coating and coating/substrate interfaces. Hence, with our choice for d, waves reflected from these interfaces will always interfere destructively.

17.6 Simple Conducting Matter

Simple conducting matter is defined by the constitutive relations $\mathbf{D} = \epsilon\mathbf{E}$, $\mathbf{B} = \mu\mathbf{H}$, and $\mathbf{j}_f = \sigma\mathbf{E}$ (Ohm's law) with constant values for ϵ, μ, and σ. We assume no sources of free charge ($\rho_f = 0$) so the Maxwell equations (17.1) and (17.2) reduce to[15]

$$\nabla\cdot\mathbf{E} = 0 \qquad \nabla\cdot\mathbf{H} = 0 \tag{17.98}$$

and

$$\nabla\times\mathbf{E} = -\mu\frac{\partial\mathbf{H}}{\partial t} \qquad \text{and} \qquad \nabla\times\mathbf{H} = \sigma\mathbf{E} + \epsilon\frac{\partial\mathbf{E}}{\partial t}. \tag{17.99}$$

Two observations are germane. First, although the conductivity σ does not appear in both equations in (17.99), the familiar trick of taking the curl of one equation and substituting into it from the other equation shows that $\mathbf{E}$ and $\mathbf{H}$ satisfy the same generalized wave equation:

$$\left(\nabla^2 - \mu\sigma\frac{\partial}{\partial t} - \mu\epsilon\frac{\partial^2}{\partial t^2}\right)\begin{Bmatrix}\mathbf{E}\\ \mathbf{H}\end{Bmatrix} = 0. \tag{17.100}$$

A second observation is that a conducting system continuously drains energy from resident electromagnetic fields. This follows from Poynting's theorem for matter (Section 15.8.1), where the rate at which the fields lose energy by doing work on the current carriers in simple conducting matter is

$$\frac{dW}{dt} = \int_V d^3r\,\mathbf{j}_f\cdot\mathbf{E} = \sigma\int_V d^3r\,|\mathbf{E}|^2. \tag{17.101}$$

The irreversible dissipation of field energy predicted by (17.101) appears as Joule heat.

17.6.1 Monochromatic Plane Waves

We seek a solution to (17.98) and (17.99) in the form of a monochromatic plane wave,

$$\mathbf{E}(\mathbf{r},t) = \mathbf{E}\exp[i(\mathbf{k}\cdot\mathbf{r} - \omega t)] \qquad \text{and} \qquad \mathbf{H}(\mathbf{r},t) = \mathbf{H}\exp[i(\mathbf{k}\cdot\mathbf{r} - \omega t)]. \tag{17.102}$$

The two divergence equations give

$$\mathbf{k}\cdot\mathbf{E} = 0 \qquad \text{and} \qquad \mathbf{k}\cdot\mathbf{H} = 0. \tag{17.103}$$

[15] An Ohmic medium does not support a surface current density $\mathbf{K}$ because $\mathbf{j} = \sigma\mathbf{E}$ is valid at all points in and on the surface of such a medium.

The two curl equations give

$$\mathbf{k} \times \mathbf{E} = \omega\mu\mathbf{H} \qquad \text{and} \qquad \mathbf{k} \times \mathbf{H} = -\omega(\epsilon + i\sigma/\omega)\mathbf{E}. \tag{17.104}$$

Comparing (17.103) and (17.104) with (17.10) with (17.11) shows that the field equations for conducting matter are identical to the field equations for transparent matter (Section 17.2) except that ϵ is replaced by a complex and frequency-dependent permittivity,[16]

$$\hat{\epsilon}(\omega) = \epsilon + i\frac{\sigma}{\omega} = \epsilon' + i\epsilon''. \tag{17.105}$$

Consequently, we mimic (17.4) to define a complex and frequency-dependent index of refraction, $\hat{n}(\omega)$, and mimic (17.14) to find the dispersion relation:

$$\mathbf{k} \cdot \mathbf{k} = \hat{k}^2 = \hat{\epsilon}(\omega)\mu\omega^2 = \mu\epsilon\omega^2 + i\mu\sigma\omega \equiv \hat{n}^2(\omega)\frac{\omega^2}{c^2}. \tag{17.106}$$

The frequency dependence of $\hat{\epsilon}(\omega)$ means that even a "simple" conductor is a dispersive medium in the sense of the first paragraph of this chapter.

The dispersion relation (17.106) requires $\mathbf{k}$ to be a complex vector. Following Section 17.3.7, we substitute $\mathbf{k} = \mathbf{q} + i\boldsymbol{\kappa}$ into (17.106) to get a generalization of (17.14) for conducting matter:

$$q^2 - \kappa^2 = \mu\epsilon\omega^2 \qquad \text{and} \qquad \mathbf{q} \cdot \boldsymbol{\kappa} = \sigma\mu\epsilon. \tag{17.107}$$

The vectors $\mathbf{q}$ and $\boldsymbol{\kappa}$ may or may not be parallel. Here, we *assume* parallelism and use a real unit vector $\hat{\mathbf{k}}$ to write

$$\mathbf{k} = \hat{k}\hat{\mathbf{k}} = \hat{n}\frac{\omega}{c}\hat{\mathbf{k}} = (n' + in'')\frac{\omega}{c}\hat{\mathbf{k}}. \tag{17.108}$$

When combined with (17.103), this choice for $\mathbf{k}$ makes the plane waves in (17.102) *transverse*, as they are in vacuum and in transparent matter. They also satisfy the analog of (17.15),

$$\mathbf{k} = \hat{n}\frac{\omega}{c}\hat{\mathbf{k}}, \qquad \hat{\mathbf{k}} \cdot \mathbf{E} = 0, \qquad \text{and} \qquad \hat{Z}\mathbf{H} = \hat{\mathbf{k}} \times \mathbf{E}, \tag{17.109}$$

with the complex and frequency-dependent wave impedance,[17]

$$\hat{Z}(\omega) = \sqrt{\frac{\mu}{\hat{\epsilon}(\omega)}}. \tag{17.110}$$

The complex nature of $\hat{Z}$ guarantees that $\mathbf{H}$ in (17.109) is not in phase with $\mathbf{E}$. This is an inductive effect associated with the Faraday's law magnetic field produced by the time-varying current density $\mathbf{j} = \sigma\mathbf{E}$ in the conducting matter.

Substituting $\mathbf{k}$ from (17.108) into the left member of (17.102) gives the electric field in the conductor as

$$\mathbf{E}(\mathbf{r}, t) = \mathbf{E}\exp[-(\omega/c)n''\hat{\mathbf{k}} \cdot \mathbf{r}]\exp\left[i\left(\frac{\omega}{c}n'\hat{\mathbf{k}} \cdot \mathbf{r} - \omega t\right)\right]. \tag{17.111}$$

The real exponential damps the amplitude of (17.111) as it propagates. Indeed, it is common to define an *absorption coefficient* α,

$$\alpha = 2n''\frac{\omega}{c}, \tag{17.112}$$

[16] This chapter uses a circumflex over a non-boldface variable to denote a complex scalar. A circumflex over a boldface variable continues to denote a real unit vector.

[17] The connection between this quantity and the complex impedance defined in Section 14.13.1 will become clear later.

to characterize the exponential decay of the wave intensity (the square of the electric field) as propagation proceeds. This is consistent with the irreversible absorption of field energy predicted by (17.101).[18]

The detailed behavior of the field emerges when we use (17.106) and (17.108) to compute the real and imaginary parts of the complex index of refraction. Specifically, n' and n'' have the same sign because

$$n'^2 - n''^2 = \mu\epsilon c^2 \qquad \text{and} \qquad 2n'n'' = \mu\sigma\omega. \tag{17.113}$$

We choose that sign as positive so (17.111) damps in the direction of propagation. Hence,

$$\begin{aligned} n' &= c\sqrt{\frac{\mu\epsilon}{2}}\left[\sqrt{1+\left(\frac{\sigma}{\omega\epsilon}\right)^2}+1\right]^{1/2} \\ n'' &= c\sqrt{\frac{\mu\epsilon}{2}}\left[\sqrt{1+\left(\frac{\sigma}{\omega\epsilon}\right)^2}-1\right]^{1/2}. \end{aligned} \tag{17.114}$$

These expressions correctly reduce to $n' = c\sqrt{\mu\epsilon}$ and $n'' = 0$ in the limit of zero conductivity. Otherwise, the factor $\sigma/\omega\epsilon$ should be familiar from our discussion of quasi-magnetostatics (Section 14.9). It is large for a "good" conductor and small for a "poor" conductor.[19] For the remainder of this chapter, we confine ourselves to the low-frequency, "good" conductor limit because only in that case is the relation $\mathbf{j}_f = \sigma\mathbf{E}$ valid with σ equal to a constant (see Section 18.5.1).

17.6.2 The Skin Depth and Refraction into a Good Conductor

Equation (17.111) diverges as $\hat{\mathbf{k}}\cdot\mathbf{r} \to -\infty$ and thus cannot be a solution of the Maxwell equations in an infinite conducting medium. On the other hand, (17.111) could well be the wave refracted into a semi-infinite conductor ($z > 0$) when a plane wave impinges upon it from a semi-infinite vacuum to its left ($z < 0$). For a good conductor, the inner square root in (17.114) approaches $\sigma/\omega\epsilon \gg 1$, and

$$n' \approx n'' \approx c\sqrt{\frac{\mu\sigma}{2\omega}} = \sqrt{\frac{\mu\sigma\omega}{2}}\frac{c}{\omega} = \frac{c/\omega}{\delta(\omega)} \gg 1. \tag{17.115}$$

The last equality in (17.115) re-introduces the *skin depth*,

$$\delta(\omega) = \sqrt{\frac{2}{\mu\sigma\omega}}, \tag{17.116}$$

which we defined in Section 14.10 as the characteristic length scale for quasistatic field penetration into a conductor. The real and imaginary parts of the wave impedance for a good conductor also have the same magnitude because

$$\hat{Z} = \sqrt{\frac{\mu}{\epsilon + i\sigma/\omega}} \approx \sqrt{\frac{\mu\omega}{i\sigma}} = \sqrt{\frac{\mu\omega}{\sigma}}e^{-i\pi/4} = \frac{1-i}{\sigma\delta}. \tag{17.117}$$

Snell's law, combined with the large value of n' implied by (17.115), suggests that the wave refracted into the conductor propagates essentially normal to the interface, independent of the angle of incidence. We leave it as an exercise for the reader to prove that this guess is correct. The corresponding fields

[18] The reader should contrast this with the damped exponential factor in (17.60), where no loss of field energy occurs. See also Section 17.3.7.

[19] In the context of low-loss dielectrics, $\sigma/\omega\epsilon$ is called the "loss tangent" because if $\hat{\epsilon} = |\hat{\epsilon}|\exp(i\theta)$, (17.105) shows that $\tan\theta = \epsilon''/\epsilon = \sigma/\omega\epsilon$.

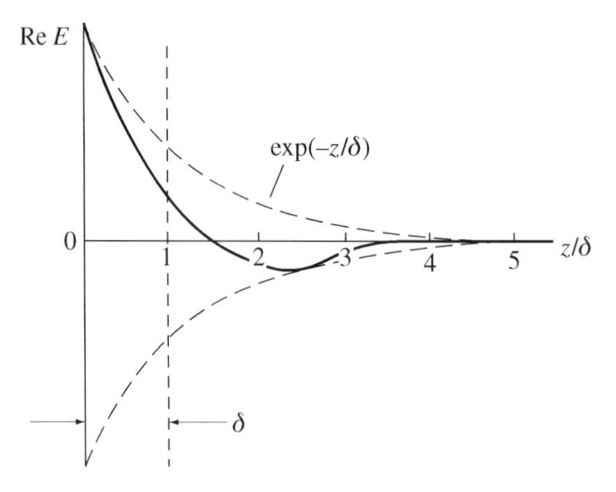

Figure 17.13: The physical electric field propagating into a sample of high conductivity matter confined to $z > 0$. The field outside the sample is not shown.

are tangential to the interface, so (17.109), (17.111), and (17.117) give the fields inside the conductor as

$$\mathbf{E}(z,t) = \mathbf{E}_{\parallel}(0)\exp(-z/\delta)\exp[i(z/\delta - \omega t)] \tag{17.118}$$

and

$$\mathbf{H}(z,t) = \frac{\sigma\delta}{1-i}\,\hat{\mathbf{z}} \times \mathbf{E}(z,t). \tag{17.119}$$

The phase velocity c/n' of this wave is typically several orders of magnitude smaller than c/n. More importantly, the equality of n' and n'' in (17.115) ensures that the main effect of a large conductivity is to damp the fields very rapidly. The graph of Re $E(z)$ in Figure 17.13 shows that the field (and thus the induced ohmic current) is confined to a distance of order δ from the conductor surface at $z = 0$. This agrees completely with the skin-depth behavior shown in Figure 14.13 for quasi-magnetostatic AC current flow in an ohmic wire.

The magnetic field (17.119) behaves similarly except that it lags the electric field by $\pi/4$ [see (17.117)] and its overall amplitude is very much larger. Actually, since the magnitude of $\mathbf{H}_{\parallel}(0)$ is approximately twice the magnitude of the tangential piece of the incident field at the surface,[20] the physics is better described if we read (17.118) and (17.119) to say that the electric field is very much smaller than the magnetic field inside a good conductor. This dominance of the magnetic field over the electric field is consistent with the fact that the electric field must disappear completely in the $\sigma \to \infty$ limit of a perfect conductor.

17.6.3 Joule Heating in a Good Conductor

The damping in Figure 17.13 occurs because the conductor irreversibly dissipates field energy into Joule heat. The rate at which this occurs is equal to the rate at which the electric field does work on the mobile charges in the conductor. Recalling (17.101), the time-averaged dissipation rate per unit volume is

$$\frac{d\langle P\rangle}{dV} = \frac{1}{2}\mathrm{Re}(\mathbf{E}\cdot\mathbf{j}) = \frac{1}{2}\sigma\,|\mathbf{E}|^2. \tag{17.120}$$

[20] At the surface of a *perfect* conductor, the tangential components of $\mathbf{H}_I$ and $\mathbf{H}_R$ at the surface are parallel and equal in magnitude while the tangential components of $\mathbf{E}_I$ and $\mathbf{E}_R$ at the surface are anti-parallel and equal in magnitude.

Combining (17.120) with (17.118) and (17.119), the loss rate per unit area of conductor surface is

$$\frac{d\langle P\rangle}{dA}=\int_0^\infty dz\frac{d\langle P\rangle}{dV}=\frac{1}{2}\sigma|\mathbf{E}_\parallel(0)|^2\int_0^\infty dz e^{-2z/\delta}=\frac{\sigma\delta}{4}|\mathbf{E}_\parallel(0)|^2=\frac{1}{2\sigma\delta}|\mathbf{H}_\parallel(0)|^2. \tag{17.121}$$

The wave frequency and constitutive constants enter (17.121) through a quantity called the *surface resistance*, R_S. This is the ohmic resistance to the induced current density $\mathbf{j}$ offered by a rectangular slab adjacent to the conductor surface with area $A = L \times L$ and thickness δ. The induced current flows parallel to the surface. Therefore,

$$R_S=\frac{L}{L\delta\sigma}=\frac{1}{\delta\sigma}=\sqrt{\frac{\mu\omega}{2\sigma}}. \tag{17.122}$$

Equation (17.121) will prove useful in Chapter 19 when we estimate ohmic losses in conducting waveguides and resonant cavities.

17.6.4 Reflection from a Good Conductor

The Fresnel equations (Section 17.3.2) remain valid for complex impedances. Since $\hat{Z} = \mu c/\hat{n}$, the reflection amplitudes in (17.34) and (17.36) generalize to

$$\hat{r}_\mathrm{p}=\frac{\hat{Z}_1\cos\theta_1-\hat{Z}_2\cos\theta_2}{\hat{Z}_1\cos\theta_1+\hat{Z}_2\cos\theta_2}=\frac{\mu_1\hat{n}_2\cos\theta_1-\mu_2\hat{n}_1\cos\theta_2}{\mu_1\hat{n}_2\cos\theta_1+\mu_2\hat{n}_1\cos\theta_2} \tag{17.123}$$

and

$$\hat{r}_\mathrm{s}=\frac{\hat{Z}_2\cos\theta_1-\hat{Z}_1\cos\theta_2}{\hat{Z}_2\cos\theta_1+\hat{Z}_1\cos\theta_2}=\frac{\mu_2\hat{n}_1\cos\theta_1-\mu_1\hat{n}_2\cos\theta_2}{\mu_2\hat{n}_1\cos\theta_1+\mu_1\hat{n}_2\cos\theta_2}. \tag{17.124}$$

The fact that $\hat{r}_\mathrm{s}$ and $\hat{r}_\mathrm{p}$ generally differ in phase implies that a linearly polarized wave reflects from a conductor with elliptical polarization.[21] Otherwise, "the complexity of what appears at first to be the simplest of problems—the reflection of a plane wave from an absorbing surface—is truly amazing".[22] Thus chastened, we limit ourselves to the case when medium 1 is a non-absorbing dielectric with $\hat{n}_1 = n_1$ and medium 2 is a good conductor.

A good conductor is very reflective at the low frequencies where our model is valid. To understand why, use $n_2' \approx n_2'' \gg n_1$ from (17.115) and $\mu_1 = \mu_2$ to evaluate (17.123) or (17.124) at normal incidence to get

$$R(\omega)=|\hat{r}(\omega)|^2=\left|\frac{\hat{n}_2-n_1}{\hat{n}_2+n_1}\right|^2\approx\frac{1-n_1/n_2'}{1+n_1/n_2'}\approx 1-\sqrt{\frac{8\epsilon_1\omega}{\sigma_2}}. \tag{17.125}$$

This is the *Hagen-Rubens relation*, which describes the high reflectivity (and thus the shiny appearance) of metals and semiconductors at visible wavelengths. Figure 17.14 shows that (17.125) provides a very good fit to low-frequency experimental reflectivity data for a typical metal.

Using $\theta_2 = 0$,[23] Figure 17.15 plots the θ_1 dependence of $R_\mathrm{s} = |\hat{r}_\mathrm{s}|^2$ and $R_\mathrm{p} = |\hat{r}_\mathrm{p}|^2$ calculated from (17.123) and (17.124) for a typical good conductor—seawater at a frequency used for satellite communication. R_p dips to a minimum value at a "pseudo-Brewster" angle of incidence, but it never goes completely to zero when $\sigma/\omega\epsilon \gg 1$. The absence of a true Brewster angle can be understood

[21] This also happens under conditions of total reflection (see Section 17.3.6).

[22] This remark appears on page 507 of Stratton (1941). See Section 17.1 of Sources, References, and Additional Reading.

[23] See the remarks following (17.117).

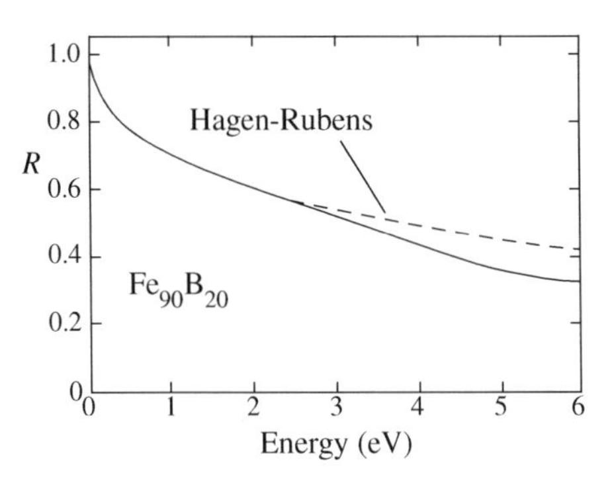

Figure 17.14: Reflectivity data for a metallic alloy. The dashed line fit to the Hagen-Rubens relation (17.125) is very good below 3 eV. Figure from Connell (1990).

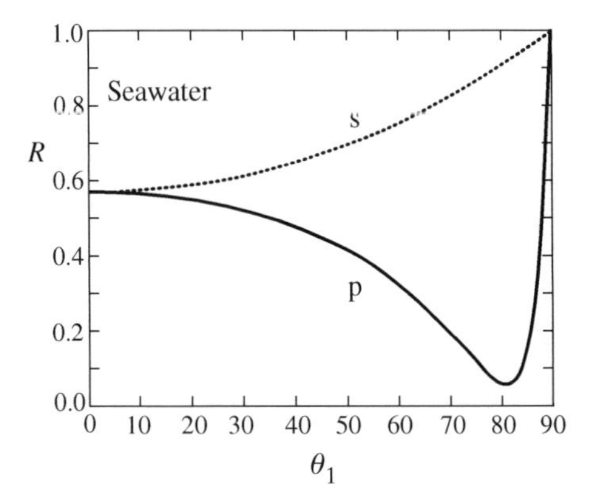

Figure 17.15: Calculated reflectivity of seawater (0°C and 3.5% salinity) at 20 GHz where $\hat{\epsilon}/\epsilon_0 \approx 22 + i32$.

qualitatively from (17.118) and (17.119). The key point is that a $\pi/4$ phase shift exists between the electric and magnetic fields just inside a good conductor. The incident wave field exhibits no such shift. Therefore, a reflected wave must be generated to make up the phase difference (at all angles of incidence) to ensure that the tangential component of the total electric field is continuous at the conductor surface.

Example 17.6 Medium 1 is a transparent dielectric and medium 2 is a good conductor. Show that the value of the p-wave reflection coefficient at the pseudo-Brewster angle (assumed to be a near-grazing angle α) can be written in terms of $\hat{Z}_2$ alone.

Solution: A good conductor has $\cos\theta_2 \approx 1$ and, at grazing incidence, $\cos\theta_1 = \cos(\pi/2 - \alpha) \approx \alpha$. Therefore, (17.123) simplifies to

$$r_{\mathrm{p}} \approx \frac{\alpha Z_1 - \hat{Z}_2}{\alpha Z_1 + \hat{Z}_2}.$$

Since $R_{\mathrm{p}} = r_{\mathrm{p}} r_{\mathrm{p}}^*$, the pseudo-Brewster angle is determined by the extremal condition

$$\frac{\partial R_{\mathrm{p}}}{\partial \alpha} = \frac{\partial}{\partial \alpha}\left[\frac{\alpha Z_1 - \hat{Z}_2}{\alpha Z_1 + \hat{Z}_2} \times \frac{\alpha Z_1 - \hat{Z}_2^*}{\alpha Z_1 + \hat{Z}_2^*}\right] = 0.$$

Carrying out the derivatives, we find

$$\left[\alpha^2 Z_1^2 - |\hat{Z}_2|^2\right](\hat{Z}_2 + \hat{Z}_2^*) = 0 \quad \Rightarrow \quad \alpha = \frac{|\hat{Z}_2|}{Z_1}.$$

Using this value of α to evaluate R_p gives the desired formula,

$$R_p = \frac{|\hat{Z}_2| - \hat{Z}_2}{|\hat{Z}_2| + \hat{Z}_2} \times \frac{|\hat{Z}_2| - \hat{Z}_2^*}{|\hat{Z}_2| + \hat{Z}_2^*} = \frac{|\hat{Z}_2| - \operatorname{Re}\hat{Z}_2}{|\hat{Z}_2| + \operatorname{Re}\hat{Z}_2}.$$

17.7 Anisotropic Matter

We conclude with an introduction to electromagnetic wave propagation in anisotropic matter. This subject is often called *crystal optics* to identify the class of materials where the scalar constitutive relation $\mathbf{D} = \epsilon\mathbf{E}$ is inadequate and a matrix constitutive relation $\mathbf{D} = \boldsymbol{\epsilon} \cdot \mathbf{E}$ is required instead. For the case of plane wave propagation in crystals, we will find that (i) the index of refraction varies with the direction of propagation; (ii) the electric field vector $\mathbf{E}$ is *not* perpendicular to the propagation vector $\mathbf{k}$; and (iii) the Poynting vector $\mathbf{S}$ is *not* parallel to $\mathbf{k}$. These results lead to optical phenomena like birefringence and practical applications like polarizing prisms and retarder wave plates.

For simplicity, we focus on lossless, non-magnetic crystals where the elements ϵ_{ij} of the dielectric matrix are real and the diagonal elements are positive in the principal axis system. Figure 17.16 shows the three possibilities. In a *cubic* crystal, all the diagonal elements are equal and the system responds to an electromagnetic wave exactly like an isotropic material. In a *uniaxial* crystal, one direction is distinguished and the two directions perpendicular to this special *optic axis* are equivalent. The dielectric properties are different in all three principal directions for a *biaxial* crystal.

17.7.1 Fresnel's Equation

Let all the field variables vary as $\exp[i(\mathbf{k}\cdot\mathbf{r} - \omega t)]$ and have real amplitudes (linear polarization). Substituting this dependence into (17.1) and (17.2) with $\rho_f = \mathbf{j}_f = 0$ gives

$$\mathbf{k}\cdot\mathbf{D} = 0 \qquad\qquad \mathbf{k}\cdot\mathbf{B} = 0 \tag{17.126}$$

and

$$\mathbf{k}\times\mathbf{E} = \omega\mathbf{B} \qquad\qquad \mathbf{k}\times\mathbf{H} = -\omega\mathbf{D}. \tag{17.127}$$

These relations show that $\mathbf{k}$, $\mathbf{D}$, and $\mathbf{B}$ are mutually perpendicular. Moreover, because $\mathbf{E}\perp\mathbf{B}$, the three vectors $\mathbf{k}$, $\mathbf{D}$, and $\mathbf{E}$ are coplanar, as shown in Figure 17.17. The fact that $\mathbf{E}\cdot\mathbf{k}\neq 0$ has many consequences. Not least, the Fresnel equations for the reflection and refraction amplitudes at a material interface (Section 17.3.2) are not valid if one or both materials are anisotropic.

It is conventional to define the *wave normal* vector $\mathbf{n}$ from

$$\mathbf{k} = \frac{\omega}{c}\mathbf{n} \tag{17.128}$$

and write

$$c\mathbf{B} = \mathbf{n}\times\mathbf{E} \qquad \text{and} \qquad c\mathbf{D} = -\mathbf{n}\times\mathbf{H}. \tag{17.129}$$

$$\begin{pmatrix} \epsilon & 0 & 0 \\ 0 & \epsilon & 0 \\ 0 & 0 & \epsilon \end{pmatrix} \qquad \begin{pmatrix} \epsilon_\perp & 0 & 0 \\ 0 & \epsilon_\perp & 0 \\ 0 & 0 & \epsilon_\| \end{pmatrix} \qquad \begin{pmatrix} \epsilon_1 & 0 & 0 \\ 0 & \epsilon_2 & 0 \\ 0 & 0 & \epsilon_3 \end{pmatrix}$$

Cubic Uniaxial Biaxial

Figure 17.16: The dielectric matrices for crystals in their principal axis systems.

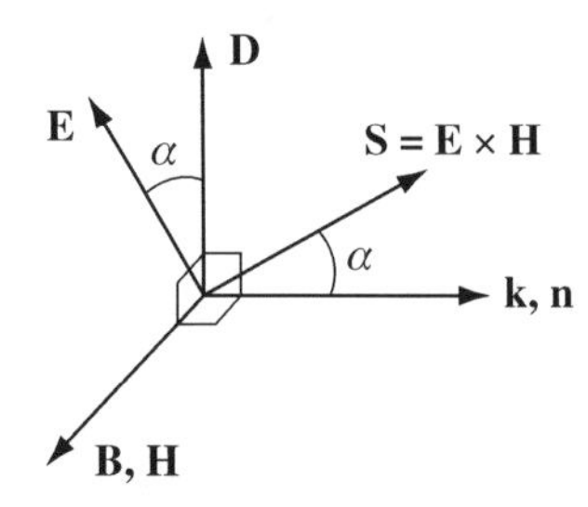

Figure 17.17: The field vectors for an anisotropic material.

Surprisingly, $\mathbf{B} = \mu_0\mathbf{H}$ and the cross product of $\mathbf{E}$ with the left member of (17.127) show that the Poynting vector is not parallel to $\mathbf{n}$. Rather, as shown in Figure 17.17, $\mathbf{S}$ is coplanar with $\mathbf{n}$, $\mathbf{E}$, and $\mathbf{D}$:

$$\mathbf{S} = \mathbf{E} \times \mathbf{H} = \sqrt{\frac{\mu_0}{\epsilon}}\left[E^2\mathbf{n} - (\mathbf{n}\cdot\mathbf{E})\mathbf{E}\right]. \tag{17.130}$$

Unlike waves in vacuum or in simple isotropic matter, the direction of wave energy flow in an anisotropic crystal is not necessarily the same as the direction of wave propagation.

The magnitude $n = |\mathbf{n}|$ of the wave normal is the index of refraction. To find it, substitute $\mathbf{H} = \sqrt{\epsilon_0/\mu_0}\mathbf{n} \times \mathbf{E}$ into the right member of (17.129) and expand the triple cross product. The result,

$$\mathbf{D} = \epsilon_0\left[n^2\mathbf{E} - (\mathbf{n}\cdot\mathbf{E})\mathbf{n}\right] = \boldsymbol{\epsilon}\cdot\mathbf{E}, \tag{17.131}$$

implies that

$$(n^2\delta_{ij} - n_i n_j - \epsilon_{ij}/\epsilon_0)E_j = 0. \tag{17.132}$$

The set of linear equations (17.132) has a solution if the determinant of the coefficients vanishes. This is one form of *Fresnel's equation*:

$$\left|n^2\delta_{ij} - n_i n_j - \epsilon_{ij}/\epsilon_0\right| = 0. \tag{17.133}$$

Using (17.128) to rewrite (17.133) makes it clear that Fresnel's equation determines the dispersion relation, $\mathbf{k}(\omega)$ or $\omega(\mathbf{k})$:

$$\left|k^2\delta_{ij} - k_i k_j - \mu_0\omega^2\epsilon_{ij}\right| = 0. \tag{17.134}$$

In practice, one evaluates (17.133) in the principal axis system where $\boldsymbol{\epsilon}$ is diagonal. For the general biaxial case (see Figure 17.16), the reader can show that the sixth-order terms in this 3×3 determinant cancel and (17.133) reduces to

$$n^2\epsilon_0^2(\epsilon_1 n_1^2 + \epsilon_2 n_2^2 + \epsilon_3 n_3^2) - \epsilon_0\left[n_1^2\epsilon_1(\epsilon_2+\epsilon_3) + n_2^2\epsilon_2(\epsilon_1+\epsilon_3) + n_3^2\epsilon_3(\epsilon_1+\epsilon_2)\right] + \epsilon_1\epsilon_2\epsilon_3 = 0. \tag{17.135}$$

Equations (17.135) and (17.128) show that the value of the index of refraction varies as a function of the wave propagation direction. Indeed, substituting

$$\mathbf{n} = n(\sin\theta\cos\phi\,\hat{\mathbf{e}}_1 + \sin\theta\sin\phi\,\hat{\mathbf{e}}_2 + \cos\theta\,\hat{\mathbf{e}}_3) \tag{17.136}$$

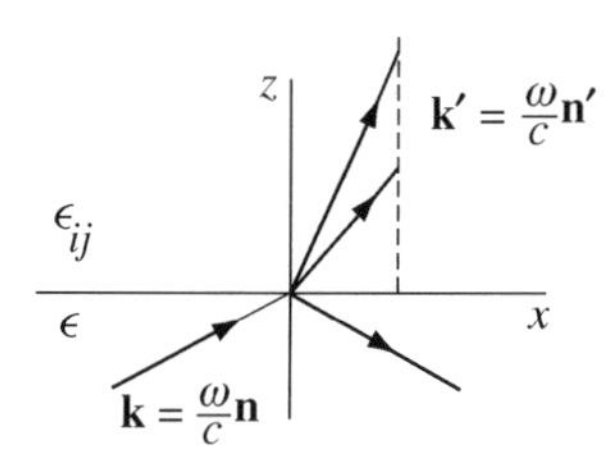

Figure 17.18: A plane wave refracts from an isotropic medium into a birefringent medium.

into (17.135) produces a quadratic equation for n^2 with real coefficients. We conclude that two waves with different indices of refraction can propagate in every direction (θ, ϕ). The two waves have orthogonal states of linear polarization and propagate at different phase speeds. This is the phenomenon of *birefringence*.

Figure 17.18 illustrates how birefringence manifests itself in the laboratory: a single optical ray incident on a transparent anisotropic dielectric appears to split into two rays inside the crystal. This happens because the anisotropy of the crystal does not change the kinematic condition that the component of the wave vector parallel to the interface must be continuous (see Section 17.3.1). From (17.128), $k_x = k'_x$ implies that $n_x = n'_x$ and, for a given incident polarization, this condition can be satisfied by two waves, each with a different perpendicular component n'_z. One ray is *ordinary* in the sense that its angle of refraction obeys Snell's law. The other ray is *extraordinary* and its angle of refraction does not obey this law. On the other hand, the observed direction of each ray follows the path of its energy flow, which is determined by the direction of the Poynting vector (17.130) rather than by the direction of **n** itself.

Application 17.3 Plane Wave Propagation in a Uniaxial Crystal

We specialize here to the uniaxial case illustrated in Figure 17.16 where $\epsilon_1 = \epsilon_2 = \epsilon_\perp$ and $\epsilon_3 = \epsilon_\parallel$. The last of these corresponds to the "optic axis" of the crystal. Bearing in mind that $n^2 = n_1^2 + n_2^2 + n_3^2$, the Fresnel equation (17.135) factors into

$$(\epsilon_0 n^2 - \epsilon_\perp)\left[\epsilon_0\epsilon_\parallel n_3^2 + \epsilon_0\epsilon_\perp(n_1^2 + n_2^2) - \epsilon_\parallel\epsilon_\perp\right] = 0. \tag{17.137}$$

By inspection, the two indices of refraction which satisfy (17.137) are

$$n^2 = \frac{\epsilon_\perp}{\epsilon_0} \qquad \text{and} \qquad \frac{n_3^2}{\epsilon_\perp} + \frac{n_1^2 + n_2^2}{\epsilon_\parallel} = \frac{1}{\epsilon_0}. \tag{17.138}$$

The solution on the left side of (17.138) is the ordinary wave and its phase speed is $c/n = 1/\sqrt{\mu_0\epsilon_\perp}$ for every direction of propagation. The solution on the right side of (17.138) is the extraordinary wave. The parameterization (17.136) gives the angle-dependent phase speed of the extraordinary ray as $c/n(\theta)$, where

$$\frac{\cos^2\theta}{\epsilon_\perp} + \frac{\sin^2\theta}{\epsilon_\parallel} = \frac{1}{n^2\epsilon_0}. \tag{17.139}$$

The two waves propagate at the same speed along the optic axis $(\theta = 0)$. ■

Sources, References, and Additional Reading

The quotation at the beginning of the chapter is taken from the Introduction to

James Clerk Maxwell, "A dynamical theory of the electromagnetic field", *Philosophical Transactions of the Royal Society of London* **155**, 459 (1865).

Section 17.1 Representative treatments of the material discussed in this chapter written for students of physics, optics, and electric engineering are, respectively,

J.A. Stratton, *Electromagnetic Theory* (McGraw-Hill, New York, 1941).

M. Born and E. Wolf, *Principles of Optics*, 6th edition (Cambridge University Press, Cambridge, 1980).

C.A. Balanis, *Advanced Engineering Electromagnetics* (Wiley, New York, 1989).

Section 17.2 Example 17.1 and Section 17.2.2 come, respectively, from

V.D. Barger and M.G. Olsson, *Classical Electricity and Magnetism* (Allyn & Bacon, Newton, MA, 1987).

R.H. Good, Jr. and T.J. Nelson, *Classical Theory of Electric and Magnetic Fields* (Academic, New York, 1971).

Section 17.3 Hertz did it first, but the broad acceptance of Maxwell's theory was hastened by the matching-condition derivation of the Fresnel equations in

Paul Drude, *Optics* (Longmans, Green and Co., London, 1902).

A good treatment of reflection and refraction at the undergraduate level and two useful papers on this subject from the pedagogical literature are

M.H. Nayfeh and M.K. Brussel, *Electricity and Magnetism* (Wiley, New York, 1985).

William T. Doyle, "Scattering approach to Fresnel's equations and Brewster's law", *American Journal of Physics* **53**, 463 (1985).

J.E. Vitela, "Electromagnetic waves in dissipative media revisited", *American Journal of Physics* **72**, 393 (2004).

The first paper on negative refraction and a non-technical review of the subject are

V.G. Veselago, "The electrodynamics of substances with simultaneously negative values of ϵ and μ", *Soviet Physics Uspekhi* **10**, 509 (1968).

J.B. Pendry, "Negative refraction", *Contemporary Physics* **45**, 191 (2004).

Figure 17.6 was adapted from

K. Zhang and D. Li, *Electromagnetic Theory for Microwaves and Optoelectronics* (Springer, Berlin, 1998). With kind permission from Springer Science & Business Media B.V.

Section 17.5 *Born and Wolf* (see Section 17.1 above) is the standard reference for electromagnetic wave propagation in layered media. Our discussion draws heavily on

P. Yeh, *Optical Waves in Layered Media* (Wiley, New York, 1988).

S.G. Lipson, H. Lipson, and D.S. Tannhauser, *Optical Physics*, 3rd edition (Cambridge University Press, Cambridge, 1995).

Airy's calculation of the transmission amplitude for the Fabry-Perot geometry (and much else) is discussed in the delightful

P. Connes, "From Newtonian fits to Wellsian heat rays: The history of multiple-beam interference", *Journal of Optics* (Paris) **17**, 5 (1986).

The contributions of two prominent physicists to the subject of radio wave propagation in layered media are on display in

A. Sommerfeld, *Partial Differential Equations in Physics* (Academic, New York, 1949), Chapter VI.

V.A. Fock, *Electromagnetic Diffraction and Propagation Problems* (Pergamon, Oxford, 1965).

The sources for the TEM and SEM micrographs in Figure 17.12 are

N. Ohnishi, Y. Nonumura, Y. Ogasaka, *et al.* "HRTEM analysis of Pt/C multilayers", *Optics for EUV, X-ray and Gamma-ray Astronomy*, edited by O. Citterio and S.L. O'Dell (SPIE, Bellingham, WA, 2004), pp. 508–517.

D. Sanvitto, A. Daraei, A. Tahraoui, *et al.* "Observation of ultrahigh quality factor in a semiconductor micro-cavity", *Applied Physics Letters* **86**, 191109 (2005). Reprinted with permission. Copyright 2005, American Institute of Physics.

S. Kinoshita and S. Yoshioka, "Structural colors in nature: The role of regularity and irregularity", *ChemPhysChem* **6**, 1442 (2005).

Section 17.6 Figure 17.14 was adapted from

G.A.N. Connell, "Optical properties of amorphous metals using a ratio reflectance method", *Applied Optics* **29**, 4560 (1990).

Example 17.6 comes from *Nayfeh and Brussel* (see Section 17.3 above). The footnote in Section 17.6.4 exclaiming the complexity of metallic reflection comes from *Stratton* (see Section 17.1 above).

Section 17.7 Our discussion follows Chapter XI of

L.D. Landau and E.M. Lifshitz, *Electrodynamics of Continuous Media* (Pergamon, Oxford, 1960).

Problems

17.1 Waves in Matter in the $\varphi = 0$ Gauge

(a) Find a gauge function that makes $\varphi(\mathbf{r}, t) = 0$ a valid choice of gauge.
(b) Derive the (generalized) inhomogeneous wave equation in matter (μ, ϵ) satisfied by the vector potential in the $\varphi = 0$ gauge. There is no free charge or free current anywhere.
(c) Find the dispersion relation for plane-wave solutions for the equation in part (b). Do waves with $\mathbf{k} \perp \mathbf{A}$ differ from waves with $\mathbf{k} \parallel \mathbf{A}$?

17.2 Faraday Rotation During Propagation For propagation along the z-axis, a medium supports left circular polarization with index of refraction n_L and right circular polarization with index of refraction n_R. If a plane wave propagating through this medium has $\mathbf{E}(z = 0, t) = \hat{\mathbf{x}}E\exp(-i\omega t)$, find the values of z where the wave is linearly polarized along the y-axis.

17.3 Optically Active Matter In the absence of free charge or free current, the Maxwell equations in optically active matter are

$$\epsilon_0 \nabla \cdot \mathbf{E} = \rho_{\text{ind}}(\mathbf{r}, t) \qquad \nabla \cdot \mathbf{B} = 0 \qquad \nabla \times \mathbf{E} = -\frac{\partial \mathbf{B}}{\partial t} \qquad \nabla \times \mathbf{B} = \mu_0 \mathbf{j}_{\text{ind}}(\mathbf{r}, t) + \frac{1}{c^2}\frac{\partial \mathbf{E}}{\partial t} = 0,$$

where $\rho_{\text{ind}}(\mathbf{r}, t) = -\nabla \cdot \mathbf{P}$ and $\mathbf{j}_{\text{ind}}(\mathbf{r}, t) = \partial \mathbf{P}/\partial t + \nabla \times \mathbf{M}$.

(a) Let $\mathbf{P} = (\epsilon - \epsilon_0)\mathbf{E}$ and $\mathbf{M} = (\mu_0^{-1} - \mu^{-1})\mathbf{B}$, but do not introduce $\mathbf{D}$ or $\mathbf{H}$. Assume plane wave behavior for all relevant quantities, e.g., $\mathbf{E}(\mathbf{r}, t) = \mathbf{E}(\mathbf{k}, \omega)\exp[i(\mathbf{k} \cdot \mathbf{r} - \omega t)]$. Find $\mathbf{j}_{\text{ind}}(\mathbf{k}, \omega)$ and $\rho_{\text{ind}}(\mathbf{k}, \omega)$. Explain why your expression for $\rho_{\text{ind}}(\mathbf{k}, \omega)$ is the most general scalar that can be constructed from $\mathbf{k}$, $\mathbf{E}$, and $\mathbf{B}$ that is linear in the fields. Explain why your expression for $\mathbf{j}_{\text{ind}}(\mathbf{k}, \omega)$ is *not* the most general vector that can be constructed from the same ingredients.
(b) Let ξ be a real constitutive constant. Explain why a term $\xi\omega\mathbf{B}$ added to $\mathbf{j}_{\text{ind}}$ in part (a) produces a completely general current density vector and show that $\epsilon\omega\mathbf{E} + \mu^{-1}\mathbf{k} \times \mathbf{B} + i\xi\omega\mathbf{B} = 0$.
(c) Show that the propagating waves in the medium of part (b) are determined by an equation of the form

$$\begin{bmatrix} a & -ib \\ ib & a \end{bmatrix} \begin{bmatrix} \hat{\mathbf{k}} \times \mathbf{E} \\ \mathbf{E} \end{bmatrix} = 0.$$

(d) Show that the solutions in part (c) are right and left circularly polarized waves that obey different dispersion relations. Find the range of allowed values of ξ. Matter that behaves this way is called *optically active*.
(e) Show that the constitutive relations $\mathbf{D} = \epsilon\mathbf{E} + \beta\mathbf{B}$ and $\mathbf{B} = \mu\mathbf{H} + \gamma\mathbf{E}$ give an equivalent description of optical activity. How are the two constants β and γ related to ξ?

17.4 Matching Conditions

(a) Explain why the matching conditions for the normal components of $\mathbf{D}$ and $\mathbf{B}$ are not needed to derive the Fresnel equations.

(b) Derive the matching conditions for the components of the Poynting vector at a flat interface between two transparent media.

17.5 Escape from a Dielectric A point source of light is embedded near the flat surface of a dielectric with index of refraction n. Treat the emitted light as a collection of plane waves (light rays) that propagate isotropically away from the source. Find the fraction of the emitted rays which could (in principle) escape from the dielectric into the vacuum space above.

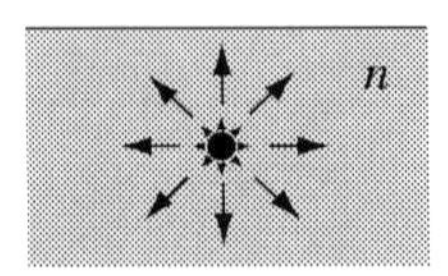

17.6 Almost Total External Reflection A plane wave in vacuum with wave vector $\mathbf{k}_I$ reflects from a non-magnetic sample into a plane wave with wave vector $\mathbf{k}_R$. At x-ray wavelengths, the index of refraction of essentially all matter is very slightly less than 1. Typically, $n \approx 1 - \delta$ with $\delta \sim 10^{-5}$. This makes the phenomenon of total *external* reflection of x-rays possible. Let $\mathbf{q} = \mathbf{k}_R - \mathbf{k}_I$ and show that the Fresnel reflection coefficients for s- and p-polarization, R_s and R_p, both vary as q^{-4} when the angle $\alpha_1 \ll 1$ but not quite small enough for total external reflection to occur.

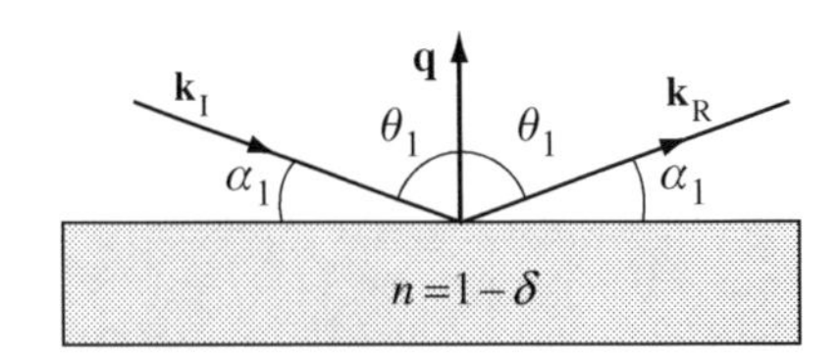

17.7 Alternate Derivation of the Fresnel Equations Let $\mathbf{k}_1 = (k_{1x}, 0, k_{1z})$ be the wave vector of a plane wave incident on the x-y plane which separates simple medium 1 from simple medium 2.

(a) Use (i) the Maxwell matching conditions for each component of the electric field and (ii) the fact that plane waves in a simple medium are transverse to prove that

$$\frac{E_z^R}{E_z^I} = \frac{k_{1z}\epsilon_2 - k_{2z}\epsilon_1}{k_{1z}\epsilon_2 + k_{2z}\epsilon_1} \qquad \frac{E_z^T}{E_z^I} = \frac{2k_{1z}\epsilon_1}{k_{1z}\epsilon_2 + k_{2z}\epsilon_1}.$$

Show that these are the Fresnel equations for p-polarization.

(b) Use a similar method and derive similar equations for H_z^R/H_z^I and H_z^T/H_z^I. Show that these are the Fresnel equations for s-polarization.

17.8 Fresnel Transmission Amplitudes Derive the Fresnel transmission amplitude formulae for non-magnetic matter:

$$\left[\frac{E_T}{E_I}\right]_{\text{TE}} = \frac{2\cos\theta_1 \sin\theta_2}{\sin(\theta_1+\theta_2)} \quad \text{and} \quad \left[\frac{E_T}{E_I}\right]_{\text{TM}} = \frac{2\cos\theta_1 \sin\theta_2}{\sin(\theta_1+\theta_2)\cos(\theta_1-\theta_2)}.$$

17.9 Guidance by Total Internal Reflection An optical fiber consist of a solid rod of material with index of refraction n_f cladded by a cylindrical shell of material with index $n_c < n_f$. Find the largest angle θ so that a wave incident from a medium with index n_a remains in the solid rod by repeated total internal reflection from the cladding layer.

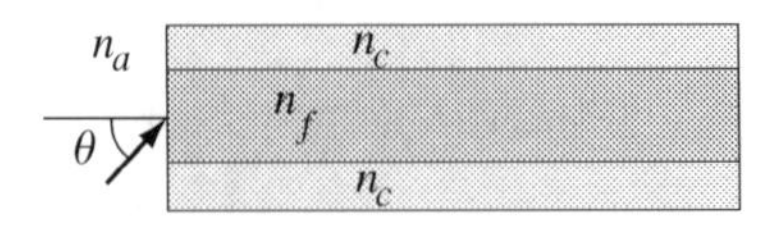

17.10 Reflection from a Metal-Coated Dielectric Slab A plane wave $\mathbf{E} = \hat{\mathbf{y}} A \exp[i(kz + \omega t)]$ propagating in vacuum in the $-z$-direction impinges at normal incidence on the front face of a (transversely infinite) slab with thickness d, index of refraction n, and magnetic permeability μ_0. The back face of the slab ($z = 0$) is coated with a very thin perfect conductor.

(a) Use a four plane-wave trial solution and show that the amplitudes of the incident wave and wave reflected back into the vacuum from the slab are equal. Show also that, when both are evaluated at the front face of the slab, the phase of this reflected wave exceeds that of the incident wave by

$$\alpha = \pi - 2\tan^{-1}\left[\frac{\tan(nkd)}{n}\right].$$

(b) Calculate the pressure exerted on the metal coating due to surface currents induced in it by the wave. Express your answer in terms of the amplitude A of the wave incident on the slab from the vacuum.

17.11 Fresnel's Problem for a Topological Insulator The optical properties of a remarkable class of materials called *topological insulators* (TI) are captured by constitutive relations which involve the fine structure constant, $\alpha = (e^2/\hbar c)/(4\pi\epsilon_0)$. With $\alpha_0 = \alpha\sqrt{\epsilon_0/\mu_0}$, the relations are

$$\mathbf{D} = \epsilon\mathbf{E} - \alpha_0\mathbf{B} \qquad\qquad \mathbf{H} = \frac{\mathbf{B}}{\mu} + \alpha_0\mathbf{E}.$$

(a) Begin with the Maxwell equations in matter with no free charge or current. Show that a monochromatic plane wave of $(\mathbf{E}, \mathbf{B})$ is a solution of these equations for a TI and find the wave speed.

(b) A plane wave with linear polarization impinges at normal incidence on the flat surface of a TI. Show that the transmitted wave remains linearly polarized with its electric field rotated by an angle θ_F. This is called *Faraday rotation* of the plane of polarization.

(c) Show that the reflected wave remains linearly polarized with its electric field rotated by an angle θ_K. This is called *Kerr rotation* of the plane of polarization.

17.12 Polarization Rotation by Reflection and Refraction A plane wave is incident on a flat interface between two transparent, non-magnetic media. Let γ_I be the angle between the incident electric field vector and the plane of incidence. The corresponding angles for the reflected and refracted field vectors are γ_R and γ_T.

(a) Use the Fresnel equations to deduce that

$$\tan\gamma_R = -\frac{\cos(\theta_1 - \theta_2)}{\cos(\theta_1 + \theta_2)}\tan\gamma_I \qquad \text{and} \qquad \tan\gamma_T = \cos(\theta_1 - \theta_2)\tan\gamma_I.$$

(b) Use these formulae to confirm the statement made in the text that, compared to the incident wave, the reflected wave is more TE and the refracted wave is more TM.

17.13 The Fresnel Rhomb A piece of glass in the shape of a rhombic prism can be used to convert linearly polarized light into circularly polarized light and vice versa. The effect is based on the phase change of totally internally reflected light.

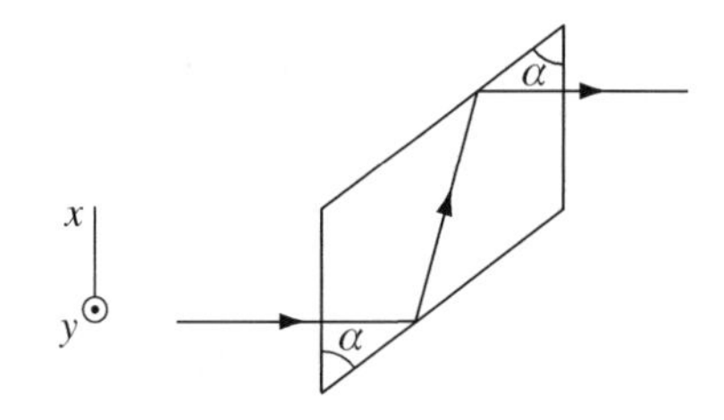

(a) Show that the "s" or $\perp$ Fresnel reflection amplitude for total internal reflection is

$$R_\perp = \left.\frac{E_R}{E_I}\right]_\perp = \frac{\cos\alpha - i\sqrt{\sin^2\alpha - 1/n^2}}{\cos\alpha + i\sqrt{\sin^2\alpha - 1/n^2}} = \exp(-i\delta_\perp).$$

(b) Show that the "p" or $\|$ Fresnel reflection amplitude for total internal reflection is

$$R_\| = \left.\frac{E_R}{E_I}\right]_\| = -\frac{\cos\alpha - in^2\sqrt{\sin^2\alpha - 1/n^2}}{\cos\alpha + in^2\sqrt{\sin^2\alpha - 1/n^2}} = \exp(-i\delta_\|).$$

(c) Let the wave incident on the first interface be $\mathbf{E}_I = E_I(\hat{\mathbf{x}} + \hat{\mathbf{y}})$. Use the foregoing to show that the polarization of the reflected wave is determined by $\delta_\perp - \delta_\|$ and that

$$\tan\left[\frac{1}{2}(\delta_\perp - \delta_\|)\right] = -\frac{\cos\alpha\sqrt{\sin^2\alpha - 1/n^2}}{\sin^2\alpha}.$$

(d) If the index of refraction of non-magnetic glass is $n = 1.5$, show that the rhomb behaves as advertised in air if the rhombic angle is $\Theta = 50.2°$ or $\Theta = 53.3°$.

17.14 Energy Transfer to an Ohmic Medium Show that the time-averaged rate at which power flows through a unit surface area of an ohmic conductor is exactly equal to the time-averaged rate of Joule heating (per unit surface area) in the bulk of the conductor.

17.15 Refraction into a Good Conductor Consider plane wave refraction from a non-conducting medium (ϵ, μ) into a conducting medium (ϵ, μ, σ). Ohmic loss requires that the refracted wave vector $\mathbf{k}_2$ be complex. The figure below shows a proposed refraction geometry where $\mathbf{k}_2 = \mathbf{q} + i\boldsymbol{\kappa}$.

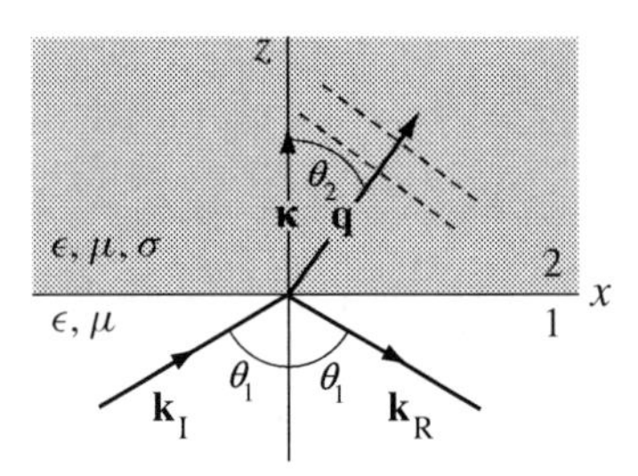

(a) Explain why $\boldsymbol{\kappa}$ points in the $+z$-direction and why the angle of incidence still equals the angle of reflection. Derive the generalization of Snell's law for this problem.
(b) Use the dispersion relation in the conductor to prove that $\theta_2 \le \theta_1$.
(c) Derive a quadratic equation for the variable κ^2. If $\delta(\omega)$ is the skin depth, find $\Lambda(\omega)$ such that

$$\kappa = \delta(\omega)^{-1}\exp\left[-\frac{1}{2}\Lambda(\omega)\right].$$

(d) Show that $\theta_2 \ll 1$ for a good conductor, independent of the angle of incidence. Neglect the displacement current for wave propagation inside a good conductor.

17.16 Phase Change for Waves Reflected from a Good Conductor Consider a TE (s-polarized) plane wave incident at angle θ_1 onto a good conductor with skin depth $\delta(\omega)$ from a transparent dielectric with index of refraction n_1. Both materials are non-magnetic. Show that the phase of the reflected wave with respect to the incident wave is approximately

$$\pi + \tan^{-1}[(\omega/c)n_1\delta(\omega)\cos\theta_1].$$

17.17 Airy's Problem Revisited Airy's problem is the transmission of a monochromatic plane wave through a transparent film (ϵ, μ) of thickness d. The text solved this problem by summing an infinite number of single-interface Fresnel reflections and transmissions. Here, we specialize to normal incidence and use the matching conditions and a five-wave analysis (a forward-going wave in the $z > d$ vacuum, a forward-going wave and a backward-going wave in the $z < 0$ vacuum, and a forward-going wave and a backward-going wave in the film) to solve the same problem.

(a) If Z_0 and Z are the wave impedances in the vacuum and in the film, show that the fraction of the incident power transmitted into the vacuum through the $z = d$ surface of the film is

$$|\mathcal{T}|^2 = \left|\frac{4ZZ_0}{(Z+Z_0)^2 - (Z-Z_0)^2 \exp(2ikd)}\right|^2.$$

(b) Show that the formula in part (a) agrees with Airy's formula (derived in the text) for the same quantity.

17.18 Radiation Pressure on a Perfect Conductor A plane wave with electric field $\mathbf{E}_{\text{inc}}(x,z) = \hat{\mathbf{y}} E_0 \exp[ik(z\sin\theta - x\cos\theta) - i\omega t]$ is incident on a perfect conductor which occupies the half-space $x < 0$. Find the pressure exerted on the conductor by (a) evaluating the Lorentz force on the currents generated on the surface of the conductor and by (b) evaluating the linear momentum change between the incident and reflected waves.

17.19 Phase Velocity of Evanescent Waves An electromagnetic wave with wave vector $\mathbf{k} = \mathbf{q} + i\boldsymbol{\kappa}$ propagates in simple matter with index of refraction n. Prove that the phase velocity of this wave is always less than c/n.

17.20 A Corner Reflector A corner reflector has two semi-infinite, perfect-conductor surfaces joined at a common edge with a right angle between the two surfaces. Prove that a right (left) circularly polarized plane wave incident on the reflector as indicated in the diagram reflects back toward the source as a right (left) circularly polarized plane wave.

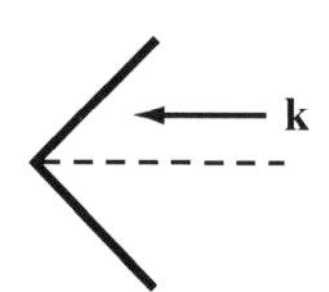

17.21 Bumpy Reflection A vacuum wave $\mathbf{E}_0(\mathbf{r},t) = \hat{\mathbf{x}} E_0 \exp[i(k_y y + k_z z - \omega t)]$ strikes a perfectly conducting surface.

(a) Write down the total electric field $\mathbf{E}$ which exists above the surface when the latter is defined by $z = 0$.
(b) To the field $\mathbf{E}$ found in part (a), add a linear combination of plane waves $\mathbf{E}'$ such that $\mathbf{E} + \mathbf{E}'$ satisfies the boundary conditions at a *corrugated* surface $z = a\sin(2\pi x/d)$ for the case when $\omega a/c \ll 1$.
(c) Show that, if the incident angle is sufficiently close to grazing, $\mathbf{E}'$ decreases exponentially from the surface into the vacuum.

17.22 Photonic Band Gap Material

(a) Derive the generalized wave equation satisfied by $\mathbf{E}(\mathbf{r},t)$ in non-magnetic matter when the permittivity is a function of position, $\epsilon(\mathbf{r})$. Specialize the equation to the case when $\epsilon(\mathbf{r}) = \epsilon(z)$ and $\mathbf{E}(\mathbf{r},t) = \hat{\mathbf{x}} E(z,t)$.
(b) Let $E(z,t) = E(z)\exp(-i\omega t)$ and let $\epsilon(z) = \epsilon_0[1 + \alpha\cos(2k_0 z)]$. Show that the Fourier components $\hat{E}(k)$ of the electric field satisfy the coupled set of linear equations

$$\left(k^2 - \frac{\omega^2}{c^2}\right)\hat{E}(k) = \frac{\omega^2\alpha}{2c^2}\left\{\hat{E}(k-2k_0) + \hat{E}(k+2k_0)\right\}.$$

(c) Suppose $\alpha \ll 1$ and focus on values of k in the immediate vicinity of k_0, i.e., $k = k_0 + q$ where $|q| \ll k_0$. Show that the Fourier components $\hat{E}(q+k_0)$ and $\hat{E}(q-k_0)$ are larger than all others and, therefore, that a 2×2 eigenvalue problem determines the dispersion relation. Hint: The wave frequency cannot differ greatly from its $\alpha = 0$ value in the limit considered.

(d) Solve the eigenvalue problem to find $\omega(k_0, q)$. Study its behavior analytically at $q = 0$. Sketch the complete dispersion curve and show that there is a range of frequencies—called a photonic band gap—where no waves occur.

17.23 Plane Wave Amplifier A plane electromagnetic wave $\mathbf{E}_I \cos(k_I z + \omega_I t)$ is incident on a perfectly reflecting mirror (solid line) that moves with constant velocity $\mathbf{v} = v\hat{\mathbf{z}}$. The reflected plane wave is $\mathbf{E}_R \cos(k_R z - \omega_R t)$.

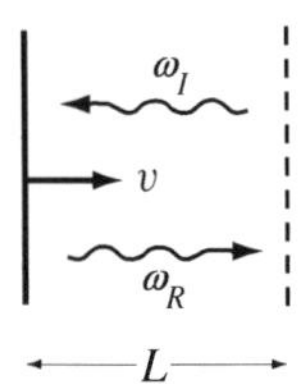

(a) Use conservation of momentum to show that the force exerted on an area A of the mirror is

$$\mathbf{F} = -A\left\{\frac{S_I + S_R}{c} + \frac{v}{c^2}(S_I - S_R)\right\}\hat{\mathbf{z}},$$

where S is the magnitude of the Poynting vector. Hint: When $v \neq 0$, the electromagnetic momentum in the volume between the mirror and any fixed reference plane (dashed line) changes in time.

(b) Use conservation of energy to show that

$$\mathbf{F} \cdot \boldsymbol{v} = A(S_I - S_R) + A(S_I + S_R)v/c.$$

(c) From (a) and (b), deduce that the reflected energy/time P_R that flows through a fixed reference plane (dashed line) exceeds the incident energy/time P_I that flows through the same plane by the ratio

$$\frac{P_R}{P_I} = \left(\frac{1 + v/c}{1 - v/c}\right)^2.$$

(d) Relate the phase of the incident and reflected waves at the surface of the mirror ($z = vt$). Use this relation to deduce that $P_R/P_I = (\omega_R/\omega_I)^2$.

17.24 Laser Beam Bent by a Magnetic Field An external magnetic field $\mathbf{B}_0$ can cause the straight-line path of a laser beam to deflect inside a *non-simple* material where the constitutive relations are $\mathbf{B} = \mu\mathbf{H}$ and $\mathbf{D} = \epsilon\mathbf{E} - i\gamma\mathbf{B} \times \mathbf{E}$. To see this, let a linearly polarized, monochromatic plane wave with electric field strength $\mathcal{E}$ enter the material at normal incidence. Assume that $\mathbf{B}_0$ lies in a plane perpendicular to the propagation vector $\mathbf{k}$ and makes an angle φ with $\mathbf{D}$.

(a) Use explicit numerical estimates to show that the typical value of $\mathbf{B}_0$ produced by a laboratory electromagnet is much larger than the magnetic field associated with, e.g., a continuous-wave argon-ion laser beam.

(b) To lowest order in $\mathbf{B}_0$, show that the time-averaged Poynting vector inside the material is

$$\langle\mathbf{S}\rangle \approx \frac{1}{2}\frac{k\mathcal{E}^2}{\mu\omega}\left\{\hat{\mathbf{k}} - \frac{\text{Im}\gamma}{\epsilon}\sin\varphi[\cos\varphi\mathbf{B}_0 - \sin\varphi\hat{\mathbf{k}} \times \mathbf{B}_0]\right\}.$$

(c) What angle of φ produces maximum deflection?

17.25 An Anisotropic Magnetic Crystal Let the half-space $z \geq 0$ be filled with a magnetic crystal where $\mathbf{H} = \boldsymbol{\mu}^{-1} \cdot \mathbf{B}$. The inverse permeability matrix $\boldsymbol{\mu}^{-1}$ is (rows and columns labeled by x, y, z)

$$\boldsymbol{\mu}^{-1} = \frac{1}{\mu_0} \begin{bmatrix} m & 0 & 0 \\ 0 & m & m' \\ 0 & m' & m \end{bmatrix}.$$

Assume that $\epsilon = \epsilon_0$ inside the crystal and that the real, dimensionless matrix elements satisfy $m > m' > 0$.

(a) Show that $\omega(k, \theta) = ck\sqrt{m - m' \sin 2\theta}$ for a wave $\mathbf{E} = \hat{\mathbf{x}}\, E_0 \exp[i(\mathbf{k} \cdot \mathbf{r} - \omega t)]$ inside the crystal when $\mathbf{k} = k(\hat{\mathbf{y}} \sin\theta + \hat{\mathbf{z}} \cos\theta)$.

(b) A plane wave $\mathbf{E} = \hat{\mathbf{x}}\, E_0 \exp[i(kz - \omega t)]$ is incident on this crystal from the vacuum ($z < 0$). Write out an explicit formula for the electric field on both sides of the refraction interface.

17.26 A Complex Dielectric Matrix The region $y < 0$ is vacuum. The region $y > 0$ is filled with material where $\mu = \mu_0$ and $D_{ij} = \epsilon_{ij} E_j$. Let α, β, and γ be real numbers and take the dielectric matrix as

$$\boldsymbol{\epsilon} = \epsilon_0 \begin{bmatrix} \alpha & i\beta & 0 \\ -i\beta & \alpha & 0 \\ 0 & 0 & \gamma \end{bmatrix}.$$

(a) Write out the electric field everywhere if a wave incident from the vacuum is $\mathbf{E} = E_0 \hat{\mathbf{x}} \exp[i\omega(y/c - t)]$.

(b) Repeat part (a) if the incident field is $\mathbf{E} = E_0 \dfrac{\hat{\mathbf{x}} + \hat{\mathbf{z}}}{\sqrt{2}} \exp[i\omega(y/c - t)]$.

18 Waves in Dispersive Matter

If we accept the electromagnetic theory of light, there is nothing left but to look for the cause of dispersion in the molecules of the medium itself.
Hendrik Lorentz (1878)

18.1 Introduction

The colored bands of a rainbow are well separated in space (dispersed) because water droplets in the atmosphere refract light with different wavelengths through different angles. Snell's law predicts this behavior because the index of refraction of water is a function of frequency. The simple conducting matter studied in Section 17.6.1 had a frequency-dependent index of refraction also. In this chapter, we argue that *all* real matter has this property of *frequency dispersion* and we discuss both its origins and consequences. Among the latter, we show that a deep connection exists between frequency dispersion and the dissipation of energy in matter. We also show that no electromagnetic information can be communicated faster than the speed of light. Otherwise, we follow tradition and use simple classical models to develop archetypes of frequency dispersion. This is perfectly adequate for a classical thermal plasma, but it is manifestly inadequate for quantum mechanical condensed matter systems. Nevertheless, with suitable caution there is much to learn from these models, even when applied to solids, liquids, and gases.

18.2 Frequency Dispersion

The frequency dispersion of the index of refraction (and other constitutive parameters) occurs because matter cannot respond instantaneously to an external perturbation. This is not a new idea. We encountered it in Section 14.13, when the inevitable time delay between voltage stimulus and current response in AC circuit theory led us to define a complex, frequency-dependent impedance, $\hat{Z}(\omega)$, as the generalization of DC resistance. For convenience, we repeat the key steps in the argument here using a time-dependent conductivity function $\sigma(\tau)$. The time-delay aspect appears when we insist that the response of the matter be *causal*. This means that the current density $\mathbf{j}(\mathbf{r}, t)$ may depend on the electric field at any time earlier than t but it cannot depend on the electric field at any time later than t. If, in addition, we confine ourselves to the *linear response* of the matter to the electric field,

$$\mathbf{j}(\mathbf{r}, t) = \int_{-\infty}^{t} dt' \sigma(t - t')\mathbf{E}(\mathbf{r}, t'). \tag{18.1}$$

In practice, $\sigma(\tau) \to 0$ as $\tau \to \infty$ because the electric field in the distant past has a negligible effect on the present-time current density. It is most convenient to extend the upper limit of integration in

(18.1) to infinity and build causality directly into the conductivity. In other words, we constrain the conductivity so

$$\sigma(\tau < 0) = 0, \tag{18.2}$$

and write

$$\mathbf{j}(\mathbf{r}, t) = \int_{-\infty}^{\infty} dt' \sigma(t - t')\mathbf{E}(\mathbf{r}, t'). \tag{18.3}$$

The right side of (18.1) is a convolution integral. This motivates us to define the Fourier transform pairs

$$\hat{\sigma}(\omega) = \int_{-\infty}^{\infty} dt\, \sigma(t)e^{i\omega t} \qquad \text{and} \qquad \sigma(t) = \frac{1}{2\pi}\int_{-\infty}^{\infty} d\omega\, \hat{\sigma}(\omega)e^{-i\omega t} \tag{18.4}$$

and write Fourier representations for the current density and electric field:

$$\mathbf{j}(\mathbf{r}, t) = \frac{1}{2\pi}\int_{-\infty}^{\infty} d\omega \hat{\mathbf{j}}(\mathbf{r}, \omega)e^{-i\omega t} \qquad \text{and} \qquad \mathbf{E}(\mathbf{r}, t) = \frac{1}{2\pi}\int_{-\infty}^{\infty} d\omega\, \hat{\mathbf{E}}(\mathbf{r}, \omega)e^{-i\omega t}. \tag{18.5}$$

These definitions permit us to apply the convolution theorem (Section 1.6.2) to the Fourier transform of (18.3) to conclude that

$$\hat{\mathbf{j}}(\mathbf{r}, \omega) = \hat{\sigma}(\omega)\hat{\mathbf{E}}(\mathbf{r}, \omega). \tag{18.6}$$

Like $\hat{Z}(\omega)$, the conductivity $\hat{\sigma}(\omega)$ is a complex, linear response function.

An important property of $\hat{\sigma}(\omega) = \sigma'(\omega) + i\sigma''(\omega)$ follows from the fact that $\mathbf{j}(\mathbf{r}, t)$ and $\mathbf{E}(\mathbf{r}, t)$ in (18.3) are both real. This implies that $\sigma(t)$ in the right member of (18.4) is also real. Therefore, equating the latter equation to its complex conjugate and changing integration variables from ω to $-\omega$ shows that

$$\hat{\sigma}(-\omega) = \hat{\sigma}^*(\omega) \qquad \Rightarrow \qquad \sigma'(\omega) = \sigma'(-\omega) \;\; \text{and} \;\; \sigma''(\omega) = -\sigma''(-\omega). \tag{18.7}$$

Generalizing this result to any equation similar to (18.3) leads us to conclude that the real (imaginary) part of any causal response function is an even (odd) function of frequency.

18.2.1 The Equivalence of Alternative Descriptions

The frequency-dependent electric and magnetic susceptibility, dielectric permittivity, and magnetic permeability are all defined like (18.6):

$$\hat{\mathbf{P}}(\mathbf{r}, \omega) = \epsilon_0\hat{\chi}(\omega)\hat{\mathbf{E}}(\mathbf{r}, \omega) \qquad \text{and} \qquad \hat{\mathbf{M}}(\mathbf{r}, \omega) = \hat{\chi}_m(\omega)\hat{\mathbf{H}}(\mathbf{r}, \omega) \tag{18.8}$$

and

$$\hat{\mathbf{D}}(\mathbf{r}, \omega) = \hat{\epsilon}(\omega)\hat{\mathbf{E}}(\mathbf{r}, \omega) \qquad \text{and} \qquad \hat{\mathbf{B}}(\mathbf{r}, \omega) = \hat{\mu}(\omega)\hat{\mathbf{H}}(\mathbf{r}, \omega). \tag{18.9}$$

The response functions in (18.8) and (18.9) are not all independent. For static fields, it is easy to distinguish the long-distance displacement of "free" charge from the short-distance displacement of "polarization" charge. This distinction becomes blurred when $\mathbf{E}(\mathbf{r}, t)$ is time-harmonic because charge oscillates back and forth in both cases. At high frequency, particularly, there is simply no way to distinguish a time-harmonic conduction current with density $\mathbf{j} = \sigma\mathbf{E}$ from a time-harmonic polarization current with density $\mathbf{j} = \partial\mathbf{P}/\partial t$. For that reason, we use the left equation in (18.8) to equate the conduction current density (18.6) to the corresponding Fourier component of the polarization current density:

$$\hat{\mathbf{j}}(\mathbf{r},\omega) = -i\omega\hat{\mathbf{P}}(\mathbf{r},\omega) = -i\omega\epsilon_0\hat{\chi}(\omega)\hat{\mathbf{E}}(\mathbf{r},\omega). \tag{18.10}$$

This relates $\hat{\sigma}(\omega)$ to $\hat{\chi}(\omega)$. Then, because

$$\hat{\mathbf{D}}(\mathbf{r},\omega) = \epsilon_0\hat{\mathbf{E}}(\mathbf{r},\omega) + \hat{\mathbf{P}}(\mathbf{r},\omega) = \hat{\epsilon}(\omega)\hat{\mathbf{E}}(\mathbf{r},\omega), \tag{18.11}$$

we discover that[1]

$$\hat{\epsilon}(\omega) = \epsilon_0 + i\frac{\hat{\sigma}(\omega)}{\omega}. \tag{18.12}$$

It is a matter of taste whether one uses a dielectric function description or a conductivity description for time-dependent electromagnetic problems in matter.

The distinction between polarization current density and magnetization current density (for linear magnets) is similarly ambiguous. This follows from the constitutive relations, the identities $\epsilon = \epsilon_0(1+\chi)$ and $\mu = \mu_0(1+\chi_m)$, and the current density formula,

$$\mathbf{j} = \frac{\partial \mathbf{P}}{\partial t} + \nabla \times \mathbf{M}. \tag{18.13}$$

Using $\nabla \times \mathbf{B} = \mu_0\mathbf{j} + c^{-2}\partial\mathbf{E}/\partial t$, it is straightforward to confirm that (18.13) can be written as

$$\mathbf{j} = \frac{\chi + \chi_m + \chi\chi_m}{\chi}\frac{\partial \mathbf{P}}{\partial t}, \tag{18.14}$$

or as

$$\mathbf{j} = \frac{\chi + \chi_m + \chi\chi_m}{\chi_m(1+\chi)}\nabla \times \mathbf{M}. \tag{18.15}$$

We conclude that time-dependent currents in matter may be described equivalently as polarization effects, magnetization effects, or any combination of the two. For condensed matter systems below optical frequencies, it is most natural to extrapolate from statics and use a dielectric permittivity $\hat{\epsilon}(\omega)$ and magnetic permeability $\hat{\mu}(\omega)$ as in (18.9). In condensed systems at optical frequencies and above, and particularly in plasma physics, it is common to set $\mathbf{M} = 0$ and build all electric *and* magnetic effects into an effective dielectric function or an effective conductivity. The ultimate choice of response functions is entirely a matter of convenience.

Example 18.1 Use the dielectric response of a medium to a free charge density $\hat{\rho}_f(\mathbf{r})\exp(-i\omega t)$ to argue that the inverse dielectric function, $\hat{\epsilon}^{-1}(\omega)$, is a causal linear response function.

Solution: Free charge produces an electric field $\hat{\mathbf{E}}_f(\mathbf{r},\omega)\exp(-i\omega t)$, where

$$\epsilon_0\nabla\cdot\hat{\mathbf{E}}_f(\mathbf{r},\omega) = \hat{\rho}_f(\mathbf{r},\omega).$$

On the other hand, dielectric theory defines a field, $\hat{\mathbf{D}}(\mathbf{r},\omega)\exp(-i\omega t)$, where

$$\nabla\cdot\hat{\mathbf{D}}(\mathbf{r},\omega) = \hat{\rho}_f(\mathbf{r},\omega).$$

Combining these with the left member of (18.9) shows that

$$\nabla\cdot\hat{\mathbf{E}}(\mathbf{r},\omega) = \hat{\epsilon}^{-1}(\omega)\nabla\cdot\hat{\mathbf{E}}_f(\mathbf{r},\omega),$$

[1] Note that (18.12) differs slightly from (17.105).

or

$$\hat{\rho}(\mathbf{r}, \omega) = \hat{\epsilon}^{-1}(\omega)\hat{\rho}_f(\mathbf{r}, \omega).$$

This equation has the same structure as (18.6), which was derived from (18.3) using the convolution theorem. Therefore, using the theorem in the opposite direction implies that an inverse dielectric function $\epsilon^{-1}(\tau)$ with the property (18.2) is also a causal linear response function. In words, the inverse dielectric function specifies how a system creates its total (free + polarization) charge by responding causally to its free charge.

18.3 Energy in Dispersive Matter

Frequency dispersion has a profound effect on the ability of a medium to store and dissipate electromagnetic energy. The key to this conclusion is the Poynting theorem in matter, which we reproduce from Section 15.8.1 as

$$\int_V d^3r \left[\mathbf{E} \cdot \frac{\partial \mathbf{D}}{\partial t} + \mathbf{H} \cdot \frac{\partial \mathbf{B}}{\partial t} \right] = - \int_V d^3r \, \mathbf{j}_{\text{ext}} \cdot \mathbf{E} - \int_V d^3r \, \nabla \cdot (\mathbf{E} \times \mathbf{H}). \tag{18.16}$$

The free current $\mathbf{j}_f$ does not appear in (18.16) if we use $\hat{\epsilon}(\omega)$ and $\hat{\mu}(\omega)$ to take account of all induced electric and magnetic effects (see Section 18.2.1). Our goal is to interpret the Poynting theorem as a balance equation for power by writing the integrand on the left side of (18.16) as the time rate of change of an electromagnetic energy density plus a term which represents the rate at which energy is lost in a unit volume of the medium by dissipative processes:

$$\mathbf{E} \cdot \frac{\partial \mathbf{D}}{\partial t} + \mathbf{H} \cdot \frac{\partial \mathbf{B}}{\partial t} = \frac{\partial u_{\text{EM}}(t)}{\partial t} + Q(t). \tag{18.17}$$

The results of the last chapter (Section 17.2.2) for simple dielectric and conducting matter suggest that $u_{\text{EM}}(t)$ will depend on the real parts of $\hat{\epsilon}(\omega)$ and $\hat{\mu}(\omega)$ and that $Q(t)$ will depend on their imaginary parts.

We begin with the first term on the left side of (18.16) and write a general, real, electric field as

$$\mathbf{E}(\mathbf{r}, t) = \frac{1}{4\pi} \int_{-\infty}^{\infty} d\omega \left[\hat{\mathbf{E}}(\mathbf{r}, \omega) \exp(-i\omega t) + \hat{\mathbf{E}}^*(\mathbf{r}, \omega) \exp(i\omega t) \right]. \tag{18.18}$$

Omitting the explicit dependence on $\mathbf{r}$, (18.18) and the left member of (18.9) combine to give

$$\mathbf{D}(t) = \frac{1}{4\pi} \int_{-\infty}^{\infty} d\omega \left[\hat{\epsilon}(\omega)\hat{\mathbf{E}}(\omega) \exp(-i\omega t) + \hat{\epsilon}^*(\omega)\hat{\mathbf{E}}^*(\omega) \exp(i\omega t) \right]. \tag{18.19}$$

Now, use $\hat{\epsilon}^*(\omega) = \hat{\epsilon}(-\omega)$, and change integration variables from ω to $-\omega$ in one term each in (18.18) and (18.19) to get[2]

$$\mathbf{E}(t) = \frac{1}{4\pi} \int_{-\infty}^{\infty} d\omega \left[\hat{\mathbf{E}}(\omega) + \hat{\mathbf{E}}^*(-\omega) \right] \exp(-i\omega t) \tag{18.20}$$

[2] See the left member of (18.7).

and

$$\mathbf{D}(t) = \frac{1}{4\pi}\int_{-\infty}^{\infty} d\omega\, \left[\hat{\mathbf{E}}(-\omega) + \hat{\mathbf{E}}^*(\omega)\right] \hat{\epsilon}^*(\omega)\exp(i\omega t). \tag{18.21}$$

Consequently,

$$\mathbf{E}\cdot\frac{\partial \mathbf{D}(t)}{\partial t} = \frac{1}{16\pi^2}\int_{-\infty}^{\infty} d\omega_1 \int_{-\infty}^{\infty} d\omega_2\, \left[\hat{\mathbf{E}}(\omega_1) + \hat{\mathbf{E}}^*(-\omega_1)\right]\cdot\left[\hat{\mathbf{E}}(-\omega_2) + \hat{\mathbf{E}}^*(\omega_2)\right] \times i\omega_2\hat{\epsilon}^*(\omega_2)\exp[-i(\omega_1-\omega_2)t]. \tag{18.22}$$

Next, change integration variables in (18.22) from ω_1 to $-\omega_2$ in one integral and from ω_2 to $-\omega_1$ in the other integral. Adding half the resulting integral to half the original integral gives

$$\mathbf{E}\cdot\frac{\partial \mathbf{D}(t)}{\partial t} = \frac{1}{32\pi^2}\int_{-\infty}^{\infty} d\omega_1 \int_{-\infty}^{\infty} d\omega_2\, \left[\hat{\mathbf{E}}(\omega_1) + \hat{\mathbf{E}}^*(-\omega_1)\right]\cdot\left[\hat{\mathbf{E}}(-\omega_2) + \hat{\mathbf{E}}^*(\omega_2)\right] \times i[\omega_2\hat{\epsilon}^*(\omega_2) - \omega_1\hat{\epsilon}(\omega_1)]\exp[-i(\omega_1-\omega_2)t]. \tag{18.23}$$

Finally, use $\epsilon''(-\omega) = -\epsilon''(\omega)$ and write out (18.23) using

$$\omega_2\hat{\epsilon}^*(\omega_2) - \omega_1\hat{\epsilon}(\omega_1) = \omega_2\epsilon'(\omega_2) - \omega_1\epsilon'(\omega_1) - i[\omega_2\epsilon''(\omega_2) + \omega_1\epsilon''(\omega_1)]. \tag{18.24}$$

The result has the anticipated form,

$$\mathbf{E}\cdot\frac{\partial \mathbf{D}}{\partial t} = \frac{\partial u_E}{\partial t} + Q_E, \tag{18.25}$$

where

$$u_E(t) = \frac{1}{32\pi^2}\int_{-\infty}^{\infty} d\omega_1 \int_{-\infty}^{\infty} d\omega_2\, \left[\hat{\mathbf{E}}(\omega_1) + \hat{\mathbf{E}}^*(-\omega_1)\right]\cdot\left[\hat{\mathbf{E}}(-\omega_2) + \hat{\mathbf{E}}^*(\omega_2)\right] \times \frac{\omega_2\epsilon'(\omega_2) - \omega_1\epsilon'(\omega_1)}{\omega_2-\omega_1}\exp[-i(\omega_1-\omega_2)t] \tag{18.26}$$

and

$$Q_E(t) = \mathbf{E}(t)\cdot\frac{1}{4\pi}\int_{-\infty}^{\infty} d\omega\, [\hat{\mathbf{E}}(\omega) + \hat{\mathbf{E}}^*(-\omega)]\omega\epsilon''(\omega)\exp(-i\omega t). \tag{18.27}$$

We analyze the second term on the left side of (18.16) similarly using $\hat{\mathbf{B}}(\omega) = \hat{\mu}(\omega)\hat{\mathbf{H}}(\omega)$. This gives

$$\mathbf{H}\cdot\frac{\partial \mathbf{B}}{\partial t} = \frac{\partial u_H}{\partial t} + Q_H, \tag{18.28}$$

where $u_H(t)$ and $Q_H(t)$ are identical to (18.26) and (18.27) with $\epsilon \to \mu$ and $\mathbf{E} \to \mathbf{H}$. This confirms (18.17) with $u_{\rm EM}(t) = u_E(t) + u_H(t)$ and $Q(t) = Q_E(t) + Q_H(t)$. On the other hand, the formulae of interest are rather complicated for fields with arbitrary time dependence. This motivates the discussion of the next section.

18.3.1 Quasi-Monochromatic Fields

The Fourier amplitude $\hat{\mathbf{E}}(\omega)$ in (18.18) encodes the entire time history of $\mathbf{E}(t)$. A similar formula with $\hat{\mathbf{B}}(\omega)$ does the same for $\mathbf{B}(t)$. Since (18.26), (18.27), and their magnetic analogs depend on these amplitudes as well, the total field energy and Joule heating rate depend on the entire time-history of the fields. Luckily, useful formulae can be derived for the physically important case of a *quasi-monochromatic* field, where both Fourier amplitudes are strongly peaked around a single frequency.[3] For such fields, the lowest-order Taylor expansions of (18.26) and (18.27) give the total energy transiently stored in the medium as

$$u_{\rm EM}(t) = \frac{1}{2}\left\{\frac{\partial}{\partial\omega}[\omega\epsilon'(\omega)]\,|\mathbf{E}(t)|^2 + \frac{\partial}{\partial\omega}[\omega\mu'(\omega)]\,|\mathbf{H}(t)|^2\right\} \tag{18.29}$$

and the rate of energy absorption as

$$Q(t) = \omega\left[\epsilon''(\omega)|\mathbf{E}(t)|^2 + \mu''(\omega)|\mathbf{H}(t)|^2\right]. \tag{18.30}$$

These expressions are valid when $\hat{\epsilon}(\omega)$ and $\hat{\mu}(\omega)$ change slowly enough to justify the neglect of their first and second derivatives in (18.30) and (18.29), respectively.

When the constitutive functions are approximated as frequency-independent constants, (18.29) reduces to the simple matter formula derived in Section 17.2.2),

$$u_{\rm EM}(t) = \frac{1}{2}\left\{\epsilon\,|\mathbf{E}(t)|^2 + \mu\,|\mathbf{H}(t)|^2\right\}. \tag{18.31}$$

Equation (18.30) is consistent with the dissipation rate (17.101) for simple conducting matter because $\mu'' = 0$ and $\epsilon'' = \sigma/\omega$ [see (17.105)]. Moreover, because the rate of heat production by electric or magnetic processes is strictly positive in lossy matter,[4] (18.30) implies that

$$\epsilon''(\omega) > 0 \qquad \text{and} \qquad \mu''(\omega) > 0. \tag{18.32}$$

There is no comparable restriction on the behavior of $\epsilon'(\omega)$ or $\mu'(\omega)$. Finally, (18.30) tells us that energy dissipation occurs at every frequency where the imaginary parts $\epsilon''(\omega)$ and $\mu''(\omega)$ are non-zero. This statement has broad significance because, as we will show in Section 18.7, if the function $\epsilon'(\omega)$ is not strictly constant, $\epsilon''(\omega)$ is guaranteed to be non-zero everywhere (albeit small at most frequencies). The same is true for the real and imaginary parts of $\hat{\mu}(\omega)$. In other words, the mere existence of frequency dispersion is sufficient to ensure that energy dissipation occurs.

18.4 Transverse and Longitudinal Waves

Plane electromagnetic waves propagate readily in dispersive matter. We study them using $\hat{\epsilon}(\omega)$ and $\hat{\mu}(\omega)$ to treat the causal response, but otherwise mimicking our treatment of plane waves in simple matter (Section 17.2.1). Therefore, we choose $\exp(i\mathbf{k}\cdot\mathbf{r})$ as the space dependence for the Fourier amplitudes in (18.9) and use the Maxwell equations in matter,

$$\nabla\times\mathbf{E} = -\frac{\partial\mathbf{B}}{\partial t} \qquad \text{and} \qquad \nabla\times\mathbf{H} = \frac{\partial\mathbf{D}}{\partial t}, \tag{18.33}$$

to deduce that

$$\mathbf{k}\times\mathbf{E} = \omega\hat{\mu}(\omega)\mathbf{H} \qquad \text{and} \qquad \mathbf{k}\times\mathbf{H} = -\omega\hat{\epsilon}(\omega)\mathbf{E}. \tag{18.34}$$

Combining the two equations in (18.34) produces the dispersion relation,

$$\mathbf{k}\times(\mathbf{k}\times\mathbf{E}) = \mathbf{k}(\mathbf{k}\cdot\mathbf{E}) - (\mathbf{k}\cdot\mathbf{k})\mathbf{E} = -\omega^2\hat{\mu}(\omega)\hat{\epsilon}(\omega)\mathbf{E}. \tag{18.35}$$

[3] This is a wave packet of the sort studied in Section 16.5 if the space dependence is a sum of plane waves. Section 18.6 analyzes the propagation properties of wave packets in a dispersive medium.

[4] A laser medium is exceptional because it *supplies* energy to the fields.

In simple matter, we set $\mathbf{k}\cdot\mathbf{E}=0$ in a formula like (18.35) to comply with the Maxwell equation $\nabla\cdot\mathbf{E}=0$ for a charge-neutral system. We proceed more cautiously now and decompose $\mathbf{E}$ into a longitudinal component $\mathbf{E}_\parallel$ parallel to $\mathbf{k}$ and a transverse component $\mathbf{E}_\perp$ perpendicular to $\mathbf{k}$:

$$\mathbf{E}=\mathbf{E}_\parallel+\mathbf{E}_\perp. \tag{18.36}$$

Because $\mathbf{k}\times\mathbf{E}_\parallel=0$ and $\mathbf{k}\cdot\mathbf{E}_\perp=0$, substituting (18.36) into (18.35) gives

$$(\mathbf{k}\cdot\mathbf{k})\mathbf{E}_\perp=\omega^2\hat{\mu}(\omega)\hat{\epsilon}(\omega)(\mathbf{E}_\parallel+\mathbf{E}_\perp). \tag{18.37}$$

Equation (18.37) defines two conditions. One is the dispersion relation for transverse waves:

$$\left[\omega^2\hat{\mu}(\omega)\hat{\epsilon}(\omega)-\mathbf{k}\cdot\mathbf{k}\right]\mathbf{E}_\perp=0. \tag{18.38}$$

Using (18.38) and (18.34), we define the complex, frequency-dependent index of refraction, $\hat{n}(\omega)$, from

$$k(\omega)=\omega\sqrt{\hat{\mu}(\omega)\hat{\epsilon}(\omega)}=\frac{\omega}{c}\hat{n}(\omega), \tag{18.39}$$

and generalize the wave impedance formula (17.110) similarly. The latter appears when we combine the left member of (18.34) with (18.39) to get

$$\hat{Z}(\omega)\mathbf{H}_\perp=\sqrt{\frac{\hat{\mu}(\omega)}{\hat{\epsilon}(\omega)}}\mathbf{H}_\perp=\hat{\mathbf{k}}\times\mathbf{E}_\perp. \tag{18.40}$$

The second condition derived from (18.37) is a dispersion relation for longitudinal waves,

$$\hat{\mu}(\omega)\hat{\epsilon}(\omega)\mathbf{E}_\parallel=0. \tag{18.41}$$

We will see in the next section that even relatively simple models of dispersive matter lead to transverse waves with a vastly richer range of behavior than we encountered for waves in simple matter. The longitudinal waves we will find are unavoidably accompanied by macroscopic accumulations of electric charge. The latter never occurs in simple matter because $\hat{\mu}(\omega)\hat{\epsilon}(\omega)=\mu\epsilon$ never vanishes in (18.41).

18.5 Classical Models for Frequency Dispersion

The epigraph to this chapter shows that physicists of the late 19th century were forced to invent models for matter to harmonize Maxwell's theory of electromagnetism with observations of frequency dispersion. Classical physics was the only tool available and it was wielded with considerable skill by Lorentz, Drude, and others of their generation to account for the optical properties of metals and insulators. Later, the same methodology was applied (with greater justification) to study electromagnetic phenomenon in plasmas in astrophysical, geophysical, and laboratory settings. Frequency dispersion arises in every case from the presence of one or more characteristic frequencies that appear naturally in the description of the matter.

We begin with classical models for matter where a propagating electromagnetic wave exerts a dominantly electric Coulomb force on the electrons and ions. The small velocities acquired by these particles ensures that the magnetic force is important only for plasmas and condensed matter systems where the magnetic field of the wave is supplemented by a strong *external* magnetic field. We treat this case because it arises very commonly, and also because it provides an introduction to the vast subject of the electrodynamics of magnetized plasmas. We assume $\hat{\mu}(\omega)=\mu_0$ except in Section 18.5.6 where we discuss an artificial material with a frequency-dependent magnetic permeability where $\mu(\omega)<0$.

18.5.1 The Drude Model of Conducting Matter

Section 9.3 introduced the classical Drude model for the steady current produced by a static electric field in a neutral system composed of free (mobile) charges. The calculated conductivity, $\sigma_0 = nq^2\tau/m$, applies to n particles per unit volume when each has charge q, mass m, and travels a mean time τ before suffering a momentum-degrading collision. A time-harmonic electric field produces a time-harmonic current and thus a complex conductivity $\hat{\sigma}(\omega)$. To find it, we generalize our previous analysis and let $\mathbf{E} \to \hat{\mathbf{E}}\exp(-i\omega t)$ in Newton's equation of motion for the velocity $\mathbf{v}$ of a typical particle:

$$m\frac{d\mathbf{v}}{dt} = q\hat{\mathbf{E}}e^{-i\omega t} - \frac{m\mathbf{v}}{\tau}. \tag{18.42}$$

Ignoring transients, a time-harmonic solution for $\mathbf{v}(t)$ is

$$\mathbf{v}(t) = \frac{q\hat{\mathbf{E}}/m}{1/\tau - i\omega}e^{-i\omega t} = \hat{\mathbf{v}}(\omega)e^{-i\omega t}. \tag{18.43}$$

The amplitude of the time-harmonic current density that develops in the system is then

$$\hat{\mathbf{j}}(\omega) = nq\hat{\mathbf{v}}(\omega) = \frac{nq^2\tau}{m}\frac{\hat{\mathbf{E}}}{1 - i\omega\tau} = \hat{\sigma}(\omega)\hat{\mathbf{E}}(\omega). \tag{18.44}$$

Equation (18.44) defines the complex, frequency-dependent, Drude conductivity as

$$\hat{\sigma}(\omega) = \frac{nq^2\tau/m}{1 - i\omega\tau} = \frac{\sigma_0}{1 - i\omega\tau}. \tag{18.45}$$

Quantum mechanical calculations give the same form for $\hat{\sigma}(\omega)$ for simple metals.

It is convenient to introduce the characteristic *plasma frequency*,

$$\omega_\mathrm{p}^2 = \frac{nq^2}{\epsilon_0 m}, \tag{18.46}$$

and substitute (18.45) into (18.12) to derive the Drude dielectric function,

$$\frac{\hat{\epsilon}(\omega)}{\epsilon_0} = \left[1 - \frac{\omega_\mathrm{p}^2\tau^2}{1 + \omega^2\tau^2}\right] + i\left[\frac{\omega_\mathrm{p}^2\tau}{\omega}\frac{1}{1 + \omega^2\tau^2}\right]. \tag{18.47}$$

We assume that the scattering time and plasma frequency obey $1/\tau \ll \omega_\mathrm{p}$, which is characteristic of metals. The low-frequency limit of (18.47) is

$$\frac{\hat{\epsilon}(\omega)}{\epsilon_0} \approx -\omega_\mathrm{p}^2\tau^2 + i\frac{\omega_\mathrm{p}^2\tau}{\omega} \qquad \omega\tau \ll 1. \tag{18.48}$$

This is the complex dielectric function studied under the name "simple conducting matter" in Section 17.6. There, we found that the reflectivity had the near-unity Hagen-Rubens form (17.125). Transverse plane waves transmitted into the medium had amplitudes that decayed exponentially as they propagated because the fields lost energy by ohmic heating. The skin depth $\delta(\omega) = \sqrt{2/\mu\sigma_0\omega}$ was the characteristic length scale for the decay.

Here, we focus on the high-frequency limit where (18.47) is purely real:

$$\frac{\epsilon(\omega)}{\epsilon_0} \approx 1 - \frac{\omega_\mathrm{p}^2}{\omega^2} \qquad \omega\tau \gg 1. \tag{18.49}$$

The dielectric function (18.49) is particularly important in what follows because it applies both to a simple metal at high frequency and to a cold, collisionless ($\tau \to \infty$) classical plasma at *all* frequencies. Figure 18.1 emphasizes that (18.49) changes sign when ω passes through ω_p.

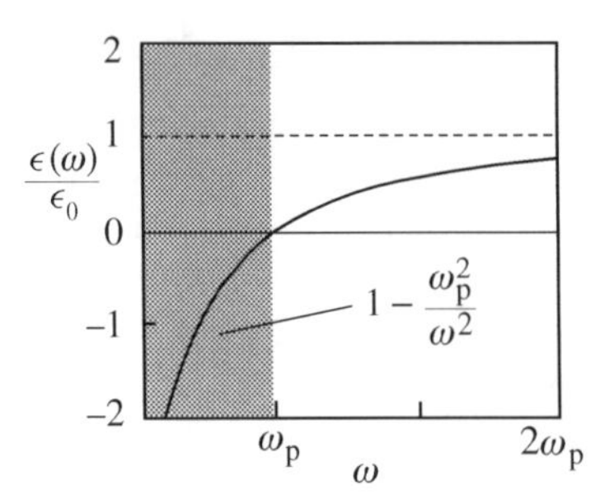

Figure 18.1: High-frequency dielectric function of the Drude model. There is no propagation in the shaded region where $\epsilon(\omega) < 0$.

18.5.2 Transverse Waves

In the high-frequency limit when (18.49) is valid, the dispersion relation (18.39) for a transverse wave ($\mathbf{E} \perp \mathbf{k}$) reads

$$\omega^2 = \omega_p^2 + c^2k^2 \qquad \omega\tau \gg 1. \tag{18.50}$$

Equation (18.50) tells us how to interpret the zero-crossing of (18.49) at $\omega = \omega_p$. When $\omega > \omega_p$, (18.50) predicts a propagating wave with a real wave vector $\mathbf{k}$. When $\omega < \omega_p$, (18.50) predicts an evanescent wave with an imaginary $\mathbf{k}$.[5] The fact that no propagation occurs when $\epsilon < 0$ (shaded region in Figure 18.1) has immediate consequences for the reflectivity of a wave that strikes a Drude medium from the vacuum. As ω increases through ω_p, the dielectric function (18.49) changes sign from negative to positive, the index of refraction in (18.39) changes from $\hat{n} = i$ to $\hat{n} = 1$, and the normal incidence reflectivity (17.44),

$$R(\omega) = \left|\frac{\hat{n}-1}{\hat{n}+1}\right|^2 = \frac{(n'-1)^2 + n''^2}{(n'+1)^2 + n''^2}, \tag{18.51}$$

drops abruptly from one to zero. The comparison between theory and experiment in Figure 18.2 shows that this abruptness is only slightly smoothed out when the full dielectric function (18.47) is used to evaluate $R(\omega)$. We speak of $\omega > \omega_p$ as the "transparency regime" of the Drude model because the fully transmitted wave is undamped [see (17.111)].

The physical origin of the behavior seen in Figure 18.2 lies with the ultimate source of the reflected and transmitted waves: the oscillating polarization of the medium. When $\omega < \omega_p$, the medium radiates two waves: a backward-propagating (reflected) wave and a forward-propagating wave which interferes destructively with the incident wave.[6] When $\omega > \omega_p$, the oscillations of the polarization suffer a change of phase and, in place of the reflected wave, the medium creates a forward-propagating (transmitted) wave. This phase change is reminiscent of the behavior of a harmonic oscillator when it is driven below and then above its natural frequency. It remains only to show that ω_p is a natural oscillation frequency of a Drude medium. This is the subject of the next section.

18.5.3 Longitudinal Waves

The dispersion relation for longitudinal waves is (18.41). A Drude medium does not support such waves at low frequency because (18.48) never vanishes. However, because the high-frequency dielectric

[5] Compare this to the behavior of the transmitted wave when passing from above to below the critical angle for total internal reflection (Section 17.3.6).

[6] This is an example of the extinction theorem. See Section 20.9.1.

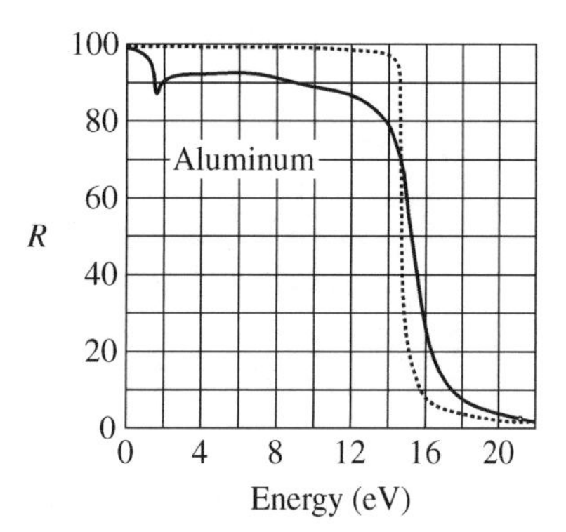

Figure 18.2: Experiment (solid curve) and Drude theory (dashed line) for the normal incidence reflectivity of Al metal. Figure from Wooten (1972).

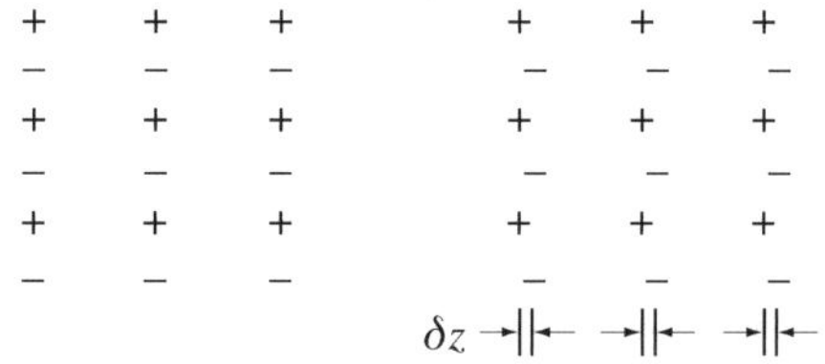

δz

Figure 18.3: A cartoon to help rationalize the phenomenon of plasma oscillations.

function (18.49) vanishes at $\omega = \omega_p$, the dispersion relation implies that a Drude medium supports $\mathbf{k} \parallel \mathbf{E}$ waves whose frequency $\omega = \omega_p$ does not depend on the wave vector $\mathbf{k}$. Two characteristics of these waves are noteworthy. First, $\mathbf{B} = 0$, which follows from the equation on the left side of (18.34) and the fact that $\mathbf{k} \times \mathbf{E}_\parallel = 0$. Second, Gauss' law reads $\mathbf{k} \cdot \mathbf{E}_\parallel = \rho/\epsilon_0$, which means that the local charge density is not zero when the wave is present. Both observations point to a phenomenon that is fundamentally electrostatic in nature. How should we understand this?

The left panel of Figure 18.3 shows a small portion of neutral matter composed of immobile positive particles and mobile negative particles. Both species have number density n in equilibrium. The right panel shows a displacement of the entire negative charge population by an amount δz. Each layer of displaced negative charge forms a parallel-plate capacitor with the layer of positive charge exposed by its motion. The electric field between the plates, $\mathbf{E} = \hat{\mathbf{z}} en\delta z/\epsilon_0$, produces a restoring force on the negative charge. Each negative layer has mass/area $M = mn\delta z$ and charge/area $Q = -en\delta z$. Therefore, the equation of motion for the layer displacement is $M\delta\ddot{z} = QE$ or

$$\delta\ddot{z} = -\frac{e^2 n}{m\epsilon_0}\delta z = -\omega_p^2 \delta z. \tag{18.52}$$

In the absence of damping, $\delta z(t)$ oscillates indefinitely at the plasma frequency (18.46). This "plasma oscillation" is a normal mode of the system. For a metal, $\omega_p \sim 10^{16}\ \text{s}^{-1}$ lies in the ultraviolet portion of the electromagnetic spectrum. For a low-density plasma in the ionosphere, $\omega_p \sim 10^6\ \text{s}^{-1}$ is a radio-frequency oscillation.

In the approximation used here, each set of "capacitor plates" oscillates at frequency ω_p independent of the other plates. This means that initial conditions may be prescribed where the $t = 0$ displacement of the negative particle plate differs from capacitor to capacitor in a wave-like manner. This construction explains why (18.41) describes waves with all possible wave vectors $\mathbf{k}$, each with the same frequency

ω_p. It also shows that these are standing waves incapable of transporting energy. More complete treatments show that truly propagating electrostatic waves occur both in real metals and in warm plasmas.[7]

Application 18.1 Charge Relaxation in an Ohmic Medium

This Application revisits the issue (Section 14.7.1) of how rapidly an ohmic medium rids itself of charge placed in its interior. The ingredients are the same as before except that we now let all quantities vary in time as $\exp(-i\omega t)$. This leads us to Fourier transform the continuity equation,

$$\nabla\cdot\mathbf{j}(\mathbf{r},t) = -\frac{\partial\rho(\mathbf{r},t)}{\partial t} \quad\Rightarrow\quad \nabla\cdot\hat{\mathbf{j}}(\mathbf{r},\omega) = i\omega\hat{\rho}(\mathbf{r},\omega), \tag{18.53}$$

and Gauss' law,

$$\epsilon_0\nabla\cdot\mathbf{E}(\mathbf{r},t) = \rho(\mathbf{r},t) \quad\Rightarrow\quad \epsilon_0\nabla\cdot\hat{\mathbf{E}}(\mathbf{r},\omega) = \hat{\rho}(\mathbf{r},\omega). \tag{18.54}$$

Ohm's law in the form (18.6) connects the right sides of (18.53) and (18.54) to give

$$[\hat{\sigma}(\omega) - i\epsilon_0\omega]\,\hat{\rho}(\mathbf{r},\omega) = 0. \tag{18.55}$$

Substituting the Drude conductivity from (18.45) into (18.55) generates a quadratic equation,

$$\omega^2 + i\omega/\tau - \omega_p^2 = 0, \tag{18.56}$$

which has solutions

$$\omega_\pm = i/2\tau \pm \sqrt{\omega_p^2 - 1/4\tau^2}. \tag{18.57}$$

The physics of (18.57) emerges when we examine the limits of a "good" conductor when σ_0 is large ($\omega_p\tau \gg 1$) and a "poor" conductor when σ_0 is small ($\omega_p\tau \ll 1$),[8] namely,

$$\omega_\pm \simeq \begin{cases} -i/2\tau \pm \omega_p & \omega_p\tau \gg 1, \\ -i/\tau, \quad -i\omega_p^2\tau & \omega_p\tau \ll 1. \end{cases} \tag{18.58}$$

For a good conductor, the upper solution in (18.58) tells us that an initial charge distribution $\rho(\mathbf{r},0)$ oscillates and damps to zero according to

$$\rho(\mathbf{r},t) = \rho(\mathbf{r},0)\exp(-t/2\tau)\cos\omega_p t. \tag{18.59}$$

The collision time τ is about 10^{-14} s for most metals. Therefore, we are justified in setting $\rho = \epsilon_0\nabla\cdot\mathbf{E} = 0$ for the analysis of electromagnetic phenomena in metals up to approximately microwave frequencies. This is the fundamental origin of our definition of a perfect conductor as a body where $\rho(\mathbf{r},t) = 0$. Of course, the charge density $\rho(\mathbf{r},0)$ does not simply disappear. It winds up on the surface of the metal by transport mechanisms that we do not discuss here.

For a poor conductor, the lower solution in (18.58) tell us that the time variation of $\rho(\mathbf{r},t)$ is the sum of two damped exponentials with different time constants. One of these is nearly the same as for a good conductor. However, because $\omega_p\tau \ll 1$, the long-time behavior is dictated by the other term:

$$\rho(\mathbf{r},t) \sim \rho(\mathbf{r},0)\exp(-\sigma_0 t/\epsilon_0). \tag{18.60}$$

This formula is the same as the quasistatic result (14.79). A typical value for $\tau_E = \epsilon_0/\sigma_0$ is about 10^{-6} s for a poor conductor such as a lightly doped semiconductor. ■

[7] See Boyd and Sanderson (2003) in Sources, References, and Additional Reading.

[8] This distinction between a "good" and a "poor" conductor is exactly the same as the one used in Section 14.7 on quasistatics.

18.5.4 The Lorentz Model for Dielectric Matter

Quantum mechanics lies at the foundation of the linear (and non-linear) response of atoms, molecules, liquids, and solids to an external electromagnetic field. This makes it all the more remarkable that a simple classical model proposed by Lorentz captures the *structure* of the complex dielectric function $\hat{\epsilon}(\omega)$ for these systems very well. The reason for this happy accident will emerge in Example 18.4 of Section 18.7. Until then, the reader should suspend disbelief and imagine a system of n independent "atoms", each composed of one electron bound to an infinitely massive nucleus by a spring with natural frequency ω_0. Dissipative processes are modeled as a drag force parameterized by a phenomenological constant Γ.

The displacement of each electron from equilibrium is assumed to satisfy the classical equation of motion

$$m\frac{d^2\mathbf{r}}{dt^2} + m\Gamma\frac{d\mathbf{r}}{dt} + m\omega_0^2\mathbf{r} = -e\mathbf{E}. \tag{18.61}$$

In steady state, both $\mathbf{r}(t)$ and $\mathbf{E}(t)$ vary as $\exp(-i\omega t)$ and the solution to (18.61) is

$$\mathbf{r}(t) = \frac{-e/m}{\omega_0^2 - \omega^2 - i\omega\Gamma}\mathbf{E}(t). \tag{18.62}$$

Lorentz identified the polarization $\mathbf{P}$ with the electric dipole moment per unit volume (see Section 6.2.2):

$$\mathbf{P}(t) = n\mathbf{p}(t) = -ne\mathbf{r}(t) = \frac{ne^2/m}{\omega_0^2 - \omega^2 - i\omega\Gamma}\mathbf{E}(t). \tag{18.63}$$

From (18.11) and the definition of the plasma frequency in (18.46), we conclude that the complex index of refraction and dielectric function for this model satisfy

$$\hat{n}^2(\omega) = \frac{\hat{\epsilon}(\omega)}{\epsilon_0} = 1 + \frac{\omega_p^2}{\omega_0^2 - \omega^2 - i\omega\Gamma}. \tag{18.64}$$

Despite our use of a classical model, there are at least three reasons to take (18.64) seriously. First, quantum mechanical perturbation theory produces a dielectric function with exactly the same form, albeit with different meanings for the adjustable parameters. Second, the resonance structure of (18.64) emerges naturally from the causality requirement built into every linear response function (see Section 18.7). Finally, experiments sensitive to the real and imaginary parts of $\hat{\epsilon}(\omega)$ look very much like those predicted by (18.64). This may be seen from Figure 18.4, which compares measured values of $\epsilon'(\omega)$ and $\epsilon''(\omega)$ for a silicon crystal with

$$\frac{\epsilon'(\omega)}{\epsilon_0} = 1 + \frac{\omega_p^2(\omega_0^2 - \omega^2)}{(\omega_0^2 - \omega^2)^2 + \omega^2\Gamma^2} \qquad \frac{\epsilon''(\omega)}{\epsilon_0} = \frac{\omega_p^2\omega\Gamma}{(\omega_0^2 - \omega^2)^2 + \omega^2\Gamma^2}. \tag{18.65}$$

Below resonance ($\omega < \omega_0$), the induced polarization oscillates in-phase with the driving field and there is an extended region where $n'(\omega) \approx \sqrt{\epsilon'(\omega)/\epsilon_0}$ increases as the frequency increases. This is called *normal dispersion* because it is consistent with the fact that a prism refracts blue light more strongly than red light. There follows a band of frequencies around ω_0 with width Γ where $n'(\omega)$ decreases sharply as the frequency increases. This region of *anomalous dispersion* is not easy to observe because, according to (18.30), the medium strongly absorbs energy from the wave in exactly the same range of frequencies. Above resonance ($\omega > \omega_0$), the induced polarization oscillates out-of-phase with the driving field and there is an extended region of normal dispersion which persists to higher frequency.

The vertical dashed lines in Figure 18.4 divide the spectral domain shown into three regions, each with characteristic behavior for the reflectivity $R(\omega)$. Using (18.51), the reader can confirm that the reflectivity is quite low in the leftmost region but rises steadily through the middle region

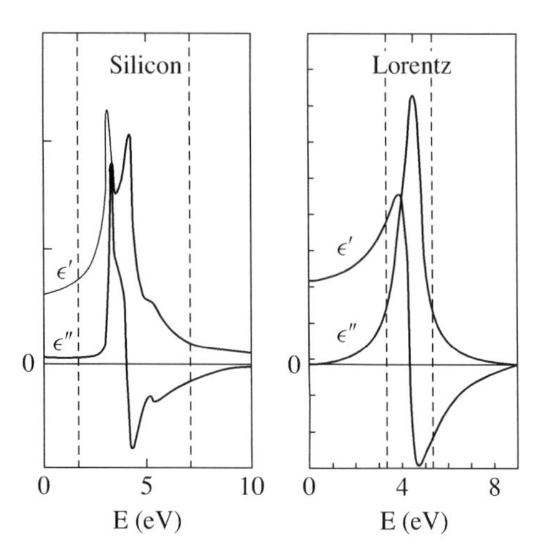

Figure 18.4: Real and imaginary parts of $\hat{\epsilon}(\omega)/\epsilon_0$ for silicon. Left panel: experiment. Right panel: Lorentz model. Vertical dashed lines are discussed in the text. Figure adapted from Wooten (1972).

where absorption occurs. The sample reflects like a metal in the rightmost region, before becoming transparent at the higher frequencies not shown in Figure 18.4.

The dielectric function for most condensed systems may be modeled by computing the total polarization as the sum of Lorentz oscillator contributions like (18.63), one from each of the various electronic and vibrational excitations of the system.[9] Figure 18.5 shows this for the case of silica glass. Metals are not excluded because we may make the Drude choice $\omega_0 = 0$ for one of the oscillators. An interesting prediction of the multiple oscillator model is that the index of refraction is slightly less than one at x-ray frequencies. This follows from (18.65) when ω far exceeds the largest excitation energy in the system, so $\epsilon''(\omega)$ is negligible and

$$n^2(\omega) \approx \frac{\epsilon'(\omega)}{\epsilon_0} \approx 1 - \frac{\omega_p^2}{\omega^2} < 1. \tag{18.66}$$

Confirmation of (18.66) comes from the observation that grazing-incidence X-rays suffer *total external reflection* from the surface of essentially all materials (cf. Section 17.3.6).

18.5.5 The Appleton Model of a Magnetized Plasma

In the middle 1920s, physicists studied the effect of the geomagnetic field on radio waves by modeling the ionosphere as a neutral collection of electrons and ions moving in a uniform magnetic field $\mathbf{B}_0 = B_0\hat{\mathbf{z}}$.[10] This medium supports transverse electromagnetic waves even if we ignore all the effects of temperature, collisions, and the motion of the relatively much heavier positive ions. Here, we derive an approximate dielectric function for this cold, magnetized plasma from the time-harmonic polarization which develops when each of n electrons per unit volume moves under the combined action of the wave's electric field and the external field $\mathbf{B}_0$. This is the Drude model of Section 18.5.1 with no collisions and a magnetic field added. The ith electron of the plasma is thus presumed to obey

[9] The low-frequency response of molecular systems is dominated by the rotation of dipoles, which plays a large role in the subject of *dielectric relaxation*.

[10] The British scientist E.V. Appleton is most closely associated with the model presented here. See, however, Sources, References, and Additional Reading.

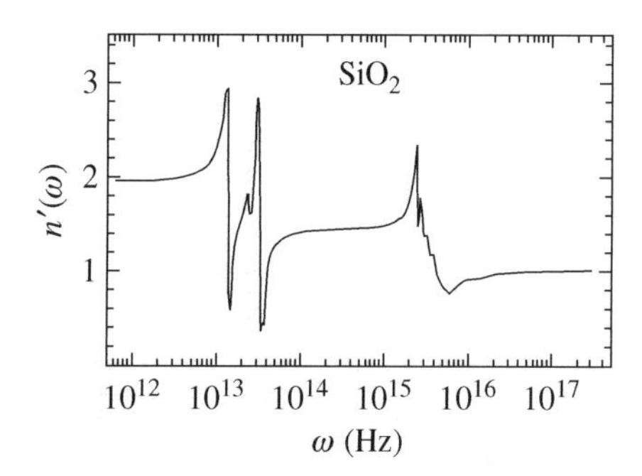

Figure 18.5: Real part of the index of refraction for SiO_2 glass. The low-frequency structure reflects vibrational excitations in the infrared. The high-frequency structure reflects electronic excitations in the ultraviolet. Figure from Fox (2001).

the classical equation of motion,

$$m\frac{d^2\mathbf{r}_i}{dt^2} = -e\left(\mathbf{E} + \frac{d\mathbf{r}_i}{dt}\times\mathbf{B}_0\right). \tag{18.67}$$

The corresponding equation of motion for the polarization $\mathbf{P} = -(e/V)\sum_i \mathbf{r}_i$ is

$$\frac{d^2\mathbf{P}}{dt^2} = \epsilon_0\omega_\text{p}^2\mathbf{E} - \omega_\text{c}\frac{d\mathbf{P}}{dt}\times\hat{\mathbf{z}}, \tag{18.68}$$

where $\omega_\text{c} = eB_0/m$ is the electron cyclotron frequency (12.12) and $\omega_p^2 = ne^2/\epsilon_0 m$ is the electron plasma frequency (18.46).

We seek a steady-state solution of (18.68) where $\mathbf{P}(t) = \mathbf{P}e^{-i\omega t}$ and $\mathbf{E}(t) = \mathbf{E}e^{-i\omega t}$. In that case, the Cartesian components of (18.68) are

$$\begin{aligned}-\omega^2 P_x &= \epsilon_0\omega_\text{p}^2 E_x + i\omega\omega_\text{c}P_y\\ -\omega^2 P_y &= \epsilon_0\omega_\text{p}^2 E_y - i\omega\omega_\text{c}P_x\\ -\omega^2 P_z &= \epsilon_0\omega_\text{p}^2 E_z.\end{aligned} \tag{18.69}$$

We use (18.69) to express each component of $\mathbf{P}$ in terms of the components of $\mathbf{E}$. Then, because $\mathbf{D} = \epsilon_0\mathbf{E} + \mathbf{P}$, we find that the scalar dielectric function in $\mathbf{D} = \epsilon\mathbf{E}$ is insufficient. Instead, we write

$$\mathbf{D} = \hat{\epsilon}\cdot\mathbf{E}, \tag{18.70}$$

where

$$\hat{\epsilon}(\omega) = \epsilon_0\begin{bmatrix} 1 - \dfrac{\omega_\text{p}^2}{\omega^2-\omega_\text{c}^2} & i\dfrac{\omega_\text{p}^2\omega_\text{c}}{\omega(\omega^2-\omega_\text{c}^2)} & 0\\ -i\dfrac{\omega_\text{p}^2\omega_\text{c}}{(\omega^2-\omega_\text{c}^2)} & 1-\dfrac{\omega_\text{p}^2}{\omega^2-\omega_\text{c}^2} & 0\\ 0 & 0 & 1-\dfrac{\omega_\text{p}^2}{\omega^2}\end{bmatrix}. \tag{18.71}$$

The matrix structure of this dielectric function identifies a magnetized plasma as an anisotropic medium of the sort discussed in Section 17.7.

We find the dispersion relation for waves in a medium where the constitutive relation is (18.70) by letting $\mathbf{E}$ and $\mathbf{B}$ vary as $\exp[i(\mathbf{k}\cdot\mathbf{r}-\omega t)]$ and combining the two Maxwell curl equations in the (by now) familiar way. The final result generalizes (18.35) to

$$(\mathbf{k}\cdot\mathbf{k})\mathbf{E} - \mathbf{k}(\mathbf{k}\cdot\mathbf{E}) = \frac{\omega^2}{c^2}\frac{\hat{\epsilon}(\omega)}{\epsilon_0}\cdot\mathbf{E}. \tag{18.72}$$

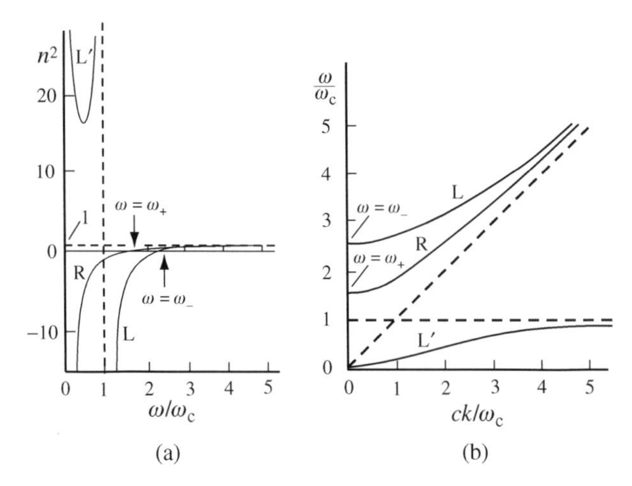

Figure 18.6: Two representations of the frequency dispersion of transverse electromagnetic waves in a magnetized plasma with $\mathbf{k} \parallel \mathbf{B}_0$ and $\omega_p = 2\omega_c$: (a) effective index of refraction vs. frequency; (b) frequency vs. wave vector. Solid lines labeled R (L) describe waves with right (left) circular polarization. Figure adapted from Sturrock (1994).

The general problem posed by (18.72) is quite complex. Here, we focus exclusively on the solutions of (18.72) where $\mathbf{k} \parallel \mathbf{B}_0$. The block diagonal structure of (18.71) gives the simplest of these as a time-harmonic electric field $\mathbf{E} = E_z\hat{\mathbf{z}}$ with $\omega = \omega_p$. This reproduces the longitudinal electrostatic wave discussed in Section 18.5.3 because the electric field drives the charged particles parallel to $\mathbf{B}_0$, which is a trajectory where the Lorentz magnetic force has no effect. More interesting are transverse waves where $\mathbf{E}$ and $\mathbf{B}$ are both perpendicular to $\mathbf{k} \parallel \mathbf{B}_0$. With $n = ck/\omega$, the system of equations to be solved is

$$\begin{pmatrix} -n^2 + 1 - \dfrac{\omega_p^2}{\omega^2 - \omega_c^2} & i\dfrac{\omega_p^2\omega_c}{\omega(\omega^2 - \omega_c^2)} \\ -i\dfrac{\omega_p^2\omega_c}{\omega(\omega^2 - \omega_c^2)} & -n^2 + 1 - \dfrac{\omega_p^2}{\omega^2 - \omega_c^2} \end{pmatrix} \begin{pmatrix} E_x \\ E_y \end{pmatrix} = 0. \tag{18.73}$$

Non-trivial solutions to (18.73) occur when the determinant of the matrix is zero. A brief calculation confirms that this is so when n^2 takes the values

$$n_\pm^2 = 1 - \frac{\omega_p^2}{\omega(\omega \pm \omega_c)}. \tag{18.74}$$

Substituting (18.74) back into (18.73) shows that the upper sign choice gives $E_x = iE_y$ (right circular polarization) and the lower sign choice gives $E_x = -iE_y$ (left circular polarization). The fact that $n_+ \neq n_-$ in (18.74) means that the phase velocities of the RCP and LCP waves are not the same. Figure 18.6 displays the dispersion information in (18.74) in two ways for $\omega_p = 2\omega_c$. Figure 18.6(a) uses R to label the curve of $n_+^2(\omega)$ and L and L′ to label the two branches of $n_-^2(\omega)$. Our experience with the Drude dielectric function (Figure 18.1) tells us to expect wave *reflection* by the medium when $n_\pm^2$ goes to zero (from above) at the *cutoff* frequencies $\omega = \omega_\pm$. We will see in a moment that wave *absorption* occurs when n_+^2 diverges to positive infinity at the *resonance* frequency $\omega = \omega_c$.

Figure 18.6(b) uses the same labeling for the branches of $\omega(k)$ with right and left circular polarization, respectively. The frequencies $\omega_\pm$ are defined by $n_+^2(\omega_+) = 0$ and $n_-^2(\omega_-) = 0$. Explicitly,

$$\frac{\omega_\pm}{\omega_c} = \frac{\mp 1 + \sqrt{1 + 4\omega_p^2/\omega_c^2}}{2}. \tag{18.75}$$

Figure 18.6(b) shows a band of frequencies between ω_c and ω_+ where no wave-like solutions exist. This "gap" in the spectrum disappears when $\omega_p < \omega_c$. The R and L modes in Figure 18.6(b) are magnetically split versions of the transverse waves (18.50) which propagate above the plasma frequency in the Drude model. The location of the cutoff frequencies $\omega_\pm$ in Figure 18.6(a) is important because undamped, propagating waves occur only when $n^2 > 0$. A circularly polarized wave incident on a magnetized plasma from the vacuum with a frequency $\omega < \omega_+$ (for LCP) or $\omega < \omega_-$ (for RCP) reflects from the plasma because a mode with $n^2(\omega) < 0$ has a purely imaginary wave vector. Such a "wave" is non-propagating and exponentially damped, exactly like the waves in Section 18.5.1 when $\omega < \omega_p$.

The L′ wave at the bottom of Figure 18.6(b) is something new. The electric field vector of this *electron cyclotron wave* rotates in the same direction and at the same speed as the cyclotron motion of the electrons induced by $\mathbf{B}_0$ (see Section 12.2.2). The phase velocity of this wave goes to zero (and the effective index of refraction diverges) as $\omega \to \omega_c$ because it resonantly transfers its energy to the electrons. At lower frequency, the name *whistler* attaches to this wave for reasons we will explain in Application 18.4. The L′ mode frequency $\omega \to 0$ as $k \to 0$, but this is *not* the shear Alfvén wave discussed in Application 17.1. The latter appears only when we supplement the electron polarization computed from (18.68) with the polarization contributed by the motion of the heavy positive ions.

Application 18.2 Reflection of Radio Waves by the Ionosphere

In 1901, Guglielmo Marconi demonstrated that radio waves could be transmitted from one side of the Atlantic Ocean to the other. To account for this surprising observation (only line-of-sight reception was thought possible), Oliver Heaviside and Arthur Kennelly independently proposed that the waves must reflect from an electrically conducting layer located about 100 kilometers above the surface of the Earth. Direct experimental evidence for the existence of the "Heaviside-Kennelly layer" (the ionosphere) was reported in 1924 by E.V. Appleton, who subsequently developed the magnetized-plasma model of the ionosphere discussed in this section.

Figure 18.7 shows the electron density n as a function of height h above the surface of the Earth in the region of the ionosphere. An electromagnetic wave with frequency ω^* launched upward from the ground experiences a steady increase in electron density as it propagates. Since $\omega_p^2 = ne^2/\epsilon_0 m$, the wave reflects when it reaches a height where the cutoff frequency in (18.75) equals ω^*. The same argument (reversed) explains why the L and R waves in Figure 18.6 are important tools in radio astronomy. There is no barrier for these waves to propagate from deep inside an astrophysical source (where the plasma density and ω_p are large) to an observer far outside the source (where the plasma density and ω_p are small).

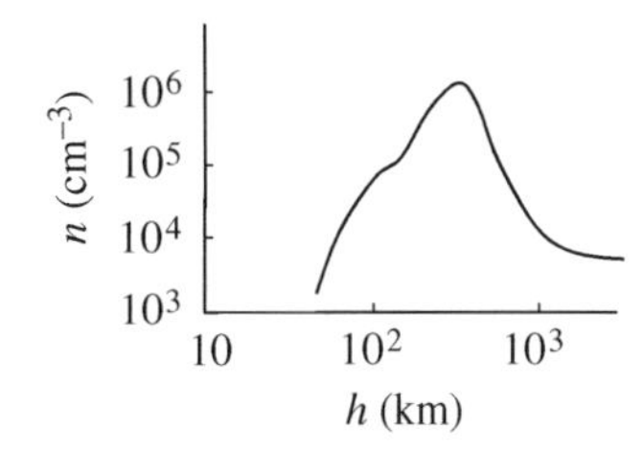

Figure 18.7: Electron density n in the ionosphere as a function of height h above the surface of the Earth.

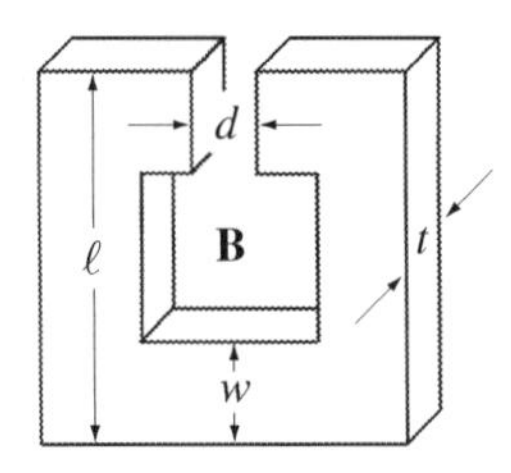

Figure 18.8: Cartoon of a metallic split-ring resonator used to make a negative-index material. The magnetic field **B** points out of the plane of the diagram.

18.5.6 The Split-Ring Model for Negative-Index Matter

We showed in Section 17.3.1 that a *negative* index of refraction results when the electric permittivity ϵ and magnetic permeability μ of a medium are both negative. This never happens for naturally occurring materials, so man-made structures are required to explore the consequences of $n = -c\sqrt{\epsilon\mu} < 0$. The Drude and Lorentz models show that it is not very difficult to make $\epsilon(\omega) < 0$ in a limited frequency band. The key is to arrange matters so $\mu(\omega) < 0$ in the same band.

Figure 18.8 is a sketch of a "split-ring resonator". Artificial crystals where these metal objects play the role of atoms in natural crystals turn out to have $n(\omega) < 0$ at infrared and higher frequencies if the size and separation of the split rings is chosen properly. At these frequencies (still below the plasma frequency), the skin depth is small enough to screen external electric and magnetic fields from the interior of the rings. All induced currents circulate on the surface and the resulting magnetic moments can be orders of magnitude larger than the magnetic moments induced in atoms and molecules. In a lumped circuit model, the circulating currents produce an effective inductance L and the electric field, which is confined to the gap which splits the ring, produces an effective capacitance C. As we will show, an oscillating magnetic field can resonantly excite an electromagnetic mode in each ring with frequency $\omega_0 = 1/\sqrt{LC}$.

The analogy with a parallel-plate capacitor permits us to estimate the gap capacitance as $C = \epsilon_0 wt/d$. A simple estimate of the inductance treats the ring as a solenoid with length t. In that case, (12.135) implies that $L = \mu_0 \ell^2/t$. Hence, the expected resonant frequency of a split ring is

$$\omega_0 = \frac{c}{\ell}\sqrt{\frac{d}{w}}. \tag{18.76}$$

To compute the magnetic permeability, $\mu(\omega)$, we recall that

$$\mathbf{B} = \mu_0(\mathbf{H} + \mathbf{M}) = \mu_0(1 + \chi_m)\mathbf{H} = \mu\mathbf{H} \tag{18.77}$$

and focus on the magnetization $\mathbf{M} = \chi_m \mathbf{H}$ induced by an external magnetic field $B\exp(-i\omega t)$ oriented perpendicular to the plane of the diagram in Figure 18.8. The magnetic moment of a single split ring is $m = \ell^2 I$, where the current induced in the ring is determined by Kirchhoff's voltage law (14.146) generalized to include the Faraday EMF induced in the ring by a flux Φ_B of external magnetic field:

$$\dot{I} + \frac{I}{LC} + \frac{1}{L}\frac{d^2\Phi_B}{dt^2} = 0. \tag{18.78}$$

For our problem, $\Phi_B = \ell^2 B\exp(-i\omega t)$. Therefore, if a is the distance between nearest-neighbor resonators in the artificial crystal we have envisioned, the frequency-dependent magnetization (magnetic moment per unit volume) of the crystal is

$$M = \frac{\ell^2 t}{a^3} \frac{\omega^2}{\omega_0^2 - \omega^2} H. \tag{18.79}$$

In this context, the dimensionless ratio $f = \ell^2 t/a^3 < 1$ is called the "filling factor" and we find

$$\frac{\mu(\omega)}{\mu_0} = 1 + f \frac{\omega^2}{\omega_0^2 - \omega^2}. \tag{18.80}$$

The magnetic permeability (18.80) develops an imaginary part when energy-loss processes are included. In that case, comparing $\mu(\omega)$ with $\epsilon(\omega)$ computed in Section 18.5.4 shows that the resonant response of a collection of metallic split rings to a time-varying magnetic field is very much like the resonant response of a collection of Lorentz oscillators to a time-varying electric field. In practice, one can engineer $\epsilon(\omega) < 0$ in the same frequency band by threading the split-ring crystal with conventional metallic wires. The final result is a man-made structure with $n(\omega) < 0$.

18.6 Wave Packets in Dispersive Matter

A wave packet in vacuum spreads out *transversely* (diffracts) as it propagates because the wave vectors of its constituent plane waves (all with the same phase speed) are not all parallel (see Section 16.5.5). By contrast, a wave packet in a dispersive medium spreads out *longitudinally* along its propagation direction (even if all the constituent wave vectors are parallel) because the phase speeds of its contributing plane waves are not all the same. We study this phenomenon quantitatively using a complex electric field $\boldsymbol{\mathcal{E}}(z, t) = \hat{\mathbf{e}} E(z, t)$ built from plane waves with real wave vectors[11]

$$\mathbf{k}(\omega) = k(\omega)\hat{\mathbf{z}} = n(\omega)\frac{\omega}{c}\hat{\mathbf{z}}. \tag{18.81}$$

If $A(\omega)$ is a weight function, the wave packet of interest is

$$E(z, t) = \int_0^\infty d\omega\, A(\omega) \exp\{i[k(\omega)z - \omega t]\}. \tag{18.82}$$

Equation (18.82) sums plane waves with different frequencies ω and uses the index of refraction $n(\omega)$ in (18.81) to distinguish one wave from another. This differs from the vacuum wave packets studied in Section 16.5 where we summed plane waves with different wave vectors and used the dispersion curve $\omega(k)$ to distinguish one wave from another. We prefer (18.82) for studying waves in matter because specifying $A(\omega)$ is equivalent (by Fourier transformation) to specifying $E(z = 0, t)$, where $z = 0$ is the entrance boundary of the dispersive medium. This is an easily measurable quantity. This may be contrasted with $E(z, t = 0)$, which is the quantity needed to determine the weight function $A(k)$ in a sum over wave vectors. The latter is more natural for an initial-value problem like the spreading of a free-field wave packet.

Typically, $A(\omega)$ is sharply peaked around a single frequency ω_0. This justifies an expansion of $k(\omega)$ in a Taylor series around ω_0. Using the variable $\Omega = \omega - \omega_0$,

$$k(\omega) = k(\omega_0 + \Omega) = k(\omega_0) + \Omega \left.\frac{dk}{d\omega}\right|_{\omega_0} + \frac{1}{2}\Omega^2 \left.\frac{d^2k}{d\omega^2}\right|_{\omega_0} + \cdots. \tag{18.83}$$

[11] We treat a complex index $\hat{n}(\omega)$ later.

If we write $k_0 = k(\omega_0)$ and similarly for its derivatives, substituting (18.83) into (18.82) factors the electric field into the product of a "carrier" plane wave and an envelope function $A(z, t)$:

$$E(z, t) = \exp\{i[k_0 z - \omega_0 t]\}\, A(z, t), \tag{18.84}$$

where

$$A(z, t) = \int_{-\omega_0}^{\infty} d\Omega\, A(\omega_0 + \Omega) \exp\left\{i\left[\Omega k_0' + \tfrac{1}{2}\Omega^2 k_0'' + \cdots\right] z - i\Omega t\right\}. \tag{18.85}$$

This way of writing $E(z, t)$ is useful because the envelope function $A(z, t)$ satisfies a simple partial differential equation. The derivatives $\partial A/\partial z$, $\partial A/\partial t$, and $\partial^2 A/\partial t^2$ all bring down factors of Ω from the exponential in (18.85) and we find without difficulty that

$$i\left(\frac{\partial}{\partial z} + k_0' \frac{\partial}{\partial t}\right) A(z, t) = \tfrac{1}{2} k_0'' \frac{\partial^2 A(z, t)}{\partial t^2}. \tag{18.86}$$

Our task is to solve (18.86) for various situations.

18.6.1 The Group Velocity Approximation

Section 16.5 identified the *group velocity*,

$$\boldsymbol{v}_g = \left.\frac{\partial \omega}{\partial \mathbf{k}}\right|_{\mathbf{k}=\mathbf{k}_0} = \left.\frac{\partial \omega}{\partial k}\right|_{k=k_0} \hat{\mathbf{k}}, \tag{18.87}$$

as the velocity of the envelope of a wave packet composed of plane waves with wave vectors clustered near $\mathbf{k}_0$.[12] This quantity appears in (18.86) because

$$k_0' = \left.\frac{\partial k}{\partial \omega}\right|_{\omega_0} = \left(\left.\frac{\partial \omega}{\partial k}\right|_{k_0}\right)^{-1} = \frac{1}{v_g}. \tag{18.88}$$

An important special case occurs when the right-hand side of (18.86) may be neglected, i.e., when $k_0' \gg k_0'' \delta\Omega$, where $\delta\Omega$ is the range of frequencies where $A(\omega)$ is large. The envelope equation (18.85) simplifies in that case to

$$\left(\frac{\partial}{\partial z} + \frac{1}{v_g}\frac{\partial}{\partial t}\right) A(z, t) = 0. \tag{18.89}$$

The solution to (18.89) is $A(z, t) = \tilde{A}(z - v_g t)$, where $\tilde{A}(x)$ is any scalar function of one variable. The corresponding electric-field wave packet (18.84) propagates undistorted along the z-axis at the group velocity v_g:

$$E(z, t) = \tilde{A}(z - v_g t) \exp\{i[k_0 z - \omega_0 t]\}. \tag{18.90}$$

Equation (18.90) is exact for simple media (where ϵ and μ are strictly constant) because $k = n\omega/c$, so $k_0'' = 0$. Moreover, the group velocity is equal to the phase velocity:

$$v_g = \frac{d\omega}{dk} = \frac{c}{n} = \frac{\omega}{k} = v_p. \tag{18.91}$$

[12] The second equality in (18.87) is valid for isotropic matter where $\omega(\mathbf{k}) = \omega(k)$.

A more interesting example is a wave packet propagating in Drude matter (Section 18.5.1). When $\omega > \omega_p$, the dispersion relation is (18.50), or

$$ck = \sqrt{\omega^2 - \omega_p^2}. \tag{18.92}$$

The group velocity and phase velocity are inversely related in this case because

$$v_g = \frac{c^2}{v_p} = c\sqrt{1 - \omega_p^2/\omega^2} < c \qquad \omega > \omega_p. \tag{18.93}$$

The quantity $k''(\omega_0)$ is never zero for this medium, but it decreases with increasing frequency. Therefore, the group velocity approximation can be used for wave propagation in a Drude conductor when the central frequency ω_0 is sufficiently large. A brief calculation shows that a quantitative condition is $\omega_0^3 \gg \delta\Omega\omega_p^2$.

It is instructive to write the group velocity in terms of the index of refraction $n(\omega)$. By direct calculation,

$$v_g = \frac{\partial \omega}{\partial k} = \frac{\partial}{\partial k}\left(\frac{ck}{n}\right) = \frac{c}{n} - \frac{ck}{n^2}\frac{\partial n}{\partial \omega}v_g. \tag{18.94}$$

Therefore,

$$v_g = \frac{c}{n + \omega \dfrac{dn}{d\omega}}. \tag{18.95}$$

When the index of refraction is complex, $v_p = c/n$ and (18.95) remain correct with n replaced by $n' = \text{Re}\,\hat{n}$. This shows that $v_g < v_p$ when the dispersion of the medium is normal $(dn'/d\omega > 0)$. On the other hand, peculiar behavior might be expected when the dispersion is anomalous $(dn/d\omega < 0)$ because the denominator of (18.95) could vanish or become negative.

Example 18.2 A transverse, quasi-monochromatic wave packet propagates in a medium with weak dispersion and negligible absorption so $u_{EM}(t)$ is given by (18.29) and $Q(t)$ in (18.30) is nearly zero. Under these conditions, show that the group velocity (18.87) is equal to the time-averaged energy velocity defined in Section 17.2 as

$$\mathbf{v}_E = \frac{\langle \mathbf{S} \rangle}{\langle u_{EM} \rangle}.$$

Solution: Because losses can be neglected, the wave impedance in (18.40) is real, and the electric and magnetic fields of the wave can be chosen as

$$\mathbf{E}(\mathbf{r}, t) = \mathbf{E}(\mathbf{r}) \cos \omega t \qquad \text{and} \qquad \mathbf{H}(\mathbf{r}, t) = \sqrt{\frac{\epsilon'}{\mu'}}\left[\hat{\mathbf{k}} \times \mathbf{E}(\mathbf{r})\right] \cos \omega t.$$

The Poynting vector averaged over one cycle is

$$\langle \mathbf{S} \rangle = \langle \mathbf{E} \times \mathbf{H} \rangle = \sqrt{\frac{\epsilon'}{\mu'}}|\mathbf{E}(\mathbf{r})|^2 \langle \cos^2 \omega t \rangle \hat{\mathbf{k}} = \frac{1}{2}\sqrt{\frac{\epsilon'}{\mu'}}|\mathbf{E}(\mathbf{r})|^2 \hat{\mathbf{k}}.$$

The corresponding time average of the electromagnetic energy density is

$$\begin{aligned}\langle u_{EM} \rangle &= \frac{1}{2}\frac{\partial}{\partial \omega}[\omega \epsilon'(\omega)]\langle \mathbf{E} \cdot \mathbf{E} \rangle + \frac{1}{2}\frac{\partial}{\partial \omega}[\omega \mu'(\omega)]\langle \mathbf{H} \cdot \mathbf{H} \rangle \\ &= \frac{1}{4\mu'}\left[2\mu'\epsilon' + \omega\mu'\frac{d\epsilon'}{d\omega} + \epsilon'\omega\frac{d\mu'}{d\omega}\right]|\mathbf{E}(\mathbf{r})|^2.\end{aligned}$$

On the other hand, because $k^2 = \omega^2\mu'\epsilon'$ when absorption is absent,

$$\frac{dk}{d\omega} = \frac{\omega}{2k}\left[2\mu'\epsilon' + \omega\mu'\frac{d\epsilon'}{d\omega} + \epsilon'\omega\frac{d\mu'}{d\omega}\right].$$

Therefore,

$$\mathbf{v}_E = \frac{\langle \mathbf{S} \rangle}{\langle u_{EM} \rangle} = \sqrt{\frac{\epsilon'}{\mu'}}\mu'\frac{\omega}{k}\frac{d\omega}{dk}\hat{\mathbf{k}} = \frac{d\omega}{dk}\hat{\mathbf{k}} = \mathbf{v}_g.$$

Application 18.3 Lorentz Model Velocities

Figure 18.9 shows the phase velocity v_p and the group velocity v_g for the Lorentz model (Section 18.5.4). The two vertical lines labeled $\sqrt{\omega_0^2 - \Gamma^2/4}$ and $\sqrt{\omega_0^2 + \omega_p^2 - \Gamma^2/4}$ bracket the region of anomalous dispersion where most of the absorption of wave energy occurs. The phase velocity exceeds the speed of light for much of the diagram. This does not contradict the special theory of relativity because v_p merely quantifies how the motion of planes of constant phase in infinite wave trains are delayed in time with respect to one another. More disconcertingly, perhaps, the group velocity v_g is *negative* in Figure 18.9 in part of the anomalous-dispersion regime, and diverges to $\pm\infty$ near the edges of the absorption band. This behavior leads some authors to conclude that the group velocity is not meaningful under these conditions.

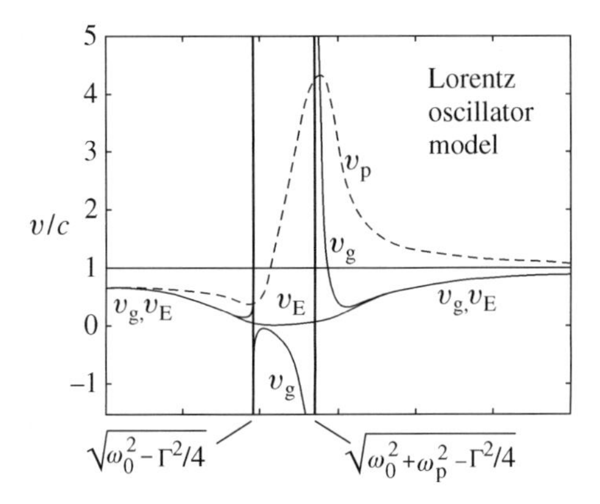

Figure 18.9: Lorentz oscillator model: phase velocity v_p (dashed curve), group velocity v_g (three-branched solid curve); energy velocity v_E (solid curve). Figure adapted from Oughstun (2006).

To the contrary, Figure 18.10 shows that a smoothly varying Gaussian wave packet can enter and exit a slab of matter with $v_g < 0$ at the same time that a Gaussian packet propagates backward through the medium as predicted by (18.90). The peak which appears at the exit face of the slab (in the second panel from the top) comes from a re-weighting of the plane waves which constitute the (exponentially small) leading edge of the original Gaussian packet. The latter impinges on the slab long before the peak does. Ultimately, *all* the distortion and "reshaping" of the original pulse shown in Figure 18.10 comes from constructive and destructive interference among the infinitely extended (in space) plane waves which contribute to the wave packet in (18.82). No contradiction of special relativity ever occurs, even when $v_g \to \infty$. We will return to the general question of signal speed in dispersive media in Section 18.7.4.

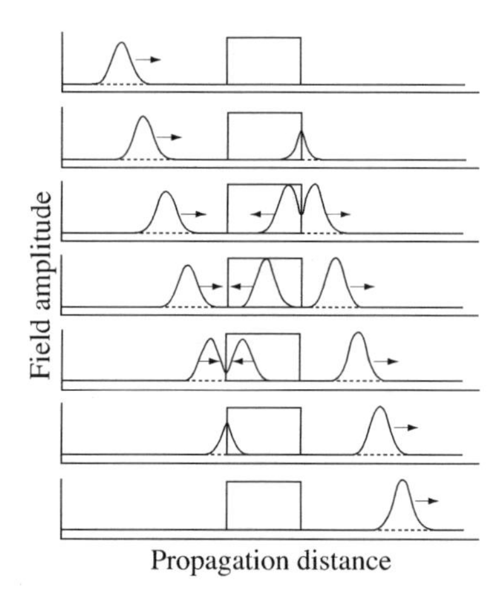

Figure 18.10: Cartoon of the propagation of a Gaussian wave packet through a slab of material with a negative group velocity. Figure from Gauthier and Boyd (2007).

Besides the phase and group velocities, Figure 18.9 shows the time average of the *energy velocity* v_E for the Lorentz model. We leave it as an exercise for the reader to show that the latter can be written in terms of the model's complex index of refraction $\hat{n} = n' + in''$ and its damping constant Γ:

$$v_E = \frac{c}{n' + 2\omega n''/\Gamma}. \tag{18.96}$$

As expected from Example 18.2, v_E coincides with the group velocity v_g when the dispersion is normal and there is negligible absorption. However, in the region of anomalous dispersion and strong damping, v_E deviates considerably from v_g and remains strictly less than c (and near zero). ■

18.6.2 Group Velocity Dispersion

The wave packet (18.90) propagates without distortion in the group velocity approximation. This approximation breaks down if the second derivative on the right side of (18.86) cannot be neglected. The name *group velocity dispersion* (GVD) is used for this second derivative because

$$k'' = \frac{d}{d\omega}\frac{dk}{d\omega} = -\frac{1}{v_g^2}\frac{dv_g}{d\omega}. \tag{18.97}$$

To study the effect of GVD on the envelope function $A(z, t)$, we define

$$\tau = t - z/v_g \qquad \text{and} \qquad A(z, t) = \psi(z, \tau). \tag{18.98}$$

Because $\partial A/\partial z = \partial\psi/\partial z + (\partial\psi/\partial\tau)(\partial\tau/\partial z)$, (18.86) simplifies to

$$i\frac{\partial\psi}{\partial z} = \frac{1}{2}k_0''\frac{\partial^2\psi}{\partial\tau^2}. \tag{18.99}$$

One way to solve (18.99) exploits the similarity of (18.99) to the time-dependent Schrödinger equation for a free particle in one space dimension. The plane wave solutions of the latter lead (after exchanging time and space variables) to plane wave solutions of (18.99) indexed by a real variable ν with

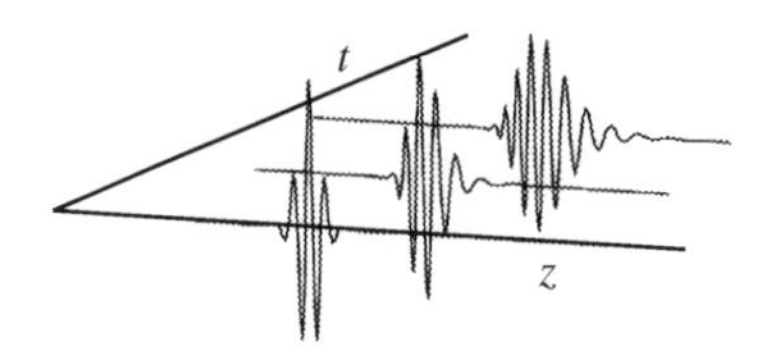

Figure 18.11: Fixed-time snapshots of Re[$E(z,t)$] from (18.104) showing the effects of group velocity dispersion on the time evolution of a wave packet.

dimensions of frequency:[13]

$$\psi_\nu(z,\tau)=\exp\left[i\left(\tfrac{1}{2}k_0''\nu^2 z-\nu\tau\right)\right]. \tag{18.100}$$

A general solution to (18.99) follows by superposing many such waves with a weight function $\hat{\psi}(\nu)$:

$$\psi(z,\tau)=\int_{-\infty}^{\infty} d\nu\,\hat{\psi}(\nu)\exp\left[i\left(\tfrac{1}{2}k_0''\nu^2 z-\nu\tau\right)\right]. \tag{18.101}$$

Setting $z=0$ (18.101) shows that $\hat{\psi}(\nu)$ is the Fourier transform of the boundary value $\psi(0,\tau)$:

$$\hat{\psi}(\nu)=\frac{1}{2\pi}\int_{-\infty}^{\infty} d\tau\,\psi(0,\tau)\exp(-i\nu\tau). \tag{18.102}$$

It is convenient (and not unrealistic) to suppose that a detector at the $z=0$ entrance to the medium records a Gaussian pulse with width (in time) ΔT:

$$\psi(0,\tau)=\exp(-\tau^2/\Delta T^2). \tag{18.103}$$

Inserting (18.103) into (18.102) gives $\hat{\psi}(\nu)=(\Delta T/\sqrt{4\pi})\exp(-\nu^2\Delta T^2/4)$. Using this in (18.101) produces an integral whose value we recorded in (16.85). With $\psi(z,\tau)$ thus in hand, the electric field given by (18.84) and (18.97) is

$$E(z,t)=\sqrt{\frac{\Delta T^2}{\Delta T^2-2ik_0''z}}\exp\left[-\frac{(t-z/v_g)^2}{\Delta T^2-2ik_0''z}\right]\exp[i(k_0 z-\omega_0 t)]. \tag{18.104}$$

Figure 18.11 shows the time evolution of the real part of (18.104) for a typical choice of parameters.

The traces in Figure 18.11 are best understood if we write the complex electric field in the form

$$E(z,t)=|E(z,t)|\exp[i\phi(z,t)]. \tag{18.105}$$

The amplitude function,

$$|E(z,t)|=\left[\frac{1}{1+\left(2k_0''z/\Delta T^2\right)^2}\right]^{1/4}\exp\left[-\frac{(t-z/v_g)^2}{\Delta T^2+\left(2k_0''z/\Delta T\right)^2}\right], \tag{18.106}$$

shows that the pulse remains a Gaussian as a function of time. However, the packet amplitude decreases and its temporal width increases as z increases. In fact, the width doubles after the pulse has propagated a distance $\Delta z\approx\Delta T^2/|k_0''|$ into the medium. Snapshots at different times show that

[13] We solved the paraxial wave equation in Section 16.5.5 using its similarity to the time-dependent Schrödinger equation for a free particle in *two* space dimensions.

the packet propagates at the group velocity v_g, but that the packet envelope stretches out and distorts more and more as time goes on. The rightmost trace in Figure 18.11 demonstrates that GVD produces the phenomenon of "chirp". This word is used to indicate that the local frequency of the wave varies as a function of position (or time). We confirm this using the phase function defined by (18.105) to compute

$$\omega(z,t) = \frac{d\phi}{dt} = \omega_0 + \frac{4k_0''z}{\Delta^4 + (2k_0''z)^2}(t - z/v_g). \tag{18.107}$$

The wave packet in Figure 18.11 has $k_0'' > 0$. Therefore, in accord with (18.107), the Fourier components with longer (shorter) wavelengths appear at the front (back) of the pulse at the latest time shown.

Example 18.3 Equation (18.106) defines the Gaussian width parameter $W^2 = \Delta T^2 + (2k_0''z/\Delta T)^2$. Derive this position-dependent width by treating the $z = 0$ packet as a superposition of sub-packets, each with a frequency bandwidth $\delta\Omega \ll \Delta\Omega$. Assume that the original packet has central frequency ω_0 and a wide but Fourier transform-limited bandwidth $\Delta\Omega \sim 1/\Delta T$.[14]

Solution: The smallness of $\delta\Omega$ ensures that each sub-packet propagates undistorted at its own group velocity. Therefore, using (18.97), the difference in time required for two sub-packets from opposite ends of the original packet to propagate a distance z is

$$\Delta t = \left| \frac{z}{v_g(\omega_0 - \Delta\Omega)} - \frac{z}{v_g(\omega_0 + \Delta\Omega)} \right| \approx \left| 2\Delta\Omega \frac{z}{v_g^2}\frac{dv_g}{d\omega} \right| \approx 2\Delta\Omega |k_0''| z \approx \frac{2|k_0''|z}{\Delta T}. \tag{18.108}$$

The delay Δt introduces a width to the wave packet which adds in quadrature with the initial width ΔT. Therefore, the total width predicted by this argument agrees with (18.106) because

$$\Delta^2 \approx \Delta T^2 + \Delta t^2 \approx \Delta T^2 + \left(2k_0''z/\Delta T\right)^2. \tag{18.109}$$

Application 18.4 Whistlers

The dispersion curves in Figure 18.6(b) describe electromagnetic waves in a magnetized electron plasma where the propagation vector **k** lies parallel to the applied magnetic field **B**. The L′ mode propagates in the frequency interval $0 < \omega < \omega_c$, where the cyclotron frequency $\omega_c = eB/m$ is about 1 MHz for electrons in the ionosphere. This interval includes the audio band, so one can imagine "listening" to ionospheric waves using a suitable transducer. This was done, unintentionally, by the physicist Heinrich Barkhausen when he served the German army in World War I by eavesdropping on enemy field communications using a telephone connected by a high-gain amplifier to a buried antenna. Besides military chatter, he frequently picked up "a very remarkable whistling" which he characterized as "an oscillation . . . of very rapidly changing frequency, beginning with the highest audible tones, passing through the entire scale, and becoming inaudible with the lowest tones".

[14] This is the time-frequency analog of the wave packets discussed in Section 16.5.3 with Fourier transform-limited spatial widths $\Delta x \sim 1/\Delta k$.

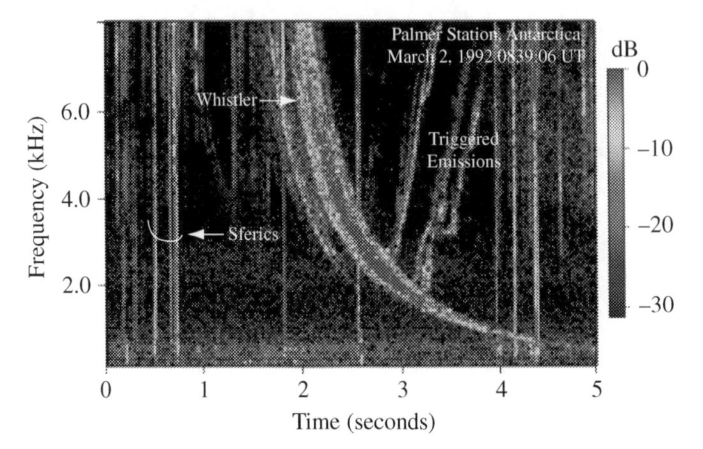

Figure 18.12: A frequency-versus-time spectrogram shows the smoothly decreasing audio tone of a "whistler". Figure from Green and Inan (2006).

Figure 18.12 is a spectrogram of a "whistler" with exactly the characteristics described by Barkhausen.[15] The modern understanding of this phenomenon supposes that a lightning strike excites a wave packet in the lower atmosphere composed of a broad (frequency) band of L′ waves. Then, as shown in Figure 18.13, the packet propagates along a terrestrial magnetic field line up into the ionosphere and then back down to the Earth where it is detected. This trajectory keeps the propagation vectors parallel to **B** and thus maintains the integrity of the L′ waves in the packet. Following Example 18.3, we treat this broad-band packet as a superposition of narrow-band packets, each propagating undistorted at its own group velocity. Because the curve of $\omega(k)$ for the L′ mode in Figure 18.6(b) has positive curvature ($d^2\omega/dk^2 > 0$) at audio frequencies, the group velocity is an increasing function of the central frequency of the sub-packets. This explains the characteristic descending tone seen in Figure 18.12 because high-frequency packets arrive before low-frequency packets.

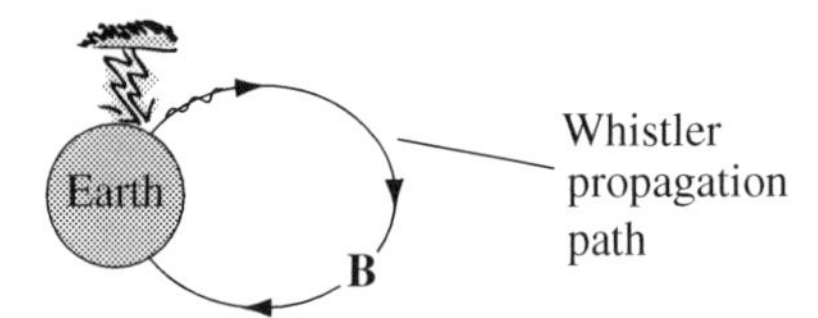

Figure 18.13: A lightning-induced packet of L′ waves propagates into the ionosphere along a magnetic field line of the Earth.

Quantitative studies presume that whistlers propagate along magnetic field lines because they are guided by "plasma ducts" (regions of anomalously high or low plasma density) which form in the immediate vicinity of a field line. Contemporary research focuses on the dynamics of duct formation and the pondermotive interaction (see Section 16.10.2) between ducts and whistler waves. Turning these arguments around, whistlers have been used as probes of the magnetosphere and to deduce the presence of lightning on other planets of the Solar System. ■

[15] Barkhausen's 1918 paper, "Two phenomena discovered with the help of a new amplifier" [*Physikalische Zeitschrift* **20**, 401 (1919)] reported both "whistling tones from the Earth" and "noise during the magnetization of iron". It is a testament to his taste in problems that studies of "Barkhausen noise" in permanent magnets are just as common today as studies of whistler phenomena.

18.7 The Consequences of Causality

The requirement of causality built into a time-dependent response function like $\chi(\mathbf{r}, t)$ (see Section 18.2) has many consequences for the complex, frequency-dependent response function $\hat{\chi}(\mathbf{r}, \omega)$ obtained from it by Fourier transformation. Prominent among these are (i) the real and imaginary parts of $\hat{\chi}(\mathbf{r}, \omega)$ can be obtained from each other; and (ii) $\hat{\chi}(\mathbf{r}, \omega)$ is an analytic function in the upper half-plane of complex ω. These mathematical properties, in turn, have direct physical consequences, including (i) frequency dispersion and energy dissipation are always present together; and (ii) no electromagnetic signal can travel faster than the speed of light in a causal medium. We discuss both below.

18.7.1 The Kramers-Kronig Relations: Derivation

Focus attention on the Fourier transform pair associated with a typical response function, say, an electric or magnetic susceptibility:

$$\hat{\chi}(\omega) = \int_{-\infty}^{\infty} dt\, \chi(t) \exp(i\omega t) \qquad \text{and} \qquad \chi(t) = \frac{1}{2\pi}\int_{-\infty}^{\infty} d\omega\, \hat{\chi}(\omega) \exp(-i\omega t). \tag{18.110}$$

Exactly as in Section 18.2, the response function $\chi(t)$ is real and causal. Reality imposes the condition $\chi(t) = \chi^*(t)$, which implies that the real part of $\hat{\chi}(\omega)$ is an even function and the imaginary part of $\hat{\chi}(\omega)$ is an odd function:

$$\chi'(-\omega) = \chi'(\omega) \qquad \text{and} \qquad \chi''(-\omega) = -\chi''(\omega). \tag{18.111}$$

Causality imposes the condition $\chi(t < 0) = 0$. If we recall the definition of the sign function from Section 1.5.3,

$$\operatorname{sgn}(x) = \begin{cases} -1 & x < 0, \\ 1 & x > 0. \end{cases} \tag{18.112}$$

we can build in causality by writing

$$\chi(t) = \frac{1}{2}[1 + \operatorname{sgn}(t)]\chi(t). \tag{18.113}$$

Equation (18.113) is the key to the calculation. Using it in the left member of (18.110) gives

$$\hat{\chi}(\omega) = \frac{1}{2}\int_{-\infty}^{\infty} dt\, \chi(t)\exp(i\omega t) + \frac{1}{2}\int_{-\infty}^{\infty} dt\, \operatorname{sgn}(t)\chi(t)\exp(i\omega t). \tag{18.114}$$

The first integral on the right-hand side of (18.114) is $\frac{1}{2}\hat{\chi}(\omega)$. Therefore, substituting the right member of (18.110) into the second integral on the right side of (18.114) gives

$$\hat{\chi}(\omega) = \frac{1}{2\pi}\int_{-\infty}^{\infty} ds\, \hat{\chi}(s) \int_{-\infty}^{\infty} dt\, \operatorname{sgn}(t)\exp[i(\omega - s)t]. \tag{18.115}$$

Finally, equate the real and imaginary parts on each side of (18.115), use (18.111), and be aware that time or frequency integrals with odd integrands vanish.[16] The final result shows that the real part of

[16] Note that $\operatorname{sgn}(t) = -\operatorname{sgn}(-t)$ in (18.112).

$\hat{\chi}(\omega)$ is completely determined by the imaginary part of $\hat{\chi}(\omega)$ and vice versa:

$$\chi'(\omega) = \frac{1}{2\pi}\int_0^\infty dt\,\cos(\omega t)\int_{-\infty}^\infty ds\,\chi''(s)\sin(st)$$
$$\chi''(\omega) = \frac{1}{2\pi}\int_0^\infty dt\,\sin(\omega t)\int_{-\infty}^\infty ds\,\chi'(s)\cos(st). \tag{18.116}$$

An unstated assumption, which we now make explicit, is that the integrals in (18.116) and (18.110) converge. This, in turn, requires that

$$\lim_{t\to\infty}\chi(t)\to 0 \text{ faster than } \frac{1}{t} \qquad \text{and} \qquad \lim_{\omega\to\infty}\hat{\chi}(\omega)\to 0 \text{ faster than } \frac{1}{\omega}. \tag{18.117}$$

Before exploring the implications of (18.116), we return to (18.115) and carry out the integration over time. This is best done using a convergence factor ϵ:

$$\int_{-\infty}^\infty dt\,\text{sgn}(t)e^{i\omega t} = \lim_{\epsilon\to 0}\int_0^\infty dt\,e^{i(\omega+i\epsilon)t} - \lim_{\epsilon\to 0}\int_{-\infty}^0 dt\,e^{i(\omega-i\epsilon)t} = \lim_{\epsilon\to 0}\left[\frac{i}{\omega-i\epsilon}+\frac{i}{\omega+i\epsilon}\right]. \tag{18.118}$$

Using (18.118), (18.115) reads

$$\hat{\chi}(\omega) = \lim_{\epsilon\to 0}\frac{i}{2\pi}\int_{-\infty}^\infty ds\left[\frac{\hat{\chi}(s)}{\omega-s+i\epsilon}+\frac{\hat{\chi}(s)}{\omega-s-i\epsilon}\right]. \tag{18.119}$$

Evaluating each term in (18.119) using the Plemelj formula,[17]

$$\lim_{\epsilon\to 0}\frac{1}{x-x_0\pm i\epsilon} = \mathcal{P}\frac{1}{x-x_0}\mp i\pi\delta(x-x_0), \tag{18.120}$$

generates the principal value integral equation

$$\hat{\chi}(\omega) = \frac{i}{\pi}\mathcal{P}\int_{-\infty}^\infty ds\,\frac{\hat{\chi}(s)}{\omega-s}. \tag{18.121}$$

Equating real and imaginary parts on either side of (18.121) gives the *Kramers-Kronig relations*,

$$\chi'(\omega) = -\frac{1}{\pi}\mathcal{P}\int_{-\infty}^\infty ds\,\frac{\chi''(s)}{\omega-s}$$
$$\chi''(\omega) = \frac{1}{\pi}\mathcal{P}\int_{-\infty}^\infty ds\,\frac{\chi'(s)}{\omega-s}. \tag{18.122}$$

The content of (18.122) is exactly the same as the content of (18.116). Both tell us that $\chi'(\omega)$ is completely determined by $\chi''(\omega)$ and $\chi''(\omega)$ is completely determined by $\chi'(\omega)$.

18.7.2 The Kramers-Kronig Relations: Meaning and Uses

The Kramers-Kronig relations are quite general and apply to essentially any physically realizable causal response function. An example is $\hat{\epsilon}(\omega)-\epsilon_0$, where the subtraction of ϵ_0 is needed to satisfy

[17] See Section 1.5.2 for discussion of the Plemelj formula and principal value integrals.

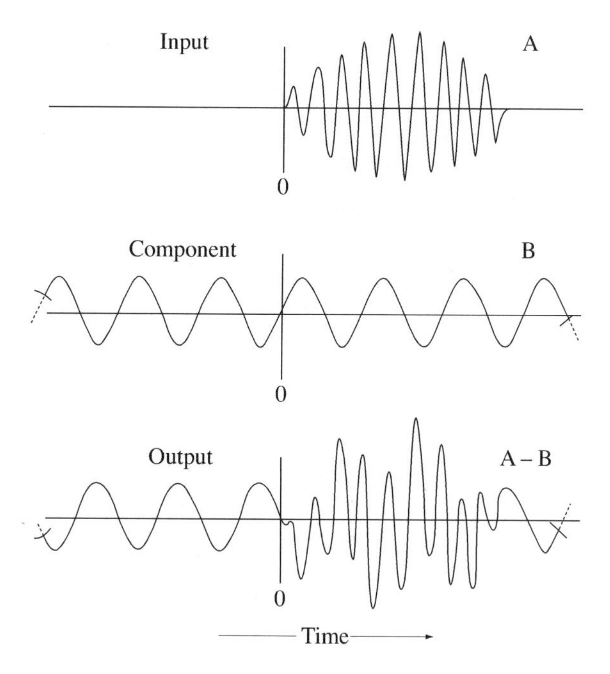

Figure 18.14: The physical origin of the connection between causality, dispersion, and absorption. Figure from Toll (1956).

(18.117). Applying (18.116) or (18.122) to this example shows that if $\epsilon'(\omega)$ is not constant anywhere, $\epsilon''(\omega)$ is non-zero everywhere (although perhaps very small at most frequencies). In light of our discussion of energy in a dispersive medium (Section 18.3), we glean the very non-trivial fact that frequency dispersion and energy dissipation are inextricably linked. The mere existence of dispersion in any interval of frequency implies that non-zero absorption occurs in every interval of frequency. Conversely, frequency dispersion occurs everywhere if energy absorption occurs anywhere.

In practice, the Kramers-Kronig relations are used to reduce the experimental labor needed to characterize the electromagnetic response of a causal medium. Thus, (18.116) or (18.122) can be used to calculate $\epsilon'(\omega)$ once $\epsilon''(\omega)$ has been deduced from, say, broadband absorption measurements. It is particularly convenient (although not obvious) that the Kramers-Kronig relations apply to the complex index of refraction, $\hat{n}(\omega) - 1$, and to the complex Fresnel reflectivity, $\hat{r}(\omega)$ (see Section 17.6.4). This makes it unsurprising that a minor industry exploits the Kramers-Kronig relations to derive the "optical constants" (not limited to optical frequencies) for essentially all materials of physical interest.

We turn now to the physical reason why causality forces the intimate relationship between dispersion and absorption revealed by the Kramers-Kronig relations. Consider the top panel of Figure 18.14. This is an amplitude-versus-time trace for an incident (stimulus) electromagnetic wave packet at a fixed observation point in a causal medium. The clock has been set so $t = 0$ corresponds to the first appearance of the packet. Suppose, for simplicity, that the medium absorbs energy only in a very narrow range of frequencies around a central frequency ω_0. This makes the imaginary part of the index of refraction, $n''(\omega)$, a very sharply peaked function. The middle panel of Figure 18.14 shows the (real part of the) plane wave component at ω_0 which dominates the absorption. The bottom panel of Figure 18.14 shows the output (or response) wave if the medium did nothing but eliminate the absorbed wave from the input wave packet. The output wave violates causality because it is not identically zero for $t < 0$. Therefore, the medium must introduce phase shifts into all the other waves in the packet to ensure destructive interference at the observation point for all $t < 0$. It is precisely the real part of the complex index of refraction, $n'(\omega)$, which quantifies the phase shift imparted by a medium to a wave with frequency ω. The final conclusion is inescapable: the real and imaginary parts of a response function must be intimately related simply to meet the demands of causality.

Example 18.4 Write the Kramers-Kronig relation for $\epsilon'(\omega)$ using positive frequencies only. Justify the Lorentz model form for this quantity in (18.65) using a narrow absorption band assumption like the one used to discuss Figure 18.14.

Solution: Begin with the Kramers-Kronig relation,

$$\frac{\epsilon'(\omega)}{\epsilon_0} = 1 - \frac{1}{\pi\epsilon_0}\mathcal{P}\int_{-\infty}^{\infty} ds \frac{\epsilon''(s)}{\omega - s}.$$

Splitting the integral into positive- and negative-frequency parts and changing variables in the latter gives

$$\frac{\epsilon'(\omega)}{\epsilon_0} = 1 + \frac{1}{\pi\epsilon_0}\mathcal{P}\int_0^{\infty} ds\epsilon''(s)\left[\frac{1}{\omega + s} - \frac{1}{\omega - s}\right] = 1 + \frac{2}{\pi\epsilon_0}\mathcal{P}\int_0^{\infty} ds \frac{s\epsilon''(s)}{s^2 - \omega^2}.$$

Suppose that absorption occurs only in the interval $(\omega_0 - \delta\omega, \omega_0 + \delta\omega)$ where $\delta\omega \ll \omega_0$. This implies that $\epsilon''(\omega)$ vanishes outside that interval and

$$\frac{\epsilon'(\omega)}{\epsilon_0} = 1 + \frac{2}{\pi\epsilon_0}\frac{1}{\omega_0^2 - \omega^2}\int_{\omega_0 - \delta\omega}^{\omega_0 + \delta\omega} ds\, s\epsilon''(s).$$

This has exactly the form of (18.65) in the limit $\Gamma \to 0$, which is consistent with the assumption of a very narrow absorption band, namely,

$$\frac{\epsilon'(\omega)}{\epsilon_0} = 1 + \frac{\omega_\text{p}^2}{\omega_0^2 - \omega^2} \quad \text{where} \quad \omega_\text{p}^2 = \frac{2}{\pi\epsilon_0}\int_{\omega_0 - \delta\omega}^{\omega_0 + \delta\omega} ds\, s\epsilon''(s).$$

18.7.3 The Analytic Properties of $\hat{\chi}(\omega)$

Causal response functions like $\hat{\chi}(\omega)$ have interesting properties when regarded as functions of the complex frequency $\omega = \omega' + i\omega''$. There are many reasons to do this, not least because it facilitates a derivation of the Kramers-Kronig relations (18.122) entirely different from the one given in the previous section. The generalization we need is straightforward because the causality condition $\chi(t < 0) = 0$ permits us to write the left member of (18.110) as

$$\hat{\chi}(\omega) = \int_0^{\infty} dt\, \chi(t)\exp(i\omega t). \tag{18.123}$$

Given the left member of (18.117), the integral (18.123) converges everywhere in the upper half of the complex ω plane where $\omega'' > 0$. The same is true of all the derivatives of $\hat{\chi}(\omega)$. This means that $\hat{\chi}(\omega)$ is *analytic* in the entire upper half ω-plane.[18]

[18] See Section 7.10 for more on analytic functions.

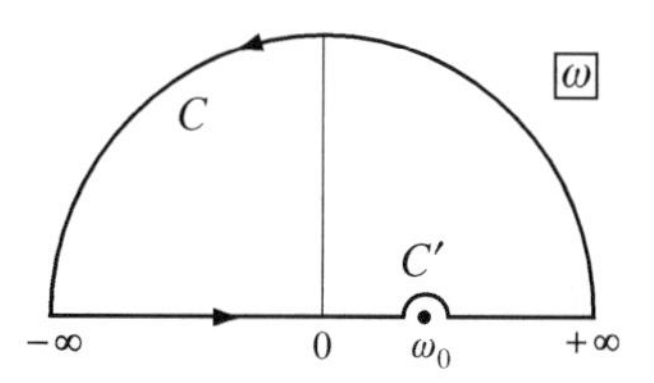

Figure 18.15: The closed integration path C in the complex ω-plane used to evaluate (18.124). The semi-circular arc C' with radius $\delta \to 0$ is part of C.

We derive the Kramers-Kronig relations using Cauchy's theorem, which states that a line integral around a closed contour in the complex plane is zero if the integrand is analytic everywhere inside the contour. Figure 18.15 shows the closed contour C we will use to deduce that

$$\int_C d\omega \frac{\hat{\chi}(\omega)}{\omega - \omega_0} = 0. \tag{18.124}$$

The key to this conclusion is that the lower portion of C traces the real axis from $\omega = -\infty$ to $\omega = +\infty$ except for a tiny detour into the upper half-plane to avoid the point $\omega = \omega_0$ on the real axis. The detour follows a semi-circular arc C' whose radius δ will go to zero at the end of the calculation.[19] The remainder of C is a semi-circular arc with radius $R \to \infty$ in the upper half-plane. Cauchy's theorem gives (18.124) because $\hat{\chi}(\omega)$ is analytic in the upper half-plane and C excludes the pole of the integrand at $\omega - \omega_0$. On the other hand, the right member of (18.117) guarantees that the integral over the large semi-circular part of C is zero by itself. Therefore, using the definition of the principal value integral in (1.104), (18.124) is equivalent to the statement that

$$\mathcal{P}\int_{-\infty}^{\infty} d\omega \frac{\hat{\chi}(\omega)}{\omega - \omega_0} + \int_{\delta\to 0} d\omega \frac{\hat{\chi}(\omega)}{\omega - \omega_0} = 0. \tag{18.125}$$

We evaluate the second integral in (18.125) using the parameterization $\omega = \omega_0 + \delta \exp(i\theta)$, where the limits of θ are chosen so the semi-circle C' is traversed from $\omega_0 - \delta$ to $\omega_0 + \delta$:

$$\int_{\delta\to 0} d\omega \frac{\hat{\chi}(\omega)}{\omega - \omega_0} = \lim_{\delta\to 0} \int_{\pi}^{0} i\delta e^{i\theta} d\theta \frac{\hat{\chi}(\omega_0 + \delta e^{i\theta})}{\delta e^{i\theta}} = -\pi i \hat{\chi}(\omega_0). \tag{18.126}$$

Combining (18.126) with (18.125) reproduces (18.121) and thus the Kramers-Kronig relations also.

18.7.4 Consistency with Special Relativity

Special relativity (Chapter 22) teaches that no signal capable of "initiating events" travels faster than the speed of light. This fact, not obviously consistent with the behavior of the group velocity in regions of anomalous dispersion (see Application 18.3), motivated Arnold Sommerfeld to ask the following question: If a wave packet with a sharp leading edge (see Figure 18.16) enters a dispersive medium at $t = 0$, how much time elapses before an observer at a distance z in the medium first detects a non-zero field amplitude?[20]

[19] When metals are considered and $\hat{\chi}(\omega) = \hat{\epsilon}(\omega) - \epsilon_0$, the contour C must avoid the origin using another small semi-circle because $\hat{\epsilon}(\omega)$ has a pole there. See (18.47).

[20] A wave packet with a sharp edge is needed so the clock can be set to $t = 0$ at the precise moment the packet enters the medium. A Gaussian wave packet cannot be used because its leading edge is never exactly zero. We note also that the energy velocity formula (18.96) was unknown in 1906 when Sommerfeld initiated his study.

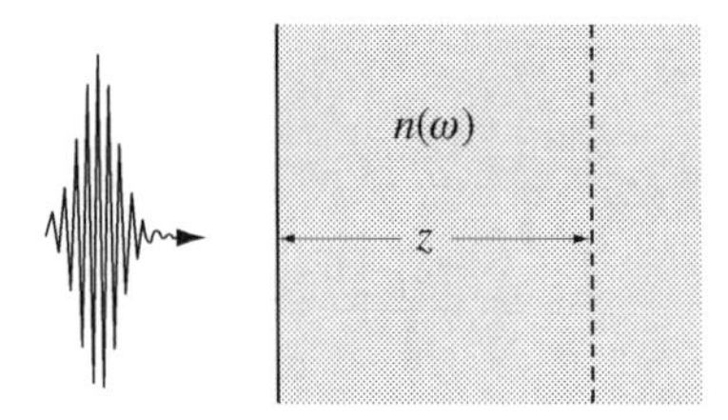

Figure 18.16: At what time does an observer at z detect the wave packet?

A physical argument observes that the packet moves at the speed of light in vacuum, but gets distorted in the medium by the addition of fields produced by the response of the medium. That response comes from the motion of charged particles in the medium which have mass and thus cannot respond instantaneously. Therefore, we expect at least some portion of the packet to arrive at the observation point in a time $t = z/c$.

Sommerfeld's calculation compares the vacuum wave packet shown in Figure 18.16,

$$E_V(z,t) = \int_{-\infty}^{\infty} d\omega\, \hat{A}(\omega) \exp\{i\omega(z/c - t)\}, \tag{18.127}$$

with the same packet after it has propagated into the medium. Using the Fresnel transmission amplitude for normal incidence (see Section 17.3.2), the latter is

$$E(z,t) = \int_{-\infty}^{\infty} d\omega \frac{2}{1+\hat{n}(\omega)} \hat{A}(\omega) \exp\{i\omega[\hat{n}(\omega)z/c - t]\}. \tag{18.128}$$

Our first observation is that (18.127) vanishes when $z > ct$ because $E_V(z,t)$ propagates undistorted at the speed of light. This fact provides information about $\hat{A}(\omega)$ in the complex ω-plane which we will use to analyze (18.128). The trick is to extend the path of integration in (18.127) to include the semi-circular contour C_1 in Figure 18.17. This does not change the value of the integral because the contribution from C_1 is zero when $z > ct$ as long as $A(\omega)$ does not increase more rapidly than the exponential decrease of $\exp[i\omega(z/c - t)]$. We conclude that the integral (18.127) vanishes when integrated over a contour which encloses the entire upper half-plane. It follows from Cauchy's theorem that $A(\omega)$ is analytic in the upper half-plane.

We analyze (18.128) similarly by extending its path of integration to include C_1 when $z > ct$. However, $\hat{n}(\omega) \to 1$ as C_1 expands to infinity because, as mentioned earlier, $\hat{n}(\omega) - 1$ is a causal response function which satisfies the Kramers-Kronig relations. Therefore, the contribution to the integral from C_1 vanishes as it did in the previous paragraph and (18.128) may be evaluated for $z > ct$ using a contour which encloses the entire upper half-plane. Invoking Cauchy's theorem shows that $E(z,t)$ is determined by the poles of the integrand of (18.128) in the upper half-plane. However, $\hat{A}(\omega)$ has no poles in the upper half-plane (see just above), and one can check that $[1 + \hat{n}(\omega)]^{-1}$ also has no poles in the upper half-plane for representative dielectric models like the one of Lorentz (Section 18.5.4). Consequently, $E(z,t) = 0$ when $z > ct$ and we confirm the prediction of special relativity that no signal can be transmitted through a dispersive medium faster than the speed of light.

The foregoing shows that all non-zero values of $E(z,t)$ come from (18.128) evaluated for $z < ct$. To perform the integration, we exploit the change of sign of the argument of the exponential factor and close the integration path using the infinitely large semi-circle C_2 in Figure 18.17. The contribution to the integral from C_2 itself is zero (as above), so Cauchy's integral formula and the pole structure of $\hat{A}(\omega)/[1 + \hat{n}(\omega)]$ in the *lower* half-plane determine the behavior $E(z,t)$. The integral cannot be done exactly so Sommerfeld used asymptotic methods and a vacuum wave packet with a step-function

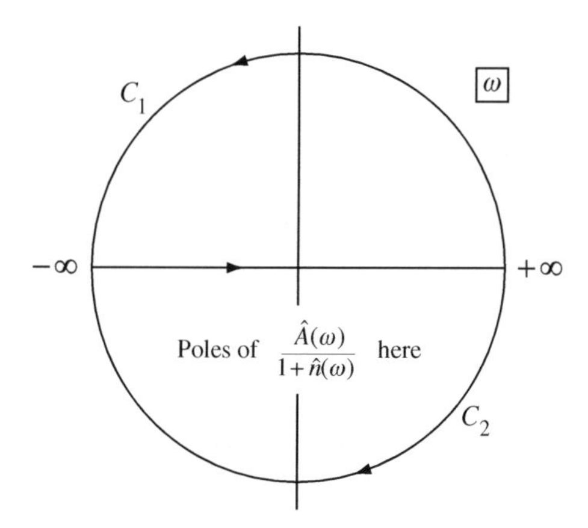

Figure 18.17: Contours in the complex ω-plane used to evaluate $E_V(z,t)$ and $E(z,t)$.

leading edge to show that the first hint of an electromagnetic field at z appears at exactly time $t = z/c$. However, the structure of this "forerunner" field differs considerably from the leading edge of the original wave packet. Follow-up work by Brillouin and others elucidated the nature of these precursor waves and the subsequent appearance of a field recognizably related to the original packet.[21]

Application 18.5 Sum Rules

The discovery of new materials demands the construction of model dielectric functions to describe their response to electromagnetic fields. Guidance for this process comes from the experience gained from earlier models (see Section 18.5) and from various *sum rules* which proposed dielectric functions must satisfy to be consistent with general principles. Two such rules follow from the positive-frequency version of the Kramers-Kronig relation derived in Example 18.4:

$$\epsilon'(\omega) = \epsilon_0 + \frac{2}{\pi}\int_0^\infty ds \frac{s\epsilon''(s)}{s^2 - \omega^2}. \tag{18.129}$$

The first sum rule constrains the value of the static dielectric constant by setting $\omega = 0$ in (18.129) to get

$$\epsilon'(0) = \epsilon_0 + \frac{2}{\pi}\int_0^\infty ds \frac{\epsilon''(s)}{s}. \tag{18.130}$$

A second sum rule follows from (18.129) and the fact that all systems possess a frequency Ω beyond which the absorption is negligible. In other words, $\epsilon''(\omega > \Omega) = 0$, so

$$\epsilon'(\omega) = \epsilon_0 + \frac{2}{\pi}\int_0^\Omega ds \frac{s\epsilon''(s)}{s^2 - \omega^2}. \tag{18.131}$$

When $\omega \gg \Omega$, we can neglect s in the denominator of (18.131) and $\epsilon'(\omega)$ approaches the Drude form (18.66). The latter statement is true because all charge carriers behave as if they were free in the

[21] See Sources, References, and Additional Reading.

presence of an extremely rapidly varying field. Therefore,

$$\epsilon_0 - \frac{2}{\pi}\int_0^{\Omega} ds \frac{s\epsilon''(s)}{\omega^2} = \epsilon_0\left[1 - \frac{\omega_p^2}{\omega^2}\right]. \tag{18.132}$$

Extending the upper limit back to infinity shows that the plasma frequency of the medium sets the scale for the rate of energy absorption integrated over all frequencies [cf. (18.30)]:

$$\int_0^{\infty} ds\, s\epsilon''(s) = \tfrac{1}{2}\pi\epsilon_0\omega_p^2. \tag{18.133}$$

Equation (18.133) is a classical version of the "f-sum rule" in quantum mechanics because ω_p^2 is proportional to the number of charge carriers in the system [see (18.46)]. ■

18.8 Spatial Dispersion

The linear response formulae discussed in Section 18.2 are far from the most general that can be written down.[22] For example, Section 18.5.5 showed that the non-isotropic response of a magnetized plasma to a propagating electromagnetic wave had the effect of replacing $\mathbf{D}(\mathbf{r}, \omega) = \hat{\epsilon}(\omega)\mathbf{E}(\mathbf{r}, \omega)$ by the matrix equation

$$D_i(\mathbf{r}, \omega) = \hat{\epsilon}_{ij}(\omega)E_j(\mathbf{r}, \omega). \tag{18.134}$$

Similarly, an isotropic medium which responds differently from point to point would lead us to write

$$\hat{\mathbf{D}}(\mathbf{r}, \omega) = \hat{\epsilon}(\mathbf{r}, \omega)\hat{\mathbf{E}}(\mathbf{r}, \omega). \tag{18.135}$$

A generalization of (18.135) recognizes that $\mathbf{P}(\mathbf{r}, \omega)$, and hence $\mathbf{D}(\mathbf{r}, \omega)$, may depend on the electric field, not just at $\mathbf{r}$ but also at other points $\mathbf{r}'$ of the medium. This is called *spatial dispersion* and the generalization of (18.135) for a homogeneous medium is

$$\hat{\mathbf{D}}(\mathbf{r}, \omega) = \int d^3r'\, \hat{\epsilon}(\mathbf{r} - \mathbf{r}', \omega)\hat{\mathbf{E}}(\mathbf{r}', \omega). \tag{18.136}$$

One way to think about (18.136) comes from changing variables to

$$\hat{\mathbf{D}}(\mathbf{r}, \omega) = \int d^3s\, \hat{\epsilon}(\mathbf{s}, \omega)\hat{\mathbf{E}}(\mathbf{r} + \mathbf{s}, \omega) \tag{18.137}$$

and noting that the dielectric response at $\mathbf{r}$ often comes from points in the immediate vicinity of $\mathbf{r}$. This justifies a Taylor expansion of $\mathbf{E}(\mathbf{r} + \mathbf{s}, \omega)$, from which we conclude that $\mathbf{D}(\mathbf{r}, \omega)$ may depend on both $\mathbf{E}(\mathbf{r}, \omega)$ and its low-order spatial derivatives:

$$\hat{\mathbf{D}}(\mathbf{r}, \omega) = \epsilon(\omega)\hat{\mathbf{E}}(\mathbf{r}, \omega) + \epsilon_k^{(1)}(\omega)\frac{\partial\hat{\mathbf{E}}(\mathbf{r}, \omega)}{\partial x_k} + \epsilon_{kj}^{(2)}(\omega)\frac{\partial^2\hat{\mathbf{E}}(\mathbf{r}, \omega)}{\partial x_k \partial x_j} + \cdots. \tag{18.138}$$

Equation (18.136) is a spatial convolution integral analogous to the temporal convolution integral in (18.1). Extending the logic used there, we Fourier transform in space (having previously transformed from time to frequency) and define a wave vector and frequency-dependent dielectric function

$$\hat{\epsilon}(\mathbf{k}, \omega) = \int d^3s\, \hat{\epsilon}(\mathbf{s}, \omega)\exp(-i\mathbf{k}\cdot\mathbf{s}). \tag{18.139}$$

[22] We do not discuss non-linear response here.

Combining (18.137) with (18.139) shows that a field $\mathbf{E}(\mathbf{r}, t) = \mathbf{E}(\mathbf{k}, \omega)\exp[i(\mathbf{k}\cdot\mathbf{r} - \omega t)]$ in a spatially dispersive medium induces a field $\mathbf{D}(\mathbf{r}, t) = \mathbf{D}(\mathbf{k}, \omega)\exp[i(\mathbf{k}\cdot\mathbf{r} - \omega t)]$ where

$$\mathbf{D}(\mathbf{k}, \omega) = \hat{\epsilon}(\mathbf{k}, \omega)\mathbf{E}(\mathbf{k}, \omega). \tag{18.140}$$

Constitutive relations like (18.140) play a large role in plasma physics and condensed matter physics.

Spatial dispersion occurs whenever a characteristic length appears in the description of a polarizable medium.[23] A good example is the screening length ℓ of a real conductor discussed in Section 5.7. In our present language, the calculation performed in that section described the static ($\omega = 0$) response of a real conductor in an approximation where the Coulomb potential of a point charge, $\varphi_0(\mathbf{r}) = q/4\pi\epsilon_0 r$, is screened by mobile charges of the opposite sign to a Yukawa-type potential,

$$\varphi(\mathbf{r}) = \frac{q}{4\pi\epsilon_0 r}\exp(-r/\ell). \tag{18.141}$$

Performing an integral like (18.139) shows that the Fourier transform partner of (18.141) is

$$\varphi(\mathbf{k}) = \frac{q/\epsilon_0}{k^2 + (1/\ell)^2}. \tag{18.142}$$

The corresponding partner of the point charge Coulomb potential is

$$\varphi_0(\mathbf{k}) = \frac{q/\epsilon_0}{k^2}. \tag{18.143}$$

In light of Example 18.1, the analog of (18.140) is $\varphi(\mathbf{k}) = \varphi_0(\mathbf{k})/\epsilon(\mathbf{k})$, and we conclude that the static but wave-vector-dependent dielectric function of this medium is

$$\epsilon(\mathbf{k}, 0) = 1 + \frac{1}{k^2\ell^2}. \tag{18.144}$$

Better approximations produce dielectric functions for conductors which are more accurate than (18.144) and which include $\omega \neq 0$ dependence more accurately than the Drude treatment of Section 18.5.1.

Sources, References, and Additional Reading

The quotation at the beginning of the chapter is taken from Lorentz' original paper (1878) on frequency-dispersive matter:

H.A. Lorentz, "Concerning the relation between the velocity of propagation of light and the density and composition of media", in *Collected Papers*, edited by P. Zeeman and A.D. Fokker (Martinus Nijhoff, The Hague, 1936), Volume II.

Section 18.1 Two good textbook treatments of the subject matter of this chapter are

L.D. Landau and E.M. Lifshitz, *Electrodynamics of Continuous Media* (Pergamon, Oxford, 1960).

R.H. Good, Jr. and T.J. Nelson, *Classical Theory of Electric and Magnetic Fields* (Academic, New York, 1971).

Two well-written monographs devoted to dispersive waves in plasmas and solids, respectively, are

D.B. Melrose and R.C. McPhedran, *Electromagnetic Processes in Dispersive Media* (Cambridge University Press, Cambridge, 1991).

M. Dressel and G. Grüner, *Electrodynamics of Solids* (Cambridge University Press, Cambridge, 2002).

[23] We recall from Section 18.5 that frequency dispersion occurs whenever a characteristic frequency appears in the description of a polarizable medium.

Section 18.2 For more on the equivalence of different formulations of linear response theory for electrodynamics, see the two books immediately preceding and

C.R. Smith and R. Inguva, "Electrodynamics in a dispersive medium: **E**, **B**, **D**, and **H**", *American Journal of Physics* **52**, 27 (1984).

P.S. Pershan, "Magneto-optical effects", *Journal of Applied Physics* **38**, 1482 (1967).

Section 18.3 Our discussion of energy and dissipation in dispersive media derives from

F. Borgnis, "On electromagnetic energy density in dispersive media", *Zeitschrift für Physik* **159**, 1 (1960).

Section 18.5 Figure 18.2 and Figure 18.4 come from *Wooten*. Figure 18.5 comes from *Fox*.

F. Wooten, *Optical Properties of Solids* (Academic, New York, 1972).

M. Fox, *Optical Properties of Solids* (Oxford University Press, Oxford, 2001). Reprinted with permission.

Application 18.1 on charge relaxation in ohmic matter was adapted from

W.M. Saslow and G. Wilkinson, "Expulsion of free electronic charge from the interior of a metal", *American Journal of Physics* **39**, 1244 (1971).

The source of Figure 18.6 and a clearly written treatment of waves in plasmas are, respectively,

P.A. Sturrock, *Plasma Physics* (Cambridge University Press, Cambridge, 1994).

T.J.M. Boyd and J.J. Sanderson, *The Physics of Plasmas* (Cambridge University Press, Cambridge, 2003).

The discovery of the dispersive properties of the ionosphere is a fascinating story. See

W.J.G. Beynon and J.A. Ratcliffe, "A discussion on the early days of ionospheric research and the theory of electric and magnetic waves in the ionosphere and magnetosphere", *Philosophical Transactions of the Royal Society of London A* **280**(1293) 1 (1975).

The principal sources for our discussion of the split-ring oscillator were

J.B. Pendry, A.J. Holden, D.J. Robbins, and W.J. Stewart, "Magnetism from conductors and enhanced nonlinear phenomena", *IEEE Transactions on Microwave Theory and Techniques* **47**, 2075 (1999).

S. Linden, C. Enkrich, G. Dolling, *et al.* "Photonic metamaterials: Magnetism at optical frequencies", *IEEE Journal of Selected Topics in Quantum Electronics* **12**, 1097 (2006).

R. Merlin, "Metamaterials and the Landau-Lifshitz permeability argument: Large permittivity begets high-frequency magnetism", *Proceedings of the National Academy of Sciences* **106**, 1693 (2009).

Section 18.6 A reference book which devotes several chapters to wave packet (pulse) propagation in dispersive media is

A.E. Siegmann, *Lasers* (University Science Books, Sausalito, CA, 1986).

Figure 18.9 and Figure 18.10 were adapted, respectively, from

K.E. Oughstun, *Electromagnetic and Optical Pulse Propagation 1: Spectral Representations in Temporally Dispersive Media* (Springer, New York, 2006). With kind permission from Springer Science & Business Media B.V.

D.J. Gauthier and R.W. Boyd, "Fast light, slow light and optical precursors: What does it all mean?", *Photonics Spectra*, January 2007, pp. 82-90.

For more on anomalous group velocities, see

K.T. McDonald, "Negative group velocity", *American Journal of Physics* **69**, 607 (2001).

The source of Figure 18.12 and a good introduction to whistler phenomena are, respectively,

J.L. Green and U.S. Inan, "Lightning effects on space plasmas and applications", in *Plasma Physics Applied*, edited by C.L. Grabbe (Transworld Research Network, Trivandrum, India, 2006), Chapter 4.

M. Hayakawa, "Whistlers", in *Handbook of Atmospheric Electrodynamics*, edited by H. Volland (CRC Press, Boca Raton, FL, 1995), Volume II, pp. 155-194.

Section 18.7 Our treatment of the Kramers-Kronig relations and Figure 18.14 (with its discussion) come from

B. Harbecke, "Application of Fourier's allied integrals to the Kramers-Kronig transformation of reflectance data", *Applied Physics A* **40**, 1151 (1986).

J.S. Toll, "Causality and the dispersion relation: Logical foundations", *Physical Review* **104**, 1760 (1956).

The classic treatise on the subject of signal velocity and forerunners, and a monograph that brings this work up to date, are

L. Brillouin, *Wave Propagation and Group Velocity* (Academic, New York, 1960).

K.E. Oughstun and G.C. Sherman, *Electromagnetic Wave Propagation in Causal Dielectrics* (Springer, New York, 1994).

A gentle introduction to spatial dispersion is

D.L. Mills, "Spatial dispersion and its effect on the properties of dielectrics", in *Polaritons*, edited by E. Burstein and F. De Martini (Pergamon, New York, 1974), pp. 147-160.

Problems

18.1 Electric Susceptibilities in Time and Frequency Find the frequency-dependent susceptibility $\hat{\chi}(\omega)$ when the temporal susceptibility $\chi(t)$ of a medium is

(a) $\chi(t) = \chi_0 \delta(t)$
(b) $\chi(t) = \chi_0 \theta(t)$
(c) $\chi(t) = \chi_0 \theta(t) \exp(-t/\tau)$
(d) $\chi(t) = \chi_0 \theta(t) \sin(\omega_0 t)$.

Hint: Delta functions appear naturally in some of these if you use a convergence factor $\lim_{\epsilon \to 0} \exp(-\epsilon t)$ and remember that $\lim_{\epsilon \to 0} \epsilon/(\omega^2 + \epsilon^2) = \pi \delta(\omega)$.

18.2 Magnetization and Conductivity

(a) A homogeneous medium is characterized by a magnetic susceptibility χ_m. Use Faraday's law to show that the magnetization current $\mathbf{J} = \nabla \times \mathbf{M}$ can be expressed in the form

$$J_i(\mathbf{r}, t) = \frac{\chi_m}{\mu_0(1 + \chi_m)} \int_{-\infty}^{t} dt' \left\{ \delta_{ij} \frac{\partial^2}{\partial x_k \partial x_k} - \frac{\partial^2}{\partial x_i \partial x_j} \right\} E_j(\mathbf{r}, t').$$

This result can be regarded as a special case of a homogeneous medium that obeys Ohm's law with a *conductivity tensor* that is non-diagonal, non-local in time, and non-local in space. That is, the total current can be written in the form

$$j_i(\mathbf{r}, t) = \int_{-\infty}^{t} dt' \int d^3 r' \sigma_{ij}(\mathbf{r} - \mathbf{r}', t - t') E_j(\mathbf{r}', t').$$

(b) Expand $\mathbf{E}(\mathbf{r}', t')$ in a Taylor series around $\mathbf{E}(\mathbf{r}, t')$ and show that $j_i(\mathbf{r}, t)$ contains a piece consistent with the form of $J_i(\mathbf{r}, t)$. Comment on the other pieces you find. What happens if the system is invariant to the symmetry operation of inversion ($\mathbf{r} \to -\mathbf{r}$)?

18.3 The Radio Operator's Friend The dielectric function of the ionosphere is $\epsilon(\omega)/\epsilon_0 = 1 - \Omega^2/\omega^2$, where Ω is a constant. Explain why a radio operator, exploiting the reflection of radio waves from the ionosphere, nearly always receives signals with $\omega > \Omega$ from distant broadcasting stations, but only occasionally or never receives signals with $\omega > \Omega$ from nearby stations.

18.4 Plane Waves of Vector Potential Show that plane wave propagation does not occur at all frequencies in a medium where the current density $\mathbf{j}$ is proportional to the vector potential: $\mu_0 \mathbf{j} = -k_0^2 \mathbf{A}$.

18.5 Plasma Sheath Let $\epsilon(\omega)/\epsilon_0 = 1 - \omega_p^2/\omega^2$ be the dielectric function of a plasma where ω_p is the plasma frequency. In a typical laboratory or astrophysical environment, any attempt to create a voltage drop $V(t) = V \cos \omega t$ across the plasma generates a region of vacuum (called the "sheath") on either side of the plasma volume as indicated in the one-dimensional sketch below.

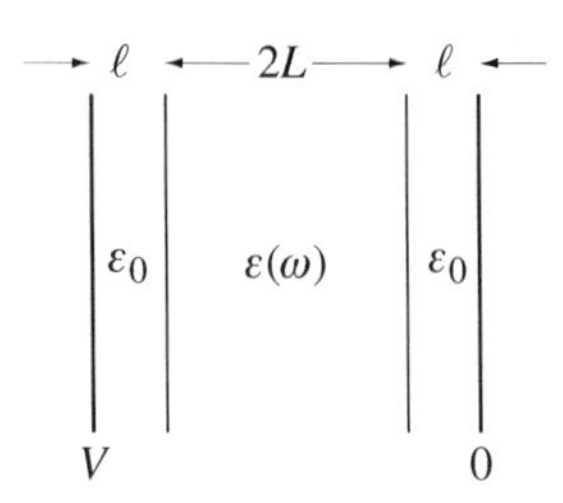

(a) Derive expressions for the uniform electric field $E_P(t) = E_P \cos \omega t$ in the plasma and for the uniform electric field $E_S(t) = E_S \cos \omega t$ in the sheath. Assume there is no free charge anywhere. Assume also that $\omega_{\rm p}$ is small enough that an *electrostatic* approximation is always valid.

(b) Plot the field amplitudes E_P and E_S on the same graph as a function of frequency. Discuss *where* the voltage drop occurs and *why* when (1) $\omega \ll \omega_{\rm p}$, (2) $\omega = \omega_{\rm p}$, and (3) $\omega \gg \omega_{\rm p}$.

(c) Make an "LC circuit" interpretation of the resonance behavior at $\omega = \omega_{\rm p}/\sqrt{1 + L/\ell}$.

18.6 Propagation in an Undamped Medium Derive the inhomogeneous wave equation satisfied by the electric field $\mathbf{E}(\mathbf{r}, t)$ in a system where $\rho(r, t) = 0$ but $\mathbf{j}(\mathbf{r}, t) \neq 0$. Show that this equation has a plane wave solution $\mathbf{E} = \mathbf{E}_0 \exp[i(\mathbf{k} \cdot \mathbf{r} - \omega t)]$ for a system of non-interacting electrons (number density n) that respond to an electric field $\mathbf{E}$ according to Newton's law. Display the dispersion relation $\omega(k)$ explicitly.

18.7 Surface Plasmon Polariton Let $\epsilon(\omega)/\epsilon_0 = 1 - \omega_{\rm p}^2/\omega^2$ be the dielectric function of the half-space $z > 0$. The half-space $z < 0$ is vacuum. Consider solutions of the appropriate wave equation inside and outside the conducting medium which are localized in the vicinity of the $z = 0$ free surface:

$$\mathbf{E}(x, z) = [\hat{\mathbf{x}} E_x^{\rm in/out} + \hat{\mathbf{z}} E_z^{\rm in/out}] \exp[i(qx - \omega t)] \exp(-\kappa_{\rm in/out}|z|).$$

(a) Relate κ to $q_\parallel$ in each medium.

(b) Use $\nabla \cdot \mathbf{D} = 0$ and the matching conditions for $\mathbf{E}$ to deduce that

$$\frac{\epsilon(\omega)}{\epsilon_0} = -\frac{\kappa_{\rm in}}{\kappa_{\rm out}} < 0.$$

(c) Derive the dispersion relation

$$q^2 = \frac{\omega^2}{c^2} \frac{\epsilon(\omega)/\epsilon_0}{1 + \epsilon(\omega)/\epsilon_0},$$

and find the physically allowed solution $\omega(q)$ for the specific choice of $\epsilon(\omega)$ given. Sketch $\omega(q)$ and investigate the limits $q \to 0$ and $q \to \infty$.

18.8 Inverse Faraday Effect An electromagnetic wave $\mathbf{E} = \delta \mathbf{E} \exp(-i\omega t)$ can induce a net magnetization in a metal. To see this, let the density and velocity of the electrons at a typical point be $n = \bar{n} + \delta n \exp(-i\omega t)$ and $\mathbf{v} = \bar{\mathbf{v}} + \delta \mathbf{v} \exp(-i\omega t)$, where $\bar{n}$ is the mean density of the electrons and $\bar{\mathbf{v}} = 0$ is the mean velocity of the electrons. The current density $\mathbf{j} = -en\mathbf{v}$ has two time-dependent pieces, one of which is $\delta \mathbf{j} = -e\bar{n}\delta \mathbf{v} = \sigma \delta \mathbf{E}$, where $\sigma = i\bar{n}e^2/m\omega$ is the collisionless Drude conductivity.

(a) Show that the time-averaged current density is $\langle \mathbf{j} \rangle = -\frac{1}{2}\mathrm{Re}\,\{e\delta n \delta \mathbf{v}^*\}$.

(b) Evaluate δn to first order in $\delta \mathbf{v}$ (using the continuity equation) and show that a piece of $\langle \mathbf{j} \rangle$ has the form $\nabla \times \mathbf{M}$ where (the plasma frequency is defined by $\omega_{\rm p}^2 = ne^2/m\epsilon_0$)

$$\mathbf{M} = \frac{i\epsilon_0 e \omega_{\rm p}^2}{4m\omega^3}(\delta \mathbf{E} \times \delta \mathbf{E}^*).$$

(c) Evaluate $\mathbf{M}$ when $\delta \mathbf{E}$ is linearly polarized. Repeat for circular polarization.

18.9 The Anomalous Skin Effect Drude's conductivity formula fails when the frequency ω is low and the mean time τ between electron collisions is large. If $\bar{v}$ is a characteristic electron speed, one says that the normal skin effect becomes *anomalous* when the mean distance between collisions $\ell = \bar{v}\tau$ exceeds the skin depth $\delta(\omega)$. To study this regime, we first write the rate of change of an ohmic current density $\mathbf{j}(t)$ as the sum of a field-driven acceleration term $d\mathbf{j}/dt|_{\rm acc} = (\sigma_0/\tau)\mathbf{E}$ and a collisional deceleration term $d\mathbf{j}/dt|_{\rm coll} = -\mathbf{j}/\tau$. This reproduces Ohm's law in the steady state $d\mathbf{j}/dt = 0$ because

$$\frac{d\mathbf{j}}{dt} = \frac{\sigma_0}{\tau}\mathbf{E} - \frac{\mathbf{j}}{\tau}.$$

(a) Approximate $d\mathbf{j}/dt$ by $\partial\mathbf{j}/\partial t$ and combine the foregoing with the Maxwell equations (neglecting the displacement current) to get a partial equation for $\mathbf{B}(\mathbf{r}, t)$ which has only first-order time derivatives:

$$\nabla^2\left[\mathbf{B} + \tau\frac{\partial\mathbf{B}}{\partial t}\right] = \mu_0\sigma_0\frac{\partial\mathbf{B}}{\partial t}.$$

Let $\mathbf{B}(z, t) = \mathbf{B}_0 e^{i(kz-\omega t)}$ and confirm that Drude's frequency-dependent conductivity emerges from your dispersion relation $k(\omega)$.

(b) Drude's conductivity formula overestimates the effect of collisions when $\ell \gg \delta$. A phenomenological way to correct this exploits the convective derivative to write

$$\frac{d\mathbf{j}}{dt} = \frac{\partial\mathbf{j}}{\partial t} - \bar{v}\frac{\partial\mathbf{j}}{\partial z}.$$

Derive a cubic equation which determines the new dispersion relation. Find $k(\omega)$ explicitly in the extreme anomalous limit (where the gradient term dominates) and show that

$$\mathbf{B}(z, t) = \mathbf{B}_0 \exp\left[(i - \sqrt{3})z/\delta^*(\omega)\right]e^{-i\omega t}.$$

The anomalous skin depth $\delta^*(\omega) = 2(\Lambda^2\bar{v}/\omega)^{1/3}$ found here describes experiments well in this regime. The constant $\Lambda^2 = m/\mu_0 ne^2$.

(c) Show that $\ell \ll \delta = \sqrt{2/\mu_0\omega\sigma_0}$ is the condition to neglect the non-local gradient term.

18.10 Energy Storage and Energy Loss

(a) Consider a medium composed of N one-dimensional, undamped Lorentz oscillators per unit volume. Show by explicit calculation that the time average of

$$u_{\rm EM}(t) = \frac{1}{2}\frac{\partial}{\partial\omega}\left[\omega\epsilon'(\omega)\right]|\mathbf{E}(t)|^2$$

is equal to the time average of the sum of the electric, kinetic, and potential energy densities of the medium.

(b) Consider a Drude medium with momentum relaxation time τ. Show by explicit calculation that the time average of

$$Q(t) = \omega\epsilon''(\omega)|\mathbf{E}(t)|^2$$

is equal to the time average of the rate at which the fields do work on the charges in a unit volume of the medium.

18.11 The Lorenz-Lorentz and Drude Formulae Let the dielectric function $\epsilon(\omega) = \epsilon_0 n^2(\omega)$ characterize a macroscopic sphere of matter composed of N electrons. If the wavelength of the incident field is large compared to the sphere radius a, it is legitimate to use a quasistatic approximation. This problem equates two expressions for the polarization $\mathbf{P}(t)$ to find $\epsilon(\omega)$.

(a) Solve a quasi-electrostatic boundary value problem to find $\mathbf{P}(t)$ when the sphere is exposed to an external electric field $\mathbf{E}_0\cos(\omega t)$.

(b) Let the sphere have polarization $\mathbf{P}\cos\omega t$. Cut out a tiny sphere (of vacuum) around any one of its electrons and show that $\mathbf{P}(t)$ produces no force on the electron. Find the dipole moment of the electron induced by the electric field in part (a) and a spring-like force (with natural frequency ω_0) which holds the electron at the center of its vacuum sphere. Sum the moments from all the electrons and equate the resulting polarization to the result of part (a) to get the Lorenz-Lorentz formula,

$$3\frac{n^2(\omega)-1}{n^2(\omega)+2}=\frac{\omega_\mathrm{p}^2}{\omega_0^2-\omega^2},$$

where $\frac{4}{3}\pi a^3 n_0 = N$ and $\omega_\mathrm{p}^2 = n_0 e^2/m\epsilon_0$.

(c) For a neutral plasma, we suppose that the N electrons are distributed in a sphere of uniform positive charge. The forces on any one electron come from (i) the electric field due to the positively charged sphere; (ii) the (quasistatic) Coulomb interaction with the other $N-1$ electrons; and (iii) the external electric field. Show that force (ii) drops out in Newton's equation of motion for the total dipole moment $\mathbf{P} = -(e/V)\sum_{k=1}^{N}\mathbf{r}_k$ where $V = \frac{4}{3}\pi a^3$. Solve the equation of motion for $\mathbf{P}$ and equate this to the result in part (a) to get the high-frequency Drude formula,

$$n^2(\omega) = 1 - \frac{\omega_\mathrm{p}^2}{\omega^2}.$$

18.12 Loss and Gain Media Consider the Lorentz-type index of refraction

$$\hat{n}^2(\omega) = 1 + \frac{f\omega_\mathrm{p}^2}{\omega_0^2-\omega^2-i\omega\Gamma}.$$

The damping constant $\Gamma > 0$ and f is called the *oscillator strength*. Assume $|f| \ll 1$.

(a) Produce an argument based on monochromatic plane wave propagation that $f > 0$ describes an absorbing medium (like a conventional dielectric) which extracts energy from the field while $f < 0$ describes a gain medium (like a population of inverted atoms in a laser cavity) which supplies energy to the field.

(b) A wave packet propagates a distance L_A through an absorbing medium with $f_A > 0$ immediately after it propagates a distance L_G through a gain medium with $f_G < 0$. Under what conditions does the packet emerge undistorted from the absorbing medium? Hint: Do not make a group velocity (or any other) approximation to the sum of monochromatic plane waves that constitutes the packet.

18.13 A Magnetic Lorentz Model The figure below shows a sample of "artificial matter" composed of infinite, parallel, filamentary wires. Each row of wires carries current in the opposite direction from the rows just above and below it. Each row is also displaced (vertically and horizontally) by a distance $a/2$ from the rows just above and below it. Each wire feels a restoring force $-ku$ (per unit length) if it moves a distance u (perpendicular to its length) from its equilibrium position shown in the diagram. Each wire (mass per unit length m) also feels a damping force $-m\gamma\dot{u}$ when it is in motion.

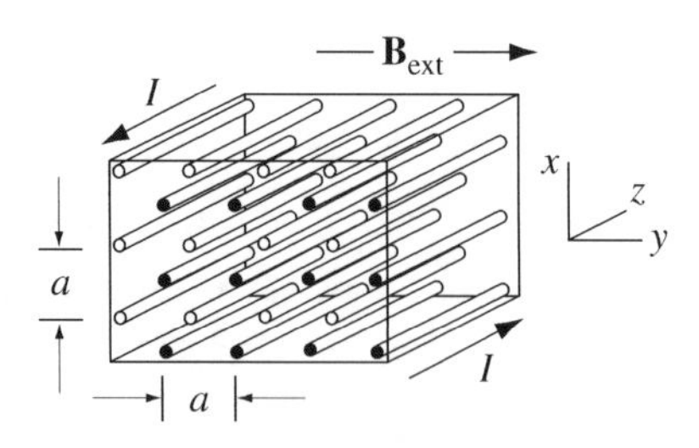

(a) Explain why the magnetic force exerted on each wire by the other wires can be neglected (compared to the force from an external magnetic field) when u is small.

(b) Let $+z$ be the direction of positive current I. Show that a uniform magnetic field $\mathbf{B} = \hat{\mathbf{y}} B \exp(-i\omega t)$ inside the sample induces a steady-state displacement of each wire in the x-direction by an amount

$$u_x(t) = \frac{IB}{m\omega^2 - k + im\gamma\omega} \exp(-i\omega t).$$

(c) The displacements in part (a) expose current sheets at the top and bottom faces (parallel to the y-z plane) of the sample. Taking this into account, show that, when exposed to an external field $\mathbf{B}_{\text{ext}} = \hat{\mathbf{y}} B_{\text{ext}} \exp(-i\omega t)$, the total field inside the sample satisfies the self-consistency equation

$$\mathbf{B} = \mathbf{B}_{\text{ext}} - \frac{\Omega_p^2}{\omega^2 - \omega_0^2 + i\gamma\omega} \mathbf{B},$$

where $\omega_0^2 = k/m$ and $\Omega_p^2 = 2\mu_0 I^2/ma^2$.

(d) Use the matching conditions for $\mathbf{H}$ to derive an expression for B/B_{ext}. Combine this information with the formula derived in part (c) to obtain an expression for the magnetic permeability $\mu(\omega)$ of the "wire matter" sample.

18.14 Energy Flow in the Lorentz Model A non-magnetic dielectric consists of N atoms per unit volume. Model the polarization of this system as the dipole moment per unit volume $\mathbf{P} = -Ne\mathbf{r}$, where $\mathbf{r}(t)$ is the displacement of each electron from its nucleus. The dynamics of each electron is modeled as a damped harmonic oscillator:

$$\frac{d^2\mathbf{r}}{dt^2} + \frac{1}{\tau}\frac{d\mathbf{r}}{dt} + \omega_0^2 \mathbf{r} = -\frac{e}{m}\mathbf{E}.$$

(a) Do not assume any particular time dependence for $\mathbf{E}$ and establish that

$$\nabla \cdot \mathbf{S} + \frac{\partial u_{\text{EM}}}{\partial t} + \frac{\partial u_{mech}}{\partial t} + N\frac{m}{\tau}\left|\frac{d\mathbf{r}}{dt}\right|^2 = 0,$$

where $\mathbf{S} = \mu_0^{-1}\mathbf{E} \times \mathbf{B}$, $u_{\text{EM}} = \frac{1}{2}\epsilon_0(\mathbf{E} \cdot \mathbf{E} + c^2\mathbf{B} \cdot \mathbf{B})$, and u_{mech} is the total mechanical (kinetic plus potential) energy density of the Lorentz oscillators. Physically interpret the last term on the left-hand side above.

(b) Now assume that a time-harmonic wave propagates through this medium. The index of refraction $\hat{n}(\omega) = \hat{n}_1(\omega) + i\hat{n}_2(\omega)$ is

$$\hat{n}^2(\omega) = 1 + \frac{\omega_{\text{p}}^2}{\omega_0^2 - \omega^2 - i\omega/\tau},$$

where $\omega_{\text{p}}^2 = Ne^2/m\epsilon_0$. Show that $\langle \mathbf{S} \rangle = \frac{1}{2}\epsilon_0 c\hat{n}_1 |E|^2 \hat{\mathbf{k}}$ is the time-averaged Poynting vector if the real and imaginary parts of $\mathbf{k}$ are parallel.

(c) Show that the energy density $\langle u \rangle = \langle u_{\text{EM}} + u_{mech} \rangle = \frac{1}{2}\epsilon_0 |E|^2 (\hat{n}_1^2 + 2\omega\tau\hat{n}_1\hat{n}_2)$ when similarly averaged so that the "energy velocity"

$$v_{\text{E}}(\omega) = \frac{\langle S \rangle}{\langle u \rangle} = \frac{c}{\hat{n}_1 + 2\omega\tau\hat{n}_2}.$$

(d) Prove that $v_{\text{E}}(\omega) < c$ for the Lorentz model.

18.15 A Paramagnetic Microwave Amplifier Let a transverse electromagnetic wave $\mathbf{H} = \hat{\mathbf{x}} H_x \exp i(ky - \omega t)$ propagate in a linear magnetic medium exposed to a static magnetic field $\mathbf{B} = B_z\hat{\mathbf{z}}$. If γ and τ are constants, experiment shows that the induced magnetization obeys

$$\frac{d\mathbf{M}}{dt} = \gamma(\mathbf{M} \times \mathbf{B}) - \frac{M}{\tau}(\hat{\mathbf{M}} - \hat{\mathbf{z}}).$$

(a) The first term on the right describes precession of the magnetization vector. The second term on the right side accounts (phenomenologically) for loss mechanisms which drive the system toward equilibrium (where $\mathbf{M}$ is aligned with the external field). Confirm this claim by computing $\mathbf{M}(t)$ when $\gamma = 0$.

(b) Let $\omega_0^2 = \gamma^2 \mu_0 B_z H_z$ and show that

$$\frac{d^2 M_x}{dt^2} + \frac{2}{\tau}\frac{dM_x}{dt} + (\omega_0^2 + \tau^{-2})M_x = \omega_0^2 \frac{M_z}{H_z} H_x.$$

Use this information to find the real and imaginary parts of the complex magnetic permeability $\hat{\mu}(\omega)$. Establish that the magnetic permeability of the system is

$$\hat{\mu}(\omega) = \mu_0 \left\{ 1 + \frac{\omega_0^2}{\omega_0^2 + 1/\tau^2 - \omega(\omega + 2i/\tau)} \frac{M_z}{H_z} \right\}.$$

(c) Derive a wave equation for this medium and relate the real and imaginary parts of k to the real and imaginary parts of $\hat{\mu}$ and to ϵ (assumed real, positive, and constant). Prove that the amplitude of the **H**-wave given above decreases (increases) as it propagates if M_z/H_z is positive (negative).

Remark: Given the result in (c), amplification of the wave occurs only if we supply energy to "pump" the system into the higher energy state with **M** anti-parallel to **H**. This is the analog of producing a "population inversion" to initiate laser action in an active medium.

18.16 Limits on the Photon Mass If the photon had a mass M, the dispersion relation for electromagnetic waves in vacuum would be

$$\omega^2 = c^2 k^2 + (Mc^2/\hbar)^2.$$

A limit on $Mc^2 \ll \hbar\omega \approx \hbar ck$ can be determined by measuring the difference in arrival times of the highest- and lowest-frequency components of a wave packet received from an astrophysical source that emits electromagnetic energy in bursts. Estimate Δt from the proposed dispersion relation. Should one collect data from a radio source or from a γ-ray source to obtain the lowest limit on M? Why?

18.17 Negative and Infinite Group Velocity An infinite slab of material with index of refraction $n(\omega)$ and group velocity $v_g < 0$ occupies the space $0 < z < a$. The rest of space is vacuum.

(a) Consider a plane wave with electric field $\mathbf{E} = \hat{\mathbf{x}} E_0 \exp[i\omega(z/c - t)]$ incident on the slab from $z < 0$. Use $n(\omega)$ to write formulae for the wave in regions $0 < z < a$ and $z > a$. Assume $n(\omega)$ is not far from unity so reflection from the slab surfaces can be ignored. State the frequency-independent transformation that permits you to derive the $z > a$ field from the $0 < z < a$ field.

(b) Make the group velocity approximation $\omega n(\omega) \approx \omega_0 n(\omega_0) + (c/v_g)(\omega - \omega_0)$ and write formulae for $E(z,t)$ in all three regions.

(c) Let $E(z < 0, t) = f(z/c - t) = \int_{-\infty}^{\infty} d\omega\, \hat{E}(z,\omega) \exp(-i\omega t)$ be a wave packet incident on the slab. Find expressions for $E(z,t)$ for $0 < z < a$ and $z > a$ using the Fourier components calculated in part (b).

(d) Choose $f(x) = E_0 \delta(x)$ and deduce that the packet exits the slab before it enters.

(e) Let the $z < 0$ field be $E_0 \exp[-(z/c - t)^2/2\tau^2] \exp[i\omega_0(z/c - t)]$ and write formulae for the electric field in all three regions.

(f) For the Gaussian packet of part (e), let $\omega_0[n(\omega_0) - 1] = -i\epsilon$, where $\epsilon = 10^{-3}$. Otherwise, choose $c = 1$, $v_g = -1/2$, $a = 50$, and $\tau = 8$. Take out the vacuum plane wave as a common factor and plot a sequence of snapshots of $E(z,t)$ in the interval $-200 < z < 250$. Choose a sufficient number of times in the interval $-150 < t < 50$ to get a clear idea of how the packet passes through the slab.

(g) Repeat part (f) with $v_g = +\infty$ and $-40 < t < 50$. Explain the physical meaning of infinite group velocity.

18.18 Parseval's Relation Let the real and imaginary parts of $\hat{\chi}(\omega) = \chi'(\omega) + i\chi''(\omega)$ satisfy the Kramers-Kronig relation.

(a) $\Delta(x)$ is an acceptable representation of the delta function $\delta(x)$ if (i) $\Delta(0)$ diverges and (ii) $\Delta(x)$ exhibits the filtering property:

$$f(0) = \int_{-\infty}^{\infty} dx\, f(x)\Delta(x).$$

Show that these two properties are satisfied if

$$\Delta = \frac{1}{\pi^2}\int_{-\infty}^{\infty} \frac{dy}{y(y-x)}.$$

Hint: Substitute $\chi'(\omega)$ into $\chi''(\omega)$ or vice versa.

(b) Use $\Delta(x)$ in part (a) to prove Parseval's relation,

$$\int_{-\infty}^{\infty} d\omega |\chi'(\omega)|^2 = \int_{-\infty}^{\infty} d\omega |\chi''(\omega)|^2.$$

18.19 A Dispersive Dielectric The polarization of a medium obeys $\mathbf{P} = \gamma\nabla\times\mathbf{E}$.

(a) Find the propagation equation for the electric field $\mathbf{E}(\mathbf{r}, t)$ in this medium.

(b) Find the dispersion relation and polarization of the plane waves that propagate in this medium.

18.20 Lorentz-Model Sum Rule The Lorentz-model dielectric function satisfies the f-sum rule (see Application 18.5),

$$\int_{0}^{\infty} d\omega\, \omega\, \mathrm{Im}\,\hat{\epsilon}(\omega) = \frac{\pi}{2}\epsilon_0\omega_{\mathrm{p}}^2.$$

Show this explicitly for the case when the damping constant Γ is small.

19 Guided and Confined Waves

I managed to illuminate the interior of a stream of water in a dark space.
Jean-Daniel Colladon (1842)

19.1 Introduction

Many contemporary technologies exploit the fact that electromagnetic waves can be guided along specified paths through space and transiently stored in low-loss enclosures. The special configurations of conductors and dielectrics used to do this are called *waveguides* and *resonant cavities*. In this chapter, we show that electromagnetic fields can be guided and stored because they adjust themselves to satisfy the required boundary (or matching) conditions at the surfaces (or internal interfaces) of a guide or cavity. The nature and characteristics of the waves are fixed by the geometry and topology of the guiding and storage structures. Besides the familiar transverse electromagnetic (TEM) waves, where **E** and **B** are both transverse to the direction of propagation, we will find transverse electric (TE) waves where only **E** is transverse and transverse magnetic (TM) waves where only **B** is transverse. By and large, our discussion focuses on the applications of waveguides and cavities to specific problems of physics. Textbooks of engineering electromagnetics discuss applications to communication and power transmission.[1]

Guided waves were discovered in 1842 by the Swiss physicist Colladon, who reported that total internal reflection could be exploited to trap light inside the parabolic streams of water produced by drilling holes in a water-filled vessel. Fifty-five years later, Hertz sought and observed meter-scale waves guided by a conducting wire. Most readers will know that two-wire (power) lines and twisted-pair (telephone) lines have been used to guide low-frequency electromagnetic waves for over a century. Coaxial cables are used at higher frequencies (into the microwave) where shielding against radiation and interference are important. Lord Rayleigh recognized the wave guiding properties of hollow metal tubes as early as 1897 and physicists played a central role in the analysis of these structures for radar purposes during World War II. For many years, hollow-tube "microwave plumbing" competed with coaxial lines for communication and other purposes in the 10-100 GHz band. Low-cost planar arrangements of conductors and dielectrics known as striplines became available in the 1970s and their compatibility with integrated-circuit technology makes them the waveguide of choice for many high-frequency applications. Dielectric structures are efficient guides for electromagnetic waves at optical frequencies and steady progress in the development of low-loss glasses and laser technology have made it possible to connect the globe with optical-fiber networks for broadband communication. The operating principle for an optical fiber comes directly from Colladon's insight.

[1] See Sources, References, and Additional Reading.

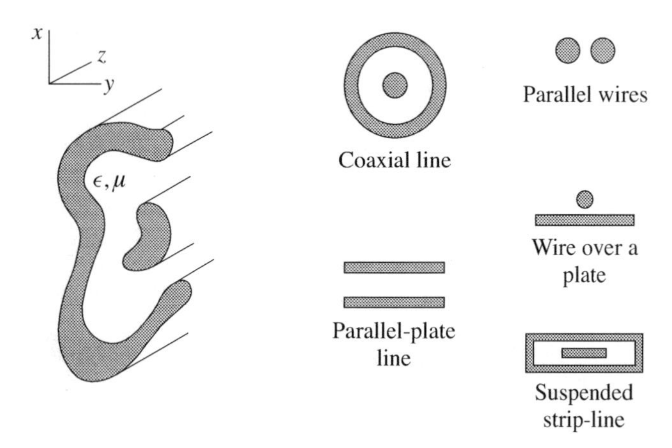

Figure 19.1: Two-conductor transmission lines. Left side: perspective view of a generic line. Right side: cross section views of typical lines. Figure from Staelin, Moregenthaler, and Kong (1994).

Resonant cavities store electromagnetic standing waves at a discrete set of frequencies. The most familiar example is the laser cavity operating in the optical and adjacent bands of the electromagnetic spectrum. The microwave klystron was invented in the late 1930s as an alternative to vacuum-tube technology. Today, it is used for high-power applications. The closely related magnetron oscillator is found in virtually every microwave oven. At lower frequencies, the tuning circuits for AM and FM radio are quasistatic resonators where the oscillating electric and magnetic fields do not overlap in space. Finally, the annular region between the surface of the Earth and the lower edge of the ionosphere turns out to form an enormous resonant cavity for very low frequency electromagnetic standing waves. The normal modes ($\sim$ 25 Hz) of this cavity are called Schumann resonances.

19.2 Transmission Lines

The vast majority of electromagnetic energy transport on the Earth's surface is performed by guided, transverse electromagnetic waves (TEM) where both **E** and **B** are transverse to the direction of propagation. The structures capable of guiding such waves are called *transmission lines*.[2] The left side of Figure 19.1 shows a generic example composed of two parallel perfect conductors whose cross sectional shapes do not vary along their lengths. The right side of Figure 19.1 shows several transmission-line geometries used in practice.

19.2.1 The Coaxial Line

The two coaxial cylinders in Figure 19.1 illustrate all the essential features of a TEM transmission line. We assume that a simple, non-conducting medium (ϵ, μ) fills the space between the inner and outer cylinders. We respect the symmetry of the problem and satisfy the perfect-conductor boundary conditions $\hat{\boldsymbol{\rho}} \times \mathbf{E}|_S = 0$ and $\hat{\boldsymbol{\rho}} \cdot \mathbf{H}|_S = 0$ on the surface S of both cylinders by guessing TEM traveling-wave solutions of the form

$$\mathbf{H} = H(\rho)\exp[i(hz - \omega t)]\hat{\boldsymbol{\phi}} \qquad \text{and} \qquad \mathbf{E} = E(\rho)\exp[i(hz - \omega t)]\hat{\boldsymbol{\rho}}. \tag{19.1}$$

[2] Transmission lines guide non-TEM waves also. We treat this more general class of waves in Section 19.4 in connection with hollow-tube waveguides.

Substituting (19.1) into the Maxwell equations (17.7) and (17.8) produces

$$\frac{\partial}{\partial\rho}\left[\rho E(\rho)\right] = 0, \qquad hE(\rho) = \omega\mu H(\rho), \qquad \text{and} \qquad hH(\rho) = \omega\epsilon E(\rho). \tag{19.2}$$

Consistency between the two rightmost equations in (19.2) requires that $\omega = h/\sqrt{\mu\epsilon}$. This dispersion relation is the same as for a uniform monochromatic plane wave in an infinite volume of the inter-cylinder medium. The additional information provided by the leftmost equation in (19.2) determines the propagating fields up to a constant I_0 with the dimensions of current:

$$\mathbf{H} = \frac{I_0}{\rho}\exp[i(hz - \omega t)]\hat{\boldsymbol{\phi}} \qquad \text{and} \qquad \mathbf{E} = \sqrt{\frac{\mu}{\epsilon}}\frac{I_0}{\rho}\exp[i(hz - \omega t)]\hat{\boldsymbol{\rho}}. \tag{19.3}$$

The TEM wave (19.3) is called *non-uniform* (see Section 17.3.7) because its amplitude is not constant on the planes of constant phase perpendicular to the z-direction of propagation. Note also that the ρ^{-1} dependencies of $\mathbf{E}$ and $\mathbf{H}$ in (19.3) are exactly what one finds, respectively, for the electric field of a cylindrical capacitor and the magnetic field of a wire carrying a steady current. We will see in the next section that this is not an accident.

The sources for the fields guided by the coaxial transmission line are the charge density $\sigma = \epsilon\hat{\mathbf{n}}\cdot\mathbf{E}|_S$ [see (5.15)] and the current density $\mathbf{K} = \hat{\mathbf{n}}\times\mathbf{H}|_S$ [see (10.39)] that appear on the surfaces of the conductors. Using (19.3) and the dispersion relation, the reader can check that the current flows in opposite directions on the inner and outer conductors and that

$$\nabla\cdot\mathbf{K} + \frac{\partial\sigma}{\partial t} = 0. \tag{19.4}$$

This surface version of the continuity equation shows that the coaxial transmission line respects conservation of charge, as it must.

19.2.2 TEM Waves

All transmission lines support TEM waves where the propagation wave vector $\mathbf{h} = h\hat{\mathbf{z}}$ can have any magnitude and the wave frequency satisfies the infinite-medium dispersion relation $\omega = h/\sqrt{\mu\epsilon}$. To prove this, we use the subscript $\perp$ to label vectors which lie in the plane perpendicular to $\mathbf{h}$ and substitute the fields $\mathbf{E} = \mathbf{E}_\perp$ and $\mathbf{H} = \mathbf{H}_\perp$ into the two Maxwell curl equations. If we separate the z-derivative from derivatives with respect to the transverse coordinates using

$$\nabla = \nabla_\perp + \hat{\mathbf{z}}\frac{\partial}{\partial z}, \tag{19.5}$$

the result is

$$\nabla_\perp\times\mathbf{E}_\perp + \hat{\mathbf{z}}\times\frac{\partial\mathbf{E}_\perp}{\partial z} = -\mu\frac{\partial\mathbf{H}_\perp}{\partial t} \qquad \text{and} \qquad \nabla_\perp\times\mathbf{H}_\perp + \hat{\mathbf{z}}\times\frac{\partial\mathbf{H}_\perp}{\partial z} = \epsilon\frac{\partial\mathbf{E}_\perp}{\partial t}. \tag{19.6}$$

The only vectors in (19.6) that point along $\hat{\mathbf{z}}$ are $\nabla_\perp\times\mathbf{E}_\perp$ and $\nabla_\perp\times\mathbf{H}_\perp$. Therefore,

$$\nabla_\perp\times\mathbf{E}_\perp = 0 \qquad \text{and} \qquad \nabla_\perp\times\mathbf{H}_\perp = 0. \tag{19.7}$$

In the absence of free charge and current embedded in the simple dielectric filling of the line, the Maxwell divergence equations read

$$\nabla_\perp\cdot\mathbf{E}_\perp = 0 \qquad \text{and} \qquad \nabla_\perp\cdot\mathbf{H}_\perp = 0. \tag{19.8}$$

Equations (19.7) and (19.8) tell us that, as far as the two variables transverse to z are concerned, the TEM electric and magnetic fields of a general two-conductor transmission line are de-coupled from one another and satisfy the equations of electrostatics and magnetostatics without sources. The

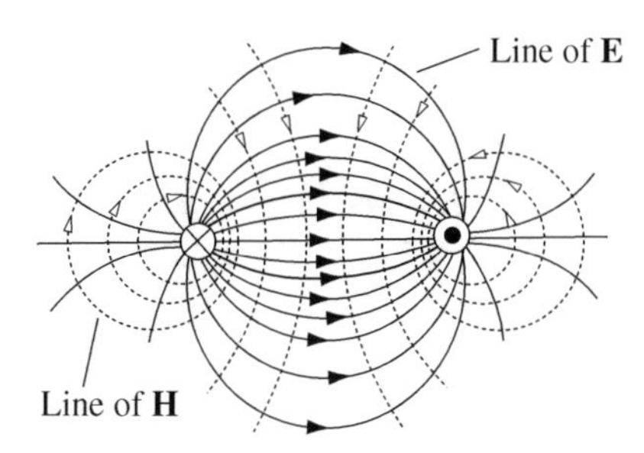

Figure 19.2: Electric field lines (solid) and magnetic field lines (dotted) in a plane transverse to the z-axis of a two-wire transmission line. The current in the right (left) wire flows out of (into) the paper.

source charge and current appear on the conductor surfaces in response to the boundary conditions $\mathbf{n} \times \mathbf{E}|_S = 0$ and $\mathbf{n} \cdot \mathbf{H}|_S = 0$, respectively. We complete the solution by looking for propagating waves where both $\mathbf{E}_\perp$ and $\mathbf{H}_\perp$ are proportional to $\exp[i(hz - \omega t)]$. Using this information in (19.6) and taking the cross product with $\hat{\mathbf{z}}$ reveals that

$$\omega = \frac{1}{\sqrt{\mu\epsilon}} h \qquad \text{and} \qquad \sqrt{\frac{\mu}{\epsilon}} \mathbf{H}_\perp = \hat{\mathbf{z}} \times \mathbf{E}_\perp. \tag{19.9}$$

Like the coaxial line studied in the previous section, a general two-conductor transmission line guides non-uniform TEM plane waves with the same dispersion relation as uniform TEM waves in an unbounded volume. This fundamental TEM mode propagates at any wavelength $2\pi/h$, with phase speed $v_\text{p} = \omega/h = 1/\sqrt{\mu\epsilon}$. The surface charge density $\sigma = \epsilon_0 \mathbf{n} \cdot \mathbf{E}|_S$ and surface current density $\mathbf{K} = \hat{\mathbf{n}} \times \mathbf{H}|_S$ similarly satisfy the surface continuity equation (19.4). We reiterate that the geometry of the guiding conductors defines electrostatic and magnetostatic problems which determine how the fields depend on the transverse variables. Figure 19.2 illustrates the field patterns for the case of two parallel wires. Note that the surface charge and surface current have opposite signs on the two conductors.

19.2.3 The Telegraph Equations

We have emphasized that the TEM wave propagation speed $v = 1/\sqrt{\mu\epsilon}$ is the same for all the transmission lines shown in Figure 19.1. This is so despite the very different spatial appearance of their propagating electric and magnetic fields. To gain some insight, we analyze the two-wire transmission line in Figure 19.2 (as a prototype) from the point of view of the voltage difference $V(z, t)$ between the wires and the current $I(z, t)$ that flows through them. Thus, let C be any path in the plane of the diagram which begins on the left wire and ends on the right wire (the curve labeled "Line of **E**" in Figure 19.2 is an example). The zero-curl condition (19.7) guarantees that the unique voltage difference is

$$V = -\int_C d\boldsymbol{\ell} \cdot \mathbf{E} = -\int_C (E_x dx + E_y dy). \tag{19.10}$$

Since $E_z = 0$ for TEM waves, $\nabla \times \mathbf{E} = -\partial \mathbf{B}/\partial t$ permits us to conclude that

$$\frac{\partial V}{\partial z} = -\frac{\partial}{\partial t} \int_C (B_x dy - B_y dx). \tag{19.11}$$

Consider now the surface S swept out by C when it translates rigidly by an amount dz. The unit normal to C (and thus also to S) satisfies $\hat{\mathbf{n}}d\ell = dy\hat{\mathbf{x}} - dx\hat{\mathbf{y}}$. Therefore, the magnetic flux through S is

$$\Phi_B = \int_S d\mathbf{S}\cdot\mathbf{B} = \int_S dz d\ell\hat{\mathbf{n}}\cdot\mathbf{B} = \int_S dz(B_x dy - B_y dx). \tag{19.12}$$

Comparing (19.11) to (19.12), we define an inductance per unit length $\mathcal{L}$ as the ratio of the magnetic flux per unit length to the current and write (19.11) as

$$\frac{\partial V}{\partial z} = -\mathcal{L}\frac{\partial I}{\partial t}. \tag{19.13}$$

We get a second relation between voltage and current as follows. First, write Ampère's law using a closed planar loop C' that encircles only the left wire (the curve labeled "Line of **H**" in Figure 19.2 is an example):

$$I = \oint_{C'} d\boldsymbol{\ell}\cdot\mathbf{H} = \oint_{C'}(H_x dx + H_y dy). \tag{19.14}$$

Second, use $H_z = 0$ and the validity of $\nabla\times\mathbf{H} = \partial\mathbf{D}/\partial t$ along the integration path to write

$$\frac{\partial I}{\partial z} = -\frac{\partial}{\partial t}\oint_{C'}(D_x dy - D_y dx). \tag{19.15}$$

Third, translate C' rigidly by dz to sweep out a surface S'. The unit normal to C' (and also to S') is $\hat{\mathbf{n}}d\ell = dy\hat{\mathbf{x}} - dx\hat{\mathbf{y}}$. Therefore, because $E_z = 0$, we define a charge per unit length along the wire, $\mathcal{Q}$, and write Gauss' law for the electric flux through S' as

$$\Phi_E = \int_{S'} d\mathbf{S}\cdot\mathbf{D} = \int_{S'} dz(D_x dy - D_y dx) = \mathcal{Q}dz. \tag{19.16}$$

Finally, define a capacitance per unit length of wire using $\mathcal{Q} = \mathcal{C}V$ and use (19.16) to write (19.15) as

$$\frac{\partial I}{\partial z} = -\mathcal{C}\frac{\partial V}{\partial t}. \tag{19.17}$$

Equations (19.13) and (19.17) are called the *telegraph equations*. They apply to any two-conductor transmission line because the geometric details of the transverse field distributions are "integrated out" by introducing the voltage and current as variables. A brief manipulation of the two telegraph equations shows that $V(z,t)$ and $I(z,t)$ both satisfy the wave equation

$$\frac{\partial^2 V}{\partial z^2} = \mathcal{L}\mathcal{C}\frac{\partial^2 V}{\partial t^2} \quad \text{and} \quad \frac{\partial^2 I}{\partial z^2} = \mathcal{L}\mathcal{C}\frac{\partial^2 I}{\partial t^2}. \tag{19.18}$$

We conclude that transmission lines support "voltage waves" and "current waves" which propagate at speed $v = 1/\sqrt{\mathcal{L}\mathcal{C}}$ down the line. This speed must equal the TEM phase speed ω/h in (19.9). Therefore, it is a characteristic feature of a transmission line that

$$\mathcal{L}\mathcal{C} = \mu\epsilon. \tag{19.19}$$

Example 19.1 Show that the telegraph equations (19.13) and (19.17) describe the electrical network shown below where every length element Δz contributes a capacitance $\mathcal{C}\Delta z$ and an inductance $\mathcal{L}\Delta z$.

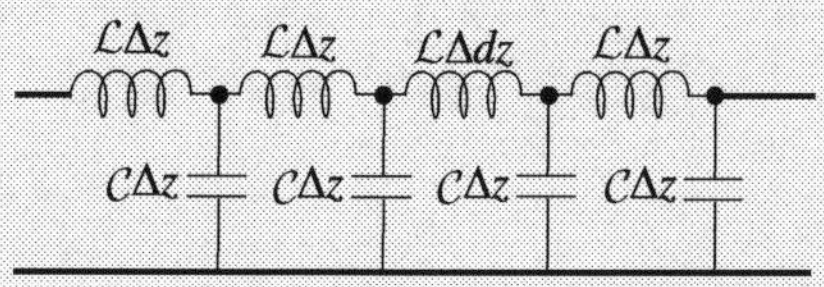

$\mathcal{L}\Delta z$ $\mathcal{L}\Delta z$ $\mathcal{L}\Delta dz$ $\mathcal{L}\Delta z$

$\mathcal{C}\Delta z$ $\mathcal{C}\Delta z$ $\mathcal{C}\Delta z$ $\mathcal{C}\Delta z$

Solution: According to AC circuit theory (Section 14.13), the voltage in this network drops by ΔV when current flows through an inductor of length Δz. The algebraic statement of this fact produces (19.13):

$$\Delta V = \frac{\partial V}{\partial z}\Delta z = -\mathcal{L}\Delta z\frac{\partial I}{\partial t}.$$

Similarly, the current decreases by ΔI when some of the charge $\Delta Q = \mathcal{C}\Delta z V$ in an element of line diverts to a shunt capacitor. This observation generates (19.17) because

$$\Delta I = \frac{\partial I}{\partial z}\Delta z = -\frac{d(\Delta Q)}{dt} = -\mathcal{C}\Delta z\frac{\partial V}{\partial t}.$$

Telegraphy and the Law of Squares

Before the telephone, the telegraph made long-distance communication possible by sending electrical signals through conducting wires. In 1851, the first telegraph line across the English channel was laid, and submarine cables below the North Sea, the Irish Sea, and the Mediterranean Sea followed soon thereafter. As cables grew longer, telegraph operators began to complain of signal retardation and distortion. Seeking help, telegraph companies approached prominent scientists of the day, including Michael Faraday and William Thomson (later Lord Kelvin). Faraday conducted a year of experiments and concluded that electric induction draws charge to the surface of the insulating material used to coat a telegraph wire. Thomson took account of the cable resistance and the cable capacitance implied by Faraday's observations to predict a "Law of Squares": the signal arrival time would increase with the square of the cable length. Experiments confirmed this prediction, thereby dampening enthusiasm for a proposed trans-Atlantic cable. Investors in the project engaged Thomson to advise them on modifications needed to salvage its technical viability by sailing on several cable-laying sea voyages as "chief cable engineer".

In 1876, Oliver Heaviside generalized Thomson's theory of telegraph signal propagation to include the effects of self-induction. To make contact with his theory, we need only consider a line resistance per unit length $\mathcal{R}$ and add a voltage drop term $-I\mathcal{R}$ to the right side of (19.13). Doing this, and combining the two telegraph equations as before gives

$$\frac{\partial^2 V}{\partial z^2} = \mathcal{L}\mathcal{C}\frac{\partial^2 V}{\partial t^2} + \mathcal{R}\mathcal{C}\frac{\partial V}{\partial t}.$$

The signal speed of modern low-loss cables is $v = 1/\sqrt{\mathcal{L}\mathcal{C}}$ because magnetic induction dominates electric resistance. The early submarine cables operated in the opposite limit where resistance dominates inductance. This situation reduces the preceding equation to Thomson's equation for the propagation of telegraph signals,

$$\frac{\partial^2 V}{\partial z^2} = \mathcal{RC}\frac{\partial V}{\partial t}.$$

This is a diffusion equation with diffusion constant $D = 1/\mathcal{RC}$. From Section 14.11 (or from dimensional analysis), we deduce that the time needed for a signal to diffuse a distance L is $T = L^2/D$. This is the Law of Squares.

19.3 Planar Conductors

The possibility of guiding waves more general than TEM, and the physics of wave guiding more generally, become clear when we revisit the problem of a plane wave incident on a perfectly conducting plane. The essential point is that an incident wave generally does not satisfy the boundary condition that the tangential component of the electric field vanish at a perfectly conducting surface,

$$\hat{\mathbf{n}} \times \mathbf{E}|_S = 0. \tag{19.20}$$

However, the incident field induces charges and/or currents on the conducting surface that are the source of a reflected (or scattered) field. It is the total field (incident plus reflected) that satisfies (19.20) and is guided by the conductor.

19.3.1 One Conducting Plane

A flat conducting surface guides electromagnetic waves parallel to itself. Consider Figure 19.3, which shows an s-polarized plane wave (see Section 17.3.2) incident on the conducting half-space $x \leq 0$. The incident electric field is tangential to the surface and thus does not satisfy (19.20). The remedy is to add a specularly reflected wave with its electric field anti-parallel to the incident electric field. With $\omega = ck_0$, the sum of the incident electric field, $\mathbf{E}_I = \hat{\mathbf{y}}E_0 \exp\left[ik_0(z\sin\theta - x\cos\theta - ct)\right]$, and the reflected electric field, $\mathbf{E}_R = -\hat{\mathbf{y}}E_0 \exp\left[ik_0(z\sin\theta + x\cos\theta - ct)\right]$, vanishes on the $x = 0$ plane:

$$\mathbf{E} = -2iE_0 \exp\left[i(k_0 z\sin\theta - \omega t)\right]\sin(k_0 x\cos\theta)\hat{\mathbf{y}}. \tag{19.21}$$

The associated magnetic field follows from Faraday's law, $\nabla \times \mathbf{E} = -\partial\mathbf{B}/\partial t$, as

$$c\mathbf{B} = 2E_0 \exp\left[i(k_0 z\sin\theta - \omega t)\right]\left\{i\sin\theta\sin(k_0 x\cos\theta)\hat{\mathbf{x}} - \cos\theta\cos(k_0 x\cos\theta)\hat{\mathbf{z}}\right\}. \tag{19.22}$$

The x-component of (19.22) vanishes on the $x = 0$ plane. This makes the normal component of $\mathbf{B}$ continuous at the conductor surface, as it must be.

The total field above the conducting surface defined by (19.21) and (19.22) is a standing wave in the x-direction and a traveling wave in the z-direction. The traveling part of the wave is not TEM. It is a transverse electric (TE) wave because only $\mathbf{E}$ is transverse to the z-direction of propagation.[3] The wave amplitude is not constant on the x-y planes of constant phase. This identifies the $x \geq 0$ field as a *non-uniform* plane wave (see Section 17.3.7) propagating in the z-direction with phase velocity

$$v_{\mathrm{p}} = \frac{\omega}{k_z} = \frac{c}{\sin\theta} > c. \tag{19.23}$$

[3] A transverse magnetic (TM) wave results if the incident plane wave is p-polarized rather than s-polarized.

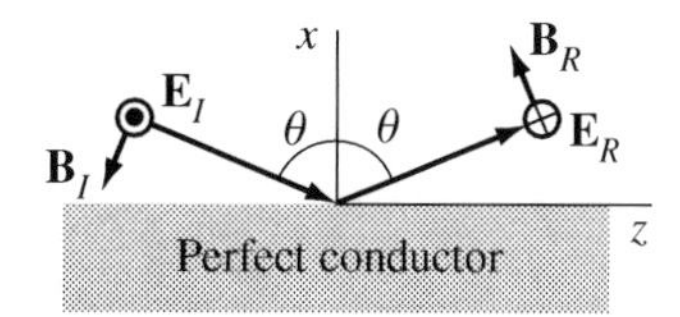

Figure 19.3: An s-polarized plane wave reflects specularly from the flat surface of a perfect conductor. The total field above the conductor is a TE wave propagating parallel to the surface.

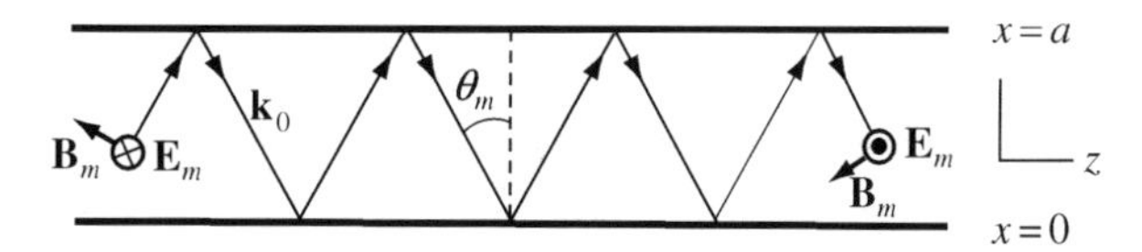

Figure 19.4: Multiple reflections of a TE mode of a parallel-plate transmission line.

The time average of the Poynting vector shows that the energy of the wave flows exclusively parallel to the interface:

$$\langle \mathbf{S} \rangle = \frac{1}{2\mu_0} \text{Re}(\mathbf{E}^* \times \mathbf{B}) = \frac{2E_0^2}{c\mu_0} \sin\theta \sin^2(k_0 x \cos\theta)\hat{\mathbf{z}}. \tag{19.24}$$

The charge density induced on the conductor is zero [see (5.15)]:

$$\sigma(z) = \epsilon_0 \hat{\mathbf{n}} \cdot \mathbf{E}|_S = \epsilon_0 E_x(x = 0) = 0. \tag{19.25}$$

The current density induced on the conductor is [see (10.39)]

$$\mathbf{K}(z, t) = \frac{1}{\mu_0}(\hat{\mathbf{n}} \times \mathbf{B})_S = \frac{1}{\mu_0}(\hat{\mathbf{x}} \times \mathbf{B})_{x=0} = \frac{2E_0}{c\mu_0} \cos\theta \exp[i(k_0 z \sin\theta - \omega t)]\hat{\mathbf{y}}. \tag{19.26}$$

The importance of the current density (19.26) cannot be overstated. It is the source of the reflected wave $\mathbf{E}_R$ in the vacuum space outside the conductor. It is also the source of a wave field *inside* the conductor which exactly annuls the incident wave in that region of space. The latter is an elementary example of the *extinction theorem*, a topic we treat more generally in Chapter 21.

19.3.2 Two Parallel Planes

A glance at (19.21) and (19.22) shows that $\hat{\mathbf{x}} \times \mathbf{E} = 0$ and $\hat{\mathbf{x}} \cdot \mathbf{B} = 0$ on planes parallel to the conductor surface defined by $x_m = m\pi/k_0 \cos\theta$, where m is a positive integer. This implies that the field is undisturbed and all boundary conditions are satisfied if a perfectly conducting sheet is inserted into every such plane. On the other hand, if we add a single conducting plane at $x = a$ to create the parallel-plate transmission line as shown in Figure 19.4, the electric field boundary condition $E_y(x = a) = 0$ applied to (19.21) quantizes the "angle of incidence" to a set of allowed values defined by

$$k_0 \cos\theta_m = m\frac{\pi}{a}. \tag{19.27}$$

The corresponding wave can be visualized as a sequence of specular reflections from the top and bottom planes as shown in Figure 19.4. The electromagnetic field in the space $x > a$ is annulled by a surface current like (19.26) which appears on the $x = a$ plane. We introduce a quantized propagation factor h_m defined by

$$h_m^2 = k_0^2 \sin^2\theta_m = k_0^2(1 - \cos^2\theta_m) = \frac{\omega^2}{c^2} - m^2\left(\frac{\pi}{a}\right)^2, \tag{19.28}$$

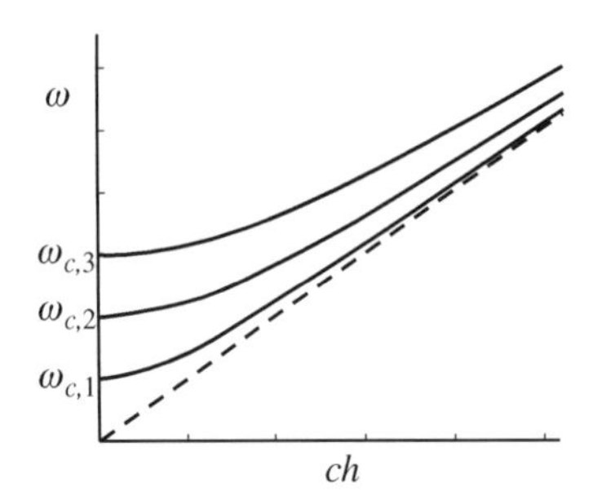

Figure 19.5: The dispersion curves $\omega(h)$ for the modes (19.29) of a parallel-plate transmission line.

and write the TE *mode functions* for this line as

$$\begin{aligned}\mathbf{E}_m &= -2iE_0 \sin\left(\frac{m\pi x}{a}\right)\exp\left[i(h_m z - \omega t)\right]\hat{\mathbf{y}} \\ c\mathbf{B}_m &= 2E_0\left\{i\sin\theta\sin\left(\frac{m\pi x}{a}\right)\hat{\mathbf{x}} - \cos\theta\cos\left(\frac{m\pi x}{a}\right)\hat{\mathbf{z}}\right\}\exp\left[i(h_m z - \omega t)\right].\end{aligned} \tag{19.29}$$

Alternatively, we can treat the propagation wave vector h as a continuous variable and plot (19.28) as shown in Figure 19.5. From this point of view, the conducting-wall boundary conditions induce frequency dispersion of the vacuum waves between the plates. This is called *structural dispersion* to distinguish it from the intrinsic dispersion of waves in matter treated in Chapter 18. Each transmission line mode (indexed by m) has its own dispersion relation,

$$\omega_m = c\sqrt{h^2 + m^2\pi^2/a^2}. \tag{19.30}$$

The hyperbolae in Figure 19.5 show that each mode propagates only when ω exceeds the characteristic *cutoff* frequency for that mode, $\omega_{c,m}$, defined by $h = 0$ in (19.30):

$$\omega_{c,m} = m\frac{\pi c}{a}. \tag{19.31}$$

Exactly as we saw for transverse wave propagation in a Drude medium above and below the plasma frequency $\omega_{\rm p}$ (Section 18.5.2), the propagation factor h is real when $\omega > \omega_c$ and imaginary when $\omega < \omega_c$. The wave is evanescent in the latter case and decays exponentially as z increases. Combining (19.31) with (19.27) shows that $\theta_m = 0$ at cutoff. From this point of view, cutoff occurs because the total field bounces back and forth in the direction normal to the conducting walls in Figure 19.4 with no advancement down the guide. The Poynting vector (19.24) confirms that energy flow ceases when propagation ceases.

The mode functions in (19.29) continue to satisfy all the relevant boundary conditions when the long open sides of the transmission line are closed by two parallel conducting planes to form a tube with a rectangular cross section. The electric field is entirely normal to the new walls and thus generates a surface charge density. The magnetic field is entirely tangential to the new walls and generates a surface current density. By their action, the waveguide entirely confines the electromagnetic field to the volume enclosed by its walls. The field is identically zero everywhere outside the walls. The same statements apply to a set of transverse magnetic (TM) mode functions. The latter appear naturally in the general discussion to follow.

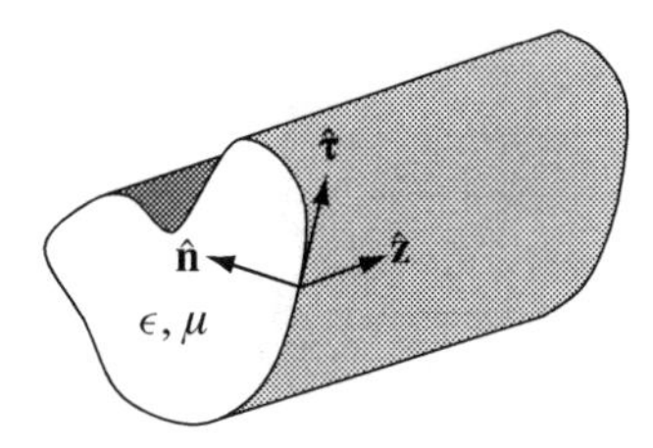

Figure 19.6: A conducting-tube waveguide. The unit normal $\hat{\mathbf{n}}$ points inward. The orthogonal unit vectors $\hat{\mathbf{t}}$ and $\hat{\mathbf{z}}$ are both tangent to the walls.

19.4 Conducting Tubes

The introduction to this chapter pointed out the importance of conducting-tube waveguides in the history of physics. The sketch in Figure 19.6 illustrates a geometry we can analyze in detail: an infinitely long tube with conducting walls and a uniform cross sectional shape. The tube is filled with simple dielectric matter (μ, ϵ). In this section, we prove that any field guided by such a structure can be decomposed into two independent fields, one where $E_z = 0$ and $H_z \neq 0$, and one where $H_z = 0$ and $E_z \neq 0$. Later sections focus on the detailed structure and characteristics of the propagating modes. If $\mathbf{r} = (\mathbf{r}_\perp, z)$, a general form for fields that propagate down the guide is

$$\begin{aligned}\mathbf{E}(\mathbf{r}, t) &= [\mathbf{E}_\perp(\mathbf{r}_\perp) + \hat{\mathbf{z}} E_z(\mathbf{r}_\perp)] \exp[i(hz - \omega t)] \\ \mathbf{H}(\mathbf{r}, t) &= [\mathbf{H}_\perp(\mathbf{r}_\perp) + \hat{\mathbf{z}} H_z(\mathbf{r}_\perp)] \exp[i(hz - \omega t)].\end{aligned} \tag{19.32}$$

The fields are harmonic, so the Maxwell divergence equations, $\nabla \cdot \mathbf{E} = 0$ and $\nabla \cdot \mathbf{H} = 0$, are satisfied automatically when we impose the Maxwell curl equations,

$$\nabla \times \mathbf{E} = i\omega\mu\mathbf{H} \qquad \text{and} \qquad \nabla \times \mathbf{H} = -i\omega\epsilon\mathbf{E}. \tag{19.33}$$

Using (19.5) to evaluate the curl, the fact that $\mathbf{E}_\perp$ and E_z do not depend on z in (19.32) shows that

$$\nabla \times \mathbf{E} = [\nabla_\perp \times \mathbf{E}_\perp + ih\hat{\mathbf{z}} \times \mathbf{E}_\perp - \hat{\mathbf{z}} \times \nabla_\perp E_z] \exp[i(hz - \omega t)]. \tag{19.34}$$

The first term in square brackets in (19.34) points along $\hat{\mathbf{z}}$. The other two terms point transverse to $\hat{\mathbf{z}}$. Therefore, substituting (19.34) and a similar expression for $\nabla \times \mathbf{H}$ into (19.33) gives

$$\begin{aligned}\nabla_\perp \times \mathbf{E}_\perp &= i\omega\mu H_z \hat{\mathbf{z}} \\ i\omega\mu\mathbf{H}_\perp - ih\hat{\mathbf{z}} \times \mathbf{E}_\perp &= -\hat{\mathbf{z}} \times \nabla_\perp E_z,\end{aligned} \tag{19.35}$$

and

$$\begin{aligned}\nabla_\perp \times \mathbf{H}_\perp &= -i\omega\epsilon E_z \hat{\mathbf{z}} \\ i\omega\epsilon\mathbf{E}_\perp + ih\hat{\mathbf{z}} \times \mathbf{H}_\perp &= \hat{\mathbf{z}} \times \nabla_\perp H_z.\end{aligned} \tag{19.36}$$

If we regard E_z and H_z as non-zero and known, (19.35) and (19.36) are four linear equations in the four unknown scalar components of $\mathbf{E}_\perp$ and $\mathbf{H}_\perp$. This proves that the transverse field components are determined by the longitudinal field components. To be more explicit, take the cross product of $\hat{\mathbf{z}}$ with the second equation in (19.36), substitute $\hat{\mathbf{z}} \times \mathbf{E}_\perp$ into the second equation in (19.35), and solve for

$\mathbf{H}_\perp$. The result can be written

$$\mathbf{H}_\perp = \frac{ih}{\gamma^2}\nabla_\perp H_z + \frac{i\omega\epsilon}{\gamma^2}\hat{\mathbf{z}} \times \nabla_\perp E_z, \tag{19.37}$$

where γ is an inverse length defined by

$$\gamma^2 = \mu\epsilon\omega^2 - h^2. \tag{19.38}$$

Next, use (19.37) to compute $\hat{\mathbf{z}} \times \mathbf{H}_\perp$, substitute this into the second equation in (19.36), and solve for $\mathbf{E}_\perp$. This gives

$$\mathbf{E}_\perp = \frac{ih}{\gamma^2}\nabla_\perp E_z - \frac{i\omega\mu}{\gamma^2}\hat{\mathbf{z}} \times \nabla_\perp H_z. \tag{19.39}$$

It remains only to determine the longitudinal components. To isolate E_z, take the curl of (19.37) and use the first equation in (19.36) to eliminate $\nabla_\perp \times \mathbf{H}_\perp$. To isolate H_z, take the curl of (19.39) and use the first equation in (19.35) to eliminate $\nabla_\perp \times \mathbf{E}_\perp$. Because E_z and H_z do not depend on z, these steps show that both functions satisfy the same two-dimensional Helmholtz equation:

$$\left[\nabla_\perp^2 + \gamma^2\right]\begin{Bmatrix} E_z \\ H_z \end{Bmatrix} = 0. \tag{19.40}$$

Collecting these results, we conclude that any Maxwell field of the form (19.32) can be decomposed into the sum of two independent fields, called transverse electric (TE) and transverse magnetic (TM), where[4]

$$\begin{array}{cc} \text{Transverse Electric (TE)} & \text{Transverse Magnetic (TM)} \\ E_z = 0 & H_z = 0 \\ \left[\nabla_\perp^2 + \gamma^2\right] H_z = 0 & \left[\nabla_\perp^2 + \gamma^2\right] E_z = 0 \\ \mathbf{H}_\perp = \dfrac{ih}{\gamma^2}\nabla_\perp H_z & \mathbf{E}_\perp = \dfrac{ih}{\gamma^2}\nabla_\perp E_z \\ \mathbf{E}_\perp = -\dfrac{\omega\mu}{h}\hat{\mathbf{z}} \times \mathbf{H}_\perp & \mathbf{H}_\perp = \dfrac{\omega\epsilon}{h}\hat{\mathbf{z}} \times \mathbf{E}_\perp . \end{array} \tag{19.41}$$

More explicit formulae for the total fields are

$$\begin{aligned} \mathbf{E}_{\mathrm{TE}} &= -\frac{i\omega\mu}{\gamma^2}\left[\hat{\mathbf{z}} \times \nabla_\perp H_z\right]\exp[i(hz - \omega t)] \\ \mathbf{H}_{\mathrm{TE}} &= \left[\frac{ih}{\gamma^2}\nabla_\perp H_z + H_z\hat{\mathbf{z}}\right]\exp[i(hz - \omega t)], \end{aligned} \tag{19.42}$$

and

$$\begin{aligned} \mathbf{E}_{\mathrm{TM}} &= \left[\frac{ih}{\gamma^2}\nabla_\perp E_z + E_z\hat{\mathbf{z}}\right]\exp[i(hz - \omega t)] \\ \mathbf{H}_{\mathrm{TM}} &= \frac{i\omega\epsilon}{\gamma^2}\left[\hat{\mathbf{z}} \times \nabla_\perp E_z\right]\exp[i(hz - \omega t)]. \end{aligned} \tag{19.43}$$

The great similarity between (19.42) and (19.43) is no accident. Recalling the wave impedance $Z = \sqrt{\mu/\epsilon}$ from (17.12), a quick check confirms that

$$Z\mathbf{H}_{\mathrm{TE}} = \mathbf{E}_{\mathrm{TM}} \qquad \text{and} \qquad \mathbf{E}_{\mathrm{TE}} = -Z\mathbf{H}_{\mathrm{TM}}. \tag{19.44}$$

[4] The next-to-last line in each column of (19.41) has been used to simplify the last line in each column derived from (19.37) and (19.39).

The fields in (19.44) reflect the duality of our starting equations and thus (ultimately) the duality of the original Maxwell equations (see Section 16.8.1).

19.4.1 No TEM Waves in Conducting Tubes

TEM waves with $E_z = H_z = 0$ do not propagate in conducting-tube waveguides. This conclusion does *not* follow from (19.41) because those formulae were derived assuming either $E_z \neq 0$ or $H_z \neq 0$. Rather, $H_z = 0$ in the first member of (19.35) tells us that $\nabla_\perp \times \mathbf{E}_\perp = 0$. This implies that $\mathbf{E}_\perp = -\nabla\varphi$. Moreover, $\nabla \cdot \mathbf{E} = 0$ applied to the first member of (19.32) guarantees that $\nabla_\perp \cdot \mathbf{E}_\perp = 0$. Therefore, $\nabla_\perp^2 \varphi = 0$, where φ is the solution of a two-dimensional electrostatic problem with a closed conducting boundary. The potential is constant on any such boundary. Therefore, by Earnshaw's theorem (Section 3.3.3), the unique solution is $\varphi = const.$ throughout the cross sectional space enclosed by the boundary. This implies that $\mathbf{E}_\perp = 0$. Exactly the same logic applied to the first member of (19.36) with $E_z = 0$ leads to $\mathbf{H}_\perp = 0$. We conclude that no waves propagate when $E_z = H_z = 0$.

The difference between a conducting-tube waveguide and the transmission lines studied in Section 19.2 (where TEM waves do propagate) is the topology of the space occupied by the waves. That space is *simply connected* for a hollow tube and *not* simply connected for a transmission line. The latter makes it possible to set up a voltage difference between two distinct conductors and avoid the pre-conditions needed to exploit Earnshaw's theorem. Indeed, by inserting a second conducting tube (with a smaller cross sectional area) into the conducting tubes studied here, we get a structure which propagates TE waves, TM waves, and a TEM wave.

19.4.2 Boundary Conditions for TE and TM Modes

The conducting-wall boundary condition $\hat{\mathbf{n}} \times \mathbf{E}|_S = 0$ transforms the two equations in (19.40) into two eigenvalue problems. One eigenvalue problem determines the propagating TE modes; the other determines the propagating TM modes. In practice, we impose the boundary conditions

$$E_z|_S = 0 \quad \text{(TM waves)} \qquad \text{and} \qquad \left.\frac{\partial H_z}{\partial n}\right|_S = 0 \quad \text{(TE waves)}, \tag{19.45}$$

because (as we will now show) these conditions *imply* that $\hat{\mathbf{n}} \times \mathbf{E}|_S = 0$.

The orthogonal triad of unit vectors sketched in Figure 19.6 are basis vectors for the electric field expansion, $\mathbf{E} = \hat{\boldsymbol{\tau}} E_\tau + \hat{\mathbf{n}} E_n + \hat{\mathbf{z}} E_z$. Using this form, the third line on the right side of (19.41) gives

$$\hat{\mathbf{n}} \times \mathbf{E}|_S = \hat{\mathbf{z}} \left.\frac{ih}{\gamma^2} \frac{\partial E_z}{\partial \tau}\right|_S - \hat{\boldsymbol{\tau}} E_z|_S \qquad \text{(TM waves)}. \tag{19.46}$$

The condition on the left side of (19.45) makes (19.46) zero because the τ-derivative is performed along the surface. Similarly, the third and fourth lines on the left side of (19.41) give

$$\hat{\mathbf{n}} \times \mathbf{E}|_S = -\hat{\mathbf{z}} \left.\frac{i\omega\mu}{\gamma^2} \frac{\partial H_z}{\partial n}\right|_S - \hat{\boldsymbol{\tau}} E_z|_S \qquad \text{(TE waves)}. \tag{19.47}$$

The condition on the right side of (19.45) makes (19.47) zero because $E_z = 0$ is true everywhere for a TE wave. We conclude that the boundary conditions in (19.45) complete the waveguide problem posed by (19.41) because $\hat{\mathbf{n}} \times \mathbf{E}|_S = 0$ is a sufficient condition to ensure a unique solution for the fields in the volume bounded by S (see Application 15.1). In later sections, we solve the two-dimensional eigenvalue problem explicitly for waveguides with rectangular and circular cross sections.

Example 19.2 Use Faraday's law to prove that the boundary condition $\hat{\mathbf{n}} \times \mathbf{E}|_S = 0$ applied to the walls of a conducting-tube waveguide implies that $\hat{\mathbf{n}} \cdot \mathbf{H}|_S = 0$ on those walls. Prove that the converse is not true.

Solution: Faraday's law is $\nabla \times \mathbf{E} = -\partial \mathbf{B}/\partial t$ and there is no essential difference between **B** and **H** for this problem. Therefore,

$$\frac{\partial}{\partial t}[\hat{\mathbf{n}} \cdot \mathbf{B}]_S = -\hat{\mathbf{n}} \cdot (\nabla \times \mathbf{E})_S = [\nabla \cdot (\hat{\mathbf{n}} \times \mathbf{E})]_S .$$

On the other hand, using the orthogonal triad of unit vectors in Figure 19.6,

$$[\nabla \cdot (\hat{\mathbf{n}} \times \mathbf{E})]_S = \left[\frac{\partial}{\partial n}\hat{\mathbf{n}} \cdot (\hat{\mathbf{n}} \times \mathbf{E})\right]_S + \left[\frac{\partial}{\partial z}\hat{\mathbf{z}} \cdot (\hat{\mathbf{n}} \times \mathbf{E})\right]_S + \left[\frac{\partial}{\partial \tau}\hat{\boldsymbol{\tau}} \cdot (\hat{\mathbf{n}} \times \mathbf{E})\right]_S .$$

The first term on the right side of this identity is zero because $\hat{\mathbf{n}}$ is perpendicular to $\hat{\mathbf{n}} \times \mathbf{E}$. Then, because $\hat{\boldsymbol{\tau}}$ and $\hat{\mathbf{z}}$ are both tangent to the boundary, the second and third terms on the right side of the identity are also zero if we impose $\hat{\mathbf{n}} \times \mathbf{E}|_S = 0$ at every point on the boundary S. This proves the assertion because $\hat{\mathbf{n}} \cdot \mathbf{B}$ varies in time and cannot be a constant on the boundary other than zero.

As for the converse, we require $\hat{\mathbf{n}} \cdot \mathbf{B}|_S = 0$, and the same algebra makes the sum of the last two terms in the identity above equal to zero. However, this could happen by some accident of the τ and z derivatives, rather than because $\hat{\mathbf{n}} \times \mathbf{E}$ vanishes everywhere on the boundary. Therefore, $\hat{\mathbf{n}} \cdot \mathbf{B}|_S = 0$ does not necessarily imply that $\hat{\mathbf{n}} \times \mathbf{E}|_S = 0$ for a hollow-tube waveguide.

19.4.3 General Properties of Modes in Conducting Tubes

In this section only, we use the function ψ to stand for both E_z and H_z. This permits us to rewrite the TE and TM problems defined by (19.40) and (19.45) as

$$\left[\nabla_\perp^2 + \gamma^2\right]\psi = 0 \qquad \text{with} \qquad \psi_{\rm TM}|_S = 0 \quad \text{or} \quad \left.\frac{\partial \psi_{\rm TE}}{\partial n}\right|_S = 0. \tag{19.48}$$

Equation (19.48) is a two-dimensional Helmholtz equation with Dirichlet or Neumann boundary conditions. The same equation and boundary conditions apply to the vertical displacements of a vibrating drumhead with either fixed (TM) or free (TE) boundary conditions. The TM case is also isomorphic to the Schrödinger problem for the wave functions and energy eigenvalues of a free particle in a two-dimensional box with hard walls. These analog problems provide useful intuition when thinking about the modal eigenfunctions ψ_i and eigenvalues γ_i of a waveguide. For example, the drumhead analogy can be used to show that the smallest value of γ_i always occurs for a TE mode. Some purely mathematical results are:

1. There are an infinite number of TE- and TM-mode eigenfunctions.
2. The eigenvalues are all real and positive.
3. The eigenfunctions can always be chosen real.
4. The eigenfunctions form a complete set of functions.
5. Eigenfunctions belonging to different eigenvalues are orthogonal.

A complete statement of proposition (5) involves the bracketed quantities in (19.32). Calling these $\mathbf{E}(\mathbf{r}_\perp)$ and $\mathbf{H}(\mathbf{r}_\perp)$, the orthogonality relations for distinct modes labeled μ and λ are integrals over the cross sectional area of the guide:

$$\int d^2 r_\perp \, \mathbf{E}_\mu \cdot \mathbf{E}_\lambda = 0 \qquad \int d^2 r_\perp \, \mathbf{H}_\mu \cdot \mathbf{H}_\lambda = 0 \qquad \int d^2 r_\perp \, \mathbf{E}_\mu \times \mathbf{H}_\lambda \cdot \hat{\mathbf{z}} = 0. \tag{19.49}$$

We leave the proof of (19.49) as an exercise. Here, we prove proposition (2) by exploiting (19.48) and an integration by parts to get a line integral over the perimeter of the guide's cross section. Bearing in mind that $\hat{\mathbf{n}}$ points inward in Figure 19.6,

$$\gamma_i^2 \int d^2r_\perp\, |\psi_i|^2 = -\int d^2r_\perp\, \psi_i^* \nabla_\perp^2 \psi_i = \int d^2r_\perp\, |\nabla_\perp \psi_i|^2 + \oint d\ell\, \psi_i^* \hat{\mathbf{n}} \cdot \nabla_\perp \psi_i. \tag{19.50}$$

The line integral in (19.50) vanishes for both TE and TM boundary conditions. Therefore, because $\gamma_i = 0$ corresponds to $\psi_i = const.$ (and thus identically zero fields), we conclude that

$$\gamma_i^2 = \frac{\int d^2r_\perp\, |\nabla_\perp \psi_i|^2}{\int d^2r_\perp\, |\psi_i|^2} > 0. \tag{19.51}$$

19.4.4 Structural Dispersion in Conducting Tubes

The wave modes of a conducting tube exhibit structural dispersion, by which we mean that the boundary condition at the tube wall makes the dispersion relation differ from $\omega(h) = h/\sqrt{\mu\epsilon}$. Indeed, the positivity of the mode eigenvalue γ_i^2 in (19.51) guarantees that (19.38) is physically meaningful for the ith mode of either TM or TE type:

$$\mu\epsilon\omega_i^2 = \left(\gamma_i^2 + h^2\right). \tag{19.52}$$

The dispersion curves sketched in Figure 19.5 also apply here, as does all the discussion in Section 19.3.2. Most important is that TE and TM waveguide modes exhibit the phenomenon of *cutoff*, where the wave vector h changes from real to imaginary and the wave (19.32) changes from propagating to evanescent when the frequency falls below the cutoff frequency,

$$\omega_{\text{c},i} = \frac{\gamma_i}{\sqrt{\mu\epsilon}} = \gamma_i \frac{c}{n}. \tag{19.53}$$

A geometrical picture of cutoff follows the argument following (19.31) for a parallel-plate transmission line. Note also, from Figure 19.5, that the (finite) number of propagating modes at a fixed value of ω increases as ω increases.

Using (19.52), a direct computation of the phase velocity of the ith propagating ($h^2 > 0$) mode gives

$$v_{\text{p},i} = \frac{\omega_i}{h} = \frac{c}{n} \frac{1}{\sqrt{1 - (\omega_{\text{c},i}/\omega_i)^2}}. \tag{19.54}$$

The corresponding group velocity is

$$v_{\text{g},i} = \frac{d\omega_i}{dh} = \frac{c}{n}\sqrt{1 - (\omega_{\text{c},i}/\omega_i)^2}. \tag{19.55}$$

Equation (19.55) shows that waves in a conducting tube exhibit group velocity dispersion of the sort studied in Section 18.6.2. This means that wave packets launched into waveguides inevitably spread and distort as they propagate down the guide. On the other hand, the connection between v_g and v_p is particularly simple for waves in conducting tubes:[5]

$$v_\text{g} v_\text{p} = \frac{c^2}{n^2} \qquad \text{and} \qquad v_\text{g} < \frac{c}{n} < v_\text{p}. \tag{19.56}$$

Figure 19.7 illustrates (19.56) geometrically using the plane wave propagation model of Figure 19.4. The planes of constant phase propagate parallel to themselves at the uniform plane wave speed c/n. An observer fixed at one point on the wall of the guide sees these planes go by at the phase speed v_p.

[5] The relation (19.56) is *not* generally valid for wave-guiding systems.

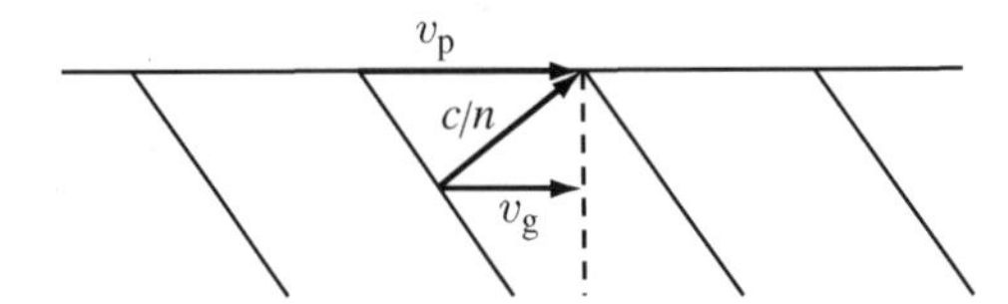

Figure 19.7: The relation between phase speed v_p, group speed v_g, and uniform plane wave speed c/n in a waveguide. Parallel lines are planes of constant phase.

The net propagation speed down the guide is the group speed v_g. The last statement requires proof, which we do in Section 19.4.7 by showing that v_g is the speed at which energy is transported down the guide.

19.4.5 Rectangular-Tube Waveguides

The simplest hollow-tube waveguide has a rectangular cross section with conducting walls at $x = 0$, $x = a$, $y = 0$, and $y = b$. Equation (19.48) reads

$$\begin{aligned} \text{TM}: \quad & \left[\frac{\partial^2}{\partial x^2} + \frac{\partial^2}{\partial y^2} + \gamma^2\right] E_z = 0 \quad \text{with} \quad E_z|_S = 0, \\ \text{TE}: \quad & \left[\frac{\partial^2}{\partial x^2} + \frac{\partial^2}{\partial y^2} + \gamma^2\right] H_z = 0 \quad \text{with} \quad \left.\frac{\partial H_z}{\partial n}\right|_S = 0. \end{aligned} \tag{19.57}$$

The eigensolutions follow immediately using separation of variables, namely,

$$\begin{aligned} E_z^{mn}(x, y) &= E \sin\left(\frac{m\pi x}{a}\right) \sin\left(\frac{n\pi y}{b}\right) \qquad m, n = 0, 1, \ldots, \\ H_z^{mn}(x, y) &= H \cos\left(\frac{m\pi x}{a}\right) \cos\left(\frac{n\pi y}{b}\right) \qquad m, n = 0, 1, \ldots, \end{aligned} \tag{19.58}$$

with the same eigenvalue spectrum for both:

$$\gamma_{mn} = \pi\sqrt{\frac{m^2}{a^2} + \frac{n^2}{b^2}}. \tag{19.59}$$

The TM fields in (19.43) vanish if $m = 0$ or $n = 0$. The TE modes in (19.42) are all non-zero except when $m = n = 0$. Indeed, the modes (19.29) of the parallel-plate transmission line are the TE_{m0} modes of a rectangular waveguide. If $a > b$, the lowest-frequency (principal) mode is TE_{10}, for which (19.53) gives the cutoff frequency,

$$\omega_c = \frac{1}{\sqrt{\mu\epsilon}} \frac{\pi}{a}. \tag{19.60}$$

In practice, ω is fixed just above ω_c so only the principal mode propagates. Combining (19.60) with the fact that it is easy to machine conducting tubes with centimeter and millimeter dimensions shows that single-mode propagation in conducting tubes occurs at microwave frequencies.

Figure 19.8 shows the instantaneous field line pattern for several low-frequency modes in a cross sectional plane of a rectangular guide. Half a period later in time, the field lines reverse direction and the surface charges (indicated by $\pm$ signs inside small circles) change sign. It is worth noting the spatial inhomogeneity of the field line density (field strength) and also the regions of space where the electric (magnetic) field lines for TM (TE) modes seem to appear or disappear. The latter are regions where the field lines have bent into or out of the direction of propagation. Figure 19.9 plots lines of current density $\mathbf{K}(x, y, z) = \hat{\mathbf{n}} \times \mathbf{H}|_S$ on the surfaces of the guide at one moment in time for a propagating TE_{10} mode.

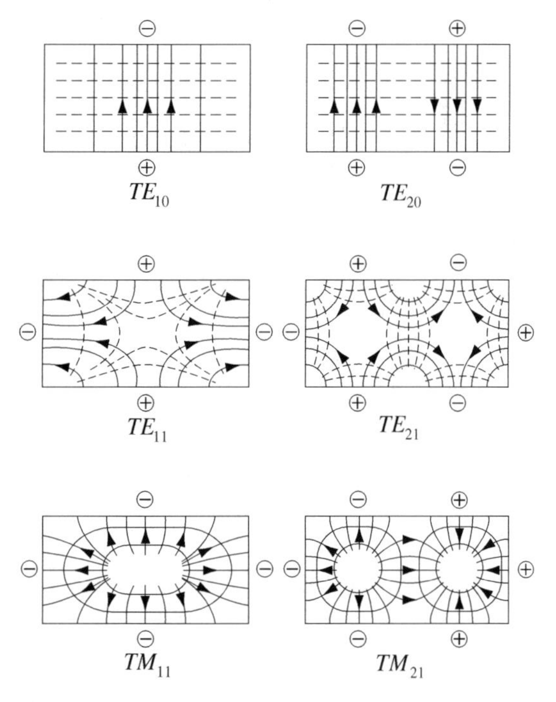

Figure 19.8: Cross sectional view of the instantaneous electric (solid) and magnetic (dashed) field line configurations for some low-frequency modes of a rectangular waveguide. Small circles indicate the sign of the charge induced on the walls. Figure from Borgnis and Papas (1958).

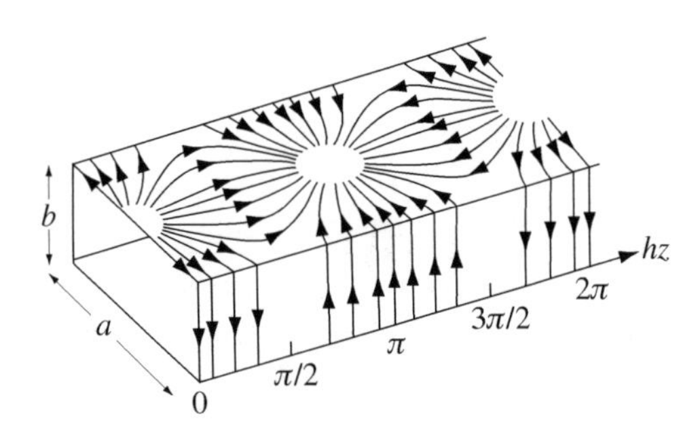

Figure 19.9: Instantaneous field line configuration of the current density **K** on the surface of a rectangular waveguide that carries a TE_{10} mode. Figure from Cheng (1989).

19.4.6 Circular-Tube Waveguides

We study the modes of a waveguide with a circular cross section by writing (19.48) in two-dimensional polar coordinates:

$$\begin{aligned} \text{TM}: \quad & \left[\frac{\partial^2}{\partial \rho^2} + \frac{1}{\rho}\frac{\partial}{\partial \rho} + \frac{1}{\rho^2}\frac{\partial^2}{\partial \phi^2} + \gamma^2\right] E_z = 0 \qquad \text{with} \qquad E_z|_S = 0, \\ \text{TE}: \quad & \left[\frac{\partial^2}{\partial \rho^2} + \frac{1}{\rho}\frac{\partial}{\partial \rho} + \frac{1}{\rho^2}\frac{\partial^2}{\partial \phi^2} + \gamma^2\right] H_z = 0 \qquad \text{with} \qquad \left.\frac{\partial H_z}{\partial n}\right|_S = 0. \end{aligned} \tag{19.61}$$

Separation of variables produces $\exp(im\phi)$ in the angular variable multiplied by Bessel functions (see Section 7.8.1) in the radial variable. Finiteness of the solution at the origin restricts us to $J_m(x)$ exclusively, so the boundary conditions for a guide of radius R are

$$J_m(\gamma^{\rm TM}R) = 0 \qquad \text{and} \qquad J'_m(\gamma^{\rm TE}R) = 0. \tag{19.62}$$

The prime on the right side of (19.62) indicates a derivative with respect to the argument. To satisfy (19.62), we exploit the $n^{\rm th}$ zero of the $m^{\rm th}$ Bessel function defined by $J_m(u_{mn}) = 0$, and the $n^{\rm th}$ zero of the derivative of the $m^{\rm th}$ Bessel function defined by $J'_m(w_{mn}) = 0$. In both cases, $n = 1$ labels the first zero, $n = 2$ labels the second zero, etc. Therefore, the eigenvalues are

$$\gamma_{mn}^{\rm TM} = \frac{u_{mn}}{R} \qquad \text{and} \qquad \gamma_{mn}^{\rm TE} = \frac{w_{mn}}{R}. \tag{19.63}$$

The mode eigenfunctions of a circular waveguide are doubly degenerate (plus/minus sign below) when $m > 0$:

$$\begin{aligned} E_z^{mn}(\rho, \phi) &= E J_m(\gamma_{mn}^{\rm TM}\rho)\exp(\pm im\phi) \qquad & m = 0, 1, \ldots \quad n = 1, 2, \ldots, \\ H_z^{mn}(\rho, \phi) &= H J_m(\gamma_{mn}^{\rm TE}\rho)\exp(+im\phi) \qquad & m = 0, 1, \ldots \quad n = 1, 2, \ldots \end{aligned} \tag{19.64}$$

The zeroes u_{mn} and w_{mn} are all irrational numbers, but they are well known and tabulated.[6] For example, the TE_{11} mode has the lowest cutoff frequency,

$$\omega_{\rm c} = \frac{1}{\sqrt{\mu\epsilon}}\frac{1.8412}{R}. \tag{19.65}$$

The numerical similarity of this frequency to (19.60) with $a = 2R$ reinforces the geometrical picture of cutoff derived from Figure 19.4. Figure 19.10 shows the instantaneous field line pattern in a cross sectional plane for the TE_{11} mode and a few other low-frequency modes.

19.4.7 The Energy Velocity

A figure of merit for any waveguide is the rate at which energy flows down the guide. A natural definition for this rate is the energy velocity,

$$\langle v_{\rm E} \rangle = \frac{\langle \mathcal{P} \rangle}{\langle U \rangle / L} = \frac{\text{time-averaged power flow through a cross section}}{\text{time-averaged energy per unit length}}. \tag{19.66}$$

The numerator of (19.66) is the integral over the guide cross section of the projection of the time-averaged Poynting vector along the guide axis. The longitudinal components of the fields in (19.32) do not contribute to this quantity because

$$\langle \mathbf{S} \rangle \cdot \hat{\mathbf{z}} = \frac{1}{2}\text{Re}\left(\mathbf{E} \times \mathbf{H}^*\right) \cdot \hat{\mathbf{z}} = \frac{1}{2}\text{Re}\left(\mathbf{E}_\perp \times \mathbf{H}_\perp^*\right) \cdot \hat{\mathbf{z}}. \tag{19.67}$$

Specializing to the TE case for illustration, the third and fourth lines of the left column of (19.41) show that (19.67) is zero for an evanescent mode where h is imaginary.[7] For a propagating mode where h is real,

$$\langle \mathcal{P}_{\rm TE} \rangle = \int d^2r_\perp \langle \mathbf{S} \rangle \cdot \hat{\mathbf{z}} = \frac{1}{2}v_{\rm p}\mu \int d^2r_\perp |\mathbf{H}_\perp|^2 = \frac{h^2 v_{\rm p}\mu}{2\gamma^4} \int d^2r_\perp |\nabla_\perp H_z|^2. \tag{19.68}$$

[6] See, e.g., M. Abramowitz and I.A. Stegun, *Handbook of Mathematical Functions* (Dover, New York, 1965).

[7] Thus, an evanescent wave does not transport energy down a waveguide. Nor does the exponential decay of its amplitude with distance imply that any dissipation of energy has occurred.

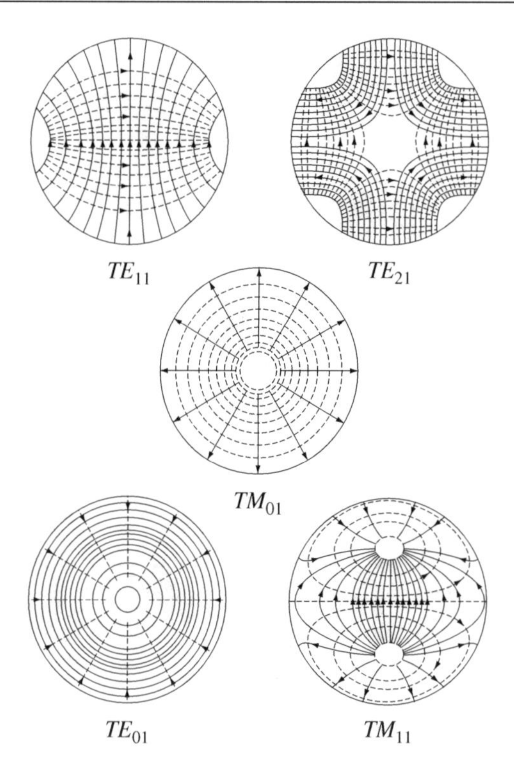

Figure 19.10: Instantaneous electric (solid) and magnetic (dashed) field line configurations for some low-frequency modes of a circular waveguide. Figure from Borgnis and Papas (1958).

Equation (19.40), the TE boundary condition in (19.45), and $v_{\rm p} = \omega/h$ reduce (19.68) to

$$\langle \mathcal{P}_{\rm TE}\rangle = \frac{h^2 v_{\rm p}\mu}{2\gamma^4}\int d^2r_\perp \left\{\nabla_\perp \cdot \left[H_z^*\nabla_\perp H_z\right] - H_z^*\nabla_\perp^2 H_z\right\} = \frac{\omega h\mu}{2\gamma^2}\int d^2r_\perp |H_z|^2. \tag{19.69}$$

Turning to the denominator of (19.66), the time-averaged total energy in a length L of guide is

$$\langle U_{\rm TE}\rangle = \frac{L}{4}\int d^2r_\perp \left[\epsilon|\mathbf{E}_\perp|^2 + \mu|\mathbf{H}|^2\right]. \tag{19.70}$$

We write $\mathbf{E}_\perp$ and $\mathbf{H} = \mathbf{H}_\perp + H_z\hat{\mathbf{z}}$ in (19.70) entirely in terms of H_z using (19.41) and the same steps used to pass from the last member of (19.68) to the last member of (19.69). Finally, the dispersion relation (19.38) allows us to write (19.70) in the form

$$\frac{\langle U_{\rm TE}\rangle}{L} = \frac{\omega^2\mu^2\epsilon}{2\gamma^2}\int d^2r_\perp |H_z|^2. \tag{19.71}$$

Forming the ratio (19.66) and using (19.56) gives the desired result: TE modes transport energy through a hollow conducting tube at the group velocity:

$$\langle v_{\rm E}\rangle = \frac{1}{\mu\epsilon v_{\rm p}} = v_{\rm g}. \tag{19.72}$$

The TM modes satisfy (19.72) also. This follows from (19.66) when we apply the duality relations (19.44) to (19.69) and (19.71) to get

$$\langle \mathcal{P}_{\rm TM}\rangle = \frac{\omega h\epsilon}{2\gamma^2}\int d^2r_\perp |E_z|^2 \tag{19.73}$$

and

$$\frac{\langle U_{\rm TM}\rangle}{L} = \frac{\omega^2\mu\epsilon^2}{2\gamma^2}\int d^2r_\perp |E_z|^2. \tag{19.74}$$

19.4.8 Energy Loss by Ohmic Heating

Waves propagating in real conducting tubes lose energy by Joule heating of the guide's imperfectly conducting walls. It is usual to write the variation of the time-averaged power flow down the guide as

$$\langle \mathcal{P}(z)\rangle = \langle P(0)\rangle \exp(-2\beta z) \tag{19.75}$$

and compute the *attenuation constant* β from the time-averaged rate of energy loss per unit length of guide, $-d\langle\mathcal{P}\rangle/dz$, because

$$\beta = -\frac{1}{2\langle\mathcal{P}\rangle}\frac{d\langle\mathcal{P}\rangle}{dz}. \tag{19.76}$$

The power lost per unit length of guide is the integral over the guide's perimeter of the power lost per unit area of conducting wall,

$$\frac{d\langle P\rangle}{dz} = \oint \frac{d\langle P\rangle}{dA}d\ell. \tag{19.77}$$

Our strategy to evaluate (19.77) has two parts. First, if $\mathbf{E}_\parallel^{\rm in}$ (parallel to the walls) dominates $\mathbf{E}_\perp^{\rm in}$ (normal to the walls) just inside the walls of a waveguide with conductivity σ and skin depth $\delta = \sqrt{2/\mu\sigma\omega}$, the results of Section 17.6.3 apply and

$$\frac{d\langle P\rangle}{dA} = \frac{\sigma\delta}{4}|\mathbf{E}_\parallel^{\rm in}|_S^2 = \frac{1}{2\sigma\delta}|\mathbf{H}_\parallel^{\rm in}|_S^2. \tag{19.78}$$

Second, we assume that the fields just outside a good ohmic conductor cannot differ greatly from the fields just outside a perfect conductor. If so, the continuity of $\mathbf{H}_\parallel$ at an ohmic surface implies that

$$\mathbf{H}_\parallel^{\rm in}(\text{imperfect})|_S \approx \mathbf{H}_\parallel^{\rm out}(\text{perfect})|_S. \tag{19.79}$$

Substituting all the above into (19.76) gives

$$\beta = \frac{\oint d\ell\,|\mathbf{H}_\parallel^{\rm out}|^2}{4\sigma\delta\langle\mathcal{P}\rangle}. \tag{19.80}$$

Rather than evaluate β exactly for any particular waveguide mode, we focus on its frequency dependence for a generic TE mode. We estimate the denominator of (19.80) using (19.69), the phase velocity (19.54), and a mean-value approximation for a guide with cross sectional area A:

$$\int d^2r_\perp |H_z^{\rm out}|^2 \approx A|H_z^{\rm out}|^2. \tag{19.81}$$

In the numerator of (19.80), $|\mathbf{H}_\parallel^{\rm out}|^2 = |H_z^{\rm out}|^2 + |\hat{\mathbf{n}}\times\mathbf{H}_\perp^{\rm out}|^2$, where $\mathbf{H}_\perp^{\rm out} = (ih/\gamma^2)\nabla_\perp H_z^{\rm out}$ from (19.41). The Helmholtz equation on the left side of (19.48) gives the estimate $\nabla_\perp H_z^{\rm out} \approx \gamma H_z^{\rm out}$. Therefore, using (19.52) and (19.53) for a guide with perimeter circumference C,

$$\oint d\ell\,|\mathbf{H}_\parallel^{\rm out}|^2 \approx C\left[\zeta + \zeta'\left(\frac{\omega^2}{\omega_c^2} - 1\right)\right]|H_z^{\rm out}|^2. \tag{19.82}$$

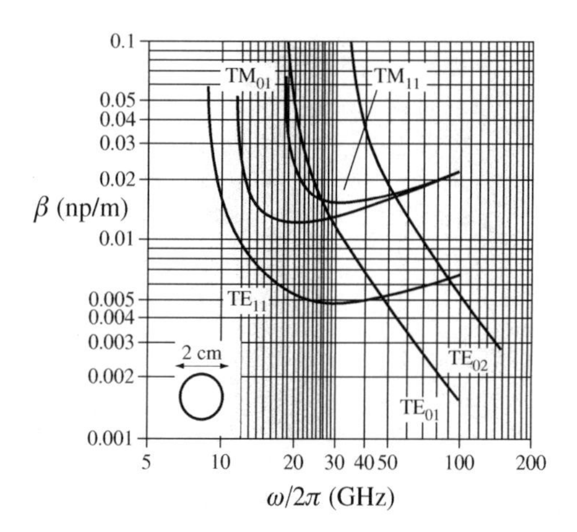

Figure 19.11: Frequency dependence of the attenuation constant (19.80) for selected modes of a circular ($R = 1$ cm) tube waveguide made of copper. Multiply the SI unit neper/m by 8.69 to get decibel/m. Figure from Inan and Inan (2000).

The dimensionless constants ζ and ζ' in (19.82) are of order one. If $\delta_c = \delta\sqrt{\omega/\omega_c}$ is the skin depth evaluated at cutoff, we conclude that

$$\beta_{\rm TE}(\omega) \approx \sqrt{\frac{\epsilon}{\mu}}\left(\frac{C}{2A}\right)\frac{1}{\sigma\delta_c}\frac{\sqrt{\omega/\omega_c}}{\sqrt{1-\omega_c^2/\omega^2}}\left[\zeta' + (\zeta-\zeta')\frac{\omega_c^2}{\omega^2}\right]. \tag{19.83}$$

A related calculation shows that $\beta_{\rm TM}(\omega)$ has the same form except that the frequency-dependent term in the square brackets in (19.83) is absent.

Figure 19.11 shows $\beta(\omega)$ for a few low-order TE and TM modes of a copper waveguide with a circular cross section. The general behavior is that $\beta(\omega)$ has a minimum. Above the minimum, the attenuation increases as $\sqrt{\omega}$ because the current induced in the walls flows in an increasingly thin (and thus more resistive) skin-depth layer. Below the minimum, the attenuation increases because (19.83) diverges as $\omega \to \omega_c$. The attenuation is not infinitely large, of course; it is simply that $\langle \mathcal{P} \rangle = 0$ below cutoff. The TE_{0n} modes are an interesting exception where $\zeta' = 0$ in (19.83) and $\beta(\omega)$ simply decreases monotonically. These modes are not lossy because their electric field lines form closed loops tangential to the tube walls (see Figure 19.10). Such modes were serious candidates for the transmission of data over long distances until the invention of optical fibers made them obsolete.

It remains only to justify our assumption that $E_\perp^{\rm in} \ll E_\parallel^{\rm in}$, despite the general fact that $E_\perp^{\rm out} \gg E_\parallel^{\rm out}$. The idea is to use $\sqrt{\mu/\epsilon}H_\parallel^{\rm out} \approx E_\perp^{\rm out}$, the refraction results from Section 17.6.3, and the continuity of the tangential component of $\mathbf{H}$ to write the tangential component of the electric field just inside the guide walls as

$$E_\parallel^{\rm in}|_S = \sqrt{\frac{\mu\omega}{\sigma}}H_\parallel^{\rm in}\Big|_S \approx \sqrt{\frac{\mu\omega}{\sigma}}H_\parallel^{\rm out}\Big|_S \approx \sqrt{\frac{\omega\epsilon}{\sigma}}E_\perp^{\rm out}\Big|_S. \tag{19.84}$$

To estimate the corresponding amplitude $E_\perp^{\rm in}|_S$, we recall that $\nabla\times\mathbf{H}^{\rm in} \approx \sigma\mathbf{E}^{\rm in}$ inside the walls of a good conductor while $\nabla\times\mathbf{H}^{\rm out} = -i\omega\epsilon\mathbf{E}^{\rm out}$ applies everywhere outside the walls [see 19.33)]. Therefore,

$$\sigma E_\perp^{\rm in}\big|_S = \hat{\mathbf{n}}\cdot\nabla\times\mathbf{H}^{\rm in}\big|_S = \hat{\mathbf{n}}\cdot\nabla\times\mathbf{H}^{\rm out}\big|_S = -i\omega\epsilon E_\perp^{\rm out}\big|_S, \tag{19.85}$$

or

$$\frac{E_\perp^{\rm in}}{E_\perp^{\rm out}}\bigg|_S = -\frac{i\omega\epsilon}{\sigma} \ll 1. \tag{19.86}$$

Comparing (19.86) to (19.84) shows that $E_{\parallel}^{\text{in}}$ is small inside the waveguide walls, but $E_{\perp}^{\text{in}}$ is even smaller. This justifies our use of (19.78).

We conclude with a remark about the tangential component of the electric field just outside the walls of a conducting waveguide with complex impedance $\hat{Z}$. To estimate this small but non-zero field, we again exploit the small difference between the fields inside the waveguide walls and the refraction wave fields (17.118) and (17.119). These refraction fields satisfy

$$\hat{\mathbf{n}} \times \mathbf{E}^{\text{in}}|_S = \hat{Z}\hat{\mathbf{n}} \times (\hat{\mathbf{n}} \times \mathbf{H}^{\text{in}})|_S. \tag{19.87}$$

Therefore, because $\mathbf{E}_{\parallel}$ and $\mathbf{H}_{\parallel}$ are continuous at S,

$$\hat{\mathbf{n}} \times \mathbf{E}^{\text{out}}|_S = \hat{Z}\hat{\mathbf{n}} \times (\hat{\mathbf{n}} \times \mathbf{H}^{\text{out}})|_S. \tag{19.88}$$

The left side of (19.88) is an estimate of the tangential electric field we seek when the perfect conductor value for $\mathbf{H}^{\text{out}}|_S$ is used on the right side. Beyond that, (19.88) is often used in the electromagnetic scattering literature as a boundary condition (in place of $\hat{\mathbf{n}} \times \mathbf{E}^{\text{out}}|_S = 0$) at the surface of an imperfect conductor. This *impedance boundary condition* is not exact because (19.87) is not exact.

Application 19.1 Slow Waves for Charged Particle Acceleration

The longitudinal electric field of the fundamental TM mode of a conducting-tube waveguide appears ideally suited to accelerating a charged particle through the tube by the Coulomb force $\mathbf{F} = q\mathbf{E}$. However, because the phase velocity of any waveguide mode is greater than the speed of light [see (19.56)], the crests and troughs of the wave pass rapidly over the particle, alternately accelerating and decelerating it, with negligible net change in the particle velocity. We show here that a modification of the guide to include a structural periodicity solves this problem by reducing the phase velocity so $v_{\text{p}} < c$. At the 2-mile, 50 GeV Stanford linear accelerator, this is done by the periodic insertion of metal irises into a circular waveguide, as shown in Figure 19.12.

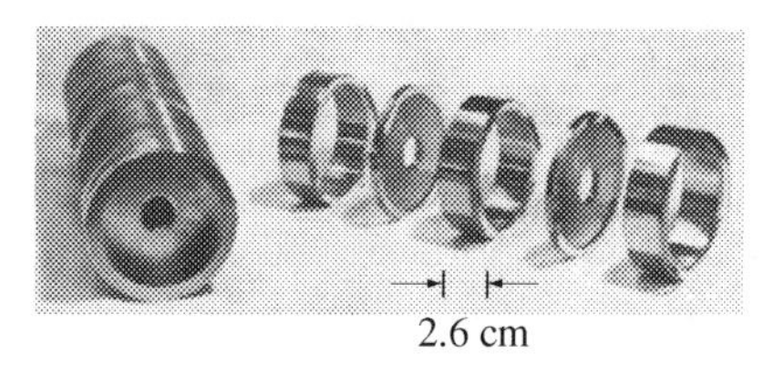

Figure 19.12: End view and exploded view of the periodic motif of the Stanford linear accelerator waveguide. Photograph from Neal (1968).

We have seen repeatedly that time-harmonic solutions to the Maxwell equations, whether in empty space or in empty tubes, ultimately derive from solutions to the scalar Helmholtz equation,

$$\left[\frac{\partial^2}{\partial x^2} + \frac{\partial^2}{\partial y^2} + \frac{\partial^2}{\partial z^2} + \frac{\omega^2}{c^2}\right]\varphi(x, y, z) = 0. \tag{19.89}$$

The difficult issue posed by the "disk-loaded" waveguide in Figure 19.12 is to find solutions to (19.89) where the associated electric field satisfies $\hat{\mathbf{n}} \times \mathbf{E}|_S = 0$, where S is the awkward-shaped interior boundary of the guide. Our strategy is to suppress the variables transverse to the z-direction of propagation and define an operator $\hat{F}(z)$ which stands for both the bracketed operator in (19.89) *and* the boundary conditions. The key point is that the latter is periodic. If the repeat distance is L, we are interested in $\varphi(z)$ when

$$\hat{F}(z)\varphi(z) = 0 \qquad \text{where} \qquad \hat{F}(z+L) = \hat{F}. \tag{19.90}$$

According to Floquet's theorem,[8] the functions $\varphi(z)$ which satisfy (19.90) can be written in terms of a periodic function $u(z) = u(z+L)$ and a parameter h:

$$\varphi(z) = \exp(ihz)u(z). \tag{19.91}$$

The periodic function $u(z)$ has a Fourier series (Section 1.6) representation. Therefore,

$$\varphi(z) = \exp(ihz)\sum_{m=-\infty}^{\infty} b_m \exp(im2\pi z/L). \tag{19.92}$$

The Maxwell fields are derived from (19.92) by some combination of space and time derivatives. Hence, the accelerating electric field of the fundamental TM mode of the guide of Figure 19.12 must have the general form

$$E_z(x,y,z,t) = \sum_{m=-\infty}^{\infty} B_m(x,y)\exp[i(h + m2\pi/L)z]\exp(-i\omega t). \tag{19.93}$$

Recall now that the electric field of the fundamental TM mode of a smooth-wall waveguide is a traveling wave $\exp[i(hz - \omega t)]$ with phase velocity $v_p = \omega/h$. By contrast, (19.93) shows that the boundary conditions imposed on the field by a periodically structured waveguide force the TM-mode field to be an infinite sum of traveling waves. Each wave has a different phase velocity,

$$v_p^m = \frac{\omega}{h_m} = \frac{\omega}{h + 2\pi m/L} = \frac{1}{1/v_p + 2\pi m/\omega L}, \tag{19.94}$$

but the same group velocity, $v_g^m = d\omega/dh_m = d\omega/dh = v_g$. The crucial observation from (19.94) is that E_z contains waves with arbitrarily small phase velocities. Hence, there is always a wave in (19.93) with a phase velocity very near the initial velocity of an injected charged particle. Continuous acceleration occurs because the particle interacts sequentially with waves in (19.93) with steadily increasing phase velocities. The energy gained by the particle is lost by the power source which maintains the field in the guide. ■

19.5 Dielectric Waveguides

Waveguides made entirely from dielectric materials are very desirable for applications where ohmic losses cannot be tolerated. The most familiar example is the *optical fiber*, which is today the dominant transmission medium for high-speed, high-capacity telecommunications. The optical fiber shown in Figure 19.13 is a dielectric waveguide with a nested-cylinder structure. The inner "core" material has an index of refraction n_1. The "cladding" material which surrounds the core has an index of refraction $n_2 < n_1$. Waves are guided down the cylindrical core with very little loss by repeated total internal reflection from the cladding layer. A buffer layer separates the cladding from a jacket which provides mechanical strength. This system is often called a "step-index" fiber because the index of refraction in the radial direction is constant except for a step discontinuity at the core/cladding interface. The more general case of a "graded-index" fiber permits a continuous variation of the index like $n(r)$.

[8] See, e.g., M.J. Gans, "A general proof of Floquet's theorem", *IEEE Transactions on Microwave Theory and Technique* **13**, 384 (1965).

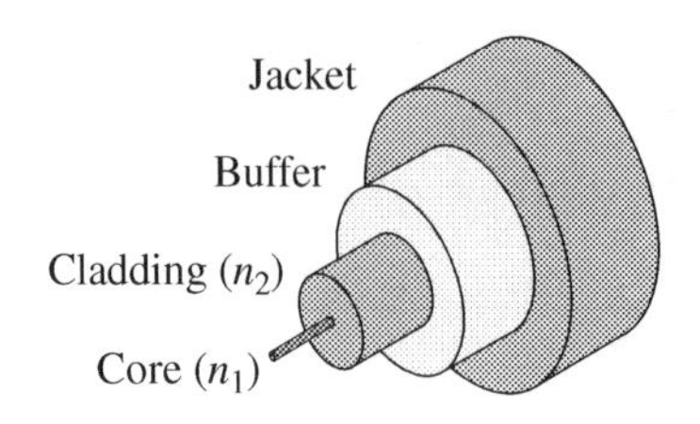

Figure 19.13: Exploded view of an optical fiber.

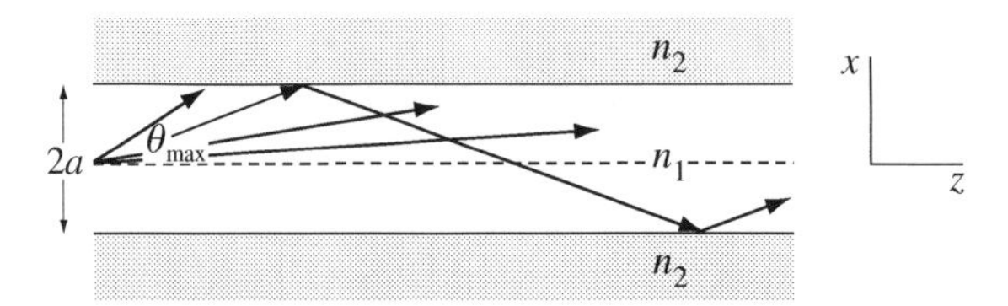

Figure 19.14: Longitudinal cross section of a typical dielectric waveguide ($n_2 < n_1$). θ_{max} is the largest angle (measured from the axis of the guide) at which total internal reflection occurs at the n_1-n_2 interface.

Dielectric waveguides differ from their perfectly conducting counterparts in three important ways: (i) the electromagnetic field is not entirely confined to the guiding volume; (ii) the number of confined modes is finite, rather than infinite; and (iii) the modes are usually not exclusively of either TE or TM type. These differences all originate from the need to impose matching (rather than boundary) conditions at dielectric interfaces.

19.5.1 Waves Guided by Total Internal Reflection

Figure 19.14 is a cross section view of an infinite slab of matter with thickness $2a$ and index of refraction n_1 embedded in an infinite medium with index of refraction $n_2 < n_1$. The physics of total internal reflection is sufficient to show that this structure is the dielectric analog of the wave-guiding parallel-plate transmission line studied in Section 19.3.2. The argument applies also to an optical fiber (whose cladding has an infinitely large outer radius) as long as the propagation wavelength is small compared to the radius of the core. For that case, we re-interpret Figure 19.14 as a longitudinal cut along a diameter of the fiber.

The rays drawn in Figure 19.14 demonstrate wave guiding in the slab due to repeated total internal reflection from the surrounding medium. The electromagnetic field is not truly confined to the n_1 material because a totally reflected wave is always accompanied by a wave which propagates along the interface in the n_2 material with an amplitude that decays exponentially with distance away from the interface (see Section 17.3.6). Similarly, the existence of a critical angle for total internal reflection tells us that only rays that propagate within a cone with opening angle $\theta_{max} = \cos^{-1}(n_2/n_1)$ from the guide axis reflect totally back into the guiding material.

Figure 19.14 is similar to Figure 19.4 in the sense that each ray represents the wave vector of a plane wave with $k = n_1\omega/c$. Another similarity is that a ray bouncing back and forth down the fiber core represents one of the dielectric guide's discrete set of allowed transverse modes. To count these modes, we use a wave-vector version of the counting argument used in Section 19.6.4, taking care

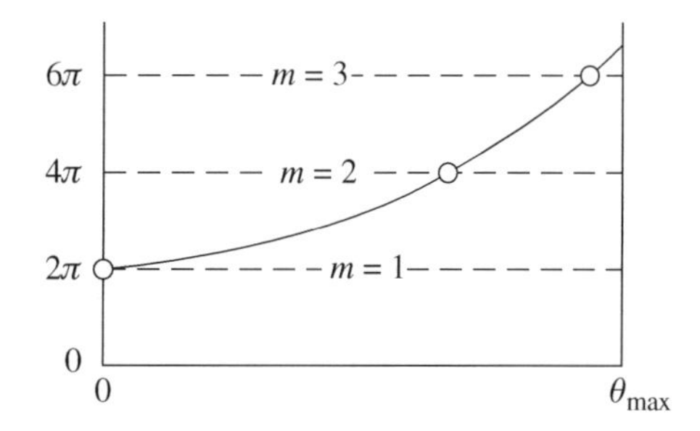

Figure 19.15: Graphical solution of (19.97) shows that only a finite number of modes propagate in a typical dielectric waveguide.

to limit the range of the transverse component of the wave vector to $0 \leq k_\perp \leq k \sin\theta_{\rm max}$. Therefore, including a factor of 2 to account for polarization, the number of modes that can be propagated in a cylindrical optical fiber with core radius a in a differential element of transverse wave vector is

$$dN = 2 \times \pi a^2 \times \frac{d^2 k_\perp}{(2\pi)^2}. \tag{19.95}$$

Since $d^2 k_\perp = 2\pi k_\perp dk_\perp$, this gives the total number of guided modes as

$$N = 2a^2 \int_0^{k\sin\theta_{\rm max}} dk_\perp k_\perp = a^2 k^2 \sin\theta_{\rm max} = a^2 k^2 \left(1 - \frac{n_2^2}{n_1^2}\right). \tag{19.96}$$

Depending on the intended application, manufacturers make different choices for a and fabricate optical fibers for both multimode operation, where $N \sim O(10^2)$, and for single-mode (two-polarization-state) operation, where $N = 2$.

The preceding geometrical optics argument neglects the fact that a true propagating mode can be constructed from the internally reflected rays in Figure 19.14 only if there is constructive interference among all the rays that travel in the same direction. Not least, this physical optics constraint distinguishes TE modes from TM modes because the Fresnel equations (Section 17.3.2) show that TE and TM waves acquire different phase shifts, $\phi_{\rm TE}$ and $\phi_{\rm TM}$, when they experience total internal reflection. These phase shifts depend on the angle θ in Figure 19.14. Therefore, because two consecutive reflections and a total *transverse* path length of $4ka\sin\theta$ occur before a ray returns to the same direction, the condition for constructive interference for one polarization is

$$4ka\sin\theta + 2\phi(\theta) = 2\pi m \qquad m = 0, 1, \ldots \tag{19.97}$$

The graphical solution of (19.97) illustrated in Figure 19.15 (for typical values of the material parameters) shows that propagating modes with $\theta_m < \theta_{\rm max}$ exist only when $m = 1, 2, 3$. This is characteristic of the general case: the core guides only a finite number of modes through the fiber.

The physical optics argument behind (19.97) does not recognize the curvature of the interface between the core and the cladding in a real optical fiber. As a result, it misses the existence of *hybrid modes* which have mixed TE and TM character. The only way to faithfully capture these modes is to solve the full Maxwell equations with dielectric matching conditions on the cylindrical interface. This is a fairly complicated task which does not provide a great deal of insight. Therefore, after setting up the general Maxwell problem in Section 19.5.2, we devote Section 19.5.3 to the guided waves of a planar slab waveguide (where no hybrid modes occur) and then give only a descriptive account of the hybrid modes of an optical fiber in Section 19.5.4.

19.5.2 Wave Fields in Inhomogeneous Dielectric Matter

This section derives a generalized Helmholtz equation for wave propagation in an inhomogeneous dielectric medium where the index of refraction varies with position. We focus on non-magnetic matter so the permittivity carries all the space variation and write the index of refraction as $n(\mathbf{r}) = c\sqrt{\mu_0\epsilon(\mathbf{r})}$. Manipulating $\nabla \times \mathbf{E} = -\partial\mathbf{B}/\partial t$ and $\nabla \times \mathbf{H} = \partial\mathbf{D}/\partial t$ in the usual way gives

$$\nabla \times \nabla \times \mathbf{E} = -\frac{n^2(\mathbf{r})}{c^2}\frac{\partial^2\mathbf{E}}{\partial t^2} = \nabla(\nabla \cdot \mathbf{E}) - \nabla^2\mathbf{E}. \tag{19.98}$$

When no free charge is present, $0 = \nabla \cdot \mathbf{D} = \nabla \cdot (\epsilon\mathbf{E}) = \epsilon\nabla \cdot \mathbf{E} + \mathbf{E} \cdot \nabla\epsilon$. Substituting this information in (19.98) produces an inhomogeneous wave equation:

$$\nabla^2\mathbf{E} - \frac{n^2(\mathbf{r})}{c^2}\frac{\partial^2\mathbf{E}}{\partial t^2} = -\nabla\left(\mathbf{E} \cdot \frac{\nabla\epsilon}{\epsilon}\right). \tag{19.99}$$

For time-harmonic fields that vary as $\exp(-i\omega t)$, (19.99) reduces to an inhomogeneous Helmholtz equation,

$$\left[\nabla^2 + n^2(\mathbf{r})\frac{\omega^2}{c^2}\right]\mathbf{E} = -\nabla\left(\mathbf{E} \cdot \frac{\nabla\epsilon}{\epsilon}\right). \tag{19.100}$$

The gradient term on the right side of (19.100) has not concerned us to this point because we have focused exclusively on systems where the dielectric function is piecewise constant. For those situations, the gradient produces a delta function on the surfaces where $n(\mathbf{r})$ is discontinuous and it is sufficient to impose the dielectric matching conditions at those surfaces. The right side of (19.100) also plays no explicit role if the variation of $\epsilon(\mathbf{r})$ is sufficiently weak.

19.5.3 Guided Modes of a Planar Waveguide

The guided modes of the dielectric slab waveguide in Figure 19.14 can be found by solving (19.100). The $\nabla\epsilon$ term is not needed because, as explained just above, we will apply the dielectric matching conditions at $x = \pm a$. We focus on TE modes and assume an electric field of the form $\mathbf{E}(x, z, t) = \hat{\mathbf{y}}E(x)\exp[i(hz - \omega t)]$. Substituting this guess into (19.100) generates the one-dimensional differential equation

$$\left[\frac{d^2}{dx^2} + \left(n^2\frac{\omega^2}{c^2} - h^2\right)\right]E(x) = 0. \tag{19.101}$$

The symmetry of the guide and (19.101) with respect to inversion through the $x = 0$ plane leads us to expect "even" solutions where $E(-x) = E(x)$ and "odd" solutions where $E(-x) = -E(x)$.

We are interested in modes that are guided by the slab and thus have the majority of their energy stored in the region $-a \le x \le a$. Therefore, we draw on our experience with total internal reflection and seek solutions which decay exponentially away from the interface into medium n_2. This leads us to define the parameters γ_1 and γ_2 so (19.101) in the slab and in the cladding take the form

$$\gamma_1^2 = n_1^2\frac{\omega^2}{c^2} - h^2 \quad \text{and} \quad \frac{d^2E_1}{dx^2} + \gamma_1^2E_1 = 0 \quad \text{in } n_1 \tag{19.102}$$

and

$$\gamma_2^2 = h^2 - n_2^2\frac{\omega^2}{c^2} \quad \text{and} \quad \frac{d^2E_2}{dx^2} - \gamma_2^2E_2 = 0 \quad \text{in } n_2. \tag{19.103}$$

The magnetic fields that accompany $\mathbf{E}_1$ and $\mathbf{E}_2$ follow from Faraday's law, $\nabla \times \mathbf{E} = -\partial \mathbf{B}/\partial t$; namely,

$$\mathbf{H} = \frac{1}{i\omega\mu}\nabla \times \mathbf{E} = \frac{1}{i\omega\mu}\left[-ih\hat{\mathbf{x}} + \hat{\mathbf{z}}\frac{\partial}{\partial x}\right] E(x)e^{i(hz-\omega t)}. \tag{19.104}$$

It remains only to find γ_1 and γ_2 in (19.102) and (19.103). These are determined uniquely and simultaneously by the matching conditions that the tangential components of $\mathbf{E}$ and $\mathbf{H}$ be continuous at $x = \pm a$. Using (19.104) and the sinusoidal and exponential solutions to (19.102) and (19.103), the even and odd components of interest are

$$E_{1y}(x) = E_1 \begin{Bmatrix} \sin(\gamma_1 x) \\ \cos(\gamma_1 x) \end{Bmatrix} \qquad H_{1z}(x) = \frac{\gamma_1 E_1}{i\omega\mu_1} \begin{Bmatrix} \cos(\gamma_1 x) \\ -\sin(\gamma_1 x) \end{Bmatrix} \tag{19.105}$$

and

$$E_{2y}(x) = E_2 \exp(-\gamma_2|x|) \qquad H_{2z}(x) = -\text{sgn}(x)\frac{\gamma_2 E_2}{i\omega\mu_2}\exp(-\gamma_2|x|). \tag{19.106}$$

The matching conditions applied to the odd-parity solutions give

$$E_1 \sin(\gamma_1 a) = E_2 e^{-\gamma_2 a} \qquad \text{and} \qquad \frac{\gamma_1}{\mu_1}E_1\cos(\gamma a) = -\frac{\gamma_2}{\mu_2}E_2 e^{-\gamma_2 a}. \tag{19.107}$$

Dividing one equation in (19.107) by the other gives a relation between γ_1 and γ_2:

$$\cot(\gamma_1 a) = -\frac{\mu_1}{\mu_2}\frac{\gamma_2}{\gamma_1}. \tag{19.108}$$

Proceeding identically for the even-parity solutions replaces (19.108) by

$$\tan(\gamma_1 a) = \frac{\mu_1}{\mu_2}\frac{\gamma_2}{\gamma_1}. \tag{19.109}$$

Finally, adding (19.102) to (19.103) gives

$$\gamma_1^2 + \gamma_2^2 = \frac{\omega^2}{c^2}\left(n_1^2 - n_2^2\right). \tag{19.110}$$

At fixed frequency, (19.110) is the equation of a circle in the γ_1-γ_2 plane. The points where this circle intersects the branches of (19.108) define the odd-parity modes guided by the slab. The points where this circle intersects the branches of (19.109) define the even-parity modes guided by the slab. The number of guided modes is finite and increases as the frequency (and hence the radius of the circle) increases.

As a numerical example, let the constitutive parameters of the guide be $\epsilon_1 = 2\epsilon_0$ and $\mu_1 = \mu_2 = \mu_0$, and fix the frequency so $\omega a = (5/4)\pi c$. These choices simplify (19.110) to

$$(\gamma_1 a)^2 + (\gamma_2 a)^2 = \left(\frac{5\pi}{4}\right)^2. \tag{19.111}$$

Figure 19.16 plots (19.111) for positive values of $\gamma_1 a$ and $\gamma_2 a$. The solid curves in the diagram marked "Odd" and "Even" are (19.108) and (19.109), respectively. The black dots indicate the solutions. With the solution values of γ_1, the equation on the far left side of (19.102) is the dispersion relation for each mode. The solution values of γ_2 inserted into (19.106) determine the degree to which each propagating mode leaks into the cladding layer. Indeed, the condition $\gamma_2 = 0$ is used to define the "cutoff" for a dielectric waveguide mode because this value implies that the wave field (19.106) is no longer guided by the slab, but extends into the cladding layer without attenuation. Substituting $\gamma_2 = 0$ into (19.108),

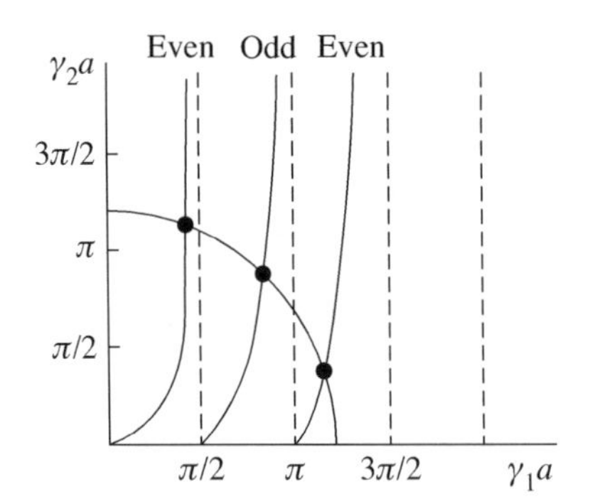

Figure 19.16: Graphical solution for the propagation constant γ_1 and evanescent decay constant γ_2 for the even- and odd-parity TE modes of a slab dielectric waveguide with thickness $2a$ and frequency $\omega = (5/4)\pi c/a$.

(19.109), and (19.110) thus determines the TE cutoff frequencies of the dielectric slab waveguide as

$$\omega_{c,m} = \frac{m\pi c}{2a\sqrt{n_1^2 - n_2^2}} \begin{cases} m = 0, 2, 4, \ldots & \text{even modes,} \\ m = 1, 3, 5, \ldots & \text{odd modes.} \end{cases} \tag{19.112}$$

The choice $m = 0$ in (19.112) shows that the lowest even-parity TE mode of the dielectric slab waveguide has zero cutoff frequency. Like the TEM mode of a parallel-plate transmission line [see (19.9)], this TE mode can propagate through the guide at arbitrarily low frequency. The next section discusses the concept of cutoff for a dielectric guide in more detail.

19.5.4 Radiation Modes and Hybrid Modes

The guided modes of the planar dielectric waveguide found in the preceding section do not form a complete set until they are supplemented by the guide's infinite continuum of *radiation modes*. The latter are nothing but the rays in Figure 19.14 which do not suffer total internal reflection because they encounter the dielectric interface at an angle $\theta > \theta_{\max}$. This is the situation for the Fabry-Perot geometry sketched in Figure 17.8, where rays zig-zag through the slab, losing power at each reflection due to refraction into the cladding material. Eventually, all the energy of the initial ray is lost to the cladding.

From the Maxwell point of view, the refracted ray is a solution of the wave equation in the cladding material that varies sinusoidally (rather than decays exponentially) in the transverse direction. This shows that the word "cutoff" is used very differently for dielectric and conducting waveguides. An above-cutoff mode in a conducting waveguide is propagating and transports time-averaged power down the guide. A below-cutoff mode in a conducting waveguide is evanescent; it neither transports time-averaged power down the guide nor dissipates energy to the environment. An above-cutoff mode in a dielectric waveguide similarly transports time-averaged power down the guide. However, a below-cutoff mode in a dielectric waveguide is radiative and transports time-averaged power transversely out of the guide.

A *hybrid mode* is a mode of a dielectric waveguide that is neither TE nor TM because its longitudinal field components E_z and H_z are simultaneously non-zero. Modes of this kind occur generically in dielectric waveguides except for special geometries where the modal field is strictly uniform in the direction transverse to both the interface normal and the direction of propagation. The planar slab waveguide analyzed in Section 19.5.3 is an example of the latter because Figure 19.14 is translationally

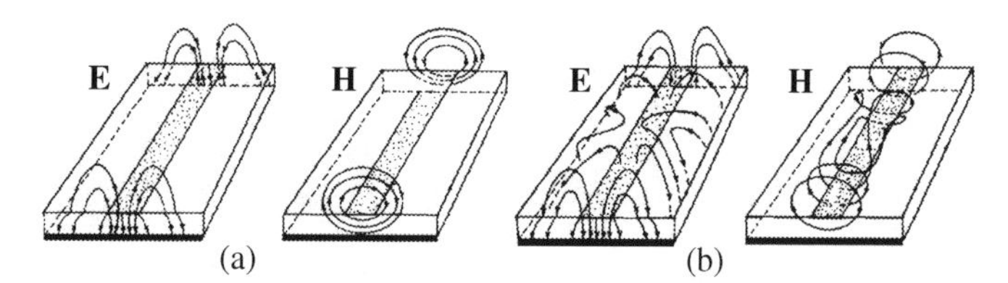

Figure 19.17: Fields lines for a microstrip where a conducting strip (speckled) and a conducting sheet (dark solid line) sandwich a slab of dielectric: (a) fundamental mode at low frequency; (b) fundamental mode at high frequency. Figure adapted from Hoffmann (1987).

invariant in the y-direction. An example where this type of invariance is lost—and the matching conditions cannot be satisfied by either TE or TM modes—is a structure commonly called a *microstrip*.

Figure 19.17 shows a microstrip used in microwave integrated circuits. It consists of a conducting strip bonded to the top face of a dielectric slab with a conducting plate bonded to the bottom of the slab. Without the dielectric, this is a two-conductor transmission line (see Section 19.2). The presence of the dielectric distorts the strictly transverse fundamental TEM mode of the transmission line into a hybrid mode with longitudinal field components. In practice, one finds that E_z and H_z are negligible (compared to the transverse field components) at low frequency and the fundamental mode is nearly indistinguishable from pure TEM [Figure 19.17(a)]. The hybrid nature of the mode becomes apparent only at higher frequencies when the wavelength becomes comparable to the characteristic dimensions of the strip [Figure 19.17(b)].

Hybrid modes occur similarly in the optical fiber in Figure 19.13 because (see Section 19.4.6) separated-variable solutions of the Helmholtz equation in cylindrical coordinates are indexed by an integer n and have the form

$$(\text{Bessel function of order } n) \times \exp(in\phi). \tag{19.113}$$

If $n = 0$, TE and TM modes exist because the field is constant in the (ϕ) direction that is perpendicular to the propagation direction and to the interface normal. This is not true for modes with $n \neq 0$, and the matching conditions can only be satisfied by hybrid modes. The name *skew ray* is used for some of these modes because the propagation path resembles a corkscrew spiraling around the symmetry axis of the fiber.

19.6 Conducting Cavities

An enclosed volume used to trap and store standing waves is called a *resonant cavity*. In this section, we study electromagnetic standing waves in a volume bounded by perfectly conducting walls.[9] Storage occurs only for waves that oscillate at one of a discrete set of frequencies because the source-free Maxwell equations and the field boundary conditions at the cavity walls define a normal-mode eigenvalue problem. Apart from the difference between vector waves and scalar waves, the normal modes of a three-dimensional electromagnetic cavity are perfectly analogous to the normal modes of a one-dimensional string or a two-dimensional drumhead. Indeed, the simplest possible electromagnetic resonator consists of two flat and perfectly parallel mirrors separated by a distance L. The normal modes are TEM waves that reflect back and forth in such a way that the round trip distance $2L$ is equal to an integer number of wavelengths. This guarantees that the electric field boundary condition is satisfied and gives the equally spaced mode frequencies as

$$\nu_m = \frac{\omega_m}{2\pi} = m\frac{c}{2L} \qquad m = 1, 2, \ldots \tag{19.114}$$

[9] We neglect the effect of small holes in the walls needed to inject waves into, and extract waves from, the cavity.

A resonator for the Gaussian beam sketched in Figure 16.11 results if the two flat mirrors are replaced by slightly curved mirrors shaped like the wavefronts of the beam far from its waist. Many laser resonators use this geometry.

19.6.1 General Properties of Cavity Modes

A general approach to finding the standing waves of a conducting resonant cavity focuses on time-harmonic and divergence-free solutions of the homogeneous wave equation. Thus, the electric field $\mathbf{E}(\mathbf{r})\exp(-i\omega t)$ of a resonant cavity formed by a volume V of simple matter (ϵ, μ) with a perfect conductor surface S satisfies

$$\begin{aligned} \left[\nabla^2 + \epsilon\mu\omega^2\right]\mathbf{E} &= 0 \qquad \text{in } V, \\ \nabla \cdot \mathbf{E} &= 0 \qquad \text{in } V, \\ \hat{\mathbf{n}} \times \mathbf{E} &= 0 \qquad \text{on } S. \end{aligned} \tag{19.115}$$

Let us choose the electric field $\mathbf{E}(\mathbf{r})$ in (19.115) to be real. The corresponding function $\mathbf{H}(\mathbf{r})$ is pure imaginary by virtue of Faraday's law for time-harmonic fields,

$$\nabla \times \mathbf{E} = i\omega\mu\mathbf{H}. \tag{19.116}$$

This means that $\mathbf{H}(\mathbf{r})\exp(-i\omega t)$ oscillates 90° out of phase with the electric field. The main challenge is to solve the Helmholtz equation in (19.115). For simple geometries, separation of variables produces a general solution. Given this, the boundary and zero-divergence conditions in (19.115) determine a complete set of eigenfunctions, $\mathbf{E}_\lambda(\mathbf{r})$, and eigenvalues, ω_λ.

Orthogonality relations follow from

$$\int_S d\mathbf{S} \cdot [\mathbf{a} \times (\nabla \times \mathbf{b}) + (\nabla \cdot \mathbf{b})\mathbf{a}] = \int_V d^3r \left[(\nabla \times \mathbf{a}) \cdot (\nabla \times \mathbf{b}) + (\nabla \cdot \mathbf{a})(\nabla \cdot \mathbf{b}) + \mathbf{a} \cdot \nabla^2\mathbf{b}\right] \tag{19.117}$$

when we set $\mathbf{a} = \mathbf{E}_\lambda$, $\mathbf{b} = \mathbf{E}_\mu$, and use all three equations in (19.115). The result is

$$\mu\epsilon\omega_\mu^2 \int_V d^3r\, \mathbf{E}_\lambda \cdot \mathbf{E}_\mu = \int_V d^3r\, (\nabla \times \mathbf{E}_\lambda) \cdot (\nabla \times \mathbf{E}_\mu). \tag{19.118}$$

Subtracting (19.118) from the same equation with λ and μ interchanged shows that

$$\int_V d^3r\, \mathbf{E}_\lambda \cdot \mathbf{E}_\mu = 0 \quad \text{and} \quad \int_V d^3r\, (\nabla \times \mathbf{E}_\lambda) \cdot (\nabla \times \mathbf{E}_\mu) = 0 \quad \text{when} \quad \omega_\lambda \neq \omega_\mu. \tag{19.119}$$

The Gram-Schmidt procedure is available for orthogonalization in the case of degeneracy.

Example 19.3 A cubical resonant cavity has perfectly conducting walls at $x = 0$, $x = a$, $y = 0$, $y = a$, $z = 0$, and $z = a$. The region $0 \leq z \leq b$ $(b < a)$ is filled with a non-magnetic dielectric with relative permittivity $\kappa = \epsilon/\epsilon_0$. Derive a transcendental equation for the frequencies of the cavity normal modes which have $E_y = E_z = 0$. Solve the equation when $\kappa = 1$.

Solution: $\nabla \cdot \mathbf{E} = \partial E_x/\partial x = 0$, so $E_x(x, y, z, t) = E_x(y, z)e^{-i\omega t}$ for harmonic modes. These satisfy

$$\begin{aligned} z < b: &\quad \left[\frac{\partial^2}{\partial y^2} + \frac{\partial^2}{\partial z^2} + \kappa\frac{\omega^2}{c^2}\right]E_x = 0 \\ z > b: &\quad \left[\frac{\partial^2}{\partial y^2} + \frac{\partial^2}{\partial z^2} + \frac{\omega^2}{c^2}\right]E_x = 0. \end{aligned}$$

The tangential part of E_x vanishes at the cavity walls. Therefore, writing $E_x(y,z) = Y(y)Z(z)$ with a separation constant λ^2, the equations to be solved are

$$Y''(y) + \lambda^2 Y(y) = 0 \qquad Y(0) = Y(a) = 0,$$

$$Z''(z) + \left(\kappa \frac{\omega^2}{c^2} - \lambda^2\right) Z(z) = 0 \qquad Z(0) = 0,\ z < b,$$

$$Z''(z) + \left(\frac{\omega^2}{c^2} - \lambda^2\right) Z(z) = 0 \qquad Z(a) = 0,\ z > b.$$

The $Y(y)$ equation (and solution) is the same for both sections of the cavity because E_x is continuous at $z = b$. If m is an integer, we find

$$z < b: \quad E_x \propto \sin\frac{m\pi y}{a} \sin[k_< z] \qquad k_<^2 = \kappa \frac{\omega^2}{c^2} - \left(\frac{m\pi}{a}\right)^2$$

$$z > b: \quad E_x \propto \sin\frac{m\pi y}{a} \sin[k_>(z-a)] \qquad k_>^2 = \frac{\omega^2}{c^2} - \left(\frac{m\pi}{a}\right)^2.$$

Continuity of E_x at $z = b$ gives the constraint $\sin[k_< b] = \sin[k_>(b-a)]$. From $\nabla \times \mathbf{E} = i\omega \mathbf{B}$, the tangential component of $\mathbf{B}$ satisfies $i\omega B_y = -\partial E_x/\partial z$. Hence,

$$z < b: \quad B_y \propto k_< \sin\frac{m\pi y}{a} \cos[k_< z]$$

$$z > b: \quad B_y \propto k_> \sin\frac{m\pi y}{a} \cos[k_>(z-a)].$$

B_y is also continuous at $z = b$. This gives the constraint $k_< \cos[k_< b] = k_> \cos[k_>(b-a)]$. Combining the two constraints produces the desired transcendental equation:

$$k_> \tan[k_< b] = k_< \tan[k_>(b-a)].$$

When $\kappa = 1$, $k_< = k_> = k$, and the transcendental equation is satisfied when the arguments of the tangent functions differ by πn, where n is an integer. This gives the $E_y = E_z = 0$ mode frequencies of a cubical box as

$$\frac{\omega^2}{c^2} = \left(\frac{m\pi}{a}\right)^2 + \left(\frac{n\pi}{a}\right)^2.$$

19.6.2 Conducting-Tube Cavities

The conducting-tube waveguide sketched in Figure 19.6 of Section 19.4 becomes a conducting-tube resonant cavity when parallel conducting plates at $z = 0$ and $z = L$ are used to cap the open ends of the waveguide. In the notation of the previous section, the new boundary conditions introduced by the end caps are

$$\mathbf{E}_\perp(z=0) = 0 \qquad \text{and} \qquad \mathbf{E}_\perp(z=L) = 0. \tag{19.120}$$

The standing-wave modes of the cavity can be derived from the propagating-wave modes of the infinite waveguide because the latter already satisfy the boundary conditions on the side walls. We satisfy the $z = 0$ condition in (19.120) by superposing a mode propagating in the $+z$-direction with an identical mode propagating in the $-z$-direction. For the TE case, subtract (19.42) from itself (with a change of sign of h) to get

$$\begin{aligned}\mathbf{E}_{\mathrm{TE}} &= \frac{2\omega\mu}{\gamma^2}\,[\hat{\mathbf{z}}\times\nabla_\perp H_z]\sin(hz)\exp(-i\omega t)\\ \mathbf{H}_{\mathrm{TE}} &= 2i\left[\frac{h}{\gamma^2}\nabla_\perp H_z\cos(hz)+H_z\sin(hz)\hat{\mathbf{z}}\right]\exp(-i\omega t).\end{aligned} \tag{19.121}$$

For the TM case, add (19.43) to itself (with a change of sign of h) to get

$$\begin{aligned}\mathbf{E}_{\mathrm{TM}} &= 2\left[-\frac{h}{\gamma^2}\nabla_\perp E_z\sin(hz)+\hat{\mathbf{z}}E_z\cos(hz)\right]\exp(-i\omega t)\\ \mathbf{H}_{\mathrm{TM}} &= \frac{2i\omega\epsilon}{\gamma^2}\,[\hat{\mathbf{z}}\times\nabla_\perp E_z]\cos(hz)\exp(-i\omega t).\end{aligned} \tag{19.122}$$

For both the TE and TM cases, the $z = L$ boundary condition in (19.120) restricts the previously continuous variable h to the values $h = p\pi/L$, where p is a non-negative integer. Using (19.52), the resulting mode eigenfrequencies are quantized by p and the waveguide mode index i to the discrete set of values

$$\omega_{ip} = \frac{1}{\sqrt{\mu\epsilon}}\sqrt{\gamma_i^2+p^2\frac{\pi^2}{L^2}} \qquad i=1,2,\ldots, \qquad p=0,1,\ldots \tag{19.123}$$

As an example, substituting (19.59) into (19.123), gives the mode frequencies of a conducting parallelepiped with dimensions $a\times b\times L$:

$$\omega_{m,n,p} = \frac{\pi}{\sqrt{\mu\epsilon}}\sqrt{\frac{m^2}{a^2}+\frac{n^2}{b^2}+\frac{p^2}{L^2}} \qquad m,n,p=0,1,\ldots \tag{19.124}$$

These cavity modes happen to be doubly degenerate because there is a TE mode and a TM mode associated with every value of γ_i for a rectangular waveguide (see Section 19.4.5). This degeneracy does not occur for cavities formed by capping the ends of waveguides with arbitrary cross sectional shapes.

19.6.3 Spherical Cavities

The electromagnetic normal modes of a spherical resonant cavity are time-harmonic, vector spherical waves that satisfy the perfect-conductor boundary condition $\hat{\mathbf{r}}\times\mathbf{E}|_S = 0$ at the cavity's walls. The candidate waves (Section 16.8) are transverse electric,

$$\mathbf{E}_{\mathrm{TE}} = -i\omega\mathbf{r}\times\nabla\hat{u} \qquad\qquad \mathbf{B}_{\mathrm{TE}} = -\nabla\times[\mathbf{r}\times\nabla\hat{u}], \tag{19.125}$$

and transverse magnetic,

$$c\mathbf{B}_{\mathrm{TM}} = -i\omega\mathbf{r}\times\nabla\hat{u} \qquad\qquad \mathbf{E}_{\mathrm{TM}} = c\nabla\times[\mathbf{r}\times\nabla\hat{u}], \tag{19.126}$$

where $\hat{u}(\mathbf{r})$ is a solution of the Helmholtz equation in spherical coordinates.

In Section 16.8, we wrote $k=\omega/c$ and used $\hat{u}(\mathbf{r}) = \sum_\ell [A_\ell h_\ell(kr)+B_\ell h_\ell^*(kr)]Y_{\ell m}(\theta,\phi)$ where $h_\ell(kr)$ and $h_\ell^*(kr)$ were outgoing wave and incoming wave spherical Hankel functions. These functions cannot be used for cavity fields because they diverge at $r=0$. Instead, we use the linearly independent *spherical Bessel functions*, $j_\ell(kr) = \frac{1}{2}[h_\ell(kr) + h_\ell^*(kr)]$, and *spherical Neumann functions*, $n_\ell(kr) = \frac{1}{2}i[h_\ell^*(kr) - h_\ell(kr)]$. The first two of each are

$$j_0(kr) = \frac{\sin(kr)}{kr} \qquad\qquad n_0(kr) = -\frac{\cos(kr)}{kr}$$

$$j_1(kr) = \frac{\sin(kr)}{(kr)^2} - \frac{\cos(kr)}{kr} \qquad\qquad n_1(kr) = -\frac{\cos(kr)}{(kr)^2} - \frac{\sin(kr)}{kr}.$$

The spherical Neumann functions diverge at the origin. Therefore, all the resonant modes of a spherical cavity derive from

$$\hat{u}_{\ell m}(\mathbf{r}) = j_\ell(kr)Y_{\ell m}(\theta, \phi). \tag{19.127}$$

Using (19.127) to evaluate (19.125), the TE electric field is

$$\mathbf{E}_{\text{TE}} = i\omega j_\ell(kr)\left[\frac{1}{\sin\theta}\frac{\partial Y_{\ell m}}{\partial\phi}\hat{\boldsymbol{\theta}} - \frac{\partial Y_{\ell m}}{\partial\theta}\hat{\boldsymbol{\phi}}\right]. \tag{19.128}$$

We compute the TM electric field in (19.126) using the vector identity

$$\nabla \times (\mathbf{r} \times \nabla u) = \mathbf{r}\nabla^2 u - 2\nabla u - r\frac{\partial}{\partial r}\nabla u \tag{19.129}$$

and the fact that (see Section 7.7)

$$\nabla^2 = \frac{1}{r^2}\frac{\partial}{\partial r}r^2\frac{\partial}{\partial r} - \frac{\hat{L}^2}{r^2} \qquad \text{where} \qquad \hat{L}^2 Y_{\ell m} = \ell(\ell+1)Y_{\ell m}. \tag{19.130}$$

A bit of algebra gives

$$\mathbf{E}_{\text{TM}} = -c\left[\hat{\mathbf{r}}\frac{\ell(\ell+1)}{r^2} + \left(\hat{\boldsymbol{\theta}}\frac{\partial}{\partial\theta} + \hat{\boldsymbol{\phi}}\frac{1}{\sin\theta}\frac{\partial}{\partial\phi}\right)\frac{1}{r}\frac{\partial}{\partial r}\right] r j_\ell(kr)Y_{\ell m}(\theta, \phi). \tag{19.131}$$

It is worth noting that the $\ell = 0$ fields vanish. This is consistent with Application 16.1.

We determine the eigenfrequencies of the spherical cavity by forcing the tangential components $(\hat{\boldsymbol{\theta}}, \hat{\boldsymbol{\phi}})$ of the electric fields (19.128) and (19.131) to be zero at the cavity walls. If the cavity has radius R, the conditions that fix $k = \omega/c$ are

$$j_\ell(k^{\text{TE}}R) = 0 \qquad \text{and} \qquad \frac{d}{dr}\left[r j_\ell(k^{\text{TM}}r)\right]_{r=R} = 0. \tag{19.132}$$

The fact that the index m in (19.127) does not appear in (19.132) means that every TE and TM mode of a spherical cavity is $(2\ell+1)$-fold degenerate. The roots needed for evaluation of (19.132) are tabulated and the lowest resonant frequency happens to belong to a TM mode with

$$\omega = 2.744\frac{c}{R}. \tag{19.133}$$

19.6.4 The Density of Modes

For many cavity problems, the precise values of the resonant frequencies are less interesting than a statistical quantity such as the number of modes per unit frequency interval. We define the *density of modes* $g(\omega)$ so

$$g(\omega)d\omega = \text{ number of normal modes with frequency between } \omega \text{ and } d\omega. \tag{19.134}$$

It is not difficult to calculate $g(\omega)$ for a cubic cavity of volume $V = L^3$, where [see (19.124)]

$$\omega = \frac{c\pi}{L}\sqrt{n^2 + m^2 + p^2} \qquad m, n, p \geq 0. \tag{19.135}$$

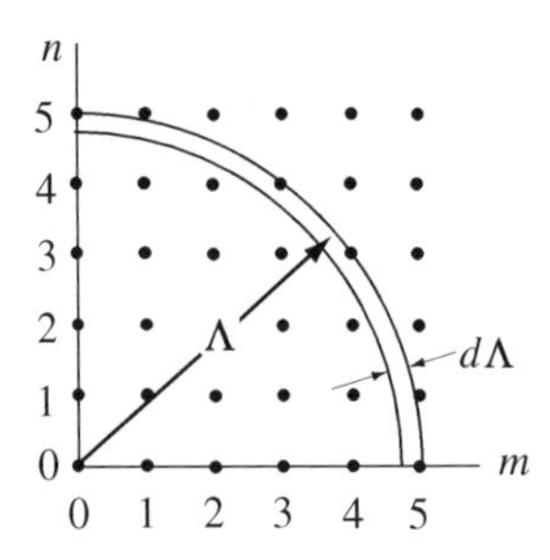

Figure 19.18: Mode counting in two dimensions.

The idea is to mark the integers on orthogonal axes labeled m, n, and p. Figure 19.18 shows the $p = 0$ plane of this space. There is one TE mode and one TM mode for every point (m, n, p) in the positive octant (except $m = n = 0$) or, equivalently, for every $1 \times 1 \times 1$ cube in the octant. This means that counting modes is equivalent to measuring volume in (m, n, p) space.

The radial distance from the origin in this space is

$$\Lambda = \sqrt{n^2 + m^2 + p^2}. \tag{19.136}$$

Therefore, when m, n, and p are real numbers, we can sensibly identify $g(\omega)d\omega$ as the volume of one octant of a spherical shell of radius Λ and thickness $d\Lambda$ (multiplied by 2 for the TE/TM degeneracy); that is,

$$g(\omega)d\omega = 2 \times \frac{1}{8} \times 4\pi\Lambda^2 d\Lambda = \frac{V\omega^2 d\omega}{\pi^2 c^3}. \tag{19.137}$$

Alternatively, we can use a delta function to count the modes directly:

$$g(\omega) = 2 \times \frac{1}{8} \times \int d\Omega \int_0^\infty d\Lambda \Lambda^2 \delta\left(\omega - c\frac{\pi}{L}\Lambda\right) = \frac{V\omega^2}{\pi^2 c^3}. \tag{19.138}$$

The key observation is that (19.137) and (19.138) remain valid when m, n, and p are integers if we restrict ourselves to frequencies $\omega \gg c/L$ so $\Delta\omega$ computed using (19.135) and adjacent integers is a good approximation to $d\omega$.[10]

The density of electromagnetic cavity modes has many uses. For example, *blackbody radiation* is the name given to electromagnetic waves that freely exchange energy with the walls of a resonant cavity held at temperature T. In 1900, Rayleigh used (19.138) to calculate $u(\nu, T)$, the average blackbody energy per unit volume and per unit interval of frequency $\nu = \omega/2\pi$. If ϵ is the average energy of a cavity normal mode, the classical equipartition theorem gives $\epsilon = kT$.[11] Then, because $g(\omega)d\omega = 2\pi g(\nu)d\nu$, we get the Rayleigh-Jeans law,

$$u(\nu, T) = \frac{2\pi}{V} g(\nu)\epsilon = \frac{8\pi\nu^2}{c^3} kT. \tag{19.139}$$

Modern physics was born when the disagreement of (19.139) with experimental measurements at high frequency led Planck, in 1900, to introduce the "quantum of energy" $h\nu$ and modify (19.139) to

$$u(\nu, T) = \frac{8\pi h\nu^3}{c^3} \frac{1}{\exp(h\nu/kT) - 1}. \tag{19.140}$$

[10] A celebrated theorem due to Weyl states that (19.138) applies to cavities of any shape if they are sufficiently large. See Sources, References, and Additional Reading.

[11] Section 19.6.7 shows that a cavity mode behaves like a one-dimensional harmonic oscillator.

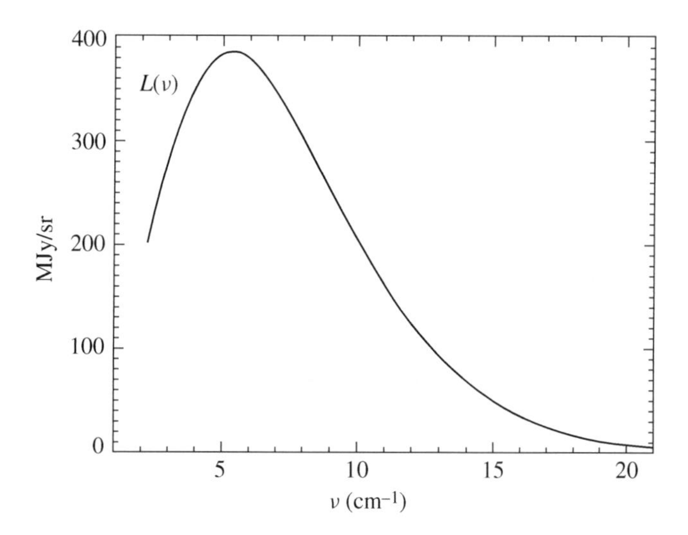

Figure 19.19: The spectral radiance, $L(\nu)$, of the cosmic microwave background radiation. The dozens of data points in the curve have uncertainties smaller than the thickness of the theoretical line drawn using the Planck distribution with $T = 2.73$ K. The non-SI unit MJy $= 10^{-23}$ W m^{-2} Hz^{-1}. Figure adapted from Fixsen *et al.* (1996).

Over a century later, the best example of blackbody radiation is the cosmic microwave background radiation which provides evidence for the Big Bang model for the origin of the Universe. Figure 19.19 is a fit to satellite-collected spectral radiance data to $L(\nu, T) = u(\nu, T)c/4\pi$ using the Planck distribution (19.140).

The density of electromagnetic modes also enters the "Golden Rule" formula,

$$\Gamma = \frac{2\pi}{\hbar}|\langle F|\mathcal{H}|I\rangle|^2 g(\omega), \tag{19.141}$$

for the rate at which an interaction Hamiltonian $\mathcal{H}$ causes a quantum system to make a radiative transition from an initial state to a final state where $E_\mathrm{F} - E_\mathrm{I} = \hbar\omega$. Using (19.138), the Golden Rule predicts spontaneous emission rates very well because empty space may be regarded as an infinitely large resonant cavity. One the other hand, when faced with very small values of Γ for microwave transitions, Purcell pointed out that spontaneous emission rates could be enhanced by arranging cavity situations where (19.138) no longer accurately predicts the density of electromagnetic modes.[12] Today, the enhancement and suppression of spontaneous emission rates by manipulation of $g(\omega)$ is an active field of research.

Application 19.2 Quantum Billiards in a Resonant Cavity

This Application demonstrates how the statistical properties of the eigenfrequencies of an electromagnetic resonant cavity have been used to to inform the subject of *quantum chaos*. Consider a billiard ball which reflects specularly from the side walls of a billiard table. For a rectangular table, infinitesimal changes in initial conditions produce infinitesimal changes in the billiard's trajectory. However, for a table with a less simple shape, infinitesimal changes in initial conditions typically produce wildly different trajectories. This is a signature of *classical chaos*. The corresponding quantum problem is a

[12] E.M. Purcell, "Spontaneous emission probabilities at radio frequencies", *Physical Review* **69**, 681 (1946).

free particle in a two-dimensional box with the shape of the billiard table. The question arises: does any signature occur in the quantum behavior when the classical motion changes from non-chaotic to chaotic?

To find the answer, it is sufficient to notice that the Schrödinger equation and boundary condition satisfied by the quantum billiard's wave function,

$$-\frac{\hbar^2}{2m}\left[\frac{\partial^2}{\partial x^2}+\frac{\partial^2}{\partial y^2}\right]\psi=\varepsilon\psi \qquad \psi|_S=0, \tag{19.142}$$

are the same as the Helmholtz equation and boundary condition in (19.57) satisfied by E_z for a TM mode in a conducting-tube waveguide. Moreover, setting $h=p\pi/L=0$ in the resonant-cavity equations (19.121) and (19.122) shows that $\mathbf{E}_{\text{TE}}=0$ and

$$\mathbf{E}_{\text{TM}}(x,y,t)=2E_z(x,y)\exp(-i\omega t)\hat{\mathbf{z}}. \tag{19.143}$$

Therefore, the eigenvalues of (19.142) are in one-to-one correspondence with the $p=0$ mode frequencies ω_{i0} in (19.123). The latter can be studied using pancake-shaped cavities where L is much smaller than the other cavity dimensions. This guarantees that the first $p=1$ mode lies at much higher frequency than very many $p=0$ modes.

Rather than the density of modes, the quantity plotted in Figure 19.20 is the distribution of the *differences* in frequency between adjacent cavity modes. The (numerical) data for a rectangular cavity follow a Poisson distribution, $P(s)\propto\exp(-as)$, which is the quantum prediction for a billiard whose classical motion is *not* chaotic. The (experimental) data for a cavity shaped like one quarter of a stadium (see inset) follow the statistics of a Gaussian orthogonal ensemble (GOE) where $P(s)\propto s\exp(-bs^2)$. This is the quantum prediction for a billiard whose classical motion is chaotic. Thus, chaotic motion on a classical billiard table manifests itself as an effective "repulsion" between the successive energy levels of a quantum billiard.

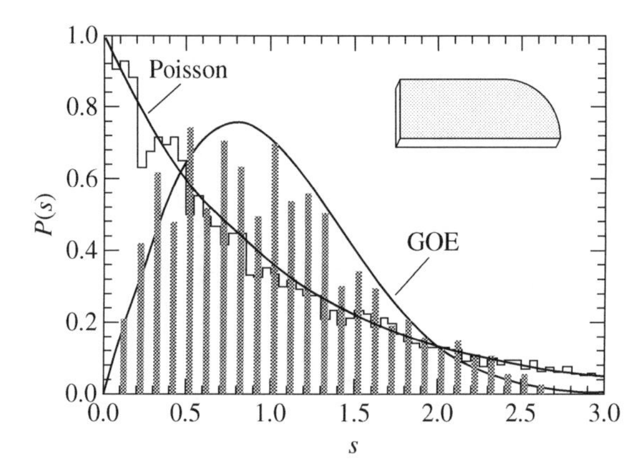

Figure 19.20: Distribution of spacings between adjacent resonant-cavity frequencies. Solid curves: predicted distributions when the classical motion is chaotic (GOE) and non-chaotic (Poisson). Histogram: numerical data for a rectangular cavity. Shaded bars: experimental data for the cavity shown in the inset. Figure adapted from Richter (1999).

19.6.5 Energy Exchange in Lossless Cavities

The total energy of an electromagnetic cavity is periodically exchanged between its electric field and its magnetic field. To prove this, we show first that the time-averaged electric energy $\langle U_E\rangle$ is equal to the time-averaged magnetic energy $\langle U_B \rangle$ for every normal mode. This, in turn, is a consequence of a time-averaged version of Poynting's theorem for the normal mode fields $\mathbf{E}_\lambda(\mathbf{r},t) = \mathbf{E}_\lambda(\mathbf{r})\exp(-i\omega_\lambda t)$ and $\mathbf{H}_\lambda(\mathbf{r},t) = \mathbf{H}_\lambda(\mathbf{r})\exp(-i\omega_\lambda t)$. The theorem in question is

$$\nabla \times (\mathbf{E}_\lambda \times \mathbf{H}^*_\lambda) = 4i\omega_\lambda \left[\langle u_E \rangle - \langle u_B \rangle\right]. \tag{19.144}$$

To derive (19.144), we need only combine the curl equations in (19.33) with the identity

$$\nabla \times (\mathbf{E} \times \mathbf{H}^*) = \mathbf{H}^* \cdot (\nabla \times \mathbf{E}) - \mathbf{E} \cdot (\nabla \times \mathbf{H}^*) \tag{19.145}$$

and use the definition of the time-averaged electric and magnetic energy densities,

$$\langle u_E \rangle = \frac{1}{4}\epsilon \mathbf{E} \cdot \mathbf{E}^* \qquad \text{and} \qquad \langle u_E \rangle = \frac{1}{4}\mu \mathbf{H} \cdot \mathbf{H}^*. \tag{19.146}$$

Integrating (19.144) over the volume of the cavity and using the divergence theorem gives

$$\int_S dA\hat{\mathbf{n}} \cdot (\mathbf{E}_\lambda \times \mathbf{H}^*_\lambda) = 4i\omega_\lambda \left[\langle U_E \rangle - \langle U_B \rangle\right]. \tag{19.147}$$

However, the left side of (19.147) vanishes because the conducting wall-boundary conditions ensure that $\hat{\mathbf{n}} \cdot (\mathbf{E}_\lambda \times \mathbf{H}^*_\lambda)|_S = \mathbf{H}^*_\lambda \cdot (\hat{\mathbf{n}} \times \mathbf{E}_\lambda)|_S = 0$. This proves that the time-averaged electric and magnetic energies are equal:

$$\langle U_E \rangle = \langle U_B \rangle. \tag{19.148}$$

The final step in the argument exploits the 90° phase difference between $\mathbf{E}_\lambda$ and $\mathbf{H}_\lambda$ pointed out in the paragraph following (19.116). This implies that the instantaneous total electric and magnetic energies in a resonant cavity are

$$\begin{aligned} U_E(t) &= \frac{\epsilon}{2}\int_V d^3r\, [\mathrm{Re}\,\mathbf{E}_\lambda(\mathbf{r},t)]^2 = 2\langle U_E \rangle \cos^2 \omega_\lambda t \\ U_B(t) &= \frac{\mu}{2}\int_V d^3r\, [\mathrm{Re}\,\mathbf{H}_\lambda(\mathbf{r},t)]^2 = 2\langle U_B \rangle \sin^2 \omega_\lambda t. \end{aligned} \tag{19.149}$$

Combining (19.148) with (19.149) shows that the total energy, $U_{\mathrm{EM}}(t) = U_E(t) + U_B(t)$, is indeed periodically exchanged between the modal electric field and the modal magnetic field.

Equation (19.149) should remind the reader of the periodic exchange of energy which occurs when an LC circuit oscillates at its resonant frequency $\omega = 1/\sqrt{LC}$. This not an accident. As Figure 19.21 shows, the passage from a circuit oscillator to a resonant cavity does not introduce new physics, just a change in geometry and an increase in resonant frequency. The latter is achieved by reducing the effective inductance (first by reducing the number of turns in the circuit solenoid from many to one and then by adding more and more one-turn solenoids in parallel) and reducing the effective capacitance (by increasing the separation between the surfaces with equal and opposite charge). The final, cylindrically symmetric resonant cavity on the far right of Figure 19.21 is typical of a two-cavity klystron. It is worth noting that the modal electric and magnetic fields largely occupy different regions of space in Figure 19.21. This is not true for the hollow-tube resonant cavities discussed in Section 19.6.2.

19.6.6 Energy Loss in Ohmic Cavities

The modes of a real resonant cavity do not oscillate forever. The mode amplitudes steadily decrease because ohmic heating in the walls drains away the field energy. A figure of merit is the quality factor Q (see Example 17.2.2), which is the mode frequency ω_0 multiplied by the ratio of the time-averaged

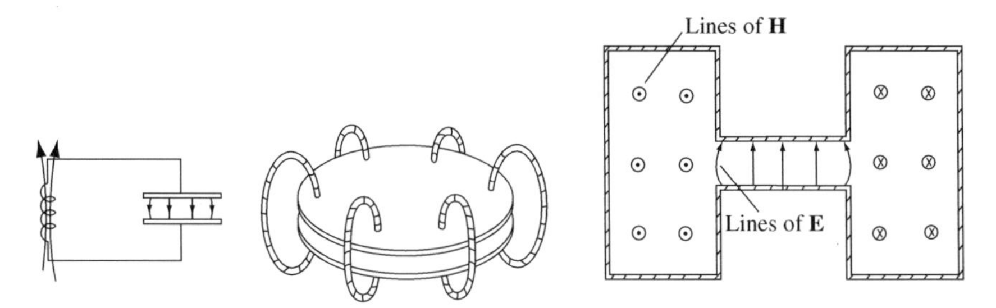

Figure 19.21: The natural frequencies increase from left to right for the resonators shown. Figure from Feynman, Leighton, and Sands (1964).

total energy to the time-averaged power dissipated:

$$Q = \omega_0 \frac{\langle U_{\rm EM} \rangle}{\langle \mathcal{P}_{\rm EM} \rangle}. \tag{19.150}$$

Using (19.148), we write $\langle U_{\rm EM} \rangle$ as twice the time-averaged magnetic energy integrated over the volume of the cavity. For $\langle \mathcal{P}_{\rm EM} \rangle$, we integrate (19.78) over the surface area of the cavity. Hence,

$$Q = \frac{2}{\delta} \times \frac{\int\limits_V dV\, |\mathbf{H}(\mathbf{r})|^2}{\int\limits_S dA\, |\mathbf{H}(\mathbf{r})|^2} \approx \frac{2}{\delta} \frac{V}{A}. \tag{19.151}$$

The estimate on the right side of (19.151) is always intuitively useful, but only quantitatively accurate for modes where the magnitude of $\mathbf{H}$ does not vary significantly over the volume of the cavity.

We get a better understanding of Q by writing (19.150) as a differential equation for the time-averaged cavity energy:

$$\frac{d\langle U_{\rm EM} \rangle}{dt} = -\frac{\omega_0}{Q} \langle U_{\rm EM} \rangle. \tag{19.152}$$

The solution to (19.152),

$$\langle U_{\rm EM}(t) \rangle = \langle U_{\rm EM}(0) \rangle \exp\left[-\frac{\omega_0 t}{Q} \right], \tag{19.153}$$

implies that $E(t)$ [and also $H(t)$] vary in time as

$$E(t) = E(0) \exp\left[-\frac{\omega_0 t}{2Q} \right] \exp\left[-i\omega_0 t \right]. \tag{19.154}$$

Frequency-domain measurements probe the Fourier transform of (19.154). Assuming that $E(t) = 0$ for $t < 0$,

$$\hat{E}(\omega) = \int_0^\infty dt\, E(t) e^{i\omega t} = \frac{E(0)}{i(\omega_0 - \omega) + \omega_0/2Q}. \tag{19.155}$$

Figure 19.22 plots the absolute square $|\hat{E}(\omega)|^2$. The line shape is Lorentzian, where the ratio of the full width at half-maximum to the resonance frequency is exactly the quality factor:

$$\frac{\Delta\omega}{\omega_0} = \frac{1}{Q}. \tag{19.156}$$

The Q factor generally differs considerably for the different modes of a resonant cavity.

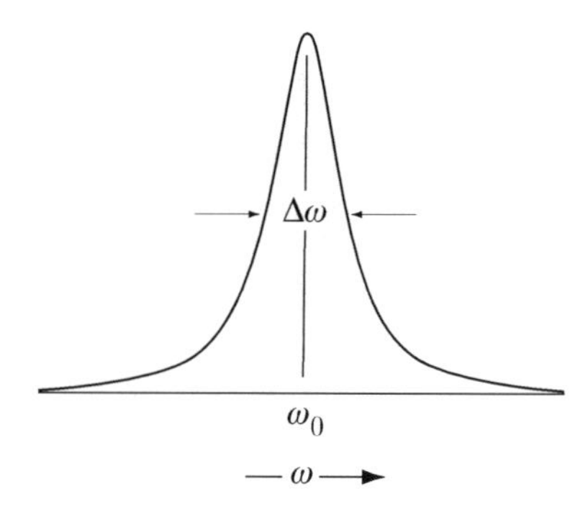

Figure 19.22: $\Delta\omega$ is the width at half-maximum of the Lorentzian line shape of $|\hat{E}(\omega)|^2$ for a typical mode of a lossy resonant cavity.

19.6.7 Excitation of Cavity Modes

A localized current source is often used to excite the electromagnetic modes of a conducting resonant cavity. Our approach to this problem uses a method (borrowed from the quantum theory of radiation) which demonstrates the dynamical similarity of the modes of a lossless resonant cavity to a collection of harmonic oscillators. Without loss of generality, we ignore free charge and adopt the Coulomb gauge where $\nabla \cdot \mathbf{A} = 0$. This permits us to neglect the scalar potential (see Section 16.2.2) and write the cavity fields as

$$\mathbf{E} = -\frac{\partial \mathbf{A}}{\partial t} \qquad \text{and} \qquad \mathbf{B} = \nabla \times \mathbf{A}. \tag{19.157}$$

Under the assumed conditions, the general time-evolution equation for the vector potential (15.9) simplifies to

$$\nabla^2 \mathbf{A} - \frac{1}{c^2}\frac{\partial^2 \mathbf{A}}{\partial t^2} = -\mu_0 \mathbf{j}. \tag{19.158}$$

The cavity-mode eigenfunctions $\mathbf{A}_\lambda(\mathbf{r})\exp(-i\omega_\lambda t)$ are the solutions to (19.158) (with $\mathbf{j} = 0$) which satisfy $\hat{\mathbf{n}} \times \mathbf{A}_\lambda|_S = 0$. This is so because (19.157) tells us that $\mathbf{E}_\lambda(\mathbf{r}) = i\omega_\lambda \mathbf{A}_\lambda(\mathbf{r})$, where $\mathbf{E}_\lambda$ is an electric field cavity mode. The key step exploits the completeness of the mode functions to write an arbitrary vector potential in the cavity as

$$\mathbf{A}(\mathbf{r}, t) = \sum_\lambda \mathbf{A}_\lambda(\mathbf{r}, t) = \sum_\lambda q_\lambda(t)\mathbf{A}_\lambda(\mathbf{r}). \tag{19.159}$$

Substituting (19.159) into (19.158) and using the first equation in (19.115) applied to $\mathbf{A}_\lambda(\mathbf{r})$ gives

$$\sum_\lambda \mathbf{A}_\lambda(\mathbf{r})\left(\ddot{q}_\lambda + \omega_\lambda^2 q_\lambda\right) = \frac{1}{\epsilon_0}\mathbf{j}(\mathbf{r}, t). \tag{19.160}$$

Finally, multiply (19.160) by A_μ, integrate over the cavity volume, and use (19.119) to get[13]

$$\ddot{q}_\mu(t) + \omega_\mu^2 q_\mu(t) = \frac{1}{\epsilon_0}\int_V d^3r\, \mathbf{j}(\mathbf{r}, t) \cdot \mathbf{A}_\mu(\mathbf{r}). \tag{19.161}$$

[13] We assume the mode functions $A(\mathbf{r})_\lambda$ are normalized in the cavity volume V.

Figure 19.23: Scanning electron micrograph of a silica dielectric resonator supported by a silicon pillar. Image from Schliesser *et al.* (2008).

This equation shows that the amplitude of the μth cavity mode behaves exactly like a driven harmonic oscillator. The "driving force" is the projection of the excitation current density onto the spatial distribution of that mode. Thus, if $\mathbf{j}(\mathbf{r}, t)$ reflects a current loop inserted into the cavity, it is not difficult to imagine that different modes will be excited as a function of the size and orientation of the loop.

Multiplying (19.161) by $\dot{q}_\lambda$ and using (19.157) and (19.159) produces a power-balance equation:

$$\frac{d}{dt}\epsilon_0\left[\frac{1}{2}\dot{q}_\mu^2(t) + \frac{1}{2}\omega_\mu^2 q_\mu^2(t)\right] + \int\limits_V d^3r\, \mathbf{j}(\mathbf{r}, t)\cdot\mathbf{E}_\mu(\mathbf{r}, t) = 0. \tag{19.162}$$

We leave it as an exercise for the reader to show that the kinetic (potential) energy term on the left side of (19.162) may be identified with the electric (magnetic) energy of the mode. From this point of view, (19.162) shows how the work done by the current source increases the total energy of the cavity mode.

19.7 Dielectric Resonators

The conducting-wall resonators studied in the previous section are useful as long as the ohmic losses in their walls can be tolerated. When this is not the case, all-dielectric structures can sometimes be used as temporary storage facilities for oscillating electromagnetic fields. We have seen an example of this already in Section 17.5.2 where the middle panel of Figure 17.12 showed an infrared laser cavity closed at either end by highly reflective mirrors made from GaAs/AlAs multilayers. Here, we consider high dielectric constant ($\kappa > 30$) materials which respond resonantly to oscillating electromagnetic fields when they are machined into simple shapes like spheres, cylinders, or disks.

In the microwave band, dielectric resonators with large quality factors ($Q > 10^4$) are used as filters and low-noise reference oscillators when small size and temperature stability are important. Figure 19.23 is a scanning electron microscope image of a more exotic, toroid-shaped dielectric resonator. The toroid was created by the selective melting and re-solidification of the perimeter of a silica (SiO_2) disk supported by a silicon pillar. The standing optical waves that propagate around this toroid's perimeter (see Application 19.3) have Q values as high as 10^8.

The normal modes of a dielectric resonator all have a finite quality factor, even if the dielectric is lossless. This is so because the fields are never completely confined to the volume of the resonator. The dielectric matching conditions guarantee that any field that lies inside the volume connects to a field (evanescent or propagating) that lies outside the volume.[14] With this complication, a complete theory of the modes can be developed along the lines developed in Section 19.6 for conducting resonant cavities.

Application 19.3 Whispering Gallery Modes

The high-Q optical modes of the micro-toroid shown in Figure 19.23 have almost all of their field strength in the immediate vicinity of the perimeter of the toroid. They propagate circumferentially at the dielectric-air interface because, in optical language, the rays trace a closed polygonal path due to repeated total internal reflection at the interface. The name *whispering gallery* modes is given to these electromagnetic waves, by analogy with sound waves (first explained by Lord Rayleigh) which propagate circumferentially along the circular base of the dome of St. Paul's Cathedral in London.

This Application exploits the similarity of the Helmholtz equation (19.100) to the Schrödinger equation to study the whispering gallery modes of a dielectric sphere. We limit ourselves to TE modes and recall from (19.125) and (19.130) that

$$\mathbf{E}(\mathbf{r}) = -\omega\hat{\mathbf{L}}u(\mathbf{r}) \qquad \text{and} \qquad \hat{\mathbf{L}} = -i\mathbf{r}\times\nabla, \tag{19.163}$$

where

$$\nabla^2 f = \frac{1}{r}\frac{\partial^2(rf)}{\partial r^2} - \frac{\hat{L}^2}{r^2}f \qquad \text{and} \qquad \hat{L}^2 Y_{\ell m} = \ell(\ell+1)Y_{\ell m}. \tag{19.164}$$

The electric field (19.163) has no radial component [see (19.128)] and we assume that the dielectric "constant" varies only radially so $\epsilon(\mathbf{r}) = \epsilon(r)$. This means that the right side of (19.100) vanishes. The key is to let $r^2 u(\mathbf{r}) = \psi(r)Y_{\ell m}(\theta,\phi)$ and substitute (19.163) and (19.164) into (19.100). Using the operator commutation relation $[\hat{L}^2, \mathbf{L}] = 0$ familiar from quantum mechanics, the result takes the form of a radial Schrödinger equation:

$$-\frac{d^2\psi}{dr^2} + \left[\frac{\omega^2}{c^2}\{1-n(r)\} + \frac{\ell(\ell+1)}{r^2}\right]\psi = \frac{\omega^2}{c^2}\psi. \tag{19.165}$$

The piecewise constant index of refraction $n(r > R) = 1$ and $n(r < R) = n > 1$ defines a uniform dielectric sphere of radius R. With this choice, the effective potential in square brackets in (19.165) has an attractive-well component for $r < R$ and a repulsive centrifugal barrier. The latter increases in magnitude as the angular momentum increases but falls off from the center of the sphere as $1/r^2$. The discontinuity of $n(r)$ requires that we impose the continuity of the tangential components of the electric and magnetic fields on the spherical shell $r = R$. From (19.163), $\mathbf{E}$ will be continuous if $\psi(r)$ is continuous. From the duality[15] of $\mathbf{H}_{\rm TE}$ and $\mathbf{E}_{\rm TM}$ [compare (19.125) to (19.126)], together with (19.131), we see that $\hat{\mathbf{r}}\times\mathbf{H}_{\rm TE}$ will be continuous if the radial derivative $\psi'(r)$ is continuous. This completes the analogy between the TE-mode problem for a dielectric sphere and the Schrödinger problem of a particle in a spherical well.

Our aim is to demonstrate the existence of whispering gallery modes localized near $r = R$. Using intuition gained from elementary quantum mechanics, Figure 19.24 shows that it is sufficient to choose $\ell \gg 1$ and locate ω just above the zero of the effective potential. The dashed line shows that most of the amplitude of $\psi(r)$ is trapped near $r = R$ by the large centrifugal barrier. The fact that this wave function "tunnels" through the barrier into the continuum at larger values of r identifies this mode as a resonance with a finite lifetime rather than a bound state with an infinite lifetime. The resonance lifetime is simply related to the quality factor Q of the mode [cf. (19.156)].

[14] Compare this with Section 19.5.4 on the radiation modes of a dielectric waveguide.

[15] See Section 16.8.1.

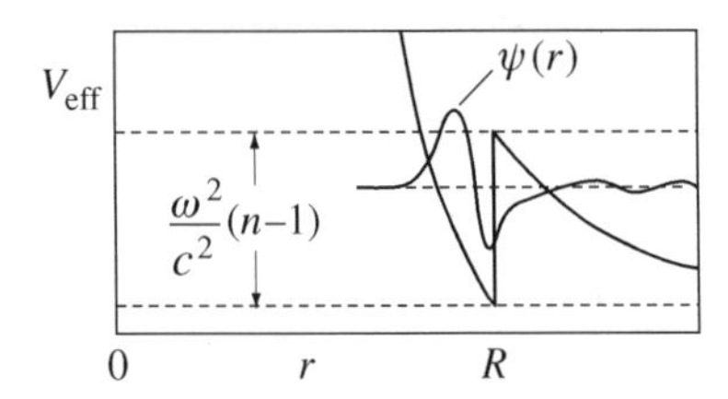

Figure 19.24: The wave function of a resonance (whispering gallery) state superimposed on the effective potential in (19.165). Figure adapted from Johnson (1993).

Sources, References, and Additional Reading

The quotation at the beginning of the chapter appears in

J.-D. Colladon, "On the reflections of a ray of light inside a parabolic liquid stream", *Comptes Rendus* **15**, 800 (July-December 1842).

Section 19.1 A comprehensive overview of the material treated in this chapter (excepting optical fibers) is

F.E. Borgnis and C.H. Papas, "Electromagnetic waveguides and resonators", in *Encyclopedia of Physics*, edited by S. Flügge (Springer, Berlin, 1958), vol. XVI, pp. 285-422.

Three good textbook treatments, in increasing order of mathematical sophistication, are

U.S. Inan and A.S. Inan, *Electromagnetic Waves* (Prentice Hall, Upper Saddle River, NJ, 2000).

K. Zhang and D. Li, *Electromagnetic Theory for Microwaves and Optoelectronics* (Springer, Berlin, 1998).

C.G. Someda, *Electromagnetic Waves*, 2nd edition (Taylor & Francis, Boca Raton, FL, 2006).

Section 19.2 The transmission line discussion of Section 19.2 benefited from the following texts. Figure 19.1 is taken from Chapter 5 of *Staelin et al.*

H.A Haus and J.R. Melcher, *Electromagnetic Fields and Energy* (Prentice Hall, Englewood Cliffs, NJ, 1989).

D.H. Staelin, A.W. Morgenthaler, and J.A. Kong, *Electromagnetic Waves* (Prentice-Hall, Englewood Cliffs, NJ, 1994).

The connection between Maxwell theory and circuit theory for wave-guiding systems was worked out independently during World War II by Julian Schwinger in the United States and Sin-itiro Tomonaga in Japan. Both later shared the Nobel Prize in Physics (with Richard Feynman) for the theory of quantum electrodynamics. For details, see

J. Mehra and K.A. Milton, *Climbing the Mountain: The Scientific Biography of Julian Schwinger* (University Press, Oxford, 2003).

S. Tomonaga, "A general theory of ultra-short wave circuits I and II", *Journal of the Physical Society of Japan* **2**, 158 (1947); *ibid.* **3**, 93 (1948).

The definitive account of Lord Kelvin's contributions to telegraphy is

C. Smith and M.N. Wise, *Energy and Empire: A Biographical Study of Lord Kelvin* (Cambridge University Press, Cambridge, 1989).

Section 19.4 Example 19.2 comes from unpublished lecture notes for a course on Electromagnetic Theory taught by Prof. Norman Kroll at Columbia University in 1955. For more on impedance boundary conditions, see

D.J. Hoppe and Y. Rahmat-Samii, *Impedance Boundary Conditions in Electromagnetics* (Taylor & Francis, Washington, DC, 1995).

A wealth of interesting waveguide and cavity physics can be found in the excellent monograph,

T.P. Wangler, *Principles of RF Linear Accelerators* (Wiley, New York, 1998).

Figure 19.8 and Figure 19.10 come from *Borgnis and Papas* (see Section 19.1 above). With kind permission from Springer Science & Business Media B.V. *Cheng, Inan and Inan* (see Section 9.1 above), and *Neal* are the sources for Figure 19.9, Figure 19.11, and Figure 19.12, respectively:

D.K. Cheng, *Field and Wave Electromagnetics*, 2nd edition (Addison-Wesley, Reading, MA, 1989). Reproduced with permission of the author and AAS.

R.B. Neal, "General description of SLAC", in *The Stanford Two-Mile Accelerator*, edited by R.B. Neal (W.A. Benjamin, New York, 1968), pp. 55-94. Reprinted with permission from the AAS.

Section 19.5 *Okamoto* is a good source for information about optical fibers. Figure 19.17 was adapted from *Hoffmann*.

K. Okamoto, *Fundamentals of Optical Waveguides*, 2nd edition (Academic, New York, 2006).

R.K. Hoffmann, *Handbook of Microwave Integrated Circuits* (Artech, Norwood, MA, 1987).

Section 19.6 Example 19.3 comes from *Harrington*. *Vahala* is an interesting account of mode density manipulation in resonant cavities. *Baltes* gives an overview of Weyl's theorem and its applications.

R.F. Harrington, *Time-Harmonic Electromagnetic Fields* (McGraw-Hill, New York, 1961), Section 4-6.

K.J. Vahala, "Optical microcavities", *Nature* **424**, 839 (2003).

H.P. Baltes, "Planck's radiation law for finite cavities and related problems", *Infrared Physics* **16**, 1 (1976).

The sources for Figure 19.19, Figure 19.21, and Figure 19.20 are, respectively,

D.J. Fixsen, E.S. Chang, J.M. Gales, *et al.*, "The cosmic microwave background spectrum from the full COBE FIRAS data set", *The Astrophysical Journal* **473**, 576 (1996). Reprinted with permission from the AAS.

R.P. Feynman, R.B. Leighton, and M. Sands, *Lectures on Physics* (Addison-Wesley, Reading, MA, 1964), Volume II, Chapter 23.

A. Richter, "Playing billiards with microwaves: Quantum manifestations of classical chaos", in *Emerging Applications of Number Theory*, edited by D.A. Hejhal, J. Friedman, M.C. Gutzwiller, and A.M. Odlyzko (Springer, Berlin, 1999), pp. 479-524. With kind permission from Springer Science & Business Media B.V.

Our treatment of cavity mode excitation (Section 19.6.7) comes from

E.U. Condon, "Electronic generation of electromagnetic oscillations", *Journal of Applied Physics* **11**, 502 (1940).

Section 19.7 For more details about dielectric resonators, the reader may consult

D. Kajfez and P. Guillon, *Dielectric Resonators* (Artech House, Dedham, MA, 1986).

Figure 19.23 comes from *Schliesser et al.* Figure 19.24 and the analogy between whispering gallery modes and resonances in quantum mechanics used in Application 19.3 come from *Johnson*.

A. Schliesser, R. Rivière, G. Anetsberger, O. Arcizet, and T.J. Kippenberg, "Resolved-sideband cooling of a micromechanical oscillator", *Nature Physics* **4**, 415 (2008).

B.R. Johnson, "Theory of morphology-dependent resonances: Shape resonances and width formulas", *Journal of the Optical Society of America A* **10**, 343 (1993).

Problems

19.1 Two-Wire Transmission Line A long transmission line consists of two identical wires embedded in a medium with permittivity ϵ and permeability μ. Let the wire separation d be large compared to the wire radius a. Calculate the capacitance per unit length C and the inductance per unit length $\mathcal{L}$. Confirm the general relation $\mathcal{L}C = \mu\epsilon$ derived in the text.

19.2 TM Wave Guided by a Flat Conductor A monochromatic plane wave in vacuum ($x > 0$) with $E_z > 0$ and $E_x < 0$ strikes a perfect conductor ($x < 0$) at an angle of incidence θ.

(a) Show that, in steady state, a non-uniform TM wave occupies the vacuum space above the conductor.
(b) Calculate the time-averaged Poynting vector everywhere.

(c) Calculate the charge density and current density induced on the surface of the conductor. Do they satisfy a continuity equation?

19.3 TEM Waves Guided by a Cone and a Plane Consider time-harmonic solutions to the Maxwell equations in vacuum where the fields are *independent* of the azimuthal angle ϕ. TEM solutions of this type also have no radial component to the fields: $E_r = B_r = 0$.

(a) Show that the conditions stated above decouple the Maxwell curl equations into two subsets, each of which describes a different type of TEM wave.
(b) Begin with the Maxwell divergence equations and find general solutions for $\mathbf{E}(r, \theta, t)$ and $\mathbf{B}(r, \theta, t)$ for each of the two TEM wave types.
(c) The figure below shows the apex of an infinite, solid conducting cone touching the conducting half-space $z < 0$. Explain why this structure can be used to guide one of the TEM wave types found above but not the other.

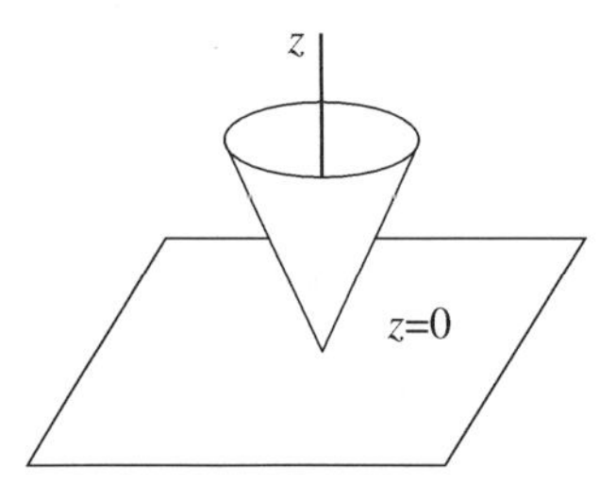

19.4 The Lowest Propagating Mode of a Waveguide The TM and TE modes of a hollow-tube waveguide are determined by the two-dimensional Helmholtz equation $[\nabla_\perp^2 + \gamma^2]\psi = 0$ with boundary conditions $\psi|_S = 0$ and $\partial\psi/\partial n|_S = 0$, respectively. The same equation and boundary conditions apply when $\psi(x, y)$ is the wave function of a free particle in a two-dimensional box with infinite or finite potential walls, and when $\psi(x, y)$ is the vibrational amplitude of a drumhead whose perimeter is held fixed or left free.

(a) Produce an argument based on the behavior of a quantum particle-in-a-box to argue that one type of mode (either TE or TM) always has the lowest (non-zero) frequency for a hollow-tube waveguide with an arbitrary cross sectional shape.
(b) Produce an argument based on the behavior of an elastic drumhead to reach the same conclusion as in part (a).

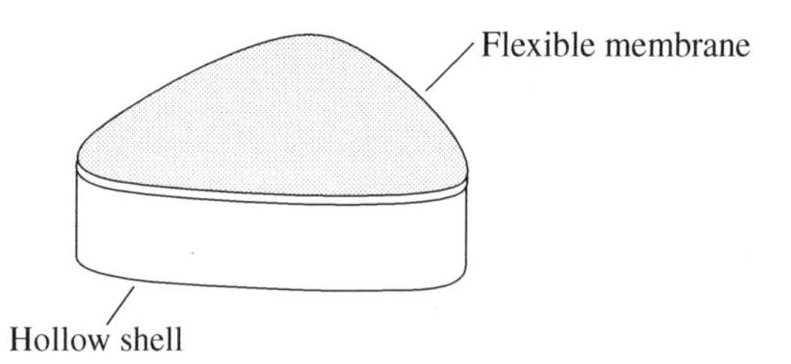

19.5 Semi-Circular Waveguide A perfectly conducting waveguide has a cross section in the shape of a semi-circle with radius R.

(a) Find the longitudinal fields E_z and B_z for the TM and TE modes, respectively. Find also the cut-off frequency for these modes.
(b) Write explicit formulae for the transverse fields for the lowest cutoff frequency found in part (a).

19.6 Whispering Gallery Modes Consider a hollow conducting tube with a circular cross section of radius R and infinite length.

(a) Find monochromatic TE ($E_z = 0$) and TM ($H_z = 0$) solutions of the Maxwell equations inside the tube which propagate around the tube circumference and thus do *not* depend on the longitudinal coordinate z.

(b) Identify the subset of waves where the electric field has no nodes inside the tube and almost all of the field energy is concentrated in a very small annular volume near the wall of the tube.

19.7 Waveguide Discontinuity Two rectangular waveguides with different major side lengths ($a_1 < a_2$) along the x-axis and equal minor sides ($b_1 = b_2$) along the y-axis are butt-joined in the $z = 0$ plane. Waveguide 1 propagates a TE_{10} mode (only) in the $+z$-direction toward waveguide 2. Find the amplitude of the various modes excited in waveguide 2 if the two guides share a corner at $x = y = 0$ and the open portion of the larger guide in the x-y plane is closed by a perfect conductor. Check the $a_1 = a_2$ limit.

19.8 A Vector-Potential Method

(a) Show that a general TM wave in a hollow-tube waveguide can be derived from a longitudinal vector potential $\mathbf{A}(\mathbf{r}, t) = \hat{\mathbf{z}}A(\mathbf{r}_\perp)\exp[i(hz - \omega t)]$ which satisfies the wave equation.

(b) Duality implies that a general TE wave can be derived from an "electric vector potential" $\tilde{\mathbf{A}}$, where $\mathbf{E}_{\text{TE}} = \nabla \times \tilde{\mathbf{A}}$. Explain why this makes perfectly good sense.

19.9 Waveguide Filters The figure below shows two circular conducting tubes in cross section. Each tube has a thin metal screen inserted at one point along its length. One screen takes the form of metal wires bent into concentric circles. The other takes the form of metal wires arranged like the spokes of a wheel. One of these tubes transmits only a low-frequency TE waveguide mode down the tube. The other transmits only a low-frequency TM waveguide mode down the tube. Explain which tube is which and why, using the fact that the fields of a general waveguide satisfy $\nabla \times \mathbf{E}_\perp = i\omega B_z \hat{\mathbf{z}}$.

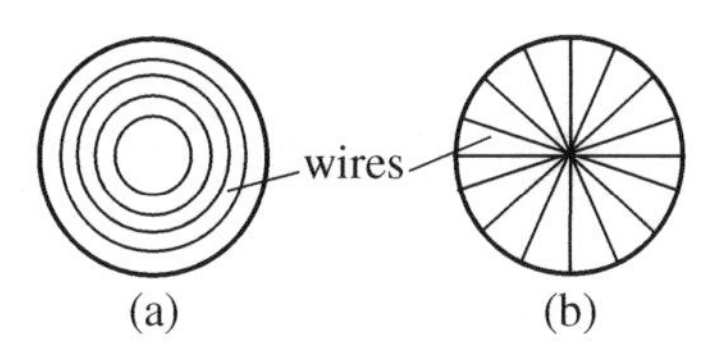

19.10 Waveguide Mode Orthogonality

(a) Suppose that $\nabla_\perp^2 \psi_p = \lambda_p \psi_p$ and $\nabla_\perp^2 \psi_q = \lambda_q \psi_q$ in a two-dimensional domain A where either $\psi|_C = 0$ or $\partial\psi/\partial n|_C = 0$ on the perimeter C of A. Use one of Green's identities to show that

$$\int_A d^2r\, \psi_p(\mathbf{r}_\perp)\psi_q(\mathbf{r}) = 0 \qquad (\lambda_p \neq \lambda_q).$$

(b) Let $\mathbf{E}_p(\mathbf{r}, t) = \mathbf{E}_p(x, y)e^{i(h_p z - \omega t)}$ and $\mathbf{B}_p(\mathbf{r}, t) = \mathbf{B}_p(x, y)e^{i(h_p z - \omega t)}$ be the fields associated with the pth mode of a cylindrical waveguide with cross sectional area A. When $p \neq q$ and both modes are TE, use the results of part (a) to show that non-degenerate modes satisfy

$$\int_A d^2r\, \mathbf{E}_p \cdot \mathbf{E}_q = 0 = \int_A d^2r\, \mathbf{B}_p \cdot \mathbf{B}_q.$$

(c) Repeat part (b) when both modes are TM.

(d) Repeat part (b) when one mode is TE and one mode is TM.

19.11 A Waveguide with a Bend A rectangular waveguide with a constant cross section and perfectly conducting walls contains a curved section as sketched below. Also indicated is a local Cartesian coordinate system where the z-axis and y-axis remain tangent and normal to the walls, respectively.

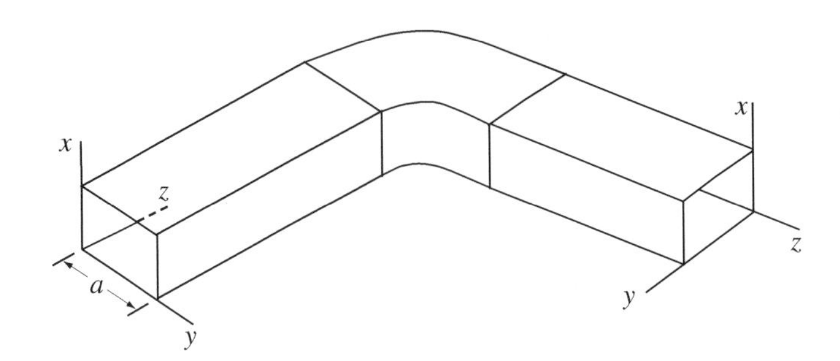

(a) The scalar function Φ satisfies $[\nabla_\perp^2 + \omega^2/c^2]\Phi(y, z) = 0$, where $\nabla_\perp^2 = \partial^2/\partial x^2 + \partial/\partial y^2$. Show that the four vacuum Maxwell equations and conducting wall-boundary conditions are satisfied by time-harmonic transverse electric (TE) modes of the form

$$\mathbf{E} = \hat{\mathbf{x}}\, i\frac{\omega}{c}\Phi \qquad\qquad c\mathbf{B} = -\hat{\mathbf{x}} \times \nabla\Phi.$$

(b) Suppose that the curvature $\kappa(z)$ of the side wall at any point on the guide satisfies $\kappa a \ll 1$ so the Laplacian operator in the local coordinate Cartesian coordinate system is well approximated by

$$\nabla^2 = \frac{\partial^2}{\partial y^2} + \frac{\partial^2}{\partial z^2} + \frac{1}{2}\kappa^2(z).$$

Separate variables in the Helmholtz equation and show that propagating modes exist in the straight portion of the guide (at least) when $\omega > \pi c/a$.

(c) Show that at least one mode exists in the curved part of the guide for $\omega < \pi c/a$. Describe the spatial characteristics of this solution. Hint: Make an analogy with the one-dimensional, time-independent Schrödinger equation.

19.12 TE and TM Modes of a Coaxial Waveguide An infinitely long coaxial waveguide is formed in the vacuum volume between two concentric, perfectly conducting cylinders with radii b and $a > b$.

(a) Find $\mathbf{E}$ and $\mathbf{B}$ for the TE and TM modes of this guide and find (but do not try to solve) the transcendental equations that determine the mode frequencies.

(b) Approximate expressions for the TE and TM mode frequencies can be found for the case when $a - b \ll \bar{\rho} = \frac{1}{2}(a + b)$. Replace the variable ρ by $\bar{\rho}$ in appropriate places in the radial part of the Helmholtz equation and redo the analysis of part (a), now including a determination of the mode frequencies.

19.13 A Baffling Waveguide The figure below shows the circular cross section of an infinitely long metallic waveguide with an infinitesimally thin, metallic baffle inserted into its otherwise hollow interior. The baffle has infinite length and a width equal to the radius R of the waveguide.

(a) Show that the baffle increases the lowest cutoff frequency for TM modes.

(b) Show that the baffle decreases the lowest cutoff frequency for TE modes.

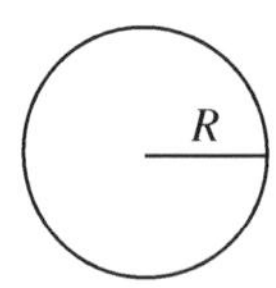

19.14 Waveguide Charge and Current

(a) Calculate the induced surface charge density $\sigma_{\rm TM}$ and the longitudinal surface current density $\mathbf{K}_{\rm TM}$ associated with the propagation of a TM mode in a perfectly conducting waveguide with a uniform cross section. Show that $\mathbf{K}_{\text{TM}} = v_p \sigma_{\text{TM}} \hat{\mathbf{z}}$ where v_p is the *phase* velocity of the wave propagating in the z-direction.

(b) Calculate the induced surface charge density σ_{TE} and the surface current density $\mathbf{K}_{\text{TE}}$ associated with the propagation of a TE mode in a perfectly conducting waveguide with a uniform cross section. Show that $\hat{\mathbf{z}} \cdot \mathbf{K}_{\text{TE}} = v_g \sigma_{\text{TE}}$ where v_g is the *group* velocity of the wave.

(c) Show that in the TE case a surface current appears in the transverse ($\hat{\boldsymbol{\tau}} = \hat{\mathbf{n}} \times \hat{\mathbf{z}}$) direction also.

(d) Show that the surface current and surface charge satisfy a "surface continuity equation" for both TE and TM modes.

19.15 Cavity Modes as Harmonic Oscillators Prove the assertion made in the text that the total electromagnetic energy of a lossless resonant cavity can be put in the form of the total mechanical energy of a collection of undamped harmonic oscillators. Identify the electric and magnetic contributions explicitly. Assume that the vector potential mode functions $\mathbf{A}_\lambda(\mathbf{r})$ are normalized in the cavity volume.

19.16 An Electromagnetic Oscillator An electromagnetic oscillator is formed when charge sloshes back and forth between two identical, perfectly conducting spheres of radius R connected by a very thin, very long, perfectly conducting rod of radius $a \ll R$ and length $l \gg R$. The net charge of the entire structure is zero. Assume that no charge accumulates on the connecting rod and that the currents that flow in the spheres are negligible.

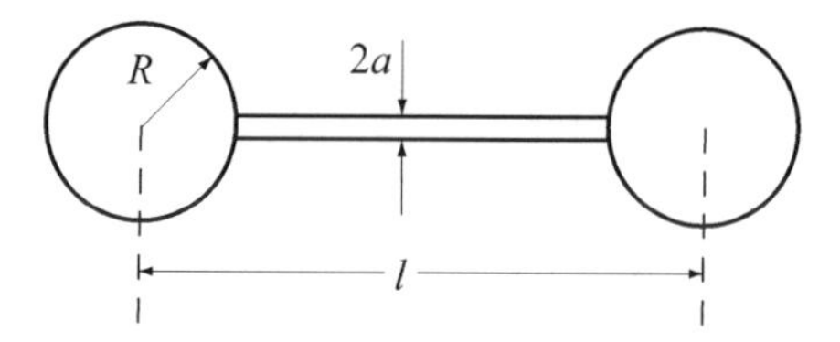

(a) Estimate the capacitance of this system.

(b) Estimate the inductance of this system.

(c) Show that the resonant frequency of the oscillator is $\omega \approx \dfrac{c}{\sqrt{Rl \ln(l/a)}}$.

19.17 A Variational Principle

(a) Show that the frequency of any mode $(\mathbf{E}, \mathbf{B})$ of a resonant cavity with volume V can be computed from

$$\frac{\omega^2}{c^2} = \frac{\int_V d^3r\, |\nabla \times \mathbf{E}|^2}{\int_V d^3r\, |\mathbf{E}|^2} = \frac{\int_V d^3r\, |\nabla \times \mathbf{B}|^2}{\int_V d^3r\, |\mathbf{B}|^2}.$$

(b) Suppose $\mathbf{E} \to \mathbf{E} + \delta\mathbf{E}$ or $\mathbf{B} \to \mathbf{B} + \delta\mathbf{B}$, where $\delta\mathbf{E}$ and $\delta\mathbf{B}$ satisfy the boundary conditions. Prove that the change in ω^2 is only second-order in $\delta\mathbf{E}$ or $\delta\mathbf{B}$. This implies that *any* choice of $\mathbf{E}(\mathbf{r})$ or $\mathbf{B}(\mathbf{r})$ in these formulae which satisfies the boundary conditions (but not the Maxwell equations) provides an upper bound on the lowest mode frequency.

(c) Obtain an estimate of the lowest TM-mode frequency in a cylindrical cavity of radius R by minimizing the frequency with respect to the variational parameter a in the choice $\mathbf{B} = (\rho + a\rho^2)\hat{\boldsymbol{\phi}}$. Compare your answer with the exact result.

19.18 An Asymmetric Two-Dimensional Resonant Cavity The two-dimensional vectors $\mathbf{k}_m$ shown below are inclined at angles $\theta_m = m\pi/3$ with respect to the positive x-axis. The vectors share a common magnitude $|\mathbf{k}_m| = k$.

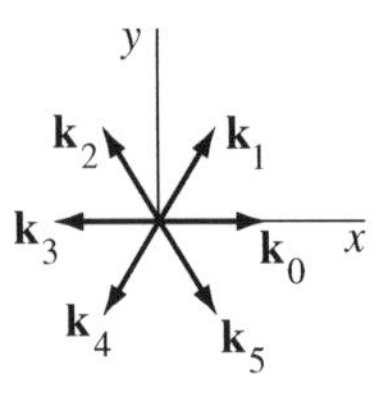

Superpose six waves with alternating amplitudes to form the scalar function

$$\psi(x, y, t) = \sum_{m=0}^{5} (-1)^m \sin(\mathbf{k}_m \cdot \mathbf{r} - ckt).$$

Draw the outline of a two-dimensional resonant cavity which supports a TM mode built from $\psi(x, y, t)$.

19.19 The Ark of the Covenant Find the numerical value of the lowest resonant frequency ω_0 and the exact half-width Γ of the resonant cavity God instructed Moses to build in Exodus 25:10-11: "Make an ark of acacia-wood: two cubits and a half shall be the length thereof, and a cubit and a half the breadth thereof, and a cubit and a half the height thereof. And thou shalt overlay it with pure gold."

19.20 Perturbation of a Cavity Resonator A small polarizable and magnetizable object inserted into a resonant cavity produces a shift in each resonance frequency by an amount $\delta\omega/\omega$. Derive an expression for this frequency shift in terms of the time-harmonic dipole moments $\mathbf{p}(t)$ and $\mathbf{m}(t)$ induced in the object by the relevant cavity mode and the values of the unperturbed mode fields $\mathbf{E}_0$ and $\mathbf{B}_0$ evaluated at the position of the object. Do this using the Boltzmann-Ehrenfest adiabatic theorem, which states that any adiabatic (quasistatic) change in the state of any oscillating system has no effect on the product of the oscillation period T and the time-averaged total energy $\langle U_{\rm EM}\rangle$. Hint: Look back at Section 6.7.3 and Section 13.7.3.

19.21 Resonant-Frequency Differences for a Cavity A perfectly conducting resonant cavity has the shape of a rectangular box where the length, width, and height are chosen as three unequal irrational numbers. Evaluate (at least) the first 10^5 resonant frequencies numerically, label them so that $\omega_1 \le \omega_2 \le \omega_3 \le \cdots$, and construct a histogram of the nearest-neighbor frequency spacings of the form

$$P(s) = \sum_{k=1}^{N} \delta(\omega_{k+1} - \omega_k - s).$$

Show that your histogram is well approximated by a Poisson distribution. Hint: Be sure your histogram takes account of *all* frequencies less than a fixed maximum $\omega_{\rm max}$ (including degeneracies) and *none* greater than $\omega_{\rm max}$.

19.22 The Panofsky-Wenzel Theorem A particle with charge q and velocity $\mathbf{v} = v\hat{\mathbf{z}}$ enters a perfectly conducting radio-frequency resonant cavity (length L) through a tiny entrance hole and then exits the cavity through an equally tiny exit hole. If v is large, we can use the impulse approximation to compute the momentum "kick" transverse to z.

(a) For a cavity with oscillation frequency ω, show that

$$\Delta\mathbf{p}_\perp = \frac{q}{v}\int_0^L dz\,[\mathbf{E} + \mathbf{v}\times\mathbf{B}]_\perp = -i\frac{q}{\omega}\int_0^L dz\,\nabla_\perp E_z.$$

Hint: A particle with velocity $\mathbf{v} = \dot{\mathbf{z}}$ experiences a change in vector potential $d\mathbf{A}/dt = \partial\mathbf{A}/\partial t + (\mathbf{v}\cdot\nabla)\mathbf{A}$.

(b) The formula derived in part (a) gives $\Delta\mathbf{p}_\perp = 0$ for TE modes. Reconcile this with the fact that both TE and TM modes produce Lorentz forces in the transverse direction.

19.23 Forces on Resonant-Cavity Walls In connection with the law of conservation of linear momentum, we showed that the electromagnetic force on a volume V can be written in the form

$$\mathbf{F} = \int_V d^3r \left[\nabla \cdot \mathbf{T} - \frac{1}{c^2} \frac{\partial \mathbf{S}}{\partial t} \right].$$

Use this formula to find the time-averaged force on each of the six perfectly conducting walls of a resonant cavity defined by $0 \le x \le a$, $0 \le y \le a$, and $0 \le z \le h$, when it is excited in a mode where $E_x = E_y = B_z = 0$ but

$$E_z = E_0 \sin \frac{\pi x}{a} \sin \frac{\pi y}{a} \exp(-i\omega t).$$

19.24 Graded Index Fiber

(a) Clearly state the conditions required for the electric field in a medium with a spatially varying index of refraction to satisfy the equation

$$\nabla^2 \mathbf{E} - \frac{n^2(\mathbf{r})}{c^2} \frac{\partial^2 \mathbf{E}}{\partial t^2} = 0.$$

(b) Let the index decrease radially away from a central axis according to $n(\rho) = n_0[1 - \alpha^2 \rho^2]$. Ensure that the foregoing applies and show that a solution $\mathbf{E} = E(\rho) \exp[i(hz - \omega t)]\hat{\boldsymbol{\phi}}$ exists where the electric field decreases very rapidly away from the central axis. This is a technique used to produce wave guiding in an optical fiber.

19.25 Interfacial Guided Waves A medium with index of refraction n_1 occupies the half-space $y < 0$. The half-space $y > 0$ is filled with a medium with index of refraction n_2. A propagating electric field bound to the $y = 0$ plane can have the form

$$\mathbf{E}(\mathbf{r}, t) = \begin{cases} \mathbf{E}_2 \exp(-\beta y) \exp[i(kx - \omega t)] & y > 0 \\ \mathbf{E}_1 \exp(\alpha y) \exp[i(kx - \omega t)] & y < 0. \end{cases}$$

(a) Find the field $\mathbf{H}(\mathbf{r}, t)$ that accompanies $\mathbf{E}(\mathbf{r}, t)$.

(b) Show that two types of solutions exist, one where $\mathbf{E}$ has only a z-component and one where $\mathbf{E}$ has no z-component. For the latter, show that the polarization of the wave is elliptical.

(c) Show that the time-averaged Poynting vector points exclusively along the x-axis for both solutions.

20 Retardation and Radiation

A portion of each of the outer [field] lines detaches itself as a self-closed line which advances independently into space.
Heinrich Hertz (1889)

20.1 Introduction

This chapter explores the electromagnetic fields produced by arbitrary but specified distributions of charge and current. After many pages of special cases and approximations, we are finally ready to solve the Maxwell equations in full generality:

$$\nabla \cdot \mathbf{E} = \frac{\rho}{\epsilon_0} \qquad \nabla \cdot \mathbf{B} = 0 \tag{20.1}$$

$$\nabla \times \mathbf{E} = -\frac{\partial \mathbf{B}}{\partial t} \qquad \nabla \times \mathbf{B} = \mu_0 \mathbf{j} + \frac{1}{c^2}\frac{\partial \mathbf{E}}{\partial t}. \tag{20.2}$$

The main conclusion is that fields produced by sources that are neither static nor quasistatic typically exhibit both *retardation* and *radiation*. Retardation refers to the fact that $\mathbf{E}(\mathbf{r}, t)$ and $\mathbf{B}(\mathbf{r}, t)$ do *not* reflect the behavior of the sources at the observation time t. Instead, they reflect the properties of the sources at an earlier (retarded) time $t - R/c$, where R is the distance between the source and the observer. Radiation is the name given to propagating fields that carry energy undiminished to points infinitely far from their sources.

Our discussion begins with the inhomogeneous wave equation—the prototype equation satisfied by the electromagnetic fields and many of the potentials of electromagnetism—which we solve using a Green function method. The physics of retardation is emphasized and, as examples, we compute the fields exactly everywhere for an infinite solenoid with a time-harmonic current and for a time-dependent point electric dipole. We derive explicit formulae for the radiation fields as integrals over the source densities and use a field line analysis to show how fields that are recognizably connected to their sources "break free" and evolve into freely propagating waves. Larmor's treatment of a slowly moving accelerated charge and a linear dipole antenna are used to illustrate radiation and its characteristics. A time-domain solution of the antenna problem provides insight complementary to the usual Fourier frequency-domain analysis. We then turn to approximate calculations and derive Cartesian and spherical multipole expansions for the radiation fields. The latter are most useful when the source size is small compared to the wavelength of the emitted radiation. The chapter concludes with the problem of plane wave refraction at a dielectric boundary analyzed from the point of view of the radiation emitted by the dielectric.

20.2 Inhomogeneous Wave Equations

The potentials and fields of electromagnetism are connected to their sources by *inhomogeneous wave equations*. For example, suppose we insert each curl equation in (20.2) into the curl of the other equation in (20.2) and use the identity

$$\nabla \times \nabla \times \mathbf{c} = \nabla(\nabla \cdot \mathbf{c}) - \nabla^2 \mathbf{c}. \tag{20.3}$$

Making use of (20.1) shows that the electric field and the magnetic field satisfy the (vector) inhomogeneous wave equations

$$\nabla^2 \mathbf{E} - \frac{1}{c^2}\frac{\partial^2 \mathbf{E}}{\partial t^2} = \frac{1}{\epsilon_0}\nabla\rho + \mu_0 \frac{\partial \mathbf{j}}{\partial t} \tag{20.4}$$

and

$$\nabla^2 \mathbf{B} - \frac{1}{c^2}\frac{\partial^2 \mathbf{B}}{\partial t^2} = -\mu_0 \nabla \times \mathbf{j}. \tag{20.5}$$

Equations (20.4) and (20.5) are more challenging to solve than their homogeneous counterparts. Moreover, any solution we find is not guaranteed to satisfy the original Maxwell equations. A similar state of affairs in Section 15.3 led us to define potential functions φ and $\mathbf{A}$ from

$$\mathbf{E} = -\nabla\varphi - \frac{\partial \mathbf{A}}{\partial t} \qquad \text{and} \qquad \mathbf{B} = \nabla \times \mathbf{A}. \tag{20.6}$$

Substituting (20.6) into (20.1) and (20.2) produced equations of motion for φ and $\mathbf{A}$ which we immediately simplified by exploiting the gauge freedom of the potentials. For example, in Section 15.3.3, the Lorenz gauge choice,

$$\nabla \cdot \mathbf{A}_L + \frac{1}{c^2}\frac{\partial \varphi_L}{\partial t} = 0, \tag{20.7}$$

led to the evolution equations

$$\nabla^2 \varphi_L - \frac{1}{c^2}\frac{\partial^2 \varphi_L}{\partial t^2} = -\rho/\epsilon_0 \tag{20.8}$$

and

$$\nabla^2 \mathbf{A}_L - \frac{1}{c^2}\frac{\partial^2 \mathbf{A}_L}{\partial t^2} = -\mu_0 \mathbf{j}. \tag{20.9}$$

Equations (20.8) and (20.9) are inhomogeneous wave equations also, but with much simpler source terms than (20.4) and (20.5). For that reason, the Lorenz gauge potentials are widely used to find the fields produced by specified distributions of free charge and current.

Two other inhomogeneous wave equations are often used to calculate the fields produced by dynamically polarized or magnetized matter. To derive them, we recall that a system with polarization $\mathbf{P}(\mathbf{r}, t)$ and magnetization $\mathbf{M}(\mathbf{r}, t)$ behaves exactly like a system with charge and current densities

$$\rho = -\nabla \cdot \mathbf{P} \qquad \text{and} \qquad \mathbf{j} = \frac{\partial \mathbf{P}}{\partial t} + \nabla \times \mathbf{M}. \tag{20.10}$$

Inserting (20.10) into (20.8) and (20.9) seems like a complication except that (20.7) is satisfied automatically if we write[1]

$$\varphi_L = -\nabla \cdot \boldsymbol{\pi}_\mathrm{e} \qquad \text{and} \qquad \mathbf{A}_L = \frac{1}{c^2}\frac{\partial \boldsymbol{\pi}_\mathrm{e}}{\partial t} + \nabla \times \boldsymbol{\pi}_\mathrm{m}. \tag{20.11}$$

[1] The similarity of (20.11) to (20.10) is not surprising in the light of the similarity of (20.7) to the continuity equation, $\nabla \cdot \mathbf{j} + \partial\rho/\partial t = 0$.

A glance back at Section 16.9 identifies $\boldsymbol{\pi}_{\rm e}(\mathbf{r},t)$ as the *electric Hertz vector* and $\boldsymbol{\pi}_{\rm m}(\mathbf{r},t)$ as the *magnetic Hertz vector*.

The reason to introduce the Hertz vectors emerges when we substitute (20.11) into (20.8) and (20.9) and do a bit of algebra using the fact that the Laplacian commutes with both the divergence and the curl. The result is that the evolution equations for $\mathbf{A}$ and φ are *both* satisfied if $\boldsymbol{\pi}_{\rm e}$ and $\boldsymbol{\pi}_{\rm m}$ themselves satisfy the inhomogeneous wave equations

$$\nabla^2\boldsymbol{\pi}_{\rm e} - \frac{1}{c^2}\frac{\partial^2\boldsymbol{\pi}_{\rm e}}{\partial t^2} = -\frac{1}{\epsilon_0}\mathbf{P} \tag{20.12}$$

and

$$\nabla^2\boldsymbol{\pi}_{\rm m} - \frac{1}{c^2}\frac{\partial^2\boldsymbol{\pi}_{\rm m}}{\partial t^2} = -\mu_0\mathbf{M}. \tag{20.13}$$

It is very advantageous for problem-solving that the polarization and magnetization appear on the right-hand sides of (20.12) and (20.13).

The electromagnetic fields follow from (20.3), (20.6), (20.11), and (20.12) as

$$\mathbf{E} = \nabla\times\nabla\times\boldsymbol{\pi}_{\rm e} - \nabla\times\frac{\partial\boldsymbol{\pi}_{\rm m}}{\partial t} - \frac{\mathbf{P}}{\epsilon_0} \qquad \text{and} \qquad \mathbf{B} = \nabla\times\nabla\times\boldsymbol{\pi}_{\rm m} + \nabla\times\frac{1}{c^2}\frac{\partial\boldsymbol{\pi}_{\rm e}}{\partial t}. \tag{20.14}$$

In practice, $\boldsymbol{\pi}_{\rm e}$ alone is sufficient to describe non-magnetic dielectric matter and $\boldsymbol{\pi}_{\rm m}$ alone is sufficient to describe non-polarizable magnetic matter. As an example, we exploit (20.12) at the end of the chapter (Section 20.9) to gain a new perspective on the problem of plane wave reflection/transmission at a vacuum/dielectric interface.

Application 20.1 The Fields of a Point Charge in Uniform Motion I

Lorentz solved the inhomogeneous wave equations (20.8) and (20.9) directly to find the fields produced by a point charge moving with constant velocity $\boldsymbol{v} = v\hat{\mathbf{z}}$ (Figure 20.1). His calculation begins with the source densities

$$\rho(\mathbf{r},t) = q\delta(x)\delta(y)\delta(z-vt) \qquad \text{and} \qquad \mathbf{j}(\mathbf{r},t) = \boldsymbol{v}\rho(\mathbf{r},t). \tag{20.15}$$

A glance at (20.9) shows that the z-directed current density in (20.15) produces, at most, a z-directed vector potential. In that case, the physical situation of uniform motion guarantees that φ and $\mathbf{A}$ can only depend on the variable $\xi = z - vt$:

$$\varphi(x,y,z-vt) \qquad\qquad A(x,y,z-vt)\hat{\mathbf{z}}. \tag{20.16}$$

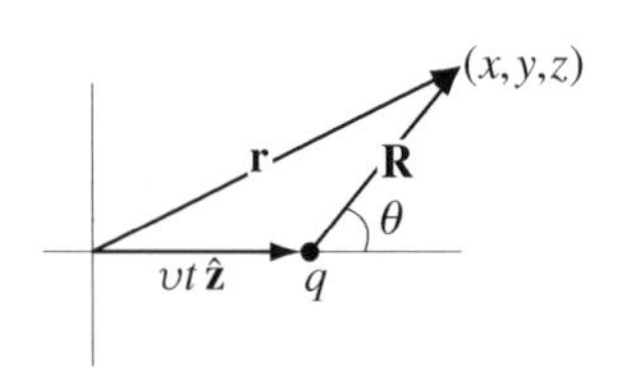

Figure 20.1: The vector $vt\hat{\mathbf{z}}$ points to a point charge moving with constant speed v. The vector $\mathbf{r}$ points to an observation point.

Setting $\beta = v/c$ and substituting $\rho(\mathbf{r},t)$ from (20.15) into (20.8) gives

$$\frac{\partial^2\varphi}{\partial x^2} + \frac{\partial^2\varphi}{\partial y^2} + (1-\beta^2)\frac{\partial^2\varphi}{\partial\xi^2} = -\frac{q}{\epsilon_0}\delta(x)\delta(y)\delta(\xi). \tag{20.17}$$

The definition $\gamma^2 = 1/(1-\beta^2)$ and a change of variable to $z' = \gamma\xi = \gamma(z - \upsilon t)$ transform (20.17) to Poisson's equation,

$$\frac{\partial^2\varphi}{\partial x^2} + \frac{\partial^2\varphi}{\partial y^2} + \frac{\partial^2\varphi}{\partial z'^2} = -\frac{\gamma q}{\epsilon_0}\delta(x)\delta(y)\delta(z'). \tag{20.18}$$

The solution to (20.18) is the familiar Coulomb potential. Therefore, in the original variables,

$$\varphi = \frac{1}{4\pi\epsilon_0}\frac{\gamma q}{\sqrt{x^2 + y^2 + \gamma^2(z - \upsilon t)^2}}. \tag{20.19}$$

A solution of the vector potential equation (20.9) is similarly

$$\mathbf{A} = \frac{\mu_0}{4\pi}\frac{\gamma q \upsilon}{\sqrt{x^2 + y^2 + \gamma^2(z - \upsilon t)^2}}\hat{\mathbf{z}}. \tag{20.20}$$

Straightforward differentiation using (20.6) shows that

$$\mathbf{E}(\mathbf{r}, t) = \frac{\gamma q}{4\pi\epsilon_0}\frac{x\hat{\mathbf{x}} + y\hat{\mathbf{y}} + (z - \upsilon t)\hat{\mathbf{z}}}{[x^2 + y^2 + \gamma^2(z - \upsilon t)^2]^{3/2}}. \tag{20.21}$$

The physical meaning of this formula becomes clearer when we define a vector $\mathbf{R}$ which points from the point charge position $\boldsymbol{\upsilon}t$ to the observation point position $\mathbf{r} = (x, y, z)$:

$$\mathbf{R} = x\hat{\mathbf{x}} + y\hat{\mathbf{y}} + (z - \upsilon t)\hat{\mathbf{z}}. \tag{20.22}$$

If $\mathbf{R}$ makes an angle θ with the direction of $\boldsymbol{\upsilon}$ (see Figure 20.1),

$$\sin^2\theta = \frac{x^2 + y^2}{x^2 + y^2 + (z - \upsilon t)^2}. \tag{20.23}$$

With this information, only a few lines of algebra are needed to confirm that

$$\mathbf{E}(\mathbf{r}, t) = \frac{q}{4\pi\epsilon_0}\frac{\hat{\mathbf{R}}}{R^2}\frac{1 - \beta^2}{(1 - \beta^2\sin^2\theta)^{3/2}}. \tag{20.24}$$

Equation (20.24) shows that a point charge in uniform motion drags an electric field along with it that is Coulomb-like except that the field magnitude varies with observation angle and the speed of the charge. When $\upsilon \ll c$, the magnitude is isotropic as in the static case. When $\upsilon \sim c$, a polar plot of the field magnitude shows that the field becomes very weak in the forward and backward directions (with respect to the direction of motion) and very strong in the transverse directions. Figure 20.2 gives a field line representation of this fact.

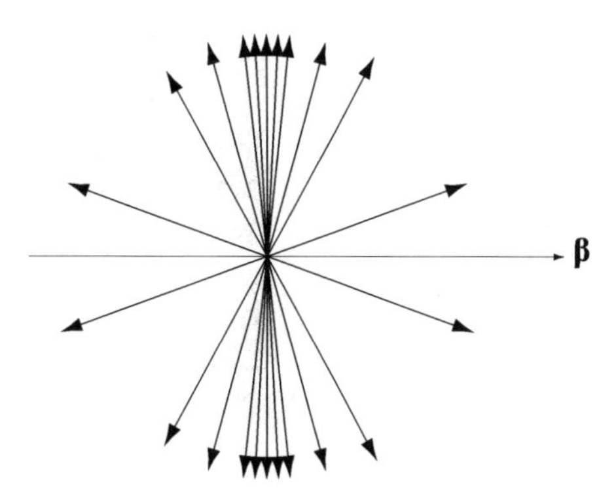

Figure 20.2: Electric field lines for a point charge in uniform motion with $\upsilon/c = 0.95$.

The reader can confirm that the right members of (20.6) and (20.20) give the magnetic field of a uniformly moving point charge as

$$\mathbf{B} = \frac{\boldsymbol{\upsilon}}{c^2} \times \mathbf{E}. \tag{20.25}$$

The magnetic field lines of (20.25) are concentric closed circles in planes perpendicular to the direction of motion of the charge. When $\upsilon \ll c$, the exact $\mathbf{B}(\mathbf{r}, t)$ in (20.25) is approximately the Biot-Savart field produced by the current density in (20.15). In that case, $\mathbf{E}$ and $\mathbf{B}$ satisfy the Ampère-Maxwell law, but not Faraday's law.[2] ■

Example 20.1 An infinitely long solenoid with radius a is concentric with the z-axis and tightly wound with n turns per unit length of wire which carries a current $I(t) = I\cos(\omega t)$.

(a) Find $\mathbf{A}(\mathbf{r}, t)$ outside the solenoid by matching solutions of the inhomogeneous wave equation (20.9) inside and outside the solenoid at $\rho = a$. Assume outgoing waves at infinity.
(b) Find the exterior fields $\mathbf{E}$ and $\mathbf{B}$ and confirm the quasi-magnetostatic result of Application 14.4 of Section 14.6.1.
(c) Calculate the time-averaged rate at which energy passes through a unit length of an imaginary cylinder that is concentric with the solenoid but has radius $\rho \gg a$.
(d) Compare the power computed in (c) with the time-averaged power per unit length supplied to the solenoid to drive its current.

Solution:

(a) The effective surface current density of the solenoid is $\mathbf{K}(t) = nI\cos(\omega t)\hat{\boldsymbol{\phi}}$. This fact and (20.9) tell us that $\mathbf{A}(\mathbf{r}, t) = A(\rho, t)\hat{\boldsymbol{\phi}}$ in cylindrical coordinates. However, as discussed in Section 1.2.7, we cannot simply write out the $\hat{\boldsymbol{\phi}}$-component of the Laplacian in (20.9). Instead, we must use (20.3) and the Coulomb gauge condition $\nabla \cdot \mathbf{A} = 0$ to write

$$\nabla \times \nabla \times \mathbf{A} + \frac{1}{c^2}\frac{\partial^2 \mathbf{A}}{\partial t^2} = \mu_0 \mathbf{j}.$$

With $A(\rho, t) = A(\rho)\exp(-i\omega t)$ and $\omega = ck$, the $\hat{\boldsymbol{\phi}}$-component of this equation for $\rho \neq a$ is

$$\rho\frac{d}{d\rho}\left(\rho\frac{dA}{d\rho}\right) + (k^2\rho^2 - 1)A = 0.$$

This is one form of Bessel's differential equation. The vector potential $\mathbf{A}$ must be finite at the origin and have outgoing-wave character as $\rho \to \infty$. Therefore, the Bessel function properties collected in Appendix C tell us that

$$A(\rho) = \begin{cases} c_1 J_1(k\rho) & \rho < a, \\ c_2 H_1^{(1)}(k\rho) & \rho > a. \end{cases}$$

The matching conditions at $\rho = a$ are $\hat{\mathbf{n}}_2 \cdot [\mathbf{A}_1 - \mathbf{A}_2] = 0$ and $\hat{\mathbf{n}}_2 \times [\mathbf{B}_1 - \mathbf{B}_2] = \mu_0 \mathbf{K}$. Using $H_1^{(1)}(x) = J_1(x) + iN_1(x)$, $(d/dx)(xJ_1) = xJ_0$, and $(d/dx)(xN_1) = xN_0$, these conditions read

$$c_1 J_1(ka) = c_2 H_1^{(1)}(ka)$$

$$c_1 k J_0(ka) - c_2 k H_0^{(1)}(ka) = \mu_0 n I.$$

[2] This is the quasi-electrostatic limit studied in Example 13.2.

Solving for c_2 and using the Bessel function Wronskian relation (C.27), we conclude that the (real) vector potential *outside* the solenoid is

$$\mathbf{A}(\rho \geq a, t) = \tfrac{1}{2}\mu_0 n I \pi a J_1(ka)\,[J_1(k\rho)\sin\omega t - N_1(k\rho)\cos\omega t]\,\hat{\boldsymbol{\phi}}.$$

(b) By direct computation, the fields $\mathbf{E} = -\partial\mathbf{A}/\partial t$ and $\mathbf{B} = \nabla \times \mathbf{A}$ outside the solenoid are

$$\mathbf{E}(\rho \geq a, t) = -\tfrac{1}{2}\mu_0 n I \pi a \omega J_1(ka)\,[N_1(k\rho)\sin\omega t + J_1(k\rho)\cos\omega t]\,\hat{\boldsymbol{\phi}}$$
$$\mathbf{B}(\rho > a, t) = \tfrac{1}{2}\mu_0 n I \pi a \omega J_1(ka)\,[J_0(k\rho)\sin\omega t - N_0(k\rho)\cos\omega t]\,\hat{\mathbf{z}}.$$

The low-frequency limit is $ka \ll 1$ and $k\rho \ll 1$. When $x \ll 1$, the Bessel functions satisfy $J_0(x) \approx 1$, $N_0(x) \approx (2/\pi)\ln(x)$, and $J_1(x) \approx x/2$. Therefore,

$$\mathbf{B}(\rho > a, t) \approx \frac{\mu_0 n I \pi \omega^2 a^2}{4c^2}\left[\sin\omega t - \frac{2}{\pi}\ln(\omega\rho/c)\cos\omega t\right]\hat{\mathbf{z}}.$$

This agrees with the calculation in Application 14.4 for the first correction to the quasi-magnetostatic limit due to the displacement current.

(c) The rate of energy flow through a large cylindrical surface concentric with the solenoid requires a calculation of the Poynting vector when $\rho \gg a$. For this purpose, we quote (from Appendix C) two Bessel function formulae which are valid when $x \gg 1$:

$$J_m(x) \longrightarrow \sqrt{\frac{2}{\pi x}}\cos\left(x - \frac{m\pi}{2} - \frac{\pi}{4}\right) \qquad N_m(x) \longrightarrow \sqrt{\frac{2}{\pi x}}\sin\left(x - \frac{m\pi}{2} - \frac{\pi}{4}\right).$$

A straightforward evaluation gives

$$\mathbf{S}(\rho \gg a, t) = \lim_{\rho\to\infty}\frac{1}{\mu_0}\mathbf{E}\times\mathbf{B} = \tfrac{1}{2}\mu_0 n^2 I^2 \pi a^2 \omega J_1^2(ka)\frac{\sin^2(k\rho - \omega t - \pi/4)}{\rho}\hat{\boldsymbol{\rho}}.$$

Therefore, the time-averaged power per unit length which flows through the walls of the imaginary cylinder is

$$\langle P \rangle = \langle \mathbf{S} \rangle \cdot 2\pi\rho\hat{\boldsymbol{\rho}} = \tfrac{1}{2}\mu_0 n^2 I^2 \pi^2 a^2 \omega J_1^2(ka).$$

Section 20.5.1 will use the word *radiation* to characterize the asymptotic ($\rho \to \infty$) fields which produce this result.

(d) The time-averaged power which maintains the current in the solenoid is the negative of the rate that work is done by the electric field on the current density which flows on the solenoid's surface. Using the electric field computed above, this power for a unit length of solenoid is

$$\langle \mathcal{P} \rangle = -n\langle I(t)\oint d\mathbf{s}\cdot\mathbf{E}(a,t)\rangle = \tfrac{1}{2}\mu_0 n^2 I^2 \pi^2 a^2 \omega J_1^2(ka).$$

The fact that $\langle P \rangle = \langle \mathcal{P} \rangle$ reflects conservation of energy for an ideal solenoid with no ohmic resistance.

20.3 Retardation

The finite propagation speed of electromagnetic waves guarantees that changes in the behavior of a source cannot be detected instantaneously by a distant (or even a nearby) observer. This phenomenon, called *retardation*, connects electromagnetic fields in a fundamental way to the time-varying distributions of charge and current that produce them. Mathematically, retardation emerges in a natural way when we solve any of the inhomogeneous wave equations derived in the previous section.[3] Accordingly, we focus on the following generic version:

$$\left[\nabla^2 - \frac{1}{c^2}\frac{\partial^2}{\partial t^2}\right]\psi(\mathbf{r},t) = -f(\mathbf{r},t). \tag{20.26}$$

The static limit of this equation is the Poisson equation studied in Chapter 8 using a Green function method. In this section, we use a time-dependent generalization of the Green function method to find solutions of (20.26). Section 20.3.3 deals explicitly with the case of sources that vary harmonically in time.

20.3.1 General Solution of the Inhomogeneous Wave Equation

The most general solution of the inhomogeneous wave equation (20.26) is constructed by manipulating the equation of motion for its Green function, $G(\mathbf{r},t\,|\,\mathbf{r}',t')$. By definition, the Green function is the wave produced at $(\mathbf{r},t)$ by a unit-strength point source which acts only at $(\mathbf{r}',t')$. Hence, the equation in question is simply a special case of (20.26) itself:

$$\left[\nabla^2 - \frac{1}{c^2}\frac{\partial^2}{\partial t^2}\right]G(\mathbf{r},t\,|\,\mathbf{r}',t') = -\delta(\mathbf{r}-\mathbf{r}')\delta(t-t'). \tag{20.27}$$

We begin by multiplying (20.27) by $\psi(\mathbf{r}',t')d^3r'dt'$, replacing the derivatives by their primed counterparts,[4] and integrating the result over a volume V and from an initial time t_1 to a final time t_2. This gives

$$\psi(\mathbf{r},t) = \frac{1}{c^2}\int_{t_1}^{t_2} dt' \int_V d^3r'\psi(\mathbf{r}',t')\frac{\partial^2}{\partial t'^2}G(\mathbf{r},t|\mathbf{r}',t') - \int_{t_1}^{t_2} dt' \int_V d^3r'\psi(\mathbf{r}',t')\nabla'^2 G(\mathbf{r},t|\mathbf{r}',t'). \tag{20.28}$$

The first term on the right side of (20.28) can be rewritten using

$$\int_{t_1}^{t_2} dt'\psi\frac{\partial^2 G}{\partial t'^2} = \int_{t_1}^{t_2} dt' G\frac{\partial^2\psi}{\partial t'^2} + \left[\psi\frac{\partial G}{\partial t'} - G\frac{\partial\psi}{\partial t'}\right]_{t_1}^{t_2}. \tag{20.29}$$

The first term on the right side of (20.29) can also be rewritten, this time using (20.26) written in primed variables. The last step is to exploit Green's second identity in the form

$$\int_{t_1}^{t_2} dt' \int_V d^3r'\left(G\,\nabla'^2\psi - \psi\,\nabla'^2 G\right) = \int_{t_1}^{t_2} dt' \int_S d\mathbf{S}'\cdot\left(G\,\nabla'\psi - \psi\,\nabla' G\right). \tag{20.30}$$

The final result is

$$\begin{aligned}
\psi(\mathbf{r},t) &= \int_{t_1}^{t_2} dt' \int_V d^3r'\,G(\mathbf{r},t|\mathbf{r}',t')f(\mathbf{r}',t') \\
&+ \int_{t_1}^{t_2} dt' \int_S d\mathbf{S}'\cdot\left[G(\mathbf{r},t|\mathbf{r}',t')\nabla'\psi(\mathbf{r}',t') - \psi(\mathbf{r}',t')\nabla' G(\mathbf{r},t|\mathbf{r}',t')\right] \\
&+ \frac{1}{c^2}\int_V d^3r'\left[\partial_{t'}G(\mathbf{r},t|\mathbf{r}',t')\psi(\mathbf{r}',t') - G(\mathbf{r},t|\mathbf{r}',t')\partial_{t'}\psi(\mathbf{r}',t')\right]\Big|_{t'=t_1}^{t'=t_2}.
\end{aligned} \tag{20.31}$$

[3] The mathematical signature of retardation derived in this section is not manifest in the solution of the inhomogeneous wave equation found in Application 20.1. See Application 23.1 of Section 23.2.4.

[4] The correctness of this step follows directly from (20.26).

This integral equation for $\psi(\mathbf{r},t)$ is recognizably a generalization of the integral equation for $\varphi(\mathbf{r})$ generated in Section 8.4 when we applied the Green function method to Poisson's equation. The first line of (20.31) is a superposition integral. The "observer" function $\psi(\mathbf{r},t)$ is synthesized by summing up the response of the wave equation to unit perturbations at different times and places weighted by the "source" function $f(\mathbf{r}',t')$. The second line in (20.31) is a boundary term in the spatial variables. Subject to later confirmation, we assume that this term vanishes for the situation of immediate interest: a spatially localized source in infinite space. The third line in (20.31) is a boundary term in the time variable.

We expect to specify ψ and $\partial_t\psi$ (at one time) because (20.26) is second-order in time. But the third line of (20.31) demands knowledge of these functions at *two* times. A similar situation occurs in the second line of (20.31) where knowledge is required of both ψ and $\nabla\psi$ on the boundary S. We resolved exactly the same issue in electrostatics by a suitable choice of spatial boundary conditions (Dirichlet or Neumann) for the Poisson Green function. Here, where we have assumed that the spatial boundary term is absent, (20.31) becomes a useful formula when we make a suitable choice of temporal "boundary conditions" for the Green function (20.27).

To see how this happens, the right side of (20.27) motivates us to seek a solution of the form

$$G(\mathbf{r},t|\mathbf{r}',t') = G(\mathbf{r}-\mathbf{r}',t-t'). \tag{20.32}$$

With the understanding that the true time and location of the perturbation can be re-inserted later, we put $\mathbf{r}'=t'=0$ and study

$$\left[\nabla^2 - \frac{1}{c^2}\frac{\partial^2}{\partial t^2}\right]G(\mathbf{r},t) = -\delta(\mathbf{r})\delta(t). \tag{20.33}$$

The delta function in (20.33) occurs at the origin of an otherwise spherically symmetric space. This implies that $G(\mathbf{r},t)=G(r,t)$ and

$$\nabla^2 G = \frac{1}{r^2}\frac{\partial}{\partial r}\left(r^2\frac{\partial G}{\partial r}\right) = \frac{2}{r}\frac{\partial G}{\partial r} + \frac{\partial^2 G}{\partial r^2}. \tag{20.34}$$

Using (20.34), (20.33) can be written in the form

$$\frac{1}{r}\left\{\frac{\partial^2}{\partial r^2} - \frac{1}{c^2}\frac{\partial^2}{\partial t^2}\right\}rG = -\delta(\mathbf{r})\delta(t). \tag{20.35}$$

The operator in curly brackets in (20.35) can be factored. Therefore, away from the origin,

$$\frac{1}{r}\left\{\frac{\partial}{\partial r} - \frac{1}{c}\frac{\partial}{\partial t}\right\}\left\{\frac{\partial}{\partial r} + \frac{1}{c}\frac{\partial}{\partial t}\right\}rG = 0 \qquad (r>0). \tag{20.36}$$

This equation has two independent solutions:

$$G_\pm(r,t) = \frac{1}{r}g_\pm(t \pm r/c), \tag{20.37}$$

where $g_+(s)$ and $g_-(s)$ are arbitrary functions of one scalar variable.

Now apply the bracketed operator on the left-hand side of (20.33) to (20.37). Using (20.34),

$$\nabla^2\left(\frac{g}{r}\right) = \nabla\cdot\left\{\frac{1}{r}\nabla g + g\nabla\left(\frac{1}{r}\right)\right\}, \tag{20.38}$$

and $\nabla^2(1/r) = -4\pi\delta(\mathbf{r})$, the chain rule leads finally to

$$\left[\nabla^2 - \frac{1}{c^2}\frac{\partial^2}{\partial t^2}\right]G_\pm(\mathbf{r},t) = -4\pi\delta(\mathbf{r})g_\pm. \tag{20.39}$$

Comparing (20.39) with (20.33) shows that $g_{\pm}(s) = \delta(s)/4\pi$. We conclude that two independent, particular solutions to (20.33) are

$$G_{\pm}(\mathbf{r}, t) = \frac{1}{4\pi r}\delta(t \pm r/c). \tag{20.40}$$

The corresponding solutions to (20.27) follow by putting $\mathbf{r} \to \mathbf{r} - \mathbf{r}'$ and $t \to t - t'$:

$$G_{\pm}(\mathbf{r}, t\,|\,\mathbf{r}', t') = \frac{1}{4\pi|\mathbf{r} - \mathbf{r}'|}\delta(t - t' \pm |\mathbf{r} - \mathbf{r}'|/c). \tag{20.41}$$

20.3.2 Advanced and Retarded Waves

The Green functions derived in the preceding section transform (20.31) from an integral equation for $\psi(\mathbf{r}, t)$ into an explicit solution of the inhomogeneous wave equation. To see this, we first write (20.31) omitting the surface integrals we expect to vanish:

$$\begin{aligned}\psi(\mathbf{r}, t) = &\int_{t_1}^{t_2} dt' \int_V d^3r'\, G(\mathbf{r}, t|\mathbf{r}', t') f(\mathbf{r}', t') \\ &+ \frac{1}{c^2}\int_V d^3r' \left[\partial_{t'} G(\mathbf{r}, t|\mathbf{r}', t')\psi(\mathbf{r}', t') - G(\mathbf{r}, t|\mathbf{r}', t')\partial_{t'}\psi(\mathbf{r}', t')\right]\Big|_{t'=t_1}^{t'=t_2}.\end{aligned} \tag{20.42}$$

The temporal behaviors of G_+ and G_- in (20.41) are the keys to simplifying (20.42). Because $R = |\mathbf{r} - \mathbf{r}'| \geq 0$ and $t_1 \leq t < t_2$, G_+ is non-zero only at the t_2 boundary $[G_+(t - t' > 0) = 0]$ and G_- is non-zero only at the t_1 boundary $[G_-(t - t' < 0) = 0]$. Therefore, substituting (20.41) into (20.42) and performing the delta function integration over t' in the first term gives two equally good particular solutions of the inhomogeneous wave equation.

The *retarded* solution makes use of G_- and defines the function $\psi_{\text{in}}(\mathbf{r}, t)$ as the integral in the second line of (20.42) evaluated at $t' = t_1$:

$$\psi_{\text{ret}}(\mathbf{r}, t) = \psi_{\text{in}}(\mathbf{r}, t) + \frac{1}{4\pi}\int_V d^3r'\, \frac{f(\mathbf{r}', t - |\mathbf{r} - \mathbf{r}'|/c)}{|\mathbf{r} - \mathbf{r}'|}. \tag{20.43}$$

The retarded observer in (20.43) feels the influence of a source $f(\mathbf{r}', t')$ that is a distance $R = |\mathbf{r} - \mathbf{r}'|$ away only if the source acted at an earlier or *retarded* time $t' = t - R/c$. The time delay reflects the finite propagation speed c of the signal and accords with our intuitive notion that cause precedes effect. Physically, $\psi_{\text{in}}(\mathbf{r}, t)$ is an "incoming wave" solution of the homogeneous wave equation that describes the physical situation at the initial time $t = t_1$ before the integral in (20.43) begins to contribute.

The *advanced* solution makes use of G_+ and defines the function $\psi_{\text{out}}(\mathbf{r}, t)$ as the integral in the second line of (20.42) evaluated at $t' = t_2$:

$$\psi_{\text{adv}}(\mathbf{r}, t) = \psi_{\text{out}}(\mathbf{r}, t) + \frac{1}{4\pi}\int_V d^3r'\, \frac{f(\mathbf{r}', t + |\mathbf{r} - \mathbf{r}'|/c)}{|\mathbf{r} - \mathbf{r}'|}. \tag{20.44}$$

The advanced observer in (20.44) feels the influence of a source $f(\mathbf{r}', t')$ that is a distance $R = |\mathbf{r} - \mathbf{r}'|$ away only if the source acted at a later or *advanced* time $t' = t + R/c$. That is, the behavior of the source at future times determines the response of the observer at the present time. $\psi_{\text{out}}(\mathbf{r}, t)$ is an "outgoing wave" solution of the homogeneous wave equation that describes the physical situation at the final time $t = t_2$ after the integral in (20.44) no longer contributes.

There is no compelling *mathematical* reason to choose between the retarded solution and the advanced solution.[5] However, when the volume $V \to \infty$ in (20.42), it is not physically possible to specify the final state conditions required by the advanced solution. For that reason, the retarded solution (20.43) is invariably used to study the fields produced by spatially localized sources in otherwise empty space. Moreover, $\psi_{\rm in} = 0$ when we specify initial conditions for the source function $f(\mathbf{r}, t)$. Since it is typical to let f be zero until $t = t_1$, the finite propagation speed of the signal guarantees that $\psi_{\rm ret}(\mathbf{r}, t)$ is zero on the infinitely distant boundary surface S in (20.31) at any finite time. This justifies our neglect of the surface integrals when passing from (20.31) to (20.42). Our final conclusion is that the practically important solution of the inhomogeneous wave equation (20.26) is

$$\psi(\mathbf{r}, t) = \int d^3r' G_-(\mathbf{r} - \mathbf{r}', t - t') f(\mathbf{r}', t') = \frac{1}{4\pi} \int d^3r' \frac{f(\mathbf{r}', t - |\mathbf{r} - \mathbf{r}'|/c)}{|\mathbf{r} - \mathbf{r}'|}. \tag{20.45}$$

Ritz vs. Einstein

The non-causal nature of the advanced potential sparked a lively exchange between Albert Einstein and the Swiss physicist Walther Ritz. They debated whether, as a matter of physical principle, the retarded potential alone must be used to discuss electromagnetic phenomena. Ritz argued that the retarded potential is chosen by Nature to distinguish the past from the future and that this choice underlies the second law of thermodynamics. Einstein asserted that the advanced and retarded potentials are equally valid for use in any finite volume of space and that the origin of the second law should be sought in the laws of probability. The two young theorists agreed to disagree in a one-paragraph paper published in 1909. Two months after the paper was published, Ritz died of tuberculosis at the age of 31. History has borne out Einstein's point of view.

20.3.3 Retarded Waves from Time-Harmonic Sources

Direct substitution of $f(\mathbf{r}, t) = f(\mathbf{r}|\omega)\exp(-i\omega t)$ into the far right side of (20.45) gives the retarded solution of the inhomogeneous wave equation for the important special case of a time-harmonic source:

$$\psi(\mathbf{r}, t) = \int d^3r' \frac{\exp(i\omega|\mathbf{r} - \mathbf{r}'|/c)}{4\pi|\mathbf{r} - \mathbf{r}'|} f(\mathbf{r}'|\omega). \tag{20.46}$$

However, the physical meaning of this retarded solution is not immediately obvious for oscillating sources which, by construction, do not turn on or turn off at any particular moment. To gain some insight, we substitute $\psi(\mathbf{r}, t) = \psi(\mathbf{r}|\omega)\exp(-i\omega t)$ into the inhomogeneous wave equation (20.26). With $\omega = ck$, the result is an inhomogeneous version of the Helmholtz equation:

$$\left[\nabla^2 + k^2\right] \psi(\mathbf{r}|\omega) = -f(\mathbf{r}|\omega). \tag{20.47}$$

Direct substitution confirms that a particular solution of this equation is

$$\psi(\mathbf{r}|\omega) = \int d^3r'\, G(\mathbf{r}, \mathbf{r}'|\omega) f(\mathbf{r}'|\omega), \tag{20.48}$$

where $G(\mathbf{r}, \mathbf{r}'|\omega)$ is the Green function for the Helmholtz equation:

$$\left[\nabla^2 + k^2\right] G(\mathbf{r}, \mathbf{r}'|\omega) = -\delta(\mathbf{r} - \mathbf{r}'). \tag{20.49}$$

A quick way to solve (20.49) recognizes that

$$G(\mathbf{r}, t) = \frac{1}{2\pi} \int_{-\infty}^{\infty} d\omega\, G(\mathbf{r}|\omega) \exp(-i\omega t) \quad \text{and} \quad \delta(t) = \frac{1}{2\pi} \int_{-\infty}^{\infty} d\omega \exp(-i\omega t) \tag{20.50}$$

[5] See Sources, References, and Additional Reading.

transform (20.33) into (20.49) with $\mathbf{r}'=0$. Therefore, the Green function we seek is the inverse Fourier transform of (20.40):

$$G_{\pm}(\mathbf{r}|\omega)=\int_{-\infty}^{\infty} dt\, G_{\pm}(\mathbf{r},t)\exp(i\omega t)=\frac{1}{4\pi r}\int_{-\infty}^{\infty} dt\,\delta(t\pm r/c)\exp(i\omega t)=\frac{\exp(\mp ikr)}{4\pi r}. \tag{20.51}$$

Back in the time domain,

$$G_{\pm}(\mathbf{r},t)=G_{\pm}(\mathbf{r}|\omega)\exp(-i\omega t)=\frac{\exp(\mp ikr-i\omega t)}{4\pi r}. \tag{20.52}$$

Equation (20.52) provides the desired physical interpretation. The retarded solution is a spherical wave which propagates *outward* from the origin to infinity. The advanced solution is a spherical wave which propagates *inward* toward the origin from infinity. To be consistent with (20.45), we substitute the outgoing wave Green function $G_{-}(\mathbf{r}|\omega)$ into (20.48) to get the retarded solution of the inhomogeneous Helmholtz equation:

$$\psi(\mathbf{r}|\omega)=\int d^3r'\,\frac{\exp(ik|\mathbf{r}-\mathbf{r}'|)}{4\pi|\mathbf{r}-\mathbf{r}'|}f(\mathbf{r}'|\omega). \tag{20.53}$$

This reproduces (20.46), as advertised. From a boundary-condition point of view, our choice of the outgoing wave solution to (20.49) is consistent with imposing the *Sommerfeld radiation condition*,

$$\lim_{r\to\infty} r\left(\frac{\partial}{\partial r}-ik\right)G=0. \tag{20.54}$$

The retarded Green function for the Helmholtz equation derived in this section plays an important role in the theory of scattering and diffraction (Chapter 21). In that context, we will refer to $G_{-}(\mathbf{r}|\omega)$ as the *free-space Green function in three dimensions* and use the notation:

$$G_0(\mathbf{r},\mathbf{r}')=\frac{\exp(ik|\mathbf{r}-\mathbf{r}'|)}{4\pi|\mathbf{r}-\mathbf{r}'|}. \tag{20.55}$$

The corresponding free space Green function in two dimensions is relevant for scattering from slits, wires, and long sharp edges. We leave it as an exercise for the reader to show that the outgoing wave solution to (20.49) in two-dimensional plane polar coordinates is

$$G_0(\boldsymbol{\rho},\boldsymbol{\rho}')=\frac{i}{4}H_0^{(1)}(k|\boldsymbol{\rho}-\boldsymbol{\rho}'|), \tag{20.56}$$

where $H_0^{(1)}(x)$ is the Hankel function of the first kind defined in Appendix C.3.

20.3.4 Retarded Potentials and Fields

The electric field, $\mathbf{E}(\mathbf{r},t)$, and the magnetic field, $\mathbf{B}(\mathbf{r},t)$, are retarded functions of the time-dependent charge density and current density that produce them. This conclusion follows immediately when we apply (20.45) to the Cartesian components of (20.4) and (20.5). However, because the explicit formulae so obtained are rather awkward (see Application 20.2 below), most practitioners begin with the *retarded potentials*. These are the particular solutions obtained when we apply (20.45) to a scalar, vector, or Hertz potential function that satisfies the inhomogeneous wave equation.

Consider first the Lorenz gauge potentials that satisfy (20.8) and (20.9). The corresponding retarded potentials are

$$\varphi_L(\mathbf{r},t)=\frac{1}{4\pi\epsilon_0}\int d^3r'\,\frac{\rho(\mathbf{r}',t-|\mathbf{r}-\mathbf{r}'|/c)}{|\mathbf{r}-\mathbf{r}'|} \tag{20.57}$$

and

$$\mathbf{A}_L(\mathbf{r},t)=\frac{\mu_0}{4\pi}\int d^3r'\,\frac{\mathbf{j}(\mathbf{r}',t-|\mathbf{r}-\mathbf{r}'|/c)}{|\mathbf{r}-\mathbf{r}'|}. \tag{20.58}$$

When the sources vary in time as $\exp(-i\omega t)$, we use (20.46) in place of (20.45) and the space parts of the time-harmonic electrodynamic potentials are

$$\varphi(\mathbf{r}|\omega)=\frac{1}{4\pi\epsilon_0}\int d^3r'\,\frac{\exp(i\omega|\mathbf{r}-\mathbf{r}'|/c)}{|\mathbf{r}-\mathbf{r}'|}\rho(\mathbf{r}'|\omega) \tag{20.59}$$

and

$$\mathbf{A}(\mathbf{r}|\omega)=\frac{\mu_0}{4\pi}\int d^3r'\,\frac{\exp(i\omega|\mathbf{r}-\mathbf{r}'|/c)}{|\mathbf{r}-\mathbf{r}'|}\mathbf{j}(\mathbf{r}'|\omega). \tag{20.60}$$

We showed in (20.11) that the Hertz vectors are potential functions for the usual scalar and vector potentials. The retarded Hertz potentials that satisfy the inhomogeneous wave equations (20.12) and (20.13) are

$$\pi_{\rm e}(\mathbf{r},t)=\frac{1}{4\pi\epsilon_0}\int d^3r'\,\frac{\mathbf{P}(\mathbf{r}',t-|\mathbf{r}-\mathbf{r}'|/c)}{|\mathbf{r}-\mathbf{r}'|} \tag{20.61}$$

and

$$\pi_{\rm m}(\mathbf{r},t)=\frac{\mu_0}{4\pi}\int d^3r'\,\frac{\mathbf{M}(\mathbf{r}',t-|\mathbf{r}-\mathbf{r}'|/c)}{|\mathbf{r}-\mathbf{r}'|}. \tag{20.62}$$

Given the meaning of retardation, it is common to say that the Lorenz and Hertz potentials "propagate at the speed of light".

It is worth noting that not all potentials are retarded. In the Coulomb gauge, the vector potential satisfies the inhomogeneous wave equation (15.17) and thus $\mathbf{A}_C(\mathbf{r},t)$ has a retarded representation. However, the Coulomb gauge scalar potential $\varphi_C(\mathbf{r},t)$ satisfies the Poisson equation, for which the relevant particular solution is the usual Coulomb integral.[6] The latter connects the scalar potential at time t to the charge density at time t. There is no retardation for this quantity, nor need there be, because the potential functions of electromagnetism are not observables. The observables are the fields, $\mathbf{B}=\nabla\times\mathbf{A}_C$ and $\mathbf{E}=-\nabla\varphi_C-\partial\mathbf{A}_C/\partial t$, which the reader can confirm (by direct calculation) are properly retarded.

Example 20.2 Beginning at $t=0$, a time-dependent but spatially uniform surface current density $\mathbf{K}(t)$ flows everywhere in the x-y plane. Show that this current generates transverse electromagnetic waves which propagate away from the plane.

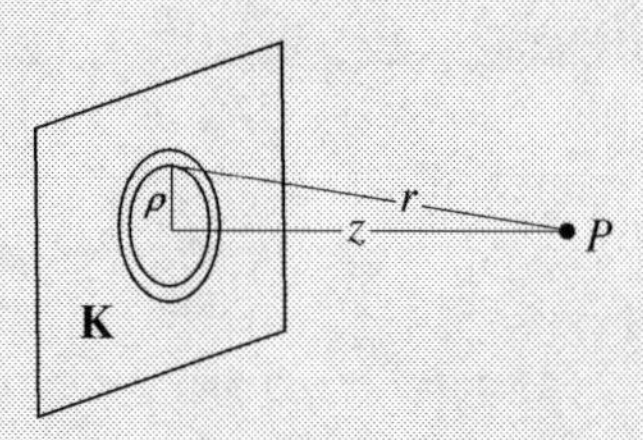

Figure 20.3: The point P lies a perpendicular distance z from a current density $\mathbf{K}(t)$ that occupies the entire x-y plane.

[6] See (15.18) of Section 15.3.2.

Solution: There is no charge and thus no scalar potential for this problem. Since $\mathbf{K}(t) = 0$ for $t < 0$, we use the Heaviside step function and write the Lorenz gauge vector potential as

$$\mathbf{A}(\mathbf{r}, t) = \frac{\mu_0}{4\pi} \int_0^{2\pi} d\phi \int_0^{\infty} d\rho \rho \frac{\mathbf{K}(t - r/c)\Theta(t - r/c)}{r}.$$

From Figure 20.3, $\rho^2 = r^2 - z^2$ so $\rho d\rho = r dr$ and

$$\mathbf{A}(\mathbf{r}, t) = \frac{\mu_0}{2} \int_{|z|}^{ct} dr\, \mathbf{K}(t - r/c)\Theta(t - r/c) = \frac{\mu_0 c}{2} \int_0^{t-|z|/c} ds\, \mathbf{K}(s)\Theta(s).$$

The electric field is

$$\mathbf{E}(z, t) = -\frac{\partial \mathbf{A}}{\partial t} = -\frac{\mu_0 c}{2}\mathbf{K}(t - |z|/c)\Theta(t - |z|/c).$$

The associated magnetic field is

$$\mathbf{B}(z, t) = \nabla \times \mathbf{A} = -(\partial A_y/\partial z)\hat{\mathbf{x}} + (\partial A_x/\partial z)\hat{\mathbf{y}},$$

or

$$\mathbf{B}(z, t) = \text{sgn}(z)\frac{\mu_0}{2}\left[\hat{\mathbf{x}}K_y(t - |z|/c) - \hat{\mathbf{y}}K_x(t - |z|/c)\right]\Theta(t - |z|/c).$$

Because $\mathbf{E}$ lies in the x-y plane, we see that

$$c\mathbf{B}(|z| - ct) = \text{sgn}(z)\hat{\mathbf{z}} \times \mathbf{E}(|z| - ct).$$

Therefore, the current sheet generates one-dimensional, transverse electromagnetic waves which propagate away from the x-y plane in the $\pm z$-direction.

Application 20.2 Schott's Formulae

The retarded Lorenz potentials lead straightforwardly to retarded expressions for the electric and magnetic fields. To derive them, let $\mathbf{R} = \mathbf{r} - \mathbf{r}'$ and $t_{\text{ret}} = t - R/c$, and introduce the notation

$$f(\mathbf{r}', t_{\text{ret}}) = f_{\text{ret}}(\mathbf{r}'). \tag{20.63}$$

The first step is to insert (20.57) and (20.58) into (20.6). This gives

$$\mathbf{E}(\mathbf{r}, t) = -\frac{1}{4\pi\epsilon_0}\nabla \int d^3r' \frac{\rho_{\text{ret}}(\mathbf{r}')}{R} - \frac{\mu_0}{4\pi}\frac{\partial}{\partial t}\int d^3r' \frac{\mathbf{j}_{\text{ret}}(\mathbf{r}')}{R} \tag{20.64}$$

and

$$\mathbf{B}(\mathbf{r}, t) = \frac{\mu_0}{4\pi}\nabla \times \int d^3r' \frac{\mathbf{j}_{\text{ret}}(\mathbf{r}')}{R}. \tag{20.65}$$

The derivatives in (20.64) and (20.65) require some care because both $\mathbf{r}$ and $\mathbf{r}'$ appear in the retarded time, $t_{\text{ret}} = t - |\mathbf{r} - \mathbf{r}'|$. This means that space derivatives of retarded-time functions are *not* the same as space derivatives of present-time functions. On the other hand, derivatives with respect to t are the same as derivatives with respect to t_{ret}. Therefore, because $\nabla t_{\text{ret}} = -\hat{\mathbf{R}}/c$,

$$\nabla\left[\frac{\rho(\mathbf{r}', t_{\text{ret}})}{R}\right] = \rho_{\text{ret}}\nabla\frac{1}{R} + \frac{1}{R}\frac{\partial \rho_{\text{ret}}}{\partial t_{\text{ret}}}\nabla t_{\text{ret}} = -\frac{\rho_{\text{ret}}\mathbf{R}}{R^3} - \frac{\mathbf{R}}{cR^2}\frac{\partial \rho_{\text{ret}}}{\partial t}. \tag{20.66}$$

Similarly,

$$\nabla \times \left[\frac{\mathbf{j}(\mathbf{r}', t_{\rm ret})}{R}\right] = -\mathbf{j}_{\rm ret} \times \nabla \frac{1}{R} + \frac{1}{R} \nabla \times \mathbf{j}_{\rm ret} = \mathbf{j}_{\rm ret} \times \frac{\mathbf{R}}{R^3} + \frac{1}{R} \nabla t_{\rm ret} \times \frac{\partial \mathbf{j}_{\rm ret}}{\partial t_{\rm ret}}. \tag{20.67}$$

Using these results to evaluate (20.64) and (20.65) produces *Schott's formulae*:[7]

$$\mathbf{E}(\mathbf{r}, t) = \frac{1}{4\pi\epsilon_0} \int d^3r' \left\{ \frac{\rho_{\rm ret}\mathbf{R}}{R^3} + \frac{\mathbf{R}}{cR^2} \frac{\partial \rho_{\rm ret}}{\partial t} - \frac{1}{c^2 R} \frac{\partial \mathbf{j}_{\rm ret}}{\partial t} \right\} \tag{20.68}$$

and

$$\mathbf{B}(\mathbf{r}, t) = \frac{\mu_0}{4\pi} \int d^3r' \left\{ \frac{\mathbf{j}_{\rm ret}}{R^3} + \frac{1}{cR^2} \frac{\partial \mathbf{j}_{\rm ret}}{\partial t} \right\} \times \mathbf{R}. \tag{20.69}$$

The "ret" subscript everywhere makes it clear that electromagnetic communication between any two points separated by a distance R requires an elapsed time of at least R/c.

We reiterate that the retardation of $\mathbf{E}(\mathbf{r}, t)$ and $\mathbf{B}(\mathbf{r}, t)$ with respect to their sources is a consequence of (20.4) and (20.5), and does *not* depend on the fact that we used the retarded Lorenz gauge potentials to derive (20.68) and (20.69). As noted earlier, the partially retarded Coulomb gauge potentials lead to exactly the same result. This is a consequence of the gauge invariance of electromagnetism. ■

20.4 The Time-Dependent Electric Dipole

A point electric dipole with moment $\mathbf{p}(t)$ is a time-varying source where the fields can be calculated exactly and analyzed in detail. If the dipole sits at the origin, (4.14) specifies its charge density as

$$\rho(\mathbf{r}, t) = -\mathbf{p}(t) \cdot \nabla \delta(\mathbf{r}). \tag{20.70}$$

The continuity equation fixes the associated current density:

$$\mathbf{j}(\mathbf{r}, t) = \dot{\mathbf{p}}(t) \delta(\mathbf{r}). \tag{20.71}$$

This is all the information we need to calculate the fields using the Schott formulae (Application 20.2). However, because the dipole potentials will be useful later for other purposes, we calculate them here as an intermediate step on the way to the fields.

The current density (20.71) determines the vector potential (20.58) as

$$\mathbf{A}(\mathbf{r}, t) = \frac{\mu_0}{4\pi} \int d^3r' \frac{\dot{\mathbf{p}}(t - |\mathbf{r} - \mathbf{r}'|/c)}{|\mathbf{r} - \mathbf{r}'|} \delta(\mathbf{r}') = \frac{\mu_0}{4\pi} \frac{\dot{\mathbf{p}}(t - r/c)}{r}. \tag{20.72}$$

We calculate the scalar potential using (20.57) and (20.70). An alternative is to use (20.72) and integrate the Lorenz gauge condition. Either way,

$$\varphi(\mathbf{r}, t) = \frac{1}{4\pi\epsilon_0} \left[\frac{\dot{\mathbf{p}}(t - r/c) \cdot \mathbf{r}}{cr^2} + \frac{\mathbf{p}(t - r/c) \cdot \mathbf{r}}{r^3} \right]. \tag{20.73}$$

[7] G.A. Schott (1912) derived (20.68) and (20.69) in the Fourier frequency domain. Contemporary authors often cite Jefimenko (1966) for the time-domain formulae derived here. See Sources, References, and Additional Reading.

Given (20.72) and (20.73), the electric and magnetic fields follow straightforwardly from (20.6). We restrict ourselves to $r > 0$ and use $f_{\rm ret} \equiv f(t - r/c)$ as in Application 20.2 so[8]

$$\mathbf{B}(\mathbf{r}, t) = -\frac{\mu_0}{4\pi}\hat{\mathbf{r}} \times \left\{\frac{\dot{\mathbf{p}}_{\rm ret}}{r^2} + \frac{\ddot{\mathbf{p}}_{\rm ret}}{cr}\right\} \tag{20.74}$$

$$\mathbf{E}(\mathbf{r}, t) = \frac{1}{4\pi\epsilon_0}\left\{\frac{3\hat{\mathbf{r}}(\hat{\mathbf{r}}\cdot\mathbf{p}_{\rm ret}) - \mathbf{p}_{\rm ret}}{r^3} + \frac{3\hat{\mathbf{r}}(\hat{\mathbf{r}}\cdot\dot{\mathbf{p}}_{\rm ret}) - \dot{\mathbf{p}}_{\rm ret}}{cr^2} + \frac{\hat{\mathbf{r}}(\hat{\mathbf{r}}\cdot\ddot{\mathbf{p}}_{\rm ret}) - \ddot{\mathbf{p}}_{\rm ret}}{c^2 r}\right\}. \tag{20.75}$$

A striking feature of these formulae is the coexistence of various terms with different algebraic dependence on the radial distance r from the source. If τ is a characteristic time for the variation of the source, the electric field formula suggests a convenient way to partition space:

$$\begin{aligned} &r \ll c\tau \qquad \text{near zone,} \\ &r \sim c\tau \qquad \text{intermediate zone,} \\ &r \gg c\tau \qquad \text{far zone.} \end{aligned} \tag{20.76}$$

The near zone is dominated by the first term in the curly brackets of both (20.74) and (20.75). The $1/r^3$ electric field has the structure of an electrostatic dipole field. The $1/r^2$ magnetic field takes account of Maxwell's displacement current but not Faraday's law. In other words, these are quasi-electrostatic (see Section 14.5) Coulomb and Biot-Savart fields calculated using the source densities (20.70) and (20.71) evaluated at the retarded time rather than the present time.

No single term dominates either the magnetic field or the electric field in the intermediate zone. The $1/r^2$ term in the latter is often called the "induction" electric field because it derives from the first term in (20.74) using Faraday's law. The dependence of this field on the time derivative $\dot{\mathbf{p}}(t)$ shows that it arises from the velocity of the elementary charges that constitute the dipole.

The far zone is dominated by the $1/r$ terms in (20.74) and (20.75). These are distinguished by their dependence on $\ddot{\mathbf{p}}(t)$, the second time derivative, showing that these fields owe their existence to the acceleration of the charges in the dipole. We will see in the next section that the $1/r$ behavior of the fields is characteristic of *radiation*. As a preview, let us calculate the rate at which energy passes through a spherical surface of radius R centered on the dipole. From Section 15.4.1, this is the integral over the surface of the radial component of the Poynting vector $\mathbf{S} = \mathbf{E} \times \mathbf{B}/\mu_0$. If the direction of $\mathbf{p}$ is fixed in space, the reader can confirm that

$$\frac{dU}{dt} = \frac{2}{3}\frac{1}{4\pi\epsilon_0}\left[\frac{d}{dt}\left\{\frac{p^2}{2R^3} + \frac{p\dot{p}}{cR^2} + \frac{\dot{p}^2}{c^2 R}\right\}_{\rm ret} + \frac{\ddot{p}^2_{\rm ret}}{c^3}\right]. \tag{20.77}$$

The near field and the intermediate field contribute exclusively to the curly brackets in (20.77). The total derivative which acts on the curly brackets implies that the time integral of this contribution to dU/dt vanishes if $\mathbf{p}(t)$ is time-harmonic and we integrate over one period. The same is true if $\mathbf{p}(t)$ turns on and off and we integrate from slightly before turn-on to slightly after turn-off. A consistent interpretation is that energy flows back and forth across any spherical surface in the near and intermediate zones for these choices for $\mathbf{p}(t)$. The last term in (20.77) is strictly positive and comes exclusively from the $1/r$ fields produced by the dipole. It is the only contribution to the rate of energy flow that remains finite as $R \to \infty$. This shows that the "acceleration" or "radiation" fields in (20.74) and (20.75), acting together, have the unique ability to carry energy undiminished all the way to infinity.

An alternative to the scalar/vector potential method used above recognizes that a point electric dipole at the origin is equivalent to a polarization $\mathbf{P}(\mathbf{r}, t) = \mathbf{p}(t)\delta(\mathbf{r})$. There is no magnetization, so the retarded Hertz potential (20.61) substituted into (20.14) gives

$$\mathbf{B}(\mathbf{r}, t) = \nabla \times \frac{\mu_0}{4\pi}\frac{\dot{\mathbf{p}}_{\rm ret}}{r} \qquad \text{and} \qquad \mathbf{E}(\mathbf{r}, t) = \nabla \times \nabla \times \frac{1}{4\pi\epsilon_0}\frac{\mathbf{p}_{\rm ret}}{r} \qquad (r > 0). \tag{20.78}$$

[8] Our notation is $\dot{\mathbf{p}}_{\rm ret} \equiv (d/dt)\mathbf{p}(t - r/c)$.

The magnetic field in (20.78) is recognizably the same as (20.74). Less obviously, the electric field in (20.78) is a very compact way to write (20.75). The Hertz potential method also reveals the dual symmetry (Section 15.2.2) between the fields of a point *electric* dipole and a point *magnetic* dipole. To see this, we note that a point magnetic dipole at the origin with moment $\mathbf{m}(t)$ is equivalent to a magnetization $\mathbf{M}(\mathbf{r}, t) = \mathbf{m}(t)\delta(\mathbf{r})$. There is no polarization in this case, so (20.62) substituted into (20.14) gives

$$\mathbf{E}(\mathbf{r}, t) = -\nabla \times \frac{\mu_0}{4\pi}\frac{\dot{\mathbf{m}}_{\text{ret}}}{r} \qquad \text{and} \qquad \mathbf{B}(\mathbf{r}, t) = \nabla \times \nabla \times \frac{\mu_0}{4\pi}\frac{\mathbf{m}_{\text{ret}}}{r} \qquad (r > 0). \tag{20.79}$$

The key observation is that (20.78) transforms into (20.79) when we let $\mathbf{p} \to \mathbf{m}/c$ and make the duality transformations $\mathbf{E} \to c\mathbf{B}$ and $\mathbf{B} \to -\mathbf{E}/c$. The same operations applied to (20.74) and (20.75), respectively, give explicit and detailed formulae for the electric and magnetic fields of a point magnetic dipole.

Example 20.3 Figure 20.4 shows $N \gg 1$ point electric dipoles arranged head-to-tail to form a ring of radius a. If $\mathbf{R} = \mathbf{r} - \mathbf{r}'$ and $d\boldsymbol{\ell}$ is an element of arc length at $\mathbf{r}'$, prove that the scalar potential of the ring vanishes and that a function $I(t)$ parameterizes the vector potential of the ring in the form

$$\mathbf{A}(\mathbf{r}, t) = \frac{\mu_0}{4\pi}\oint d\boldsymbol{\ell}\frac{I(t - R/c)}{R}.$$

Comment on the physical connection between $I(t)$ and the moment $p(t)$ of each electric dipole.

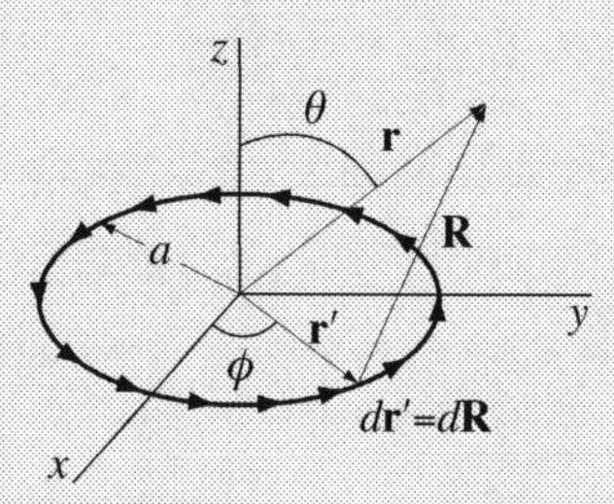

Figure 20.4: A ring formed from N point electric dipoles arranged head-to-tail.

Solution: Let $\varphi_d(\mathbf{r}, t)$ be the point dipole scalar potential (20.73). By superposition, the scalar potential of the ring is

$$\varphi(\mathbf{r}, t) = \frac{N}{2\pi a}\oint d\ell\, \varphi_d(\mathbf{R}, t).$$

Now, $\mathbf{p}(t) = p(t)\hat{\boldsymbol{\phi}}$ for each dipole and $d\mathbf{R} = d\boldsymbol{\ell} = d\ell\hat{\boldsymbol{\phi}}$. Therefore, if

$$\psi(R) = -\frac{1}{4\pi\epsilon_0}\frac{p(t - R/c)}{R},$$

a brief calculation confirms that $d\mathbf{R} \cdot \nabla_{\mathbf{R}}\psi = d\ell\, \varphi_d$. Consequently,

$$\varphi(\mathbf{r}, t) = \frac{N}{2\pi a}\oint d\mathbf{R} \cdot \nabla\psi = 0.$$

The scalar potential vanishes because the charge density of each dipole is canceled by the charge density of the dipoles that immediately precede and follow it.

Let $\mathbf{A}_d(\mathbf{r}, t)$ be the point dipole vector potential (20.72). Because $\mathbf{p}(t)$ and $d\boldsymbol{\ell}$ both point in the $\hat{\boldsymbol{\phi}}$ direction, superposition and the choice

$$I(t) = \frac{N}{2\pi a}\frac{dp}{dt}$$

produce a vector potential with the suggested form,

$$\mathbf{A}(\mathbf{r}, t) = \frac{N}{2\pi a}\oint d\boldsymbol{\ell}\, \mathbf{A}_d(\mathbf{R}, t) = \frac{\mu_0}{4\pi}\oint d\boldsymbol{\ell}\, \frac{I(t - R/c)}{R}.$$

Dimensional analysis shows that $I(t)$ is an electric current. To get the precise form, note that $V = 2\pi a S$ is the volume of a circular toroid with cross sectional area $S \to 0$ which closely encloses the ring. This volume has uniform polarization $P = Np/V$ and $I(t) = j_P(t)S$ is the current associated with the polarization current density $j_P = dP/dt$ (see Section 14.2.2).

20.5 Radiation

Radiation is the name given to electromagnetic fields that transport energy undiminished from their sources to observation points infinitely far away. For physically realizable sources of finite spatial extent, this occurs only if the radiation fields somehow "detach" from their sources and propagate to infinity as electromagnetic waves. In Section 16.1, detached waves of this kind were called *free fields*. In this section, we define radiation fields precisely and use an oscillating point electric dipole to illustrate the process of radiation field detachment. We derive expressions for the angular distribution of radiated power (in both the time and frequency domains) and derive Larmor's formula for the total power radiated by a slowly moving but accelerating point charge.

20.5.1 The Definition of Radiation

By definition, a compact source *radiates* into a differential element of solid angle $d\Omega$ if $\mathbf{r}$ lies on the spherical surface element $r^2 d\Omega$ and the Poynting vector $\mathbf{S} = \mathbf{E} \times \mathbf{B}/\mu_0$ gives a non-zero value for

$$dP(t) = \lim_{r\to\infty} \hat{\mathbf{r}} \cdot \mathbf{S}(\mathbf{r}, t) r^2 d\Omega. \tag{20.80}$$

This definition implies that $\mathbf{S} \propto \hat{\mathbf{r}}/r^2$ as $r \to \infty$. The corresponding radiation electric and magnetic fields, $\mathbf{E}_{\rm rad}(\mathbf{r}, t)$ and $\mathbf{B}_{\rm rad}(\mathbf{r}, t)$ must each vary as $1/r$ in the same limit. Therefore, (20.80) predicts an angular distribution of radiated power given by

$$\frac{dP}{d\Omega} = r^2 \hat{\mathbf{r}} \cdot (\mathbf{E}_{\rm rad} \times \mathbf{B}_{\rm rad})/\mu_0. \tag{20.81}$$

The total power radiated to infinity, $P(t)$, is the integral of (20.80) over all angles. In other words, $P(t)$ is the flux of the Poynting vector through a spherical surface A at infinity:

$$P(t) = \lim_{r\to\infty} \int_A d\mathbf{A} \cdot \mathbf{S}. \tag{20.82}$$

An alternative to (20.82) follows from Poynting's theorem (Section 15.4.1),

$$\int_A d\mathbf{A} \cdot \mathbf{S} + \int_V d^3r\, \mathbf{j} \cdot \mathbf{E} = -\int_V d^3r\, \frac{\partial}{\partial t}\frac{1}{2}\epsilon_0 \left[\mathbf{E} \cdot \mathbf{E} + c^2 \mathbf{B} \cdot \mathbf{B}\right] = -\frac{dU_{\rm EM}}{dt}. \tag{20.83}$$

Comparing (20.82) to (20.83) shows that

$$-\int d^3r\, \mathbf{j}\cdot\mathbf{E} = P(t) + \frac{dU_{\rm EM}}{dt}. \tag{20.84}$$

Hence, the total radiated power $P(t)$ can be calculated from the integral on the left side of (20.84) if we agree to drop all total time derivative terms. The important special case when $\mathbf{j}(\mathbf{r},t) = \mathbf{j}(\mathbf{r}|\omega)\exp(-i\omega t)$ simplifies matters because the quantity of interest is the power (20.82) averaged over one period. Performing this time average on (20.84) eliminates the total time derivative. Using the time-averaging theorem (Section 1.6.3), we conclude that the power radiated into the far zone is equal to the power required to maintain the source current:

$$\langle P\rangle = -\frac{1}{2}\mathrm{Re}\int_V d^3r\, \mathbf{j}\cdot\mathbf{E}^*. \tag{20.85}$$

In the literature, the term "induced EMF method" is often used in connection with (20.85).

Example 20.1 at the end of Section 20.2 demonstrated explicitly that (20.85) and the time average of (20.82) give the same answer for the radiation produced by an infinitely long cylindrical solenoid with a sinusoidally time-varying current. We note in passing that this two-dimensional current source requires a slightly different definition of radiation. Part (c) of Example 20.1 showed that the asymptotic electric and magnetic fields of the solenoid vary as $1/\sqrt{\rho}$. These are two-dimensional *cylindrical waves* rather than three-dimensional spherical waves and the associated Poynting vector varies as $1/\rho$. Therefore, the power radiated into an element of polar angle $d\phi$ is

$$\frac{dP}{d\phi} = \rho\hat{\boldsymbol{\rho}}\cdot(\mathbf{E}_{\rm rad}\times\mathbf{B}_{\rm rad})/\mu_0. \tag{20.86}$$

The same remarks apply to the radiation produced by a long, straight wire which carries a time-varying current.

20.5.2 The Birth of Radiation

In this section, we develop some intuition about the radiation process by using the fields of a time-dependent electric dipole derived in Section 20.4 to study how radiation fields detach from their sources and propagate to infinity as free electromagnetic waves. Our field line analysis follows the original discussion given by Heinrich Hertz—the first person to purposefully produce and detect electromagnetic waves in the laboratory—in his book *Electric Waves* (1893). From this point of view, radiation is created by a topological process called *field line reconnection*.

Hertz' analysis of the radiation process exploits an alternative form of the point dipole electric field (20.75) obtained by substituting the vector potential (20.72) into the Ampère-Maxwell law,

$$\frac{1}{c^2}\frac{\partial\mathbf{E}}{\partial t} = \nabla\times\mathbf{B} = \nabla(\nabla\cdot\mathbf{A}) - \nabla^2\mathbf{A}. \tag{20.87}$$

For a z-oriented dipole, the result is

$$\frac{1}{c^2}\frac{\partial\mathbf{E}}{\partial t} = \frac{\mu_0}{4\pi}\left[\nabla\frac{d}{dz}\frac{\dot{p}}{r} - \hat{\mathbf{z}}\nabla^2\frac{\dot{p}}{r}\right]_{\rm ret}. \tag{20.88}$$

Integration over time and evaluation of the gradient and Laplacian in cylindrical coordinates gives

$$\mathbf{E} = \frac{1}{4\pi\epsilon_0}\left[\hat{\boldsymbol{\rho}}\frac{\partial}{\partial z}\frac{\partial}{\partial\rho}\frac{p}{r} - \hat{\mathbf{z}}\frac{1}{\rho}\frac{\partial}{\partial\rho}\left(\rho\frac{\partial}{\partial\rho}\frac{p}{r}\right)\right]_{\rm ret}. \tag{20.89}$$

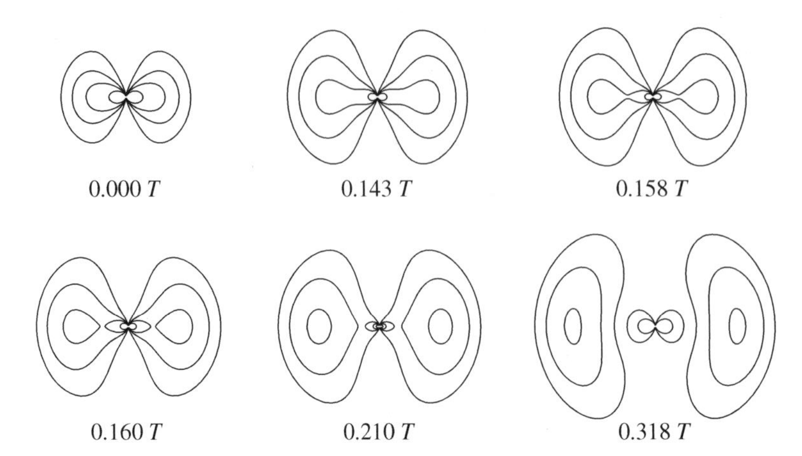

Figure 20.5: Electric field lines for a vertically oriented, time-harmonic point dipole located at the center of each panel. The indicated times are fractions of the oscillation period T. Field direction arrows are omitted for clarity.

The definition

$$R(t) \equiv -\rho \frac{\partial}{\partial \rho} \frac{p(t - r/c)}{r} \tag{20.90}$$

puts (20.89) into the desired final form,

$$\mathbf{E} = \frac{1}{4\pi\epsilon_0 \rho} \left[\frac{dR}{d\rho} \hat{\mathbf{z}} - \frac{dR}{dz} \hat{\boldsymbol{\rho}} \right]. \tag{20.91}$$

The virtue of (20.91) is that $\nabla R = (\partial R/\partial \rho)\hat{\boldsymbol{\rho}} + (\partial R/\partial z)\hat{\mathbf{z}}$. This means that $\mathbf{E} \cdot \nabla R = 0$ and the set of curves $R(\rho, z) = const.$ are everywhere tangent to electric field lines. Therefore, except for their direction, we are at liberty to regard the level curves of $R(\rho, z)$ *as* electric field lines.

Figure 20.5 shows a series of snapshots of the field lines associated with (20.91) when $R(t)$ in (20.90) is evaluating using the time-harmonic dipole moment

$$p(t - r/c) = p_0 \cos(kr - \omega t + \phi). \tag{20.92}$$

We have chosen the phase ϕ in (20.92) so $t = 0$ for the panel in the upper left corner of the figure. The field line pattern in this first panel is recognizably similar to a static dipole field (Figure 3.2). The remaining panels indicate the time as a fraction of the period $T = 2\pi/\omega$. A notable feature of the top row of panels is that the field line loop closest to the dipole shrinks as time goes on. The other three loops expand during the same time interval. These dynamical features may remind the reader of the ebb and flow of electromagnetic energy in the near and intermediate zones of the dipole discussed in the paragraph following (20.77).

The last two panels in the top row of Figure 20.5 show that the innermost of the three expanding loops distorts and pinches inward. A closed loop of electric field line has formed by the time of the first panel of the second row. Figure 20.6 gives an expanded (and schematic) view of how this occurs. The pinching-in process terminates when the field lines intersect at a point where $\mathbf{E} = 0$. Infinitesimally later, the topology of the field lines changes and a closed loop of electric field appears. The last two panels of Figure 20.5 show that the middle and outer loops pinch off, one at a time, and thereby launch two more concentric closed loops of $\mathbf{E}$. As these loops propagate away from the dipole, new field lines attached to the dipole are created. The entire cycle repeats indefinitely, as near fields evolve into intermediate fields and then break off to form fields which propagate freely to infinity. This is how radiation is born.

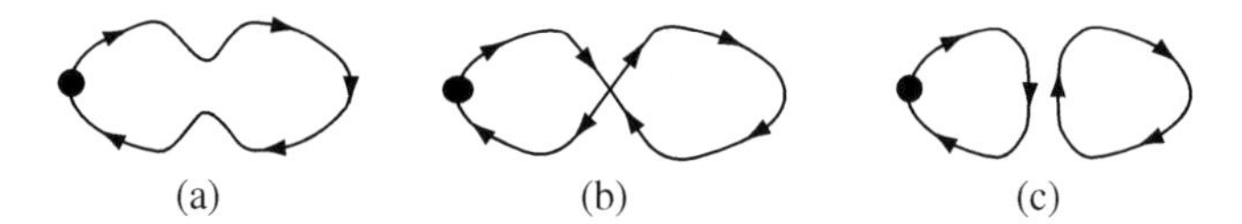

Figure 20.6: Schematic view of field line reconnection between $t = 0.158T$ and $t = 0.160T$ in Figure 20.5. Arrows indicate the direction of $\mathbf{E}$. (a) A dipole loop attached to its source (black dot) begins to pinch in; (b) $\mathbf{E} = 0$ at the point where the loop intersects itself; (c) field line reconnection creates a detached loop which propagates away from the source and a loop which remains connected to (and shrinks back toward) the source.

20.5.3 Radiation Fields in the Time Domain

In this section, we derive explicit formulae for the radiation fields, $\mathbf{E}_{\rm rad}(\mathbf{r}, t)$ and $\mathbf{B}_{\rm rad}(\mathbf{r}, t)$, produced by a specified (but otherwise arbitrary) time-dependent current distribution and the angular distribution of power radiated by that distribution. Consistent with (20.81), we do this by keeping only the leading $1/r$ contribution to the total fields computed from $\mathbf{B} = \nabla \times \mathbf{A}$ and $\mathbf{E} = -\nabla\varphi - \partial\mathbf{A}/\partial t$.

For convenience, we repeat the Lorenz gauge expression (20.58) for the retarded vector potential:

$$\mathbf{A}(\mathbf{r}, t) = \frac{\mu_0}{4\pi}\int d^3r' \frac{\mathbf{j}(\mathbf{r}', t - |\mathbf{r} - \mathbf{r}'|/c)}{|\mathbf{r} - \mathbf{r}'|}. \tag{20.93}$$

For a current density with characteristic size L, radiation appears when $L/r \ll 1$. Therefore, because $r' \leq L$ in (20.93), we consider the expansion

$$|\mathbf{r} - \mathbf{r}'| = r\sqrt{1 - 2\frac{\hat{\mathbf{r}}\cdot\mathbf{r}'}{r} + \left(\frac{r'}{r}\right)^2} = r\left\{1 - \left(\frac{r'}{r}\right)(\hat{\mathbf{r}}\cdot\hat{\mathbf{r}}') + \frac{1}{2}\left(\frac{r'}{r}\right)^2\left[1 - (\hat{\mathbf{r}}\cdot\hat{\mathbf{r}}')^2\right]\right\} + \cdots. \tag{20.94}$$

The approximation $|\mathbf{r} - \mathbf{r}'| \approx r$ is sufficient in the denominator of (20.93) because we are looking for derivatives of $\mathbf{A}$ that vary exactly as $1/r$.

It is necessary to retain the second term in the curly brackets in (20.94) when we approximate the numerator of (20.93). This is so because (i) the fields would have no angular dependence otherwise and (ii) the quantity $\hat{\mathbf{r}}\cdot\mathbf{r}'/c \approx L/c$ is the time needed for an electromagnetic signal to propagate across the source. This correction to the retarded time $t - r/c$ can be significant when the source is large or its time variation is rapid. This contrasts with the third term in curly brackets in (20.94), which produces a correction to $t - r/c$ that can be made as small as we like by choosing r large enough. Putting all this together, we conclude that the part of (20.93) that contributes to radiation is

$$\mathbf{A}_{\rm rad}(\mathbf{r}, t) = \frac{\mu_0}{4\pi r}\int d^3r'\, \mathbf{j}(\mathbf{r}', t - r/c + \hat{\mathbf{r}}\cdot\mathbf{r}'/c). \tag{20.95}$$

The corresponding radiation scalar potential is

$$\varphi_{\rm rad}(\mathbf{r}, t) = \frac{1}{4\pi\epsilon_0 r}\int d^3r'\, \rho(\mathbf{r}', t - r/c + \hat{\mathbf{r}}\cdot\mathbf{r}'/c). \tag{20.96}$$

We compute $\mathbf{B}_{\rm rad} = \nabla \times \mathbf{A}_{\rm rad}$ using (20.95) and the chain rule. For the latter, it is convenient to define

$$t^\star = t - r/c + \hat{\mathbf{r}}\cdot\mathbf{r}'/c \tag{20.97}$$

and use the fact that

$$\nabla t^\star = -\frac{1}{c}\nabla r + \frac{1}{c}\nabla(\hat{\mathbf{r}}\cdot\mathbf{r}') = -\frac{\hat{\mathbf{r}}}{c} - \frac{\hat{\mathbf{r}}}{c} \times \left(\hat{\mathbf{r}} \times \frac{\mathbf{r}'}{r}\right) \approx -\frac{\hat{\mathbf{r}}}{c}. \tag{20.98}$$

The approximation in (20.98) is valid because the omitted factor is higher-order in $1/r$ and thus irrelevant to radiation. The final result for the radiation magnetic field is

$$\mathbf{B}_{\rm rad}(\mathbf{r},t) = -\frac{\mu_0}{4\pi c}\frac{\hat{\mathbf{r}}}{r} \times \int d^3r' \frac{\partial}{\partial t}\mathbf{j}(\mathbf{r}', t - r/c + \hat{\mathbf{r}}\cdot\mathbf{r}'/c). \tag{20.99}$$

The computation of $\mathbf{E}_{\rm rad} = -\nabla\varphi_{\rm rad} - \partial\mathbf{A}_{\rm rad}/\partial t$ using (20.95) and (20.96) is similar in the sense that we only retain terms that vary as $1/r$. We begin with (20.98) and the fact that derivatives with respect to $t^\star$ are the same as derivatives with respect to t to deduce that

$$-\nabla\rho(\mathbf{r}', t^\star) = -\frac{\partial\rho(\mathbf{r}', t^\star)}{\partial t^\star}\nabla t^\star = \frac{\partial\rho(\mathbf{r}', t^*)}{\partial t}\frac{\hat{\mathbf{r}}}{c}. \tag{20.100}$$

This gives the intermediate result

$$\mathbf{E}_{\rm rad}(\mathbf{r},t) = \frac{\hat{\mathbf{r}}}{4\pi\epsilon_0 cr}\int d^3r' \frac{\partial\rho(\mathbf{r}', t^\star)}{\partial t} - \frac{\mu_0}{4\pi r}\int d^3r' \frac{\partial\mathbf{j}(\mathbf{r}', t^\star)}{\partial t}. \tag{20.101}$$

A brief side-calculation permits us to eliminate the charge density in favor of the current density in (20.101). The first step uses (20.97), (20.98), and the continuity equation to write

$$\frac{\partial\,\rho(\mathbf{r}', t^\star)}{\partial t} = \frac{\partial\,\rho(\mathbf{r}', t^\star)}{\partial t^\star} = \left[\frac{\partial}{\partial t'}\rho(\mathbf{r}', t')\right]_{t'=t^\star} = -\left[\nabla'\cdot\mathbf{j}(\mathbf{r}', t')\right]_{t'=t^\star}. \tag{20.102}$$

The second step uses $\nabla' t^\star = \hat{\mathbf{r}}/c$ and the chain rule to get

$$\nabla'\cdot\mathbf{j}(\mathbf{r}', t^\star) = \left[\nabla'\cdot\mathbf{j}(\mathbf{r}', t')\right]_{t'=t^\star} + \frac{\partial\,\mathbf{j}(\mathbf{r}', t^\star)}{\partial t^\star}\cdot\nabla' t^\star = \left[\nabla'\cdot\mathbf{j}(\mathbf{r}', t')\right]_{t'=t^\star} + \frac{\hat{\mathbf{r}}}{c}\cdot\frac{\partial\,\mathbf{j}(\mathbf{r}', t^\star)}{\partial t}. \tag{20.103}$$

Using (20.103) to eliminate the rightmost member of (20.102) gives the desired result:

$$\frac{\partial\rho(\mathbf{r}', t^\star)}{\partial t} = -\nabla'\cdot\mathbf{j}(\mathbf{r}', t^\star) + \frac{\hat{\mathbf{r}}}{c}\cdot\frac{\partial\mathbf{j}(\mathbf{r}', t^\star)}{\partial t}. \tag{20.104}$$

Substituting (20.104) into (20.101) expresses $\mathbf{E}_{\rm rad}(\mathbf{r},t)$ as the sum of three integrals over all space. The integral of $\nabla'\cdot\mathbf{j}(\mathbf{r}', t^\star)$ vanishes (after an integration by parts) for a localized current distribution. The two others combine using $\hat{\mathbf{r}}(\hat{\mathbf{r}}\cdot\mathbf{a}) - \mathbf{a} = \hat{\mathbf{r}}\times(\hat{\mathbf{r}}\times\mathbf{a})$ to give the desired formula for the radiation electric field:

$$\mathbf{E}_{\rm rad}(\mathbf{r},t) = \hat{\mathbf{r}}\times\left[\frac{\mu_0}{4\pi}\frac{\hat{\mathbf{r}}}{r}\times\int d^3r' \frac{\partial}{\partial t}\mathbf{j}(\mathbf{r}', t - r/c + \hat{\mathbf{r}}\cdot\mathbf{r}'/c)\right]. \tag{20.105}$$

20.5.4 Summary of Radiation-Zone Results

The radiation produced by a localized current distribution, $\mathbf{j}(\mathbf{r},t)$, can be determined entirely from the radiation vector potential,

$$\mathbf{A}_{\rm rad}(\mathbf{r},t) = \frac{\mu_0}{4\pi r}\int d^3r'\,\mathbf{j}(\mathbf{r}', t - r/c + \hat{\mathbf{r}}\cdot\mathbf{r}'/c). \tag{20.106}$$

The radiation fields themselves are

$$c\mathbf{B}_{\rm rad}(\mathbf{r},t) = -\hat{\mathbf{r}}\times\frac{\partial\mathbf{A}_{\rm rad}(\mathbf{r},t)}{\partial t} \tag{20.107}$$

and

$$\mathbf{E}_{\rm rad}(\mathbf{r},t) = -\hat{\mathbf{r}}\times c\mathbf{B}_{\rm rad}. \tag{20.108}$$

We have already emphasized the $1/r$ dependence of $\mathbf{E}_{\rm rad}$ and $\mathbf{B}_{\rm rad}$. This implies that the Poynting vector $\mathbf{S} = \mathbf{E}_{\rm rad}\times\mathbf{B}_{\rm rad}/\mu_0$ transports energy undiminished to radial distances arbitrarily far from the source. Otherwise, we draw attention to three consequences of (20.106), (20.107), and (20.108).

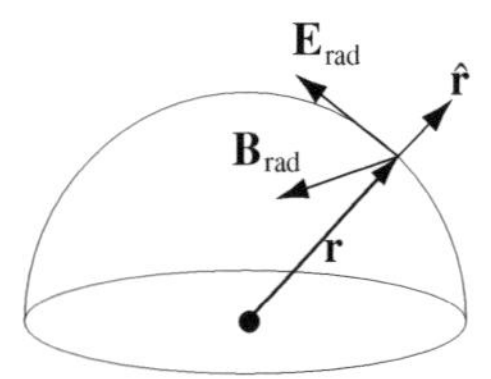

Figure 20.7: The vectors $\mathbf{E}_{\rm rad}$, $\mathbf{B}_{\rm rad}$, and $\hat{\mathbf{r}}$ are mutually orthogonal in the radiation zone far from the source (black dot).

First, the field magnitudes satisfy $|\mathbf{E}_{\rm rad}| = c|\mathbf{B}_{\rm rad}|$. Second, the vectors $\hat{\mathbf{r}}$, $\mathbf{E}_{\rm rad}$, and $\mathbf{B}_{\rm rad}$ form the right-handed orthogonal triad shown in Figure 20.7:

$$\hat{\mathbf{r}} \times \mathbf{E}_{\rm rad} = c\mathbf{B}_{\rm rad} \qquad \text{and} \qquad \mathbf{E}_{\rm rad} \times c\mathbf{B}_{\rm rad} = \hat{\mathbf{r}}|\mathbf{E}_{\rm rad}|^2. \tag{20.109}$$

Finally, as r increases and the wave front flattens out, the *local* characteristics of the radiation fields (20.107) and (20.108) increasingly resemble the characteristics of a transverse electromagnetic plane wave with a propagation vector which points from the source to the observer.

We conclude with a general formula for the angular distribution of radiated power derived by substituting (20.109) into (20.81) and using (20.107) and (20.108):

$$\frac{dP}{d\Omega} = \frac{r^2}{c\mu_0}|\mathbf{E}_{\rm rad}|^2 = \frac{1}{c\mu_0}\left|\mathbf{r} \times \frac{\partial \mathbf{A}_{\rm rad}(\mathbf{r}, t)}{\partial t}\right|^2. \tag{20.110}$$

The total radiated power is the integral of (20.110) over all angles.

20.5.5 Larmor's Formula

A famous formula due to Joseph Larmor predicts the total power radiated by an accelerating but slowly moving charged particle. This result is important because moving charged particles are the constituents of every real current density. For a single point charge q with velocity $\mathbf{v}(t) = \dot{\mathbf{r}}_0(t)$,

$$\mathbf{j}(\mathbf{r}, t) = q\mathbf{v}(t)\delta[\mathbf{r} - \mathbf{r}_0(t)]. \tag{20.111}$$

When a sum of terms like (20.111) is used to evaluate (20.106), the time derivative in (20.110) shows that radiation *cannot* occur unless at least some of the charged particles accelerate for at least some of the time. This is consistent with the absence of radiation fields for the problem of a charged particle moving with uniform velocity (Section 20.1).

It is important to appreciate that the mere presence of accelerating particles does not *guarantee* that radiation exists. The fields produced by different accelerating particles can interfere destructively and conspire to produce no radiation at all. An example is a uniformly charged spherical shell whose radius oscillates periodically in time. Regardless of the distribution of charge over the surface of the shell, $\mathbf{j} = j\hat{\mathbf{r}}$, so (20.107) and (20.108) imply that the radiation fields are identically zero. Other classes of time-dependent but radiation-less current distributions are known as well.

Let us use (20.111) to evaluate the vector potential (20.106). If the particle speed $v \ll c$, it is a good approximation to write

$$\mathbf{j}(\mathbf{r}, t - r/c + \hat{\mathbf{r}} \cdot \mathbf{r}_0/c) \approx \mathbf{j}(\mathbf{r}, t - r/c) \tag{20.112}$$

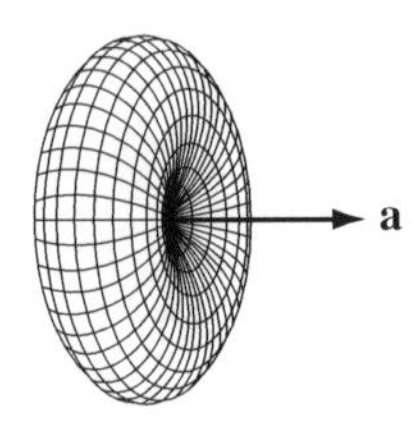

Figure 20.8: The angular distribution of power radiated by a slowly moving point particle with instantaneous acceleration **a**. The latter is approximately equal to $\mathbf{a}_{\rm ret}$ when $v \ll c$.

because the first correction is of order v/c. Using (20.112) to evaluate (20.105) and defining $\mathbf{a}_\perp$ as the component of the particle acceleration **a** perpendicular to $\hat{\mathbf{r}}$, we find[9]

$$\mathbf{E}_{\rm rad} = \frac{\mu_0 q}{4\pi r}\hat{\mathbf{r}} \times [\hat{\mathbf{r}} \times \mathbf{a}(t - r/c)] = -\frac{\mu_0 q}{4\pi r}\mathbf{a}_\perp(t - r/c). \tag{20.113}$$

If $\hat{\mathbf{r}} \cdot \hat{\mathbf{a}} = \cos\theta$, the angular distribution of radiated power from (20.110) is[10]

$$\frac{dP}{d\Omega} = \frac{\mu_0 q^2 |\mathbf{a}_{\rm ret}|^2}{16\pi^2 c}\sin^2\theta. \tag{20.114}$$

Figure 20.8 shows that the radiation pattern resembles a doughnut, with no radiation emitted either parallel or anti-parallel to $\mathbf{a}_{\rm ret}$. Acceleration and deceleration produce exactly the same pattern and the total radiated power is given by *Larmor's formula*:

$$P = \int d\Omega \frac{dP}{d\Omega} = \frac{1}{4\pi\epsilon_0}\frac{2q^2|\mathbf{a}_{\rm ret}|^2}{3c^3}. \tag{20.115}$$

A quick application of (20.115) is to the emission of *cyclotron radiation* by a particle with charge q and mass m moving non-relativistically in a uniform magnetic field **B**. As discussed in Section 12.2.2, such a particle follows a helical trajectory. The motion parallel to **B** is uniform. The motion perpendicular to **B** is circular with uniform speed v. The circular (cyclotron) motion is maintained by a centripetal acceleration v^2/R where $R = mv/qB$ is the orbit radius. With this data, we conclude from (20.115) that energy radiates from the circling charge at a rate

$$P = \frac{q^4 v^2 B^2}{6\pi\epsilon_0 m^2 c^3}. \tag{20.116}$$

20.5.6 Radiation Fields in the Frequency Domain

Many authors focus exclusively on the radiation produced by a time-harmonic current density. They do this because Fourier's theorem guarantees that the harmonic content of any current density $\mathbf{j}(\mathbf{r}, t)$ and radiation vector potential $\mathbf{A}_{\rm rad}(\mathbf{r}, t)$ is revealed by writing

$$\begin{bmatrix} \mathbf{j}(\mathbf{r}, t) \\ \mathbf{A}_{\rm rad}(\mathbf{r}, t) \end{bmatrix} = \frac{1}{2\pi}\int_{-\infty}^{\infty} d\omega \begin{bmatrix} \mathbf{j}(\mathbf{r}|\omega) \\ \mathbf{A}_{\rm rad}(\mathbf{r}|\omega) \end{bmatrix} \exp(-i\omega t). \tag{20.117}$$

If we substitute the top member of (20.117) into (20.106) and let $\mathbf{k} = (\omega/c)\hat{\mathbf{r}}$, the bottom member of (20.117) permits us to deduce that

$$\mathbf{A}_{\rm rad}(\mathbf{r}|\omega) = \frac{\mu_0}{4\pi}\frac{e^{ikr}}{r}\int d^3r'\, \mathbf{j}(\mathbf{r}'|\omega)e^{-i\mathbf{k}\cdot\mathbf{r}'} \equiv \frac{\mu_0}{4\pi}\frac{e^{ikr}}{r}\hat{\mathbf{j}}(\mathbf{k}|\omega). \tag{20.118}$$

The function $\hat{\mathbf{j}}(\mathbf{k}|\omega)$ is the Fourier transform (over space and time) of the source current density.

[9] We choose the position of the particle as $r = 0$. The (slow) motion of this origin is not problematical for observations made in the radiation zone.

[10] As in Section 20.4, the subscript "ret" refers to the retarded time $t - r/c$.

Using (20.107) and (20.108), the time-harmonic fields associated with the vector potential (20.118) are

$$c\mathbf{B}_{\text{rad}}(\mathbf{r}\,|\,\omega) = i\omega\hat{\mathbf{r}} \times \mathbf{A}_{\text{rad}}(\mathbf{r}\,|\,\omega) \qquad \mathbf{E}_{\text{rad}}(\mathbf{r}\,|\,\omega) = -i\omega\hat{\mathbf{r}} \times [\hat{\mathbf{r}} \times \mathbf{A}_{\text{rad}}(\mathbf{r}\,|\,\omega)]\,. \tag{20.119}$$

The corresponding time-averaged angular distribution of radiated power follows from (20.110) and the time-averaging theorem (see Section 1.6.3):

$$\left\langle \frac{dP}{d\Omega} \right\rangle = \frac{\omega^2}{2c\mu_0}\,|\mathbf{r} \times \mathbf{A}_{\text{rad}}(\mathbf{r}\,|\,\omega)|^2 = \frac{\mu_0\omega^2}{32\pi^2 c}|\hat{\mathbf{r}} \times \hat{\mathbf{j}}(\mathbf{k}\,|\,\omega)|^2 = \frac{\mu_0\omega^2}{32\pi^2 c}|\hat{\mathbf{j}}_\perp(\mathbf{k}\,|\,\omega)|^2. \tag{20.120}$$

This shows that the frequencies of the waves radiated by a current source $\mathbf{j}(\mathbf{r}, t)$ in the propagation direction $\mathbf{k}$ are the same as the frequencies present in the transverse components of the Fourier transform $\hat{\mathbf{j}}(\mathbf{k}, \omega)$.

Equation (20.120) shows that a current density $\mathbf{j}(\mathbf{r}, t)$ produces *no* radiation if all its transverse Fourier components $\hat{\mathbf{j}}_\perp(\mathbf{k}\,|\,\omega = ck)$ vanish. Moreover, we will prove in Section 23.4.2 that (20.120) is essentially the angular-resolved *frequency spectrum* of the radiation produced by a current density $\mathbf{j}(\mathbf{r}, t)$ with Fourier transform $\hat{\mathbf{j}}(\mathbf{k}\,|\,\omega)$. This is the amount of energy radiated into a unit solid angle in a unit interval of frequency. An example is the non-relativistic cyclotron emission studied in Section 20.5.5, where the radiation occurs exclusively at the cyclotron frequency $\omega_c = |q|B/m$.

20.6 Thin-Wire Antennas

Every radio listener knows that a thin wire conductor of almost any shape acts as an antenna for the reception of electromagnetic waves broadcast from distant sources. Incident waves induce a time-dependent current on the antenna's surface which drives a voltage across the antenna's load resistance. This voltage encodes the information content of the broadcast. Conversely, an external voltage applied to its input terminals drives a time-varying current in a broadcast antenna. This makes a thin-wire antenna a prototype radiating system.

20.6.1 The Dipole Antenna: Frequency Domain

Figure 20.9 shows a straight, thin-wire, linear antenna driven at its center point by an input voltage. This is called a *dipole antenna*. The current which actually flows in this antenna is determined by an integral equation which guarantees that the tangential component of the total electric field vanishes at the antenna surface.[11] In this section, we suppose that the driving voltage varies in time as $\exp(-i\omega t)$ and focus on the limit when the radius of the wire goes to zero. In that case, a first approximation to the current which flows in a time-harmonic dipole antenna is the standing wave ($\omega = ck$)

$$I(z, t) = I_0 \sin k(d - |z|)e^{-i\omega t} \qquad -d \le z \le d. \tag{20.121}$$

We use $\mathbf{j}(\mathbf{r}')d^3r' \to I(z')dz'$ to evaluate (20.118) with the antenna current (20.121). Because $\hat{\mathbf{z}} \cdot \hat{\mathbf{r}} = \cos\theta$ is effectively constant in the radiation zone, and

$$\int dz \exp(az)\sin(bz + c) = \frac{\exp(az)}{a^2 + b^2}\left[a\sin(bz + c) - b\cos(bz + c)\right], \tag{20.122}$$

[11] Approximate analytic solutions to this classic boundary value problem in antenna theory have given way to accurate numerical solutions. See, e.g., S.J. Orfanidis, *Electromagnetic Waves and Antennas*, Chapter 21. Available at http://www.ece.rutgers.edu/~orfanidi/ewa/.

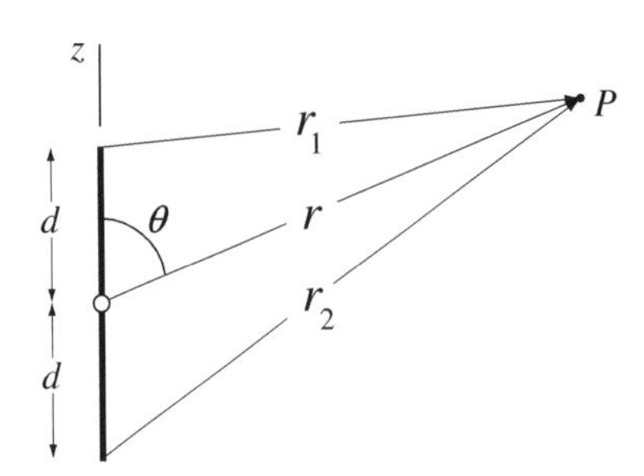

Figure 20.9: A linear (dipole) antenna of length $2d$ driven by a voltage source at its midpoint (open circle). P labels an observation point.

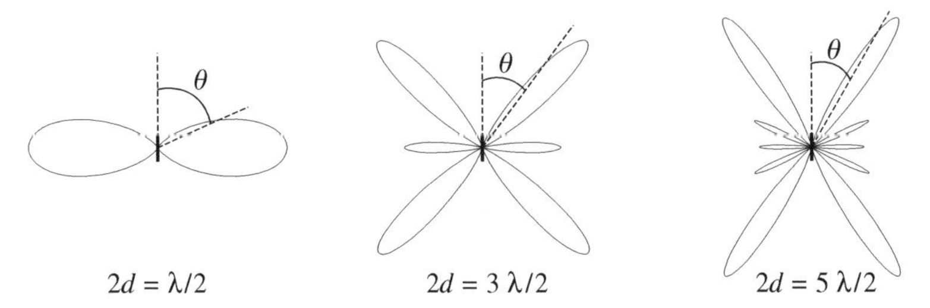

Figure 20.10: Angular distribution of radiation produced by the dipole antenna in Figure 20.9 when $2d = \lambda/2$, $2d = 3\lambda/2$, and $2d = 5\lambda/2$. The figures are rotationally symmetric around the vertical axis of the antenna.

a few lines of algebra are sufficient to confirm that

$$\mathbf{A}_{\text{rad}}(\mathbf{r}, |\omega) = \frac{\mu_0 I_0}{2\pi} \frac{\exp(ikr)}{kr} \left[\frac{\cos(kd\cos\theta) - \cos kd}{\sin^2\theta} \right] \hat{\mathbf{z}}. \tag{20.123}$$

The time-averaged angular distribution of radiated power (20.120) is[12]

$$\left\langle \frac{dP}{d\Omega} \right\rangle = \frac{\mu_0 c I_0^2}{8\pi^2} \left[\frac{\cos(kd\cos\theta) - \cos kd}{\sin\theta} \right]^2. \tag{20.124}$$

Figure 20.10 illustrates (20.124) for three choices of antenna length: $2d = \lambda/2$, $2d = 3\lambda/2$, and $2d = 5\lambda/2$. The number of distinct lobes is equal to the number of radiation half-wavelengths m that "fit" into the length of the antenna. The principal radiation lobe becomes narrower and lies closer and closer to the antenna axis as $m \to \infty$.

The integration of (20.124) over all angles cannot be reduced to elementary functions. Nevertheless, the dependence of $\langle dP/d\Omega \rangle$ on I_0^2 guarantees that we can define a *radiation resistance* R_{rad} such that[13]

$$\langle P \rangle = \frac{1}{2} I_0^2 R_{\text{rad}}. \tag{20.125}$$

[12] The angular distribution of power radiated when an antenna operates in "broadcast" mode is the same as the angular distribution of power absorbed when the same antenna operates in "receiving" mode. See Sources, References, and Further Reading.

[13] In practical antenna theory, the radiation resistance (the real part of the complex impedance) is defined by $\langle P \rangle = \frac{1}{2}|I_T|^2 R_{\text{rad}}$ where $|I_T|$ is the magnitude of the current which flows through the input terminals of the antenna. The input impedance of an antenna is discussed clearly in Section 9.3 of D.H. Staelin, A.W. Morgenthaler, and J.A. Kong, *Electromagnetic Waves* (Prentice Hall, Englewood Cliffs, NJ, 1994).

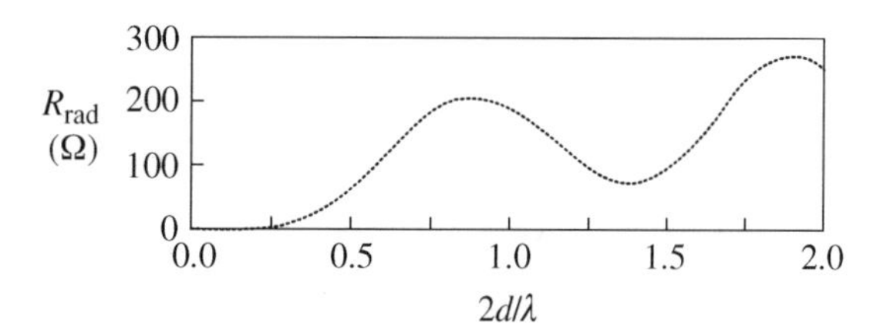

Figure 20.11: The radiation resistance of a dipole antenna of total length $2d$ that produces radiation with wavelength λ. Figure adapted from Balanis (2005).

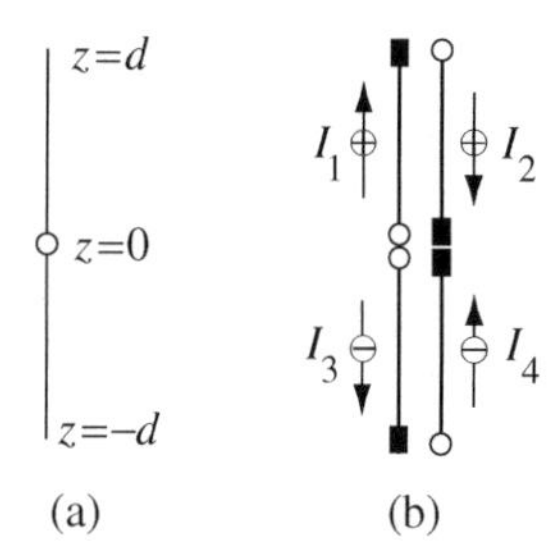

Figure 20.12: (a) A dipole antenna; (b) four linear end-fed antennas. For each, a wave of current launches from the open circle and is absorbed at the closed square. Plus and minus signs identify the charge driven in the direction of the arrows. Figure adapted from Smith (2001).

Figure 20.11 shows that R_{rad} predicted by (20.125) and (20.124) oscillates around a mean value which increases roughly as $\ln(2d/\lambda)$. The local maxima occur when $2d$ is slightly less than an integer number of wavelengths. Because current flows into and out of the center point, the condition $2d = n\lambda$ corresponds to situations where equal lengths of antenna carry current in opposite directions.

20.6.2 The Dipole Antenna: Time Domain

The frequency-domain analysis of the previous section does not reveal the process by which an antenna radiates. To gain this insight, we follow Smith (2001) and switch to the time domain. The key step is to represent the standing wave (20.121) on the dipole antenna of Figure 20.12(a) as a sum of four traveling waves, each of which propagates an identical current $I_S(t)$ at speed c. We will show below that a suitable choice of $I_S(t)$ reproduces the vector potential (20.123). More importantly, we will discover the time-history of the radiation experienced by an observer.

Figure 20.12(b) illustrates four traveling waves, each confined to its own linear, end-fed antenna of length d. Waves 1 and 3 launch from $z = 0$ at $t = 0$. Wave 1 drives positive charge upward until it is absorbed at $z = d$. Wave 3 drives negative charge downward until it is absorbed at $z = -d$. No charge accumulates at the ends of the original dipole antenna because, at $t = d/c$, wave 2 launches positive charge downward from $z = d$ and wave 4 launches negative charge upward from $z = -d$. Both are absorbed at $z = 0$ at $t = 2d/c$.

We focus first on the radiation produced by wave 1 due to its current:

$$I_1(z,t) = I_S(t - z/c) \qquad 0 \le z \le d. \tag{20.126}$$

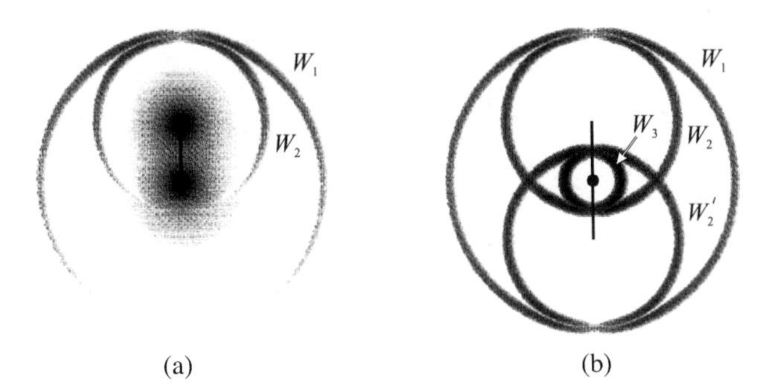

Figure 20.13: Logarithmic grayscale plots of $|\mathbf{E}_{\rm rad}(\mathbf{r}, t)|$ at a fixed time produced by (a) a traveling wave of current created and then absorbed at opposite ends of a single end-fed antenna; (b) four traveling waves of current superposed to produce a standing wave of current on the dipole antenna of Figure 20.12(a). The various spherical wave fronts W_k are discussed in the text. Figure adapted from Smith (2001).

If r and θ are defined with respect to the open circle at the end of antenna 1 as they are in Figure 20.9, the radiation vector potential (20.106) generated by this current satisfies

$$\frac{\partial \mathbf{A}_1}{\partial t} = \hat{\mathbf{z}}\frac{\mu_0}{4\pi r}\int_0^d dz' \frac{d}{dt} I_S[t - r/c - (z'/c)(1 - \cos\theta)]. \tag{20.127}$$

The integral (20.127) is straightforward because the chain rule relates dI_S/dt to dI_S/dz'. The result is

$$\frac{\partial \mathbf{A}_1}{\partial t} = \hat{\mathbf{z}}\frac{\mu_0}{4\pi r}\frac{c}{1 - \cos\theta}\left[I_S(t - r/c) - I_S(t - r/c - (d/c)(1 - \cos\theta))\right]. \tag{20.128}$$

Using (20.107) and (20.108), we conclude that the electric field radiated by the current I_1 in Figure 20.12 is

$$\mathbf{E}_1(\mathbf{r}, t) = \frac{\mu_0 c}{4\pi}\frac{\sin\theta}{1 - \cos\theta}\left[\frac{I_S(t - r/c)}{r} - \frac{I_S(t - d/c - (r - d\cos\theta)/c)}{r}\right]\hat{\boldsymbol{\theta}}. \tag{20.129}$$

Equation (20.129) is valid for any choice of the time-dependent current $I_S(t)$. We follow Smith (2001) and choose a Gaussian pulse of the form $I_S(t) = I_0 \exp[-(4ct/d)^2]$. Figure 20.13(a) plots the magnitude of $|\mathbf{E}_1(\mathbf{r}, t)|$ from (20.129) at time $t = 2.5d/c$. The grayscale intensity is logarithmic. The large ring W_1 comes from the first term in the square brackets in (20.129). It is a spherical wave which propagates at the speed of light away from the $z = 0$ launch point of the wave. The small ring W_2 comes from the second term in (20.129). This is also a spherical wave, but it is delayed by a time d/c and propagates at the speed of light away from the $z = d$ absorption point of the wave. The latter assignment follows from the retardation factor in the second term and the fact that the distance r_1 in Figure 20.9 is

$$r_1 \approx r - d\cos\theta \qquad\qquad r \gg d. \tag{20.130}$$

We note in passing that the electrically neutral traveling wave exposes negative charge when it leaves the launch point and deposits positive charge when it reaches the absorption point. These charges are the source of the static electric field seen in the immediate vicinity of the end points in Figure 20.13(a).

Our interpretation of Figure 20.13(a) must be consistent with the fact that radiation is produced by the acceleration of charged particles (Section 20.5.5). Electrons on the surface of the antenna are transiently accelerated by a guided electromagnetic wave (associated with the current pulse) which propagates down the wire at the speed of light. However, the numerical results of Smith (2001) show that the field produced by the charge induced when the wave reaches any particular segment of the wire is effectively canceled by the field produced by the charge induced when the wave propagates to the immediately adjacent segment. The only places where field cancellation does not occur are the launch point $z = 0$ at $t = 0$ and the reflection points $z = \pm d$ at $t = d/c$. It is important to appreciate that the moving charges that constitute the current pulse itself are *not* relevant here.[14]

The radiation electric fields produced by the traveling waves 2, 3, and 4 in Figure 20.12(b) are calculated similarly. Care must be taken because the sign of the moving charge, the launch times, the launch points, and the absorption points are not the same for each. When these fields are added to (20.129), we get the total radiation electric field:

$$\mathbf{E}_{\rm tot}(\mathbf{r}, t) = \frac{\mu_0 c}{2\pi \sin\theta} \left\{ \frac{I_S(t - r/c)}{r} + \frac{I_S(t - r/c - 2d/c)}{r} - \frac{I_S[t - d/c - (r - d\cos\theta)/c]}{r} - \frac{I_S[t - d/c - (r + d\cos\theta)/c]}{r} \right\} \hat{\boldsymbol{\theta}}. \tag{20.131}$$

Figure 20.13(b) is a grayscale plot of $|\mathbf{E}_{\rm tot}(\mathbf{r}, t)|$ from (20.131) at time $t = 2.5d/c$ for the current choice $I_S(t) = I_0 \exp[-(4ct/d)^2]$. This is the same Gaussian pulse used in Figure 20.13(a). The ring W_1 comes from the launch of waves 1 and 3 at $t = 0$ from the center of the antenna. Rings W_2 and W_2' come from the absorption of these waves at $t = d/c$ and the simultaneous launch of waves 2 and 4 from the ends of the antenna.[15] The ring W_3 comes from the absorption of waves 2 and 4 at $t = 2d/c$ at the center of the antenna. We conclude that an observer experiences the radiation produced by this pulsed dipole antenna as a sequence of four signals (separated in time), one from each of the four spherical wave shells indicated in Figure 20.13(b).

It remains only to choose $I_S(t)$ properly to connect the radiation electric field (20.131) to the field produced by the time-harmonic current (20.121). To do this, we note that $\mathbf{A}_{\rm rad}(\mathbf{r}, t) = A_{\rm rad}(\mathbf{r}|\omega)\exp(-i\omega t)\hat{\mathbf{z}}$ and use (20.107) and (20.108) to write

$$\mathbf{E}_{\rm rad}(\mathbf{r}, t) = \hat{\mathbf{r}} \times \left[\hat{\mathbf{r}} \times \frac{\partial \mathbf{A}_{\rm rad}(\mathbf{r}, t)}{\partial t} \right] = \frac{\partial A_{\rm rad}(\mathbf{r}, t)}{\partial t} \sin\theta \, \hat{\boldsymbol{\theta}}. \tag{20.132}$$

With a few lines of algebra, the reader can confirm that the electric field computed using (20.132) with the vector potential (20.123) is exactly the same as the electric field (20.131) with the particular choice

$$I_S(t) = \frac{i}{2} I_0 \exp[-i(kd + \omega t)]. \tag{20.133}$$

If we approximate the time-harmonic current (20.133) as a time-series of Gaussian current pulses with alternating signs, it is not difficult to see that the discrete bursts of radiation represented by the rings in Figure 20.13 evolve into continuous radiation emission from the center and end points of the antenna. The constructive and destructive interference of these waves is responsible for the lobe structure of the emission pattern in Figure 20.10.

20.6.3 Antenna Arrays

Any collection of radiating elements where the positions of the individual radiators are chosen on purpose is called an *antenna array*. Interference effects guarantee that the radiation pattern produced

[14] These electrons move at the drift velocity $v_d \ll c$. See Section 9.3.

[15] The distance r_2 in Figure 20.9 is approximately $r + d\cos\theta$ when $r \gg d$. The latter distance appears in the last term in (20.131).

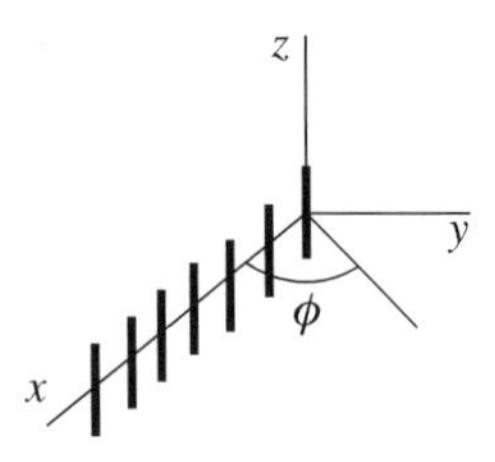

Figure 20.14: A line of parallel and equally spaced linear antennas.

by an array depends sensitively on the spatial arrangement of the individual radiators. This makes it possible to "engineer" the pattern in various ways, including (but not limited to) focusing the bulk of the radiated power into a relatively small angular range.

As an example, Figure 20.14 shows N linear antennas sitting on the x-axis at positions $x = ma$ where $m = 0, 1, \ldots, N-1$. All the antennas point in the z-direction and the current in each antenna is

$$I(z,t) = I_0 \cos(kz)\cos\omega t \qquad -\lambda/4 \leq z \leq \lambda/4. \tag{20.134}$$

Our task is to find the angular distribution of power radiated by this array.

The current density of the array is a sum of N terms, one for each antenna. For the mth antenna, we write the integration variable in (20.118) as $\mathbf{r}' = ma\hat{\mathbf{x}} + z''\hat{\mathbf{z}}$ and use the azimuthal angle ϕ in Figure 20.14 to locate the observation point in the x-y plane. Since $\omega = ck = 2\pi c/\lambda$,

$$\mathbf{A}_{\text{rad}}(\omega) = \hat{\mathbf{z}}\frac{\mu_0 I_0}{4\pi}\frac{\exp(ikr)}{r}\left[\int_{-\lambda/4}^{\lambda/4} dz''\cos kz''\right]\sum_{m=0}^{N-1} e^{-ikma\cos\phi}. \tag{20.135}$$

Evaluating the integral and performing the sum (a geometric series) gives

$$\mathbf{A}_{\text{rad}}(\omega) = \hat{\mathbf{z}}\frac{\mu_0 I_0}{2\pi}\frac{\exp(ikr)}{kr}e^{i(N-1)\frac{1}{2}ka\cos\phi}\frac{\sin\left(\frac{1}{2}kNa\cos\phi\right)}{\sin\left(\frac{1}{2}ka\cos\phi\right)}. \tag{20.136}$$

It is conventional to define an *array factor*,

$$\mathcal{F}(\cos\phi) = \frac{\sin\left(\frac{1}{2}kNa\cos\phi\right)}{\sin\left(\frac{1}{2}ka\cos\phi\right)}, \tag{20.137}$$

and use (20.120) to write the time-averaged distribution of power radiated into the x-y plane as

$$\left\langle\frac{dP}{d\phi}\right\rangle = \frac{\mu_0 c I_0^2}{8\pi^2}\left|\mathcal{F}(\cos\phi)\right|^2. \tag{20.138}$$

Figure 20.15 is a graph of the absolute value of the normalized array factor, $|\mathcal{F}(\cos\phi)|/N$, for the case $N = 7$. Compared to a single z-oriented antenna (which radiates isotropically in the x-y plane), the array factor introduces a strong angular dependence to the radiated power.

The important features of Figure 20.15 are the position, relative amplitude, and angular width of the main peak. Of these, only the position ($\cos\phi = 0$ or $\phi = \pi/2$) is not a function of N. For simplicity, let $\lambda = a$ so

$$\mathcal{F}(\cos\phi) = \frac{\sin(N\pi\cos\phi)}{\sin(\pi\cos\phi)}. \tag{20.139}$$

By l'Hospital's rule, $|\mathcal{F}(0)| = N$, and the first zero occurs when $\cos\phi = 1/N$. This angle can be made arbitrarily close to the y-axis by making N arbitrarily large. The next (secondary) maximum occurs

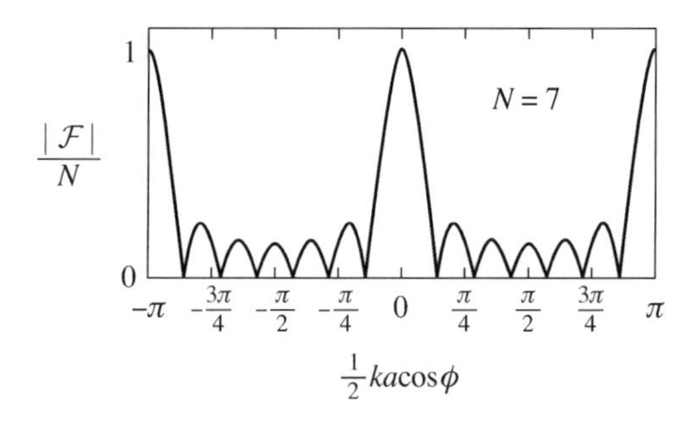

Figure 20.15: The absolute value of the normalized array factor $|\mathcal{F}|/N$ for $N = 7$.

when $\cos\phi = 3/2N$. When $N \gg 1$, the radiated power at the secondary maximum is proportional to

$$|\mathcal{F}(3/2N)|^2 = \frac{\sin^2(3\pi/2)}{\sin^2(3\pi/2N)} \approx \left(\frac{2N}{3\pi}\right)^2 \approx 0.045|\mathcal{F}(0)|^2. \tag{20.140}$$

This shows that the majority of the radiation emitted by the array can be confined to a narrow beam along the $\pm y$-axes. This is called the *broadside* direction with respect to the line of antennas.

A new feature appears if we introduce a constant phase difference between the currents in successive antennas. In other words, we let $\cos\omega t \to \cos(\omega t + m\delta)$ in (20.134) so the summand in the sum over antennas in (20.135) becomes $\exp[-im(ka\cos\phi + \delta)]$. This changes the array factor from (20.137) to

$$\mathcal{F}(\cos\phi) = \frac{\sin[\frac{1}{2}N(ka\cos\phi + \delta)]}{\sin[\frac{1}{2}(ka\cos\phi + \delta)]}. \tag{20.141}$$

As a result, the position of the main radiation beam rotates from $\phi = \pi/2$ to an angle $\phi^\star$ where $ka\cos\phi^\star + \delta = 0$. This shows that the angular position of the principal radiation beam can be "steered" by changing the value of the phase δ. For that reason, the name *phased-array antenna* is used for the entire arrangement.

The footnote to (20.124) implies that all the results of this section apply equally well to the receiving characteristics of an antenna array. This fact was essential to the 1967 discovery of the pulsar in the Crab nebula by Antony Hewish and Jocelyn Bell. The phased array of 2048 radio antennas used by them had the ability to resolve distant radio sources with unprecedentedly small size.

20.7 Cartesian Multipole Radiation

Radiation fields can rarely be computed exactly for physically interesting current densities. This motivates us to develop an approximation scheme that reliably approximates radiation fields when the source size is sufficiently small or when its time variation is sufficiently slow. To set the stage, recall from (20.107) and (20.108) that $\mathbf{B}_{\text{rad}}(\mathbf{r},t)$ and $\mathbf{E}_{\text{rad}}(\mathbf{r},t)$ are completely determined by the time derivative of the radiation vector potential (20.106). However, rather than work with $\mathbf{A}_{\text{rad}}(\mathbf{r},t)$ directly, it is somewhat more convenient to work with the *radiation vector*, $\boldsymbol{\alpha}(\mathbf{r},t)$, defined by

$$\frac{\partial \mathbf{A}_{\text{rad}}(\mathbf{r},t)}{\partial t} = \frac{\mu_0}{4\pi r}\boldsymbol{\alpha}(\mathbf{r},t), \tag{20.142}$$

where

$$\boldsymbol{\alpha}(\mathbf{r},t) = \frac{\partial}{\partial t}\int d^3r'\,\mathbf{j}(\mathbf{r}', t - r/c + \hat{\mathbf{r}}\cdot\mathbf{r}'/c). \tag{20.143}$$

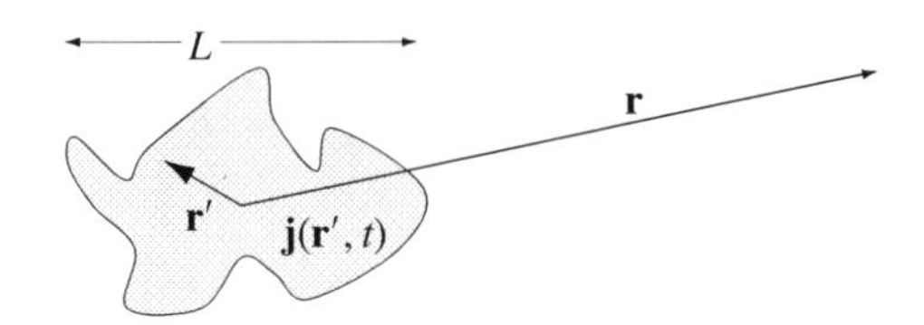

Figure 20.16: To calculate fields in the radiation zone ($r \gg L$), a low-order multipole expansion is valid when $\mathbf{j}(\mathbf{r}, t)$ varies sufficiently slowly in time.

In this language, the angular distribution of radiated power (20.110) is

$$\frac{dP}{d\Omega} = \frac{\mu_0}{16\pi^2 c}\,|\hat{\mathbf{r}} \times \boldsymbol{\alpha}(\mathbf{r}, t)|^2. \tag{20.144}$$

The key point is that $\hat{\mathbf{r}} \cdot \mathbf{r}'/c \simeq L/c$ is the time needed for an electromagnetic signal to propagate across the source (see Figure 20.16). Consequently, a rapidly converging series results if we Taylor expand the current density in (20.143) around the time $t - r/c$ for a spatially small or slowly varying source. The series in question generates a *Cartesian multipole expansion* for $\boldsymbol{\alpha}(\mathbf{r}, t)$, and thus for the radiation fields themselves. Similar to the electrostatic and magnetostatic Cartesian multipole expansions studied earlier in this book, an approximation based on a low-order truncation of this expansion is often sufficient to extract the most important physics.

The expansion of interest is

$$\begin{aligned}\mathbf{j}(\mathbf{r}', t - r/c + \hat{\mathbf{r}} \cdot \mathbf{r}'/c) \approx \mathbf{j}(\mathbf{r}', t - r/c) &+ \frac{\hat{\mathbf{r}} \cdot \mathbf{r}'}{c}\frac{\partial}{\partial t}\mathbf{j}(\mathbf{r}', t - r/c) \\ &+ \frac{1}{2}\left(\frac{\hat{\mathbf{r}} \cdot \mathbf{r}'}{c}\right)^2 \frac{\partial^2}{\partial t^2}\mathbf{j}(\mathbf{r}', t - r/c) + \cdots.\end{aligned} \tag{20.145}$$

The meaning of (20.145) becomes clear if the source current varies maximally over a characteristic time T. In that case, the estimates $\partial j/\partial t \sim j/T$ and $\partial^2 j/\partial t^2 \sim j/T^2$ are valid and we see that (20.145) is an expansion in powers of

$$\frac{L}{cT} \equiv \frac{\text{source size}}{\text{distance travelled by light in time } T}. \tag{20.146}$$

For harmonic time dependence with wavelength $\lambda = 2\pi/k$, we have $\omega = ck$ and $T = 2\pi/\omega$. This identifies the expansion parameter in (20.146) as L/λ. Therefore, truncation of (20.145) after a small number of terms is most valid for long wavelength or slow time variations of the driving source current density. These low-order terms produce radiation fields driven by temporal changes of the electric dipole moment $\mathbf{p}(t)$, the magnetic dipole moment $\mathbf{m}(t)$, and the electric quadrupole tensor $\mathbf{Q}(t)$ of the source current density. More precisely, we will demonstrate in the next three subsections that

$$\boldsymbol{\alpha}(\mathbf{r}, t) = \frac{d^2}{dt^2}\mathbf{p}(t - r/c) + \frac{1}{c}\frac{d^2}{dt^2}\mathbf{m}(t - r/c) \times \hat{\mathbf{r}} + \frac{1}{c}\frac{d^3}{dt^3}\mathbf{Q}(t - r/c) \cdot \hat{\mathbf{r}} + \cdots. \tag{20.147}$$

20.7.1 Electric Dipole Radiation

The term *electric dipole radiation* is used when only the first term on the right-hand side of (20.145) is used to evaluate (20.143) and (20.144). To see why, we focus attention on[16]

$$\boldsymbol{\alpha}_{\mathrm{E1}}(\mathbf{r}, t) = \frac{d}{dt}\int d^3r'\,\mathbf{j}(\mathbf{r}', t - r/c). \tag{20.148}$$

[16] The notation "Eℓ" is shorthand for *electric* multipole of order 2^ℓ. Thus E1 stands for electric dipole, E2 for electric quadrupole, etc. M1 and M2 similarly stand for *magnetic* dipole and quadrupole.

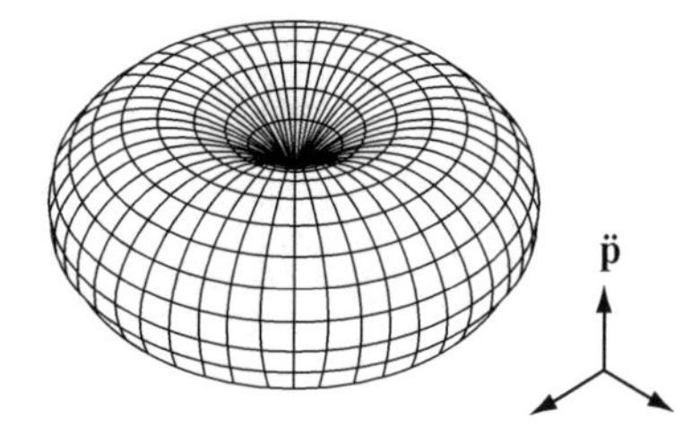

Figure 20.17: Radiation pattern produced by a vertically oriented electric dipole located at the center of the diagram.

We encountered this integral previously in magnetostatics (Section 11.1.2), where its value was *zero* as a consequence of the steady-current condition $\nabla \cdot \mathbf{j} = 0$. Repeating this calculation in the present case gives a non-zero value for (20.148) because the current density is constrained only by the continuity equation,

$$\nabla \cdot \mathbf{j} + \frac{\partial \rho}{\partial t} = 0. \tag{20.149}$$

An alternative and equivalent way to evaluate the integral uses (20.111) for the current density and $\rho(\mathbf{r}, t) = \sum_k q_k \delta[\mathbf{r} - \mathbf{r}_k(t)]$. In detail,

$$\int d^3r'\, \mathbf{j}(\mathbf{r}', t) = \sum_k q_k \boldsymbol{v}_k(t) = \frac{d}{dt} \sum_k q_k \mathbf{r}_k(t) = \frac{d}{dt} \int d^3r'\, \rho(\mathbf{r}', t)\mathbf{r}' = \frac{d}{dt}\mathbf{p}(t). \tag{20.150}$$

Using (20.150) to evaluate (20.148) shows that the radiation from a slowly time-varying current distribution is dominated by the second time derivative of the electric dipole moment of the distribution:

$$\alpha_{\mathrm{E1}}(\mathbf{r}, t) = \frac{d^2}{dt^2}\mathbf{p}(t - r/c) \equiv \ddot{\mathbf{p}}_{\mathrm{ret}}. \tag{20.151}$$

Substituting (20.151) into (20.107) and (20.108) gives the dipole radiation fields

$$\mathbf{B}_{\mathrm{E1}} = -\frac{\mu_0}{4\pi c} \frac{\hat{\mathbf{r}} \times \ddot{\mathbf{p}}_{\mathrm{ret}}}{r} \tag{20.152}$$

and

$$\mathbf{E}_{\mathrm{E1}} = -\hat{\mathbf{r}} \times c\mathbf{B}_{\mathrm{E1}} = \frac{\mu_0}{4\pi} \frac{\hat{\mathbf{r}}(\hat{\mathbf{r}} \cdot \ddot{\mathbf{p}}_{\mathrm{ret}}) - \ddot{\mathbf{p}}_{\mathrm{ret}}}{r}. \tag{20.153}$$

These are identical to the last terms in the formulae (20.74) and (20.75) for the magnetic and electric fields of a point electric dipole. This is consistent with our understanding from electrostatics and magnetostatics that suitably chosen point multipole sources reproduce the effect of a spatially extended source for observation points that lie sufficiently far from the sources.

Substituting (20.151) into (20.144) gives the angular distribution of dipole radiation as

$$\left(\frac{dP}{d\Omega}\right)_{\mathrm{E1}} = \frac{\mu_0}{16\pi^2 c} |\hat{\mathbf{r}} \times \ddot{\mathbf{p}}_{\mathrm{ret}}|^2. \tag{20.154}$$

If the direction of $\mathbf{p}(t)$ is fixed in space, (20.154) predicts that the radiated power varies as $\sin^2\theta$ where θ is the angle between $\mathbf{p}$ and $\hat{\mathbf{r}}$ (see Figure 20.17). There is no emission along the dipole axis because radiation fields are transverse and the rotational symmetry of a dipole implies that no transverse direction can be singled out. The maximum emission occurs in the plane perpendicular to

p that passes through the source. The subscript "ret" reminds us that the angular distribution of power is evaluated at fixed values of the retarded time $t - r/c$ when the fields left the source.

The total radiated power is the integral of (20.154) over all angles:

$$P_{\rm E1}(t) = \frac{\mu_0}{16\pi^2 c}|\ddot{\mathbf{p}}_{\rm ret}|^2 \int d\Omega \sin^2\theta = \frac{\mu_0}{6\pi c}|\ddot{\mathbf{p}}_{\rm ret}|^2. \tag{20.155}$$

A superficially less simple way to get (20.155) has the virtue that it generalizes to more complicated situations in a way that direct integration over angles does not. The first step is to write

$$|\hat{\mathbf{r}} \times \ddot{\mathbf{p}}_{\rm ret}|^2 = |\ddot{\mathbf{p}}_{\rm ret}|^2 - |\hat{\mathbf{r}} \cdot \ddot{\mathbf{p}}_{\rm ret}|^2. \tag{20.156}$$

The integral over the first term in (20.156) gives a multiplicative factor of 4π. Replacing $\ddot{\mathbf{p}}_{\rm ret}$ by $\mathbf{b}$, the integral over the second term is

$$\int d\Omega |\hat{\mathbf{r}} \cdot \mathbf{b}|^2 = b_i b_j^* \int d\Omega \hat{r}_i \hat{r}_j = b_i b_j^* I_{ij}. \tag{20.157}$$

Since $\hat{r}_i = r_i/r$, symmetry implies that $I_{ij} = I\delta_{ij}$. Then, because $\sum_i \hat{r}_i \hat{r}_i = 1$, we find that $I = 4\pi/3$. Combining both terms in (20.156) gives

$$P_{\rm E1}(t) = \frac{\mu_0}{6\pi c}|\ddot{\mathbf{p}}_{\rm ret}|^2 = \frac{1}{4\pi\epsilon_0}\frac{2|\ddot{\mathbf{p}}_{\rm ret}|^2}{3c^3}. \tag{20.158}$$

This agrees with (20.155) and the last term in (20.77), as it should.

The alert reader will have noticed the similarity of Figure 20.17 to Figure 20.8. This is no accident, because (20.158) becomes the Larmor formula (20.115) if we set $\mathbf{p} = q\mathbf{r}$, where $\mathbf{r}$ is the displacement of a charge q from a fixed origin. This choice is consistent with the definition of the electric dipole moment in (20.150). From this point of view, the cyclotron radiation discussed at the end of Section 20.5.5 is interpreted as the radiation produced by a *rotating* dipole moment.

A rotating dipole moment is a special case of a time-harmonic dipole moment $\mathbf{p}(t) = \mathbf{p}\exp(-i\omega t)$, where $\mathbf{p}$ is complex. With $\omega = ck$, the radiation fields in such a case are the real parts of

$$\mathbf{B}_{\rm E1} = \frac{\mu_0}{4\pi}\frac{\omega^2}{c}(\hat{\mathbf{r}} \times \mathbf{p})\frac{e^{i(kr-\omega t)}}{r} \tag{20.159}$$

$$\mathbf{E}_{\rm E1} = -\frac{\mu_0}{4\pi}\omega^2[\hat{\mathbf{r}} \times (\hat{\mathbf{r}} \times \mathbf{p})]\frac{e^{i(kr-\omega t)}}{r}. \tag{20.160}$$

The quantity in square brackets in (20.160) gives the direction of the radiation electric field and thus, by our convention (Section 16.4), fixes the polarization of the emitted electric dipole radiation. Since $\ddot{\mathbf{p}} = -\omega^2\mathbf{p}$ for a time-harmonic source, using (20.151) to evaluate (20.144) gives the time-averaged angular distribution of power radiated by a time-harmonic electric dipole as

$$\left\langle \frac{dP}{d\Omega} \right\rangle_{\rm E1} = \frac{ck^4}{32\pi^2\epsilon_0}\left[(\hat{\mathbf{r}} \times \mathbf{p}) \cdot (\hat{\mathbf{r}} \times \mathbf{p}^*)\right]. \tag{20.161}$$

The corresponding total radiated power obtained by time-averaging (20.158) is

$$\left\langle \frac{dU}{dt} \right\rangle_{\rm E1} = \frac{\mu_0\omega^4}{12\pi c}\mathbf{p} \cdot \mathbf{p}^*. \tag{20.162}$$

Application 20.3 The Classical Zeeman Effect

In 1896, Zeeman observed that the D emission line of sodium splits into three components when the source atoms are placed between the poles of a magnet. Lorentz provided an immediate explanation by taking account of an external magnetic field in the equation of motion for the displacement $\mathbf{s}(t)$ of a particle with charge q and mass m which oscillates with frequency ω_0 around a stationary equilibrium point.[17]

Assuming that the entire system is charge-neutral, Lorentz' idea was to study the radiation produced by the dipole moment $\mathbf{p}(t) = q\mathbf{s}(t)$. The equation of motion is

$$\ddot{\mathbf{s}} + \omega_0^2 \mathbf{s} = \frac{q}{m}(\dot{\mathbf{s}} \times \mathbf{B}). \tag{20.163}$$

If $\mathbf{s}(t) = \mathbf{s}\exp(-i\omega t)$, $\mathbf{B} = B\hat{\mathbf{z}}$, and $\omega_L = qB/2m$ is the Larmor frequency (see Section 12.2.3), the Cartesian components of $\mathbf{s}$ satisfy

$$\begin{bmatrix} \omega^2 - \omega_0^2 & 2\omega_L\omega & 0 \\ -2i\omega_L\omega & \omega^2 - \omega_0^2 & 0 \\ 0 & & \omega^2 - \omega_0^2 \end{bmatrix} \begin{bmatrix} s_x \\ s_y \\ s_z \end{bmatrix} = 0. \tag{20.164}$$

It is straightforward to check that, when $\omega_L \ll \omega$, the three solutions of (20.164) are

$$\begin{aligned} \mathbf{s}_0(t) &= a_0\hat{\mathbf{z}}e^{-i\omega_0 t} \\ \mathbf{s}_+(t) &= a_+(\hat{\mathbf{x}} + i\hat{\mathbf{y}})e^{-i(\omega_0+\omega_L)t} \\ \mathbf{s}_-(t) &= a_-(\hat{\mathbf{x}} - i\hat{\mathbf{y}})e^{-i(\omega_0-\omega_L)t}. \end{aligned} \tag{20.165}$$

The dipole moment $\mathbf{p}_0(t) = q\mathbf{s}_0(t)$ oscillates along the z-axis. The corresponding radiation fields (20.159) and (20.160) are linearly polarized and propagate with frequency ω_0. The dipole moments $\mathbf{p}_+(t) = q\mathbf{s}_+(t)$ and $\mathbf{p}_-(t) = q\mathbf{s}_-(t)$ rotate in the x-y plane (in opposite directions) and produce radiation at frequencies $\omega_0 + \omega_L$ and $\omega_0 - \omega_L$, respectively. The unit vector identities

$$\hat{\mathbf{x}} = \sin\theta\cos\phi\,\hat{\mathbf{r}} + \cos\theta\cos\phi\,\hat{\boldsymbol{\theta}} - \sin\phi\,\hat{\boldsymbol{\phi}} \tag{20.166}$$

$$\hat{\mathbf{y}} = \sin\theta\sin\phi\,\hat{\mathbf{r}} + \cos\theta\sin\phi\,\hat{\boldsymbol{\theta}} + \cos\phi\,\hat{\boldsymbol{\phi}} \tag{20.167}$$

imply that

$$\hat{\mathbf{r}} \times \mathbf{p}_\pm \propto e^{\pm i\phi}(\cos\theta\,\hat{\boldsymbol{\phi}} \mp i\,\hat{\boldsymbol{\theta}}). \tag{20.168}$$

Applying this result to (20.160) and (20.161) shows that the $\omega_0 \pm \omega_L$ radiation fields have opposite elliptical polarization (circular for emission along the z-axis) and angular distributions that vary as $1 + \cos^2\theta$.

From a quantitative analysis of the line splitting and the polarization, Zeeman and Lorentz concluded that q was negative and that $|q|/m$ was 2000 times larger than the charge-to-mass ratio known for hydrogen ions. Today, we identify their particle as the electron and draw the same conclusions about the spectrum using quantum mechanics rather than classical mechanics. ■

Example 20.4 Use (20.84) to derive Larmor's formula (20.158) by assuming that the localized sources of $\mathbf{E}(\mathbf{r}, t)$ vary slowly in time. Begin with the electromagnetic potentials and expand the current density to second order and the charge density to third order.

[17] Zeeman and Lorentz shared the 1902 Nobel Prize in "recognition of the extraordinary service they rendered by their researches into the influence of magnetism upon radiation phenomena".

Solution: Let $\mathbf{R} = \mathbf{r} - \mathbf{r}'$ and suppress the spatial argument of the charge and current densities. The suggested expansion for the latter is $\mathbf{j}(t - R/c) \approx \mathbf{j}(t) - (R/c)\dot{\mathbf{j}}(t)$. Therefore, using (20.150),

$$\mathbf{A} = \frac{\mu_0}{4\pi}\int d^3r' \frac{\mathbf{j}(t - R/c)}{R} = \frac{\mu_0}{4\pi}\int d^3r' \left[\frac{\mathbf{j}}{R} - \frac{1}{c}\frac{\partial \mathbf{j}}{\partial t}\right] = \frac{\mu_0}{4\pi}\int d^3r' \frac{\mathbf{j}}{R} - \frac{\mu_0}{4\pi c}\ddot{\mathbf{p}}.$$

A third-order expansion of the charge density gives

$$\varphi = \frac{1}{4\pi\epsilon_0}\int d^3r' \left[\frac{\rho}{R} - \frac{1}{c}\frac{\partial \rho}{\partial t} + \frac{1}{2c^2}R\frac{\partial^2\rho}{\partial t^2} - \frac{1}{6c^3}R^2\frac{\partial^3\rho}{\partial t^3}\right] = \varphi_1 + \varphi_2 + \varphi_3 + \varphi_4.$$

By conservation of charge,

$$\varphi_2 = -\frac{1}{4\pi\epsilon_0 c}\frac{d}{dt}\int d^3r'\,\rho = 0.$$

Otherwise, $\mathbf{E} = -\nabla\varphi - \partial\mathbf{A}/\partial t$, integration by parts, and (20.149) show that

$$-\int d^3r\,\mathbf{j}\cdot\mathbf{E} = \int d^3r\,\mathbf{j}\cdot[\nabla\varphi + \dot{\mathbf{A}}] = \int d^3r\left[\nabla\cdot(\varphi\mathbf{j}) - \varphi\nabla\cdot\mathbf{j} + \mathbf{j}\cdot\frac{\partial\mathbf{A}}{\partial t}\right]$$
$$= \int d^3r\,[\varphi\dot{\rho} + \mathbf{j}\cdot\dot{\mathbf{A}}].$$

For our purposes, it is somewhat simpler to separate the φ_4 term and calculate

$$-\int d^3r\,\mathbf{j}\cdot\mathbf{E} = \int d^3r\,[\dot{\rho}(\varphi_1 + \varphi_3) + \mathbf{j}\cdot(\nabla\varphi_4 + \dot{\mathbf{A}})],$$

where $\dot{\rho}$ and $\mathbf{j}$ are evaluated at time t. Specifically, because $\nabla R^2 = 2\mathbf{R}$ and $\mathbf{p}(t) = \int d^3r'\,\mathbf{r}'\rho(\mathbf{r}', t)$ is the electric dipole moment, conservation of charge implies that

$$\nabla\varphi_4 = -\frac{1}{4\pi\epsilon_0}\frac{1}{3c^3}\frac{d^3}{dt^3}\int d^3r'\,(\mathbf{r} - \mathbf{r}')\rho(\mathbf{r}', t) = \frac{1}{4\pi\epsilon_0}\frac{1}{3c^3}\dddot{\mathbf{p}}.$$

This result, (20.150), and the expressions for φ_1, φ_3, and $\dot{\mathbf{A}}$ above lead without difficulty to

$$-\int d^3r\,\mathbf{j}\cdot\mathbf{E} = -\frac{\mu_0}{4\pi}\frac{2}{3c}\dot{\mathbf{p}}\cdot\dddot{\mathbf{p}}$$
$$+\frac{d}{dt}\int d^3r\int d^3r'\frac{1}{4\pi\epsilon_0}\left[\frac{\rho(\mathbf{r})\rho(\mathbf{r}')}{2R} + \frac{1}{4c^2}\dot{\rho}(\mathbf{r})\dot{\rho}(\mathbf{r}')R + \frac{\mathbf{j}(\mathbf{r})\cdot\mathbf{j}(\mathbf{r}')}{2c^2R}\right].$$

The final step uses $-\dot{\mathbf{p}}\cdot\dddot{\mathbf{p}} = |\ddot{\mathbf{p}}|^2 - (d/dt)(\dot{\mathbf{p}}\cdot\ddot{\mathbf{p}})$ to conclude that

$$-\int d^3r\,\mathbf{j}\cdot\mathbf{E} = \frac{\mu_0}{6\pi c}|\ddot{\mathbf{p}}|^2 + \text{terms which are total time derivatives.}$$

Comparing this to (20.84) shows that the total time derivative terms contribute to the stored electromagnetic energy. What remains is Larmor's formula (20.158) because $\mathbf{p}_{\rm ret} \approx \mathbf{p}(t)$ when the sources vary slowly in time.

20.7.2 Magnetic Dipole Radiation

If the electric dipole moment $\mathbf{p}(t)$ happens to vanish, or merely if $\ddot{\mathbf{p}}(t) = 0$ [see (20.151)], the dominant contribution to long-wavelength radiation comes from the second term on the right side of (20.145). Not obviously, the analysis of this term simplifies if we use the identity

$$(\mathbf{r}'\cdot\hat{\mathbf{r}})\mathbf{j} = (\mathbf{r}'\times\mathbf{j})\times\hat{\mathbf{r}} + (\mathbf{j}\cdot\hat{\mathbf{r}})\mathbf{r}' \tag{20.169}$$

to rewrite $(\mathbf{r}'\cdot\hat{\mathbf{r}})\mathbf{j}$ as the sum of two terms, one that is symmetric and one that is anti-symmetric with respect to the interchange of $\mathbf{j}$ and $\mathbf{r}'$. The decomposition we want is

$$(\mathbf{r}'\cdot\hat{\mathbf{r}})\mathbf{j} = \underbrace{\tfrac{1}{2}(\mathbf{r}'\times\mathbf{j})\times\hat{\mathbf{r}}}_{\text{anti-symmetric}} + \underbrace{\tfrac{1}{2}[(\mathbf{r}'\cdot\hat{\mathbf{r}})\mathbf{j} + (\hat{\mathbf{r}}\cdot\mathbf{j})\mathbf{r}']}_{\text{symmetric}}. \tag{20.170}$$

The anti-symmetric piece of (20.170) produces *magnetic dipole radiation*. The symmetric piece of (20.170) produces *electric quadrupole radiation*. For the first of these, we recall from magnetostatics (Section 11.2) that the magnetic dipole moment of a current distribution is

$$\mathbf{m} = \tfrac{1}{2}\int d^3r'\,\mathbf{r}'\times\mathbf{j}. \tag{20.171}$$

Using (20.171) in (20.170), and using the latter in (20.145) to evaluate (20.143) produces[18]

$$\boldsymbol{\alpha}_{\rm M1}(\mathbf{r},t) = \frac{1}{c}\ddot{\mathbf{m}}(t-r/c)\times\hat{\mathbf{r}}. \tag{20.172}$$

The radiation fields generated by (20.172) using (20.107) and (20.108) are

$$\mathbf{B}_{\rm M1} = \frac{\mu_0}{4\pi c^2}\frac{\hat{\mathbf{r}}(\hat{\mathbf{r}}\cdot\ddot{\mathbf{m}}_{\rm ret}) - \ddot{\mathbf{m}}_{\rm ret}}{r} \tag{20.173}$$

and

$$\mathbf{E}_{\rm M1} = \frac{\mu_0}{4\pi c}\frac{\hat{\mathbf{r}}\times\ddot{\mathbf{m}}_{\rm ret}}{r}. \tag{20.174}$$

If $\mathbf{m}(t) = \mathbf{m}\exp(-i\omega t)$ and $\omega = ck$, the fields are the real parts of

$$\mathbf{B}_{\rm M1} = \frac{\mu_0}{4\pi}k^2[\hat{\mathbf{r}}\times(\mathbf{m}\times\hat{\mathbf{r}})]\frac{e^{i(kr-\omega t)}}{r} \tag{20.175}$$

$$\mathbf{E}_{\rm M1} = -\frac{\mu_0}{4\pi}\frac{\omega^2}{c}(\hat{\mathbf{r}}\times\mathbf{m})\frac{e^{i(kr-\omega t)}}{r}. \tag{20.176}$$

We make two remarks about these fields. First, the replacement $\mathbf{p}\to\mathbf{m}/c$ in the *electric* dipole radiation fields of the previous subsection generates the *magnetic* dipole radiation fields of the present subsection if we use the *duality* transformation (see Section 15.2.2),

$$\mathbf{B}_{\rm E1}(\mathbf{r},t)\to -\mathbf{E}_{\rm M1}(\mathbf{r},t)/c \qquad \mathbf{E}_{\rm E1}(\mathbf{r},t)\to c\mathbf{B}_{\rm M1}(\mathbf{r},t). \tag{20.177}$$

Second, if $\mathbf{m} = m\hat{\mathbf{z}}$, the identity $\hat{\mathbf{z}} = \cos\theta\,\hat{\mathbf{r}} - \sin\theta\,\hat{\boldsymbol{\theta}}$ implies that

$$\mathbf{B}_{\rm M1}(r,\theta) \propto -\sin\theta\frac{e^{i(kr-\omega t)}}{r}\hat{\boldsymbol{\theta}} \qquad \mathbf{E}_{\rm M1}(r,\theta)\propto \sin\theta\frac{e^{i(kr-\omega t)}}{r}\hat{\boldsymbol{\phi}}. \tag{20.178}$$

These magnetic dipole radiation fields are the asymptotic ($r\to\infty$) part of the spherical TE waves one would compute in Section 16.8.1 from the $\ell = 1$ solution of the homogeneous scalar wave equation in spherical coordinates (see Section 16.8.2). The electric dipole fields in Section 20.7.1 are the asymptotic part of the corresponding spherical TM waves.

The angular distribution of power from (20.110) and (20.172) is

$$\left(\frac{dP}{d\Omega}\right)_{\rm M1} = \frac{\mu_0}{16\pi^2c^3}|\hat{\mathbf{r}}\times\ddot{\mathbf{m}}_{\rm ret}|^2, \tag{20.179}$$

[18] See footnote at the beginning of Section 20.7.1 for the M1 notation.

with its time-averaged counterpart for harmonic fields,

$$\left\langle \frac{dP}{d\Omega} \right\rangle_{\text{M1}} = \frac{\mu_0 c k^4}{32\pi^2} (\hat{\mathbf{r}} \times \mathbf{m}) \cdot (\hat{\mathbf{r}} \times \mathbf{m}^*). \tag{20.180}$$

As expected from the duality (20.177), the formulae (20.179) and (20.180) are exactly (20.154) and (20.161) with $\mathbf{p} \to \mathbf{m}/c$. By the same argument, the electric dipole formulae (20.158) and (20.162) imply that the total power radiated to infinity by a magnetic dipole source is

$$P_{\text{M1}}(t) = \frac{\mu_0}{4\pi} \frac{2|\ddot{\mathbf{m}}_{\text{ret}}|^2}{3c^3}, \tag{20.181}$$

or, for time-harmonic fields,

$$\langle P_{\text{M1}} \rangle = \frac{\mu_0 \omega^4}{12\pi c^3} \mathbf{m} \cdot \mathbf{m}^*. \tag{20.182}$$

Electromagnetic duality also implies that the radiation pattern produced by a vertical magnetic dipole source is identical to the radiation pattern produced by a vertical electric dipole source (Figure 20.17). On the other hand, comparison of (20.160) with (20.176) shows that the polarization of the radiation is different in the two cases.

Our discussion of M1 radiation assumes that no E1 radiation is present. If an electromagnetic source has both $\ddot{\mathbf{p}} \neq 0$ and $\ddot{\mathbf{m}} \neq 0$, the total radiation is a superposition of both electric and magnetic dipole fields. This produces interference effects in the angular distribution of power. On the other hand, conservation of energy guarantees that the total radiated power does *not* show the effect of interference. P_{tot} is simply the arithmetic sum of (20.158) and (20.181). With some care, this fact can be used to estimate the relative power emitted into E1 and M1 radiation by a single source.

Consider a model for a rotating nucleus where the charge and current are related by $\mathbf{j} = \rho \mathbf{v}$. Because

$$\mathbf{m} = \frac{1}{2} \int d^3 r \, \mathbf{r} \times \mathbf{j} \qquad \text{and} \qquad \mathbf{p} = \int d^3 r \, \mathbf{r} \rho, \tag{20.183}$$

an order-of-magnitude estimate for the ratio of the magnetic dipole power to the electric dipole power is

$$\frac{P_{\text{M1}}}{P_{\text{E1}}} \sim \left(\frac{m}{cp} \right)^2 \sim \left(\frac{j}{c\rho} \right)^2 \sim \left(\frac{v}{c} \right)^2. \tag{20.184}$$

This shows that electric dipole radiation dominates magnetic dipole radiation when the charged particle velocities are non-relativistic. When the charge and current densities vary in time as $\exp(-i\omega t)$, the continuity equation (20.149) estimate $j/L \sim \omega\rho$ (L is the system size) gives the same answer in the long-wavelength limit when the multipole expansion applies:

$$\frac{\langle P_{\text{M1}} \rangle}{\langle P_{\text{E1}} \rangle} \sim \left(\frac{m}{cp} \right)^2 \sim \left(\frac{j}{c\rho} \right)^2 \sim \left(\frac{L}{\lambda} \right)^2. \tag{20.185}$$

An exception to the rule of E1 dominance is a perfectly conducting sphere immersed in a long-wavelength, monochromatic plane wave. If a is the sphere radius and $\mathbf{E}_0 = c\mathbf{B}_0$ are the field amplitudes, the electric dipole moment is $\mathbf{p} = 4\pi\epsilon_0 a^3 \mathbf{E}_0$ (see Example 5.1 of Section 5.2.1) and the magnetic dipole moment is $\mathbf{m} = -2\pi a^3 \mathbf{B}_0/\mu_0$ (see Example 13.4 of Section 13.6.6). The E1 and M1 channels radiate comparably in this case because $m/cp = \frac{1}{2}$. The general argument responsible for (20.185) does not apply because the electric moment $\mathbf{p}$ comes from a charge density ρ_E induced by $\mathbf{E}_0$ and the magnetic moment $\mathbf{m}$ comes from a current density $\mathbf{j}_B$ induced by $\mathbf{B}_0$. The continuity equation ensures that the densities ρ_B and $\mathbf{j}_E$ exist, but they contribute negligibly to the dipole moments.

Example 20.5

(a) Show that a magnetic moment $\mathbf{m}(t)$ radiates angular momentum at the rate

$$\frac{d\mathbf{L}}{dt} = -\frac{\mu_0}{4\pi}\frac{2}{3c^3}(\dot{\mathbf{m}}_{\rm ret} \times \ddot{\mathbf{m}}_{\rm ret}).$$

(b) Apply this formula to a pulsar modeled as a rotating neutron star with a frozen-in magnetic dipole moment $\mathbf{m}_0$ that is misaligned with the z-axis of rotation (Figure 20.18). Compare the time average of dL_z/dt to the time average of dU/dt.

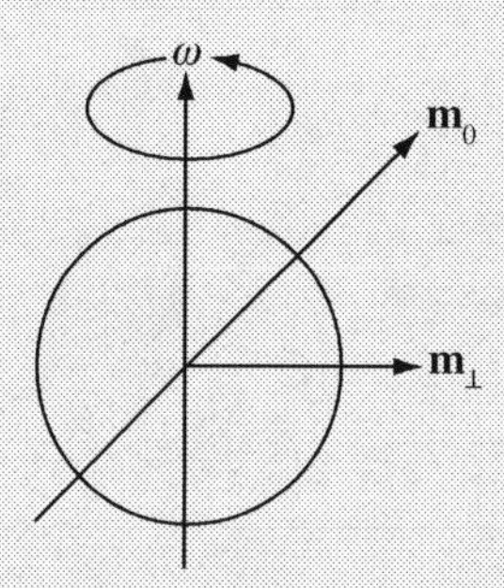

Figure 20.18: A pulsar modeled as a rotating neutron star. Only the component of the frozen-in magnetic moment $\mathbf{m}$ perpendicular to the rotation axis changes in time.

Solution:

(a) Far outside itself, we can replace any extended source having a magnetic dipole moment $\mathbf{m}(t)$ by a point source with the same moment. In Section 15.6.1, we calculated the rate of change of angular momentum in a volume V due to fields that pass through its bounding surface S to be

$$\frac{d\mathbf{L}}{dt} = \epsilon_0 \int_S d\mathbf{S} \cdot \{\mathbf{E}(\mathbf{r} \times \mathbf{E}) + c^2\mathbf{B}(\mathbf{r} \times \mathbf{B})\} + \tfrac{1}{2}\epsilon_0 \int_S d\mathbf{S} \times \mathbf{r}(E^2 + c^2B^2).$$

We choose S as a sphere centered on $\mathbf{m}(t)$ so $d\mathbf{S} = \hat{\mathbf{r}}R^2 d\Omega$. This immediately eliminates the second integral above. Following the logic of (20.80), it is necessary that the quantities $\mathbf{B}(\mathbf{r} \times \mathbf{B})$ and $\mathbf{E}(\mathbf{r} \times \mathbf{E})$ both decay as $1/R^2$ for the integral to be finite in the $R \to \infty$ limit. Moreover, the factors $d\mathbf{S} \cdot \mathbf{E}_{\rm rad}$ and $d\mathbf{S} \cdot \mathbf{B}_{\rm rad}$ vanish because radiation fields are transverse to $\hat{\mathbf{r}}$. Combining all this information tells us that, if $\mathbf{E}_2$ and $\mathbf{B}_2$ are *non-radiation* fields which decay as $1/R^2$,

$$\frac{d\mathbf{L}}{dt} = \epsilon_0 \int_S d\mathbf{S} \cdot \{\mathbf{E}_2(\mathbf{r} \times \mathbf{E}_{\rm rad}) + c^2\mathbf{B}_2(\mathbf{r} \times \mathbf{B}_{\rm rad})\} = \epsilon_0 \int_S (d\mathbf{S} \cdot \mathbf{E}_2)cr\mathbf{B}_{\rm rad} - \frac{1}{c\mu_0}\int_S (d\mathbf{S} \cdot \mathbf{B}_2)r\mathbf{E}_{\rm rad}.$$

The second equality above is a consequence of (20.108) and (20.109). $\mathbf{E}_2$ and $\mathbf{B}_2$ are the "intermediate-zone" fields of a point magnetic dipole analogous to the intermediate-zone fields of a point electric dipole in (20.74) and (20.75). At points away from the source, we can use $\mathbf{p} \to \mathbf{m}/c$ and the duality transformation (20.177) to read off the magnetic dipole fields from the electric dipole fields:

$$\mathbf{E}_2 = \frac{\mu_0}{4\pi}\hat{\mathbf{r}} \times \frac{\dot{\mathbf{m}}_{\rm ret}}{r^2} \qquad \mathbf{B}_2 = \frac{\mu_0}{4\pi c}\frac{3\hat{\mathbf{r}}(\hat{\mathbf{r}} \cdot \dot{\mathbf{m}}_{\rm ret}) - \dot{\mathbf{m}}_{\rm ret}}{r^2}.$$

$\mathbf{E}_2$ makes no contribution because $d\mathbf{S} \cdot \mathbf{E}_2 = 0$. Therefore, reading $\mathbf{E}_{\rm rad}$ from (20.174),

$$\frac{d\mathbf{L}}{dt} = -\frac{\mu_0}{8\pi^2 c^3}\int d\Omega(\hat{\mathbf{r}} \cdot \dot{\mathbf{m}}_{\rm ret})(\hat{\mathbf{r}} \times \ddot{\mathbf{m}}_{\rm ret}).$$

The integral to be done here appeared previously in (20.157). The answer is quoted on the left side of (20.197) below. Therefore,

$$\frac{d\mathbf{L}}{dt} = -\frac{\mu_0}{4\pi}\frac{2}{3c^3}(\dot{\mathbf{m}}_{\rm ret} \times \ddot{\mathbf{m}}_{\rm ret}).$$

(b) For the pulsar, the component of $\mathbf{m}_0$ parallel to the rotation axis does not change in time. The perpendicular component $\mathbf{m}_\perp$ rotates in a plane with frequency ω (see Figure 20.18). If the angular velocity of the pulsar is $\boldsymbol{\omega} = \omega\hat{\mathbf{z}}$, we may choose

$$\mathbf{m}_\perp(t) = \mathbf{m}_\perp e^{-i\omega t} = m_\perp(\hat{\mathbf{x}} + i\hat{\mathbf{y}})e^{-i\omega t}.$$

Therefore, ignoring the sign that tells us that the system *loses* angular momentum to radiation,

$$\left\langle\frac{d\mathbf{L}}{dt}\right\rangle = \frac{\mu_0}{6\pi c^3}\frac{1}{2}\mathrm{Re}\left\{\dot{\mathbf{m}}_\perp \times \ddot{\mathbf{m}}_\perp^*\right\} = \frac{\mu_0}{4\pi}\frac{2m_\perp^2\omega^3}{3c^3}\hat{\mathbf{z}}.$$

Equation (20.182) gives the time-averaged energy loss rate as

$$\left\langle\frac{dU}{dt}\right\rangle = \frac{\mu_0}{4\pi}\frac{2m_\perp^2\omega^4}{3c^3} = \omega\left\langle\frac{dL}{dt}\right\rangle.$$

20.7.3 Electric Quadrupole Radiation

The term *electric quadrupole radiation* is used when the symmetric piece of (20.170) is used from (20.145) to evaluate the angular distribution of radiated power from (20.143) and (20.144). Our focus is now the vector[19]

$$\boldsymbol{\alpha}_{\rm E2}(\mathbf{r}, t) = \frac{1}{2c}\frac{d^2}{dt^2}\int d^3r'\,[(\mathbf{r}' \cdot \hat{\mathbf{r}})\mathbf{j} + (\hat{\mathbf{r}} \cdot \mathbf{j})\mathbf{r}']. \tag{20.186}$$

As in (20.150), we use $\mathbf{j}(\mathbf{r}, t) = \sum_k q_k\boldsymbol{v}_k\delta(\mathbf{r} - \mathbf{r}_k)$ and $\rho(\mathbf{r}, t) = \sum_k q_k\delta(\mathbf{r} - \mathbf{r}_k)$ to write (20.186) in the form

$$\boldsymbol{\alpha}_{\rm E2}(\mathbf{r}, t) = \frac{1}{2c}\frac{d^2}{dt^2}\sum_k q_k[\boldsymbol{v}_k(\mathbf{r}_k \cdot \hat{\mathbf{r}}) + \mathbf{r}_k(\boldsymbol{v}_k \cdot \hat{\mathbf{r}})] = \frac{1}{2c}\frac{d^3}{dt^3}\sum_k q_k\mathbf{r}_k(\mathbf{r}_k \cdot \hat{\mathbf{r}}). \tag{20.187}$$

Switching back to continuous variables gives $\boldsymbol{\alpha}_{\rm E2}$ in terms of the electric quadrupole moment tensor $\mathbf{Q}$ of the charge distribution (see Section 4.4):

$$\boldsymbol{\alpha}_{\rm E2}(\mathbf{r}, t) = \frac{1}{c}\frac{d^3}{dt^3}\frac{1}{2}\left[\int d^3r'\,\rho(\mathbf{r}', t - r/c)\mathbf{r}'\mathbf{r}'\right] \cdot \hat{\mathbf{r}} = \frac{1}{c}\dddot{\mathbf{Q}}(t - r/c) \cdot \hat{\mathbf{r}}. \tag{20.188}$$

Substituting (20.188) into (20.107), (20.108), and (20.110) generates the fields and the angular distribution of power for electric quadrupole radiation. In detail,

$$\mathbf{B}_{\rm E2} = -\frac{\mu_0}{4\pi c^2}\frac{\hat{\mathbf{r}} \times (\dddot{\mathbf{Q}}_{\rm ret} \cdot \hat{\mathbf{r}})}{r} \tag{20.189}$$

$$\mathbf{E}_{\rm E2} = \frac{\mu_0}{4\pi c}\frac{(\hat{\mathbf{r}} \cdot \dddot{\mathbf{Q}}_{\rm ret} \cdot \hat{\mathbf{r}})\hat{\mathbf{r}} - \dddot{\mathbf{Q}}_{\rm ret} \cdot \hat{\mathbf{r}}}{r} \tag{20.190}$$

$$\left(\frac{dP}{d\Omega}\right)_{\rm E2} = \frac{\mu_0}{16\pi^2c^3}|\hat{\mathbf{r}} \times (\dddot{\mathbf{Q}}_{\rm ret} \cdot \hat{\mathbf{r}})|^2. \tag{20.191}$$

[19] See footnote at the beginning of Section 20.7.1 for the E2 notation.

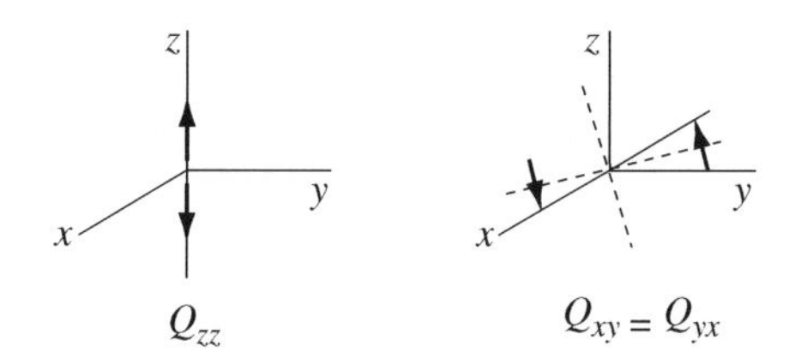

Figure 20.19: A point quadrupole is formed when two oppositely oriented dipoles are displaced toward one another and meet at the origin. Left: "axial" quadrupole where the dipoles point along $\hat{\mathbf{z}}$ and are displaced along $\hat{\mathbf{z}}$. Right: "lateral" quadrupole where the dipoles point along a line in the x-y plane and are displaced in the x-y plane at right angles to their orientation.

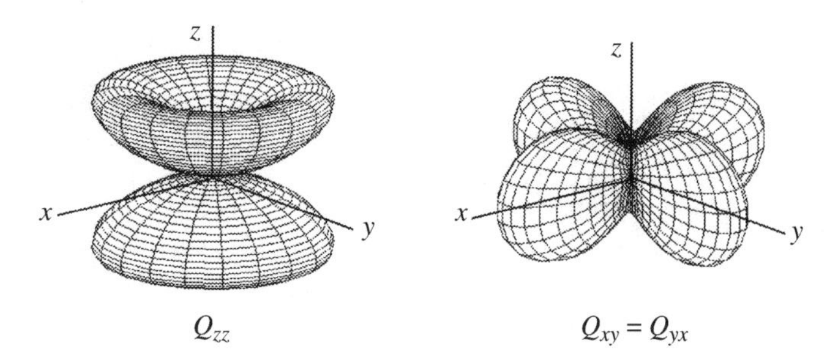

Figure 20.20: Angular distribution of power from point-like radiators. Left: axial quadrupole Q_{zz}. Right: lateral quadrupole $Q_{xy} = Q_{yx}$.

The angular distribution of quadrupole radiation depends on the details of $\mathbf{Q}$. For example, a point electric quadrupole can be regarded as the superposition of two identical but oppositely oriented point electric dipoles (see Section 4.2.1). The left side of Figure 20.19 is a cartoon of an "axial" quadrupole where only the component Q_{zz} is non-zero. In dyadic notation (see Section 1.8),

$$\mathbf{Q} = \hat{\mathbf{z}} Q_{zz} \hat{\mathbf{z}}. \tag{20.192}$$

Therefore, since $\hat{\mathbf{z}} = \hat{\mathbf{r}} \cos\theta - \hat{\boldsymbol{\theta}} \sin\theta$,

$$\hat{\mathbf{r}} \times (\mathbf{Q} \cdot \hat{\mathbf{r}}) = (\hat{\mathbf{r}} \times \hat{\mathbf{z}}) Q_{zz} (\hat{\mathbf{z}} \cdot \hat{\mathbf{r}}) = -Q_{zz} \sin\theta \cos\theta \hat{\boldsymbol{\phi}}. \tag{20.193}$$

The right side of Figure 20.19 shows a "lateral" quadrupole where the quadrupole dyadic reads

$$\mathbf{Q} = \hat{\mathbf{x}} Q_{xy} \hat{\mathbf{y}} + \hat{\mathbf{y}} Q_{yx} \hat{\mathbf{x}}. \tag{20.194}$$

Because $Q_{xy} = Q_{yx}$ [see (20.188)], the reader can confirm that

$$\hat{\mathbf{r}} \times (\mathbf{Q} \cdot \hat{\mathbf{r}}) = Q_{xy} \sin\theta [\cos\theta \sin(2\phi) \hat{\boldsymbol{\phi}} - \cos(2\phi) \hat{\boldsymbol{\theta}}]. \tag{20.195}$$

Figure 20.20 shows the angular distributions of radiation for these two cases calculated using (20.191).

The total power radiated by a quadrupole source is best computed by writing out (20.191) in Cartesian components:

$$|\hat{\mathbf{r}} \times (\ddot{\mathbf{Q}} \cdot \hat{\mathbf{r}})|^2 = \hat{r}_m \hat{r}_j \ddot{Q}_{im} \ddot{Q}_{ij} - \hat{r}_t \hat{r}_m \hat{r}_k \hat{r}_p \ddot{Q}_{km} \ddot{Q}_{tp}. \tag{20.196}$$

The formulae needed to integrate (20.196) over all angles are

$$\int d\Omega \, \hat{r}_m \hat{r}_j = \frac{4\pi}{3} \delta_{mj} \qquad \int d\Omega \, \hat{r}_t \hat{r}_m \hat{r}_k \hat{r}_p = \frac{4\pi}{15} (\delta_{tm} \delta_{kp} + \delta_{tk} \delta_{mp} + \delta_{tp} \delta_{mk}). \tag{20.197}$$

The integral on the left side of (20.197) is not new; we evaluated it following (20.157). The integral on the right side of (20.197) is similar.[20]

Collecting this information together and remembering that repeated indices are summed over x, y, and z, we find without difficulty that

$$P_{\rm E2}(t) = \int d\Omega \left(\frac{dP}{d\Omega}\right)_{\rm E2} = \frac{\mu_0}{20\pi c^3}\left[\dddot{Q}_{im}\dddot{Q}_{im} - \tfrac{1}{3}\dddot{Q}_{ii}\dddot{Q}_{mm}\right]_{\rm ret}. \tag{20.198}$$

The appearance in (20.198) of the square of the trace[21] of $\dddot{\mathbf{Q}}$ reminds us that a traceless electric quadrupole tensor $\boldsymbol{\Theta}$ can be defined with components (Section 4.4.1)

$$\Theta_{ij} = \tfrac{1}{2}\int d^3r\, \rho(\mathbf{r})(3r_ir_j - r^2\delta_{ij}) = 3Q_{ij} - Q_{kk}\delta_{ij}. \tag{20.199}$$

Substituting (20.199) into (20.198) gives the more compact formula

$$P_{\rm E2}(t) = \frac{\mu_0}{180\pi c^3}\left[\dddot{\Theta}_{im}\dddot{\Theta}_{im}\right]_{\rm ret}. \tag{20.200}$$

Some care is needed when performing time averages of $dP/d\Omega$ and the total power when a quadrupole source varies periodically in time. The time-averaging theorem (1.139) is not immediately useful when, as can happen, the time dependencies of the individual elements of $\mathbf{Q}$ are not exactly the same. In such cases, the time average must be performed explicitly. On the other hand, when $\dddot{\Theta} = i\omega^3\Theta$, we can use the ratio of (20.200) to (20.162) to estimate the relative importance of E1 and E2 radiated power when a single source radiates both simultaneously. The result,

$$\frac{\langle P_{\rm E2}\rangle}{\langle P_{\rm E1}\rangle} \sim \left(\frac{L}{\lambda}\right)^2, \tag{20.201}$$

shows that E1 radiation always dominates E2 radiation at long wavelength. The discussion surrounding (20.184) should be consulted to estimate the relative importance of $E2$ to $M1$ radiation.

Example 20.6 Show that quadrupole radiation dominates the long-wavelength emission from an isolated group of N moving particles when all the particles have the same charge-to-mass ratio q/m.

Solution: Quadrupole radiation dominates at long wavelength if there is no dipole radiation. There will be no electric dipole radiation if $\ddot{\mathbf{p}}(t) = 0$ and no magnetic dipole radiation if $\ddot{\mathbf{m}}(t) = 0$. As in (20.150), the total electric dipole moment is

$$\mathbf{p} = \sum_{k=1}^{N} q_k\mathbf{r}_k = \frac{q}{m}\sum_{k=1}^{N} m_k\mathbf{r}_k.$$

Therefore, $\ddot{\mathbf{p}} = (q/m)\sum_k m_k\ddot{\mathbf{r}}_k$. The sum is proportional to the acceleration of the center of mass, which is zero for an isolated system. The total magnetic dipole moment is

$$\mathbf{m} = \frac{1}{2}\sum_{k=1}^{N} q_k(\mathbf{r}_k\times\boldsymbol{v}_k) = \frac{q}{2m}\sum_{k=1}^{N} q_k(\mathbf{r}_k\times m_k\boldsymbol{v}_k) = \frac{q}{2m}\sum_{k=1}^{N}\mathbf{L}_k.$$

The sum is equal to the total orbital angular momentum, which is constant for an isolated system. Therefore, $\ddot{\mathbf{m}} = 0$.

[20] The delta function structure follows from the symmetry of the integral with respect to exchanging any two indices. The numerical pre-factor follows because the integral equals 4π when we set $t = m$, $k = p$, and sum over t and p.

[21] ${\rm Tr}\mathbf{A} = \sum_k A_{kk}$.

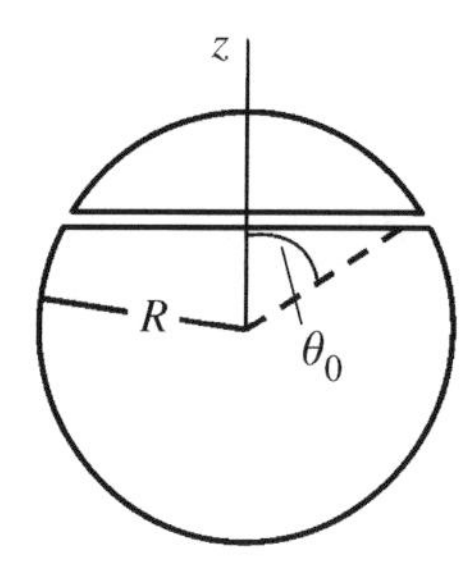

Figure 20.21: A perfectly conducting sphere with a gap at a fixed latitude. A voltage $V\cos\omega t$ is applied across the gap.

20.8 Spherical Multipole Radiation

In this section, we use outgoing spherical waves to construct a multipole decomposition for electromagnetic fields radiated by a compact source of time-dependent charge and current. For rotationally invariant sources, this representation is the natural generalization of the spherical expansions derived in Section 4.6 and Section 11.4.2 for electrostatic and magnetostatic sources, respectively. An important step is to find the spherical multipole moments that relate the amplitudes of the various outgoing multipole fields to the charge density $\rho(\mathbf{r}, t)$ and current density $\mathbf{j}(\mathbf{r}, t)$. We restrict ourselves throughout to time-harmonic sources that vary as $\exp(-i\omega t)$. The formidable algebra required to treat the case of full spherical symmetry motivates us to begin with an example with azimuthal symmetry. We then pass to the general case and derive expressions for the fields and the multipole moments. We compute the energy and angular momentum radiated by a single multipole field and indicate (in brief) how the general formalism applies to radiative transitions in atoms.

We learned in Section 16.8 that any solution of the free-space Maxwell equations can be written as a linear combination of a transverse electric (TE) wave (where $\mathbf{r}\cdot\mathbf{E} = 0$) and a transverse magnetic (TM) wave (where $\mathbf{r}\cdot\mathbf{B} = 0$). If $u(\mathbf{r}, t)$ and $w(\mathbf{r}, t)$ are arbitrary solutions of the scalar wave equation, the solutions found previously were

$$\begin{aligned}\mathbf{E} &= \mathbf{r}\times\nabla\frac{\partial u}{\partial t} + c\nabla\times(\mathbf{r}\times\nabla w)\\ c\mathbf{B} &= \mathbf{r}\times\nabla\frac{\partial w}{\partial t} - c\nabla\times(\mathbf{r}\times\nabla u).\end{aligned}\tag{20.202}$$

In this section, we restrict ourselves to harmonic waves and let $\partial/\partial t \to -i\omega$ in (20.202). The space parts of u and w then satisfy the Helmholtz equation with $\omega = ck$:

$$\left[\nabla^2 + k^2\right]\psi(\mathbf{r}) = 0.\tag{20.203}$$

For radiation problems, we will always choose the outgoing wave solutions to (20.203).

20.8.1 A Slotted Conducting Sphere

Figure 20.21 shows a conducting sphere of radius R separated into two parts by a narrow gap at a fixed latitude. Our interest is the radiation produced by the sphere when a voltage $V\cos\omega t$ is applied across the gap. Away from the gap, the tangential component of the electric field must vanish at the conductor surface. Just at the gap, we impose the electric field $\mathbf{E}(\theta) = (V/R)\delta(\theta-\theta_0)\hat{\boldsymbol{\theta}}$. By symmetry, the electric field produced by the sphere does not depend on ϕ and does not have a component in the $\hat{\boldsymbol{\phi}}$ direction. A glance at the curl operator in spherical coordinates shows that this will be true if $u(\mathbf{r}, t) = 0$ in (20.202). In other words, the fields produced by the slotted sphere are purely transverse magnetic (TM).

We showed in Section 16.8.2 that each elementary, outgoing wave solution of (20.203) is the product of a spherical Hankel function, $h_\ell^{(1)}(kr)$, and a spherical harmonic, $Y_{\ell m}(\theta, \phi) = Y_{\ell m}(\Omega)$. Here, the spherical harmonic may be replaced by a Legendre polynomial, $P_\ell(\theta)$. Therefore, with a minus sign inserted for convenience, the general solution for $w(\mathbf{r})$ is

$$w(r, \theta) = -\sum_{\ell=0}^{\infty} A_\ell h_\ell^{(1)}(kr) P_\ell(\cos\theta). \tag{20.204}$$

Substituting (20.204) into (20.202), we use the fact that

$$L^2 = -\frac{1}{\sin\theta}\frac{\partial}{\partial\theta}\left(\sin\theta\frac{\partial}{\partial\theta}\right) - \frac{1}{\sin^2\theta}\frac{\partial^2}{\partial^2\phi} \tag{20.205}$$

is the square of the angular momentum operator (with $\hbar = 1$) in quantum mechanics, and $L^2 P_\ell(\cos\theta) = \ell(\ell+1)P_\ell(\cos\theta)$, to deduce that

$$\begin{aligned} cB_\phi &= i\omega \sum_{\ell=1}^{\infty} A_\ell h_\ell^{(1)}(kr)\frac{d}{d\theta}P_\ell(\cos\theta) \\ E_r &= c\sum_{\ell=1}^{\infty} A_\ell \frac{\ell(\ell+1)}{r} h_\ell^{(1)}(kr) P_\ell(\cos\theta) \\ E_\theta &= c\sum_{\ell=1}^{\infty} A_\ell \frac{1}{r}\frac{d}{dr}\left[r h_\ell^{(1)}(kr)\right]\frac{d}{d\theta}P_\ell(\cos\theta). \end{aligned} \tag{20.206}$$

The identity (C.14) makes the orthogonality relation (C.16) useful for finding A_ℓ when we apply the boundary condition $E_\theta(R,\theta) = (V/R)\delta(\theta-\theta_0)$ to (20.206). The result is

$$A_\ell = -\frac{V}{c}\frac{2\ell+1}{2\ell(\ell+1)}\frac{\sin\theta_0 P_\ell^1(\cos\theta_0)}{\dfrac{d}{dr}\left[r h_\ell^{(1)}(kr)\right]_{r=R}}. \tag{20.207}$$

Inserting (20.207) into (20.206) gives fields which are exact at all distances. However, to find the angular distribution of radiated power, we need only one of the transverse fields, E_θ or H_ϕ, in the $r \to \infty$ limit. This, in turn, requires the asymptotic formula

$$\lim_{r\to\infty} h_\ell^{(1)}(kr) = \frac{1}{kr}\exp\{i[kr - \tfrac{1}{2}(\ell+1)\pi]\}. \tag{20.208}$$

Figure 20.22 shows the corresponding time-averaged angular distribution of power for $\lambda = 4R$ and $\lambda = R/2$. The first of these is the beginning of the long-wavelength regime and the radiation pattern exhibits the broadside behavior expected for dipole emission (see Figure 20.17). Many multipoles contribute to the shorter-wavelength case shown where the emission is nearly isotropic, albeit with significant lobes near $\theta = 0$ and $\theta = \pi$.

20.8.2 Multipole Expansions for $\mathbf{E}(\mathbf{r}, t)$ and $\mathbf{B}(\mathbf{r}, t)$

In this section, we derive a spherical multipole expansion for a general radiation field produced by arbitrary time-harmonic sources. We change notation slightly from the previous section and write the functions $u(\mathbf{r}, t)$ and $w(\mathbf{r}, t)$ that appear in (20.202) in the form

$$\omega u(\mathbf{r}, t) = \Lambda^{\mathrm{M}}(\mathbf{r})\exp(-i\omega t) \qquad \text{and} \qquad \omega w(\mathbf{r}, t) = \Lambda^{\mathrm{E}}(\mathbf{r})\exp(-i\omega t). \tag{20.209}$$

The assumed time dependence implies that $\Lambda^{\mathrm{M}}(\mathbf{r})$ and $\Lambda^{\mathrm{E}}(\mathbf{r})$ must be taken from among the solutions of the scalar Helmholtz equation (20.203). We conform with the physics literature of multipole fields

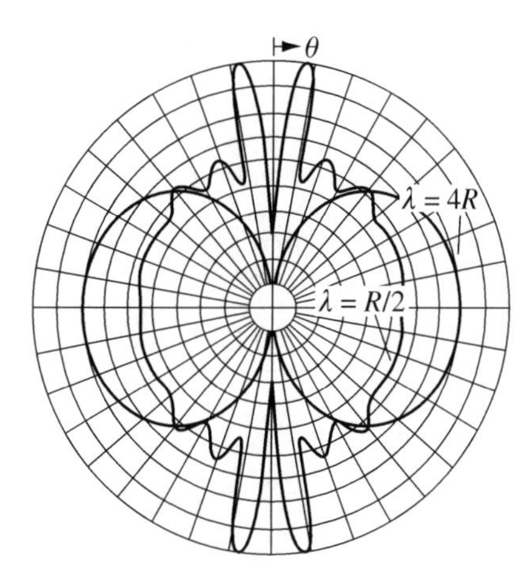

Figure 20.22: The time-averaged angular distribution of power radiated by the slotted sphere in Figure 20.21 with $\theta_0 = \pi/2$. Figure from Harrington (1961).

if we again introduce the quantum mechanical operator for angular momentum (with $\hbar = 1$),

$$\mathbf{L} = -i\mathbf{r} \times \nabla, \tag{20.210}$$

and write (20.202) in the form

$$\begin{aligned} \mathbf{E} &= \mathbf{L}\Lambda^{\mathrm{M}} + \frac{i}{k}\nabla \times \mathbf{L}\Lambda^{\mathrm{E}} \\ c\mathbf{B} &= \mathbf{L}\Lambda^{\mathrm{E}} - \frac{i}{k}\nabla \times \mathbf{L}\Lambda^{\mathrm{M}}. \end{aligned} \tag{20.211}$$

The reader can check that the fields in (20.211) satisfy $\nabla \cdot \mathbf{E} = 0$, $\nabla \cdot \mathbf{B} = 0$, and the two Maxwell curl equations,[22]

$$\mathbf{E} = \frac{i}{k}\nabla \times c\mathbf{B} \qquad \text{and} \qquad c\mathbf{B} = -\frac{i}{k}\nabla \times \mathbf{E}. \tag{20.212}$$

The key step is to expand $\Lambda^{\mathrm{M}}(\mathbf{r})$ and $\Lambda^{\mathrm{E}}(\mathbf{r})$ in (20.211) as sums of the elementary outgoing wave solutions of the Helmholtz equation (20.203) written in spherical coordinates. As noted just before (20.204), each of these elementary solutions is a product of a spherical Hankel function, $h_\ell^{(1)}(kr)$, and a spherical harmonic, $Y_{\ell m}(\theta, \phi) = Y_{\ell m}(\Omega)$. Therefore, if $\Lambda^{\mathrm{M}}_{\ell m}$ and $\Lambda^{\mathrm{E}}_{\ell m}$ are expansion coefficients with the dimensions of electric field, the fields in (20.211) take the spherical multipole form

$$c\mathbf{B}(\mathbf{r}) = \sum_{\ell=1}^{\infty}\sum_{m=-\ell}^{\ell} \Lambda^{\mathrm{E}}_{\ell m}\mathbf{L}h_\ell^{(1)}(kr)Y_{\ell m}(\Omega) - \frac{i}{k}\sum_{\ell=1}^{\infty}\sum_{m=-\ell}^{\ell} \Lambda^{\mathrm{M}}_{\ell m}\nabla \times \mathbf{L}h_\ell^{(1)}(kr)Y_{\ell m}(\Omega) \tag{20.213}$$

and

$$\mathbf{E}(\mathbf{r}) = \sum_{\ell=1}^{\infty}\sum_{m=-\ell}^{\ell} \Lambda^{\mathrm{M}}_{\ell m}\mathbf{L}h_\ell^{(1)}(kr)Y_{\ell m}(\Omega) + \frac{i}{k}\sum_{\ell=1}^{\infty}\sum_{m=-\ell}^{\ell} \Lambda^{\mathrm{E}}_{\ell m}\nabla \times \mathbf{L}h_\ell^{(1)}(kr)Y_{\ell m}(\Omega). \tag{20.214}$$

There are no $\ell = 0$ terms in (20.213) and (20.214) because $\mathbf{L}$ acts only on the angular variables and Y_{00} is a constant.[23] The *magnetic multipole moment* $\Lambda^{\mathrm{M}}_{\ell m}$ fixes the amplitude of each TE wave and the

[22] This requires the use of (20.203) and the operator identities $\nabla \cdot \mathbf{L} = 0$ and $\nabla \times (\nabla \times \mathbf{L}) = -\nabla^2\mathbf{L}$.

[23] This is consistent with Application 16.1 of Section 16.8.2.

electric multipole moment $\Lambda^{\rm E}_{\ell m}$ fixes the amplitude of each TM wave. The reason for these names will emerge in the next section.

20.8.3 Explicit Formulae for the Multipole Moments

Our task now is to connect the expansion coefficients $\Lambda^{\rm M}_{\ell m}$ and $\Lambda^{\rm E}_{\ell m}$ to the charge and current densities of the source. For this purpose, we take the scalar product of $\mathbf{r}$ with the two equations in (20.211). Because $\mathbf{r}\cdot\mathbf{L}=0$ and $\mathbf{r}\cdot\nabla\times\mathbf{L}=iL^2$ are operator identities, we find

$$\mathbf{r}\cdot c\mathbf{B}=\frac{1}{k}L^2\Lambda^{\rm M} \qquad \text{and} \qquad \mathbf{r}\cdot\mathbf{E}=-\frac{1}{k}L^2\Lambda^{\rm E}. \tag{20.215}$$

Equation (20.215) implies that $\mathbf{r}\cdot\mathbf{E}$ and $\mathbf{r}\cdot\mathbf{B}$ can be expanded in elementary solutions of the Helmholtz equation, just like Λ^M and $\Lambda^{\rm E}$. Indeed, because $L^2Y_{\ell m}(\Omega)=\ell(\ell+1)Y_{\ell m}(\Omega)$, the expansion coefficients are related by

$$\Lambda^{\rm M}_{\ell m}=\frac{k}{\ell(\ell+1)}(\mathbf{r}\cdot c\mathbf{B})_{\ell m} \qquad \text{and} \qquad \Lambda^{\rm E}_{\ell m}=-\frac{k}{\ell(\ell+1)}(\mathbf{r}\cdot\mathbf{E})_{\ell m}. \tag{20.216}$$

In light of (20.216), we focus on the connection between the expansion coefficients, $(\mathbf{r}\cdot\mathbf{B})_{\ell m}$ and $(\mathbf{r}\cdot\mathbf{E})_{\ell m}$, and the charge density $\rho(\mathbf{r},t)$ and current density $\mathbf{j}(\mathbf{r},t)$. This is not difficult to discover because $\mathbf{r}\cdot\mathbf{E}$ and $\mathbf{r}\cdot\mathbf{B}$ both satisfy inhomogeneous Helmholtz equations derived by specializing the inhomogeneous wave equations (20.4) and (20.5) to time-harmonic fields and using the identity

$$\mathbf{r}\cdot(\nabla^2\mathbf{F})=\nabla^2(\mathbf{r}\cdot\mathbf{F})-2\nabla\cdot\mathbf{F}. \tag{20.217}$$

These steps produce

$$\left[\nabla^2+k^2\right](\mathbf{r}\cdot\mathbf{B})=-\mu_0\,\mathbf{r}\cdot\nabla\times\mathbf{j} \tag{20.218}$$

and

$$\left[\nabla^2+k^2\right](\mathbf{r}\cdot\mathbf{E})=\frac{1}{\epsilon_0}\left[2\rho+\mathbf{r}\cdot\nabla\rho\right]-i\omega\mu_0\,\mathbf{r}\cdot\mathbf{j}. \tag{20.219}$$

We found the retarded solution to the inhomogeneous Helmholtz equation in Section 20.3.3. If

$$G_0(\mathbf{r},\mathbf{r}')=\frac{\exp(ik|\mathbf{r}-\mathbf{r}'|)}{4\pi|\mathbf{r}-\mathbf{r}'|} \tag{20.220}$$

is the free-space Green function, two applications of (20.53) give

$$\mathbf{r}\cdot\mathbf{B}=\mu_0\int d^3r'\,G_0(\mathbf{r},\mathbf{r}')\left[\mathbf{r}'\cdot\nabla'\times\mathbf{j}(\mathbf{r}')\right] \tag{20.221}$$

and

$$\mathbf{r}\cdot\mathbf{E}=\frac{1}{\epsilon_0 c}\int d^3r'\,G_0(\mathbf{r},\mathbf{r}')\left[ik\mathbf{r}'\cdot\mathbf{j}(\mathbf{r}')-c(2+\mathbf{r}'\cdot\nabla')\rho(\mathbf{r}')\right]. \tag{20.222}$$

We now write a spherical expansion for $G_0(\mathbf{r},\mathbf{r}')$ by combining (C.54) with the spherical harmonic addition theorem (4.79) to get[24]

$$G_0(\mathbf{r},\mathbf{r}')=ik\sum_{\ell=0}^{\infty}\sum_{m=-\ell}^{\ell}j_\ell(kr_<)h^{(1)}_\ell(kr_>)Y^*_{\ell m}(\Omega_<)Y_{\ell m}(\Omega_>). \tag{20.223}$$

[24] The inequality $r'<r$ is always true because we enclose the charge and current in a sphere and compute the fields outside that sphere only.

Spherical expansions for $\mathbf{r}\cdot\mathbf{B}$ and $\mathbf{r}\cdot\mathbf{E}$ follow by substituting (20.223) into (20.221) and (20.222). The coefficients $(\mathbf{r}\cdot\mathbf{B})_{\ell m}$ and $(\mathbf{r}\cdot\mathbf{E})_{\ell m}$ are the pre-factors that multiply $h_\ell^{(1)}(kr)Y_{\ell m}(\Omega)$ in each expression. Using these in (20.216) gives explicit expressions for the spherical multipole moments which appear in (20.213) and (20.215):

$$\Lambda^{\rm M}_{\ell m} = \frac{ick^2\mu_0}{\ell(\ell+1)}\int d^3r\; j_\ell(kr)Y^*_{\ell m}(\Omega)\,[\mathbf{r}\cdot\nabla\times\mathbf{j}(\mathbf{r})] \tag{20.224}$$

and

$$\Lambda^{\rm E}_{\ell m} = -\frac{ik^2}{\epsilon_0 c\,\ell(\ell+1)}\int d^3r\; j_\ell(kr)Y^*_{\ell m}(\Omega)\,[ik\mathbf{r}\cdot\mathbf{j}(\mathbf{r}) - c(2+\mathbf{r}\cdot\nabla)\rho(\mathbf{r})]\,. \tag{20.225}$$

The origin of the names "magnetic" and "electric" for the moments $\Lambda^{\rm M}_{\ell m}$ and $\Lambda^{\rm E}_{\ell m}$ becomes clear when we assume that the source has linear size a and pass to the long-wavelength ($ka \ll 1$) limit. Using $\lim_{x\to 0} j_\ell(x) = x^\ell/(2\ell+1)!!$ from (C.46), we find[25]

$$\lim_{ka\to 0}\Lambda^{\rm M}_{\ell m} = \frac{ick^{\ell+2}\mu_0}{\ell(2\ell+1)!!}M_{\ell m} \qquad\text{and}\qquad \lim_{ka\to 0}\Lambda^{\rm E}_{\ell m} = -\frac{ik^{\ell+2}}{4\pi\epsilon_0\ell(2\ell-1)!!}A_{\ell m}. \tag{20.226}$$

$\Lambda^{\rm M}_{\ell m}$ is proportional to the magnetostatic moment $M_{\ell m}$ defined in (11.66) and $\Lambda^{\rm E}_{\ell m}$ is proportional to the electrostatic moment $A_{\ell m}$ defined in (4.87). We conclude that slowly time-varying currents produce long-wavelength TE waves and slowly time-varying charges produce long-wavelength TM waves.

20.8.4 Electric Multipole Radiation of Energy

We consider here the radiation of energy from a single electric multipole source. From (20.213) and (20.214), the exact fields of interest are

$$c\mathbf{B} = \Lambda^{\rm E}_{\ell m}\mathbf{L}h_\ell^{(1)}(kr)Y_{\ell m}(\Omega) \qquad\text{and}\qquad \mathbf{E} = \Lambda^{\rm E}_{\ell m}\frac{i}{k}\nabla\times\mathbf{L}h_\ell^{(1)}(kr)Y_{\ell m}(\Omega). \tag{20.227}$$

Equation (20.227) simplifies when $r\to\infty$ because $\lim_{x\to\infty} h_\ell^{(1)}(x) = (-i)^{\ell+1}\exp(ix)/x$ [see (C.49)] and $\mathbf{L}$ operates only on the angular variables. The fields that result are tangential and fall off as $1/r$, as expected for radiation fields:

$$c\mathbf{B}_{\rm rad} = \Lambda^{\rm E}_{\ell m}(-i)^{\ell+1}\frac{\exp(ikr)}{kr}\mathbf{L}Y_{\ell m}(\Omega) \qquad\text{and}\qquad \mathbf{E}_{\rm rad} = c\mathbf{B}_{\rm rad}\times\hat{\mathbf{r}}. \tag{20.228}$$

Using (20.120), the time-averaged angular distribution of power radiated by the multipole field (20.228) is

$$\left\langle\frac{dP}{d\Omega}\right\rangle = \frac{r^2}{2\mu_0 c}\,|c\mathbf{B}_{\rm rad}|^2 = \frac{1}{2\mu_0 ck^2}\left|\Lambda^{\rm E}_{\ell m}\right|^2|\mathbf{L}Y_{\ell m}(\Omega)|^2\,. \tag{20.229}$$

We state without proof that the vector $\mathbf{L}Y_{\ell m}$ in (20.228) and (20.229) is a linear combination of the form[26]

$$\mathbf{L}Y_{\ell m} = \hat{\mathbf{z}}A_0Y_{\ell m} + (\hat{\mathbf{x}}+i\hat{\mathbf{y}})A_1Y_{\ell,m+1} + (\hat{\mathbf{x}}-i\hat{\mathbf{y}})A_{-1}Y_{\ell,m-1}, \tag{20.230}$$

where A_0, A_1, and A_{-1} are functions of ℓ and m. $\mathbf{L}Y_{\ell m}$ is also one member of a set of spherical tensors called *vector spherical harmonics*. The most sophisticated treatments of multipole radiation exploit

[25] The electric moment requires an integration by parts and the convention that $(-1)!! = 1$.

[26] See, for example, D.M. Brink and G.R. Satchler, *Angular Momentum*, 2nd edition (Clarendon Press, Oxford, 1968), Section 4.10.2.

the rotational properties of these tensors. Here, we note only that the angular distribution (20.229) is a different function of the polar angle θ (only) for each value of ℓ and $|m|$.[27]

The total radiated power is the integral of (20.229) over all angles. Because $\mathbf{L}$ is a Hermitian operator,

$$\int d\Omega\, [\mathbf{L}Y_{\ell m}]^* \cdot [\mathbf{L}Y_{\ell m}] = \int d\Omega \left[L^2 Y_{\ell m}\right]^* Y_{\ell m} = \ell(\ell+1). \tag{20.231}$$

Therefore, the rate at which an (ℓm) electric multipole radiates energy to infinity is

$$\left\langle \frac{dU}{dt} \right\rangle = \frac{\ell(\ell+1)}{2\mu_0 c k^2} \left|\Lambda^{\mathrm{E}}_{\ell m}\right|^2. \tag{20.232}$$

Substitution of Λ^{E} from (20.226) into (20.232) shows that the radiated power is proportional to $\omega^{2\ell+2}$ when the radiation wavelength is large compared to the source size. The radiated power decreases rapidly as the ℓ-order increases. The ω^4 and ω^6 dependence of the radiated power predicted for dipole ($\ell = 1$) and quadrupole ($\ell = 2$) sources agrees with the Cartesian multipole analysis of Section 20.7.

The results for the power radiated by a single magnetic multipole are similar. The vector spherical harmonic in (20.229) again gives the radiation pattern, but the polarization is rotated by 90°. When several multipoles radiate simultaneously, the reader can confirm that interference between multipoles occurs in the angular distribution of power, but not in the total radiated power.

20.8.5 Electric Multipole Radiation of Angular Momentum

The rate at which a single electric multiple source radiates angular momentum is fixed by the conservation law derived in Section 15.6.1. Equation (15.77) gives the rate at which electromagnetic fields carry angular momentum through a surface S which completely encloses the sources:

$$\frac{d\mathbf{L}}{dt} = \epsilon_0 \int_S d\mathbf{S} \cdot \left\{\mathbf{E}(\mathbf{r} \times \mathbf{E}) + c^2 \mathbf{B}(\mathbf{r} \times \mathbf{B})\right\} + \tfrac{1}{2}\epsilon_0 \int_S d\mathbf{S} \times \mathbf{r}(E^2 + c^2 B^2). \tag{20.233}$$

If S is an enormous sphere, the last integral in (20.233) vanishes because $d\mathbf{S} \times \mathbf{r} = 0$. For an electric multipole, the term proportional to $d\mathbf{S} \cdot \mathbf{B}$ also vanishes because $\mathbf{B}$ in (20.227) is transverse magnetic and has no radial component. Therefore, the time-averaging theorem (Section 1.6.3) gives

$$\left\langle \frac{d\mathbf{L}}{dt} \right\rangle = \tfrac{1}{2}\epsilon_0 \mathrm{Re} \int_S dS\, (\hat{\mathbf{r}} \cdot \mathbf{E}^*)(\mathbf{r} \times \mathbf{E}). \tag{20.234}$$

We have not written $\mathbf{E}_{\mathrm{rad}}$ in (20.234) because (20.228) shows that this field is tangential and thus makes no contribution to the radiated angular momentum. Accordingly, we return to the exact fields (20.227) and use the operator identity $\mathbf{r} \cdot \nabla \times \mathbf{L} = iL^2$ to find that the radial piece of the electric field is

$$\hat{\mathbf{r}} \cdot \mathbf{E} = -\frac{\ell(\ell+1)}{kr} \Lambda^{\mathrm{E}}_{\ell m} h^{(1)}_\ell(kr) Y_{\ell m}(\Omega). \tag{20.235}$$

The $x \to \infty$ form of $h^{(1)}_\ell(x)$ following (20.227) shows that the asymptotic behavior of (20.235) is $1/r^2$, which is just what we need to compensate the extra factor of $\mathbf{r}$ in the integrand of (20.234). We will also need the identity $\mathbf{r} \times (\nabla \times \mathbf{L}) = -\mathbf{L}(1 + \mathbf{r} \cdot \nabla)$, which shows that

$$\mathbf{r} \times \mathbf{E} = -\Lambda^{\mathrm{E}}_{\ell m} \frac{i}{k} \left(1 + r\frac{\partial}{\partial r}\right) h^{(1)}_\ell(kr) \mathbf{L} Y_{\ell m}(\Omega). \tag{20.236}$$

[27] Jackson (1999) gives polar plots of (20.229) for $\ell = 1$ and $\ell = 2$.

The final result follows by substituting (20.235) and (20.236) into (20.234), and using the asymptotic form of $h_\ell^{(1)}(kr)$ everywhere:

$$\left\langle \frac{d\mathbf{L}}{dt} \right\rangle = \frac{\epsilon_0 \ell(\ell+1)}{2k^3} \left| \Lambda^{\rm E}_{\ell m} \right|^2 \mathrm{Re} \int d\Omega\, Y^*_{\ell m} \mathbf{L} Y_{\ell m}. \tag{20.237}$$

We evaluate (20.237) using (20.230), the orthogonality of the spherical harmonics, and the familiar fact that $L_z Y_{\ell m} = m Y_{\ell m}$. The result is that only the z-component of (20.237) is non-zero:

$$\left\langle \frac{dL_z}{dt} \right\rangle = \frac{\epsilon_0 \ell(\ell+1)}{2k^3} \left| \Lambda^{\rm E}_{\ell m} \right|^2 m. \tag{20.238}$$

The explicit multiplicative factor of m in (20.238) shows that each of the $2\ell+1$ components of an ℓ-order electric multipole source radiates a different amount of (z-component of) angular momentum. Moreover, using the remarks following (20.230), we see that there is a one-to-one correspondence between the absolute magnitude of L_z radiated by an (ℓm) electric multipole and the angular pattern of power radiated by that multipole. The results for the angular momentum radiated by a single magnetic multipole are similar. However, interference occurs when several multipoles radiate simultaneously, e.g., the x- and y-components of (20.237) are no longer zero. This is so even if the simultaneously radiating multipoles belong to the same value of ℓ.

20.8.6 Radiation from Atoms and Nuclei

The main use of the classical multipole formalism developed in this section is to analyze the radiation emitted by atoms and nuclei. A first hint that our classical theory might describe radiation from a quantum system comes when we combine (20.232) with (20.238). The result is[28]

$$\frac{1}{m}\left\langle \frac{dL_z}{dt} \right\rangle = \frac{1}{\omega}\left\langle \frac{dU}{dt} \right\rangle. \tag{20.239}$$

Equation (20.239) has an immediate and familiar "photon" interpretation: if an (ℓm) electric multipole radiates energy in units of $\hbar\omega$, the z-component of the angular momentum must radiate in units of $m\hbar$. Beyond this, the transition to quantum mechanics obliges us to replace certain functions in our theory by their operator counterparts. The most important of these is the current density which appears in the radiation vector potential (20.118) and in the multipole moments (20.224) and (20.225).[29]

Textbooks of quantum mechanics define the current density operator for a point particle of charge q and mass m at position $\mathbf{s}$ as

$$\mathbf{j}_{\rm op}(\mathbf{r},\mathbf{s}) = \frac{q\hbar}{2mi}\left[\nabla_{\mathbf{s}}\delta(\mathbf{r}-\mathbf{s}) + \delta(\mathbf{r}-\mathbf{s})\nabla_{\mathbf{s}}\right]. \tag{20.240}$$

Radiation from an atom is associated with the transition of an electron from an initial state with wave function $\psi_I(\mathbf{s})$ to a final state with wave function $\psi_F(\mathbf{s})$. If the energy difference between the two states is $\hbar\omega$, the effective current density needed to describe this transition in our classical theory is the matrix element

$$\mathbf{j}(\mathbf{r}|\omega) = \int d^3s\, \psi^*_F(\mathbf{s})\mathbf{j}_{\rm op}(\mathbf{r},\mathbf{s})\psi_I(\mathbf{s}) = \frac{q\hbar}{2mi}\left[\psi^*_F(\mathbf{r})\nabla\psi_I(\mathbf{r}) - \psi_I(\mathbf{r})\nabla\psi^*_F(\mathbf{r})\right]. \tag{20.241}$$

Substituting the spatial Fourier transform of (20.241) into the next-to-rightmost member of (20.120) gives exactly the same angular distribution of radiated power as predicted by quantum mechanics for the $\psi_I \to \psi_F$ transition.[30]

[28] Compare with the last equation in Example 20.5.

[29] The charge density in (20.225) can be eliminated in favor of $\mathbf{j}$ using the continuity equation.

[30] See, e.g., Chapter 13 of G. Baym, *Lectures on Quantum Mechanics* (Benjamin/Cummings, Menlo Park, CA, 1969).

To include spin, we supplement the electron orbital current density (20.241) with a spin magnetization current density $\mathbf{j}_S(\mathbf{r}) = \nabla \times \mathbf{M}_S$ and use the Pauli spin matrix, $\boldsymbol{\sigma}$, to generalize the definition of spin magnetization offered in Section 13.2.1 to

$$\mathbf{M}_S = \frac{e\hbar}{2m}\langle\psi_F|\boldsymbol{\sigma}|\psi_I\rangle. \tag{20.242}$$

The matrix element notation in (20.242) indicates a sum over the two values of spin-$\frac{1}{2}$.

We now remind the reader that not all current distributions produce radiation [see the remarks following (20.111) and (20.120)]. The distribution (20.241) must have this property for some choices of ψ_I and ψ_F because (by construction) it reproduces the transition matrix element (and thus the selection rules) of the quantum theory. The angular momentum selection rules follow by conservation of angular momentum because the angular momentum carried away by the relevant multipole radiation field (see Section 20.8.5) must account for the difference in intrinsic angular momentum associated with $\psi_I(\mathbf{r})$ and $\psi_F(\mathbf{r})$. The parity selection rules can be deduced similarly because, like the initial and final state wave functions, the electric and magnetic multipole fields have definite parity.[31]

20.9 Radiation in Matter

Except for neutron stars, all matter consists of atoms (or ions and free electrons) distributed more or less densely in otherwise empty space. The bound or free electrons (and even the ions) can be set into accelerated motion in various ways. As a result, radiation is easily produced inside matter and easily propagates in matter before (possibly) exiting into the vacuum. The differences between radiation in matter and radiation in vacuum come from the influence of the index of refraction, $n(\omega)$, and its frequency dispersion. For example, the field vectors $\mathbf{E}_{\text{rad}}$ and $\mathbf{B}_{\text{rad}}$ need not be transverse to $\mathbf{r}$ when a time-dependent current density produces radiation in a medium which supports longitudinal waves. A low-density plasma is an example (see Section 18.5.3).

Unusual effects also occur when the velocity of a charged particle in a medium exceeds the phase speed of electromagnetic waves in that medium (Cherenkov radiation). We will discuss this effect separately in Chapter 23. In this section, we limit ourselves to a non-dispersive dielectric and demonstrate the power of the Hertz vector method by showing explicitly that the reflected and transmitted waves of elementary refraction theory are produced by the mutual interference of all the radiation fields produced when an incident wave enters and polarizes a dielectric medium.

20.9.1 The Ewald-Oseen Extinction Theorem

Figure 20.23 shows a plane wave in vacuum approaching a linear dielectric half-space at normal incidence. We have analyzed this situation previously using the Fresnel equations (Section 17.3.2) to find the amplitude of the reflected and transmitted waves. The wave equation in matter (Section 17.2) shows that the phase speed of the transmitted wave is c/n, where $n = \sqrt{1+\chi}$ is the index of refraction. In this section, we derive both results simultaneously by solving an inhomogeneous wave equation which has the electric polarization $\mathbf{P}(\mathbf{r}, t)$ as its source term. We will gain new insight by interpreting the solution as an example of the *Ewald-Oseen extinction theorem*. This is the statement that the time-dependent polarization of the dielectric medium induced by the incident wave is the source of a retarded field which plays several roles. Inside the medium, it cancels the incident wave and replaces it by the transmitted wave. Outside the medium, it is exactly the reflected wave.

[31] See Sources, References, and Additional Reading.

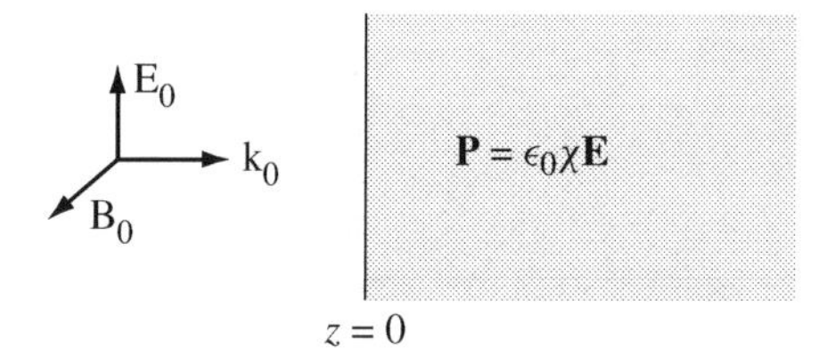

Figure 20.23: The half-space $z > 0$ is a linear dielectric medium. The dielectric develops a polarization $\mathbf{P} = \epsilon_0 \chi \mathbf{E}$ when a plane wave enters it at normal incidence from the vacuum half-space $z < 0$.

To prove these statements, we begin with the inhomogeneous wave equation satisfied by the electric Hertz vector, $\boldsymbol{\pi}_{\rm e}(\mathbf{r}, t)$ (see Section 20.2):

$$\nabla^2 \boldsymbol{\pi}_{\rm e} - \frac{1}{c^2}\frac{\partial^2 \boldsymbol{\pi}_{\rm e}}{\partial t^2} = -\frac{1}{\epsilon_0}\mathbf{P}. \tag{20.243}$$

Also from Section 20.2, we recall that $\boldsymbol{\pi}_{\rm e}$ determines the electric field in a non-magnetic medium from

$$\mathbf{E} = \nabla \times \nabla \times \boldsymbol{\pi}_{\rm e} - \frac{\mathbf{P}}{\epsilon_0}. \tag{20.244}$$

Matching conditions do not play a role if we seek a solution to (20.243) that is valid for *all* values of z in Figure 20.23.

Our strategy is to write $\boldsymbol{\pi}_{\rm e}$ as the sum of a homogeneous solution, $\boldsymbol{\pi}_0$, and a particular solution, $\boldsymbol{\pi}_{\rm p}$:

$$\boldsymbol{\pi}_{\rm e}(\mathbf{r}, t) = \boldsymbol{\pi}_0(\mathbf{r}, t) + \boldsymbol{\pi}_{\rm p}(\mathbf{r}, t). \tag{20.245}$$

The homogeneous ($\mathbf{P} = 0$) solution accounts for the incident wave propagating in the positive z-direction with wave vector $k_0 = \omega/c$ and amplitude $\mathbf{E}_0 \perp \hat{\mathbf{z}}$. Accordingly,

$$\mathbf{E}_0 \exp[i(k_0 z - \omega t)] = \nabla \times \nabla \times \boldsymbol{\pi}_0. \tag{20.246}$$

The particular solution accounts for the field produced by the polarization of the medium. This is the retarded integral (20.61), which we repeat here for convenience:

$$\boldsymbol{\pi}_{\rm p}(\mathbf{r}, t) = \frac{1}{4\pi\epsilon_0}\int d^3r' \, \frac{\mathbf{P}(\mathbf{r}', t - |\mathbf{r} - \mathbf{r}'|/c)}{|\mathbf{r} - \mathbf{r}'|}. \tag{20.247}$$

To find the physical solution of interest, we evaluate (20.247) for a plane wave-type polarization, $\mathbf{P}(\mathbf{r}, t) = \mathbf{P}\exp[i(kz - \omega t)]$, and require consistency between the total electric field inside the medium and the polarization which contributes to it. The first step leads to

$$\boldsymbol{\pi}_{\rm p}(\mathbf{r}, t) = \frac{\mathbf{P}\exp[i(kz - \omega t)]}{4\pi\epsilon_0}\int_0^\infty dz' \exp[ik(z' - z)] \int_{-\infty}^{\infty} dx' \int_{-\infty}^{\infty} dy' \frac{\exp(ik_0|\mathbf{r} - \mathbf{r}'|)}{|\mathbf{r} - \mathbf{r}'|}. \tag{20.248}$$

The integrand of the x' and y' integrals is azimuthally symmetric and thus produces a factor of 2π when performed in cylindrical coordinates. Changing variables to $R^2 = \rho^2 + |z' - z|^2$ and using a convergence factor λ to regularize the final integral over R gives

$$2\pi \int_0^\infty d\rho \rho \frac{\exp(ik_0 R)}{R} = \lim_{\lambda \to 0} 2\pi \int_{|z'-z|}^{\infty} dR \exp(ik_0 R)\exp(-\lambda R) = \frac{2\pi i}{k_0}\exp[i(k_0|z' - z|)]. \tag{20.249}$$

Therefore,

$$\boldsymbol{\pi}_{\mathrm{p}}(z,t) = i\frac{\mathbf{P}\exp[i(kz-\omega t)]}{2\epsilon_0 k_0}\int_0^{\infty} dz' \exp[ik(z'-z)]\exp(ik_0|z-z'|). \tag{20.250}$$

The integral over z' in (20.250) must be done separately for $z<0$ and for $z>0$. We find

$$\boldsymbol{\pi}_{\mathrm{p}}(z,t) = -\frac{\mathbf{P}\exp(-i\omega t)}{2\epsilon_0 k_0}\begin{cases} \dfrac{1}{k+k_0}\exp(-ik_0 z) & z<0, \\ \dfrac{2k_0}{k_0^2-k^2}\exp(ikz)+\dfrac{1}{k-k_0}\exp(ik_0 z) & z>0. \end{cases} \tag{20.251}$$

Nominally, the electric field follows from (20.244), (20.245), (20.246), and (20.251). However, it is convenient to use (20.3) and (20.12) to eliminate the double curl and Laplacian operators acting on $\boldsymbol{\pi}_{\mathrm{p}}$. Moreover, $\nabla\cdot\boldsymbol{\pi}_{\mathrm{p}}=0$ because $\mathbf{P}\cdot\hat{\mathbf{z}}=0$. Therefore,

$$\mathbf{E}(z,t) = \mathbf{E}_0\exp[i(k_0 z-\omega t)] + k_0^2\boldsymbol{\pi}_{\mathrm{p}}(z,t). \tag{20.252}$$

Substituting (20.251) into (20.252) for $z<0$ gives

$$\mathbf{E}(z<0,t) = \mathbf{E}_0\exp[i(k_0 z-\omega t)] - \frac{\mathbf{P}}{2\epsilon_0}\frac{k_0}{k+k_0}\exp[-i(k_0 z+\omega t)]. \tag{20.253}$$

Doing the same for $z>0$ gives

$$\mathbf{E}(z>0,t) = \mathbf{E}_0\exp[i(k_0 z-\omega t)] - \frac{\mathbf{P}}{2\epsilon_0}\frac{k_0}{k-k_0}\exp[i(k_0 z-\omega t)] - \frac{\mathbf{P}}{\epsilon_0}\frac{k_0^2}{k_0^2-k^2}\exp[i(kz-\omega t)]. \tag{20.254}$$

Focus now on the dielectric half-space and set $\mathbf{E}(z>0,t)=\mathbf{E}\exp[i(kz-\omega t)]$ on the left side of (20.254) and $\mathbf{P}=\epsilon_0\chi\mathbf{E}$ on the right side of (20.254). The requirement that the coefficients of $\exp(ik_0 z)$ and $\exp(ikz)$ agree on both sides of this equation generates two self-consistency conditions,

$$1 = -\chi\frac{k_0^2}{k_0^2-k^2} \qquad \text{and} \qquad \mathbf{E}_0 = \frac{1}{2}\chi\frac{k_0}{k-k_0}\mathbf{E}. \tag{20.255}$$

The left side of (20.255) determines the propagation wave vector in the medium as

$$k = \sqrt{1+\chi}\,k_0 = nk_0. \tag{20.256}$$

The right side of (20.255) determines the polarization amplitude as

$$\mathbf{P} = \epsilon_0\chi\mathbf{E} = 2\epsilon_0(n-1)\mathbf{E}_0. \tag{20.257}$$

Substituting (20.256) and (20.257) into (20.253) gives the final result for the field in the vacuum as

$$\mathbf{E}(z<0,t) = \mathbf{E}_0\exp[i(k_0 z-\omega t)] - \left(\frac{n-1}{n+1}\right)\mathbf{E}_0\exp[-i(k_0 z+\omega t)]. \tag{20.258}$$

Making the same substitutions in (20.254) gives the final result for the field in the dielectric:

$$\mathbf{E}(z>0,t) = \left(\frac{2}{n+1}\right)\mathbf{E}_0\exp[i(nk_0 z-\omega t)]. \tag{20.259}$$

The electric fields (20.258) and (20.259) illustrate the Ewald-Oseen extinction theorem, which states that there are three parts to the field produced by the dynamic polarization induced in a dielectric by an incident electromagnetic wave. One part is a "reflected" wave which propagates in the vacuum with speed c. Another part annuls the incident wave inside the medium by destructive interference. The third part is a "transmitted" wave which propagates in the medium with phase speed c/n. The amplitudes of the reflected and transmitted waves are exactly those predicted by a conventional Fresnel analysis which matches solutions of different wave equations at $z = 0$. The sheer implausibility of our conclusion motivates us to repeat it. The time-dependent polarization induced in a dielectric medium by an incident plane launches retarded waves from every point in the medium. These waves have just the right amplitude and phase for their superposition to annihilate the incident wave and create both the usual transmitted and reflected waves of Fresnel theory.

Sources, References, and Additional Reading

The quotation at the beginning of the chapter is taken from

H. Hertz, *Electric Waves* (Macmillan, London, 1893), Chapter IX. Reproduced with permission of Palgrave Macmillan.

Section 20.1 Three good treatments of retardation and radiation are

R.H. Good, Jr. and T.J. Nelson, *Classical Theory of Electric and Magnetic Fields* (Academic, New York, 1971).

E.J. Konopinski, *Electromagnetic Fields and Relativistic Particles* (McGraw-Hill, New York, 1981).

J. Schwinger, L.L. DeRaad, Jr., K.A. Milton, and W.-Y. Tsai, *Classical Electrodynamics* (Perseus, Reading, MA, 1998).

Section 20.2 The sources for Application 20.1 and Example 20.1 are, respectively,

H.A. Lorentz, *The Theory of Electrons*, 2nd edition (Dover, New York, 1952), Section 26.

J.D. Templin, "Exact solution to the field equations in the case of an ideal, infinite solenoid", *American Journal of Physics* **63**, 916 (1995).

Section 20.3 The phrase "arrow of time" is often used for the apparent breaking of time-reversal symmetry in thermodynamic and electrodynamic phenomena. An illuminating one-dimensional model calculation is

A.D. Boozer, "Retarded potentials and the radiative arrow of time", *European Journal of Physics* **28**, 1131 (2007).

Hansen and Yaghjian discuss a "causality trick" where the advanced solution can be useful (for numerical reasons) to solve propagation problems when V is finite. *Wheeler and Feynman* famously studied radiation in the $V \to \infty$ limit using a linear combination of $\psi_{\rm ret}$ and $\psi_{\rm adv}$ with an "absorbing boundary condition" at infinity. As it turns out, the requirements of their model do not accord with contemporary cosmological ideas.

T.B. Hansen and A.D. Yaghjian, *Plane Wave Theory of Time-Domain Fields* (IEEE Press, New York, 1999).

J.A. Wheeler and R.P. Feynman, "Interaction of the absorber as the mechanism for radiation", *Reviews of Modern Physics* **17**, 157 (1945).

The one-page Ritz-Einstein paper mentioned at the end of Section 20.3.2 is

W. Ritz and A. Einstein, "On the current state of the radiation problem", *Physikalische Zeitschrift* **10**, 323 (1909).

The original, frequency-domain derivation of the Schott formulae appears in Chapter II of

G.A. Schott, *Electromagnetic Radiation* (Cambridge University Press, Cambridge, 1912).

The time-domain version of the Schott formulae derived in the text appears in, e.g.,

O.D. Jefimenko, *Electricity and Magnetism* (Appleton-Century-Crofts, New York, 1966).

P.D. Clemmow and J.P. Dougherty, *Electrodynamics of Particles and Plasmas* (Addison-Wesley, Reading, MA, 1969).

Section 20.4 Our discussion of (20.77) follows *Mandel*. Example 20.3 was taken from *Nevesskii*.

L. Mandel, "Energy flow from an atomic dipole in classical electrodynamics", *Journal of the Optical Society of America* **62**, 1101 (1972).

N. Ye. Nevesskii, "Electromagnetic fields of current structures", *Electrical Technology* **4**, 141 (1994).

Section 20.5 The "induced EMF" method to calculate the power radiated by a time-dependent current source was introduced by the French-American polymath Léon Brillouin during a period when he lectured on radio science at the École Supérieure d'Electricité. He subsequently made many contributions to electromagnetic theory, optics, quantum mechanics, solid state theory, and information theory.

L. Brillouin, "On the origin of radiation resistance", *Radioelectricité* **3**, 147 (1922).

Our treatment of Hertz' analysis of the "birth of radiation" benefitted from

G. Scharf, *From Electrostatics to Optics* (Springer, Berlin, 1994), Section 3.4.

For an appreciation of the life and work of Joseph Larmor, see

A. Warwick, "Frequency, theorem and formula: Remembering Joseph Larmor in electromagnetic theory", *Notes and Records of the Royal Society of London* **47**, 49 (1993).

Section 20.6 Our frequency-domain analysis, including Figure 20.11, comes from *Balanis*. Our time-domain discussion, including Figure 20.12 and Figure 20.13, comes from *Smith*.

C. Balanis, *Antenna Theory: Analysis and Design*, 3rd edition (Wiley, New York, 2005).

G.S. Smith, "Teaching antenna radiation from a time-domain perspective", *American Journal of Physics* **69**, 288 (2001). Reprinted with permission. © 2001 American Association of Physics Teachers.

de Hoop uses electromagnetic theory to prove the equivalence of the broadcast and receiving patterns of an antenna. *Rohlfs* presents an alternative proof based on thermodynamic considerations. *Smith* explains the physics.

A.T. de Hoop, "A reciprocity relation between the transmitting and receiving properties of an antenna", *Applied Scientific Research* **19**, 90 (1968).

K. Rohlfs, *Tools of Radio Astronomy* (Springer, Berlin, 1986), Section 4.5.3.

G.S. Smith, "Teaching antenna reception and scattering from a time-domain perspective", *American Journal of Physics* **70**, 829 (2002).

The phased antenna array used to detect the first pulsar is described in the article which earned Antony Hewish a share of the 1974 Nobel Prize:

A. Hewish, S.J. Bell, J.D.H. Pilkington, P.F. Scott, and R.A. Collins, "Observation of a rapidly pulsating radio source", *Nature* **217**, 709 (1968).

Section 20.7 Application 20.3, Example 20.4, and Example 20.6 come, respectively, from

E. Collett, "The description of polarization in classical physics", *American Journal of Physics* **36**, 713 (1968).

J. Schwinger, L.L. DeRaad, Jr., K.A. Milton, and W.-Y. Tsai, *Classical Electrodynamics* (Perseus, Reading, MA, 1998), Chapter 33.

L.D. Landau and E.M. Lifshitz, *The Classical Theory of Fields*, 2nd edition (Addison-Wesley, Reading, MA, 1962), Section 71.

Two quite different approaches to the complete Cartesian multipole expansion are

C. Vrejoiu, "Electromagnetic multipoles in Cartesian coordinates", *Journal of Physics A* **35**, 9911 (2002).

N. Kemmer, "Elementary derivation of general multipole moments for classical charge and current distributions" in *Quanta: Essays in Theoretical Physics Dedicated to Gregor Wentzel*, edited by P.G.O. Freund, C.J. Goebel, and Y. Nambu (University Press, Chicago, IL, 1970), pp. 1-12.

Section 20.8 The slotted-sphere problem discussed in Section 20.8.1 comes from *Karr*. The accompanying Figure 20.22 was reproduced from *Harrington*.

P.R. Karr, "Radiation properties of a spherical antenna as a function of the location of the driving force", *Journal of Research of the National Bureau of Standards* **46**, 422 (1951).

R.F. Harrington, *Time-Harmonic Electromagnetic Fields* (McGraw-Hill, New York, 1961).

We follow *Gray* for our general discussion of spherical multipole radiation. *Blatt and Weisskopf* and *Konopinski* (see Section 20.1 above) treat this subject using vector spherical harmonics. This is the method of choice for rotationally invariant quantum systems. *Jackson* discusses the radiation from atoms and molecules from this point of view.

C.G. Gray, "Multipole expansions of electromagnetic fields using Debye potentials", *American Journal of Physics* **46**, 169 (1978).

J.M. Blatt and V.F. Weisskopf, *Theoretical Nuclear Physics* (Wiley, New York, 1952).
J.D. Jackson, *Classical Electrodynamics*, 3rd edition (Wiley, New York, 1999).

Section 20.9 Two complementary treatments of radiation in dispersive matter are

V.L. Ginzburg, *Theoretical Physics and Astrophysics* (Pergamon, Oxford, 1979).
D.B. Melrose and R.C. McPhedran, *Electromagnetic Processes in Dispersive Matter* (Cambridge University Press, Cambridge, 1991).

Our Hertz vector approach to the extinction theorem was inspired by

R.K. Wangsness, "Effect of matter on the phase velocity of an electromagnetic wave", *American Journal of Physics* **49**, 950 (1981).
J.J. Sein, "Solutions to time-harmonic Maxwell equations with a Hertz vector", *American Journal of Physics* **57**, 834 (1989).

Problems

20.1 Poynting Flux above a Thunderstorm A lighting strike associated with a thunderstorm acts very much like a broadband antenna. Explain why data from airplane-borne electric and magnetic field sensors flown immediately above (in the near zone of) such storms reveal Poynting fluxes in the $\hat{\boldsymbol{\rho}}$ direction (with respect to the $\hat{\mathbf{z}}$-direction of the vertical lighting strike) which increase linearly with frequency between 100 Hz and 10 kHz. Hint: Thunderclouds rise to about 50,000 feet above sea level.

20.2 Fields from an Alternating Current in an Ohmic Wire An infinitely long straight wire on the z-axis has a circular cross section and obeys $\mathbf{j}(\omega) = \sigma_0 \mathbf{E}(\omega)$ for all $\rho \leq a$. After initial transients, one finds the charge density $\rho(\mathbf{r}, t) \equiv 0$ and the current $I(t) = I_0 \cos \omega t$ everywhere inside the wire.

(a) Solve an appropriate Helmholtz equation and find the exact $\mathbf{E}(\mathbf{r}, t)$ inside the wire. Express the amplitude of the field in terms of I_0.
(b) Solve an appropriate Helmholtz equation and find $\mathbf{E}$ and $\mathbf{B}$ exactly outside the wire.
(c) Use Poynting's theorem to show that the normal component of the time-averaged Poynting vector $\langle \mathbf{S} \rangle$ evaluated on any cylindrical surface concentric with the wire always points toward the z-axis.
(d) Use the Poynting vector to calculate the rate at which energy is lost to ohmic heating per unit length of wire.
(e) Use the Poynting vector to calculate the rate at which energy is lost to radiation per unit length of wire. How is this result consistent with conservation of energy and the answer to part (c)?

20.3 Free-Space Green Function in Two Dimensions Confirm by direct substitution into the defining equation that the free-space Green function for the Helmholtz equation in two-dimensional plane polar coordinates $\boldsymbol{\rho} = (\rho, \phi)$ is

$$G_0(\boldsymbol{\rho}, \boldsymbol{\rho}') = \frac{i}{4} H_0^{(1)}(k|\boldsymbol{\rho} - \boldsymbol{\rho}'|),$$

where $H_0^{(1)}(x)$ is the zero-order Hankel function of the first kind.

20.4 The Method of Descent The Green function $G(x, y, z, t > 0) = \delta(t - r/c)/4\pi r$ is a solution of the inhomogeneous wave equation in three space dimensions with the source term $-\delta(x)\delta(y)\delta(z)\delta(t)$.

(a) Show that $G_2(x, y, t) = \int_{-\infty}^{\infty} dz\, G(x, y, z, t)$ is a solution of the inhomogeneous wave equation in two space dimensions with the source term $-\delta(x)\delta(y)\delta(t)$.
(b) Evaluate the integral in part (a) to find $G_2(\rho, t)$ explicitly, where $\rho^2 = x^2 + y^2$.

20.5 Retarded Fields from Non-Retarded Potentials The scalar and vector potentials in the Coulomb gauge are

$$\varphi_C(\mathbf{r}, t) = \frac{1}{4\pi\epsilon_0} \int d^3r' \frac{\rho(\mathbf{r}', t)}{|\mathbf{r} - \mathbf{r}'|} \qquad \text{and} \qquad \mathbf{A}_C(\mathbf{r}, t) = \frac{\mu_0}{4\pi} \int d^3r' \frac{\mathbf{j}_\perp(\mathbf{r}, t - |\mathbf{r} - \mathbf{r}'|/c)}{|\mathbf{r} - \mathbf{r}'|},$$

where

$$\mathbf{j}_{\perp}(\mathbf{r}, t) = \nabla \times \frac{1}{4\pi} \int d^3r' \frac{\nabla' \times \mathbf{j}(\mathbf{r}', t)}{|\mathbf{r} - \mathbf{r}'|}.$$

(a) The scalar potential $\varphi_C(\mathbf{r}, t)$ is not retarded because it depends on $\rho(\mathbf{r}', t)$. The vector potential $\mathbf{A}_C(\mathbf{r}, t)$ looks retarded, because it depends on $\mathbf{j}_{\perp}(\mathbf{r}', t - |\mathbf{r} - \mathbf{r}'|/c)$, but it is not. Use the properties of the transverse current density, $\mathbf{j}_{\perp}$, to explain why at least some contributions to $\mathbf{A}_C$ are not retarded.

(b) Show by direct calculation that $\mathbf{A}_C(\mathbf{r}, t)$ nevertheless produces a properly retarded magnetic field, $\mathbf{B}(\mathbf{r}, t)$. Hint: Use an extra integration over t' and a delta function $\delta(t' - t + |\mathbf{r} - \mathbf{r}'|/c)$ to impose the retardation of $\mathbf{j}_{\perp}$.

(c) Exploit the Ampère-Maxwell equation to derive a retarded expression for $\mathbf{E}(\mathbf{r}, t)$.

20.6 Radiation from a Magnetized Electron Gas A classical electron gas with number density n_0 exhibits a Maxwell velocity distribution at temperature T. In the presence of a uniform magnetic field $\mathbf{B}_0$, the gas emits radiation at a wavelength which is much larger than the mean separation between the electrons. Find the radiated power per unit volume.

20.7 Energy Flow from a Point Electric Dipole A point electric dipole with moment $\mathbf{p}(t)$ has a fixed position in space. Show that the rate at which energy flows through a spherical surface of radius R centered at the dipole is

$$\frac{dU}{dt} = \frac{2}{3}\frac{1}{4\pi\epsilon_0}\left[\frac{d}{dt}\left\{\frac{p^2}{2R^3} + \frac{p\dot{p}}{cR^2} + \frac{\dot{p}^2}{c^2R}\right\}_{\text{ret}} + \frac{\ddot{p}^2_{\text{ret}}}{c^3}\right].$$

20.8 A Point Charge Blinks On A charge density $\rho(\mathbf{r}, t) = q(t)\delta(\mathbf{r})$ where $q(t) = 0$ for $t < 0$ and $q(t) = q$ for $t > \tau$.

(a) Calculate $\mathbf{E}(\mathbf{r}, t)$ and $\mathbf{B}(\mathbf{r}, t)$ using symmetry and elementary methods.

(b) Calculate $\mathbf{E}(\mathbf{r}, t)$ and $\mathbf{B}(\mathbf{r}, t)$ from the Coulomb gauge scalar and vector potentials.

(c) Calculate $\mathbf{E}(\mathbf{r}, t)$ and $\mathbf{B}(\mathbf{r}, t)$ from the Lorenz gauge scalar and vector potentials. As an intermediate step, you will prove that

$$\int d^3s \, \frac{\delta(\tau - s/c)}{|\mathbf{r} - \mathbf{s}|} = 4\pi\left[\frac{c^3\tau^2}{r}\Theta(r - c\tau)\Theta(\tau) + c^2\tau\Theta(c\tau - r)\right].$$

Hint: Use an extra integration over t' and a delta function $\delta(t' - t + |\mathbf{r} - \mathbf{r}'|/c)$ to impose retardation of the vector potential in part (c).

20.9 The Birth of Radiation The text shows that the electric field of a point electric dipole at the origin with moment $\mathbf{p}(t) = p(t)\hat{\mathbf{z}}$ produces an electric field (away from the source) in cylindrical coordinates of the form

$$\mathbf{E}(\rho, z, t) = \frac{1}{4\pi\epsilon_0}\left[\frac{1}{\rho}\frac{dR}{d\rho}\hat{\mathbf{z}} - \frac{1}{\rho}\frac{dR}{dz}\hat{\boldsymbol{\rho}}\right],$$

where

$$R(\rho, z, t) = -\rho\frac{\partial}{\partial\rho}\left[\frac{p(t - r/c)}{r}\right].$$

The family of curves $R(\rho, z) = \textit{const.} = R_0$ may be interpreted as electrical field lines.

(a) Show that the points (ρ, z) of null (zero) electric field where the field lines pinch off and detach from the source at time t satisfy

$$z = 0 \qquad\qquad \frac{R_0}{\rho} + \frac{\ddot{p}(t - \rho/c)}{c^2} = 0.$$

(b) Assume that $p(t) = p_0 \cos \omega t$ and show that only those field lines with $|R_0| < 2\omega p_0/c\sqrt{3}$ can detach from the source. Check that the detachment-point solutions correspond to $\rho > 0$.

20.10 An Electrically Short Antenna Consider the time-harmonic dipole antenna with imposed current $I(z) = I_0 \sin(kd - k|z|)$ discussed in Section 20.6.1 of the text.

(a) Evaluate the time-averaged angular distribution of power in the limit when the radiation wavelength is very large compared to the length of the antenna. Such an antenna is called "electrically short".
(b) Compute the exact electric dipole moment of the antenna. Take the long-wavelength limit and use this moment to evaluate the time-averaged angular distribution of power radiated by a point electric dipole. Compare with part (a).
(c) In the literature, $\langle dP/d\Omega \rangle$ is often quoted using the current at the input point, $I(0) = I(z = 0)$, rather than the peak current I_0. Show that doing this changes the explicit wavelength dependence into the dependence found for $\langle dP/d\Omega \rangle$ for a time-harmonic point electric dipole modeled as two point charges $\pm q(t)$ at $z = \pm d$ connected by a *spatially uniform* current $I(0)$.

20.11 The Time-Domain Electric Field of a Dipole Antenna Derive the expression (20.131) for $\mathbf{E}_{\rm tot}(\mathbf{r}, t)$, the total, time-domain electric field produced by a dipole antenna synthesized from four traveling current waves $I_S(t)$.

20.12 Radiation Recoil

(a) Explain why a localized (and entirely classical) source of charge and current does not recoil when it emits dipole radiation.
(b) Is recoil ever possible for a classical radiation source? If not, explain why not. If so, give an example.

20.13 Non-Radiating Sources Suppose $\mathbf{f}(\mathbf{r})$ is a localized vector function and $\mathbf{j}(\mathbf{r}, t) = \mathbf{j}(\mathbf{r}|\omega) \exp(-i\omega t)$ is a time-harmonic current density where

$$i\omega\mu_0 \mathbf{j}(\mathbf{r}|\omega) = \nabla \times [\nabla \times \mathbf{f}(\mathbf{r})] - \frac{\omega^2}{c^2}\mathbf{f}(\mathbf{r}).$$

Prove that $\mathbf{j}(\mathbf{r}, t)$ does not radiate and find the physical meaning of $\mathbf{f}(\mathbf{r}, t)$.

20.14 Lorentz Reciprocity A current density $\mathbf{j}_1(\mathbf{r}) \exp(-i\omega t)$ produces fields $\mathbf{E}_1(\mathbf{r}) \exp(-i\omega t)$ and $\mathbf{B}_1(\mathbf{r}) \times \exp(-i\omega t)$. A second current density $\mathbf{j}_2 \exp(-i\omega t)$ produces fields $\mathbf{E}_2(\mathbf{r}) \exp(-i\omega t)$ and $\mathbf{B}_2(\mathbf{r}) \times \exp(-i\omega t)$.

(a) If V is a volume bounded by a surface S, prove the *Lorentz reciprocity theorem*:

$$\mu_0 \int_V d^3r\, (\mathbf{E}_2 \cdot \mathbf{j}_1 - \mathbf{E}_1 \cdot \mathbf{j}_2) = \int_S d\mathbf{S} \cdot (\mathbf{E}_1 \times \mathbf{B}_2 - \mathbf{E}_2 \times \mathbf{B}_1).$$

(b) Specialize to a spherical volume which both encloses the current sources and is large enough that all the fields on S can be taken to be radiation fields. Prove that

$$\int_V d^3r\, \mathbf{E}_2 \cdot \mathbf{j}_1 = \int_V d^3r\, \mathbf{E}_1 \cdot \mathbf{j}_2.$$

(c) Specialize further to time-harmonic point electric dipole sources located at $\mathbf{r}_1$ and $\mathbf{r}_2$. If $\mathbf{p}_k$ is the dipole moment of the $k^{\rm th}$ dipole, prove that

$$\mathbf{p}_1 \cdot \mathbf{E}_2(\mathbf{r}_1) = \mathbf{p}_2 \cdot \mathbf{E}_1(\mathbf{r}_2).$$

Lorentz reciprocity is used to prove that the angular distribution of power radiated by an antenna in broadcast mode is identical to the angular distribution of power absorbed by the same antenna in receiving mode.

20.15 Radiation from a Phased Array A current distribution consists of N identical sources. The kth source is identical to the first source except for a rigid translation by an amount $\mathbf{R}_k$ ($k = 1, 2, \ldots, N$). The sources oscillate at the same frequency ω but have different phases δ_k. That is,

$$\mathbf{j}_k \propto \exp[-i(\omega t + \delta_k)].$$

(a) Show that the angular distribution of radiated power can be written as the product of two factors: one is the angular distribution for $N = 1$; the other depends on $\mathbf{R}_k$ and δ_k but not the structure of the sources.
(b) The planes of two square loops (each with side length a) are centered on (and lie perpendicular to) the z-axis at $z = \pm a/2$. The loop edges are parallel to the x and y coordinate axes. Find the angular distribution of power, $dP/d\Omega$, in the x-z plane if the current at all points in both loops is $I \cos \omega t$. Make a polar plot of the angular distribution for $\omega a/c = 2\pi$ and $\omega a/c \ll 1$. Identify the multipole character of the radiation in the latter case.
(c) Repeat part (b) when the current in the upper loop is $I \cos \omega t$ and the current in the lower loop is $-I \cos \omega t$.

20.16 Radiation from a Square Loop A square loop of wire in the x-y plane is centered at the origin with its edges (each of length $2a$) parallel to the axes. Current flows counterclockwise around the loop as viewed from the positive z-axis. The time-dependence of the current is

$$I(t) = \begin{cases} 0 & t < 0, \\ I_0 t/\tau & 0 < t < \tau, \\ I_0 & \tau < t, \end{cases}$$

where $\tau > 2a/c$.

(a) Show that the radiation vector $\boldsymbol{\alpha}(\mathbf{r}, t) = \hat{\mathbf{y}}\,\alpha(r, \theta, t)$ when $\mathbf{r} = r\cos\theta\,\hat{\mathbf{z}} + r\sin\theta\,\hat{\mathbf{x}}$.
(b) Compute $\alpha(t)$ and make a careful graph of this function for fixed r and θ.

20.17 Linear Antenna Radiation Let the current density in a linear antenna of length h be

$$\mathbf{j}(\mathbf{r}, t) = \begin{cases} I(z, t)\delta(x)\delta(y)\hat{\mathbf{z}} & -h \le z \le h, \\ 0 & |z| > h. \end{cases}$$

(a) Find $\mathbf{E}_{\text{rad}}(\mathbf{r}, t)$ for the current $I(z, t) = A\,\delta(t)$. Your answer will have two terms. Determine the apparent origin of each term and give an argument for the time delay between the two. Make a polar plot centered on the antenna and regard each ray as a time axis in units of t/τ where $h = c\tau$. For each of the two terms above, draw a closed, dashed curve which indicates when the signals arrive at each angle. At a few representative angles, draw solid dots on the dashed curves to indicate the relative magnitude of $\mathbf{E}_{\text{rad}}$ at that angle.
(b) Repeat part (a) for the traveling-wave current $I(z, t) = \bar{A}\,\delta(t - z/c)$.
(c) Show that a uniform current $I(z, t) = A\,\exp(-i\omega t)$ radiates total power P where

$$P \sim \begin{cases} \omega^2\tau^2 & \omega\tau \ll 1, \\ \omega\tau & \omega\tau \gg 1. \end{cases}$$

(d) Show that the traveling-wave current $I(z, t) = \bar{A}\,\exp i(kz - \omega t)$ radiates total power $\bar{P}$ where

$$\bar{P} \sim \begin{cases} \omega^2\tau^2 & \omega\tau \ll 1, \\ \ln(\omega\tau) & \omega\tau \gg 1. \end{cases}$$

Some useful integrals are

$$\int dx \frac{\sin^2 x}{x} = \frac{1}{2}\ln x - \frac{1}{2}\text{Ci}(2x) \qquad \int dx \frac{\sin^2 x}{x^2} = -\frac{1}{2x} + \frac{\cos 2x}{2x} + \text{Si}(2x),$$

where $\text{Ci}(u)$ is the *cosine integral* and $\text{Si}(u)$ is the *sine integral*.

20.18 Radiation from a Filamentary Current

(a) Find the potentials and the fields produced by a current $I(t) = I_0\Theta(t)$ that turns on abruptly at $t = 0$ in a neutral, filamentary wire coincident with the entire z-axis.
(b) Show that the electric and magnetic fields approach their expected values as $t \to \infty$.

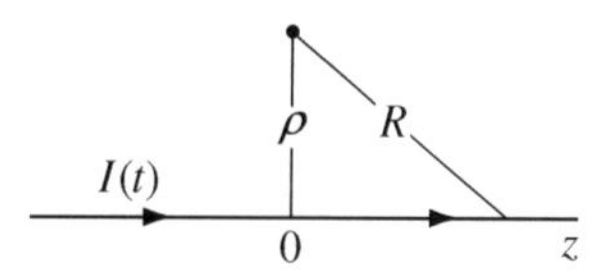

20.19 Crossed and Oscillating Electric Dipoles Two point electric dipoles are crossed in the x-y plane as shown below (at left). Both oscillate at frequency ω but with a $\pi/2$ phase difference.

(a) Sketch the time evolution of the instantaneous angular distribution of power in the x-y plane at several representative times between $t = 0$ and $t = 2\pi/\omega$.
(b) Show that the radiation emitted along the z-axis is circularly polarized. Is the polarization the same for $+z$ as it is for $-z$?
(c) Show that two crossed dipoles oscillating at frequency ω with no phase difference between them also emit circular polarized radiation along the z-axis if they are displaced by $\lambda/4$ as shown below (at right). Is the polarization the same for z as it is for $-z$?

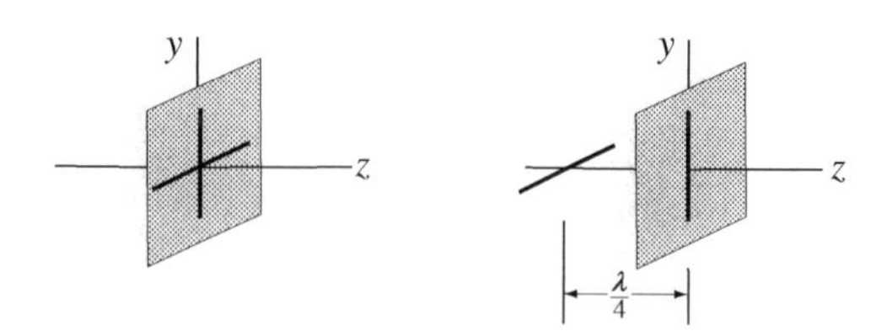

20.20 An Uncharged Rotor Two equal and opposite charges are attached to the ends of a rod of length s. The rod rotates counterclockwise in the x-y plane with angular speed $\omega = ck$. The electric dipole moment of the system at $t = 0$ has the value $\mathbf{p}_0 = qs\hat{\mathbf{x}}$.

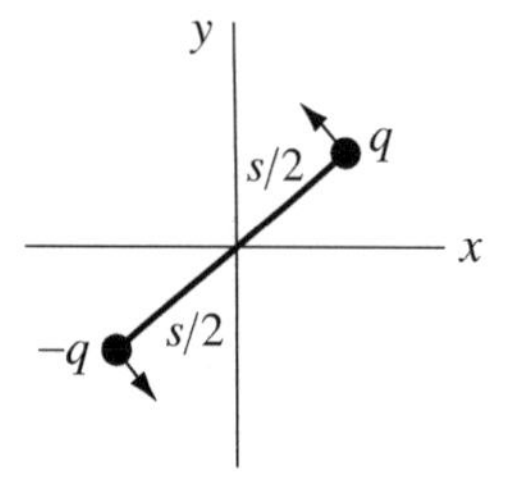

(a) Show that the electric field in the radiation zone is

$$\mathbf{E}_{\text{rad}}(r, \theta, \phi, t) = \frac{k^2 p_0}{4\pi\epsilon_0}\left(\cos\theta\hat{\boldsymbol{\theta}} + i\hat{\boldsymbol{\phi}}\right)\frac{e^{i(kr-\omega t+\phi)}}{r}.$$

Explain why the observer's azimuthal angle ϕ appears in the phase.
(b) Write out the (real) electric field on the positive x-, y-, and z-axes. Identify the state of polarization observed in each case and make a physical argument why each might be expected.
(c) Find the time-averaged rate at which energy is radiated per unit solid angle and the total rate at which energy is radiated to infinity.

20.21 Pulsar Radiation The pulsar SGR1806-20 has a period $T = 7.5$ sec and a "spindown" rate of $|\dot{T}| = 10^{-11}$. Estimate the maximum magnetic field strength $|\mathbf{B}|$ at the surface of this pulsar by assuming that the rotational kinetic energy ($M = 3 \times 10^{30}$ kg and $R = 10^3$ m) is dissipated by magnetic dipole radiation.

20.22 Neutron Radiation Apply a magnetic field $\mathbf{B} = B\hat{\mathbf{z}}$ to a neutron with its spin (assign a "classical" magnitude $S = \hbar/2$ to the spin) oriented initially along the $+y$-axis. Show that, because the precessing magnetic moment $\mathbf{m}$ radiates energy, the (polar) orientation angle of the moment decays to zero as

$$\Theta(t) \approx \cos^{-1}\left[\tanh\frac{t}{\tau}\right] \qquad \text{where} \qquad \frac{1}{\tau} = \frac{8m^5 B^3}{3\pi\epsilon_0 c^5 \hbar^4}.$$

Hint: Neglect $d\Theta/dt$ compared to the precession frequency when you compute the radiated power. Derive a condition on the parameters of the problem which justifies this approximation.

20.23 Radiation Interference A compact source of radiation is characterized by an electric dipole moment $\mathbf{p}(t) = \mathbf{p}_0 \cos\omega t$ and a magnetic dipole moment $\mathbf{m}(t) = \mathbf{m}_0 \cos\omega t$. Assume that the vectors $\mathbf{p}_0$ and $\mathbf{m}_0$ are real.

(a) Show that the time-averaged angular distribution of power generally exhibits interference between the two types of dipole radiation. Under what conditions is there no interference?
(b) Show that the time-averaged total power emitted by the source does not exhibit interference.

20.24 Wire Radiation A very long (infinite) wire with a very small cross sectional area A is coincident with the z-axis. The wire carries a current $I(t) = I\exp(-i\omega t)$.

(a) Use the Poynting vector to determine the dependence of the radiation fields on the variable ρ (cylindrical coordinates) which measures the distance perpendicular to the wire.
(b) The radiation magnetic field from a time-dependent point electric dipole $p(t)$ at the origin is

$$c\mathbf{B}_{\text{rad}}(r,t) = -\frac{\mu_0}{4\pi}\frac{\hat{\mathbf{r}}}{r} \times \frac{d^2\mathbf{p}(t - r/c)}{dt^2}.$$

Explain why the radiation magnetic field from the wire can be regarded as a superposition of fields like the one above.
(c) Carry out the superposition and confirm the answer you got in part (a). Simplify the integral you get by assuming that $z \ll \rho$ (even if it appears not to be justified) and then justify the approximation afterwards.

20.25 A Charged Rotor Two identical point charges q are fixed to the ends of a rod of length 2ℓ which rotates with constant angular velocity $\frac{1}{2}\omega$ in the x-y plane about an axis perpendicular to the rod and through its center.

(a) Calculate the electric dipole moment $\mathbf{p}(t)$. Is there electric dipole radiation?
(b) Calculate the magnetic dipole moment $\mathbf{m}(t)$. Is there magnetic dipole radiation?
(c) Show that the electric quadrupole moment is $Q(t) = \frac{1}{2}q\ell^2 \begin{bmatrix} 1+\cos\omega t & \sin\omega t & 0 \\ \sin\omega t & 1-\cos\omega t & 0 \\ 0 & 0 & 0 \end{bmatrix}$.

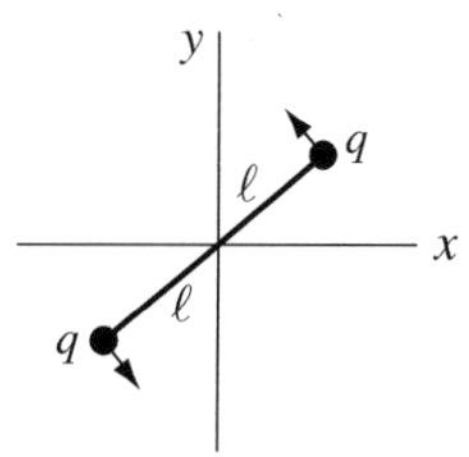

(d) Show that the time-averaged angular distribution of radiated power is

$$\left\langle \frac{dP}{d\Omega} \right\rangle = \frac{\mu_0}{4\pi} \frac{q^2\omega^6\ell^4}{32\pi c^3} \left(1 - \cos^4\theta\right),$$

where θ is the polar angle measured from the z-axis.

20.26 Rotating-Triangle Radiation Three identical point charges q are glued to the corners of an equilateral triangle that lies in the x-y plane. The charges rotate with constant angular velocity ω around the z-axis, which passes through the center of the triangle. Find the angular distribution of electric dipole, magnetic dipole, and electric quadrupole radiation (treated separately) produced by this source.

20.27 Collision Radiation Two point particles with masses m_1 and m_2 and charges q_1 and q_2 move slowly toward one another. For what choices of mass and charge is this motion *not* accompanied by dipole radiation?

20.28 Radiation of Linear Momentum

(a) Compare the conservation laws for energy and linear momentum for a spherical volume. Use this to define an angular distribution of the rate at which electromagnetic waves radiate linear momentum, $d\mathbf{P}_{\rm EM}/dt\,d\Omega$, by analogy with the definition of the angular distribution of the rate at which electromagnetic waves radiate energy, $dP/d\Omega$.

(b) Express $dP/d\Omega$ and $d\mathbf{P}_{\rm EM}/dt\,d\Omega$ entirely in terms of $\mathbf{E}_{\rm rad}$.

(c) Show that the relation between the two quantities computed in part (b) is consistent with the mechanical properties of electromagnetic plane waves.

20.29 Angular Momentum of Electric Dipole Radiation A collection of N charges lies inside a volume V. With respect to a fixed origin, the angular momentum of the charges and the electromagnetic fields they produce within V is

$$\mathbf{L} = \epsilon_0 \int_V d^3r \; \{\mathbf{r} \times (\mathbf{E} \times \mathbf{B})\} + \sum_{i=1}^{N} \mathbf{r}_i \times \mathbf{p}_i.$$

(a) Let S be the surface which bounds V. The text used a general conservation law to establish that

$$\frac{d\mathbf{L}}{dt} = \epsilon_0 \int_S d\mathbf{S} \cdot \{c\mathbf{B}(\mathbf{r} \times c\mathbf{B}) + \mathbf{E}(\mathbf{r} \times \mathbf{E})\} + \frac{1}{2}\epsilon_0 \int_S d\mathbf{S} \times \mathbf{r}\left\{E^2 + c^2B^2\right\}.$$

Derive this expression using the expression for $\mathbf{L}$ stated just above, the Maxwell equations, and the Lorentz force law.

(b) Suppose that the N charges generate a time-dependent electric dipole moment $\mathbf{p}(t)$. Compute $d\mathbf{L}/dt$ when V is chosen as a spherical volume (centered at the origin) so large that only the radiation fields due to $\mathbf{p}(t)$ have any significant magnitude on S. Does the answer surprise you?

(c) Now use the *exact* fields associated with $\mathbf{p}(t)$ and recalculate $d\mathbf{L}/dt$ keeping only those terms which survive in the limit when the sphere radius goes to infinity. Show that

$$\frac{d\mathbf{L}}{dt} = \frac{\mu_0}{4\pi} \frac{1}{2\pi c} \int_S d\Omega \left([\ddot{\mathbf{p}}]_{\rm ret} \times \hat{\mathbf{r}}\right)\left(\hat{\mathbf{r}} \cdot [\dot{\mathbf{p}}]_{\rm ret}\right),$$

where $[\mathbf{p}]_{\rm ret} = \mathbf{p}(t - r/c)$.

(d) Carry out the integral and confirm that

$$\frac{d\mathbf{L}}{dt} = \frac{\mu_0}{6\pi c}[\ddot{\mathbf{p}} \times \dot{\mathbf{p}}]_{\text{ret}}.$$

20.30 Dipole Moment of the Slotted Sphere Find the electric dipole moment of the slotted sphere discussed in Section 20.8.1.

21 Scattering and Diffraction

The solution of this problem presents mathematical difficulties which arise from the necessity of taking into account the geometrical shape of the obstacles on which the wave is falling.
Vladimir Fock (1948)

21.1 Introduction

An incident electromagnetic wave is said to *scatter* or *diffract* from a sample of matter when the field produced by the sample *cannot* be described using Fresnel's theory of reflection and refraction from a flat interface (Section 17.3). In this chapter, we focus on the class of problems where this occurs because the wavelength of the incident monochromatic field is *not small* compared to the curvature of a material boundary. From a Fresnel point of view, the total field in these cases results from the interference of many different "reflected" and "refracted" waves propagating in different directions. We will encounter other points of view as we proceed. Figure 21.1 shows some typical geometries of interest. There is no universal naming practice, but many authors say that "scattering" occurs from objects with smooth boundaries and "diffraction" occurs from objects with sharp edges.

The physics that produces scattering and diffraction is identical to the physics that produces the Fresnel equations. An incident electromagnetic wave sets the charged particles of a medium into motion. Each accelerated charge produces a retarded field which is felt by, and thus affects the motion of, every other charge in the medium. The motion of every charge and the field it produces must be consistent with the total field each charge experiences. The sum of the fields produced by all the particles of the medium is called the "scattered field" and the total field at any point (inside or outside the object) is the sum of the incident field and the scattered field:

$$\mathbf{E} = \mathbf{E}_{\rm inc} + \mathbf{E}_{\rm scatt}. \tag{21.1}$$

The same phenomenon viewed from a macroscopic perspective treats the moving charged particles as an induced current. This current is the source of $\mathbf{E}_{\rm scatt}$. Bearing in mind the constitutive relation of the matter, the actual current density is the one which ensures that (21.1) satisfies the boundary (or matching) condition at the surface of the scattering medium. The complexities of the problem all arise from the shape of the scatterer.

The proper treatment of scattering and diffraction has consistently attracted the attention of mathematical physicists of the first rank. The names Fresnel, Helmholtz, Rayleigh, and Kirchhoff are closely associated with a scalar theory developed for acoustics and optics. After Maxwell folded optics into electromagnetism, vector field calculations of lasting influence were performed by Debye, Mie, and Sommerfeld. The problems of radar engaged the quantum pioneers Fock and Schwinger, both of whom made profound contributions to the classical theory of electromagnetic scattering and diffraction. New insights appear today in contexts as diverse as the rainbow, laser propagation, and near-field microscopy.

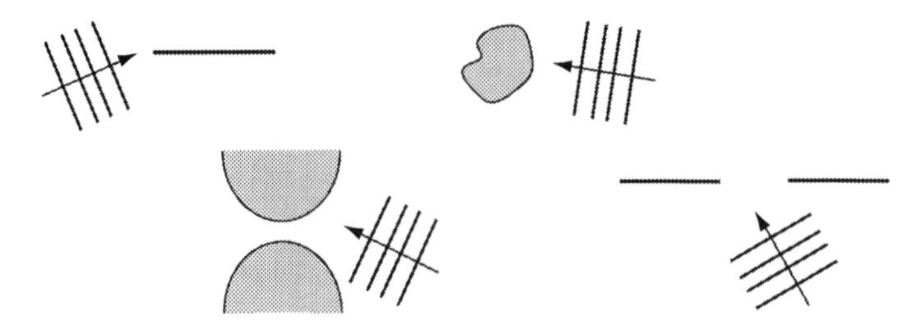

Figure 21.1: Four situations where a plane wave approaches an object with a sharp edge or a surface curvature which is large compared to the incident wavelength. Figure adapted from Toraldo di Francia (1953).

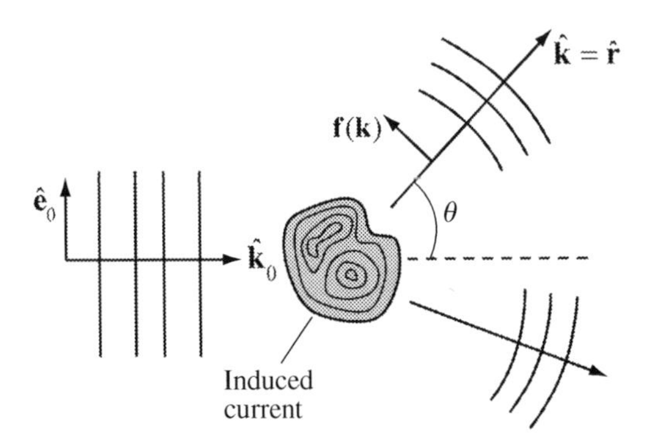

Figure 21.2: A plane wave with polarization vector $\hat{\mathbf{e}}_0$ propagating in the $\hat{\mathbf{k}}_0$ direction induces currents in a sample of matter. The scattering amplitude $\mathbf{f}(\mathbf{k})$ transverse to the local plane wave propagation direction $\hat{\mathbf{k}}$ characterizes the strength of the field radiated in that direction.

The diversity of scattering and diffraction problems, and the diversity of the mathematical techniques developed to treat them, is very great. For that reason, the aim of this chapter is less to teach quantitative scattering theory than to expose the reader to some of the basic ideas and to present a selection of representative results. To that end, we begin with scattering from small objects and thereby make immediate contact with the theory of electric and magnetic dipole radiation. Higher-order contributions become increasingly important as the size of the scattering object increases and an exact partial wave analysis is possible for cylinders and spheres. Most real scattering problems cannot be solved exactly, so we describe the Born and physical optics approximations as examples of what can be done without extensive numerical work. Our approach to diffraction focuses on the classic problem of a plane wave incident on an aperture in a plane screen. This leads us to discuss the electromagnetic version of optical concepts like Huygens' principle and Babinet's principle. The example of wave transmission through a sub-wavelength aperture serves to demonstrate that scattering and diffraction are far from exhausted as topics for contemporary research.

21.2 The Scattering Cross Section

The cartoon in Figure 21.2 shows a conducting or dielectric object which scatters an incident plane wave with polarization vector $\hat{\mathbf{e}}_0$ and propagation vector $\hat{\mathbf{k}}_0$. The figure of merit for this process is called the *differential cross section* for scattering. Up to a factor of r^2, this is the radial component of the time-averaged Poynting vector of the radiation field $\mathbf{E}_{\text{rad}}$ (the long-distance part of $\mathbf{E}_{\text{scatt}}$) divided by the magnitude of the time-averaged Poynting vector of the incident field $\mathbf{E}_{\text{inc}}$:

$$\frac{d\sigma_{\text{scatt}}}{d\Omega} = \frac{\text{scattered power radiated into a unit solid angle}}{\text{incident power per unit area}} = \frac{r^2\hat{\mathbf{r}}\cdot\langle\mathbf{S}_{\text{rad}}\rangle}{|\langle\mathbf{S}_{\text{inc}}\rangle|}. \tag{21.2}$$

A convenient form for (21.2) follows from the general characteristics of $\mathbf{E}_{\text{rad}}$ outlined in Section 20.5 for a compact, three-dimensional, time-harmonic source. First, it is an outgoing spherical wave with radial dependence $\exp(ikr)/r$. Second, the spherical wave front flattens out (locally) into a plane wave front with propagation vector $\mathbf{k} = k\hat{\mathbf{r}}$. Third, the field direction is transverse to the propagation direction. A *scattering amplitude* $\mathbf{f}(\mathbf{k}) \perp \mathbf{k}$ is commonly used to describe the vector and angular behavior of the scattered radiation field. Combining all this information permits us to write the asymptotic ($r \to \infty$) total field (21.1) in the form

$$\lim_{r\to\infty} \mathbf{E} = \mathbf{E}_{\text{inc}} + \mathbf{E}_{\text{rad}} = E_0\left[\hat{\mathbf{e}}_0 e^{i\mathbf{k}_0\cdot\mathbf{r}} + \frac{e^{ikr}}{r}\mathbf{f}(\mathbf{k})\right]e^{-i\omega t}. \tag{21.3}$$

All the wave fields in (21.3) propagate in vacuum, so $k = k_0 = \omega/c$.

Comparing (21.2) with (20.81) shows that the differential cross section is a normalized version of the time-averaged angular distribution of radiated power. Making use of (20.120) and (21.3),

$$\frac{d\sigma_{\text{scatt}}}{d\Omega} = \frac{\langle dP/d\Omega\rangle}{\frac{1}{2}\epsilon_0 c E_0^2} = r^2\frac{|\mathbf{E}_{\text{rad}}|^2}{|\mathbf{E}_0|^2} = |\mathbf{f}(\mathbf{k})|^2. \tag{21.4}$$

If our interest is a scattered electric field with a particular polarization $\boldsymbol{\epsilon}$, (21.4) generalizes to[1]

$$\left.\frac{d\sigma_{\text{scatt}}}{d\Omega}\right|_{\boldsymbol{\epsilon}} = r^2\frac{|\boldsymbol{\epsilon}^*\cdot\mathbf{E}_{\text{rad}}|^2}{|\mathbf{E}_0|^2} = |\boldsymbol{\epsilon}^*\cdot\mathbf{f}(\mathbf{k})|^2. \tag{21.5}$$

The results of Section 20.5.6 relate $\mathbf{E}_{\text{rad}}$ to the time-harmonic current density $\mathbf{j}(\mathbf{r}|\omega)\exp(-i\omega t)$ induced in the scatterer by the incident field:

$$\mathbf{E}_{\text{rad}} = -\frac{ik}{4\pi\epsilon_0 c}\hat{\mathbf{k}}\times\left[\hat{\mathbf{k}}\times\int d^3r'\,\mathbf{j}(\mathbf{r}'|\omega)\exp(-i\mathbf{k}\cdot\mathbf{r}')\right]\frac{\exp[i(kr-\omega t)]}{r}. \tag{21.6}$$

Then, because $|\hat{\mathbf{k}}\times\hat{\mathbf{k}}\times\mathbf{v}|^2 = |\hat{\mathbf{k}}\times\mathbf{v}|^2$ for any vector $\mathbf{v}$, substituting (21.6) into (21.4) gives

$$\frac{d\sigma_{\text{scatt}}}{d\Omega} = \left(\frac{k}{4\pi\epsilon_0 E_0 c}\right)^2\left|\hat{\mathbf{k}}\times\int d^3r'\,\mathbf{j}(\mathbf{r}'|\omega)\exp(-i\mathbf{k}\cdot\mathbf{r}')\right|^2. \tag{21.7}$$

The integral over all space of the differential cross section is called the *total cross section* for scattering:

$$\sigma_{\text{scatt}} = \int d\Omega\frac{d\sigma_{\text{scatt}}}{d\Omega}. \tag{21.8}$$

The dimensions of σ_{scatt} are length squared, which makes it convenient to compare (21.8) with the geometric cross section, σ_{geom}, defined as the projected area of the scattering object intercepted by the incident plane wave.

21.3 Thomson Scattering

Thomson scattering occurs when an electromagnetic plane wave interacts with a single free electron. Classically, we model the electron as a point particle with charge $-e$ and mass m which responds

[1] Let $\boldsymbol{\epsilon}_k$ be a complete set of polarization vectors. Then $|\mathbf{E}_{\text{rad}}|^2 = |\sum_k(\boldsymbol{\epsilon}_k^*\cdot\mathbf{E}_{\text{rad}})\boldsymbol{\epsilon}_k|^2 = \sum_k|\boldsymbol{\epsilon}_k^*\cdot\mathbf{E}_{\text{rad}}|^2$.

to $\mathbf{E}_{\rm inc} = \hat{\mathbf{e}}_0 E_0 \exp[i(\mathbf{k}_0 \cdot \mathbf{r} - \omega t)]$ and $c\mathbf{B}_{\rm inc} = \hat{\mathbf{k}}_0 \times \mathbf{E}_{\rm inc}$. We showed in Section 16.10.1 that the magnetic Lorentz force may be neglected compared to the electric Coulomb force when the field strength is weak. Therefore, with $\hat{\mathbf{e}}_0 \parallel \hat{\mathbf{z}}$, the equation of motion for the trajectory $\mathbf{r}_0(t)$ of the electron is

$$m\ddot{\mathbf{r}}_0 = -eE_0 \exp[i(\mathbf{k}_0 \cdot \mathbf{r}_0 - \omega t)]\hat{\mathbf{e}}_0. \tag{21.9}$$

This motion produces the time-harmonic current density

$$\mathbf{j}(\mathbf{r}, t) = -e\dot{z}_0(t)\delta(\mathbf{r} - \mathbf{r}_0)\hat{\mathbf{e}}_0 = \frac{ie^2 E_0}{m\omega} \exp[i(\mathbf{k}_0 \cdot \mathbf{r}_0 - \omega t)]\delta(\mathbf{r} - \mathbf{r}_0)\hat{\mathbf{e}}_0. \tag{21.10}$$

Because $\hat{\mathbf{e}}_0 \cdot \hat{\mathbf{e}}_0^* = 1$, using (21.10) with $\mathbf{r}_0 = 0$ to evaluate (21.7) gives the Thomson scattering cross section,[2]

$$\frac{d\sigma_{\rm Thom}}{d\Omega} = \left(\frac{e^2}{4\pi\epsilon_0 mc^2}\right)^2 |\hat{\mathbf{k}} \times \hat{\mathbf{e}}_0|^2 \equiv r_e^2 \left(1 - |\hat{\mathbf{k}} \cdot \hat{\mathbf{e}}_0|^2\right). \tag{21.11}$$

This formula is valid for all choices of $\hat{\mathbf{e}}_0$, whether real (for linear polarization) or complex (for circular or elliptical polarization).

The magnitude of the frequency-independent Thomson cross section is set by a length called the *classical electron radius*,

$$r_e = \frac{e^2}{4\pi\epsilon_0 mc^2} \approx 2.82 \times 10^{-15} \text{ m}. \tag{21.12}$$

This is the radius of a charged sphere whose Coulomb self-energy is equal to the rest energy of the electron. Alternately, $r_e = \alpha\lambda_c = \alpha^2 a_{\rm B}$, where $a_{\rm B} \approx 5.3 \times 10^{-11}$ m is the Bohr radius, $\alpha \approx 1/137$ is the fine structure constant, and $\lambda_c = \alpha a_{\rm B}$ is the Compton wavelength of the electron.

The absence of scattering along the direction of the electric field ($\mathbf{k} \parallel \hat{\mathbf{e}}_0$) and the angular dependence of (21.11) are reminiscent of the behavior of dipole radiation shown in Figure 20.17 of Section 20.7.1. This is not an accident. From (21.10), the motion of the oscillating point charge produces an electric dipole moment,

$$\mathbf{p}(t) = -ez(t) = -\frac{e^2 E_0}{m\omega^2}\hat{\mathbf{e}}_0 \exp(-i\omega t) = \mathbf{p}\exp(-i\omega t), \tag{21.13}$$

which in turn produces an electric dipole radiation field (20.160),

$$\mathbf{E}_{\rm rad} = -\frac{\mu_0}{4\pi}\omega^2[\hat{\mathbf{k}} \times (\hat{\mathbf{k}} \times \mathbf{p})]\frac{e^{i(kr-\omega t)}}{r}. \tag{21.14}$$

Comparing (21.14) to (21.3) gives $\mathbf{f} = (\mu_0 e^2/4\pi m)[\hat{\mathbf{e}}_0 - (\hat{\mathbf{k}} \cdot \hat{\mathbf{e}}_0)\hat{\mathbf{k}}]$, and inserting this into the far right-hand side of (21.4) reproduces the Thomson cross section (21.11).

It is useful to introduce the coordinate system shown in Figure 21.3 where $\hat{\mathbf{e}}_1 = \hat{\mathbf{e}}_\perp$ and $\hat{\mathbf{e}}_2 = \hat{\mathbf{e}}_\parallel$ are orthogonal unit vectors which lie perpendicular and parallel to the *scattering plane* defined by $\hat{\mathbf{k}}_0$ and $\hat{\mathbf{k}}$. For example, suppose we substitute first $\mathbf{p} = p\hat{\mathbf{e}}_\perp$ and then $\mathbf{p} = p\hat{\mathbf{e}}_\parallel$ into (21.14). Using Figure 21.3, we deduce that $\mathbf{E}_{\rm rad}$ shares the polarization of the incident wave in the sense of being polarized either perpendicular or parallel to the scattering plane.

Equation (21.11) is the cross section for scattering when the incident plane wave has fixed polarization $\hat{\mathbf{e}}_0$. To find the cross section for an *unpolarized* incident wave (a random mixture of waves with

[2] The time-independent choice $\mathbf{r}_0 = 0$ implies that the displacement of the electron away from the origin may be neglected in (21.7). This is a long-wavelength, low-velocity approximation.

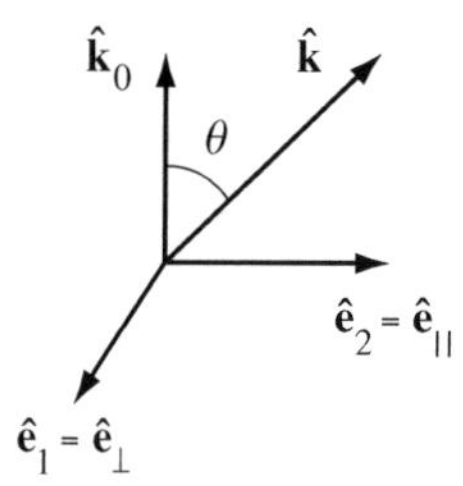

Figure 21.3: Orthogonal linear polarization vectors $\hat{\mathbf{e}}_1$ and $\hat{\mathbf{e}}_2$. The latter lies in the scattering plane defined by $\mathbf{k}_0$ and $\mathbf{k}$.

any two orthogonal polarization vectors, $\hat{\mathbf{e}}_1$ and $\hat{\mathbf{e}}_2$), we perform a statistical average of (21.11) over the two polarizations:

$$\left.\frac{d\sigma_{\text{Thom}}}{d\Omega}\right|_{\text{unpol}} = \frac{1}{2}\sum_{m=1}^{2} r_e^2(1 - |\hat{\mathbf{k}} \cdot \hat{\mathbf{e}}_m|^2). \tag{21.15}$$

Using the polarization vectors defined in Figure 21.3 to evaluate the contributions to (21.15), we find immediately that

$$\frac{d\sigma_\perp}{d\Omega} = r_e^2 \qquad \text{and} \qquad \frac{d\sigma_{||}}{d\Omega} = r_e^2\cos^2\theta. \tag{21.16}$$

Therefore,

$$\left.\frac{d\sigma_{\text{Thom}}}{d\Omega}\right|_{\text{unpol}} = \frac{1}{2}\left[\frac{d\sigma_\perp}{d\Omega} + \frac{d\sigma_{||}}{d\Omega}\right] = \frac{1}{2}r_e^2\left(1 + \cos^2\theta\right). \tag{21.17}$$

Unlike (21.11), the cross section (21.17) for scattering unpolarized waves is non-zero at every scattering angle. The total *Thomson cross section* is the integral of (21.17) over all these angles:

$$\sigma_{\text{Thom}} = \int d\Omega \left.\frac{d\sigma_{\text{Thom}}}{d\Omega}\right|_{\text{unpol}} = \frac{8\pi}{3}r_e^2. \tag{21.18}$$

The information in (21.16) also leads naturally to a definition for the *degree of polarization* of the scattered radiation. This is

$$\Pi(\theta) = \frac{\dfrac{d\sigma_\perp}{d\Omega} - \dfrac{d\sigma_{||}}{d\Omega}}{\dfrac{d\sigma_\perp}{d\Omega} + \dfrac{d\sigma_{||}}{d\Omega}} = \frac{\sin^2\theta}{1 + \cos^2\theta}. \tag{21.19}$$

The plots of (21.17) and (21.19) in Figure 21.4 show that the electric dipole scattering of unpolarized waves peaks in the forward ($\theta = 0$) and backward ($\theta = \pi$) directions and that the radiation is 100% linearly polarized for scattering at right angles ($\theta = \pi/2$) to the direction of incidence.

Example 21.1 Evaluate (21.15) in a coordinate-system-independent way using the fact that $\hat{\mathbf{e}}_1$, $\hat{\mathbf{e}}_2$, and $\hat{\mathbf{k}}_0$ form an orthonormal triad [see (16.48)].

Solution: The stated information means that the vectors in question satisfy a completeness relation. We write the latter in both abstract and component form as

$$|\hat{\mathbf{e}}_1\rangle\langle\hat{\mathbf{e}}_1| + |\hat{\mathbf{e}}_2\rangle\langle\hat{\mathbf{e}}_2| + |\hat{\mathbf{k}}_0\rangle\langle\hat{\mathbf{k}}_0| = 1 \qquad \text{or} \qquad \hat{e}_{1i}\hat{e}^*_{1j} + \hat{e}_{2i}\hat{e}^*_{2j} + \hat{k}_{0i}\hat{k}_{0j} = \delta_{ij}.$$

Using the second of these,

$$\sum_{m=1}^{2} |\hat{\mathbf{k}}\cdot\hat{\mathbf{e}}_m|^2 = \hat{k}_i\hat{k}_j\hat{e}_{mi}\hat{e}^*_{mj} = \hat{k}_i\hat{k}_j\left(\delta_{ij} - \hat{k}_{0i}\hat{k}_{0j}\right) = 1 - (\hat{k}\cdot\hat{k}_0)^2.$$

Therefore, because $\hat{\mathbf{k}}_0 \cdot \hat{\mathbf{k}} = \cos\theta$, (21.15) becomes

$$\left.\frac{d\sigma_{\text{Thom}}}{d\Omega}\right|_{\text{unpol}} = \frac{1}{2}r_e^2\left[2 - \left(1-\cos^2\theta\right)\right] = \frac{1}{2}r_e^2\left(1+\cos^2\theta\right).$$

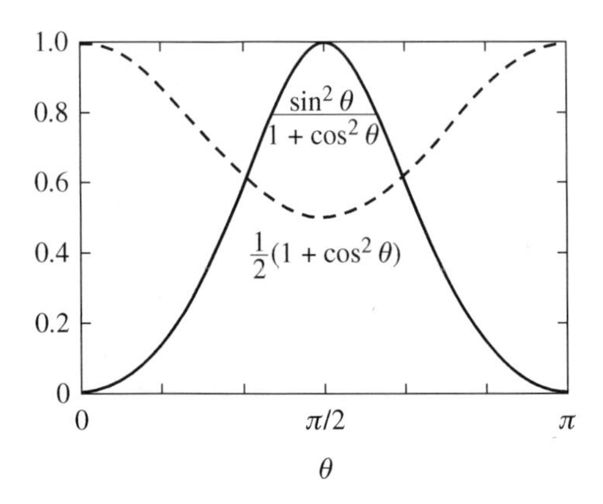

Figure 21.4: Angular dependence of the cross section (dashed curve) and degree of polarization (solid curve) for Thomson (and Rayleigh) scattering.

Example 21.2 Show that the cross section for monochromatic plane wave scattering from a collection of N electrons with fixed positions $\mathbf{r}_k$ is proportional to the absolute square of the spatial Fourier transform of the electron density $n(\mathbf{r}) = \sum_{k=1}^{N}\delta(\mathbf{r}-\mathbf{r}_k)$.

Solution: In the presence of a time-harmonic plane wave, the current density for all N electrons is the sum of terms like (21.10), except that the particles are located at positions $\mathbf{r}_k$ rather than at the origin. The space part of this density is

$$\mathbf{j}(\mathbf{r}|\omega) = -\frac{ie^2E_0}{m\omega}\exp(i\mathbf{k}_0\cdot\mathbf{r})\hat{\mathbf{e}}_0\sum_{k=1}^{N}\delta(\mathbf{r}-\mathbf{r}_k).$$

Substituting the foregoing into (21.7) shows that the cross section for scattering from the electron ensemble is proportional to the cross section for scattering from a single electron:

$$\left.\frac{d\sigma}{d\Omega}\right|_{\text{ensemble}} = \frac{d\sigma_{\text{Thom}}}{d\Omega}\times\left|\sum_{k=1}^{N}\exp[i(\mathbf{k}_0-\mathbf{k})\cdot\mathbf{r}_k]\right|^2.$$

On the other hand, the Fourier transform of the electron density is

$$n(\mathbf{q}) = \int d^3r\, n(\mathbf{r})\exp(-i\mathbf{q}\cdot\mathbf{r}) = \sum_{k=1}^{N}\int d^3r\,\delta(\mathbf{r}-\mathbf{r}_k)\exp(-i\mathbf{q}\cdot\mathbf{r})$$

$$= \sum_{k=1}^{N}\exp(-i\mathbf{q}\cdot\mathbf{r}_k) = n^*(-\mathbf{q}).$$

If we define a *scattering wave vector* $\mathbf{q} = \mathbf{k}_0 - \mathbf{k}$, the two preceding equations give the advertised result:

$$\left.\frac{d\sigma}{d\Omega}\right|_{\text{ensemble}} = \frac{d\sigma_{\text{Thom}}}{d\Omega} \times |n(\mathbf{q})|^2 = \frac{d\sigma_{\text{Thom}}}{d\Omega} \times \sum_{k=1}^{N}\sum_{j=1}^{N} \exp[i\mathbf{q}\cdot(\mathbf{r}_k - \mathbf{r}_j)].$$

This cross section formula underpins the theory of x-ray scattering, and the double sum is called a *form factor* in that context. X-ray scattering is sensibly regarded as an energy-conserving, elastic process for valence electrons whose binding energies are small compared to the energy of an incident x-ray. This implies that $|\mathbf{k}_0| = |\mathbf{k}|$, in which case Figure 21.2 shows that $q = 2k\sin(\frac{1}{2}\theta)$.

Application 21.1 The Polarization of Cosmic Microwave Radiation

Figure 16.5 shows that the background microwave radiation received from the cosmos is linearly polarized. The origin of this polarization is Thomson scattering and a small spatial anisotropy in the temperature of the radiation itself. The cartoons in Figure 21.5 tell the story by focusing on the radiation seen by an observer on the positive z-axis. Figure 21.5(a) shows linearly polarized plane waves incident on an electron at the origin from the $\pm x$ and $\pm y$ directions. Because Thomson scattering preserves polarization and is maximal at right angles to the incident wave polarization (see Figure 20.17), the waves incident from the $\pm x$ directions scatter into y-polarized waves on the z-axis and the waves incident from the $\pm y$ directions scatter into x-polarized waves on the z-axis. Thus, the observed radiation is unpolarized. This conclusion does not change if we add incoming z-polarized waves from the $\pm x$ and $\pm y$ directions (so all the incident radiation is unpolarized) because their cross section for scattering into the z-direction is zero.

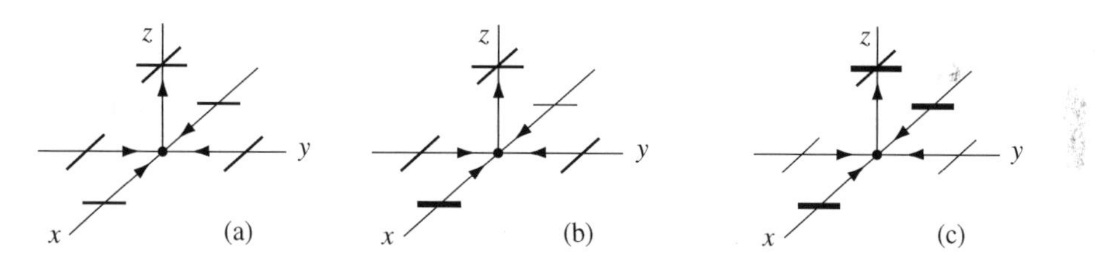

Figure 21.5: Cartoon of linearly polarized planes waves incident from $\pm x$ and $\pm y$ that scatter from an electron at the origin into the $+z$-direction of observation. The horizontal bars indicate the direction of polarization of each wave. The thickness of the bars indicates the intensity of the wave. The spatial distribution of the incident wave intensity is (a) isotropic; (b) dipole anisotropic; and (c) quadrupole anisotropic in the rest frame of the Universe. The observed radiation is (a) unpolarized; (b) unpolarized; and (c) linearly polarized. Figure adapted from Dodelson (2003).

Figure 21.5(b) introduces a dipole anisotropy by making the wave amplitude bigger (smaller) than the average for plane waves incident from the $+x$ ($-x$) direction. A dipole anisotropy in the blackbody radiation temperature would have this effect. The radiation seen from the $+z$-axis is unpolarized because the equal (unequal) intensity waves incident from the $\pm y$ ($\pm x$) axes scatter into the $+z$ direction preserving both intensity and polarization to give the mean intensity. Finally, Figure 21.5(c) introduces a quadrupole anisotropy by making the wave amplitude bigger (smaller) than the average for plane waves incident from the $\pm x$ ($\pm y$) directions. This produces the desired result: a net linear polarization of the waves scattered into the direction of the observer. This is consistent with independent measurements which confirm that the large-scale spatial isotropy of the blackbody temperature of the microwave radiation is perturbed by a small quadrupole anisotropy. ■

21.4 Rayleigh Scattering

Rayleigh scattering occurs when an electromagnetic plane wave impinges on a small dielectric or conducting object. By small, we mean that all characteristic linear dimensions of the object are small compared to the wavelength $\lambda = 2\pi/k_0$ of the plane wave. Under these conditions, we showed in Section 20.7 that the first few terms of a Cartesian multipole expansion of the radiation fields are sufficient to extract the physics. The incident **E** and **B** fields are nearly constant over the object's volume and the far-zone field is dominated by the radiation produced by the time-harmonic electric and magnetic dipole moments induced in the object. We compute the cross section from (21.4) by adding the magnetic dipole electric field (20.176) to the electric dipole electric field (21.14):

$$\frac{d\sigma_{\text{Ray}}}{d\Omega} = \left(\frac{k_0^2}{4\pi\epsilon_0 E_0 c}\right)^2 \left|\hat{\mathbf{k}} \times \mathbf{m} + \hat{\mathbf{k}} \times (\hat{\mathbf{k}} \times c\mathbf{p})\right|^2. \tag{21.20}$$

Let a be the largest linear dimension of the scattering object. The condition $k_0 a \ll 1$ implies that the plane wave fields are nearly uniform over its volume and a quasistatic approximation is sufficient to calculate the induced moment(s). The simplest example is a non-magnetic dielectric object with electric polarizability α where the electric dipole moment is

$$\mathbf{p} = \alpha\epsilon_0 E_0 \hat{\mathbf{e}}_0. \tag{21.21}$$

The corresponding cross section is

$$\frac{d\sigma_{\text{Ray}}}{d\Omega} = \left(\frac{k_0^2\alpha}{4\pi}\right)^2 \left(1 - |\hat{\mathbf{k}} \cdot \hat{\mathbf{e}}_0|^2\right). \tag{21.22}$$

On dimensional grounds alone, the polarizability scales with the object volume, which is of the order of a^3. Therefore, the total Rayleigh scattering cross section, σ_{Ray}, obtained by integrating (21.22) over all angles, is much smaller than the geometric cross section of the object:

$$\sigma_{\text{Ray}} \sim a^2(k_0 a)^4 \ll a^2. \tag{21.23}$$

The electric dipole Rayleigh cross section (21.22) and the Thomson cross section (21.11) have exactly the same angular dependence. Therefore, the discussion leading to the $1 + \cos^2\theta$ dependence in (21.17) for an unpolarized incident wave remains valid for electric dipole Rayleigh scattering. The same is true for the discussion leading to the degree of polarization (21.19). This means that Figure 21.4 applies to Rayleigh scattering as well as to Thomson scattering. Indeed, the strong linear polarization observed for skylight is a powerful indicator that dipole scattering dominates the interaction of the Sun's rays with the atmosphere.

21.4.1 Atmospheric Color

The striking λ^{-4} dependence of (21.22) is the origin of the blue color of the daylight sky and the red color of the setting Sun.[3] Looking away from the Sun during the day, we see sunlight scattered into our eyes by the molecules of the atmosphere. Because $\lambda_{\text{blue}} < \lambda_{\text{red}}$, (21.23) implies that blue light is the dominant component of the scattered light. We see red light looking directly at the Sun at sunset because the blue light has been scattered out of our line of sight. This argument is appealing precisely because it is so simple. On the other hand, it glosses over some subtleties.[4]

One subtlety is that the incoherent scattering argument presented above is not obviously relevant when (as is usual) $N \gg 1$ molecules lie inside a typical volume $V \sim \lambda^3$ of the visible sky. It could be

[3] See the cover of this book.

[4] See Sources, References, and Additional Reading for more details.

that destructive interference wipes out the incoherent scattering intensity altogether. A proper coherent calculation requires that we perform the double sum over N quoted at the end of Example 21.2. There are two ways to do this, provided some information is supplied about the spatial correlations of the molecules within the volume. The first method (due to Rayleigh) is a brute-force sum over the molecular positions in V. The second method (due to Smoluchowski and Einstein) divides V into smaller sub-volumes and rearranges the sum to show that fluctuations (deviations) from the mean number of molecules in each sub-volume are essential to avoid destructive interference. The two methods are equivalent and give the same answer, namely, that the incoherent scattered intensity survives: the scattering intensity from all the molecules in V is simply N times the scattering intensity from a single molecule.

Another subtlety is that the Sun radiates as a $T = 6000\,\text{K}$ blackbody with a spectral radiance which decreases with wavelength in the visible.[5] Under these conditions, the λ^{-4} argument more naturally predicts a violet sky than a blue sky. This is not what you see because the human brain assigns a color to incoming light based on the stimulus it receives from photoreceptive cells in the eye called cones. It happens that there are three types of cones, each with its own sensitivity to different visible wavelengths. An average observer reports that the daylight sky is blue because that color (rather than violet) represents the combined response of all three cone types to the spectral radiance of the Sun weighted by the Rayleigh λ^{-4} factor.

21.5 Two Exactly Solvable Problems

The scattering cross section can be computed exactly for objects with very simple shapes. In this section, we derive the exact cross section for scattering from a *conducting cylinder* and discuss the exact cross section for scattering from a *dielectric sphere*. We do this (i) to discover some generic features of scattering from extended distributions of matter; (ii) for later use in assessing the quality of approximate calculations; and (iii) because cylinders and spheres are good models for wires and raindrops (among other things).

The calculations are possible because a plane wave can be expressed as a sum of elementary solutions of the Helmholtz equation with either cylindrical or spherical symmetry. If the scattered field (produced by the object) is written as a similar sum with unknown expansion coefficients, the coefficients can be determined from the condition that the total field satisfies appropriate boundary or matching conditions at the surface of the object. The final solutions take the form of infinite sums of special functions which are amenable to numerical evaluation and thus to the identification of trends over large ranges of the parameters. Special analytic techniques valid for limited ranges of the parameters provide insight into the mechanisms of scattering. Indeed, the latter are practically essential for interpreting the results of numerical scattering calculations for both conducting and dielectric bodies.

21.5.1 Scattering from a Conducting Cylinder

Figure 21.6 shows a plane wave propagating in the $+x$-direction toward a perfectly conducting cylinder of radius a aligned with the z-axis. The polarization (electric field) vector lies in the y-z plane and it is sufficient to compute the scattered field separately for $\mathbf{E} \parallel \hat{\mathbf{z}}$ and $\mathbf{E} \perp \hat{\mathbf{z}}$. The results for these two cases can be superposed to treat the general case. Polarization parallel to the cylinder axis induces surface currents in the $\hat{\mathbf{z}}$-direction only. The associated vector potential points along $\hat{\mathbf{z}}$ also, so $B_z = (\nabla \times \mathbf{A})_z = 0$. On the other hand, $E_z = -\partial_z \varphi - \partial_t A_z \neq 0$. This is the only field component needed because $cB_\phi = -E_z$ in the radiation zone where the cross section is defined. Polarization

[5] Figure 19.19 is a graph of spectral radiance at $T = 2.73\,\text{K}$.

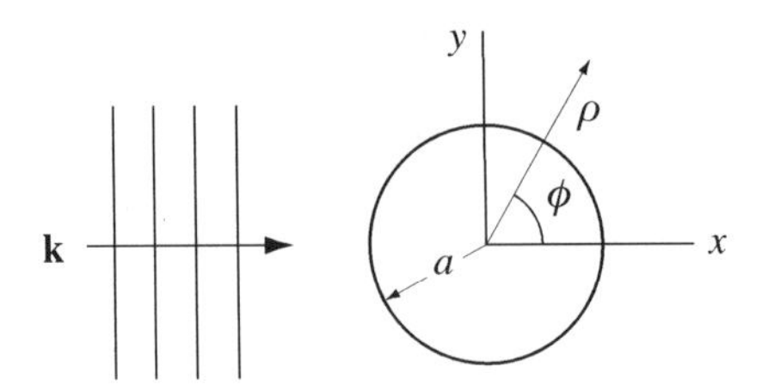

Figure 21.6: A plane wave approaches an infinite cylinder at normal incidence.

perpendicular to the cylinder axis induces surface currents in the $\hat{\boldsymbol{\phi}}$ direction. The corresponding circumferential vector potential has only x- and y-components. Then, because translational invariance guarantees that no physical quantity can depend on the z-coordinate, $E_z = -\partial_z \varphi - \partial_t A_z = 0$. The magnetic field component of interest in this case is B_z, because $E_\phi = cB_z$ in the radiation zone.

A monochromatic plane wave with frequency ω induces time-harmonic currents and thus a time-harmonic scattered field. Outside the cylinder, the electric and magnetic fields satisfy the wave equation. Therefore, the space parts of E_z and B_z satisfy the Helmholtz equation (see Section 16.6),

$$\left[\nabla^2 + \frac{\omega^2}{c^2}\right] u = 0. \tag{21.24}$$

Writing out (21.24) in cylindrical coordinates, we drop the z-derivatives and let $k = \omega/c$ to get

$$\frac{1}{\rho}\frac{\partial}{\partial \rho}\left(\rho \frac{\partial u}{\partial \rho}\right) + \frac{1}{\rho^2}\frac{\partial^2 u}{\partial \phi^2} + k^2 u = 0. \tag{21.25}$$

Separation of variables assumes that $u(\rho, \phi) = R(\rho)G(\phi)$. Inserting this into (21.25) and choosing p^2 as a separation constant gives $G(\phi) = \exp(\pm ip\phi)$ and shows that $R(\rho)$ satisfies Bessel's differential equation with $x = k\rho$:

$$\frac{d^2 R}{dx^2} + \frac{1}{x}\frac{dR}{dx} + \left(1 - \frac{p^2}{x^2}\right) R = 0. \tag{21.26}$$

The full angular range of ϕ is relevant, so $p = m$ is a non-negative integer. In Section 7.8.1, we discussed the linearly independent Bessel function solutions $J_m(x)$ and $N_m(x)$. Here, the scattered field must behave like an outgoing cylindrical wave when $\rho \to \infty$. Therefore, the relevant solution is the Hankel function, $H_m^{(1)}(x) = J_m(x) + iN_m(x)$, which has the asymptotic form

$$\lim_{x\to\infty} H_m^{(1)}(x) = \sqrt{\frac{2}{\pi x}} \exp\{i[x - (2m+1)\pi/4]\}. \tag{21.27}$$

We consider $\|$ polarization first and write the scattered field, $E_z(\text{scatt})$, as a sum of elementary solutions $H_m^{(1)}(k\rho)\exp(im\phi)$ with expansion coefficients A_m. The incident field is $E_z(\text{inc}) = E_0 \exp(ikx)$. Because $x = \rho\cos\phi$ in Figure 21.6, we can use the two-dimensional plane wave expansion (C.52),

$$E_z(\text{inc}) = E_0 \exp(ik\rho\cos\phi) = E_0 \sum_{m=-\infty}^{\infty} i^m J_m(k\rho)\exp(im\phi), \tag{21.28}$$

and write the total field outside the cylinder as

$$E_z = E_z(\text{inc}) + E_z(\text{scatt}) = E_0 \sum_{m=-\infty}^{\infty} \left[i^m J_m(k\rho) + A_m H_m^{(1)}(k\rho)\right]\exp(im\phi). \tag{21.29}$$

The boundary condition $E_z(\rho = a) = 0$ uniquely determines the A_m. Therefore,

$$E_z(\rho, \phi) = E_0 \exp(ik\rho \cos\phi) - E_0 \sum_{m=-\infty}^{\infty} i^m \frac{J_m(ka)}{H_m^{(1)}(ka)} H_m^{(1)}(k\rho) \exp(im\phi). \tag{21.30}$$

The $\perp$ polarization case is similar except that $cB_z(\text{inc}) = E_0 \exp(ikx)$ and the condition that the tangential component of the electric field vanish at the cylinder boundary reads

$$E_\phi(\rho = a) = \frac{c}{k} \frac{\partial B_z}{\partial \rho}\bigg|_{\rho=a} = 0. \tag{21.31}$$

Using a prime to indicate a derivative with respect to the argument, the final result for $\perp$ polarization is

$$cB_z(\rho, \phi) = E_0 \exp(ik\rho \cos\phi) - E_0 \sum_{m=-\infty}^{\infty} i^m \frac{J'_m(ka)}{H_m'^{(1)}(ka)} H_m^{(1)}(k\rho) \exp(im\phi). \tag{21.32}$$

For this two-dimensional problem, the definition of the differential scattering cross section differs slightly from that given earlier. In place of (21.7), we have

$$\frac{d\sigma}{d\phi} = \frac{\rho \hat{\boldsymbol{\rho}} \cdot \langle \mathbf{S}_{\text{rad}} \rangle}{\frac{1}{2} E_0^2 \epsilon_0 c}, \tag{21.33}$$

where the time-averaged Poynting vector in the radiation zone is (cf. Section 20.5.1)

$$\langle \mathbf{S}_{\text{rad}} \rangle_\parallel = \frac{\hat{\boldsymbol{\rho}}}{2} \epsilon_0 c |E_z(\text{rad})|^2 \quad \text{and} \quad \langle \mathbf{S}_{\text{rad}} \rangle_\perp = \frac{\hat{\boldsymbol{\rho}}}{2} \frac{c}{\mu_0} |B_z(\text{rad})|^2. \tag{21.34}$$

The radiation field is the $\rho \to \infty$ part of the scattered field. Therefore, we use the asymptotic formula (21.27) in (21.30) and (21.32), insert the radiation fields into (21.34), and evaluate the cross section expression (21.33) for parallel and perpendicular polarization. The result of these steps is

$$\begin{aligned} \frac{d\sigma_\parallel}{d\phi} &= \frac{2}{\pi k} \left| \sum_{m=-\infty}^{\infty} \frac{J_m(ka)}{H_m^{(1)}(ka)} \exp(im\phi) \right|^2 \\ \frac{d\sigma_\perp}{d\phi} &= \frac{2}{\pi k} \left| \sum_{m=-\infty}^{\infty} \frac{J'_m(ka)}{H_m'^{(1)}(ka)} \exp(im\phi) \right|^2. \end{aligned} \tag{21.35}$$

These cross sections have dimensions of length (rather than area) because they measure the relative power radiated to infinity per unit length of cylinder, i.e., the power radiated through the boundary of a large circle coaxial with the cylinder.

Integrating each expression in (21.35) over ϕ gives the total scattering cross sections,

$$\sigma_\parallel = \frac{4}{k} \sum_{m=-\infty}^{\infty} \frac{J_m^2(ka)}{J_m^2(ka) + N_m^2(ka)} \quad \text{and} \quad \sigma_\perp = \frac{4}{k} \sum_{m=-\infty}^{\infty} \frac{J_m'^2(ka)}{J_m'^2(ka) + N_m'^2(ka)}. \tag{21.36}$$

Figure 21.7 plots $\sigma_\parallel$ and $\sigma_\perp$ as a function of ka. Our first observation is that both cross sections approach a common value when $ka \to \infty$. This is not surprising because geometrical optics becomes valid in the short-wavelength limit. However, instead of approaching the geometric value $\sigma_{\text{geom}} = 2a$,

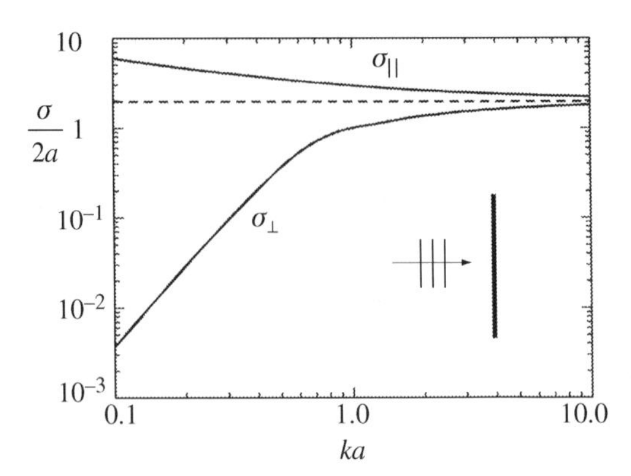

Figure 21.7: The total scattering cross section for plane wave scattering from an infinitely long conducting cylinder (radius a) at normal incidence. $\sigma_{\|}$ and $\sigma_{\perp}$ correspond to **E** parallel and perpendicular to the cylinder axis, respectively. The curves are normalized by the geometric cross section, $\sigma_{\text{geom}} = 2a$.

the limiting value seen in Figure 21.7 is actually *twice* this value. This is called the *extinction paradox* and we will find an explanation for it in Section 21.7.

The low-frequency, long-wavelength behaviors of $\sigma_{\|}$ and $\sigma_{\perp}$ are strikingly different:

$$\lim_{ka\to 0} \sigma_{\|} = \frac{\pi^2}{k \ln^2\left(\dfrac{1}{ka}\right)} \qquad \text{and} \qquad \lim_{ka\to 0} \sigma_{\perp} = \frac{3}{4}\pi^2 a(ka)^3. \tag{21.37}$$

The k^3 behavior of $\sigma_{\perp}$ is the signature of Rayleigh scattering (Section 21.4) in two dimensions. An electric field oriented perpendicular to the cylinder axis drives an electric dipole whose size cannot exceed the diameter of the cylinder. By contrast, no natural length scale limits the surface currents induced along the length of the cylinder when an electric field is oriented parallel to the cylinder axis. That being said, the $1/k$ divergence of $\sigma_{\|}$ in (21.37) is an artifact of the assumed infinite length of the cylinder. For a cylinder with length ℓ, our two-dimensional results apply for observation points (away from the ends) where $a^2/\lambda \ll \rho \ll \ell^2/\lambda$. A reasonable approximation for the cross section in this regime is[6]

$$\sigma_{\|}(\ell) \approx \frac{k\ell^2}{\pi}\sigma_{\|}(\infty). \tag{21.38}$$

Equation (21.38) decreases as ka decreases and crosses over smoothly to three-dimensional Rayleigh scattering when $r \gg \ell^2/\lambda$ and the finite size of the cylinder becomes apparent.

An application of these results is shown in Figure 21.8 where a planar and parallel array of perfectly conducting wires acts as a polarizer for plane electromagnetic waves. When the wire spacing d satisfies $a \ll d \ll \lambda$, waves polarized parallel to the wires are reflected and waves polarized perpendicular to the wires are transmitted. The transmission of $\perp$ waves follows immediately from the vanishingly small interaction between a long-wavelength $\perp$ wave and each wire predicted by Figure 21.7. Long-wavelength $\|$ waves are reflected because, in the limit considered, the longitudinal currents set up in the grid are indistinguishable from the currents set up when a plane wave strikes (and is reflected from) a flat conducting plate.

[6] See Ruck (1970) in Sources, References, and Additional Reading.

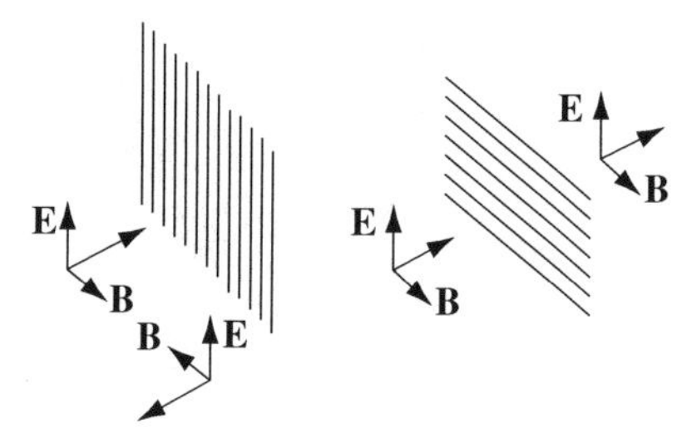

Figure 21.8: A wire-grid polarizer reflects (transmits) waves with **E** parallel (perpendicular) to the direction of the wires in the grid.

21.5.2 Mie Scattering from a Dielectric Sphere

The solution to the electromagnetic problem of plane wave scattering by a dielectric sphere is usually attributed to Gustav Mie.[7] His method was to expand the incident field and the scattered field in an infinite set of vector spherical waves and use the dielectric matching conditions at the surface of the sphere to determine the expansion coefficients. In this section, we outline only the basic steps of this calculation and focus on the results and their interpretation. The key observation is that the incident wave breaks the symmetry of the sphere and makes the matching conditions impossible to satisfy unless the electromagnetic field is written as a linear combination of TE and TM vector spherical waves.

Assume a time dependence $\exp(-i\omega t)$ and put $\omega = ck$. We showed in Section 16.8 that the Maxwell equations have solutions of the form

$$\begin{aligned} \mathbf{E} &= \nabla \times (\mathbf{r}u) - \frac{i}{k}\nabla \times [\nabla \times (\mathbf{r}w)] \\ c\mathbf{B} &= -\nabla \times (\mathbf{r}w) - \frac{i}{k}\nabla \times [\nabla \times (\mathbf{r}u)], \end{aligned} \tag{21.39}$$

where $u(\mathbf{r})$ and $w(\mathbf{r})$ are solutions of the scalar Helmholtz equation,

$$[\nabla^2 + k^2]\psi(\mathbf{r}) = 0. \tag{21.40}$$

We also showed that the separated-variable solutions to (21.40) in spherical coordinates are spherical Bessel radial functions multiplied by spherical harmonic angular functions. However, for the sake of consistency with the majority of the optical, chemical, and atmospheric literature of Mie scattering, we will write sines and cosines for the ϕ dependence and associated Legendre polynomials for the θ dependence [cf. (C.18)].

Beginning with the plane wave identity (C.53), it is an exercise in the use of recurrence relations to show that $\mathbf{E}_{\text{inc}} = E_0 \exp(ikz)\hat{\mathbf{x}}$ can be written in the form (21.39) with

$$\left\{ \begin{array}{c} u_{\text{inc}} \\ w_{\text{inc}} \end{array} \right\} = E_0 \sum_{\ell=1}^{\infty} i^{\ell} \frac{2\ell+1}{\ell(\ell+1)} j_\ell(kr) P_\ell^1(\cos\theta) \left\{ \begin{array}{c} \sin\phi \\ \cos\phi \end{array} \right\}. \tag{21.41}$$

The representation of the scattered field *inside* the sphere also has the form (21.39) with the expansion (21.41) except that $k \to nk$ (where n is the index of refraction of the sphere) and each term in the sum for u_{in} (w_{in}) includes an expansion coefficient c_ℓ (d_ℓ). The representation of the scattered field *outside* the sphere is again (21.39) with the expansion (21.41) except that the spherical Hankel function $h_\ell^{(1)}(kr)$ replaces the spherical Bessel function $j_\ell(kr)$ and each term in the sum for u_{out} (w_{out})

[7] Mie published his solution in 1908. An identical solution was published by Ludvig Lorenz in 1890. See the box at the end of the section for more about Lorenz.

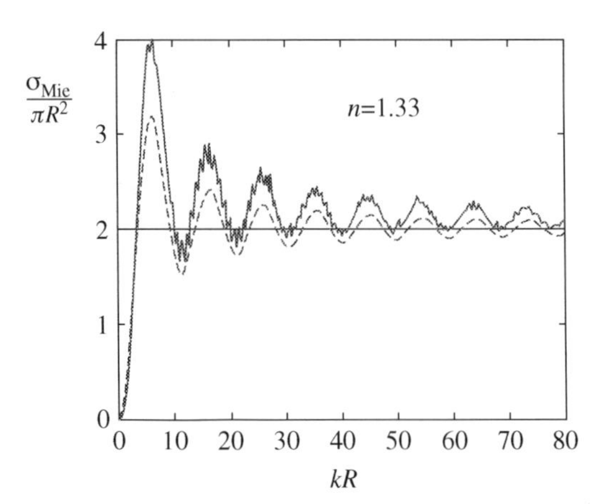

Figure 21.9: The cross section for plane wave scattering from a dielectric sphere with the index of refraction of water: numerical evaluation of exact Mie theory (solid curve); analytic approximation (21.45) (dashed curve). The curves are normalized by the geometric cross section, $\sigma_{\rm geom} = \pi R^2$.

includes an expansion coefficient a_ℓ (b_ℓ). The Hankel function ensures that the exterior field behaves asymptotically like the outgoing spherical wave in (21.3).

The (complex) expansion coefficients a_ℓ, b_ℓ, c_ℓ, and d_ℓ are determined by imposing the four matching conditions at $r = R$:

$$\hat{\mathbf{r}} \times (\mathbf{E}_{\rm inc} + \mathbf{E}_{\rm out}) = \hat{\mathbf{r}} \times \mathbf{E}_{\rm in} \qquad \text{and} \qquad \hat{\mathbf{r}} \times (\mathbf{H}_{\rm inc} + \mathbf{H}_{\rm out}) = \hat{\mathbf{r}} \times \mathbf{H}_{\rm in}. \tag{21.42}$$

The coefficients turn out to be quite complicated, even for non-magnetic spheres. Here, we use a prime to denote differentiation with respect to the argument and display only

$$\begin{aligned} a_\ell &= -\frac{j_\ell(nkR)[xj_\ell(x)]'_{x=kR} - j_\ell(kR)[xj_\ell(x)]'_{x=nkR}}{j_\ell(nkR)[xh_\ell^{(1)}(x)]'_{x=kR} - h_\ell^{(1)}(kR)[xj_\ell(x)]'_{x=nkR}} \\ b_\ell &= -\frac{j_\ell(kR)[xj_\ell(x)]'_{x=nkR} - n^2 j_\ell(nkR)[xj_\ell(x)]'_{x=kR}}{h_\ell^{(1)}(kR)[xj_\ell(x)]'_{x=nkR} - n^2 j_\ell(nkR)[xh_\ell^{(1)}(x)]'_{x=kR}}. \end{aligned} \tag{21.43}$$

The total scattering cross section is

$$\sigma_{\rm Mie} = \frac{2\pi}{k^2}\sum_{\ell=1}^{\infty}(2\ell+1)\left\{|a_\ell|^2 + |b_\ell|^2\right\}. \tag{21.44}$$

The solid curve in Figure 21.9 is a numerical evaluation of the Mie cross section (21.44) for a sphere with the dielectric constant of water ($n = 1.33$). The dominant features are a sharp rise to a global maximum followed by a quasi-periodic damped oscillation toward an asymptotic value equal to twice the geometric cross section, $\sigma_{\rm geom} = \pi R^2$. The latter is another of example of the "extinction paradox" mentioned in connection with Figure 21.7. The oscillations come from interfering plane wave rays which accumulate different amounts of phase as they pass through the sphere at different impact parameters. Specifically, if $\Delta = 2R(k_{\rm in} - k) = 2kR(n-1)$ is the difference in phase accumulated by a plane wave traversing the sphere diameter and a plane wave propagating in vacuum, the dashed curve in Figure 21.9 is the prediction of an approximate cross section formula we will derive in Application 21.2 of Section 21.7.1:

$$\sigma_{\rm approx} = 2\pi R^2\left[1 - \frac{2}{\Delta}\sin\Delta + \frac{2}{\Delta^2}(1-\cos\Delta)\right]. \tag{21.45}$$

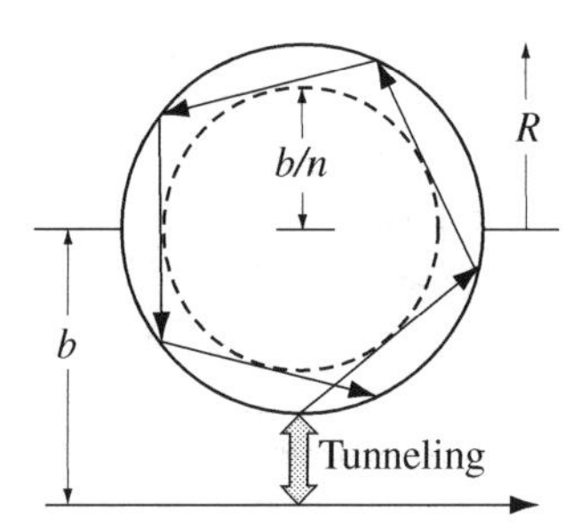

Figure 21.10: Ray-optic representation of the coupling between an incident wave and a normal mode of a spherical dielectric resonator. The radial boundaries are the inner and outer turning points of the effective potential in Figure 19.24 at the energy of the mode. Figure after Guimarães and Nussenzveig (1992).

The high-frequency, non-periodic "ripples" which decorate the damped oscillation in Figure 21.9 persist out to arbitrarily large values of kR and are extraordinarily sensitive to changes in the index n and the sphere radius R. Each ripple is a resonance in the cross section which occurs when the incident wave transiently couples to one of the electromagnetic normal modes of the dielectric sphere. To understand this, we recall from Section 19.7 that these modes can be described as virtual bound states trapped by the centrifugal barrier of the ℓ-dependent effective potential sketched in Figure 19.24. By analogy with the quantum theory of potential scattering, we identify $\ell = bk$ as the angular momentum of a ray with impact parameter b. If $b < R$, the ray simply refracts into the sphere in the usual way. If $b \gg R$, there is virtually no interaction between the ray and the sphere. In between, there is a range of impact parameters where the ray can tunnel through the barrier and resonantly excite a normal mode.[8] Figure 21.10 represents such a mode by a ray which propagates inside the sphere in the annular volume defined by the inner turning point of the potential at $r = b/n$ and the outer turning point at $r = R$. Eventually, the ray tunnels back out and rejoins the original ray.

The Great Dane

Eighteen years before Gustav Mie published his analysis of what we call "Mie" scattering, the entire problem was solved and published by the Danish physicist Ludvig Lorenz (1829-1891). Remarkably, this is only one of several instances where Lorenz' precedence in the solution of an electromagnetic problem went unrecognized by his peers and forgotten by history. For most of the last century, he was remembered only for the Lorentz-Lorenz equation, which generalizes the Clausius-Mossotti relation of static dielectric theory and relates the index of refraction $n(\omega)$ of a gas of molecules with density N to the molecular polarizability, $\alpha(\omega)$:

$$\frac{n^2(\omega) - 1}{n^2(\omega) + 2} = \frac{N\alpha(\omega)}{3\epsilon_0}.$$

The pairing of Lorenz with the Dutch physicist Hendrik Lorentz is particulary ironic because the familiar electromagnetic gauge choice,

$$\nabla \cdot \mathbf{A} + \frac{1}{c^2}\frac{\partial \varphi}{\partial t} = 0,$$

has long been incorrectly attributed to Lorentz. In fact, this gauge constraint first appeared in 1867 when Lorenz introduced the concept of the retarded potential into electrodynamics in a paper entitled "On the identity of the vibrations of light with electrical currents". Lorenz generalized a

[8] The "tunneling" wave is an evanescent wave of the sort discussed in Section 17.3.7.

quasistatic theory of Kirchhoff to include retardation and developed a theory which (in retrospect) was practically identical to Maxwell's. Lorenz went to some trouble to insist that the concept of the aether was superfluous.

A third example of neglect dates from 1860, when Lorenz extended Fresnel's theory of refraction to account for a thin "transition layer" between two bulk media where the dielectric properties interpolate smoothly between the two bulk values. The elliptical polarization of the reflected light induced by the presence of this layer became the diagnostic tool of "ellipsometry" 40 years later when Drude re-derived Lorenz' results using Maxwell theory.

It is puzzling that Lorenz' work was forgotten so quickly. His most important papers were quickly translated from Danish and all of his work could be read in German, French, or English by 1896, when his collected works were published. On the other hand, Lorenz was an autodidact who did not correspond much with other physicists. He did not travel widely and spent his entire career as a physics teacher for army cadets at a military high school outside Copenhagen. For a man born in the Danish city of Elsinore (the setting for Shakespeare's *Hamlet*), it would be altogether fitting if the ghost of his achievements walked again among contemporary scientists.

21.6 Two Approximation Schemes

Modern computers can generate essentially exact solutions to most electromagnetic scattering problems. Nevertheless, one always gains intuition from less exact methods if they are based on physically well-motivated approximations. This section treats two such schemes. The *Born approximation* applies to objects which are only weakly polarizable, magnetizable, or conducting, but where the ratio of the object size to the wavelength (a/λ) can have any value. The *physical optics approximation* applies to highly conducting objects in the limit when $a/\lambda \gg 1$.

21.6.1 The Born Approximation

The Born approximation presumes that the total field $\mathbf{E}$ inside a weakly dielectric or weakly conducting scattering medium does not differ significantly from the incident field $\mathbf{E}_{\rm inc}$. If we use the language of polarization (rather than conductivity), this means that the current induced in the object by a monochromatic plane wave is

$$\mathbf{j} = \frac{\partial \mathbf{P}}{\partial t} = -i\omega\epsilon_0\chi\mathbf{E} \approx -i\omega\epsilon_0\chi\mathbf{E}_{\rm inc} = -i\omega\epsilon_0\chi\hat{\mathbf{e}}_0 E_0 \exp[(i\mathbf{k}_0\cdot\mathbf{r} - \omega t)]. \tag{21.46}$$

Substituting (21.46) into (21.7) gives the approximate cross section

$$\left.\frac{d\sigma_{\rm scatt}}{d\Omega}\right|_{\rm Born} = \left(\frac{k_0^2}{4\pi}\right)^2 |\hat{\mathbf{k}}\times\hat{\mathbf{e}}_0|^2 \left|\int d^3r'\, \chi(\mathbf{r},\omega)\exp[i(\mathbf{k}_0 - \mathbf{k})\cdot\mathbf{r}']\right|^2. \tag{21.47}$$

The angular factor $|\hat{\mathbf{k}}\times\hat{\mathbf{e}}_0|^2$ in (21.47) is familiar from (21.11) and all the remarks made there about averaging this quantity for an unpolarized incident wave apply here also.

A low-frequency application of (21.47) applies to a dielectric object with volume V and uniform electric susceptibility χ. When $k_0 a \ll 1$, the exponential may be neglected in the integral and (21.47) reduces to

$$\left.\frac{d\sigma_{\rm scatt}}{d\Omega}\right|_{\rm Born} = \left(\frac{k_0^2}{4\pi}\right)^2 (V\chi)^2 |\hat{\mathbf{k}}\times\hat{\mathbf{e}}_0|^2. \tag{21.48}$$

The polarizability of this object is $\alpha = V\chi$. Therefore, (21.48) reproduces the Rayleigh scattering formula (21.22), as expected.

A high-frequency application of (21.47) applies to a Drude plasma where (see Section 18.5.1)

$$\chi(\omega) = \frac{\epsilon(\omega)}{\epsilon_0} - 1 = \left(1 - \frac{\omega_p^2}{\omega^2}\right) - 1 = -\frac{nq^2}{\omega^2 m\epsilon_0}. \tag{21.49}$$

Generalizing to a particle density $n(\mathbf{r})$ which varies with position, substituting (21.49) into (21.47) gives

$$\left.\frac{d\sigma_{\text{scatt}}}{d\Omega}\right|_{\text{Born}} = \left(\frac{q^2}{4\pi\epsilon_0 mc^2}\right)^2 |\hat{\mathbf{k}} \times \hat{\mathbf{e}}_0|^2 \left|\int d^3r'\, n(\mathbf{r}')\exp[i(\mathbf{k}_0 - \mathbf{k})\cdot\mathbf{r}']\right|^2. \tag{21.50}$$

If the mobile particles of the plasma are electrons, where $q = -e$, (21.50) reproduces the Thomson scattering results of Section 21.3 multiplied by a form factor that reproduces the results of Example 21.2.

Example 21.3 Find the Born approximation to the differential cross section for scattering from a uniform and lossless dielectric sphere with radius R. Plot the cross section for an unpolarized incident beam with $k_0 R = 15/2$.

Solution: The susceptibility χ is constant inside the sphere and zero outside the sphere. We make no assumptions about the magnitude of $k_0 R$. With $\mathbf{q} = \mathbf{k} - \mathbf{k}_0$, (21.47) becomes

$$\left.\frac{d\sigma_{\text{scatt}}}{d\Omega}\right|_{\text{Born}} = \left(\frac{k_0^2\chi}{4\pi}\right)^2 |\hat{\mathbf{k}} \times \hat{\mathbf{e}}_0|^2 \left|\int d^3r' \exp(i\mathbf{q}\cdot\mathbf{r}')\right|^2.$$

The integral to be evaluated is

$$I = \int d^3r' \exp(i\mathbf{q}\cdot\mathbf{r}') = 2\pi\int_{-1}^{1} du \int_0^R dr' r'^2 \exp(iqru) = \frac{4\pi}{q}\int_0^R dr' r' \sin(qr') = 4\pi R^3 \frac{j_1(qR)}{qR},$$

where $j_1(x) = \sin(x)/x^2 - \cos(x)/x$ is the spherical Bessel function of order one. The scattering is elastic, so the remark at the end of Example 21.2 tells us that $q = 2k_0\sin(\theta/2)$. Finally, (21.17) shows that $|\hat{\mathbf{k}} \times \hat{\mathbf{e}}_0|^2 = \frac{1}{2}(1 + \cos^2\theta)$ for an unpolarized incident wave. We conclude that the Born cross section for the sphere is

$$\left.\frac{d\sigma_{\text{scatt}}}{d\Omega}\right|_{\text{Born}} = \frac{R^2}{4}(k_0R)^2\chi^2 \frac{1+\cos^2\theta}{2} \frac{j_1^2[2k_0R\sin(\theta/2)]}{\sin^2(\theta/2)}.$$

Figure 21.11 compares the Born approximation (solid line) to the differential scattering cross section with the exact Mie cross section (dashed curve) for a dielectric sphere with index of refraction $n = 1.05$ and radius $R = (15/4\pi)\lambda$. The majority of the scattering is near the forward direction, as one might expect from a sphere whose dielectric properties are so weak that the field inside the sphere is indistinguishable from the incident field. The zeroes of the Born cross section (which come from the zeroes of the spherical Bessel function) are replaced by local minima in

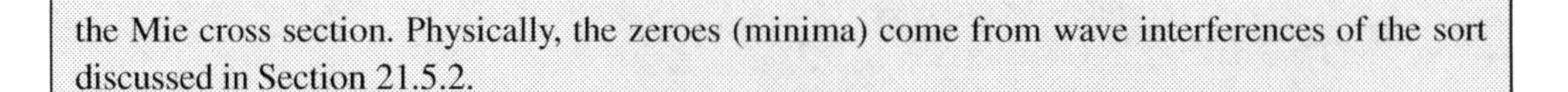

the Mie cross section. Physically, the zeroes (minima) come from wave interferences of the sort discussed in Section 21.5.2.

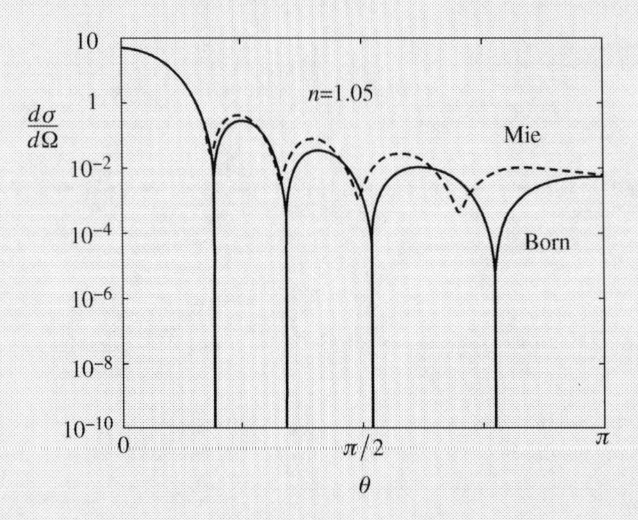

Figure 21.11: Differential cross section (in arbitrary units) for scattering from a uniform dielectric sphere with index of refraction $n = 1.05$ and radius $R = (15/4\pi)\lambda$. The exact Mie result is the dashed curve. The Born approximation is the solid curve.

21.6.2 The Physical Optics Approximation

Geometrical optics is the natural starting point for scattering situations where the wavelength of the incident wave is very small compared to the characteristic feature size of a target object. Such objects are called *electrically large*. For a plane wave incident on a conductor, the frequency-independent physics of geometrical optics is simply specular reflection of a parallel set of rays from the illuminated portions of the target object. The points on the surface shadowed from the incident wave by other parts of the body play no role.

The physical optics approximation is a physically motivated correction to geometrical optics which simplifies the radiation fields computed from Maxwell's equations. The key idea, due to Macdonald (1912), is to assign a current density $\mathbf{K}$ to every illuminated surface point as if that point were part of a flat and infinite conducting plane oriented tangent to the surface at that point. Zero surface current density is assigned to surface points which are shadowed. From Section 19.3.1, we know that specular reflection of a plane wave incident on a conducting plane annuls the tangential component of the surface electric field and doubles the tangential component of the surface magnetic field. Hence, the physical optics approximation for the current density induced on the surface of a conducting object by the incident wave is

$$\mu_0 \mathbf{K}(\mathbf{r}, t) = \begin{cases} 2\hat{\mathbf{n}} \times \mathbf{B}_{\text{inc}}(\mathbf{r}, t) & \text{at illuminated surface points} \\ 0 & \text{at shadowed surface points.} \end{cases} \tag{21.51}$$

If S is the illuminated part of the surface and $\mathbf{B}_{\text{inc}} = \mathbf{B}_0 \exp[i(\mathbf{k}_0 \cdot \mathbf{r} - \omega t)]$ is the magnetic field of the incident plane wave, substituting (21.51) into the vector potential expression (20.118) to evaluate

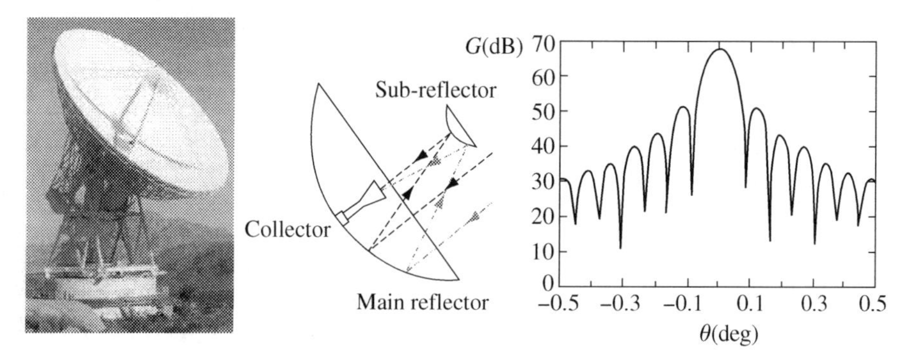

Figure 21.12: Left panel: a parabolic reflector antenna. Middle panel: geometrical optics ray tracing of an incoming bundle of parallel rays. Right panel: physical optics prediction for the gain, $G(\theta)$, which is the angular distribution of power received by this reflector, normalized by the total power received. The vertical scale (decibels) is logarithmic. Figure adapted from Imbriale (2003).

the electric field in (20.119) gives

$$\mathbf{E}_{\rm rad}(\mathbf{r}, t) \approx \frac{i\omega}{2\pi}\frac{\exp[i(kr-\omega t)]}{r}\hat{\mathbf{k}} \times \left\{ \hat{\mathbf{k}} \times \int_S dS' \, (\hat{\mathbf{n}}' \times \mathbf{B}_0) \exp[i(\mathbf{k}_0 - \mathbf{k}) \cdot \mathbf{r}'] \right\}. \tag{21.52}$$

A common use of (21.52) is to calculate the angular distribution of radiation transmitted or received (they are the same) by an electrically large antenna.[9] For example, the reflector-type radio antenna shown in the leftmost panel of Figure 21.12 is part of the NASA Deep Space Network. The middle panel of the figure shows the geometrical optics prediction that parallel rays received from space are focused by the main parabolic reflector onto a sub-reflector which reflects them onto a collector. The rightmost panel shows the physical optics prediction for the gain function (the angular distribution normalized by the total power) at 7.2 GHz for the particular 32 m diameter antenna shown in the leftmost panel. The polar angle θ is measured with respect to the symmetry axis of the main reflector paraboloid. Because the vertical scale (decibels) is logarithmic, the physical optics gain function confirms the geometrical optics expectation that most of the received power is confined to a very small angular range around the symmetry axis. Discussion of the side lobes in $G(\theta)$ is best postponed until Section 21.8 when we take up the subject of diffraction.

21.7 The Total Cross Section

We have so far neglected the possibility that the scatterer in Figure 21.2 could *absorb* some of the energy carried by the incident monochromatic plane wave. To account for this, we invoke conservation of energy and evaluate the Poynting theorem for a volume V which includes the scatterer:

$$\int_V d^3r \, \nabla \cdot \mathbf{S} + \frac{d}{dt} U_{\rm EM} = - \int_V d^3r \, \mathbf{j} \cdot \mathbf{E}. \tag{21.53}$$

[9] See the footnote which accompanies (20.124).

The total energy term in (21.53) drops out after averaging over one period of a time-harmonic field. Therefore,

$$\int_S dA\,\hat{\mathbf{n}}\cdot\langle\mathbf{S}\rangle = -\int_V d^3r\,\langle\mathbf{j}\cdot\mathbf{E}\rangle. \tag{21.54}$$

The right side of (21.54) is the rate at which field energy is absorbed by the particles of the scatterer. This motivates us to define the total cross section for absorption, $\sigma_{\rm abs}$, as the ratio of the time-averaged absorption rate to the time-averaged rate at which the incident beam supplies energy per unit area:

$$\sigma_{\rm abs} = \frac{1}{|\langle\mathbf{S}_{\rm inc}\rangle|}\int_V d^3r\,\langle\mathbf{j}\cdot\mathbf{E}\rangle = -\frac{1}{|\langle\mathbf{S}_{\rm inc}\rangle|}\int_S dA\,\hat{\mathbf{n}}\cdot\langle\mathbf{S}\rangle. \tag{21.55}$$

The total cross section is the sum of the cross sections for scattering and absorption:

$$\sigma_{\rm tot} = \sigma_{\rm scatt} + \sigma_{\rm abs}. \tag{21.56}$$

21.7.1 The Optical Theorem

The *optical theorem* is an exact relation that relates the total cross section (21.56) to the amplitude of the electromagnetic field scattered in the forward direction (the propagation direction of the incident wave). To derive it, we choose the closed integration surface S in (21.55) as a sphere (centered on the scatterer) whose surface lies entirely in the radiation zone. Then, because the flux of the incident plane wave in (21.3) is $|\langle\mathbf{S}_{\rm inc}\rangle| = \frac{1}{2}\epsilon_0 cE_0^2$, we can rewrite the leftmost and rightmost members of (21.55) as

$$-\sigma_{\rm abs} = \frac{1}{E_0^2}{\rm Re}\int d\Omega\, r^2\hat{\mathbf{r}}\cdot\left[(\mathbf{E}_{\rm inc}+\mathbf{E}_{\rm rad})^*\times c(\mathbf{B}_{\rm inc}+\mathbf{B}_{\rm rad})\right]. \tag{21.57}$$

The integral in (21.57) which involves $\mathbf{E}^*_{\rm inc}\times\mathbf{B}_{\rm inc}$ is zero because $\mathbf{S}_{\rm inc}$ is a constant vector. From (21.4), the integral which involves $\mathbf{E}^*_{\rm rad}\times\mathbf{B}_{\rm rad}$ is $\sigma_{\rm scatt}$. The cross terms which remain can be evaluated using (21.3) and $c\mathbf{B} = \hat{\mathbf{k}}_0\times\mathbf{E}_{\rm inc} + \hat{\mathbf{r}}\times\mathbf{E}_{\rm rad}$.

Because $\hat{\mathbf{r}}\cdot\mathbf{f} = 0$, the remarks just above render (21.57) in the form

$$\begin{aligned}-\sigma_{\rm abs} = \sigma_{\rm scatt} &+ {\rm Re}\int d\Omega\,(\hat{\mathbf{e}}_0^*\cdot\mathbf{f})\exp(-i\mathbf{k}_0\cdot\mathbf{r})\,r\exp(ikr)\\ &+{\rm Re}\int d\Omega\left[(\hat{\mathbf{r}}\cdot\hat{\mathbf{k}}_0)(\hat{\mathbf{e}}_0\cdot\mathbf{f}^*) - (\hat{\mathbf{r}}\cdot\hat{\mathbf{e}}_0)(\hat{\mathbf{k}}_0\cdot\mathbf{f}^*)\right]\exp(i\mathbf{k}_0\cdot\mathbf{r})\,r\exp(-ikr).\end{aligned} \tag{21.58}$$

We evaluate the integrals in (21.58) using the asymptotic identity:

$$\lim_{k_0r\to\infty}\exp(i\mathbf{k}_0\cdot\mathbf{r}) = 2\pi i\left[\frac{\exp(-ik_0r)}{k_0r}\delta(\hat{\mathbf{r}}+\hat{\mathbf{k}}_0) - \frac{\exp(ik_0r)}{k_0r}\delta(\hat{\mathbf{r}}-\hat{\mathbf{k}}_0)\right]. \tag{21.59}$$

The reader can derive (21.59) by substituting the spherical harmonic addition theorem (C.23) into the plane wave expansion formula (C.53) to get

$$\exp(i\mathbf{k}_0\cdot\mathbf{r}) = 4\pi\sum_{\ell=0}^{\infty} i^\ell j_\ell(k_0r)\sum_{m=-\ell}^{\ell}Y^*_{\ell m}(\hat{\mathbf{k}}_0)Y_{\ell m}(\hat{\mathbf{r}}). \tag{21.60}$$

After inserting the large-argument form of the spherical Bessel function (C.46) into (21.60), the inversion symmetry (C.24) and completeness (C.20) of the spherical harmonics produces (21.59).

Substituting (21.59) into (21.58) eliminates one integral because the factor $\hat{\mathbf{r}}\cdot\hat{\mathbf{e}}_0$ vanishes when $\hat{\mathbf{r}} = \pm\hat{\mathbf{k}}_0$. Otherwise, we note that $\hat{\mathbf{r}} = \hat{\mathbf{k}}_0$ is the $\theta = 0$ forward direction and $\hat{\mathbf{r}} = -\hat{\mathbf{k}}_0$ is the $\theta = \pi$

backward direction. Therefore, because $k = k_0$, the remaining integrals and (21.56) reduce (21.58) to

$$-\sigma_{\text{tot}} = -\frac{4\pi}{k}\text{Re}\left[\frac{\hat{\mathbf{e}}_0^* \cdot \mathbf{f}(\mathbf{k}_0) - \hat{\mathbf{e}}_0 \cdot \mathbf{f}^*(\mathbf{k}_0)}{2i} - \frac{\hat{\mathbf{e}}_0^* \cdot \mathbf{f}(-\mathbf{k}_0)e^{i2kr} + \hat{\mathbf{e}}_0 \cdot \mathbf{f}^*(-\mathbf{k}_0)e^{-i2kr}}{2i}\right]. \quad (21.61)$$

The first (second) term in square brackets in (21.61) is purely real (imaginary). Therefore, (21.61) reduces to our desired result, the optical (or forward scattering) theorem:

$$\sigma_{\text{tot}} = \frac{4\pi}{k}\text{Im}\left[\hat{\mathbf{e}}_0^* \cdot \mathbf{f}(\mathbf{k}_0)\right]. \quad (21.62)$$

Because (21.62) was derived from (21.55), conservation of energy is the physical principle which underlies the optical theorem. The flow of energy is determined by the interference between the incident wave and the wave created by the excitation of the scatterer. This interference happens at all points in space, but only the amplitude of the radiated wave in the forward direction appears in the final statement of the theorem.

Two points are worth noting. First, our proof of (21.62) assumed an incident (plane) wave with infinite extent in the transverse direction. The theorem does not apply to a beam-like wave if its transverse width is smaller than the size of the scatterer. Second, the theorem is exact and therefore easy to violate when approximate expressions are used for σ_{tot} or $\mathbf{f}(\mathbf{k}_0)$. An example is the Thomson scattering calculation performed in Section 21.3. We found $\sigma_{\text{scatt}} = (8\pi/3)r_e^2$, yet $\hat{\mathbf{e}}_0^* \cdot \mathbf{f}(\mathbf{k}_0)$ was purely real. This is not consistent with (21.62).

To identify the problem, we use (21.55) and (21.56) and write the theorem in the form

$$\frac{4\pi}{k}\text{Im}\left[\hat{\mathbf{e}}_0^* \cdot \mathbf{f}(\mathbf{k}_0)\right] = \sigma_{\text{scatt}} + \frac{1}{2|\langle \mathbf{S}_{\text{inc}}\rangle|}\text{Re}\left[\int d^3r\, \mathbf{j}^* \cdot \mathbf{E}_{\text{inc}} + \int d^3r\, \mathbf{j}^* \cdot \mathbf{E}_{\text{rad}}\right]. \quad (21.63)$$

Equation (21.63) splits σ_{abs} into two terms. For Thomson scattering, the contribution $\int d^3r\, \mathbf{j}^* \cdot \mathbf{E}_{\text{inc}} = 0$ because a single electron has no internal dynamics to dissipate energy. On the other hand, $\int d^3r\, \mathbf{j}^* \cdot \mathbf{E}_{\text{rad}} \neq 0$ because the radiation field does work on the electron. This is the phenomenon of *radiation reaction*, which we neglected when writing the equation of motion (21.9) but which contributes an imaginary part to $\mathbf{f}(\mathbf{k}_0)$ when it is taken into account.[10] This omission is justified as far as the particle dynamics is concerned. It is *not* justified when we study conservation of energy as reflected in the optical theorem.

Application 21.2 Approximate Mie Scattering

This Application uses the optical theorem to derive the approximate cross section formula (21.45) for wave scattering from a sphere with index of refraction n. The idea is to replace the asymptotic electric field (21.3) by a scalar wave with the same form and mimic the partial-wave expansion of the scattering amplitude $f(\theta)$ which appears, e.g., in the quantum theory of potential scattering. This expansion uses a set of phase shifts, δ_ℓ, to parameterize the scattering:

$$f(\theta) = \frac{1}{2ik}\sum_{\ell=0}^{\infty}(2\ell+1)P_\ell(\cos\theta)\left[\exp(i2\delta_\ell) - 1\right]. \quad (21.64)$$

Following the ray optics discussion at the end of Section 21.5.2, we use $\ell = bk$ to replace the angular momentum index in (21.64) by the impact parameter $b = R\cos\alpha$ shown in Figure 21.13. A straight-line ray with this impact parameter accumulates a phase shift $2\delta_\ell = \Delta\sin\alpha$ with respect to a wave in vacuum where $\Delta = 2R(k_{\text{in}} - k_{\text{out}}) = 2kR(n-1)$.

[10] See Section 23.6.

We evaluate (21.64) in a continuum approximation where

$$\sum_{\ell=0}^{\infty} \to k \int_0^R db = kR \int_0^{\pi/2} d\alpha \sin\alpha. \tag{21.65}$$

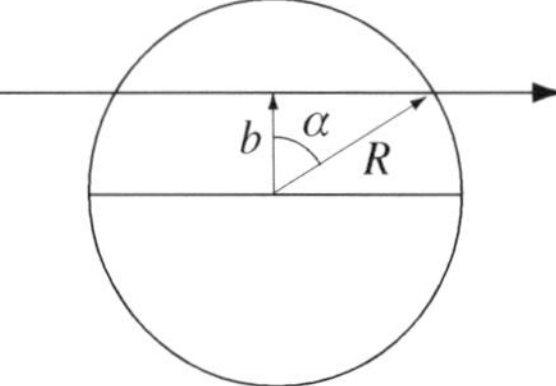

Figure 21.13: A ray passes through a dielectric sphere with impact parameter b.

Therefore, because $P_\ell(1) = 1$ and $2\ell + 1 \approx 2kR\cos\alpha$, the forward scattering amplitude becomes

$$f(0) = -ikR^2 \int_0^{\pi/2} d\alpha \sin\alpha \cos\alpha \left[\exp(i\Delta \sin\alpha) - 1\right]. \tag{21.66}$$

The substitution $x = \sin\alpha$ and an integration by parts yields

$$f(0) = ikR^2 \left[\frac{1}{2} - \frac{\exp(i\Delta)}{i\Delta} - \frac{\exp(i\Delta) - 1}{\Delta^2}\right]. \tag{21.67}$$

Replacing $\hat{\mathbf{e}}_0^* \cdot \mathbf{f}$ by f in the optical theorem (21.62) gives the cross section (21.45):

$$\sigma_{\text{approx}} = 2\pi R^2 \left[1 - \frac{2}{\Delta}\sin\Delta + \frac{2}{\Delta^2}(1 - \cos\Delta)\right]. \tag{21.68}$$

■

21.7.2 The Extinction Paradox

Consider an object of linear size a which does not absorb energy. In the short-wavelength limit when $ka \gg 1$, our experience with optics suggests that $\sigma_{\text{tot}} = \sigma_{\text{scatt}}$ will approach the geometric cross section σ_{geom}.[11] However, both Figure 21.7 for a conducting cylinder and Figure 21.9 for a dielectric sphere show, instead, that σ_{scatt} approaches *twice* this value. These are examples of a more general high-frequency result, called the *extinction paradox*, which is valid even when $\sigma_{\text{abs}} \neq 0$, namely, $\sigma_{\text{tot}} \to 2\sigma_{\text{geom}}$ when $ka \gg 1$.

We resolve this paradox (at least qualitatively) by paying careful attention to the scattered field in (21.1). The key is the Ewald-Oseen extinction theorem (see Section 20.9.1), which can be interpreted as dividing $\mathbf{E}_{\text{scatt}}$ into three parts. The first part annihilates the incident wave inside the object; the second part forms the entire wave field inside the object; the final part superposes with the incident field to form the entire wave field outside the object. With these in mind, we consider two objects: one perfectly conducting and one perfectly absorbing. In both cases, the short-wavelength limit $ka \gg 1$ implies that a dark and sharply defined shadow forms behind the object.

For the perfect conductor, $\sigma_{\text{abs}} = 0$ and the entire effect comes from σ_{scatt}. One contribution to σ_{scatt} comes from the incident wave power which is reflected from the illuminated face of the object. By definition, this piece has magnitude σ_{geom}. Another contribution comes from the power in the

[11] See the last line of Section 21.2 for the definition of σ_{geom}.

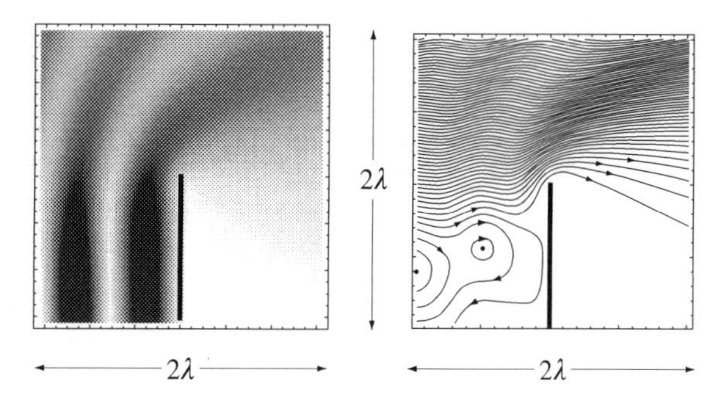

Figure 21.14: A plane wave incident from the left diffracts at the edge of a conducting half-plane (vertical black line). The electric field points out of the plane of the diagram. Left panel: electric field intensity; right panel: lines of the Poynting vector. Both panels are two wavelengths on a side. Figure from Berry (2001).

scattered wave which forms the shadow. This also has magnitude σ_{geom} because complete destructive interference requires that $\mathbf{E}_{\text{scatt}} = -\mathbf{E}_{\text{inc}}$ in the shadow region. The final result is $\sigma_{\text{scatt}} = 2\sigma_{\text{geom}}$. For the perfect absorber, $\sigma_{\text{abs}} = \sigma_{\text{geom}}$ because the field created inside the object now contains all the power which was reflected in the previous case. On the other hand, $\sigma_{\text{scatt}} = \sigma_{\text{geom}}$ because the formation of the shadow still requires that the sources inside the sphere annihilate the incident field in the shadow region. This gives $\sigma_{\text{tot}} = 2\sigma_{\text{geom}}$ as a consequence of (21.56).

So far, we have relied on the presence of a sharp shadow (and thus the large-object limit $ka \gg 1$) to explain the extinction paradox. However, numerical studies by Berg *et al.* (2011) show that $\sigma_{\text{tot}} \to 2\sigma_{\text{geom}}$ even for scattering from semi-transparent dielectric spheres where no shadow forms because ka is *not* large. If we generalize the definition of Δ in Application 21.2 to

$$\Delta = 2ka\,\text{Re}(n-1), \tag{21.69}$$

a summary of the numerical results for a sphere is

$$\lim_{\Delta \gg 1} \sigma_{\text{tot}} = 2\sigma_{\text{geom}}. \tag{21.70}$$

Some insight into (21.70) comes from studying the integration over angles of $d\sigma_{\text{tot}}/d\Omega$ for spheres with a complex index of refraction. When this integration is carried out over the illuminated part of the sphere's surface *alone*, the result is very nearly $2\sigma_{\text{geom}}$ for all values of Δ. Continuing the integration over the non-illuminated part of the surface drives the integral away from this value *except* when $\Delta \gg 1$. In that case, σ_{tot} remains equal to $2\sigma_{\text{geom}}$ because either (i) the contributions from every part of the non-illuminated surface are negligibly small (the shadow-forming conductor case) or (ii) the contributions from different parts of the non-illuminated surface vary rapidly in phase and cancel one another out (the non-shadow-forming dielectric case). It remains an open question to prove (21.70) rigorously for an arbitrary scatterer.

21.8 Diffraction by a Planar Aperture

Wave scattering from an object with a sharp edge is called *diffraction*. The first solution of the Maxwell equations for a true diffraction problem appeared in 1896, when Sommerfeld published his *tour de force* analysis of a plane wave incident on a semi-infinite conducting plane. Figure 21.14 illustrates some features of Sommerfeld's solution. The vertical dark line at the center of the each panel is the half-plane, which extends infinitely in three directions: downward, toward the reader, and away from the reader. The plane wave propagates from left to right with the electric vector $\mathbf{E}$ oriented perpendicular to the page. The shading in the left panel is darkest (lightest) in regions of highest

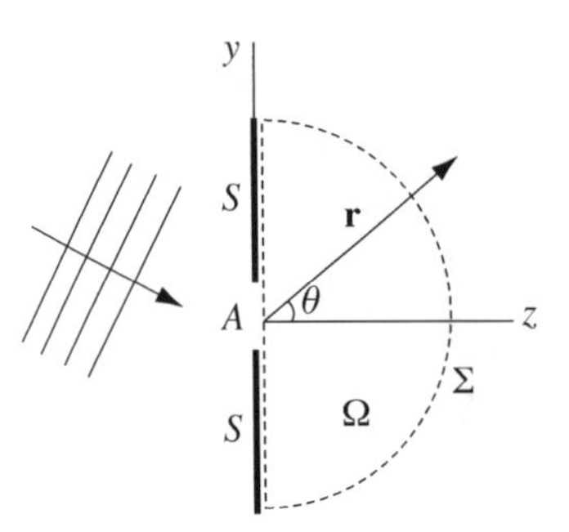

Figure 21.15: A monochromatic plane wave incident on an aperture (A) cut out of a plane screen (S). The dashed curve Σ is the surface of the volume Ω.

(lowest) electric field strength. The appearance of non-zero field intensity in the geometric shadow of the screen is one of the characteristic signatures of diffraction by a sharp edge.

The right panel of Figure 21.14 shows streamlines of the Poynting vector $\mathbf{S} = \mathbf{E} \times \mathbf{H}$ during edge diffraction. Several of these paths for energy flow bend around the edge and into the geometrical shadow. Other lines of $\mathbf{S}$ illustrate the reflection of the incident wave from the conducting surface. The small closed circle on the illuminated side of the plane is centered at one of many isolated points (actually lines perpendicular to the page) where the electric field vanishes. This may be compared to the parallel lines (or planes perpendicular to the page) of zero electric field which occur in the standing wave which forms when a plane wave falls with normal incidence on an infinite (rather than semi-infinite) conducting plane (see Section 19.3.1).

Broadly viewed, edge diffraction by a semi-infinite plane is a special case of electromagnetic diffraction from an aperture cut out of a flat screen. This more general problem—with screens made from different materials and single or multiple apertures shaped like circles, rectangles, or long narrow slits—has long attracted the attention of physicists, microscopists, and astronomers. We follow tradition and analyze this situation first using scalar diffraction theory. We then solve the problem in a manner that is fully consistent with the Maxwell equations.

21.8.1 Scalar Diffraction Theory

Figure 21.15 shows a monochromatic plane wave incident on an aperture (A) cut out of a planar screen (S) at $z = 0$. Scalar diffraction theory treats the plane wave as a scalar field and seeks the unique solution to the wave equation which propagates into the $z > 0$ half-space subject to suitable boundary conditions on the screen and at infinity. The guess $u(\mathbf{r})\exp(-i\omega t)$ implies that $u(\mathbf{r})$ satisfies Helmholtz' equation (21.24), and the Green function method is particularly well suited to our task. If $k_0 = \omega/c$, the Green function of interest satisfies

$$\left[\nabla^2 + k_0^2\right] G(\mathbf{r}, \mathbf{r}') = -\delta(\mathbf{r} - \mathbf{r}'). \tag{21.71}$$

The key step uses $u(\mathbf{r})$ and $G(\mathbf{r}', \mathbf{r})$ to write out Green's second identity (1.80) for the volume Ω enclosed by the surface Σ indicated in Figure 21.15. Ω comprises the entire $z > 0$ half-plane when we let the radius of the hemisphere go to infinity.

The identity in question reads

$$\int_\Omega d^3r' \,[u(\mathbf{r}')\nabla'^2 G(\mathbf{r}', \mathbf{r}) - G(\mathbf{r}', \mathbf{r})\nabla'^2 u(\mathbf{r}')] = \int_\Sigma d\mathbf{S}' \cdot [u(\mathbf{r}')\nabla' G(\mathbf{r}', \mathbf{r}) - G(\mathbf{r}', \mathbf{r})\nabla' u(\mathbf{r}')]. \tag{21.72}$$

For $\mathbf{r} \in \Omega$, (21.24) and (21.71) reduce (21.72) to

$$u(\mathbf{r}) = \int_{\Sigma} d\mathbf{S}' \cdot \left[G(\mathbf{r}', \mathbf{r})\nabla' u(\mathbf{r}') - u(\mathbf{r}')\nabla' G(\mathbf{r}', \mathbf{r})\right]. \tag{21.73}$$

Our previous experience with Green functions suggests that we impose boundary conditions on (21.71) so one of the two terms in (21.73) vanishes. We choose a Dirichlet condition on the planar surface $z = 0^+$ and an outgoing spherical wave condition at infinity:

$$G(\mathbf{r}', \mathbf{r}) = 0 \quad \text{when} \quad z' = 0^+$$

and (21.74)

$$r'\left(\partial G/\partial r' - ik_0 G\right) = 0 \quad \text{when} \quad r' \to \infty.$$

The second of these guarantees that the integral over the large hemispherical part of Σ vanishes. Therefore, because the outward normal to Ω along the screen is $-\hat{\mathbf{z}}$, (21.73) simplifies to

$$u(\mathbf{r}) = \int_{z'=0^+} dS' u(\mathbf{r}') \frac{\partial}{\partial z'} G(\mathbf{r}', \mathbf{r}). \tag{21.75}$$

We now need an explicit expression for $G(\mathbf{r}', \mathbf{r})$. A good starting point is the free-space Green function (20.53), which satisfies both (21.71) and the outgoing wave condition (see Section 20.3.3):

$$G_0(\mathbf{r}, \mathbf{r}') = \frac{\exp\left(ik_0|\mathbf{r} - \mathbf{r}'|\right)}{4\pi|\mathbf{r} - \mathbf{r}'|}. \tag{21.76}$$

Using (21.76) and the method of images (Section 8.3), it is straightforward to construct a Green function which vanishes when $z' = 0^+$, in accordance with (21.74). Because $\mathbf{R} = (x' - x)\hat{\mathbf{x}} + (y' - y)\hat{\mathbf{y}} + (z' - z)\hat{\mathbf{z}}$ and $\mathbf{R}^\star = (x' - x)\hat{\mathbf{x}} + (y' - y)\hat{\mathbf{y}} + (z' + z)\hat{\mathbf{z}}$ are image points with respect to $z' = 0$, the Green function we seek is

$$G(\mathbf{r}', \mathbf{r}) = \frac{\exp(ik_0 R)}{4\pi R} - \frac{\exp(ik_0 R^\star)}{4\pi R^\star}. \tag{21.77}$$

Now, the z'-derivative of (21.77) evaluated at $z' = 0$ is proportional to the z-derivative of (21.76) evaluated at $z' = 0$. Therefore, if $s^2 = (x' - x)^2 + (y' - y)^2 + z^2$, using (21.77) to evaluate (21.75) gives what is often called the *Rayleigh-Sommerfeld diffraction integral*:

$$u(\mathbf{r}) = -2\int_{z'=0} dS'\, u(\mathbf{r}')\hat{\mathbf{z}} \cdot \nabla G_0 = -\frac{1}{2\pi}\int_{z'=0} dS'\, u(\mathbf{r}')\frac{\partial}{\partial z}\left[\frac{\exp(ik_0 s)}{s}\right]. \tag{21.78}$$

The integral equation (21.78) relates the unique solution of the scalar Helmholtz equation (21.24) in the $z > 0$ half-space to its boundary values $u(x, y, 0)$. Of course, the latter are generally not known.

The name Kirchhoff is associated with an approximation which transforms (21.78) into a formula for $u(\mathbf{r})$. Let $u_0(\mathbf{r})$ be the *incident* plane wave. In Figure 21.15, the idea is to set $u(x, y, 0) = 0$ on the screen and $u(x, y, 0) = u_0(x, y, 0)$ in the aperture. We thereby restrict the integration in (21.78) to the aperture. This is a high-frequency, small-wavelength approximation which performs best when the majority of the field in the aperture is far from the perturbation induced by the boundary of the aperture, in other words, when $\lambda/a \ll 1$ where λ is the wavelength and a is the smallest characteristic dimension of the aperture. This is the usual domain of optics, which was the class of problems of interest to Kirchhoff.

The name Fraunhofer is associated with the far-field limit of (21.78) where $k_0 r \gg 1$. In that case, the z-derivative in (21.78) may be approximated by the multiplicative factor ik_0 and the rightmost integral in (21.78) becomes a mathematical expression of *Huygens' principle*: a wave incident on an aperture propagates into the far field as if every element of the aperture is the source of a spherical wave with an amplitude and phase given by the incident wave. If, in addition, the aperture size is small and we restrict ourselves to observation points not far from the z-axis, it is common to define a variable $\mathbf{p} = k_0\hat{\mathbf{r}}$, use the far-field limit of ∇G_0 [see (21.97)], and ignore the angular factor $\hat{\mathbf{z}} \cdot \hat{\mathbf{r}}$ to write

$$u(\mathbf{r}) \approx -\frac{ik_0}{2\pi}\frac{\exp(ik_0 r)}{r} \int_{\text{aperture}} dS' u_0(\mathbf{r}') \exp(-i\mathbf{p} \cdot \mathbf{r}'). \tag{21.79}$$

The subject of Fourier optics exploits the conclusion from (21.79) that the far-zone field field $u(x, y, z)$ is proportional to the two-dimensional Fourier transform of the aperture field $u_0(x, y, 0)$.

Further progress requires that we relate the scalar field $u(\mathbf{r})$ to a specific electromagnetic quantity. The Cartesian components of $\mathbf{E}(\mathbf{r})$ and $\mathbf{B}(\mathbf{r})$ are natural choices because each one individually satisfies (21.24) in the $z > 0$ half-space. Unfortunately, the fields obtained by solving the integral equation (21.78) exactly for each component do not generally satisfy the Maxwell equations.[12] This fact notwithstanding, widespread practice simply compares $|u(\mathbf{r})|^2$ with the measured intensity of the field. Figure 21.16 is an example which compares experiment (left side) with the prediction of (21.78) using Kirchhoff's approximation (right side) for a plane wave that diffracts from a circular aperture (radius a) in a plane screen at normal incidence.

The three panels plot the field intensity in planes parallel to the screen for different values of $F = a^2/\lambda z$. When F is an integer and $\lambda/a \ll 1$ (as is the case here) the diffraction pattern in each plane has F maxima and $F - 1$ minima. These features arise from the constructive and destructive interference of waves that arrive at the observation plane from different points in the aperture. The same interference produces oscillating intensity maxima and minima on the symmetry axis ($r = 0$) as a function of F. The agreement between experiment and scalar diffraction theory in Figure 21.16 is remarkable and various explanations of this fact have been offered. All of these draw on the vector theory of electromagnetic diffraction, to which we turn next.[13]

21.8.2 Vector Diffraction Theory

Vector diffraction theory treats the incident field in Figure 21.15 as an electromagnetic plane wave and seeks the unique solution to the Maxwell equations which propagates into the $z > 0$ half-space subject to suitable boundary conditions on the screen and at infinity. There is more than one way to do this. One approach, indicated schematically on the left side of Figure 21.17, exploits the angular spectrum of plane waves (Section 16.6.1) to represent the field at every point in the aperture as an appropriate linear combination of plane waves. The interference of these waves in the near zone (close to the aperture) evolves, in the far zone, to a radiation field determined by the single plane which propagates in the direction $\mathbf{r}$ of observation.

The right side of Figure 21.17 illustrates a different approach to the vector diffraction problem. Here, one generalizes the Huygens' principle idea mentioned in the paragraph preceding (21.79) and regards each point in the aperture as the source of a single *vector* spherical wave. The field at all distances reflects the interference among all the sources. The plane wave and spherical wave points of view provide complementary intuition about the phenomenon of diffraction. In this section, we begin with the plane wave method and describe how it is used to calculate the electric field diffracted by a planar aperture. However, rather than using it to analyze any particular aperture geometry, we sum over

[12] The Sommerfeld half-plane problem (Figure 21.14) has enough symmetry that complete knowledge of one particular field component is sufficient (through the Maxwell equations) to determine all the other components. In that case, $u(\mathbf{r})$ may be identified with that one particular component.

[13] We discuss the apparent success of scalar diffraction theory seen here in Section 21.8.3 below.

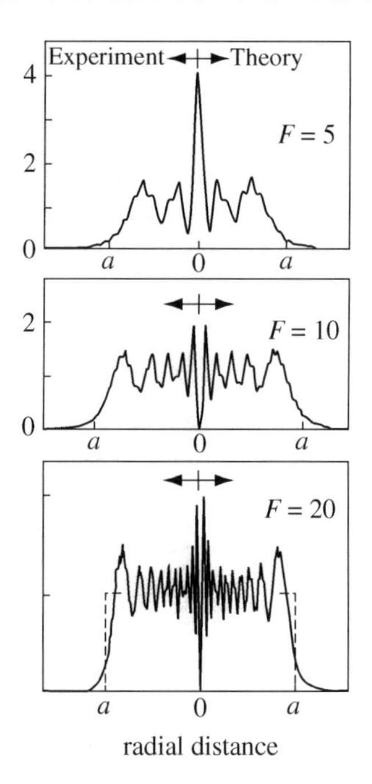

Figure 21.16: Normal-incidence diffraction from a circular aperture of radius a for different values of $F = a^2/\lambda z$. Left side: measured field intensity. Right side: $|u(\rho, z)|^2$ calculated using (21.78) and Kirchhoff's approximation. The dashed rectangle is the geometrical optics approximation. Figure from Siegman (1986).

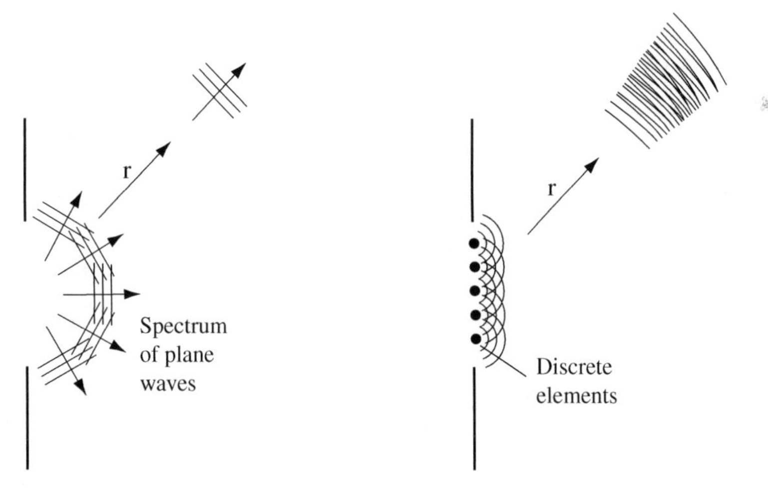

Figure 21.17: Two representations of the field diffracted by an aperture. Left: angular spectrum of plane waves. Right: Huygens' effective sources. Figure from Smith (1997).

the spectrum of plane waves to derive an expression for the diffracted electric field which manifestly illustrates Huygens' principle. Using the latter, we rationalize the success of scalar diffraction theory in Figure 21.16 and calculate the vector fields diffracted by a circular aperture.

Assume that all fields vary as $\exp(-i\omega t)$ and let $\omega = ck_0$. The angular-spectrum approach to diffraction from a planar aperture at $z = 0$ superposes plane wave solutions of the vector Helmholtz equation,

$$(\nabla^2 + k_0^2)\mathbf{E} = 0, \tag{21.80}$$

in such a way that the sum converges for $z \geq 0$. For a representative plane wave, $\exp(i\mathbf{k}\cdot\mathbf{r})$, these two conditions constrain the wave vector $\mathbf{k} = k_x\hat{\mathbf{x}} + k_y\hat{\mathbf{y}} + k_z\hat{\mathbf{z}}$, so

$$k_z = \begin{cases} \sqrt{k_0^2 - k_x^2 - k_y^2} & k_x^2 + k_y^2 \leq k_0^2, \\ i\sqrt{k_x^2 + k_y^2 - k_0^2} & k_x^2 + k_y^2 \geq k_0^2. \end{cases} \tag{21.81}$$

The plane waves with k_z real in (21.81) propagate with undiminished amplitude as z increases; the waves with k_z imaginary are evanescent and decay exponentially as z increases. In all that follows, it is important to interpret every appearance of the variable k_z (either explicitly or as a component of $\mathbf{k}$) as a shorthand for (21.81).

It is straightforward to construct a sum over plane waves for $\mathbf{E}(x, y, z \geq 0)$ which satisfies $\nabla\cdot\mathbf{E} = 0$ automatically. The key is to make the expansion coefficients the Cartesian components of a two-dimensional vector function $\boldsymbol{\mathcal{E}}(k_x, k_y) = \mathcal{E}_x(k_x, k_y)\hat{\mathbf{x}} + \mathcal{E}_y(k_x, k_y)\hat{\mathbf{y}}$. Doing this, an expression with manifestly zero divergence is

$$\mathbf{E}(x, y, z \geq 0) = \frac{1}{(2\pi)^2}\int\limits_{-\infty}^{\infty} dk_x \int\limits_{-\infty}^{\infty} dk_y \left[\boldsymbol{\mathcal{E}} - \frac{1}{k_z}(\mathbf{k}\cdot\boldsymbol{\mathcal{E}})\,\hat{\mathbf{z}}\right]\exp(i\mathbf{k}\cdot\mathbf{r}). \tag{21.82}$$

It is essential to the completeness of the Fourier representation that (21.82) includes the evanescent waves in (21.81). It does not matter that these waves never reach the far zone. What matters is that their presence in (21.82) influences the amplitudes of the propagating waves in the sum which *do* reach the radiation zone. Now, let $\mathbf{k}_\perp = k_x\hat{\mathbf{x}} + k_y\hat{\mathbf{y}}$ and $\mathbf{r}_\perp = x\hat{\mathbf{x}} + y\hat{\mathbf{y}}$. Equation (21.82) shows that the x- and y-components of the electric field evaluated at $z = 0$ are

$$E_{x,y}(\mathbf{r}_\perp, z = 0) = \frac{1}{2\pi^2}\int d^2k_\perp \mathcal{E}_{x,y}(\mathbf{k}_\perp)\exp(i\mathbf{k}_\perp\cdot\mathbf{r}_\perp). \tag{21.83}$$

Equation (21.83) is a two-dimensional Fourier transform. Therefore, the expansion coefficients we seek are determined by the inverse Fourier transform. Dropping the explicit $z = 0$ on the left side of (21.83), the result is

$$\mathcal{E}_{x,y}(\mathbf{k}_\perp) = \int d^2r'_\perp E_{x,y}(\mathbf{r}'_\perp)\exp(-i\mathbf{k}_\perp\cdot\mathbf{r}'_\perp). \tag{21.84}$$

Substituting (21.84) into (21.82) expresses $\mathbf{E}(x, y, z \geq 0)$ in terms of its transverse components evaluated at $z = 0$. If the solution of this integral equation can be found (or approximated), the associated magnetic field follows from $\nabla\times\mathbf{E} = i\omega\mathbf{B}$.

Practical diffraction calculations using the angular-spectrum method require (i) a guess for the (generally unknown) values of $E_{x,y}(x, y, z = 0)$; (ii) evaluation of the integral (21.84) to find the expansion coefficients; and (iii) evaluation of the field integral (21.82). The reader may wish to explore this approach for specific aperture geometries. Our choice is to bypass such calculations and, instead, transform the sum over plane waves (21.82) into a sum over spherical waves. The calculation proceeds most smoothly if we replace (21.83) by

$$\hat{\mathbf{z}}\times\mathbf{E}(x, y, z = 0) = \frac{1}{(2\pi)^2}\int d^2k_\perp\,[\hat{\mathbf{z}}\times\boldsymbol{\mathcal{E}}(\mathbf{k}_\perp)]\exp(i\mathbf{k}_\perp\cdot\mathbf{r}). \tag{21.85}$$

The associated inverse Fourier transform replaces (21.84):

$$\hat{\mathbf{z}} \times \mathcal{E}(\mathbf{k}_\perp) = \int d^2r'_\perp \, [\hat{\mathbf{z}} \times \mathbf{E}(\mathbf{r}'_\perp)] \exp(-i\mathbf{k}_\perp \cdot \mathbf{r}'_\perp). \tag{21.86}$$

Our strategy is to exploit (21.86) and a brief, explicit calculation confirms that (21.82) is identical to

$$\mathbf{E}(x, y, z \geq 0) = \frac{i}{(2\pi)^2} \nabla \times \int d^2k_\perp \left[\frac{\hat{\mathbf{z}} \times \mathcal{E}(\mathbf{k}_\perp)}{k_z} \right] \exp(i\mathbf{k} \cdot \mathbf{r}). \tag{21.87}$$

Substituting (21.86) into (21.87) gives the key intermediate result,

$$\mathbf{E}(\mathbf{r}_\perp, z \geq 0) = \nabla \times 2 \int d^2r'_\perp \, [\hat{\mathbf{z}} \times \mathbf{E}(\mathbf{r}'_\perp)] \left\{ \frac{i}{8\pi^2} \int \frac{d^2k_\perp}{k_z} \exp[i\mathbf{k}_\perp \cdot (\mathbf{r} - \mathbf{r}') + ik_z z] \right\}. \tag{21.88}$$

The quantity in curly braces in (21.88) may be rewritten using *Weyl's identity*,[14]

$$\frac{\exp(ik_0 r)}{4\pi r} = \frac{i}{8\pi^2} \int \frac{d^2k_\perp}{k_z} \exp(i\mathbf{k}_\perp \cdot \mathbf{r}_\perp + ik_z|z|). \tag{21.89}$$

Therefore, because the left side of (21.89) is the free-space Green function (21.76), (21.88) becomes *Smythe's formula* for the diffracted electric field,[15]

$$\mathbf{E}(\mathbf{r}_\perp, z \geq 0) = \nabla \times 2 \int_{z'=0} d^2r'_\perp \, [\hat{\mathbf{z}} \times \mathbf{E}(\mathbf{r}'_\perp)] G_0(\mathbf{r}, \mathbf{r}'_\perp). \tag{21.90}$$

Equation (21.90) is the vector analog of the scalar integral equation (21.78). The associated magnetic field follows from $\nabla \times \mathbf{E} = i\omega\mathbf{B}$. A precisely analogous calculation gives

$$\mathbf{B}(\mathbf{r}_\perp, z \geq 0) = \nabla \times 2 \int_{z'=0} d^2r'_\perp \, [\hat{\mathbf{z}} \times \mathbf{B}(\mathbf{r}'_\perp)] G_0(\mathbf{r}, \mathbf{r}'_\perp), \tag{21.91}$$

with an electric field given by $\nabla \times \mathbf{B} = -i\omega\mathbf{E}/c^2$.

A glance back at Application 16.2 of Section 16.9.2 shows that (21.90) and (21.91) are superpositions of the fields (16.183) and (16.184) with harmonic time dependence and different choices for the direction of the constant vector $\hat{\mathbf{s}}$ in the plane of the screen. Taken together, the results of this section show that it is sufficient to know the tangential components of $\mathbf{E}(x, y, z = 0)$ *or* $\mathbf{B}(x, y, z = 0)$ to compute both fields in the $z \geq 0$ half-space. This is consistent with the uniqueness theorem for time-dependent fields (Application 15.1 of Section 15.4.1). It remains only to explore the approximations and limits needed to evaluate (21.90) and (21.91) for practical calculations of the fields diffracted by an aperture in a planar screen.

21.8.3 The Kirchhoff Approximation

The boundary values of $\hat{\mathbf{z}} \times \mathbf{E}$ and $\hat{\mathbf{z}} \times \mathbf{B}$ in (21.90) and (21.91) are generally not known. To make progress, the *Kirchhoff approximation* for this vector problem puts $\mathbf{F} = \mathbf{E}$ in (21.90), $\mathbf{F} = \mathbf{B}$ in (21.91),

[14] The Weyl identity, which makes use of (21.81), can be established by (i) Fourier transforming (21.71) with $\mathbf{r}' = 0$; (ii) solving the resulting algebraic equation for $G(\mathbf{k})$; and (iii) performing the k_z integration of the inverse Fourier transform by contour integration. For a method which avoids the complex plane, see A.S. Marathay, "Fourier transform of the Green function for the Helmholtz equation", *Journal of the Optical Society of America* **65**, 964 (1975).

[15] See W.R. Smythe, "The double current sheet in diffraction", *Physical Review* **72**, 1066 (1947), and Section 12.18 of W.R. Smythe, *Static and Dynamic Electricity*, 3rd edition (McGraw Hill, New York, 1969).

and supposes that

$$\begin{aligned}\hat{\mathbf{z}} \times \mathbf{F} &\approx 0 \quad \text{on the screen } S \\ \hat{\mathbf{z}} \times \mathbf{F} &\approx \hat{\mathbf{z}} \times \mathbf{F}_{\text{inc}} \quad \text{in the aperture } A.\end{aligned} \tag{21.92}$$

This approximation is used for screens of all types in the high-frequency (optical) limit when the wavelength is small compared to the aperture size. As mentioned earlier, this is the regime where the perturbation to the field in the aperture by the aperture boundary should be negligible.[16]

The first line of the Kirchhoff approximation in (21.92) limits the domain of integration in (21.90) and (21.91) to the aperture. This and the spherical wave character of $G_0(\mathbf{r}, \mathbf{r}')$ in (21.76) show that (21.89) and (21.90) reflect the Huygens' principle point of view illustrated on the right side of Figure 21.17: each point in the aperture is the origin of a spherical wave with a phase and amplitude determined by the incident field.

It is worth noting that, whether one uses $\hat{\mathbf{n}} \times \mathbf{E}$ data or $\hat{\mathbf{n}} \times \mathbf{B}$ data, one of the two Kirchhoff boundary conditions is always *exact* when the screen is an infinitely thin perfect conductor. In that case,

$$\begin{aligned}\hat{\mathbf{z}} \times \mathbf{E} &= 0 \quad \text{on the screen } S \\ \hat{\mathbf{z}} \times \mathbf{B} &= \hat{\mathbf{z}} \times \mathbf{B}_{\text{inc}} \quad \text{in the aperture } A.\end{aligned} \tag{21.93}$$

The first condition is familiar from all our previous work with perfect conductors. The second follows from the surface current density matching condition, $\hat{\mathbf{n}}_2 \times [\mathbf{B}_1 - \mathbf{B}_2] = \mu_0 \mathbf{K}$, and the invariance of the screen with respect to reflection in the $z = 0$ plane. The latter implies that $\hat{\mathbf{z}} \times \mathbf{B}_{\text{ind}}(0^+) = -\hat{\mathbf{z}} \times \mathbf{B}_{\text{ind}}(0^-)$, where $\mathbf{B}_{\text{ind}}$ is the field produced by the currents induced in the screen by the incident field. We conclude that $\hat{\mathbf{z}} \times \mathbf{B}_{\text{ind}}(0^\pm) \propto \mathbf{K}$. The current density is zero in the aperture, so only the incident plane wave contributes to $\hat{\mathbf{z}} \times \mathbf{B}(0^+)$.

We now return to Figure 21.16. The Kirchhoff approximation to scalar diffraction theory was used to compute the "theory" half of that figure and our vector diffraction theory will help rationalize the agreement between that theory and experiment. The first step is to perform the curl operation in (21.90). This puts Smythe's formula in the form

$$\mathbf{E}(\mathbf{r}_\perp, z \geq 0) = -2 \int\limits_{z'=0} d^2 r'_\perp \, [\hat{\mathbf{z}} \times \mathbf{E}(\mathbf{r}'_\perp)] \times \nabla G_0(\mathbf{r}, \mathbf{r}'_\perp). \tag{21.94}$$

The second step writes out the cross products in (21.94) to get

$$\mathbf{E}(\mathbf{r}) = -2 \int\limits_{z'=0} d^2 r'_\perp \, \mathbf{E} \, \hat{\mathbf{z}} \cdot \nabla G_0 + 2\hat{\mathbf{z}} \int\limits_{z'=0} d^2 r'_\perp \, \mathbf{E} \cdot \nabla G_0. \tag{21.95}$$

Comparing (21.95) to (21.78) shows that the latter reproduces the former exactly for the field components $E_x(\mathbf{r})$ and $E_y(\mathbf{r})$, but fails to do this for $E_z(\mathbf{r})$. Therefore, the scalar theory should predict experimental field intensities well when E_z is negligible. This is true when the angle of incidence is near normal and the observation points are not far from the symmetry axis (except when $z < \lambda$). These conditions happen to be satisfied for the data shown in Figure 21.16.

21.8.4 Fraunhofer Diffraction

The term *Fraunhofer diffraction* is used when the distance from the aperture to the observation point is large compared to the wavelength and large compared to the size of the aperture. The exact electric

[16] The Kirchhoff approximation can be used to study scattering from a two-dimensional scatterer like a disk or a strip by regarding the plane of the scatterer as "mostly aperture".

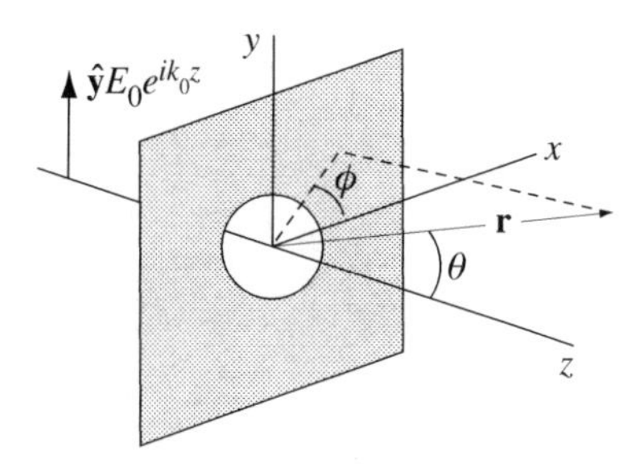

Figure 21.18: Diffraction of a plane wave by a circular aperture.

field expression (21.94) simplifies considerably in this limit. If $\mathbf{R} = \mathbf{r} - \mathbf{r}'$, a brief calculation confirms that the gradient of the free-space Green function (21.76) is

$$\nabla G_0 = (ik_0 R - 1)\frac{G_0(R)}{R}\hat{\mathbf{R}}. \tag{21.96}$$

The stated conditions imply that $k_0 R \gg 1$ and $r \gg r'$. Therefore,

$$\nabla G_0 \approx ik_0 \hat{\mathbf{r}} \frac{\exp(ik_0 r)}{4\pi r} \exp(-ik_0 \hat{\mathbf{r}} \cdot \mathbf{r}'). \tag{21.97}$$

Substituting (21.97) into (21.94) gives the Fraunhofer limit of Smythe's formula as

$$\mathbf{E}_{\rm rad}(\mathbf{r}_\perp, z \geq 0) = ik_0 \frac{\exp(ik_0 r)}{2\pi r} \hat{\mathbf{r}} \times \int\limits_{z'=0} d^2 r'_\perp \left[\hat{\mathbf{z}} \times \mathbf{E}(\mathbf{r}'_\perp)\right] \exp(-ik_0 \hat{\mathbf{r}} \cdot \mathbf{r}'_\perp). \tag{21.98}$$

As expected for a radiation field, $\mathbf{E}_{\rm rad}$ is transverse to $\hat{\mathbf{r}}$ and decreases as $1/r$. The asymptotic ($r \to \infty$) magnetic field calculated from $\mathbf{B} = (\nabla \times \mathbf{E})/i\omega$ also has the anticipated radiation zone form,

$$c\mathbf{B}_{\rm rad} = \hat{\mathbf{r}} \times \mathbf{E}_{\rm rad}. \tag{21.99}$$

As an example, we consider the Kirchhoff approximation to (21.98) when a normal-incidence plane wave with $\mathbf{E}_{\rm inc} = E_0 \exp(ik_0 z)\hat{\mathbf{y}}$ strikes a conducting screen with a circular aperture. The screen in Figure 21.18 occupies $z = 0$ and the origin-centered aperture has radius a. The z-axis is normal to the aperture, so $\mathbf{r} = r \sin\theta \cos\phi\, \hat{\mathbf{x}} + r \sin\theta \sin\phi\, \hat{\mathbf{y}} + r \cos\theta\, \hat{\mathbf{z}}$ defines a spherical coordinate system. In the plane of the aperture ($\theta = \pi/2$), we write $\mathbf{r}_\perp = \rho \cos\phi\, \hat{\mathbf{x}} + \rho \sin\phi\, \hat{\mathbf{y}}$. Using these and the formulae in Section 1.2.3,

$$\hat{\mathbf{r}} \times [\hat{\mathbf{z}} \times \mathbf{E}_{\rm inc}]_{z=0} = -E_0(\sin\phi \hat{\boldsymbol{\theta}} + \cos\phi \cos\theta \hat{\boldsymbol{\phi}}) \quad \text{and} \quad \hat{\mathbf{r}} \cdot \mathbf{r}' = \rho' \sin\theta \cos(\phi - \phi'). \tag{21.100}$$

Therefore,

$$\mathbf{E} = -ik_0 E_0 \frac{\exp(ik_0 r)}{2\pi r} (\sin\phi \hat{\boldsymbol{\theta}} + \cos\phi \cos\theta \hat{\boldsymbol{\phi}}) \int\limits_0^a d\rho' \rho' \int\limits_0^{2\pi} d\phi' \exp[-ik_0 \rho' \sin\theta \cos(\phi - \phi')]. \tag{21.101}$$

The symmetry of a circle implies that all the ϕ dependence of the diffracted field is carried by the vector pre-factor in (21.101). Therefore, we can set $\phi = 0$ inside the ϕ integral and use

$$\int\limits_0^{2\pi} d\phi' \exp[-ik_0 \rho' \sin\theta \cos\phi'] = 4 \int\limits_0^{\pi/2} d\phi' \cos[k_0 \rho' \sin\theta \cos\phi']. \tag{21.102}$$

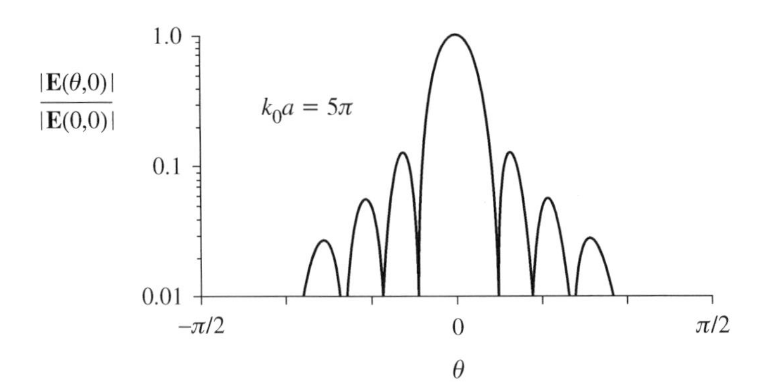

Figure 21.19: Angular dependence of the electric field magnitude diffracted by a circular aperture with $k_0a = 5\pi$. Figure from Smith (1997).

The final result takes a convenient form if we combine (21.102) with an integral representation of the zero-order Bessel function,

$$J_0(u) = \frac{2}{\pi}\int_0^{\pi/2} d\phi \cos(u\cos\phi), \tag{21.103}$$

and the definite integral,

$$\int_0^z du\, u J_0(u) = zJ_1(z), \tag{21.104}$$

namely,

$$\mathbf{E}(r,\theta,\phi) = -\frac{i}{2}E_0(k_0a)^2\frac{\exp(ik_0r)}{k_0r}\left[\frac{2J_1(k_0a\sin\theta)}{k_0a\sin\theta}\right](\sin\phi\hat{\boldsymbol{\theta}} + \cos\phi\cos\theta\hat{\boldsymbol{\phi}}). \tag{21.105}$$

Figure 21.19 plots the normalized magnitude of the diffracted electric field (21.105) in the $\phi = 0$ plane for an electrically large aperture with $k_0a = 5\pi$. The vertical scale is logarithmic, so the vast majority of the diffracted field is contained in the central lobe around the forward ($\theta = 0$) direction. The number of side lobes and interference zeroes increases as k_0a increases because they are determined by the zeroes of J_1.

The secondary maxima and minima occur in Figure 21.19 for the same reason they occur in Figure 21.16: constructive and destructive interference among Huygens' wavelets emanating from different points in the aperture. Indeed, a scalar diffraction calculation leads to the same integral that appears in (21.101). For this particular problem, the full Maxwell formalism adds only the vector post-factor in (21.105). On the other hand, the alert reader will have noticed the distinct similarity between the circular aperture diffraction pattern in Figure 21.19 and the radiation pattern for the circular reflector shown in the rightmost panel of Figure 21.12. This is so because the plane represented by the straight line in the middle panel of Figure 21.12 may be regarded as the "aperture" in a "screen" which happens to be made of vacuum. The field reflected from the concave dish onto this circular aperture serves as the "incident" field in a Kirchhoff/Fraunhofer approximation to the field radiated by the reflector.

We conclude with a brief historical remark. Lord Rayleigh famously used the scalar version of (21.105) to determine the resolving power of a telescope or microscope with a circular aperture. He suggested that two point sources are just resolvable if their angular separation θ is such that the first

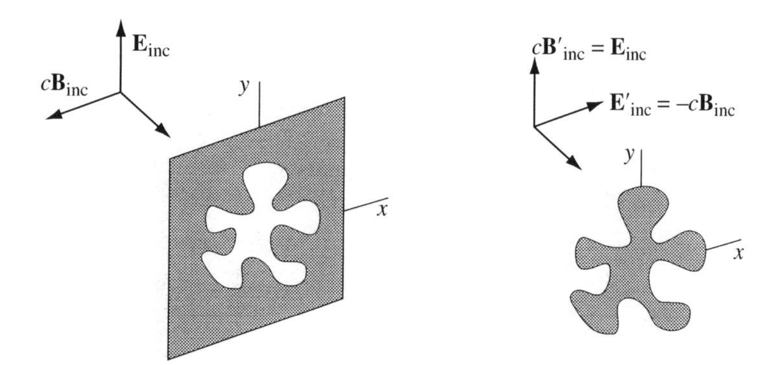

Figure 21.20: Plane wave scattering from an aperture in a conducting screen (left panel) and from a complementary conducting plate (right panel).

minimum of the diffraction pattern of one source coincides with the central maximum of the diffraction pattern of the other source. Since $x_1 = 3.832$ is the first zero of $J_1(x)$, we set $k_0 a \sin\theta = x_1$ and use $\lambda = 2\pi/k_0$ to write the *Rayleigh criterion* as

$$\sin\theta = 0.6\,\frac{\lambda}{a}. \tag{21.106}$$

21.9 Generalized Optical Principles

Before Maxwell, the laws of optics were developed using a scalar field (the light intensity or its square root) as the fundamental quantity. After Maxwell, the optical laws were generalized to be consistent with the vector field theory of electromagnetism. The previous section presented an example: a Huygens-like principle for the electric and magnetic fields diffracted by an aperture cut out of an infinite planar sheet. In this section, we specialize to an aperture cut out of a *conducting* sheet and prove an electromagnetic version of Babinet's principle. We then return to Huygens' principle and prove a vector version of the principle for the electric and magnetic fields diffracted by an *arbitrary* scatterer.

21.9.1 Babinet's Principle for Vector Fields

Babinet's principle elegantly relates the scattering produced by the two *complementary* objects shown in Figure 21.20. One is a flat, infinitely large, infinitesimally thin, and perfectly conducting sheet with an aperture cut out. The other is a flat, infinitesimally thin, and perfectly conducting plate with the exact size and shape of the aperture. The incident plane waves in the two cases are not identical. Rather,

$$\mathbf{E}'_{\rm inc} = -c\mathbf{B}_{\rm inc} \qquad \text{and} \qquad c\mathbf{B}'_{\rm inc} = \mathbf{E}_{\rm inc}. \tag{21.107}$$

The total fields in each case are the sum of the incident fields shown and the field scattered by each conductor. Thus,

$$\mathbf{B} = \mathbf{B}_{\rm inc} + \mathbf{B}_{\rm scatt} \qquad \text{and} \qquad \mathbf{E} = \mathbf{E}_{\rm inc} + \mathbf{E}_{\rm scatt}, \tag{21.108}$$

and similarly for the primed fields. Babinet's principle asserts that the electromagnetic fields of these complementary scattering problems are related by

$$\mathbf{E}'_{\rm scatt} = c\mathbf{B} \qquad \text{and} \qquad c\mathbf{B}'_{\rm scatt} = -\mathbf{E}. \tag{21.109}$$

We prove (21.109) by writing out the time-harmonic ($\omega = ck_0$) Maxwell equations and the boundary conditions satisfied by $(\mathbf{E}, \mathbf{B})$ and by $(\mathbf{E}'_{\text{scatt}}, \mathbf{B}'_{\text{scatt}})$. For the aperture problem, the total field in the $z > 0$ half-space satisfies

$$\nabla \times \mathbf{E} = ik_0(c\mathbf{B}) \qquad \text{and} \qquad \nabla \times (c\mathbf{B}) = -ik_0\mathbf{E}. \tag{21.110}$$

We repeat from (21.93) the boundary conditions for the total field at $z = 0$:

$$\begin{aligned} \hat{\mathbf{z}} \times \mathbf{E} &= 0 && \text{off the aperture} \\ \hat{\mathbf{z}} \times \mathbf{B} &= \hat{\mathbf{z}} \times \mathbf{B}_{\text{inc}} && \text{on the aperture.} \end{aligned} \tag{21.111}$$

For the plate problem, the scattered field in the $z > 0$ half-space satisfies

$$\nabla \times \mathbf{E}'_{\text{scatt}} = ik_0(c\mathbf{B}'_{\text{scatt}}) \qquad \text{and} \qquad \nabla \times (c\mathbf{B}'_{\text{scatt}}) = -ik_0\mathbf{E}'_{\text{scatt}}. \tag{21.112}$$

The boundary conditions for the scattered field at $z = 0$ are

$$\begin{aligned} \hat{\mathbf{z}} \times (c\mathbf{B}'_{\text{scatt}}) &= 0 && \text{off the plate} \\ \hat{\mathbf{z}} \times \mathbf{E}'_{\text{scatt}} &= \hat{\mathbf{z}} \times (c\mathbf{B}_{\text{inc}}) && \text{on the plate.} \end{aligned} \tag{21.113}$$

The first condition in (21.113) is true for the same reason that the second condition in (21.111) is true. The second condition in (21.113) follows from $\hat{\mathbf{z}} \times (\mathbf{E}'_{\text{scatt}} + \mathbf{E}'_{\text{inc}}) = 0$ and (21.107). This is all we need. Babinet's principle (21.109) is the statement that (21.112) and (21.113) transform to (21.110) and (21.111) when we make the duality substitutions $\mathbf{E}'_{\text{scatt}} \to c\mathbf{B}$ and $c\mathbf{B}'_{\text{scatt}} \to -\mathbf{E}$.[17] We emphasize that Babinet's principle (21.109) is an *exact* result that applies to both the near-field and far-field limits of the scattered fields.

A straightforward application of Babinet's principle relates the diffraction pattern produced by a circular hole in a conducting sheet to the diffraction pattern produced by a conducting disk which just fills the hole. Because $(\hat{\mathbf{k}}, \mathbf{E}_{\text{rad}}, \mathbf{B}_{\text{rad}})$ forms a right-handed orthogonal triad of vectors and $|\mathbf{E}| = c|\mathbf{B}|$ in the radiation zone, the cross product of $\hat{\mathbf{k}}$ with (21.109) evaluated in the radiation zone gives

$$\begin{aligned} \hat{\mathbf{k}} \times \mathbf{E}_{\text{rad}}(\text{disk}) &= \hat{\mathbf{k}} \times c\mathbf{B}_{\text{rad}}(\text{hole}) && \Longrightarrow && c\mathbf{B}_{\text{rad}}(\text{disk}) = -\mathbf{E}_{\text{rad}}(\text{hole}) \\ \hat{\mathbf{k}} \times c\mathbf{B}_{\text{rad}}(\text{disk}) &= -\hat{\mathbf{k}} \times \mathbf{E}_{\text{rad}}(\text{hole}) && \Longrightarrow && \mathbf{E}_{\text{rad}}(\text{disk}) = c\mathbf{B}_{\text{rad}}(\text{hole}). \end{aligned} \tag{21.114}$$

Consequently,

$$|\mathbf{E}_{\text{rad}}(\text{disk})| = |\mathbf{E}_{\text{rad}}(\text{hole})| \qquad \text{and} \qquad |\mathbf{B}_{\text{rad}}(\text{disk})| = |\mathbf{B}_{\text{rad}}(\text{hole})|. \tag{21.115}$$

This tells us that, apart from the difference in incident field polarization implied by (21.107), the far-field diffraction patterns produced by a circular aperture and a conducting disk are the same. For normal incidence, that pattern is Figure 21.19 when the Kirchhoff approximation is valid. The latter replaces the true field by the incident field in the hole, so this will be true only for electrically large apertures where the perturbation from the hole's perimeter can be neglected.

Application 21.3 Sub-Wavelength Apertures and Near-Field Optics

This Application uses Babinet's principle and the physics of the angular-spectrum representation to analyze the diffraction of a plane wave by a sub-wavelength aperture ($a \ll \lambda$) in a conducting sheet. In the *far field*, we will confirm the expectation from Section 21.8.3 that Kirchhoff's approximation fails for an electrically small aperture. In the *near field*, we will learn how to "beat the diffraction limit" set by the Rayleigh criterion (21.106).

[17] See Section 15.2.2 for the duality of free electromagnetic fields.

Our statement about the far field is true because Babinet's principle (see Section 21.9.1) guarantees that the fields transmitted through the aperture into $z > 0$ are related by duality to the fields diffracted into $z > 0$ by a perfectly conducting, two-dimensional object which just fills the hole. But according to Section 21.4, an electrically small object produces Rayleigh scattering with its characteristic λ^{-4} variation of the cross section. This disagrees with the Kirchhoff prediction that $\sigma_{\text{scatt}} \propto \lambda^{-2}$ because the quantity in square brackets in (21.105) approaches one when $k_0 a \ll 1$. In words, the transmission of a plane wave through a very small aperture in a perfect conductor is *very much less* than the Kirchhoff approximation predicts.

The Babinet argument implies that the differential cross section for scattering by a small aperture in a conducting sheet is given by (21.20) if we can compute the electric and magnetic dipole moments induced in the apertured conductor when a long-wavelength plane wave strikes it. Using a quasistatic approximation appropriate to $\lambda \gg a$, we leave it as an exercise for the reader to show that these moments can be calculated directly using the aperture geometry, or indirectly using Babinet's principle and the dipole moments calculated for the flat, conducting object shaped like the aperture.[18] Here, we content ourselves with a qualitative discussion for a plane wave incident on a circular aperture from an arbitrary angle of incidence, as shown in Figure 21.15.

Consider the polarization where $\mathbf{B}_0$ lies parallel to the screen. The conductor boundary condition $\hat{\mathbf{n}} \cdot \mathbf{B}|_S = 0$ ensures that the field line pattern very near the hole looks something like the left panel of Figure 21.21. In the far field, the magnetic field line pattern for $z > 0$ is the same as that produced by a magnetic dipole at the center of the hole oriented anti-parallel to the long-wavelength field incident from $z < 0$. The electric field for this polarization has components both parallel to the screen and perpendicular to the screen. However, the conductor boundary condition $\hat{\mathbf{n}} \times \mathbf{E}|_S = 0$ ensures that the field line pattern very near the hole looks something like the right panel of Figure 21.21. In the far field, the electric field line pattern for $z > 0$ is the same as that produced by an electric dipole at the center of the hole oriented parallel to the long-wavelength field incident from $z < 0$. The other polarization has $\mathbf{E}_0$ parallel to the screen and produces only a magnetic dipole response in the far field.

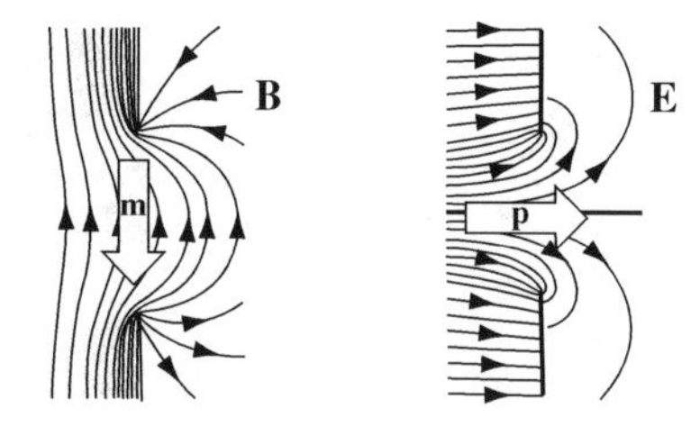

Figure 21.21: Uniform fields induce dipole moments in a metal sheet with an aperture. Left panel: magnetic dipole moment **m**. Right panel: electric dipole moment **p**. Figure adapted from Drezet, Woehl, and Huant (2002).

There is more to learn about the *near field* in the immediate vicinity of a very small aperture than Figure 21.21 suggests. To see this, we exploit the angular-spectrum representation (21.82) of the field diffracted into $z > 0$ by the aperture. The restriction (21.81) is crucial. If a plane wave with $k_x^2 + k_y^2 > \omega^2/c^2$ contributes to the transmitted field, k_z is a positive imaginary number and the plane wave in question does not propagate into the far zone. It is *evanescent* and decays exponentially as z increases.[19] The key observation is that evanescent waves contribute more and more to

[18] For an "elementary" derivation of the dipole moments induced in a conducting disk, see R. Friedberg, "The electrostatics and magnetostatics of a conducting disk", *American Journal of Physics* **61**, 1084 (1993).

[19] Section 17.3.7 reviews the properties of evanescent waves.

the sum (21.82) as the aperture size becomes smaller. This can be understood from our discussion of the complementarity inequality, $\Delta x \Delta k_x \geq \frac{1}{2}$, which connects the spatial size of a wave packet in the x-direction to the spread in wave vectors needed for its Fourier synthesis (see Section 16.5.3). For the present problem, we identify $\Delta x \Delta y$ with the area of an electrically small aperture and conclude that the angular spectrum of any wave it transmits must contain many evanescent waves with large values of k_x and k_y.

The mere fact that far fewer propagating waves contribute to the radiation zone of an electrically small aperture compared to an electrically large aperture is one way to understand the much reduced intensity of Rayleigh scattering compared to Kirchhoff scattering. Similarly, the Rayleigh criterion (21.106) applies only to images formed from propagating waves. It says nothing about the resolution possible for an object which lies so close to a sub-wavelength aperture that it is illuminated principally by evanescent waves. In that case, the main issue is the exponential decay of the waves and a spatial resolution $d \ll \lambda$ can be achieved simply by locating the object within a distance d of the aperture. This is the operating principle behind the field of *near-field optics*. ■

21.9.2 Huygens' Principle for Vector Fields

Let the surface S in Figure 21.22 completely enclose an arbitrary scattering or diffracting object. We take this to mean that the volume Ω contains one or more sources of a time-harmonic electromagnetic field. In this section, we generalize the results of Section 21.8.2 and prove that the scattered or diffracted fields in the volume V outside S can be represented by a set of spherical wave sources distributed over S. The source strengths are determined by the tangential components of $\mathbf{E}$ and $\mathbf{B}$ on S. This is a generalized *Huygens' principle* for electromagnetic fields.

Among the several ways to proceed, we avoid dyadic Green functions and excessive algebra by exploiting a fairly unfamiliar identity of vector calculus.[20] If $\psi(\mathbf{r}')$ is a scalar function, $\mathbf{E}(\mathbf{r}')$ is a vector function, and $\hat{\mathbf{n}}'$ is the unit normal to S that points *into* V,

$$\int_V d^3r' \left[\psi \nabla' \times (\nabla' \times \mathbf{E}) + \mathbf{E}\nabla'^2\psi + (\nabla' \cdot \mathbf{E})\nabla'\psi\right]$$

$$= -\int_S dS' \left[\psi\, \hat{\mathbf{n}}' \times (\nabla' \times \mathbf{E}) + (\hat{\mathbf{n}}' \times \mathbf{E}) \times \nabla'\psi + (\hat{\mathbf{n}}' \cdot \mathbf{E})\nabla'\psi\right]. \tag{21.116}$$

For our application, the vector $\mathbf{E}(\mathbf{r}')$ is the space part of a time-harmonic electric field with frequency $\omega = ck_0$. In V, this quantity satisfies $\nabla' \cdot \mathbf{E} = 0$, $\nabla' \times \mathbf{E} = i\omega\mathbf{B}$, and $\nabla' \times \nabla' \times \mathbf{E} = k_0^2\mathbf{E}$. The scalar $\psi(\mathbf{r}') = G_0(\mathbf{r}, \mathbf{r}')$ is the free-space Green function (21.76). The latter satisfies (21.71) and goes to zero when either argument goes to infinity. Substituting all this information into (21.116) and choosing $\mathbf{r}$ to lie in the complementary volume Ω gives

$$\mathbf{E}(\mathbf{r} \in \Omega) = 0. \tag{21.117}$$

Choosing $\mathbf{r} \in V$ gives a formula where only integrals over the surface S appear:

$$\mathbf{E}(\mathbf{r} \in V) = i\omega \int_S dS' \,[\hat{\mathbf{n}}' \times \mathbf{B}(\mathbf{r}')] G_0(\mathbf{r}, \mathbf{r}') + \int_S dS' \,[\hat{\mathbf{n}}' \times \mathbf{E}(\mathbf{r}')] \times \nabla' G_0(\mathbf{r}, \mathbf{r}')$$
$$+ \int_S dS' \,[\hat{\mathbf{n}}' \cdot \mathbf{E}(\mathbf{r}')] \nabla' G_0(\mathbf{r}, \mathbf{r}'). \tag{21.118}$$

[20] A one-page paper with a straightforward proof of (21.116) is H. Unz, "Scalar-vector analog of Green's theorem", *IRE Transactions on Antennas and Propagation*, **6**, 300 (1958).

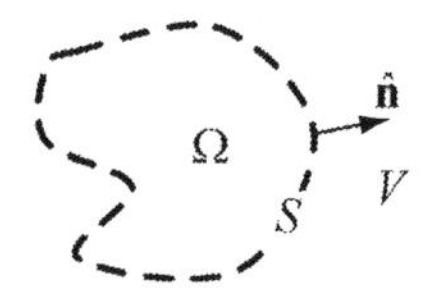

Figure 21.22: Geometry used to discuss Huygens' principle for electromagnetic fields.

A similar calculation replaces $\mathbf{E}$ by $\mathbf{B}$ in (21.116) and uses $\nabla' \cdot \mathbf{B} = 0$, $\nabla' \times \mathbf{B} = -i\omega\mathbf{E}/c^2$, and $\nabla' \times \nabla' \times \mathbf{B} = k_0^2\mathbf{B}$. The result is $\mathbf{B}(\mathbf{r} \in \Omega) = 0$ and

$$\begin{aligned}\mathbf{B}(\mathbf{r} \in V) = -i(k_0/c) \int_S dS' \, [\hat{\mathbf{n}}' \times \mathbf{E}(\mathbf{r}')]G_0(\mathbf{r}, \mathbf{r}') + \int_S dS' \, [\hat{\mathbf{n}}' \times \mathbf{B}(\mathbf{r}')] \times \nabla' G_0(\mathbf{r}, \mathbf{r}') \\ + \int_S dS' \, [\hat{\mathbf{n}}' \cdot \mathbf{B}(\mathbf{r}')]\nabla' G_0(\mathbf{r}, \mathbf{r}').\end{aligned} \tag{21.119}$$

Equations (21.118) and (21.119) are called the *Stratton-Chu formulae*. They are important for situations where the compact volume Ω contains the charge and current and it is necessary to know the details of the fields in the much larger volume V. A typical calculation uses a numerical method to find the solution in Ω and on S, and then substitutes the latter into the Stratton-Chu formulae to get the solution in V. For conceptual purposes, we transform (21.118) and (21.119) as follows: (i) use $\nabla' G_0 = -\nabla G_0$ to bring the gradient operators outside the integrals; (ii) take the curl of each equation; and (iii) use a Maxwell curl equation to eliminate the curl on the left side of each. These operations generate what are called the *Franz formulae*:

$$\begin{aligned}\mathbf{E}(\mathbf{r} \in V) &= \nabla \times \int_S dS' \, [\hat{\mathbf{n}}' \times \mathbf{E}(\mathbf{r}')]G_0(\mathbf{r}, \mathbf{r}') + \frac{ic^2}{\omega} \nabla \times \nabla \times \int_S dS' \, [\hat{\mathbf{n}}' \times \mathbf{B}(\mathbf{r}')]G_0(\mathbf{r}, \mathbf{r}') \\ \mathbf{B}(\mathbf{r} \in V) &= \nabla \times \int_S dS' \, [\hat{\mathbf{n}}' \times \mathbf{B}(\mathbf{r}')]G_0(\mathbf{r}, \mathbf{r}') - \frac{i}{\omega} \nabla \times \nabla \times \int_S dS' \, [\hat{\mathbf{n}}' \times \mathbf{E}(\mathbf{r}')]G_0(\mathbf{r}, \mathbf{r}').\end{aligned} \tag{21.120}$$

The Franz formulae are interesting, not least because they require more information than strictly should be necessary. Each involves both $\hat{\mathbf{n}} \times \mathbf{E}|_S$ *and* $\hat{\mathbf{n}} \times \mathbf{B}|_S$, while the uniqueness theorem (mentioned at the end of Section 21.8.2) tells us that only $\hat{\mathbf{n}} \times \mathbf{E}|_S$ *or* $\hat{\mathbf{n}} \times \mathbf{B}|_S$ should be needed. That being said, the only geometry known where an explicit formula exists to compute the fields in a volume when only one tangential field component is known on the boundary is the planar aperture to which (21.90) and (21.91) apply. For all other geometries, both surface quantities must be known or approximated. An example is the physical optics expression (21.52). This may be derived by substituting $\hat{\mathbf{n}} \times \mathbf{E}|_S = 0$ and (21.51) into the right side of (21.120) and then passing to the radiation zone using (21.97) to evaluate the interior curl operation and $\nabla \to i\mathbf{k}$ to evaluate the exterior curl operation.

The physics represented by the Franz formulae becomes clear when we recall (from Section 20.4) the fields produced by an origin-centered point electric dipole with moment $\mathbf{p}(t)$,

$$\mathbf{B}(\mathbf{r}, t) = \nabla \times \frac{\mu_0}{4\pi} \frac{\dot{\mathbf{p}}_{\text{ret}}}{r} \quad \text{and} \quad \mathbf{E}(\mathbf{r}, t) = \nabla \times \nabla \times \frac{1}{4\pi\epsilon_0} \frac{\mathbf{p}_{\text{ret}}}{r}, \tag{21.121}$$

and the fields produced by an origin-centered point magnetic dipole with moment $\mathbf{m}(t)$,

$$\mathbf{E}(\mathbf{r}, t) = -\nabla \times \frac{\mu_0}{4\pi} \frac{\dot{\mathbf{m}}_{\text{ret}}}{r} \quad \text{and} \quad \mathbf{B}(\mathbf{r}, t) = \nabla \times \nabla \times \frac{\mu_0}{4\pi} \frac{\mathbf{m}_{\text{ret}}}{r}. \tag{21.122}$$

We are concerned with time-harmonic sources, so $\mathbf{p}_{\text{ret}} = \mathbf{p}(t - r/c) = \mathbf{p}\exp[i(k_0 r - \omega t)]$ and $\mathbf{m}_{\text{ret}} = \mathbf{m}(t - r/c) = \mathbf{m}\exp[i(k_0 r - \omega t)]$. In that case, a direct comparison of (21.121) and (21.122) with (21.120) shows that the latter are exactly the fields of a surface S endowed with point electric and magnetic dipoles with areal densities[21]

$$\frac{d\mathbf{p}}{dS} = \frac{i}{\omega\mu_0}[\hat{\mathbf{n}} \times \mathbf{B}]_S \qquad \text{and} \qquad \frac{d\mathbf{m}}{dS} = -\frac{i}{\omega\mu_0}[\hat{\mathbf{n}} \times \mathbf{E}]_S. \tag{21.123}$$

The presence of the free-space Green functions in (21.120) embodies the Huygens' principle idea that the fields in V may be thought of as produced by effective spherical wave sources distributed over S.

Example 21.4 Show that (21.120) reduces to (21.90) and (21.91) when the volume V in Figure 21.22 is the $z > 0$ half-space. Hint: Compare the fields at $\mathbf{r} = (x, y, z > 0)$ with the fields at the image point $\bar{\mathbf{r}} = (x, y, -z)$.

Solution: The given V identifies the surface S as the $z = 0$ plane and the normal $\hat{\mathbf{n}}' = \hat{\mathbf{z}}$. Using the hint and (21.117), the left side of (21.120) vanishes when the right side is evaluated at the image point $\bar{\mathbf{r}}$, which lies in Ω. Accordingly, the electric field equation reads

$$0 = \nabla \times \int_S dS'\, [\hat{\mathbf{z}} \times \mathbf{E}(\mathbf{r}')]G_0(\bar{\mathbf{r}}, \mathbf{r}') + \frac{ic^2}{\omega}\nabla \times \nabla \times \int_S dS'\, [\hat{\mathbf{z}} \times \mathbf{B}(\mathbf{r}')]G_0(\bar{\mathbf{r}}, \mathbf{r}').$$

Each curl operation refers to the observation point. Therefore, the z-derivative of the gradient operator introduces a minus sign when it acts on $G_0(\bar{\mathbf{r}}, \mathbf{r}')$. Moreover, $G_0(\bar{\mathbf{r}}, \mathbf{r}') = G_0(\mathbf{r}, \mathbf{r}')$ when $z' = 0$. Therefore, the z-component of the first term of the equation just above has the same sign as the corresponding component of (21.120) while the x- and y-components have the opposite sign. Similarly, the z-component of the second term of the null equation has the opposite sign as the corresponding component of (21.120) while the x- and y-components have the same sign. Therefore, adding the z-components of the two equations and subtracting the x- and y-components of the two equations gives the advertised result,

$$\mathbf{E}(x, y, z > 0) = \nabla \times 2\int_S dS'\, [\hat{\mathbf{z}} \times \mathbf{E}(\mathbf{r}')]G_0(\mathbf{r}, \mathbf{r}').$$

A similar argument reduces the magnetic field equation in (21.120) to (21.91).

Sources, References, and Additional Reading

The quotation at the beginning of the chapter is taken from

V. Fock, "New methods in diffraction theory" *Philosophical Magazine* **39**, 149 (1948). Reproduced by permission of Taylor & Francis Ltd.

Section 21.1 *Smith* and *Landau and Lifshitz* treat the subject of this chapter very differently, but with equal clarity and insight. Our debt to the former will be obvious. *Ishimaru* presents many different approaches to scattering and diffraction. *Newton* compares electromagnetic scattering to particle scattering and quantum mechanical wave scattering. The latter two books treat polarization effects in detail.

G.S. Smith, *An Introduction to Classical Electromagnetic Radiation* (Cambridge University Press, Cambridge, 1997).

[21] Textbooks of engineering electromagnetism often relate $[\hat{\mathbf{n}} \times \mathbf{B}]_S$ to a current of electric charge and $[\hat{\mathbf{n}} \times \mathbf{E}]_S$ to a current of (fictitious) magnetic charge.

L.D. Landau and E.M. Lifshitz, *Electrodynamics of Continuous Media* (Pergamon, Oxford, 1960).

A. Ishimaru, *Electromagnetic Wave Propagation, Radiation, and Scattering* (Prentice-Hall, Upper Saddle River, NJ, 1991).

R.G. Newton, *Scattering Theory of Waves and Particles* (McGraw-Hill, New York, 1966).

Two expert-level treatments of scattering and diffraction are

G.T. Ruck, D.E. Barrick, W.D. Stuart, and C.K. Krichbaum, *Radar Cross Section Handbook* (Plenum, New York, 1970), Volume 1.

M. Nieto-Vesperinas, *Scattering and Diffraction in Physical Optics* (Wiley, New York, 1991).

Figure 21.1 was adapted from

G. Toraldo di Francia, *Electromagnetic Waves* (Wiley-Interscience, New York, 1953). Reproduced with kind permission of Maria Luisa Dalla Chiara.

Section 21.3 Application 21.1 is based on a discussion in

S. Dodelson, *Modern Cosmology* (Academic, Amsterdam, 2003), Chapter 10. © 2003 Elsevier.

Section 21.4 *Smith* discusses dipole scattering from small particles, its role in the human perception of atmospheric color, and the contributions of Rayleigh, Smoluchowski, and Einstein to the theory in

G.S. Smith, "Human color vision and the unsaturated blue color of the daytime sky", *American Journal of Physics* **73**, 590 (2005).

G.S. Smith, "Summing the molecular contributions to skylight", *American Journal of Physics* **76**, 816 (2008).

Section 21.5 Our discussion of scattering from an infinite conducting cylinder follows *Egyes*. The classic reference on Mie scattering from dielectric spheres is *van de Hulst*. The treatise by *Grandy* brings his discussion up to date. *Ruck et al.* (see Section 21.1 above) review the literature of scattering from conducting spheres and cylinders in great detail.

L. Eyges, *The Classical Electromagnetic Field* (Dover, New York, 1972).

H.C. van de Hulst, *Light Scattering by Small Particles* (Wiley, New York, 1957).

W.T. Grandy, Jr., *Scattering of Waves by Large Spheres* (Cambridge University Press, Cambridge, 2000).

Figure 21.9 was produced by MiePlot, a code at wwww.philiplaven.com/mieplot.htm. Figure 21.10 comes from

L.G. Guimarães and H.M. Nussenzveig, "Theory of Mie resonances and ripple fluctuations", *Optics Communications* **89**, 363 (1992). With permission from Elsevier.

For an appreciation of Ludvig Lorenz, see

O. Keller, "Optical works of L.V. Lorenz", in *Progress in Optics*, vol. 43 (Elsevier, Amsterdam, 2002), Chapter 3.

Section 21.6 Figure 21.11 was produced using MiePlot (see Section 21.5 above). *Macdonald* gave the first discussion of the physical optics approximation. Figure 21.12 was adapted from *Imbriale*:

H.M. Macdonald, "The effect produced by an obstacle on a train of electric waves", *Philosophical Transactions of the Royal Society of London A* **212**, 299 (1912).

W.A. Imbriale, *Large Antennas of the Deep Space Network* (Wiley-Interscience, New York, 2003), Chapter 9.

Section 21.7 *Newton* gives a short history of the optical theorem. Our proof is due to *Saxon* in the form given by *Mishchenko et al.* The approximation derived in Application 21.2 is due to *van de Hulst* (see Section 21.5 above) in the form given by *Loudec et al.*

R.G. Newton, "Optical theorem and beyond", *American Journal of Physics* **44**, 639 (1976).

D.S. Saxon, "Tensor scattering matrix for the electromagnetic field", *Physical Review* **100**, 1771 (1955).

M.I. Mishchenko, L.D. Travis, and A.A. Lacis, *Scattering, Absorption, and Emission of Light by Small Particles* (Cambridge University Press, Cambridge, 2002).

K. Louedec, S. Dagoret-Campagne, and M. Urban, "Ramsauer approach to Mie scattering of light on spherical particles", *Physica Scripta* **80**, 35403 (2009).

Van de Hulst (see Section 21.5 above) and *Brillouin* give differing explanations of the extinction paradox. Our discussion follows *Berg et al.*

L. Brillouin, "The scattering cross section of spheres for electromagnetic waves", *Journal of Applied Physics* **20**, 1110 (1949).

M.J. Berg, C.M. Sorensen, and A. Chakrabarti, "A new explanation of the extinction paradox", *Journal of Quantitative Spectroscopy & Radiative Transfer* **112**, 1170 (2011).

Section 21.8 The textbook by *Sommerfeld* describes his exact solution to the electromagnetic diffraction problem for a conducting half-plane. Figure 21.14 comes from *Berry*.

A. Sommerfeld, *Optics* (Academic, New York 1954).

M. Berry, "Geometry of phase and polarization singularities, illustrated by edge diffraction and the tides", in *The Second International Conference on Singular Optics*, edited by M.S. Soskin and M.V. Vasnetsov, Proceedings of the SPIE, vol. 4403 (2001), pp. 1-12.

Most optics textbooks (including *Sommerfeld* just above) discuss scalar diffraction theory in more or less detail. *Goodman* is the standard reference for the role of the Fourier transform in far-field diffraction. Figure 21.16 comes from *Siegman*.

J. Goodman, *Introduction to Fourier Optics*, 3rd edition (Roberts and Company, Greenwood Village, CO, 2005).

A.E. Siegman, *Lasers* (University Science Books, Sausalito, CA, 1986), Section 18.4.

Our treatment of vector diffraction theory follows *Smith* (see Section 21.1 above), which is also the source of Figure 21.17 and Figure 21.19.

Section 21.9 Our proof of Babinet's principle is due to *Smith* (see Section 21.1). *Bethe* is the author of the classic paper on diffraction by small apertures. *Hecht et al.* provide a very readable overview of the field of near field optics. Figure 21.21 was adapted from *Drezet et al.*

H.A. Bethe, "Theory of diffraction by small holes", *Physical Review* **66**, 163 (1944).

B. Hecht, B. Sick, U.P. Wild, *et al.*, "Scanning near-field optical microscopy with aperture probes: Fundamentals and applications", *Journal of Chemical Physics* **112**, 7761 (2000).

A. Drezet, J.C. Woehl, and S. Huant, "Diffraction by a small aperture in a conical geometry", *Physical Review E* **65**, 46611 (2002).

Our discussion of Huygens' principle and the Franz formulae benefitted from *Tai*. Example 21.4 comes from the mathematically rigorous monograph by *Jones*.

C.-T. Tai, "Kirchhoff theory: Scalar, vector, or dyadic?", *IEEE Transactions on Antennas and Propagation* **20**, 114 (1972).

D.S. Jones, *The Theory of Electromagnetism* (Macmillan, New York, 1964).

Problems

21.1 Scattering from a Bound Electron Find the total scattering cross section when a circularly polarized wave scatters from an electron bound to a point in space by a spring with spring constant k. Assume that the amplitude of the incident wave is not large.

21.2 Scattering from a Hydrogen Atom Let $\mathbf{q} = \mathbf{k}_0 - \mathbf{k}$ be the scattering vector defined in Example 1.2. If a_B is the Bohr radius, show that the cross section for plane wave scattering from a hydrogen atom is proportional to the factor $[1 + (qa_B/2)^2]^{-4}$.

21.3 Double Scattering A long-wavelength, left circularly polarized, monochromatic plane wave scatters into the direction $\hat{\mathbf{k}}_1$ from a uniform dielectric sphere with radius a and polarizability α. The scattered wave travels a distance $r_1 \gg a$ and scatters from an identical sphere into the direction $\hat{\mathbf{k}}_2$. Find the twice-scattered electric field at a distance $r_2 \gg a$ from the second sphere. Express your answer using polarization vectors which are (i) transverse to $\hat{\mathbf{k}}_2$ and (ii) parallel and perpendicular to the plane of the diagram.

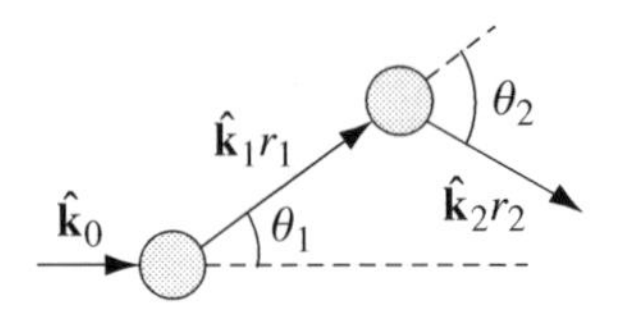

21.4 Rayleigh Scattering à la Rayleigh In one of his papers devoted to the color of skylight, Lord Rayleigh used physical reasoning and dimensional analysis to deduce the wavelength dependence of the intensity of light scattered by a particle in the atmosphere. Invent Rayleigh's argument, beginning with his assumption that the ratio of the scattered field amplitude to the incident field amplitude *could* depend on (i) the volume of the particle, (ii) the distance from the particle to the observation point, (iii) the wavelength of the scattered light, and (iv) the speed of light.

21.5 Rayleigh Scattering from a Conducting Sphere

(a) Place a perfectly conducting sphere with radius a in a uniform electric field $\mathbf{E}_0$ and let an origin-centered electric dipole field represent the field produced by the sphere. Use this information to deduce that $\mathbf{p} = 4\pi\epsilon_0 a^3\mathbf{E}_0$ is the dipole moment induced in the sphere.
(b) Place the sphere in a uniform magnetic field $\mathbf{B}_0$ and let an origin-centered magnetic dipole field represent the field produced by the sphere. Use this information to deduce that $\mathbf{m} = -(2\pi a^3/\mu_0)\mathbf{B}_0$ is the dipole moment induced in the sphere.
(c) Let θ be the angle between the incident wave vector $\mathbf{k}_0$ and the scattered wave vector $\mathbf{k}$. If $ka \ll 1$, show that the differential cross section for scattering of an unpolarized plane wave by the perfectly conducting sphere is

$$\left\langle \frac{d\sigma}{d\Omega} \right\rangle_{\text{unpol}} = a^2(ka)^4 \left[\frac{5}{8}(1+\cos^2\theta) - \cos\theta \right].$$

(d) Compare the answer in (c) with the cross section when the incident plane wave is circularly polarized.

21.6 Scattering from a Molecular Rotor A linearly polarized, monochromatic plane wave scatters from a polar molecule by exerting a torque that sets the molecule into motion. Treat the molecule as an electric dipole with moment $\mathbf{p}$ and moment of inertia I. Ignore terms quadratic in the (very slow) angular velocity $\boldsymbol{\Omega}$ of the molecule and average over all orientations of $\mathbf{p}$ to show that the total scattering cross section is $\sigma_{\text{scatt}} = \mu_0^2 p^4/9\pi I^2$. Hint: A rotating dipole moment satisfies $\dot{\mathbf{p}} = \boldsymbol{\Omega} \times \mathbf{p}$.

21.7 Preservation of Polarization I A linearly polarized plane wave with electric field amplitude $\mathbf{E}_0$ is incident on a small, perfectly conducting sphere. Use the dipole moment information provided in Problem 21.5 and find the angle between the scattering wave vector $\mathbf{k}$ and the incident wave vector $\mathbf{k}_0$ where the radiated electric field points in the same direction as $\mathbf{E}_0$.

21.8 Scattering and Absorption by an Ohmic Sphere A low-frequency, plane electromagnetic wave Rayleigh scatters from a sphere with radius a and conductivity σ. Assume that the skin depth $\delta \gg a$.

(a) Find the electric dipole moment induced in the sphere by the incident wave.
(b) Calculate the absorption cross section of the sphere.
(c) Show that the optical theorem is satisfied by the absorption cross section (alone) and the electric dipole scattering amplitude.
(d) Rationalize the result of part (c) with the fact that the scattering cross section is not zero.

21.9 Scattering from a Dielectric Cylinder The symmetry axis of an infinitely long dielectric cylinder with radius a and permittivity ϵ coincides with the z-axis. A monochromatic wave with wave vector $\mathbf{k}_0$ is normally incident on the cylinder as shown below. Find the electric field everywhere if the incident wave is polarized in the z-direction.

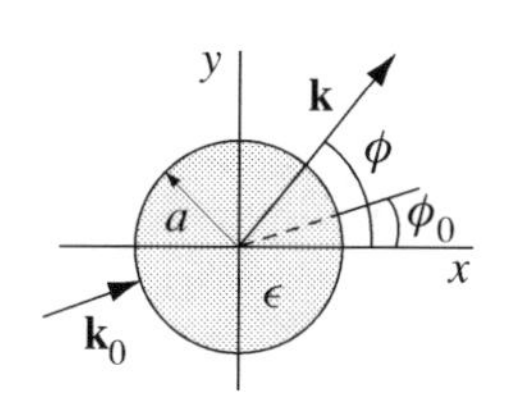

21.10 Preservation of Polarization II A monochromatic plane wave with electric field amplitude $\mathbf{E}_0$ is incident on a perfectly conducting object with an arbitrary shape. Prove that the electric field radiated in the backward direction is parallel to $\mathbf{E}_0$ in the physical optics approximation.

21.11 Scattering from a Conducting Strip A thin and infinitely long, perfectly conducting strip occupies the area $0 \leq x \leq w$ of the $y = 0$ plane. A monochromatic plane wave polarized along $+z$ scatters from the strip as shown below. Assume specular reflection and let $\omega = ck = ck_0$.

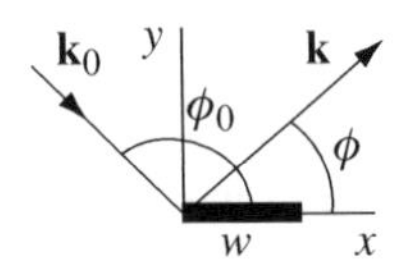

(a) Find the physical optics surface current density.

(b) Perform the z' integration in the exact vector potential produced by $\mathbf{K}_{\text{PO}}(\mathbf{r}, t)$ and use the Hankel function identity,

$$i\pi H_0^{(1)}(x) = \int_{-\infty}^{\infty} d\eta \frac{\exp\left(i\sqrt{x^2+\eta^2}\right)}{\sqrt{x^2+\eta^2}},$$

to show that

$$\mathbf{A}(\mathbf{r}, t) = \hat{\mathbf{z}}\frac{i\mu_0 E_0}{2Z_0}\sin\phi_0 \int_0^w dx' H_0^{(1)}(k|\boldsymbol{\rho} - \boldsymbol{\rho}'|)\exp(-ikx'\cos\phi_0)\exp(-i\omega t).$$

(c) Evaluate $\mathbf{A}(\rho, \phi)$ in the far zone and show that the two-dimensional differential cross section for scattering is

$$\frac{d\sigma}{d\phi} = \frac{2}{\pi k}\sin^2\phi_0 \frac{\sin^2\left[\frac{1}{2}kw(\cos\phi + \cos\phi_0)\right]}{(\cos\phi + \cos\phi_0)^2}.$$

Hint: Do not completely neglect the dependence of $|\boldsymbol{\rho} - \boldsymbol{\rho}'|$ on x' when approximating a phase factor in the far zone.

21.12 Physical Optics Backscattering

(a) Let S be the illuminated portion of a conductor. If $\hat{\mathbf{n}}$ is the local unit normal vector to S and $\mathbf{k} = k_0\hat{\mathbf{k}}$ is the propagation direction of the backscattered wave, show that the cross section for backscattering in the physical optics approximation is

$$\sigma_{\text{R}} = \frac{k_0^2}{4\pi^2}\left|\int_S dS'\,\hat{\mathbf{k}}\cdot\hat{\mathbf{n}}'\exp(-2i\mathbf{k}\cdot\mathbf{r}')\right|^2.$$

(b) Specialize to a flat, rectangular plate with negligible thickness that lies in the $z = 0$ plane. The side parallel to the x-axis has length a and the side parallel to the y-axis has length b. If the center of the plate coincides with the origin of spherical coordinates and the plate area is A, show that

$$\sigma_{\text{R}} = \frac{A^2}{\lambda^2}\cos^2\theta\left[\frac{\sin(k_0 a\sin\theta\cos\phi)}{k_0 a\sin\theta\cos\phi} \times \frac{\sin(k_0 b\sin\theta\sin\phi)}{k_0 b\sin\theta\sin\phi}\right]^2.$$

21.13 Born Scattering from a Dielectric Cube A plane wave $\mathbf{E}_0\exp[i(\mathbf{k}_0\cdot\mathbf{r} - \omega t)]$ scatters from a dielectric cube with volume $V = a^3$ and electric susceptibility $\chi \ll 1$. Two cube edges align with $\mathbf{k}_0$ and $\mathbf{E}_0$.

(a) Calculate the differential scattering cross section in the Born approximation.

(b) Show that $\sigma_{\text{Born}} \approx \frac{1}{4}k^2a^4\chi^2$ when $ka \gg 1$. Hint: The near-forward direction dominates the scattering when $ka \gg 1$.

(c) The weak scattering assumed by the Born approximation implies that $|\mathbf{E}_{\text{rad}}|/|\mathbf{E}_0| \ll 1$ for all $\mathbf{q}$, even when $r \approx a$. Deduce from this that the $ka \gg 1$ result of part (b) is valid only when $\sigma_{\text{Born}} \ll \chi a^2$.

21.14 Scattering from a Short Conducting Wire A monochromatic plane wave scatters from a perfectly conducting wire where a $\ll h$. Assume that both the propagation vector and the electric field of the incident wave lie in the y-z plane as shown below.

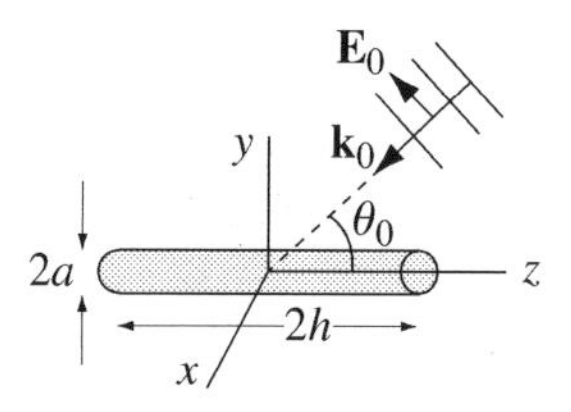

(a) In the Rayleigh limit when $k_0h \ll 1$, the scattering is dominated by a z-directed electric dipole moment $\mathbf{p} = \epsilon_0\alpha(\hat{\mathbf{z}} \cdot \mathbf{E}_0)\hat{\mathbf{z}}$. Assume an induced surface current density,

$$\mathbf{K}(z) = \frac{I_0}{2\pi a}\left[1 - (z/h)^2\right]\hat{\mathbf{z}},$$

and determine the parameter I_0 by imposing the perfect-conductor condition, $[\mathbf{E}_{\text{scatt}} + \mathbf{E}_0] \cdot \hat{\mathbf{z}} = 0$, at a single point: the origin of coordinates at the center of the wire. Specifically, calculate $\mathbf{E}_{\text{scatt}}$ from a near-zone expansion of the retarded potentials and show that

$$I_0 = \frac{-i\pi\omega h^2}{\{[\ln(2h/a) - 1][1 - \frac{1}{2}(kh)^2] + i\frac{2}{9}(kh)^3\}}\hat{\mathbf{z}} \cdot \mathbf{E}_0.$$

(b) Drop the terms of order $(k_0h)^2$ and $(k_0h)^3$ from the expression for I_0, compute the induced electric dipole moment, and use this to find the total scattering cross section.

(c) Show that the terms dropped in part (b) must be retained in the radiation zone forward scattering amplitude to satisfy the optical theorem with the cross section computed in part (b).

21.15 Absorption Cross Section for a Microscopic Object Show that $\sigma_{\text{abs}} = (\omega/c)\text{Im}\alpha$ is the frequency-dependent absorption cross section for a microscopic object (atom, molecule, or nucleus) with polarizability α.

21.16 Absorption Sum Rule for a Lorentz Oscillator A monochromatic plane wave scatters from a Lorentz atom where a bound electron obeys the classical equation of motion $\ddot{\mathbf{r}} + \gamma\dot{\mathbf{r}} + \omega_0^2\mathbf{r} = 0$. Assume that the electron displacement and damping are both very small. If $r_e = e^2/4\pi\epsilon_0 mc^2$ is the classical electron radius, show that the integrated total absorption cross section is independent of the damping constant:

$$\int_0^\infty d\omega\, \sigma_{\text{abs}}(\omega) = 2\pi^2 r_e c.$$

21.17 The Optical Theorem in Two Dimensions

(a) Integrate the differential cross sections derived in the text to find the total scattering cross sections $\sigma_\parallel$ and $\sigma_\perp$ for an infinitely long and perfectly conducting cylinder with $\mathbf{E}_{\text{inc}}$ oriented, respectively, parallel and perpendicular to the cylinder axis.

(b) In two dimensions, the scattering amplitude $\mathbf{f}(\mathbf{k}) = \mathbf{f}(k, \phi)$ is defined by the asymptotic electric field

$$\mathbf{E}_{\text{rad}}(\rho, \phi) = E_0\sqrt{\frac{i}{k}}\mathbf{f}(\mathbf{k})\frac{\exp(ik\rho)}{\sqrt{\rho}}.$$

Use the results of part (a) for either polarization to confirm that the optical theorem in two dimensions is

$$\sigma_{\rm tot} = \frac{\sqrt{8\pi}}{k}\,{\rm Im}[\hat{\mathbf{e}}_0^* \cdot \mathbf{f}(k, 0)].$$

21.18 The Optical Theorem for Pedestrians A unit-amplitude, monochromatic plane wave of a scalar field $\psi(\mathbf{r})$ scatters from an origin-centered obstacle of finite size. Apart from a factor of $\exp(-i\omega t)$, the field takes the asymptotic form

$$\psi(r, \theta, \phi) \approx \exp(ikz) + \frac{\exp(ikr)}{r} f(\theta, \phi).$$

Focus on the almost-forward direction where $\theta \ll 1$ and suppose that a flat screen with a radius $R \gg z/k$ collects the energy of the wave at a distance $z \gg R$ from the origin. Show that

$$\int\limits_{\rm screen} dS\,|\psi|^2 \approx \pi R^2 - \frac{4\pi}{k}{\rm Im} f(0),$$

where $f(0)$ is the scattering amplitude evaluated on the z-axis. Use this result to argue that

$$\sigma_{\rm tot} = \sigma_{\rm scatt} + \sigma_{\rm abs} = \frac{4\pi}{k}{\rm Im} f(0).$$

21.19 Total Cross Section Sum Rule An incident plane wave $\hat{\mathbf{e}}_0 E_0 \exp[i(\mathbf{k}_0 \cdot \mathbf{r} - \omega t)]$ scatters from a target with amplitude $\mathbf{f}(\mathbf{k})$. One can prove that $\mathbf{f}(\mathbf{k}_0) \cdot \hat{\mathbf{e}}_0^*/k^2$ is a causal response function of the sort discussed in Section 18.7. Use this information to prove the wavelength sum rule,

$$\lim_{\lambda \to \infty} {\rm Re}\,[\mathbf{f}(\lambda, \mathbf{k}_0) \cdot \hat{\mathbf{e}}_0^*] = \frac{1}{\pi \lambda^2} \int_0^\infty d\lambda'\, \sigma_{\rm tot}(\lambda').$$

21.20 The Index of Refraction Let $\mathbf{E}_{\rm inc} = \mathbf{e}_0 E_0 \exp[i(kz - \omega t)]$ be the electric field of a plane wave propagating in a homogeneous dielectric medium. The wave vector $k = nk_0 = n\omega/c$, where n is the index of refraction of the medium. Suppose that the number density of scatters increases from N to $N + \delta N$ in a thin layer of the medium between $z = 0$ and $z = \delta t$. Because δt is infinitesimal, $\mathbf{E}_{\rm inc}$ scatters once from every extra atom in the layer. Therefore, if the atomic scattering amplitude is $\mathbf{f}(\theta, \phi)$, the extra electric field produced at the distant observation point $z \gg \delta t$ is

$$\mathbf{E}_{\rm rad}(z) = \delta N E_0 \delta t \int\limits_{\rm layer} d^2 r\, \frac{\exp(ikR)}{R} \mathbf{f}(\theta, \phi).$$

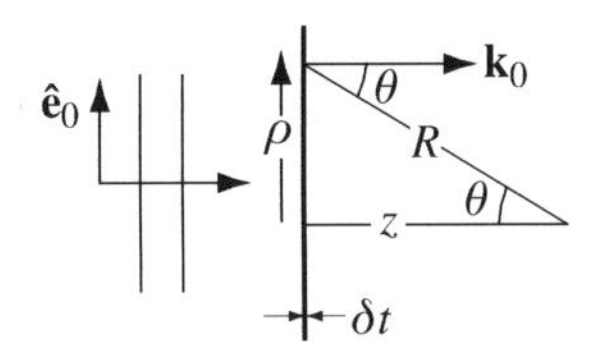

(a) Change variables to $\eta = R/z$, integrate by parts, and compare the original integral to the new integral in the limit $kz \gg 1$. Note that $\mathbf{f}(0) \equiv \mathbf{f}(\theta = 0, \phi)$ does not depend on ϕ and establish that

$$\mathbf{E}_{\rm rad}(z) = \frac{2\pi i}{k} \delta N E_0 \delta t \exp(ikz)\,\mathbf{f}(0) \qquad kz \gg 1.$$

(b) Construct $\mathbf{E}(z) = \mathbf{E}_{\rm inc}(z) + \mathbf{E}_{\rm rad}(z)$ from the results of part (a) and argue that your expression remains valid at $z = \delta t$. Derive from this fact an expression for $\delta k/\delta N$, the change in wave vector induced by the density perturbation. Integrate and conclude that the index of refraction of the unperturbed medium satisfies

$$n^2 = 1 + \frac{4\pi N}{k_0^2}\hat{\mathbf{e}}_0^* \cdot \mathbf{f}(0).$$

21.21 Radiation Pressure from Scattering An object scatters an incident plane wave with $\mathbf{E}_{\rm inc}(\mathbf{r}, t) = \hat{\mathbf{e}}_0 E_0 \exp[i(\mathbf{k}_0 \cdot \mathbf{r} - \omega t)]$. Use the Maxwell stress tensor formalism to show that the time-averaged force on the object can be written in terms of the incident wave intensity $I_{\rm inc}$, the total cross section $\sigma_{\rm tot}$, and the differential cross section for scattering $d\sigma_{\rm scatt}/d\Omega$ as

$$\langle \mathbf{F} \rangle = \frac{I_{\rm inc}}{c}\left[\sigma_{\rm tot}\hat{\mathbf{k}}_0 - \int d\Omega\, \hat{\mathbf{r}} \frac{d\sigma_{\rm scatt}}{d\Omega}\right].$$

The projection of this force on the direction $\hat{\mathbf{k}}_0$ is often called the radiation pressure due to scattering. Hint: Integrate the stress tensor over the surface of an enormous sphere in the radiation zone.

21.22 A Backscatter Theorem Theorem: a monochromatic plane wave incident on a body with $\epsilon(\mathbf{r}) = \mu(\mathbf{r})$ produces zero scattered field intensity in the far zone in the backward direction if the direction of incidence is an axis of symmetry where rotation by 90° leaves the body unchanged. To prove this,

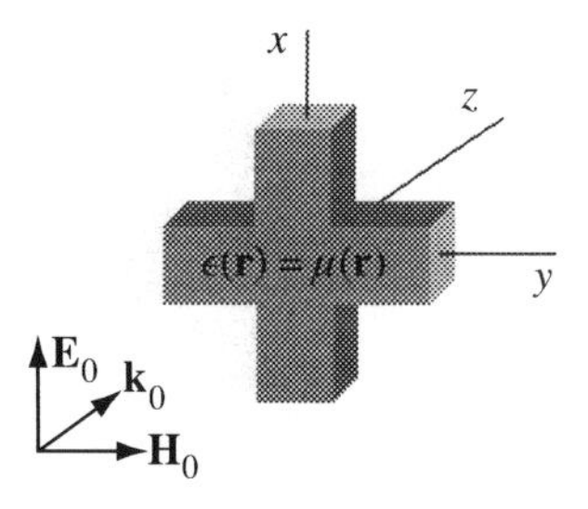

(a) Begin with the Maxwell equations for matter with a spatially varying permittivity and permeability. Show that $\mathbf{E}(\mathbf{r})$ and $\mathbf{H}(\mathbf{r})$ satisfy the same generalized wave equation when $\epsilon(\mathbf{r}) = \mu(\mathbf{r})$.

(b) Let $\mathbf{E}^{\rm scatt}$ be the exact electric field produced by the body. Show that

$$\mathbf{H}^{\rm scatt}(x, y, z) = -E_y^{\rm scatt}(y, -x, z)\hat{\mathbf{x}} + E_x^{\rm scatt}(y, -x, z)\hat{\mathbf{y}} + E_z^{\rm scatt}(y, -x, z)\hat{\mathbf{z}}.$$

(c) Use the results of part (a), part (b), and a consideration of the Poynting vector in the radiation zone to prove the theorem.

21.23 The Angular Spectrum of Plane Waves in Two Dimensions Consider time-harmonic electromagnetic fields in the domain $z \geq 0$ of the form

$$\mathbf{E}(x, z \geq 0, t) = \mathbf{E}(x, z, \omega)\exp(-i\omega t) \qquad \mathbf{B}(x, z \geq 0, t) = \mathbf{B}(x, z, \omega)\exp(-i\omega t).$$

(a) Let $\hat{\mathbf{y}} \cdot \mathbf{E}_{TE}(x, z = 0, t = 0) = \bar{E}_y(x)$. Determine the scalar function $\Lambda_{TE}(k_x)$ and the vector function $\boldsymbol{\Gamma}_{TE}(k_x)$ so that

$$\mathbf{E}_{TE}(x, z, \omega) = \hat{\mathbf{y}} \int\limits_{-\infty}^{\infty} \frac{dk_x}{2\pi}\, \Lambda_{TE}(k_x)\exp(i\mathbf{k} \cdot \mathbf{r})$$

$$\mathbf{B}_{TE}(x, z, \omega) = \int\limits_{-\infty}^{\infty} \frac{dk_x}{2\pi c}\left[\frac{k_x}{k_0}\hat{\mathbf{z}} - \frac{k_z}{k_0}\hat{\mathbf{x}}\right] \boldsymbol{\Lambda}_{TE}(k_x)\exp(i\mathbf{k} \cdot \mathbf{r})$$

solve Maxwell's equations in free space. The wave vector $\mathbf{k} = \hat{\mathbf{x}}\, k_x + \hat{\mathbf{z}}\, k_z$ is two-dimensional. Explain why there is no integral over k_z.

(b) Let $\hat{\mathbf{x}} \cdot \mathbf{E}_{TM}(x, z = 0, t = 0) = \bar{E}_x(x)$. Determine the scalar function $\Lambda_{TM}(k_x)$ and the vector function $\Gamma_{TM}(k_x)$ so that

$$\mathbf{B}_{TM}(x, z, \omega) = \hat{\mathbf{y}} \int_{-\infty}^{\infty} \frac{dk_x}{2\pi c} \frac{k_0}{k_z} \Lambda_{TM}(k_x) \exp(i\mathbf{k} \cdot \mathbf{r})$$

$$\mathbf{E}_{TM}(x, z, \omega) = \int_{-\infty}^{\infty} \frac{dk_x}{2\pi} \left[\hat{\mathbf{x}} - \frac{k_x}{k_z} \hat{\mathbf{z}} \right] \Lambda_{TM}(k_x) \exp(i\mathbf{k} \cdot \mathbf{r})$$

solve Maxwell's equations in free space. The factor $k_0 = \omega/c$.

(c) Represent a general $z \geq 0$ field as $\mathbf{E} = \mathbf{E}_{TE} + \mathbf{E}_{TM}$ and $\mathbf{B} = \mathbf{B}_{TE} + \mathbf{B}_{TM}$. If $\mathbf{S}$ is the Poynting vector, show that the time-averaged power transmitted down the z-axis is

$$P_z = \int_{-\infty}^{\infty} dx < \mathbf{S} \cdot \hat{\mathbf{z}} > = \frac{1}{4\pi} \sqrt{\frac{\epsilon_0}{\mu_0}} \int_{-k_0}^{k_0} dk_x \left\{ |\Lambda_{TM}|^2 \frac{k_0}{k_z} + |\Lambda_{TE}|^2 \frac{k_z}{k_0} \right\}.$$

What is the physical origin of the limits of integration on the k_x integral?

21.24 Weyl's Identity This problem outlines a contour integration method to prove that

$$\frac{\exp(ik_0 r)}{4\pi r} = \frac{i}{8\pi^2} \int \frac{d^2 k_\perp}{k_z} \exp(i\mathbf{k}_\perp \cdot \mathbf{r}_\perp + ik_z|z|),$$

where

$$k_z = \begin{cases} \sqrt{k_0^2 - k_x^2 - k_y^2} & k_x^2 + k_y^2 \leq k_0^2, \\ i\sqrt{k_x^2 + k_y^2 - k_0^2} & k_x^2 + k_y^2 \geq k_0^2. \end{cases}$$

See Problem 8.23 for another method.

(a) The left side of the Weyl identity is the free-space Green function, $G_0(\mathbf{r})$, which satisfies $(\nabla^2 + k_0^2)G_0(\mathbf{r}) = -\delta(\mathbf{r})$. Fourier transform this differential equation and show that

$$G_0(\mathbf{r}) = \frac{1}{(2\pi)^3} \int d^3k \, \frac{\exp(i\mathbf{k} \cdot \mathbf{r})}{k^2 - k_0^2}.$$

(b) Use contour integration to perform the integral over k_z in part (a). Assume that k_0 has a small positive imaginary part to establish the location of the poles and to decide how to close the contour.

21.25 Radiation from an Open Waveguide The $a \times b$ rectangular aperture of an infinite conducting plane is illuminated by the TE_{10} mode of a rectangular waveguide with the same cross sectional shape as the aperture. Evaluate the radiated electric field at the point (r, θ, ϕ) with respect to an origin at the center of the aperture using Kirchoff's approximation to the Fraunhofer-Smythe formula (21.98).

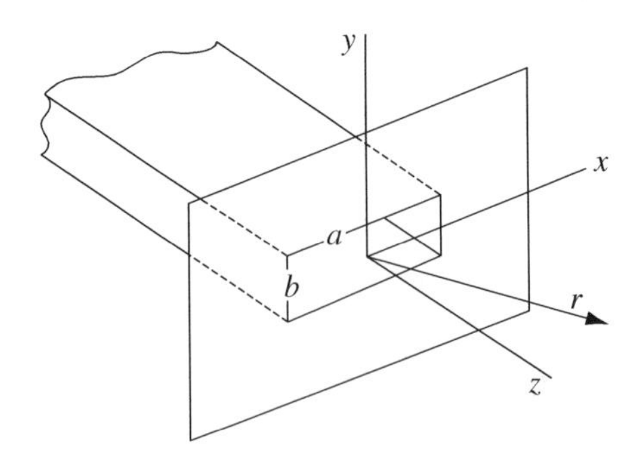

21.26 Diffraction from a Slit A plane wave propagating in the $+x$-direction with electric field $\mathbf{E}_0$ strikes a thin metal screen at $x = 0$ and diffracts from a long and narrow horizontal slit (width a) cut out of the screen. The scattering vector $\mathbf{k}$ lies in the x-y plane (perpendicular to the long direction of the slit) at an angle ϕ from the forward direction. Evaluate Smythe's formula (21.90) using Kirchoff's approximation and the far-field limit of the free-space Green function in two dimensions, $G_0^{(2)}(\boldsymbol{\rho}, \boldsymbol{\rho}') = (i/4)H_0^{(1)}(k|\boldsymbol{\rho} - \boldsymbol{\rho}'|)$, to show that the diffracted electric field is

$$\mathbf{E}(\rho, \phi) = -\frac{a}{2}\mathbf{k} \times (\hat{\mathbf{x}} \times \mathbf{E}_0)\sqrt{\frac{2}{\pi k\rho}} \exp[i(k\rho - \pi/4)]\frac{\sin(\frac{1}{2}ka\sin\phi)}{\frac{1}{2}ka\sin\phi}.$$

21.27 Diffraction of a Beam by a Large Aperture

(a) Consider the electric field diffracted by a circular aperture of radius a using a Kirchoff approximation where $\mathbf{E}_{\text{inc}} = E_0 \exp(-\rho^2/w^2)\hat{\mathbf{y}}$ in the plane of the aperture. Show that the far-zone field still has a Gaussian profile when the beam waist $w \ll a$.

(b) Repeat the calculation for a square aperture with side length a in the same limit and compare to part (a).

21.28 Effective Aperture Dipoles I Let $z = 0$ be a perfect conductor except for an aperture whose size is very small compared to the wavelength of a plane wave incident from $z < 0$. Perform a multipole expansion of the far-zone limit of Smythe's formula and find the effective electric and magnetic dipole moments of the aperture in the terms of $\mathbf{E}_\parallel$, the component of the exact electric field in the plane of the aperture. Are your results consistent with Figure 21.21?

21.29 Effective Aperture Dipoles II A monochromatic plane wave with fields $\mathbf{E}_0$ and $\mathbf{B}_0$ scatters from a thin conducting disk of radius a. In the long-wavelength limit, the scattered field is described by electric and magnetic dipole radiation fields with moments

$$\mathbf{p}_{\text{d}} = -\frac{16}{3}a^3\epsilon_0\hat{\mathbf{n}} \times (\hat{\mathbf{n}} \times \mathbf{E}_0) \qquad \text{and} \qquad \mathbf{m}_{\text{d}} = -\frac{8}{3\mu_0}a^3(\hat{\mathbf{n}} \cdot \mathbf{B}_0)\hat{\mathbf{n}}.$$

The unit vector $\hat{\mathbf{n}}$ points in the direction of the incident wave propagation vector when the latter is normal to the plane of the disk. Use Babinet's principle to deduce the effective dipole moments which characterize the diffracted field when a circular hole of radius a in a flat conducting plane is illuminated by a plane wave with aperture fields $\mathbf{E}_{\text{a}}$ and $\mathbf{B}_{\text{a}}$. Do not assume normal incidence for the diffraction problem.

21.30 Kirchhoff's Approximation for Complementary Scatterers A monochromatic plane wave polarized along $\hat{\mathbf{y}}$ is normally incident from $z < 0$ onto a two-dimensional conducting scatterer confined to the $z = 0$ plane. Use Kirchoff's approximation but do *not* use the Fraunhofer approximation.

(a) Let the scatterer be a conducting disk of radius a. Find $\mathbf{E}_{\text{disk}}(0, 0, z > 0)$.

(b) Let the scatterer be an infinite conducting sheet with a circular aperture of radius a centered on the z-axis. Find $\mathbf{E}_{\text{aperture}}(0, 0, z > 0)$.

(c) Confirm that $\mathbf{E}_{\text{aperture}}(0, 0, z > 0) = \mathbf{E}_{\text{inc}} - \mathbf{E}_{\text{disk}}(0, 0, z > 0)$. Explain why Babinet's principle is *not* the reason this is true.

22 Special Relativity

The special theory of relativity owes its origin to Maxwell's equations of the electromagnetic field.
Albert Einstein (1949)

22.1 Introduction

Special relativity is the theory of how different observers, moving at constant velocity with respect to one another, report their experience of the same physical event.[1] This description is completely accurate, but it conceals the fact that special relativity radically altered physicists' conceptions of space and time. It also obscures the deep connection between special relativity and electromagnetism, a connection Albert Einstein chose to emphasize in the opening paragraph of his ground-breaking paper on the subject (1905):

It is well known that Maxwell's electrodynamics—as usually understood at the present time—when applied to moving bodies, leads to asymmetries which do not appear to be inherent in the phenomena. Take, for example, the reciprocal action of a magnet and a conductor. The observable phenomenon here depends only on the relative motion of the conductor and the magnet, whereas the customary view draws a sharp distinction between the two cases in which either the one or the other of these bodies is in motion.

The issue that concerned Einstein was the perceived difference between "transformer" EMF and "motional" EMF when a conductor and a magnet move relative to one another (see Section 14.4.1). From the point of view of the conductor, the moving magnet produces an electric field at every point in space, including within the body of the conductor, where it induces a current. From the point of view of the magnet, no electric field appears anywhere and the magnetic Lorentz force is responsible for the current flow. Einstein found the unquestioning acceptance of this asymmetry intolerable for the description of a single physical phenomenon. He was also concerned with a paradox that had vexed him since the age of 16:[2]

If I pursue a beam of light with the velocity c, I should observe such a beam of light as a spatially oscillatory electromagnetic field at rest. However, there seems to be no such thing, whether on the basis of experience or according to Maxwell's equations.

Einstein's contemplation of these matters led him to the concept of the *relativity of simultaneity* and to a theory of space and time which forced him to abandon Newtonian dynamics. Indeed, special

[1] *General* relativity addresses the same issue for observers whose relative motion is completely arbitrary.

[2] See "Autobiographical notes", in *Albert Einstein: Philosopher-Scientist*, 3rd edition, edited by P.A. Schilpp (Open Court, La Salle, IL, 1969), Volume I.

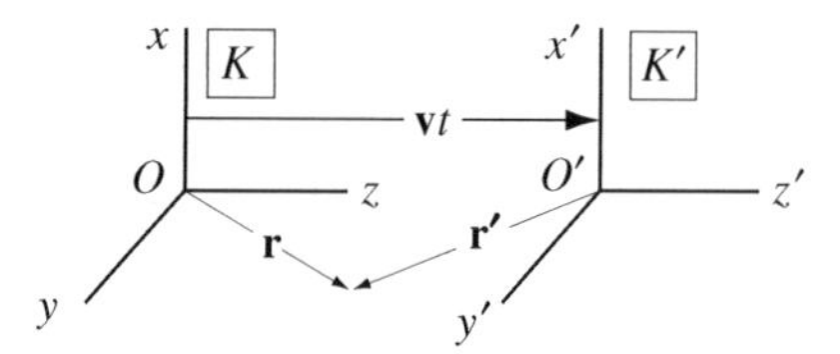

Figure 22.1: Two reference frames in a "standard configuration" where the coordinate axes are oriented identically, the origins coincide at $t = t' = 0$, and their relative motion occurs along one axis. The positions vectors are drawn in the Galilean limit where $\mathbf{r}' = \mathbf{r} - \mathbf{v}t$.

relativity is a conceptually subtle subject precisely because the relativity of simultaneity does not comport with intuitive notions based on Newton's laws. Some of these subtleties are the source of "paradoxes" which confound the inexpert and the unwary (and even the expert and the wary). Paradoxes have great value for those who wish to delve deeply into the subject, but they do not figure prominently in our presentation of special relativity. Our focus is to acquaint the reader with the language, principal consequences, and contemporary applications of the theory, particularly as they bear on classical electromagnetism.

Our discussion begins with the physical postulates of relativity, the Lorentz transformation, and some of the simpler consequences of the Lorentz transformation. We treat the kinematics and dynamics of point particles quite briefly and do not discuss spin at all. A central topic is the transformation laws for electromagnetic quantities like charge density, current density, the electromagnetic potentials, and the electromagnetic fields. Using these, we revisit the physics of moving point charges and plane electromagnetic waves. We then introduce the concept of the Lorentz tensor and derive manifestly covariant representations for the Maxwell equations and for the conservation laws of electrodynamics. An application of the latter to an isolated electromagnetic pulse provides insight into this system that is otherwise difficult to deduce. The chapter concludes with some consequences of special relativity for dielectric and magnetic matter in motion.

22.2 Galileo's Relativity

Before special relativity was formulated, the fundamental laws of physics were understood to obey Galileo's principle of relativity. The key concept is the *reference frame*, which we define as an oriented system of coordinates in three-dimensional space equipped with rulers and clocks to perform measurements of position and time. The latter permit us to define an *event* as an occurrence at a fixed point in space and time (x, y, z, t). Newton gave special attention to a class of frames where objects move with constant velocity if they are not acted on by external forces. These are called *inertial* frames and it follows that every inertial frame moves with constant velocity with respect to any other inertial frame. Newton also emphasized the universal nature of time, in the sense that the clocks in all inertial frames tick at the same rate, independent of all external influences. Among other things, the concept of universal (or absolute) time implies that two events judged to be simultaneous in one inertial frame are presumed to be simultaneous in all other inertial frames.

22.2.1 Particle Motion

Galileo's relativity principle states that *the laws of bodily motion are the same in all inertial frames*. Consider, for example, the inertial frames K and K' shown in Figure 22.1. The frames are arranged in a "standard configuration" where their coordinate axes are similarly oriented, their origins O and O' coincide in space when $t = t' = 0$, and their relative motion occurs with constant speed v along the parallel axes z and z'. Often, we will say that K is the "laboratory frame". In this frame, Newton's second law reads

$$m\frac{d^2\mathbf{r}}{dt^2} = \mathbf{F}. \tag{22.1}$$

Consider now the frame K'. Newton's assumption of universal time guarantees that $t' = t$. Combining this with the vector addition indicated in Figure 22.1 gives the complete Galilean transformation as

$$\mathbf{r}' = \mathbf{r} - \mathbf{v}t \qquad \text{and} \qquad t' = t. \tag{22.2}$$

Because $\mathbf{v}$ is constant (by assumption), (22.2) implies that $d^2\mathbf{r}'/dt'^2 = d^2\mathbf{r}/dt^2$. Therefore, if the mass does not depend on velocity, the rule $\mathbf{F}' = \mathbf{F}$ brings Newton's law into accord with Galileo's relativity principle because (22.1) transforms to

$$m\frac{d^2\mathbf{r}'}{dt'^2} = \mathbf{F}'. \tag{22.3}$$

22.2.2 Wave Motion

Unlike single-particle motion, sound waves and water waves do *not* behave identically in all inertial frames. This is the common observation of, say, a body surfer who rides along with a wave rather than allowing it to wash over him. To see this explicitly, consider a scalar field which satisfies the wave equation in frame K with speed c:

$$\left[\nabla^2 - \frac{1}{c^2}\frac{\partial^2}{\partial t^2}\right] f(\mathbf{r}, t) = 0. \tag{22.4}$$

The wave $f(\mathbf{r}, t)$ in frame K becomes a wave $f'(\mathbf{r}', t')$ in frame K' and its time evolution is determined by the wave operator in (22.4) transformed into primed variables. Using (22.2), the chain rule gives the derivatives we need as

$$\frac{\partial}{\partial \mathbf{r}} = \frac{\partial \mathbf{r}'}{\partial \mathbf{r}} \cdot \frac{\partial}{\partial \mathbf{r}'} + \frac{\partial t'}{\partial \mathbf{r}}\frac{\partial}{\partial t'} = \frac{\partial}{\partial \mathbf{r}'} \tag{22.5}$$

and

$$\frac{\partial}{\partial t} = \frac{\partial \mathbf{r}'}{\partial t} \cdot \frac{\partial}{\partial \mathbf{r}'} + \frac{\partial t'}{\partial t}\frac{\partial}{\partial t'} = \frac{\partial}{\partial t'} - \mathbf{v} \cdot \frac{\partial}{\partial \mathbf{r}'}. \tag{22.6}$$

Using (22.5) and (22.6) to compute the second derivatives in (22.4) gives the propagation equation in K' as

$$\left[\nabla'^2 - \frac{1}{c^2}\frac{\partial^2}{\partial t'^2} + \frac{2}{c^2}(\mathbf{v} \cdot \nabla')\frac{\partial}{\partial t'} - \frac{1}{c^2}(\mathbf{v} \cdot \nabla')^2\right] f'(\mathbf{r}', t') = 0. \tag{22.7}$$

Comparing (22.4) to (22.7) shows that the Galilean transformation (22.2) does *not* preserve the form of the wave equation (as it does Newton's second law) because classical waves propagate *relative* to any uniform motion of the host medium (water, air, etc.). For example, let $\mathbf{v} = v\hat{\mathbf{z}}$ and consider waves propagating in the $+z$-direction. If $g(s)$ is an arbitrary function of one variable, direct substitution confirms that a plane wave solution to (22.7) is

$$f'(x', y', z', t') = h(z' - ct' + vt'). \tag{22.8}$$

If $v = c$, the solution (22.8) tells us that an observer at rest in frame K' sees no wave propagation at all, only a static displacement of the particles of the medium. Wave motion is not invariant to a Galilean transformation.

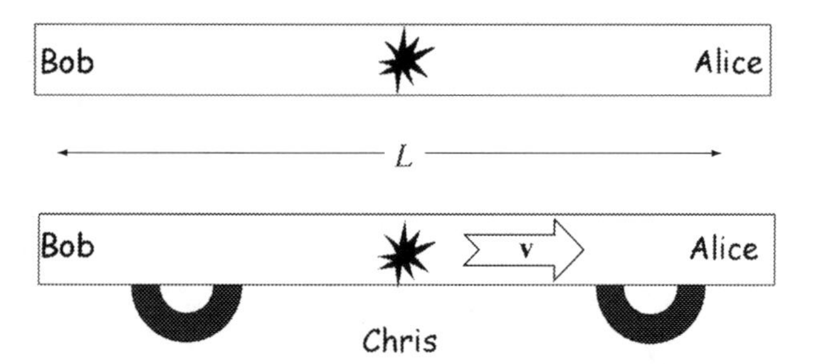

Figure 22.2: Top: Alice and Bob use a flash of light from a source at the midpoint between them to synchronize their clocks. Bottom: Chris sees Alice, Bob, and the light source in uniform motion and concludes that their clocks are not synchronized.

22.3 Einstein's Relativity

Special relativity appeared at a time when scientists were struggling to understand the *absence* of Galilean relative-motion effects (see Section 22.2.2) in light-propagation experiments. Einstein's beam-of-light thought-experiment (quoted in the Introduction) makes this point explicitly. Most proposed solutions either questioned the validity of Maxwell's equations or postulated special properties for the "aether", the presumptive host medium for light waves. Einstein resolved the conceptual issues associated with the electrodynamics of moving bodies by rejecting the universal validity of Newton's laws and embracing the universal validity of Maxwell's laws. His famous and highly readable 1905 paper on the subject frames the solution using two postulates:

I. The laws of physics take the same form in every inertial frame.
II. The speed of light in vacuum is the same in every inertial frame.

Postulate I is a generalization of Galileo's relativity principle to include Maxwell's laws of electrodynamics. Postulate II explained the failure to detect relative motion between light and the aether by the simple expedient of making the aether superfluous. We will show below that Einstein's two postulates (and the tacit assumptions that empty space is isotropic and spatially homogeneous) are sufficient to construct the entire edifice of special relativity.[3] Part of the program is to discover the transformation laws that preserve the forms of the wave equation and the Maxwell equations, and to discover the dynamical law of motion that replaces Newton's laws. An equally important part of the program is to discover the physical consequences of the postulates. We begin with the most important of the physical consequences.

22.3.1 The Relativity of Simultaneity

Special relativity destroyed Newton's concept of absolute time, and with it the previously unquestioned concept of universal temporal simultaneity for all observers. To make this clear, Figure 22.2 presents a thought-experiment where Alice and Bob wish to synchronize their clocks. They do this by agreeing to start their clocks when each observes a flash of light from a source located at the midpoint between them. In this way, each reasonably concludes that their flash observations were simultaneous events.

Now consider the same scenario from the point of view of an observer, Chris, who sees Alice, Bob, and the light source all moving uniformly to the right with speed v. Because Chris knows that light

[3] Einstein's actual supposition was that the speed of light is "independent of the state of motion of the emitting body". Our postulate II is a consequence of this statement and postulate I.

travels at speed c, he recognizes that the light reaches Bob sooner than it reaches Alice, and concludes that Bob starts his clock before Alice starts hers. If Chris saw Alice and Bob moving in the opposite direction, he would conclude that Alice starts her clock before Bob. Quantitatively, Chris computes that the two clocks are out of synchronization by an amount

$$\Delta T = \frac{L/2}{c-v} - \frac{L/2}{c+v} = \frac{L}{c}\frac{v/c}{1-v^2/c^2}. \tag{22.9}$$

This number is very small if $v \ll c$, and thus goes unnoticed for all but the most precise human applications.

We learn from this thought-experiment that two inertial observers will not necessarily agree that two events are simultaneous, nor even that one event precedes the other! This conclusion has profound consequences for the concept of causality (see Section 22.4.3) and for Newton's concept of absolute time. The latter, in particular, is revealed to be an approximation which breaks down when the relative velocity between frames approaches the speed of light. The tale of Alice and Bob makes clear that time is the personal, experiential property of every inertial observer.

Application 22.1 The Global Positioning System

The Global Positioning System is a network of 24 Earth-orbiting satellites used to determine the position of a receiver on the Earth's surface with a spatial resolution of 1 m^2. Each satellite carries a highly stable atomic clock synchronized (before launch) to all the other clocks. In orbit, each satellite broadcasts a synchronous radio-frequency signal every millisecond, tagged with the position and time of the transmission. The satellite network is distributed in space so any receiver on the surface of the Earth is always in direct line of sight of four satellites.

Let (x, y, z, t) be the event where an observer simultaneously receives the transmission data (x_k, y_k, z_k, t_k) from each of four satellites $(k = 1, 2, 3, 4)$. Because the speed of light is the same for all inertial observers, the observer's position, $\mathbf{r} = (x, y, z)$ is determined by solving the four simultaneous one-way signal propagation equations

$$|\mathbf{r} - \mathbf{r}_k| = c(t - t_k) \qquad k = 1, 2, 3, 4. \tag{22.10}$$

The magnitude of the speed of light ($c = 299\,792\,458$ m/s) implies that a timing error of 1 ns leads to a positioning error of the order of 30 cm.

Equation (22.10) is correct only if the four transmitting satellites are inertial with respect to the observer. To make this as true as possible, a combination of special and general relativistic corrections are applied to take account of the centripetal acceleration of the satellites, the effect of the Earth's gravitational field, the eccentricity of the satellite orbits, and the rotation of the Earth during the transit time of the signals.[4] It is even possible to use the GPS to test special relativity itself. ■

22.4 The Lorentz Transformation

The renunciation of absolute time required by the relativity of simultaneity (Section 22.3.1) implies that the Galilean transformation (22.2) between two inertial reference frames cannot be exact. On the other hand, everyday experience shows that Galileo's relativity works very well if the relative speed v between the frames is very small compared to the speed of light. In this section, we reconcile these statements using two observations of a single optical event to deduce the exact transformation law between inertial frames.

[4] See Ashby and Spilker, Jr. (1995) in Sources, References, and Additional Reading.

The different perception of time by different inertial observers leads us to treat space and time on an equal footing and to locate events in a venue called *space-time*. The most general transformation law between two inertial frames K and K' in space-time is

$$x' = x'(x, y, z, t) \quad y' = y'(x, y, z, t) \quad z' = z'(x, y, z, t) \quad t' = t'(x, y, z, t'). \tag{22.11}$$

The functions in (22.11) are very restricted if we assume that the properties of space are homogeneous and do not vary from point to point or as a function of time. In particular, the infinitesimal displacement

$$dx' = \frac{\partial x'}{\partial x}dx + \frac{\partial x'}{\partial y}dy + \frac{\partial x'}{\partial z}dz + \frac{\partial x'}{\partial t}dt \tag{22.12}$$

cannot be an explicit function of (x, y, z, t). This tells us that the partial derivatives in (22.12) are constants. The same is true for the three other functions in (22.11). Therefore, the transformation laws are linear functions of their arguments. It is a preview of future notation when we let r_μ (with $\mu = 1, 2, 3, 4$) stand for x, y, z, ct and write our deduction for the transformation law to this point in the form

$$r'_\mu = L_{\mu\nu} r_\nu + a_\mu. \tag{22.13}$$

22.4.1 Boosting the Standard Configuration

The right side of (22.13) contains 20 parameters. This number drops to a handful if we assume that (i) the coordinate axes in frame K are aligned with their counterparts in frame K'; (ii) the origins of the two frames coincide when $t = t' = 0$; and (iii) the velocity vector which "boosts" frame K to frame K' is $\mathbf{v} = v\hat{\mathbf{z}}$. This returns us to the "standard configuration" of Figure 22.1, where the most general transformation law consistent with the rotational invariance of isotropic free space is

$$x' = Cx \qquad y' = Cy \qquad z' = Az + Bt \qquad t' = Dz + Et. \tag{22.14}$$

Our task is to determine the constants A, B, C, D, and E. We make a few remarks about more general Lorentz transformations at the end of the section.

Our first deduction is that $B = -vA$. This follows from (22.14) because the standard configuration requires $z' = 0$ to coincide with $z = vt$. By symmetry, $x = Cx'$ and $y = Cy'$ supplement (22.14) because C cannot depend on the direction of motion of one frame with respect to the other. $C = 1$ is the only reasonable conclusion. We now adopt Einstein's original method and consider a point source of light that emits a spherical wave at $t = 0$ from the origin of K. Such a wave propagates radially at speed c and reaches the observation point (x, y, z) at a time t such that

$$x^2 + y^2 + z^2 - c^2t^2 = 0. \tag{22.15}$$

According to postulate II of Section 22.3, the same event viewed from frame K' of Figure 22.1 satisfies

$$x'^2 + y'^2 + z'^2 - c^2t'^2 = 0. \tag{22.16}$$

Our strategy is to substitute (22.14) into (22.16) (with $B = -vA$ and $C = 1$) and insist that the result reproduce (22.15). This procedure generates three constraints on the coefficients:

$$A^2 - c^2D^2 = 1 \tag{22.17}$$

$$E^2 - \frac{v^2}{c^2}A^2 = 1 \tag{22.18}$$

$$vA^2 + c^2DE = 0. \tag{22.19}$$

The remaining steps are straightforward. First, we use (22.17) and (22.19) to eliminate D and derive a relation between E^2 and A^2. Combining this relation with (22.18) to eliminate A^2 produces

$E^2 = 1/(1 - v^2/c^2)$. With this information, (22.18) shows that $A^2 = E^2$. In fact, $A = E = 1/\sqrt{1 - v^2/c^2}$, because we must recover the Galilean limit (22.2) as $c \to \infty$. Finally, we insert these values into (22.19) to get $D = -v/c^2\sqrt{1 - v^2/c^2}$. Therefore, the Lorentz transformation from inertial frame K to inertial frame K' in Figure 22.1 is

$$x' = x \qquad y' = y \qquad z' = \frac{z - vt}{\sqrt{1 - \dfrac{v^2}{c^2}}} \qquad t' = \frac{t - vz/c^2}{\sqrt{1 - \dfrac{v^2}{c^2}}}. \tag{22.20}$$

The conclusion $v < c$ follows immediately from (22.20) because the real numbers (x, y, z, t) must transform into the real numbers (x', y', z', t'). In words, *the speed of light is greater than the speed which can be achieved by any inertial frame*. By extension, no material particle or object at rest in an inertial frame can be accelerated to the speed of light.

It is standard practice in special relativity to define the symbols

$$\beta = \frac{v}{c} < 1 \qquad \text{and} \qquad \gamma = \frac{1}{\sqrt{1 - \beta^2}} > 1. \tag{22.21}$$

Using (22.21), we write (22.20) in the form

$$x' = x \qquad y' = y \qquad z' = \gamma(z - \beta ct) \qquad ct' = \gamma(ct - \beta z). \tag{22.22}$$

We derive the reverse transformation from K' to K by solving the linear equations in (22.22) for the functions $z(z', t')$ and $t(z', t')$. By symmetry, we get the same answer by exchanging the primed and unprimed variables in (22.22) and letting $v \to -v$:

$$x = x' \qquad y = y' \qquad z = \gamma(z' + \beta ct') \qquad ct = \gamma(ct' + \beta z'). \tag{22.23}$$

Example 22.1 Frame A moves at constant velocity $s\hat{\mathbf{z}}$ with respect to frame B. Frame B moves at constant velocity $w\hat{\mathbf{z}}$ with respect to frame C. Find the velocity $\mathbf{u} = u\hat{\mathbf{z}}$ at which frame A moves with respect to frame C. Figure 22.3 shows all three frames in standard configurations.

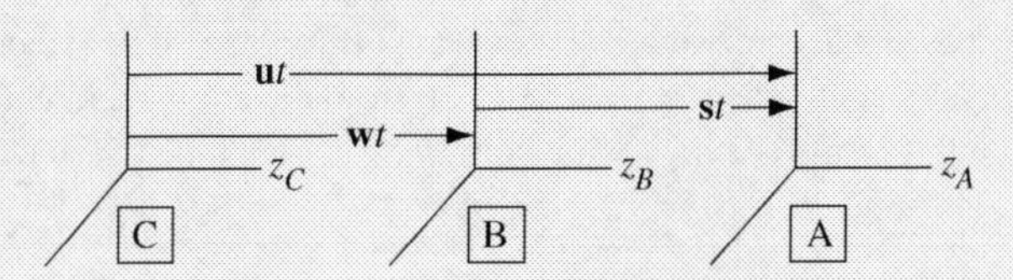

Figure 22.3: Three standard-configuration inertial frames moving uniformly with respect to one another.

Solution: Using (22.22) twice,

$$z_A = \gamma(s)[z_B - st_B] \qquad t_A = \gamma(s)(t_B - sz_B/c^2)$$

and

$$z_B = \gamma(w)[z_C - wt_C] \qquad t_B = \gamma(w)(t_C - wz_C/c^2).$$

By direct substitution,

$$z_A = \gamma(s)\gamma(w)(z_C - wt_C - st_C + swz_C/c^2) = \gamma(s)\gamma(w)(1 + sw/c^2)\left[z_C - \frac{s + w}{1 + sw/c^2}t_C\right].$$

Comparing the square brackets just above to the square brackets in the first two equations suggests that the composition law for parallel boost velocities is

$$u = \frac{s + w}{1 + \dfrac{sw}{c^2}}.$$

If so, consistency demands that the factor in front of the square brackets is $\gamma(u)$. A bit of algebra confirms this because the quantity $1/\gamma^2(u)$ is equal to

$$1 - u^2/c^2 = 1 - \frac{1}{c^2}\left(\frac{s + w}{1 + sw/c^2}\right)^2 = \frac{1 + s^2w^2/c^4 - s^2/c^2 - w^2/c^2}{(1 + sw/c^2)^2}$$

$$= \left[\frac{1}{\gamma(s)\gamma(w)(1 + sw/c^2)}\right]^2.$$

Very similar manipulations show that the same value of u correctly generates t_A from x_C and t_C. Finally, we note that the identity proved by the preceding equation is often used in problem-solving, particularly when it is written in the form

$$\gamma(u) = \gamma(s)\gamma(w)\left(1 + \frac{sw}{c^2}\right).$$

22.4.2 Time Dilation and Length Contraction

The mixing of space and time implied by the Lorentz transformation produces a variety of non-intuitive predictions. Consider, for example, two arbitrary events, (x_1, y_1, z_1, t_1) and (x_2, y_2, z_2, t_2), and the difference variables, $\Delta x = x_1 - x_2$, $\Delta y = y_1 - y_2$, $\Delta z = z_1 - z_2$, and $\Delta t = t_1 - t_2$. For the geometry of Figure 22.1, the linearity of (22.22) and (22.23) imply that

$$\Delta z' = \gamma(\Delta z - \beta c \Delta t) \qquad c\Delta t' = \gamma(c\Delta t - \beta \Delta z) \tag{22.24}$$

and

$$\Delta z = \gamma(\Delta z' + \beta c \Delta t') \qquad c\Delta t = \gamma(c\Delta t' + \beta \Delta z'). \tag{22.25}$$

The phenomenon of *time dilation* reveals itself when we identify the two events as two readings of a clock at rest in K'. The clock does not move in this frame, so $\Delta z' = 0$. We make no attempt to measure Δz, but the elapsed time in K' is $\Delta t' = T'$. Therefore, using (22.25),

$$T = \Delta t = \gamma \Delta t' = \frac{T'}{\sqrt{1 - \dfrac{v^2}{c^2}}} > T'. \tag{22.26}$$

The observer in the laboratory reports a longer elapsed time than does the observer in the moving frame. He/she concludes that "clocks in uniform motion tick more slowly than stationary clocks". Experimental evidence for time dilation comes from the radioactive decay of cosmic ray muons. These particles carry their own clock in the sense that a non-zero muon lifetime guarantees that an initial population of muons decays exponentially as time goes on. Experiments that compared the downward flux of velocity-selected muons at the top of a mountain to the downward flux of muons at sea level found that many more muons survived than predicted by the known muon lifetime, $\tau' \approx 2.2$ μsec.[5]

[5] See, for example, D.H. Frisch and J.H. Smith, "Measurement of relativistic time dilation using μ-mesons", *American Journal of Physics* **31**, 342 (1963).

Instead, the measured flux was consistent with a muon population

$$N(t) = N(0)\exp(-t/\tau), \tag{22.27}$$

with $\tau = \gamma\tau'$ and γ computed from (22.21) and the measured mean speed of the muons. The effective lifetime $\tau \gg \tau'$ is exactly what we predict if the time-dilation formula (22.26) is correct.

The phenomenon of *length contraction* reveals itself when we identify the two events introduced at the beginning of this section as sightings of the two end points of a rod at rest in K'. The time lapse $\Delta t'$ needed to measure the length $L' = \Delta z'$ is irrelevant in the rest frame of the rod. By contrast, the only sensible way to measure the "length" of a moving rod is to perform the sightings of its end points simultaneously in the lab frame ($\Delta t = 0$) when we establish that $L = \Delta z$. Using (22.24), we conclude that $L' = \Delta z' = \gamma\Delta z = \gamma L$. Therefore,

$$L = \Delta z = \frac{1}{\gamma}\Delta z' = L'\sqrt{1 - \frac{v^2}{c^2}} < L'. \tag{22.28}$$

The observer in the laboratory reports a shorter length than does the observer in the moving frame. He/she concludes that "moving rods contract in their direction of motion". Because $x = x'$ and $y = y'$ in (22.20), there is no length contraction in the direction *transverse* to the direction of motion.

Experimental evidence for length (or Lorentz) contraction is less direct than for time dilation. One example is the behavior of a particle beam in a linear accelerator. The particles in such a beam inevitably have a small velocity component transverse to the direction of acceleration. However, the total transverse spread of the beam at the end of the accelerator is never very large. This can be understood by transforming to the rest frame of a typical relativistic particle in the beam. From this perspective, the apparent length of the accelerator is greatly contracted compared to its laboratory length.[6] The beam cannot spread very much if the distance traveled by every particle is very small.

It is important to appreciate that length contraction is a *kinematic* effect in the sense that no forces of any kind are involved. Length contraction does *not* occur because motion induces some sort of longitudinal compressive stress. After all, an observer in the rest frame of the rod detects no motion at all. Rather, the operational definition of length given above identifies the inescapable relativity of simultaneity as the ultimate source of the contraction effect.

Application 22.2 Heavy Ion Collisions and the Quark Gluon Plasma

Current theories of the immediate aftermath of the Big Bang place all matter in a highly excited state called a quark-gluon plasma (QGP). As the Universe expanded and cooled, it is suggested that the plasma underwent a phase transition to a state where the quarks and gluons condensed into baryons and mesons. The nuclei of ordinary matter formed later from the condensation of protons and neutrons. A glimpse into this exotic physics can now be gained in the laboratory because recent experiments strongly suggest that a QGP forms (transiently) when two heavy atomic nuclei are collided at relativistic energies. As we will see, length contraction plays an important role in QGP formation.

[6] Wangler (2008) estimates that the Stanford linear accelerator appears only 0.25 m long to a particle accelerated from 40 Mev to 40 GeV over the 3 km laboratory length of the device.

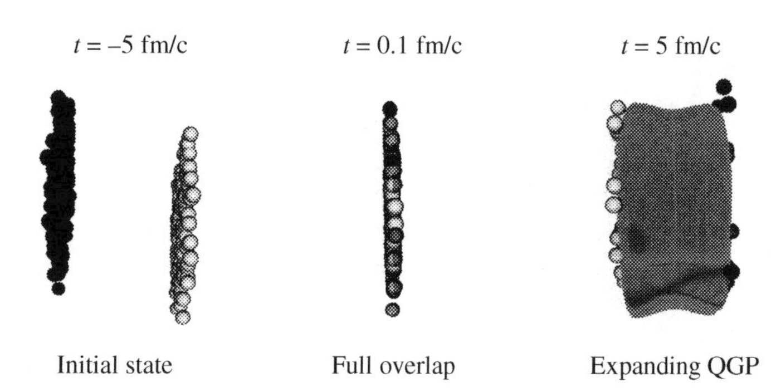

Figure 22.4: Simulation of the collision of two Au nuclei at 200 GeV per nucleon pair. Left: Lorentz-contracted nuclei approach one another. Middle: the nuclei overlap and create a region of very high energy density. Right: a quark-gluon plasma expands into the volume between the separating nuclei. Images courtesy of Hannah Petersen (Duke University) and the MADAI collaboration.

Figure 22.4 shows three snapshots from a relativistic nuclear fluid-dynamics computer simulation of the collision of two Au nuclei with velocities of 99.9995% of the speed of light. The left panel shows that the approaching nuclei are Lorentz contracted ($\gamma \sim 100$) into pancakes in the center-of-mass frame. The middle panel shows the region of maximum nuclei overlap where the energy density is highest. The right panel shows the situation after the QGP has formed and it expands and cools. After the temperature has fallen below the critical temperature, the plasma recombines into hadrons which can be measured by particle detectors. The overall time scale of the collision process is about 30 fm/c, where 1 fm/c $= 3.33 \times 10^{-24}$ s.

It takes about 0.14 fm/c for the two length-contracted nuclei to reach full overlap. The quarks and gluons in the two nuclei equilibrate and the QGP forms about 0.6 fm/c after the onset of the collision. However, if Lorentz contraction of the nuclei did not occur, it would take at least 7 fm/c to reach full overlap and the equilibration time would be approximately 14 fm/c. This is more than one order of magnitude larger than what is observed when the results of simulations like the one shown in Figure 22.4 are compared to experimental data. The Lorentz contraction of the colliding nuclei is also important for the angular distribution of emitted particles. In both experiment and the simulations, this emission is confined to the reaction plane (defined by the beam direction and the impact parameter). When non-relativistic, spherical nuclei collide, the emission of particles occurs primarily perpendicular to the reaction plane. ■

22.4.3 The Invariant Interval

A relativistic (or Lorentz) *invariant* quantity takes the same numerical value in every inertial frame. Invariants play a special role in relativity, beginning with Einstein's postulate (Section 22.3) that the speed of light is a relativistic invariant. Another relativistic invariant is electric charge. There is no experimental evidence that the charge of an electron (or proton or neutron) depends on its speed. In this section, we introduce a third invariant called the *interval* and use it to distinguish past events from future events and causes from effects.

Using the variables defined at the beginning of Section 22.4.2, the square of the *interval* between the two events is defined as

$$(\Delta s)^2 = (\Delta x)^2 + (\Delta y)^2 + (\Delta z)^2 - (c\Delta t)^2. \tag{22.29}$$

The interval combines a distance in space, $d = \sqrt{(\Delta x)^2 + (\Delta y)^2 + (\Delta z)^2}$, with a distance (or lapse) in time, Δt, into a single quantity. Like d, the interval is invariant to rotations and translations in space.

Like Δt, the interval is invariant to translations in time. Most importantly, Δs is invariant to a Lorentz transformation. To prove this, we use the standard configuration (Figure 22.1) and (22.24) to write the interval evaluated in K' in terms of the coordinates defined in K. This gives

$$(\Delta s')^2 = (\Delta x)^2 + (\Delta y)^2 + \gamma^2(\Delta z - \beta c\Delta t)^2 - \gamma^2(c\Delta t - \beta \Delta z)^2, \tag{22.30}$$

or

$$(\Delta s')^2 = (\Delta x)^2 + (\Delta y)^2 + \gamma^2(1 - \beta^2)\left[\Delta z^2 - c^2\Delta t^2\right]. \tag{22.31}$$

However, because $\gamma^2(1 - \beta^2) = 1$,

$$(\Delta s')^2 = (\Delta x)^2 + (\Delta y)^2 + (\Delta z)^2 - (c\Delta t)^2 = (\Delta s)^2. \tag{22.32}$$

This proves that $(\Delta s)^2$ takes the same value in all inertial frames.

Special relativity exploits the invariance of the interval in various ways. A first observation is that $(\Delta s)^2$ can be positive, negative, or zero. The three cases differ in the nature of the "separation" between the events:

$$\begin{aligned} &(\Delta s)^2 > 0 \quad \text{space-like separation} \\ &(\Delta s)^2 = 0 \quad \text{null separation} \\ &(\Delta s)^2 < 0 \quad \text{time-like separation.} \end{aligned} \tag{22.33}$$

A pair of events with *null separation* can be connected by a signal traveling at the speed of light. An example is the null separation between the origin and every point on the expanding spherical wave front described by (22.15). For a pair of events with a *space-like separation* $[(\Delta s)^2 > 0]$, the distance in space is greater than the distance $c\Delta t$ that can be covered by a light beam in the time Δt. For such events, it is always possible to perform a Lorentz transformation to an inertial frame where the event pair are simultaneous. If we call the latter frame K', and locate both events on the z' axis,

$$(\Delta s')^2 = (\Delta z')^2 - (c\Delta t')^2 = (\Delta z')^2. \tag{22.34}$$

The last equality in (22.34) follows from the $\Delta t' = 0$ condition for simultaneity in K' and shows why the label "space-like" is used for this case. We deduce from (22.24) that

$$\Delta t' = 0 \quad \Rightarrow \quad \beta = \frac{c\Delta t}{\Delta z}. \tag{22.35}$$

However, $|c\Delta t/\Delta z| < 1$ because (22.29) is space-like and $\Delta x = \Delta y = 0$. This shows that the boost required to make $\Delta t' = 0$ has $\beta < 1$, which is indeed physically realizable. A similar demonstration shows that a pair of events with a *time-like separation* $[(\Delta s)^2 < 0]$ can always be made to occur at a single point in space ($\Delta z' = 0$). For such events, the distance in space is less than the distance $c\Delta t$ that can be covered by a light beam in the time Δt.

We are now in a position to reconcile the concept of causality with the relativity of simultaneity. Figure 22.5 is a space-time or "Minkowski" diagram where the x- and y-axes are represented by a single axis ρ where $\rho^2 = x^2 + y^2$. An event labeled O occupies the origin of space-time. The two "light cones" drawn in Figure 22.5 are defined by the equation $\rho^2 + z^2 = c^2t^2$. Therefore, the interval (22.29) between O and any event on the surface of either cone is zero. From (22.33), the corresponding interval is space-like for events which lie *outside* both cones and time-like for events which lie *inside* either cone.

The event labeled S in Figure 22.5 is space-like with respect to O. This event, and all other events outside the light cones, are "absolutely distant" from O because their Euclidean distance from the origin can never be reduced to zero without violating the condition $(\Delta s)^2 > 0$ for a space-like interval. Moreover, these events cannot be said to be earlier or later than the event at O because the time interval between them can have different signs for different observers. For example, if $\Delta t > 0$, we can make

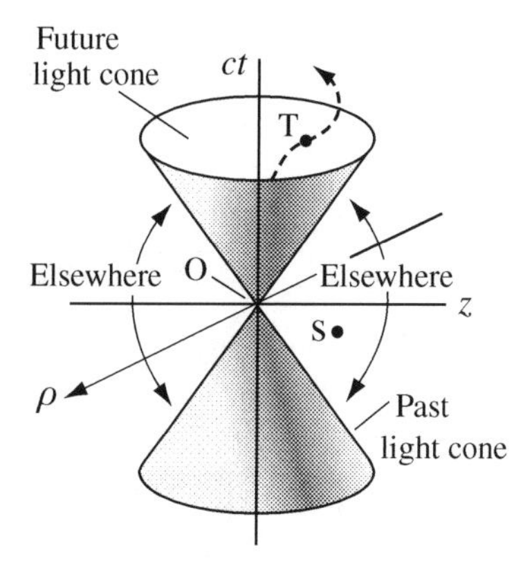

Figure 22.5: A Minkowski diagram where past and future light cones separate space-time into space-like intervals and time-like intervals with respect to an event O at the origin. The dashed curve is the "world line" of a particle moving with non-uniform velocity.

$\Delta t' < 0$ in (22.24) by choosing the boost speed so $|c\Delta t/\Delta z| < \beta < 1$. The possibility of this sign inversion implies that a cause-and-effect relationship cannot exist between two space-like events. This is consistent with the impossibility of transforming these events to the same point in space, which would be needed to compare their clocks.

The event labeled T in Figure 22.5 is time-like with respect to O. We say that T lies in the "future light cone" of O because it occurs *later* in time than O in all inertial frames. An event which lies inside the complementary "past light cone" of O occurs *earlier* in time than O in all inertial frames. These statements are true because $\Delta t'$ in (22.24) has the same sign as Δt for all time-like intervals. This is true, in turn, because the criteria for it *not* to be true is $|\beta| > |c\Delta t/\Delta z|$, which is impossible because $(\Delta s)^2 < 0$ implies that $|c\Delta t/\Delta z| > 1$ when $\Delta x = \Delta y = 0$ in (22.29). We conclude that causality is a meaningful concept for events with a time-like separation.

22.4.4 Proper Time

The *proper time* is an invariant measure of the motion of a particle along its trajectory in space-time. A definition for proper time follows naturally if we define the "world line" of a particle as the locus of points in space-time which describes the trajectory in question. The dashed curve in Figure 22.5 is a typical world line for a particle with non-uniform velocity $\mathbf{u}(t) = d\mathbf{r}/dt$. All world lines lie inside the light cone because the particle speed is always less than the speed of light.

Focus now on the interval between two points on the world line which lie infinitesimally close to each other. Using (22.30), this is the time-like quantity

$$(ds)^2 = (d\mathbf{r})^2 - (cdt)^2 = -(cdt)^2\left[1 - \frac{u^2(t)}{c^2}\right]. \tag{22.36}$$

Dividing (22.36) by the speed of light produces another invariant. This leads us to define a differential element of the invariant proper time in an inertial frame K as

$$d\tau = \sqrt{-\frac{(ds)^2}{c^2}} = \sqrt{1 - \frac{u^2(t)}{c^2}}dt = \frac{dt}{\gamma(u)}. \tag{22.37}$$

The invariance of $d\tau$ means that (22.37) has the same numerical value in any other inertial frame K' where $u' \neq u$, $t' \neq t$, and $dt' \neq dt$. This fact will provide a natural way to define Lorentz invariant time derivatives in what follows.

We close with two points. First, the last equality in (22.37) generalizes the meaning of γ in (22.21) so the argument can be a particle speed rather than merely the speed of a Lorentz boost from one inertial frame to another. Second, the definition of $d\tau$ tells us that the "proper time" is the time measured by a clock in its own rest frame.

22.4.5 Boosting a General Configuration

The most general Lorentz transformation between two inertial frames differs from the standard configuration transformation (22.22) in two ways.[7] First, the boost velocity $\mathbf{v}$ need not lie along one of the coordinate axes as it does in the standard configuration of Figure 22.1. Second, the Cartesian axes of frame K' may not be aligned with the Cartesian axes of frame K. For a boost (without a rotation), we decompose $\mathbf{r}$ into its components $\mathbf{r}_\parallel$ and $\mathbf{r}_\perp$ which lie parallel and perpendicular to $\boldsymbol{\beta} = \mathbf{v}/c$. Using these variables, the Lorentz transformation and its inverse take the form

$$\begin{aligned} \mathbf{r}'_\perp &= \mathbf{r}_\perp & \qquad \mathbf{r}_\perp &= \mathbf{r}'_\perp \\ \mathbf{r}'_\parallel &= \gamma(\mathbf{r}_\parallel - \boldsymbol{\beta} ct) & \qquad \mathbf{r}_\parallel &= \gamma(\mathbf{r}'_\parallel + \boldsymbol{\beta} ct') \\ ct' &= \gamma(ct - \boldsymbol{\beta}\cdot\mathbf{r}_\parallel) & \qquad ct &= \gamma(ct' + \boldsymbol{\beta}\cdot\mathbf{r}'_\parallel). \end{aligned} \tag{22.38}$$

For a rotation (without a boost), the Euler angle transformation law between the rotated variables (x, y, z) and the un-rotated variables (x', y', z') is derived in every textbook of classical mechanics and in every discussion of the quantum theory of angular momentum.

The general case when boosts and rotations occur together lies beyond the needs of this book. Here, we simply state without proof that a general Lorentz transformation between a frame K and a frame K' can be decomposed into either (i) a pure boost followed by a pure rotation or (ii) a pure rotation followed by a pure boost. However, the boost velocities and the rotation angles in the two decompositions are not generally the same. A related observation is that successive Lorentz boosts in different directions are equivalent to a single boost accompanied by a pure rotation. The latter fact is used to understand the phenomenon of *Thomas precession* in atomic physics.[8]

22.5 Four-Vectors

Einstein's first postulate of relativity (Section 22.3) states that the mathematical laws of physics have the same form in every inertial frame. In this section, we introduce the *four-vector* as the first step toward a formalism that will make this form-invariance (or covariance) self-evident. Our immediate aim is the physics that can be learned directly from the manipulation of individual four-vectors. Later, we will exploit the four-vector (and its generalizations) to facilitate covariance and streamline calculations.

Let $\mathbf{a} = (a_1, a_2, a_3)$ and $\mathbf{b} = (b_1, b_2, b_3)$ be three-vectors in Euclidean space. The scalar product of two three-vectors is invariant to translations and rotations of the coordinate system. In other words, if K and K' are two such systems,

$$\mathbf{a}\cdot\mathbf{b} = a_k b_k = a'_k b'_k = \mathbf{a}'\cdot\mathbf{b}'. \tag{22.39}$$

The invariance of the norm $\sqrt{\mathbf{a}\cdot\mathbf{a}}$ is a special case of (22.39).

We denote a four-vector in Minkowski space by

$$\vec{a} = (a_1, a_2, a_3, a_4). \tag{22.40}$$

[7] We continue to restrict ourselves to *homogeneous* transformations where $a_\mu = 0$ in (22.13).

[8] See, for example, G.P. Fisher, " The Thomas precession", *American Journal of Physics* **40**, 1772 (1972).

To justify calling (22.40) a "vector", we similarly require the scalar product of two four-vectors be invariant to translations, rotations, and Lorentz boosts from one inertial frame to another. In other words,[9]

$$\vec{a} \cdot \vec{b} = a_\mu b_\mu = a'_\mu b'_\mu = \vec{a}' \cdot \vec{b}'. \tag{22.41}$$

A special case is the invariance of

$$a_\mu a_\mu = a'_\mu a'_\mu. \tag{22.42}$$

By analogy with (22.33), it is common to say that a four-vector is null, space-like, or time-like depending on whether (22.42) is zero, positive, or negative.

The prototype of a four-vector in special relativity is the space-time coordinate,

$$\vec{r} = (x, y, z, ict) = (\mathbf{r}, ict). \tag{22.43}$$

The fourth component of (22.43) is a pure imaginary number.[10] This choice ensures that (22.42) produces the appropriate minus sign when we write (22.15) as $\vec{r} \cdot \vec{r} = 0$ and the invariant interval (22.29) as $(\Delta s)^2 = \Delta\vec{r} \cdot \Delta\vec{r}$. It was precisely the assumed invariance of these quantities which led us to the standard-configuration Lorentz transformation (22.22) and its inverse (22.23). Using (22.43), matrix representations for these transformations are

$$r'_\mu = \left[\frac{\partial r'_\mu}{\partial r_\nu}\right] r_\nu = L_{\mu\nu} r_\nu \qquad \text{and} \qquad r_\mu = \left[\frac{\partial r_\mu}{\partial r'_\nu}\right] r'_\nu = L^{-1}_{\mu\nu} r'_\nu, \tag{22.44}$$

where

$$\mathbf{L} = \begin{bmatrix} 1 & 0 & 0 & 0 \\ 0 & 1 & 0 & 0 \\ 0 & 0 & \gamma & i\beta\gamma \\ 0 & 0 & -i\beta\gamma & \gamma \end{bmatrix} \qquad \mathbf{L}^{-1} = \begin{bmatrix} 1 & 0 & 0 & 0 \\ 0 & 1 & 0 & 0 \\ 0 & 0 & \gamma & -i\beta\gamma \\ 0 & 0 & i\beta\gamma & \gamma \end{bmatrix}. \tag{22.45}$$

Inspection of (22.45) shows that $\mathbf{L}$ is an *orthogonal* matrix where $\mathbf{L}^{\mathrm{T}} = \mathbf{L}^{-1}$. Hence,

$$L_{\mu\lambda} L_{\nu\lambda} = \delta_{\mu\nu}. \tag{22.46}$$

The determinant of these transformation matrices is one:

$$|\mathbf{L}| = |\mathbf{L}^{-1}| = 1. \tag{22.47}$$

By definition, the components of an arbitrary four-vector $\vec{a}$ transform exactly like $\vec{r}$. Hence, a_1, a_2, and a_3 are often called the "space components" of $\vec{a}$, and a_4 is called the "time component" of $\vec{a}$. More precisely, $\vec{a}$ is a four-vector if

$$a'_\mu = L_{\mu\nu} a_\nu. \tag{22.48}$$

Referring back to (22.38), (22.48) is equivalent to

$$\begin{aligned} \mathbf{a}'_\perp &= \mathbf{a}_\perp & \qquad \mathbf{a}_\perp &= \mathbf{a}'_\perp \\ \mathbf{a}'_\parallel &= \gamma(\mathbf{a}_\parallel + i\boldsymbol{\beta} a_4) & \qquad \mathbf{a}_\parallel &= \gamma(\mathbf{a}'_\parallel - i\boldsymbol{\beta} a'_4) \\ a'_4 &= \gamma(a_4 - i\boldsymbol{\beta}\cdot\mathbf{a}_\parallel) & \qquad a_4 &= \gamma(a'_4 + i\boldsymbol{\beta}\cdot\mathbf{a}'_\parallel). \end{aligned} \tag{22.49}$$

[9] It is customary to use a repeated Greek index $(\mu, \nu, \sigma, \ldots)$ to sum over the four components of a four-vector and a repeated Latin index $(i, j, k, \ldots)$ to sum over the three components of a three-vector.

[10] Furry (1969), Veltman (1994), and 't Hooft (2001) all note the virtues of an imaginary time component when discussing special relativity only. Appendix D discusses the use of four-vectors with all real components and a diagonal metric tensor to evaluate the scalar product.

We leave it as an exercise for the reader to check that the transformation rules (22.49) guarantee that the scalar product of two four-vectors is invariant as indicated in (22.41).

22.5.1 The Four-Velocity and Four-Acceleration

The *four-velocity* $\vec{U}$ is another prototype four-vector. In this section, we use $\vec{U}$ to illustrate the construction logic and general usefulness of all four-vectors. The creative task is to identify a quantity that is closely related to the three-velocity $\mathbf{u} = d\mathbf{r}/dt$ of a particle, yet has the Lorentz transformation properties of the four-vector $\vec{r}$ defined in (22.43). The natural solution is to divide the four-vector $d\vec{r}$ by a differential element of proper time, which is a Lorentz invariant scalar (see Section 22.4.4). This prescription gives

$$\vec{U} = \frac{d\vec{r}}{d\tau} = \gamma(u)\frac{d}{dt}(\mathbf{r}, ict) = \gamma(u)\left(\frac{d\mathbf{r}}{dt}, ic\right) = \gamma(u)(\mathbf{u}, ic) \equiv (\mathbf{U}, U_4). \tag{22.50}$$

Regardless of its three-velocity, (22.50) shows that $\vec{U}$ is a time-like four-vector because it defines the Lorentz invariant scalar

$$\vec{U}\cdot\vec{U} = \mathbf{U}\cdot\mathbf{U} + U_4^2 = \frac{\mathbf{u}\cdot\mathbf{u} - c^2}{1 - u^2/c^2} = -c^2. \tag{22.51}$$

This is sensible because the discussion in Section 22.4.3 implies that we can always find an inertial frame where $\mathbf{u}$ is (instantaneously) zero.

It is instructive to Lorentz transform the four-velocity (22.50) to deduce the more complicated transformation law obeyed by the three-velocity. For that purpose, let $\mathbf{u}'$ be the velocity of a particle in the inertial frame K' in Figure 22.1. A particle at rest in this frame has speed $v = \beta c$ along the z-axis when viewed from the lab frame K. The four-vector $\vec{U}$ transforms like (22.49), so

$$U_1' = U_1 \qquad U_2' = U_2 \qquad U_3' = \gamma(v)(U_3 + i\beta U_4) \qquad U_4' = \gamma(v)(U_4 - i\beta U_3). \tag{22.52}$$

Dividing the components of $\mathbf{U}'$ in (22.52) by U_4' gives

$$\frac{U_1'}{U_4'} = \frac{U_1}{\gamma(v)(U_4 - i\beta U_3)} \qquad \frac{U_2'}{U_4'} = \frac{U_2}{\gamma(v)(U_4 - i\beta U_3)} \qquad \frac{U_3'}{U_4'} = \frac{U_3 + i\beta U_4}{U_4 - i\beta U_3}. \tag{22.53}$$

Finally, use the identity $\mathbf{U}/U_4 = \mathbf{u}/ic$ to eliminate all the four-velocity components in (22.53) in favor of three-velocity components. The result is the desired transformation law,

$$u_x' = \frac{u_x}{\gamma(v)(1 - u_z v/c^2)} \qquad u_y' = \frac{u_y}{\gamma(v)(1 - u_z v/c^2)} \qquad u_z' = \frac{u_z - v}{1 - u_z v/c^2}. \tag{22.54}$$

The inverse of (22.54) is the same formula with $\mathbf{u}$ and $\mathbf{u}'$ interchanged and $v \to -v$:

$$u_x = \frac{u_x'}{\gamma(v)(1 + u_z' v/c^2)} \qquad u_y = \frac{u_y'}{\gamma(v)(1 + u_z' v/c^2)} \qquad u_z = \frac{u_z' + v}{1 + u_z' v/c^2}. \tag{22.55}$$

As it must, (22.55) reduces to the Galilean velocity addition formula, $\mathbf{u} = \mathbf{u}' + \mathbf{v}$, when $v/c \ll 1$. The special case $u_x' = u_y' = 0$ merits attention also. First, it confirms our earlier conclusion that c is a limiting velocity because $u_z \to c$ when the boost velocity $v \to c$.[11] Second, we can return to Figure 22.3 of Example 22.1 and re-interpret the inertial reference frame A as a "particle" which moves with *uniform* speed $u_z' = s$ with respect to an inertial frame $K' = B$. Because the latter moves with uniform speed $v = w$ with respect to the lab frame $K = C$, the far left member of (22.55) reproduces the law for parallel boost velocities derived in the Example.

[11] This conclusion does not change when u_x' and u_y' are non-zero.

A natural definition for the *four-acceleration* of a particle is

$$\vec{\mathcal{A}} = \frac{d\vec{U}}{d\tau} = \gamma(u)\frac{d}{dt}\frac{(\mathbf{u}, ic)}{\sqrt{1-u^2/c^2}} \equiv (\boldsymbol{\mathcal{A}}, \mathcal{A}_4). \tag{22.56}$$

With $\mathbf{a} = d\mathbf{u}/dt$, the time derivative in (22.56) gives

$$\boldsymbol{\mathcal{A}} = \frac{\mathbf{a}}{1-u^2/c^2} + \frac{\mathbf{u}(\mathbf{u}\cdot\mathbf{a})/c^2}{(1-u^2/c^2)^2} \qquad \text{and} \qquad \mathcal{A}_4 = \frac{i(\mathbf{u}\cdot\mathbf{a})/c}{(1-u^2/c^2)^2}. \tag{22.57}$$

An interesting property of $\vec{\mathcal{A}}$ is its orthogonality to the four-velocity $\vec{U}$. This follows immediately from $\vec{U}\cdot\vec{U} = -c^2$ in (22.51) because

$$\vec{U}\cdot\vec{\mathcal{A}} = \vec{U}\cdot\frac{d\vec{U}}{d\tau} = \frac{1}{2}\frac{d}{d\tau}\left(\vec{U}\cdot\vec{U}\right) = 0. \tag{22.58}$$

22.5.2 The Four-Momentum and Energy

The *four-momentum* $\vec{p}$ plays a central role in relativistic particle dynamics. Given the four-velocity in (22.50), we define $\vec{p}$ using a scalar $\mathcal{E}$ and a three-vector $\mathbf{p}$:

$$\vec{p} = m\vec{U} = m(\mathbf{U}, U_4) = (\mathbf{p}, i\mathcal{E}/c). \tag{22.59}$$

The mass m in (22.59) must be a Lorentz invariant scalar if we require $\vec{p}$ to be a four-vector like $\vec{U}$.[12] Our main task is to identify $\mathcal{E}$ and $\mathbf{p}$, and extract the physics from the component equations

$$\mathcal{E}/c = -imU_4 = \gamma(u)mc \qquad \text{and} \qquad \mathbf{p} = m\mathbf{U} = \gamma(u)m\mathbf{u}. \tag{22.60}$$

The meaning of $\mathcal{E}$ becomes clear when we Taylor expand $\gamma(u)mc$ in (22.60) for $u \ll c$ to get

$$\mathcal{E} = \frac{mc^2}{\sqrt{1-u^2/c^2}} = mc^2 + \frac{1}{2}mu^2 + \frac{3}{8}m\frac{u^4}{c^2} + \cdots. \tag{22.61}$$

The second term on the far right side of (22.61) is the familiar low-velocity kinetic energy. The first term is a constant which may sensibly be called the *rest energy*. The *total energy* is $\mathcal{E}$ and the impossibility of accelerating a massive particle to speed c appears here as the impossibility of accelerating a particle to infinite energy. The exact kinetic energy is

$$T = \mathcal{E} - mc^2 = mc^2\left[\frac{1}{\sqrt{1-u^2/c^2}} - 1\right]. \tag{22.62}$$

The meaning of $\mathbf{p}$ in (22.60) emerges similarly from a Taylor expansion of $\gamma(u)m\mathbf{u}$ for $u \ll c$. The result,

$$\mathbf{p} = \frac{m\mathbf{u}}{\sqrt{1-u^2/c^2}} = m\mathbf{u}\left[1 + \frac{1}{2}\frac{u^2}{c^2} + \frac{3}{8}\frac{u^4}{c^4} + \cdots\right], \tag{22.63}$$

shows that $\mathbf{p}$ reduces to the ordinary Newtonian linear momentum, $m\mathbf{u}$, when the particle speed is very small compared to the speed of light.

Using (22.51), the Lorentz invariant length of the energy-momentum four-vector (22.59) is

$$\vec{p}\cdot\vec{p} = m^2\vec{U}\cdot\vec{U} = -m^2c^2. \tag{22.64}$$

[12] The numerical value of m is what older textbooks call the rest mass, m_0. We do not introduce the latter quantity in this book.

On the other hand, evaluating the norm of $\vec{p}$ using its own components gives

$$\vec{p}\cdot\vec{p} = \mathbf{p}\cdot\mathbf{p} - \frac{\mathcal{E}^2}{c^2} = p^2 - \frac{\mathcal{E}^2}{c^2}. \tag{22.65}$$

Combining (22.64) with (22.65) expresses the total energy in terms of the mass and three-momentum of a particle:

$$\mathcal{E} = \sqrt{c^2p^2 + m^2c^4}. \tag{22.66}$$

Eliminating $\gamma(u)$ from the left and right sides of (22.60) relates the three-velocity to the three-momentum and the total energy:

$$\mathbf{u} = \frac{c^2\mathbf{p}}{\mathcal{E}}. \tag{22.67}$$

Experiment shows that (22.66) and (22.67) remain valid for zero-mass particles. The $m = 0$ limits of these formulae are

$$\mathcal{E} = cp \qquad \text{and} \qquad \mathbf{u} = c\frac{\mathbf{p}}{p}. \tag{22.68}$$

The energy-momentum four-vector (22.59) in this case is $\vec{p} = (\mathbf{p}, ip)$. This shows that there is no frame of reference where $\mathbf{p} = 0$ unless $\mathcal{E} = 0$, in which case the "particle" does not exist. Photons have no rest frame.

Finally, with the relativistic momentum (22.63) in hand, it is natural to suppose that the equation of motion for a particle with charge q and mass m in an arbitrary electromagnetic field is

$$\frac{d\mathbf{p}}{dt} = \frac{d}{dt}\left[\frac{m\mathbf{u}}{\sqrt{1-u^2/c^2}}\right] = q(\mathbf{E} + \mathbf{u}\times\mathbf{B}). \tag{22.69}$$

We will confirm the correctness of (22.69) in Example 22.5 below. We will also confirm that the time rate of change of the work done on the particle by the Coulomb force is equal to the time rate of change of the kinetic energy of the particle:

$$\frac{dT}{dt} = \frac{d}{dt}\left(\mathcal{E} - mc^2\right) = \frac{d\mathcal{E}}{dt} = \mathbf{u}\cdot\frac{d\mathbf{p}}{dt} = q\mathbf{u}\cdot\mathbf{E}. \tag{22.70}$$

This is a familiar statement of the conservation of power.

Application 22.3 Charged Particle Motion in a Plane Wave

In this Application, we integrate the relativistic equation of motion (22.69) exactly for a particle with charge q and mass m in the field of a monochromatic plane wave. The results confirm the non-relativistic conclusion from Section 16.10.1 that a single plane wave cannot impart a net acceleration to a charged particle. On the other hand, the method we present can be used to study relativistic charged particle motion in more complex electromagnetic wave fields (like the focus of a laser beam) where net particle acceleration can be achieved.

Consider an elliptically polarized plane wave propagating in the $+z$-direction. If $\varphi = kz - \omega t$ is the phase and $\omega = ck$ is the wave frequency, the fields are

$$\mathbf{E} = \hat{\mathbf{x}}E_1\cos\varphi - \hat{\mathbf{y}}E_2\sin\varphi \quad \text{and} \quad c\mathbf{B} = \hat{\mathbf{x}}E_2\sin\varphi + \hat{\mathbf{y}}E_1\cos\varphi. \tag{22.71}$$

The assumed initial conditions for the particle are $\mathbf{r}(0) = \mathbf{u}(0) = 0$. Using (22.71), the Cartesian components of (22.69) are

$$\frac{dp_x}{dt} = qE_1\cos\varphi\left(1 - \frac{u_z}{c}\right) \qquad \frac{dp_y}{dt} = qE_2\sin\varphi\left(\frac{u_z}{c} - 1\right) \qquad \frac{dp_z}{dt} = q\frac{\mathbf{u}}{c}\cdot\mathbf{E}. \tag{22.72}$$

The time derivative of the phase at the position of the particle is

$$\frac{d\varphi}{dt} = \left(\frac{u_z}{c} - 1\right)\omega. \tag{22.73}$$

Therefore, the first first two equations in (22.72) can be rewritten as

$$\frac{dp_x}{dt} = -\frac{qE_1}{\omega}\cos\varphi\frac{d\varphi}{dt} \qquad \text{and} \qquad \frac{dp_y}{dt} = \frac{qE_2}{\omega}\sin\varphi\frac{d\varphi}{dt}. \tag{22.74}$$

Integrating (22.74) and imposing the initial conditions gives

$$p_x = -\frac{qE_1}{\omega}\sin\varphi \qquad \text{and} \qquad p_y = \frac{qE_2}{\omega}(1-\cos\varphi). \tag{22.75}$$

Comparing the last equation of (22.72) to (22.70) shows that we can integrate the p_z equation in (22.72) also. The initial condition for the total energy is $\mathcal{E}(0) = mc^2$. Therefore, this integration gives

$$cp_z = \mathcal{E} - mc^2. \tag{22.76}$$

Squaring both sides of (22.76) and using (22.66) gives the desired final result:

$$p_z = \frac{p_x^2 + p_y^2}{2mc} = \frac{1}{2mc}\left\{\left[\frac{qE_1}{\omega}\sin\varphi\right]^2 + \left[\frac{qE_2}{\omega}(1-\cos\varphi)\right]^2\right\}. \tag{22.77}$$

The first step to find the trajectory equations inserts (22.76) into (22.67):

$$\mathbf{u} = \frac{c\mathbf{p}}{p_z + mc}. \tag{22.78}$$

Next, combine (22.73) with (22.78) to get

$$\frac{d\mathbf{r}}{d\varphi} = \frac{d\mathbf{r}}{dt}\frac{dt}{d\varphi} = \frac{\mathbf{u}}{\omega(u_z/c - 1)} = \frac{c^2\mathbf{p}}{\omega(u_z - c)(p_z + mc)} = -\frac{\mathbf{p}}{\omega m}. \tag{22.79}$$

Finally, integrate (22.79) using (22.75) and (22.77). Imposing the initial conditions gives the particle trajectory in terms of the phase, $\varphi = kz - \omega t$:

$$\begin{aligned} x &= \frac{qE_1}{m\omega^2}(1-\cos\varphi) \\ y &= \frac{qE_2}{m\omega^2}(\sin\varphi - \varphi) \\ z &= \frac{q^2}{8m^2\omega^3 c}\left[(E_1^2 - E_2^2)\sin(2\varphi) - (2E_1^2 + 6E_2^2)\varphi + 8E_2^2\sin\varphi\right]. \end{aligned} \tag{22.80}$$

Equation (22.80) predicts that the charged particle moves in a closed orbit around an origin which drifts at constant velocity. The non-relativistic special case discussed in Section 16.10.1 corresponds to $E_2 = 0$ and $kd \ll 1$, where $d = qE_1/m\omega^2$. In that case, the replacement $\varphi \to \omega t$ in (22.80) reproduces both the equations of motion discussed in that section and the figure-eight particle trajectories shown in Figure 16.16. ■

22.6 Electromagnetic Quantities

Special relativity explains why different inertial observers interpret the same electromagnetic phenomenon in different ways. The most elegant discussions of this subject exploit the manifestly covariant formulation of electromagnetism (see Section 22.7). In this section, we offer a more pedestrian approach which, if only by contrast, will make the power, elegance, and economy of the covariant approach all the more impressive.

Our first task is to discover how derivatives with respect to space and time transform under a Lorentz transformation. Using (22.38), the answer is

$$\begin{aligned}
\frac{\partial}{\partial \mathbf{r}'_{\parallel}} &= \frac{\partial \mathbf{r}_{\parallel}}{\partial \mathbf{r}'_{\parallel}} \cdot \frac{\partial}{\partial \mathbf{r}_{\parallel}} + \frac{\partial \mathbf{r}_{\perp}}{\partial \mathbf{r}'_{\parallel}} \cdot \frac{\partial}{\partial \mathbf{r}_{\perp}} + \frac{\partial t}{\partial \mathbf{r}'_{\parallel}} \frac{\partial}{\partial t} = \gamma \left(\frac{\partial}{\partial \mathbf{r}_{\parallel}} + \frac{\boldsymbol{\beta}}{c} \frac{\partial}{\partial t} \right) \\
\frac{\partial}{\partial \mathbf{r}'_{\perp}} &= \frac{\partial}{\partial \mathbf{r}_{\perp}} \\
\frac{\partial}{\partial t'} &= \frac{\partial \mathbf{r}_{\parallel}}{\partial t'} \cdot \frac{\partial}{\partial \mathbf{r}_{\parallel}} + \frac{\partial t}{\partial t'} \frac{\partial}{\partial t} = \gamma \left(c\boldsymbol{\beta} \cdot \frac{\partial}{\partial \mathbf{r}_{\parallel}} + \frac{\partial}{\partial t} \right).
\end{aligned} \tag{22.81}$$

Because $\nabla = \partial/\partial \mathbf{r}$, comparing (22.81) with (22.49) shows that we are fully justified in defining the four-vector operator

$$\vec{\nabla} = \left(\nabla, \frac{\partial}{\partial(ict)} \right). \tag{22.82}$$

The invariant length associated with (22.82) is the wave equation operator

$$\vec{\nabla} \cdot \vec{\nabla} = \nabla^2 - \frac{1}{c^2} \frac{\partial^2}{\partial t^2}. \tag{22.83}$$

Equation (22.83) implies that waves which propagate at the speed of light in one inertial frame propagate at the speed of light in all inertial frames. This is an important consistency check because we derived the Lorentz transformation assuming precisely this behavior for a spherical wave of light (Section 22.4.1).

22.6.1 The Continuity Equation

Einstein's first postulate states that the laws of electromagnetism are valid in every inertial frame. An example is the conservation of charge, which we represent using the continuity equation,

$$\nabla \cdot \mathbf{j} + \frac{\partial \rho}{\partial t} = 0. \tag{22.84}$$

Because $\nabla \cdot \mathbf{j}$ is the scalar product of two three-vectors, and the gradient and time derivative are components of the four-vector (22.82), it is natural to inquire whether (22.84) is the scalar product (22.41) of two four-vectors. Indeed, if we guess

$$\vec{j} = (\mathbf{j}, ic\rho), \tag{22.85}$$

(22.84) assumes the frame-independent form

$$\vec{\nabla} \cdot \vec{j} = 0. \tag{22.86}$$

The left-hand side of this equation—the scalar product of the four-gradient (22.82) with a four-vector—is called a *four-divergence*.

To confirm that (22.85) is a four-vector, consider a bit of electric charge $dq' = \rho' dx' dy' dz'$ at rest in the inertial frame K' of Figure 22.1. When viewed from the lab frame K, the charge is $dq = \rho dx dy dz$ and $dz = dz'/\gamma$ because the volume element is contracted like (22.28) along the direction of motion. On the other hand, the invariance of electric charge mentioned at the beginning of Section 22.4.3 requires that $dq = dq'$. To make this so, the relation between the charge densities observed in K and K' must be $\rho = \gamma \rho'$:

$$dq = \rho dx dy dz = [\gamma \rho'] dx' dy' \left[\frac{dz'}{\gamma} \right] = \rho' dx' dy' dz' = dq'. \tag{22.87}$$

The "charge density dilation" [cf. (22.26)] implied by (22.87) follows immediately if (22.85) is indeed a four-vector with its law of transformation:

$$\mathbf{j}'_{\parallel} = \gamma(\mathbf{j}_{\parallel} - \rho\mathbf{v}) \qquad \mathbf{j}'_{\perp} = \mathbf{j}_{\perp} \qquad \rho' = \gamma(\rho - \mathbf{v}\cdot\mathbf{j}/c^2). \tag{22.88}$$

Specifically, since $\mathbf{j} = \rho\mathbf{v}$ in the laboratory frame K, the last member of (22.88) confirms that

$$\rho' = \gamma(\rho - jv/c^2) = \gamma(1-\beta^2)\rho = \frac{\rho}{\gamma}. \tag{22.89}$$

Example 22.2 Use delta function representations for the charge density and current density of a collection of point particles to prove that $\vec{j} = (\mathbf{j}, ic\rho)$ is a four-vector. Hint: Introduce a delta function and an integral over time.

Solution: A collection of particles with charges q_k and positions $\mathbf{r}_k(t)$ have charge density and current density

$$\rho(\mathbf{r}, t) = \sum_k q_k\delta[\mathbf{r} - \mathbf{r}_k(t)] \qquad \text{and} \qquad \mathbf{j}(\mathbf{r}, t) = \sum_k q_k\dot{\mathbf{r}}_k(t)\delta[\mathbf{r} - \mathbf{r}_k(t)].$$

Let $\vec{r} = (x, y, z, ict)$ and define $\vec{r}_k(t) = [x_k(t), y_k(t), z_k(t), ict_k(t)]$ where $t_k(t) = t$. Using these, the candidate four-vector,

$$\vec{j}(\vec{r}) = \sum_k q_k \frac{d\vec{r}_k(t)}{dt}\delta[\mathbf{r} - \mathbf{r}_k(t)],$$

is the same as $\vec{j} = (\mathbf{j}, ic\rho)$. To prove that $\vec{j}(\vec{r})$ is indeed a four-vector, introduce the four-dimensional delta function

$$\delta^4[\vec{r} - \vec{r}_k(t)] = \delta(x - x_k)\delta(y - y_k)\delta(z - z_k)\delta(ict - ict_k)$$

and rewrite the candidate vector as an integral over a time variable s:

$$\vec{j}(\vec{r}) = \int ds \sum_k q_k\delta^4[\vec{r} - \vec{r}_k(s)]\frac{d\vec{r}_k(s)}{ds}.$$

The differentials ds cancel and can be replaced by the proper time differential $d\tau$:

$$\vec{j}(\vec{r}) = \int d\tau \sum_k q_k\delta^4[\vec{r} - \vec{r}_k(\tau)]\frac{d\vec{r}_k(\tau)}{d\tau}.$$

On the right-hand side, $\vec{r}_k(\tau)$ is a four-vector and $d\tau$ and q_k are Lorentz scalars. Therefore, $\vec{j}(\vec{r})$ is a four-vector if $\delta^4[\vec{r} - \vec{r}_k(\tau)]$ is a Lorentz scalar. Two facts make this so. First, if K and K' are two inertial frames, we know from (1.118) that

$$|\mathbf{J}(\vec{r}, \vec{r}')|\delta^4[\vec{r} - \vec{r}_k(\tau)] = \delta^4[\vec{r}' - \vec{r}'_k(\tau)].$$

Second, (22.44) and (22.47) show that the Jacobian matrix $\mathbf{J}$ is exactly the unit determinant Lorentz transformation matrix $\mathbf{L}$.

Example 22.3 (a) A charge distribution is characterized by an electric dipole moment $\mathbf{p}'$ in its own rest frame. Find the electric dipole moment $\mathbf{p}$ reported by a laboratory observer who sees the distribution move with constant velocity $\mathbf{v}$.[13] (b) A current distribution with magnetic dipole moment $\mathbf{m}'$ moves with constant velocity $\mathbf{v}$ in the laboratory. Show that an observer in the laboratory perceives that the distribution has an electric dipole moment $\mathbf{p} = \mathbf{v}\times\mathbf{m}'/c^2$.

[13] The electric dipole moment $\mathbf{p}$ in this Example should not be confused with the linear momentum three-vector $\mathbf{p}$ defined in Section 22.5.2.

Solution:

(a) The inverse of (22.88) expresses lab-frame quantities in terms of rest-frame quantities. Moreover, like the volume element in (22.87), all lengths parallel to the boost direction suffer a Lorentz contraction. Therefore, because $\mathbf{j}' = 0$, the dipole moment in the laboratory is

$$\mathbf{p} = \int d^3r\, \mathbf{r}\rho = \int \frac{d^3r'}{\gamma}\left[\mathbf{r}'_\perp + \frac{\mathbf{r}'_\parallel}{\gamma}\right]\gamma\rho' = \mathbf{p}'_\perp + \frac{\mathbf{p}'_\parallel}{\gamma}.$$

(b) If $\hat{\boldsymbol{\beta}} = \mathbf{v}/v$, the quantity in square brackets immediately above says that $\mathbf{r} = \mathbf{r}' + (\gamma^{-1} - 1)\hat{\boldsymbol{\beta}}\hat{\boldsymbol{\beta}} \cdot \mathbf{r}'$. Hence, because $\rho' = 0$ in this case,

$$p_\ell = \int d^3r\, \rho r_\ell = \int \frac{d^3r'}{\gamma}\frac{\gamma v_k j'_k}{c^2}\left[r'_\ell + \left(\frac{1}{\gamma} - 1\right)\beta_\ell\beta_j r'_j\right].$$

The integral needed here is not new. From (11.11),

$$\int d^3r'\, j'_k r'_\ell = \epsilon_{ki\ell} m'_i,$$

where the rest-frame magnetic dipole moment is

$$\mathbf{m}' = \frac{1}{2}\int d^3r'\, \mathbf{r}' \times \mathbf{j}'.$$

Therefore, because $\mathbf{v}$ and $\boldsymbol{\beta}$ are parallel,

$$p_\ell = \frac{v_k}{c^2}\epsilon_{ki\ell} m'_i \quad \Rightarrow \quad \mathbf{p} = \frac{\mathbf{v}}{c^2} \times \mathbf{m}'.$$

In the lab frame, a uniformly moving magnetic dipole moment appears to acquire an electric dipole moment. A uniformly moving electric dipole moment similarly appears to acquire a magnetic dipole moment $\mathbf{m} = \mathbf{p} \times \mathbf{v}$ (see the last paragraph of Section 14.2.2). Unfortunately, one cannot deduce this merely by mimicking the method used here because the magnetic moment definition $\mathbf{m} = 1/2\int d^3r\mathbf{r} \times \mathbf{j}$ assumes $\nabla \cdot \mathbf{j} = 0$ (see Section 11.2), which is not true for a moving electric dipole.

22.6.2 Lorenz Gauge Potentials

The gauge freedom enjoyed by the scalar potential $\varphi(\mathbf{r}, t)$ and the vector potential $\mathbf{A}(\mathbf{r}, t)$ imply that these quantities possess no intrinsic transformation properties when we change inertial frames. However, they acquire quite specific transformation properties if we choose a gauge constraint that is preserved by a Lorentz transformation. Using (22.84) as a model, the Lorenz gauge condition,

$$\nabla \cdot \mathbf{A} + \frac{1}{c^2}\frac{\partial \varphi}{\partial t} = 0, \tag{22.90}$$

has this property if we can define the four-vector

$$\vec{A} = (\mathbf{A}, i\varphi/c) \tag{22.91}$$

and use (22.82) to write (22.90) as an invariant four-divergence:

$$\vec{\nabla} \cdot \vec{A} = 0. \tag{22.92}$$

To confirm that (22.91) is indeed a four-vector, we recall that the electromagnetic potentials in the Lorenz gauge satisfy the inhomogeneous wave equations (see Section 15.3.3)

$$\left[\nabla^2 - \frac{1}{c^2}\frac{\partial^2}{\partial t^2}\right]\varphi = -\rho/\epsilon_0 \qquad \text{and} \qquad \left[\nabla^2 - \frac{1}{c^2}\frac{\partial^2}{\partial t^2}\right]\mathbf{A} = -\mu_0\mathbf{j}. \tag{22.93}$$

With the judicious insertion of a factor of i on both sides of the scalar potential equation, a row-vector representation of both equations in (22.93) is

$$\left[\nabla^2 - \frac{1}{c^2}\frac{\partial^2}{\partial t^2}\right](\mathbf{A}, i\varphi/c) = -\mu_0(\mathbf{j}, ic\rho). \tag{22.94}$$

Combining (22.94) with the Lorentz invariance of the wave operator (22.83) and the four-current character of (22.85) shows that the transformation properties on the left and right sides of (22.94) will not be the same unless (22.91) is indeed a four-vector.

22.6.3 Field Transformation Laws

Observers in different inertial frames can disagree about the origin of a given electromagnetic phenomenon because the concepts of "electric field" and "magnetic field" are intrinsically observer-dependent. This conclusion—which resolves at a stroke the "asymmetry" pointed out by Einstein in the paragraph quoted at the beginning of the chapter—is not surprising once we accept the observer-dependent meaning of charge density and current density implied by (22.88). Indeed, if $\mathbf{c} = \mathbf{c}_\parallel + \mathbf{c}_\perp$ partitions a three-vector into components parallel and perpendicular to the velocity $\mathbf{v} = c\boldsymbol{\beta}$ at which a frame K' moves with respect to a frame K (see Figure 22.1), we will show in this section that

$$\mathbf{E}'_\parallel = \mathbf{E}_\parallel \qquad \mathbf{E}'_\perp = \gamma(\mathbf{E} + \boldsymbol{\beta} \times c\mathbf{B})_\perp \qquad \mathbf{E}_\perp = \gamma(\mathbf{E}' - \boldsymbol{\beta} \times c\mathbf{B}')_\perp \tag{22.95}$$

$$\mathbf{B}'_\parallel = \mathbf{B}_\parallel \qquad c\mathbf{B}'_\perp = \gamma(c\mathbf{B} - \boldsymbol{\beta} \times \mathbf{E})_\perp \qquad c\mathbf{B}_\perp = \gamma(c\mathbf{B}' + \boldsymbol{\beta} \times \mathbf{E}')_\perp. \tag{22.96}$$

These transformation laws show that the pre-relativity concepts of a purely electric field and a purely magnetic field do not survive the transition to Einstein's relativity. A boost from one inertial frame to another generally mixes the fields together. We note in passing that a field null ($\mathbf{E} = \mathbf{B} = 0$) in one frame is a field null in every frame.

Our strategy to deduce (22.95) and (22.96) exploits

$$\mathbf{B} = \nabla \times \mathbf{A}, \qquad \mathbf{E} = -\nabla\varphi - \frac{\partial \mathbf{A}}{\partial t}, \tag{22.97}$$

and the transformation rule (22.49) for the four-vectors $\vec{\nabla}$ in (22.82) and $\vec{A}$ in (22.91).

We begin with the magnetic field, where the key equation is

$$\mathbf{B}' = \nabla' \times \mathbf{A}' = (\nabla'_\parallel + \nabla'_\perp) \times (\mathbf{A}'_\parallel + \mathbf{A}'_\perp). \tag{22.98}$$

Because $\nabla'_\parallel \times \mathbf{A}'_\parallel = 0$, the parallel and perpendicular components of $\mathbf{B}'$ are

$$\mathbf{B}'_\parallel = (\nabla' \times \mathbf{A}')_\parallel = \nabla'_\perp \times \mathbf{A}'_\perp \tag{22.99}$$

and

$$\mathbf{B}'_\perp = (\nabla' \times \mathbf{A}')_\perp = \nabla'_\parallel \times \mathbf{A}'_\perp + \nabla'_\perp \times \mathbf{A}'_\parallel. \tag{22.100}$$

The three-vectors $\nabla'_\perp$ and $\mathbf{A}'_\perp$ are both transverse components of the space part of a four-vector. Therefore, both are invariant under a Lorentz transformation [see (22.49)] and we conclude from (22.99) that

$$\mathbf{B}'_\parallel = \mathbf{B}_\parallel. \tag{22.101}$$

A bit more work is needed to evaluate (22.100) because $\nabla'_\parallel$ and $\mathbf{A}'_\parallel$ are not Lorentz invariant. Substituting from (22.49) gives

$$\mathbf{B}'_\perp = \gamma\left(\nabla_\parallel + \frac{\mathbf{v}}{c^2}\frac{\partial}{\partial t}\right) \times \mathbf{A}_\perp + \nabla_\perp \times \gamma\left(\mathbf{A}_\parallel - \frac{\mathbf{v}}{c^2}\varphi\right), \tag{22.102}$$

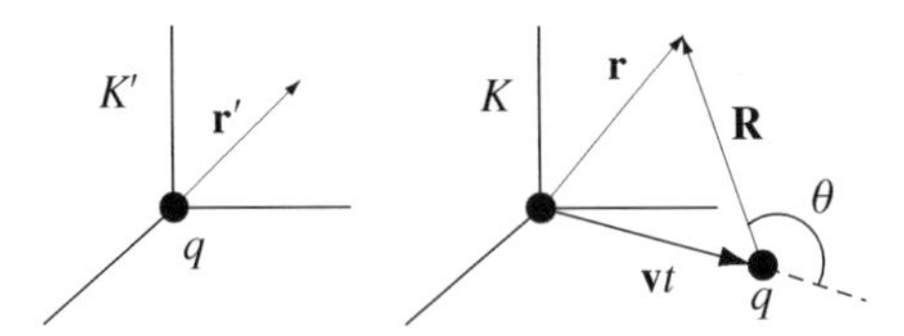

Figure 22.6: A point charge at rest at the origin of frame K' is observed to move with uniform velocity $\mathbf{v} = c\boldsymbol{\beta}$ in the laboratory frame K.

which may be rearranged to

$$\mathbf{B}'_\perp = \gamma\left(\nabla_\parallel \times \mathbf{A}_\perp + \nabla_\perp \times \mathbf{A}_\parallel\right) + \frac{\gamma}{c^2}\left[\mathbf{v} \times \frac{\partial \mathbf{A}_\perp}{\partial t} - \nabla_\perp \times (\mathbf{v}\varphi)\right]. \tag{22.103}$$

The structure of (22.100) and the fact that $\mathbf{v}$ is a constant vector permit us to rewrite (22.103) as

$$\mathbf{B}'_\perp = \gamma \mathbf{B}_\perp - \frac{\gamma}{c^2}\mathbf{v} \times \left(-\frac{\partial \mathbf{A}_\perp}{\partial t} - \nabla_\perp \varphi\right). \tag{22.104}$$

Finally, the definition of $\mathbf{E}$ in (22.97) simplifies (22.104) to

$$\mathbf{B}'_\perp = \gamma\left(\mathbf{B}_\perp - \frac{\mathbf{v}}{c^2} \times \mathbf{E}_\perp\right) = \gamma\left(\mathbf{B} - \frac{\mathbf{v}}{c^2} \times \mathbf{E}\right)_\perp. \tag{22.105}$$

This is the expression on the right side of (22.96).

The electric field formulae in (22.95) require even more algebra to prove. Using (22.49) and the four-vector definitions in (22.82) and (22.91), we transform $\mathbf{E}'_\parallel = -\nabla'_\parallel \varphi' - \partial \mathbf{A}'_\parallel/\partial t'$ to

$$\mathbf{E}'_\parallel = -\gamma\left(\nabla_\parallel + \frac{\mathbf{v}}{c^2}\frac{\partial}{\partial t}\right)\gamma(\varphi - \mathbf{v}\cdot\mathbf{A}) - \gamma\left(\frac{\partial}{\partial t} + \mathbf{v}\cdot\nabla\right)\gamma\left(\mathbf{A}_\parallel - \frac{\mathbf{v}}{c^2}\varphi\right). \tag{22.106}$$

The right side of (22.106) generates eight terms. Two of these are $\pm(\mathbf{v}/c^2)\partial\varphi/\partial t$, which cancel. Because $\mathbf{v}$ is a constant vector, the terms $(\mathbf{v}\cdot\nabla)\mathbf{A}_\parallel$ and $-\nabla_\parallel(\mathbf{v}\cdot\mathbf{A})$ cancel also. The four terms which remain can be manipulated to read

$$\mathbf{E}_\parallel = -\gamma^2\left[\nabla_\parallel \varphi + \frac{\partial \mathbf{A}_\parallel}{\partial t}\right](1 - v^2/c^2). \tag{22.107}$$

This confirms the left side of (22.95) because $\gamma^2(1-\beta^2) = 1$ and

$$\mathbf{E}'_\parallel = -\nabla_\parallel \varphi - \frac{\partial \mathbf{A}_\parallel}{\partial t} = \mathbf{E}_\parallel. \tag{22.108}$$

We leave the derivation of the transformation law for $\mathbf{E}'_\perp$ [the right side of (22.95)] as an exercise for the reader.

22.6.4 A Point Charge in Uniform Motion

The electromagnetic field produced by a point charge in uniform motion can be calculated by transforming the field produced by the charge in its own rest frame K' to the laboratory frame K. Accordingly, the left panel of Figure 22.6 shows q at rest at the origin of K'. The right panel shows q moving with constant velocity $\mathbf{v}$ in the laboratory frame K. The observation point is called $\mathbf{r}$ in K and $\mathbf{r}'$ in K'.

The field of the point charge in its own frame is entirely electrostatic:

$$\mathbf{E}' = \frac{q'}{4\pi\epsilon_0}\frac{\mathbf{r}'}{r'^3} = \frac{q'}{4\pi\epsilon_0}\left[\frac{\mathbf{r}'_\parallel}{(\mathbf{r}'_\parallel\cdot\mathbf{r}'_\parallel + \mathbf{r}'_\perp\cdot\mathbf{r}'_\perp)^{3/2}} + \frac{\mathbf{r}'_\perp}{(\mathbf{r}'_\parallel\cdot\mathbf{r}'_\parallel + \mathbf{r}'_\perp\cdot\mathbf{r}'_\perp)^{3/2}}\right] = \mathbf{E}'_\parallel + \mathbf{E}'_\perp. \quad (22.109)$$

This means that the rest-frame magnetic field $\mathbf{B}' = 0$. The key assumption is that frame K' moves with velocity $\mathbf{v}$ with respect to frame K. Therefore, the transformations on the far right sides of (22.95) and (22.96) tell us that the laboratory fields are

$$\mathbf{E} = \gamma\mathbf{E}'_\perp + \mathbf{E}'_\parallel \qquad \text{and} \qquad c\mathbf{B} = \boldsymbol{\beta}\times\gamma\mathbf{E}'_\perp. \quad (22.110)$$

The fields in (22.110) are functions of the rest-frame variables. However, the Lorentz transformation permits us to write them as functions of the laboratory-frame variables. We do this using (22.38) and the invariance of electric charge ($q' = q$). The resulting laboratory electric field is

$$\mathbf{E} = \frac{q}{4\pi\epsilon_0}\frac{\gamma\mathbf{r}'_\perp + \mathbf{r}'_\parallel}{(\mathbf{r}'_\parallel\cdot\mathbf{r}'_\parallel + \mathbf{r}'_\perp\cdot\mathbf{r}'_\perp)^{3/2}} = \frac{q}{4\pi\epsilon_0}\frac{\gamma\mathbf{r}_\perp + \gamma(\mathbf{r}_\parallel - \mathbf{v}t)}{[\gamma(\mathbf{r}_\parallel - \mathbf{v}t)\cdot\gamma(\mathbf{r}_\parallel - \mathbf{v}t) + \mathbf{r}_\perp\cdot\mathbf{r}_\perp]^{3/2}}. \quad (22.111)$$

Equation (22.111) simplifies using the K-frame vector $\mathbf{R} = \mathbf{r} - \mathbf{v}t$ (see Figure 22.6) which points from the instantaneous position of the charge to the observation point. $\mathbf{R}$ also defines an angle θ such that $\hat{\mathbf{v}}\cdot\hat{\mathbf{R}} = \cos\theta$. Then, because $\mathbf{R}_\parallel = \mathbf{r}_\parallel - \mathbf{v}t$ and $\mathbf{R}_\perp = \mathbf{r}_\perp$, we find

$$\mathbf{E} = \frac{q}{4\pi\epsilon_0}\frac{\gamma\mathbf{R}}{(\gamma^2R_\parallel^2 + R_\perp^2)^{3/2}} = \frac{q}{4\pi\epsilon_0}\frac{\gamma\mathbf{R}}{\gamma^2R^3(\cos^2\theta + \sin^2\theta/\gamma^2)^{3/2}}. \quad (22.112)$$

A final simplification exploits $\gamma^2(1-\beta^2) = 1$ to write (22.112) in the form

$$\mathbf{E} = \frac{q}{4\pi\epsilon_0}\frac{\hat{\mathbf{R}}}{R^2}\frac{1-\beta^2}{(1-\beta^2\sin^2\theta)^{3/2}}. \quad (22.113)$$

The corresponding magnetic field from (22.110) is

$$\mathbf{B} = \frac{\mathbf{v}}{c^2}\times\gamma\mathbf{E}'_\perp = \frac{\mathbf{v}}{c^2}\times\mathbf{E}_\perp = \frac{\mathbf{v}}{c^2}\times\mathbf{E}. \quad (22.114)$$

These fields agree with those found by a different method and discussed in Application 20.1 at the end of Section 20.2. They will appear again in the next chapter as a special case of the Liénard-Wiechert fields of a point charge in arbitrary motion.

22.6.5 Plane Waves

Special relativity provides interesting insight into various common optical phenomena. For example, let a monochromatic plane wave propagate in vacuum with speed $c = \omega/k$ in a reference frame K. The electromagnetic fields in this frame are

$$\mathbf{E}(\mathbf{r},t) = \mathbf{E}_0\exp[i(\mathbf{k}\cdot\mathbf{r} - \omega t)] \qquad \text{and} \qquad c\mathbf{B} = \hat{\mathbf{k}}\times\mathbf{E}. \quad (22.115)$$

The Lorentz invariance of the wave operator (Section 22.6) implies that the plane wave (22.115) has exactly the same *form* when observed in a frame K' which moves with uniform speed $\mathbf{v}$ with respect to K:

$$\mathbf{E}'(\mathbf{r}',t') = \mathbf{E}'_0\exp[i(\mathbf{k}'\cdot\mathbf{r}' - \omega' t')] \qquad \text{and} \qquad c\mathbf{B}' = \hat{\mathbf{k}}'\times\mathbf{E}'. \quad (22.116)$$

Our deduction that special relativity transforms one plane wave into another plane wave suggests an alternative to writing down (22.116) directly: simply Lorentz transform (22.115) from K to the moving frame K'. We begin with the field amplitudes and use (22.95) to relate the components of $\mathbf{E}'_0$ to the components of $\mathbf{E}_0$ and $\mathbf{B}_0$. Assuming this is done, the electric field (22.115) becomes

$$\mathbf{E}'(\mathbf{r}, t) = \mathbf{E}'_0 \exp[i(\mathbf{k} \cdot \mathbf{r} - \omega t)]. \tag{22.117}$$

Next, use (22.38) to eliminate the variables $\mathbf{r}$ and t in (22.117) in favor of the variables $\mathbf{r}'$ and t'. This gives

$$\mathbf{E}(\mathbf{r}', t') = \mathbf{E}'_0 \exp\left[i\mathbf{k} \cdot \{\mathbf{r}'_\perp + \gamma(\mathbf{r}'_\parallel + \boldsymbol{\beta} ct')\} - i\omega\gamma(t' + \boldsymbol{\beta} \cdot \mathbf{r}'_\parallel/c)\right]. \tag{22.118}$$

Comparing (22.118) to (22.116) produces the final result that

$$\mathbf{k}'_\perp = \mathbf{k}_\perp \qquad \mathbf{k}'_\parallel = \gamma(\mathbf{k}_\parallel - \boldsymbol{\beta}\omega/c) \qquad \omega' = \gamma(\omega - \mathbf{v} \cdot \mathbf{k}_\parallel). \tag{22.119}$$

Equation (22.119) is significant because comparison with (22.49) shows that the frequency and wave vector of a plane wave are the components of a four-vector,

$$\vec{k} = (\mathbf{k}, i\omega/c). \tag{22.120}$$

An immediate consequence of (22.120) is that the *phase* of a plane wave is a Lorentz invariant scalar. This is so because the phase can be written as the scalar product (22.41) of two four-vectors:

$$\phi(\mathbf{r}, t) = \mathbf{k} \cdot \mathbf{r} - \omega t = \vec{k} \cdot \vec{r}. \tag{22.121}$$

The invariant length of (22.120) is zero ($\vec{k} \cdot \vec{k} = 0$) because $\omega = c|\mathbf{k}|$ for vacuum waves. Otherwise, the transformation properties of $\vec{k}$ generate two well-known optical effects when there is relative motion between a source of waves and a detector of waves: the *Doppler effect* and *stellar aberration*. The Doppler effect refers to the fact that the frequency of the observed waves differs from the frequency of the emitted waves. Stellar aberration refers to the shift in the apparent position of a star because the direction of the observed light's wave vector differs from the direction of the emitted light's wave vector. Application 22.4 discusses some aspects of the Doppler effect. We leave aberration as a topic for the reader to explore as an exercise.

Application 22.4 Reflection from a Moving Mirror

In his original paper on relativity, Einstein discussed the reflection of a monochromatic plane wave from a moving mirror. Here, we consider normal-incidence reflection from a large mirror in the x-y plane which moves with velocity $\mathbf{v} = v\hat{\mathbf{z}}$ (Figure 22.7). In the laboratory, the incident wave fields with wave vector $\mathbf{k}_i = k_i\hat{\mathbf{z}}$ are

$$\mathbf{E}_i = \hat{\mathbf{x}}E_i \exp\left[i(\mathbf{k}_i \cdot \mathbf{r} - \omega_i t)\right] \qquad \text{and} \qquad c\mathbf{B}_i = \hat{\mathbf{z}} \times \mathbf{E}_i. \tag{22.122}$$

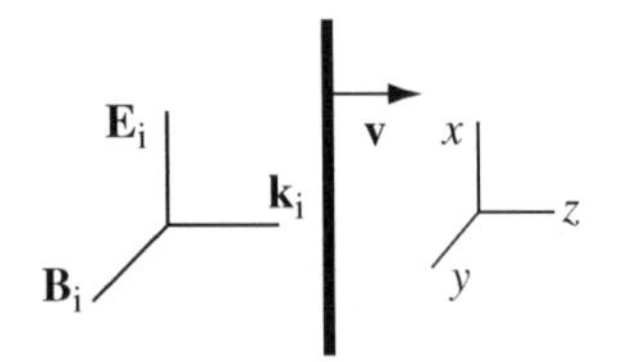

Figure 22.7: A plane wave reflects at normal incidence from a plane mirror moving with velocity **v**.

The incident wave fields in the rest frame of the mirror are

$$\mathbf{E}'_\mathrm{i} = \hat{\mathbf{x}} E'_\mathrm{i} \exp\left[i(\mathbf{k}'_\mathrm{i} \cdot \mathbf{r}' - \omega'_\mathrm{i} t')\right] \qquad \text{and} \qquad c\mathbf{B}'_\mathrm{i} = \hat{\mathbf{z}} \times \mathbf{E}'_\mathrm{i}. \tag{22.123}$$

Using (22.95) and (22.119), the wave parameters in (22.123) are related to those in (22.122) by

$$\begin{aligned}
\omega'_\mathrm{i} &= \gamma(\omega_\mathrm{i} - \mathbf{v}\cdot\mathbf{k}_\mathrm{i}) = \gamma(1-\beta)\omega_\mathrm{i} = \sqrt{\frac{1-\beta}{1+\beta}}\,\omega_\mathrm{i} = ck'_\mathrm{i} \\
\mathbf{k}'_\mathrm{i} &= \gamma(\mathbf{k}_\mathrm{i} - \mathbf{v}\omega_\mathrm{i}/c^2) = \gamma(1-\beta)k_\mathrm{i}\hat{\mathbf{z}} = \sqrt{\frac{1-\beta}{1+\beta}}\,k_\mathrm{i}\hat{\mathbf{z}} = k'_\mathrm{i}\hat{\mathbf{z}} \\
E'_\mathrm{i} &= \gamma(E_\mathrm{i} - vB_\mathrm{i}) = \gamma(1-\beta)E_\mathrm{i} = \sqrt{\frac{1-\beta}{1+\beta}}\,E_\mathrm{i}.
\end{aligned} \tag{22.124}$$

The factor γ in the frequency formula is a relativistic correction to the Doppler effect formula of Newtonian physics. The correction is small when $v \ll c$, but it produces the entire *transverse Doppler effect* ($\omega'_i = \gamma\omega_i$) when $\mathbf{v}\cdot\mathbf{k}_\mathrm{i} = 0$. This transverse effect has no counterpart in Newtonian physics.

In the rest frame of the mirror, the reflected fields have wave vector $\mathbf{k}'_\mathrm{r} = -\mathbf{k}'_\mathrm{i}$ and oscillate at frequency $\omega'_\mathrm{r} = \omega'_\mathrm{i}$. The electric field amplitude changes sign upon reflection. Therefore, with $E'_\mathrm{r} = -E'_\mathrm{i}$, the wave fields in the rest frame of the mirror are

$$\mathbf{E}'_\mathrm{r} = \hat{\mathbf{x}} E'_\mathrm{r} \exp\left[i(\mathbf{k}'_\mathrm{r} \cdot \mathbf{r}' - \omega'_\mathrm{r} t')\right] \qquad \text{and} \qquad c\mathbf{B}'_\mathrm{r} = -\hat{\mathbf{z}} \times \mathbf{E}'_\mathrm{r}. \tag{22.125}$$

The wave vector $\mathbf{k}'_\mathrm{r}$ is anti-parallel to $\mathbf{v}$. Therefore, the transformation back to the laboratory frame produces a plane wave with fields

$$\mathbf{E}_\mathrm{r} = \hat{\mathbf{x}} E_\mathrm{r} \exp\left[i(\mathbf{k}_\mathrm{r} \cdot \mathbf{r} - \omega_\mathrm{r} t)\right] \qquad \text{and} \qquad c\mathbf{B}_\mathrm{r} = -\hat{\mathbf{z}} \times \mathbf{E}_\mathrm{r}, \tag{22.126}$$

where

$$\begin{aligned}
\omega_\mathrm{r} &= \gamma(\omega'_\mathrm{r} + \mathbf{v}\cdot\mathbf{k}'_\mathrm{r}) = \gamma ck'_\mathrm{i}(1-\beta) = \left(\frac{1-\beta}{1+\beta}\right)\omega_\mathrm{i} < \omega_\mathrm{i} \\
\mathbf{k}_\mathrm{r} &= \gamma(\mathbf{k}'_\mathrm{r} + \mathbf{v}\omega'_\mathrm{r}/c^2) = -\gamma k'_\mathrm{i}(1-\beta)\hat{\mathbf{z}} = -\left(\frac{1-\beta}{1+\beta}\right)\mathbf{k}_\mathrm{i} \\
E_\mathrm{r} &= \gamma(E'_\mathrm{r} - vB'_\mathrm{r}) = \gamma(1-\beta)E'_\mathrm{r} = -\left(\frac{1-\beta}{1+\beta}\right)E_\mathrm{i}.
\end{aligned} \tag{22.127}$$

The first line of (22.127) shows that the reflected wave frequency is "red shifted" compared to the incident wave frequency for the receding mirror situation shown in Figure 22.7. A "blue shift" ($\omega_\mathrm{r} > \omega_\mathrm{i}$) occurs when the mirror approaches the incident wave and $\beta \to -\beta$ in (22.127). The last line of (22.127) shows that the energy and momentum densities of the reflected wave decrease and increase similarly (compared to the incident wave) because they are proportional to the square of the field amplitude (see Section 16.3.4):

$$u_\mathrm{EM} = \epsilon_0|\mathbf{E}|^2 \qquad\qquad c\mathbf{g} = u_\mathrm{EM}\hat{\mathbf{k}}. \tag{22.128}$$

The force of radiation pressure (Section 17.4) mediates the exchange of energy and momentum between the plane wave and the moving mirror. No violation of conservation of energy or linear momentum occurs because an external agent maintains the constant speed of the mirror. ■

Example 22.4 Let $U_{\rm EM}$ and $\mathbf{P}_{\rm EM}$ be the total energy and total linear momentum of the portion of a monochromatic plane wave that lies inside a finite volume V which moves with the wave. Prove that $U_{\rm EM}/\omega$ is a Lorentz invariant.

Solution: Let (22.122) be the plane wave in the lab frame K. The first and last lines of (22.124) show that the wave frequency and electric field amplitude in a standard configuration frame K' which moves with velocity $\mathbf{v} = v\hat{\mathbf{z}}$ are

$$\omega' = \gamma(1-\beta)\omega \qquad \text{and} \qquad E'_{\rm i} = \gamma(1-\beta)E_{\rm i}.$$

Using the leftmost equation in (22.128), the energy density in K' is

$$u'_{\rm EM} = \epsilon_0 E'^2_{\rm i} = \epsilon_0\gamma^2(1-\beta)^2E^2_{\rm i} = \gamma^2(1-\beta)^2u_{\rm EM}.$$

We are interested in the total energy $U_{\rm EM}$ contained in a finite volume of space which moves with the wave at the speed of light. No Lorentz transformation can bring this volume to rest. Therefore, we temporarily reduce the speed of the volume and suppose that it moves with velocity $u\hat{\mathbf{z}}$ when viewed from K and velocity $u'\hat{\mathbf{z}}$ when viewed from K' (see Figure 22.8). If V_0 is the rest volume, length contraction implies that observers in K and K' measure the size of the moving volume to be

$$V = V_0\sqrt{1-u^2/c^2} = V_0/\gamma(u) \qquad \text{and} \qquad V' = V_0\sqrt{1-u'^2/c^2} = V_0/\gamma(u').$$

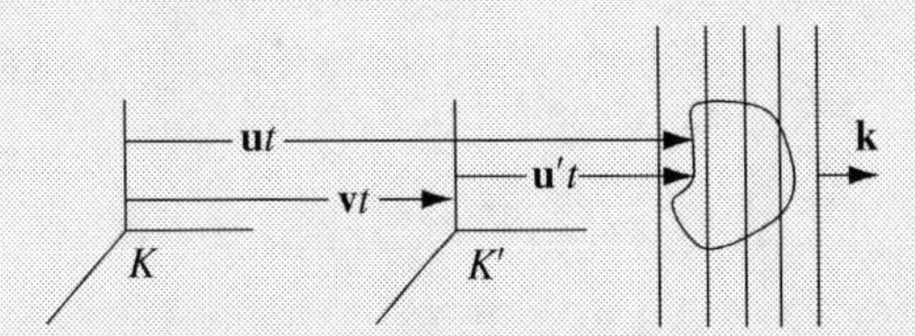

Figure 22.8: A finite and arbitrarily shaped volume of space moves with velocity $u\hat{\mathbf{z}}$ when viewed from a frame K and with velocity $u'\hat{\mathbf{z}}$ when viewed from a frame K'. K' moves with velocity $v\hat{\mathbf{z}}$ with respect to K.

However, comparison with Figure 22.8 and the last equation in Example 22.1 of Section 22.4.1 shows that

$$\gamma(u) = \gamma(v)\gamma(u')(1 + vu'/c^2).$$

Noting that $\gamma = \gamma(v)$, we take $u' \to c$ at the end to get

$$V' = \frac{\gamma(u)}{\gamma(u')}V = \gamma(1 + vu'/c^2)V = \gamma(1 + vu'/c^2)V \to \gamma(1+\beta)V.$$

Combining all the above demonstrates the suggested invariance because

$$\frac{U'_{\rm EM}}{\omega'} = \frac{u'_{\rm EM}V'}{\omega'} = \frac{\gamma^2(1-\beta)^2u_{\rm EM}\gamma(1+\beta)V}{\gamma(1-\beta)\omega} = \frac{u_{\rm EM}V}{\omega} = \frac{U_{\rm EM}}{\omega}.$$

A similar result holds in quantum theory for a "box of photons" because $U_{\rm EM} = N\hbar\omega$ where N is the number of photons contained in the box.

22.7 Covariant Electrodynamics

The equations of electrodynamics take the same form in every inertial frame. Unfortunately, it is quite tedious to demonstrate this *covariance* explicitly using the transformation rules derived earlier for partial derivatives, charge and current density, and the electric and magnetic fields. This motivates the search for a representation of the Maxwell equations which makes covariance more obvious.

The clue we need to construct a transparently covariant electrodynamics comes from the fact that $\nabla \cdot \mathbf{B} = 0$ and $\nabla \cdot \mathbf{E} = \rho/\epsilon_0$ are scalar equations and the two curl-type Maxwell equations are vector equations. This structure guarantees that the Maxwell equations are form-invariant to rotations and translations in space because these are the defining characteristic of scalars and vectors. The *components* of ∇, $\mathbf{E}$, $\mathbf{B}$, and $\mathbf{j}$ change when rotations and translations are performed, but their three-vector character does not change. Therefore, to show that electrodynamics is similarly covariant, it is sufficient to show that the fundamental equations of the subject can be written entirely in terms of *Lorentz tensors* whose components change under a Lorentz boost, but whose essential tensor character does not change. This is called writing Maxwell's theory in manifestly covariant form.

22.7.1 Lorentz Tensors

Lorentz tensors are defined in complete analogy with the rotational Cartesian tensors discussed in Section 1.8. Thus, a Lorentz tensor of rank 0 is what we have previously called a Lorentz scalar: a one-component quantity which is invariant to a change of inertial frame:

$$c' = c. \tag{22.129}$$

A Lorentz tensor of rank 1 is what we have previously called a four-vector: an object whose four components transform according to the Lorentz transformation matrix (22.45):

$$a'_\mu = L_{\mu\nu} a_\nu. \tag{22.130}$$

A Lorentz tensor of rank 2 is an object whose 16 components transform according to the rule

$$s'_{\mu\nu} = L_{\mu\alpha} L_{\nu\beta} s_{\alpha\beta}. \tag{22.131}$$

Lorentz tensors of higher rank are defined similarly.

Two theorems are noteworthy for the role they play in the manipulation of Lorentz tensors. Consider a rank 1 tensor b_μ and a rank 2 tensor $W_{\mu\nu}$. An example of the *contraction theorem* states that multiplying the two together and contracting (summing over) the common index μ produces an object d_ν which is a rank 1 tensor:

$$b_\mu W_{\mu\nu} = d_\nu. \tag{22.132}$$

Conversely, suppose that d_ν in (22.132) is a specified rank 1 tensor and b_μ is an *arbitrary* rank 1 tensor. In that case, the *quotient theorem* states that $W_{\mu\nu}$ is a rank 2 tensor.

Finally, three expressions derived earlier in the chapter will serve to illustrate the use of Lorentz tensors to achieve manifest covariance. These are the continuity equation (22.86), the Lorenz gauge condition (22.92), and the inhomogeneous wave equation for the Lorenz gauge potentials (22.94):

$$\partial_\mu j_\mu = 0, \qquad \partial_\mu A_\mu = 0, \qquad \text{and} \qquad \partial_\mu \partial_\mu A_\nu = -\mu_0 j_\nu. \tag{22.133}$$

The first two of these are covariant because their structure is "zero-rank tensor = zero-rank tensor". The last is covariant because its structure is "first-rank tensor = first-rank tensor".

22.7.2 The Maxwell Equations

The path to writing the Maxwell equations in manifestly covariant form begins with the observation that the four-gradient $\partial_\mu = (\partial_1, \partial_2, \partial_3, \partial_4) = (\nabla, \partial/\partial(ict))$ from (22.82) and the four-potential $A_\mu = (A_1, A_2, A_3, A_4) = (\mathbf{A}, i\varphi/c)$ from (22.91) are sufficient to write out the two equations in (22.97) explicitly. The magnetic field components are

$$\begin{aligned} B_x &= \frac{\partial}{\partial y}A_z - \frac{\partial}{\partial z}A_y &= \partial_2 A_3 - \partial_3 A_2 \\ B_y &= \frac{\partial}{\partial z}A_x - \frac{\partial}{\partial x}A_y &= \partial_3 A_1 - \partial_1 A_3 \\ B_z &= \frac{\partial}{\partial x}A_y - \frac{\partial}{\partial y}A_x &= \partial_1 A_2 - \partial_2 A_1 \end{aligned} \tag{22.134}$$

and the electric field components are

$$\begin{aligned} \frac{iE_x}{c} &= \frac{\partial A_x}{\partial(ict)} - \frac{\partial}{\partial x}\left(\frac{i\varphi}{c}\right) &= \partial_4 A_1 - \partial_1 A_4 \\ \frac{iE_y}{c} &= \frac{\partial A_y}{\partial(ict)} - \frac{\partial}{\partial y}\left(\frac{i\varphi}{c}\right) &= \partial_4 A_2 - \partial_2 A_4 \\ \frac{iE_z}{c} &= \frac{\partial A_z}{\partial(ict)} - \frac{\partial}{\partial z}\left(\frac{i\varphi}{c}\right) &= \partial_4 A_3 - \partial_3 A_4. \end{aligned} \tag{22.135}$$

Equations (22.134) and (22.135) show that the Cartesian components of **E** and **B** are components of a second-rank Lorentz tensor with the form of a "generalized curl":

$$F_{\mu\nu} = \partial_\mu A_\nu - \partial_\nu A_\mu. \tag{22.136}$$

The *electromagnetic field-strength tensor* $F_{\mu\nu}$ has only 6 (rather than 16) independent components because it is asymmetric ($F_{\mu\nu} = -F_{\nu\mu}$) and the diagonal elements ($\mu = \nu$) are zero. In matrix form,

$$\mathbf{F} = \begin{bmatrix} 0 & B_z & -B_y & -iE_x/c \\ -B_z & 0 & B_x & -iE_y/c \\ B_y & -B_x & 0 & -iE_z/c \\ iE_x/c & iE_y/c & iE_z/c & 0 \end{bmatrix}, \tag{22.137}$$

and we see that

$$E_k = icF_{k4} \qquad \text{and} \qquad B_k = \tfrac{1}{2}\epsilon_{k\ell m}F_{\ell m}. \tag{22.138}$$

The reader can confirm that the tensor transformation rule (22.131) applied to $F_{\mu\nu}$ reproduces the field transformation formulae (22.95) and (22.96) derived earlier. Also noteworthy is the Lorentz invariant scalar,

$$F_{\alpha\beta}F_{\alpha\beta} = 2(\mathbf{B}\cdot\mathbf{B} - \mathbf{E}\cdot\mathbf{E}/c^2). \tag{22.139}$$

For present purposes, the importance of $F_{\mu\nu}$ is that the two inhomogeneous Maxwell equations are contained in the manifestly covariant equation

$$\partial_\nu F_{\mu\nu} = \mu_0 j_\mu. \tag{22.140}$$

This equation is of the type "first-rank tensor = first-rank tensor" because the indicated sum over the (second) index of the second-rank field tensor $F_{\mu\nu}$ leaves an object which transforms like a first-rank tensor. The $\mu = 4$ component of (22.140) written out in (22.141) is Gauss' law, $\nabla\cdot\mathbf{E} = \rho/\epsilon_0$, because $j_4 = ic\rho$ and $\mu_0\epsilon_0 c^2 = 1$:

$$i\mu_0 c\rho = \partial_\mu F_{4\mu} = \frac{\partial}{\partial x}\left(\frac{iE_x}{c}\right) + \frac{\partial}{\partial y}\left(\frac{iE_y}{c}\right) + \frac{\partial}{\partial z}\left(\frac{iE_z}{c}\right) + 0. \tag{22.141}$$

Similarly, the $\mu = 1, 2, 3$ components of (22.140) are the x-, y-, z components of the Ampère-Maxwell equation, $\nabla \times \mathbf{B} = \mu_0 \mathbf{j} + c^{-2}\partial \mathbf{E}/\partial t$. For example, $\mu = 1$ gives

$$\mu_0 j_x = \partial_\mu F_{1\mu} = 0 + \frac{\partial}{\partial y} B_z + \frac{\partial}{\partial z}(-B_y) + \frac{\partial}{\partial(ict)}\left(-\frac{iE_x}{c}\right), \tag{22.142}$$

or

$$\mu_0 j_x = [\nabla \times \mathbf{B}]_x - \frac{1}{c^2}\frac{\partial E_x}{\partial t}. \tag{22.143}$$

The homogeneous Maxwell equations can be written in terms of $F_{\mu\nu}$ also. To see this, consider a third-rank Lorentz tensor expression where a cyclic permutation of the indices (no summation) relates one term to the next:

$$\partial_\lambda F_{\mu\nu} + \partial_\mu F_{\nu\lambda} + \partial_\nu F_{\lambda\mu} = 0. \tag{22.144}$$

Because $F_{\mu\mu} = 0$ and $F_{\lambda\mu} = -F_{\mu\lambda}$, the left-hand side of (22.144) is zero if any two indices are equal. However, setting $\lambda = 1$, $\mu = 2$, and $\nu = 3$ in (22.144) gives

$$0 = \partial_1 F_{23} + \partial_2 F_{31} + \partial_3 F_{12} = \frac{\partial B_x}{\partial x} + \frac{\partial B_y}{\partial y} + \frac{\partial B_z}{\partial z} = \nabla \cdot \mathbf{B}. \tag{22.145}$$

Similarly, setting $\lambda = 4$, $\mu = 1$, and $\nu = 2$ in (22.144) gives

$$0 = \partial_4 F_{12} + \partial_1 F_{24} + \partial_2 F_{41} = \frac{\partial B_z}{\partial(ict)} + \frac{\partial}{\partial x}\left(-\frac{iE_y}{c}\right) + \frac{\partial}{\partial y}\left(\frac{iE_x}{c}\right) = -i\left[\frac{\partial \mathbf{B}}{\partial t} + \nabla \times \mathbf{E}\right]_z. \tag{22.146}$$

The remaining choices for the indices in (22.144) generate the other Cartesian components of Faraday's law.

The inelegant appearance of (22.144) provides an opportunity to point out that (22.137) is not the only way to embed $\mathbf{E}$ and $\mathbf{B}$ in a second-rank Lorentz tensor. This is so because the Lorentz transformation formulae (22.95) and (22.96) are invariant to the duality transformation $\mathbf{B} \to -\mathbf{E}/c$ and $\mathbf{E}/c \to \mathbf{B}$ (see Section 15.2.2). As a result, the same discrete symmetry operation applied to the elements of $F_{\mu\nu}$ produces an independent second-rank Lorentz tensor we will call $G_{\mu\nu}$. The matrix form of this *dual tensor* is

$$\mathbf{G} = \begin{bmatrix} 0 & -E_z/c & E_y/c & -iB_x \\ E_z/c & 0 & -E_x/c & -iB_y \\ -E_y/c & E_x/c & 0 & -iB_z \\ iB_x & iB_y & iB_z & 0 \end{bmatrix}. \tag{22.147}$$

The reason for defining $G_{\mu\nu}$ is compelling, and straightforward to confirm: the homogeneous Maxwell equations $\nabla \cdot \mathbf{B} = 0$ and $\nabla \times \mathbf{E} = -\partial \mathbf{B}/\partial t$ are contained in the single manifestly covariant equation

$$\partial_\nu G_{\mu\nu} = 0. \tag{22.148}$$

To facilitate calculations, it is useful to know that

$$F_{\mu\nu} G_{\mu\nu} = -4\mathbf{E} \cdot \mathbf{B}/c \tag{22.149}$$

is a Lorentz scalar. Finally,

$$G_{\mu\nu} = \tfrac{1}{2} i \epsilon_{\mu\nu\alpha\beta} F_{\alpha\beta}, \tag{22.150}$$

where $\epsilon_{\mu\nu\alpha\beta}$ is a four-dimensional generalization of the Levi-Cività symbol defined in (1.34).[14]

[14] Thus, $\epsilon_{\mu\nu\alpha\beta}$ takes the value $+1$ if $(\mu\nu\alpha\beta)$ is an even permutation of (1234), it takes the value -1 if $(\mu\nu\alpha\beta)$ is an odd permutation of (1234), and it takes the value 0 if any two indices are equal.

Example 22.5 Derive a manifestly covariant equation of motion for a object with charge q and mass m in an arbitrary electromagnetic field. Use it to confirm (22.69) and (22.70).

Solution: Newton's equation of motion for a mass point m with charge q and linear momentum $m\mathbf{u}$ is

$$\frac{d}{dt}m\mathbf{u} = q(\mathbf{E} + \mathbf{u} \times \mathbf{B}) \qquad \text{(non-relativistic)}.$$

A covariant generalization may be expected to replace the ordinary time interval dt by the proper time interval $d\tau$ from (22.37), the three-velocity $\mathbf{u}$ by the four-velocity U_μ from (22.50), and the field vectors $\mathbf{E}$ and $\mathbf{B}$ by the field tensor $F_{\mu\nu}$ from (22.137). Because the four-momentum from (22.59) is $p_\mu = mU_\mu$, the first-rank Lorentz tensor equation which results is

$$\frac{dp_\mu}{d\tau} = qU_\nu F_{\mu\nu}.$$

Using the definition of $\mathbf{p}$ in (22.63), the space components of this equation written in three-vector form give the equation of motion for the point charge in any inertial frame as

$$\frac{d\mathbf{p}}{dt} = \frac{d}{dt}\left[\frac{m\mathbf{u}}{\sqrt{1-u^2/c^2}}\right] = q(\mathbf{E} + \mathbf{u} \times \mathbf{B}) \qquad \text{(relativistic)}.$$

Using (22.62) to add zero to the time component of our proposed covariant equation gives

$$\frac{dT}{dt} = \frac{d}{dt}\left(\mathcal{E} - mc^2\right) = \frac{d\mathcal{E}}{dt} = q\mathbf{u} \cdot \mathbf{E}.$$

These equations confirm the guesses (22.69) and (22.70) made in Section 22.5.2.

22.7.3 Conservation Laws

Covariant notation provides a powerful way to express and organize the conservation laws of electromagnetism. In this section, we use results from earlier in this chapter to write the conservation laws for energy, linear momentum, and angular momentum in manifestly covariant form. The following section uses this representation to extract new physics about free fields which are non-zero only in finite regions of space. Chapter 24 revisits conservation laws from a Lagrangian point of view.

Energy and Linear Momentum

We begin with the expressions for j_μ in (22.85) and $F_{\mu\nu}$ in (22.137). A brief computation confirms that the Coulomb-Lorentz force density, $\mathbf{f} = \rho\mathbf{E} + \mathbf{j} \times \mathbf{B}$, is the space part of a force density four-vector defined by[15]

$$f_\mu = j_\nu F_{\mu\nu} = (\mathbf{f}, i\mathbf{j} \cdot \mathbf{E}/c). \tag{22.151}$$

Our strategy is to mimic Section 15.5.1 and try to express the first-rank tensor f_μ as the four-divergence of a suitably defined second-rank tensor by eliminating the sources in favor of the fields.

The first step substitutes (22.140) into (22.151) to get

$$\mu_0 f_\mu = F_{\mu\nu}\partial_\sigma F_{\nu\sigma} = \partial_\sigma(F_{\mu\nu}F_{\nu\sigma}) - F_{\nu\sigma}\partial_\sigma F_{\mu\nu}. \tag{22.152}$$

[15] To simplify notation, we write $\mathbf{f}$ in place of the symbol $\mathbf{f}_{\text{mech}}$ defined just before (15.54).

Next, use the anti-symmetry of $F_{\mu\nu}$ and an exchange of the dummy indices ν and σ to rewrite the last term in (22.152) as

$$F_{\nu\sigma}\partial_\sigma F_{\mu\nu} = F_{\sigma\nu}\partial_\sigma F_{\nu\mu} = F_{\nu\sigma}\partial_\nu F_{\sigma\mu}. \tag{22.153}$$

The first and last members of (22.153), and the covariant form (22.144) of the homogeneous Maxwell equations, show that

$$F_{\nu\sigma}\partial_\sigma F_{\mu\nu} = \tfrac{1}{2}F_{\nu\sigma}\left(\partial_\sigma F_{\mu\nu} + \partial_\nu F_{\sigma\mu}\right) = -\tfrac{1}{2}F_{\nu\sigma}\partial_\mu F_{\nu\sigma} = -\tfrac{1}{4}\partial_\mu\left(F_{\nu\sigma}F_{\nu\sigma}\right). \tag{22.154}$$

Substituting (22.154) into (22.152) gives

$$\mu_0 f_\mu = -\partial_\sigma(F_{\mu\nu}F_{\sigma\nu}) + \tfrac{1}{4}\partial_\mu(F_{\alpha\beta}F_{\alpha\beta}). \tag{22.155}$$

It is interesting to note that the invariant (22.139) appears in the last term in (22.155). Otherwise, we achieve our goal because by defining the symmetric, second-rank, *electromagnetic stress-energy tensor*,

$$\Theta_{\mu\sigma} = \frac{1}{\mu_0}\left[F_{\mu\nu}F_{\sigma\nu} - \tfrac{1}{4}\delta_{\mu\sigma}F_{\alpha\beta}F_{\alpha\beta}\right] = \Theta_{\sigma\mu}, \tag{22.156}$$

(22.155) takes the desired form:

$$\partial_\sigma\Theta_{\sigma\mu} = -f_\mu. \tag{22.157}$$

An immediate consequence of (22.151) and (22.157) is that a *free* electromagnetic field (a propagating field detached from its sources) has a divergence-free stress-energy tensor[16]

$$\partial_\sigma\Theta_{\sigma\mu} = 0. \tag{22.158}$$

Let us display the elements of $\Theta_{\mu\sigma}$ in one inertial frame. Using (22.139), and our convention that a Latin index like k runs over the space components only, Θ_{44} simplifies to the negative of the electromagnetic density $u_{\rm EM}$ defined in (15.32):

$$\Theta_{44} = \frac{1}{\mu_0}\left[F_{4k}F_{4k} - \tfrac{1}{2}\left(\mathbf{B}\cdot\mathbf{B} - \mathbf{E}\cdot\mathbf{E}/c^2\right)\right] = -\tfrac{1}{2}\epsilon_0\left[\mathbf{E}\cdot\mathbf{E} + c^2\mathbf{B}\cdot\mathbf{B}\right] = -u_{\rm EM}. \tag{22.159}$$

The off-diagonal elements Θ_{4k} are proportional to the Cartesian components of the electromagnetic linear momentum density $\mathbf{g}$ defined in (15.50):

$$\Theta_{4k} = \frac{1}{\mu_0}F_{4j}F_{kj} = \frac{i}{\mu_0 c}(\mathbf{E}\times\mathbf{B})_k = icg_k. \tag{22.160}$$

A bit of algebra confirms that the space-space components Θ_{ij} are the negative of the components of Maxwell stress tensor T_{ij} defined in (15.47):

$$\Theta_{ij} = -T_{ij} = -\epsilon_0\left[E_iE_j + c^2B_iB_j - \tfrac{1}{2}\delta_{ij}(E^2 + c^2B^2)\right]. \tag{22.161}$$

Putting all this together gives the matrix of $\Theta_{\mu\sigma}$ as

$$\boldsymbol{\Theta} = \begin{bmatrix} -T_{xx} & -T_{xy} & -T_{xz} & icg_x \\ -T_{xy} & -T_{yy} & -T_{yz} & icg_y \\ -T_{zx} & -T_{yz} & -T_{zz} & icg_z \\ icg_x & icg_y & icg_z & -u_{\rm EM} \end{bmatrix}. \tag{22.162}$$

[16] If we neglect gravity, a divergence-free condition like (22.158) applies to the *total* stress-energy tensor of any closed system composed of matter and electromagnetic fields. This tensor is $\boldsymbol{\Theta}_{\rm tot} = \boldsymbol{\Theta}_{\rm em} + \boldsymbol{\Theta}_{\rm mat}$, where $\boldsymbol{\Theta}_{\rm em}$ is the electromagnetic stress-energy tensor discussed in this section and $\boldsymbol{\Theta}_{\rm mat}$ is the stress-energy tensor contributed by the matter.

The electromagnetic stress-energy tensor is traceless because

$$\Theta_{\mu\mu} = -u_{\rm EM} - \epsilon_0 \sum_{k=1}^{3} \left[E_k^2 + c^2 B_k^2 - \tfrac{1}{2}(E^2 + c^2 B^2) \right] = 0. \qquad (22.163)$$

With the representation (22.162) in hand, it is straightforward to confirm that (22.157) contains two conservation laws in differential form. Equation (22.151) and $\mathbf{S} = c^2\mathbf{g}$ (see Section 15.4.1) show that the time component of (22.157) is the Poynting's theorem statement (15.33) of the conservation of energy:

$$-\mathbf{j}\cdot\mathbf{E} = \nabla\cdot\mathbf{S} + \frac{\partial u_{\rm EM}}{\partial t}. \qquad (22.164)$$

Similarly, if $\partial_k T_{kj} \equiv [\nabla\cdot\mathbf{T}]_j$, the space components of (22.157) give the law (15.54) for the conservation of linear momentum:

$$-\mathbf{f} = \nabla\cdot(-\mathbf{T}) + \frac{\partial\mathbf{g}}{\partial t}. \qquad (22.165)$$

Angular Momentum and Center of Energy

Given the Lorentz force density four-vector (22.151), it is natural to study the properties of a second-rank Lorentz torque density tensor defined by

$$\mathcal{N}_{\mu\nu} = r_\mu f_\nu - r_\nu f_\mu. \qquad (22.166)$$

The structure of this anti-symmetric tensor in one inertial frame is

$$\mathcal{N} = \begin{bmatrix} 0 & (\mathbf{r}\times\mathbf{f})_z & -(\mathbf{r}\times\mathbf{f})_y & \mathcal{N}_{14} \\ -(\mathbf{r}\times\mathbf{f})_z & 0 & (\mathbf{r}\times\mathbf{f})_x & \mathcal{N}_{24} \\ (\mathbf{r}\times\mathbf{f})_y & -(\mathbf{r}\times\mathbf{f})_x & 0 & \mathcal{N}_{34} \\ \mathcal{N}_{41} & \mathcal{N}_{42} & \mathcal{N}_{43} & 0 \end{bmatrix}, \qquad (22.167)$$

where $\mathbf{r}\times\mathbf{f}$ is the mechanical torque density (15.65) and

$$\mathcal{N}_{k4} = -\mathcal{N}_{4k} = i\left[r_k(\mathbf{j}\cdot\mathbf{E})/c - ctf_k\right]. \qquad (22.168)$$

Analogous to (22.157), it is possible to write the second-rank torque density tensor as the four-divergence of a third-rank Lorentz tensor:

$$M_{\sigma\mu\nu} = \Theta_{\sigma\mu} r_\nu - \Theta_{\sigma\nu} r_\mu = -M_{\rho\nu\mu}. \qquad (22.169)$$

In detail, we use (22.157), (22.169), and the symmetry of the stress-energy tensor ($\Theta_{\mu\nu} = \Theta_{\nu\mu}$) to write (22.166) in the form

$$\mathcal{N}_{\mu\nu} = r_\nu\partial_\sigma\Theta_{\sigma\mu} - r_\mu\partial_\sigma\Theta_{\sigma\nu} = \partial_\sigma M_{\sigma\mu\nu} - (\Theta_{\nu\mu} - \Theta_{\mu\nu}) = \partial_\sigma M_{\sigma\mu\nu}. \qquad (22.170)$$

The anti-symmetry of $M_{\rho\nu\mu}$ with respect to its last two indices implies that only 24 of its $4^3 = 64$ components are independent. We focus first on the 12 components $M_{\sigma ij}$ where $ij = 12, 23, 31$. Of these, the reader can check that the nine with $\sigma = 1, 2, 3$ are exactly the components of the second-rank tensor of angular momentum current density, $\mathbf{M} = \mathbf{T}\times\mathbf{r}$, defined in Section 15.6.1. The three components M_{4ij} are similarly the components of the vector density of field angular momentum, $-ic\mathbf{r}\times\mathbf{g}$. With this information, it is not difficult to confirm that the three independent components of (22.170) with $\mu\nu = 12, 23, 31$ can be collected into the continuity-like equation for angular momentum written previously in (15.72):

$$\frac{\partial}{\partial t}(\mathbf{r}\times\mathbf{g}) + \nabla\cdot(\mathbf{T}\times\mathbf{r}) = -\mathbf{r}\times\mathbf{f}. \qquad (22.171)$$

The three independent components of (22.170) that remain have $\mu\nu = 14, 24, 34$. Using (22.168), these can be collected into the vector equation[17]

$$\frac{\partial}{\partial t}\left(\mathbf{r}u_{\text{EM}} - c^2 t\mathbf{g}\right) + c^2\nabla\cdot(\mathbf{gr}+\mathbf{T}) = c^2 t\mathbf{f} - (\mathbf{j}\cdot\mathbf{E})\mathbf{r}. \tag{22.172}$$

Because $\mathbf{j}\cdot\mathbf{E} = \mathbf{j}\cdot\mathbf{f}$ (the Lorentz magnetic force does no work on charged particles) and the Poynting vector $\mathbf{S} = c^2\mathbf{g}$, (22.172) is the same as the previously derived (15.82). The latter was used in Section 15.7 to define the center of energy for sources and fields confined to a volume where the divergence term in (22.172) disappears after integrating over a volume. The next section revisits this question from a covariant perspective.

22.7.4 Particle-Like Properties of Free Fields

Free electromagnetic fields behave like relativistic particles in the sense that their energy and linear momentum transform like the energy-momentum four-vector of a particle. In this section, we prove this for (i) any finite volume of the radiation field produce by a localized source and (ii) any localized electromagnetic wave packet. Both are *free* in the sense that their fields are detached from their sources and propagate at the speed of light.

Monochromatic Radiation Fields

The portion of the radiation field produced by a localized current source that lies inside a finite volume V increasingly resembles the propagating electromagnetic field of a plane wave when V is chosen farther and farther away from the source (see Section 20.5.4). Here, we apply this observation to a monochromatic radiation field of frequency ω and let U_{EM} be the total electromagnetic energy contained in V. These choices make Example 22.4 applicable and we conclude that the ratio U_{EM}/ω is a Lorentz invariant scalar.

For the single, transverse plane wave of interest, the rightmost equation in (22.128) and $\omega = ck$ give the total linear momentum contained in V as

$$\mathbf{P}_{\text{EM}} = \mathbf{g}V = \frac{u_{\text{EM}}}{c}V\hat{\mathbf{k}} = \frac{u_{\text{EM}}}{\omega}V\mathbf{k} = \left(\frac{U_{\text{EM}}}{\omega}\right)\hat{\mathbf{k}}. \tag{22.173}$$

Moreover,

$$i\frac{U_{\text{EM}}}{c} = \left(\frac{U_{\text{EM}}}{\omega}\right)i\frac{\omega}{c}. \tag{22.174}$$

Therefore, because U_{EM}/ω is a Lorentz scalar and $(\mathbf{k}, i\omega/c)$ is a four-vector, (22.173) and (22.174) tell us that $(\mathbf{P}_{\text{EM}}, iU_{\text{EM}}/c)$ is a four-vector also. Comparison with the four-vector $(\mathbf{p}, i\mathcal{E}/c)$ for a relativistic particle with linear momentum $\mathbf{p}$ and total energy $\mathcal{E}$ [see (22.59)] establishes the similarity between a volume of propagating radiation and a relativistic particle. In quantum theory, the same result follows by treating radiation as a collection of relativistic particles with zero mass (photons).

Localized Wave Packets

A localized wave packet or *electromagnetic pulse* is an electromagnetic wave with the property that $\mathbf{E}(\mathbf{r}, t)$ and $\mathbf{B}(\mathbf{r}, t)$ fall to negligible values outside a finite volume of space. Pulses of this kind are created by turning on and off a flashlight or a laser. In what follows, we use a manifestly covariant method to prove that $(\mathbf{P}_{\text{EM}}, iU_{\text{EM}}/c)$ is a four-vector when $\mathbf{P}_{\text{EM}}$ and U_{EM} are the total linear momentum and total energy of the pulse. We begin with a bit of mathematics.

[17] The dyadic expression $\nabla\cdot(\mathbf{gr}) = \nabla_k g_k \mathbf{r}$.

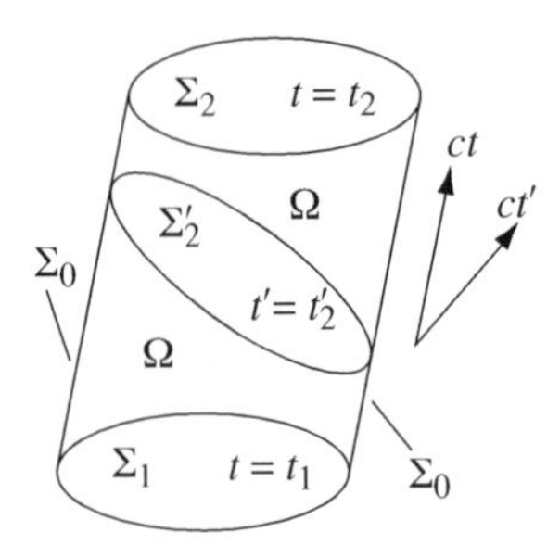

Figure 22.9: A four-dimensional volume Ω in Minkowski space. The ellipses are "endcaps" which represent three-dimensional volumes at fixed times.

The left member of (22.175) is the usual statement of the divergence theorem for a vector $\mathbf{Z}$. V is a three-dimensional Euclidean volume and $d\mathbf{S}$ is a vector which points outward from V and has the magnitude of a differential element of the two-dimensional surface S which bounds V:

$$\int_V d^3r\, \partial_k Z_k = \int_S dS_k Z_k \qquad\qquad \int_\Omega d^4r\, \partial_\mu Z_\mu = \int_\Sigma dS_\mu Z_\mu. \tag{22.175}$$

The right member of (22.175) is the divergence theorem for a four-vector Z_μ. Ω is a four-dimensional Minkowski volume and dS_μ is a four-vector which points outward from Ω and has the magnitude of a differential element of the three-dimensional "surface" Σ which bounds Ω.[18]

Figure 22.9 is a cartoon of a four-dimensional "volume" Ω in Minkowski space where time flows in the direction perpendicular to the elliptical planes, each of which represents three-dimensional Euclidean space at a fixed time in some inertial frame. The surface of Ω is composed of a tubular part Σ_0 and two endcaps, Σ_1 and Σ_2. We assume that an electromagnetic pulse lies entirely *inside* Ω and that its sources (j_μ) lie entirely *outside* Ω. In light of (22.158), this means that

$$\partial_\nu \Theta_{\mu\nu} = 0 \qquad \text{in } \Omega. \tag{22.176}$$

Now, let a_μ be a *constant* four-vector and integrate the four-divergence of $b_\nu = a_\mu \Theta_{\mu\nu}$ over Ω. This integral is zero by (22.176). Therefore, when we apply the four-divergence theorem in (22.175), the assumed localized nature of the fields of the pulse eliminates the integral over Σ_0. Accordingly,

$$0 = \int_\Omega d^4r\, \partial_\nu b_\nu = \int_{\Sigma_0} dS_\nu b_\nu + \int_{\Sigma_1} dS_\nu b_\nu + \int_{\Sigma_2} dS_\nu b_\nu = \int_{\Sigma_1} dS_\nu b_\nu + \int_{\Sigma_2} dS_\nu b_\nu. \tag{22.177}$$

A key observation is that the integrand $dS_\nu b_\nu$ in (22.177) is a Lorentz invariant scalar. Therefore, the Σ_1 and Σ_2 integrals can be evaluated in any desired frame of reference. For example, let us evaluate both integrals in a frame K whose time coordinate ct is shown in Figure 22.9. The diagram makes clear that the endcaps correspond to $t = t_1$ and $t = t_2$ and that $dS_\mu = (0, 0, 0, \pm i\, d^3r)$ for the two caps. Hence, (22.177) is equivalent to

$$\int_{t=t_1} d^3r\, b_4 = \int_{t=t_2} d^3r\, b_4. \tag{22.178}$$

Since a_μ is arbitrary and does not depend on space or time, we learn from (22.178) that the three-dimensional volume integral

$$\int d^3r\, \Theta_{4\mu} \quad \text{is a time-independent constant.} \tag{22.179}$$

[18] See C. Møller, *The Theory of Relativity* (Clarendon, Oxford, 1952) for a detailed proof.

A glance back at (22.162) for $\mu = 1, 2, 3, 4$ shows that the constants in questions are the total field energy and the total field linear momentum:

$$\int d^3r\,\Theta_{44} = -U_{\rm EM} \qquad \text{and} \qquad \int d^3r\,\Theta_{4i} = icP_{{\rm EM},i}. \tag{22.180}$$

We now return to (22.177) and modify the previous calculation by evaluating the Σ_2 integral in a frame K' with the coordinate ct' shown in Figure 22.9. This replaces (22.178) by the statement that a particular volume integral takes the same numerical value in two different inertial frames:

$$\int\limits_{t=t_1} d^3r\,b_4 = \int\limits_{t'=t'_2} d^3r'\,b'_4. \tag{22.181}$$

In other words,

$$a_\mu \int d^3r\,\Theta_{4\mu} \quad \text{is a Lorentz invariant scalar.} \tag{22.182}$$

Then, because a_μ is an arbitrary four-vector, and the entire expression in (22.182) is a Lorentz scalar, the quotient theorem (22.132) tells us that the integral is itself a four-vector. Using (22.162) as before, the four-vector in question is

$$P_\mu = -\frac{i}{c}\int d^3r\,\Theta_{4\mu} = (\mathbf{P}_{\rm EM}, iU_{\rm EM}/c). \tag{22.183}$$

As noted in the paragraph following (22.174), (22.183) has exactly the same structure as the energy-momentum four-vector of a relativistic particle, $p_\mu = (\mathbf{p}, i\mathcal{E}/c)$.

Let us investigate the similarity between an electromagnetic pulse and a relativistic particle a bit more. When there are no sources in Ω, the torque density tensor $\mathcal{N}_{\mu\nu} = 0$ in (22.166) and (22.170) implies that

$$\partial_\sigma M_{\sigma\mu\nu} = 0 \qquad \text{in } \Omega. \tag{22.184}$$

This equation is similar to (22.176) and the same method used to derive (22.179) from (22.176) implies that

$$\int d^3r\,M_{4\mu\nu} \quad \text{is a time-independent constant.} \tag{22.185}$$

From the discussion of (22.171) and (22.172), the conserved quantities implied by (22.185) are the total field angular momentum of the pulse,

$$\mathbf{L}_{\rm EM} = \int d^3r\,\mathbf{r}\times\mathbf{g}, \tag{22.186}$$

and

$$\int d^3r\,(\mathbf{r}u_{\rm em} - c^2t\mathbf{g}) = \int\limits_V d^3r\,\mathbf{r}u_{\rm em} - c^2\mathbf{P}_{\rm EM}. \tag{22.187}$$

Following the example of Section 15.7, we define the *center of energy* as

$$\mathbf{R}_{\rm E} = \frac{\int\limits_V d^3r\,\mathbf{r}u_{\rm EM}}{\int\limits_V d^3r\,u_{\rm EM}}. \tag{22.188}$$

The denominator of (22.188) is the total field energy $U_{\rm EM}$. Therefore, (22.187) can be put in the form

$$\mathbf{R}_{\rm E}U_{\rm EM} - c^2t\mathbf{P}_{\rm EM} = \text{constant vector.} \tag{22.189}$$

From (22.179), we know that $U_{\rm EM}$ and $\mathbf{P}_{\rm EM}$ are constants of the motion of the pulse. This suggests that we take the time derivative of (22.189) and define the velocity of the center of energy as

$$\mathbf{v}_{\rm E} = \frac{d\mathbf{R}_{\rm E}}{dt} = \frac{c^2\mathbf{P}_{\rm EM}}{U_{\rm EM}}. \tag{22.190}$$

Equation (22.190) has the same structure as (22.67) for a relativistic particle. Putting (22.190) together with (22.183), we conclude that a propagating "pulse" of electromagnetic radiation and a relativistic particle are really very similar, at least as far as energy and momentum are concerned.

22.8 Matter in Uniform Motion

The extension of Einstein's theory of relativity to material media was achieved by Hermann Minkowski in 1908. His theory is worth our attention, *not* because the acceleration of bodies of matter to relativistic speeds is conceivable in principle, but because the low-velocity limit of the theory provides a reliable and unambiguous basis for predicting the fields produced when such bodies move through external electromagnetic fields at non-relativistic speeds. We begin by reminding the reader that the extension of vacuum electromagnetism to polarizable, magnetizable, and conducting matter (Section 2.4) uses the polarization **P** and magnetization **M** to define the auxiliary fields

$$\mathbf{D} = \epsilon_0\mathbf{E} + \mathbf{P} \qquad \text{and} \qquad \mathbf{B} = \mu_0(\mathbf{H} + \mathbf{M}). \tag{22.191}$$

With (22.191) in hand, the Maxwell equations in matter supplement the homogeneous Maxwell equations

$$\nabla\cdot\mathbf{B} = 0 \qquad \text{and} \qquad \nabla\times\mathbf{E} = -\frac{\partial\mathbf{B}}{\partial t}, \tag{22.192}$$

with inhomogeneous equations where the free charge density ρ_f and free current density $\mathbf{j}_f$ are the field sources:

$$\nabla\cdot\mathbf{D} = \rho_f \qquad \text{and} \qquad \nabla\times\mathbf{H} = \mathbf{j}_f + \frac{\partial\mathbf{D}}{\partial t}. \tag{22.193}$$

22.8.1 Minkowski Electrodynamics

Minkowski *assumed* that (22.192) and (22.193) are valid in every inertial frame. Of course, (22.148) is already a manifestly covariant expression of (22.192). Otherwise, (i) the substitutions $\mathbf{D}\to\epsilon_0\mathbf{E}$ and $\mathbf{H}\to\mathbf{B}/\mu_0$ convert the field terms in (22.193) to the field terms in the corresponding vacuum Maxwell equations; and (ii) free charge and current density transform like total charge and current density. Therefore, with the stated substitutions, (22.140) is a manifestly covariant expression of (22.193). Making the same substitutions in (22.95) and (22.96) gives the desired transformation laws for **D** and **H**. We write these here with the rest-frame values on the right-hand side:

$$\mathbf{D}_\parallel = \mathbf{D}'_\parallel \qquad \mathbf{D}_\perp = \gamma\left[\mathbf{D}' - \frac{\mathbf{v}}{c^2}\times\mathbf{H}'\right]_\perp \tag{22.194}$$

and

$$\mathbf{H}_\parallel = \mathbf{H}'_\parallel \qquad \mathbf{H}_\perp = \gamma[\mathbf{H}' + \mathbf{v}\times\mathbf{D}']_\perp. \tag{22.195}$$

The transformation rules for polarization and magnetization follow immediately from (22.191) as

$$\mathbf{P}_\parallel = \mathbf{P}'_\parallel \qquad \mathbf{P}_\perp = \gamma\left[\mathbf{P}' + \frac{\mathbf{v}}{c^2}\times\mathbf{M}'\right]_\perp \tag{22.196}$$

and

$$\mathbf{M}_{\parallel} = \mathbf{M}'_{\parallel} \qquad\qquad \mathbf{M}_{\perp} = \gamma[\mathbf{M}' - \mathbf{v} \times \mathbf{P}']_{\perp}. \tag{22.197}$$

We note in passing that (22.197) reproduces the result from Section 14.2.2 that a body with rest-frame values $\mathbf{M}' = 0$ and $\mathbf{P}' \neq 0$ is observed to have magnetization $\mathbf{M} = \mathbf{P}' \times \mathbf{v}$ when it moves with uniform velocity $\mathbf{v}$ in the laboratory.

It remains only to derive laboratory-frame constitutive equations for matter in uniform motion. For this purpose, we assume that $\mathbf{D}' = \epsilon\mathbf{E}'$ and $\mathbf{B}' = \mu\mathbf{H}'$ in the rest frame and eliminate all the primed variables using the transformation laws. The result of this algebra is

$$\begin{aligned} \mathbf{D} + \frac{\mathbf{v}}{c^2} \times \mathbf{H} &= \epsilon(\mathbf{E} + \mathbf{v} \times \mathbf{B}) \\ \mathbf{B} - \frac{\mathbf{v}}{c^2} \times \mathbf{E} &= \mu(\mathbf{H} - \mathbf{v} \times \mathbf{D}). \end{aligned} \tag{22.198}$$

Alternatively, we can use (22.191) and the electric and magnetic susceptibilities defined by $\chi\epsilon_0 = \epsilon - \epsilon_0$ and $\chi_m\mu_0 = \mu - \mu_0$ to write (22.198) in the form

$$\begin{aligned} \mathbf{P} &= \epsilon_0\chi(\mathbf{E} + \mathbf{v} \times \mathbf{B}) + \frac{\mathbf{v}}{c^2} \times \mathbf{M} \\ \mathbf{M} &= \chi_m(\mathbf{H} - \mathbf{v} \times \mathbf{D}) - \mathbf{v} \times \mathbf{P}. \end{aligned} \tag{22.199}$$

Substituting each equation in (22.198) into the other eliminates $\mathbf{B}$ from the first equation and $\mathbf{D}$ from the second. The resulting fully relativistic constitutive relations have the form $\mathbf{D}[\mathbf{E}, \mathbf{H}]$ and $\mathbf{B}[\mathbf{E}, \mathbf{H}]$. For uncharged conducting matter which obeys Ohm's law in its rest frame, the reader can show that the total current density in the laboratory is

$$\mathbf{j} = \gamma\sigma(\mathbf{E} + \mathbf{v} \times \mathbf{B}). \tag{22.200}$$

Finally, moving matter modifies the interface conditions we derived from the Maxwell equations. From (2.49) and (2.57), the matching conditions in the rest frame K' are

$$\begin{aligned} \hat{\mathbf{n}}'_2 \cdot [\mathbf{D}'_1 - \mathbf{D}'_2] &= \sigma'_f & \qquad \hat{\mathbf{n}}'_2 \cdot [\mathbf{B}'_1 - \mathbf{B}'_2] &= 0 \\ \hat{\mathbf{n}}'_2 \times [\mathbf{H}'_1 - \mathbf{H}'_2] &= \mathbf{K}'_f & \qquad \hat{\mathbf{n}}'_2 \times [\mathbf{E}'_1 - \mathbf{E}'_2] &= 0. \end{aligned} \tag{22.201}$$

The matching conditions in the laboratory frame K follow when we rewrite (22.201) using the transformation laws for the fields, for the surface charge and current densities, and for the unit normal to the interface. The latter two are not very familiar and the calculation is rather tedious.[19] Therefore, we merely quote the results when the interface moves with velocity $\mathbf{v}$ as viewed from K. The normal components of $\mathbf{B}$ and $\mathbf{D}$ remain continuous and [cf. (2.50)]

$$\begin{aligned} \hat{\mathbf{n}}_2 \times [\mathbf{E}_1 - \mathbf{E}_2] &= (\hat{\mathbf{n}}_2 \cdot \mathbf{v})[\mathbf{B}_1 - \mathbf{B}_2] \\ \hat{\mathbf{n}}_2 \times [\mathbf{H}_1 - \mathbf{H}_2] &= -(\hat{\mathbf{n}}_2 \cdot \mathbf{v})[\mathbf{D}_1 - \mathbf{D}_2]. \end{aligned} \tag{22.202}$$

There is no correction to the original matching conditions when an interface moves parallel to itself.

[19] See Section 1.9, Section 3.4, and Section 5.1 of Van Bladel (1984).

22.8.2 Slowly Moving Matter

We specialize now to the low-velocity limit when $v \ll c$. This leads us to drop the terms proportional to v^2/c^2 in the constitutive relations $\mathbf{D}[\mathbf{E}, \mathbf{H}]$ and $\mathbf{B}[\mathbf{E}, \mathbf{H}]$ mentioned after (22.199). The result is

$$\begin{aligned} \mathbf{D} &= \epsilon \mathbf{E} + \left\{ \mu\epsilon - \frac{1}{c^2} \right\} \mathbf{v} \times \mathbf{H} \\ \mathbf{B} &= \mu \mathbf{H} + \left\{ \frac{1}{c^2} - \mu\epsilon \right\} \mathbf{v} \times \mathbf{E}. \end{aligned} \tag{22.203}$$

The low-velocity form of (22.199) is (22.203) with $\mathbf{D} \to \mathbf{P}$, $\epsilon\mathbf{E} \to \epsilon_0 \chi \mathbf{E}$, $\mathbf{B} \to \mathbf{M}$, and $\mu\mathbf{H} \to \mu_0 \chi_m \mathbf{H}$. Now, dielectric bodies tend to be only weakly magnetic and magnetic bodies tend to be only weakly dielectric. Therefore, when the motion of a polarizable body through an electric field induces a magnetic field [as predicted by (22.197)], it is often possible to approximate (22.203) by

$$\mathbf{D} = \epsilon \mathbf{E} \qquad \text{and} \qquad \mathbf{B} = \mu \mathbf{H} + \left\{ \frac{1}{c^2} - \mu\epsilon \right\} \mathbf{v} \times \mathbf{E}. \tag{22.204}$$

Similarly, and more commonly, when the motion of a magnetizable body through a magnetic field induces an electric field [as predicted by (22.196)], it is common to approximate (22.203) by

$$\mathbf{B} = \mu \mathbf{H} \qquad \text{and} \qquad \mathbf{D} = \epsilon \mathbf{E} + \left\{ \mu\epsilon - \frac{1}{c^2} \right\} \mathbf{v} \times \mathbf{H}. \tag{22.205}$$

Example 22.6 A non-magnetic, dielectric cylinder moves parallel to its symmetry axis at constant velocity $\mathbf{v} = v\hat{\mathbf{z}}$ in a uniform magnetic field $\mathbf{B} = -B_0\hat{\mathbf{y}}$. Find the electric field everywhere and the polarization of the cylinder. Neglect end effects.

Solution: The magnetic field is time-independent in every inertial frame. This reduces the right member of (22.192) to $\nabla \times \mathbf{E} = 0$, from which we deduce that $\mathbf{E} = -\nabla\varphi$. Using this in the right member of (22.205) with $\mu = \mu_0$ gives

$$\nabla \cdot \mathbf{D} = \rho_f = \epsilon \nabla \cdot \mathbf{E} + (\epsilon - \epsilon_0) \nabla \cdot (\mathbf{v} \times \mathbf{B}),$$

or

$$\nabla^2 \varphi = -\frac{\rho_f}{\epsilon} + \frac{\epsilon - \epsilon_0}{\epsilon} \nabla \cdot (\mathbf{v} \times \mathbf{B}).$$

Both terms on the right side of the preceding equation are zero because (i) there is no free volume charge and (ii) the velocity and magnetic field are uniform. Therefore, $\nabla^2\varphi = 0$ inside and outside the cylinder. From the right member of (22.205) and the absence of free surface charge, the matching conditions at the cylinder surface, $\rho = a$, are

$$\varphi_{\text{in}} = \varphi_{\text{out}}$$

and

$$0 = \sigma_f = \hat{\boldsymbol{\rho}} \cdot (\mathbf{D}_{\text{out}} - \mathbf{D}_{\text{in}}) = -\left[\epsilon_0 \frac{\partial \varphi_{\text{out}}}{\partial \rho} - \epsilon \frac{\partial \varphi_{\text{in}}}{\partial \rho} \right]_{\rho=a} - (\epsilon - \epsilon_0) \hat{\boldsymbol{\rho}} \cdot (\mathbf{v} \times \mathbf{B}).$$

Because $\hat{\boldsymbol{\rho}} = \cos\phi \hat{\mathbf{x}} + \sin\phi \hat{\mathbf{y}}$, the latter is

$$\left[\epsilon \frac{\partial \varphi_{\text{in}}}{\partial \rho} - \epsilon_0 \frac{\partial \varphi_{\text{out}}}{\partial \rho} \right]_{\rho=a} = (\epsilon - \epsilon_0) v B_0 \cos\phi.$$

Our experience with potential theory (Section 7.9) permits us to write down the solution by inspection:

$$\varphi(\rho,\phi) = \frac{\epsilon-\epsilon_0}{\epsilon+\epsilon_0} v B_0 \cos\phi \times \begin{cases} \rho & \rho \leq a, \\ \dfrac{a^2}{\rho} & \rho \geq a. \end{cases}$$

The associated electric field is

$$\mathbf{E}_{\text{in}} = -\frac{\epsilon-\epsilon_0}{\epsilon-\epsilon_0} v B_0 \hat{\mathbf{x}}$$

$$\mathbf{E}_{\text{out}} = \frac{\epsilon-\epsilon_0}{\epsilon+\epsilon_0} v B_0 \frac{a^2}{\rho^2} \left[\cos\phi\hat{\boldsymbol{\rho}} + \sin\phi\hat{\boldsymbol{\phi}}\right].$$

Finally, we extract $\mathbf{P}$ by combining $\mathbf{D} = \epsilon_0\mathbf{E} + \mathbf{P}$ with the right member of (22.205) with $\mu = \mu_0$. The result is

$$\mathbf{P} = (\epsilon - \epsilon_0)(\mathbf{E}_{\text{in}} + \mathbf{v} \times \mathbf{B}) = 2\epsilon_0 v B_0 \frac{\epsilon-\epsilon_0}{\epsilon+\epsilon_0} \hat{\mathbf{x}}.$$

Application 22.5 The Electric Field outside a Pulsar

A *pulsar* is an astrophysical object that emits polarized radiation in bursts with a period that can range from 1 msec to 10 sec. The simplest model of a pulsar is a rapidly rotating neutron star which emits magnetic dipole radiation in a narrow beam. As shown in Figure 22.10, the rotation axis and the magnetic dipole axis are not aligned. Distant observers detect pulses because a rotating beam produces a "lighthouse effect". The interior and exterior of a pulsar are thought to be good conductors, but the origin of the pulsar's magnetic dipole field is not known.

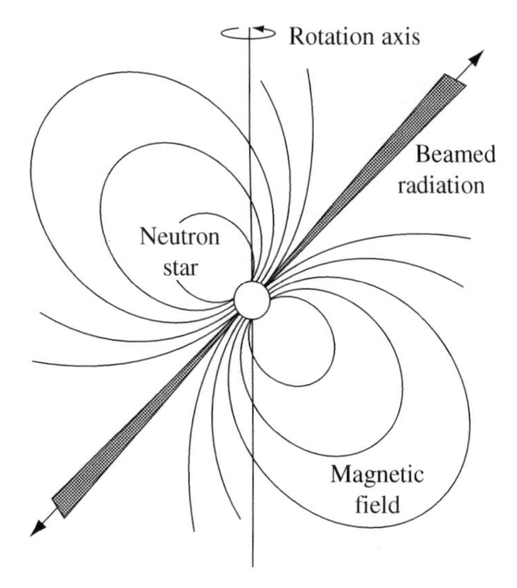

Figure 22.10: A cartoon model of a pulsar as a rotating, magnetized, neutron star.

In this Application, we model a pulsar as a sphere with radius R and uniform magnetization $\mathbf{M}'$. For simplicity, we align $\mathbf{M}'$ with the angular velocity vector $\boldsymbol{\omega} = \omega\hat{\mathbf{z}}$ and suppose that only the interior of the sphere is a good conductor. Our goal is to show that, in the vacuum space outside of itself, the pulsar produces an *electric quadrupole* field in addition to its magnetic dipole field. Inside the pulsar, we assume that the magnitude of the linear velocity $\mathbf{v} = \omega \times \mathbf{r}$ is everywhere small compared to the speed of light. Therefore, we will consistently ignore the factor $\gamma = 1/\sqrt{1 - v^2/c^2}$ when we transform fields from the non-rotating rest frame of the sphere to the laboratory frame where the sphere rotates.

The sphere is a conductor, so the electric vectors in their rest frame vanish everywhere:

$$\mathbf{E}' = \mathbf{D}' = \mathbf{P}' = 0. \tag{22.206}$$

A sphere with volume V and uniform magnetization $\mathbf{M}' = M_0\hat{\mathbf{z}}$ was the subject of Application 13.1. The fields we found were

$$\mathbf{H}' = \begin{cases} -\dfrac{1}{3}\mathbf{M}' & r < R, \\[2ex] \dfrac{V}{4\pi}\left\{\dfrac{3(\hat{\mathbf{r}} \cdot \mathbf{M}')\hat{\mathbf{r}} - \mathbf{M}'}{r^3}\right\} & r > R, \end{cases} \tag{22.207}$$

and

$$\mathbf{B}' = \begin{cases} \dfrac{2}{3}\mu_0\mathbf{M}' & r < R, \\[1ex] \mu_0\mathbf{H}' & r > R. \end{cases} \tag{22.208}$$

It is not immediately clear how to transform these fields to the laboratory frame because every point of a rotating sphere experiences centripetal acceleration. Fortunately, when terms of order v^2/c^2 can be neglected, the general relativistic formulae which transform fields in a rotating frame K' to fields in a non-rotating frame K reduce to the special relativistic formulae (22.95), (22.194), and (22.196) with $\gamma = 1$.[20] Therefore, in terms of the cylindrical coordinate $\rho = r\sin\theta$, the electric vectors in the laboratory are

$$\begin{aligned} \mathbf{E}_{\text{in}} &= -\mathbf{v} \times \mathbf{B}'_{\text{in}} = -\frac{2}{3}\mu_0\omega M_0\rho\hat{\boldsymbol{\rho}} \\ \mathbf{D}_{\text{in}} &= -\frac{\mathbf{v}}{c^2} \times \mathbf{H}'_{\text{in}} = \frac{\omega}{3c^2}M_0\rho\hat{\boldsymbol{\rho}} \\ \mathbf{P}_{\text{in}} &= \frac{\mathbf{v}}{c^2} \times \mathbf{M}' = \frac{\omega}{c^2}M_0\rho\hat{\boldsymbol{\rho}}. \end{aligned} \tag{22.209}$$

The physical origin of the inward-directed field $\mathbf{E}_{\text{in}}$ becomes clear when we note from (22.96) that $\mathbf{B}_{\text{in}} = \mathbf{B}'_{\text{in}}$. Therefore, the first line of (22.209) tells us that charge redistributes inside the pulsar until mechanical equilibrium is established and $\mathbf{E}_{\text{in}} + \mathbf{v} \times \mathbf{B}_{\text{in}} = 0$. Although the sphere is a conductor, a non-zero electric field develops so the electric Coulomb force balances the magnetic Lorentz force at every interior point. This electric field is not uniform in space, but satisfies $\mathbf{E}_{\text{in}} \cdot \mathbf{B}_{\text{in}} = 0$ everywhere inside the pulsar.

The electric field outside the pulsar satisfies $\nabla \cdot \mathbf{E}_{\text{out}} = 0$ and $\nabla \times \mathbf{E}_{\text{out}} = 0$. The first condition follows from the assumed absence of charge outside the pulsar. The second follows because the alignment of the rotation axis with the magnetic dipole axis guarantees that $\mathbf{B}_{\text{out}}$ is equal to the time-independent field $\mathbf{B}'_{\text{out}}$ in (22.208). We conclude that $\nabla^2\varphi_{\text{out}} = 0$ where $\mathbf{E}_{\text{out}} = -\nabla\varphi_{\text{out}}$. The boundary conditions are $\varphi_{\text{out}} \to 0$ as $r \to \infty$ and continuity of the tangential component of the electric field at the sphere boundary $r = R$ [see (22.202)]. Because $\hat{\boldsymbol{\rho}} \cdot \hat{\boldsymbol{\theta}} = \cos\theta$, the latter condition reads

$$\hat{\boldsymbol{\theta}} \cdot \mathbf{E}_{\text{out}}(R,\theta) = \hat{\boldsymbol{\theta}} \cdot \mathbf{E}_{\text{in}}(R,\theta) = -\frac{2}{3}\mu_0\omega M_0 R\sin\theta\cos\theta. \tag{22.210}$$

[20] See Ridgley (1998) in Sources, References, and Additional Reading.

We now draw on our knowledge of the solutions of Laplace's equation for problems with azimuthal symmetry (Section 7.6). The unique potential outside the pulsar is

$$\varphi_{\text{out}} = -\frac{2}{9}\mu_0\omega M_0 R^5 \frac{P_2(\cos\theta)}{r^3} = -\frac{1}{9}\mu_0\omega M_0 R^5 \frac{3\cos^2\theta - 1}{r^3}, \tag{22.211}$$

because

$$\mathbf{E}_{\text{out}} = -\nabla\varphi_{\text{out}} = -\frac{\mu_0\omega M_0 R^5}{3r^4}\left[(3\cos^2\theta - 1)\hat{\mathbf{r}} + 2\sin\theta\cos\theta\hat{\boldsymbol{\theta}}\right] \tag{22.212}$$

satisfies the boundary condition (22.210). As advertised, the exterior potential (22.211) and field (22.212) are those of an electric quadrupole. This conclusion survives in more realistic models of the electrodynamics of pulsars.

The reader may wish to check that the sources of $\mathbf{E}_{\text{out}}$ are volume and surface densities of free charge (which derive from the divergence of **D**) and volume and surface densities of polarization charge (which derive from the divergence of **P**). The total charge of the pulsar is zero. It is also worth noting that (22.211) implies that a potential difference exists between two points on the sphere with different latitudes. Hence, a current would flow through a wire whose ends touched the sphere at these points. This shows that the steady-state electrodynamics of a pulsar is closely related to the electrodynamics of the Faraday disk generator studied in Application 14.2. The main difference is that a pulsar rotates through its own magnetic field rather than through an external magnetic field. ■

Sources, References, and Additional Reading

The quotation at the beginning of the chapter is taken from "Autobiographical notes", Einstein's contribution to

P.A. Schilpp, *Albert Einstein, Philosopher-Scientist* (Library of Living Philosophers, Evanton, IL, 1949).

Section 22.1 The quotation in the first paragraph is the opening three sentences from the English translation of Einstein's original article [*Annalen der Physik* **17**, 891 (1905)] on special relativity:

A. Einstein, "On the electrodynamics of moving bodies", *The Collected Papers of Albert Einstein*, translated by A. Beck and P. Havas (University Press, Princeton, 1989), Volume 2.

Textbooks of electrodynamics with good discussions of special relativity include

L.D. Landau and E.M. Lifshitz, *The Classical Theory of Fields*, 2nd edition (Addison-Wesley, Reading, MA, 1962).

E.J. Konopinski, *Electromagnetic Fields and Relativistic Particles* (McGraw-Hill, New York, 1981).

J.D. Jackson, *Classical Electrodynamics*, 3rd edition (Wiley, New York, 1999).

There are a great many monographs devoted exclusively to relativity. This chapter benefitted particularly from

C. Møller, *The Theory of Relativity* (Clarendon Press, Oxford, 1952).

W.G.V. Rosser, *An Introduction to the Theory of Relativity* (Butterworths, London, 1964).

V.A. Ugarov, *Special Relativity* (Mir, Moscow, 1979).

Section 22.2 A discussion of both Galilean relativity and special relativity from the point of view of fundamental symmetries appears in

N. Doughty, *Lagrangian Interaction* (Westview, New York, 1990).

Section 22.3 *Miller* is the gold standard for the history of the development of special relativity. Among other things, he provides section-by-section commentary on Einstein's 1905 paper. The thought-provoking book by *Galison* contains no equations.

A.I. Miller, *Albert Einstein's Special Theory of Relativity* (Addison-Wesley, Reading, MA, 1981).

P. Galison, *Einstein's Clocks and Poincaré's Maps* (Norton, New York, 2003).

Figure 22.2 and our deduction of the relativity of simultaneity come from

D.F. Comstock, "The principle of relativity", *Science* **31**, 767 (1910).

The physics of the Global Positioning System is reviewed in

N. Ashby and J.J. Spilker, Jr., "Introduction to relativistic effects on the global positioning system", in *Global Positioning System: Theory and Applications*, edited by B.W. Parkinson and J.J. Spilker, Jr. (American Institute of Aeronautics and Astronautics, Washington, DC, 1995).

Section 22.4 A critical appraisal of different derivations of the Lorentz transformation appears in

J.R. Lucas and P.E. Hodgson, *Spacetime and Electromagnetism* (Clarendon Press, Oxford, 1990).

The source of the length-contraction estimate in the footnote at the end of Section 22.4.2 is

T. P. Wangler, *RF Linear Accelerators*, 2nd edition (Wiley-VCH, Weinheim, Germany, 2008), Chapter 7.

Prof. Steffen Bass (Duke University) kindly provided the prose for Application 22.2. His associate, Dr. Hannah Petersen, provided the simulation results shown in Figure 22.4. The MADAI collaboration develops visualization tools for very large scale simulations. For an overview of the physics, see

M.J. Tannenbaum, "Recent results in relativistic heavy ion collisions: From a new state of matter to the perfect fluid", *Reports on Progress in Physics* **69**, 2005 (2006).

A fascinating topic we do not treat is discussed with great clarity in

V.F. Weisskopf, "The visual appearance of rapidly moving objects", *Physics Today* **13**(9), 24 (1960).

Section 22.5 *Lorentz et al.* includes the 1908 article where Minkowski introduced the idea of four-dimensional space-time. *Pauli* is the English translation of the still-authoritative encyclopedia article he wrote in 1921 at the age of 21.

H.A. Lorentz, A. Einstein, H. Minkowski, and H. Weyl, *The Principle of Relativity* (Dover, New York, 1952).

W. Pauli, *Theory of Relativity* (Dover, New York, 1958).

Authors who endorse the use of *ict* when treating special relativity only include

W.H. Furry, "Examples of momentum distributions in the electromagnetic field and in matter", *American Journal of Physics* **37**, 621 (1969).

M. Veltman, *Diagrammatica: The Path to Feynman Rules* (Cambridge University Press, Cambridge, 1994), Chapter 1 and Appendix F.

G. 't Hooft, *Introduction to General Relativity* (Rinton Press, Princeton, N.J., 2001).

Application 22.3 was taken from

S. Acharya and A.C. Saxena, "The exact solution of the relativistic equation of motion of a charged particle driven by an elliptically polarized electromagnetic wave", *IEEE Transactions on Plasma Science* **21**, 257 (1993).

Section 22.6 A particularly thorough discussion of relativistic electrodynamics at the advanced undergraduate level is

W.G.V. Rosser, *Classical Electromagnetism via Relativity* (Plenum, New York, 1968).

Examples 22.2, 22.3, and 22.4 come from *Weinberg*, *Fisher* and *Hnizdo*, and *Becker*, respectively. *Leonhardt and Piwnicki* provide a novel perspective on relativistic effects in optics.

S. Weinberg, *Gravitation and Cosmology* (Wiley, New York, 1972), Chapter 2.

G. P. Fisher, "The electric dipole moment of a moving magnetic dipole", *American Journal of Physics* **39**, 1528 (1971).

V. Hnizdo, "Magnetic dipole moment of a moving electric dipole", *American Journal of Physics* **80**, 645 (2012).

R. Becker, *Electromagnetic Fields and Interactions* (Dover, New York, 1982), Volume I, Section 84.

U. Leonhardt and P. Piwnicki, "Light in moving media", *Contemporary Physics* **41**, 308 (2000).

Section 22.7 *Konopinski* (see Section 22.1 above) was the main source for this section. A recent textbook which introduces covariant methods early and uses them often is

C. A. Brau, *Modern Problems in Classical Electrodynamics* (University Press, Oxford, 2004).

Section 22.7.4 was adapted from *Møller* (see Section 22.1 above) and *Pauli* (see Section 22.5 above). We do not treat the angular momentum of particles. A good discussion of this topic and the related issues of Thomas precession and a covariant equation for spin precession can be found in *Jackson* (see Section 22.1 above).

Section 22.8 Our discussion of Minkowski's theory of matter in uniform motion benefitted from the treatment in *Landau and Lifshitz* (see Section 22.1 above). The transformation laws needed to derive the matching conditions at a moving interface may be found in *Van Bladel*. The title of this excellent book should not deter the reader from consulting this goldmine of problems where special relativity is used as a practical tool. Example 22.6 comes from *Jefimenko*.

J. Van Bladel, *Relativity and Engineering* (Springer, Berlin, 1984).
O.D. Jefimenko, *Electricity and Magnetism* (Appleton-Century-Crofts, New York, 1966).

Application 22.5 combines elements of

C.T. Ridgely, "Applying relativistic electrodynamics to a rotating medium", *American Journal of Physics* **66**, 114 (1998).
A. Baños, Jr. and R. K. Golden, "The electrodynamic field of a rotating uniformly magnetized sphere", *Journal of Applied Physics* **23**, 1294 (1952).
F.C. Michel and H. Li, "Electrodynamics of neutron stars", *Physics Reports* **318**, 227 (1999).

Problems

22.1 Low-Velocity Limit Let Δz and Δt be the difference between the space coordinates and the time coordinates of a pair of events. Show that, for at least some pairs of events, a Lorentz transformation of these differences *does not* reduce to a Galilean transformation in the limit of very low boost speed. Are the events in question space-like or time-like?

22.2 Linearity of the Lorentz Transformation The text exploits the homogeneity of space to conclude that the Lorentz transformation must be linear. Some authors state that this conclusion also follows if we demand that uniform rectilinear motion in K corresponds to uniform rectilinear motion in K'. Show, to the contrary, that the same property is a consequence of the non-linear transformation law

$$x_i' = \frac{A_{ij}x_j + b_i}{c_j x_j + d} \qquad i = 1, 2, 3, 4.$$

Examine the points which transform to infinity and, using this information, invent a physical argument which forces $c_j = 0$. Hint: It is convenient to set $x_4 = ct$ so $v_4 = dx_4/dt = c$.

22.3 Velocity Addition Let $\mathbf{u} = d\mathbf{r}/dt$ be the velocity of a particle observed in an inertial frame K. The same quantity observed in an inertial frame K' moving with velocity $\mathbf{v}$ with respect to K is $\mathbf{u}' = d\mathbf{r}'/dt'$.

(a) Use the transformation properties of dt, $\mathbf{r}_\parallel$, and $\mathbf{r}_\perp$ directly to derive the velocity addition rule,

$$\mathbf{u}_\parallel = \frac{d\mathbf{r}_\parallel}{dt} = \frac{\mathbf{u}_\parallel' + \mathbf{v}}{1 + \dfrac{\mathbf{v}\cdot\mathbf{u}_\parallel'}{c^2}} \qquad \text{and} \qquad \mathbf{u}_\perp = \frac{d\mathbf{r}_\perp}{dt} = \frac{\mathbf{u}_\perp'}{\gamma(v)\left[1 + \dfrac{\mathbf{v}\cdot\mathbf{u}_\parallel'}{c^2}\right]}.$$

(b) Let $\mathbf{v}$ define a polar axis with polar coordinates $\mathbf{u} = (u, \theta)$ and $\mathbf{u}' = (u', \theta')$ for the particle velocities as measured in K and K'. Write the transformation laws in part (a) in the form $u = u(u', \theta')$ and $\theta = \theta(u', \theta')$.

(c) Use the results of (b) to show that $u \to c$ when $v \to c$.

22.4 Invariance of the Scalar Product Let $\vec{a} = (\mathbf{a}, a_4)$ and $\vec{b} = (\mathbf{b}, b_4)$ be two four-vectors. Show that the scalar product $\vec{a}\cdot\vec{b} = \mathbf{a}\cdot\mathbf{b} + a_4 b_4$ is a Lorentz invariant scalar. It will be convenient to write $\mathbf{a} = \mathbf{a}_\parallel + \mathbf{a}_\perp$ and similarly for $\mathbf{b}$.

22.5 The Quotient Theorem for a Four-Vector Let $\vec{A}$ be a four-vector and let (B_1, B_2, B_3, B_4) be an ordered set of four variables with unknown properties. Prove that this set constitutes a four-vector $\vec{B}$ if $A_\mu B_\mu$ is a Lorentz invariant scalar for *any* choice of $\vec{A}$.

22.6 Transformation Law for $\mathbf{E}_\perp$ Use the transformation properties of the four-vectors $\vec{A}$ and $\vec{\nabla}$ directly to prove that

$$\mathbf{E}'_\perp = \gamma[\mathbf{E} + \mathbf{v} \times \mathbf{B}]_\perp.$$

22.7 Transformation of Force A cylindrical column of electrons has uniform charge density ρ_0 and radius a.

(a) Find the force on an electron at a radius $r < a$.
(b) A moving observer sees the column as a beam of electrons, each moving with uniform speed $\mathbf{v}$. What force does this observer report is felt by an electron in the beam at a radius $r < a$?

22.8 A Charged, Current-Carrying Wire The charged particles of an infinitely long and filamentary wire produce a linear charge density λ' and a current I'. Let the vector $\mathbf{I}' = I'\hat{\mathbf{z}}$ indicate the direction of current flow.

(a) What is the current $\mathbf{I}$ and linear charge density λ measured in the laboratory frame when the wire moves with velocity $\mathbf{v} = v\hat{\mathbf{z}}$ in that frame?
(b) Show that, by an appropriate choice of inertial frame, one can reduce the magnetic field of the wire to zero *or* one can reduce the electric field of the wire to zero. In the first case, find the linear charge density responsible for the remaining electric field. In the second case, find the current responsible for the remaining magnetic field. Why is only one of these choices possible?

22.9 Poynting in the Wrong Direction? A static charge distribution generates an electric field in the frame K'. Set the distribution into motion with velocity $\mathbf{v} = v\hat{\mathbf{x}}$ as viewed from the laboratory frame K.

(a) Let $u_{\rm EM}$ be the electromagnetic energy density. Show that $\partial u_{\rm EM}/\partial t' = 0$. Use this fact to show that $\nabla \cdot \mathbf{S}_0 + \partial u_{\rm EM}/\partial t = 0$ in K where $\mathbf{S}_0 = u_{\rm EM}\mathbf{v}$.
(b) Show that the Poynting vector, $\mathbf{S} = \mathbf{E} \times \mathbf{B}/\mu_0$, is not equal to $\mathbf{S}_0$, but that $\nabla \cdot \mathbf{S} = \nabla \cdot \mathbf{S}_0$ at points in space that do not lie on the moving charge distribution.

22.10 Boost the Electromagnetic Field For some event, observer A measures $\mathbf{E} = (\alpha, 0, 0)$ and $\mathbf{B} = (\alpha/c, 0, 2\alpha/c)$ and observer B measures $\mathbf{E}' = (E'_x, \alpha, 0)$ and $\mathbf{B}' = (\alpha/c, B'_y, \alpha/c)$. Observer C moves with velocity $v\hat{\mathbf{x}}$ with respect to observer B. Find (a) the fields $\mathbf{E}'$ and $\mathbf{B}'$ measured by observer B; and (b) the fields $\mathbf{E}''$ and $\mathbf{B}''$ measured by observer C.

22.11 Covariance of the Maxwell Equations Establish the covariance of the four vacuum Maxwell equations the "hard way" by using the transformation properties of the derivatives, the fields, and the charge and current density. Let the boost be in an arbitrary direction $\boldsymbol{\beta} = \mathbf{v}/c$. Hint: Each equation does not immediately transform into itself.

22.12 Transformations of E and B

(a) Use the transformation laws for $\mathbf{E}$ and $\mathbf{B}$ to show that $\mathbf{E} \cdot \mathbf{B}$ is a Lorentz invariant.
(b) Find the boost velocity $\mathbf{v}$ from K to K' so that the field is purely electric or purely magnetic in K' if $\mathbf{E} \perp \mathbf{B}$ and $E \neq cB$ in K.
(c) Static fields $\mathbf{E} = E_0\hat{\mathbf{y}}$ and $c\mathbf{B} = E_0(\hat{\mathbf{y}}\cos\theta + \hat{\mathbf{z}}\sin\theta)$ exist in some inertial frame K. Find the boost velocities from K to K' which make $\mathbf{E}'$ either parallel or anti-parallel to $\mathbf{B}'$ in K'. For both cases, find the θ dependence of the magnitudes of $\mathbf{E}'$, $\mathbf{B}'$, and the boost velocity.

22.13 Covariant Charge and Current Density Show by explicit calculation that the formulae for the charge and current densities of a collection of point charges,

$$\rho(\mathbf{r}, t) = \sum_{k=1}^{N} q_k \delta(\mathbf{r} - \mathbf{r}_k(t)) \quad \text{and} \quad \mathbf{j}(\mathbf{r}, t) = \sum_{k=1}^{N} q_k \mathbf{v}_k(t)\, \delta(\mathbf{r} - \mathbf{r}_k(t)),$$

have exactly the same form when we boost from frame K to frame K' by a velocity $\mathbf{v}_0$.

Hint: Show first that $\delta[\mathbf{r}' - \mathbf{r}'_k(t')] = \gamma(1 - \mathbf{v}_0 \cdot \mathbf{v}_k/c^2)\delta[\mathbf{r} - \mathbf{r}_k(t)]$ by evaluating the Jacobian determinant in the volume element transformation $d^3R = |\mathbf{J}(\mathbf{R}, \mathbf{r}')|d^3r'$ where $\mathbf{R} = \mathbf{r} - \mathbf{r}_k(t)$.

22.14 A Relativistic Particle in a Constant Electric Field A point particle with charge q and mass m moves in response to a uniform electric field $\mathbf{E} = E\hat{\mathbf{z}}$. The initial energy, linear momentum, and velocity are $\mathcal{E}_0$, p_0, and $\mathbf{u}(0) = u_0\hat{\mathbf{y}}$. Find $\mathbf{r}(t)$ and show that eliminating t gives the particle trajectory

$$z = \frac{\mathcal{E}_0}{qE}\cosh\left(\frac{qEy}{cp_0}\right).$$

Check the non-relativistic limit.

22.15 A Charged Particle in Uniform Motion Revisited

(a) Boost from the laboratory frame K to the rest frame K' to find the vector potential $\mathbf{A}(\mathbf{r}, t)$ and the scalar potential $\varphi(\mathbf{r}, t)$ for a charged particle q which moves with constant velocity $\mathbf{v}$ when viewed from the laboratory.

(b) Find $\mathbf{E}$ and $\mathbf{B}$ in the laboratory frame using the potentials computed in part (a).

(c) Let $\mathbf{r} = \mathbf{r}_\perp + \mathbf{r}_\parallel$ be the decomposition of the position vector into components perpendicular and parallel to $\mathbf{v}$. In the limit when $v \to c$, show that the fields you computed in part (b) reduce to

$$\mathbf{E}(\mathbf{r}, t) = \frac{1}{2\pi\epsilon_0}\frac{\mathbf{r}_\perp}{r_\perp^2}q\delta(ct - r_\parallel) \qquad \text{and} \qquad \mathbf{B}(\mathbf{r}, t) = \frac{\mu_0}{2\pi}\frac{\mathbf{v}\times\mathbf{r}_\perp}{r_\perp^2}q\delta(ct - r_\parallel)\,.$$

Interpret your result in terms of the Lorentz contraction.

(d) Compare and contrast the fields in part (c) with the radiation fields produced by a compact time-dependent source.

22.16 The Four-Potential

(a) Use the four-vector character of $\vec{\nabla}$ and the Lorentz transformation laws for $\mathbf{B}$ and $\mathbf{E}$ to deduce that the scalar and vector potentials form a four-vector $(\mathbf{A}, i\varphi/c)$. This reverses the methodology followed in the text.

(b) Show that the spherical equipotentials of a stationary point charge distort to ellipsoids when the charge is in uniform rectilinear motion with velocity $\mathbf{v}$. Plot a cross sectional view of one equipotential on the same graph for $v = 0.3c$, $0.5c$, $0.7c$, and $0.9c$.

22.17 A Moving Current Loop

(a) A small current loop moves with constant velocity $\mathbf{v}_0$ as viewed in the laboratory frame. Find the vector potential $\mathbf{A}(\mathbf{r})$ and the scalar potential $\varphi(\mathbf{r})$ in the lab frame. It may be convenient to introduce the vector $\mathbf{R} = \mathbf{r} - \mathbf{v}_0 t$.

(b) Take the limit $v_0 \ll c$ in your formulae and deduce that the moving loop possesses both a magnetic dipole moment and an electric dipole moment.

22.18 Transformation of Dipole Moments Use the inertial-frame transformation laws for the polarization $\mathbf{P}$ and magnetization $\mathbf{M}$ to deduce the electric dipole moment $\mathbf{p}$ and magnetic dipole moment $\mathbf{m}$ observed in the laboratory for a body that moves with velocity $\mathbf{v}$. Assume that the body has non-zero values for both these moments in its own rest frame.

22.19 TE and TM Modes of a Waveguide Show that a TE (TM) waveguide mode remains a TE (TM) waveguide mode for a Lorentz boost along the propagation direction of a straight waveguide.

22.20 Stellar Aberration The position of a star in the heavens is determined by the direction of the propagation vector of the light it emits. Let inertial frame K' move with velocity $\mathbf{v}$ with respect to K. If $\mathbf{k}\cdot\mathbf{v} = kv\cos\theta$

and $\mathbf{k}' \cdot \mathbf{v} = k'v\cos\theta'$, show that

$$\tan\theta = \frac{\sin\theta'}{\gamma(\cos\theta' + \beta)}.$$

22.21 Reflection from a Rotating Mirror Surprisingly good astronomical mirrors can be constructed from a *rotating* dish of liquid mercury. Consider a ray of light incident on such a mirror at an arbitrary angle. Treat the mirror surface as locally flat but moving with a linear velocity $\mathbf{v}$ at the point of reflection. Show that, when observed in the lab frame, the reflected ray suffers no Doppler shift of its wavelength and no Doppler aberration of its angle of reflection.

22.22 Reflection from a Moving Mirror Revisited A plane wave in vacuum with electric field $\mathbf{E}_0 = \hat{\mathbf{x}}E_0\exp[i(k_0z - \omega_0 t)]$ reflects at normal incidence from a mirror which moves at constant speed $v\hat{\mathbf{z}}$. Derive the results found in Application 22.4 for the frequency and electric field amplitude of the reflected wave using a method which never leaves the laboratory frame.

22.23 Transformation of Phase and Group Velocity

(a) Consider a wave with dispersion relation $\omega(\mathbf{k})$. Show that the group velocity $\mathbf{u} = \nabla_{\mathbf{k}}\omega$ transforms under a Lorentz transformation exactly like a particle velocity.

(b) Show that the phase velocity $\mathbf{u}_p = (\omega/k)\hat{\mathbf{k}}$ transforms under a Lorentz transformation exactly like a particle velocity when $\omega = c|\mathbf{k}|$.

22.24 The Invariance of $U_{\rm EM}/\omega$ Revisited A flat-screen photodetector continuously absorbs energy from a plane wave (frequency ω) when the wave strikes the screen (area A) at normal incidence. Use the lab frame (where the detector is at rest) and a frame moving uniformly in the direction of the wave to study the total energy $U_{\rm EM}$ absorbed by the detector in a finite time interval. Deduce thereby that the ratio $U_{\rm EM}/\omega$ is a Lorentz invariant scalar.

22.25 Conservation of Energy-Momentum The scalar $p_\mu p_\mu = -m^2c^2$ is Lorentz invariant because it is the magnitude of a four-vector. Use the covariant equation of motion for a point charge in an electromagnetic field to prove, independently, that $p_\mu p_\mu$ is a constant, independent of proper time.

22.26 Gauge Freedom and Lorentz Invariance The scalar and vector potentials satisfy the homogeneous wave equation in free space. Often, we choose $\varphi = 0$ and require that the polarization vector $\mathbf{e}$ of the vector potential be transverse to its wave vector: $\mathbf{e}\cdot\mathbf{k} = 0$. This is not a Lorentz invariant requirement. However, consider a plane wave solution $A_\nu = e_\nu\exp(ik_\sigma r_\sigma)$ of the wave equation, $\partial_\mu\partial_\mu A_\nu = 0$ which satisfies the Lorenz gauge constraint, $\partial_\mu A_\mu = 0$.

(a) Use covariant notation and show that a change of gauge does not affect the field tensor $F_{\mu\nu}$.

(b) Suppose that a Lorentz boost from A_ν to A'_ν destroys the transverse property of the polarization, i.e., $\mathbf{k}'\cdot\mathbf{A}' \neq 0$. Show that a plane wave as a gauge function is permissible and sufficient to ensure that $\tilde{A}_\nu = A'_\nu + \partial_\nu\Lambda$ is transverse and satisfies the Lorentz gauge constraint.

22.27 Covariant Properties of a Plane Wave For a monochromatic plane wave, the vectors $(\mathbf{E}, \mathbf{B}, \mathbf{k})$ form a right-handed orthogonal triad. Prove that this is a Lorentz invariant statement because $G_{\mu\nu}F_{\mu\nu}$ is a Lorentz scalar and $k_\mu F_{\mu\nu}$ is a four-vector.

22.28 A Stress-Energy Invariant Evaluate the Lorentz invariant $\Theta_{\mu\nu}\Theta_{\mu\nu}$ in an arbitrary inertial frame. Identify a type of electromagnetic field where this invariant is zero.

22.29 Diagonalize the Stress-Energy Tensor At a single space-time point, it is always possible to orient the space axes so $E_z = B_z = 0$ and $\mathbf{E}\cdot\mathbf{B} = EB\cos\theta$.

(a) Under these conditions, diagonalize $\Theta_{\mu\nu}$ and show that the two distinct eigenvalues are

$$\lambda = \pm\frac{1}{4\mu_0}\sqrt{(F_{\mu\nu}F_{\mu\nu})^2 + (F_{\mu\nu}G_{\mu\nu})^2} = \pm\frac{1}{2}\epsilon_0\sqrt{E^4 + c^4B^4 + 2E^2B^2c^2\cos(2\theta)}.$$

(b) Show that part (a) implies that the electromagnetic energy density at the space-time point in question is either zero or not less than $|\lambda|$ in every inertial frame.

22.30 Stress-Energy Tensor for Matter A set of particles has charges q_k, masses m_k, and positions $\mathbf{r}_k(t)$. Let $\vec{s} = (x, y, z, ict)$ be the space-time four-vector and define $\vec{r}_k = (x_k, y_k, z_k, ict) = (\mathbf{r}_k, ict)$. The components of the stress-energy tensor for this system are the sum of the density and current density of energy-momentum of the individual particles:

$$\Theta^{\rm mat}_{\alpha\beta}(\mathbf{s}, t) = \sum_k p_{k,\alpha} \frac{dr_{k,\beta}}{dt} \delta[\mathbf{s} - \mathbf{r}_k(t)].$$

(a) Prove that $\Theta^{\rm mat}_{\alpha\beta} = \Theta^{\rm mat}_{\beta\alpha}$.

(b) Prove that $\partial_\beta \Theta^{\rm mat}_{\alpha\beta} = j_\nu F_{\alpha\nu}$. This divergence is the negative of the divergence of the electromagnetic stress-energy tensor. Therefore, $\nabla \cdot (\boldsymbol{\Theta} + \boldsymbol{\Theta}^{\rm mat}) = 0$. Hint: Begin with the space divergence $\partial_i \Theta^{\rm mat}_{\alpha i}$.

23 Fields from Moving Charges

I have been looking for a tolerably simple way of expressing the radiation at a distance from an electron.
Oliver Heaviside (1904)

23.1 Introduction

The electromagnetic fields produced by point charges in motion play some role in practically every sub-discipline of physics. The key issues are not new because retardation and radiation were the main subjects of Chapter 20. Indeed, all the topics studied in this chapter could have been treated immediately after Section 20.3.4 when we wrote down the retarded integrals for the electromagnetic potentials in the Lorenz gauge. The value added by delaying the discussion until now is that the methods and insights of special relativity simplify calculations and help build intuition.

The first section below derives the potentials and fields produced by a point charge that moves along a specified trajectory. Subsequent sections look into the details for simple trajectories with and without particle acceleration. We will be particulary interested in the changes that occur when the particle speed increases from non-relativistic to ultra-relativistic values. The experimentally important frequency spectrum of emitted power emerges when we Fourier analyze the time dependence of the emitted fields. The emission of radiation implies energy loss by a moving particle and thus some perturbation of its trajectory. We treat this problem using the concept of *radiation reaction*. The chapter concludes with a brief introduction to Cherenkov radiation.

23.2 The Liénard-Wiechert Problem

Figure 23.1 shows the trajectory $\mathbf{r}_0(t)$ of a point charge q. The instantaneous velocity of the charge is $\mathbf{v}(t) = d\mathbf{r}_0(t)/dt$. Our task is to compute the exact electromagnetic fields associated with this moving charge. Following the original approach of Liénard and Wiechert (circa 1900), we calculate the electromagnetic potentials first and then pass to the fields using

$$\mathbf{E} = -\nabla\varphi - \frac{\partial \mathbf{A}}{\partial t} \qquad \text{and} \qquad \mathbf{B} = \nabla \times \mathbf{A}. \tag{23.1}$$

The potentials are sufficiently important that we discuss them from three points of view: an explicit evaluation of retarded integrals, a heuristic argument, and a derivation based on relativistic covariance.

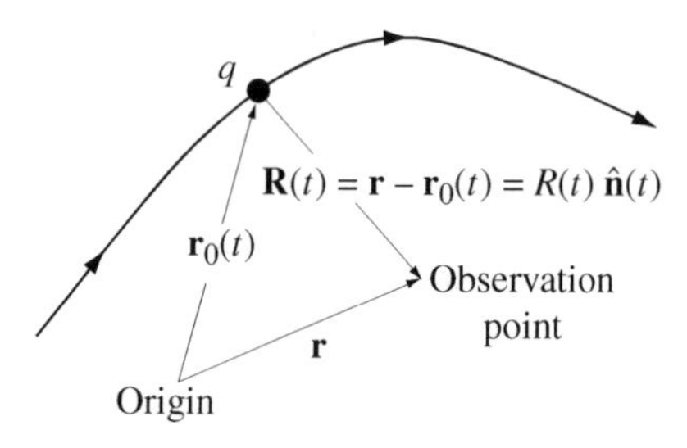

Figure 23.1: The trajectory $\mathbf{r}_0(t)$ of a point charge q.

23.2.1 The Liénard-Wiechert Potentials

The retarded electromagnetic potentials in the Lorenz gauge are (see Section 20.3.4)

$$\varphi(\mathbf{r}, t) = \frac{1}{4\pi\epsilon_0}\int d^3r' \,\frac{\rho(\mathbf{r}', t - |\mathbf{r} - \mathbf{r}'|/c)}{|\mathbf{r} - \mathbf{r}'|} \tag{23.2}$$

and

$$\mathbf{A}(\mathbf{r}, t) = \frac{\mu_0}{4\pi}\int d^3r' \,\frac{\mathbf{j}(\mathbf{r}', t - |\mathbf{r} - \mathbf{r}'|/c)}{|\mathbf{r} - \mathbf{r}'|}. \tag{23.3}$$

The charge density and current density of the moving point charge in Figure 23.1 are

$$\rho(\mathbf{r}, t) = q\delta(\mathbf{r} - \mathbf{r}_0(t)) \qquad \text{and} \qquad \mathbf{j}(\mathbf{r}, t) = q\mathbf{v}(t)\delta(\mathbf{r} - \mathbf{r}_0(t)). \tag{23.4}$$

We will also need the length $R(t)$ and a unit vector $\hat{\mathbf{n}}(t)$ defined by

$$\mathbf{R}(t) = \mathbf{r} - \mathbf{r}_0(t) = R(t)\hat{\mathbf{n}}(t). \tag{23.5}$$

Substituting (23.4) into (23.2) and (23.3) produces some challenging integrals. To make progress, we focus on the scalar potential and temporarily complicate (23.2) by introducing an integration over a dummy variable t' and a delta function to enforce the time-retardation of the charge density:

$$\varphi(\mathbf{r}, t) = \frac{1}{4\pi\epsilon_0}\int d^3r' \int dt' \frac{\rho(\mathbf{r}', t')}{|\mathbf{r} - \mathbf{r}'|}\delta(t' - t + |\mathbf{r} - \mathbf{r}'|/c). \tag{23.6}$$

Inserting the charge density in (23.4) into (23.6) and performing the space integral leaves

$$\varphi(\mathbf{r}, t) = \frac{q}{4\pi\epsilon_0}\int dt' \frac{\delta(t' - t + R(t')/c)}{R(t')}. \tag{23.7}$$

We evaluate (23.7) using the identity (see Section 1.5.1)

$$\delta[g(x)] = \sum_n \frac{1}{|g'(x_n)|}\delta(x - x_n) \qquad \text{where} \qquad g(x_n) = 0,\;\; g'(x_n) \neq 0. \tag{23.8}$$

When the particle speed $v < c$, the argument of the delta function in (23.7) is zero at one unique time. This is the *retarded time*, $t_{\rm ret}$, defined by

$$t_{\rm ret} = t - \frac{R(t_{\rm ret})}{c}. \tag{23.9}$$

If $\boldsymbol{\beta} = \mathbf{v}/c$, the derivative that appears in the denominator in (23.8) is the Doppler-like factor

$$g(t') = \frac{d}{dt'}\left[t' - t + R(t')/c\right] = 1 + \frac{1}{c}\frac{d}{dt'}\sqrt{\mathbf{R}(t') \cdot \mathbf{R}(t')} = 1 - \boldsymbol{\beta}(t') \cdot \hat{\mathbf{n}}(t') > 0. \tag{23.10}$$

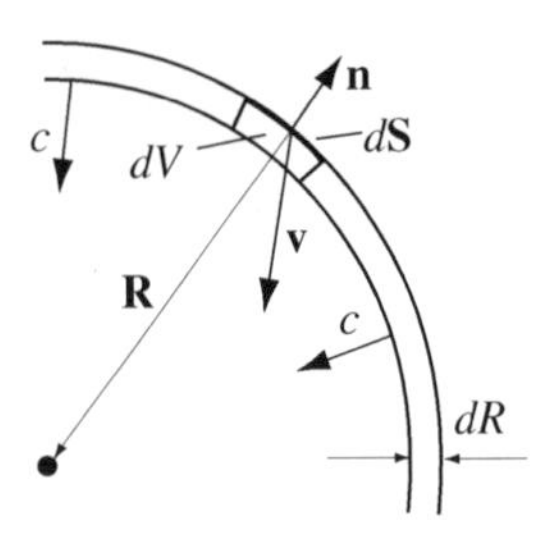

Figure 23.2: Concentric circular arcs are snapshots of an information-collecting shell as it collapses onto an observation point (black dot) at the speed of light. The charge in the volume V moves at the velocity $\mathbf{v}$.

This derivative is strictly positive because $v < c$. Using (23.8) and (23.10) to evaluate (23.7) gives the Liénard-Wiechert scalar potential,

$$\varphi(\mathbf{r}, t) = \frac{1}{4\pi\epsilon_0}\left[\frac{q}{|R(t)g(t)|}\right]_{t=t_{\rm ret}} = \frac{1}{4\pi\epsilon_0}\left[\frac{q}{|R - \boldsymbol{\beta}\cdot\mathbf{R}|}\right]_{\rm ret}. \tag{23.11}$$

The subscript "ret" indicates that all quantities inside the brackets are evaluated at the retarded time. An entirely similar evaluation of (23.3) using the current density in (23.4) gives the Liénard-Wiechert vector potential,

$$\mathbf{A}(\mathbf{r}, t) = \frac{\mu_0}{4\pi}\left[\frac{q\mathbf{v}(t)}{|R(t)g(t)|}\right]_{t=t_{\rm ret}} = \frac{\mu_0}{4\pi}\left[\frac{q\mathbf{v}}{|R - \boldsymbol{\beta}\cdot\mathbf{R}|}\right]_{\rm ret}. \tag{23.12}$$

23.2.2 A Heuristic Discussion

The Liénard-Wiechert potentials (23.11) and (23.12) differ in two ways from the electromagnetic potentials derived in Example 14.3 for a slowly moving point charge,

$$\varphi(\mathbf{r}, t) = \frac{1}{4\pi\epsilon_0}\frac{q}{|\mathbf{r} - \mathbf{r}_0(t)|} \qquad \text{and} \qquad \mathbf{A}(\mathbf{r}, t) = \frac{\mu_0}{4\pi}\frac{q\mathbf{v}(t)}{|\mathbf{r} - \mathbf{r}_0(t)|}. \tag{23.13}$$

The first difference is that the quasistatic potentials in (23.13) are evaluated at the present time, while the exact potentials are evaluated at the retarded time. This is straightforward to understand: electromagnetic information is transmitted at the speed of light rather than instantaneously.

The second difference between the Liénard-Wiechert potentials and the quasistatic potentials is the appearance of the factor $g(t_{\rm ret}) = 1 - \boldsymbol{\beta}(t_{\rm ret})\cdot\hat{\mathbf{n}}(t_{\rm ret})$ in the denominator of the exact potentials. This is a more subtle effect and, as suggested just before (23.10), amounts to a Doppler-like correction associated with the motion of the charge. The argument is best made using an "information-collecting shell" interpretation of the retarded scalar potential (23.2). A very similar argument applies to the retarded vector potential (23.3). The idea is that a source point $\mathbf{r}'$ contributes to the integral when it lies on the surface of a spherical shell of radius $R = |\mathbf{r} - \mathbf{r}'|$ centered at the observation point $\mathbf{r}$. However, the integrand involves the charge density at the retarded time, $t_{\rm ret} = t - |\mathbf{r} - \mathbf{r}'|/c$, rather than at the time t of observation. Therefore, the contribution to $\varphi(\mathbf{r}, t)$ from $\mathbf{r}'$ is counted correctly if the shell collapses radially at speed c and passes through $\mathbf{r}'$ at precisely $t_{\rm ret}$.

Focus now on the volume element labeled dV in Figure 23.2. If the charge distribution is stationary, the information-collecting shell encounters a charge $dq = \rho_{\rm ret}dV = \rho_{\rm ret}R^2dRd\Omega$ as it collapses a radial distance dR toward the observation point. However, the density of charge in dV is not static if the particles that carry the charge move. Qualitatively, the collapsing shell encounters a charge $dq > \rho_{\rm ret}dV$ when the particles have a net *inward* radial velocity and a charge $dq < \rho_{\rm ret}dV$ when the particles have a net *outward* radial velocity.

To be quantitative, let the charge in dV have velocity $\mathbf{v}$ and note that this motion causes a current $dI = [\rho\mathbf{v}]_{\text{ret}} \cdot d\mathbf{S}$ to flow through the outward surface element $d\mathbf{S}$ in Figure 23.2. It takes a time dR/c for the sphere to sweep through dV, so the actual charge encountered is

$$dq = \rho_{\text{ret}}dV - dI\frac{dR}{c} = \left[\rho dV - \rho\boldsymbol{\beta}\cdot\hat{\mathbf{R}}R^2dRd\Omega\right]_{\text{ret}} = \left[1 - \boldsymbol{\beta}\cdot\hat{\mathbf{R}}\right]_{\text{ret}}\rho_{\text{ret}}dV. \tag{23.14}$$

Equation (23.14) shows that the integral (23.2) can be evaluated by replacing $\rho_{\text{ret}}dV$ in the integrand by $dq/\left[1 - \boldsymbol{\beta}\cdot\hat{\mathbf{R}}\right]_{\text{ret}}$. Performing the integral for a point charge reproduces the Liénard-Wiechert scalar potential (23.11).

23.2.3 A Covariant Derivation

According to Hermann Minkowski—the inventor of the four-dimensional formulation of Einstein's special theory of relativity—the Liénard-Wiechert potentials provided "perhaps the most striking example" of the advantages afforded by his approach.[1] To see this, let $R^2 = (x - x_0)^2 + (y - y_0)^2 + (z - z_0)^2$ be the distance in (23.9) between an observer at the present space-time point $(\mathbf{r}, ict)$ and a moving point charge at the retarded space-time point $(\mathbf{r}_0, ict_{\text{ret}})$.

Just at the retarded time, perform a Lorentz transformation from the laboratory frame to a frame where the velocity $\mathbf{v}$ of the charge is momentarily zero. In that frame, the electromagnetic potentials at $\mathbf{r}(t)$ are

$$\varphi = \frac{1}{4\pi\epsilon_0}\frac{q}{R} \qquad \text{and} \qquad \mathbf{A} = 0. \tag{23.15}$$

Our goal is to find a manifestly covariant expression for the four-potential $(\mathbf{A}, i\varphi/c)$ which reduces to (23.15) in the rest frame of the charge. Evaluating A_ν in a frame where $\mathbf{v}$ is *not* zero should yield the Liénard-Wiechert potentials. For this to happen, the four-velocity of the particle, $U_\nu = \gamma(\mathbf{v})(\mathbf{v}, ic)$, must enter our expression for A_ν. Another four-vector which should enter is the difference between $(\mathbf{r}, ict)$ and $(\mathbf{r}_0, ict_{\text{ret}})$:

$$R_\nu = (\mathbf{r} - \mathbf{r}_0, ic(t - t_{\text{ret}})). \tag{23.16}$$

A crucial observation is that the two events that define R_ν are connected by a signal traveling at the speed of light. This means that the events have *null separation* (see Section 22.4.3) and R_ν has zero length. In other words,

$$R_\nu R_\nu = R^2 - c^2(t - t_{\text{ret}})^2 = 0 \quad \Rightarrow \quad R_\nu = (\mathbf{R}, iR). \tag{23.17}$$

The charge q is a relativistic invariant. Therefore, the only four-vector $A_\nu = (\mathbf{A}, i\varphi/c)$ which reproduces (23.15) in the rest frame where $U_\mu = (0, ic)$ is[2]

$$A_\nu = -\frac{q}{4\pi\epsilon_0}\frac{U_\nu}{cR_\sigma U_\sigma}. \tag{23.18}$$

Writing out the components of (23.18) gives

$$\varphi = -\frac{1}{4\pi\epsilon_0}\frac{qc}{\mathbf{v}\cdot\mathbf{R} - c^2(t - t_{\text{ret}})} \qquad \text{and} \qquad \mathbf{A} = -\frac{1}{4\pi\epsilon_0}\frac{q\mathbf{v}/c}{\mathbf{v}\cdot\mathbf{R} - c^2(t - t_{\text{ret}})}. \tag{23.19}$$

[1] See Minkowski's article, "Space and time", in H.A. Lorentz, A. Einstein, and H. Minkowski, *The Principle of Relativity* (Dover, New York, 1952), pp. 75-91.

[2] Because $R_\sigma R_\sigma = 0$, $R_\sigma U_\sigma$ is the only scalar available for the denominator of (23.17).

However, because the event $(\mathbf{r}_0, ict_{\rm ret})$ involves the retarded time, $\mathbf{v}$ and $\mathbf{R}$ must all be evaluated at that time. This information and $R = c(t - t_{\rm ret})$ reduce (23.19) to the Liénard-Wiechert expressions,

$$\varphi(\mathbf{r}, t) = \frac{1}{4\pi\epsilon_0}\left[\frac{q}{R - \boldsymbol{\beta}\cdot\mathbf{R}}\right]_{\rm ret} \quad \text{and} \quad \mathbf{A}(\mathbf{r}, t) = \frac{\mu_0}{4\pi}\left[\frac{q\mathbf{v}}{R - \boldsymbol{\beta}\cdot\mathbf{R}}\right]_{\rm ret}. \tag{23.20}$$

The reader may wonder why the steps used here to derive (23.20) are valid when the particle velocity is not uniform. The answer is that the transformation and evaluation are performed instantaneously and apply only at the moment when the particle has the non-zero velocity of interest. In other words, the calculation in this section exploits a sequence of independent Lorentz transformations, each performed at a different point and at a different time along the particle trajectory.

23.2.4 The Liénard-Wiechert Fields

Evaluating (23.1) with the Liénard-Wiechert potentials generates the electric and magnetic fields produced by a point charge with a specified trajectory $\mathbf{r}_0(t)$. This can be done using (23.20), but the calculation is simpler if we use the scalar potential (23.7) and the corresponding integral for the vector potential. The electric field is

$$\mathbf{E}(\mathbf{r}, t) = -\frac{q}{4\pi\epsilon_0}\nabla\int dt'\,\frac{\delta(t' - t + R(t')/c)}{R(t')} - \frac{\mu_0 q}{4\pi}\frac{\partial}{\partial t}\int dt'\,\frac{\mathbf{v}(t')\delta(t' - t + R(t')/c)}{R(t')}. \tag{23.21}$$

The definition of R in (23.5) shows that $\nabla R = \hat{\mathbf{n}}$ and $\nabla(1/R) = -\mathbf{R}/R^3$. The first of these, and the chain rule, give

$$\nabla\delta(t' - t + R(t')/c) = -\frac{\partial}{\partial t}\delta(t' - t + R(t')/c)\frac{\hat{\mathbf{n}}}{c}. \tag{23.22}$$

Therefore,

$$\mathbf{E}(\mathbf{r}, t) = \frac{q}{4\pi\epsilon_0}\left[\int dt'\delta(t' - t + R(t')/c)\frac{\hat{\mathbf{n}}}{R^2} + \frac{\partial}{\partial t}\int dt'\delta(t' - t + R(t')/c)\frac{\hat{\mathbf{n}} - \boldsymbol{\beta}}{cR}\right]. \tag{23.23}$$

Both integrals in (23.23) are like the integral (23.7). Therefore, the logic which led to (23.11) applies here as well. We conclude from this that

$$\mathbf{E}(\mathbf{r}, t) = \frac{q}{4\pi\epsilon_0}\left[\frac{\hat{\mathbf{n}}}{gR^2}\right]_{\rm ret} + \frac{q}{4\pi\epsilon_0}\frac{d}{dt}\left[\frac{\hat{\mathbf{n}} - \boldsymbol{\beta}}{gcR}\right]_{\rm ret}, \tag{23.24}$$

where $g = 1 - \boldsymbol{\beta}\cdot\hat{\mathbf{n}}$ is the Doppler factor defined in (23.10) and discussed in Section 23.2.2. A very similar calculation gives the corresponding magnetic field as

$$\mathbf{B}(\mathbf{r}, t) = \frac{\mu_0 q}{4\pi}\left[\frac{\mathbf{v}\times\hat{\mathbf{n}}}{gR^2}\right]_{\rm ret} + \frac{\mu_0 q}{4\pi}\frac{d}{dt}\left[\frac{\mathbf{v}\times\hat{\mathbf{n}}}{gcR}\right]_{\rm ret}. \tag{23.25}$$

The fields (23.24) and (23.25) are exact, but not particularly useful in the form given. To make progress, we need a few derivative relations. First, combine (23.9) with [see (23.10)]

$$\frac{dR}{dt} = -\hat{\mathbf{n}}\cdot c\boldsymbol{\beta} \tag{23.26}$$

to get

$$\frac{dt}{dt_{\rm ret}} = 1 + \frac{1}{c}\frac{dR_{\rm ret}}{dt_{\rm ret}} = [1 - \boldsymbol{\beta}\cdot\hat{\mathbf{n}}]_{\rm ret} = g_{\rm ret}. \tag{23.27}$$

Second, differentiate $\hat{\mathbf{n}} = \mathbf{R}/R$ directly and use (23.26) to deduce that

$$\frac{d\hat{\mathbf{n}}}{dt} = \frac{c}{R}\hat{\mathbf{n}}\times(\hat{\mathbf{n}}\times\boldsymbol{\beta}). \tag{23.28}$$

Finally, use the definition in (23.27) to get

$$\frac{dg}{dt} = -\left[\frac{d\hat{\mathbf{n}}}{dt}\cdot\boldsymbol{\beta} + \hat{\mathbf{n}}\cdot\frac{d\boldsymbol{\beta}}{dt}\right]. \tag{23.29}$$

Return now to the electric field (23.24) and use (23.27) to replace d/dt by $d/dt_{\rm ret}$. All variables are now retarded, and carrying out the indicated derivatives produces

$$\mathbf{E}(\mathbf{r},t) = \frac{q}{4\pi\epsilon_0}\left[\frac{\hat{\mathbf{n}}}{gR^2} + \frac{1}{cgR}\left(\frac{d\hat{\mathbf{n}}}{dt} - \frac{d\boldsymbol{\beta}}{dt}\right) - \frac{(\hat{\mathbf{n}}-\boldsymbol{\beta})}{cg^2R^2}\left(\frac{dg}{dt}R + g\frac{dR}{dt}\right)\right]_{\rm ret}. \tag{23.30}$$

Apart from the dimensionless acceleration $\dot{\boldsymbol{\beta}} = d\boldsymbol{\beta}/dt$, the time derivatives in (23.30) are given by (23.26), (23.28), and (23.29). Using these, only a bit of algebra is needed to find the *Liénard-Wiechert electric field*:

$$\mathbf{E} = \frac{q}{4\pi\epsilon_0}\left[\frac{(\hat{\mathbf{n}}-\boldsymbol{\beta})(1-\beta^2)}{g^3R^2} + \frac{\hat{\mathbf{n}}\times\left\{(\hat{\mathbf{n}}-\boldsymbol{\beta})\times\dot{\boldsymbol{\beta}}\right\}}{cg^3R}\right]_{\rm ret} = \mathbf{E}_{\rm v} + \mathbf{E}_{\rm a}. \tag{23.31}$$

The second equality in (23.31) decomposes the electric field into a velocity field, $\mathbf{E}_{\rm v}$, and an acceleration field, $\mathbf{E}_a$. The velocity field does not contain $\dot{\boldsymbol{\beta}}$ and varies in space as $1/R^2$. This field remains "attached" to the source charge in the sense of the near field in Figure 20.5. The acceleration field goes to zero as $\dot{\boldsymbol{\beta}} \to 0$ and varies in space as $1/R$. This field propagates to infinity (see below).

The reader can carry out very similar steps beginning with (23.25) and confirm that the Liénard-Wiechert magnetic field is

$$\mathbf{B} = \frac{\mu_0 q}{4\pi}\left[\frac{(\mathbf{v}\times\hat{\mathbf{n}})(1-\beta^2)}{g^3R^2} + \frac{(\boldsymbol{\beta}\times\hat{\mathbf{n}})(\dot{\boldsymbol{\beta}}\cdot\hat{\mathbf{n}}) + g\dot{\boldsymbol{\beta}}\times\hat{\mathbf{n}}}{g^3R}\right]_{\rm ret} = \mathbf{B}_{\rm v} + \mathbf{B}_{\rm a}. \tag{23.32}$$

Like the electric field, (23.32) decomposes naturally into a velocity magnetic field, $\mathbf{B}_{\rm v}$, and an acceleration magnetic field, $\mathbf{B}_{\rm a}$. Comparing (23.31) to (23.32) demonstrates that

$$c\mathbf{B}_{\rm v} = \hat{\mathbf{n}}_{\rm ret}\times\mathbf{E}_{\rm v} \qquad \text{and} \qquad c\mathbf{B}_{\rm a} = \hat{\mathbf{n}}_{\rm ret}\times\mathbf{E}_{\rm a}. \tag{23.33}$$

Adding the two equations in (23.33) shows that the total Liénard-Wiechert fields satisfy

$$c\mathbf{B} = \hat{\mathbf{n}}_{\rm ret}\times\mathbf{E}. \tag{23.34}$$

We deduce from (23.33) that $\mathbf{B}\cdot\mathbf{E} = 0$, $\mathbf{B}\cdot\hat{\mathbf{n}}_{\rm ret} = 0$, and $c|\mathbf{B}| \le |\mathbf{E}|$. These relations hold for the total fields and (separately) for the velocity and acceleration fields. On the other hand, it is straightforward to check that $\mathbf{E}_{\rm v}\cdot\hat{\mathbf{n}}_{\rm ret} \neq 0$. We leave it as an exercise for the reader to show that the separation of the Liénard-Wiechert electromagnetic field into a velocity part and an acceleration part is relativistically invariant.

The acceleration field is particularly important because $(\hat{\mathbf{n}}_{\rm ret}, \mathbf{E}_{\rm a}, \mathbf{B}_{\rm a})$ is an orthogonal triad of vectors with $c|\mathbf{B}| = |\mathbf{E}|$:

$$\mathbf{E}_{\rm a} = c\mathbf{B}_{\rm a}\times\hat{\mathbf{n}}_{\rm ret} \qquad \mathbf{E}_{\rm a}\cdot\hat{\mathbf{n}}_{\rm ret} = 0 \qquad \mathbf{B}_{\rm a}\cdot\hat{\mathbf{n}}_{\rm ret} = 0. \tag{23.35}$$

Combining (23.35) with the $1/R$ spatial variation of $\mathbf{E}_{\rm a}$ and $\mathbf{B}_{\rm a}$ shows that the acceleration field of a moving charged particle has all the characteristics of a radiation field (see Section 20.5). This is consistent with our earlier conclusion (Section 20.5.5) that particle acceleration is necessary (but not sufficient) to produce radiation.

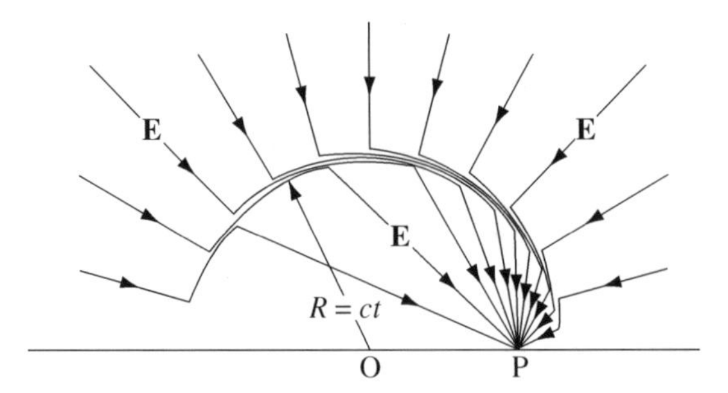

Figure 23.3: Electric field lines for a point (negative) charge that is abruptly accelerated from rest (at point O) to at state of uniform speed at point P. Figure from Purcell (1965).

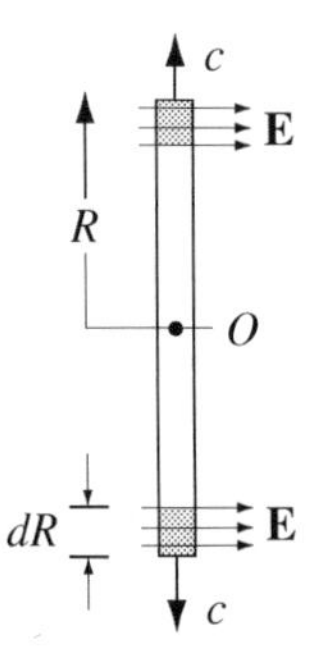

Figure 23.4: Side view of an annular ring whose radius R expands at the speed of light. The annulus of thickness dR (shaded) collects flux from the part of the Liénard-Wiechert electric field in Figure 23.3 that is transverse to the radius vector $\mathbf{R}$.

Interesting qualitative physics lies behind (23.35) and the observation that an accelerating charge generates electromagnetic field vectors with components that are *transverse* to $\hat{\mathbf{n}}_{\rm ret}$. To understand this, we follow J.J. Thomson (the discover of the electron) and plot the electric field produced by a point charge which undergoes a short burst of acceleration. Figure 23.3 shows the field lines of $\mathbf{E}$ for a point charge which was at rest at point O until it was abruptly accelerated to the velocity $\beta = 0.2$ at time $t = 0$. Thereafter, the charge moved at constant velocity and reached the point P at time t.

Everywhere *outside* a sphere of radius $R = t/c$ centered at O, the field lines point radially inward to O. This is expected because the "news" of the abrupt acceleration has not yet reached these distant points. The field in this outer region comes from the velocity term in (23.31). Everywhere *inside* the same sphere, the field lines point directly toward the charge. This is also expected because a point charge moving at constant speed drags a Coulomb-like field along with it (see Section 20.1). This is also a velocity field. However, because all field lines are continuous, the two patterns must knit together in the thin spherical shell that separates them. The field in this region is, by necessity, nearly tangential to the shell and comes primarily from the acceleration term in (23.31). This agrees with the middle equation in (23.35) because the unit vector $\hat{\mathbf{n}}_{\rm ret}$ points radially outward from O. A similar graphical analysis can be carried out for the magnetic field.[3]

The R-dependence of the acceleration electric field in (23.31) can be understood using the expanding annulus shown in side view in Figure 23.4. The annulus is centered at the point O in Figure 23.3 and is concentric with the particle trajectory. The radius R of the annulus increases at the speed

[3] See Ohanian (1980) in Sources, References, and Further Reading.

of light and the annular width dR is large enough to capture all the flux from the field lines in Figure 23.3 that are transverse to the vector $\mathbf{R}$. This group of field lines also moves outward at the speed of light. The total electric flux captured by the ring is $E_a 2\pi R dR$ and, by construction, this value does not change as R increases. Therefore, $E_a(R) \propto 1/R$. We conclude that an impulsive acceleration event causes a charged particle to launch a pulse of radiation (a kink in its electric field line pattern) which propagates away from the retarded position of the charge at the speed of light. The reader can compare this with the time-domain analysis of antenna radiation presented in Section 20.6.2.

Example 23.1 A linearly polarized, time-harmonic plane wave propagating in the $+z$-direction with polarization $\hat{\boldsymbol{\epsilon}}_i$ and electric field $\mathbf{E}_i$ scatters from a moving electron. This Thomson scattering event (see Section 21.3) leaves the electron with velocity $\boldsymbol{\beta}$ and acceleration $\dot{\boldsymbol{\beta}}$. A convenient measure of the polarization of the scattered field is

$$p(\hat{\boldsymbol{\epsilon}}_i, \hat{\boldsymbol{\epsilon}}_f) = \frac{|\hat{\boldsymbol{\epsilon}}_f \cdot \mathbf{E}_f|^2}{|\hat{\boldsymbol{\epsilon}}_i \cdot \mathbf{E}_f|^2}.$$

The Liénard-Wiechert fields and the relativistic equations of motion for a point charge make it possible to write p in a form which does not explicitly contain the acceleration $\dot{\boldsymbol{\beta}}$. Consider backscattering only and find p as a function of $\boldsymbol{\beta} = (\beta, \theta, \phi)$ when $\hat{\boldsymbol{\epsilon}}_i = \hat{\mathbf{x}}$ and $\hat{\boldsymbol{\epsilon}}_f = \hat{\mathbf{y}}$. Use this result to identify situations when $p = 0$ and when $p \gg 1$. Polarization analysis of this kind is used in diagnostic studies of relativistic electron beams.

Solution: The backscattered electric field $\mathbf{E}_f$ in the far zone is the acceleration electric field in (23.31) with $\hat{\mathbf{n}}_{\rm ret} \approx -\hat{\mathbf{z}}$. To eliminate $\dot{\boldsymbol{\beta}}$ from $\mathbf{E}_f$, we write the relativistic equations of motion for a particle with charge q (Section 22.5.2):

$$\frac{d}{dt}(\gamma m \mathbf{v}) = q(\mathbf{E}_i + \mathbf{v} \times \mathbf{B}_i) \qquad \text{and} \qquad \frac{d}{dt}(\gamma m c^2) = q\mathbf{E}_i \cdot \mathbf{v}.$$

Combining these two with the incoming plane wave relation, $c\mathbf{B}_i = \hat{\mathbf{z}} \times \mathbf{E}_i$, leads without difficulty to

$$\dot{\boldsymbol{\beta}} = \frac{q}{\gamma m c}\left[(1 - \boldsymbol{\beta} \cdot \hat{\mathbf{z}})\mathbf{E}_i + (\boldsymbol{\beta} \cdot \mathbf{E}_i)(\hat{\mathbf{z}} - \boldsymbol{\beta})\right].$$

Using this formula for $\dot{\boldsymbol{\beta}}$ and $g = [1 - \hat{\mathbf{n}} \cdot \boldsymbol{\beta}]_{\rm ret} \approx 1 + \beta_z$, the projection of the backscattered $\mathbf{E}_f$ onto any unit vector $\hat{\boldsymbol{\epsilon}} \perp \hat{\mathbf{z}}$ is

$$\hat{\boldsymbol{\epsilon}} \cdot \mathbf{E}_f = \frac{\mu_0 q^2}{4\pi \gamma m g^3 R}\left[2(\boldsymbol{\beta} \cdot \mathbf{E}_i)(\hat{\boldsymbol{\epsilon}} \cdot \boldsymbol{\beta}) + (\beta_z^2 - 1)(\hat{\boldsymbol{\epsilon}} \cdot \mathbf{E}_i)\right].$$

Therefore,

$$p(\hat{\mathbf{x}}, \hat{\mathbf{y}}) = \frac{|\hat{\mathbf{y}} \cdot \mathbf{E}_f|^2}{|\hat{\mathbf{x}} \cdot \mathbf{E}_f|^2} = \left(\frac{2\beta^2 \sin^2\theta \sin\phi \cos\phi}{2\beta^2 \sin^2\theta \cos^2\phi + \beta^2 \cos^2\theta - 1}\right)^2.$$

This formula says that the polarization of the backscattered radiation is the same as the polarization of the incident radiation if the scattering event leaves the charge at rest ($\beta = 0$) or if $\boldsymbol{\beta}$ is aligned exactly with the x-, y-, or z-axes. Conversely, because

$$\lim_{\beta \to 1} p = \tan^2(2\phi),$$

the field backscattered by an ultra-relativistic ($v \approx c$) electron is nearly cross-polarized with respect to the incident field when $\mathbf{v} = (v_x, \pm v_x, 0)$.

Application 23.1 The Fields of a Point Charge in Uniform Motion II

The electromagnetic fields produced by a point charge with constant velocity were calculated in Application 20.1 (by solving the inhomogeneous wave equation) and again in Section 22.6.4 (by Lorentz transformation of the field of a static point charge). This situation produces no radiation, but the phenomenon of retardation did not play an obvious role in either calculation. We shed light on this issue here by transforming the retarded-time formulae (23.31) and (23.33) into observation-time formulae for the special case when $\dot{\boldsymbol{\beta}} = 0$.

The point charge q in Figure 23.5 moves uniformly with velocity $\mathbf{v} = c\boldsymbol{\beta}$. The black dot labels the position of the charge at the time t when the fields are observed at the point C. The point labeled A is the position of the charge at the retarded time, $t_{\rm ret}$. We begin with the electric field and, because $\dot{\boldsymbol{\beta}} = 0$, only the first (velocity) term in (23.31) requires our attention:

$$\mathbf{E}_{\rm v} = \frac{q}{4\pi\epsilon_0}\left[\frac{(\hat{\mathbf{n}} - \boldsymbol{\beta})(1-\beta^2)}{g^3 R^2}\right]_{\rm ret}. \tag{23.36}$$

Our first task is to prove that

$$[\hat{\mathbf{n}} - \boldsymbol{\beta}]_{\rm ret} = \frac{\mathbf{R}}{R_{\rm ret}}. \tag{23.37}$$

This requires the implicit equation (23.9) for the retarded time, which we use to compute the vector $\Delta\mathbf{s}$ which points from the retarded-time position of the charge to the observation-time position in Figure 23.5:

$$\Delta\mathbf{s} = \mathbf{v}(t - t_{\rm ret}) = \boldsymbol{\beta} R_{\rm ret}. \tag{23.38}$$

On the other hand, the geometry of Figure 23.5 shows that

$$\boldsymbol{\beta} R_{\rm ret} + \mathbf{R} = \mathbf{R}_{\rm ret} = R_{\rm ret}\hat{\mathbf{n}}_{\rm ret}. \tag{23.39}$$

Combining (23.38) with (23.39) gives the advertised formula (23.37). Inserting the latter into (23.36) gives

$$\mathbf{E}_{\rm v}(\mathbf{r}, t) = \frac{q}{4\pi\epsilon_0}\frac{\mathbf{R}(1-\beta^2)}{[gR]^3_{\rm ret}}. \tag{23.40}$$

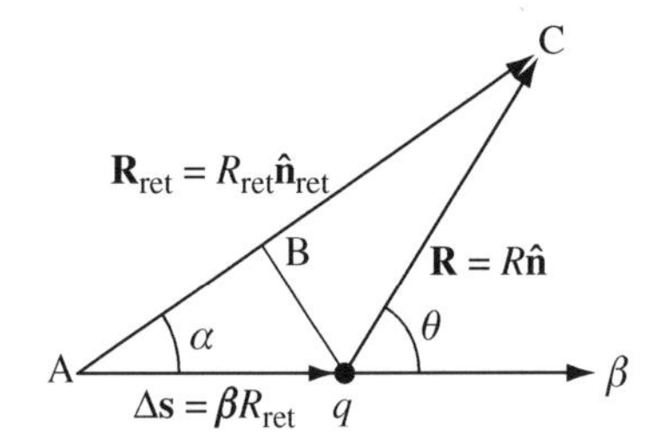

Figure 23.5: A point charge q with uniform velocity $\mathbf{v} = c\boldsymbol{\beta}$. The location of the charge at the retarded time is labeled A. The observation point is labeled C.

To make further progress, we note that the distance $\overline{AB} = \beta(\mathbf{R}_{\rm ret} \cdot \hat{\mathbf{n}}_{\rm ret})$ is the projection of the vector $\beta\mathbf{R}_{\rm ret}$ onto the direction $\hat{\mathbf{n}}_{\rm ret}$. In that case, the definition of $g_{\rm ret}$ from (23.27) tells us that the distance

$$\overline{BC} = R_{\rm ret}(1 - \hat{\mathbf{n}}_{\rm ret} \cdot \boldsymbol{\beta}) = [gR]_{\rm ret}. \tag{23.41}$$

Finally, the geometry of Figure 23.5 shows that

$$\overline{BC}^2 + \beta^2 R^2_{\rm ret}\sin^2\alpha = R^2 \qquad \text{and} \qquad R_{\rm ret}\sin\alpha = R\sin\theta. \tag{23.42}$$

Together, (23.41) and (23.42) give $[gR] = R(1-\beta^2\sin^2\theta)^{1/2}$. Substituting this into (23.40) gives the desired final result,

$$\mathbf{E}_v(\mathbf{r},t) = \frac{q}{4\pi\epsilon_0}\frac{\mathbf{R}}{R^3}\frac{1-\beta^2}{(1-\beta^2\sin^2\theta)^{3/2}}. \tag{23.43}$$

This is the electric field we found in our earlier treatments of this problem. Using (23.37) to evaluate (23.33) gives a magnetic field expression we have also seen before:

$$\mathbf{B}_v = \frac{\hat{\mathbf{n}}_{\rm ret}}{c}\times\mathbf{E}_v = \frac{1}{c}\left[\boldsymbol{\beta}+\frac{\mathbf{R}}{R_{\rm ret}}\right]\times\mathbf{E}_v = \frac{\mathbf{v}}{c^2}\times\mathbf{E}_v. \tag{23.44}$$

23.2.5 The Heaviside-Feynman Fields

Heaviside (1904) and Feynman (1963) published formulae for the Liénard-Wiechert fields which differ entirely from (23.31) and (23.33). Three identities are needed to reproduce their results. The first follows from (23.27) and (23.9):

$$\frac{1}{g_{\rm ret}} = \frac{dt_{\rm ret}}{dt} = 1-\frac{1}{c}\frac{dR_{\rm ret}}{dt}. \tag{23.45}$$

The second uses (23.45) to replace $d/dt_{\rm ret}$ by d/dt:

$$\boldsymbol{\beta}_{\rm ret} = -\frac{1}{c}\frac{d\mathbf{R}_{\rm ret}}{dt_{\rm ret}} = -\frac{g_{\rm ret}}{c}\frac{d\mathbf{R}_{\rm ret}}{dt}. \tag{23.46}$$

Finally, (23.28) is valid using either all present-time variables or all retarded-time variables. Choose the latter but use (23.45) to replace the retarded-time derivative with a present-time derivative. This leads immediately to the third identity:

$$\hat{\mathbf{n}}_{\rm ret}\times\frac{d\hat{\mathbf{n}}_{\rm ret}}{dt} = \left[\frac{c}{gR}\boldsymbol{\beta}\times\hat{\mathbf{n}}\right]_{\rm ret}. \tag{23.47}$$

Now, return to the electric field in the form (23.24). Using (23.45) and (23.46) to eliminate the factors $[1/g]_{\rm ret}$ and $[\boldsymbol{\beta}/g]_{\rm ret}$ gives

$$\mathbf{E}(\mathbf{r},t) = \frac{q}{4\pi\epsilon_0}\left\{\frac{\hat{\mathbf{n}}_{\rm ret}}{R^2_{\rm ret}}\left(1-\frac{1}{c}\frac{dR_{\rm ret}}{dt}\right)+\frac{1}{c}\frac{d}{dt}\left[\frac{\hat{\mathbf{n}}_{\rm ret}}{R_{\rm ret}}\left(1-\frac{1}{c}\frac{dR_{\rm ret}}{dt}\right)+\frac{1}{cR_{\rm ret}}\frac{d\mathbf{R}_{\rm ret}}{dt}\right]\right\}. \tag{23.48}$$

Writing $\mathbf{R} = R\hat{\mathbf{n}}$ in the last term, the derivatives in (23.48) produce some internal cancellation and, ultimately, the *Heaviside-Feynman formula* for the electric field of a point charge with a specified trajectory:

$$\mathbf{E}(\mathbf{r},t) = \frac{q}{4\pi\epsilon_0}\left\{\left[\frac{\hat{\mathbf{n}}}{R^2}\right]_{\rm ret}+\frac{R_{\rm ret}}{c}\frac{d}{dt}\left[\frac{\hat{\mathbf{n}}}{R^2}\right]_{\rm ret}+\frac{1}{c^2}\frac{d^2\hat{\mathbf{n}}_{\rm ret}}{dt^2}\right\}. \tag{23.49}$$

For the magnetic field, we use (23.47) in both terms of (23.25) to get

$$\mathbf{B}(\mathbf{r},t) = \frac{\mu_0 q}{4\pi}\left\{\left[\frac{\hat{\mathbf{n}}}{R}\right]_{\rm ret}\times\frac{d\hat{\mathbf{n}}_{\rm ret}}{dt}+\frac{\hat{\mathbf{n}}_{\rm ret}}{c}\times\frac{d^2\hat{\mathbf{n}}_{\rm ret}}{dt^2}\right\}. \tag{23.50}$$

Comparing (23.50) to (23.49) confirms (23.33):

$$c\mathbf{B}(\mathbf{r},t) = \hat{\mathbf{n}}_{\rm ret}\times\mathbf{E}(\mathbf{r},t). \tag{23.51}$$

Heaviside and Feynman pointed out many interesting features of (23.49) and (23.50). For example, using (23.9) and (23.27) in (23.49) shows that

$$\mathbf{E}(\mathbf{r},t)=\frac{q}{4\pi\epsilon_0}\left\{\left[\frac{\hat{\mathbf{n}}}{R^2}\right]_{\rm ret}+(t-t_{\rm ret})\frac{d}{dt_{\rm ret}}\left[\frac{\hat{\mathbf{n}}}{R^2}\right]_{\rm ret}\frac{1}{g_{\rm ret}}+\frac{1}{c^2}\frac{d^2\hat{\mathbf{n}}_{\rm ret}}{dt^2}\right\}. \tag{23.52}$$

On the other hand, a Taylor series expansion gives

$$\frac{\hat{\mathbf{n}}}{R^2}(t)=\frac{\hat{\mathbf{n}}}{R^2}(t_{\rm ret}+R_{\rm ret}/c)\approx\left[\frac{\hat{\mathbf{n}}}{R^2}\right]_{\rm ret}+(t-t_{\rm ret})\frac{d}{dt_{\rm ret}}\left[\frac{\hat{\mathbf{n}}}{R^2}\right]_{\rm ret}+\cdots. \tag{23.53}$$

If the particle speed is non-relativistic (so $\beta\ll 1$ and $g_{\rm ret}\approx 1$) *and* the particle trajectory is such that the higher-order derivatives in the Taylor expansion can be neglected, comparing (23.53) to (23.52) shows that

$$\mathbf{E}(\mathbf{r},t)\approx\frac{q}{4\pi\epsilon_0}\frac{\hat{\mathbf{n}}}{R^2}+\frac{\mu_0 q}{4\pi}\frac{d^2\hat{\mathbf{n}}_{\rm ret}}{dt^2}. \tag{23.54}$$

In words, the observed electric field is approximately the quasistatic (instantaneous) Coulomb field plus a second derivative (in time). The latter must be the radiation electric field because our approximations do not preclude acceleration of the charge. We will exploit this fact to analyze synchrotron radiation in Section 23.5.

23.3 Radiation in the Time Domain

The energy radiated to infinity by a moving charged particle is determined by the acceleration fields $\mathbf{E}_{\rm a}(\mathbf{r},t)$ and $\mathbf{B}_{\rm a}(\mathbf{r},t)$ in (23.31) and (23.32). These are radiation fields because they fall off as $1/R$ and form an orthogonal triad with the retarded line-of-sight unit vector, $\hat{\mathbf{n}}_{\rm ret}$. The associated Poynting vector,

$$\mathbf{S}(t)=\frac{1}{\mu_0}\mathbf{E}_{\rm a}\times\mathbf{B}_{\rm a}=\epsilon_0 cE_{\rm a}^2\,\hat{\mathbf{n}}_{\rm ret}=\epsilon_0 c\left(\frac{q}{4\pi\epsilon_0}\right)^2\left|\frac{\hat{\mathbf{n}}\times\left[(\hat{\mathbf{n}}-\boldsymbol{\beta})\times\dot{\boldsymbol{\beta}}\right]}{cg^3R}\right|^2_{\rm ret}\hat{\mathbf{n}}_{\rm ret}, \tag{23.55}$$

determines the rate at which energy flows through a solid angle $d\Omega$ of a distant enclosing sphere of radius R:

$$\frac{dP(t)}{d\Omega}=\frac{dU}{dtd\Omega}=R^2\mathbf{S}(t)\cdot\hat{\mathbf{n}}_{\rm ret}. \tag{23.56}$$

A subtlety associated with (23.56) is that the energy detected in a time interval dt measured by a distant observer is generally not equal to the energy emitted in a time interval $dt_{\rm ret}$ measured by a nearby observer. Quantitatively, the detection rate $dU_{\rm rad}/dt$ differs from the emission rate $dU_{\rm rad}/dt_{\rm ret}$ by a factor of $dt/dt_{\rm ret}=(1-\boldsymbol{\beta}\cdot\hat{\mathbf{n}})_{\rm ret}=g_{\rm ret}$ [see (23.27)]. The presence of the factor $g_{\rm ret}$ shows that the physics is closely related to the phenomenon described in Figure 23.2 and is most pronounced for relativistic particles. The emission rate is more fundamental – it is a property of the charge itself – so we multiply (23.56) by $g_{\rm ret}$ and focus on the angular distribution of emitted power,

$$\frac{dP(t_{\rm ret})}{d\Omega}=\frac{dU}{dt_{\rm ret}d\Omega}=g_{\rm ret}R^2\mathbf{S}(t)\cdot\hat{\mathbf{n}}_{\rm ret}=\frac{q^2}{16\pi^2\epsilon_0 c}\left.\frac{\left|\hat{\mathbf{n}}\times\left[(\hat{\mathbf{n}}-\boldsymbol{\beta})\times\dot{\boldsymbol{\beta}}\right]\right|^2}{(1-\hat{\mathbf{n}}\cdot\boldsymbol{\beta})^5}\right|_{\rm ret}. \tag{23.57}$$

The leftmost and rightmost members of (23.57) are now both functions of the emission time, which we may consider the proper time measured by the particle itself. Much of the rest of this chapter is devoted to a variety of limiting cases, special cases, and applications of (23.57). We will sometimes drop the subscript "ret" in what follows, but the reader should assume that all variables are evaluated at the retarded time unless stated otherwise.

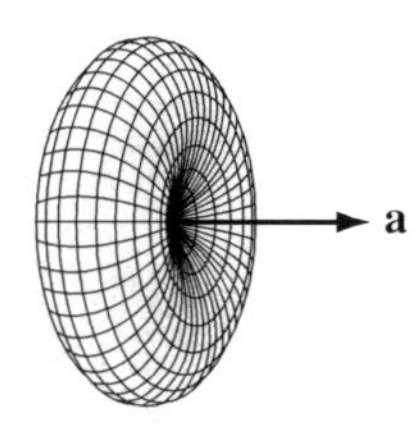

Figure 23.6: Angular distribution of radiation emitted by a slowly moving ($v \ll c$) point charge with acceleration **a**.

23.3.1 Non-Relativistic Motion

When the particle acceleration $\mathbf{a} = c\dot{\boldsymbol{\beta}}$ does not vanish, the $\boldsymbol{\beta} = 0$ limit of (23.57) is a good approximation to the angular distribution of radiation emitted by a slowly moving ($v \ll c$) point charge. We choose the position of the charge as the origin of a spherical coordinate system and align the z-axis with the instantaneous acceleration **a**. In that case, $\hat{\mathbf{n}} = \hat{\mathbf{r}}$ and (23.57) simplifies to a formula derived in Section 20.5.5:

$$\frac{dP}{d\Omega} = \frac{\mu_0 q^2}{16\pi^2 c}|\hat{\mathbf{n}} \times (\hat{\mathbf{n}} \times \mathbf{a})| = \frac{\mu_0 q^2}{16\pi^2 c}|\hat{\mathbf{r}} \times \mathbf{a}|^2 = \frac{\mu_0 q^2 a^2}{16\pi^2 c}\sin^2\theta. \tag{23.58}$$

Figure 23.6 repeats Figure 20.8 and shows the donut-shaped emission pattern predicted by (23.58). There is no radiation along $\pm\mathbf{a}$ and the emission is largest in the "broadside" direction perpendicular to **a**. Section 20.7.1 offered an interpretation of (23.58) as a dipole radiation formula.

23.3.2 Acceleration || Velocity

Equation (23.57) simplifies considerably when the acceleration **a** is parallel to the velocity **v**. The radiation pattern has azimuthal symmetry around their common direction and thus depends only on the angle θ defined by $\hat{\mathbf{n}} \cdot \boldsymbol{\beta} = (v/c)\cos\theta$. Remembering that all quantities refer to the (retarded) time of emission, the angular distribution of emitted radiation is

$$\left.\frac{dP}{d\Omega}\right|_{\parallel} = \frac{\mu_0 q^2}{16\pi^2 c}\frac{a^2\sin^2\theta}{(1-\beta\cos\theta)^5}. \tag{23.59}$$

The dependence of (23.59) on a^2 shows that the same radiation pattern occurs whether the charge accelerates or decelerates. The German word *bremsstrahlung* for "braking radiation" is often used for the deceleration case. Figure 23.7 shows that the radiated power vanishes when $\theta = 0$, and takes the same value, independent of β, when $\theta = \pm\pi/2$. Otherwise, the diagram demonstrates that the $\beta = 0$ broadside lobes progressively narrow and bend toward the forward direction as β increases.

The most interesting case is the *ultra-relativistic limit* where $\beta \approx 1$ and the denominator of (23.59) nearly vanishes when $\theta = 0$. This leads us to combine $\cos\theta \approx 1 - \frac{1}{2}\theta^2$ with

$$1 - \beta = \frac{1-\beta^2}{1+\beta} = \frac{1}{\gamma^2(1+\beta)} \approx \frac{1}{2\gamma^2} \qquad (\beta \approx 1) \tag{23.60}$$

to get

$$1 - \beta\cos\theta \approx \frac{1+\gamma^2\theta^2}{2\gamma^2} \qquad (\beta \approx 1, \theta \ll 1). \tag{23.61}$$

Substituting (23.61) in (23.59) gives the ultra-relativistic expression

$$\left.\frac{dP}{d\Omega}\right|_{\parallel} \approx \frac{2\mu_0 q^2 a^2}{\pi^2 c}\frac{\gamma^8(\gamma\theta)^2}{[1+(\gamma\theta)^2]^5} \qquad (\gamma \gg 1, \theta \ll 1). \tag{23.62}$$

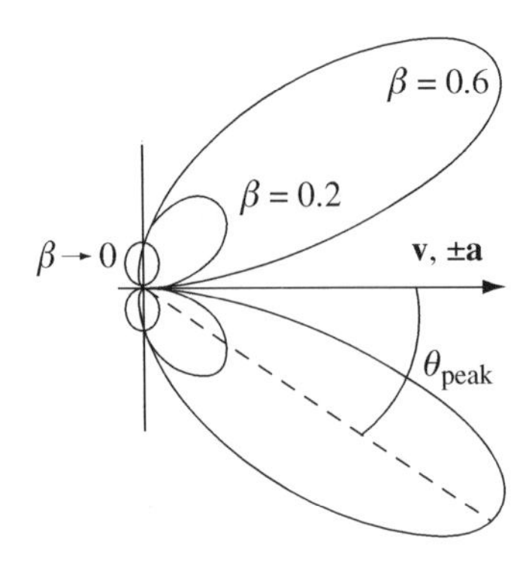

Figure 23.7: Angular distribution of radiation from a moving charge with $\mathbf{v} \parallel \mathbf{a}$ for several values of $\beta = v/c$. The diagram is rotationally symmetric around the $\mathbf{v}$ axis.

Maximizing (23.62) as a function of $\gamma\theta$ shows that the $\gamma \gg 1$ angular distribution is dominated by two intense (γ^8) peaks at $\theta = \pm\theta_{\text{peak}} \sim \pm 1/2\gamma$. Both peaks have an angular width of $\Delta\theta \sim 1/\gamma$. In other words, when $\beta \approx 1$, almost all the radiated power is emitted very near $\theta = 0$. This is the mathematical limit of the trend seen in Figure 23.7.

A highly relativistic and accelerating charged particle "beams" its radiation into the near-forward direction as a consequence of the Doppler effect. To see this, consider a plane wave emitted in the laboratory with wave vector $\mathbf{k}$ and frequency $\omega = ck$. Because $(\mathbf{k}, i\omega/c)$ is a four-vector, the transformation law (22.49) relates these plane wave variables to the corresponding primed variables in an inertial frame that moves with the instantaneous velocity $\mathbf{v}$ of the moving charge:

$$\mathbf{k}_\parallel = \gamma(\mathbf{k}'_\parallel + \boldsymbol{\beta}\omega'/c) \qquad \text{and} \qquad \mathbf{k}_\perp = \mathbf{k}'_\perp. \tag{23.63}$$

Using these, we deduce that

$$\cot\theta = \frac{k_\parallel}{k_\perp} = \frac{\gamma k' \cos\theta' + \gamma\beta k'}{k' \sin\theta'} = \gamma\left(\cot\theta' + \frac{\beta}{\sin\theta'}\right). \tag{23.64}$$

According to (23.64), a plane wave emitted broadside ($\theta' = \pi/2$) in the rest frame of the charge propagates in the laboratory frame at the angle $\theta = \tan^{-1}(1/\gamma\beta)$. When $\beta \approx 1$ and $\gamma \gg 1$, $\theta \sim 1/\gamma$, as predicted by (23.62).

23.3.3 Acceleration ⊥ Velocity

Figure 23.8 shows a point charge moving along a circular arc in the x-z plane. The charge has velocity $\mathbf{v}$ and acceleration $\mathbf{a} \perp \mathbf{v}$ at the moment when it passes through the origin of coordinates O. The angular distribution of radiation (23.57) for this situation is more complex than the $\mathbf{a} \parallel \mathbf{v}$ case (Section 23.3.2) because both the polar angle θ and the azimuthal angle ϕ are needed for a complete description.

Substituting $\mathbf{v} = v\hat{\mathbf{z}}$, $\mathbf{a} = a\hat{\mathbf{x}}$, and $\hat{\mathbf{n}}_{\text{ret}} \approx \sin\theta\cos\phi\hat{\mathbf{x}} + \sin\theta\sin\phi\hat{\mathbf{y}} + \cos\theta\hat{\mathbf{z}}$ into (23.57) leads (after some algebra) to

$$\left.\frac{dP}{d\Omega}\right|_\perp = \frac{\mu_0 q^2 a^2}{16\pi^2 c}\frac{1}{(1-\beta\cos\theta)^3}\left[1 - \frac{\sin^2\theta\cos^2\phi}{\gamma^2(1-\beta\cos\theta)^2}\right]. \tag{23.65}$$

The non-relativistic limit ($\beta \to 0$) of (23.65) has the angular dependence $1 - \sin^2\theta\cos^2\phi$. This distribution, shown on the left side of Figure 23.9, is exactly the low-velocity *cyclotron radiation*

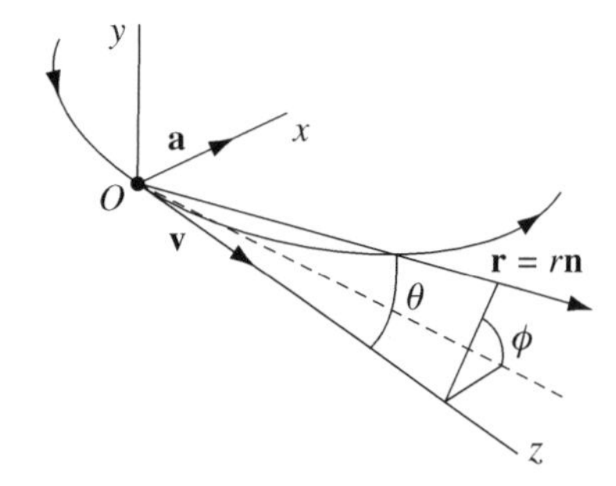

Figure 23.8: A point charge moves along a curved trajectory in the x-z plane.

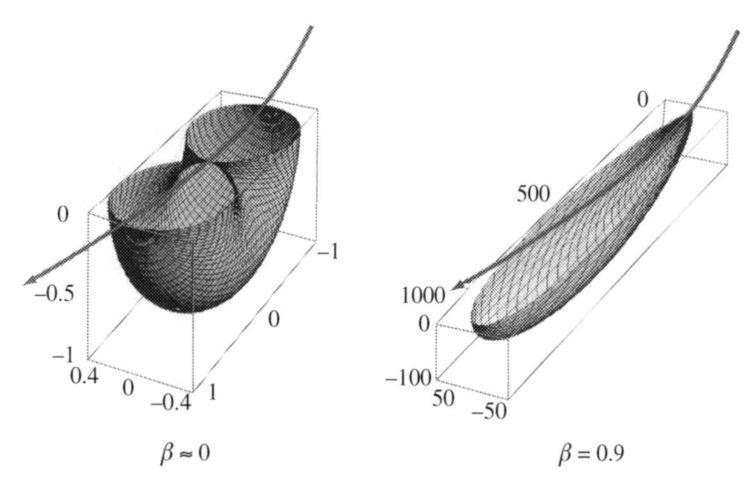

Figure 23.9: Angular distribution of radiation from a point charge moving along a circular arc. Left side: cyclotron radiation for $\beta \ll 1$. Only half the pattern is shown for clarity. Right side: synchrotron radiation for $\beta \approx 1$. Note the large change in scale between the two. Figure courtesy of Prof. Dr. Klaus Wille (Technical University of Dortmund)

pattern shown in Figure 23.6.[4] Once again, the interesting limit of (23.65) is $\beta \to 1$ because the denominator nearly vanishes when $\theta = 0$. This is the $\gamma \gg 1$ ultra-relativistic regime of *synchrotron radiation* where (23.61) is valid and the angular distribution simplifies to

$$\left.\frac{dP}{d\Omega}\right|_{\perp} \approx \frac{\mu_0 q^2 a^2}{16\pi^2 c} \frac{8\gamma^6}{(1+\gamma^2\theta^2)^3}\left[1 - \frac{4\gamma^2\theta^2\cos^2\phi}{(1+\gamma^2\theta^2)^2}\right] \qquad (\gamma \gg 1). \qquad (23.66)$$

The quantity in square brackets in (23.66) is equal to one at $\theta = 0$. Therefore, the fact that $\gamma \gg 1$ implies that the emitted radiation is concentrated in a narrow beam ($\Delta\theta \sim 1/\gamma$) pointed in the forward ($\theta = 0$) direction of particle motion. This beaming is apparent on the right side of Figure 23.9 and [as explained in the paragraph that contains (23.63) and (23.64)] is best understood as a consequence of the relativistic Doppler effect. The same physics explains why the $\beta \ll 1$ emission into the backward hemisphere virtually disappears when $\beta \to 1$. A tiny vestige of this emission survives, but it is too weak to see on the scale of the right side of Figure 23.9. The reader can also check that (23.65) is identically zero only when $\phi = 0$ and $\theta = \pm\theta_0$, where $\cos\theta_0 = \beta$.

[4] Note that the angle θ is defined differently in (23.58) and (23.65).

23.3.4 Larmor's Formula Generalized

The total amount of energy radiated by a moving charge with $\beta \approx 1$ is much greater than the total energy radiated when $\beta \ll 1$. The truth of this statement is apparent from Figure 23.7 and Figure 23.9. Our task here is to prove and rationalize it. The first step is to recall from Section 20.5.5 that integrating (23.58) over all angles yields Larmor's formula for the instantaneous rate at which a slowly moving particle radiates energy:

$$P = \frac{1}{4\pi\epsilon_0}\frac{2q^2|\mathbf{a}_{\rm ret}|^2}{3c^3}. \tag{23.67}$$

The generalization of (23.67) to situations where the charge is *not* moving slowly follows similarly by integrating the exact angular distribution of emitted power (23.57) over all angles. The integration is straightforward, but tedious, and gives

$$P = \frac{1}{4\pi\epsilon_0}\frac{2q^2}{3c^3}\gamma^6\left[a^2 - \left(\frac{\mathbf{v}\times\mathbf{a}}{c}\right)^2\right]_{\rm ret}. \tag{23.68}$$

A quite different way to derive (23.68) exploits the fact that the emitted power $P(t)$ is a Lorentz scalar. To prove this, we use the fact that both $(\mathbf{r}, ict)$ and $(\mathbf{P}_{\rm rad}, iU_{\rm rad}/c)$ are four-vectors.[5] The argument begins in an inertial frame K' where the moving charge is instantaneously at rest and the emission pattern looks like the left panel of Figure 23.9. The symmetry of this pattern (and the fact that no preferred direction can be defined) guarantees that $d\mathbf{P}'_{\rm rad} = 0$ is the momentum radiated in a time dt'. Similarly, $d\mathbf{r}' = 0$ because the velocity $\mathbf{v}' = d\mathbf{r}'/dt' = 0$. Using the transformation law (22.49), a Lorentz boost to a frame K where the charge has instantaneous velocity $\mathbf{v}$ shows that the radiated power is indeed the same in the two frames:

$$P = \frac{dU_{\rm rad}}{dt} = \frac{\gamma(dU'_{\rm rad} + \mathbf{v}\cdot d\mathbf{P}'_{\rm rad})}{\gamma(dt' + \mathbf{v}\cdot d\mathbf{r}'/c^2)} = \frac{dU'_{\rm rad}}{dt'} = P'. \tag{23.69}$$

A similar boost of the space-like components of $(\mathbf{P}_{\rm rad}, iU_{\rm rad}/c)$ shows that the rate at which a moving charge radiates linear momentum is proportional to the rate at which it radiates energy:

$$\frac{d\mathbf{P}_{\rm rad}}{dt} = \frac{\gamma(d\mathbf{P}'_{\rm rad} + \mathbf{v}\,dU'_{\rm rad}/c^2)}{\gamma(dt' + \mathbf{v}\cdot d\mathbf{r}'/c^2)} = \frac{\mathbf{v}}{c^2}\frac{dU'_{\rm rad}}{dt'} = \frac{\mathbf{v}}{c^2}\frac{dU_{\rm rad}}{dt}. \tag{23.70}$$

Now, Larmor's formula (23.67) is *exact* in the rest frame of the charge K'. Therefore, we follow the logic used in Section 23.2.3 to derive the Liénard-Wiechert potentials, and seek a manifestly covariant expression for the power which reduces to (23.67) when evaluated in K'. A natural choice replaces the three-acceleration $\mathbf{a}$ by the four-acceleration $\mathcal{A}_\mu = dU_\mu/d\tau$, where $U_\mu = \gamma(\mathbf{v}, ic)$ is the four-velocity and $\tau = t/\gamma$ is the proper time. For later use, we recall the momentum-energy four-vector, $p_\mu = mU_\mu = (\mathbf{p}, i\mathcal{E}/c)$, and write the equivalent expressions:

$$P = \frac{1}{4\pi\epsilon_0}\frac{2q^2}{3c^3}\mathcal{A}_\mu\mathcal{A}_\mu = \frac{1}{4\pi\epsilon_0}\frac{2q^2}{3m^2c^3}\frac{dp_\mu}{d\tau}\frac{dp_\mu}{d\tau}. \tag{23.71}$$

The remaining task is to evaluate (23.71) in an inertial frame where the instantaneous particle velocity is $\boldsymbol{\beta}$. Using $\mathcal{E} = \gamma mc^2$ and $\mathbf{p} = \gamma m\mathbf{v}$, the scalar product of interest is

$$\frac{dp_\mu}{d\tau}\frac{dp_\mu}{d\tau} = \frac{d\mathbf{p}}{d\tau}\cdot\frac{d\mathbf{p}}{d\tau} - \frac{1}{c^2}\left(\frac{d\mathcal{E}}{d\tau}\right)^2 = (mc\gamma)^2\left[\frac{d(\gamma\boldsymbol{\beta})}{dt}\cdot\frac{d(\gamma\boldsymbol{\beta})}{dt} - \left(\frac{d\gamma}{dt}\right)^2\right]. \tag{23.72}$$

[5] See Section 22.7.4 and Teitelboim (1970) in Sources, References, and Additional Reading.

Moreover,

$$\frac{d\gamma}{dt} = \gamma^3 \boldsymbol{\beta} \cdot \dot{\boldsymbol{\beta}} \qquad \text{and} \qquad \frac{d(\gamma \boldsymbol{\beta})}{dt} = \gamma \dot{\boldsymbol{\beta}} + \gamma^3 (\boldsymbol{\beta} \cdot \dot{\boldsymbol{\beta}}) \boldsymbol{\beta}. \tag{23.73}$$

After inserting (23.73) into (23.72), the identity $\gamma^2 = 1/(1-\beta^2)$ and a few lines of algebra are sufficient to show that (23.71) is equivalent to

$$P = \frac{1}{4\pi\epsilon_0} \frac{2q^2}{3c} \gamma^6 \left[\dot{\beta}^2 - (\boldsymbol{\beta} \times \dot{\boldsymbol{\beta}})^2 \right]_{\rm ret}. \tag{23.74}$$

This expression is the same as (23.68) and correctly reduces to (23.67) in the (instantaneous) proper frame of reference where $\beta = 0$ and $\gamma = 1$.

For fixed magnitudes of **v** and **a**, (23.74) predicts that linear acceleration ($\mathbf{a} \parallel \mathbf{v}$) produces radiation at a rate $P_\parallel \propto \gamma^6 a^2$ while centripetal acceleration ($\mathbf{a} \perp \mathbf{v}$) produces radiation at a rate $P_\perp \propto \gamma^4 a^2$. The factors of γ^6 and γ^4 account for the large size of the emission lobes in Figure 23.7 and Figure 23.9. Our derivation of (23.74) shows that they arise from the Lorentz transformation properties of energy and time.

Example 23.2

(a) A unit length of a linear accelerator increases the energy of a charged particle by an amount $d\mathcal{E}/dx$. Show that the rate at which the particle radiates energy is

$$P_\parallel = \frac{q^2 c}{6\pi\epsilon_0} \frac{1}{(mc^2)^2} \left(\frac{dp}{dt} \right)^2 = \frac{q^2 c}{6\pi\epsilon_0} \frac{1}{(mc^2)^2} \left(\frac{d\mathcal{E}}{dx} \right)^2.$$

(b) A circular accelerator with radius R raises the energy of a charged particle to an energy $\mathcal{E}$. Show that the rate at which the particle radiates energy is

$$P_\perp = \frac{q^2 c}{6\pi\epsilon_0} \frac{\gamma^2}{(mc^2)^2} \left(\frac{dp}{dt} \right)^2 = \frac{q^2 c}{6\pi\epsilon_0} \frac{\beta^4}{(mc^2)^4} \frac{\mathcal{E}^4}{R^2}.$$

Solution:

(a) For linear motion, $(d\mathbf{p}/dt) \cdot (d\mathbf{p}/dt) = (dp/dt)^2$. Moreover, $\mathcal{E}^2 = c^2 p^2 + m^2 c^4$ and $\boldsymbol{\beta} = c\mathbf{p}/\mathcal{E}$ (see Section 22.5.2) imply that

$$dp = \frac{\mathcal{E}}{c^2 p} d\mathcal{E} = \frac{d\mathcal{E}}{v}.$$

Using this information in the leftmost equation in (23.72), together with $d\tau = dt/\gamma$ and $\gamma^2(1-\beta^2)$ gives

$$\frac{dp_\mu}{d\tau} \frac{dp_\mu}{d\tau} = \gamma^2 \left[\left(\frac{dp}{dt} \right)^2 - \frac{1}{c^2} \left(\frac{d\mathcal{E}}{dt} \right)^2 \right] = \left(\frac{dp}{dt} \right)^2 = \frac{1}{v^2} \left(\frac{d\mathcal{E}}{dt} \right)^2.$$

Finally, because $dx = v dt$ is valid in any frame, (23.71) gives the advertised result,

$$P_\parallel = \frac{q^2 c}{6\pi\epsilon_0} \frac{1}{(mc^2)^2} \left(\frac{d\mathcal{E}}{dx} \right)^2.$$

(b) For circular motion, the particle speed does not change, so $\mathbf{p} = \gamma m \mathbf{v}$ implies that $d\mathbf{p}/dt = \gamma m d\mathbf{v}/dt$. Therefore, (23.68) says that

$$P_\perp = \frac{q^2}{6\pi\epsilon_0 c^3} \gamma^6 a^2 (1-\beta^2) = \frac{q^2 c}{6\pi\epsilon_0} \frac{\gamma^2}{(mc^2)^2} \left(\frac{dp}{dt} \right)^2.$$

On the other hand, the centripetal acceleration is $a = v^2/R$ and the total energy is $\mathcal{E} = \gamma mc^2$. Therefore,

$$P_\perp = \frac{q^2}{6\pi\epsilon_0 c^3}\gamma^4 a^2 = \frac{q^2 c}{6\pi\epsilon_0}\frac{(\gamma\beta)^4}{R^2} = \frac{q^2 c}{6\pi\epsilon_0}\frac{\beta^4}{(mc^2)^4}\frac{\mathcal{E}^4}{R^2}.$$

Remark: The paragraph following (23.74) compares two situations with the same magnitude of imposed acceleration. The present Example is more relevant in practice because it compares two situations with the same magnitude of applied force (change in momentum). The fact that $P_\perp/P_\parallel = \gamma^2$ for the same value of dp/dt implies that much more machine power is needed to overcome relativistic radiation losses in a circular accelerator than in a linear accelerator. This is so because a linear machine accelerates particles rather quickly to near the speed of light. Thereafter, very little acceleration (and thus radiation loss) occurs and the applied "acceleration force" works mainly to increases the total energy of the particle.

By contrast, the particles in a circular machine undergo continuous centripetal acceleration (and radiation loss) merely to retain their circular orbits. Circular machines are popular nevertheless because, unlike in a linear machine, particles pass through each accelerator element many times and multiple opportunities exist for particles to collide with one another. Radiation losses are reduced in a circular machine by increasing the orbit radius R, but the amount of real estate (and money) consumed becomes prohibitive as the energy $\mathcal{E}$ increases.

23.4 Radiation in the Frequency Domain

The history of physics abounds with important discoveries based on the spectral analysis of electromagnetic radiation. The discrete spectra of atoms and the continuous spectrum of the cosmic microwave background are familiar examples. In this section, we study the continuous frequency spectrum of radiation produced by a charged particle that follows an arbitrary (but specified) trajectory. The key step is to use Fourier's theorem to resolve the time dependence of the radiated electric field into its frequency components.

23.4.1 The Angle-Resolved Frequency Spectrum

Consider a time-dependent electric field $\mathbf{E}(t)$ and its frequency-dependent Fourier transform $\hat{\mathbf{E}}(\omega)$. The two are related by (Section 1.6)

$$\mathbf{E}(t) = \frac{1}{2\pi}\int_{-\infty}^{\infty} d\omega\, \hat{\mathbf{E}}(\omega)\exp(-i\omega t) \qquad \hat{\mathbf{E}}(\omega) = \int_{-\infty}^{\infty} dt\, \mathbf{E}(t)\exp(i\omega t). \tag{23.75}$$

From (23.55) and (23.56), the amount of energy radiated *per unit time* into a unit solid angle at a large distance r from the moving charge is

$$\frac{dP(t)}{d\Omega} = \frac{d^2 U_{\text{rad}}}{dt d\Omega} = \epsilon_0 c r^2 |\mathbf{E}_{\text{rad}}(t)|^2. \tag{23.76}$$

The key result we will establish is that the amount of energy radiated *per unit frequency* into a unit solid angle is proportional to $|\hat{\mathbf{E}}_{\text{rad}}(\omega)|^2$.

The proof begins by integrating (23.76) over all time to get the total energy radiated into a unit solid angle:

$$\frac{dU_{\rm rad}}{d\Omega} = \int_{-\infty}^{\infty} dt \, \frac{dP(t)}{d\Omega} = \epsilon_0 c r^2 \int_{-\infty}^{\infty} dt \, |\mathbf{E}_{\rm rad}(t)|^2. \tag{23.77}$$

Next, use Parseval's theorem (Section 1.6.1) to deduce that

$$\int_{-\infty}^{\infty} dt \, |\mathbf{E}_{\rm rad}(t)|^2 = \frac{1}{2\pi} \int_{-\infty}^{\infty} d\omega \, |\hat{\mathbf{E}}_{\rm rad}(\omega)|^2. \tag{23.78}$$

The reality condition $\mathbf{E}(t) = \mathbf{E}^*(t)$ applied to (23.75) implies that $\hat{\mathbf{E}}(\omega) = \hat{\mathbf{E}}^*(-\omega)$. Therefore, (23.78) transforms (23.77) to

$$\frac{dU_{\rm rad}}{d\Omega} = \frac{\epsilon_0 c}{\pi} r^2 \int_{0}^{\infty} d\omega \, |\hat{\mathbf{E}}_{\rm rad}(\omega)|^2. \tag{23.79}$$

Finally, the integral of $d^2U_{\rm rad}/d\omega d\Omega$ over all positive frequencies must also be the total energy radiated into a unit solid angle:

$$\frac{dU_{\rm rad}}{d\Omega} = \int_{0}^{\infty} d\omega \frac{d^2U_{\rm rad}}{d\omega d\Omega} \equiv \int_{0}^{\infty} d\omega \frac{dI(\omega)}{d\Omega}. \tag{23.80}$$

Comparing (23.80) to (23.79) gives the advertised result for the angle-resolved spectrum of radiated energy:

$$\frac{dI(\omega)}{d\Omega} = \frac{d^2U_{\rm rad}}{d\omega d\Omega} = \frac{\epsilon_0 c}{\pi} r^2 |\hat{\mathbf{E}}_{\rm rad}(\omega)|^2. \tag{23.81}$$

The integral of (23.81) over all angles is the frequency spectrum of the total radiated energy,

$$I(\omega) = \int d\Omega \, \frac{dI(\omega)}{d\Omega}. \tag{23.82}$$

A slightly different analysis is most natural when the trajectory of the charged particle is periodic and the radiated electric field satisfies $\mathbf{E}_{\rm rad}(\mathbf{r}, t) = \mathbf{E}_{\rm rad}(\mathbf{r}, t + T)$. In this case, the spectrum is discrete rather than continuous and the only frequencies that occur are harmonics (integer multiples) of the fundamental frequency $\omega_0 = 2\pi/T$. In place of (23.75) we write

$$\mathbf{E}(\mathbf{r}, t) = \sum_{m=-\infty}^{\infty} \hat{\mathbf{E}}_m(\mathbf{r}) \exp(-im\omega_0 t) \qquad \hat{\mathbf{E}}_m(\mathbf{r}) = \frac{\omega_0}{2\pi} \int_{0}^{2\pi/\omega_0} dt \, \mathbf{E}(\mathbf{r}, t) \exp(im\omega_0 t), \tag{23.83}$$

and in place of (23.77) we study the angular distribution of the average power radiated into the various harmonics during one period of the motion:

$$\left\langle \frac{dP}{d\Omega} \right\rangle = \frac{1}{T} \int_{0}^{T} dt \, \frac{dP}{d\Omega} = \sum_{m=1}^{\infty} \frac{dP_m}{d\Omega}. \tag{23.84}$$

We will learn in Section 23.5 how relativistic retardation effects redistribute spectral weight from the fundamental frequency (which dominates for slowly moving particles) to higher harmonics.

23.4.2 The Spectrum of an Arbitrary Current Density

As a prelude to the point charge problem, we derive here a formula for the radiation spectrum associated with an arbitrary time-dependent current density, $\mathbf{j}(\mathbf{r}, t)$. The ingredients come from Section 20.5.6, namely, a relation between the frequency-domain electric field and the frequency-domain vector potential,

$$\mathbf{E}_{\rm rad}(\mathbf{r}|\omega) = -i\omega\hat{\mathbf{r}} \times [\hat{\mathbf{r}} \times \mathbf{A}_{\rm rad}(\mathbf{r}|\omega)]\,, \tag{23.85}$$

and a relation between $\mathbf{A}_{\rm rad}(\mathbf{r}|\omega)$ and the space-and-time Fourier transform of the current density,

$$\mathbf{A}_{\rm rad}(\mathbf{r}\,|\,\omega) = \frac{\mu_0}{4\pi}\frac{e^{ikr}}{r}\int d^3r'\,\mathbf{j}(\mathbf{r}'\,|\,\omega)e^{-i\mathbf{k}\cdot\mathbf{r}'} \equiv \frac{\mu_0}{4\pi}\frac{e^{ikr}}{r}\hat{\mathbf{j}}(\mathbf{k}\,|\,\omega). \tag{23.86}$$

Substituting these into (23.81) gives the angular frequency spectrum as [cf. (20.120)]

$$\frac{dI(\omega)}{d\Omega} = \frac{\epsilon_0 c\,\omega^2}{\pi}r^2|\hat{\mathbf{r}} \times \mathbf{A}_{\rm rad}(\omega)|^2 = \frac{\mu_0\omega^2}{16\pi^3 c}|\hat{\mathbf{r}} \times \hat{\mathbf{j}}(\mathbf{k}|\omega)|^2. \tag{23.87}$$

This expression specifies the amount of energy radiated into a unit solid angle in the direction of $\mathbf{k} = (\omega/c)\hat{\mathbf{r}}$ in a unit frequency interval centered on ω.

23.4.3 The Spectrum of a Moving Point Charge

The current density for a point charge moving on the trajectory $\mathbf{r}_0(t)$ with velocity $\mathbf{v}(t) = \dot{\mathbf{r}}_0(t)$ is $\mathbf{j}(\mathbf{r}, t) = q\mathbf{v}_0(t)\delta(\mathbf{r} - \mathbf{r}_0(t))$. The Fourier transform of this quantity in space and time is

$$\hat{\mathbf{j}}(\mathbf{k}|\omega) = \int_{-\infty}^{\infty} dt \int d^3r\,\mathbf{j}(\mathbf{r}, t)\exp[-i(\mathbf{k}\cdot\mathbf{r} - \omega t)] = q\int_{-\infty}^{\infty} dt\,\mathbf{v}(t)\exp[-i(\mathbf{k}\cdot\mathbf{r}_0(t) - \omega t)]. \tag{23.88}$$

Substituting (23.88) into (23.87) gives the radiation spectrum produced by a moving point charge:

$$\frac{dI(\omega)}{d\Omega} = \frac{\mu_0 q^2\omega^2}{16\pi^3 c}\left|\hat{\mathbf{r}} \times \int_{-\infty}^{\infty} dt\,\mathbf{v}(t)\exp[-i(\mathbf{k}\cdot\mathbf{r}_0(t) - \omega t)]\right|^2. \tag{23.89}$$

The special case $\mathbf{v}$ = constant provides a consistency check because a particle that never accelerates also never radiates (see Section 23.1). Taking $\mathbf{v}$ out of the integral in (23.89) leaves an integral which is proportional to the delta function, $\delta(\omega[1 - \hat{\mathbf{r}}\cdot\boldsymbol{\beta}])$. This correctly gives $dI/d\omega = 0$ because $\beta < 1$ and the argument of the delta function is never zero (except at $\omega = 0$ which is not germane to radiation). Another special case is the dipole limit where $\mathbf{k}\cdot\mathbf{r}_0(t) \ll 1$ and (23.89) simplifies to a formula that requires only the Fourier transform of the particle velocity $\mathbf{v}(t)$ or acceleration $\mathbf{a}(t)$:

$$\frac{dI(\omega)}{d\Omega} = \frac{\mu_0 q^2\omega^2}{16\pi^3 c}\,|\hat{\mathbf{r}} \times \hat{\mathbf{v}}(\omega)|^2 = \frac{\mu_0 q^2}{16\pi^3 c}\,|\hat{\mathbf{r}} \times \hat{\mathbf{a}}(\omega)|^2\,. \tag{23.90}$$

The reader may wish to derive (23.90) directly from the non-relativistic Larmor formula (23.58).

Returning to (23.89), subtleties can arise in connection with the behavior of $\mathbf{v}(t)$ when $t \to \pm\infty$. This will become clear in the next section when we derive it again by a different method. The key is to be alert to the convergence of the integral, particulary when simple models for the trajectory function are used. Finally, we leave it as an exercise for the reader to show that the case of a periodic trajectory where (23.84) applies leads to an analogous formula for the average power radiated into the

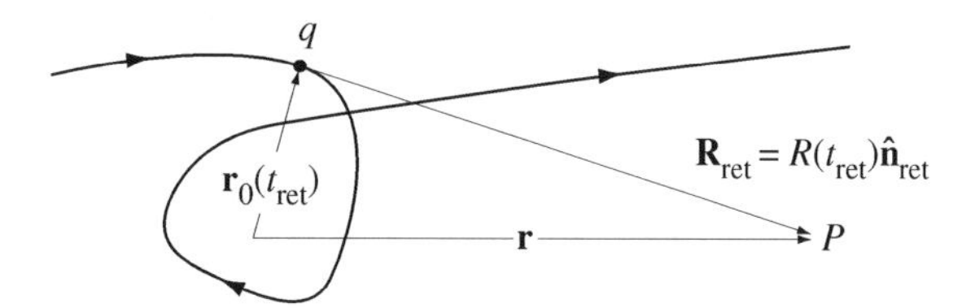

Figure 23.10: The approximations $\hat{\mathbf{n}}_{\rm ret} \approx \hat{\mathbf{r}}$ and $R(t_{\rm ret}) \approx r - \mathbf{r}_0 \cdot \hat{\mathbf{r}}$ apply when the observation point P is very far from the regions of particle acceleration.

mth harmonic:

$$\frac{dP_m}{d\Omega} = \frac{\mu_0 q^2 m^2 \omega_0^4}{32\pi^4 c}\left|\hat{\mathbf{r}} \times \int\limits_0^{2\pi/\omega_0} dt\, \mathbf{v}(t)\exp\{-im\omega_0[\hat{\mathbf{r}}\cdot\mathbf{r}_0(t)/c - t]\}\right|^2. \tag{23.91}$$

23.4.4 The Liénard-Wiechert Spectrum

It is instructive to re-derive (23.89) using the Liénard-Wiechert acceleration electric field in (23.31). This field is evaluated at the retarded time, so we use (23.9) and (23.27) to change the integration variable in (23.75) from the observation time t to the retarded time $t_{\rm ret}$. Doing this, the angular distribution of radiation (23.81) takes the form

$$\frac{dI(\omega)}{d\Omega} = \frac{\mu_0 c q^2}{16\pi^3}\left|\int\limits_{-\infty}^{\infty} dt_{\rm ret}\frac{\hat{\mathbf{n}}\times\left[(\hat{\mathbf{n}}-\boldsymbol{\beta})\times\dot{\boldsymbol{\beta}}\right]}{(1-\boldsymbol{\beta}\cdot\hat{\mathbf{n}})^2}\exp\{i\omega[t_{\rm ret} + R(t_{\rm ret})/c]\}\right|^2. \tag{23.92}$$

In practice, the observation point is almost always very far from the regions of space where $\dot{\boldsymbol{\beta}} \neq 0$. If so, we are justified in writing (see Figure 23.10)

$$R(t_{\rm ret}) = \sqrt{r^2 - 2\mathbf{r}_0(t_{\rm ret})\cdot\mathbf{r} + r_0^2(t_{\rm ret})} \approx r - \hat{\mathbf{r}}\cdot\mathbf{r}_0(t_{\rm ret}). \tag{23.93}$$

Substituting (23.93) into (23.92) and making the change of variable $t_{\rm ret} \to \tau$ gives

$$\frac{dI(\omega)}{d\Omega} = \frac{\mu_0 c q^2}{16\pi^3}\left|\int\limits_{-\infty}^{\infty} d\tau\frac{\hat{\mathbf{n}}\times\left[(\hat{\mathbf{n}}-\boldsymbol{\beta})\times\dot{\boldsymbol{\beta}}\right]}{(1-\boldsymbol{\beta}\cdot\hat{\mathbf{n}})^2}\exp\{i\omega[\tau - \hat{\mathbf{n}}\cdot\mathbf{r}_0(\tau)/c]\}\right|^2. \tag{23.94}$$

Finally, because $\mathbf{R} = R\hat{\mathbf{n}}$, our radiation zone approximation is consistent with $\hat{\mathbf{n}}_{\rm ret} \approx \hat{\mathbf{r}}$. It is with this understanding that we use (23.94) and its variations to follow. An important virtue of the widely used formula (23.94) is that the integrand vanishes whenever the particle acceleration vanishes. This is always the case for scattering and open-orbit problems where the charge feels a field of force for only a limited amount of time.

To make contact with (23.89), we recall that $\hat{\mathbf{n}} = \hat{\mathbf{r}}$ is a constant vector in our approximation. In that case, the identity

$$\frac{\hat{\mathbf{n}}\times\left[(\hat{\mathbf{n}}-\boldsymbol{\beta})\times\dot{\boldsymbol{\beta}}\right]}{(1-\boldsymbol{\beta}\cdot\hat{\mathbf{n}})^2} = \frac{d}{d\tau}\left[\frac{\hat{\mathbf{n}}\times(\hat{\mathbf{n}}\times\boldsymbol{\beta})}{1-\hat{\mathbf{n}}\cdot\boldsymbol{\beta}}\right] \tag{23.95}$$

permits us to integrate (23.94) by parts. With $k = \omega/c$ and $\phi(\tau) = \omega\tau - \mathbf{k}\cdot\mathbf{r}_0(\tau)$, the result is

$$\frac{dI(\omega)}{d\Omega} = \frac{\mu_0 c q^2}{16\pi^3}\left|\frac{\hat{\mathbf{n}}\times(\hat{\mathbf{n}}\times\boldsymbol{\beta})}{1-\hat{\mathbf{n}}\cdot\boldsymbol{\beta}}\exp(i\phi)\Big|_{\tau=-\infty}^{\tau=\infty} - i\omega\int\limits_{-\infty}^{\infty} d\tau\,[\hat{\mathbf{n}}\times(\hat{\mathbf{n}}\times\boldsymbol{\beta})]\exp(i\phi)\right|^2. \tag{23.96}$$

When the integrated term vanishes, e.g., when the particle velocity vanishes at $\tau = \pm\infty$, (23.96) simplifies to

$$\frac{dI(\omega)}{d\Omega} = \frac{\mu_0 c q^2 \omega^2}{16\pi^3} \left| \int_{-\infty}^{\infty} d\tau \; [\hat{\mathbf{n}} \times (\hat{\mathbf{n}} \times \boldsymbol{\beta})] \exp[-i(\mathbf{k} \cdot \mathbf{r}_0(\tau) - \omega\tau)] \right|^2 . \tag{23.97}$$

This expression is the same as (23.89) because $\hat{\mathbf{n}} \approx \hat{\mathbf{r}}$ is a constant vector which comes outside the integral and $|\hat{\mathbf{r}} \times (\hat{\mathbf{r}} \times \mathbf{s})|^2 = |\hat{\mathbf{r}} \times \mathbf{s}|^2$ for any vector $\mathbf{s}$.

Application 23.2 A Collinear Acceleration Burst

This Application calculates the radiation spectrum when a point charge q with uniform velocity $\mathbf{v}_0$ accelerates briefly in the direction parallel to its motion. This is the scenario analyzed graphically in Figure 23.3. To simplify the calculation, we assume a square-pulse acceleration burst (left panel of Figure 23.11) during which the particle velocity changes from $\mathbf{v}_0$ to $\mathbf{v}_f = \mathbf{v}_0 + \Delta\mathbf{v}$. In the interval $0 \leq \tau \leq \Delta t$ when the integrand in (23.94) is not zero, we approximate the trajectory function as $\mathbf{r}_0(\tau) \approx \bar{\mathbf{v}}\tau$, where $\bar{\mathbf{v}} = \frac{1}{2}(\mathbf{v}_0 + \mathbf{v}_f)$. The estimates $\boldsymbol{\beta} \approx \bar{\boldsymbol{\beta}} = \bar{\mathbf{v}}/c$ and $\dot{\boldsymbol{\beta}} \approx \Delta\mathbf{v}/c\Delta t$ apply also. Therefore,

$$\frac{dI(\omega)}{d\Omega} \approx \frac{\mu_0 q^2}{16\pi^3 c} \left| \frac{\hat{\mathbf{n}} \times (\hat{\mathbf{n}} \times \Delta\mathbf{v})}{(1 - \bar{\boldsymbol{\beta}} \cdot \hat{\mathbf{n}})^2 \Delta t} \int_0^{\Delta t} d\tau \exp[i\omega(1 - \hat{\mathbf{n}} \cdot \bar{\boldsymbol{\beta}})\tau] \right|^2 . \tag{23.98}$$

Evaluating the integral in (23.98) with $\hat{\mathbf{n}} \cdot \bar{\mathbf{v}} = \bar{v}\cos\theta$ gives the final result,

$$\frac{dI(\omega)}{d\Omega} \approx \frac{\mu_0 q^2}{16\pi^3 c} \frac{(\Delta v)^2 \sin^2\theta}{(1 - \bar{\beta}\cos\theta)^4} \frac{\sin^2 x}{x^2} \quad \text{where} \quad x = \frac{1}{2}\omega\Delta t(1 - \bar{\beta}\cos\theta). \tag{23.99}$$

The explicit variation with θ on the left side of (23.99) is very similar (if not quite identical) to the angular distribution of power $dP/d\Omega|_{\parallel}$ computed in Section 23.3.2. The variable x carries all the frequency dependence and some additional angle dependence through the function $\sin^2 x/x^2$ plotted in the right panel of Figure 23.11. This function is largest at $x = 0$, decreases monotonically to zero at $x = \pi$, and is negligibly small when $x > \pi$. The same general behavior is found when the square pulse in Figure 23.11 is replaced by a more realistic model for the acceleration burst. It is enough that $dI/d\Omega$ goes to zero very rapidly when x exceeds some characteristic value.

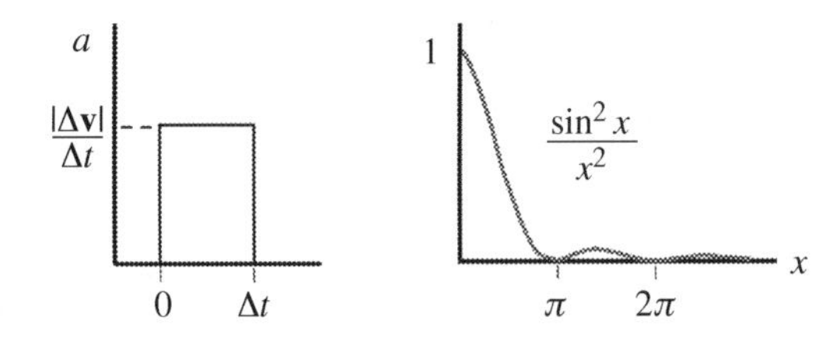

Figure 23.11: Left panel: the model acceleration burst. Right panel: the function that appears in (23.99) with $x = \frac{1}{2}\omega\Delta t(1 - \bar{\beta}\cos\theta)$.

In the low-velocity limit, $x = \frac{1}{2}\omega\Delta t$ and the frequency dependence of (23.99) is the same for every value of θ. From Figure 23.11, radiation emitted in the interval $0 \leq \tau \leq \Delta t$ has almost all of its spectral weight in the frequency range $0 \leq \omega \leq \omega_C$, where $\omega_C = 2\pi/\Delta t$. The inverse relation between Δt and the characteristic frequency ω_C is an example of the complementarity between Fourier transform pair variables encountered earlier in connection with wave packets (see Section 16.5.3).

The relativistic limit is more interesting because the frequency spectrum is *not* the same at all angles of observation. When $\gamma \gg 1$, we already know from Section 23.3.2 that the emission is most intense in the neighborhood of $\theta \approx \pm 1/2\gamma$. The approximation (23.61) is valid in this limit, so

$$x \approx \frac{\omega \Delta t}{4\bar{\gamma}^2}\left(1 + \bar{\gamma}^2\theta^2\right). \tag{23.100}$$

Setting (23.100) equal to π shows that the radiation spectrum in the ultra-relativistic limit extends from zero frequency to

$$\omega_C \approx \frac{4\pi}{\Delta t}\frac{\bar{\gamma}^2}{1+\bar{\gamma}^2\theta^2}. \tag{23.101}$$

In the angular region where maximum emission occurs, (23.101) predicts that the ultra-relativistic spectrum extends to a frequency that is $\bar{\gamma}^2$ larger than the corresponding non-relativistic cutoff frequency. ■

23.4.5 The Heaviside-Feynman Spectrum

The Heaviside-Feynman expression for the electric field of a moving point charge (see Section 23.2.5) leads to a formula for the angle- and frequency-resolved spectrum of radiation which differs from both (23.89) and (23.92). The radiation part of this field is

$$\mathbf{E}_{\rm rad}(t) = \frac{\mu_0 q}{4\pi}\frac{d^2\hat{\mathbf{n}}_{\rm ret}}{dt^2}. \tag{23.102}$$

Substituting (23.102) into (23.81), integrating by parts twice, and discarding the $t = \pm\infty$ boundary terms gives

$$\frac{dI(\omega)}{d\Omega} = \frac{\mu_0 q^2 r^2}{16\pi^3 c}\left|\int\limits_{-\infty}^{\infty} dt\, \frac{d^2\hat{\mathbf{n}}_{\rm ret}}{dt^2}\exp(i\omega t)\right|^2 = \frac{\mu_0 q^2 r^2 \omega^4}{16\pi^3 c}\left|\int\limits_{-\infty}^{\infty} dt\, \hat{\mathbf{n}}_{\rm ret}\exp(i\omega t)\right|^2. \tag{23.103}$$

This expression relates the radiation spectrum to the Fourier transform of the unit vector that points from the (retarded) position of the moving charge to the observation point (see Figure 23.10). We will make use of (23.103) in the next section.

23.5 Synchrotron Radiation

We showed in Section 23.3.3 that a charged particle moving on a circular arc near the speed of light emits *synchrotron radiation* in the form of a narrow and intense beam directed tangent to the arc (see Figure 23.9). This implies that a fixed observer sees a brief flash or pulse of radiation every time the particle moves directly toward him/her. In this section, we use the Liénard-Wiechert electric field to study the polarization and time evolution of such a pulse. Later, we use the Heaviside-Feynman electric field to study the transition from cyclotron radiation to synchrotron radiation and the frequency spectrum of the latter.

The extra attention to these topics is well warranted. Besides its importance to circular accelerators for particle physics, the synchrotron radiation produced by the ~75 electron storage ring "light sources" around the world is used for a wide range of spectroscopic and imaging purposes by physicists, chemists, biologists, and materials scientists. Moreover, the detection of astrophysical synchrotron radiation is a powerful indicator of the presence of magnetic fields and particle acceleration mechanisms near pulsars and black holes.

23.5.1 The Electric Field Pulse

The properties of a typical pulse of synchrotron radiation are most easily studied using a scaling form of its electric field. To derive this form, we begin with the acceleration part of the Liénard-Wiechert electric field produced by the generic moving particle in Figure 23.10:

$$\mathbf{E}_{\text{rad}}(\mathbf{r}, t) = \frac{q}{4\pi\epsilon_0}\left[\frac{\hat{\mathbf{n}} \times \left\{(\hat{\mathbf{n}} - \boldsymbol{\beta}) \times \dot{\boldsymbol{\beta}}\right\}}{cg^3 R}\right]_{\text{ret}}. \tag{23.104}$$

The subscript "ret" reminds us that the trajectory function $\mathbf{r}_0(t)$ in $\mathbf{R}(t) = \mathbf{r} - \mathbf{r}_0(t)$ and all the other quantities inside the square brackets are evaluated at the retarded time t_{ret} defined by

$$t_{\text{ret}} = t - R(t_{\text{ret}})/c. \tag{23.105}$$

Specialize now to a point charge moving relativistically in a circle of radius a with frequency ω_0. Our goal is to write the most intense part of (23.104) using suitably scaled time and observation-angle variables. Accordingly, return to Figure 23.8 and assume that the charge passes the origin of coordinates at times $t_{\text{ret}} = 0, 2\pi/\omega_0, \ldots$ The trajectory function in that case is

$$\mathbf{r}_0(t_{\text{ret}}) = a[1 - \cos(\omega_0 t_{\text{ret}})]\hat{\mathbf{x}} + a\sin(\omega_0 t_{\text{ret}})\hat{\mathbf{z}}. \tag{23.106}$$

Our interest is the ultra-relativistic limit when $\gamma = 1/\sqrt{1-\beta^2} \gg 1$ and

$$\beta = 1 - \epsilon \approx 1 - \frac{1}{2\gamma^2}. \tag{23.107}$$

The corresponding ultra-relativistic emission pattern (23.66) depends only weakly on the angle ϕ. Therefore, we locate the observer in the y-z plane ($\phi = \pi/2$) of Figure 23.8 and write

$$\mathbf{r} = r\sin\theta\,\hat{\mathbf{y}} + r\cos\theta\,\hat{\mathbf{z}}. \tag{23.108}$$

The first order of business is to evaluate $R(t_{\text{ret}}) = |\mathbf{r} - \mathbf{r}_0(t_{\text{ret}})|$ in the radiation zone where $r \gg a$. Using (23.106), we find

$$R(t_{\text{ret}}) = \sqrt{r^2 - 2ar\cos\theta\sin(\omega_0 t_{\text{ret}}) + 2a^2[1 - \cos(\omega_0 t_{\text{ret}})]} \approx r - a\cos\theta\sin(\omega_0 t_{\text{ret}}). \tag{23.109}$$

The further approximation $R(t_{\text{ret}}) \approx r$ is sufficient for direct use in the denominator of (23.104). However, to compute g_{ret}, we insert (23.109) into (23.105) and use the definition (23.27) to get

$$g_{\text{ret}} = dt/dt_{\text{ret}} = 1 - \beta\cos\theta\cos(\omega_0 t_{\text{ret}}). \tag{23.110}$$

Because $\beta = a\omega_0/c$, (23.110) is consistent with $g_{\text{ret}} = (1 - \boldsymbol{\beta}\cdot\hat{\mathbf{n}})_{\text{ret}}$ [see (23.27)] provided we adopt the radiation-zone approximation for $\hat{\mathbf{n}}_{\text{ret}}$ made just before (23.65), namely,

$$\hat{\mathbf{n}}_{\text{ret}} \approx \hat{\mathbf{n}} = \hat{\mathbf{r}} = \sin\theta\,\hat{\mathbf{y}} + \cos\theta\,\hat{\mathbf{z}}. \tag{23.111}$$

Now, the right side of Figure 23.9 makes clear that the radiation emitted by an ultra-relativistic orbiting charge most nearly resembles the pencil-like beam of a rotating searchlight or lighthouse. This means that a distant observer located very close to the z-axis ($\theta \ll 1$) is periodically illuminated by brief pulses of radiation in the immediate vicinity of $t_{\text{ret}} = 0, 2\pi/\omega_0, 4\pi/\omega_0, \ldots$ Therefore, we specialize to $t_{\text{ret}} \approx 0$ and Taylor expand (23.110) in both θ and t_{ret} to get

$$g_{\text{ret}} \approx \epsilon + \tfrac{1}{2}\theta^2 + \tfrac{1}{2}\omega_0^2 t_{\text{ret}}^2 + \cdots. \tag{23.112}$$

Motivated by this expansion, we define scaled angle and retarded time variables

$$\Theta = \frac{\theta}{\sqrt{2\epsilon}} = \theta\gamma \qquad \text{and} \qquad \tau_{\rm ret} = t_{\rm ret}\frac{\omega_0\gamma}{\sqrt{1+\Theta^2}}, \tag{23.113}$$

and write (23.112) in the form

$$g_{\rm ret} \approx \epsilon\left(1+\Theta^2\right)\left(1+\tau_{\rm ret}^2\right). \tag{23.114}$$

For the numerator of (23.104), we use $t_{\rm ret} \approx 0$ to neglect the $\hat{\mathbf{x}}$-component of (23.106) and differentiate the $\hat{\mathbf{z}}$-component to compute $\boldsymbol{\beta}_{\rm ret}$ and $\dot{\boldsymbol{\beta}}_{\rm ret}$. Using these and (23.111), a few lines of algebra confirm that

$$\hat{\mathbf{n}} \times [(\hat{\mathbf{n}} - \boldsymbol{\beta}_{\rm ret}) \times \dot{\boldsymbol{\beta}}_{\rm ret}] \approx \omega_0\beta\left\{[\beta\cos\theta - \cos(\omega_0 t_{\rm ret})]\hat{\mathbf{x}} + \sin\theta\sin(\omega_0 t_{\rm ret})(\hat{\mathbf{x}} \times \hat{\mathbf{n}})\right\}. \tag{23.115}$$

The final steps are to expand (23.115) to second order in θ and $t_{\rm ret}$, replace the latter by the scaling variables in (23.113), and substitute the result into (23.104). Because $\hat{\mathbf{x}} \times \hat{\mathbf{n}} = -\hat{\boldsymbol{\theta}}$ when $\phi = \pi/2$ [see (1.23)], we obtain the desired scaling form for the electric field pulse near $t_{\rm ret} = 0$:

$$\mathbf{E}(\mathbf{r},t) \approx -\frac{q}{4\pi\epsilon_0}\frac{\omega_0}{cr}\frac{1}{(1+\Theta^2)^3}\frac{1}{(1+\tau_{\rm ret}^2)^3}\frac{1}{\epsilon^2}\left[(1-\tau_{\rm ret}^2)(1+\Theta^2)\hat{\mathbf{x}} + 2\tau_{\rm ret}\Theta\sqrt{1+\Theta^2}\,\hat{\boldsymbol{\theta}}\right]. \tag{23.116}$$

23.5.2 Polarization and Temporal Shape

Synchrotron radiation is polarized predominantly in the plane of the orbiting particle. This follows immediately from (23.116) and the fact that $\Theta = \theta\gamma$ is the (scaled) angle of observation above that plane. In Figure 23.8, the radiation is polarized *entirely* along $\hat{\mathbf{x}}$ for $\Theta = 0$ observations where the emission is most intense. A smaller out-of-plane component appears when $\Theta \neq 0$. We will make this statement more quantitative when we turn to the frequency spectrum in Section 23.5.4.

To discover the temporal *shape* of a synchrotron radiation pulse, we must express the electric field (23.116) as a function of observer time rather than retarded time. To that end, use (23.109) to expand (23.105) to second order in θ and to third order in $t_{\rm ret}$:

$$t - r/c = t_{\rm ret} - (a/c)\left(1 - \tfrac{1}{2}\theta^2 + \cdots\right)\left(\omega_0 t_{\rm ret} - \tfrac{1}{6}\omega_0^3 t_{\rm ret}^3 + \cdots\right). \tag{23.117}$$

The approximation $\omega_0 a/c = \beta \approx 1$, (23.107), and (23.113) permit us to write (23.117) as

$$t - r/c \approx \left(1 - \beta + \frac{1}{2}\theta^2\right)t_{\rm ret} + \frac{1}{6}\omega_0^2 t_{\rm ret}^3 \approx \frac{\sqrt{2}}{\omega_0}\epsilon^{3/2}(1+\Theta^2)^{3/2}\left(\tau_{\rm ret} + \tfrac{1}{3}\tau_{\rm ret}^3\right). \tag{23.118}$$

With a scaled observer time τ defined as[6]

$$\tau = (t - r/c)\frac{\omega_0}{\sqrt{2}}\epsilon^{-3/2}(1+\Theta^2)^{-3/2}, \tag{23.119}$$

(23.118) simplifies to the cubic equation

$$\tau \approx \tau_{\rm ret} + \tfrac{1}{3}\tau_{\rm ret}^3. \tag{23.120}$$

It remains only to solve (23.120) to find $\tau_{\rm ret}$ as a function of τ and insert the latter into (23.116) to express (23.116) as a function of scaled observer time. We will do this exactly in Section 23.5.4. Here, we proceed approximately and note from (23.120) that $\tau_{\rm ret} \approx \tau$ when $|\tau_{\rm ret}| \ll 1$ and $\tau_{\rm ret} \approx (3\tau)^{1/3}$ when $|\tau_{\rm ret}| \gg 1$. Knitting these two limiting cases together produces the qualitative curve of $\tau_{\rm ret}(\tau)$ sketched in the left panel of Figure 23.12. The right panel of Figure 23.12 shows the corresponding behavior of $E_x(\tau)$ and $E_\theta(\tau)$ for $\Theta = 1$. This choice of polar angle corresponds to the "edge" of the

[6] Notice that the definitions of the scaled variables τ and $\tau_{\rm ret}$ in terms of t and $t_{\rm ret}$ are not exactly the same.

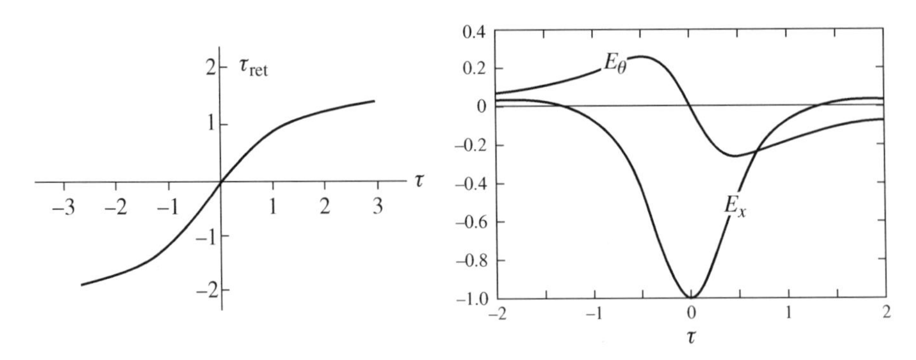

Figure 23.12: Left panel: approximate solution for the function $\tau_{ret}(\tau)$ that satisfies (23.120). Right panel: the electric field from a pulse of synchrotron radiation as a function of scaled observer time. The observation point is $\Theta = 1$ in the y-z plane of Figure 23.8.

synchrotron beam where E_θ is relatively large but the total emission intensity is an order of magnitude smaller than its value at the center of the beam ($\Theta = 0$), where $E_\theta = 0$ and E_x is largest.

23.5.3 The Cyclotron-Synchrotron Transition

In this section and the next, we use the Heaviside-Feynman electric field (23.102) to gain further insight into synchrotron radiation. For example, because $\mathbf{E}_{\rm rad} \propto \ddot{\hat{\mathbf{n}}}_{\rm ret}$, we see immediately that the polarization lies entirely in the x-z plane of Figure 23.8 when $\theta = 0$ and acquires an out-of-plane component when $\theta \neq 0$. The latter is always very small in the relativistic limit of synchrotron radiation because the emission is negligible when $\theta > 1/\gamma \ll 1$.

An especially interesting issue is the evolution of the radiation field as a function of the velocity of the circulating charge. Using (23.106), (23.108), and (23.109), $\hat{\mathbf{n}}_{\rm ret} = [\mathbf{r} - \mathbf{r}_0(t_{\rm ret})]/R_{\rm ret}$ in the radiation zone is

$$\hat{\mathbf{n}}_{\rm ret} = \frac{a[\cos(\omega_0 t_{\rm ret}) - 1]\hat{\mathbf{x}} + r\sin\theta\hat{\mathbf{y}} + [r\cos\theta - a\sin(\omega_0 t_{\rm ret})]\hat{\mathbf{z}}}{r - a\cos\theta\sin(\omega_0 t_{\rm ret})}. \tag{23.121}$$

If we limit ourselves to an observer in the $\theta = 0$ plane, the z-component of (23.121) has no time dependence and does not contribute to the electric field (23.102). This is consistent with the fact that all radiation fields are transverse. Moreover, it is sufficient to retain only the factor r in the denominator of the x-component of $\hat{\mathbf{n}}_{\rm ret}$. Substituting the latter into (23.102) and using $\beta = a\omega_0/c$, the radiation electric field can be written in the form

$$\mathbf{E}_{\rm rad} = -\frac{\mu_0 q}{4\pi r}\frac{c}{\omega_0}\frac{d^2\Lambda}{dt^2}\hat{\mathbf{x}} \quad \text{where} \quad \Lambda = 1 - \beta\cos(\omega_0 t_{\rm ret}). \tag{23.122}$$

A graphical method due to Feynman *et al.* (1963) reveals the behavior of $\mathbf{E}_{\rm rad}$ as a function of the observer's time. The idea is to pair (23.122) with (23.105) because the latter relates the retarded time to the observer time. In light of (23.109), the $\theta = 0$ limit of (23.105) is

$$t - r/c = t_{\rm ret} - (a/c)\sin(\omega_0 t_{\rm ret}). \tag{23.123}$$

If we reset the clock of the fixed observer by r/c, (23.123) is equivalent to

$$\omega_0 t = \omega_0 t_{\rm ret} - \beta\sin(\omega_0 t_{\rm ret}). \tag{23.124}$$

Two observations complete the story. First, the equations

$$\Lambda = 1 - \beta\cos(\omega_0 t_{\rm ret}) \quad \text{and} \quad \omega_0 t = \omega_0 t_{\rm ret} - \beta\sin(\omega_0 t) \tag{23.125}$$

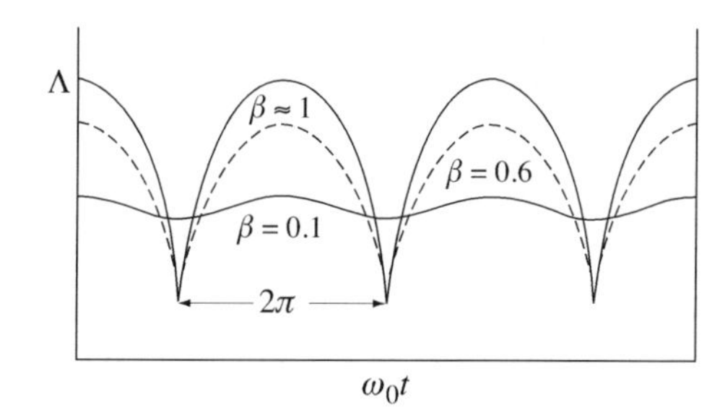

Figure 23.13: The curvature of $\Lambda(\omega_0 t)$ is proportional to the radiation electric field $E(t)$ observed at a fixed point in the plane of a charge moving uniformly in a circle with frequency ω_0. The three curves correspond to three values of the particle speed, $\upsilon = c\beta$.

are a parametric representation (with parameter $t_{\rm ret}$) of a planar curve $\Lambda(\omega_0 t)$ called a *curtate cycloid.* This is the curve traced out by a dot painted at a radius $\beta < 1$ on a wheel of unit radius which rolls with angular velocity ω_0. Second, the curvature (second derivative with respect to t) of $\Lambda(\omega_0 t)$ is proportional to the electric field $E_x(t)$.

Figure 23.13 shows plots of $\Lambda(\omega_0 t)$ for three values of $\beta = \upsilon/c$. The non-relativistic ($\beta = 0.1$) curve is a low-amplitude sinusoid with frequency ω_0. The second derivative is also a low-amplitude sinusoid with frequency ω_0. Invoking (23.122), we conclude that this situation corresponds to cyclotron radiation, where the electric field oscillates at the fundamental frequency ω_0. The amplitude of the cyclotron radiation electric field modulates only slightly over each period of the charge's motion. A glance back at the fairly isotropic emission pattern on the left side of Figure 23.9 makes this very plausible.

By contrast, the ultra-relativistic ($\beta \approx 1$) curve is a full cycloid with sharp cusps which repeat with a period $T = 2\pi/\omega_0$. The corresponding curvature function implies that the electric field consists of a sequence of very short and very intense pulses observed once per revolution of the charge. This is synchrotron radiation, which must have a very broad frequency spectrum because many terms in a Fourier series are needed to represent the second derivative of a function with cusps. The shape of the moderately relativistic ($\beta = 0.6$) curve in Figure 23.13 suggests how the transition from cyclotron radiation to synchrotron radiation occurs as β increases.

23.5.4 The Continuous Spectrum

A single charge moving relativistically on a circular arc emits synchrotron radiation over a wide range of frequencies. Qualitatively, this broadband character of synchrotron radiation may be inferred from the fact that the dominant E_x component in Figure 23.12 has a pulse width of $\tau \sim 1$. Then, combining (23.107) and (23.119) with the reciprocal Fourier relation between variances in time and frequency (see Section 16.5.3), we predict that there is significant emission of radiation with frequencies between ω_0 and $T^{-1} \sim \omega_0 \gamma^3$. The latter is very much larger than the fundamental frequency ω_0 in the ultra-relativistic limit when $\gamma \gg 1$.

Alternatively, we can exploit the sweeping-searchlight picture of synchrotron radiation implied by the right side of Figure 23.9 and let $t(1)$ and $t(2)$ be the times when a distant observer first sees the leading and trailing edges of the emission beam. This radiation was emitted at times $t_{\rm ret}(1)$ and $t_{\rm ret}(2)$, respectively. Now, we learned in the paragraph following (23.66) that the angular width of the beam is $\Delta\theta \sim 1/\gamma$. Therefore, using (23.27) and (23.107), the time $\Delta t = t(2) - t(1)$ required for the beam to sweep past the distant observer is

$$\Delta t = \Delta t_{\rm ret}\frac{dt}{dt_{\rm ret}} = \frac{\Delta\theta}{\omega_0}(1 - \boldsymbol{\beta}\cdot\hat{\mathbf{n}})_{\rm ret} \sim \frac{1}{\gamma\omega_0}(1-\beta) \sim \frac{1}{\omega_0\gamma^3}. \tag{23.126}$$

Schott (1912) and Schwinger (1949) derived analytic expressions for the frequency spectrum of synchrotron radiation.[7] The exact result is a sequence of delta functions at integer multiples of the fundamental frequency ω_0. A continuous approximation which passes smoothly through the intensity predicted for each harmonic can be computed by evaluating the function (23.97). The latter used the Liénard-Wiechert electric field (23.104) as input. Here, we exploit the methodology of Section 23.5.3 and derive the same result using the Heaviside-Feynman spectrum function (23.103).

Following Section 23.5.1, we focus on small polar angles near $\theta = 0$ and a single synchrotron pulse emitted near $t_{\rm ret} = 0$. This permits us to expand (23.121) to first order in a/r, first order in θ, and second order in $t_{\rm ret}$. Keeping only the time-dependent terms gives

$$\hat{\mathbf{n}}_{\rm ret} \approx \frac{a}{r}\left[-\tfrac{1}{2}\omega_0^2 t_{\rm ret}^2 \hat{\mathbf{x}} + \theta\omega_0 t_{\rm ret}\hat{\mathbf{y}}\right]. \tag{23.127}$$

At this point, it is convenient to introduce the variables

$$\omega^\star = \tfrac{3}{2}\gamma^3\omega_0 \qquad \eta = 3\gamma^3(1+\gamma^2\theta^2)^{-3/2}\omega_0 t \qquad \xi = \frac{\omega}{2\omega^\star}(1+\gamma^2\theta^2)^{3/2}. \tag{23.128}$$

The first of these is the characteristic frequency derived qualitatively just above. The second and third satisfy $\omega_0 t = \xi\eta$. Consequently, inserting (23.127) and (23.128) into (23.103) and using $\beta = a\omega_0/c \approx 1$ gives the frequency spectrum in the form

$$\frac{dI(\omega)}{d\Omega} = \frac{1}{4\pi\epsilon_0}\frac{q^2}{4\pi^2 c}\frac{\omega^2}{\omega_0^2}\left|\xi\int_{-\infty}^{\infty} d\eta\,\left[\tfrac{1}{2}\omega_0^2 t_{\rm ret}^2\hat{\mathbf{x}} + \theta\omega_0 t_{\rm ret}\hat{\mathbf{y}}\right]\exp(i\xi\eta)\right|^2. \tag{23.129}$$

The integration in (23.133) requires the variable $t_{\rm ret}$ expressed as a function of t (or η). We did this in Section 23.5.2 (for the relevant case when both θ and $t_{\rm ret}$ are near zero) and obtained the cubic equation on the far left side of (23.118). Dropping the factor r/c as explained following (23.123), this equation reads

$$(\omega_0 t_{\rm ret})^3 + 3\gamma^{-2}(1+\gamma^2\theta^2)(\omega_0 t_{\rm ret}) - 6\omega_0 t = 0. \tag{23.130}$$

The exact (real) solution to (23.130) is

$$\omega_0 t_{\rm ret} = \gamma^{-1}\sqrt{1+\gamma^2\theta^2}\left[\left(\sqrt{\eta^2+1}+\eta\right)^{1/3} - \left(\sqrt{\eta^2+1}-\eta\right)^{1/3}\right]. \tag{23.131}$$

Substituting (23.131) into (23.129) and using[8]

$$\begin{aligned}
&\int_0^\infty d\eta\,\left[\left(\sqrt{\eta^2+1}+\eta\right)^{1/3} - \left(\sqrt{\eta^2+1}-\eta\right)^{1/3}\right]^2\cos(\xi\eta) = \frac{2}{\sqrt{3}}\frac{1}{\xi}K_{2/3}(\xi)\\
&\int_0^\infty d\eta\,\left[\left(\sqrt{\eta^2+1}+\eta\right)^{1/3} - \left(\sqrt{\eta^2+1}-\eta\right)^{1/3}\right]\sin(\xi\eta) = -\frac{1}{\sqrt{3}}\frac{1}{\xi}K_{1/3}(\xi)
\end{aligned} \tag{23.132}$$

gives the angular spectrum for a single burst of synchrotron radiation in terms of modified Bessel functions. Using $\Theta = \gamma\theta$ from (23.113), this distribution is

$$\frac{dI(\omega)}{d\Omega} = \frac{1}{4\pi\epsilon_0}\frac{3q^2\gamma^2}{4\pi^2 c}\left(\frac{\omega}{\omega^\star}\right)^2(1+\Theta^2)\left[(1+\Theta^2)K_{2/3}^2(\xi) + \Theta^2 K_{1/3}^2(\xi)\right]. \tag{23.133}$$

[7] See Sources, References, and Additional Reading.

[8] The integrals in (23.132) are derived from closely related integrals in Section 3.775 of I.S. Gradeshteyn and I.M. Ryzhik, *Tables of Integrals, Series, and Products* (Academic, New York, 1980). See Wang (1993) for the details.

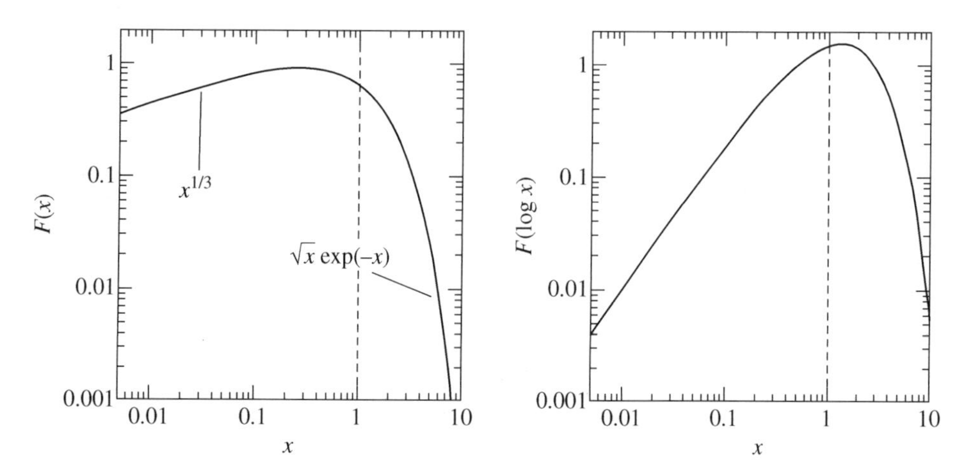

Figure 23.14: The angle-integrated spectrum of synchrotron radiation as a function of $x = \omega/\omega^\star$. Left panel: log-log plot of $F(x)$. Right panel: log-log plot of $F(\log x)$. Figure from Condon and Ransom (2010).

The $K^2_{2/3}$ term in (23.133) comes from E_x in Figure 23.12. This is the component of the radiated field parallel to the plane of the particle motion. The $K^2_{1/3}$ term comes from E_θ in Figure 23.12. This is the radiated electric field component perpendicular to the plane of the particle motion.

The frequency spectrum of synchrotron radiation is the integral of (23.133) over all angles. The result is[9]

$$I(\omega) = \frac{1}{4\pi\epsilon_0}\frac{\sqrt{3}q^2\gamma}{c}F(\omega/\omega^\star) \qquad \text{where} \qquad F(x) = x\int_x^\infty d\xi\, K_{5/3}(\xi). \tag{23.134}$$

The left side of Figure 23.14 shows the universal function $F(x)$ defined in (23.134) on a log-log scale. This common way of plotting the spectrum function clearly shows its power-law variation when $\omega \ll \omega^\star$. We have also indicated the exponential decay of the spectrum when $\omega \gg \omega^\star$. However, because $I(\omega)$ is the power emitted per unit frequency rather than the power emitted per unit log (frequency), this log-log representation obscures the fact that most of the power is radiated very near $\omega = \omega^\star$. We remedy this with the graph of $F(\log x)$ on the right side of Figure 23.14. This figure makes it more plausible that the total power emitted with $\omega < \omega^\star$ is exactly equal to the total power emitted with $\omega > \omega^\star$. Another interesting quantity is the ratio of the power radiated with the field polarized parallel to the plane of the particle trajectory to the power radiated with the field polarized perpendicular to the particle trajectory. A detailed calculation shows that this ratio of powers is

$$\frac{P_\parallel}{P_\perp} = 7, \tag{23.135}$$

compared to the non-relativistic limit of cyclotron radiation where $P_\parallel/P_\perp = 3$. In practice, polarization information of this kind is used to identify synchrotron radiation in the non-thermal emission observed from nebulae and supernovas.

23.5.5 The Discrete Spectrum

The spectrum function (23.134) applies to a single pulse of synchrotron radiation like the one shown in Figure 23.12. Such a pulse is produced whenever a relativistic charge traverses even a very small segment of circular arc. For a fixed observer, a relativistic charge moving in a circular orbit produces

[9] See Wiedemann (2003) for the details.

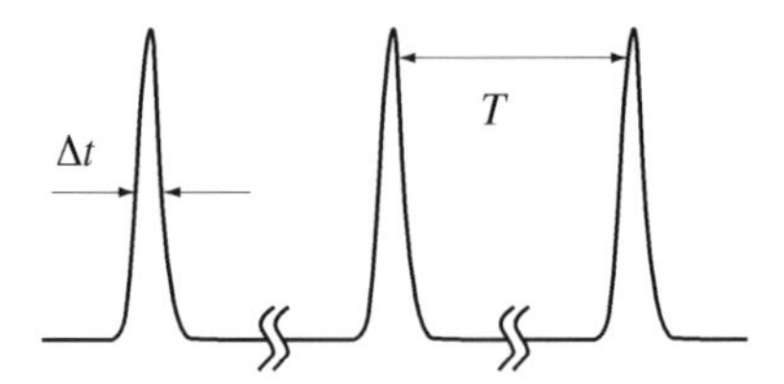

Figure 23.15: A periodic array of electric field pulses of synchrotron radiation produced by a charge moving relativistically in a circle with period T.

a sequence of these pulses separated by the period T of the orbit. Figure 23.15 is a cartoon time-trace for this situation, which may be regarded as the second derivative of the $\beta \approx 1$ curve in Figure 23.13.

The goal of this section is to show that the frequency spectrum of the function sketched in Figure 23.15 is discrete and to relate it to the continuous function (23.134). The key to the calculation is that the total electric field at any time can be written

$$E_{\text{tot}}(t) = \sum_{k--\infty}^{\infty} E(t - kT), \tag{23.136}$$

where $E(t)$ is the electric field of a single pulse. The expression (23.136) is intuitively reasonable when $\Delta t \ll T$ and the pulses in Figure 23.15 do not overlap in time. However, a moment's reflection shows that the strict periodicity of the motion (and hence of the electric field) implies that (23.136) is correct even for non-relativistic speeds when $\Delta t \sim T$.

We now define the function[10]

$$\Delta_T(t) = \sum_{k=-\infty}^{\infty} \delta(t - kT) \tag{23.137}$$

and use it to write (23.136) in the form

$$E_{\text{tot}} = \int_{\infty}^{\infty} dt' \sum_{k=-\infty}^{\infty} \delta(t - t' - kT)E(t') = \int_{-\infty}^{\infty} dt' \Delta_T(t - t')E(t'). \tag{23.138}$$

The rightmost member of (23.138) is a convolution integral. Therefore, the convolution theorem (Section 1.6.2) guarantees that the Fourier transforms of E_{tot}, $\Delta_T(t)$, and $E(t)$ satisfy

$$\hat{E}_{\text{tot}}(\omega) = \hat{\Delta}_T(\omega)\hat{E}(\omega). \tag{23.139}$$

On the other hand, using $T = 2\pi/\omega_0$ and the Fourier series derived in Example 1.6, the function $\hat{\Delta}_T(\omega)$ is

$$\hat{\Delta}_T(\omega) = \int_{-\infty}^{\infty} dt \sum_{k=-\infty}^{\infty} \delta(t - kT)\exp(i\omega t) = \sum_{k=-\infty}^{\infty} \exp(i\omega kT) = \omega_0 \sum_{n=-\infty}^{\infty} \delta(\omega - n\omega_0). \tag{23.140}$$

Substituting (23.140) into (23.139) gives the spectrum of the periodic train of electric field pulses in Figure 23.15 as

$$\hat{E}_{\text{tot}}(\omega) = \omega_0 \sum_{n=-\infty}^{\infty} \hat{E}(n\omega_0)\delta(\omega - n\omega_0). \tag{23.141}$$

[10] The names "Dirac comb" and "Shah function" are sometimes used for (23.137).

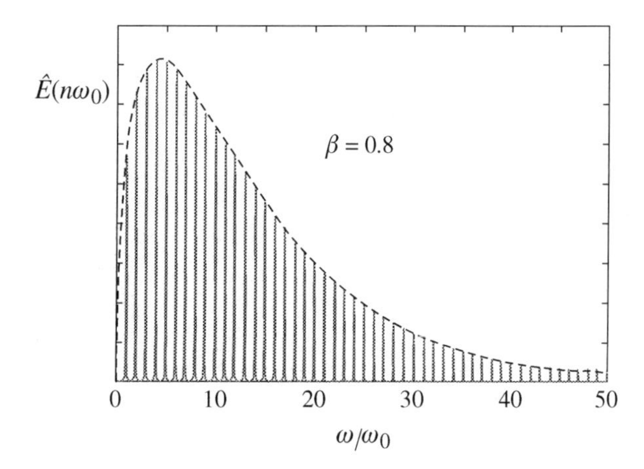

Figure 23.16: The weights $\hat{E}(n\omega_0)$ of the delta functions in the spectrum function (23.141). The envelope of these weights defines the continuous function $\hat{E}(\omega)$ (dashed curve).

Equation (23.141) confirms our statement that the radiation spectrum is discrete for a point charge moving in uniform circular motion. The frequencies that occur are harmonics of the fundamental orbital frequency ω_0. The relative weight of each harmonic is determined by the Fourier transform of the single-pulse function evaluated at the harmonic frequency. This is shown in Figure 23.16. Similar remarks apply to the spectrum of radiated power $I(\omega) \propto |\hat{E}_{\text{tot}}(\omega)|^2$.

Example 23.3 N identical point particles, each with charge q, are spaced uniformly around a ring of radius a. All move with the same constant speed v around the ring. Describe the frequency spectrum of the emitted radiation. Explain why the radiation disappears in the limit when $N \to \infty$ and $q \to 0$ while Nq remains finite. Do the conclusions change if the charges are distributed randomly around the ring?

Solution: The orbit period $T = 2\pi/\omega_0 = 2\pi a/v$ is common to all the charges. This makes T/N the period of motion for the entire configuration and we predict from (23.83) that radiation occurs only at frequencies which are integer multiples of $N\omega_0$. The suppression of all the lower harmonics is an interference effect. The lowest radiation harmonic is pushed to higher and higher frequency as N increases and, in the proposed limit, we expect no radiation at all. What remains is the static electric field produced by a ring with total charge Nq and the static magnetic field produced by a ring with steady current $I = Nq\omega_0/2\pi$. These conclusions fail when the charges are randomly distributed because the period of motion reverts back to $2\pi/\omega_0$. Even when $N \to \infty$, the resulting current density is not time-independent and radiation should be expected.

23.6 Radiation Reaction

The emission of radiation irreversibly changes the trajectory of a charged particle. A vivid example was sketched by Niels Bohr in his first paper on the quantum theory of atomic spectra:[11]

[11] See Sources, References, and Additional Reading.

The inadequacy of classical electrodynamics to account for the properties of atoms from a model like Rutherford's appears very clearly ... if we consider the effect of the energy of radiation. The electron no longer describes stationary orbits. The whole system loses energy ... and the electron approaches the nucleus describing orbits of smaller and smaller dimensions.

The electromagnetic force responsible for the spiralling motion described by Bohr comes from the electron's radiation field and is called *radiation reaction*. Similarly, a source that radiates linear momentum non-isotropically must recoil to conserve total linear momentum.

In this section, we derive the electromagnetic force which a radiating system exerts on itself. The magnitude of this force is generally negligible for macroscopic distributions of current. This contrasts with the effect of radiation reaction on the microscopic current distribution envisioned by Bohr. Indeed, Example 23.2 at the end of this section estimates the lifetime of an excited atomic electron from the effect of radiation reaction on the arguably "better" classical atomic model of a point charge bound harmonically to a fixed nucleus. Along the way, we derive the famous Lorentz-Abraham equation of motion for an electron for the main purpose of replacing it by a more physically acceptable alternative.

23.6.1 Background

The concept of radiation reaction was developed around 1900 by Planck, Abraham, and Lorentz. A major motivation was to derive an equation of motion for the electron, which these classical physicists regarded as an electromagnetic object. The advent of quantum theory mooted these efforts, but various aspects of the problem have continued to attract interest for over a century.

Our first task is to understand why it has been possible to ignore radiation reaction up to this point. An instructive case is an electron which emits synchrotron radiation as it moves in a circle with speed $v \approx c$ and period $T = 2\pi R/v$ under the influence of a uniform magnetic field B. The Lorentz force provides the centripetal acceleration, so $\gamma m v^2/R = evB$. Using part (b) of Example 23.2, the energy lost by the electron during one complete orbit is

$$\Delta E_{\rm rad} = P_\perp T = \frac{e^2 c(\gamma\beta)^4}{6\pi\epsilon_0 R^2}\frac{2\pi R}{v} \approx \frac{e^2\gamma^4}{3\epsilon_0 c}\frac{v}{R} = \frac{e^3\gamma^3 B}{3\epsilon_0 cm}. \tag{23.142}$$

Comparing (23.142) to the electron's total energy, γmc^2, we conclude that radiation reaction is important for this problem if

$$\gamma^2 B \gtrsim \frac{3\epsilon_0 m^2 c^3}{e^3} = \frac{3}{4\pi\alpha} B_{\rm c} \approx 10^{11}\ {\rm T}, \tag{23.143}$$

where $\alpha = e^2/4\pi\epsilon_0\hbar c \approx 1/137$ is the fine structure constant and $B_{\rm c} = m^2c^2/e\hbar$ is the critical magnetic field for vacuum polarization in quantum electrodynamics (see Section 2.5.2). To avoid the appearance of strong-field quantum effects, it turns out we should combine (23.143) with two additional restrictions:[12]

$$B \lesssim 10^7\ {\rm T} \qquad \text{and} \qquad \gamma \gtrsim 200. \tag{23.144}$$

It is not easy to satisfy (23.143) and (23.144) simultaneously. Therefore, radiation reaction can be safely ignored for most problems where charges move in external magnetic fields. Circular accelerators and storage rings are important exceptions, but a considerable effort is made in those cases to preserve the integrity of the charged particle orbits by using external sources to replenish the energy loss per orbit (23.142).

[12] See R.W. Nelson and I. Wasserman, "Synchrotron radiation with radiation reaction", *The Astrophysical Journal* **371**, 265 (1991).

For the hydrogen atom which motivated Bohr, the electric field of the proton generates the centripetal acceleration of the electron. Glancing back at (23.142), it is not difficult to calculate the energy lost by the electron during one complete orbit at the Bohr radius a_B. The electron speed in the ground state is $v = \alpha c$, so

$$\Delta E_{\text{rad}} = \frac{e^2 c(\gamma\beta)^4}{6\pi\epsilon_0 R^2}\frac{2\pi R}{v} \approx \frac{e^2}{3\epsilon_0}\frac{\alpha^3}{a_B} = \frac{8\pi}{3}\alpha^3\,\text{Ry}, \tag{23.145}$$

where $\text{Ry} = e^2/8\pi\epsilon_0 a_B$ is the Rydberg energy. The binding energy of the electron is 1 Ry, so (23.145) implies that radiation reaction has a very small effect per orbit. Nevertheless, with no source of energy to replace (23.145), the electron is doomed to spiral into the proton, albeit after a very large number of orbit periods. A general conclusion is that radiation reaction may be ignored except for situations (like orbit stability) when even extremely small trajectory perturbations cannot be tolerated.

23.6.2 The Force on a Slowly Varying Source

Let **E** and **B** be the retarded fields produced by a spatially extended and radiating source with charge and current densities ρ and $\mathbf{j}$. The radiation reaction force $\mathbf{F}_{\text{RR}} = \int d^3r\,(\rho\mathbf{E} + \mathbf{j}\times\mathbf{B})$ accounts for the energy lost by radiation because it degrades the current density by performing work on it. The existence of this force does not contradict our previous understanding that static and quasistatic sources do not exert forces on themselves because these fields are not retarded. In this section, we focus on sources that vary slowly enough in time that a low-order multipole expansion is sufficient to find the radiation reaction.

A trick introduced by Dirac (1938) makes it straightforward to identify the terms in the multipole expansion which contribute to $\mathbf{F}_{\text{RR}}$. The idea is to compare the retarded potentials in (23.2) and (23.3) with their *advanced* counterparts (see Section 20.3.3). The retarded fields radiate outgoing waves and decrease the system energy. The advanced fields radiate incoming waves and increase the system energy. Therefore, we need only isolate the terms in the multipole expansions which distinguish the advanced solution from the retarded solution and use the retarded one. The other terms cannot contribute a net force on the source.

The fields responsible for $\mathbf{F}_{\text{RR}}$ lie firmly in the near zone. Hence, the multipole expansion needed is precisely the one performed in Example 20.4 at the end of Section 20.7.1 to derive Larmor's radiation formula. If a plus/minus subscript identifies an advanced/retarded potential, the low-order terms derived there imply that

$$\mathbf{A}_\pm(\mathbf{r},t) = \frac{\mu_0}{4\pi}\int d^3r'\,\frac{\mathbf{j}(\mathbf{r}',t)}{|\mathbf{r}-\mathbf{r}'|} \pm \frac{\mu_0}{4\pi c}\frac{d}{dt}\int d^3r'\,\mathbf{j}(\mathbf{r}',t) + \cdots \tag{23.146}$$

and

$$\begin{aligned}\phi_\pm(\mathbf{r},t) = \frac{1}{4\pi\epsilon_0}\Bigg[\int d^3r'\,\frac{\rho(\mathbf{r}',t)}{|\mathbf{r}-\mathbf{r}'|} &\pm \frac{1}{c}\frac{d}{dt}\int d^3r'\,\rho(\mathbf{r},t) \\ &+ \frac{1}{2c^2}\frac{d^2}{dt^2}\int d^3r'\,\rho(\mathbf{r}',t)|\mathbf{r}-\mathbf{r}'| \\ &\pm \frac{1}{6c^3}\frac{d^3}{dt^3}\int d^3r'\,\rho(\mathbf{r}',t)|\mathbf{r}-\mathbf{r}'|^2 + \cdots\Bigg].\end{aligned} \tag{23.147}$$

A third-order expansion of the scalar potential is needed because the time derivative of the total charge in (23.147) vanishes by conservation of charge. If $\mathbf{p}(t)$ is the electric dipole moment of the source, Dirac's argument instructs us to read off the radiation reaction potentials from (23.146) and (23.147) as

$$\mathbf{A}_{\rm RR}(\mathbf{r},t) = -\frac{\mu_0}{4\pi c}\frac{d}{dt}\int d^3r'\,\mathbf{j}(\mathbf{r}',t) = -\frac{\mu_0}{4\pi c}\frac{d^2\mathbf{p}(t)}{dt^2} \tag{23.148}$$

and

$$\phi_{\rm RR}(\mathbf{r},t) = -\frac{\mu_0}{24\pi c}\frac{d^3}{dt^3}\int d^3r'\,\rho(\mathbf{r}',t)|\mathbf{r}-\mathbf{r}'|^2. \tag{23.149}$$

The radiation reaction fields follow from $\mathbf{B}_{\rm RR} = \nabla\times\mathbf{A}_{\rm RR}$, $\mathbf{E}_{\rm RR} = -\nabla\phi_{\rm RR} - \partial\mathbf{A}_{\rm RR}/\partial t$, and $\nabla|\mathbf{r}-\mathbf{r}'|^2 = 2(\mathbf{r}-\mathbf{r}')$:

$$\mathbf{E}_{\rm RR}(t) = \frac{\mu_0}{6\pi c}\frac{d^3\mathbf{p}(t)}{dt^3} \qquad \text{and} \qquad \mathbf{B}_{\rm RR} = 0. \tag{23.150}$$

Using (23.150), the retardation-induced force which acts back on, say, a collection of radiating particles with charges q_k is

$$\mathbf{F}_{\rm RR}(t) = \sum_k q_k\mathbf{E}_{\rm RR}(t) = \sum_k \frac{\mu_0 q_k}{6\pi c}\frac{d^3\mathbf{p}(t)}{dt^3}. \tag{23.151}$$

An interesting feature of (23.151) is that the *net* reaction force on a neutral source is *zero* despite the non-zero force felt by each of its constituent charges. Higher-order terms in the expansions (23.146) and (23.147) generate magnetic dipole and electric quadrupole contributions to the radiation reaction.

We complete the connection to Example 20.4 by computing the rate at which the reaction force (23.151) does work on the particles of the source. Because $\mathbf{p}(t) = \sum_k q_k\mathbf{r}_k(t)$ is the electric dipole moment,

$$\frac{dW}{dt} = \sum_k q_k\mathbf{E}_{\rm rr}\cdot\mathbf{v}_k = \frac{\mu_0}{6\pi c}\dot{\mathbf{p}}\cdot\dddot{\mathbf{p}} = \frac{\mu_0}{6\pi c}\left[\frac{d}{dt}(\dot{\mathbf{p}}\cdot\ddot{\mathbf{p}}) - |\ddot{\mathbf{p}}|^2\right]. \tag{23.152}$$

An interesting quantity is the *average* of (23.152) over a time interval T:

$$\langle\dot{W}\rangle = \frac{\mu_0}{6\pi c}\frac{1}{T}\int_0^T dt\left[\frac{d}{dt}(\dot{\mathbf{p}}\cdot\ddot{\mathbf{p}}) - |\ddot{\mathbf{p}}|^2\right]. \tag{23.153}$$

If the total derivative term in (23.153) vanishes for any reason, e.g., $\mathbf{p}(t)$ is periodic with period T, (23.153) simplifies to

$$\langle\dot{W}\rangle = -\frac{\mu_0}{6\pi c}\langle|\ddot{\mathbf{p}}|^2\rangle. \tag{23.154}$$

The work (23.154) is exactly the negative of the average of the rate (20.158) at which energy is carried to infinity by electric dipole radiation (see Section 20.7.1). This shows how conservation of energy works (on average) for one important class of charge/current distributions.

23.6.3 The Lorentz-Abraham Equation

Let us apply the results of the previous section to find the equation of motion for a radiating point particle with charge e and mass m. For simplicity, assume that an external force $\mathbf{F}_{\rm ext}$ causes the particle to accelerate and move slowly through the origin of coordinates at time $t=0$. In that case, the motion-induced electric dipole moment $\mathbf{p} = q\mathbf{r}(t)$ permits us to evaluate the reaction force $\mathbf{F}_{\rm RR}$ in (23.151) and add it to $\mathbf{F}_{\rm ext}$ in Newton's second law. The *Lorentz-Abraham equation* that results is

$$m\mathbf{a} = \mathbf{F}_{\rm ext} + m\tau_0\dot{\mathbf{a}}, \tag{23.155}$$

where τ_0 is (approximately) the time required for light to travel a distance equal to the classical electron radius r_e:

$$\tau_0 = \frac{\mu_0 e^2}{6\pi mc} = \frac{2}{3}\frac{r_e}{c}. \tag{23.156}$$

The radiation reaction force in (23.155) is proportional to a time derivative of acceleration. This is unusual in physics and has many unusual consequences. Not least, (23.155) predicts changes in acceleration when $\mathbf{F}_{\rm ext} = 0$.

The general solution of (23.155) is

$$\mathbf{a}(t) = \exp(t/\tau_0)\left[\mathbf{b} - \frac{1}{m\tau_0}\int_{-\infty}^{t} dt'\,\mathbf{F}_{\rm ext}(t')\exp(-t'/\tau_0)\right], \tag{23.157}$$

where $\mathbf{b}$ is a constant vector. This constant does not occur in a typical Newton's law trajectory problem because the initial conditions $\mathbf{r}(0)$ and $\dot{\mathbf{r}}(0)$ specify the solution uniquely for an equation which is second order in time. Equation (23.155) written in terms of $\mathbf{r}(t)$ is third-order in time and extra information must be supplied. As an example, let a constant external force turn on at $t = 0$ by choosing $\mathbf{F}_{\rm ext}(t) = \mathbf{f}\Theta(t)$. Inserting this into (23.157) gives

$$\mathbf{a}(t) = \exp(t/\tau_0)\left\{\mathbf{b} - \frac{\mathbf{f}}{m}\left[1 - \exp(-t/\tau_0)\right]\Theta(t)\right\}. \tag{23.158}$$

This solution is pathological. Unless $\mathbf{b}$ has a very special value, (23.158) predicts a "runaway" solution where $\mathbf{a}(t) \sim \mathbf{b}\exp(t/\tau_0)$. This is surely unphysical and we are entitled to ask: what has gone wrong and what can be done to correct it?

The problems with (23.155) reach back to the observation in Section 3.6 that the electrostatic self-energy of a point charge is infinite. The relevance of this becomes clear when we use Example 14.3 at the end of Section 14.5 to write the quasistatic vector potential when a point charge with uniform velocity $\mathbf{v}$ moves slowly through $\mathbf{r} = 0$ at $t = 0$:

$$\mathbf{A}(\mathbf{r}, t = 0) = \frac{\mu_0 e\mathbf{v}}{4\pi r}. \tag{23.159}$$

Any change in the velocity $\delta\mathbf{v}$ induces a change in the vector potential $\delta\mathbf{A}$. There is a corresponding change in the magnetic field and, by Lenz' law, a force from the Faraday's law electric field acts back on the charge to oppose the change. Put another way, the force is a consequence of conservation of total linear momentum for a point charge and its velocity and acceleration fields. Either way, there is a reaction force

$$\mathbf{F}_{\rm L} = -e\frac{\partial\mathbf{A}}{\partial t} = -\left[\frac{e^2}{4\pi\epsilon_0 c^2 r}\frac{d\mathbf{v}}{dt}\right]_{r=0}. \tag{23.160}$$

The force (23.160) diverges at the position of the point charge. To make sense of this, we introduce a cutoff at the classical electron radius and write $\mathbf{F}_{\rm L}$ in terms of an electromagnetic energy $U_{\rm em}$ and an equivalent electromagnetic mass $m_{\rm em}$:

$$\mathbf{F}_{\rm L} \approx -\frac{e^2}{4\pi\epsilon_0 r_e}\frac{1}{c^2}\frac{d\mathbf{v}}{dt} \equiv -\frac{U_{\rm em}}{c^2}\frac{d\mathbf{v}}{dt} \equiv -m_{\rm em}\mathbf{a}. \tag{23.161}$$

This force (23.161) supplements $\mathbf{F}_{\rm ext} + \mathbf{F}_{\rm RR}$ when we use Newton's second law to find the acceleration of our particle. Presumably, the latter has some "bare" mass $m_0 > 0$ of non-electromagnetic origin. Therefore, the equation of motion for the particle is

$$m_0\mathbf{a} = \mathbf{F}_{\rm ext} + m\tau_0\dot{\mathbf{a}} - m_{\rm em}\mathbf{a}. \tag{23.162}$$

Comparing (23.162) to (23.155) shows that we should interpret $m = m_0 + m_{\rm em}$ in the latter as a *renormalized* mass. Moreover, $\mathbf{F}_L$ did not show up when we used Dirac's trick to compute $\mathbf{F}_{\rm RR}$ in (23.151) because $\mathbf{F}_L$ contributes to the inertia of the charge rather than to a net force on it. By contrast, the entire right side of (23.162) shows up explicitly when one computes the Coulomb-Lorentz self-force exerted on a spherical source of characteristic size r_e using the *retarded* potentials in (23.146) and (23.147) with all the terms in the multipole expansion retained.[13]

In the next section, we connect the ideas of cutoff and mass renormalization to the problem of runaway solutions. First, however, it is instructive to generalize (23.155) to situations where the point charge moves at relativistic speeds. If $\mathcal{F}_\mu$ is an external four-force, the fact that $\mathcal{A}_\mu = \dot{U}_\mu$ is the four-acceleration suggests the covariant equation of motion,[14]

$$m\mathcal{A}_\mu = \mathcal{F}_\mu + m\tau_0\dot{\mathcal{A}}_\mu \qquad \text{(wrong)}. \tag{23.163}$$

Unfortunately, (23.163) cannot be correct because $\mathcal{A}_\mu U_\mu = 0$ [see (22.58)] implies that any true four-force (of whatever origin) must satisfy $\mathcal{F}_\mu U_\mu = 0$ and a brief calculation confirms that $U_\mu \dot{\mathcal{A}}_\mu \neq 0$. On the other hand, any term we add to the right side of (23.163) must vanish when the particle speed vanishes. This is necessary to recover (23.155) in that limit. Therefore, we replace $\dot{\mathcal{A}}_\mu$ in (23.163) by $\dot{\mathcal{A}}_\mu + pU_\mu$ where p is a Lorentz scalar. Applying the condition $(\dot{\mathcal{A}}_\mu + pU_\mu)U_\mu = 0$ and $U_\mu U_\mu = -c^2$ [see (22.51)] determines p and we get the *Lorentz-Abraham-Dirac* equation:[15]

$$m\mathcal{A}_\mu = \mathcal{F}_\mu + m\tau_0\left(\delta_{\mu\nu} + \frac{U_\mu U_\nu}{c^2}\right)\dot{\mathcal{A}}_\nu = \mathcal{F}_\mu + m\tau_0\left(\dot{\mathcal{A}}_\mu - \frac{\mathcal{A}_\nu\mathcal{A}_\nu}{c^2}U_\mu\right). \tag{23.164}$$

Dirac (1938) derived (23.164) by computing the flux of electromagnetic energy-momentum through an infinitesimally narrow tube surrounding the world line of a point charge. He proposed to eliminate runaway solutions by imposing the condition that $\mathcal{A}_\mu \to 0$ as $t \to \infty$.

23.6.4 The Landau-Lifshitz Equation

It is possible to derive an effective equation of motion for a point charge which includes radiation reaction but which does not have runaway solutions like (23.158). The key observation is that the Lorentz-Abraham equation (23.155) does not have universal validity. To see this, we note that all derivations of (23.155) implicitly let some length scale $R \to 0$ (either the particle size or the distance to a point particle). On the other hand, the discussion surrounding (23.162) shows that mass renormalization introduces an electromagnetic energy $e^2/4\pi\epsilon_0 R$ which cannot exceed mc^2 if the bare mass m_0 is to be positive.[16] Therefore, $R > c\tau_0$ [see (23.156)] is the shortest length scale where the theory can be valid and (23.155) will have cutoff-induced corrections of the order of $m\tau_0\ddot{\mathbf{a}}R/c \approx m\tau_0^2\ddot{\mathbf{a}}$. In other words, if $T = a/\dot{a}$ is a typical time over which the acceleration changes, a physically more meaningful way to write the Lorentz-Abraham equation is

$$\mathbf{a} = \frac{1}{m}\mathbf{F}_{\rm ext} + \tau_0\dot{\mathbf{a}} + O(\tau_0^2/T^2). \tag{23.165}$$

Our discussion shows that (23.155) becomes unphysical when the last term in (23.165) cannot be neglected. The implication that the radiation reaction force $\mathbf{F}_{\rm RR}$ must always remain small compared to $\mathbf{F}_{\rm ext}$ motivated Landau and Lifshitz (1962) to replace (23.165) by its first iterate. Their calculation

[13] This is the method presented by Jackson (1999).

[14] A dot over a Lorentz tensor means a derivative with respect to *proper* time.

[15] The second equality in (23.164) is true because $\mathcal{A}_\mu U_\mu = 0$ implies that $\dot{\mathcal{A}}_\mu U_\mu + \mathcal{A}_\mu \dot{U}_\mu = 0$.

[16] We do not follow authors who permit $m_{\rm em} \to \infty$ and $m_0 \to -\infty$.

begins with the observation that the leading-order version of (23.165) is

$$\mathbf{a} = \frac{1}{m}\mathbf{F}_{\rm ext} + O(\tau_0/T). \tag{23.166}$$

Taking the time derivative of (23.166) and substituting back into (23.165) gives

$$\mathbf{a} = \frac{1}{m}\mathbf{F}_{\rm ext} + \frac{\tau_0}{m}\dot{\mathbf{F}}_{\rm ext} + O(\tau_0^2/T^2). \tag{23.167}$$

This equation is just as accurate as (23.165) but, because there is no $\dot{\mathbf{a}}$ term, it has no runaway solutions or other pathologies of the sort which plague (23.165). These conclusions are supported by more rigorous mathematical analysis than we have presented here.[17]

We conclude that the trajectory of a slowly moving point charge which includes the effect of radiation reaction is best described using the *Landau-Lifshitz equation*,

$$m\dot{\mathbf{v}} = \mathbf{F}_{\rm ext} + \tau_0\dot{\mathbf{F}}_{\rm ext} = \mathbf{F}_{\rm ext} + \tau_0\frac{\partial \mathbf{F}_{\rm ext}}{\partial t} + \tau_0(\mathbf{v}\cdot\nabla)\mathbf{F}_{\rm ext}. \tag{23.168}$$

The second equality in (23.168) recognizes that the total time variation of $\mathbf{F}_{\rm ext}(\mathbf{r}, t)$ experienced by a moving particle is the convective derivative defined in Section 1.3.3. We leave it as an exercise for the reader to derive the covariant analog of (23.168) beginning with the Lorentz-Abraham-Dirac equation (23.164).

Example 23.4 Solve the Landau-Lifshitz equation for a non-relativistic electron in an excited state of an atom modeled as a harmonic oscillator with natural frequency ω_0 damped by radiation reaction. Discuss the resonance characteristics of the spectrum of emitted radiation.

Solution: Let $\mathbf{r}(t)$ be the displacement of the oscillator from equilibrium. The external force is $\mathbf{F}_{\rm ext} = -m\omega_0^2\mathbf{r}$ for an electron of mass m. With this choice, the Landau-Lifshitz equation (23.168) reads

$$m\ddot{\mathbf{r}} = -m\omega_0^2\mathbf{r} - m\tau_0\omega_0^2\dot{\mathbf{r}}.$$

The guess $\mathbf{r}(t) = \mathbf{r}_0\exp(-i\Omega t)$ is a solution if $\Omega^2 + i\tau_0\omega_0^2\Omega - \omega_0^2 = 0$ and we deduce that

$$\Omega = -i\tfrac{1}{2}\omega_0^2\tau_0 \pm \omega_0\sqrt{1-(\omega_0\tau_0/2)^2} \approx -i\tfrac{1}{2}\omega_0^2\tau_0 \pm \omega_0 + O(\tau_0^2).$$

We cannot retain the square root in the equation above because (23.168) is reliable only to first order in τ_0. Therefore, with $\Gamma = \omega_0^2\tau_0$,

$$\mathbf{r}(t) = \mathbf{r}_0\exp(-i\Omega t)\Theta(t) = \mathbf{r}_0\exp(-\tfrac{1}{2}\Gamma t)\exp(-i\omega_0 t)\Theta(t).$$

The spectrum of emitted radiation (23.90) requires the Fourier transform of the electron velocity $\mathbf{v}(t) = \dot{\mathbf{r}}(t)$. Specifically

$$\hat{\mathbf{v}}(\omega) = -i\Omega\mathbf{r}_0\int_0^\infty dt\,\exp[-i(\Omega-\omega)t] = \frac{\Omega}{\omega-\Omega}\mathbf{r}_0 \approx \frac{\omega_0 - i\Gamma/2}{\omega-\omega_0+i\Gamma/2}\mathbf{r}_0,$$

and the spectrum function is

$$\frac{dI(\omega)}{d\Omega} = \frac{\mu_0 q^2\omega^2}{16\pi^3 c}|\hat{\mathbf{v}}(\omega)|^2 \propto \frac{\omega_0^2 + (\Gamma/2)^2}{(\omega-\omega_0)^2 + (\Gamma/2)^2}.$$

[17] See H. Spohn, *Dynamics of Charged Particles and Their Radiation Field* (Cambridge University Press, Cambridge, 2004).

This a called a Lorentzian line shape. Its intensity is largest at the *resonance frequency* ω_0 and falls to half its maximum value at the frequencies $\omega_0 \pm \Gamma/2$. For that reason, $\Gamma = \omega_0^2\tau_0$ is called the *radiative linewidth*. Lorentzian line shapes are the rule when radiative decays occur in Nature.

23.7 Cherenkov Radiation

A charged particle moving at constant speed $v > c/n$ through a medium with index of refraction n emits a characteristic blue light called *Cherenkov radiation*. From a microscopic point of view, it is clear from Application 23.1 that the moving particle itself does not emit the radiation. The source is rather the time-dependent polarization of the medium induced by the motion of the particle. On the other hand, a particle moving at *any* speed through a medium induces transient accelerations of the particles of the medium. Therefore, subtle interference must be at work to restrict the generation of Cherenkov radiation to situations where the particle speed exceeds the phase velocity of light in the medium.

Notwithstanding these remarks, it is convenient from a macroscopic point of view to regard the moving charge as the source of the radiation. This makes the following elementary discussion sufficient to reveal its basic characteristics and practical importance. Figure 23.17(a) shows a point charge (black dot) moving at constant speed $v > c/n$ through a dielectric medium. The small open circles represent the position of the charge at earlier, equally spaced moments in time. The large circles indicate the outer limit of the spherical wave fronts emitted by the particle at those times. The geometry shows that the expanding radiation front is the surface of a cone which is tangent to all the spherical fronts and which has its apex at the position of the charge. The phrase "Mach cone" is often used because the Cherenkov wave front is analogous to the conical wave front formed behind an airplane when it flies at supersonic speeds. The geometry shows that the cone angle θ_C is

$$\sin\theta_C = \frac{c/n}{v}. \tag{23.169}$$

Particle physicists have long exploited (23.169) to measure the speed of charged particles passing through a detector. When combined with an independent measurement of momentum, (22.66) and (22.67) permit the mass and energy of the particle to be determined. More recently, arrays of imaging Cherenkov telescopes have been used by astrophysicists to turn the Earth's atmosphere into a detector for very high-energy gamma rays. The gamma rays create a secondary shower of high-energy charged particles when they strike the atmosphere and the telescopes detect the Cherenkov light produced by the shower.

23.7.1 Potentials and Fields

An analytic approach to Cherenkov radiation recognizes that the retarded potentials (23.2) and (23.3) remain valid as long as we replace ϵ_0, μ_0, c, and $\boldsymbol{\beta} = \mathbf{v}/c$ by ϵ, μ,

$$c_n = \frac{c}{n}, \qquad \text{and} \qquad \boldsymbol{\beta}_n = \frac{\mathbf{v}}{c_n} \tag{23.170}$$

wherever they occur. Therefore, the modified Líenard-Wiechert potentials (23.11) and (23.12) appropriate to the Cherenkov problem are[18]

$$\varphi(\mathbf{r}, t) = \frac{1}{4\pi\epsilon}\left[\frac{q}{R - \boldsymbol{\beta}_n \cdot \mathbf{R}}\right]_{\text{ret}} \qquad \text{and} \qquad \mathbf{A}(\mathbf{r}, t) = \frac{\mu}{4\pi}\left[\frac{q\mathbf{v}}{R - \boldsymbol{\beta}_n \cdot \mathbf{R}}\right]_{\text{ret}}. \tag{23.171}$$

[18] This problem seems ideally suited for solution by Lorentz transformation from the rest frame of the charge. Unfortunately, the calculation is less simple than when the charge moves in vacuum. See Nag and Sayied (1956) in Sources, References, and Additional Reading.

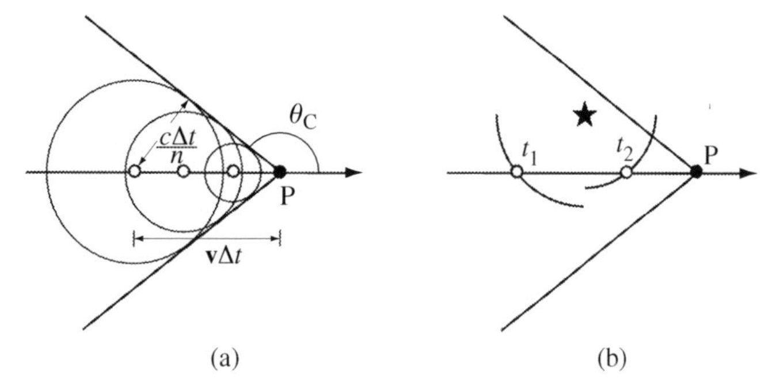

Figure 23.17: The Cherenkov effect for a charged particle (black dot) moving at constant speed. (a) Spherical waves are emitted at previous positions of the particle (small open circles) and expand at speed c/n. (b) When the particle speed $v > c/n$, two information-collecting shells collapse onto every observation point (star) inside the Mach cone at the same observation time. Figure from Ohanian (1995).

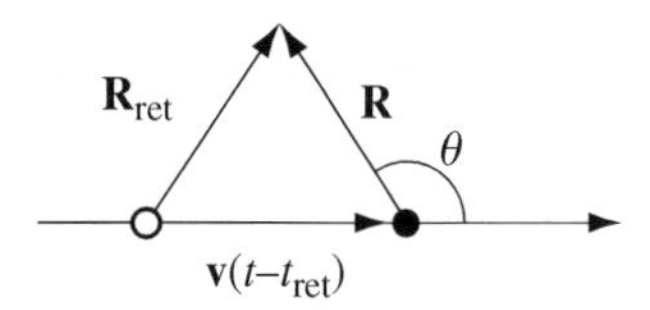

Figure 23.18: The present-time position (black dot) and retarded-time position (small white circle) of a charge moving with constant velocity $\mathbf{v}$. $\mathbf{R}$ and $\mathbf{R}_{\rm ret}$ point to the observation point.

The interesting feature of these equations is that the equation that determines the retarded time,

$$t - t_{\rm ret} = \frac{R(t_{\rm ret})}{c_n}, \tag{23.172}$$

has *two* solutions when the observation point lies inside the Mach cone and *no* solutions when it lies outside the cone. We can demonstrate this graphically using the concept of the "information-collecting shell" introduced in Section 23.2.2.

Figure 23.17(b) shows an observation point (star) inside the Mach cone of a charge moving at constant velocity. Also shown are two views of an information-collecting shell which collapses at speed c_n and arrives at the star at time t. The moving charge enters the volume enclosed by the shell at time $t_1 < t$ and exits that volume at time t_2 where $t > t_2 > t_1$. Both of these are legitimate "retarded times" when $v > c_n$. By contrast, the trajectory $\mathbf{r}_0(t)$ of the charge never enters or exits the volume when the observation point lies outside the Mach cone. There is no retarded time and the field is zero at such points. To be more quantitative, we refer to Figure 23.18 and note that the square of $R_{\rm ret} = |\mathbf{R} + \mathbf{v}(t - t_{\rm ret})|$ is a quadratic equation for $t - t_{\rm ret}$ with the solutions

$$t - t_{\rm ret} = \frac{-\mathbf{v}\cdot\mathbf{R} \pm \sqrt{(\mathbf{v}\cdot\mathbf{R})^2 - (v^2 - c_n^2)R^2}}{v^2 - c_n^2} \geq 0. \tag{23.173}$$

If $v > c_n$, the positivity condition on the far right side of (23.173) imposes two conditions:

$$\mathbf{v}\cdot\mathbf{R} < 0 \qquad \text{and} \qquad (\mathbf{v}\cdot\mathbf{R})^2 > (v^2 - c_n^2)R^2. \tag{23.174}$$

With the definition of θ_C in (23.169), the left and right sides of (23.174) respectively imply that solutions to (23.173) exist if

$$\theta > \frac{\pi}{2} \qquad \text{and} \qquad \sin\theta \leq \sin\theta_C = 1/\beta_n. \tag{23.175}$$

The conditions (23.175) define the volume inside the Mach cone in Figure 23.17(a).

We leave it as an exercise for the reader to confirm that the two retarded solutions in (23.173) make identical contributions to the denominators in (23.171). In that case,

$$\varphi(\mathbf{r},t) = \frac{1}{2\pi\epsilon}\frac{q}{R(1-\beta_n^2\sin^2\theta)^{1/2}}\Theta(\cos\theta_{\rm C}-\cos\theta) \tag{23.176}$$

and

$$\mathbf{A}(\mathbf{r},t) = \frac{\mu}{2\pi\epsilon}\frac{q\mathbf{v}}{R(1-\beta_n^2\sin^2\theta)^{1/2}}\Theta(\cos\theta_{\rm C}-\cos\theta). \tag{23.177}$$

Now, if the black dot in Figure 23.18 is the origin of polar coordinates,

$$\nabla\Theta(\cos\theta_{\rm C}-\cos\theta) = \frac{\sin\theta}{R}\delta(\cos\theta_{\rm C}-\cos\theta)\hat{\boldsymbol{\theta}}. \tag{23.178}$$

Also, if $R_\parallel = \mathbf{R}\cdot\hat{\mathbf{v}}$, we have $d(\cos\theta)/dt = (\sin^2\theta/R)\dot{R}_\parallel = -(\upsilon\sin^2\theta)/R$, so

$$\frac{\partial}{\partial t}\Theta(\cos\theta_{\rm C}-\cos\theta) = \frac{\upsilon\sin^2\theta}{R}\delta(\cos\theta_{\rm C}-\cos\theta). \tag{23.179}$$

Therefore, the electromagnetic fields $\mathbf{E} = -\nabla\varphi - \partial\mathbf{A}/\partial t$ and $\mathbf{B} = \nabla\times\mathbf{A}$ are

$$\begin{aligned}\mathbf{E} = &-\frac{q}{2\pi\epsilon}\frac{(\beta_n^2-1)\hat{\mathbf{R}}}{R^2(1-\beta_n^2\sin^2\theta)^{3/2}}\Theta(\cos\theta_{\rm C}-\cos\theta)\\ &+\frac{q}{2\pi\epsilon}\frac{\sqrt{\beta_n^2-1}\,\hat{\mathbf{R}}}{R^2\beta_n(1-\beta_n^2\sin^2\theta)^{1/2}}\delta(\cos\theta_{\rm C}-\cos\theta)\end{aligned} \tag{23.180}$$

and

$$\mathbf{B} = \frac{\upsilon}{c_n^2}\sin\theta(-\hat{\boldsymbol{\theta}}\times\mathbf{E}). \tag{23.181}$$

The fields (23.180) have a number of peculiar characteristics. One of these is that the electric field at observation points inside the Mach cone points directly *toward* the moving charge rather than away from it. This would contradict the integral form of Gauss' law for a point charge except that the (singular) electric field on the cone itself points *away* from the charge. Gauss' law is properly satisfied when both contributions to the electric field are considered. The mere fact that the fields are singular at $\theta = \theta_{\rm C}$ is troubling until we learn (see below) that the frequency spectrum of Cherenkov radiation is continuous. Combining this with the fact (see Chapter 18) that the index of refraction of any real medium is a function of frequency tells us that we should have let $n = n(\omega)$ from the start. Roughly speaking, frequency dispersion combined with (23.169) implies that there is a different $\theta_{\rm C}$ for every frequency. This blurring of the cone edge is enough to smooth out the delta function singularity into a large and narrow maximum which constitutes the Cherenkov "signal".

23.7.2 The Frequency Spectrum

A stationary observer located outside the Mach cone sees no electromagnetic field until the motion of the charge causes the cone surface to sweep over the observer's position. The abrupt appearance of the short-lived Cherenkov "signal" (see just above) implies (through the reciprocal relation between Δt and $\Delta\omega$ in a Fourier transform) that the spectral width of the observed radiation is large. If we set ourselves the goal to find this spectrum function, Section 23.4.1 directs us to focus on the Fourier transform $\hat{\mathbf{E}}(\omega)$ of the electric field $\mathbf{E}(t)$.

To calculate $\mathbf{E}(t)$, it is convenient to fix the origin of a polar coordinate system at the position of the charge q. This origin moves as the charge moves, so the polar coordinates of a stationary observer change as a function of time. Let us assume that the surface of the Mach cone passes over the observer

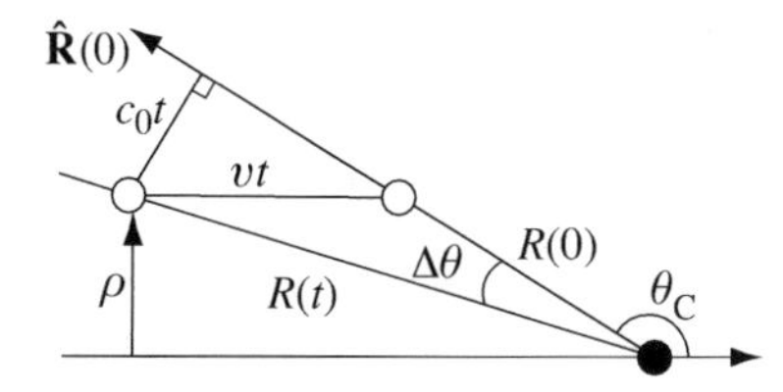

Figure 23.19: The polar coordinates of a stationary laboratory observer change as a function of time (small white circles) when they are referred to coordinate axes centered at the position of a moving point charge (black dot). Figure adapted from Smith (1997).

at $t = 0$. In that case, Figure 23.19 shows that the polar coordinates of the observer are $(R(0), \theta_C)$. A short time later, the observer finds him/herself inside the cone with polar coordinates $(R(t), \theta(t))$ where $\theta(t) = \theta_C + \Delta\theta$. This information, and the assumptions that $R(0) \approx R(t) \gg vt$ and $\Delta\theta \ll 1$, will permit us to write the Cherenkov electric field (23.180) as an explicit function of time.

Using (23.169) and $\beta_n = v/c_n = nv/c$, we infer from the geometry of Figure 23.19 that

$$\Delta\theta \approx vt\frac{\sin\theta_C}{R(0)} \tag{23.182}$$

and

$$\cos\theta_C - \cos\theta(t) = \cos\theta_C - \cos(\theta_C + \Delta\theta) \approx \sin\theta_C\Delta\theta = \frac{c_n t}{\beta_n R(0)}. \tag{23.183}$$

Then, because $\beta_n^2 \sin^2\theta_C = 1$,

$$1 - \beta_n^2 \sin^2\theta(t) \approx -2\beta_n^2 \cos\theta_C \sin\theta_C \Delta\theta = \frac{2c_n t\sqrt{1-\beta_n^2}}{R(0)}. \tag{23.184}$$

Substituting these results into (23.180) gives the electric field as

$$\mathbf{E}(t) \approx \frac{q(\beta_n^2 - 1)^{1/4}}{(2c_n)^{3/2}\pi\epsilon\sqrt{R(0)}}\left[\frac{\delta(t)}{\sqrt{t}} - \frac{\Theta(t)}{2t^{3/2}}\right]\hat{\mathbf{R}}(0). \tag{23.185}$$

The Fourier transform $\hat{\mathbf{E}}(\omega)$ follows by substituting (23.185) into the rightmost integral in (23.75). Integrating the delta function by parts, we find

$$\hat{\mathbf{E}}(\omega) = -\frac{i\omega q(\beta_n^2-1)^{1/4}}{(2c_n)^{3/2}\pi\epsilon\sqrt{R(0)}}\hat{\mathbf{R}}(0)\int_0^\infty dt\, t^{-1/2}\exp(i\omega t) = \frac{q(\beta_n^2-1)^{1/4}\omega^{1/2}}{4\sqrt{\pi}\epsilon c_n^{3/2}\sqrt{R(0)}}(1-i)\hat{\mathbf{R}}(0). \tag{23.186}$$

The fact that $\mathbf{E}(t)$ and $\hat{\mathbf{E}}(\omega)$ vary as $1/\sqrt{R(0)}$ tells us that we are dealing with a two-dimensional cylindrical wave rather than a three-dimensional spherical wave.[19] This motivates us to alter slightly the analysis of Section 23.4.1 and evaluate the energy radiated per unit *area* per unit frequency [cf. (23.80)]:

$$\frac{d^2U_{\rm rad}}{d\omega dA} = \frac{\epsilon c_n}{\pi}|\hat{\mathbf{E}}(\omega)|^2. \tag{23.187}$$

The literature of Cherenkov radiation typically transforms (23.187) into a spectral distribution per unit path length ℓ. This the energy per unit frequency which passes through a unit length of a cylinder of

[19] See the discussion surrounding (20.86).

radius ρ centered on the trajectory of the moving particle. A glance back at Figure 23.19 shows that this quantity is

$$\frac{d^2U_{\rm rad}}{d\omega d\ell} = 2\pi\rho[\hat{\boldsymbol{\rho}}\cdot(-\hat{\boldsymbol{\theta}})]\frac{d^2U_{\rm rad}}{d\omega dA}. \tag{23.188}$$

Using (23.186) to evaluate (23.187) and (23.188) gives the final result,

$$\frac{d^2U_{\rm rad}}{d\omega d\ell} = -2\pi R(0)\sin\theta_C\cos\theta_C\frac{d^2U_{\rm rad}}{d\omega dA} = \frac{\mu q^2}{4\pi}\left(1-\frac{c^2}{v^2n^2}\right)\omega. \tag{23.189}$$

Cherenkov radiation appears blue to the naked eye because (23.189) is an increasing function of frequency.[20] A more reliable statement requires the generalization of (23.189) to a medium with frequency-dependent index of refraction $n(\omega)$:

$$\frac{d^2U_{\rm rad}}{d\omega d\ell} = \frac{\mu(\omega)q^2}{4\pi}\left(1-\frac{c^2}{v^2n^2(\omega)}\right)\omega. \tag{23.190}$$

Consider the graph of $n(\omega)$ for SiO_2 glass in Figure 18.5. This curve, which is typical of transparent dielectrics, confirms our conclusion that Cherenkov light is blue because $n(\omega)$ does not vary much in the visible part of the spectrum. On the other hand, Figure 18.5 predicts that there will be *no* Cherenkov emission at frequencies beyond the near-ultraviolet because $n(\omega) < 1$ there and it is not possible to satisfy (23.169) with a real angle θ_C. This is confirmed by experiment.

Sources, References, and Additional Reading

The quotation at the beginning of the chapter is the first sentence from

O. Heaviside, "The radiation from an electron moving in an elliptic, or any other orbit", *Nature* **69**, 342 (1904).

Section 23.1 Good textbook treatments of the fields produced by moving charges are

L.D. Landau and E.M. Lifshitz, *The Classical Theory of Fields*, 2nd edition (Addison-Wesley, Reading, MA, 1962).

J. D. Jackson, *Classical Electrodynamics*, 3rd edition (Wiley, New York, 1999).

E. J. Konopinski, *Electromagnetic Fields and Relativistic Particles* (McGraw-Hill, New York, 1981).

G. S. Smith, *Classical Electromagnetic Radiation* (Cambridge University Press, Cambridge, 1997).

Section 23.2 To the best of our knowledge, the delta function and information-gathering shell approaches to the Liénard-Wiechert potentials appeared first in *Oppenheimer* and *Abraham*, respectively. *Sommerfeld* provided the details for Minkowski's covariant method.

J. R. Oppenheimer, *Lectures on Electrodynamics* (Gordon & Breach, New York, 1970).

M. Abraham, *Elektromagnetische Theorie der Strahlung* (Teubner, Leipzig, 1905).

A. Sommerfeld, "On the theory of relativity II: Four-dimensional vector analysis" (in German), *Annalen der Physik* **33**, 649 (1910).

Equation (23.49) for the electric field of a moving point charge was first discussed by *Heaviside* (in the paper cited above as the source of the quotation at the beginning of this chapter). It was rediscovered years later by *Feynman et al.*:

R.P. Feynman, R.B. Leighton, and M. Sands, *The Feynman Lectures on Physics* (Addison-Wesley, Reading, MA, 1963), Volume I, Chapters 28 and 34; Volume II, Chapter 21.

[20] See the last paragraph of Section 21.4.1 for the role of the human eye in suppressing violet.

Figure 23.3 comes from *Purcell*. *Thomson* is the original source for the electric field line argument. *Ohanian* discusses the magnetic field.

E. Purcell, *Electricity and Magnetism* (McGraw-Hill, New York, 1965), Chapter 5.

J.J. Thomson, *Electricity and Matter* (Charles Scribner's Sons, New York, 1904).

H.C. Ohanian, "Electromagnetic radiation fields: A simple approach via field lines", *American Journal of Physics* **48**, 170 (1980).

Example 23.1 was adapted from

S. Ruschin, "Free electron laser diagnosis by polarization analysis of Thomson backscattered light", *Journal of Applied Physics* **69**, 1923 (1991).

Section 23.3 *Teitelboim* proves that $(\mathbf{P}_{\rm EM}, iU_{\rm EM}/c)$ is a four-vector when the acceleration field of a moving charged particle is integrated over any finite volume. Example 23.2 was adapted from *Jackson* (see Section 23.1 above) and *Panofsky*.

C. Teitelboim, "Splitting of the Maxwell tensor: Radiation reaction without advanced fields", *Physical Review D* **1**, 1572 (1970).

W.K.H. Panofsky, "Special relativity in engineering", in *Some Strangeness in the Proportion*, edited by H. Woolf (Addison-Wesley, New York, 1980), pp. 94-105.

Section 23.4 *Jackson* and *Konopinski* (see Section 23.1 above) have particularly good treatments of the Liénard-Wiechert spectrum. The Heaviside-Feynman spectrum is derived in

C. Wang, "Concise expression of a classical radiation spectrum", *Physical Review E* **47**, 4358 (1993).

Section 23.5 *Margaritondo* uses qualitative arguments and minimal mathematics to derive the basic properties of synchrotron radiation. The frequency spectrum was derived by *Schott* and is the subject of a famous article by *Schwinger*. *Wiedemann* provides mathematical details we omit.

G. Margaritondo, "A primer in synchrotron radiation: Everything you always wanted to know about sex (synchrotron emission of x-rays) but were afraid to ask", *Journal of Synchrotron Radiation* **2**, 148 (1995).

G.A. Schott, *Electromagnetic Radiation* (Cambridge University Press, Cambridge, 1912).

J Schwinger, "On the classical radiation of accelerated electrons", *Physical Review* **75**, 1912 (1949).

H. Wiedemann, *Synchrotron Radiation* (Springer, Berlin, 2003).

Our scaling form of the electric field comes from unpublished lecture notes by Prof. Michael Cohen (University of Pennsylvania). The graphical treatment of synchrotron radiation appears in Chapter 34 of Volume I of *Feynman et al.* (see Section 23.2 above). The derivation of $dI/d\Omega$ for synchrotron radiation from the Heaviside-Feynman electric field is due to *Wang* (see Section 23.4 above). Figure 23.14 and Section 23.5.5 come from

J.J. Condon and S.M. Ransom, *Essential Radio Astronomy*, 2010. www.cv.nrao.edu/course/ astr534/ERA.shtml

Section 23.6 *Dirac* stimulated an avalanche of research on radiation reaction. Our discussion follows *Landau and Lifshitz* (see Section 23.1 above) and unpublished lecture notes by Prof. Eric Poisson (University of Guelph).

P.A.M. Dirac, "Classical theory of radiating electrons", *Proceedings of the Royal Society of London A* **167**, 148 (1938).

E. Poisson, *Electromagnetic Theory*, www.physics.uoguelph.ca/poisson/research/em.pdf

The quotation from Bohr comes from

N. Bohr, "On the constitution of atoms and molecules", *Philosophical Magazine* **26**, 1 (1913).

For more on the "classical electron", see *Jackson* (see Section 23.1 above) and

F. Rohrlich, *Classical Charged Particles* (Addison-Wesley, Reading, MA, 1965).

A.D. Yaghjian, *Relativistic Dynamics of a Charged Sphere*, 2nd edition (Springer, New York, 2006).

Section 23.7 This section follows the excellent treatment of Cherenkov radiation in *Smith* (see Section 23.1 above), which is also the source of Figure 23.19. Figure 23.17 comes from *Ohanian*. *Levich* treats the effects of frequency dispersion and *Nag and Sayied* use special relativity and Minkowski's electrodynamics of moving media to find the fields and the frequency spectrum.

H.C. Ohanian, *Classical Electrodynamics* (Allyn & Bacon, New York, 1995).

B.G. Levich, *Theoretical Physics* (North-Holland, Amsterdam, 1971), Volume 2, Section 39.

B.D. Nag and A.M. Sayied, "Electrodynamics of moving media and the theory of the Cherenkov effect", *Proceedings of the Royal Society of London* A **235**, 544 (1956).

Problems

23.1 Smith-Purcell and Undulator Radiation

(a) The side view to the left below shows an electron with velocity $\mathbf{v} = v\hat{\mathbf{z}}$ skimming over a diffraction grating composed of a periodic array of metal strips with periodicity L. Explain why radiation is produced. Use a constructive interference argument to show that the radiation is observed at an angle θ above the horizontal at the wavelengths $\lambda_n = (c/v - \cos\theta)L/n$ where n is a positive integer. This is called Smith-Purcell radiation.

(b) The side view to the right below shows an electron with *average* velocity $\mathbf{v} = v\hat{\mathbf{z}}$ skimming over the surface of a periodic array of permanent magnets with periodicity L. Explain why radiation is produced. Why does the radiation observed at an angle θ above the horizontal occur at the same wavelengths as in part (a)? This is called undulator radiation.

Hint: You may assume that the electron speed is relativistic.

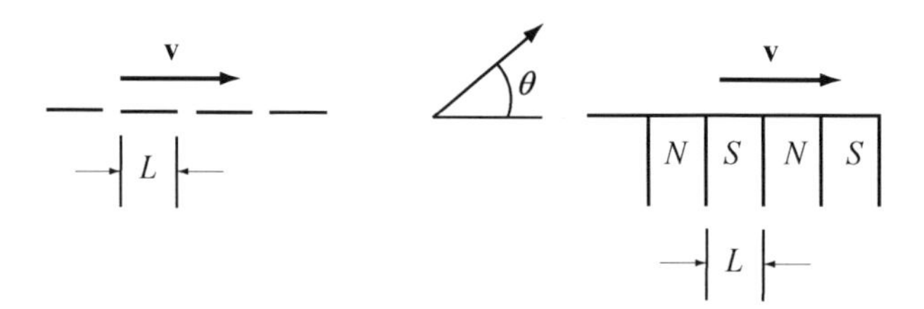

23.2 Gauss' Law for a Moving Charge Show by direct integration (in the laboratory frame) that the electric field of a point charge q moving with constant speed v in the x-direction satisfies Gauss' law in integral form.

23.3 The Retarded Time A point charge q moves along a specified trajectory $\mathbf{r}_0(t)$ with velocity $\mathbf{v}(t) = \dot{\mathbf{r}}_0(t)$. For each choice of t, show that the equation $t_{\rm ret} = t - |\mathbf{r} - \mathbf{r}_0(t_{\rm ret})|/c$ has exactly one solution for the retarded time $t_{\rm ret}$, provided $|\mathbf{v}(t)| < c$.

23.4 The Direction of the Velocity Field Prove that the "velocity" part of the Liénard-Wiechert electric field points to the observer from the "anticipated position" of the moving point charge. The latter is the position the charge *would* have moved to *if* it retained the velocity $\mathbf{v}_{\rm ret}$ from $t = t_{\rm ret}$ to the present time of observation.

23.5 Inverting the Retarded Field This problem reconstructs the trajectory $\mathbf{r}_0(t)$ of a charged particle from the fields produced by the particle at observation points where the magnetic field does not vanish.

(a) Use $c\mathbf{B} = \hat{\mathbf{n}}_{\rm ret} \times \mathbf{E}$ to deduce that

$$\hat{\mathbf{n}}_{\rm ret} \cdot \mathbf{E} = \frac{q}{|q|}\sqrt{E^2 - c^2B^2}.$$

Hint: The Liénard-Wiechert electric field determines the factor $q/|q|$.

(b) Use the information in part (a) to show that

$$\hat{\mathbf{n}}_{\rm ret} = \frac{\mathbf{E} \times \mathbf{B} + \mathbf{E}(q/|q|)\sqrt{E^2 - c^2B^2}}{E^2}.$$

(c) Evaluate the formula for $c\mathbf{B}$ in part (a) using the Heaviside-Feynman electric field, solve for $R_{\rm ret}$, and use the result of part (b) to conclude that

$$\mathbf{R}_{\rm ret} = \frac{q\,\hat{\mathbf{n}}_{\rm ret}\mathbf{B}\cdot(\hat{\mathbf{n}}\times\dot{\mathbf{n}})_{\rm ret}}{4\pi\epsilon_0 c^2 B^2 - (q/c)\mathbf{B}\cdot(\hat{\mathbf{n}}\times\ddot{\mathbf{n}})_{\rm ret}}.$$

(d) Part (b) and (c) express $\mathbf{R} = R\hat{\mathbf{n}} = \mathbf{r} - \mathbf{r}_0$ at the retarded time entirely in terms of the fields at the present time. Show that we also have enough information to express $t_{\rm ret}$ entirely in terms of present-time quantities.

23.6 The Covariant Liénard-Wiechert Field

(a) Calculate $F_{\mu\nu} = \partial_\mu A_\nu - \partial_\nu A_\mu$ from the Liénard-Wiechert potential A_μ. Express the result in terms of R_μ, the four-velocity U_μ, and the four-acceleration $\partial U_\mu/\partial\tau$. Hint: It will be useful to evaluate $\partial_\mu(R_\sigma R_\sigma)$ in order to show that $\partial_\mu \tau = R_\mu/(R_\sigma U_\sigma)$.
(b) Show that the splitting of $F_{\mu\nu}$ into a velocity part and an acceleration part is relativistically invariant.
(c) Find $\mathbf{E}$ and $\mathbf{B}$ when $\mathbf{v} = 0$ (but $\dot{\mathbf{v}} \neq 0$) by evaluating appropriate components of $F_{\mu\nu}$ in the rest frame of the charge.

23.7 *N* Charges Moving in a Circle I N identical, equally-spaced point particles, each with charge q, move in a circle of radius a. Each particle moves with the same constant speed v around the ring. Show that the Liénard-Wiechert electric field is *static* everywhere on the symmetry axis.

23.8 Energy Loss from Gyro-Radiation A particle with charge q and mass m moves in a uniform external magnetic field $\mathbf{B}$. Find the total rate at which the particle loses energy by radiation when its motion is relativistic.

23.9 The Path of Minimum Radiation A non-relativistic charged particle begins at rest, moves in a straight line, and then comes back to rest. The total journey of distance d takes a time T to complete.

(a) Find the total amount of energy radiated if the particle accelerates at a constant rate for the first half of its journey and decelerates at a constant rate for the second half of its journey.
(b) Find the acceleration function $a(t)$ so the journey radiates the least amount of energy.

23.10 Radiation Energy Loss from Coulomb Repulsion A non-relativistic particle with charge e and mass m begins at infinity with initial speed v_0 and collides head-on with a fixed field of force. The force is Coulombic with potential $V(r) = Ze^2/r$. Integrate Larmor's formula to show that the total energy lost by the particle to radiation is $\Delta E = 2mv_0^5/45\pi\epsilon_0 Zc^3$.

23.11 Frequency of Dipole Radiation A non-relativistic particle with charge q follows a trajectory $\mathbf{r}_0(t) = R[\hat{\mathbf{x}}\cos(\omega_1 t)\cos(\omega_2 t) + \hat{\mathbf{y}}\sin(\omega_2 t)]$. Identify the frequencies at which dipole radiation occurs.

23.12 Larmor's Formula with Fields Displayed Write Larmor's formula for the power radiated by a charged particle in a manifestly covariant form which explicitly displays the electromagnetic field experienced by the particle. Evaluate this formula in an inertial frame where the particle has velocity $c\boldsymbol{\beta}$.

23.13 Emission Rates by Lorentz Transformation An electron enters and exits a capacitor with parallel-plate separation d through two small holes. The electron velocity $v\hat{\mathbf{z}}$ is parallel to the capacitor electric field $\mathbf{E}$ and the change in the electron velocity is small. Calculate the total energy $\Delta U'_{\rm EM}$ and linear momentum $\Delta P'_{\rm EM}$ radiated by the electron in its (momentary) rest frame and Lorentz transform to the laboratory frame to find $\Delta U_{\rm EM}$ and $\Delta P_{\rm EM}$.

23.14 Emission Rates by Explicit Integration

(a) Use direct integration to show that a radiating charge q emits energy at the rate

$$\frac{dU_{EM}}{dt} = \int_A d\mathbf{A}\cdot g\,\mathbf{S} = \frac{2\gamma^6}{3c^3}\frac{q^2}{4\pi\epsilon_0}\left[|\mathbf{a}|^2 - |\mathbf{a}\times\boldsymbol{\beta}|^2\right].$$

In this expression, $\mathbf{S}$ is the Poynting vector, $\boldsymbol{\beta} = \mathbf{v}/c$, $\mathbf{a} = \dot{\mathbf{v}}$, $g = 1 - \hat{\mathbf{n}} \cdot \boldsymbol{\beta}$, and $\gamma = 1/\sqrt{1-\beta^2}$. The integration is carried out over the surface A of an enormously large sphere. Hint: Show first that

$$I = \int \frac{d\Omega}{(1 - \hat{\mathbf{n}} \cdot \boldsymbol{\beta})^3} = \frac{4\pi}{(1-\beta^2)^2}$$

and then show that

$$J_i = \int \frac{n_i d\Omega}{(1 - \hat{\mathbf{n}} \cdot \boldsymbol{\beta})^4} = \frac{1}{3}\frac{\partial I}{\partial \beta_i} \qquad \text{and} \qquad K_{ij} = \int \frac{n_i n_j d\Omega}{(1 - \hat{\mathbf{n}} \cdot \boldsymbol{\beta})^5} = \frac{1}{4}\frac{\partial J_i}{\partial \beta_j}.$$

(b) Use direct integration to show that a radiating charge q emits linear momentum at the rate

$$\frac{d\mathbf{P}_{\text{EM}}}{dt} = -\int_A d\mathbf{A} \cdot g\mathbf{T} = \frac{\boldsymbol{\beta}}{c}\frac{dU_{\text{EM}}}{dt}.$$

In this expression, $T_{ij} = \epsilon_0 \left\{ E_i E_j + c^2 B_i B_j - \frac{1}{2}\delta_{ij}(E^2 + B^2) \right\}$ is the electromagnetic stress tensor. Hint: Show first that

$$M = \int \frac{d\Omega}{(1 - \hat{\mathbf{n}} \cdot \boldsymbol{\beta})^2} = \frac{4\pi}{1-\beta^2},$$

and use successive differentiation to derive other integrals you need.

23.15 The Radiated Power Spectrum of a Linear Oscillator A point charge oscillates with the trajectory $\mathbf{r}_0(t) = a\cos(\omega_0 t)\hat{\mathbf{z}}$. Find the angular distribution of power radiated into the mth harmonic, $dP_m/d\Omega$, during one period of the motion. Which harmonics dominate in the non-relativistic limit? Two Bessel function identities will be useful:

$$J_m(x) = \frac{i^m}{2\pi}\int_0^{2\pi} d\phi \, \exp[i(m\phi - x\cos\phi)] \qquad \text{and} \qquad J_{m+1}(x) + J_{m-1}(x) = (2m/x)J_m(x).$$

23.16 The Radiation Spectrum of Beta Decay Model the beta decay reaction $n \to p + e + \bar{\nu}_e$ as the abrupt creation of an electron at $t = 0$ with constant velocity $\mathbf{v} = c\boldsymbol{\beta}$.

(a) Find the angular distribution of energy radiated per unit frequency, $dI/d\Omega$. Use θ for the angle between $\mathbf{v}$ and the observation point. Hint: Consider the use of a convergence factor if you encounter an ill-behaved integral.

(b) Show that the total energy radiated per unit frequency is

$$I(\omega) = \frac{\mu_0 q^2 c}{4\pi^2}\left[\frac{1}{\beta}\ln\left(\frac{1+\beta}{1-\beta}\right) - 2\right].$$

(c) The fact that $dI/d\Omega$ and $I(\omega)$ are independent of frequency implies that the total amount of energy radiated is infinite. How must our model of beta decay be corrected to eliminate this unphysical behavior?

23.17 Energy Loss and Electric Field Spectrum A charged particle produces an electric field $\mathbf{E}(\mathbf{r}, t)$ as it passes by an isolated atom. Model a bound electron in the atom as a damped harmonic oscillator with natural frequency ω_0 and damping constant Γ. Assume that the electric field induces small-amplitude, non-relativistic motion of the bound electron near the origin of coordinates. If $\hat{\mathbf{E}}(\omega)$ is the Fourier transform of $\mathbf{E}(\mathbf{r} = 0, t)$, show that the energy transferred to the bound electron from the field is

$$\Delta E = \frac{e^2}{m\pi}\int_0^{\infty} d\omega \, |\hat{\mathbf{E}}(\omega)|^2 \frac{\omega^2 \Gamma}{(\omega^2 - \omega_0^2)^2 + \omega^2\Gamma^2}.$$

What is ΔE as $\Gamma \to 0$?

23.18 Angular Distribution of Radiated Frequency Harmonics A particle with charge q follows a periodic trajectory where $\mathbf{r}_0(t) = \mathbf{r}_0(t + T)$. If $\omega_0 = 2\pi/T$, prove that the angular distribution of the average power radiated into the mth harmonic during one period of the motion is

$$\frac{dP_m}{d\Omega} = \frac{\mu_0 q^2 m^2 \omega_0^4}{32\pi^4 c}\left|\hat{\mathbf{r}} \times \int_0^{2\pi/\omega_0} dt\, \mathbf{v}(t) \exp\{-im\omega_0[\hat{\mathbf{r}}\cdot\mathbf{r}_0(t)/c - t]\}\right|^2 .$$

23.19 ***N* Charges Moving in a Circle II** N identical point particles, each with charge q, move in a circle of radius a. The angular position of the jth particle is $\phi_j(t) = \omega_0(t - t_0) + \theta_j$.

(a) Prove that the power spectrum satisfies

$$\left.\frac{dP_m}{d\Omega}\right|_{\text{N}} = \left|\sum_{j=1}^{N} \exp(-im\theta_j)\right|^2 \left.\frac{dP_m}{d\Omega}\right|_1 .$$

(b) Specialize to the case where the charges are equally spaced around the circle at all times. Use part (a) to prove that this configuration radiates only at the frequencies $kN\omega_0$ where k is a positive integer.
(c) Show that $dP_m/d\Omega|_N \sim N$ when the charges are distributed randomly around the ring.

23.20 **Covariant Radiation of Energy-Momentum** Construct a covariant expression for the rate at which a moving charged particle loses total energy-momentum $P_\mu = (\mathbf{P}_{\text{rad}}, iU_{\text{rad}}/c)$. Evaluate your expression in an arbitrary inertial frame as a check.

23.21 **Lorentz Transformation of $dP/d\Omega$** Show that the angular distribution of power emitted by a moving point charge transforms like

$$\frac{dP}{d\Omega} = \gamma^2(1+\beta\cos\theta')^3\frac{dP'}{d\Omega'} = \frac{1}{\gamma^4(1+\beta\cos\theta)^3}\frac{dP'}{d\Omega'}.$$

Hint: Use the fact that $(\mathbf{P}_{\text{rad}}, iU_{\text{rad}}/c)$ is a four-vector for a finite volume of radiation.

23.22 **Cyclotron Motion with Radiation Reaction** A non-relativistic particle with charge q performs circular cyclotron motion in a uniform magnetic field $\mathbf{B} = B\hat{\mathbf{z}}$. Include the radiation reaction force $m\tau_0\ddot{\mathbf{v}}$ in the equation of motion and solve it assuming that the motion remains approximately circular. Find the time constant for the decay of the particle velocity when the reaction is weak.

23.23 **Radiation Pressure Due to Radiation Reaction** An electron is scattered by an electromagnetic plane wave $\mathbf{E}_0 \exp[i(\mathbf{k}\cdot\mathbf{r} - \omega t)]$. Show that radiation reaction induces a small, time-averaged, self-Lorentz force on the electron which may be interpreted as radiation pressure exerted by the wave. Express the force is terms of the Thomson scattering cross section σ_{T}.

23.24 **Angular Momentum Decay by Radiation Reaction** Consider a classical atom where a particle with mass m and charge q orbits a fixed charge Zq non-relativistically. If L_z is the component of the angular momentum $\mathbf{L}$ that characterizes the orbit, show that the radiation reaction force $m\tau_0\ddot{\mathbf{v}}$ causes the energy and angular momentum of the "atom" to decay together such that

$$\frac{dL_z}{dE} = \frac{1}{2}\frac{L_z}{|E|}.$$

Hint: The force of radiation reaction is much smaller than the force between q and Zq.

23.25 **Covariant Landau-Lifshitz Equation** An electromagnetic field $F_{\mu\nu}$ accelerates a relativistic particle with charge e and mass m. Show that the covariant Landau-Lifshitz equation for this situation is

$$m\dot{U}_\mu = eU_\alpha F_{\mu\alpha} - \frac{e^2\tau_0}{m}U_\alpha F_{\alpha\beta}F_{\mu\beta} + e\tau_0 U_\beta U_\alpha \partial_\alpha F_{\mu\beta} - \frac{e^2\tau_0}{c^2}(F_{\beta\alpha}U_\alpha)^2 U_\mu .$$

24 Lagrangian and Hamiltonian Methods

Larmor had an intense, almost mystical devotion to the principle of least action ... To [him] it was the ultimate natural principle—the mainspring of the Universe.
Arthur Eddington (1942)

24.1 Introduction

This chapter provides an introduction to the use of Lagrangian and Hamiltonian methods in classical electrodynamics. Our goal is to demonstrate that the powerful variational methods developed to derive the equations of motion and conservation laws for conventional mechanical systems can be extended to describe electrodynamics. By its nature, the material in this chapter is rather formal and most of our attention focuses on deriving the Maxwell equations and the Coulomb-Lorentz force law from a single Lagrangian or Hamiltonian. The new physics we will encounter bears principally on the gauge invariance of the theory. At the Lagrangian level, we will show that gauge invariance implies conservation of charge and vice versa. At the Hamiltonian level, we will show that electrodynamics is an example of a *constrained* dynamical system and that the maintenance of the constraints exploits gauge invariance in an essential way.

Our main theoretical tool is *Hamilton's principle* of stationary action. Originally conceived in the context of geometrical optics—and then extended to include mechanical systems—Hamilton's principle determines the equations of motion for any system where generalized coordinates can be sensibly defined. In the most familiar examples, a small number of degrees of freedom are sufficient to characterize the system of interest. The method becomes applicable to Maxwell's theory when we regard the electromagnetic field and/or the electromagnetic potentials at every point in space as independent generalized coordinates. These are treated on an equal footing with the coordinates of mobile charged particles. Among other things, the connection between conservation laws and the invariance properties of the Lagrangian known for simple mechanical systems survives the generalization to Maxwell's field theory.

This chapter passes freely back and forth between non-covariant and covariant notation. The former helps fix ideas and makes immediate contact with most readers' previous experience with variational methods. The latter is the natural language of the relativistic field theories developed in the second half of the 20th century as generalizations of electromagnetic field theory.

24.2 Hamilton's Principle

Let a closed mechanical system be described by a set of generalized coordinates and generalized velocities,

$$q_1(t), q_2(t), \dots, q_N(t) \qquad \text{and} \qquad \dot{q}_1(t), \dot{q}_2(t), \dots, \dot{q}_N(t). \tag{24.1}$$

Classical mechanics teaches us that a scalar function $L[q_k(t), \dot{q}_k(t)]$ called the *Lagrangian*[1] determines the dynamical behavior of each generalized coordinate via Lagrange's equations,

$$\frac{d}{dt}\left(\frac{\partial L}{\partial \dot{q}_k}\right) = \frac{\partial L}{\partial q_k}. \tag{24.2}$$

For example, the Lagrangian for many mechanical systems is the kinetic energy minus the potential energy because (24.2) written out for

$$L = T - V = \tfrac{1}{2}\sum_k m_k \dot{q}_k^2 - V(q_1, q_2, \ldots) \tag{24.3}$$

reproduces Newton's equations of motion,

$$m\ddot{q}_k = -\frac{\partial V}{\partial q_k}. \tag{24.4}$$

One way to derive (24.2) focuses on a functional of the generalized coordinates called the *action*,

$$S[q_k(t)] = \int_{t_1}^{t_2} dt\, L[q_k(t), \dot{q}_k(t)]. \tag{24.5}$$

Hamilton's principle states that, among all the trajectories $q_k(t)$ which take fixed values at t_1 and t_2, the particular trajectory that occurs in Nature is the one which produces a stationary (usually minimum) value for $S[q_k(t)]$. A mathematical expression of Hamilton's principle follows when we permit *variations* of the trajectory $q_k(t) \to q_k(t) + \delta q_k(t)$ subject to the restriction that $\delta q_k(t_1) = \delta q_k(t_2) = 0$. These variations induce variations of the velocities, $\delta \dot{q}_k$, and a corresponding variation of the action:

$$\delta S = \int_{t_1}^{t_2} dt\, \delta L = \int_{t_1}^{t_2} dt\, \{L[q_k(t) + \delta q_k(t), \dot{q}_k(t) + \delta \dot{q}_k(t)] - L[q_k(t), \dot{q}_k(t)]\}. \tag{24.6}$$

Expanding (24.6) to first order in the infinitesimal variations gives

$$\delta S = \int_{t_1}^{t_2} dt \left[\frac{\partial L}{\partial q_k}\delta q_k + \frac{\partial L}{\partial \dot{q}_k}\delta \dot{q}_k\right]. \tag{24.7}$$

Substituting the identity (see Example 24.1)

$$\delta \dot{q}_k = \delta\left(\frac{dq_k}{dt}\right) = \frac{d}{dt}(\delta q_k) \tag{24.8}$$

into (24.7) and integrating by parts, we find

$$\delta S = \int_{t_1}^{t_2} dt \left[\frac{\partial L}{\partial q_k} - \frac{d}{dt}\frac{\partial L}{\partial \dot{q}_k}\right]\delta q_k + \left.\frac{\partial L}{\partial \dot{q}_k}\delta q_k\right|_{t_1}^{t_2}. \tag{24.9}$$

The stationary action condition identified by Hamilton's principle is $\delta S = 0$. Moreover, the last term of (24.9) vanishes from the assumed end-point restrictions on $q_k(t)$. Consequently,

$$\int_{t_1}^{t_2} dt \left[\frac{\partial L}{\partial q_k} - \frac{d}{dt}\frac{\partial L}{\partial \dot{q}_k}\right]\delta q_k = 0. \tag{24.10}$$

The key observation is that the variations δq_k are arbitrary. Therefore, (24.10) vanishes only if the bracketed quantity in the integrand vanishes. This establishes Lagrange's equations (24.2).

[1] The Lagrangian for an isolated system can be chosen to not depend on time explicitly.

Two points are worth noting. First, the Lagrangian is not unique. As an example, let $\Lambda[q_m(t), t]$ be any scalar function of time and the generalized coordinates. The Lagrangian $L[q_m(t), \dot{q}_m(t)]$ produces exactly the same dynamics as the Lagrangian

$$L[q_m(t), \dot{q}_m(t)] + \frac{d\Lambda[q_m(t), t]}{dt}, \tag{24.11}$$

because the total derivative term does not alter Lagrange's equations. Specifically, the change in the left side of (24.2),

$$\frac{d}{dt}\frac{\partial \dot{\Lambda}}{\partial \dot{q}_k} = \frac{d}{dt}\frac{\partial}{\partial \dot{q}_k}\left(\frac{\partial \Lambda}{\partial q_m}\dot{q}_m + \frac{\partial \Lambda}{\partial t}\right) = \frac{d}{dt}\frac{\partial \Lambda}{\partial q_k} = \frac{\partial \dot{\Lambda}}{\partial q_k}, \tag{24.12}$$

is equal to the change in the right side of (24.2).

Second, it will be useful for later work if we define a set of *canonical momenta*,

$$p_k = \frac{\partial L}{\partial \dot{q}_k}, \tag{24.13}$$

and write (24.2) in the form

$$\frac{dp_k}{dt} = \frac{\partial L}{\partial q_k}. \tag{24.14}$$

Equation (24.14) shows that a particular canonical momentum is a constant of the motion if the Lagrangian does not depend explicitly on the corresponding generalized coordinate.

Example 24.1 Prove (24.8).

Solution: Let the infinitesimal ϵ and the function $\eta_k(t)$ parameterize the variation $\delta q_k(t) = \epsilon\eta_k(t)$. If $q'_k(t)$ stands for the generalized coordinate $q_k(t)$ after the variation is performed,

$$\delta q_k(t) = q'_k(t) - q_k(t) = \epsilon\eta_k(t).$$

The derivative of interest is

$$\frac{d}{dt}(\delta q_k) = \frac{d}{dt}(\epsilon\eta_k) = \epsilon\dot{\eta}_k.$$

On the other hand,

$$\delta\left(\frac{dq_k}{dt}\right) = \frac{dq'_k}{dt} - \frac{dq_k}{dt} = \epsilon\dot{\eta}_k.$$

The rightmost members of the two expressions just above are equal. Therefore,

$$\frac{d}{dt}(\delta q_k) = \delta\left(\frac{dq_k}{dt}\right) = \delta\dot{q}_k.$$

24.3 Lagrangian Description

We establish in this section that the Lagrangian for a closed system of non-relativistic point particles and the electromagnetic fields they produce is

$$L_{\text{EM}} = \tfrac{1}{2}\sum_k m_k v_k^2 + \int d^3r\,(\mathbf{j}\cdot\mathbf{A} - \rho\varphi) + \tfrac{1}{2}\epsilon_0 \int d^3r\,(\mathbf{E}\cdot\mathbf{E} - c^2\mathbf{B}\cdot\mathbf{B}). \tag{24.15}$$

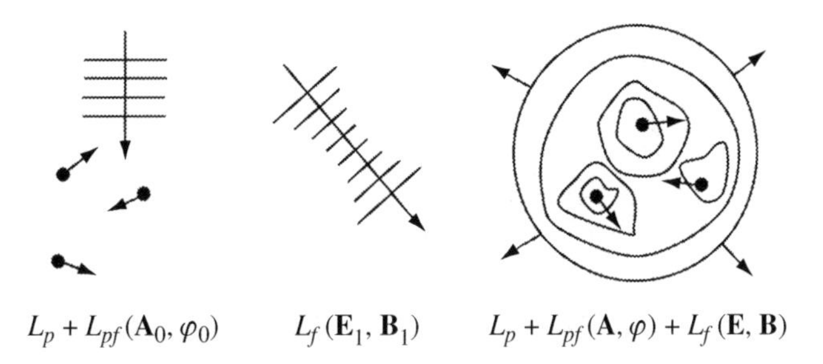

Figure 24.1: A tale of three Lagrangians. Left panel: particle trajectories are determined by the Coulomb-Lorentz force exerted by an external field with specified potentials $(\mathbf{A}_0, \varphi_0)$. Middle panel: a field $(\mathbf{E}_1, \mathbf{B}_1)$ evolves according to the source-free Maxwell equations. Right panel: a field $(\mathbf{E}, \mathbf{B})$ evolves according to the Maxwell-equations with the particles as the source of charge and current. The Coulomb-Lorentz force exerted by this field acts back on the particles and determines their trajectories.

The crucial test is that the Lagrange equations associated with (24.15) correctly reproduce the Maxwell equations for the fields and Newton's second law with the Coulomb-Lorentz force law for the particles.

Two strategies are used to establish (24.15). One approach acknowledges that we already know the equations of motion. Hence, it may be possible to derive $L_{\rm EM}$ simply by "working backwards". This exercise is worthwhile because (i) it is not obvious that $L_{\rm EM}$ exists and (ii) knowledge of the Lagrangian produces insights not immediately apparent from the equations of motion. An alternative approach aims to deduce the field-dependent terms in (24.15) from nearly first principles. The main assumption is that each term in the corresponding action (24.5) is a Lorentz invariant scalar. Thereafter, the "construction" process is guided by (i) respect for the other symmetries of the problem; (ii) an aesthetic desire that $L_{\rm EM}$ be as simple as possible; and (iii) an appeal to minimal experimental data. The last of these recognizes that "the form of the [electromagnetic] action . . . cannot be fixed on the basis of general considerations alone".[2]

Once it is known, $L_{\rm EM}$ provides a direct route to the equations of motion and the conservation laws. The experience gained from this approach to $L_{\rm EM}$ proves useful when the time comes to construct Lagrangians for problems where the dynamics is not known. In this section, we incorporate elements of both the "working backward" and deductive approaches.

24.3.1 A Look Ahead

The mathematics of the Lagrangian method has the potential to obscure the physics. To combat this, we rationalize the logic followed in the next three subsections and state the final results now. The key point is that $L_{\rm EM}$ in (24.15) is the sum of three terms. These are a particle Lagrangian L_p, a particle-field interaction Lagrangian $L_{pf}(\mathbf{A}, \varphi)$, and a field Lagrangian $L_f(\mathbf{E}, \mathbf{B})$:

$$L_{\rm EM} = L_p + L_{pf}(\mathbf{A}, \varphi) + L_f(\mathbf{E}, \mathbf{B}). \tag{24.16}$$

Because L_p is simply the Lagrangian for a free particle, we include particle-field interactions and derive the auxiliary Lagrangian $L_0 = L_p + L_{pf}$ first. We then derive the free-field Lagrangian L_f. The sum of these is the total Lagrangian, $L_{\rm EM} = L_0 + L_f$. Figure 24.1 summarizes the (possibly) surprising result we will find that L_0, L_f, and $L_{\rm EM}$ each describe a distinct and physically realizable situation.

[2] L.D. Landau and E.M. Lifshitz, *The Classical Theory of Fields*, 2nd edition (Addison-Wisley, Reading, MA, 1962), Section 16.

The left panel of Figure 24.1 indicates the physics described by the auxiliary Lagrangian $L_0 = L_p + L_{pf}(\mathbf{A}_0, \varphi_0)$. We show in Section 24.3.2 that Lagrange's equations derived from L_0 determine the particle trajectories from Newton's second law and the Coulomb-Lorentz force exerted on the particles by an external field (illustrated by a plane wave in the panel) described by the potentials $(\mathbf{A}_0, \varphi_0)$. The time dependence of these external potential functions must be specified ahead of time. This Lagrangian completely neglects the fields produced by the particles themselves. For that reason, L_0 may be relevant to situations where the particle density is very low.

The middle panel of Figure 24.1 indicates the physics described by the free-field Lagrangian $L_f(\mathbf{E}_1, \mathbf{B}_1)$. There are no particles present and the time dependence of the field is not specified ahead of time. Instead, the fields evolve according to Lagrange's equations for L_f, which turn out to be the source-free Maxwell equations (see Section 24.3.3). The example sketched in Figure 24.1 is a beam-like solution of these equations (see Section 16.7).

The right panel of Figure 24.1 indicates the physics described by the total Lagrangian $L_{\rm EM}$ in (24.16). This a closed system and Lagrange's equations derived in Section 24.3.5 from $L_{\rm EM}$ describe the field dynamics and particle dynamics simultaneously and self-consistently. The potentials $(\mathbf{A}, \varphi)$ are the source of the Coulomb-Lorentz force on the particles and the particles produce the charge density and current density in the Maxwell equations which determine how the fields $\mathbf{E} = \nabla\varphi - \partial\mathbf{A}/\partial t$ and $\mathbf{B} = \nabla \times \mathbf{A}$ evolve in time. To include the effect of an external field on the particle motion (e.g., the field in left panel), we need only add $\mathbf{A}_0$ to $\mathbf{A}$ and φ_0 to φ in L_{pf}.

Comparing (24.15) and (24.16) shows that L_p is the non-relativistic kinetic energy of the particles. The second term, L_{pf}, couples the scalar potential and the vector potential, respectively, to the charge and current densities,

$$\rho(\mathbf{r}, t) = \sum_k e_k \delta[\mathbf{r} - \mathbf{r}_k(t)] \qquad \text{and} \qquad \mathbf{j}(\mathbf{r}, t) = \sum_k e_k \mathbf{v}_k \delta[\mathbf{r} - \mathbf{r}_k(t)]. \tag{24.17}$$

The scalar potential term in (24.15) is reminiscent of the negative of a quasi-electrostatic potential energy and the entire interaction term L_{pf} would have the $T{-}V$ structure of a mechanical Lagrangian if the magnetic vector potential term could be interpreted as a form of kinetic energy. Conversely, a similar interpretation of the minus sign in the field Lagrangian L_f would associate electric energy with kinetic energy and magnetic energy with potential energy. A more correct inference is that the field dynamically exchanges electric energy and magnetic energy in the same way that a simple mechanical system dynamically exchanges kinetic energy and potential energy.

In the sections to follow, we derive the constituent pieces of (24.15) and confirm that the corresponding Lagrange equations produce the behavior summarized above. Later in the chapter, we relate the electromagnetic conservation laws for charge, energy-momentum, and angular momentum to the gauge invariance, translational invariance, and rotational invariance of the total Lagrangian.

24.3.2 A Charged Particle in a Specified Field

In this section, we show that the Lagrangian $L_0(\mathbf{r}, \mathbf{v})$ for a single particle with charge e and mass m moving with velocity $\mathbf{v}(t) = \dot{\mathbf{r}}(t)$ in a specified electromagnetic field is the sum of the particle and particle-field interaction terms in (24.16). Our method is to work backwards from the Coulomb-Lorentz equation of motion,

$$m\frac{d\mathbf{v}}{dt} = e[\mathbf{E}(\mathbf{r}, t) + \mathbf{v} \times \mathbf{B}(\mathbf{r}, t)]. \tag{24.18}$$

More precisely, we deduce L_0 after manipulating (24.18) into the form of Lagrange's equation (24.2) for this problem:

$$\frac{d}{dt}\left(\frac{\partial L_0}{\partial \mathbf{v}}\right) = \frac{\partial L_0}{\partial \mathbf{r}} = \nabla L_0. \tag{24.19}$$

The gradient operator on the right side of (24.19) suggests that we write the Maxwell fields in terms of the scalar potential $\varphi(\mathbf{r}, t)$ and vector potential $\mathbf{A}(\mathbf{r}, t)$:

$$\mathbf{B} = \nabla \times \mathbf{A} \qquad \mathbf{E} = -\nabla\varphi - \frac{\partial \mathbf{A}}{\partial t}. \tag{24.20}$$

The velocity $\mathbf{v}(t)$ is not a function of position, so (24.20) and the vector identity

$$\nabla(\mathbf{v}\cdot\mathbf{A}) = \mathbf{v}\times(\nabla\times\mathbf{A}) + \mathbf{A}\times(\nabla\times\mathbf{v}) + (\mathbf{v}\cdot\nabla)\mathbf{A} + (\mathbf{A}\cdot\nabla)\mathbf{v} \tag{24.21}$$

transform (24.18) to

$$m\frac{d\mathbf{v}}{dt} = -e\left[\nabla\varphi + \frac{\partial \mathbf{A}}{\partial t} + (\mathbf{v}\cdot\nabla)\mathbf{A} - \nabla(\mathbf{v}\cdot\mathbf{A})\right]. \tag{24.22}$$

We recall now from Section 1.3.3 that the term $(\mathbf{v}\cdot\nabla)\mathbf{A}$ in (24.22) accounts for the variation of $\mathbf{A}$ due to the motion of the charged particle in the definition of the total (convective) time derivative:

$$\frac{d\mathbf{A}}{dt} = \frac{\partial \mathbf{A}}{\partial t} + (\mathbf{v}\cdot\nabla)\mathbf{A}. \tag{24.23}$$

Inserting (24.23) into (24.22) and rearranging terms gives

$$\frac{d}{dt}(m\mathbf{v} + e\mathbf{A}) = e\nabla(\mathbf{v}\cdot\mathbf{A} - \varphi). \tag{24.24}$$

Comparing (24.24) to (24.19) shows that the latter produces the former if the Lagrangian for a charged particle in an external electromagnetic field is

$$L_0(\mathbf{r}, \mathbf{v}) = \tfrac{1}{2}mv^2 + e\mathbf{v}\cdot\mathbf{A}(\mathbf{r}, t) - e\varphi(\mathbf{r}, t). \tag{24.25}$$

Substituting (24.17) into the second term of (24.15) and comparing to (24.16) confirms our statement that (24.25) is the sum $L_0 = L_p + L_{pf}$.

We conclude with two remarks. First, with L_0 in hand, (24.13) identifies

$$\mathbf{p} = \frac{\partial L_0}{\partial \mathbf{v}} = m\mathbf{v} + e\mathbf{A} \tag{24.26}$$

as the canonical momentum for this problem. This formalizes the interpretation of $e\mathbf{A}$ given in Section 15.5.3 as the linear momentum stored in an electromagnetic field which can be exchanged with the kinetic momentum $m\mathbf{v}$ of a charged particle when external forces [the right side of (24.24)] act. If we write Newton's second law as $\mathbf{F} = m\dot{\mathbf{v}}$, the reader will recognize (24.24) as the instantaneous rest-frame ($\mathbf{v} = 0$) expression (15.38) from our discussion of conservation of total linear momentum in Section 15.5.

Second, a "working backward" argument similar to the one used in this section can be used to generalize (24.25) to the case of particles moving at relativistic speeds. We leave it as an exercise for the reader to show that Lagrange's equation (24.19) evaluated using

$$L_0(\mathbf{r}, \mathbf{v}) = -\frac{mc^2}{\gamma} + e\mathbf{v}\cdot\mathbf{A}(\mathbf{r}, t) - e\varphi(\mathbf{r}, t) \tag{24.27}$$

produces the relativistic equation of motion (22.69) deduced in Section 22.5.2. It is interesting to note that the first term on the right side of (24.27) is *not* the relativistic kinetic energy, $T = mc^2(\gamma - 1)$.

Example 24.2 A neutral and point-like particle possesses an electric dipole moment $\mathbf{p}(t)$ and a magnetic dipole moment $\mathbf{m}$.

(a) Use (24.15) to evaluate $L_0 = L_p + L_{pf}$ in (24.16) and show that a suitable Lagrangian for this object moving in specified electric and magnetic fields $\mathbf{E}(\mathbf{r}, t)$ and $\mathbf{B}(\mathbf{r}, t)$ is

$$L = \tfrac{1}{2}mv^2 + \mathbf{m} \cdot \mathbf{B} + \mathbf{p} \cdot (\mathbf{E} + \mathbf{v} \times \mathbf{B}).$$

(b) Show that the force on the particle predicted by L is the same as the one found in Section 15.9.1:

$$\mathbf{F} = \frac{d}{dt}(\mathbf{p} \times \mathbf{B}) + \nabla\left[\mathbf{p} \cdot (\mathbf{E} + \mathbf{v} \times \mathbf{B})\right] + \nabla(\mathbf{m} \cdot \mathbf{B}).$$

Hint: A moving polarization $\mathbf{P}$ produces a magnetization $\mathbf{M} = \mathbf{P} \times \mathbf{v}$ (see Section 14.2.2).

Solution:

(a) The Lagrangian

$$L_{pf} = \int d^3r\, \mathcal{L}_{pf} = \int d^3r\, (\mathbf{j} \cdot \mathbf{A} - \rho\varphi)$$

requires the charge density ρ and current density $\mathbf{j}$ of an electric and magnetic point dipole with trajectory $\mathbf{r}_0(t)$. Using the hint,

$$\rho = -\nabla \cdot \mathbf{P} = -\nabla \cdot [\mathbf{p}\delta(\mathbf{r} - \mathbf{r}_0)]$$

and

$$\mathbf{j} = \frac{\partial \mathbf{P}}{\partial t} + \nabla \times \mathbf{M} = \frac{\partial}{\partial t}[\mathbf{p}\delta(\mathbf{r} - \mathbf{r}_0)] + \nabla \times [\mathbf{m}\delta(\mathbf{r} - \mathbf{r}_0) + \mathbf{p}\delta(\mathbf{r} - \mathbf{r}_0) \times \mathbf{v}].$$

Because $\mathbf{v}(t) = \dot{\mathbf{r}}_0(t)$, writing out the divergence and curl operations gives

$$\rho = -\mathbf{p} \cdot \nabla\delta(\mathbf{r} - \mathbf{r}_0)$$

and

$$\mathbf{j} = \dot{\mathbf{p}}\delta(\mathbf{r} - \mathbf{r}_0) - \mathbf{m} \times \nabla\delta(\mathbf{r} - \mathbf{r}_0) - \mathbf{v}(\mathbf{p} \cdot \nabla)\delta(\mathbf{r} - \mathbf{r}_0).$$

Now, substitute ρ and $\mathbf{j}$ into the Lagrangian L_{pf} and integrate by parts to eliminate terms like $\nabla\delta(\mathbf{r} - \mathbf{r}_0)$. This localizes every function of position at $\mathbf{r} = \mathbf{r}_0$. Because $\mathbf{B} = \nabla \times \mathbf{A}$ and $\mathbf{p} \cdot (\mathbf{v} \times \mathbf{B}) = (\mathbf{p} \cdot \nabla)(\mathbf{v} \cdot \mathbf{A}) - (\mathbf{v} \cdot \nabla)(\mathbf{p} \cdot \mathbf{A})$, we find

$$L_{pf} = \dot{\mathbf{p}} \cdot \mathbf{A} + \mathbf{m} \cdot \mathbf{B} + \mathbf{p} \cdot (\mathbf{v} \times \mathbf{B}) - \mathbf{p} \cdot \nabla\varphi + (\mathbf{v} \cdot \nabla)(\mathbf{p} \cdot \mathbf{A}).$$

Finally, using $\mathbf{E} = -\nabla\varphi - \partial\mathbf{A}/\partial t$, $\dot{\mathbf{A}} = \partial\mathbf{A}/\partial t + (\mathbf{v} \cdot \nabla)\mathbf{A}$, and the low-velocity Lagrangian $L_p = \frac{1}{2}mv^2$, the sum $L = L_p + L_{pf}$ takes the form,

$$L = \frac{1}{2}mv^2 + \frac{d}{dt}(\mathbf{p} \cdot \mathbf{A}) + \mathbf{m} \cdot \mathbf{B} + \mathbf{p} \cdot (\mathbf{v} \times \mathbf{B}) + \mathbf{p} \cdot \mathbf{E}.$$

This differs from the proposed Lagrangian by a total time derivative. From (24.11) and (24.12), both Lagrangians produce the same equation of motion for the particle.

(b) Dropping the total time derivative term from L, the left side of Lagrange's equation (24.2) is

$$\frac{d\mathbf{p}}{dt} = \frac{d}{dt}\frac{\partial L}{\partial \mathbf{v}} = \frac{d}{dt}[m\mathbf{v} + \mathbf{B} \times \mathbf{p}].$$

The right side of (24.2) is

$$\frac{\partial L}{\partial \mathbf{r}} = \nabla \left[\mathbf{p} \cdot (\mathbf{E} + \mathbf{v} \times \mathbf{B}) + \mathbf{m} \cdot \mathbf{B} \right].$$

Setting these two expressions equal gives the advertised force from Newton's second law,

$$\frac{d}{dt} m\dot{\mathbf{v}} = \frac{d}{dt} (\mathbf{p} \times \mathbf{B}) + \nabla \left[\mathbf{p} \cdot (\mathbf{E} + \mathbf{v} \times \mathbf{B}) \right] + \nabla (\mathbf{m} \cdot \mathbf{B}).$$

24.3.3 The Free Electromagnetic Field

The conservation laws developed in Chapter 15 established that an electromagnetic field behaves like a mechanical system capable of exchanging energy, linear momentum, and angular momentum with charged particles. A typical "motion" of this system is a redistribution of the fields in space and time as dictated by the Maxwell equations. A Lagrangian description becomes possible when we generalize the idea of a "dynamical variable" to include the Cartesian components of the electromagnetic field at every point in space.

Consider a free electromagnetic field with no charged particles present. If the discrete index k runs over x, y, and z and the continuous index $\mathbf{r}$ runs over every point in space, the preceding suggests that $E_k(\mathbf{r}, t)$ represents $3 \times \infty$ dynamically independent electric field variables. Similarly, $B_k(\mathbf{r}, t)$ represents $3 \times \infty$ dynamically independent magnetic field variables. From this point of view, the field-only part of the Lagrangian can be at most a combination of these variables summed over both the discrete and continuous indices. This is exactly what we find in the last term of (24.15) and designated in the last term of (24.16) as the *free-field* Lagrangian,

$$L_f = \tfrac{1}{2}\epsilon_0 \int d^3r \, (\mathbf{E} \cdot \mathbf{E} - c^2 \mathbf{B} \cdot \mathbf{B}) = \int d^3r \, \mathcal{L}_f. \tag{24.28}$$

The rightmost member of (24.28) defines the free-field *Lagrangian density* $\mathcal{L}_f$ and a deductive approach to this quantity aims to establish the particular form shown in the middle member of (24.28) from first principles. For guidance, we use the fact that Lagrange's equations derived from this Lagrangian should produce the Maxwell equations for a free electromagnetic field.

The absence of sources suggests that $\mathcal{L}_f$ depends on $\mathbf{E}$ and $\mathbf{B}$ alone. Otherwise, the form of $\mathcal{L}_f$ is constrained by three facts. First, we expect the generalized coordinates and velocities to be linear functions of $\mathbf{E}$ and $\mathbf{B}$ (see Section 24.3.4 below). Second, Lagrange's equation (24.2) involves only first derivatives of the coordinates and velocities. Third, the Maxwell equations are linear in the fields. These observations imply that $\mathcal{L}_f$ can be at most quadratic in the field amplitudes. Moreover, $\mathcal{L}_f$ must be a scalar. Therefore, if a, b, and g are constants, we conclude that the Lagrangian density must have the form

$$\mathcal{L}_f = a\mathbf{E} \cdot \mathbf{E} + g\mathbf{E} \cdot \mathbf{B} + b\mathbf{B} \cdot \mathbf{B}. \tag{24.29}$$

The fact that the Lagrangian density $\mathcal{L}_f$ and the energy density u_{EM} have the same dimensions permits us to infer that $a \propto \epsilon_0$, $g \propto \epsilon_0 c$, and $b \propto \epsilon_0 c^2$.

An issue with (24.29) is that its three terms do not transform identically under the parity operation where $\mathbf{r} \to -\mathbf{r}$. A glance back at Section 15.2.1 shows that $\mathbf{E}$ is odd under parity and $\mathbf{B}$ is even under parity. Therefore, we have the choice in (24.29) of discarding the $\mathbf{E} \cdot \mathbf{B}$ term ($g = 0$) or discarding both the $\mathbf{E} \cdot \mathbf{E}$ and $\mathbf{B} \cdot \mathbf{B}$ terms ($a = b = 0$). We rule out the latter by the additional requirement that $\mathcal{L}_f$ apply to electrostatic ($\mathbf{B} = 0$) and magnetostatic ($\mathbf{E} = 0$) situations separately. Therefore, $g = 0$ and we conclude that there is no $\mathbf{E} \cdot \mathbf{B}$ term in the free-field Lagrangian density.

This is as far as we can go using non-relativistic heuristics. The best way to check that the specific values $a = \epsilon_0/2$ and $b = -\epsilon_0 c^2/2$ quoted in (24.28) are correct is to confirm that the corresponding

Lagrange equations produce the free-field Maxwell equations. We skip this step here because we will perform the analogous calculation in Section 24.3.5 using the complete Lagrangian (24.15). The results we get will confirm the correctness of (24.28) *a posteriori*.

24.3.4 The Lagrange Equations for Fields

It is necessary to generalize Lagrange's equations (24.2) before we can derive the Maxwell equations from the total electromagnetic Lagrangian (24.15). This is so because the Lagrangian offers $\mathbf{E}(\mathbf{r}, t)$, $\mathbf{B}(\mathbf{r}, t)$, $\mathbf{A}(\mathbf{r}, t)$, and $\varphi(\mathbf{r}, t)$ as possible generalized coordinates and these are functions of time *and* space rather than functions of time alone. In this section, we make a choice of coordinates and then derive the generalized Lagrange equation needed to discover how they evolve in time.

One modus of Lagrangian electrodynamics uses all four of the functions displayed above as independent generalized coordinates. We leave this approach as an exercise for the reader. Here, we proceed differently and retain only the vector potential and the scalar potential as generalized coordinates. Our main reason for doing this is familiar from previous work. The representation of the fields in terms of the potentials in (24.20) implies that the homogeneous Maxwell equations

$$\nabla \cdot \mathbf{B} = 0 \qquad \text{and} \qquad \nabla \times \mathbf{E} + \frac{\partial \mathbf{B}}{\partial t} = 0 \tag{24.30}$$

are satisfied automatically and require no further attention. This permits us to focus on using the Lagrangian formalism to derive the inhomogeneous Maxwell equations: Gauss' law, and the Ampère-Maxwell law. It also means we must write the Lagrangian (24.15) entirely in terms of the potentials:

$$L_{\text{EM}} = \tfrac{1}{2}\sum_k m_k v_k^2 + \int d^3r\,(\mathbf{j}\cdot\mathbf{A} - \rho\varphi) + \tfrac{1}{2}\epsilon_0 \int d^3r\,\left[(\nabla\varphi + \partial\mathbf{A}/\partial t)^2 - c^2(\nabla\times\mathbf{A})^2\right]. \tag{24.31}$$

Apart from the integration over space, the action associated with (24.31) differs from (24.5) because the Lagrangian contains not only the generalized velocity $\partial\mathbf{A}/\partial t$ but also the derivatives $\nabla\varphi$ and $\nabla\times\mathbf{A}$. This suggests we should derive Lagrange's equations for an action which includes space derivatives of the generalized coordinates:

$$S = \int_{t_1}^{t_2} dt \int d^3r\,\mathcal{L}\left[q_k(\mathbf{r}, t), \dot{q}_k(\mathbf{r}, t), \partial_i q_k(\mathbf{r}, t)\right]. \tag{24.32}$$

The variation of (24.32) generalizes (24.7) to

$$\delta S = \int_{t_1}^{t_2} dt \int d^3r \left[\frac{\partial\mathcal{L}}{\partial q_k}\delta q_k + \frac{\partial\mathcal{L}}{\partial\dot{q}_k}\delta\dot{q}_k + \frac{\partial\mathcal{L}}{\partial(\partial_i q_k)}\delta(\partial_i q_k)\right]. \tag{24.33}$$

Using $\delta(\partial_i q_k) = \partial_i(\delta q_k)$, the integrand of (24.33) may be written as

$$\frac{\partial\mathcal{L}}{\partial q_k}\delta q_k + \frac{d}{dt}\left(\frac{\partial\mathcal{L}}{\partial\dot{q}_k}\delta q_k\right) - \frac{d}{dt}\left(\frac{\partial\mathcal{L}}{\partial\dot{q}_k}\right)\delta q_k + \partial_i\left\{\frac{\partial\mathcal{L}}{\partial(\partial_i q_k)}\delta q_k\right\} - \partial_i\left\{\frac{\partial\mathcal{L}}{\partial(\partial_i q_k)}\right\}\delta q_k. \tag{24.34}$$

Following Section 24.2, we eliminate the integral over the second and fourth (total derivative) terms in (24.34) by constraining the variations of the position-dependent generalized coordinates so

$$\begin{aligned} &\delta q_k(\mathbf{r}, t_1) = \delta q_k(\mathbf{r}, t_2) = 0 \\ &\delta q_k(\mathbf{r}, t) = 0 \quad \text{when } \mathbf{r}\to\infty. \end{aligned} \tag{24.35}$$

The terms that remain produce the net variation

$$\delta S = \int_{t_1}^{t_2} dt \int d^3r \left[\frac{\partial\mathcal{L}}{\partial q_k} - \frac{d}{dt}\left(\frac{\partial\mathcal{L}}{\partial\dot{q}_k}\right) - \partial_i\left\{\frac{\partial\mathcal{L}}{\partial(\partial_i q_k)}\right\}\right]\delta q_k. \tag{24.36}$$

Again following Section 24.2, the variation δq_k is arbitrary, so $\delta S = 0$ is guaranteed only if

$$\frac{d}{dt}\left(\frac{\partial \mathcal{L}}{\partial \dot{q}_k}\right) = \frac{\partial \mathcal{L}}{\partial q_k} - \partial_i \frac{\partial \mathcal{L}}{\partial(\partial_i q_k)}. \tag{24.37}$$

24.3.5 The Coulomb-Lorentz and Maxwell Equations

We are now in a position to derive the equations of motion for both the particles and the fields from the total electromagnetic Lagrangian $L_{\rm EM}$ in (24.31). The first step notes that the free-field part of $L_{\rm EM}$ is independent of the positions and velocities of the charged particles. Therefore, Lagrange's equation for each coordinate $\mathbf{r}_k$ is exactly the same as the equation of motion derived from the Lagrangian $L_0(\mathbf{r}_k, \mathbf{v}_k)$ recorded in (24.27). Transcribing from (24.24) gives

$$\frac{d}{dt}\left[m_k \mathbf{v}_k + e_k \mathbf{A}(\mathbf{r}_k)\right] = e_k \nabla_k \left[\mathbf{v}_k \cdot \mathbf{A}(\mathbf{r}_k) - \varphi(\mathbf{r}_k)\right], \tag{24.38}$$

and following the steps backward from (24.24) to (24.18) gives the Coulomb-Lorentz equations of motion for the charged particles:

$$m_k \frac{d\mathbf{v}_k}{dt} = e_k[\mathbf{E}(\mathbf{r}_k, t) + \mathbf{v}_k \times \mathbf{B}(\mathbf{r}_k, t)]. \tag{24.39}$$

Turning to the fields, we remind the reader that our use of the potentials in (24.31) establishes the homogeneous Maxwell equations in (24.30) as identities based on (24.20). Therefore, if we begin with the generalized coordinate $\mathbf{A}(\mathbf{r}, t)$, the left side of (24.37) is the time derivative of a position-dependent canonical momentum density $\boldsymbol{\pi}(\mathbf{r}, t)$ defined by (24.13) and the generalized velocity $\dot{\mathbf{A}} \equiv \partial \mathbf{A}/\partial t$. The latter appears only in the free-field part of the total Lagrangian density $\mathcal{L}_{\rm EM}$ and we find

$$\boldsymbol{\pi} = \frac{\partial \mathcal{L}_{\rm EM}}{\partial \dot{\mathbf{A}}} = \epsilon_0 \left(\nabla\varphi + \dot{\mathbf{A}}\right) = -\epsilon_0 \mathbf{E}. \tag{24.40}$$

The corresponding Lagrange equation for each Cartesian component of $\boldsymbol{\pi}$ is

$$\frac{d\pi_k}{dt} = \frac{\partial \mathcal{L}_{\rm EM}}{\partial A_k} - \partial_i \frac{\partial \mathcal{L}_{\rm EM}}{\partial(\partial_i A_k)}. \tag{24.41}$$

However, $\partial \mathcal{L}_{\rm EM}/\partial A_k = (\partial/\partial A_k)(\mathbf{j}\cdot\mathbf{A}) = j_k$ and

$$\frac{\partial}{\partial(\partial_i A_k)}\left[\frac{1}{2\mu_0}(\nabla\times\mathbf{A})^2\right] = \frac{1}{\mu_0}\frac{\partial}{\partial(\partial_i A_k)}(B_j \epsilon_{j\ell m}\partial_\ell A_m) = \frac{1}{\mu_0} B_j \epsilon_{jik}. \tag{24.42}$$

Therefore, (24.41) reads

$$\frac{d\pi_k}{dt} = j_k - \frac{1}{\mu_0}\epsilon_{kij}\partial_i B_j. \tag{24.43}$$

Using (24.40), we identify (24.43) as the Ampère-Maxwell law,

$$\nabla\times\mathbf{B} = \mu_0 \mathbf{j} + \frac{1}{c^2}\frac{\partial \mathbf{E}}{\partial t}. \tag{24.44}$$

The generalized coordinate $\varphi(\mathbf{r}, t)$ is distinguished by the fact that its "velocity" $\dot{\varphi}(\mathbf{r}, \mathbf{t})$ does not appear anywhere in the Lagrangian (24.31). This means that the associated canonical momentum vanishes,

$$\pi_0(\mathbf{r}, t) = \frac{\partial \mathcal{L}_{\rm EM}}{\partial \dot{\varphi}} = 0, \tag{24.45}$$

and the corresponding Lagrange equation is

$$0 = \frac{\partial \mathcal{L}_{\rm EM}}{\partial \varphi} - \partial_i \frac{\partial \mathcal{L}_{\rm EM}}{\partial(\partial_i \varphi)}. \tag{24.46}$$

However, $\partial \mathcal{L}/\partial\varphi = (\partial/\partial\varphi)(-\rho\varphi) = -\rho$ and

$$\frac{\partial}{\partial(\partial_i\varphi)}\left[\frac{\epsilon_0}{2}(\nabla\varphi + \dot{\mathbf{A}})^2\right] = -\epsilon_0 \frac{\partial}{\partial(\partial_i\varphi)}\left[E_k(\partial_k\varphi + \dot{A}_k)\right] = -\epsilon_0 E_i. \tag{24.47}$$

Therefore, (24.46) reads

$$0 = -\rho + \epsilon_0 \partial_i E_i. \tag{24.48}$$

This is Gauss' law,

$$\nabla \cdot \mathbf{E} = \rho/\epsilon_0. \tag{24.49}$$

Several remarks are in order. First, our evaluation of Lagrange's equations (24.37) produced first-order (Maxwell) equations for the fields rather than second-order equations for the potentials equivalent to those derived in Section 15.3. This happened because we substituted from (24.20) at an early stage of each calculation. Second, the consequence of (24.45) that there is no true equation of motion for $\varphi(\mathbf{r}, t)$ is consistent with the "initial condition" status afforded to Gauss' law in (2.89) at the end of Chapter 2.

Finally, it is rather remarkable that a complete description of charged particles interacting with their fields follows from the Lagrangian $L_{\rm EM}$ in (24.31). After all, we "derived" $L_{\rm EM}$ as the sum of a Lagrangian L_0 for charged particles moving in an external field and a Lagrangian L_f for free fields in the absence of charges. This implies that the matter-field interaction term $e(\mathbf{v} \cdot \mathbf{A} - \varphi)$ which accounts for the effect of an electromagnetic field on charged particles *also* accounts for the creation of electromagnetic fields by charged particles.

24.3.6 Covariant Formulation

Covariant forms for the total electromagnetic action and its Lagrange equation follow straightforwardly from the work we have done so far. For the action, we replace the non-relativistic kinetic energy in (24.15) by the relativistic particle Lagrangian in (24.27) and exploit the invariants $j_\mu A_\mu = \mathbf{j} \cdot \mathbf{A} - \rho\varphi$ and $F_{\mu\nu} F_{\mu\nu} = 2(\mathbf{B} \cdot \mathbf{B} - \mathbf{E} \cdot \mathbf{E}/c^2)$ [see (22.139)]. The action is the time-integral of the total Lagrangian, so

$$S = \int dt \left\{ -\frac{mc^2}{\gamma} + \int d^3 r \left[j_\mu A_\mu - \frac{1}{4\mu_0} F_{\mu\nu} F_{\mu\nu} \right] \right\}. \tag{24.50}$$

The principle of relativity tells us that Hamilton's principle must produce the same physics in every inertial frame. An immediate consequence is that the action S must be a Lorentz invariant scalar. The expression (24.50) satisfies this requirement because both the proper time $d\tau = dt/\gamma$ and the four-dimensional volume element $d^4 r = dxdydzd(ct)$ are Lorentz invariant scalars. Indeed, this argument and the Lorentz scalar nature of $j_\mu A_\mu$ and $F_{\mu\nu} F_{\mu\nu}$ are central to the "deductive" approach to deriving $L_{\rm EM}$ mentioned in Section 24.3.1.

The Lagrange equations follow from (24.50) by a direct application of Hamilton's principle. Alternatively, the structure of the four-gradient $\partial_\mu = [\nabla, \partial/\partial(ict)]$ shows that the equation of motion (24.37) for the generalized coordinate A_α can be written in the manifestly covariant form

$$\partial_\mu \frac{\partial \mathcal{L}}{\partial(\partial_\mu A_\alpha)} = \frac{\partial \mathcal{L}}{\partial A_\alpha}. \tag{24.51}$$

A Brief History of Electromagnetic Lagrangians

The action principle was embraced by many 19th-century scientists as a central organizing principle for the mechanistic world. Therefore, it is not surprising that Maxwell devotes an entire chapter of his *Treatise on Electricity and Magnetism* (1873) to an account of the equations of Lagrange and Hamilton. On the other hand, his actual use of these equations to develop his theory is rather sparing. It was left to Helmholtz to challenge his contemporaries to discover a single Lagrangian from which one could derive both Maxwell's system of field equations and the Coulomb-Lorentz force law for moving charges. Helmholtz' own solution to this problem was complicated and has largely been forgotten.

An important step forward was made by Lorentz in 1892. Although he used virtual-work arguments, the translation of his work into Lagrangian language by Darrigol (2000) permits us to conclude that Lorentz was the first person to write the Lagrangian L_0 in (24.25) and use it to derive the force $\mathbf{F} = q(\mathbf{E} + \mathbf{v} \times \mathbf{B})$. Lorentz also (essentially) wrote the free-field Lagrangian (24.28) and used it to derive Faraday's law by treating $\mathbf{E}$ as a generalized coordinate. On the other hand, he did not derive either Gauss' law or the Ampère-Maxwell law from a dynamical principle.

The Lagrangian $L_{\rm EM}$ in (24.15) and the idea that one should use the vector and scalar potentials alone as generalized coordinates is due to Karl Schwarzchild (1903). The content of our Section 24.3.5 is essentially his derivation of the Maxwell equations and the Coulomb-Lorentz force law from the corresponding Lagrange equations. Today, Schwarzchild is best remembered as the first person to obtain an exact solution (the non-rotating black hole) of Einstein's equations of general relativity.

24.4 Invariance and Conservation Laws

Textbooks of classical mechanics typically point out that the total energy of a mechanical system is conserved if its Lagrangian is not an explicit function of time. In a similar way, conservation of linear momentum and conservation of angular momentum are related to the invariance of the Lagrangian with respect to uniform translations and rotations in space. In this section, we generalize this idea and show how the conservation laws of electromagnetism result when Hamilton's principle of stationary action (Section 24.2) is applied to infinitesimal variations of variables that reflect the fundamental symmetries of the theory.

24.4.1 Conservation of Charge

The conservation of electric charge is a consequence of the gauge invariance of electrodynamics and vice versa. To see this, consider the Lagrangian for a charged particle in an electromagnetic field, $L_0 = \frac{1}{2}mv^2 + e\mathbf{v} \cdot \mathbf{A} - e\varphi$ (see Section 24.3.2). The presence of the electromagnetic potentials tells us that L_0 is not gauge invariant. If $\Lambda(\mathbf{r}, t)$ is an arbitrary scalar function, a gauge change to $\mathbf{A}' = \mathbf{A} + \nabla\Lambda$ and $\varphi' = \varphi - \partial\Lambda/\partial t$ changes the Lagrangian from L_0 to [cf. (24.23]

$$L_0' = L_0 + e\left(\mathbf{v} \cdot \nabla\Lambda + \frac{\partial\Lambda}{\partial t}\right) = L_0 + e\frac{d\Lambda}{dt}. \tag{24.52}$$

On the other hand, we know from (24.11) and (24.12) that L_0' and L_0 produce exactly the same Lagrange equation (24.2). This confirms what we already know: the Coulomb-Lorentz force law in (24.18) depends only on the gauge invariant electric and magnetic fields.

From the perspective of Hamilton's principle (Section 24.2), the gauge invariance of Lagrange's equation implies that the stationary action condition $\delta S = 0$ is valid when we treat the change of gauge as a variation. In other words, if $\delta\Lambda$ is an infinitesimal gauge function, (24.52) tells us that the generalized coordinate variations $\mathbf{A} \to \mathbf{A} + \nabla\delta\Lambda$ and $\varphi \to \varphi - \partial(\delta\Lambda)/\partial t$ applied to the Lagrangian for a collection of charged particles produces a variation in the action,

$$\delta S = \int_{t_1}^{t_2} dt \sum_k e_k \left\{\mathbf{v}_k \cdot \nabla[\delta\Lambda(\mathbf{r}_k, t)] + \frac{\partial[\delta\Lambda(\mathbf{r}_k, t)]}{\partial t}\right\}. \tag{24.53}$$

Using (24.17) to rewrite (24.53) gives the condition for gauge invariance as

$$\delta S = \int_{t_1}^{t_2} dt \int d^3r \left\{ \mathbf{j} \cdot \nabla(\delta\Lambda) + \rho \frac{\partial(\delta\Lambda)}{\partial t} \right\} = 0. \tag{24.54}$$

The key step is to rewrite (24.54) as

$$\delta S = \int_{t_1}^{t_2} dt \int d^3r \left[\nabla \cdot (\mathbf{j}\delta\Lambda) + \frac{\partial}{\partial t}(\rho\delta\Lambda) \right] - \int_{t_1}^{t_2} dt \int d^3r \left(\nabla \cdot \mathbf{j} + \frac{\partial \rho}{\partial t} \right) \delta\Lambda = 0. \tag{24.55}$$

The first space-time integral in (24.55) vanishes from the end-point restrictions on the variations of the generalized coordinates φ and $\mathbf{A}$ stated in (24.35). Therefore, because $\delta\Lambda$ is otherwise arbitrary, the gauge invariance of electrodynamics has led us to the continuity equation for electric charge,

$$\nabla \cdot \mathbf{j} + \frac{\partial \rho}{\partial t} = 0. \tag{24.56}$$

Running this argument backwards for a localized distribution demonstrates that gauge invariance is a consequence of the conservation of charge.[3]

24.4.2 Noether's Theorem

A celebrated theorem due to Noether makes a general connection between conserved quantities and the invariance of the action with respect to variations of a continuous variable. The setting for the theorem is a variation of the action which supplements the functional variations we have allowed so far with infinitesimal variations of the space-time coordinates. Our interest is the covariant electromagnetic action (24.50). Therefore, we vary an action with the general form

$$S = \frac{1}{c} \int d^4r\, \mathcal{L}[r_\alpha, A_\alpha(r_\mu), \partial_\beta A_\alpha(r_\mu)]. \tag{24.57}$$

An infinitesimal transformation of the coordinates is

$$r_\mu \to r'_\mu = r_\mu + \delta r_\mu. \tag{24.58}$$

The four-potential and its derivatives vary in two ways. The first is an explicit functional variation performed without any change of coordinates. This we have done previously. The second is an implicit functional variation induced by the coordinate transformation (24.58). If we use the notation $\delta_0 A_\alpha$ for the first type of variation and δA for the total variation,

$$A_\alpha \to A_\alpha + \delta A_\alpha = A_\alpha + \delta_0 A_\alpha + \frac{\partial A_\alpha}{\partial r_\mu} \delta r_\mu. \tag{24.59}$$

The action varies in two ways also. One comes from the variation of the Lagrangian density with no change of coordinates. The other comes from the variation of the integration volume element induced by (24.58):

$$\delta S = \delta \int d^4r\, \mathcal{L} = \int d^4r\, \delta_0 \mathcal{L} + \int \delta(d^4r)\, \mathcal{L}. \tag{24.60}$$

The variation of the Lagrangian density in (24.60) is

$$\delta_0 \mathcal{L} = \frac{\partial \mathcal{L}}{\partial r_\mu} \delta r_\mu + \frac{\partial \mathcal{L}}{\partial A_\alpha} \delta_0 A_\alpha + \frac{\partial \mathcal{L}}{\partial(\partial_\mu A_\alpha)} \delta_0(\partial_\mu A_\alpha). \tag{24.61}$$

[3] The boxed material in Section 15.3.1 presents Wigner's qualitative argument for this conclusion.

The identity $\delta_0(\partial_\mu A_\alpha) = \partial_\mu(\delta_0 A_\alpha)$ implies that (24.61) is the same as

$$\delta_0 \mathcal{L} = \frac{\partial \mathcal{L}}{\partial r_\mu}\delta r_\mu + \frac{\partial \mathcal{L}}{\partial A_\alpha}\delta_0 A_\alpha + \partial_\mu\left[\frac{\partial \mathcal{L}}{\partial(\partial_\mu A_\alpha)}\delta_0 A_\alpha\right] - \partial_\mu\left[\frac{\partial \mathcal{L}}{\partial(\partial_\mu A_\alpha)}\right]\delta_0 A_\alpha. \tag{24.62}$$

The variation of the four-dimensional volume element in (24.61) is

$$\delta(d^4r) = d^4r' - d^4r = d^4r\,J - d^4r = d^4r\,\partial_\mu(\delta r_\mu), \tag{24.63}$$

because the Jacobian of the coordinate transformation (24.58) is (see Section 1.4.1)

$$J = \left|\det\left(\frac{\partial r'_\mu}{\partial r_\nu}\right)\right| = |\det(\delta_{\mu\nu} + \partial_\nu \delta r_\mu)| = 1 + \partial_\mu(\delta r_\mu) + O(\delta r_\mu^2). \tag{24.64}$$

Substituting (24.62) and (24.63) into (24.60) gives the intermediate result

$$\delta S = \int d^4r\left\{\partial_\mu(\mathcal{L}\delta r_\mu) + \left[\frac{\partial \mathcal{L}}{\partial A_\alpha} - \partial_\mu\left\{\frac{\partial \mathcal{L}}{\partial(\partial_\mu A_\alpha)}\right\}\right]\delta_0 A_\alpha + \partial_\mu\left[\frac{\partial \mathcal{L}}{\partial(\partial_\mu A_\alpha)}\delta_0 A_\alpha\right]\right\}. \tag{24.65}$$

At this point, we pause and apply Hamilton's principle (Section 24.2) to the variation (24.65) with $\delta r_\mu = 0$. The first term in the integrand vanishes identically and the integral of the last, total derivative term vanishes from the end-point restriction on the variations of $\delta_0 A_\alpha$ discussed earlier. Combining $\delta S = 0$ with the arbitrariness of $\delta_0 A_\alpha$ leads to the conclusion that

$$\frac{\partial \mathcal{L}}{\partial A_\alpha} - \partial_\mu\left\{\frac{\partial \mathcal{L}}{\partial(\partial_\mu A_\alpha)}\right\} = 0. \tag{24.66}$$

This is Lagrange's equation (24.51). Henceforth, we assume that the four-potential satisfies this equation. Therefore, (24.65) reduces to

$$\delta S = \int d^4r\,\partial_\mu\left[\mathcal{L}\delta r_\mu + \partial_\mu\left(\frac{\partial \mathcal{L}}{\partial(\partial_\mu A_\alpha)}\delta_0 A_\alpha\right)\right]. \tag{24.67}$$

Noether's theorem considers variations which leave the action invariant such as spatial and temporal symmetry operations. This means that $\delta S = 0$ and we derive from (24.67) a conservation law in the form of a generalized equation of continuity:

$$\partial_\mu\left[\mathcal{L}\delta r_\mu + \partial_\mu\left(\frac{\partial \mathcal{L}}{\partial(\partial_\mu A_\alpha)}\delta_0 A_\alpha\right)\right] = 0. \tag{24.68}$$

An alternative form uses (24.59) to replace $\delta_0 A_\alpha$ by δA_α in (24.68). The result is

$$\partial_\mu\left[\left(\mathcal{L}\delta_{\mu\nu} - \frac{\partial \mathcal{L}}{\partial(\partial_\mu A_\alpha)}\partial_\nu A_\alpha\right)\delta r_\nu + \frac{\partial \mathcal{L}}{\partial(\partial_\mu A_\alpha)}\delta A_\alpha\right] = 0. \tag{24.69}$$

24.4.3 Conservation of Energy-Momentum

Relativistic energy-momentum is conserved for an isolated system that is invariant to uniform translations in space-time. This follows from Noether's theorem with the choice $\delta r_\mu = \delta\epsilon_\mu$ where $\delta\epsilon_\mu$ is a *constant* four-vector. The explicit functional change in $A_\alpha(r_\nu)$ produced by this translation without any change of coordinates is [cf. (12.122)]

$$\delta_0 A_\alpha(r_\nu) = A_\alpha(r_\nu - \delta\epsilon_\nu) - A_\alpha(r_\nu) = -\frac{\partial A_\alpha}{\partial r_\nu}\delta\epsilon_\nu. \tag{24.70}$$

Using (24.59), the total variation $\delta A_\alpha = 0$. In other words, the translated function at the translated coordinate position is the same as the original function at the original coordinate position. With this information, the arbitrary nature of the translation vector $\delta\epsilon_\mu$ implies that (24.69) reduces to

$$\partial_\mu\left(\delta_{\mu\nu}\mathcal{L} - \frac{\partial\mathcal{L}}{\partial(\partial_\mu A_\alpha)}\partial_\nu A_\alpha\right) = 0. \tag{24.71}$$

It is tempting to evaluate the conservation law (24.71) using the total Lagrangian density $\mathcal{L}_{\text{EM}}$. However, the presence of the $j_\mu A_\mu$ term in (24.50) makes this impermissible because this interaction piece of the Lagrangian density facilitates the exchange of energy-momentum between the field-only system described by (24.72) and the matter system described by the four-current $j_\mu(x_\nu)$. A more general conservation law would result if we took account of the matter in the Lagrangian and permitted variations of this Lagrangian induced by variations of the matter variables.

Let us evaluate (24.71) using only the free-field Lagrangian density,

$$\mathcal{L}_f = -\frac{1}{4\mu_0}F_{\sigma\beta}F_{\sigma\beta}. \tag{24.72}$$

The first step is to note that $F_{\sigma\beta} = \partial_\sigma A_\beta - \partial_\beta A_\sigma$ and $F_{\sigma\beta} = -F_{\beta\sigma}$ imply that

$$\frac{\partial\mathcal{L}_f}{\partial(\partial_\mu A_\alpha)}\partial_\nu A_\alpha = -\frac{1}{4\mu_0}\left[2F_{\sigma\beta}\frac{\partial F_{\sigma\beta}}{\partial(\partial_\mu A_\alpha)}\right]\partial_\nu A_\alpha = -\frac{1}{\mu_0}F_{\mu\alpha}\partial_\nu A_\alpha. \tag{24.73}$$

Therefore, using $\partial_\nu A_\alpha = F_{\nu\alpha} + \partial_\alpha A_\nu$,

$$-\partial_\mu\left(\frac{\partial\mathcal{L}_f}{\partial(\partial_\mu A_\alpha)}\partial_\nu A_\alpha\right) = \frac{1}{\mu_0}\left[\partial_\mu(F_{\mu\alpha}F_{\nu\alpha}) + (\partial_\mu F_{\mu\alpha})\partial_\alpha A_\nu + F_{\mu\alpha}\partial_\mu\partial_\alpha A_\nu\right]. \tag{24.74}$$

The last term in (24.74) vanishes because $F_{\mu\alpha} = -F_{\alpha\mu}$. The next-to-last term vanishes also from the $j_\mu = 0$ limit of the inhomogeneous Maxwell equations [see (22.140)],

$$\partial_\nu F_{\mu\nu} = 0. \tag{24.75}$$

Using (24.72) for the first term of (24.71) and (24.74) for the second term, we arrive finally at the desired conservation law, namely,

$$\partial_\mu\Theta_{\mu\nu} = 0, \tag{24.76}$$

where

$$\mu_0\Theta_{\mu\nu} = F_{\mu\sigma}F_{\nu\sigma} - \tfrac{1}{4}\delta_{\mu\nu}F_{\alpha\beta}F_{\alpha\beta}. \tag{24.77}$$

This agrees with the results of Section 22.7.3 where we used (24.77) to define the electromagnetic stress-energy tensor and found the energy-momentum conservation law $\partial_\mu\Theta_{\mu\nu} = -j_\alpha F_{\nu\alpha}$ for a system with sources.

24.4.4 Conservation of Angular Momentum

We showed in Section 22.7.4 that uniform motion of the center of energy and a statement of the conservation of electromagnetic angular momentum are contained in the covariant conservation law,

$$\partial_\sigma M_{\sigma\mu\nu} = 0, \tag{24.78}$$

where

$$M_{\sigma\mu\nu} = \Theta_{\sigma\mu}r_\nu - \Theta_{\sigma\nu}r_\mu = -M_{\rho\nu\mu}. \tag{24.79}$$

These results are a consequence of rotational invariance. Therefore, we expect (24.78) to follow from Noether's theorem when the variations in (24.68) or (24.69) correspond to an infinitesimal rotation in space.

Let $\delta\boldsymbol{\omega}$ be an infinitesimal vector whose components are the Euler angles of a rigid rotation. The change in a position vector due to this rotation is

$$\delta\mathbf{r} = \delta\boldsymbol{\omega} \times \mathbf{r}. \tag{24.80}$$

An equivalent representation uses the anti-symmetric object $\delta\Omega_{ij} = -\delta\Omega_{ji}$ defined by

$$\delta r_i = \epsilon_{ijk}\delta\omega_j r_k \equiv \delta\Omega_{ik} r_k. \tag{24.81}$$

If we choose $\delta r_4 = 0$, a four-dimensional representation of (24.81) is[4]

$$\delta r_\nu = \delta\Omega_{\mu\nu} r_\nu. \tag{24.82}$$

The same rotation (without a change of coordinates) produces a change in $\mathbf{A}(\mathbf{r})$ given by[5]

$$\delta_0\mathbf{A}(\mathbf{r}) = -(\delta\mathbf{r}\cdot\nabla)\cdot\mathbf{A}(\mathbf{r}) + \delta\boldsymbol{\omega}\times\mathbf{A}(\mathbf{r}), \tag{24.83}$$

or

$$\delta_0 A_\alpha = -\partial_\nu A_\alpha \delta r_\nu + \delta\Omega_{\alpha\beta} A_\beta. \tag{24.84}$$

Using (24.84) in (24.59) gives the total variation including rotation of the coordinates as

$$\delta A_\alpha = \delta\Omega_{\alpha\beta} A_\beta. \tag{24.85}$$

Equation (24.85) agrees with (24.82), as expected, because r_ν and A_β transform the same way.

The conservation law (24.69) written out using (24.82), (24.85), and the derivative of the free-field Lagrangian density computed in (24.73) is

$$\frac{1}{\mu_0}\delta\Omega_{\nu\sigma}\partial_\mu\left[\left(-\frac{1}{4}\delta_{\mu\nu}F_{\sigma\beta}F_{\sigma\beta} + F_{\mu\alpha}\partial_\nu A_\alpha\right)r_\sigma - F_{\mu\nu}A_\sigma\right] = 0. \tag{24.86}$$

It is not difficult to rewrite (24.86) using $\partial_\nu A_\alpha = F_{\nu\alpha} + \partial_\alpha A_\nu$ and the definition of $\Theta_{\mu\nu}$ in (24.77):

$$\delta\Omega_{\mu\nu}\partial_\mu\left[\Theta_{\mu\sigma}r_\nu + \mu_0 F_{\mu\alpha}(\partial_\alpha A_\nu)r_\sigma - \mu_0 F_{\mu\nu}A_\sigma\right] = 0. \tag{24.87}$$

Now write (24.87) again with the indices ν and σ exchanged, and add the result to (24.87). Because $\delta\Omega_{\sigma\nu} = -\delta\Omega_{\nu\sigma}$, we find

$$\delta\Omega_{\nu\sigma}\partial_\mu\left(\Theta_{\mu\nu}r_\sigma - \Theta_{\mu\sigma}r_\nu\right) = \delta\Omega_{\nu\sigma}\partial_\mu\left[F_{\mu\alpha}(\partial_\alpha A_\nu)r_\sigma - F_{\mu\alpha}(\partial_\alpha A_\sigma) - F_{\mu\nu}A_\sigma + F_{\mu\sigma}A_\nu\right]. \tag{24.88}$$

Equation (24.75) and the anti-symmetry of $F_{\mu\nu}$ are all the reader needs to show that the four-divergence on the right side of (24.88) is zero. Therefore, because $\delta\Omega_{\nu\sigma}$ is arbitrary, we conclude that

$$\partial_\mu\left(\Theta_{\mu\nu}r_\sigma - \Theta_{\mu\sigma}r_\nu\right) = 0. \tag{24.89}$$

This is the anticipated angular momentum conservation law (24.78).

24.5 Hamiltonian Description

Hamilton's equations of motion are completely equivalent to Lagrange's equations of motion for mechanical systems of the sort usually studied in classical mechanics. In this section, we

[4] The only property of $\delta\Omega_{\mu\nu}$ used below is $\delta\Omega_{\mu\nu} = -\delta\Omega_{\nu\mu}$.

[5] Equation (24.83) is the possibly more familiar expression $\delta_0\mathbf{A}(\mathbf{r}) = \mathbf{R}\cdot\mathbf{A}(\mathbf{R}^{-1}\cdot\mathbf{r}) - \mathbf{A}(\mathbf{r})$ evaluated with a rotation matrix $\mathbf{R}$ which describes the infinitesimal rotation $\delta\boldsymbol{\omega}$. See, e.g., Section 17.1 of O. Keller, *Quantum Theory of Near-Field Electrodynamics* (Springer, Berlin, 2011).

show that the conventional Hamiltonian method is not entirely satisfactory when applied to classical electrodynamics. There is no problem using Hamilton's equations to derive either the inhomogeneous Maxwell equations or the Coulomb-Lorentz equations for the charged particles. Inconsistency appears only when we write Hamilton's equation for the electromagnetic scalar potential. We outline Dirac's solution to this problem and thereby highlight the central roles of gauge invariance and gauge fixing in constructing a consistent Hamiltonian formulation of electrodynamics.

We begin with the Lagrangian $L(q_k, \dot{q}_k, t)$ introduced in Section 24.2, generalized to include an explicit dependence on time. The time derivative of this object is

$$\frac{dL}{dt} = \frac{\partial L}{\partial t} + \sum_k \left(\frac{\partial L}{\partial q_k}\dot{q}_k + \frac{\partial L}{\partial \dot{q}_k}\ddot{q}_k \right). \tag{24.90}$$

The definition (24.13) of the canonical momentum,

$$p_k = \frac{\partial L}{\partial \dot{q}_k}, \tag{24.91}$$

simplifies Lagrange's equations (24.2) to

$$\frac{\partial L}{\partial q_k} = \frac{d}{dt}\left(\frac{\partial L}{\partial \dot{q}_k}\right) = \frac{dp_k}{dt}. \tag{24.92}$$

The two preceding equations put (24.90) in the form

$$\frac{dL}{dt} = \frac{\partial L}{\partial t} + \sum_k (\dot{p}_k\dot{q}_k + p_k\ddot{q}_k), \tag{24.93}$$

and motivate us to define the *Hamiltonian* function:

$$H(q_k, p_k) = \sum_k p_k\dot{q}_k - L(q_k, \dot{q}_k). \tag{24.94}$$

Using (24.94) to rewrite (24.93) gives

$$\frac{dH}{dt} = -\frac{\partial L}{\partial t}. \tag{24.95}$$

Equation (24.95) shows that the Hamiltonian is a conserved quantity whenever the Lagrangian is not an explicit function of time.

The notation on the left side of (24.94) reflects our expectation that the Legendre transformation (see Section 5.6.2) on the right side of (24.94) produces a function of the coordinates and momenta only. To confirm this, we differentiate H and use (24.91) and (24.92). The result is

$$dH = \sum_k \left(p_k d\dot{q}_k + \dot{q}_k dp_k - \frac{\partial L}{\partial q_k}dq_k - \frac{\partial L}{\partial \dot{q}_k}d\dot{q}_k \right) = \sum_k (\dot{q}_k dp_k - \dot{p}_k dq_k). \tag{24.96}$$

With this information, the rules of calculus tell us that

$$dH = \sum_k \left(\frac{\partial H}{\partial p_k}dp_k + \frac{\partial H}{\partial q_k}dq_k \right). \tag{24.97}$$

Comparing (24.97) to (24.96) produces *Hamilton's equations*:

$$\dot{q}_k = \frac{\partial H}{\partial p_k} \qquad \dot{p}_k = -\frac{\partial H}{\partial q_k}. \tag{24.98}$$

These equations are first-order in time in contrast to the second-order Lagrange equations.

Example 24.3 Derive the Hamiltonian for a relativistic point charge in an external electromagnetic field. Find the non-relativistic limit.

Solution: Since $1/\gamma = \sqrt{1 - \mathbf{v}\cdot\mathbf{v}/c^2}$, the Lagrangian (24.27) for this situation is

$$L_0(\mathbf{r}, \mathbf{v}) = -mc^2\sqrt{1 - \frac{\mathbf{v}\cdot\mathbf{v}}{c^2}} + e\mathbf{v}\cdot\mathbf{A}(\mathbf{r}, t) - e\varphi(\mathbf{r}, t).$$

The corresponding canonical momentum is

$$\mathbf{p} = \frac{\partial L_0}{\partial \mathbf{v}} = \gamma m\mathbf{v} + e\mathbf{A}.$$

Therefore, the Hamiltonian is

$$H = \mathbf{p}\cdot\mathbf{v} - L = (\gamma m\mathbf{v} + e\mathbf{A})\cdot\mathbf{v} + \frac{mc^2}{\gamma} - e\mathbf{v}\cdot\mathbf{A} + e\varphi = \gamma mc^2 + e\varphi.$$

Finally, we use the identity $\gamma mc = \sqrt{(\gamma m\mathbf{v})^2 + m^2c^2}$ to write H as a function of $\mathbf{p}$ rather than $\mathbf{v}$:

$$H = c\sqrt{(\gamma m\mathbf{v})^2 + m^2c^2} + e\varphi = c\sqrt{(\mathbf{p} - e\mathbf{A})^2 + m^2c^2} + e\varphi.$$

The non-relativistic limit of the canonical momentum is $\mathbf{p} = m\mathbf{v} + e\mathbf{A}$. Therefore, the corresponding non-relativistic Hamiltonian is

$$H = c\left\{m^2c^2\left[1 + \frac{1}{m^2c^2}(\mathbf{p} - e\mathbf{A})^2\right]^{1/2}\right\} + e\varphi \approx mc^2 + \frac{(\mathbf{p} - e\mathbf{A})^2}{2m} + e\varphi.$$

24.5.1 The Coulomb-Lorentz Equations

The Hamiltonian for a set of charged particles coupled to their electromagnetic fields is constructed directly from the corresponding Lagrangian. For convenience, we recall the latter from Section 24.3.5 and write it here:

$$L_{\text{EM}} = \tfrac{1}{2}\sum_k m_k v_k^2 + \int d^3r\,(\mathbf{j}\cdot\mathbf{A} - \rho\varphi) + \tfrac{1}{2}\epsilon_0\int d^3r\,\left[(\nabla\varphi + \partial\mathbf{A}/\partial t)^2 - c^2(\nabla\times\mathbf{A})^2\right]. \tag{24.99}$$

We also recall from (24.26), (24.40), and (24.45) that L_{EM} and its Lagrangian density $\mathcal{L}_{\text{EM}}$ define the canonical momentum associated with the particle velocity $\mathbf{v}_k$, and the canonical momentum densities associated with the generalized velocities $\dot{\mathbf{A}}$ and $\dot{\varphi}$, as

$$\mathbf{p}_k = \frac{\partial L_{\text{EM}}}{\partial \mathbf{v}_k} = m\mathbf{v}_k + e\mathbf{A} \qquad \boldsymbol{\pi} = \frac{\partial \mathcal{L}_{\text{EM}}}{\partial \dot{\mathbf{A}}} = -\epsilon_0\mathbf{E} \qquad \pi_0 = \frac{\partial \mathcal{L}_{\text{EM}}}{\partial \dot{\varphi}} = 0. \tag{24.100}$$

Using all three terms in (24.100), the definition (24.94) of the Hamiltonian generalizes to

$$H = \sum_k \mathbf{p}_k\cdot\mathbf{v}_k + \int d^3r\,\left(\boldsymbol{\pi}\cdot\dot{\mathbf{A}} + \pi_0\dot{\varphi}\right) - L_{\text{EM}}. \tag{24.101}$$

Substituting (24.99) and (24.100) into (24.101) gives the intermediate result

$$\begin{aligned} H &= \sum_k (m_k \mathbf{v}_k + e_k \mathbf{A}) \cdot \mathbf{v}_k + \int d^3 r\, \boldsymbol{\pi} \cdot \dot{\mathbf{A}} \\ &- \sum_k \left[\frac{1}{2} m_k v_k^2 + e_k \mathbf{v}_k \cdot \mathbf{A}(\mathbf{r}_k) - e_k \varphi(\mathbf{r}_k) \right] \\ &- \frac{1}{2}\epsilon_0 \int d^3 r \left[\left(\nabla\varphi + \dot{\mathbf{A}}\right)^2 - c^2 (\nabla \times \mathbf{A})^2 \right]. \end{aligned} \tag{24.102}$$

The Hamiltonian formalism requires that we express (24.102) as a function of the canonical momenta. Using (24.100) and $\mathbf{E} = -\nabla\varphi - \dot{\mathbf{A}}$, we find without difficulty that

$$\begin{aligned} H &= \sum_k \frac{1}{2m_k} [\mathbf{p}_k - e_k \mathbf{A}(\mathbf{r}_k)]^2 + \sum_k e_k \varphi(\mathbf{r}_k) \\ &- \int d^3 r\, \boldsymbol{\pi} \cdot \nabla\varphi + \frac{1}{2}\epsilon_0 \int d^3 r \left[\left(\frac{\boldsymbol{\pi}}{\epsilon_0}\right)^2 + c^2 (\nabla \times \mathbf{A})^2 \right]. \end{aligned} \tag{24.103}$$

The Hamiltonian (24.103) is best suited to evaluate Hamilton's equations (24.98) for the particle degrees of freedom. The results are

$$\frac{d\mathbf{r}_k}{dt} = \frac{\partial H}{\partial \mathbf{p}_k} = \frac{\mathbf{p}_k - e_k \mathbf{A}(\mathbf{r}_k)}{m_k} = \mathbf{v}_k \tag{24.104}$$

and

$$\frac{d\mathbf{p}_k}{dt} = -\frac{\partial H}{\partial \mathbf{r}_k} = \frac{e_k}{m_k} [(\mathbf{p}_k - e_k \mathbf{A}) \cdot \nabla]\mathbf{A} - e_k \nabla\varphi = e_k \nabla(\mathbf{v}_k \cdot \mathbf{A}) - e_k \nabla\varphi. \tag{24.105}$$

Comparing (24.105) to (24.24) and working backward to (24.18) shows that Hamilton's equation of motion reproduces the expected Coulomb-Lorentz expression,

$$m\frac{d\mathbf{v}}{dt} = e[\mathbf{E}(\mathbf{r}, t) + \mathbf{v} \times \mathbf{B}(\mathbf{r}, t)]. \tag{24.106}$$

24.5.2 Hamilton's Equations For Fields

The structure of the Hamiltonian (24.103) and our experience with the Lagrangian approach to field dynamics in Section 24.3.4 motivates us to generalize Hamilton's equations (24.98) to situations where the Hamiltonian is a space integral over a Hamiltonian density with the general form

$$\mathcal{H}\left[q_k(\mathbf{r}, t), \partial_i q_k(\mathbf{r}, t), p_k(\mathbf{r}, t), \partial_i p_k(\mathbf{r}, t)\right]. \tag{24.107}$$

Our strategy is to apply Hamilton's principle (Section 24.2) and set to zero the variation of the action

$$0 = \delta S = \int dt\, d^3 r\, \delta\mathcal{L} = \int dt\, d^3 r\, \delta[p_k \dot{q}_k - \mathcal{H}] = \int dt\, d^3 r\, [\dot{q}_k \delta p_k + p_k \delta\dot{q}_k - \delta\mathcal{H}]. \tag{24.108}$$

Using

$$p_k \delta\dot{q}_k = p_k \frac{d(\delta q_k)}{dt} = \frac{d}{dt}(p_k \delta q_k) - \dot{p}_k \delta q_k, \tag{24.109}$$

and our (by now) usual argument that the total derivative in (24.109) vanishes in the action integral from the end-point restrictions (24.35) imposed on δq, we find

$$0 = \int dt\, d^3 r\, [\dot{q}_k \delta p_k - \dot{p}_k \delta q_k - \delta\mathcal{H}]. \tag{24.110}$$

The variation of the Hamiltonian density in (24.110) is

$$\delta\mathcal{H} = \frac{\partial\mathcal{H}}{\partial q_k}\delta q_k + \frac{\partial\mathcal{H}}{\partial(\partial_i q_k)}\delta(\partial_i q_k) + \frac{\partial\mathcal{H}}{\partial p_k}\delta p_k + \frac{\partial\mathcal{H}}{\partial(\partial_i p_k)}\delta(\partial_i p_k). \tag{24.111}$$

Manipulating the second and fourth terms on the right side of (24.111) in a manner similar to (24.109), and following the steps from (24.33) to (24.36) puts (24.110) in the form

$$\begin{aligned} 0 &= \int dt\, d^3r \left[\dot{q}_k - \frac{\partial\mathcal{H}}{\partial p_k} + \partial_i\left\{\frac{\partial\mathcal{H}}{\partial(\partial_i p_k)}\right\}\right]\delta p_k \\ &\quad - \int dt\, d^3r \left[\dot{p}_k + \frac{\partial\mathcal{H}}{\partial p_k} - \partial_i\left\{\frac{\partial\mathcal{H}}{\partial(\partial_i q_k)}\right\}\right]\delta q_k. \end{aligned} \tag{24.112}$$

The variations δq_k and δp_k are independent, so (24.112) is satisfied only if the field coordinates and momenta satisfy the generalized Hamilton's equations

$$\begin{aligned} \dot{p}_k &= -\frac{\partial\mathcal{H}}{\partial q_k} + \partial_i\frac{\partial\mathcal{H}}{\partial(\partial_i q_k)} \\ \dot{q}_k &= \frac{\partial\mathcal{H}}{\partial p_k} - \partial_i\frac{\partial\mathcal{H}}{\partial(\partial_i p_k)}. \end{aligned} \tag{24.113}$$

24.5.3 The Maxwell Equations

Our use of the vector and scalar potentials in (24.20) establishes the homogeneous Maxwell equations as identities:

$$\nabla\cdot\mathbf{B} = 0 \qquad \text{and} \qquad \nabla\times\mathbf{E} + \frac{\partial\mathbf{B}}{\partial t} = 0. \tag{24.114}$$

To make further progress, and exploit the generalized Hamilton equations (24.113), it is convenient to replace the sum over $e_k\varphi(\mathbf{r}_k)$ in the Hamiltonian (24.103) by an integral over space using the charge density $\rho(\mathbf{r}, t)$ in (24.17). The latter is localized, so we can integrate $\boldsymbol{\pi}\cdot\nabla\varphi$ by parts and discard the surface term. These steps transform (24.103) to the Hamiltonian we will use for the remainder of this chapter:

$$\begin{aligned} H &= \sum_k \frac{1}{2m_k}[\mathbf{p}_k - e_k\mathbf{A}(\mathbf{r}_k)]^2 + \int d^3r\, \varphi(\nabla\cdot\boldsymbol{\pi} + \rho) \\ &\quad + \frac{1}{2}\epsilon_0 \int d^3r \left[\left(\frac{\boldsymbol{\pi}}{\epsilon_0}\right)^2 + c^2(\nabla\times\mathbf{A})^2\right]. \end{aligned} \tag{24.115}$$

The two inhomogeneous Maxwell equations derive from the first Hamilton equation in (24.113). Consider the momentum $\boldsymbol{\pi}$ conjugate to the generalized coordinate $\mathbf{A}$ in (24.100). Using (24.115),

$$\frac{\partial\pi_j}{\partial t} = -\frac{\partial}{\partial A_j}\sum_k \frac{1}{2m_k}[\mathbf{p}_k - e_k\mathbf{A}(\mathbf{r})]^2\delta(\mathbf{r} - \mathbf{r}_k) + \frac{1}{2}\epsilon_0 c^2\partial_i\frac{\partial}{\partial(\partial_k A_j)}(\nabla\times\mathbf{A})^2. \tag{24.116}$$

After carrying out the derivatives, the leftmost member of (24.100) and the definition of the current density in (24.17) simplify (24.116) to

$$\mu_0\frac{\partial\boldsymbol{\pi}}{\partial t} = \mu_0\mathbf{j} - \nabla\times(\nabla\times\mathbf{A}). \tag{24.117}$$

The substitutions $\boldsymbol{\pi} = -\epsilon_0\mathbf{E}$ and $\mathbf{B} = \nabla\times\mathbf{A}$ identify (24.117) as the Ampère-Maxwell law,

$$-\frac{1}{c^2}\frac{\partial\mathbf{E}}{\partial t} = \mu_0\mathbf{j} - \nabla\times\mathbf{B}. \tag{24.118}$$

Gauss' law follows similarly from the momentum density $\pi_0 = 0$ conjugate to the generalized coordinate φ in (24.100). Again using $\boldsymbol{\pi} = -\epsilon_0 \mathbf{E}$,

$$0 = \frac{\partial \pi_0}{\partial t} = -\frac{\partial}{\partial \varphi}\left[\varphi(\nabla \cdot \boldsymbol{\pi} + \rho)\right] = -\epsilon_0 \nabla \cdot \mathbf{E} + \rho. \tag{24.119}$$

Consider now the second Hamilton equation in (24.113). For the generalized coordinate $\mathbf{A}$,

$$\frac{\partial \mathbf{A}}{\partial t} = \frac{\partial \mathcal{H}}{\partial \boldsymbol{\pi}} - \partial_i \frac{\partial \mathcal{H}}{\partial(\partial_i \boldsymbol{\pi})} = \frac{\boldsymbol{\pi}}{\epsilon_0} - \nabla \varphi. \tag{24.120}$$

Because $\mathbf{E} = -\boldsymbol{\pi}/\epsilon_0$, (24.120) is simply a restatement of $\mathbf{E} = -\nabla\varphi - \partial \mathbf{A}/\partial t$. However, for the generalized coordinate φ, the Hamilton equation of motion is

$$\frac{\partial \varphi}{\partial t} = \frac{\partial \mathcal{H}}{\partial \pi_0} = 0. \tag{24.121}$$

This result is suspicious. How can the scalar potential not change with time? At the very least, (24.121) is inconsistent with the Maxwell equations derived from the first of Hamilton's equations.

We will see in the next section that the origin and removal of this inconsistency is closely related to issues of gauge invariance and gauge fixing. Indeed, the issue we raise here does not appear in most treatments of the Hamiltonian form of Maxwell's theory precisely because a specific choice of gauge is made at an early stage.[6]

24.5.4 Dirac's Method of Constraints

Equation (24.121) is *not* correct. The reason becomes clear when we carry out the total derivative on the left side of Lagrange's equation (24.2) explicitly to get

$$\frac{\partial^2 L}{\partial \dot{q}_k \partial \dot{q}_m}\ddot{q}_m = \frac{\partial L}{\partial q_k} - \frac{\partial^2 L}{\partial \dot{q}_k \partial q_m}\dot{q}_m. \tag{24.122}$$

The goal of Lagrangian dynamics is to solve (24.122) for the accelerations $\ddot{q}_m$. This is straightforward for mechanics problems of the usual sort because the matrix

$$W_{km} = \frac{\partial^2 L}{\partial \dot{q}_k \partial \dot{q}_m} = \frac{\partial p_k}{\partial \dot{q}_m} \tag{24.123}$$

has an inverse and the change of variables from $L(q_k, \dot{q}_k)$ to $H(q_k, p_k)$ is one-to-one. This means that the velocities can be expressed as unique functions of the coordinates and the momenta. The momenta are all independent and no irregularities arise.

However, if $\det \mathbf{W} = 0$, the inverse matrix does not exist and not all of the conjugate momenta are independent. The Lagrangian is said to be *singular* and there exist so-called *Lagrangian constraints* among the variables (q_k, p_k). A consistent way to pass from the Lagrangian formalism to the Hamiltonian formalism in the singular case was given by Dirac (1964). In this section, we outline his method far enough to expose the roles of gauge invariance and gauge fixing in the theory.

The Lagrangian of classical electrodynamics is singular because the vanishing of the canonical momentum (24.45) associated with the scalar potential,

$$\pi_0(\mathbf{r}, t) = \frac{\partial \mathcal{L}_{\rm EM}}{\partial \dot{\varphi}} = 0, \tag{24.124}$$

implies that the matrix elements $\partial \pi_0/\partial \dot{q}_m$ on the far right side of (24.123) are all zero. This makes $\pi_0(\mathbf{r}, t) = 0$ a Lagrangian constraint in precisely the sense indicated above. This is called a *primary*

[6] The usual context for this discussion is the quantization of classical electrodynamics. See Sources, References, and Additional Reading.

constraint to indicate that it arises directly from the Lagrangian definition of conjugate momentum. It follows that (24.119) is not a true equation of motion but simply a requirement that the primary constraint be preserved in time. Gauss' law, $\nabla \cdot \boldsymbol{\pi} + \rho = 0$, is thus an independent, so-called *secondary* constraint. Could there be further constraints? To check, we demand that our secondary constraint be preserved in time as well. Using the Ampère-Maxwell law (24.118), we find

$$0 = \frac{\partial}{\partial t}(\nabla \cdot \boldsymbol{\pi} + \rho) = -\epsilon_0 \nabla \cdot \frac{\partial \mathbf{E}}{\partial t} + \frac{\partial \rho}{\partial t} = \nabla \cdot \mathbf{j} + \frac{\partial \rho}{\partial t}. \tag{24.125}$$

Equation (24.125) is *not* a new constraint because the continuity equation $\nabla \cdot \mathbf{j} + \partial \rho / \partial t = 0$ follows by direct computation from (24.17) (cf. Application 2.1).

The primary and secondary constraints we have identified restrict the variations δq_k and δp_k used in (24.108) to derive Hamilton's equations. Let us use the method of Lagrange multipliers (Section 1.10) to enforce the constraints. This leads us to define an *extended* action,

$$S_E(q_k, p_k, u, w) = \int dt\, d^3r \left[p_k \dot{q}_k - \mathcal{H} - u\pi_0 - w(\nabla \cdot \boldsymbol{\pi} + \rho) \right], \tag{24.126}$$

and apply Hamilton's principle ($\delta S_E = 0$) with the variations δq_k and δp_k performed independently and without restriction. For reasons which will become clear below, we permit extra variational freedom in (24.126) by choosing $u = u(t)$ and $w = w(\mathbf{r})$ as arbitrary Lagrange multiplier *functions* rather than as simple constants. Moreover, comparing (24.126) to (24.108) shows that we can retain the latter equation if we replace the Hamiltonian density $\mathcal{H}$ by an *extended* Hamiltonian density,

$$\mathcal{H}_E = \mathcal{H} + u(t)\pi_0 + w(\mathbf{r})(\nabla \cdot \boldsymbol{\pi} + \rho). \tag{24.127}$$

Using $\mathcal{H}_E$ in place of $\mathcal{H}$ in (24.113), the reader can verify that the equations of motion for $\mathbf{r}$ (24.104), $\boldsymbol{\pi}$ (24.117), and π_0 (24.119) do not change. Thus, we reproduce the inhomogeneous Maxwell equations derived in the previous section. The equations of motion which *do* change are[7]

$$\frac{d\mathbf{p}_k}{dt} = e_k \left[\nabla(\mathbf{v}_k \cdot \mathbf{A}) - \nabla(\varphi + w) \right], \tag{24.128}$$

$$\frac{\partial \mathbf{A}}{\partial t} = \frac{\boldsymbol{\pi}}{\epsilon_0} - \nabla(\varphi + w), \tag{24.129}$$

and

$$\frac{d\varphi}{dt} = u. \tag{24.130}$$

Equation (24.128) correctly describes charged particle motion despite the presence of the arbitrary function $w(\mathbf{r})$. To see this, use $\mathbf{p}_k = m_k \mathbf{v}_k + e_k \mathbf{A}$ and (24.129) to eliminate $\nabla(\varphi + w)$ in favor of $\boldsymbol{\pi}/\epsilon_0 - \dot{\mathbf{A}} = -\mathbf{E} - \dot{\mathbf{A}}$. The result,

$$m_k \frac{d\mathbf{v}_k}{dt} = e_k \nabla(\mathbf{v}_k \cdot \mathbf{A}) + e_k \left[\frac{\partial \mathbf{A}}{\partial t} - \frac{d\mathbf{A}}{dt} \right] + e_k \mathbf{E}, \tag{24.131}$$

combined with (24.23) is the Coulomb-Lorentz force law in the form (24.22).

By contrast, the general solutions to the equations of motion for $\mathbf{A}(\mathbf{r}, t)$ and $\varphi(\mathbf{r}, t)$ in (24.129) and (24.130) depend explicitly on the arbitrary functions $u(t)$ and $w(\mathbf{r})$. The meaning of this becomes clear when we ask for the change in the vector and scalar potentials induced by these Lagrange functions after an infinitesimal time δt:

$$\delta \mathbf{A} = -\nabla w\, \delta t \qquad \text{and} \qquad \delta \varphi = u\, \delta t. \tag{24.132}$$

[7] To derive (24.128), it is simplest to write the $w(\mathbf{r})\rho(\mathbf{r})$ term in (24.127) as a discrete sum like the $e_k \varphi(\mathbf{r}_k)$ term in (24.103).

However, because $u = u(t)$ and $w = w(\mathbf{r})$, the changes in (24.132) are indistinguishable from the changes

$$\Delta \mathbf{A} = -\nabla \Lambda \qquad \text{and} \qquad \Delta \varphi = +\frac{\partial \Lambda}{\partial t} \tag{24.133}$$

induced by a gauge transformation with the gauge function

$$\Lambda(\mathbf{r}, t) = \left[w(\mathbf{r}) + \int_{\infty}^{t} ds\, u(s) \right] \delta t. \tag{24.134}$$

Hence, the only effect of the Lagrange multiplier functions is to impose a continuous sequence of gauge transformations at every space-time point. These gauge changes have no effect on the observables $\mathbf{E}(\mathbf{r}, t)$ and $\mathbf{B}(\mathbf{r}, t)$.

24.5.5 A Heuristic Approach to Gauge Fixing

The explicit gauge freedom of Maxwell's theory on display in (24.129) and (24.130) is interesting as a matter of principle but inconvenient for practical calculations. Dirac and others developed a *gauge fixing* machinery which chooses the Lagrange multiplier functions in a manner consistent with any particular choice of gauge. For the most common choices of gauge (Coulomb and Lorenz), this formalism reproduces (and therefore justifies) gauge fixing schemes which had been developed long before. We will not pursue this approach here, despite its implications for the canonical quantization of electrodynamics.[8]

Our strategy in this section is to proceed heuristically and identify the true dynamical variables from among the set $(\varphi, \pi_0, \mathbf{A}, \boldsymbol{\pi})$ by requiring that their equations of motion not depend on $u(t)$ or $w(\mathbf{r})$. This procedure avoids the Lagrange multiplier functions (rather than making specific choices for them) but amounts to fixing a gauge nonetheless because no arbitrary functions appear in the final dynamics.

Our first step is to discard the scalar potential φ because (24.129) shows that it depends unavoidably on $u(t)$.[9] This forces us to discard the corresponding conjugate momentum π_0 as well. Of course, the fact that $\pi_0 = \dot{\pi}_0 = 0$ would induce us to do this anyway. To proceed further, we recall from Helmholtz' theorem (Section 1.9) that a well-behaved vector field like the vector potential can be decomposed as $\mathbf{A} = \mathbf{A}_\perp + \mathbf{A}_\parallel$ where $\nabla \cdot \mathbf{A}_\perp = 0$ and $\nabla \times \mathbf{A}_\parallel = 0$. Using this fact, the equations of motion for $\mathbf{A}$ (24.129) and $\boldsymbol{\pi}$ (24.117) separate into

$$\frac{\partial \mathbf{A}_\perp}{\partial t} = \frac{\boldsymbol{\pi}_\perp}{\epsilon_0} \qquad \text{and} \qquad \mu_0 \frac{\partial \boldsymbol{\pi}_\perp}{\partial t} = \mu_0 \mathbf{j}_\perp - \nabla \times \nabla \times \mathbf{A}_\perp, \tag{24.135}$$

and

$$\frac{\partial \mathbf{A}_\parallel}{\partial t} = \frac{\boldsymbol{\pi}_\parallel}{\epsilon_0} - \nabla(\varphi + w) \qquad \text{and} \qquad \frac{\partial \boldsymbol{\pi}_\parallel}{\partial t} = \mathbf{j}_\parallel. \tag{24.136}$$

The presence of $w(\mathbf{r})$ in (24.136) directs us to discard $\mathbf{A}_\parallel$ and thus its conjugate momentum $\boldsymbol{\pi}_\parallel$ from our set of dynamical variables. This leaves the canonical pair $(\mathbf{A}_\perp, \boldsymbol{\pi}_\perp)$ as the only "true" degrees of freedom. As advertised, no arbitrary functions appear in (24.135).

To see that this choice is sufficient, note first from (24.20) that $\mathbf{B}_\perp = \nabla \times \mathbf{A}_\perp$ while

$$\mathbf{E}_\perp = -\frac{\partial \mathbf{A}_\perp}{\partial t} \qquad \text{and} \qquad \mathbf{E}_\parallel = -\nabla \varphi - \frac{\partial \mathbf{A}_\parallel}{\partial t}. \tag{24.137}$$

[8] Chapter 8 of Weinberg (1995) gives a very clear account of the "Dirac bracket" method and its application to electrodynamics.

[9] A "discarded" variable is not set to zero. It simply has no independent equation of motion.

Using these, the Maxwell equations take the form

$$\begin{aligned} \nabla \times \mathbf{E}_\perp &= -\frac{\partial \mathbf{B}_\perp}{\partial t} & \qquad \nabla \cdot \mathbf{E}_\parallel &= \rho/\epsilon_0 \\ \nabla \times \mathbf{B}_\perp &= \mu_0 \mathbf{j}_\perp + \frac{1}{c^2}\frac{\partial \mathbf{E}_\perp}{\partial t} & \qquad \nabla \cdot \mathbf{B}_\parallel &= 0. \end{aligned} \tag{24.138}$$

The equations on the left side of (24.138) confirm that $\mathbf{A}_\perp$ and $\boldsymbol{\pi}_\perp$ (or $\mathbf{E}_\perp$ and $\mathbf{B}_\perp$) are sufficient to compute the time evolution of the fields. The equations on the right of (24.138) are satisfied for all time if they are satisfied initially.

The procedure outlined above amounts to fixing the Coulomb gauge. To see this, we need only write out the usual expression for this gauge, $\nabla \cdot \mathbf{A} = 0$, in the language of the Helmholtz theorem. That is, $\nabla \cdot \mathbf{A} = \nabla \cdot (\mathbf{A}_\perp + \mathbf{A}_\parallel) = \nabla \cdot \mathbf{A}_\parallel = 0$. The last equality, combined with the definition $\nabla \times \mathbf{A}_\parallel = 0$, is equivalent to the constraint

$$\mathbf{A}_\parallel = 0. \tag{24.139}$$

Suppose we impose (24.139) as an initial condition on the dynamics. Our previous experience with constraints directs us to demand that (24.139) be preserved in time. This leads to $\dot{\mathbf{A}}_\parallel = 0$ as a second independent constraint on the initial data. Because (24.136) is first order in time, this is sufficient to guarantee that $\mathbf{A}_\parallel$ remains zero at all later times. Combining these facts with the expression for $\mathbf{E}_\parallel$ in (24.137) and Gauss' law in (24.138) yields Poisson's equation $\nabla^2 \varphi = -\rho/\epsilon_0$. This fixes the scalar potential at every point in space-time as

$$\varphi(\mathbf{r}, t) = \frac{1}{4\pi\epsilon_0}\int d^3r' \, \frac{\rho(\mathbf{r}', t)}{|\mathbf{r} - \mathbf{r}'|}. \tag{24.140}$$

The Coulomb gauge assignments for $\mathbf{A}_\parallel$ (24.139) and φ (24.140) eliminates them (and their conjugate momenta) as dynamical variables and thus establishes the equivalency of this gauge choice with the heuristic argument given above.

We conclude by writing the Hamiltonian (24.115) in its Coulomb gauge form. The key step is to recognize that $\boldsymbol{\pi} = -\epsilon_0 \mathbf{E}$, combined with the Coulomb gauge formulae $\mathbf{E} = -\nabla\varphi$ and $\nabla^2\varphi = -\rho/\epsilon_0$, imply that $\nabla \cdot \boldsymbol{\pi} + \rho = 0$. This yields the intermediate result

$$\begin{aligned} H_C = &\sum_k \frac{1}{2m_k}[\mathbf{p}_k - e_k \mathbf{A}(\mathbf{r}_k)]^2 \\ &+ \frac{1}{2\epsilon_0}\int d^3r \, \left(\boldsymbol{\pi}_\perp + \boldsymbol{\pi}_\parallel\right)^2 + \frac{1}{2\mu_0}\int d^3r \, (\nabla \times \mathbf{A}_\perp)^2. \end{aligned} \tag{24.141}$$

However,

$$\int d^3r \, \boldsymbol{\pi}_\perp \cdot \boldsymbol{\pi}_\parallel = \epsilon_0 \int d^3r \, \boldsymbol{\pi}_\perp \cdot \nabla\varphi = -\epsilon_0 \int d^3r \, \varphi \nabla \cdot \boldsymbol{\pi}_\perp = 0 \tag{24.142}$$

and

$$\frac{1}{2\epsilon_0}\int d^3r \, \boldsymbol{\pi}_\parallel \cdot \boldsymbol{\pi}_\parallel = \frac{1}{2}\epsilon_0 \int d^3r \, \mathbf{E}_\parallel \cdot \nabla\varphi = \frac{1}{2}\int d^3r \, \rho \, \varphi. \tag{24.143}$$

Combining these results with (24.140) simplifies (24.141) to

$$\begin{aligned} H_C = &\sum_k \frac{1}{2m_k}[\mathbf{p}_k - e_k \mathbf{A}_\perp]^2 + \frac{1}{8\pi\epsilon_0}\int d^3r \int d^3r' \, \frac{\rho(\mathbf{r}, t)\rho(\mathbf{r}', t)}{|\mathbf{r} - \mathbf{r}'|} \\ &+ \frac{1}{2}\epsilon_0 \int d^3r \left[\left(\frac{\boldsymbol{\pi}_\perp}{\epsilon_0}\right)^2 + c^2 (\nabla \times \mathbf{A}_\perp)^2\right]. \end{aligned} \tag{24.144}$$

The Hamiltonian (24.144) is widely applied to problems in atomic, molecular, and condensed matter physics because the Coulomb binding energy term is already present in the matter-only Hamiltonian.

It is not so easy to identify the "true" degrees of freedom for other choices of gauge. This is true for electrodynamics, and also for more complex field theories which share the constrained dynamics character of electrodynamics. For these cases, Dirac's systematic approach to gauge fixing is always available.

Sources, References, and Additional Reading

The quotation at the beginning of the chapter is taken from

A.S. Eddington, *Royal Society Obituaries* **4**, 197 (1942).

Section 24.1 *Yourgrau and Mandelstam* is a superb introduction to variational methods. *Podolsky and Kunz* and *Konopinski* are intermediate-level textbooks that discuss Lagrangian and Hamiltonian methods in electrodynamics with particular clarity.

W. Yourgrau and S. Mandelstam, *Variational Principles in Dynamics and Quantum Theory*, 3rd edition (Dover, New York, 1979).

B. Podolsky and K.S. Kunz, *Fundamentals of Electrodynamics* (Marcel Dekker, New York, 1969).

E.J. Konopinski, *Electromagnetic Fields and Relativistic Particles* (McGraw-Hill, New York, 1981).

Section 24.2 Hamilton's principle is discussed in these textbooks of classical mechanics:

H. Goldstein, *Classical Mechanics* (Addison-Wesley, Cambridge, MA, 1950).

M.G. Calkin, *Lagrangian and Hamiltonian Mechanics* (World Scientific, Singapore, 1996).

Section 24.3 *Doughty* is an interesting general reference for Lagrangian dynamics. The non-relativistic approach taken by *Schwinger et al.* is idiosyncratic but consistently fascinating and enlightening. Our deduction of the Maxwell equations follows *Goldstein* (see Section 24.2 above).

N.A. Doughty, *Lagrangian Interaction* (Addison-Wesley, Reading, MA, 1990).

J. Schwinger, L.L. DeRaad, Jr., K.A. Milton, and W.-Y. Tsai, *Classical Electrodynamics* (Perseus, Reading, MA, 1998).

The boxed material on the history of Lagrangians in electrodynamics is based on Chapter 8 and Appendix 9 of *Darrigol*. *Schwarzschild* was the first paper to display the modern Lagrangian for electrodynamics.

O. Darrigol, *Electrodynamics from Ampère to Einstein* (University Press, Oxford, 2000).

K. Schwarzschild, "On electrodynamics", *Nachrichten von der Gesellschaft der Wissenschaften zu Götingen, Mathematisch-Physikalische Klasse*, 126 (1903).

Section 24.4 Our presentation of Noether's theorem benefitted from the discussion in *Konopinski* (see Section 24.1) as well as

A.O. Barut, *Electrodynamics and Classical Field Theory of Fields and Particles* (Macmillan, New York, 1964).

F. Rohrlich, *Classical Charged Particles* (Addison-Wesley, Reading, MA, 1965).

Section 24.5 *Schwinger et al.* (see Section 24.3 above) and *Yourgrau and Mandelstam* (see Section 24.1 above) give complementary and non-covariant discussions of the Hamiltonian formulation of electrodynamics. *Dirac* and *Weinberg* discuss electrodynamics as a constrained classical field theory before turning to its quantization:

P.A.M. Dirac, *Lectures on Quantum Mechanics* (Belfer Graduate School of Yeshiva University, New York, 1964).

S. Weinberg, *The Quantum Theory of Fields* (Cambridge University Press, Cambridge, 1995), Volume I.

The Hamiltonian formulation of classical electrodynamics is given in both the Coulomb gauge and the Lorenz gauge in

E.C.G. Sudarshan and N. Mukunda, *Classical Dynamics: A Modern Perspective* (Wiley, New York, 1974).

W. Heitler, *The Quantum Theory of Radiation*, 3rd edition (Dover, New York, 1984).

C. Cohen-Tannoudji, J. Dupont-Roc, and G. Grynberg, *Photons and Atoms* (Wiley-Interscience, New York, 1997).

Problems

24.1 Working Backward The success of the "working backward" method used in the text to find the Lagrangian for a charged particle in an external electromagnetic field is not obvious. Helmholtz proved in 1887 that a Lagrangian description exists for a general position- and velocity-dependent force with Cartesian components $F_k = F_k(\mathbf{r}, \dot{\mathbf{r}}, t)$ if

$$\frac{\partial F_i}{\partial \dot{r}_j} = -\frac{\partial F_j}{\partial \dot{r}_i} \qquad \text{and} \qquad \frac{\partial F_i}{\partial r_j} - \frac{\partial F_j}{\partial r_i} = \frac{1}{2}\frac{d}{dt}\left(\frac{\partial F_i}{\partial \dot{r}_j} - \frac{\partial F_j}{\partial \dot{r}_i}\right).$$

(a) Show that the first Helmholtz condition implies that $\partial^2 F_i/\partial \dot{r}_j \partial \dot{r}_k = 0$. Integrate this equation explicitly and deduce that $\mathbf{F}$ has the Coulomb-Lorentz form

$$F_i(\mathbf{r}, \dot{\mathbf{r}}, t) = P_i\,(\mathbf{r}, t) + \epsilon_{ijk}\dot{r}_j Q_k\,(\mathbf{r}, t)\,,$$

where $\mathbf{P}(\mathbf{r}, t)$ and $\mathbf{Q}(\mathbf{r}, t)$ are arbitrary vector functions of the coordinates and time only.

(b) Show that the second Helmholtz condition implies that $\mathbf{P}$ and $\mathbf{Q}$ satisfy the homogeneous "Maxwell" equations

$$\nabla \cdot \mathbf{Q} = 0 \qquad \text{and} \qquad \nabla \times \mathbf{P} + \frac{\partial \mathbf{Q}}{\partial t} = 0.$$

Hint: The convective derivative will play a role here.

(c) The equation of motion for a particle subject to no external forces but in a reference frame that rotates with angular velocity $\boldsymbol{\omega}$ with respect to an inertial reference frame is

$$m\ddot{\mathbf{r}} = m\mathbf{r} \times \dot{\boldsymbol{\omega}} + 2m\dot{\mathbf{r}} \times \boldsymbol{\omega} + m\boldsymbol{\omega} \times (\mathbf{r} \times \boldsymbol{\omega})\,.$$

The second and third terms on the right-hand side are the familiar Coriolis and centrifugal forces. Show that the total force for this problem satisfies both Helmholtz relations. Identify the electric-like field $\mathbf{P}$ and the magnetic-like field $\mathbf{Q}$. Confirm by explicit calculation that these fields satisfy the homogeneous "Maxwell" equations.

24.2 An Effective Nuclear Force An effective force between two nucleons used to analyze scattering data can be derived from a non-relativistic Hamiltonian that depends only on the relative coordinate $r = |\mathbf{r}_1 - \mathbf{r}_2|$ between the nucleons and the corresponding radial momentum p:

$$H\,(r, p) = \frac{p^2}{2m} + g\,(r) + p^2 f\,(r)\,.$$

(a) Treat r and p as canonically conjugate and derive the corresponding Lagrangian $L = L\,(r, \dot{r})$.

(b) Show that the effective Newton's law force between the nucleons depends on $\ddot{r}$ and $\dot{r}$ as well as $f(r)$ and $g(r)$.

24.3 Relativistic Lagrangian Show that the Lagrangian $L(\mathbf{r}, \mathbf{v}) = -mc^2/\gamma + e\mathbf{v} \cdot \mathbf{A}(\mathbf{r}, t) - e\varphi(\mathbf{r}, t)$ predicts the correct relativistic equation of motion for a point particle with mass m and charge e.

24.4 A Relativistic Particle Coupled to a Scalar Field The action for a relativistic point particle coupled by a strength g to a space-time-dependent Lorentz scalar field $\varphi(x)$ is

$$S = -mc\int ds - g\int ds\,\varphi(\mathbf{r}(s)).$$

Find the equation of motion for the particle. How does the force on the particle differ from the Coulomb force of an electric field?

24.5 The Clausius and Darwin Lagrangians In 1880, Rudolf Clausius proposed a Lagrangian for a collection of charged particles moving in an external electromagnetic field. If $\mathbf{r}_{\alpha\beta} = \mathbf{r}_\alpha - \mathbf{r}_\beta$, the *Clausius Lagrangian* is

$$L_C = \frac{1}{2}\sum_{\alpha=1}^{N} m_\alpha v_\alpha^2 - \frac{1}{8\pi\epsilon_0}\sum_{\alpha=1}^{N}\sum_{\beta\neq\alpha}^{N}\frac{q_\alpha q_\beta}{r_{\alpha\beta}}\left(1 - \frac{\mathbf{v}_\alpha\cdot\mathbf{v}_\beta}{c^2}\right).$$

(a) Show that L_C follows from the Lagrangian L_0 for a collection of charges in an external field if one uses the static forms for the scalar potential and vector potential. What choice of gauge is implied by this construction?

(b) The Clausius approximation does not provide a consistent description of Maxwell's electrodynamics to $O(v^2/c^2)$. To do better, Charles Darwin (grandson of the biologist) chose the Coulomb gauge so the neglect of retardation in the scalar potential is exact. In this gauge, the vector potential is calculated using the transverse current density $\mathbf{j}_\perp(\mathbf{r}, t)$ rather than the total current density $\mathbf{j}(\mathbf{r}, t)$ (See Section 15.3.2). Neglect retardation (why?) and expand the vector potential and the relativistic particle Lagrangian to $O(v^2/c^2)$ and derive thereby the *Darwin Lagrangian*:

$$L_D = \frac{1}{2}\sum_{\alpha=1}^{N} m_\alpha v_\alpha^2\left(1 + \frac{v_\alpha^2}{4c^2}\right) - \frac{1}{8\pi\epsilon_0}\sum_{\alpha=1}^{N}\sum_{\beta\neq\alpha}^{N}\frac{q_\alpha q_\beta}{r_{\alpha\beta}}\left[1 - \frac{\mathbf{v}_\alpha\cdot\mathbf{v}_\beta + (\mathbf{v}_\alpha\cdot\hat{\mathbf{r}}_{\alpha\beta})(\mathbf{v}_\beta\cdot\hat{\mathbf{r}}_{\alpha\beta})}{2c^2}\right].$$

Hint: If $\nabla_k = \partial/\partial r_k$,

$$\frac{1}{4\pi}\int d^3r \frac{1}{|\mathbf{r}-\mathbf{r}_\alpha|}\nabla_k\nabla_m\frac{1}{|\mathbf{r}-\mathbf{r}_\beta|} = \frac{1}{2}\left[\frac{(\mathbf{r}_\alpha - \mathbf{r}_\beta)_k(\mathbf{r}_\alpha - \mathbf{r}_\beta)_m}{r_{\alpha\beta}^3} - \frac{\delta_{km}}{r_{\alpha\beta}}\right].$$

24.6 Equivalent Lagrangians

(a) The Lagrangians L and $L + d\Lambda/dt$ produce the same Lagrange equations if $\Lambda = \Lambda(q_k, t)$. What happens if $\Lambda = \Lambda(\dot{q}_k, q_k, t)$?

(b) Not all equivalent Lagrangians differ by a total time derivative. Find a Lagrangian which yields the same equations of motion as $L = \dot{x}\dot{y} - xy$ but does not differ from it by a total time derivative.

(c) Show that there is a class of vector functions $\boldsymbol{\zeta}\,[\psi(\mathbf{r}, t)]$ of a field variable $\psi\,(\mathbf{r}, t)$ where the equations of motion derived from a Lagrangian density $\mathcal{L} = \mathcal{L}[\psi(\mathbf{r}, t), \dot{\psi}(\mathbf{r}, t)]$ do not change if $\mathcal{L} \to \mathcal{L} + \nabla\cdot\boldsymbol{\zeta}\,[\psi(\mathbf{r}, t)]$.

24.7 Practice with Lagrangian Densities Find the equation of motion for the scalar field $\phi(\mathbf{r}, t)$ if the Lagrangian density is

(a) $\mathcal{L} = \frac{1}{2}\dot{\phi}\phi - \frac{1}{2}|\nabla\phi|^2$

(b) $\mathcal{L} = \frac{1}{2}(\partial_\mu\phi)(\partial_\mu\phi) - \frac{1}{2}\sigma\phi^2$.

24.8 One-Dimensional Massive Scalar Field A one-dimensional field theory with scalar potential $\varphi\,(x, t)$ is characterized by the action

$$S = \frac{1}{2}\int\int dtdx\left[\frac{1}{c^2}\left(\frac{\partial\varphi}{\partial t}\right)^2 - \left(\frac{\partial\varphi}{\partial x}\right)^2 - m^2\varphi^2\right].$$

Find the equation of motion for $\varphi\,(x, t)$ by both Lagrangian and Hamiltonian methods.

24.9 Proca Electrodynamics The free-field Lagrangian density,

$$\mathcal{L}_P = \frac{1}{2}\epsilon_0\left[\left(\nabla\varphi + \frac{\partial\mathbf{A}}{\partial t}\right)^2 - c^2(\nabla\times\mathbf{A})^2\right] - \frac{1}{2\mu_0\ell^2}\left[\mathbf{A}^2 - (\varphi/c)^2\right],$$

with $\ell = \hbar/mc$, was introduced by Alexandre Proca in 1936 as an alternative to Dirac's theory of the positron. Today, it serves as a model for electrodynamics with a photon with mass m when matter-field coupling is added to get the total Lagrangian density, $\mathcal{L} = \mathbf{j} \cdot \mathbf{A} - \rho\varphi + \mathcal{L}_P$.

(a) Find the effect of the Proca mass term on the Maxwell equations.
(b) Show that the Proca model violates gauge invariance because a particular choice of gauge must be made to guarantee conservation of charge.
(c) Find the scalar potential for a static point charge q in the Proca model.

24.10 Podolsky Electrodynamics The Lagrangian density,

$$\mathcal{L}_P = j_\mu A_\mu - \frac{1}{4\mu_0} F_{\sigma\beta} F_{\sigma\beta} - \frac{a^2}{2\mu_0} (\partial_\lambda F_{\beta\lambda})(\partial_\rho F_{\beta\rho}),$$

was introduced by Boris Podolsky in 1942 as a generalization of Maxwell theory which preserves the linear character of the field equations yet avoids certain unwanted divergences.

(a) Apply Hamilton's principle and derive generalized Lagrange equations appropriate to a Lagrangian density of the form $\mathcal{L} = \mathcal{L}(A_\alpha, \partial_\beta A_\alpha, \partial_\mu \partial_\nu A_\alpha)$.
(b) Apply the result of part (a) to $\mathcal{L}_P$ and show that the equations of motion are

$$\left[1 - a^2 \partial_\mu \partial_\mu\right] \partial_\lambda F_{\lambda\alpha} = 0.$$

(c) Write out the $a \neq 0$ generalizations of Gauss' law and the Ampère-Maxwell law in three-vector form.

24.11 Chern-Simons Electrodynamics A model for an electrodynamics which respects gauge invariance but violates Lorentz invariance supplements the usual Maxwell Lagrangian with terms drawn from a four-vector $\vec{d} = (\mathbf{d}, i d_0)$:

$$L_{\text{CS}} = \int d^3 r \left[\rho\varphi - \mathbf{j} \cdot \mathbf{A} + \tfrac{1}{2} \left\{\epsilon_0 \left(\mathbf{E}^2 - c^2 \mathbf{B}^2\right) - \varphi\, (\mathbf{d} \cdot \mathbf{B}/c) + \mathbf{d} \cdot (\mathbf{A} \times \mathbf{E}/c) + d_0 \mathbf{A} \cdot \mathbf{B}\right\}\right].$$

(a) Find the restrictions that must be imposed on $\vec{d}$ to ensure that a gauge transformation does not alter the dynamics.
(b) Assume that $\vec{d}$ is a *constant* four-vector. Is this consistent with your answer to part (a)? Find the Chern-Simons Maxwell equations which replace the usual Maxwell equations. Confirm that the theory is gauge invariant but does not respect Lorentz invariance.

24.12 First-Order Lagrangian Treat the ten scalar functions in the set $(\varphi, \mathbf{A}, \mathbf{E}, \mathbf{B})$ as independent generalized coordinates in the Lagrangian density

$$\mathcal{L}(\varphi, \mathbf{A}, \mathbf{E}, \mathbf{B}) = \mathbf{j} \cdot \mathbf{A} - \rho\varphi - \tfrac{1}{2}\epsilon_0(\mathbf{E}^2 - c^2\mathbf{B}^2) - \epsilon_0 \mathbf{E} \cdot \left(\nabla\varphi + \dot{\mathbf{A}}\right) - \epsilon_0 c^2 \mathbf{B} \cdot (\nabla \times \mathbf{A}).$$

(a) Show that the Lagrange equations produce all four Maxwell equations directly.
(b) Identify the primary constraints associated with $\mathcal{L}$.

24.13 Primary Hamiltonian A system with Lagrangian $L = L(q, \dot{q})$ and canonical momentum $p = \partial L/\partial \dot{q}$ exhibits a primary constraint $\Psi(p, q) = 0$. If $u(p, q)$ is an arbitrary function, show that the Lagrange equation and the primary constraint together imply Hamilton's equations for a "primary Hamiltonian",

$$H_P = H + u\Psi.$$

Hint: compare the total differentials of H and Ψ.

24.14 Gauge Fixing à la Fermi Let $\Omega = \nabla \cdot \mathbf{A} + (1/c^2)\partial\varphi/\partial t$. Fermi showed that one can use a Lagrange parameter λ to impose the Lorenz gauge condition $\Omega = 0$ using the Lagrangian density

$$\mathcal{L} = \mathbf{j} \cdot \mathbf{A} - \rho\varphi + \frac{1}{2}\epsilon_0 \left(\nabla\varphi + \frac{\partial \mathbf{A}}{\partial t}\right)^2 - \frac{1}{2\mu_0}(\nabla \times \mathbf{A})^2 - \frac{\lambda}{2\mu_0}\Omega^2.$$

(a) Show that the Lagrange equations for the potentials are modified inhomogeneous wave equations.
(b) Find the corresponding modified inhomogeneous Maxwell equations.
(c) Manipulate the equations in (a) and use conservation of charge to deduce that Ω satisfies a homogeneous wave equation.
(d) Find initial conditions for the wave equation in (c) which guarantee that the Lorenz gauge condition is always satisfied and that $\mathbf{E}$ and $\mathbf{B}$ in (b) are the physical electric and magnetic fields.

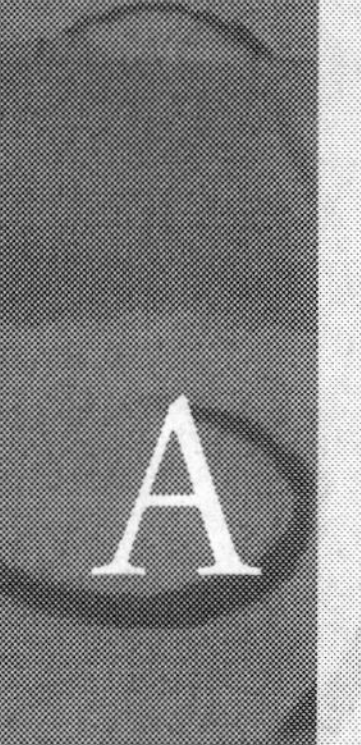

Appendix A
List of Important Symbols

Symbol	Meaning	Defined in Section
$\mathbf{A}$	vector potential	10.5
$\vec{A}$, A_μ	four-potential	22.6.2
$\vec{\mathcal{A}}$, $\mathcal{A}_\mu$	four-acceleration	22.5.1
$A_{\ell m}$	exterior spherical electric multipole moments	4.6.1
α	polarizability	6.6.1
$\boldsymbol{\alpha}$	radiation vector	20.7
$\mathbf{B}$	magnetic field	10.1
$B_{\ell m}$	interior spherical electric multipole moments	4.6.2
$\boldsymbol{\beta}$	dimensionless velocity $\boldsymbol{v}/c$	22.4.1
β	waveguide attenuation coefficient	19.4.8
C, C_{ij}	capacitance; coefficients of capacitance	5.4
c	speed of light	2.6
χ	electric susceptibility	6.5
χ_m	magnetic susceptibility	13.6
$\mathbf{D}$	auxiliary (electric) field	6.4.1
D	particle diffusion constant	9.9
$\mathcal{D}$	magnetic diffusion constant	14.11
δ	skin depth	17.6.2
δ_{ij}, $\delta(x)$	Kronecker and Dirac delta functions	1.2.5, 1.5
e	electron charge magnitude	2.1.1
$\mathbf{E}$	electric field	3.1
$\mathcal{E}$	EMF (electromotive force)	9.7
$\mathcal{E}$	relativistic total energy	22.5.2
$\hat{\mathbf{e}}_1$, $\hat{\mathbf{e}}_2$	polarization basis vectors	16.4.1
ϵ	dielectric permittivity	6.5
ϵ_{ijk}	Levi-Cività symbol	1.2.5
f_μ	four-force density	22.7.3
$F_{\mu\nu}$	electromagnetic field tensor	22.7.2
G_0	free-space Green function	20.3.3
$G_{\mu\nu}$	electromagnetic field dual tensor	22.7.2

Symbol	Meaning	Defined in Section
$\mathbf{g}$	electromagnetic momentum density	15.5.2
g	g-factor	11.2.2
$g(t)$	relativistic Doppler factor	23.2.1
$g_{\mu\nu}$	metric tensor	D.1
Γ	Lorentz model damping factor	18.5.4
γ	relativistic factor $(1 - v^2/c^2)^{-1/2}$	22.4.1
$\mathbf{H}$	auxiliary (magnetic) field	13.5
$H, \mathcal{H}$	Hamiltonian and Hamiltonian density	24.5, 24.5.2
$H_m^{(1,2)}(x)$	Hankel functions	C.3
$h_m^{(1,2)}(x)$	spherical Hankel functions	C.3.2
I	electric current	2.1.2
I	light intensity	16.3.5
$I(\omega)$	frequency spectrum of radiated energy	23.4.1
$\frac{dI}{d\Omega}$	angular spectrum of radiated energy	23.4.1
$I_m(x)$	modified Bessel function	C.3.1
$J_m(x)$	Bessel function	C.3
$\mathbf{j}$	current density	2.1.2
$\mathbf{j}_f, \mathbf{j}_\mathrm{M}$	free and magnetization current density	13.2
$j_m(x)$	spherical Bessel function	C.3.2
$\vec{j}, j_\mu$	four-current density	22.6.1
$\mathbf{K}$	surface current density	2.1.2
$K_m(x)$	modified Bessel function	C.3.1
$\mathbf{k}$	wave vector	16.3.1
κ	dielectric constant	6.5
κ_m	relative magnetic permeability	13.6
$\mathbf{L}_\mathrm{mech}$	total mechanical angular momentum	15.6.1
$\mathbf{L}_\mathrm{EM}$	total electromagnetic angular momentum	15.6.1
L	self-inductance	12.8
$L, \mathcal{L}$	Lagrangian and Lagrangian density	24.2, 24.3.3
$L_{\mu\nu}$	Lorentz transformation matrix	22.5
$\Lambda_{\ell m}^{\mathrm{E,M}}$	electric and magnetic spherical multipole moments	20.8.2
λ	linear charge density	3.3
λ	wavelength	16.3.4
$\mathbf{M}$	magnetization	13.2
$\mathbf{M}$	angular momentum current density	15.6.1
M_{ij}	mutual inductances	12.8.2
$M_{\ell m}$	spherical magnetic multipole moments	11.4.2
$\mathbf{m}$	magnetic dipole moment	11.2
$m_{ip\ldots q}^{(n)}$	Cartesian magnetic multipole moments	11.4.1
μ	chemical potential	5.7
μ	magnetic permeability	13.6
μ_B	Bohr magneton	11.2.2
$\mathbf{N}$	electric torque	3.1
$\mathbf{N}$	magnetic torque	10.1
$\mathcal{N}_{\mu\nu}$	Lorentz torque density tensor	22.7.3
$N_m(x)$	Neumann function	C.3

Symbol	Meaning	Defined in Section
n	number density of particles	5.7
n	index of refraction	17.2
$n_m(x)$	spherical Neumann function	C.3.2
$\hat{\mathbf{n}}$	outward normal vector	1.4.2
Ω_S	solid angle	3.4.4
ω_c	cyclotron frequency	12.2.2.
ω_L	Larmor frequency	12.2.3
ω_p	plasma frequency	18.5.1
$\mathbf{P}$	electric polarization	6.2
$\mathbf{P}_{\text{mech}}$	total mechanical linear momentum	15.5
$\mathbf{P}_{\text{EM}}$	total electromagnetic linear momentum	15.5.2
P	power radiated	20.5.1
$P_\ell(x)$	Legendre polynomial	C.1
$\mathcal{P}$	power dissipated	14.8
P_{ij}	coefficients of potential	5.4.4
$\dfrac{dP}{d\Omega}$	angular distribution of radiated power	20.5.1
$\mathbf{p}$	electric dipole moment	4.1.1
$\mathbf{p}$	particle linear momentum	22.5.2
$\vec{p}$, p_μ	four-momentum	22.5.2
Φ_E, Φ_B	electric and magnetic flux	3.3.4, 10.1.2
φ	electric scalar potential	3.3
ϕ	phase of a plane wave	16.3.2
$\boldsymbol{\pi}$	canonical momentum density	24.5.1
$\boldsymbol{\pi}_e$, $\boldsymbol{\pi}_m$	electric and magnetic Hertz vector	16.9.1, 16.9.2
Π	degree of polarization	21.3
ψ	magnetic scalar potential	10.4
Q	electric charge	2.1.1
Q_{ij}	electric quadrupole moment tensor	4.1.1
$\mathcal{Q}$	quality factor	14.13.2, 17.2.2
R	electric resistance	9.5
R	reflection coefficient	17.3.4
$\mathbf{R}_E$	center of energy	15.7
r	reflection amplitude	17.3.2
r_e	classical electron radius	21.3
$\vec{r}$, r_μ	space-time four-vector	22.5
ρ	electric charge density	2.1.1
ρ	resistivity	9.5
ρ_f, ρ_P	free and polarization charge density	6.2
ρ^*	fictitious magnetic charge density	13.4
$\mathbf{S}$	Poynting vector	15.4.1
S	action	24.2
s_k	Stokes parameters	16.4.5
Δs	invariant interval	22.4.3
σ	surface charge density	2.1.1
σ	conductivity	9.3
σ^*	fictitious magnetic surface charge density	13.4

Symbol	Meaning	Defined in Section
σ_{tot}	total cross section	21.7
σ_{abs}	absorption cross section	21.7
σ_{scatt}	scattering cross section	21.2
$\dfrac{d\sigma_{\text{scatt}}}{d\Omega}$	differential scattering cross section	21.2
T	transmission coefficient	17.3.4
T	relativistic kinetic energy	22.5.2
$T_{ij}(\mathbf{E})$	electric stress tensor in vacuum	3.7
$T_{ij}(\mathbf{B})$	magnetic stress tensor in vacuum	12.5
T_{ij}	electromagnetic stress tensor in vacuum	15.5.1
$\mathcal{T}_{ij}(\mathbf{D})$	electric stress tensor in matter	6.8.5
$\mathcal{T}_{ij}(\mathbf{H})$	magnetic stress tensor in matter	13.8.5
$\mathcal{T}_{ij}$	electromagnetic stress tensor in matter	15.8.2
t	transmission amplitude	17.3.2
t_{ret}	retarded time	23.2.1
$\Theta(x)$	Heaviside step function	1.5.3
Θ_{ij}	traceless electric quadrupole moment tensor	4.4.1
$\Theta_{\mu\nu}$	stress-energy tensor	22.73
$T^{(\ell)}_{ij\cdots m}$	traceless Cartesian electric multipole moments	4.7
$\boldsymbol{\tau}$	surface dipole moment density	4.3
τ	collision time	9.3
τ	proper time	22.4.4
τ_0	radiation reaction time constant	23.6.3
τ_E	quasi-electrostatic time constant	14.7.1
τ_M	quasi-magnetostatic time constant	14.9
$\vec{U}, U_\mu$	four-velocity	22.5.1
U_E	electrostatic total energy	3.6
U_B	magnetostatic total energy	12.6
U_{EM}	electromagnetic total energy	15.4.1
u_{EM}	electromagnetic energy density	15.4.1
V	voltage	9.7.1
V_E	electrostatic potential energy	3.5
$\hat{V}_B$	magnetostatic potential energy for fixed currents	12.7
v_{d}	drift velocity	9.3
v_{g}	group velocity	18.6.1
v_{E}	energy velocity	17.2.1
v_{p}	phase velocity	16.3.2
W	mechanical work	3.6
W	work function	3.6.2
x^μ, x_μ	contravariant and covariant four-vectors	D.1
$Y_{\ell m}$	spherical harmonics	C.2
Z	intrinsic impedance	17.2
Z_{wave}	wave impedance	17.2.1

B

Appendix B
Gaussian Units

B.1 Definition

Gaussian units employ the centimeter (cm), gram (g), and second (s) as independent *base units* for length, mass, and time. Unlike the SI system discussed in Section 2.6, no fourth base unit is introduced to describe current, charge, or anything else. The electric field is defined exactly as in SI, so we transcribe (2.74) and write the force between two point charges as

$$F_e = k_e \frac{q_1 q_2}{r^2} = q_1 E. \tag{B.1}$$

The magnetic field is defined using the force between two current-carrying wires, as we did for SI. However, **E** and **B** are assigned the *same* dimensions by replacing (2.75) with

$$F_m = k_m \frac{2 I_1 I_2 L_2}{\rho} = \frac{I_1 L_2 B}{c}. \tag{B.2}$$

Otherwise, the Gaussian system[1] makes the choices $k_e = 1$ and $k_m = 1/c^2$. Therefore, a dimensionally correct Lorentz force law (2.76) and the Maxwell equations (2.77) and (2.78) now read

$$\mathbf{F} = q\left(\mathbf{E} + \frac{\boldsymbol{v}}{c} \times \mathbf{B}\right) \tag{B.3}$$

and

$$\nabla \cdot \mathbf{E} = 4\pi\rho \qquad \nabla \cdot \mathbf{B} = 0 \tag{B.4}$$

$$\nabla \times \mathbf{E} = -\frac{1}{c}\frac{\partial \mathbf{B}}{\partial t} \qquad \nabla \times \mathbf{B} = \frac{4\pi}{c}\mathbf{j} + \frac{1}{c}\frac{\partial \mathbf{E}}{\partial t}. \tag{B.5}$$

The Gaussian system relates the potentials to the fields by

$$\mathbf{B} = \nabla \times \mathbf{A} \qquad \mathbf{E} = -\nabla\varphi - \frac{1}{c}\frac{\partial \mathbf{A}}{\partial t}. \tag{B.6}$$

[1] Different choices for k_e, k_m, and the definition of **B** distinguish the CGS Gaussian system from two other CGS systems: electrostatic units (esu) and electromagnetic units (emu). The Gaussian system is often called "mixed" because it uses esu units for ρ, **j**, and **E** and emu units for **B**. See Brown (1984) cited at the end of Chapter 2 for the details, none of which affect the discussion here.

The Poynting vector and electromagnetic energy density are similarly

$$\mathbf{S} = \frac{c}{4\pi}(\mathbf{E} \times \mathbf{B}) \qquad u_{\text{EM}} = \frac{1}{8\pi}\left(|\mathbf{E}|^2 + |\mathbf{B}|^2\right). \tag{B.7}$$

In the presence of matter,

$$\mathbf{D} = \mathbf{E} + 4\pi\mathbf{P} \qquad \mathbf{H} = \mathbf{B} - 4\pi\mathbf{M}, \tag{B.8}$$

and the various constitutive relations are

$$\mathbf{D} = \epsilon\mathbf{E} \qquad \mathbf{B} = \mu\mathbf{H} \qquad \mathbf{j}_f = \sigma\mathbf{E} \qquad \mathbf{P} = \chi\mathbf{E} \qquad \mathbf{M} = \chi_m\mathbf{H}. \tag{B.9}$$

Therefore,

$$\epsilon = 1 + 4\pi\chi \qquad \mu = 1 + 4\pi\chi_m, \tag{B.10}$$

and **E**, **D**, **B**, **H**, **P**, and **M** all have the same dimensions in the Gaussian system.[2]

B.2 Coping with Conversion

Two issues arise when we convert from Gaussian units to SI units and vice versa.[3] First, we must transform numerical values of physical quantities in one unit system to numerical values in the other system. Second, we must rewrite equations and definitions in one unit system as equations and definitions in the other unit system. Table B.1 addresses the first question by listing the units and numerical equivalents in both systems for a collection of electromagnetic quantities.

An effective way to deal with converting equations and definitions between unit systems begins by noting that variables with dimensions of position, time, mass, force, or energy require no conversion factors apart from those indicated in Table B.1. Otherwise, recalling that γ is the gyromagnetic ratio,

$$\frac{j_{\text{SI}}}{j_{\text{G}}} = \frac{\rho_{\text{SI}}}{\rho_{\text{G}}} = \frac{I_{\text{SI}}}{I_{\text{G}}} = \frac{Q_{\text{SI}}}{Q_{\text{G}}} = \frac{P_{\text{SI}}}{P_{\text{G}}} = \frac{E_{\text{G}}}{E_{\text{SI}}} = \frac{\varphi_{\text{G}}}{\varphi_{\text{SI}}} = \sqrt{4\pi\epsilon_0} \tag{B.11}$$

$$\frac{M_{\text{SI}}}{M_{\text{G}}} = \frac{\gamma_{\text{SI}}}{\gamma_{\text{G}}} = \frac{B_{\text{G}}}{B_{\text{SI}}} = \sqrt{\frac{4\pi}{\mu_0}} \tag{B.12}$$

$$\frac{D_{\text{G}}}{D_{\text{SI}}} = \sqrt{\frac{4\pi}{\epsilon_0}} \qquad \frac{H_{\text{G}}}{H_{\text{SI}}} = \sqrt{4\pi\mu_0} \qquad \frac{\chi_{\text{SI}}}{\chi_{\text{G}}} = 4\pi. \tag{B.13}$$

Beginning with an equation or definition in one unit system, use the foregoing to solve for each variable in terms of the same variable in the other system. Substitute these into the starting expression and use $\mu_0\epsilon_0 c^2 = 1$ to simplify what remains.

As an example of Gaussian-to-SI conversion, let us check the assertion in (B.5) that the Ampère–Maxwell equation is

$$\nabla \times \mathbf{B}_{\text{G}} = \frac{4\pi}{c}\mathbf{j}_{\text{G}} + \frac{1}{c}\frac{\partial \mathbf{E}_{\text{G}}}{\partial t}. \tag{B.14}$$

Using (B.11) and (B.12), (B.14) transforms to

$$\nabla \times \sqrt{\frac{4\pi}{\mu_0}}\mathbf{B}_{\text{SI}} = \frac{4\pi}{c}\frac{1}{\sqrt{4\pi\epsilon_0}}\mathbf{j}_{\text{SI}} + \frac{1}{c}\frac{\partial}{\partial t}\sqrt{4\pi\epsilon_0}\mathbf{E}_{\text{SI}} \tag{B.15}$$

[2] Nevertheless, different names are given to the units for many of these quantities.

[3] This section follows the excellent treatment of A.S. Arrott, "Magnetism in SI units and Gaussian units", in *Ultrathin Magnetic Films I*, edited by J.A.C. Bland and B.Heinrich (Springer, Berlin, 1994), pp. 7-19. Arrott points out some subtleties with unit conversion for permanent magnets we do not discuss here.

Table B.1. SI-Gaussian unit conversion for electromagnetic quantities.

Quantity	SI	Gaussian
length d	1 meter (m)	10^2 centimeter (cm)
mass m	1 kilogram (kg)	10^3 gram (g)
time t	1 second (s)	1 second (s)
force $\mathbf{F}$	1 newton (N)	10^5 dyne (dyn)
energy U	1 joule (J)	10^7 erg (erg)
capacitance C	1 farad (F)	9×10^{11} statfarad
charge Q	1 coulomb (C)	3×10^9 statcoulomb
charge density ρ	1 C/m^3	3×10^3 statcoulomb/cm^3
resistance R	1 ohm (Ω)	$\frac{1}{9} \times 10^{-11}$ statohm
conductivity σ	$1\ (\Omega \cdot \text{m})^{-1}$	9×10^9 (statohm·cm)$^{-1}$
current I	1 ampere (A)	3×10^9 statampere $= 10^{-1}$ abampere
$\mathbf{j}$	1 A/m^2	3×10^5 statampere/cm^2
$\mathbf{D}$	C/m^2	$12\pi \times 10^5$ statampere/cm
$\mathbf{E}$	1 volt per meter (V/m)	$\frac{1}{3} \times 10^{-4}$ statvolt/cm
L	1 henry (H)	$\frac{1}{9} \times 10^{-11}$ stathenry
$\mathbf{H}$	1 A/m	$4\pi \times 10^{-3}$ oersted (Oe)
Φ_B	1 weber (Wb)	10^8 maxwell (Mx)
$\mathbf{B}$	1 Wb/m^2 = 1 tesla (T)	10^4 gauss (G)
$\mathbf{M}$	1 A/m	10^{-3} Oe
$\mathbf{P}$	C/m^2	3×10^5 statvolt/cm
φ	1 volt (V)	$\frac{1}{300}$ statvolt

or

$$\nabla \times \mathbf{B}_{\text{SI}} = \frac{1}{c}\sqrt{\frac{\mu_0}{\epsilon_0}}\mathbf{j}_{\text{SI}} + \sqrt{\frac{\mu_0\epsilon_0}{c}}\frac{\partial \mathbf{E}_{\text{SI}}}{\partial t}. \tag{B.16}$$

Since $\mu_0\epsilon_0 c^2 = 1$, we get the correct SI expression,

$$\nabla \times \mathbf{B}_{\text{SI}} = \mu_0 \mathbf{j}_{\text{SI}} + \frac{1}{c^2}\frac{\partial \mathbf{E}_{\text{SI}}}{\partial t}. \tag{B.17}$$

As an example of SI-to-Gaussian conversion, consider the electric field (6.17) inside a sphere with uniform polarization **P**:

$$\mathbf{E}_{\text{SI}} = -\frac{\mathbf{P}_{\text{SI}}}{3\epsilon_0}. \tag{B.18}$$

Using (B.11), (B.18) transforms to

$$\frac{\mathbf{E}_{\text{G}}}{\sqrt{4\pi\epsilon_0}} = -\frac{1}{3\epsilon_0}\sqrt{4\pi\epsilon_0}\,\mathbf{P}_{\text{G}}, \tag{B.19}$$

or

$$\mathbf{E}_{\text{G}} = -\frac{4\pi}{3}\mathbf{P}_{\text{G}}. \tag{B.20}$$

Our final example is the transformation of the magnetic moment expression

$$\mathbf{m}_{\text{SI}} = \frac{1}{2}\int d^3r\, \mathbf{r} \times \mathbf{j}_{\text{SI}}. \tag{B.21}$$

The moment $\mathbf{m}$ does not appear in (B.11), (B.12), or (B.13). However, for conversion purposes, we regard it as derived from the volume integral of a distribution of magnetization (see Section 13.2.3):

$$\mathbf{m} = \int d^3r\, \mathbf{M}. \tag{B.22}$$

We can now use (B.12) to conclude that $\mathbf{m}_{\rm SI} = \mathbf{m}_{\rm G}\sqrt{4\pi/\mu_0}$. Therefore, (B.21) transforms to

$$\sqrt{\frac{4\pi}{\mu_0}}\mathbf{m}_{\rm G} = \frac{1}{2}\int d^3r\, \mathbf{r} \times \sqrt{4\pi\epsilon_0}\mathbf{j}_{\rm G}, \tag{B.23}$$

or

$$\mathbf{m}_{\rm G} = \frac{1}{2c}\int d^3r\, \mathbf{r} \times \mathbf{j}_{\rm G}. \tag{B.24}$$

Appendix C
Special Functions

C.1 Legendre Polynomials

The real-valued Legendre polynomials, $P_\ell(x)$, are defined by the generating function

$$\frac{1}{\sqrt{1-2xt+t^2}} \equiv \sum_{\ell=0}^{\infty} t^\ell P_\ell(x) \qquad |x| \le 1,\ 0 < t < 1. \tag{C.1}$$

Alternatively, consider the differential equation

$$(1-x^2)\frac{d^2P(x)}{dx^2} - 2x\frac{dP(x)}{dx} + \ell(\ell+1)P(x) = 0. \tag{C.2}$$

The $P_\ell(x)$ are the eigenfunctions of the Sturm-Liouville eigenvalue problem defined by (C.2) on the interval $-1 \le x \le 1$ with the boundary conditions that $P(1)$ and $P(-1)$ are finite. The index ℓ is a non-negative integer. The polynomials are orthogonal,

$$\int_{-1}^{1} dx\, P_\ell(x) P_m(x) = \frac{2}{2\ell+1}\delta_{\ell m}, \tag{C.3}$$

and complete,

$$\sum_{\ell=0}^{\infty} \left(\ell + \tfrac{1}{2}\right) P_\ell(x) P_\ell(x') = \delta(x-x'). \tag{C.4}$$

Using the *Rodriguez formula*,

$$P_\ell(x) = \frac{1}{2^\ell \ell!}\frac{d^\ell}{dx^\ell}(x^2-1)^\ell, \tag{C.5}$$

we find

$$\begin{aligned}
P_0(x) &= 1 \\
P_1(x) &= x \\
P_2(x) &= \frac{1}{2}(3x^2-1) \\
P_3(x) &= \frac{1}{2}(5x^3-3x) \\
P_4(x) &= \frac{1}{8}(35x^4-30x^2+3).
\end{aligned} \tag{C.6}$$

The parity of the polynomials is $P_\ell(-x) = (-1)^\ell P_\ell(x)$. A few special values are $P_\ell(1) = 1$, $P_\ell(-1) = (-1)^\ell$, and

$$P_\ell(0) = \begin{cases} 0 & \text{odd } \ell, \\ \dfrac{(\ell-1)!!}{(-2)^{\ell/2}(\ell/2)!} & \text{even } \ell \geq 2. \end{cases} \tag{C.7}$$

Three useful recurrence relations satisfied by the Legendre polynomials and their derivatives are

$$(2\ell+1)P_\ell(x) = P'_{\ell+1}(x) - P'_{\ell-1}(x) \tag{C.8}$$

$$(2\ell+1)xP_\ell(x) = (\ell+1)P_{\ell+1}(x) + \ell P_{\ell-1}(x) \tag{C.9}$$

$$P_\ell(x) = P'_{\ell+1}(x) - 2xP'_\ell(x) + P'_{\ell-1}(x). \tag{C.10}$$

C.1.1 Associated Legendre Polynomials

Legendre's differential equation is

$$(1-x^2)\frac{d^2P(x)}{dx^2} - 2x\frac{dP(x)}{dx} + \left\{\ell(\ell+1) - \frac{m^2}{1-x^2}\right\}P(x) = 0. \tag{C.11}$$

For integer values of $|m| \leq \ell$, the associated Legendre polynomials $P_\ell^m(x)$ are the complete and orthonormal eigenfunctions of the Sturm-Liouville eigenvalue problem defined by (C.11) on the interval $-1 \leq x \leq 1$ with the boundary conditions that $P(1)$ and $P(-1)$ are finite. An explicit formula is

$$P_\ell^m(x) = (-1)^m(1-x^2)^{m/2}\frac{d^m}{dx^m}P_\ell(x), \tag{C.12}$$

from which we deduce that

$$P_\ell^0(x) = P_\ell(x), \qquad P_\ell^1(0) = P'_\ell(0), \qquad \text{and} \qquad P_\ell^m(1) = 0. \tag{C.13}$$

Three useful identities are

$$\frac{d}{d\theta}P_\ell(\cos\theta) = P_\ell^1(\cos\theta) \tag{C.14}$$

$$P_\ell^1(\cos\theta) = (-1)^{\ell+1}P_\ell^1(-\cos\theta) \tag{C.15}$$

$$\int_0^\pi d\theta\,\sin\theta\,P_\ell^m(\cos\theta)P_{\ell'}^m(\cos\theta) = \frac{2}{2\ell+1}\frac{(\ell+m)!}{(\ell-m)!}\delta_{\ell\ell'}. \tag{C.16}$$

C.2 Spherical Harmonics

Consider the partial differential equation

$$\frac{1}{\sin\theta}\frac{\partial}{\partial\theta}\left(\sin\theta\frac{\partial Y}{\partial\theta}\right) + \frac{1}{\sin^2\theta}\frac{\partial^2 Y}{\partial\phi^2} = -\ell(\ell+1)Y. \tag{C.17}$$

The complex-valued spherical harmonics $Y_{\ell m}(\theta,\phi)$ are the eigenfunctions of the eigenvalue problem defined by (C.17) on the sphere with the boundary conditions that $Y(\theta,\phi)$ be finite and single-valued. With integer values of $|m| \leq \ell$, these eigenfunctions can be written in terms of the associated Legendre polynomials (see Section C.1.1) as

$$Y_{\ell m}(\theta,\phi) = \sqrt{\frac{2\ell+1}{4\pi}\frac{(\ell-m)!}{(\ell+m)!}}P_\ell^m(\cos\theta)e^{im\phi} \qquad m \geq 0, \tag{C.18}$$

with $Y_{\ell,-m}(\theta,\phi) = (-1)^m Y^*_{\ell m}(\theta,\phi)$. The spherical harmonics are both orthonormal,

$$\int_0^{2\pi} d\phi \int_0^{\pi} d\theta \, \sin\theta \, Y_{\ell m}(\theta,\phi) Y^*_{\ell' m'}(\theta,\phi) = \delta_{\ell\ell'}\delta_{mm'} \tag{C.19}$$

and complete,

$$\sum_{\ell=0}^{\infty} \sum_{m=-\ell}^{\ell} Y_{\ell m}(\theta,\phi) Y^*_{\ell m}(\theta',\phi') = \delta(\cos\theta - \cos\theta')\delta(\phi - \phi'). \tag{C.20}$$

With $x = r\sin\theta\cos\phi$, $y = r\sin\theta\sin\phi$, and $z = r\cos\theta$, the spherical harmonic $Y_{\ell m}(x, y, z)$ is a homogeneous polynomial in x, y, and z of degree ℓ. If $A_{\ell m} = \sqrt{(2\ell+1)(\ell+m)!(\ell-m)!/4\pi}$,

$$r^\ell Y_\ell^m(x,y,z) = A_{\ell m} \sum_{p,q,s=1}^{\infty} \frac{1}{p!q!s!} \left(-\frac{x+iy}{2}\right)^p \left(\frac{x-iy}{2}\right)^q z^s \, \delta_{p+q+s,\ell} \, \delta_{p-q,m}. \tag{C.21}$$

The first few spherical harmonics are

$$\begin{aligned}
Y_{00}(\theta,\phi) &= \frac{1}{\sqrt{4\pi}} \\
Y_{10}(\theta,\phi) &= \sqrt{\frac{3}{4\pi}}\cos\theta = \sqrt{\frac{3}{4\pi}}\frac{z}{r} \\
Y_{1\pm1}(\theta,\phi) &= \mp\sqrt{\frac{3}{8\pi}}\sin\theta\exp(\pm i\phi) = \mp\sqrt{\frac{3}{8\pi}}\frac{x \pm iy}{r} \\
Y_{20}(\theta,\phi) &= \sqrt{\frac{5}{16\pi}}\left\{3\cos^2\theta - 1\right\} = \sqrt{\frac{5}{16\pi}}\frac{3z^2 - r^2}{r^2} \\
Y_{2\pm1}(\theta,\phi) &= \mp\sqrt{\frac{15}{8\pi}}\sin\theta\cos\theta\exp(\pm i\phi) = \mp\sqrt{\frac{15}{8\pi}}\frac{z(x \pm iy)}{r^2} \\
Y_{2\pm2}(\theta,\phi) &= \sqrt{\frac{15}{32\pi}}\sin^2\theta\exp(\pm 2i\phi) = \sqrt{\frac{15}{32\pi}}\frac{(x \pm iy)^2}{r^2}.
\end{aligned} \tag{C.22}$$

If $\mathbf{r} = (r,\theta,\phi)$ and $\mathbf{r}' = (r',\theta',\phi')$ are defined with respect to a fixed set of coordinate axes in space, the *spherical harmonic addition theorem* states that

$$P_\ell(\hat{\mathbf{r}} \cdot \hat{\mathbf{r}}') = \frac{4\pi}{2\ell+1} \sum_{m=-\ell}^{+\ell} Y^*_{\ell m}(\theta',\phi') Y_{\ell m}(\theta,\phi). \tag{C.23}$$

Finally, the spherical harmonics transform under space inversion as

$$Y_{\ell m}(-\hat{\mathbf{r}}) = (-1)^\ell Y_{\ell m}(\hat{\mathbf{r}}). \tag{C.24}$$

C.3 Bessel Functions

Bessel's differential equation is

$$\frac{d^2R}{dx^2} + \frac{1}{x}\frac{dR}{dx} + \left[1 - \frac{m^2}{x^2}\right]R = 0. \tag{C.25}$$

We restrict ourselves to $x \geq 0$ and integer values for m. Linearly independent solutions of (C.25) are the Bessel function $J_m(x) = (-1)^m J_{-m}(x)$ and the Neumann function

$$N_m(x) = \frac{J_m(x)\cos m\pi - J_{-m}(x)}{\sin m\pi}. \tag{C.26}$$

The associated Wronskian is

$$\left[J_m(x)N_m'(x) - N_m(x)J_m'(x)\right] = \frac{2}{\pi x}. \tag{C.27}$$

A generating function for $J_m(x)$ is

$$\exp\left\{\frac{x}{2}\left(t - \frac{1}{t}\right)\right\} = \sum_{m=0}^{\infty} J_m(x)t^m. \tag{C.28}$$

Noting Euler's constant, $\gamma = 0.5772\ldots$, we write the small-argument forms,

$$\lim_{x\to 0} J_m(x) = \frac{1}{m!}\left(\frac{x}{2}\right)^m \tag{C.29}$$

and

$$\lim_{x\to 0} N_m(x) = \begin{cases} -\dfrac{(m-1)!}{\pi}\left(\dfrac{x}{2}\right)^{-m} & m \neq 0, \\ \dfrac{2}{\pi}\left[\ln\left(\dfrac{x}{2}\right) + \gamma\right] & m = 0. \end{cases} \tag{C.30}$$

When $x \sim m$, crossover begins to the asymptotic values

$$\begin{aligned} \lim_{x\to\infty} J_m(x) &= \sqrt{\frac{2}{\pi x}}\cos\left(x - \frac{m\pi}{2} - \frac{\pi}{4}\right) \\ \lim_{x\to\infty} N_m(x) &= \sqrt{\frac{2}{\pi x}}\sin\left(x - \frac{m\pi}{2} - \frac{\pi}{4}\right). \end{aligned} \tag{C.31}$$

The trigonometric behavior in (C.31) suggests the definition of the *Hankel functions*:

$$\begin{aligned} H_m^{(1)}(x) &= J_m(x) + iN_m(x) \\ H_m^{(2)}(x) &= J_m(x) - iN_m(x). \end{aligned} \tag{C.32}$$

$H_m^{(1)}(x)$ behaves asymptotically like an outgoing wave. $H_m^{(2)}(x)$ behaves asymptotically like an incoming wave.

The nth zero of the Bessel function $J_m(x)$ is defined by $J_m(\alpha_{mn}) = 0$. For a finite interval, $0 \le \rho \le R$, these appear in the completeness relation,

$$\sum_{n=1}^{\infty} \frac{2}{R^2} \frac{J_m\left(\alpha_{mn}\dfrac{\rho}{R}\right) J_m\left(\alpha_{mn}\dfrac{\rho'}{R}\right)}{J_{m+1}^2(\alpha_{mn})} = \frac{1}{\rho}\delta(\rho - \rho') \qquad 0 \le \rho, \rho' < R, \tag{C.33}$$

and the orthogonality relation,

$$\int_0^R d\rho\rho J_m\left(\alpha_{mn}\frac{\rho}{R}\right) J_m\left(\alpha_{mp}\frac{\rho}{R}\right) = \delta_{np}\frac{R^2}{2}J_{m+1}^2(\alpha_{mn}). \tag{C.34}$$

When $R \to \infty$, the completeness relation (C.33) evolves to

$$\int_0^\infty dkkJ_m(k\rho)J_m(k\rho') = \frac{1}{\rho}\delta(\rho - \rho') \qquad \rho, \rho' > 0. \tag{C.35}$$

Useful in the same limit are the Bessel function analogs of Fourier transform pairs:

$$f(\rho) = \int_0^\infty dkkJ_m(k\rho)g(k) \qquad g(\rho) = \int_0^\infty d\rho\rho J_m(k\rho)f(\rho). \tag{C.36}$$

Finally, we draw attention to two integral representations:

$$J_m(x) = \frac{1}{2\pi i^m} \int_0^{2\pi} d\phi \exp[i(x\cos\phi + m\phi)]$$
$$H_0^{(1)}(x) = -\frac{i}{\pi} \int_{-\infty}^{\infty} \frac{ds}{\sqrt{x^2+s^2}} \exp\left(i\sqrt{x^2+s^2}\right). \quad \text{(C.37)}$$

C.3.1 Modified Bessel Functions

Real-valued Bessel functions of pure imaginary argument are defined by

$$I_m(x) = i^{-m} J_m(ix)$$
$$K_m(x) = \frac{\pi}{2} i^{m+1} \left[J_m(ix) + iN_m(ix)\right]. \quad \text{(C.38)}$$

The associated Wronskian is

$$[K_m(x)I_m'(x) - I_m(x)K_m'(x)] = \frac{1}{x}. \quad \text{(C.39)}$$

The limiting and asymptotic values follow immediately from (C.29) through (C.31) as

$$\lim_{x\to 0} I_m(x) = \frac{1}{m!}\left(\frac{x}{2}\right)^m \quad \text{(C.40)}$$

$$\lim_{x\to 0} K_m(x) = \begin{cases} \dfrac{(m-1)!}{2}\left(\dfrac{x}{2}\right)^{-m} & m \neq 0, \\ -\left[\ln\left(\dfrac{x}{2}\right) + \gamma\right] & m = 0, \end{cases} \quad \text{(C.41)}$$

$$\lim_{x\to\infty} I_m(x) = \sqrt{\frac{1}{2\pi x}}\, e^x$$
$$\lim_{x\to\infty} K_m(x) = \sqrt{\frac{\pi}{2x}}\, e^{-x}. \quad \text{(C.42)}$$

C.3.2 Spherical Bessel Functions

Real-valued spherical Bessel functions and spherical Neumann functions are defined by

$$j_m(x) = \sqrt{\frac{\pi}{2x}} J_{m+\frac{1}{2}}(x) \qquad n_m(x) = \sqrt{\frac{\pi}{2x}} N_{m+\frac{1}{2}}(x). \quad \text{(C.43)}$$

The associated Wronskian is

$$[j_m(x)n_m'(x) - n_m(x)j_m'(x)] = \frac{1}{x^2}. \quad \text{(C.44)}$$

Explicit formulae are

$$j_m(x) = (-x)^m \left(\frac{1}{x}\frac{d}{dx}\right)^m \frac{\sin x}{x}$$
$$n_m(x) = -(-x)^m \left(\frac{1}{x}\frac{d}{dx}\right)^m \frac{\cos x}{x}. \quad \text{(C.45)}$$

When the argument is small,

$$\lim_{x\to 0} j_m(x) = \frac{x^m}{(2m+1)!!} \qquad \lim_{x\to 0} n_m(x) = -\frac{(2m-1)!!}{x^{m+1}}. \quad \text{(C.46)}$$

When the argument is large,

$$\lim_{x\to\infty} j_m(x) = \frac{1}{x}\sin(x - m\pi/2)$$
$$\lim_{x\to\infty} n_m(x) = -\frac{1}{x}\cos(x - m\pi/2). \tag{C.47}$$

Spherical Hankel functions are defined by

$$h_m^{(1)}(x) = j_m(x) + i n_m(x) \qquad h_m^{(2)}(x) = j_m(x) - i n_m(x). \tag{C.48}$$

Their limiting and asymptotic behaviors follow immediately from (C.46) and (C.47). For example,

$$\lim_{x\to\infty} h_m^{(1)}(x) = (-i)^{\ell+1}\frac{\exp(ix)}{x} \quad \text{and} \quad \lim_{x\to\infty} h_m^{(2)}(x) = (-i)^{\ell+1}\frac{\exp(-ix)}{x}. \tag{C.49}$$

The nth zero of the spherical Bessel function $j_m(x)$ is defined by $j_m(\gamma_{mn}) = 0$. For a finite interval, $0 \le \rho \le R$, these appear in the orthogonality relation

$$\int_0^R d\rho\rho^2 j_m(\gamma_{mn}\rho/R) j_m(\gamma_{mk}\rho/R) = \delta_{nk}\frac{1}{2}R^2 j_{m+1}^2(\gamma_{mn}). \tag{C.50}$$

On the real line, the $j_m(x)$ are complete because

$$\frac{2}{\pi}\int_0^\infty dk k^2 j_m(kr) j_m(kr') = \frac{1}{r^2}\delta(r - r'). \tag{C.51}$$

C.3.3 Miscellaneous Identities

Plane wave expansions in two and three dimensions are

$$\exp(i\mathbf{k}\cdot\boldsymbol{\rho}) = \exp(ik\rho\cos\phi) = \sum_{m=-\infty}^{\infty} i^m J_m(k\rho)\exp(im\phi) \tag{C.52}$$

$$\exp(i\mathbf{k}\cdot\mathbf{r}) = \exp(ikr\cos\theta) = \sum_{\ell=0}^{\infty}(2\ell+1)i^\ell j_\ell(kr)P_\ell(\cos\theta). \tag{C.53}$$

Let $r_<$ ($r_>$) be the lesser (greater) of r and r' and let $\cos\Theta = \hat{\mathbf{r}}\cdot\hat{\mathbf{r}}'$. Then,

$$\frac{\exp(ik|\mathbf{r}-\mathbf{r}'|)}{|\mathbf{r}-\mathbf{r}'|} = ik\sum_{\ell=0}^{\infty}(2\ell+1) j_\ell(kr_<) h_\ell^{(1)}(kr_>) P_\ell(\cos\Theta) \tag{C.54}$$

$$\frac{2}{\pi}\int_0^\infty dk j_\ell(kr) j_\ell(kr') = \frac{1}{2\ell+1}\frac{r_<^\ell}{r_>^{\ell+1}}. \tag{C.55}$$

Appendix D

Managing Minus Signs in Special Relativity

D.1 The Metric Tensor

Special relativity flows from the assumption that the quantity

$$(ds)^2 = (dx)^2 + (dy)^2 + (dz)^2 - (cdt)^2 \tag{D.1}$$

takes the same value in all inertial frames. The main text exploits the imaginary number i to handle the minus sign in (D.1). An alternative approach represents the space-time four-vector $d\vec{x}$ in two ways: a *contravariant* form with superscripted components,

$$dx^\mu = (dx^0, dx^1, dx^2, dx^3) = (cdt, dx, dy, dz), \tag{D.2}$$

and a *covariant* form with subscripted components,[1]

$$dx_\mu = (-dx_0, dx_1, dx_2, dx_3) = (-cdt, dx, dy, dz). \tag{D.3}$$

With these definitions and the Einstein summation convention, (D.1) is the scalar product,

$$(ds)^2 = d\vec{x} \cdot d\vec{x} = dx_\mu dx^\mu = dx^\mu dx_\mu. \tag{D.4}$$

The contraction of a repeated index *always* occurs between an upper index and a lower index.

Calculations are facilitated by the introduction of a *metric tensor*, $g_{\mu\nu}$. The metric tensor for the flat space-time of special relativity has the matrix representation

$$g_{\mu\nu} = \begin{pmatrix} -1 & 0 & 0 & 0 \\ 0 & 1 & 0 & 0 \\ 0 & 0 & 1 & 0 \\ 0 & 0 & 0 & 1 \end{pmatrix}. \tag{D.5}$$

We use $g_{\mu\nu}$ to "lower an index" and convert a contravariant vector into a covariant vector:

$$x_\mu = g_{\mu\nu} x^\nu. \tag{D.6}$$

[1] We follow N.A. Doughty, *Lagrangian Interaction* (Addison-Wesley, Reading, MA, 1990) and present the "East Coast metric" used by most general relativists and string theorists. The "West Coast metric" used by most particle physicists makes the time component positive and the space components negative in (D.3).

Combining (D.6) with (D.4) permits us to write $(ds)^2$ using contravariant components only

$$(ds)^2 = x^\mu g_{\mu\nu} x^\nu = g_{\mu\nu} x^\mu x^\nu. \tag{D.7}$$

More generally, the scalar product of two arbitrary four-vectors is

$$\vec{a} \cdot \vec{b} = a^\mu b_\mu = a^\mu g_{\mu\nu} b^\nu = g_{\mu\nu} a^\mu b^\nu. \tag{D.8}$$

The inverse of $g_{\mu\nu}$ is written $g^{\mu\nu}$ and has the matrix representation

$$g^{\mu\nu} = \begin{pmatrix} -1 & 0 & 0 & 0 \\ 0 & 1 & 0 & 0 \\ 0 & 0 & 1 & 0 \\ 0 & 0 & 0 & 1 \end{pmatrix}. \tag{D.9}$$

Our rules for contraction dictate that the inverse relation produces a Kronecker delta with one index "up" and one index "down", namely,

$$g_{\alpha\mu} g^{\mu\beta} = \delta_\alpha^{\ \beta}. \tag{D.10}$$

The right side of (D.10) reflects a convention: whether up or down, the left index labels the rows and the right index labels the columns of the matrix representation. We use $g^{\mu\nu}$ to "raise an index" and convert a covariant vector into a contravariant vector:

$$x^\mu = g^{\mu\nu} x_\nu. \tag{D.11}$$

Using (D.11), the scalar product (D.8) can also be written using covariant components only

$$\vec{a} \cdot \vec{b} = a^\nu b_\nu = a_\mu g^{\mu\nu} b_\nu = g^{\mu\nu} a_\mu b_\nu. \tag{D.12}$$

Finally, (D.10) facilitates an index manipulation trick used very often in calculations:

$$a_\mu b^\mu = g_{\mu\nu} a^\nu g^{\mu\alpha} b_\alpha = a^\nu g_{\nu\mu} g^{\mu\alpha} b_\alpha = a^\nu \delta_\nu^{\ \alpha} b_\alpha = a^\nu b_\nu. \tag{D.13}$$

We note in passing that the scalar products (D.8) and (D.12) remain valid in general relativity except that the components of the metric tensor generally depend on space and time.

D.2 The Lorentz Transformation

The Lorentz transformation (22.49) describes a boost of the standard configuration along the z-axis for a four-vector (a_1, a_2, a_2, a_4). Here, the four-vector is (a_0, a_1, a_2, a_3), so we boost along the x-axis to keep the matrix of the transformation block diagonal. When the frame K' moves with velocity $c\beta\hat{\mathbf{x}}$ when viewed from the frame K, the transformation rules are

$$\begin{aligned} x'^0 &= \gamma(x^0 - \beta x^1) & x'^1 &= \gamma(x^1 - \beta x^0) & x'^2 &= x^2 & x'^3 &= x^3 \\ x^0 &= \gamma(x'^0 + \beta x'^1) & x^1 &= \gamma(x'^1 + \beta x'^0) & x^2 &= x'^2 & x^3 &= x'^3. \end{aligned} \tag{D.14}$$

Hence, the contravariant components of a four-vector transform like

$$a'^\mu = \frac{\partial x'^\mu}{\partial x^\nu} a^\nu = L^\mu_{\ \nu} a^\nu \qquad \text{and} \qquad a^\mu = \frac{\partial x^\mu}{\partial x'^\nu} a'^\nu = [L^{-1}]^\mu_{\ \nu} a'^\nu. \tag{D.15}$$

Using the leftmost equation in (D.15), the rightmost equality in the Lorentz invariance statement,

$$a'_\sigma a'^\sigma = a'_\sigma \frac{\partial x'^\sigma}{\partial x^\nu} a^\nu = a_\nu a^\nu, \tag{D.16}$$

implies that

$$a'_\sigma \frac{\partial x'^\sigma}{\partial x^\nu} = a_\nu \;\Rightarrow\; a'_\sigma \frac{\partial x'^\sigma}{\partial x^\nu} \frac{\partial x^\nu}{\partial x'^\mu} = a_\nu \frac{\partial x^\nu}{\partial x'^\mu} \;\Rightarrow\; a'_\sigma \delta^\sigma_{\ \mu} = a_\nu \frac{\partial x^\nu}{\partial x'^\mu}. \tag{D.17}$$

Therefore, the transformation rules for the covariant components of a four-vector are

$$a'_\mu = a_\nu \frac{\partial x^\nu}{\partial x'^\mu} = a_\nu L^\nu{}_\mu \qquad \text{and} \qquad a_\mu = a'_\nu \frac{\partial x'^\nu}{\partial x^\mu} = a'_\nu [L^{-1}]^\nu{}_\mu, \tag{D.18}$$

where

$$\mathbf{L} = \begin{bmatrix} \gamma & -\gamma\beta & 0 & 0 \\ -\gamma\beta & \gamma & 0 & 0 \\ 0 & 0 & 1 & 0 \\ 0 & 0 & 0 & 1 \end{bmatrix} \qquad \mathbf{L}^{-1} = \begin{bmatrix} \gamma & \gamma\beta & 0 & 0 \\ \gamma\beta & \gamma & 0 & 0 \\ 0 & 0 & 1 & 0 \\ 0 & 0 & 0 & 1 \end{bmatrix}. \tag{D.19}$$

A brief computation confirms that $\mathbf{L}$, $\mathbf{L}^{-1}$, and the metric tensor matrix $\mathbf{g}$ satisfy

$$\mathbf{L}\mathbf{g} = \mathbf{L}^{\mathrm{T}}\mathbf{g} = \mathbf{g}\mathbf{L}^{-1}. \tag{D.20}$$

The chain rule and comparison with (D.18) show that derivatives with respect to the *contravariant* components of $\vec{x}$ transform like the *covariant* components of a four-vector:

$$\frac{\partial}{\partial x'^\mu} = \frac{\partial x^\nu}{\partial x'^\mu}\frac{\partial}{\partial x^\nu} \quad \text{or} \quad \partial'_\mu = \frac{\partial x^\nu}{\partial x'^\mu}\partial_\nu. \tag{D.21}$$

Hence, the covariant and contravariant forms of the gradient four-vector differ from the space-time vectors (D.2) and (D.3) by the sign of their time components,

$$\partial_\mu = \left(\frac{\partial}{\partial(ct)}, \nabla\right) \qquad \text{and} \qquad \partial^\mu = g^{\mu\nu}\partial_\nu = \left(-\frac{\partial}{\partial(ct)}, \nabla\right). \tag{D.22}$$

Accordingly,

$$\partial_\mu a^\mu = \frac{1}{c}\frac{\partial a^0}{\partial t} + \nabla\cdot\mathbf{a} \qquad \text{and} \qquad \partial_\mu \partial^\mu = \nabla^2 - \frac{1}{c^2}\frac{\partial^2}{\partial t^2}. \tag{D.23}$$

Finally, we note the generalization of the foregoing to second-rank Lorentz tensors which have contravariant, covariant, or mixed character:

$$A'^{\mu\nu} = L^\mu{}_\alpha L^\nu{}_\beta A^{\alpha\beta} \qquad B'_{\mu\nu} = B_{\alpha\beta}[L^{-1}]^\alpha{}_\mu [L^{-1}]^\beta{}_\nu \qquad C'^\mu{}_\nu = L^\mu{}_\gamma C^\gamma{}_\sigma [L^{-1}]^\sigma{}_\nu. \tag{D.24}$$

D.3 Other Lorentz Tensors

The contravariant four-vectors for velocity and energy-momentum are

$$U^\mu = \gamma(c, \mathbf{u}) \qquad \text{and} \qquad p^\mu = mU^\mu = (\mathcal{E}/c, \mathbf{p}). \tag{D.25}$$

The corresponding Lorentz invariant scalars are

$$U^\mu U_\mu = -c^2 \qquad \text{and} \qquad p^\mu p_\mu = p^2 - \frac{\mathcal{E}^2}{c^2} = -m^2c^2. \tag{D.26}$$

The contravariant four-vectors for the electromagnetic potential and charge-current density are

$$A^\mu = (\varphi/c, \mathbf{A}) \qquad \text{and} \qquad j^\mu = (c\rho, \mathbf{j}). \tag{D.27}$$

Using (D.22), the invariant Lorenz gauge condition and continuity equation are

$$\partial_\mu A^\mu = 0 \qquad \text{and} \qquad \partial_\mu j^\mu = 0. \tag{D.28}$$

The contravariant form of the anti-symmetric electromagnetic field tensor is

$$F^{\mu\nu} = \partial^\mu A^\nu - \partial^\nu A^\mu, \tag{D.29}$$

and it follows from (D.22), (D.27), and (D.29) that

$$F^{\mu\nu} = \begin{pmatrix} 0 & E_x/c & E_y/c & E_z/c \\ -E_x/c & 0 & B_z & -B_y \\ -E_y/c & -B_z & 0 & B_x \\ -E_z/c & B_y & -B_x & 0 \end{pmatrix}. \tag{D.30}$$

The covariant form of the field tensor is

$$F_{\alpha\beta} = g_{\alpha\mu} g_{\beta\nu} F^{\mu\nu}, \tag{D.31}$$

and the Maxwell equations take the form

$$\partial_\nu F^{\mu\nu} = \mu_0 j^\mu \qquad \text{and} \qquad \partial_\sigma F_{\mu\nu} + \partial_\mu F_{\nu\sigma} + \partial_\nu F_{\sigma\mu} = 0. \tag{D.32}$$

In this notation, the equation of motion of a particle with charge q in an electromagnetic field is

$$\frac{dp^\mu}{d\tau} = qU_\nu F^{\mu\nu}. \tag{D.33}$$

The symmetric, electromagnetic stress-energy tensor,

$$\Theta^{\mu\nu} = \frac{1}{\mu_0}\left(F^{\mu\alpha} F^{\nu\sigma} g_{\sigma\alpha} - \tfrac{1}{4} g^{\mu\nu} F_{\alpha\beta} F^{\alpha\beta}\right) = \frac{1}{\mu_0}\left(F^{\mu\alpha} F^{\nu}_{\ \alpha} - \tfrac{1}{4} g^{\mu\nu} F_{\alpha\beta} F^{\alpha\beta}\right), \tag{D.34}$$

has the block matrix representation

$$\mathbf{\Theta} = \begin{pmatrix} u_{\text{EM}} & c\mathbf{g}_{\text{EM}} \\ c\mathbf{g}_{\text{EM}} & -\mathbf{T} \end{pmatrix}. \tag{D.35}$$

This tensor satisfies the energy-momentum conservation law,

$$\partial_\nu \Theta^{\mu\nu} = F^{\mu\alpha} j_\alpha. \tag{D.36}$$

D.3.1 Manipulating Indices

Index raising and index lowering occur frequently in covariant calculations. To illustrate this, let us confirm (D.36) by direct evaluation of the covariant derivative. For convenience, let $\mu_0 = 1$ and use the rightmost member of (D.34) to write

$$\partial_\nu \Theta^{\mu\nu} = \partial_\nu (F^{\mu\alpha} F^{\nu}_{\ \alpha} - \tfrac{1}{4} g^{\mu\nu} F_{\gamma\beta} F^{\gamma\beta}). \tag{D.37}$$

From the product rule,

$$\partial_\nu \Theta^{\mu\nu} = F^{\mu\alpha}(\partial_\nu F^{\nu}_{\ \alpha}) + F^{\nu}_{\ \alpha}(\partial_\nu F^{\mu\alpha}) - \tfrac{1}{4} g^{\mu\nu}(\partial_\nu F_{\gamma\beta}) F^{\gamma\beta} - \tfrac{1}{4} g^{\mu\nu}(\partial_\nu F^{\gamma\beta}) F_{\gamma\beta}. \tag{D.38}$$

The inhomogeneous Maxwell equation on the left side of (D.32), the index-raising property of $g^{\mu\nu}$, and the exchange of upper and lower repeated indices illustrated by (D.13) transform (D.38) to

$$\partial_\nu \Theta^{\mu\nu} = F^{\mu\alpha} j_\alpha + F_{\nu\alpha}(\partial^\nu F^{\mu\alpha}) - \tfrac{1}{4}(\partial^\mu F_{\gamma\beta}) F^{\gamma\beta} - \tfrac{1}{4}(\partial^\mu F^{\gamma\beta}) F_{\gamma\beta}. \tag{D.39}$$

Now replace the repeated index pair $(\gamma\beta)$ by $(\nu\alpha)$ in the last two terms of (D.39) and exchange upper and lower indices in one of the terms to get

$$\partial_\nu \Theta^{\mu\nu} = F^{\mu\alpha} j_\alpha + \tfrac{1}{2}\left[2F_{\nu\alpha}(\partial^\nu F^{\mu\alpha}) - (\partial^\mu F^{\nu\alpha}) F_{\nu\alpha}\right]. \tag{D.40}$$

Writing $2F_{\nu\alpha}(\partial^\nu F^{\mu\alpha}) = F_{\nu\alpha}(\partial^\nu F^{\mu\alpha}) + F_{\nu\alpha}(\partial^\nu F^{\mu\alpha})$, exchanging the dummy indices α and ν in the second term (only), and using $F_{\nu\alpha} = -F_{\alpha\nu}$, we deduce that

$$\partial_\nu \Theta^{\mu\nu} = F^{\mu\alpha} j_\alpha + \tfrac{1}{2} F_{\nu\alpha}(\partial^\nu F^{\mu\alpha} - \partial^\alpha F^{\mu\nu} - \partial^\mu F^{\nu\alpha}). \tag{D.41}$$

The asymmetry $F^{\mu\alpha} = -F^{\alpha\mu}$ transforms (D.41) to

$$\partial_\nu \Theta^{\mu\nu} = F^{\mu\alpha} j_\alpha - \tfrac{1}{2} F_{\nu\alpha}(\partial^\nu F^{\alpha\mu} + \partial^\alpha F^{\mu\nu} + \partial^\mu F^{\nu\alpha}). \tag{D.42}$$

The parenthetical term on the right side of (D.42) vanishes by virtue of the homogeneous Maxwell equation on the right side of (D.32). Therefore,

$$\partial_\nu \Theta^{\mu\nu} = F^{\mu\alpha} j_\alpha, \tag{D.43}$$

which is the equation (D.36) we wished to confirm.

Index

Abraham-Lorentz equation, 902
Abraham-Minkowski controversy, 526
absorption, 575, 588, 629, 644, 651
  coefficient, 608
acceleration field of a moving charge, 875
acceleration four-vector, 836
accelerator, particle, 686, 830, 885
action
  definition, 917
  Lorentz invariance of, 919
  total electromagnetic, 926
addition theorem, spherical harmonics, 108, 955
adiabatic invariance, 376
advanced Green function, 722
Aepinus, F., 30
aether, 82, 586, 790, 825
Airy's formula, 603
Airy, G., 602
Alfvén waves, 587
aluminum, reflectivity of, 633
Ampère's formula, 34
Ampère's law, 307
Ampère's theorem, 345
Ampère, A.-M., 365
Ampère-Maxwell law, 36, 456
Ampèrian molecular current, 409
analytic function theory, 221
anapole moment, 348
angle, solid, 71, 319
angular distribution of radiation from
  a point charge in circular motion, 883
  a relativistic source, 880
  a slotted sphere, 756
  a slowly moving charge, 736
  a specified current density, 737
  a wire antenna, 738
  an antenna array, 742
  an electric multipole source, 759
  an oscillating electric dipole, 746
  an oscillating electric quadrupole, 752
  an oscillating magnetic dipole, 750
angular momentum
  and magnetic moments, 340
  conservation of, 516, 854, 930
  current density of, 517
  of a paraxial beam, 563
  of electromagnetic fields, 516
  operator, 6, 756
  radiation of, 750
angular spectrum of plane waves, 558, 801
anisotropic matter, waves in, 613
anode, 140
anomalous dispersion, 635
antenna, 737
  arrays, 741
  dipole, 737, 739
  phased-array, 743
aperture, diffraction by an, 797
Appleton model for a magnetized plasma, 636
Appleton, E.V, 639
approximation
  Born, 790
  Fraunhofer, 799
  Kirchhoff, 799, 803
  paraxial, 559, 561
  physical optics, 792
atmospheric color, 782
attenuation in conducting-tube waveguides, 684
auxiliary field **D**
  defined, 44
  matching condition, 45
auxiliary field **H**
  defined, 44
  matching condition, 45
axial vector, 21
azimuthal symmetry
  potential problems with, 209

Babinet's principle for vector fields, 807
Barkhausen, H., 647
battery, 283
beam waist, 560
beam-like waves, 558
Bessel functions, 216, 955
  modified, 957
  spherical, 957
betatron, 461
biaxial crystal, 613
Big Bang, 699
Biot-Savart law, 35, 304
birefringence, 615
blue sky, Rayleigh explanation of, 782
Bohr magneton, 341
boost, Lorentz, 827, 834
Born approximation, 790
bound charge, 159
boundary conditions
  and uniqueness, 199, 509
  conducting waveguide, 677
  Dirichlet, 199
  impedance, 686
  Kirchhoff, 798, 804
  magnetic field of the solar corona, 386

mixed, 199, 252
Neumann, 199, 252, 278
temporal, 721
Bragg mirror, 605
bremsstrahlung, 881
Brewster's angle, 594

canonical momentum, 918
capacitance
cross, 137
matrix, 136
of a circular disk, 135
self, 134
capacitor, 140
fringing fields, 225
with fixed charge, 169
with fixed potential, 172
carrier wave, 556
Cartesian coordinates, 2
Cartesian multipole radiation, 743
Cartesian symmetry, potential problems with, 203
cathode, 140
Cauchy's theorem, 652
causality, 486, 624, 649, 651, 832
Cavendish, H., 126
cavity resonator
chaos in, 699
closed tube, 695
conducting, 693
density of modes, 697
energy exchange, 700
spherical, 696
center of energy, 519, 854, 857, 930
center of energy theorem, 520
chaos, in a resonant cavity, 699
charge
absence of magnetic, 48
and gauge invariance, 927
bound, 159
conservation of electric, 32, 501
electric, 30
invariance of, 831
inversion, 79
polarization, 159
relaxation in an ohmic medium, 473, 634
space, 274
charge density, 30
at a perfect conductor surface, 130
at a real metal surface, 40
at the surface of a conducting disk, 131
fictitious magnetic, 415
force on, 58
in crystalline silicon, 38
macroscopic versus microscopic, 44
of a point electric dipole, 95
on the surface of a current-carrying wire, 285
polarization, 118, 159
singularity at a sharp corner or edge, 219
torque on, 58
charged particle motion, 366
and Larmor's formula, 735
and strong focusing, 356
in a cylindrical electron lens, 218
in a disk-loaded waveguide, 686
in a plane wave, 572, 838
in a plasma, 459
in a synchrotron, 891
in a uniform magnetic field, 366
in crossed fields, 368
in ohmic matter, 275
in the Earth's magnetic field, 377
in time-harmonic fields, 573
Lagrangian for, 920
with radiation reaction, 899
Cherenkov radiation, 906
Child-Langmuir law, 275
chirp, 646
circuit theory
AC, 486
DC, 284
circular current loop, magnetic field of, 304
circular-tube waveguides, 681
classical electron radius, 778, 903
Clausius-Mossotti formula, 176
closure relation, 202
coaxial transmission line, 667
Colladon, J.-D., 666
color
of the daylight sky, 782
of the setting sun, 782
complementarity, 554
complementary objects, diffraction theory, 807
completeness of
Bessel functions, 956
complex exponentials, 12
Legendre polynomials, 107, 953
orthonormal sets of functions, 202
spherical harmonics, 108, 955
complex
dielectric function, 608
index of refraction, 608
permittivity, 608
wave impedance, 608
complex logarithm potential, 260
conducting matter
dielectric function, 608
Drude model for, 631
skin depth, 609
waves in, 607
conducting-tube waveguide, 675
absence of TEM waves, 677
boundary conditions, 677
modes, 678
conductivity
frequency-dependent, 625
static Drude, 275
conductors
boundary condition, 130
energy of a collection of, 142
force on, 143
perfect, 126, 431
permeability of, 431
real, 149
spatial dispersion, 657
surface charge density, 130
surface current density, 311
confined waves, 666
conformal mapping, 224
conservation laws
and Lagrangian invariance, 927
in covariant form, 852
in matter, 522
in vacuum, 501
conservation of
angular momentum, 516, 854, 930
charge, 32, 501, 927
energy
in dispersive matter, 627
in simple matter, 523
in vacuum, 507
energy-momentum, 852, 929
linear momentum
in matter, 524
in vacuum, 511
conservative force, 61
constitutive relations, 45
dielectric matter, 166
frequency-dispersive matter, 626
magnetic matter, 421
matter in uniform motion, 859
simple conducting matter, 607
contact resistance, 279
continuity equation, 32, 272, 456, 501, 505, 668
and gauge invariance, 928
covariant, 840
continuous symmetries, 503
contraction theorem for tensors, 849
contravariant four-vector, 959
convective derivative, 8, 458
conversion between unit systems, 950
convolution theorem, 16
coordinate four-vector, 835

cosmic microwave background
  polarization, 545, 781
  spectral radiance, 699
Coulomb blockade, 142
Coulomb gauge, 321, 505, 538, 939
Coulomb's law, 33, 48, 59
  for magnetism, 435
Coulomb-Lorentz force, 29, 37, 456, 920
  from Hamilton's equations, 933
  from Lagrange's equations, 925
covariance, relativistic, 834, 848
covariant
  conservation laws, 852
  electrodynamics, 848
  equations of particle dynamics, 852
  four-vector, 959
  Larmor formula, 884
  Liénard-Wiechert potentials, 873
  Maxwell equations, 849
Crab nebulae, 743
critical angle, 595
cross section
  absorption, 793
  scattering
    differential (2D), 785
    differential (3D), 776
    total, 777
  total, 793
crystal optics, 613
current
  Ampèrian molecular, 409
  bound, 408
  displacement, 456
  free, 408
  in matter, 275
  in vacuum, 273
  polarization, 458
  sheet, 309
  sources, 287
  steady, 272, 302
current density, 31, 272
  at a perfect conductor surface, 311
  convection, 273
  force on, 301
  four-vector, 840
  magnetization, 410
  of a point magnetic dipole, 343
  of electromagnetic angular momentum, 517
  of electromagnetic energy, 508
  of electromagnetic linear momentum, 514
  orbital magnetization, 409
  polarization, 458
  spin magnetization, 409
  torque on, 301
current, electric, 31
cutoff in a
  conducting waveguide, 674
  dielectric waveguide, 691
  magnetized plasma, 638
cyclotron
  frequency, 366
  radiation, 882
  radius, 367
cylindrical coordinates, 2
cylindrical symmetry
  potential problems with, 215

Debye-Hückel, 291
delta function
  one dimension, 11
  three dimensions, 14
demagnetization field, 418
density of modes, 697
diamagnet, 407, 422
dielectric constant, 167, 175
  of a plasma, 459
  of a polar liquid, 212
dielectric function for
  conducting matter, 631
  dielectric matter, 635
  dispersive matter, 626
  magnetized plasma, 637
  negative-index matter, 640
  silicon, 636
dielectric matter, 158
  constitutive relation, 166
  energy of, 178
  forces on, 184
  linear, 167
  Lorentz model for, 635
  response to fixed fields, 172
  response to free charge, 169
  short range forces in, 187
  simple, 167
  waves in, 584
dielectric permittivity, 167
  frequency-dependent, 625
dielectric waveguides, 687
dielectrophoresis, 185
diffraction, 775
  Babinet's principle, 807
  by a planar aperture, 797
  by a sub-wavelength aperture, 808
  Fraunhofer
    from a circular aperture, 805
    of scalar fields, 799
    of vector fields, 804
  free space, 556
  Huygens' principle, 810
  scalar theory, 798
  Smythe's formula, 803
  Sommerfeld solution for a half-plane, 797
  vector theory, 800
diffusion
  analogy with electrostatics, 220
  in matter, 291
  magnetic, 481
dimensional analysis, 181, 243, 291, 395, 672, 730
dipole
  electric, 92
  magnetic, 337
dipole antenna
  frequency-domain analysis, 737
  time-domain analysis, 739
dipole field
  electric, 96
  magnetic, 343
  scattered, 778
  time-dependent, 728
dipole force
  electric, 96
    time-dependent, 527, 574
  magnetic, 373
dipole moment
  effective, of an aperture, 808
  electric, 92
  magnetic, 338
    conducting sphere, 432
    magnetized matter, 411
    of a current loop, 339
    spin, 340
dipole potential
  electric scalar, 92
  magnetic scalar, 339
  magnetic vector, 338
dipole-dipole interaction
  electric, 98
  magnetic, 378
Dirac's method of constraints, 936
Dirac, P., 901
direct integration
  Dirichlet Green function, 256
Dirichlet boundary conditions, 199
Dirichlet Green function, 251, 252
  direct integration, 256
  eigenfunction expansion, 254
  magic rule, 253
  method of splitting, 258
discontinuity
  of macroscopic fields, 42
  of potential at a dipole layer, 100
discrete symmetries, 502
disk generator, Faraday, 466
dispersion, 651
  anomalous, 635
  frequency, 624
    classical models for, 630

normal, 635
spatial, 656
structural, 674, 679
dispersion relation, 555, 586
dispersive matter
conservation of energy, 627
wave packets in, 641
waves in, 624
displacement current, 36, 456
divergence theorem
for a four-vector, 856
for a three-vector, 9
domain, magnetic, 443
Doppler effect, 847
double-curl equation, 323
double layer, 292
drift velocity, 275, 372
drift, $\mathbf{E} \times \mathbf{B}$, 368
drift-diffusion equation, 291
Drude model
for frequency-dependent conductivity, 631
for static conductivity, 275
duality, 49, 503, 566, 677, 808, 851
dyadic, 513

Earnshaw's theorem, 63
Eddington, A., 916
eddy current, 483
induced force, 484, 492
induced ohmic loss, 484
eigenfunction expansion
of an arbitrary function, 202
of Dirichlet Green function, 254
Einstein relation, 291
Einstein summation convention, 4
Einstein, A., 536, 723, 782, 822
electric charge, 30
and gauge invariance, 927
bound, 159
conservation of, 32, 501
free, 159
invariance of, 831
electric current, 31
electric dipole, 92
force on a, 96
layer, 98
moment, 92
point, 95
potential, 92
potential energy, 97
time-dependent, 727
torque, 97
electric dipole moment,
of a conducting sphere, 129
of polarized matter, 159
electric dipole radiation, 744
electric field **E**
local, 177
matching condition, 42
near a sharp corner or edge, 219
of a charged cylinder, 68
of a charged line segment, 65
of a charged ring, 59
of a charged sheet, 69
of a charged sphere, 68
of a point charge in uniform motion, 716, 844
of a point electric dipole, 96
of an electric dipole, 92
of polarized matter, 162
outside a current-carrying wire, 286
electric flux, 65
electric force, 58
between point charges, 33
from variation of potential energy, 74
on a dielectric interface, 188
on a dielectric sub-volume, 186
on a dipole, 96
on a quadrupole, 104
on an embedded dielectric, 189
on an isolated body, 184
electric Hertz vector, 570, 716
electric multipole expansion
azimuthal, 112
Cartesian
primitive, 90
traceless, 116
spherical, 109
electric multipoles, 90
electric polarization, 158
electric quadrupole, 102
electric dipole moment, 92
electric quadrupole radiation, 752
electric stress tensor, 81
for a simple dielectric, 188
electric susceptibility, 167
frequency-dependent, 625
electric torque, 58
electro-kinetic momentum, 515
electrocardiography, 290
electromagnet, iron core, 423
electromagnetic
angular momentum density, 516
dual tensor, 851
energy density, 508
field-strength tensor, 850, 961
induction, 462
linear momentum density, 513
potentials, 503
stress-energy tensor, 853, 962
electromagnetic fields
of a charge in uniform motion, 878
free, 536, 853, 855
from relativistic charges, 870
general properties, 501
non-classical, 47
of charge in arbitrary motion, 874
quasistatic, 455
electromotive force (EMF), 282, 462, 464
electron microscope, 358
electrostatic
analogy with diffusion, 220
energy of a system of conductors, 142
field, 58
induction, 126
lens, 217
potential, 60
complex, 221
multipole expansion, 90
near a sharp corner or edge, 219
of a charged line segment, 65
of a charged ring, 211
of a conducting sphere, 126
of a current source, 287
of a dipole layer, 99
of a line dipole, 260
of an electric dipole, 92
of an electric quadrupole, 102
of polarized matter, 162
potential theory
Laplace's equation, 198
Poisson's equation, 236
electrostatic energy
potential, 74
total, 76
electrostatics, 58
history of, 33
EMF, 464
energy
conservation of, 507, 523, 852
in matter, 588
current density, 508
density, 508
electric dipole, 97
electric dipole-dipole interaction, 98
electrostatic, potential, 74
electrostatic, total, 76
exchange in lossless cavities, 700
in dispersive matter, 627
magnetic dipole, 377
magnetic dipole-dipole interaction, 378
magnetic potential, 389
magnetic total, 384
minimization (Thomson's theorem), 128
of a system of conductors, 142
of a wave packet, 552

energy (*cont.*)
  of dielectric matter, 178
  of electrostatic interaction, 79
  of magnetic matter, 433
  relativistic, 837
  transport, 593
  transport by radiation, 730
  velocity
    of a plane wave in matter, 544
energy loss
  and radiation reaction, 899
  by a point particle, relativistic, 884
  by electric dipole radiation, 746
  by electric multipole radiation, 759
  by electric quadrupole radiation, 754
  by Joule heating, 280
  by magnetic dipole radiation, 750
  by particle accelerators, 885
  by the classical Bohr atom, 900
  in a conducting-tube waveguide, 684
  in resistive wires, 510
  in resonant cavities, 701
  in simple conducting matter, 607
energy velocity
  in a conducting-tube waveguide, 682
  in Lorentz matter, 644
envelope of a wave packet, 556
equipartition theorem, 698
equipotential surface, 63
Euler, L., 58
evanescent plane wave, 558, 598
events, 823
  separation in space-time, 832
Ewald-Oseen extinction theorem, 762
extinction paradox, 796
extinction theorem, 632, 673, 762
extraordinary ray, 615

f-sum rule, 655
Fabry-Perot geometry, 602
far zone, 728
Faraday disk generator, 466
Faraday EMF, 464
Faraday's cage, 207
Faraday's law, 35, 460
Faraday, M., 35, 427, 671
ferroelectric, 167
ferromagnet, 407
  hard and soft, 443
  permeability, 430
Fick's law, 291
field concept, 34
field lines
  charged line segment, 66
  electric, 63
  electric dipole, 93
  for an accelerated point charge, 876
  magnetic, 302
  magnetic dipole, 338
  point charge in a uniform field, 64, 73
  refraction of, 173
  topology of magnetic, 325
fission, nuclear, 113
flip coil, 464
Floquet's theorem, 687
flux
  electric, 65
  magnetic, 302
flux rule, 464
flux theorem, 10, 462
Fock, V., 775
focusing
  by electrostatic fields, 217
  by magnetostatic fields, 356, 359
  strong, 356
force
  Coulomb-Lorentz, 29, 37, 920
  electromagnetic on a classical atom, 527
  electromagnetic on isolated matter, 526
  electrostatic, 58
  magnetostatic, 301
  on a charged surface, 71
  on a conductor, 143
  on a magnetic dipole, 373
  on a polarizable particle, 574
  on an electric dipole, 96
  on dielectric matter, 184
  on magnetic matter, 435
  on particles in free fields, 571
  pondermotive, 573
force density four-vector, 852
form factor, 780
four-point resistance probe, 288
four-vector
  charge-current density, 840
  contravariant, 959
  covariant, 959
  energy-momentum, 837
  force density, 852
  frequency-wave vector, 846
  general, 834
  scalar-vector potential, 842
  space-time coordinate, 835
  velocity, 836
Fourier analysis, 15
Fourier transform, 554
Fourier-Bessel series, 216
Franklin, B., 30
Franz formulae, 811
Fraunhofer diffraction
  of scalar fields, 799
  of vector fields, 804
free fields, 536
  particle-like properties, 855
  radiation, 730
free-space diffraction, 556
free-space Green function, 724
frequency
  cyclotron, 366, 459
  Larmor, 367, 381
frequency dispersion, 624
  Appleton model for, 636
  classical models for, 630
  Drude model for, 631
  Lorentz model for, 635
  split-ring model for, 640
frequency spectrum
  of Cherenkov radiation, 908
  of Heaviside-Feynman radiation, 891
  of Liénard-Wiechert radiation, 889
  of radiation from an arbitrary current distribution, 888
  of synchrotron radiation, 895
Fresnel equations, 590, 762
Furry, W., 835

g-factor, 341
Galilean transformation, 824
Galvani, L., 31
gauge
  Coulomb, 321, 505, 538, 725
  fixing, 938
  invariance, 321, 504
    and conservation of charge, 927
    and Dirac's method of constraints, 937
  Lorenz, 507, 537, 715
Gauss' law, 34, 68
Gaussian beam, 559
Gaussian units, 949
Gaussian wave packet, 554, 646
generalized coordinates, 916
generalized optical principles, 807
Gilbert, W., 30, 34, 407
Ginzburg, V., 523
global positioning system (GPS), 826
going off the axis, 211
Golden Rule, 699
gradient force
  on a magnetic dipole, 373
  on an electric dipole, 96
Green function
  Dirichlet, 251, 252
    direct integration, 256
    eigenfunction expansion, 254

magic rule, 253
method of splitting, 258
for free space, 724, 799
cylindrical representation, 256
for Poisson's equation, 250
for the exterior of a hollow tube, 259
for the Helmholtz equation, 724
for the wave equation, 720
magic rule, 253
Neumann, 252
scalar diffraction theory, 798
Green's identities, 9
Green's reciprocity relation, 75, 137
Green, G., 236, 251
grounding a conductor, 135
group velocity, 555, 642
and the index of refraction, 643
dispersion, 645
in a conducting-tube waveguide, 679
in Drude matter, 643
in Lorentz matter, 644
negative, 644
guided waves, 666
Guoy phase, 561
gyromagnetic ratio, 341

Hagens-Rubens relation, 611
hairy ball theorem, 568
Hall effect, quantum, 389
Hamilton's equations
for fields, 934
for particles, 933
Hamilton's principle, 916
Hamiltonian
total electromagnetic, 934, 935, 939
treatment of electrodynamics, 931
Hamiltonian density, 934, 937
Hankel functions, 956
Heaviside, O., 639, 671, 870
Heaviside-Feynman fields, 879
heavy ion collisions, 830
helicity, 327, 565
Helmholtz coil, 315
Helmholtz equation, 557, 705
Green function for the, 724
Helmholtz theorem, 22
Helmholtz, H., 926
Hertz potentials, 725
Hertz vector
electric, 570, 716
magnetic, 569, 716
Hertz, H., 33, 714, 731
heuristic derivation
of the Liénard-Wiechert potentials, 872
of the Maxwell equations, 51
hidden momentum, 521
homopolar generator, Faraday, 466
Huygens' principle, 799, 804, 810
hyperfine interaction energy, 419
hysteresis, magnetic, 443, 446

image
dipole, 240
force, 238
method for
a conducting cylinder, 248
a conducting sphere, 245, 247
a dielectric cylinder, 249
dielectric boundaries, 240
magnetic matter, 428
multiple conducting planes, 242
one conducting plane, 237
potential, 238
potential states, 239
impedance
boundary condition, 686
in circuit theory, 487
matching, 592
of the vacuum, 585
wave, 586
complex, 608
index manipulations, 962
index of refraction, 585, 762
complex, 608
negative, 590
of silicon dioxide, 637
induced EMF method, 731
inductance, 394
mutual, 396
self, 395
induction
electromagnetic, 462
electrostatic, 126
inertial frame, 823
Infeld, L., 536
information-collecting shell, 872
inhomogeneous plane wave, 598
intensity, 544
interface matching conditions, 42
interfacial wave, 596
intermediate zone, 728
intrinsic impedance, 585
invariance
adiabatic, 376
gauge, 321, 504
rotational, 52, 68, 308, 503, 827, 834
translational, 52, 68, 308, 503, 834
invariant interval, 831
inverse distance, expansion in spherical harmonics, 109
inverse-square force law, 33
inversion, method of, 246
inversion, space, 18, 21, 52, 502, 690
Ioffe-Pritchard geometry, 377
ionosphere, 633, 639, 666
irrotational current sources, 304

Jacobian determinant, 9
Jeans, J., 197
Jefimenko, O.D., 726
Joule heating, 280, 484, 610, 684

Kennelly, A., 639
Kirchhoff approximation
for scalar field diffraction, 799
for vector field diffraction, 803
Kirchhoff's laws, 284
klystron, 666
Kramers-Krönig relations, 649
Kronecker symbol, 4

Lagrange multipliers, 24
Lagrange's equations
covariant form, 926
for fields, 924
for particles, 917
Lagrangian, 916
approach to the conservation laws, 927
for a free field, 923
for a moving point dipole, 922
for a non-relativistic charge in a field, 920
for a relativistic charge in a field, 921
singular, 936
total electromagnetic, 918, 925
treatment of electrodynamics, 918
Lagrangian density, 923
Landau-Lifshitz equation, 904
Langmuir-Child law, 275
Laplace's equation, 174, 197
analytic function theory, 221
and multipole theory, 113
in image theory, 237, 239
in magnetostatics, 312, 417
separation of variables, 201
uniqueness of solutions, 199
Larmor frequency, 367, 747
Larmor precession, 381
Larmor's formula, 735, 746, 747
relativistic, 884
Larmor's theorem, 367
Larmor, J., 367
law of squares, 671
left-handed matter, 590
Legendre functions, 209

Legendre polynomials, 107, 953
  associated, 954
Legendre transformation, 145, 932
length contraction, 830
lens, electrostatic, 217
Lenz' law, 36, 464
Levi-Cività symbol, 4
Liénard-Wiechert
  fields, 874
  potentials, 871
light cone, 832
light, speed of, 51
linear momentum
  conservation of, 511, 524
  electromagnetic, 513
  electromagnetic, in matter, 526
linewidth, radiative, 906
liquid drop model, 113
local field, 177
longitudinal waves
  in a Drude medium, 632
  in dispersive matter, 629
Lord Rayleigh, 705
Lorentz averaging, 39
Lorentz force, 29, 37, 365, 366, 456
Lorentz invariant scalar, 831
Lorentz model
  for dielectric matter, 635
  for magnetization, 411
  for polarization, 160
  wave velocities, 644
Lorentz reciprocity, 769
Lorentz tensors, 849
Lorentz transformation
  of a static Coulomb field, 844
  of electromagnetic fields, 843
  of four-vectors, 835, 961
  of magnetization, 858
  of plane wave fields, 845
  of polarization, 858
  of space-time coordinates, 834
  standard configuration, 827
Lorentz, H., 624, 927
Lorentz-Abraham equation, 902
Lorentzian line shape, 702, 906
Lorenz gauge, 507, 537
Lorenz, L., 507, 789

macroscopic sources and fields, 44
macroscopic vs. microscopic, 38
magnetic anisotropy, 379
magnetic bacteria, 379
magnetic bottle, 377
magnetic charge
  absence of physical, 48
  fictitious, 415, 436, 443, 445
  real, 419
magnetic diffusion, 481
magnetic dipole, 337
  field lines, 338
  force on a, 373
  layers, 345
  moment, 338
  point, 343
  potential energy, 377
  scalar potential, 339
  torque, 378
  vector potential, 338
magnetic dipole moment
  adiabatic invariance of, 376
  of a conducting sphere, 432
  of a current loop, 339
  of a magnetized body, 411
  of a permeable sphere, 422
  of the proton, 419
  orbital, 340
  spin, 341
magnetic dipole radiation, 748
magnetic domain, 443
magnetic energy
  potential, 389
  total, 384
magnetic field **B**
  axially symmetric, 357
  matching condition, 42
  of a current line, 308
  of a current ring, 313
  of a current segment, 307
  of a current sheet, 309
  of a current-carrying wire, 323
  of a magnetic dipole, 338
  of a point magnetic dipole, 343
  of a solenoid, 305
  of a torus winding, 311
  of a uniformly magnetized sphere, 417
  of magnetized matter, 412
  of the Earth, 339
  of the solar corona, 386
magnetic flux, 302, 388, 461
magnetic force, 301, 365
  between steady currents, 34, 368
  from variation of potential energy, 391
  on a current sheet, 311
  on a dipole, 373
  on a magnetic interface, 441
  on a magnetic sub-volume, 438
  on an embedded magnet, 440
  on an isolated body, 436
magnetic helicity, 327
magnetic Hertz vector, 569, 716
magnetic hysteresis, 443, 446
magnetic matter, 407
  constitutive relation, 421
  energy of, 433
  linear, 421
  permanent, 443
  simple, 421
  waves in, 584
magnetic mirror, 375
magnetic monopole, 49, 302, 337
magnetic multipole expansion
  for scalar potential
    axial, 357
    azimuthal, 351
    spherical, 349
  for vector potential
    Cartesian, 336, 347
    interior, 353
    spherical, 351
magnetic multipoles, 336
magnetic permeability, 421
  for a split-ring resonator, 641
  frequency-dependent, 625
magnetic plasma
  Appleton model for a, 636
magnetic pressure, 382
magnetic quadrupole, 356
magnetic reconnection, 325
magnetic resonance imaging, 316
magnetic response
  to a fixed field, 425
  to free current, 423
magnetic scalar potential, 312
  and the method of images, 428
  multi-valued nature of, 318
  multipole expansion of, 349
  of a current loop, 314
  of a magnetic dipole, 339
  of magnetized matter, 415
  topological aspects of, 317
magnetic shielding
  AC, 480
  DC, 428
magnetic stress tensor, 381
  for a simple magnet, 439
magnetic susceptibility, 421
  frequency-dependent, 625
magnetic tension, 382
magnetic torque, 301, 365
  on a dipole, 378
magnetic trapping, 377
magnetic virial theorem, 383
magnetic work, 366, 371
magnetization **M**
  as a sum of point dipoles, 413
  energy to create, 434
  Lorentz model for, 411
  magnetic field produced by, 412
  non-uniqueness of, 412
  of the vacuum, 46
  orbital, 408, 409
  spin, 408

total, 410
magnetostatics, 301
history of, 34
magnetron, 666
Marconi, G., 639
mass renormalization, 904
matching conditions
at a moving interface, 43, 859
at an electric dipole layer, 100
for **B**, 42, 310
for **D**, 45, 166
for **E**, 42, 70
for **H**, 45, 420
for electrostatic potential, 62, 197
for magnetic scalar potential, 417
in ohmic matter, 277
Maxwell equations, 849
electrostatic, 58
from Hamilton's equations, 935
from Lagrange's equations, 925
heuristic derivation, 51
in matter, 43
in vacuum, 33
macroscopic vs. macroscopic, 39
magnetostatic, 301
Maxwell inequalities, 138
Maxwell stress tensor, 513
Maxwell, J.C., 33, 90, 584
metal cluster ionization potential, 79
metal surface, charge density at a, 40
metallic alloy, reflectivity of a, 612
method of images for
a dielectric cylinder, 249
a conducting cylinder, 248
a conducting sphere, 245, 247
dielectric boundaries, 240
magnetic matter, 428
multiple conducting planes, 242
one conducting plane, 237
method of inversion, 246
metric tensor in special relativity, 959
microscopic vs. macroscopic, 38
microstrip, 693
Mie scattering, 787
approximate, 795
Minkowski diagram, 832
Minkowski electrodynamics, 858
Minkowski, H., 835, 858
mirror, magnetic, 375
mixed boundary conditions, 199, 252
mobility, 292
modes
density of, 697
excitation of cavity, 703
for a parallel-plate transmission line, 673
in conducting cavities, 694
in conducting-tube cavities
TE and TM, 695
in conducting-tube waveguides, 678
in dielectric waveguides, 687
in spherical cavities
TE and TM, 696
of a dielectric waveguide
hybrid, 692
radiation, 692
moment
anapole, 348
electric dipole, 92
electric quadrupole, 102
magnetic dipole, 372
momentum
canonical, 918
conservation of, 511, 524, 852, 929
electromagnetic, 513
relativistic, 837
momentum density
canonical, 925
electromagnetic, 513
momentum four-vector, 837
momentum, hidden, 521
monochromatic plane waves, 543
monopole, magnetic, 49, 302, 337
motional EMF, 464
multilayer, wave propagation in a, 604
multipole expansion
for electrostatic potential
azimuthal, 112
spherical, 109
for magnetic scalar potential
axial, 357
azimuthal, 351
spherical, 349
for radiation fields
Cartesian, 743
spherical, 755
for vector potential
Cartesian, 336, 347
interior, 353
spherical, 351
multipole moments
electromagnetic, 758
electrostatic
spherical, 110, 112
magnetostatic
azimuthal, 351
Cartesian, 347
spherical, 349
multipole radiation
Cartesian, 743
from atoms and nuclei, 761
spherical, 755
multipoles
electric, 90
magnetic, 336
mutual inductance, 396

near zone, 728
near-field optics, 810
negative
group velocity, 644
index matter, 590
split-ring model for, 640
negative refraction, 590
Nernst-Planck equation, 292
network circuits, 490
Neumann boundary conditions, 199, 278
Neumann Green function, 252
Noether's theorem, 928
non-uniform
plane wave, 598, 672
TEM wave, 668
normal dispersion, 635
nuclear quadrupole moments, 105

Ohm's law, 36, 275, 340, 473, 510, 607
in a moving medium, 859
Onsager, L., 177, 212
optical fiber, 687
optical theorem, 794
optical tweezers, 574
orbital magnetization, 409
orbital, magnetic moment, 340
ordinary ray, 615
orthogonality
of Bessel functions, 217, 956
of cavity modes, 694
of complex exponentials, 217
of complex vectors, 548
of Legendre polynomials, 107, 953
of sinusoids, 204
of spherical Bessel functions, 958
of spherical harmonics, 108, 213, 955
of waveguide modes, 678
orthonormal functions, 202

p-polarization, 590
parallel-plate
capacitor, 141
transmission line, 673
paramagnet, 407, 422
paraxial approximation, 559, 561
paraxial beam, angular momentum of, 563
paraxial waves, 562
parity, 94, 348, 502, 762
Parseval's theorem, 16
Pauli, W., 455, 835

Peierls, R., 29
perfect
absorber, 796
conductor, 126, 431
diamagnet, 430
ferromagnet, 430
lens, 590
permanent magnet matter, 443
permeability, 421
of a ferromagnet, 430
of a perfect conductor, 431
of a superconductor, 430
permittivity, 167
complex, 608
phase of a wave, 541
phase velocity
and charged particle acceleration, 686
and Cherenkov radiation, 906
in a conducting-tube waveguide, 679
in a good conductor, 610
in a magnetized plasma, 639
in Lorentz matter, 644
in negative-index matter, 590
of a plane wave
in matter, 586
in vacuum, 541
of an Alfvén wave, 587
of an evanescent wave, 598
physical optics approximation, 792
pickup coil, 465
pinch effect, 371
Planck distribution, 698
Planck, M., 501
plane of incidence, 589
plane wave expansions, 958
plane waves
angular spectrum of, 558
evanescent, 558, 598
in anisotropic matter, 613
in conducting matter, 607
in simple matter, 584
in special relativity, 845
in vacuum, 539
mechanical properties of, 542
monochromatic, 543
non-uniform, 672
standing, 539
transverse, 539
plasma
frequency, 631
oscillation, 633
waves in a magnetized, 636
Poincaré sphere, 550
point electric dipole, 95
point magnetic dipole, 343
Poisson's equation, 236
Green function for, 250
particular solution, 236
uniqueness of solutions, 199
vector, 321
Poisson's formula, 162
Poisson-Boltzmann equation, 262
polar symmetry, potential problems with, 218
polar vector, 21
polarizability, 129, 175
polarization
by reflection, 594
by Thomson/Rayleigh scattering, 779
circular, 547
ellipse, 545
elliptical, 549
linear, 546
of an electromagnetic wave, 545
of synchrotron radiation, 893
of the cosmic microwave background, 545, 781
polarization **P**
as a sum of point dipoles, 164
electric field produced by, 162
energy to create, 181
Lorentz model of, 160
modern theory of, 160
of a conductor, 126
of a dielectric, 158
of the vacuum, 46
polarization charge, 159
at an interface, 170
density, 118
polarization current, 458
in ice, 459
pondermotive force, 573
potential
electromagnetic, 503
electrostatic, 60
complex, 221
matching condition for, 62
multipole expansion, 90
near a sharp corner or edge, 219
of a charged line segment, 65
of a charged ring, 211
of a conducting sphere, 126
of a current source, 287
of a dipole layer, 99
of a line dipole, 260
of an electric dipole, 92
of an electric quadrupole, 102
of polarized matter, 162
four-vector, 842
magnetic scalar, 312
of a current loop, 314
of a magnetic dipole, 339
momentum, 515
scalar
Coulomb gauge, 505
in special relativity, 842
Lorenz gauge, 724
of a point charge in arbitrary motion, 872
of a point charge in uniform motion, 716
of a time-dependent electric dipole, 727
vector, 320
Coulomb gauge, 506
for radiation, 734
gauge freedom, 504
in special relativity, 842
Lorenz gauge, 724
of a charge in uniform motion, 716
of a current ring, 324
of a current carrying wire, 322, 323
of a magnetic dipole, 338
of a magnetic dipole layer, 345
of a point charge in arbitrary motion, 872
of a point magnetic monopole, 344
of a time-dependent electric dipole, 727
of magnetized matter, 412
physical significance, 514
potential energy
and Green's reciprocity, 75
barrier in an ion channel, 244
electrostatic, 74
force from variation of, 74, 391
landscape in matter, 126, 158
magnetostatic, 389
of a magnetic dipole, 377
of an electric dipole, 97
of magnetic matter, 433
potential theory
for a simple magnet, 426
for magnetic matter, 416
for simple dielectrics, 174
ohmic matter, 276
Poisson's equation, 236
uniqueness of solutions, 199
power dissipated
by a conducting-tube waveguide, 684
by a resonant cavity, 701
by an ohmic medium, 474
in circuit theory, 486
power radiated
by a point particle, relativistic, 884
by a slowly moving point charge, 736

by a specified current source, 737
by a wire antenna, 738
by an oscillating electric dipole, 746
by an oscillating electric multipole, 760
by an oscillating electric quadrupole, 754
by an oscillating magnetic dipole, 750
by at time-harmonic solenoid, 718
by particle accelerators, 885
in the frequency domain, 886
in the general case, 730
power transported
by a conducting-tube waveguide, 682
Poynting vector
and the definition of radiation, 730
at an interface, 593
field lines
for a resistive wire, 510
for a plane wave, 544
in a conducting-tube waveguide, 682
in matter, 523
in negative-index matter, 590
in vacuum, 508
of an evanescent wave, 598
uniqueness, 511
Poynting's theorem, 507
in matter, 523
in special relativity, 854
time-averaged, 700
precession, Larmor, 381
precession, Thomas, 834
pressure, magnetic, 382
pressure, radiation, 599, 847
principal value integral, 13
proper time, 833
pseudovector, 21
pulsar, 743
Purcell, E.M, 158

Q, quality factor of
a dielectric resonator, 704
a lossy medium, 588
a resonant cavity, 702
a resonant circuit, 489
quadrupole mass spectrometer, 469
quadrupole moment tensor
of an ellipsoid, 106
of nuclei, 105
primitive, 91, 102
traceless, 103
quadrupole, electric, 102
force and torque on, 104
quantum electrodynamics, 46
quantum Hall effect, 389
quark confinement, classical model for, 180
quark-gluon plasma, 830
quasi-electrostatics
in poor conductors, 473
in vacuum, 468
quasi-magnetostatics
in good conductors, 475
in vacuum, 471
quasi-monochromatic fields, 628
quotient theorem, 849

Rabi, I., 336
radiation, 714
angular distribution of, 730
birth of, 731
blackbody, 698
damping, 899
definition of, 730
from
a current sheet, 725
a magnetic dipole, 748
a point charge in circular motion, 883
a relativistic source, 880
a slotted sphere, 756
a slowly moving charge, 736
a specified current density, 737
a time-harmonic solenoid, 718
a wire antenna, 738
an antenna array, 742
an electric dipole, 744
an electric multipole, 759
an electric quadrupole, 752
atoms and nuclei, 761
cyclotron motion, 882
synchrotron motion, 882, 891
the cosmic microwave background, 699
Hertz analysis of, 731
in matter, 762
in the frequency domain, 736, 886
in the time domain, 733, 880
multipole
Cartesian, 743
spherical, 755
of angular momentum, 751, 760
pressure, 599, 847
reaction, 795, 899
resistance, 739
vector, 743
zone
summary of results, 734
radiation condition, Sommerfeld, 724
radiative linewidth, 906
Rayleigh criterion, 806
Rayleigh distance, 560
Rayleigh scattering
three dimensions, 782
two dimensions, 786
Rayleigh, Lord, 782
Rayleigh-Jeans law, 698
reciprocity
electric, 75
Lorentz, 769
magnetic, 387
reconnection
electic field line, 732
magnetic field line, 325
rectangular-tube waveguides, 680
red sun, Rayleigh explanation of, 782
reference frame, 823
reflection
amplitude, 593
coefficient, 593
from a good conductor, 611
from a moving mirror, 846
from a planar boundary, 588
of radio waves by the ionosphere, 639
polarization by, 594
total internal, 595
reflectivity, 632
of a metallic alloy, 612
of aluminum, 633
of seawater, 612
refraction
from a planar boundary, 588
index of, 762
into a good conductor, 609
of magnetic field lines, 425
refrigerator magnet, 445
relativistic covariance, 834, 848
relativistic invariance of
electric charge, 831
proper time, 833
radiated power, 884
the action, 919
the four-vector scalar product, 835
the interval, 831
the phase of a plane wave, 846
the speed of light, 831
the wave operator, 840
relativistic transformation of
a static Coulomb field, 844
electromagnetic fields, 843
four-vectors, 961
magnetization, 858
plane wave fields, 845
polarization, 858
space-time coordinates, 827, 834, 835

relativity of simultaneity, 825
relativity, special, 653, 822
remanent magnetization, 443
resistance
  contact, 279
  electrical, 277
  four-point probe, 288
resistivity, 278
resonant cavities, 666
  chaos in, 699
  closed tube, 695
  conducting, 693
  density of modes, 697
  energy exchange, 700
  Q-factor of, 702
  spherical, 696
response functions
  analyticity of, 652
  causal, 624, 649
resting potential of a cell, 291
retardation, 714, 719
retarded Green function, 722
retarded potentials, 724
retarded time, 714
right-hand rule, 10, 36, 97, 307
Ritz, W., 723
runaway solutions, 903
Rutherford, Ernest, 480

s-polarization, 590
scalar potential
  Coulomb gauge, 505
  electrostatic, 60
    complex, 221
    matching condition for, 62
    multipole expansion, 90
    near a sharp corner or edge, 219
    of a charged line segment, 65
    of a charged ring, 211
    of a conducting sphere, 126
    of a current source, 287
    of a dipole layer, 99
    of a line dipole, 260
    of an electric dipole, 92
    of an electric quadrupole, 102
    of polarized matter, 162
  in special relativity, 842
  Lorenz gauge, 724
  magnetic, 312
    and the method of images, 428
    multi-valued nature of, 318
    multipole expansion of, 349
    of a current loop, 314
    of magnetized matter, 415
  of a point charge in arbitrary motion, 872
  of a point charge in uniform motion, 716
  of a time-dependent electric dipole, 727
scalar product of two four-vectors, 835
scattering, 775
  amplitude, 777
  and the blue sky, 782
  and the red sun, 782
  cross section
    differential (2D), 785
    differential (3D), 776
    total, 777
  form factor, 780
  from a conducting cylinder, 783
  from a dielectic sphere, 787
  long wavelength, 777, 782
  Mie, 787
    approximate, 795
  plane, 778
  Rayleigh, 782
  Thomson, 777
  wave vector, 780
  x-ray, 780
Schott's formulae, 726
Schumann resonances, 666
Schwarzchild, K., 927
screening length, 149, 291, 657
screening, electrostatic, 133
seawater, reflectivity of, 612
self-inductance, 395
separation of variables
  Helmholtz equation, 567
  Laplace equation, 201
    azimuthal symmetry, 209
    Cartesian symmetry, 203
    cylindrical symmetry, 215
    polar coordinates, 218
    spherical symmetry, 212
shielding
  electrostatic, 133
  magnetic AC, 480
  magnetic DC, 428
SI units, 50
sign function, 14
silicon dioxide, index of refraction of, 637
silicon, dielectric function of, 636
simple dielectric matter
  defined, 167
  waves in, 584
simple magnetic matter
  defined, 421
  waves in, 584
singular behavior
  of **E** at a sharp corner or edge, 219
  of a point electric dipole, 95
  of a point magnetic dipole, 343
skin depth, 609
skin effect, 477
Stanford linear accelerator, 686
Smoluchowski, M., 782
Smythe's diffraction formula, 803
Snell's law, 589
solenoid, 305
  time-dependent, 718
  toroidal, 312, 349, 397
solid angle, 71, 319
Sommerfeld radiation condition, 724
Sommerfeld, A., 58, 653
space charge, 274
space inversion, 18, 21, 52, 502, 690
space-time, 827
spatial dispersion, 656
special relativity, 653, 822
speed of light, 51
spherical
  Bessel functions, 957
  cavity resonator, 696
  coordinates, 3
  harmonics, 108, 213, 954
  multipole radiation, 755
  symmetry, potential problems with, 212
  waves, 565
spin magnetic moment, 340
spin magnetization, 408
split-ring model for negative-index matter, 640
splitting method for Dirichlet Green function, 258
spontaneous emission, 699
standard configuration, 823
standing wave, 539, 634, 672, 695, 737, 798
steady-current condition, 272, 284, 302, 337, 439, 475
stellar aberration, 846
step function, 13
Stokes
  parameters, 549
  relations, 603
  theorem, 10
Stratton-Chu formulae, 811
stress tensor
  electric, 81
  electromagnetic (Maxwell), 513
  magnetic, 381
stress-energy tensor, 853
strong focusing, 356
structural dispersion, 674, 679
sum rules, 655
superconductor
  compared to a perfect conductor, 407, 432
  perfect diamagnetism, 432
  permeability of a, 430

zero resistance, 388
surface charge density
defined, 30
of a perfect conductor, 130
surface current density
defined, 31
of a perfect conductor, 311
surface wave, 596
susceptibility
electric, 167
magnetic, 421
symmetry
continuous, 503
discrete, 502
dual, 503, 566
in electromagnetism, 501
symmetry arguments
to find **A**, 322
to find **B**, 307
synchrotron radiation, 882, 891
frequency spectrum, 895
polarization, 893
pulse shape, 893
transition from cyclotron radiation, 894
Système International (SI) units, 50

TE (transverse electric) modes
in conducting-tube cavities, 695
in spherical cavities, 696
TE (transverse electric) waves
guided by a transmission line, 673
in a circular waveguide, 681
in a conducting-tube waveguide, 676
in a dielectric waveguide, 690
in a rectangular waveguide, 680
in an optical fiber, 689
in Fresnel theory, 590
in vacuum, 566
telegraph equations, 669
TEM (transverse electromagnetic) waves, 668
absence in hollow-tube waveguides, 677
guided by a coaxial transmission line, 667
non-uniform, 668
tensor
contraction theorem, 849
decomposition of a second rank, 338
demagnetization, 418
dual, 851
electric stress, 81
electromagnetic field strength, 850
electromagnetic stress-energy, 853
Lorentz, transformation properties, 849
magnetic stress, 381
Maxwell stress, 513
metric, in special relativity, 959
moment of inertia, 379
quadrupole moment
primitive, 91, 102
traceless, 103
quotient theorem, 849
rotational, definition, 20
torque density, 854
theorem
Ampère's, 345
Cauchy's, 652
center of energy, 520
convolution, 16
divergence, 9
Earnshaw's, 63
equipartition, 698
extinction, 762
Floquet's, 687
hairy ball, 568
Helmholtz, 22
Larmor's, 367
magnetic virial, 383
Noether's, 928
optical, 794
Parseval's, 16
Poynting's, 507
Stokes, 10
Thomson's (electrostatics), 128
Thomson's (magnetostatics), 303
time-averaging, 17
uniqueness, 199, 509
Whittaker's, 539
Thomas precession, 834
Thomas-Fermi, 291
Thomson scattering, 777
Thomson's
formula, 488
jumping ring, 492
problem, 78
theorem
of electrostatics, 128
of magnetostatics, 303
Thomson, W. (Lord Kelvin), 35, 301, 671
't Hooft, G., 835
time dilation, 829
time reversal, 502
time-averaging theorem, 17
TM (transverse magnetic) modes
in conducting-tube cavities, 695
in spherical cavities, 696
TM (transverse magnetic) waves
in a circular waveguide, 681
in a conducting-tube waveguide, 676
in a rectangular waveguide, 680
in an optical fiber, 689
in Fresnel theory, 590
in vacuum, 566
topology and the magnetic scalar potential, 317
topology of magnetic field lines, 325
toroidal solenoid, 312, 349, 397
torque
electric, 58
on a dipole, 97
on a quadrupole, 104
Lorentz tensor of, 854
magnetic, 301, 365
on a dipole, 378
mechanical, 517
total energy
electrostatic, 76
magnetostatic, 384
of a plane wave, 544
of a relativistic particle, 837
of a wave packet, 552
of dielectric matter, 179
of magnetic matter, 433
of the electromagnetic field, 508
total internal reflection, 595, 666, 688
transformation
Galilean, 824
Lorentz
of a static Coulomb field, 844
of electromagnetic fields, 843
of four-vectors, 835, 961
of magnetization, 858
of plane wave fields, 845
of polarization, 858
of space-time coordinates, 827, 834
standard configuration, 827
transformer EMF, 464
transmission amplitude, 594
transmission coefficient, 594
transmission line, 667
coaxial, 667
parallel-plate, 673
TEM waves guided by a, 667
transverse electric (TE) modes
in conducting-tube cavities, 695
in spherical cavities, 696
transverse electric (TE) waves
guided by a transmission line, 673
in a circular waveguide, 681
in a conducting-tube waveguide, 676
in a dielectric waveguide, 690
in a rectangular waveguide, 680
in an optical fiber, 689

in Fresnel theory, 590
in vacuum, 566
transverse magnetic (TM) modes
in conducting-tube cavities, 695
in spherical cavities, 696
transverse magnetic (TM) waves
in a circular waveguide, 681
in a conducting-tube waveguide, 676
in a rectangular waveguide, 680
in an optical fiber, 689
in Fresnel theory, 590
in vacuum, 566
transverse waves
in a Drude medium, 632
in a magnetized plasma, 638
in dispersive matter, 629
in simple matter, 584
in vacuum, 539
triode, 140
two-dimensional potential theory problems, 221

uniaxial crystal, 613
waves in a, 615
uniqueness theorem
for time-dependent fields, 509
Laplace's equation, 199
Poisson's equation, 199
Unisphere, 213
units
conversion, 950
Gaussian, 949
Système International (SI), 50

vacuum diode, 273
vacuum polarization, 46
vacuum tube, 140
variational principle
electrostatic, 226
for the action, 916
vector Poisson equation, 321
vector potential, 320
Coulomb gauge, 506
for radiation, 734
gauge freedom, 504
in special relativity, 842
Lorenz gauge, 724
multipole expansion
Cartesian, 336, 347
interior, 353
spherical, 351
of a charge in uniform motion, 716
of a current line, 322
of a current ring, 324
of a current-carrying wire, 322, 323
of a magnetic dipole, 338
of a magnetic dipole layer, 345
of a point charge in arbitrary motion, 872
of a point magnetic monopole, 344
of a time-dependent electric dipole, 727
of magnetized matter, 412
physical significance, 514
velocity
energy
in a conducting-tube waveguide, 682
in Lorentz matter, 644
of a plane wave in matter, 544
group
and the index of refraction, 643
approximation, 642
in a conducting-tube waveguide, 679
in Drude matter, 643
in Lorentz matter, 644
negative, 644
of a wave packet, 555
phase
and Cherenkov radiation, 906
in a conducting-tube waveguide, 679
in a good conductor, 610
in a magnetized plasma, 639
in charged particle acceleration, 686
in Lorentz matter, 644
in negative index matter, 590
of an Alfvén wave, 587
of an evanescent wave, 598
velocity field of a moving charge, 875
velocity four-vector, 836
Veltman, M., 835
Volta, A., 31, 272
voltage, 283
voltaic cell, 283

waist, Gaussian beam, 560
wave equation
covariant form of, 840
for **E** and **B**, 537
for the electric Hertz vector, 570
for the electromagnetic potentials, 537
for the magnetic Hertz vector, 569
Green function for the, 720
inhomogeneous, 715, 720
wave impedance, 586
wave normal, 613
wave packet
consistency with special relativity, 653
dispersion relation, 555
envelope, 556
Gaussian, 554, 646
in dispersive matter, 641
particle-like properties of a, 855
scalar, 553
synthesis, 552
wave vector, 540
waveguide
absence of TEM waves, 677
boundary conditions for, 677
conducting tube, 675
cutoff in a, 674
dielectric, 687
disk-loaded, 686
energy loss in, 684
energy velocity in, 682
general mode properties, 678
particle acceleration in, 686
phase and group velocities in, 679
structural dispersion in, 679
TE and TM waves in, 676
waves
Alfvén, 587
beam-like, 558
confined by
a cavity resonator, 693
a dielectric resonator, 704
guided by
a transmission line, 667
conducting tubes, 675
planar conductors, 672
in a Drude medium, 632
in a magnetized plasma, 638
in a multilayer, 604
in anisotropic matter, 613
in dispersive matter, 624
in simple conducting matter, 607
in simple matter, 584
in vacuum, 536
interfacial, 596
longitudinal, 629
paraxial, 562
partially polarized, 551
plane, 539
polarization of, 545
slow, 686
spherical, 565
surface, 596
TE (transverse electric)
guided by a transmission line, 673
in a circular waveguide, 681
in a conducting-tube waveguide, 676
in a dielectric waveguide, 690
in a rectangular waveguide, 680
in an optical fiber, 689

in Fresnel theory, 590
in vacuum, 566
TEM (transverse electromagnetic), 668
TM (transverse magnetic)
in a circular waveguide, 681
in a conducting-tube waveguide, 676
in a rectangular waveguide, 680
in an optical fiber, 689
in Fresnel theory, 590
in vacuum, 566
transverse
in dispersive matter, 629
in simple matter, 584
in vacuum, 539
unpolarized, 551
whispering gallery modes, 705
whistlers, 647
Whittaker's theorem, 539, 570
Wigner, E., 1, 505
world line, 833
Wronskian
Bessel functions, 956
modified Bessel functions, 257, 957
spherical Bessel functions, 957

x-ray scattering, 780

Zeeman effect, 747

Grad, Div, Curl, and Laplacian

CARTESIAN $\quad d\boldsymbol{\ell} = dx\hat{\mathbf{x}} + dy\hat{\mathbf{y}} + dz\hat{\mathbf{z}} \qquad d^3r = dxdydz$

$$\nabla\psi = \frac{\partial\psi}{\partial x}\hat{\mathbf{x}} + \frac{\partial\psi}{\partial y}\hat{\mathbf{y}} + \frac{\partial\psi}{\partial z}\hat{\mathbf{z}}$$

$$\nabla\cdot\mathbf{A} = \frac{\partial A_x}{\partial x} + \frac{\partial A_y}{\partial y} + \frac{\partial A_z}{\partial z}$$

$$\nabla\times\mathbf{A} = \left(\frac{\partial A_z}{\partial y} - \frac{\partial A_y}{\partial z}\right)\hat{\mathbf{x}} + \left(\frac{\partial A_x}{\partial z} - \frac{\partial A_z}{\partial x}\right)\hat{\mathbf{y}} + \left(\frac{\partial A_y}{\partial x} - \frac{\partial A_x}{\partial y}\right)\hat{\mathbf{z}}$$

$$\nabla^2\psi = \frac{\partial^2\psi}{\partial x^2} + \frac{\partial^2\psi}{\partial y^2} + \frac{\partial^2\psi}{\partial z^2}$$

CYLINDRICAL $\quad d\boldsymbol{\ell} = d\rho\hat{\boldsymbol{\rho}} + \rho d\phi\hat{\boldsymbol{\phi}} + dz\hat{\mathbf{z}} \qquad d^3r = \rho d\rho d\phi dz$

$$\nabla\psi = \frac{\partial\psi}{\partial\rho}\hat{\boldsymbol{\rho}} + \frac{1}{\rho}\frac{\partial\psi}{\partial\phi}\hat{\boldsymbol{\phi}} + \frac{\partial\psi}{\partial z}\hat{\mathbf{z}}$$

$$\nabla\cdot\mathbf{A} = \frac{1}{\rho}\frac{\partial}{\partial\rho}(\rho A_\rho) + \frac{1}{\rho}\frac{\partial A_\phi}{\partial\phi} + \frac{\partial A_z}{\partial z}$$

$$\nabla\times\mathbf{A} = \left(\frac{1}{\rho}\frac{\partial A_z}{\partial\phi} - \frac{\partial A_\phi}{\partial z}\right)\hat{\boldsymbol{\rho}} + \left(\frac{\partial A_\rho}{\partial z} - \frac{\partial A_z}{\partial\rho}\right)\hat{\boldsymbol{\phi}} + \frac{1}{\rho}\left[\frac{\partial}{\partial\rho}(\rho A_\phi) - \frac{\partial A_\rho}{\partial\phi}\right]\hat{\mathbf{z}}$$

$$\nabla^2\psi = \frac{1}{\rho}\frac{\partial}{\partial\rho}\left(\rho\frac{\partial\psi}{\partial\rho}\right) + \frac{1}{\rho^2}\frac{\partial^2\psi}{\partial\phi^2} + \frac{\partial^2\psi}{\partial z^2}$$

SPHERICAL $\quad d\boldsymbol{\ell} = dr\hat{\mathbf{r}} + rd\theta\hat{\boldsymbol{\theta}} + r\sin\theta d\phi\hat{\boldsymbol{\phi}} \qquad d^3r = r^2\sin\theta drd\theta d\phi$

$$\nabla\psi = \frac{\partial\psi}{\partial r}\hat{\mathbf{r}} + \frac{1}{r}\frac{\partial\psi}{\partial\theta}\hat{\boldsymbol{\theta}} + \frac{1}{r\sin\theta}\frac{\partial\psi}{\partial\phi}\hat{\boldsymbol{\phi}}$$

$$\nabla\cdot\mathbf{A} = \frac{1}{r^2}\frac{\partial}{\partial r}(r^2A_r) + \frac{1}{r\sin\theta}\frac{\partial}{\partial\theta}(\sin\theta A_\theta) + \frac{1}{r\sin\theta}\frac{\partial A_\phi}{\partial\phi}$$

$$\nabla\times\mathbf{A} = \frac{1}{r\sin\theta}\left[\frac{\partial}{\partial\theta}(\sin\theta A_\phi) - \frac{\partial A_\theta}{\partial\phi}\right]\hat{\mathbf{r}} + \left[\frac{1}{r\sin\theta}\frac{\partial A_r}{\partial\phi} - \frac{1}{r}\frac{\partial}{\partial r}(rA_\phi)\right]\hat{\boldsymbol{\theta}} + \frac{1}{r}\left[\frac{\partial}{\partial r}(rA_\theta) - \frac{\partial A_r}{\partial\theta}\right]\hat{\boldsymbol{\phi}}$$

$$\nabla^2\psi = \frac{1}{r^2}\frac{\partial}{\partial r}\left(r^2\frac{\partial\psi}{\partial r}\right) + \frac{1}{r^2\sin\theta}\frac{\partial}{\partial\theta}\left(\sin\theta\frac{\partial\psi}{\partial\theta}\right) + \frac{1}{r^2\sin^2\theta}\frac{\partial^2\psi}{\partial\phi^2}$$

Vector Identities

$$\mathbf{a} \cdot (\mathbf{b} \times \mathbf{c}) = \mathbf{b} \cdot (\mathbf{c} \times \mathbf{a}) = \mathbf{c} \cdot (\mathbf{a} \times \mathbf{b})$$

$$\mathbf{a} \times (\mathbf{b} \times \mathbf{c}) = (\mathbf{a} \cdot \mathbf{c})\mathbf{b} - (\mathbf{a} \cdot \mathbf{b})\mathbf{c}$$

$$(\mathbf{a} \times \mathbf{b}) \cdot (\mathbf{c} \times \mathbf{d}) = (\mathbf{a} \cdot \mathbf{c})(\mathbf{b} \cdot \mathbf{d}) - (\mathbf{a} \cdot \mathbf{d})(\mathbf{b} \cdot \mathbf{c})$$

$$\nabla \times \nabla \psi = 0$$

$$\nabla \cdot (\nabla \times \mathbf{a}) = 0$$

$$\nabla \times (\nabla \times \mathbf{a}) = \nabla(\nabla \cdot \mathbf{a}) - \nabla^2 \mathbf{a}$$

$$\nabla \cdot (\psi \mathbf{a}) = \mathbf{a} \cdot \nabla \psi + \psi \nabla \cdot \mathbf{a}$$

$$\nabla \times (\psi \mathbf{a}) = \nabla \psi \times \mathbf{a} + \psi \nabla \times \mathbf{a}$$

$$\nabla(\mathbf{a} \cdot \mathbf{b}) = (\mathbf{a} \cdot \nabla)\mathbf{b} + (\mathbf{b} \cdot \nabla)\mathbf{a} + \mathbf{a} \times (\nabla \times \mathbf{b}) + \mathbf{b} \times (\nabla \times \mathbf{a})$$

$$\nabla \cdot (\mathbf{a} \times \mathbf{b}) = \mathbf{b} \cdot (\nabla \times \mathbf{a}) - \mathbf{a} \cdot (\nabla \times \mathbf{b})$$

$$\nabla \times (\mathbf{a} \times \mathbf{b}) = \mathbf{a}(\nabla \cdot \mathbf{b}) - \mathbf{b}(\nabla \cdot \mathbf{a}) + (\mathbf{b} \cdot \nabla)\mathbf{a} - (\mathbf{a} \cdot \nabla)\mathbf{b}$$

Integral Identities

$$\int_V d^3 r \, \nabla \cdot \mathbf{A} = \int_S dS \, \hat{\mathbf{n}} \cdot \mathbf{A}$$

$$\int_V d^3 r \, \nabla \psi = \int_S dS \, \hat{\mathbf{n}} \psi$$

$$\int_V d^3 r \, \nabla \times \mathbf{A} = \int_S dS \, \hat{\mathbf{n}} \times \mathbf{A}$$

$$\int_S dS \, \hat{\mathbf{n}} \cdot \nabla \times \mathbf{A} = \oint_C d\boldsymbol{\ell} \cdot \mathbf{A}$$

$$\int_S dS \, \hat{\mathbf{n}} \times \nabla \psi = \oint_C d\boldsymbol{\ell} \, \psi$$

Miscellaneous

$$\epsilon_{ijk}\epsilon_{i\ell m} = \delta_{j\ell}\delta_{km} - \delta_{jm}\delta_{k\ell}$$

$$\nabla^2\left(\frac{1}{r}\right) = -4\pi\,\delta(\mathbf{r})$$

$$\frac{\partial}{\partial r_k}\frac{\partial}{\partial r_m}\left(\frac{1}{r}\right) = \frac{3r_k r_m - r^2\delta_{km}}{r^5} - \frac{4\pi}{3}\delta_{km}\delta(\mathbf{r})$$

$$\delta[g(x)] = \sum_n \frac{1}{|g'(x_n)|}\delta(x - x_n) \qquad g(x_n) = 0, \qquad g'(x_n) \neq 0$$

$$\delta(x) = \frac{1}{2\pi}\int_{-\infty}^{\infty} dk\ \exp(ikx)$$

$$f(\mathbf{r}, t) = \frac{1}{(2\pi)^4}\int d^3k \int_{-\infty}^{\infty} d\omega\,\hat{f}(\mathbf{k}|\omega)\exp[i(\mathbf{k}\cdot\mathbf{r} - \omega t)]$$

$$\hat{f}(\mathbf{k}|\omega) = \int d^3r \int_{-\infty}^{\infty} dt\ f(\mathbf{r}, t)\exp[-i(\mathbf{k}\cdot\mathbf{r} - \omega t)]$$

$$\mathbf{C}(\mathbf{r}) = \nabla\times\frac{1}{4\pi}\int d^3r'\,\frac{\nabla'\times\mathbf{C}(\mathbf{r}')}{|\mathbf{r}-\mathbf{r}'|} - \nabla\frac{1}{4\pi}\int d^3r'\,\frac{\nabla'\cdot\mathbf{C}(\mathbf{r}')}{|\mathbf{r}-\mathbf{r}'|}$$

Unit Vector Relations

$$\hat{\boldsymbol{\rho}} = \hat{\mathbf{x}}\cos\phi + \hat{\mathbf{y}}\sin\phi \qquad \hat{\mathbf{x}} = \hat{\boldsymbol{\rho}}\cos\phi - \hat{\boldsymbol{\phi}}\sin\phi$$

$$\hat{\boldsymbol{\phi}} = -\hat{\mathbf{x}}\sin\phi + \hat{\mathbf{y}}\cos\phi \qquad \hat{\mathbf{y}} = \hat{\boldsymbol{\rho}}\sin\phi + \hat{\boldsymbol{\phi}}\cos\phi$$

$$\hat{\mathbf{r}} = \hat{\mathbf{x}}\sin\theta\cos\phi + \hat{\mathbf{y}}\sin\theta\sin\phi + \hat{\mathbf{z}}\cos\theta \qquad \hat{\mathbf{x}} = \hat{\mathbf{r}}\sin\theta\cos\phi + \hat{\boldsymbol{\theta}}\cos\theta\cos\phi - \hat{\boldsymbol{\phi}}\sin\phi$$

$$\hat{\boldsymbol{\theta}} = \hat{\mathbf{x}}\cos\theta\cos\phi + \hat{\mathbf{y}}\cos\theta\sin\phi - \hat{\mathbf{z}}\sin\theta \qquad \hat{\mathbf{y}} = \hat{\mathbf{r}}\sin\theta\sin\phi + \hat{\boldsymbol{\theta}}\cos\theta\sin\phi + \hat{\boldsymbol{\phi}}\cos\phi$$

$$\hat{\boldsymbol{\phi}} = -\hat{\mathbf{x}}\sin\phi + \hat{\mathbf{y}}\cos\phi \qquad \hat{\mathbf{z}} = \hat{\mathbf{r}}\cos\theta - \hat{\boldsymbol{\theta}}\sin\theta$$

Maxwell Equations

$$\nabla \cdot \mathbf{E} = \frac{\rho}{\epsilon_0} \qquad \nabla \cdot \mathbf{D} = \rho_f$$

$$\nabla \times \mathbf{E} = -\frac{\partial \mathbf{B}}{\partial t} \qquad \nabla \times \mathbf{E} = -\frac{\partial \mathbf{B}}{\partial t}$$

$$\nabla \cdot \mathbf{B} = 0 \qquad \nabla \cdot \mathbf{B} = 0$$

$$\nabla \times \mathbf{B} = \mu_0 \mathbf{j} + \frac{1}{c^2}\frac{\partial \mathbf{E}}{\partial t} \qquad \nabla \times \mathbf{H} = \mathbf{j}_f + \frac{\partial \mathbf{D}}{\partial t}$$

Auxiliary Quantities

$$\mathbf{D} = \epsilon_0 \mathbf{E} + \mathbf{P} = \epsilon \mathbf{E} \qquad \mathbf{P} = \epsilon_0 \chi \mathbf{E}$$

$$\mathbf{H} = \frac{1}{\mu_0}\mathbf{B} - \mathbf{M} = \frac{1}{\mu}\mathbf{B} \qquad \mathbf{M} = \chi_m \mathbf{H}$$

$$\mathbf{j} = \mathbf{j}_f + \frac{\partial \mathbf{P}}{\partial t} + \nabla \times \mathbf{M} \qquad \rho = \rho_f - \nabla \cdot \mathbf{P}$$

$$\mathbf{E} = -\nabla \varphi - \frac{\partial \mathbf{A}}{\partial t} \qquad \mathbf{B} = \nabla \times \mathbf{A}$$

Matching Conditions

$$\hat{\mathbf{n}}_2 \cdot [\mathbf{E}_1 - \mathbf{E}_2] = \frac{\sigma}{\epsilon_0}$$

$$\hat{\mathbf{n}}_2 \cdot [\mathbf{B}_1 - \mathbf{B}_2] = 0$$

$$\hat{\mathbf{n}}_2 \times [\mathbf{E}_1 - \mathbf{E}_2] = 0$$

$$\hat{\mathbf{n}}_2 \times [\mathbf{B}_1 - \mathbf{B}_2] = \mu_0 \mathbf{K}$$

$$\varphi_1 - \varphi_2 = 0$$

$$\mathbf{A}_1 - \mathbf{A}_2 = 0$$